CW01183393

Modern Soft Tissue Pathology

Tumors and Non-Neoplastic Conditions

Second Edition

Markku Miettinen, MD, PhD, is Senior Clinician and Head of Surgical Pathology at the National Cancer Institute, National Institutes of Health, Bethesda, Maryland, USA

Modern Soft Tissue Pathology

Tumors and Non-Neoplastic Conditions

Second Edition

Edited by
Markku Miettinen
National Cancer Institute, National Institutes of Health, Bethesda, Maryland, USA

CAMBRIDGE UNIVERSITY PRESS

CAMBRIDGE
UNIVERSITY PRESS

University Printing House, Cambridge CB2 8BS, United Kingdom

Cambridge University Press is part of the University of Cambridge.

It furthers the University's mission by disseminating knowledge in the pursuit of education, learning and research at the highest international levels of excellence.

www.cambridge.org
Information on this title: www.cambridge.org/9781107567276

© Cambridge University Press 2016

This publication is in copyright. Subject to statutory exception and to the provisions of relevant collective licensing agreements, no reproduction of any part may take place without the written permission of Cambridge University Press.

First published: 2010
Second edition 2016

Printed in the United Kingdom by Clays, St Ives plc

A catalogue record for this publication is available from the British Library

ISBN 978-1-107-56727-6 Mixed Media
ISBN 978-1-107-07060-8 Hardback
ISBN 978-1-107-70750-4 Cambridge Books Online

Additional resources for this publication at www.cambridge.org/9781107567276

Cambridge University Press has no responsibility for the persistence or accuracy of URLs for external or third-party internet websites referred to in this publication, and does not guarantee that any content on such websites is, or will remain, accurate or appropriate.

....

Every effort has been made in preparing this book to provide accurate and up-to-date information which is in accord with accepted standards and practice at the time of publication. Although case histories are drawn from actual cases, every effort has been made to disguise the identities of the individuals involved. Nevertheless, the authors, editors and publishers can make no warranties that the information contained herein is totally free from error, not least because clinical standards are constantly changing through research and regulation. The authors, editors and publishers therefore disclaim all liability for direct or consequential damages resulting from the use of material contained in this book. Readers are strongly advised to pay careful attention to information provided by the manufacturer of any drugs or equipment that they plan to use.

Contents

Contributors vii
Preface and Acknowledgements ix

1. **Overview of soft tissue tumors** 1
   Markku Miettinen

2. **Radiologic evaluation of soft tissue tumors** 11
   Mark D. Murphey and Mark J. Kransdorf

3. **Immunohistochemistry of soft tissue tumors** 41
   Markku Miettinen

4. **Genetics of soft tissue tumors** 92
   Julia A. Bridge and Marilu Nelson

5. **Molecular genetics of soft tissue tumors** 115
   Jerzy Lasota

6. **Fibroblast biology, fasciitis, retroperitoneal fibrosis, and keloids** 181
   Markku Miettinen

7. **Fibromas and benign fibrous histiocytomas** 204
   Markku Miettinen

8. **Fibromatoses** 233
   Markku Miettinen

9. **Benign fibroblastic and myofibroblastic proliferations in children** 249
   Markku Miettinen

10. **Childhood fibroblastic and myofibroblastic proliferations of variable biologic potential** 277
    Markku Miettinen

11. **Myxomas and ossifying fibromyxoid tumor** 299
    Markku Miettinen

12. **Solitary fibrous tumor, hemangiopericytoma, and related tumors** 324
    Markku Miettinen

13. **Fibroblastic and myofibroblastic neoplasms with malignant potential** 336
    Markku Miettinen

14. **Lipoma variants and conditions simulating lipomatous tumors** 379
    Markku Miettinen

15. **Atypical lipomatous tumors and liposarcomas** 416
    Markku Miettinen

16. **Smooth muscle tumors** 443
    Markku Miettinen

17. **Gastrointestinal stromal tumor (GIST)** 474
    Markku Miettinen

18. **Stromal tumors and tumor-like lesions of the female genital tract** 491
    Markku Miettinen

19. **Angiomyolipoma and other perivascular epithelioid cell tumors (PEComas)** 513
    Markku Miettinen

20. **Rhabdomyomas and rhabdomyosarcomas** 527
    Erin R. Rudzinski, Vinay Prasad, Kadria Sayed, Faridali Ramji, and David M. Parham

21. **Hemangiomas, lymphangiomas, and reactive vascular proliferations** 553
    Markku Miettinen

22. **Hemangioendotheliomas, angiosarcomas, and Kaposi's sarcoma** 593
    Markku Miettinen

23. **Glomus tumor, glomangiopericytoma, myopericytoma, and juxtaglomerular tumor** 624
    Markku Miettinen

24. **Nerve sheath tumors** 637
    Markku Miettinen

25. **Neuroectodermal tumors: melanocytic, glial, and meningeal neoplasms** 694
    Markku Miettinen

26. **Paragangliomas** 723
    Markku Miettinen

27. **Primary soft tissue tumors with epithelial differentiation** 744
    Markku Miettinen

Contents

28. **Malignant mesothelioma and other mesothelial proliferations** 777
    Markku Miettinen

29. **Merkel cell carcinoma and metastatic and sarcomatoid carcinomas involving soft tissue** 798
    Markku Miettinen

30. **Cartilage- and bone-forming tumors** 825
    Julie C. Fanburg-Smith and Mark D. Murphey

31. **Small round cell tumors** 855
    Nick Shillingford, Hiroyuki Shimada and David M. Parham

32. **Alveolar soft part sarcoma** 895
    Markku Miettinen

33. **Pathology of synovia and tendons** 903
    Markku Miettinen

34. **Miscellaneous tumor-like lesions, and histiocytic and foreign body reactions** 926
    Markku Miettinen

35. **Lymphoid, myeloid, histiocytic, and dendritic cell proliferations in soft tissues** 941
    Markku Miettinen

36. **Cytology of soft tissue lesions** 975
    Matjaž Šebenik and Živa Pohar-Marinšek

37. **Surgical management of soft tissue sarcoma: histologic type and grade guide surgical planning and integration of multimodality therapy** 1012
    Charlotte Ariyan and Samuel Singer

38. **Medical oncology of soft tissue sarcomas** 1024
    Robert G. Maki

Index 1036

Contributors

Charlotte Ariyan, MD, PhD
Department of Surgery
Memorial Sloan-Kettering Cancer Center
New York, New York

Julia A. Bridge, MD, FACMG
Departments of Pathology and Microbiology,
Pediatrics, and Orthopedic Surgery
University of Nebraska Medical Center
Omaha, Nebraska

Julie C. Fanburg-Smith, MD*
Professor of Pathology
Uniformed Services University of the
Health Sciences (USUHS)
Bethesda, Maryland

Mark J. Kransdorf, MD
Department of Radiologic Pathology
Armed Forces Institute of Pathology
Washington, DC
Department of Radiology
Mayo Clinic
Jacksonville, Florida

Jerzy Lasota, MD, PhD*
Department of Soft Tissue Pathology
Armed Forces Institute of Pathology
Washington, DC

Robert G. Maki, MD, PhD
Department of Medicine
Memorial Sloan-Kettering Cancer Center
New York, New York

Markku Miettinen, MD, PhD (Editor)
Senior Clinician and Head of Surgical Pathology,
National Cancer Institute, National Institutes of Health,
Bethesda, Maryland

Mark D. Murphey, MD*
Department of Radiologic Pathology
Armed Forces Institute of Pathology
Washington, DC
Departments of Radiology and Nuclear Medicine
Uniformed Services University of the Health Sciences
Bethesda, Maryland
Department of Radiology
University of Maryland School of Medicine
Baltimore, Maryland

Marilu Nelson, MS, CLsp(CG, MG)
Department of Pathology and Microbiology
Nebraska Medical Center
Omaha, Nebraska

David M. Parham, MD
Department of Pathology
University of Southern California Keck
School of Medicine
Los Angeles, California

Živa Pohar-Marinšek, MD, PhD
Department of Cytopathology
Institute of Oncology
Ljubljana, Slovenia

Vinay Prasad, MD
Department of Pathology
Ohio State University
Department of Laboratory Medicine
Nationwide Children's Hospital
Columbus, Ohio

Faridali Ramji, MD, FRCFC
Departments of Radiological Sciences and
Pediatric Radiology
University of Oklahoma Health Sciences Center
The Children's Hospital,
Oklahoma University Medical Center
Oklahoma City, Oklahoma

Erin R. Rudzinski, MD
Department of Laboratories
Seattle Children's Hospital
Seattle, Washington

List of contributors

Kadria Sayed, MD
Department of Pathology
University of Arkansas Medical Sciences
Arkansas Children's Hospital
Little Rock, Arkansas

Matjaž Šebenik, MD
Department of Pathology
John F. Kennedy Hospital
Fort Lauderdale, Florida

Nick Shillingford
Pediatric Pathologist
Children's Hospital Los Angeles
Los Angeles, California

Hiroyuki Shimada, MD, PhD
Professor of Clinical Pathology
Children's Hospital Los Angeles
Los Angeles, California

Samuel Singer, MD
Department of Surgery
Memorial Sloan-Kettering Cancer Center
New York, New York

Disclaimer: The opinions and assertions of the authors associated with the Armed Forces Institute of Pathology, contained herein, are the expressed views of the authors and do not necessarily reflect the view of the Armed Forces Institute of Pathology or the US Department of Defense.

Preface and Acknowledgements

The second edition brings updates with new tumor entities, changes due to new classification and terminology, and lot of advances in tumor genetics and immunohistochemistry. Even if some updates increased volume, removing less important material whenever indicated by changes of practice allowed us to keep the book size equal to that of the 1st edition.

I thank those authors who updated the chapters on the rapidly evolving topics. Drs. Julia Bridge and Jerzy Lasota brought a great wealth of new knowledge on cytogenetics and molecular genetics helping others to follow this field – a growing necessity for all pathologists. Drs. David Parham, Hiroyuki Shimada and associates updated the complex fields of rhabdomyosarcoma and small round cell tumors. Drs. Julie Fanburg-Smith and Mark Murphey integrated radiology in the knowledge of cartilaginous and osseous lesions in the way already long practiced in bone pathology. Dr. Robert Maki updated the fast moving field of sarcoma oncology in an admirably concise way. I also thank the staff of Cambridge University Press for their help during the preparation of this edition. The guidance and encouragement of Jade Scard and Nisha Doshi is gratefully acknowledged. Anne Kenton performed meticulous proofreading and reference editing, for which I am very thankful. I also thank my wife Birgit for all kinds of help and support during this project.

Chapter 1

Overview of soft tissue tumors

Markku Miettinen
National Cancer Institute, National Institutes of Health

Definition

Soft tissue tumors are generally defined as tumors of connective tissues, including nonosseous sarcomas, benign mesenchymal tumors, and tumor-like proliferations. These tumors are usually considered to include nonosseous tumors of the extremities, trunk wall, intra-abdominal and intrathoracic space, and head and neck, although definitions vary. Generally excluded from this definition are nonmesenchymal tumors of the skin, cutaneous melanoma, most primary epithelial tumors, and brain and bone tumors.

In this book, soft tissue is understood broadly to include any important tumors of nonbony tissues of the extremities, trunk wall, retroperitoneum, mediastinum, and head and neck, except organ-specific tumors. Gastrointestinal stromal tumors are included because of their clinical importance and common occurrence as metastatic abdominal masses. Cutaneous nevi and primary cutaneous melanoma, and intracranial nerve sheath tumors have been excluded. Metastatic epithelial tumors and metastatic melanoma are included because of their practical importance (Table 1.1). As a subspecialty, soft tissue pathology intersects many other subspecialties of pathology.

Classification

The purpose of classification is to group similar tumors to create an understanding of tumor biology and behavior for developing treatment and follow-up strategies. The study of properly classified tumors also aids in the discovery of pathogenesis, devising biology-based treatments, and perhaps preventing tumors.

Soft tissue tumors continue to be classified according to the cell type that they resemble or have been thought to resemble. The clinical correlations that have already been obtained by the present classification are so numerous that the basis of this classification will probably remain, although cytogenetic, molecular genetic, and gene expression studies continue to refine the classification system. Reaching an ideal classification for this complex group of tumors, one that would be at the same time simple, highly reproducible, and clinically most informative, is not an easy task.

Table 1.1 Summary of soft tissue tumors by definition

Primary tumors of different locations
    Extremities
    Trunk wall
    External genitalia
    Body cavities, including retroperitoneum and mediastinum
    Head and neck
    Mesenchymal tumors of the gastrointestinal tract
    Other organ-based connective tissue tumors

Tumors of different histogenetic categories
    Mesenchymal tumors
        Benign mesenchymal tumors (e.g., lipoma)
        Malignant mesenchymal tumors (sarcomas)
    Neuroectodermal tumors (e.g., neurofibroma)
        Benign: neurofibroma, schwannoma
        Malignant neuroectodermal tumors

Primary and metastatic carcinomas

Primary and metastatic malignant melanoma

Hematolymphatic neoplasms

Miscellaneous tumors and reactive conditions

By tumor type, soft tissue tumors comprise a diverse group of benign, malignant, and borderline malignant (intermediate malignant) tumors. Most of them arise from (or show differentiation toward) mesenchymal cells, but some are of neuroectodermal (e.g., Schwann cell tumors), epithelial (metastatic carcinomas), or hematolymphatic (extranodal lymphoid and histiocytic infiltrates and lymphomas) origin (Table 1.2). The generally accepted basis for soft tissue tumor classification is the World Health Organization (WHO) system, last published in 2013.[1]

Malignant mesenchymoma is a sarcoma that displays differentiation toward more than one specific line, except the fibroblastic one. This designation is rarely applied today, because most tumors formerly classified as mesenchymoma are now more preferably diagnosed as liposarcomas or nerve sheath tumors with heterologous differentiation.

Benign tumors generally show the greatest similarity to their normal cell counterparts. For example, lipoma is histologically

Table 1.2 Simplified chart of the major types of primary soft tissue tumors grouped according to the cell types that they resemble[a]

Cell type	Benign tumor	Malignant tumor
Fibroblast, including myofibroblast	Fibroma, myxoma	Fibrosarcoma, malignant fibrous histiocytoma
Adipocyte	Lipoma	Liposarcoma
Smooth muscle cell	Leiomyoma	Leiomyosarcoma
Skeletal muscle cell	Rhabdomyoma	Rhabdomyosarcoma
Endothelial cell	Hemangioma	Angiosarcoma, Kaposi's sarcoma
Schwann cell	Schwannoma, neurofibroma	Some malignant peripheral nerve sheath tumors
Cartilage cell	Chondroma	Chondrosarcoma
Interstitial cell of Cajal of intestines		Gastrointestinal stromal tumors, a spectrum from benign to malignant
Histiocyte	Juvenile xanthogranuloma Rosai–Dorfman disease?	Histiocytic sarcoma (true histiocytic lymphoma)
Lymphoid cells	Benign lymphoid hyperplasia	Extranodal lymphomas in soft tissues
No known normal cell or benign counterparts		Ewing family tumors, synovial sarcoma, epithelioid sarcoma
		Alveolar soft part sarcoma

[a]Intermediate categories between benign and malignant tumors are excluded for simplicity.

indistinguishable from normal adipose tissue, and leiomyoma cells greatly resemble normal smooth muscle cells.

Sarcoma cells show varying resemblance to normal cell types, depending on their degree of differentiation. For example, well-differentiated liposarcoma cells greatly resemble fat cells, whereas pleomorphic ones contain more limited numbers of cells with specific fat differentiation. Some sarcoma types, such as synovial, epithelioid, and alveolar soft part sarcoma, have no normal cell counterparts.

Histogenesis

According to current understanding, most tumors are derived from multipotential precursor cells (stem cells) that are preprogrammed to differentiate into various mature cell types. The tumors are thought not to derive from mature cells, such as skeletal muscle and mature adipocytes, because such cells are terminally differentiated and incapable of cellular division.

The preprogrammed nature of many stem cells explains why some sarcomas closely resemble their mature cell types. For example, the cells of leiomyosarcoma closely resemble smooth muscle cells; however, some tumors contain cellular components that have no resemblance to normal cell types in that location. For example, metaplastic or neoplastic cartilage components can be present in different sarcomas. Similarly, the origin of benign rhabdomyoma as a polyp in the vaginal mucosa and rhabdomyosarcoma of the urinary bladder cannot be understood based on the normal cell types present in these locations.

The tissue origin of soft tissue stem cells is not fully clear, but it seems likely that many of them come from the local, organ-specific pools of stem cells. New data indicate that some soft tissue components are replenished from stem cells of bone marrow origin; these cells also could be the origin of some soft tissue tumors. For example, some regenerating skeletal muscle cells have been shown to have their origins in bone marrow,[2] and a portion of endothelial progenitor cells are of bone marrow stromal origin.[3]

The histogenesis of many sarcomas with no known normal cell counterparts (e.g., synovial sarcoma, alveolar soft part sarcoma) could reflect the unique genetic makeup that has created unprecedented tumor phenotypes that are not comparable to those of any normal tissue.

Epidemiology

Approximately 11 930 people are estimated to have a soft tissue sarcoma diagnosed in the United States in 2015, and 4870 people (41%) are expected to die of these tumors, according to American Cancer Society estimates.[4] The total incidence of sarcomas is higher and might be close to double if organ-based tumors are included, however. In cancer statistics, these tumors customarily are pooled with carcinomas of different organs. Overall, sarcomas are slightly more common in men, although some types (leiomyosarcomas) occur more often in women.

Sarcomas are rare tumors, and non-organ-based sarcomas constitute only 0.7% of all cancers. Their incidence is only about 5% of that of the most common carcinomas (i.e., prostate, breast, and lung), and one half of that of brain tumors and leukemias.[5] The relative rarity of soft tissue sarcomas might be explained by mesenchymal cells being located behind protective epithelial barriers that take most of the carcinogenic hits.

Based on these data, the incidence of sarcomas in the United States is approximately 4.0 per 100 000. This finding is similar to that obtained in the survey of epidemiology and end results (SEERS) program based on a sample population of the United States, where the overall incidence of soft tissue sarcomas was approximately 4 per 100 000, if Kaposi's sarcoma is excluded. According to the SEERS data, the incidence of soft tissue sarcomas has increased from the 1960s, although some studies have attributed this increase solely to the Kaposi's sarcoma epidemic.[6]

There might be global differences in sarcoma incidence, according to data from different cancer registries. For example, the incidence per 100 000 was only 0.8 in Osaka, Japan; 1.4 in Bombay, India; and 2.4 in Shanghai, China.[7] These figures are less than those reported for the United States and Europe, where the incidence is between 3 and 4 per 100 000. The apparent variances in incidence could result from differences in diagnosis, coding, and classification, however.

Like most other cancers, most soft tissue sarcomas occur in older adults, who have higher age-specific incidence of these tumors.[7,8] Important subgroups of soft tissue tumors, however, occur predominantly or exclusively in children (e.g., neuroblastoma, embryonal rhabdomyosarcoma, angiomatoid fibrous histiocytoma) and young adults (e.g., Ewing family tumors, alveolar rhabdomyosarcoma, and synovial sarcoma).

The incidence of benign soft tissue tumors is impossible to determine accurately, because benign tumors are underrepresented in hospital materials and usually are not included in tumor registries. As surgical specimens, however, benign soft tissue tumors outnumber their malignant counterparts by a margin of at least 100:1. In major teaching centers and tertiary hospitals with active musculoskeletal tumor surgery, however, the ratio might be closer to 10 or 20:1 because of the relative enrichment of malignant tumors, especially liposarcomas.

Etiology

The etiology of soft tissue sarcomas is relatively poorly understood, and known causes apply to only a small percentage of these tumors, much less than for many other cancers that are more clearly related to environmental carcinogenesis, for example lung cancer.

The most important known etiologic factors for soft tissue sarcomas include ionizing radiation, oncogenic viruses, and chemicals. The role of trauma is disputable, although anecdotal cases seem to support it.

All tumors are thought to arise as a result of acquired genetic alterations leading to abnormal quality or quantity of proteins that control cellular proliferation and differentiation. Certain etiologic environmental factors, radiation, certain viruses, and chemicals are known to be capable of causing genetic alterations that can lead to tumorigenesis.

A small percentage of sarcomas arise from host factors. Among them, the most important are hereditary genetic alterations (tumor syndromes), of which the most common by far is neurofibromatosis type 1 (Chapters 5 and 24). Rarely, other host factors are involved, such as immunosuppression and chronic lymphedema.

Radiation

Radiation-induced sarcomas develop in a small minority of patients (<1%) who have undergone therapeutic irradiation, typically 5 to 10 or more years after the radiation. Such postirradiation sarcomas most commonly include undifferentiated tumor types such as fibrosarcoma, malignant fibrous histiocytoma (MFH), osteosarcoma, and, rarely, angiosarcoma and malignant peripheral nerve sheath tumor (MPNST). The most common locations of postirradiation sarcoma are breast and chest wall in women irradiated for breast carcinoma, and the pelvis and lower abdominal wall in patients irradiated for gynecological or urological cancer.[9–11]

Thorotrast, colloidal thorium-oxide-containing radioactive radiologic contrast medium, was used until the 1940s, especially for angiographic studies. This material is permanently deposited to the reticuloendothelial cell system, especially in the liver, and some patients subsequently developed angiosarcoma of the liver or, more commonly, hepatic carcinoma or leukemia.[12]

Viruses

Oncogenic viruses might introduce new genomic material, which encodes for oncogenic proteins that disrupt the regulation of cellular proliferation. These genes are read, and the host makes proteins that alter cell cycle regulation or otherwise promote the viral infection. Two DNA viruses of the herpesvirus family have been linked to specific types of soft tissue sarcomas: human herpesvirus 8 (HHV8) to Kaposi's sarcoma,[13] and Epstein–Barr virus (EBV) to certain leiomyosarcomas.[14] In both instances, the virus–sarcoma connection is more common in immunosuppressed patients.

Most, if not all, Kaposi's sarcomas contain HHV8 sequences. This gamma herpesvirus is parenterally or sexually transmitted and is thought to be etiologically significant for the development of Kaposi's sarcoma, which explains its epidemic nature in HIV-infected populations and higher incidence in populations with higher prevalence of HHV8 infection.[13]

The EBV-associated leiomyosarcomas occur in immunodeficient or immunosuppressed patients, especially in children with HIV infection. Some have been seen in patients under chronic medically induced immunosuppression.[14]

Chemicals

Epidemiologic studies have linked phenoxyacetic acid herbicides to increased incidence of peripheral soft tissue sarcomas in some studies, although others have not confirmed this association. Dioxin contaminants have been suspected as the base of the carcinogenicity of these herbicide preparations.[5]

Sarcomas have developed around permanently retained metal objects, such as shrapnel and implanted surgical devices. These tumors have been mainly angiosarcomas and MFHs in the small numbers of reported cases.[15] Experimental studies support the capability of implanted metal or plastic objects to cause sarcomas. Several types of plastics and metals implanted long term in tissues were shown to induce local sarcoma formation, most commonly MFH or fibrosarcoma in rats. Proliferative mesenchymal lesions, possibly representing preneoplastic changes, were also observed.[16]

Epidemiologic studies have failed to show the association of soft tissue sarcomas with smoking, alcohol use, and organic solvent exposure.[5,8]

Hepatic angiosarcoma is an exceptional sarcoma that is associated with specific chemical agents more commonly than any other sarcoma. According to a large epidemiologic study, such factors were present in the history of approximately 25% of patients with these sarcomas; these factors included vinyl chloride (used in plastic manufacturing), inorganic arsenic (used as a pesticide or historically as a syphilis medicine), thorotrast, and androgenic anabolic steroids (the latter used medicinally or for doping purposes).[17]

Host factors

Immunosuppression is known to be associated with sarcomas having a viral connection, but could also cause other sarcomas. Hereditary or acquired (infection-associated or iatrogenic) lymphedema is a rare cause of extremity-based angiosarcomas, of which postmastectomy angiosarcoma in the lymphedematous arm is the most common example.

Grading

Grading is an arbitrary estimate of the degree of malignancy. The grading of soft tissue sarcomas by histologic parameters is to provide guidance for prognostic prediction and treatment, especially in relation to the patient's need for adjuvant therapy. Important other factors independent of grade are tumor size, completeness of the surgical excision, and the overall clinical situation.

Low-grade sarcomas are locally aggressive, but have a very low metastatic potential and usually a good prognosis. Consequently, they are usually treated with a wide surgical excision whenever possible. High-grade sarcomas have a high risk for both local recurrence and metastasis, and chemotherapy is the main mode of treatment for some high-grade sarcomas, such as childhood rhabdomyosarcoma and Ewing family tumors. Grading is also part of the current staging systems.

None of the grading systems replaces histologic typing, which is very important whenever type can be specified. In fact, the most widely used grading systems include histologic type as a grading variable. Several excellent reviews on sarcoma grading are recommended to the reader.[18–21]

Grading systems

The most extensively documented and widely used grading system for adult soft tissue sarcomas is the one developed by the French Federation of Cancer Centers (FFCC). This system has evolved over the past 25 years and has been specifically validated for spindle cell sarcomas.[21–24] The FFCC system uses the parameters of tumor differentiation, necrosis, and mitotic activity; the last two parameters were employed differently in the previously introduced grading system of the National Cancer Institute.[24] A comparative study has suggested that the FFCC system might result in a more informative grade, a higher number of high-grade tumors, a lesser number of intermediate-grade tumors, and a higher prognostic predictive value.[25]

The current FFCC system is based on a points score obtained as a sum of three factors: differentiation, mitotic rate, and tumor necrosis (Table 1.3). Each soft tissue sarcoma type has a differentiation score assigned, based on the histologic type (Table 1.4).

Table 1.3 Grading system of the French Federation of Cancer Centers, based on Coindre[21]

Tumor differentiation, according to Table 1.4	1–3
Well-differentiated tumors of defined histogenetic types	1
Moderately differentiated tumors of defined histologic types	2
Poorly differentiated tumors and undefined histogenetic types	3
Mitotic count	
0–9/10 HPF[a]	1
10–19/10 HPF	2
20 or more/10 HPF	3
Tumor necrosis[a]	
None	0
<50%	1
≥50%	2
Histologic grade	Sum of the above scores
1	2 or 3
2	4 or 5
3	6, 7, or 8

This grading system formulates the overall grade from total points from scores from tumor differentiation, mitotic rate, and tumor necrosis.
[a] High-power field (HPF) defined as 0.1734 mm^2.

The pediatric oncology grading system is applicable to nonrhabdomyosarcomatous soft tissue sarcomas of children. This system relies significantly on the histologic type as the basis of the grade, especially for the low-grade (grade 1) and high-grade (grade 3) tumors. It also incorporates necrosis and mitotic activity as grading parameters for some histologic types.[26] This system is summarized in Table 1.5.

A four-tiered system, dividing both low- and high-grade tumors into two grades, has also been suggested.[27] Another proposed grading system was based mainly on mitotic activity,[28] and one system included tumor size, vascular invasion, and microscopic necrosis as prognostic parameters.[29]

Limitations of grading

Tumor grading applies best to an excision specimen. Limited sampling (e.g., needle biopsy) can give a minimum grade only. Preoperative treatments, such as radiation, chemotherapy, and embolization, often induce tumor necrosis and variable tumor abolition, making grading inapplicable. Grading of tumors that often vary remarkably little from case to case seems to add limited value to the histogenetic diagnosis (e.g., extraskeletal myxoid chondrosarcoma and alveolar soft part sarcoma); these tumors therefore are often considered ungradable. Some grading parameters include a subjective element. For example, the determination of well-differentiated versus conventional or poorly differentiated examples of various tumors can be subjective (Table 1.4).

Table 1.4 Tumor differentiation score according to the updated version of the French Federation of Cancer Centers grading system

Differentiation score 1
  Well-differentiated fibro-, lipo-, or leiomyosarcoma

Differentiation score 2
  Conventional fibrosarcoma
  Myxoid sarcomas (MFH, liposarcoma, chondrosarcoma)
  Undifferentiated pleomorphic sarcoma
  Conventional leiomyosarcoma
  Well-differentiated or conventional angiosarcoma
  Conventional MPNST

Differentiation score 3
  Poorly differentiated fibrosarcoma
  MFH with a nonstoriform pattern
  Round cell liposarcoma
  Pleomorphic sarcomas (liposarcoma, leiomyosarcoma)
  Rhabdomyosarcoma (embryonal, alveolar, pleomorphic)
  Poorly differentiated and epithelioid angiosarcoma
  Triton tumor, epithelioid MPNST
  Extraskeletal mesenchymal chondrosarcoma
  Osteosarcoma
  Ewing family tumors/primitive neuroectodermal turmor (PNET)
  Synovial sarcoma
  Clear cell sarcoma
  Epithelioid sarcoma
  Alveolar soft part sarcoma
  Malignant rhabdoid tumor
  Undifferentiated sarcoma

Modified from Coindre.[21]

Table 1.5 The Pediatric Oncology Group grading system for nonrhabdomyosarcomatous soft tissue sarcomas of children

Grade 1
  Dermatofibrosarcoma protuberans, deep
  Infantile fibrosarcoma, well-differentiated (children not >4 years)
  Infantile hemangiopericytoma, well-differentiated
  Well-differentiated and myxoid liposarcoma
  Well-differentiated MPNST
  Extraskeletal myxoid chondrosarcoma
  Angiomatoid (malignant) fibrous histiocytoma

Grade 2
  Sarcomas not included in grades 1 and 3 with <15% of necrosis with no more than 5 mitoses/10 HPF
  No marked atypia, no markedly high cellularity
  Includes noninfantile fibrosarcomas, poorly differentiated infantile fibrosarcomas, leiomyosarcomas, and MPNSTs filling the previous criteria

Grade 3
  Round cell and pleomorphic liposarcoma
  Mesenchymal chondrosarcoma
  Extraskeletal osteosarcoma
  Malignant triton tumor
  Alveolar soft part sarcoma
  Sarcomas not included in grade 1 with >15% of necrosis, or with over 5 mitoses/10 HPF. Marked atypia and cellularity can also result in assignment into grade 3.

Modified and adapted from Parham et al.[26]

The impact of histologic grading is diluted by the nearly automatic high-grade assignment for some tumor types, because all examples of such entities would result in a high-grade scoring, and because of the lack of grading principles for some tumor types. This was pointed out by the Association of Directors of Anatomic and Surgical Pathology.[30]

Even the best grading systems include an element of subjectivity in the assessment of tumor differentiation, mitosis counting, and evaluation of the amount of necrosis. The reliability of grading is probably greater for the more common tumor types, and conversely, the assessment of grading systems for rare tumor types often is based on a very limited number of cases. Ideally, grading for all sarcoma types should be based on studies comparing large numbers of sarcomas of different parameters within a single histologic diagnosis group.

Staging

The stage is an estimate for the extent or dissemination of a tumor; the current system includes tumor grade as a component. Stage is an important characterization of a tumor for treatment formulation, cooperative clinical trials, and clinicopathologic studies of tumor behavior. The stage is based on clinical and radiologic evaluation of the tumor. The reader is referred to an illustrative review on sarcoma staging systems.[31]

The most widely used staging system is the Union Internationale Contre le Cancer-TNM (UICC-TNM) system.[32] Its current version has merged with the American Joint Committee of Cancer (AJCC) staging system.[33] These two identical systems classify tumors from stages I to IV, in which a low stage represents a small local tumor and stage IV describes metastatic disease (Table 1.6). All low-grade tumors are stage IA or IB, depending on the tumor size. Nonmetastatic high-grade tumors are divided into stages II and III, in which the latter stage is assigned to deep large tumors (>5 cm). This system excludes visceral sarcomas and certain cutaneous tumors such as Kaposi's sarcoma and dermatofibrosarcoma protuberans. Angiosarcoma has been excluded because its common multifocal nature makes the evaluation of tumor size and metastasis problematic.

The current UICC-TNM staging system was developed based on previous systems, especially the one originally suggested by the Task Force on Soft Tissue Sarcoma of the American Joint Committee for Cancer Staging and End Results Reporting (AJCC). This system incorporated histologic grade into the final stage and was based on evaluation of 1215 cases of 13 types of soft tissue sarcomas, mainly from the extremities. The study documented the value of staging in predicting survival.[34]

Table 1.6 Summary of the current TNM or the American Joint Committee for Cancer staging system for soft tissue sarcomas

Stage	Histologic grade (G)	Primary tumor	Lymph node status (N)	Distant metastasis (M)
I–IV	Low or high	T1 or T2	Negative/Positive	Absent/Present
IA	Low	T1a or T1b	Negative	Absent
IB	Low	T2a or T2b	Negative	Absent
IIA	High	T1a or T1b	Negative	Absent
IIB	High	T2a	Negative	Absent
III	High	T2b	Negative	Absent
IV	Any	Any	Positive	Absent
	Any	Any	Negative or positive	Present
Grade (G)	An arbitrary determination based on current grading systems. In a three-tier system, intermediate grade (G 2 or 3) is merged with high grade.			
T	Maximum diameter of tumor	T1 = 5 cm or less		T2 = >5 cm
Tumors of each T group are subclassified based on depth or anatomic location or both:				
	a = Superficial tumors of the trunk and extremities not invading in the superficial fascia b = Deep tumors invading, permeating, or located below the superficial fascia, or tumors in intra-abdominal, retroperitoneal, and intrathoracic location			

The surgical staging system developed by Enneking et al.[35] applies mainly to extremity sarcomas. Like that of the UICC-TNM, this system also incorporates grade as a factor, but it subdivides the stages by the tumor's relationship to the compartments of the extremities, instead of tumor size. In this system, low-grade tumors are stage 1, high-grade nonmetastatic tumors stage 2, and metastatic tumors stage 3. Further summary of the radiologic aspects of staging systems is found in Chapter 2.

Many investigators have realized that accurate prognostication must go beyond histologic typing, grading, and staging. Additional informative prognostic parameters include tumor size, tumor depth, anatomic site, and patient age. Based on these parameters and the histologic type, nomograms have been developed to predict potential mortality from sarcoma.[36–39]

Evaluation of soft tissue tumor specimens

The nature of specimens varies, depending on whether punch or needle biopsies are employed, or if incisional biopsies, piecemeal excisions, and complete resection specimens are used.

The trend toward minimally invasive diagnostic procedures has led to increasing use of small specimens, such as needle biopsies. Diagnostic specimens from internal sites, such as intra-abdominal and intrathoracic tumors, are commonly needle biopsies. Ultrasound or other radiologically guided procedures have increased the accuracy of lesion sampling.

Success in the definitive diagnosis and typing of tumors varies with needle biopsies. Abdominal tumors that can often be diagnosed reliably on needle biopsy include diffuse large cell lymphoma, well-differentiated liposarcoma, leiomyosarcoma, gastrointestinal stromal tumor, schwannoma, and solitary fibrous tumor. A small biopsy cannot rule out a malignant or high-grade component, however, and also can underestimate the potential of the tumor. Definitive diagnosis of reactive conditions and low-grade lymphomas is often impossible.

Radiologic correlation can enhance the information value of a small specimen by providing additional parameters, such as tumor configuration, relation to surrounding structures, and even tissue composition (e.g., fat and fluid). Magnetic resonance imaging (MRI) studies can help to identify a dedifferentiated liposarcoma, based on the presence of an integral fatty component in a spindle cell or pleomorphic sarcoma.

Open biopsy and resection specimens

Ideally, all tumor specimens should be received fresh immediately from surgery without fixation, because this increases the options for special studies and scientific evaluation. This is not necessary with needle biopsies, however, because the material is limited and optimal fixation can be best assured if the biopsy is immediately placed in the fixative.

Table 1.7 lists the steps that can be taken for the comprehensive analysis of a soft tissue tumor. These steps depend on the clinical environment and the scope of the studies planned in the future. High-quality clinicopathologic evaluation with thorough gross description, evaluation of margins, and histologic sampling must be performed in all cases.

If the diagnosis is unknown when the specimen is received, frozen section is useful for triage purposes, to guide the pathologist in the optimal selection of special studies. In some centers, frozen section is also used as a primary diagnostic mode with many tumors and can be highly accurate with an experienced pathology staff.

Grossing

Small specimens should usually be inked universally, and large specimens selectively, in the areas closest to the tumor. The specimen should then be sliced with 5–10 mm intervals and the tumor measured in three dimensions. The margins are usually best evaluated by sections perpendicular to the specimen surface closest to the tumor. Possible satellite nodules outside the main tumor mass should also be recorded, and the percentage of gross necrosis estimated. Other features to be recorded include color and consistency, as well as the presence of any hemorrhage, calcification, ossification, cysts, and grossly different tumor components. Representative sections should be documented and sampled for microscopy, and small tumors should be submitted entirely (1–2 cm or less). A minimal sampling should include one section per each 1 cm of tumor diameter. A diagnostic pitfall is missing a lipomatous component in dedifferentiated liposarcoma; therefore, surrounding fat should be included in any sarcoma sampling, especially one that is retroperitoneal.

Although smaller tumors can be submitted for tissue processing on the same day that the specimen is received, larger tumors and all lipomatous tumors should be sliced and allowed to fix overnight to improve tissue processing and quality of tissue sections.

Gross photography

Gross photography is an excellent permanent tumor documentation. The ideal photographic documentation includes intact and sliced tumor with overview and close-up views, some of them with a metric scale. Digital photographs can also assist grossing, and annotations can be made on the microscopic sampling.

Reporting

As suggested by the Association of the Directors of Anatomic and Surgical Pathology,[30] the surgical pathology report should accurately document the tumor site, histologic type, grade, tumor size, status or margins, percentage of necrosis, lymph node status, and several other factors (Table 1.8).

Table 1.7 Steps and parameters in a comprehensive analysis of a soft tissue tumor specimen on gross examination

1. Gross photography (preferably fresh tissue, possibly also fixed)
2. Evaluation and inking of margins
3. Gross description and tumor measurements
4. Sampling of tumor and margins for histology (perpendicular recommended to document distance from the margin)
5. Frozen section for diagnostic or triaging purposes
6. Sampling of fresh tissue for further analysis[a]
   - 6.1. Frozen tissue procurement (frozen, in special preservatives) for RNA, DNA, FISH, proteomics, chemical and immunohistochemical analysis. Formalin fixation of adjacent tissues for morphologic documentation of the selections
   - 6.2. Submission of material for cytogenetics or assessment of chemosensitivity for initiation of a short-term cell culture and a possible continuous cell line
   - 6.3. Sampling in special fixatives (alcohol, Carnoy's, B5) for further studies
   - 6.4. Sampling for glutaraldehyde fixation for electron microscopy

[a] Fresh tissue sampling should be performed as soon as possible, preferably first.

Table 1.8 Suggested parameters to be included in the surgical pathology report of a sarcoma, as suggested by the Association of the Directors of Anatomic and Surgical Pathology[30]

Final report
1 Tumor site, type of biopsy or excision
2 Depth of the tumor (subcutis, fascia, skeletal muscle)
3 Tumor type, possible variant
4 Grade, if possible
5 Tumor size (maximum diameter in cm), plus possible presence of satellite nodules
6 Status of margins (minimal distance to margins) and lymph node status if present
7 Microscopic quantitation of necrosis
8 Vascular invasion (if present)

Addendum report or reports (if studies cannot be completed by the issue of final report)
1 Immunohistochemistry
2 Electron microscopy
3 Cytogenetics

Tissue procurement for special studies

Tissue procurement for special studies is an important part of specimen handling, especially in academic centers, and can usually be accommodated easily without interfering with the diagnostic sampling and evaluation of the margins. It not only relates to scientific studies, but also allows the optimal use of advanced diagnostic modalities.

Frozen tissue

Freezing of tissue is clinically indicated to help perform molecular genetic assays more easily or reliably, and it is also scientifically indicated to build knowledge of genetic and biochemical changes in tumors, compared with normal tissue. Frozen tissue is required for optimal and more effective analysis of nucleic acids. High-molecular-weight DNA and RNA usually can be obtained only from fresh and not from formaldehyde-fixed tissue. Similarly, native proteins can be reliably obtained only from fresh or frozen tissue for proteomics, biochemical microanalysis of the spectrum of cell signaling, and other functionally important proteins.

Freezing in liquid nitrogen is optimal, but long-term storage in a −70 °C freezer or liquid nitrogen are both adequate. Well-organized storage compartments and inventory systems are required for optimal retrieval. Liquid nitrogen storage has the advantage of being independent of electric power, which protects the tissue bank from power outages. For liquid nitrogen tanks, automatic refilling systems are available.

Ideally, aliquots of both tumor and normal tissue should be sampled separately. The best way to store the tissue is to freeze small pieces separately in a liquid nitrogen bath and transfer them to the cryovial (allowing the nitrogen to evaporate in −20 °C cryostate to prevent the vial's cap popping). Such separately frozen "pearls" of tissue can be easily poured from the cryovial and used one at a time or as needed.

Cytogenetics and cell culture

Cytogenetic studies are diagnostically indicated in tumors with specific translocations or other chromosomal morphologic changes (see Chapter 4). They are also indicated to increase scientific knowledge of previously uncharacterized tumor types.

Cytogenetic specimens can be sent to the laboratory in a culture medium and should be preserved in a sterile manner. A very thin slice of well-preserved, non-necrotic tumor tissue should be submitted. Fine-needle aspirates are also suitable.

Short-term cultures needed for karyotyping can be successfully obtained for malignant tumors, but benign tumors can be difficult to grow. Such cultures can also be used for *in vitro* testing of tumor chemosensitivity.

Long-term cultures and establishing a cell line from a tumor are more challenging, and the success rate is only modest, even with highly experienced investigators. Long-term cultures offer priceless dynamic models for investigating cell biologic, biochemical, and pharmacological characteristics of the tumor, however.

Special fixatives

Fixation of specimens in special fixatives is often indicated for optimal evaluation of tumors that might be expected to be diagnostically difficult. The tissue aliquots should be small, not exceeding the thickness of a standard coin, to allow the penetration of the fixative prior to autolysis.

Alcohol (absolute ethanol)-fixed tissue can be saved for further studies, such as RNA, DNA, and protein extraction, or they can be embedded in paraffin for tissue section-based studies. Alcohol-fixed tissue can be advantageous for immunohistochemical analysis of some antigens. It can also be suitable for Western blot analysis of proteins and obtaining high-molecular-weight nucleic acids.

Carnoy's fixative is a modified alcohol fixation, added with glacial acetic acid in a ratio of 1:4. Methacarn is an alcoholic fixative using methanol instead of ethanol.

Heavy-metal-containing formalin fixative (B5 solution containing mercury chloride) yields superior nuclear detail and is often used for hematopoietic neoplasia. Similarly, zinc salts can be used as a less toxic and more environmentally friendly alternative. Tissues processed with these fixatives, however, are not generally suitable for molecular genetic studies.

A small aliquot of viable tumor, or if necessary, samples of several different areas, should be sampled in 2.5% buffered glutaraldehyde for electron microscopy. The easiest way to prepare a sample is first to cut a thin slice, section this further into a rod, and then mince the rod into cubes not exceeding 1 mm at the thickest point.

Role of electron microscopy

Electron microscopy is very rarely mandatory for diagnosis, but it can have diagnostic potential and therefore should be considered for tissue procurement program. The processed tissue can be saved for future studies, if analysis is not needed immediately for diagnosis. The most useful application of electron microscopy is for those soft tissue tumors with highly distinctive ultrastructural features (Table 1.9).

Ultrastructural details at magnifications ranging from 3000 to 50 000 can give valuable diagnostic information about selected soft tissue tumors.[40,41] For many groups of tumors, such as lymphomas, melanoma, and undifferentiated tumors, immunohistochemistry has mostly replaced electron microscopy as a diagnostic method; for others it is used only infrequently because of its labor-intensive nature and cost.

Although glutaraldehyde fixation is optimal for preserving cytoplasmic details, fixation in buffered formalin is also adequate. Cytoplasmic texture and many membranous structures deteriorate during routine formalin fixation, however, and they can be lost during tissue processing and paraffin embedding.

Table 1.9 Frequent distinctive electron microscopic findings in selected soft tissue tumors

Tumor type	Diagnostic features
Fibroblastic neoplasms	Variable myofibroblastic differentiation found in subsets of tumor cells
Desmoid, fibrosarcoma, MFH	Myofibroblasts with scattered bundles of actin filaments
Smooth muscle tumors, glomus tumor	Cytoplasmic bundles of actin filaments, attachment plaques, basal lamina
Rhabdomyosarcoma	Ribosome myosin complexes, collections of thin and thick filaments, possible sarcomeres and Z-bands
Angiosarcoma	Weibel–Palade bodies (predominantly in well-differentiated tumor cells)
Schwannoma	Spindle cells, complex interdigitating cell processes and prominent basal laminas
Perineurial cell tumors	Spindle cell with long cytoplasmic processes, prominent intermediate filaments, frequent pinocytic vesicles, basal laminas
Melanoma	Melanosomes. Can be sparse and difficult to find.
Paraganglioma	Dense core granules of variable size and morphology, typically abundant
Neuroendocrine carcinoma	Membrane-bound dense core granules of 100–400 nm in diameter, typically abundant, but could be sparse in high-grade tumors
Rhabdoid tumor, sarcomas with rhabdoid cytologic features	Spherical collections of perinuclear collections of cytoplasmic intermediate filaments displacing the cytoplasmic organelles
Alveolar soft part sarcoma	Cytoplasmic rhomboid crystals with 70 Angstrom periodicity
Mesothelioma, differentiated	Long, slender microvilli that are typically 15 times longer than their width
Dendritic reticulum cell sarcoma	Desmosomes, elongated cell processes

General tips for diagnosis making

The more one knows of the possible diagnostic entities, the better one is equipped to identify various tumor entities. Knowledge on tumor entities and their sites of occurrence can help significantly to narrow down the differential diagnosis.

One should remember that metastatic melanoma and poorly differentiated carcinomas often involve soft tissue tumors and both can simulate various sarcomas so that these diagnoses need to be also considered in the differential diagnosis.

Six immunohistochemical markers: CD34, desmin, epithelial membrane antigen, keratin cocktail AE1/AE3, S100 protein, and smooth muscle actin can be very helpful in initial screening of an unknown tumor. A detailed approach on immunohistochemistry is given in Chapter 3.

Review of radiological studies can be useful in many ways. They can help with interpreting the significance of findings in small specimens, and in some cases, radiology can give clues to tumor histogenesis (presence of fat, close association of blood vessels or nerves).

References

1. Fletcher CDM, Bridge JA, Hogendoorn PCV, Mertens F. *WHO Classification of Tumours of Soft Tissue and Bone.* International Agency for Research on Cancer: Lyon; 2013.
2. Ferrari G, Cusella-De Angelis G, Coletta M, *et al.* Muscle regeneration by bone marrow-derived myogenic progenitors. *Science* 1998;279:1528–1530.
3. Asahara T, Masuda H, Takahashi T, *et al.* Bone marrow origin of endothelial progenitor cells responsible for postnatal vasculogenesis in physiological and pathological neovascularization. *Circ Res* 1999;85:221–228.
4. American Cancer Society on-line. www.cancer.org.
5. Zahm SH, Fraumeni JF. The epidemiology of soft tissue sarcoma. *Semin Oncol* 1997;24:504–514.
6. Ross JA, Severson RK, Davis S, Brooks JJ. Trends in the incidence of soft tissue sarcomas in the United States from 1973 through 1987. *Cancer* 1993;72:486–490.
7. Clemente C, Orazi A, Rilke F. The Italian registry of soft tissue tumors. *Appl Pathol* 1988;6:221–240.
8. Olsson H. A review of the epidemiology of soft tissue sarcoma. *Acta Orthop Scand (Suppl 285)* 1999;70:8–19.
9. Laskin WB, Silverman TA, Enzinger FM. Postirradiation soft tissue sarcomas: an analysis of 53 cases. *Cancer* 1988;62:2230–2240.
10. Wiklund TA, Blomqvist CP, Räty J, *et al.* Postirradiation sarcoma: analysis of a nationwide cancer registry material. *Cancer* 1991;68:524–531.
11. Mark RJ, Poen J, Tran LM, *et al.* Postirradiation sarcomas. A single-institution study and review of the literature. *Cancer* 1994;73:2653–2662.

12. Stover BJ. Effect of thorotrast in humans. *Health Phys* 1983;44 (Suppl 1):253–257.
13. Boshoff C, Chang Y. Kaposi's sarcoma-associated herpesvirus: a new DNA tumor virus. *Annu Rev Med* 2001;52:453–470.
14. Hsu JL, Glaser SL. Epstein–Barr virus-associated malignancies: epidemiologic patterns and etiologic implications. *Crit Rev Oncol Hematol* 2000;34: 27–53.
15. Jennings TA, Peterson L, Axiotis CA, *et al*. Angiosarcoma associated with foreign body material: a report of three cases. *Cancer* 1988;62:2436–2444.
16. Kirkpatrick CJ, Alves A, Köhler H, *et al*. Biomaterial-induced sarcoma: a novel model to study preneoplastic change. *Am J Pathol* 2000;156:1455–1467.
17. Falk H, Thomas LB, Popper H, Ishak KG. Hepatic angiosarcoma associated with androgenic-anabolic steroids. *Lancet* 1979;2:1120–1123.
18. Kilpatrick SE. Histologic prognostication in soft tissue sarcomas: grading versus subtyping or both? A comprehensive review of the literature with proposed practical guidelines. *Ann Diagn Pathol* 1999;3:48–61.
19. Oliveira AM, Nascimento AG. Grading in soft tissue tumors: principles and problems. *Skeletal Radiol* 2001;30: 543–559.
20. Deyrup AT, Weiss SW. Grading of soft tissue sarcomas: the challenge of providing information in an imprecise world. *Histopathology* 2006;48: 42–50.
21. Coindre JM. Grading of soft tissue sarcomas: from histological to molecular assessment. *Pathology* 2014;46:113–120.
22. Trojani M, Contesso G, Coindre JM, *et al*. Soft-tissue sarcomas of adults: study of pathological prognostic variables and definition of a histopathological grading system. *Int J Cancer* 1984;33:37–42.
23. Coindre JM, Bui NB, Bonichon F, de Mascarel I, Trojani M. Histopathologic grading in spindle cell soft tissue sarcomas. *Cancer* 1988;61:2305–2309.
24. Costa J, Wesley RA, Glatstein E, Rosenberg SA. The grading of soft tissue sarcomas: results of a clinicohistopathologic correlation in a series of 163 cases. *Cancer* 1984;53: 530–541.
25. Guillou L, Coindre JM, Bonichon F, *et al*. Comparative study of the National Cancer Institute and French Federation of Cancer Centers Sarcoma Group grading systems in a population of 410 adult patients with soft tissue sarcoma. *J Clin Oncol* 1997;15:350–362.
26. Parham DM, Webber BL, Jenkins JJ, Cantor AB, Maurer HM. Nonrhabdomyosarcomatous soft tissue sarcomas of childhood: formulation of a simplified system for grading. *Mod Pathol* 1995;8:705–710.
27. Markhede G, Angervall L, Stener B. A multivariate analysis of the prognosis after surgical treatment of malignant soft tissue tumors. *Cancer* 1982;49:1721–1733.
28. Myhre-Jensen O, Kaae S, Madsen EH, Sneppen O. Histopathological grading in soft tissue tumours: relation to survival in 261 surgically treated patients. *Acta Pathol Microbiol Scand A* 1983;91:145–150.
29. Gustafson P, Akerman M, Alvegord TA, *et al*. Prognostic information in soft tissue sarcoma using tumour size, vascular invasion and microscopic tumour necrosis – the SIN-system. *Eur J Cancer* 2003;39:1568–1576.
30. Association of the Directors of Anatomic and Surgical Pathology. Recommendations for the reporting of soft tissue sarcomas. *Mod Pathol* 1998;11:1257–1260.
31. Peabody TD, Gibbs CP, Simon MA. Evaluation and staging of musculoskeletal neoplasms. *J Bone Joint Surg* 1998;80A:1204–1218.
32. Sobin LH, Wittekind C. *TNM Classification of Malignant Tumors*. UICC: Wiley-Liss; 2002.
33. Greene FL, Page D, Morrow M, *et al*. (eds.) *AJCC Cancer Staging Manual*, 6th edn. New York: Springer; 2002.
34. Russell WO, Cohen J, Enzinger F, *et al*. A clinical and pathological staging system for soft tissue sarcomas. *Cancer* 1977;40:1562–1570.
35. Enneking WF, Spanier SS, Goodman MA. A system for the surgical staging of musculoskeletal tumors. *Orthop Rel Res* 1980;153:106–120.
36. Kattan MW, Laung DHY, Brennan MF. Postoperative nomogram for 12-year sarcoma-specific death. *J Clin Oncol* 2002;20:791–796.
37. Mariani L, Miceli R, Kattan MW, *et al*. Validation and adaptation of a nomogram for predicting the survival of patients with extremity soft tissue sarcoma using a three-grade system. *Cancer* 2005;103:402–408.
38. Eilber FC, Brennan MF, Eilber FR, *et al*. Validation of the postoperative nomogram for 12-year sarcoma-specific mortality. *Cancer* 2004;101: 2270–2275.
39. Dalal KM, Kattan MW, Antonescu CR, Brennan MF, Singer S. Subtype specific prognostic nomogram for patients with primary liposarcoma of the retroperitoneum, extremity, or trunk. *Ann Surg* 2006;244:381–391.
40. Erlandson RA, Woodruff JM. Role of electron microscopy in the evaluation of soft tissue neoplasms, with emphasis on spindle cell and pleomorphic tumors. *Hum Pathol* 1998;29:1372–1381.
41. Ordonez MG, Mackay B. Electron microscopy in tumor diagnosis: indications for its use in the immunohistochemical era. *Hum Pathol* 1998;29:1403–1411.

Chapter 2

Radiologic evaluation of soft tissue tumors

Mark D. Murphey
Armed Forces Institute of Pathology, Uniformed Services University of the Health Sciences, University of Maryland School of Medicine

Mark J. Kransdorf
Armed Forces Institute of Pathology, Mayo Clinic

The goals of imaging of soft tissue tumors are threefold: (1) lesion detection, (2) identifying a specific diagnosis or reasonable differential diagnosis, and (3) lesion staging. The radiologic evaluation of soft tissue tumors to achieve these goals has markedly evolved, improved, and expanded with the advent of computerized tomography (CT) and particularly magnetic resonance imaging (MRI). CT and more recently MRI allow lesion detection and staging by delineating anatomic extent in essentially all cases and relatively specific diagnosis in approximately 25% to 50% of soft tissue tumors.[1–7] We would suggest that evaluation of soft tissue tumors is now similar to that of bone tumors in that pathologic diagnosis should incorporate the imaging findings in most cases. This is particularly true in large tumors in which only a small amount of tissue might be available for histologic review initially and there is doubt about whether the tissue is a true representation of the entire lesion. As in bone tumors, a close working relationship among three groups of physicians is needed: the pathologist, radiologist, and orthopedic oncologist. The purpose of this chapter is to provide a framework for the use of radiologic evaluation of soft tissue tumors. Although this approach reviews multiple imaging modalities, the authors emphasize MRI because it is generally considered the optimal radiologic tool in evaluation of soft tissue tumors.

The annual incidence of benign soft tissue tumors has been estimated at 300 per 100 000 people, leading to an estimated 750 000 to 800 000 lesions in the United States.[8,9] Thus, benign soft tissue tumors are relatively common lesions and outnumber malignant soft tissue tumors by a ratio of 100 to 150 to 1.[8,9] Soft tissue sarcomas occur relatively infrequently, although increasing in incidence, representing approximately 1% of all malignant tumors.[10,11] The American Cancer Society figures of 2008 estimate 10 390 soft tissue malignancies annually in the United States, an increase of 28% since the year 2000.[12] These malignancies are estimated to account for 3680 deaths, a decrease of 16% since the year 2000.[12,13]

Radiologic evaluation
Radiographs

In our opinion, the radiologic evaluation of a soft tissue tumor should *always* begin with radiographs. Although not helpful in a high percentage of lesions, certain features can be diagnostic and are actually more difficult to appreciate on advanced imaging (CT and MRI). Unfortunately, in this day and age of high-tech imaging, this simple inexpensive step is frequently forgotten, an omission that can lead to disastrous results. Radiographs can reveal that an apparent soft tissue mass is related to an underlying osseous lesion such as an osteochondroma or post-traumatic deformity. Calcification of a soft tissue mass significantly restricts reasonable diagnostic considerations and can be pathognomonic in hemangioma, synovial chondromatosis, tumoral calcinosis, or heterotopic bone formation (myositis ossificans; Figure 2.1).[8,14] Although calcification can be nonspecific in appearance on radiographs, nonspecific mineralization frequently is associated with extraskeletal chondrosarcoma or osteosarcoma and synovial sarcoma. Finally, radiographs allow detection of underlying bone involvement, as evidenced by the presence of periosteal reaction or cortical destruction and marrow invasion.[8,14] In the authors' experience, the common soft tissue sarcomas to reveal bone involvement are synovial sarcoma and the fibrous malignancies (malignant fibrous histiocytoma [MFH] and fibrosarcoma).

Nuclear medicine

Scintigraphic evaluation does not play a primary role in evaluation of soft tissue tumors. Both benign and malignant soft tissue tumors, particularly more vascularized lesions, often reveal mild increased uptake of radionuclide on bone scintigraphy, owing to increased blood flow. In addition, lesions with mineralization can show more extensive radiotracer activity resulting from the increased turnover of calcium and phosphate. Although evaluation has included only small numbers of patients, gallium scanning has been advocated to distinguish benign peripheral

---

Modern Soft Tissue Pathology, Second Edition, ed. Markku Miettinen. Published by Cambridge University Press. © Cambridge University Press 2016.

Chapter 2: Radiologic evaluation of soft tissue tumors

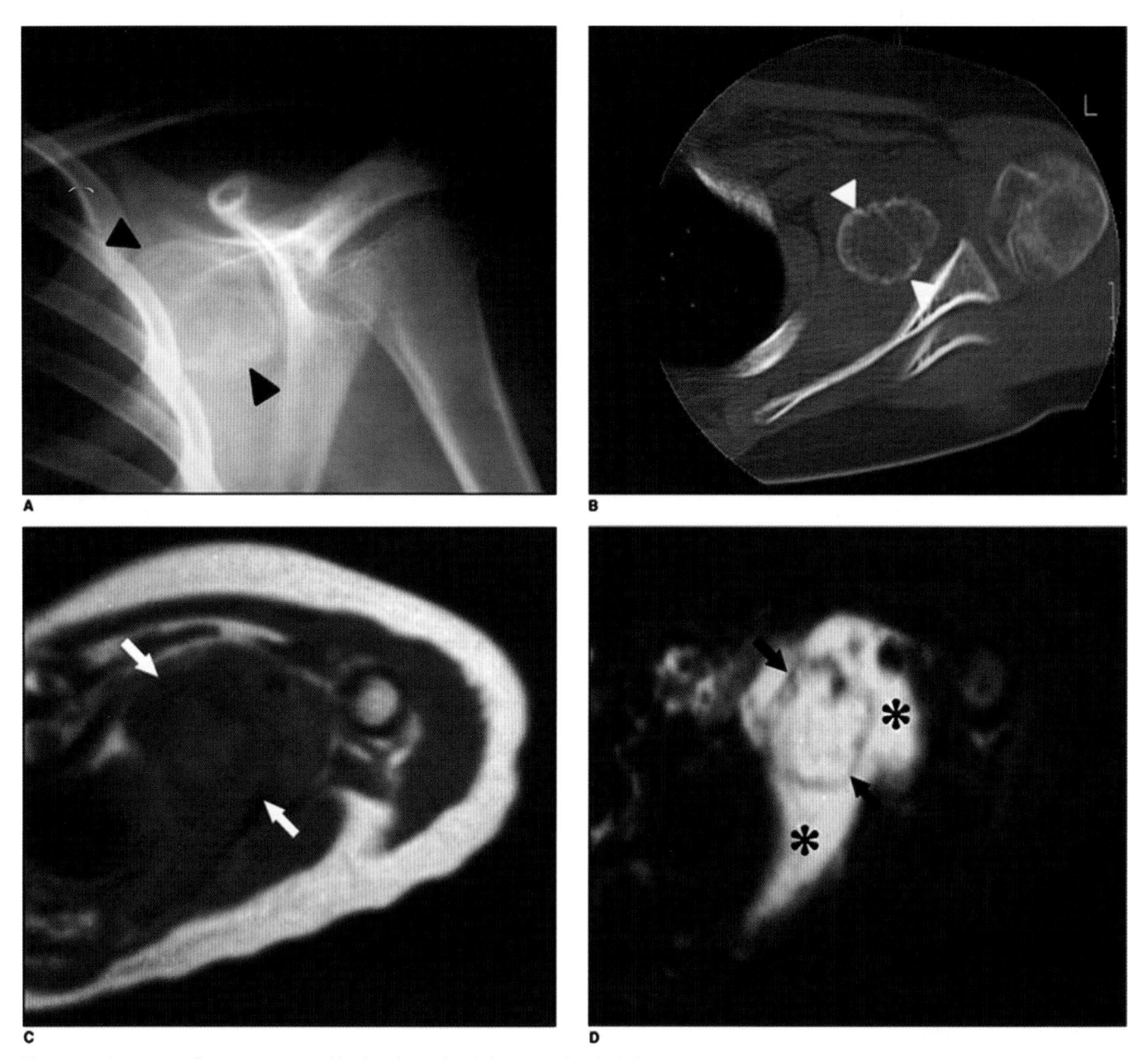

Figure 2.1 Myositis ossificans in a 10-year-old girl. Radiograph (**A**) shows peripheral calcification (arrowheads) overlying the scapula. CT (**B**) reveals characteristic peripheral rim appearance of the calcification (arrowheads), reflecting the "zonal phenomenon" of maturation and separation from the scapula. Axial T1- (**C**) and T2- weighted (**D**) MR images show the mass (arrows), although the margins appear infiltrative owing to surrounding edema (*) and pathognomonic calcification and relationship to the scapula is not as well seen as on the CT (**B**). (**E**) The zonal phenomenon is identically demonstrated on the photomicrograph (hematoxylin and eosin stain, 10 ×) with more mature bone on the periphery (between arrows) and less mature areas centrally (*)

nerve sheath tumors (BPNSTs) from malignant peripheral nerve sheath tumors (MPNSTs).[15–17] Specifically, prominent gallium uptake is seen in MPNSTs in contradistinction to BPNSTs (neurilemmoma or neurofibroma), which demonstrate limited or no uptake.[15–17] FDG (fluorine-18 fluoro-2-deoxy-D-glucose) positron emission tomography (PET) has received more attention recently in assessing soft tissue masses by measuring the avidity of glucose turnover (quantitated by the standardized uptake value [SUV]).[18–21] FDG PET has been used in attempts to distinguish benign from malignant tumors (SUV greater than 2–3), to evaluate the response of neoplasms to treatment (radiation or chemotherapy), and to evaluate possible neoplasm recurrence following surgical resection.[18–21] FDG PET images are also frequently performed with CT (PET-CT fusion) to improve the anatomic detail. As more experience with FDG PET emerges, however, a significant overlap between benign and malignant processes is becoming apparent (Figure 2.2). Thallium has also been used to assess response to therapy (radiation therapy or chemotherapy) and to evaluate possible recurrent tumors following surgical resection.

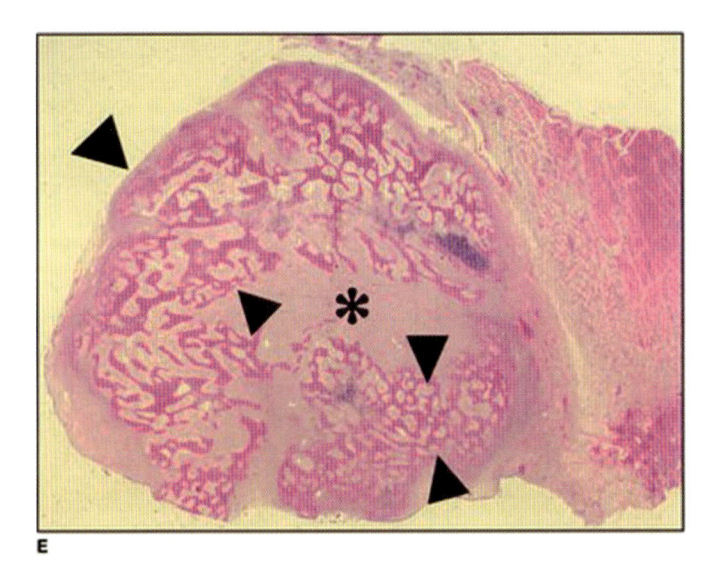

Figure 2.1 (cont.)

Angiography

In the past, angiographic evaluation of soft tissue tumors, particularly sarcomatous lesions, was relatively common to assess the degree of vascularity and serve as a vascular road map during surgery. In addition, angiography was often used to evaluate the effect of preoperative therapy depicted, because decreased vascularity usually is a result of hemorrhage and/or necrosis. However, angiography has largely been supplanted by other imaging modalities, such as CT or MR angiography (CTA or MRA).[22–24] Embolization of soft tissue neoplasms can be performed by means of angiographic access as a method to decrease blood loss at surgical resection in lesions with prominent vascularity. Angiomatous lesions, particularly when diffuse or extensive (angiomatosis), can be embolized as the sole method of treatment.

CT, MRI, and ultrasonography

The advantages of these three-dimensional imaging modalities compared with radiographs in evaluation of soft tissue tumors is primarily a function of their superior contrast resolution. Although not universally accepted, in the authors' opinion and experience, MRI is generally far superior to CT in radiologic evaluation of soft tissue tumors owing to its marked improvement in contrast resolution and multiplanar capabilities (ability to image in any plane desired; see Figure 2.1).[1–7,25–35] MRI might be contraindicated and CT preferred for a variety of reasons, including severe claustrophobia, cardiac pacemaker, and metallic foreign bodies (including recent surgery with clips in anatomically sensitive areas such as the brain). CT is also preferable in certain anatomic regions such as the periscapular area or chest or abdominal wall, where motion artifact can be problematic with MRI (Figure 2.1). In addition, CT is the imaging modality of choice to both detect and characterize calcification (chondroid or osteoid) in soft tissue tumors in cases for which radiographs are inadequate owing to the subtle nature of the mineralization or obscuration by complex anatomy, particularly overlying osseous structures (i.e., pelvis; Figure 2.1). Although some studies suggest that CT and MRI are equivalent for identification of bone involvement by soft tissue tumors, in the authors' opinion, CT remains preferable for this assessment, particularly when subtle or affecting a small osseous structure such as the fibula or scapula.

Placement of a marker over the soft tissue lesion is helpful with both CT and MRI. This is particularly important for superficial lesions such as lipomas, which can be obscured by the surrounding subcutaneous fat (Figure 2.3). Patient positioning is also important with both modalities, particularly again in superficial lesions, with care being taken that the lesion is not compressed. This can require prone positioning in patients with superficial paraspinal lesions, buttock, or posterior extremity masses.

Although an extensive review of technical factors to optimize MRI in evaluation of soft tissue tumors is beyond the scope of this chapter, a few basic concepts are essential. As a general rule it is essential to obtain an MRI in two orthogonal planes. The axial plane is usually optimal for evaluation, and both T1- and T2-weighted MRIs should be obtained. Technical factors and the appearance of different musculoskeletal tissues are listed in Table 2.1 for standard T1- and T2-weighted MRI. In general, T1-weighted images are used for anatomic delineation of structures (spatial resolution), whereas T2-weighted sequences are used to distinguish various types of tissue such as tumor versus normal or a fluid collection (contrast resolution). A second plane of imaging should also be performed either coronally (best for masses located medially or laterally in a compartment) or sagittally (best for masses located anteriorly or posteriorly in a compartment). Fat suppression T2-weighted sequences (essentially subtracting fat, making its signal null and black) are often employed in the second plane to increase the conspicuity of signal (high signal intensity – white in appearance) of tumors.[36–39]

Intravenous (IV) contrast material can be administered for either CT (iodinated material) or MRI (paramagnetic substance – gadolinium) in an attempt to improve the contrast resolution in evaluation of soft tissue tumors. Contrast is generally much more important for CT in differentiating soft tissue tumors from surrounding muscle owing to its inferior contrast resolution. In contradistinction, lesion detection and delineation are typically evaluated easily on MRI without IV contrast. An exception to this is encountered with neoplasms that are highly necrotic, hemorrhagic, or myxomatous (Figure 2.4). In these cases, IV contrast can be very informative in identification of areas of enhancing solid and cellular tissue (Figure 2.4).[40–51] This becomes vitally important in directing biopsy to those areas that harbor diagnostic tissue for histologic evaluation (as opposed to nondiagnostic regions of hemorrhage, necrosis, and nonspecific myxoid tissue). The use of IV contrast with MRI should be approached carefully because it is not without hazards, particularly with the advent

Chapter 2: Radiologic evaluation of soft tissue tumors

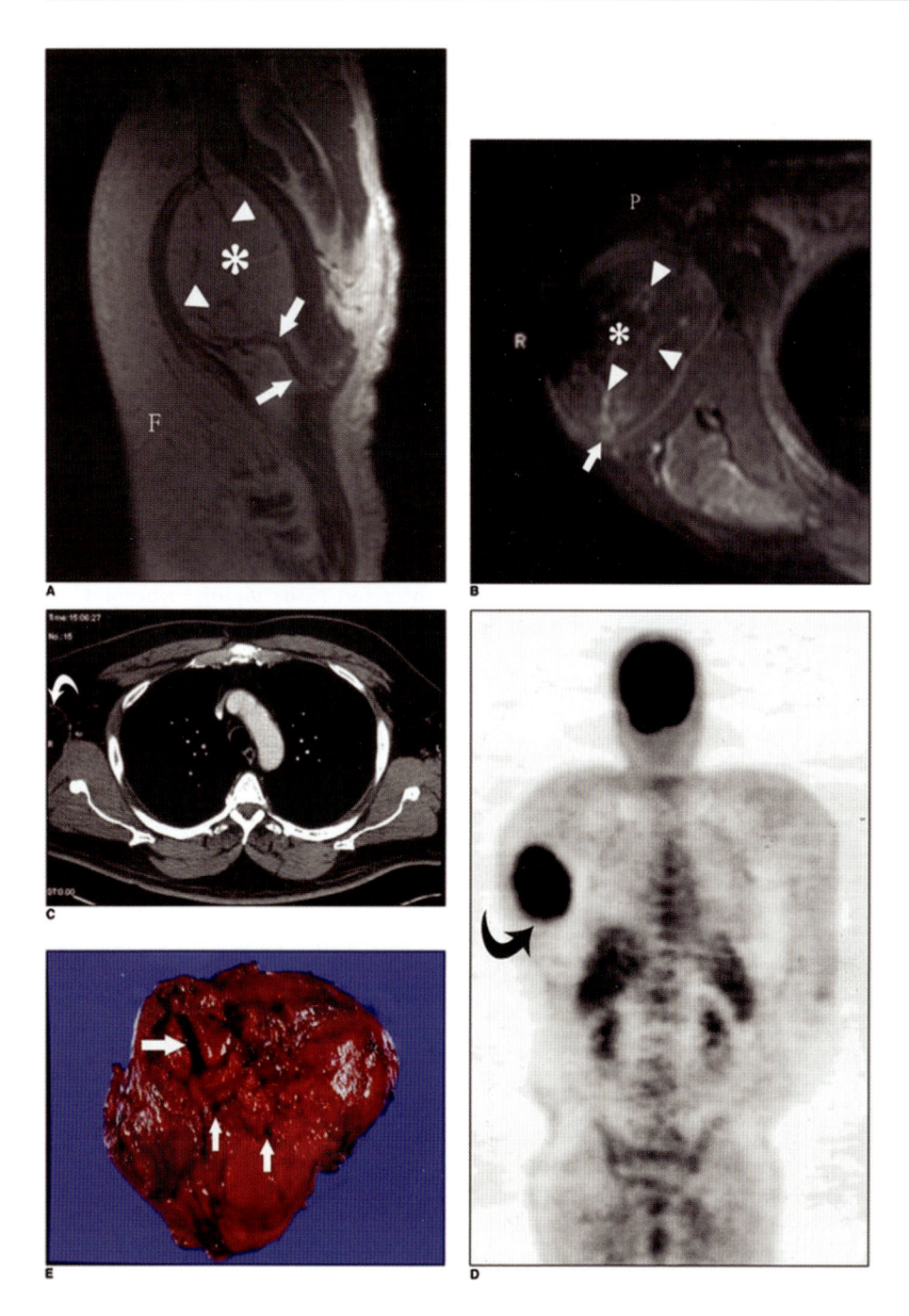

Figure 2.2 Hibernoma in a 63-year-old man with an upper arm soft tissue mass and PET imaging. (**A, B**) Sagittal T1-weighted (**A**) and axial fat-suppressed T2-weighted (**B**) MR images show a soft tissue mass (*) very similar in intrinsic appearance to to subcutaneous fat (F). Curvilinear nonadipose components (arrowheads) could be seen in a well-differentiated liposarcoma, although the serpentine nature suggests that these are vascular channels and there is a feeding vascular pedicle (arrows). (**C**) CT of the chest also reveals the lipomatous soft tissue mass (white curved arrow). (**D**) PET imaging reveals marked increased uptake of radionuclide (black curved arrow). (**E**) Photograph of sectioned gross specimen demonstrated brown fat and vascular channels (arrows) corresponding to the imaging appearance. These vascular channels and degree of PET avidity would not be seen in well-differentiated liposarcoma and are pathognomonic for hibernoma. This case emphasizes that PET measures glucose turnover, not benignity versus malignancy, as this benign lesion (hibernoma) has more intense activity than the low-grade malignant tumor in the differential diagnosis (well-differentiated liposarcoma) in this instance.

of nephrogenic systemic fibrosis associated with the use of gadolinium products in patients with chronic renal insufficiency.

Ultrasonography (US) is another imaging modality that can be used in the evaluation of soft tissue tumors. Advantages of sonography include its low cost, real-time scanning, lack of ionizing radiation, and no need for IV contrast. Lesions located subcutaneously are often best evaluated by US because of their superficial position.[52,53] Perhaps the most important use of US is in distinguishing a solid from a cystic mass.[54,55] This distinction is very important in differential diagnosis, and US is quite adept at identifying truly cystic structures, which are anechoic with posterior acoustic enhancement (Figure 2.5). These cystic lesions include ganglion, synovial cyst, bursa, and abscess (Figure 2.5). In contradistinction, US demonstrates the solid consistency with internal echoes of other soft tissue tumors, including myxomas and myxoid neoplasms (i.e., myxoid liposarcoma) which on CT and or standard MR sequences can simulate a cyst owing to the high intrinsic water content (Figure 2.6). Doppler US assessment can also be applied to evaluate lesion vascularity and response of neoplasms to preoperative chemotherapy and radiation therapy by depicting progressive necrosis and decrease in lesion size.[56]

Despite the many advantages of US, this modality also has many disadvantages. US is very operator dependent, and

Chapter 2: Radiologic evaluation of soft tissue tumors

Table 2.1 MRI signal intensity of various tissues

Tissue type	T1 signal intensity	T2 signal intensity
Fat	High	High
Bone marrow (yellow)	High	Intermediate
Tumor[a]	Intermediate	High
Muscle	Intermediate	Low
Hyaline cartilage	Intermediate	High
Water	Very low	Very high
Tendons/ligaments	Very low	Very low
Cortex	Very low	Very low
Fibrocartilage	Very low	Very low
Fibrous tissue	Low to intermediate	Variable[b]
Blood[c]	Variable	Variable

High = bright (white); intermediate = gray; low = dark (black).
[a] Tumor – majority of lesions.
[b] Highly collagenized low signal, more cellular high signal.
[c] Dependent on components of methemoglobin, hemosiderin, oxyhemoglobin, deoxyhemoglobin, frequently heterogeneous with areas of high signal on both T1- and T2-weighting except hemosiderin, which is low signal on all pulse sequences.

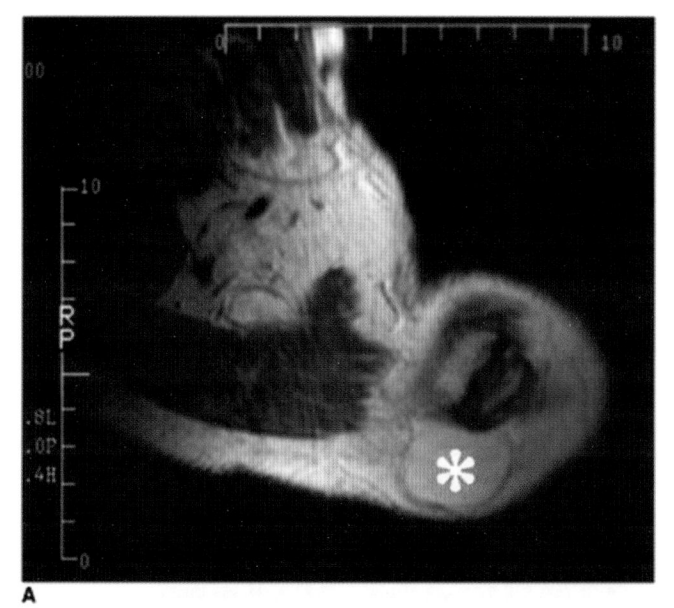

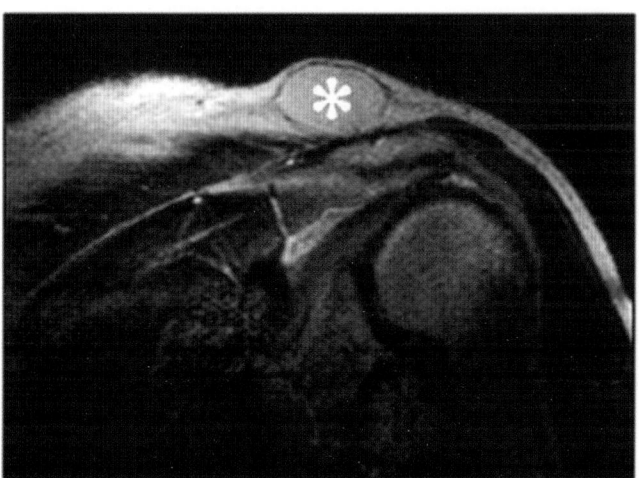

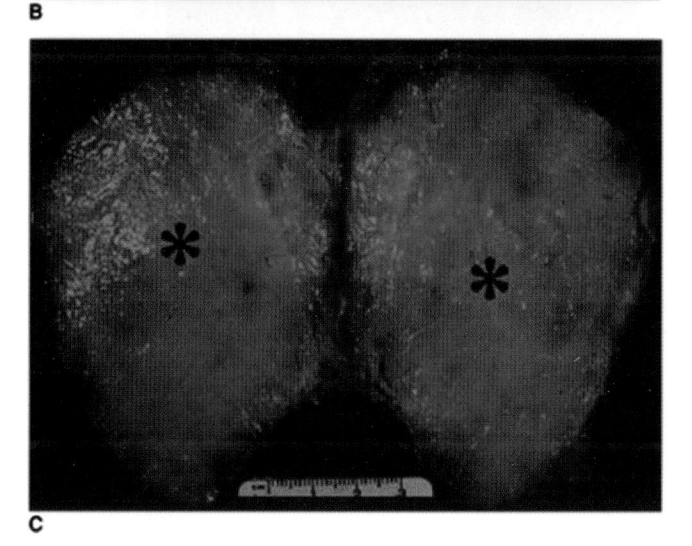

Figure 2.3 Subcutaneous lipoma of the shoulder in a 50-year-old woman. Axial T1- (**A**) and coronal T2-weighted (**B**) MR images show a subcutaneous lipoma (*) isointense to subcutaneous fat on both pulse sequences. Distinction from the surrounding fat is seen as a low signal peripheral pseudocapsule. (**C**) Photograph of gross specimen of a lipoma with homogeneous adipose tissue (*).

relatively unusual examinations, such as musculoskeletal tumors, often require physician scanning and a commitment of significant time and expertise. Additionally, the field of view can be limited, making large lesions more difficult to assess entirely compared with CT and MRI. Although US is quite adept at evaluating superficial lesions, deep-seated lesions, particularly in complex areas of anatomy such as the pelvis, can be obscured by overlying osseous structures. The contrast resolution of sonography is also inferior to that of MRI, particularly in detecting adipose tissue within lesions accurately and confidently.

Staging

There are several staging systems (see Table 2.2 and Chapter 1)[57–59] that are used in the evaluation of soft tissue tumors. All of these systems, however, share the need for a close working relationship among the orthopedic oncologist, musculoskeletal radiologist, and pathologist to stage lesions and guide treatment options appropriately. Limb salvage is the surgical treatment of choice for most soft tissue sarcomas.[60–63] Much of the information necessary, particularly the extent of the lesion and involvement of adjacent structures, for staging soft tissue tumors is obtained by imaging. Important features to assess include lesion extent (does the tumor cross a major fascial plane to involve more than one compartment?), size (is the lesion less than or greater than 5 cm?), and determining involvement of adjacent bone, joint, or neurovascular structures (Figure 2.7).[64] Historically, the Enneking staging system (also referred to as the

15

Chapter 2: Radiologic evaluation of soft tissue tumors

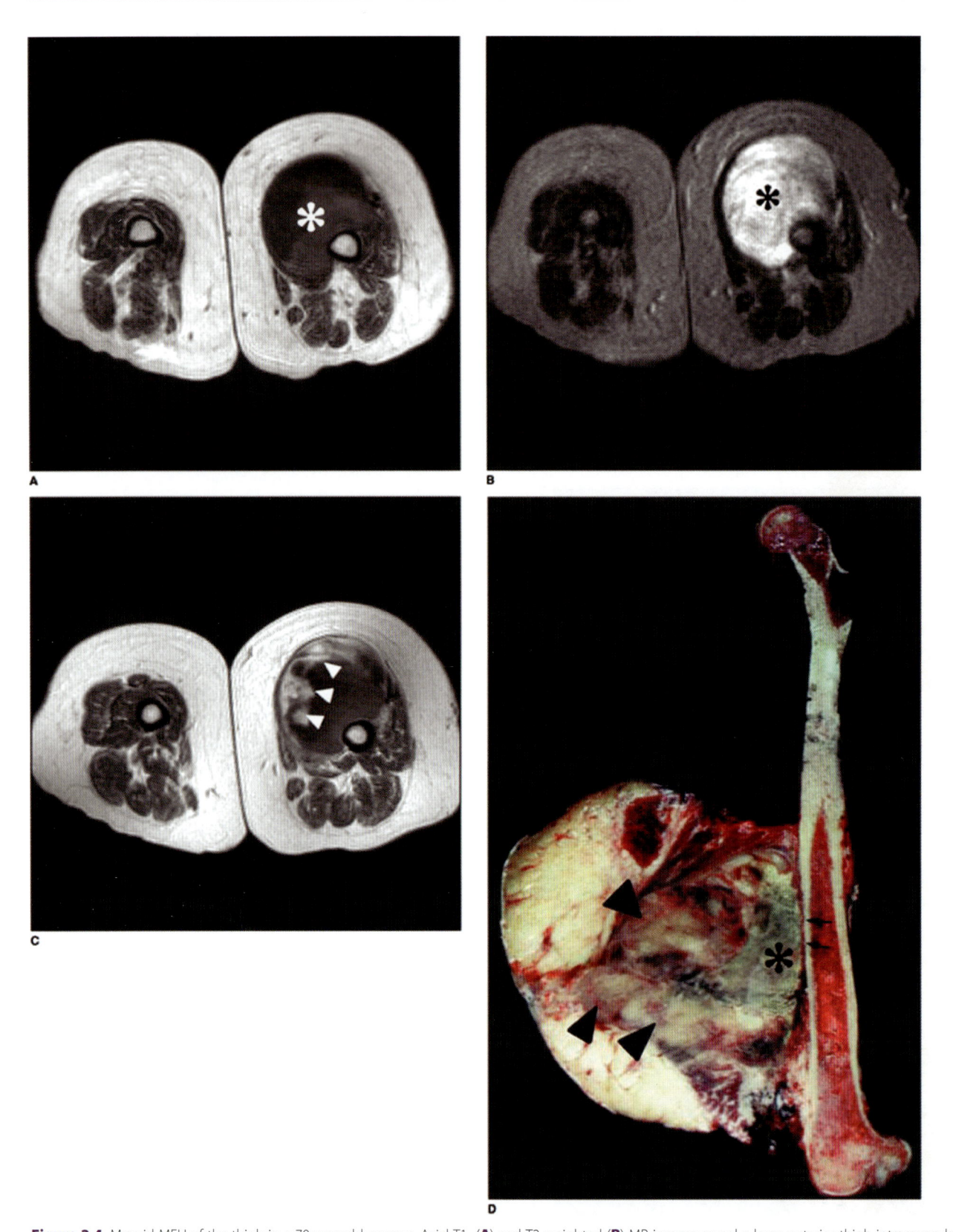

Figure 2.4 Myxoid MFH of the thigh in a 70-year-old woman. Axial T1- (**A**) and T2-weighted (**B**) MR images reveal a large anterior thigh intramuscular soft tissue mass (*). There is intermediate signal on T1-weighting (**A**) and high signal on T2-weighting (**B**). Postcontrast axial T1-weighting MR image (**C**) reveals nodular peripheral enhancement *(arrowheads)* in areas of more solid tissue. (**D**) Photograph of the gross specimen shows the higher water content myxoid central region (*) and solid periphery *(arrowheads)* and adjacent femur *(arrows)*. Biopsy should be directed toward these solid areas that harbor diagnostic pathologic tissue as opposed to more myxoid areas, and only the postcontrast image allows this distinction.

Musculoskeletal Tumor Society Staging System; Table 2.2) was used by orthopedic oncologists and is still referred to because it is better related to the type of surgical resection performed (wide versus radical excision).[50] Again, the multiplanar capabilities and superior contrast resolution of MRI make this modality the technique of choice in the assessment of most soft tissue tumors (particularly deep-seated lesions).

Lesion specificity by imaging

In the authors' opinion and experience, the two most important radiologic features in evaluation of soft tissue tumors for diagnosis are *lesion location* and *intrinsic imaging characteristics* (including lesion size, morphology, shape, and extent).[65] Patient age (Table 2.3) can also be a factor in limiting the differential diagnosis, but is much less important compared with evaluation of bone neoplasms (Figure 2.8).

Lesion location

Specific anatomic location (e.g., thigh) of various soft tissue tumors is available with cross-sectional imaging, although in the authors' opinion this information is less helpful compared with the compartment involved: subcutaneous, intermuscular, intramuscular, intra-articular/periarticular, and the identification of multiple or extensive lesions. This is analogous to evaluation of bone tumors (e.g., epiphysis, metaphysis, or diaphysis) and provides general differential diagnoses, since various neoplasms are more likely to occur in certain compartments. Common benign and malignant lesions in these various compartments are listed in Table 2.4. There are exceptions to this concept, such as a lesion deep to the scapular tip, which almost invariably represents elastofibroma.

Subcutaneous masses are extremely common clinically; however, they are relatively infrequently evaluated by imaging owing to the "ease" of clinical evaluation. In contradistinction, deep-seated soft tissue masses (i.e., intermuscular, intramuscular, or intra-articular) are usually imaged because of inadequate clinical assessment (Figure 2.3).

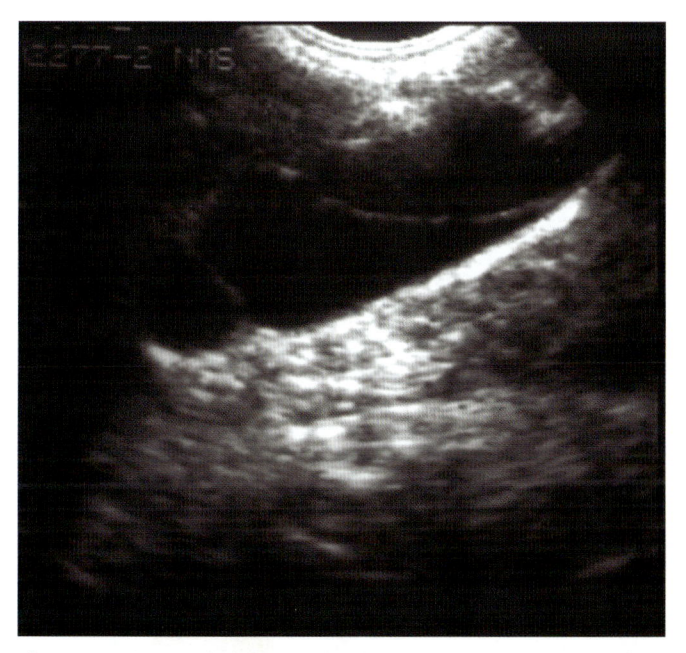

Figure 2.5 Ultrasound of a popliteal (Baker) cyst in the posterior knee of a 50-year-old woman. Sonogram shows an anechoic mass with posterior acoustic enhancement typical of a fluid-filled mass. Several thin septations are also seen.

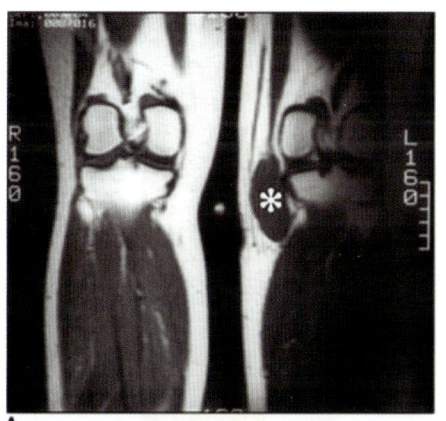

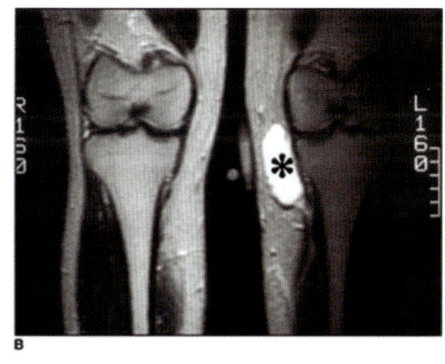

Figure 2.6 Myxoid liposarcoma medial to the knee in a 45-year-old woman simulating a cyst. Coronal T1-weighted (**A**) and T2-weighted (**B**) MR images show a soft tissue mass (*). There is homogeneous low signal on T1-weighting and high signal on T2-weighting, simulating a cyst, although the unusual location should raise doubts as to this diagnosis. The sonogram (**C**) clearly demonstrates the hypoechoic but solid (not anechoic – see Figure 2.5) internal consistency of the mass. (**D**) Photograph of resected myxoid liposarcoma demonstrates the high water content (*) accounting for the MR appearance.

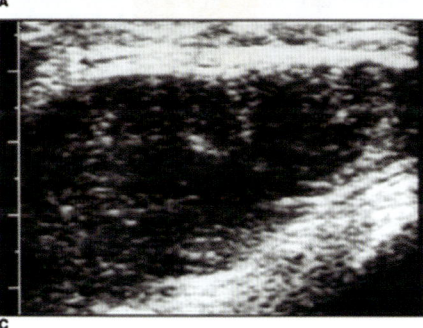

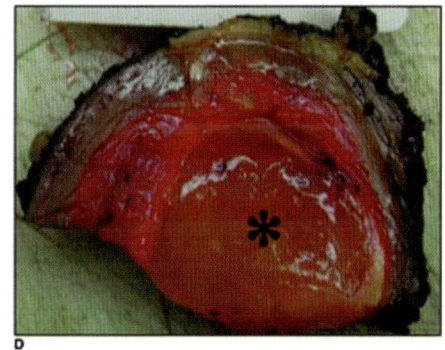

Chapter 2: Radiologic evaluation of soft tissue tumors

Table 2.2 Enneking staging of sarcomas of soft tissue and bone

Stage	Grade	Extent	Metastasis
IA	G1	T1	M0
IB	G1	T2	M0
IIA	G2	T1	M0
IIB	G2	T2	M0
III	G1–G2	T1	M1
	G1–G2	T2	M1

Histologic Grade: G1: low risk of metastasis, <25%, G2: high risk of metastasis, >25%.
Site: T1: intracompartmental, T2: extracompartmental.
Metastasis: M0: no regional or distant metastases, M1: regional or distant metastases.
Modified from Seeger et al.[50]

Intermuscular soft tissue masses are usually surrounded by a rim of fat ("split-fat" sign) with the surrounding musculature draped around the lesion (Figure 2.9). This is best depicted by MRI, but is also seen on US and CT. This situation occurs because the deep intermuscular tissue is primarily composed of fat, and soft tissue masses arising in this location maintain a rim of adipose tissue as they enlarge. In large lesions the intermuscular location is often best determined by evaluating the superior and inferior aspect of the tumor.

Soft tissue masses centered in muscle (intramuscular) replace the normal muscle texture signal intensity on MRI (Figure 2.10). These lesions reveal surrounding muscle and typically lack a fat rim unless they involve an entire compartment.

Soft tissue masses that arise within or immediately adjacent to a joint have a relatively limited differential diagnosis and are usually benign. Synovial sarcoma is the most frequent malignancy, although it rarely originates in a joint (5–10% of cases). Intra-articular lesions often diffusely involve the joint, as seen

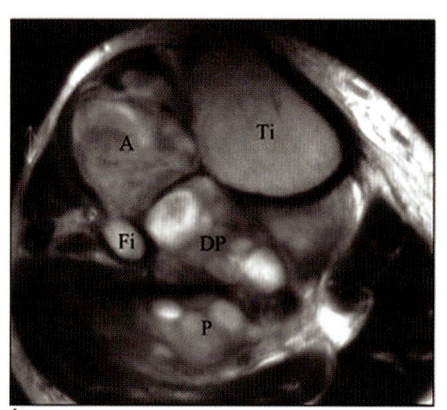

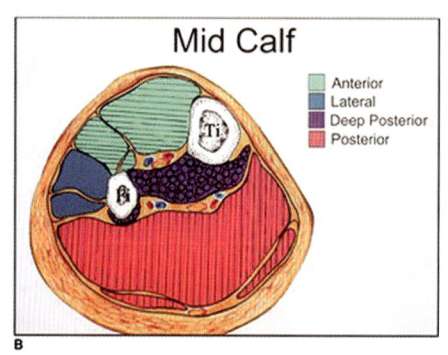

Figure 2.7 Synovial sarcoma of the calf in a 30-year-old man with involvement of multiple compartments and the neurovascular bundle leading to amputation as opposed to limb salvage. The axial T2-weighted MR image (**A**) shows the heterogeneous intermediate to high signal intensity soft tissue mass with involvement of the anterior [A], deep posterior [DP], and posterior [P] compartments of the calf and intervening neurovascular bundles on the MRI as well as the anatomic drawing (**B**) and photograph of the axially sectioned gross specimen (**C**). (Ti = tibia; Fi = fibula. Drawing by Dr. Annie Frazier, AFIP).

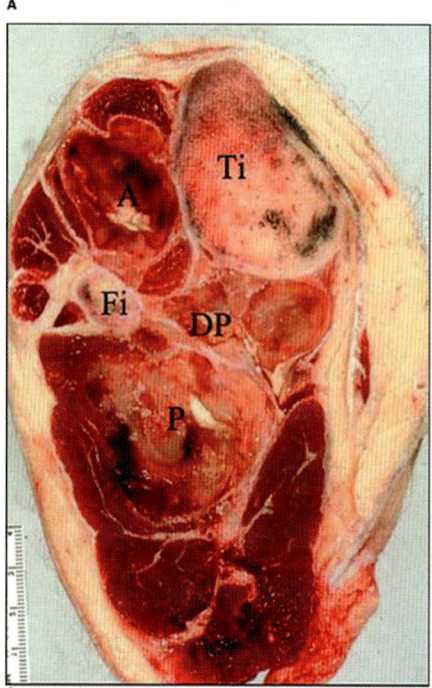

Table 2.3 Soft tissue tumor differential diagnosis by age[a]

CHILD (<16 YEARS)	
Benign	**Malignant**
Hemangioma	Fibrosarcoma/MFH
Fibromatosis	Synovial sarcoma
Fibrous histiocytoma	Rhabdomyosarcoma
Granuloma annulare	MPNST

YOUNG ADULT (16–45 YEARS)	
Benign	**Malignant**
Ganglion	MFH/Fibrosarcoma
Fibrous histiocytoma	Liposarcoma
nodular fasciitis	Dermatofibrosarcoma protuberans (DFSP)
Neurogenic neoplasm (neurilemmoma, neurofibroma)	Synovial sarcoma
Lipoma	MPNST
Hemangioma	

OLDER ADULT (46 YEARS AND OLDER)	
Benign	**Malignant**
Ganglion	MFH/fibrosarcoma
Lipoma	Liposarcoma
Neurogenic neoplasm	Leiomyosarcoma
Fibrous histiocytoma	MPNST
Nodular fasciitis	DFSP
Myxoma	

[a] From pathologic referral series AFIP. References (145, 146) in order of decreasing frequency.

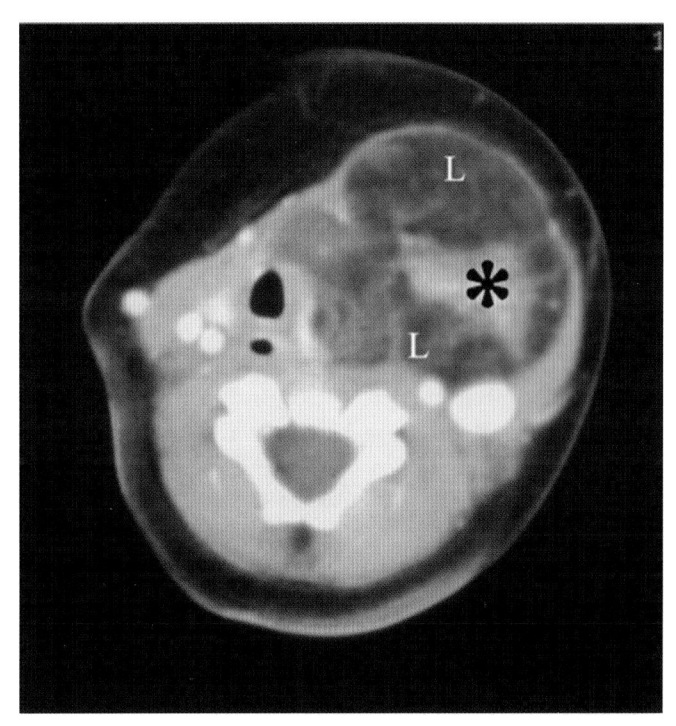

Figure 2.8 Lipoblastoma of the neck in a 12-month-old boy with an enlarging soft tissue mass. CT of the neck shows a heterogeneous soft tissue mass with both lipomatous (L) and solid components (*). In an adult this imaging appearance suggests the diagnosis of well-differentiated liposarcoma with possible dedifferentiation; however, in a young child this is an untenable diagnosis. Factoring the patient age along with imaging appearance allows the accurate diagnosis of lipoblastoma.

with pigmented villonodular synovitis (PVNS, Figure 2.11), synovial chondromatosis, and lipoma arborescens.

The detection of multiple soft tissue masses (or extensive/diffuse involvement) also markedly restricts the differential diagnosis and is generally associated with benign lesions, in contradistinction to other organ systems, where this finding suggests malignancy. Lipomas are multiple in 5% to 15% of patients. The fibromatoses and angiomatous lesions are multifocal in up to 20% of patients. Patients with type one neurofibromatosis (NF1) invariably reveal multiple neurogenic tumors (Figure 2.12), and a lesion that enlarges rapidly should be viewed as representing malignant degeneration to MPNST. Myxomas can be multiple in association with fibrous dysplasia (Mazabraud syndrome). Finally, metastases can be multifocal, particularly to the subcutaneous tissue (melanoma and breast), or rarely to muscle (bronchogenic).

Lesion intrinsic characteristics

Relatively specific diagnosis by imaging characteristics has been reported in between 20% and 50% of soft tissue tumors.[1–7] In the authors' opinion, this percentage will gradually increase with more experience and description of specific characteristics, particularly on MRI, although likely not beyond approximately 70% to 80% of soft tissue masses. The imaging characteristics that allow specific diagnosis include lesion location (e.g., lesion deep to scapular tip for elastofibroma; Figure 2.13), lesion shape or morphology (e.g., fusiform shape in neurogenic tumor with entering and exiting nerve), and intrinsic appearance (e.g., presence of fat for lipomatous lesions). Soft tissue masses that can frequently be diagnosed with imaging alone include lipomatous lesions, angiomatous lesions, neurogenic tumors, elastofibroma, intra-articular masses (PVNS, synovial chondromatosis), cystic lesions (ganglion, synovial cysts), and occasionally fibrous masses such as fibromatosis and nodular fasciitis.

Soft tissue masses with specific imaging appearances

Lipomatous lesions

Soft tissue tumors containing significant amounts of adipose tissue are usually easily detected on CT (low attenuation or density) or MRI (high signal on T1-weighting and intermediate signal on T2-weighting) because of their intrinsic appearance (at least focally), which is identical to subcutaneous fat. These lesions include lipoma, lipoblastoma, hibernoma, liposarcoma,

Chapter 2: Radiologic evaluation of soft tissue tumors

Table 2.4 Soft tissue mass evaluation: lesion location

Subcutaneous	Angiomatous lesions Benign fibrous histiocytoma Dermatofibrosarcoma protuberans Granuloma annulare Leiomyosarcoma Lipoma Lymphoma MFH/fibrosarcoma Metastasis (particularly melanoma) Myxoma Nodular fasciitis Skin appendage tumors
Intermuscular	Fibromatosis Ganglion Leiomyosarcoma Nodular fasciitis Neurogenic tumors Synovial cyst Synovial sarcoma Lipoma Liposarcoma (particularly myxoid type)
Intramuscular	Nodular fasciitis Angiomatous lesions Lipoma MFH/fibrosarcoma Myxoma Liposarcoma (well-differentiated/pleomorphic) Leiomyosarcoma/rhabdomyosarcoma Soft tissue Ewing sarcoma/PNET
Intra-articular/juxtaarticular	Lipoma arborescens Pigmented villonodular synovitis Synovial chondromatosis Giant cell tumor of tendon sheath Synovial cyst/bursa/ganglion Synovial hemangioma Tumoral calcinosis Synovial sarcoma

and hemangioma. Most of these lesions have a diagnostic appearance on imaging.

Imaging of lipomas often (11–22% of lesions) reveals a mass with homogeneous adipose tissue reflecting its monotonous pathologic appearance.[66–75] Intramuscular and intermuscular lesions are easily differentiated from the surrounding tissue (Figure 2.14).[66–68] Subcutaneous lipomas can be difficult to detect, however, because of their location in surrounding adipose tissue, with which they have an identical appearance (Figure 2.3). Identification of a thin surrounding pseudocapsule is necessary to distinguish these lesions from the background of subcutaneous fat (Figure 2.3). This capsule is of low signal intensity on all MR pulse sequences and of muscle density on CT. Nonlipomatous mesenchymal components in lipomas are occasionally seen, typically as small inconspicuous septa (37–49% of lipomas; Figure 2.14). A more complex

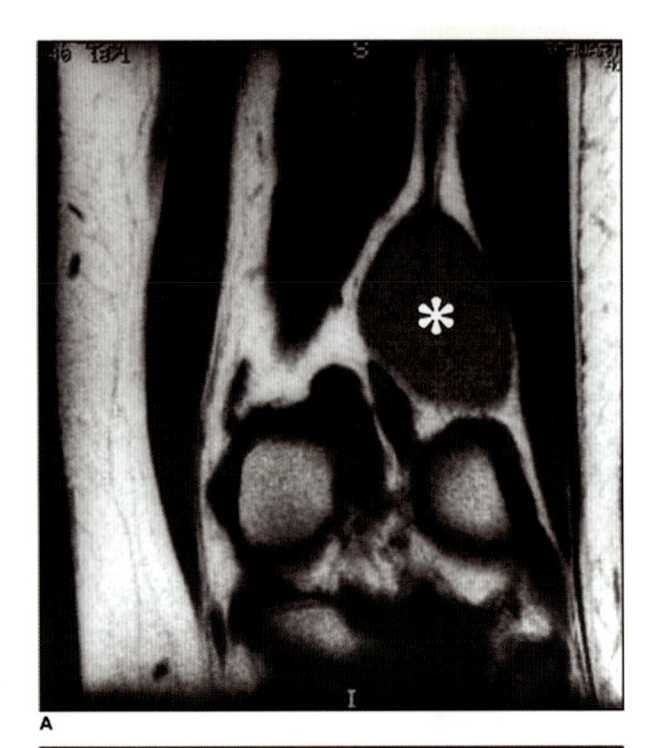

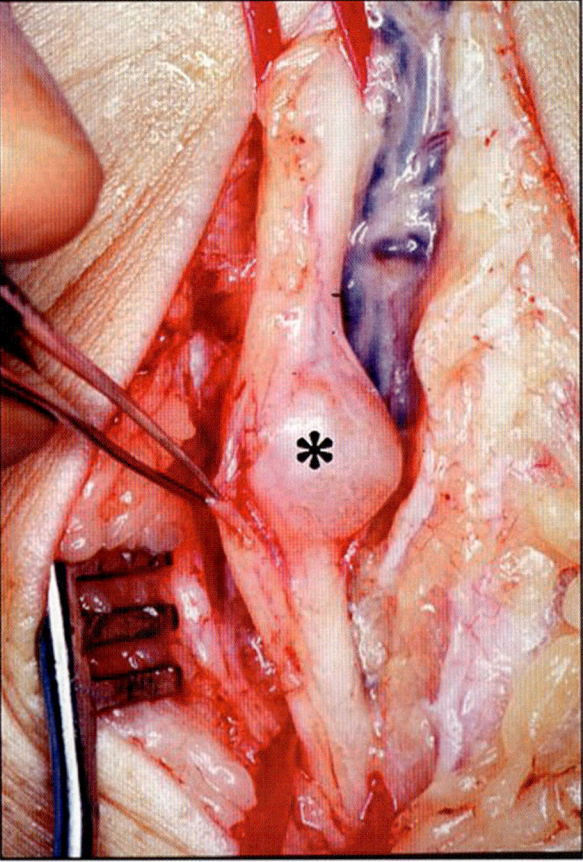

Figure 2.9 Neurilemmoma of the popliteal fossa in a 49-year-old woman. Coronal T1-weighted MR image (**A**) shows linear tube representing the sciatic nerve entering the mass (*), which is surrounded by fat (split-fat sign). Intraoperative photograph (**B**) reveals the entering and exiting nerve relationship to the mass (*) and the cause of the MRI appearance.

Chapter 2: Radiologic evaluation of soft tissue tumors

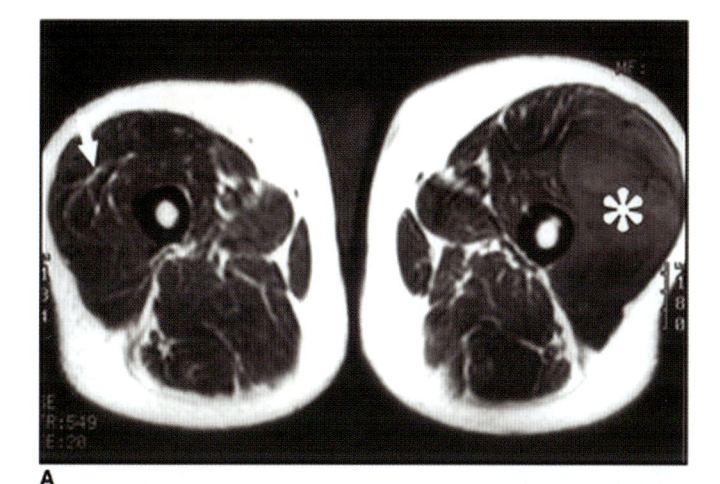

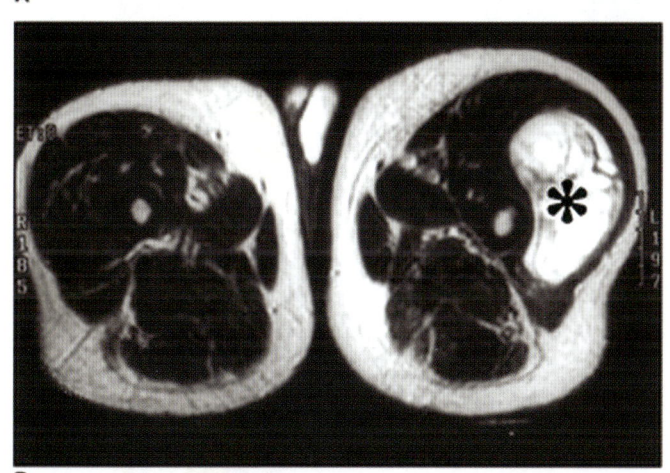

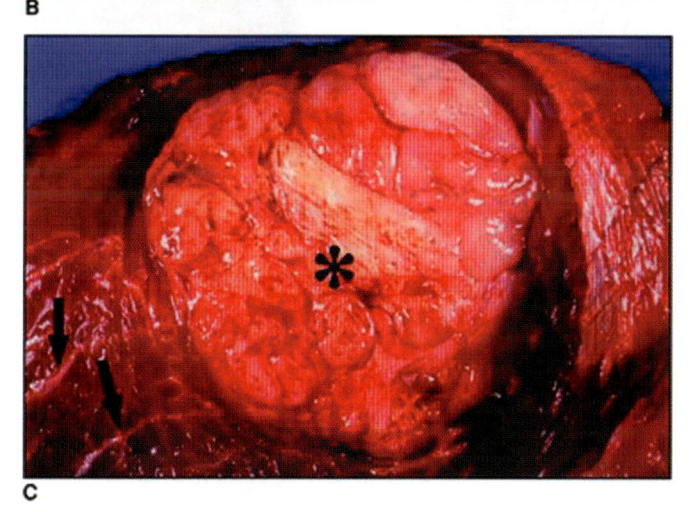

Figure 2.10 Intramuscular malignant fibrous histiocytoma (MFH) of the thigh in a 60-year-old man. Axial T1-weighted (**A**) and T2-weighted (**B**) MR images show the soft tissue mass (*) replacing the normal muscle texture seen on the contralateral side *(arrow)*. Photograph of a gross specimen (**C**) similarly reveals a large intramuscular MFH (*) replacing the normal muscle seen adjacent to the lesion (arrows).

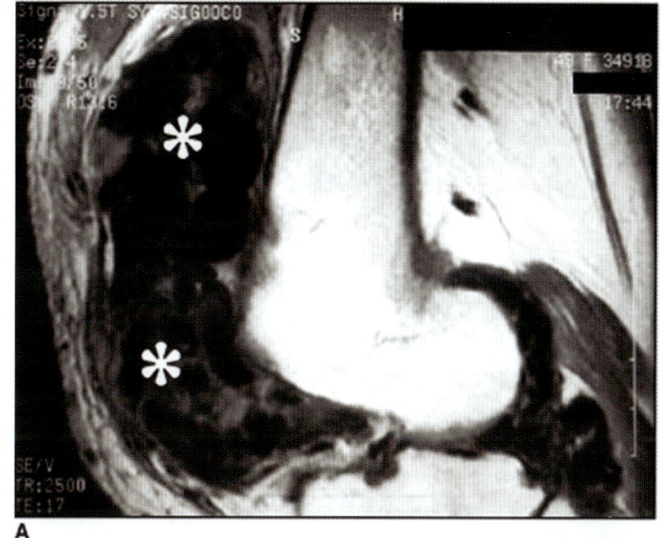

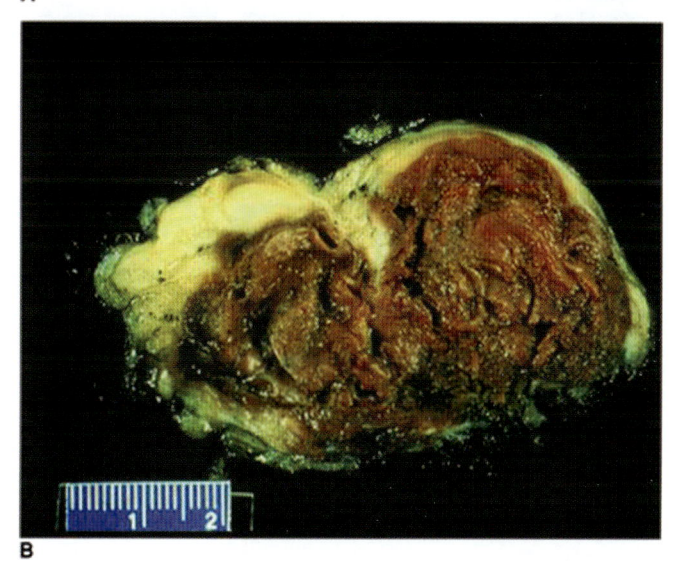

Figure 2.11 Intra-articular pigmented villonodular synovitis of the knee in a 34-year-old woman with joint pain. Sagittal T2-weighted MR image (**A**) shows extensive abnormal tissue in the knee joint (*) remaining low signal intensity as a result of hemosiderin deposition. Subtle bone erosion is also seen. Photograph of the gross specimen (**B**) reveals typical brown discoloration of the mass caused by the hemosiderin deposition, which also results in the characteristic low signal intensity on MRI.

imaging appearance of lipomas with significant nonlipomatous components is seen in 28% to 31% of lesions, and these are difficult to distinguish radiologically from well-differentiated liposarcomas. Contrast enhancement is only minimal in lipomas, usually in the nonlipomatous components. Lipomas can be multiple in 5% to 15% of cases and also can be infiltrative in cases of lipomatosis.[69–73,76]

Lipoblastomas and the diffusely infiltrative lipoblastomatosis usually reveal a predominantly fat-containing mass.[77,78] Particularly in very young patients, however, myxoid components may predominate, with only small elements of adipose tissue.[77,78] These largely nonlipomatous lesions can simulate liposarcoma by imaging, although the young age of the patient should make this diagnosis untenable (Figure 2.8).

Reflective of its pathology (composed of brown fat), the imaging of hibernoma images in most patients reveal tissue

21

Chapter 2: Radiologic evaluation of soft tissue tumors

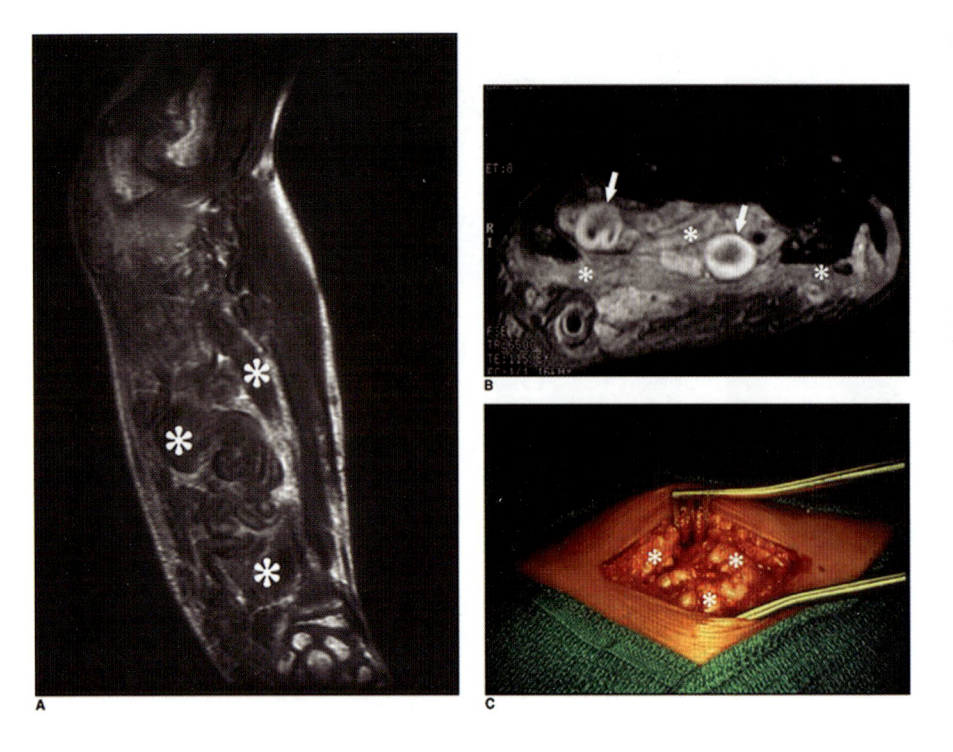

Figure 2.12 Extensive soft tissue mass involving the forearm in a patient with neurofibromatosis type 1 representing a plexiform neurofibroma. Sagittal T1-weighted (**A**) and axial T2-weighted (**B**) MR images show an extensively infiltrating soft tissue mass (*) of the forearm. On the T2-weighted MR image (**B**) there is low signal intensity centrally and high signal intensity peripherally ("target" sign – arrows) typical of these lesions. Photograph of the intraoperative appearance (**C**) of a plexiform neurofibroma also revealing the diffusely infiltrating serpentine soft tissue mass (*), which has been described as a "bag of worms" appearance.

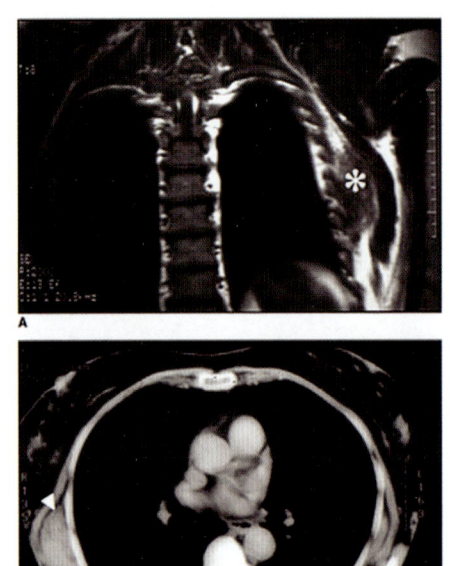

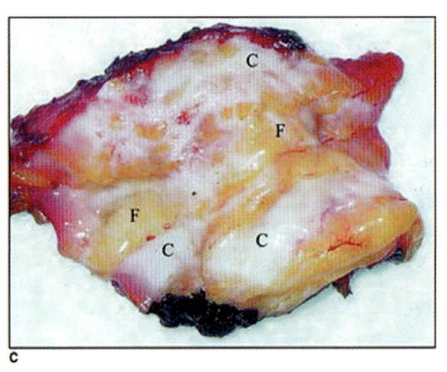

Figure 2.13 Elastofibroma in several different patients. Coronal T1-weighted MR image (**A**) shows a chest wall mass (*) with some intermixed higher signal intensity tissue representing fat. CT (**B**) reveals a soft tissue mass deep to the right scapular tip (arrowheads) and a smaller elastofibroma lesion on the left. Photograph of the gross specimen (**C**) demonstrates intermixed collagenized regions [C] and fatty elements [F].

similar but not identical to fat.[79–81] These lesions are invariably more vascular than lipomas and can reveal prominent enhancement after IV contrast by CT or MRI.[70] In addition and in contradisdinction from well-differentiated liposarcoma, a feeding vascular pedicle and internal serpentine vascular structures are frequently identified (Figure 2.2). Recently, PET imaging has shown intense radionuclide uptake in hibernoma that in the authors' opinion also allows differentiation of these lesions from well-differentiated liposarcoma (Figure 2.2).

Liposarcomas show a varying imaging appearance depending on their histologic subtype.[82–89] Well-differentiated liposarcomas typically reveal extensive areas identical to subcutaneous fat (>75% of the lesion volume) by CT and MRI (Figure 2.15).[82,83] In fact, differentiation from lipoma, similar to pathologic evaluation, can be difficult. The majority of well-differentiated liposarcomas (91–96% of lesions), however, demonstrate significant nonlipomatous components, particularly prominent septa both in number and thickness (96% of

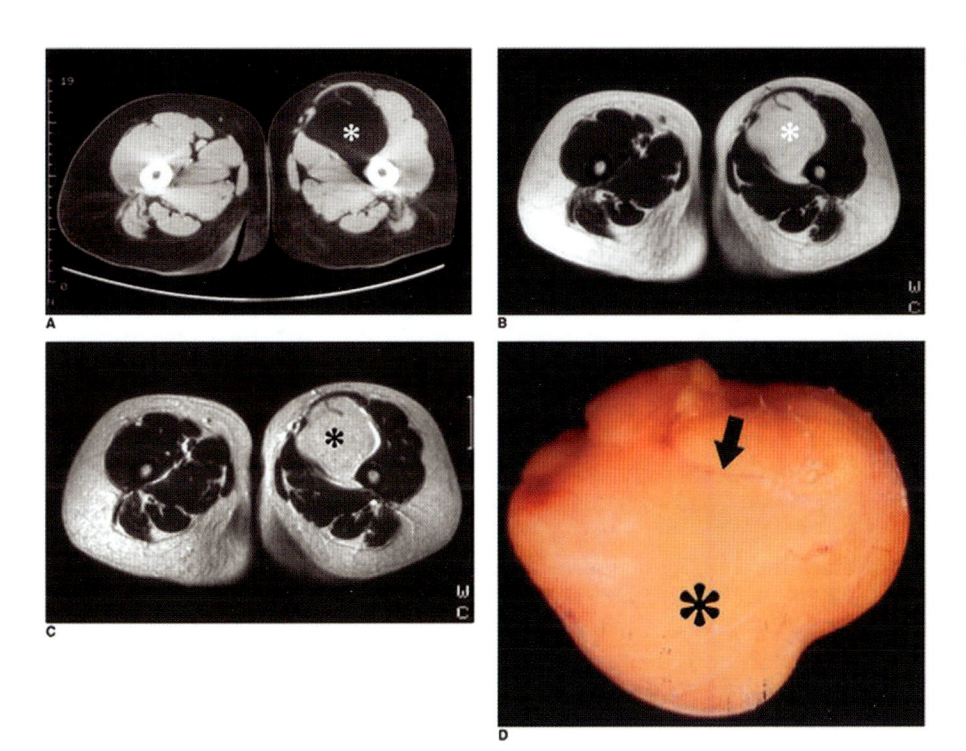

Chapter 2: Radiologic evaluation of soft tissue tumors

Figure 2.14 Lipoma of the thigh in a 54-year-old woman. All imaging studies including CT (**A**), T1-weighted MR (**B**), and T2-weighted MR (**C**) show the soft tissue mass (*) with intrinsic characteristics identical to subcutaneous fat. A single thin, small fibrous septation is seen anteriorly. Photograph of gross specimen (**D**) demonstrates the monotonous fat homogeneous adipose tissue (*) and thin septa *(arrow)*.

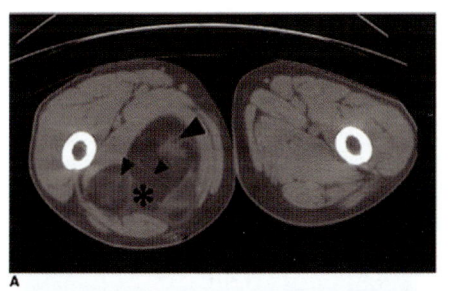

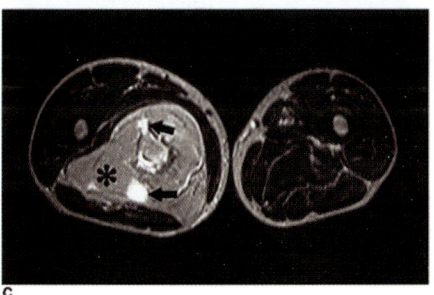

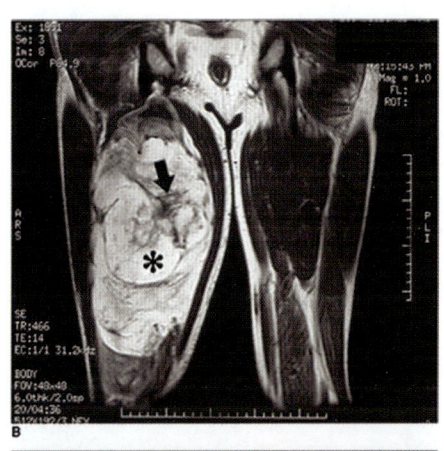

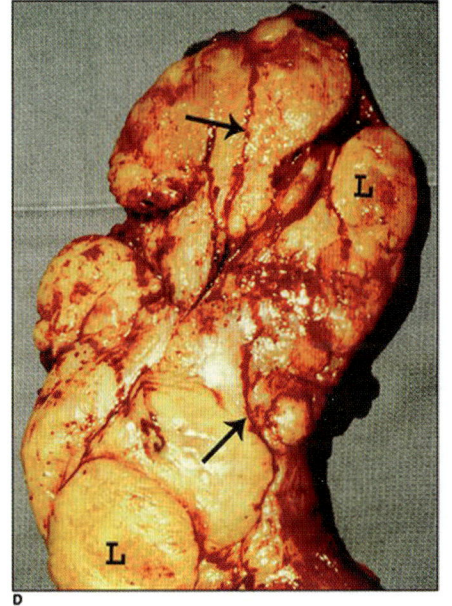

Figure 2.15 Well-differentiated liposarcoma of the thigh in a 68-year-old man. CT (**A**) shows a heterogeneous mass in the posterior thigh with areas of fat (*) and multiple septae and regions of soft tissue density *(arrowheads)*. Coronal T1-weighted (**B**) and axial T2-weighted (**C**) MR images also reveal the mass to contain large areas of fat (*) and prominent nonlipomatous nodules and septae *(arrows)*. Photograph of gross specimen (**D**) demonstrates the lipomatous component [L] and prominent thick septa *(arrows)*.

23

cases) or focal nodular regions (83% of cases) that allow distinction from lipoma.[82,83] Only 4% to 9% of well-differentiated liposarcomas simulate lipomas more closely, with thin septa in an otherwise diffusely fat-containing lesion. Focal areas of metaplastic mineralization (calcification or ossification) have been reported in up to 10% of cases.[22] Higher-grade liposarcomas (myxoid/round cell and pleomorphic), reflecting more anaplasia, reveal a less lipomatous imaging appearance.[84,85] In the authors' experience, however, focal areas of fat (usually <10% of the tumor by volume) are seen in the majority (approximately 90% of cases) of lesions by MRI (superior to CT), suggesting the diagnosis (Figure 2.16).[86] Myxoid liposarcomas are usually intermuscular lesions, and their high water content histologically is reflected in much of the lesion, showing a "cyst-like" appearance. In the authors' experience, 5% to 10% of cases of myxoid liposarcoma suggest an entirely cystic lesion on CT and MRI, although IV contrast demonstrates more diffuse or peripheral nodular enhancement, and US also reveals evidence of a solid, not cystic, mass (Figure 2.5).[86–89] Round cell components within a myxoid liposarcoma are sometimes suggested by focal, often nodular, areas that are not as high signal intensity on T2-weighted MR images and that enhance diffusely. Areas suggesting possible round cell components should be biopsied because of the effect on treatment and prognosis. Pleomorphic liposarcomas, although high-grade lesions, also usually demonstrate focal areas of identifiable fat, particularly by MRI (62–75%).[82, 87–89]

Dedifferentiated liposarcomas usually arise in an otherwise well-differentiated liposarcoma and are most common in the retroperitoneum.[84] These lesions are often quite large, with only a relatively small focus of dedifferentiation. Imaging by CT or MRI typically depicts this small focus of nonlipomatous solid tissue as distinctly different from the surrounding adipose tissue (Figure 2.17).[87–91] These imaging features allow the biopsy to be directed to the dedifferentiated focus as well as to lipomatous regions. Similarly, biopsy, either percutaneous or open, can be directed by imaging in other high-grade liposarcomas (myxoid/round cell, or pleomorphic) to more lipomatous or nonlipomatous solid areas of tumor and away from nondiagnostic regions of hemorrhage or necrosis. This is extremely helpful, particularly in large heterogeneous tumors, to improve the ease and accuracy of pathologic

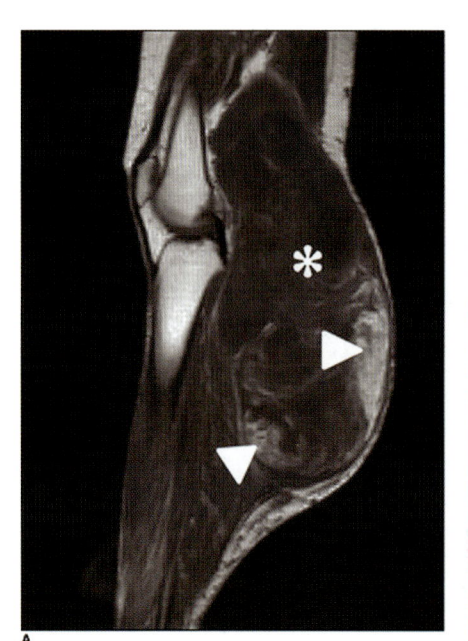

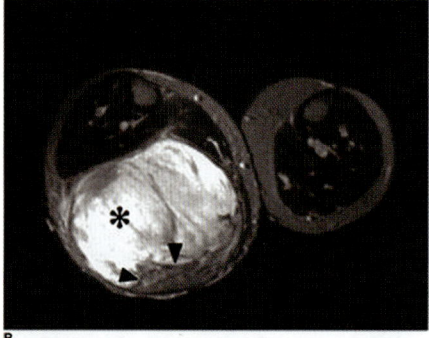

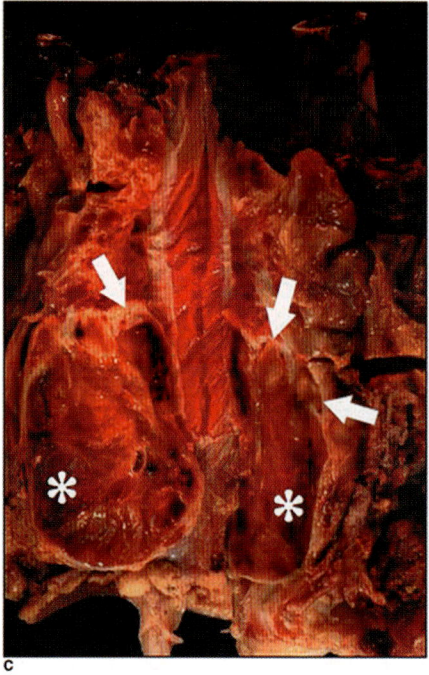

Figure 2.16 Myxoid liposarcoma of the popliteal fossa in a 60-year-old woman. Sagittal T1-weighted (**A**) and axial T2-weighted (**B**) MR images show a large soft tissue mass with a high water content myxoid component (*) and other areas containing fat *(arrowheads)*. The myxoid regions are low signal on T1-weighting and high signal on T2-weighting, whereas the fat areas are isointense on both pulse sequences to subcutaneous adipose tissue. Photograph of sectioned gross specimen (**C**) demonstrates the prominent high water myxoid component (*) and the small lipomatous component *(arrows)*.

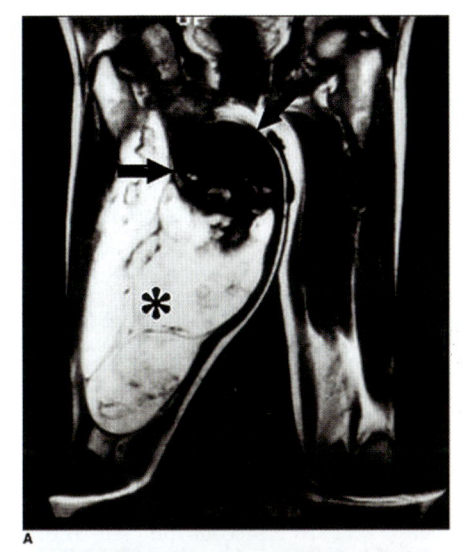

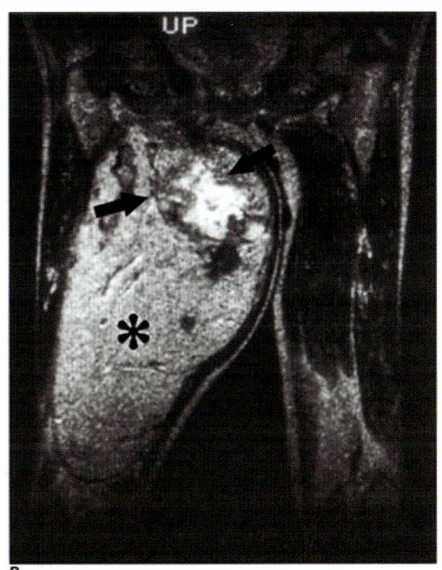

Figure 2.17 Dedifferentiated liposarcoma of the thigh in a 65-year-old woman with two differing solid components of tumor requiring biopsy. Coronal T1-weighted (**A**) and T2-weighted (**B**) MR images show a large soft tissue mass in the thigh with two solid components. The larger inferior component (*) is composed of adipose tissue with thick septa consistent with well-differentiated liposarcoma. The smaller superior component differs in signal intensity *(arrows)*. Photograph of the sectioned gross specimen (**C**) reveals these features as well, with the well-differentiated liposarcomatous components (left *) and the high-grade component (right *). Biopsy of only the superior component would have led to inappropriate initial treatment.

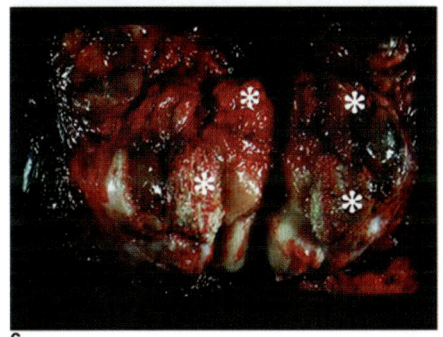

diagnosis and certainly has implications in patient management and prognosis.

Angiomatous lesions

Angiomatous lesions include hemangioma, lymphangioma, angiomatosis (and angiomatous syndromes), and the more aggressive vascular neoplasms: hemangioendothelioma, hemangiopericytoma, and angiosarcoma. Hemangioma is the most common angiomatous lesion, accounting for 7% of all benign soft tissue tumors.[8,22] Hemangioma represents the most common neoplasm in infancy and childhood. It has been estimated that 1.5% of the general population and 10% of children under the age of 1 year have a hemangioma.[92] This discussion uses the broader definition of hemangioma to represent any lesion composed of normal or abnormal vascular channels or spaces, with the acknowledgment that many investigators consider these lesions vascular malformations. Imaging of a hemangioma depends significantly on the histologic type of lesions, specifically, the size of the vascular channels composing most of the lesion (capillary, cavernous, arteriovenous, or venous). Radiographs can be normal, reveal a nonspecific soft tissue mass, or show characteristic calcification in the form of phleboliths (particularly in cavernous lesions, 30–50%).[22,93] Nonspecific calcification can also be seen. Bone scintigraphy frequently is normal or reveals only mild increase uptake of radionuclide. Angiography commonly shows enlarged feeding vessels and staining. Venous hemangiomas can require venography or direct puncture with injection of contrast for detection. US examination demonstrates large vascular channels and spaces (cavernous and venous lesions) with low vascular resistance and forward flow in both systole and diastole.[94,95] Conversely, high-flow arteriovenous lesions reveal a high vascular resistance arterial Doppler signal.

CT evaluation requires IV contrast for optimal detection of enhancing serpentine vascular channels and spaces (Figure 2.18).[95,96] Areas of calcification are usually easily detected on radiographs. Very small phleboliths or location in a complex area of anatomy with overlying osseous structures (e.g., the pelvis) can obscure these diagnostic calcifications on radiographs, however, as opposed to CT, which allows identification.

MRI of hemangioma is frequently diagnostic in appearance. In fact, in our opinion and experience, hemangiomas often reveal pathognomonic characteristics that make biopsy unnecessary unless deemed a requirement prior to surgical resection. Cavernous hemangiomas are most frequently evaluated radiologically because they present as nonspecific palpable deep soft tissue masses and are usually intramuscular.[92,97–100] MRI signal intensity is typically very heterogeneous on all pulse sequences, related to the intermixture of fat and vascular components (Figure 2.19). Recognition of the morphologic

Chapter 2: Radiologic evaluation of soft tissue tumors

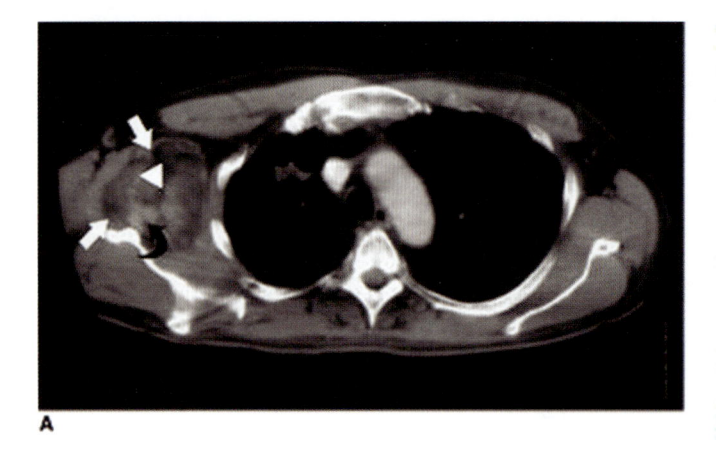

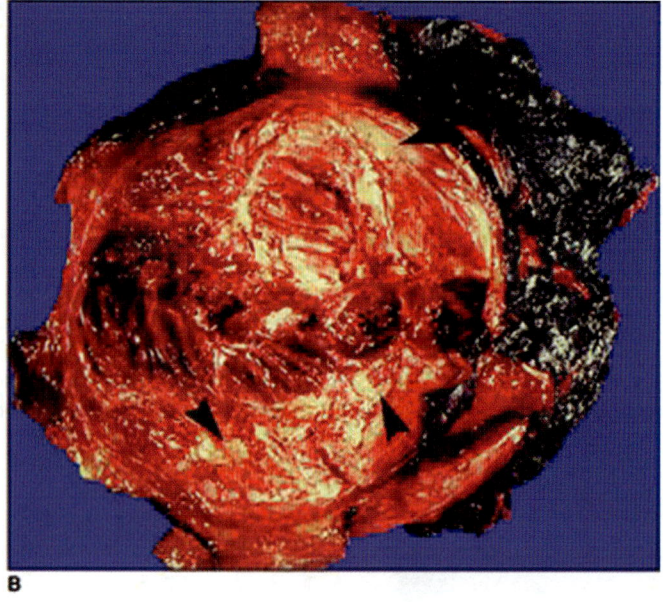

Figure 2.18 Intramuscular cavernous hemangioma about the shoulder in a 43-year-old man. CT (**A**) after intravenous contrast shows the heterogeneous mass with serpentine *(arrowhead)* and circular *(curved arrow)* enhancing regions representing cavernous vascular structures and low-density periphery *(arrows)* caused by secondary fat atrophy of muscle. Photograph of sectioned gross specimen (**B**) reveals the large cavernous vascular spaces and the surrounding muscle fat atrophy *(arrowheads)*.

shape of the vascular elements with serpentine vascular channels and spaces, as well as the fat overgrowth in adjacent muscle, are the pathognomonic features of these lesions by imaging (Figure 2.19). In the authors' experience, the presence of fat in muscle adjacent to the vascular channels and spaces represents atrophy resulting from chronic ischemia.[93] This essentially represents a vascular steal phenomenon, and of interest is a clinical history of pain with exercise in these patients, likely corresponding to this occurrence as well. In the authors' experience, visualization of one or both of these MRI features is present in between 90% and 95% of intramuscular hemangiomas.[101]

The vascular channels or spaces in cavernous hemangiomas are low to intermediate intensity on T1-weighting and become very high signal on T2-weighted MRI, reflecting the slow blood flow in these lesions (Figure 2.19). The vascular channels and spaces appear circular when seen *en face*, as opposed to linear and often serpentine when depicted longitudinally. This often creates multiple circular foci of high signal on T2-weighted MRI, often referred to as the "spot and dot" appearance (Figure 2.19). Phleboliths remain low signal intensity on all MRI pulse sequences, but are better identified by radiographs or CT (Figure 2.19).

Arteriovenous hemangiomas (often referred to as *arteriovenous malformations*) demonstrate high or low flow in the serpentine vascular channels and are generally devoid of larger cavernous spaces, which affects the imaging appearance. High-flow lesions reveal persistent low signal intensity on all MRI pulse sequences owing to the lack of mobile protons in rapidly moving blood. Doppler US shows these vascular structures as high-resistance vessels.[92] Low-flow arteriovenous hemangiomas show similar intrinsic characteristics on MRI and US as seen with cavernous lesions. Perhaps the most important feature in distinguishing high-flow arteriovenous lesions is in the recognition of the potential for bleeding during biopsy or surgical resection. It can be impossible to obtain vascular stasis after biopsy of deep-seated lesions such as in the retroperitoneum, and cases of death by exsanguination have been reported.

Capillary hemangiomas are probably the least common to be imaged, although clinically they are the most frequently encountered (1 in 200 births).[8,22] These lesions are often located in the superficial subcutaneous tissues and are easily evaluated clinically. The diagnosis is usually obvious and is related to the apparent bluish skin discoloration or "strawberry" nevus of a patient in the first few years of life. In addition, 75% to 90% of juvenile capillary hemangiomas involute by age seven, and therefore patients are not typically surgical candidates.[8,22,93] For these reasons, as with other subcutaneous lesions, imaging is infrequently performed. Of note, the imaging appearance of these hemangiomas is often nonspecific, reflecting their underlying histology. In the authors' opinion and experience, the high degree of cellularity and small size of the vascular channels accounts for the nonspecific imaging appearance of a soft tissue mass seen on CT and MRI in capillary hemangiomas (Figure 2.20). Fat overgrowth is usually not identified in these superficial lesions. The extent of the capillary hemangioma and effect on adjacent structures is well evaluated by imaging, which fortunately is the clinical question because the diagnosis is generally already known, as previously discussed.

Hemangiomas involving the synovium are rare, accounting for fewer than 1% of all lesions.[8] The imaging appearance is

Chapter 2: Radiologic evaluation of soft tissue tumors

identical to that already described for other hemangiomas. In addition, however, repetitive episodes of hemarthrosis can result in associated synovitis and subsequent joint destruction. The appearance of the synovial thickening can simulate that of hemophiliac arthropathy, except that synovial hemangioma is a monoarticular disease. On MRI the synovial thickening remains low signal intensity on all pulse sequences, identical to pigmented PVNS because of associated hemosiderin deposition.

Lymphangiomas (or lymphatic malformations) are generally cavernous lesions present at birth or discovered in the first two years of life (50–90%) affecting the head, neck, or axilla.[8,22] Imaging by CT, MRI, or US reveals these large cystic spaces as either unilocular but more frequently multilocular (Figure 2.21). Admixture with a hemangiomatous component or hemorrhage can be associated with a more complex appearance resulting from bleeding and recognized on imaging as fluid levels.

Angiomatosis and angiomatous syndromes (e.g., Maffucci, Osler–Weber–Rendu, Klippel–Trenaunay–Weber) represent diffuse infiltration of either one tissue plane extensively (bone, subcutaneous, muscle, or viscera) or more commonly multiple tissue planes by hemangiomatous or lymphangiomatous lesions (Figure 2.22). Individual lesions appear identical to those previously described for solitary lesions but are much more extensive and infiltrative. Imaging is frequently performed to evaluate visceral involvement, which when present is associated with a worsened prognosis.[22,93]

The more aggressive vascular lesions hemangioendothelioma, hemangiopericytoma, and angiosarcoma are uncommon lesions and frequently show a nonspecific imaging appearance. Identification of a soft tissue mass with recognizable areas of serpentine vascular channels with high flow should suggest the diagnosis of a more aggressive vascular neoplasm (Figure 2.23). Unlike hemangiomas, these lesions do not demonstrate fat overgrowth, are not composed of only vascular channels/spaces, and are usually centered in an intermuscular location (not intramuscular, as with hemangiomas). Angiosarcomas arising in chronic lymphedema can reveal diagnostic imaging features of subcutaneous edema, extremity enlargement, and development of a focal plaque-like skin and subcutaneous mass representing the malignant transformation. Interestingly, most other soft tissue sarcomas do not generally reveal identifiable vascular channels on MRI. In the authors' opinion this reflects the small size of the vascular channels feeding most soft tissue sarcomas below the spatial resolution of MRI. The differential diagnosis of a soft tissue lesion with prominent identifiable vascular channels includes synovial sarcoma (unusual pattern), alveolar soft part sarcoma, hibernoma, rhabdomyosarcoma, and extraskeletal Ewing sarcoma or primitive neuroectodermal tumor (PNET), in addition to the aggressive vascular lesions.[22]

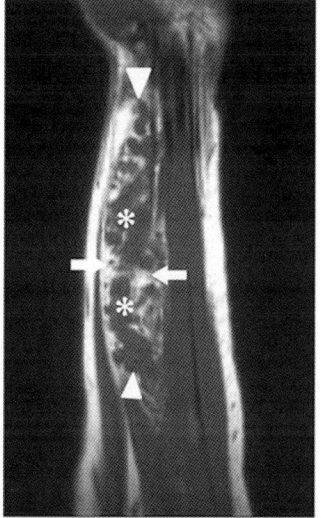

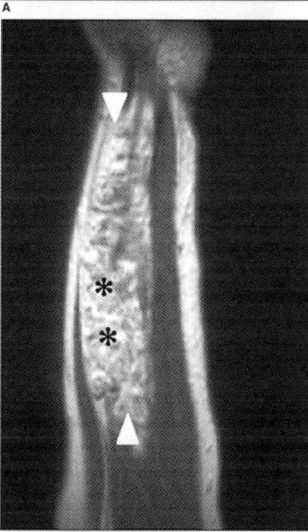

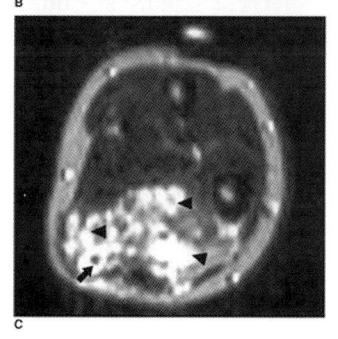

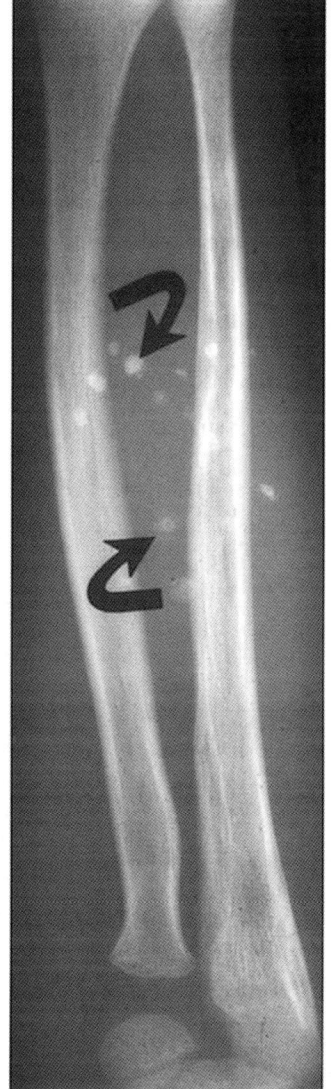

Figure 2.19 Intramuscular cavernous hemangioma of the forearm in a 15-year-old girl. Sagittal T1-weighted MR images before (**A**) and after (**B**) intravenous contrast show a large heterogeneous volar compartment mass (*large arrowheads*) composed of low signal serpentine vascular channels and spaces (white *) with associated fat overgrowth (*arrows*). Enhancement of these vascular areas are seen after contrast (B – black *). The T2-weighted MR image (**C**) reveals high signal in the slow-flowing circular vascular channels and spaces (*small arrowheads*), whereas fat becomes lower signal and phlebolith is low signal (*arrow*), resulting in the "spot and dot" appearance. Radiograph of the forearm (**D**) more obviously demonstrates the phleboliths (*curved arrows*) characteristic of this diagnosis.

Neurogenic tumors

Neurogenic tumors include traumatic neuroma, Morton neuroma, neurilemmoma (schwannoma), neurofibroma, and

Chapter 2: Radiologic evaluation of soft tissue tumors

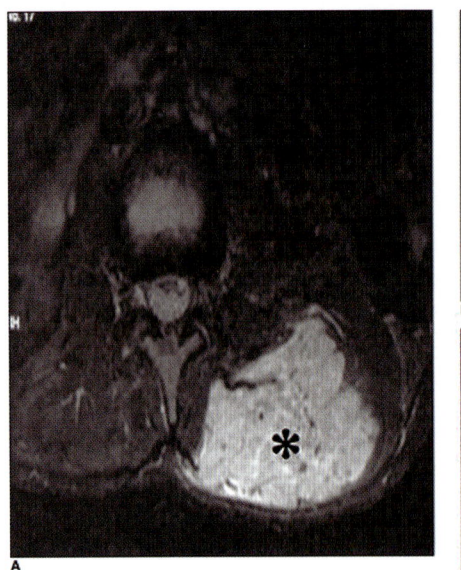

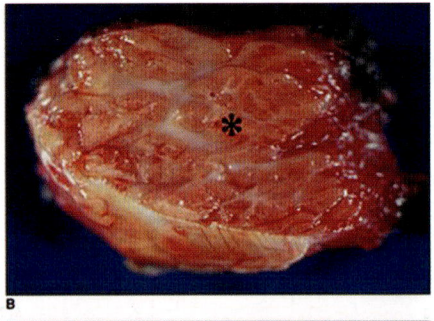

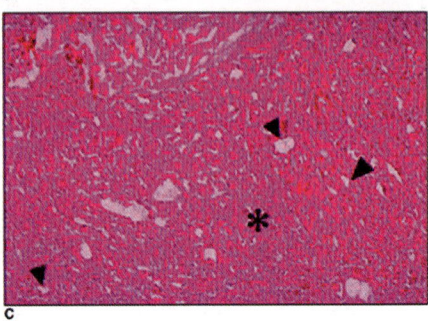

Figure 2.20 Intramuscular capillary hemangioma of the paraspinal musculature in a 6-year-old girl. Axial T2-weighted MR image (**A**) shows a large heterogeneous soft tissue mass (*) with nonspecific features. There are no recognizable vascular channels/spaces or fat overgrowth typical of cavernous hemangiomas as seen in Figure 2.19. Photograph of sectioned gross specimen (**B**) and photomicrograph (**C**) (hematoxylin and eosin stain; 175 ×) also reveal the soft tissue mass (*) without large vascular spaces. Histology demonstrates a largely cellular lesion with only small capillary-sized vessels *(arrowheads)* that cannot be discerned by imaging, leading to the nonspecific radiologic appearance.

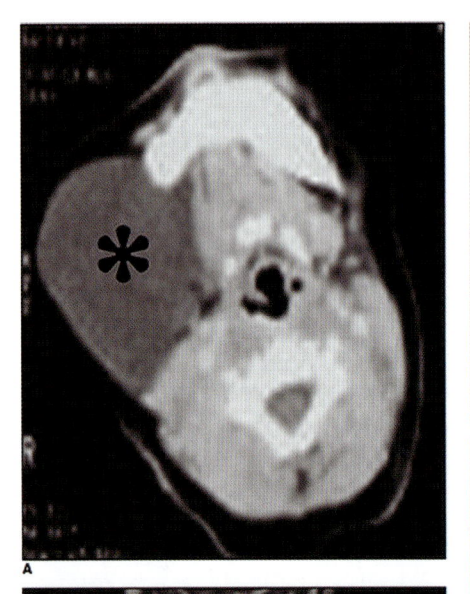

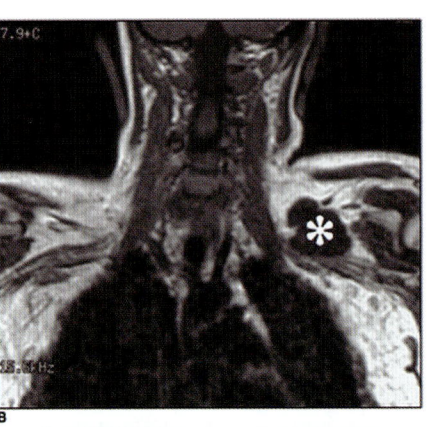

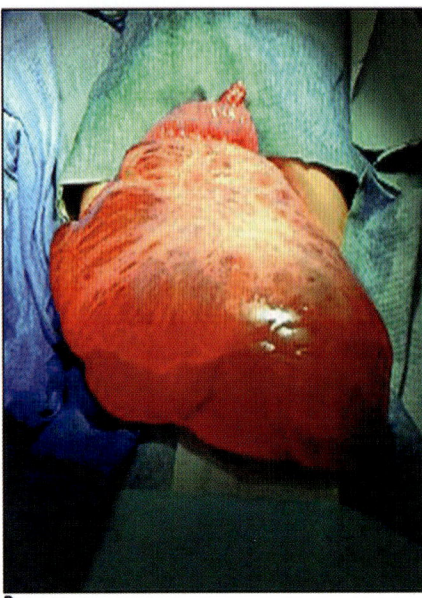

Figure 2.21 Lymphangioma (lymphatic malformation) of the neck in several different patients. CT (**A**) shows a low attenuation high water content soft tissue mass (*). Coronal T1-weighted (**B**) and T2-weighted (**C**) MR images also reveal similar features of the soft tissue mass (*) with homogeneous high signal intensity on the T2-weighted image. Photograph (**D**) of a gross specimen demonstrates the multilocular cystic morphology of a lymphangioma.

Chapter 2: Radiologic evaluation of soft tissue tumors

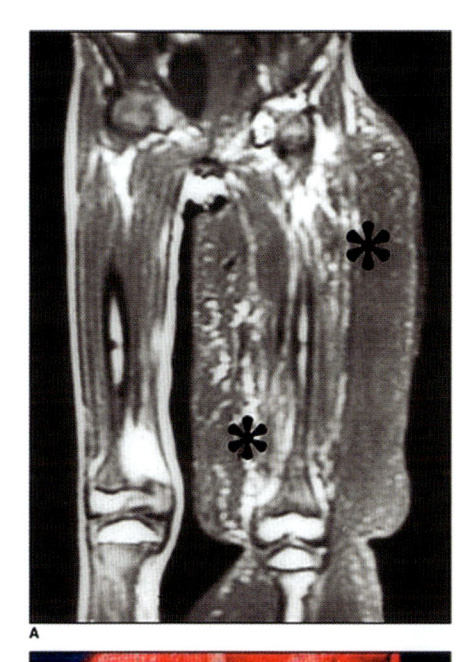

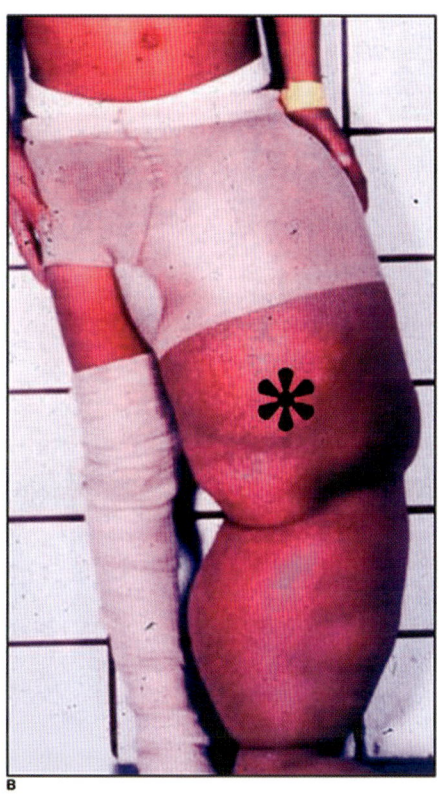

Figure 2.22 Angiomatosis is largely lymphangiomatosis of the entire lower extremity, resulting in elephantiasis. Coronal T1-weighted (**A**) MR image shows extensively enlarged and infiltrated lower extremity (*). Identical features are seen in the clinical photograph (**B**) and the sectioned gross specimen photograph (**C**) with extensively enlarged and infiltrated lower extremity (*).

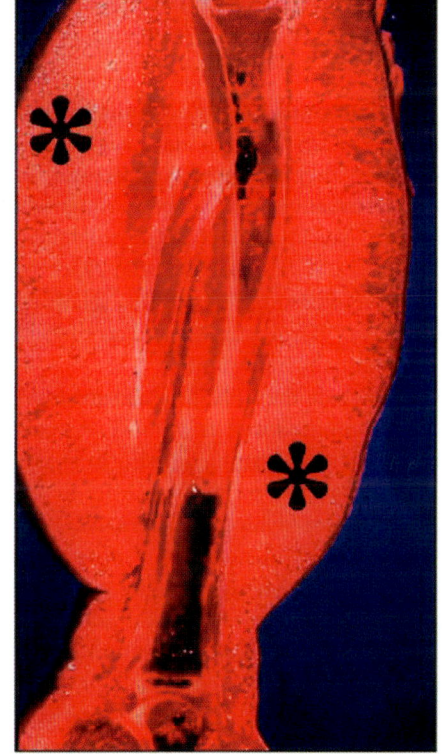

MPNST. Advanced imaging (CT, US, or MRI) of these lesions is frequently diagnostic. Important features are typically either location (Morton or traumatic neuroma) or characteristic morphology and relationship to a known nerve distribution (BPNST or MPNST).

Radiographs of neurogenic tumors are usually normal or reveal only a nonspecific soft tissue mass. Calcification of the lesions is unusual. Bone overgrowth is rarely associated with neurogenic neoplasms, particularly in patients with NF1.[102]

Similar to radiographs, bone scintigraphy and angiography usually reveal nonspecific features in evaluation of neurogenic tumors. PNSTs generally show mild increased radionuclide uptake on bone scintigraphy. Angiography reveals a variable degree of staining with neurogenic neoplasms, although MPNST typically shows more prominent vascularity.[22,102]

Advanced imaging (US, CT, or MRI) more clearly depicts the location and intrinsic characteristics of these lesions. Neurogenic tumors are often fusiform, representing the

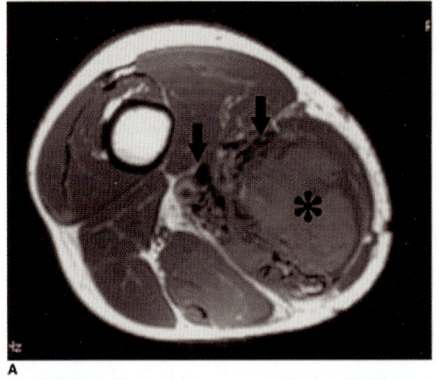

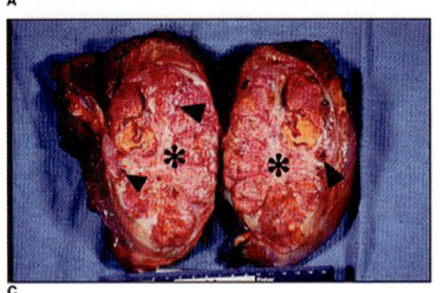

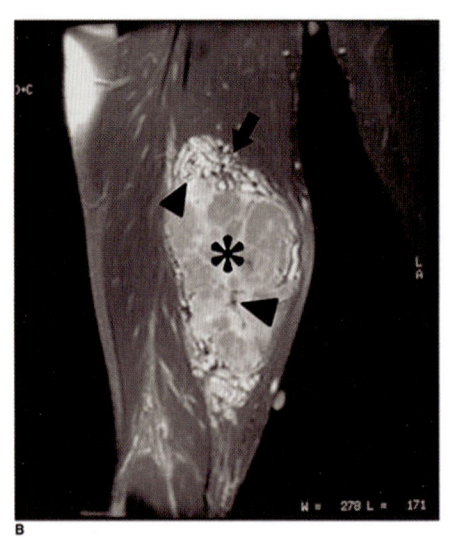

Figure 2.23 Hemangiopericytoma of the thigh in a 25-year-old man. Axial T1-weighted (**A**) and coronal T2-weighted (**B**) MR images show a large soft tissue mass (*) with nonspecific intrinsic signal characteristics. There is intermediate signal intensity on T1-weighting and intermediate to high signal intensity on T2-weighting. Serpentine high-flow vascular structures are noted both as feeding vessels *(arrows)* and internal vascularity *(arrowheads)*, suggesting the diagnosis, and no fat overgrowth or slow-flow vascular channels are seen, as expected for a hemangioma. Identical features are seen in the photograph of the sectioned and bivalved gross specimen (**C**) of the soft tissue mass (*) and vascular structures (arrowheads).

entering and exiting nerve at the lesion margins (Figures 2.9 and 2.12). This morphologic appearance can be seen well on US, CT, or MRI and should always suggest the diagnosis of a neurogenic tumor because it is unusual in other soft tissue masses.[102–107] MRI and US are often superior to CT because of their improved contrast and spatial resolution (Figures 2.2, 2.9, and 2.12).

Traumatic neuromas are usually seen in association with amputation (traumatic or surgical) and are typically painful. A palpable mass is also sometimes present. Imaging of traumatic neuromas of larger nerve trunks typically reveals a thickened tubular structure ending in a bulbous mass *(terminal neuroma)*.[108,109] No exiting nerve is seen in cases related to amputation. In contradistinction, traumatic neuromas related to chronic irritation (spindle neuromas) reveal focal nerve thickening with both the entering and exiting nerve. Traumatic neuromas of small superficial nerves can show entirely nonspecific imaging findings of only a round or oval soft tissue mass. US might be more adept at detecting the associated nerve in these superficial lesions or those involving small nerves.

Morton neuromas are usually diagnosed because of their characteristic location. These nonneoplastic masses are related to the plantar digital nerve between the third and fourth metatarsals, or less commonly, the second and third metatarsals.[8,22] MRI and US are the best radiologic methods to detect Morton neuroma and demonstrate small focal masses centered in the neurovascular bundle location (Figure 2.24).[110–112] The small round or oval mass represents perineural fibrosis of the plantar nerve, which typically cannot be seen well normally owing to its small size. Morton neuromas that are smaller than 5 mm can be difficult to detect on US.[111] Lesions are low to intermediate signal on T1-weighted MRI and unlike many other tumors do not typically show high signal on T2-weighting. For this reason they are more conspicuous on T1-weighted MRI. Lesions usually enhance, often intensely following IV MR contrast. Associated intermetatarsal bursal fluid is also a common feature. It has recently been reported that asymptomatic Morton neuromas may be relatively common (30% in a group of 70 volunteers), and as always, clinical correlation to symptoms is important.[112]

Neurogenic neoplasms include the BPNST neurilemmoma (schwannoma) and neurofibroma, as well as MPNST. As previously discussed, these lesions demonstrate a fusiform shape, with the entering and exiting nerve best seen on MRI or US particularly when large nerves are involved (Figures 2.9, 2.12, and 2.25). Other imaging features of neurogenic neoplasms include low attenuation on CT, target sign (MRI), fascicular sign (MRI), split-fat sign (CT, MRI; Figure 2.9), and associated muscle atrophy (CT, MRI).[102–107] The low attenuation (density) of these lesions on CT is likely related to a high water content or the lipid contained in the myelin.[103] The target sign on T2-weighted MRI represents low signal intensity centrally (fibrous tissue with higher collagen content) and high signal intensity peripherally (more myxoid areas with higher water content) (Figure 2.12).[102,105] This sign is suggestive of a neurogenic neoplasm and is more common with neurofibroma. The fascicular sign is seen on MRI (best on proton density or T2-weighted images) as multiple ring-like structures in the lesion.[102] In the authors' opinion, this corresponds to the fascicular morphology seen pathologically in neurogenic neoplasms. The *split-fat sign* represents a rim of fat around neurogenic neoplasms as a result of these lesions arising from an intermuscular location about the neurovascular bundle.[102]

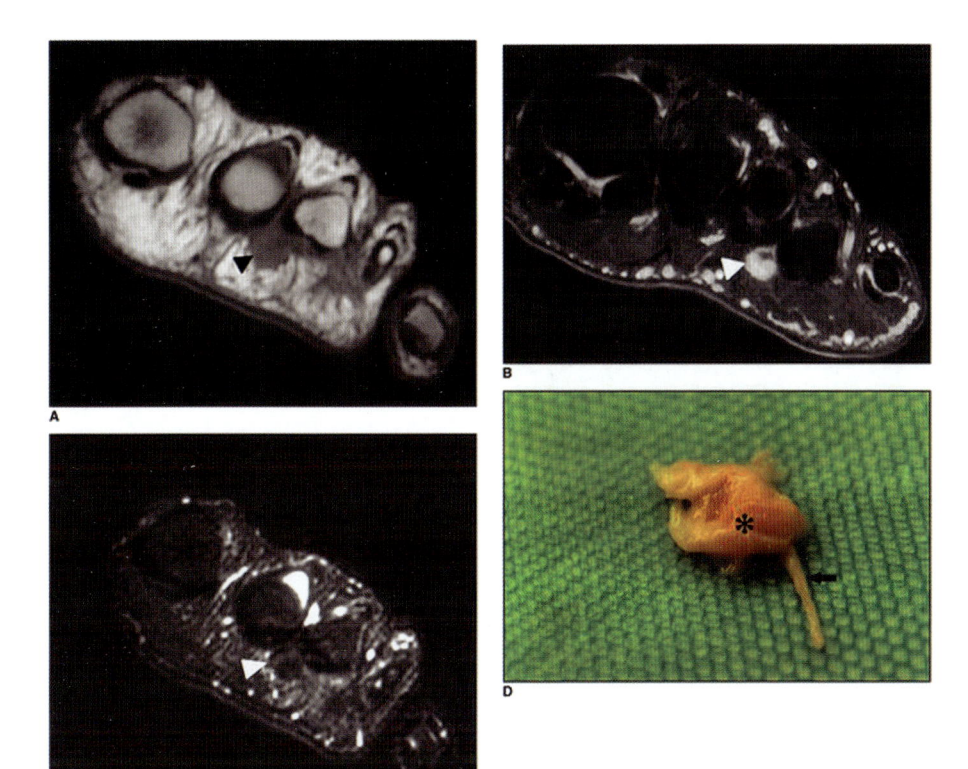

Figure 2.24 Morton neuroma of the foot in several patients. Coronal T1-weighted precontrast (**A**), postcontrast (**B**), and T2-weighted (**C**) MR images show a small intermetatarsal soft tissue mass *(arrowheads)*. The lesion is intermediate in signal intensity on T1-weighting, diffusely enhances with contrast, and is low signal intensity on T2-weighting. There is a small associated intermetatarsal bursa. Photograph (**D**) of the gross specimen reveals the Morton neuroma (*) encasing the plantar digital nerve *(arrow)*.

The neurovascular bundle is normally surrounded by fat, and as these neoplasms slowly enlarge, a rim of fat is maintained about the lesion, creating this imaging appearance, best seen by MRI followed by CT. Finally, associated fat atrophy of surrounding muscle is seen in up to 23% of cases related to nerve involvement and is relatively uncommon in association with other soft tissue masses.[104]

Differentiation of neurilemmoma (nerve eccentric to the mass) and neurofibroma should be possible (nerve inseparable from the mass) by the relationship of the soft tissue mass to the entering and exiting nerve. The authors agree with Cerofolini and coworkers that these lesions can be differentiated in nearly 70% of cases, particularly in lesions affecting large nerves.[106]

Neurofibromas can also be multiple or plexiform in association with NF1 (Figure 2.12). The imaging of plexiform neurofibromas identically reflects their pathologic appearance of a serpentine "bag of worms."[8,22,102] There is diffuse nerve thickening and nodularity, and the target sign is frequently seen on T2-weighted MRI. Diffuse neurofibromas typically involve the subcutaneous tissue and demonstrate a reticulated branching appearance along the connective tissue septa or plaque-like growth on MRI or CT.[113,114]

The distinction of BPNST from MPNST by imaging is often difficult or impossible. Imaging features that suggest MPNST include size greater than 5 cm, infiltrative margins, central necrosis, rapid growth, and increased uptake on gallium nuclear medicine studies (Figures 2.9 and 2.25).[22,102] In addition, MPNSTs less commonly reveal the target sign, fascicular sign, and split-fat sign, reflecting their higher degree of anaplasia and more aggressive growth. Neurilemmomas are one of the few benign neoplasms that reveal central necrosis on imaging; these lesions are often termed *ancient schwannomas*.[92] Recently FDG PET and PET methionine studies have been advocated as modalities that could allow more accurate distinction of these lesions with significantly higher SUV numbers in MPNST arising in NF1 patients.[115]

"Cystic" masses

Soft tissue masses that typically have an imaging appearance of cysts include synovial cyst, bursal fluid collection, ganglion, perilabral or meniscal cyst, hematoma, abscess, and myxomatous neoplasms. These lesions all can have a similar intrinsic appearance on noncontrast CT or MRI. CT shows low attenuation, and MRI reveals low signal intensity on T1-weighting and very high signal on T2-weighting. US demonstrates an anechoic lesion with posterior acoustic enhancement in truly cystic masses (Figure 2.5). Clinical history and lesion location are very helpful in distinguishing these lesions.

Synovial cysts represent an extension of the joint, usually a result of chronic effusion from mechanical internal derangement or arthropathy (particularly rheumatoid arthritis).[116,117] These lesions are lined by synovium, and the most common site is the popliteal cyst (Baker cyst). A *Baker cyst* is typically easily diagnosed by its location and by identifying the narrow neck of communication to the joint from the

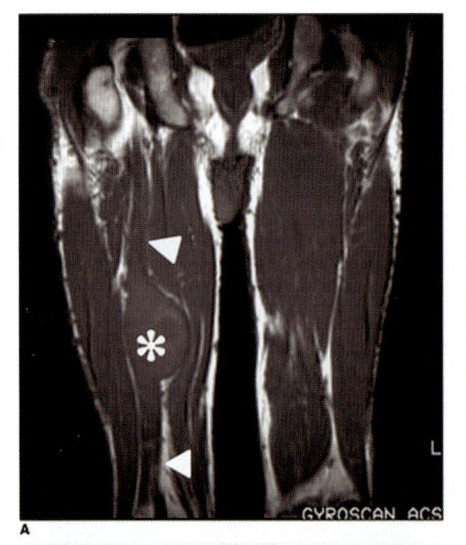

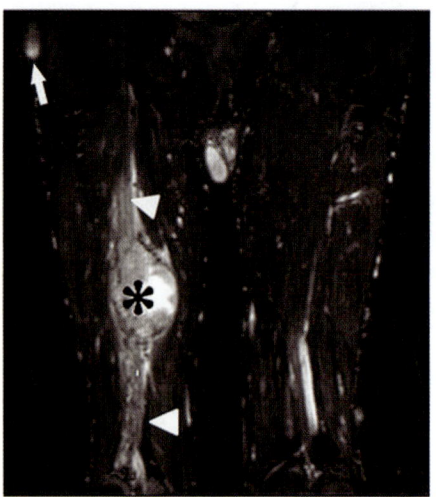

Figure 2.25 MPNST arising in a plexiform neurofibroma of the sciatic nerve in a 42-year-old man with type I neurofibromatosis and recent rapidly enlarging thigh mass. Coronal T1-weighted (**A**) and fat-suppressed (STIR) T2-weighted (**B**) MR images show a fusiform mass (*) with marked thickening of the entering and exiting sciatic nerve plexiform neurofibroma (arrowheads). A second subcutaneous neurofibroma is seen superiorly (arrow). Gross specimen (**C**) reveals an identical appearance, with the thickened plexiform nerve including the sciatic branches (curved arrows) and the MPNST.

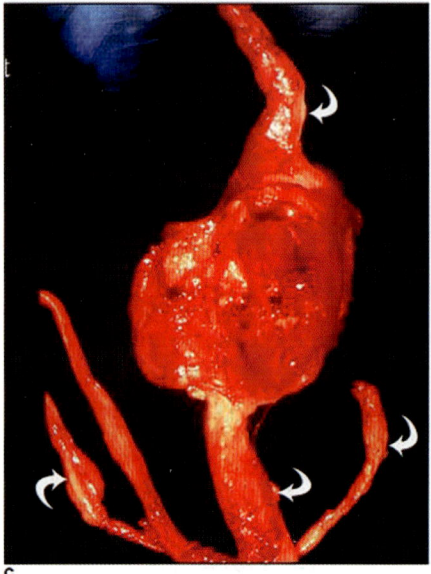

gastrocnemius-semimembranosus bursa (Figure 2.26). Septations, osteochondral fragments, and nodules of synovitis may be seen in synovial cysts. Popliteal cysts can rupture and dissect (complicated Baker cyst) into the calf, clinically simulating deep venous thrombosis. Imaging reveals this dissection as fusiform and the often prominent resultant surrounding edema caused by tissue irritation. The shape and surrounding edema usually allow distinction from other causes of popliteal soft tissue masses, despite the fact that complicated Baker cysts sometimes show solid regions resulting from hemorrhage.

Ganglia are not true cysts but are lined by fibrous tissue. These lesions more frequently involve the wrist (70%) and less commonly affect the foot, ankle, knee, shoulder, and hip.[118–120] Ganglia are typically near but not in joints, although more recently intra-articular lesions have been recognized.[119] Ganglia are usually small (1–3 cm) lesions that show fluid characteristics on CT, MRI, and US and may be multilocular (Figure 2.27). US is often the quickest and least expensive method to image and diagnose a ganglion (Figure 2.27). Recently, a more complex appearance on US has been reported in a ganglion about the hand and wrist compared with the typical anechoic morphology.[121]

Focal fluid collections adjacent to menisci and labral cartilage have been referred to as ganglia and juxtaarticular myxomas in the past. It is now recognized, however, that most of these lesions in fact result from tears of these fibrocartilaginous structures, with fluid extending through these defects and accumulating in adjacent soft tissues, thus creating the "cyst." These perilabral or meniscal cysts are thus a result of trauma and do not represent ganglia or myxomas.[122–124] This has treatment implications, in that these lesions are likely to recur if the underlying cartilaginous damage is not repaired. Although CT and US can identify these "cysts," MRI is the modality of choice owing to its ability to detect the associated internal joint derangement. The cyst can be complex and not purely fluid, containing material related to associated debris.

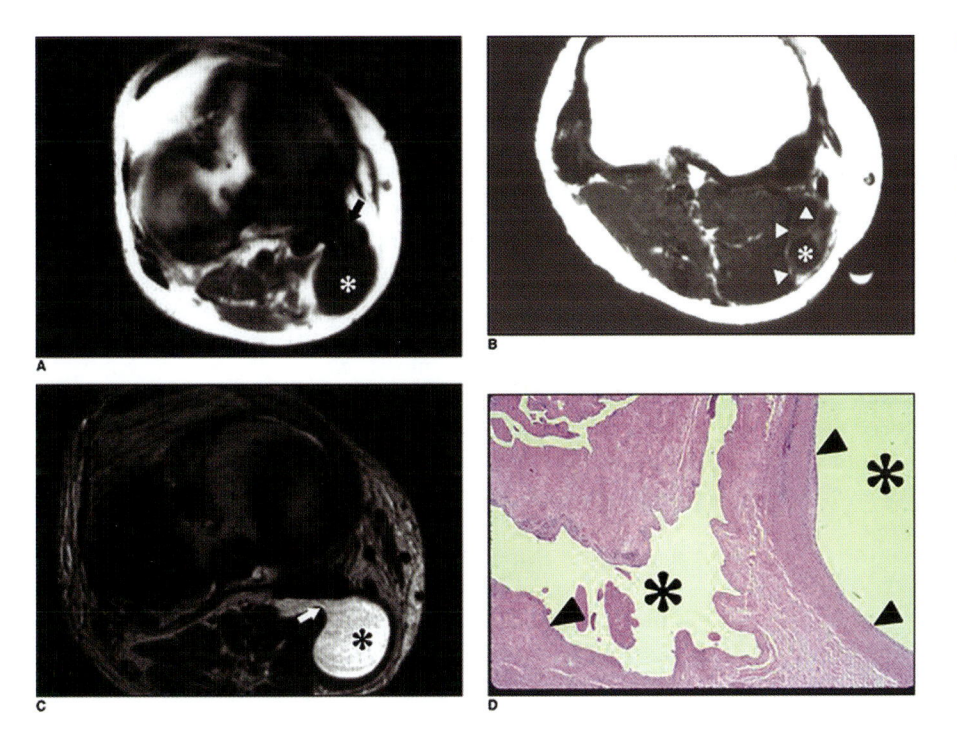

Chapter 2: Radiologic evaluation of soft tissue tumors

Figure 2.26 Popliteal cyst in a 40-year-old man. Axial T1-weighted MR images before (**A**) and after intravenous contrast (**B**) and T2-weighted image (**C**) show a homogeneous cystic mass (low intensity on T1-weighting and very high signal on T2-weighting) (*) and a neck extending back toward the joint *(arrows)*. After contrast, thin peripheral and septal enhancement is seen *(arrowheads)*, confirming the cystic consistency of the lesion. Photomicrograph (**D**, hematoxylin and eosin stain; 150 ×) demonstrates multilocular appearance (*) and septa lined by synovial cells *(arrowheads)*.

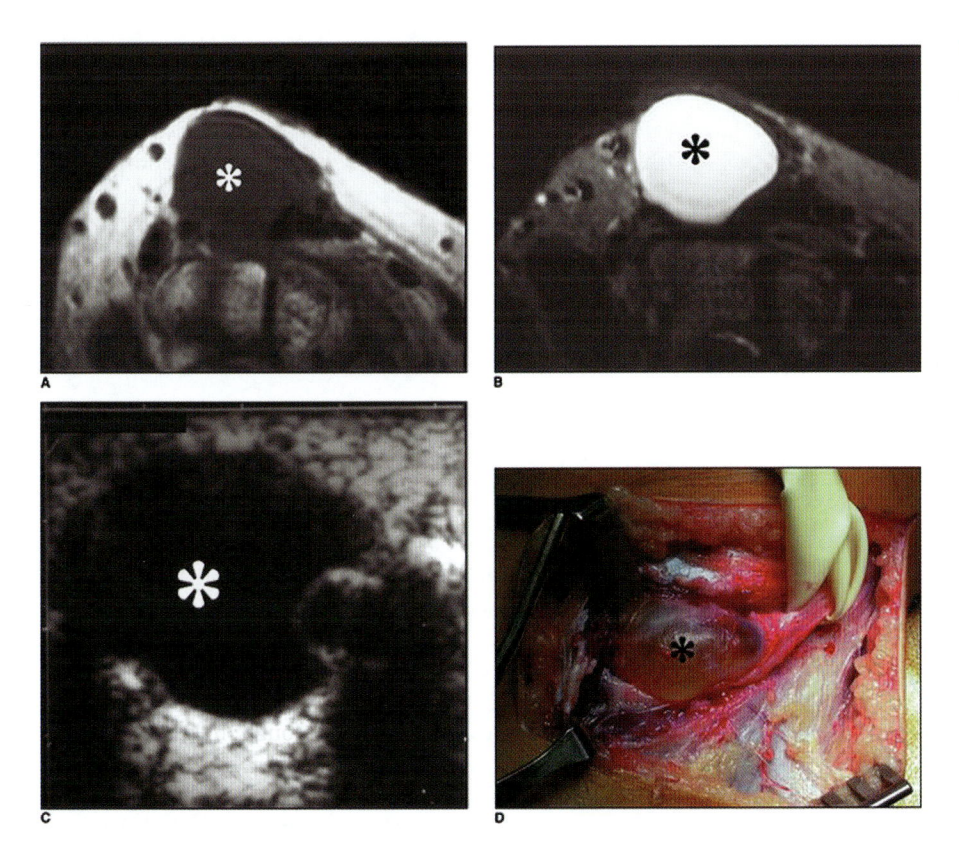

Figure 2.27 Ganglion of the wrist in several different patients. Axial T1-weighted (**A**) and T2-weighted (**B**) MR images show a high water content soft tissue mass (*) dorsal to the wrist. Ultrasound (**C**) reveals a homogeneous anechoic soft tissue mass (*). Intraoperatve photograph (**D**) demonstrates the multilocular cystic morphology of a ganglion (*).

Extrinsic erosion of bone rarely occurs, particularly with lesions of the hip and acetabular labral tears.

Liquified hematoma and abscess formation also can cause "cystic" soft tissue masses.[22] Clinical history and associated inflammation are helpful in the diagnosis. Imaging typically reveals a much thicker and more irregular wall (but not nodular) around these lesions compared with other cystic masses. Chronic hematomas often contain prominent hemosiderin in the lesion wall, which can be detected with a gradient echo MR sequence, making it very suggestive of the diagnosis. Abscess

33

formation often reveals linear extension, representing a sinus tract and an intermediate signal intensity wall on MR (penumbra or gray-wall sign), both features being very indicative of this diagnosis.

Myxomatous neoplasms include myxoma, neurogenic neoplasm, liposarcoma, MFH (or myxofibrosarcoma), and extraskeletal chondrosarcoma.[8,22,125,126] These lesions can simulate cystic masses on noncontrast CT and MRI, reflecting their intrinsic high water content. After IV contrast peripheral nodules of mild diffuse enhancement are apparent, distinguishing these myxoid neoplasms from a ganglion, synovial cyst, abscess, or hematoma, which demonstrate peripheral and/or septal enhancement (Figure 2.26). US can also distinguish these lesions because myxomatous lesions, which do have a high water content, are still solid lesions and are hypoechoic but not anechoic as are cystic masses (Figures 2.5 and 2.27).

Additional lesions with specific imaging appearances

Elastofibroma represents a reactive process that can usually be diagnosed by its specific location: a soft tissue mass deep to the scapular tip (Figure 2.13). Elastofibroma is a common lesion that can be recognized on 2% of CT chest examinations.[127] Small streaks of fat can be seen on both CT and MRI, and lesions typically show intermediate to low signal intensity on all MR pulse sequences.[127-129]

PVNS represents a benign proliferative lesion of the synovium. Both the diffuse intra-articular form and the localized and usually extra-articular form (giant cell tumor of tendon sheath [GCTTS]) often show characteristic imaging findings.[130-132]

Radiographs of GCTTS can show only a nonspecific soft tissue mass most commonly involving the hand or wrist (65–89% of lesions).[131,133] Extrinsic erosion of underlying bone is apparent in approximately 15% to 20% of cases.[131,133] Calcification has been reported in 6% of lesions.[131] Radiographs of the diffuse form of PVNS typically reveal only evidence of effusion or a nonspecific soft tissue mass. Extrinsic erosion of bone has been described in 50% of cases, affecting both sides of the joint.[130-132] This finding depends on the joint involved and specifically the size of the articulation. Small-capacity joints such as the hip (the second most commonly involved joint) almost invariably (93%) demonstrate erosion, whereas the knee (the most frequently affected joint) with its large size reveals erosion in only 26% of cases.[130-132]

Hemosiderin deposition in PVNS often allows more definitive imaging diagnosis on CT or more commonly MRI (Figure 2.11). CT reveals extrinsic erosion of bone and the accompanying soft tissue mass can have increased attenuation (density) resulting from the hemosiderin. MRI shows lesion extent and the hemosiderin effect more clearly, causing marked decreased signal intensity on T2-weighting. This low signal intensity is even more pronounced on gradient echo MR sequences ("blooming" artifact; Figure 2.11). Focal areas of effusion are common and reveal high signal intensity on T2-weighted MRI. The degree of hemosiderin deposition is usually prominent in the diffuse form of PVNS, although GCTTS often reveals a more variable degree of pigment deposition. This variation is reflected in the MRI appearance, and lesions with little or no hemosiderin, particularly GCTTS, sometimes show nonspecific higher signal intensity on T2-weighted MRI.

Synovial chondromatosis also represents a synovially based process that frequently shows a characteristic imaging appearance. Radiographs classically reveal multifocal (often innumerable) areas of calcification (70–95% of cases) frequently with a typical chondroid appearance (rings and arcs) scattered throughout the articulation.[134-138] Enchondral ossification can also occur in these cartilaginous nodules. Similar to PVNS, other nonspecific radiographic changes include a soft tissue mass, effusion, extrinsic erosion of bone, and secondary osteoarthritis.

CT is the best imaging modality to detect calcification in the areas of chondroid synovial neoplasia, particularly if they are small or involve complex areas of anatomy such as the hip (Figure 2.28). The calcifications are often very similar in size and shape. MRI shows diffuse synovial abnormality with intermediate signal on T1-weighting. The high water content of hyaline cartilage (75–80%) shows high signal on T2-weighted MRI, similar to effusion.[22,134-138] Areas of calcification remain low signal on all MRI pulse sequences; however, small calcifications are often difficult or impossible to detect on MRI compared with CT or radiographs (Figure 2.28). Synovial chondromatosis usually affects joints (knee in 50–65% of cases), but tendon sheaths (tenosynovial chondromatosis) or bursa (bursal chondromatosis) may also be affected, with identical imaging appearances.[8,22]

There are multiple various forms of fibromatosis that can involve the musculoskeletal system, including aggressive infantile, extra-abdominal desmoid, palmar, and myofibromatosis.[8,22,139-143] Although an extensive discussion of these lesions individually is beyond the scope of this chapter, these lesions as a group often show characteristic MRI findings that suggest the diagnosis. The fibromatoses commonly affect the foot, hand, shoulder (Figure 2.29), chest wall, and paraspinal areas.[8,22] Highly collagenized lesions frequently reveal prominent areas of diffuse low signal intensity on T2-weighted MRI (Figure 2.29). Conversely, less collagenized and more cellular lesions can show nonspecific high signal on T2-weighted MRI. Even the latter lesions often show collagenized low signal intensity bands.[139,140,143,144] (Figure 2.29). In addition, the growth pattern frequently reveals tails of extension along fascial planes (Figure 2.29), which is an unusual growth pattern in other neoplasms, except nodular fasciitis and dermatofibrosarcoma protuberans.[140] In the authors' opinion and experience, fibromatoses with predominant low signal intensity on T2-weighted MRI are less likely to recur or grow

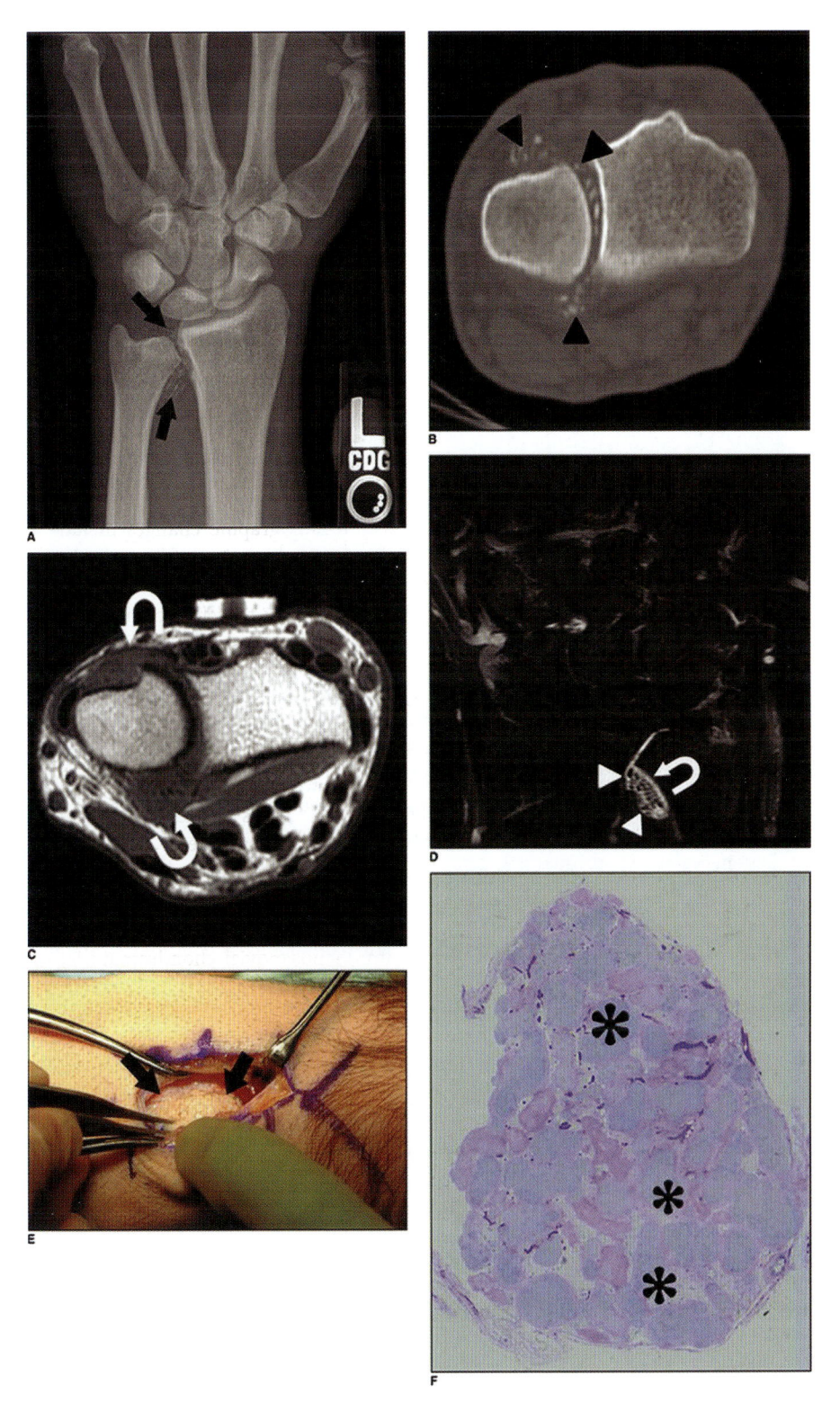

Chapter 2: Radiologic evaluation of soft tissue tumors

Figure 2.28 Synovial chondromatosis of the radioulnar joint in a 25-year-old man with wrist pain. Hand radiograph (**A**) shows subtle calcifications (*small arrows*) within the radioulnar joint. CT (**B**) easily reveals multiple small calcifications (*large arrowheads*) within the widened radioulnar joint. Axial T1-weighted (**C**) and T2-weighted (**D**) MR images show distension of the radioulnar joint with fluid (*curved arrows*) and multiple small filling defects (*small arrowheads*) representing the calcifications on CT. Intraoperative photograph (**E**) demonstrates distended radioulnar joint with multiple small osteochondral nodules (*large arrows*). The multilobular growth and cartilaginous morphology (*) is well shown on the photomicrograph (**F**, hematoxylin and eosin stain; 15 ×)

significantly. In patients treated with radiation, MRI can assess tumor response to the therapy. Good response to radiation is reflected as increasing collagenization, pathologically and progressively increasing low signal intensity (by MRI), and decreased mass size.

Soft tissue masses with nonspecific imaging appearance

Soft tissue masses with a nonspecific imaging appearance should be assessed for extent and staging.[145,146] Imaging

35

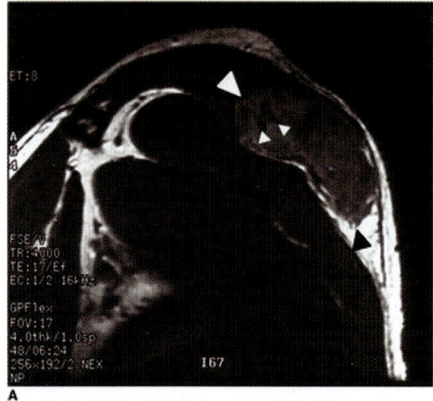

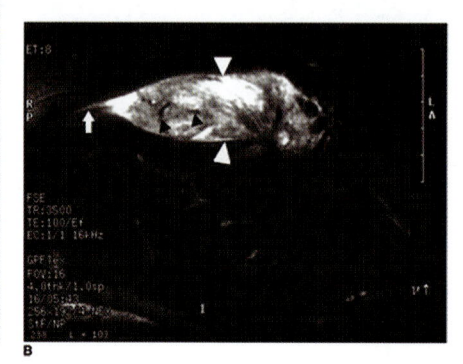

Figure 2.29 Extra-abdominal desmoid of the shoulder in a 32-year-old man. Sagittal T1-weighted (**A**) and coronal T2-weighted (**B**) MR images show a large mass *(large arrowheads)* about the shoulder. Intrinsic signal characteristics of low to intermediate signal intensity on T1-weighting and high signal on T2-weighting are nonspecific. Low signal intensity bands *(small arrowheads)*, fascial extension *(arrow)*, and lesion location suggest the diagnosis. Photograph of a sectioned gross specimen (**C**) from a different patient with extra-abdominal desmoid demonstrating the soft tissue mass (*) and prominent fascial extension *(curved arrows)*.

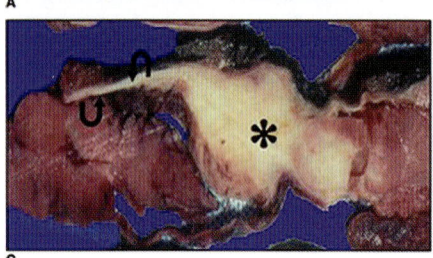

features of small size (<5 cm), defined margins, homogeneity, and lack of neurovascular encasement suggest a benign process.[147–150] In contradistinction, large size (>5 cm), ill-defined margins, heterogeneity, and neurovascular or bone involvement suggest a malignant (aggressive) process.[147–150] In the authors' opinion, however, differentiation of benign versus malignant cannot be made with enough confidence to alter the need for a biopsy of a soft tissue mass that is solid but nonspecific by intrinsic imaging. Benign lesions that can reveal aggressive characteristics simulating malignancy include hematoma, fibromatosis, reactive lymph nodes, abscess, and myositis ossificans. Malignant lesions that at times reveal indolent features simulating benign disease include synovial sarcoma and myxoid liposarcoma.[22]

Conclusion

The role of imaging in evaluation of soft tissue tumors has markedly improved because of the advent of CT, US, and more recently, MRI. Indeed, the last modality has particularly improved the goal of imaging, which includes lesion detection, characterization, and staging. Relatively specific histologic diagnosis can be made by imaging characteristics in 25% to 50% of lesions. The authors' opinion is that this will gradually increase over time with continuing experience, description of more specific intrinsic imaging characteristics, and confidence in these radiologic appearances. Soft tissue masses that often reveal pathognomonic imaging appearance include lipomatous lesions, angiomatous lesions, neurogenic tumors, "cystic" masses, elastofibroma, PVNS, synovial chondromatosis, and the fibromatoses. In the evaluation of soft tissue tumors with a nonspecific imaging appearance, the extent of the lesion remains vital in radiologic assessment, but distinction of benign from malignant lesions is fraught with uncertainty. The authors strongly believe that these nonspecific soft tissue masses require biopsy to direct definitive treatment. Clinical evaluation and treatment of soft tissue masses must emphasize a team approach, incorporating the combined skills of radiologists, pathologists, and orthopedic oncologists with the ultimate goal of improving patient management and outcome.

Acknowledgements

The authors gratefully acknowledge the support of Janice Danqing Liu for manuscript preparation. In addition, the authors would like to thank the residents – without whom this project would not have been possible – who attend the Armed Forces Institute of Pathology's radiologic pathology courses, for their contribution to our series of patients.

References

1. Weekes RG, McLeod RA, Reiman HM, Pritchard DJ. CT of soft-tissue neoplasms. *AJR Am J Roentgenol* 1985;144:355–360.
2. Sundaram M, McGuire MH, Herbold DR. Magnetic resonance imaging of soft tissue masses: an evaluation of fifty-three histologically proven tumors. *Magn Reson Imaging* 1988;6:237–248.
3. Petasnick JP, Turner DA, Charters JR, Gitelis S, Zacharias CE. Soft-tissue masses of the locomotor system: comparison of MR imaging with CT. *Radiology* 1986;160:125–133.

4. Totty WG, Murphy WA, Lee JK. Soft-tissue tumors: MR imaging. *Radiology* 1986;160:135–141.

5. Kransdorf MJ, Jelinek JS, Moser R, et al. Soft-tissue masses: diagnosis using MR imaging. *AJR Am J Roentgenol* 1989;153:541–547.

6. Berquist TH, Ehman RL, King BF, Hodgman CG, Ilstrup DM. Value of MR imaging in differentiating benign from malignant soft-tissue masses: study of 95 lesions. *AJR Am J Roentgenol* 1990;155:1251–1255.

7. Crim JR, Seeger LL, Yao L, Chandnani V, Eckardt JJ. Diagnosis of soft-tissue masses with MR imaging: can benign masses be differentiated from malignant ones? *Radiology* 1992;185:581–586.

8. Enzinger FM, Weiss SW. General considerations. In *Soft Tissue Tumors*. Weiss SW, Goldblum JR (eds.) St. Louis: CV Mosby; 2008: 1995.

9. Mettlin C, Priore R, Rao U, et al. Results of the national soft-tissue sarcoma registry. *J Surg Oncol* 1982;19:224–227.

10. Angervall L, Kindblom LG. Principles for pathologic-anatomic diagnosis and classification of soft-tissue sarcomas. *Clin Orthop Relat Res* 1993;(289):9–18.

11. Baldursson G, Agnarsson BA, Benediktsdottir KR, Hrafnkelsson J. Soft tissue sarcomas in Iceland 1955–1988. Analysis of survival and prognostic factors. *Acta Oncol* 1991;30:563–568.

12. Jemal A, Siegel R, Ward E, et al. Cancer statistics, 2008. *CA Cancer J Clin* 2008;58:71–96.

13. Greenlee RT, Murray T, Bolden S, Wingo PA. Cancer statistics, 2000. *CA Cancer J Clin* 2000;50:7–33.

14. Kransdorf MJ, Murphey MD. Radiologic evaluation of soft-tissue masses: a current perspective. *AJR Am J Roentgenol* 2000;175:575–587.

15. Levine E, Huntrakoon M, Wetzel LH. Malignant nerve-sheath neoplasms in neurofibromatosis: distinction from benign tumors by using imaging techniques. *AJR Am J Roentgenol* 1987;149:1059–1064.

16. Hammond JA, Driedger AA. Detection of malignant change in neurofibromatosis (von Recklinghausen's disease) by gallium-67 scanning. *Can Med Assoc J* 1978;119:352–353.

17. Kaplan IL, Swayne LC, Baydin JA. Uptake of Ga-67 citrate in a benign neurofibroma. *Clin Nucl Med* 1989;14:224.

18. Aoki J, Watanabe H, Shinozaki T, et al. FDG PET of primary benign and malignant bone tumors: standardized uptake value in 52 lesions 1. *Radiology* 2001;219:774–777.

19. Feldman F, van Heertum R, Manos C. 18-FDG PET scanning of benign and malignant musculoskeletal lesions. *Skeletal Radiol* 2003;32:201–208.

20. Ioannidis JP, Lau J. 18F-FDG PET for the diagnosis and grading of soft-tissue sarcoma: a meta-analysis. *J Nucl Med* 2003;44:717–724.

21. Tateishi U, Yamaguchi U, Seki K, et al. Bone and soft-tissue sarcoma: preoperative staging with fluorine 18 fluorodeoxyglucose PET/CT and conventional imaging. *Radiology* 2007; 245:839.

22. Kransdorf MJ, Murphey MD. *Imaging of Soft Tissue Tumors*. Philadelphia: WB Saunders; 1997.

23. Feydy A, Anract P, Tomeno B, Chevrot A, Drape JL. Assessment of vascular invasion by musculoskeletal tumors of the limbs: use of contrast-enhanced MR angiography. *Radiology* 2006;238: 611–621.

24. Swan JS, Grist TM, Sproat IA, et al. Musculoskeletal neoplasms: preoperative evaluation with MR angiography. *Radiology* 1995;194: 519–524.

25. Panicek DM, Gatsonis C, Rosenthal DI, et al. CT and MR imaging in the local staging of primary malignant musculoskeletal neoplasms: report of the Radiology Diagnostic Oncology Group. *Radiology* 1997;202:237–246.

26. Dalinka MK, Zlatkin MB, Chao P, Kricun ME, Kressel HY. The use of magnetic resonance imaging in the evaluation of bone and soft-tissue tumors. *Radiol Clin North Am* 1990;28: 461–470.

27. Pettersson H, Gillespy T, 3rd, Hamlin DJ, et al. Primary musculoskeletal tumors: examination with MR imaging compared with conventional modalities. *Radiology* 1987;164: 237–241.

28. Tehranzadeh J, Mnaymneh W, Ghavam C, Morillo G, Murphy BJ. Comparison of CT and MR imaging in musculoskeletal neoplasms. *J Comput Assist Tomogr* 1989;13:466–472.

29. Aisen AM, Martel W, Braunstein EM, et al. MRI and CT evaluation of primary bone and soft-tissue tumors. *AJR Am J Roentgenol* 1984;146: 749–756.

30. Chang AE, Matory YL, Dwyer AJ, et al. Magnetic resonance imaging versus computed tomography in the evaluation of soft tissue tumors of the extremities. *Ann Surg* 1987;205:340–348.

31. Demas BE, Heelan RT, Lane J, et al. Soft-tissue sarcomas of the extremities: comparison of MR and CT in determining the extent of disease. *AJR Am J Roentgenol* 1988;150:615–620.

32. Hudson TM, Hamlin DJ, Enneking WF, Pettersson H. Magnetic resonance imaging of bone and soft tissue tumors: early experience in 31 patients compared with computed tomography. *Skeletal Radiol* 1985;13:134–146.

33. Weekes RG, Berquist TH, McLeod RA, Zimmer WD. Magnetic resonance imaging of soft-tissue tumors: comparison with computed tomography. *Magn Reson Imaging* 1985;3:345–352.

34. Bloem JL, Taminiau AH, Eulderink F, Hermans J, Pauwels EK. Radiologic staging of primary bone sarcoma: MR imaging, scintigraphy, angiography, and CT correlated with pathologic examination. *Radiology* 1988;169: 805–810.

35. Rubin DA, Kneeland JB. MR imaging of the musculoskeletal system: technical considerations for enhancing image quality and diagnostic yield. *AJR J Roentgenol* 1994;163:1155–1163.

36. Mirowitz SA. Fast scanning and fat-suppression MR imaging of musculoskeletal disorders. *AJR Am J Roentgenol* 1993;161:1147–1157.

37. Fujimoto H, Murakami K, Ichikawa T, et al. MRI of soft-tissue lesions: opposed-phase T2*-weighted gradient-echo images. *J Comput Assist Tomogr* 1993;17:418–424.

38. Shuman WP, Baron RL, Peters MJ, Tazioli PK. Comparison of STIR and spin-echo MR imaging at 1.5 T in 90 lesions of the chest, liver, and pelvis. *AJR Am J Roentgenol* 1989;152:853–859.

39. Dwyer AJ, Frank JA, Sank VJ, et al. Short-Ti inversion-recovery pulse sequence: analysis and initial experience

in cancer imaging. *Radiology* 1988;168:827–836.

40. Beltran J, Chandnani V, McGhee RA, Jr., Kursunoglu-Brahme S. Gadopentetate dimeglumine-enhanced MR imaging of the musculoskeletal system. *AJR Am J Roentgenol* 1991;156:457–466.

41. Verstraete KL, De Deene Y, Roels H, et al. Benign and malignant musculoskeletal lesions: dynamic contrast-enhanced MR imaging-parametric "first-pass" images depict tissue vascularization and perfusion. *Radiology* 1994;192:835–843.

42. Benedikt RA, Jelinek JS, Kransdorf MJ, Moser RP, Berrey BH. MR imaging of soft-tissue masses: role of gadopentetate dimeglumine. *J Magn Reson Imaging* 1994;4:485–490.

43. Takebayashi S, Sugiyama M, Nagase M, Matsubara S. Severe adverse reaction to IV gadopentetate dimeglumine. *AJR Am J Roentgenol* 1993;160:659.

44. Tardy B, Guy C, Barral G, et al. Anaphylactic shock induced by intravenous gadopentetate dimeglumine. *Lancet* 1992;339:494.

45. Tishler S, Hoffman JC, Jr. Anaphylactoid reactions to i.v. gadopentetate dimeglumine. *AJNR Am J Neuroradiol* 1990;11:17–23.

46. Omohundro JE, Elderbrook MK, Ringer TV. Laryngospasm after administration of gadopentetate dimeglumine. *J Magn Reson Imaging* 1992;2:729–730.

47. Shellock FG, Hahn HP, Mink JH, Itskovich E. Adverse reaction to intravenous gadoteridol. *Radiology* 1993;189:151–152.

48. Jordan RM, Mintz RD. Fatal reaction to gadopentetate dimeglumine. *AJR Am J Roentgenol* 1995;164:743–744.

49. Harkens KL, Moore TE, Yuh WT, et al. Gadolinium-enhanced MRI of soft tissue masses. *Australas Radiol* 1993;37:30–34.

50. Seeger LL, Widoff BE, Bassett LW, Rosen G, Eckardt JJ. Preoperative evaluation of osteosarcoma: value of gadopentetate dimeglumine-enhanced MR imaging. *AJR Am J Roentgenol* 1991;157:347–351.

51. Kransdorf MJ, Murphey MD. The use of gadolinium in the MR evaluation of soft tissue tumors. *Semin Ultrasound CT MR* 1997;18:251–268.

52. Lin J, Fessell DP, Jacobson JA, Weadock WJ, Hayes CW. An illustrated tutorial of musculoskeletal sonography: part I, introduction and general principles. *AJR Am J Roentgenol* 2000;175:637–645.

53. Fornage BD, Tassin GB. Sonographic appearances of superficial soft tissue lipomas. *J Clin Ultrasound* 1991;19:215–220.

54. Lin J, Jacobson JA, Fessell DP, Weadock WJ, Hayes CW. An illustrated tutorial of musculoskeletal sonography: part 4, musculoskeletal masses, sonographically guided interventions, and miscellaneous topics. *AJR Am J Roentgenol* 2000;175:1711–1719.

55. Loyer EM, DuBrow RA, David CL, Coan JD, Eftekhari F. Imaging of superficial soft-tissue infections: sonographic findings in cases of cellulitis and abscess. *AJR Am J Roentgenol* 1996;166:149–152.

56. Choi H, Varma DG, Fornage BD, Kim EE, Johnston DA. Soft-tissue sarcoma: MR imaging vs sonography for detection of local recurrence after surgery. *AJR Am J Roentgenol* 1991;157:353–358.

57. Enneking WF, Spanier SS, Goodman MA. A system for the surgical staging of musculoskeletal sarcoma. *Clin Orthop Relat Res* 1980;(153):106–120.

58. Hajdu SI. *Pathology of Soft Tissue Tumors*. Philadelphia: Lea & Febiger; 1979.

59. Russell WO, Cohen J, Edmonson JH, et al. Staging system for soft tissue sarcoma. *Semin Oncol* 1981;8:156–159.

60. McDonald DJ. Limb-salvage surgery for treatment of sarcomas of the extremities. *AJR Am J Roentgenol* 1994;163:509–513; discussion 514–506.

61. Myhre-Jensen O. A consecutive 7-year series of 1331 benign soft tissue tumours. Clinicopathologic data: comparison with sarcomas. *Acta Orthop Scand* 1981;52:287–293.

62. Rydholm A. Management of patients with soft-tissue tumors: strategy developed at a regional oncology center. *Acta Orthop Scand Suppl* 1983;203:13–77.

63. Peabody TD, Simon MA. Principles of staging of soft-tissue sarcomas. *Clin Orthop Relat Res* 1993;(289):19–31.

64. Anderson MW, Temple HT, Dussault RG, Kaplan PA. Compartmental anatomy: relevance to staging and biopsy of musculoskeletal tumors. *AJR Am J Roentgenol* 1999;173:1663–1671.

65. Robinson E, Bleakney RR, Ferguson PC, O'Sullivan B. Oncodiagnosis panel: 2007: multidisciplinary management of soft-tissue sarcoma. *Radiographics* 2008;28:2069–2086.

66. Osment LS. Cutaneous lipomas and lipomatosis. *Surg Gynecol Obstet* 1968;127:129–132.

67. Leffert RD. Lipomas of the upper extremity. *J Bone Joint Surg Am* 1972;54:1262–1266.

68. Rydholm A, Berg NO. Size, site and clinical incidence of lipoma. Factors in the differential diagnosis of lipoma and sarcoma. *Acta Orthop Scand* 1983;54:929–934.

69. Dooms GC, Hricak H, Sollitto RA, Higgins CB. Lipomatous tumors and tumors with fatty component: MR imaging potential and comparison of MR and CT results. *Radiology* 1985;157:479–483.

70. Kransdorf MJ, Moser RP, Jr., Meis JM, Meyer CA. Fat-containing soft-tissue masses of the extremities. *Radiographics* 1991;11:81–106.

71. Hunter JC, Johnston WH, Genant HK. Computed tomography evaluation of fatty tumors of the somatic soft tissues: clinical utility and radiologic-pathologic correlation. *Skeletal Radiol* 1979;4:79–91.

72. Dolph JL, Demuth RJ, Miller SH. Familial multiple lipomatosis. *Plast Reconstr Surg* 1980;66:620–622.

73. Barkhof F, Melkert P, Meyer S, Blomjous CE. Derangement of adipose tissue: a case report of multicentric retroperitoneal liposarcomas, retroperitoneal lipomatosis and multiple subcutaneous lipomas. *Eur J Surg Oncol* 1991;17:547–550.

74. Bancroft LW, Kransdorf MJ, Peterson JJ, O'Connor MI. Benign fatty tumors: classification, clinical course, imaging appearance, and treatment. *Skeletal Radiol* 2006;35:719–733.

75. Murphey MD, Carroll JF, Flemming DJ, et al. From the archives of the AFIP: benign musculoskeletal lipomatous lesions. *Radiographics* 2004;24:1433–1466.

76. Leffell DJ, Braverman IM. Familial multiple lipomatosis: report of a case and a review of the literature. *J Am Acad Dermatol* 1986;15:275–279.

77. Chung EB, Enzinger FM. Benign lipoblastomatosis: an analysis of 35 cases. *Cancer* 1973;32:482–492.

78. Jimenez JF. Lipoblastoma in infancy and childhood. *J Surg Oncol* 1986;32:238–244.

79. Lateur L, Van Ongeval C, Samson I, Van Damme B, Baert AL. Case report 842: benign hibernoma. *Skeletal Radiol* 1994;23:306–309.

80. Seynaeve P, Mortelmans L, Kockx M, Van Hoye M, Mathijs R. Case report 813: hibernoma of the left thigh. *Skeletal Radiol* 1994;23:137–138.

81. Ritchie DA, Aniq H, Davies AM, Mangham DC, Helliwell TR. Hibernoma-correlation of histopathology and magnetic-resonance-imaging features in 10 cases. *Skeletal Radiol* 2006;35:579–589.

82. Jelinek JS, Kransdorf MJ, Shmookler BM, Aboulafia AJ, Malawer MM. Liposarcoma of the extremities: MR and CT findings in the histologic subtypes. *Radiology* 1993;186:455–459.

83. London J, Kim EE, Wallace S, et al. MR imaging of liposarcomas: correlation of MR features and histology. *J Comput Assist Tomogr* 1989;13:832–835.

84. Kransdorf MJ, Meis JM, Jelinek JS. Dedifferentiated liposarcoma of the extremities: imaging findings in four patients. *AJR Am J Roentgenol* 1993;161:127–130.

85. Sundaram M, Baran G, Merenda G, McDonald DJ. Myxoid liposarcoma: magnetic resonance imaging appearances with clinical and histological correlation. *Skeletal Radiol* 1990;19:359–362.

86. Murphey MD, Flemming DJ, Jelinek JS, et al. Imaging of higher grade liposarcoma with pathologic correlation. *Radiology* 1997;208:332.

87. Kransdorf MJ, Bancroft LW, Peterson JJ, et al. Imaging of fatty tumors: distinction of lipoma and well-differentiated liposarcoma. *Radiology* 2002;224:99–104.

88. Murphey MD, Arcara LK, Fanburg-Smith J. From the archives of the AFIP: imaging of musculoskeletal liposarcoma with radiologic-pathologic correlation. *Radiographics* 2005;25:1371–1395.

89. Ohguri T, Aoki T, Hisaoka M, et al. Differential diagnosis of benign peripheral lipoma from well-differentiated liposarcoma on MR imaging: is comparison of margins and internal characteristics useful? *AJR Am J Roentgenol* 2003;180:1689–1694.

90. Sung MS, Kang HS, Suh JS, et al. Myxoid liposarcoma: appearance at MR imaging with histologic correlation. *Radiographics* 2000;20:1007–1019.

91. Tateishi U, Hasegawa T, Beppu Y, et al. Prognostic significance of MRI findings in patients with myxoid-round cell liposarcoma. *AJR Am J Roentgenol* 2004;182:725–731.

92. Vilanova JC, Barcelo J, Smirniotopoulos JG, et al. Hemangioma from head to toe: MR imaging with pathologic correlation. *Radiographics* 2004;24:367–385.

93. Murphey MD, Fairbairn KJ, Parman LM, et al. From the archives of the AFIP. Musculoskeletal angiomatous lesions: radiologic-pathologic correlation. *Radiographics* 1995;15:893–917.

94. Derchi LE, Balconi G, De Flaviis L, Oliva A, Rosso F. Sonographic appearances of hemangiomas of skeletal muscle. *J Ultrasound Med* 1989;8:263–267.

95. Greenspan A, McGahan JP, Vogelsang P, Szabo RM. Imaging strategies in the evaluation of soft-tissue hemangiomas of the extremities: correlation of the findings of plain radiography, angiography, CT, MRI, and ultrasonography in 12 histologically proven cases. *Skeletal Radiol* 1992;21:11–18.

96. Hawnaur JM, Whitehouse RW, Jenkins JP, Isherwood I. Musculoskeletal haemangiomas: comparison of MRI with CT. *Skeletal Radiol* 1990;19:251–258.

97. Buetow PC, Kransdorf MJ, Moser RP, Jr., Jelinek JS, Berrey BH. Radiologic appearance of intramuscular hemangioma with emphasis on MR imaging. *AJR Am J Roentgenol* 1990;154:563–567.

98. Yuh WT, Kathol MH, Sein MA, Ehara S, Chiu L. Hemangiomas of skeletal muscle: MR findings in five patients. *AJR Am J Roentgenol* 1987;149:765–768.

99. Cohen EK, Kressel HY, Perosio T, et al. MR imaging of soft-tissue hemangiomas: correlation with pathologic findings. *AJR Am J Roentgenol* 1988;150:1079–1081.

100. Nelson MC, Stull MA, Teitelbaum GP, et al. Magnetic resonance imaging of peripheral soft tissue hemangiomas. *Skeletal Radiol* 1990;19:477–482.

101. McRae GA, Murphey MD, Temple HT, Torop AH, Fanburg-Smith J. Imaging of soft tissue hemangioma with pathologic correlation. *Radiology* 1997;205:449.

102. Murphey MD, Smith WS, Smith SE, Kransdorf MJ, Temple HT. From the archives of the AFIP. Imaging of musculoskeletal neurogenic tumors: radiologic-pathologic correlation. *Radiographics* 1999;19:1253–1280.

103. Suh JS, Abenoza P, Galloway HR, Everson LI, Griffiths HJ. Peripheral (extracranial) nerve tumors: correlation of MR imaging and histologic findings. *Radiology* 1992;183:341–346.

104. Kumar AJ, Kuhajda FP, Martinez CR, et al. Computed tomography of extracranial nerve sheath tumors with pathological correlation. *J Comput Assist Tomogr* 1983;7:857–865.

105. Stull MA, Moser RP, Jr., Kransdorf MJ, Bogumill GP, Nelson MC. Magnetic resonance appearance of peripheral nerve sheath tumors. *Skeletal Radiol* 1991;20:9–14.

106. Cerofolini E, Landi A, DeSantis G, et al. MR of benign peripheral nerve sheath tumors. *J Comput Assist Tomogr* 1991;15:593–597.

107. Cohen LM, Schwartz AM, Rockoff SD. Benign schwannomas: pathologic basis for CT inhomogeneities. *AJR Am J Roentgenol* 1983;147:141–143.

108. Singson RD, Feldman F, Slipman CW, et al. Postamputation neuromas and other symptomatic stump abnormalities: detection with CT. *Radiology* 1987;162:743–745.

109. Boutin RD, Pathria MN, Resnick D. Disorders in the stumps of amputee patients: MR imaging. *AJR Am J Roentgenol* 1998;171:497–501.

110. Zanetti M, Ledermann T, Zollinger H, Hodler J. Efficacy of MR imaging in patients suspected of having Morton's neuroma. *AJR Am J Roentgenol* 1997;168:529–532.

111. Redd RA, Peters VJ, Emery SF, Branch HM, Rifkin MD. Morton neuroma: sonographic evaluation. *Radiology* 1989;171:415–417.

112. Zanetti M, Strehle JK, Zollinger H, Hodler J. Morton neuroma and fluid in

113. Hassell DS, Bancroft LW, Kransdorf MJ, et al. Imaging appearance of diffuse neurofibroma. *AJR Am J Roentgenol* 2008;190:582–588.

114. Huang GS, Huang CW, Lee HS, et al. Diffuse neurofibroma of the arm: MR characteristics. *AJR Am J Roentgenol* 2005;184:1711–1712.

115. Bredella MA, Torriani M, Hornicek F, et al. Value of PET in the assessment of patients with neurofibromatosis type 1. *AJR Am J Roentgenol* 2007;189:928–935.

116. Feldman F, Singson RD, Staron RB. Magnetic resonance imaging of para-articular and ectopic ganglia. *Skeletal Radiol* 1989;18:353–358.

117. Schwimmer M, Edelstein G, Heiken JP, Gilula LA. Synovial cysts of the knee: CT evaluation. *Radiology* 1985;154:175–177.

118. Haller J, Resnick D, Greenway G, et al. Juxtaacetabular ganglionic (or synovial) cysts: CT and MR features. *J Comput Assist Tomogr* 1989;13:976–983.

119. Recht MP, Applegate G, Kaplan P, et al. The MR appearance of cruciate ganglion cysts: a report of 16 cases. *Skeletal Radiol* 1994;23:597–600.

120. De Flaviis L, Nessi R, Del Bo P, Calori G, Balconi G. High-resolution ultrasonography of wrist ganglia. *J Clin Ultrasound* 1987;15:17–22.

121. Teefey SA, Dahiya N, Middleton WD, Gelberman RH, Boyer MI. Ganglia of the hand and wrist: a sonographic analysis. *AJR Am J Roentgenol* 2008;191:716–720.

122. Burk DL, Jr., Dalinka MK, Kanal E, et al. Meniscal and ganglion cysts of the knee: MR evaluation. *AJR Am J Roentgenol* 1988;150:331–336.

123. Schuldt DR, Wolfe RD. Clinical and arthrographic findings in meniscal cysts. *Radiology* 1980;134:49–52.

124. Tyson LL, Daughters TC, Jr., Ryu RK, Crues JV, 3rd. MRI appearance of meniscal cysts. *Skeletal Radiol* 1995;24:421–424.

125. Sundaram M, McDonald DJ, Merenda G. Intramuscular myxoma: a rare but important association with fibrous dysplasia of bone. *AJR Am J Roentgenol* 1989;153:107–108.

126. Wirth WA, Leavitt D, Enzinger FM. Multiple intramuscular myxomas. Another extraskeletal manifestation of fibrous dysplasia. *Cancer* 1971;27:321–340.

127. Brandser EA, Goree JC, El-Khoury GY. Elastofibroma dorsi: prevalence in an elderly patient population as revealed by CT. *AJR Am J Roentgenol* 1998;171:977–980.

128. Bui-Mansfield LT, Chew FS, Stanton CA. Elastofibroma dorsi of the chest wall. *AJR Am J Roentgenol* 2000;175:244.

129. Kransdorf MJ, Meis JM, Montgomery E. Elastofibroma: MR and CT appearance with radiologic-pathologic correlation. *AJR Am J Roentgenol* 1992;159:575–579.

130. Cotten A, Flipo RM, Chastanet P, et al. Pigmented villonodular synovitis of the hip: review of radiographic features in 58 patients. *Skeletal Radiol* 1995;24:1–6.

131. Jelinek JS, Kransdorf MJ, Shmookler BM, Aboulafia AA, Malawer MM. Giant cell tumor of the tendon sheath: MR findings in nine cases. *AJR Am J Roentgenol* 1994;162:919–922.

132. Jelinek JS, Kransdorf MJ, Utz JA, et al. Imaging of pigmented villonodular synovitis with emphasis on MR imaging. *AJR Am J Roentgenol* 1989;152:337–342.

133. Murphey MD, Rhee JH, Lewis RB, et al. Pigmented villonodular synovitis: radiologic-pathologic correlation. *Radiographics* 2008;28:1493–1518.

134. Milgram JW. Synovial osteochondromatosis: a histopathological study of thirty cases. *J Bone Joint Surg Am* 1977;59:792–801.

135. Blandino A, Salvi L, Chirico G, et al. Synovial osteochondromatosis of the ankle: MR findings. *Clin Imaging* 1992;16:34–36.

136. Sundaram M, McGuire MH, Fletcher J, et al. Magnetic resonance imaging of lesions of synovial origin. *Skeletal Radiol* 1986;15:110–116.

137. Kramer J, Recht M, Deely DM, et al. MR appearance of idiopathic synovial osteochondromatosis. *J Comput Assist Tomogr* 1993;17:772–776.

138. Murphey MD, Vidal JA, Fanburg-Smith JC, Gajewski DA. Imaging of synovial chondromatosis with radiologic-pathologic correlation. *Radiographics* 2007;27:1465–1488.

139. Rock MG, Pritchard DJ, Reiman HM, Soule EH, Brewster RC. Extra-abdominal desmoid tumors. *J Bone Joint Surg Am* 1984;66:1369–1374.

140. Sundaram M, Duffrin H, McGuire MH, Vas W. Synchronous multicentric desmoid tumors (aggressive fibromatosis) of the extremities. *Skeletal Radiol* 1988;17:16–19.

141. Quinn SF, Erickson SJ, Dee PM, et al. MR imaging in fibromatosis: results in 26 patients with pathologic correlation. *AJR Am J Roentgenol* 1991;156:539–542.

142. Sundaram M, McGuire MH, Schajowicz F. Soft-tissue masses: histologic basis for decreased signal (short T2) on T2-weighted MR images. *AJR Am J Roentgenol* 1987;148:1247–1250.

143. Robbin MR, Murphey MD, Temple HT, Kransdorf MJ, Choi JJ. Imaging of musculoskeletal fibromatosis. *Radiographics* 2001;21:585–600.

144. Kransdorf MJ, Jelinek JS, Moser RP, Jr., et al. Magnetic resonance appearance of fibromatosis: a report of 14 cases and review of the literature. *Skeletal Radiol* 1990;19:495–499.

145. Kransdorf MJ. Malignant soft-tissue tumors in a large referral population: distribution of diagnoses by age, sex, and location. *AJR Am J Roentgenol* 1995;164:129–134.

146. Kransdorf MJ. Benign soft-tissue tumors in a large referral population: distribution of specific diagnoses by age, sex, and location. *AJR Am J Roentgenol* 1995;164:395–402.

147. Beltran J, Simon DC, Katz W, Weis LD. Increased MR signal intensity in skeletal muscle adjacent to malignant tumors: pathologic correlation and clinical relevance. *Radiology* 1987;162:251–255.

148. Hanna SL, Fletcher BD, Parham DM, Bugg MF. Muscle edema in musculoskeletal tumors: MR imaging characteristics and clinical significance. *J Magn Reson Imaging* 1991;1:441–449.

149. Mirowitz SA, Totty WG, Lee JK. Characterization of musculoskeletal masses using dynamic Gd-DTPA enhanced spin-echo MRI. *J Comput Assist Tomogr* 1992;16:120–125.

150. Deschepper A, Ramon FA, Degryse HR. Statistical analysis of MRI parameters predicting malignancy in 141 soft tissue masses. *Rofo* 1992;156:587–591.

Chapter 3

Immunohistochemistry of soft tissue tumors

Markku Miettinen
National Cancer Institute, National Institutes of Health

This chapter outlines the biologic background and the tissue distribution of the most significant immunohistochemical markers for soft tissue tumors. Tumor-type-specific applications are detailed in Chapters 6 to 35. Markers for metastatic carcinomas are presented in Chapter 29, and those for lymphohematopoietic tumors in Chapter 35. Some very specialized "one-tumor markers" are discussed only with their specific applications in tumor-specific Chapters 6 to 35. Several reviews are suggested for further reading.[1-6]

The markers discussed in this chapter are grouped under subheadings of endothelial and multispecific, muscle cell, neural and neuroendocrine, nerve sheath tumor/melanoma, histiocytic markers, keratins, other epithelial and mesothelial markers, and other markers. These markers and their main diagnostic targets have been listed in Table 3.1. Markers that have proved nonspecific or impractical and are no longer recommended are listed in Table 3.2.

Very few markers are totally specific for one cell or tumor type, and no cell cycle or proliferation marker can separate benign and malignant tumors. The advances in epitope retrieval and more complete automation have improved consistency and reduced interlaboratory variation. Literature still shows variance in results, probably related to differences in reagents, laboratory procedures, and analyzed tissues.

The most useful markers include cell surface and cell membrane antigens such as receptors, structural proteins, secretory products, cytoplasmic antigens, and nuclear antigens (cell cycle and transcriptional regulators; Table 3.3). Many membrane antigens are classified as *leukocyte* antigens with a cluster of differentiation (CD) number, and their expression in mesenchymal cells is sometimes diagnostically useful.

Use of well-documented specific antibodies, good detection systems, knowledge of the distribution of various antigens in different tumors, and experience in interpretation are needed for successful application of immunohistochemistry. Diagnostic immunohistochemistry is an adjunct to diagnosis and has to be interpreted in the context of histology and clinicopathologic situation.

Endothelial or multispecific markers

Antigens of endothelial cells help to confirm endothelial cell differentiation of tumors such as angiosarcoma and Kaposi's sarcoma. None of these antigens is entirely endothelial cell specific, and their use in combination is recommended in problem cases. The most useful endothelial markers have been summarized in Table 3.4.

CD31 (platelet endothelium cell adhesion molecule 1, PECAM-1)

This transmembrane glycoprotein of 130 kD has six extracellular immunoglobulin-like loops and homology to other cell adhesion molecules, such as the carcinoembryonic antigen. CD31 is constitutively expressed in most endothelial cells and associates internally with catenin family proteins.[7,8] In hematopoietic cells, CD31 is present in a subset of myeloblasts, platelets, megakaryocytes, sinus histiocytes of lymph nodes, some lymphoid cells, plasma cells, and histiomonocytic cells.[9,10] A distinctive membrane staining pattern is typically observed in positive cells. Platelet aggregates and areas of thrombosis and hemorrhage are positive. Inflammatory infiltrates often contain positive cells, which should not be confused with tumor cell reactivity (Figure 3.1).

Most data published on CD31 in pathology literature are based on the use of monoclonal antibody JC/70. CD31 is nearly consistently present in benign and malignant vascular tumors, including hemangiomas, epithelioid hemangioendothelioma, angiosarcomas, and Kaposi's sarcoma.[9-12] Many angiosarcomas show a membrane staining pattern, which is more distinct in epithelioid vascular tumors. Although positivity in carcinoma or mesothelioma has been reported occasionally, this appears to be rare (<1%).[13] No significant reactivity has been reported in melanoma and nonvascular sarcomas.

This marker, along with ERG (see below) is probably closest to the immunohistochemical gold standard for the definition of an endothelial neoplasm; without a positive result for CD31 it might be difficult to verify a tumor as an angio-

Modern Soft Tissue Pathology, Second Edition, ed. Markku Miettinen. Published by Cambridge University Press. © Cambridge University Press 2016.

Chapter 3: Immunohistochemistry of soft tissue tumors

Table 3.1 Main groups of immunohistochemical cell differentiation markers, with selected single markers and their most important applications in soft tissue tumors

Group of markers	Diagnostic use and additional positive cell types/tumors
Endothelial markers	
CD31	Angiosarcoma, hemangioendotheliomas, Kaposi's sarcoma, histiocytes, and primitive hematopoietic cells
CD34	Kaposi's sarcoma GIST, many fibroblastic tumors, bone marrow precursor cells, angiosarcoma, epithelioid sarcoma (50% of the two latter positive)
von Willebrand factor	Epithelioid hemangioendothelioma, some angiosarcomas
VEGFR3	Kaposi's sarcoma, lymphatic vascular tumors, many angiosarcomas and hemangiomas
Podoplanin	Kaposi's sarcoma, many angiosarcomas, lymphatic vascular tumors, mesothelioma, many carcinomas
ERG*	Angiosarcoma, hemangioendotheliomas, prostate cancer, some cartilage tumors
Muscle cell markers	
Actins: common muscle actin	Smooth and skeletal muscle tumors, myofibroblastic tumors
Smooth muscle actin	Smooth muscle and myofibroblastic tumors
Desmin	Smooth and skeletal muscle tumors, desmoplastic small round cell tumor, aggressive angiomyxoma, angiomatoid fibrous histiocytoma, tenosynovial giant cell tumor (focal)
Heavy caldesmon	Smooth muscle and its tumors, myoepithelia, gastrointestinal stromal tumors (50–60%)
Calponin	Smooth muscle, myofibroblasts, myoepithelia, synovial sarcoma (often)
MyoD1*, myogenin*	Rhabdomyosarcoma, fetal and regenerative skeletal muscle
Smooth muscle myosin	Smooth muscle and myofibroblastic tumors
Neural and neuroendocrine specific markers	
Synaptophysin	Neuroblastoma, paraganglioma, neuroendocrine carcinoma
Chromogranin	Paraganglioma, neuroendocrine carcinoma (especially low-grade)
Neurofilament proteins	Neuroblastoma, paraganglioma, Merkel cell carcinoma
Alpha-internexin	Essentially same as neurofilaments
S100 protein and other multispecific neural markers	
S100 protein	Melanocytic, schwannian, chondroid, Langerhans cell tumors
Sox10*	Melanocytic, schwannian, and myoepithelial tumors. Malignant peripheral nerve sheath tumors and myoepitheliomas are often negative
Melanoma markers other than S100 protein	
HMB45	Melanoma, clear cell sarcoma, angiomyolipoma/PEComa family tumors (PEComas)
Tyrosinase	Nevi, melanoma
Melan A	Nevi, melanoma, angiomyolipoma/PEComa (variably)
Microphthalmia*	Nevi, melanoma, osteoclastic giant cells, neurothekeoma
Histiocytic markers	
Lysozyme	Histiocytes, myelomonocytic cells
Factor XIIIa	Histiocytes, especially dendritic ones
CD68	Histiocytes and any lysosome-rich cells of any derivation, melanoma, paraganglioma, schwannoma, granular cell tumor
CD163	Histiocytes, histiocytic sarcoma, juvenile xanthogranuloma
Keratins	Carcinomas, synovial and epithelioid sarcoma, chordoma, adamantinoma, and myoepithelioma. Sporadically and focally in many other sarcomas and metastatic melanoma (see separate tables for mesenchymal tumors)

Table 3.1 (cont.)

Group of markers	Diagnostic use and additional positive cell types/tumors
Other epithelial and mesothelial markers with variable specificities	
EMA/MUC1	Epithelial tumors in general, synovial sarcoma, perineurioma, meningioma, epithelioid sarcoma, chordoma
B72.3	Many adenocarcinomas, epithelioid angiosarcoma
Cadherins	Complex distribution in epithelial and some nonepithelial tumors
Calretinin	Mesothelioma, some carcinomas, synovial sarcoma, desmoid fibromatosis, and some other myofibroblastic tumors (often)
Desmoplakins	Epithelial tumors in general, meningioma, Ewing sarcoma
Ep-CAM (Ber-Ep4, MOC31)	Synovial sarcoma, most adenocarcinomas, rare mesotheliomas
HBME-1	Mesothelioma, some adenocarcinomas, synovial sarcoma, chordoma, chondrosarcoma
Mesothelin	Synovial sarcoma, mesothelioma, ovarian serous carcinoma, pancreatic adenocarcinoma, and some other adenocarcinomas
Wilms' tumor protein (WT)*	Small round cell desmoplastic tumor, mesothelioma, ovarian serous carcinoma and related tumors, female genital stromal tumors, uterine and pelvic leiomyomas in women
Other important tumor type markers	
ALK	Inflammatory myofibroblastic tumor, large cell anaplastic lymphoma (variable)
Basement membrane proteins	Schwann cell tumors, angiosarcoma
CD10	Endometrial stromal sarcoma, many fibroblastic tumors
CD99	Ewing sarcoma, widespread in different tumors
CD117 (KIT)	Gastrointestinal stromal tumor (GIST), mast cell neoplasms, angiosarcoma, Ewing sarcoma, neuroblastoma, seminoma/dysgerminoma, some melanomas and clear cell sarcomas, adenoid cystic carcinoma, few other carcinomas (for example, thymic carcinoma)
DOG1/Ano 1	GIST, many carcinomas, occasionally in synovial sarcoma and leiomyomas
Estrogen and progesterone receptors	Carcinomas of breast, endometrium, and ovary (some). Most uterine and retroperitoneal leiomyomas and genital stromal tumors (in women), some uterine leiomyosarcomas angiomyofibroblastoma, aggressive angiomyxoma
Glial fibrillary acidic protein	Glial tumors, schwannomas, myoepithelioma
Inhibin	Granular cell tumor, sex cord stromal tumors, adrenal cortical carcinoma
MDM2*	Atypical lipomatous tumor and dedifferentiated liposarcoma (some other sarcomas)
MUC4	Low-grade fibromyxoid sarcoma, sclerosing epithelioid fibrosarcoma, synovial sarcoma (focal)
RB1*	Loss in spindle cell lipoma
SDHB	Loss in some gastric GISTs and paragangliomas (SDH-deficient tumors)
SATB2*	Osteosarcoma, osteoid material, colorectal carcinoma
SMARCB1/INI1*	Loss in epithelioid sarcoma, rhabdoid tumors, and some extraskeletal myxoid chondrosarcoma, relative loss in synovial sarcomas
STAT6*	Solitary fibrous tumor
Vimentin	Widespread in mesenchymal tumors, many poorly differentiated carcinomas highlighted as negative, but some are variably positive

* Nuclear labeling

Chapter 3: Immunohistochemistry of soft tissue tumors

Table 3.2 Markers no longer recommended for clinical use in the typing of soft tissue tumors because of insufficient specificity or lack of high quality antibodies

Alpha-1-antichymotrypsin and alpha-1-antitrypsin	Histiocytes, many tumors of any lineage, very low histiocytic specificity
Bcl2	Solitary fibrous tumor, synovial sarcoma, many others
CD56 (NCAM)	Neuroendocrine carcinomas, rhabdomyosarcoma, synovial sarcoma, many other sarcomas
CD57	Schwann cell tumors, synovial sarcoma, cartilaginous tumors, neuroendocrine tumors
CD63	Melanoma, many mesenchymal tumors, and some carcinomas
Lysozyme (muramidase)	Identifies histiocytes but also shows epithelial positivity
Myoglobin	Rhabdomyosarcoma (differentiated). Problems of specificity of polyclonal antibodies
Nerve growth factor receptor, low affinity (p75)	Nerve sheath tumors, dermatofibrosarcoma protuberans
Neuron-specific enolase	Neural and melanocytic tumors, smooth muscle, myofibroblasts
Sarcomeric actin	Insufficiently specific for skeletal muscle neoplasms
Ulex europaeus lectin	Recognizes endothelia and their neoplasms but in addition a wide variety of different types of carcinomas

Table 3.3 Examples of classes of proteins by location or function, used as cell-type or diagnostic markers in immunohistochemistry of soft tissue tumors

Cell membrane proteins	Organelle-specific	Cytosolic proteins	Cytoskeletal proteins	Nuclear proteins
CD10	Lysosomes	S100 protein	Intermediate filament proteins	Transcriptional regulators
CD31		Alpha-1 antitrypsin		
CD34	CD68		Vimentin	Estrogen receptor
CD117 (KIT)			Keratins	Progesterone receptor
Ep-CAM (Ber-Ep4, MOC31)			Desmin	Myogenic regulators
EMA	Mitochondria		GFAP	ERG
	SDHB		Neurofilaments	Sox10
				Brachyury
	Melanosomes		Microfilament proteins	
	HMB45			Other nuclear proteins
	Tyrosinase		Actins, myosins	Ki67 analogs
				RB1
				INI1/SMARCB1
	Cytoplasmic granules and vesicles: Chromogranin Synaptophysin			

sarcoma. In addition, CD31 is variably expressed in blasts of extramedullary myeloid tumors.

CD34 (hematopoietic progenitor cell antigen)

This glycosylated transmembrane glycoprotein of 115 kD is a sialomucin of an unknown, possibly receptor-related function, but no ligands have been identified.[14] CD34 is considered a progenitor cell marker for hematopoietic and some nonhematopoietic cell lineages, including fibroblastic and epithelial cells.[15] It is expressed in early hematopoietic blasts, virtually all endothelial cells, and subsets of fibroblasts. In the skin, the positive fibroblasts are located especially periadnexally and perivascularly, and in different organs they are commonly seen in septal structures and around blood vessels.[16,17] Superficial stroma of the uterine cervix and vagina are also positive.

Chapter 3: Immunohistochemistry of soft tissue tumors

Table 3.4 Overview of endothelial cell markers

Name of antigen	Alternative name/designation	Distribution in normal tissues	Distribution in tumors
CD31	PECAM1	All endothelia	Most angiosarcomas Kaposi's sarcoma
CD34	Hematopoietic progenitor cell antigen	All endothelia	50% of angiosarcomas All Kaposi's sarcomas
von Willebrand factor	Factor VIII-related antigen	Most endothelia, hepatic sinusoids	
Vascular endothelial growth factor receptor 3 (VEGFR3)	FLT4	Lymphatics, fenestrated endothelia, neovascular endothelia	Angiosarcomas (variably) Kaposi's sarcoma
Podoplanin	D2–40 antibody	Lymphatic endothelia	Kaposi's sarcoma Many angiosarcomas Many nonvascular tumors
ERG	ETS-related transcription factor (nuclear)	Endothelia, some cartilage cells	Angiosarcoma, hemangioendothelioma, 40% of prostatic adenocarcinomas, chondromas, some chondrosarcomas, epithelioid sarcoma (often)

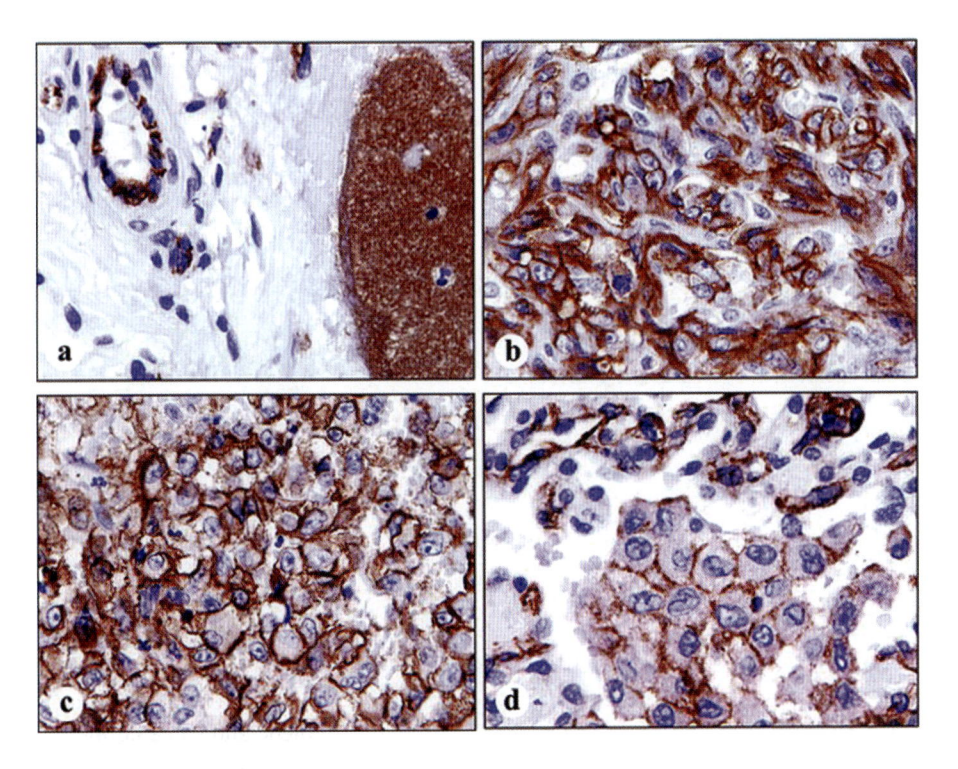

Figure 3.1 Examples of CD31 immunoreactivity. (**a**) Most normal endothelial cells are positive, as are platelet thrombi (right). (**b,c**) Angiosarcomas typically show a membrane-staining pattern. (**d**) Membrane positivity for CD31 is commonly seen in histiocytes, which should not be misconstrued as evidence for endothelial differentiation.

Positive spindle cells usually show an apparently cytoplasmic staining pattern, whereas a distinct membrane staining is seen especially in large cytoplasmic cells.

Two monoclonal antibodies to CD34 are commonly used in formalin-fixed and paraffin-embedded tissue: QBEND10 and My10/HPCA1, both following enzyme pretreatment or heat-induced epitope retrieval. CD34 is useful in the evaluation of vascular and spindle cell tumors, especially fibroblastic tumors. The differential diagnostic applications have been listed in Table 3.5.

Kaposi's sarcoma is consistently CD34 positive (Figure 3.2). Different types of angiosarcomas and epithelioid

Chapter 3: Immunohistochemistry of soft tissue tumors

Table 3.5 Differential diagnostic applications of CD34 in immunohistochemical analysis of soft tissue tumors: examples of tumors with contrasting patterns of immunoreactivity

CD34-positive tumors	Typically CD34-negative tumors
Dermatofibrosarcoma protuberans	Cellular fibrous histiocytoma, benign fibrous histiocytoma, scar tissue
Solitary fibrous tumor	Desmoid, fibrosarcoma, sarcomatous mesothelioma, synovial sarcoma
Myxoid spindle cell lipoma	Myxoid liposarcoma
Hemangiopericytoma	Meningioma (<10% positive), synovial sarcoma
Kaposi's sarcoma (spindle cells)	Spindle cell hemangioma (spindle cells)
Gastrointestinal stromal tumors (70%)	Abdominal and gastrointestinal leiomyomas and schwannomas
	Leiomyosarcoma (20–30% variably positive)
Angiosarcoma (50%)	Metastatic carcinoma, primary squamous cell carcinoma
Epithelioid sarcoma (50%)	
Neurofibroma (an often prominent CD34-positive component)	Schwannoma (few CD34-positive cells)
	Malignant peripheral nerve sheath tumor

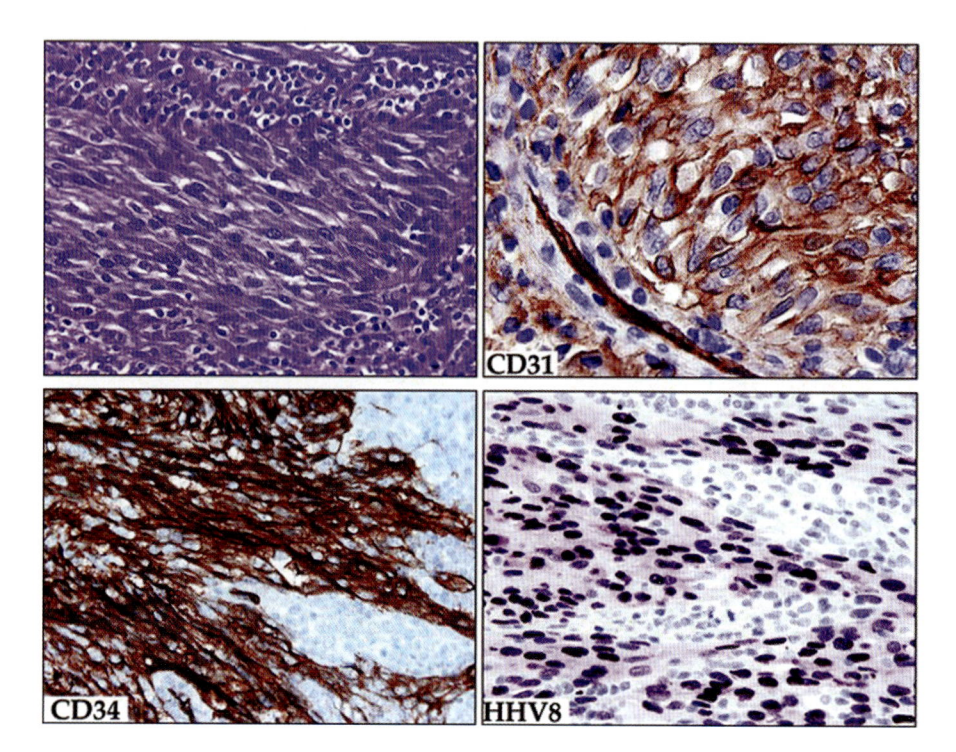

Figure 3.2 Positivity for both CD31 and CD34 is typical of Kaposi's sarcoma, and nuclear HHV8 positivity confirms the diagnosis. This example shows a fascicular spindle cell pattern that should not be confused with tumors such as leiomyosarcoma.

hemangioendotheliomas are variably positive in approximately 50% of cases.[18–20]

Several spindle cell fibroblastic tumors are consistently positive for CD34; these include dermatofibrosarcoma protuberans.[21–23] Tumors having fibrosarcomatous transformation are variably and inconsistently CD34 positive, however.[24] CD34 is useful in the differential diagnosis of dermatofibrosarcoma (DFSP) and benign fibrous histiocytoma, because the latter is almost always negative, except for possible focal positivity in tumor periphery.

Solitary fibrous tumors and hemangiopericytomas of different locations are equally positive,[16,25,26] but desmoplastic mesotheliomas[27] and synovial sarcomas are negative. A small subset of true meningiomas is variably CD34 positive, and this is more common in fibroblastic variants.

Of tumors of adipose tissue, spindle cell lipomas are positive, and many well-differentiated and some dedifferentiated liposarcomas contain CD34-positive tumor cells, whereas myxoid liposarcomas are negative. Similar CD34 positivity can also be detected in some myxofibrosarcomas and undifferentiated sarcomas (malignant fibrous histiocytoma; MFH).[28]

A CD34-positive fibroblastic component is typically present in neurofibromas, often in a netlike pattern.[29,30] Loss of the CD34-positive cell population can be a helpful feature in evaluating the transformation of neurofibroma into a malignant peripheral nerve sheath tumor.

Gastrointestinal stromal tumors (GISTs) are CD34 positive in 70% to 80% of cases. More consistently positive are spindle cell GISTs and esophageal, gastric, and rectal GISTs, whereas small intestinal and epithelioid gastric GISTs express CD34 in 50% of the cases; the positivity does not appear to vary by tumor malignancy. Gastrointestinal leiomyomas and schwannomas are negative. Typical leiomyosarcomas, however, especially the retroperitoneal and uterine ones, are variably CD34 positive in 20% to 30% of cases.[31,32]

Other positive tumors include epithelioid sarcoma (50%), which is diagnostically useful considering the very rare CD34 positivity of squamous and other carcinomas.[20] Approximately 30% of glomus tumors are variably positive. In hematopoietic tumors, CD34 is essentially limited to primitive leukemias and blastic extramedullary myeloid tumors.[16]

von Willebrand factor (vWF, Factor VIII-related antigen)

This structurally complex polymeric glycoprotein, von Willebrand factor (vWF), is composed of monomers of 270 kD, noncovalently bound to Factor VIII, and acts in the coagulation cascade.[33] It is synthesized and expressed in megakaryocytes and vascular endothelial cells, which also secrete vWF into the circulating blood and subendothelial matrix. Vascular endothelia of different types are positive, except some capillaries; liver sinusoids and lymphatics can be negative. Immunostaining often reveals granular cytoplasmic positivity, reflecting the subcellular location in the Weibel–Palade bodies of endothelial cells.[34]

vWF was the first marker to be used for the identification of neoplastic endothelial cells.[34] Although consistently present in benign endothelia, this antigen is inconsistently expressed in transformed endothelia and angiosarcomas, although its endothelial specificity is high. It is more often present in epithelioid hemangioendotheliomas (Figure 3.3) and well-differentiated angiosarcomas, often only focally.[11,34,35]

vWF is best used together with other endothelial markers, especially CD31. Necrotic and hemorrhagic tissue shows a high biologic background owing to the content of this antigen in the serum; this can make the staining uninterpretable.

Podoplanin (D2–40)

Podoplanin is a membrane protein present in renal glomerular podocytes and is also expressed in lymphatic vessels, mesothelial cells, dendritic reticulum cells, and some epithelial cells, for example in the sebaceous glands. The most widely used podoplanin antibody is monoclonal antibody D2-40.[36–40]

Among tumors, podoplanin is expressed in lymphangioma, Kaposi's sarcoma, angiosarcoma, and a subset of benign and malignant vascular tumors sometimes considered as displaying lymphatic endothelial-like differentiation (Figure 3.4). Podoplanin is also expressed in mesothelioma, seminoma, skin adnexal tumors, dendritic reticulum cell

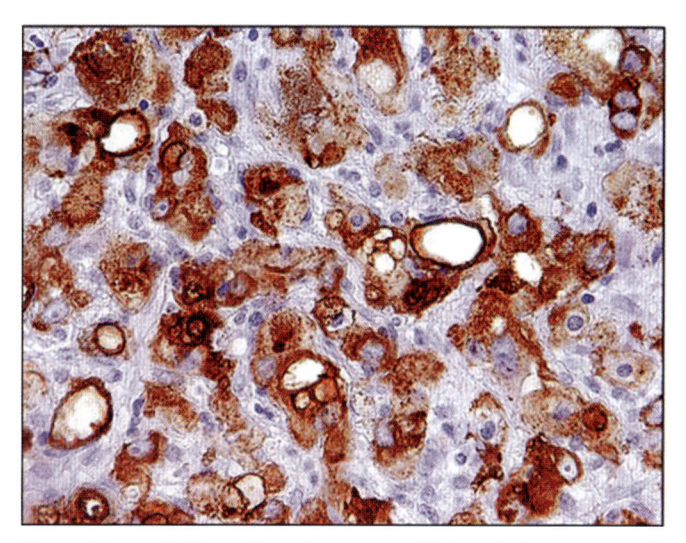

Figure 3.3 von Willebrand factor immunoreactivity is typically present in well-differentiated vascular neoplasms, such as low-grade hemangioendotheliomas. Shown here is an epithelioid hemangioendothelioma.

sarcoma, and schwannoma. Great caution is needed when using this marker, because of its multispecific nature. Podoplanin is best used in special circumstances with differential diagnostic entities having contrasting patterns of podoplanin expression (Table 3.6).[36–48]

Vascular endothelial growth factor receptor 3 (VEGFR3)

Vascular endothelial growth factor receptor type 3 (VEGFR3, formerly FLT4) is a transmembrane receptor tyrosine kinase specific for subsets of endothelia and trophoblast. The receptor is constitutively expressed in lymphatics and certain capillary endothelia, especially in the fenestrated endothelia in the renal glomeruli, endocrine organs, and the nasal mucosa. Its ligands are vascular endothelial growth factors C and D (VEGFC and VEGFD). The growth factors themselves are produced by different cell types; VEGFD is expressed by the cells of the dispersed endocrine system.[49]

VEGFR3 is expressed in 100% of Kaposi's sarcomas. A majority of angiosarcomas (80%) from different sites are positive, but those having epithelioid cytology, including epithelioid hemangioendotheliomas, are less often positive (40–50%). Endothelia of neovascular capillaries in carcinomas and sarcomas are often positive as well. Limited review of nonvascular tumors has shown them to be negative, suggesting that this marker can be useful in the separation of malignant vascular tumors from their nonvascular counterparts.[49–51]

ERG

ERG (abrreviated from ETS-related gene) is an ETS (E26 transformation-specific) family transcription factor. It is constitutively expressed in endothelial cell nuclei and also serves as an endothelial marker for vascular tumors, highlighting the nuclei of hemangiomas, hemangioendotheliomas, and angiosarcomas. It is one of the best, if not the best, endothelial cell markers. In addition, ERG is expressed in 40% of prostatic adenocarcinomas, apparently reflecting the subset with ERG-gene rearrangements[52,53]

Additional soft tissue tumors often positive for ERG include epithelioid sarcomas, where positivity has been reported with antibodies to N-terminus, but not carboxyterminus.[53,54] ERG is also expressed in cartilage and can be detected in many cartilaginous tumors, among them chondroma of soft parts, well-differentiated chondrosarcoma, chondroblastic osteosarcoma, and extraskeletal myxoid chondrosarcoma (ESMC; variably). However, enchondromas were negative when studied.[55] Also, tumors containing ERG expression-activating translocations, such as EWSR1-ERG in Ewing sarcoma, express nuclear ERG.[56]

The Freund's leukemia integration site (Fli-1) gene encodes for a transcriptional regulator protein consistently expressed in the nuclei of endothelial cells, lymphocytes, and Ewing sarcoma cells. The Fli-1 gene is a partner in the t(11;22) translocation in Ewing sarcoma; although Fli-1 is also a potential endothelial marker, its reactivity with lymphoid tissue makes it a less practical marker.[57,58]

Muscle cell markers

Smooth muscle, skeletal muscle, and myofibroblastic differentiation can be evaluated with several markers (Table 3.7). Many of them are cytoskeletal-associated proteins. *Actins* are a family of microfilament (diameter 6 nm) proteins. Desmin is the intermediate filament (diameter 10 nm) protein typical of muscle cells, and calponin and heavy caldesmon are the actin-binding, cytoskeleton-associated proteins of smooth muscle cells.

Transcriptional regulators of skeletal muscle, MyoD1 and myogenin, are specific nuclear markers for skeletal muscle differentiation: rhabdomyosarcoma and other tumors with rhabdomyoblastic differentiation.

A diagnostic pitfall for all muscle cell markers is the presence of entrapped or reactive muscle cells or myofibroblasts between the tumor cells. These components should not be misidentified as part of the tumor.

Table 3.6 Applications of podoplanin in tumor diagnosis: tumors with contrasting patterns of expression as potential diagnostic applications

Positive	Negative
Lymphatic vessels, lymphangioma	Blood vessels, endothelia, hemangioma
Kaposi's sarcoma	Leiomyosarcoma
Chondrosarcoma	Chordoma
Seminoma	Embryonal carcinoma
Schwannoma	Neurofibroma
Adrenal cortical lesion	Metastatic renal carcinoma Pheochromocytoma
Malignant mesothelioma	Pulmonary adenocarcinoma
Skin adnexal carcinomas	Metastatic carcinomas to skin

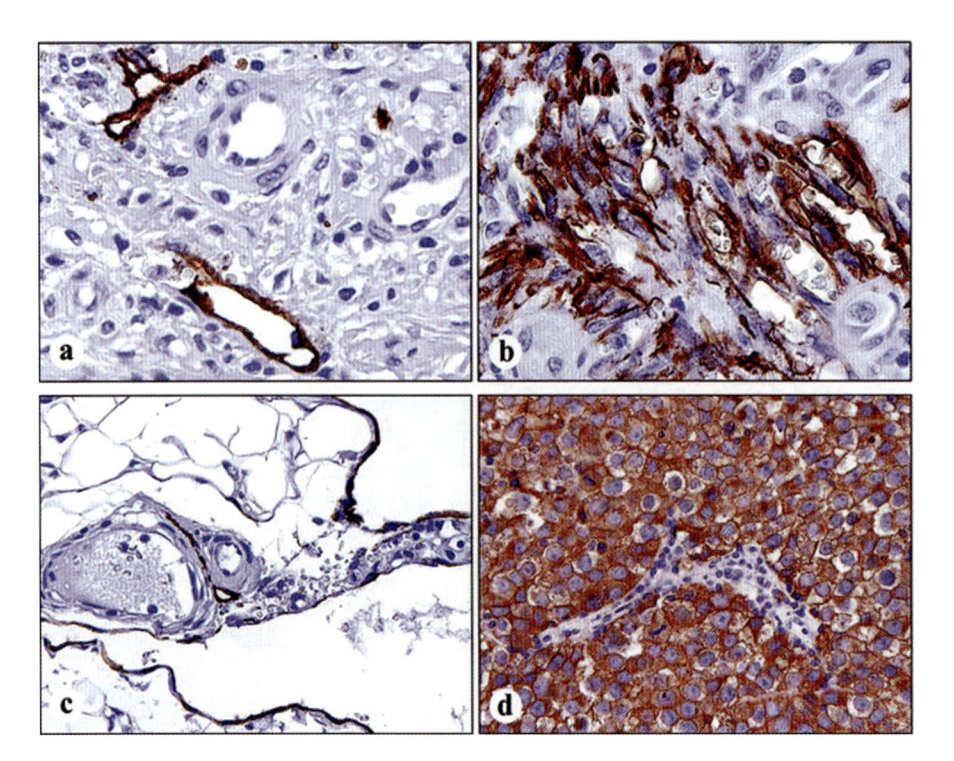

Figure 3.4 Examples of podoplanin immunoreactivity (D2–40 antibody). (**a**) In the skin, lymphatic vessels are positive, whereas blood vessels, recognizable here as vessels with pericytes, are negative. (**b**) Kaposi's sarcoma is strongly positive. (**c**) Normal mesothelial cells are positive. (**d**) Seminoma shows strong podoplanin immunoreactivity.

Table 3.7 Overview of the most widely used muscle cell markers and their application in muscle cell and other tumors

Marker	Antibody clone	Normal distribution	Tumors
Muscle-specific actin	HHF-35	Smooth and skeletal muscle Myoepithelia	Myofibroblastic (nodular fasciitis) Leiomyoma, leiomyosarcoma Rhabdomyoma, rhabdomyosarcoma Some myoepithelial cell tumors
Alpha smooth muscle actin	1A4	Smooth muscle, myoepithelia	Myofibroblastic (nodular fasciitis) Leiomyoma, leiomyosarcoma Rare cells in rhabdomyosarcoma Some myoepithelial cell tumors
Desmin	D33	Smooth muscle (most) Skeletal muscle Some mesothelial cells (reactive) Myoid cells of lymph nodes and seminiferous tubules	Some myofibroblastic tumors Leiomyoma, leiomyosarcoma (70%) Rhabdomyosarcoma Angiomatoid fibrous histiocytoma (50%) GIST (rarely) Desmoplastic small round cell tumor Aggressive angiomyxoma Angiomyofibroblastoma Tenosynovial giant cell tumor (rare cells), ossifying fibromyxoid tumor (rarely)
Calponin		Myofibroblasts, smooth muscle	Myofibroblastic and smooth muscle tumors Synovial sarcoma
Heavy caldesmon		Smooth muscle	Leiomyomas, leiomyosarcoma GIST (often)
Smooth muscle myosin		Smooth muscle	Smooth muscle tumors (myofibroblastic tumors negative)
MyoD1, myogenin		Skeletal muscle (fetal and reactive)	Rhabdomyosarcoma, heterologous skeletal muscle differentiation in any tumor

Actins

The actin microfilament (diameter 6 nm) cytoskeleton, ubiquitous in all types of cells, is abundant and bundled into myofilaments in muscle cells. There are at least six different and highly homologous actin isoforms (43 kD) originally discovered by two-dimensional gel electrophoresis and found to have different cell-type specificities. There are three alpha actins specific for smooth, skeletal, and cardiac muscle, respectively, a smooth-muscle-specific gamma actin and a ubiquitous beta actin.[59] Actin cytoskeleton is not only related to cell contraction and motility, but is also linked to numerous membrane and cytoplasmic cytoskeletal- and cytoskeleton-associated proteins with complex interactions and functions.[60]

Muscle-specific actin (HHF-35)

This antibody, HHF-35, identifies smooth and skeletal muscle-specific alpha and gamma actins and has been widely used in diagnostic immunohistochemistry. In normal tissues, it reacts with subsets of myoepithelial cells of complex glands, vascular and parenchymal smooth muscle, pericytes, and skeletal muscle cells.[61,62]

Skeletal muscle tumors are typically positive for HHF-35, including all types of rhabdomyosarcomas. The previous idea, that HHF-35 is more sensitive than desmin, was probably based on historical difficulties in the detection of desmin in formalin-fixed tissue.

Smooth muscle tumors, leiomyomas, leiomyosarcomas, and glomus tumors are consistently positive, and hemangiopericytomas are typically negative for HHF-35, representing a clinically useful contrasting pattern.[62] Leiomyosarcomas can show higher numbers of HHF-35-positive cells than those observed in desmin immunostaining.

Some myofibroblastic lesions, such as nodular fasciitis and tendon sheath fibroma, are strongly HHF-35 positive, a finding that should not lead to a misdiagnosis of leiomyosarcoma.[63] Other myofibroblastic tumors, for example, desmoids, have only scattered HHF-35-positive cells. Myofibroblasts in the desmoplastic stroma of carcinomas are more often positive for HHF-35 than desmin. Malignant fibroblastic tumors (e.g., fibrosarcomas, MFHs) can contain HHF-35-positive cells, reflecting a neoplastic or reactive myofibroblastic component.[62,64]

Alpha smooth muscle actin (SMA)

Antibodies specific to this actin subset (e.g., the widely used 1A4) have reactivities in normal tissues similar to those of HHF-35, except that they do not react with skeletal muscle cells. In neoplastic smooth muscle and myofibroblastic tissues, they are relatively similar to HHF-35. For example, leiomyomas, leiomyosarcomas, and glomus tumors are positive and hemangiopericytomas negative for SMA.[65–67] SMA antibodies generally do not stain rhabdomyosarcomas, although some tumors are focally positive.[68]

The best-documented SMA- (and mannitol salt agar; MSA)-positive carcinomas are perhaps myoepithelial carcinomas, which have been reported positive in 50% and 31%, respectively, for these markers.[69] It is not well understood whether the SMA-positive cells in other spindle cell carcinomas represent true myoid differentiation or myofibroblastic interstitial components, or if they are a reflection of multipotential mesenchymal differentiation or metaplasia.

Smooth muscle myosin and transgelin

Patterns of immunoreactivity resemble those of smooth muscle actin. Both smooth muscle and some myofibroblastic tumors, such as nodular fasciitis, are positive.[70] Transgelin has been proposed as a more sensitive marker to detect poorly differentiated leiomyosarcomas.[71]

Desmin

Biology and normal distribution

Desmin, a mostly muscle-specific intermediate filament protein (Mw 53 kD), binds the myofilaments together as bundles in both smooth and skeletal muscle and is one of the oldest markers for myogenic sarcomas. Desmin is present in some but not all vascular and all parenchymal smooth muscle cells and in cardiac and skeletal muscle cells.[72] Aortic vascular smooth muscle has been specifically reported as desmin negative,[73] whereas the smooth muscle cells of most small veins and arteries are positive. Desmin is also present in subsets of interstitial reticulum cells in the paracortex of lymph nodes.[74]

Desmin positivity in reactive mesothelial cells is a diagnostic pitfall that can potentially lead to a misdiagnosis of rhabdomyosarcoma. Transient desmin expression has been described in fetal mesothelial cells,[75] and reactive adult mesothelial cells also can be desmin positive.[75–77] Desmin presence in reactive mesothelial cells in effusions has been confirmed by Western blot.[77]

Desmin in soft tissue tumors

In most tumors, desmin positivity reflects muscle cell differentiation, but some myofibroblastic tumors and others of unknown histogenesis are also positive, a finding that suggests muscle cell lineage implications for desmin positivity in some cases.

All rhabdomyomas and nearly all rhabdomyosarcomas are positive, and this includes differentiated rhabdomyoblasts and varying numbers of small undifferentiated cells.[64,72] Desmin is also useful for highlighting the heterologous skeletal muscle components in triton tumor, endometrial carcinosarcoma, and dedifferentiated liposarcoma.

Of smooth muscle tumors, practically all leiomyomas and 70% to 80% of leiomyosarcomas are positive. The negative subset is not simply an artifact of formalin-fixed tissue, because lack of desmin expression has also been observed in frozen sections of ultrastructurally documented leiomyosarcomas.[65] Whether the desmin negativity reflects loss of antigen in less-differentiated cells or origin from a desmin-negative subset of smooth muscle is unknown.

In the gastrointestinal tract, desmin is useful in separating leiomyomas from GISTs, because the latter (CD117 [KIT]-positive GISTs) are usually negative. Fewer than 5% of GISTs are desmin positive, usually only focally.

Some myofibroblasts in fibrous tumors are desmin positive, for example, subsets of tumor cells in desmoids. Some specific female genital stromal tumors, angiomyofibroblastoma[78] and aggressive angiomyxoma,[79] are typically desmin positive, although they are not thought to be tumors of ordinary smooth muscle but rather myofibroblastic tumors. Desmin-positive tumor cells can also be present in typical myxoid[80] and sometimes in well-differentiated liposarcomas and in pleomorphic MFH; in the last they probably represent desmin-positive myofibroblasts.

Approximately 50% of angiomatoid (malignant) fibrous histiocytomas,[81,82] and 30% to 50% of alveolar soft part sarcomas and some ossifying fibromyxoid tumors are also desmin positive.[83,84] In desmoplastic small round cell tumors, desmin is typically coexpressed with keratins, often showing a dot-like cytoplasmic pattern.[85] Some Ewing family tumors express desmin focally, although these tumors should be separated from desmoplastic small round cell tumors.[86,87]

A few malignant mesotheliomas have been desmin positive, similar to some reactive mesothelial cells.[76,77,88,89] Desmin has also been commonly detected in the blastemal component of Wilms' tumors, which have been negative for other muscle cell markers.[90] Large desmin-positive cells with dendritic processes have been identified in tenosynovial giant cell tumors, where they could represent the true neoplastic component hidden behind an extensive histiocytic component.[91] Examples of desmin-positive tumors are show in Figure 3.5.

Heavy caldesmon (HCD)

The heavy molecular weight isoform of caldesmon (HCD, 34 kD) is an actin, Ca^{2+}, and calmodulin-binding cytoskeleton-associated protein involved in the regulation of smooth muscle contraction. This protein is highly expressed in smooth muscle and myoepithelial cells, but usually not in myofibroblasts.[92–94]

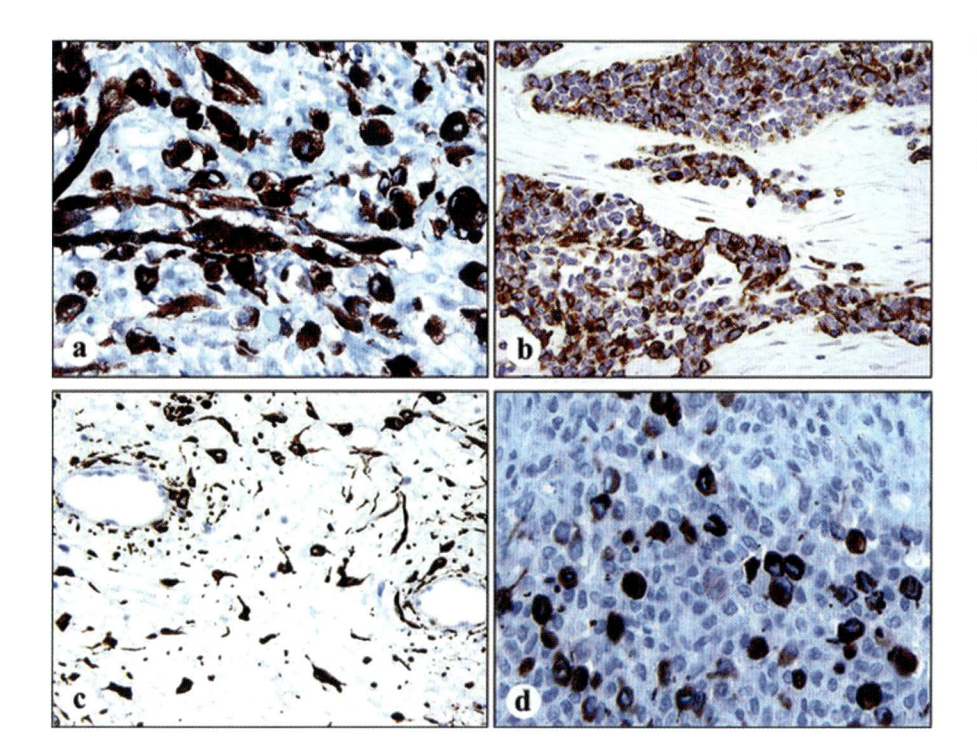

Figure 3.5 (**a**) Rhabdomyosarcoma cells are positive for desmin. (**b**) Cytoplasmic or dot-like desmin positivity is typical of desmoplastic small round cell tumor. (**c**) Aggressive angiomyxoma cells. (**d**) Some cells in tenosynovial giant cell tumors are desmin positive.

HCD is consistently present in leiomyomas, most glomus tumors, and nearly all leiomyosarcomas, but poorly differentiated (dedifferentiated) areas of the last category can be negative.[95–98] HCD is absent from rhabdomyosarcomas.[95,96] Interestingly, most GISTs are also positive, indicating traits of smooth muscle differentiation in these tumors.[96]

HCD can be used as a supplemental marker to diagnose smooth muscle tumors and to separate them from myofibroblastic tumors that are typically although not always negative. Examples of HCD and calponin immunoreactive tumors are shown in Figure 3.6.

Calponin

Calponin is an F actin- and tropomyosin-binding, cytoskeleton-associated protein considered important in the regulation of smooth muscle contraction. It is highly expressed in smooth muscle, myoepithelial cells, and myofibroblasts.[99] The desmoplastic stroma of many carcinomas is positive, as are tumor cells in leiomyomas (Figure 3.6) and leiomyosarcomas.[94] Most GISTs are negative. Myofibroblastic tumors such as nodular fasciitis are often positive. Synovial sarcomas often show positivity in spindle cell components.[96]

Myogenic regulatory factors MyoD1 (Myf3 antibody) and myogenin (Myf4 antibody)

These transcriptional regulators (transcription factors) of the myogenic determination/myogenic factor (MyoD/Myf) family are DNA-binding nuclear proteins with c-myc homologous helix-loop-helix (HLH) regions.[100–102] They regulate expression of the skeletal muscle-specific proteins and are lineage specific. MyoD1 (46 kD) determines the commitment to skeletal muscle differentiation, and myogenin (32 kD) is responsible for the terminal differentiation of myotubes and maintenance of skeletal muscle phenotype.[102] Both proteins are expressed in the nuclei of fetal (Figure 3.7) and regenerating cells, but not in normal adult skeletal muscle cells or other mesenchymal cells. Only nuclear positivity should be considered, and cytoplasmic positivity should be disregarded.[100–108] Both MyoD1 and myogenin are expressed in many different rhabdomyosarcomas (90%), and only nuclear positivity should be counted as positive (Figure 3.7). Their reactivities in rhabdomyosarcoma vary and therefore both markers should be used together.

Expression is typically higher in less-differentiated rhabdomyosarcoma cells, and these markers can be undetectable in differentiating rhabdomyoblasts in postchemotherapy specimens.[104–107] Alveolar rhabdomyosarcomas are often more strongly myogenin positive than are embryonal rhabdomyosarcomas.[108,109] Many embryonal rhabdomyosarcomas are extensively positive for MyoD1, but show limited if any myogenin positivity. Both markers appear highly specific for skeletal muscle tumors and have not been found expressed in other small round cell tumors and nonrhabdomyosarcoma spindle cell sarcomas with very rare exceptions, such as Wilms' tumor with myogenous differentiation.[106]

Expression of MyoD1/myogenin in reactive skeletal muscle in the periphery of any muscle-infiltrating tumor should not lead to a misdiagnosis of rhabdomyosarcoma. Cytoplasmic MyoD1 positivity can be present in nonmuscle tumors, for

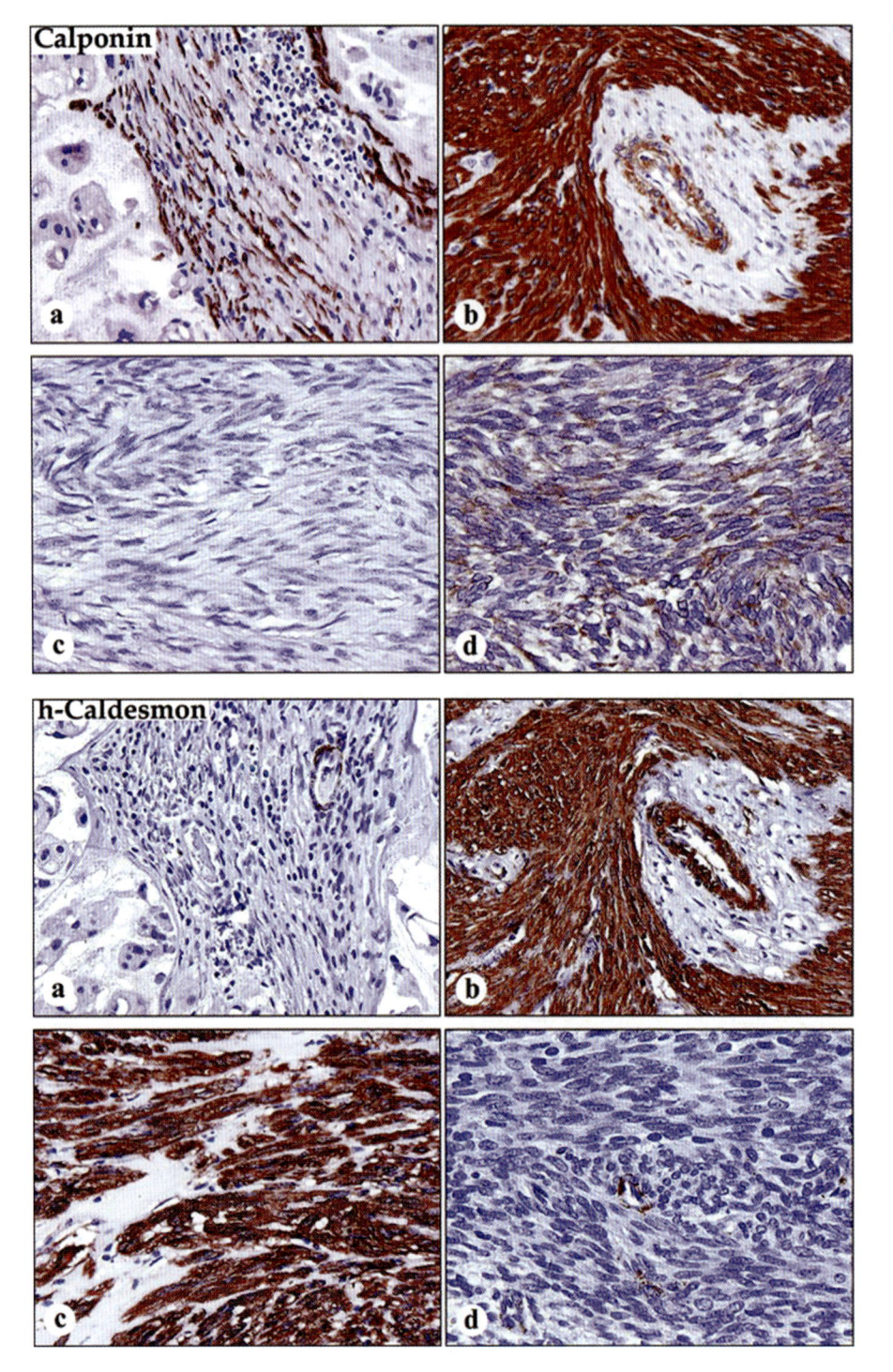

Figure 3.6 Examples of tissue distribution of calponin and heavy caldesmon. (**a**) Non-neoplastic myofibroblasts are positive for calponin and negative for caldesmon; only vascular smooth muscle is positive. (**b**) Leiomyoma cells are positive for both calponin and caldesmon. (**c**) Gastrointestinal stromal tumor is negative for calponin but positive for caldesmon. (**d**) Monophasic synovial sarcoma cells are positive for calponin but negative for caldesmon.

example in alveolar soft part sarcoma, in which this pattern of immunoreactivity has an unknown significance.[110,111]

Myoglobin

This oxygen-transporting hemoprotein is present in both skeletal and cardiac muscle.[112–114] Antibodies currently available are polyclonal, and common immunohistochemical background problems reduce the value of these reagents. Acquired phagocytosed myoglobin can have positive staining in histiocytes, which can be a diagnostic pitfall.[115] Myoglobin antibodies are no longer recommended for tumor-type diagnosis.

Neural and neuroendocrine specific markers

Chromogranin and synaptophysin are important in the diagnosis of paraganglioma, and primary and metastatic neuroendocrine carcinoma. Synaptophysin and NB84 can be used to

Table 3.8 Applications of neural and neuroendocrine markers in soft tissue tumors

	Synaptophysin	Chromogranin	Neurofilament proteins	Keratins
Neuroblastoma	+	+/−	+	−
Paraganglioma	+	+	−/(+)	−
Low-grade neuroendocrine carcinomas (carcinoids)	+	+	−/+[a]	+
Merkel cell carcinoma	+/−	−(+)	+	+
High-grade neuroendocrine carcinomas	−/+	+/−	+/(−)	+
Adenocarcinomas	−	−	−	+

[a] Neurofilament positivity in these tumors limited to NF68.

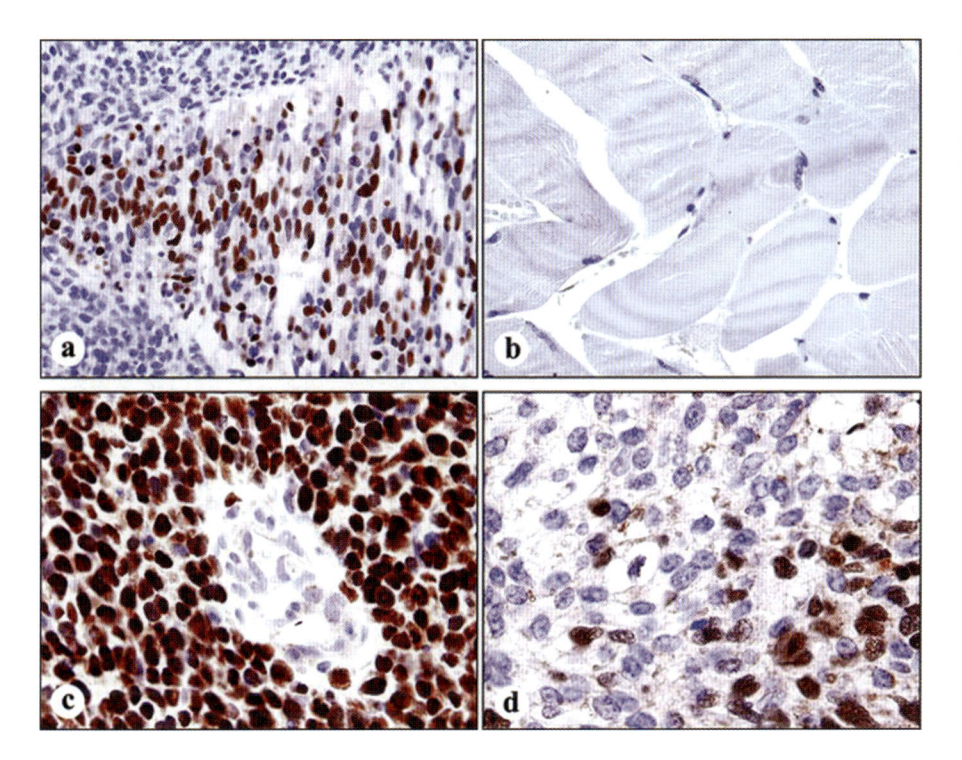

Figure 3.7 (a,b) MyoD1 immunostaining in fetal mesenchyme shows nuclear positivity in the whole somite developing into skeletal muscle, but adult skeletal muscle is negative. (c,d) Embryonal rhabdomyosarcoma cells show strong nuclear positivity that varies from extensive to focal.

separate neuroblastomas from other small round cell tumors. The most important applications have been listed in Table 3.8.

Synaptophysin

Synaptophysin is a primary marker, perhaps the first-line marker, for neuroendocrine differentiation. This acidic, 38-kD transmembrane glycoprotein of the small presynaptic vesicles is a membrane channel protein.[116,117] In peripheral tissues, it is expressed in neural and neuroendocrine cells, such as ganglion cells, axons, paraganglia, and most cells of the dispersed neuroendocrine system.[118,119]

Neuroblastomas, including most of the poorly differentiated ones, are positive for synaptophysin, the latter ones often only focally (Figure 3.8). Well-differentiated examples are more consistently positive, and in ganglioneuroma, the ganglion cells and stromal axons are positive.[120,121] Synaptophysin is also detectable in medulloblastoma,[122] central PNET, and in most esthesioneuroblastomas.[123]

The Ewing family of tumors, or peripheral PNETs, sometimes have been reported as synaptophysin positive, and positivity for this and other neuroendocrine markers has been considered evidence for the diagnosis of PNET, as opposed to a typical Ewing sarcoma.[124] Typical Ewing sarcomas are usually synaptophysin negative or only rarely positive.[125,126] In the author's experience, synaptophysin is rarely immunohistochemically demonstrable in any type of Ewing tumor. One series concluded that PNET-like differentiation with neuroendocrine markers does not have adverse prognostic significance.[127]

Paragangliomas of all types contain synaptophysin in the chief cells in 100% of cases.[128–130] Both primary and metastatic low- and high-grade epithelial neuroendocrine tumors (carcinoids and small cell carcinomas, including Merkel cell

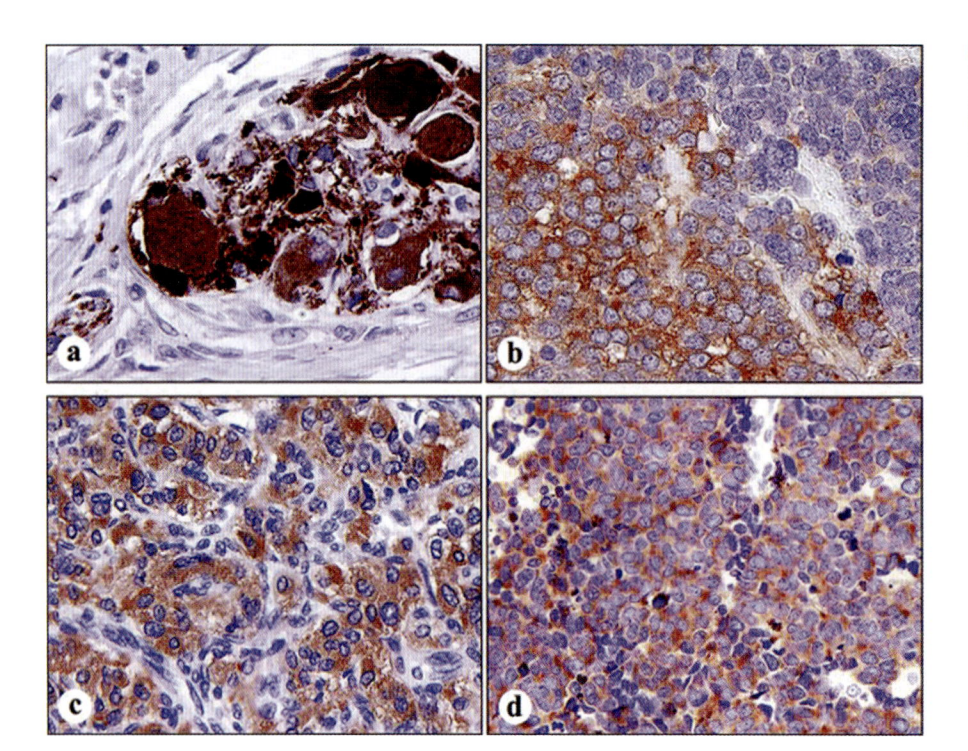

Figure 3.8 Cytoplasmic synaptophysin positivity: (**a**) Autonomic ganglion and a nerve (lower left corner). (**b**) Primitive neuroblastoma. (**c**) Carotid body paraganglioma. (**d**) Metastatic pulmonary small cell carcinoma.

carcinoma) are almost invariably synaptophysin positive, although high-grade tumors can display limited positivity; this is especially true in paraffin-embedded tissue.[128,129]

Synaptophysin is particularly helpful in distinguishing spindled neuroendocrine tumors (e.g., carcinoid, medullary thyroid carcinoma) from mesenchymal spindle cell neoplasms. High-grade neuroendocrine carcinomas, such as small cell carcinoma, tend to be more consistently positive for synaptophysin than for chromogranin.

Synaptophysin positivity has been reported in ESMCs; one study demonstrated that 87% of cases were positive. Neuroendocrine granules were also demonstrated by electron microscopy and verified by immunoelectron microscopy. Neuron-specific enolase was demonstrated in all cases. Although the significance of these observations is currently unclear, they have been suggested to indicate neural/neuroendocrine differentiation in ESMC.[131] In addition, many sarcomas of different lineages and metastatic malignant melanomas can occasionally express synaptophysin.[132]

Chromogranin A

Chromogranin A is a primary neuroendocrine marker, like synaptophysin, although somewhat less sensitive in poorly differentiated tumors. Chromogranin A is an acidic, glutamic-acid-rich calcium-binding soluble protein of the dense core granules of neural and neuroendocrine cells.[133] It is widely used as an immunohistochemical marker, and most data are based on this chromogranin. Other chromogranins (secretogranins) are less well-known as immunohistochemical markers because of the lack of established reagents. Chromogranin A is present in paraganglia, pheochromocytes (chromaffin cells) of the adrenal medulla, and epithelial neuroendocrine cells in the respiratory and gastrointestinal tracts, thyroid, and pituitary.[134]

Chromogranin has a limited distribution in soft tissue tumors. It is expressed in most neuroblastomas, but not in some primitive examples.[121,135] Paragangliomas, pheochromocytomas, and metastatic neuroendocrine carcinomas are all chromogranin positive.[130,134] Low-grade tumors such as carcinoids are more consistently positive, whereas high-grade tumors, such as Merkel cell carcinoma, often show limited positivity (25% of tumor cells). Small carcinomas can display even fewer positive cells (>1%), reflecting their low content of dense core granules.[134]

Chomogranin A has a very limited distribution in soft tissue sarcomas. Glandular malignant peripheral nerve sheath tumors (MPNSTs) were reported chromogranin positive in five of eight cases, and neuropeptides (i.e., serotonin, pancreatic polypeptide, and gastrin) were also detected, indicating neuroendocrine epithelial differentiation in these exceedingly rare tumors, which histologically could have colon-carcinoma-like columnar epithelial differentiation.[136] The Ewing family of tumors (PNETs especially) have been found to be almost invariably chromogranin negative, although they might have other neuroendocrine markers.[124–126] Secretogranin II, a protein of the chromogranin family, has been detected in Ewing sarcoma.[137]

Neurofilament proteins

The intermediate filament of the cytoskeleton of neurons and their axons contains neurofilament (NF) proteins as a major component. Aberrant organization or accumulation of NFs is

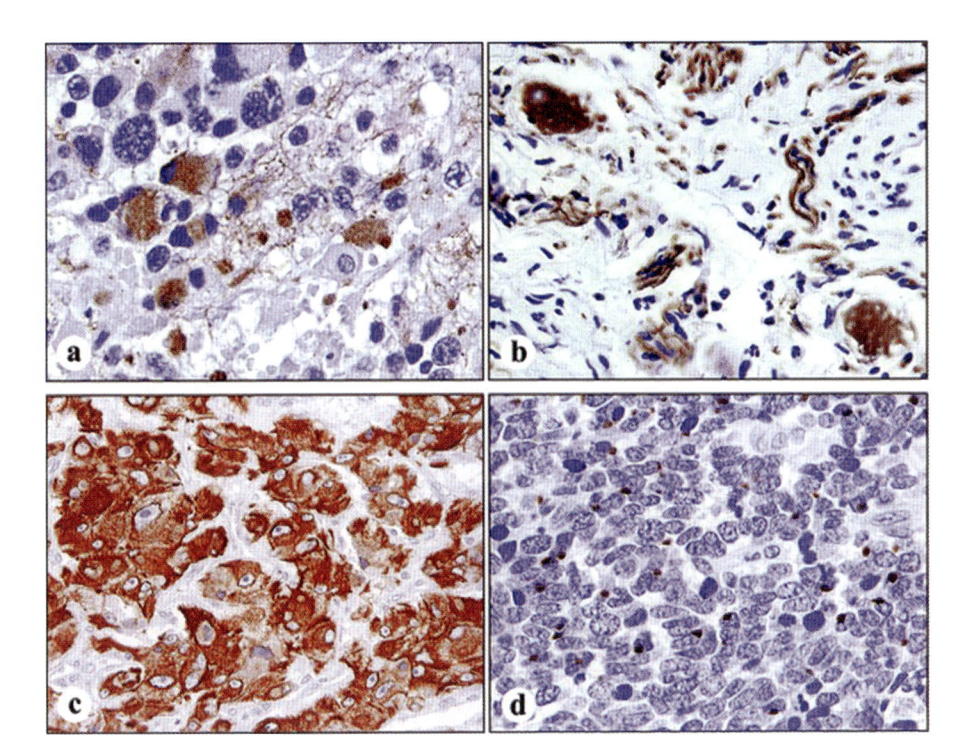

Figure 3.9 Neurofilament-positive tumors. (a) Larger ganglion-like cells in neuroblastoma. (b) Axons and ganglion cells in ganglioneuroma. (c) NF68-positive chief cells in retroperitoneal paraganglioma. (d) Dot-like paranuclear positivity in Merkel cell carcinoma.

probably central to the pathogenesis of certain neurodegenerative diseases.[138] There are three NF forms: the low-molecular-weight NF68 (68 kD), medium-weight NF160 (160 kD), and high-molecular-weight NF200 (200 kD). These are also referred to as NFL, NFM, and NFH, respectively. These forms are differentially expressed in different types of neurons and are developmentally regulated.[139]

Neurofilaments have a very restricted distribution in normal tissues. They are present in addition to neurons in the adrenal medulla, but not in normal epithelial or neuroendocrine cells.

Neurofilaments in soft tissue tumors

In soft tissue tumors, NF in Merkel cell carcinoma, often in a perinuclear dot-like pattern, has a great diagnostic value; most other carcinomas are negative.[140] Neuroblastomas are almost invariably NF positive, based on frozen-section studies. All subunits have been detected, but NF68 is expressed most frequently. In paraffin sections, however, only differentiated cells are positive (Figure 3.9).

Adrenal pheochromocytomas are consistently and globally NF positive, similar to paragangliomas of head and neck. These tumors mostly contain NF68, however, so that antibodies that react with this must be used.[141,142]

Epithelial neuroendocrine tumors with NF68-positive cells other than Merkel cell carcinoma include pancreatic islet cell tumors[143] and bronchial carcinoids.[144]

NF proteins have been reported in different sarcomas, including Ewing sarcoma,[145] metastatic epithelioid sarcoma,[146] poorly differentiated synovial sarcoma,[147] and embryonal rhabdomyosarcoma. Embryonal rhabdomyosarcoma contained NF68-positive cells.[148] The diagnostic and biologic significance of these observations is unknown.

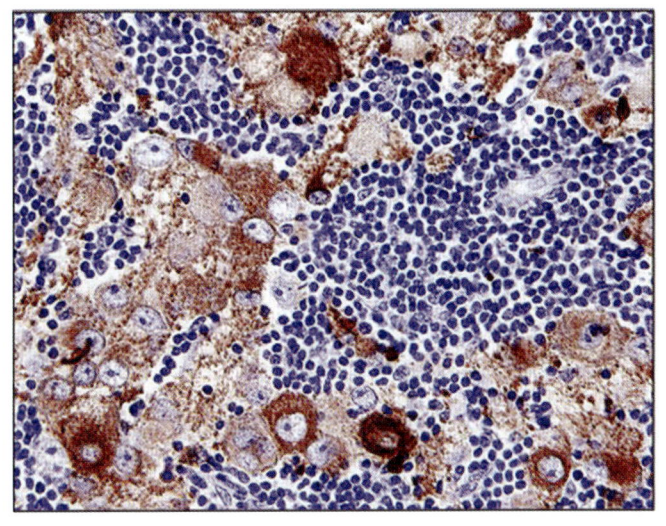

Figure 3.10 Lymph node metastasis of differentiating neuroblastoma is positive for NB84.

Axon presence can often be demonstrated with NF antibodies in neurofibroma and ganglioneuroma, whereas axons are generally not present in schwannomas (except possibly focally in the periphery at the nerve junction). This finding could have some differential diagnostic value.

Neuroblastoma marker (NB84)

The monoclonal antibody NB84 raised to antigen isolated from neuroblastoma cells detects a biochemically uncharacterized 57-kD protein.[149] Immunohistochemically it shows a cytoplasmic staining pattern (Figure 3.10). The antibody reacts with neuroblastomas consistently, irrespective of their level of

Table 3.9 Distribution of S100 protein in non-neoplastic and neoplastic soft tissues: selected most important applications

Normal cell	Tumor
Melanocyte	Nevi, malignant melanoma, all types
Schwann cell	Schwannoma, neurofibroma nerve sheath myxoma
Cartilage	Chondroma, chondrosarcoma Extraskeletal myxoid chondrosarcoma <30%
Adipocyte	Liposarcoma (variable)
Regenerating skeletal muscle	Rhabdomyosarcoma (variable)
Myoepithelial cells	Myoepithelioma/mixed tumor
Langerhans cell/ interdigitating reticulum cell	Rosai–Dorfman disease Langerhans cell histiocytosis Interdigitating reticulum cell sarcoma
Tumors with unknown normal cell counterpart	Ossifying fibromyxoid tumor (>90%) Synovial sarcoma (20–30%)

differentiation.[149–151] Approximately one-third of Ewing sarcomas also show positive cells, whereas positivity in other small round cell tumors is limited, but has occasionally been reported in rhabdomyosarcoma, Wilms' tumor, and small cell osteosarcoma.[150,151]

S100 protein and other multispecific neural markers

S100 protein, low-affinity nerve growth factor receptor p75, CD56 (NCAM), and CD57 (Leu7) have schwannian or neural specificities, but all of these markers react with many other cell types as well.

Although S100 protein has a moderately wide distribution in normal and neoplastic tissues, it remains a clinically useful marker because of its contrasting patterns of reactivity in morphologically similar tumors, and because it has been well studied. Table 3.9 lists the diagnostic applications of S100 protein.

Nerve growth factor receptor p75 is broadly distributed in neural and mesenchymal tissues, but its ability to separate dermatofibrosarcoma protuberans (positive) from dermatofibroma (negative) could be a diagnostic application.

CD56 and CD57, leukocyte antigens of natural killer cell lineage, are also expressed in many other tissues, including neural and neuroendocrine ones. They have limited applications in soft tissue tumors. Sox10 is a new useful marker for Schwann cell and melanocytic lineages, although it is also expressed in myoepithelial cells and their tumors, and basal-cell-type breast cancers.

S100 protein

S100 protein is a small, acidic, calcium-binding protein (Mw 21 kD) that belongs to the family of EF-hand proteins. It was originally isolated in bovine brain extract and so named by its solubility in saturated ammonium sulfate; most other proteins precipitate.[152] S100 protein has multiple extra- and intracellular functions related to regulation of Ca^{2+} homeostasis, enzymatic activities, and cell proliferation and differentiation.[153] There are two types of subunits: S100A and S100B, which are differentially expressed. The immunostaining patterns of the commonly used polyclonal antisera are thought to reflect the distribution of S100B.[154]

S100 protein is expressed in a variety of cell types that include melanocytes, Schwann cells and sustentacular cells of paraganglia, chondrocytes, adipocytes, myoepithelial cells of various glands, Langerhans cells of the skin, and the related interdigitating reticulum cells. Positive cells typically show both cytoplasmic and nuclear staining.[155]

Although S100 protein has many specific uses, it is also one of the most important everyday markers available in diagnostic immunohistochemistry. Examples of S100-protein-positive cell types are shown in Figure 3.11.

Melanocytes and their tumors

Normal and neoplastic melanocytes are consistently positive, and S100 protein is present in virtually all benign and malignant melanocytic tumors.[156–159] It is expressed in primary as well as metastatic melanomas; only 1% to 2% of the latter are negative. Melanoma is by far the most common malignant and strongly S100-protein-positive tumor, and this diagnosis must always be ruled out whenever such an epithelioid or spindled mesenchymal-appearing tumor is examined. Despite its lack of tumor specificity, S100 protein remains an important screening marker in the identification of metastatic melanoma.

Clear cell sarcoma of tendons and aponeurosis is a S100-protein-positive tumor that is distinct from melanoma,[159,160] although it also is HMB45 positive.

Schwann cells and their tumors

Schwann cells are the strongly S100-protein-positive components in peripheral nerves. Benign Schwann cell tumors, including the cellular Antoni-A areas of conventional and cellular schwannomas are strongly positive. S100 protein immunohistochemistry is often useful in the differential diagnosis from smooth muscle tumors and neurogenic sarcomas, which are positive much less often. Neurofibromas have a relatively smaller number of S100-protein-positive cells, but some elements in these tumors, especially the Meissner-like bodies and tumor cells with epithelioid morphology, are strongly positive.[156,159]

Almost all granular cell tumors are positive, with the notable exception of those in the newborn; newborn cells represent a distinct histogenetically unrelated tumor category.

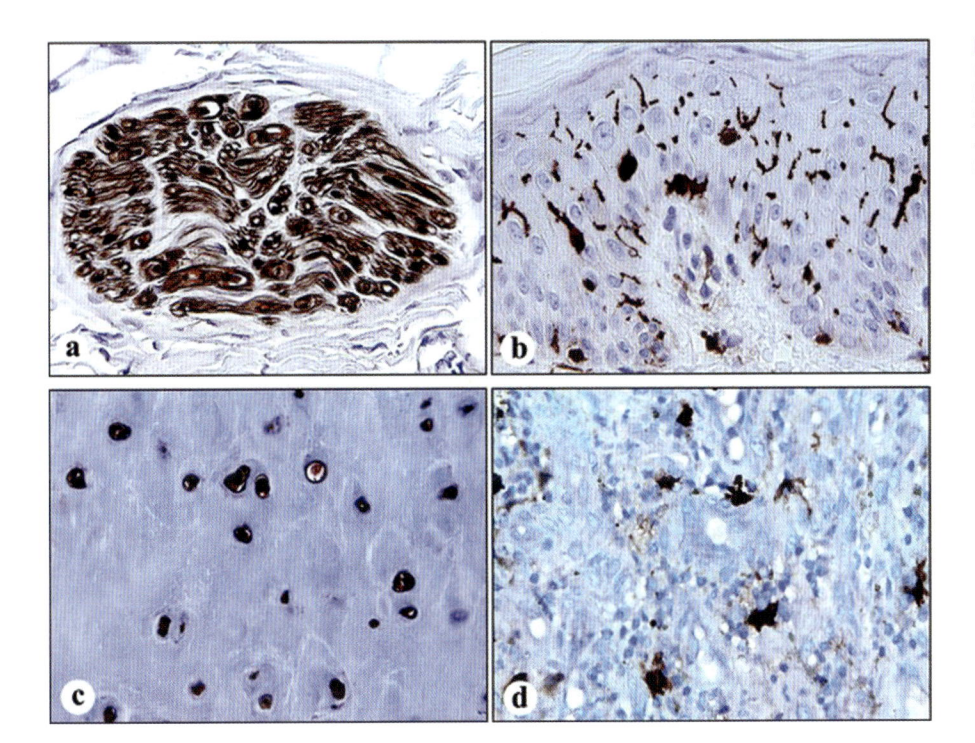

Figure 3.11 S100 protein positivity in non-neoplastic tissues. (**a**) Peripheral nerves. (**b**) Epidermal Langerhans cells. (**c**) Cartilage cells. (**d**) Scattered dendritic cells (antigen-presenting cells) are a common finding in many tumors and reactive conditions.

Rare other granular cell neoplasms occur in adults; these are fibroblastic tumors (see Chapter 7).

MPNSTs are S100 protein positive inconsistently. In published series they have shown positivity in 50% to 60% of cases.[161,162] In the author's experience, S100 protein positivity in MPNST has the following two aspects: first, in many tumors, pre-existent neurofibroma elements are seen as S100-protein-positive cells intermingling with the S100-protein-negative tumor cells. Second, approximately one-third of MPNSTs have schwannian or melanocytic-like differentiation and are S100 protein positive; the latter group includes most epithelioid MPNSTs.

Normal paraganglia and most paragangliomas contain a pericompartmental sheath of slender, S100-protein-positive Schwann cell-like cells, called *sustentacular cells*.[163,164] Their numbers vary in paragangliomas, and low numbers or no such cells are found more commonly in malignant than benign tumors.[163,164]

Cartilage and bone tumors

Normal cartilage cells are S100 protein positive, but osteoblasts are negative. Well-differentiated cartilaginous tumors, such as chondromas and low-grade hyaline cartilage-type chondrosarcomas, are positive. Islands of differentiated cartilage are highlighted as S100 protein positive in mesenchymal chondrosarcoma.[165,166] SMCs are positive in 30% of cases, with some variation; S100 protein has limited value in their diagnosis. Chordomas are often, but not always S100 protein positive, similar to their ancestral notochord cells.[167]

Fat and its tumors

Normal adipocytes are variably positive, and S100 protein has been suggested as useful in highlighting lipoblasts in liposarcomas and in revealing adipocyte differentiation in poorly differentiated myxoid (round cell) liposarcomas, and in differentiating between myxoid liposarcoma and myxoid MFH, the latter being negative.[168,169] In the author's experience, S100 protein positivity is variable in all types of liposarcomas, especially the pleomorphic ones, and dedifferentiated tumors are usually negative. S100 protein can be useful in the diagnosis of poorly differentiated myxoid/round cell liposarcomas, however.

Other mesenchymal tumors with S100 protein positivity

Ossifying fibromyxoid tumor is a distinctive soft tissue tumor of unknown histogenesis that demonstrates S100 protein positivity in most cases.[170] Approximately 20% to 30% of synovial sarcomas contain S100-protein-positive cells, some of them extensively so. Spindle cell and epithelial areas in both biphasic and monophasic tumors can show positivity.[171]

Although generally negative for S100 protein, many other sarcomas and benign mesenchymal tumors can have sporadic S100-protein-positive tumor cells. Rhabdomyosarcomas can be S100 protein positive; this could reflect the fact that regenerative, although not normal, skeletal muscle also can be S100 protein positive.[172] KIT-positive GISTs, especially those of the small intestine, are S100 protein positive in 10% of cases, and these tumors can also coexpress smooth muscle actins.

Epithelial cells and carcinomas

Among benign epithelial cells, myoepithelial cells of sweat glands, salivary glands, and other complex epithelia are typically S100 protein positive.[155]

S100 protein positivity is usually seen in myoepithelial and mixed tumors and occurs relatively infrequently in carcinomas. In some studies, however, 12% to 42% of primary and metastatic carcinomas of the breast and other sites had S100-protein-positive tumor cells.[173,174] Carcinomas reported as being S100 protein positive with a 50% to 60% frequency include carcinomas of eccrine sweat glands, salivary glands, and breast, and some report metastatic renal, endometrial, and ovarian carcinomas positive, with a frequency of 66% to 87%.[174]

Histiocyte-related cells

The cutaneous Langerhans cells[175] and their homolog in lymphoid tissues, the interdigitating reticulum cells, are S100 protein positive,[176] and varying numbers of similar cells with dendritic processes are seen in almost all reactive and neoplastic soft tissues. These cells serve as a good universal internal control for S100 protein immunostaining and should not be mistaken for positive tumor cells. The large epithelioid histiocytes in Rosai–Dorfman disease (sinus histiocytosis with massive lymphadenopathy) are strongly S100 positive,[177,178] which is diagnostically helpful. Langerhans cell histiocytosis cells in bone and nodal lesions are strongly positive.[155]

Interdigitating reticulum cell sarcomas are extremely rare S100-protein-positive malignant hematopoietic tumors. This diagnosis is often considered in the case of a nodal lesion with no apparent primary location. In the author's experience, most tumors initially suspected as such prove to be metastatic malignant melanomas, and interdigitating reticulum cell sarcoma should be a diagnosis that is made only after careful exclusion of melanoma, supported by other more specific markers (e.g., CD1a or CD45).

Sox10

Sox10 is a SOX-family transcription factor expressed in melanocytes, Schwann cells, and myoepithelial cells. In contrast to S100 protein, it is not expressed in Langerhans cells and dendritic S100-protein-positive cells.[179–183]

Sox10 is similar to S100 protein in its nearly uniform reactivity with benign and malignant melanocytic tumors and benign Schwann cell neoplasms. Like S100 protein, it also is only variably expressed in MPNSTs often only labeling pre-existing neurofibroma elements.

Sox10 is also expressed in myoepithelial tumors of soft tissue although it is inconsistently present in malignant myoepitheliomas. It is not expressed in other soft tissue tumors. Sox10 is not expressed in some other S100-protein-positive cell and tumor types, such as Langerhans and cartilaginous ones, and in this context it is more melanocyte/schwannian lineage-specific than S100 protein.[183]

Sox10 can be useful in the differential diagnosis between schwannian nerve sheath tumors vs perineurioma and meningioma, which are negative. It also helps to distinguish melanocytic tumors from PEComas, as the latter are Sox10 negative.[183]

Melanoma markers other than S100 protein

HMB45, tyrosinase, melan A (MART-1), and microphthalmia transcription factor (MITF) are the most important melanoma markers. All of them have additional specificities that are sometimes diagnostically useful (Table 3.10). CD63, which also

Table 3.10 Summary of melanocyte-specific markers and their reactivities

	HMB45	Tyrosinase	Melan A	Microphthalmia transcription factor
Normal melanocytes	−	+/−	+	+
Activated melanocytes	+	+	+	+
Cellular blue nevus	+	+	+	+
Primary melanoma	+	+	+	+
Metastatic melanoma	+/(−)	+/(−)	+/(−)	+/(−)
Desmoplastic melanoma	−	−	−	−
Clear cell sarcoma	+	+/−	+/−	+/−
Angiomyolipoma	+	−/(+)	+	−/+
Lymphangiomyoma	+	−	−/+	−/+
Adrenal cortical carcinoma	−	−	+	−
Metastatic carcinomas, other	−	−	−	−
Osteoclast-like giant cells	−	−	−	+

+ Almost always positive; +/− = usually positive, minority of cases negative; −/+ = usually negative, minority of cases positive; − = almost invariably negative.

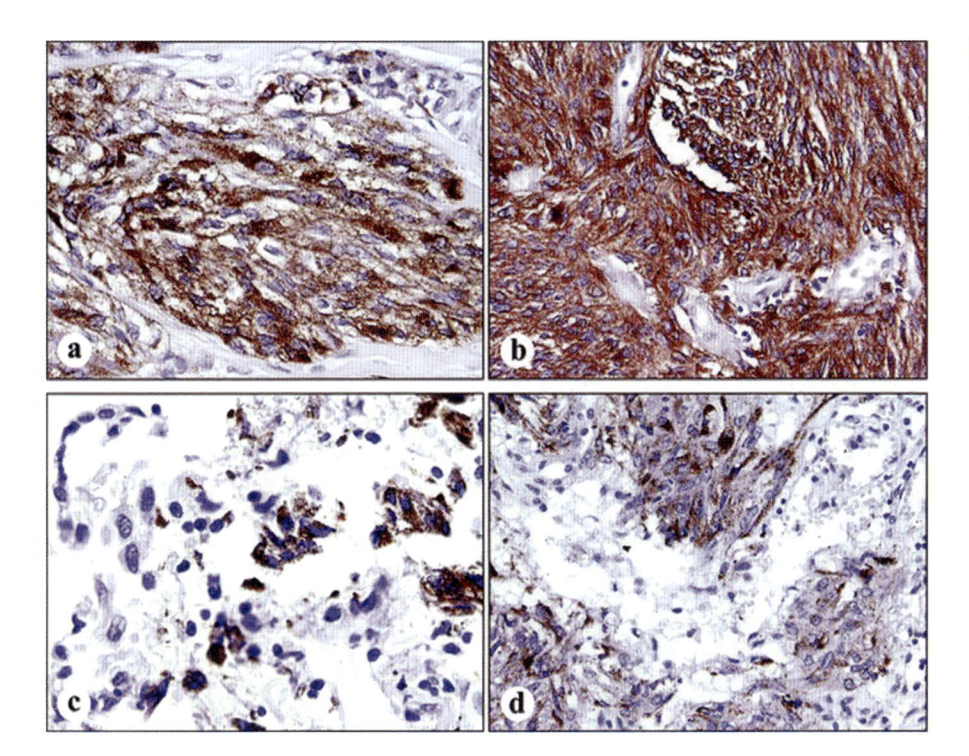

Figure 3.12 (a) Granular cytoplasmic positivity for HMB45 in metastatic melanoma. (b) Cellular blue nevus is positive. (c,d) Lymphangiomyoma and angiomyolipoma are focally positive for HMB45.

labels melanocytic tumors, is less specific and therefore impractical.

HMB45 (human melanoma, black)

This monoclonal antibody was raised to pigmented metastatic melanoma by Gown *et al.* and so named after the immunogen, human melanoma, black (HMB).[184] It recognizes an oncofetal 100-kD glycoprotein gp100 present in the immature but not mature melanosomes and is thought to be a melanosome-specific marker.[185] Fetal and activated junctional adult melanocytes are positive, whereas the resting adult melanocytes and intradermal nevus cells and melanophages are negative. Blue nevi of conventional and cellular types and primary melanomas are consistently positive.

Several studies have shown primary melanomas to be positive in 90% to 100% of cases, but spindle cell melanomas can be negative.[186,187] Metastatic melanomas are positive, with a slightly lower frequency, varying from 70% to 90%.[186–189] The number of positive cells varies and can be limited to only a few cells. Desmoplastic melanomas are HMB45 negative in more than 90% of cases.[189,190]

Clear cell sarcoma of tendons and aponeuroses is a tumor with melanocytic differentiation, and it is consistently HMB45 positive.[160] Sometimes this tumor has a stronger reactivity for HMB45 than for S100. Generally, MPNSTs are negative for HMB45, which is useful in separating them from metastatic melanoma; however, rare pigmented nerve sheath tumors with a melanocytic component can be HMB45 positive. Of these, pigmented neurofibroma has scattered HMB45-positive cells,[191] and psammomatous melanotic schwannomas are also positive.

Angiomyolipoma and the closely related lymphangiomyoma (tosis) are unique nonmelanocytic renal, hepatic, retroperitoneal, or pulmonary tumors that contain HMB45-positive cells in most cases, although the number of positive cells varies.[191–194] These cells often have epithelioid morphology and show granular cytoplasmic immunostaining. Clear cell ("sugar") tumor of the lung and some similar extrarenal tumors, designated as perivascular epithelioid cell tumors, PEComas, or myomelanocytic tumors, are HMB45-positive, angiomyolipoma-related tumors (Figure 3.12) that variably coexpress smooth muscle actins and desmin.[195]

HMB45 is thought to be a highly specific lineage marker for melanocytic differentiation, but it does not distinguish malignant from benign lesions. A spectrum of carcinomas, sarcomas, and lymphomas has been consistently negative. The reported positivity in primary and metastatic carcinomas is now thought to be based on impure antibody preparations in most cases.[185]

Tyrosinase

This 75-kD enzyme glycoprotein catalyzes the two earliest steps of tyrosine incorporation into melanin pigment.[196] It has been identified as a target for melanoma immunotherapy, for which monoclonal antibodies were initially raised.[197] More recently, these antibodies have been applied for diagnostic immunohistochemistry.

Based on several studies, tyrosinase has been suggested as an excellent marker for metastatic melanoma, which shows diffuse cytoplasmic staining in 80% to 90% of cases.[188,189,198–200] Desmoplastic melanomas are almost invariably negative, however.[189,200] In cutaneous pigmented nevi,

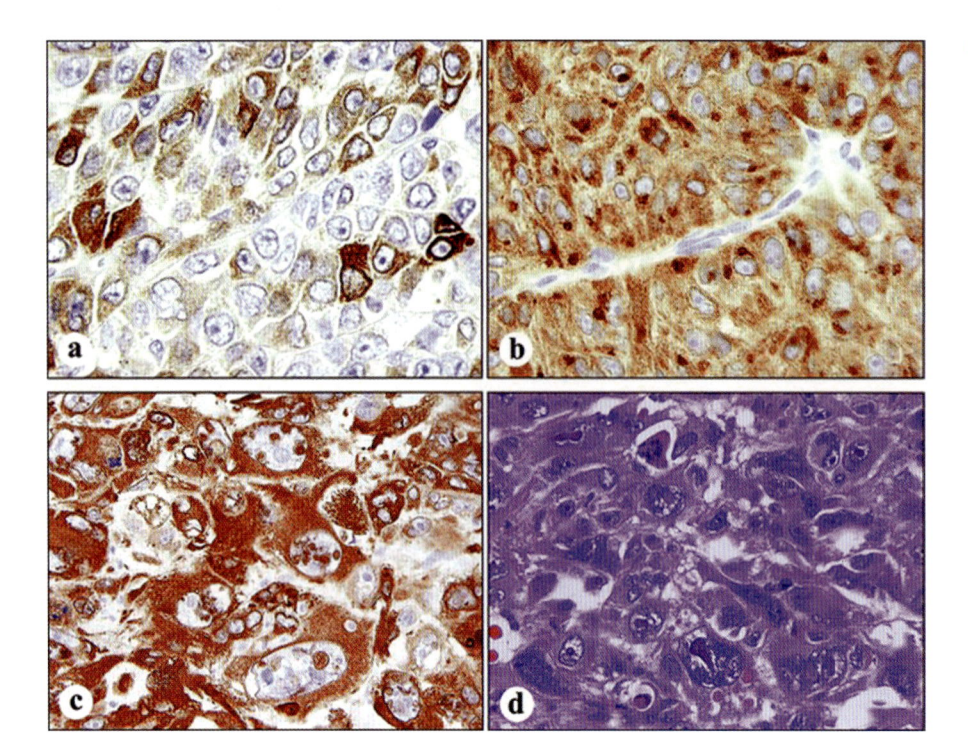

Figure 3.13 Tyrosinase in metastatic melanoma. (**a**) Focal positivity. (**b**) Extensive cytoplasmic and perinuclear positivity. (**c,d**) Pleomorphic melanoma metastasis is strongly positive.

superficial components are more tyrosinase positive than deeper components. Clear cell sarcomas are also often positive,[189] whereas carcinomas, other sarcomas, and lymphomas have been consistently negative.[188] Renal angiomyolipoma is only rarely positive for this melanocytic marker.[201]

Tyrosinase is also a sensitive marker for metastatic melanoma, because a larger number of tumor cells tends to be positive (Figure 3.13).

Melanoma antigen recognized by T-lymphocytes or melan A

This antigen, melanoma antigen recognized by T-lymphocytes (MART-1), was originally identified as a target for T-cell immunologic response to melanoma. The gene was cloned and simultaneously identified as coding for MART-1[202] and melan A.[203] The antigen is closely related to the gp100 recognized by HMB45 monoclonal antibody and has been identified as an antigen recognized by autologous T-lymphocytes in an immune response leading to tumor lysis.[202] The melan A protein is expressed in all normal melanocytes and adrenal cortical cells (Figure 3.14).[202,204,205]

Melan A is present in normal melanocytes and nevi and has been shown to be a useful marker for melanoma. Based on two large series, melan A is present in all primary melanomas and 81% to 88% of metastatic conventional melanomas, but desmoplastic melanomas have been generally negative.[189–205] Mucosal melanomas have also been positive.[206]

Angiomyolipomas are positive for melan A, typically more homogeneously than for HMB45, and melan A transcript has also been demonstrated.[207] Lymphangiomyomas have been variably positive.[208]

Melan A positivity has also been detected in steroid cell tumors, including adrenal cortical adenomas and carcinomas, and in ovarian and testicular Leydig cell tumors with A103 antibody to melan A. Because carcinomas of other organs have been melan A negative, this marker has been suggested as useful in the identification of carcinomas of adrenal cortical origin.[209] The common tendency of the adrenal cortex and its tumors for endogenous avidin-biotin binding should be considered as a pitfall, causing false-positive staining.

Microphthalmia transcription factor

This nuclear protein is a transcriptional regulator for key melanocyte genes, such as tyrosinase.[210,211] It is also expressed and functionally important in osteoclasts. MITF-deficient mice do not develop bone marrow space and suffer from osteopetrosis.[212]

MITF is expressed in melanocytes, primary melanomas, and most metastatic malignant melanomas, typically with most tumor cells showing nuclear positivity.[189,213,214] Clear cell sarcomas are often but not always positive. MITF can be a useful adjunct for the diagnosis of metastatic melanomas. Some histiocytes, especially osteoclastic giant cells and epithelioid histiocytes granulomas, are also positive, however (Figure 3.15). The fibrohistiocytic tumor, neurothekeoma, is also positive.

Desmoplastic melanomas are essentially negative for MITF,[189,200,215,216] although one study showed 50% of these variants to be positive.[216] This antigen has been variably detected in angiomyolipoma from 20% to 70%.[201] Potential positivity in mononuclear histiocytes and leiomyosarcoma necessitates caution in its diagnostic use.[189,200]

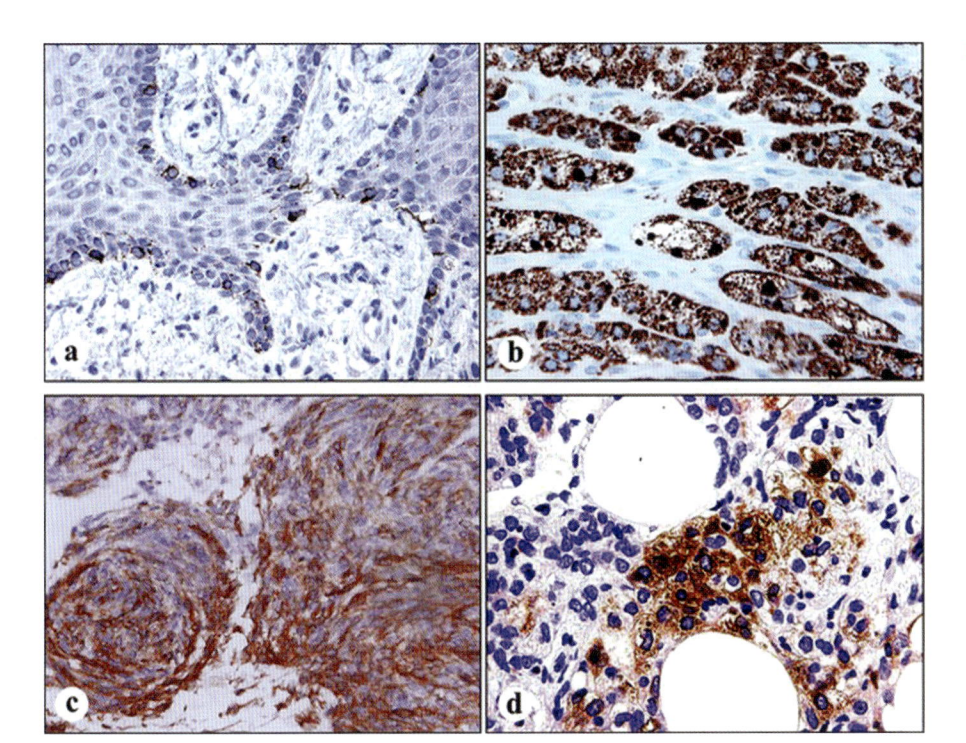

Figure 3.14 Melan-A-positive tissues. (a) Normal epidermal basal layer melanocytes. (b) Adrenal cortical cells. (c) Cellular blue nevus. (d) Angiomyolipoma of kidney.

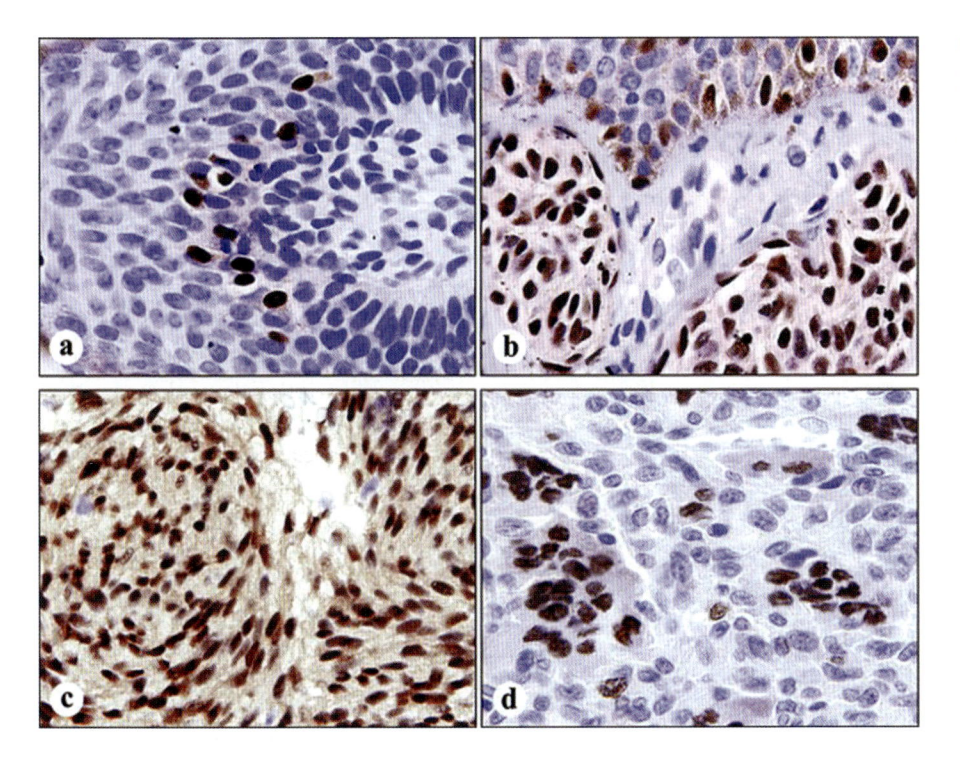

Figure 3.15 (a,b) Melanocytes in normal skin and a nevus show uniform nuclear positivity for microphthalmia transcription factor. (c) Cellular blue nevus and (d) osteoclast nuclei and those of some mononuclear xanthomatous histiocytes are also positive.

KBA62 and PNL2

KBA62 and PNL2 are two new markers that recognize a majority of melanomas. Of these KBA62 has a low specificity. PNL2, while relatively specific for melanoma but failing to recognize subtypes such as desmoplastic or spindle cell melanoma, also reacts with PEComas.[217,218]

Histiocytic markers

These markers identify subsets of myeloid and histiomonocytic cells and true histiocytic tumors, which are very few: extramedullary myeloid tumors, very rare histiocytic sarcomas (formerly true histiocytic lymphomas), reticulohistiocytomas, and juvenile xanthogranuloma. Hardly any of the histiocytic

Table 3.11 Summary of reactivity for histiocytic markers in tumor cells in various lesions[a]

	Histiocytic specificity	Positive tumors
Factor XIIIa	High	Juvenile xanthogranuloma Infiltrating histiocytes in fibrous histiocytomas and many sarcomas
CD68	Low	Histiocytic, schwannian, melanocytic tumors Granular cell tumor Some carcinomas
CD163	High	True histiocytic tumors: Histiocytic sarcoma Reticulohistiocytoma Juvenile xanthogranuloma Mononuclear histiocytes positive Osteoclasts negative

[a] All markers also detect tumor-infiltrating histiocytes or dendritic histiocytes.

markers are truly histiocyte specific, and some "histiocytic" antigens are widely expressed. The histiocytic markers discussed here are Factor XIIIa, CD68, and CD163. The last seems to be the most specific marker. Table 3.11 lists diagnostic applications.

Factor XIIIa (FXIIIa)

This enzyme protein, a protransglutaminase, operates in the late stages of blood coagulation in the formation of the fibrin clot.[219] This immunohistochemical marker selectively identifies subsets of dendritic histiocytes in lymph nodes and skin, where these cells have been called *dermal dendrocytes* and were once thought to be the origin for Kaposi's sarcoma spindle cells.[220,221]

FXIIIa immunostaining is mainly used in dermatopathologic lesions, and researchers have now realized that the FXIIIa-positive cells mostly represent non-neoplastic infiltrating cells. These cells are prominent in fibrous histiocytomas, but scant in dermatofibrosarcoma protuberans, which could have some diagnostic value.[22] FXIII-positive stromal histiocytes are abundant in MFHs, but also can be present in other tumors, such as leiomyosarcoma; therefore their demonstration does not have a significant value in tumor typing.[222] The lesional cells of juvenile xanthogranuloma are positive, and in this instance the positivity reflects true histiomonocytic differentiation.[223,224]

CD68

This 110-kD lysosomal protein can be detected in formalin-fixed and paraffin-embedded tissue with the KP1 and PG-M1 antibodies.[225–227] CD68 is a lysosomal and not histiocyte-specific marker, and this explains its expression in many nonhistiocytic lysosome-rich cells and tumors, such as granular cell tumor and Schwann cell neoplasms, metastatic melanoma, and paraganglioma.[228–230] In epithelial tumors, positivity has been noted especially in renal adenocarcinoma.[222,230,231] Although some malignant fibrous histiocytomas show tumor cell immunoreactivity,[232] this finding does not indicate histiocytic differentiation, and it has no specific diagnostic value toward the definition of MFH, in view of the usual staining in definitely nonhistiocytic cells.

CD68 is highly expressed in histiomonocytic cells, such as tumor-infiltrating mono- and multinucleated histiocytes, including osteoclasts. A cytoplasmic staining pattern is observed, and its demonstration could have some value in their evaluation. Juvenile xanthogranuloma is a CD68-positive true histiocytic tumor.

CD163

This antigen, an erythrocyte scavenger receptor, is expressed in histiocytes in a cytokine-dependent manner.[233,234] In the author's experience, it is an excellent histiocytic marker in paraffin-embedded and formalin-fixed tissue. Most mononuclear histiocytes are positive in a membrane and cytoplasmic pattern. Most epithelioid cells of granulomas and multinucleated histiocytic giant cells, such as osteoclasts, are negative, however. CD163 is more histiocyte-specific than CD68; melanomas and granular cell tumors are CD163 negative, whereas they are usually CD68 positive.[235,236]

True histiocytic neoplasms typically positive for CD163 include juvenile xanthogranuloma, reticulohistiocytoma, histiocytic sarcoma, and some leukemias with monocyte/macrophage lineage differentiation. In contrast, benign and malignant fibrous histiocytomas and tenosynovial giant cell tumors are rich in infiltrating CD163-positive histiocytes, but the neoplastic component – sometimes overshadowed by extensive histiocytic infiltration – is negative (Figure 3.16). CD163 is currently the best histiocytic marker in the diagnosis of true histiocytic tumors.

Keratins (cytokeratins) and epithelial membrane antigen

Keratins form the intermediate filament (10 nm) cytoskeleton typical of but not totally restricted to epithelial cells and their tumors. Epithelial keratins (cytokeratins, soft keratins) form a group of proteins of two multigene families. There are additional keratins restricted to hair and nail. The following discussion centers on epithelial (soft, non-hair) keratins.

The functional significance of keratins includes support of structural integrity, but keratins also interact with many cytoskeletal and other proteins and participate in cell signaling and regulation of apoptosis.[237]

The individual epithelial keratins are relevant to pathologists, because antibodies to most of them are now applicable in formalin-fixed tissue, and diagnostic patterns in their

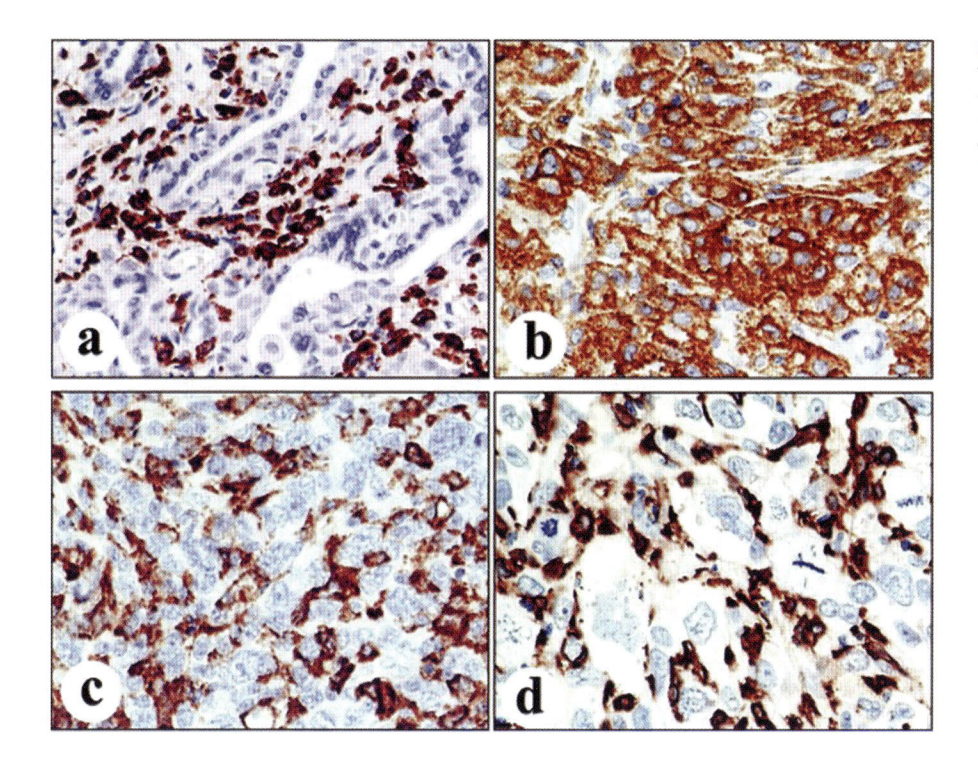

Figure 3.16 (**a**) CD163 positivity in placental villous histiocytes. (**b**) Juvenile xanthogranuloma cells are strongly CD163 positive. (**c,d**) Malignant melanoma and malignant fibrous histiocytoma contain large numbers of CD163-positive histiocytes, whereas the tumor cells are negative.

distribution have emerged. Knowledge of the individual keratins also makes it easier to understand the results obtained with keratin cocktails. Although keratins are filaments, positivity is typically seen as diffuse cytoplasmic immunostaining. Their filamentous nature can often be seen in immunofluorescence staining, especially in cytologic specimens, or by confocal microscopy.

History and terminology

The 20 different epithelial keratins were originally identified biochemically by two-dimensional (2D) gel electrophoresis by Franke et al.[238] and numbered by Moll et al.[239,240] and Sun et al.[241–243] In the 2D-system, the keratins are first solubilized with high molar urea. Then they are separated first by isoelectric properties (isoelectric focusing or non-equilibrium pH-gradient electrophoresis [NEPHGE]), and thereafter by molecular weights under denaturing conditions (SDS-polyacrylamide gel electrophoresis [SDS-PAGE]). Keratins were identified in most carcinomas but not in mesenchymal cells in general.[239,241] Mass spectrometric proteomic analysis has been used to identify individual keratins separated by 2D gel electrophoresis.[244]

In soft tissues, keratins were first identified by polyclonal epidermal keratin antibodies in synovial sarcoma,[245] epithelioid sarcoma,[246] chordoma,[247] and normal myometrium.[248] These results have been subsequently confirmed with monoclonal antibodies and some by Western blotting, and detailed analysis of keratin subsets has been performed on many tumor types.

Keratins (especially keratins 8 and 18) have also been reported in many other sarcomas, complicating the separation of epithelial and mesenchymal tumors by keratin immunohistochemistry.[249] The imunohistochemical and biochemical demonstration of keratins in fetal mesenchymal cells[75] and cultured fibroblasts[250,251] forms a rational basis for their presence in various mesenchymal tumors.

The term *keratin* is recommended over *cytokeratin*, and the gene names consist of letters KRT, followed by the number.

Overview of keratins

The individual epithelial keratins (recommended abbreviation: K, followed by the number), and their main normal distribution are summarized in Table 3.12. Their distribution in carcinomas, mesenchymal cells, and soft tissue tumors are summarized in Tables 3.13 to 3.15.

Keratins of type I (type A, with acidic isoelectric points) are numbered 9 to 20. They constitute a spectrum from low- (K19: 40 kD) to high-molecular-weight keratins (K9: 58 kD),[239–241] and, with the exception of K18, are encoded by genes clustered in 17q, suggesting evolution by gene duplication.[252,253] Keratins of type II (type B, with basic isoelectric properties) are numbered 1 to 8. They also constitute a spectrum from low (K8: 52 kD) to higher molecular weights (K1: 67 kD), and together with K18 are encoded by a cluster of genes in 12q.[253,254]

Keratin filaments are formed as heteropolymers of at least one type I and at least one type II keratin. The pair forming type II keratin is typically 8 kD larger than the corresponding type I keratin. Based on this principle, all epithelial cells contain at least two different keratins. For example, hepatocytes contain two keratins (K8 and K18), whereas epidermal keratinocytes have a complex constellation of 6 to 10 keratin polypeptides.[239,241,242]

Chapter 3: Immunohistochemistry of soft tissue tumors

Table 3.12 Summary of the epithelial keratins and their most important distribution patterns[a]

Type II (basic, B) keratins			Type I (acidic, A) keratins	
Name	**Size**	**Distribution pattern**	**Name**	**Size**
		Soft epithelia of palms and soles	K9	64 kD
K1	67 kD	Keratinizing epithelia of skin Some endothelia (K1 only)	K10	56–57 kD
		Keratinizing epithelia of skin	K11	56 kD
K2	65 kD	Some keratinizing epithelia		
K3	62 kD	Restricted to corneal epithelium	K12	54 kD
K4	60 kD	Cells other than basal layer of internal squamous epithelia urothelial cells	K13	52 kD
K5	58 kD	Basal cells of glandular and some squamous epithelia, including myoepithelial cells, mesothelial cells Upper layers of some squamous epithelia Some skin adnexal epithelia Especially hair shaft basal cells	K14 K15	50 kD 50 kD
K6	56 kD	Hyperproliferative epidermis, some skin adnexa (sweat glands, subsets of hair shaft epithelia) Myoepithelia, basal cells of complex glandular epithelia, subsets of hair shaft epithelia	K16 K17	48 kD 46 kD
K7	54 kD	Respiratory and upper gastrointestinal epithelia, mesothelia, some basal cells, complex glandular epithelia, some endothelial cells		
K8	52 kD	Most types of simple epithelia + mesothelium Superficial layer of urothelia (umbrella cells) Some endothelial cells (K18 only) Most types of simple epithelia and mesothelia. Absent in hepatocytes, some renal tubules and normal thyroid follicles Complex distribution in internal squamous epithelia Absent in normal epidermis Lower gastrointestinal tract epithelia, urothelia, Merkel cells of skin, and lip mucosa	K18 K19 K20	45 kD 40 kD 46 kD

[a] Keratins typically expressed as a pair are shown on the same line, and those with less known pairing patterns are shown singly on a line.

Table 3.13 Immunohistochemical distribution of keratins in normal mesenchymal cells

	K8	K18	K19	K7	K1
Myometrium, other parenchymal smooth muscle	(+)	(+)	(+)	−	−
Vascular smooth muscle	(+)	(+)	(+)	(+)	−
Capillary endothelia	−	(+)	−	(+)	(+)
Meningothelial cells	−	+	−	−	−
Interstitial reticulum cells of lymphoid tissue	+	+	−	−	−

+ = consistently positive; (+) = variably positive; − = negative.

The lowest-molecular-weight keratins in each group are typically expressed in simple, nonstratified, noncomplex epithelia (e.g., gastrointestinal epithelia and mesothelia), whereas the others are present in complex glandular, ductal, and stratified squamous and transitional cell epithelia. The keratins of simple epithelia are often called *low-molecular-weight keratins*. This is partly misleading, because type II keratins of the lowest molecular weight have higher molecular weights than some of the higher-molecular-weight type I keratins.

The keratins are discussed here in their order of relative importance. Keratins of simple epithelia are the ones most commonly detected in carcinomas and sarcomas, but extensive

Table 3.14 Expected results of immunostainings for different keratin polypeptides in adenocarcinomas and selected other carcinomas

Metastatic adenocarcinoma from	K7	K8	K13	K14	K17	K18	K19	K20
Breast	+	+	−/(+)	−/+	−/(+)	+	+	−/(+)
Lung	+	+	−/(+)	−/(+)	−/(+)	+	+	−/(+)
Stomach	+/(−)	+	−	−	−	+	+	+/−
Pancreas	+	+	−	+/−	+/−	+	+	+/−
Colon	−/(+)	+	−	−	−	+	+	+/(−)
Kidney	−/(+)	+	−	−	−	+	+/−	−
Prostate	−/(+)	+	−	−	−	+	+	−(+)
Squamous cell carcinoma	+/−	+/−	+/−	+	+	+	+	−
Transitional cell carcinoma	+	+	+/−	−/+	+/−	+	+	+/−
Small cell carcinoma, lung origin	+	+	−	−	−	+	+	−
Merkel cell carcinoma	−/(+)	+	−	−	−	+	+	+

Note that results represent typical cases, and aberrations are possible. − = negative; + = positive; +/(−) = generally positive, rarely negative; +/− = variable (approximately equally often positive as negative); −/+ = positive in a minority of cases; −/(+) = generally negative, rarely positive, usually focally. Based on references[221–225, 230], and author's unpublished results.

Table 3.15 Expected results from immunostainings for different keratin polypeptides in selected soft tissue tumors

	K7	K8	K13	K14[a]	K17	K18	K19	K20
Malignant mesothelioma, epithelial	+	+	−	+/−	−	+	+	−
Synovial sarcoma, biphasic, epithelia	+	+	−/+	+	−/+	+	+	−/+
Synovial sarcoma, monophasic	+	+	−	−/(+)	−	+	+	−
Epithelioid sarcoma	−/(+)	+	−	−/+	−	+	+/−	−
Epithelioid hemangioendothelioma	+/−	−/(+)	−	−	−	+	−	−
Epithelioid angiosarcoma	−/+	−/+	−	−	−	+/−	−/(+)	−
Angiosarcoma, nonepithelioid	−/(+)	−/(+)	−	−	−	−/+	−	−
Ewing family of tumors	−	(+)	−	−	−	(+)	−	−
Desmoplastic small round cell tumor	−	+	−	−	−	+	+/−	−
Chordoma	−/(+)	+	−	−	−	+	+	−(+)
Leiomyosarcoma	−	−/+	−	−	−	−/+	−	−
Melanoma metastasis	−	−/+	−	−	−	−/+	−	−

Note that results represent typical cases, and aberrations are possible. − = negative; + = positive; +/(−) generally positive, rarely negative; +/− = variable (approximately equally often positive as negative); −/+ = positive in a minority of cases −/(+) = generally negative, rarely positive, usually focally. Based on references[230–235, 250–253, 256], and author's unpublished results.
[a] Results largely also apply to K5, except that mesotheliomas tend to be more often positive for the latter.

studies have shown stratified and even keratinizing epithelial keratins to a surprising degree in mesenchymal cells and tumors. Examples of keratin immunoreactivity are shown in Figure 3.17.

Keratins of simple, nonstratified epithelia: K8 and K18

K8 (52 kD) and K18 (45 kD), the evolutionarily oldest keratins, are already present at the earliest embryonal stages. They are widely expressed and generally codistributed in practically all of the simple epithelia of all organs, including mesothelia.[239,255] In normal undisturbed hepatocytes and some renal tubular cells they are the only keratins present. In many complex glandular epithelia, they are preferentially expressed in the luminal epithelial cells, but are absent in most normal squamous epithelia. In the epidermis, they are present only in Merkel cells. In the urothelium, they are limited to the uppermost cell layer, the umbrella cells.[256,257]

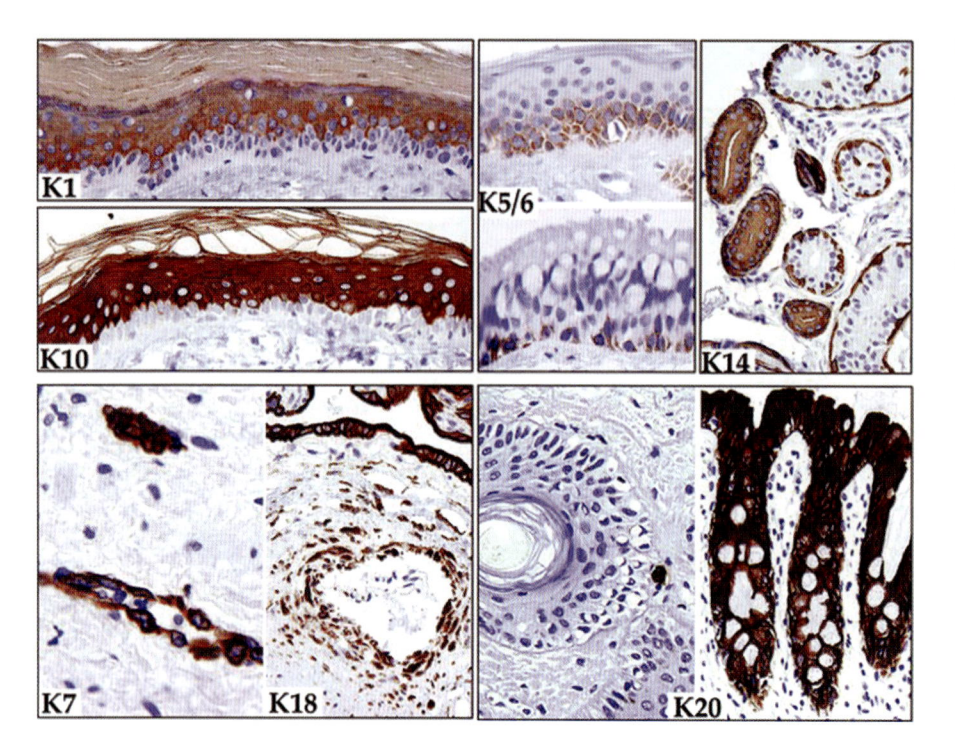

Figure 3.17 Keratins 1 and 10 are essentially limited to keratinizing squamous epithelia. Keratin 5/6 and 14 antibodies typically recognize basal layers of squamous and respiratory epithelia. Keratin 7 positivity in vascular endothelia is a common finding. Keratin 18 is often detected in vascular smooth muscle cells. Here also placental trophoblasts are positive. Keratin 20 is present in rare hair-shaft-associated cells representing Merkel cells, and in intestinal epithelia.

K8 and K18 are present in most nonsquamous cell carcinomas, absent in most well-differentiated squamous cell carcinomas of skin, and variably expressed in poorly differentiated cutaneous and many internal organ squamous cell carcinomas.[258,259]

K8 and K18 in mesenchymal cells and soft tissue tumors

In mesenchymal cells, reactivity to K8 or K18 antibodies or both has been demonstrated in smooth muscle cells,[260-262] where their presence has also been demonstrated by Western blotting.[262] These keratins can also be expressed in developing myocardium, whereas they are absent from the adult heart.[75,263] Myometrial smooth muscle often shows dot-like immunoreactivity, and gastrointestinal, prostatic, and vascular smooth muscle cells can also be positive.

K18 (but not K8) is expressed in some capillary endothelial cells and meningothelial cells.[264-266] Both K8 and K18 are present in interstitial reticulum cells of lymph nodes, a subset of paracortical spindle cells.[267]

Expression of K8 and K18 in virally transformed fibroblasts has been shown by multiple methods. Such expression is thought to indicate a relaxed control of K8 and K18 expression in transformed mesenchymal cells and could explain the presence of these keratins in some sarcomas and other nonepithelial tumors.[250,251] Nearly ubiquitous expression of keratin 8 and 18 transcripts has been demonstrated by reverse transcription polymerase chain reaction (RT-PCR) in various nonepithelial tissues.[268] The cellular correlation of these studies is difficult to ascertain, however. Keratin expression in smooth muscle and endothelial cells is probably responsible for this finding.

In epithelial soft tissue tumors, K8 and K18 are expressed in the epithelia of biphasic synovial sarcoma, and in scattered cells in monophasic and variably in poorly differentiated synovial sarcomas.[269,270] Epithelioid sarcoma[271-273] and normal notochord and chordoma[274-276] are usually positive, although one study found K18 in only one half of the chordomas.[275] Merkel cell carcinoma and most metastatic carcinomas are positive.[277]

K8 and K18 are present in reactive submesothelial spindle cells, with other simple epithelial keratins.[278] Some angiosarcomas are positive for keratins K8 and K18 and the antibodies that recognize them.[265,279] Epithelioid hemangioendotheliomas are typically K18 positive, but negative for K8, and over 50% of angiosarcomas are K18 and K8 positive; this probably represents neoexpression in transformed cells, because normal endothelial cells are K8 negative.[264,265]

Leiomyosarcomas are positive for K8 or K18 or the antibody cocktails recognizing them in 20% to 40% of cases.[280,281] Other sarcomas, such as MFH,[282-284] MPNST,[269] and rhabdomyosarcoma, can also be focally positive.[148,172] GISTs, especially the malignant variety, are K18 positive in 20% to 30% of cases, but they are less commonly and usually only focally K8 positive.[285]

Meningiomas commonly contain K18-positive cells. They generally show only limited K7 and K8 expression. Secretory meningiomas have a complex keratin expression pattern in pseudoglandular structures, however.[266]

The Ewing family of tumors commonly have scattered K8- and K18-positive cells, often in a starry sky-like random pattern, and keratin positivity has been verified in molecular genetically confirmed cases by a AE1/AE3 keratin cocktail.[145,286]

Other nonepithelial malignant tumors with K8 or K18 expression include 10% to 20% of metastatic melanomas[287] and occasional B-cell lymphomas, mostly those with plasmacytic differentiation,[288,289] and large cell anaplastic lymphomas.[290]

K18 is an important marker in poorly differentiated and sarcomatoid carcinomas, where it can be the most prominent and best detectable keratin type.

Keratins of simple, nonstratified epithelia: K19

K19 (40 kD) is the keratin with the lowest molecular weight. K19 is normally present in most simple epithelia, including mesothelia, but normal hepatocytes and some renal tubular cells are negative.[239] K19 is widely expressed in complex glandular epithelia, especially in the luminal cells. K19 is absent in the epidermis, but it is present in skin adnexa and basal cells of internal squamous epithelia of the esophagus and cervix, and in other layers with a complex, variable distribution pattern.[291]

K19 in mesenchymal cell and soft tissue tumors

In mesenchymal cells, K19 can be expressed in myometrium, which often shows dot-like cytoplasmic positivity. Leiomyosarcomas are generally negative, however.[269] Submesothelial spindle cells are positive, and endothelia are negative.

Most types of primary and metastatic carcinomas are positive, except some hepatocellular, renal, and adrenal cortical carcinomas.[258]

The epithelial cells in biphasic synovial sarcoma and small numbers of spindle cells in most monophasic synovial sarcomas are positive. K19 can be demonstrated in 70% to 80% of epithelioid sarcomas and virtually all chordomas.[269,270,273,276] K19 is very rarely present in malignant epithelioid vascular tumors, and therefore antibodies to it or multispecific antibodies that do not recognize K18 or K19 (e.g., AE1 antibody) are useful in the differential diagnosis of malignant epithelioid vascular tumors and carcinomas.[265]

Keratins of simple, nonstratified epithelia: K7

K7 (54 kD) is a simple epithelial keratin with a moderately wide distribution in simple and complex glandular epithelia, such as breast, respiratory tract, upper gastrointestinal tract, mesothelia, and urogenital tract epithelia.[239,292,293] In breast epithelia, it shows a predominantly luminal distribution, whereas in the prostate it marks selectively basal cells. K7 is absent in colorectal epithelium (except for neuroendocrine cells). In squamous epithelia, K7 is absent in the epidermis, but is variably present in internal squamous epithelia of the esophagus and larynx.

K7 in soft tissue tumors

Carcinomas generally show a distribution similar to that seen in normal tissues. Typically positive are adenocarcinomas of breast, lung, hepatobiliary and urogenital tracts, uterus, and ovaries. Typically negative are colorectal, prostatic, and hepatocellular carcinomas, although minor subsets of these tumors are positive.[294-297] Merkel cell carcinomas are usually negative but can be focally positive.[298]

In mesenchymal cells, K7 is often present in endothelia of venules and lymphatics (Figure 3.17). In vascular tumors, K7 is present in lymphangioma, epithelioid hemangioendothelioma (50%), and in 10% to 20% of angiosarcomas. K7 also seems to be present in some smooth muscle cells.

In sarcomas, K7 is strongly expressed in the epithelial cells of biphasic synovial sarcoma, and it is probably the most prevalent keratin present in monophasic synovial sarcoma with positive foci in most cases. K7 has been found useful for separating poorly differentiated synovial sarcoma from MPNST and Ewing family tumors, which are negative.[269,299]

Epithelioid sarcomas show limited K7 reactivity, usually focal, in 30% of cases. Chordomas are focally positive in 10% to 20% of cases.[276]

Keratins of simple, nonstratified epithelia: K20

K20 (46 kD), originally discovered and named *protein IT*, is expressed in gastrointestinal epithelia, urothelia, and Merkel cells and Merkel cell carcinomas of skin.[240,300] Biphasic synovial sarcoma epithelial cells are often focally positive,[270] but monophasic tumors and epithelioid sarcomas are negative.[301] Metastatic carcinomas of colorectal, urothelial, and, to lesser degree, other gastrointestinal origin are positive, whereas only a minority of breast, prostatic, and renal origin show positive cells, usually focally.[296,297,300,302]

Keratins of basal cells of complex epithelia: K5 and K14

K5 (58 kD) and K14 (50 kD) are present in basal cells of squamous and other complex epithelia, including the myoepithelial cells and tumors derived thereof.[239,241,303,304] Although normally present only in the basal cells of squamous epithelia, these keratins are typically present in most cells of squamous cell carcinomas. In addition, they are variably expressed in subsets of adenocarcinomas and ductal carcinomas, which are clinically aggressive tumors.[305,306] K5 is typical of mesothelioma, often detected with K5/6 antibody, and has been considered one of the most useful markers in the differential diagnosis between mesothelioma and adenocarcinoma, because the latter tumors are usually negative.[307,308] One study of paraffin-embedded carcinomas suggested that K14 antibody could be useful in demonstrating squamous differentiation of carcinomas. Focal positivity in ductal adenocarcinomas of various organs is possible, however.[304]

In soft tissue tumors, K14 is typically extensively expressed in the epithelia of most biphasic synovial sarcomas, but is rarely present in monophasic and poorly differentiated tumors.[270] It is also present in 50% of epithelioid sarcomas, usually focally. In our experience, K14 expression in mesothelioma is usually less prominent than that of K5.

K5 (often studied with an antibody that reacts with both K5 and K6) is less prominently expressed in synovial sarcoma than K14; however, foci of positive cells are seen in most biphasic tumors and in about one half of the monophasic examples.

Keratins of basal cells of complex epithelia: K17

K17 (46 kD) is expressed in myoepithelia, the basal cells of many glandular complex epithelia (e.g., respiratory epithelia), and in subsets of hair shaft epithelia.[309,310] Although absent in normal squamous epithelia, K17 can be detectable in squamous hyperplasia.

In soft tissue tumors, K17 is often present focally in the glandular epithelia of biphasic synovial sarcoma, but is absent in monophasic tumors; it is almost always absent in epithelioid sarcoma as well. Squamous carcinomas are usually positive, and some adenocarcinomas, especially those of pancreatic origin, show variable positivity.[311,312]

Keratins of internal squamous epithelia and urothelium: K4 and K13

K4 (60 kD) and K13 (52 kD) are present and coexpressed in all but basal cells of internal squamous epithelia (e.g., mouth, esophagus, cervix, vagina) and in the urothelium.[310,313,314] Differentiated urothelial and squamous carcinomas of internal squamous epithelia are typically positive, whereas this marker is generally lost in the poorly differentiated carcinomas. Focal positivity is rare in adenocarcinomas of different organs.[313,314]

In soft tissue tumors, focal K13 reactivity occurs in the epithelia of some biphasic synovial sarcomas, but K4 has been rarely demonstrable, in the author's experience.

Keratins of keratinizing stratified epithelia: K1

K1 is the highest-molecular-weight keratin (67 kD). In normal epithelia it is selectively expressed in keratinizing squamous epithelia, and in the epidermis except in the basal layer. Keratinizing squamous carcinomas show variable positivity, but nonkeratinizing squamous and other carcinomas are negative for K1.[258,277,315]

K1 has also been reported as a kininogen receptor in endothelial cells.[316] It can be immunohistochemically demonstrated in many but not all normal endothelia. Some vascular tumors, such as epithelioid hemangioendotheliomas, are consistently positive. Variably positive are hemangiomas and angiosarcomas, the epithelioid one being more commonly positive. Focal expression also can be seen in the spindle cell components of biphasic synovial sarcoma, epithelioid sarcoma, and schwannoma.[315]

Keratins of keratinizing stratified epithelia: K10

Of the other keratins of stratified epithelia, experience is more extensive with K10, to which antibodies reactive in formalin-fixed tissue have been available for a longer time. This keratin is quite specific for squamous differentiation in carcinomas.[317] In soft tissue tumors, reactivity is limited to rare cells in those biphasic synovial sarcomas that have squamous differentiation, and to subsets of keratinizing squamous cell carcinomas.[317] Although expected to be the pair for K1, K10 cannot be equally demonstrated in endothelial and mesenchymal tumors.[315]

Other keratins

K2, K9, K11 are the other high-molecular-weight keratins. They are thought to be limited to keratinizing squamous epithelia. Of these, K9 is essentially limited to the epithelia of soles and palms.[239,241,242,318] These keratins are absent in synovial and epithelioid sarcoma.[270,301]

K6 (56 kD) and K16 (48 kD) are present in hyperproliferative keratinocytes of the skin and in subsets of hair shaft epithelia.[241,242] These keratins are expressed only sporadically in the epithelia of biphasic synovial sarcoma and are not present in epithelioid sarcoma. Distribution in carcinomas is essentially limited to those with squamous differentiation.

K15 (50 kD) is present in certain skin adnexa. K15 is also recognized by a monoclonal antibody to CD8, which cross-reacts with this keratin.[319] Expression is limited in soft tissue tumors.[301]

K3 (62 kD) and K12 (54 kD) are a keratin pair thought to be restricted to the corneal epithelium of the eye.[241,242] There are no data on soft tissue tumors.

Multispecific keratin antibodies and keratin cocktails

Patterns of reactivities of some commonly used monoclonal keratin antibodies and cocktails in soft tissue tumors are summarized in Table 3.16.

Monoclonal antibodies AE1 to AE3 used as a cocktail react with most keratins except K18 and K20.[241] This cocktail shows positivity with nearly all synovial sarcomas, epithelioid sarcomas, and chordomas. Vascular endothelia of some venules and lymphatics are also positive, probably reflecting their reactivity with K7. Almost all carcinomas are positive, except some hepatocellular and adrenocortical carcinomas.

Monoclonal antibody CAM5.2 reacts with K8 based on immunoblotting,[263] and its patterns of reactivity seem to

Table 3.16 Patterns of keratin reactivities of selected monoclonal antibodies

Monoclonal antibody	Keratins recognized
AE1	K9–K17, K19
AE3	K1–K8
CAM5.2	K8
34βE12	K1, K5, K10, K14
MNF11	K8, K18, K19

correspond to those of other K8 antibodies. For example, it does not react with normal endothelia. Despite what is commonly thought, CAM5.2 does not recognize K18.

Most synovial sarcomas, epithelioid sarcomas, and chordomas are positive. Most carcinomas are positive, except well-differentiated and some moderately differentiated squamous cell carcinomas, whereas the poorly differentiated ones commonly reacquire simple epithelial keratins.

Monoclonal antibodies PKK1 and MNF11 react with combinations of simple epithelial keratins 8, 18, and 19 (said to react with all of them; therefore most synovial sarcomas, epithelioid sarcomas, chordomas, and carcinomas are expected to be positive in most cases).

Monoclonal antibody 34βE12 reacts with a combination of keratins 1, 5, 10, and 14. Based on this, it shows reactivity with epidermis and other squamous epithelia, myoepithelial cells, and basal cells of various ductal epithelia (e.g., prostatic basal epithelia). In soft tissue tumors, biphasic synovial sarcoma shows variable epithelial positivity. Monophasic synovial sarcomas and epithelioid sarcomas commonly show focal reactivity. Chordoma is also variably positive.

Monoclonal antibodies that react with both keratin 5 and 6 (keratin 5/6) are often used in lieu of K5 antibodies. They practically reflect K5 content, since K6 has a relatively limited expression in hyperproliferative keratinocytes and some complex (stratified) epithelial carcinomas.

Epithelial membrane antigen

Epithelial membrane antigen (EMA) is a large, highly glycosylated, integral transmembrane glycoprotein weighing more than 200 kD. Its protein has been recognized as a MUC1 gene product, and it belongs to the family of polymorphic epithelial mucins. Its functions range from protective surface coat and antimicrobial barrier to regulation of cell signaling, often inhibitory in nature.[320,321] EMA was initially referred to as *human milk fat globule protein* (HMFG), based on the immunogen used to generate the first antibodies.[322]

EMA is expressed in many epithelial cell membranes. EMA is present in the luminal aspect of most glandular and secretory epithelia. In nonepithelial cells, EMA is also detectable in perineurial and meningeal cells, and in plasma cells (Figure 3.18). Positive staining can appear in the membrane or be apparently cytoplasmic, and many carcinomas overexpress EMA, often with predominantly nonmembranous distribution.[323–325]

EMA-positive soft tissue tumors include synovial sarcoma (glandular lumina in biphasic tumors, spindle cell foci in monophasic tumors), epithelioid sarcoma, chordoma, and metastatic carcinomas, except hepatocellular and adrenocortical carcinomas. Detection of EMA is often useful in the diagnosis of meningioma (most cases are positive), and is very important in the recognition of many types of perineurial cell tumors, perineuriomas.[326–329]

EMA can be focally detectable in some other sarcomas such as low-grade fibromyxoid sarcoma, leiomyosarcoma, and rarely in angiosarcoma so that it cannot be considered specific for perineurial tumors.[265,280,281] It is also often expressed in fibroblastic tumors such as low-grade fibromyxoid sarcoma[330] and benign epithelioid fibrous histiocytoma.[331]

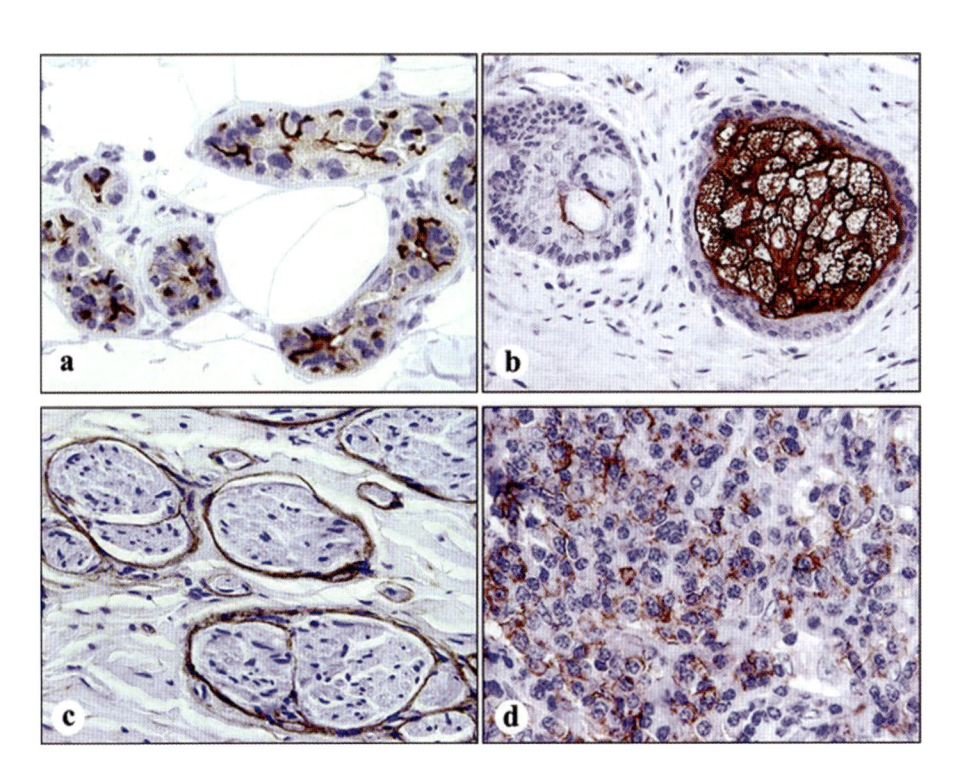

Figure 3.18 Epithelial membrane antigen (EMA, MUC1) immunoreactivity in normal cell types. (**a**) Luminal positivity in sweat gland ducts. (**b**) Strong positivity in sebaceous glands and luminal positivity in hair shaft epithelia. (**c**) Perineurial cells are strongly positive. (**d**) Plasma cells show membrane positivity for EMA.

Other markers

Anaplastic lymphoma kinase (ALK)

The anaplastic lymphoma kinase (ALK) gene located in 2p23 is rearranged, and ALK-fusion genes are overexpressed in large cell anaplastic lymphoma and in inflammatory myofibroblastic tumors (IMT), in which the plump spindle cells often demonstrate strong cytoplasmic ALK positivity.[332,333] In the malignant epithelioid variant (epithelioid myofibroblastic sarcoma), ALK positivity is specifically perinuclear.[334] ALK positivity can also occur in other sarcomas such as rhabdomyosarcoma and MPNST,[335] and has been detected in epithelioid fibrous histiocytoma of the skin. The latter may also have ALK gene rearrangements.[336]

Cadherins

This is a family of large (120–140 kD) calcium-dependent cell adhesion molecules that maintain the integrity of epithelial and some nonepithelial tissues by homotypic cell adhesion. E-cadherin is expressed in most normal epithelia, N-cadherin in neural tissues and mesothelia, P-cadherin in placental trophoblast, and VE-cadherin in vascular endothelia and perineurial cells.[337,338]

In soft tissue tumors, E-cadherin-positive elements occur in epithelial cells in biphasic synovial sarcoma, but rarely in epithelioid sarcoma. This cadherin is commonly present in clear cell sarcoma and metastatic melanoma.[339,340] One study found VE-cadherin in five of seven epithelioid sarcomas.[341]

Malignant mesotheliomas are more commonly positive for N-cadherin and usually negative for E-cadherin, whereas many carcinomas have the opposite patterns.[342]

Our experience suggests the following diagnostically useful, contrasting patterns in E- and N-cadherin expression: epithelioid sarcomas are negative for E-cadherin in 90% of cases, whereas squamous carcinomas are always positive. Chordoma is positive for N-cadherin, but conventional and myxoid chondrosarcomas are negative.[342]

CD10, common acute lymphoblastic leukemia antigen

The common acute lymphoblastic leukemia antigen (CALLA) is a cell surface aminopeptidase. It is expressed in follicular lymphocytes, some epithelial surfaces (small intestine), and has been used as a marker for common acute lymphoblastic lymphomas and follicular lymphomas. In mesenchymal tissues, subsets of fibroblasts and endometrial stromal cells are positive (Figure 3.20), and benign nerve sheath tumors are negative.[343,344]

Endometrial stromal sarcomas and normal endometrial stroma are usually CD10 positive, but uterine smooth muscle tumors are negative, and this feature can be useful in the differential diagnosis of endometrial stromal sarcoma and cellular smooth muscle tumors, and possibly also for diagnosis of endometriosis.[345,346] In addition, many fibroblastic neoplasms are CD10 positive, including both benign (fibrous histiocytoma) and malignant tumors (fibrosarcoma, MFH) (Figure 3.20).[347,348] A large series of sarcomas of different

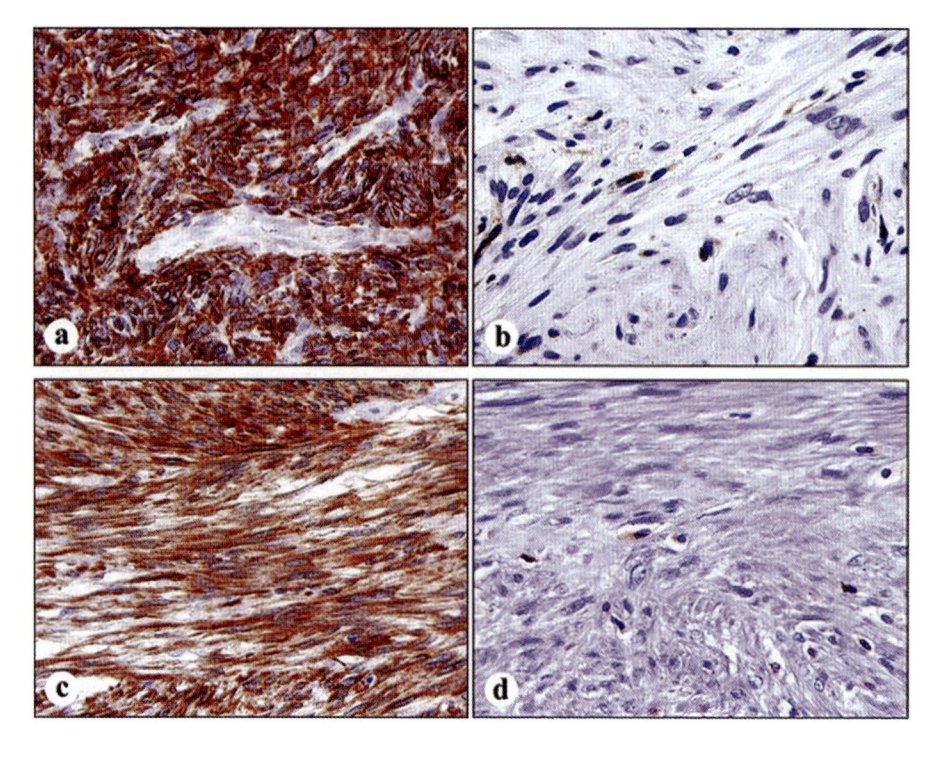

Figure 3.19 Bcl2 immunoreactivity in tumors. (**a**) Solitary fibrous tumor is strongly positive. (**b**) Myofibroma is negative; a few cells are positive. (**c, d**) Gastrointestinal stromal tumor is strongly positive, whereas esophageal leiomyoma is negative.

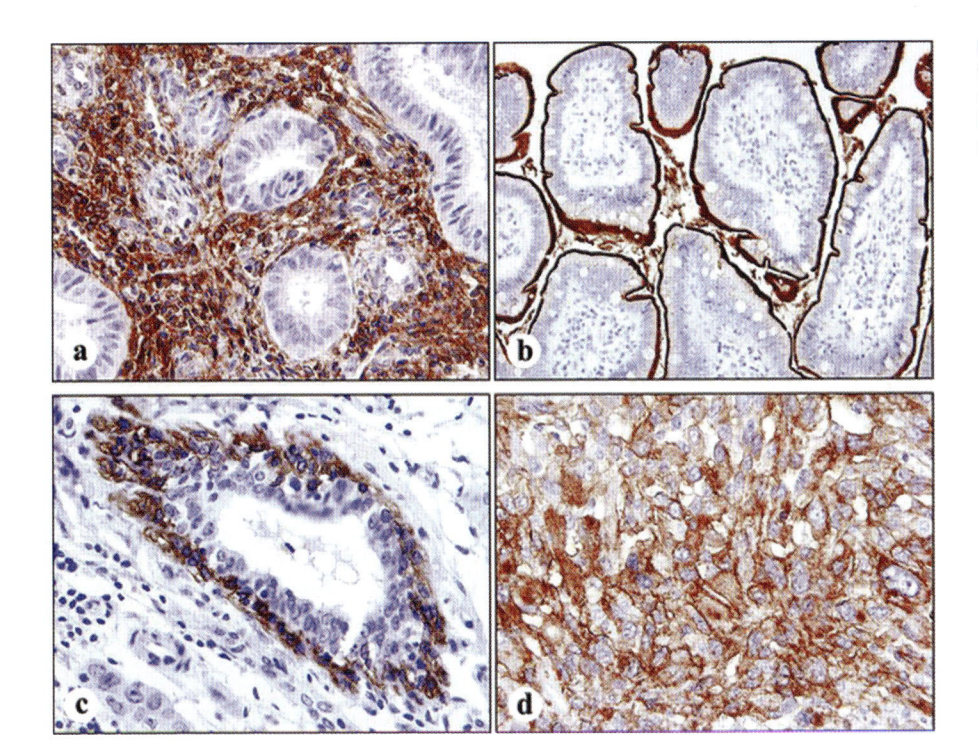

Chapter 3: Immunohistochemistry of soft tissue tumors

Figure 3.20 Examples of CD10-positive cells and tissues. (**a**) Endometrial stroma. (**b**) Exterior surface of small intestinal villi. (**c**) Myoepithelial cells of breast ducts. (**d**) Pleomorphic undifferentiated sarcoma (MFH).

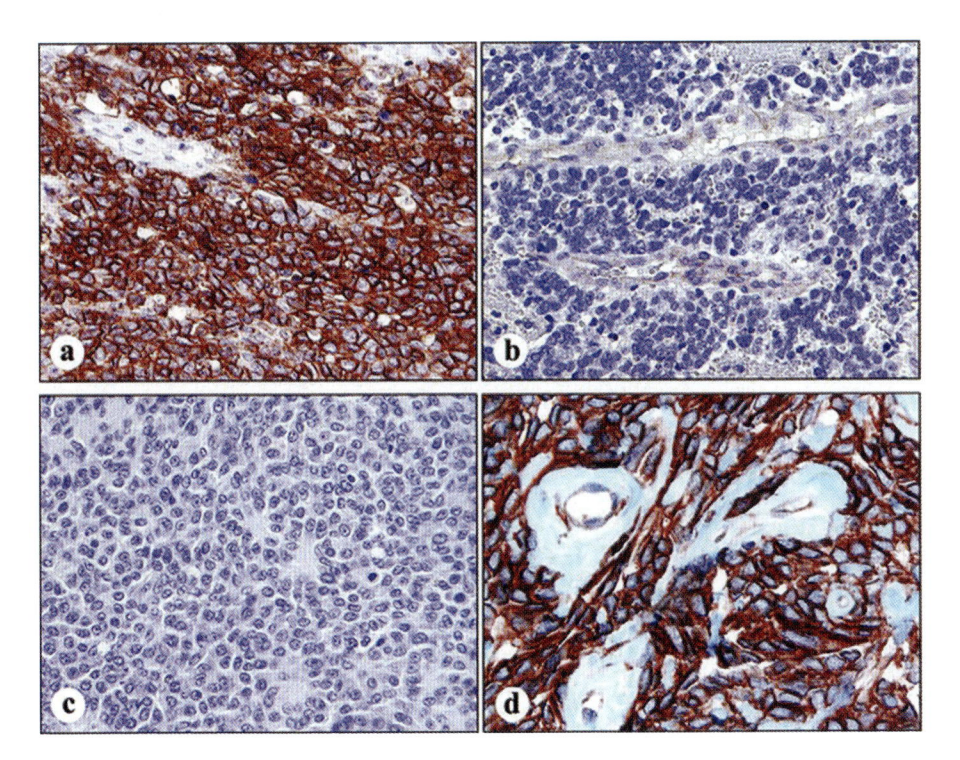

Figure 3.21 (**a**) Membrane positivity for CD99 is a typical although not a specific feature of the Ewing family tumors. (**b,c**) CD99 negativity in small round cell tumors, neuroblastoma, and a malignant melanocytic neoplasm is significant evidence against the diagnosis of Ewing sarcoma. (**d**) Among fibroblastic tumors, solitary fibrous tumor is strongly CD99 positive.

lineages showed positivity in nearly 50% of cases, indicating that the lineage-specificity of CD10 for sarcomas is low, at least for malignant tumors.[348]

CD99

This antigen, also known as mic-2 gene product and protein p30/32, was originally discovered in T-lymphoblasts. CD99 is widely expressed in different tissues at a low level, but may be diagnostically useful in Ewing sarcoma and related tumors (PNET), which typically show a strong membrane staining. Neuroblastoma and other small round cell tumors are typically negative, however (Figure 3.21) T-lymphoblastic lymphomas are also strongly positive.[349–351] Strong immunoreactivity can also be seen in other tumors such as neuroendocrine

71

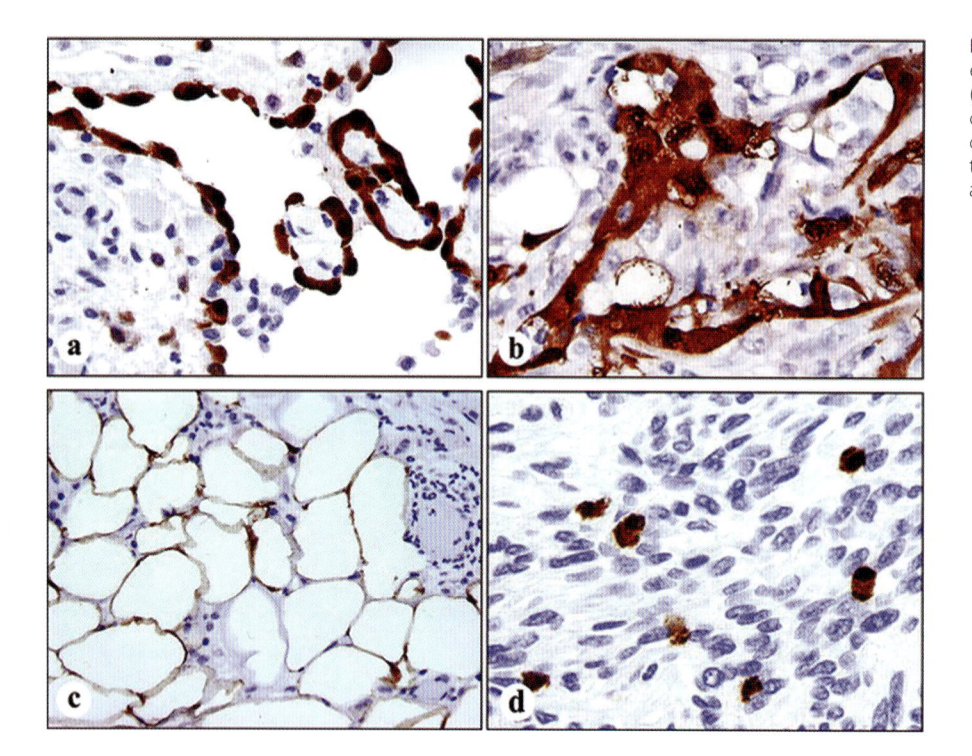

Figure 3.22 Nuclear and cytoplasmic calretinin expression in selected normal tissues and tumor. (**a**) Mesothelial cells show strong nuclear and cytoplasmic positivity. (**b**) Malignant mesothelioma cells are typically calretinin positive. (**c**) Adipose tissue is positive. (**d**) Calretinin-positive mast cells in a cellular leiomyoma.

carcinomas, synovial sarcoma, hemangiopericytoma/solitary fibrous tumor, other fibroblastic tumors, and meningioma.[352,353] A positive result does not specifically define Ewing sarcoma; however, a negative result is evidence against Ewing sarcoma. More specific tests such as EWSR1 gene rearrangement and more specific fusion studies are therefore recommended for confirmation. Transcription factor NKX2.2, a transcriptional target of EWSR1-FLI1 fusion protein has been proposed a more specific marker for Ewing sarcoma.[354]

Calretinin

This calcium-binding protein belongs to the EF-hand proteins that are related to S100 protein and regulate intracellular calcium levels. Calretinin is expressed in normal mesothelial cells, mast cells, and fat cells. These cells show cytoplasmic and often also nuclear positivity (Figure 3.22). Calretinin has proved a useful although not fully specific marker for malignant mesothelioma. Most mesotheliomas of all types are positive, whereas most adenocarcinomas are negative. Small cell and squamous carcinomas are often positive.[355]

Among soft tissue tumors, many synovial sarcomas are calretinin positive. In biphasic tumors, positivity in the spindle cell component is more common, but it can also occur in the epithelial component. Monophasic and poorly differentiated synovial sarcomas often contain calretinin-positive tumor cells. Calretinin positivity is rare in other sarcomas, but can occur in some MPNSTs. Calretinin therefore could be of some use in the differential diagnosis of synovial sarcoma.[356]

A subset of myofibroblastic neoplasms also contain calretinin-positive lesional cells, especially desmoid fibromatosis and proliferative fasciitis/myositis.[357] Paralleling its expression in fat cells, lipomatous tumors are variably calretinin positive. Inconsistent calretinin immunoreactivity limits its application as a marker for lipomatous tumors.[358]

Collagen IV and laminin

These basement membrane proteins are present around the basement membrane/basal lamina-containing mesenchymal cell types: smooth and skeletal muscle and Schwann cells (Figure 3.23 a and b) and also on the basal aspect of basal epithelial cells and vascular endothelia. In soft tissue tumors, schwannomas with complex cell processes show apparently diffuse positivity, whereas well-differentiated smooth and skeletal muscle tumors and glomus tumors typically show pericellular positivity in a network-like pattern.[359–362]

Because of their lack of cell-type specificity, antibodies to basement membrane proteins are rarely used diagnostically; however, their distinct staining patterns make them potential alternative markers (Figure 3.23 c and d).

Estrogen and progesterone receptors

Estrogen and progesterone receptors (ER, PR) are nuclear transcriptional regulator proteins for a number of hormone-dependent genes. ER and PR are expressed in female sex-hormone-sensitive epithelia of the breast and endometrium, and in female genital stromal cells, including uterine cervical stroma and myometrium. The following applies to ER alpha unless stated otherwise.

In soft tissue tumors, ER positivity occurs in uterine and abdominal extrauterine leiomyomas in women (Figure 3.24). Peritoneal leiomyomatosis lesions are also positive. Uterine

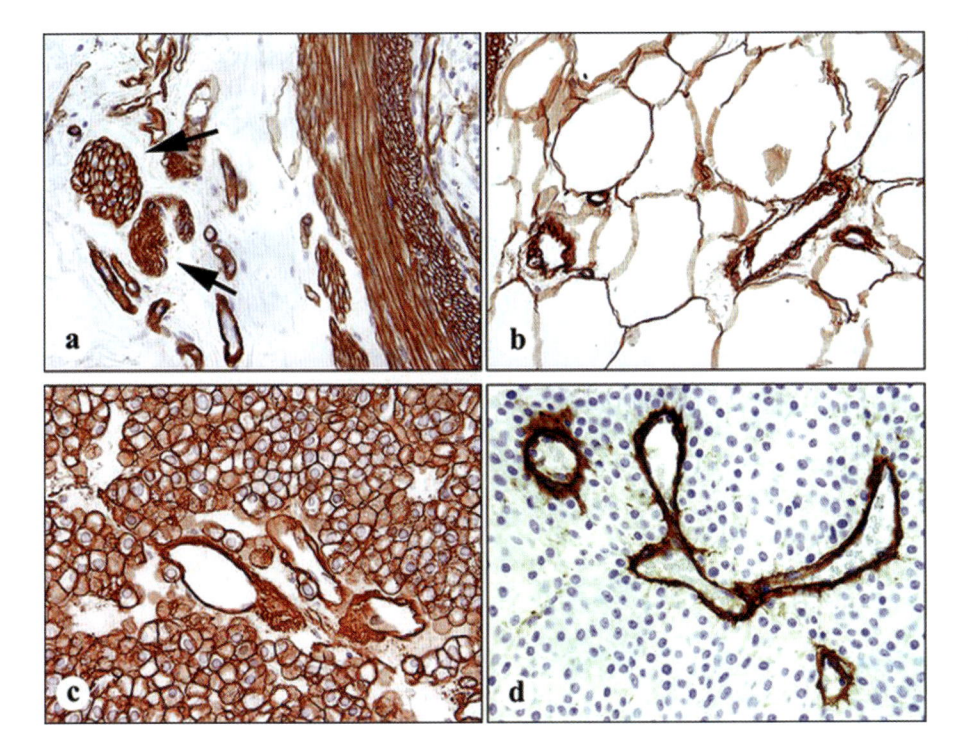

Figure 3.23 Collagen IV immunostaining patterns in examples of basement-membrane-positive and -negative tissues and tumors. (**a**) Smooth muscle is positive showing a pericellular net-like pattern in cross-sectioned fibers. Also positive are nerves (arrows) and blood vessels. (**b**) Fat cells and blood vessels show prominent basement membranes, whereas fibrous stroma is negative. (**c,d**) Glomus tumor cells are surrounded by collagen-IV-positive material, whereas hidradenoma is negative (except for epitheliostromal interface).

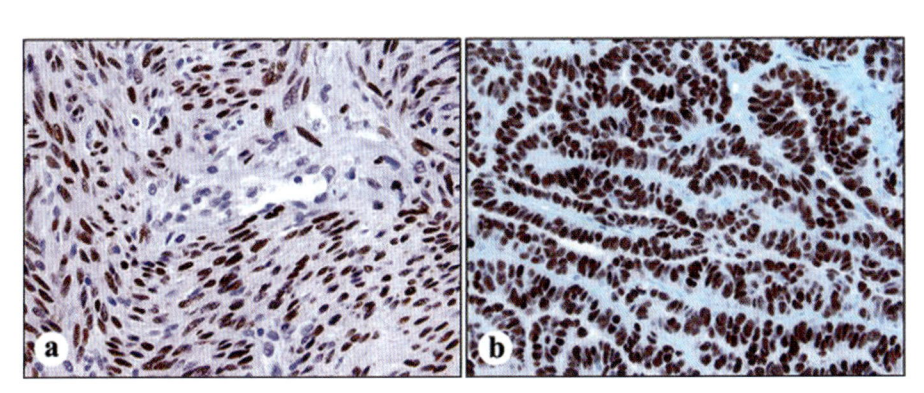

Figure 3.24 (**a**) Most retroperitoneal leiomyomas in women show nuclear positivity for estrogen receptor, similar to uterine leiomyomas. (**b**) Metastatic papillary endometrioid carcinoma with strong nuclear estrogen receptor positivity.

and analogous extrauterine leiomyosarcomas can also be ER positive with a lower frequency. ER positivity therefore does not solely define benign smooth muscle tumors. Ultimately, correlation with other parameters, such as atypia and mitotic activity, is needed to determine the biologic potential of ER-positive smooth muscle tumors in women.[363–366]

Other ER-positive mesenchymal tumors are female genital stromal tumors such as endometrial stromal sarcoma, aggressive angiomyxoma of the pelvis, and angiomyofibroblastoma of the vulva. Evaluation of ER and PR can be useful in the diagnosis of these entities. PR can be present in a broader spectrum of tumors than ER and therefore can be less specific for genital stromal tumors.

ER positivity of a metastatic carcinoma can point to mammary (or uterine/ovarian) origin. In a large study, 66% of ductal and 88% of lobular mammary carcinomas were positive, whereas all 500 pulmonary and gastrointestinal carcinomas were negative.[367]

In addition, ER beta has a broader distribution, including expression in desmoid fibromatosis and many vascular tumors.[368,369]

Glial fibrillary acidic protein

The intermediate filaments of glial cells, especially astrocytes and some ependymal cells, contain glial fibrillary acidic protein (GFAP, Mw 51 kD). Some myoepithelial cells of salivary gland, breast, and skin also express GFAP, but otherwise this protein has a restricted distribution.[370] Although most peripheral nerve Schwann cells are negative, increased GFAP expression has been observed in reactive Schwann cells in axonal neuropathies.[371] GFAP is strongly expressed in most astrocytic and ependymal glial tumors and is a marker for myxopapillary and other soft tissue ependymomas.

In the peripheral nervous system, it is present in 30% to 50% of schwannomas, and in true nerve sheath myxomas, but it is generally absent in MPNSTs.[372–375] Most gastrointestinal schwannomas are positive, whereas GISTs are negative, representing a useful discriminating parameter.[285]

Myoepitheliomas and mixed tumors of salivary glands and skin adnexa often contain GFAP-positive epithelial cells, consistent with their myoepithelial-like differentiation.[370]

INI1 (SMARC1 protein, BAF47)

INI1 (integrase interactor 1) also known as SMARCB1 (SWI/SNF [Switch/sugar non-fermenting] matrix-associated actin-dependent regulator of chromatin subfamily B member) is a nuclear chromatin remodeling and organizer protein. BAF47 means BRG-associated factor 47, and BRG1 is another chromatin remodeling protein in the SWI/SNF complex. INI1 is known as a tumor suppressor and its loss is observed in a small number of tumor types. Biallelic genomic losses or loss-of-function mutations are responsible for loss of expression in these tumors.[376,377]

For soft tissue tumors, INI1 loss is a good diagnostic marker for epithelioid sarcoma, extrarenal rhabdoid tumors, and atypical teratoid rhabdoid tumor (ATRT) of the brain and this loss can be practically considered a definitional feature for these tumor types.[378–380] INI1 is also commonly lost in ESMC,[381] myoepithelial carcinomas of soft tissue,[377] and malignant epithelioid peripheral nerve sheath tumor.[382] It is also mutated in some nerve sheath tumors in schwannomatosis, but not totally lost.[383] In addition, reduced expression without necessarily a total loss of INI1 is a consistent feature for synovial sarcoma that can be diagnostically helpful.[384] In synovial sarcoma, INI1 loss is a consequence of the SS18 gene rearrangement, so that to some degree, INI1 is a potential surrogate marker for the synovial sarcoma translocation.[385]

INI1 is also lost in renal medullary carcinoma associated with sickle cell disease and in a small number of carcinomas reported in the sinonasal and gastrointestinal tracts. These carcinomas are typically associated with a rhabdoid cytology.[377,386] When evaluating INI1-immunohistochemistry, one has to observe the presence of an internal control (retained expression in non-neoplastic cells) to verify technical validity of the immunostain (Figure 3.25).

KIT, or CD117, stem cell factor/mast cell growth factor receptor

This transmembrane protein is a growth factor receptor, a type III receptor tyrosine kinase with five immunoglobulin-like extracellular loops. CD117, or KIT, was originally discovered as the feline Hardy–Zuckerman sarcoma retroviral oncoprotein.[387–389]

KIT is constitutionally expressed in early hematopoietic cells, mast cells, germ cells, melanocytes, and interstitial cells of Cajal, the gastrointestinal pacemaker and neuroregulatory cells within and around the myenteric plexus. KIT-positive

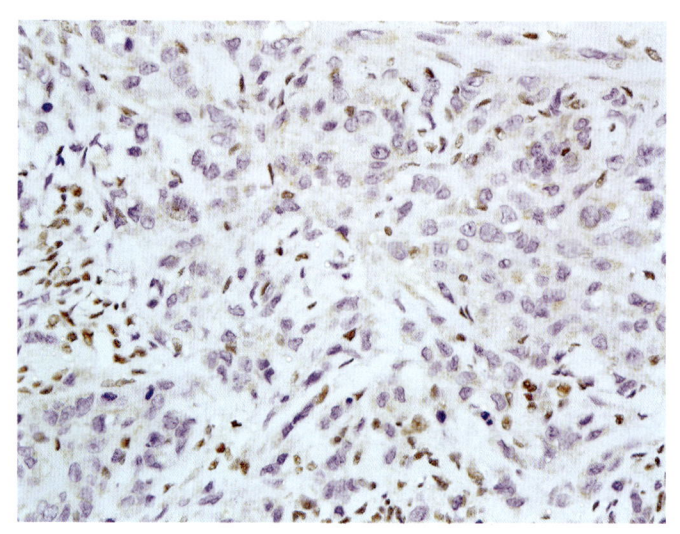

Figure 3.25 INI1 (SMARCB1) loss is detected here in the tumor cells of epithelioid sarcoma. Note preserved nuclear positivity in non-neoplastic cells, which is a necessary internal control for test validity.

epithelial cells include breast lobular epithelium and some basal cells of skin and hair shafts (Figure 3.26). The positive staining usually appears as distinctive membrane staining in normal cells and as diffuse cytoplasmic, Golgi zone, or membrane staining in neoplastic cells.[389–392]

GISTs with Cajal-cell-like differentiation are almost always KIT positive, and the positivity is now considered a major definitional feature for GIST that separates the GISTs from true smooth muscle tumors and schwannomas.[393–395] Mast cell tumors, mastocytoma, urticaria pigmentosa, and extramedullary myeloid tumors are positive.[389,396] Seminomas are positive, but other germ cell tumors are typically negative.[390]

Approximately 50% of angiosarcomas, Ewing sarcomas, and clear cell sarcomas, and 20% to 45% of melanomas are also KIT positive.[389,397] KIT expression has been noted to decrease in melanoma progression. Nevertheless, up to 50% of metastatic melanomas are KIT positive.[398] Most other soft tissue tumors are negative, but sporadic cells in other tumors and neovascular endothelia can be positive.

DOG1/Ano1, another antigen expressed in Cajal cells, is a new supplemental marker for GIST and is discussed in Chapter 17.

MUC4

MUC4 protein is typically expressed in low-grade fibromyxoid sarcoma and its variants, including a majority of cases showing sclerosing epithelioid fibrosarcoma-like morphology. It is also expressed in synovial sarcoma and many carcinomas, but is generally absent in other spindle cell neoplasms and is therefore a useful marker for low-grade fibromyxoid sarcoma.[399,400]

SATB2

This protein is a nuclear transcription factor expressed in osteoblasts and important in bone matrix formation. SATB2

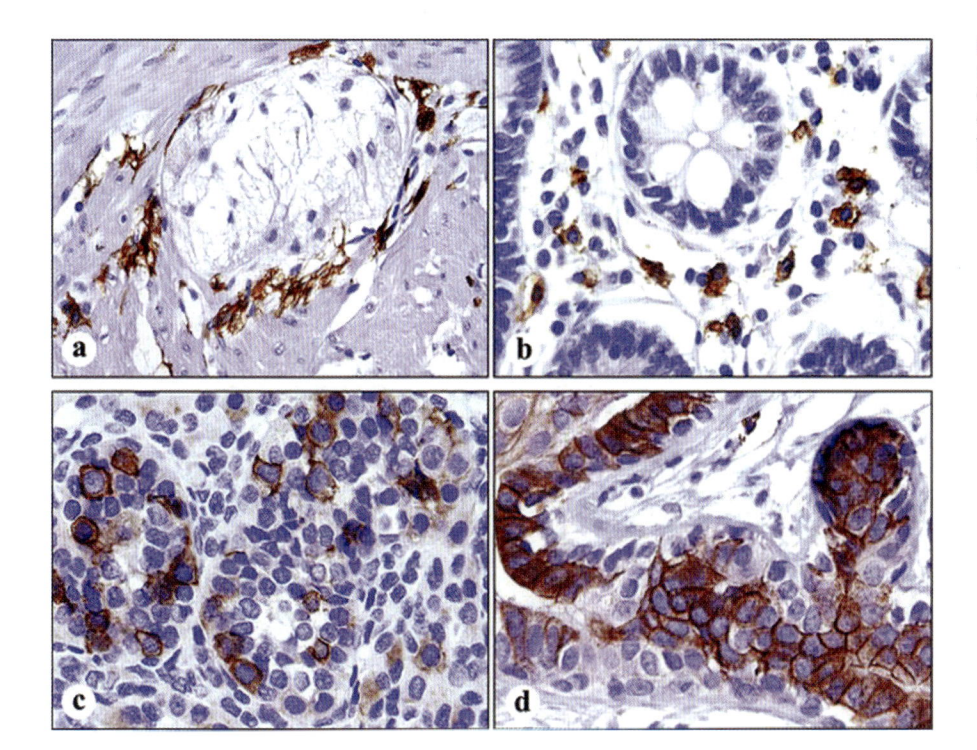

Figure 3.26 Examples of KIT expression in non-neoplastic tissues. (**a**) Gastrointestinal interstitial cells of Cajal around the myenteric plexus. (**b**) Mast cells in colonic mucosa. (**c**) Germ cells, here seen in the prepubertal testis. (**d**) Ductal epithelial cells of breast.

is expressed in osteosarcoma matrix, but has also been detected in giant cell tumors of bone. It is expressed in many gastrointestinal carcinomas, especially colorectal carcinomas. Because experience is limited, caution is advisable in interpreting this marker.[401,402]

Vimentin

The intermediate filaments in most mesenchymal cells (i.e., fibroblasts, endothelial cells, cartilage, histiocytes, and lymphoid cells) are predominantly or exclusively composed of vimentin (Mw 57 kD), which is the name given by Franke et al. to this intermediate filament protein. Among mesenchymal cells, vimentin is specifically absent in most mature smooth and skeletal muscle cells. Certain epithelial cells (i.e., thyroid follicle cells, endometrial glands, mesothelia, and ovarian surface epithelial cells) also contain vimentin, often in the basal aspect of the cytoplasm.[403,404]

Vimentin is widely expressed in mesenchymal and neuroectodermal tumors, including most sarcomas and malignant melanoma, but it is absent from some leiomyomas. Many renal, thyroid, ovarian, and endometrial carcinomas, and small portions of other carcinomas, are variably positive. Poorly differentiated carcinomas are variably and sarcomatoid carcinomas typically vimentin positive.[404]

Vimentin has limited application in the differential diagnosis of soft tissue tumors because of its widespread tissue distribution, but its absence in the tumor cells of many poorly differentiated carcinomas (with the presence of internal control) can be diagnostically useful. Because vimentin can be consistently demonstrated after heat-induced epitope retrieval virtually independent of fixation parameters, it has lost its past use as a marker for antigenic preservation.[405]

Wilms' tumor protein

Wilms' tumor protein (WT1), a nuclear protein, is a transcriptional regulator (transcription factor) and is expressed in developing glomerular epithelia of the kidney, mesothelial cells, and müllerian epithelia.[406]

Positive tumors with nuclear staining include Wilms' tumor and (intra-abdominal) desmoplastic small round cell tumor (DSRCT); other small round cell tumors are negative. However, for the diagnosis of DSRCT, antibodies to WT1 have been used that react with the carboxyterminus, because the aminoterminus is eliminated in the translocation (Figure 3.27).[407,408]

In epithelial tumors, WT1 is expressed in most malignant mesotheliomas, but is often lost in sarcomatoid examples.[409] WT1 is selectively expressed in serous carcinomas and leiomyomas of müllerian origin. WT1 can be useful in evaluating the müllerian epithelial origin of metastatic carcinomas, or in the diagnosis of primary peritoneal müllerian serous carcinomas, and in separating ovarian carcinomas from those from the breast (breast carcinomas are negative).[410,411]

Cell cycle markers
Ki67 and analogs

Ki67 (MKI67 gene product) is a cell cycle stage-related nuclear protein identified with the Ki67 monoclonal antibody. It is selectively expressed in cells that have entered into the cell

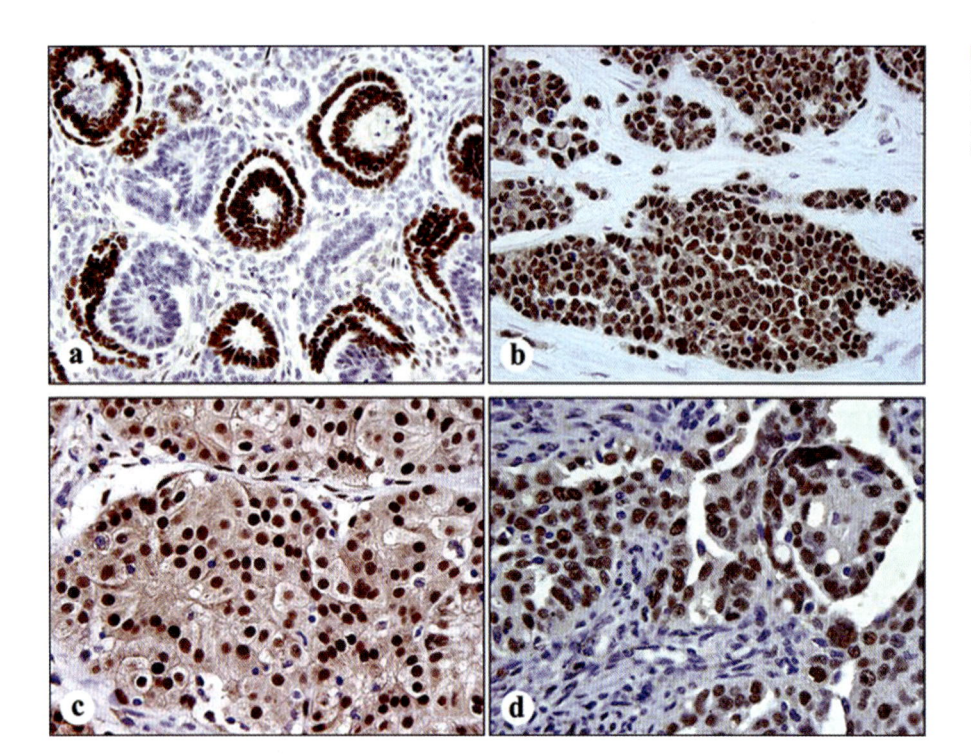

Figure 3.27 Nuclear WT1 expression. (**a**) Developing renal glomeruli are positive. (**b**) Desmoplastic small round cell tumor. (**c**) Malignant mesothelioma. (**d**) Serous ovarian carcinoma shows nuclear positivity.

cycle and are in phase other than G0. While original antibodies reacted only in frozen sections, Ki-67-analog antibodies such as MIB1 work well in formalin-fixed tissue.[412,413]

In sarcomas, a higher score has generally correlated with tumor grade and proliferative fraction by flow cytometry, and some studies have concluded that higher MIB1 scores correlate with tumor aggressiveness.[414–419] Cutoff values between 10% and 20% of positive nuclei have been suggested to give independent prognostic information predictive of metastasis and tumor-related death in high-grade sarcomas of the extremities.[415–419] Because benign processes, such as nodular fasciitis and germinal centers of lymphoid tissue, can have high Ki67-scores, 30–50% or more, Ki67-labeling cannot be used as a discriminator between benign and malignant conditions outside histologic classification.[420] Ki67-labeling may help identify mitotically active areas and thus indirectly assist determination of maximum mitotic rates. Examples of various MIB1 scores are shown in Figure 3.28.

p53 (TP53)

This cell cycle regulator is a nuclear protein of 53 kD. It is often referred to as p53, although the official protein and gene name is TP53. TP53 has been studied extensively, and there is a general assumption that most normal and benign cells do not show detectable expression, and that its detected expression is abnormal. Some studies have documented that immunohistochemically detected p53 expression has prognostic value in soft tissue sarcomas,[421,422] but some studies, for example those on MFH, have found no prognostic significance.[423] There is some correlation between the p53 expression and sarcoma grade.[424] Immunoreactivity for p53 has been detected in some benign soft tissue tumors, so that the presence of this marker cannot be used to diagnose malignancy.[425,426] We refer to Chapter 5 for a discussion of the TP53 gene and germline mutation syndromes, and Chapters 5 and 15 for MDM2.

Retinoblastoma gene product (RB1)

Retinoblastoma gene product p110RB (or simply RB), is a nuclear protein that acts as an inhibitory cell cycle regulator. Phosphorylation of RB protein by cell cycle proteins (cyclin kinase complexes) releases its inhibitory function and activates cell proliferation. Loss of RB1 protein as a consequence of genomic losses is the cause of hereditary bilateral retinoblastoma.[427,428] Somatic loss of RB protein occurs in some soft tissue sarcomas, such as leiomyosarcoma, osteosarcoma, and MFH. Loss of RB protein can be detected immunohistochemically, but there are concerns about its reproducible detection in paraffin-embedded tissue.[429] One study suggested that RB protein is more commonly lost in embryonal versus alveolar rhabdomyosarcoma, and therefore it has been suggested as a potential diagnostic marker for this subtype.[430] RB1-loss also seems to be a common feature in spindle cell lipoma (Chapter 14).

Cyclin-dependent kinase inhibitors

Genomic losses or loss of cyclin-dependent kinase inhibitors leading to loss of their expression, including proteins p16 (CKND2), p21 (WAF1), and p27 (KIP), is a feature of many sarcomas. For example, low expression of p27 has been found as a marker for poor prognosis in synovial sarcoma.[431] Also,

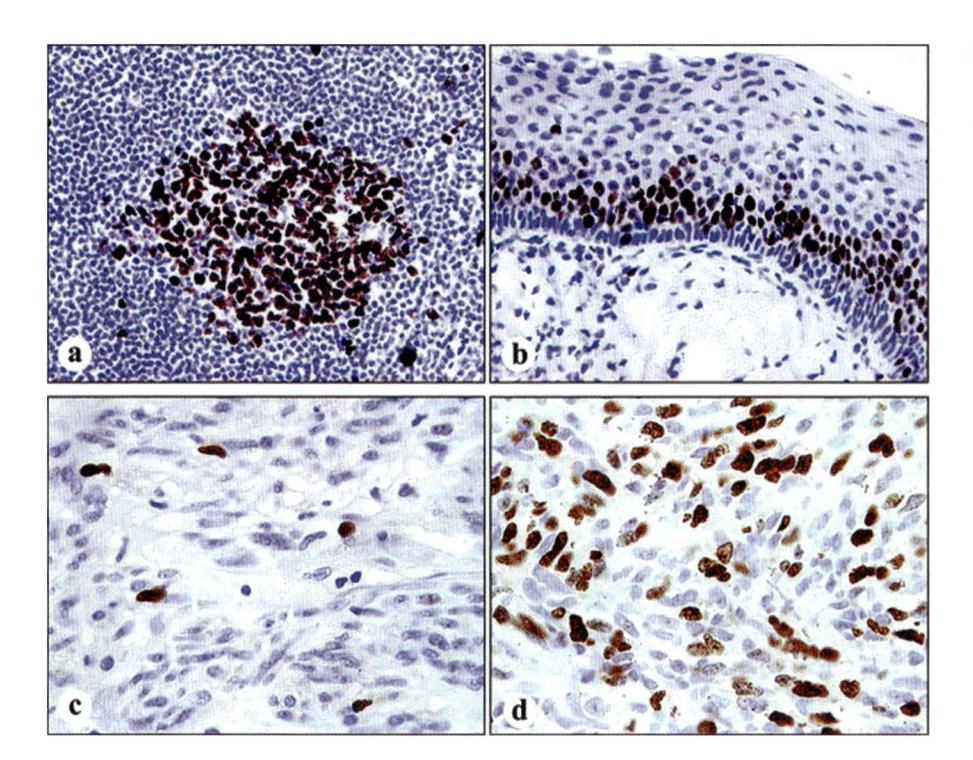

Figure 3.28 (**a,b**) In normal tissues, nuclear MIB1 positivity occurs in germinal centers and epidermal basal cells. (**c,d**) Low and high MIB1 indices in mitotically inactive and active gastrointestinal stromal tumors.

homozygous loss of p16 has been commonly observed in MPNSTs, as opposed to neurofibromas.[432,433] In some instances, malignant tumors have higher p16 expression than benign tumors. For example, uterine leiomyosarcomas are more commonly p16 positive than leiomyomas, and atypical lipomatous tumors are commonly p16 positive, in contrast to ordinary lipomas.[434–436] In general, diagnostic applications of these proteins are not fully developed.

References

Reviews, general

1. Fisher C. Immunohistochemistry in diagnosis of soft tissue tumors. *Histopathology* 2011;58:1001–1012.
2. Hornick JL. Novel uses of immunohistochemistry in the diagnosis and classification of soft tissue tumors. *Mod Pathol* 2014;27 (Suppl 1):S47–S63.
3. Miettinen M. Immunohistochemistry of soft-tissue tumours. Review with emphasis on 10 markers. *Histopathology* 2014;64:101–118.
4. Ordonez NG. Immunohistochemical endothelial markers: a review. *Adv Anat Pathol* 2012;19:281–295.
5. Chu PG, Weiss LM. Keratin expression in human tissues and neoplasms. *Histopathology* 2002;40:403–439.
6. Nagle RB. Intermediate filaments: a review of the basic biology. *Am J Surg Pathol* 1988;12 Suppl 1: 4–16.

Endothelial and multispecific markers

7. Newman PJ, Berndt MC, Gorski J, et al. PECAM-1 (CD31) cloning and relation to adhesion molecules of the immunoglobulin gene superfamily. *Science* 1990;247:1219–1222.
8. Ilan N, Cheug L, Pinter E, Madri JA. Platelet-endothelial cell adhesion molecule-1 (CD31), a scaffolding molecule for selected catenin family members whose binding is mediated by different tyrosine and serine/threonine phosphorylation. *J Biol Chem* 2000;275:21435–21443.
9. Kuzu I, Bicknell R, Harris AL, et al. Heterogeneity of vascular endothelial cells with relevance to diagnosis of vascular tumours. *J Clin Pathol* 1992;45:143–148.
10. McKenney JK, Weiss SW, Folpe AL. CD31 expression in intratumoral macrophages: a potential diagnostic pitfall. *Am J Surg Pathol* 2001;25:1167–1173.
11. de Young BR, Wick MR, Fitzgibbon JF, Sirgi KE, Swanson PE. CD31: an immunospecific marker for endothelial differentiation in human neoplasms. *Appl Immunohistochem* 1993;1:97–100.
12. Miettinen M, Lindenmayer AE, Chaubal A. Endothelial cell markers CD31, CD34, and BNH9 antibody to H- and Y-antigens: evaluation of their specificity and sensitivity in the diagnosis of vascular tumors and comparison with von Willebrand's factor. *Mod Pathol* 1994;7:82–90.
13. de Young BR, Frierson HF Jr, Ly MN, Smith D, Swanson PE. CD31 immunoreactivity in carcinomas and mesotheliomas. *Am J Clin Pathol* 1998;110:374–377.
14. Lanza F, Healy L, Sutherland DR. Structural and functional features of the CD34 antigen: an update. *J Biol Regul Homeostat Agents* 2001;15:1–13.
15. Sidney LE, Branch MJ, Dynphy SE, Dua HS, Hopkinson A. Concise review: evidence for CD34 as a common marker for diverse progenitors. *Stem Cells* 2014;32:1380–1389.

16. van de Rijn M, Rouse RV. CD34 – a review. *Appl Immunohistochem* 1994;2:71–80.
17. Nickoloff BJ. The human progenitor cell antigen (CD34) is localized on endothelial cells, dermal dendritic cells, and perifollicular cells in formalin-fixed normal skin, and on proliferating endothelial cells and stromal spindle-shaped cells in Kaposi's sarcoma. *Arch Dermatol* 1991;127:523–529.
18. Ramani P, Bradley NJ, Fletcher CDM. QBEND/10, a new monoclonal antibody to endothelium: assessment of its diagnostic utility in paraffin sections. *Histopathology* 1990;17:237–242.
19. Sankey EA, More L, Dhillon AP. QBEnd/10. A new immunostain for the routine diagnosis of Kaposi's sarcoma. *J Pathol* 1990;161:267–271.
20. Traweek ST, Kandalaft PL, Mehta P, Battifora H. The human hematopoietic progenitor cell antigen (CD34) in vascular neoplasia. *Am J Clin Pathol* 1991;96:25–31.
21. Aiba S, Tabata N, Ishii H, Ootani H, Tagami H. Dermatofibrosarcoma protuberans is a unique fibrohistiocytic tumor expressing CD34. *Br J Dermatol* 1992;127:79–84.
22. Abenoza P, Lillemoe T. CD34 and factor XIIIa in the differential diagnosis of dermatofibroma and dermatofibrosarcoma protuberans. *Am J Dermatopathol* 1993;15:429–434.
23. Cohen PR, Rapini RP, Farhood AI. Expression of the human hematopoietic progenitor cell antigen CD34 in vascular and spindle cell tumors. *J Cutan Pathol* 1993;20:15–20.
24. Goldblum JR. CD34 positivity in fibrosarcomas which arise in dermatofibrosarcoma protuberans. *Arch Pathol Lab Med* 1995;119:238–241.
25. Westra WH, Gerald WL, Rosai J. Solitary fibrous tumor. Consistent CD34 immunoreactivity and occurrence in the orbit. *Am J Surg Pathol* 1994;18:992–998.
26. Hanau CA, Miettinen M. Solitary fibrous tumor. Histological and immunohistochemical spectrum of benign and malignant variants presenting at different sites. *Hum Pathol* 1995;26:440–449.
27. Flint A, Weiss SW. CD34 and keratin expression distinguishes solitary fibrous tumor (fibrous mesothelioma) of pleura from desmoplastic mesothelioma. *Hum Pathol* 1995;26:428–431.
28. Suster S, Fisher C. Immunoreactivity for the human hematopoietic progenitor cell antigen (CD34) in lipomatous tumors. *Am J Surg Pathol* 1997;21:195–200.
29. Weiss SW, Nickoloff BJ. CD34 is expressed by a distinctive cell population in peripheral nerve, nerve sheath tumors, and related lesions. *Am J Surg Pathol* 1993;17:1039–1045.
30. Chaubal A, Paetau A, Zoltick P, Miettinen M. CD34 immunoreactivity in nervous system tumors. *Acta Neuropathol* 1994;88:454–458.
31. Miettinen M, Sobin LH, Sarlomo-Rikala, M. Immunohistochemical spectrum of GISTs at different sites and their differential diagnosis with other tumors with a reference to CD117 (KIT). *Mod Pathol* 2000;13:1134–1142.
32. Rizeq MN, van de Rijn M, Hendrickson MR, Rouse RV. A comparative immunohistochemical study of uterine smooth muscle neoplasms with emphasis on the epithelioid variant. *Hum Pathol* 1994;25:671–677.
33. Ruggeri ZM. Structure and function of von Willebrand factor. *Thromb Haemost* 1999;82:576–584.
34. Burgdorf WHC, Mukai K, Rosai J. Immunohistochemical identification of factor VIII-related antigen in endothelial cells of cutaneous lesions of alleged vascular nature. *Am J Clin Pathol* 1981;75:167–171.
35. Leader M, Collins M, Patel J, Henry K. Staining for Factor VIII related antigen and Ulex europaeus agglutinin I (UEA I) in 230 tumors. An assessment of their specificity for angiosarcoma and Kaposi's sarcoma. *Histopathology* 1986;10:1153–1162.
36. Breiteneder-Geleff S, Soleiman A, Kowalski H, *et al.* Angiosarcomas express mixed endothelial phenotypes of blood and lymphatic capillaries: podoplanin as a specific marker for lymphatic endothelium. *Am J Pathol* 1999;154:385–394.
37. Kahn HJ, Bailey D, Marks A. Monoclonal antibody D2-40, a new marker for lymphatic endothelium, reacts with Kaposi sarcoma and a subset of angiosarcomas. *Mod Pathol* 2002;15:434–440.
38. Gomaa AH, Yaar M, Bhawan J. Cutaneous immunoreactivity of D2-40 antibody beyond the lymphatics. *Am J Dermatopathol* 2007;29:18–21.
39. Ordonez NG. Podoplanin: a novel diagnostic immunohistochemical marker. *Adv Anat Pathol* 2006;13:83–88.
40. Kalof AN, Cooper K. D2-40 immunohistochemistry – so far! *Adv Anat Pathol* 2009;16:62–64.
41. Chu AY, Litzky LA, Pasha TL, Acs G, Zhang PJ. Utility of D2-40, a novel mesothelial marker, in the diagnosis of malignant mesothelioma. *Mod Pathol* 2005;18:105–110.
42. Hinterberger M, Reinecke T, Storz M, *et al.* D2–40 and calretinin: a tissue microarray analysis of 341 malignant mesotheliomas with emphasis on sarcomatoid differentiation. *Mod Pathol* 2007;20:248–255.
43. Lau SK, Chu PG, Weiss LM. D2–40 immunohistochemistry in the differential diagnosis of seminoma and embryonal carcinoma: a comparative immunohistochemical study with KIT (CD117) and CD30. *Mod Pathol* 2007;20:320–325.
44. Huse JT, Pasha TL, Zhang PJ. D2-40 functions as an effective chondroid marker distinguishing true chondroid tumors from chordoma. *Acta Neuropathol* 2007;113:87–94.
45. Liang H, Wu H, Giorgadze TA, *et al.* Podoplanin is a highly sensitive and specific marker to distinguish primary skin adnexal carcinomas from adenocarcinomas metastatic to skin. *Am J Surg Pathol* 2007;31:304–310.
46. Shibahara J, Kashima T, Kikuchi Y, Kunita A, Fukyyama M. Podoplanin is expressed in subsets of tumors of the central nervous system. *Virchows Arch* 2006;448:493–499.
47. Jokinen CH, Dadras SS, Goldblum JR, *et al.* Diagnostic implications of podoplanin expression in peripheral nerve sheath tumors. *Am J Clin Pathol* 2008;129:886–893.
48. Browning L, Bailey D, Parker A. D2-40 is a sensitive and specific marker in differentiating primary adrenal cortical tumours from both metastatic clear cell renal carcinoma and pheochromocytoma. *J Clin Pathol* 2008;61:293–296.
49. Partanen TA, Arola J, Saaristo A, *et al.* VEGF-C and VEGF-D expression in

neuroendocrine cells and their receptor, VEGFR-3, in fenestrated blood vessels in human tissues. *FASEB J* 2000;14:2087–2096.

50. Jussila L, Valtola R, Partanen TA, et al. Lymphatic endothelium and Kaposi's sarcoma spindle cells detected by antibodies against the vascular endothelial growth factor receptor-3. *Cancer Res* 1998;58:1599–1604.

51. Partanen TA, Alitalo K, Miettinen M. Lack of lymphatic vascular specificity of vascular endothelial growth factor receptor 3 in 185 vascular tumors. *Cancer* 1999;86:2406–2412.

52. Miettinen M, Wang ZF, Paetau A, et al. ERG transcription factor as an immunohistochemical marker for vascular endothelial tumors and prostatic carcinoma. *Am J Surg Pathol* 2011;35:432–441.

53. Miettinen M, Wang Z, Sarlomo-Rikala M, et al. ERG expression in epithelioid sarcoma: a diagnostic pitfall. *Am J Surg Pathol* 2013;37:1580–1585.

54. Stockman DL, Hornick JL, Deavers MT, et al. ERG and FLI1 protein expression in epithelioid sarcoma. *Mod Pathol* 2014;27:496–501.

55. Shon W, Folpe AL, Fritchie KJ. ERG expression in chondrogenic bone and soft tissue tumours. *J Clin Pathol* 2015;68:125–129.

56. Wang WL, Patel NR, Caragea M, et al. Expression of ERG, an Ets family transcription factor, identifies ERG-rearranged Ewing sarcoma. *Mod Pathol* 2012;25:1378–1383.

57. Rossi S, Orvieto E, Furlanetto A, et al. Utility of the immunohistochemical detection of FLI-1 expression in round cell and vascular neoplasms using a monoclonal antibody. *Mod Pathol* 2004;17:547–552.

58. Folpe AL, Chand EM, Goldblum JR, Weiss SW. Expression of Fli-1, a nuclear transcription factor, distinguishes vascular neoplasms from potential mimics. *Am J Surg Pathol* 2001;25:1061–1066.

Muscle cell markers

59. Vandekerckhove J, Weber K. At least six different actins are expressed in higher mammals: an analysis based on the amino acid sequence of the amino terminal tryptic peptide. *J Mol Biol* 1978;126:783–802.

60. Small JV. Structure-function relationships in smooth muscle. The missing links. *Bioessays* 1995;17:785–792.

61. Tsukada T, McNutt MA, Ross R, Gown AM. HHF35, a muscle actin-specific monoclonal antibody. II. Reactivity in normal, reactive and neoplastic human tissues. *Am J Pathol* 1987;127:389–402.

62. Miettinen M. Antibody specific for muscle actin in the diagnosis and classification of soft tissue tumors. *Am J Pathol* 1988;130:205–215.

63. Montgomery EA, Meis JM. Nodular fasciitis. Its morphologic spectrum and immunohistochemical profile. *Am J Surg Pathol* 1991;15:942–948.

64. Rangdaeng S, Truong LD. Comparative immunohistochemical staining for desmin and muscle-specific actin: a study of 576 cases. *Am J Clin Pathol* 1991;96:32–45.

65. Schurch W, Skalli O, Seemayer TA, Gabbiani G. Intermediate filament proteins and actin isoforms as markers for soft tissue tumor differentiation and origin. I. Smooth muscle tumors. *Am J Pathol* 1987;128:91–103.

66. Schurch W, Skalli O, Lagace R, Seemayer TA, Gabbiani G. Intermediate filament proteins and actin isoforms as markers for soft tissue tumor differentiation and origin. III. Hemangiopericytomas and glomus tumors. *Am J Pathol* 1990;136:771–786.

67. Roholl PJM, Elbers HRJ, Prinsen I, et al. Distribution of actin isoforms in sarcomas: an immunohistochemical study. *Hum Pathol* 1990;21:1269–1274.

68. Skalli O, Gabbiani G, Babai F, et al. Intermediate filament proteins and actin isoforms as markers for soft tissue tumor differentiation and origin. II. Rhabdomyosarcomas. *Am J Pathol* 1988;130:515–531.

69. Savera AT, Sloman A, Huvos AG, Klimstra DS. Myoepithelial carcinoma of the salivary glands: a clinicopathologic study of 25 patients. *Am J Surg Pathol* 2000;24:761–774.

70. Perez-Montiel MD, Plaza JA, Domininguez-Malagon H, Suster S. Differential expression of smooth muscle myosin, smooth muscle actin, h-caldesmon and calponin in the differential diagnosis of myofibroblastic and smooth muscle lesions of skin and soft tissue. *Am J Dermatopathol* 2006;28:105–111.

71. Robin YM, Penel N, Pérot G, et al. Transgelin is a novel marker of smooth muscle differentiation that improves diagnostic accuracy of leiomyosarcomas: a comparative immunohistochemical reappraisal of myogenic markers in 900 soft tissue tumors. *Mod Pathol.* 2013;26:502–510.

72. Franke WW, Schmid E, Schiller DL, et al. Differentiation-related patterns of expression of proteins of intermediate-size filaments in tissues and cultured cells. *Cold Spring Harb Symp Quant Biol* 1982;46 (Pt 1):431–453.

73. Frank ED, Warren L. Aortic smooth muscle cells contain vimentin instead of desmin. *Proc Natl Acad Sci USA* 1981;78:2020–2024.

74. Toccanier-Pelte MF, Skalli O, Kapanci Y, Gabbiani G. Characterization of stromal cells with myoid features in lymph nodes and spleen in normal and pathologic conditions. *Am J Pathol* 1987;129:109–118.

75. van Muijen GNP, Ruiter DJ, Warnaar SO. Coexpression of intermediate filament polypeptides in human fetal and adult tissues. *Lab Invest* 1987;57:359–369.

76. Hurlimann J. Desmin and neural markers in mesothelial cells and mesotheliomas. *Hum Pathol* 1994;25:753–757.

77. Ferrandez-Izquierdo A, Navarro-Fos S, Gonzalez-Devesa M, Gil-Benso R, Llombart-Bosch A. Immunocytochemical typification of mesothelial cells in effusions: in vivo and in vitro models. *Diagn Cytopathol* 1994;10:256–262.

78. Fletcher CDM, Tsang WYW, Fisher C, Lee KC, Chan JKC. Angiomyofibroblastoma of the vulva. A benign neoplasm distinct from aggressive angiomyxoma. *Am J Surg Pathol* 1992;16:373–382.

79. Fetsch JF, Laskin WB, Lefkowitz M, Kindblom LG, Meis-Kindblom JM. Aggressive angiomyxoma: a clinicopathologic study of 29 female patients. *Cancer* 1996;78:79–90.

80. Gibas Z, Miettinen M, Limon J, et al. Cytogenetic and immunohistochemical profile of myxoid liposarcoma. *Am J Clin Pathol* 1995;103:20–26.

81. Fletcher CD. Angiomatoid "malignant fibrous histiocytoma": an immunohistochemical study indicative of myoid differentiation. *Hum Pathol* 1991;22:563–568.

82. Fanburg-Smith JC, Miettinen M. Angiomatoid "malignant" fibrous histiocytoma: a clinicopathologic study of 158 cases and further exploration of the myoid phenotype. *Hum Pathol* 1999;30:1336–1343.

83. Persson S, Willems JS, Kindblom LG, Angervall L. Alveolar soft part sarcoma. An immunohistochemical, cytologic and electron-microscopic study and a quantitative DNA analysis. *Virchows Arch A Pathol Anat Histopathol* 1988;412:499–513.

84. Miettinen M, Ekfors T. Alveolar soft part sarcoma. Immunohistochemical evidence for muscle cell differentiation. *Am J Clin Pathol* 1990;93:32–38.

85. Gerald WL, Miller HK, Battifora H, et al. Intra-abdominal desmoplastic small, round cell tumor: report of 19 cases of a distinctive type of high-grade polyphenotypic malignancy affecting young individuals. *Am J Surg Pathol* 1991;15:499–513.

86. Parham DM, Dias P, Kelly DR, Rutledge JC, Houghton PJ. Desmin-positivity in primitive neuroectodermal tumors of childhood. *Am J Surg Pathol* 1992;16:483–492.

87. Thorner P. Intra-abdominal polyphenotypic tumor. *Pediatr Pathol Lab Med* 1996;16:161–169.

88. Truong LD, Rangdaeng S, Cagle P, et al. The diagnostic utility of desmin: a study of 584 cases and review of the literature. *Am J Clin Pathol* 1990;93:305–314.

89. Mayall FG, Goddard H, Gibbs AR. Intermediate filament expression in mesotheliomas: leiomyoid mesotheliomas are not uncommon. *Histopathology* 1992;21:453–457.

90. Folpe AL, Patterson K, Gown AM. Antibodies to desmin identify the blastemal component of nephroblastoma. *Mod Pathol* 1997;10:896–900.

91. Folpe AL, Weiss SW, Fletcher CD, Gown AM. Tenosynovial giant cell tumors: evidence of a desmin-positive dendritic cell subpopulation. *Mod Pathol* 1998;11:939–944.

92. Takahashi K, Hiwada K, Kokubu T. Isolation and characterization of a 34,000 dalton calmodulin- and F-actin binding protein from chicken gizzard smooth muscle. *Biochem Biophys Res Commun* 1986;141:20–26.

93. Sobue K, Sellers JR. Caldesmon, a novel regulatory protein in smooth muscle and nonmuscle actomyosin systems. *J Biol Chem* 1990;266:12115–12118.

94. Lazard D, Sastre X, Frid MG, et al. Expression of smooth muscle-specific proteins in myoepithelium and stromal myofibroblasts of normal and malignant human breast tissue. *Proc Natl Acad Sci USA* 1993;90:999–1003.

95. Watanabe K, Kusakabe T, Hoshi N, Saito A, Suzuki T. h-Caldesmon in leiomyosarcoma and tumors with smooth muscle cell-like differentiation: its specific expression in the smooth muscle cell tumor. *Hum Pathol* 1999;30:392–396.

96. Miettinen M, Sarlomo-Rikala M, Kovatich AJ, Lasota J. Calponin and h-caldesmon in soft tissue tumors: consistent h-caldesmon immunoreactivity in gastrointestinal stromal tumors indicates traits of smooth muscle differentiation. *Mod Pathol* 1999;12:756–762.

97. Ceballos KM, Nielsen GP, Selig MK, O'Connell JX. Is anti-h-caldesmon useful for distinguishing smooth muscle and myofibroblastic tumors?: an immunohistochemical study. *Am J Clin Pathol* 2001;14:746–753.

98. Hisaoka M, Wei-Qi S, Jian W, Morio T, Hashimoto H. Specific but variable expression of h-caldesmon in leiomyosarcomas: an immunohistochemical reassessment of a novel myogenic marker. *Appl Immunohistochem Mo1 Morphol* 2001;9:302–308.

99. Gimona M, Herzog M, Vandekerckhove J, Small JV. Smooth muscle specific expression of calponin. *FEBS Lett* 1990;274:159–162.

100. Weintraub H. The MyoD family and myogenesis: redundancy, networks and thresholds. *Cell* 1993;75:1241–1244.

101. Dias P, Dilling M, Hougton P. The molecular basis of skeletal muscle differentiation. *Semin Diagn Pathol* 1994;11:3–14.

102. Rudnicki MA, Jaenisch R. The MyoD family of transcription factors and skeletal myogenesis. *Bioessays* 1995;17:203–209.

103. Parham DM, Dias P, Bertorini T, et al. Immunohistochemical analysis of the distribution of MyoD1 in muscle biopsies of primary myopathies and neurogenic atrophy. *Acta Neuropathol* 1994;87:605–611.

104. Dias P, Parham DM, Shapiro DN, Tapscott SJ, Houghton PJ. Monoclonal antibodies to the myogenic regulatory protein MyoD1 epitope mapping and diagnostic utility. *Cancer Res* 1992;52:6431–6439.

105. Tallini G, Parham DM, Dias P, et al. Myogenic regulatory protein expression in adult soft tissue sarcomas: a sensitive and specific marker of skeletal muscle differentiation. *Am J Pathol* 1994;144:693–701.

106. Wang NP, Marx J, McNutt MA, Gown AM. Expression of myogenic regulatory proteins (myogenin and MyoD1) in small blue round cell tumors of childhood. *Am J Pathol* 1995;147:1799–1810.

107. Cui S, Hano H, Harada T, et al. Evaluation of new monoclonal anti-MyoD1 and anti-myogenin antibodies for the diagnosis of rhabdomyosarcoma. *Pathol Int* 1999;49:62–68.

108. Kumar S, Perlman E, Harris CA, Raffeld M, Tsokos M. Myogenin is a specific marker for rhabdomyosarcoma: an immunohistochemical study in paraffin-embedded tissues. *Mod Pathol* 2000;13:988–993.

109. Dias P, Chen B, Dilday B, et al. Strong immunostaining for myogenin in rhabdomyosarcoma is significantly associated with tumors of the alveolar subclass. *Am J Pathol* 2000:156: 399–408.

110. Wang NP, Bacchi CE, Jiang JJ, McNutt MA, Gown AM. Does alveolar soft-part sarcoma exhibit skeletal muscle differentiation? An immunocytochemical and biochemical study of myogenic regulatory protein expression. *Mod Pathol* 1996;9:496–506.

111. Gomez JA, Amin MB, Ro JY, et al. Immunohistochemical profile of myogenin and Myo D1 does not support skeletal muscle lineage in alveolar soft part sarcoma. A study of 19 tumors. *Arch Pathol Lab Med* 1999;123:503–507.

112. Mukai K, Rosai J, Hallaway BE. Localization of myoglobin in normal and neoplastic human skeletal muscle cells using an immunoperoxidase

method. *Am J Surg Pathol* 1979;3:373–376.

113. Corson JM, Pinkus GS. Intracellular myoglobin – a specific marker for skeletal muscle differentiation in soft tissue sarcomas. *Am J Pathol* 1980;103:384–389.

114. Brooks JJ. Immunohistochemistry of soft tissue tumors: myoglobin as a tumor marker for rhabdomyosarcoma. *Cancer* 1982;50:1757–1763.

115. Eusebi V, Bondi A, Rosai J. Immunohistochemical localization of myoglobin in nonmuscular cells. *Am J Surg Pathol* 1984;8:51–55.

Neural and neuroendocrine markers
Synaptophysin

116. Wiedenmann B, Franke WW. Identification and localization of synaptophysin, an integral membrane glycoprotein of Mr 38000 characteristic of presynaptic vesicles. *Cell* 1985;41:1017–1028.

117. Thomas L, Hartung K, Langosch D, et al. Identification of synaptophysin as a hexameric channel protein of the synaptic vesicle membrane. *Science* 1988;242:1050–1052.

118. Wiedenmann B, Franke WW, Kuhn C, Moll R, Gould VE. Synaptophysin: a marker protein for neuroendocrine cells and neoplasms. *Proc Natl Acad Sci USA* 1986;83:3500–3504.

119. Wiedenmann B, Huttner WB. Synaptophysin and chromogranins/secretogranins: widespread constituents of distinct types of neuroendocrine vesicles and new tools in tumor diagnosis. *Virchows Arch B Cell Pathol Incl Mol Pathol* 1989;58:95–121.

120. Miettinen M, Rapola J. Synaptophysin: an immunohistochemical marker for childhood neuroblastoma. *Acta Pathol Microbiol Scand A* 1987;95:167–170.

121. Hachitanda Y, Tsuneyoshi M, Enjoji M. Expression of pan-neuroendocrine proteins in 53 neuroblastic tumors. *Arch Pathol Lab Med* 1989;113:381–384.

122. Schwechheimer K, Wiedenmann B, Franke WW. Synaptophysin: a reliable marker for medulloblastomas. *Virchows Arch A Pathol Anat Histopathol* 1987;411:53–59.

123. Frierson HF, Ross GW, Mills SE, Frankfurter A. Olfactory neuroblastoma: additional immunohistochemical characterization. *Am J Clin Pathol* 1990;94:547–553.

124. Cavazzana AO, Ninfo V, Roberts J, Triche TJ. Peripheral neuroepithelioma: a light microscopic, immunocytochemical, and ultrastructural study. *Mod Pathol* 1992;5:71–78.

125. Ladanyi M, Heinemann FS, Huvos AG, et al. Neural differentiation in small round cell tumors of bone and soft tissue with the translocation t(11l22)(q24;q12): an immunohistochemical study of 11 cases. *Hum Pathol* 1990;21:1245–1251.

126. Amann G, Zoubek A, Salzer-Kuntschik M, Windhager R, Kovar H. Relation of neurological marker expression and EWS gene fusion types in MIC2/CD99-positive tumors of the Ewing family. *Hum Pathol* 1999;30:1058–1064.

127. Parham DM, Hijazi Y, Steinberg SM, et al. Neuroectodermal differentiation in Ewing's sarcoma family of tumors does not predict tumor behavior. *Hum Pathol* 1999;30:911–918.

128. Gould VE, Wiedenmann B, Lee I, et al. Synaptophysin expression in neuroendocrine neoplasms as determined by immunocytochemistry. *Am J Pathol* 1987;126:243–257.

129. Miettinen M. Synaptophysin and neurofilament proteins as markers for neuroendocrine tumors. *Arch Pathol Lab Med* 1987;111:813–818.

130. Johnson TL, Zarbo RJ, Lloyd RV, Crissman JD. Paragangliomas of the head and neck: immunohistochemical neuroendocrine and intermediate filament typing. *Mod Pathol* 1988;1:216–223.

131. Goh YW, Spagnolo DV, Platten M, et al. Extraskeletal myxoid chondrosarcoma: a light microscopic, immunohistochemical, ultrastructural, and immunoultrastructural study indicating neuroendocrine differentiation. *Histopathology* 2001;39:514–524.

132. Banerjee SS, Menasce LP, Eyden BP, Brain AN. Malignant melanoma showing ganglioneuroblastic differentiation: report of a unique case. *Am J Surg Pathol* 1999;23:582–588.

Chromogranin A

133. O'Connor DT, Mahata SK, Taupenot L, et al. Chromogranin A in human disease. *Adv Exp Med Biol* 2000;482:377–388.

134. Lloyd RV. Immunohistochemical localization of chromogranin in normal and neoplastic endocrine tissues. *Pathol Annu* 1987;22(Part 2):69–90.

135. Molenaar WM, Baker DL, Pleasure D, Lee VMY, Trojanowski JQ. The neuroendocrine and neural profiles of neuroblastomas, ganglioneuroblastomas, and ganglioneuromas. *Am J Pathol* 1990;136:375–382.

136. Christensen WN, Strong EW, Bains MS, Woodruff JM. Neuroendocrine differentiation in the glandular peripheral nerve sheath tumor: pathologic distinction from the biphasic synovial sarcoma with glands. *Am J Surg Pathol* 1988;12:417–426.

137. Pagani A, Fischer-Colbrie R, Sanfilippo B, et al. Secretogranin II expression in Ewing's sarcomas and primitive neuroectodermal tumors. *Diagn Mol Pathol* 1992;1:165–172.

Neurofilament and related proteins

138. Lee MK, Cleveland DW. Neuronal intermediate filaments. *Annu Rev Neurosci* 1996;19:187–217.

139. Dahl D. Immunohistochemical differences between neurofilaments in perikarya, dendrites and axons: immunofluorescence study with antisera raised to neurofilament polypeptides (200K, 150K, 70K) isolated by anion exchange chromatography. *Exp Cell Res* 1983;149:397–408.

140. Gould VE, Moll R, Moll I, Lee I, Franke WW. Neuroendocrine (Merkel) cells of the skin: hyperplasias, dysplasias, and neoplasms. *Lab Invest* 1985;52:334–353.

141. Mukai M, Torikata C, Iri H, et al. Expression of neurofilament triplet proteins in human neural tumors: an immunohistochemical study of paraganglioma, ganglioneuroma, ganglioneuroblastoma and neuroblastoma. *Am J Pathol* 1986;122:28–35.

142. Miettinen M, Lehto VP, Virtanen I. Immunofluorescence microscopic evaluation of the intermediate filament expression of the adrenal cortex and

medulla and their tumors. *Am J Pathol* 1985;118:360–366.

143. Miettinen M, Lehto VP, Dahl D, Virtanen I. Varying expression of cytokeratin and neurofilaments in neuroendocrine tumors of human gastrointestinal tract. *Lab Invest* 1985;52:429–436.

144. Lehto VP, Miettinen M, Virtanen I. A dual expression of cytokeratin and neurofilaments in bronchial carcinoid cells. *Int J Cancer* 1985;35:421–425.

145. Moll R, Lee I, Gould VE, et al. Immunocytochemical analysis of Ewing's tumors. Patterns of expression of intermediate filaments and desmosomal proteins indicate cell-type heterogeneity and pluripotential differentiation. *Am J Pathol* 1987;27:288–304.

146. Gerharz CD, Moll R, Meister P, Knuth A, Gabbert H. Cytoskeletal heterogeneity of an epithelioid sarcoma with expression of vimentin, cytokeratins, and neurofilaments. *Am J Surg Pathol* 1990;14:274–283.

147. Folpe AL, Gown AM. Poorly differentiated synovial sarcoma: immunohistochemical distinction from primitive neuroectodermal tumors and high grade malignant peripheral nerve sheath tumors. *Am J Surg Pathol* 1998;22:673–682.

148. Miettinen M, Rapola J. Immunohistochemical spectrum of rhabdomyosarcoma and rhabdomyosarcoma-like tumors: expression of cytokeratin and the 68 kD neurofilament protein. *Am J Surg Pathol* 1989;13:120–132.

NB 84

149. Thomas JO, Nijjar J, Turley H, Micklem H, Gatter KC. NB84: a new monoclonal antibody for the recognition of neuroblastoma in routinely processed material. *J Pathol* 1991;163:69–75.

150. Miettinen M, Chatten J, Paetau A, Stevenson AJ. Monoclonal antibody NB84 in the differential diagnosis of neuroblastoma and other small round cell tumors. *Am J Surg Pathol* 1998;22:327–332.

151. Folpe AL, Patterson K, Gown AM. Antineuroblastoma antibody NB-84 also identifies a significant subset of other small blue round cell tumors. *Appl Imunohistochem* 1997;5:239–245.

S100 protein

152. Moore BW. A soluble protein characteristic of the nervous system. *Biochem Biophys Res Commun* 1965;19:739–744.

153. Donato R. S100: a multigenic family of calcium-modulated proteins of the EF-hand type with intracellular and extracellular functional roles. *Int J Biochem Cell Biol* 2001;33:638–668.

154. Takahashi K, Isobe T, Ohtsuki Y, et al. Immunhistochemical study on the distribution of alpha and beta subunits of the S-100 protein in human neoplasms and normal tissues. *Virchows Arch B Cell Pathol Incl Mol Pathol* 1984;45:385–396.

155. Nakajima T, Watanabe S, Sato Y, et al. An immunoperoxidase study of S-100 protein distribution in normal and neoplastic tissues. *Am J Surg Pathol* 1982;6:715–727.

156. Stefansson K, Wollmann R, Jerkovic M. S100 protein in soft tissue tumors derived from Schwann cells and melanocytes. *Am J Pathol* 1982;106:261–268.

157. Kahn HJ, Marks A, Thom H, Baumal R. Role of antibody to S100 protein in diagnostic pathology. *Am J Clin Pathol* 1983;79:341–347.

158. Cochran AJ, Lu HF, Li PX, Saxton R, Wen DR. S-100 protein remains a practical marker for melanocytic and other tumours. *Melanoma Res* 1993;3:325–330.

159. Weiss SW, Langloss JM, Enzinger FM. Value of S-100 protein in the diagnosis of soft tissue tumors with particular reference to benign and malignant Schwann cell tumors. *Lab Invest* 1983;49:299–308.

160. Swanson PE, Wick MR. Clear cell sarcoma: an immunohistochemical analysis of six cases and comparison with other epithelioid neoplasms of soft tissue. *Arch Pathol Lab Med* 1989;113:55–60.

161. Daimaru Y, Hashimoto H, Enjoji M. Malignant peripheral nerve sheath tumors (malignant schwannomas): an immunohistochemical study of 29 cases. *Am J Surg Pathol* 1985;9:434–444.

162. Wick MR, Swanson PE, Scheithauer BW, Manivel JC. Malignant peripheral nerve sheath tumors: an immunohistochemical study of 62 cases. *Am J Clin Pathol* 1987;87:425–433.

163. Kliewer KE, Wen DR, Cancilla PA, Cochran AJ. Paragangliomas. Assessment of prognosis by histologic, immunohistochemical, and ultrastructural techniques. *Hum Pathol* 1989;20:29–39.

164. Achilles E, Padberg BC, Holl K, Klöppel G, Schröder S. Immunocytochemistry of paragangliomas – value of staining for S-100 protein and glial fibrillary acid protein in diagnosis and prognosis. *Histopathology* 1991;18:453–458.

165. Nakamura Y, Becker LE, Marks A. S-100 protein in tumors of cartilage and bone. *Cancer* 1983;52:1820–1824.

166. Okajima K, Honda I, Kitagawa T. Immunohistochemical distribution of S-100 protein in tumors and tumor like lesions of bone and cartilage. *Cancer* 1988;61:792–799.

167. Nakamura Y, Becker LE, Marks A. S100 protein in human chordoma and human and rabbit notochord. *Arch Pathol* 1983;107:118–120.

168. Hashimoto H, Daimaru Y, Enjoji M. S-100 protein distribution in liposarcoma. An immunoperoxidase study with special reference to the distinction of liposarcoma from myxoid malignant fibrous histiocytoma. *Virchows Arch A Pathol Anat Histopathol* 1984;405:1–10.

169. dei Tos A, Wadden C, Fletcher CDM. S-100 protein staining in liposarcoma. Its diagnostic utility in the high-grade myxoid (round-cell) variant. *Appl Immunohistochem* 1996;4:95–101.

170. Enzinger FM, Weiss SW, Liang CY. Ossifying fibromyxoid tumor of soft parts. A clinicopathological analysis of 59 cases. *Am J Surg Pathol* 1989;13:817–827.

171. Guillou L, Wadden C, Kraus MD, Dei Tos AP, Fletcher CDM. S-100 protein reactivity in synovial sarcomas: a potentially frequent diagnostic pitfall. Immunohistochemical analysis of 100 cases. *Appl Immunohistochem* 1996;4:167–175.

172. Coindre JM, de Mascarel A, Trojani M, de Mascarel I, Pages A. Immunohistochemical study of rhabdomyosarcoma: unexpected staining with S100 protein and cytokeratin. *J Pathol* 1988;155:127–132.

173. Drier JK, Swanson PE, Cherwitz DL, Wick MR. S100 protein immunoreactivity in poorly differentiated carcinomas. Immunohistochemical comparison

174. Herrera GA, Turbat-Herrera EA, Lott RL. S-100 protein expression by primary and metastatic carcinomas. *Am J Clin Pathol* 1988;89:168–176.

175. Cocchia D, Michetti F, Donato R. Immunochemical and immunocytochemical localization of S-100 antigen in normal human skin. *Nature* 1981;294:85–87.

176. Takahashi K, Yamaguchi H, Ishizeki J, Nakajima T, Nakazato Y. Immunohistochemical and immunoelectron microscopic localization of S-100 protein in the interdigitating reticulum cells of the human lymph node. *Virchows Arch B Cell Pathol Incl Mol Pathol* 1981;37:125–135.

177. Aoyama K, Terashima K, Imai Y, *et al*. Sinus histiocytosis with massive lymphadenopathy. A histogenetic analysis of histiocytes found in the fourth Japanese case. *Acta Pathol Jpn* 1984;34:375–388.

178. Eisen RN, Rosai J. Immunohistochemical characterization of sinus histiocytosis with massive lymphadenopathy (Rosai-Dorfman disease). *Semin Diagn Pathol* 1990;7:74–82.

Sox10

179. Nonaka D, Chiriboga L, Rubin BP. Sox10: a pan-schwannian and melanocytic marker. *Am J Surg Pathol* 2008;32:1291–1298.

180. Heerema MG, Suurmeijer AJ. Sox10 immunohistochemistry allows the pathologist to differentiate between prototypical granular cell tumors and other granular cell lesions. *Histopathology* 2012;61:997–999.

181. Karamchandani JR, Nielsen TO, van de Rijn M, West RB. Sox10 and S100 in the diagnosis of soft-tissue neoplasms. *Appl Immunohistochem Mol Morphol* 2012;20:445–450.

182. Naujokas A, Charli-Joseph Y, Ruben BS, *et al*. SOX-10 expression in cutaneous myoepitheliomas and mixed tumors. *J Cutan Pathol* 2014;41:353–363.

183. Miettinen M, McCue PA, Sarlomo-Rikala M, *et al*. Sox10–a marker for not only schwannian and melanocytic neoplasms but also myoepithelial cell tumors of soft tissue: a systematic analysis of 5134 tumors. *Am J Surg Pathol* 2015;39:826–835.

Melanocytic markers other than S100 protein

184. Gown AM, Vogel AM, Hoak D, Gough F, McNutt MA. Monoclonal antibodies specific for melanocytic tumors distinguish subpopulations of melanocytes. *Am J Pathol* 1985;123:195–203.

185. Bacchi CE, Bonetti F, Pea M, Martignoni G, Gown AM. HMB-45: a review. *Appl Immunohistochem* 1996;4:73–85.

186. Wick MR, Swanson PE, Rocamora A. Recognition of malignant melanoma by monoclonal antibody HMB-45. An immunohistochemical study of 200 paraffin-embedded cutaneous tumors. *J Cutan Pathol* 1988;15:201–207.

187. Ordonez NG, Ji XL, Hickey RC. Comparison of HMB-45 monoclonal antibody and S100 protein in the immunohistochemical diagnosis of melanoma. *Am J Clin Pathol* 1988;90:385–390.

188. Kaufmann O, Koch S, Burghardt J, Audring H, Dietel M. Tyrosinase, Melan-A and KBA62 as markers for the immunohistochemical identification of metastatic amelanotic melanomas. *Mod Pathol* 1998;11:740–746.

189. Miettinen M, Fernandez M, Franssila K, *et al*. Microphthalmia transcription factor in the immunohistochemical diagnosis of metastatic melanoma. Comparison with four other melanoma markers. *Am J Surg Pathol* 2001;25:205–211.

190. Longacre TA, Egbert BM, Rouse RV. Desmoplastic and spindle-cell malignant melanoma. An immunohistochemical study. *Am J Surg Pathol* 1996;20:1489–1500.

191. Fetsch JF, Michal M, Miettinen M. Pigmented (melanotic) neurofibroma: a clinicopathologic and immunohistochemical analysis of 19 lesions from 17 patients. *Am J Surg Pathol* 2000;24:331–343.

192. Pea M, Bonetti F, Zamboni G, *et al*. Melanocyte marker HMB-45 is regularly expressed in angiomyolipoma of the kidney. *Pathology* 1991;23:185–188.

193. Ashfaq R, Weinberg A, Albores-Saavedra J. Renal angiomyolipomas and HMB-45 reactivity. *Cancer* 1993;71:3091–3097.

194. Chan JK, Tsang WY, Pau M, *et al*. Lymphangiomyomatosis and angiomyolipoma: closely related entities characterized by hamartomatous proliferation of HMB-45 positive smooth muscle. *Histopathology* 1993;22:445–455.

195. Bonetti F, Pea M, Martignoni G, *et al*. Clear cell ("sugar") tumor of the lung is a lesion strictly related to angiomyolipoma: the concept of a family of lesions characterized by the presence of the perivascular epithelioid cells (PEC). *Pathology* 1994;26:230–236.

196. Kwon BS. Pigmentation genes: the tyrosinase gene family and the Pmel 17 family. *J Invest Dermatol* 1993;100 (2 Suppl):134S–140S.

197. Chen YT, Stockert E, Tsang S, Coplan KA, Old LJ. Immunophenotyping of melanomas for tyrosinase: implications for vaccine development. *Proc Natl Acad Sci USA* 1995;92:8125–8129.

198. Hofbauer GF, Kamarashev J, Geertsen R, Boni R, Dummer R. Tyrosinase immunoreactivity in formalin-fixed, paraffin embedded primary and metastatic melanoma: frequency and distribution. *J Cutan Pathol* 1998;25:204–209.

199. Jungbluth AA, Iversen K, Coplan K, *et al*. T311: an anti-tyrosinase monoclonal antibody for the detection of melanocytic lesions in paraffin-embedded tissues. *Pathol Res Pract* 2000;196:235–242.

200. Busam KJ, Iversen K, Coplan KC, Jungbluth AA. Analysis of microphthalmia transcription factor expression in normal tissues and tumors, and comparison of its expression with S-100 protein, gp100, and tyrosinase in desmoplastic malignant melanoma. *Am J Surg Pathol* 2001;25:197–204.

201. Zavala-Pompa A, Folpe AL, Jimenez RE, *et al*. Immunohistochemical study of microphthalmia transcription factor and tyrosinase in angiomyolipoma of the kidney, renal cell carcinoma, and renal and retroperitoneal sarcomas. *Am J Surg Pathol* 2001;25:65–70.

202. Kawakami Y, Eliyahu S, Delgado CH, *et al*. Cloning of the gene coding for a shared human melanoma antigen recognized by autologous T-cells

infiltrating into tumor. *Proc Natl Acad Sci USA* 1994;91:3515–3519.

203. Coulie PG, Brichard V, van Pel A, et al. A new gene coding for a differentiation antigen recognized by autologous cytolytic T lymphocytes on HLA-A2 melanomas. *J Exp Med* 1994;180:35–42.

204. Kawakami Y, Eliyahu S, Delgado CH, et al. Identification of a human melanoma antigen recognized by tumor-infiltrating lymphocytes associated with in vivo tumor rejection. *Proc Natl Acad Sci USA* 1994;91:6458–6462.

205. Jungbluth AA, Busam KJ, Gerald WL, et al. A103: an anti melan-A monoclonal antibody for the detection of malignant melanoma in paraffin-embedded tissues. *Am J Surg Pathol* 1998;22:595–602.

206. Orosz Z. Melan-A/MART-1 expression in various melanocytic lesions and in non-melanocytic soft tissue tumours. *Histopathology* 1999;34:517–525.

207. Jungbluth AA, Iversen K, Coplan K, et al. Expression of melanocyte-associated markers gp-100 and Melan-A/MART-1 in angiomyolipomas. An immunohistochemical and RT-PCR analysis. *Virchows Arch* 1999;434:429–435.

208. Fetsch PA, Fetsch JF, Marincola FM, et al. Comparison of melanoma antigen recognized by T-cells (MART-1) to HMB-45: additional evidence to support a common lineage for angiomyolipoma, lymphangiomyomatosis and clear cell sugar tumor. *Mod Pathol* 1998;11:699–703.

209. Busam KJ, Iversen K, Coplan KA, et al. Immunoreactivity for A103, an antibody to melan-A (Mart-1), in adrenocortical and other steroid tumors. *Am J Surg Pathol* 1998;22:57–63.

210. Yasumoto K, Yokoyama K, Shibata K, Tomita Y, Shibahara S. Microphthalmia-associated transcription factor as a regulator for melanocyte-specific transcription of the human tyrosinase gene. *Mol Cell Biol* 1994;14:8058–8070.

211. Hemesath TJ, Steingrimsson E, McGill G, et al. Microphthalmia, a critical factor in melanocyte development, defines a discrete transcription factor family. *Genes Dev* 1994;8:2770–2780.

212. Weilbacher KN, Hershey CL, Takemoto CM, et al. Age-resolving osteopetrosis: a rat model implicating microphthalmia and the related transcription factor TEF3. *J Exp Med* 1998;187:775–785.

213. King R, Weilbaecher KN, McGill G, et al. Microphthalmia transcription factor. A sensitive and specific melanocytic marker for melanoma diagnosis. *Am J Pathol* 1999;155:731–738.

214. King R, Googe PB, Weilbaecher KN, Mihm MC, Fisher DE. Microphthalmia transcription factor expression in cutaneous benign, malignant melanocytic, and nonmelanocytic tumors. *Am J Surg Pathol* 2001;25:51–57.

215. Granter SR, Weilbaecher KN, Quigley C, Fletcher CDM, Fisher DE. Microphthalmia transcription factor. Not a sensitive or specific marker for the diagnosis of desmoplastic melanoma and spindle cell (non-desmoplastic) melanoma. *Am J Dermatopathol* 2001;23:185–189.

216. Koch MB, Shih IM, Weiss SW, Folpe AL. Microphthalmia transcription factor and melanoma cell adhesion molecule expression distinguish desmoplastic/spindle cell melanoma from morphologic mimics. *Am J Surg Pathol* 2001;25:58–64.

217. Busam KJ, Kucukgöl D, Sato E, et al. Immunohistochemical analysis of novel monoclonal antibody PNL2 and comparison with other melanocyte differentiation markers. *Am J Surg Pathol* 2005;29:400–406.

218. Aung PP, Sarlomo-Rikala M, Lasota J, et al. KBA62 and PNL2: 2 new melanoma markers-immunohistochemical analysis of 1563 tumors including metastatic, desmoplastic, and mucosal melanomas and their mimics. *Am J Surg Pathol* 2012;36:265–272.

Histiocytic markers

219. Ichinose A. Physiopathology and regulation of factor XIII. *Thromb Haemost* 2001;86:57–65.

220. Nemes Z, Thomazy V, Adany L, Muszbek L. Identification of histiocytic reticulum cells by the immunohistochemical domonstration of factor XIII (F-XIIIa) in human lymph nodes. *J Pathol* 1986;149:121–132.

221. Nickoloff BJ, Griffiths CEM. Factor XIIIa-expressing dermal dendrocytes in AIDS-associated cutaneous Kaposi's sarcomas. *Science* 1989;243:1736–1737.

222. Nemes Z, Thomazy V. Factor XIIIa and the classic histiocytic markers in malignant fibrous histiocytoma: a comparative immunohistochemical study. *Hum Pathol* 1988;19:822–829.

223. Misery L, Boucheron S, Claudy AL. Factor XIIIa expression in juvenile xanthogranuloma. *Acta Dermatovenerol* 1994;74:43–44.

224. Kraus MD, Haley JC, Ruiz R, et al. "Juvenile" xanthogranuloma. An immunophenotypic study with reappraisal of histogenesis. *Am J Dermatopathol* 2001;23:104–111.

225. Pulford KAF, Rigney EM, Jones M, et al. KP1: a new monoclonal antibody detecting a monocyte/macrophage associated antigen in routinely processed tissue sections. *J Clin Pathol* 1989;42:414–421.

226. Weiss LM, Arber DA, Chang KL. CD68: a review. *Appl Immunohistochem* 1994;2:2–8.

227. Warnke RA, Pulford KAF, Pallesen G, et al. Diagnosis of myelomonocytic and macrophage neoplasms in routinely processed tissue biopsies with monoclonal antibody KP1. *Am J Pathol* 1989;135:1089–1095.

228. Tsang WY, Chan JK. KP1 (CD68) staining of granular cell neoplasms: is KP1 a marker for lysosomes rather than the histiocytic lineage? *Histopathology* 1992;21:84–86.

229. dei Tos A, Doglioni C, Laurino L, Fletcher CDM. KP1 (CD68) expression in benign neural tumours. Further evidence of its low specificity as a histiocytic/myeloid marker. *Histopathology* 1993;23:185–187.

230. McHugh M, Miettinen M. CD68: its limited specificity for histiocytic tumors. *Appl Immunohistochem* 1994;2:186–190.

231. Cassidy M, Loftus B, Whelan A, et al. KP-1: not a specific marker. Staining of 137 sarcomas, 48 lymphomas, 28 carcinomas, 7 malignant melanomas and 8 cystosarcoma phyllodes. *Virchows Arch* 1994;424:635–640.

232. Binder SW, Said JW, Shintaku IP, Pinkus GS. A histiocyte-specific marker in the diagnosis of malignant fibrous histiocytoma: use of monoclonal

antibody KP-1 (CD68). *Am J Clin Pathol* 1992;97:759–763.

233. Buechler C, Ritter M, Orso E, et al. Regulation of scavenger receptor CD163 expression in human monocytes and macrophages by pro- and antiinflammatory stimuli. *J Leukoc Biol* 2000;67:97–103.

234. Fabriek BO, Dijkstra CD, Van Den Berg TK. The macrophage scavenger receptor CD163. *Immunobiology* 2005;210:153–160.

235. Nguyen TT, Schwartz EJ, West RB, et al. Expression of CD163 (hemoglobin scavenger receptor) in normal tissues, lymphomas, carcinomas, and sarcomas is largely restricted to the monocyte/macrophage lineage. *Am J Surg Pathol* 2005;29:617–624.

236. Lau SK, Chu PG, Weiss LM. CD163: a specific marker of macrophages in paraffin-embedded tissue samples. *Am J Clin Pathol* 2004;122:794–801.

Keratins

237. Moll R, Divo M, Langbein L. The human keratins: biology and pathology. *Histochem Cell Biol* 2008;129:705–733.

238. Franke WW, Schiller DL, Moll R, et al. Diversity of cytokeratins. Differentiation specific expression of cytokeratin polypeptides in epithelial cells and tissues. *J Mol Biol* 1981;153:933–959.

239. Moll R, Franke WW, Schiller DL, Geiger B, Krepler R. The catalog of human cytokeratins: patterns of expression in normal epithelia, tumors and cultured cells. *Cell* 1982;31:11–24.

240. Moll R, Schiller DL, Franke WW. Identification of protein IT of the intestinal cytoskeleton as a novel type I cytokeratin with unusual properties and expression patterns. *J Cell Biol* 1990;111:567–580.

241. Sun T-T, Eichner R, Schermer A, et al. Classification, expression and possible mechanisms of evolution of mammalian epithelial keratins: a unifying model. In *Cancer Cell I/The Transformed Phenotype*. Levine AJ, van de Voude GF, Topp WC, Watson JD (eds.) Cold Spring Harbor, NY: Cold Spring Harbor Laboratory; 1985:169–176.

242. Cooper D, Schermer A, Sun TT. Biology of disease. Classification of human epithelia and their neoplasms using monoclonal antibodies to keratins: strategies, applications and limitations. *Lab Invest* 1985;52:243–256.

243. Eichner R, Bonitz P, Sun TT. Classification of epidermal keratins according to their immunoreactivity, isoelectric point, and mode of expression. *J Cell Biol* 1984;98:1388–1396.

244. Alfonso P, Nunez A, Mazdoz-Gurpide J, et al. Proteomic expression analysis of colorectal cancer by two-dimensional differential gel electrophoresis. *Proteomics* 2005;5:2602–2611.

245. Miettinen M, Lehto VP, Virtanen I. Keratin in the epithelial-like cells of classical biphasic synovial sarcoma. *Virchows Arch B Cell Pathol Incl Mol Pathol* 1982;40:157–161.

246. Chase DR, Enzinger FM, Weiss SW, Langloss JM. Keratin in epithelioid sarcoma. An immunohistochemical study. *Am J Surg Pathol* 1984;8:435–441.

247. Miettinen M, Lehto VP, Dahl D, Virtanen I. Differential diagnosis of chordoma, chondroid, and ependymal tumors as aided by anti-intermediate filament antibodies. *Am J Pathol* 1983;112:160–169.

248. Huitfeldt HS, Brandtzaeg P. Various keratin antibodies produce immunohistochemical staining of human myocardium and myometrium. *Histochemistry* 1985;83:381–389.

249. Miettinen M. Keratin immunohistochemistry: update of applications and pitfalls. *Pathol Annu* 1993;24(Part 2):113–143.

250. von Koskull H, Virtanen I. Induction of cytokeratin expression in human mesenchymal cells. *J Cell Physiol* 1987;133:321–329.

251. Knapp AC, Franke WW. Spontaneous losses of control of cytokeratin gene expression in transformed, non-epithelial human cells occurring at different levels of regulation. *Cell* 1989;59:67–79.

252. Rosenberg M, Ray Chaudhury A, Shows TB, LeBeau MM, Fuchs E. A group of type I keratin genes on human chromosome 17: characterization and expression. *Mol Cell Biol* 1988;8:722–736.

253. Romano V, Bosco P, Rocchi M, et al. Chromosomal assignments of human type I and type II cytokeratin genes to different chromosomes. *Cytogenet Cell Genet* 1988;48:148–151.

254. Rosenberg M, Fuchs E, Le Beau MM. Three epidermal and one simple epithelial type II keratin genes map to chromosome 12. *Cytogenet Cell Genet* 1991;57:33–38.

255. Blobel GA, Moll R, Franke WW, Kayser KW, Gould VE. The intermediate filament cytoskeleton of malignant mesotheliomas and its diagnostic significance. *Am J Pathol* 1985;121:235–247.

256. Moll R, Achstetter T, Becht E, et al. Cytokeratins in normal and malignant transitional epithelium. Maintenance of expression of urothelial differentiation features in transitional cell carcinomas and bladder carcinoma cell culture lines. *Am J Pathol* 1990;132:123–144.

257. Schaafsma HE, Ramaekers FC, van Muijen GN, et al. Distribution of cytokeratin polypeptides in human transitional cell carcinomas, with special emphasis on changing expression patterns during tumor progression. *Am J Pathol* 1990;136:329–343.

258. Moll R. Cytokeratins as markers of differentiation in the diagnosis of epithelial tumors. *Subcell Biochem* 1998;31:205–262.

259. Markey AC, Lane EB, Churchill LJ, McDonald DM, Leigh IM. Expression of simple epithelial keratins 8 and 18 in epidermal neoplasia. *J Invest Dermatol* 1991;97:763–770.

260. Brown DC, Theaker JM, Banks PM, Gatter KC, Mason DY. Cytokeratin expression in smooth muscle and smooth muscle tumours. *Histopathology* 1987;11:477–486.

261. Norton AJ, Thomas JA, Isaacson PG. Cytokeratin-specific monoclonal antibodies are reactive with tumours of smooth muscle derivation. An immunocytochemical and biochemical study using antibodies to intermediate filament cytoskeletal proteins. *Histopathology* 1987;11:487–499.

262. Gown AM, Boyd HC, Chang Y, et al. Smooth muscle cells can express cytokeratins of "simple" epithelium. Immunocytochemical and biochemical studies in vitro and in vivo. *Am J Pathol* 1988;132:223–232.

263. Kuruc N, Franke WW. Transient coexpression of desmin and cytokeratins 8 and 18 in developing

myocardial cells of some vertebrate species. *Differentiation* 1986;38:177–193.

264. Jahn L, Fouquet B, Rohe K, Franke WW. Cytokeratins in certain endothelial and smooth muscle cells of two taxonomically distant vertebrate species, *Xenopus laevis* and man. *Differentiation* 1987;36:234–254.

265. Miettinen M, Fetsch JF. Distribution of keratins in normal endothelial cells and a spectrum of vascular tumors: implications in tumor diagnosis. *Hum Pathol* 2000;31:1062–1067.

266. Miettinen M, Paetau A. Mapping of the keratin polypeptides in meningiomas of different types: an immunohistochemical analysis of 463 cases. *Hum Pathol* 2002;33:590–598.

267. Franke WW, Moll R. Cytoskeletal components of lymphoid organs. I. Synthesis of cytokeratins 8 and 18 and desmin in subpopulations of extrafollicular reticulum cells of human lymph nodes, tonsils and spleen. *Differentiation* 1987;36:145–163.

268. Traweek ST, Liu J, Battifora H. Keratin gene expression in non-epithelial tissues: detection with polymerase chain reaction. *Am J Pathol* 1993;142:1111–1118.

269. Miettinen M. Keratin subsets in spindle cell sarcomas. Keratins are widespread but synovial sarcoma contains a distinctive keratin polypeptide pattern and desmoplakins. *Am J Pathol* 1991;138:505–513.

270. Miettinen M, Limon J, Niezabitowski A, Lasota J. Patterns of keratin polypeptides in 110 biphasic, monophasic and poorly differentiated synovial sarcomas. *Virchows Arch* 2000;437:275–283.

271. Manivel JC, Wick MR, Dehner LP, Sibley RK. Epithelioid sarcoma. An immunohistochemical study. *Am J Clin Pathol* 1987;87:319–326.

272. Daimaru Y, Hashimoto H, Tsuneyoshi M, Enjoji M. Epithelial profile of epithelioid sarcoma. An immunohistochemical analysis of six cases. *Cancer* 1987;59:34–41.

273. Miettinen M, Fanburg-Smith JC, Virolainen M, Shmookler BM, Fetsch JF. Epithelioid sarcoma: an immunohistochemical analysis of 112 classical and variant cases and a discussion of the differential diagnosis. *Hum Pathol* 1999;30:934–942.

274. Heikinheimo K, Persson S, Kindblom LG, Morgan PR, Virtanen I. Expression of different cytokeratin subclasses in human chordoma. *J Pathol* 1991;164:145–150.

275. Naka T, Iwamoto Y, Shinohara N, et al. Cytokeratin subtyping in chordoma and the fetal notochord: an immunohistochemical analysis of aberrant expression. *Mod Pathol* 1997;10:545–551.

276. O'Hara BJ, Paetau A, Miettinen M. Keratin subsets and monoclonal antibody HBME-1 in chordoma: immunohistochemical differential diagnosis between tumors simulating chordoma. *Hum Pathol* 1998;29:119–126.

277. Chu PG, Weiss LM. Keratin expression in human tissues and neoplasms. *Histopathology* 2002;40:403–439.

278. Bolen JW, Hammar SP, McNutt MA. Reactive and neoplastic serosal tissue: a light microscopic, ultrastructural, and immunocytochemical study. *Am J Surg Pathol* 1986;10:34–47.

279. Gray MH, Rosenberg AE, Dickersin GR, Bhan AK. Cyto keratin expression in epithelioid vascular neoplasms. *Hum Pathol* 1990;21:212–217.

280. Miettinen M. Immunoreactivity for cytokeratin and epithelial membrane antigen in leiomyosarcoma. *Arch Pathol Lab Med* 1988;112:637–640.

281. Iwata J, Fletcher CD. Immunohistochemical detection of cytokeratin and epithelial membrane antigen in leiomyosarcoma: a systematic study of 100 cases. *Pathol Int* 2000;50:7–14.

282. Miettinen M, Soini Y. Malignant fibrous histiocytoma. Heterogeneous patterns of intermediate filament proteins by immunohistochemistry. *Arch Pathol Lab Med* 1989;113:1363–1366.

283. Rosenberg AE, O'Connell JX, Dickersin GR, Bhan AK. Expression of epithelial markers in malignant fibrous histiocytoma of the musculoskeletal system: an immunohistochemical and electron microscopic study. *Hum Pathol* 1993;24:284–293.

284. Litzky LA, Brooks JJ. Cytokeratin immunoreactivity in malignant fibrous histiocytoma and spindle cell tumors: comparison between frozen and paraffin-embedded tissues. *Mod Pathol* 1992;5:30–34.

285. Sarlomo-Rikala M, Tsujimura T, Lendahl U, Miettinen M. Patterns of nestin and other intermediate filament expression distinguish between gastrointestinal stromal tumors, leiomyomas, and schwannomas. *APMIS* 2002;110:499–507.

286. Gu M, Antonescu CR, Guiter G, et al. Cytokeratin immunoreactivity in Ewing's sarcoma: prevalence in 50 cases confirmed by molecular diagnostic studies. *Am J Surg Pathol* 2000;24:410–416.

287. Zarbo RJ, Gown AM, Nagle RB, Visscher DW, Crissman JD. Anomalous cytokeratin expression in malignant melanoma: one- and two-dimensional western blot analysis and immunohistochemical survey of 100 melanomas. *Mod Pathol* 1990;3:494–501.

288. Wotherspoon AC, Norton AJ, Isaacson PG. Immunoreactive cytokeratins in plasmacytomas. *Histopathology* 1989;14:141–150.

289. Lasota J, Hyjek E, Koo C, Blonski J, Miettinen M. Cytokeratin-positive B-cell lymphomas: verification by polymerase chain reaction. *Am J Surg Pathol* 1996;20:346–354.

290. Gustmann C, Altmannsberger M, Osborn M, Griesser H, Feller AC. Cytokeratin expression and vimentin content in large cell anaplastic lymphomas and other non-Hodgkin's lymphomas. *Am J Pathol* 1991;138:1413–1422.

291. Bartek J, Vojtesek B, Staskova Z, et al. A series of 14 new monoclonal antibodies to keratins: characterization and value in diagnostic histopathology. *J Pathol* 1991;164:215–224.

292. Ramaekers F, Huysmans A, Schaart G, Moesker O, Vooijs P. Tissue distribution of keratin 7 as monitored by a monoclonal antibody. *Exp Cell Res* 1987;170:235–249.

293. van Niekerk CC, Jap PH, Ramaekers FC, van de Molengraft F, Poels LG. Immunohistochemical demonstration of keratin 7 in routinely fixed paraffin-embedded human tissues. *J Pathol* 1991;165:145–152.

294. Osborn M, van Lessen G, Weber K, Kloppel G, Altmannsberger M. Differential diagnosis of gastrointestinal carcinomas by using monoclonal antibodies specific for individual

295. Ramaekers F, van Niekerk C, Poels L, et al. Use of monoclonal antibodies to keratin 7 in the differential diagnosis of adenocarcinomas. *Am J Pathol* 1990;136:641–655.

296. Wang NP, Zee S, Zarbo RJ, Bacchi CE, Gown AM. Coordinate expression of cytokeratins 7 and 20 defines unique subsets of carcinomas. *Appl Immunohistochem* 1995;3:99–107.

297. Chu P, Wu E, Weiss LM. Cytokeratin 7 and cytokeratin 20 expression in epithelial neoplasms: a survey of 435 cases. *Mod Pathol* 2000;13:962–972.

298. Jensen K, Kohler S, Rouse RV. Cytokeratin staining in Merkel cell carcinoma: an immunohistochemical study of cytokeratins 5/6, 7, 17, and 20. *Appl Immunohistochem Mol Morphol* 2000;8:310–315.

299. Folpe AL, Schmid RA, Chapman D, Gown AM. Poorly differentiated synovial sarcoma: immunohistochemical distinction from primitive neuroectodermal tumors and high-grade malignant peripheral nerve sheath tumors. *Am J Surg Pathol* 1998;22:673–682.

300. Moll R, Lowe A, Laufer J, Franke WW. Cytokeratin 20 in human carcinomas. A new histodiagnostic marker detected by monoclonal antibodies. *Am J Pathol* 1992;140:427–447.

301. Laskin WB, Miettinen M. Epithelioid sarcoma: new insights based on extended immunohistochemical analysis. *Arch Pathol Lab Med* 2003;127:1161–1168.

302. Miettinen M. Keratin 20: immunohistochemical marker for gastrointestinal, urothelial, and Merkel cell carcinomas. *Mod Pathol* 1995;8:384–388.

303. Purkis PE, Steel JB, Mackenzie IC, et al. Antibody markers of basal cells in complex epithelia. *J Cell Sci* 1990;97:39–50.

304. Chu PG, Luda MH, Weiss LM. Cytokeratin 14 expression in epithelial neoplasms: a survey of 435 cases with emphasis on its value in differentiating squamous carcinomas from other epithelial tumours. *Histopathology* 2001;39:9–16.

305. Wetzels RHW, Kuijpers HJH, Lane EB, et al. Basal cell-specific and hyperproliferation-related keratins in human breast cancer. *Am J Pathol* 1991;138:751–763.

306. Malzahn K, Mitze M, Thoenes M, Moll R. Biological and prognostic significance of stratified epithelial cytokeratins in infiltrating ductal breast carcinomas. *Virchows Arch* 1988;433:119–129.

307. Moll R, Dhouailly D, Sun TT. Expression of keratin 5 as a distinctive feature of epithelial and biphasic mesotheliomas: an immunohistochemical study using monoclonal antibody AE14. *Virchows Arch B Cell Pathol Incl Mol Pathol* 1989;58:129–145.

308. Ordonez NG. Value of cytokeratin 5/6 immunostaining in distinguishing epithelial mesothelioma of the pleura from lung adenocarcinoma. *Am J Surg Pathol* 1998;22:1215–1221.

309. Troyanovsky SM, Guelstein VI, Tchipysheva TA, Krutovskikh VA, Bannikov GA. Patterns of expression of keratin 17 in human epithelia: dependency on cell position. *J Cell Sci* 1989;93:419–426.

310. Troyanovsky SM, Leube RE, Franke WW. Characterization of the human gene encoding cytokeratin 17 and its expression pattern. *Eur J Cell Biol* 1992;59:127–137.

311. Miettinen M, Nobel MP, Tuma BT, Kovatich AJ. Keratin 17. Immunohistochemical mapping of its distribution in human epithelial tumors and its potential applications. *Appl Immunohistochem* 1997;5:152–159.

312. Goldstein NS, Bassi D, Uzieblo A. WT1 is an integral component of an antibody panel to distinguish pancreaticobiliary and some ovarian epithelial neoplasms. *Am J Clin Pathol* 2001;116:246–252.

313. Moll R, Krepler R, Franke WW. Complex cytokeratin polypeptide patterns observed in certain human carcinomas. *Differentiation* 1983;23:256–269.

314. van Muijen GNP, Ruiter DJ, Franke WW. Cell-type heterogeneity of cytokeratin expression in complex epithelia and carcinomas as demonstrated by monoclonal antibodies specific for cytokeratins 4 and 13. *Exp Cell Res* 1986;62:97–113.

315. Remotti F, Fetsch JF, Miettinen M. Keratin 1 expression in endothelia and mesenchymal tumors: immunohistochemical analysis of normal and neoplastic tissues. *Hum Pathol* 2001;32:873–879.

316. Hasan AAK, Zisman T, Schmaier AH. Identification of cytokeratin 1 as a binding protein and presentation receptor for kininogens on endothelial cells. *Proc Natl Acad Sci USA* 1998;95:3615–3620.

317. Ivanyi D, Ansink A, Groeneweld E, et al. New monoclonal antibodies recognizing epidermal differentiation-associated keratins in formalin-fixed, paraffin-embedded tissue: keratin 10 expression in carcinoma of the vulva. *J Pathol* 1989;159:7–12.

318. Knapp AC, Franke WW, Heid H, et al. Cytokeratin No. 9, an epidermal type I keratin characteristic of a special program of keratinocyte differentiation displaying body site specificity. *J Cell Biol* 1986;103:657–667.

319. Jih DM, Lyle S, Elenitsas R, Elder DE, Cotsarelis G. Cyto keratin 15 expression in trichoepitheliomas and a subset of basal cell carcinomas suggests they originate from hair follicle stem cells. *J Cutan Pathol* 1999;26:113–118.

EMA (MUC1)

320. Gendler SJ. MUC1, the renaissance molecule. *J Mammary Gland Biol Neoplasia* 2001;6:339–353.

321. Brayman M, Thathiah A, Carson DD. MUC1: a multifunctional cell surface component of reproductive tissue epithelia. *Reprod Biol Endocrinol* 2004;2:4.

322. Sasaki M, Peterson JA, Wara WM, Ceriani RL. Human mammary epithelial antigens (HME-Ags) in the circulation of nude mice implanted with a breast tumor and non-breast tumors. *Cancer* 1981;48:2204–2210.

323. Sloane JP, Ormerod MG. Distribution of epithelial membrane antigen in normal and neoplastic tissues and its value in diagnostic tumor pathology. *Cancer* 1981;47:1786–1795.

324. Pinkus GS, Kurtin PJ. Epithelial membrane antigen: a diagnostic discriminant in surgical pathology; immunohistochemical profile in epithelial, mesenchymal, and hematopoietic neoplasms using paraffin sections and monoclonal antibodies. *Hum Pathol* 1985;16:929–940.

325. Heyderman E, Strudley I, Powell G, et al. A new monoclonal antibody to

epithelial membrane antigen (EMA)-E29: a comparison of its immunocytochemical reactivity with polyclonal anti-EMA antibodies and with another monoclonal antibody, HMFG-2. *Br J Cancer* 1985;52:355–361.

326. Swanson PE, Manivel JC, Scheithauer BW, Wick MR. Epithelial membrane antigen reactivity in mesenchymal neoplasms: an immunohistochemical study of 306 soft tissue sarcomas. *Surg Pathol* 1989;2:313–322.

327. Ariza A, Bilbao JM, Rosai J. Immunohistochemical detection of epithelial membrane antigen in normal perineurial cells and perineurioma. *Am J Surg Pathol* 1988;12:678–683.

328. Theaker JM, Fletcher CDM. Epithelial membrane antigen expression by the perineurial cell: further studies on peripheral nerve lesions. *Histopathology* 1989;14:581–588.

329. Fetsch JF, Miettinen M. Sclerosing perineurioma: a clinicopathologic study of 19 cases of a distinctive soft tissue lesion with a predilection for the fingers and palms of young adults. *Am J Surg Pathol* 1997;21:1433–1442.

330. Guillou L, Benhattar J, Gengler C, et al. Translocation-positive low-grade fibromyxoid sarcoma: clinicopathologic and molecular analysis of a series expanding the morphologic spectrum and suggesting a potential relationship with sclerosing epithelioid fibrosarcoma: a study from the French Sarcoma Group. *Am J Surg Pathol* 2007;31:1387–1402.

331. Doyle LA, Fletcher CD. EMA positivity in epithelioid fibrous histiocytoma: a potential diagnostic pitfall. *J Cutan Pathol* 2011;38:697–703.

ALK

332. Coffin CM, Patel A, Perkins S, et al. ALK1 and p80 expression and chromosomal rearrangements involving 2p23 in inflammatory myofibroblastic tumor. *Mod Pathol* 2001;14:569–576.

333. Cook JR, Dehner LP, Collins MH, et al. Anaplastic lymphoma kinase (ALK) expression in the inflammatory myofibroblastic tumor: a comparative immunohistochemical study. *Am J Surg Pathol* 2001;25:1364–1371.

334. Mariño-Enríquez A, Wang WL, Roy A, et al. Epithelioid inflammatory myofibroblastic sarcoma: an aggressive intra-abdominal variant of inflammatory myofibroblastic tumor with nuclear membrane or perinuclear ALK. *Am J Surg Pathol* 2011;35:135–144.

335. Cessna MH, Zhou H, Sanger WG, et al. Expression of ALK1 and p80 in inflammatory myofibroblastic tumor and its mesenchymal mimics: a study of 135 cases. *Mod Pathol* 2002;15:931–938.

336. Doyle LA, Mariño-Enriquez A, Fletcher CD, Hornick JL. ALK rearrangement and overexpression in epithelioid fibrous histiocytoma. *Mod Pathol* 2015;28:904–912.

Cadherins

337. Angst RD, Marcozzi C, Magee AI. The cadherin superfamily. *J Cell Sci* 2001;114:625–626.

338. Smith MEF, Pignatelli M. The molecular histology of neoplasia: the role of the cadherin/catenin complex. *Histopathology* 1997;31:107–111.

339. Sato H, Hasegawa T, Abe Y, Skai H, Hirohashi S. Expression of E-cadherin in bone and soft tissue sarcoma: a possible role in epithelial differentiation. *Hum Pathol* 1999;30:1344–1349.

340. Danen EH, de Vries TJ, Morandini R, et al. E-cadherin expression in human melanoma. *Melanoma Res* 1996;6:127–131.

341. Smith ME, Brown JI, Fisher C. Epithelioid sarcoma: presence of vascular-endothelial cadherin and lack of epithelial cadherin. *Histopathology* 1998;33:425–431.

342. Laskin WB, Miettinen M. Epithelial-type and neural-type cadherin expression in malignant noncarcinomatous neoplasms with epithelioid features that involve the soft tissues. *Arch Pathol Lab Med* 2002;126:425–431.

CD10

343. Mechtersheimer G, Moller P. Expression of the common acute lymphoblastic leukemia antigen (CD10) in mesenchymal tumors. *Am J Pathol* 1989;134:961–965.

344. Chu P, Arber DA. Paraffin-section detection of CD10 in 505 nonhematopoietic neoplasms: frequent expression in renal cell carcinoma and endometrial stromal sarcoma. *Am J Clin Pathol* 2000;113:374–382.

345. Chu PG, Arber DA, Weiss LM, Chang KL. Utility of CD10 in distinguishing between endometrial stromal sarcoma and uterine smooth muscle tumors: an immunohistochemical comparison of 34 cases. *Mod Pathol* 2001;14:465–471.

346. McCluggage WG, Sumathi VP, Maxwell P. CD10 is a sensitive and diagnostically useful immunohistochemical marker of normal endometrial stroma and of endometrial stromal neoplasms. *Histopathology* 2001;39:273–278.

347. Kanitakis J, Bourchany D, Claudy A. Expression of the CD10 antigen (neutral endopeptidase) by mesenchymal tumors of the skin. *Anticancer Res* 2000;20:3539–3544.

348. Deniz K, Çoban G, Okten T. Anti-CD10 (56C6) expression in soft tissue sarcomas. *Pathol Res Pract* 2012;208:281–285.

CD99

349. Ambros IM, Ambros PF, Strehl J, et al. MIC2 is a specific marker for Ewing's sarcoma and peripheral neuroectodermal tumors: evidence for a common histogenesis of Ewing's sarcoma and peripheral primitive neuroectodermal tumors from MIC2 expression and common chromosome aberration. *Cancer* 1991;67:1886–1893.

350. Fellinger EJ, Garin-Chesa P, Triche TJ, Huvos AG, Rettig WJ. Immunohistochemical analysis of Ewing's sarcoma cell surface antigen p30/32MIC2 *Am J Pathol* 1991;1139:317–325.

351. Stevenson AJ, Chatten J, Bertoni F, Miettinen M. CD99 (p30/32-MIC2) neuroectodermal/Ewing sarcoma antigen as an immunohistochemical marker. Review of more than 600 tumors and the literature experience. *Appl Immunohistochem* 1994;2:231–240.

352. Renshaw AA. O13 (CD99) in spindle cell tumors. Reactivity with hemangiopericytoma, solitary fibrous tumor, synovial sarcoma, and meningioma but rarely with sarcomatoid mesothelioma. *Appl Immunohistochem* 1995;3:250–256.

353. Diwan AH, Skelton HG III, Horenstein MG, et al. Dermatofibrosarcoma protuberans and giant cell fibroblastoma exhibit CD99 positivity. *J Cutan Pathol* 2008;35:547–550.

354. Shibuya R, Matsuyama A, Nakamoto M, et al. The combination of CD99 and NKX2.2, a transcriptional target of EWSR1-FLI1, is highly specific for the diagnosis of Ewing sarcoma. *Virchows Arch* 2014;465:599–605.

Calretinin

355. Dei Tos AP, Doglioni C. Calretinin: a novel tool for diagnostic immunohistochemistry. *Adv Anat Pathol* 1998;5:61–66.

356. Miettinen M, Limon J, Niezabitowski A, Lasota J. Calretinin and other mesothelioma markers in synovial sarcoma: analysis of antigenic similarities and differences with malignant mesothelioma. *Am J Surg Pathol* 2001;25:610–617.

357. Barak S, Wang Z, Miettinen M. Immunoreactivity for calretinin and keratins in desmoid fibromatosis and other myofibroblastic tumors: a diagnostic pitfall. *Am J Surg Pathol* 2012;36:1404–1409.

358. Cates JM, Coffing BN, Harris BT, Black CC. Calretinin expression in tumors of adipose tissue. *Hum Pathol* 2006;37:312–321.

Collagen IV and laminin

359. Miettinen M, Foidart JM, Ekblom P. Immunohistological demonstration of laminin, the major glycoprotein of basement membranes, as an aid in the diagnosis of soft tissue tumors. *Am J Clin Pathol* 1983;79:306–311.

360. Autio-Harmainen H, Apaja-Sarkkinen M, Martikainen J, Taipale A, Rapola J. Production of basement membrane laminin and type IV collagen by tumors of striated muscle: an immunohistochemical study of rhabdomyosarcomas of different histologic types and a benign vaginal rhabdomyoma. *Hum Pathol* 1986;17:1218–1224.

361. Ogawa K, Oguchi M, Yamabe H, Nakashima Y, Hamashima Y. Distribution of collagen type IV in soft tissue tumors: an immunohistochemical study. *Cancer* 1986;58:269–277.

362. Leong ASY, Vinyuvat S, Suthipintawong C, Leong FJ. Patterns of basal lamina immunostaining in soft tissue and bony tumors. *Appl Immunohistochem* 1997;5:1–7.

Estrogen receptors

363. Billings SD, Folpe AI, Weiss SW. Do leiomyomas of deep soft tissue exist?: an analysis of highly differentiated smooth muscle tumors of deep soft tissue supporting two distinct subtypes. *Am J Surg Pathol* 2001;25:1134–1142.

364. Paal E, Miettinen M. Retroperitoneal leiomyomas: a clinicopathologic and immunohistochemical study of 56 cases with a comparison to retroperitoneal leiomyosarcomas. *Am J Surg Pathol* 2001;25:1355–1363.

365. Rao UN, Finkelstein SD, Jones MW. Comparative immunohistochemical and molecular analysis of uterine and extrauterine leiomyosarcomas. *Mod Pathol* 1999;12:1001–1009.

366. Kelley TW, Borden EC, Goldblum JR. Estrogen and progesterone receptor expression in uterine and extrauterine leiomyosarcomas: an immunohistochemical study. *Appl Immunohistochem Mol Morphol* 2004;12:338–341.

367. Deamant FD, Pombo MT, Battifora H. Estrogen receptor immunohistochemistry as a predictor of site of origin in metastatic breast cancer. *Appl Immunohistochem* 1993;1:188–192.

368. Deyrup AT, Tretiakova M, Montag AG. Estrogen receptor-beta expression in extra abdominal fibromatoses: an analysis of 40 cases. *Cancer* 2006;106:208–213.

369. Deyrup AT, Tretiakova M, Khramtsov A, Montag AG. Exstrogen receptor beta expression in vascular neoplasia: an analysis of 53 benign and malignant cases. *Mod Pathol* 2004;17:1372–1377.

Glial fibrillary acidic protein

370. Achstatter T, Moll R, Anderson A, et al. Expression of glial filament protein (GFP) in nerve sheaths and non-neural cells re-examined using monoclonal antibodies, with special emphasis on the co-expression of GFP and cytokeratins in epithelial cells of human salivary gland and pleomorphic adenomas. *Differentiation* 1986;31:206–227.

371. Mancardi GL, Cadoni A, Tabaton M, et al. Schwann cell GFAP expression increases in axonal neuropathies. *J Neurol Sci* 1991;102:177–183.

372. Memoli VA, Brown EF, Gould VE. Glial fibrillary acidic protein (GFAP) immunoreactivity in peripheral nerve sheath tumors. *Ultrastruct Pathol* 1984;7:269–275.

373. Gould VE, Moll R, Moll I, et al. The intermediate filament complement of the spectrum of nerve sheath neoplasms. *Lab Invest* 1986;55:463–474.

374. Kawahara E, Oda Y, Ooi A, et al. Expression of glial fibrillary acidic protein (GFAP) in peripheral nerve sheath tumors. A comparative study of immunoreactivity of GFAP, vimentin, S100-protein and neurofilament in 38 schwannomas and 18 neurofibromas. *Am J Surg Pathol* 1988;12:115–120.

375. Gray MH, Rosenberg AE, Dickersin GR, Bhan AK. Glial fibrillary acidic protein and keratin expression by benign and malignant nerve sheath tumors. *Hum Pathol* 1989;20:1089–1096.

INI1

376. Masliah-Planchon J, Bièche I, Guinebretière JM, Bourdeaut F, Delattre O. SWI/SNF chromatin remodeling and human malignancies. *Annu Rev Pathol* 2015;10:145–171.

377. Hollmann TJ, Hornick JL. INI1-deficient tumors; diagnostic features and molecular genetics. *Am J Surg Pathol* 2011:35:e47–e63

378. Hornick JL, Dal Cin P, Fletcher CD. Loss of INI1 expression is characteristic of both conventional and proximal-type epithelioid sarcoma. *Am J Surg Pathol* 2009;33:542–550.

379. Sigauke E, Rakheja D, Maddox DL, et al. Absence of expression of SMARCB1/INI1 in malignant rhabdoid tumors of the central nervous system, kidneys and soft tissue: an immunohistochemical study with implications for diagnosis. *Mod Pathol* 2006;19:717–725.

380. Hoot AC, Russo P, Judkins AR, Perlman EJ, Biegel JA. Immunohistochemical analysis of hSNF5/INI1 distinguishes renal and extra-renal malignant rhabdoid tumors from other pediatric soft tissue tumors. *Am J Surg Pathol* 2004;28:1485–1491.

381. Kohashi K, Oda Y, Yamamoto H, et al. SMARCB1/INI1 protein expression in round cell soft tissue sarcomas associated with chromosomal translocations involving EWS: a special reference to SMARCB1/INI1 negative variant extraskeletal myxoid chondrosarcoma. *Am J Surg Pathol* 2008;32:1168–1174.

382. Jo VY, Fletcher CD. Epithelioid malignant peripheral nerve sheath tumor: clinicopathologic analysis of 63

cases. *Am J Surg Pathol* 2015;39:673–682.

383. Rizzo D, Fréneaux P, Brisse H, et al. SMARCB1 deficiency in tumors from the peripheral nervous system: a link between schwannomas and rhabdoid tumors? *Am J Surg Pathol* 2012;36:964–972.

384. Arnold MA, Arnold CA, Li G, et al. A unique pattern of INI1 immunohistochemistry distinguishes synovial sarcoma from its histologic mimics. *Hum Pathol* 2013;44:881–887.

385. Kadoch C, Crabtree GR. Reversible disruption of mSWI/SNF (BAF) complexes by the SS18-SSX oncogenic fusion in synovial sarcoma. *Cell* 2013;153:71–85.

386. Agaimy A. The expanding family of SMARCB1(INI1)-deficient neoplasia: implications of phenotypic, biological, and molecular heterogeneity. *Adv Anat Pathol* 2014;21:394–410.

KIT (CD117)

387. Besmer P, Murphy JE, George PC, et al. A new acute transforming feline retrovirus and relationship of its oncogene v-kit with the protein kinase gene family. *Nature* 1986;320:415–421.

388. Kitamura Y, Hirota S, Nishida T. Molecular pathology of c-kit proto-oncogene and development of gastrointestinal stromal tumors. *Ann Chir Gynaecol* 1998;87:282–286.

389. Miettinen M, Lasota J. KIT (CD117): a review on expression in normal and neoplastic tissues, and mutations and their clinicopathologic correlation. *Appl Immunohistochem Mol Morphol* 2005;13:205–220.

390. Tsuura Y, Hiraki H, Watanabe K, et al. Preferential localization of c-kit product in tissue mast cells, basal cells of skin, epithelial cells of breast, small cell lung carcinoma and seminoma/dysgerminoma in human: immunohistochemical study of formalin-fixed, paraffin-embedded tissues. *Virchows Arch* 1994;424:135–141.

391. Lammie A, Drobnjak M, Gerald W, et al. Expression of c-kit and kit ligand proteins in normal human tissues. *J Histochem Cytochem* 1994;42:1417–1425.

392. Arber DA, Tamayo R, Weiss LM. Paraffin section detection of the c-kit gene product (CD117) in human tissues: value in the diagnosis of mast cell disorders. *Hum Pathol* 1998;29:498–504.

393. Kindblom LG, Remotti HE, Aldenborg F, Meis-Kindblom JM. Gastrointestinal pacemaker cell tumor (GIPACT): gastrointestinal stromal tumors show phenotypic characteristics of the interstitial cells of Cajal. *Am J Pathol* 1998;152:1259–1269.

394. Sarlomo-Rikala M, Kovatich AJ, Barusevicius A, Miettinen M. CD117: a sensitive marker for gastrointestinal stromal tumors that is more specific than CD34. *Mod Pathol* 1998;11:728–734.

395. Sircar K, Hewlett BR, Huizinga JD, et al. Interstitial cells of Cajal as precursors for gastrointestinal stromal tumors. *Am J Surg Pathol* 1999;23:377–389.

396. Chen J, Yanuck RR III, Abbondanzo SL, Chu WS, Aguilera NS. c-kit (CD117) reactivity in extramedullary myeloid tumor/granulocytic sarcoma. *Arch Pathol Lab Med* 2001;125:1448–1452.

397. Miettinen M, Sarlomo-Rikala M, Lasota J. KIT expression in angiosarcomas and in fetal endothelial cells. Lack of c-kit mutations in exon 11 and 17 of c-kit. *Mod Pathol* 2000;13:536–541.

398. Montone KT, van Belle P, Elenitsas R, Elder DE. Protooncogene c-kit expression in malignant melanoma: protein loss with tumor progression. *Mod Pathol* 1997;10:939–944.

MUC4

399. Doyle LA, Möller E, Dal Cin P, et al. MUC4 is a highly sensitive and specific marker for low-grade fibromyxoid sarcoma. *Am J Surg Pathol* 2011;35:733–741.

400. Doyle LA, Wang WL, Dal Cin P, et al. MUC4 is a sensitive and extremely useful marker for sclerosing epithelioid fibrosarcoma: association with FUS gene rearrangement. *Am J Surg Pathol* 2012;36:1444–1451.

SATB2

401. Conner JR, Hornick JL. SATB2 is a novel marker of osteoblastic differentiation in bone and soft tissue tumors. *Histopathology* 2013;63:36–49.

402. Ordonez NG. SATB2 is a novel marker of osteoblastic differentiation and colorectal carcinoma. *Adv Anat Pathol* 2014;21:63–67.

Vimentin

403. Franke WW, Schmid E, Osborn M, Weber K. Different intermediate-sized filaments distinguished by immunofluorescence microscopy. *Proc Natl Acad Sci USA* 1978;75:5034–5038.

404. Azumi N, Battifora H. The distribution of vimentin and keratin in epithelial and nonepithelial neoplasms. A comprehensive immunohistochemical study on formalin- and alcohol-fixed tumors. *Am J Clin Pathol* 1987;88:286–296.

405. Battifora H. Assessment of antigen damage in immunohistochemistry. The vimentin internal control. *Am J Clin Pathol* 1991;96:669–671.

WT1

406. Scharnhorst V, Van Der Eb AJ, Jochemsen AG. WT1 proteins: functions in growth and differentiation. *Gene* 2001;273:141–161.

407. Ordonez NG. Desmoplastic small round cell tumor. II: an ultrastructural and immunohistochemical study with emphasis on new immunohistochemical markers. *Am J Surg Pathol* 1998;22:1314–1327.

408. Barnoud R, Sabourin J, Pasquier D, et al. Immunohistochemical expression of WT1 by desmoplastic small round cell tumor: a comparative study with other small round cell tumors. *Am J Surg Pathol* 2000;24:830–836.

409. Amin KM, Litzky LA, Smythe WR, et al. Wilms' tumor 1 susceptibility (WT1) gene products are selectively expressed in malignant mesothelioma. *Am J Pathol* 1995;146:344–356.

410. Shimizu M, Toki T, Takagi Y, Konishi I, Fujii S. Immunohistochemical detection of the Wilms' tumor gene (WT1) in epithelial ovarian tumors. *Int J Gynecol Pathol* 2000;19:158–163.

411. Tornos C, Soslow R, Chen S, et al. Expression of WT1, CA125, and GCDFP-15 as useful markers in the differential diagnosis of primary ovarian carcinomas versus metastatic breast cancer to the ovary. *Am J Surg Pathol* 2005;29:1482–1489.

Cell cycle markers

412. Gerdes J, Li L, Schlueter C, et al. Immunobiochemical and molecular biologic characterization of the cell proliferation-associated nuclear antigen that is defined by monoclonal antibody Ki-67. *Am J Pathol* 1991:138;867–873.

413. Key G, Becker MH, Baron B, et al. New Ki-67 equivalent murine monoclonal antibodies (MIB 1–3) generated against bacterially expressed parts of the Ki-67 cDNA containing three 62 base pair repetitive elements encoding for the Ki-67 epitope. *Lab Invest* 1993;68:629–636.

414. Swanson SA, Brooks JJ. Proliferation markers Ki-67 and p105 in soft tissue lesions. Correlation with DNA flow cytometric characteristics. *Am J Pathol* 1990;137:1491–1500.

415. Huuhtanen RL, Blomqvist CP, Wiklund TA, et al. Comparison of the Ki-67-score and S-phase fraction as prognostic variables in soft-tissue sarcoma. *Br J Cancer* 1999;79:945–951.

416. Heslin MJ, Cordon-Cardo C, Lewis JJ, Woodruff JM, Brennan MF. Ki-67 detected by MIB-1 predicts distant metastasis and tumor mortality in primary, high-grade extremity sarcomas. *Cancer* 1998;83:490–497.

417. Hoos A, Stojadinovic A, Mastorides S, et al. High Ki-67 proliferative index predicts disease-specific survival in patients with high-risk soft tissue sarcomas. *Cancer* 2001;92:869–874.

418. Hasegawa T. Histological grading and MIB-1 labeling index of soft tissue sarcomas. *Pathol Int* 2007;57:121–125.

419. Seinen JM, Jönsson M, Bendahl PO, et al. Prognostic value of proliferation in pleomorphic soft tissue sarcomas: a new look at an old measure. *Hum Pathol* 2012;43;2247–2254.

420. Lin XY, Wang L, Zhang Y, Dai SD, Wang EH. Variable Ki67 proliferative index in 65 cases of nodular fasciitis, compared with fibrosarcoma and fibromatosis. *Diagn Pathol* 2013;8:50.

421. Wurl P, Taubert H, Meye A, et al. Prognostic value of immunohistochemistry for p53 in primary soft-tissue sarcomas: a multivariate analysis of five antibodies. *J Cancer Res Clin Oncol* 1997;123:502–508.

422. Antonescu CR, Leuang DH, Dudas M, et al. Alterations of cell cycle regulators in localized synovial sarcoma: a multifactorial study with prognostic implications. *Am J Pathol* 2000;156:977–983.

423. Yang P, Hirose T, Hasegawa T, et al. Prognostic implication of the p53 protein and Ki-67 antigen immunohistochemistry in malignant fibrous histiocytoma. *Cancer* 1995;76:618–625.

424. Sabah M, Cummins R, Leader M, Kay E. Immunoreactivity of p53, MDM2, p21 (WAF/CIP1), Bcl-2, and Bax in soft tissue sarcomas: correlation with histologic grade. *Appl Immunohistochem Mol Morphol* 2007;15:64–69.

425. Soini Y, Vähäkangas K, Nuorva K, Kamel D, Lane DP, Pääkkö P. p53 immunohistochemistry in malignant fibrous histiocytomas and other mesenchymal tumours. *J Pathol* 1992;168:29–33.

426. Dei Tos AP, Doglioni C, Laurino L, Barbareschi M, Fletcher CD. p53 protein expression in non-neoplastic lesions and benign and malignant neoplasms of soft tissues. *Histopathology* 1993;22:45–50.

427. Giordano A, Kaiser HE. The retinoblastoma gene: its role in cell cycle and cancer. *In Vivo* 1996;10:223–227.

428. Hamel PA, Gallie BL, Phillips RA. The retinoblastoma protein and cell cycle regulation. *Trends Genet* 1992;8:180–185.

429. Wang J, Coltrera MD, Gown AM. Abnormalities of p53 and p110RB tumor suppressor gene expression in human soft tissue tumors: correlations with cell proliferation and tumor grade. *Mod Pathol* 1995;8:837–842.

430. Kohashi K, Oda Y, Yamamoto H, et al. Alterations of RB1 gene in embryonal and alveolar rhabdomyosarcoma: special reference to utility of pRB immunoreactivity in differential diagnosis of rhabdomyosarcoma subtype. *J Cancer Res Clin Oncol* 2008;134:1097–1103.

431. Kawauchi S, Goto Y, Liu XP, et al. Low expression of p27(Kip1), a cyclin-dependent kinase inhibitor, is a marker of poor prognosis in synovial sarcoma. *Cancer* 2001;91:1005–1012.

432. Kourea HP, Orlow I, Scheithauer BW, Cordon-Cardo C, Woodruff JM. Deletions of the INK4A gene occur in malignant peripheral nerve sheath tumors but not in neurofibromas. *Am J Pathol* 1999;155:1855–1860.

433. Nielsen GP, Stemmer-Rachamimov AO, Ino Y, et al. Malignant transformation of neurofibromas in neurofibromatosis 1 is associated with CDKN2A/p16 inactivation. *Am J Pathol* 1999;155:1879–1884.

434. O'Neill CJ, McBride HA, Connolly LE, McCluggage WG. Uterine leiomyosarcomas are characterized by high p16, p53 and MIB1 expression in comparison with usual leiomyomas, leiomyoma variants and smooth muscle tumours of uncertain potential. *Histopathology* 2007;50:851–858.

435. Atkins KA, Arronte N, Darus CJ, Rice LW. The use of p16 in enhancing the histologic classification of uterine smooth muscle tumors. *Am J Surg Pathol* 2008;32:98–102.

436. Gonzalez RS, McClain CM, Chamberlain BK, Coffin CM, Cates JM. Cyclin-dependent kinase inhibitor 2A (p16) distinguishes well-differentiated liposarcoma from lipoma. *Histopathology* 2013;62:1109–1111.

Chapter 4: Genetics of soft tissue tumors

Julia A. Bridge
University of Nebraska Medical Center

Marilu Nelson
Nebraska Medical Center

The evolution of mesenchymal tumor classification schemes has coincided with cytogenetic and molecular advances. Increasing recognition of the specific genetic abnormalities inherent in these tumors and the growing use of cytogenetic and molecular genetic procedures have aided in the formulation of a diagnosis and the resolution of cellular origin. Several of the genetic markers are also of prognostic value, and the importance of molecular testing for guiding targeted therapeutic strategies in mesenchymal neoplasia is expanding.

The objective of this chapter is to review recurrent or tumor-specific genetic events in mesenchymal neoplasms and discuss the cytogenetic and molecular cytogenetic approaches commonly used in clinical practice to identify them.

Genetic events and molecular pathologic approaches

Common genetic approaches used to identify mesenchymal tumor-specific abnormalities include: (1) conventional cytogenetic; (2) molecular cytogenetic (fluorescence *in situ* hybridization [FISH]); (3) reverse transcription polymerase chain reaction (RT-PCR) analyses; and more recently, (4) next-generation sequencing (to include panels designed for sarcoma testing). Historically, many of the genetic abnormalities that have come to be recognized as tumor specific in sarcomas were first identified by conventional cytogenetic analysis.[1] In turn, the initial cytogenetic evidence facilitated the cloning of many candidate genes involved in the pathogenesis of mesenchymal tumors. Cytogenetic analysis has provided clinicians with a valuable tool to supplement their diagnostic armamentarium. The addition of molecular cytogenetic (FISH) and molecular approaches (e.g. RT-PCR, next-generation sequencing) has further enhanced the sensitivity and accuracy of detecting nonrandom chromosomal imbalances or structural rearrangements in sarcomas, including assessment in formalin-fixed, paraffin-embedded tissues. Molecular assays that are also diagnostically and prognostically informative in mesenchymal neoplasms include microarrays and database software for genomic (DNA copy number changes [imbalances] and copy neutral loss of heterozygosity [cnLOH]) and transcriptomic (RNA, gene expression patterns) comprehensive assessments or profiles.

Conventional cytogenetic analysis

Acquired or somatic chromosomal abnormalities play an important role in sarcomagenesis. Modifications of earlier conventional cytogenetic protocols coupled with increasing molecular cytogenetic techniques continue to advance the field of solid tumor cytogenetics.

Cell culture and chromosome banding

Specimen requirements

Tissue submitted for cytogenetic analysis must be fresh (not frozen or fixed in formalin) because living, dividing cells are required (Figure 4.1). A mesenchymal tumor sample submitted for cytogenetic analysis should be representative of the neoplastic process and preferably be part of the specimen submitted for pathologic study. Ideally a 1 to 2 cm^3 (approximately 0.5–1.0 g sample is provided for analysis; however, small biopsy specimens or fine-needle aspirates (<500 mg) can also be analyzed successfully (although prolonged culture might be needed to produce sufficient numbers of cells for examination).[2] Notably, a limited sample size might be more restrictive for a few neoplasms, such as benign adipose tissue or cartilaginous tumors, which often have a low cell density per volume unit. Although certain culture techniques can be employed in an attempt to reduce potential overgrowth of non-neoplastic components and optimize cell viability, it remains imperative that tissue selected for cytogenetic analysis is composed of predominantly tumor with minimal necrosis.[3-6] Efforts to perform cytogenetic analysis are worthwhile even when material is limited, because the presence of even a single cell exhibiting a tumor-specific chromosomal abnormality provides strong diagnostic support.

Specimen transport

Specimens can be successfully cultured following transportation; this can require 24 to 72 hours for delivery. Whenever possible, however, transport time should be kept to a minimum, preferably with receipt of the sample on the same day as the surgical procedure. Care should also be taken to protect the specimen from extreme changes in temperature. If transportation is delayed, the specimen should be placed in

---

Modern Soft Tissue Pathology, Second Edition, ed. Markku Miettinen. Published by Cambridge University Press. © Cambridge University Press 2016.

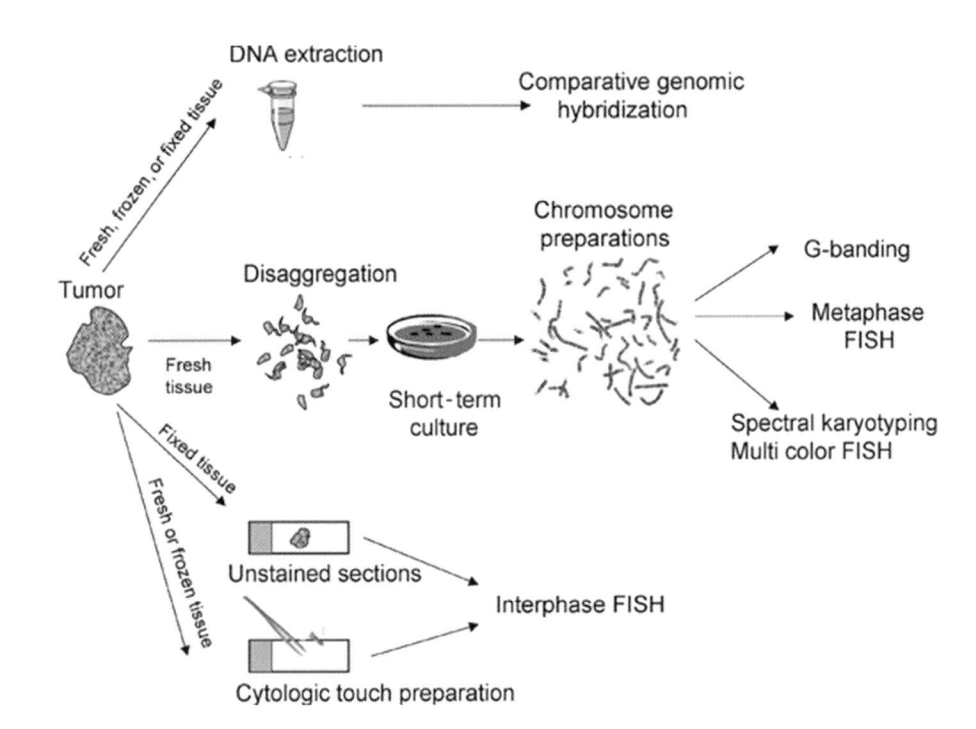

Figure 4.1 Basic specimen requirements: Fresh tissue is required for cell culture and subsequent G-banded metaphase cell preparations or metaphase FISH procedures. DNA can be extracted from fresh, frozen, or formalin-fixed, paraffin-embedded tissue for cytogenomic array platforms, and interphase FISH studies can be conducted on cytologic or formalin-fixed, paraffin-embedded tissue section preparations.

supplemented tissue culture medium and kept at room temperature or stored at 4 °C.

Various media, such as RPMI 1640, MEM or Ham's, or sterile isotonic solutions can be used for specimen transport. Supplementation of transport media with a minimal amount of fetal bovine serum (FBS; ~5%) and antibiotics (~1%) is beneficial.[3–5] A fungicide also can be added if deemed necessary. If the specimen is to be transported immediately and media are not available, placement of the tissue onto sterile saline-soaked gauze is acceptable. The tissue should not be allowed to remain in saline for extended periods, however.

Sample preparation

On receipt in the cytogenetic laboratory, the tissue should be rinsed and cleared of blood and necrotic cellular debris. Balanced salt solutions can be used as a rinse solution to assist in dissolving extracellular mucin when applicable with yield of a tighter cell pellet after centrifugation.[5] A small portion of the tissue should be immediately snap frozen and stored in the event that FISH or other molecular assays are subsequently required.

Disaggregation of solid tissues into single cells and small cell clusters is necessary before culture initiation. For some neoplasms, the tissue is readily disaggregated by simple mechanical means such as mincing the tissue with scalpels or curved scissors, and additional enzymatic digestion is not needed. A potential advantage of this approach is minimization of stromal contamination, because stromal cells are often locked in fibrous connective tissues.[4,6]

Owing to the increased efficiency, enzymatic digestion in combination with mechanical disaggregation has proved to be preferable when processing most soft tissue tumors.

Although there are several options available as to the type, concentration, and duration of proteolytic enzyme that can be used, crude collagenase is the most widely used because of several distinct advantages of this enzyme (i.e., it causes minimal cell damage, effectively breaks down connective fibrous tissue, and is unaffected by the presence of serum). Moreover, crude collagenase not only dissociates several types of collagen, but also other types of intercellular substances owing to the presence of contaminating proteases.[4] The duration of specimen incubation in an enzyme digest depends on the type and concentration of enzyme as well as the cellular and connective tissue density of the individual specimen, but could range from 20 minutes for certain neural tumors to 16 hours or longer for cartilaginous or osseous neoplasms.

Culture initiation and maintenance

After enzymatic digestion, the cell suspension should be examined under a phase microscope to ensure that sufficient dissociation has occurred. It is important to also take note of the viability of the tissue. A trypan blue stain can be performed on a small portion of the cell suspension to ascertain the percentage of viable cells and help to determine the most accurate dilution to use for culture initiation. If cell counts are performed, a typical dilution can be approximately 10^6 cells/5 mL.[5] Sarcomas generally have a higher plating efficiency than epithelial neoplasms.

There are several medium options available for the culture of mesenchymal tumors. The most commonly used is RPMI 1640. Standard protocols require supplementation of the medium with 10% to 20% FBS, 1% penicillin/streptomycin (pen/strep) and 1% L-glutamine. HEPES also can be added at a final concentration of 1% to help maintain a balanced pH.[3–5]

For tissue culture purposes, mesenchymal tumors are commonly inoculated into flasks (T-12.5 or T-25) or onto 22 mm^2 cover slips for *in situ* cultures. Cells propagated as a suspension culture also can be grown in a flask or petri dish, or in a 15-mL conical tube. Cell cultures, incubated at 37 °C in a moist 5% CO_2 environment, should be monitored daily under an inverted microscope. Media are changed as needed to eliminate blood and cellular debris and to nourish the dividing neoplastic cells. If nonviable or deteriorating cells remain in the culture, they can produce proteases that will harm the surrounding viable tumor cells.[3,6]

Harvest and banding of chromosome preparations

The length of time that a soft tissue tumor is cultured to attain satisfactory karyotypic findings varies and depends on the histopathologic type, grade of tumor, tumor cellularity, and size of specimen submitted for analysis. A short-term culture usually results in a sufficient number of mitoses in 3 to 6 days. Lengthy culture times should be avoided because undesired overgrowth by normal fibroblasts is more likely to occur.[3]

An alternative to tissue culture is a direct or same-day harvest. With this technique, endemic dividing cells are arrested after a 1- to 12-hour incubation in colchicine and culture medium. A direct harvest captures metaphase cells entering cell division shortly after the sample was obtained. This approach can circumvent the possibility of introducing additional chromosomal aberrations and polyploidization with culture and therefore provide a more accurate representation of the *in vivo* state. Effusions (e.g., ascites and pleural fluids) and fine-needle aspirations of some tumors, such as the small round blue cell neoplasm, can have a high mitotic index and therefore can be suitable for direct preparations or overnight culturing.[3,4] Direct harvests also yield results quickly (within 24 hours), a factor of great importance when the patient is clinically unstable. Limitations of direct harvest include constraints by the *in vivo* mitotic index or the generation of morphologically inferior chromosomes.

When an optimal number of mitoses are observed, the proliferating cells are arrested in mid-division with the addition of colchicine to block mitotic spindle formation.[3–6] Cells grown in culture flasks are trypsinized into a suspension culture, whereas cells grown on cover slips remain attached throughout the harvest procedure. Following incubation in colchicine, the cells are exposed to a hypotonic solution for approximately 20 minutes to swell the cells and enhance the chromosome preparations. Subsequently, the cells are fixed in a 3:1 methanol:glacial acetic acid solution.

Superlative chromosome morphology and metaphase cell spreading can be achieved through the use of a self-contained drying chamber to regulate the seasonal changes in the ambient temperature and humidity. Giemsa stain, introduced in the 1970s, was one of the first methods used to produce G-banded chromosomes and is still used today; however, care must be taken not to destroy the chromosome morphology with the required trypsin digestion.[7,8] Giemsa–trypsin–Wright (GTW) banding, described by Yunis in the 1980s, is now preferred by many laboratories owing to its ease of use, retention of chromatin morphology that can be at greater risk to overdigestion such as double minutes, and its simple removability for subsequent studies such as FISH.[9]

Chromosome analysis

Microscopy and automated image analysis systems

G-banded metaphase cell coordinates are obtained by manual or automated slide scanning under low magnification (10× lens objective) followed by detailed analysis of individual chromosomes under higher magnification (63× or 100×). If possible, a minimum of 15 to 20 banded metaphase cells are analyzed, with documentation of the aberrations observed. Analysis is conducted on cells of varying morphology from several different cultures or culture vessels to avoid selection. Complete characterization of multiple cell clones might require the analysis of additional metaphase cells.

Today, most cytogenetic laboratories engage automated image analysis systems to generate digital karyotypes and obviate the need for darkroom facilities required by traditional photomicrography. Additional advantages of automated systems for cytogenetic and FISH analyses include image reproducibility, storage reliability, and data communication. Image analysis software facilitates examination of metaphase chromosomes through automatic classification of the chromosomes based on length, localization of the centromere, and banding pattern. Cases with extensive chromosomal overlapping or structural abnormalities represent a caveat of these systems, however, because the software might not recognize the chromosome, and thus correct karyotypic assignment requires the skills of an experienced technologist.

Karyotype

A normal chromosomal complement is composed of 22 pairs of autosomal chromosomes and two sex chromosomes aligned by size, centromere location, and banding pattern. Numerical and structural abnormalities of a clonal nature acquired through the neoplastic process are presented in an abnormal karyotype. An *aberrant clone* is defined as two or more cells that share the same structural abnormality or numerical gain, or three or more cells that exhibit the same numerical loss.[10] There are three basic terms used by cytogeneticists to describe clonal evolution. The term *mainline* (ml) refers to the most common clone and is strictly a quantitative measure. The most basic and simplistic clone present in a sample is referred to as the *stemline* (sl), whereas clones that show increased complexity and appear to evolve from the stemline are called *sidelines* (sdl).[10]

Nomenclature

Karyotype descriptions follow nomenclature rules that were established at a conference held in Denver, CO in 1960.[11] Although technologic advances make periodic revisions necessary, the basic principles established in Denver, and later expanded at a 1971 conference in Paris, prevail in the

International System for Human Cytogenetic Nomenclature.[10,12] The primary constriction of a chromosome, the *centromere*, divides the chromosome into an upper (short) arm and lower (long) arm designated "p" and "q." Each arm is divided into regions, bands, and sub-bands of lighter and darker staining intensity. For example, "1p36.3," referring to the short arm of chromosome 1, region 3, band 6, sub-band 3, is an example of the nomenclature required for specifying breakpoints in structurally aberrant, rearranged, or deleted chromosomes, and for defining the location of genes on chromosomes. The number of alternating light and dark bands detectable with G-banding varies with the level of chromosomal contraction in each metaphase cell. Preparations from mesenchymal tumors typically range from 400 to 600 bands per haploid set, resulting in approximately 5 to 10×10^6 base pairs of DNA per band.

When one is describing the chromosomal complement of a neoplasm, the total number of chromosomes is listed first, followed by the sex chromosomes and the chromosomal abnormalities in numerical order; abbreviations are used to denote each structural abnormality. Symbols and aspects of nomenclature likely to be encountered in the most articles covering cytogenetic and molecular cytogenetic changes in mesenchymal neoplasms are presented in Table 4.1. For more information regarding these areas, the reader should consult the *International System for Human Cytogenetic Nomenclature*.[10]

Advantages and limitations of conventional cytogenetic analysis

Perhaps the greatest drawback of cytogenetic analysis is the requirement for fresh tissue. Successful cytogenetic analysis depends on achieving tumor cell growth in culture. This specimen prerequisite is distinctly limiting, because for some hospitals, specimens are received in the cutting room already fixed in formalin or the pathologist might prematurely place it in formalin. To use the karyotyping approach, therefore, the pathologist must plan ahead and submit a viable portion of the specimen prior to fixation. Complex karyotypes and suboptimal metaphase cell morphology encountered in the analysis of some soft tissue tumors can hamper evaluation. A significant strength of cytogenetic analysis is that it provides a global assessment of both numerical and structural abnormalities in a single assay, including both primary and secondary anomalies. Moreover, in contrast to FISH or RT-PCR, knowledge of the anticipated anomaly or histologic diagnosis is not necessary. Additional advantages and limitations are summarized in Table 4.2.[13,14]

Molecular cytogenetic analysis

A revolutionary tool in the analysis and characterization of chromosomes and chromosomal abnormalities has been the development of FISH techniques. Initially, FISH involved the detection of common numerical abnormalities through the use of centromere-specific repetitive-DNA probes.[15] Additional DNA FISH probe options and the subsequent development of alternative, novel FISH approaches such as multicolor FISH (M-FISH),[16] spectral karyotyping (SKY),[17] combined binary ratio FISH (COBRA-FISH),[18] metaphase-based comparative genomic hybridization (CGH),[19] and cytogenomic arrays (aCGH and single nucleotide polymorphism array [SNP array])[20,21] have led to powerful genome-wide applications that increase the diagnostic and prognostic capabilities of molecular cytogenetics. These molecular cytogenetic approaches, driven by advancing technical unions between molecular diagnostics and standard cytogenetics, have enhanced our knowledge regarding genetic alterations in mesenchymal neoplasia.

Fluorescence *in situ* hybridization

Hybridization refers to the binding or annealing of complementary DNA or RNA sequences that serve as probes. For the purposes of this chapter, the discussion is confined to DNA-based probes most commonly labeled with fluorescent dyes. By way of this approach, specific nucleic acid sequences can be detected in morphologically preserved chromosomes (metaphase FISH) or in nuclei from cell suspensions or cytologic touch preparations of fresh or frozen tumor as well as fixed, paraffin-embedded tissue (interphase FISH). FISH is the only form of analysis that provides cellular localization of DNA sequences in a heterogeneous population, including detection of low-level mosaicism. FISH detects regions of interest usually larger than those assessed by RT-PCR, but smaller than the band-level resolution of conventional cytogenetic analysis. The overall resolution of interphase FISH is approximately 50 to 100 kb compared with an overall resolution of 10 Mb for routine cytogenetic analysis, 2 to 3 Mb for M-FISH and SKY, 2 to 10 Mb for metaphase-based CGH, and 40 to 100 kb for cytogenomic arrays (aCGH and SNP arrays).[22,23]

Specimen requirements

Although FISH can be applied to the same preparations as conventional cytogenetic studies (metaphase cells), a distinct advantage of this technique is the ability to analyze nonproliferating or interphase cells. Interphase nuclei obtained from cytological preparations (e.g., smears, touch imprints, or cytospins) or fresh tissue samples following enzymatic disaggregation, as well as from archival, formalin-fixed, paraffin-embedded tissue are acceptable. Although FISH of paraffin-embedded tissue sections allows interpretation with histomorphologic correlation (i.e., tissue architecture is preserved), the thickness of the tissue section required for optimal hybridization (4–5 μm) can affect the interpretation of results owing to nuclear truncation. A spurious consequence of nuclear truncation can be an underestimation of probe copy number because the sectioned or partial nuclei lack a complete DNA complement. For the most accurate assessment of subtle aneuploidy or structural changes, analysis of whole or intact

Chapter 4: Genetics of soft tissue tumors

Table 4.1 Cytogenetic and molecular cytogenetic abbreviations (nomenclature)

Cytogenetic abbreviations	
del	Deletion or loss of a chromosome segment Example: del(8)(q24) represents a deletion of band 24 located on the long arm of chromosome 8
der	Structurally abnormal chromosome created by the rearrangement of two or more chromosomes or more than one rearrangement in the same chromosome
dup	Duplication, a replicated or duplicated chromosome segment next to itself; if in the same orientation = *direct* duplication (*dir dup*); if inverted = *inverted* duplication (*inv dup*)
i	Isochromosome, a symmetric chromosome composed of duplicated long or short arms formed after misdivision of the centromere in a transverse plane Example: i(3)(p10) refers to a duplication of the short arm of chromosome 3 with loss of the long arm
ins	Insertion, a chromosomal segment from one chromosome is inserted into a nonhomologous chromosome; analogous to a duplication, an insertion can be direct or inverted Example: dir ins(10;12)(q22;q13q14) refers to a direct insertion of the chromosomal segment of a long arm of chromosome 12 (q13–14) into chromosome 10 at q22
inv	Inversion, a segment of a chromosome is reversed 180 degrees; a *paracentric* inversion does not involve the centromere in the inversion, a *pericentric* inversion does (a break in each chromosomal arm is necessary) Example: inv(12)(p12q13) is a pericentric inversion involving the centromere of chromosome 12
p	Short arm of a chromosome (from French *petit*)
q	Long arm of a chromosome (letter following *p*)
t	Translocation, a *reciprocal* translocation is an exchange of chromosomal material between two or more nonhomologous chromosomes (can be balanced or unbalanced); a *Robertsonian* translocation involves acrocentric chromosomes with fusion at the centromere and loss of the short arms and satellites Example: t(11;22)(q24;q12) denotes a reciprocal translocation involving breaks at band q24 on the long arm of chromosome 11 and at band q12 on the long arm of chromosome 22 with exchange of the segments distal to those breakpoints
+ (plus sign)	Added chromosomes (+7) or chromosomal segment (7q+)
− (minus sign)	Lost (deleted) chromosome (−9) or chromosomal segment (9q−)
Molecular cytogenetic abbreviations	
arr	Microarray Example: arr[hg19](8)x3,(12)x3 represents single copy gain of chromosomes 8 and 12 with specification of the genome build used (i.e. [hg19])
ish	*In situ* hybridization on metaphase cell preparations Example: 47,XY,+r.ish der(12)(wcp12+) Metaphase FISH with a whole chromosome paint (wcp) probe for chromosome 12 indicates that the ring chromosome in this karyotypically abnormal cell is derived from chromosome 12 material
nuc ish	*In situ* hybridization on nuclear or interphase cell preparations
amp	Amplified signal Example: nuc ish(D12Z1 × 2),amp(MDM2)[200] Interphase FISH indicating two copies of the chromosome 12 centromere-specific probe were observed in 200 nuclei and that the corresponding quantity of *MDM2* locus signals in these same nuclei were too numerous to be counted accurately (i.e., amplified)
con	Connected signals Example: nuc ish 2q35(PAX3 × 2),13q14(FOXO1 × 2)(PAX3 con FOXO1 × 2) Interphase FISH of a tumor cell indicating that two *PAX3* and two *FOXO1* signals are connected (signals adjacent to each other) suggesting the presence of a t(2;13)(q35;q14) translocation (dual fusion)
sep	Separated signals (signals are split) Example: nuc ish 16p11(FUS × 4)(FUS sep FUS × 1)[180/200] Interphase FISH with a *FUS* break-apart probe indicating that the *FUS* signals on one chromosome 16 homolog are separated, consistent with a rearrangement of the *FUS* locus. The 180/200 in brackets denotes the observation of this result in 180 of the 200 cells analyzed

Table 4.2 Conventional cytogenetic analysis advantages

Advantages	Limitations
Provides global information in a single assay • includes primary and secondary anomalies • knowledge of anticipated anomaly or histologic diagnosis not necessary	Requires fresh tissue • although direct preparations can be performed, cell culture is typically required (1–10 days)
Variants undetectable by interphase FISH or RT-PCR may be uncovered.	• might encounter complex karyotypes with suboptimal morphology • submicroscopic or cryptic rearrangements may result in a false-negative result
Diagnostically useful • sensitive and specific • can be performed on fine-needle aspirates	Normal karyotypes sometimes observed following therapy-induced tumor necrosis or overgrowth of normal supporting stromal cells

nuclei as obtained from a cytologic preparation or disaggregation and release of cells from a thick (50 μm) paraffin-embedded tissue section might be preferable.

Probes and hybridization

Chromosomal probes frequently used in clinical practice to examine soft tissue tumors can be divided into three categories: (1) centromere-specific (repetitive sequence) probes; (2) locus-specific (unique sequence) probes; and (3) "paint" or whole-chromosome probes (WCP). Centromere-specific probes are composed of pericentromeric tandemly repeated 171 bp monomers or alpha-satellite sequences. Distinct centromere-specific probes have been developed for nearly every chromosome, based on dissimilarities in the alpha-satellite sequences of the different chromosomes.[24] Centromeric probes are predominantly used for chromosome enumeration (i.e., determination of aneuploidy). Locus-specific probes are single-copy probes homologous to specific targets (15 to >500 kb in size) and are of particular value in assessing oncogenes or tumor-suppressor genes. WCPs, generated from microdissected chromosomes or DNA libraries established from chromosome flow sorting, are composed of probe mixtures with homology at multiple sites along the target chromosomes.[25,26] These latter probes are nearly exclusively reserved for metaphase cell analysis when further characterization of structural chromosomal abnormalities is required.

A variety of quality-controlled DNA probes intended for clinical purposes are commercially manufactured and sold as analyte-specific reagents (ASRs). For most cytogenetic laboratories, FISH analysis is limited to probes that are commercially available. The number of commercially available probes designed specifically for the study of mesenchymal neoplasms is limited. A dual-color, break-apart probe specific for the assessment of the *EWSR1* gene locus at 22q12 was one of the first probes commercially developed for the examination of translocations in soft tissue tumors. Rearrangements of *EWSR1* are seen in >95% of Ewing sarcomas/peripheral primitive neuroectodermal tumors (pPNET), >99% of desmoplastic small round cell tumors, >90% of clear cell sarcomas, ~75% of extraskeletal myxoid chondrosarcomas, and ~5% of myxoid or round cell liposarcomas, among others.[27–31] Other examples of dual-color, break-apart FISH probes marketed for the detection of mesenchymal-tumor-associated chromosomal translocations include the *SS18* locus at 18q11 for synovial sarcoma; the *DDIT3* locus at 12q13 for myxoid or round cell liposarcoma; the *FUS* locus at 16p11 for myxoid or round cell liposarcoma, angiomatoid fibrous histiocytoma, and low-grade fibromyxoid sarcoma; the *FOXO1* locus at 13q14 for alveolar rhabdomyosarcoma; the *ALK* locus at 2p23 for inflammatory myofibroblastic tumor; and the *ETV6* locus at 12p13 for infantile fibrosarcoma (rearrangements also seen in congenital mesoblastic nephroma).

Some laboratories also elect to custom design probes for the assessment of specific research questions or for routine clinical use in analyzing rearrangements that lack commercial probes. As a consequence of the Human Genome Project, abundant genomic DNA fragments have been characterized and mapped to specific chromosomal regions, resulting in additional options for probe development. Yeast artificial chromosome (YAC; 2 Mb is the approximate maximum probe length [mpl] that can be cloned in this vector), bacterial artificial chromosome (BAC; mpl = 300 kb), P-1 derived artificial chromosome (PAC; mpl = 200 kb), cosmid (mpl = 45 kb), bacteriophage λ (mpl = 25 kb), and plasmids (mpl = 20 kb) are readily available cloning vectors for the generation of custom-designed probes, although YAC, BAC, and PAC are most practical given the more robust probe size. Custom-designed or laboratory-developed probes are used exclusively in-house and not sold to other laboratories. Laboratory-developed probes are not currently regulated by the FDA, and therefore clinical laboratories using such probes must verify or establish, for each specific use of each probe, the performance specifications for applicable performance characteristics (e.g., accuracy, precision, analytical sensitivity and specificity).[32] Resources for guidance on validation processes, ongoing interlaboratory FISH quality assurance comparison programs and proficiency testing for cytogenetic and molecular cytogenetic platforms are available through the College of American Pathologists (CAP)/ACMG Laboratory Accreditation Program.[33]

Chapter 4: Genetics of soft tissue tumors

Table 4.3 Probe selection strategies for interphase cell analysis

Analysis for:	Probe strategy
Gene amplification (e.g., *MYCN*, *MDM2*)	Couple a gene locus-specific, unique sequence probe with the corresponding chromosomal centromere-specific probe • the latter representing a control for ploidy status
Deletion (e.g., *RB*, *TP53*, *SMARCB1* [*INI1*])	Couple a gene locus-specific, unique sequence probe with either the corresponding chromosomal centromere-specific probe or another reference probe (e.g., a separate unique sequence probe located on the same chromosome) • the latter two options representing controls for ploidy status
Translocation – assessing for rearrangement of one of the translocation gene partners involved (e.g., *EWSR1*)	Dual-color, break-apart probe sets represent a mixture of one probe located just proximal (centromeric) to the gene locus or breakpoint of interest and another probe that is located just distal (telomeric)
Translocation – assessing for rearrangement of two translocation gene partners on one of the two derivative chromosomes (fusion gene)	Dual-color, breakpoint flanking probe sets represent a mixture of one probe located centromeric and another probe telomeric to each translocation gene partner
Translocation – assessing for rearrangement of two translocation gene partners on both of the two derivative chromosomes (dual fusion)	Dual-color, breakpoint spanning probe sets represent a mixture of two probes each spanning a translocation gene partner

The general processing of a specimen for FISH includes: (1) specimen fixation; (2) specimen pretreatment; (3) probe and target denaturation and hybridization; and (4) microscopy. Cytological preparations (e.g., smears, touch imprints, and cytospins) are fixed briefly in a 3:1 methanol:glacial acetic acid (Carnoy's fixative) following air drying. For paraffin-embedded tissue, neutral buffered formalin is superior to Bouin's fixative with respect to FISH.[34] Immediate specimen fixation is important because target preservation deteriorates with delays. A prolonged formalin fixation should be avoided, however, because it can produce a highly cross-linked (protein–protein and protein–nucleic-acid) network or barrier, adversely affecting the entry or penetration of FISH probes.[35]

Most soft tissue tumor specimens necessitate a pretreatment to eliminate contaminating proteins or extracellular matrix components for optimal hybridization. For formalin-fixed tissues, pretreatment reduces cross-linking and increases probe permeability with exposure of the target nucleotide sequences. Inclusion of proteolytic enzymes such as pepsin, trypsin, or proteinase K in pretreatment protocols requires proper calculation of the duration of contact, but greatly facilitates successful hybridization. Following pretreatment, the specimen and probe are denatured through the use of formamide at high temperatures and then hybridized together under optimal incubation conditions, where they anneal to form a duplex founded on complementary base pairing. Stringent washes are used to eliminate nonspecific binding and visualization of the hybridized probe to the target DNA is carried out by fluorescence microscopy.[36] In a diagnostic setting, a comprehensible protocol for interpreting FISH results is essential and should include the assorted criteria for detecting abnormalities in individual tests.

Probe strategies

The specific clinical question to be addressed and the type of tissue available for study (metaphase versus interphase cell analysis) are important factors in determining which probe (s) to use in practice. Clinically, FISH performed on metaphase cell preparations plays an important role in deciphering complex chromosomal rearrangements and in verifying the presence or absence of cryptic (submicroscopic) changes. In research, metaphase FISH is useful for positional cloning of chromosomal breakpoints and associated candidate disease genes. A rapid determination of whether specific genes, loci, or regions are present, or if deletions, amplifications, or other structural rearrangements (e.g., translocations) have occurred in a mesenchymal neoplasm, can be achieved by interphase FISH. Probe selection strategies commonly used in soft tissue tumor interphase cell analysis are summarized in Table 4.3.

Interpreting probe signal patterns in interphase FISH can be challenging and requires practice. Certain pitfalls can be avoided, however. For example, evaluation for amplification of an oncogene or loss of a tumor-suppressor gene locus should include not only the unique-sequence probe for that oncogene or tumor-suppressor gene locus but also a second probe (the corresponding chromosomal centromere-specific probe or a second reference probe [e.g., a separate unique sequence probe located on the same chromosome]) as a control for ploidy status. Omission of a ploidy control could result in a false interpretation. For example, 6 to 13 probe signals for the *MYCN* locus (2p24) are identified in red in Figure 4.2A, a copy number that is increased over the normal diploid number. In these same cells with the elevated *MYCN* copy number, however, reside three to seven green ploidy control probe signals (centromeric alpha-satellite sequence probe

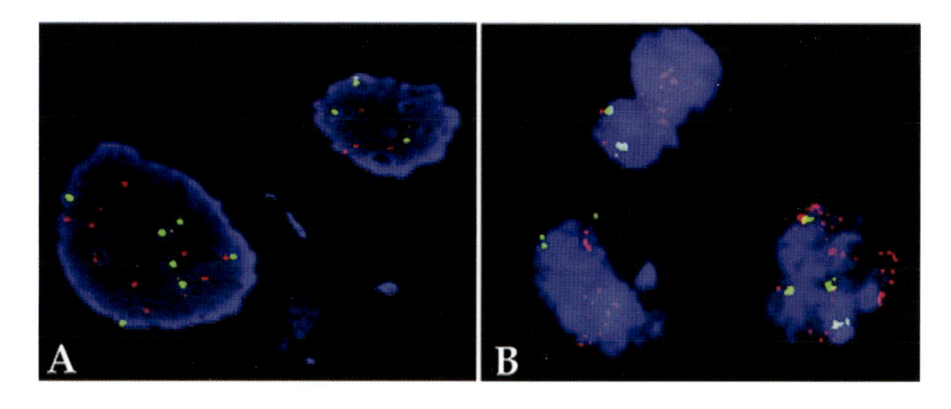

Figure 4.2 (**A**) Relative duplication of MYCN in a neuroblastoma (MYCN probe signal copy number in red to CEP2 signal copy number in green is a ratio of ~2). Conventional cytogenetic analysis of this neoplasm revealed near-triploid and near-hexaploid chromosomal complements, a pattern associated with a favorable prognosis. (**B**) Amplification of MYCN in a different neuroblastoma; a pattern associated with an unfavorable prognosis. Amplification of MYCN as assessed by FISH has been defined by some groups as greater than a fourfold increase of the MYCN probe signal copy number as compared with the reference probe copy number (e.g., CEP2). (From Ambros IM, Speleman F, Ambros PF. Nervous system: peripheral neuroblastic tumours (neuroblastoma, ganglioneuroblastoma, ganglioneuroma). *Atlas Genet Cytogenet Oncol Haematol* 2009;13(1):84–89. http://atlasgeneticsoncology.org/Tumors/neurob5002.html)

specific to chromosome 2 [CEP2]), a finding that is consistent with a relative duplication (*MYCN* to CEP2 ratio of ~2), not amplification of the *MYCN*.[37] Inclusion of the ploidy control probe allows for discrimination between gene duplication (where the chromosome is aneusomic, increasing gene copy number) and gene amplification (when an excess of gene copies over chromosome copies is observed; an increase in gene ratio to >2).[38] Because *MYCN* amplification (Figure 4.2B) defines a high-risk neuroblastoma subgroup requiring intensive therapy, accurate quantification of this gene plays an important role in the care of these patients.

Two commercially available probe strategies commonly used to diagnose rearrangements of specific gene regions include the application of either dual-color, break-apart probes or dual-color, dual-fusion probes. A break-apart probe design employs two differentially labeled DNA probes that flank a specific gene locus or breakpoint region. A split of the two normally adjacent probes is indicative of a rearrangement. Break-apart probes are optimal for assessing rearrangements of a gene that could have more than one translocation partner (e.g., *FOXO1* [*FKHR*] localized to 13q14 and involved in the 2;13 and 1;13 alveolar-rhabdomyosarcoma-associated translocations, Figure 4.3A). If the gene of interest is rearranged in more than one histopathologic entity, such as the *EWSR1* gene in both Ewing sarcoma and desmoplastic small round cell tumor, a break-apart probe will not distinguish them, however.[39] In contrast, dual-fusion probes are composed of two independent fluorophore-labeled DNA probes that span both rearranged loci of a specific translocation, allowing for the identification of pathognomonic rearrangements. In normal cells hybridized with a dual-color, dual-fusion probe, two separate signals are observed for each differentially labeled probe. In cells harboring the translocation of interest, the probes on each derivative chromosome are juxtaposed, however, producing two fusion signals (Figure 4.3B). A third strategy, using differentially labeled probes directed to only one side of each of the two chromosomal breakpoints involved in a translocation, results in a single fusion signal representing one of the two derivative chromosomes involved in the translocation (Figure 4.3C). This latter strategy is not as specific, owing to a higher false-positive rate.[40]

Technical variations

Combining immunofluorescence labeling with FISH is a valuable technical alternative that permits the simultaneous study of immunophenotypic markers and chromosomal abnormalities present in tumor cells (sometimes referred to as the FICTION method [fluorescence immunophenotyping, and interphase cytogenetics as a tool for the investigation of neoplasms] or ImmunoFISH [ImFISH]).[41,42] A modified ImFISH technique has also been described whereby a conventional chromogenic immunohistochemical technique is performed in concert with interphase FISH and both fluorescent and bright-field microscopic captured images are superimposed for analysis. Demonstration of chromosomal alterations in cells of specific phenotypes within a heterogeneous population can be achieved using these methodologic variations, as illustrated by the demonstration of loss of *NF1* restricted to S100-protein-immunoreactive Schwann cells in plexiform neurofibromas.[43] Moreover, the spectrum of presentations associated with certain diagnoses has also expanded through the use of this combined approach. For example, some lesions originally described as adamantinomas prior to the availability of genetic characterization were found to harbor the 11;22 translocation characteristic of Ewing sarcoma in the nuclei of cytokeratin-immunoreactive cells, delineating a novel histological variant, *adamantinoma-like Ewing sarcoma*.[44]

Chromogenic *in situ* hybridization (CISH) is an alternative to FISH.[45] The probes are detected through an immunohistochemistry-like peroxidase reaction, and visualization of the red, green, or brown enzyme precipitates is accomplished on a bright-field microscope.[46] The CISH assay is remarkably stable and can be archived for several years. Verification of the histopathology is easily achieved because CISH slides are counterstained with hematoxylin and eosin, and analysis of signals can be readily performed with a 40× objective magnification (i.e., oil immersion lens not required).

Conversely, CISH analysis with multicolor probe combinations is more challenging than FISH, and the number of probes available in the CISH format is comparatively few. Consequently, most CISH assays are aimed at aneuploidy or amplification detection.

Conventional karyotyping is limited by its ability to decipher complex structural aberrations fully and marker chromosomes accurately. To facilitate karyotyping, particularly analysis of cytogenetically complex neoplasms exhibiting multiple chromosomal rearrangements and marker chromosomes, universal chromosome painting techniques (M-FISH, SKY, and COBRA-FISH) have been developed.[47] Both M-FISH and SKY utilize mixtures of spectrally distinct fluorochromes such that no two probes have the same labeling combination – a combinatorial labeling process that "paints" each of the 24 human chromosomes in a unique color. The principal difference between the two techniques relates to image acquisition. SKY analysis relies on the capture of a single image acquired through a customized multiband optical filter,[17] whereas M-FISH captures the five spectrally distinct fluorochromes separately through single-band filters and collates them into a single image.[16] COBRA-FISH is based on the simultaneous use of combinatorial labeling and ratio labeling.[18] Ratio labeling uses combinations of fluorochromes in different ratios to distinguish between probes. Only four fluorophores instead of five are needed to produce color discrimination of 24 targets, a feature distinguishing this method from M-FISH and SKY. Furthermore, multiplicity can be increased to 48 by introduction of a fifth fluorophore, thereby promoting color identification of each of the p and q chromosomal arms.[48] These approaches are well-suited to soft tissue tumors because the complexity of some karyotypes can mask the primary or tumor-specific anomaly of interest (Figure 4.4). Limitations of these techniques include their: (1) metaphase cell prerequisite, (2) inability to detect intrachromosomal or intrachromosomal arm aberrations (e.g., inversions, duplications) or to specify chromosomal region/band involvement, and (3) relatively low resolution (1–2 Mb).

Gene amplifications and deletions frequently contribute to sarcomagenesis. CGH is a technique that was designed for detecting genomic imbalances by cohybridizing differentially labeled test and reference genomic DNAs to normal metaphase chromosomes and examining the fluorescence ratios along the length of the chromosomes for DNA copy number variations.[49] Balanced translocations, representing tumor-specific markers for many soft tissue tumors, are not identifiable by CGH. Chromosomal CGH is also limited by its low resolution (5–10 Mb for the detection of copy number loss and gain, and 2 Mb for amplification).[50] More recently, array-based CGH has

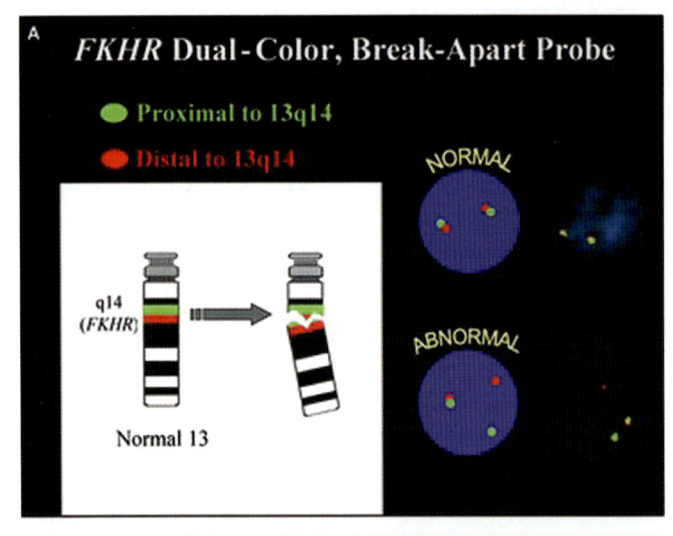

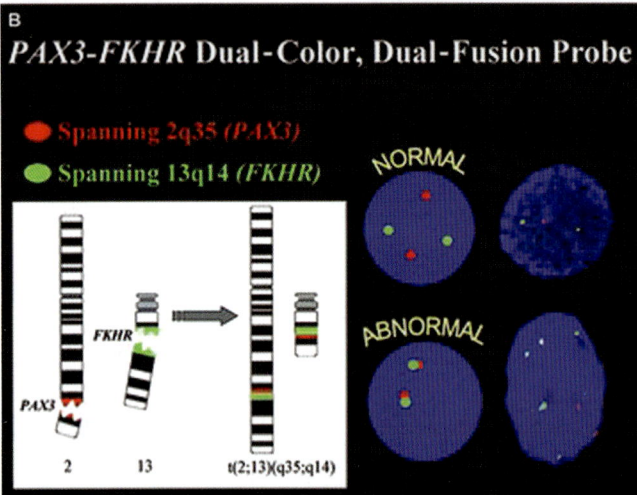

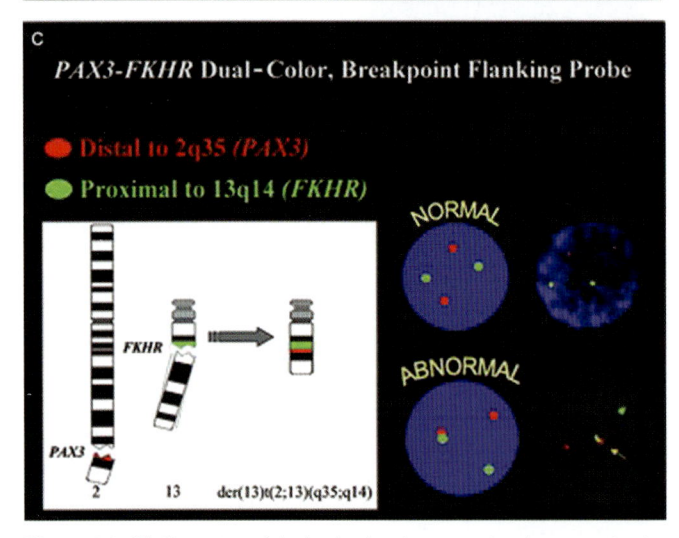

Figure 4.3 (**A**) Illustration of dual-color, break-apart probe demonstrating in the abnormal cell a split of one set of orange and green normally adjacent signals, indicative of a arrangement of the *FOXO1* gene locus. (**B**) Illustration of dual-color, dual-fusion probe demonstrating two separate orange and green signals in the normal cell in contrast to multiple fused orange and green signals (indicative of *PAX3/FOXO1* fusions), as well as individual orange and green signals in the abnormal cell. Note that there can be variations of abnormal patterns. The variant pattern in this example is related to aneuploidy. (**C**) Illustration of dual-color, breakpoint flanking probe demonstrating a single fusion of orange and green signals (arrow) in the abnormal cell consistent with the presence of the derivative 13 of the t(2;13)(q35;q14).

replaced the chromosomal targets of standard CGH with well-defined genomic clones or oligonucleotides and SNPs, a technological advance that has greatly improved the resolution.[51,52] Cytogenomic array platforms are particularly useful for analysis of DNA sequence copy number changes in dedifferentiated, undifferentiated, or high-grade sarcomas in which high-quality metaphase preparations are often difficult to produce, and complex karyotypes with numerous markers, double minutes, and homogeneously stained chromosomal regions are common. The analysis of a large number of individual genes in a single specimen can be accomplished by studying gene dosage when tumor genomic DNA is tested or gene expression patterns when tumor RNA is tested.[53,54]

Cytogenomic array platforms also contribute to the identification of chromosomal regions harboring candidate tumor-suppressor genes or oncogenes in specific soft tissue tumors, including regions that previously have not been identified by other means. Examples include copy number alterations of the *CDKN2A*, *SMARCB1*, *SOX6*, and *PTEN* loci in Ewing sarcoma and amplification of *IGF1R* within a 15q25–26 amplicon in rhabdomyosarcoma.[55–57] Identification of these aberrations aid in further elucidating the molecular pathogenesis of soft tissue tumors and also offer promising targets for therapeutic strategies.[55–57]

Clinically, gene amplification has diagnostic and prognostic usefulness and is a mechanism of acquired drug resistance.[58] Well-differentiated and dedifferentiated liposarcomas are characterized by the presence of supernumerary ring or giant marker chromosomes (Figure 4.5) composed of 12q13–15 amplicons (Figure 4.6). Identification of amplification of the *MDM2* and *CDK4* genes corresponding to these amplicons by FISH (Figure 4.7) or real-time polymerase chain reaction (quantitative PCR [Q-PCR]) is a valuable adjunct in the differential diagnosis of adipose tissue tumors.[59] Because *MDM2* and *CDK4* genomic copy number change is associated with changes in the expression of these genes, assessment of MDM2 and CDK4 expression using immunohistochemistry is a possible diagnostic alternative,[60] although some studies have suggested that it might not be as specific as FISH.[59,61]

Chromothripsis is a recently discovered phenomenon represented by local "shattering" of chromosomes that is thought to arise through a single catastrophic event of chromosome breakage and inaccurate reassembly and is identifiable by cytogenomic array techniques as well as by sequencing.[62,63] The result is genetic complexity to include formation of gene fusions, disruption of tumor suppressors, and amplification of oncogenes. The occurrence and the outcomes of

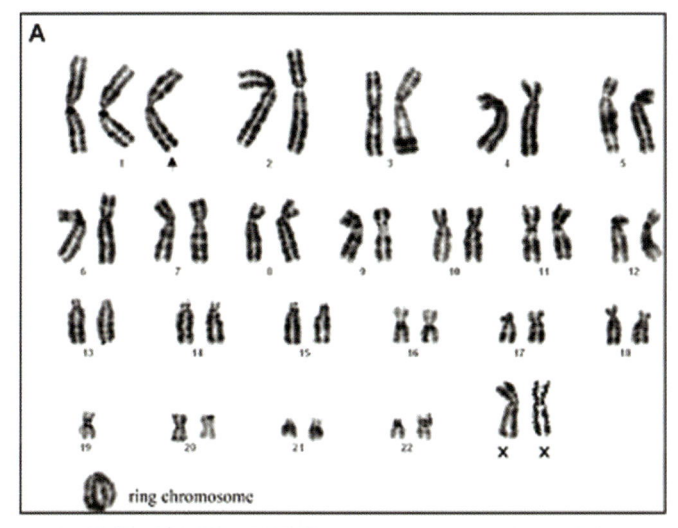

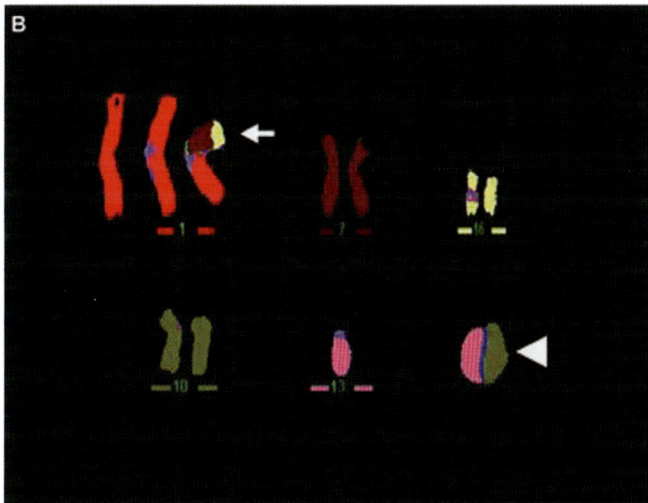

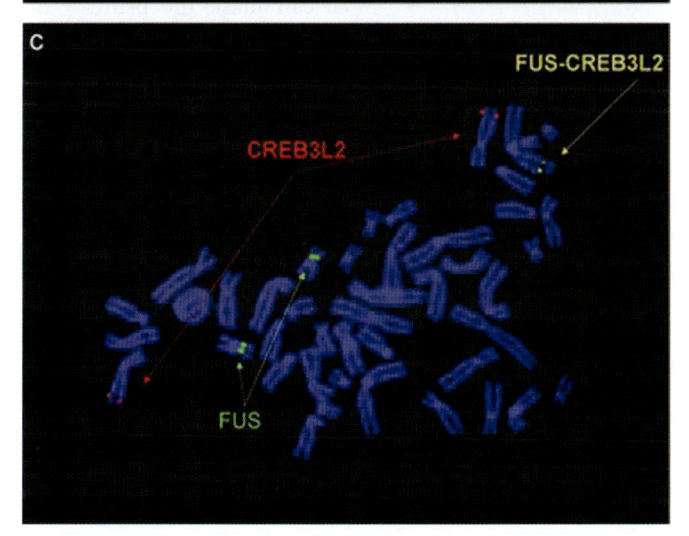

Figure 4.4 (**A**) Conventional cytogenetic analysis of a low-grade fibromyxoid sarcoma revealed the presence of an abnormal clone characterized by the presence of a supernumerary ring chromosome (composition could not be determined with G-banding) and an extra chromosome 1 with additional material of unknown origin on the short arm (arrow). Loss of chromosome 19 was a random anomaly in this cell. (**B**) M-FISH analysis revealed that the supernumerary ring chromosome was composed of chromosome 10 and 13 material (arrowhead) and that the characteristic low-grade fibromyxoid sarcoma 7;16 translocation was a component of a complex rearrangement involving chromosome 1 [der(1)t(1;7)(p12;p11.1)t(7;16)(q33;p11.2)] (arrow). (**C**) Metaphase FISH analysis with custom-designed probes spanning the gene loci of FUS (green signals) and CREB3L2 (red signals) confirmed the presence of a FUS/CREB3L2 fusion (yellow arrow) on the +der(1).

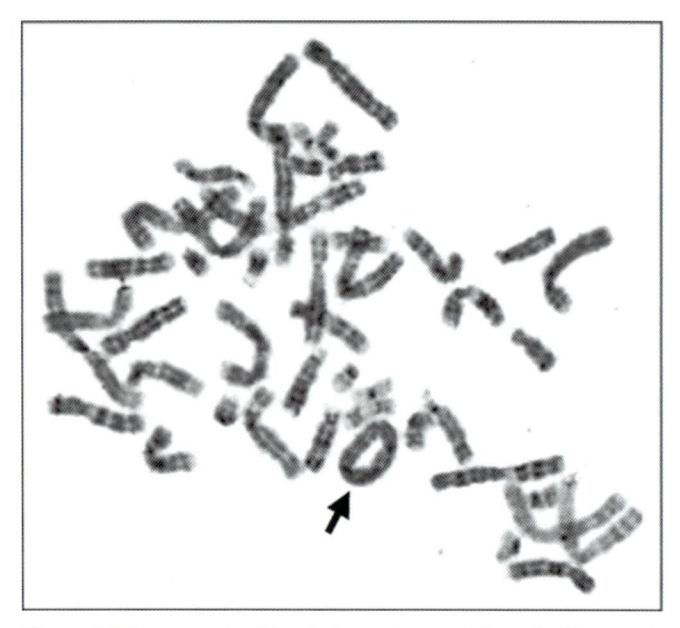

Figure 4.5 Representative G-banded metaphase cell of a well-differentiated liposarcoma exhibiting a supernumerary ring chromosome (arrow).

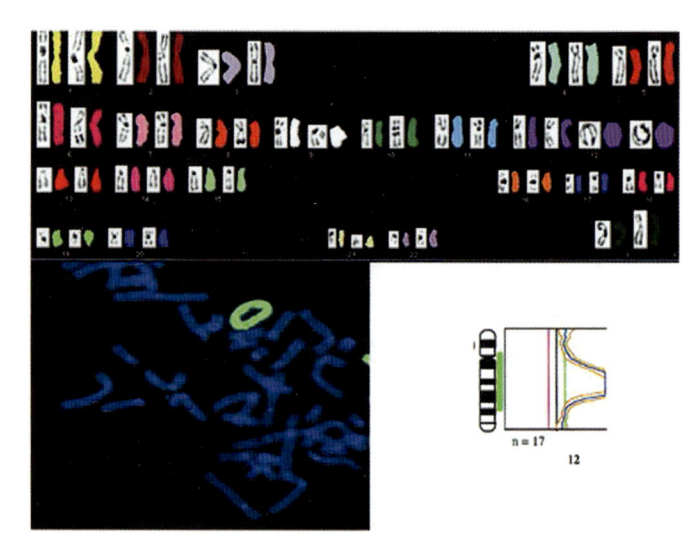

Figure 4.6 (Upper and lower left) SKY and FISH analysis with a chromosome 12 paint probe (green) demonstrate that the ring chromosomes are composed of chromosome 12 material in this well-differentiated liposarcoma (lower right). CGH analysis further defines the anomaly as a 12q13–15 amplicon.

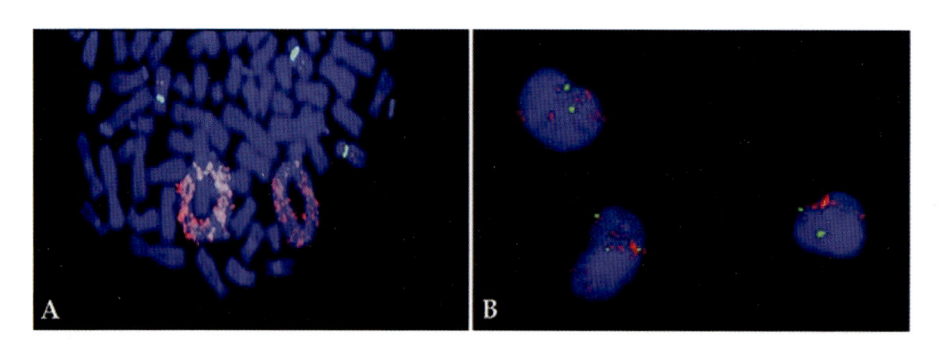

Figure 4.7 (**A**) Metaphase FISH analysis conducted with a unique sequence probe for the MDM2 locus in orange coupled with a ploidy control probe (CEP12) in green demonstrates MDM2 amplification within the ring chromosomes of this well-differentiated liposarcoma. In addition, there are four normal chromosome 12 homologs. (**B**) Interphase FISH analysis on a cytologic touch preparation with the same probe combination also illustrates amplification of the MDM2 locus in another well-differentiated liposarcoma.

chromothripsis have been reported in both bone and soft tissue tumors.[64] For example, the presence of chromothripsis is associated with a poor prognosis in neuroblastoma.[65]

Advantages and limitations of molecular cytogenetic analysis

Usually it is wise to have more than one genetic diagnostic modality available, to be ready to confirm unexpected or discrepant results by two independent techniques.[27] Interphase FISH is a rapid (same-day or overnight procedure) and effective alternative to RT-PCR analysis for the assessment of soft tissue tumor translocation events. In contrast to conventional cytogenetic analysis, FISH, a highly sensitive and reliable technology, does not require actively dividing cells (fresh tissue), making it a valuable tool not only for routine patient care but also for retrospective studies of large numbers of cases using archival fixed tissues. Conversely, unlike conventional cytogenetics, which is capable of providing a comprehensive appraisal of both numerical and structural abnormalities in a soft tissue tumor in a single assay, interphase FISH is a targeted approach requiring a priori knowledge of the anomaly of interest or suspected aberration. Moreover, for some diagnostic quandaries, multiple FISH assays might be required to reach a definitive conclusion.[14] A summary of additional advantages and limitations of molecular cytogenetic approaches is provided in Table 4.4.

Chromosomal abnormalities in soft tissue tumors

Conventional cytogenetics provided the first chromosomal marker for malignancy in 1960.[66] This was to be known as the *Philadelphia chromosome*. Additional chromosomal changes associated with various hematologic disorders followed in the 1970s with the advent of chromosomal banding. The chromosome changes in soft tissue tumors were not realized until relatively late in the era of cancer cytogenetics. The description in the early 1980s of a specific change in Ewing sarcoma catalyzed studies of other tumors of mesenchymal origin.[67,68]

Table 4.4 Molecular cytogenetic analysis

Advantages	Limitations
Can be performed on metaphase or interphase cell preparations (fresh, frozen, or fixed samples) • can localize anomaly within specific cells or tissue types	Targeted approach; not screening tool (generally requires a priori knowledge of anomaly of interest or suspected aberration) Exceptions include • multicolor banding techniques (M-FISH, SKY, COBRA-FISH) capable of a comprehensive assessment of structural chromosomal alterations on metaphase cell preparations • cytogenomic array platforms capable of a comprehensive assessment of genomic imbalances (and with SNP array analysis, cnLOH) in a neoplasm (extracted DNA)
Can provide results when • tissue is insufficient or unsatisfactory for cytogenetic analysis • conventional cytogenetics has failed to yield results • cryptic rearrangements are present	Relatively low resolution when contrasting with other molecular approaches capable of detecting single base changes
Diagnostically useful • sensitive and specific	• requires fluorescence microscopy • interpretation can be challenging when analyzing aneuploid/polyploid samples or suboptimal specimens (weak probe hybridization or penetration, background fluorescence or autofluorescence)
Rapid turn-around time	Limited selection of commercially available probes

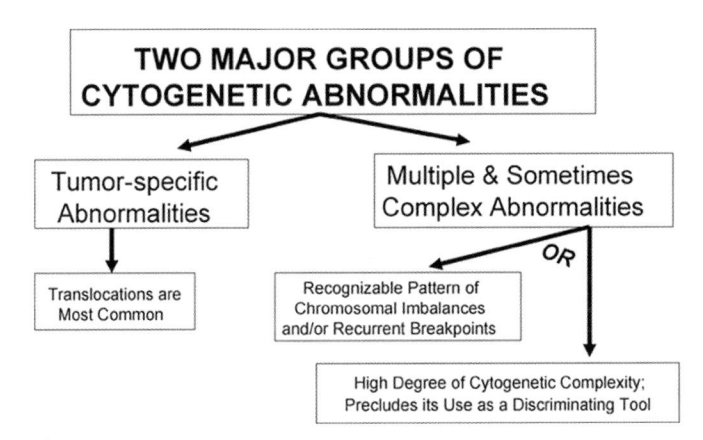

Figure 4.8 Cytogenetic abnormality groups.

Simplistically, cytogenetic abnormalities in mesenchymal neoplasms can be divided into two major groups (Figure 4.8): (1) A significant number of sarcomas are characterized by a relatively simple karyotype dominated by a recurrent anomaly, usually a defining translocation (Table 4.5).[30,31,69-75] These translocations result in the production of chimeric genes encoding for abnormal, oncogenic proteins that are central to the causation of these tumors. Tumor-specific chromosomal anomalies serve as valuable diagnostic aids, particularly in the differential diagnosis of those sarcomas of a confusing nature, such as poorly differentiated sarcomas, small round cell neoplasms, or sarcomas arising in unusual anatomic sites or exhibiting other atypical clinicopathologic presentations.[27,31,76-79] (2) A second major group of sarcomas is associated with multiple and sometimes complex chromosomal changes. Although more difficult to appreciate the diagnostic value of these karyotypic changes, a pattern of chromosomal imbalances or recurrent breakpoints can be recognizable for some neoplasms, such as embryonal rhabdomyosarcoma and malignant peripheral nerve sheath tumor, among others.[80-85] These aberrant patterns, when viewed in association with other clinicohistopathologic features, can contribute to accurate nosology. Regrettably, for most sarcomas in this group, the high degree of cytogenetic complexity (including large numbers of unidentifiable marker chromosomes and intratumoral heterogeneity) precludes its use as a discriminating tool.

Biological considerations
Relatively simple karyotypes (translocations)

Translocations, or exchange of chromosomal material between two or more nonhomologous chromosomes, are encountered as the most common tumor-specific anomalies in mesenchymal neoplasms. Approximately one-third of all sarcomas exhibit a nonrandom chromosomal translocation or associated fusion gene.[86] Translocations can be reciprocal (Figure 4.9A) or nonreciprocal/unbalanced (Figure 4.9B). A tumor-defining translocation is considered a *primary* chromosomal abnormality. Because it is often present as the sole karyotypic aberration, it is presumed to be the initiating oncogenic event (driver mutation). Moreover, these defining translocations are present throughout the clinical course, underscoring a role in

Chapter 4: Genetics of soft tissue tumors

Table 4.5 Characteristic and variant chromosomal translocations and associated fusion genes in benign and malignant soft tissue tumors

Neoplasm	Translocation	Fusion gene(s)
Alveolar rhabdomyosarcoma	t(2;13)(q35;q14) t(1;13)(p36;q14) t(X;2)(q13;q35) t(2;2)(q35;p23) t(8;13)(p12;q13)	PAX3-FOXO1 PAX7-FOXO1 PAX3-FOXO4 PAX3-NCOA1 FOXO1-FGFR1
Alveolar soft part sarcoma	der(17)t(X;17)(p11;q25)	ASPSCR1-TFE3
Aggressive angiomyxoma	12q15 rearrangements	HMGA2
Aneurysmal bone cyst	t(16;17)(q22;p13) t(1;17)(p34.1–34.3;p13) t(3;17)(q21;p13) t(9;17)(q22;p11–12) t(17;17)(p13;q12)	CDH11-USP6 THRAP3-USP6 CNBP -USP6 OMD-USP6 COL1A1-USP6
Angiomatoid fibrous histiocytoma	t(12;22)(q13;q12) t(2;22)(q33;q12) t(12;16)(q13;p11)	EWSR1-ATF1 EWSR1-CREB1 FUS-ATF1
Benign metastasizing leiomyoma	6p21	HMGA1
Chondroid lipoma	t(11;16)(q13;p12–13)	C11orf95-MKL2
Chondromyxoid fibroma	inv(6)(p25q13) 6q25 rearrangements	?
Clear cell sarcoma	t(12;22)(q13;q12) t(2;22)(q33;q12)	EWSR1-ATF1 EWSR1-CREB1
Congenital/infantile fibrosarcoma	t(12;15)(p13;q25)	ETV6-NTRK3
Congenital/spindle cell rhabdomyosarcoma	t(6;8)(p12;q11.2) t(8;11)(q11.2;p15)	SRF-NCOA2 TEAD1-NCOA2
Dermatofibrosarcoma protuberans and giant cell fibroblastoma	t(17;22)(q21;q13)	COL1A1-PDGFB (often within a ring chromosome)
Desmoplastic fibroblastoma	t(2;11)(q31;q12) t(11;17)(q12;p11.2)	Deregulated expression of FOSL1
Desmoplastic small round cell tumor	t(11;22)(p13;q12)	EWSR1-WT1
Endometrial stromal sarcoma	t(7;17)(p15;q21) t(10;17)(q22;p13)	JAZF1-SUZ12 YWHAE-NUTM2A/B
Epithelioid hemangioendothelioma	t(1;3)(p36;q25) t(X;11)(p11.2;q13)	WWTR1-CAMTA1 YAP1-TFE3
Epithelioid hemangioma	ZFP36 (19q13.1) FOSB (19q13.3) WWTR1 (3q23–24)	ZFP36-FOSB WWTR1-FOSB
Ewing sarcoma/pPNET/Ewing family tumors (EFT)-like	t(11;22)(q24;q12) t(21;22)(q22;q12) t(7;22)(q22;q12) t(17;22)(q21;q12) t(2;22)(q36;q12) inv(22)(q12q12) t(20;22)(q13;q12) t(6;22)(p21;q12) t(16;21)(p11;q22) t(2;16)(q36;p11) t(4;19)(q35;q13) t(10;19)(q26.3;q13) inv(X)(p11.2p11.4)	EWSR1-FLI1 EWSR1-ERG EWSR1-ETV1 EWSR1-ETV4 EWSR1-SP3 EWSR1-PATZ1 EWSR1-NFATC2 EWSR1-POU5F1 FUS-ERG FUS-FEV CIC-DUX4 CIC-DUX4 BCOR-CCNB3
Extrapleural solitary fibrous tumor	12q13 rearrangements	NAB2-STAT6

Table 4.5 (cont.)

Neoplasm	Translocation	Fusion gene(s)
Extraskeletal mesenchymal chondrosarcoma	inv(8)(q13q21)	*HEY1-NCOA2*
Extraskeletal myxoid chondrosarcoma	t(9;22)(q22;q12) t(9;17)(q22;q11) t(9;15)(q22;q21) t(3;9)(q12;q22)	*EWSR1-NR4A3* *TAF15-NR4A3* *TCF12-NR4A3* *TFG-NR4A3*
Hibernoma	11q13–21 rearrangements	*MEN1* and/or *AIP* homozygous or hemizygous loss
Inflammatory myofibroblastic tumor	t(1;2)(q22;p23) t(2;19)(p23;p13) t(2;17)(p23;q23) t(2;2)(p23;q13) t(2;2)(p23;q35) t(2;11)(p23;p15) t(2;4)(p23;q21) inv(2)(p23;q35) t(2;12)(p23;p11)	*TPM3-ALK* *TPM4-ALK* *CLTC-ALK* *RANBP2-ALK* *ATIC-ALK* *CARS-ALK* *SEC31A –ALK* *ATIC-ALK* *PPFIBP1-ALK*
Leiomyoma (uterus)	t(12;14)(q15;q24) 6p21	*HMGA2-RAD51B* *HMGA1*
Lipoblastoma	8q12 rearrangements	*PLAG1*
Lipoma, conventional	12q15 rearrangements 6p21 rearrangements	*HMGA2* *HMGA1*
Low-grade fibromyxoid sarcoma and hyalinizing spindle cell tumor with giant rosettes	t(7;16)(q33;p11) t(11;16)(p13;p11)	*FUS-CREB3L2* *FUS-CREB3L1*
Myoepithelioma/myoepithelial carcinoma/mixed tumor	t(1;22)(q23;q12) t(6;22)(p21;q12) t(19;22)(q13;q12) 16p11.2 rearrangement	*EWSR1-PBX1* *EWSR1-POU5F1* *EWSR1-ZNF444* *FUS-?*
Myolipoma of soft tissue		*HMGA2*
Myxoid/round cell liposarcoma	t(12;16)(q13;p11) t(12;22)(q13;q12)	*FUS-CHOP* *EWSR1-CHOP*
Myxoinflammatory fibroblastic sarcoma, hemosiderotic fibrolipomatous tumor, and pleomorphic hyalinizing angiectatic tumor	t(1;10)(p22;q24) with amplified 3p11–12	Involves *TGFBR3* and *MGEA5* without detectable fusion, transcriptional upregulation of *FGF8* and amplification of the *VGLL3* locus
Nodular fasciitis	t(17;22)(p13;q13.1)	*MYH9-USP6*
Ossifying fibromyxoid tumor	6p21 rearrangement	*PHF1*
Pericytoma	t(7;12)(p22;q13)	*ACTB-GLI1*
Primary pulmonary myxoid sarcoma	t(2;22)(q34;q12)	*EWSR1-CREB1*
Pseudomyogenic hemangioendothelioma	t(7;19)(q22;q13)	*SERPINE1-FOSB*
Sclerosing epithelioid fibrosarcoma	t(7;16)(q33;p11) – identified in LGFMS with SEF-like foci	*FUS* rearrangement has been detected in a minority of "pure" SEF cases. In contrast, *EWSR1* and *CREB3L1* rearrangements in pure SEF
Soft tissue angiofibroma	t(5;8)(p15;q13)	*AHRR-NCOA2*
Soft tissue chondroma	12q13–15 rearrangements	*HMGA2*
Synovial sarcoma	t(X;18)(p11.2;q11.2) t(X;20)(p11.2;q13.3)	*SYT-SSX1* *SYT-SSX2* *SYT-SSX4* *SS18L1-SSX1*
Tenosynovial giant cell tumor	t(1;2)(p13;q37)	*CSF1-COL6A3*

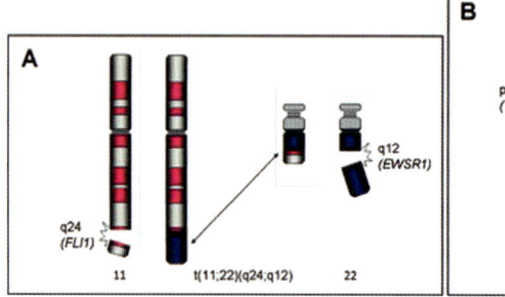

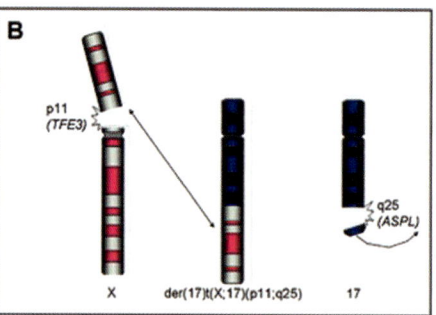

Figure 4.9 (A) Schematic illustrating a balanced 11;22 translocation characteristic of Ewing sarcoma. (B) The X;17 translocation characteristic of alveolar soft part sarcoma is frequently unbalanced, exhibiting only the derivative 17 [der(17)t(X;17)(p11;q25)].

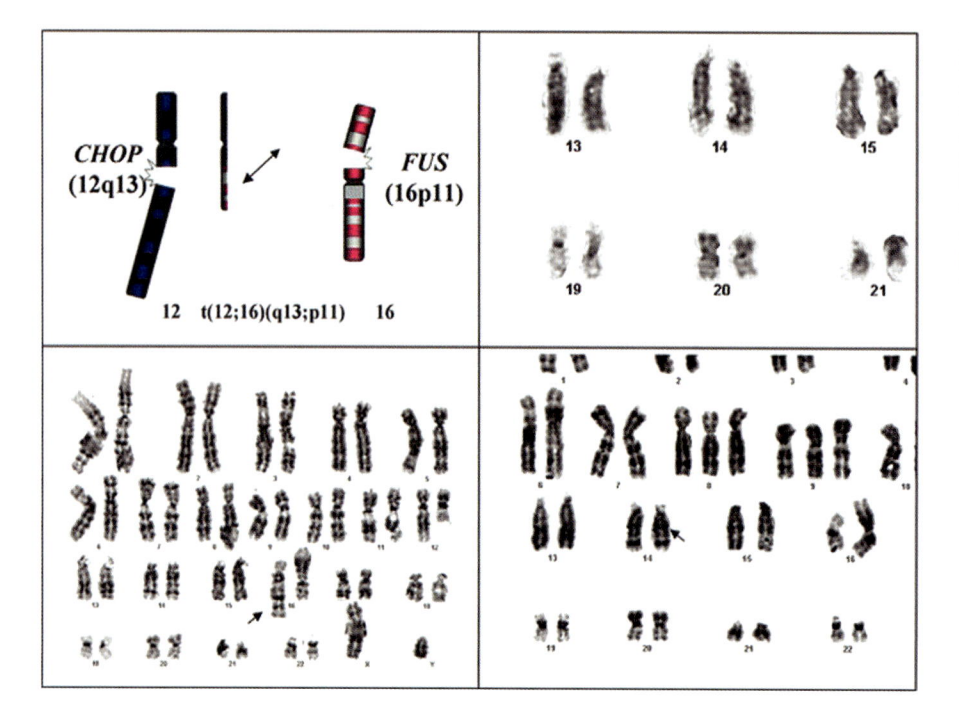

Figure 4.10 (Upper left and right) Schematic and G-banded karyotype illustrating the primary 12;16 chromosomal translocation in myxoid liposarcoma. (Bottom left) The common structural secondary chromosomal anomaly, der(16)t(1;16) (arrow), is seen in addition to the primary 12;16 translocation in this myxoid liposarcoma. (Bottom right) The common numerical secondary chromosomal anomaly, +8 (arrow), is seen in addition to the primary 12;16 translocation in a different myxoid liposarcoma.

neoplastic maintenance as well. Dominantly acting oncoproteins associated with chromosomal translocations in soft tissue tumors have multiple functions and might or might not require additional mutations for cell transformation.[87–89] In contrast, *secondary* chromosomal abnormalities can be consistent in a particular neoplasm but are also observed in other histologic tumor types, thereby lacking the specificity of the primary change. Secondary changes or additional genetic mutations are also thought to be essential in cancer development or contributory to tumor progression, but little is known about this group of changes in sarcomas. Two of the most common secondary changes in sarcomas characterized by a tumor-specific chromosomal abnormality (e.g., Ewing sarcoma, synovial sarcoma, or myxoid liposarcoma) are the presence of extra copies of chromosome 8 and an unbalanced 1;16 translocation, resulting in relative gain of 1q and loss of 16q material (Figure 4.10). Gain of chromosome 8 appears to be significantly overrepresented in large tumors (>5 cm), suggesting that tumors with this genetic abnormality have an increased growth rate, and the der(16)t(1;16) could represent an early event in clonal evolution that portends an adverse prognosis.[90–95]

In soft tissue tumors as well as other malignancies, translocations are almost always somatic and mechanistically result in disruption or misregulation of normal gene function. The following are possible genetic consequences of translocations: (1) the translocation breakpoint(s) can occur within the gene(s), destroying gene function; (2) the translocation leads to juxtapositioning of inactive oncogenes within transcriptionally active chromosomal regions or regulatory sequences (e.g., immunoglobulin or T-cell receptor genes), resulting in overexpression of an oncogene (e.g., MYC overexpression associated with Burkitt's lymphoma translocations); and (3) the translocation leads to fusion of two genes, one from each translocation partner, resulting in a hybrid gene that encodes for an abnormal protein not found in normal cells and that may be tumor specific.

For sarcomas, chromosomal translocations most commonly result in the production of a highly specific, novel chimeric gene. Functionally there are three types of resultant fusion proteins: (1) chimeric transcription factors that cause transcriptional deregulation (examples include the fusion proteins of Ewing and synovial sarcoma, and myxoid liposarcoma, among others); (2) chimeric tyrosine kinases that elicit deregulation of kinase signaling pathways (e.g., inflammatory myofibroblastic tumor, congenital/infantile fibrosarcoma); and (3) chimeric autocrine growth factors that also result in kinase signaling pathway deregulation (e.g., dermatofibrosarcoma protuberans, tenosynovial giant cell tumor). Of note, constitutive activation of specific kinases by oncogenic mutations is an additional mechanism of genetic deregulation of kinase signaling. This mechanism is exemplified in gastrointestinal stromal tumors (GIST) as *KIT* or *PDGFRA* mutations.[96–98]

Both cytogenetic and molecular genetic variant translocations exist for several of the translocation-characterized sarcomas, and new cytogenetic and molecular genetic variants continue to be discovered and defined. Therefore, depending on the type of molecular pathologic approach used, these less common variant translocations and gene fusions might not be identified.

Cytogenetic variant translocations occur as the result of rearrangement of one consistent gene with differing chromosomal translocation partners (Figure 4.11). For example, 85% to 90% of Ewing sarcomas exhibit an 11;22 translocation, resulting in fusion of the *EWSR1* gene to *FLI1* (*ETS* gene family member).[99] Less common cytogenetic variants, however, have also been identified and are characterized by the fusion of the *EWSR1* gene with other members of the *ETS* family of transcription factors, including *ERG* (21q22), *ETV1* (7p22), *ETV4* (17q21), *FEV* (2q35–36), *PATZ1* (22q12) and *SP3* (2q31).[100–105] Rarely, the *FUS* gene substitutes for *EWSR1* in Ewing sarcoma-associated translocations [e.g., *FUS-ERG*, t(16;21)(p11;q22) and *FUS-FEV*, t(2;16)(q36;p11)].[106,107]

Molecular variants are often the result of genomic breakpoint differences that lead to distinct fusion product exon combinations.[14,31] For example, in Ewing sarcoma, the chromosomal translocation breakpoints can arise within *EWSR1* introns 7 to 10 and within introns 3 to 9 of the *ETS* gene family, enabling the generation of several possible *EWSR1-FLI1* chimeric transcripts (Figure 4.12). The two most frequent exon combinations in Ewing sarcoma-associated *EWSR1-FLI1* fusion transcripts include fusion of *EWSR1* exon

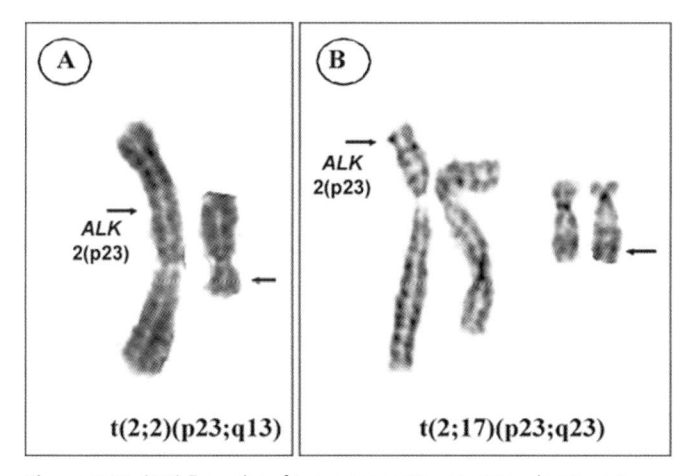

Figure 4.11 (A,B) Examples of two cytogenetic variant translocations in an inflammatory myofibroblastic tumor. Note that both involve the 2p23 locus of the *ALK* gene (the consistently involved gene in this neoplasm), but differ by their chromosomal translocation partners.

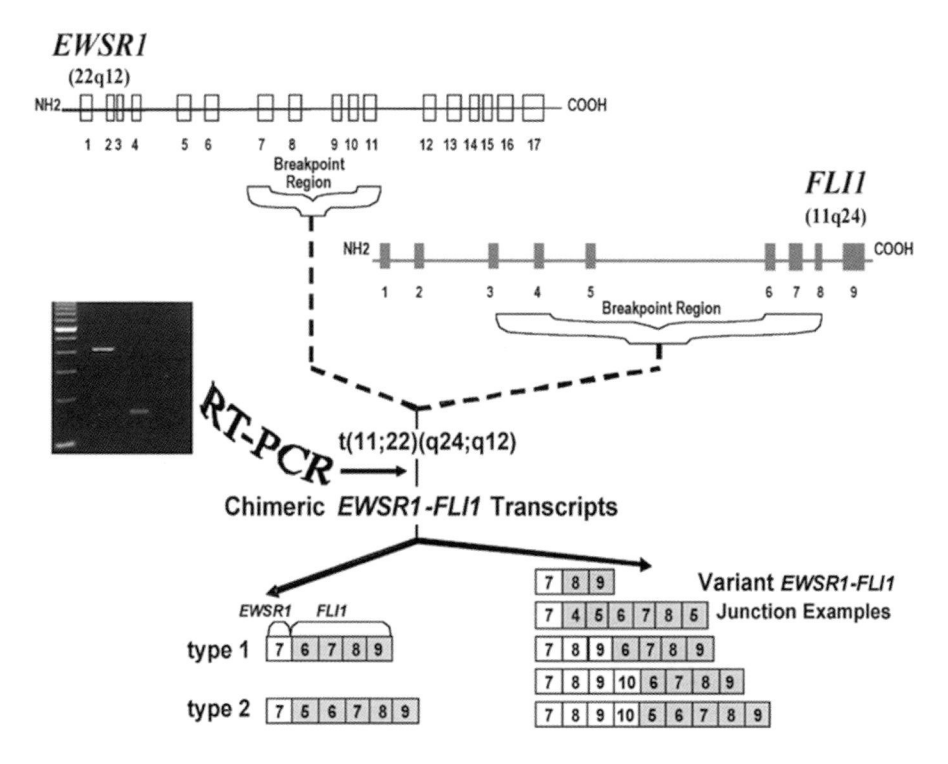

Figure 4.12 The genomic breakpoints of *EWSR1* and *FLI1* are heterogeneous, resulting in various types of in-frame *EWSR1-FLI1* chimeric transcripts. The two main types, fusion of *EWSR1* exon 7 to *FLI1* exon 6 (type 1) and fusion of *EWSR1* exon 7 to *FLI1* exon 5 (type 2), account for 85% to 90% of *EWSR1-FLI1* fusions.

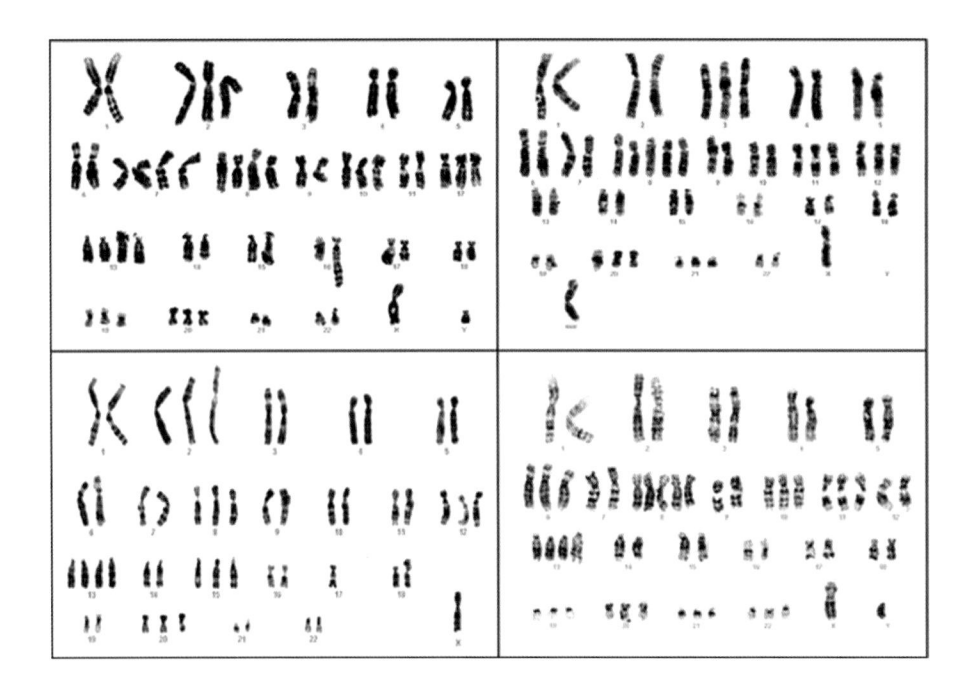

Figure 4.13 Embryonal rhabdomyosarcoma exhibits a recurrent chromosomal pattern (gain of all or portions of chromosomes 2, 7, 8, 11, 12, 13, and/or 20, with or without loss of 22) as illustrated in these four different cases.

7 to *FLI1* exon 6 (type 1) and fusion of *EWSR1* exon 7 to *FLI1* exon 5 (type 2). Type 1 and type 2 molecular variants can readily be detected by their unique RT-PCR product band size. The identity of less common or unexpected product band sizes should be confirmed by using additional approaches such as direct sequencing or digestion with specific restriction endonucleases. Cytogenetic analysis does not distinguish between molecular variants. FISH analysis can be useful for detecting rare cytogenetic variants that primer sets are not commonly designed for.

Multiple, sometimes complex abnormalities

A second (larger) group of sarcomas is associated with multiple and sometimes complex abnormalities. Although this group does not feature the diagnostically useful and sometimes prognostically informative tumor-specific chromosomal translocations or specific activating or inactivating mutations, a recurrent or recognizable pattern of karyotypic aberrations can contribute to the classification of a subset of these sarcomas. For example, embryonal rhabdomyosarcoma exhibits a recurrent and familiar pattern of chromosomal imbalances (gain of all or portions of chromosomes 2, 7, 8, 11, 12, 13, and/or 20, with or without loss of 22) that can be useful diagnostically when viewed in conjunction with other clinicohistopathologic features (Figure 4.13).[80,81] Another case in point is the nonrandom cytogenetic pattern of monosomies of chromosomes 14 and 22 and structural aberrations (typically resulting in loss) of 1p and 9p in GISTs.[108] These regions of chromosomal loss occurring in addition to *KIT* or *PDGFRA* mutations presumably correspond with loss of tumor-suppressor genes and are significantly associated with GIST malignancy, response to treatment, and intestinal versus nonintestinal localization.[109,110]

A distinct advantage of cytogenetic analysis is that primary or characteristic chromosomal aberrations are present in all tumor cells and are expressed throughout the clinical course. These alterations are not lost as a neoplasm becomes less differentiated or metastasizes. Similar to well-differentiated liposarcoma, dedifferentiated liposarcoma karyotypes feature supernumerary ring or giant marker chromosomes composed chiefly of amplified 12q13–15 material, signifying a kinship to, if not in fact derivation of, these tumors from well-differentiated liposarcomas.[111] Unlike well-differentiated liposarcoma, however, the ring or marker chromosome(s) in dedifferentiated liposarcoma is typically accompanied by additional complex changes. Notably *MDM2*, and to a variable extent other 12q13–15 localized oncogenes (e.g., *CDK4*, *HMGA2*, and *CHOP* among others), are amplified within the WDL/DDL-associated ring or marker chromosomes. Cytogenetic or molecular demonstration of chromosome 12 comprised supernumerary ring/marker chromosomes or *MDM2/CDK4* amplification can serve to distinguish WDL/DDL from benign adipose tissue tumors and other poorly differentiated or undifferentiated sarcomas, respectively.[59,111-116]

A subset of benign and malignant soft tissue tumors that cytogenetically exhibit recurrent regions of loss and gain or amplification are listed in Table 4.6. Several of these entities feature loss of recognized tumor-suppressor gene loci and are associated with inherited genetic disorders.[30] For example, neurofibromas and malignant peripheral nerve sheath tumors can arise sporadically or as part of the spectrum of neurofibromatosis type 1 (NF1), an autosomal dominant disorder caused by germline inactivating mutations of the *NF1* gene.[117] The *NF1* gene is a large tumor-suppressor gene, spanning approximately 335 kb of genomic DNA on 17q11.2 with 60 exons encoding the neurofibromin protein. Mutations/loss

Table 4.6 Recurrent chromosomal imbalances in benign and malignant soft tissue tumors

Neoplasm	Cytogenetic abnormality	Associated molecular event
Angiosarcoma of soft tissue	8q24 amplification	High-level amplification of *MYC* is consistent hallmark of radiation-induced lymphedema-associated angiosarcoma
Cellular angiofibroma	−13q, −16q	?
Desmoid-type fibromatosis	+8, +20 del(5)(q21–22)	? *APC* loss
Ectomesenchymoma	+2, +8, +11, +12, +13, +20	?
Embryonal rhabdomyosarcoma	+2, +7, +8, +11, +12, +13, +20, −22 Loss or UPD of 11p15.5	? *IGF2*, *H19*, *CDKN1C* and *HOTS*
Epithelioid sarcoma	−22, t or del(22)(q11.2)	*SMARCB1* loss
Gastrointestinal stromal tumor	−14, −22 deletions of 1p, 9p, 9q, 10, 11p, and 13q and gains/amplifications of 5p, 3q, 8q, and 17q are associated with malignant behavior	? *CDKN2A/B* loss
Intimal sarcoma	Gain or amplification of 12q12–15 and 4q12	*CDK4*, *TSPAN31*, *MDM2*, *GLI1* and *PDGFRA*, *KIT*, *CHC2* respectively
Malignant peripheral nerve sheath tumor	−17, del(17)(q) del(9)(p)	*NF1* loss *CDKN2A* loss
Mammary-type myofibroblastoma	−13q, −16q	?
Neurofibroma (NF1 associated and sporadic)	−17 or del(17)(q)	*NF1* loss
Neuroblastoma	del(1)(p36) del(11)(q23) + del(3)(p26) dmin, HSR	? ? *MYCN* amplification
Palmar/plantar fibromatosis	+7, +8	?
PEComa	−16p	*TSC2* loss
Perineurioma	−22, t or del(22)(q)	*NF2* loss
Rhabdoid tumor (extrarenal)	−22, t or del(22)(q11.2)	*SMARCB1* loss
Spindle cell lipoma/pleomorphic lipoma	−13 or 13q, −16 or 16q22–qter	?
Schwannoma (NF2 associated and sporadic)	−22 or del(22)(q)	*NF2*, *SMARCB1* loss
Well-differentiated/dedifferentiated liposarcoma	+ring/giant marker chromosome (12q13–15 amplicon)	*MDM2*, *CDK4*, *HMGA2* amplification

of *NF1* in both familial and sporadic neurofibromas and malignant peripheral nerve sheath tumors follow the "two-hit" model put forward first by Knudson with respect to retinoblastoma.[118] In neurofibromas arising in NF1 patients, one allele of the *NF1* gene is constitutionally inactivated, while the other allele is subsequently inactivated ("second hit") at the somatic level. In sporadic neurofibromas, both alleles of *NF1* are inactivated at the somatic level.[119,120] Loss of 17q is rarely evidenced by gross cytogenetic analysis in neurofibromas, but is frequently observed as a chromosomal abnormality within the typically complex karyotypes of malignant peripheral nerve sheath tumors.[82–84] In addition, cytogenetic aberrations resulting in loss of 1p, 9p, and 22q are also common in malignant peripheral nerve sheath tumors. These aberrations can be associated with loss of other tumor-suppressor genes such as *CDKN2A* and *CDKN2B*. Over and above *NF1* mutations, mutations of *CDKN2A* or *TP53* appear to be essential for the progression from neurofibroma to malignant peripheral nerve sheath tumor.[121–128] Furthermore, recurrent gain of 7p15–p21 and 17q22–qter, to include genes such as *TOP2A* and *BIRC5*, as detected by FISH and microarray-based comparative genomic hybridization (aCGH) analyses, is associated with poor overall survival for patients with malignant peripheral nerve sheath tumors.[129–132]

Overall, for most sarcomas demonstrating multiple chromosomal anomalies (predominantly high-grade pleomorphic sarcomas), little to no diagnostic utility is the more common outcome because of the high degree of karyotypic complexity. The karyotypic abnormalities in these neoplasms are usually unbalanced and correspond to numerous gene copy number changes. In contrast to sarcomas characterized by tumor-specific translocations, *TP53* mutations are more common in cytogenetically complex sarcomas, but their presence does not usually impart the same prognostic significance.[133–137]

References

1. Sandberg AA, Bridge JA. *The Cytogenetics of Bone and Soft Tissue Tumors.* Austin: RG Landes, CRC Press; 1994.
2. Fletcher JA, Kozakewich HP, Hoffer FA, *et al.* Diagnostic relevance of clonal cytogenetic aberrations in malignant soft-tissue tumors. *N Engl J Med* 1991;324:436–442.
3. Dal Cin P. Metaphase harvest and cytogenetic analysis of malignant hematological specimens. In *Current Protocols in Human Genetics*. 2003: Chapter 10: Unit 10.2.
4. Mandahl N. Methods in solid tumour cytogenetics. In *Human Cytogenetics. A Practical Approach. Volume II. Malignancy and Acquired Abnormalities*, 3rd edn. Rooney DE (eds.) Oxford: Oxford University Press; 2001: 165–203.
5. Thompson FH. Cytogenetic methods and findings in human solid tumors. In *The AGT Cytogenetics Laboratory Manual*, 3rd edn. Brach MJ, Knutsen T, Spurbeck J (eds.) Philadelphia: Lippincott-Raven; 1997: 375–430.
6. Freshney RI. Culture of tumor cells. In *Culture of Animal Cells: A Manual of Basic Technique*, 5th edn. Hoboken, New Jersey: John Wiley & Sons, Inc.; 2005: 421–433.
7. Wang HC, Federoff S. Banding in human chromosomes treated with trypsin. *Nat New Biol* 1972;235:52–53.
8. Seabright M. A rapid banding technique for human chromosomes. *Lancet* 1971;2:971–972.
9. Yunis JJ. New chromosome techniques in the study of human neoplasia. *Hum Pathol* 1981;12:540–549.
10. Shaffer LG, McGowan-Jordan J, Schmid M. (eds.) *ISCN (2013) An International System for Human Cytogenetic Nomenclature.* Basel: S. Karger; 2013.
11. Lejeune J, Levan A, Böök JA, *et al*. A proposed standard system on nomenclature of human mitotic chromosomes. *Lancet* 1960;275:1063–1065.
12. Paris Conference (1971): standardization in human cytogenetics. *Birth Defects Orig Artic Ser* 1972;8:1–46.
13. Bridge JA. Advantages and limitations of cytogenetic, molecular cytogenetic, and molecular diagnostic testing in mesenchymal neoplasms. *J Orthop Sci* 2008;13:273–282.
14. Bridge JA, Cushman-Vokoun AM. Molecular diagnostics of soft tissue tumors. *Arch Pathol Lab Med* 2011;135(5):588-601.
15. Pinkel D, Gray J, Trask B, *et al*. Cytogenetic analysis by in situ hybridization with fluorescently labeled nucleic acid probes. *Cold Spring Harbor Symp Quant Biol* 1986;51:151–157.
16. Speicher MR, Gwyn Ballard S, Ward DC. Karyotyping human chromosomes by combinatorial multi-fluor FISH. *Nat Genet* 1996;12:368–375.
17. Schröck E, du Manoir S, Veldman T, *et al*. Multicolor spectral karyotyping of human chromosomes. *Science* 1996;273:494–497.
18. Tanke HJ, Wiegant J, van Gijlswijk RP, *et al*. New strategy for multi-colour fluorescence in situ hybridization: COBRA: Combined binary ratio labelling. *Eur J Hum Genet* 1999;7:2–11.
19. Kallioniemi A, Kallioniemi OP, Sudar D, *et al*. Comparative genomic hybridization for molecular cytogenetic analysis of solid tumors. *Science* 1992;258:818–821.
20. Pinkel D, Segraves R, Sudar D, *et al*. High resolution analysis of DNA copy number variation using comparative genomic hybridization to microarrays. *Nat Genet* 1998;20:207–211.
21. Solinas-Toldo S, Lampel S, Stilgenbauer S, *et al*. Matrix-based comparative genomic hybridization: biochips to screen for genomic imbalances. *Genes Chromosomes Cancer* 1997;20:399–407.
22. Mundle SD, Koska RJ. Fluorescence in situ hybridization: a major milestone in luminous cytogenetics. In *Molecular Diagnostics: For the Clinical Laboratorian*, 2nd edn. Coleman WB, Tsongalis GJ (eds.) Totowa, NJ: Humana Press; 2006: 196.
23. Michelson DJ, Shevell MI, Sherr EH, *et al*. Evidence report: genetic and metabolic testing on children with global developmental delay: report of the Quality Standards Subcommittee of the American Academy of Neurology and the Practice Committee of the Child Neurology Society. *Neurology* 2011;77(17):1629–1635.
24. Jabs EW, Wolf SF, Migeon BR. Characterization of a cloned DNA sequence that is present at centromeres of all human autosomes and the X chromosome and shows polymorphic variation. *Proc Natl Acad Sci USA* 1984;81:4882–4888.
25. Guan, XY, Meltzer P, Trent J. Rapid generation of whole chromosome painting probes (WCPs) by chromosome microdissection. *Genomics* 1994;22:101–107.
26. Lichter P, Ledbetter SA, Ledbetter DH, Ward DC. Fluorescence in situ hybridization with ALU and L1 polymerase chain reaction probes for rapid characterization of human chromosomes in hybrid cell lines. *Proc Natl Acad Sci USA* 1990;85:9138–9142.
27. Ladanyi M, Bridge JA. Contribution of molecular genetic data to the classification of sarcomas. *Hum Pathol* 2000;31:532–538.

28. Sandberg AA, Bridge JA. Updates on the cytogenetics and molecular genetics of bone and soft tissue tumors: clear cell sarcoma (malignant melanoma of soft parts). *Cancer Genet Cytogenet* 2001;130:1–7.
29. Patel RM, Downs-Kelly E, Weiss SW, et al. Dual-color, break-apart fluorescence in situ hybridization for EWS gene rearrangement distinguishes clear cell sarcoma of soft tissue from malignant melanoma. *Mod Pathol* 2005;18:1585–1590.
30. Fletcher CDM, Bridge JA, Hogendoorn PCW, Mertens F. *Classification of Tumours of Soft Tissue and Bone*, 4th edn. Lyon, France: IARC Press; 2013.
31. Bridge JA. The role of cytogenetics and molecular diagnostics in the diagnosis of soft-tissue tumors. *Mod Pathol* 2014;27:580-597.
32. https://www.acmg.net/Pages/ACMG_Activities/stds-2002/e.htm (accessed January 2016).
33. http://www.cap.org/apps/cap.portal?_nfpb=true&cntvwrPtlt_actionOverride=/portlets/contentViewer/show&cntvwrPtlt{actionForm.contentReference}=committees/cytogenetics/cytogenetics_index.html&_pageLabel=cntvwr (accessed January 2016).
34. Weiss LM, Chen YY. Effects of different fixatives on detection of nucleic acids from paraffin-embedded tissues by in situ hybridization using oligonucleotide probes. *J Histochem Cytochem* 1991;30:1237–1241.
35. Watters AD, Bartlett JM. Fluorescence in situ hybridization in paraffin tissue sections: pretreatment protocol. *Mol Biotechnol* 2002;21:217–220.
36. Kearney L, Hammond DW. Molecular cytogenetic technologies. In *Human Cytogenetics. A Practical Approach. Volume II. Malignancy and Acquired Abnormalities*, 3rd edn. Rooney DE, (ed.) Oxford: Oxford University Press; 2001: 129–163.
37. Schwab M, Shimada H, Joshi V, Brodeur GM. Neuroblastic tumours of adrenal gland and sympathetic nervous system. In *World Health Organization Classification of Tumours: Pathology & Genetics. Tumours of the Nervous System*. Kleinhues P, Cavenee WK (eds.) Lyon: IARC Press; 2000: 159.
38. Fluorescence in situ hybridization: technical overview. In *Molecular Diagnosis of Cancer: Methods and Protocols*, 2nd edn. Roulston JE, Bartlett JMS (eds.) Totowa, NJ: Humana Press; 2004: 77–87.
39. Sandberg AA. Cytogenetics and molecular genetics of bone and soft-tissue tumors. *Am J Med Genet* 2002;115:189–193.
40. Dewald GW. Interphase FISH studies of chronic myeloid leukemia. *Methods Mol Biol* 2002;204:311–342.
41. Weber-Matthiesen K, Winkemann M, Müller-Hermelink A, Schlegelberger B, Grote W. Simultaneous fluorescence immunophenotyping and interphase cytogenetics: a contribution to the characterization of tumor cells. *J Histochem Cytochem* 1992;40:171–175.
42. Duval C, de Tayrac M, Sanschagrin F, et al. ImmunoFISH is a reliable technique for the assessment of 1p and 19q status in oligodendrogliomas. *PLoS One* 2014;9(6): e100342.
43. Perry A, Roth KA, Banerjee R, Fuller CE, Gutmann DH. NF1 deletions in S-100 protein-positive and negative cells of sporadic and neurofibromatosis 1 (NF1)-associated plexiform neurofibromas and malignant peripheral nerve sheath tumors. *Am J Pathol* 2001;159:57–61.
44. Bridge JA, Fidler ME, Neff JR, et al. Adamantinoma-like Ewing's sarcoma: genomic confirmation, phenotypic drift. *Am J Surg Pathol* 1999;23:159–165.
45. Summersgill B, Clark J, Shipley J. Fluorescence and chromogenic in situ hybridization to detect genetic aberrations in formalin-fixed paraffin embedded material, including tissue microarrays. *Nat Protoc* 2008;3:220–234.
46. Dandachi N, Dietze O, Hauser-Kronberger C. Chromogenic in situ hybridization: a novel approach to a practical and sensitive method for the detection of HER2 oncogene in archival human breast carcinoma. *Lab Invest* 2002;82(8):1007–1014.
47. Speicher MR, Carter NP. The new cytogenetics: blurring the boundaries with molecular biology. *Nat Rev Genet* 2005;6:782–792.
48. Wiegant J, Bezrookove V, Rosenberg C, et al. Differentially painting human chromosome arms with combined binary ratio-labeling fluorescence in situ hybridization. *Genome Res* 2000;10(6):861–865.
49. Kallioniemi A, Kallioniemi OP, Sudar D, et al. Comparative genomic hybridization for molecular cytogenetic analysis of solid tumors. *Science* 1992;258:818.
50. Kallioniemi OP, Kallioniemi A, Piper J, et al. Optimizing comparative genomic hybridization for analysis of DNA sequence copy number changes in solid tumors. *Genes Chromosomes Cancer* 1994;10:231–243.
51. Pinkel D, Albertson DG. Comparative genomic hybridization. *Annu Rev Genomics Hum Genet* 2005;6:331.
52. Cowell JK, Hawthorn L. The application of microarray technology to the analysis of the cancer genome. *Curr Mol Med* 2007;7:103–120.
53. Nielsen TO. Microarray analysis of sarcomas. *Adv Anat Pathol* 2006 Jul;13(4):166–173.
54. West RB, van de Rijn M. The role of microarray technologies in the study of soft tissue tumours. *Histopathology* 2006;48:22–31.
55. Lynn M, Wang Y, Slater J, et al. High-resolution genome-wide copy-number analyses identify localized copy-number alterations in Ewing sarcoma. *Diagn Mol Pathol* 2013;22(2):76–84.
56. Jahromi MS, Putnam AR, Druzgal C, et al. Molecular inversion probe analysis detects novel copy number alterations in Ewing sarcoma. *Cancer Genet* 2012;205(7-8):391–404.
57. Bridge JA, Liu J, Qualman SJ, et al. Genomic gains and losses are similar in genetic and histologic subsets of rhabdomyosarcoma, whereas amplification predominates in embryonal with anaplasia and alveolar subtypes. *Genes Chromosomes Cancer* 2002;33:310–321.
58. Albertson DG. Gene amplification in cancer. *Trends Genet* 2006;22:447–455.
59. Sirvent N, Coindre JM, Maire G, et al. Detection of MDM2-CDK4 amplification by fluorescence in situ hybridization in 200 paraffin-embedded tumor samples: utility in diagnosing adipocytic lesions and comparison with immunohistochemistry and real-time PCR. *Am J Surg Pathol* 2007;31:1476–1489.
60. Binh MB, Sastre-Garau X, Guillou L, et al. MDM2 and CDK4 immunostainings are useful adjuncts in diagnosing well-differentiated and

dedifferentiated liposarcoma subtypes: a comparative analysis of 559 soft tissue neoplasms with genetic data. *Am J Surg Pathol* 2005;29:1340–1347.

61. Weaver J, Goldblum JR, Turner S, et al. Detection of MDM2 gene amplification or protein expression distinguishes sclerosing mesenteritis and retroperitoneal fibrosis from inflammatory well-differentiated liposarcoma. *Mod Pathol* 2009;22:66–70.

62. Forment JV, Kaidi A, Jackson SP. Chromothripsis and cancer: causes and consequences of chromosome shattering. *Nat Rev Cancer* 2012;12(10):663–670.

63. Cai H, Kumar N, Bagheri HC, et al. Chromothripsis-like patterns are recurring but heterogeneously distributed features in a survey of 22,347 cancer genome screens. *BMC Genomics* 2014;15:82.

64. Stephens PJ, Greenman CD, Fu B, et al. Massive genomic rearrangement acquired in a single catastrophic event during cancer development. *Cell* 2011;144:27–40.

65. Molenaar JJ, Koster J, Zwijnenburg DA, et al. Sequencing of neuroblastoma identifies chromothripsis and defects in neuritogenesis genes. *Nature* 2012;483(7391):589–593.

66. Nowell P, Hungerford D. A minute chromosome in chronic granulocytic leukemia. *Science* 1960;1332:1947.

67. Aurias A, Rimbaut C, Buffe D, Dubousset J, Mazabraud A. Translocation of chromosome 22 in Ewing's sarcoma. *CR Seances Acad Sci III* 1983;296:1105–1107.

68. Delattre O, Zucman J, Melot T, et al. The Ewing family of tumors: a subgroup of small-round-cell tumors defined by specific chimeric transcripts. *N Engl J Med* 1994;331:294–299.

69. Ieremia E, Thway K. Myxoinflammatory fibroblastic sarcoma: morphologic and genetic updates. *Arch Pathol Lab Med* 2014;138(10):1406–1411.

70. Wachtel M, Dettling M, Koscielniak E, et al. Gene expression signatures identify rhabdomyosarcoma subtypes and detect a novel t(2;2)(q35;p23) translocation fusing PAX3 to NCOA1. *Cancer Res* 2004;64(16):5539–5545.

71. Sumegi J, Streblow R, Frayer RW, et al. Recurrent t(2;2) and t(2;8) translocations in rhabdomyosarcoma without the canonical PAX-FOXO1 fuse PAX3 to members of the nuclear receptor transcriptional coactivator family. *Genes Chromosomes Cancer* 2010;49(3):224–236.

72. Mosquera JM, Sboner A, Zhang L, et al. Recurrent NCOA2 gene rearrangements in congenital/infantile spindle cell rhabdomyosarcoma. *Genes Chromosomes Cancer* 2013;52(6):538–550.

73. Antonescu CR, Chen HW, Zhang L, et al. ZFP36-FOSB fusion defines a subset of epithelioid hemangioma with atypical features. *Genes Chromosomes Cancer* 2014;53(11):951–959.

74. Walther C, Tayebwa J, Lilljebjörn H, et al. A novel SERPINE1-FOSB fusion gene results in transcriptional up-regulation of FOSB in pseudomyogenic haemangioendothelioma. *J Pathol* 2014;232(5):534–540.

75. Thway K, Nicholson AG, Lawson K, et al. Primary pulmonary myxoid sarcoma with EWSR1-CREB1 fusion: a new tumor entity. *Am J Surg Pathol* 2011;35(11):1722–1732.

76. Fletcher CD, Busam KJ. The clinical role of molecular genetics in soft tissue tumor pathology. *Cancer Metastasis Rev* 1997;16:207–227.

77. Antonescu CR. The role of genetic testing in soft tissue sarcoma. *Histopathology* 2006;48:13–21.

78. Ladanyi M, Antonescu CR, Dal Cin P. Cytogenetic and molecular genetic pathology of soft tissue tumors. In *Enzinger and Weiss's Soft Tissue Tumors*. Weiss SW, Goldblum JR (eds.) Philadelphia: Mosby Elsevier Inc.; 2008: 73–102.

79. Bridge JA. Contribution of cytogenetics to the management of poorly differentiated sarcomas. *Ultrastruct Pathol* 2008;32:63–71.

80. Polito P, Dal Cin P, Sciot R, et al. Embryonal rhabdomyosarcoma with only numerical chromosome changes. Case report and review of the literature. *Cancer Genet Cytogenet* 1999;109:161–165.

81. Bridge JA, Liu J, Weibolt V, et al. Novel genomic imbalances in embryonal rhabdomyosarcoma revealed by comparative genomic hybridization and fluorescence in situ hybridization: an intergroup rhabdomyosarcoma study. *Genes Chromosomes Cancer* 2000;27:337–344.

82. Plaat BE, Molenaar WM, Mastik MF, et al. Computer-assisted cytogenetic analysis of 51 malignant peripheral nerve sheath tumors: sporadic versus neurofibromatosis-type-1-associated malignant schwannomas. *Int J Cancer* 1999;83:171–178.

83. Mertens F, Dal Cin P, De Wever I, et al. Cytogenetic characterization of peripheral nerve sheath tumours: a report of the CHAMP study group. *J Pathol* 2000;190:31–38.

84. Bridge RS Jr, Bridge JA, Neff JR, et al. Recurrent chromosomal imbalances and structurally abnormal breakpoints within complex karyotypes of malignant peripheral nerve sheath tumour and malignant triton tumour: a cytogenetic and molecular cytogenetic study. *J Clin Pathol* 2004;57:1172–1178.

85. Bowen JM, Cates JM, Kash S, et al. Genomic imbalances in benign metastasizing leiomyoma: characterization by conventional karyotypic, fluorescence in situ hybridization, and whole genome SNP array analysis. *Cancer Genet* 2012;205(5):249–54.

86. Borden EC, Baker LH, Bell RS, et al. Soft tissue sarcomas of adults: state of the translational science. *Clin Cancer Res* 2003;9:1941–1956.

87. Deneen B, Denny CT. Loss of p16 pathways stabilizes EWS/FLI1 expression and complements EWS/FLI1 mediated transformation. *Oncogene* 2001;20:6731–6741.

88. Lessnick SL, Dacwag CS, Golub TR. The Ewing's sarcoma oncoprotein EWS/FLI induces a p53-dependent growth arrest in primary human fibroblasts. *Cancer Cell* 2002;1:393–401.

89. Xia SJ, Barr FG. Chromosome translocations in sarcomas and the emergence of oncogenic transcription factors. *Eur J Cancer* 2005;41:2513–2527.

90. Mugneret F, Lizard S, Aurias A, Turc-Carel C. Chromosomes in Ewing's sarcoma. II. Nonrandom additional changes, trisomy 8 and der(16)t(1;16). *Cancer Genet Cytogenet* 1988;32:239–245.

91. Höglund M, Gisselsson D, Mandahl N, Mitelman F. Ewing tumours and synovial sarcomas have critical features of karyotype evolution in common with

92. Skytting BT, Szymanska J, Aalto Y, et al. Clinical importance of genomic imbalances in synovial sarcoma evaluated by comparative genomic hybridization. *Cancer Genet Cytogenet* 1999;115:39–46.
93. Hattinger CM, Rumpler S, Ambros IM, et al. Demonstration of the translocation der(16)t(1;16)(q12;q11.2) in interphase nuclei of Ewing tumors. *Genes Chromosomes Cancer* 1996;17:141–150.
94. Stark B, Mor C, Jeison M, et al. Additional chromosome 1q aberrations and der(16)t(1;16), correlation to the phenotypic expression and clinical behavior of the Ewing family of tumors. *J Neurooncol* 1997;31:3–8.
95. Sandberg AA, Bridge JA. Updates on cytogenetics and molecular genetics of bone and soft tissue tumors: Ewing sarcoma and peripheral primitive neuroectodermal tumors. *Cancer Genet Cytogenet* 2000;123:1–26.
96. Hirota S, Isozaki K, Moriyama Y, et al. Gain-of-function mutations of c-kit in human gastrointestinal stromal tumors. *Science* 1998;279:577–580.
97. Hirota S, Ohashi A, Nishida T, et al. Gain-of-function mutations of platelet-derived growth factor receptor alpha gene in gastrointestinal stromal tumors. *Gastroenterology* 2003;125:660–667.
98. Lasota J, Dansonka-Mieszkowska A, Sobin LH, Miettinen M. A great majority of GISTs with PDGFRA mutations represent gastric tumors of low or no malignant potential. *Lab Invest* 2004;84:874–883.
99. Delattre O, Zucman J, Plougastel B, et al. Gene fusion with an ETS DNA-binding domain caused by chromosome translocation in human tumours. *Nature* 1992;359:162–165.
100. Zucman J, Melot T, Desmaze C, et al. Combinatorial generation of variable fusion proteins in the Ewing family of tumours. *EMBO J* 1993;12:4481–4487.
101. Jeon IS, Davis JN, Braun BS, et al. A variant Ewing's sarcoma translocation (7;22) fuses the EWS gene to the ETS gene ETV1. *Oncogene* 1995;10:1229–1234.
102. Kaneko Y, Yoshida K, Handa M, et al. Fusion of an ETS-family gene, EIAF, to EWS by t(17;22)(q12;q12) chromosome translocation in an undifferentiated sarcoma of infancy. *Genes Chromosomes Cancer* 1996;15:115–121.
103. Peter M, Couturier J, Pacquement H, et al. A new member of the ETS family fused to EWS in Ewing tumors. *Oncogene* 1997;14:1159–1164.
104. Mastrangelo T, Modena P, Tornielli S, et al. A novel zinc finger gene is fused to EWS in small round cell tumor. *Oncogene* 2000;19:3799–3804.
105. Wang L, Bhargava R, Zheng T, et al. Undifferentiated small round cell sarcomas with rare EWS gene fusions: identification of a novel EWS-SP3 fusion and of additional cases with the EWS-ETV1 and EWS-FEV fusions. *J Mol Diagn* 2007;9:498–509.
106. Shing DC, McMullan DJ, Roberts P, et al. FUS/ERG gene fusions in Ewing's tumors. *Cancer Res* 2003;63:4568–4576.
107. Ng TL, O'Sullivan MJ, Pallen CJ, et al. Ewing sarcoma with novel translocation t(2;16) producing an in-frame fusion of FUS and FEV. *J Mol Diagn* 2007;9:459–463.
108. Sandberg AA, Bridge JA. Updates on the cytogenetics and molecular genetics of bone and soft tissue tumor: gastrointestinal stromal tumors. *Cancer Genet Cytogenet* 2002;135:1–22.
109. López-Guerrero JA, Noguera R, Llombart-Bosch A. GIST: particular aspects related to cell cultures, xenografts, and cytogenetics. *Semin Diagn Pathol* 2006;23:103–110.
110. Wozniak A, Sciot R, Guillou L, et al. Array CGH analysis in primary gastrointestinal stromal tumors: cytogenetic profile correlates with anatomic site and tumor aggressiveness, irrespective of mutational status. *Genes Chromosomes Cancer* 2007;46:261–276.
111. Sandberg AA. Updates on the cytogenetics and molecular genetics of bone and soft tissue tumors: liposarcoma. *Cancer Genet Cytogenet* 2004;155:1–24.
112. Meis-Kindblom JM, Sjogren H, Kindblom LG, et al. Cytogenetic and molecular genetic analyses of liposarcoma and its soft tissue simulators: recognition of new variants and differential diagnosis. *Virchows Arch* 2001;439:141–151.
113. Segura-Sanchez J, Gonzalez-Campora R, Pareja-Megia MJ, et al. Chromosome-12 copy number alterations and MDM2, CDK4 and TP53 expression in soft tissue liposarcoma. *Anticancer Res* 2006;26:4937–4942.
114. Shimada S, Ishizawa T, Ishizawa K, et al. The value of MDM2 and CDK4 amplification levels using real-time polymerase chain reaction for the differential diagnosis of liposarcomas and their histologic mimickers. *Hum Pathol* 2006;37:1123–1129.
115. Weaver J, Downs-Kelly E, Goldblum JR, et al. Fluorescence in situ hybridization for MDM2 gene amplification as a diagnostic tool in lipomatous neoplasms. *Mod Pathol* 2008;21:943–949.
116. Le Guellec S, Chibon F, Ouali M, et al. Are peripheral purely undifferentiated pleomorphic sarcomas with MDM2 amplification dedifferentiated liposarcomas? *Am J Surg Pathol* 2014;38(3):293-304.
117. DeClue, JE, Papageorge AG, Fletcher JA, et al. Abnormal regulation of mammalian p21ras contributes to malignant tumor growth in von Recklinghausen (type 1) neurofibromatosis. *Cell* 1992;69:265–273.
118. Knudson AG. Mutation and statistical study of retinoblastoma. *Proc Natl Acad Sci USA* 1971;68:820–823.
119. Serra E, Puig S, Otero D, et al. Confirmation of a double hit model for the NF1 gene in benign neurofibromas. *Am J Hum Genet* 1997;61:512–519.
120. Storlazzi CT, Von Steyern FV, Domanski HA, Mandahl N, Mertens F. Biallelic somatic inactivation of the NF1 gene through chromosomal translocations in a sporadic neurofibroma. *Int J Cancer* 2005;117:1055–1057.
121. Birindelli S, Perrone F, Oggionni M, et al. Rb and TP53 pathway alterations in sporadic and NF1-related malignant peripheral nerve sheath tumors. *Lab Invest* 2001;81:833–844.
122. Legius E, Dierick H, Wu R, et al. TP53 mutations are frequent in malignant NF1 tumors. *Genes Chromosomes Cancer* 1994;10:250–255.
123. Kourea HP, Orlow I, Scheithauer BW, Cordon-Cardo C, Woodruff JM. Deletions of the INK4A gene occur in malignant peripheral nerve sheath tumors but not in neurofibromas. *Am J Pathol* 1999;155:1855–1860.
124. Berner JM, Sorlie T, Mertens F, et al. Chromosome band 9p21 is frequently

altered in malignant peripheral nerve sheath tumors: studies of CDKN2A and other genes of the pRB pathway. *Genes Chromosomes Cancer* 1999;26:151–160.

125. Nielsen GP, Stemmer-Rachamimov AO, Ino Y, et al. Malignant transformation of neurofibromas in neurofibromatosis 1 is associated with CDKN2A/p16 inactivation. *Am J Pathol* 1999;155:1879–1884.

126. Perrone F, Tabano S, Colombo F, et al. p15INK4b, p14ARF, and p16INK4a inactivation in sporadic and neurofibromatosis type 1-related malignant peripheral nerve sheath tumors. *Clin Cancer Res* 2003;9:4132–4138.

127. Agesen TH, Førenes VA, Molenaar WM, et al. Expression patterns of cell cycle components in sporadic and neurofibromatosis type 1-related malignant peripheral nerve sheath tumors. *J Neuropathol Exp Neurol* 2005;64:74–81.

128. Carroll SL, Stonecypher MS. Tumor suppressor mutations and growth factor signaling in the pathogenesis of NF1-associated peripheral nerve sheath tumors: II. The role of dysregulated growth factor signaling. *J Neuropathol Exp Neurol* 2005;64:1–9.

129. Schmidt H, Wurl P, Taubert H, et al. Genomic imbalances of 7p and 17q in malignant peripheral nerve sheath tumors are clinically relevant. *Genes Chromosomes Cancer* 1999;25:205–211.

130. Skotheim RI, Kallioniemi A, Bjerkhagen B, et al. Topoisomerase-II alpha is upregulated in malignant peripheral nerve sheath tumors and associated with clinical outcome. *J Clin Oncol* 2003;21:4586–4591.

131. Storlazzi CT, Brekke HR, Mandahl N, et al. Identification of a novel amplicon at distal 17q containing the BIRC5/SURVIVIN gene in malignant peripheral nerve sheath tumours. *J Pathol* 2006; 209:492–500.

132. Kresse SH, Skårn M, Ohnstad HO, et al. DNA copy number changes in high-grade malignant peripheral nerve sheath tumors by array CGH. *Mol Cancer* 2008;7:48.

133. Huang HY, Illei PB, Zhao Z, et al. Ewing sarcomas with p53 mutations or p16/p14ARF homozygous deletions: a highly aggressive subset associated with poor chemoresponse. *J Clin Oncol* 2005;23:548–558.

134. Antonescu CR, Tschernyavsky SJ, Decuseara R, et al. Prognostic impact of P53 status, TLS-CHOP fusion transcript structure, and histological grade in myxoid liposarcoma: a molecular and clinicopathologic study of 82 cases. *Clin Cancer Res* 2001;7:3977–3987.

135. Drobnjak M, Latres E, Pollack D, et al. Prognostic implications of p53 nuclear overexpression and high proliferation index of Ki-67 in adult soft-tissue sarcomas. *J Natl Cancer Inst* 1994;86:549–554.

136. Wurl P, Taubert H, Meye A, et al. Prognostic value of immunohistochemistry for p53 in primary soft-tissue sarcomas: a multivariate analysis of five antibodies. *J Cancer Res Clin Oncol* 1997;123:502–508.

137. Cordon-Cardo C, Latres E, Drobnjak M, et al. Molecular abnormalities of mdm2 and p53 genes in adult soft tissue sarcomas. *Cancer Res* 1994;54: 794–799.

Chapter 5

Molecular genetics of soft tissue tumors

Jerzy Lasota
Armed Forces Institute of Pathology, Washington, DC

It is now widely accepted that cancer is caused by an accumulation of complementary genetic and epigenetic changes. These changes lead to the destabilization of cellular growth control and promote uncontrolled clonal proliferation and tumor development. Over the last decades, significant genetic and epigenetic alterations affecting the function of *oncogenes*, *tumor-suppressor genes*, and *microRNA (miRNA) genes* have been identified. The gain of function of oncogenes, loss of function of tumor-suppressor genes, and deregulation of miRNA genes are among the major molecular events in the development of heritable and sporadic human cancer.

The identification of genetic and epigenetic abnormalities in the cancer genome and the understanding of their functional consequences are leading to the development of new diagnostic approaches and better treatment strategies, targeting the affected gene products and their signaling pathways. The ongoing "genetic revolution" continues to deliver data that is impacting the surgical pathology of soft tissue tumors. Detection of translocations and fusion gene products has been used as a disease-specific marker to improve the diagnosis and prognosis of soft tissue tumors. Other genetic changes, such as the amplification of oncogenes and deletion of tumor-suppressor genes, have been correlated with poor clinical outcomes. Finally, specific genetic and epigenetic mutations have been incorporated in the molecular staging of diseases, including the screening of peripheral blood, bone marrow, or other fluids for minimal residual disease or micrometastases. Although integration of genetics and epigenetics data into surgical pathology of soft tissue tumors has significantly expanded, it is expected that utilization of a wider spectrum of new technologies, including next-generation sequencing (NGS), will define new approaches to sarcoma diagnosis, prognosis, and treatment. The incorporation of new information into diagnostic pathology requires a critical evaluation of genetic research data, however.

This chapter briefly reviews several topics concerning the nature and detection of genetic and epigenetic mutations in soft tissue tumors, whereas specific applications are discussed in tumor-specific chapters.

Oncogenes

Proto-oncogenes are human genes that are involved in cell growth and differentiation, often as growth factor receptors. Some of these genes show sequence homology to known retroviral oncogenes. Gain-of-function genetic mutations can change proto-oncogenes into active oncogenes, which are presumed to lead to uncontrolled cell growth and the development of cancer. Gene transfer assays were the first to provide strong evidence of the existence of oncogenes in human cancer. In these assays, human cancer DNA transferred *in vitro* into mouse recipient cells induced neoplastic transformation. Several oncogenes were identified following this strategy.[1] However, progress in cancer cytogenetics and positional cloning of genetic alterations, such as translocations, prompted the discovery of a great number of human oncogenes.[2]

The first human oncogene, HRAS, was discovered, cloned, and sequenced following a gene transfer experiment in which DNA extracted from the human bladder carcinoma cell line was transferred to mouse NIH3T3 cells and induced neoplastic transformation.[3,4] Comparison of normal HRAS proto-oncogene and oncogene sequences revealed a G-to-T mutation at codon 12, leading to the change of one amino acid being the sole difference in the encoded protein. This proved that subtle molecular changes can alter gene functions and can be responsible for neoplastic transformation.[5] In subsequent studies, other members of the RAS family, namely KRAS and NRAS, have been identified.[6]

Human RAS genes encode distinct but closely related proteins of 188–189 amino acids, HRAS, KRAS 4A, KRAS 4B, and NRAS. RAS proteins play a central role in signal transduction, regulating cell proliferation, and apoptotic cell death. These proteins belong to the short G protein (guanine nucleotide-binding protein) family and are involved in second messenger cascades, functioning as "molecular switches" between an inactive guanosine diphosphate (GDP) and an active guanosine triphosphate (GTP) bound state.[7]

Although gain-of-function RAS mutations involving few mutational "hot spots" (amino acid residues G12, G13, and Q61) are among the most common genetic changes identified in human cancer, the incidence of RAS mutations varies widely between different types of cancer.[8] In soft tissue tumors, mutational activation of RAS is a rather uncommon event. RAS mutations have been reported in a subset of high-grade sarcomas (malignant fibrous histiocytomas)[9,10] and occasionally in other tumors, including angiosarcoma, leiomyosarcoma, liposarcoma, and embryonal rhabdomyosarcoma.[10–14]

Activation of RAS proteins, located on the inner surface of the cell membrane, occurs through the signaling initiated by the growth factor binding to the membrane tyrosine kinase receptors. The activated RAS-GTP protein, in turn, starts signaling cascades that lead to cell growth and differentiation.[15]

Because of the high prevalence of RAS mutations in human tumors, it has been suggested for years that blocking of the pathologically activated RAS signaling pathway could be part of rational cancer therapy.[16,17] These studies are still at the research level and have not reached clinical trials.

Tumor-suppressor genes

The concept of *tumor-suppressor genes* was developed in the early 1970s by Alfred Knudson, who formulated the "two-hit hypothesis" explaining the genetic mechanism of cancer development in hereditary retinoblastoma. Based on epidemiologic studies, Knudson assumed that both alleles of the hypothetic "retinoblastoma gene" had to be mutationally inactivated for tumor formation.[18,19] The Knudson hypothesis was validated by the studies on retinoblastoma in children with inherited constitutional deletion at chromosome 13q14.2, in which a second copy of RB1 was somatically mutated.[20-22] However, most children with retinoblastoma do not have a family history of the disease. In such cases, RB1 mutations are developed primarily during spermatogenesis.[23,24] The second hit frequently results in loss of heterozygosity at the RB locus and can arise from allelic loss, mitotic recombination, or the loss of the whole chromosome 13.[25,26] In addition to genomic mutations, extensive abnormal hypermethylation of the RB1 promoter CpG islands provide an epigenetic mechanism for RB1 inactivation.[27] Also, other epigenetic mechanisms, including interactions with miRNAs and hyperphosphorylation at a post-translational level, could compromise RB1 functions.[28,29] *In vitro* experiments have demonstrated that the introduction of a functional RB1 gene suppresses tumorigenicity of a tumor cell line with an inactivated endogenous RB1 gene. Thus, RB1 is a prototype of the tumor-suppressor gene, which can inhibit tumor formation, acting as a negative regulator of cell proliferation.[21,22]

The RB1 gene encodes for a 105-kD phosphoprotein, which, together with p107 and p130, belongs to the RB family of nuclear proteins. This gene spans over 200 kb of genomic DNA and consists of 27 exons. Members of this family form part of a signal transduction pathway called the RB pathway, which is important in cell cycle regulation and has a role in growth suppression, differentiation, and apoptosis in different cell types.[20-22] The loss-of-function RB1 mutations and defects in RB1 pathways have been reported in many human cancers, including sarcomas.[30,31] Nevertheless, the frequency of mutational inactivation of the RB1 gene in soft tissue sarcomas appears to be relatively low.[32,33]

Following the discovery of RB1, a number of other genes have been identified, whose loss of function affects different cellular processes and can promote a malignant phenotype *in vitro*. Further, it has been documented that somatic inactivation of these genes (tumor-suppressor genes) can play an important role in the development of sporadic tumors. Some of the tumor-suppressor genes, such as RB1, have turned out to be cell cycle regulators, while others are involved in DNA mismatch repair and cell apoptosis. Regardless, functions of many tumor-suppressor genes are not well characterized.[34]

Epimutations

There is strong evidence that oncogene activation followed by tumor-suppressor gene inactivation is the first step in soft tissue sarcoma tumorigenesis. In contrast, initial tumor-suppressor gene alteration in carcinomas is followed by oncogene activation. These genomic alterations are often accompanied or shadowed by epigenetic changes (epimutations).[35] Epimutations can affect the CpG island methylation pattern, chromatin organization, and histone protein modifications, disrupt genomic imprinting, and subsequently cause the inactivation of tumor-suppressor genes or activation of oncogenes. The association between epimutations and tumor initiation, progression, and metastatic dissemination is well documented.[36] The term *epimutation (epigenetic changes)* refers to the heritable modifications of phenotypes that result from genetic changes without alterations in the DNA sequence.[37] However, in the field of cancer research, the word epimutations is often used to describe epigenetic changes driven by DNA mutations.[38] Thus, the separation of epimutations into two different categories – primary, which occur in the absence of any DNA sequence change, and secondary, which follow DNA mutation – has been suggested.[39]

MicroRNA genes

The human genome project completed in April 2003 identified approximately 24 000 protein-coding genes. However, the sequences of these genes only account for less than 2% of the human genome. High-throughput cDNA sequencing tasks have shown that the human genome is widely transcribed.[40] Successively, a large number of regulatory elements, like promotor/enhancer regions, transcription binding sites, and different non-protein coding RNAs, were identified, including a class of evolutionarily conserved miRNAs.[41-43]

MicroRNAs are small (approximately 20 nucleotides in size) noncoding RNAs that function as gene regulators. MiRNAs fine-tune the expression of target genes through mRNA degradation or translational repression. In general, these single-stranded molecules bind through the "seed sequence," defined by two to eight 5′ nucleotides to the complementary binding sites within the targeted transcript's 3′ untranslated region. However, in some cases, the nucleotide sequence context outside the "seed sequence" influences target recognition.[44] Additionally, single nucleotide polymorphisms and mutations can affect miRNA functions.[45] Further restraint of biological functions of a given miRNA could be related to naturally occurring variations in the sequence length, creating isomiRNAs, microRNA isoforms.[46]

MiRNAs have been shown to play a role in a variety of physiological cellular processes such as differentiation, development, proliferation, metabolism and cell-cycle control, and cell death control. Moreover, the miRNA expression pattern is often tissue specific.[47–49] A single miRNA can interact with the transcripts of hundreds of different genes. Also, a specific mRNA can be targeted by multiple miRNAs. Thus, miRNAs have the capacity to regulate multiple pathways and the deregulation of their functions has widespread and complex biological effects. Most of the miRNA coding sequences were found in introns (intragenic regions) and, occasionally, in exons. However, some genes encoding miRNAs have their own promotors and are located in intergenic regions.[50] Approximately 2000 human miRNAs have been identified, but this number might change with the progression of new studies and incorporation of next-generation deep sequencing data (www.miRbase.org).

Genes coding miRNAs are often located in chromosomal regions subjected to genetic mutations, such as amplifications, deletions, or translocations. Changes in the gene copy number of miRNA genes (common to several types of tumors or unique to a specific tumor type) lead to an increase or decrease in miRNA expression. Distinguished miRNA expression signatures have been linked to tumor-specific translocations.[51,52] Besides genetic mutations, epigenetic mechanisms such as promotor hypermethylation can change miRNA expression in cancer.[53] Furthermore, changes in the tumor environment, leading to hypoxia, can influence miRNA expression.[54]

MiRNAs strongly associated with cancer, often called "oncomir," have been reported to act, depending on the genetic and epigenetic context, as oncogenes or tumor-suppressor genes, either through direct interaction or by indirectly targeting genes involved in the pathways controlled by oncogenes or tumor-suppressor genes.[55] Aberrant miRNA expressions were reported in various types of cancer and have been associated with tumor initiation and progression.[56,57] Tumor-specific miRNA expression profiles have been identified in several sarcomas. Sarcoma-specific miRNA profiles could be used for tumor diagnosis and classification.[56,58,59] Additionally, the profiling of miRNA at different stages of tumor development offers new biomarkers associated with progression and metastasis.[60,61] The detection of circulating miRNAs in serum and plasma has recently emerged as a new cancer diagnostic and prognostic biomarker. A number of studies have shown significant differences between the expression profiles of circulating miRNA in cancer patients and healthy individuals. Also, increased or decreased levels of serum/plasma tumor-specific miRNA have been shown to be predictive of earlier relapse or disease-free survival.[62,63] In soft tissue tumors, the detection of circulating miRNAs was reported in rhabdomyosarcoma (RMS) and malignant peripheral nerve sheath tumor (MPNST) patients.[61] Serum levels of circulating muscle-specific miRNAs (especially miR-206) have been shown to be higher in RMS patients than in patients with non-RMS tumors and have been used as a diagnostic marker for rhabdomyosarcoma.[64] Sporadic MPNSTs could be distinguished from MPNSTs associated with neurofibromatosis type 1 (NF1) based on the expression profile of three (miR-801, -214, and -24) serum-circulating miRNAs.[65] A recently established, sarcoma-specific miRNA database (www.oncomir.umn.edu) provides detailed information about miRNA expression in different types of soft tissue tumors.[66]

There is growing evidence based on animal studies that miRNAs could be used as therapeutic agents.[67] Thus, miRNA expression profiling has diagnostic, prognostic, and potentially therapeutic significance. Therefore, standardized procedures, sample preparation, platforms, and bioinformatics should be established to ensure data reproducibility.[68]

Inherited cancer syndromes

Genetic studies on families with inherited tumors resulted in the discovery of genes affected by germline mutations. Activation or inactivation of these genes plays a crucial role in the development of inherited cancer syndromes. Furthermore, the modification of their functions by somatic, genetic, or epigenetic mutations was indicated in the development of sporadic tumors. Syndromes of inherited predisposition to soft tissue tumors and genes critical for their development are listed in Table 5.1.

Genetic alterations in soft tissue tumors
Fusion genes

Fusion genes are formed due to chromosomal structural rearrangements, such as translocation, inversion, and interstitial deletions. In soft tissue tumors, translocations are common primary cytogenetic aberrations resulting in abnormal hybrid chromosomes that contain DNA from two or more chromosomes. Examples of chromosome translocations and other rearrangements leading to fusion gene formation or pathologic activation of normal genes identified in soft tissue tumors are listed in Table 5.2. A reciprocal translocation generates two translocated chromosomes by a reciprocal exchange of chromosomal material. A nonreciprocal translocation refers to a situation in which loss of translocated genomic material occurs and only one translocated chromosome is present. Translocations and other chromosomal rearrangements variably lead to the recombination of regulatory or coding sequences of different genes. Replacement of regulatory sequences can result in the overexpression of a normal protein, which can cause oncogenic transformation. Recombination (fusion) of coding sequences of different genes subsequently leads to the expression of a chimeric fusion gene protein. Fusion gene proteins act as oncoproteins interacting with specific and multiple pathways and promoting uncontrolled cell growth and tumor formation.

Chapter 5: Molecular genetics of soft tissue tumors

Table 5.1 Examples of hereditary/familial cancer syndromes with predisposition to soft tissue tumors

Cancer syndrome (*OMIM number*)	Type of inheritance	Affected gene/locus (cytogenetic map location)	Soft tissue tumor
Basal cell nevus syndrome/Gorlin syndrome (109400)	AD	PTCH1 (9q22.32)	Cardiac fibroma, ovarian fibroma[69,70]
Beckwith–Wiedemann syndrome (130650)	SP/AD (complex)	NSD1 (5q35.2–q35.3) H19 (11p15.5) KCNQ1OT1 (11p15.5) CDKN1C (11p15.4)	Wilms' tumor, neuroblastoma, rhabdomyosarcoma[71,72]
Blue rubber bleb nevus syndrome/bean syndrome (112200)	AD	Unknown	Cavernous hemangioma in different body parts including skin, GI, CNS, liver, other organs[73]
Carney complex type I (160980) Carney complex type II (605244)	AD AD	PRKAR1A (17q24.2) Unknown (2p16)	Myxoma, melanotic schwannoma[74]
Costello syndrome (218040)	SP	HRAS (11p15.5)	Embryonal rhabdomyosarcoma, neuroblastoma[75]
Familial adenomatous polyposis/Gardner syndrome and hereditary desmoid disease (175100, 135290, 175100)	AD	APC (5q22.2)	Desmoid[76–78]
Familial atypical multiple malignant melanoma with multiple neurofibromas (155600)	AD?	Multiple genes indicated including CDKN2A (9p21.3)	Malignant melanoma, multiple neurofibromas[79–81]
Familial GIST syndrome (606764)	AD	KIT (4q12) PDGFRA (4q12)	Gastrointestinal stromal tumor[82,83]
Familial multiple lipomas (151900)	AD	HMGA2 (12q14.3)	Lipomas[84]
Familial retinoblastoma (180200)	AD	RB1 (13q14.2)	Retinoblastoma, osteosarcoma[85,86]
Familial Wilms' tumor, WAGR syndrome, Denys–Drash syndrome (194070, 194072, 194080)	AD	WT1 (11p13), other loci	Wilms' tumor[87–90]
Glomuvenous malformations (138000)	AD or SP	GLMN (1p22.1)	Glomuvenous malformation[91]
Hereditary leiomyomatosis and renal cell cancer (150800)	AD	FH (1q42.1)	Leiomyomas[92,93]
Hyaline fibromatosis syndrome (228600)	AR	ANTXR2 (4q21.21)	Hyaline fibromatosis[94]
Li–Fraumeni familial cancer syndrome (151623)	AD	TP53 (17p13.1) CHEK2 (22q12.1)	Soft tissue and bone sarcomas including rhabdomyosarcoma[95,96]
Multiple enchondromatosis Maffucci type/Maffucci syndrome (614569)	SP	IDH1 (2q33.3) IDH2 (15q26.1)	Enchondromas and hemangiomas[97]
Neurofibromatosis type 1 (162200)	AD	NF1 (17q11.2)	Neurofibroma, MPNST, paraganglioma, GIST[98]
Neurofibromatosis type 2 (101000)	AD	NF2 (22q12.2)	Meningioma, schwannoma[99]
Nijmegen breakage syndrome (251260)	AR	NBN (8q21.3)	Perianal rhabdomyosarcoma[100]
Paraganglioma/pheochromocytoma syndrome type 1–4 (115310, 168000, 605373, 601650)	AD	SDHD (11q23.1) Type 1 SDHAF2 (11q12.2) Type 2 SDHC (1q23.3) Type 3 SDHB (1p36.13) Type 4	Paraganglioma, pheochromocytoma[101,102]

Table 5.1 (cont.)

Cancer syndrome (OMIM number)	Type of inheritance	Affected gene/locus (cytogenetic map location)	Soft tissue tumor
Paraganglioma and GIST, familial (606864)	AD	SDHB (1p36.13) SDHC (1q23.3) SDHD (11q23.1)	Paraganglioma, GIST[103,104]
PTEN hamartoma tumor syndromes: Bannayan–Riley–Ruvalcaba syndrome (153480) Cowden syndrome (158350) Proteus syndrome (176920)	AD AD SP	PTEN (10q23.31) PTEN (10q23.31) PTEN (10q23.31) AKT1 (14q32.33)	Lipomas, hemangiomata[105] Fibromas, lipomas, meningioma[105] Lipomas[106]
Rhabdoid predisposition syndrome 1 (609322) Rhabdoid predisposition syndrome 2 (613325)	AD AD	SMARCB1 (22q11.23) SMARCA4 (19p13.2)	Malignant rhabdoid tumor including "small cell carcinoma of the ovary"[107–110]
Rubinstein–Taybi syndrome (180849)	AD	CREBBP (16p13.3) EP300 (22q13.2)	Keloid, neuroblastoma, leiomyosarcoma, rhabdomyosarcoma[111–113]
Schwannomatosis (162091)	AD	SMARCB1 (22q11.23)	Multiple intracranial, spinal, or peripheral schwannomas, without vestibular nerve involvement[114]
Tuberous sclerosis complex type 1 (191100) Tuberous sclerosis complex type 2 (613254)	AD AD	TSC1 (9q34.13) TSC2 (16p13.3)	Angiomyolipomas, pulmonary lymphangioleiomyomatosis, angiofibromas[115,116] Facial angiofibromas, ungual fibromata, heart rhabdomyoma, chordoma[115,116]
von Hippel–Lindau syndrome (193300)	AD	VHL (3p25.3)	Hemangioblastoma, hemangioma[117,118]
Werner syndrome/adult progeria (277700)	AR	RECQL2 (8p12)	Sarcomas e.g. fibrosarcoma, malignant fibrous histiocytoma, leiomyosarcoma[119]

Abbreviations: OMIM – Online Mendelian Inheritance in Man (www.omim.org); AD – autosomal dominant; AR – autosomal recessive, SP – sporadic; Gene symbols as recommended by HUGO Gene Nomenclature Committee (www. genenames.org).

In the past, fusion genes were cloned and characterized following the detection of chromosome rearrangements by chromosome banding studies. Subsequently, fluorescence in situ hybridization (FISH) technologies with specific probes have been employed and have allowed for more effective pinpointing of the breakpoint regions. In clinical practice, an interphase FISH based on a break-apart rearrangement probe has been widely used to detect rearrangements leading to fusion gene formation. This technique, if well structured, can produce data in 24 hours and could be used on formalin-fixed paraffin-embedded (FFPE) tissues. A break-apart probe consists of two large DNA fragments flanking the genomic breakpoint of the main fusion gene partner. After hybridization, the DNA fragments are juxtaposed in normal cells, but become separated in cells with gene rearrangement. This strategy does not allow for the detection of the second fusion gene partner or the type of fusion. Identification of both fusion gene partners requires a more sophisticated FISH technique employing multiple probes. However, the use of probes representing both fusion partners is practically difficult due to the involvement of multiple gene partners in some cases and the lack of commercially available probes.[2,166]

The invention of the polymerase chain reaction (PCR) amplification technique, particularly reverse transcription PCR (RT-PCR) assays, offered a fast and simple tool for identifying known fusion gene transcripts in RNA obtained from either frozen or FFPE tissues. The main reason for searching for fusion genes at the RNA level was the fact that fusion breaks are often dispersed in the large introns, and thus, practically impossible to amplify from DNA using the standard PCR approach.[261] The advantage of RT-PCR compared to break-apart probe interphase FISH is an ability to identify both fusion partners; however, the structural heterogeneity of fusion gene transcripts because of the various translocation breakpoints might require the use of a multiplex PCR to detect all fusion variants.[261] RACE (rapid amplification of cDNA ends), an RT-PCR-based technique, was successfully employed to amplify chimeric transcripts expressed by the fusion genes with only one known partner.[262] In some cases, the structural diversity of fusion gene products can translate into functional

Table 5.2 Examples of translocations leading to pathologic fusion gene formation or gene overexpression identified in soft tissue tumors

Soft tissue tumor	Cytogenetic aberration	Fusion gene/rearranged gene/deregulated gene
Aggressive angiomyxoma[120–123]	t(8;12)(p12;q15) t(11;12)(q23;q15) t(12;21)(q15;q21.1).	HMGA2 HMGA2 HMGA2
Alveolar soft part sarcoma[124,125]	t(X;17)(p11;q25)	ASPSCR1-TFE3
Angiofibroma[126,127]	t(5;8)(p15;q13) t(7;8;14)(q11;q13;q31)	AHRR-NCOA2 GTF2I-NCOA2
Angiomatoid fibrous histiocytoma[128–131]	t(2;22)(q33;q12) t(12;22)(q13;q12) t(12;16)(q13;p11)	EWSR1-CREB1 EWSR1-ATF1 FUS-ATF1
Angiosarcoma[132]	Identified by NGS	CEP85L-ROS1
Chondroid lipoma[133]	t(11;16)(q13;p13)	C11orf95-MKL2
Chondroma[134,135]	12q14–15 rearrangements t(3;12)(q27;q15)	HMGA2 HMGA2-LPP
Chondrosarcoma, extraskeletal myxoid[136–142]	t(9;22)(q22;q12) t(9;17)(q22;q11) t(9;15)(q22;q21) t(9;3)(q22;q11–q12) t(9;16)(q22;11.2)	EWSR1-NR4A3 TAF15-NR4A3 TCF12-NR4A3 TFG-NR4A3 FUS-NR4A3
Chondrosarcoma, mesenchymal[143–144]	t(8;8)(q13;q21) or del(8)(q13q21) t(1;5)(q42;q32)	HEY1-NCOA2 IRF2BP2-CDX1
Clear cell sarcoma[145–150]	t(12;22)(q13;q12) t(2;22)(q33;q12)	EWSR1-ATF1 EWSR1-CREB1
Collagenous fibroma (desmoplastic fibroblastoma)[151,152]	t(2;11)(q31;q12)	FOSL1
Dermatofibrosarcoma protuberans Giant cell fibroblastoma[153–158]	r(17;22)(q21;q13)/t(17;22)(q21;q13) t(5;8)(q14.3;p21.2) in t(17;22) negative cases	COL1A1-PDGFB VCAN-PTK2B
Desmoplastic small round cell tumor[159–163]	t(11;22)(p13;q12) t(21;22)(q22;q12)	EWSR1-WT1 EWSR1-ERG
Endometrial stromal sarcoma[164–170]	t(6;10;10)(p21;q22;p11) t(7;17)(p15;q21) t(X;17)(p11.2;q21.33) t(6;7)(p21;p15) t(1;6)(p34;p21) t(10;17)(q23;p13) t(10;17)(q22;p13) t(X;22)(p11;q13)	EPC1-PHF1 JAZF1-SUZ12 MBTD1-CXorf67 JAZF1-PHF1 MEAF6-PHF1 YWHAE-NUTM2A YWHAE-NUTM2B ZC3H7B-BCOR
Epithelioid sarcoma of the ovary[166,171]	?t(12;12)(q23;q24) ?t(12;12)(q13;q22) t(1;22)(p36;q11)	CMKLR1-HNF1A ERBB3-CRADD SMARCB1-WASF2
Ewing sarcoma/PNET[166,172–186]	t(11;22)(q24;q12) t(21;22)(q22;q12) t(7;22)(p22;q12) t(17;22)(q12;q12) t(2;22)(q33;q12) r(20;22)(q13;q12), t(20;22)(q13;q12) t(2;22)(q31;q12) inv(22)(q12q12); 22q12 rearrangement t(4;22)(q31;q12) t(16;21)(p11;q22) t(2;16)(q35;p11)	EWSR1-FLI1 EWSR1-ERG EWSR1-ETV1 EWSR1-ETV4 EWSR1-FEV EWSR1-NFATC2 EWSR1-SP3 EWSR1-ZNF278 EWSR1-SMARCA5 FUS-ERG FUS-FEV

Table 5.2 (cont.)

Soft tissue tumor	Cytogenetic aberration	Fusion gene/rearranged gene/deregulated gene
Ewing sarcoma-like[187–190]	t(4;19)(q35;q13) or t(10;19)(q26.3;q13)	CIC-DUX4
	t(X;19)(q13;q13.3)	CIC-FOXO4
	inv(X)(p11.4p11.22)	BCOR-CCNB3
Fibrosarcoma, infantile[191,192]	t(12;15)(p13;q25)	ETV6-NTRK3
Hemangioendothelioma, epithelioid[193–195]	t(1;3)(q36.3;q25)	WWTR1-CAMTA1
	t(X;11)(p11.2;q22.1)	TFE3-YAP1
Hemangioendothelioma, pseudomyogenic[196,197]	t(7;19)(q22;q13)	SERPINE1-FOSB
Hemosiderotic fibrolipomatous tumor[198]	t(1;10)(p22;q24)	TGFBR3-MGEA5
Inflammatory myofibroblastic tumor[199–207]	2p23 rearrangements	ALK
	t(1;2)(q25;p23)	TPM3-ALK
	t(2;19)(p23;p13.1)	TPM4-ALK
	t(2;17)(p23;q23)	CLTC-ALK
	t(2;11)(p23;p15)	CARS-ALK
	inv(2)(p23;q35)	ATIC-ALK
	t(2;12)(p23;q11)	DCTN1-ALK
	t(2;12)(p23;q13)	PPFIBP1-ALK
	inv(2)(p23;q13)	RANBP2-ALK
	t(2;4)(p23;q21)	SEC31A-ALK
	Identified by NGS	LNIN-ALK
	Identified by NGS	PRKAR1A-ALK
	Identified by NGS	TFG-ALK
	inv(7)(p21q22)	CUTL1-AGR3
	Identified by NGS	TFG-ROS1
	Identified by NGS	YWHAE-ROS1
	Identified by NGS	NAB2-PDGFRB
Leiomyoma, uterine[208–213]	t(7;12)(q31;q14)	HMGA2-COG5
	t(12;14)(q15;q24)	HMGA2-RAD51L1
	t(12;14)(q15;q11)	HMGA2-CCNB1IP1
	t(12;8)(q15;q22-23)	HMGA2-COX6C
	t(12;3)(q13–15;q27–28)	HMGA2-LPP
	t(12;13)(q15;q12)	HMGA2-LHFP
	12q15 rearrangement	HMGA2-RTV-H-related sequences
	12q15 rearrangement	HMGA2-ALDH2
	t(10;17)(q22;q21)	KAT6B-KAT2A
Leiomyoma, retroperitoneal[214]	t(10;17)(q22;q21)	KAT6B-KANSL1
Lipoblastoma[215–218]	8q12 rearrangements	PLAG1
	t(7;8)(q22;q12)	COL1A2-PLAG1
	t(8;14)(q12;q24)	PLAG1-RAD51L1
	8q12:8q24.1 rearrangement	HAS2-PLAG1
Lipoma[219–224]	12q14–15 rearrangements	HMGA2
	6p23–21 rearrangements	HMGA1
	t(5;12)(q33;q14)	EBF1-RPSAP52
	t(2;12)(q37;q14)	HMGA2- CXCR7
	t(5;12)(q33;q14)	HMGA2-EBF1
	t(12;13)(q14;q13)	HMGA2-LHFP
	t(3;12)(q28;q14)	HMGA2-LPP
	t(9;12)(q22;q14)	HMGA2-NFIB
	t(1;12)(p32;q14)	HMGA2-PPAP2B
Liposarcoma, myxoid/round cell[225–228]	t(12;16)(q13;p11)	FUS-DDIT3
	t(12;22)(q13;q12)	EWSR1-DDIT3

Chapter 5: Molecular genetics of soft tissue tumors

Table 5.2 (cont.)

Soft tissue tumor	Cytogenetic aberration	Fusion gene/rearranged gene/deregulated gene
Liposarcoma, dedifferentiated[143,166,229]	t(9;12)(q33;q15) ?t(12)(q14;q14) t(9;12)(q33;q21) ?t(12)(q15;q21) t(9;12)(q33;q15) t(5;14)(p13;q32)	CNOT2-ASTN2 CTDSP2-FAM19A2 NR6A1-TRHDE NUP107-LGR5 NUP107-PAPPA RCOR1-WDR70
Low-grade fibromyxoid sarcoma/hyalinizing spindle cell tumor with giant rosettes/sclerosing epithelioid fibrosarcoma[230-234]	t(7;16)(q34;p11) t(11;16)(p11;p11)	FUS-CREB3L2 FUS-CREB3L1
Myoepithelioma, soft tissue[166,235-239]	t(12;22)(q13;q12) t(1;22)(q23;q12) t(6;22)(p21;q12) t(19;22)(q13;q12) 8q12 rearrangements 16p11 rearrangements	EWSR1-ATF1 EWSR1-PBX1 EWSR1-POU5F1 EWSR1-ZNF444 PLAG1 FUS
Myofibroblastic sarcoma, intermediate grade[171]	t(X;6)(p11;p24)	RREB1-TFE3
Myxoinflammatory fibroblastic sarcoma[198]	t(1;10)(p22;q24)	TGFBR3-MGEA5
Nodular fasciitis[240]	t(17;22)(p13;q13.1)	MYH9-USP6
Ossifying fibromyxoid tumor[241-243]	6p21 rearrangements 12q24 and 6p21 rearrangements 10p11 and 6p21 rearrangements t(X;22)(p11;q13)	EP400-PHF1 EPC1-PHF1 MEAF6-PHF1 ZC3H7B-BCOR
PEComa[244]	t(X;1)(p11;p34)	SFPQ-TFE3
Pericytoma[245]	t(7;12)(p22;q13)	ACTB-GLI1
Pleomorphic hyalinizing angiectatic tumor[246]	11p22 and 10q24 rearrangements	TGFBR3-MGEA5 MGEA5 rearrangements
Primary pulmonary myxoid sarcoma[247]	22q12 and 2q33 rearrangements	EWSR1-CREB1
Rhabdomyosarcoma, alveolar[166,248-251]	t(8;13;9)(p11;q14;q32) t(2;13)(q36;q14) t(X;2)(q13;q35) t(2;2)(p23;q36) t(2;8)(q36;q13) t(1;13)(p36;q14)	FOXO1A-FGFR1 PAX3-FOXO1A PAX3-FOXO4 PAX3-NCOA1 PAX3-NCOA2 PAX7-FOXO1A
Rhabdomyosarcoma, spindle cell[166,252]	t(6;8)(p21;q13) t(8;11)(q13;p15)	SRF-NCOA2 TEAD1-NCOA2
Solitary fibrous tumor[253,254]	inv(12)(q13q13)	NAB2-STAT6
Synovial sarcoma[255-257]	t(X;18)(p11;q11) t(X;18)(p11;q11) t(X;18)(p11;q11) t(X;20)(p11;q13.3)	SS18-SSX1 SS18-SSX2 SS18-SSX4 SS18L1-SSX1
Tenosynovial giant cell tumor/pigmented villonodular synovitis[258-260]	t(1;2)(p13;q37) 1p13 and 1q21 rearrangements 1p13 rearrangements	CSF1-COL6A3 CSF1-S100A10 CSF1-rearrangements

Abbreviations: NGS – next-generation sequencing. Gene symbols as recommended by HUGO Gene Nomenclature Committee (www.genenames.org).

differences and might have a prognostic significance.[263–267] RT-PCR evaluation of FFPE tissue for fusion gene products can be hampered by RNA degradation. This could be overcome by the use of a nested PCR amplification, though the employment of such a strategy substantially increases the risk of cross-contamination among samples and can create PCR amplification artifacts. Alternatively, real-time PCR, free of cross-contamination risk, could be used to detect specific and multiple gene fusion transcripts.[268]

Occasionally, RT-PCR-based assays fail to detect fusion gene transcripts in tumors with FISH documented gene rearrangement, indicating gene fusion. These tumors might carry common fusion genes with unusual fusion points or rare fusion gene variants involving different partners. Also, fusion gene transcripts have been detected in tumors without specific chromosome translocations, which could be a result of cytogenetically undetectable cryptic rearrangements.[269,270] A good example of such a rearrangement is the case of a subtle chromosome aberration, an inversion of a 5′ portion of EWSR1 and insertion into chromosome 21, which created an EWSR1-ERG gene fusion typical of the t(22;21) chromosome translocation reported in Ewing sarcoma.[270] A cryptic rearrangement can be a part of complex cytogenetic aberrations unrecognized by standard G-banding and seen only when multicolored karyotyping is employed. Such a complex chromosomal rearrangement, including t(X;18) and t(5;19) in the marker chromosomes, as well as ins(6;18) and t(16;20), has been reported in a case of synovial sarcoma with SS18-SSX2 fusion gene transcripts.[271]

Several studies unexpectedly reported detection of specific fusion gene transcripts in tumors characterized by different fusion genes. These controversial reports were often based solely on PCR assays. For example, the (X;18)(p11;q11) translocation leading to SS18-SSX fusion genes is a genetic hallmark of synovial sarcoma.[255–257] In one study, however, SS18-SSX fusion transcripts were found in a subset of MPNSTs.[272] Because a highly sensitive nested PCR assay was used in this study, concerns of false-positive results from PCR cross-contamination have been raised. The lack of cytogenetic support for the occurrence of this translocation in MPNST also casts doubt on these PCR findings.[273,274]

PCR-based assays are highly sensitive and can detect genetic alterations in a subset of cells undergoing clonal evolution. In one study, a small number of soft tissue tumors yielded a strong signal for the expected fusion gene products and detectable weak signals for unexpected fusion gene transcripts.[275] This might suggest that in some tumors a secondary genetic aberration could lead to the formation of a clone characterized by more than one fusion gene. Thus, in some cases, detection of unexpected fusion gene transcripts might reflect chromosomal instability and clonal evolution. Such a hypothesis seems to be supported by the identification of cryptic balanced translocations leading to EWSR1-ERG and EWSR1-FLI1 fusions in Wilms' tumors by high-resolution molecular cytogenetic analysis.[276] The finding of the reciprocal t(11;22)(q24;q12) translocation in a short culture from a mesenchymal chondrosarcoma of the bone could also reflect a similar phenomenon.[277] Introduction of ultra-deep NGS might help to better understand the frequency and nature of clonal heterogeneity.

The unexpected PCR products of fusion gene amplification could also originate from normal cells. The JAZF1-SUZ12 fusion gene transcripts, commonly seen in human endometrial stromal tumors because of the t(7;17) translocation, have been shown to mimic RNA from physiologically regulated trans-splicing between precursor messenger RNAs for JAZF1 and SUZ12. Identical sequences have been shown for both chimeric RNA and trans-spliced mRNA. This finding raises the concern that other fusion gene transcripts could mimic normal products resulting from the trans-splicing of pre-mRNA. Detection of such products using highly sensitive PCR-based assays can infer the presence of gene rearrangements in fusion-gene-free samples.[278] Finally, histologic misclassification might contribute to some sporadic and unexpected genetic findings.

Global gene expression profiling has been successfully used to search for new fusion gene transcripts.[143,279] The custom-made oligomicroarray with all known sarcoma fusion gene transcripts was designed to detect all possible exon–exon breakpoints between fusion genes in a single analysis.[280,281] Other methods, like the RT-PCR-microarray pipeline or the GeneChip exon array screening, have been developed to search for known or new fusion gene transcripts.[282,283] However, an advance in so-called NGS technologies suggests that gene expression profiling is unlikely to be developed into commonly used clinical assay.

Next-generation sequencing of whole-genome tumor DNA has been successfully used to identify unknown fusion genes, for example, recently NAB2-STAT6 fusion gene, in solitary fibrous tumor.[253] However, the majority of the NGS-based studies searching for new fusion genes employed a transcriptome sequencing (RNA-seq) strategy.[166] The advantage of the RNA-based NGS strategy is the ability to identify fusion gene transcript variants and the level of fusion gene transcript expression. Yet, chromosomal rearrangements causing replacement of regulatory sequences cannot be detected using this approach; such gene fusions represent a relatively small fraction of all known gene fusions in soft tissue tumors. Moreover, chromosomal rearrangements silencing involved genes may not be detected by the NGS RNA-seq.[284]

Although fusion genes and their transcripts were once considered to be tumor specific, there is growing evidence that in some instances identical fusion genes occur in morphologically different soft tissue tumors. Furthermore, the same translocations and fusion genes could be found in tumors of other than mesenchymal cellular lineages. Table 5.3 shows multiple examples of similar fusion genes identified in different tumors. Nevertheless, the detection of fusion genes and their transcripts still remains diagnostically important when analyzed in a morphologic and clinical context. Also, understanding of the fusion gene pathological

Chapter 5: Molecular genetics of soft tissue tumors

Table 5.3 A wide variety of pathologic fusion genes identified in tumors derived from different lineages

Gene	Fusion partner	Tumor	Lineage
ALK	ATIC, CLTCL1, MSN, MYH9, NPM1, RNF213, TFG, TPM3, TPM4	Anaplastic large cell lymphoma[288–295]	H
	ELM4	Breast cancer[296]	E
	C2orf44, ELM4	Colorectal cancer[296,297]	E
	CLTC, NPM1, SEC31A, SQSTM1	Diffuse large B-cell lymphoma[298–301]	H
	TPM4	Esophageal squamous cell carcinoma[302]	E
	ATIC, CARS, CLTC, DCTN1, LNIN, PPFIBP1, PRKAR1A, RANBP2, SEC31A, TFG, TPM3, TPM4	Inflammatory myofibroblastic tumor[199–205]	M
	VCL	Medullary carcinoma of kidney[303]	E
	RANBP2	Myeloproliferative disorders[304]	H
	EML4, KIF5B, KLC1, TFG	Non-small-cell lung cancer[295,305–307]	E
	FN1	Ovarian stromal sarcoma[308]	M
	TPM3	Systematic histiocytosis[309]	H
BCOR	RARA	Acute promyelocytic leukemia[310]	H
	ZC3H7B	Endometrial stromal sarcoma[170]	M
	CCNB3	Ewing sarcoma-like[189]	M
CIC	DUX4, FOXO4	Ewing sarcoma-like[187,188,190]	M
ETV6	NTRK3	Acute leukemia[311,312]	H
	NTRK3	Congenital mesoblastic nephroma/infantile fibrosarcoma[191,192]	M
	NTRK3	Secretory breast carcinoma[313]	E
	NTRK3	Mammary analog secretory carcinoma of salivary gland[314]	E
	NTRK3	Radiation-associated thyroid cancer[315]	E
EWSR1	ATF1, CREB1	Angiomatoid fibrous histiocytoma[128,130]	M
	ATF1	Clear cell odontogenic carcinoma[316]	E
	ATF1, CREB1	Clear cell sarcoma[145–149]	M
	ATF1, CREB1	Clear cell sarcoma-like tumor of GI tract[150]	M
	ERG, WT1	Desmoplastic small round cell tumor[159–163]	M
	FLI1, ERG, ETV1, ETV4, FEV, ZNF278, SMARCA5, SP3	Ewing sarcoma[162–182,186]	M
	NFATC2	Ewing sarcoma-like malignant round cell tumor of bone[185]	M
	NFATC1	Hemangioma, bone[317]	M
	POU5F1	Hidradenoma, skin[318]	E
	ATF1	Hyalinizing clear cell carcinoma, salivary gland[319,320]	E
	CREB3L1, CREB3L2	Low grade fibromyxoid sarcoma	M
		Sclerosing epithelioid fibrosarcoma[230–234]	M
	CREM	Malignant melanoma[132]	M
	YY1	Mesothelioma[321]	E
	ATF1, ZNF393, PBX1, POU5F1, ZNF44	Myoepithelioma, bone and soft tissue[235–237,239]	E/M
	NR4A3	Myxoid chondrosarcoma, extraskeletal[136,137]	M
	DDIT3	Myxoid liposarcoma[225–228]	M
	CREB1	Primary pulmonary myxoid sarcoma[247]	M
	POU5F1	Undifferentiated bone sarcoma[322]	M
FUS	ERG	Acute myeloid leukemia[323]	H
	ATF1	Angiomatoid fibrous histiocytoma[128,129]	M
	ERG, FEV	Ewing sarcoma[183,184]	M
	CREB3L2, CREB3L1	Low-grade fibromyxoid sarcoma[232,233]	M
	NR4A3	Myxoid chondrosarcoma, extraskeletal[142]	M
	DDIT3	Myxoid liposarcoma[225–229]	M
	ZNF393, POU5F1	Myoepithelioma, bone[239,324]	M
HMGA2	Unknown	Aggressive angiomyxoma[120-123]	M
	CXCR7, EBF1, LHFP, LPP, NFIB, PPAP2B	Lipoma[222–224]	M
	FHIT	Pleomorphic adenoma parotid gland[325]	E
	LPP	Pulmonary chondroid hamartoma[326]	M
	ALDH2, CCNB1IP1, COX6C, COG5, RAD51L1	Uterine leiomyoma[208–213]	M

Table 5.3 (cont.)

Gene	Fusion partner	Tumor	Lineage
NCOA2	ETV6, KAT6A	Acute/biphenotypic leukemia[327]	H
	PAX3	Alveolar rhabdomyosarcoma[247–250]	M
	AHRR, GTF2I	Angiofibroma[126,127]	M
	HEY1	Mesenchymal chondrosarcoma[143]	M
	SRF, TEAD1	Spindle cell rhabdomyosarcoma[251]	M
PHF1	EPC1, JAZF1, MEAF6	Endometrial stromal sarcoma[167,170]	M
	EP400	Ossifying fibromyxoid tumor[241,242]	M
PLAG1	COL1A2, HAS2, RAD51L1	Lipoblastoma[215–218]	M
	LIFR	Myoepithelioma, soft tissue[238]	E/M
	CHCHD7, CTNNB1, LIFR, TCEA1	Pleomorphic adenoma salivary gland[328]	E
ROS1	TFG, YWHAE	Inflammatory myofibroblastic tumor[206]	M
	CEP85L	Angiosarcoma[132]	M
	CCDC6, CD74, EZR, KDELR2, LRIG3, SLC34A2, SDC4, TPM3	Cholangiocarcinoma	E
		Gastric adenocarcinoma	E
		Non-small-cell lung cancer[329]	E
TFE3	ASPSCR1	Alveolar soft part sarcoma[124,125]	M
	YAP1	Epithelioid hemangioendothelioma[195]	M
	IGFBP7	Perivascular epithelioid cell tumor (PEComa)[244]	M
	ASPSCR1, CLTC, IGFBP7, NONO, PRCC	Xp11-renal cell carcinoma[330]	E
USP6 (TRE17)	CDH11, COL1A1, OMD, THRAP3, ZNF9,	Aneurysmal bone cyst[331]	M
	MYH9	Nodular fasciitis[240,332]	M

Abbreviations: E – epithelioid, H – hematopoietic, M – mesenchymal
Gene symbols as recommended by HUGO Gene Nomenclature Committee (www.genenames.org).

signal transduction pathways provides a biological rationale for specific targeted therapies.[285–287]

The categories of the fusion genes associated with sarcomas are often defined by the oncogenic function of the fusion gene proteins.

Fusion genes involving RNA- and DNA-binding proteins: so-called FET proteins

The family of RNA- and DNA-binding proteins was named the FET protein family after genes identified in Ewing/PNET sarcoma-associated translocations: **F**US (fused in sarcoma), **E**WSR1 (Ewing sarcoma breakpoint region 1), and **T**AF15, a TATA-binding protein-associated factor 15. FET proteins are primarily located in the nucleus and highly expressed in almost all human fetal and adult tissues. Members of this family are characterized by a putative RNA-binding domain at the C-terminus and a glutamine-rich N-terminus and were implicated in central cellular processes, such as the regulation of gene expression, maintenance of genomic integrity, protein–RNA binding, transcription, and miRNA processing. Nearly half of fusion gene proteins identified in sarcomas contain a portion of the FET gene family protein.[333,334]

EWSR1 and FUS (previously also known as TLS) are the two most common FET family members involved in gene fusions in soft tissue tumors. Fusion gene EWSR1-FLI1 is the molecular event corresponding to the Ewing sarcoma-specific t(11;22)(q24;q12) translocation, the first chromosome translocation characterized in soft tissue tumors.[172,174,175,335]

FLI1, a EWSR1 fusion partner, belongs to the ETS family of transcription factors. This gene family is defined by a specific DNA-binding domain and flanking protein–protein interaction domains and has been indicated in different types of cancer. The ETS gene family members (ERG, ETV1, and ETV4) are commonly involved in the fusion with 5′ untranslated region of transmembrane protease, serine 2 (TMPRSS2). The androgen-responsive promoter elements of TMPRSS2 mediate oncogenic overexpression of ETS genes, which seems to be critical for prostate cancer development.[336–339]

Products of FET-ETS fusion consist of an N-terminal portion of FET and a DNA-binding domain of ETS. These chimeric proteins act as aberrant transcription factors, targeting a network of genes normally modulated by the ETS family transcription factors. Furthermore, FET-ETS oncoproteins impact RNA splicing or affect other proteins by disturbing their ability to form functional complexes.[340,341]

In the Ewing sarcoma/PNET family of tumors, EWSR1-FLI1 gene fusion is the most common and occurs in 85% to 95% of cases. The second most common is the EWSR1-ERG gene fusion, detected in approximately 5% to 10% of analyzed cases. However, other rare fusions of EWSR1 and members of the ETS family, ETV1, ETV4, FEV, or other genes, such as ZNF278 (zinc finger protein 278), SMARCA5

(SWI/SNF-related, matrix-associated, actin-dependent regulator of chromatin, subfamily A, member 5), and SP3 (transcription factor SP3), have also been reported.[176–181,186]

In other than Ewing sarcoma/PNET soft tissue tumors, EWSR1 was found to be a fusion partner to the ATF-1 and to the CREB1 in angiomatoid fibrous histiocytoma,[128–131] clear cell sarcoma,[145] clear cell sarcoma-like tumor in gastrointestinal tract,[150] and primary pulmonary myxoid sarcoma,[321] to the CREB3L1 and CREB3L2 in low-grade fibromyxoid sarcoma/sclerosing epithelioid fibrosarcoma,[230–234] to the DDIT3 (C/EBP homologous protein[CHOP]) gene in myxoid liposarcoma,[225] to the NR4A3 (previously called CHN or TEC) in extraskeletal myxoid chondrosarcoma,[136] to the PBX1, POU5F1, and to the genes encoding zinc finger proteins 393 and 444 in myoepithelioma,[235–239] and to the WT1 gene in desmoplastic small round cell tumor (DSRCT).[159] Yet, EWSR1 rearrangement is not specific for soft tissue tumors. EWSR1-ATF1 fusion was reported in clear cell odontogenic carcinoma and hyalinizing clear cell carcinoma of the salivary gland,[316] while EWSR1-POU5F1 fusion occurred in skin hidradenoma.[318]

cAMP-dependent transcription factor-1 (cATF1) and cAMP-response element-binding protein (CREB1) belong to the ATF/CREB family of transcription factors. The DNA binding of these proteins is mediated by means of its basic leucine zipper (bZIP) domain.[342] EWSR1-ATF1 and EWSR1-CREB1 proteins consist of strong EWSR1 transcriptional activation domains fused to the ATF1 or CREB1 DNA-binding domain. These chimeric proteins escape cAMP control and ultimately dysregulate the expression of various genes normally controlled by cAMP.[343] CREB3L2 (cAMP response element-binding protein 3 like 2) is a member of the old astrocyte specifically induced substance (OASIS) DNA binding and basic leucine zipper dimerization (bZIP) family of transcription factors. These transcription factors can dimerize with each other, but not with transcription factors belonging to other bZIP families.[344]

Wilms' tumor-suppressor gene (WT1), a Krüppel-like transcription factor encodes four Cys2His2 zinc fingers in the C-terminal and N-terminal domains, the latter containing both transcriptional activation and repression domains.[345] Although this gene is biallelically inactivated in a subset of Wilms' tumors, in DSRCT it is altered by reciprocal t(11;22)(p13;q12) translocation involving the EWSR1 gene. In this oncogenic translocation, the N-terminal domain of the EWSR gene is fused to the carboxyterminus (last three zinc fingers) of the WT1 gene. The EWSR1-WT1 fusion gene product is represented by two isoforms, EWS/WT1(−KTS) and EWS/WT1(+KTS). The latter is created due to alternative splicing with insertion of three amino acids: lysine, threonine, and serine (KTS) in the linker region between zinc fingers 3 and 4.[162] Introduction of EWS/WT1(−KTS) into NIH3T3 cells causes their tumorigenic transformation, whereas EWS/WT1(+KTS) shows no transformative potential.[346] However, the latter contributes to the DSRCT invasive phenotype by induction of LRRC15 expression.[347]

Nuclear receptor subfamily 4, group A, member 3 (NR4A3), initially designated as neuron-derived orphan receptor-1 (NOR1), belongs to the steroid/thyroid nuclear receptor gene superfamily.[348] The EWSR1-NR4A3 gene fusion protein, also called a EWS-CHN chimeric protein, consists of the EWSR1 transcriptional activation domain fused to the NR4A3 DNA-binding domain. This chimeric protein is acting as a potent transcriptional activator;[136] however, it can also affect pre-mRNA splicing. Modifications of RNA metabolism, leading to the instability of pre-mRNA splicing, might contribute to the process of oncogenesis.[349] In approximately 25% of extraskeletal myxoid chondrosarcoma, EWSR1 is substituted in the fusion with TAF15 (also known as TAFII68) or TCF12.[137–139] EWSR1 and TAF15 share an extensive sequence homology,[140] whereas TCF12 encodes the helix-loop-helix transcription factor 4.[350] TRK-fused gene (TFG) has been identified as a fusion partner of NR4A3 in one case.[141] TFG encodes a predicted 400-amino acid protein with a putative N-terminal coiled-coil region and was found by screening for genes that have similar N-terminal regions to EWSR1 and FUS.[351] TFG is an activating partner in oncogenic translocations involving neurotrophic tyrosine kinase-1 receptor (NTRK1) in thyroid papillary carcinoma and ALK in anaplastic large cell lymphoma.[293,352]

Pre-B-cell leukemia transcription factor 1 (PBX1) is involved in t(1;19)(q23;p13.3), one of the most common translocations in childhood and adult acute lymphoblastic leukemia (ALLs). This translocation usually results in TCF3 (previously E2A) and PBX1 gene fusion, in which the E2A DNA-binding domain is replaced by a putative DNA-binding domain of PBX1.[353] The NIH3T3 cells, transfected with TCF-PBX1 cDNAs encoding the fusion proteins, were able to cause malignant tumors in nude mice.[354] EWSR1-PBX1 gene fusion has been identified in a myoepithelioma that had a balanced translocation t(1;22)(q23;q12) as the sole karyotypic change. Based on the structural similarities between EWSR1-PBX1 and other EWSR1 or PBX1 gene fusions, it appears that the EWSR1-PBX1 chimeric protein could have oncogenic activity.[235]

POU5F1 is a sequence-specific transcription factor that is essential for keeping germ cells and embryonic stem cells in an immature and pluripotent status. The identified transcript in an undifferentiated sarcoma is predicted to encode a chimeric protein consisting of the EWSR1 amino-terminal domain and the POU5F1 carboxy-terminal domain. EWSR1-POU5F1 acts as an oncogenic transcription factor.[322] It has been suggested that the expression of this chimeric protein might contribute to undifferentiated and immature phenotypes; however, the presence of a similar EWSR1-POU5F1 gene fusion in two types of epithelial tumors, hidradenoma and mucoepidermoid carcinoma of the salivary glands, undermines this hypothesis.[318,322]

EWSR1-DDIT3 (also known as EWS-CHOP) fusion was found in myxoid liposarcomas with t(12;22)(q13;q12), although a t(12;16)(q13;p11) chromosome translocation leading to FUS-DDIT3 (FUS-CHOP) gene fusion is a molecular hallmark for this tumor.[355,356] In both cases, the N-terminus of FUS or EWSR1 is fused to DDIT3.[225,228] The expression of FUS-DDIT3 or EWSR1-DDIT3 in murine cells revealed transforming activity. The oncogenic effect was related to the FUS or EWSR1 component of the fusion protein and could not be fully substituted by fusion of other activators with the DDIT3.[357,358] In nude mice, FUS-DDIT3-induced tumors were analogous to human myxoid liposarcomas.[357]

FUS-ERG or FUS-FEV fusion genes have been identified in Ewing sarcomas.[184] Thus, EWSR1 and FUS are interchangeable when forming fusions with ETS transcription factors. Also, the same t(16;21)(p11;q22) chromosome translocation leading to the FUS and ERG gene fusion was reported in a subset of acute myeloid leukemia.[323,359]

FUS (for fusion), previously called TLS (for translocated in liposarcoma), encodes a nuclear RNA-binding protein highly homologous to EWSR1.[360] DDIT3 (DNA damage-inducible transcript 3) is a member of the leucine zipper transcription factor family[226,227] and regulates adipocyte differentiation and growth arrest.[361,362] FUS-DDIT3 fusion consists of the N-terminus of FUS and DNA binding and leucine zipper dimerization domain of DDIT3.[226,227] The FUS-DDIT3 oncoprotein can inhibit adipocyte differentiation.[363]

FUS gene is also a fusion partner to the ATF1 and the CREB1 genes in angiomatoid fibrous histiocytoma[131] and to CREB3L1 and CREB3L2 in low-grade fibromyxoid sarcoma.[232,234] CREB3L1 proteins belong to the same basic leucine-zipper family of transcription factors and display extensive sequence homology in their DNA-binding domains. Thus, FUS-CREB3L1 is thought to promote tumorigenesis through pathways similar to that of FUS-CREB3L2.[233] In the FUS-CREB3L2 chimeric protein, the CREB3L2 bZIP-encoding domain is controlled by the FUS promoter, which causes deregulation of genes normally controlled by CREB3L2.[364]

Fusion genes involving receptor tyrosine kinase genes

Tyrosine kinases (TKs) are important regulators of signal transduction pathways. More than 90 genes in the human genome encode either receptor or cytoplasmic nonreceptor TKs.[365] Receptor TKs are cell surface receptors for growth factors, cytokines, and hormones. Their activation regulates many cellular functions, especially cell proliferation and apoptosis. Structural or functional alterations of receptor TKs result in dominant oncogenic activity, causing malignant transformation in hematopoietic, mesenchymal, and epithelial cells.[366] In soft tissue tumors, several receptor TKs, such as ALK, KIT, and NTRK3, have been shown to act as powerful oncogenes. Targeting the pathologically activated receptor TKs with TK inhibitors is a new targeted therapy in the treatment of sarcoma.[287,367,368]

Anaplastic lymphoma kinase (ALK) gene

The anaplastic lymphoma kinase (ALK) gene was first identified as a fusion partner of the NPM1 (nucleophosmin) gene in t(2;5)(p23;q35), the main chromosomal translocation in anaplastic large cell lymphoma.[292] The ALK gene, mapped to 2p23, encodes for a transmembrane receptor TK protein and has a structure typical of receptor TK, with a large extracellular domain, a transmembrane segment, and a TK domain. NPM, mapped to 5q35, encodes for a 38-kD nucleolar phosphoprotein.[292]

The NPM-ALK fusion protein consists of an N-terminus of NPM, thus containing an oligomerization motif and a cytoplasmic tyrosine kinase domain of ALK. The NPM-mediated oligomerization leads to the constitutive activation of a NPM-ALK TK function. A strong NPM promoter results in a high level of expression of NPM-ALK.[369,370] This chimeric protein can transform rodent fibroblasts and bone marrow cells.[371,372] Retrovirus-mediated gene transfer of NPM-ALK also causes lymphoid malignancy in mice.[373] Furthermore, NPM-ALK transgenic mice spontaneously develop T-cell lymphomas and plasma cell tumors.[374]

In inflammatory myofibroblastic tumors (IMT), TPM3 and TPM4 are the most common translocation partners for ALK, although several other partner genes have been reported. Similar ALK fusion genes and chimeric proteins have been identified in different types of carcinoma, hematopoetic disorder, and sarcoma (examples are listed in Table 5.2). All ALK chimeric proteins, encoded by fusion genes, are structurally similar to the NPM-ALK protein. They consist of an N-terminal dimerization or an oligomerization domain of a strongly expressed ALK fusion partner and of a C-terminal TK domain of ALK. This causes the consistent activation of ALK TK activity. The effects of ALK fusion protein variants, however, vary with regard to cell proliferation rate, colony formation in soft agar, invasion, migration through the endothelial barrier, and tumorigenicity. These differences could be due to differential activation of various signaling pathways by the ALK gene fusion protein variants. The mechanisms of ALK-mediated oncogenesis are not completely understood, although there is strong evidence that ALK fusion chimeric proteins activate multiple signaling pathways, including RAS/RAF/MEK/ERK cell proliferation and JAK/STAT cell survival pathways.[375,376]

A multitargeted TK inhibitor crizotinib has been successfully used for treatment of ALK-rearranged tumors.[377] IMTs with an ALK fusion gene showed response to crizotinib treatment, while no response to crizotinib was seen in the ALK gene fusion negative case.[378,379] A recent study showed that some ALK fusion gene negative IMTs are driven by oncogenic gene fusions involving ROS1, a receptor TK that is closely related to ALK.[206,329]

Neurotrophic tyrosine kinase receptor type 3 (NTRK3) gene

NTRK3 (neurotrophic tyrosine kinase receptor type 3) belongs to the family of nerve growth factor receptors, which play an important role in regulating the development of both the central and the peripheral nervous systems.[380]

ETV6-NTRK3 fusion gene, formed by t(12;15)(p13;q25), was identified in infantile fibrosarcoma and congenital mesoblastic nephroma. Thus, morphologic and genetic similarities of these tumors suggest that they belong to a single neoplastic entity.[191,192]

ETV6, ETS variant gene 6 (also known as TEL), is a member of the ETS family of transcription factors. ETV6 alterations, including fusion gene formations with multiple partners, deletions, and point mutations, and other possible alterations at the promoter level, have been reported in human leukemias and are thought to contribute to leukemogenic processes.[311,312] The ETV6-NTRK3 fusion gene has also been found in secretory breast carcinoma, a rare subtype of infiltrating ductal carcinoma and secretory carcinoma of the salivary gland.[313,314] Retroviral transfer of ETV6-NTRK3 into murine mammary epithelial cells leads to tumor formation in nude mice, confirming that the ETV6-NTRK3 transfection is compatible with epithelial tumorigenesis.[313] The presence of similar gene fusions in tumors derived from different cell lineages, including mesenchymal, epithelioid, and hematopoietic, suggests that the oncogenic activation of receptor RTKs and deregulation of RTKs signaling pathways could represent a universal oncogenic mechanism.[369,381]

The ETV6-NTRK3 chimeric transcript consists of a helix-loop-helix dimerization domain of ETV6 fused to the protein tyrosine kinase domain of NTRK3. The ETV6-NTRK3 oncoprotein expressed in NIH3T3 cells deregulates NTRK3 signaling pathways.[382,383] This chimeric protein functions as a constitutively active TK, activating both the RAS/MAPK and PI3K/AKT pathways.[384]

Midostaurin (PKC412), a multitarget TK inhibitor, suppresses ETV6-NTRK3 downstream signaling, leading to inhibition of cell proliferation and induction of apoptosis. This molecule might be used for the treatment of patients with tumors driven by ETV6-NTRK3 oncoprotein.[385]

Fusion genes involving chromatin-remodeling genes

SS18-SSX fusion gene in synovial sarcoma and a variety of fusion genes identified in endometrial stromal sarcoma and ossifying fibromyxoid tumor belong to this category (Table 5.2).

Chromosomal translocation t(X;18)(p16;q11) was first described in 1986 as a cytogenetic observation.[386] t(X;18) and its rare variants are synovial sarcoma specific and have not been reported in other cancers.[387–389] This translocation involves the synovial sarcoma translocation chromosome 18 gene (SS18) and one of the synovial sarcoma X breakpoint genes (SSX), leading to the functional SS18-SSX fusion gene.[390]

The SSX gene family consists of nine members,[391] of which SSX1 and SSX2 are common fusion partners with SS18.[255,392] A few cases with SS18-SSX4 fusions have also been reported, however.[256,393] SSX1 and SSX2 genomic sequences and their encoded proteins are highly homologous. Thus, SS18-SSX1 and SS18-SSX2 fusion junctions are almost identical. Heterogeneity within the fusion junctions is rare, although variants of nonrecurrent fusions with different junctions and with the insertions of additional genetic material have been described.[255,392,394]

Expression of SSX genes is restricted to the testis and therefore the corresponding proteins are referred to as cancer/testis antigens.[255,395,396] The SSX proteins possess two transcriptional repressor domains, a Krüppel-associated box and a SSX repressor domain. These proteins function as transcriptional corepressors and are associated with the polycomb complex. Both SS18 and SSX products and SS18-SSX chimeric proteins are localized in the nucleus but lack DNA-binding domains. The SS18 gene (previously called SYT) is ubiquitously expressed in a wide range of human tissues and encodes for the protein that contains a transcriptional activating domain rich in glycine, proline, glutamine, and tyrosine (QPGY domain). The SS18 protein interacts with a putative transcriptional factor, AF10, an acetyltransferase p300, a component of histone deacetylase complex mSin3A, and a component of SWI/SNF chromatin remodeling complexes BRM and BRG1.[397–402] Both SS18 protein and SS18-SSX oncoprotein show high-affinity binding to the core subunits of SWI/SNF. Also, expression of the SS18-SSX chimeric protein induces depletion of the SMARCB1 subunit from SWI/SNF. SMARCB1 and BRG1 inactivation has been implicated in the regulation of cellular proliferation.[403] Thus, expression of SS18-SSX hampers normal SWI/SNF functions.[404–406]

The tumor suppressor early growth response 1 (EGR1) is repressed by the SS18-SSX oncoprotein through a direct association with the EGR1 promoter. This SS18-SSX binding correlates with the trimethylation of Lys (27) of histone H3 (H3K27-M3) and the recruitment of the so-called polycomb group proteins to this promoter. Thus, polycomb-mediated epigenetic gene repression is another mechanism of oncogenesis in the synovial sarcoma.[407]

In general, primary synovial sarcoma is characterized by a "flat" genomic copy number profile. However, recurrent and metastatic lesions show additional genetic changes. Mutations affecting APC and CTNNB1 were identified in a subset of cases, suggesting activation of the Wnt pathways.[408] In fact, SS18-SSX itself can induce nuclear beta-catenin accumulation and activation of Wnt-beta-catenin signaling.[409] The gene expression profiling studies revealed a high expression of RTK in synovial sarcoma, including PDGFRA, EGFR, and ERBB2.[410] Although PTEN and PIK3CA mutations were found only in a few cases, activation of the AKT/mTOR pathway has been documented by several studies.[411] Thus, inhibition of the Wnt-beta-catenin and AKT-mTOR pathways appears to be a therapeutic target in synovial sarcoma.[412]

COL1A1-PDGFB, a fusion gene involving growth factor

A t(17;22)(q22;q13.1) translocation or supernumerary ring chromosome derived from t(17;22) are characteristic features of dermatofibrosarcoma protuberans (DFSP).[413] As a result of this translocation, the COL1A1 gene, mapped to 17q22, encoding component of type I collagen, and the platelet-derived growth factor, beta chain (PDGFB) gene, a human homolog of the v-sis oncogene, are rearranged and fused.[153] The COL1A1-PDGFB fusion gene results in excessive production of the chimeric COL1A1-PDGFB fusion protein and leads to autocrine growth stimulation through the PDGF receptor. The overexpression of PDGFB increases the growth rate, is transforming for transfected cells, and is tumorigenic in nude mice.[154,155] Imatinib mesylate, a tyrosine kinase inhibitor, inhibits the growth of DFSP cell cultures and mouse tumors, apparently by its ability to inhibit PDGFR.[414] Clinical trial evidence suggests some effectiveness of imatinib mesylate in the treatment of recurrent or metastatic DFSP.[415,416]

Other types of fusion genes
PAX-FKHR fusion genes

Alveolar rhabdomyosarcoma is characterized by two tumor-specific chromosomal translocations: t(2;13)(q35;q14) and t(1;13)(p36;q14). These translocations result in fusions of the PAX3 or PAX7 genes, mapped to 2q35 and 1p36, with the FKHR (FOXO1A) gene mapped to 13q14.[417,418]

PAX3 and PAX7 genes are members of the PAX (paired box) transcription factor gene family and show high structural and sequence similarities. PAX genes are involved in embryogenesis and the expression of murine PAX3 and PAX7 homologs was documented during the development of nervous and muscular systems.[418–420] PAX3 and PAX7 regulate functions of the muscle progenitor cells, satellite cells, which lie under the basal lamina of muscle fibers. PAX genes control the entry of these cells into the myogenic program by the activation of the myogenic determination gene, MYOD1. Also, PAX7 is essential for the survival of satellite cells.[421] Loss-of-function PAX3 mutations have been implicated in congenital neural and muscular anomalies, both in mouse (splotch mouse) and human (Waardenburg type I to III and craniofacial-deafness-hand) malformation syndromes.[422–425]

FOXO1A (previously called FKHR [forkhead-related]) belongs to one of the FOX gene subfamilies, which encode transcription factors with a highly conserved DNA-binding motif, related to the *Drosophila* region-specific homeotic gene "forkhead."[426,427] FOXO1A and other subfamily members, such as FOXO4 and FOXO3A (respectively known as MLLT7 or AFX and FKHRL1), are conserved beyond species and regulated by an insulin signaling pathway.[428–430] Forkhead transcription factors encoded by these genes have a role in metabolic cell proliferation and apoptosis signaling pathways.[431] Besides FOXO1A involvement in PAX gene fusions, two other members, FOXO3A and FOXO4, are fused to the KMT2A (previously known as MLL) gene in some of the acute myeloid leukemias.[432,433]

PAX3-FOXO1A and PAX7-FOXO1A fusion genes contain highly homologous PAX3- and PAX7-derived N-terminal DNA-binding domains fused to the activating, COOH-terminal domain of the FOXO1A gene. Expression studies in alveolar rhabdomyosarcomas showed that fusion PAX proteins are overexpressed when compared with wild-type PAX. Overexpression of PAX7-FOXO1A results from a secondary genetic event, the amplification of the PAX7-FOXO1A fusion gene. Compared with the wild-type PAX3, the PAX3-FOXO1A is thought to result in a higher transcriptional rate of the pathologic fusion gene.[434–437] Furthermore, this fusion protein is resistant to pathways controlling FOXO1A expression.[438]

The PAX3-FOXO1A fusion has been shown to inhibit myogenic differentiation of murine myoblasts and to induce oncogenic transformation of NIH3T3.[439–441] It exerts growth suppression and cell death in immortalized, non-transformed cells, however.[442] This phenomenon has been seen for other types of oncoproteins, including EWSR1-FLI1 in Ewing sarcoma. Growth suppression and cell death can be attenuated by the absence of tumor suppressors, however, as seen in p53, p16, and p19, or in activation of the RAS signaling pathway.[443,444] PAX3-FOXO1A mice with disrupted RB and p53 pathways developed ARMS with a higher frequency.[445]

Oncogenic PAX3-FOXO1A and PAX7-FOXO1A fusion proteins act as strong transcriptional activators of genes with PAX3 and PAX7 DNA-binding sites. The expression profile of cell lines with an ectopically expressed PAX3-FOXO1A has resulted in the identification of a significant number of downstream targets of the PAX3 fusion oncoprotein, although their contribution to tumor development is not clear. Also, downregulation or upregulation of specific genes depends on the type of cell line in which the fusion gene was ectopically expressed. It is thus unclear if any given ectopic cell culture system mimics gene expression of alveolar rhabdomyosarcoma. Nevertheless, targets identified by multiple studies included among others P-cadherin (CDH3), cannabinoid receptor 1 (CNR1), CXC motif, chemokine receptor type 4 (CXCR4), fibroblast growth factor receptor 4 (FGFR4) mesenchymal-epithelial transition factor (MET), and MYCN protooncogene.[446,447] Note, data obtained from the gene expression profiling of tumor samples must be taken cautiously because of a low concordance between such studies. CNR1 was also identified as a top gene of the alveolar rhabdomyosarcoma gene expression profile.[448] This gene is regulated by both wild-type and fusion PAX3 proteins.[446] MET and CXCR4 encode cell surface receptors. Activation of MET and CXCR4 signaling pathways could impact ARMS metastatic behavior.[449] In addition, many genes identified in gene transfer studies are known targets of STAT3, which is involved in a protein–protein interaction with PAX3-FOXO1A.[450]

Amplification and overexpression of the miR-17–92 cluster of miRNAs has been identified in PAX7-FOXO1A, but not in PAX3-FOXO1A tumors and might contribute to a more aggressive phenotype.[451]

ASPSCR1-TFE3 fusion gene

A t(X;17)(p11.2;q25) chromosomal translocation is typical of alveolar soft part sarcoma. This translocation fuses the ASPSCR1 gene (at 17q25) in-frame to the TFE3 gene (at Xp11.2).[124,125] The ASPSCR1-TFE3 chimeric proteins consist of the N-terminus of ASPSCR1 and basic helix-loop-helix, leucine zipper DNA binding, and multimerization domains of TFE3. A high level of ASPSCR1-TFE3 expression from constitutive activation of the ASPL promoter causes transcriptional deregulation of downstream target genes.[452]

ASPSCR1 (alveolar soft part sarcoma chromosome region, candidate 1), alternatively termed ASPL (alveolar soft part sarcoma locus) or TUG (Tether-containing UBX domain for GLUT4 [glucose transporter type 4]), encodes a putative protein of 553 amino acids and is ubiquitously expressed in normal adult tissues, though heart and skeletal muscles display the highest ASPSCR1 expression.[452] ASPSCR1 functions as a tether interacting with GLUT4 (currently called SLC2A4) and cellular/organellar membranes.[453]

TFE3 (transcription factor for immunoglobulin heavy-chain enhancer 3) is a member of the microphthalmia transcription factor/transcription factor E (MITF-TFE) family of basic helix-loop-helix/leucine-zipper transcription factor genes. TFE3 interacts with transcriptional regulators such as E2F3 or SMAD3 and plays a role in cell growth and proliferation.[454]

In papillary renal cell carcinoma, TFE3 and TFEB (T-cell transportation factor EB), another member of the MITF gene family, form fusion genes with several partners, including PRCC, SFPQ, NONO, CLTC, and ASPSCR1. Thus, some papillary renal cell carcinomas share the translocation with alveolar soft part sarcoma.[452,455–459]

ASPL-, PSF-, and NONO-TFE3 chimeric proteins bind to MET and strongly activate its promoter, resulting in MET autophosphorylation and activation of downstream signaling in the presence of the MET ligand, hepatocyte growth factor. Aberrant transcriptional upregulation of MET by oncogenic TFE3 fusion proteins is another mechanism for activating MET signaling.[460] A selective inhibition of the MET receptor tyrosine kinase signaling pathway with ARQ 197 in tumors bearing fusion genes with TFE3 is currently under clinical testing.[461]

Oncogene amplifications

DNA amplification is defined as the gain of genetic material in a defined region, often called an *amplicon*. In general, the multiplication of genes located in the amplicon leads to the overexpression of their protein products. Relatively high gene expressions can occur without an increased copy number, however, and a lack of or low gene expression can be found in tumors with multiple gene copy numbers. The latter could be related to epigenetic upregulation, whereas the former could result from a nonproductive gain of genetic material. Amplified DNA is associated with two cytogenetically detectable abnormalities, extrachromosomal double minutes (dmins) and intrachromosomal, homogenous staining regions (hsrs). *Dmins* and *hsrs* are interchangeable forms of amplified DNA. Hsrs can become detached and are excised from the chromosome to form dmins. The latter can be integrated into another chromosome and form hsrs.[2]

The gain of genetic material/gene amplification can be a primary, early molecular event or a secondary genetic change acquired during tumor progression. The gain/amplification of PAX7-FOXO1A in alveolar rhabdomyosarcoma and COL1A1-PDGFB in DFSP with fibrosarcomatous transformation are such secondary changes, elevating the oncogenic potential of the fusion gene.[434,462,463]

Amplicons can be demonstrated by a spectrum of cytogenetic and molecular genetic techniques. These include classic karyotyping, FISH, brightfield *in situ* hybridization (BISH), comparative genomic hybridization (CGH), array CGH and single nucleotide polymorphism (SNP) arrays, and real-time PCR (quantitative PCR) assays,[464–468] although NGS, which provides the unique opportunity to evaluate DNA sequence and copy number variations of multiple genes/loci simultaneously, would most likely replace these technologies in both research and clinical practice.[469]

The MYC gene was the first cellular oncogene shown to be amplified in human cancer.[470,471] The MYC gene family (helix-loop-helix leucine zipper class of transcription factors) includes MYC (8q24), MYCN (2p24), and MYCL (1p34). All MYC family genes stimulate cell growth and proliferation, prevent terminal cell differentiation, and have been indicated in the development of various types of tumors when deregulated.[472,473] Amplification of MYC was found in primary aggressive and metastatic tumors, including high-grade sarcomas.[474,475] High level MYC amplification is a distinctive feature of secondary angiosarcomas developed after irradiation and/or chronic lymphedema.[476] Other radiation-induced sarcomas, including undifferentiated pleomorphic sarcoma and leiomyosarcoma, show low-level MYC amplifications and atypical vascular lesions lack MYC amplification.[477] More recently, MYC amplification was also reported in a subset of primary hepatic and cutaneous angiosarcomas.[478,479] MYC amplification upregulates the miR-17–92 cluster and subsequently downregulates THBS1, a potent endogenous inhibitor of angiogenesis.[478]

High-level MYCN amplification and overexpression occurs in childhood neuroblastoma cell lines and aggressive neuroblastomas with a poor clinical outcome.[480,481] The evaluation of MYCN amplification status is considered clinically important for the design of optimal therapy in neuroblastoma.[482] The ALK locus, which is located 13.2 megabases (Mb) centromeric to the MYCN locus, has been shown to be a target of copy number gain in neuroblastoma. However, ALK and MYCN loci were amplified in separated amplicons. Similarly to MYCN, the ALK copy number status is highly associated with malignant phenotypes and adverse clinical outcomes. Also, the identification of oncogenic ALK kinase mutations strongly implicates ALK, not MYCN, as the critical player in neuroblastoma pathogenesis.[483–487] A mutated ALK protein was hyperphosphorylated and displayed constitutive kinase activity. The cells transfected with ALK mutants showed a transforming capacity *in vitro* and were able to develop tumors in nude mice. Furthermore, the knockdown of the mutant ALK expression caused reduced cell proliferation and apoptosis.[483-487] Thus, inhibition of the pathologically activated ALK receptor TK with ALK-specific kinase inhibitors appears as an attractive therapeutic target.[488,489] Therefore, therapeutic strategies targeting ALK oncoproteins are already in early phase of clinical testing.[490] ALK gene amplification was also identified in some leiomyosarcomas, rhabdomyosarcomas, and MFHs, so alteration of this gene could contribute to the development and/or progression of other soft tissue tumors.[491] MYCN amplification and overexpression have been identified in both embryonal and alveolar rhabdomyosarcoma. The copy number and expression levels are significantly higher in the alveolar type, however. MYCN amplification has also indicated an adverse clinical outcome in alveolar rhabdomyosarcoma.[492]

MYCL amplification has not been seen in sarcoma except for a case of highly malignant, atypical EWSR1-FLI1 Ewing sarcoma.[493]

Amplification and rearrangement of the 12q13–15 region often occurs in benign and malignant soft tissue tumors.[2,494] Several genes, including GLI, MDM2, DDIT3, CDK4, TSPAN31 (previously SAS), and HMGA2, have been mapped in this region. The molecular structure of the amplicon is complex and still not completely understood.[495] Two amplification units, one including CDK4/SAS and another MDM2, have been identified, although others could exist.[496-499] These units can either be coamplified or amplified separately.[496] In some cases, other genes (DDIT3, GLI, RAP1B, LRP1, and IFNG) located in the 12q13–15 region are coamplified, but not consistently expressed.[497]

The murine double minute (MDM2) gene is a human homolog of the evolutionarily highly conserved gene, initially identified in the 3T3DM mouse cell line.[498] The MDM2 gene encodes for a 90-kD zinc finger protein with a p53 binding site.[500] MDM2 regulates p53 either by inhibiting its nuclear transactivating function or its cytoplasmic degradation.[501] MDM2 also interacts with the retinoblastoma (RB) protein and is its critical negative regulator by promoting RB protein degradation.[502,503] The overexpression of the MDM2 oncoprotein destabilizes p53 and RB pathways and promotes cell survival and cell cycle progression, leading to deregulated cell proliferation.[503] An SNP in the MDM2 promoter, SNP309, increases the affinity of the transcriptional activator Sp1, resulting in higher levels of MDM2 RNA and protein and the disruption of the p53 pathway. This polymorphism accelerates tumor formation in both hereditary and sporadic cancers, including soft tissue sarcomas.[504-506]

Among fatty tumors, the amplification of 12q14–15 was seen almost exclusively in atypical lipomas and well-differentiated liposarcomas, with lipomas being characterized by simple structural chromosome aberrations often involving HMGA2 at 12q15.[507,508] Supernumerary ring and giant rod marker chromosomes are characteristic cytogenetic features of these tumors and carry amplified 12q13–21 sequences with multiple copies of MDM2.[509,510] Based on CGH studies, over-representation of 12q13–21 seems to be useful in separating lipoma-like liposarcomas from lipomas.[511] The lack of MDM2 immunoreactivity in lipomas confirms the diagnostic significance of this observation.[512] Moderate gains of 12q are not always associated with a malignant phenotype, however, intermediary forms could exist between lipomas and well-differentiated liposarcoma. Benign tumors showing gain of 12q14–15 with extra copies of MDM2 and CDK4 were reported to have no expression of these genes, suggesting that this amplification had no functional consequences.[513] Evaluation of MDM2 copy numbers by real-time PCR and FISH-based assays has been found to be useful for the differential diagnosis of well-differentiated liposarcomas and their histologic mimics.[514-516] MDM2 amplification is also a highly sensitive marker of dedifferentiated liposarcoma.[515,517] However, the value of MDM2 amplification analysis in differential diagnosis of dedifferentiated liposarcoma remains controversial because MDM2 amplification can occur in other types of pleomorphic sarcoma.[518] For example, in intimal sarcomas of large vessels and of the heart, amplification of MDM2 is a driving genetic alteration.[519,520]

Nutlins (-1, -2 and -3) are potent selective MDM2 inhibitors. These small molecules bind to the p53-binding pocket of MDM2 and reactivate the p53 pathway, leading to cell cycle arrest, apoptosis, and growth inhibition of human tumor xenografts in nude mice.[521] In general, Nutlins require retention of wild-type p53 and functional p53 signaling to be effective in cancer cells.[522] However, Nutlin-3 seems to disrupt p73-MDM2 binding and enhance the stability of p73, a p53 homolog transactivating proapoptotic gene. Thus, Nutlin-3 is effective in tumors with an inactivated p53 pathway and in tumors containing wild-type p53.[522,523] Also, it induces apoptosis in liposarcoma cells with MDM2 amplification.[524] Activation of the p53 pathway by antagonizing its negative regulator, MDM2, is a promising new sarcoma therapy.[287,525]

Gain-of-function mutations

Gastrointestinal stromal tumor (GIST) is the archetypical oncogenic-mutation-associated tumor driven by a KIT or PDGFRA gain-of-function mutation.

KIT proto-oncogene, the human homolog of v-kit (segment of Hardy–Zuckerman 4 feline sarcoma virus), is mapped to 4q11–q12. It encodes for a transmembrane glycoprotein of the type III receptor tyrosine kinase family.[526] Activation of KIT by its ligand, stem cell factor α, activates signal transduction pathways that include RAS/MAP kinase, RAC/RHO-JNK, PI3K/AKT, and SFK/STAT signaling networks.[527] KIT-tyrosine kinase activity is regulated by the juxtamembrane domain, which inhibits KIT kinase activity in the absence of a ligand.[528]

Gain-of-function KIT mutations typically occur in sporadic and familial GISTs and play an essential role in pathogenesis.[529,530] Most of these mutations (deletions/deletion-insertions, missense point mutations, duplications, and in rare cases, insertions) cluster in the 5′ part of the juxtamembrane domain. Less frequently, mutations involve extracellular and TK domains. These mutations lead to constitutional activation of KIT TK and promote cell survival and proliferation.[531,532] The expression of the mutant KIT elicits a transforming ability in cell lines.[529,533] Subjects with germline KIT mutations develop ICC hyperplasia and multiple GISTs.[530,534] Also, multiple GISTs have developed in transgenic mice with germline (inherited) gain-of-function KIT mutations.[535,536]

Imatinib mesylate (Gleevec/Glivec, http://www.novartis.com), which specifically inhibits ABL, KIT, and platelet-derived growth factor receptor alpha (PDGFRA) receptor TK, has been successfully used in the treatment of clinically advanced, unresectable, and metastatic GISTs. The inhibition of oncogenic KIT activity in GISTs using TK inhibitors is the first example of targeted therapy in sarcomas.[537,538] Although many patients benefit from this treatment, resistance often develops because of so-called acquired secondary KIT mutations, genomic amplification of KIT, or other incompletely defined molecular mechanisms.[539–541] Considering recent studies on intratumor heterogeneity, resistance to imatinib mesylate in GIST could develop due to the selection of imatinib-resistant clones harboring secondary KIT mutations.[542] Other multitargeted TK inhibitors have been used for treatment of imatinib-resistant GISTs.[543–545]

The PDGFRA gene is localized on chromosome 4q12 in the vicinity of the KIT locus. Both genes show high homology and have probably evolved from a duplication of a common ancestral gene.[546,547] Gain-of-function PDGFRA mutations occur in a subset of sporadic GISTs and in familial GISTs, especially those of gastric origin.[548,549] The most common mutation is the Asp842Val substitution in the second TK domain; however, other mutations clustering in the vicinity of codon 842 have been reported. Mutations in PDGFRA juxtamembrane domain are relatively rare.[534] In general, Asp842Val PDGFRA mutants are imatinib resistant.[541]

Loss-of-function mutations

Neurofibromatosis type 1 tumor-suppressor gene

NF1 is an autosomal dominant disorder caused by genetic alterations of the NF1 gene, leading to the loss of a functional NF1 protein.[98] The NF1 gene is considered a recessive tumor-suppressor gene in the sense of the Knudson's two-hit hypothesis. This gene maps to chromosome 17q11.2 and encodes a 320-kD protein, termed *neurofibromin,* which was originally demonstrated to function as a negative regulator of RAS.[550–552] Presence of the GTPase-activating protein (GAP) domain, identified within neurofibromin, is consistent with the proposed role of neurofibromin as a RAS-GAP protein.[98,553]

Loss of the NF1 protein function results in increased RAS activity and stimulates cell growth by activating PI3K/AKT/mTOR signaling and increasing RAF kinase and MEK signaling.[554] Reconstitution of the NF1 GAP-related domain in NF1-deficient cells restores normal RAS activity and reverses the growth advantage that results from NF1 inactivation.[555] Various RAS pathway inhibitors have been tested in preclinical models and clinical trials for treatment of NF1-related complications.[556,557] Although neurofibromin primarily regulates cell growth by modulating RAS activity, it also positively regulates the intracellular cAMP pathway.[558] Reduction of neurofibromin expression decreases cAMP levels affecting cell survival.[559]

Several genetically engineered mouse (GEM) models of NF1-associated tumors have been developed. GEM strains are cancer prone and have developed different types of NF1-associated tumors, including optic glioma and malignant glioma, cutaneous neurofibroma and plexiform neurofibroma, MPNST pheochromocytomas, and leukemias. These mouse models provided platforms for critical experiments on new therapeutic targets, prior to human clinical trials. In the last decade, several clinical trials with drugs targeting pathways deregulated by NF1 deficiency have been tested.[560,561]

NF1 is characterized by multiple neurofibromas. Schwann cells have been determined to be the cells of origin for neurofibroma. Although the recruitment and role of the mast cells and fibroblasts in the formation of neurofibromas in NF1 is not fully understood, emerging evidence points to it being driven by KIT expression in mast cells as crucial contributors to neurofibroma tumorigenesis.[562,563] This has led to mouse studies and, subsequently, human clinical trials with imatinib mesylate, a potent KIT receptor TK inhibitor.[564] Thus, understanding of the tumor microenvironment has an important implication for the NF1-associated tumor therapeutic strategies.

Germline NF1 mutations have been extensively documented in NF1 patients. Most of these mutations directly or indirectly yielded premature termination codons;[565] however, other mutants have also been reported. At least 5% to 20% of NF1 patients have a submicroscopic deletion spanning the entire NF1 gene.[566] The *NF1 microdeletion syndrome* is often characterized by a more severe NF1 phenotype. In particular, patients with NF1 gene microdeletion often show variable facial dysmorphism, mental retardation, developmental delay, and an excessive number of neurofibromas.[567]

A loss of heterozygosity at the NF1 locus and inactivating mutations in the NF1 gene have been shown in different types of NF1-associated malignancies.[568–574] Some studies, however, have shown a low frequency of NF1 mutations. This could be related to the cellular heterogeneity of the samples composed of neoplastic cells carrying mutated NF1 and non-neoplastic cells with NF1 in the germline configuration. Furthermore, the large size of the NF1 gene, which spans over 350 kb of genomic DNA and contains 60 exons, and the presence of several pseudogenes makes standard PCR-based screening for mutations difficult, especially in partially degraded DNA extracted from FFPE tissues.[575] The relatively high frequency of large deletions, undetectable by PCR assays, and common NF1 splicing errors, detectable only in well-preserved RNA, might contribute to this problem as well.[576] The recent application of multiple ligation-dependent probe amplification (MLPA) analysis and NGS-based methods for the molecular testing of NF1 are expected to overcome some of these problems.[577,578]

NF1 mutation data are available on the Human Gene Mutation Database (HGMD, htpp://www.uwcm.ac.uk/uwcm/mg/search/120231.html) and the Human Genome Organization (HUGO, htpp://www.nf.org/nf1gene/nf1.gene.home.html).

Neurofibromatosis type 2 tumor-suppressor gene

Neurofibromatosis type 2 (NF2) is an autosomal dominant disorder caused by loss-of-function mutations of the NF2 gene mapped to chromosome 22q12.2. Bilateral vestibular schwannomas and other hereditary brain tumors are typical diagnostic features of NF2. In these tumors, both NF2 alleles are altered by deletions or point mutations, as predicted by Knudson's two-hit hypothesis of recessive tumor-suppressor gene inactivation.[99]

Genetic alterations of NF2 consist of nonsense mutations, splice donor site mutations, deletions, and insertions, causing frameshifts and STOP codons, and subsequently leading to nonfunctional or truncated protein. Initially, inactivating mutations of NF2 were detected in approximately 50% of screened NF2 patients.[579,580] Subsequently, large deletions, duplications, and insertions affecting several exons were identified, raising the detection frequency of constitutional NF2 mutations.[581,582] However, the combination of multiple screening techniques facilitated a mutation-detection rate of 100% for the 21 inherited cases. Genetic profiles of somatic and constitutional mutations differ significantly. Somatic NF2 mutations are skewed toward frameshift (>50%), when compared with constitutional changes that are primarily nonsense and splice site.[583,584] In meningiomas, somatic mutations show a tendency to affect the 5' part of the gene, where exons 2 and 3 are considered crucial for NF2 protein tumor-suppressor functioning.[583]

The NF2 gene encodes a 595-amino acid protein called *merlin* or *schwannomin*.[585,586] Merlin is a member of the band 4.1 superfamily of cytoskeleton-associated proteins. This family consists of several membrane-associated cytoplasmic proteins, including protein 4.1, talin, ezrin, radixin, and moesin (ERM) proteins, several protein tyrosine phosphatases, and nonmuscle myosins.[587] Although merlin shares a structural similarity with ERM proteins, its cellular function is unique. Expression of merlin but not ERM proteins in NIH3T3 cells inhibits cell growth and RAS transformation.[588,589] Mice with a heterozygous knockout of NF2 develop a spectrum of malignant tumors, but there is no similar effect from the knockout of other genes encoding ERM proteins.[590,591] Merlin has been indicated as a negative regulator in multiple signaling pathways, whose activation leads to increased proliferation and survival. At the membrane, it regulates integrins mediating the Rac/PAK/JNK, mTORC1, and FAK/Src pathways, RTKs, such as EGFR, and HER2 mediating Ras/Raf/MEK/ERK, PI3K/AKT, and the Wnt/beta-catenin pathways. At the nucleus, it suppresses the E3 ubiquitin ligase CRL4^{DCAF1}.[592–594] Inhibition of pathways regulated by merlin encompass the targeted therapy of NF2-associated tumors. Recently, lapatinib (a dual inhibitor of EGFR and HER2) and everolimus (inhibitor of mTOR complex 1) have been used for treatment of progressive vestibular schwannomas in NF2 patients.[595,596]

The monosomy of chromosome 22, LOH at NF2 locus, and inactivating NF2 mutations have been documented in different types of soft tissue tumors, including mesotheliomas,[597,598] schwannomas,[599–601] and perineuriomas.[602] Reduced expression or absence of merlin is the functional consequence of mutational NF2 inactivation, and this has been shown in schwannomas and meningiomas.[603–605] Also, increased proteolytic degradation of merlin by calpain was detected in meningiomas and schwannomas without NF2 gene alterations.[606] This could explain the functional inactivation of the NF2 protein in tumors without allelic losses or mutations.

NF2 mutation databases are available online in the HGMD (htpp://www.uwcm.ac.uk/uwcm/mg/search/120232.html) and HUGO (htpp://neuro-www2mgh.harvard.edu/nf2) databases.

SMARCB1 tumor-suppressor gene

SMARCB1, also known as SNF5 or INI1, encodes the BAF47 protein, a core component of the SWI/SNF complexes.[607,608] These large protein complexes are critical regulators of transcription through ATP-dependent chromatin remodeling.[609] SWI/SNF complexes can carry out different cellular functions and impact different cellular processes, depending upon the configuration of the complex components.[607,608] SMARCB1 loss-of-function mutations indicating involvement of SWI/SNF complexes in tumorigenesis were first identified in malignant rhabdoid tumor (MRT), an early childhood cancer.[610,611] Monosomy of chromosome 22 and the loss of 22q11.2 are characteristic cytogenetic features of MRT.[612–614] Biallelic inactivation of SMARCB1 through genetic mutations, such as whole-gene deletions, frameshift mutations, nonsense and splice-site mutations, is fully consistent with Knudson's two-hit hypothesis of tumor-suppressor gene inactivation and has been documented in >90% of sporadic tumors.[614] In general, MRTs have a normal genome, with SMARCB1 alteration being the only detectable genetic change.[615] Thus, disruption of the SMARCB1 epigenetic function in this tumor largely substitutes the genomic instability often seen in other sarcomas.[616]

SMARCB1 mutations were identified in constitutional DNA (germline mutations) from families with a predisposition to develop extrarenal MRTs and tumors of the CNS, including choroid plexus carcinoma, medulloblastoma, and central primitive neuroectodermal tumor.[617–620] SMARCB1-germline mutations (splice-site, missense, and 3' untranslated region mutations, rarely seen in MRT) have been described in familial schwannomatosis (multiple intracranial, spinal, or peripheral schwannomas, without involvement of the vestibular nerve), a condition frequently associated with NF2 germline mutations.[621–623] The biallelic inactivation of SMARCB1 and NF2 genes (a four-hit mechanism) has been implicated in schwannomatosis-related tumorigenesis.[624,625] SMARCB1 mutations were also identified in families with multiple meningiomas of the falx cerebri of the cranium.[625] More recently, loss-of-function germline mutations in SMARCE1, which encodes BAF57, another component of SWI/SNF complexes, have been reported in multiple meningiomas of the spine.[626]

Although initial studies suggested that SMARCB1 inactivation is specific for MRT, subsequently SMARCB1 mutations and loss of the BAF47 protein expression has been reported in other soft tissue tumors, including epithelioid sarcoma,[627,628] EWSR1-NR4A3 fusion gene negative extraskeletal myxoid chondrosarcomas,[629] epithelioid MPNSTs,[628,630] and poorly differentiated chordomas.[631] Lack of BAF47 expression was detected in renal medullary carcinomas[632] and sinonasal carcinomas,[633] indicating a broader role of SMARCB1 in carcinogenesis. Some of these tumors, despite different pathogenesis, display prominent rhabdoid features, so the lack of BAF47 expression might be associated with a rhabdoid cell phenotype.[629] Recent studies employing NGS strategies have identified mutations in genes encoding different components of SWI/SNF complexes in a large number (>20%) of human cancers.[634]

In mice, homozygous inactivation of SNF5 resulted in early embryonic lethality, whereas animals haploinsufficient for SNF5 developed tumors strikingly similar to human MRTs.[635] In the conditional mouse model, induced inactivation of SNF5 led to 100% of the mice developing lymphomas or MRTs with a rapid median onset, indicating a critical role of SNF5 inactivation in cancer development.[636] Furthermore, inactivation of p16(INK4a) or RB does not accelerate tumor formation in SNF5 conditional mice, whereas the mutation of p53 leads to a dramatic acceleration of tumor formation.[637] Thus, alterations in the cellular functions of SWI/SNF complexes are critical in cancer initiation and progression, despite the genetic mechanisms underlying these processes not being completely elucidated yet.

Complex genetic and epigenetic mutations

High-grade spindle cell and pleomorphic sarcomas (leiomyosarcoma, liposarcoma, malignant fibrous histiocytoma, and rhabdomyosarcoma) and other pleomorphic, undifferentiated sarcomas often associated with radiation exposure carry complex genetic mutations and display intratumor heterogeneity.[638,639] Chaotic chromosomal rearrangements involving multiple loci are common and cytogenetic classification of these aneuploid tumors is practically impossible.[640] More recent studies employing new technologies offer hope for progress in this area. One study based on array-CGH and transcriptome analysis reported alteration of RB1 and loss of RBL2 (retinoblastoma-like protein 2), deletion of the PTEN tumor-suppressor gene, and DKK1 down or upregulation as major recurrent genetic mutations in soft tissue sarcomas with complex genomics. Alterations of PTEN and DKK1 functions could suggest involvement of the Wnt canonical pathway in this subset of tumors.[641] Another study based on integrative analysis of DNA sequence, copy number, and mRNA expression reported p53 mutations in 17% of pleomorphic liposarcomas.[642] However, clear discriminations between leiomyosarcoma and undifferentiated pleomorphic sarcoma was not possible based on gene expression profiling.[643] Molecular profiling of undifferentiated, pleomorphic sarcomas with NGS could lead to a breakthrough by identifying multiple genetic mutations and altered signaling pathways that are pivotal for tailoring targeted therapy.[644]

Genetic pathways altered in sarcomas
Cell cycle regulators

The cell cycle is regulated by a complex net of molecular signaling pathways, with the p16-cyclin D1/CDK4-pRB pathway and the p53 pathway playing pivotal roles in this process. Abnormal expression of cell cycle regulators leads to uncontrolled and unlimited cellular proliferation, and causes cancer.

p16-cyclin D1/CDK4-RB pathway

The RB tumor-suppressor gene encodes Rb protein that is involved in the regulation of cell cycle progression from the first gap phase (G_1) to the synthesis phase (S). The p16-cyclin D1/CDK4-RB pathway is controlled by G_1 cyclins (cyclin D1, E) and cyclin-dependent kinases (CDK4/6, CDK2). The activity of the latter is modulated by cyclin-dependent kinase inhibitors, such as p16, p21, and p27. Hypophosphorylated RB, an active form of the protein, binds to transcription regulators and prevents the cell from progressing through the G_1 phase of the cell cycle. Absence of RB or phosphorylated RB allows cell cycle progression into the late G_1 phase and the rest of the cycle. Loss or alteration of RB affects the RB pathway and leads to abnormal cell proliferation.[645,646]

The cyclin-dependent kinase inhibitor (CDKN) locus mapped to chromosome 9p21 harbors genes encoding three cell cycle inhibitory proteins: p15(INK4b) encoded by CDKN2b, p16(INK4a) encoded by CDKN2a, and p14ARF, encoded by an alternative reading frame of CDKN2a. A spectrum of genetic mutations, mostly deletions, affecting chromosome 9p21 and the CDKN2a/2b locus has been described in a variety of human cancers, including sarcomas.[647,648] Also, methylation of the 5' CpG islands of the CDKN2a promoter region has been reported.[649] Both the loss of CDKN2a and hypermethylation of the CDKN2a promoter region indicate poor clinical outcomes of cancer, as shown in leiomyosarcoma and GIST.[650,651] In alveolar rhabdomyosarcoma, the PAX3-FOXO1A oncoprotein might require a loss of p16(INK4a) to promote its oncogenic function.[652,653] This early step, coupled with MYCN amplification and telomere stabilization, is thought to be one of the crucial events in alveolar rhabdomyosarcoma tumorigenesis.[654]

Another member of the so-called INK4 protein family, p15(INK4b) is encoded by CDKN2b, which is located adjacent to CDKN2a on chromosome 9p21.[655] CDKN2b has been shown to be codeleted with CDKN2a in cancer cell lines and in primary tumors,[655,656] but mice deficient in p15(INK4b) do not develop cancer.[648] In contrast, the deletions of p16(INK4a) and p19(ARF) or p19(ARF) alone give rise to different types of cancers, including sarcomas.[657,658] Mice deficient for all three proteins were more tumor prone, however, and developed a wider spectrum of tumors than the CDKN2a mutant mice did, with a preponderance of skin tumors and soft tissue sarcomas. Also, CDKN2ab-/- mouse embryonic fibroblasts (MEFs) were substantially more sensitive to oncogenic transformation than CDKN2a mutant MEFs.[659]

p16(INK4a) inhibits phosphorylation of RB and induces a G_1 cell cycle arrest through the interaction with CDK4 and CDK6. The p19(ARF) protein binds to the MDM2 protein and promotes its rapid degradation. Degradation of MDM2 leads to the stabilization and accumulation of p53. The functional consequence of this interaction is the restoration of the G_1 cell cycle arrest. Thus, genetic mutations involving chromosome 9p21 region and the CDKN2a gene simultaneously affect the p16-cyclin D1/CDK4-pRB pathway and ARF-MDM2-p53 cell-cycle-related suppression pathways.[645,660]

Recently developed selective CDK4/6 inhibitors have been entered into clinical trials and have shown antitumor activity in different types of cancer, including sarcoma.[661]

p53 tumor-suppressor gene pathway

The human p53 gene (TP53) has been mapped to chromosome 17p13.1 and encodes a 393-amino acid protein, which acts as a sequence-specific transcription factor.[662] The p53 gene was named after the corresponding 53-kD cellular protein isolated from SV40 virus-transformed cells that bound to the large T antigen of SV40 T.[663,664] Because of an internal promoter and alternative splicing events, the p53 transcriptional expression pattern is complex, with isoforms encoding 28- to 53-kD proteins. The interplay between full-length p53 and p53 isoforms mediates p53 functions.[665]

Activated p53 can engage many targets and either promote or inhibit their transcription.[666] Based on expression profiling studies, p53 more often represses than transactivates the targeted genes.[667,668] p53 transcriptionally repressed genes include several cell cycle regulatory genes that act at each phase of the cell cycle.[668]

Early studies showed that introduction of wild-type p53 into the cell line without the endogenous p53 caused growth arrest or induced apoptosis.[669,670] Also, p53 activity was essential for radiation-induced death in thymocytes and chemotherapy-induced apoptosis in fibroblasts expressing deregulated oncogenes in p53 knockout mice.[671-673] p53 transgenic mice carrying multiple copies of a p53 mutant and a normal wild-type p53 allele developed osteosarcomas, pulmonary adenocarcinomas, and lymphomas. Furthermore, mice with a homozygous null p53 genotype developed tumors and died earlier than did mice with a heterozygous null/wild-type genotype.[674-677] CDKN1a (p21) is one of the p53-transactivated targets. Activated p21 inhibits cyclin-dependent kinase activity of cyclin-CDK2 and cyclin-CDK4 complexes, preventing phosphorylation of cyclin-dependent kinase substrates (RB protein) and blocking cell cycle progression. A cell arrested in G1 undergoes repair of DNA damage before replication. Unsuccessful repair of DNA damage directs a cell into apoptosis.[678] Thus, the terms *guardian of the genome* or *cellular gatekeeper* have been applied for growth and division to stress the importance of the p53 function in cell cycle control.[679]

Inactivation of the p53 protein is the most frequently occurring single-gene event in human cancer.[680] Several mechanisms can lead to the alteration of the p53 expression. Genomic mutation can cause the deletion of p53 or expression of either a truncated or dysfunctional protein.[681,682] Most p53 mutations represent missense mutations clustering in the part of the gene encoding the DNA-binding domain. The most commonly mutated codons, 175, 245, 248, 249, and 273, have been designated as mutational hotspots.[680]

Li–Fraumeni syndrome (LFS OMIM #151623), an autosomal dominant cancer predisposition syndrome characterized by an increased frequency of early-onset cancer, is associated with germline mutation in the p53 gene.[683,684] Children and young adults from LFS families developed a spectrum of tumors, including soft tissue and bone sarcomas, brain tumors, adenocortical carcinomas, acute leukemias, and premenopausal breast cancers. Members of some of the families who do not fulfill all the classic epidemiologic criteria of LFS are often diagnosed with a Li–Fraumeni-like syndrome (LFL). Molecular studies have revealed germline p53 mutations in most LFS and in some LFL families.[685,686] In addition, LOH at the p53 locus was documented in approximately 50% of the tumors from LFS families.[687] The lack of p53 mutations in some of the LFS or LFL families could suggest epigenetic inactivation or that another tumor-suppressor gene is involved in those cases. Heterozygous checkpoint kinase 2 gene (CHK2) germline mutations have been identified in the LFS and LFL families.[688,689] CHK2 acts as a cell cycle regulator required for DNA damage and replication checkpoints.[690-692] CHK2 protein regulates the cell cycle, interacting with ATM and p53.[690,693,694] Mutations in the ATM-CHK2-p53 pathway might allow cell proliferation, survival, increased genomic instability, and tumor progression.[695]

In soft tissue sarcomas, the genomic integrity of p53 is preserved and the frequency of p53 mutations is relatively low, ranging from 5% to 20%, depending on the type of tumor.[696] However, other genetic and epigenetic mechanisms, which indirectly suppress p53 function, have been identified. Overexpression owing to amplification of p53-negative regulators, human MDM2 or human MDM4, which promotes p53 degradation, was identified in a number of tumors, including well-differentiated liposarcoma, dedifferentiated liposarcomas, Ewing sarcoma, and synovial sarcomas, among others.[697] Recent studies based on next-generation technologies revealed amplification of MDM2 in human sarcomas more frequently than previously estimated.[698] Use of an MDM2 and MDM4 antagonist to reactivate the p53 pathway, an attractive therapeutic strategy for cancer, has yet to be validated clinically.[699]

p53 interacts with more than 400 proteins, including pathologic fusion gene products. Knowledge of these interactions is limited, although some progress has been achieved in recent years. In Ewing sarcoma, EWS-FLI1 oncoprotein silences p53 through either the Notch signaling pathway or through the formation of a protein complex involving EWS-FLI1 and p53.[700,701] In synovial sarcoma, the SS18-SSX1 oncoprotein acts as a positive regulator of HDM2 stability. Therefore, SS18-SSX1 expression promotes the ubiquitination and degradation of p53 in an HDM2-dependent manner.[702] Kaposi's sarcoma-associated herpesvirus, also known as HHV8, encodes viral interferon regulatory factor 3, which downregulates p53 functions through the inhibition of the upstream regulatory ATM pathway or by directly impairing p53 oligomerization and DNA-binding ability.[703] Thus, modification of the protein–protein interactions involving p53 offers an accurate therapeutic intervention.[704]

The stability and transcriptional activity of p53 are regulated through multiple post-translational modifications, such as phosphorylation, acetylation, and ubiquitination.[705] The epigenetic mechanism, such as p53 acetylation[706] or methylation,[707] can enable the p53-mediated stress response. Many miRNAs, for example miR-15a–16-1 cluster, miR-24, miR-34a, miR-124, miR-125b, miR-129, miR-137, miR-188, miR-195, miR-449, and members of let-7 family, have been implicated in the regulation of key players in cell cycle control pathways.[708–712] Reduced expression of miRNAs found in different types of human cancers further indicates the importance of miRNAs as a new generation of tumor suppressors.[713,714]

Growth factor signaling pathways

Growth factors, usually proteins or steroid hormones, act as signaling molecules. They activate receptor (RTK) and non-receptor (PTK) protein tyrosine kinases and induce intracellular signaling pathways that regulate a spectrum of biological processes, such as cell proliferation, activation, differentiation, and migration. Pathologic activation of growth factor signaling has been associated with cellular transformation and tumorigenesis. In sarcomas, genetic mutations in growth factor and growth factor receptor genes such as ALK, KIT, NTRK3, PDGFRA, PDGFRB, and VEGF/VEGFR often underline abnormal activation of growth factor signaling pathways. However, the activation of many signaling pathways is not associated with genetic mutations, such as the pathologic activation of the insulin-like growth factor 1 receptor (IGF1R) pathway frequently implicated in a variety of sarcomas, including rhabdomyosarcoma, leiomyosarcoma, osteosarcoma, synovial sarcoma, and Ewing sarcoma.[715]

IGF1R is one of the components of the insulin growth factor (IGF) axis, a cellular network of two ligands, IGF1 and IGF2, and three receptors, IGF1R, IGF2R, and INSR (insulin receptor). IGF1R and INSR share significant sequence homology and can form hybrid receptors. IGF1R is activated by IGF1 and IGF2 with high affinity and by insulin with low affinity, whereas INSR binds insulin with high affinity. Both IGFs and insulin can activate the IGF1R/INSR hybrid receptor. IGF activities are controlled by insulin-like growth-factor-binding proteins (IGFBP).[716] Activation of IGF1R and INSR leads to autophosphorylation, then tyrosine phosphorylation of the substrates, and finally to the activation of two main signaling pathways: the PI3K/AKT/mTOR pathway and the RAS/MAPK pathway. The MAPK pathway increases cellular proliferation, while the PI3K pathway inhibits apoptosis and stimulates protein synthesis.[717] Downstream activation of the PI3K/AKT/mTOR pathway is often a point of convergence for different growth factor signaling pathways and a molecular target for the treatment of sarcoma.[718,719]

Deregulation of the IGF axis has been linked to the loss of transcriptional suppression by altered tumor-suppressor genes, such as p53, or transcriptional enhancement by onco-proteins.[720] In osteosarcoma and rhabdomyosarcoma cell lines, a lack of the wild-type p53 strongly enhances IGF1R activity, while the EWS-WT1 oncoprotein transactivates the IGF1R promotor in DSRCT.[721] Similarly, EWS-FLI1 activates the IGF/IGF1R pathway directly or via miRNA regulatory functions in Ewing sarcoma.[722–724] Thus, the spectrum of different molecular events could activate the IGF axis in a given tumor and in different tumors. Therefore, inhibition of IGF/IGF1R signaling remains an attractive therapeutic approach.[725]

HGF/MET is another common growth factor signaling pathway often indicated in human cancer, including a variety of sarcomas.[715] Ink4a/Arf knockout mice with deregulation of MET signaling due to HGF overexpression have been shown to develop embryonal rhabdomyosarcomas.[726] MET, also known as hepatocyte growth factor receptor (HGFR), is an RTK whose signaling is critical for cell proliferation, motility, and migration. Its ligand, the hepatocyte growth factor (HGF), a heparin-binding protein, is normally produced by different types of mesenchymal cells. HGF stimulation induces phosphorylation of MET tyrosine residues and, subsequently, the activation of RAS/ERK, PI3K/AKT, and SRC downstream signaling pathways.[727] Aberrant MET activation occurs via genetic alterations, including gain-of-function mutations, translocations, and gene amplifications. In the absence of genetic mutations, overexpression of MET is a consequence of transcriptional upregulation.[728] In clear cell sarcoma, c-MET is directly activated by MITF, a transcriptional target of the EWS-ATF1 oncoprotein.[729] Similarly, c-MET is a transcriptional target gene for PAX3-FOXO1 and ASPL-TFE3 oncoproteins in alveolar rhabdomyosarcoma and alveolar soft part sarcoma, respectively.[730,731] Also, high levels of MET, documented in advanced tumors, are associated with hypoxia. The latter activates MET via the transcription factor hypoxia inducible factor 1α (HIF1α).[732] A number of therapeutic agents targeting HGF or MET have been developed and are now being evaluated in clinical trials.[733]

Over the past decades, a significant number of inhibitors targeting activated growth factor signaling pathways in sarcomas have been developed and investigated. A few examples of soft tissue tumors with growth factor signaling pathways altered by genetic or epigenetic mutations and recently developed therapeutic agents are listed in Table 5.4.

APC inactivation and beta-catenin activation in the Wnt pathway

Familial adenomatous polyposis of the colon (FAP) is an inherited disease that typically shows itself at an early adult age with colon polyposis and a predisposition for developing colon cancer and other extraintestinal malignancies.[744,745] An increased frequency of desmoid tumors is one of the clinical features of FAP.[746,747] Presence of germline APC mutations in hereditary desmoid disease[748] and familial infiltrative fibromatosis (FIF) have suggested that FAP and FIF could be different clinical manifestations of the same genetic syndrome. Adenomatous polyposis with desmoids has also been referred to as *Gardner syndrome*.[749]

FAP locus and the adenomatous polyposis coli gene (APC) map to the chromosome 5q21. Germline mutations of the APC gene have been documented in a majority of FAP cases.[744,745,750,751] As predicted by Knudson's two-hit hypothesis, somatic APC mutations occur in desmoid tumors in FAP patients.[752] Germline mutations in APC are either nucleotide substitutions, creating nonsense codons, or small deletions/insertions, leading to frameshifts. Somatic mutations also include a loss of heterozygosity.[744,745,750–752]

Table 5.4 Examples of growth factor signaling pathways activated in sarcomas and examples of targeted therapies developed to inhibit these pathways

Sarcoma	Growth factor signaling pathway	Mechanism of activation	Targeted therapy (clinical trials, xenograft, *in vitro* experiments)
Alveolar soft part sarcoma	HGF/MET	MET is transcriptional target of ASPL-TFE3 oncoprotein[730]	Tivantinib/ARQ197[734]
Clear cell sarcoma	HGF/MET	MET is transcriptional target of MITF activated by EWS-ATF1 oncoprotein[729]	Tivantinib/ARQ197[734]
Congenital fibrosarcoma	NTF3/NTRK3	NTRK3 is constitutively activated by the fusion with ETV6[384]	Crizotinib[735]
Dermatofibrosarcoma protuberans	PDGFs/PDGFRB	PDGFB is constitutively activated by the fusion with COL1A1[155]	Imatinib[415]
Ewing sarcoma	IGF/IGF1R	IGF1 is upregulated and IGFBP is repressed by EWS-FLI1 oncoprotein[722–724]	R1507 IGF1R Mab[736]
Gastrointestinal stromal tumor	SCF/KIT	KIT is constitutively activated by gain-of-function mutations[529]	Imatinib[537]
Gastrointestinal stromal tumor	PDGFs/PDGFRA	PDGFRA is constitutively activated by gain-of-function mutations[548]	Imatinib[545]
Inflamatory myofibroblastic tumor	PTN/ALK	ALK fusion chimeric proteins activate multiple signaling pathways, including RAS/RAF/MEK/ERK cell proliferation and JAK/STAT cell survival pathways[737]	Crizotinib[377]
MPNST	IGF/IGF1R	IGF1R is overexpressed due to gene amplification[738]	Cabozantinib[739]
Rhabdomyosarcoma, alveolar	HGF/MET	MET is transcriptional target of PAX3-FOXO1 oncoprotein[731]	SU11274[740]
Rhabdomyosarcoma, alveolar	IGF/IGF1R	IGF1R is transcriptional target of PAX3-FOXO1 oncoprotein[741]	R1507 IGF1R Mab[736]
Synovial sarcoma	IGF/IGF1R	IGF2 transcription is enhanced by SS18-SSX1 and SS18-SSX2 oncoproteins[742]	R1507 IGF1R Mab[736]
Tenosynovial giant cell tumor	CSF1/CSF1R	CSF1 is constitutively activated by the fusion with COL6A3[258]	Imatinib[743]

Unlike some other tumor-suppressor genes, the APC gene can act in a fashion that does not entirely follow Knudson's classic two-hit hypothesis for tumorigenesis, which assumes that the two "hits," resulting in a loss of tumor-suppressor function, are independent mutation events. In the case of APC inactivation, both the position and type of the second hit depend on the localization and type of the first germline mutation. This nonrandom distribution of somatic hits has been interpreted as the result of selection for more advantageous mutations during tumor formation.[753,754] The site of the germline APC mutation can also determine the severity of the disease; germline mutations affecting the 3' end of APC gene result in a severe desmoid phenotype.[755] Furthermore, a 50% decrease in the expression of one APC allele can lead to the development of FAP.[756,757]

The APC protein is a component of the Wnt signaling pathway, which plays a pivotal role in several embryonal and adult cellular processes and has been implicated in many types of cancer.[758] The transcription factor beta-catenin is the key effector of this pathway, stimulating proliferation through modulating the expression of specific target genes. Normally, the APC protein interacts with beta-catenin and induces its degradation. Mutational inactivation of APC leads to beta-catenin overexpression and pathologic signaling through the Wnt pathway.[758,759]

FAP-associated desmoid tumors are caused by germline APC mutations followed by somatic inactivation of the wild-type APC allele, whereas sporadic desmoid tumors are usually characterized by oncogenic activation of beta-catenin. The constitutive activation of beta-catenin occurs through mutations that affect phosphorylation sites within exon 3 and cause protein stabilization.[760–762] Certain types of beta-catenin mutation can indicate an increased risk of disease recurrence.[763] APC mutations have also been documented but less commonly in sporadic desmoid tumors.[764]

Targeting key components of the Wnt/beta-catenin signaling pathway is a new approach in anticancer drug development.[765,766]

Dysfunction of tricarboxylic acid cycle (Krebs cycle)

In 1956, Otto Warburg observed that cancer cells exhibited high rates of glycolysis (aerobic glycolysis), even in the presence of oxygen, and hypothesized that cancer is caused by metabolic alterations.[767] The enhanced aerobic glycolysis exhibited by cancer cells has been extensively studied and used to develop diagnostic and therapeutic tools. Subsequently, the discovery of oncogenes and tumor-suppressor genes caused interest to wane in the investigation of the role of so-called "Warburg effect" in human cancer. However, in the last decade, findings of loss- or gain-of-function mutations in genes encoding for tricarboxylic acid (TCA) cycle enzymes, fumarate hydratase (FH), isocitrate dehydrogenase (IDH), and succinate dehydrogenase (SDH), have restored interest in Warburg's hypothesis.[768–771]

SDH and FH are considered tumor suppressors, as their inactivation follows Knudson's classic two-hit hypothesis of inherited germline loss-of-function mutations in one allele and acquired loss-of-function somatic mutations in the tumor cells. In contrast, IDH mutations are somatic and affect only one tumor allele. Therefore, they appear to be oncogenic gain-of-function mutations. Mutational alteration of SDH, FH, and IDH genes result in the pathologic accumulation of succinate, fumarate, and D-2-hydroxyglutarate, oncometabolites implicated in cellular transformation and oncogenesis.[768–771]

SDH is a heterotetramer mitochondrial enzyme complex comprising four subunits, A, B, C, and D, encoded by SDHA, SDHB, SDHC, and SDHD nuclear genes. SDHA flavination and assembly of the tetramer requires the presence of SDH5 (SDHAF2). In the TCA cycle, SDHA is responsible for conversion of succinate to fumarate. SDHB is an iron sulfur protein that participates in the electron transport chain for the oxidation of ubiquinone to ubiquinol. SDHC and SDHD are membrane-anchoring subunits.[772] SDH deficiency was identified in gastrointestinal stromal tumors, paragangliomas, pituitary adenomas, and renal cell carcinomas, and was associated with loss-of-function mutations. Although, in some cases, no genetic mutations could be identified and the mechanism of inactivation might have been related to epigenetic silencing.[773–775]

Fumarate hydratase (FH) gene encodes two FH isoforms, mitochondrial and cytoplasmic. FH catalyzes conversion of fumarate to malate in the TCA cycle. Recessive FH mutations have been linked to severe encephalopathy and early death, while dominant FH mutations predispose to the development of multiple cutaneous and uterine leiomyomas, hereditary leiomyomatosis, and renal cell cancer. Mutant cells show reduced mitochondrial FH activity and lack cytoplasmic FH activity. The latter suggests that FH tumor-suppressor gene function is associated with cytoplasmic isoform.[776]

IDH is an enzyme that catalyzes decarboxylation of isocitrate into alpha-ketoglutarate. Three isoforms encoded by different genes, IDH1, IDH2, and IDH3, have been identified. Somatic mutations in IDH1 and IDH2 were found in different types of human cancer including low-grade glioma, secondary glioblastoma, chondrosarcoma, cholangiocarcinoma, and acute myeloid leukemia.[777,778]

Clonality assessment with HUMARA assay

The analysis of the clonal nature of a lesion is rarely diagnostically important in soft tissue tumors, but it is often used as an investigational measure to examine the nature of a cellular proliferation, on the assumption that clonal proliferations are neoplastic. Consistent nonrandom chromosomal aberrations, such as translocations, large deletions, amplifications, numerical changes, and molecular genetic changes, such as single base pair substitutions, deletions, insertions of a few nucleotides, and fusion gene transcripts, are markers confirming the clonal nature of a given lesion. These markers can be specific for a particular tumor type or a particular case.

Tumor-specific markers are not always easy to identify, however, and other, more universal approaches must be applied to confirm the clonal nature of the proliferating tumor cells. According to Lyon's hypothesis, one of the copies of the X chromosome is inactivated in each somatic cell of an adult female.[779] Inactivation occurs randomly at an early stage of embryogenesis and results in a cellular mosaic pattern with either a maternal or paternal chromosome X inactivated.[780] Polyclonal cells extracted from normal tissue contain equal numbers of maternally and paternally derived X chromosomes. In contrast, monoclonal proliferation has only one type of inactivated chromosome X of paternal or maternal origin, transmitted from its progenitor cell. The nature of these assays means that they can be performed in tissue from female patients only.

Various polymorphic genes on the X chromosome have been used in clonality studies based on Lyon's hypothesis. Early assays evaluated the expression pattern of glucose-6-phosphate dehydrogenase (G6PD) isoenzyme[781,782] or examined the methylation pattern of the chromosome X polymorphic loci (phosphoglycerate kinase [PGK] gene, hypoxanthine phosphoribosyltransferase [HPRT] gene, and hypervariable DXS255 locus [M27β]) using methylation-sensitive restriction endonucleases.[783–785] The low informativeness of polymorphisms and incomplete methylation were major limitations of these clonality assays.[786]

The most widely used clonality assay (human androgen receptor [HUMARA]) is based on amplification of the short tandem trinucleotide (CAG) repeat (STR) polymorphism, identified in the coding region of the first exon of human androgen receptor gene and closely located to methylation-sensitive restriction enzyme sites (Hhp II and Hha I). Tumor and normal tissue DNA samples obtained from the same subject are cleaved with methylation-sensitive restriction endonucleases and amplified by PCR. The pattern of PCR amplification products differs among monoclonal and polyclonal samples. A valid interpretation is impossible without the appropriate normal tissue controls, however.[787]

The HUMARA test is considered most reliable and informative based on consistent methylation pattern and high rate of allelic polymorphism.[788,789] An expression (mRNA)-based HUMARA has been developed, but the usefulness of this test might be limited by the expression of the androgen receptor in different tumors.[790] Several benign soft tissue lesions have been evaluated for clonality using the HUMARA test, and in some tumors, discordant results have been reported. Technical problems, including altered methylation status at the HpaII sites[791] and a large nontumor or normal cell component in analyzed samples, might have contributed to these discrepancies. The latter most likely is responsible for the incorrect conclusion that giant cell tumor of the tendon sheath is a polyclonal non-neoplastic lesion,[792] whereas other studies have confirmed its neoplastic nature and documented the COL6A3-CSF1 fusion gene in tumor cells.[258,259] Thus, HUMARA should be interpreted with caution, especially when the analyzed lesion is composed of a heterogeneous cell population. Examples of HUMARA tests on soft tissue and related lesions are shown in Table 5.5.[793–815]

Table 5.5 Examples of soft tissue and related lesions evaluated for clonality status using the human androgen receptor assay (HUMARA)

Lesion	Result of HUMARA
Angiomyolipoma	Monoclonal[793,794] Monoclonal smooth muscle and blood vessel[795,796] Polyclonal adipose tissue[795,796]
Chester–Erdheim disease (non-Langerhans cell histiocytosis)	Monoclonal[797,798] Polyclonal[799]
Chordoma	Polyclonal[800]
Desmoid fibromatosis	Monoclonal[791,801,802]
Dermatofibroma	Monoclonal[803] Monoclonal histiocytoid cells[804] Polyclonal fibroblastic cells[804]
Giant cell tumor of tendon sheath	ᵃPolyclonal[792]
Histiocytosis X	Monoclonal[805]
Kaposi's sarcoma	Monoclonal but multiple KS lesions in a given patient can represent different clones[806,807] Polyclonal[808]
Leiomyomatosis, disseminated peritoneal	Monoclonal[809]
Leiomyomatosis, intravenous	Monoclonal[810]
Melanocytic nevi	Monoclonal[811] Polyclonal[812]
Nodular fasciitis	Polyclonal[813]
Palmar fibromatosis	Polyclonal[814]
Sclerosing hemangioma of the lung	Monoclonal[815]

ᵃ Incorrect conclusion.

Examples of molecular genetic techniques

This section presents examples of molecular genetic techniques in the context of specific applications and the required cell/tissue material necessary for optimal performance. Genetic alterations in soft tissue tumors vary from simple single nucleotide changes (point mutations) to complex, large structural genetic abnormalities. Identification of these mutations provides an important adjunct to standard histopathological diagnosis and allows for proper tumor classification. In some

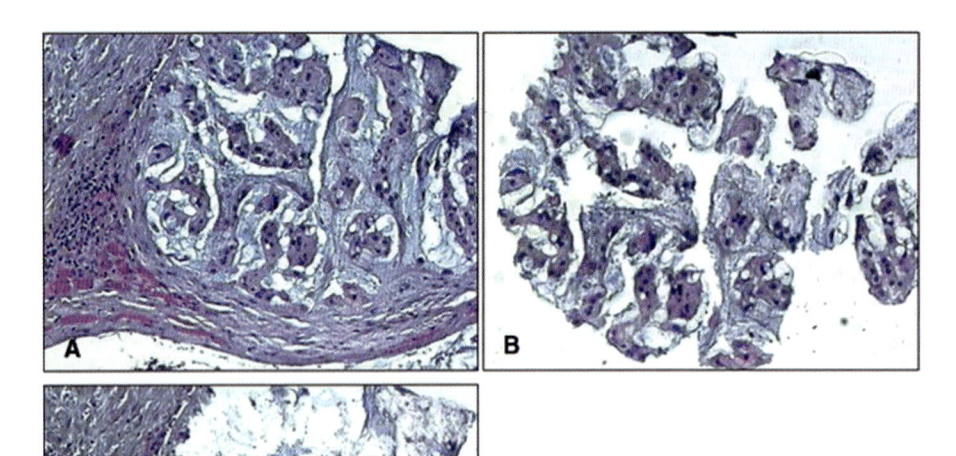

Figure 5.1 Example of laser-based capture microdissection of tumor tissue. (**A**) Sample before dissection. (**B**) Captured tumor tissue. (**C**) Sample after dissection.

cases, a mutation profile of tumor DNA has a prognostic value and may indicate sensitivity or resistance to targeted anticancer therapy. Over the past few decades, a wide variety of molecular genetic assays and testing platforms have been developed. Some of these techniques are useful for clinical testing, whereas others remain within the domain of scientific studies.[261]

Sample preparation for molecular genetic studies requires careful evaluation of the tissue material. Cross-contamination of tumor samples with normal tissue and vice versa can influence the results of the molecular genetic assays and lead to false-negative or false-positive results. Morphologic verification of frozen or FFPE tissue before submitting for the nucleic acid extraction should always be performed. Normal tissue, surrounding or trapped inside the tumor, must be precisely dissected out, using a laser-based capture microdissection technique if necessary. An example of such a procedure is shown in Figure 5.1.

DNA and *RNA* extraction from FFPE tissues is particularly important in the molecular testing and research of soft tissue tumors. Despite progress in understanding the need to perform molecular genetic testing, tumor samples are not consistently freeze preserved, and therefore the molecular genetic studies must rely on nucleic acids recovered from FFPE tissues. The research of new genetic markers also requires testing well-characterized cases, with long-term follow-up studies. This cannot be easily achieved in prospective studies and analysis of FFPE tumors retrieved from the archival files is very helpful. Also, recent advances in targeted therapies increased the demands for tumor

Table 5.6 Fixatives and preservation of DNA for PCR amplification[817–819]

Fixative	PCR amplification
Acetone	Very good
Alcohol	Very good
Formalin	Good
B-5	Unsatisfactory
Bouin's	Unsatisfactory
Carnoy's	Variable (contradictory results reported)
Zenker's	Unsatisfactory

genotyping. Partially degraded, but amplifiable DNA or RNA can be recovered from FFPE tissues. Different extraction procedures have been developed,[816–818] including commercially available kits and robotic systems (www.promega.com, www.qiagen.com, www.lifetechnologies.com). The quality of the recovered nucleic acids can vary from sample to sample, however, and it depends on several factors, including the type of fixative. The effect of different fixatives on DNA preservation and PCR amplification is shown in Table 5.6. Formalin is the most frequently used formaldehyde-based fixative in pathology. Use of low pH (unbuffered) formalin significantly intensifies degradation of nucleic acids. Also, long-term storage of already fixed tissue may increase DNA and RNA fragmentation.[818–821] In general, if successful DNA or RNA extraction is achieved, fragments below 200 base pairs are relatively easy targets for

PCR amplification. In some cases, however, neither DNA nor RNA can be extracted successfully. Suboptimal PCR amplification of low-quantity and -quality templates can generate false results with a relatively high frequency.[822,823] These artifacts could mimic single nucleotide variations and, in the context of tumor clonal heterogeneity and increased use of NGS, might be difficult to distinguish from true mutations.[824,825] Simultaneous use of different mutation detection techniques and repeated analysis of independent amplification products can help to identify such artifacts.[824]

PCR amplification is a technique that allows automated, enzymatic *in vitro* synthesis of millions of copies of target DNA sequences for subsequent sequence analysis. The method was invented by Kary Mullis,[826] for which he received the 1993 Nobel Prize in Chemistry, and was initially applied to amplify the human beta-globin gene for the prenatal diagnosis of sickle cell anemia.[827–829] A standard PCR contains template (double-stranded DNA), primers (oligonucleotides complementary to the sequences flanking the target of amplification), enzyme (Taq DNA polymerase), mixture of nucleotides (dNTPs-A,C,G,T), and reaction buffer. Amplification of the target sequence occurs through cycling (approximately 25–40 cycles) and each cycle consists of three steps. In the first step, the template DNA is denatured (separated into single strands) by heating the reaction to 98 °C. In the second, annealing step, primers complementary to the opposite strands of DNA bind to the specific sites. In the third elongation step, Taq DNA polymerase initiates synthesis of the target sequence directionally from the 5' end. Since the PCR technique was introduced, many new applications and modifications of this technique have been developed. PCR-based techniques have had an enormous impact on basic molecular biology and have contributed to the rapid development of molecular diagnostic pathology. *Multiplex PCR* allows for the amplification of several target sequences simultaneously. In soft tissue tumor pathology, multiplex PCR amplification could be a method of choice in screening for variant fusion gene transcripts or for transcripts of different fusion genes.[830] *Nested PCR* is a modified PCR technique that allows for effective amplification of low copy number templates. In this technique, PCR products from the first, standard PCR amplification are used as a template for the second (nested) PCR amplification. Although the nested PCR has been successfully used to amplify target sequences from severely degraded DNA and RNA obtained from FFPE tissues[831] and could be used for a wide spectrum of applications in diagnostic molecular pathology, use of this technique carries a serious danger of random contamination, which is difficult to monitor. The possibility of random contamination is one of the major concerns preventing the introduction of a nested PCR technique into diagnostic pathology. To avoid cross-contamination, precautions, including use of disposable labware, gloves, and lab coats at all stages and extensive cleaning of all work areas after each experiment, are required. Physical isolation of the template preparation area from the PCR preparatory and PCR product amplification analysis areas is also crucial. In addition, the use of multiple negative controls at all stages of nested PCR can help to identify and eliminate random contamination.[261,831,832]

Reverse transcription polymerase chain reaction (RT-PCR) allows for the amplification of messenger RNA (mRNA). First, mRNA is converted to complementary DNA (cDNA) using reverse transcriptase (RT). Next, the cDNA is used as a template for PCR amplification as in regular PCR. RT-PCR can detect small quantities of transcripts undetectable by less-sensitive methods. RT-PCR amplification of the transcripts expressed at a low level can be difficult to achieve, and negative results should always be evaluated in the context of the ability to amplify other transcripts. Real-time PCR can be helpful in such cases because of highly specific detection and direct quantification of the amplified sequences. In clinical settings, fusion gene transcripts resulting from the chromosome translocations in STTs are common targets for RT-PCR amplification.[261]

Rapid amplification of cDNA ends (RACE) is an RT-PCR-based technique used to amplify the mRNA sequence when only a portion of the sequence is known.[833] 5' RACE and 3' RACE allow amplification of an unknown sequence located 5' and 3' to a known sequence, respectively.[834,835] This technique has been frequently used to amplify and identify unknown gene fusion partners of EWSR1, ALK1, and other genes involved with multiple partner genes.

Allele-specific PCR (AS-PSR) is a simple, cost-effective technique developed to determine known polymorphisms or common cancer mutations without Sanger sequencing of PCR products. AS-PCR is based on the principle that extension of the primer will occur only when its 3' end is perfectly complementary to the template. Thus, in this method, specific primers are designed to amplify either a polymorphic/mutant allele or wild-type allele.[836] In the last decade, several other alternatives to Sanger sequencing, including single-base extension and pyrosequencing, have been developed and successfully used to identify common cancer mutations.[837,838] However, some of these methods might be replaced by the molecular testing, based on the NGS technology.

Analyses of PCR amplification products vary and depend on the primary goal of amplification. A spectrum of molecular techniques, including different types of gel and capillary electrophoreses, radioactive or nonradioactive hybridization with specific probes, and sequencing can be used to evaluate and identify PCR products.[261] Examples of different strategies for PCR products analysis are shown in Table 5.7 and Figure 5.2. Although direct sequencing of PCR products seems to be the gold standard when searching for mutations, it can be labor intensive and relatively expensive, when large numbers (thousands) of samples are analyzed. Screening of the PCR products with a high-throughput system is necessary to separate mutant from wild-type cases.[839] Denaturing gradient gel electrophoresis (DGGE) and single-strand conformation polymorphism (SSCP) assay are two conventional methods commonly used to prescreen PCR products for mutations. DGGE is based on differential melting of double-stranded DNA molecules during electrophoresis in the gel, with an increasing concentration of a denaturing agent. The melting behavior is sequence dependent and allows separation of the fragments differing by a single nucleotide.[840,841] SSCP assay is based on different, sequence-related mobilities of denatured single-stranded DNA molecules in nondenaturing polyacrylamide gel. Single nucleotide substitutions, losses, or insertions of a few nucleotides change the conformation of the single-stranded DNA molecule and change its migration, compared with the wild-type, during the electrophoresis.[842,843] Denaturing high-performance liquid chromatography (DHPLC) has been used to analyze DNA fragments and has been applied to mutation analysis.[844] For DHPLC analysis, PCR products are denatured by heating and then allowed to reanneal. Reanealed fragments form either homoduplexes, if the target DNA matches the normal DNA fragment, or heteroduplexes, if the target DNA contains a mutation. The heteroduplexes have a different melting temperature than the homoduplexes and generate a distinctive chromatographic pattern. Major DHPLC advantages include high sensitivity and specificity in mutation detection, a high throughput (time of single analysis ~5 min), and the ability to evaluate relatively large (up to 1.5 kb) DNA fragments.[845,846]

Real-time PCR, also referred to as *quantitative-PCR (Q-PCR)*, is a rapid temperature cycling and fluorescence-detection-based technique allowing the amplification and reliable quantification of PCR products. An oligonucleotide fluorogenic probe 5' labeled with a reporter dye and 3' labeled with a quencher dye anneals specifically to the template between forward and reverse primers. During the PCR amplification, Taq polymerase cleaves the probe between reporter

Table 5.7 Examples of different strategies for conventional PCR product analysis

Agarose/polyacrylamide gel electrophoresis (separation based on size)	Blotting after separation Hybridization with probe Purification Sequencing
SSCP analysis (separation based on structure)	Blotting after separation Hybridization with probe
DDGE analysis (separation based on structure)	Blotting after separation Hybridization with probe
Capillary electrophoresis (separation based on size)	
DHPLC (separation based on structure)	Purification Sequencing
Blotting without separation	Hybridization with probe
Purification	Cloning followed by sequencing
	Direct sequencing of PCR products

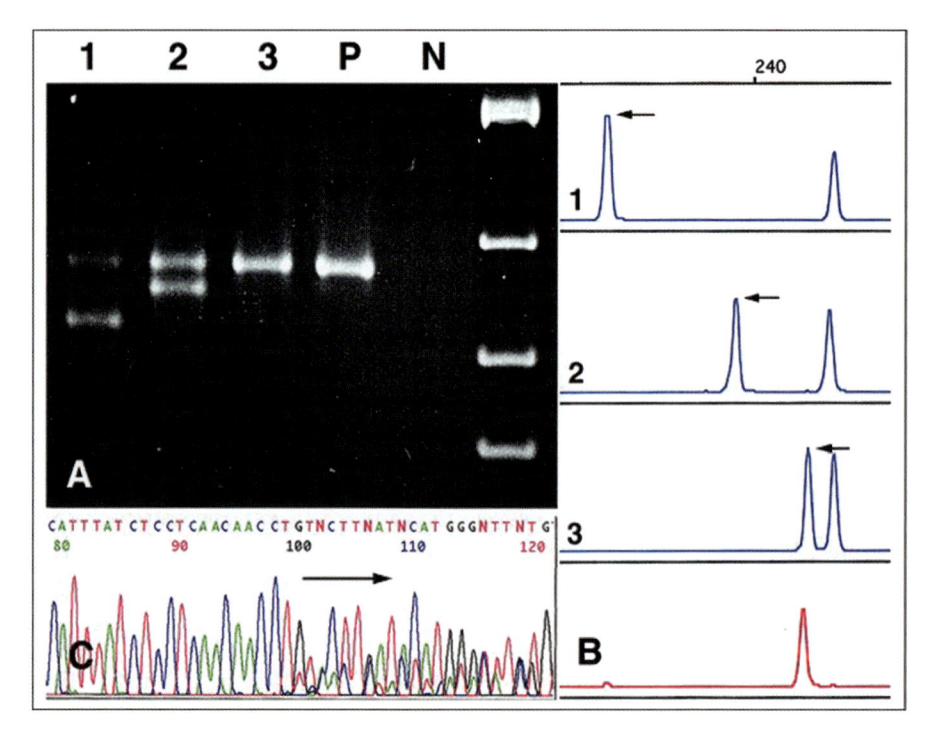

Figure 5.2 Example of PCR amplification product analysis using classical, standard gel electrophoresis, capillary gel electrophoresis, and direct sequencing. (**A**) Standard nondenaturing polyacrylamide gel electrophoresis of KIT PCR products. PCR products amplified from GIST samples are numbered 1 to 3; P and N represent placental DNA and negative control, respectively. Smaller bands seen in lines 1 and 2 indicate the presence of deletions. (**B**) More sensitive capillary gel electrophoresis has detected deletions (arrows on smaller peaks) in all three samples. (**C**) Direct sequencing of the PCR products. (A deletion-related shift of the sequences is indicated by the horizontal arrow.)

and quencher and releases a fluorescent reporter, owing to its 5′–3′ nucleotidase activity. PCR products are measured in real time at each cycle during the extension phase of amplification by quantification of the fluorescence, which increases proportionally to the amount of the amplified product.[847] Monitoring the fluorescence during temperature changes allows for the identification of the PCR products by their melting temperatures and serves as an indirect measure of sequence alterations. Multiple targets can be amplified at the same time using different oligonucleotide probes labeled with different fluorescent dyes. Simple, rapid real-time PCR-based clinical assays detecting simultaneously multiple soft tissue tumor-specific gene fusion products have been developed.[848,849] The method allows for a quantitative and reproducible detection of low copy number DNA or RNA targets and has been shown to be useful in monitoring minimal residual disease.[850,851] Because the quantification of the PCR product occurs in real time, there is no need for post-PCR processing of the PCR products. This substantially reduces the chance of cross-contamination of the samples.

PCR amplification of microsatellite markers has been used to detect deletions causing a loss of heterozygosity (LOH) and has been used extensively in genetic studies on tumor-suppressor genes. Microsatellite markers consist of several repeats of two to seven nucleotides dispersed throughout the genome, mostly outside of the coding sequences. Many of these repeats are highly polymorphic in size, which means that the allele inherited from the father differs substantially from the one inherited from the mother. The status of polymorphic microsatellite markers is evaluated by comparing capillary gel electrophoresis of PCR products amplified from normal and tumor DNA obtained from the same subject.[852] An example of an LOH study is shown in Figure 5.3.

Multiplex ligation-dependent probe amplification (MLPA) is an efficient, low-cost technique developed to detect genomic copy number variations (https://www.mlpa.com).This method is very useful for detecting aberrations that are too small to be identified by FISH and too big to be identified by standard PCR amplification and sequencing. In typical MLPA reaction, multiple probes consisting of two oligonucleotides are hybridized to their genomic targets. Then adjacent oligonucleotides with embedded 5′ or 3′ primer sequences are ligated and PCR amplified by a single set of primers. Because only ligated probes are amplified, the number of PCR products corresponds to the number of targets in the given sample. Subsequently, PCR amplification products are analyzed by capillary electrophoresis and the peak pattern of tumor and reference samples is compared.

Comparative genomic hybridization (*CGH*) is a method to study gains and losses of DNA copy numbers, but balanced translocations or polyploidy cannot be detected. Differentially labeled tumor DNA and normal reference DNA are hybridized simultaneously to normal human metaphase chromosomes. The intensity of the hybridization is measured, and the ratio between reference, normal DNA (labeled red), and tumor DNA (labeled green) indicates the DNA copy number changes.[853–855] Standard CGH has been modified by introducing array technology. In *array CGH* (*aCGH*), normal human metaphase chromosomes are replaced by high-density arrays with thousands of DNA or cDNA samples, which have been mapped to a specific region of the genome. aCGH therefore offers a much higher resolution than standard CGH.[856,857] Because CGH can be performed on DNA extracted from FFPE tissue, it is a powerful screening procedure, allowing detection of total gains and losses in the archival cases. CGH databases are available online (http://www.ncbi.nlm.nih.gov/sky, http://www.helsinki.fi/~lgl_www/CMG.html). Combining the data obtained by different investigators can be limited by the technical differences between CGH studies, however.[858] Although the prognostic value of CGH data has not yet been clearly established for soft tissue tumors, analysis of CGH data with clinical outcome might help to identify patterns of gains and losses correlating with malignant or benign clinical outcome, as documented in GISTs.[859]

cDNA microarrays, also known as *DNA chips* or *gene arrays*, represent a generation of assays developed to analyze large numbers of genes or transcripts simultaneously. Gene sequences on the chip are represented by homologous cDNAs, oligonucleotides, or by peptide nucleic acid (PNA) probes. Hundreds of thousands of probes attached to the chip are hybridized with RNA from tumor and normal tissue. Comparison of the hybridization signal between the normal and tumor RNA shows upregulated and downregulated genes.[860,861] The gene expression profiles are expected to define new diagnostic and prognostic markers and be useful in defining possible therapeutic targets.[862] Gene expression profiling has been used to identify novel fusion genes, including PAX3-NCOA1 in alveolar rhabdomyosarcoma,[251] CSPG2-PTK2B in DFSP,[158] and COL6A3-CSF1 in tenosynovial giant cell tumor.[258] However, application of this technology to the pathology of soft tissue tumors is still limited.[863,864] Common obstacles include a lack of normal RNA counterparts to the tumor RNA and the lack of availability of fresh-frozen tissue necessary to extract large amounts of RNA. The nature of soft tissue tumors, including the complex cytologic composition of

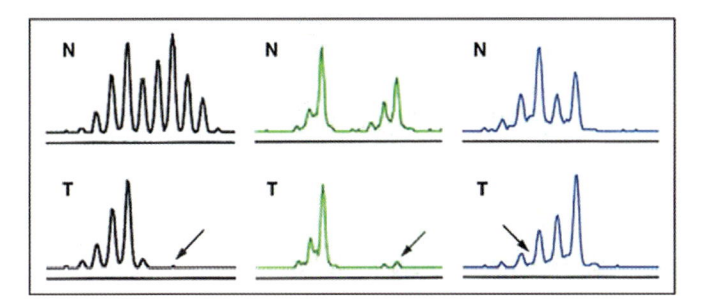

Figure 5.3 Examples of LOH detected in tumor DNA using PCR amplification of microsatellite markers and capillary gel electrophoresis. N and T refer to normal and tumor DNA. (Arrows indicate lost alleles in tumors.)

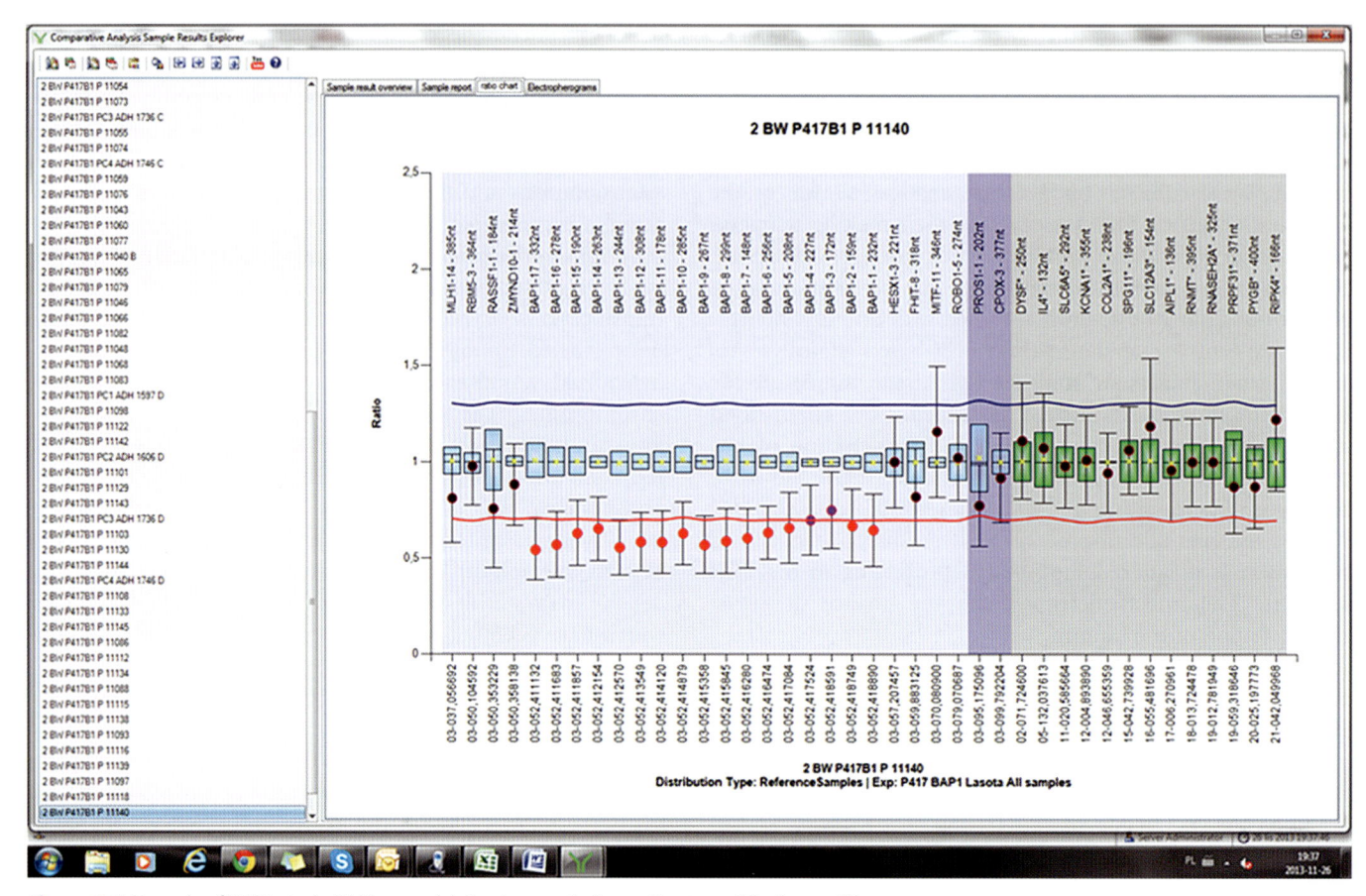

Chapter 5: Molecular genetics of soft tissue tumors

Figure 5.4 Example of MLPA study. BAP1 gene deletion in mesothelioma. Courtesy of Dr. Bartosz Wasag.

the lesions, contributes to technical problems and possible misinterpretations.

Nucleic acid sequencing is a method used to identify the order of nucleotide bases (dNTPs), adenine (A), guanine (G), cytosine (C), and thymine (T), in an analyzed DNA molecule. The chain-termination method, also referred to as dideoxy sequencing, was developed by Edward Sanger in 1975. This method with some modifications has been used for almost four decades and is considered the gold standard in DNA sequencing.[865] Dideoxy sequencing is based on DNA chain elongation with dNTPs enriched by 1% of labeled dideoxynucleotides (ddNTPs), molecules that terminate DNA chain elongation because they cannot form a phosphodiester bond with the next deoxynucleotide. Labeled terminators such as ddATP, ddGTP, ddCTP, and ddTTP are detected following electrophoretic separation of randomly terminated DNA extension products and sequence of nucleotide is identified in a given DNA molecule.

Pyrosequencing (sequencing by synthesis) is a method of DNA sequencing determining the order of nucleotides by monitoring DNA synthesis. Immobile single-strand DNA is hybridized to a sequencing primer and incubated with the enzymes. The A, C, G, and T nucleotides are sequentially added and removed from the reaction. Incorporation of the complementary to the first unpaired base nucleotide produces chemiluminescent signals, which allows the sequence to be determined. This technique is slightly more sensitive than Sanger sequencing. However, pyrosequencing reads of DNA sequences are shorter than those obtained by the Sanger method. Several pyrosequencing platforms with clinical applications have been developed and are commercially available.[838]

Although DNA fragments up to 1 kb in length can be read by the Sanger sequencing method, daily throughput of the Sanger automated instrument is limited to 115 kb. For example, to sequence an entire human genome using this method on one machine would take around 60 years and cost 5 to 30 million US dollars.[866,867] Therefore, growing demands for a faster generation of sequencing data are leading to development of high-throughput, low-cost NGS technologies.

Although *next-generation sequencing* (NGS) was primarily developed as a research tool, it is rapidly becoming an important tool in clinical molecular oncology.[868] Even though NGS platforms are unique and based on different chemistry, the general sequencing strategy employed by these systems remains the same and includes specific preparation of the template, sequencing, and imaging and bioinformatics data analysis. In general, double-stranded DNA is converted into a "sequencing library." This requires fragmentation of DNA molecules, selection of the sequenceable fragments, and ligation of the synthetic DNAs, which serve as primers for amplification or sequencing to the ends of the selected fragments. Also, template preparation separates/immobilizes DNA fragments by attaching them to the solid surface or to the bids. The majority of NGS platforms require initial template amplification. This step, however, can introduce errors into sequencing and theoretically reduce the detection of low-level variant clones. The latter problems could be bypassed by employing a single-molecule sequencing approach; however, single-molecule sequencing platforms are still at the developmental stage. Although NGS platforms differ substantially in their chemistry and nucleotide sequence detection mechanisms, a majority of them rely on the "sequencing by synthesis" strategy, employing DNA polymerase or the ligase enzyme to synthesize new DNA fragments from the library DNA fragments. Completed NGS reactions can generate a huge volume of data ranging from mega- to gigabases. Analysis of such data could be time-consuming and may require special knowledge of bioinformatics. The growing number of NGS applications highlights the development of a foundation for a new generation of personalized genome-based medicine. It is believed that some NGS applications may replace other molecular techniques, including microarray analysis (Figure 5.5).

In soft tissue pathology, the NGS strategy was successfully used by several studies. For example, an NGS-based study identified novel PTPRB and PLCG1 mutations in angiosarcomas[869] and a novel EWSR1-CREB3L1 gene fusion in sclerosing epithelioid fibrosarcoma.[870] Also, targeted NGS studies of cancer-related genes in soft tissue sarcomas offered a comprehensive genetic profile of analyzed tumors, which could be used for risk stratification and tailoring rational, personalized therapy.[871,872] Ultra-deep targeted NGS (UDT-sequencing) is well suited to study low copy-number targets in the cases of clonal selection and identify subclonal somatic mutations in heterogeneous samples.[873] More recently, anchored multiplex PCR (AMP), a rapid method for gene rearrangement detection by NGS has been developed.[874]

NGS research has revealed a larger than expected complexity of intertumor and intratumor heterogeneity.[875] Apparently primary tumors often harbor "private genetic mutations" limited to specific subclones, which is in addition to the driver genetic mutation presented in all cells. Over time, these mutations may become secondary driver mutations and provide a growth or survival advantage to the host clone.[876] It is now believed that a great majority of malignant tumors by the time of diagnosis consist of multiple distinct clones. These clones might be difficult to detect though, especially in tumors driven by simple balanced translocations.[877] Since the same gene may be affected by either genetic or epigenetic mutations, or a combination of both, identification of driving mutations might require the application of different molecular technologies.[875] The UDT-sequencing of tumor samples collected over the course of the disease could elucidate heterogeneity in a given tumor.[873]

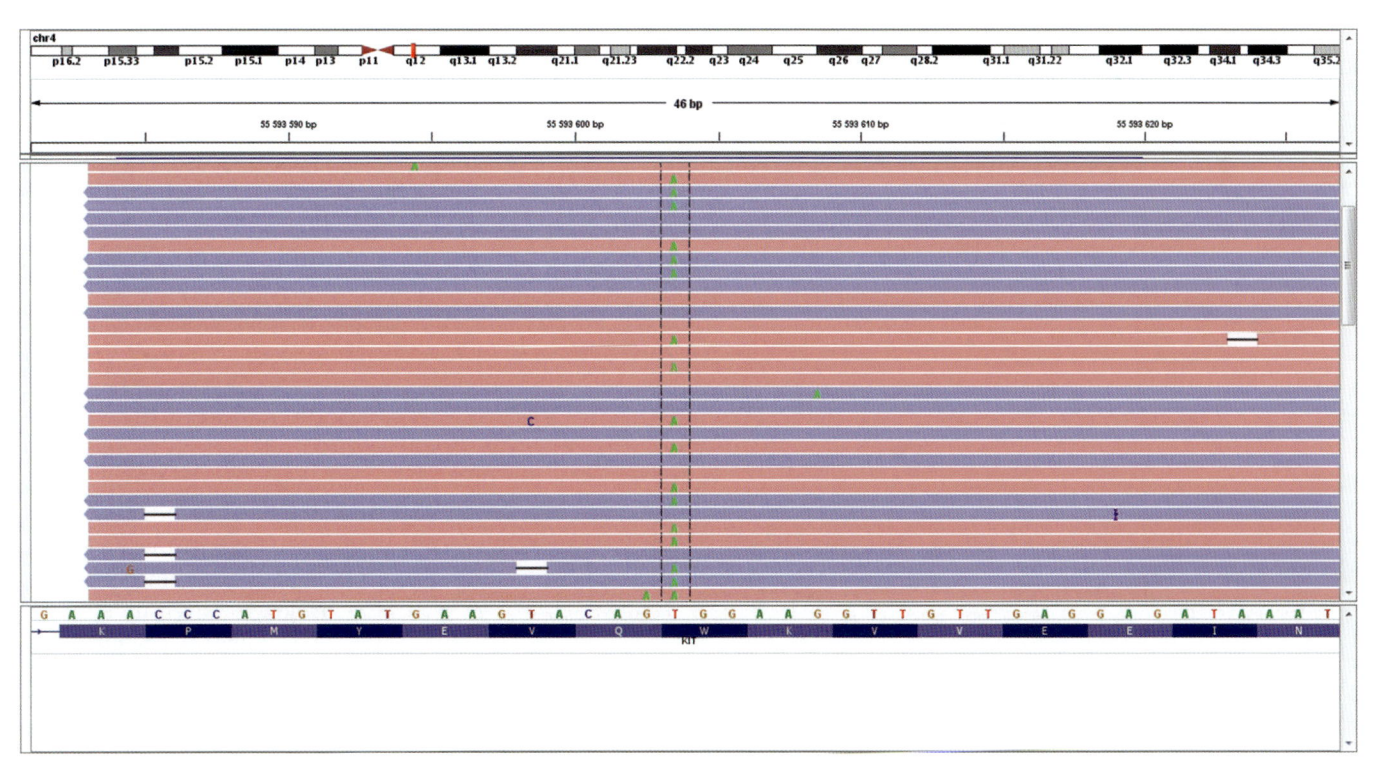

Figure 5.5 Example of next-generation sequencing of KIT gene. Integrative Genomics Viewer visualization of KITc.1669T>C (W557R) oncogenic mutation in gastrointestinal stromal tumor. Courtesy of Dr. Artur Kowalik.

Genes and their names

According to the Human Genome Project, approximately 24 000 genes will be gradually identified in the human genome.[878] Thus, uniform genetic nomenclature, which will ensure proper communication among molecular biologists, pathologists, and clinicians, is extremely important. Unfortunately, some of the genes are recognized by more than one name, while with others the same name may be given to different genes. This can be confusing for those who did not follow the history of genetic discoveries and do not know that some genes were cloned simultaneously by different groups of investigators or were found to be altered in different diseases and therefore initially were called by different names. These names circulate in scientific literature and will continue to be used for some time before uniform genetic nomenclature is widely accepted. *Gene names and symbols used in this chapter are recommended by HUGO Gene Nomenclature Committee (www.genenames.org).*

Glossary

Gene symbol	Gene name	Cytogenetic location	OMIM#	Gene ID
ACTB	ACTIN, BETA	7p22.1	102630	60
AGR3	ANTERIOR GRADIENT 3	7p21.1	609482	155465
AHRR	ARYLHYDROCARBON RECEPTOR REPRESSOR	5p15.33	606517	57491
AKT1	V-AKT MURINE THYMOMA VIRAL ONCOGENE HOMOLOG 1	14q32.33	164730	207
ALDH2	ALDEHYDE DEHYDROGENASE 2 FAMILY	12q24.12	100650	217
ALK	ANAPLASTIC LYMPHOMA KINASE	2p23.2	105590	238
ANTXR2	ANTHRAX TOXIN RECEPTOR 2	4q21.21	608041	118429
APC	APC GENE	5q22.2	611731	324
ASPSCR1	ALVEOLAR SOFT PART SARCOMA CHROMOSOME REGION, CANDIDATE 1	17q25.3	606236	79058
ASTN2	ASTROTACTIN 2	9q33.1	612856	23245
ATF1	ACTIVATING TRANSCRIPTION FACTOR 1	12q13.12	123803	466
ATIC	5-AMINOIMIDAZOLE-4-CARBOXAMIDE RIBONUCLEOTIDE FORMYLTRANSFERASE/IMP CYCLOHYDROLASE	2q35	601731	471
ATM	ATAXIA-TELANGIECTASIA MUTATED GENE	11q22.3	607585	472
BAP1	BRCA1-ASSOCIATED PROTEIN 1	3p21.1	603089	8314
BCOR	BCL6 COREPRESSOR	Xp11.4	300485	54880
C2orf44	CHROMOSOME 2 OPEN READING FRAME 44	2p23.3	616234	80304
C11orf95	CHROMOSOME 11 OPEN READING FRAME 95	11q13.1	615699	65998
CAMTA1	CALMODULIN-BINDING TRANSCRIPTION ACTIVATOR 1	1p36.31-p36.23	611501	23261
CARS	CYSTEINYL-tRNA SYNTHETASE	11p15.4	123859	833
CCDC6	COILED-COIL DOMAIN-CONTAINING PROTEIN 6	10q21.2	601985	8030
CCNB1IP1	CYCLIN B1 INTERACTING PROTEIN 1	14q11.2	608249	57820
CCNB3	CYCLIN B3	Xp11.22	300456	85417
CCND1	CYCLIN D1	11q13.3	168461	595
CCNE1	CYCLIN E1	19q12	123837	898
CD74	CD74 ANTIGEN	5q32	142790	972
CDH3	CADHERIN 3	16q22.1	114021	1001
CDH11	CADHERIN 11	16q21	600023	1009
CDK2	CYCLIN-DEPENDENT KINASE 2	12q13.2	116953	1017

(cont.)

Gene symbol	Gene name	Cytogenetic location	OMIM#	Gene ID
CDK4	CYCLIN-DEPENDENT KINASE 4	12q14.1	123829	1019
CDK6	CYCLIN-DEPENDENT KINASE 6	7q21.2	603368	1021
CDKN1C	CYCLIN-DEPENDENT KINASE INHIBITOR 1C	11p15.4	600856	1028
CDKN2A	CYCLIN-DEPENDENT KINASE INHIBITOR 2A	9p21.3	600160	1031
CDX1	CAUDAL-TYPE HOMEOBOX TRANSCRIPTION FACTOR 1	5q32	600746	1044
CEP85L	Centrosomal protein 85kD-like	6q22.31	NA	387119
CHCHD7	COILED-COIL-HELIX-COILED-COIL-HELIX DOMAIN-CONTAINING PROTEIN 7	8q12.1	611238	79145
CHEK2	CHECKPOINT KINASE 2, S. POMBE, HOMOLOG OF	22q12.1	604373	11200
CIC	CAPICUA, DROSOPHILA, HOMOLOG OF	19q13.2	612082	23152
CLTC	CLATHRIN, HEAVY POLYPEPTIDE	17q23.1	118955	1213
CLTCL1	CLATHRIN, HEAVY POLYPEPTIDE-LIKE 1	22q11.21	601273	8218
CMKLR1	CHEMOKINE-LIKE RECEPTOR 1	12q23.3	602351	1240
CNOT2	CCR4-NOT TRANSCRIPTION COMPLEX, SUBUNIT 2	12q15	604909	4848
CNR1	CANNABINOID RECEPTOR 1	6q15	114610	1268
COG5	COMPONENT OF OLIGOMERIC GOLGI COMPLEX 5	7q22.3	606821	10466
COL1A1	COLLAGEN, TYPE I, ALPHA-1	17q21.33	120150	1277
COL1A2	COLLAGEN, TYPE I, ALPHA-2	7q21.3	120160	1278
COL6A3	COLLAGEN, TYPE VI, ALPHA-3	2q37.3	120250	1293
COX6C	CYTOCHROME c OXIDASE, SUBUNIT VIc	8q22.2	124090	1345
CRADD	CASP2 AND RIPK1 DOMAIN-CONTAINING ADAPTOR WITH DEATH DOMAIN	12q22	603454	8738
CREB1	cAMP RESPONSE ELEMENT-BINDING PROTEIN 1	2q33.3	123810	1385
CREB3L1	cAMP RESPONSE ELEMENT-BINDING PROTEIN 3-LIKE 1	11p11.2	616215	90993
CREB3L2	cAMP RESPONSE ELEMENT-BINDING PROTEIN 3-LIKE 2	7q33	608834	64764
CREBBP	CREB-BINDING PROTEIN	16p13.3	600140	1387
CREM	cAMP RESPONSE ELEMENT MODULATOR	10p11.21	123812	1390
CSF1	COLONY-STIMULATING FACTOR 1	1p13.3	120420	1435
CTDSP2	CTD SMALL PHOSPHATASE 2	12q14.1	608711	10106
CTNNB1	CATENIN, BETA-1	3p22.1	116806	1499
CUTL1	CUT-LIKE 1	7q22.1	116896	1523
CXCR4	CHEMOKINE, CXC MOTIF, RECEPTOR 4	2q22.1	162643	7852
CXCR7	CHEMOKINE, CXC MOTIF, RECEPTOR 7	2q37.3	610376	57007
CXorf67	Chromosome X open reading frame 67	Xp11.22	NA	340602
DCTN1	DYNACTIN 1	2p13.1	601143	1639
DDIT3	DNA DAMAGE-INDUCIBLE TRANSCRIPT 3	12q13.3	126337	1649
DKK1	DICKKOPF, XENOPUS, HOMOLOG OF, 1	10q21.1	605189	22943
DUX4	DOUBLE HOMEOBOX PROTEIN 4	4q35	606009	100288687
E2F3	E2F TRANSCRIPTION FACTOR 3	6p22.3	600427	1871
EBF1	EARLY B-CELL FACTOR 1	5q33.3	164343	1879

Chapter 5: Molecular genetics of soft tissue tumors

(cont.)

Gene symbol	Gene name	Cytogenetic location	OMIM#	Gene ID
EGFR	EPIDERMAL GROWTH FACTOR RECEPTOR	7p11.2	131550	1956
EGR1	early growth response 1	5q31.2	128990	1958
EML4	ECHINODERM MICROTUBULE ASSOCIATED PROTEIN LIKE-4	2p21	607442	27436
EP300	E1A-BINDING PROTEIN, 300-KD	22q13.2	602700	2033
EP400	E1A-BINDING PROTEIN, 400-KD	12q24.33	606265	57634
EPC1	ENHANCER OF POLYCOMB, DROSOPHILA, HOMOLOG OF, 1	10p11.22	610999	80314
ERBB2	V-ERB-B2 AVIAN ERYTHROBLASTIC LEUKEMIA VIRAL ONCOGENE HOMOLOG 2	17q12	164870	2064
ERBB3	V-ERB-B2 AVIAN ERYTHROBLASTIC LEUKEMIA VIRAL ONCOGENE HOMOLOG 3	12q13.2	190151	2065
ERG	V-ETS AVIAN ERYTHROBLASTOSIS VIRUS E26 ONCOGENE HOMOLOG	21q22.2	165080	2078
ETV1	ETS VARIANT GENE 1	7p21.2	600541	2115
ETV4	ETS VARIANT GENE 4	17q21.31	600711	2118
ETV6	ETS VARIANT GENE 6	12p13.2	600618	2120
EWSR1	EWING SARCOMA BREAKPOINT REGION 1	22q12.2	133450	2130
EZR	EZRIN	6q25.3	123900	7430
FAM19A2	family with sequence similarity 19 (chemokine (C-C motif)-like), member A2	12q14.1	NA	338811
FEV	FIFTH EWING SARCOMA VARIANT	2q35	607150	54738
FGFR1	FIBROBLAST GROWTH FACTOR RECEPTOR 1	8p11.23-p11.22	136350	2260
FGFR4	FIBROBLAST GROWTH FACTOR RECEPTOR 4	5q35.2	134935	2264
FH	FUMARATE HYDRATASE	1q43	136850	2271
FHIT	FRAGILE HISTIDINE TRIAD GENE	3p14.2	601153	2272
FLI1	FRIEND LEUKEMIA VIRUS INTEGRATION 1	11q24.3	193067	2313
FN1	FIBRONECTIN 1	2q35	135600	2335
FOSB	V-FOS FBJ MURINE OSTEOSARCOMA VIRAL ONCOGENE HOMOLOG B	19q13.32	164772	2354
FOSL1	FOS-LIKE ANTIGEN 1	11q13.1	136515	8061
FOXO1A	FORKHEAD BOX O1A	13q14.11	136533	2308
FOXO3A	FORKHEAD BOX O3A	6q21	602681	2309
FOXO4	FORKHEAD BOX O4	Xq13.1	300033	4303
FUS	FUSED IN SARCOMA	16p11.2	137070	2521
GLI	GLIOMA-ASSOCIATED ONCOGENE HOMOLOG	12q13.3	165220	2735
GLMN	GLOMULIN	1p22.1	601749	11146
GTF2I	GENERAL TRANSCRIPTION FACTOR II-I	7q11.23	601679	2969
H19	H19, IMPRINTED MATERNALLY EXPRESSED NONCODING TRANSCRIPT	11p15.5	103280	283120
HAS2	HYALURONAN SYNTHASE 2	8q24.13	601636	3037
HEY1	HAIRY/ENHANCER OF SPLIT-RELATED WITH YRPW MOTIF 1	8q21	602953	23462
HGF	HEPATOCYTE GROWTH FACTOR	7q21.11	142409	3082
HIF1A	HYPOXIA-INDUCIBLE FACTOR 1, ALPHA SUBUNIT	14q23.2	603348	3091
HMGA1	HIGH MOBILITY GROUP AT-HOOK 1	6q21.31	600701	3159

(cont.)

Gene symbol	Gene name	Cytogenetic location	OMIM#	Gene ID
HMGA2	HIGH MOBILITY GROUP AT-HOOK 2	12q14.3	600698	8091
HNF1A	HNF1 HOMEOBOX A	12q24.31	142410	6927
HPRT1	HYPOXANTHINE GUANINE PHOSPHORIBOSYLTRANSFERASE 1	Xq26.2-q26.3	308000	3251
HRAS	V-HA-RAS HARVEY RAT SARCOMA VIRAL ONCOGENE HOMOLOG	11p15.5	190020	3265
IDH1	ISOCITRATE DEHYDROGENASE 1	2q34	147700	3417
IDH2	ISOCITRATE DEHYDROGENASE 2	15q26.1	147650	3418
IDH3A	ISOCITRATE DEHYDROGENASE 3, ALPHA SUBUNIT	15q25.1	601149	3419
IDH3B	ISOCITRATE DEHYDROGENASE 3, BETA SUBUNIT	20p13	604526	3420
IFNG	INTERFERON, GAMMA	12q15	147570	3458
IGF1	INSULIN-LIKE GROWTH FACTOR I	12q23.2	147440	3479
IGF1R	INSULIN-LIKE GROWTH FACTOR I RECEPTOR	15q26.3	147370	3480
IGF2	INSULIN-LIKE GROWTH FACTOR II	11p15.5	147470	3481
IGF2R	INSULIN-LIKE GROWTH FACTOR II RECEPTOR	6q25.3	147280	3482
IGFBP7	INSULIN-LIKE GROWTH FACTOR-BINDING PROTEIN 7	4q12	602867	3490
INSR	INSULIN RECEPTOR	19p13.2	147670	3643
IRF2BP2	INTERFERON REGULATORY FACTOR 2 BINDING PROTEIN 2	1q42.3	615332	359948
JAZF1	JUXTAPOSED WITH ANOTHER ZINC FINGER GENE 1	7p15.2-p15.1	606246	221895
KANSL1	KAT8 REGULATORY NSL COMPLEX, SUBUNIT 1	17q31.31	612452	284058
KAT2A	LYSINE ACETYLTRANSFERASE 2A	17q21.2	602301	2648
KAT6A	LYSINE ACETYLTRANSFERASE 6A	8p11.21	601408	7994
KAT6B	LYSINE ACETYLTRANSFERASE 6B	10q22.2	605880	23522
KCNQ1OT1	KCNQ1-OVERLAPPING TRANSCRIPT 1	11p15.5	604115	10984
KDELR2	KDEL ENDOPLASMIC RETICULUM PROTEIN RETENTION RECEPTOR 2	7p22.1	609024	11014
KIF5B	KINESIN FAMILY MEMBER 5B	10p11.22	602809	3799
KIT	V-KIT HARDY-ZUCKERMAN 4 FELINE SARCOMA VIRAL ONCOGENE HOMOLOG	4q12	164920	16590
KITLG	KIT LIGAND	12q21.32	184745	4254
KLC1	KINESIN LIGHT CHAIN 1	14q32.33	600025	3831
KMT2A	LYSINE-SPECIFIC METHYLTRANSFERASE 2A	11q23.3	159555	4297
KRAS	V-KI-RAS2 KIRSTEN RAT SARCOMA VIRAL ONCOGENE HOMOLOG	12p12.1	190070	3845
LGR5	LEUCINE-RICH REPEAT-CONTAINING G PROTEIN-COUPLED RECEPTOR 5	12q21.1	606667	8549
LHFP	LIPOMA HMGIC FUSION PARTNER	13q13.3-q14.1	606710	10186
LIFR	LEUKEMIA INHIBITORY FACTOR RECEPTOR	5p13.1	151443	3977
LMNA	LAMIN A/C	1q22	150330	4000
LPP	LIM DOMAIN-CONTAINING PREFERRED TRANSLOCATION PARTNER IN LIPOMA	3q27-q28	600700	4026
LRIG3	LEUCINE-RICH REPEATS- AND IMMUNOGLOBULIN-LIKE DOMAINS-CONTAINING PROTEIN 3	12q14.1	608870	121227
LRP1	LOW DENSITY LIPOPROTEIN RECEPTOR-RELATED PROTEIN 1	12q13.3	107770	4035
LRRC15	LEUCINE RICH REPEAT CONTAINING 15	3q29	NA	131578

Chapter 5: Molecular genetics of soft tissue tumors

(cont.)

Gene symbol	Gene name	Cytogenetic location	OMIM#	Gene ID
MBTD1	MBT DOMAIN CONTAINING 1	17q21.33	NA	54799
MDM2	MOUSE DOUBLE MINUTE 2 HOMOLOG	12q15	164785	4193
MEAF6	MYST/ESA1-ASSOCIATED FACTOR 6	1p34.3	611001	64769
MET	MET PROTOONCOGENE	7q31.2	164860	4233
MGEA5	MENINGIOMA-EXPRESSED ANTIGEN 5	10q24.32	604039	10724
MKL2	MYOCARDIN-LIKE 2	16p13.12	609463	57496
MSN	MOESIN	Xq12	309845	4478
MTOR	MECHANISTIC TARGET OF RAPAMYCIN	1p36.22	601231	2475
MYC	V-MYC AVIAN MYELOCYTOMATOSIS VIRAL ONCOGENE HOMOLOG	8q24.21	190080	4609
MYCL	V-MYC AVIAN MYELOCYTOMATOSIS VIRAL ONCOGENE HOMOLOG, LUNG CARCINOMA-DERIVED	1p34.2	164850	4610
MYCN	V-MYC AVIAN MYELOCYTOMATOSIS VIRAL-RELATED ONCOGENE, NEUROBLASTOMA-DERIVED	2p24.3	164840	4613
MYH9	MYOSIN, HEAVY CHAIN 9, NONMUSCLE	22q12.3	160775	4627
MYOD1	MYOGENIC DIFFERENTIATION ANTIGEN 1	11p15.1	159970	17927
NAB2	NGFIA-BINDING PROTEIN 2	12q13.3	602381	4665
NBN	NIBRIN	8q21.3	602667	4683
NCOA1	NUCLEAR RECEPTOR COACTIVATOR 1	2p23.3	602691	4648
NCOA2	NUCLEAR RECEPTOR COACTIVATOR 2	8q13.3	601993	10499
NF1	NEUROFIBROMIN 1	17q11.2	162200	4763
NF2	NEUROFIBROMIN 2	22q12.2	607379	4771
NFATC1	NUCLEAR FACTOR OF ACTIVATED T CELLS, CYTOPLASMIC, CALCINEURIN-DEPENDENT 1	18q23	600489	4772
NFATC2	NUCLEAR FACTOR OF ACTIVATED T CELLS, CYTOPLASMIC, CALCINEURIN-DEPENDENT 2	20q13.2	600490	4773
NFIB	NUCLEAR FACTOR I/B	9p23-p22	600728	4781
NONO	NON-POU DOMAIN-CONTAINING OCTAMER-BINDING PROTEIN	Xq13.1	300084	4841
NPM1	NUCLEOPHOSMIN/NUCLEOPLASMIN FAMILY, MEMBER 1	5q35.1	164040	4869
NR4A3	NUCLEAR RECEPTOR SUBFAMILY 4, GROUP A, MEMBER 3	9q22.3-q31.1	600542	8013
NR6A1	NUCLEAR RECEPTOR SUBFAMILY 6, GROUP A, MEMBER 1	9q33.3	602778	2649
NRAS	NEUROBLASTOMA RAS VIRAL ONCOGENE HOMOLOG	1p13.2	164790	4893
NSD1	NUCLEAR RECEPTOR-BINDING Su-var, ENHANCER OF ZESTE, AND TRITHORAX DOMAIN PROTEIN 1	5q35.2-q35.3	606681	64324
NTF3	NEUROTROPHIN 3	12p13.31	162660	4908
NTRK1	NEUROTROPHIC TYROSINE KINASE, RECEPTOR, TYPE 1	1q23.1	191315	4914
NTRK3	NEUROTROPHIC TYROSINE KINASE, RECEPTOR, TYPE 3	15q25.3	191316	4916
NUP107	NUCLEOPORIN, 107-KD	12q15	607617	57122
NUTM2A	NUT family member 2A	10q23.2	NA	728118
NUTM2B	NUT family member 2B	10q22.3	NA	729262
OMD	OSTEOMODULIN	9q22.31	NA	4958
PAPPA	PREGNANCY-ASSOCIATED PLASMA PROTEIN A	9q33.1	176385	5069

Chapter 5: Molecular genetics of soft tissue tumors

(cont.)

Gene symbol	Gene name	Cytogenetic location	OMIM#	Gene ID
PAX3	PAIRED BOX GENE 3	2q36.1	606597	5077
PAX7	PAIRED BOX GENE 7	1p36.13	167410	5081
PBX1	PRE-B-CELL LEUKEMIA TRANSCRIPTION FACTOR 1	1q23.3	176310	5087
PDGFB	PLATELET-DERIVED GROWTH FACTOR, BETA POLYPEPTIDE	22q13.1	190040	5155
PDGFRA	PLATELET-DERIVED GROWTH FACTOR RECEPTOR, ALPHA	4q12	173490	5156
PDGFRB	PLATELET-DERIVED GROWTH FACTOR RECEPTOR, BETA	5q32	173410	5159
PGK1	PHOSPHOGLYCERATE KINASE 1	Xq21.1	311800	5230
PHF1	PHD FINGER PROTEIN 1	6p21.32	602881	5252
PLAG1	PLEOMORPHIC ADENOMA GENE 1	8q12.1	603026	5324
PLCG1	PHOSPHOLIPASE C, GAMMA-1	20q12	172420	5335
POU5F1	POU DOMAIN, CLASS 5, TRANSCRIPTION FACTOR 1	6p21.33	164177	5460
PPAP2B	PHOSPHATIDIC ACID PHOSPHATASE TYPE 2B	1p32.2	607125	8613
PPFIBP1	PROTEIN-TYROSINE PHOSPHATASE, RECEPTOR-TYPE, F POLYPEPTIDE-INTERACTING PROTEIN-BINDING PROTEIN 1	12p12.1	603141	8496
PRCC	PAPILLARY RENAL CELL CARCINOMA TRANSLOCATION-ASSOCIATED GENE	1q23.1	179755	5546
PRKAR1A	PROTEIN KINASE, cAMP-DEPENDENT, REGULATORY, TYPE I, ALPHA	17q24.2	188830	5573
PTCH1	PATCHED, DROSOPHILA, HOMOLOG OF, 1	9p22.32	601309	5727
PTEN	PHOSPHATASE AND TENSIN HOMOLOG	10q23.31	601728	5728
PTK2	PROTEIN-TYROSINE KINASE, CYTOPLASMIC	8q24.3	600758	5747
PTK2B	PROTEIN-TYROSINE KINASE 2, BETA	8p21.2	601212	2185
PTPRB	PROTEIN-TYROSINE PHOSPHATASE, RECEPTOR-TYPE, BETA	12q15	176882	5787
RAD51L1	RAD51, S. CEREVISIAE, HOMOLOG OF, B	14q24.1	602948	5890
RANBP2	RAN-BINDING PROTEIN 2	2q12.3	601181	5903
RAP1B	RAS-RELATED PROTEIN RAP1B	12q15	179530	5908
RARA	RETINOIC ACID RECEPTOR, ALPHA	17q21.2	180240	5914
RB1	RB1 GENE	13q14.2	614041	5925
RBL2	RETINOBLASTOMA-LIKE 2	16q12.2	180203	5934
RCOR1	REST COREPRESSOR	14q32.31	607675	23186
RECQL2	RECQ PROTEIN-LIKE 2	8p12	604611	7486
RHO	RHODOPSIN	3q22.1	180380	6010
RNF213	RING FINGER PROTEIN 213	17q25.3	613768	57674
ROS1	V-ROS AVIAN UR2 SARCOMA VIRUS ONCOGENE HOMOLOG 1	6q22.1	165020	6098
RPSAP52	RIBOSOMAL PROTEIN SA PSEUDOGENE 52	12q14.3	NA	204010
RREB1	RAS-RESPONSIVE ELEMENT BINDING PROTEIN 1	6p24.3	602209	6239
S100A10	S100 CALCIUM-BINDING PROTEIN A10	1q21.3	114085	6281
SDC4	SYNDECAN 4	20q13.12	600017	6385
SDHA	SUCCINATE DEHYDROGENASE COMPLEX, SUBUNIT A, FLAVOPROTEIN	5p15.33	600857	6389
SDHAF2	SUCCINATE DEHYDROGENASE COMPLEX ASSEMBLY FACTOR 2	11q12.2	613019	54949

Chapter 5: Molecular genetics of soft tissue tumors

(cont.)

Gene symbol	Gene name	Cytogenetic location	OMIM#	Gene ID
SDHB	SUCCINATE DEHYDROGENASE COMPLEX, SUBUNIT B, IRON SULFUR PROTEIN	1p36.13	185470	6390
SDHC	SUCCINATE DEHYDROGENASE COMPLEX, SUBUNIT C, INTEGRAL MEMBRANE PROTEIN, 15-KD	1q23.3	602413	6391
SDHD	SUCCINATE DEHYDROGENASE COMPLEX, SUBUNIT D, INTEGRAL MEMBRANE PROTEIN	11q23.1	602690	6392
SEC31A	SEC31, YEAST, HOMOLOG OF, A	4q21.22	610257	22872
SERPINE1	SERPIN PEPTIDASE INHIBITOR, CLADE E, MEMBER 1	7q22.1	NA	5054
SFPQ	SPLICING FACTOR, PROLINE- AND GLUTAMINE-RICH	1p34.3	605199	6421
SLC2A4	SOLUTE CARRIER FAMILY 2 (FACILITATED GLUCOSE TRANSPORTER), MEMBER 4	17p13.1	138190	6517
SLC34A2	SOLUTE CARRIER FAMILY 34 (SODIUM/PHOSPHATE COTRANSPORTER), MEMBER 2	4p15.2	604217	10568
SMAD3	MOTHERS AGAINST DECAPENTAPLEGIC, DROSOPHILA, HOMOLOG OF, 3	15q22.33	603109	4088
SMARCA4	SWI/SNF-RELATED, MATRIX-ASSOCIATED, ACTIN-DEPENDENT REGULATOR OF CHROMATIN, SUBFAMILY A, MEMBER 4	19p13.2	603254	6597
SMARCA5	SWI/SNF-RELATED, MATRIX-ASSOCIATED, ACTIN-DEPENDENT REGULATOR OF CHROMATIN, SUBFAMILY A, MEMBER 5	4q31.21	603375	8467
SMARCB1	SWI/SNF-RELATED, MATRIX-ASSOCIATED, ACTIN-DEPENDENT REGULATOR OF CHROMATIN, SUBFAMILY B, MEMBER 1	22q11.23	601607	6598
SP3	TRANSCRIPTION FACTOR Sp3	2q31.1	601804	6670
SQSTM1	SEQUESTOSOME 1	5q35.3	601530	8878
SRC	V-SRC AVIAN SARCOMA (SCHMIDT-RUPPIN A-2) VIRAL ONCOGEN	20q11.23	190090	6714
SRF	SERUM RESPONSE FACTOR	6p21.1	600589	6722
SS18	SYNOVIAL SARCOMA TRANSLOCATION, CHROMOSOME 18	18q11.2	600192	6760
SS18L1	SS18-LIKE GENE 1	20q13.33	606472	26039
SSX1	SARCOMA, SYNOVIAL, X BREAKPOINT 1	Xp11.23	312820	6756
SSX2	SARCOMA, SYNOVIAL, X BREAKPOINT 2	Xp11.23	300192	6757
SSX4	SYNOVIAL SARCOMA, X BREAKPOINT 4	Xp11.23	300326	6759
STAT6	SIGNAL TRANSDUCER AND ACTIVATOR OF TRANSCRIPTION 6	12q13.3	601512	6778
SUZ12	SUPPRESSOR OF ZESTE 12, DROSOPHILA, HOMOLOG OF	17q11.2	606245	23512
TAF15	TAF15 RNA POLYMERASE II, TATA BOX-BINDING PROTEIN-ASSOCIATED FACTOR, 68-KD	17q12	601574	8148
TCEA1	TRANSCRIPTION ELONGATION FACTOR A, 1	8q11.23	601425	6917
TCF3	TRANSCRIPTION FACTOR 3	19p13.3	147141	6929
TCF12	TRANSCRIPTION FACTOR 12	15q21.3	600480	6938
TEAD1	TEA DOMAIN FAMILY MEMBER 1	11p15.3-p15.2	189967	7003
TFE3	TRANSCRIPTION FACTOR FOR IMMUNOGLOBULIN HEAVY-CHAIN ENHANCER 3	Xp11.23	314310	7030
TFG	TRK-FUSED GENE	3q12.2	602498	10342
TGFBR3	TRANSFORMING GROWTH FACTOR-BETA RECEPTOR, TYPE III	1p22.1	600742	7049

(cont.)

Gene symbol	Gene name	Cytogenetic location	OMIM#	Gene ID
THBS1	THROMBOSPONDIN I	15q14	188060	7057
THRAP3	THYROID HORMONE RECEPTOR-ASSOCIATED PROTEIN 3	1p34.3	603809	9967
TMPRSS2	TRANSMEMBRANE PROTEASE, SERINE 2	21q22.3	602060	7113
TP53	TUMOR PROTEIN p53	17p13.1	191170	7157
TPM3	TROPOMYOSIN 3	1q21.3	191030	7170
TPM4	TROPOMYOSIN 4	19p13.12	600317	7171
TRHDE	THYROTROPIN-RELEASING HORMONE-DEGRADING ECTOENZYME	12q21.1	606950	29953
TSC1	TSC1 GENE	9q34.13	605284	7248
TSC2	TSC2 GENE	16p13.3	191092	7249
TSPAN31	TETRASPANIN 31	12q14.1	181035	6302
USP6	UBIQUITIN-SPECIFIC PROTEASE 6	17p13.2	604334	9098
VCAN	VERSICAN	5q14.2-q14.3	118661	1462
VCL	VINCULIN	10q22.2	193065	7414
VHL	VHL GENE	3p25.3	608537	7428
WASF2	WASP PROTEIN FAMILY, MEMBER 2	1p36.11	605875	10163
WDR70	WD REPEAT DOMAIN 70	5p13.2	NA	55100
WT1	WT1 GENE	11p13	607102	7490
WWTR1	WW DOMAIN-CONTAINING TRANSCRIPTION REGULATOR 1	3q25.1	607392	25937
YAP1	YES-ASSOCIATED PROTEIN 1, 65-KD	11q22.1-q22.2	606608	10413
YWHAE	YROSINE 3-MONOOXYGENASE/TRYPTOPHAN 5-MONOOXYGENASE ACTIVATION PROTEIN, EPSILON ISOFORM	17p13.3	605066	7531
YY1	TRANSCRIPTION FACTOR YY1	14q32.2	600013	7528
ZC3H7B	ZINC FINGER CCCH-TYPE CONTAINING 7B	22q13.2	NA	23264
ZNF9	ZINC FINGER PROTEIN 9	3q21.3	116955	7555
ZNF44	ZINC FINGER PROTEIN 44	19p13.2	194542	51710
ZNF278	ZINC FINGER PROTEIN 278	22q12.2	605165	23598
ZNF393	ZINC FINGER PROTEIN 393	1p34.1	609602	128209
ZNF444	ZINC FINGER PROTEIN 444	19q13.43	607874	55311

References

Oncogenes

1. Cooper GM. Oncogenes, 2nd edn. Boston: Jones and Bartlett Publishers International; 1995.
2. Heim S, Mitelman F. Cancer Cytogenetics. Chromosomal and Molecular Genetic Aberrations of Tumor Cells. New York: Wiley-Liss; 1995.
3. Der CJ, Krontiris TG, Cooper GM. Transforming genes of human bladder and lung carcinoma cell lines are homologous to the ras genes of Harvey and Kirsten sarcoma viruses. Proc Natl Acad Sci USA 1982;79:3637–3640.
4. Parada LF, Tabin CJ, Shih C, Weinberg RA. Human EJ bladder carcinoma oncogene is homologue of Harvey sarcoma virus ras gene. Nature 1982;297:474–478.
5. Reddy EP, Reynolds RK, Santos E, Barbacid M. A point mutation is responsible for the acquisition of transforming properties by the T24 human bladder carcinoma oncogene. Nature 1982;300:149–152.
6. Barbacid M. ras genes. Annu Rev Biochem 1987;56:779–827.
7. Quilliam LA, Rebhun JF, Castro AF. A growing family of guanine nucleotide exchange factors is responsible for activation of Ras-family GTPases. Prog Nucleic Acid Res Mol Biol 2002;71:391–444.
8. Bos JL. Ras oncogenes in human cancer: a review. Cancer Res 1989;49:4682–4689.

9. Bohle RM, Brettreich S, Repp R, et al. Single somatic ras gene point mutation in soft tissue malignant fibrous histiocytomas. *Am J Pathol* 1996;148:731–738.

10. Yoo J, Robinson RA, Lee JY. H-ras and K-ras gene mutations in primary human soft tissue sarcomas: concomitant mutations of the ras genes. *Mod Pathol* 1999;12:775–780.

11. Hill MA, Gong C, Casey TJ, et al. Detection of K-ras mutations in resected primary leiomyosarcoma. *Cancer Epidemiol Biomarkers Prev* 1997;6:1095–1100.

12. Marion MJ, Froment O, Trepo C. Activation of Ki-ras gene by point mutation in human liver angiosarcoma associated with vinyl chloride exposure. *Mol Carcinog* 1991;4:450–454.

13. Przygodzki RM, Finkelstein SD, Keohavong P, et al. Sporadic and thorotrast-induced angiosarcomas of the liver manifest frequent and multiple point mutations in K-ras-2. *Lab Invest* 1997;76:153–159.

14. Stratton MR, Fisher C, Gusterson BA, Cooper CS. Detection of point mutations in N-ras and K-ras genes of human embryonal rhabdomyosarcomas using oligonucleotide probes and the polymerase chain reaction. *Cancer Res* 1989;49:6324–6327.

15. Rebollo A, Martinez-A C. Ras proteins: recent advances and new functions. *Blood* 1999;94:2971–2980.

16. Reuter CW, Morgan MA, Bergmann L. Targeting the Ras signaling pathway: a rational, mechanism-based treatment for hematologic malignancies? *Blood* 2000;96:1655–1669.

17. Adjei AA. Blocking oncogenic Ras signaling for cancer therapy. *J Natl Cancer Inst* 2001;93:1062–1074.

Tumor-suppressor genes

18. Knudson AJ. Mutation and cancer: statistical study of retinoblastoma. *Proc Natl Acad Sci USA* 1971;68:820–823.

19. Knudson AJ, Hethcote HW, Brown BW. Mutation and childhood cancer: a probabilistic model for the incidence of retinoblastoma. *Proc Natl Acad Sci USA* 1975;72:5116–5120.

20. Friend SH, Bernards R, Rogelj S, et al. A human DNA segment with properties of the gene that predisposes to retinoblastoma and osteosarcoma. *Nature* 1986;323:643–646.

21. Fung Y-KT, Murphree AL, T'Ang A, et al. Structural evidence for the authenticity of the human retinoblastoma gene. *Science* 1987;236:1657–1661.

22. Hong FD, Huang H-JS, To H, et al. Structure of the human retinoblastoma gene. *Proc Natl Acad Sci USA* 1989;86:5502–5506.

23. Dryja TP, Mukai S, Petersen R, et al. Parental origin of mutations of the retinoblastoma gene. *Nature* 1989;339:556–558.

24. Zhu XP, Dunn JM, Phillips RA, et al. Preferential germline mutation of the paternal allele in retinoblastoma. *Nature* 1989;340:312–313.

25. Horowitz JM, Yandell DW, Park S-H, et al. Point mutational inactivation of the retinoblastoma antioncogene. *Science* 1989;243:937–940.

26. Lohmann DR. RB1 mutations in retinoblastoma. *Hum Mutat* 1999;14:283–288.

27. Stirzaker C, Millar DS, Paul CL, et al. Extensive DNA methylation spanning the Rb promoter in retinoblastoma tumors. *Cancer Res* 1997;57:2229–2237.

28. Polsky D, Mastorides S, Kim D, et al. Altered patterns of RB expression define groups of soft tissue sarcoma patients with distinct biological and clinical behavior. *Histol Histopathol* 2006;21:743–752.

29. Volinia S, Calin GA, Liu CG, et al. A microRNA expression signature of human solid tumors defines cancer gene targets. *Proc Natl Acad Sci USA* 2006;103:2257–2261.

30. Dei Tos AP, Maestro R, Doglioni C, et al. Tumor suppressor genes and related molecules in leiomyosarcoma. *Am J Pathol* 1996;148:1037–1045.

31. Cohen JA, Geradts J. Loss of RB and MTS1/CDKN2 (p16) expression in human sarcomas. *Hum Pathol* 1997;28:893–898.

32. Stratton MR, Williams S, Fisher C, et al. Structural alterations of the RB1 gene in human soft tissue tumours. *Br J Cancer* 1989;60:202–205.

33. Wunder JS, Czitrom AA, Kandel R, Andrulis IL. Analysis of alterations in the retinoblastoma gene and tumor grade in bone and soft-tissue sarcomas. *J Natl Cancer Inst* 1991;83:194–200.

34. Fisher DE. *Tumor Suppressor Genes in Human Cancer.* Totowa, NJ: Humana Press; 2001.

Epimutations

35. Croce CM. Oncogenes and cancer. *N Engl J Med* 2008;358:502–511.

36. Roy DM, Walsh LA, Chan TA. Driver mutations of cancer epigenomes. *Protein Cell* 2014;5:265–296.

37. Holliday R. The inheritance of epigenetic defects. *Science* 1987;238:163–170.

38. Oey H, Whitelaw E. On the meaning of the word "epimutation." *Trends Genet* 2014;30:519–520.

39. Horsthemke B. Epimutations in human disease. *Curr Top Microbiol Immunol* 2006;310:45–59.

MicroRNAs (miRNAs)

40. Frith MC, Pheasant M, Mattick JS. The amazing complexity of the human transcriptome. *Eur J Hum Genet* 2005;13:894–897.

41. ENCODE Project Consortium. An integrated encyclopedia of DNA elements in the human genome. *Nature* 2012;489:57–74.

42. Sana J, Faltejskova P, Svoboda M, Slaby O. Novel classes of non-coding RNAs and cancer. *J Transl Med* 2012;10:103.

43. Taft RJ, Pang KC, Mercer TR, Dinger M, Mattick JS. Non-coding RNAs: regulators of disease. *J Pathol* 2010;220:126–139.

44. Sun G, Li H, Rossi JJ. Sequence context outside the target region influences the effectiveness of miR-223 target sites in the RhoB 3'UTR. *Nucleic Acids Res* 2010;38:239–252.

45. Gottwein E, Cai X, Cullen BR. A novel assay for viral microRNA function identifies a single nucleotide polymorphism that affects Drosha processing. *J Virol* 2006;80:5321–5326.

46. Ebhardt HA, Fedynak A, Fahlman RP. Naturally occurring variations in sequence length creates microRNA isoforms that differ in argonaute effector complex specificity. *Silence* 2010;1:12.

47. Carleton M, Cleary MA, Linsley PS. MicroRNAs and cell cycle regulation. *Cell Cycle* 2007;6:2127–1232.

48. Harfe BD. MicroRNAs in vertebrate development. *Curr Opin Genet Dev* 2005;15:410–415.

49. Boehm M, Slack FJ. MicroRNA control of lifespan and metabolism. *Cell Cycle* 2006;5:837–840.

50. Rodriguez A, Griffiths-Jones S, Ashurst JL, Bradley A. Identification of mammalian microRNA host genes and transcription units. *Genome Res* 2004;14:1902–1910.

51. Calin GA, Sevignani C, Dumitru CD, et al. Human microRNA genes are frequently located at fragile sites and genomic regions involved in cancers. *Proc Natl Acad Sci USA* 2004;101:2999–3004.

52. Rossi S, Sevignani C, Nnadi SC, Siracusa LD, Calin GA. Cancer-associated genomic regions (CAGRs) and noncoding RNAs: bioinformatics and therapeutic implications. *Mamm Genome* 2008;19:526–540.

53. Baer C, Claus R, Plass C. Genome-wide epigenetic regulation of miRNAs in cancer. *Cancer Res* 2013;73:473–477.

54. Kulshreshtha R, Ferracin M, Wojcik SE, et al. A microRNA signature of hypoxia. *Mol Cell Biol* 2007;27:1859–1867.

55. Zhang B, Pan X, Cobb GP, Anderson TA. microRNAs as oncogenes and tumor suppressors. *Dev Biol* 2007;302:1–12.

56. Lu J, Getz G, Miska EA, et al. MicroRNA expression profiles classify human cancers. *Nature* 2005;435:834–838.

57. Calin GA, Croce CM. MicroRNA signatures in human cancers. *Nat Rev Cancer* 2006;6:857–866.

58. Subramanian S, Lui WO, Lee CH, et al. MicroRNA expression signature of human sarcomas. *Oncogene* 2008;27:2015–2026.

59. Renner M, Czwan E, Hartmann W, et al. MicroRNA profiling of primary high-grade soft tissue sarcomas. *Genes Chromosomes Cancer* 2012;51:982–996.

60. Dela Cruz F, Matushansky I. MicroRNAs in chromosomal translocation-associated solid tumors: learning from sarcomas. *Discov Med* 2011;12:307–317.

61. Fujiwara T, Kunisada T, Takeda K, et al. MicroRNAs in soft tissue sarcomas: overview of the accumulating evidence and importance as novel biomarkers. *Biomed Res Int* 2014;2014:592868.

62. Cortez MA, Calin GA. MicroRNA identification in plasma and serum: a new tool to diagnose and monitor diseases. *Expert Opin Biol Ther* 2009;9:703–711.

63. Sachdeva M, Mito JK, Lee CL, et al. MicroRNA-182 drives metastasis of primary sarcomas by targeting multiple genes. *J Clin Invest* 2014;124:4305–4319.

64. Miyachi M, Tsuchiya K, Yoshida H, et al. Circulating muscle-specific microRNA, miR-206, as a potential diagnostic marker for rhabdomyosarcoma. *Biochem Biophys Res Commun* 2010;400:89–93.

65. Weng Y, Chen Y, Chen J, Liu Y, Bao T. Identification of serum microRNAs in genome-wide serum microRNA expression profiles as novel noninvasive biomarkers for malignant peripheral nerve sheath tumor diagnosis. *Med Oncol* 2013;30:531.

66. Sarver AL, Phalak R, Thayanithy V, Subramanian S. S-MED: sarcoma microRNA expression database. *Lab Invest* 2010;90:753–761.

67. Taylor MA, Schiemann WP. Therapeutic opportunities for targeting microRNAs in cancer. *Mol Cell Ther* 2014;2:1–13.

68. Chugh P, Dittmer DP. Potential pitfalls in microRNA profiling. *Wiley Interdiscip Rev RNA* 2012;3:601–616.

Inherited cancer syndromes
Table 5.1

69. Kimonis VE, Goldstein AM, Pastakia B, et al. Clinical manifestations in 105 persons with nevoid basal cell carcinoma syndrome. *Am J Med Genet* 1997;69:299–308.

70. Fujii K, Miyashita T. Gorlin syndrome (nevoid basal cell carcinoma syndrome): update and literature review. *Pediatr Int* 2014;56:667–674.

71. Weksberg R, Shuman C, Beckwith JB. Beckwith-Wiedemann syndrome. *Eur J Hum Genet* 2010;18:8–14.

72. Wiedemann, H-R. Tumours and hemihypertrophy associated with Wiedemann-Beckwith syndrome. (Letter) *Eur J Pediatr* 1983;141:129.

73. Jin XL, Wang ZH, Xiao XB, Huang LS, Zhao XY. Blue rubber bleb nevus syndrome: a case report and literature review. *World J Gastroenterol* 2014;20:17254–17259.

74. Espiard S, Bertherat J. Carney complex. *Front Horm Res* 2013;41:50–62.

75. Estep AL, Tidyman WE, Teitell MA, Cotter PD, Rauen KA. HRAS mutations in Costello syndrome: detection of constitutional activating mutations in codon 12 and 13 and loss of wild-type allele in malignancy. *Am J Med Genet A* 2006;140:8–16.

76. Nishisho I, Nakamura Y, Miyoshi Y, et al. Mutations of chromosome 5q21 genes in FAP and colorectal cancer patients. *Science* 1991;253:665–669.

77. Eccles DM, Van Der Luijt R, Breukel C, et al. Hereditary desmoid disease due to a frameshift mutation at codon 1924 of the APC gene. *Am J Hum Genet* 1996;59:1193–1201.

78. Scott RJ, Froggatt NJ, Trembath RC, et al. Familial infiltrative fibromatosis (desmoid tumours) (MIM135290) caused by a recurrent 3′ APC gene mutation. *Hum Mol Genet* 1996;5:1921–1924.

79. Hussussian CJ, Struewing JP, Goldstein AM, et al. Germline p16 mutations in familial melanoma. *Nat Genet* 1994;8:15–21.

80. Randerson-Moor JA, Harland M, Williams S, et al. A germline deletion of p14(ARF) but not CDKN2A in a melanoma-neural system tumour syndrome family. *Hum Mol Genet* 2001;10:55–62.

81. Vanneste R, Smith E, Graham G. Multiple neurofibromas as the presenting feature of familial atypical multiple malignant melanoma (FAMMM) syndrome. *Am J Med Genet A* 2013;161A:1425–1431.

82. Nishida T, Hirota S, Taniguchi M, et al. Familial gastrointestinal stromal

tumours with germline mutation of the KIT gene. *Nat Genet* 1998;19:323–324.

83. Chompret A, Kannengiesser C, Barrois M, et al. PDGFRA germline mutation in a family with multiple cases of gastrointestinal stromal tumor. *Gastroenterology* 2004;126:318–321.

84. Ligon AH, Moore SD, Parisi MA, et al. Constitutional rearrangement of the architectural factor HMGA2: a novel human phenotype including overgrowth and lipomas. *Am J Hum Genet* 2005;76:340–348.

85. Dryja TP, Mukai S, Petersen R, et al. Parental origin of mutations of the retinoblastoma gene. *Nature* 1989;339:556–558.

86. Zhu XP, Dunn JM, Phillips RA, et al. Preferential germline mutation of the paternal allele in retinoblastoma. *Nature* 1989;340:312–313.

87. Barbaux S, Niaudet P, Gubler MC, et al. Donor splice-site mutations in WT1 are responsible for Frasier syndrome. *Nat Genet* 1997;17:467–470.

88. Pelletier J, Bruening W, Kashtan CE, et al. Germinal mutations in the Wilms' tumor suppressor gene are associated with abnormal urogenital development in Denys-Drash syndrome. *Cell* 1991;67:437–447.

89. van Heyningen V, Bickmore WA, Seawright A, et al. Role for the Wilms tumor gene in genital development? *Proc Natl Acad Sci USA* 1990;87:5383–5386.

90. Ruteshouser EC, Huff V. Familial Wilms tumor. *Am J Med Genet C Semin Med Genet* 2004;129C:29–34.

91. Brouillard P, Boon LM, Mulliken JB, et al. Mutations in a novel factor, glomulin, are responsible for glomuvenous malformations ("glomangiomas"). *Am J Hum Genet* 2002;70:866–874.

92. Kiuru M, Launonen V. Hereditary leiomyomatosis and renal cell cancer (HLRCC). *Curr Mol Med* 2004;4(8):869–875.

93. Gardie B, Remenieras A, Kattygnarath D, et al. Novel FH mutations in families with hereditary leiomyomatosis and renal cell cancer (HLRCC) and patients with isolated type 2 papillary renal cell carcinoma. *J Med Genet* 2011;48:226–234.

94. Denadai R, Raposo-Amaral CE, Bertola D, et al. Identification of 2 novel ANTXR2 mutations in patients with hyaline fibromatosis syndrome and proposal of a modified grading system. *Am J Med Genet A* 2012;158A:732–742.

95. Li FP, Fraumeni JF. Rhabdomyosarcoma in children: epidemiologic study and identification of a cancer family syndrome. *J Natl Cancer Inst* 1969;43:1365–1373.

96. Li FP, Fraumeni JF. Soft tissue sarcomas, breast cancer and other neoplasms: a familial syndrome? *Ann Int Med* 1969;71:747–752.

97. Amary MF, Damato S, Halai D, et al. Ollier disease and Maffucci syndrome are caused by somatic mosaic mutations of IDH1 and IDH2. *Nat Genet* 2011;43:1262–1265.

98. Rasmussen SA, Friedman JM. NF1 gene and neurofibromatosis 1. *Am J Epidemiol* 2000;151:33–40.

99. Gutmann DH. Molecular insights into neurofibromatosis 2. *Neurobiol Dis* 1997;3:247–261.

100. Meyer S, Kingston H, Taylor AM, et al. Rhabdomyosarcoma in Nijmegen breakage syndrome: strong association with perianal primary site. *Cancer Genet Cytogenet* 2004;154:169–174.

101. Astuti D, Latif F, Dallol A, et al. Gene mutations in the succinate dehydrogenase subunit SDHB cause susceptibility to familial pheochromocytoma and to familial paraganglioma. *Am J Hum Genet* 2001;69:49–54.

102. Müller U. Pathological mechanisms and parent-of-origin effects in hereditary paraganglioma/pheochromocytoma (PGL/PCC). *Neurogenetics* 2011;12:175–181.

103. Pasini B, McWhinney SR, Bei T, et al. Clinical and molecular genetics of patients with the Carney–Stratakis syndrome and germline mutations of the genes coding for the succinate dehydrogenase subunits SDHB, SDHC, and SDHD. *Eur J Hum Genet* 2008;16:79–88.

104. Miettinen M, Lasota J. Succinate dehydrogenase deficient gastrointestinal stromal tumors (GISTs): a review. *Int J Biochem Cell Biol* 2014;53:514–519.

105. Blumenthal GM, Dennis PA. PTEN hamartoma tumor syndromes. *Eur J Hum Genet* 2008;16:1289–1300.

106. Cohen MM Jr. Proteus syndrome review: molecular, clinical, and pathologic features. *Clin Genet* 2014;85:111–119.

107. Taylor MD, Gokgoz N, Andrulis IL, et al. Familial posterior fossa brain tumors of infancy secondary to germline mutation of the hSNF5 gene. *Am J Hum Genet* 2000;66:1403–1406.

108. Foulkes WD, Clarke BA, Hasselblatt M, et al. No small surprise: small cell carcinoma of the ovary, hypercalcaemic type, is a malignant rhabdoid tumour. *J Pathol* 2014;233:209–214.

109. Jelinic P, Mueller JJ, Olvera N, et al. Recurrent SMARCA4 mutations in small cell carcinoma of the ovary. *Nat Genet* 2014;46:424–426.

110. Ramos P, Karnezis AN, Craig DW, et al. Small cell carcinoma of the ovary, hypercalcemic type, displays frequent inactivating germline and somatic mutations in SMARCA4. *Nat Genet* 2014;46:427–429.

111. Miller RW, Rubinstein JH. Tumors in Rubinstein–Taybi syndrome. *Am J Med Genet* 1995;56:112–115.

112. van de Kar AL, Houge G, Shaw AC, et al. Keloids in Rubinstein-Taybi syndrome: a clinical study. *Br J Dermatol* 2014;171:615–621.

113. Negri G, Milani D, Colapietro P, et al. Clinical and molecular characterization of Rubinstein-Taybi syndrome patients carrying distinct novel mutations of the EP300 gene. *Clin Genet* 2015;87:148–154.

114. Hadfield KD, Newman WG, Bowers NL, et al. Molecular characterisation of SMARCB1 and NF2 in familial and sporadic schwannomatosis. *J Med Genet* 2008;45:332–339.

115. Crino PB, Nathanson KL, Henske EP. The tuberous sclerosis complex. *N Engl J Med* 2006;355:1345–1356.

116. Curatolo P, Bombardieri R, Jozwiak S. Tuberous sclerosis. *Lancet* 2008;372:657–668.

117. O' Brien FJ, Danapal M, Jairam S, et al. Manifestations of Von Hippel Lindau syndrome: a retrospective national review. *QJM* 2014;107:291–296.

118. Wanebo JE, Lonser RR, Glenn GM, Oldfield EH. The natural history of hemangioblastomas of the central nervous system in patients with von

Hippel-Lindau disease. *J Neurosurg* 2003;98:82–94.

119. Yamamoto K, Imakiire A, Miyagawa N, Kasahara T. A report of two cases of Werner's syndrome and review of the literature. *J Orthop Surg (Hong Kong)* 2003;11:224–233.

Genetic alterations in soft tissue tumors

Fusion genes

Table 5.2

120. Nucci MR, Weremowicz S, Neskey DM, et al. Chromosomal translocation t(8;12) induces aberrant HMGIC expression in aggressive angiomyxoma of the vulva. *Genes Chromosomes Cancer* 2001;32:172–176.

121. Micci F, Panagopoulos I, Bjerkehagen B, Heim S. Deregulation of HMGA2 in an aggressive angiomyxoma with t(11;12)(q23;q15). *Virchows Arch* 2006;448:838–842.

122. Rabban JT, Dal Cin P, Oliva E. HMGA2 rearrangement in a case of vulvar aggressive angiomyxoma. *Int J Gynecol Pathol* 2006;25:403–407.

123. Rawlinson NJ, West WW, Nelson M, Bridge JA. Aggressive angiomyxoma with t(12;21) and HMGA2 rearrangement: report of a case and review of the literature. *Cancer Genet Cytogenet* 2008;181:119–124.

124. Joyama S, Ueda T, Shimizu K, et al. Chromosome rearrangement at 17q25 and Xp11.2 in alveolar soft-part sarcoma: a case report and review of the literature. *Cancer* 1999;86:1246–1250.

125. Ladanyi M, Lui MY, Antonescu CR, et al. The der(17)t(X;17)(p11;q25) of human alveolar soft part sarcoma fuses the TFE3 transcription factor gene to ASPL, a novel gene at 17q25. *Oncogene* 2001;20:48–57.

126. Jin Y, Möller E, Nord KH, et al. Fusion of the AHRR and NCOA2 genes through a recurrent translocation t(5;8)(p15;q13) in soft tissue angiofibroma results in upregulation of aryl hydrocarbon receptor target genes. *Genes Chromosomes Cancer* 2012;51:510–520.

127. Arbajian E, Magnusson L, Mertens F, et al. A novel GTF2I/NCOA2 fusion gene emphasizes the role of NCOA2 in soft tissue angiofibroma development. *Genes Chromosomes Cancer* 2013;52:330–331.

128. Waters BL, Panagopoulos I, Allen EF. Genetic characterization of angiomatoid fibrous histiocytoma identifies fusion of the FUS and ATF-1 genes induced by a chromosomal translocation involving bands 12q13 and 16p11. *Cancer Genet Cytogenet* 2000;121:109–116.

129. Raddaoui E, Donner LR, Panagopoulos I. Fusion of the FUS and ATF1 genes in a large deep-seated angiomatoid fibrous histiocytoma. *Diagn Mol Pathol* 2002;11:157–162.

130. Antonescu CR, Dal Cin P, Nafa K, et al. EWSR1-CREB1 is the predominant gene fusion in angiomatoid fibrous histiocytoma. *Genes Chromosomes Cancer* 2007;46:1051–1060.

131. Hallor KH, Micci F, Meis-Kindblom JM, et al. Fusion genes in angiomatoid fibrous histiocytoma. *Cancer Lett* 2007;251:158–163.

132. Giacomini CP, Sun S, Varma S, et al. Breakpoint analysis of transcriptional and genomic profiles uncovers novel gene fusions spanning multiple human cancer types. *PLoS Genet* 2013;9: e1003464.

133. Huang D, Sumegi J, Dal Cin P, et al. C11orf95-MKL2 is the resulting fusion oncogene of t(11;16)(q13;p13) in chondroid lipoma. *Genes Chromosomes Cancer* 2010;49:810–818.

134. Tallini G, Dorfman H, Brys P, et al. Correlation between clinicopathological features and karyotype in 100 cartilaginous and chordoid tumours: a report from the Chromosomes and Morphology (CHAMP) Collaborative Study Group. *J Pathol* 2002;196:194–203.

135. Dahlén A, Mertens F, Rydholm A, et al. Fusion, disruption, and expression of HMGA2 in bone and soft tissue chondromas. *Mod Pathol* 2003;16:1132–1140.

136. Labelle Y, Bussières J, Courjal F, Goldring MB. The EWS/TEC fusion protein encoded by the t(9;22) chromosomal translocation in human chondrosarcomas is a highly potent transcriptional activator. *Oncogene* 1999;18:3303–3308.

137. Attwooll C, Tariq M, Harris M, et al. Identification of a novel fusion gene involving hTAFII68 and CHN from a t(9;17)(q22;q11.2) translocation in an extraskeletal myxoid chondrosarcoma. *Oncogene* 1999;18:7599–7601.

138. Sjögren H, Meis-Kindblom J, Kindblom LG, Åman P, Stenman G. Fusion of the EWS-related gene TAF2N to TEC in extraskeletal myxoid chondrosarcoma. *Cancer Res* 1999;59:5064–5067.

139. Sjögren H, Wedell B, Kindblom JM, Kindblom LG, Stenman G. Fusion of the basic helix-loop-helix protein TCF12 to TEC in extraskeletal myxoid chondrosarcoma with translocation t(9;15)(q22;q21). *Cancer Res* 2000;60:6832–6835.

140. Morohoshi F, Arai K, Takahashi EI, Tanigami A, Ohki M. Cloning and mapping of a human RBP56 gene encoding a putative RNA binding protein similar to FUS/TLS and EWS proteins. *Genomics* 1996;38:51–57.

141. Hisaoka M, Ishida T, Imamura T, Hashimoto H. TFG is a novel fusion partner of NOR1 in extraskeletal myxoid chondrosarcoma. *Genes Chromosomes Cancer* 2004;40:325–328.

142. Broehm CJ, Wu J, Gullapalli RR, Bocklage T. Extraskeletal myxoid chondrosarcoma with a t(9;16)(q22;p11.2) resulting in a NR4A3-FUS fusion. *Cancer Genet* 2014;207:276–280.

143. Wang L, Motoi T, Khanin R, et al. Identification of a novel, recurrent HEY1-NCOA2 fusion in mesenchymal chondrosarcoma based on a genome-wide screen of exon-level expression data. *Genes Chromosomes Cancer* 2012;51:127–139.

144. Nyquist KB, Panagopoulos I, Thorsen J, et al. Whole-transcriptome sequencing identifies novel IRF2BP2-CDX1 fusion gene brought about by translocation t(1;5)(q42;q32) in mesenchymal chondrosarcoma. *PLoS One* 2012;7: e49705.

145. Zucman J, Delattre O, Desmaze C, et al. EWS and ATF-1 gene fusion induced by t(12;22) translocation in malignant melanoma of soft parts. *Nat Genet* 1993;4:341–345.

146. Panagopoulos I, Mertens F, Debiec-Rychter M, et al. Molecular genetic characterization of the EWS/ATF1 fusion gene in clear cell sarcoma of tendons and aponeuroses. *Int J Cancer* 2002;99:560–567.

147. Speleman F, Delattre O, Peter M, et al. Malignant melanoma of the soft parts (clear-cell sarcoma): conformation of EWS and ATF-1 gene fusion caused by

a t(11;22) translocation. *Mod Pathol* 1997;10:496–499.

148. Antonescu CR, Tschernyavsky SJ, Woodruff JM, et al. Molecular diagnosis of clear cell sarcoma: detection of EWS-ATF1 and MITF-M transcripts and histopathological and ultrastructural analysis of 12 cases. *J Mol Diagn* 2002;4:44–52.

149. Covinsky M, Gong S, Rajaram V, Perry A, Pfeifer J. EWS-ATF1 fusion transcripts in gastrointestinal tumors previously diagnosed as malignant melanoma. *Hum Pathol* 2005;36:74–81.

150. Antonescu CR, Nafa K, Segal NH, Dal Cin P, Ladanyi M. EWS-CREB1: a recurrent variant fusion in clear cell sarcoma–association with gastrointestinal location and absence of melanocytic differentiation. *Clin Cancer Res* 2006;12:5356–5362.

151. Sciot R, Samson I, van den Berghe H, Van Damme B, Dal Cin P. Collagenous fibroma (desmoplastic fibroblastoma): genetic link with fibroma of tendon sheath? *Mod Pathol* 1999;12:565–568.

152. Macchia G, Trombetta D, Möller E, et al. FOSL1 as a candidate target gene for 11q12 rearrangements in desmoplastic fibroblastoma. *Lab Invest* 2012;92:735–743.

153. Simon MP, Pedeutour F, Sirvent N, et al. Deregulation of the platelet-derived growth factor B-chain gene via fusion with collagen gene COL1A1 in dermatofibrosarcoma protuberans and giant-cell fibroblastoma. *Nat Genet* 1997;15:95–98.

154. Shimizu A, O'Brien KP, Sjöblom T, et al. The dermatofibrosarcoma protuberans-associated collagen type Iα1/platelet-derived growth factor (PDGF) B-chain fusion gene generates a transforming protein that is processed to functional PDGF-BB. *Cancer Res* 1999;59:3719–3723.

155. Simon MP, Navarro M, Roux D, Pouysségur J. Structural and functional analysis of a chimeric protein COL1A1-PDGFB generated by the translocation t(17;22)(q22;q13.1) in dermatofibrosarcoma protuberans (DP). *Oncogene* 2001;20:2965–2975.

156. O'Brien KP, Seroussi E, Dal Cin P, et al. Various regions within the alpha-helical domain of the COL1A1 gene are fused to the second exon of the PDGFB gene in dermatofibrosarcoma protuberans and giant cell fibroblastomas. *Genes Chromosomes Cancer* 1998;23:187–193.

157. Sirvent N, Maire G, Pedeutour F. Genetics of dermatofibrosarcoma protuberans family of tumors: from ring chromosomes to tyrosine kinase inhibitor treatment. *Genes Chromosomes Cancer* 2003;37:1–19.

158. Bianchini L, Maire G, Guillot B, et al. Complex t(5;8) involving the CSPG2 and PTK2B genes in a case of dermatofibrosarcoma protuberans without the COL1A1-PDGFB fusion. *Virchows Arch* 2008;452:689–696.

159. Ladanyi M, Gerald WL. Fusion of the EWS and WT1 genes in the desmoplastic small round cell tumor. *Cancer Res* 1994;54:2837–2840.

160. Gerald WL, Rosai J, Ladanyi M. Characterization of the genomic breakpoint and chimeric transcripts in the EWS-WT1 gene fusion of desmoplastic small round cell tumor. *Proc Natl Acad Sci USA* 1995;14:1028–1032.

161. Gerald WL, Ladanyi M, de Alava E, et al. Clinical, pathologic, and molecular spectrum of tumors associated with t(11;22)(p13;q12): desmoplastic small-round-cell tumor and its variants. *J Clin Oncol* 1998;16:3028–3036.

162. Gerald WL, Haber DA. The EWS-WT1 gene fusion in desmoplastic small round cell tumor. *Semin Cancer Biol* 2005;15:197–205.

163. Ordi J, de Alava E, Torné A, et al. Intraabdominal desmoplastic small round cell tumor with EWS/ERG fusion transcript. *Am J Surg Pathol* 1998;22:1026–1032.

164. Koontz JI, Soreng AL, Nucci M, et al. Frequent fusion of the JAZF1 and JJAZ1 genes in endometrial stromal tumors. *Proc Natl Acad Sci* 2001;98:6348–6353.

165. Micci F, Panagopoulos I, Bjerkehagen B, Heim S. Consistent rearrangement of chromosomal band 6p21 with generation of fusion genes JAZF1/PHF1 and EPC1/PHF1 in endometrial stromal sarcomas. *Cancer Res* 2006;66:107–112.

166. Mertens F, Tayebwa J. Evolving techniques for gene fusion detection in soft tissue tumours. *Histopathology* 2014;64:151–162.

167. Micci F, Gorunova L, Gatius S, et al. MEAF6/PHF1 is a recurrent gene fusion in endometrial stromal sarcoma. *Cancer Lett* 2014;347:75–78.

168. Dewaele B, Przybyl J, Quattrone A, et al. Identification of a novel, recurrent MBTD1-CXorf67 fusion in low-grade endometrial stromal sarcoma. *Int J Cancer* 2014;134:1112–1122.

169. Lee CH, Nucci MR. Endometrial stromal sarcoma - the new genetic paradigm. *Histopathology* 2015;67: 1–19.

170. Panagopoulos I, Thorsen J, Gorunova L, et al. Fusion of the ZC3H7B and BCOR genes in endometrial stromal sarcomas carrying an X;22-translocation. *Genes Chromosomes Cancer* 2013;52:610–618.

171. McPherson A, Hormozdiari F, Zayed A, et al. deFuse: an algorithm for gene fusion discovery in tumor RNA-Seq data. *PLoS Comput Biol* 2011;7: e1001138

172. Aurias A, Rimbaut C, Buffe D, Zucker JM, Mazabraud A. Translocation involving chromosome 22 in Ewing's sarcoma: a cytogenetic study of four fresh tumors. *Cancer Genet Cytogenet* 1984;12:21–25.

173. Whang-Peng J, Triche TJ, Knutsen T, et al. Chromosome translocation in peripheral neuroepithelioma. *N Engl J Med* 1984;311:584–585.

174. Delattre O, Zucman J, Plougastel B, et al. Gene fusion with an ETS DNA binding domain caused by chromosome translocation in human tumors. *Nature* 1992;359:162–165.

175. Zucman J, Delattre O, Desmaze C, et al. Cloning and characterization of the Ewing's sarcoma and peripheral neuroepithelioma t(11;22) translocation breakpoints. *Genes Chromosomes Cancer* 1992;5:271–277.

176. Zucman J, Melot T, Desmaze C, et al. Combinatorial generation of variable fusion proteins in Ewing family of tumors. *EMBO J* 1993;12:4481–4487.

177. Sorensen PH, Lessnick SL, Lopez-Terrada D, et al. A second Ewing's sarcoma translocation, t(21;22), fuses the EWS gene to another ETS-family transcription factor, ERG. *Nat Genet* 1994;6:146–151.

178. Jeon IS, Davis JN, Braun BS, et al. A variant Ewing's sarcoma translocation t(7;22) fuses the EWS gene to the ETS gene ETV1. *Oncogene* 1995;10:1229–1234.

179. Kaneko Y, Yoshida K, Handa M, et al. Fusion of an ETS-family gene, EIAF, to EWS by t(17;22)(q12;q12) chromosome translocation in an undifferentiated sarcoma of infancy. Genes Chromosomes Cancer 1996;15:115–121.

180. Peter M, Couturier J, Pacquement H, et al. A new member of the ETS family fused to EWS in Ewing tumors. Oncogene 1997;14:1159–1164.

181. Mastrangelo T, Modena P, Tornielli S, et al. A novel zinc finger gene is fused to EWS in small round cell tumor. Oncogene 2000;19:3799–3804.

182. Wang L, Bhargava R, Zheng T, et al. Undifferentiated small round cell sarcomas with rare EWS gene fusions: identification of the novel EWS-SP3 fusion and of additional cases with the EWS-ETV1 and EWS-FEV fusions. J Mol Diagn 2007;9:498–509.

183. Ng TL, O'Sullivan MJ, Pallen CJ, et al. Ewing sarcoma with novel translocation t(2;16) producing an in-frame fusion of FUS and FEV. J Mol Diagn 2007;9:459–463.

184. Shing DC, McMullan DJ, Roberts P, et al. FUS/ERG gene fusions in Ewing's tumors. Cancer Res 2003;63:4568–4576.

185. Szuhai K, Ijszenga M, de Jong D, et al. The NFATc2 gene is involved in a novel cloned translocation in a Ewing sarcoma variant that couples its function in immunology to oncology. Clin Cancer Res 2009;15:2259–2268.

186. Sumegi J, Nishio J, Nelson M, et al. A novel t(4;22)(q31;q12) produces an EWSR1-SMARCA5 fusion in extraskeletal Ewing sarcoma/primitive neuroectodermal tumor. Mod Pathol 2011;24:333–342.

187. Choi EY, Thomas DG, McHugh JB, et al. Undifferentiated small round cell sarcoma with t(4;19)(q35;q13.1) CIC-DUX4 fusion: a novel highly aggressive soft tissue tumor with distinctive histopathology. Am J Surg Pathol 2013;37:1379–1386.

188. Italiano A, Sung YS, Zhang L, et al. High prevalence of CIC fusion with double-homeobox (DUX4) transcription factors in EWSR1-negative undifferentiated small blue round cell sarcomas. Genes Chromosomes Cancer 2012;51:207–218.

189. Pierron G, Tirode F, Lucchesi C, et al. A new subtype of bone sarcoma defined by BCOR-CCNB3 gene fusion. Nat Genet 2012;44:461–466.

190. Sugita S, Arai Y, Tonooka A, et al. A novel CIC-FOXO4 gene fusion in undifferentiated small round cell sarcoma: a genetically distinct variant of Ewing-like sarcoma. Am J Surg Pathol 2014;38:1571–1576.

191. Knezevich SR, McFadden DE, Tao W, Lim JF, Sorensen PH. A novel ETV6-NTRK3 gene fusion in congenital fibrosarcoma. Nat Genet 1998;18:184–187.

192. Rubin BP, Chen CJ, Morgan TW, et al. Congenital mesoblastic nephroma t(12;15) is associated with ETV6-NTRK3 gene fusion: cytogenetic and molecular relationship to congenital (infantile) fibrosarcoma. Am J Pathol 1998;153:1451–1458.

193. Errani C, Zhang L, Sung YS, et al. A novel WWTR1-CAMTA1 gene fusion is a consistent abnormality in epithelioid hemangioendothelioma of different anatomic sites. A novel WWTR1-CAMTA1 gene fusion is a consistent abnormality in epithelioid hemangioendothelioma of different anatomic sites. Genes Chromosomes Cancer 2011;50:644–653.

194. Tanas MR, Sboner A, Oliveira AM, et al. Identification of a disease-defining gene fusion in epithelioid hemangioendothelioma. Sci Transl Med 2011;3:98ra82.

195. Antonescu CR, Le Loarer F, Mosquera JM, et al. Novel YAP1-TFE3 fusion defines a distinct subset of epithelioid hemangioendothelioma. Genes Chromosomes Cancer 2013;52:775–784.

196. Trombetta D, Magnusson L, von Steyern FV, et al. Translocation t(7;19)(q22;q13) - a recurrent chromosome aberration in pseudomyogenic hemangioendothelioma? Cancer Genet 2011;204:211–215.

197. Walther C, Tayebwa J, Lilljebjörn H, et al. A novel SERPINE1-FOSB fusion gene results in transcriptional up-regulation of FOSB in pseudomyogenic haemangioendothelioma. J Pathol 2014;232:534–540.

198. Antonescu CR, Zhang L, Nielsen GP, et al. Consistent t(1;10) with rearrangements of TGFBR3 and MGEA5 in both myxoinflammatory fibroblastic sarcoma and hemosiderotic fibrolipomatous tumor. Genes Chromosomes Cancer 2011;50:757–764.

199. Cools J, Wlodarska I, Somers R, et al. Identification of novel fusion partners of ALK, the anaplastic lymphoma kinase, in anaplastic large-cell lymphoma and inflammatory myofibroblastic tumor. Genes Chromosomes Cancer 2002;34:354–362.

200. Bridge JA, Kanamori M, Ma Z, et al. Fusion of the ALK gene to the clathrin heavy chain gene, CLTC, in inflammatory myofibroblastic tumor. Am J Pathol 2001;159:411–415.

201. Debiec-Rychter M, Marynen P, Hagemeijer A, Pauwels P. ALK-ATIC fusion in urinary bladder inflammatory myofibroblastic tumor. Genes Chromosomes Cancer 2003;38:187–190.

202. Ma Z, Hill DA, Collins MH, et al. Fusion of ALK to the Ran-binding protein 2 (RANBP2) gene in inflammatory myofibroblastic tumor. Genes Chromosomes Cancer 2003;37:98–105.

203. Panagopoulos I, Nilsson T, Domanski HA, et al. Fusion of the SEC31L1 and ALK genes in an inflammatory myofibroblastic tumor. Int J Cancer 2006;118:1181–1186.

204. Lawrence B, Perez-Atayde A, Hibbard MK, et al. TPM3-ALK and TPM4-ALK oncogenes in inflammatory myofibroblastic tumors. Am J Pathol 2000;157:377–384.

205. Takeuchi K, Soda M, Togashi Y, et al. Pulmonary inflammatory myofibroblastic tumor expressing a novel fusion, PPFIBP1-ALK: reappraisal of anti-ALK immunohistochemistry as a tool for novel ALK fusion identification. Clin Cancer Res 2011;17:3341–3348.

206. Lovly CM, Gupta A, Lipson D, et al. Inflammatory myofibroblastic tumors harbor multiple potentially actionable kinase fusions. Cancer Discov 2014;4:889–895.

207. Antonescu CR, Suurmeijer AJ, Zhang L, et al. Molecular characterization of inflammatory myofibroblastic tumors with frequent ALK and ROS1 gene fusions and rare novel RET rearrangement. Am J Surg Pathol 2015;39:957–967.

208. Kazmierczak B, Pohnke Y, Bullerdiek J. Fusion transcripts between HMGIC gene and RTVL-H-related sequences in mesenchymal tumors without

208. cytogenetic aberrations. *Genomics* 1996;38:223–226.

209. Schoenmakers EF, Huysmans C, van de Ven WJ. Allelic knockout of novel splice variants of human recombination repair gene RAD51B in t(12;14) uterine leiomyomas. *Cancer Res* 1999;59:19–23.

210. Kurose K, Mine N, Doi D, et al. Novel gene fusion COX6C at 8q22-23 to HMGIC at 12q15 in a uterine leiomyoma. *Genes Chromosomes Cancer* 2000;27:303–307.

211. Mine N, Kurose K, Konishi H, et al. Fusion of a sequence from HEI10 (14q11) to the HMGIC gene at 12q15 in a uterine leiomyoma. *Jpn J Cancer Res* 2001;92:135–139.

212. Moore SD, Herrick SR, Ince TA, et al. Uterine leiomyomata with t(10;17) disrupt the histone acetyltransferase MORF. *Cancer Res* 2004;64:5570–5577.

213. Velagaleti GV, Tonk VS, Hakim NM, et al. Fusion of HMGA2 to COG5 in uterine leiomyoma. *Cancer Genet Cytogenet* 2010;202:11–16.

214. Panagopoulos I, Gorunova L, Bjerkehagen B, Heim S. Novel KAT6B-KANSL1 fusion gene identified by RNA sequencing in retroperitoneal leiomyoma with t(10;17)(q22;q21). *PLoS One* 2015;10:e0117010.

215. Astrom A, D'Amore ES, Sainati L, et al. Evidence of involvement of the PLAG1 gene in lipoblastomas. *Int J Oncol* 2000;16:1107–1110.

216. Gisselsson D, Hibbard MK, Dal Cin P, et al. PLAG1 fusion oncogenes in lipoblastoma. *Cancer Res* 2000;60:4869–4872.

217. Sciot R, De Wever I, Debiec-Rychter M. Lipoblastoma in a 23-year-old male: distinction from atypical lipomatous tumor using cytogenetic and fluorescence in-situ hybridization analysis. *Virchows Arch* 2003;442:468–471.

218. Deen M, Ebrahim S, Schloff D, Mohamed AN. A novel PLAG1-RAD51L1 gene fusion resulting from a t(8;14)(q12;q24) in a case of lipoblastoma. *Cancer Genet* 2013;206:233–237.

219. Petit MM, Mols R, Schoenmakers EF, Mandahl N, Van de Ven WJ. LPP, the preferred fusion partner gene of HMGIC in lipomas, is a novel member of the LIM protein gene family. *Genomics* 1996;36:118–129.

220. Petit MM, Schoenmakers EF, Huysmans C, et al. LHFP, a novel translocation partner gene of HMGIC in a lipoma, is a member of a new family of LHFP-like genes. *Genomics* 1999;57:438–441.

221. Kazmierczak B, Dal Cin P, Wanschura S, , et al. Cloning and molecular characterization of part of a new gene fused to HMGIC in mesenchymal tumors. *Am J Pathol* 1998;152:431–435.

222. Broberg K, Zhang M, Strömbeck B, et al. Fusion of RDC1 with HMGA2 in lipomas as the result of chromosome aberrations involving 2q35-37 and 12q13-15. *Int J Oncol* 2002;21:321–326.

223. Nilsson M, Panagopoulos I, Mertens F, Mandahl N. Fusion of the HMGA2 and NFIB genes in lipoma. *Virchows Arch* 2005;447:855–858.

224. Bianchini L, Birtwisle L, Saâda E, et al. Identification of PPAP2B as a novel recurrent translocation partner gene of HMGA2 in lipomas. *Genes Chromosomes Cancer* 2013;52:580–590.

225. Panagopoulos I, Höglund M, Mertens F, et al. Fusion of EWS and CHOP genes in myxoid liposarcoma. *Oncogene* 1996;12:489–494.

226. Crozat A, Aman P, Mandahl N, Ron D. Fusion of CHOP to a novel RNA-binding protein in human myxoid liposarcoma. *Nature* 1993;363:640–644.

227. Rabbitts TH, Forster A, Larson R, Nathan P. Fusion of the dominant negative transcription regulator CHOP with a novel gene FUS by translocation t(12;16) in malignant liposarcoma. *Nat Genet* 1993;4:175–180.

228. Dal Cin P, Sciot R, Panagopoulos I, et al. Additional evidence of a variant translocation t(12;22) with EWS/CHOP fusion in myxoid liposarcoma: clinicopathological features. *J Pathol* 1997;182:437–441.

229. Taylor BS, DeCarolis PL, Angeles CV, et al. Frequent alterations and epigenetic silencing of differentiation pathway genes in structurally rearranged liposarcomas. *Cancer Discov* 2011;1:587–597.

230. Storlazzi CT, Mertens F, Nascimento A, et al. Fusion of the FUS and BBF2H7 genes in low grade fibromyxoid sarcoma. *Hum Mol Genet* 2003;12:2349–2358.

231. Reid R, de Silva MV, Paterson L, Ryan E, Fisher C. Low-grade fibromyxoid sarcoma and hyalinizing spindle cell tumor with giant rosettes share a common t(7;16)(q34;p11). *Am J Surg Pathol* 2003;27:1229–1236.

232. Panagopoulos I, Storlazzi CT, Fletcher CD, et al. The chimeric FUS/CREB3L2 gene is specific for low-grade fibromyxoid sarcoma. *Genes Chromosomes Cancer* 2004;40:218–228.

233. Mertens F, Fletcher CD, Antonescu CR, et al. Clinicopathologic and molecular genetic characterization of low-grade fibromyxoid sarcoma, and cloning of a novel FUS/CREB3L1 fusion gene. *Lab Invest* 2005;85:408–415.

234. Guillou L, Benhattar J, Gengler C, et al. Translocation-positive low-grade fibromyxoid sarcoma: clinicopathologic and molecular analysis of a series expanding the morphologic spectrum and suggesting potential relationship to sclerosing epithelioid fibrosarcoma: a study from the French Sarcoma Group. *Am J Surg Pathol* 2007;31:1387–1402.

235. Brandal P, Panagopoulos I, Bjerkehagen B, et al. Detection of a t(1;22)(q23;q12) translocation leading to an EWSR1-PBX1 fusion gene in a myoepithelioma. *Genes Chromosomes Cancer* 2008;47:558–564.

236. Antonescu CR, Zhang L, Chang NE, et al. EWSR1-POU5F1 fusion in soft tissue myoepithelial tumors. A molecular analysis of sixty-six cases, including soft tissue, bone, and visceral lesions, showing common involvement of the EWSR1 gene. *Genes Chromosomes Cancer* 2010;49:1114–1124.

237. Brandal P, Panagopoulos I, Bjerkehagen B, Heim S. t(19;22)(q13;q12) Translocation leading to the novel fusion gene EWSR1-ZNF444 in soft tissue myoepithelial carcinoma. *Genes Chromosomes Cancer* 2009;48:1051–1056.

238. Antonescu CR, Zhang L, Shao SY, et al. Frequent PLAG1 gene rearrangements in skin and soft tissue myoepithelioma with ductal differentiation. *Genes Chromosomes Cancer* 2013;52:675–682.

239. Huang SC, Chen HW, Zhang L, et al. Novel FUS-KLF17 and EWSR1-KLF17 fusions in myoepithelial tumors. *Genes Chromosomes Cancer* 2015;54:267–275.

240. Erickson-Johnson MR, Chou MM, Evers BR, et al. Nodular fasciitis: a novel model of transient neoplasia

induced by MYH9-USP6 gene fusion. *Lab Invest* 2011;91:1427–1433.

241. Gebre-Medhin S, Nord KH, Möller E, et al. Recurrent rearrangement of the PHF1 gene in ossifying fibromyxoid tumors. *Am J Pathol* 2012;181:1069–1077.

242. Endo M, Kohashi K, Yamamoto H, et al. Ossifying fibromyxoid tumor presenting EP400-PHF1 fusion gene. *Hum Pathol* 2013;44:2603–2608.

243. Antonescu CR, Sung YS, Chen CL, et al. Novel ZC3H7B-BCOR, MEAF6-PHF1, and EPC1-PHF1 fusions in ossifying fibromyxoid tumors: molecular characterization shows genetic overlap with endometrial stromal sarcoma. *Genes Chromosomes Cancer* 2014;53:183–193.

244. Tanaka M, Kato K, Gomi K, et al. Perivascular epithelioid cell tumor with SFPQ/PSF-TFE3 gene fusion in a patient with advanced neuroblastoma. *Am J Surg Pathol* 2009;33:1416–1420.

245. Dahlén A, Mertens F, Mandahl N, Panagopoulos I. Molecular genetic characterization of the genomic ACTB-GLI fusion in pericytoma with t(7;12). *Biochem Biophys Res Commun* 2004;325:1318–1323.

246. Carter JM, Sukov WR, Montgomery E, et al. TGFBR3 and MGEA5 rearrangements in pleomorphic hyalinizing angiectatic tumors and the spectrum of related neoplasms. *Am J Surg Pathol* 2014;38:1182–1992.

247. Thway K, Nicholson AG, Lawson K, et al. Primary pulmonary myxoid sarcoma with EWSR1-CREB1 fusion: a new tumor entity. *Am J Surg Pathol* 2011;35:1722–1732.

248. Barr FG, Galili N, Holick J, et al. Rearrangement of the PAX3 paired box gene in the paediatric solid tumour alveolar rhabdomyosarcoma. *Nat Genet* 1993;3:113–117.

249. Davis RJ, D'Cruz CM, Lovell MA, Biegel JA, Barr FG. Fusion of PAX7 to FKHR by the variant t(1;3)(p36;q14) translocation in alveolar rhabdomyosarcoma. *Cancer Res* 1994;54:2869–2872.

250. Barr FG, Qualman SJ, Macris MH, et al. Genetic heterogeneity in the alveolar rhabdomyosarcoma subset without typical gene fusions. *Cancer Res* 2002;62:4704–4710.

251. Wachtel M, Dettling M, Koscielniak E, et al. Gene expression signatures identify rhabdomyosarcoma subtypes and detect a novel t(2;2)(q35;p23) translocation fusing PAX3 to NCOA1. *Cancer Res* 2004;64:5539–5545.

252. Mosquera JM, Sboner A, Zhang L, et al. Recurrent NCOA2 gene rearrangements in congenital/infantile spindle cell rhabdomyosarcoma. *Genes Chromosomes Cancer* 2013;52:538–550.

253. Robinson DR, Wu YM, Kalyana-Sundaram S, et al. Identification of recurrent NAB2-STAT6 gene fusions in solitary fibrous tumor by integrative sequencing. *Nat Genet* 2013;45:180–185.

254. Mohajeri A, Tayebwa J, Collin A, et al. Comprehensive genetic analysis identifies a pathognomonic NAB2/STAT6 fusion gene, nonrandom secondary genomic imbalances, and a characteristic gene expression profile in solitary fibrous tumor. *Genes Chromosomes Cancer* 2013;52:873–886.

255. Crew AJ, Clark J, Fisher C, et al. Fusion of SYT to two genes, SSX1 and SSX2, encoding proteins with homology to the Kruppel-associated box in human synovial sarcoma. *EMBO J* 1995;14:2333–2340.

256. Skytting B, Nilsson G, Brodin B, et al. A novel fusion gene, SYT-SSX4, in synovial sarcoma. *J Natl Cancer Inst* 1999;91:974–975.

257. Storlazzi CT, Mertens F, Mandahl N, et al. A novel fusion gene, SS18L1/SSX1, in synovial sarcoma. *Genes Chromosomes Cancer* 2003;37:195–200.

258. West RB, Rubin BP, Miller MA, et al. A landscape effect in tenosynovial giant-cell tumor from activation of CSF1 expression by a translocation in a minority of tumor cells. *Proc Natl Acad Sci USA* 2006;103:690–695.

259. Möller E, Mandahl N, Mertens F, Panagopoulos I. Molecular identification of COL6A3-CSF1 fusion transcripts in tenosynovial giant cell tumors. *Genes Chromosomes Cancer* 2008;47:21–25.

260. Panagopoulos I, Brandal P, Gorunova L, Bjerkehagen B, Heim S. Novel CSF1-S100A10 fusion gene and CSF1 transcript identified by RNA sequencing in tenosynovial giant cell tumors. *Int J Oncol* 2014;44:1425–1432.

261. Pfeifer JD. *Molecular Genetic Testing In Surgical Pathology*. Philadelphia: Lippincott Williams & Wilkins; 2006.

262. Yeku O, Frohman MA. Rapid amplification of cDNA ends (RACE). *Methods Mol Biol* 2011;703:107–122.

263. Ladanyi M. The emerging molecular genetics of sarcoma translocation. *Diagn Mol Pathol* 1995;4:162–173.

264. Rabbitts TH. Chromosomal translocation master genes, mouse models and experimental therapeutics. *Oncogene* 2001;20:5763–5777.

265. Xia SJ, Barr FG. Chromosome translocations in sarcomas and the emergence of oncogenic transcription factors. *Eur J Cancer* 2005;41:2513–2527.

266. Slater O, Shipley J. Clinical relevance of molecular genetics to paediatric sarcomas. *J Clin Pathol* 2007;60:1187–1194.

267. Oda Y, Tsuneyoshi M. Recent advances in the molecular pathology of soft tissue sarcoma: implications for diagnosis, patient prognosis, and molecular target therapy in the future. *Cancer Sci* 2009;100:200–208.

268. Peter M, Gilbert E, Delattre O. A multiplex real-time pcr assay for the detection of gene fusions observed in solid tumors. *Lab Invest* 2001;81:905–912.

269. Geurts van Kessel A, de Bruijn D, Hermsen L, et al. Masked t(X;18)(p11;q11) in a biphasic synovial sarcoma revealed by FISH and RT-PCR. *Genes Chromosomes Cancer* 1998;23:198–201.

270. Kaneko Y, Kobayashi H, Hanada M, Satake N, Maseki N. EWS-ERG fusion transcript produced by chromosomal insertion in a Ewing sarcoma. *Genes Chromosomes Cancer* 1997;18:228–231.

271. Lestou VS, O'Connell JX, Robichaud M, et al. Cryptic t(X;18), ins(6;18), and SYT-SSX2 gene fusion in a case of intraneural monophasic synovial sarcoma. *Cancer Genet Cytogenet* 2002;138:153–156.

272. O'Sullivan MJ, Kyriakos M, Zhu X, et al. Malignant peripheral nerve sheath tumors with t(X;18): a pathologic and molecular genetic study. *Mod Pathol* 2000;13:1253–1263.

273. Ladanyi M, Woodruff JM, Scheithauer BW, et al. Re: O'Sullivan MJ, Kyriakos M, Zhu X, Wick MR, Swanson PE, Dehner LP, Humphrey PA, Pfeifer JD: Malignant peripheral nerve sheath tumors with

t(X;18): a pathologic and molecular genetic study. *Mod Pathol* 2000;13:1336–1346.

274. Tamborini E, Agus V, Perrone F, *et al.* Lack of SYT-SSX fusion transcripts in malignant peripheral nerve sheath tumors on RT-PCR analysis of 34 archival cases. *Lab Invest* 2002;82:609–618.

275. Fritsch MK, Bridge JA, Schuster AE, Perlman EJ, Argani P. Performance characteristics of a reverse transcriptase-polymerase chain reaction assay for the detection of tumor-specific fusion transcripts from archival tissue. *Pediatr Dev Pathol* 2003;6:43–53.

276. Stewénius Y, Jin Y, Ora I, *et al.* High-resolution molecular cytogenetic analysis of Wilms tumors highlights diagnostic difficulties among small round cell kidney tumors. *Genes Chromosomes Cancer* 2008;47:845–852.

277. Sainati L, Scapinello A, Montaldi A, *et al.* A mesenchymal chondrosarcoma of a child with the reciprocal translocation (11;22)(q24;q12). *Cancer Genet Cytogenet* 1993;71:144–147.

278. Li H, Wang J, Mor G, Sklar J. A neoplastic gene fusion mimics trans-splicing of RNAs in normal human cells. *Science* 2008;321:1357–1361.

279. Beck AH, West RB, van de Rijn M. Gene expression profiling for the investigation of soft tissue sarcoma pathogenesis and the identification of diagnostic, prognostic, and predictive biomarkers. *Virchows Arch* 2010;456:141–151.

280. Løvf M, Thomassen GO, Mertens F, *et al.* Assessment of fusion gene status in sarcomas using a custom made fusion gene microarray. *PLoS One* 2013;8:e70649.

281. Skotheim RI, Thomassen GO, Eken M, *et al.* A universal assay for detection of oncogenic fusion transcripts by oligo microarray analysis. *Mol Cancer* 2009;8:5.

282. Xiong FF, Li BS, Zhang CX, *et al.* A pipeline with multiplex reverse transcription polymerase chain reaction and microarray for screening of chromosomal translocations in leukemia. *Biomed Res Int* 2013;2013:135086.

283. Wada Y, Matsuura M, Sugawara M, *et al.* Development of detection method for novel fusion gene using GeneChip exon array. *J Clin Bioinforma* 2014;4:3.

284. Agerstam H, Lilljebjörn H, Lassen C, *et al.* Fusion gene-mediated truncation of RUNX1 as a potential mechanism underlying disease progression in the 8p11 myeloproliferative syndrome. *Genes Chromosomes Cancer* 2007;46:635–643.

285. Frith AE, Hirbe AC, Van Tine BA. Novel pathways and molecular targets for the treatment of sarcoma. *Curr Oncol Rep* 2013;15:378–385.

286. Aragon-Ching JB, Maki RG. Treatment of adult soft tissue sarcoma: old concepts, new insights, and potential for drug discovery. *Cancer Invest* 2012;30:300–308.

287. Martín Liberal J, Lagares-Tena L, Sáinz-Jaspeado M, *et al.* Targeted therapies in sarcomas: challenging the challenge. *Sarcoma* 2012;2012:626094.

Table 5.3

288. Wlodarska I, De Wolf-Peeters C, Falini B, *et al.* The cryptic inv(2)(p23q35) defines a new molecular genetic subtype of ALK-positive anaplastic large-cell lymphoma. *Blood* 1998;92:2688–2695.

289. Touriol C, Greenland C, Lamant L, *et al.* Further demonstration of the diversity of chromosomal changes involving 2p23 in ALK-positive lymphoma: 2 cases expressing ALK kinase fused to CLTCL (clathrin chain polypeptide-like). *Blood* 2000;95:3204–3207.

290. Tort F, Pinyol M, Pulford K, *et al.* Molecular characterization of a new ALK translocation involving moesin (MSN-ALK) in anaplastic large cell lymphoma. *Lab Invest* 2001;81:419–426.

291. Lamant L, Gascoyne RD, Duplantier MM, *et al.* Non-muscle myosin heavy chain (MYH9): a new partner fused to ALK in anaplastic large cell lymphoma. *Genes Chromosomes Cancer* 2003;37:427–432.

292. Morris SW, Kirstein MN, Valentine MB, *et al.* Fusion of a kinase gene, ALK to a nucleolar protein gene, NPM, in non-Hodgkin's lymphoma. *Science* 1994;263:1281–1284.

293. Hernández L, Pinyol M, Hernández S, *et al.* TRK-fused gene (TFG) is a new partner of ALK in anaplastic large cell lymphoma producing two structurally different TFG-ALK translocations. *Blood* 1999;94:3265–3268.

294. Lamant L, Dastugue N, Pulford K, Delsol G, Mariamé B. A new fusion gene TPM3-ALK in anaplastic large cell lymphoma created by a (1;2)(q25;p23) translocation. *Blood* 1999;93:3088–3095.

295. Liang X, Meech SJ, Odom LF, *et al.* Assessment of t(2;5)(p23;q35) translocation and variants in pediatric ALK+ anaplastic large cell lymphoma. *Am J Clin Pathol* 2004;121:496–506.

296. Lin E, Li L, Guan Y, *et al.* Exon array profiling detects EML4-ALK fusion in breast, colorectal, and non-small cell lung cancers. *Mol Cancer Res* 2009;7:1466–1476.

297. Lipson D, Capelletti M, Yelensky R, *et al.* Identification of new ALK and RET gene fusions from colorectal and lung cancer biopsies. *Nat Med* 2012;18:382–384.

298. Chikatsu N, Kojima H, Suzukawa K, *et al.* ALK+, CD30-, CD20- large B-cell lymphoma containing anaplastic lymphoma kinase (ALK) fused to clathrin heavy chain gene (CLTC). *Mod Pathol* 2003;16:828–832.

299. Adam P, Katzenberger T, Seeberger H, *et al.* A case of a diffuse large B-cell lymphoma of plasmablastic type associated with the t(2;5)(p23;q35) chromosome translocation. *Am J Surg Pathol* 2003;27:1473–1476.

300. Van Roosbroeck K, Cools J, Dierickx D, *et al.* ALK-positive large B-cell lymphomas with cryptic SEC31A-ALK and NPM1-ALK fusions. *Haematologica* 2010;95:509–513.

301. Takeuchi K, Soda M, Togashi Y, *et al.* Identification of a novel fusion, SQSTM1-ALK, in ALK-positive large B-cell lymphoma. *Haematologica* 2011;96:464–467.

302. Jazii FR, Najafi Z, Malekzadeh R, *et al.* Identification of squamous cell carcinoma associated proteins by proteomics and loss of beta tropomyosin expression in esophageal cancer. *World J Gastroenterol* 2006;12:7104–7112.

303. Debelenko LV, Raimondi SC, Daw N, *et al.* Renal cell carcinoma with novel VCL-ALK fusion: new representative of ALK-associated tumor spectrum. *Mod Pathol* 2011;24:430–442.

304. Röttgers S, Gombert M, Teigler-Schlegel A, et al. ALK fusion genes in children with atypical myeloproliferative leukemia. *Leukemia* 2010;24:1197–1200.

305. Takeuchi K, Choi YL, Togashi Y, et al. KIF5B-ALK, a novel fusion oncokinase identified by an immunohistochemistry-based diagnostic system for ALK-positive lung cancer. *Clin Cancer Res* 2009;15:3143–3149.

306. Togashi Y, Soda M, Sakata S, et al. KLC1-ALK: a novel fusion in lung cancer identified using a formalin-fixed paraffin-embedded tissue only. *PLoS One* 2012;7:e31323.

307. Rikova K, Guo A, Zeng Q, et al. Global survey of phosphotyrosine signaling identifies oncogenic kinases in lung cancer. *Cell* 2007;131:1190–1203.

308. Ren H, Tan ZP, Zhu X, et al. Identification of anaplastic lymphoma kinase as a potential therapeutic target in ovarian cancer. *Cancer Res* 2012;72:3312–3323.

309. Chan JK, Lamant L, Algar E, et al. ALK+ histiocytosis: a novel type of systemic histiocytic proliferative disorder of early infancy. *Blood* 2008;112:2965–2968.

310. Yamamoto Y, Tsuzuki S, Tsuzuki M, et al. BCOR as a novel fusion partner of retinoic acid receptor alpha in a t(X;17)(p11;q12) variant of acute promyelocytic leukemia. *Blood* 2010;116:4274–4283.

311. Bohlander SK. ETV6: a versatile player in leukemogenesis. *Semin Cancer Biol* 2005;15:162–174.

312. Eguchi M, Eguchi-Ishimae M, Tojo A, et al. Fusion of ETV6 to neurotrophin-3 receptor TRKC in acute myeloid leukemia with t(12;15)(p13;q25). *Blood* 1999;93:1355–1363.

313. Tognon C, Knezevich SR, Huntsman D, et al. Expression of the ETV6-NTRK3 gene fusion as a primary event in human secretory breast carcinoma. *Cancer Cell* 2002;2:367–376.

314. Skálová A, Vanecek T, Sima R, et al. Mammary analogue secretory carcinoma of salivary glands, containing the ETV6-NTRK3 fusion gene: a hitherto undescribed salivary gland tumor entity. *Am J Surg Pathol* 2010;34:599–608.

315. Leeman-Neill RJ, Kelly LM, Liu P, et al. ETV6-NTRK3 is a common chromosomal rearrangement in radiation-associated thyroid cancer. *Cancer* 2014;120:799–807.

316. Bilodeau EA, Weinreb I, Antonescu CR, et al. Clear cell odontogenic carcinomas show EWSR1 rearrangements: a novel finding and a biological link to salivary clear cell carcinomas. *Am J Surg Pathol* 2013;37:1001–1005.

317. Arbajian E, Magnusson L, Brosjö O, et al. A benign vascular tumor with a new fusion gene: EWSR1-NFATC1 in hemangioma of the bone. *Am J Surg Pathol* 2013;37:613–616.

318. Möller E, Stenman G, Mandahl N, et al. POU5F1, encoding a key regulator of stem cell pluripotency, is fused to EWSR1 in hidradenoma of the skin and mucoepidermoid carcinoma of the salivary glands. *J Pathol* 2008;215:78–86.

319. Antonescu CR, Katabi N, Zhang L, et al. EWSR1-ATF1 fusion is a novel and consistent finding in hyalinizing clear-cell carcinoma of salivary gland. *Genes Chromosomes Cancer* 2011;50:559–570.

320. Skálová A, Weinreb I, Hyrcza M, et al. Clear cell myoepithelial carcinoma of salivary glands showing EWSR1 rearrangement: molecular analysis of 94 salivary gland carcinomas with prominent clear cell component. *Am J Surg Pathol* 2015;39:338–348.

321. Panagopoulos I, Thorsen J, Gorunova L, et al. RNA sequencing identifies fusion of the EWSR1 and YY1 genes in mesothelioma with t(14;22)(q32;q12). *Genes Chromosomes Cancer* 2013;52:733–740.

322. Yamaguchi S, Yamazaki Y, Ishikawa Y, et al. EWSR1 is fused to POU5F1 in a bone tumor with translocation t(6;22)(p21;q12). *Genes Chromosomes Cancer* 2005;43:217–222.

323. Panagopoulos I, Aman P, Fioretos T, et al. Fusion of the FUS gene with ERG in acute myeloid leukemia with t(16;21)(p11;q22). *Genes Chromosomes Cancer* 1994;11:256–262.

324. Puls F, Arbajian E, Magnusson L, et al. Myoepithelioma of bone with a novel FUS-POU5F1 fusion gene. *Histopathology* 2014;65:917–922.

325. Geurts JM, Schoenmakers EF, Röijer E, Stenman G, Van de Ven WJ. Expression of reciprocal hybrid transcripts of HMGIC and FHIT in a pleomorphic adenoma of the parotid gland. *Cancer Res* 1997;57:13–17.

326. Rogalla P, Lemke I, Kazmierczak B, Bullerdiek J. An identical HMGIC-LPP fusion transcript is consistently expressed in pulmonary chondroid hamartomas with t(3;12)(q27-28;q14-15). *Genes Chromosomes Cancer* 2000;29:363–366.

327. Strehl S, Nebral K, König M, et al. ETV6-NCOA2: a novel fusion gene in acute leukemia associated with coexpression of T-lymphoid and myeloid markers and frequent NOTCH1 mutations. *Clin Cancer Res* 2008;14:977–983.

328. Asp J, Persson F, Kost-Alimova M, Stenman G. CHCHD7-PLAG1 and TCEA1-PLAG1 gene fusions resulting from cryptic, intrachromosomal 8q rearrangements in pleomorphic salivary gland adenomas. *Genes Chromosomes Cancer* 2006;45:820–828.

329. Davies KD, Doebele RC. Molecular pathways: ROS1 fusion proteins in cancer. *Clin Cancer Res* 2013;19:4040–4045.

330. Ross H, Argani P. Xp11 translocation renal cell carcinoma. *Pathology* 2010;42:369–373.

331. Oliveira AM, Perez-Atayde AR, Dal Cin P, et al. Aneurysmal bone cyst variant translocations upregulate USP6 transcription by promoter swapping with the ZNF9, COL1A1, TRAP150, and OMD genes. *Oncogene* 2005;24:3419–3426.

332. Oliveira AM, Chou MM. USP6-induced neoplasms: the biologic spectrum of aneurysmal bone cyst and nodular fasciitis. *Hum Pathol* 2014;45:1–11.

Fusion genes involving RNA- and DNA-binding proteins: so-called FET proteins

333. Law WJ, Cann KL, Hicks GG. TLS, EWS and TAF15: a model for transcriptional integration of gene expression. *Brief Funct Genomic Proteomic* 2006;5:8–14.

334. Kovar H. Kovar H. Dr. Jekyll and Mr. Hyde: the two faces of the FUS/EWS/TAF15 protein family. *Sarcoma* 2011;2011:837474.

335. Turc-Carel C, Philip I, Berger M-P, Philip T, Lenoir GM. Chromosomal translocations 11;22 in cell lines of Ewing's sarcoma. *N Engl J Med* 1983;309:497–498.

336. Tomlins SA, Rhodes DR, Perner S, et al. Recurrent fusion of TMPRSS2 and ETS transcription factor genes in prostate cancer. *Science* 2005;310:644–648.

337. Wang J, Cai Y, Ren C, Ittmann M. Expression of variant TMPRSS2/ERG fusion messenger RNAs is associated with aggressive prostate cancer. *Cancer Res* 2006;66:8347–8351.

338. Tomlins SA, Mehra R, Rhodes DR, et al. TMPRSS2:ETV4 gene fusions define a third molecular subtype of prostate cancer. *Cancer Res* 2006;66:3396–3400.

339. Kumar-Sinha C, Tomlins SA, Chinnaiyan AM. Recurrent gene fusions in prostate cancer. *Nat Rev Cancer* 2008;8:497–511.

340. Truong AH, Ben-David Y. The role of Fli-1 in normal cell function and malignant transformation. *Oncogene* 2000;19:6482–6489.

341. Arvand A, Denny CT. Biology of EWS/ETS fusions in Ewing's family tumors. *Oncogene* 2001;20:5747–5754.

342. Hai T, Hartman MG. The molecular biology and nomenclature of the activating transcription factor/cAMP responsive element binding family of transcription factors: activating transcription factor proteins and homeostasis. *Gene* 2001;273:1–11.

343. Brown AD, Lopez-Terrada D, Denny C, Lee KA. Promoters containing ATF-binding sites are de-regulated in cells that express the EWS/ATF1 oncogene. *Oncogene* 1995;10:1749–1756.

344. Panagopoulos I, Möller E, Dahlén A, et al. Characterization of the native CREB3L2 transcription factor and the FUS/CREB3L2 chimera. *Genes Chromosomes Cancer* 2007;46:181–191.

345. Lee SB, Haber DA. Wilms tumor and the WT1 gene. *Exp Cell Res* 2001;264:74–99.

346. Kim J, Lee K, Pelletier J. The desmoplastic small round cell tumor t(11;22) translocation produces EWS/WT1 isoforms with differing oncogenic properties. *Oncogene* 1998;16:1973–1979.

347. Reynolds PA, Smolen GA, Palmer RE, et al. Identification of a DNA-binding site and transcriptional target for the EWS-WT1(+KTS) oncoprotein. *Genes Dev* 2003;17:2094–2107.

348. Ohkura N, Hijikuro M, Yamamoto A, Miki K. Molecular cloning of a novel thyroid/steroid receptor superfamily gene from cultured rat neuronal cells. *Biochem Biophys Res Commun* 1994;205:1959–1965.

349. Ohkura N, Yaguchi H, Tsukada T, Yamaguchi K. The EWS/NOR1 fusion gene product gains a novel activity affecting pre-mRNA splicing. *J Biol Chem* 2002;277:535–543.

350. Gan TI, Rowen L, Nesbitt R, et al. Genomic organization of human TCF12 gene and spliced mRNA variants producing isoforms of transcription factor HTF4. *Cytogenet Genome Res* 2002;98:245–248.

351. Mencinger M, Panagopoulos I, Andreasson P, et al. Characterization and chromosomal mapping of the human TFG gene involved in thyroid carcinoma. *Genomics* 1997;41:327–331.

352. Greco A, Mariani C, Miranda C, et al. The DNA rearrangement that generates the TRK-T3 oncogene involves a novel gene on chromosome 3 whose product has a potential coiled-coil domain. *Mol Cell Biol* 1995;15:6118–6127.

353. Hunger SP, Galili N, Carroll AJ, et al. The t(1;19)(q23;p13) results in consistent fusion of E2A and PBX1 coding sequences in acute lymphoblastic leukemias. *Blood* 1991;77:687–693.

354. Kamps MP, Murre C, Sun XH, Baltimore D. A new homeobox gene contributes the DNA binding domain of the t(1;19) translocation protein in pre-B ALL. *Cell* 1990;60:547–555.

355. Turc-Carel C, Limon J, Dal Cin P, et al. Cytogenetic studies of adipose tissue tumors. II. Recurrent reciprocal translocation t(12;16)(q13;p11) in myxoid liposarcomas. *Cancer Genet Cytogenet* 1986;23:291–299.

356. Åman P, Ron D, Mandahl N, et al. Rearrangement of the transcription factor gene CHOP in myxoid liposarcomas with t(12;16)(q13;p11). *Genes Chromosomes Cancer* 1992;5:278–285.

357. Kuroda M, Ishida T, Takanashi M, et al. Oncogenic transformation and inhibition of adipocytic conversion of preadipocytes by TLS/FUS-CHOP type II chimeric protein. *Am J Pathol* 1997;151:735–744.

358. Zinszner H, Albalat R, Ron D. A novel effector domain from the RNA-binding protein TLS or EWS is required for oncogenic transformation by CHOP. *Genes Dev* 1994;8:2513–2526.

359. Ichikawa H, Shimizu K, Hayashi Y, Ohki M. An RNA-binding protein gene, TLS/FUS, is fused to erg in human myeloid leukemia with t(16;21) chromosomal translocation. *Cancer Res* 1994;54:2865–2868.

360. Åman P, Panagopoulos I, Lassen C, et al. Expression patterns of the human sarcoma-associated genes FUS and EWS and the genomic structure of FUS. *Genomics* 1996;37:1–8.

361. Ron D, Brasier AR, McGehee RE Jr, Habener JF. Tumor necrosis factor-induced reversal of adipocytic phenotype of 3T3-L1 cells is preceded by a loss of nuclear CCAAT/enhancer binding protein (C/EBP). *J Clin Invest* 1992;89:223–233.

362. Ron D, Habener JF. CHOP a novel developmentally regulated nuclear protein that dimerizes with transcription factors C/EBP and LAP and functions as a dominant negative inhibitor of gene transcription. *Genes Dev* 1992;6:439–453.

363. Adelmant G, Gilbert JD, Freytag SO. Human translocation liposarcoma-CCAAT/enhancer binding protein (C/EBP) homologous protein (TLS-CHOP) oncoprotein prevents adipocyte differentiation by directly interfering with C/EBPbeta function. *J Biol Chem* 1998;273:15574–15581.

364. Panagopoulos I, Möller E, Dahlén A, et al. Characterization of the native CREB3L2 transcription factor and the FUS/CREB3L2 chimera. *Genes Chromosomes Cancer* 2007;46:181–191.

Fusion genes involving receptor tyrosine kinase genes

365. Robinson DR, Wu YM, Lin SF. The protein tyrosine kinase family of the human genome. *Oncogene* 2000;19:5548–5557.

366. Blume-Jensen P, Hunter T. Oncogenic kinase signalling. *Nature* 2001;411:355–365.

367. Demetri GD. Targeting c-kit mutations in solid tumors: scientific rationale and novel therapeutic options. *Semin Oncol* 2001;28(5 Suppl 17):19–26.

368. Zhang J, Hochwald SN. Targeting receptor tyrosine kinases in solid tumors. *Surg Oncol Clin N Am* 2013;22:685–703.

Anaplastic lymphoma kinase (ALK) gene

369. Ladanyi M. Aberrant ALK tyrosine kinase signaling. Different cellular lineages, common oncogenic mechanisms? *Am J Pathol* 2000;157:341–345.

370. Duyster J, Bai R-Y, Morris SW. Translocations involving anaplastic lymphoma kinase (ALK). *Oncogene* 2001;20:5623–5637.

371. Bai RY, Dieter P, Peschel C, Morris SW, Duyster J. Nucleophosmin-anaplastic lymphoma kinase of large-cell anaplastic lymphoma is a constitutively active tyrosine kinase that utilizes phospholipase C-gamma to mediate its mitogenicity. *Mol Cell Biol* 1998;18:6951–6961.

372. Bai RY, Ouyang T, Miething C, *et al.* Nucleophosmin-anaplastic lymphoma kinase associated with anaplastic large-cell lymphoma activates the phosphatidylinositol 3-kinase/Akt antiapoptotic signaling pathway. *Blood* 2000;96:4319–4327.

373. Kuefer MU, Look AT, Pulford K, *et al.* Retrovirus-mediated gene transfer of NPM-ALK causes lymphoid malignancy in mice. *Blood* 1997;90:2901–2910.

374. Chiarle R, Gong JZ, Guasparri I, *et al.* NPM-ALK transgenic mice spontaneously develop T-cell lymphomas and plasma cell tumors. *Blood* 2003;101:1919–1927.

375. Armstrong F, Duplantier MM, Trempat P, *et al.* Differential effects of X-ALK fusion proteins on proliferation, transformation, and invasion properties of NIH3T3 cells. *Oncogene* 2004;23:6071–6082.

376. Roskoski R Jr. Anaplastic lymphoma kinase (ALK): structure, oncogenic activation, and pharmacological inhibition. *Pharmacol Res* 2013;68:68–94.

377. Ou SH. Crizotinib: a novel and first-in-class multitargeted tyrosine kinase inhibitor for the treatment of anaplastic lymphoma kinase rearranged non-small cell lung cancer and beyond. *Drug Des Devel Ther* 2011;5:471–485.

378. Butrynski JE, D'Adamo DR, Hornick JL, *et al.* Crizotinib in ALK-rearranged inflammatory myofibroblastic tumor. *N Engl J Med* 2010;363:1727–1733.

379. Jacob SV, Reith JD, Kojima AY, *et al.* An unusual case of systemic inflammatory myofibroblastic tumor with successful treatment with ALK-inhibitor. *Case Rep Pathol* 2014;2014:470340.

Neurotrophic tyrosine kinase receptor type 3 (NTRK3) gene

380. Bibel M, Barde YA. Neurotrophins: key regulators of cell fate and cell shape in the vertebrate nervous system. *Genes Dev* 2000;14:2919–2937.

381. Lannon CL, Sorensen PH. ETV6-NTRK3: a chimeric protein tyrosine kinase with transformation activity in multiple cell lineages. *Semin Cancer Biol* 2005;15:215–223.

382. Wai DH, Knezevich SR, Lucas T, *et al.* The ETV6-NTRK3 gene fusion encodes a chimeric protein tyrosine kinase that transforms NIH3T3 cells. *Oncogene* 2000;19:906–915.

383. Tognon C, Garnett M, Kenward E, *et al.* The chimeric protein tyrosine kinase ETV6-NTRK3 requires both Ras-Erk1/2 and PI3-kinase-Akt signaling for fibroblast transformation. *Cancer Res* 2001;61:8909–8916.

384. Lannon CL, Martin MJ, Tognon CE, *et al.* A highly conserved NTRK3 C-terminal sequence in the ETV6-NTRK3 oncoprotein binds the phosphotyrosine binding domain of insulin receptor substrate-1: an essential interaction for transformation. *J Biol Chem* 2004;279:6225–6234.

385. Chi HT, Ly BT, Kano Y, *et al.* ETV6-NTRK3 as a therapeutic target of small molecule inhibitor PKC412. *Biochem Biophys Res Commun* 2012;429:87–92.

Fusion genes involving chromatin-remodeling genes

386. Limon J, Dal Cin P, Sandberg AA. Translocations involving the X chromosome in solid tumors: presentation of of two sarcomas with t(X;18)(q13;p11). *Cancer Genet Cytogenet* 1986;23:87–91.

387. Turc-Carel C, Dal Cin P, Limon J, *et al.* Involvment of chromosome X in primary cytogenetic change in human neoplasia: nonrandom translocation in synovial sarcoma. *Proc Natl Acad Sci USA* 1987;84:1981–1985.

388. Dal Cin P, Rao U, Jani-Sait S, Karakousis C, Sandberg AA. Chromosomes in the diagnosis of soft tissue tumors. I. Synovial sarcoma. *Mod Pathol* 1992;5:357–362.

389. dos Santos NR, de Bruijn DR, van Kessel AG. Molecular mechanisms underlying human synovial sarcoma development. *Genes Chromosomes Cancer* 2001;30:1–14.

390. Clark AJ, Rocques PJ, Crew AJ, *et al.* Identification of novel genes, SYT and SSX, involved in the t(X;18)(p11.2;q11.2) translocation found in human synovial sarcoma. *Nat Genet* 1994;7:502–508.

391. Güre AO, Wei IJ, Old LJ, Chen YT. The SSX gene family: characterization of 9 complete genes. *Int J Cancer* 2002;101:448–453.

392. Fligman I, Lonardo F, Jhanwar SC, *et al.* Molecular diagnosis of synovial sarcoma and characterization of a variant SYT-SSX2 fusion transcript. *Am J Pathol* 1995;147:1592–1599.

393. Mancuso T, Mezzelani A, Riva C, *et al.* Analysis of SYT-SSX fusion transcripts and bcl-2 expression and phosphorylation status in synovial sarcoma. *Lab Invest* 2000;80:805–813.

394. Safar A, Wickert R, Nelson M, Neff JR, Bridge JA. Characterization of a variant SYT-SSX1 synovial sarcoma fusion transcript. *Diagn Mol Pathol* 1998;7:283–287.

395. Hendricks KB, Shanahan F, Lees E. Role for BRG1 in cell cycle control and tumor suppression. *Mol Cell Biol* 2004;24:362–376.

396. Gure AO, Türeci Ö, Sahin U, *et al.* SSX: a multigene family with several members transcribed in normal testis and human cancer. *Int J Cancer* 1997;72:965–971.

397. de Bruijn DR, dos Santos NR, Thijssen J, *et al.* The synovial sarcoma associated protein SYT interacts with the acute leukemia associated protein AF10. *Oncogene* 2001;20:3281–3289.

398. Eid JE, Kung AL, Scully R, Livingston DM. p300 Interacts with the nuclear proto-oncoprotein SYT as part of the active control of cell adhesion. *Cell* 2000;102:839–848.

399. Ishida M, Tanaka S, Ohki M, Ohta T. Transcriptional coactivator activity of SYT is negatively regulated by BRM and Brg1. *Genes Cells* 2004;9:419–428.

400. Ito T, Ouchida M, Ito S, et al. SYT, a partner of SYT-SSX oncoprotein in synovial sarcomas, interacts with mSin3A, a component of histone deacetylase complex. *Lab Invest* 2004;84:1484–1490.

401. Nagai M, Tanaka S, Tsuda M, et al. Analysis of transforming activity of human synovial sarcoma-associated chimeric protein SYT-SSX1 bound to chromatin remodeling factor hBRM/hSNF1 alpha. *Proc Natl Acad Sci USA* 2001;98:3843–3848.

402. Thaete C, Brett D, Monaghan P, et al. Functional domains of SYT and SYT-SSX synovial sarcoma translocation proteins and co-localization with the SNF protein BRM in the nucleus. *Hum Mol Genet* 1999;8:585–591.

403. Tureci O, Sahin U, Schobert I, et al. The SSX-2 gene, which is involved in the t(X;18) translocation of synovial sarcomas, codes for the human tumor antigen HOM-MEL-40. *Cancer Res* 1996;56:4766–4772.

404. de Bruijn DR, Nap JP, van Kessel AG. The (epi)genetics of human synovial sarcoma. *Genes Chromosomes Cancer* 2007;46:107–117.

405. de Bruijn DR, Allander SV, van Dijk AH, et al. The synovial sarcoma-associated SS18-SSX2 fusion protein induces epigenetic gene (de)regulation. *Cancer Res* 2006;66:9474–9482.

406. Ito T, Ouchida M, Morimoto Y, et al. Significant growth suppression of synovial sarcomas by the histone deacetylase inhibitor FK228 in vitro and in vivo. *Cancer Lett* 2005;224:311–319.

407. Lubieniecka JM, de Bruijn DR, Su L, et al. Histone deacetylase inhibitors reverse SS18-SSX-mediated polycomb silencing of the tumor suppressor early growth response 1 in synovial sarcoma. *Cancer Res* 2008;68:4303–4310.

408. Subramaniam MM, Calabuig-Fariñas S, Pellin A, Llombart-Bosch A. Mutation analysis of E-cadherin, beta-catenin and APC genes in synovial sarcoma. *Histopathology* 2010;57:482–486.

409. Barham W, Frump AL, Sherrill TP, et al. Targeting the Wnt pathway in synovial sarcoma models. *Cancer Discov* 2013;3:1286–1301.

410. Friedrichs N, Trautmann M, Endl E, et al. Phosphatidylinositol-3'-kinase/AKT signaling is essential in synovial sarcoma. *Int J Cancer* 2011;129:1564–1575.

411. Setsu N, Kohashi K, Fushimi F, et al. Prognostic impact of the activation status of the Akt/mTOR pathway in synovial sarcoma. *Cancer* 2013;119:3504–3513.

412. Nielsen TO, Poulin NM, Ladanyi M. Synovial sarcoma: recent discoveries as a roadmap to new avenues for therapy. *Cancer Discov* 2015;5:124–134.

COL1A1-PDGFB, a fusion gene involving growth factor

413. Pedeutour F, Simon MP, Minoletti F, et al. Translocation, t(17;22)(q22;q13), in dermatofibrosarcoma protuberans: a new tumor-associated chromosome rearrangement. *Cytogenet Cell Genet* 1996;72:171–174.

414. Sjoblom T, Shimizu A, O'Brien KP, et al. Growth inhibition of dermatofibrosarcoma protuberans tumors by the platelet-derived growth factor receptor antagonist STI571 through induction of apoptosis. *Cancer Res* 2001;61:5778–5783.

415. Abrams TA, Schuetze SM. Targeted therapy for dermatofibrosarcoma protuberans. *Curr Oncol Rep* 2006;8:291–296.

416. Llombart B, Sanmartín O, López-Guerrero JA, et al. Dermatofibrosarcoma protuberans: clinical, pathological, and genetic (COL1A1-PDGFB) study with therapeutic implications. *Histopathology* 2009;54:860–872.

Other types of fusion genes
PAX-FKHR fusion genes

417. Barr FG. Gene fusions involving PAX and FOX family members in alveolar rhabdomyosarcoma. *Oncogene* 2001;20:5736–5746.

418. Tremblay P, Gruss P. Pax: genes for mice and men. *Pharmacol Ther* 1994;61:205–226.

419. Underhill DA. Genetic and biochemical diversity in the Pax gene family. *Biochem Cell Biol* 2000;78:629–638.

420. Chi N, Epstein JA. Getting your Pax straight: Pax proteins in development and disease. *Trends Genet* 2002;18:41–47.

421. Buckingham M, Relaix F. The role of Pax genes in the development of tissues and organs: Pax3 and Pax7 regulate muscle progenitor cell functions. *Annu Rev Cell Dev Biol* 2007;23:645–673.

422. Epstein DJ, Vekemans M, Gros P. Splotch (Sp-2H), a mutation affecting development of the mouse neural tube, shows a deletion within the paired homeodomain of Pax-3. *Cell* 1991;67:767–774.

423. Baldwin CT, Hoth CF, Amos JA, da-Silva EO, Milunsky A. An exonic mutation in the HuP2 paired domain gene causes Waardenburg's syndrome. *Nature* 1992;355:637–638.

424. Tassabehji M, Read AP, Newton VE, et al. Mutations in the PAX3 gene causing Waardenburg syndrome type 1 and type 2. *Nat Genet* 1993;3:26–30.

425. Asher JH Jr, Sommer A, Morell R, Friedman TB. Missense mutation in the paired domain of PAX3 causes craniofacial-deafness-hand syndrome. *Hum Mutat* 1996;7:30–35.

426. Kaufmann E, Knochel W. Five years on the wings of fork head. *Mech Dev* 1996;57:3–20.

427. Katoh M, Katoh M. Human FOX gene family [review]. *Int J Oncol* 2004;25:1495–1500.

428. Durham SK, Suwanichkul A, Scheimann AO, et al. FKHR binds the insulin response element in the insulin-like growth factor binding protein-1 promoter. *Endocrinology* 1999;140:3140–3146.

429. Guo S, Rena G, Cichy S, et al. Phosphorylation of serine 256 by protein kinase B disrupts transactivation by FKHR and mediates effects of insulin on insulin-like growth factor-binding protein-1 promoter activity through a conserved insulin response sequence. *J Biol Chem* 1999;274:17184–17192.

430. Brunet A, Bonni A, Zigmond MJ, et al. Akt promotes cell survival by phosphorylating and inhibiting a

Forkhead transcription factor. *Cell* 1999;96:857–868.

431. Accili D, Arden KC. FoxOs at the crossroads of cellular metabolism, differentiation, and transformation. *Cell* 2004;117:421–426.

432. Hillion J, Le Coniat M, Jonveaux P, Berger R, Bernard OA. AF6q21, a novel partner of the MLL gene in t(6;11)(q21;q23), defines a forkhead transcriptional factor subfamily. *Blood* 1997;90:3714–3719.

433. Parry P, Wei Y, Evans G. Cloning and characterization of the t(X;11) breakpoint from a leukemic cell line identify a new member of the forkhead gene family. *Genes Chromosomes Cancer* 1994;11:79–84.

434. Barr FG, Nauta LE, Davis RJ, et al. In vivo amplification of the PAX3-FKHR and PAX7-FKHR fusion genes in alveolar rhabdomyosarcoma. *Hum Mol Genet* 1996;5:15–21.

435. Davis RJ, Barr FG. Fusion genes resulting from alternative chromosomal translocations are overexpressed by gene-specific mechanisms in alveolar rhabdomyosarcoma. *Proc Natl Acad Sci USA* 1997;94:8047–8051.

436. Weber-Hall S, McManus A, Anderson J, et al. Novel formation and amplification of the PAX7-FKHR fusion gene in a case of alveolar rhabdomyosarcoma. *Genes Chromosomes Cancer* 1996;17:7–13.

437. Fitzgerald JC, Scherr AM, Barr FG. Structural analysis of PAX 7 rearrangements in alveolar rhabdomyosarcoma. *Cancer Genet Cytogenet* 2000;117:37–40.

438. del Peso L, González VM, Hernández R, Barr FG, Núñez G. Regulation of the forkhead transcription factor FKHR, but not the PAX3-FKHR fusion protein, by the serine/threonine kinase Akt. *Oncogene* 1999;18:7328–7333.

439. Fredericks WJ, Galili N, Mukhopadhyay S, et al. The PAX3-FKHR fusion protein created by the t(2;13) translocation in alveolar rhabdomyosarcomas is a more potent transcriptional activator than PAX3. *Mol Cell Biol* 1995;15:1522–1535.

440. Scheidler S, Fredericks WJ, Rauscher FJ III, Barr FG, Vogt PK. The hybrid PAX3-FKHR fusion protein of alveolar rhabdomyosarcoma transforms fibroblasts in culture. *Proc Natl Acad Sci USA* 1996;93:9805–9809.

441. Lam PY, Sublett JE, Hollenbach AD, Roussel MF. The oncogenic potential of the Pax3-FKHR fusion protein requires the Pax3 homeodomain recognition helix but not the Pax3 paired-box DNA binding domain. *Mol Cell Biol* 1999;19:594–601.

442. Xia SJ, Barr FG. Analysis of the transforming and growth suppressive activities of the PAX3-FKHR oncoprotein. *Oncogene* 2004;23:6864–6871.

443. Deneen B, Denny CT. Loss of p16 pathways stabilizes EWS/FLI1 expression and complements EWS/FLI1 mediated transformation. *Oncogene* 2001;20:6731–6741.

444. Ren YX, Finckenstein FG, Abdueva DA, et al. Mouse mesenchymal stem cells expressing PAX-FKHR form alveolar rhabdomyosarcomas by cooperating with secondary mutations. *Cancer Res* 2008;68:6587–6597.

445. Keller C, Arenkiel BR, Coffin CM, et al.. Alveolar rhabdomyosarcomas in conditional Pax3:Fkhr mice: cooperativity of Ink4a/ARF and Trp53 loss of function. *Genes Dev* 2004;18:2614–2626.

446. Begum S, Emami N, Cheung A, et al. Cell-type-specific regulation of distinct sets of gene targets by Pax3 and Pax3/FKHR. *Oncogene* 2005;24:1860–1872.

447. Marshall AD, Grosveld GC. Alveolar rhabdomyosarcoma-The molecular drivers of PAX3/7-FOXO1-induced tumorigenesis. *Skelet Muscle* 2012;2:25.

448. Wachtel M, Dettling M, Koscielniak E, et al. Gene expression signatures identify rhabdomyosarcoma subtypes and detect a novel t(2;2)(q35;p23) translocation fusing PAX3 to NCOA1. *Cancer Res* 2004;64:5539–5545.

449. Jankowski K, Kucia M, Wysoczynski M, et al. Both hepatocyte growth factor (HGF) and stromal-derived factor-1 regulate the metastatic behavior of human rhabdomyosarcoma cells, but only HGF enhances their resistance to radiochemotherapy. *Cancer Res* 2003;63:7926–7935.

450. Nabarro S, Himoudi N, Papanastasiou A, et al. Coordinated oncogenic transformation and inhibition of host immune responses by the PAX3-FKHR fusion oncoprotein. *J Exp Med* 2005;202:1399–1410.

451. Reichek JL, Duan F, Smith LM, et al. Genomic and clinical analysis of amplification of the 13q31 chromosomal region in alveolar rhabdomyosarcoma: a report from the Children's Oncology Group. *Clin Cancer Res* 2011;17:1463–1473.

ASPL-TFE3 fusion gene

452. Heimann P, el Housni H, Ogur G, et al. Fusion of a novel gene, RCC17, to the TFE3 gene in t(X;17)(p11.2;q25.3)-bearing papillary renal cell carcinomas. *Cancer Res* 2001;61:4130–4135.

453. Alexandru G, Graumann J, Smith GT, et al. UBXD7 binds multiple ubiquitin ligases and implicates p97 in HIF1alpha turnover. *Cell* 2008;134:804–816.

454. Martina JA, Diab HI, Li H, Puertollano R. Novel roles for the MiTF/TFE family of transcription factors in organelle biogenesis, nutrient sensing, and energy homeostasis. *Cell Mol Life Sci* 2014;71:2483–2497.

455. Sidhar SK, Clark J, Gill S, et al. The t(X;1)(p11.2;q21.2) translocation in papillary renal cell carcinoma fuses a novel gene PRCC to the TFE3 transcription factor gene. *Hum Mol Genet* 1996;5:1333–1338.

456. Clark J, Lu YJ, Sidhar SK, et al. Fusion of splicing factor genes PSF and NonO (p54nrb) to the TFE3 gene in papillary renal cell carcinoma. *Oncogene* 1997;15:2233–2239.

457. Argani P, Lui MY, Couturier J, et al. A novel CLTC-TFE3 gene fusion in pediatric renal adenocarcinoma with t(X;17)(p11.2;q23). *Oncogene* 2003;22:5374–5378.

458. Davis IJ, Hsi BL, Arroyo JD, et al. Cloning of an alpha-TFEB fusion in renal tumors harboring the t(6;11)(p21;q13) chromosome translocation. *Proc Natl Acad Sci USA* 2003;100:6051–6056.

459. Argani P, Antonescu CR, Illei PB, et al. Primary renal neoplasms with the ASPL-TFE3 gene fusion of alveolar soft part sarcoma: a distinctive tumor entity previously included among renal cell carcinomas of children and adolescents. *Am J Pathol* 2001;159:179–192.

460. Tsuda M, Davis IJ, Argani P, et al. TFE3 fusions activate MET signaling by transcriptional up-regulation, defining another class of tumors as candidates for therapeutic MET inhibition. *Cancer Res* 2007;67:919–929.

461. Mitton B, Federman N. Alveolar soft part sarcoma: molecular pathogenesis and implications for novel targeted therapies. *Sarcoma* 2012;2012:428789.

Oncogene amplifications

462. Kiuru-Kuhlefelt S, El-Rifai W, Fanburg-Smith J, et al. Concomitant DNA copy number amplification at 17q and 22q in dermatofibrosarcoma protuberans. *Cytogenet Cell Genet* 2001;92:192–195.

463. Abbott JJ, Erickson-Johnson M, Wang X, Nascimento AG, Oliveira AM. Gains of COL1A1-PDGFB genomic copies occur in fibrosarcomatous transformation of dermatofibrosarcoma protuberans. *Mod Pathol* 2006;19:1512–1518.

464. Nitta H, Grogan TM. Bright-field in situ hybridization methods to discover gene amplifications and rearrangements in clinical samples. *Methods Mol Biol* 2013;986:341–352.

465. Pinkel D, Albertson DG. Array comparative genomic hybridization and its applications in cancer. *Nat Genet* 2005;37 Suppl: S11-17.

466. Schwab M. Oncogene amplification in solid tumors. *Semin Cancer Biol* 1999;9:319–325.

467. Sirvent N, Coindre JM, Maire G, et al. Detection of MDM2-CDK4 amplification by fluorescence in situ hybridization in 200 paraffinembedded tumor samples: utility in diagnosing adipocytic lesions and comparison with immunohistochemistry and real-time PCR. *Am J Surg Pathol* 2007;31:1476–1489.

468. Thorner PS, Ho M, Chilton-MacNeill S, Zielenska M. Use of chromogenic in situ hybridization to identify MYCN gene copy number in neuroblastoma using routine tissue sections. *Am J Surg Pathol* 2006;30:635–642.

469. Abel HJ, Duncavage EJ. Detection of structural DNA variation from next generation sequencing data: a review of informatic approaches. *Cancer Genet* 2013;206:432–440.

470. Collins S, Groudine M. Amplification of endogenous mycrelated sequences in a human myeloid leukaemia cell line. *Nature* 1982;298:679–681.

471. Dalla-Favera R, Wong-Staal F, Gallo RC. Onc gene amplification in promyelocytic leukaemia cell line HL-60 and primary leukaemic cells of the same patient. *Nature* 1982;299:61–63.

472. Grandori C, Cowley SM, James LP, Eisenman RN. The Myc/Max/Mad network and the transcriptional control of cell behavior. *Annu Rev Cell Dev Biol* 2000;16:653–699.

473. Oster SK, Ho CS, Soucie EL, Penn LZ. The myc oncogene: MarvelouslY Complex. *Adv Cancer Res* 2002;84:81–154.

474. Yokota J, Tsunetsugu-Yokota Y, Battifora H, Le Fevre C, Cline MJ. Alterations of myc, myb, and ras(Ha) proto-oncogenes in cancers are frequent and show clinical correlation. *Science* 1986;231:261–265.

475. Barrios C, Castresana JS, Ruiz J, Kreicbergs A. Amplification of the c-myc proto-oncogene in soft tissue sarcomas. *Oncology* 1994;51:13–17.

476. Manner J, Radlwimmer B, Hohenberger P, et al. MYC high level gene amplification is a distinctive feature of angiosarcomas after irradiation or chronic lymphedema. *Am J Pathol* 2010;176:34–39.

477. Käcker C, Marx A, Mössinger K, et al. High frequency of MYC gene amplification is a common feature of radiation-induced sarcomas. Further results from EORTC STBSG TL 01/01. *Genes Chromosomes Cancer* 2013;52:93–98.

478. Italiano A, Thomas R, Breen M, et al. The miR-17–92 cluster and its target THBS1 are differentially expressed in angiosarcomas dependent on MYC amplification. *Genes Chromosomes Cancer* 2012;51:569–578.

479. Shon W, Sukov WR, Jenkins SM, Folpe AL. MYC amplification and overexpression in primary cutaneous angiosarcoma: a fluorescence in-situ hybridization and immunohistochemical study. *Mod Pathol* 2014;27:509–515.

480. Brodeur GM, Seeger RC, Schwab M, Varmus HE, Bishop JM. Amplification of N-myc in untreated human neuroblastomas correlates with advanced disease stage. *Science* 1984;224:1121–1124.

481. Seeger RC, Brodeur GM, Sather H, et al. Association of multiple copies of the N-myc oncogene with rapid progression of neuroblastomas. *N Engl J Med* 1985;313:1111–1116.

482. Brodeur GM, Azar C, Brother M, et al. Neuroblastoma: effect of genetic factors on prognosis and treatment. *Cancer* 1992;70:1685–1694.

483. Chen Y, Takita J, Choi YL, et al. Oncogenic mutations of ALK kinase in neuroblastoma. *Nature* 2008;455:971–974.

484. George RE, Sanda T, Hanna M, et al. Activating mutations in ALK provide a therapeutic target in neuroblastoma. *Nature* 2008;455:975–978.

485. Janoueix-Lerosey I, Lequin D, Brugiéres L, et al. Somatic and germline activating mutations of the ALK kinase receptor in neuroblastoma. *Nature* 2008;455:967–970.

486. Mossé YP, Laudenslager M, Longo L, et al. Identification of ALK as a major familial neuroblastoma predisposition gene. *Nature* 2008;455:930–935.

487. Carén H, Abel F, Kogner P, Martinsson T. High incidence of DNA mutations and gene amplifications of the ALK gene in advanced sporadic neuroblastoma tumours. *Biochem J* 2008;416:153–159.

488. Christensen JG, Zou HY, Arango ME, et al. Cytoreductive antitumor activity of PF-2341066, a novel inhibitor of anaplastic lymphoma kinase and c-Met, in experimental models of anaplastic large-cell lymphoma. *Mol Cancer Ther* 2007;6:3314–3322.

489. Li R, Morris SW. Development of anaplastic lymphoma kinase (ALK) small-molecule inhibitors for cancer therapy. *Med Res Rev* 2008;28:372–412.

490. Barone G, Anderson J, Pearson AD, Petrie K, Chesler L. New strategies in neuroblastoma: therapeutic targeting of MYCN and ALK. *Clin Cancer Res* 2013;19:5814–5821.

491. Li XQ, Hisaoka M, Shi DR, Zhu XZ, Hashimoto H. Expression of anaplastic lymphoma kinase in soft tissue tumors: an immunohistochemical and molecular study of 249 cases. *Hum Pathol* 2004;35:711–721.

492. Williamson D, Lu YJ, Gordon T, et al. Relationship between MYCN copy number and expression in rhabdomyosarcomas and correlation with adverse prognosis in the alveolar subtype. *J Clin Oncol* 2005;23:880–888.

493. Ozaki T, Nakagawa Y, Yoshida A, et al. Amplification of MYCL in atypical Ewing tumor. Analysis of metaphase and microarray comparative genomic hybridization. *Cancer Genomics Proteomics* 2004;1:275–282.

494. Ragazzini P, Gamberi G, Pazzaglia L, et al. Amplification of CDK4, MDM2, SAS and GLI genes in leiomyosarcoma, alveolar and embryonal rhabdomyosarcoma. *Histol Histopathol* 2004;19:401–411.

495. Wolf M, Aaltonen LA, Szymanska J, et al. Complexity of 12q13–22 amplicon in liposarcoma: microsatellite repeat analysis. *Genes Chromosomes Cancer* 1997;18:66–70.

496. Elkahloun AG, Bittner M, Hoskins K, Gemmill R, Meltzer PS. Molecular cytogenetic characterization and physical mapping of 12q13–15 amplification in human cancer. *Genes Chromosomes Cancer* 1996;17:205–214.

497. Reifenberger G, Ichimura K, Reinferberger J, et al. Refined mapping of 12q13–15 amplicons in human malignant gliomas suggests CDK4/SAS and MDM2 as independent amplification targets. *Cancer Res* 1996;56:5141–5145.

498. Fakharzadeh SS, Trusko SP, George DL. Tumorigenic potential associated with enhanced expression of a gene that is amplified in a mouse tumor cell line. *EMBO J* 1991;10:1565–1569.

499. Berner JM, Forus A, Elkahloun A, et al. Separate amplified regions encompassing CDK4 and MDM2 in human sarcomas. *Genes Chromosomes Cancer* 1996;17:254–259.

500. Kussie P, Gorina S, Marechal V, et al. Structure of the MDM2 oncoprotein bound to the p53 tumor suppressor transactivation domain. *Science* 1996;274:921–922.

501. Buschmann T, Fuchs SY, Lee CG, Pan ZQ, Ronai Z. SUMO-1 modification of Mdm2 prevents its selfubiquitination and increases Mdm2 ability to ubiquitinate p53. *Cell* 2000;101:753–762.

502. Xiao ZX, Chen J, Levine AJ, et al. Interaction between the retinoblastoma protein and the oncoprotein MDM2. *Nature* 1995;375:694–698.

503. Sdek P, Ying H, Chang DL, et al. MDM2 promotes proteasome-dependent ubiquitin-independent degradation of retinoblastoma protein. *Mol Cell* 2005;20:699–708.

504. Bond GL, Hu W, Bond EE, et al. A single nucleotide polymorphism in the MDM2 promoter attenuates the p53 tumor suppressor pathway and accelerates tumor formation in humans. *Cell* 2004;119:591–602.

505. Bougeard G, Baert-Desurmont S, Tournier I, et al. Impact of the MDM2 SNP309 and p53 Arg72Pro polymorphism on age of tumour onset in Li-Fraumeni syndrome. *J Med Genet* 2006;43:531–533.

506. Ruijs MW, Schmidt MK, Nevanlinna H, et al. The single-nucleotide polymorphism 309 in the MDM2 gene contributes to the Li-Fraumeni syndrome and related phenotypes. *Eur J Hum Genet* 2007;15:110–114.

507. Fedele M, Battista S, Manfioletti G, et al. Role of the high mobility group A proteins in human lipomas. *Carcinogenesis* 2001;22:1583–1591.

508. Nakayama T, Toguchida J, Wadayama B, et al. MDM2 gene amplification in bone and soft tissue tumors: association with tumor progression in differentiated adipose tissue tumors. *Int J Cancer* 1995;64:342–346.

509. Pedeutour F, Forus A, Coindre JM, et al. Structure of the supernumerary ring and giant rod chromosomes in adipose tissue tumors. *Genes Chromosomes Cancer* 1999;24:30–41.

510. Suijkerbuijk RF, Olde Weghuis DE, Van Den Berg M, et al. Comparative genomic hybridization as a tool to define two distinct chromosome 12-derived amplification units in well-differentiated liposarcomas. *Genes Chromosomes Cancer* 1994;9:292–295.

511. Szymanska J, Virolainen M, Tarkkanen M, et al. Overrepresentation of 1q21–23 and 12q13–21 in lipoma-like liposarcomas but not in benign lipomas: a comparative genomic hybridization study. *Cancer Genet Cytogenet* 1997;99:14–18.

512. Pilotti S, Della Torre G, Lavarino C, et al. Distinct mdm2/p53 expression patterns in liposarcoma subgroups: implication for different pathogenetic mechanisms. *J Pathol* 1997;181:14–24.

513. Italiano A, Cardot N, Dupré F, et al. Gains and complex rearrangements of the 12q13–15 chromosomal region in ordinary lipomas: the "missing link" between lipomas and liposarcomas? *Int J Cancer* 2007;121:308–315.

514. Hostein I, Pelmus M, Aurias A, et al. Evaluation of MDM2 and CDK4 amplification by real-time PCR on paraffin wax-embedded material: a potential tool for the diagnosis of atypical lipomatous tumours/well-differentiated liposarcomas. *J Pathol* 2004;202:95–102.

515. Binh MB, Sastre-Garau X, Guillou L, et al. MDM2 and CDK4 immunostainings are useful adjuncts in diagnosing well-differentiated and dedifferentiated liposarcoma subtypes: a comparative analysis of 559 soft tissue neoplasms with genetic data. *Am J Surg Pathol* 2005;29:1340–1347.

516. Weaver J, Downs-Kelly E, Goldblum JR, et al. Fluorescence in situ hybridization for MDM2 gene amplification as a diagnostic tool in lipomatous neoplasms. *Mod Pathol* 2008;21:943–949.

517. Coindre JM, Pédeutour F, Aurias A. Well-differentiated and dedifferentiated liposarcomas. *Virchows Arch* 2010;456:167–179.

518. Goldblum JR. An approach to pleomorphic sarcomas: can we subclassify, and does it matter? *Mod Pathol* 2014;27(Suppl 1):S39–46.

519. Bode-Lesniewska B, Zhao J, Speel EJ, et al. Gains of 12q13–14 and overexpression of mdm2 are frequent findings in intimal sarcomas of the pulmonary artery. *Virchows Arch* 2001;438:57–65.

520. Neuville A, Collin F, Bruneval P, et al. Intimal sarcoma is the most frequent primary cardiac sarcoma: clinicopathologic and molecular retrospective analysis of 100 primary cardiac sarcomas. *Am J Surg Pathol* 2014;38:461–469.

521. Vassilev LT, Vu BT, Graves B, et al. In vivo activation of the p53 pathway by small-molecule antagonists of MDM2. *Science* 2004;303:844–848.

522. Tovar C, Rosinski J, Filipovic Z, et al. Small-molecule MDM2 antagonists reveal aberrant p53 signaling in cancer: implications for therapy. *Proc Natl Acad Sci USA* 2006;103:1888–1893.

523. Lau LM, Nugent JK, Zhao X, Irwin MS. HDM2 antagonist Nutlin-3 disrupts

p73-HDM2 binding and enhances p73 function. *Oncogene* 2008;27:997–1003.

524. Müller CR, Paulsen EB, Noordhuis P, *et al.* Potential for treatment of liposarcomas with the MDM2 antagonist Nutlin-3A. *Int J Cancer* 2007;121:199–205.

525. Vassilev LT. MDM2 inhibitors for cancer therapy. *Trends Mol Med* 2007;13:23–31.

Gain-of-function mutations

526. Taylor ML, Metcalfe DD. KIT signaling trunsduction. *Hematol Oncol Clin North Am* 2000;14:517–535.

527. Fletcher JA. Role of KIT and platelet-derived growth factor receptors as oncoproteins. *Semin Oncol* 2004;31 (Suppl 6):4–11.

528. Mol CD, Dougan DR, Schneider TR, *et al.* Structural basis for the autoinhibition and STI-571 inhibition of c-Kit tyrosine kinase. *J Biol Chem* 2004;279:31655–31663.

529. Hirota S, Isozaki K, Moriyama Y, *et al.* Gain-of-function mutations of c-kit in human gastrointestinal stromal tumors. *Science* 1998;279:577–580.

530. Nishida T, Hirota S, Taniguchi M, *et al.* Familial gastrointestinal stromal tumours with germline mutation of the KIT gene. *Nat Genet* 1998;19:323–324.

531. Ma Y, Cunningham M, Wang X, *et al.* Inhibition of spontaneous receptor phosphorylation by residues in putative alpha-helix in the KIT intracellular juxtamembrane region. *J Biol Chem* 1999;274:13399–13402.

532. Chan PM, Ilangumaran S, La Rose J, Chakrabartty A, Rottapel R. Autoinhibition of the kit receptor tyrosine kinase by the cytosolic juxtamembrane region. *Mol Cell Biol* 2003;23:3067–3078.

533. Nakahara M, Isozaki K, Hirota S, *et al.* A novel gain-of-function mutation of c-kit gene in gastrointestinal stromal tumors. *Gastroenterology* 1998;115:1090–1095.

534. Lasota J, Miettinen M. Histopathology. Clinical significance of oncogenic KIT and PDGFRA mutations in gastrointestinal stromal tumours. *Histopathology* 2008;53:245–266.

535. Sommer G, Agosti V, Ehlers I, *et al.* Gastrointestinal stromal tumors in a mouse model by targeted mutation of the Kit receptor tyrosine kinase. *Proc Natl Acad Sci USA* 2003;100:6706–6711.

536. Rubin BP, Antonescu CR, Scott-Browne JP, *et al.* A knock-in mouse model of gastrointestinal stromal tumor harboring Kit K641E. *Cancer Res* 2005;65:6631–6639.

537. Joensuu H, Roberts PJ, Sarlomo-Rikala M, *et al.* Effect of the tyrosine kinase inhibitor STI571 in a patient with metastatic gastrointestinal stromal tumor. *N Engl J Med* 2001;344:1052–1056.

538. van Oosterom AT, Judson I, Verweij J, *et al.* Safety and efficacy of imatinib (STI571) in metastatic gastrointestinal stromal tumours: a phase I study. *Lancet* 2001;358:1421–1423.

539. Chen LL, Trent JC, Wu EF, *et al.* A missense mutation in KIT domain 1 correlates with imatinib resistance in gastrointestinal stromal tumors. *Cancer Res* 2004;64:5913–5919.

540. Debiec-Rychter M, Cools J, Dumez H, *et al.* Mechanisms of resistance to imatinib mesylate in gastrointestinal stromal tumors and activity of the PKC412 inhibitor against imatinib-resistant mutants. *Gastroenterology* 2005;128:270–279.

541. Heinrich MC, Corless CL, Blanke CD, *et al.* Molecular correlates of imatinib resistance in gastrointestinal stromal tumors. *J Clin Oncol* 2006;24:4764–4774.

542. Fisher R, Pusztai L, Swanton C. Cancer heterogeneity: implications for targeted therapeutics. *Br J Cancer* 2013;108:479–485.

543. Faivre S, Delbaldo C, Vera K, *et al.* Safety, pharmacokinetic, and antitumor activity of SU11248, a novel oral multitarget tyrosine kinase inhibitor, in patients with cancer. *J Clin Oncol* 2006;24:25–35.

544. Joensuu H. Second-line therapies for the treatment of gastrointestinal stromal tumor. *Curr Opin Oncol* 2007;19:353–358.

545. Maki RG. Recent advances in therapy for gastrointestinal stromal tumors. *Curr Oncol Rep* 2007;9:165–169.

546. Roberts WM, Look AT, Roussel MF, Sherr CJ. Tandem linkage of human CSF-1 receptor (c-fms) and PDGF receptor genes. *Cell* 1989;55:655–661.

547. Stenman G, Eriksson A, Claesson-Welsh L. Human PDGFA receptor gene maps to the same region on chromosome 4 as the KIT oncogene. *Genes Chromosomes Cancer* 1989;1:155–158.

548. Heinrich MC, Corless CL, Duensing A, *et al.* PDGFRA activating mutations in gastrointestinal stromal tumors. *Science* 2003;299:708–710.

549. Chompret A, Kannengiesser C, Barrois M, *et al.* PDGFRA germline mutation in a family with multiple cases of gastrointestinal stromal tumor. *Gastroenterology* 2004;126:318–321.

Loss-of-function mutations

Neurofibromatosis type 1 gene (NF1)

550. Cawthon RM, Weiss R, Xu GF, *et al.* A major segment of the neurofibromatosis type 1 gene: cDNA sequence, genomic structure, and point mutations. *Cell* 1990;62:193–201.

551. Wallace MR, Marchuk DA, Anderson LB, *et al.* Type 1 neurofibromatosis gene: identification of a large transcript disrupted in three NF1 patients. *Science* 1990;249:181–186.

552. Xu GF, Lin B, Tanaka K, *et al.* The catalytic domain of the neurofibromatosis type 1 gene product stimulates ras GTPase and complements ira mutants of S. cerevisiae. *Cell* 1990;63:835–841.

553. Martin GA, Viskochil D, Bollag G, *et al.* The GAP-related domain of the neurofibromatosis type 1 gene product interacts with ras p21. *Cell* 1990;63:843–849.

554. Bollag G, Clapp DW, Shih S, *et al.* Loss of NF1 results in activation of the Ras signaling pathway and leads to aberrant growth in haematopoietic cells. *Nat Genet* 1996;12:144–148.

555. Hiatt KK, Ingram DA, Zhang Y, Bollag G, Clapp DW. Neurofibromin GTPase-activating protein-related domains restore normal growth in Nf12/2 cells. *J Biol Chem* 2001;276:7240–7245.

556. Dilworth JT, Kraniak JM, Wojtkowiak JW, *et al.* Molecular targets for emerging anti-tumor therapies for neurofibromatosis type 1. *Biochem Pharmacol* 2006;72:1485–1492.

557. Parada LF, Kwon CH, Zhu Y. Modeling neurofibromatosis type 1 tumors in the mouse for therapeutic intervention. *Cold Spring Harb Symp Quant Biol* 2005;70:173–176.

558. Tong J, Hannan F, Zhu Y, Bernards A, Zhong Y. Neurofibromin regulates G protein-stimulated adenylyl cyclase activity. *Nat Neurosci* 2002;5:95–96.

559. Brown JA, Gianino SM, Gutmann DH. Defective cAMP generation underlies the sensitivity of CNS neurons to neurofibromatosis-1 heterozygosity. *J Neurosci* 2010;30:5579–5589.

560. Cichowski J, Shih TS, Schmitt E, et al. Mouse models of tumor development in neurofibromatosis type 1. *Science* 1999;286:2172–2176.

561. Lin AL, Gutmann DH. Advances in the treatment of neurofibromatosis-associated tumours. *Nat Rev Clin Oncol* 2013;10:616–624.

562. Gottfried ON, Viskochil DH, Fults DW, Couldwell WT. Molecular, genetic, and cellular pathogenesis of neurofibromas and surgical implications. *Neurosurgery* 2006;58:1–16.

563. Le LQ, Parada LF. Tumor microenvironment and neurofibromatosis type I: connecting the GAPs. *Oncogene* 2007;26:4609–4616.

564. Robertson KA, Nalepa G, Yang FC, et al. Imatinib mesylate for plexiform neurofibromas in patients with neurofibromatosis type 1: a phase 2 trial. *Lancet Oncol* 2012;13:1218–1224.

565. Fahsold R, Hoffmeyer S, Mischung C, et al. Minor lesion mutational spectrum of the entire NF1 gene does not explain its high mutability but points to a functional domain upstream of the GAP-related domain. *Am J Hum Genet* 2000;66:790–818.

566. Dorschner MO, Sybert VP, Weaver M, Pletcher BA, Stephens K. NF1 microdeletion breakpoints are clustered at flanking repetitive sequences. *Hum Mol Genet* 2000;9:35–46.

567. Venturin M, Guarnieri P, Natacci F, et al. Mental retardation and cardiovascular malformations in NF1 microdeleted patients point to candidate genes in 17q11.2. *J Med Genet* 2004;41:35–41.

568. Skuse GR, Kosciolek BA, Rowley PT. Molecular genetic analysis of tumors in von Recklinghausen neurofibromatosis: loss of heterozygosity for chromosome 17. *Genes Chromosomes Cancer* 1989;1:36–41.

569. Xu W, Mulligan LM, Ponder MA, et al. Loss of NF1 alleles in phaeochromocytomas from patients with type I neurofibromatosis. *Genes Chromosomes Cancer* 1992;4:337–342.

570. Legius E, Marchuk DA, Collins FS, Glover TW. Somatic deletion of the neurofibromatosis type 1 gene in neurofibrosarcoma supports a tumour suppressor gene hypothesis. *Nat Genet* 1993;3:122–126.

571. Colman SD, Williams CA, Wallace RW. Benign neurofibromas in type 1 neurofibromatosis (NF1) show somatic deletions of the NF1 gene. *Nat Genet* 1995;11:90–92.

572. Kluwe L, Friedrich RE, Mautner VF. Allelic loss of the NF1 gene in NF1-associated plexiform neurofibromas. *Cancer Genet Cytogenet* 1999;113:65–69.

573. Gutzmer R, Herbst RA, Mommert S, et al. Allelic loss at the neurofibromatosis type 1 (NF1) gene locus is frequent in desmoplastic neurotropic melanoma. *Hum Genet* 2000;107:357–361.

574. Kluwe L, Hagel C, Tatagiba M, et al. Loss of NF1 alleles distinguish sporadic from NF1-associated pilocytic astrocytomas. *J Neuropathol Exp Neurol* 2001;60:917–920.

575. Viskochil DH. Gene structure and function. In *Neurofibromatosis Type 1: From Genotype to Phenotype.* Uphadhyaya M, Cooper DN (eds.) Oxford: BIOS Scientific Publishers; 1998: 39–56.

576. Messiaen LM, Callens T, Mortier G, et al. Exhaustive mutation analysis of the NF1 gene allows identification of 95% of mutations and reveals a high frequency of unusual splicing defects. *Hum Mutat* 2000;15:541–555.

577. Nemethova M, Bolcekova A, Ilencikova D, et al. Thirty-nine novel neurofibromatosis 1 (NF1) gene mutations identified in Slovak patients. *Ann Hum Genet* 2013;77:364–379.

578. Maruoka R, Takenouchi T, Torii C, et al. The use of next-generation sequencing in molecular diagnosis of neurofibromatosis type 1: a validation study. *Genet Test Mol Biomarkers* 2014;18:722–735.

Neurofibromatosis type 2 gene (NF2)

579. Merel P, Hoang-Xuan K, Sanson M, et al. Screening for germ-line mutations in the NF2 gene. *Genes Chromosomes Cancer* 1995;12:117–127.

580. Ruttledge MH, Andermann AA, Phelan CM, et al. Type of mutation in the neurofibromatosis type 2 gene (NF2) frequently determines severity of disease. *Am J Hum Genet* 1996;59:331–342.

581. Zucman-Rossi J, Legoix P, Der Sarkissian H, et al. NF2 gene in neurofibromatosis type 2 patients. *Hum Mol Genet* 1998;7:2095–2101.

582. Kluwe L, Nygren AO, Errami A, et al. Screening for large mutations of the NF2 gene. *Genes Chromosomes Cancer* 2005;42:384–391.

583. Baser ME, Contributors to the International NF2 Mutation Database. The distribution of constitutional and somatic mutations in the neurofibromatosis 2 gene. *Hum Mutat* 2006;27:297–306.

584. Ahronowitz I, Xin W, Kiely R, et al. Mutational spectrum of the NF2 gene: a meta-analysis of 12 years of research and diagnostic laboratory findings. *Hum Mutat* 2007;28:1–12.

585. Rouleau GA, Merel P, Lutchman M, et al. Alteration in a new gene encoding a putative membrane-organizing protein causes neurofibromatosis type 2. *Nature* 1993;363:515–521.

586. Trofatter JA, MacCollin MM, Rutter JL, et al. A novel moesin-, ezrin-, radixin-like gene is a candidate for the neurofibromatosis 2 tumor suppressor. *Cell* 1993;72:791–800.

587. McCartney BM, Fehon RG. The ERM family proteins and their roles in cell-cell interactions. In *Cytoskeletal-Membrane Interactions and Signal Transduction.* Cowijn P, Klymkowsky MW (eds.) Austin, TX: R. G. Landes Bioscience; 1997: 200–210.

588. Lutchman M, Rouleau GA. The neurofibromatosis type 2 gene product, schwannomin, suppresses growth of NIH 3T3 cells. *Cancer Res* 1995;55:2270–2274.

589. Tikoo A, Varga M, Ramesh V, Gusella J, Maruta H. An anti-Ras function of neurofibromatosis type 2 gene product (NF2/Merlin). *J Biol Chem* 1994;269:23387–23390.

590. Giovannini M, Robanus-Maandag E, Niwa-Kawakita M, et al. Schwann cell hyperplasia and tumors in transgenic mice expressing a naturally occurring mutant NF2 protein. *Genes Dev* 1999;13:978–986.

591. McClatchey AI, Saotome I, Mercer K, et al. Mice heterozygous for a mutation at the Nf2 tumor suppressor locus develop a range of highly metastatic tumors. *Genes Dev* 1998;12:1121–1133.

592. Scoles DR. The merlin interacting proteins reveal multiple targets for NF2 therapy. *Biochim Biophys Acta* 2008;1785:32–54.

593. Zhou L, Hanemann CO. Merlin, a multi-suppressor from cell membrane to the nucleus. *FEBS Lett* 2012;586:1403–1408.

594. Cooper J, Giancotti FG. Molecular insights into NF2/Merlin tumor suppressor function. *FEBS Lett* 2014;588:2743–2752.

595. Karajannis MA, Legault G, Hagiwara M, et al. Phase II trial of lapatinib in adult and pediatric patients with neurofibromatosis type 2 and progressive vestibular schwannomas. *Neurooncol* 2012;14:1163–1170.

596. Karajannis MA, Legault G, Hagiwara M, et al. Phase II study of everolimus in children and adults with neurofibromatosis type 2 and progressive vestibular schwannomas. *Neurooncol* 2014;16:292–297.

597. Bianchi AB, Mitsunaga SI, Cheng JQ, et al. High frequency of inactivating mutations in the neurofibromatosis type 2 gene (NF2) in primary malignant mesotheliomas. *Proc Natl Acad Sci USA* 1995;92:10854–10858.

598. Cheng JQ, Lee WC, Klein MA, et al. Frequent mutations of NF2 and allelic loss from chromosome band 22q12 in malignant mesothelioma: evidence for a two-hit mechanism of NF2 inactivation. *Genes Chromosomes Cancer* 1999;24:238–242.

599. Bijlsma EK, Merel P, Bosch DA, et al. Analysis of mutations in the SCH gene in schwannomas. *Genes Chromosomes Cancer* 1994;11:7–14.

600. Twist EC, Ruttledge MH, Rousseau M, et al. The neurofibromatosis type 2 gene is inactivated in schwannomas. *Hum Mol Genet* 1994;3:147–151.

601. Jacoby LB, MacCollin M, Barone R, Ramesh V, Gusella JF. Frequency and distribution of NF2 mutations in schwannomas. *Genes Chromosomes Cancer* 1996;17:45–55.

602. Lasota J Fetsch JF, Wozniak A, et al. The neurofibromatosis type 2 gene is mutated in perineural cell tumors: A molecular genetic study of eight cases. *Am J Pathol* 2001;158:1223–1229.

603. Stemmer-Rachamimov AO, Xu L, Gonzalez-Agosti C, et al. Universal absence of merlin, but not other ERM family members, in schwannomas. *Am J Pathol* 1997;151:1649–1654.

604. Gutmann DH, Giordano MJ, Fishback AS, Guha A. Loss of merlin expression in sporadic meningiomas, ependymomas and schwannomas. *Neurology* 1997;49:267–270.

605. Lee JH, Sundaram V, Stein DJ, et al. Reduced expression of schwannomin/merlin in human sporadic meningiomas. *Neurosurgery* 1997;40:578–587.

606. Kimura Y, Koga H, Araki N, et al. The involvement of calpain-dependent proteolysis of the tumor suppressor NF2 (merlin) in schannomas and meningiomas. *Nat Med* 1998;4:915–922.

SMARCB1 tumor-suppressor gene

607. Biegel JA, Busse TM, Weissman BE. SWI/SNF chromatin remodeling complexes and cancer. *Am J Med Genet C Semin Med Genet* 2014;166C:350–366.

608. Masliah-Planchon J, Bièche I, Guinebretière JM, Bourdeaut F, Delattre O. SWI/SNF chromatin remodeling and human malignancies. *Annu Rev Pathol* 2015;10:145–171.

609. Peterson CL, Herskowitz I. Characterization of the yeast SWI1, SWI2, and SWI3 genes, which encode a global activator of transcription. *Cell* 1992;68:573–583.

610. Beckwith JB, Palmer NF. Histopathology and prognosis of Wilms tumors: results from the First National Wilms' Tumor Study. *Cancer* 1978;41:1937–1948.

611. Parham DM, Weeks DA, Beckwith JB. The clinicopathologic spectrum of putative extrarenal rhabdoid tumors: an analysis of 42 cases studied with immunohistochemistry or electron microscopy. *Am J Surg Pathol* 1994;18:1010–1029.

612. Biegel JA, Rorke LB, Packer RJ, Emanuel BS. Monosomy 22 in rhabdoid or atypical tumors of the brain. *J Neurosurg* 1990;73:710–714.

613. Biegel JA, Burk CD, Parmiter AH, Emanuel BS. Molecular analysis of partial deletion of 22q in a central nervous system rhabdoid tumor. *Genes Chromosomes Cancer* 1992;5:104–108.

614. Biegel JA, Allen CS, Kawasaki K, et al. Narrowing the critical region for the rhabdoid tumor locus in 22q11. *Genes Chromosomes Cancer* 1996;16:94–105.

615. Lee RS, Stewart C, Carter SL, et al. A remarkably simple genome underlies highly malignant pediatric rhabdoid cancers. *J Clin Invest* 2012;122:2983–2988.

616. McKenna ES, Sansam CG, Cho YJ, et al. Loss of the epigenetic tumor suppressor SNF5 leads to cancer without genomic instability. *Mol Cell Biol* 2008;28:6223–6233.

617. Versteege I, Sevenet N, Lange J, et al. Truncating mutations of hSNF5/INI1 in aggressive paediatric cancer. *Nature* 1998;394:203–206.

618. Sevenet N, Lellouch-Tubiana A, Schofield D, et al. Spectrum of hSNF5/INI1 somatic mutations in human cancer and genotype-phenotype correlations. *Hum Mol Genet* 1999;8:2359–2368.

619. Sevenet N, Sheridan E, Amram D, et al. Constitutional mutations of the hSNF5/INI1 gene predispose to a variety of cancers. *Am J Hum Genet* 1999;65:1342–1348.

620. Taylor MD, Gokgoz N, Andrulis IL, et al. Familial posterior fossa brain tumors of infancy secondary to germline mutation of the hSNF5 gene. *Am J Hum Genet* 2000;66:1403–1406.

621. Hulsebos TJ, Plomp AS, Wolterman RA, et al. Germline mutation of INI1/SMARCB1 in familial schwannomatosis. *Am J Hum Genet* 2007;80:805–810.

622. Rousseau G, Noguchi T, Bourdon V, Sobol H, Olschwang S. SMARCB1/INI1 germline mutations contribute to 10%

of sporadic schwannomatosis. *BMC Neurol* 2011;11:9.

623. Smith MJ, Wallace AJ, Bowers NL, et al. Frequency of SMARCB1 mutations in familial and sporadic schwannomatosis. *Neurogenetics* 2012;13:141–145.

624. Sestini R, Bacci C, Provenzano A, Genuardi M, Papi L. Evidence of a four-hit mechanism involving SMARCB1 and NF2 in schwannomatosis-associated schwannomas. *Hum Mutat* 2008;29:227–231.

625. van den Munckhof P, Christiaans I, Kenter SB, Baas F, Hulsebos TJ. Germline SMARCB1 mutation predisposes to multiple meningiomas and schwannomas with preferential location of cranial meningiomas at the falx cerebri. *Neurogenetics* 2012;13:1–7.

626. Smith MJ, O'Sullivan J, Bhaskar SS, et al. Loss-of-function mutations in SMARCE1 cause an inherited disorder of multiple spinal meningiomas. *Nat Genet* 2013;45:295–298.

627. Modena P, Lualdi E, Facchinetti F, et al. SMARCB1/INI1 tumor suppressor gene is frequently inactivated in epithelioid sarcomas. *Cancer Res* 2005;65:4012–4019.

628. Hornick JL, Dal Cin P, Fletcher CD. Loss of INI1 expression is characteristic of both conventional and proximal-type epithelioid sarcoma. *Am J Surg Pathol* 2009;33:542–550.

629. Kohashi K, Oda Y, Yamamoto H, et al. SMARCB1/INI1 protein expression in round cell soft tissue sarcomas associated with chromosomal translocations involving EWS: a special reference to SMARCB1/INI1 negative variant extraskeletal myxoid chondrosarcoma. *Am J Surg Pathol* 2008;32:1168–1174.

630. Carter JM, O'Hara C, Dundas G, et al. Epithelioid malignant peripheral nerve sheath tumor arising in a schwannoma, in a patient with "neuroblastoma-like" schwannomatosis and a novel germline SMARCB1 mutation. *Am J Surg Pathol* 2012;36:154–160.

631. Mobley BC, McKenney JK, Bangs CD, et al. Loss of SMARCB1/INI1 expression in poorly differentiated chordomas. *Acta Neuropathol* 2010;120:745–753.

632. Cheng JX, Tretiakova M, Gong C, et al. Renal medullary carcinoma: rhabdoid features and the absence of INI1 expression as markers of aggressive behavior. *Mod Pathol* 2008;21:647–652.

633. Agaimy A, Koch M, Lell M, et al. SMARCB1(INI1)-deficient sinonasal basaloid carcinoma: a novel member of the expanding family of SMARCB1-deficient neoplasms. *Am J Surg Pathol* 2014;38:1274–1281.

634. Kadoch C, Hargreaves DC, Hodges C, et al. Proteomic and bioinformatic analysis of mammalian SWI/SNF complexes identifies extensive roles in human malignancy. *Nat Genet* 2013;45:592–601.

635. Roberts CW, Galusha SA, McMenamin ME, Fletcher CD, Orkin SH. Haploinsufficiency of Snf5 (integrase interactor 1) predisposes to malignant rhabdoid tumors in mice. *Proc Natl Acad Sci USA* 2000;97:13796–13800.

636. Roberts CW, Leroux MM, Fleming MD, Orkin SH. Highly penetrant, rapid tumorigenesis through conditional inversion of the tumor suppressor gene Snf5. *Cancer Cell* 2002;2:415–425.

637. Isakoff MS, Sansam CG, Tamayo P, et al. Inactivation of the Snf5 tumor suppressor stimulates cell cycle progression and cooperates with p53 loss in oncogenic transformation. *Proc Natl Acad Sci USA* 2005;102:17745–17750.

Complex genetic and epigenetic mutations

638. Orndal C, Rydholm A, Willén H, Mitelman F, Mandahl N. Cytogenetic intratumor heterogeneity in soft tissue tumors. *Cancer Genet Cytogenet* 1994;78:127–137.

639. Wang R, Lu YJ, Fisher C, Bridge JA, Shipley J. Characterization of chromosome aberrations associated with soft-tissue leiomyosarcomas by twenty-four-color karyotyping and comparative genomic hybridization analysis. *Genes Chromosomes Cancer* 2001;31:54–64.

640. Mertens F, Fletcher CD, Dal Cin P, et al. Cytogenetic analysis of 46 pleomorphic soft tissue sarcomas and correlation with morphologic and clinical features: a report of the CHAMP Study Group. Chromosomes and MorPhology. *Genes Chromosomes Cancer* 1998;22:16–25.

641. Gibault L, Pérot G, Chibon F, et al. New insights in sarcoma oncogenesis: a comprehensive analysis of a large series of 160 soft tissue sarcomas with complex genomics. *J Pathol* 2011;223:64–71.

642. Barretina J, Taylor BS, Banerji S, et al. Subtype-specific genomic alterations define new targets for soft-tissue sarcoma therapy. *Nat Genet* 2010;42:715–721.

643. Villacis RA, Silveira SM, Barros-Filho MC, et al. Gene expression profiling in leiomyosarcomas and undifferentiated pleomorphic sarcomas: SRC as a new diagnostic marker. *PLoS One* 2014;9:e102281.

644. Jour G, Scarborough JD, Jones RL, et al. Molecular profiling of soft tissue sarcomas using next-generation sequencing: a pilot study toward precision therapeutics. *Hum Pathol* 2014;45:1563–1571.

Genetic pathways altered in sarcomas

Cell cycle regulators

p16-cyclin D1/CDK4-pRb pathway

645. Sherr CJ. Cancer cell cycles. *Science* 1996;274:1672–1677.

646. Classon M, Salama S, Gorka C, et al. Combinatorial roles for pRB, p107, and p130 in E2F-mediated cell cycle control. *Proc Natl Acad Sci USA* 2000;97:10820–10825.

647. Ruas M, Peters G. The p16INK4a/CDKN2A tumor suppressor and its relatives. *Biochim Biophys Acta* 1998;1378:F115–F177.

648. Orlow I, Drobnjak M, Zhang ZF, et al. Alterations of INK4A and INK4B genes in adult soft tissue sarcomas: effect on survival. *J Natl Cancer Inst* 1999;91:73–79.

649. Merlo A, Herman JG, Mao L, et al. 5-prime CpG island methylation is associated with transcriptional silencing of the tumour suppressor p16/CDKN2/MTS1 in human cancers. *Nat Med* 1995;1:686–692.

650. Kawaguchi K, Oda Y, Saito T, et al. Mechanisms of inactivation of the p16INK4a gene in leiomyosarcoma of soft tissue: decreased p16 expression correlates with promotor methylation and poor prognosis. *J Pathol* 2003;201:487–495.

651. Schneider-Stock R, Boltze C, Lasota J, et al. Loss of p16 protein defines high-risk patients with gastrointestinal

stromal tumors: a tissue microarray study. *Clin Cancer Res* 2005;11:638–645.

652. Obana K, Yang HW, Piao HY, *et al.* Aberrations of p16INK4A, p14ARF, and p15INK4B genes in pediatric solid tumors. *Int J Oncol* 2003;23:1151–1157.

653. Linardic CM, Naini S, Herndon JE 2nd, *et al.* The PAX3-FKHR fusion gene of rhabdomyosarcoma cooperates with loss of p16INK4A to promote bypass of cellular senescence. *Cancer Res* 2007;67: 6691–6699.

654. Naini S, Etheridge KT, Adam SJ, *et al.* Defining the cooperative genetic changes that temporally drive alveolar rhabdomyosarcoma. *Cancer Res* 2008;68:9583–9588.

655. Hannon GJ, Beach D. p15(INK4B) is a potential effector of TGF-beta-induced cell cycle arrest. *Nature* 1994;371:257–261.

656. Nabori T, Miura K, Wu DJ, *et al.* Deletions of the cyclin-dependent kinase-4 inhibitor gene in multiple human cancers. *Nature* 1994;368:753–756.

657. Serrano M, Lee H, Chin L, *et al.* Role of the INK4a locus in tumor supression and cell mortality. *Cell* 1996;85:27–37.

658. Kamijo T, Zindy F, Roussel MF, *et al.* Tumor suppression at the mouse INK4a locus mediated by the alternative reading frame product p19ARF. *Cell* 1997;91:649–659.

659. Krimpenfort P, Ijpenberg A, Song JY, *et al.* p15Ink4b is a critical tumour suppressor in the absence of p16Ink4a. *Nature* 2007;448:943–946.

660. Cordon-Cardo C. Mutation of cell cycle regulators: Biological and clinical implications for human neoplasia. *Am J Pathol* 1995;147:545–560.

661. Dickson MA, Tap WD, Keohan ML, *et al.* Phase II trial of the CDK4 inhibitor PD0332991 in patients with advanced CDK4-amplified well-differentiated or dedifferentiated liposarcoma. *J Clin Oncol* 2013;31:2024–2028.

p53 tumor-suppressor gene pathway

662. Isobe M, Emanuel BS, Givol D, Oren M, Croce CM. Localization of gene for human p53 antigen to band 17p13. *Nature* 1986;320:84–85.

663. Lane DP, Crawford LV. T antigen is bound to a host protein in SV40-transformed cells. *Nature* 1979;278:261–263.

664. Linzer DI, Levine AJ. Characterization of a 54K dalton cellular SV40 tumor antigen present in SV40-transformed cells and uninfected embryonal carcinoma cells. *Cell* 1979;17:43–52.

665. Bourdon JC, Fernandes K, Murray-Zmijewski F, *et al.* p53 isoforms can regulate p53 transcriptional activity. *Genes Dev* 2005;19:2122–2137.

666. Laptenko O, Prives C. Transcriptional regulation by p53: one protein, many possibilities. *Cell Death Differ* 2006;13:951–961.

667. Kho PS, Wang Z, Zhuang L, *et al.* p53-regulated transcriptional program associated with genotoxic stress-induced apoptosis. *J Biol Chem* 2004;279:21183–21192.

668. Spurgers KB, Gold DL, Coombes KR, *et al.* Identification of cell cycle regulatory genes as principal targets of p53-mediated transcriptional repression. *J Biol Chem* 2006;281:25134–25142.

669. Finlay CA, Hinds PW, Levine AJ. The p53 proto-oncogene can act as a suppressor of transformation. *Cell* 1989;57:1083–1093.

670. Eliyahu D, Michalovitz D, Eliyahu S, Pinhasi-Kimhi O, Oren M. Wild-type p53 can inhibit oncogene-mediated focus formation. *Proc Natl Acad Sci USA* 1989;86:8763–8767.

671. Clarke AR, Purdie CA, Harrison DJ, *et al.* Thymocyte apoptosis induced by p53-dependent and independent pathways. *Nature* 1993;362:849–852.

672. Lowe SW, Schmitt EM, Smith SW, Osborne BA, Jacks T. p53 is required for radiation-induced apoptosis in mouse thymocytes. *Nature* 1993;362:847–849.

673. Lowe SW, Ruley HE, Jacks T, Housman DE. p53-dependent apoptosis modulates the cytotoxicity of anticancer agents. *Cell* 1993;74:957–967.

674. Lavigueur A, Maltby V, Mock D, *et al.* High incidence of lung, bone, and lymphoid tumors in transgenic mice overexpressing mutant alleles of the p53 oncogene. *Mol Cell Biol* 1989;9:3982–3991.

675. Donehower LA, Harvey M, Slagle BL, *et al.* Mice deficient for p53 are developmentally normal but susceptible to spontaneous tumours. *Nature* 1992;356:215–221.

676. Harvey M, McArthur MJ, Montgomery Jr CA, *et al.* Spontaneous and carcinogen-induced tumorigenesis in p53-deficient mice. *Nat Genet* 1993;5:225–229.

677. Jacks T, Remington L, Williams BO, *et al.* Tumor spectrum analysis in p53-deficient mice. *Curr Biol* 1994;4:1–7.

678. Oren M. Regulation of the p53 tumor suppressor protein. *J Biol Chem* 1999;274:36031–36034.

679. Levine AJ. p53, the cellular gatekeeper for growth and devision. *Cell* 1997;88:323–331.

680. Hollstein M, Shomer B, Greenblatt M, *et al.* Somatic point mutations in the p53 gene of human tumors and cell lines: updated compilation. *Nucleic Acids Res* 1996;24:141–146.

681. Baker SJ, Fearon ER, Nigro JM, *et al.* Chromosome 17 deletions and p53 gene mutations in colorectal carcinomas. *Science* 1989;244:217–221.

682. Nigro JM, Baker SJ, Preisinger AC, *et al.* Mutations in the p53 gene occur in diverse human tumour types. *Nature* 1989;342:705–708.

683. Li FP, Fraumeni JF. Rhabdomyosarcoma in children; epidemiologic study and identification of a cancer family syndrome. *J Natl Cancer Inst* 1969;43:1365–1373.

684. Li FP, Fraumeni JF. Soft tissue sarcomas, breast cancer and other neoplasms: a familial syndrome? *Ann Int Med* 1969;71:747–752.

685. Malkin D, Li FP, Strong LC, *et al.* Germ line p53 mutations in a familial syndrome of breast cancer, sarcomas, and other neoplasms. *Science* 1990;250:1233–1238.

686. Varley JM, Evans DGR, Birch JM. Li-Fraumeni syndrome – a molecular and clinical review. *Br J Cancer* 1997;76:1–14.

687. Varley JM, Thorncroft M, McGown G, *et al.* A detailed study of loss of heterozygosity on chromosome 17 in tumours from Li-Fraumeni patients carrying a mutation to the TP53 gene. *Oncogene* 1997;14:865–871.

688. Bell DW, Varley JM, Szydlo TE, *et al.* Heterozygous germ line hCHK2 mutations in Li-Fraumeni syndrome. *Science* 1999;286:2528–2531.

689. Vahteristo P, Tamminen A, Karvinen P, et al. p53, CHK2 and CHK1 genes in Finnish families with Li–Fraumeni syndrome: further evidence of CHK2 in inherited cancer predisposition. *Cancer Res* 2001;61:5718–5722.

690. Matsuoka S, Huang M, Elledge SJ. Linkage of ATM to cell cycle regulation by the Chk2 protein kinase. *Science* 1998;282:1893–1897.

691. Blasina A, de Weyer IV, Laus MC, et al. A human homologue of the checkpoint kinase Cds1 directly inhibits Cdc25 phosphatase. *Curr Biol* 1999;14:1–10.

692. Chaturvedi P, Eng WK, Zhu Y, et al. Mammalian Chk2 is a downstream effector of the ATM-dependent DNA damage checkpoint pathway. *Oncogene* 1999;18:4047–4054.

693. Brown AL, Lee C-H, Schwarz JK, et al. A human Cda1-related kinase that functions downstream of ATM protein in the cellular response to DNA damage. *Proc Natl Acad Sci USA* 1999;96:3745–3750.

694. Chehab NH, Malikzay A, Appel M, Halazonetis TD. Chk2/hCds1 functions as a DNA damage checkpoint in G-1 by stabilizing p53. *Genes Dev* 2000;14:278–288.

695. Bartkova J, Horejsí Z, Koed K, et al. DNA damage response as a candidate anti-cancer barrier in early human tumorigenesis. *Nature* 2005;434:864–870.

696. Neilsen PM, Pishas KI, Callen DF, Thomas DM. Targeting the p53 pathway in Ewing sarcoma. *Sarcoma* 2011;2011:746939.

697. Ito M, Barys L, O'Reilly T, et al. Comprehensive mapping of p53 pathway alterations reveals an apparent role for both SNP309 and MDM2 amplification in sarcomagenesis. *Clin Cancer Res* 2011;17:416–426.

698. Taylor BS, Barretina J, Maki RG, et al. Advances in sarcoma genomics and new therapeutic targets. *Nat Rev Cancer* 2011;11:541–557.

699. Zhan C, Lu W. Peptide activators of the p53 tumor suppressor. *Curr Pharm Des* 2011;17:603–609.

700. Ban J, Bennani-Baiti IM, Kauer M, et al. EWS-FLI1 suppresses NOTCH-activated p53 in Ewing's sarcoma. *Cancer Res* 2008;68:7100–7109.

701. Li Y, Tanaka K, Fan X, et al. Inhibition of the transcriptional function of p53 by EWS-Fli1 chimeric protein in Ewing Family Tumors. *Cancer Lett* 2010;294:57–65.

702. D'Arcy P, Maruwge W, Ryan BA, Brodin B. The oncoprotein SS18-SSX1 promotes p53 ubiquitination and degradation by enhancing HDM2 stability. *Mol Cancer Res* 2008;6:127–138.

703. Baresova P, Musilova J, Pitha PM, Lubyova B. p53 tumor suppressor protein stability and transcriptional activity are targeted by Kaposi's sarcoma-associated herpesvirus-encoded viral interferon regulatory factor 3. *Mol Cell Biol* 2014;34:386–399.

704. van der Ent W, Jochemsen AG, Teunisse AF, et al. Ewing sarcoma inhibition by disruption of EWSR1-FLI1 transcriptional activity and reactivation of p53. *J Pathol* 2014;233:415–424.

705. Brooks CL, Gu W. Ubiquitination, phosphorylation and acetylation: the molecular basis for p53 regulation. *Curr Opin Cell Biol* 2003;15:164–171.

706. Tang Y, Zhao W, Chen Y, Zhao Y, Gu W. Acetylation is indispensable for p53 activation. *Cell* 2008;133:612–626.

707. Scoumanne A, Chen X. Protein methylation: a new mechanism of p53 tumor suppressor regulation. *Histol Histopathol* 2008;23:1143–1149.

708. Lal A, Kim HH, Abdelmohsen K, et al. p16(INK4a) translation suppressed by miR-24. *PLoS One* 2008;3:e1864.

709. Pierson J, Hostager B, Fan R, Vibhakar R. Regulation of cyclin dependent kinase 6 by microRNA 124 in medulloblastoma. *J Neurooncol* 2008;90:1–7.

710. Johnson CD, Esquela-Kerscher A, Stefani G, et al. The let-7 microRNA represses cell proliferation pathways in human cells. *Cancer Res* 2007;67:7713–7722.

711. Sun F, Fu H, Liu Q, et al. Downregulation of CCND1 and CDK6 by miR-34a induces cell cycle arrest. *FEBS Lett* 2008;582:1564–1568.

712. Wu J, Qian J, Li C, et al. miR-129 regulates cell proliferation by downregulating Cdk6 expression. *Cell Cycle* 2010;9:1809–1818.

713. Esquela-Kerscher A, Slack FJ. Oncomirs: microRNAs with a role in cancer. *Nat Rev Cancer* 2006;6:259–269.

714. Lynam-Lennon N, Maher SG, Reynolds JV. The roles of microRNA in cancer and apoptosis. *Biol Rev Camb Philos Soc* 2009;84:55–71.

Growth factor signaling pathways

715. Helman LJ, Meltzer P. Mechanisms of sarcoma development. *Nat Rev Cancer* 2003;3:685–694.

716. Foulstone E, Prince S, Zaccheo O, et al. Insulin-like growth factor ligands, receptors, and binding proteins in cancer. *J Pathol* 2005;205:145–153.

717. Pollak M. Insulin-like growth factor-related signaling and cancer development. *Recent Results Cancer Res* 2007;174:49–53.

718. Polivka J Jr, Janku F. Molecular targets for cancer therapy in the PI3K/AKT/mTOR pathway. *Pharmacol Ther* 2014;142:164–175.

719. Forscher C, Mita M, Figlin R. Targeted therapy for sarcomas. *Biologics* 2014;8:91–105.

720. Werner H. Tumor suppressors govern insulin-like growth factor signaling pathways: implications in metabolism and cancer. *Oncogene* 2012;31:2703–2714.

721. Idelman G, Glaser T, Roberts CT Jr, Werner H. WT1-p53 interactions in insulin-like growth factor-I receptor gene regulation. *J Biol Chem* 2003;278:3474–3482.

722. Cironi L, Riggi N, Provero P, et al. IGF1 is a common target gene of Ewing's sarcoma fusion proteins in mesenchymal progenitor cells. *PLoS One* 2008;3:e2634.

723. Prieur A, Tirode F, Cohen P, Delattre O. EWS/FLI-1 silencing and gene profiling of Ewing cells reveal downstream oncogenic pathways and a crucial role for repression of insulin-like growth factor binding protein 3. *Mol Cell Biol* 2004;24:7275–7283.

724. McKinsey EL, Parrish JK, Irwin AE, et al. A novel oncogenic mechanism in Ewing sarcoma involving IGF pathway targeting by EWS/Fli1-regulated microRNAs. *Oncogene* 2011;30:4910–4920.

725. Heidegger I, Pircher A, Klocker H, Massoner P. Targeting the insulin-like growth factor network in cancer therapy. *Cancer Biol Ther* 2011;11:701–707.

726. Sharp R, Recio JA, Jhappan C, et al. Synergism between INK4a/ARF inactivation and aberrant HGF/SF

signaling in rhabdomyosarcomagenesis. *Nat Med* 2002;8:1276–1280.

727. Trusolino L, Bertotti A, Comoglio PM. MET signalling: principles and functions in development, organ regeneration and cancer. *Nat Rev Mol Cell Biol* 2010;11:834–848.

728. Danilkovitch-Miagkova A, Zbar B. Dysregulation of Met receptor tyrosine kinase activity in invasive tumors. *J Clin Invest* 2002;109:863–867.

729. Davis IJ, Kim JJ, Ozsolak F, et al. Oncogenic MITF dysregulation in clear cell sarcoma: defining the MiT family of human cancers. *Cancer Cell* 2006;9:473–484.

730. Tsuda M, Davis IJ, Argani P, et al. TFE3 fusions activate MET signaling by transcriptional up-regulation, defining another class of tumors as candidates for therapeutic MET inhibition. *Cancer Res* 2007;67:919–929.

731. Ginsberg JP, Davis RJ, Bennicelli JL, Nauta LE, Barr FG. Up-regulation of MET but not neural cell adhesion molecule expression by the PAX3-FKHR fusion protein in alveolar rhabdomyosarcoma. *Cancer Res* 1998;58:3542–3546.

732. Kitajima Y, Ide T, Ohtsuka T, Miyazaki K. Induction of hepatocyte growth factor activator gene expression under hypoxia activates the hepatocyte growth factor/c-Met system via hypoxia inducible factor-1 in pancreatic cancer. *Cancer Sci* 2008;99:1341–1347.

733. Yan S, Nakagawa T. The current state of molecularly targeted drugs targeting HGF/Met. *Jpn J Clin Oncol* 2014;44:9-12.

Table 5.4

734. Wagner AJ, Goldberg JM, Dubois SG, et al. Tivantinib (ARQ 197), a selective inhibitor of MET, in patients with microphthalmia transcription factor-associated tumors: results of a multicenter phase 2 trial. *Cancer* 2012;118:5894–5902.

735. Taipale M, Krykbaeva I, Whitesell L, et al. Chaperones as thermodynamic sensors of drug-target interactions reveal kinase inhibitor specificities in living cells. *Nat Biotechnol* 2013;31:630–637.

736. Pappo AS, Vassal G, Crowley JJ, et al. A phase 2 trial of R1507, a monoclonal antibody to the insulin-like growth factor-1 receptor (IGF-1R), in patients with recurrent or refractory rhabdomyosarcoma, osteosarcoma, synovial sarcoma, and other soft tissue sarcomas: results of a Sarcoma Alliance for Research Through Collaboration study. *Cancer* 2014;120:2448–2456.

737. Hallberg B, Palmer RH. Mechanistic insight into ALK receptor tyrosine kinase in human cancer biology. *Nat Rev Cancer* 2013;13:685–700.

738. Yang J, Ylipää A, Sun Y, et al. Genomic and molecular characterization of malignant peripheral nerve sheath tumor identifies the IGF1R pathway as a primary target for treatment. *Clin Cancer Res* 2011;17:7563–7573.

739. Torres KE, Zhu QS, Bill K, et al. Activated MET is a molecular prognosticator and potential therapeutic target for malignant peripheral nerve sheath tumors. *Clin Cancer Res* 2011;17:3943–3955.

740. Hou J, Dong J, Sun L, et al. Inhibition of phosphorylated c-Met in rhabdomyosarcoma cell lines by a small molecule inhibitor SU11274. *J Transl Med* 2011;9:64.

741. Ayalon D, Glaser T, Werner H. Transcriptional regulation of IGF-I receptor gene expression by the PAX3-FKHR oncoprotein. *Growth Horm IGF Res* 2001;11:289–297.

742. Friedrichs N, Küchler J, Endl E, et al. Insulin-like growth factor-1 receptor acts as a growth regulator in synovial sarcoma. *J Pathol* 2008;216:428–439.

743. Cassier PA, Gelderblom H, Stacchiotti S, et al. Efficacy of imatinib mesylate for the treatment of locally advanced and/or metastatic tenosynovial giant cell tumor/pigmented villonodular synovitis. *Cancer* 2012;118:1649–1655.

APC inactivation and beta-catenin activation in the Wnt pathway

744. Kinzler KW, Nilbert MC, Su L-K, et al. Identification of FAP locus genes from chromosome 5q21. *Science* 1991;253:661–665.

745. Nishisho I, Nakamura Y, Miyoshi Y, et al. Mutations of chromosome 5q21 genes in FAP and colorectal cancer patients. *Science* 1991;253:665–669.

746. Klemmer S, Pascoe L, DeCosse J. Occurrence of desmoids in patients with familial adenomatous polyposis of the colon. *Am J Med Genet* 1987;28:385–392.

747. Clark SK, Neale KF, Landgrebe JC, Phillips RKS. Desmoid tumours complicating familial adenomatous polyposis. *Br J Surg* 1999;86:1185–1189.

748. Eccles DM, Van Der Luijt R, Breukel C, et al. Hereditary desmoid disease due to a frameshift mutation at codon 1924 of the APC gene. *Am J Hum Genet* 1996;59:1193–1201.

749. Scott RJ, Froggatt NJ, Trembath RC, et al. Familial infiltrative fibromatosis (desmoid tumours) (MIM135290) caused by a recurrent 3′ APC gene mutation. *Hum Mol Genet* 1996;5:1921–1924.

750. Groden J, Thliveris A, Samowitz W, et al. Identification and characterization of the familial adenomatous polyposis coli gene. *Cell* 1991;66:589–600.

751. Joslyn G, Carlson M, Thliveris A, et al. Identification of deletion mutation and three new genes at the familial polyposis locus. *Cell* 1991;66:601–613.

752. Sen-Gupta S, Van Der Luijt R, Bowles LV, Meera Khan P, Delhanty JDA. Somatic mutation of APC gene in desmoid tumour in familial adenomatous polyposis. *Lancet* 1993;342:552–553.

753. Lamlum H, Ilyas M, Rowan A, et al. The type of somatic mutation at APC in familial adenomatous polyposis is determined by the site of the germline mutation: a new facet to Knudson's 'two-hit' hypothesis. *Nat Med* 1999;5:1071–1075.

754. Crabtree M, Sieber OM, Lipton L, et al. Refining the relation between 'first hits' and 'second hits' at the APC locus: the 'loose fit' model and evidence for differences in somatic mutation spectra among patients. *Oncogene* 2003;22:4257–4265.

755. Couture J, Mitri A, Lagace R, et al. A germline mutation at the extreme 3′ end of the APC gene results in a severe desmoid phenotype and is associated with overexpression of beta-catenin in the desmoid tumor. *Clin Genet* 2000;57:205–212.

756. Laken SJ, Papadopoulos N, Petersen GM, et al. Analysis of masked mutations in familial adenomatous polyposis. *Proc Natl Acad Sci USA* 1999;96:2322–2326.

757. Yan H, Dobbie Z, Gruber SB, et al. Small changes in expression affect

757. predisposition to tumorigenesis. *Nat Genet* 2002;30:25–36.

758. Polakis P. The many ways of Wnt in cancer. *Curr Opin Genet Dev* 2007;17:45–51.

759. Segditsas S, Tomlinson I. Colorectal cancer and genetic alterations in the Wnt pathway. *Oncogene* 2006;25:7531–7537.

760. Miyoshi Y, Iwao K, Nawa G, et al. Frequent mutations in the beta-catenin gene in desmoid tumors from patients without familial adenomatous polyposis. *Oncol Res* 1998;10:591–594.

761. Tejpar S, Michils G, Denys H, et al. Analysis of Wnt/beta catenin signalling in desmoid tumors. *Acta Gastroenterol Belg* 2005;68:5–9.

762. Kotiligam D, Lazar AJ, Pollock RE, Lev D. Desmoid tumor: a disease opportune for molecular insights. *Histol Histopathol* 2008;23:117–126.

763. Lazar AJ, Tuvin D, Hajibashi S, et al. Specific mutations in the beta-catenin gene (CTNNB1) correlate with local recurrence in sporadic desmoid tumors. *Am J Pathol* 2008;173:1518–1527.

764. Alman BA, Li C, Pajerski ME, Diaz-Cano S, Wolfe HJ. Increased beta-catenin protein and somatic APC mutations in sporadic aggressive fibromatoses (desmoid tumors). *Am J Pathol* 1997;151:329–334.

765. Luu HH, Zhang R, Haydon RC, et al. Wnt/beta-catenin signaling pathway as a novel cancer drug target. *Curr Cancer Drug Targets* 2004;4:653–671.

766. Takahashi-Yanaga F, Sasaguri T. The Wnt/beta-catenin signaling pathway as a target in drug discovery. *J Pharmacol Sci* 2007;104:293–302.

Dysfunction of the tricarboxylic acid cycle (Krebs cycle)

767. Warburg O. On the origin of cancer cells. *Science* 1956;123:309–314.

768. Frezza C, Pollard PJ, Gottlieb E. Inborn and acquired metabolic defects in cancer. *J Mol Med (Berl)* 2011;89:213–220.

769. Yang M, Soga T, Pollard PJ. Oncometabolites: linking altered metabolism with cancer. *J Clin Invest* 2013;123:3652–3658.

770. Adam J, Yang M, Soga T, Pollard PJ. Rare insights into cancer biology. *Oncogene* 2014;33:2547–2556.

771. Desideri E, Vegliante R, Ciriolo MR. Mitochondrial dysfunctions in cancer: genetic defects and oncogenic signaling impinging on TCA cycle activity. *Cancer Lett* 2015;356:217–223.

772. Rutter J, Winge DR, Schiffman JD. Succinate dehydrogenase - assembly, regulation and role in human disease. *Mitochondrion* 2010;10:393–401.

773. Pasini B, Stratakis CA. SDH mutations in tumorigenesis and inherited endocrine tumours: lesson from the phaeochromocytoma-paraganglioma syndromes. *J Intern Med* 2009;266:19–42.

774. Barletta JA, Hornick JL. Succinate dehydrogenase-deficient tumors: diagnostic advances and clinical implications. *Adv Anat Pathol* 2012;19:193–203.

775. Miettinen M, Lasota J. Succinate dehydrogenase deficient gastrointestinal stromal tumors (GISTs): a review. *Int J Biochem Cell Biol* 2014;53:514–519.

776. Tomlinson IP, Alam NA, Rowan AJ, et al. Germline mutations in FH predispose to dominantly inherited uterine fibroids, skin leiomyomata and papillary renal cell cancer. *Nat Genet* 2002;30:406–410.

777. Dang L, White DW, Gross S, et al. Cancer-associated IDH1 mutations produce 2-hydroxyglutarate. *Nature* 2009;462:739–744.

778. Ward PS, Patel J, Wise DR, et al. The common feature of leukemia-associated IDH1 and IDH2 mutations is a neomorphic enzyme activity converting alpha-ketoglutarate to 2-hydroxyglutarate. *Cancer Cell* 2010;17:225–234.

Clonality assessment with HUMARA assay

779. Lyon MF. Gene action in the X-chromosome of the mouse (*Mus musculus L.*) *Nature* 1961;190:372–373.

780. Lyon MF. The William Allan Memorial Award address: X-chromosome inactivation and the location and expression of X-linked genes. *Am J Hum Genet* 1988;42:8–16.

781. Beutler E, Yeh M, Fairbanks VF. Normal human female as a mosaic of X-chromosome activity: studies using the gene for G6PD deficiency as a marker. *Proc Natl Acad Sci USA* 1962;48:9–16.

782. Fialkow PJ. Clonal origin of human tumors. *Biochem Biophys Acta* 1976;458:283–321.

783. Boyd Y, Fraser NJ. Methylation patterns at the hypervariable X-chromosome locus DXS255 (M27β): correlation with X-inactivation status. *Genomics* 1990;7:182–187.

784. Keith DH, Singer-Sam J, Riggs AD. Active X chromosome DNA is unmethylated at eight CCGG sites clustered in a guanine-plus-cytosine-rich island at the 5′ end of the gene for phosphoglycerate kinase. *Mol Cell Biol* 1986;6:4122–4125.

785. Vogelstein B, Fearon ER, Hamilton SR, Feinberg AP. Use of restriction fragment length polymorphisms to determine the clonal origin of human tumors. *Science* 1985;227:642–645.

786. Fey MF, Liechti-Gallati S, von Rohr A, et al. Clonality and X-inactivation patterns in hematopoietic cell populations detected by the highly informative M27β DNA probe. *Blood* 1994;83:931–938.

787. Diaz-Cano SJ. Designing a molecular analysis of clonality in tumors. *J Pathol* 2000;191:343–344.

788. Allen RC, Zoghbi HY, Moseley AB, Rosenblatt HM, Belmont JW. Methylation of HpaII and HhaI sites near the polymorphic CAG repeat in the human androgen-receptor gene correlates with X chromosome inactivation. *Am J Hum Genet* 1992;51:1229–1239.

789. Busque L, Gilliland DG. Clonal evolution in acute myeloid leukemia. *Blood* 1993;82:337–342.

790. Busque L, Zhu J, DeHart D, et al. An expression-based clonality assay at the human androgen receptor locus (HUMARA) on chromosome X. *Nucleic Acids Res* 1994;22:697–698.

791. Li M, Cordon-Cardo C, Gerald WL, Rosai J. Desmoid fibromatosis is a clonal process. *Hum Pathol* 1996;27:939–943.

792. Vogrincic GS, O'Connell JX, Gilks CB. Giant cell tumor of tendon sheath is a polyclonal cellular proliferation. *Hum Pathol* 1997;28:815–819.

793. Paradis V, Laurendeau I, Vieillefond A, et al. Clonal analysis of renal sporadic angiomyolipomas. *Hum Pathol* 1998;29:1063–1067.

794. Flemming P, Lehmann U, Becker T, Klempnauer J, Kreipe H. Common and epithelioid variants of hepatic angiomyolipoma exhibit clonal growth and share a distinctive immunophenotype. *Hepatology* 2000;32:213–217.

795. Saxena A, Alport EC, Custead S, Skinnider LF. Molecular analysis of clonality of sporadic angiomyolipoma. *J Pathol* 1999;189:79–84.

796. Tang LH, Hui P, Garcia-Tsao G, Salem RR, Jain D. Multiple angiomyolipomata of the liver: a case report. *Mod Pathol* 2002;15:167–171.

797. Chetritt J, Paradis V, Dargere D, *et al.* Chester-Erdheim disease: a neoplastic disorder. *Hum Pathol* 1999;30:1093–1096.

798. Dickson BC, Pethe V, Chung CT, *et al.* Systemic Erdheim–Chester disease. *Virchows Arch* 2008;452:221–227.

799. Al-Quran S, Reith J, Bradley J, Rimsza L. Erdheim-Chester disease: case report, PCR-based analysis of clonality, and review of literature. *Mod Pathol* 2002;15:666–672.

800. Klingler L, Trammell R, Allan DG, Butler MG, Schwartz HS. Clonality studies in sacral chordoma. *Cancer Genet Cytogenet* 2006;171:68–71.

801. Lucas DR, Shroyer KR, McCarthy PJ, *et al.* Desmoid tumor is a clonal cellular proliferation: PCR amplification of HUMARA for analysis of patterns of X-chromosome inactivation. *Am J Surg Pathol* 1997;21:306–311.

802. Middleton SB, Frayling IM, Phillips RK. Desmoids in familial adenomatous polyposis are monoclonal proliferations. *Br J Cancer* 2000;82:827–832.

803. Chen TC, Kuo T, Chan HL. Dermatofibroma is a clonal proliferative disease. *J Cutan Pathol* 2000;27:36–39.

804. Hui P, Glusac EJ, Sinard JH, Perkins AS. Clonal analysis of cutaneous fibrous histiocytoma (dermatofibroma). *J Cutan Pathol* 2002;29:385–389.

805. Willman CL, Busque L, Griffith BB, *et al.* Langerhans'-cell histiocytosis (histiocytosis X): a clonal proliferative disease. *N Engl J Med* 1994;331:154–160.

806. Rabkin CS, Bedi G, Musaba E, *et al.* AIDS-related Kaposi's sarcoma is a clonal neoplasm. *Clin Cancer Res* 1995;1:257–260.

807. Gill PS, Tsai YC, Rao AP, *et al.* Evidence for multiclonality in multicentric Kaposi's sarcoma. *Proc Natl Acad Sci USA* 1998;95:8257–8261.

808. Delabesse E, Oksenhendler E, Lebbé C, *et al.* Molecular analysis of clonality in Kaposi's sarcoma. *J Clin Pathol* 1997;50:664–668.

809. Quade BJ, McLachlin CM, Soto-Wright V, *et al.* Disseminated peritoneal leiomyomatosis: clonality analysis by X chromosome inactivation and cytogenetics of a clinically benign smooth muscle proliferation. *Am J Pathol* 1997;150:2153–2166.

810. Quade BJ, Dal Cin P, Neskey DM, Weremowicz S, Morton CC. Intravenous leiomyomatosis: molecular and cytogenetic analysis of a case. *Mod Pathol* 2002;15:351–356.

811. Indsto JO, Cachia AR, Kefford RF, Mann GJ. X inactivation, DNA deletion, and microsatellite instability in common acquired melanocytic nevi. *Clin Cancer Res* 2001;7:4054–4059.

812. Sanz Esponera J. Genetic alterations in the differential diagnosis of melanocytic diseases. *Ann R Acad Nac Med (Madr)* 2000;117:815–824.

813. Koizumi H, Mikami M, Doi M, Tadokoro M. Clonality analysis of nodular fasciitis by HUMARA-methylation-specific PCR. *Histopathology* 2005;47:320–321.

814. Wang L, Zhu HG. Clonal analysis of palmar fibromatosis: a study whether palmar fibromatosis is a real tumor. *J Transl Med* 2006;4:21.

815. Niho S, Suzuki K, Yokose T, *et al.* Monoclonality of both pale cells and cuboidal cells of sclerosing hemangioma of the lung. *Am J Pathol* 1998;152:1065–1069.

Examples of molecular genetic techniques

816. Mies C. Molecular biology analysis of paraffin-embedded tissues. *Hum Pathol* 1994;25:555–560.

817. Lewis F, Maughan NJ, Smith V, Hillan KJ, Quirke P. Unlocking the archive-gene expression in paraffin-embedded tissue. *J Pathol* 2001;195:66–71.

818. Jackson DP, Lewis FA, Taylor GR, Boylston AW, Quirke P. Tissue extraction of DNA and RNA and analysis by the polymerase chain reaction. *J Clin Pathol* 1990;43:499–504.

819. Greer CE, Peterson SL, Kiviat NB, Manos MM. PCR amplification from paraffin-embedded tissues: effects of fixative and fixation time. *Am J Clin Pathol* 1991;95:117–124.

820. Shibata D. The polymerase chain reaction and the molecular genetic analysis of tissue biopsies. In *Diagnostic Molecular Pathology: A Practical Approach*, Vol II. Herrington CS, McGee JOD (eds.) Oxford, England: IRL Press; 1992: 85–111.

821. Ludyga N, Grünwald B, Azimzadeh O, *et al.* Nucleic acids from long-term preserved FFPE tissues are suitable for downstream analyses. *Virchows Arch* 2012;460:131–140.

822. Williams C, Ponten F, Moberg C, *et al.* A high frequency of sequence alterations is due to formalin fixation of archival specimens. *Am J Pathol* 1999;155:1467–1471.

823. Sieben NL, ter Haar NT, Cornelisse CJ, Fleuren GJ, Cleton-Jansen AM. PCR artifacts in LOH and MSI analysis of microdissected tumor cells. *Hum Pathol* 2000;31:1414–1419.

824. Do H, Dobrovic A. Sequence artifacts in DNA from formalin-fixed tissues: causes and strategies for minimization. *Clin Chem* 2015;61:64–71.

825. Wong SQ, Li J, Tan AY, *et al.* Sequence artefacts in a prospective series of formalin-fixed tumours tested for mutations in hotspot regions by massively parallel sequencing. *BMC Med Genomics* 2014;7:23.

826. Mullis KB, Faloona F. Specific synthesis of DNA in vitro via a polymerase-catalyzed chain reaction. *Methods Enzymol* 1987;155:335–350.

827. Saiki R, Scharf S, Faloona F, *et al.* Enzymatic amplification of beta-globin genomic sequences and restriction site analysis for diagnosis of sickle cell anemia. *Science* 1985;230:1350–1354.

828. Saiki RK, Bugawan TL, Horn GT, Mullis KB, Erlich HA. Analysis of enzymatic amplificatically amplified beta-globin and HLA-DQ alpha DNA with allele-specific oligonucleotide probes. *Nature* 1986;324:163–166.

829. Embury SH, Scharf SJ, Saiki RK, *et al.* Rapid prenatal diagnosis of sickle cell anemia by a new method of DNA

analysis. *N Engl J Med* 1987;316:656–661.

830. Downing JR, Khandekar A, Shurtleff SA, et al. Multiplex RT-PCR assay for the differential diagnosis of alveolar rhabdomyosarcoma and Ewing's sarcoma. *Am J Pathol* 1995;46:626–634.

831. Lasota J, Miettinen M. Absence of Kaposi's sarcoma-associated virus (human herpesvirus-8) sequences in angiosarcoma. *Virchows Arch* 1999;434:51–56.

832. Lasota J, Jasinski M, Debiec-Rychter M, et al. Detection of the SYT-SSX fusion transcripts in formaldehyde-fixed, paraffin-embedded tissue: a reverse transcription polymerase chain reaction amplification assay useful in the diagnosis of synovial sarcoma. *Mod Pathol* 1998;11:626–633.

833. Frohman MA, Dush MK, Martin GR. Rapid production of full-length cDNAs from rare transcripts: amplification using a single gene-specific oligonucleotide primer. *Proc Natl Acad Sci USA* 1988;85:8998–9002.

834. Scotto-Lavino E, Du G, Frohman MA. 5′ end cDNA amplification using classic RACE. *Nat Protoc* 2006;1:2555–2562.

835. Scotto-Lavino E, Du G, Frohman MA. 3′ end cDNA amplification using classic RACE. *Nat Protoc* 2006;1:2742–2745.

836. Gaudet M, Fara AG, Beritognolo I, Sabatti M. Allele-specific PCR in SNP genotyping. *Methods Mol Biol* 2009;578:415–424.

837. Syvänen AC. From gels to chips: "minisequencing" primer extension for analysis of point mutations and single nucleotide polymorphisms. *Hum Mutat* 1999;13:1–10.

838. Ahmadian A, Ehn M, Hober S. Pyrosequencing: history, biochemistry and future. *Clin Chim Acta* 2006;363:83–94.

839. Cotton, RGH. Slowly but surely towards better scanning for mutations. *Trends Genet* 1997;13:43–46.

840. Fischer SG, Lerman LS. DNA fragments differing by single base-pair substitutions are separated in denaturing gradient gels: correspondence with melting theory. *Proc Natl Acad Sci USA* 1983;80:1579–1583.

841. Fodde R, Losekoot M. Mutation detection by denaturing gradient gel electrophoresis (DGGE). *Hum Mutat* 1994; 3:83–94.

842. Orita M, Iwahana H, Kanazawa H, Hayashi K, Sekiya T. Detection of polymorphism of human DNA by gel electrophoresis as single-strand conformation polymorphisms. *Proc Natl Acad Sci USA* 1989;86:2766–2770.

843. Hayashi K. PCR-SSCP: a method for detection of mutations. *Genet Anal Tech Appl* 1992;9:73–79.

844. Oefner PJ, Underhill PA. Comparative DNA sequencing by denaturing high-performance liquid chromatography (DHPLC). *Am J Hum Genet* 1995;57: A266.

845. Liu W, Smith DI, Rechtzigel KJ, Thibodeau SN, James CD. Denaturing high performance liquid chromatography (DHPLC) used in the detection of germline and somatic mutations. *Nucleic Acids Res* 1998;26:1396–1400.

846. Han SS, Cooper DN, Upadhyaya MN. Evaluation of denaturing high performance liquid chromatography (DHPLC) for the mutational analysis of the neurofibromatosis type 1 (NF1) gene. *Hum Genet* 2001;109:487–497.

847. Livak KJ, Flood SJ, Marmaro J, Giusti W, Deetz K. Oligonucleotides with fluorescent dyes at opposite ends provide a quenched probe system useful for detecting PCR product and nucleic acid hybridization. *PCR Methods Appl* 1995;4:357–362.

848. Peter M, Gilbert E, Delattre O. A multiplex real-time pcr assay for the detection of gene fusions observed in solid tumors. *Lab Invest* 2001;81:905–912.

849. Bijwaard KE, Fetsch JF, Przygodzki R, Taubenberger JK, Lichy JH. Detection of SYT-SSX fusion transcripts in archival synovial sarcomas by real-time reverse transcriptase-polymerase chain reaction. *J Mol Diagn* 2002;4:59–64.

850. Pongers-Willemse MJ, Verhagen OJ, Tibbe GJ, et al. Real-time PCR for the detection of minimal residual disease in acute lymphoblastic leukemia using junctional region specific TaqMan probes. *Leukemia* 1998;12:2006–2014.

851. Preudhomme C, Revillion F, Merlat A, et al. Detection of BCR-ABL transcripts in chronic myeloid leukemia (CML) using a 'real time' quantitative RT-PCR assay. *Leukemia* 1999;13:957–964.

852. Oda S, Oki E, Maehara Y, Sugimachi K. Precise assessment of microsatellite instability using high resolution fluorescent microsatellite analysis. *Nucleic Acids Res* 1997;25:3415–3420.

853. Kallioniemi A, Kallioniemi O-P, Sudar D, et al. Comparative genomic hybridization for molecular cytogenetic analysis of solid tumors. *Science* 1992; 258:818–821.

854. Kallioniemi O-P, Kallioniemi A, Piper J, et al. Optimizing comparative genomic hybridization for analysis of DNA sequence copy number changes in solid tumors. *Gene Chromosomes Cancer* 1994;10:231–243.

855. du Manoir S, Speicher MR, Joos S, et al. Detection of complete and partial chromosome gains and losses by comparative genomic in situ hybridization. *Hum Genet* 1993;90:590–610.

856. Oostlander AE, Meijer GA, Ylstra B. Microarray-based comparative genomic hybridization and its applications in human genetics. *Clin Genet* 2004;66:488–495.

857. Lockwood WW, Chari R, Chi B, Lam WL. Recent advances in array comparative genomic hybridization technologies and their applications in human genetics. *Eur J Hum Genet* 2006;14:139–148.

858. Knuutila S, Autio K, Aalto Y. On line access to CGH data of DNA sequence copy number changes. *Am J Pathol* 2000;157:689–690.

859. El-Rifai W, Sarlomo-Rikala M, Andersson LC, Knuutila S, Miettinen M. DNA sequence copy number changes in gastrointestinal stromal tumors: tumor progression and prognostic significance. *Cancer Res* 2000;60:3899–3903.

860. Duggan DJ, Bittner M, Yidong C, Meltzer P, Trent JM. Expression profiling using cDNA microarrays. *Nature Genet* 1999;21:10–14.

861. Lockhard DJ, Winzeler EA. Genomics, gene expression and DNA arrays. *Nature* 2000;405:827–836.

862. Golub TR, Slonim DK, Tamayo P, et al. Molecular classification of cancer: class discovery and class prediction by gene expression monitoring. *Science* 1999;286:531–536.

863. Khan J, Simon R, Bittner M, et al. Gene expression profiling of alveolar

864. Nielsen TO. Microarray analysis of sarcomas. *Adv Anat Pathol* 2006;13:166–173.

865. Sanger F, Nicklen S, Coulson AR. DNA sequencing with chain-terminating inhibitors. *Proc Natl Acad Sci USA* 1977;74:5463–5467.

866. Bennett ST, Barnes C, Cox A, Davies L, Brown C. Toward the 1,000 dollars human genome. *Pharmacogenomics* 2005;6:373–382.

867. Hert DG, Fredlake CP, Barron AE. Advantages and limitations of next generation sequencing technologies: a comparison of electrophoresis and non-electrophoresis methods. *Electrophoresis* 2008;29:4618–4626.

868. Koboldt DC, Steinberg KM, Larson DE, Wilson RK, Mardis ER. The next-generation sequencing revolution and its impact on genomics. *Cell* 2013;155:27–38.

869. Behjati S, Tarpey PS, Sheldon H, et al. Recurrent PTPRB and PLCG1 mutations in angiosarcoma. *Nat Genet* 2014;46:376–379.

870. Stockman DL, Ali SM, He J, Ross JS, Meis JM. Sclerosing epithelioid fibrosarcoma presenting as intraabdominal sarcomatosis with a novel EWSR1-CREB3L1 gene fusion. *Hum Pathol* 2014;45:2173–2178.

871. Jour G, Scarborough JD, Jones RL, et al. Molecular profiling of soft tissue sarcomas using next-generation sequencing: a pilot study toward precision therapeutics. *Hum Pathol* 2014;45:1563–1571.

872. Shukla N, Ameur N, Yilmaz I, et al. Oncogene mutation profiling of pediatric solid tumors reveals significant subsets of embryonal rhabdomyosarcoma and neuroblastoma with mutated genes in growth signaling pathways. *Clin Cancer Res* 2012;18:748–757.

873. Harismendy O, Schwab RB, Bao L, et al. Detection of low prevalence somatic mutations in solid tumors with ultra-deep targeted sequencing. *Genome Biol* 2011;12:R124.

874. Zheng Z, Liebers M, Zhelyazkova B, et al. Anchored multiplex PCR for targeted next-generation sequencing. *Nat Med* 2014;20:1479–1484.

875. Burrell RA, McGranahan N, Bartek J, Swanton C. The causes and consequences of genetic heterogeneity in cancer evolution. *Nature* 2013;501:338–345.

876. Ashworth A, Lord CJ, Reis-Filho JS. Genetic interactions in cancer progression and treatment. *Cell* 2011;145:30–38.

877. Roche-Lestienne C, Soenen-Cornu V, Grardel-Duflos N, et al. Several types of mutations of the Abl gene can be found in chronic myeloid leukemia patients resistant to STI571, and they can pre-exist to the onset of treatment. *Blood* 2002;100:1014–1018.

Genes and their names

878. Lander ES, Linton LM, Birren B, et al. Initial sequencing and analysis of the human genome. *Nature* 2001;409:860–921.

Chapter 6

Fibroblast biology, fasciitis, retroperitoneal fibrosis, and keloids

Markku Miettinen
National Cancer Institute, National Institutes of Health

Preceded by an introduction to fibroblastic neoplasms and fibroblast biology, this chapter discusses lesions that include the word *fasciitis* in their name and two putative reactive lesions, retroperitoneal fibrosis and keloids. All of these are composed of fibroblasts and myofibroblasts; the latter are fibroblasts that typically contain smooth muscle actin and have contractile properties, originally described in wound healing.

Fasciitis is a clinicopathologically heterogeneous group that contains several distinct entities. Eosinophilic and necrotizing fasciitis are inflammatory and infectious conditions, respectively. They are unrelated to other lesions termed *fasciitis* but are included here because of similar terminology.

Benign fibroblastic tumors and tumor-like lesions form a large and heterogeneous group of non-neoplastic and neoplastic tumor entities. Some of these have a predilection for children, and some occur only in adults. Benign fibroblastic proliferations in this book are divided arbitrarily into four groups: (1) fasciitis and related lesions, (2) benign fibrous neoplasms (fibromas and related tumors), (3) fibromatoses, and (4) myxomas and other myxoid lesions. (Fibrous proliferations of children are presented in two separate chapters encompassing benign tumors and those of variable biologic potential.)

Lack of hard evidence makes it difficult to separate reactive and neoplastic conditions in some instances. Likewise, some lesions thought to be neoplasms could actually be reactive. In some instances, the commonly used designations are actually misleading about the true nature of the lesion; for example, nuchal fibroma, penile fibromatosis, and fibromatosis colli may all be reactive processes.

Fibroblastic neoplasms of variable biologic potential include tumors that are mostly benign (solitary fibrous tumor/hemangiopericytoma) but include a definite subset of malignant variants. There are also low-grade tumors, such as dermatofibrosarcoma protuberans and acral myxoinflammatory fibrosarcoma, that have the potential to recur but a very low, if any, risk of metastasis. Myxofibrosarcoma and malignant fibrous histiocytoma (MFH; undifferentiated pleomorphic sarcoma) are malignant fibroblastic neoplasms that have a spectrum ranging from relatively indolent tumors to high-grade sarcomas (see Chapter 13).

Biology of fibroblasts

Fibroblast biology is a large research area that is relevant not only to fibroblastic tumors but also to many other tumor groups, especially vascular and nerve sheath tumors, because their morphogenesis is governed by some of the same growth factors, growth factor receptors, and matrix interactions that are involved in fibroblast biology.[1,2] Currently the biologic information and morphologic knowledge of fibroblastic tumors intersect only sporadically, indicating that translation of biologic knowledge to the pathology of tumors is only beginning. Many fibroblastic tumor entities are known by morphologic parameters only, without our having any insight into their biology. The purpose of this section is to provide some framework for fibroblast biology, to help to understand biologic discoveries on fibroblast tumors and further develop the connections between fibroblast biology and fibroblastic tumors.

Definition of fibroblasts and their subsets

Fibroblasts are collagen-producing mesenchymal cells that are capable of forming the collagenous matrix of dense connective tissue. Fibroblasts are derived from primitive mesenchyme and can be identified as spindle-shaped, vimentin-positive mesenchymal cells in a wide variety of connective tissues of the body. Fibroblasts include a replication-capable stem-cell-like subpopulation that is activated during tissue repair, and neoplasia.

Two specific subsets of fibroblasts can be identified based on morphology and antigen expression: myofibroblasts and the CD34-positive fibroblasts.

Myofibroblasts

Most reactive fibroblastic lesions, stroma of malignant tumors, and most fibroblastic neoplasms contain myofibroblasts. Gabbiani and Majno originally identified these derivatives of fibroblasts ultrastructurally as cells with features intermediate between fibroblasts and smooth muscle cells and functionally as the cells responsible for scar contraction.[3] Like fibroblasts, myofibroblasts contain an ultrastucturally prominent rough endoplasmic reticulum, and they have some cytoplasmic

---

Modern Soft Tissue Pathology, Second Edition, ed. Markku Miettinen. Published by Cambridge University Press. © Cambridge University Press 2016.

bundles of actin microfilaments and membrane-attached actin filaments (attachment plaques), but to lesser degree than seen in true smooth muscle cells. Unlike smooth muscle cells, they have an incomplete and only focally developed basement membrane. Immunohistochemically myofibroblasts span a broad spectrum, including subsets that are positive for vimentin only; vimentin and actin; or vimentin, actin, and desmin.[4]

CD34-positive fibroblasts

The CD34-positive fibroblasts were originally described by Nickoloff around skin adnexa and blood vessels.[5] The possible relationship of these cells with the CD34-positive bone marrow stem cells was originally suspected but has not been demonstrated convincingly. CD34-positive fibroblasts are present perivascularly and in connective tissue septa throughout the body. These cells are candidates for the histogenetic origin of some CD34-positive fibroblastic tumors, such as solitary fibrous tumor and dermatofibrosarcoma protuberans.

Collagens

Collagens are a large, complex group of extracellular matrix-forming peptides. Currently, 28 collagen types are known, but only a few of these are major components of the fibrillary extracellular matrix formed by fibroblasts. Other collagens are fibril-associated collagen with interrupted triple helices (FACIT-collagens), network-forming collagens, and basement membrane collagens.[6,7]

Collagen type I is the most abundant. This protein is deposited by fibroblasts to form polymers that are assembled into collagen fibrils and collagen fibers. Collagen type III is the predominant fibrillary collagen in newly formed fibroblasts, such as those in granulation tissue. Collagen type II is the predominant collagen in cartilaginous matrix.

Mutations in fibrillary collagen genes (especially those that encode for collagen types I, II, III, and XI) are known causes for diseases such as Ehlers–Danlos syndrome (hyperelastic collagen) and osteogenesis imperfecta.[8]

Fibroblast growth factors and their receptors

The proliferation and differentiation of fibroblasts is regulated by the interaction of several fibroblast growth factors and their receptors with other growth factors, such as proteins of the transforming growth factor beta (TGF β) family. Interaction with the extracellular matrix also regulates fibroblast behavior.[1,2,9,10]

In non-neoplastic conditions, the fibroblastic proliferation can be activated by inflammation (through growth factors/cytokines) or by another type of tissue injury.

Molecular mechanisms of fibroblastic tumors

The molecular mechanisms of fibroblastic neoplasia are still understood very incompletely. Pathologic activation of signal transduction pathways is being revealed in some tumor types, however, for example, in intramuscular myxoma. Fibromatosis could be a consequence of pathologic activation of Wnt signaling. Dysfunction of capillary morphogenesis protein seems to be the cause of juvenile hyaline fibromatosis.

Under certain conditions, fibroblasts are capable of secreting abundant proteoglycan or glycoprotein matrix. Such secretory activity of fibroblasts contributes to the histogenesis of myxoid neoplasms, which are essentially fibroblastic tumors, based on evidence of the fibroblastic and myofibroblastic nature of the tumor cells.

Nodular fasciitis

Nodular fasciitis is the designation for a tumor-like, cellular, sometimes pseudosarcomatous myofibroblastic spindle cell proliferation. Although originally uniformly considered a reactive lesion, recent reports of clonal cytogenetic changes and a specific translocation suggest that at least some cases represent clonal neoplastic proliferations. Considering the histologic variation, however, nodular fasciitis could be biologically and genetically a heterogeneous group.

The first reports of this entity used the term *subcutaneous pseudosarcomatous fibromatosis* (fasciitis)[11] and pseudosarcomatous fasciitis;[12] the term *nodular fasciitis* came to use in the early 1960s.[13]

Intravascular fasciitis is a microscopic variant with intravascular involvement. Periosteal fasciitis is the designation for a nodular fasciitis lesion apposed to a bone. Ossifying fasciitis or panniculitis ossificans refers to lesions combining features of nodular fasciitis and metaplastic bone formation. Cranial fasciitis refers to nodular fasciitis in the scalp of infants, associated with lytic change of adjacent immature bone susceptible to such change. This entity is discussed in Chapter 9.

Clinical features

Nodular fasciitis is most common in young adults, and half of all cases occur between the ages of 20 and 40 years. Almost 20% of cases are seen in children, in whom a substantial male predominance is noted. Only 3% of the cases are diagnosed in adults over 65 years of age. In adults, there is equal incidence in both sexes (Figure 6.1).

Nearly two-thirds of the cases occur in the upper extremities or the head and neck, and the two most common sites are the forearm and the arm (Figure 6.2). Occurrence in the trunk wall is common, but only 15% of cases occur in the lower extremities. In the head, neck, and trunk, nodular fasciitis is relatively more common in children, although the median ages in all major sites vary within a narrow range (28–33 years).

The tumor usually measures 2 to 3 cm in diameter and is centered in the subcutis and often attached to the fascia. Size of more than 5 cm is exceptional. Extension to the dermis is possible, especially in the head and neck, and purely dermal variants have been noted on occasion. Some examples involve skeletal muscle, and a small minority (1–2%) are purely intramuscular.[11–17]

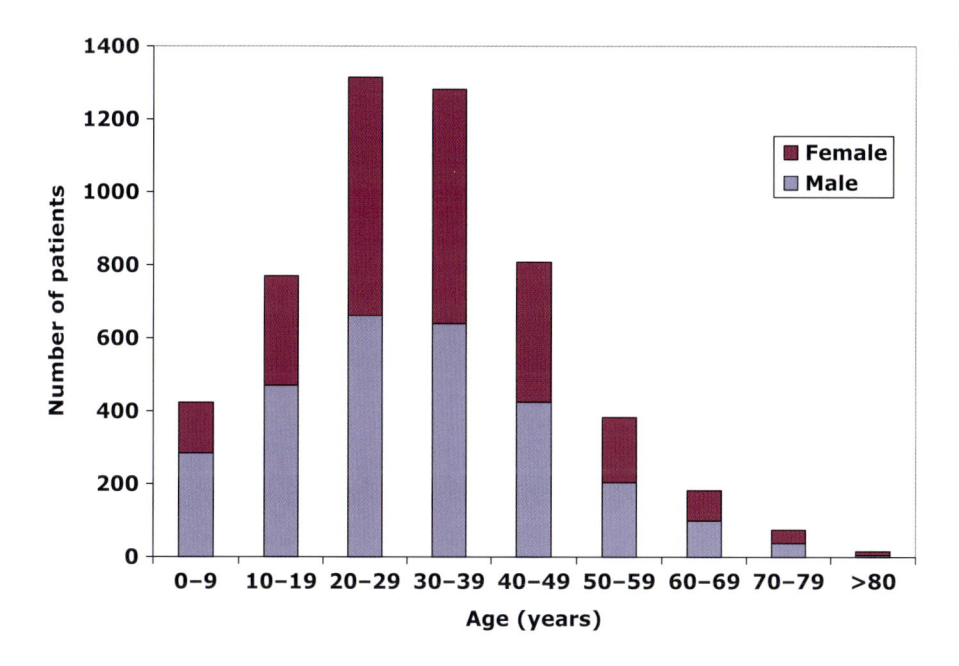

Figure 6.1 Age and sex distribution of 5254 patients with nodular fasciitis.

Nodular fasciitis often appears as a rapidly growing, painless mass. According to one study, median duration was 2 months, and 85% of cases had a duration <4 months.[4] Exceptionally, there is a long history of a lesion, lasting for several years. Some patients report an antecedent trauma, and historically nodular fasciitis has been thought to be a reactive process. Local excision is sufficient, although on rare occasions, nodular fasciitis does recur. In some authors' experience, however, all recurrent examples have represented other entities, including (myofibroblastic) sarcomas.[16]

Pathology

Grossly, nodular fasciitis typically forms a well-circumscribed mass surrounded by a dense collagenous band, but it is not truly encapsulated. On sectioning, the lesion varies from mucoid to firm, having a tan, often discolored, cut surface with focal hemorrhage. Some lesions contain cysts (Figure 6.3). Some variants, especially the so-called repair variant (granulation-tissue-like variant), commonly have infiltrative, spiculated borders to subcutaneous fat (Figure 6.4).

The architecture is slightly storiform or shows a vague fascicular pattern with the cells separated by varying numbers of collagen fibers, which often increase toward the periphery and on maturation of the lesion. Microcystic change, tissue cleavages, and focal mucoid degeneration are characteristic features. Extravasated erythrocytes, lymphocytes, and scattered osteoclast-like giant cells with three to seven nuclei are often sprinkled between the myofibroblasts, giving a heterogeneous appearance (Figure 6.5). The histological features of nodular fasciitis vary from myxoid to variably fibrous, which is possibly related to the age of the lesion. It has been suggested that nodular fasciitis evolves from a myxoid early stage to a cellular phase and later involutes to a fibrous phase (Figure 6.6), sometimes with progressive cystic change.[17]

Cytologically, the myofibroblasts of nodular fasciitis often have "a tissue culture appearance," which refers to the resemblance of these myofibroblasts with an epithelioid appearance to cultured fibroblasts spreading onto the surface of the culture dish. These cells have large hypochromatic (never hyperchromatic) nuclei with an open chromatin, a delicate nucleolus, and angulated cytoplasmic processes. Numerous mitoses can be present, especially in the highly cellular variants. Such variants can be truly pseudosarcomatous, but they still have, at least focally, characteristic features, such as focal clefting, osteoclast-like giant cells, and peripheral maturation (Figure 6.7).

The repair (granulation-tissue-like) variant has a prominent capillary pattern, often in the lesional periphery, and they typically infiltrate as spicules into the surrounding subcutaneous fat (Figure 6.8). A variant with spicules of maturing bone (and cartilage) has been named *fasciitis ossificans*. This lesion might be alternatively viewed as a form of myositis ossificans arising in the subcutis.[18]

Intramuscular nodular fasciitis is histologically similar to the subcutaneous variety, but often contains regenerative/degenerative skeletal muscle cells (sarcoplasmic giant cells) in the lesional periphery. Some lesions have a permeative growth with large numbers of entrapped skeletal muscle cells (Figure 6.9). This variant has to be separated from desmoid fibromatosis, which is more uniformly cellular and solid, lacking microcystic and mucoid changes.

Electron microscopic studies on nodular fasciitis have shown myofibroblastic differentiation, characterized by spindle cells with focal cytoplasmic and membrane-associated actin bundles and focal basement membranes.[19]

Immunohistochemically, nodular fasciitis cells are positive for vimentin and usually for smooth muscle actin, muscle actin, CD10 (Figure 6.10), and smooth muscle myosin.[20,21]

Figure 6.2 Anatomic distribution of 5263 cases of nodular fasciitis. (**A**) General view. (**B**) Detailed view of the most commonly involved sites.

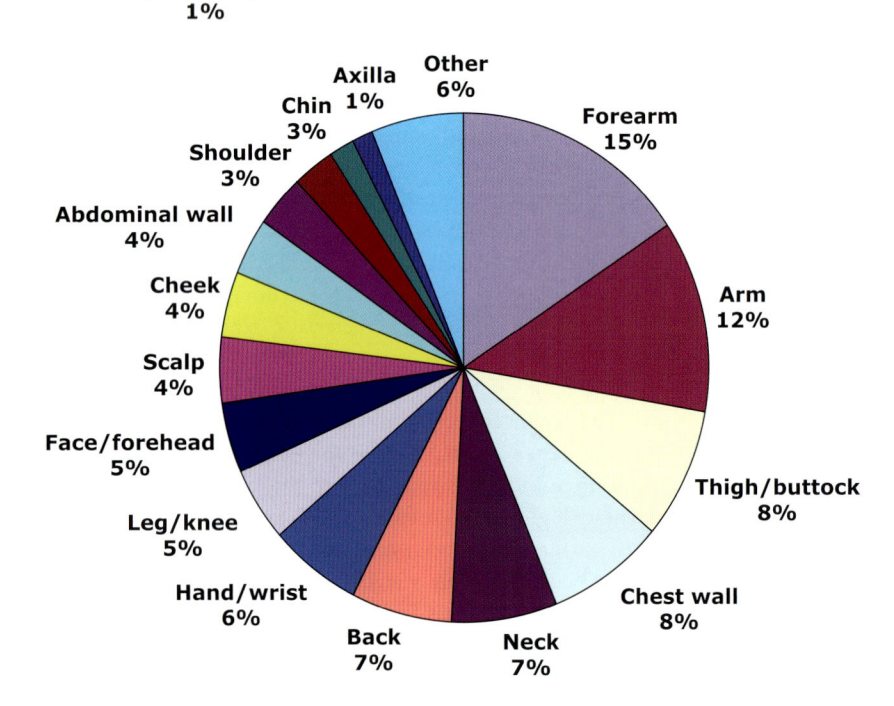

The tumor cells are negative for BCL2, desmin, heavy caldesmon, CD34, S100 protein, and keratin cocktail AE1/AE3. Scattered dendritic (antigen-presenting) cells are positive for S100 protein, and variable numbers of histiocytes and possibly some myofibroblasts are CD68 positive. The number of true histiocytes labeled for CD163 is typically smaller than the number of CD68-positive cells.

Differential diagnosis

Some carcinomas, notably a variant thyroid papillary carcinoma, can contain nodular-fasciitis-like stroma; therefore thorough sampling is required to rule out the presence of carcinoma components before nodular fasciitis is diagnosed in parenchymal organs, such as the thyroid.[22]

The highly cellular variant can be distinguished from sarcoma by lack of nuclear atypia, sharp circumscription, tendency to peripheral maturation (increased amount of collagen fibers), and features typical of fasciitis, such as focal microcystic/mucoid degeneration, and occurrence of osteoclastic giant cells, at least focally.

Desmoid and low-grade fibrosarcomas/myofibroblastic sarcomas can contain areas resembling nodular fasciitis, and definitive differential diagnosis therefore might not be possible from limited material, such as a needle biopsy specimen. Areas of relatively uniform myofibroblastic proliferation with moderate to abundant collagen and prominent blood vessels are typical of desmoid fibromatosis, and the presence of enlarged or pleomorphic nuclei with atypia should alert one to the possibility of a (low-grade) sarcoma. Such atypia should be sought out, especially with large lesions and those occurring in older people. The presence of any significant atypia should raise serious questions about the diagnosis of nodular fasciitis.

Genetics

Typical of nodular fasciitis is the fusion translocation involving MYH9-USP6 genes. This translocation pairs the promoter of the myosin heavy chain gene with the coding sequences of ubiquitin-specific peptidase 6, which corresponds to a t(17;22) translocation.[23] Other clonal gene rearrangements at 3q21 and 15q22 have been reported, including a reciprocal translocation of t(13;15).[24,25] Rearrangements of segments of chromosome 15 involving region 15q22–25 have also been reported; this region includes genes for fibroblast growth factor 7 (FGF7), and neurotropin kinase C (NTRK3).[26,27] These findings support the concept that nodular fasciitis is a self-limiting or transient neoplasia, and that the observed gene rearrangements could be pathogenetically important.

Intravascular fasciitis

Nodular-fasciitis-like lesions with dominant intravascular components are designated as intravascular fasciitis, as originally reported by Patchefsky and Enzinger in 1981.[28]

Clinical features

Intravascular fasciitis is a rare condition, with a frequency of only 2.2% of that of nodular fasciitis, based on Armed Forces Institute of Pathology (AFIP) data. It typically occurs in children and young adults and is more common in children than is nodular fasciitis. The overall median age of patients is 27 years, with a significant male predominance (Figure 6.11A).

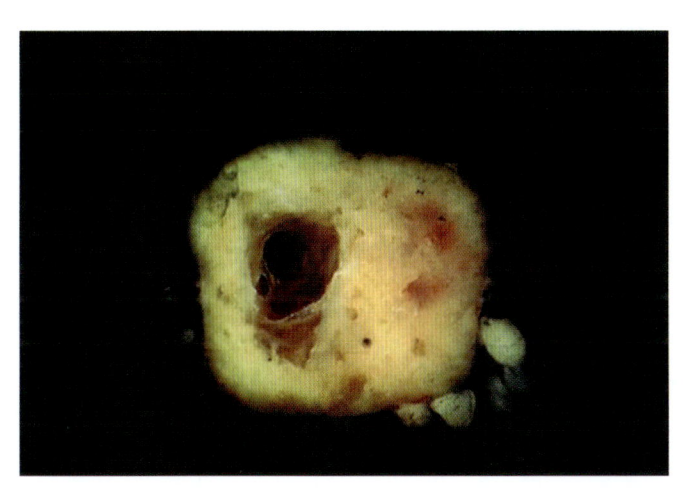

Figure 6.3 Grossly, this example of nodular fasciitis forms a whitish, fleshy mass of approximately 2 cm, with a central cyst.

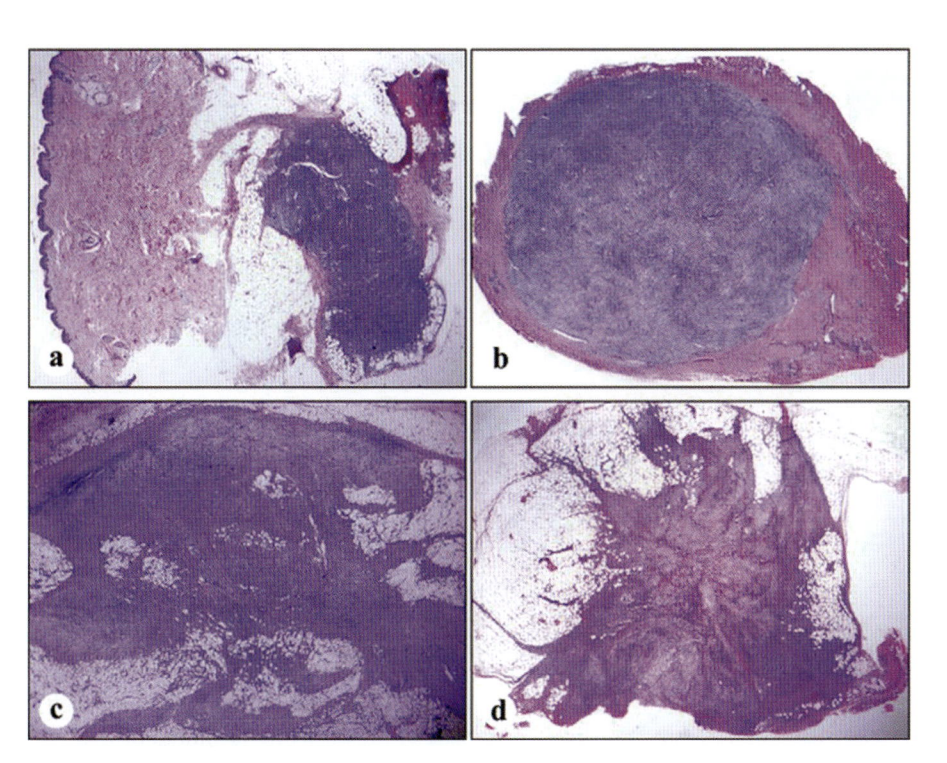

Figure 6.4 Low-magnification appearance of nodular fasciitis. (**a,b**) Two examples of variably circumscribed lesions. (**c,d**) Two examples of the infiltrative ("repair") variant of nodular fasciitis.

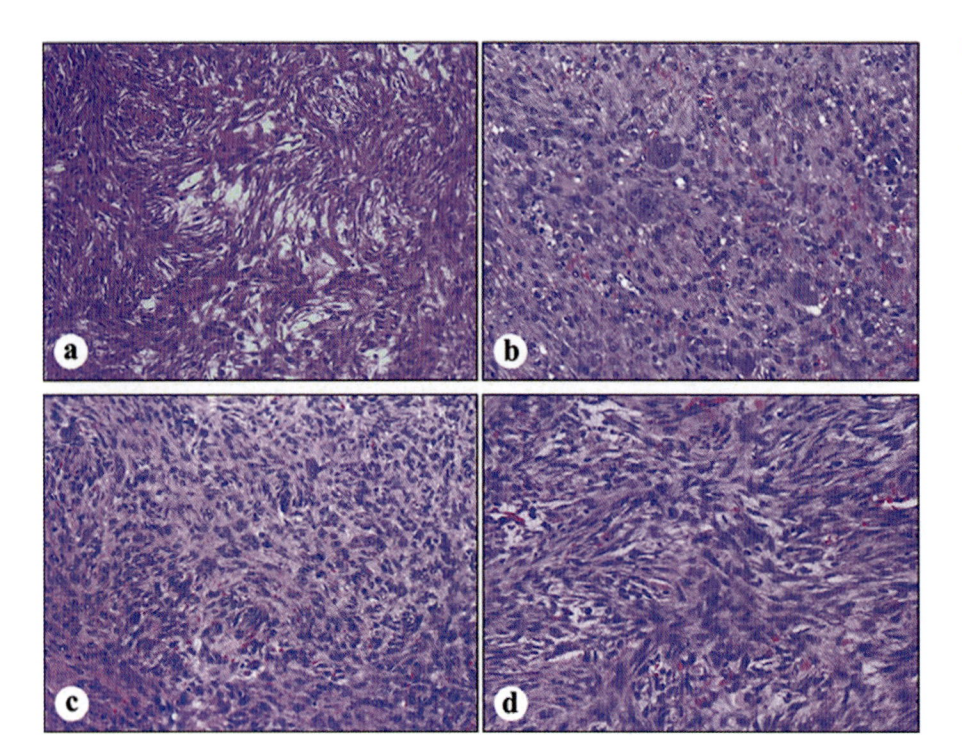

Chapter 6: Fibroblast biology, fasciitis, retroperitoneal fibrosis, and keloids

Figure 6.5 Typical histologic features of moderately cellular variants of nodular fasciitis. Microcystic degeneration and clefts (a,d), moderate deposition of collagen (a–d), osteoclast-like giant cells (b,c), extravasated erythrocytes (d), and occasional mitotic activity (c,d).

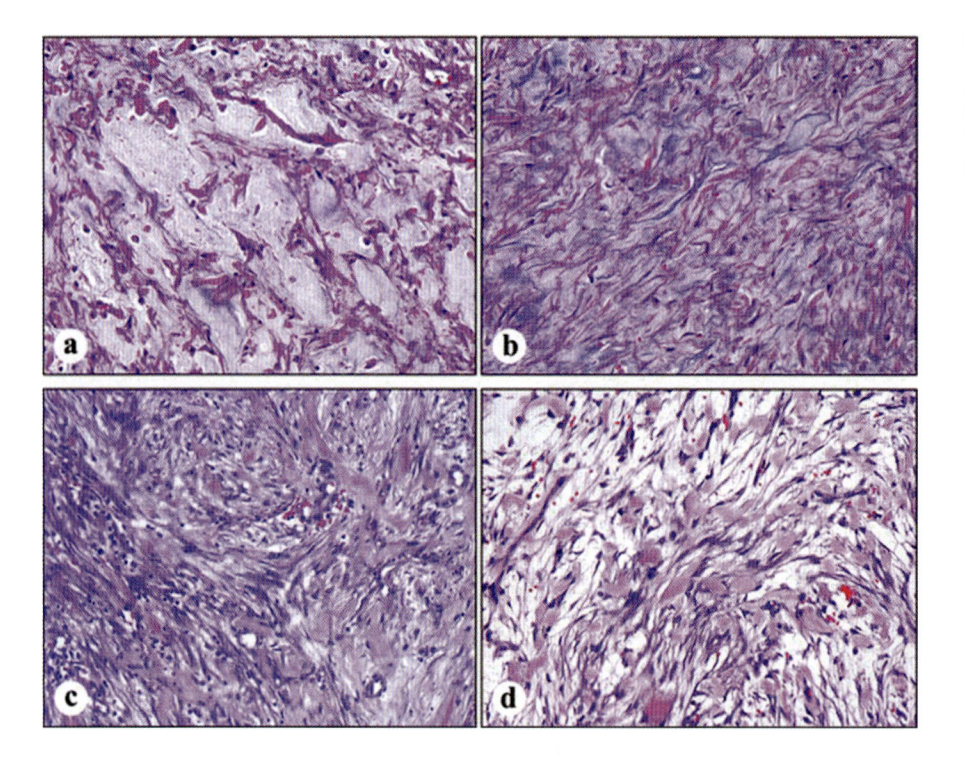

Figure 6.6 Variants of nodular fasciitis that could be related to the duration of the lesion. (a,b) Myxoid "early" example with prominent deposition of extracellular mucin. (c,d) A fibrous "late" example with relatively low cellularity and prominent collagen deposition.

The anatomic distribution differs from nodular fasciitis in that the head and neck are the most common sites, followed by upper extremities, whereas the trunk and lower extremities are unusual sites (Figure 6.11B). Local excision is curative.

Pathology

Intravascular fasciitis forms a small, circumscribed intravascular mass usually <2 cm (Figure 6.12). Some cases have a multinodular, plexiform pattern caused by involvement of multiple

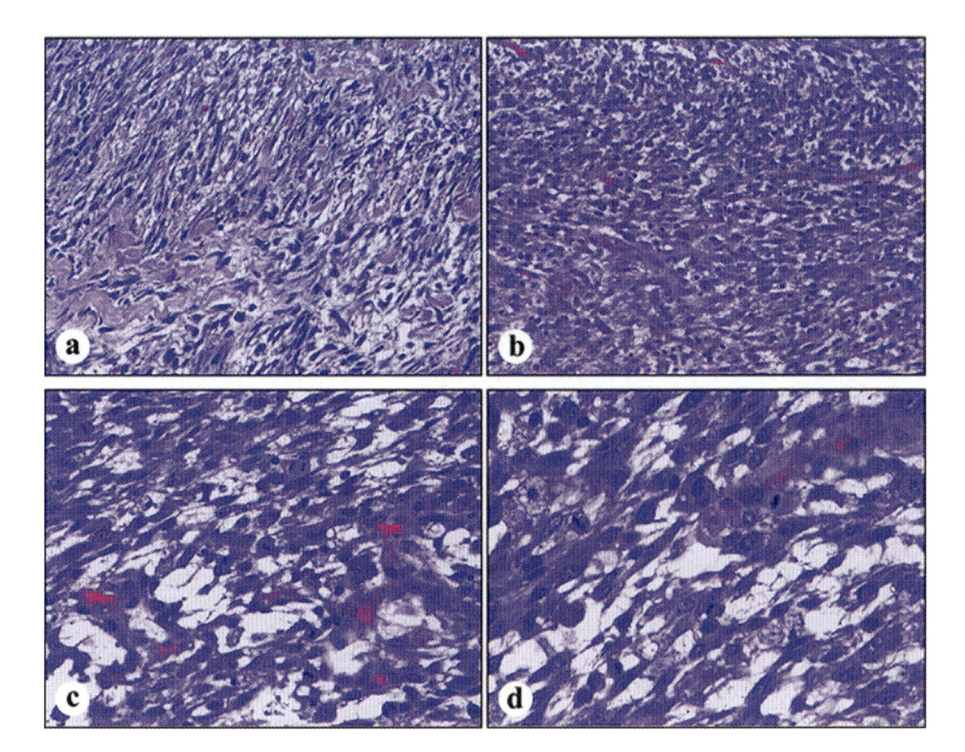

Figure 6.7 Extremely cellular "pseudosarcomatous" example of nodular fasciitis with numerous mitotic figures. (a) Focal collagenous maturation is present in the lesion periphery. (b,c) The tumor is highly cellular but contains cleavage spaces, extracellular mucin, and scattered osteoclastic giant cells typical of nodular fasciitis. Numerous mitoses are present (c,d).

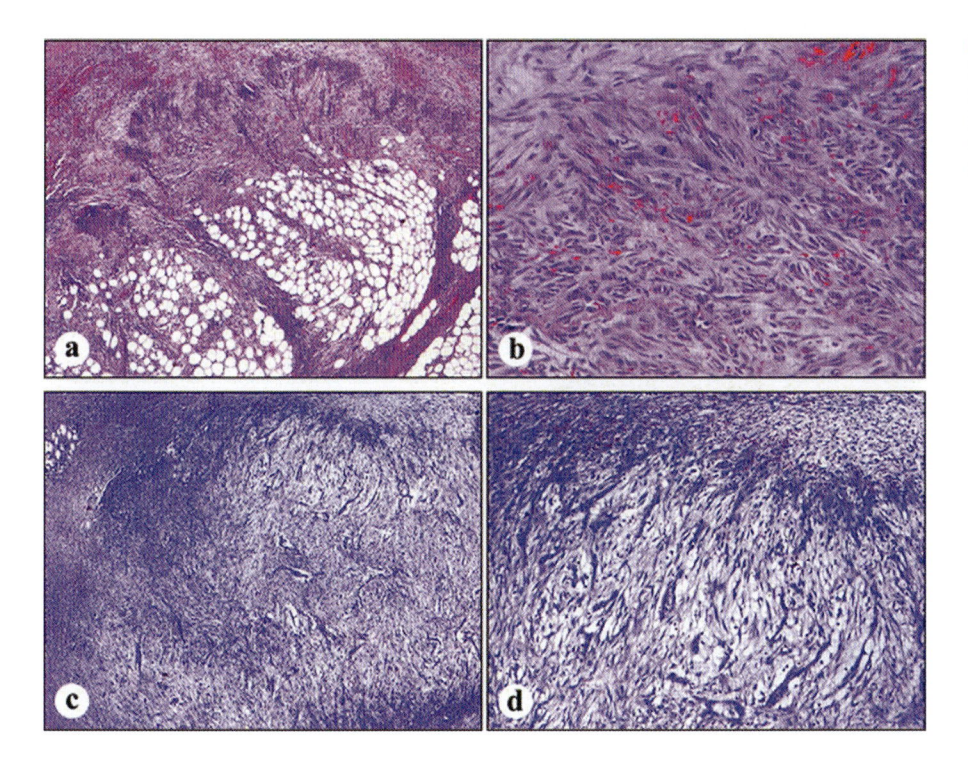

Figure 6.8 Examples of repair variant of nodular fasciitis with granulation-tissue-like morphology. Note the poorly circumscribed nature of the process, prominent arcades or newly formed capillaries, fat infiltration, and collagenous maturation.

vascular profiles representing one tortuous vessel or multiple vessels, usually veins. In a minority of cases, arterial segments are also involved. The histological appearance varies from cases identical to nodular fasciitis to more compact examples with near-to fascicular structure, often with numerous osteoclast-like giant cells (Figure 6.13). Central necrosis can be present in the intravascular growth, and some examples have focal cartilaginous metaplasia.

In the author's experience, the immunohistochemical features are identical to nodular fasciitis, having prominent

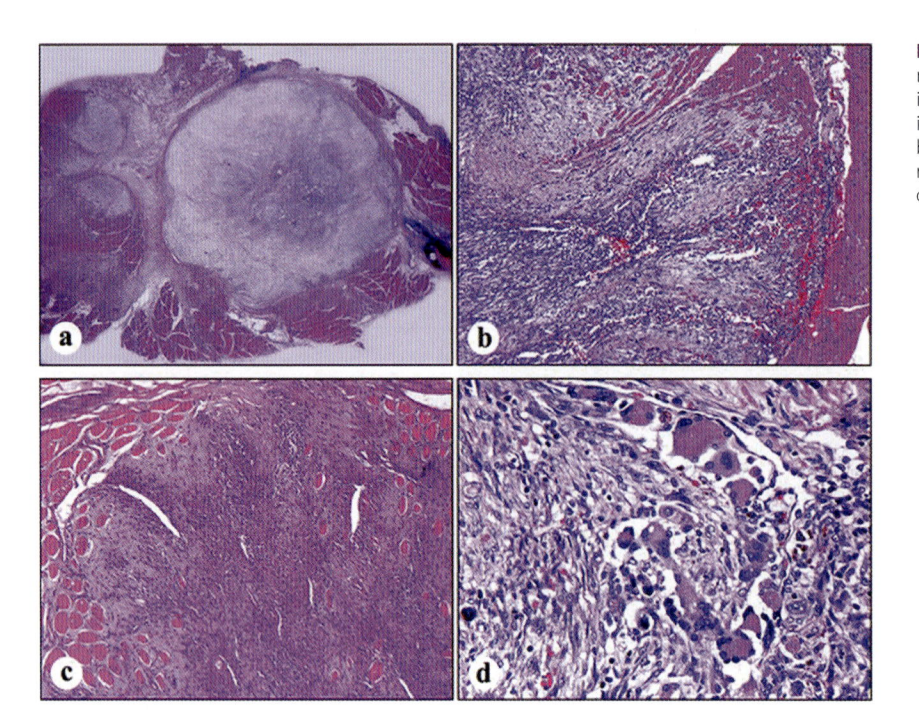

Chapter 6: Fibroblast biology, fasciitis, retroperitoneal fibrosis, and keloids

Figure 6.9 Intramuscular nodular fasciitis. (**a**) Low magnification reveals a sharply demarcated intramuscular nodule with two small satellite nodules to its left. (**b,c**) The lesion periphery shows an infiltrative border. (**d**) High magnification shows multinucleated regenerative skeletal muscle cells (sarcoplasmic giant cells) intermingling with lesional myofibroblasts.

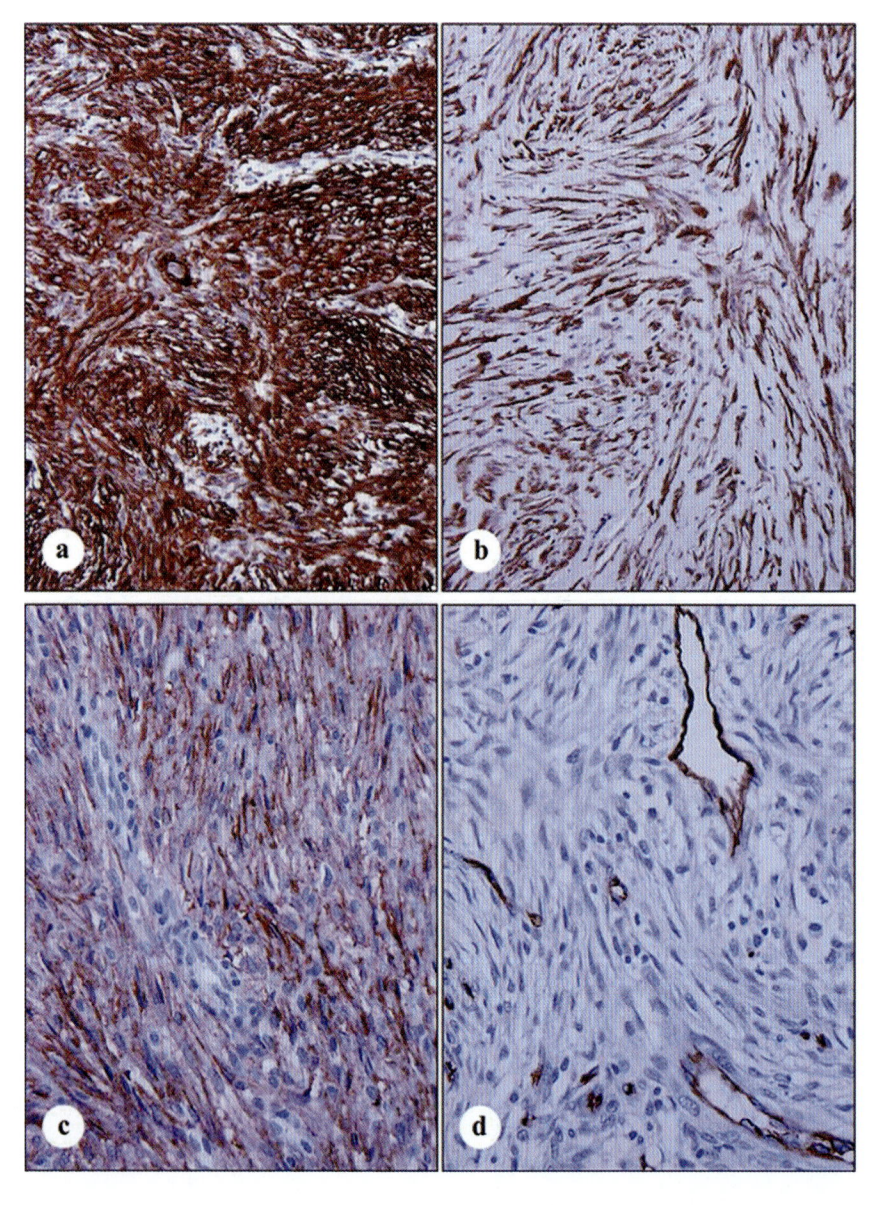

Figure 6.10 Immunohistochemical features typical of nodular fasciitis. (**a,b**) Variably prominent smooth muscle actin positivity. (**c**) CD10 positivity in the spindle cells. (**d**) CD34 is seen in the endothelial cells only.

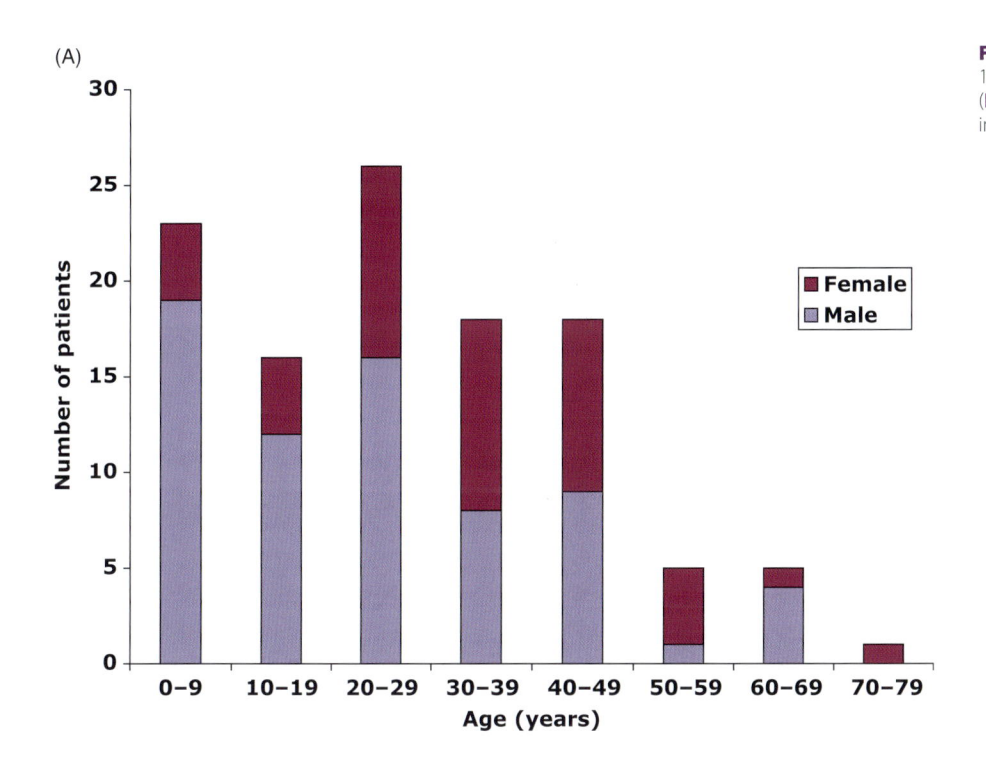

Figure 6.11 (A) Age and sex distribution of 112 patients with intravascular fasciitis. (B) Anatomic distribution of 111 cases of intravascular fasciitis.

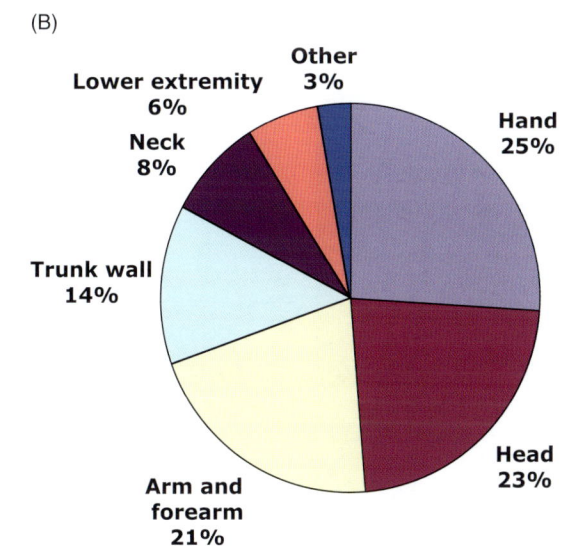

smooth muscle actin positivity, whereas the tumor cells are negative for CD34, desmin, keratin cocktail, and S100 protein.

Penile myointimoma

This benign, intravascular myofibroblastic proliferation, described by Fetsch et al. in 2000,[29] can be considered as related to intravascular fasciitis. This condition involves the corpus spongiosum of the glans penis of male patients of various ages (reported between 2 and 61 years),[29,30] with an overall median age of 31 years. The tumor typically presents as a small, usually <1-cm nodule of short duration, but occasionally there is a history of several months. Intralesional or marginal excision appears to be curative, and an incompletely excised lesion often undergoes spontaneous resolution, supporting the practice of conservative treatment.

Histologically, myointimoma typically shows a plexiform architecture at a low magnification by its involvement of multiple vascular profiles. The involved vessels contain a uniform proliferation of stellate myofibroblasts in a myxoid matrix, without significant mitotic activity or atypia (Figure 6.14). There is some resemblance to the myofibroblastic bundles of myofibroma. This lesion represents a more monotonous and cytologically uniform cellular proliferation than does intravascular fasciitis.

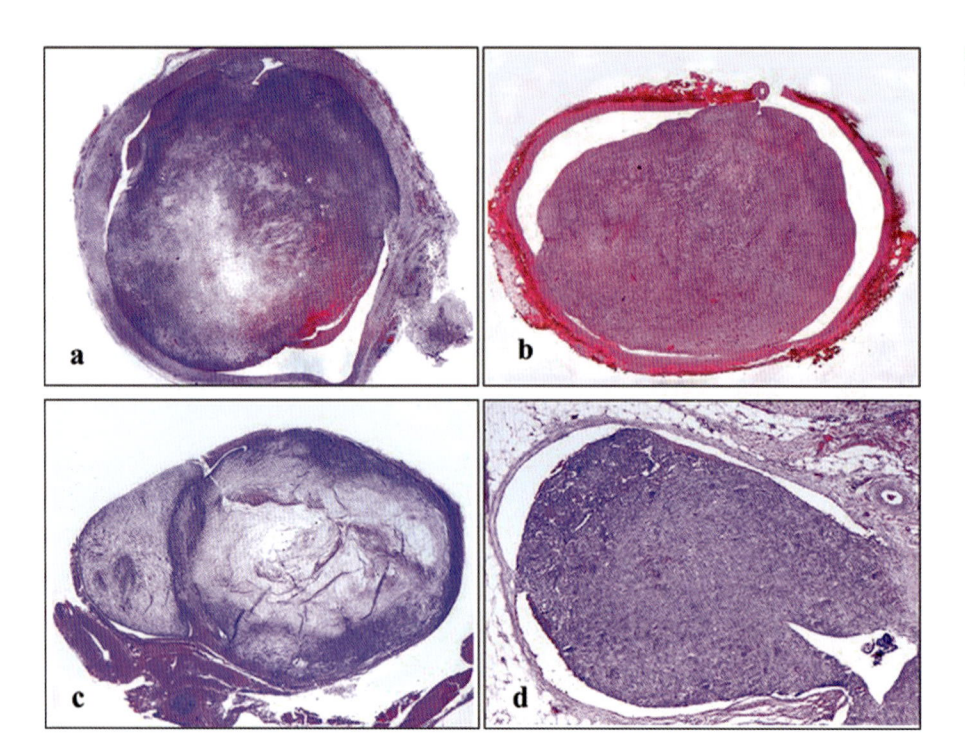

Figure 6.12 Four low-magnification examples of intravascular fasciitis growing inside small veins.

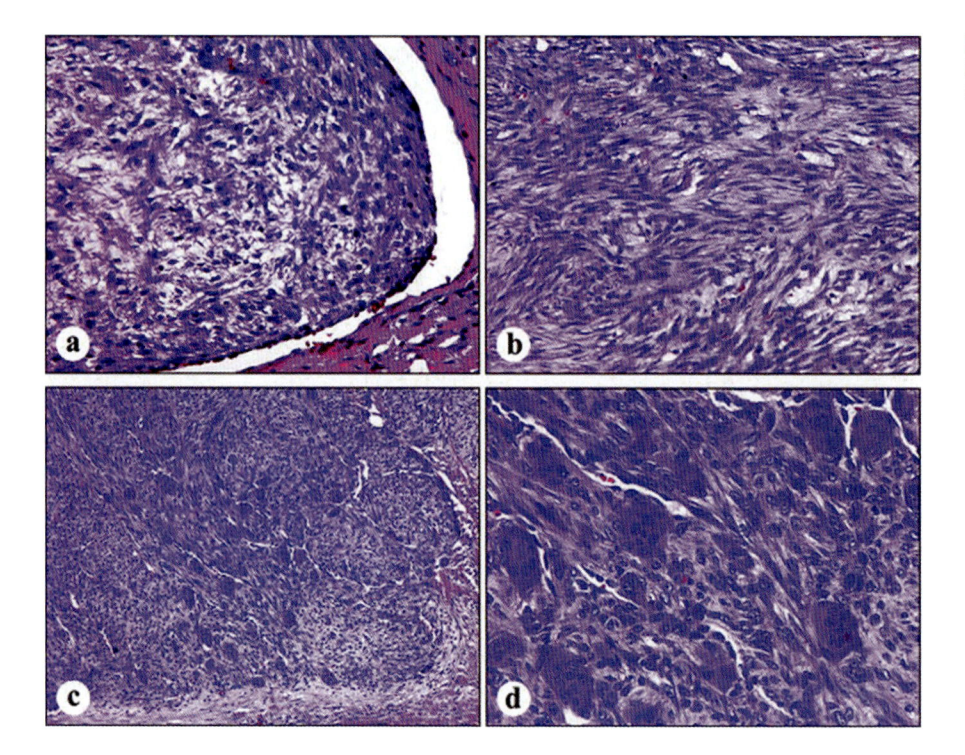

Figure 6.13 Intravascular fasciitis can be similar to typical nodular fasciitis (**a,b**), but some examples are rich in osteoclastic giant cells (**c,d**).

Immunohistochemically, the intravascular myofibroblastic proliferation in myointimoma is positive for smooth muscle actin, muscle actin, and calponin, and is negative for CD34 and S100 protein and desmin. The last marker highlights vascular smooth muscle typically surrounding the intravascular proliferation (Figure 6.15).

Proliferative fasciitis

The term *proliferative fasciitis* was coined by Chung and Enzinger in 1975[31] for subcutaneous fasciitis-like lesions with distinctive ganglion cell-like fibroblasts, which were similar to intramuscular tumefactions earlier reported as proliferative myositis.[32,33]

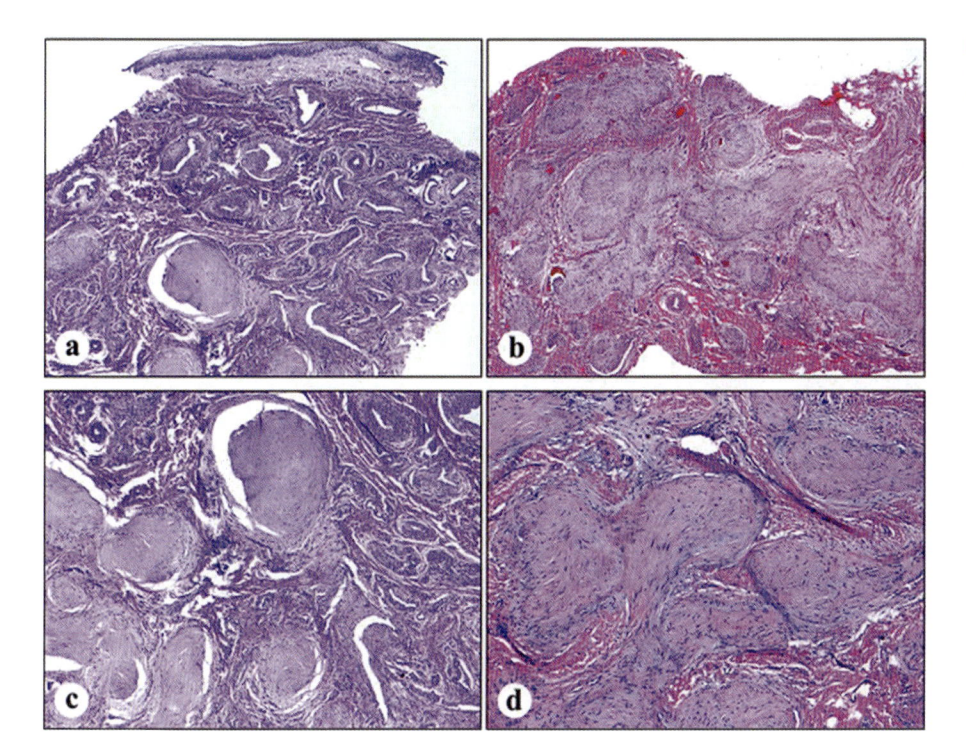

Figure 6.14 Penile myointimoma has a plexiform architecture, with the lesion being composed of multiple profiles of intravascular bundles of cytologically uniform myofibroblasts. The sharp confinement of the bundles reflects their intravascular location.

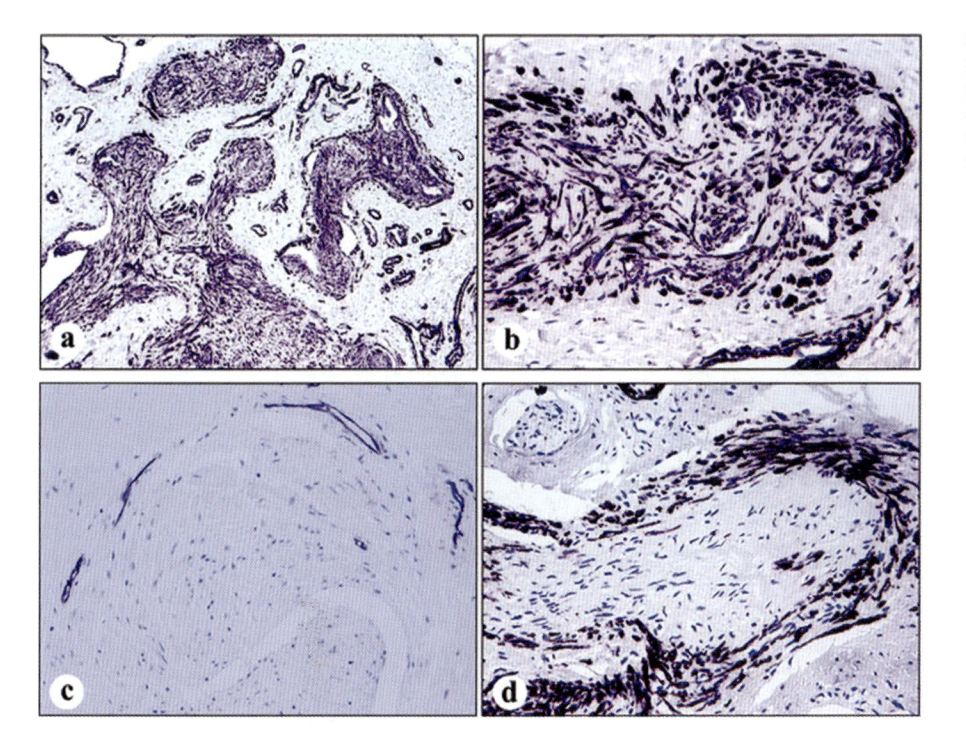

Figure 6.15 Immunohistochemical features of penile myointimoma. (**a,b**) The myofibroblasts are positive for smooth muscle actin. (**c**) Only the endothelial cells are CD34 positive. (**d**) The surrounding vascular smooth muscle elements only are desmin positive.

Clinical features

Proliferative fasciitis is much less common than nodular fasciitis, with a frequency of 6% of that of nodular fasciitis, based on AFIP files. In contrast to nodular fasciitis, proliferative fasciitis occurs predominantly in adults with a mild male predominance; 70% of patients are between 31 and 70 years of age, with a median age of 53 years (Figure 6.16A). Rare occurrence in children (7% of all cases) has been documented. The most commonly involved sites are the thigh, forearm, arm, leg, chest wall, and back, with occurrence in the distal extremities and head and neck being rare (Figure 6.16B). Approximately one-third of patients have a history of recent local trauma. The lesion grows rapidly and is often tender. Local excision is sufficient, with no tendency to recur.[31–35]

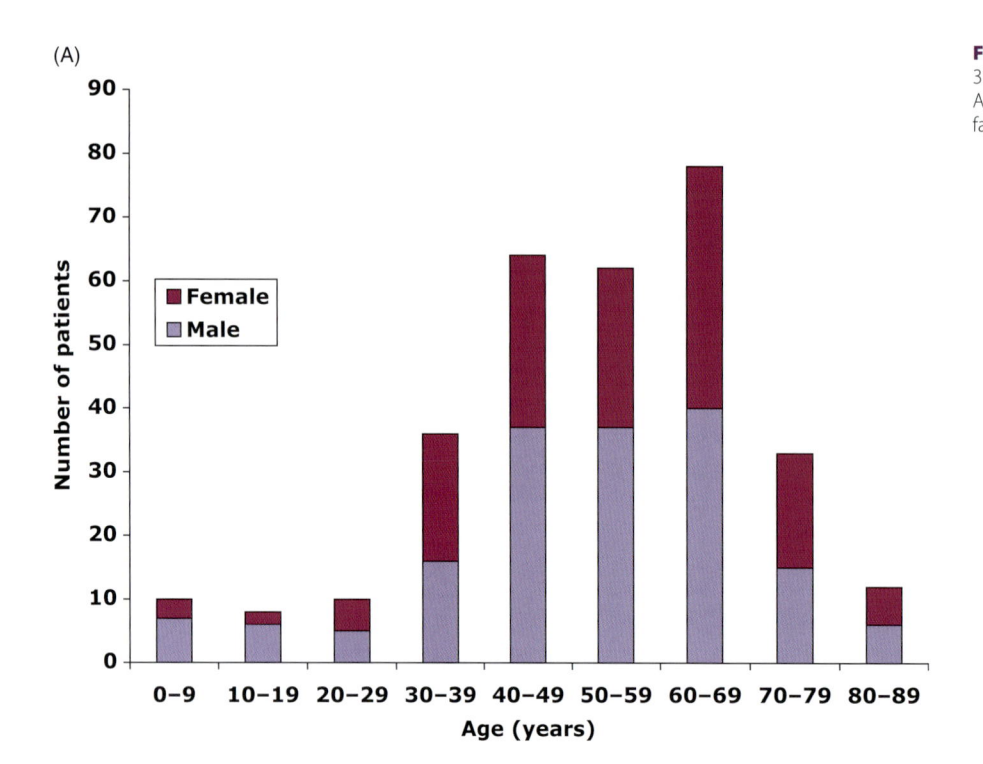

Chapter 6: Fibroblast biology, fasciitis, retroperitoneal fibrosis, and keloids

Figure 6.16 (A) Age and sex distribution of 313 patients with proliferative fasciitis. (B) Anatomic distribution of 303 cases of proliferative fasciitis.

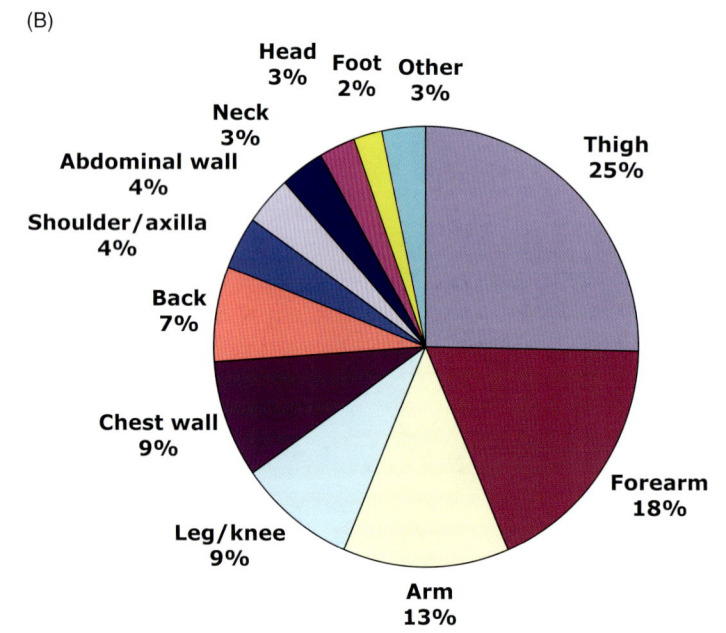

Pathology

Proliferative fasciitis forms a 1- to 4-cm poorly circumscribed, infiltrative, often elongated, oblong, or wedge-shaped mass expanding a septum of subcutaneous fat that is sometimes attached to the fascia (Figure 6.17). Grossly, the tumor is yellowish gray to tan, and often the margin is a ragged or spiculated, with the lesion typically infiltrating the fat.

Histologically, the lesion can be solid or contain a central, elongated cystic space, a "tract" that may be surrounded by arcades or neovascular capillaries. The cellular component is an admixture of bland spindle cells and ganglion-cell-like fibroblasts, and interstitial hemorrhage is often present (Figure 6.18). These cells are large, with abundant, basophilic bubbly cytoplasm and large nuclei with open chromatin and often with a prominent eosinophilic nucleolus (Figures 6.18 and 6.19). They can be binucleated or trinucleated, and forms resembling Reed–Sternberg cells are also sometimes present. The number of ganglion-like cells varies from a few scattered cells to most of the lesional cells. Mitotic activity varies from none to moderate (with several mitoses being present in a

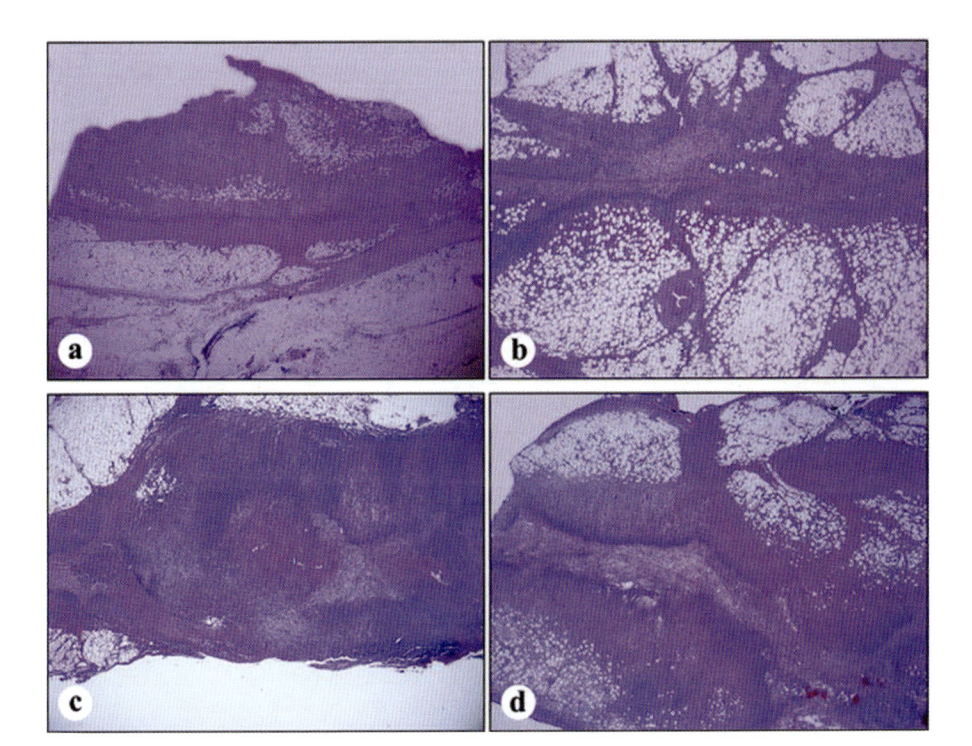

Chapter 6: Fibroblast biology, fasciitis, retroperitoneal fibrosis, and keloids

Figure 6.17 At low magnification, proliferative fasciitis lesions are typically poorly circumscribed and form elongated, sometimes centrally cystic tracts in the subcutaneous fat.

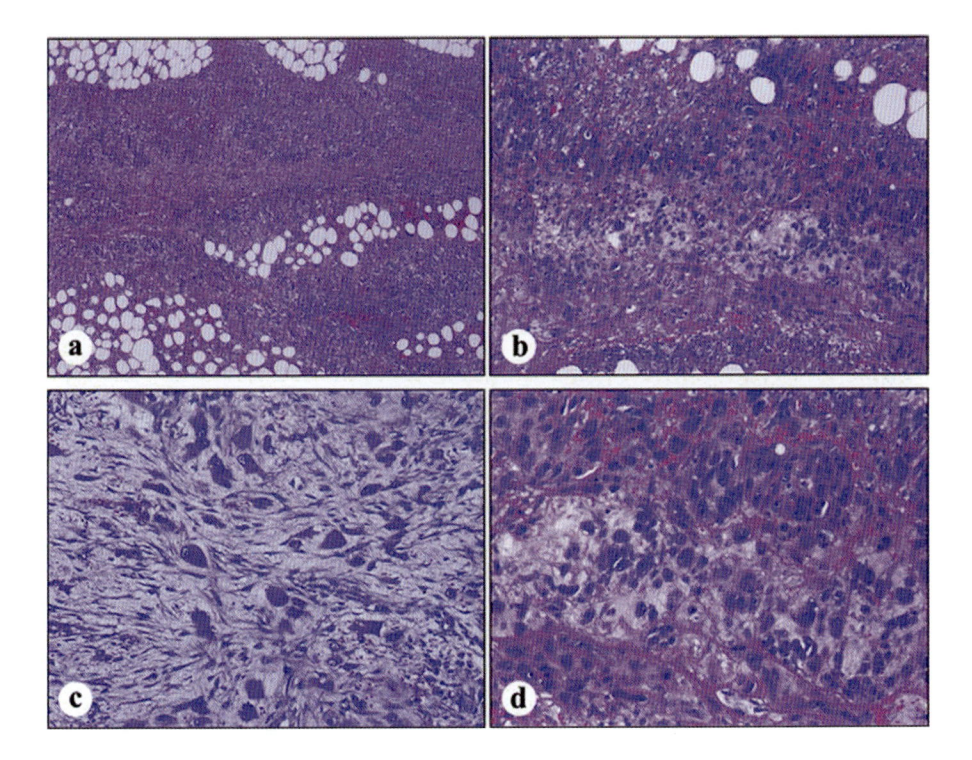

Figure 6.18 Proliferative fasciitis showing a central cavity rich in ganglion-like cells (a,b,d). A spindle cell component is present in case (c).

single HPF in some cases) and in rare cases, atypically appearing mitoses occur.

In some cases, notably lesions in children, proliferative fasciitis can be alarmingly cellular, with mitotic activity. When predominantly composed of ganglion-like cells, it should not be confused with rhabdomyosarcoma. Mitotic activity is also sometimes present in these cells (Figure 6.19).

Immunohistochemically the spindle cells and ganglion-like cells are negative for CD34 and desmin, but they can be positive for muscle actins and smooth muscle actin. The ganglion-like cells are typically negative for actins. However, they are often positive for keratin and calretinin, which should not lead to confusion with epithelial neoplasms such as malignant mesothelioma.[36]

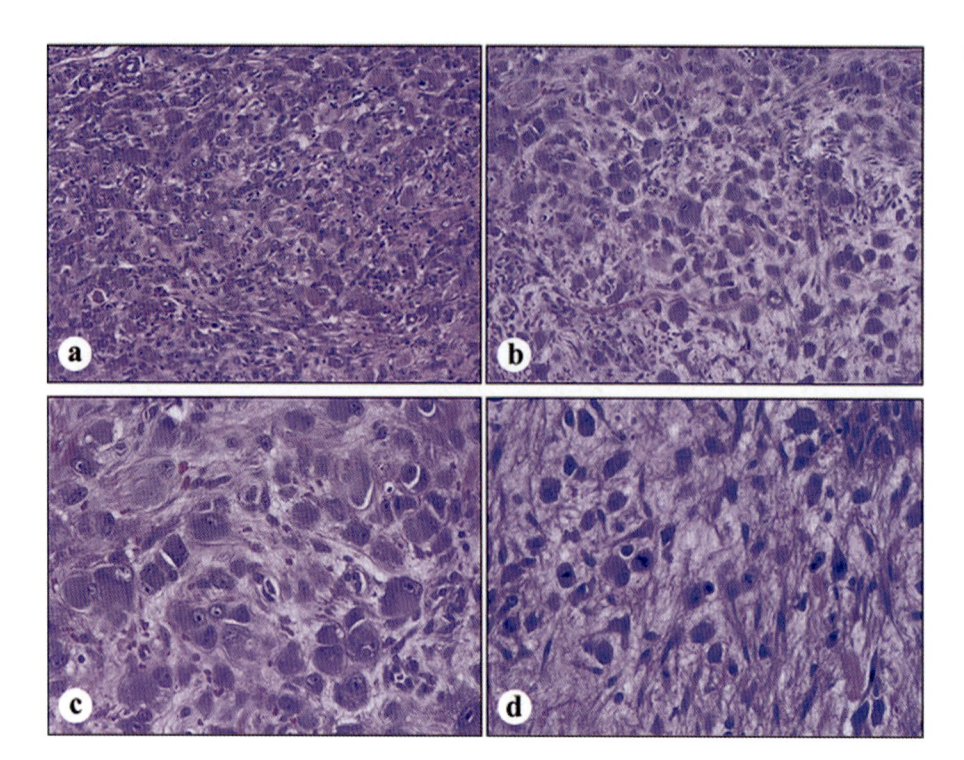

Figure 6.19 (a–c) Some proliferative fasciitis lesions, especially those in children, can be composed predominantly of ganglion-cell-like fibroblasts. (d) Two mitotic figures are seen to the left.

The differential diagnosis includes true histiocytic proliferations, and in some cases, metastatic carcinoma. In some cases, immunohistochemistry (CD163 for histiocytes; keratins and EMA for carcinomas) helps in the differential diagnosis.

Proliferative myositis

Proliferative myositis was reported by Kern *et al.* in 1960[32] as a putative reactive post-traumatic intramuscular lesion.

Clinical features

Proliferative myositis is the designation for an intramuscular lesion that shows histologic features similar to proliferative fasciitis. The lesion probably represents a reparative process of unknown origin and is benign and self-limited.

Similar to proliferative fasciitis, proliferative myositis occurs predominantly in middle-aged adults and has a mild male predominance. The most common sites of presentation are the muscles of the neck, shoulder, and upper trunk. Occurrence in the tongue has also been reported. The lesion presents as a rapidly growing, painless mass, and as such simulates a malignant tumor. Local excision is sufficient, and there is no tendency to recur.[32–34]

Pathology

Grossly the lesions are typically small, poorly delineated, pale, fibrous foci involving the skeletal muscle tissue and surrounding muscle fibers and fascia. Histologically the lesion typically expands the space between the muscle fibers and widens the connective tissue septa. In evolving lesions, foci of muscle cell degeneration and lymphoid and histiocytic infiltration are present. The early phase of the lesion, which is more cellular, shows components essentially similar to proliferative fasciitis, with spindle cells and large cytoplasmic ganglion-like cells (Figure 6.20); the latter have been shown to be modified fibroblasts by electron microscopy.[37,38] Long-standing lesions can be predominantly fibrous (Figure 6.20c). Mitoses are inconspicuous. Foci of metaplastic bone and cartilage can be present (Figure 6.20d). A resolved lesion sometimes develops into a paucicellular diagnostically nonspecific scar.

Differential diagnosis

The histologically unusual, large ganglion-like cells and cellular spindle cell component should not be confused with sarcoma; in fact, the gross appearance does not resemble a tumor. Skeletal muscle infiltration of low-grade sarcomas can resemble the histologic pattern of proliferative myositis in a biopsy specimen, however. In the differential diagnosis, one should pay attention to cellular atypia and radiologic or surgical evidence of a dominant tumor mass (almost always seen with true sarcomas).

Focal myositis is the designation for a nonspecific fibroinflammatory lesion of skeletal muscle. This condition can represent a reparative phase of a muscle rupture. Histologically there is a mixture of evolving fibrosis with lymphoid infiltration and degenerative skeletal muscle fibers. No ganglion-like cells are present, in contrast to proliferative myositis.

Genetics

One report has shown clonal chromosomal abnormalities in proliferative fasciitis, and another report has shown this in

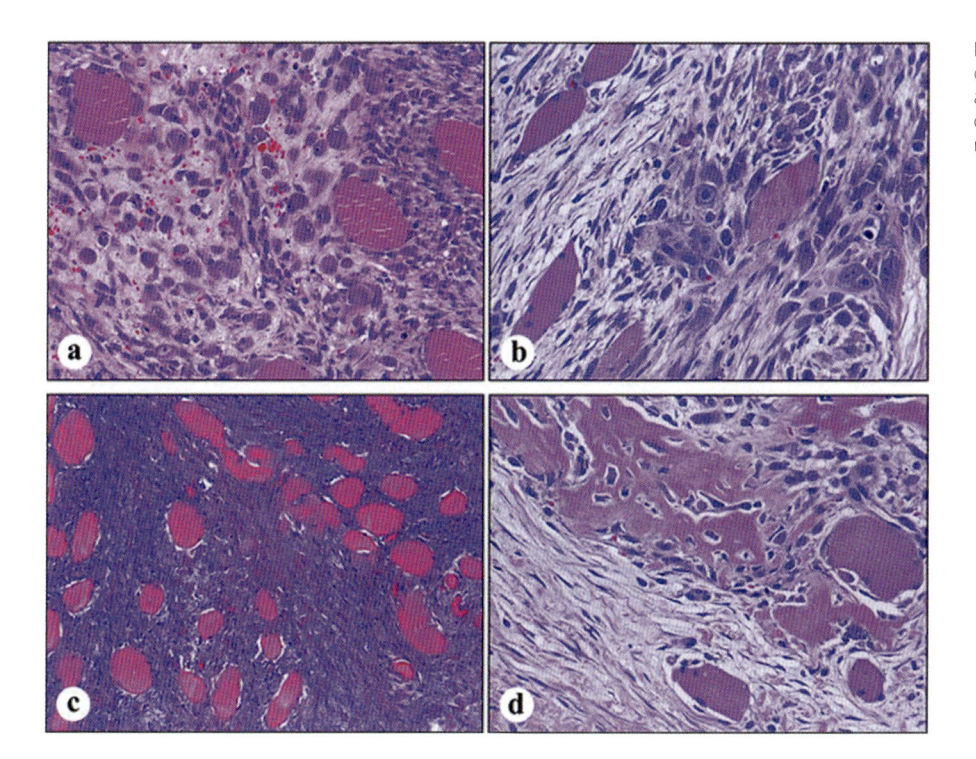

Figure 6.20 (a,b) Proliferative myositis lesion contains numerous ganglion cell-like fibroblasts and infiltrates in the skeletal muscle. (c) Fibrosis dominates in long-standing lesions. (d) Foci of metaplastic bone are present in some cases.

myositis. The former report showed trisomy of chromosome 2 as the sole abnormality,[39] and the latter had a translocation t(6;14)(q23;q32) without numeric chromosomal aberrations.[40]

Ischemic fasciitis (atypical decubital fibroplasia)

Clinical features

This presumably reactive, pseudoneoplastic fibroblastic proliferation was described almost simultaneously by two previously cited reports by Montgomery *et al.* and Perosio and Weiss.[41,42] The process probably represents a faulty reparative fibrovascular situation in an ischemic environment. It typically occurs in older, sometimes bedridden patients with a median patient age of close to 80 years, and there is a mild male predominance (Figure 6.21A). This condition also can occur in elderly patients without an underlying condition, suggesting that it could be an age-specific histologic reaction, perhaps related to fragility of atrophic tissues. Conversely, occurrence in young patients with underlying disorders (melorheostosis and long-term use of crutches) has also been reported.[43] Ischemic fasciitis has a predilection for soft tissues over bony prominences, and the most common locations are hip, shoulder, back, and chest wall (Figure 6.21B).

Pathology

The lesion bulges underneath an overlying hyperemic skin, but the skin is rarely ulcerated. The lesion is subcutis based but can extend to the fascia and periosteum, typically ranging from 1 to 8 cm in diameter. Larger lesions are often centrally cystic, containing fibrinous material.

Histologically there is vague lobulation by necrotic subcutaneous fat infiltrated by atypical fibromyxoid, zones of vascular proliferation, and central pools of fibrin (Figures 6.22 and 6.23). The lesion contains scattered, atypical, bipolar spindled or epithelioid fibroblasts with abundant basophilic cytoplasm and smudged nuclei with prominent nucleoli. These cells have some resemblance to the ganglion-like cells of proliferative fasciitis, and they can have some mitotic activity, even atypical-appearing mitoses (Figure 6.23d). Prominent hemangioma-like clusters and zones of newly formed capillaries bordering necrosis are typical of this condition and could reflect the fact that tissue ischemia is a powerful angiogenic stimulus (Figures 6.22 and 6.23). The stromal cells can be positive for CD34 and muscle actins but negative for desmin and S100 protein.

Although thought to be non-neoplastic, these lesions can show alarming histologic features and could therefore be confused with neoplasms such as myxoid MFH. The presence of a fibrinous cavity, arcades of capillary proliferation, and lack of nuclear pleomorphism, as well as the clinical setting, could help in this distinction.

Eosinophilic fasciitis

Eosinophilic fasciitis is a rare, diffuse fascial inflammatory disease that is considered to be related to autoimmune collagen vascular diseases, such as scleroderma, although it is considered a separate clinical entity in which fascial pathology predominates. It is also known by the eponym *Shulman syndrome*, for the author of the first report on this condition.[44] In contrast to the forms of fasciitis discussed before, it does not form a localized mass, but diffusely involves fascial planes.

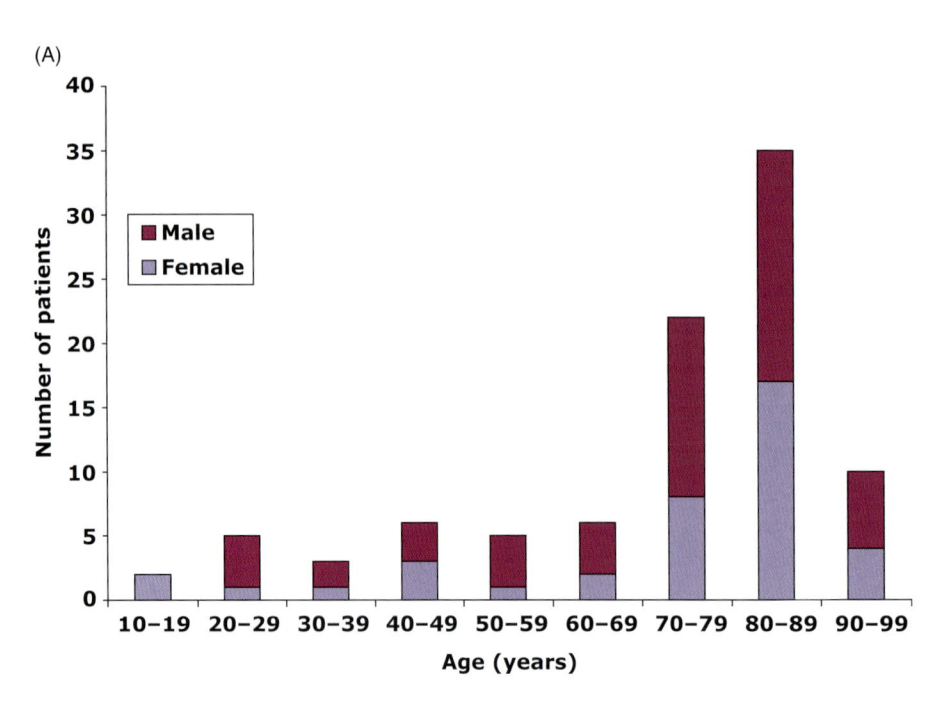

Figure 6.21 (A) Age and sex distribution of 94 patients with ischemic fasciitis. (B) Anatomic location of 93 cases of ischemic fasciitis.

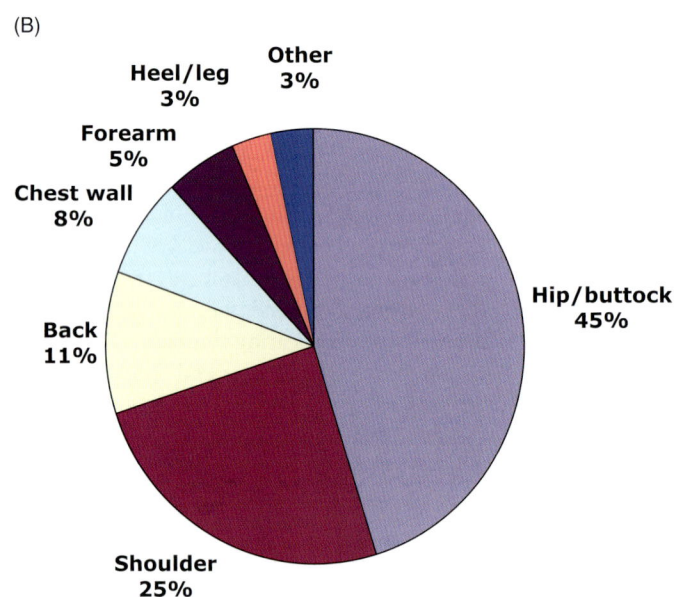

Other reasons for tissue eosinophilia, such as systemic parasitic infestation, have to be ruled out.

Clinical features

Eosinophilic fasciitis occurs mainly in middle-aged adults, presenting with painful swellings in legs and arms, peripheral blood eosinophilia, hypergammaglobulinemia, and elevated erythrocyte sedimentation rate. Patients often develop motion limitations as a consequence of fascial fibrosis, sometimes including flexion contractures of extremities. Other clinical signs include skin puckering or "orange-peel" appearance and loss of skin retractability.[44–48]

The condition is idiopathic, but acute onset after strenuous exercise has led to the suggestion that autoimmune reaction to injured fascia could be a causal factor. This syndrome has clinicopathological similarities to the eosinophilia-myalgia syndrome caused by contaminated L-tryptophan preparations in the 1980s.[48]

Proper biopsy for this condition must include fascial tissue, and therefore conventional dermatologic punch biopsies are generally insufficient. Systemic corticosteroid treatment is usually effective.

Pathology

Histologic features vary according to the stage of disease. The typical changes include fascial thickening with edema and fibrin deposition, and focal, often band-like capillary

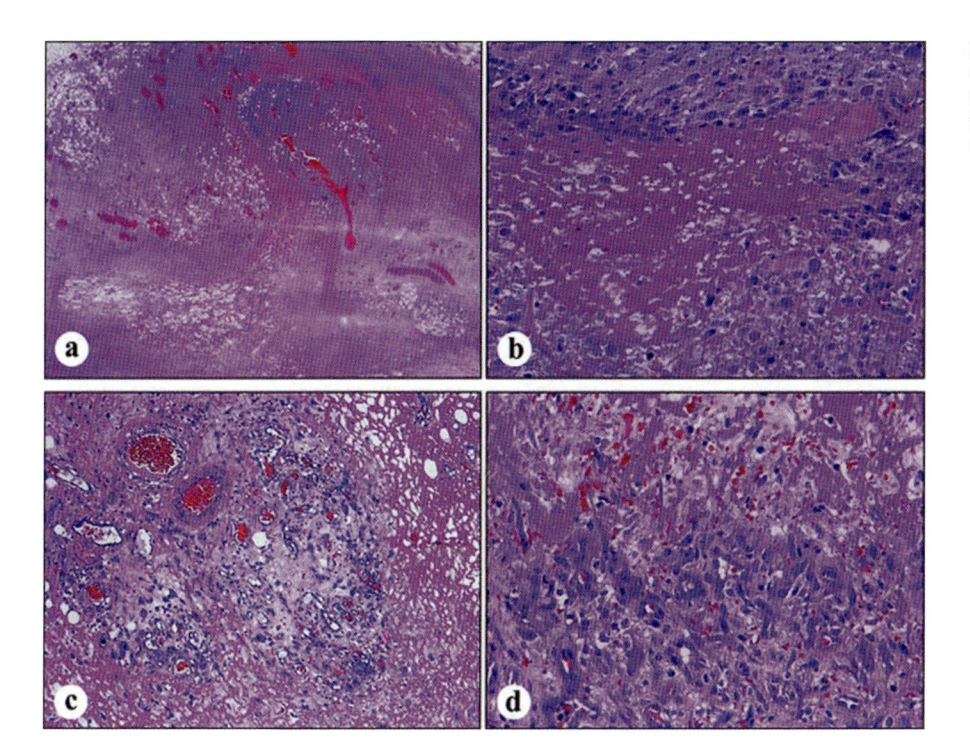

Figure 6.22 (a–c). Ischemic fasciitis lesion containing a large central fibrin-filled cavity surrounded by slightly atypical epithelioid fibroblasts. (c,d) Presence of arcades of prominent capillary proliferation is a typical feature.

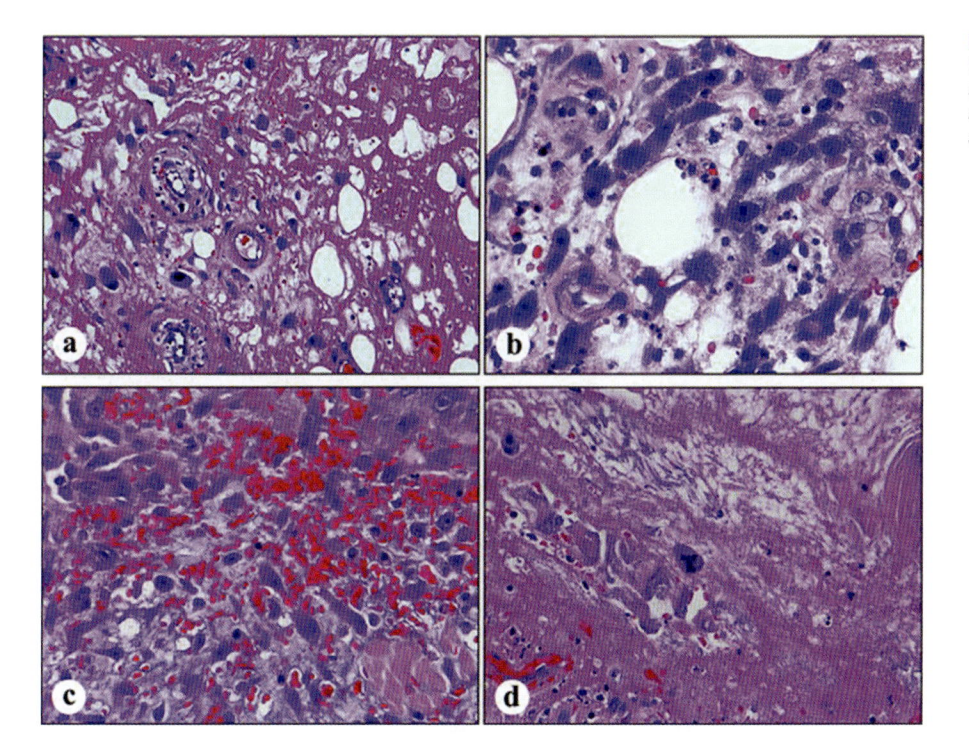

Figure 6.23 (a–c) The atypical fibroblasts often have basophilic moderately abundant cytoplasm and moderately prominent nucleoli. (d) Mitotic activity can be present, including occasional atypical-appearing mitotic figures.

neovascularization and infiltration of lymphocytes, plasma cells, histiocytes, and variably eosinophils. Pronounced fibrosis can develop in later stages of the lesion (Figure 6.24). In some forms of this disorder, however, tissue eosinophilia is focal or absent, possibly as a result of corticosteroid treatment. Perineurial inflammation and skeletal muscle involvement can occur. Chronic inflammation can extend to the septa of the subcutis. Loss of skin adnexa can occur, similar to scleroderma, but dermal fibrosis is not a feature.

Necrotizing fasciitis

Necrotizing fasciitis is a life-threatening, invasive septic infection involving deep soft tissues and caused by various

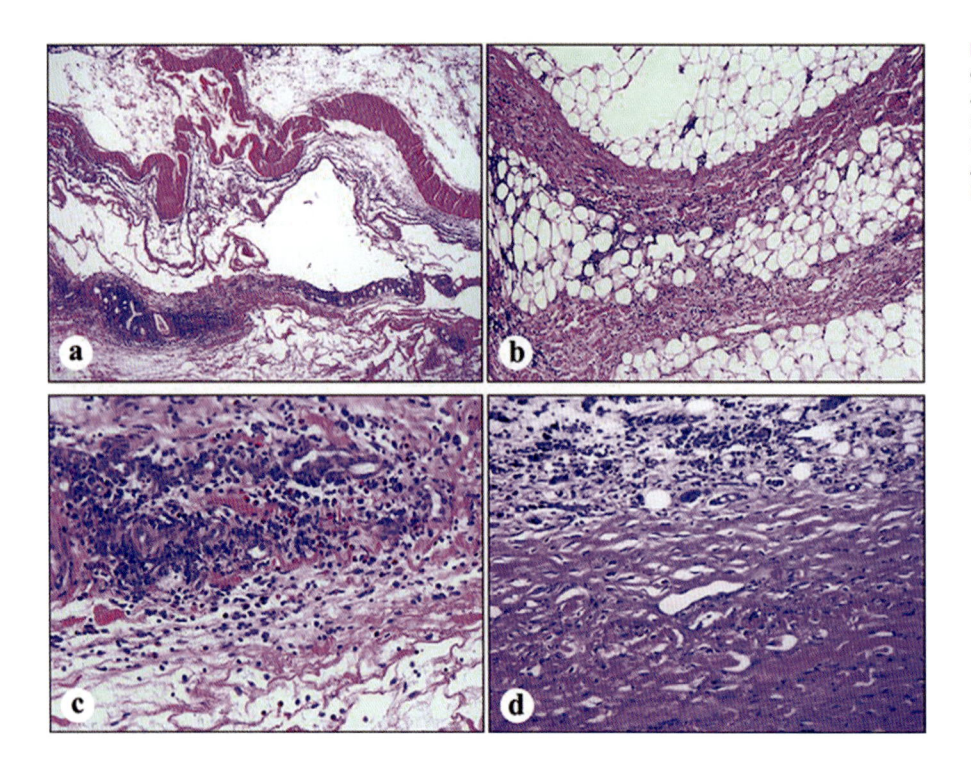

Figure 6.24 (a,b) Low magnification of eosinophilic fasciitis shows thickening of the fascia and fat septa with inflammatory infiltration. (c,d) High magnification shows mixed inflammatory infiltrate rich in eosinophils in edematous fascia, and prominent fibrosis can often be found in older lesions.

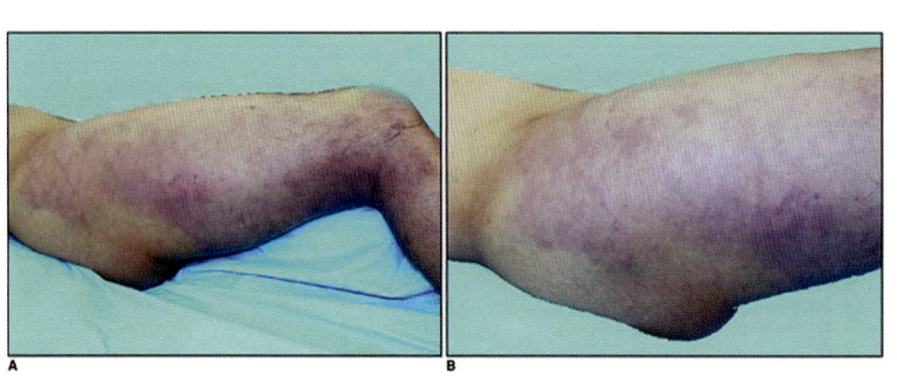

Figure 6.25 Necrotizing fasciitis can appear clinically as diffuse swelling and bluish discoloration of the involved area. In this patient, the lesion extended from below the knee into the proximal thigh, leading to the loss of the entire lower extremity after attempted fasciectomy and débridement.

microorganisms. Like eosinophilic fasciitis, it does not form localized mass lesions.

Clinical features

This condition occurs in a wide age range from newborn to the elderly, and it is more common in men. It is most common in middle-aged subjects, but there are smaller peaks of occurrence in newborns and the elderly. Predisposing factors include malnutrition, diabetes, medically induced (or rarely AIDS-associated) immunosuppression, and trauma. Sometimes this infection develops as a surgical complication. A more localized infection, such as dental infection or cellulites, can also lead into necotizing fasciitis.[49,50]

Necrotizing fasciitis occurs in a variety of body sites, most commonly in the abdominal wall (children) and lower extremities, especially in adults. Necrotizing fasciitis involving the scrotum and surrounding tissues is referred to as *Fournier's gangrene*. Other sites, such as scalp and chest wall, can also be involved. Septic shock commonly ensues, and mortality rates are high, despite systemic antibiotics, surgical débridement, and possibly amputation. Antibiotic resistance often contributes to the therapeutic failure. A mortality rate of 30% was cited in a recent review.[50]

The microbiologic background is heterogeneous, but it most commonly includes streptococci (especially group A) and *Staphylococcus aureus*. Other bacteria implicated in this disease are enterococci and *Clostridium difficile*, or the etiology can be polymicrobial. Fungal organisms, such as *Candida, Cryptococcus, Histoplasma*, or others, are also possible culprits. Gas gangrene caused by *Clostridium perfringens* also causes an invasive infection of deep soft tissues, but is often classified separately from necrotizing fasciitis.

Pathology

Clinically the involved area is swollen and erythematous (Figure 6.25). The subcutaneous tissue, fascia, and skeletal muscle are edematous with red discoloration, often with a necrotic appearance.

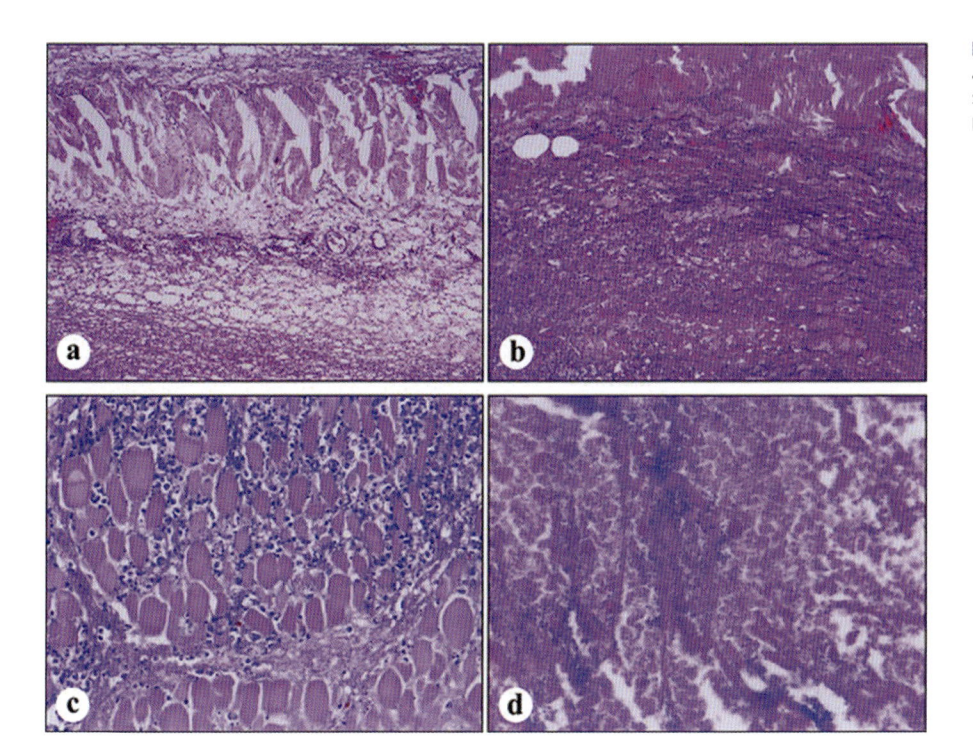

Figure 6.26 (a–c) Necrotizing fasciitis shows acute inflammatory infiltration in the fascia and skeletal muscle. (d) Numerous coccobacteria are present in the Gram's stain.

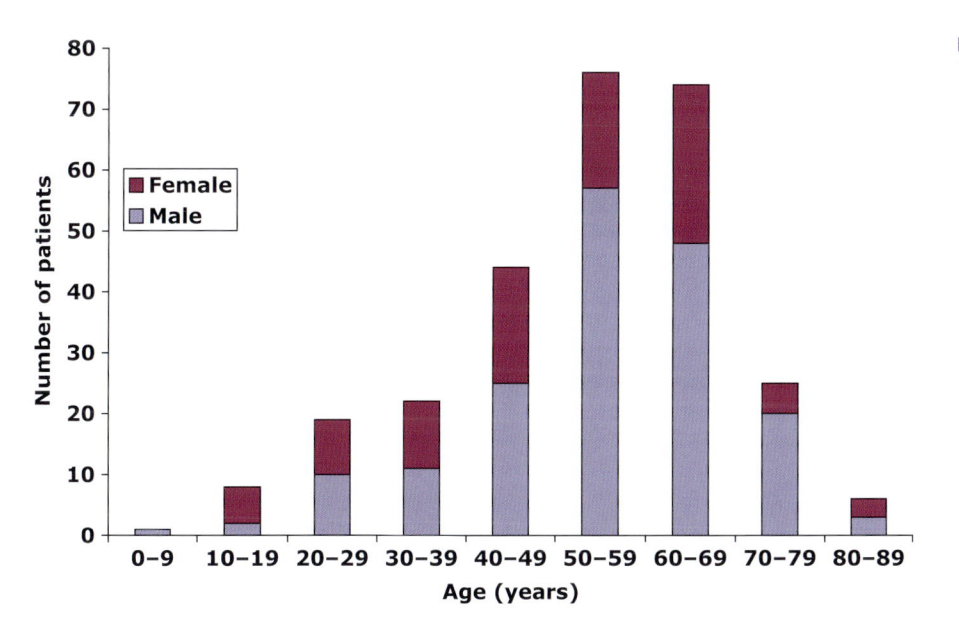

Figure 6.27 Age and sex distribution of 167 patients with retroperitoneal fibrosis.

Histologic findings are nonspecific and usually include massive neutrophil infiltration and necrosis of subcutaneous fat, fascia, and skeletal muscle (Figure 6.26). Gram's stain or other bacterial stains can show confluent bacterial colonies in the involved tissue.

Retroperitoneal fibrosis (idiopathic retroperitoneal fibrosis, Ormond's disease)

Retroperitoneal fibrosis is also known by the eponym, for J. K. Ormond, a urologist who described the first patients in 1948.[51] The term *idiopathic* sometimes precedes the disease name. Because different specific (aortic atherosclerosis/aneurysm, collagen vascular diseases) and unknown causes can be associated with similar pathology, however, *idiopathic* is often omitted from the diagnostic name. Sclerosing malignancies such as lymphoma, carcinoma, carcinoid syndrome, and specific infections, such as histoplasmosis, must be ruled out before this diagnosis is made.

Clinical features

This condition usually occurs in middle-aged and older adults, with a 3:1 male predominance, but it has been also reported in children (Figure 6.27).[51–55] Systemic symptoms, such as fever,

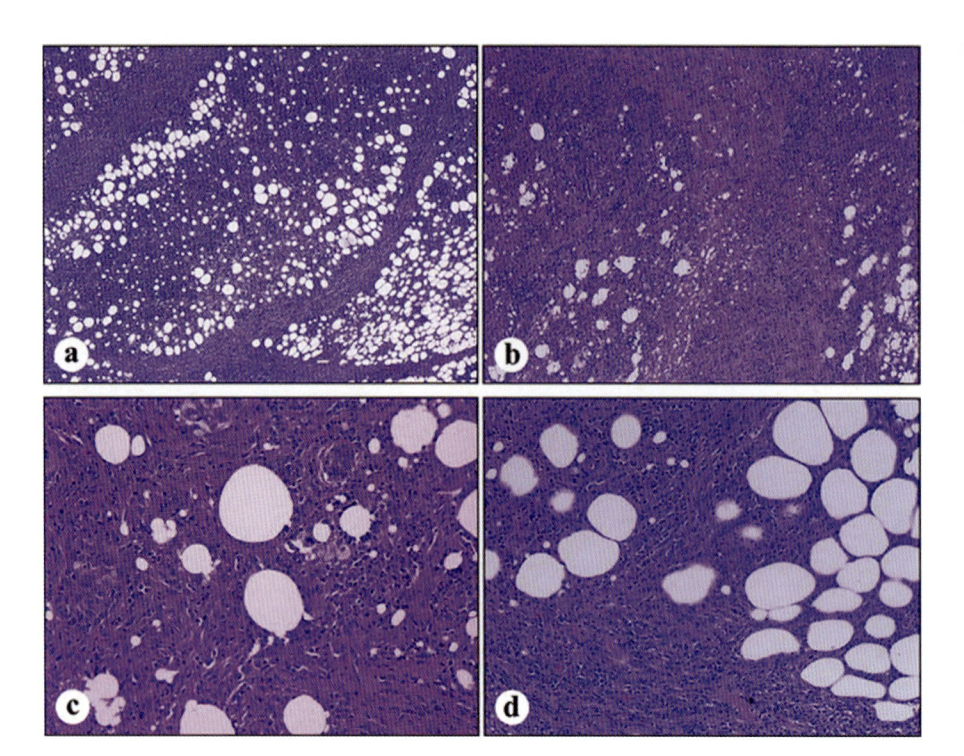

Figure 6.28 Retroperitoneal fibrosis with predominantly inflammatory features. Note vague lobulation, permeative growth in the fat, and fat necrosis, and lymphoid infiltration with germinal centers, xanthoma cells, eosinophils, and plasma cells.

elevated erythrocyte sedimentation rate, and lumbar back or abdominal pain, are common. Clinically distinctive is ureteral obstruction, which prior to modern diagnostic modalities often led to renal failure. Etiology is unknown, but an autoimmune cause has been suspected. Similar lesions have been associated with causative factors, drugs (e.g., methysergide used for migraine treatment), and therapeutic radiation. Retroperitoneal fibrosis is often considered among collagen vascular diseases, and some cases are associated with systemic lupus erythematosus or rheumatoid arthritis.

The process often begins as a plaque-like inflammatory cuffing around the aorta. Later it draws the ureters toward the midline, ultimately encasing them, which has the potential to cause obstructive renal damage. By comparison, retroperitoneal tumors often displace the ureters laterally.

Inflammatory aneurysm of the aorta and perianeurysmal retroperitoneal fibrosis are thought to be part of the spectrum of retroperitoneal fibrosis, and immunologic reaction to atherosclerotic aortic lesions has been proposed as a cause. Idiopathic mediastinal fibrosis probably represents a similar disease in a thoracic location and can cause superior vena cava obstruction. Mediastinal fibrosis may also occur in younger adults.[56]

The most important treatment modalities include surgical relief of ureteral obstruction, systemic corticosteroids and other immunosuppressive treatments, tamoxifen, and aortic aneurysm repair in some cases. Because the lesions typically grow between and around the great vessels and ureters, surgical excision often is not possible. These lesions more commonly are biopsied for diagnosis only.

Pathology

In the abdomen, the lesion often presents as a seemingly circumscribed mass on the sacral promontorium, but grows between the great vessels. The process often forms a mat-like extension along the ureters up to the kidneys, extending from 2 cm to 5 cm deep (anteroposterior dimension). The lesional tissue is typically biopsied as single or multiple plaque-like strips or nodules.

Histologically there is a variably dense, fibrosclerosing process infiltrating fat (Figure 6.28). The fibrous component contains fibroblasts and myofibroblasts, but the fibrous tissue is typically less cellular than that seen in the desmoplastic stroma of carcinomas (Figure 6.29). The inflammatory component includes lipophages and other histiocytes, plasma cells, eosinophils, and clusters of lymphocytes, often with focal germinal center formation. In older lesions, fibrosis predominates, and cellular infiltration is less conspicuous.[57]

Imunohistochemically the lymphoid component contains B-cells forming germinal centers and scattered within the patches of diffuse lymphoid infiltrates, among which T-cells predominate. The plasma cell population is polyclonal, containing comparable numbers of kappa and lambda light-chain-positive cells. A conspicuous SMA-positive myofibroblastic population is typically present. Also, the lesions are rich in histiocytes,[58] and the number of CD163-positive histiocytes is typically far larger than expected from histologic review.

Differential diagnosis

Caution is needed in interpretation of small biopsies, because the diagnosis of idiopathic retroperitoneal fibrosis requires exclusion of specific causes such as infections (e.g., histoplasmosis,

Chapter 6: Fibroblast biology, fasciitis, retroperitoneal fibrosis, and keloids

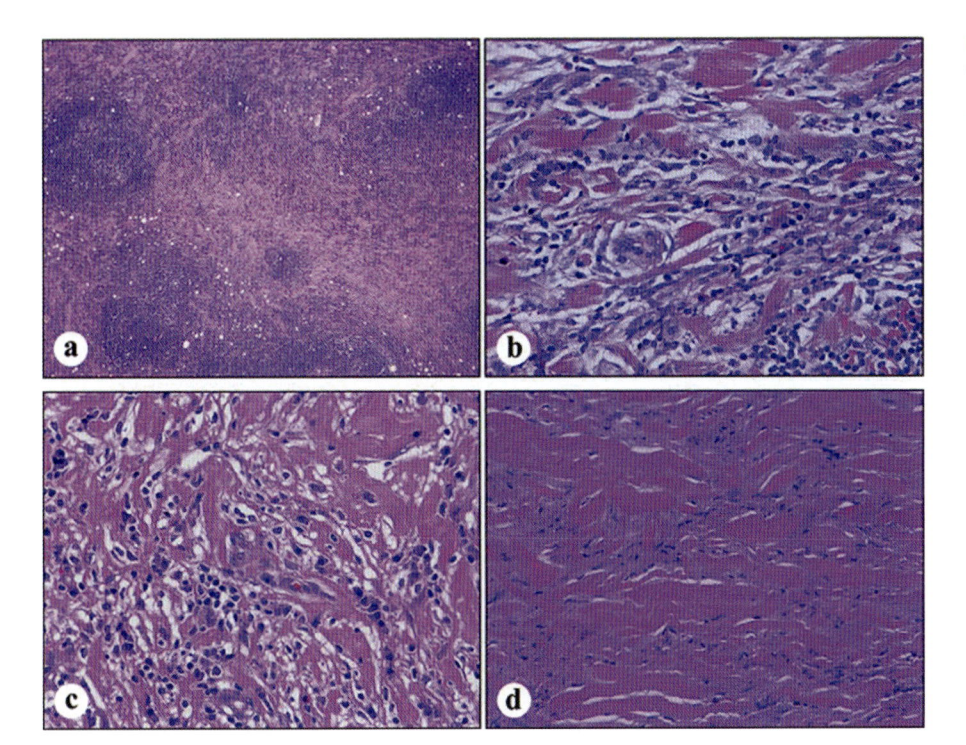

Figure 6.29 (a–d) Retroperitoneal fibrosis with evolving fibrosis. Inflammatory background with mixed lymphocytic, plasma cell, and eosinophilic infiltrate coexists with evolving fibrosis. (c,d) In a late phase, fibrosis predominates over inflammatory component.

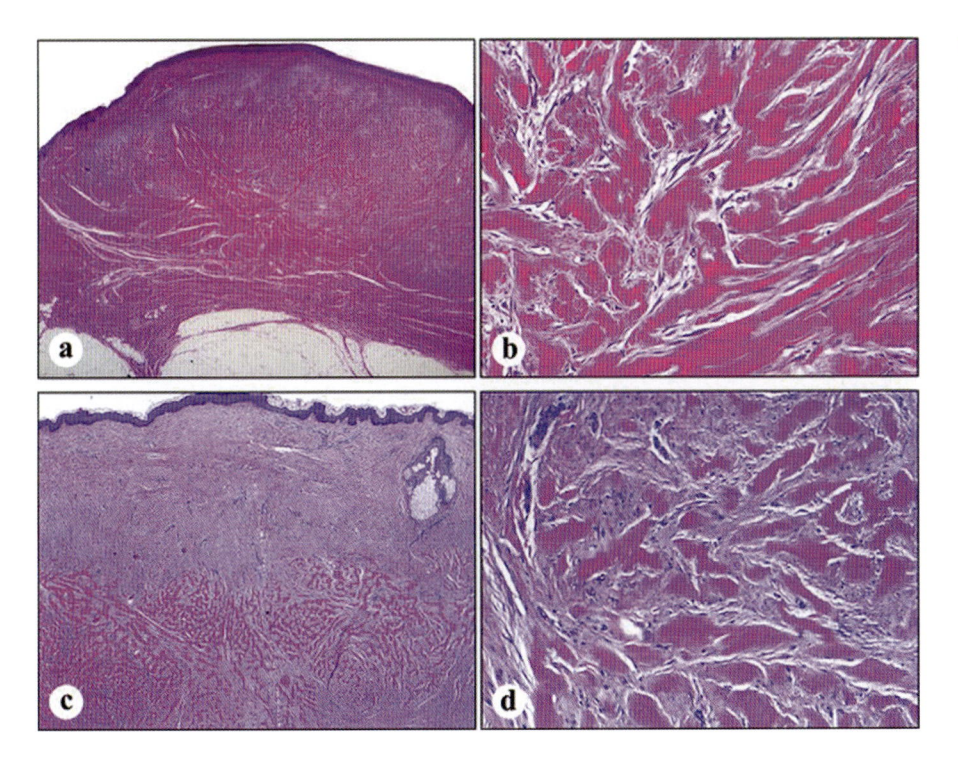

Figure 6.30 A keloid can form a slightly elevated band-like dermal lesion or a dome-shaped nodule. Keloids contain thick, glassy collagen fibers surrounded by scant myofibroblasts.

actinomycosis) and sclerosing neoplasms, and clinical or radiologic correlation is necessary. Sclerosing (follicular) lymphoma, mesothelioma, or metastatic carcinoma (prostatic, sclerosing carcinoid) must be ruled out; keratin immunostaining is recommended, especially if there is any suspicion of an epithelial component. Other markers such as CD3 and CD20 are often indicated to evaluate lymphoid populations and rule out lymphoma. There is some overlap with the IgG4 sclerosing disease. However, most cases of retroperitoneal fibrosis do not meet the stringent criteria of IgG4 sclerosing disease in terms of the relative content of IgG4-plasma cells (>40% of IgG4-positive plasma cells of all IgG-positive plasma cells).[59]

The diffuse nature, lack of atypical cellular components, and cellular heterogeneity differentiate this condition from fibromatosis, sclerosing (inflammatory) liposarcoma, and inflammatory myofibroblastic tumor.

Keloids

Clinical features

A *keloid* is a reactive, hyperplastic fibrous overgrowth representing a form of hypertrophic scar. Keloids typically present in surgical and other scars and sites of recent trauma, such as in body-piercing sites; they represent a pathologic wound-healing process with excessive collagen formation. Keloids predominantly occur in young adults and are more common in blacks. Chronic inflammation is thought to be etiologically significant, probably by means of local excess of fibrosis, promoting cytokines such as TGF-β. Recurrence following excision is common, probably reflecting the persistence of the causative stimulus.[60,61]

Pathology

Grossly, keloid specimens are firm and resilient to rubbery on sectioning, sometimes having a trabeculated or whorled surface. Histologically, keloids are composed of thick, deeply eosinophilic "waxy" collagen fibers surrounded by scattered hyperplastic fibroblasts or myofibroblasts (Figure 6.30); this component separates a keloid from a hypertrophic scar. The vascular pattern can be prominent with focal perivascular lymphoplasmacytic infiltration, but inflammation is not always present.

References

Introduction to fibroblastic tumors and fibroblast biology

1. McKeehan WL, Wang F, Kan M. The heparan sulfate-fibroblast growth factor family: diversity of structure and function. *Prog Nucleic Acid Res Mol Biol* 1998;59:135–176.
2. Eckes B, Kessler D, Aumailley M, Krieg T. Interactions of fibroblasts with the extracellular matrix: implications for the understanding of fibrosis. *Springer Semin Immunopathol* 1999;21:415–429.
3. Schurch W, Seemayer TA, Gabbiani G. The myofibroblast: a quarter century after its discovery. *Am J Surg Pathol* 1998;22:141–147.
4. Skalli O, Schurch W, Seemayer T, et al. Myofibroblasts from diverse pathologic settings are heterogeneous in their content of actin isoforms and intermediate filament proteins. *Lab Invest* 1989;60:275–285.
5. Nickoloff BJ. CD34. The human progenitor cell antigen (CD34) is localized on endothelial cells, dermal dendritic cells, and perifollicular cells in formalin-fixed normal skin, and on proliferating endothelial cells and stromal spindle-shaped cells in Kaposi's sarcoma. *Arch Dermatol* 1991;127:523–529.
6. Burgeson RE, Nimni ME. Collagen types: Molecular structure and tissue distribution. *Clin Orthop Relat Res* 1992;282:250–272.
7. Shaw LM, Oslen BR. FACIT collagens: diverse molecular bridges in extracellular matrices. *Trends Biochem Sci* 1991;16:191–194.
8. Kuivaniemi H, Tromp G, Prockop DJ. Mutations in fibrillar collagens (types I, II, III, and XI), fibril-associated collagen (type IX), and network-forming collagen (type X) cause a spectrum of diseases of bone, cartilage, and blood vessels. *Hum Mutat* 1997;9:300–315.
9. Ornitz DM, Itoh N. Fibroblast growth factors. *Genome Biol* 2001;2:1–12.
10. Eswarakumar VP, Lax I, Schlesinger J. Cellular signaling by fibroblast growth factor receptors. *Cytokine Growth Factor Rev* 2005;16:139–149.

Nodular fasciitis

11. Konwaler BE, Keasbey L, Kaplan L. Subcutaneous pseudosarcomatous fibromatosis (fasciitis). *Am J Clin Pathol* 1955;25:241–252.
12. Culberson JD, Enterline HT. Pseudosarcomatous fasciitis: a distinctive clinical-pathologic entity: report of five cases. *Ann Surg* 1960;151:235–240.
13. Price EB Jr, Silliphant WM, Shuman R. Nodular fasciitis: a clinicopathologic analysis of 65 cases. *Am J Clin Pathol* 1961;35:122–136.
14. Allen PW. Nodular fasciitis. *Pathology* 1972;4:9–26.
15. Dahl I, Jarlstedt J. Nodular fasciitis in the head and neck: a clinicopathological study on 18 cases. *Acta Otolaryngol* 1980;90:152–159.
16. Bernstein KE, Lattes R. Nodular (pseudosarcomatous) fasciitis, a nonrecurrent lesion: clinicopathologic study of 134 cases. *Cancer* 1982;49:1668–1678.
17. Shimizu S, Hashimoto H, Enjoji M. Nodular fasciitis: an analysis of 250 patients. *Pathology* 1984;16:161–166.
18. Kwittken J, Branche M. Fasciitis ossificans. *Am J Clin Pathol* 1969;51:251–255.
19. Wirman JA. Nodular fasciitis, a lesion of myofibroblasts: an ultrastructural study. *Cancer* 1976;38:2378–2389.
20. Montgomery EA, Meis JM. Nodular fasciitis: its morphologic spectrum and immunohistochemical profile. *Am J Surg Pathol* 1991;15:942–948.
21. Perez-Montiel MD, Plaza JA, Dominguez-Malagon H, Suster S. Differential expression of smooth muscle myosin, smooth muscle actin, h-caldesmon, and calponin in the diagnosis of myofibroblastic and smooth muscle lesions of skin and soft tissue. *Am J Dermatopathol* 2006;28:105–111.
22. Chan JK, Carcangiu ML, Rosai J. Papillary carcinoma of thyroid with exuberant nodular fasciitis-like stroma: report of three cases. *Am J Clin Pathol* 1991;95:309–314.
23. Erickson-Johnson MR, Chou MM, Evers BR, et al. Nodular fasciitis: a novel model of transient neoplasia induced by MYH9-USP6 gene fusion. *Lab Invest* 2011;91:1427–1433.
24. Sawyer JR, Sammartino G, Baker GF, Bell JM. Clonal chromosome aberrations in a case of nodular fasciitis. *Cancer Genet Cytogenet* 1994;76:154–156.
25. Weibolt VM, Buresh CJ, Roberts CA, et al. Involvement of 3q21 in nodular fasciitis. *Cancer Genet Gytogenet* 1998;106:177–179.
26. Donner LR, Silva T, Dobin SM. Clonal rearrangement of 15p11.2, 16p11.2, 16p13.3 in a case of nodular fasctiitis: additional evidence favoring nodular fasciitis as a benign neoplasm and not a reactive tumefaction. *Cancer Genet Cytogenet* 2002;139:138–140.
27. Velagaleti GV, Tapper JK, Panova NE, Miettinen M, Gatalica Z. Cytogenetic

findings in a case of nodular fasciitis of subclavicular region. *Cancer Genet Cytogenet* 2003;141:160–163.

Intravascular fasciitis and penile myointimoma

28. Patchefsky AS, Enzinger FM. Intravascular fasciitis: a report of 17 cases. *Am J Surg Pathol* 1981;5:29–36.
29. Fetsch JF, Brinsko RW, Davis CJ, Mostofi FK, Sesterhenn IA. A distinctive myointimal proliferation ("myointimoma") involving the corpus spongiosum of the glans penis: a clinicopathologic and immunohistochemical analysis of 10 cases. *Am J Surg Pathol* 2000;24:1524–1530.
30. McKenney JK, Collins MH, Carretero AP, et al. Penile myointimoma in children and adolescents: a clinicopathologic study of 5 cases supporting a distinct entity. *Am J Surg Pathol* 2007;31:1622–1626.

Proliferative fasciitis and proliferative myositis

31. Chung EM, Enzinger FM. Proliferative fasciitis. *Cancer* 1975;36:1450–1458.
32. Kern WH. Proliferative myositis: a pseudosarcomatous reaction to injury. *Arch Pathol* 1960;69:209–216.
33. Enzinger FM, Dulcey F. Proliferative myositis: report of 33 cases. *Cancer* 1967;20:2213–2223.
34. Kitano M, Iwasaki H, Enjoji M. Proliferative fasciitis: a variant of nodular fasciitis. *Acta Pathol Jpn* 1977;27:485–493.
35. Meis JM, Enzinger FM. Proliferative fasciitis and myositis in childhood. *Am J Surg Pathol* 1992;16:364–372.
36. Barak S, Wang Z, Miettinen M. Imunoreactivity for calretinin and keratins in desmoid fibromatosis and other myofibroblastic tumors: a diagnostic pitfall. *Am J Surg Pathol* 2012;36:1404–1409.
37. Lundgren L, Kindblom LG, Willems J, Falkmer U, Angervall L. Proliferative myositis and fasciitis: a light and electron microscopic, cytologic, DNA-cytometric and immunohistochemical study. *APMIS* 1992;100:437–448.
38. Rose AG. An electron microscopic study of the giant cells in proliferative myositis. *Cancer* 1974;33:1543–1547.
39. Dembinski A, Bridge JA, Neff JR, Berger C, Sandberg AA. Trisomy 2 in proliferative fasciitis. *Cancer Genet Cytogenet* 1992;60:27–30.
40. McComb EN, Neff JR, Johansson SL, Nelon M, Bridge JA. Chromosomal anomalies in a case of proliferative myositis. *Cancer Genet Cytogenet* 1997;98:142–144.

Ischemic fasciitis (atypical decubital fibroplasia)

41. Montgomery EA, Meis JM, Mitchell MS, Enzinger FM. Atypical decubital fibroplasia: a distinctive fibroblastic pseudotumor occurring in debilitated patients. *Am J Surg Pathol* 1992;16:708–715.
42. Perosio PM, Weiss SW. Ischemic fasciitis: a juxtaskeletal fibroblastic proliferation with a predilection for elderly patients. *Mod Pathol* 1993;6:69–72.
43. Yamamoto M, Ishida T, Machinami R. Atypical decubital fibroplasia in a young patient with melorheostosis. *Pathol Int* 198; 48:160–163.

Eosinophilic fasciitis and related conditions

44. Shulman LE. Diffuse fasciitis with eosinophilia: a new syndrome? *Trans Am Assoc Physicians* 1975;88:70–86.
45. Barnes L, Rodnan GP, Medsker TA, Short D. Eosinophilic fasciitis: a pathologic study of twenty cases. *Am J Pathol* 1979;96:493–517.
46. Shulman LE. Diffuse fasciitis with hypergammaglobulinemia and eosinophilia: a new syndrome. *J Rheumatol* 1984;11:569–570.
47. Doyle JA, Ginsburg WW. Eosinophilic fasciitis. *Med Clin North Am* 1989;73:1157–1166.
48. Varga J, Kahari VM. Eosinophilia-myalgia syndrome, eosinophilic fasciitis, and related fibrosing disorders. *Curr Opin Rheumatol* 1997;9:562–570.

Necrotizing fasciitis

49. Seal DV. Necrotizing fasciitis. *Curr Opin Infect Dis* 2001;14:127–132.
50. Young MH, Aronoff DM, Engleberg NC. Necrotizing fasciitis: pathogenesis and treatment. *Expert Rev Anti Infect Ther* 2005;3:279–294.

Retroperitoneal fibrosis and related disorders

51. Ormond JK. Idiopathic retroperitoneal fibrosis. *JAMA* 1960;174:1561–1568.
52. Mitchinson MJ. Retroperitoneal fibrosis revisited. *Arch Pathol Lab Med* 1986;110:784–786.
53. Gilkeson GL, Allen NB. Retroperitoneal fibrosis: a true connective tissue disease. *Rheum Dis Clin North Am* 1996;22:23–38.
54. van Bommel EF. Retroperitoneal fibrosis. *Neth J Med* 2002;60:231–242.
55. Miller OF, Smith LJ, Ferrara EX, McAleer IM, Kaplan GW. Presentation of idiopathic retroperitoneal fibrosis in the pediatric population. *J Pediatr Surg* 2003;38:1685–1688.
56. Flieder DB, Suster S, Moran CA. Idiopathic fibroinflammatory (fibrosing/sclerosing) lesions of the mediastinum: a study of 30 cases with emphasis on morphologic heterogeneity. *Mod Pathol* 1999;12:257–264.
57. Dehner LP, Coffin CM. Idiopathic fibrosclerotic disorders and other inflammatory pseudotumors. *Semin Diagn Pathol* 1998;15:161–173.
58. Hughes D, Buckley PJ. Idiopathic retroperitoneal fibrosis is a macrophage-rich process. Implications for its pathogenesis and treatment. *Am J Surg Pathol* 1993;17:482–490.
59. Yachoui R, Sehgal R, Carmichael B. Idiopathic retroperitoneal fibrosis: clinicopathologic features and outcome analysis. *Clin Rheumatol* 2016;35:401–407.

Keloids

60. Niessen FB, Spauwen PH, Shalkwijk J, Kon M. On the nature of hypertrophic scars and keloids: a review. *Plast Reconstr Surg* 1999;104:1435–1458.
61. Tuan TL, Nichter LS. The molecular basis of keloid and hypertrophic scar formation. *Mol Med Today* 1998;4:19–24.

Chapter 7: Fibromas and benign fibrous histiocytomas

Markku Miettinen
National Cancer Institute, National Institutes of Health

Fibromas and benign fibrous histiocytomas (BFHs) are fibroblastic-myofibroblastic growths. Most but perhaps not all of these are true neoplasms. Fibrocartilaginous pseudotumor is included because of its anatomic juxtaposition with nuchal fibroma. Nuchal fibroma and irritation fibroma of the oral cavity are other probably reactive processes. Fibromas typical of children are discussed in Chapter 9. Giant cell angiofibroma is discussed with solitary fibrous tumor (Chapter 12).

Elastofibroma
Clinical features

Originally described by Järvi *et al.* in 1961 as "elastofibroma dorsi," this clinicopathologically distinctive fibrous proliferation is rich in thick, convoluted elastic fibers. Elastofibroma occurs in middle-aged and older adults exclusively underneath the lower end of scapula below the dorsal muscles in the posterior chest wall.[1] Armed Forces Institute of Pathology (AFIP) cases show a mild male predominance, especially in the older age groups (Figure 7.1). Autopsy and radiologic studies have suggested that incidental microscopic elastofibromas might be common (10–20%).[2] Half of the patients are asymptomatic, and bilateral tumors occur in 20% of patients. Behavior is benign, and recurrences are very rare after a simple excision.[3,4] Although initial studies suggested reactive origin, possibly by chronic minor trauma, recent findings showing clonal chromosomal changes suggest a neoplastic nature.

Lesions containing abundant elastic fibers and designated as elastofibromas were reported in the gastric wall associated with elastofibroma of the back.[5] Otherwise there is no evidence that occasional lesions reported as elastofibromas outside the scapular region are comparable with true elastofibromas.

Pathology

Clinically, elastofibroma is best seen as a subscapular mass after elevation of the arm, and the intraoperative view is one of a variably demarcated deep mass (Figure 7.2a,b).

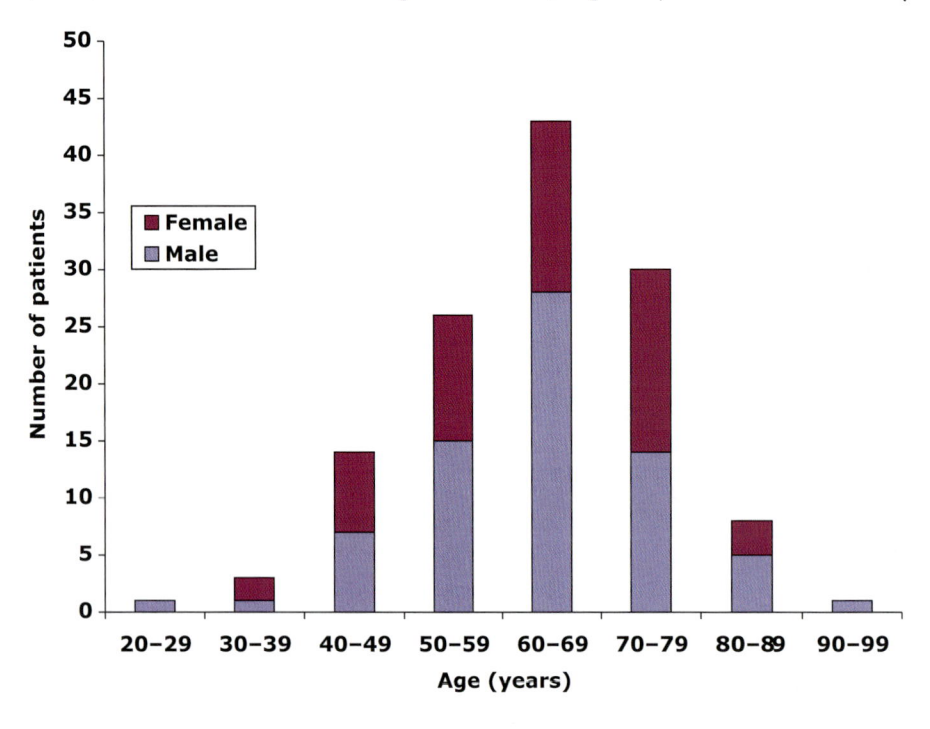

Figure 7.1 Age and sex distribution of 122 elastofibromas. Note predilection for older middle age.

Modern Soft Tissue Pathology, Second Edition, ed. Markku Miettinen. Published by Cambridge University Press. © Cambridge University Press 2016.

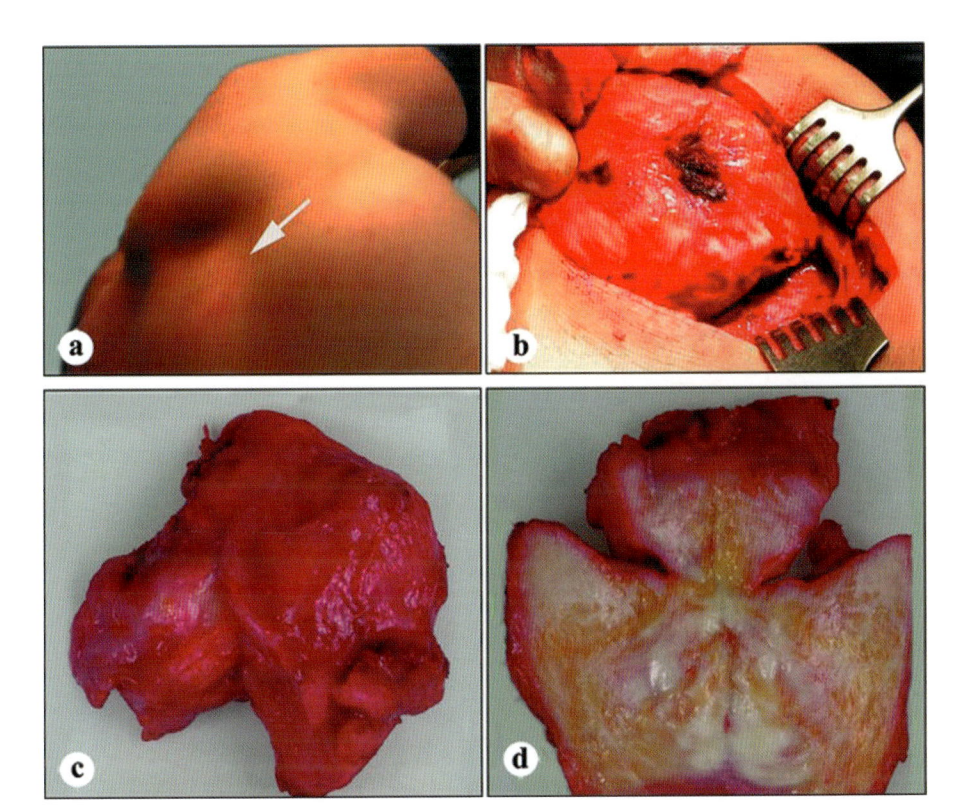

Chapter 7: Fibromas and benign fibrous histiocytomas

Figure 7.2 (**a**) Elastofibroma is clinically best visible as a subscapular mass after elevating the arm. (Arrow points to upper edge of the tumor.) (**b**) Intraoperative view shows a demarcated mass. (**c, d**) Grossly, elastofibroma forms a fibrous mass internally, showing an admixture of yellow fatty and gray fibrous elements in a striped pattern.

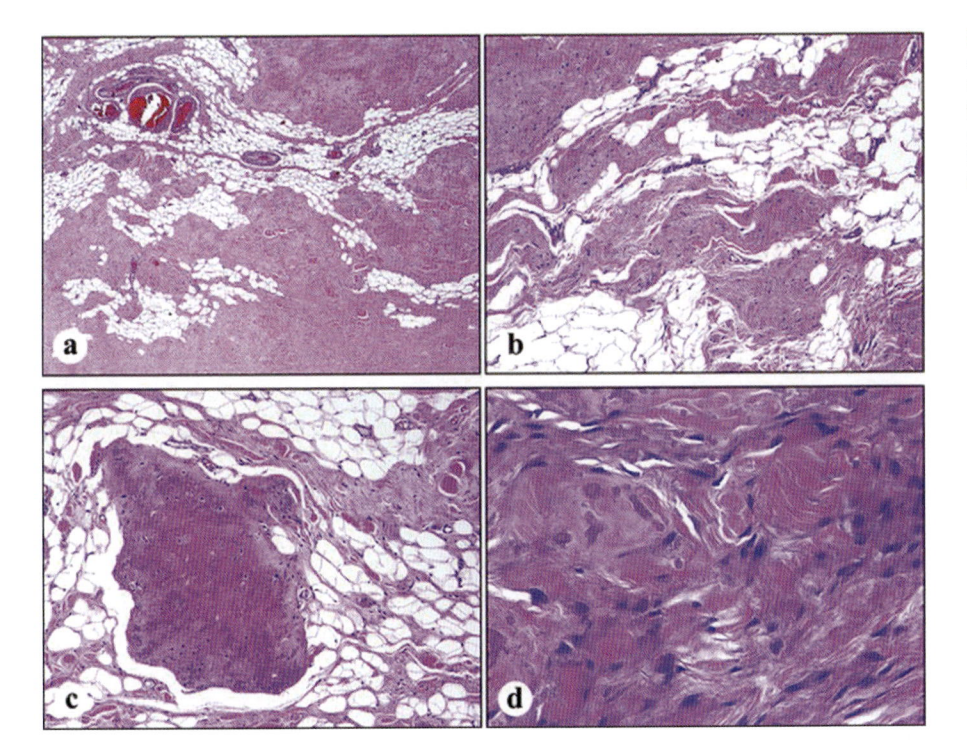

Figure 7.3 (**a–c**) Low-magnification view of elastofibroma shows a paucicellular mass with alternating fibrous and fatty elements. (**c**) The fibrous elements can have a slightly cartilaginous appearance. (**d**) Higher magnification of a cellular focus reveals bland fibroblasts and a few elastic fibers in this view.

Elastofibroma commonly measures 3 to 10 cm and is firm to rubbery. On sectioning it often shows alternating yellow and gray zones corresponding to fatty and fibrous components (Figure 7.2c,d). The gross appearance varies according to the proportions of these components.

Microscopically, elastofibroma shows alternating streaks of fat and paucicellular fibrous component containing bland spindle cells in a dense collagenous matrix (Figure 7.3). The collagenous matrix contains numerous, thick, convoluted elastic fibers that can appear fragmented and beaded, resembling

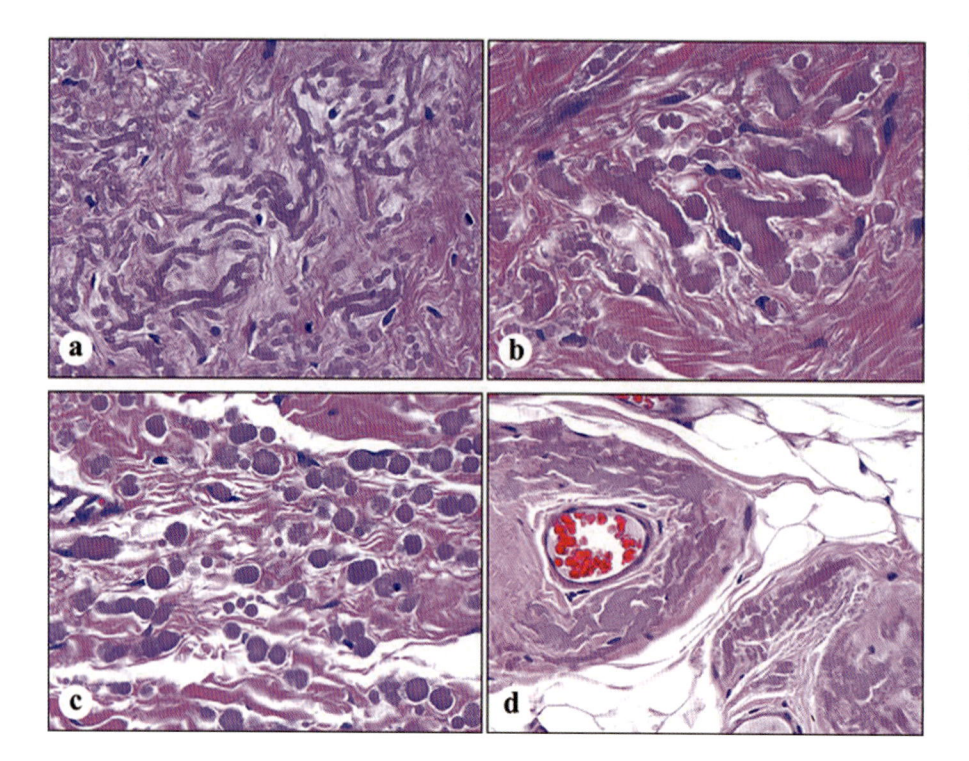

Figure 7.4 (a,b) Elastic fibers in elastofibroma can form cords resembling fungal hyphae. (c) Cross-sectional view of the fibers, which appear as rounded bodies. (d) The fibers can also form amorphous, perivascular collections reminiscent of amyloid deposition.

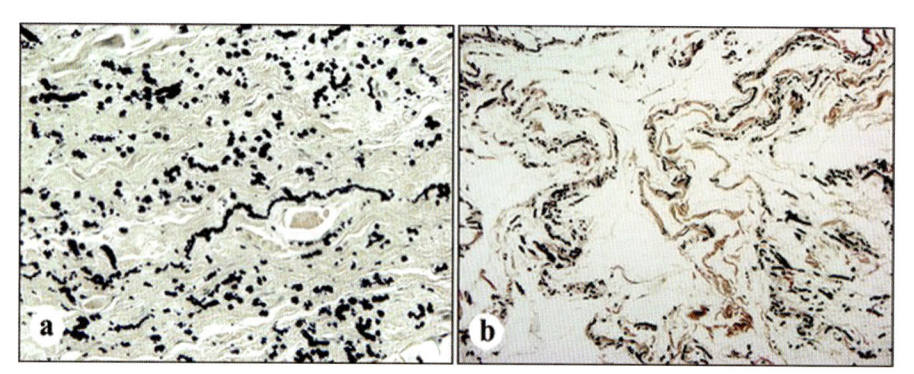

Figure 7.5 Elastin stain shows numerous elastic fibers both in the solid and fatty areas.

fungal hyphae or amyloid (Figure 7.4). The elastic fibers are highlighted with elastin staining, are often more numerous than appears in the hematoxylin and eosin (HE) stain, and are also present in fat septa (Figure 7.5). The tumor cells have been found positive for collagen type II, which is normally found only in hyaline cartilage and fibrocartilage.[6] The fibrous component also has been found to have CD34 positivity.[7]

The presentation and histologic appearance of elastofibroma are so characteristic that it can hardly be confused with other tumors. Desmoid tumors and sarcomas (malignant fibrous histiocytoma, MFH) can occur in the same location, however, and should be distinguished by their much greater cellularity, and cellular atypia in the case of the latter. Elastofibroma should not be confused with a fibrolipoma or atypical lipomatous tumor; the adipocytic component shows no atypia.

Genetics

Complex chromosomal changes, including heterogeneous nonrecurrent translocations, have been found in each of the five karyotyped cases, and this has been regarded as chromosomal instability.[8–10] Recurrent chromosomal rearrangements have been found in 1p and 7q.[9–10] Comparative genomic hybridization has revealed copy number changes, the most common of which was a gain of Xq.[11] The specific significance of these findings awaits further studies, however. Analysis of human androgen receptor (HUMARA) locus has revealed nonrandom X-chromosome inactivation, supporting the clonal nature of tumor cells.[7]

Nuchal-type fibroma

Clinical features

This designation, of nuchal-type fibroma, refers to a paucicellular tumor-like accumulation of collagen, predominantly presenting in the posterior neck. Enzinger and Weiss reported the original description of this fibroma.[12] Because similar lesions can present in extranuchal sites, the designation of *nuchal-type fibroma* has been proposed.[13]

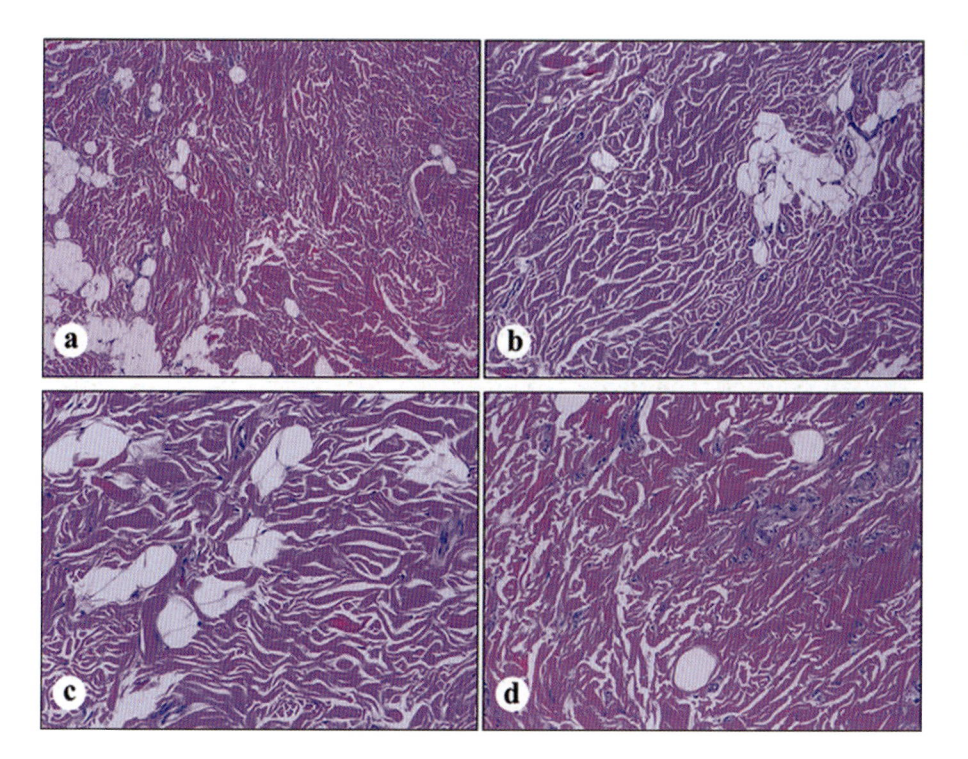

Figure 7.6 (a–c) Nuchal fibroma is paucicellular and contains dense, coarse collagen with vague lobular arrangement and foci of entrapped subcutaneous fat. (d) Traumatic neuroma-like clusters of small nerves are often present.

Nuchal-type fibroma presents in a wide age range, but its frequency peaks at the sixth decade. There is a strong male predominance.[12–15] More than 70% of the lesions present in the posterior neck, but similar lesions also can occur in the back, shoulder, face, and occasionally elsewhere. The etiology is unknown, but there is an association with diabetes; diabetic scleredema, a cutaneous swelling in the posterior neck, can include cases of nuchal fibroma.[16] Cases earlier reported in children in connection with Gardner syndrome could represent the recently described Gardner fibroma[17,18] (Chapter 9). Recurrences have been reported following simple excision, but it is unknown whether this is a true recurrence or persistence of the triggering factors. Nuchal-type fibroma might represent a reactive condition to unknown stimuli, and there is no evidence that it is a true neoplasm.

Pathology

Nuchal-type fibroma appears as a firm, poorly delineated, cutaneous tumor-like mass that measures 1 cm to 6 cm in diameter (average 3 cm). Histologically, it is composed of paucicellular accumulation of lobulated bundles of dermal collagen extending to the subcutis and intermingling with scant residual fat. The lesional fibroblasts are few and inconspicuous, and the overall appearance is that of an exaggerated nuchal dermis. Small, apparently proliferating nerve twigs resembling a traumatic neuroma are often present (Figure 7.6). In rare instances, the lesion extends to skeletal muscle. Scant elastic fibers can be highlighted with special stains, but there are no clumped fibers similar to those seen in elastofibroma.

Gardner fibroma (occurring with the Gardner syndrome variant of familial adenomatous polyposis [FAP]) often can be distinguished from nuchal fibroma by its slightly higher cellularity, by its nuclear positivity for beta-catenin in particular, and by its occurrence in children. Some examples can be histologically inseparable from nuchal fibroma and distinguished from it only by clinical connection with FAP, however. Fibrolipoma forms a distinct, pseudocapsulated mass with a substantial mature lipomatous component.

Nuchal fibrocartilaginous pseudotumor
Clinical features

This rare but probably under-reported tumor-like lesion represents acquired fibrocartilaginous metaplasia of the nuchal ligament. Some patients have a history of past cervical trauma, such as whiplash injury, suggesting post-traumatic etiology. Presentation is in adult life, predominantly in young adults and early middle age (median age: 39 years). Seven of the reported cases have been in women and four in men. The lesion is sometimes tender and is located in the posterior neck, overlying the posterior aspect of the lower cervical vertebrae. Follow-up studies have shown no recurrences after simple excision.[19–21]

Pathology

The nodule measures 1 cm to 3 cm in diameter and is located in the deep soft tissue posterior to the cervical vertebrae, involving the nuchal ligament. Microscopically it is a poorly delineated nodular fibrocartilaginous proliferation with linear arrays of chondrocytes within a fibrocartilage-type dense matrix, surrounded by dense fibrosis (Figure 7.7).

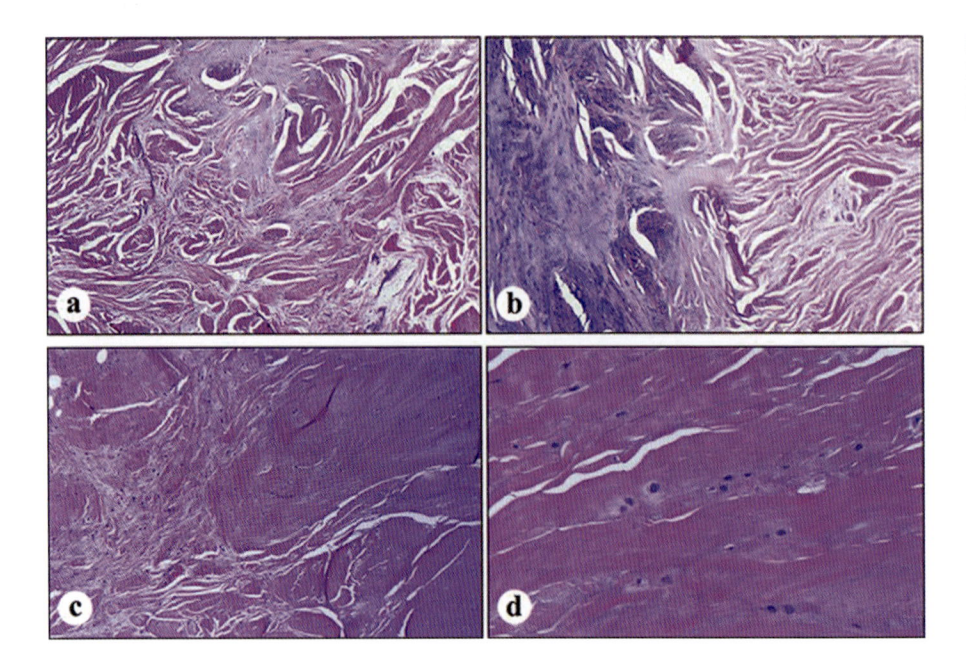

Figure 7.7 Nuchal fibrocartilaginous pseudotumor is composed of reparative fibrocartilaginous metaplasia of the nuchal ligament.

A CD34-positive component has been reported,[21] but these lesions have been negative for desmin and SMA.

Collagenous fibroma (desmoplastic fibroblastoma)

Clinical features

Originally reported by Evans in 1995,[22] *collagenous fibroma* is a grossly and microscopically distinctive, relatively rare, benign fibrous tumor that predominantly occurs in men between the fifth and seventh decades (70%) and is rare in children and adolescents (Figure 7.8A). It occurs in a wide variety of peripheral soft tissue sites, usually the subcutis, but one-third involves skeletal muscle, and some examples are purely intramuscular. The most common locations are the arm (24%), shoulder (19%), back, upper extremities, and feet (Figure 7.8B). The behavior is benign, and none of the published series has demonstrated recurrences in long-term follow-up.[22–25]

Pathology

The tumor usually measures 2 cm to 4 cm at the widest part, but examples >10 cm and as large as 20 cm have occurred. Grossly, it appears as a well-circumscribed, oval, elongated, or disk-shaped mass that can be grossly lobulated. On sectioning, the tumor is firm, pearl gray, and homogeneous, often with a cartilage-like consistency on sectioning (Figure 7.9a–c). Radiologically it appears as a well-circumscribed, often elongated mass (Figure 7.9d).

Microscopically, a collagenous fibroma is an apparently circumscribed mass that can be lobulated and can extend into vascular lumina (Figure 7.10) and skeletal muscle. It is paucicellular and composed of scattered, unevenly spaced spindled or stellate-shaped fibroblasts and myofibroblasts in a dense, amorphous or irregularly bundled, deeply eosinophilic, sometimes focally or rarely more extensively myxoid, collagenous matrix (Figure 7.11). There is no sign of necrosis, atypia, or mitotic activity, beyond rare regular mitotic figures. Some tumors have nearly acellular collagenous areas. Despite the apparent gross circumscription, the lesion microscopically infiltrates the subcutaneous fat in 70% of cases and extends into the skeletal muscle in 25% of cases. The vessels are usually inconspicuous, with small lumens and thin walls. Intramuscular and intravascular extension has no apparent clinical significance.

Immunohistochemically, the tumor cells are variably, usually focally positive for SMA and occasionally focally for keratins (numbers 7 and 18), as is sometimes seen in other myofibroblastic lesions. The tumor cells are positive for vimentin, and this immunostaining often highlights the stellate cell processes better than does HE staining (Figure 7.11d). The tumor cells are negative for desmin, EMA, S100 protein, and CD34.

Differential diagnosis

Intramuscular examples of this tumor should not be confused with desmoid. Lower cellularity, predominance of amorphous collagenous matrix, and inconspicuous vasculature separate collagenous fibroma from desmoid fibromatosis. Fibroma of tendon sheath is hypervascular, usually located in the fingers or hand, and also is more prominently SMA positive than collagenous fibroma. Much lower overall cellularity and vascularity, as well as lack of atypia, separates collagenous fibroma from a low-grade fibrosarcoma.

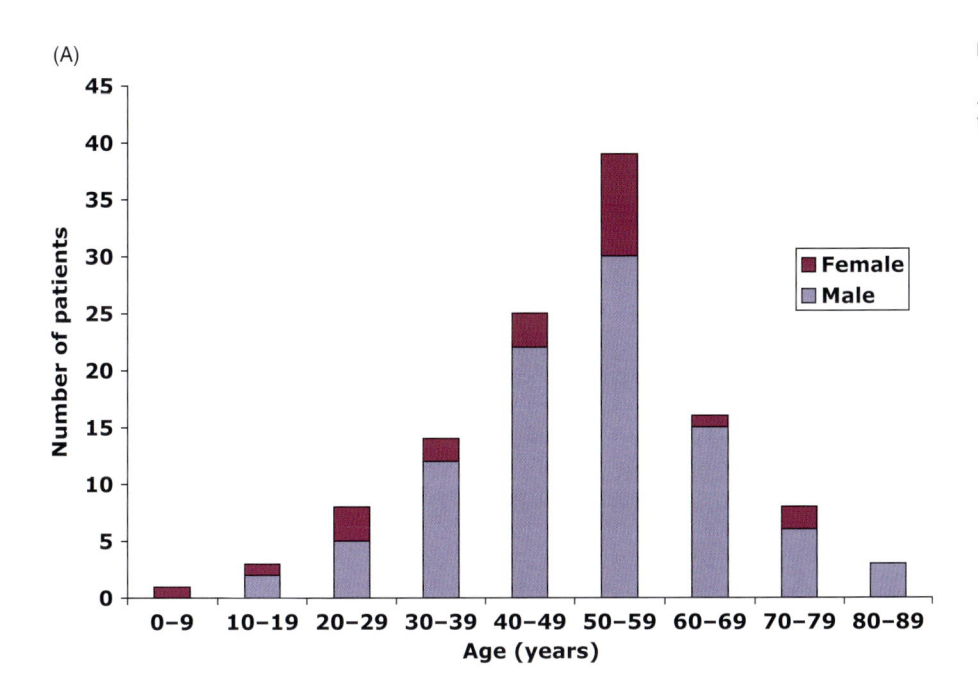

Figure 7.8 (A) Age and sex distribution of 117 patients with collagenous fibroma. (B) Anatomic distribution of 117 cases of collagenous fibroma.

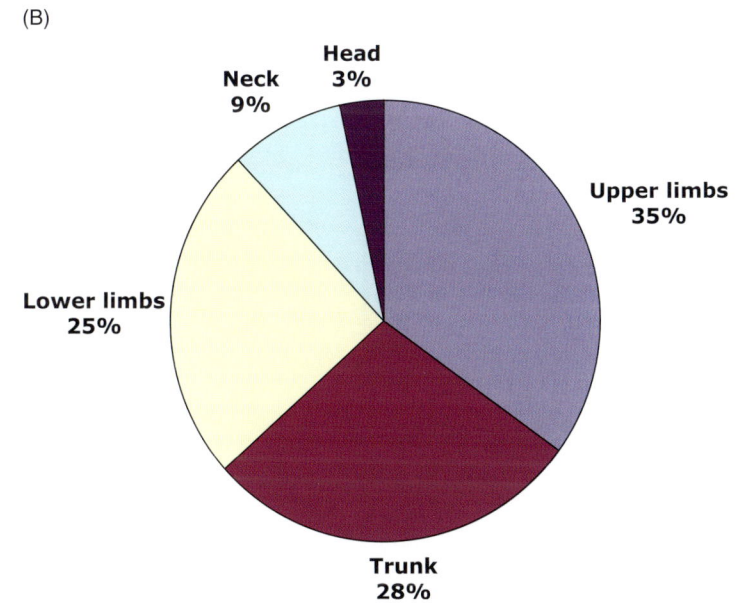

Genetics

Cytogenetic studies have confirmed t(2;11)(q31;q12) translocation as a recurrent chromosomal change.[26] Overexpression of FOS-like antigen 1 (FOSL1) seems to be associated with the 11q gene rearrangements, although no fusion genes have been detected so far.[27]

Fibroma of the tendon sheath

Originally reported by Chung and Enzinger in 1979, *fibroma of tendon sheath* is a relatively rare myofibroblastic tumor with a predilection for the fingers and hand.[28]

Clinical features

This tumor occurs predominantly in young and middle-aged adults, with a strong male predominance. Examples in children have also been reported (Figure 7.12A). The lesion typically occurs as a small painless mass in the subcutis of fingers and hands, closely associated with tendon sheaths but has also been found in the feet, knee, and shoulder region (Figures 7.12B and 7.13). Simple excision is curative, and the tumor has no noticeable tendency to recur.[28–36]

Pathology

This tumor is usually small, 1 cm to 2 cm in diameter, and appears well-circumscribed, sometimes lobulated.

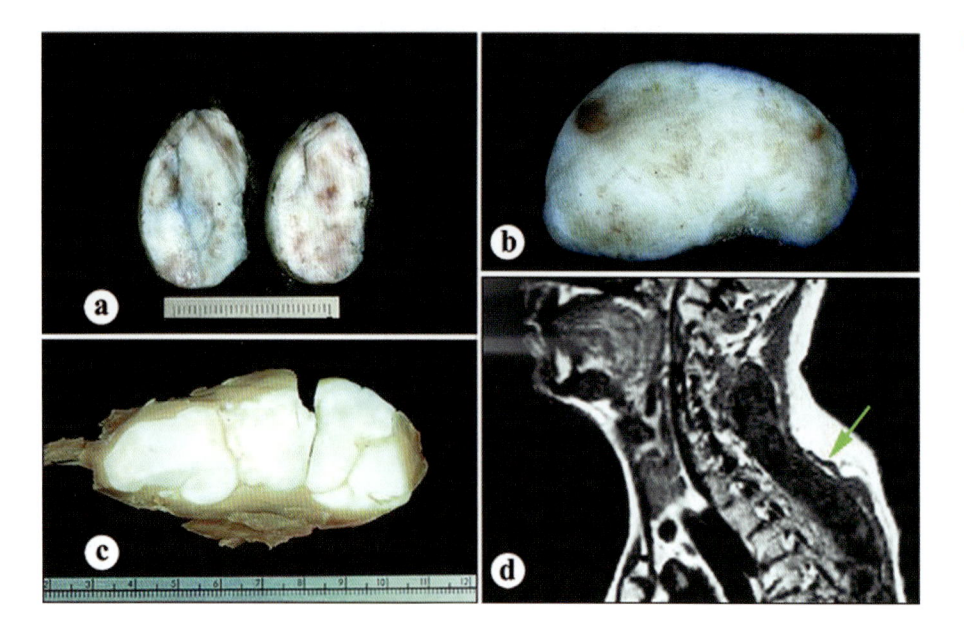

Chapter 7: Fibromas and benign fibrous histiocytomas

Figure 7.9 (a,b) Grossly collagenous fibroma forms an oval mass, which is solid, gray-white, and glistening on sectioning. (c) The tumor can also form an intramuscular, lobulated mass. (d) MRI scan of an unusually large, intramuscular, sausage-shaped, collagenous fibroma of 20 cm involving the posterior neck and upper back. (Arrow points to midpoint of the tumor.)

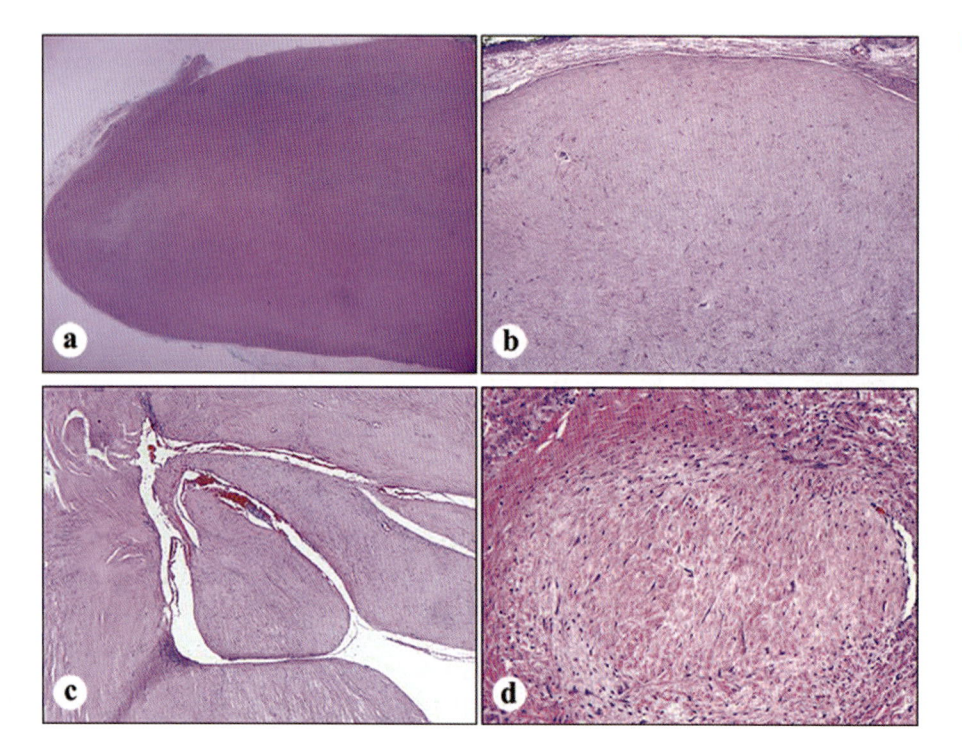

Figure 7.10 Low-magnification histology of collagenous fibroma. (a,b) Atypical example forms a well-demarcated, densely collagenous mass. (c) In some cases, the tumor shows microscopic lobulation. (d) An unusual example shows microscopic involvement of a vessel wall.

Histologically it is composed of uniform slender spindled fibroblasts and myofibroblasts in a variably myxoid to densely fibrous, typically highly vascular stroma. Slit-like blood vessels are often present, especially in the tumor periphery (Figures 7.14 and 7.15). A pleomorphic example has been reported but it is uncertain if this is fully comparable with the typical variants.[35] Immunohistochemically typical is muscle actin positivity and negativity for CD34, desmin, and S100 protein.

Differential diagnosis

This entity sometimes is difficult to distinguish histologically from nodular fasciitis and collagenous fibroma. The former has a more heterogeneous composition, and the latter has stellate fibroblasts. Although fibroma of tendon sheath has been suggested as a hyalinized variant of tenosynovial giant cell tumor, evidence for such a relationship remains weak.[37]

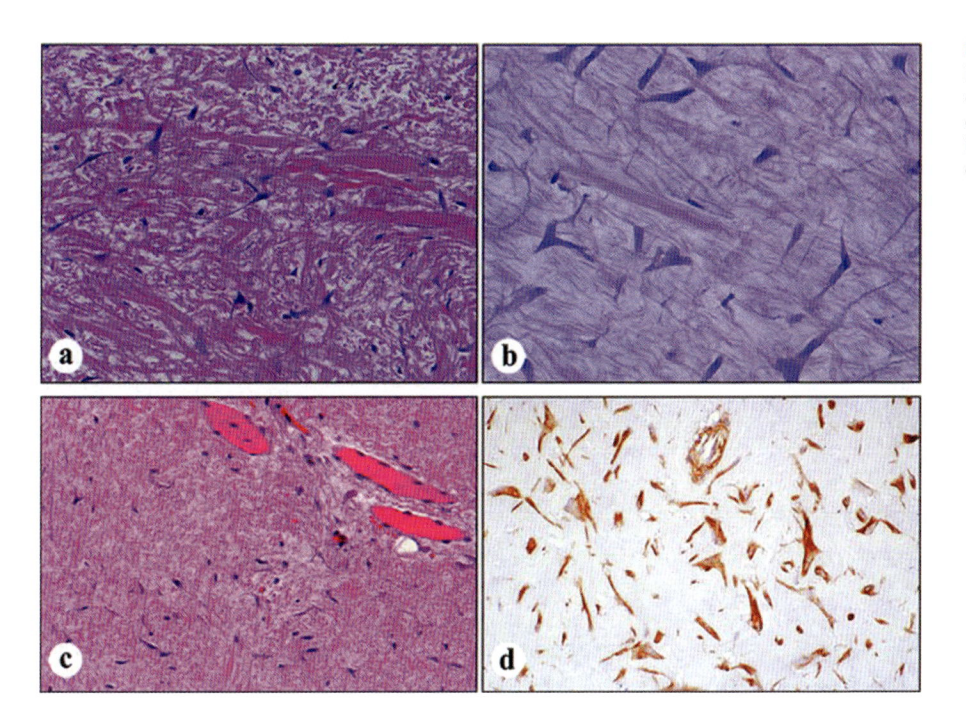

Figure 7.11 (a,b) Collagenous fibroma is histologically composed of stellate fibroblasts in a densely collagenous background. (c) The tumor can infiltrate skeletal muscle. (d) Vimentin immunostain highlights the cytoplasmic processes of the stellate fibroblasts.

Genetics

Cytogenetic analysis has revealed translocation t(2;11) in one case.[38] The translocation is similar to the one known in desmoplastic fibroblastoma, and may reflect the fact that prior to description of desmoplastic fibroblastoma this entity was commonly classified as fibroma of tendon sheath.

Sclerotic fibroma of skin (circumscribed storiform collagenoma, "plywood fibroma")

Sclerotic fibroma is a designation for an uncommon skin tumor that clinically manifests as a small, dome-shaped dermal lesion. This tumor occurs in adults of either sex with no site predilection, usually sporadically, and rarely in connection with Cowden's disease (multiple hamartoma syndrome), in which multiple lesions can occur. Local excision is curative.[39–43]

On histologic analysis, the tumor forms a circumscribed, often spherical, paucicellular dermal nodule composed of inconspicuous fibroblasts in a dense collagenous background with a storiform organization, often with focal pericellular cleavages (Figure 7.16). The tumor cells are often positive for CD34.[44,45]

The differential diagnosis includes sclerotic BFH (dermatofibroma) and dermatofibrosarcoma protuberans (DFSP). The former can be recognized by the presence, typically, of fibrous histiocytoma in some areas (especially the periphery of the lesion), and the latter by a more extensive lesion with fat infiltration. Some dermatofibromas undergo sclerosis, showing areas reminiscent of sclerotic fibroma.[46] Examples having morphology overlapping with pleomorphic fibroma also have been reported.[47]

Pleomorphic fibroma of the skin

Clinical features

First reported by Kamino *et al.* in 1989,[48] *pleomorphic fibroma* is an uncommon skin tumor that presents as a small (5–15 mm) cutaneous polypoid or dome-shaped lesion covered by intact epidermis. This tumor occurs in young and middle-aged adults in various sites, mostly in the proximal extremities and trunk. The tumor grows slowly, and patients often have a long history of the lesion prior to surgery. There seems to be no significant tendency for recurrence following a simple excision, although recurrence can develop after an incomplete excision.[48]

Pathology

Histologically, pleomorphic fibroma often forms an exophytic mass (Figure 7.17a). It is paucicellular and contains scattered, mildly to moderately atypical fibroblasts in a dense collagenous background (Figures 7.17 and 7.18). The tumor cells often have abundant cytoplasm with multiple dendritic processes. The nuclei are large, irregularly shaped, and mildly hyperchromatic, and multinucleation is common. Stromal cleavages similar to those in sclerotic fibroma and multinucleated cells with nuclear arrangements similar to those in pleomorphic lipoma can occur (Figure 7.18d) so that association with pleomorphic/atypical lipomatous tumor has been discussed.[49] Occasional mitoses, even atypical ones, can be present. These tumor cells have been reported as CD34 positive and S100 protein negative.[50,51]

Atypical fibroxanthoma has a higher cellularity, significant nuclear atypia, and often a higher mitotic activity, with numerous atypical mitoses. Positivity for CD34 and

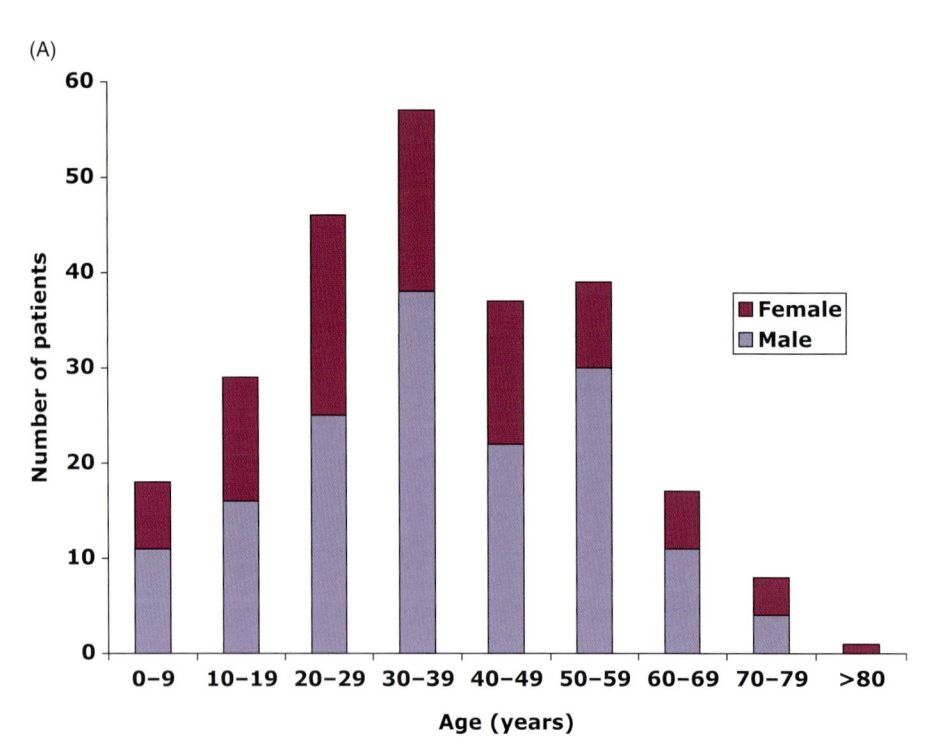

Figure 7.12 (A) Age and sex distribution of 252 civilian patients with fibromas of tendon sheath. (B) Anatomic distribution of 361 fibromas of tendon sheath.

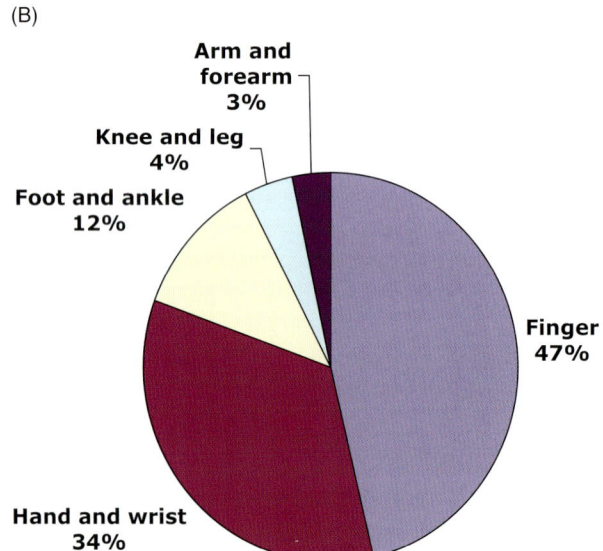

paucicellularity differentiate this fibroma from atypical BFH with giant (monster) cells.[51]

Dermatomyofibroma

Described by Kamino et al. in 1992, *dermatomyofibroma* is the designation for a small, <1- to 2-cm, plaque-like lesion that typically occurs in young adults in the arm and shoulder region.[52,53] This apparently rare tumor is probably under-recognized. Histologically it is composed of uniform, horizontally oriented, spindled myofibroblasts and fibroblasts admixed with thin collagen fibers and entrapping cutaneous adnexa and resembling superficial variants of fibromatosis (Figure 7.19). Foamy histiocytes and giant cells are not present, and inflammatory cells are scant, differing from most variants of BFA. Dermatomyofibroma is positive for muscle actins but negative for desmin.

Irritation fibroma of the oral cavity

This very common lesion is one of the most frequently made diagnoses in oral pathology and constituted 25% of all surgical biopsies in one university dental clinic practice.[54] It is thought to be an acquired, probably post-traumatic scar-like formation. This fibroma seems to occur slightly more commonly in men and have a predilection for middle age, although it can occur at any age (Figure 7.20A). It usually occurs in the buccal mucosa, but often also in the lower lip, or gingiva and tongue; it is rare elsewhere in the oral cavity (Figure 7.20B).

This fibroma appears grossly as a firm, dome-shaped or shallow sessile mucosal polypoid elevation. Histologic analysis shows a densely collagenous, slightly hypervascular paucicellular stroma containing scattered fibroblasts, and focal lymphoplasmacytic infiltration is often present (Figure 7.21).

Giant cell fibroma of the oral cavity
Clinical features
This relatively common lesion was found to compose 1.0% to 2.7% of all biopsies in two university dental clinic practices.[54,55] It has a predilection for young adults, with a mild 1.3:1 female predominance (Figure 7.22A). The most common locations are the gingivae, occurring almost twice as frequently in the mandibular as in the maxillary gingivae (Figure 7.22B). Other common locations include the tongue, palate, and buccal mucosa, whereas occurrence in the lip is uncommon and in the floor of the mouth exceptional. The lesion forms a small, <1 cm (average: 0.4 cm), sessile, dome-shaped or pedunculated mucosal elevation. Recurrence has been reported rarely after a local excision.[55–60]

Pathology
This fibroma forms a small polypoid mucosal lesion. Histologically it is typically covered by a thickened, parakeratotic epithelium. The stroma is densely collagenous and contains scattered, oval, fibroblast-like cells, many of which are multinucleated with nuclei variably arranged, sometimes in a horseshoe pattern. Numerous dilated capillaries are typical (Figure 7.23). On immunohistochemical analysis, the fibroma cells are positive for vimentin and negative for S100 protein and keratins.[60]

Palisaded myofibroblastoma
This very rare tumor, reported simultaneously by Weiss et al.[61] and Suster and Rosai,[62] occurs in lymph nodes, predominantly in the inguinal region. It is histologically distinctive for nuclear palisading, resembling Verocay bodies of schwannoma. One of the original reports named it as an "intranodal hemorrhagic spindle cell tumor with amianthoid fibers."[62]

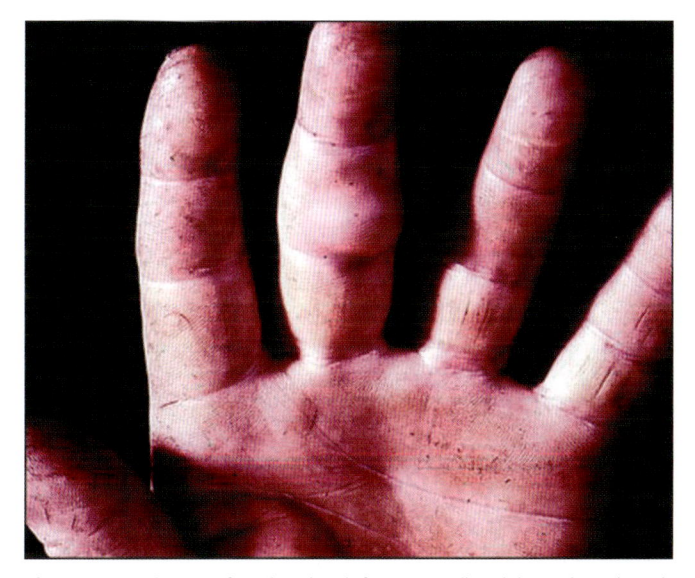

Figure 7.13 Fibroma of tendon sheath forms a small nodule on the volar side of the middle finger.

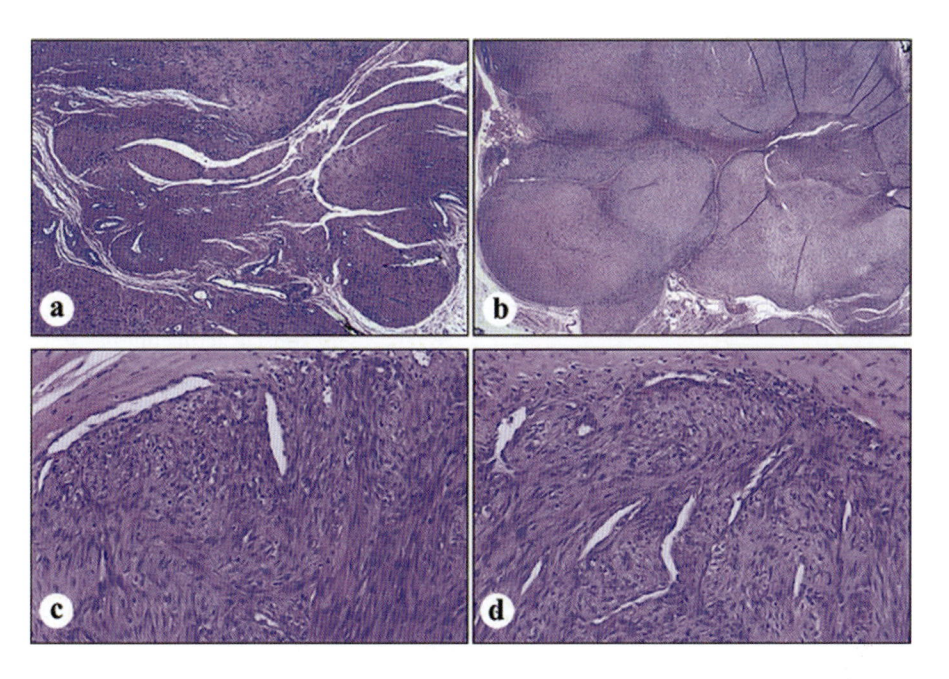

Figure 7.14 Fibroma of the tendon sheath is lobulated and sharply circumscribed, containing numerous slit-like vessels and uniform myofibroblasts.

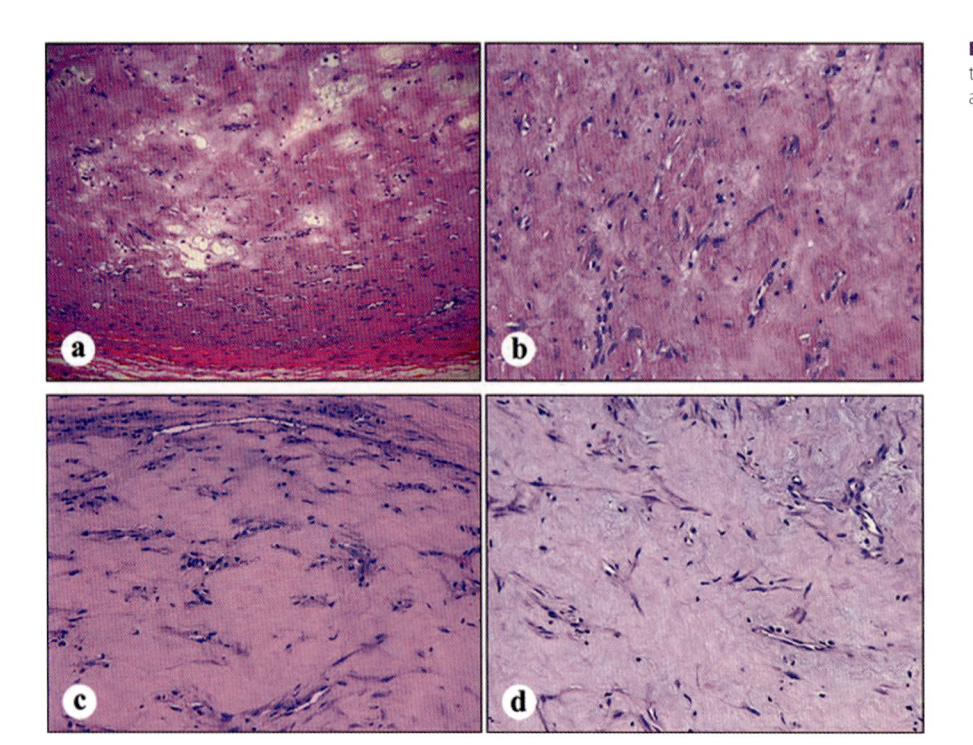

Figure 7.15 Less-cellular variants of fibroma of tendon sheath with fibrous or fibromyxoid stroma and a prominent capillary pattern.

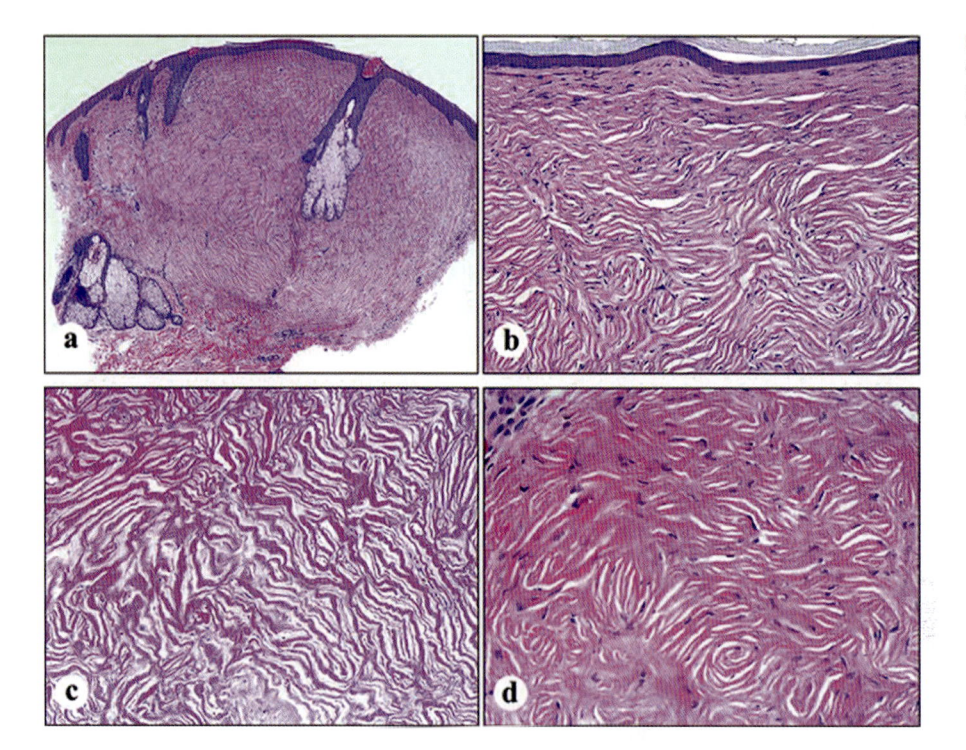

Figure 7.16 Sclerotic fibroma of the skin forms a circumscribed dermal nodule. The tumor is relatively paucicellular, collagenous, and typically contains slit-like spaces in the collagen.

Clinical features

Palisaded myofibroblastoma occurs in middle-aged adults, with a 2:1 male predominance. Almost all cases have been reported in inguinal lymph nodes, and occasional ones in the submandibular region and axilla, where the tumor forms a painless mass.[61–69] Predominant location in the inguinal region has been suggested to reflect the myofibroblastic population specific to this region.[70]

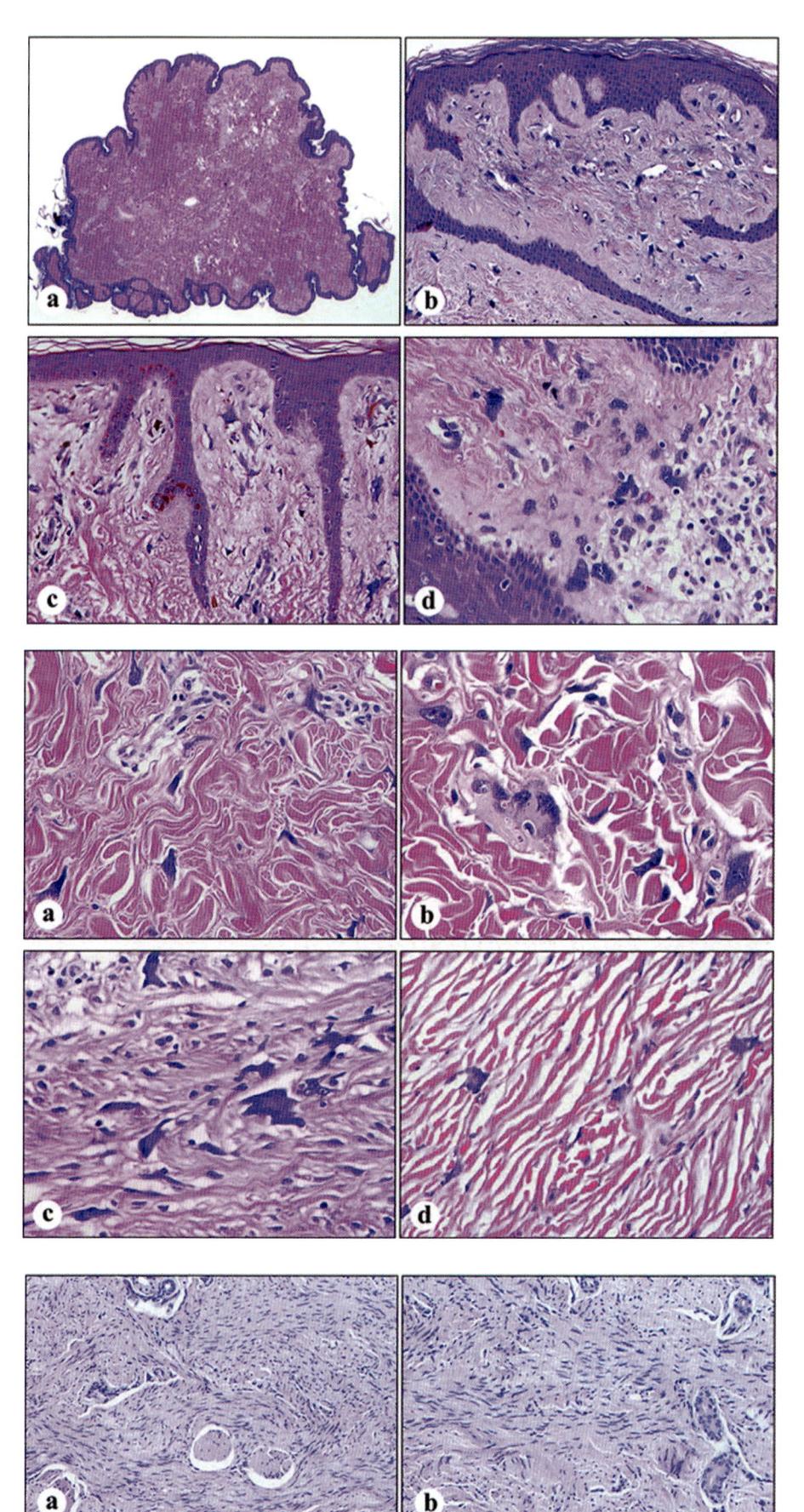

Chapter 7: Fibromas and benign fibrous histiocytomas

Figure 7.17 Pleomorphic fibroma typically forms a polypoid dermal lesion containing scattered fibroblasts with nuclear atypia in a dense collagenous background.

Figure 7.18 (a–d) Examples of higher magnifications of pleomorphic fibroma. Note multinucleated cells in (b–d) Slit-like spaces in collagen can resemble those seen in sclerotic fibroma (d).

Figure 7.19 Dermatomyofibroma is composed of a band of spindle fibroblasts in the dermis with a pattern slightly resembling fibromatosis. The lesion entraps pilar smooth muscle and epithelial skin adnexa.

215

Figure 7.20 (A) Age and sex distribution of 380 oral irritation fibromas. (B) Anatomic distribution of 408 oral irritation fibromas.

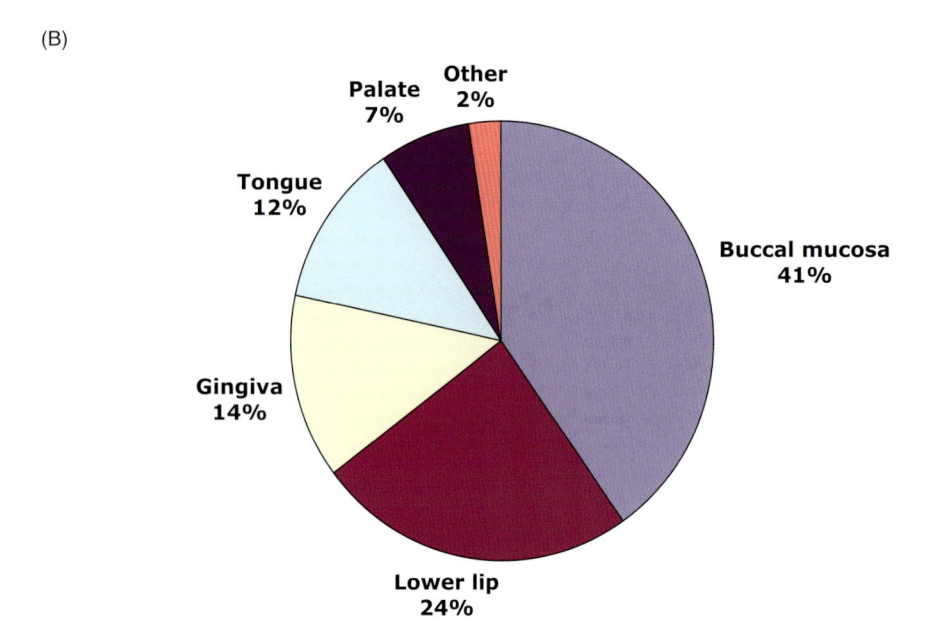

Although the tumor is benign, there is at least one report of recurrence[71] and one report of multicentricity.[72]

Pathology

Grossly, the lesion forms a gray-brown mass typically rimmed by residual lymph node tissue (Figure 7.24). In most cases, marginal sinus can be identified as a sign of a true lymph node. Histologically distinctive are thick bundles of collagen seen as strands or ovoid profiles, depending on the plane of section, surrounded by nuclear palisades of spindle cells, often perpendicular to the collagen (Figures 7.25 and 7.26). In some cases, the collagen fibers are abundant throughout, whereas in others they are seen only at the interface of the tumor and lymph node, where a thick collagen band is often also present. Gamna–Gandy bodies can arise from the collagen fibers. The thick collagen fibers might only be a focal feature, with a prominent perivascular hyalinization surrounded by the cellular component.

Some tumors have a diffuse, variably cellular, spindle cell component somewhat resembling the appearance of fibromatosis, whereas some cases have an alternating pattern of collagenous and cellular zones (Figure 7.26d). The spindle cells usually have a myofibroblastic appearance. Nuclei can appear

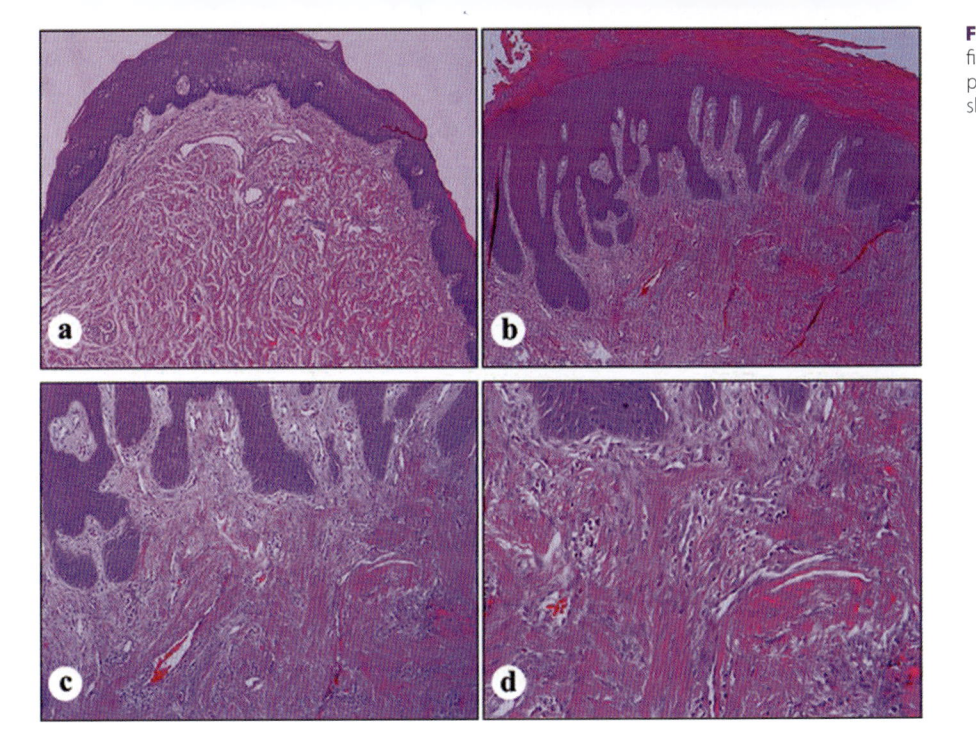

Figure 7.21 Histologic features of oral irritation fibroma. Note a plaque-like, fibrosclerosing paucicellular process, with moderate numbers of slit-like vessels.

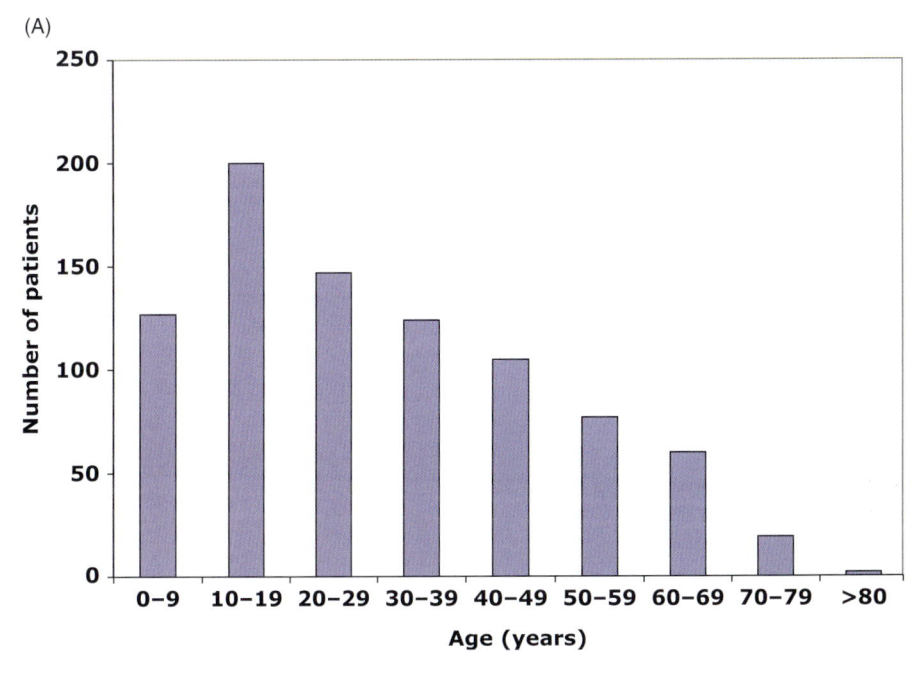

Figure 7.22 (**A**) Age distribution of oral giant cell fibroma in eight series. (**B**) Anatomic location of 652 oral giant cell fibromas in three series.

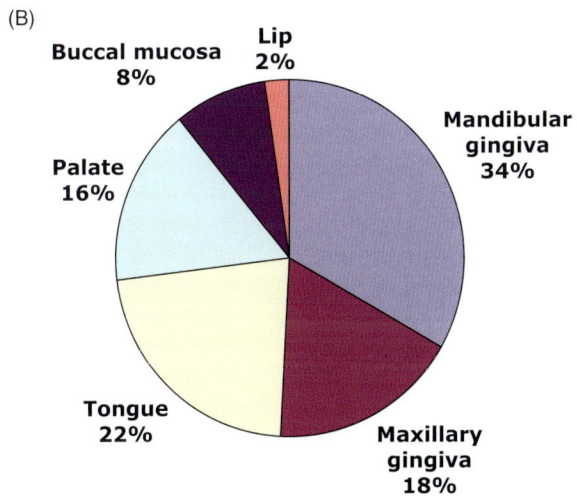

217

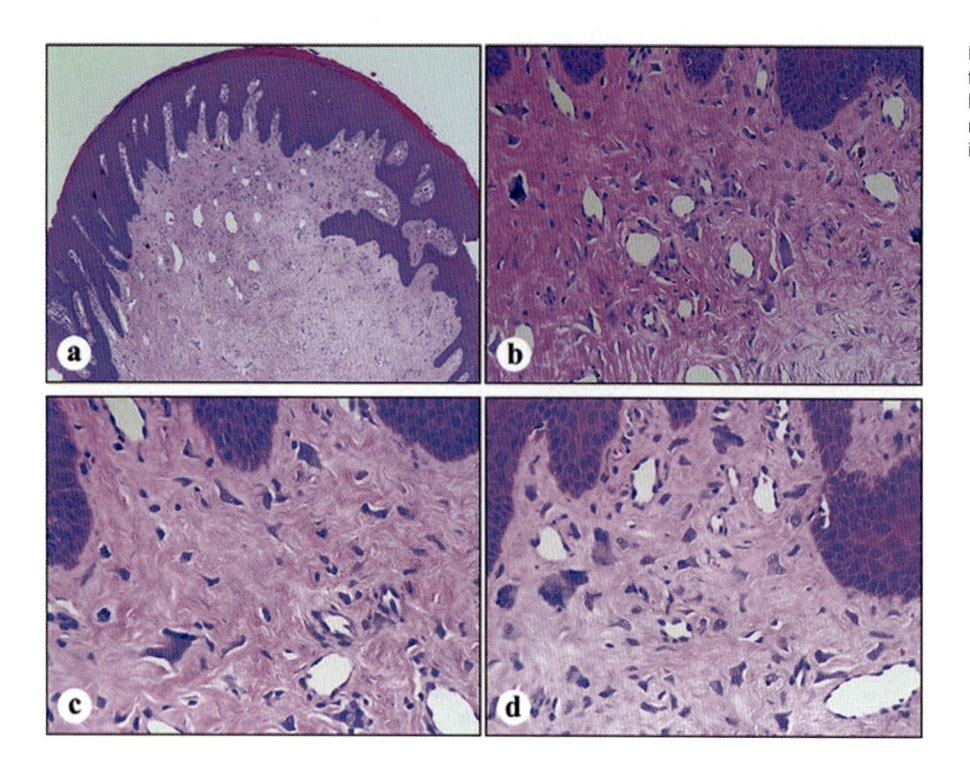

Chapter 7: Fibromas and benign fibrous histiocytomas

Figure 7.23 Giant cell fibroma of the oral mucosa forms a polypoid lesion typically covered by hyperkeratotic epithelium. The stroma contains mononuclear cells and multinucleated giant cells in a dense collagenous background.

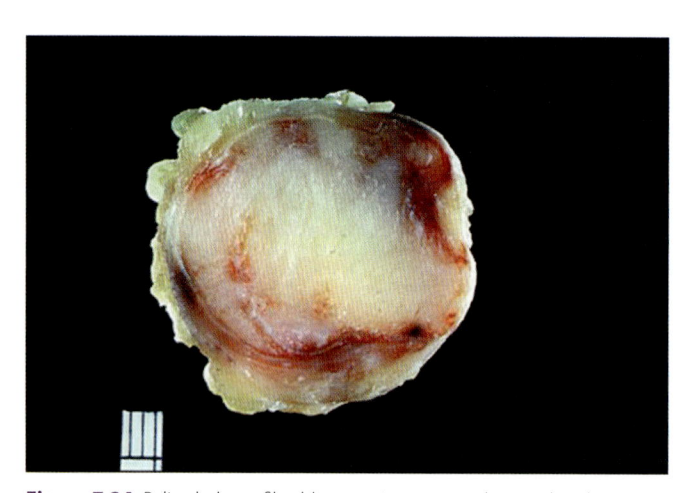

Figure 7.24 Palisaded myofibroblastoma is seen grossly as a sharply circumscribed nodule, and a small peripheral crescent of the remaining lymph node is seen especially on the lower left aspect of the tumor. Scale elements are millimeters.

hyperchromatic but are uniform. Some spindle cells have perinuclear eosinophilic globules similar to those present in digital fibroma. Neither nuclear atypia nor mitotic activity is present. Interstitial hemorrhage is prominent in some tumors, and metaplastic bone is occasionally present.

According to one study, the crystalloid "amianthoid" collagen fibers are actually abnormally thick collagen fibers, composed of multiple coalesced fibers consisting of collagen types I and III and not typical amianthoid fibers.[73]

On immunohistochemical analysis, the tumor cells are positive for vimentin, SMA, collagen IV, and sometimes for desmin, but they are negative for CD34 and S100 protein.

Cyclin D1 overexpression is a common feature.[74] Typical of this tumor is strong nuclear expression of beta-catenin, which is associated with beta-catenin stabilizing mutations and as a consequence activates cyclin D1 expression.[75]

Differential diagnosis

Some examples, especially those with prominent hemorrhage and few hyaline cords, resemble Kaposi's sarcoma. A tendency for a higher collagen content and lack of the hyaline globules typical of Kaposi's sarcoma are characteristic of palisaded myofibroblastoma. Immunohistochemical positivity for CD34 and HHV8 are additional features seen only in Kaposi's sarcoma and not in palisaded myofibroblastoma.

Myofibroblastoma of the breast
Clinical features

Originally described by Wargotz et al., this tumor occurs in adults, with a slight female predominance.[76] The median age in the series was 64 years (range, 40–85 years). The tumor forms a firm, circumscribed, relatively small nodule. Follow-up information indicates that this tumor neither recurs nor metastasizes and is therefore amenable to a simple excision.[76–78]

Pathology

Grossly this tumor is typically a well-circumscribed, rubbery, gray-tan lesion and measures 1 cm to 4 cm, with a mean diameter of 2.3 cm in the largest series.[76]

Histologically, the tumor is quite cellular and has a pushing, noninfiltrative border. It is composed of plump, uniform,

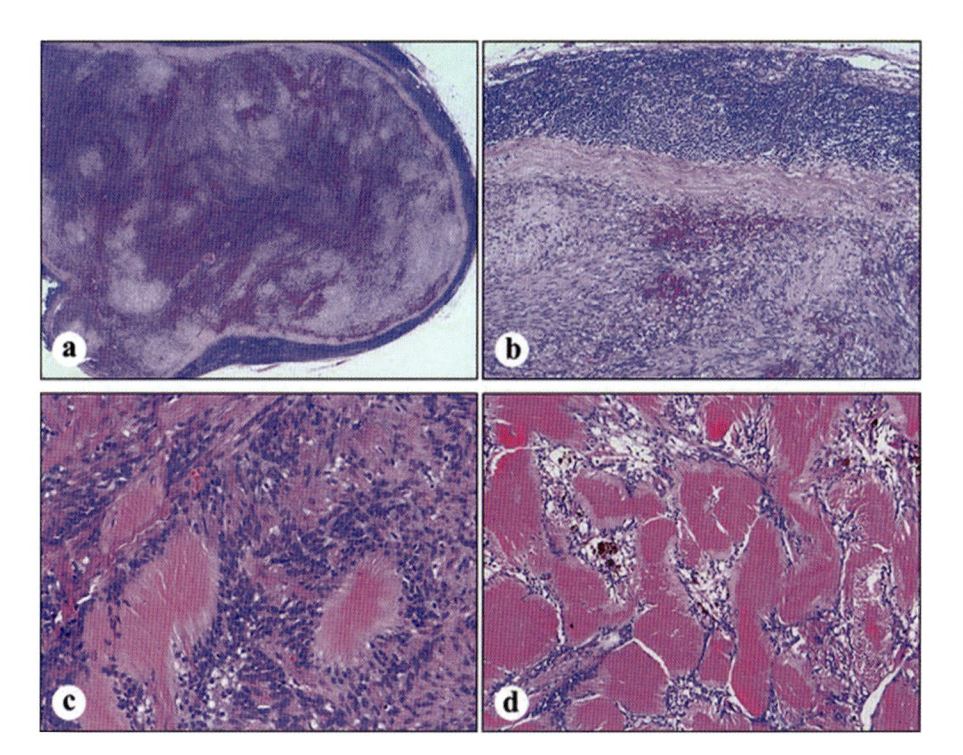

Chapter 7: Fibromas and benign fibrous histiocytomas

Figure 7.25 (**a**) Palisaded myofibroblastoma forms a sharply circumscribed nodule located in a lymph node with extensive hemorrhage. (**b**) The tumor is often delineated from the nodal elements by a thick, collagenous, capsule-like band. (**c,d**) The number of collagen bundles varies, and they often surround spindle-shaped tumor cells in various arrangements.

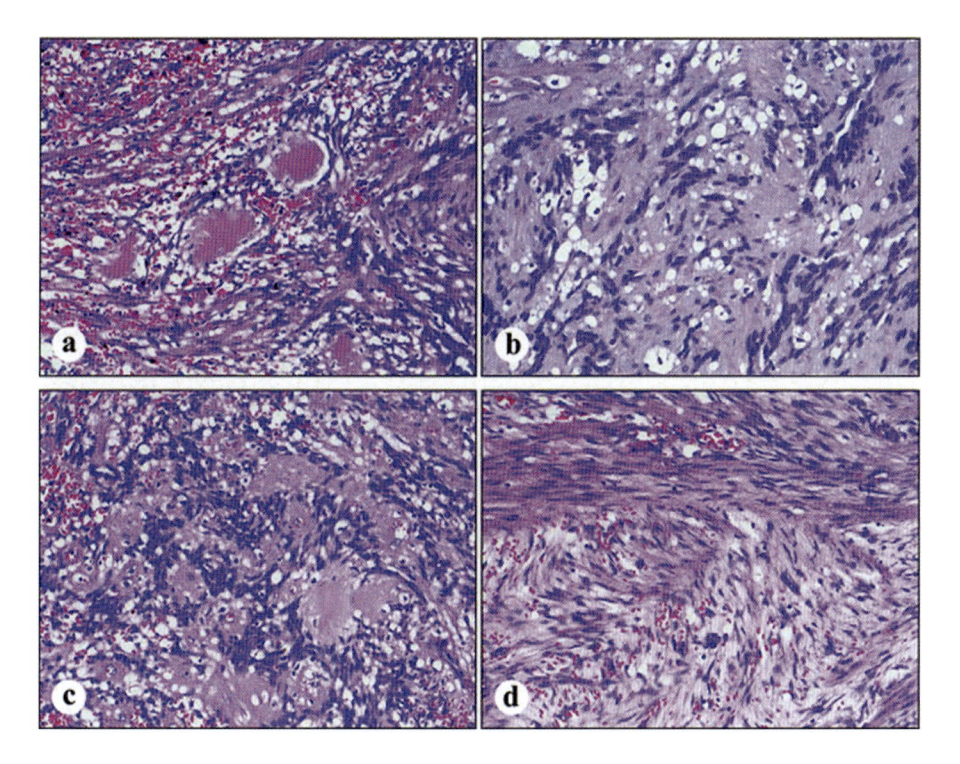

Figure 7.26 (**a**) Interstitial hemorrhage is typical of palisaded myofibroblastoma. (**b,c**) Myxoid stromal and trabecular and palisaded nuclear arrangements. (**d**) Fibromatosis-like cellular areas are occasionally present.

spindled or sometimes oval to epithelioid myofibroblasts separated by broad streaks of collagen (Figure 7.27). Nuclear grooves are often seen, and mild pleomorphism may occur. Mitotic activity is low. Immunohistochemically the tumor cells are positive for actin, CD34, estrogen and progesterone receptors, and variably for desmin, and negative for S100 protein and keratins.[77–80]

The epithelioid variants should not be confused with invasive carcinoma, a danger especially present when the tumor is associated with lobular carcinoma *in situ*.[81,82] Alternating cellular areas and collagen, as opposed to uniform fibrous matrix, separate myofibroblastoma from fibromatosis of breast.

Mammary-type myofibroblastoma is the term used to describe myofibroblastoma-like tumors occurring outside the

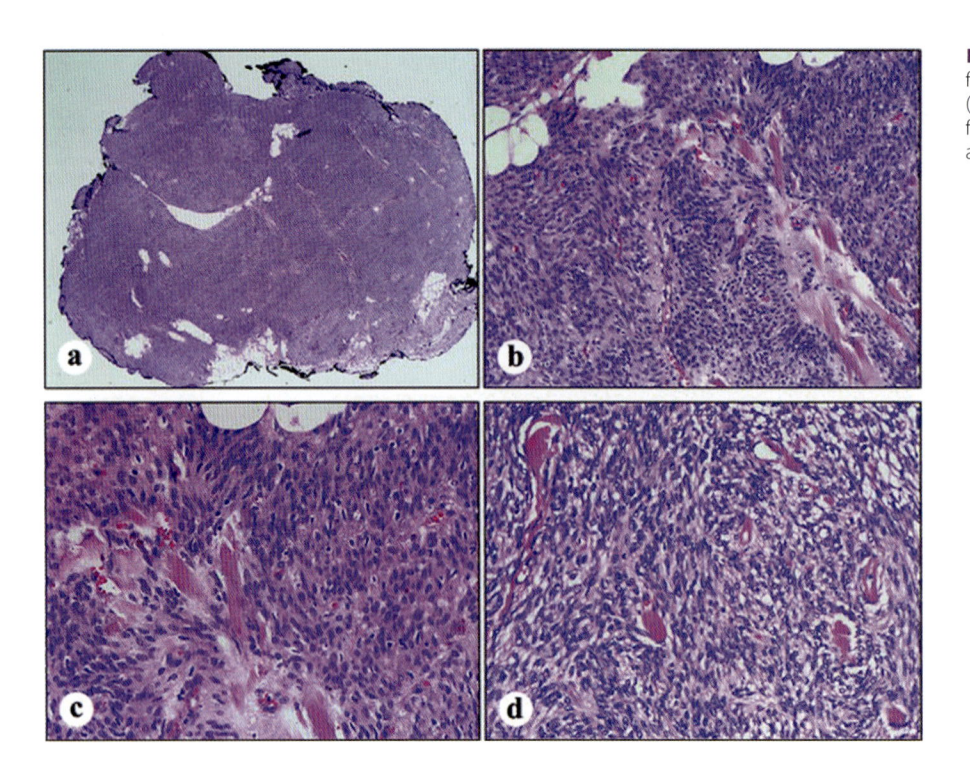

Figure 7.27 (a) Myofibroblastoma of the breast forms a circumscribed nodule that infiltrates in fat. (b–d) The tumor cells are uniform spindle cells with focal palisading. Fat cells and thick collagen fibers are also evident.

breast. These tumors have occurred at different peripheral sites, mostly in subcutaneous tissue.[83] They contain fat, spindle cells, and are CD34 positive, showing resemblance to spindle cell lipoma. Indeed, genomic losses in 13q have also been reported, suggesting that the extramammary myofibroblastoma-like lesions are related to spindle cell lipoma.[84]

Benign fibrous histiocytoma (dermatofibroma)

BFH, dermatofibroma, previously also called *sclerosing hemangioma, histiocytoma*, and *subepidermal nodular fibrosis* is one of the most common benign soft tissue tumors. Although most authors consider BFHs to be tumors,[85] some have entertained the possibility that they might be local responses to injury by inflammation or trauma.[86]

Clinical features

BFHs occur in a wide age range, but are rare in the first decade and most common in young adults between the ages of 20 and 40 years. Some studies have found a moderate female predominance. For example, a large series of 290 cases showed more than 2:1 female predominance.[87]

The most common locations are the extremities, especially the leg and thigh, but BFH also occurs in the upper extremity, trunk wall, neck, and head, including the face.[88] Various subtypes presented here have essentially similar patterns of occurrence, except that some have slight gender or site predilections, as detailed later.

BFHs mainly involve the dermis, but they can also extend into the subcutis and superficial skeletal muscle, especially in the head and neck area. Such tumors are sometimes designated *deep fibrous histiocytomas*. In the author's experience, however, virtually all deep intramuscular tumors historically designated as BFH variants are actually other tumors, such as solitary fibrous tumors and low-grade fibromyxoid sarcomas.

Small typical variants most commonly present as slightly elevated, pigmented nodules, but an early lesion can also be flat or dimpled. Larger tumors can form dome-shaped, sessile or polypoid, exophytic masses.[89] Some variants (especially the aneurysmal ones) clinically resemble a melanoma or Kaposi's sarcoma because of reddish-brown discoloration from hemorrhage.

Occurrences of multiple BFHs have been reported on rare occasions, mostly in immunosuppressed patients, such as patients with HIV/AIDS, or those treated with corticosteroids for autoimmune disease, such as lupus erythematosus. Multiple lesions can form local eruptions, but in some patients these occurred in separate sites.[90]

Recurrences and metastases

Although benign, BFH can recur unless completely excised. The frequency of recurrence of conventional variants is low, probably <10%.[85] Examples in the face are prone to recur perhaps because surgery in this site tends to be more conservative.[88] Cellular variants have a higher recurrence rate, approximately 25%. Complete excision of the cellular variant therefore seems prudent.[91] On rare occasions, regional lymph node metastases have been reported in the cellular variants.

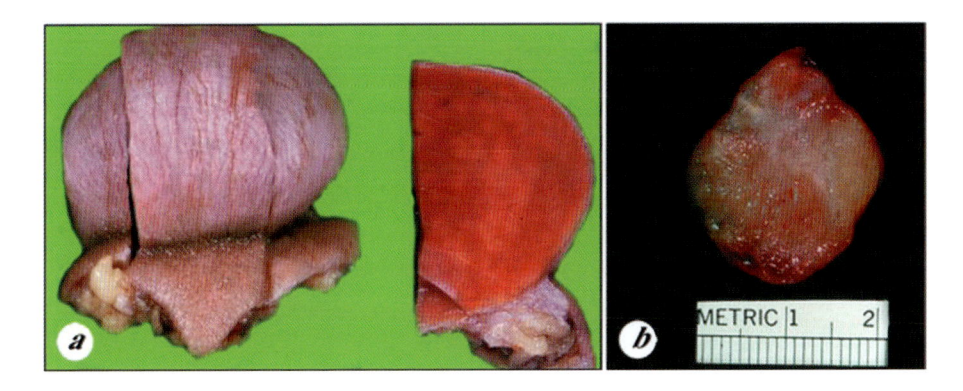

Figure 7.28 Gross appearances of two benign fibrous histiocytomas. (**a**) A polypoid tumor showing bright orange-brown tissue on sectioning. (**b**) An unfixed specimen of benign fibrous histiocytoma shows a fleshy, pale tan cut surface.

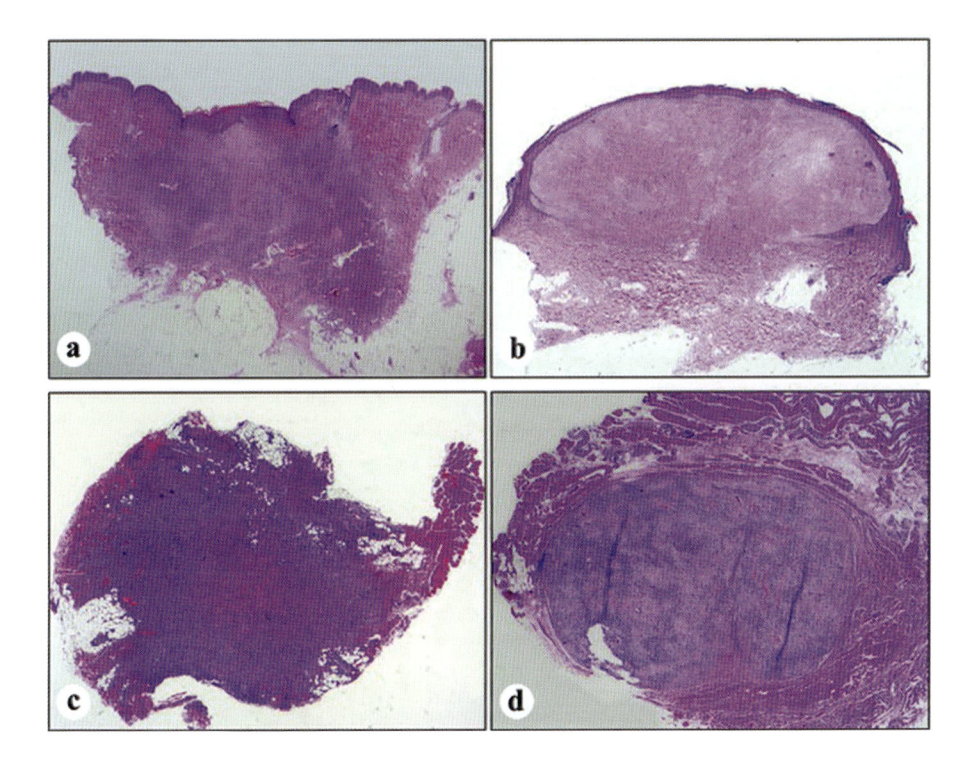

Figure 7.29 Different configurations of benign fibrous histiocytoma on low magnification. (**a**) Dimpled. (**b**) Dome-shaped with an epithelial collar. (**c**) Subcutaneous fat-infiltrating variant adjacent to superficial skeletal muscle of the neck. (**d**) Intramuscular example in the neck.

Rare pulmonary metastases of cellular BFH have also been reported, and most of such cases have been young male adults <40 years of age. In many cases, the metastases have been multiple and have occurred <1 to 23 years after the primary tumor. The course has been variable with initial reports indicating rather indolent nature.[92,93] However, a large more recent series showed 50% mortality. The deaths occurred 2 to 13 years after lung metastases.[94] In some patients, pulmonary BFH has occurred without an apparent primary tumor, potentially raising the question of pulmonary origin. While most metastasized BFHs have been of the aneurysmal or cellular variant, some have been atypical FH.[92–94]

Pathology of conventional BFH

Grossly, these lesions vary. Smaller examples are hyperpigmented nodules or plaques, the latter often showing a central dimpled depression. Larger tumors can form sessile dome-shaped or polypoid masses. The cut surface is often yellow to brown because of extensive lipid and hemosiderin content (Figure 7.28). Tumors located in the head and neck can extend into or infiltrate subcutaneous fat and superficial skeletal muscle (Figure 7.29).

The typical variant is composed of fibroblasts admixed with histiocytes (sometimes xanthomatous or multinucleated) admixed with streaks of entrapped collagen fibers. Some tumors are microscopically storiform, with whorl-like patterns (Figure 7.30). Hemosiderin pigmentation is common, especially as a focal finding (Figure 7.31).

Most lesions are moderately vascular and many show perivascular hyalinization and superficial fibrosclerosis. Palisaded nuclei seen in some cases can be prominent enough to resemble a schwannoma (Figure 7.32a,b). Palisaded variants are often located in the fingers.[95] Some examples contain Touton-type giant cells with a ring-like nuclear arrangement (Figure 7.32c,d). The periphery of BFH is typically ragged and

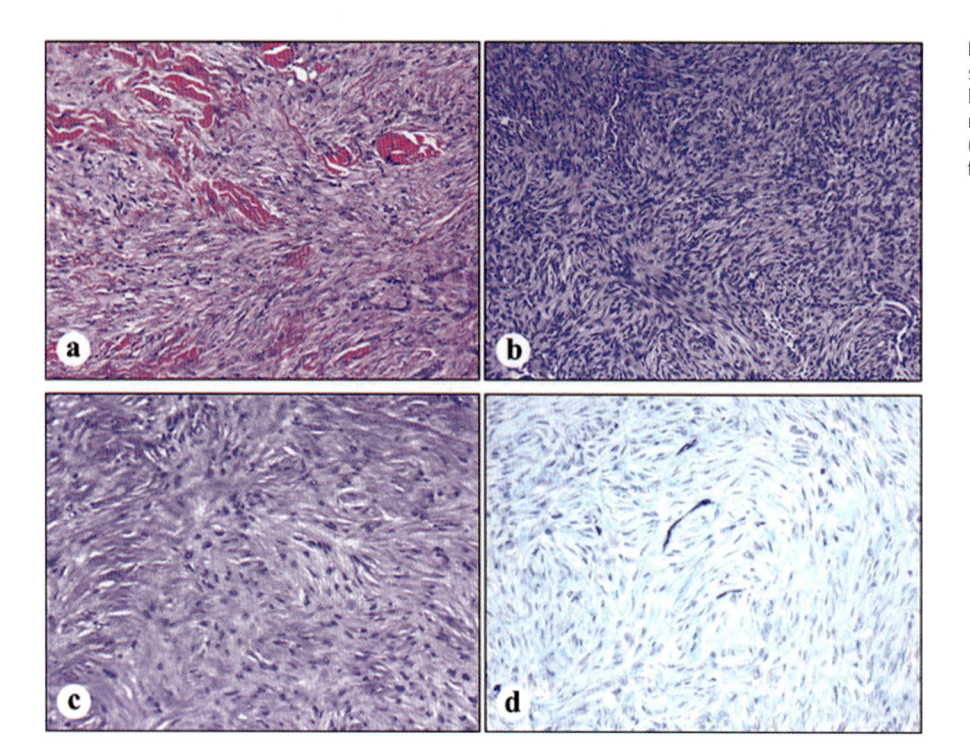

Figure 7.30 (a,b) Collagen trapping and storiform growth patterns in benign fibrous histiocytoma. (c) This storiform tumor has some resemblance to dermatofibrosarcoma protuberans. (d) The tumor is immunohistochemically negative for CD34.

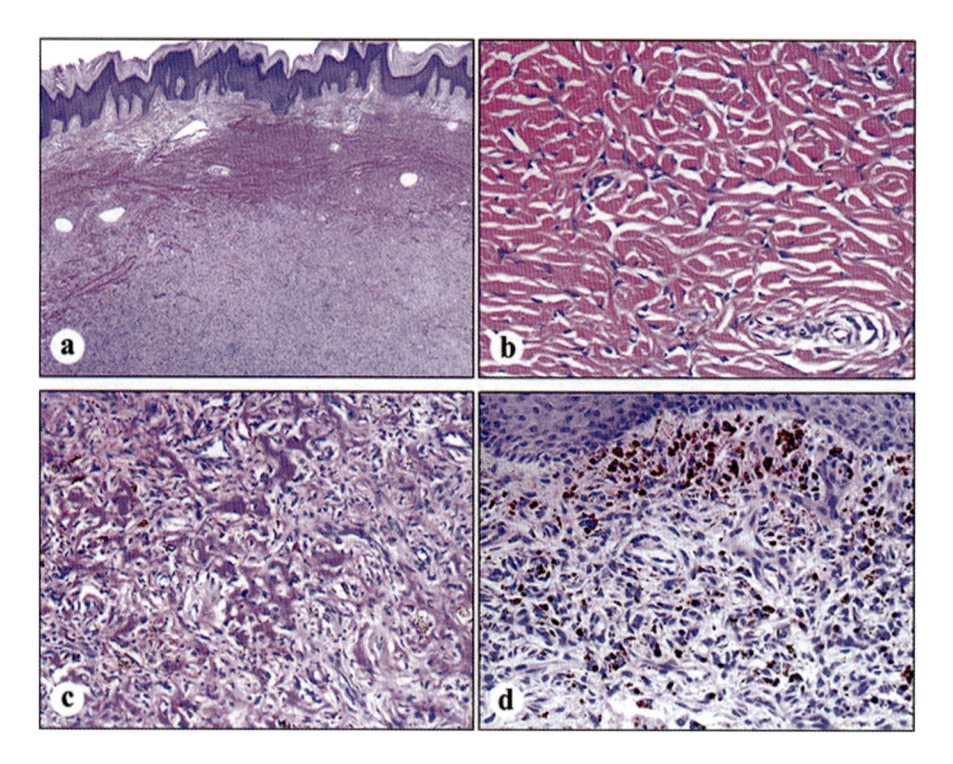

Figure 7.31 Sclerosing and pigmented examples of benign fibrous histiocytoma. (a–c) are from the same tumor, the upper part of which shows sclerosis; the lower part has pigmentation. (d) Abundant hemosiderin pigment is present in histiocytes as clumped aggregates.

infiltrative, and the margin to the subcutis varies from a penetrating to an infiltrative quality; the latter can be extensive, especially in cellular variants (see later discussion).

Immunohistochemistry

BFH is almost invariably negative for CD34 (Figure 7.33a, b), although this marker can be present at the periphery of the lesion, with the positivity probably contributed by adjacent nonlesional fibroblasts.[96–98] BFHs are also negative for desmin, EMA, keratins, and S100 protein. The coexpression of proliferative markers preferentially seen in the lesional fibroblasts but not histiocytes has been interpreted to support the idea that the fibroblasts are the proliferative component.[99]

Chapter 7: Fibromas and benign fibrous histiocytomas

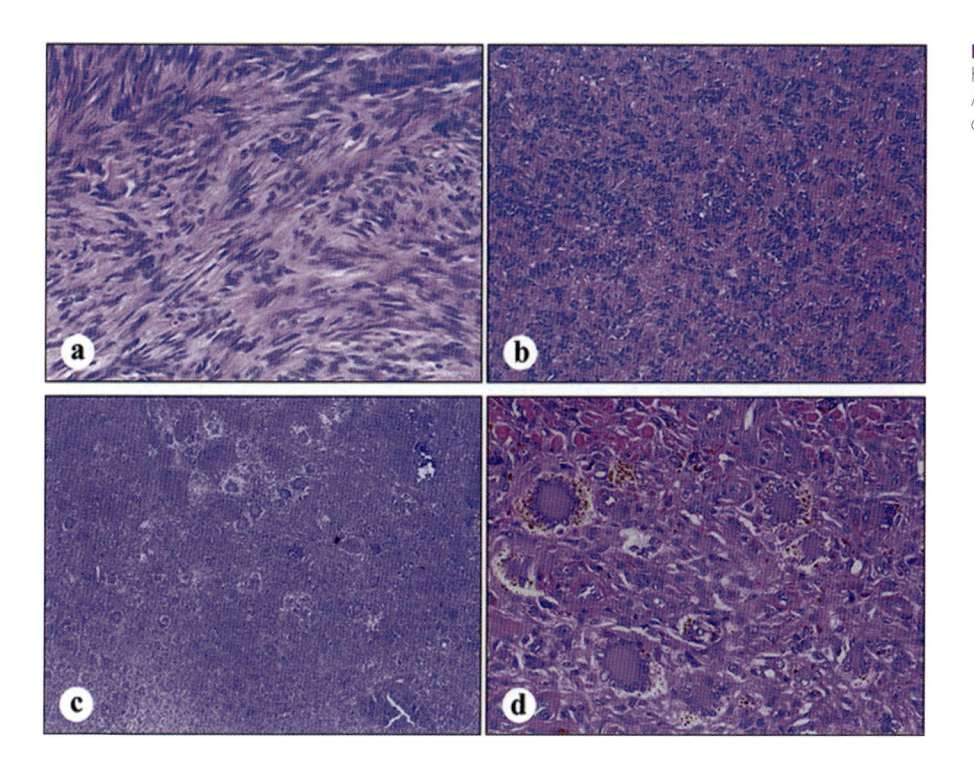

Figure 7.32 Variants of benign fibrous histiocytoma. (**a,b**) Nuclear palisading. (**c,d**) A tumor rich in multinucleated Touton-like giant cells with hemosiderin pigment.

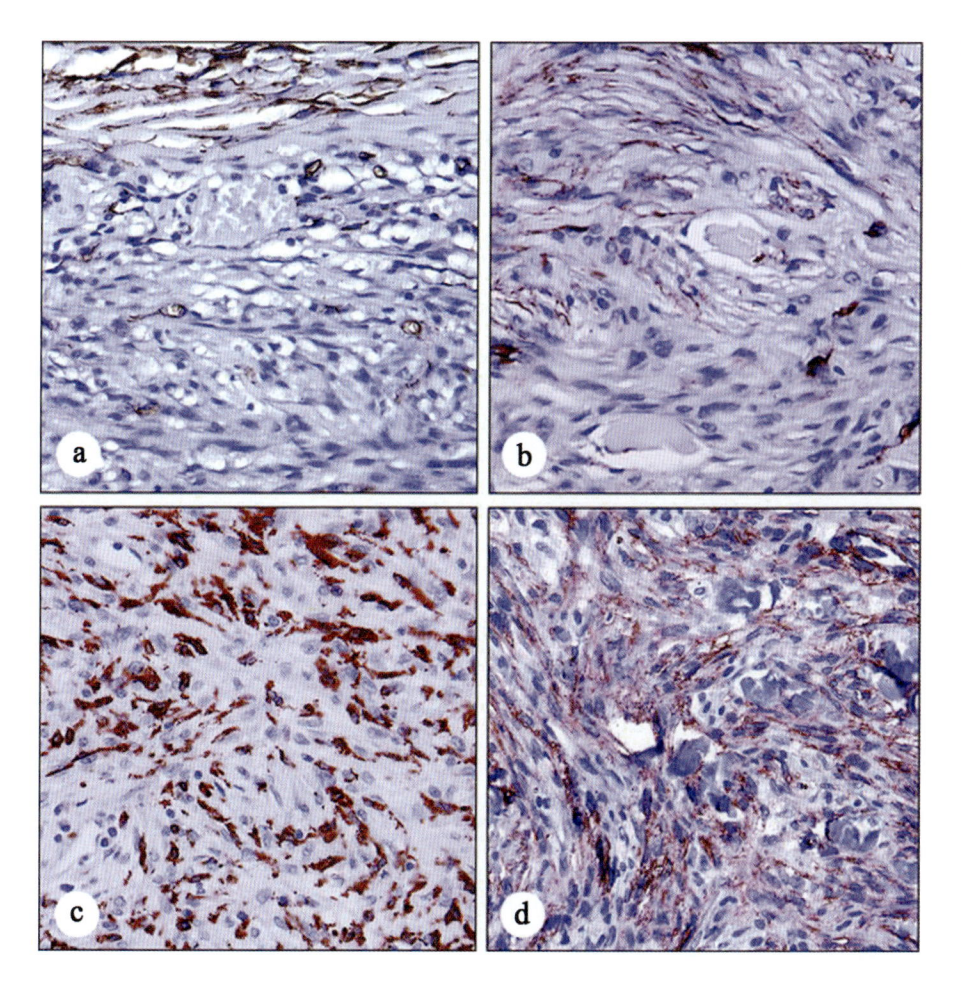

Figure 7.33 Immunohistochemical features of benign fibrous histiocytoma. (**a,b**) CD34 positivity is often seen in the tumor periphery, whereas the positivity in the central areas is essentially limited to endothelial cells. (**c**) CD163-positive histiocytes are often numerous, similar to the appearance seen in FXIIIa immunostain. Note that the lesional fibroblasts are negative for this histiocytic marker. (**d**) Most benign fibrous histiocytomas are CD10 positive.

BHFs are variably positive for alpha smooth muscle actin and CD10, and often contain high numbers of sprinkled histiocytes positive for markers such as Factor XIIIa, CD68, and CD163. In BFH, however, the main neoplastic population is negative for histiocytic markers (Figure 7.33c), although antibodies to CD68 might label some nonhistiocytic cells also.[96–98]

Differential diagnosis

When infiltrating the subcutis, BFH often has a pushing margin or infiltrates into the fat only focally, in contrast to the diffuse fat infiltration commonly seen in DFSP. The lesional cells in BFH are typically negative for CD34, in contrast to DFSP.[96–98] DFSP is also more uniform and composed of more slender and delicate cells than the typically cytologically heterogeneous, more coarsely built and collagenous BFH.

BFH and nodular fasciitis can overlap in some cases, for example in their storiform cellular pattern. The latter is typically based in the subcutis and contains several features not typical of BFH: a looser texture with cleavages and microcystic degeneration, mucoid stroma, and osteoclastic giant cells.

Historically, larger intramuscular tumors sometimes designated as deep fibrous histiocytomas usually represented other entities, notably solitary fibrous tumors and sometimes even low-grade fibromyxoid sarcomas. Most tumors previously designated as BFHs of the orbit are now histologically classified as solitary fibrous tumors and these tumors are also CD34 positive.[100]

Plexiform fibrohistiocytic tumor and neurothekeoma are distinctive multinodular childhood lesions. The former is usually subcutaneous, showing a tentacle-like growth pattern, whereas the latter is a dermal lesion typically having a compartmental micronodular pattern. These tumors are discussed in Chapter 10.

Juvenile xanthogranuloma is a true histiocytic lesion positive for CD163 and arising from precursors of bone marrow origin. It is typically seen in children and young adults and often contains Touton giant cells, with peripheral nuclei arranged in a wreath-like pattern, and eosinophils (Chapter 35).

Variants of BFH

Several histologic variants of BFH have been identified. These variants show only minor clinical differences from typical BFHs, the most important being increased recurrence rate in cellular and atypical variants. They collectively constitute no more than 10% of all BFHs. The most important are: (1) aneurysmal, (2) cellular, (3) epithelioid, (4) lipidized, (5) atypical fibrous histiocytoma (with "monster cells"), and (6) granular cell variant. These variants are each discussed in detail here.

Aneurysmal (cellular) fibrous histiocytoma

This relatively rare cellular variant (2% of all BFHs) occurs in a wide age range, with a mild female predominance and a predilection for proximal parts of the extremities.[101–104] The tumor size varies from <0.5 cm to 4 cm, and grossly there is a blue or dark surface and cystic consistency, with some examples clinically simulating Kaposi's sarcoma or melanoma. Intratumoral hemorrhage can cause worrisome lesional enlargement, prompting medical attention. This variant can also form a large, exophytic sessile polypoid cutaneous mass. There is a 10% to 20% recurrence rate, and rare metastases have been reported.[94,104]

Histologically, this variant is cellular and at low magnification is characterized by cysts and hemorrhagic spaces, often containing extravasated erythrocytes and xanthoma cells. The main cellular component is a plump, slightly epithelioid or polygonal component admixed with variably prominent histiocytes (Figure 7.34). There is often vague storiform

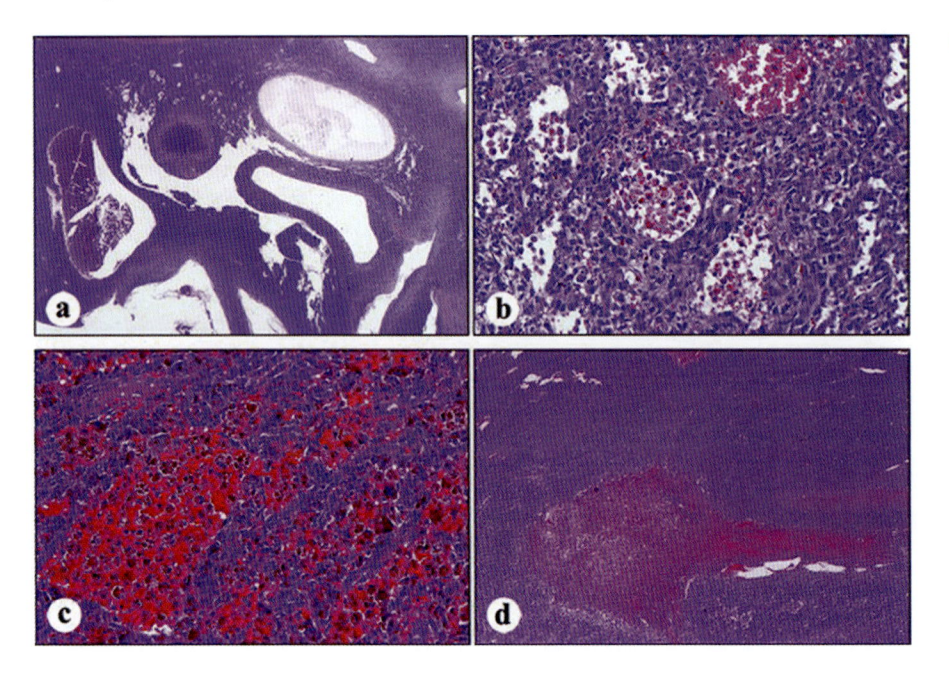

Figure 7.34 Aneurysmal fibrous histiocytoma contains cystic, often blood- or sometimes siderophage-filled spaces surrounded by spindled or slightly epithelioid tumor cells.

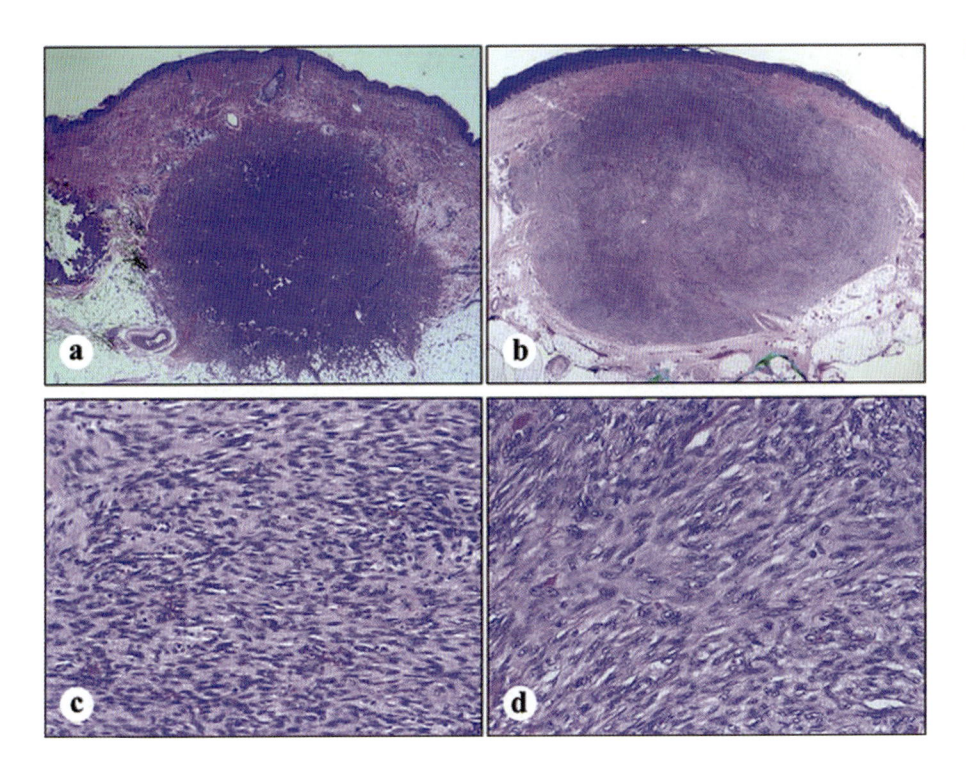

Figure 7.35 Cellular fibrous histiocytoma. Two examples with low and high magnifications. Note the predominantly subcutaneous location of a, with both dermal and subcutaneous components in b, fat infiltration in a, and a tendency to fascicular pattern in c and d.

architecture, and a hemangiopericytoma-like vascular pattern can be present.[103] Moderate mitotic activity (5 mitoses per 10 HPFs) is not unusual. This variant should not be confused with angiomatoid fibrous histiocytoma, a childhood tumor composed of densely cellular sheets of round to spindled cells. Aneurysmal fibrous histiocytoma can be distinguished from Kaposi's sarcoma by its having a more heterogeneous cellular composition, and by further immunohistochemical analysis (i.e., negative for CD34 and HHV8). Melanocytic markers are also absent.

Cellular fibrous histiocytoma

Cellular fibrous histiocytoma is the designation for densely cellular benign fibrous histiocytoma composed of fascicular sheets of spindle cells without significant atypia.[91] Clinical presentation is similar to that of other fibrous histiocytomas. The lesions usually measure 0.5 cm to 2.5 cm in maximum diameter, have a higher mitotic rate than typical fibrous histiocytomas (up to 10 mitoses per 10 HPFs), and often extend to the subcutis (Figure 7.35). A zone of central necrosis is commonly present, and fat infiltration can be prominent (Figure 7.36). Recurrence is more common than in other variants of BFH, occurring in up to 25% of patients. Immunohistochemically, the tumors do not differ from other fibrous histiocytomas, except that SMA positivity might be more prominent, and they are typically negative for CD34, in contrast to DFSP.

Epithelioid fibrous histiocytoma

The epithelioid variant typically forms a small, dome-shaped or polypoid lesion of 0.5 cm to 2 cm (median: <1 cm), often surrounded by an epithelial collar (Figure 7.37a,b). It occurs in a wide age range, with the median age being 43 years in the collective data from the largest series, most commonly in the lower extremities.[105–108] Histologically, the epithelioid variant is composed of uniform polygonal epithelioid cells with abundant cytoplasm and distinct borders, with a variably collagenous background (Figure 7.37c,d). Reported immunohistochemical findings include positivity for EMA and ALK. The latter is associated with ALK-gene rearrangements.[109,110]

The epithelioid variant of BFH should not be confused with Spitz nevus, which usually shows nevoid type clusters in the periphery and often has a spindle cell component; S100 immunostaining is helpful in making this distinction, because epithelioid BFHs are negative.

Epithelioid BFH should also be separated from true histiocytic proliferations with epithelioid morphology, such as solitary reticulohistiocytoma (solitary epithelioid histiocytoma). Whereas epithelioid BFH contains high numbers of infiltrating histiocytes, the large tumor cells are negative for histiocytic markers such as CD163. In reticulohistiocytoma, the epithelioid cells are positive.[111]

Lipidized fibrous histiocytoma

The lipidized variant occurs in slightly older persons than does BFH on average – median age, 50–53 years – and one study showed a male predominance and more common occurrence in the ankles and legs.[112] Based on a recent comparative study, however, these tendencies might not be greater than that for BFHs in general.[113] Tumor size varies, but examples of this variant can be larger than other BFHs; some are >5 cm. Histologically, there are abundant lipid-laden histiocytes,

Chapter 7: Fibromas and benign fibrous histiocytomas

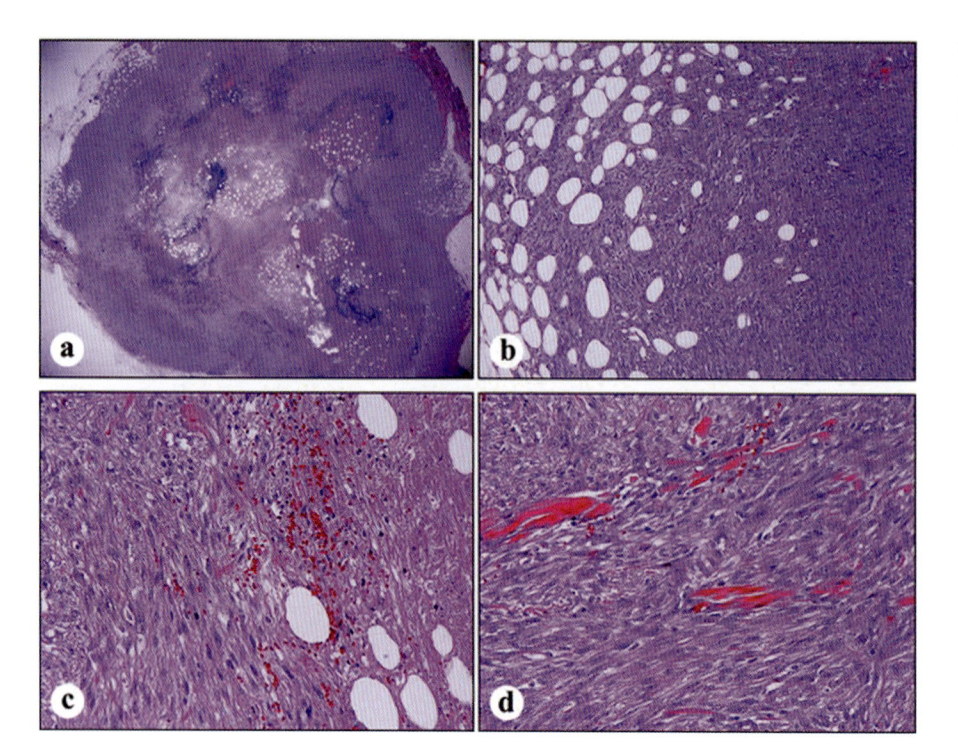

Figure 7.36 (a–c) Cellular fibrous histiocytoma showing fat infiltration and central coagulative necrosis. (d) The viable areas show collagen trapping and uniform cytology consistent with cellular fibrous histiocytoma.

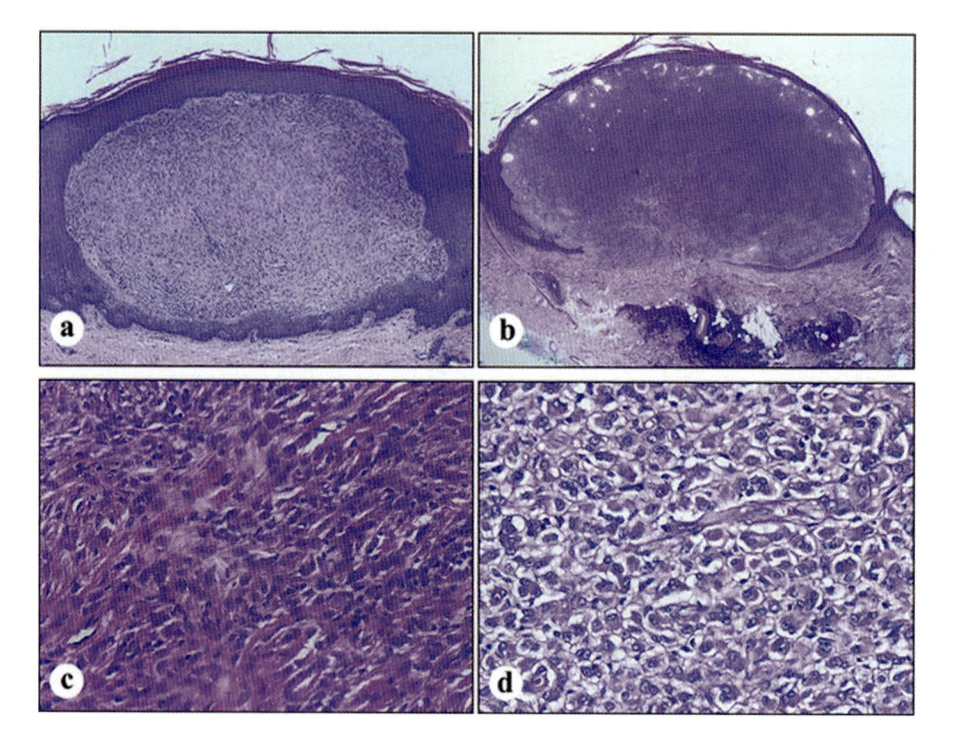

Figure 7.37 (a,b) Epithelioid fibrous histiocytoma typically forms a polypoid nodule, often surrounded by an epithelial collar. (c,d) The tumor cells have a moderate amount of epithelioid cytoplasm and are embedded in a collagenous matrix.

xanthoma cells, and often a rich vascular pattern with hyalinized vessel walls. The fibroblastic tumor cells are rather inconspicuous (Figure 7.38). There does not seem to be any significant association with hyperlipidemia.[112,113]

In contrast, xanthoma is a true histiocytic proliferation that frequently occurs in connection with hyperlipidemia, and often in tendons or bursae. It often contains cholesterol crystals and lacks a fibroblastic myofibroblastic neoplastic component.

Atypical fibrous histiocytoma

Atypical fibrous histiocytoma, pseudosarcomatous fibrous histiocytoma, or *dermatofibroma with monster cells* are the designations used for BFHs with variable, usually focal cytologic

Chapter 7: Fibromas and benign fibrous histiocytomas

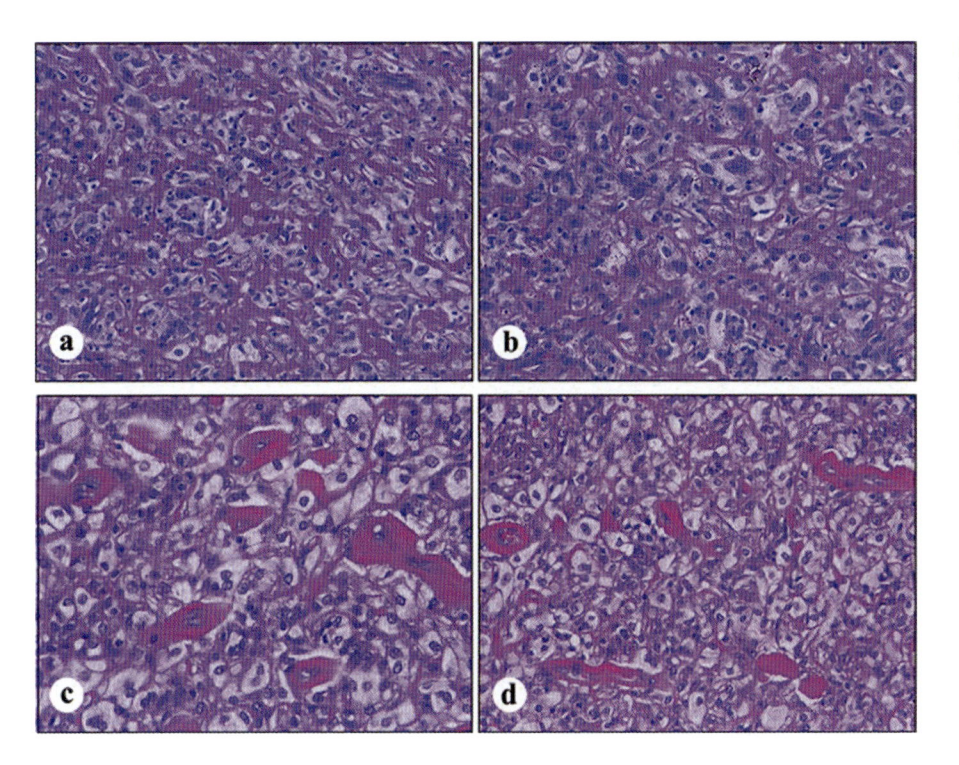

Figure 7.38 Lipidized fibrous histiocytoma contains pale cytoplasmic, lipid-laden cells, some of which also have cytoplasmic hemosiderin. Perivascular and variable stromal sclerosis are typical of this variant.

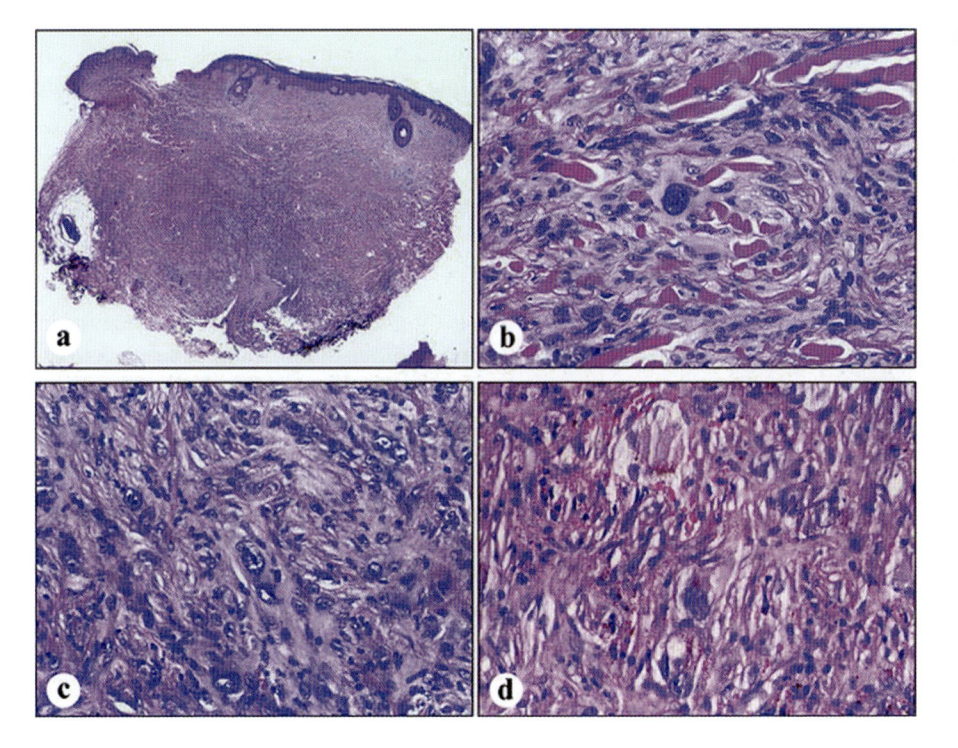

Figure 7.39 Atypical fibrous histiocytoma. (a) Low magnification of this case is similar to an ordinary variant of BFH with predominantly dermal involvement. (b,c) Higher magnification reveals scattered cells with abundant cytoplasm and markedly enlarged nuclei. (d) Another example shows large atypical cells, some of which are multinucleated with prominent nucleoli.

atypia and the potential for atypical mitoses.[114–118] These tumors are clinically similar to other variants. They occur in a wide age range, with a predilection for the extremities. The scattered atypical cells have large nuclei, often with prominent nucleoli, and some also have abundant cytoplasm, but mitotic activity is very low (Figure 7.39). Although such lesions have demonstrated benign behavior, rare metastases have been reported to other sites of skin, colon, and lung. Some atypical BFHs have features of aneurysmal fibrous histiocytoma.[118] Complete excision and follow-up is advisable. Some forms of atypical fibrous histiocytoma are CD34 positive, and even focal keratin positivity is sometimes present. Tumors recently designated as superficial CD34-positive fibroblastic tumors may also have been previously included under "atypical fibrous histiocytoma."[119] Examples with ALK and CD30 immunoreactivity and ALK gene rearrangements have been reported.

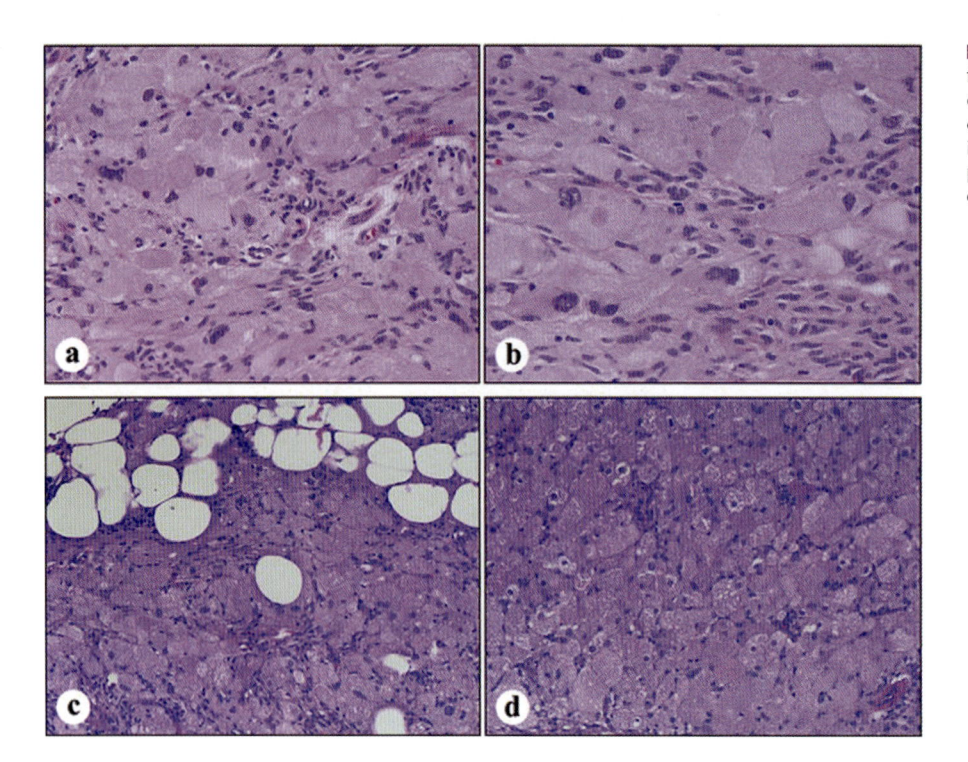

Figure 7.40 (a,b) Two examples of granular cell fibrous histiocytoma (primitive non-neural granular cell tumors of the skin). These tumors have cytoplasmic granularity not dissimilar to that seen in ordinary granular cell tumor, but they are S100 protein negative. (c) Subcutaneous fat infiltration is common in this variant.

These should not be confused with anaplastic large cell lymphoma variants.[120] Atypical fibrous histiocytoma differs conceptually from atypical fibroxanthoma (AFX) by the presence of widespread, often MFH-like cytologic atypia in the latter.

Granular cell fibrous histiocytoma and related lesions

The granular cell variant is rare. Based on a small number of reported cases, this variant occurs in young adults with an apparent male predominance and seems to have a predilection for the shoulder and back. It forms papules or nodules up to 2 cm to 3 cm, often being larger than conventional BFHs.[121–123] Histologically, the tumor is composed of cells with abundant granular cytoplasm quite similar to the ordinary (almost invariably S100-protein-positive) granular cell tumor, but the tumor cells are negative for S100 protein, in contrast to the ordinary granular cell tumor (Figure 7.40). Focal muscle actin positivity has been reported. The tumor can also contain areas resembling conventional BFH. Behavior has been benign in one study with follow-up.[122]

Somewhat similar, S100-protein-negative cutaneous (dermal) tumors were also reported as "primitive non-neural granular cell tumors of skin."[124,125] Most of these cases occurred in the neck, back, or shoulder, and most were <1 cm (range, 0.2–2.8 cm). Most patients were <40 years old. There can be significant mitotic activity equal to 2 to 12/10 HPFs; nuclear atypia with pleomorphism occurs in some cases. Local recurrences were not reported with limited follow-up, but one patient had a regional lymph node metastasis with no distant metastases, and a benign clinical course following a local excision without adjuvant therapy.[125]

Genetics

BFHs have been proved monoclonal by HUMARA analysis in one study,[126] although another study showed heterogeneity, with a polyclonal pattern in some and monoclonal in others.[127] Considering the large numbers of infiltrating non-neoplastic cells, polyclonal results are not unexpected, and one should not rule out the presence of a monoclonal component.

Cytogenetic study of 13 cases revealed heterogeneous, clonal chromosomal changes with near-diploid karyotypes in 38% of cases, more commonly in the cellular fibrous histiocytomas (three of four cases) than the conventional examples (two of eight cases).[128] A t(12;19)(p12;q13) translocation was reported as the only cytogenetic change in one case of recurrent aneurysmal BFH of the leg.[129]

References

Elastofibroma

1. Järvi OH, Saxén EA, Hopsu-Havu VK, Vartiovaara JJ, Vaissalo VT. Elastofibroma: a degenerative pseudotumor. *Cancer* 1969;23:42–63.
2. Järvi OH, Länsimies PH. Subclinical elastofibromas in the scapular region in an autopsy series. *Acta Pathol Microbiol Scand A* 1975;83:87–108.
3. Nagamine N, Nohary Y, Ito E. Elastofibromas in Okinawa: a clinicopathologic study of 170 cases. *Cancer* 1982;50:1794–1805.
4. Lococo F, Cesario A, Mattei F, et al. Elastofibroma dorsi: clinicopathological analysis of 71 cases. *Thorac Cardiovasc Surg* 2013;61:215–222.
5. Enjoji M, Sumiyoshi K, Sueyoshi K. Elastofibromatous lesion of the stomach in a patient with elastofibroma dorsi. *Am J Surg Pathol* 1985;9:233–237.

6. Madri JA, Dise CA, LiVolsi VA, Merino MJ, Bibro MC. Elastofibroma dorsi: an immunohistochemical study of collagen content. *Hum Pathol* 1981;12:186–190.

7. Hisaoka M, Hashimoto H. Elastofibroma: clonal fibrous proliferation with predominant CD34-positive cells. *Virchows Arch* 2006;448:195–199.

8. McComb EN, Feely MG, Neff JR, et al. Cytogenetic instability, predominantly involving chromosome 1, is characteristic of elastofibroma. *Cancer Genet Cytogenet* 2001;126:68–72.

9. Vanni R, Marras S, Faa G, et al. Chromosome instability in elastofibroma. *Cancer Genet Cytogenet* 1999;111:182–183.

10. Batstone P, Forsyth L, Goodlad J. Clonal chromosome aberrations secondary to chromosome instability in elastofibroma. *Cancer Genet Cytogenet* 2001;128:46–47.

11. Nishio JN, Iwasaki H, Ohjimi Y, et al. Gain of Xq detected by comparative genomic hybridization in elastofibroma. *Int J Mol Med* 2001;10:277–280.

Nuchal fibroma and nuchal fibrocartilaginous pseudotumor

12. Enzinger FM, Weiss SW. *Soft Tissue Tumors*, 2nd edn. St. Louis; CV Mosby; 1983.

13. Michal M, Fetsch JF, Hes O, Miettinen M. Nuchal-type fibroma: a clinicopathologic study of 52 cases. *Cancer* 1999;85:156–163.

14. Balachandran K, Allen PW, MacCormac LB. Nuchal fibroma: a clinicopathologic study of nine cases. *Am J Surg Pathol* 1995;19:313–317.

15. Abraham Z, Rosenbaum M, Rosner I, et al. Nuchal fibroma. *J Dermatol* 1997;24:262–265.

16. Banney LA, Weedon D, Muir JB. Nuchal fibroma associated with scleredema, diabetes mellitus and organic solvent exposure. *Australas J Dermatol* 2000;41:39–41.

17. Diwan AH, Graves ED, King JA, Horenstein MG. Nuchal-type fibroma in two related patients with Gardner's syndrome. *Am J Surg Pathol* 2000;24:1563–1567.

18. Dawes CL, LaHei ER, Tobias V, Kern I, Stening W. Nuchal fibroma should be recognized as a new extracolonic manifestation of Gardner-variant familial adenomatous polyposis. *Aust N Z J Surg* 2000;70:824–826.

19. O'Connell JX, Janzen DL, Hughes TR. Nuchal fibrocartilaginous pseudotumor: a distinctive soft tissue lesion associated with prior neck injury. *Am J Surg Pathol* 1997;21:836–840.

20. Laskin WB, Fetsch JF, Miettinen M. Nuchal fibrocartilaginous pseudotumor: a clinicopathologic study of five cases and review of the literature. *Mod Pathol* 1999;12:663–668.

21. Zamecnik M, Michal M. Nuchal fibrocartilaginous pseudotumor: immunohistochemical and ultra structural study of two cases. *Pathol Int* 2001;51:723–728.

Collagenous fibroma

22. Evans HL. Desmoplastic fibroblastoma. A report of seven cases. *Am J Surg Pathol* 1995;19:1077–1081.

23. Nielsen GP, O'Connell JX, Dickersin GR, Rosenberg AE. Collagenous fibroma (desmoplastic fibroblastoma): a report of seven cases. *Mod Pathol* 1996;9:781–785.

24. Hasegawa T, Shimoda T, Hirohashi S, Hizawa K, Sano T. Collagenous fibroma (desmoplastic fibroblastoma): report of four cases and review of the literature. *Arch Pathol Lab Med* 1998;122:455–460.

25. Miettinen M, Fetsch JF. Collagenous fibroma (desmoplastic fibroblastoma). A clinicopathological analysis of 63 cases of a distinctive soft tissue lesion with stellate-shaped fibroblasts. *Hum Pathol* 1998;28:676–682.

26. Bernal K, Nelson M, Neff JR, Nielsen SM, Bridge JA. Translocation (2;11)(q31;q12) is recurrent in collagenous fibroma (desmoplastic fibroblastoma). *Cancer Genet Cytogenet* 2004;149:161–163.

27. Macchia G, Trombetta D, Möller E, et al. FOSL1 as a candidate target gene for 11q12 rearrangements in desmoplastic fibroblastoma. *Lab Invest* 2012;92:735–743.

Fibroma of tendon sheath

28. Chung EB, Enzinger FM. Fibroma of tendon sheath. *Cancer* 1979;44:1945–1954.

29. Smith PS, Pieterse AS, McClure J. Fibroma of tendon sheath. *J Clin Pathol* 1992;35:842–848.

30. Cooper PH. Fibroma of tendon sheath. *J Am Acad Dermatol* 1984;11:625–628.

31. Lundgren LG, Kindblom LG. Fibroma of tendon sheath: a light and electron microscopic study of 6 cases. *Acta Pathol Microbiol Immunol Scand [A]*. 1984;92:401–409.

32. Hashimoto H, Tseneyoshi M, Daimaru Y, Ushijama M, Enjoji M. Fibroma of tendon sheath: a tumor of myofibroblasts: a clinicopathologic study of 18 cases. *Acta Pathol Jpn* 1985;35:1099–1107.

33. Humphreys S, McKee PH, Fletcher CD. Fibroma of tendon sheath: a clinicopathologic study. *J Cutan Pathol* 1986;13:331–338.

34. Pulitzer DR, Martin PC Reed RJ. Fibroma of tendon sheath: a clinicopathologic study of 32 cases. *Am J Surg Pathol* 1989;13:472–479.

35. Lamovec J, Brock M, Voncina D. Pleomorphic fibroma of tendon sheath. *Am J Surg Pathol* 1991;15:1202–1205.

36. Millon SJ, Bush DC, Garbes AD. Fibroma of tendon sheath in the hand. *J Hand Surg (Am)* 1994;19:788–793.

37. Satti MB. Tendon sheath tumours: a pathological study of the relationship between giant cell tumor of tendon sheath and fibroma of tendon sheath. *Histopathology* 1992;20:213–220.

38. Dal Cin P, Sciot R, Se Smet L, Van Den Berghe H. Translocation 2;11 in a fibroma of tendon sheath. *Histopathology* 1988;32:433–435.

Sclerotic fibroma of skin (circumscribed storiform collagenoma) and pleomorphic fibroma

39. Weary PE, Gorlin RJ, Gentry WC Jr, Comer JF, Greer KE. Multiple hamartoma syndrome (Cowden's disease). *Arch Dermatol* 1972;106:682–690.

40. Rapini RP, Golitz LS. Sclerotic fibromas of the skin. *Am J Acad Dermatol* 1989;20:266–271.

41. Lo WL, Wong CK. Solitary sclerotic fibroma. *J Cutan Pathol* 1990;17:269–273.

42. Metcalf JS, Maize JC, le Boit PW. Circumscribed storiform collagenoma (sclerosing fibroma). *Am J Dermatopathol* 1991;13:122–129.

43. Requena L, Gutierrez J, Sanchez Yus E. Multiple sclerotic fibromas. A cutaneous marker of Cowden's disease. *J Cutan Pathol* 1992;19:346–351.

44. High WA, Stewart D, Essary LR, et al. Sclerotic fibroma-like change in various neoplastic and inflammatory skin lesions: is sclerotic fibroma a distinct entity? *J Cutan Pathol* 2004;31: 373–378.

45. Mahmood MN, Salama ME, Chaffins M, et al. Solitary sclerotic fibroma of skin: a possible link with pleomorphic fibroma with immunophenotypic expression for O13 (CD99) and CD34. *J Cutan Pathol* 2003;30:631–636.

46. Hanft VN, Shea CR, McNutt NS, et al. Expression of CD34 in sclerotic ("plywood") fibromas. *Am J Dermatopathol* 2000;22:17–21.

47. Martin-Lopez R, Feal-Cortizas C, Fraga J. Pleomorphic sclerotic fibroma. *Dermatology* 1999;198:69–72.

48. Kamino H, Lee JY, Berke A. Pleomorphic fibroma of the skin: a benign neoplasm with cytologic atypia: a clinicopathologic study of eight cases. *Am J Surg Pathol* 1989;13:107–113.

49. Layfield LJ, Fain JS. Pleomorphic fibroma of skin. A case report and immunohistochemical study. *Arch Pathol Lab Med* 1991;115:1046–1049.

50. Garcia-Doval I, Casas L, Toribio J. Pleomorphic fibroma of the skin, a form of sclerotic fibroma: an immunohistochemical study. *Clin Exp Dermatol* 1998;23:22–24.

51. Rudolph P, Schubert C, Zelger BG, Zelger B, Parwaresch R. Differential expression of CD34 and Ki-M1p in pleomorphic fibroma and dermatofibroma with monster cells. *Am J Dermatopathol* 1999;21:414–419.

52. Kamino H, Reddy VB, Gero M, Greco MA. Dermatomyofibroma. A benign cutaneous, plaque-like proliferation of fibroblasts and myofibroblasts in young adults. *J Cutan Pathol* 1992;19: 85–93.

53. Mentzel T, Calonje E, Fletcher CD. Dermatomyofibroma: additional observations on a distinctive cutaneous myofibroblastic tumour with emphasis on differential diagnosis. *Br J Dermatol* 1993;129:69–73.

Giant cell fibroma and irritation fibroma of oral cavity

54. Weathers DR, Callihan MD. Giant-cell fibroma. *Oral Surg Oral Med Oral Pathol* 1974;37:374–384.

55. Houston GD. The giant cell fibroma: a review of 464 cases. *Oral Surg Oral Med Oral Pathol* 1982;53:582–587.

56. Reibel J. Oral fibrous hyperplasias containing stellate and multinucleated cells. *Scand J Dent Res* 1982;90:217–226.

57. Bouquot JE, Gundlach KK. Oral exophytic lesions in 23,616 white Americans over 35 years of age. *Oral Surg Oral Med Oral Pathol* 1986;62:284–291.

58. Bakos LH. The giant cell fibroma: a review of 116 cases. *Ann Dent* 1992;51:32–35.

59. Odell EW, Lock C, Lombardi TL. Phenotypic characterization of stellate and giant cells in giant cell fibroma by immunohistochemistry. *J Oral Pathol Med* 1994;23:284–287.

60. Magnusson BC, Rasmusson LG. The giant cell fibroma: a review of 103 cases with immunohistochemical findings. *Acta Odont Scand* 1995;53:293–296.

Palisaded myofibroblastoma

61. Weiss SW, Gnepp DR, Bratthauer GL. Palisaded myofibroblastoma: a benign mesenchymal tumor of lymph node. *Am J Surg Pathol* 1989;13:341–346.

62. Suster S, Rosai J. Intranodal hemorrhagic spindle-cell tumor with "amianthoid" fibers: report of six cases of a distinctive mesenchymal neoplasm of the inguinal region that simulates Kaposi's sarcoma. *Am J Surg Pathol* 1989;13:347–357.

63. Michal M, Chlumska A, Povysilova V. Intranodal "amianthoid" myofibroblastoma: report of six cases. Immunohistochemical and electron microscopical study. *Pathol Res Pract* 1992;188:199–204.

64. Rossi A, Bulgarini A, Rondanelli E, Incensati R. Intranodal palisaded myofibroblastoma: report of three new cases. *Tumori* 1995;81:464–468.

65. Eyden BP, Harris M, Greywoode GI, Christensen L, Banerjee SS. Intranodal myofibroblastoma: report of a case. *Ultrastruct Pathol* 1996;20:79–88.

66. Hisaoka M, Hashimoto H, Daimaru Y. Intranodal palisaded myofibroblastoma with so-called amianthoid fibers: a report of two cases with a review of the literature. *Pathol Int* 1998;48:307–312.

67. Basu A, Harvey DR. Palisaded myofibroblastoma: an uncommon tumour of lymph nodes. *Eur J Surg Oncol* 1998;24:609.

68. Fletcher CD, Stirling RW. Intranodal myofibroblastoma presenting in the submandibular region: evidence of a broader clinical and histological spectrum. *Histopathology* 1990;16:287–293.

69. D'Antonio A, Addesso M, Amico P, Fragetta F. Axillary intranodal palisaded myofibroblastoma: report of a case associated with chronic mastitis. *BMJ Case Rep* 2014;2014 doi:10.1136/bcr-2014-205877.

70. Bigotti G, Coli A, Mottolese M, Di Filippo F. Selective localization of palisaded myofibroblastoma with amianthoid fibers. *J Clin Pathol* 1991;44;761–764.

71. Creager AJ, Garwacki CP. Recurrent intranodal palisaded myofibroblastoma with metaplastic bone formation. *Arch Pathol Lab Med* 1999; 123:433–436.

72. Lioe TF, Allen DC, Bell JC. A case of multicentric intranodal palisaded myofibroblastoma. *Histopathology* 1994;24:173–175.

73. Skalova A, Michal M, Chlumska A, Leivo I. Collagen composition and ultrastructure of the so-called amianthoid fibres in palisaded myofibroblastoma: ultrastructural and immunohistochemical study. *J Pathol* 1992;167:335–340.

74. Kleist B, Poetsch M, Schmoll J. Intranodal palisaded myofibroblastoma with overexpression of cyclin D1. *Arch Pathol Lab Med* 2003;127:1040–1043.

75. Laskin WB, Lasota JP, Fetsch JF, et al. Intranodal palisading myofibroblastoma: another mesenchymal neoplasm with CTNNB1 (B-catenin gene) mutations. Clinicopathological, immunohistochemical, and molecular genetic study of 18 cases. *Am J Surg Pathol* 2015;39:197–205.

Myofibroblastoma of breast

76. Wargotz ES, Weiss SW, Norris HJ. Myofibroblastoma of the breast. Sixteen cases of a distinctive benign mesenchymal tumor. *Am J Surg Pathol* 1987;11:493–502.

77. Lee AH, Sworn MJ, Theaker JM, Fletcher CD. Myofibroblastoma of breast: an immunohistochemical study. *Histopathology* 1993;22:75–78.

78. Lazaro-Santander R, Garcia-Prats MD, Nieto S, et al. Myofibroblastoma of the breast with diverse histological features. *Virchows Arch* 1999;434:547–550.

79. Eyden BP, Shanks JH, Ioachim E, et al. Myofibroblastoma of breast:

evidence favoring smooth-muscle rather than myofibroblastic differentiation. *Ultrastruct Pathol* 1999;23:249–257.

80. Magro G, Bisceglia M, Michal M. Expression of steroid hormone receptors, their regulated proteins, and bcl-2 protein in myofibroblastoma of the breast. *Histopathology* 2000;36:515–521.

81. Magro G, Vecchio GM, Michal M, Eusebi V. Atypical epithelioid cell myofibroblastoma of the breast with multinodular growth pattern: a potential pitfall of malignancy. *Pathol Res Pract* 2013;209:463–466.

82. Arafah MA, Ginter PS, D'Alfonso TM, Hoda SA. Epithelioid mammary myofibroblastoma mimicking invasive lobular carcinoma. *Int J Surg Pathol* 2015;23:284–288.

83. McMenamin ME, Fletcher CD. Mammary-type myofibroblastoma of soft tissue: a tumor closely related to spindle cell lipoma. *Am J Surg Pathol* 2001;25:1022–1029.

84. Pauwels P, Sciot R, Croiset F, *et al*. Myofibroblastoma of the breast: genetic link with spindle cell lipoma. *J Pathol* 2000;191:282–285.

Benign fibrous histiocytoma and its variants

85. Calonje E, Fletcher CDM. Cutaneous fibrohistiocytic tumors: an update. *Adv Anat Pathol* 1994;1:2–15.

86. Zelger B, Zelger BG, Burgdorf WHC. Dermatofibroma – a critical evaluation. *Int J Surg Pathol* 2004;12:333–344.

87. Gonzalez S, Duatte I. Benign fibrous histiocytoma of the skin: a morphologic study of 290 cases. *Pathol Res Pract* 1982;174:379–391.

88. Mentzel T, Kutzner H, Rütten A, Hügel H. Benign fibrous histiocytoma (dermatofibroma) of the face: clinicopathological and immunohistochemical study of 34 cases associated with an aggressive clinical course. *Am J Dermatopathol* 2001;23:419–426.

89. Requena L, Farina MC, Fuente C, *et al*. Giant dermatofibroma: a little known clinical variant of dermatofibroma. *J Am Acad Dermatol* 1994;30: 714–718.

90. Kanitakis J, Carbonnel E, Delmonte S, *et al*. Multiple eruptive dermatofibromas with HIV infection: case report and literature review. *J Cutan Pathol* 2000;27:54–56.

91. Calonje E, Mentzel T, Fletcher CD. Cellular benign fibrous histiocytoma: clinicopathologic analysis of 74 cases of a distinctive variant of cutaneous fibrous histiocytoma. *Am J Surg Pathol* 1994;18:668–676.

92. Colome-Grimmer MI, Evans HL. Metastasizing cellular dermatofibroma: a report of two cases. *Am J Surg Pathol* 1996;20:1361–1367.

93. Bisceglia M, Attino V, Bacchi CE. Metastasizing "benign" fibrous histiocytoma of the skin: a report of two additional cases and review of the literature. *Adv Anat Pathol* 2006;13:89–96.

94. Doyle LA, Fletcher CD. Metastasizing "benign" cutaneous fibrous histiocytoma: a clinicopathologic analysis of 16 cases. *Am J Surg Pathol* 2013;37:484–495.

95. Schwob VS, Santa Cruz DJ. Palisading cutaneous fibrous histiocytoma. *J Cutan Pathol* 1986;13:403–407.

96. Prieto VG, Reed JA, Shea CR. Immunohistochemistry of dermatofibromas and benign fibrous histiocytomas. *J Cutan Pathol* 1995;22:336–341.

97. Lee KJ, Yang JM, Lee ES, Lee DY, Jang KT. CD10 is expressed in dermatofibromas. *Br J Dermatol* 2005;155:622–623.

98. Abenoza P, Lillemoe T. CD34 and factor XIIIa in the differential diagnosis of dermatofibroma and dermatofibrosarcoma protuberans. *Am J Dermatopathol* 1993;15:429–434.

99. Li DF, Iwasaki H, Kikuchi M, Ichiki M, Ogata K. Dermatofibroma: superficial fibrous proliferation with reactive histiocytes. A multiple immunostaining analysis. *Cancer* 1994;74:66–73.

100. Furusato E, Valenzuela IA, Fanburg-Smith JC, *et al*. Orbital solitary fibrous tumor: encompassing the terminology for hemangiopericytoma, giant cell angiofibroma, and fibrous histiocytoma of the orbit: reappraisal of 41 cases. *Hum Pathol* 2011;42:120–128.

101. Santa-Cruz DJ, Kyriakos M. Aneurysmal ("angiomatoid") fibrous histiocytoma of the skin. *Cancer* 1981;47:2053–2061.

102. Calonje E, Fletcher CDM. Aneurysmal benign fibrous histiocytoma: clinicopathological analysis of 40 cases of a tumour frequently misdiagnosed as a vascular neoplasm. *Histopathology* 1995;26:323–331.

103. Zelger BW, Zelger BG, Steiner H, Ofner D. Aneurysmal and hemangiopericytoma-like fibrous histiocytoma. *J Clin Pathol* 1996;49:313–318.

104. Sheehan KM, Leader MB, Sexton S, Cunningham F, Leen E. Recurrent aneurysmal fibrous histiocytoma. *J Clin Pathol* 2004;57:312–314.

105. Wilson Jones E, Cerio R, Smith NP. Epithelioid fibrous histiocytoma: a new entity. *Br J Dermatol* 1989;120:185–195.

106. Mehregan AH, Mehregan DR, Broecker A. Epithelioid cell histiocytoma. *J Am Acad Dermatol* 1992;26:243–246.

107. Glusac EJ, Barr RJ, Everett MA, Pitha J, Santa Cruz DJ. Epithelioid cell histiocytoma: a report of 10 cases including a cellular variant. *Am J Surg Pathol* 1994;18:583–590.

108. Singh Gomez C, Calonje E, Fletcher CDM. Epithelioid benign fibrous histiocytoma of skin: clinicopathological analysis of 20 cases of a poorly known variant. *Histopathology* 1994;24:123–129.

109. Doyle LA, Fletcher CD. EMA positivity in epithelioid fibrous histiocytoma: a potential diagnostic pitfall. *J Cutan Pathol* 2011;38:697–703.

110. Doyle LA, Mariño-Enriquez A, Fletcher CD, Hornick JL. ALK rearrangement and overexpression in epithelioid fibrous histiocytoma. *Mod Pathol* 2015;28:904–912.

111. Miettinen M, Fetsch JF. Solitary reticulohistiocytoma (solitary epithelioid histiocytoma): a clinicopathologic and immunohistochemical study of 44 cases. *Am J Surg Pathol* 2006;30:521–528.

112. Iwata J, Fletcher CD. Lipidized fibrous histiocytoma: clinicopathologic analysis of 22 cases. *Am J Dermatopathol* 2000; 22:126–134.

113. Wagamon K, Somach SC, Bass J, *et al*. Lipidized dermatofibromas and their relationship to serum lipids. *J Am Acad Dermatol* 2006;54:494–498.

114. Fukamizu H, Oku T, Inoue K, *et al*. Atypical pseudosarcomatous cutaneous

115. Leyva WH, Santa Cruz DJ. Atypical cutaneous fibrous histiocytoma. *Am J Dermatopathol* 1986;8:467–471.
116. Tamada S, Ackerman AB. Dermatofibroma with monster cells. *Am J Dermatopathol* 1987;9:380–387.
117. Beham A, Fletcher CDM. Atypical "pseudosarcomatous" variant of cutaneous benign fibrous histiocytoma: report of eight cases. *Histopathology* 1990;17:167–169.
118. Kaddu S, McMenamin ME, Fletcher CD. Atypical fibrous histiocytoma of the skin: clinicopathologic analysis of 59 cases with evidence of infrequent metastasis. *Am J Surg Pathol* 2002;26:35–46.
119. Carter JM, Weiss SW, Linos K, DiCaudo DJ, Folpe AL. Superficial CD34-positive fibroblastic tumor: report of 18 cases of a distinctive low-grade mesenchymal neoplasm of intermediate (borderline) malignancy. *Mod Pathol* 2014;27:294–302.
120. Szablewski V, Laurent-Roussel S, Rethers L, et al. Atypical fibrous histiocytoma of the skin with CD30 and p80/ALK1 positivity and ALK gene rearrangement. *J Cutan Pathol* 2014;41:715–719.
121. Val-Bernal JF, Mira C. Dermatofibroma with granular cells. *J Cutan Pathol* 1996;23:562–565.
122. Zelger BG, Steiner H, Kutzner H, Rütten A, Zelger B. Granular cell dermatofibroma. *Histopathology* 1997;31:258–262.
123. Soyer HP, Metze D, Kerl H. Granular cell dermatofibroma. *Am J Surg Pathol* 1997;19:168–173.
124. LeBoit PE, Barr RJ, Burall S, et al. Primitive granular-cell tumor and other cutaneous granular-cell neoplasms of apparent non-neural origin. *Am J Surg Pathol* 1991;15:48–58.
125. Lazar AF, Fletcher CDM. Primitive nonneural granular cell tumors of skin. Clinicopathologic analysis of 13 cases. *Am J Surg Pathol* 2005;29:927–934.
126. Chen TC, Kuo T, Chan HL. Dermatofibroma is a clonal proliferative disease. *J Cutan Pathol* 2000;27:36–39.
127. Hui P, Glusac EJ, Sinard JH, Perkins AS. Clonal analysis of cutaneous fibrous histiocytoma (dermatofibroma). *J Cutan Pathol* 2002;29:385–389.
128. Vanni R, Fletcher CD, Sciot R, et al. Cytogenetic evidence of clonality in cutaneous benign fibrous histiocytomas: a report of the CHAMP study group. *Histopathology* 2000;37:212–217.
129. Botrus G, Sciot R, Debiec, Rychter M. Cutaneous aneurysmal fibrous histiocytoma with a t(12;19)(p12;q13) as the sole cytogenetic abnormality. *Cancer Genet Cytogenet* 2006;164:155–158.

Chapter 8

Fibromatoses

Markku Miettinen
National Cancer Institute, National Institutes of Health

The designation *fibromatosis* refers to a specific group of locally recurring and potentially aggressive but nonmetastasizing fibroblastic and myofibroblastic tumors. These include palmar and plantar fibromatosis, which are often together referred to as *superficial fibromatoses*. Penile fibromatosis (Peyronie's disease) is often attached to this group, although it is not certain whether it is truly comparable with other fibromatoses.

Deep, fascia-associated or intramuscular fibromatoses are called *desmoid tumors*, synonymously known as *aggressive fibromatoses* and *musculoaponeurotic fibromatoses*. All fibromatoses are composed of fibroblasts with a substantial myofibroblastic component. General features and clonal chromosomal changes suggest that both superficial and deep fibromatoses are clonal, neoplastic processes. Although there have been misgivings about the continuing use of the term *desmoid fibromatosis* (or briefly, *desmoid*), the existence of different types of fibromatoses (especially in childhood) justifies this as a specifying designation.

Some fibromatoses, such as juvenile fibromatosis and lipofibromatosis, occur exclusively in children. Several childhood conditions called fibromatosis are not true fibromatoses, but reactive processes (fibromatosis colli), peculiar hyperplastic processes (gingival fibromatosis), or aberrations in morphogenesis (juvenile hyaline fibromatosis). These childhood lesions termed fibromatoses are discussed in Chapter 9.

Palmar and plantar fibromatosis
Clinical features

Palmar fibromatosis is also known as *Dupuytren's contracture*, named for an early nineteenth-century French surgeon who described this condition. Palmar fibromatosis is common, especially in men of northern and eastern European, Scottish, and Irish descent.[1,2] A random sample from the Icelandic population revealed clinical signs of palmar fibromatosis in 19% of men and 4% of women, and increasing incidence by age, but only a fraction of these patients were symptomatic.[2] The median age of palmar fibromatosis patients is in the sixth decade, although it can occur at almost any age after infancy. It is rare in premenopausal women and very rare in children, however (Figure 8.1).

Based on epidemiologic studies, clinical associations and risk factors include heavy smoking, manual labor, low body mass index, elevated fasting blood glucose level/diabetes, and seizures. The nature of these associations is unclear, but free oxygen radicals have been hypothesized to promote this fibromatosis. Some patients have concurrent plantar or penile fibromatosis, clinodactyly (valgus position) of the fifth finger, and tendency to form keloids.[3–5]

Palmar fibromatosis manifests as multiple discontinuous nodules and string-like formations in the palmar fascia on the volar, especially ulnar, side of the hand, especially in the fourth and fifth fingers, where the fibromatosis usually starts. Advanced lesions can result in flexion contractures of the fingers and hand, and in rare cases, involvement of extensor fasciae can cause extension contracture of fingers. In 10% to 50% patients, both hands are involved.[3–7]

Knuckle pads are small (usually 1–3 mm), periarticular, subdermal nodules on the dorsolateral side of proximal interphalangeal or metacarpopharangeal joints of fingers. They are histologically similar to palmar fibromatosis, with which they can coexist, and in fact represent palmar fibromatosis lesions in a dorsolateral location. Knuckle pads do not cause contractures, however.[8]

Plantar fibromatosis is also named as *Ledderhose disease*, for a German surgeon who described this condition in the end of the nineteenth century. Plantar fibromatosis occurs more often in young adults than does palmar fibromatosis, with a peak incidence in the third decade, with a female predominance.[5,9,10] Occurrence in childhood is rare but has been reported from the age of 3 years, with an 8% relative frequency in the pediatric population in the Armed Forces Institute of Pathology (AFIP) consultation series (Figure 8.2).[5] Familial cases in multiple siblings with apparent autosomal dominant transmission have been reported, and familial tumors can occur at a younger age than the sporadic ones.[11,12]

Plantar fibromatosis forms single or multiple nodules in the plantar fascia of the foot, typically starting from the medial aspect of the plantar arch. The lesions are often asymptomatic, but they can become painful and interfere with footwear and walking. The nodules vary from microscopic to 2 cm in diameter.

Modern Soft Tissue Pathology, Second Edition, ed. Markku Miettinen. Published by Cambridge University Press. © Cambridge University Press 2016.

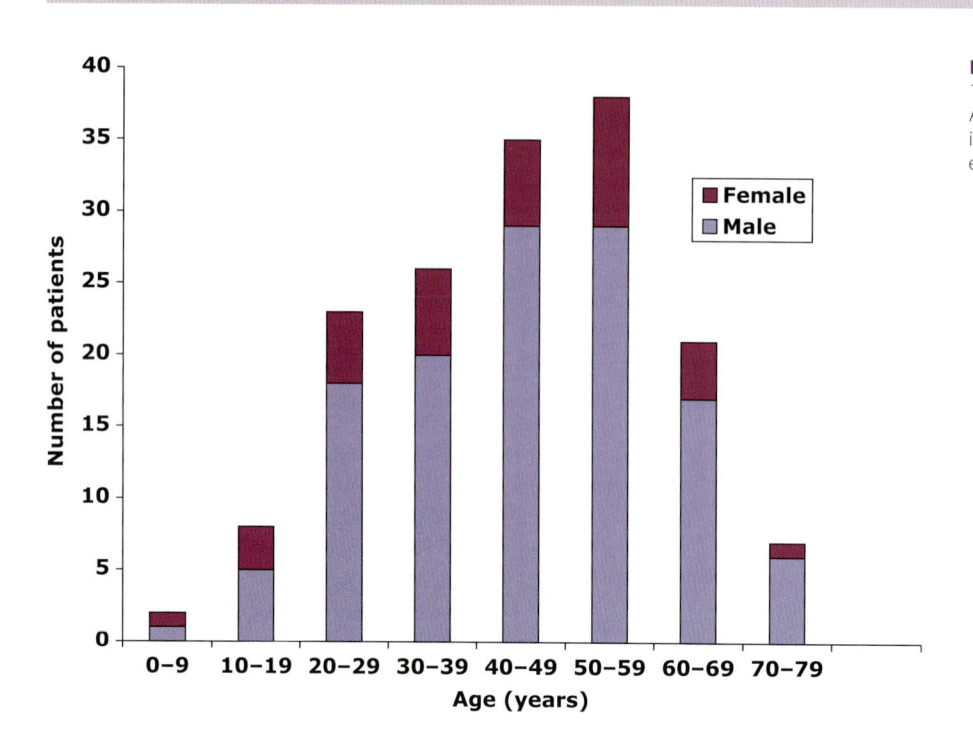

Figure 8.1 Age and sex distribution of 160 patients with palmar fibromatosis based on AFIP data on all cases. If civilian patients only are included, there is a lesser male predominance, especially for the middle-aged groups.

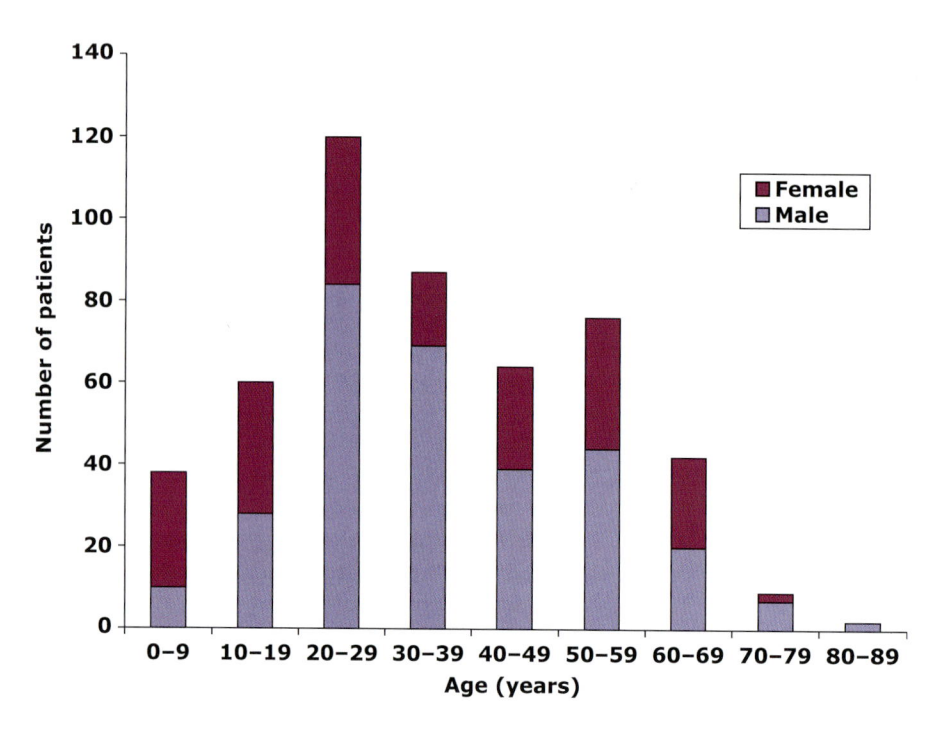

Figure 8.2 Age and sex distribution of 498 patients with plantar fibromatosis (AFIP data). If only civilian patients are included, there is a lesser male predominance in the middle-aged groups.

Some patients have synchronous plantar and palmar fibromatoses, and these patients in particular can have bilateral lesions.

Treatment

Treatment of plantar and palmar fibromatosis focuses on symptom control in the foot and correction of contractures in the hand. Selective excisions, debulking, and segmental fasciectomies are generally used. In advanced cases, more extensive fasciectomy and skin transplantation is sometimes necessary. Nonsurgical options include corticosteroid (triamcinolone acetonide) injections, especially for palmar lesions, and orthosis devices apposing the plantar nodules.[3,4,9,10]

Pathology

Grossly the lesions are firm to rubbery, gray-white fascial nodules varying from <1 mm up to 1.5 to 2 cm in diameter, but in plantar fibromatosis, coalescent nodules can form larger masses, perhaps >5 cm. Palmar and plantar fibromatoses are

Chapter 8: Fibromatoses

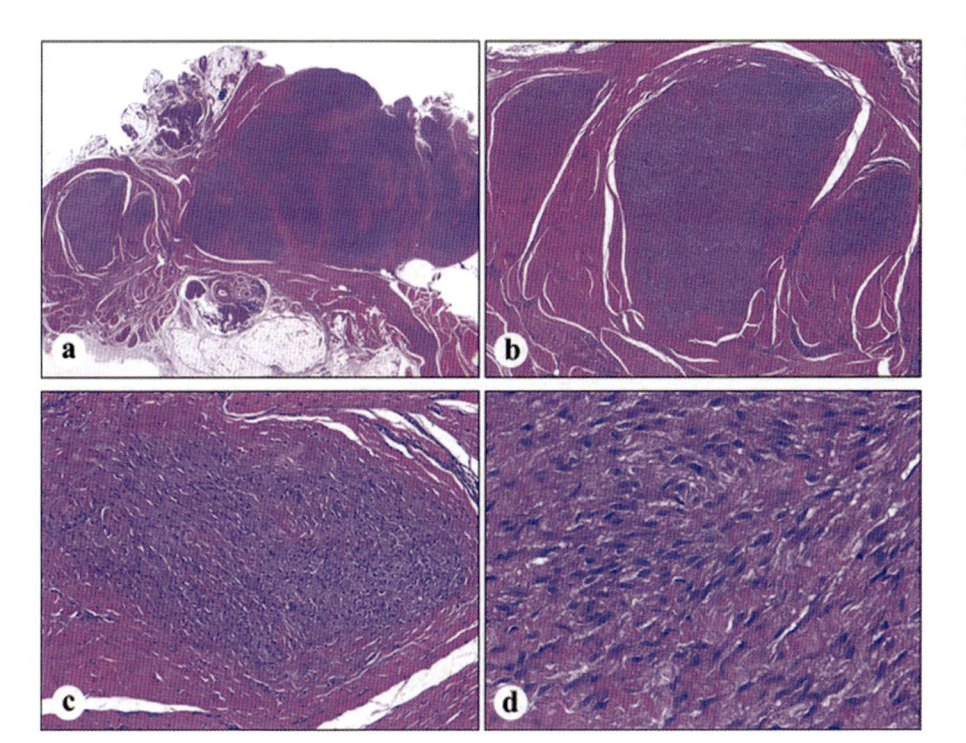

Figure 8.3 (a,b) Palmar fibromatosis lesion involves the palmar fascia in a multinodular manner. (c,d) The nodules may be circumscribed or coalescent, and are composed of uniform spindled myofibroblasts with no atypia in a variably collagenous stroma.

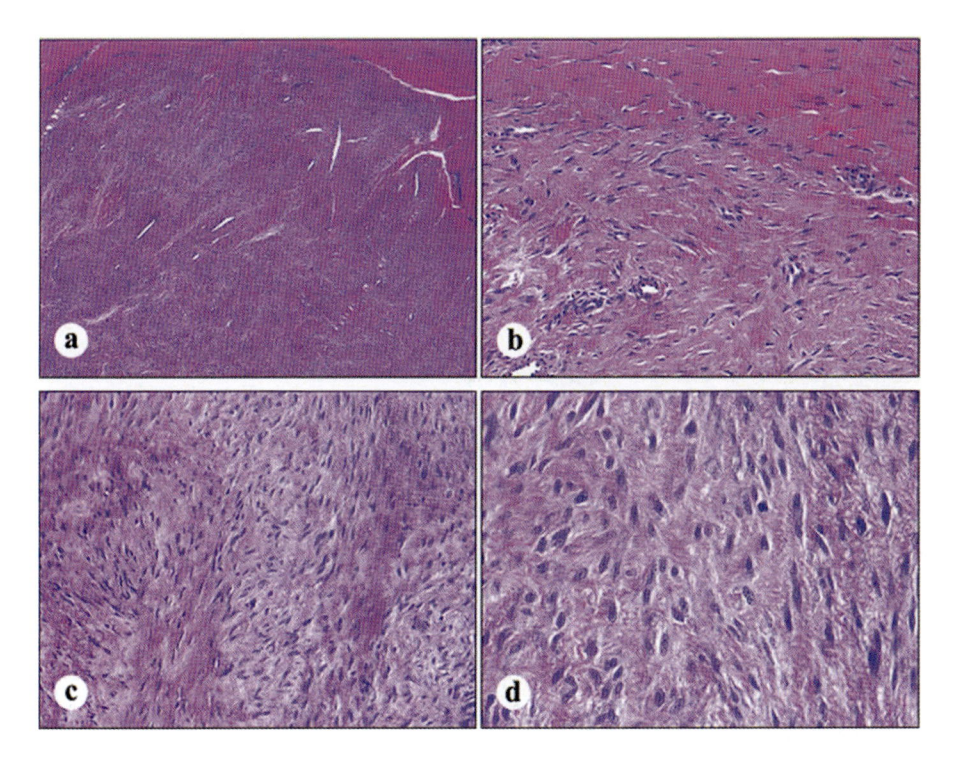

Figure 8.4 The cellularity and collagen content in palmar fibromatosis varies, probably in part by lesional age. On a higher magnification the myofibroblasts are uniform, with delicate nucleoli.

virtually indistinguishable histologically. Palmar lesions are sometimes divided into cellular (proliferative), involutional, and residual phases, in order of decreasing cellularity.[13] Larger nodules often show involution with central fibrosis (Figures 8.3 and 8.4).

Plantar fibromatosis often forms multiple, separate or conjoined, serpiginous nodules that are usually sharply demarcated, but sometimes have a ragged margin to the fascia (Figure 8.5). The cellular nodules are composed of uniform, spindled fibroblasts and myofibroblasts that are often oriented parallel to each other in a dense, collagenous background. When cut cross-sectionally, the lesional cells have a round appearance, sometimes with pericellular halos (Figure 8.6a,b). Profiles of blood vessels with prominent endothelia and smooth muscle layer are present inside and around the nodules, with some vessels having wide, gaping lumens (Figure 8.6c,d). Moderate

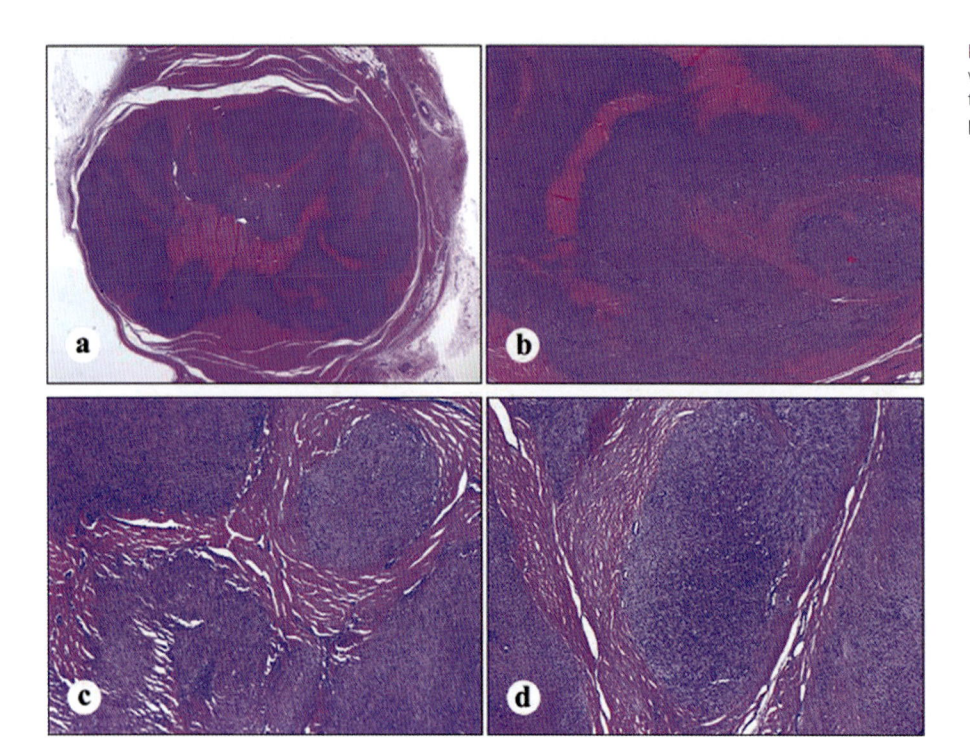

Figure 8.5 Plantar fibromatosis forms well-demarcated, sometimes coalescent nodules that contain serpiginous areas of myofibroblastic proliferation in the dense fascial collagen.

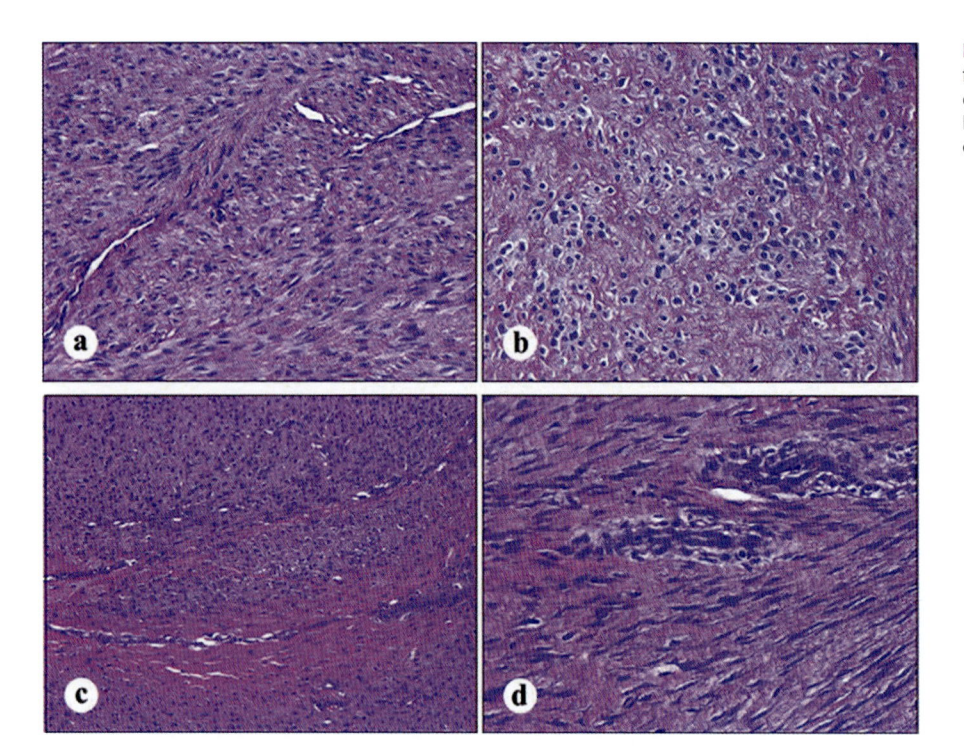

Figure 8.6 (a,b) The myofibroblasts of plantar fibromatosis can appear spindled or epithelioid, depending on the plane of sectioning. (c,d) Prominent blood vessels are often present around or sometimes within the cellular nodules.

mitotic activity (commonly 3–5 mitoses per 10 HPF, occasionally >10 mitoses per 10 HPFs) is sometimes present, especially in childhood lesions, but atypical mitoses do not occur (Figure 8.7a,b). Multinucleated giant cells, often with nuclei in a horseshoe pattern, are often seen in plantar fibromatosis, although they are rarely a prominent finding (Figure 8.7d).[5,14] Scattered mast cells and focal hemorrhage also can be present.

Immunohistochemically, variable positivity for smooth muscle actin is an expression of the myofibroblastic phenotype.[5,15] The lesional cells are negative for CD34, desmin, keratins, and S100 protein. Some cases of palmar and plantar fibromatosis show nuclear beta-catenin positivity, generally seen in a minority of cells, in contrast with desmoids.[16,17]

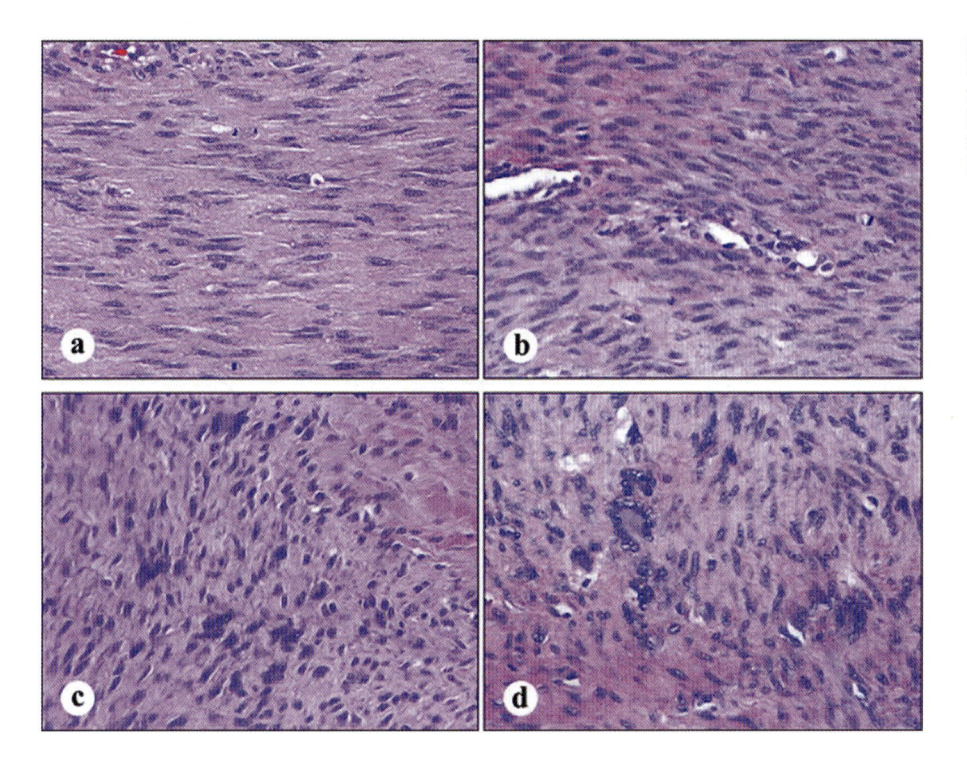

Figure 8.7 (a,b) Considerable mitotic activity can be found in plantar fibromatosis, and each of these two fields contains two mitotic figures. (c,d) Multinucleated giant cells are often present, and their nuclear orientation varies from random to a horseshoe pattern.

Differential diagnosis

Virtually no other lesion type arises precisely within the palmar or plantar fascia, and fibrosarcomas practically never occur in this location. Synovial sarcoma can present in the plantar area, but usually is located below the plantar fascia and never centered around it. Synovial sarcoma is typically more homogeneous and more highly cellular than plantar fibromatosis and lacks the homogeneous collagenous background, although it can have collagenous areas. Immunohistochemical demonstration of epithelial markers is helpful in problem cases. Epithelioid sarcoma often involves fascial structures, although this is very rare in the plantar foot. It contains epithelioid cells, often in a sclerosing stroma, typically with eosinophilic cytoplasm and demonstrating keratin and often CD34 immunoreactivity. Fibrous histiocytomas are skin-based lesions that practically never involve the plantar fascia and usually show a storiform pattern and a heterogeneous cellular composition.

Genetics

Cytogenetic studies of palmar fibromatosis have shown trisomies of chromosomes 7 and 8.[18] One study of 26 palmar fibromatoses showed heterogeneity but no recurrent changes in 69% of cases, including different translocations.[19] Another study found trisomy 8 to be the only chromosomal aberration in 2 of 21 palmar fibromases.[20]

In plantar fibromatosis, the reported cytogenetic changes include trisomies of chromosomes 8 and 14, although their frequency is not high,[20,21] and a reciprocal translocation t(2;7)(p13;p13) was reported in a tumor cell subpopulation in one case.[22] The human androgen receptor (HUMARA) test has failed to verify the monoclonal nature of palmar fibromatosis,[23] but this does not disprove its monoclonal nature, considering the potential for false-negative results with this test. Beta-catenin mutations have not been found in palmar/plantar fibromatosis, in contrast to desmoids,[16] and nuclear expression, typical of desmoids, is rare or absent in palmar and plantar fibromatoses.[16,17]

Penile fibromatosis (Peyronie's disease, plastic induration of the penis)

Peyronie's disease is clinically well-known but morphologically poorly understood, partly because histologic specimens are rare because the treatment is mostly conservative. There is some relationship with palmar-plantar fibromatosis, but unrelated fibroinflammatory processes may also have been included in this entity.

Clinical features

Peyronie's disease was named after the surgeon to King Louis XV of France. This fibromatosis-like process involves the inner part of the penile fascia (tunica albuginea) around the corpora cavernosa. An indurated penile plaque forms and causes a curvature on the side of plaque on erection, sometimes with pain or erectile dysfunction. The plaque is more commonly on the dorsal aspect, but it can be on either side.[24] Based on one epidemiologic study, this condition is not rare: there was a 3.2% prevalence rate in a questionnaire survey in Cologne, Germany.[25] In the same study, the mean age of patients with Peyronie's disease was 55 years, and the age range was from 26 to 69 years.

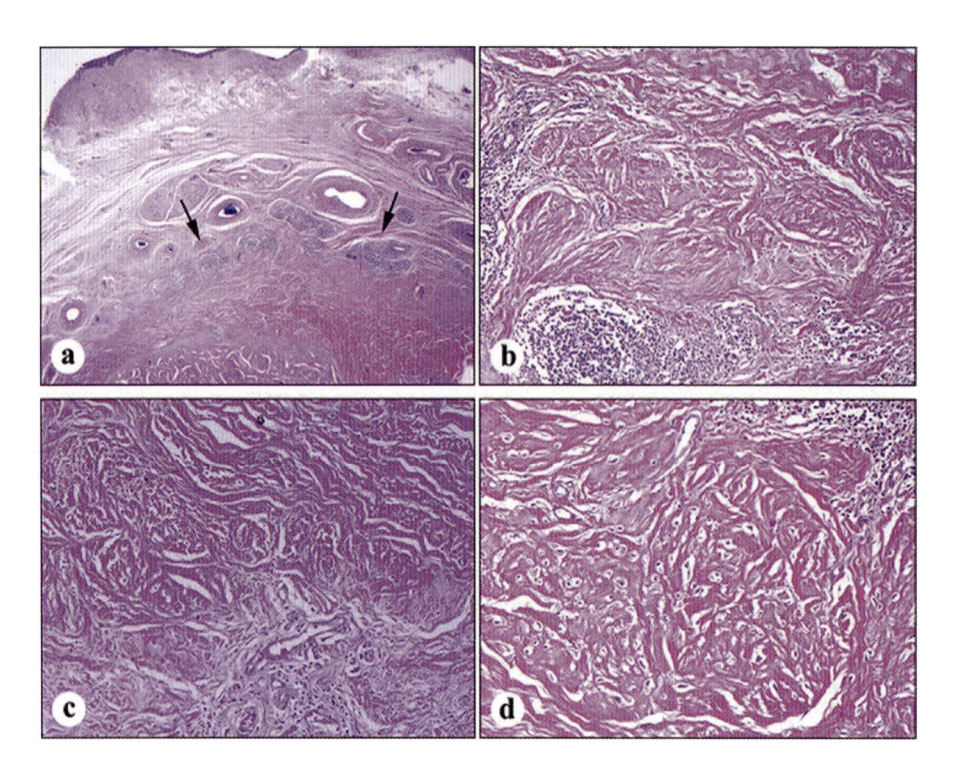

Figure 8.8 (a) Low magnification of an autopsy specimen of the penile shaft of a patient with Peyronie's disease shows marked fibrous thickening of tunica albuginea, the outer border of which contains patchy lymphoid infiltration (arrows). (b–d) The thickened tunica contains a paucicellular fibrosclerosing thickening with focal lymphoplasmacytic infiltrate.

The pathogenesis of this condition is unknown, but it has been thought to be related to a fibroinflammatory reaction to a subtle trauma, and subsequent effect of transforming growth factor beta. Treatment is usually conservative but can include surgery to excise the plaque, and the use of a penile prosthesis.

Pathology

Based on an older study, a Peyronie lesion varies from <1 cm to >5 cm, on average 2 cm.[26] A typical excision specimen is a firm, rubbery, sometimes calcified plaque-like lesion. The lesion primarily involves the tunica albuginea, but some excised specimens include elements of corpus cavernosum. Histologic features of penile fibromatosis can include a lymphoid reaction and formation of perivascular fibrosis progressing into a scar-like formation (Figure 8.8), sometimes with metaplastic bone.[26,27] Many lesions of Peyronie's disease histologically resemble palmar and plantar fibromatosis.

Caution is necessary to rule out sclerosing primary or metastatic carcinoma and epithelioid sarcoma; the presence of atypical epithelial cells is diagnostic and is aided by keratin immunohistochemistry. In addition, unrelated conditions, such as trauma (e.g., hematoma, fascial rupture) can cause scar-like indurations that clinically resemble penile fibromatosis.

Desmoid fibromatosis (desmoid tumor, aggressive fibromatosis, musculoaponeurotic fibromatosis)

Desmoid fibromatosis refers to deep, fascial, intramuscular, intra-abdominal or intrathoracic fibromatoses. These are synonymously known by the names shown in the heading, and sometimes referred to only as fibromatoses.[28] For clarity, the term *desmoid fibromatosis* (or briefly, desmoid) is used here.

By site of occurrence, desmoids are often clinicopathologically divided into (1) abdominal wall desmoids (sometimes referred to as *abdominal desmoids*, although this latter designation is also used in the literature of intra-abdominal desmoids), (2) intra-abdominal (mesenteric, pelvic), and (3) extra-abdominal (non-abdominal wall) desmoids, although this designation is sometimes collectively used for all desmoids presenting outside the abdominal cavity. These three groups have clinical differences, which merit their separate discussion, but pathologically and genetically desmoids are thought to be essentially similar, and their pathology and genetics are therefore discussed together.

The overall incidence of desmoids has been estimated as 2.5 to 4 per million based on a Finnish population-based study.[29] A minority of desmoids (9–16% in two series[30,31]) occurs in connection with familial adenomatous polyposis (FAP) syndrome, a tumor susceptibility syndrome including intestinal adenomatous polyposis, desmoids, osteomas, and epidermal inclusion cysts. Previously, the eponym *Gardner syndrome* was used when referring to FAP with desmoids, especially when associated with osteomas and epidermal cysts. Although FAP-associated desmoids more often occur intra-abdominally, they also can involve the abdominal wall and occasionally other peripheral sites.[32]

Desmoid tumors can be locally aggressive and recur, but they do not metastasize. Nevertheless, fatal complications are possible in large desmoids or those involving vital structures in

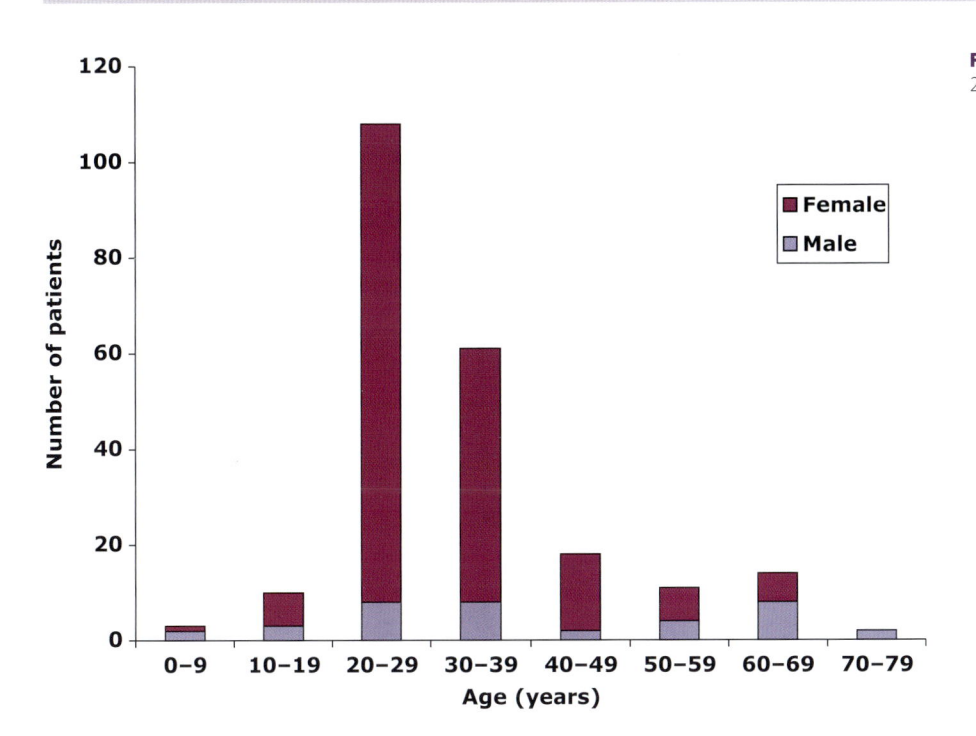

Figure 8.9 Age and sex distribution of 227 patients with abdominal wall desmoids.

locations such as the neck and thorax. Mesenteric desmoids, especially the intra-abdominal type in FAP, can lead to fatalities, particularly when related to surgical and infectious complications, especially in earlier times.[33]

Abdominal wall desmoid

These tumors, sometimes slightly misleadingly called *abdominal desmoids*, compose 30% to 40% of all desmoids and have a significant (3–9:1) female predominance (Figure 8.9), as reported in two separate studies.[29,34]

Abdominal wall desmoids usually present in young women of childbearing age between 20 and 40 years, are rare in men, and measure 5 cm on average, but can reach a size >20 cm. The midline abdominal wall inside the rectus abdominis muscle sheath is the most common location, but other abdominal wall muscles are occasionally involved. Many abdominal wall desmoids develop in scars, especially those following cesarean sections or other surgical trauma, and pregnancy and other hormonal influences are known to promote their growth.[35]

Most abdominal wall desmoids are easily cured by surgery, and two studies indicated that they have a 10% to 20% recurrence rate, lower than that in extra-abdominal desmoids.[29,34] Some studies, however, indicate that site is not a factor in determining recurrence rates of desmoids.[36]

Extra-abdominal desmoid

Extra-abdominal desmoids (outside the abdomen and abdominal wall) comprise more than 50% of all desmoids and occur in all ages >1 year, most commonly in young adults between the ages 20 and 40 years (Figure 8.11A).

Two larger studies showed a marked 3:1 female predominance.[37,38] A Finnish study showed a decrease of female predominance by age, with an even gender ratio observed in older adults.[29] Desmoids are rare in the first decade, but in the author's experience compose a minor portion (20–25%) of tumors sometimes designated as infantile fibromatosis by the age at which they occur.

The most common sites for extra-abdominal desmoids are the shoulder girdle (23%), chest wall (19%), thigh (14%), and neck (10%). Approximately 5% to 7% occur in the back, arm, forearm, buttocks, and knee, with rare examples in the head and distal extremities (Figure 8.11B).[37–40] Some chest wall desmoids extend into the pleural cavity and mediastinum, and others present primarily in the thoracic cavity or involve the breast; such tumors have been designated as fibromatosis of the breast. These tumors seem pathogenetically comparable to other extra-abdominal desmoids by their involvement of the beta-catenin pathway.[41]

Trauma preceding formation of the mass was reported in 19% of cases in one study, and the mass was painful in one-third of patients.[35] Isolated cases have been reported associated with silicone implants of the breast, sometimes associated with implant rupture.[42]

Tumor size varies from a small, <1- to 2-cm nodule to masses >20 cm; the larger tumors are especially seen in the shoulder girdle, chest wall, and inside the thoracic cavity. Recurrence rates reported in the largest and most recent series have varied between 20% and 30%,[37,38] although rates as high as 45% have been reported recently.[36] Repeated recurrences can follow incomplete excision, but none of the long-term follow-up studies has revealed distant metastases.

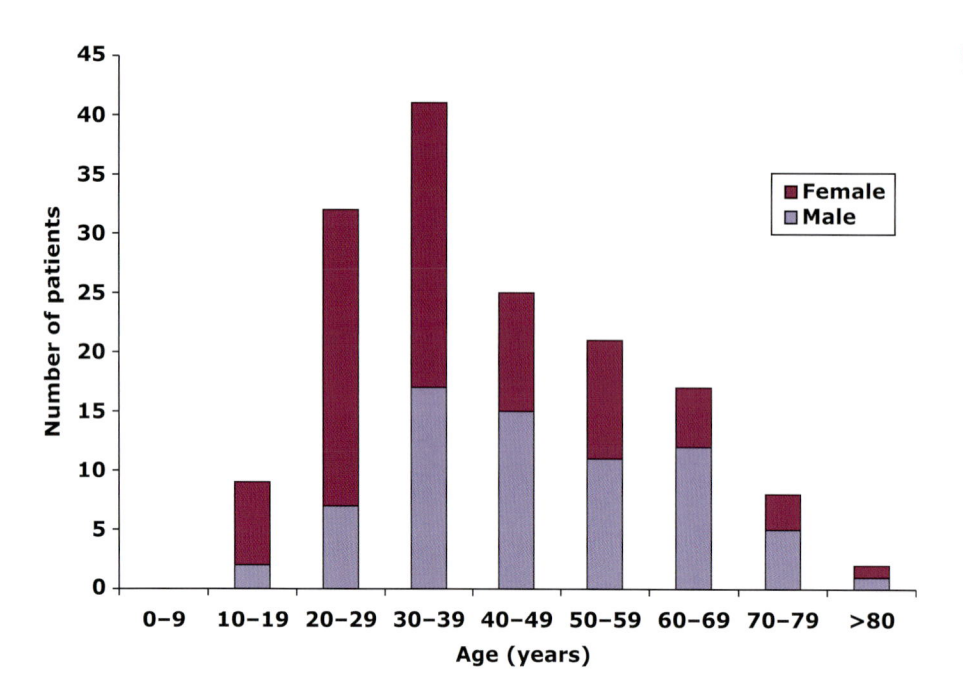

Figure 8.10 Age and sex distribution of 156 patients with mesenteric desmoids.

Intra-abdominal desmoid

Most commonly, intra-abdominal desmoids (<10% of all desmoids) occur in young adults in or around the small bowel mesentery (Figure 8.10). Many are associated with FAP, whose phenotypic variant, Gardner syndrome, includes adenomatous polyposis, desmoids, skull osteomas, and epidermal inclusion cysts. In the largest retrospective series, evidence of FAP was present in 15% of patients.[43] This might be a low estimate of the association of mesenteric desmoid and FAP, considering the retrospective nature of the study. It has been estimated that FAP patients have a cumulative 20% to 25% lifetime risk for developing a desmoid, and their risk for developing a desmoid is nearly 1000-fold higher than that of the normal population.[32,44] Mesenteric desmoids in FAP often develop 1 to 2 years after other abdominal surgery, such as prophylactic colectomy, and surgical trauma seems to be a significant triggering factor for this group of desmoids.[32,44]

Mesenteric desmoids can be relatively asymptomatic, but they can form a palpable mass, and some cause intestinal obstruction, bleeding, or perforation. Tumor size varies from relatively small nodules to large, >10- to 20-cm complex tumors that can involve a single segment of bowel or encase multiple loops of small bowel, making them impossible to excise. Some examples involve the stomach and colon. Those patients who have FAP have a higher risk of recurrence and complications, including some tumor-related mortality, mainly because of intestinal obstruction, infections by tumor fistulation, and postsurgical short bowel syndrome.[32,44,45] In one study, however, tumors unassociated with FAP had a surprisingly low recurrence rate (10%), and many patients whose tumors were only biopsied or partially excised did well.[45]

Pelvic desmoids especially occur in young women. These tumors can reach a large size, necessitating surgery to relieve large vessels and other vital structures, and recurrences are common. There seems to be no significant association with FAP.[46]

Multiple desmoids

There are two main scenarios for multiple desmoids: *syndromic* and *sporadic*. Syndromic desmoids are generally associated with FAP; no other well-defined genetic syndromes have been reported in association with desmoids. Nevertheless, familial occurrence of a multicentric desmoid unassociated with FAP was reported in a 43-year-old woman and her young adult son and daughter. Two of these tumors were extra-abdominal, and one was in the abdominal wall.[47]

Multiple desmoids occurring in different parts of the same extremity probably represent regional recurrences of the original tumor, a phenomenon sometimes referred to as *tumor migration*. In a large study of extra-abdominal desmoids, 10% of the cases were multicentric, and 20% demonstrated such tumor migration.[40] In one case, an abdominal wall desmoid was followed by a desmoid in the breast 3 years later in a 41-year-old woman apparently without evidence of FAP.[48]

Treatment

Complete excision with negative margins is the main mode of treatment in abdominal wall and extra-abdominal desmoids, especially when this is possible without structural sacrifice. An incomplete correlation between a positive margin and tumor recurrence has been pointed out, however. Structurally, sacrifice is therefore not generally favored in sensitive sites. Adjuvant postoperative radiation therapy has been used in

Figure 8.11 (A) Age and sex and (B) site distribution of extra-abdominal desmoids.

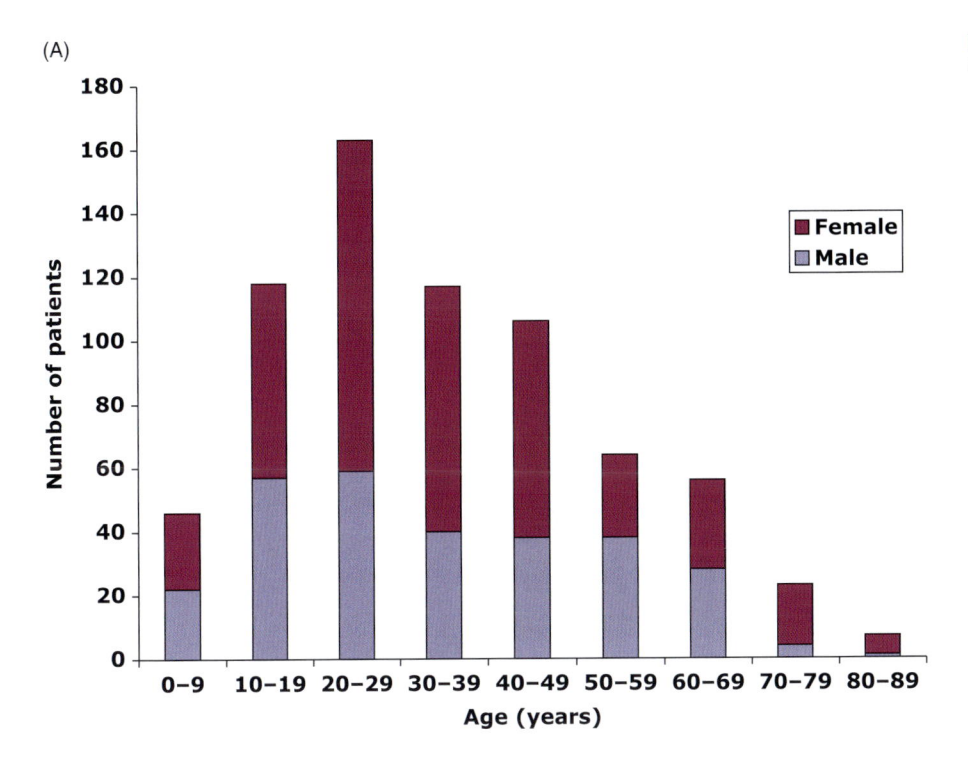

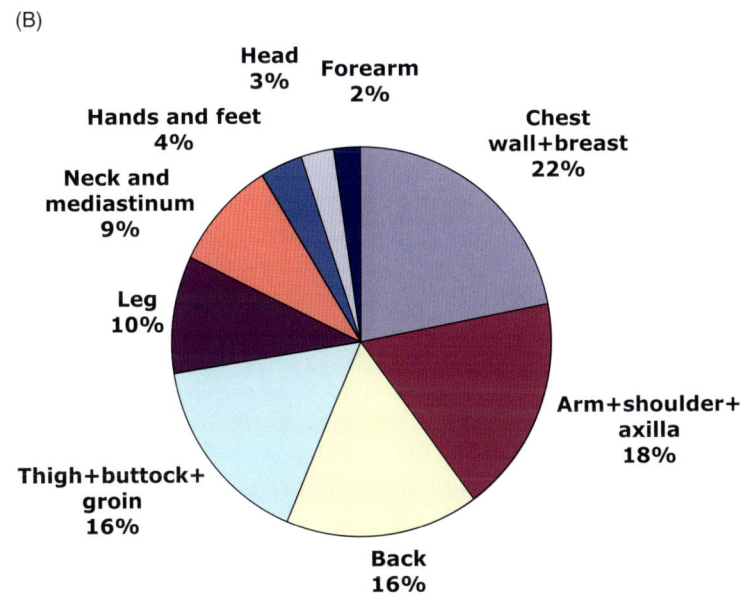

some cases, but its benefit has been difficult to characterize objectively without randomized trial settings.[37–39,48,49]

Intra-abdominal desmoids are now generally treated less aggressively because of their relatively indolent nature and because surgery can promote further tumor growth and cause complications, such as short bowel syndrome from extensive intestinal resections. The potential for spontaneous regression, as occasionally reported, is thought to allow for a more conservative treatment strategy.[44]

Medical therapy has been used as the primary treatment modality for intra-abdominal desmoids (especially in FAP), and for large extra-abdominal desmoids or those involving central neurovascular plexuses, formerly necessitating amputations. Agents used have included nonsteroidal anti-inflammatory drugs (e.g., sulindac), antiestrogens/estrogen modulators (tamoxifen and related agents), and conventional (especially low-dose) chemotherapies. Assessing the efficacy of these medical treatments is difficult because of potential spontaneous tumor regression and lack of control patients. In the future, randomized trials might resolve the value of such treatments in desmoids more conclusively.[39,44]

During the past 5 years, clinical trials have been conducted on imatinib mesylate, a KIT and platelet-derived growth factor receptor alpha (PDGFRA) receptor tyrosine kinase inhibitor

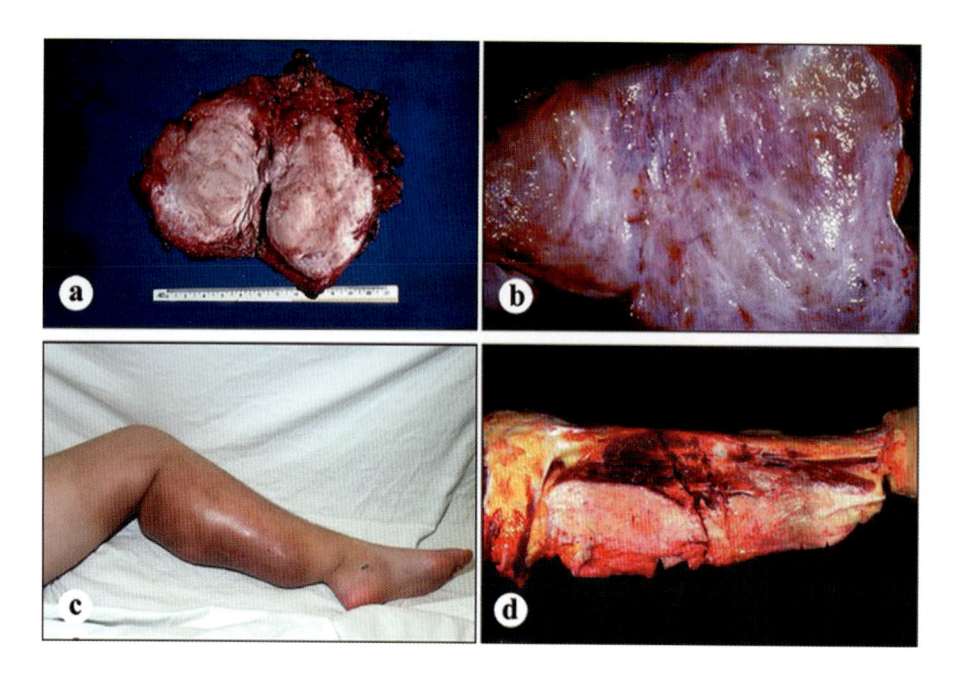

Figure 8.12 Gross and clinical images of desmoid. (**a**) An intramuscular example removed with a margin of normal skeletal muscle. (**b**) The cut surface of the desmoid is gray-white and glistening with a trabecular appearance. (**c**) An example of an aggressive desmoid involving nearly the entire leg. Amputation was the only means of treatment to restore the function of this painful leg. (**d**) This desmoid extensively involves the muscle compartments of the posterior aspect of the calf.

more commonly used in gastrointestinal stromal tumors. One study showed partial response or stable disease in nearly 40% of patients, but long-term experience is limited. In that study, desmoids were found not to express KIT and PDGFRA, but they had PDGFRB expression similar to that of fibroblasts, with low phosphorylation levels of the latter receptor and no evidence of mutations. Plasma PDGFRB levels had an inverse correlation with treatment failure.[50] These results suggest a pathogenetic role for PDGFRB in desmoids.

Pathology

On gross inspection, desmoids vary greatly in size, from small nodules <1 cm to bulky tumors >20 cm to 30 cm, but they usually measure 3 cm to 15 cm in diameter when excised (Figure 8.12). They are typically located inside the muscle or attached to the fascia, sometimes with a grossly spiculated periphery owing to infiltrative intramuscular growth. A small percentage of desmoids, however, especially those of small size, involve subcutis and fascia, forming only masses with tentacle-like, spiculated extensions.

On sectioning, the tissue is typically gray-white, firm, and resilient, with a coarsely trabeculated surface (Figure 8.12a,b). Some lesions show focal blood color on sectioning, reflecting the well-vascularized nature of the lesions. Focal myxoid change can occur, especially in mesenteric tumors.

Microscopically, desmoids show variably formed, longitudinally oriented fascicles of spindled fibroblasts and myofibroblasts in a prominently collagenous background (Figure 8.13). The collagenous matrix can contain thick keloid-like collagen fibers (Figure 8.14). In some cases, however, the fascicular structure is not well developed. The lesions have a prominent, evenly spaced vascular pattern composed of mildly gaping vessels, often with prominent endothelial and myopericytic cells (Figure 8.15). Atrophic skeletal muscle fibers (Figure 8.16a–c) are often entrapped in the periphery of intramuscular lesions; multinucleated regenerative skeletal muscle cells (sarcoplasmic giant cells) should not be confused with atypical tumor cells.[51] Subcutaneous fat infiltration also occurs (Figure 8.16d).

The cell borders are often poorly defined and merge with the abundant extracellular collagen. In some examples, however, lesional fibroblasts have an epithelioid or dendritic character with well-defined cytoplasmic borders (Figure 8.17). The nuclei are oval, mildly enlarged, with little if any variation, and have one or more delicate nucleoli. Mitotic activity is typically low, but can be higher in some cases, up to 2–3 or more per 10 HPFs. Elevated mitotic frequency alone therefore is not grounds for classifying a tumor with overall desmoid characteristics as a fibrosarcoma. Atypical mitoses are not present in desmoids, however.

Mesenteric desmoids often have a focal perivascular or more generalized fibromyxoid appearance. Otherwise, these tumors are similar to other desmoids (Figure 8.18).

Immunohistochemistry

Nuclear positivity for beta-catenin occurs in 70% to 80% of all desmoids (Figure 8.19a), but this finding is not specific, because other fibroblastic tumors, such as solitary fibrous tumors, also can be positive.[23,52,53] Desmoids, like many other fibroblastic and nonfibroblastic tumors, also show variable cytoplasmic beta-catenin positivity, and this cytoplasmic positivity is not specific for desmoids. Overexpression of cyclin D1 has been found to be associated with overexpression of beta-catenin.[54]

Desmoid tumors are positive for vimentin and variably focally for alpha smooth muscle actin (Figure 8.19b) and

Chapter 8: Fibromatoses

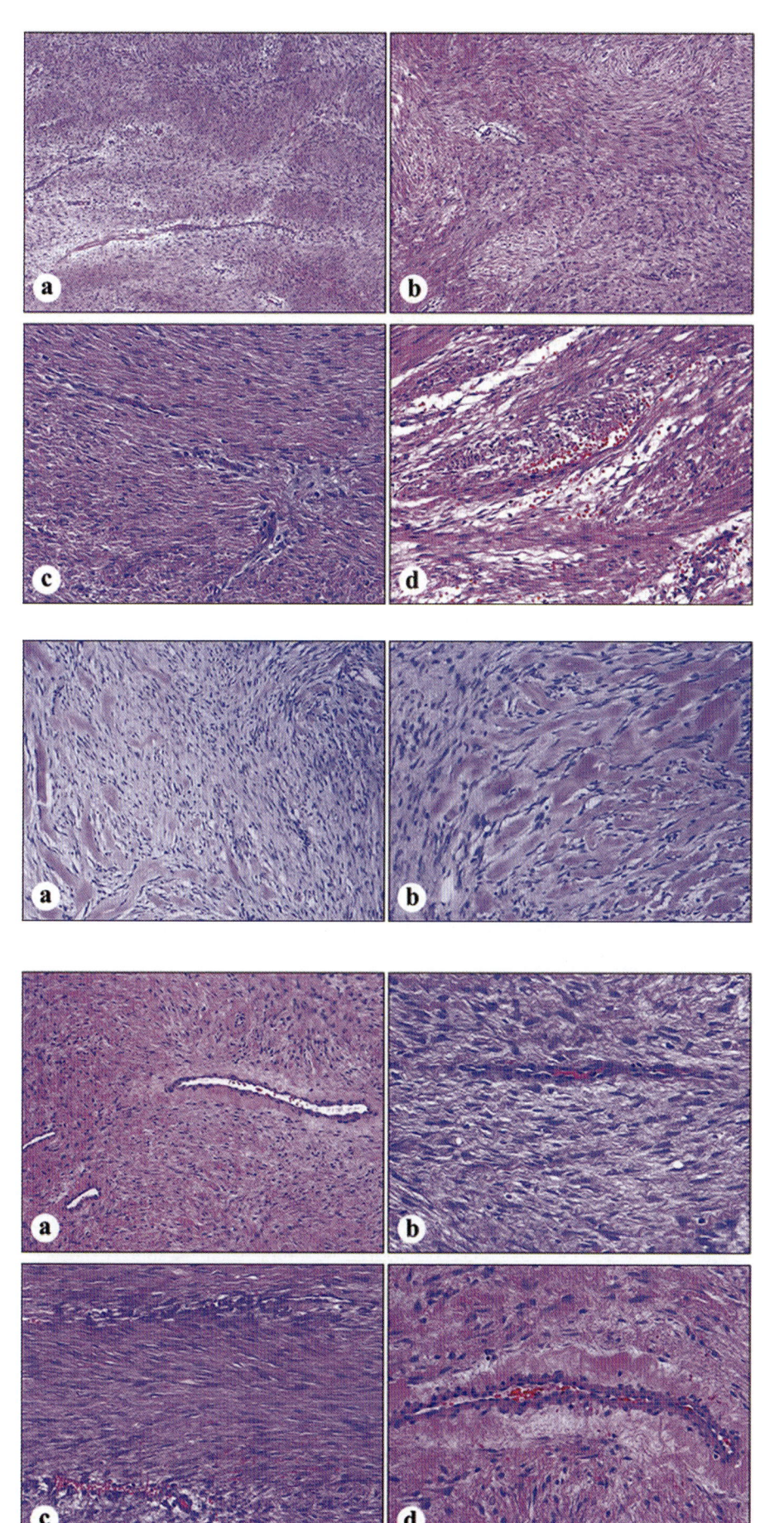

Figure 8.13 Low- to medium-power magnification shows the variably developed fascicular structure, relative paucicellularity, focal myxoid change, and high collagen content in desmoid.

Figure 8.14 Keloid-like collagen fibers are an occasional finding in desmoids.

Figure 8.15 Desmoids typically contain numerous mildly dilated blood vessels with variably prominent endothelial cells and pericytes.

243

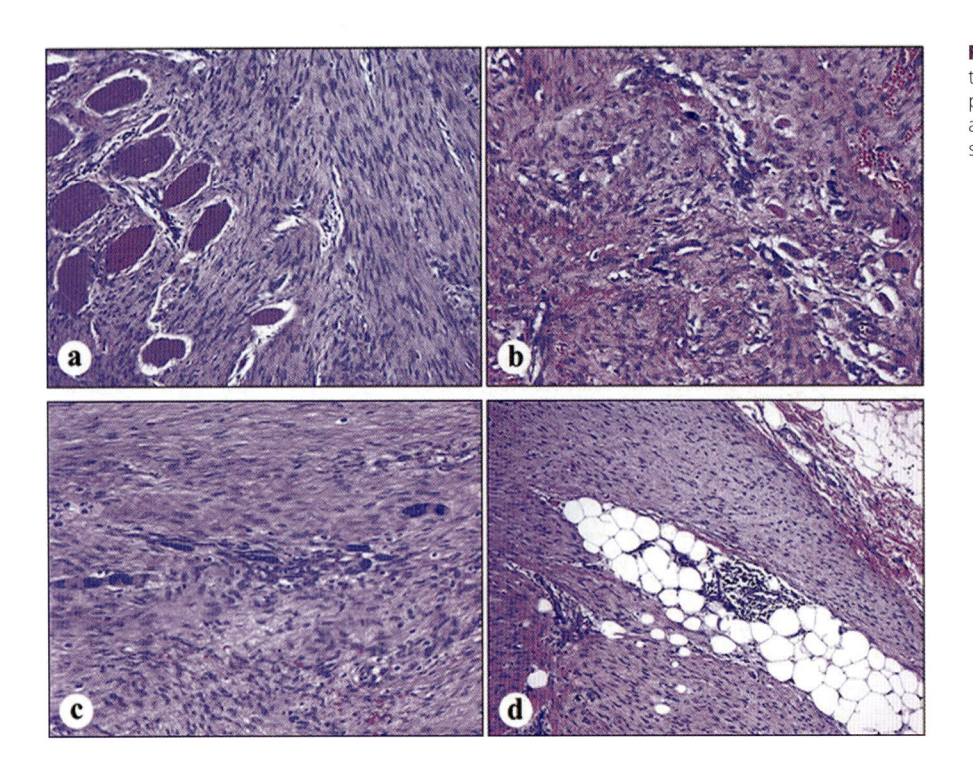

Figure 8.16 (a–d) Skeletal muscle infiltration is typical of desmoid. (c) In some cases, pseudoatypical regenerating skeletal muscle cells are present. (d) Some desmoids infiltrate subcutaneous fat.

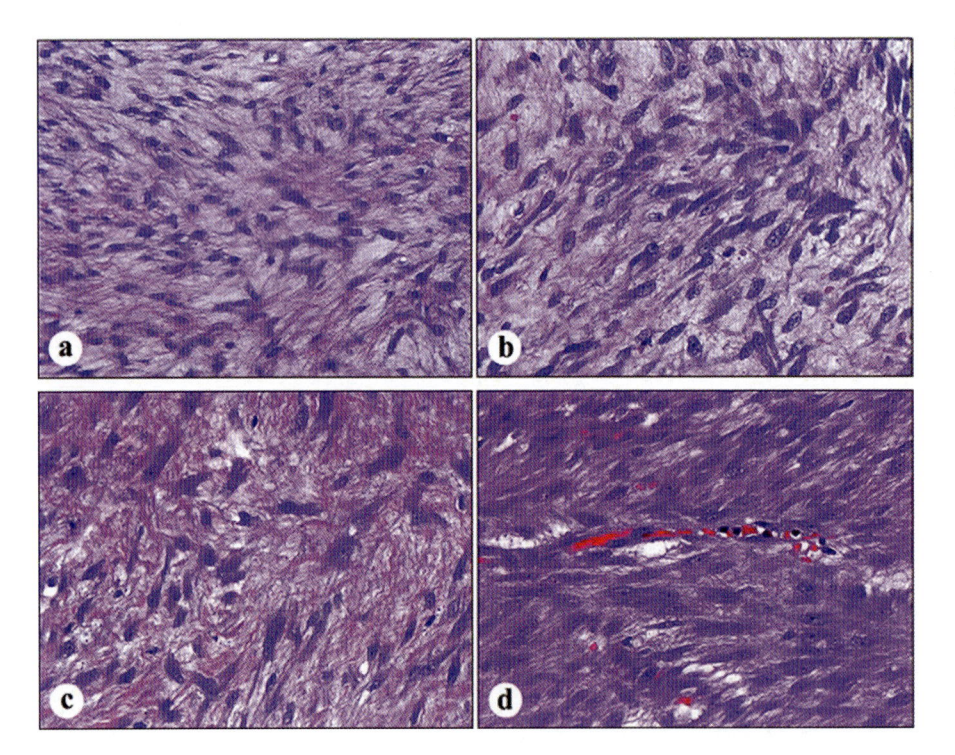

Figure 8.17 High magnifications of desmoid cells show a variable amount of cytoplasm, and some cells have distinct cytoplasm with epithelioid features and dendritic cytoplasmic processes. Delicate to slightly prominent nucleoli are often present.

desmin, but they are negative for BCL2, CD34, and S100 protein. Although negative for conventional estrogen receptor (estrogen receptor alpha), desmoid tumors have been found to express estrogen receptor beta.[55] This finding might reconcile the previously apparent discrepancy between positive radioimmunoassays for estrogen receptor and negative immunohistochemistry for estrogen receptor alpha.

Differential diagnosis

Distinguishing a desmoid from scar tissue can be difficult in a small biopsy, and if scar tissue is present at the margins, exclusion of a desmoid can be impossible. Small biopsies of desmoids can sometimes be difficult to distinguish from nodular fasciitis. A more homogeneous appearance, desmoid-type gaping vessels, and lack of components (e.g., scattered

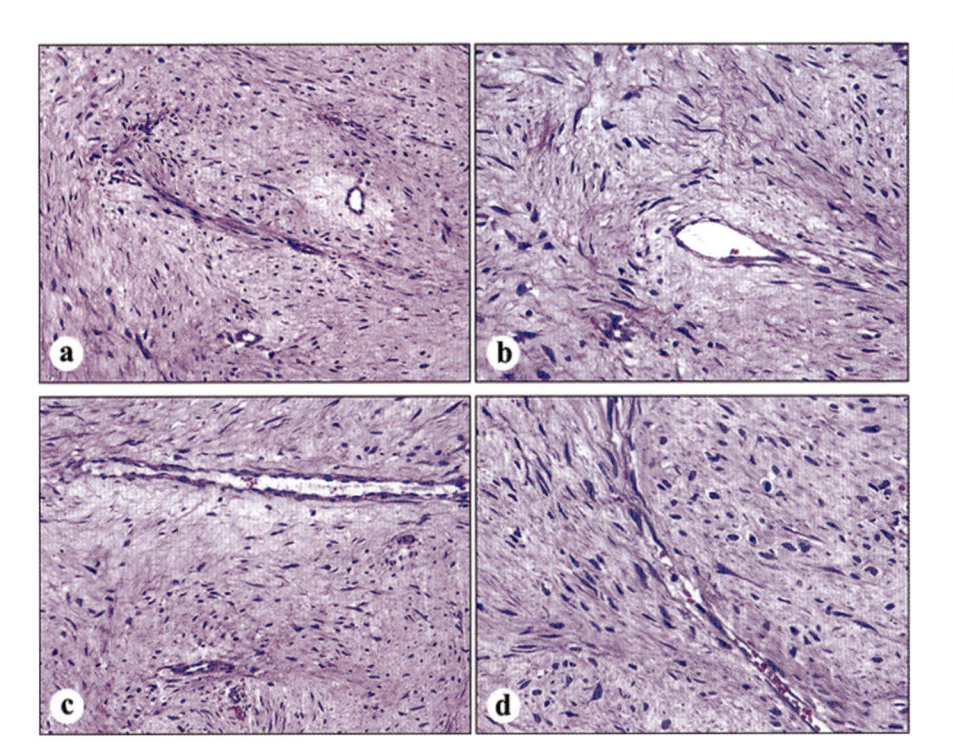

Figure 8.18 Mesenteric desmoid often shows an edematous stroma, and these tumors can appear less cellular than other desmoids.

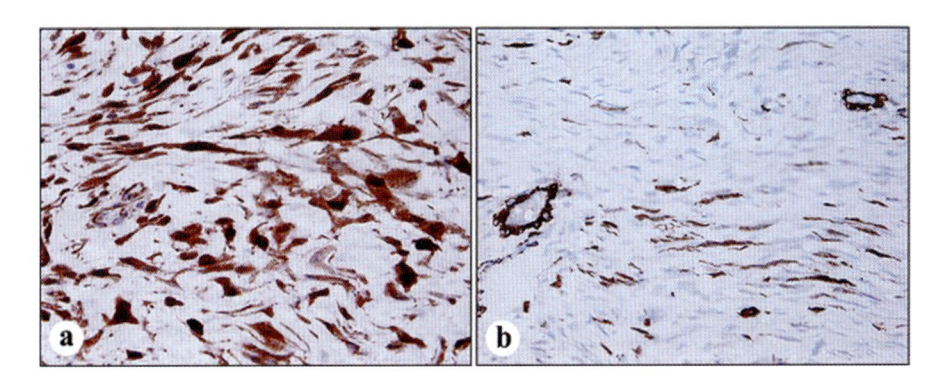

Figure 8.19 (a) Nuclear positivity for beta-catenin is a typical although not totally diagnostic feature of desmoids. Note that cytoplasmic positivity is also present, but this is not specific for desmoid. (b) Focal, variable smooth muscle actin positivity is a common finding in desmoids.

lymphohistiocytic infiltrate, mucoid degeneration, and osteoclast-like giant cells) support the diagnosis of a desmoid. Highly cellular desmoids, especially with mitotic activity, have to be separated from fibrosarcoma. Uniform cytologic features, lack of atypia, and high amount of extracellular collagen support the diagnosis of a desmoid. Solitary fibrous tumors demonstrate higher cellularity, hemangiopericytoma-like vessels, and CD34 immunoreactivity, all of which allow these tumors to be distinguished from desmoid.

Pathogenesis of desmoids

Much of the knowledge about desmoid pathognesis has evolved from studies on FAP patients with desmoids. Especially in the earlier literature, the constellation of polyposis, desmoid tumor, skull osteomas, and epidermal inclusion cysts was designated as Gardner syndrome. Currently, however, this is considered a phenotypic variant of FAP. Because of almost inevitable development of colon carcinoma by midlife, preventive colectomy or proctocolectomy is usually performed in early adulthood. By this preventive treatment, mortality to colon cancer has decreased; however, this then exposes the patients to other complications, especially development of intra-abdominal desmoid tumors and small intestinal or duodenal and ampullary carcinomas.[52,56]

The key event in FAP pathogenesis is constitutional activation of the wingless Int (Wnt) signaling pathway through loss of function of the adenomatous polyposis (APC) protein. This occurs through a combination of loss-of-function germline mutations or allelic losses in one copy of the APC tumor-suppressor gene (at 5q21–22) and loss-of-function somatic mutation of the other gene copy, resembling the classic tumor-suppressor gene model.[57,58] Location of mutation can modulate phenotype and disease severity.[59] The germline mutations are frameshift mutations that most commonly include microdeletions, and less commonly, point mutations and insertions.[60] Inactivation of APC protein leads to defective degradation and abnormal nuclear accumulation of beta-catenin, which probably modulates transcriptional regulation of various genes related to cell proliferation.[61] A minority (≥20–30%) of sporadic desmoids show a combination of mutations and allelic losses in APC gene leading to production of defective or no APC protein.[41,62,63]

A more common disease mechanism in sporadic desmoids is the occurrence of somatic beta-catenin gene (CTNNB1) mutations that cause increased half-life of that protein, leading to its abnormal accumulation in the cytoplasm and nuclei, with consequences similar to the loss of APC function. Sporadic desmoids showed somatic CTNNB1 mutations in 40% to 50% of cases in four different studies.[41,54,63,64] The most common CTNNB1 mutations in desmoid are T41A (45%) and S45F (25%). The latter mutation may be associated with a greater potential for recurrence and seems to be more common in extra-abdominal desmoids.[65,66] However, some studies have not confirmed the prognostic association.[67]

Other genetic changes in desmoids

Several studies have demonstrated a clonal X-chromosome inactivation pattern in sporadic[68–70] and FAP-associated desmoids[71] by HUMARA analysis, supporting the concept that desmoid formation is a monoclonal process.

The only recurrent clonal cytogenetic anomalies in desmoids, in addition to loss of 5q material related to the involvement of the APC-locus,[72] have been trisomies of chromosomes 8 and 20.[73–76] The trisomies are relatively rare in cytogenetic studies but are more commonly detected by FISH. This could be a consequence of their presence in only some of the tumor cells, and of their being a negative selection factor in short-term culture for cytogenetics.[73,77] Although it has been suspected that trisomy 8 is a risk factor for recurrence of desmoids,[74] others have not supported this impression.[78] A minority of desmoids has shown complex chromosomal changes.[79,80]

References
Palmar and plantar fibromatosis

1. Ross DC. Epidemiology of Dupuytren's disease. *Hand Clin* 1999;15:53–62.
2. Gudmundsson KG, Arngrimsson R, Sigfusson N, Bjornsson A, Jonsson T. Epidemiology of Dupuytren's disease: clinical, serological, and social assessment. The Reykjavik study. *J Clin Epidemiol* 2000;53:291–296.
3. Benson LS, Williams CS, Kahle M. Dupuytren's contracture. *J Am Acad Orthop Surg* 1998;6:24–35.
4. Rayan GM. Dupuytren disease: anatomy, pathology, presentation, and treatment. *J Bone Joint Surg Am* 2007;89:189–198.
5. Fetsch JF, Laskin WB, Miettinen M. Palmar-plantar fibromatosis in children and preadolescents: a clinicopathological study of 56 cases with newly recognized demographics and extended follow-up information. *Am J Surg Pathol* 2005;29:1095–1105.
6. Allen PW. The fibromatoses: a clinicopathologic classification based on 140 cases. *Am J Surg Pathol* 1977;1:255–270.
7. Ushijima M, Tsuneyoshi M, Enjoji M. Dupuytren type fibromatoses: a clinicopathologic study of 62 cases. *Acta Pathol Jpn* 1984;34:991–1001.
8. Hueston JT. Some observations on knuckle pads. *J Hand Surg Br* 1984;9:75–78.
9. Aluisio FV, Mair SD, Hall RL. Plantar fibromatosis: treatment of primary and recurrent lesions and factors associated with recurrence. *Foot Ankle Int* 1996;17:672–678.
10. Zgonis T, Jolly GP, Polyzois V, Kanuck DM, Stamatis ED. Plantar fibromatosis. *Clin Podiatr Med Surg North Am* 2005;22:11–18.
11. Chen KT, van Dyne DA. Familial plantar fibromatosis. *J Surg Oncol* 1985;29:240–241.
12. Graells Estrada J, Garcia Fernandez D, Badina Tottoella F, Moreno Carzano A. Familial plantar fibromatosis. *Clin Exp Dermatol* 2003;28:669–670.
13. Luck JV. Dupuytren's contracture: a new concept of the pathogenesis correlated with surgical management. *J Bone Joint Surg Am* 1959;41:635–664.
14. Evans HL. Multinucleated giant cell in plantar fibromatosis. *Am J Surg Pathol* 2002;26:244–248.
15. de Palma L, Santucci A, Gigante A, Di Giulio A, Carloni S. Plantar fibromatosis: an immunohistochemical and ultrastructural study. *Foot Ankle Int* 1999;20:253–257.
16. Montgomery E, Lee JH, Abraham SC, Wu TT. Superficial fibromatoses are genetically distinct from deep fibromatosis. *Mod Pathol* 2001;14:695–701.
17. Bhattacharya B, Dilworth HP, Iacobuzio-Donahue C, et al. Nuclear beta-catenin expression distinguishes deep fibromatosis from other benign and malignant fibroblastic and myofibroblastic lesions. *Am J Surg Pathol* 2005;29:653–659.
18. Dal Cin P, De Smet L, Sciot R, van Damme B, Van Den Berghe H. Trisomy 7 and trisomy 8 in dividing and nondividing tumor cells in Dupuytren's disease. *Cancer Genet Cytogenet* 1999;108:137–140.
19. Casalone R, Mazzola D, Meroni E, et al. Cytogenetic and interphase cytogenetic analyses reveal chromosome instability but no clonal trisomy 8 in Dupuytren contracture. *Cancer Genet Cytogenet* 1997;99:73–76.
20. De Wever I, Dal Cin P, Fletcher CDM, et al. Cytogenetic, clinical, and morphologic correlations in 78 cases of fibromatosis: a report from the CHAMP study group. *Mod Pathol* 2000;13:1080–1085.
21. Breiner JA, Nelson M, Bredthauer BD, Neff JR, Bridge JA. Trisomy 8 and trisomy 14 in plantar fibromatosis. *Cancer Genet Cytogenet* 1999;108:176–177.
22. Sawyer JR, Sammartino G, Godden N, Nicholas RW. A clonal reciprocal t(2;7)(p13;p13) in plantar fibromatosis. *Cancer Genet Cytogenet* 2005;158:67–69.
23. Chansky HA, Trumble TE, Conrad EU, et al. Evidence for a polyclonal etiology

of palmar fibromatosis. *J Hand Surg* 1999;24:339–344.

Peyronie's disease (penile fibromatosis)

24. Sommer F, Schwarzer U, Wassmer G, *et al*. Epidemiology of Peyronie's disease. *Int J Impot Res* 2002;14:379–383.
25. Hellstrom WJ, Bivalaqua TJ. Peyronie's disease: etiology, medical and surgical therapy. *J Androl* 2000;21:347–354.
26. Smith BH. Peyronie's disease. *Am J Clin Pathol* 1966;43:670–678.
27. Davis CJ Jr. The microscopic pathology of Peyronie's disease. *J Urol* 1997;157:282–284.

Desmoid fibromatosis: clinicopathologic

28. Ferenc T, Sygut J, Kopczynski J, *et al*. Aggressive fibromatosis (desmoid tumors): definition, occurrence, pathology, diagnostic problems, clinical behavior, genetic background. *Pol J Pathol* 2006;57:5–15.
29. Reitamo JJ, Häyry P, Nykyri E, Saxen E. The desmoid tumor. I. Incidence, sex-, age- and anatomical distribution in the Finnish population. *Am J Clin Pathol* 1982;77:665–673.
30. Lev D, Kotilingam D, Wei C, *et al*. Optimizing treatment of desmoid tumors. *J Clin Oncol* 2007;25:1785–1791.
31. Fallen T, Wilson M, Morlan B, Lindor NM. Desmoid tumors – a characterization of patients seen at Mayo clinic 1976–1999. *Fam Cancer* 2006;5:191–194.
32. Heiskanen I, Järvinen H. Occurrence of desmoid tumours in familial adenomatous polyposis and results of treatment. *Int J Colorectal Dis* 1996;11:157–162.
33. Posner MC, Shiu MH, Newsome JL, *et al*. The desmoid tumor. Not a benign disease. *Arch Surg* 1989;124:191–196.
34. Brasfield RD. Das Gupta TK. Desmoid tumors of the anterior abdominal wall. *Surgery* 1969;65:241–246.
35. Häyry P, Reitamo JJ, Totterman S, Hopfner-Hallikainen D, Sivula A. The desmoid tumor. II. Analysis of factors possibly contributing to the etiology and growth behavior. *Am J Clin Pathol* 1982;77:674–680.
36. Pignatti G, Barbanti-Brodano G, Ferrari D, *et al*. Extra-abdominal desmoid tumor: a study of 83 cases. *Clin Orthop Relat Res* 2000;375:207–213.
37. Gronchi A, Casali PG, Mariani L, *et al*. Quality of surgery and outcome in extraabdominal aggressive fibromatosis: a series of patients surgically treated at a single institution. *J Clin Oncol* 2003;21:1390–1397.
38. Merchant NB, Lewis JJ, Woodruff JM, Leung DH, Brennan MF. Extremity and trunk desmoid tumors. *Cancer* 1999;86: 2045–2052.
39. Hosalkar HS, Fox EJ, Delaney T, *et al*. Desmoid tumors and current status of management. *Orthop Clin North Am* 2006;37:53–63.
40. Rock MG, Pritchard DJ, Reiman HM, Soule EH, Brewster AR. Extra-abdominal desmoid tumors. *J Bone Joint Surg Am* 1984;66:1369–1374.
41. Abraham SC, Reynolds C, Lee JH, *et al*. Fibromatosis of the breast and mutations involving the APC/beta-catenin pathway. *Hum Pathol* 2002;33:39–46.
42. Aaron AD, O'Mara JW, Legendre KE, *et al*. Chest wall fibromatosis associated with silicone implants. *Surg Oncol* 1996;5:93–99.
43. Burke AP, Sobin LH, Shekitka KM, Federspiel BH, Helwig EB. Intra-abdominal fibromatosis: a pathologic analysis of 130 tumors with comparison of clinical subgroups. *Am J Surg Pathol* 1990;14:335–341.
44. Sturt NJH, Clark SK. Current ideas in desmoid tumors. *Fam Cancer* 2000;5:275–285.
45. Burke AP, Sobin LH, Shekitka KM. Mesenteric fibromatosis: a follow-up study. *Arch Pathol Lab Med* 1990;114:832–835.
46. Mariani A, Nascimento AG, Webb MJ, Sim FH, Podratz KC. Surgical management of desmoid tumors of the female pelvis. *J Am Coll Surg* 2000;191:175–183.
47. Zayid I, Dihmis C. Familial multicentric fibromatosis – desmoids. *Cancer* 1969;24:786–795.
48. Reis-Filho JS, Milanezi F, Pope ZB, Fillus-Neto J, Schmitt FC. Primary fibromatosis of the breast in a patient with multiple desmoid tumors. *Pathol Res Pract* 2001;197:775–779.
49. Biermann JS. Desmoid tumors. *Curr Treat Options Oncol* 2000;1:262–266.
50. Heinrich MC, McArthur GA, Demetri GD, *et al*. Clinical and molecular studies of the effect of imatinib on advanced aggressive fibromatosis (desmoid tumor). *J Clin Oncol* 2006;24:1195–1203.
51. Enzinger FM, Shiraki M. Musculo-aponeurotic fibromatosis of the shoulder girdle (extra-abdominal desmoid): analysis of 30 cases followed-up for 10 or more years. *Cancer* 1967;20:1131–1140.
52. Ng TL, Gown AM, Barry TS, *et al*. Nuclear beta-catenin in mesenchymal tumors. *Mod Pathol* 2005;18:68–74.
53. Carlson JW, Fletcher CDM. Imunohistochemistry for β-catenin in the differential diagnosis of spindle cell lesions: analysis of a series and review of the literature. *Histopathology* 2007;51:509–514.
54. Saito T, Oda Y, Tanaka K, *et al*. β-catenin nuclear expression correlates with cyclin D1 overexpression in sporadic desmoid tumors. *J Pathol* 2001;195:222–228.
55. Deyrup A, Tretiakova M, Montag AG. Estrogen receptor-β expression in extraabdominal fibromatoses: an analysis of 40 cases. *Cancer* 2006;106:208–213.

Desmoid fibromatosis: genetics

56. Galiatsatos P, Foulkes WD. Familial adenomatous polyposis. *Am J Gastroenterol* 2006;101: 385–398.
57. Okamoto M, Sato C, Kohno Y, *et al*. Molecular nature of chromosome 5q loss in colorectal tumors and desmoids from patients with familial adenomatous polyposis. *Hum Genet* 1990;85:595–599.
58. Miyaki M, Konishi M, Kikuchi-Yanoshita R, *et al*. Coexistence of somatic and germ-line mutations of APC gene in desmoid tumors from patients with familial adenomatous plyposis. *Cancer Res* 1993;53:5079–5082.
59. Couture J, Mitri A, Lagace R, *et al*. A germline mutation at the extreme 3′ end of the APC gene results in a severe desmoid phenotype and is associated with overexpression of beta-catenin in the desmoid tumor. *Clin Genet* 2000;57:205–212.
60. Latchford A, Volikos E, Johnson V, *et al*. APC mutations in FAP-associated desmoid tumours are non-random but not "just right." *Hum Mol Genet* 2007;16:78–82.

61. Li C, Bapat B, Alman BA. Adenomatous polyposis coli gene mutation alters proliferation through its beta-catenin regulatory function in aggressive fibromatosis (desmoid tumor). *Am J Pathol* 1998;153:709–714.

62. Alman BA, Li C, Pajerski ME, Diaz-Cano S, Wolfe HJ. Increased beta-catenin protein and somatic APC mutations in sporadic aggressive fibromatosis. *Am J Pathol* 1997;151:329–334.

63. Tejpar S, Nollet F, Li C, et al. Predominance of beta-catenin mutations and beta-catenin dysregulation in sporadic aggressive fibromatosis (desmoid tumor). *Oncogene* 1999;18:6615–6620.

64. Miyoshi Y, Iwao K, Nawa G, Ochi T, Nakamura Y. Frequent mutations in the beta-catenin gene in desmoid tumors from patients without familial adenomatous polyposis. *Oncol Res* 1998;10:591–594.

65. Lazar AJ, Tuvin D, Hajibashi S, et al. Specific mutations in the beta-catenin gene (CTNNB1) correlate with local recurrence in sporadic desmoid tumors. *Am J Pathol* 2008;173:1518–1527.

66. Colombo C, Miceli R, Lazar AJ, et al. CTNNB1 45F mutation is a molecular prognosticator of increased postoperative primary desmoid tumor recurrence: an independent, multicenter validation study. *Cancer* 2013;119(20):3696–3702.

67. Mullen JT, DeLaney TF, Rosenberg AE, et al. β-Catenin mutation status and outcomes in sporadic desmoid tumors. *Oncologist* 2013;18:1043–1049.

68. Li M, Cordon-Cardo C, Gerald WL, Rosai J. Desmoid fibromatosis is a clonal process. *Hum Pathol* 1996;27:939–943.

69. Alman BA, Pajerski ME, Diaz-Cano S, Corboy K, Wolfe HJ. Aggressive fibromatosis (desmoid tumor) is a monoclonal disorder. *Diagn Mol Pathol* 1997;6:98–101.

70. Lucas DR, Schroyer KR, McCarthy PJ, et al. Desmoid tumor is a clonal cellular proliferation: PCR-amplification of HUMARA for analysis of patterns of X-chromosome inactivation. *Am J Surg Pathol* 1997;21:306–311.

71. Middleton SB, Frayling IM, Phillips RKS. Desmoids in familial adenomatous polyposis are monoclonal proliferations. *Br J Cancer* 2000;82:827–832.

72. Yoshida MA, Ikeuchi T, Iwama T, et al. Chromosome changes in desmoid tumors developed in patients with familial adenomatous polyposis. *Jpn J Cancer Res* 1991;82:916–921.

73. Dal Cin P, Sciot R, Aly MS, et al. Some desmoid tumors are characterized by trisomy 8. *Genes Chromosomes Cancer* 1994;10:131–135.

74. Fletcher JA, Naeem R, Xiao S, Corson JM. Chromosome aberrations in desmoid tumors: trisomy 8 may be a predictor of recurrence. *Cancer Genet Cytogenet* 1995;79:139–143.

75. Mertens F, Willen H, Rydholm A, et al. Trisomy 20 is a primary chromosome aberration in desmoid tumors. *Int J Cancer* 1995;63:527–529.

76. Qi H, Dal Cin P, Hernandez JM, et al. Trisomies 8 and 20 in desmoid tumors. *Cancer Genet Cytogenet* 1996;92:147–149.

77. Kouho H, Aoki T, Hisaoka M, Hashimoto H. Clinicopathological and interphase cytogenetic analysis of desmoid tumors. *Histopathology* 1997;31:336–341.

78. De Wever I, Dal Cin P, Fletcher CD, et al. Cytogenetic, clinical, and morphologic correlations in 78 cases of fibromatosis: a report from the CHAMP study group. *Mod Pathol* 2000;13:1080–1085.

79. Bridge JA, Sreekantaiah C, Mouron B, et al. Clonal chromosomal abnormalities in desmoid tumors. *Cancer* 1992;69:430–436.

80. Karlsson I, Mandahl N, Heim S, et al. Complex chromosome rearrangements in an extraabdominal desmoid tumor. *Cancer Genet Cytogenet* 1988;34:241–245.

Chapter 9
Benign fibroblastic and myofibroblastic proliferations in children

Markku Miettinen
National Cancer Institute, National Institutes of Health

Benign fibroblastic proliferations typical of children have here been divided into three categories: fasciitis and pseudotumors, fibromas and fibromatoses, and other benign fibrous proliferations. Among these are included a total of 15 different clinicopathologic entities, which will be discussed here. These entities span a wide clinicopathologic and morphologic spectrum from non-neoplastic to neoplastic, with some examples of indeterminate conditions. Common to them all, however, is the potential for recurrence only, and none for metastasis. Some of these tumors also occur in adults (e.g., calcifying fibrous tumor, calcifying aponeurotic fibroma, myofibroma). Conversely, many fibrous proliferations typically seen in adults also occur in children (e.g., nodular fasciitis, palmar, plantar, and desmoid fibromatoses) – these are discussed in Chapter 8. Fibroblastic and myofibroblastic lesions of childhood with variable biologic potential are discussed in Chapter 10. Desmoid tumors are discussed in Chapter 8.

The terminology of some childhood fibrous tumors does not match the lesion type: some names are clearly misnomers and probably will be adjusted in the future. For example, the term *fibromatosis* has been historically applied to purely reactive, reparative processes such as fibromatosis colli, in addition to their main use for lesions with recurrence potential, such as desmoids. Similarly, some conditions classified as fibromas might be closer to fibromatosis in some respects. For example, calcifying aponeurotic fibroma tends to have diffuse infiltrative borders and significant potential for recurrence. The meaning of the term *hamartoma* has remained somewhat obscure; for example, *fibrous hamartoma of infancy* is a condition that could be a benign neoplasm. Mutations in capillary morphogenesis 2 protein have been discovered as the probable pathogenesis of the rare juvenile hyaline fibromatosis.

Cranial fasciitis
Clinical features

Cranial fasciitis is the designation for a nodular fasciitis-like lesion presenting in the scalp of infants, typically associated with focal lytic change in the skull at the site of the lesion. It presents as a rapidly growing scalp or facial mass in infants and young children, usually before the age of 2 years (Figure 9.1). This condition is congenital in a minority of

Figure 9.1 Age and sex distribution of 28 children with cranial fasciitis.

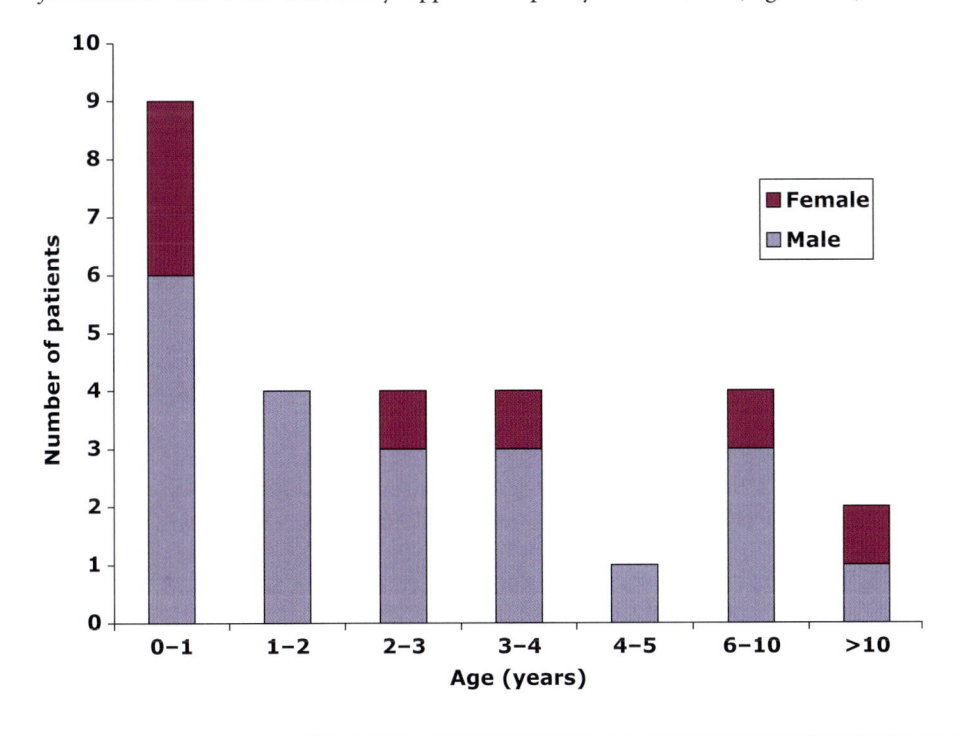

Modern Soft Tissue Pathology, Second Edition, ed. Markku Miettinen. Published by Cambridge University Press. © Cambridge University Press 2016.

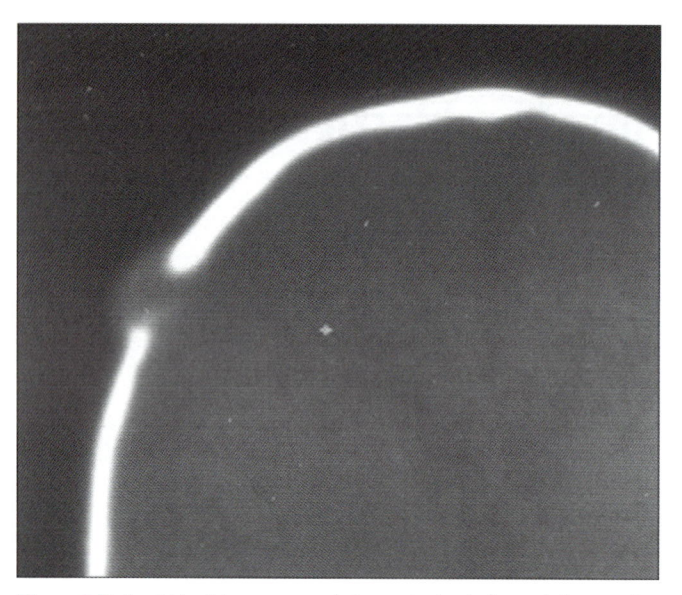

Figure 9.2 Cranial fasciitis appears as a lytic mass in the skull on radiologic studies.

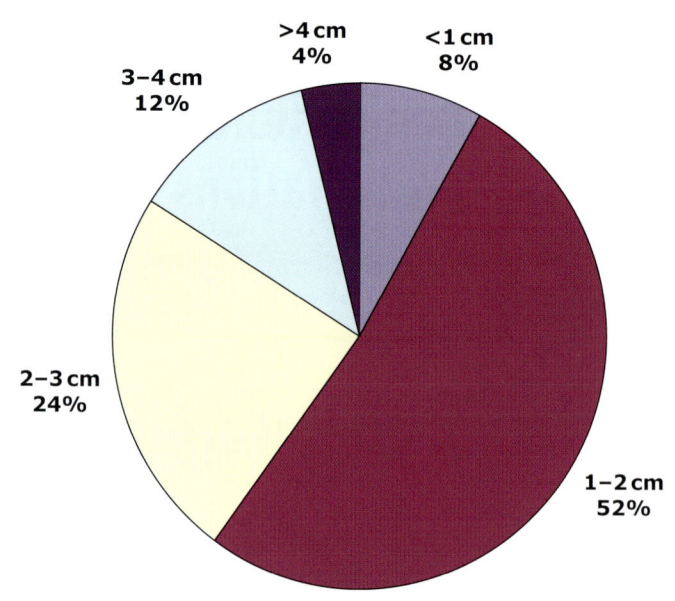

Figure 9.3 Size distribution of 27 cases of cranial fasciitis.

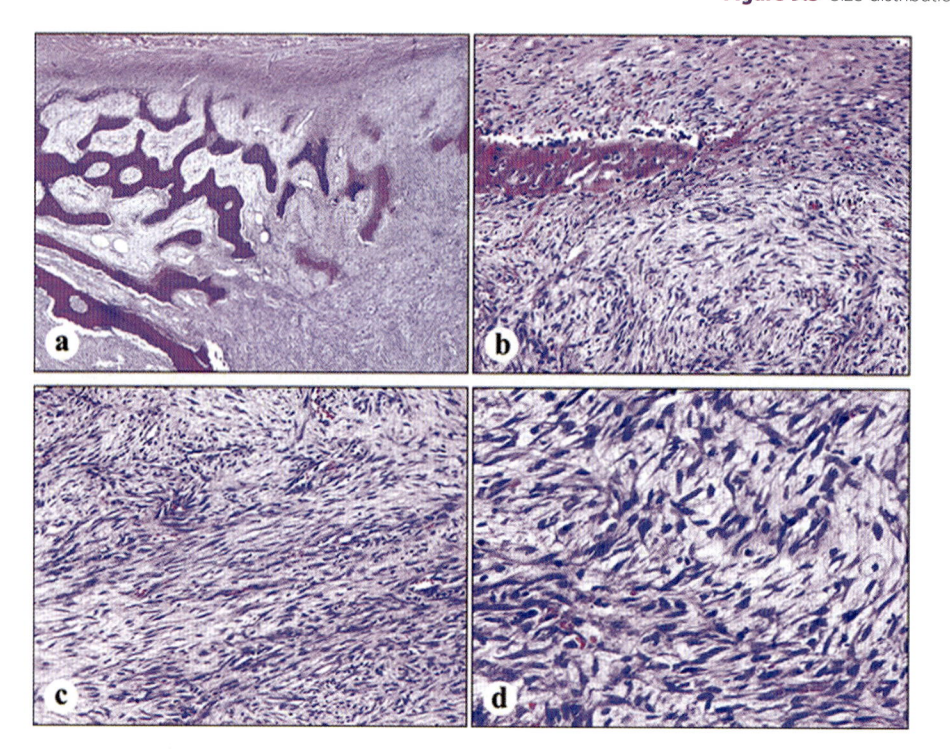

Figure 9.4 (a,b) Cranial fasciitis lesions can contain spicules of bone related to their bone involvement. (c,d) Histologically there is fibromyxoid myofibroblastic proliferation resembling nodular fasciitis.

cases.[1–3] Single cases in adults forming an intracranial mass have been reported.[4]

The lesion variably erodes the outer table of skull and adjacent facial bone can permeate the entire skull bone (Figure 9.2), and can even extend into the dura and leptomeninges. Local, tissue-conserving excision is considered sufficient because of the self-limiting nature of the process; there seems to be no potential for recurrence. The bone erosion characteristic of cranial fasciitis is probably a consequence of interaction of a nodular fasciitis lesion with immature fetal bone, which is more susceptible to focal resorption by macrophages and osteoclasts than mature bone. Whether some cases reported as cranial fasciitis, such as one with massive intracranial involvement, represent this tumor type is doubtful.[5]

Pathology

Grossly, cranial fasciitis varies from a mildly mucoid to rubbery, and white to pink mass usually measuring 1 cm to 3 cm (Figure 9.3); in the author's experience all large masses >5 cm represent other entities, such as desmoid fibromatosis or myofibroma.

The lesions are histologically similar to nodular fasciitis, except for their osseous involvement and the presence of bone spicules in some cases (Figure 9.4). They vary from highly myxoid and vascular to those with evolving collagenous background with features identical to different phases of nodular fasciitis. Mitotic activity can be present (generally up to 5 mitoses per 10 HPFs). Like nodular fasciitis, cranial fasciitis is immunohistochemically positive for vimentin and smooth muscle actin, and negative for desmin and S100 protein.

Differential diagnosis

Nodular fasciitis in the scalp of older children and adults without lytic bone change should simply be designated as nodular fasciitis, and the diagnosis of cranial fasciitis should be used instead of nodular fasciitis in small children with lytic change in the skull bone. Lesions that contain beta-catenin-positive nuclei and have features similar to fibromatosis, are better classified as desmoid fibromatosis.

Cranial fasciitis must be differentiated from desmoid tumor, which rarely occurs in the head and neck of small children. Desmoids are more highly cellular and typically have a prominent vascular pattern, often with conspicuous pericytes. Myofibroma contains pale, staining clusters of myofibroblasts and an immature hemangiopericytoma-like vascular component; this tumor is usually subcutaneous, but it can involve bone as well. Myxomas of the jaw bones are lytic, intraosseous mass lesions in the jaw bones containing scattered tumor cells in mucoid matrix. An inflammatory myofibroblastic tumor contains an admixture of spindled cells with abundant amphophilic cytoplasm, lymphocytes, and plasma cells.

Fibromatosis colli
Clinical features

This reactive, reparative condition presenting in newborns was referred to as *muscular torticollis* and *sternomastoid tumor* in earlier literature. It is usually thought to be a consequence of intrauterine or perinatal trauma to the sternocleidomastoid muscle. This assumption is partly based on coexistence of known birth traumas, such as paresthesia of the arm *(Erb's palsy)*. Fibromatosis colli develops during the first two months of life, and appears as a relatively small (usually 2–3 cm) mass in any part of the sternocleidomastoid muscle, more often in its lower portion, and more often on the right side.[6] Torticollis leaning into the side of the lesion can develop. Torticollis is not synonymous with fibromatosis colli, however, because it can also result from several unrelated orthopedic and neurologic conditions.

Because the harmless nature and spontaneous resolution of these lesions is well recognized, excisional surgery is avoided, and treatment is oriented toward surgical correction of the torticollis position when needed, instead of mass removal.

Pathology

Histologic specimens are rarely received because of generally conservative treatment. The tumor is grossly a whitish, solid mass (Figure 9.5). Histologically there is an evolving collagenous, myofibroblastic scar-like process diffusely infiltrating skeletal muscle and separating single muscle fibers or clusters of them. Both degenerating and regenerating muscle fibers are present, including multinucleated forms (Figure 9.6). Smooth muscle actin-positive myofibroblasts can be detected immunohistochemically.

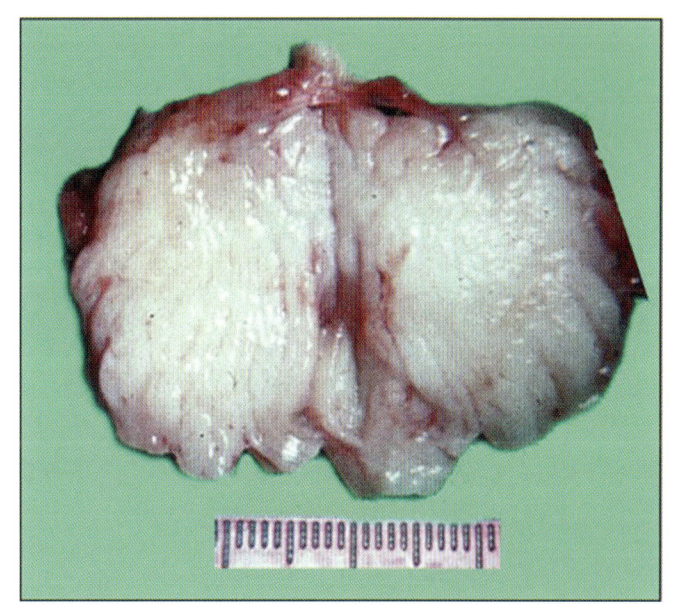

Figure 9.5 Grossly, a fibromatosis colli lesion appears as a grayish, fleshy mass. Red discoloration represents residual skeletal muscle.

Fine-needle aspiration biopsy has been advocated for the confirmation of the diagnosis and to rule out a true neoplasm when indicated. The aspirates typically show fibroblasts, myofibroblasts, and fragments of degenerated skeletal muscle cells.[7,8]

Fibromatosis colli, a reparative process, should be distinguished from desmoid and myofibromatosis involving the neck. The histologic composition is more heterogeneous than that of desmoids and myofibromatosis, which contain a more solid, more richly collagenous, moderately cellular, myofibroblastic proliferation with "biphasic" features in the latter, combining myofibroblastic nests and hemangiopericytoma-like vascular elements.

Fibrous umbilical polyp
Clinical features

This clinically distinctive, small polypoid tumor described by Vargas in 2001 occurs in the umbilicus nearly exclusively in male infants (>90% male predominance); one half of all cases are diagnosed before the age of 8 months (range of age at diagnosis: 3–18 months).[9] It can represent an abnormal reparative process of the umbilicus, which is the termination point of the omphalomesenteric duct remnant. There have been no recurrences following a simple excision.

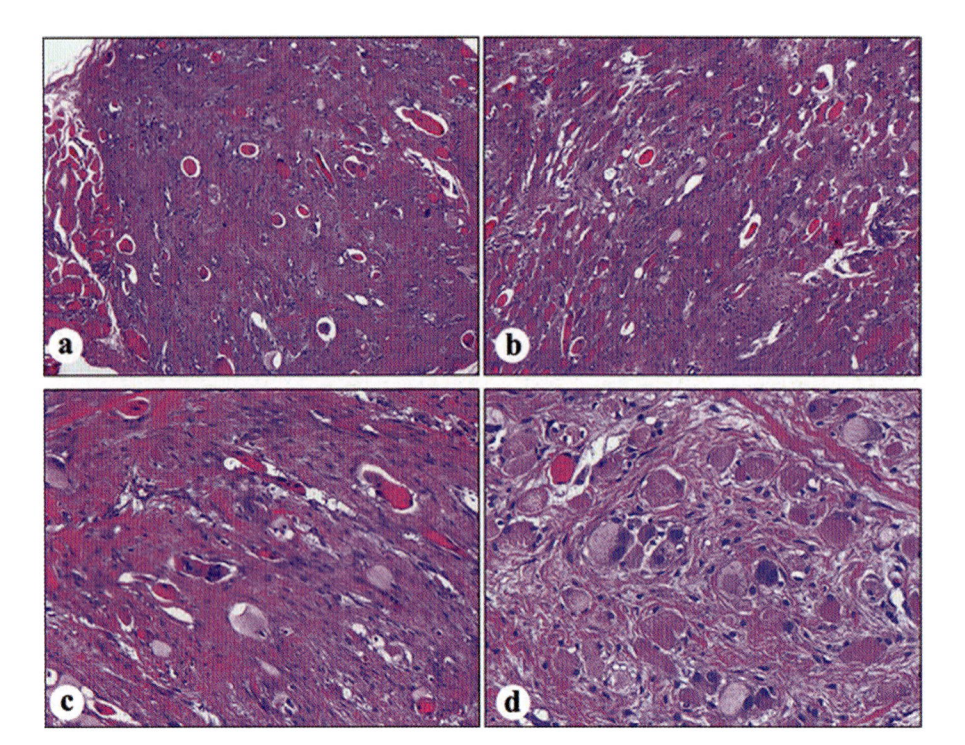

Figure 9.6 (a–c) A fibromatosis colli lesion contains infiltrates in the sternocleidomastoid muscle and is composed of alternating fibrous scarring and skeletal muscle cells. (d) Some skeletal muscle cells show degenerative vacuolization, and regenerative satellite cell proliferation is focally present.

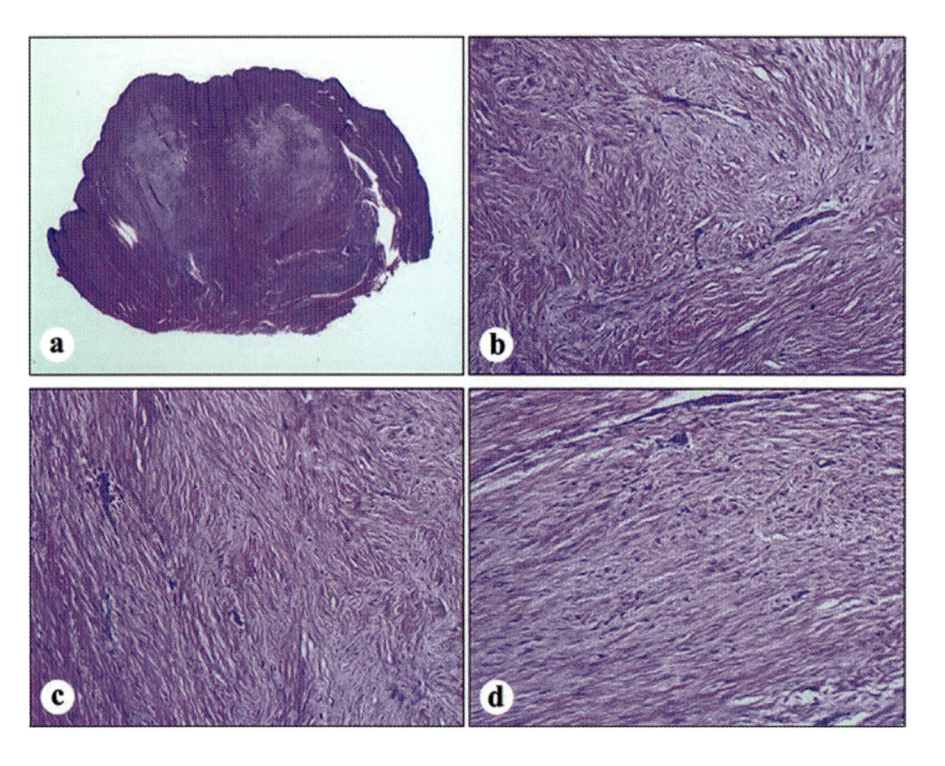

Figure 9.7 Fibrous umbilical polyp forms a dome-shaped elevation of the umbilicus. Histologically there is a relatively paucicellular, scar-like fibrous process with a mild vascular accentuation.

Pathology

The polyp forms a dome-shaped or pedunculated umbilical elevation measuring 4 mm to 12 mm (median, 6 mm).

Histologically, it can involve both dermis and deeper tissues, containing relatively paucicellular, fibrocollagenous, or fibromyxoid tissue, with some resemblance to scar, fasciitis, or fibromatosis (Figure 9.7). Some lesions include fibroblasts with ample cytoplasm and epithelioid features resembling similar cells in desmoids, or slightly resembling ganglion cells. Blood vessels are generally not prominent. No epithelial elements, keloidal collagen, or inflammatory components are present, separating this lesion from pilonidal disease and keloid. This polyp is different from the umbilical polyps sometimes seen in older children and young adults. The polyp cells are often positive for smooth muscle actin, and occasionally focally for desmin, and are negative for CD34, S100 protein, epithelial membrane antigen (EMA), and keratins.[9]

Gingival fibromatosis

Gingival fibromatosis often refers to hereditary gingival fibromatosis (HGF) that occurs as an isolated finding or as a component of several different syndromes, common to which is gingival hyperplasia of childhood onset. This condition was previously known under names such as congenital macrogingivae, elephantiasis gingivae, gigantism of the gingiva, and hypertrophic gingivitis; it is not a true fibromatosis but most likely a connective tissue hyperplasia.[10–12]

Syndrome-associated HGF includes manifestations such as hypertrichosis, mental retardation, epilepsy, corneal dystrophy, microphthalmia, and deafness. Most HGF syndromes have autosomal dominant transmission, but autosomal recessive transmission occurs in some syndromes associated with HGF.[13]

The term *gingival fibromatosis* has also been used for unrelated reactive conditions, the most common of which is iatrogenic gingival hyperplasia, associated with drugs such as diphenylhydantoin, cyclosporine, and calcium channel blockers. This condition combines chronic inflammation and fibrosis, and the reader is referred to other sources for a more complete discussion. Gingival fibromatosis-like enlargement of gums, previously also referred to as gingival fibromatosis, occurs in juvenile hyaline fibromatosis.

Clinical features

The gingival hyperplasia in HGF syndromes starts in childhood, most commonly at the eruption of permanent teeth, but some forms are present already at the eruption of deciduous teeth or before. The pathologic growth of gingiva is tooth dependent and could be halted by extraction of teeth, a treatment historically used in the most severe forms of this disease. Gingival hyperplasia can be localized, involving one gingival region, or it can be generalized, involving all regions of mandibular and maxillary gingivae (Figure 9.8). The lesions can cause both aesthetic and functional problems. The latter problems include blocked tooth eruption, orthodontic problems, and failure of jaw closure in severe cases. Gingivectomy has been used as treatment, and this must sometimes be repeated on recurrence of gingival growth.

Pathology

Histologic findings are nonspecific. Gingival tissue is thickened, and histologically there is increased extracellular collagenous matrix, typically consisting of thick collagen bundles radiating into different directions (Figure 9.9). Fibroblasts are scattered, bland, and rather inconspicuous, with virtually no mitotic activity. Focal chronic inflammation is sometimes present. Rete pegs of surface epithelium are elongated. Significant epithelial hyperplasia occurs with associated inflammation, and the surface can be focally ulcerated. There is no indication that the histologic changes are specific to the different forms of isolated or syndrome-associated HGFs. Some researchers have noted heterogeneity in terms of the presence of different types of myofibroblasts in different forms of gingival fibromatosis.[14] Juvenile hyaline fibromatosis can also cause gingival overgrowth, but is typically associated with multiple cutaneous nodules and has a different pathogenesis.

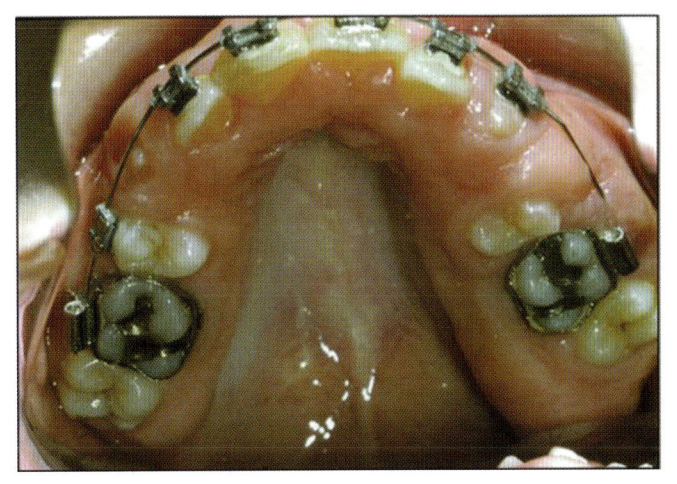

Figure 9.8 Gingival fibromatosis can cause extensive, diffuse enlargement of gingival soft tissue. (Courtesy of Dr. Robert Foss, Washington, DC.)

Genetics

The genetic background of two forms of isolated HGF syndromes with autosomal dominant transmission has been discovered. HGF in a Brazilian family was found to have involvement of 2p21, identified as the HGF1 locus, where SOS1 (son of sevenless-1) gene was found to be involved by a frameshift mutation leading to a stop codon.[15] Another locus in 5q13–q22, designated as GINGF2, was identified in a Chinese family with a form of HGF of infancy onset, with calcium/calmodulin-dependent protein kinase IV (CAMK4) as the candidate gene.[16] If the latter gene is involved, then this might explain a connection of gingival fibromatosis-like proliferation reported with calcium channel blockers.

Calcifying fibrous (pseudo) tumor

Originally reported by Rosenthal *et al.* in 1998[17] as "childhood fibrous tumor with psammoma bodies," this tumor was named a *calcifying fibrous pseudotumor* (CAFP) by Fetsch *et al.*;[18] the designation *pseudo-* was omitted in the World Health Organization (WHO) classification of 2001.[19] Paucicellular fibrous tissue suggests a reactive nature, although isolated reports of recurrence raises the possibility of a true fibroblastic neoplasm.

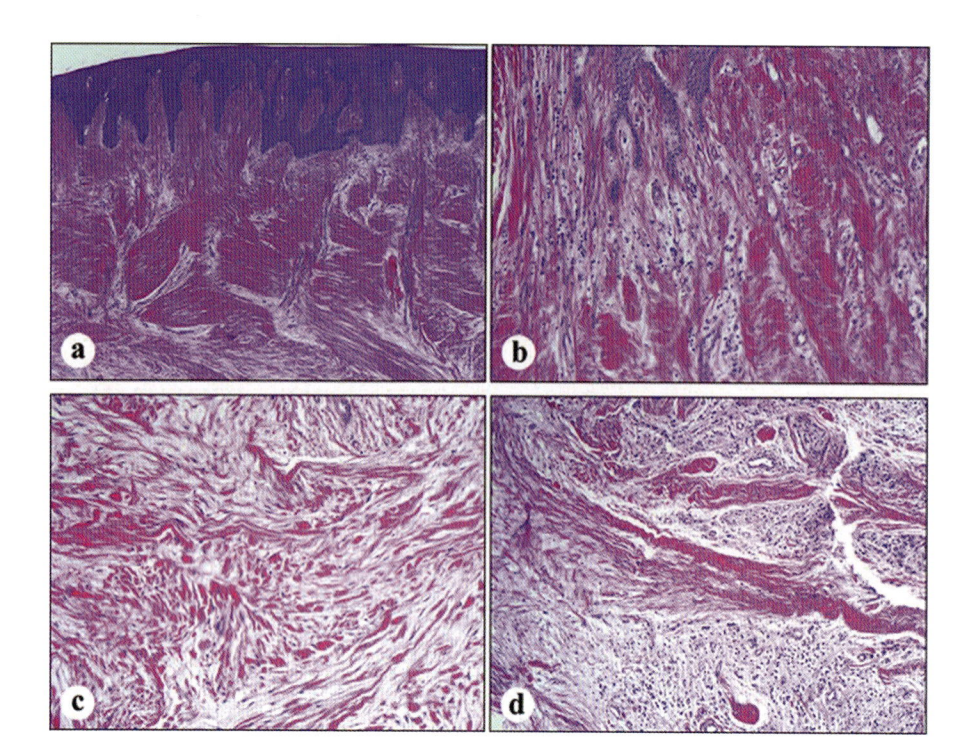

Figure 9.9 (**a–c**) Histologically, gingival fibromatosis shows a band-like fibrous enlargement of the gingival stroma. (**d**) The stroma shows focal myxoid change and can have foci of lymphoplasmacytic infiltration.

Clinical features

This histologically distinctive fibrous tumor occurs predominantly in children and young adults in peripheral soft tissues and body cavities with equal sex distribution, but the tumor also occurs in older adults, especially in the visceral locations. Some patients have had a long history (>10 years) of tumor. Among the peripheral sites are extremities, trunk, and neck, where the tumor has a predilection for deep soft tissues and has measured from 2.5 cm to 15 cm in diameter (median size, 3.5 cm).[18,20] Available series indicate that visceral cases outnumber those in peripheral soft tissues.[20–27] The abdominal, pleural, and scrotal tumors can be multiple, and their size varies greatly, ranging from <1 cm to >5 cm. Occurrence has been reported in the gastric wall.[28] The author has also seen examples involving the intestinal wall, predominantly in the outer layers. Local recurrence after simple excision is rare.[27]

Pathology

Grossly, CAFP appears circumscribed and sometimes lobulated. It is solid, firm, textureless, and gray-white on sectioning, with possible gritty calcifications. Histologically the tumor is composed of thick, partly amorphous and partly fibrillary collagen, scattered psammomatous or amorphous, irregularly stellate calcifications, and focal lymphoplasmacytic infiltrate, sometimes including larger lymphoid foci (Figures 9.10 and 9.11). Some histologically similar tumors without calcifications have also been included in this entity. There is no cytologic atypia or mitotic activity. Immunohistochemically the lesional fibroblasts are positive for CD34 and sometimes focally for desmin and smooth muscle actin, but they are negative for ALK1 and S100 protein.[29–31]

CAFP is consistently ALK negative and is not related to inflammatory myofibroblastic tumor.[30,31] CAFP is much less cellular than is inflammatory myofibroblastic tumor; the latter does not usually feature psammomatous calcifications. No cytogenetic data have been reported to date (March 2015).

Infantile digital fibroma/fibromatosis (recurrent digital fibroma, inclusion body fibromatosis)

It has recently been suggested that this rare tumor, originally reported as a "recurrent digital fibrous tumor" by Reye in 1965,[32] be renamed as *inclusion body fibromatosis* in the WHO classification, with the thought that lesions with similar inclusions can occur in adults elsewhere (e.g., in the breast).[19] Because proof of the identification of such lesions with the digital fibroma is weak, the designation "infantile digital fibroma" is used in this book.

Clinical features

This superficial fibromatosis-like lesion typically occurs in infants and small children of either sex on the extensor surfaces of the second to the fifth fingers and toes; the lesion almost never occurs in the thumb and is rare in the big toe. Some of these tumors are congenital, and some children have multiple lesions. The pathogenesis of this tumor is unknown, and attempts to isolate viral organisms, inspired by the cellular inclusions, have been unsuccessful. The lesion appears as an

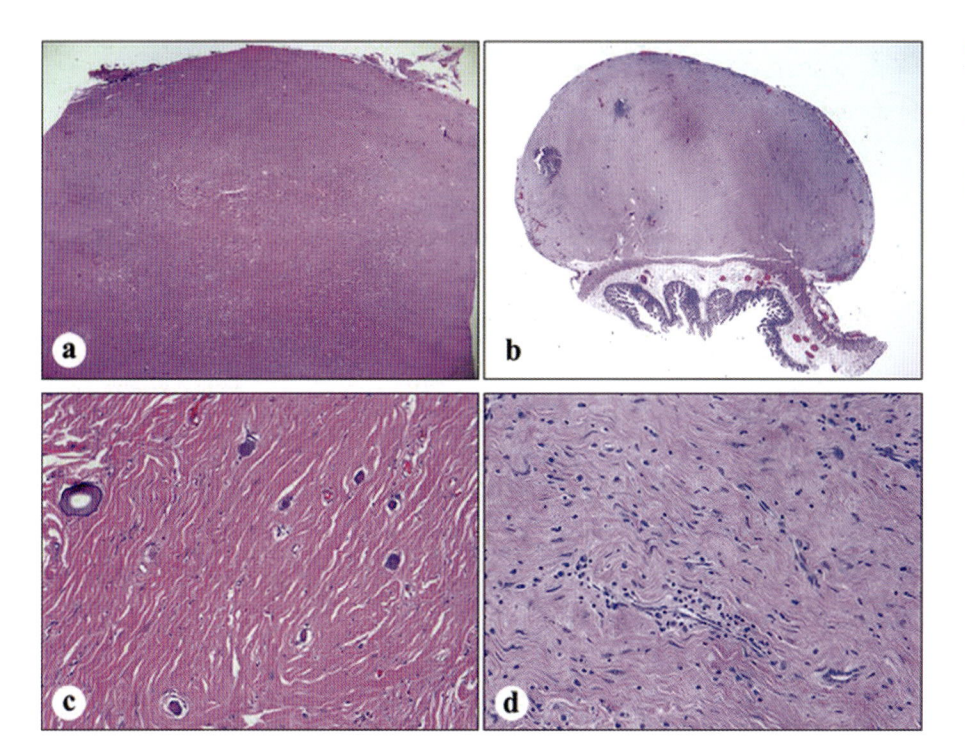

Figure 9.10 (**a**) Calcifying fibrous pseudotumor forms a dense, homogeneous-appearing paucicellular mass. (**b**) This tumor can also involve intestinal wall, as here seen in the external aspect of the small intestine. (**c**) Varying numbers of psammomatous calcifications are present. (**d**) Focal lymphoplasmacytic infiltration is typically present.

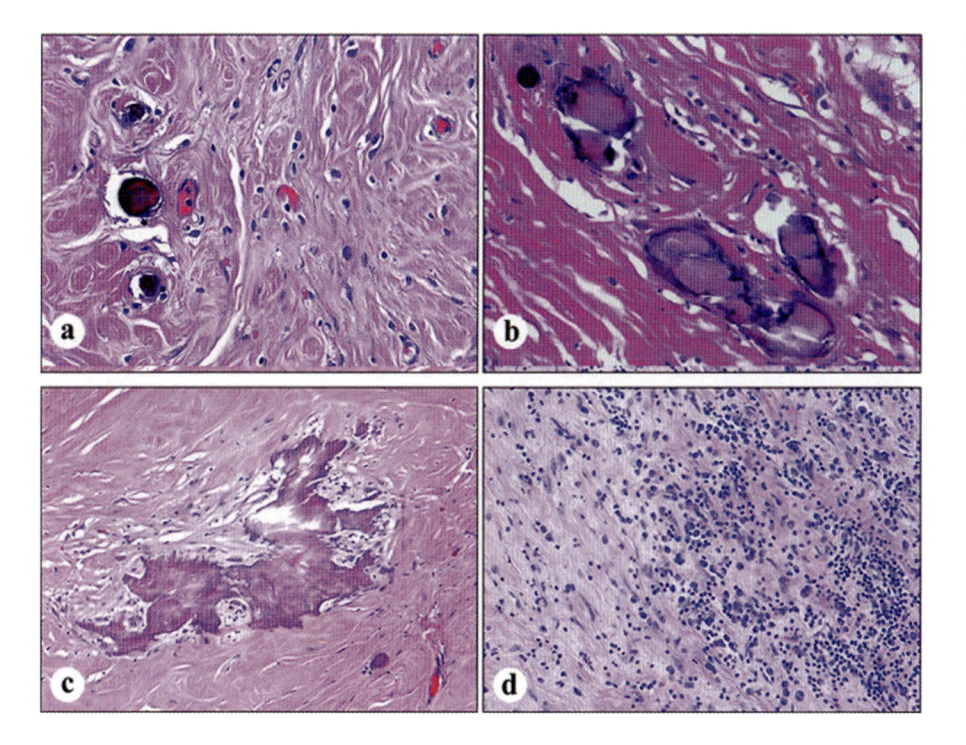

Figure 9.11 (**a–c**) The configuration of the calcifications in calcifying fibrous pseudotumor varies from round to oval or irregularly shaped. (**d**) More extensive inflammatory infiltration can be focally present.

elevated, smooth, contoured nodule underneath the sometimes stretched and erythematous skin of fingers or toes (Figure 9.12). The tumor clinically can resemble an infection.[32–38]

Local recurrences are common following an incomplete excision. Nevertheless, aggressive surgery is not advocated because of the nondestructive nature of recurrences and their tendency to spontaneous regression in long-term follow-up. Familial occurrence, with limb and other malformations, has been reported.[39]

Pathology

Grossly, infantile digital fibroma forms a poorly demarcated, gray-white nodular lesion. Histologically it usually involves both the dermis and subcutis, sometimes extending into the fascia. It is composed of relatively small, irregular, crisscrossing fascicles composed of uniform, bland fibroblasts and myofibroblasts in a collagenous background. Entrapped eccrine ducts with a regenerative appearance are a common finding.

Nuclei are relatively hypochromatic, with a delicate nucleolus. The presence of eosinophilic, rounded cytoplasmic, often perinuclear inclusions is typical of this tumor. The number of inclusions and their prominence vary widely (Figures. 9.13 and 9.14). The inclusions have been shown to contain actin microfilaments, as demonstrated by immunohistochemistry and heavy meromyosin binding.[40,41] The globules have been shown to be actin positive especially after "demasking" with potassium hydroxide treatment.[42,43] The lesional cells are typically positive for desmin.[38] Ultrastructural studies have revealed myofibroblastic differentiation and the filamentous nature of the inclusions.[43]

Fibrous tumors with cytoplasmic inclusions similar to those of infantile digital fibromatosis have been sporadically reported in adults in nondigital sites. These include occurrence in fibroma of the arm, in phyllodes tumor and other fibroepithelial tumors of the breast,[44–47] and in the peculiar scleroderma-like fibrous lesions that developed in conjunction with the epidemic toxic oil syndrome, especially reported from Spain.[48] It is by no means certain that these lesions are part of the same entity as infantile digital fibroma.

Differential diagnosis

Although the lesion might resemble desmoid fibromatosis at first glance, the superficial location, distinctive eosinophilic cytoplasmic globules, and lack of prominent vessels are differentiating features.

Calcifying (juvenile) aponeurotic fibroma
Clinical features

Calcifying aponeurotic fibroma is a very rare, distinctive fibrous lesion of childhood, originally termed *juvenile aponeurotic fibroma* or *calcifying fibroma* by Keasbey in 1953.[49] This tumor

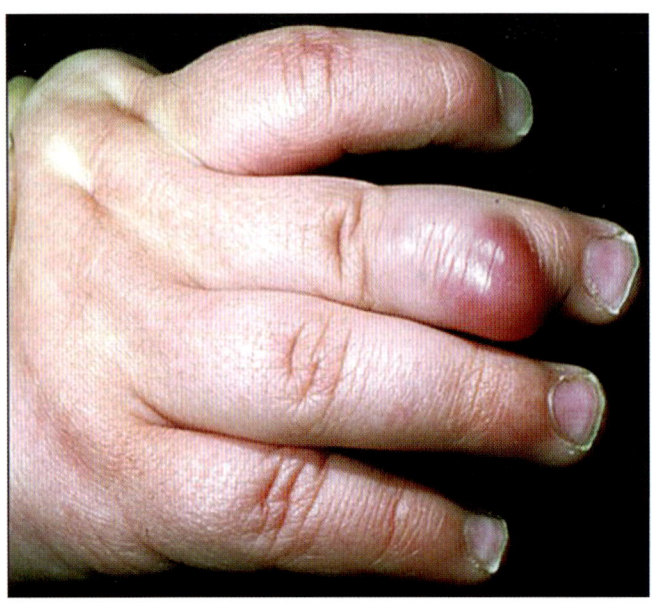

Figure 9.12 Digital fibroma forms an erythematous mass on the dorsum of the distal middle finger.

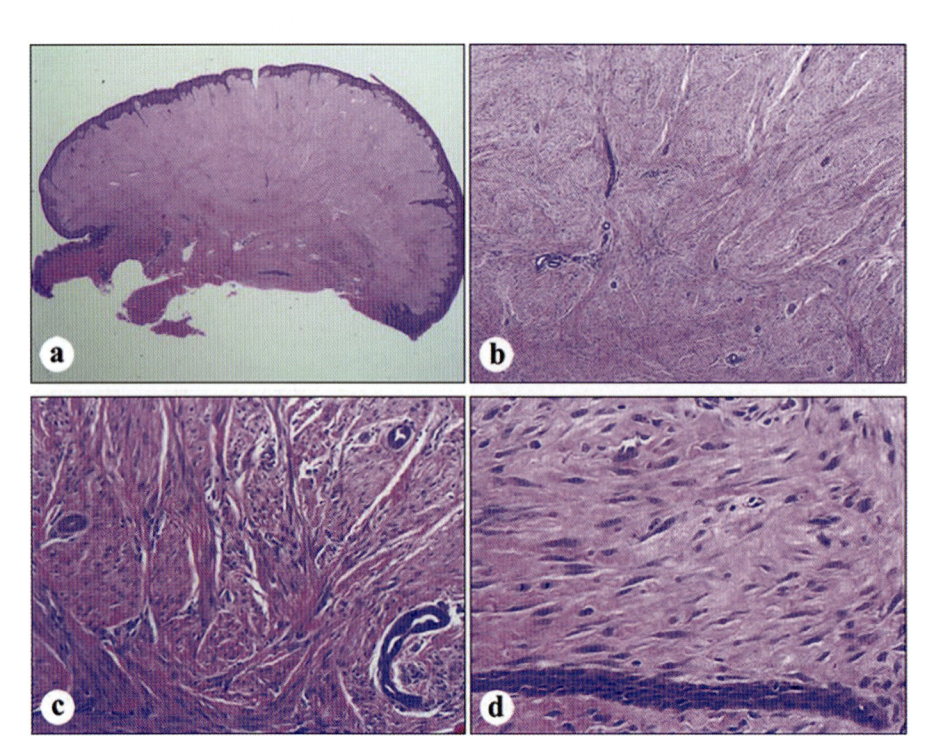

Figure 9.13 (a) Digital fibroma forms a polypoid, dome-shaped cutaneous mass. (b) The tumor is composed of intersecting fascicles of myofibroblastic proliferation that trap sweat glands. (c) The lesion is homogeneous and has only moderate cellularity. (d) High magnification shows numerous perinuclear, eosinophilic cytoplasmic globules that are typical and virtually diagnostic of this entity. Note also a hyperplastic, entrapped sweat gland duct in the lower aspect; these are common findings.

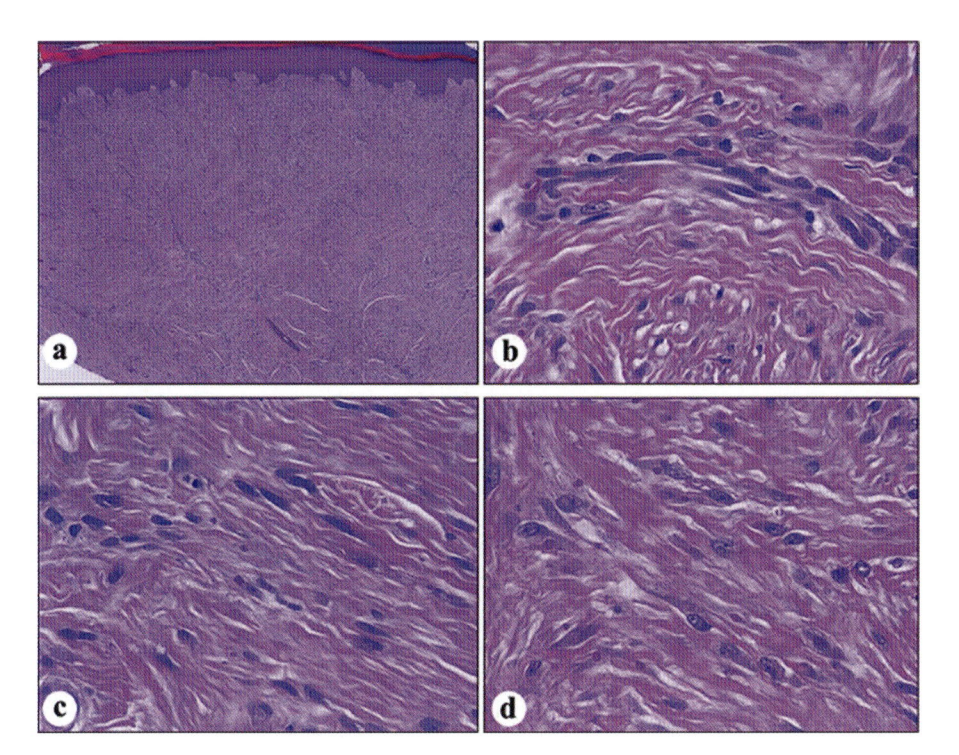

Figure 9.14 Another example of digital fibroma. (**a**) The tumor involves the dermis. (**b–d**) The tumor cells are set in a dense fibrous matrix and often contain the typical eosinophilic cytoplasmic globules.

occurs almost exclusively in children and young adults, usually in the hand.[50,51] Nearly one half occur in the first decade, with the median age of 11 years according to the AFIP material (Figure 9.15). Occasional cases are seen in older subjects, mostly representing late recurrences of childhood tumors, and there is a 2:1 male predominance.[52] The tumor forms a superficial soft tissue mass, with 80% of the lesions occurring in the distal extremities, especially in the palmar side of the hand and fingers, and 20% in the proximal parts of the extremities and trunk.[52] It has a significant tendency to recur locally, and in some cases, recurrences have developed over the span of 20 years or more. Whether malignant transformation occurs is questionable; histologic evidence has been offered, but metastases have never been documented. Because the recurrences are not destructive, and lesional maturation or regression can occur over time, preservation of function is a priority over radical excision in sensitive sites such as the hand.

Pathology

Grossly, the lesion has irregular contours and is usually a relatively small, rubbery-to-firm gray-white mass with possible calcified foci, with a median size of 1 cm to 3 cm, rarely exceeding 4 cm to 5 cm. Histologically, small tumors involve the dermis, and larger ones infiltrate into subcutaneous fat. Most cases contain distinctive round or oval fibrocartilaginous foci (Figures 9.16 and 9.17). These foci may be calcified and surrounded by osteoclastic giant cells; they can undergo cystic change and dissolution in an old lesion. Nodular foci of epithelioid fibroblasts typically surrounding these foci can precede their formation. Between these nodules there is bland, fibromatosis-like spindled fibroblastic proliferation that is the prevailing component in some parts of the lesions (Figure 9.18). In some cases, the spindled component is highly cellular, but mitotic activity and atypia are low (<2 in 10 HPFs). In such cases, the presence of epithelioid foci with calcification elsewhere in the tumor establishes the diagnosis. The predominance of the fibrous component in lesions of small children and cartilaginous components in older patients suggests that the cartilaginous component increases during the evolution of the lesion. No genetic data have been reported.

Lipofibromatosis
Clinical features

This rare pediatric tumor, first reported by Fetsch et al.,[53] typically occurs in the extremities of infants and children, with a 2:1 male predominance and a predilection for the hands and feet, the fingers being the most common site (>20%). Less commonly it occurs in the proximal extremities, knee region, thigh, arm, and relatively rarely (10%) in the trunk wall, head, and neck.[54] One-half of the lesions are present at birth, and the first surgery is typically before the age of 2 years. The tumor has a high tendency to persist and recur locally following an incomplete excision but does not metastasize. Patients with recurrences have been treated more than 10 years after the primary excision.

The tumor involves the deep dermis and subcutis, and sometimes also skeletal muscle; it measures 1 cm to 7 cm (median, 2 cm). Because this tumor has the potential for recurrence only, relatively conservative management is advocated for sensitive sites such as the hand, whereas complete excision is preferable in more feasible sites such as the proximal

Figure 9.15 (A) Age, sex, and (B) anatomic distribution of calcifying aponeurotic fibroma.

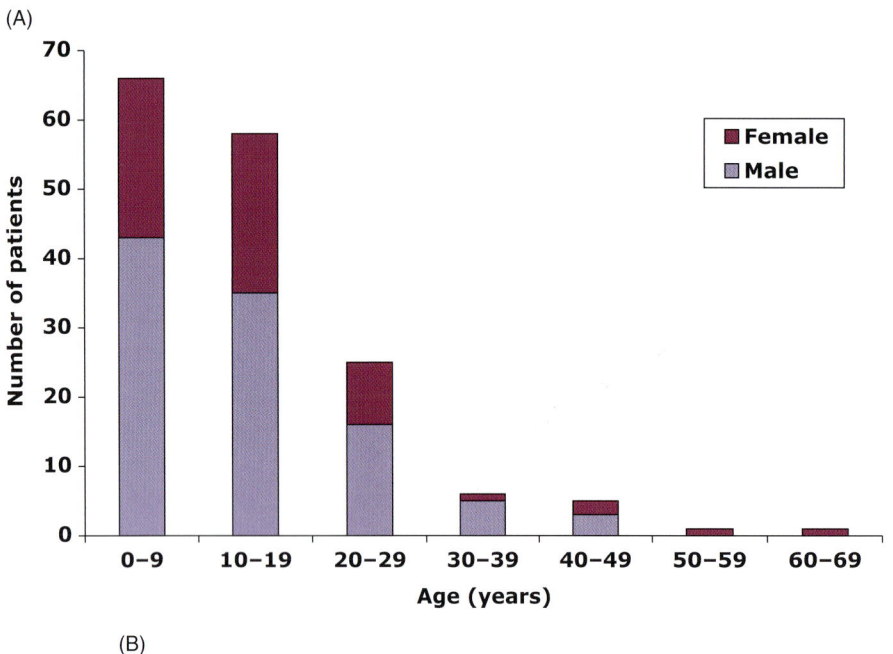

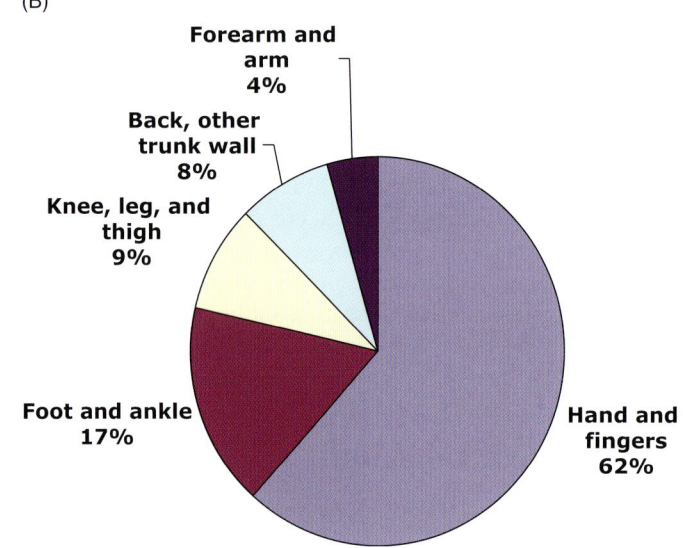

extremities and trunk wall. In some earlier cases, aggressive treatment, such as amputating the hand or foot, was performed, but this should not be necessary in view of the low biologic potential of this tumor.

Pathology

On gross inspection, the lesion is fibrolipomatous, having an integral, generally prominent fatty component with visible fibrous white-to-tan streaks. A similar combination of fatty and fibrous components is also evident in MRI studies. The lesions are poorly marginated, often infiltrating around blood vessels, nerves, skin adnexa, and sometimes into skeletal muscle.

Microscopically the tumor is composed of a dominant fatty component interspersed by fibromatosis-like streaks traversing through the fat, often forming cellular, septal-like structures in the fat, enhancing its lobulation. Dermal involvement is also possible (Figures 9.19 and 9.20). The fatty component is mostly mature but can contain focally immature features such as numerous small capillaries and focal myxoid changes. The fibrous streaks are composed of uniform, at best only mildly atypical, oval to spindled fibroblasts with a moderate, occasionally extensive collagenous matrix. Mitotic activity is often absent, and if present, hardly exceeds 5 mitoses per 50 HPFs. Small, univacuolated fat cells, probably representing evolving primitive fat, are often present at the interphase of the fat and fibrous elements.

Immunohistochemically the spindle cell components are partly positive for smooth muscle actin, CD34, CD99, and often for BCL2. S100 protein positivity occurs in the mature and immature fatty component and occasionally in some

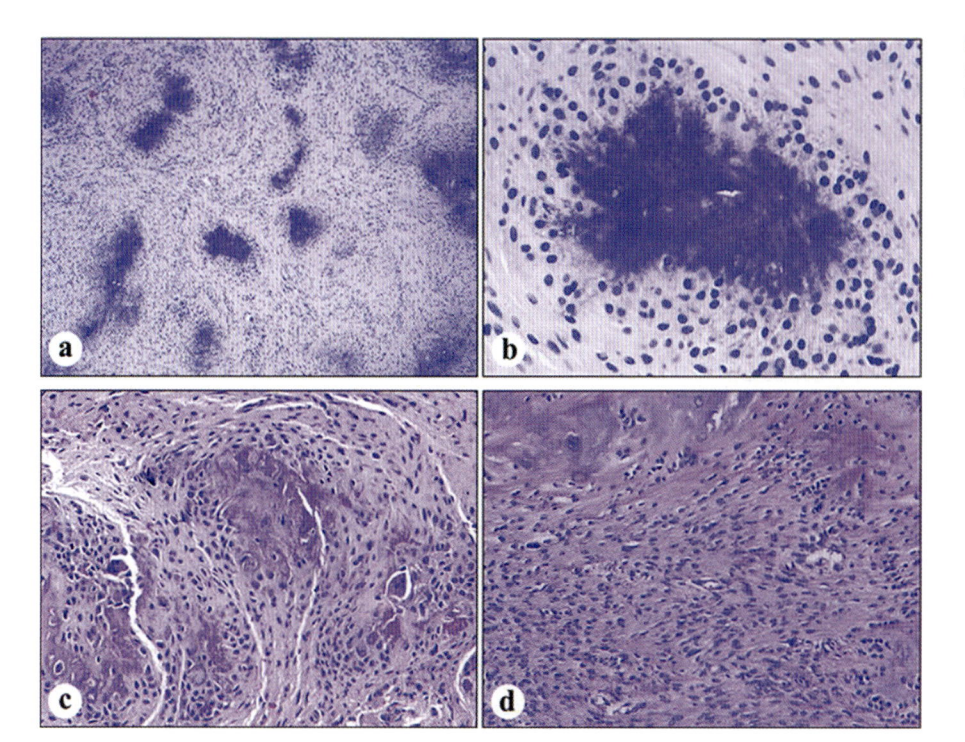

Chapter 9: Benign fibroblastic and myofibroblastic proliferations in children

Figure 9.16 Calcifying aponeurotic fibroma contains foci of calcifications surrounded by epithelioid fibrocartilage-like cells.

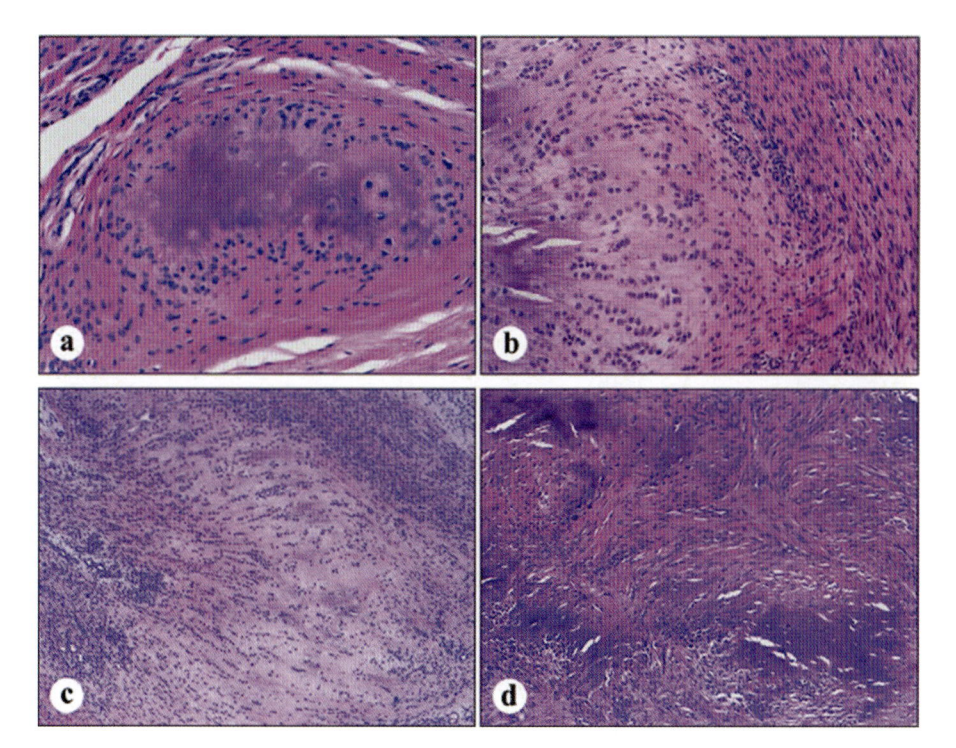

Figure 9.17 The tinctorial quality of the calcification varies, and the fibrocartilage elements can show nuclear palisading.

spindle cells. Scattered HMB45-, tyrosinase-, and melan-A-positive dendritic melanocytes have been present in some lesions. All components are negative for desmin and AE1/AE3 keratin cocktail.[53]

A three-way chromosomal translocation t(4;9;6)(q21;g22;q2?4) has been reported in one case.[55]

Differential diagnosis

Previously, these tumors were classified as *juvenile fibromatosis* or *fibrous hamartoma of infancy*. There is a resemblance to fibrous hamartoma of infancy, except that no primitive round cell component is present. Juvenile diffuse fibromatosis is a rare condition characterized by oval fibroblastoid cells, often

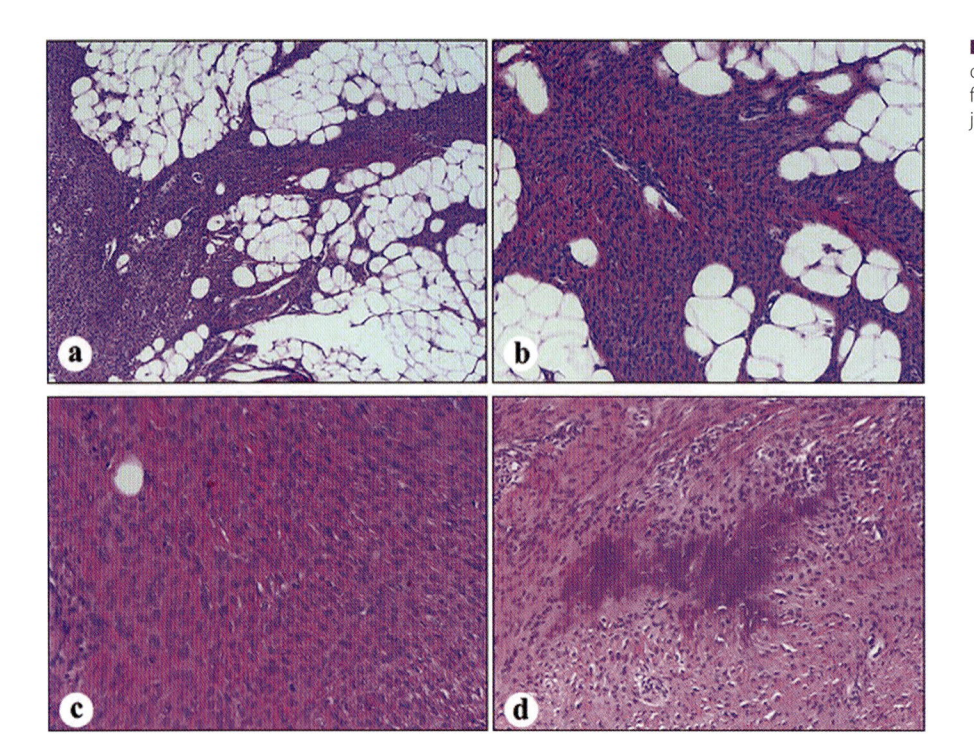

Figure 9.18 (a–c) Fibromatosis-like components can be prominent in calcifying aponeurotic fibroma. (d) Focal calcification is present juxtaposed to a fibromatous element.

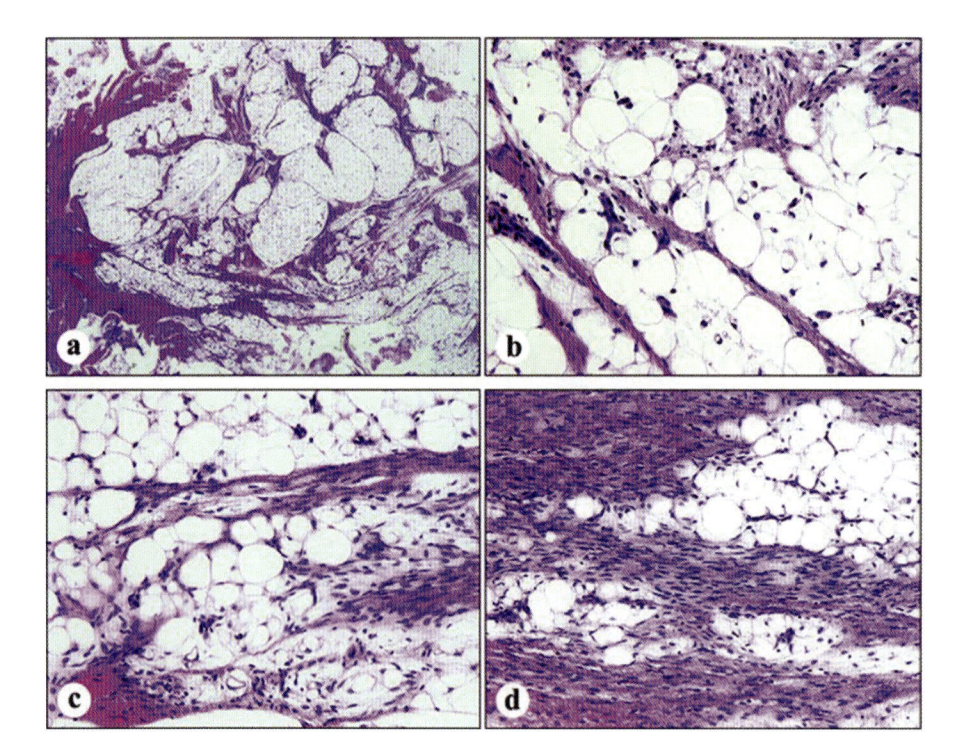

Figure 9.19 (a–c) Lipofibromatosis lesion contains streaks of moderately cellular fibroblastic foci and intervening fat somewhat resembling fibrous hamartoma of infancy. The tumor typically infiltrates in skeletal muscle and can have lipoblast-like fat cells (d).

with diffuse skeletal muscle infiltration. Desmoid fibromatosis contains dominant spindle cell fascicles with well-developed, mildly gaping vessels. Lipoblastoma lacks the interspersed cellular fibrous elements seen in lipofibromatosis.

Infantile diffuse fibromatosis

The terms *infantile* and *juvenile fibromatosis* have been used in the past in a generic sense to denote the young age of the patient, without necessarily identifying a specific tumor type.

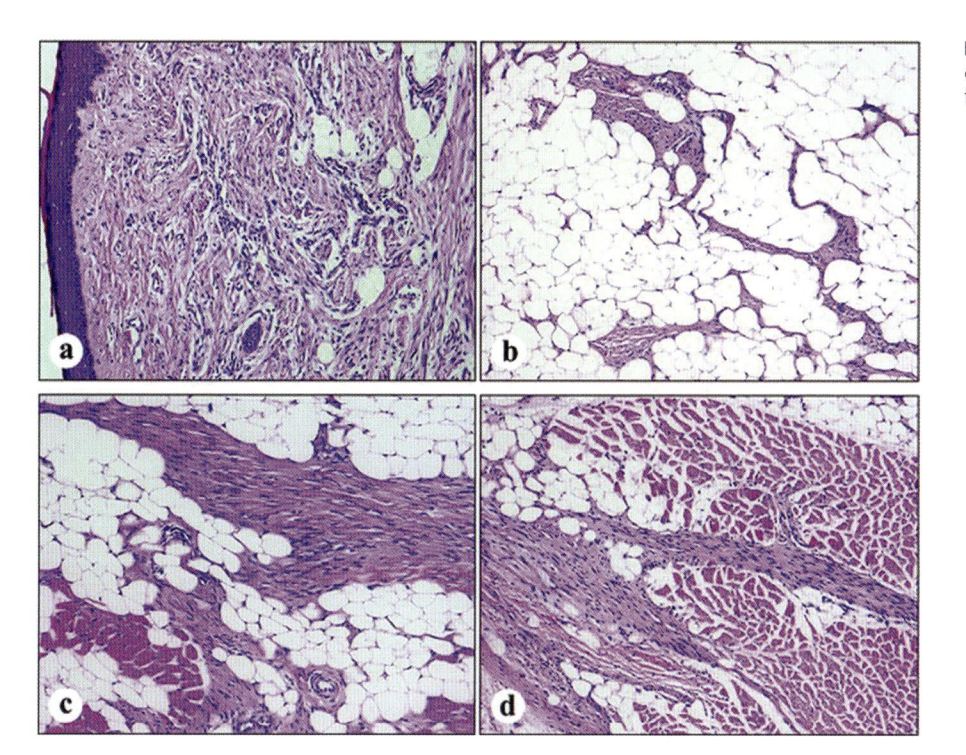

Figure 9.20 (**a**) Lipofibromatosis involving dermis. (**b–d**) The alternating streaks of fat and fibromatous streaks are typical.

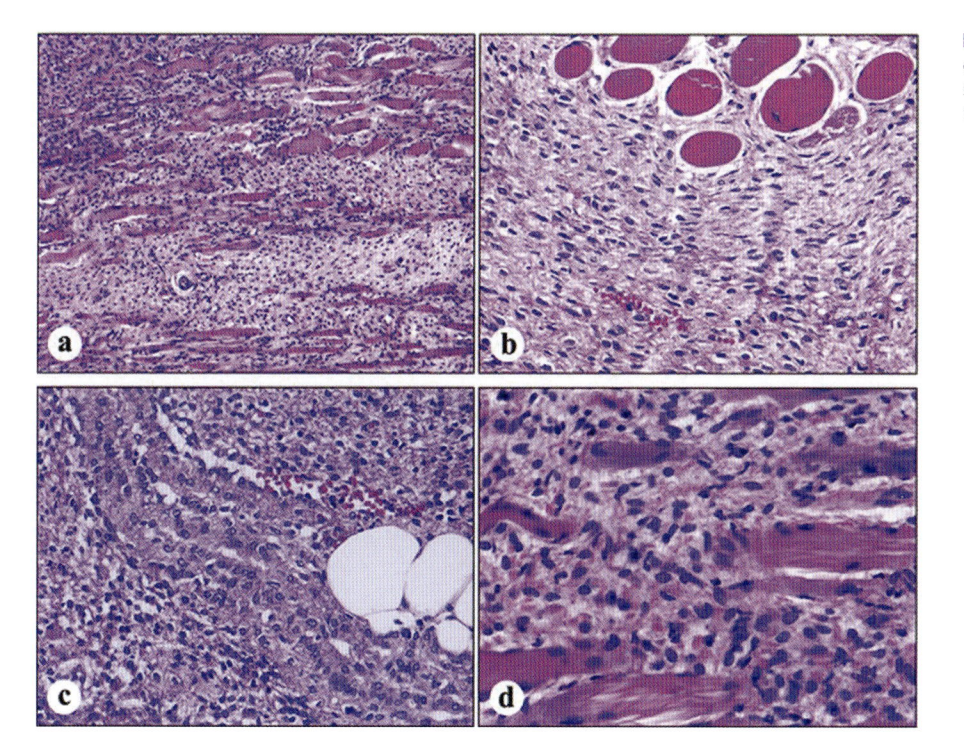

Figure 9.21 Infantile diffuse fibromatosis. Round cells infiltrate diffusely in the skeletal muscle and subcutaneous fat. There is no significant atypia or mitotic activity.

Thus, typical desmoids and lipofibromatosis might have fallen under this designation. The description here refers to a very rare type of fibromatosis that seems to occur exclusively in small children.

This very rare fibromatosis, unrelated to desmoids, typically forms intramuscular masses at various body sites in the extremities, trunk, and head and neck, including the tongue. Local recurrences occur, but there is no metastatic potential. Documentation of this entity is limited.[56]

Histologically the lesions are composed of oval to round, fibroblast-like cells that diffusely infiltrate in fat and skeletal muscle (Figure 9.21). There is no nuclear atypia, and mitotic activity is inconspicuous. These tumors often recur, but long-term follow-up data are not available.

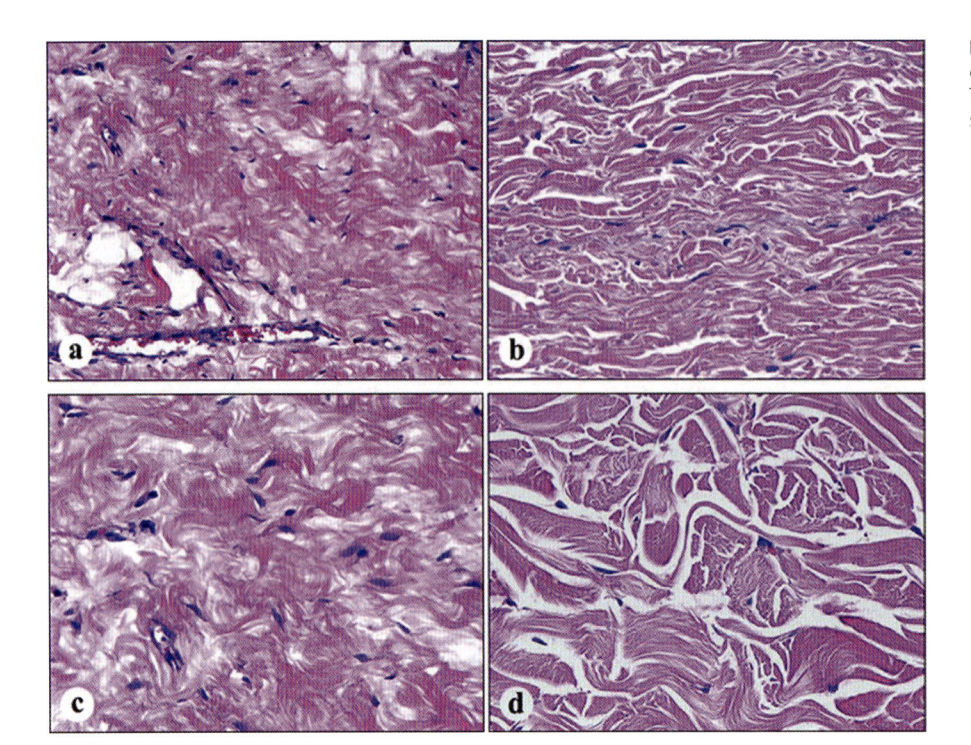

Figure 9.22 (a–d) Gardner fibroma is composed of scattered fibroblasts in a collagenous matrix. (d) The collagen might show lobulation, similar to that seen in nuchal fibroma.

Gardner fibroma

Gardner fibroma is a newly described fibroma type occurring with familial adenomatous polyposis (FAP, Gardner syndrome), and it has been considered a desmoid precursor. The difference between nuchal fibroma (Chapter 7) and Gardner fibroma is subtle and must be supported by clinical and genetic correlation.

Clinical features

Described by Wehrli et al.,[57] this fibroma typically occurs in children and young adults with a median age of 5 years, with the majority of these tumors (78%) being detected during the first decade, and 93% by the age of 20. The oldest patient in the largest series was 36 years old.[58] There is a mild male predominance and a high association with Gardner syndrome (70%). This fibroma therefore has been considered a sentinel lesion for Gardner syndrome; it can be the first manifestation of the disease in young children.[58] It typically forms an ill-defined, plaque-like mass, most commonly in the trunk wall, with a predilection for paraspinal soft tissues. It also occurs in other trunk-based locations in the back, axilla, chest or abdominal wall, and flank.[58,59] Some patients have multiple lesions, and familial occurrence is not unexpected, as is the case with Gardner syndrome in general. Some patients have developed desmoid in an area previously affected by Gardner fibroma.[60] It is unclear, however, whether this means that Gardner fibroma is a precursor lesion for desmoid or if desmoid arises in surgical scar formation.

Pathology

Grossly, Gardner fibroma forms a rubbery, poorly circumscribed, whitish mass measuring from <1 cm to >10 cm (average: 4 cm). Histologically the lesion is relatively paucicellular, with a dense collagenous background and prominent mildly dilated vessels resembling those found in desmoid tumors (Figure 9.22). The process can infiltrate subcutaneous fat and nerves. There are sparse, bland fibroblasts with small nuclei, and neither atypia nor mitotic activity is encountered. Immunohistochemically, the lesional cells often (>60%) show nuclear positivity for beta-catenin, similar to desmoids (Figure 9.23). They have also been reported positive for cyclin D1 and c-Myc.[58]

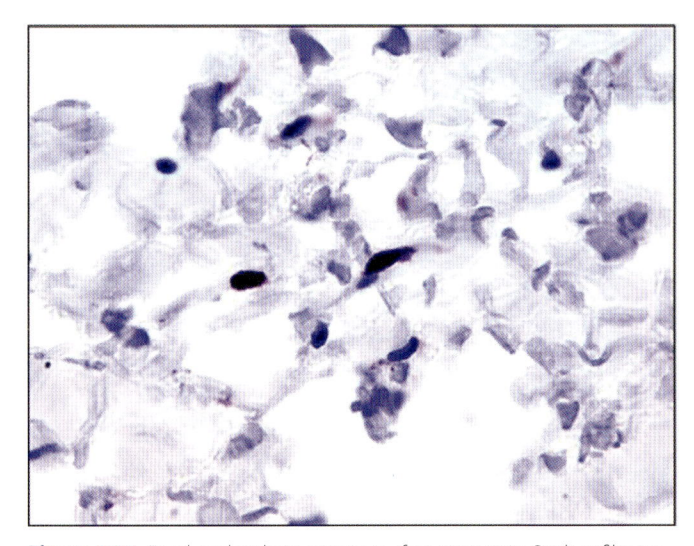

Figure 9.23 Focal nuclear beta-catenin is often present in Gardner fibroma.

Differential diagnosis

The histological features of nuchal-type fibroma can be somewhat similar to Gardner fibroma, although nuchal fibroma generally has a lower cellularity and often has fatty and traumatic neuroma-like components.

Juvenile nasopharyngeal angiofibroma
Clinical features

This histologically distinctive tumor presents in the nasopharynx nearly exclusively in young male patients between the ages of 10 and 20 years. It is site and gender specific. Some series have reported occasional cases in older men and women, but a histologic relationship with the juvenile tumor has not been proved. The author has identified only one potential female patient in the AFIP files, a 12-year-old girl. Based on reported series, this tumor has a predilection for Caucasians.[61–66]

The tumor typically presents with unilateral nasal obstruction, episodes of epistaxis that can be massive, nasal discharge, and sometimes serous otitis media. Larger tumors with extranasal spread can cause facial asymmetry, proptosis, and headaches. Most tumors are sporadic, but some occur in patients with FAP, in whom a 25-fold excess of incidence has been estimated, prompting consideration of nasopharyngeal angiofibroma as a part of the spectrum of FAP-associated tumors.[67] Some series have shown no association with FAP.[68]

Angiofibroma originates in the posterior nasal cavity, probably from its lateral roof at the superior aspect of the sphenopalatine foramen. It often bulges into the soft palate, extending into the maxillary and sphenoid sinuses and sometimes into the orbit and medial cranial fossa.

Treatment is usually surgical, and choice of strategy is based on radiologic staging, especially CT scans with contrast. Endoscopic intranasal surgery is usually successful in smaller tumors, but open surgery is used for larger tumors and those extending into the sinuses and skull base.

Preoperative embolization is often used to reduce intraoperative bleeding. For large tumors, especially those with intracranial spread, radiation therapy (and rarely even systemic chemotherapy) has been used in some centers. Hormonal treatment has not been advocated, despite the apparent androgen dependency of this tumor.

Recurrence rates have been approximately 30% in recent studies[63,64] and >50% in older reports,[61] but the tumor does not metastasize. Rare fatal complications such as meningitis, pneumonia, and exsanguination have occurred, especially with intracranial disease. Malignant transformation has been reported in at least four patients, all of them probably radiation-related cases (postradiation sarcoma), with the transformation being diagnosed 6 months to 18 years after radiation.[69–72]

Pathology

On gross inspection, the tumor typically measures 1 cm to 3 cm, and is gray-white, solid and rubbery, lined by smooth mucosa, sometimes having focal ulceration. Several lobular components can be present, representing extension to different parts of the nasal cavity, paranasal sinuses, and adjacent structures. On sectioning, it can appear porous and focally hemorrhagic because of dilated vessels within.

Nasopharyngeal angiofibroma, especially when relatively small, forms a polypoid lesion that can extend into and ulcerate mucosa (Figure 9.24). In some cases, the tumor is covered by a

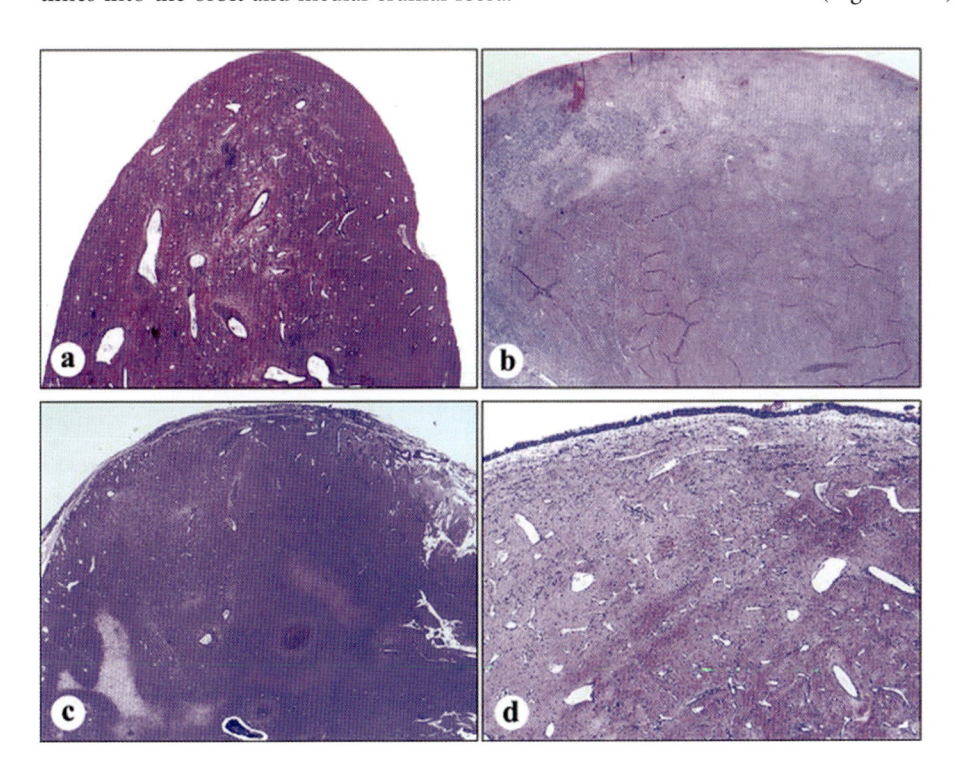

Figure 9.24 (**a**) Configuration of a small nasopharyngeal angiofibroma presenting as a polypoid lesion. (**b**) A nonspecific superficial zone with ulceration covers the luminal aspect of the tumor. (**c,d**) These two tumors extend close to the mucosa, forming sharply demarcated masses containing multiple vascular profiles.

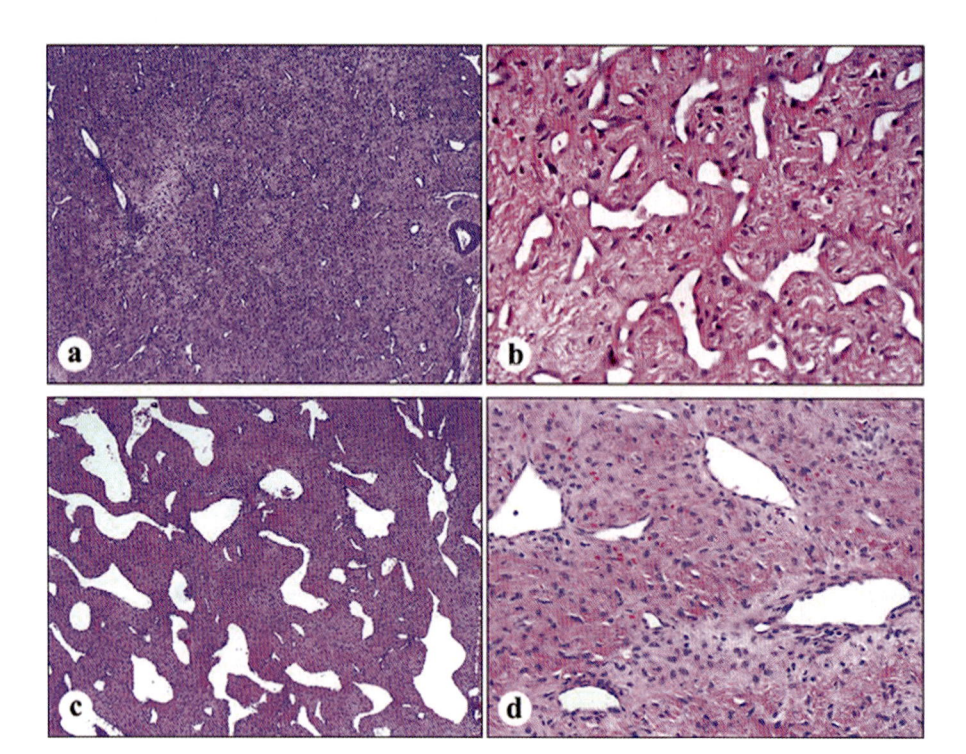

Figure 9.25 Nasopharyngeal angiofibromas have a hemangiopericytoma-like abundant vascular pattern.

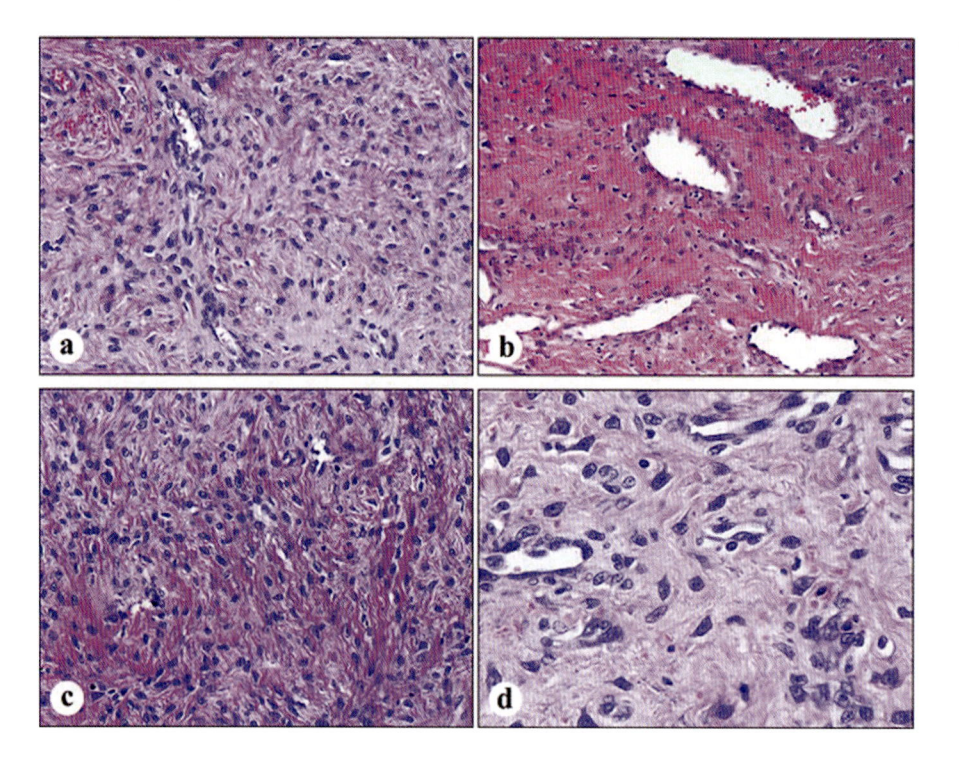

Figure 9.26 A higher magnification shows spindled to polygonal cells containing uniform nuclei with delicate nucleoli. A coarse collagenous matrix is a consistent finding.

layer of nonspecific granulation tissue and fibrosis (Figure 9.24b). Histologically, this tumor is distinctive due to numerous, irregularly shaped, gaping or narrow blood vessels, with an attenuated or no smooth muscle layer and a thin endothelial layer with no atypia (Figure 9.25). The vessels often have a staghorn shape, resembling that of solitary fibrous tumor or hemangiopericytoma. Between the vessels there is a dense, collagenous, fibrous stroma containing scattered plump, epithelioid or stellate fibroblasts having occasional mild nuclear atypia but low mitotic activity (Figure 9.26). The fibroblasts often have an amphophilic cytoplasm. Stromal cellularity varies, and hyalinized, nearly acellular areas can be present.

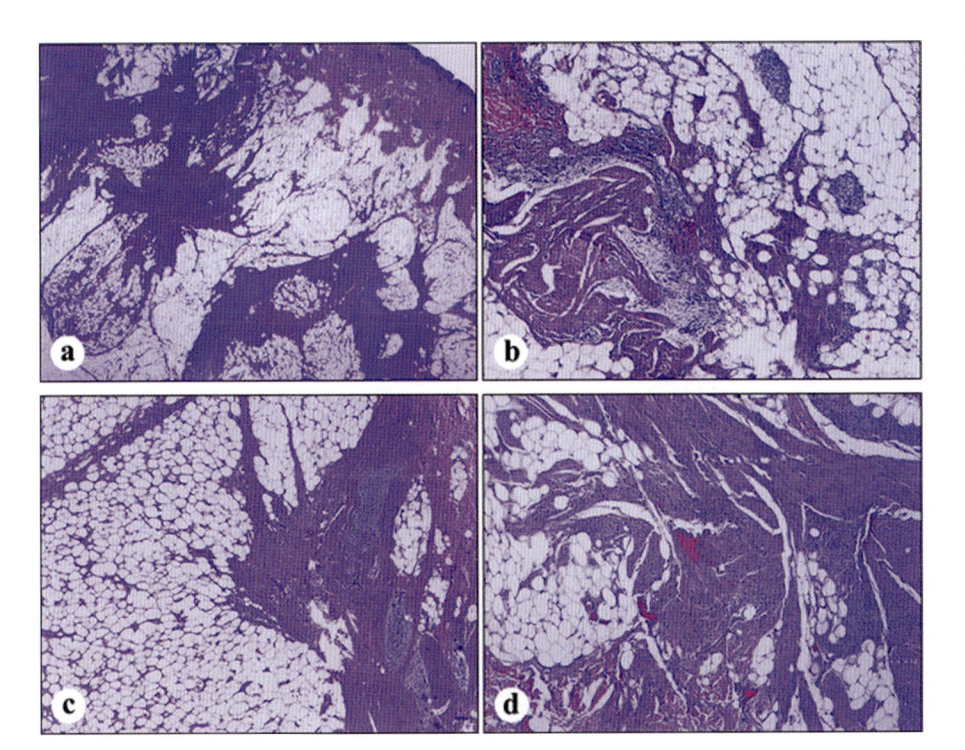

Figure 9.27 Low magnification reveals three components in fibrous hamartoma of infancy: fat, fibromatosis-like element, and small spherules of more primitive cells. Their relative proportions vary in different tumors and in various parts of the same tumor.

Malignant transformation, probably representing postradiation sarcoma in each case, has been a spindle cell sarcoma classified as a fibrosarcoma in three cases, and a pleomorphic sarcoma classified as malignant fibrous histiocytoma (MFH) in one case.

Immunohistochemically, the stromal cells are positive for vimentin and generally negative for smooth muscle actin and desmin; ultrastructurally they resemble fibroblasts. Occasional myofibroblasts have been considered a regressive component.[73,74] The stromal cells express beta-catenin in the nuclei, and sporadic tumors seem to have a high incidence for somatic beta-catenin mutation.[75] Androgen receptor and estrogen receptor beta are also expressed by the stromal cells.[76–78] Cytogenetic studies have not been reported so far.

Differential diagnosis

Sinonasal hemangiopericytoma is smooth muscle actin positive and solitary fibrous tumor, CD34 positive; both are more highly cellular, consisting of rounded and spindled tumor cells, respectively. Choanal polyps have an edematous, paucicellular stroma, where the blood vessels are dilated, thin-walled capillaries; scattered inflammatory cells are often present. Biopsies of enlarged turbinates should not be confused with angiofibroma; they are angiomyomatous, consisting of coarse vessels with well-developed vascular smooth muscle elements.

Fibrous hamartoma of infancy

Originally reported by Enzinger in 1965, fibrous hamartoma of infancy is a rare fibroblastic tumor with triphasic histology, typically occurring in a truncal location in small children.[79]

Clinical features

This histologically distinctive lesion typically presents in the trunk wall of children younger than 2 years old, with a 2:1 male predominance; it can be congenital. The lesion, commonly 3 cm to 4 cm, is subcutaneous and most commonly located in the anterior axilla or chest wall. Rarely, it presents in the proximal extremities but practically never in the hands, fingers, or toes.[79–84] Simple excision is curative, and incomplete excision does not seem to carry a significant risk for recurrence, indicating that conservative excision is sufficient.[85]

Pathology

On gross inspection, a typical fibrous hamartoma of infancy is a poorly circumscribed lesion, merging with the surrounding fat and usually measuring 1 cm to 5 cm. Histologically, the lesion consists of mature adipose tissue interspersed with curving streaks of variably cellular fibroblasts in a collagenous background. Distinctive and unique to this lesion are rounded clusters of oval fibroblasts, which are often seen at the interface of fat and fibrous tissue (Figure 9.27). This constellation of components is virtually diagnostic for fibrous hamartoma of infancy. Sometimes (the author's estimate: 20%) the lesion is entirely composed of cellular fibrous elements with wavy collagen, and such lesions can resemble pseudoangiomatous stroma of the breast or sometimes a neurofibroma (Figure 9.28). There are no true Schwann cell elements, however, and the lesion is negative for S100 protein.

Somewhat similar tumors in distal parts of extremities lack the clusters of oval fibroblasts and have been classified as lipofibromatosis.

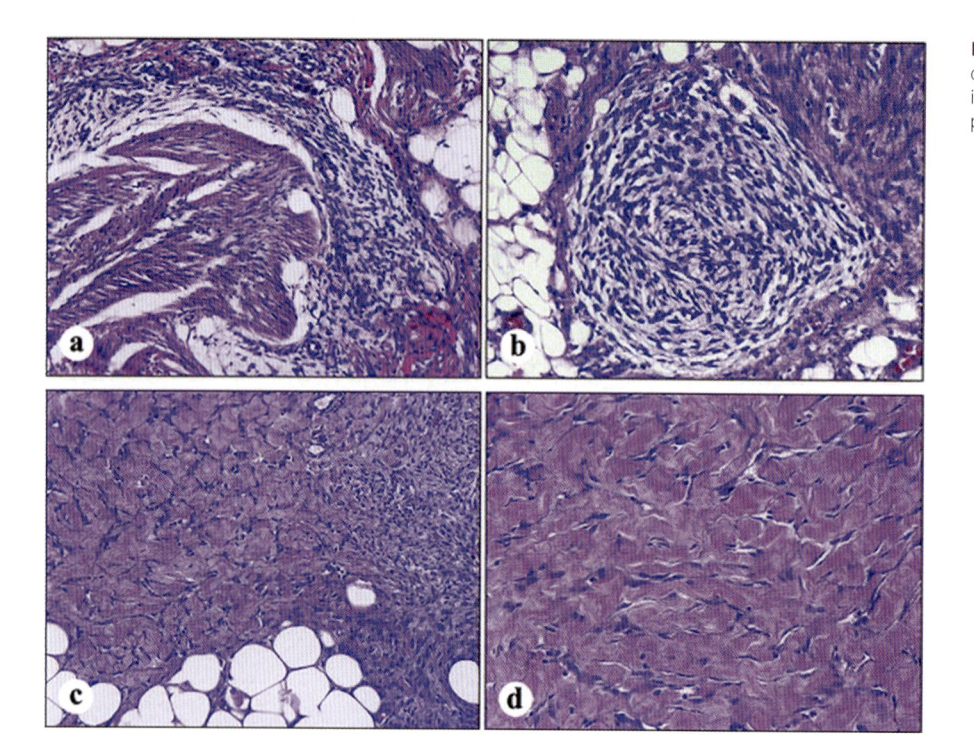

Figure 9.28 (a,b) Higher magnification of the different elements of fibrous hamartoma of infancy. (c,d) Some cases contain a prominent pseudoangiomatous fibrous element.

Genetics

Cytogenetic studies in one case showed a reciprocal t(2;3)(q31;q21) translocation as the only abnormality.[86]

Myofibroma and myofibromatosis

The designation of *infantile myofibromatosis* was introduced by Chung and Enzinger in 1981 for pediatric tumors previously often designated as *congenital fibromatosis*.[87]

Clinical features

Originally reported in infants and children,[87,88] the clinicopathologic spectrum of myofibroma has been expanded to cover similar lesions in adults of all ages.[89–91] The designation of *myofibromatosis* is applied to multiple lesions, which when disseminated were historically referred to alternatively as *congenital fibromatosis, congenital fibrosarcoma,* or *generalized fibromatosis*. Multiple lesions seem to be more common in female infants and are either regional or sometimes involve multiple superficial sites without representing a viscerally disseminated disease. Presentation as a solitary nodule is far more common and is referred to as *myofibroma* or *solitary myofibroma*. In adults, the designation of adult myofibroma has been applied. The age distribution is shown in Figure 9.29.

Most examples are sporadic, but rare familial occurrence in an autosomal dominant pattern has been reported.[92] Some cases have occurred in siblings of different generations of one family.[87,91] Myofibroma is the most common fibrous tumor of infancy.[93]

Myofibroma can occur in a wide variety of peripheral soft tissue locations, and the multiple lesions can widely involve the superficial soft tissues, internal organs, bone, and bone marrow (Figure 9.30). The head and neck is the most commonly involved region in all ages, but in adults, occurrence in the lower extremities is much more common than in children. Extremity lesions do not seem to have a preference for any particular region, except that occurrence in the forearm seems to be uncommon.

Peripheral examples usually involve the dermis and subcutis; they sometimes extend into skeletal muscle.[87] A large series reported occurrence in the oral region, most commonly in the mandible, lips, cheeks, and tongue, where the tumor sometimes presents as a polypoid intraoral projection.[91]

The behavior of peripheral soft tissue myofibroma is benign, even when multifocal. Rare cases with extensive internal organ involvement in newborns can be fatal, however, and a 20% mortality rate for the multicentric disease was reported in one original series.[87] Other widely disseminated fatal cases have been reported, with lesions involving the heart, lungs, and liver.[94] This condition also can lead to intrauterine fetal demise.

Local recurrence rate of solitary myofibroma in the largest series was 7% to 13%, with all recurrences being curable with local re-excision.[87] A relatively conservative excision is therefore preferred, especially in sensitive sites such as the face and other regions in the head. In many cases, the tumor undergoes spontaneous regression and maturation. This has been attributed to a tendency to tumor cell apoptosis.[95] Older patients sometimes have a long history of a stable lesion before the excision.

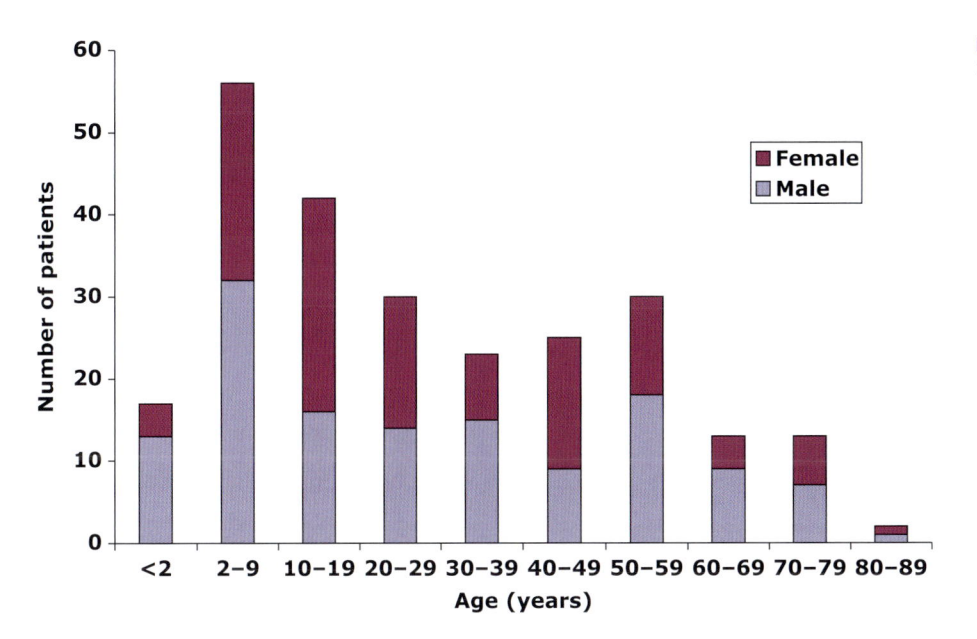

Figure 9.29 Age and sex distribution of 240 patients with myofibromas.

Pathology

Solitary myofibroma usually measures 1 cm to 5 cm and forms a sharply demarcated but unencapsulated mass, often with a histologically infiltrative margin. Grossly, the nodules can be lobulated, and on sectioning they are firm or gritty and vary from white to gray-pink to tan, often with hemorrhagic discoloration.

Microscopically, myofibroma has a biphasic pattern with two distinctive components: (1) a central, highly vascular and cellular, hemangiopericytoma-like component with plump, oval intervascular cells, and (2) scattered or crisscrossing clusters and bundles of spindled, pale-staining, cytologically bland myofibroblasts often most conspicuously located in the tumor periphery (Figure 9.31). When the latter component is not present in the biopsy, the diagnosis can be difficult, whereas tumors lacking the former component can be recognized more easily. In many cases the central, immature part of the lesion has an extensive, typically sharply demarcated area of coagulative necrosis, sometimes with calcification, hyalinization, and cystic change (Figure 9.32).

The more differentiated, eosinophilic myofibroblastic component often forms distinctive, histologically paler and more paucicellular foci and swirling nests of spindle cells separated by collagen fibers (Figure 9.32d). In some cases, this component is seen as the only preserved component in extensively centrally necrotic lesions, and in others it can compose most of the lesion. This component can also show intimal vascular involvement, often outside of the confines of the main lesion; this has no adverse clinical significance (Figure 9.32c).

In small children, the cellular, immature areas of myofibromas sometimes predominate (Figure 9.33). These tumors can have a high mitotic rate (8 per 10 HPF), although usually it is less (<5 per 10 HPF). Nuclear pleomorphism is not a feature of this tumor.

Solitary cutaneous lesions and tumors in adults are generally similar to those in children, except that they often have a more prominent, mature myofibroblastic component and a less prominent, often only focal, vascular hemangiopericytoma-like component (Figure 9.33). The same predominance of the mature component can also be seen in some disseminated lesions; for example, in bone marrow, the myofibroblastic clusters might be the predominant element.

In myofibromas of adults, the differentiated myofibroblastic components often dominate (Figure 9.34). A variant of cutaneous/subcutaneous myofibroma, often seen in adults, has the myofibroblastic component placed as intravascular plugs filling dilated lumina. Although the differentiated myofibroblastic component predominates, a primitive vascular element is also focally present (Figure 9.35).

On immunohistochemical analysis, the differentiated myofibroblastic component is positive for alpha smooth muscle actin, and negative for desmin and S100 protein. CD34 is often seen in the immature hemangiopericytoma-like component.

Genetic data are scant, but one case of an interstitial deletion of 6q12q15 has been reported.[96]

Differential diagnosis

Highly cellular variants of myofibroma can resemble infantile fibrosarcoma, but usually differ from them by the biphasic composition by vascular and myofibroblastic

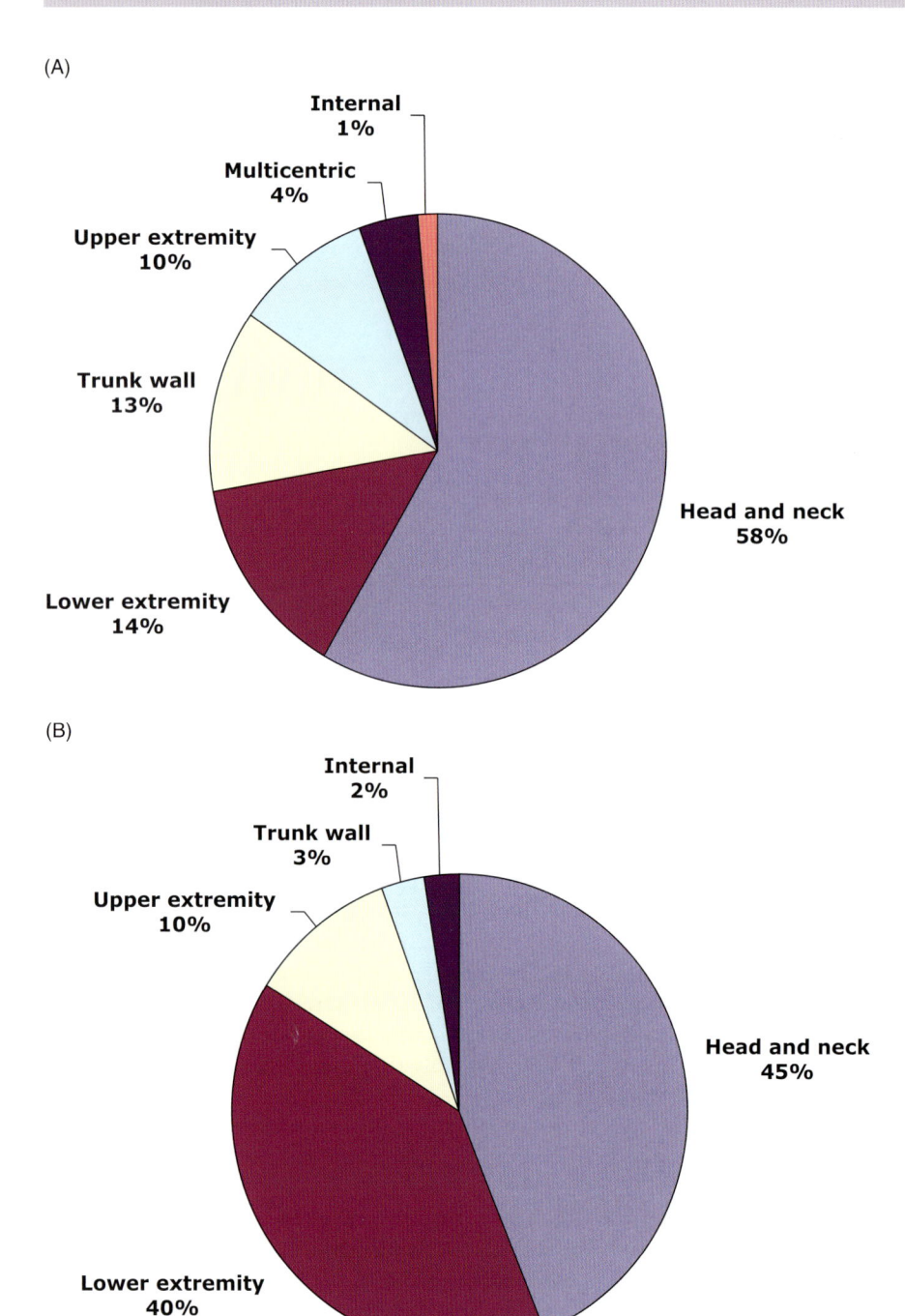

Figure 9.30 (**A**) Anatomic distribution of 72 myofibromas in the first decade and (**B**) distribution of 126 myofibromas in adults >20 years.

(A)
- Internal 1%
- Multicentric 4%
- Upper extremity 10%
- Trunk wall 13%
- Lower extremity 14%
- Head and neck 58%

(B)
- Internal 2%
- Trunk wall 3%
- Upper extremity 10%
- Lower extremity 40%
- Head and neck 45%

components. If the myofibroblastic nodules are not present, then molecular genetic testing for infantile fibrosarcoma or synovial sarcoma could be useful. Infiltrative growth and apparent perineurial (vascular) invasion should also not be confused with malignancy.[97] Myofibromas were historically classified as fibromatosis (e.g., under the designation congenital fibromatosis), from which they differ by lack of a uniform fibroblastic proliferation with collagen formation.

Most tumors previously classified as infantile hemangiopericytomas are myofibromas, and careful search of these tumors usually reveals the myofibroblastic foci typical of myofibroma.[98] The highly cellular myofibromas in small children can invoke the differential diagnosis of infantile fibrosarcoma. The presence of the myofibroblastic clusters, even focally and in subtle forms, is a strong diagnostic clue.

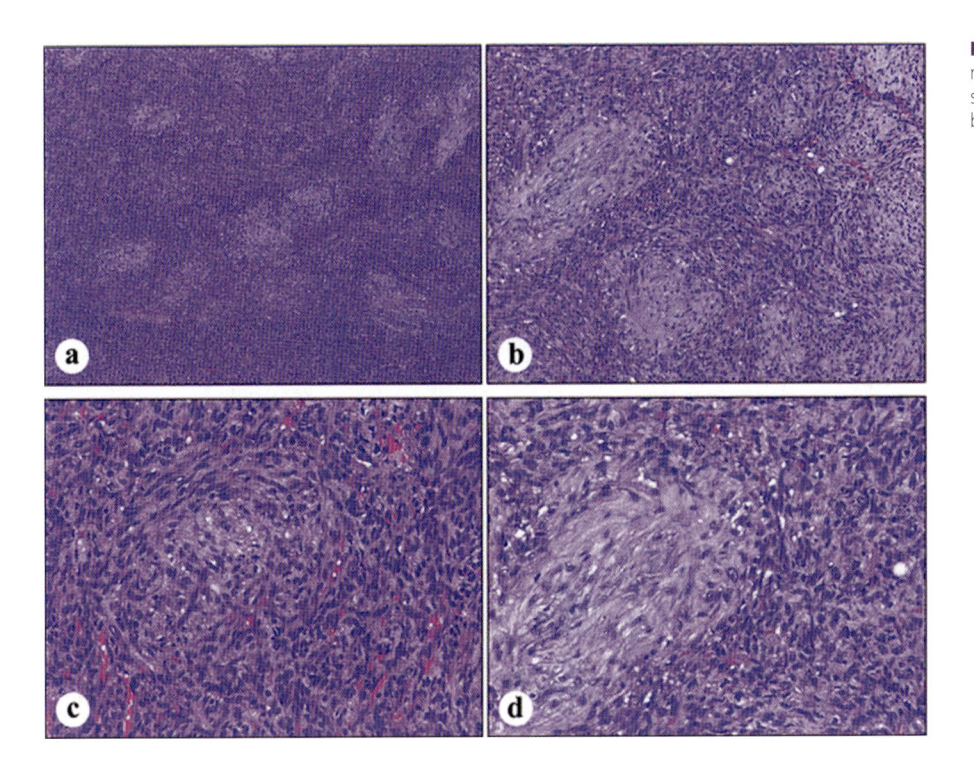

Figure 9.31 A typical example of infantile myofibroma shows a biphasic pattern with pale staining foci of differentiated myofibroblasts and basophilic, highly vascular element.

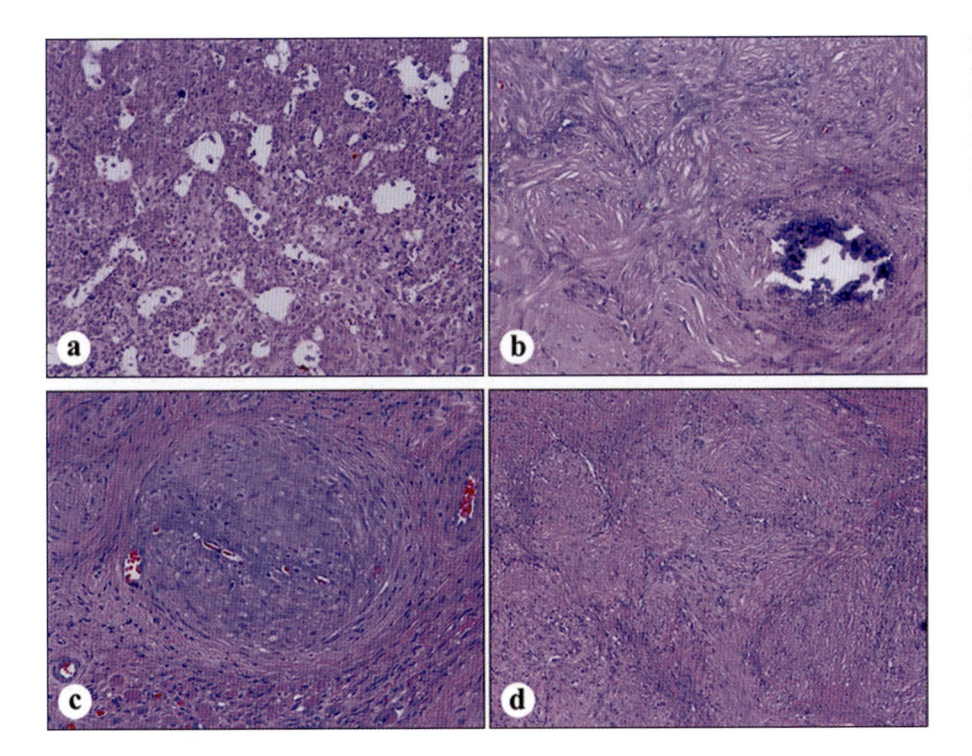

Figure 9.32 (**a**) Central coagulative necrosis is common in the primitive vascular component of infantile myofibroma. (**b**) Focal calcifications can occur in the myofibroblastic component. (**c**) The myofibroblastic foci can be intravascular. (**d**) In some cases, the differentiated myofibroblastic clusters predominate.

Juvenile hyaline fibromatosis
Clinical features

This rare hereditary syndrome (also known as juvenile hyalin fibromatosis, hyalinosis fibrotica multiplex, systemic hyalinosis, Murray–Puretic–Drescher syndrome, and infantile systemic hyalinosis) is a disease with mendelian autosomal recessive inheritance. It mainly affects children of phenotypically normal carrier parents, who have been consanguineous in some cases.[99–105]

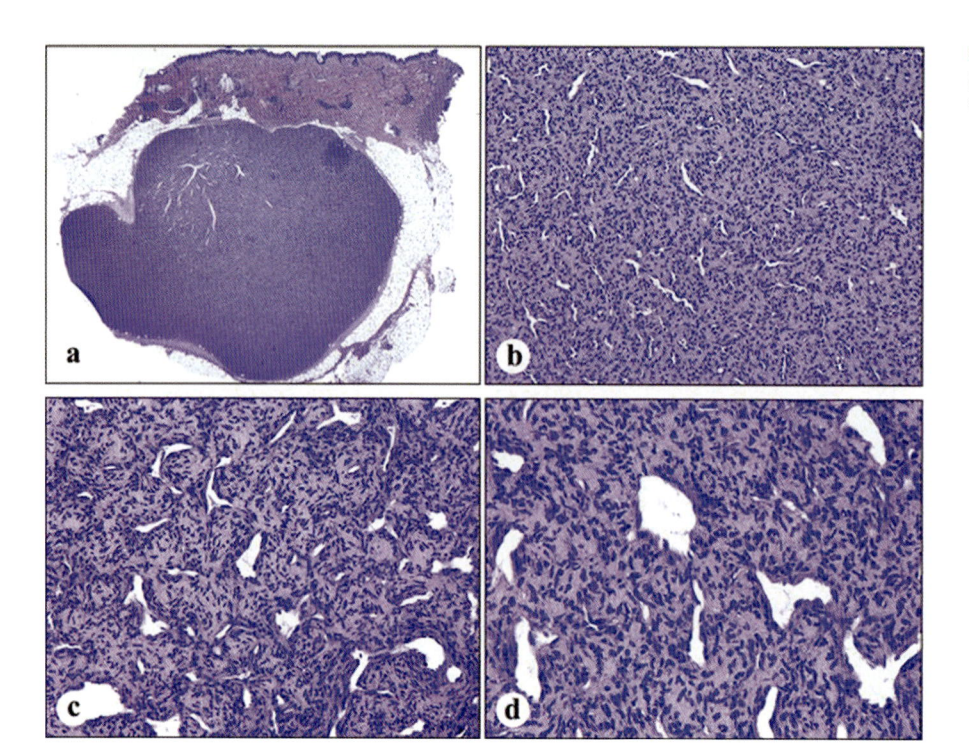

Chapter 9: Benign fibroblastic and myofibroblastic proliferations in children

Figure 9.33 Some infantile myofibromas are predominantly composed of primitive vascular component in a hemangiopericytoma-like pattern.

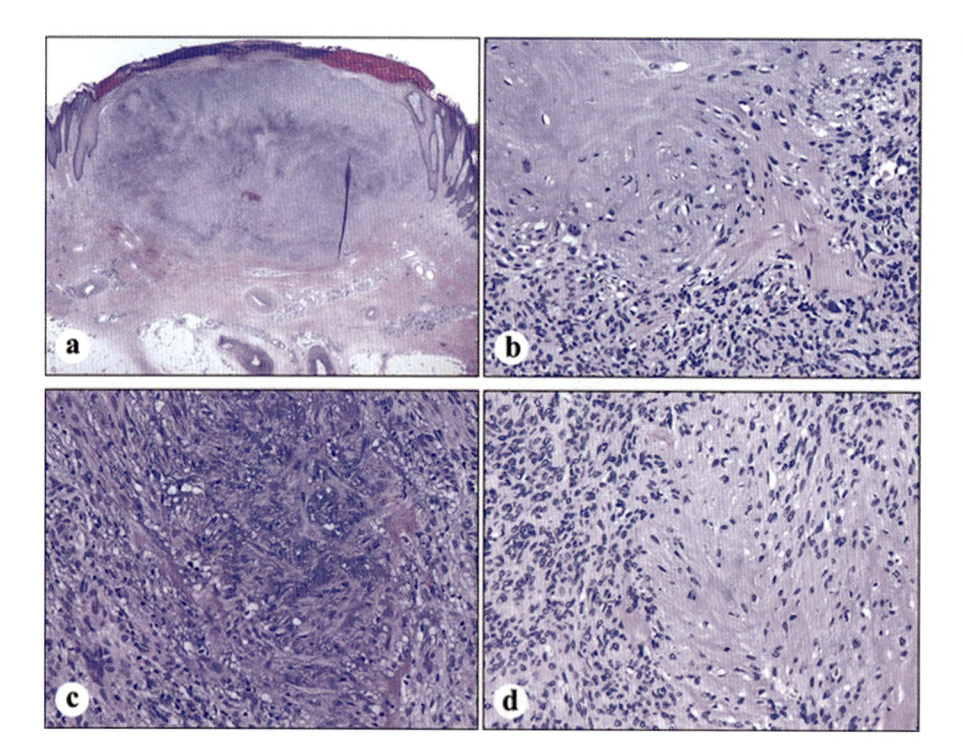

Figure 9.34 Adult myofibroma forming a cutaneous mass. This tumor shows the typical alternating pattern of differentiated myofibroblastic clusters and the primitive vascular component.

This syndrome manifests in young children as slowly growing multiple subcutaneous fibrous nodules in the head and neck, back, or extremities (Figure 9.36). Associated changes include fibrous gingival hyperplasia, flexion contractures of joints, and osteolytic lesions. The fibrous tumors involve the skin and subcutis, and when located in the scalp and face, they can reach disfiguring proportions. Because of the occurrence of multiple skin tumors, this syndrome has sometimes been clinically misinterpreted as neurofibromatosis, especially when the associated changes have not been prominent. Infantile

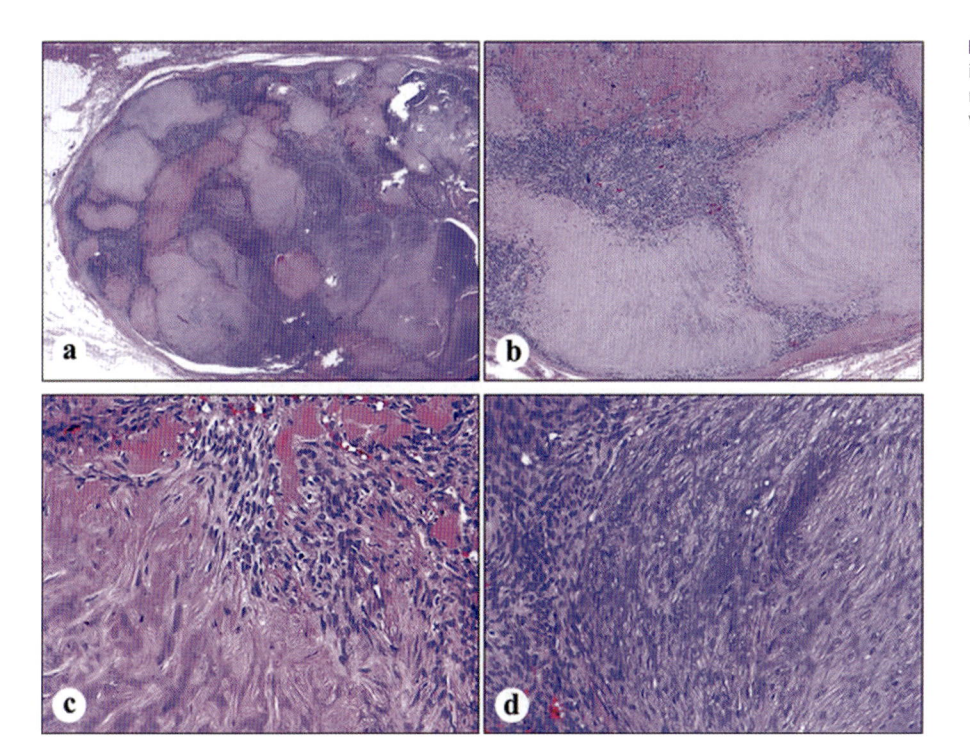

Figure 9.35 Adult myofibroma often forms an intravascular mass. The differentiated myofibroblastic foci predominate, but the primitive vascular element is focally present.

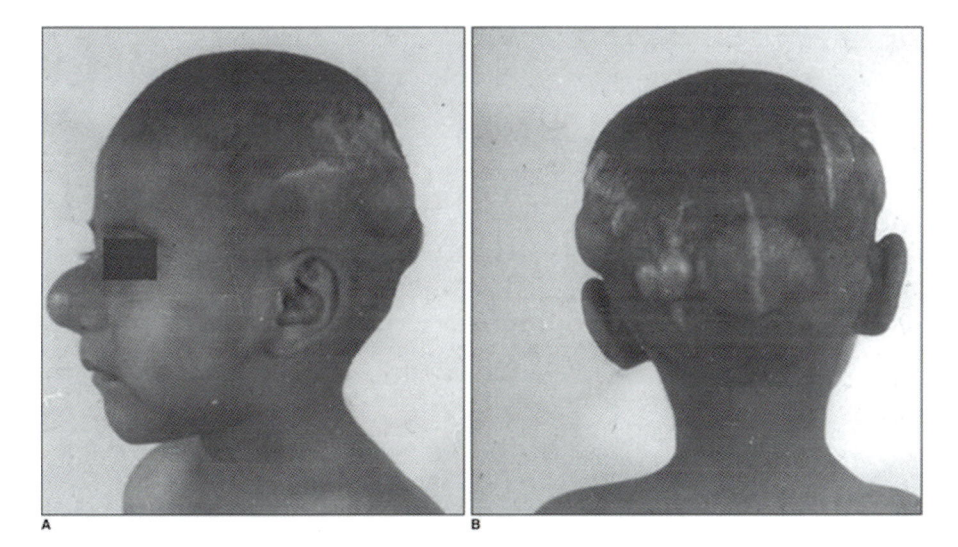

Figure 9.36 Juvenile hyaline fibromatosis has formed multiple tumor-like masses in the head of a young boy; the process also involves the nose. Many nodules show scars from previous surgery.

systemic hyalinosis is a more severe, early-onset variant of the same syndrome. This form also has visceral involvement, and patients with it have a poor prognosis.[106]

Pathology

Histologically the nodules are composed of bland fibroblasts and poorly formed capillaries separated by abundant hyaline-myxoid eosinophilic, collagenous matrix (Figure 9.37). The key change is accumulation of extracellular myxohyaline mucopolysaccharide material in connective tissue caused by an inborn error of morphogenesis.

Although clinically this lesion might resemble neurofibromatosis type 1 because of the multiplicity of the lesions, the histologic features of fibroblasts in the myxohyaline matrix differ markedly from neurofibroma.

Genetics

Germline loss-of-function mutations in ANTXR2 (anthrax toxin receptor) gene encoding for capillary morphogenetic

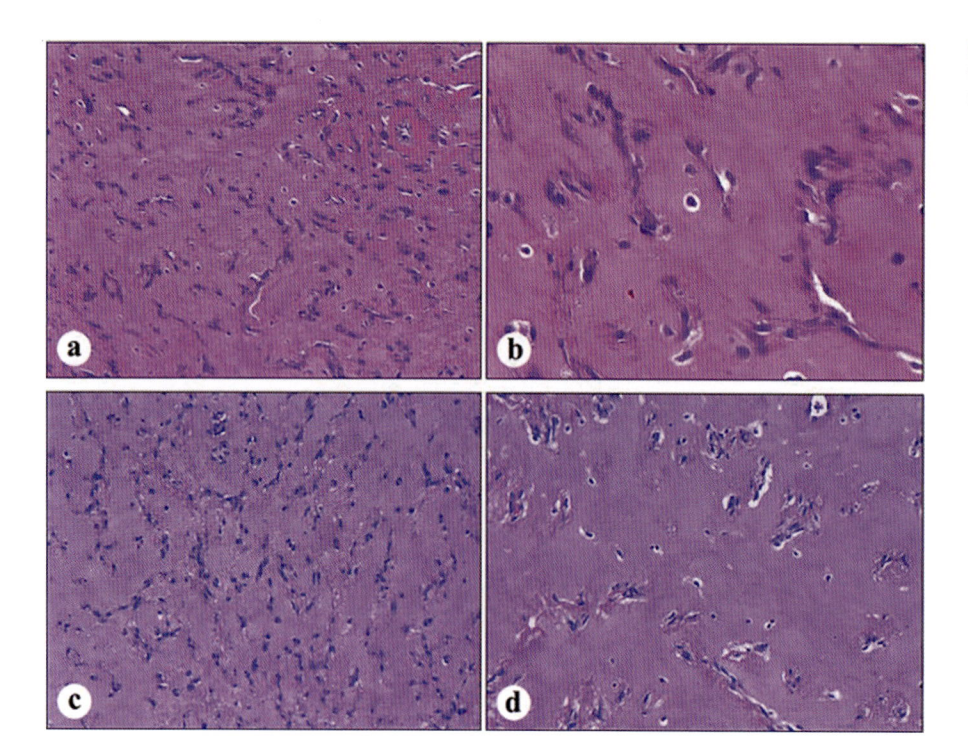

Figure 9.37 Two examples of juvenile hyaline fibromatosis show deposition of abundant hyaline material between variably formed capillary vessels.

protein 2 (at 4q21) have been found in both juvenile hyaline fibromatosis and infantile systemic hyalinosis and are considered causative changes.[107,108] The mutations lead into abnormal cell matrix interactions and subsequent accumulation of the myxohyaline material.[107,108]

Cerebriform fibrous proliferation in Proteus syndrome

Clinical features

Proteus syndrome is a rare dysmorphogenetic syndrome with variable manifestations, named by Wiedemann *et al.* after the Greek god Proteus, known for his changing appearance.[109] Severe examples are diagnosed in early childhood and milder variants in older children and young adults. It is a partial gigantism syndrome, the multiple expressions of which most commonly include macrodactyly, macrocephaly and cranial exostoses, hemihypertrophy, and scoliosis.[109–112] Associated tumors include epidermal nevi (especially collagen nevi), lipomas or lipomatosis, and hemangiomas. Epithelial neoplasms unusual in children (i.e., ovarian or thyroid) have also been reported.[113]

Pathology

Proteus syndrome is relevant to fibroblastic tumors by its superficial soft tissue lesions in sites such as the foot, where

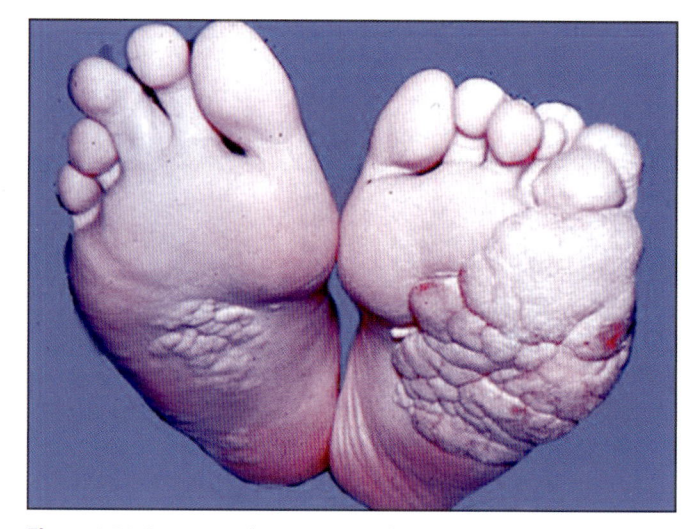

Figure 9.38 Proteus syndrome patients often have gyriform lesions on the foot soles.

macrodactylia can occur, and complex, gyriform, tumor-like formations can develop in the plantar aspect (Figure 9.38).

Histologically these formations consist of verruciform, hyperkeratotic epidermis and dense, hypocellular to nearly acellular, collagenous fibrous thickening of the dermis. Fibrocollagenous lesions without gyriform features can be designated as collagen nevi (Figure 9.39).

Activating AKT1 mutations (c. 49G to A, p. Glu17Lys) have been found responsible for most cases of Proteus

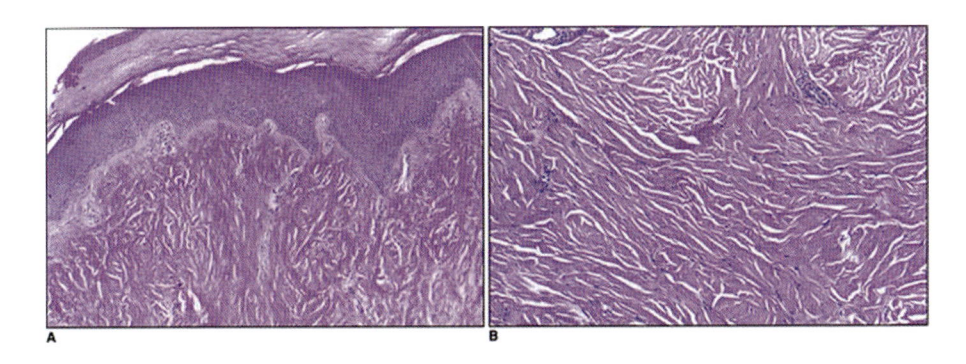

Figure 9.39 Cerebriform fibrous proliferation in Proteus syndrome shows dense, paucicellular collagenous fibrosis and papilliform epidermal change on histologic analysis.

syndrome. These mutations have been somatic and present only in the involved regions indicating somatic mosaicism.[114] Patients with previously reported PTEN mutations probably have PTEN hamartoma-tumor syndrome.[115]

References

Cranial fasciitis

1. Lauer DH, Enzinger FM. Cranial fasciitis in childhood. *Cancer* 1980;45:401–406.
2. Patterson JW, Moran SL, Konerding H. Cranial fasciitis. *Arch Dermatol* 1989;125:674–678.
3. Sarangarajan R, Dehner LP. Cranial and extracranial fasciitis of childhood: a clinicopathologic and immunohistochemical study. *Hum Pathol* 1999;30:87–92.
4. Rapana A, Iaccarino C, Bellotti A, et al. Exclusively intracranial and cranial fasciitis of the adult age. *Clin Neurol Neurosurg* 2002;105:35–38.
5. Sayama T, Morioka T, Baba T, Ikezaki K, Fukui M. Cranial fasciitis with massive intracranial extension. *Childs Nerve Syst* 1995;11:242–245.

Fibromatosis colli

6. MacDonald D. Sternomastoid tumor and muscular torticollis. *J Bone Joint Surg Br* 1969;51:432–443.
7. Wakely PE Jr, Price WG, Frable WJ. Sternomastoid tumor of infancy (fibromatosis colli): diagnosis by aspiration cytology. *Mod Pathol* 1989;2:378–381.
8. Sauer F, Sehner L, Freng A. Cytologic features of fibromatosis colli of infancy. *Acta Cytol* 1997;41:633–635.

Fibrous umbilical polyp

9. Vargas SO. Fibrous umbilical polyp: a distinct fasciitis-like proliferation of early childhood with a marked male predominance. *Am J Surg Pathol* 2001;25:1438–1442.

Gingival fibromatosis

10. Clark D. Gingival fibromatosis and its related syndromes: a review. *J Can Dent Assoc* 1987;53:137–140.
11. Coletta RD, Graner E. Hereditary gingival fibromatosis: a systematic review. *J Periodontol* 2006;77:753–766.
12. Doufexi A, Mina M, Ioannidou E. Gingival overgrowth in children: epidemiology, pathogenesis, and complications: a literature review. *J Periodontol* 2005;76:3–10.
13. Hart TC, Pallos D, Bozzo L, et al. Evidence of genetic heterogeneity for hereditary gingival fibromatosis. *J Dent Res* 2000;79:1758–1764.
14. Bitu CC, Sobral LM, Kellermann MG, et al. Heterogeneous presence of myofibroblasts in hereditary gingival fibromatosis. *J Clin Periodontol* 2006;33:393–400.
15. Hart TC, Zhang Y, Gorry MC, et al. A mutation in the SOS1 gene causes hereditary gingival fibromatosis. *Am J Hum Genet* 2002;70:943–954.
16. Xiao S, Bu L, Zhu L, et al. A new locus for hereditary gingival fibromatosis (GINGF2) maps to 5q13-q22. *Genomics* 2001;74:180–185.

Calcifying fibrous (pseudo) tumor

17. Rosenthal NS, Abdul-Karim FW. Childhood fibrous tumor with psammoma bodies: clinicopathologic features in two cases. *Arch Pathol Lab Med* 1988;112:798–800.
18. Fetsch JF, Montgomery EA, Meis JM. Calcifying fibrous pseudotumor. *Am J Surg Pathol* 1993;17:502–508.
19. Fletcher CD, Mertens F, Unni KK, et al. WHO Classification of Tumours. *Pathology and Genetics: Tumours of Soft Tissue and Bone.* Lyon: IRC Press; 2002.
20. Nascimento AF, Ruiz R, Hornick JL, Fletcher CD. Calcifying fibrous 'pseudotumor': clinicopathologic study of 15 cases and analysis of its relationship to inflammatory myofibroblastic tumor. *Int J Surg Pathol* 2002;10:189–196.
21. Pinkard NB, Wilson RW, Lawless N, et al. Calcifying fibrous pseudotumor of pleura: a report of three cases of a newly described entity involving the pleura. *Am J Clin Pathol* 1996;105:189–194.
22. Hainaut P, Lesage V, Weynand B, Coche E, Noirhomme P. Calcifying fibrous pseudotumor (CFPT): a patient presenting with multiple pleural lesions. *Acta Clin Belg* 1999;54:162–164.
23. Kocova L, Michal M, Sulc M, Zamecnik M. Calcifying fibrous pseudotumour of visceral peritoneum. *Histopathology* 1997;31:181–184.
24. Soyer T, Ciftci AO, Gucer S, Orhan D, Senocak ME. Calcifying fibrous pseudotumor of lung: a previously unreported entity. *J Pediatr Surg* 2004;39:1729–1730.
25. Dumont P, de Muret A, Skrobala D, Robin P, Toumieux B. Calcifying fibrous pseudotumor of the mediastinum. *Ann Thorac Surg* 1997;63:543–544.
26. Chon SH, Lee CB, Oh YH. Calcifying fibrous pseudotumor causing thoracic outlet syndrome. *Eur J Cardiothorac Surg* 2005;27:353–355.
27. Maeda A, Kawabata K, Kusuzaki K. Rapid recurrence of calcifying fibrous pseudotumor (a case report). *Anticancer Res* 2002;22:1795–1797.

28. Agaimy A, Bihl MP, Tornillo L, et al. Calcifying fibrous tumor of the stomach: clinicopathologic and molecular study of seven cases with literature review and reappraisal of the histogenesis. *Am J Surg Pathol* 2010;34:271–278.

29. Ben-Izhak O, Itin L, Feuchtwanger Z, Lifschitz-Mercer B, Czernobilsky B. Calcifying fibrous pseudotumor of mesentery presenting with acute peritonitis: case report with immunohistochemical study and review of literature. *Int J Surg Pathol* 2001;9:249–253.

30. Sigel JE, Smith TA, Reith JD, Goldblum JR. Immunohistochemical analysis of anaplastic lymphoma kinase expression in deep soft tissue calcifying fibrous pseudotumor: evidence of a late sclerosing stage of inflammatory myofibroblastic tumor? *Ann Diagn Pathol* 2001;5:10–14.

31. Hill KA, Gonzalez-Crussi F, Chou PM. Calcifying fibrous pseudotumor versus inflammatory myofibroblastic tumor: a histological and immunohistochemical comparison. *Mod Pathol* 2001;14:784–790.

Recurrent digital fibroma

32. Reye RDK. Recurring digital fibrous tumor of childhood. *Arch Pathol* 1965;80:228–231.

33. Allen PW. Recurring digital fibrous tumor of childhood. *Pathology* 1972;4:215–223.

34. Beckett JH, Jacobs AH. Recurring digital fibrous tumor of childhood: a review. *Pediatrics* 1977;59:401–406.

35. Bhawan J, Bacchetta C, Joris I, Majno G. A myofibroblastic tumor: infantile digital fibroma (recurrent digital fibrous tumor of childhood). *Am J Pathol* 1979;94:19–36.

36. Ishii N, Matsui K, Ichiyama S, Takahashi Y, Nakajima H. A case of infantile digital fibromatosis showing spontaneous regression. *Br J Dermatol* 1989;121:129–133.

37. Rimareix F, Bardot J, Andrac L, et al. Infantile digital fibroma: report on eleven cases. *Eur J Pediatr Surg* 1997;7:345–348.

38. Laskin WB, Miettinen M, Fetsch JF. Infantile digital fibroma/fibromatosis: a clinicopathological and immunohistochemical study of 69 tumors from 57 patients with long-term follow-up. *Am J Surg Pathol* 2009; 33:1–13.

39. Breuning MH, Oranje AP, Langemeijer RA, et al. Recurrent digital fibroma, focal dermal hypoplasia, and limb malformations. *Am J Med Genet* 2000;94:91–101.

40. Iwasaki H, Kikuchi M, Ohtsuki I, et al. Infantile digital fibromatosis: identification of actin filaments in cytoplasmic inclusions by heavy meromyosin binding. *Cancer* 1983;52:1653–1661.

41. Fringer B, Thais H, Bohm N, Altmannsberger M, Osborn M. Identification of actin microfilaments in the intracytoplasmic inclusions present in recurring infantile digital fibromatosis (Reye tumor). *Pediatr Pathol* 1986;6:311–324.

42. Mukai M, Torikata C, Iri H, et al. Immunohistochemical identification of aggregated actin filaments in formalin-fixed, paraffin-embedded sections, I. A study of infantile digital fibromatosis by a new pretreatment. *Am J Surg Pathol* 1992;16:110–115.

43. Hyashi T, Tsuda N, Chowdhury PR, et al. Infantile digital fibromatosis: a study of the development and regression of cytoplasmic inclusion bodies. *Mod Pathol* 1995;8:548–552.

44. Sarma DP, Hoffmann EO. Infantile digital fibroma-like tumor in an adult. *Arch Dermatol* 1980;116:578–579.

45. Viale G, Doglioni C, Iuzzolino P, et al. Infantile digital fibromatosis-like tumour (inclusion body fibromatosis) of adulthood: report of two cases with ultrastructural and immunocytochemical findings. *Histopathology* 1988;12:415–424.

46. Bittesini L, Dei Tos AP, Doglioni C, et al. Fibroepithelial tumor of the breast with digital fibroma-like inclusions in the stromal component: case report with immunocytochemical and ultrastructural analysis. *Am J Surg Pathol* 1994;18:296–301.

47. Hiraoka N, Mukai M, Hosoda Y, Hata J. Phyllodes tumor of the breast containing the intracytoplasmic inclusion bodies identical with infantile digital fibromatosis. *Am J Surg Pathol* 1994;18:506–511.

48. Navas-Palacios JJ, Conde-Zurita JM. Inclusion body myofibroblasts other than those seen in recurrent digital fibroma of childhood. *Ultrastruct Pathol* 1984;7:109–121.

Juvenile aponeurotic fibroma

49. Keasbey LE. Juvenile aponeurotic fibroma (calcifying fibroma): a distinctive tumor arising in the palms and soles of young children. Calcifying aponeurotic fibroma. *Cancer* 1953;6:338–346.

50. Goldman RL. The cartilage analogy of fibromatosis (aponeurotic fibroma): further observations based on seven new cases. *Cancer* 1970;26:1325–1331.

51. Allen PW, Enzinger FM. Juvenile aponeurotic fibroma. *Cancer* 1970;26:857–867.

52. Fetsch JF, Miettinen M. Calcifying aponeurotic fibroma: a clinicopathologic study of 22 cases arising in uncommon sites. *Hum Pathol* 1998;29:1504–1510.

Lipofibromatosis and diffuse infantile fibromatosis

53. Fetsch JF, Miettinen M, Laskin WB, Michal M, Enzinger FM. A clinicopathologic study of 45 pediatric soft tissue tumors with an admixture of adipose tissue and fibroblastic elements, and a proposal for classification as lipofibromatosis. *Am J Surg Pathol* 2000;24:1491–1500.

54. Herrmann BW, Dehner LP, Forsen JW Jr. Lipofibromatosis presenting as a pediatric neck mass. *Int J Pediatr Otolaryngol* 2004;68:1545–1549.

55. Kenney B, Richkind KE, Friedlaender G, Zambrano E. Chromosomal rearrangements in lipofibromatosis. *Cancer Genet Cytogenet* 2007;179:136–139.

56. Enzinger FM. Fibrous tumors of infancy. In *Tumors of Bone and Soft Tissue*. Chicago: Year Book Medical Publishers; 1965: 231–268.

Gardner fibroma

57. Wehrli BM, Weiss SW, Yandow S. Gardner-associated fibromas (GAF) in young patients: a distinct fibrous lesion that identifies unsuspected Gardner syndrome and risk for fibromatosis. *Am J Surg Pathol* 2001;25:645–651.

58. Coffin CM, Hornick JL, Zhou H, Fletcher CDM. Gardner fibroma: a clinicopathologic and immunohistochemical analysis of 45 patients with 57 fibromas. *Am J Surg Pathol* 2006;31:410–416.

59. Michal M, Boudova L, Mukensnabl P. Gardner's syndrome associated fibromas. *Pathol Int* 2004;54:523–526.
60. Diwan AH, Graves ED, King JA, Horenstein MG. Nuchal-type fibroma in two related patients with Gardner's syndrome. *Am J Surg Pathol* 2000;24:1563–1567.

Juvenile angiofibroma

61. Sternberg SS. Pathology of juvenile nasopharyngeal angiofibroma: a lesion of adolescent males. *Cancer* 1954;7:15–28.
62. Neel HB 3rd, Whicker JH, Devine KD, Weiland LH. Juvenile angiofibroma: review of 120 cases. *Am J Surg* 1973;126:547–556.
63. Economou TS, Abemayor E, Ward PH. Juvenile nasopharyngeal angiofibroma: an update of the UCLA experience, 1960–1985. *Laryngoscope* 1988;98:170–175.
64. Ungkanont K, Byers RM, Weber RS, et al. Juvenile nasopharyngeal angiofibroma: an update of therapeutic management. *Head Neck* 1996;18:60–66.
65. Enepekides DJ. Recent advances in the treatment of juvenile angiofibroma. *Curr Opin Otolaryngol Head Neck Surg* 2004;12:495–499.
66. Onerci M, Ogretmenoglu O, Yucel T. Juvenile nasopharyngeal angiofibroma: a revised staging system. *Rhinology* 2006;44:39–45.
67. Giardiello FM, Hamilton SR, Krush AJ, et al. Nasopharyngeal angiofibroma in patients with familial adenomatous polyposis. *Gastroenterology* 1993;105:1550–1552.
68. Klockars T, Renkonen S, Leivo I, Hagstrom J, Makitie AA. Juvenile nasopharyngeal angiofibroma: no evidence for inheritance or association with familial adenomatous polyposis. *Fam Cancer* 2010;9:401–403.
69. Batsakis JG, Klopp CT, Newman W. Fibrosarcoma arising in a juvenile nasopharyngeal angiofibroma following extensive radiation therapy. *Am Surg* 1955;21:786–793.
70. Chen KT, Bauer FW. Sarcomatous transformation of nasopharyngeal angiofibroma. *Cancer* 1982;49:369 371.
71. Spagnolo DV, Papadimitriou JM, Archer M. Postirradiation malignant fibrous histiocytoma arising in juvenile nasopharyngeal angiofibroma and producing alpha-1-antitrypsin. *Histopathology* 1984;8:339–352.
72. Makek MS, Andrews JC, Fisch U. Malignant transformation of a nasopharyngeal angiofibroma. *Laryngoscope* 1989;99(10 Pt 1):1088–1092.
73. Beham A, Fletcher CD, Kainz J, Schmid C, Humer U. Nasopharyngeal angiofibroma: an immunohistochemical study of 32 cases. *Virchows Arch A Pathol Anat Histopathol* 1993;423:281–285.
74. Beham A, Kainz J, Stammberger H, Aubock L, Beham-Schmid C. Immunohistochemical and electron microscopical characterization of stromal cells in nasopharyngeal angiofibromas. *Eur Arch Otorhinolaryngol* 1997;254:196–199.
75. Abraham SC, Montgomery EA, Giardiello FM, Wu TT. Frequent beta-catenin mutations in juvenile nasopharyngeal angiofibromas. *Am J Pathol* 2001;158:1073–1078.
76. Hwang HC, Mills SE, Patterson K, Gown AM. Expression of androgen receptors in nasopharyngeal angiofibroma: an immunohistochemical study of 24 cases. *Mod Pathol* 1998;11:1122–1126.
77. Zhang PJ, Weber R, Liang HH, Pasha TL, LiVolsi VA. Growth factors and receptors in juvenile nasopharyngeal angiofibroma and nasal polyps: an immunohistochemical study. *Arch Pathol Lab Med* 2003;127:1480–1484.
78. Montag AG, Tretiakova M, Richardson M. Steroid hormone receptor expression in nasopharyngeal angiofibromas. *Am J Clin Pathol* 2006;125:832–837.

Fibrous hamartoma of infancy

79. Enzinger FM. Fibrous hamartoma of infancy. *Cancer* 1965;18:241–248.
80. Efem SE, Ekpo MD. Clinicopathological features of untreated fibrous hamartoma of infancy. *J Clin Pathol* 1993;46:522–524.
81. Popek EJ, Montgomery EA, Fourcroy JL. Fibrous hamartoma of infancy in the genital region: findings in 15 cases. *J Urol* 1994;151:990–993.
82. Sotelo-Avila C, Bale PM. Subdermal fibrous hamartoma of infancy: pathology of 40 cases and differential diagnosis. *Pediatr Pathol* 1994;14:39–52.
83. Dickey GE, Sotelo-Avila C. Fibrous hamartoma of infancy: current review. *Pediatr Dev Pathol* 1999;2:236–243.
84. Michal M, Mukensnabl P, Chlumska A, Kodet R. Fibrous hamartoma of infancy: a study of eight cases with immunohistochemical and electron microscopic findings. *Pathol Res Pract* 1992;188:1049–1053.
85. Carretto E, Dall'Igna P, Alaggio R, et al. Fibrous hamartoma of infancy: an Italian multi-institutional experience. *J Am Acad Dermatol* 2006;54:800–803.
86. Lakshminarayanan R, Konia T, Welborn J. Fibrous hamartoma of infancy: a case report with associated cytogenetic findings. *Arch Pathol Lab Med* 2005;129:520–522.

Myofibroma and myofibromatosis

87. Chung EB, Enzinger FM. Infantile myofibromatosis: a review of 59 cases with localized and generalized involvement. *Cancer* 1981;48:1807–1818.
88. Briselli MF, Soule EH, Gilchrist GS. Congenital fibromatosis: report of 18 cases of solitary and 4 cases of multiple tumors. *Mayo Clin Proc* 1980;55:554–562.
89. Daimaru Y, Hashimoto H, Enjoji M. Myofibromatosis in adults (adult counterpart of infantile myofibromatosis). *Am J Surg Pathol* 1989;13:859–865.
90. Smith KJ, Skelton HG, Barrett TL, Lupton GP, Graham JH. Cutaneous myofibroma. *Mod Pathol* 1989;2:603–609.
91. Foss RD, Ellis GL. Myofibromas and myofibromatosis of the oral region: a clinicopathologic analysis of 79 cases. *Oral Surg Oral Med Oral Pathol Oral Radiol Endod* 2000;89:57–65.
92. Jennings T, Duray PH, Collins FS, et al. Infantile myofibromatosis: evidence for an autosomal-dominant disorder. *Am J Surg Pathol* 1984;8:529–538.
93. Wiswell TE, Davis J, Cunningham BE, Solenberger L, Thomas PJ. Infantile myofibromatosis: the most common fibrous tumor of infancy. *J Pediatr Surg* 1988;23:315–318.
94. Coffin CM, Neilson KA, Ingels S, Frank-Gerszberg R, Dehner LP. Congenital generalized fibromatosis: a disseminated angiocentric myofibromatosis. *Pediatr Pathol Lab Med* 1995;15:571–587.

95. Fukasawa Y, Ishikura H, Takada A, et al. Massive apoptosis in infantile myofibromatosis: a putative mechanism for tumor regression. *Am J Pathol* 1995;144:480–485.

96. Stenman G, Nadal N, Persson S, Gunterberg B, Angervall L. del(6) (q12q15) as the sole cytogenetic abnormality in a case of solitary infantile myofibromatosis. *Oncol Rep* 1999;6:1101–1104.

97. Linos K, Carter JM, Gardner JM, et al. Myofibromas with atypical features: expanding the morphologic spectrum of a benign entity. *Am J Surg Pathol* 2014;38:1649–1654.

98. Mentzel T, Calonje E, Nascimento AG, Fletcher CDM. Infantile hemangiopericytoma versus infantile myofibromatosis: study of a series suggesting a continuous spectrum of infantile myofibroblastic lesions. *Am J Surg Pathol* 1994;18:922–930.

Juvenile hyaline fibromatosis

99. Kitano Y, Horiki M, Aoki T, Sagami S. Two cases of juvenile hyaline fibromatosis: some histological, electron microscopic, and tissue culture observations. *Arch Dermatol* 1972;106:877–883.

100. Ishikawa H, Mori S. Systemic hyalinosis or fibromatosis hyalinica multiplex juvenilis as a congenital syndrome: a new entity based on the inborn error of acid mucopolysaccharide metabolism in connective tissue cells? *Acta Derm Venereol* 1973;53:185–191.

101. Remberger K, Krieg T, Kunze D, Weinmann HM, Hubner G. Fibromatosis hyalinica multiplex (juvenile hyalin fibromatosis): light microscopic, electron microscopic, immunohistochemical, and biochemical findings. *Cancer* 1985;56:614–624.

102. Fayad MN, Yacoub A, Salman S, Khudr A, Der Kaloustian VM. Juvenile hyaline fibromatosis: two new patients and review of the literature. *Am J Med Genet* 1987;26:123–131.

103. Mancini GM, Stojanov L, Willemsen R, et al. Juvenile hyaline fibromatosis: clinical heterogeneity in three patients. *Dermatology* 1999;198:18–25.

104. Senzaki H, Kiyozuka Y, Uemura Y, et al. Juvenile hyaline fibromatosis: a report of two unrelated adult sibling cases and a literature review. *Pathol Int* 1998;48:230–236.

105. Haleem A, Al-Hindi HN, Joboury MA, Husseini HA, Ajlan AA. Juvenile hyaline fibromatosis: morphologic, immunohistochemical, and ultrastructural study of three siblings. *Am J Dermatopathol* 2002;24:218–224.

106. Landing BH, Nadorra R. Infantile systemic hyalinosis: report of four cases of a disease, fatal in infancy, apparently different from juvenile systemic hyalinosis. *Pediatr Pathol* 1986;6:55–79.

107. Dowling O, Difeo A, Ramirez MC, et al. Mutations in capillary morphogenesis gene-2 result in the allelic disorders juvenile hyaline fibromatosis and infantile systemic hyalinosis. *Am J Hum Genet* 2003;73:957–966.

108. Hanks S, Adams S, Douglas J, et al. Mutations in the gene encoding capillary morphogenesis protein 2 cause juvenile hyaline fibromatosis and infantile systemic hyalinosis. *Am J Hum Genet* 2003;73:791–800.

Cerebriform fibrous proliferation in Proteus syndrome

109. Wiedemann HR, Burgio GR, Aldenhof P, et al. The proteus syndrome. Partial gigantism of the hands and/or feet, nevi, hemihypertrophy, subcutaneous tumors, macrocephaly, or other skull anomalies, and possible accelerated growth and visceral affections. *Eur J Pediatr* 1983;140:5–12.

110. Biesecker LG, Happle R, Mulliken JB, et al. Proteus syndrome: diagnostic criteria, differential diagnosis, and patient evaluation. *Am J Med Genet* 1999;84:389–395.

111. Nguyen T, Turner JT, Olsen C, Biesecker LG, Darling TN. Cutaneous manifestations of Proteus syndrome: correlations with general clinical severity. *Arch Dermatol* 2004;140:1001–1002.

112. Twede JV, Turner JT, Biesecker LG, Darling TN. Evolution of skin lesions in Proteus syndrome. *J Am Acad Dermatol* 2005;52:834–838.

113. Gordon PL, Wilroy LS, Lasater OE, Cohen MM Jr. Neoplasms in Proteus syndrome. *Am J Med Genet* 1995;57:74–78.

114. Lindhurst MJ, Sapp JC, Teer JK, et al. A mosaic activating mutation in AKT1 associated with Proteus syndrome. *N Engl J Med* 2011;365:611–619.

115. Lofeld A, McLellan NJ, Cole T, et al. Epidermal naevus in Proteus syndrome showing loss of heterozygosity for an inherited PTEN mutation. *Br J Dermatol* 2006;154:1194–1198.

Chapter 10

Childhood fibroblastic and myofibroblastic proliferations of variable biologic potential

Markku Miettinen
National Cancer Institute, National Institutes of Health

The fibroblastic and myofibroblastic lesions of childhood with variable biologic potential covered in this chapter include neurothekeoma, plexiform fibrohistiocytic tumor (PFHT), angiomatoid fibrous histiocytoma (AFH), inflammatory myofibroblastic tumor (IMT), and infantile fibrosarcoma. Although these tumor types typically occur in children, many of them also occur in younger and older adults at a lower frequency.

Neurothekeoma is a benign myofibroblastic tumor separate from true nerve sheath myxoma. It is included here because of the rare occurrence of atypical variants and its resemblance to PFHT. All other lesions have the potential mainly for local recurrence; however, they also have a variable but usually low risk for metastasis.

Understanding of the molecular genetics of all of these tumors has improved because of the discovery of tumor-specific fusion translocations in AFH, IMT, and infantile fibrosarcoma. These gene rearrangements are diagnostic markers, and the corresponding gene products probably play a pathogenetic role.

Other borderline to malignant fibroblastic lesions that are more typical of adults can also occur in children, for example, low-grade fibromyxoid sarcoma and dermatofibrosarcoma protuberans (DFSP). These tumors, including giant cell fibroblastoma, the juvenile variant of DFSP, are discussed in Chapter 13.

Neurothekeoma

Originally described by Gallager and Helwig in 1980 and then thought to be a nerve sheath tumor,[1] *neurothekeoma* has recently been verified conclusively as a fibroblastic-myofibroblastic neoplasm that is unrelated to nerve sheath myxoma and therefore should be separated from it.[2] The original description of neurothekeoma contained a minor component of nerve sheath myxomas (because these tumors are far less common than neurothekeomas), and similarly, the early reports on nerve sheath myxomas probably contained examples of myxoid neurothekeomas, because at that time immunohistochemical studies were not available for conclusive separation of these entities.

Immunohistochemical studies have indicated that cellular neurothekeomas consistently lack S100 protein positivity, in contrast to most of the myxoid variants.[3,4] Today, these S100-protein-positive myxoid tumors are recognized as nerve sheath myxomas, an unrelated entity (Chapter 24). A small number of true myxoid neurothekeomas, S100-protein-negative neoplasms, still remain. Histologic differential diagnosis between these entities is discussed below in a separate section. Recent studies have rejected the concept that neurothekeoma and nerve sheath myxoma are part of a single entity and also have dispelled the notion that transitional forms between these tumors exist.[2,5] Because neurothekeoma is a myofibroblastic rather than a nerve sheath neoplasm, the designation *multinodular (myxoid) myofibroblastoma* is probably histogenetically more accurate.[5]

Clinical features

Neurothekeoma has a predilection for children and young adults, with a median age from 17 to 21 years and a 1.8:1 female predominance.[5,6] It is less common in older adults, and occurrence after the age of 60 years is rare (Figure 10.1). This tumor is most commonly located in the head (especially the nose and face), shoulder, and arm. It is less common in the lower extremity, trunk, and neck, and distinctly uncommon in the hands and feet (Figure 10.2). There is a predilection for people with fair skin.[5]

Neurothekeoma usually forms a small (often <1 cm) polypoid or dome-shaped skin lesion that rarely exceeds 2 cm in diameter. Occurrence of multiple lesions is rare. The tumor grows slowly, and the mean tumor duration has been 7 to 19 months in the two large series reported, with occasional patients having a history of the tumor for several years.[5,6] There is clinical resemblance to sebaceous cyst, nevus, and skin adnexal tumor, and occasionally to basal cell carcinoma or melanoma.

Neurothekeoma has a relatively low (10%) local recurrence rate after a marginal or incomplete excision, but metastases have not been reported. The recurrence rate seems to be higher in the myxoid variant, but analysis of small numbers of examples with nuclear atypia has not revealed any greater tendency for recurrence.[5,6] In view of the low risk of recurrence, relatively conservative management (local excision) without need to re-excise after a positive margin is appropriate, especially in sensitive sites such as the face, for nonatypical

Modern Soft Tissue Pathology, Second Edition, ed. Markku Miettinen. Published by Cambridge University Press. © Cambridge University Press 2016.

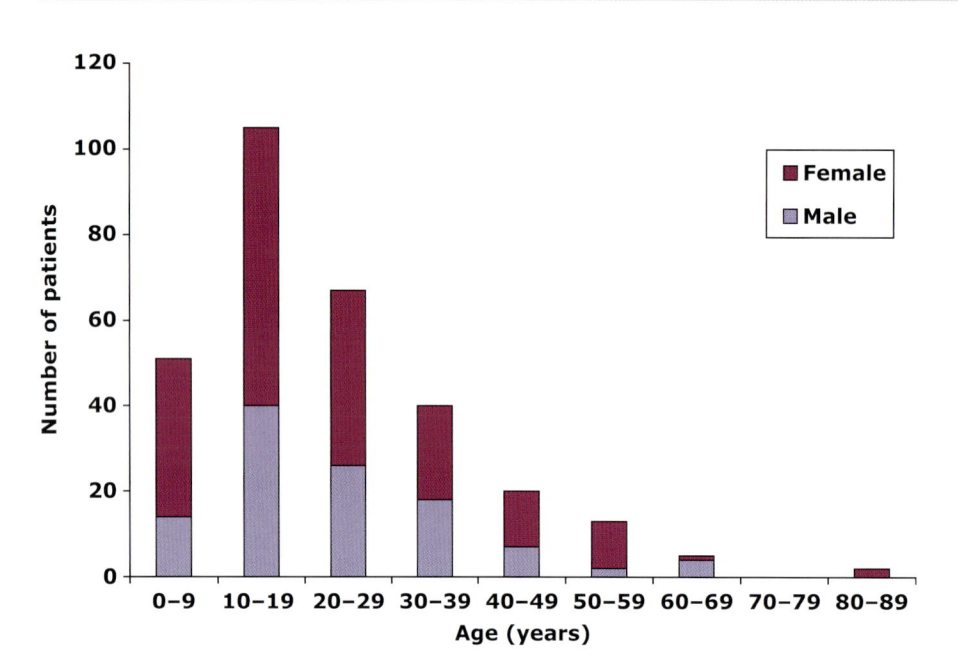

Figure 10.1 Age and sex distribution of 311 patients with neurothekeomas in two series.[5,6]

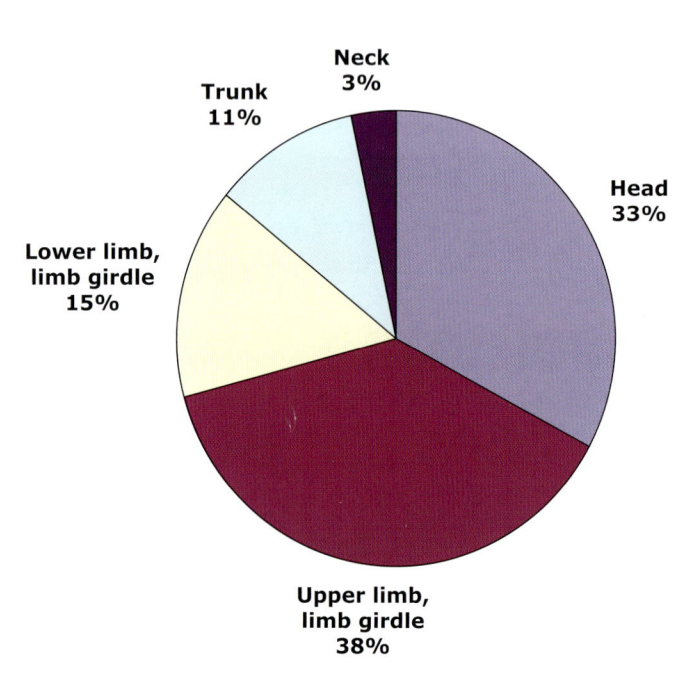

Figure 10.2 Anatomic distribution of 307 cases of neurothekeomas in two series.[5,6]

variants. A complete excision with negative margins should be performed for atypical variants, however.

Pathology

Grossly, cellular neurothekeoma presents as a small cutaneous or superficial subcutaneous nodule, which can be exophytic and is rubbery or hard on sectioning.

Microscopically, the lesion forms an unencapsulated, poorly circumscribed nodule that almost always involves the dermis and can leave an uninvolved zone below the epidermal junction or can extend into it (Figure 10.3). Most of these tumors, at least focally, involve the subcutis, and a few cases, especially those in the head and neck, extend into the superficial skeletal muscle.

Architecturally, the tumor is typically composed of multiple, variably well-defined or partly coalescent, round or ovoid nodules composed of spindled or epithelioid cells separated by variable amounts of myxoid matrix, often in the periphery of the nodules (Figure 10.4). The two ends of the spectrum include tumors with no myxoid matrix (Figure 10.5), and tumors with a prominent myxoid matrix with no solid components, creating a resemblance to nerve sheath myxoma. Between the nodules, there are thick collagenous bands or septa. Some tumors also contain areas of sheet-like growth. A plexiform pattern with ample amount of collagenous matrix is present in some cases. Additionally, there is often a fascicular cellular component composed of spindle cells (Figure 10.6).

The tumor nuclei are ovoid, vesicular, with open chromatin and a small nucleolus, and the cytoplasm is often granular and varies from pale to eosinophilic, often with poorly emarginated cell borders, probably depending on fixation (Figure 10.5). Mitotic activity is usually low, rarely exceeding 5 mitoses per 50 HPFs (with a conventional, non-wide-field microscope). Occasional osteoclasts are present within the cellular nodules in one-third of cases, and scattered lymphocytes and plasma cells are sometimes evident. Variants having nuclear atypia with focal pleomorphism occur, and occasional atypical mitoses can be present (Figure 10.6).[5,6]

Immunohistochemistry

The tumor cells are consistently positive for vimentin and CD10, and most cases also show microphthalmia transcription factor (MITF)-positive nuclei. Immunoreactivity for CD63

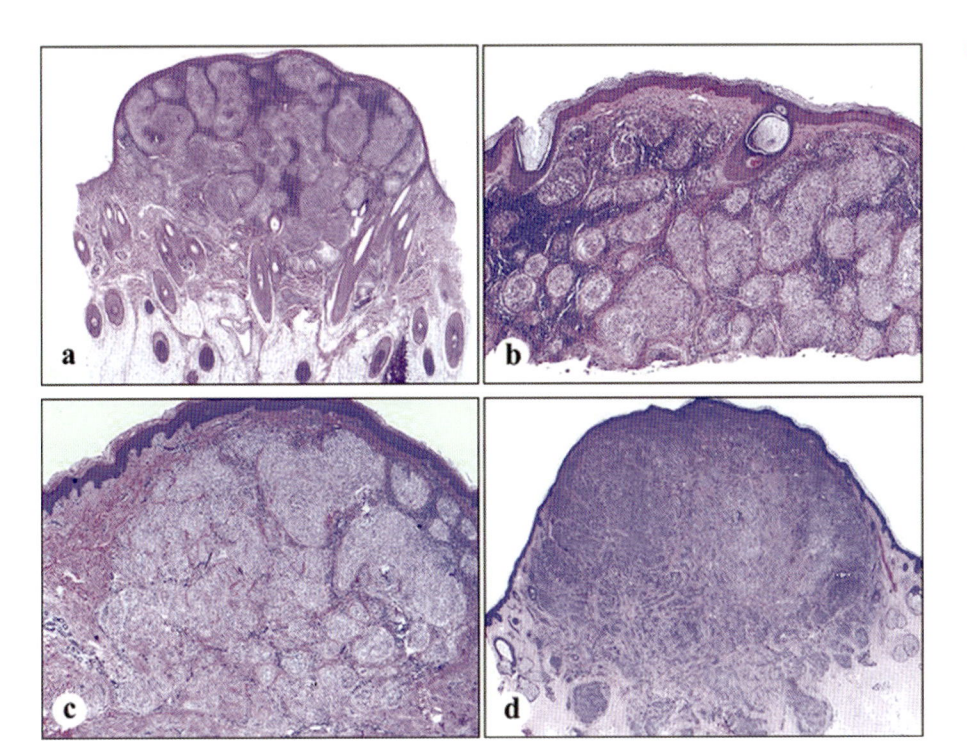

Figure 10.3 Low-magnification view of four different neurothekeomas shows dome-shaped or sessile polypoid cutaneous nodules. Essentially dermal distribution of tumors and their multinodular character is clearly apparent. By their composition, all tumors in this picture can be characterized as "cellular."

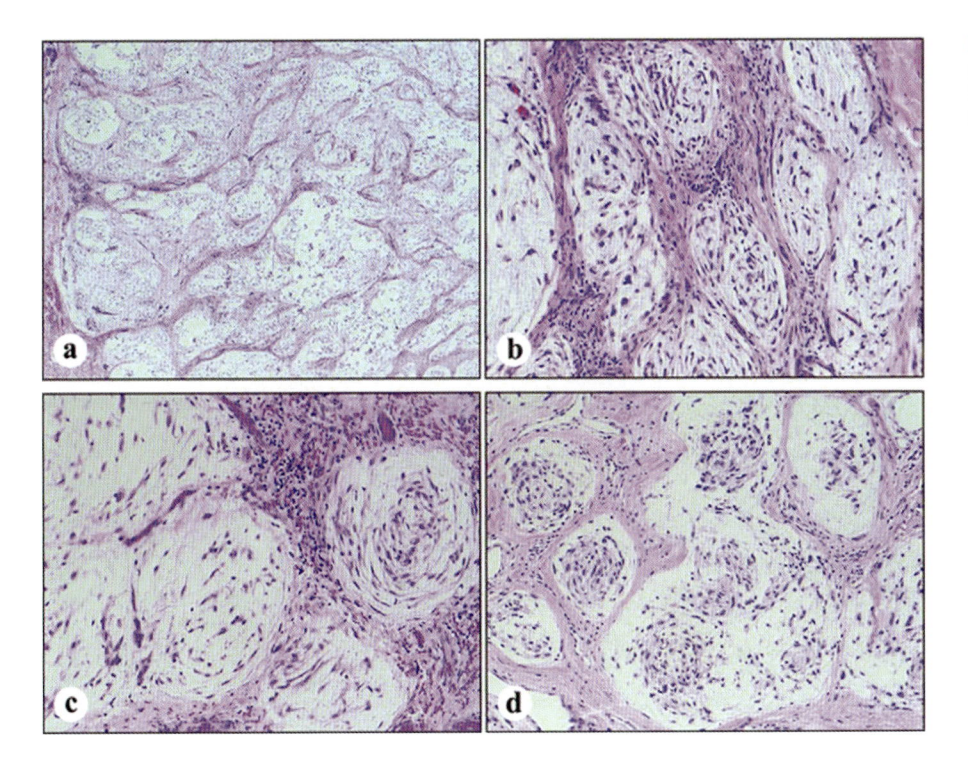

Figure 10.4 Examples of myxoid or mixed neurothekeomas also show multinodularity. The nodules often contain a more solid central component and a peripheral myxoid component.

(NK1C3), CD68, CD99, NSE, and PGP9.5 (in 60–100% of cases) is characteristic, but nonspecific.[3–7] There is focal to extensive positivity for SMA (40–50%) and calponin, indicating myofibroblastic differentiation. The tumor cells are consistently negative for S100 protein, although varying numbers of positive antigen-presenting dendritic cells are present (Figure 10.7). Common positivity for KBA62, an antigen also present in melanomas, should not lead to confusion with melanocytic lesions.[8] This tumor is also negative for GFAP, NGFR (p75), CD34, CD117 (KIT), CD163, and desmin. Rare

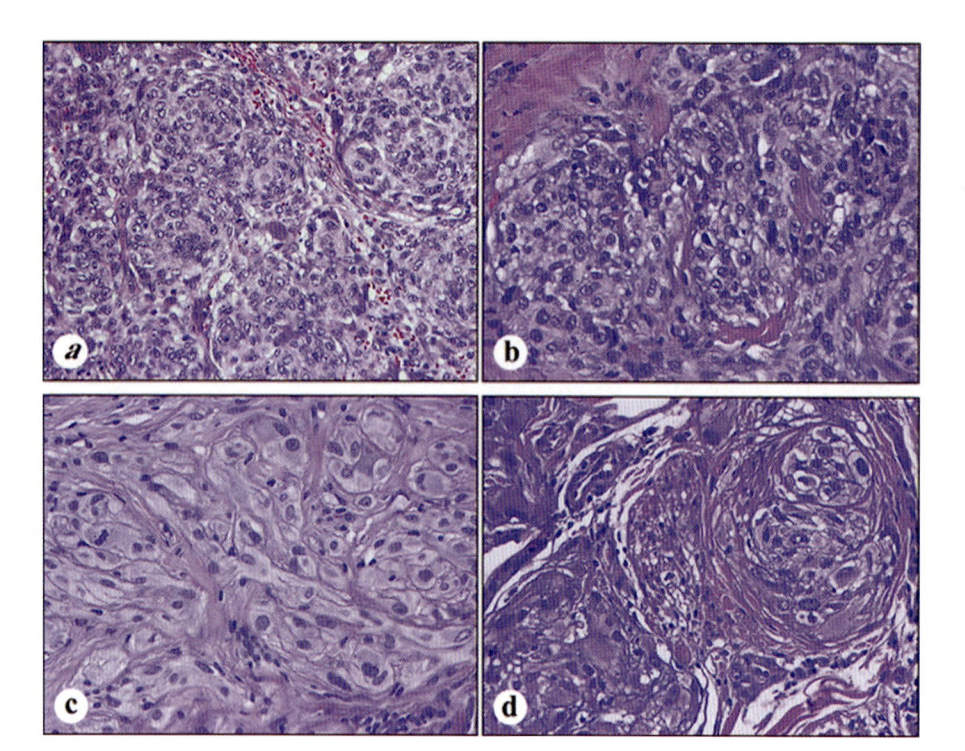

Figure 10.5 View of a single nodule in cellular neurothekeoma reveals cells that are spindled to epithelioid, with a moderate amount of cytoplasm. (**a**) Osteoclastic giant cells are present. (**c**) Well-delineated cytoplasm and crisp nuclear morphology in this case are probably a result of B5 fixation.

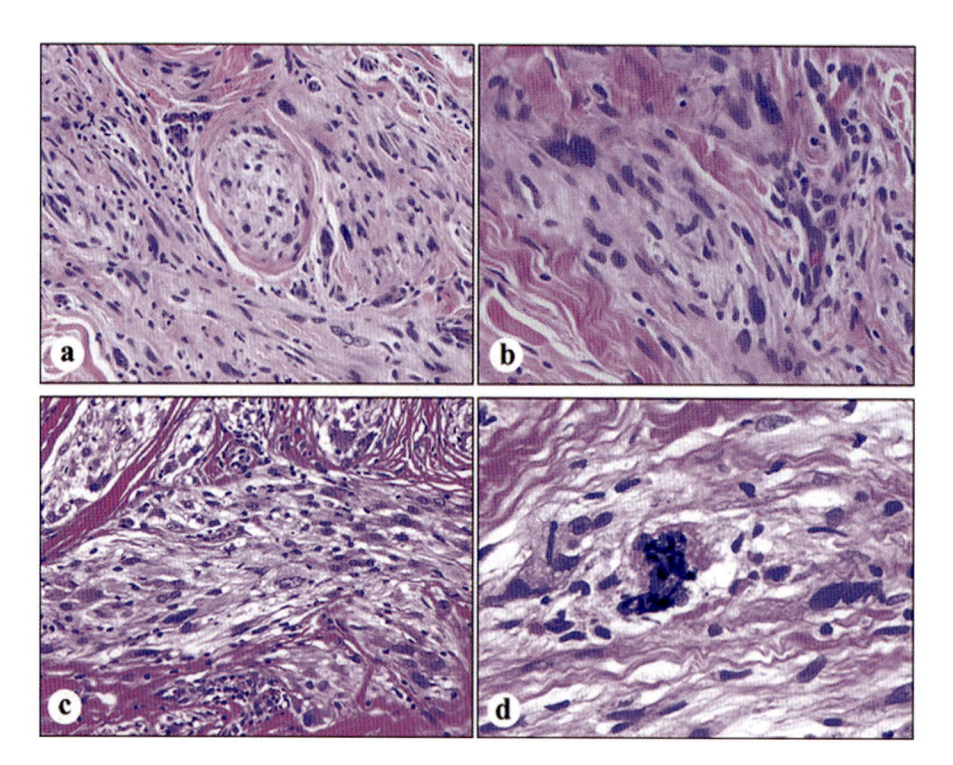

Figure 10.6 Examples of cellular atypia and focal fascicular growth in neurothekeoma. (**a–c**) Focal atypia and fascicular growth pattern. (**d**) Atypical mitoses are an exceptional finding.

HMB45 and melan A positivity has been detected. Based on this antigenic profile, neurothekeoma is unrelated to true nerve sheath myxoma, and there is no evidence for schwannian or melanocytic differentiation. Gene expression array studies also support distinction of neurothekeoma from nerve sheath tumors.[9]

Differential diagnosis

Nerve sheath myxoma differs from myxoid neurothekeoma by a higher contrast between the myxoid nodules and thick, acellular septa, and trabecular and syncytial cellular arrangements within the cellular nodules, as seen in juxtaposed

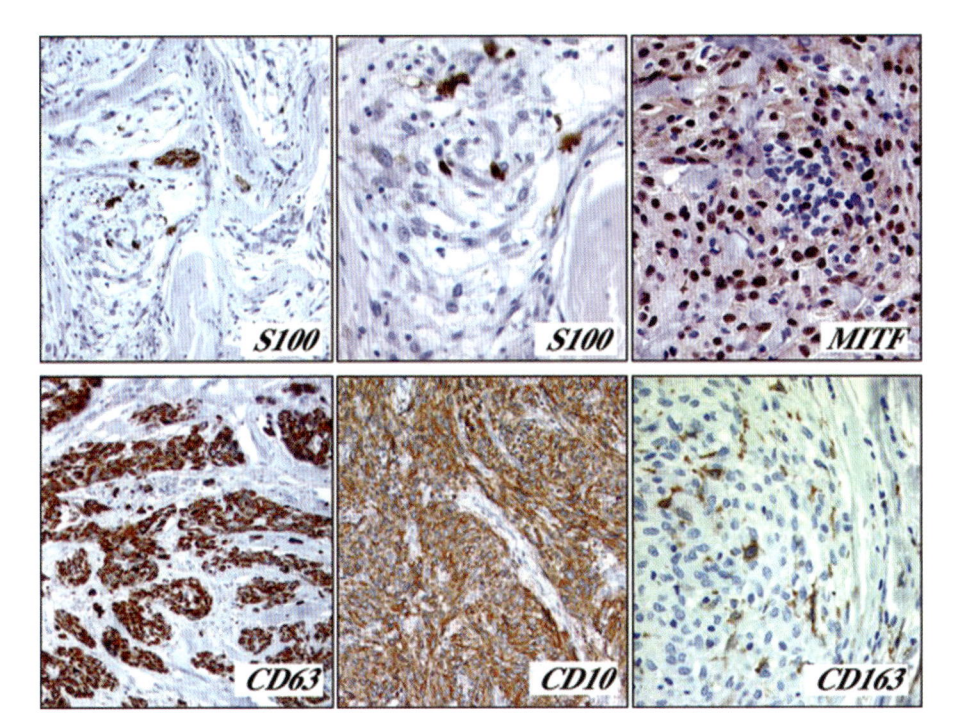

Figure 10.7 Immunohistochemical features of cellular neurothekeoma. The tumors are uniformly negative for S100 protein, in contrast to nerve sheath myxomas. However, scattered S100-protein-positive dendritic cells are often present. Typical positive immunohistochemical features include nuclear reactivity for microphthalmia transcription factor, and cytoplasmic positivity for CD63 (NKIC3), and CD10. Histiocyte-specific marker CD163 highlights only infiltrating histiocytes, whereas the tumor cells are negative.

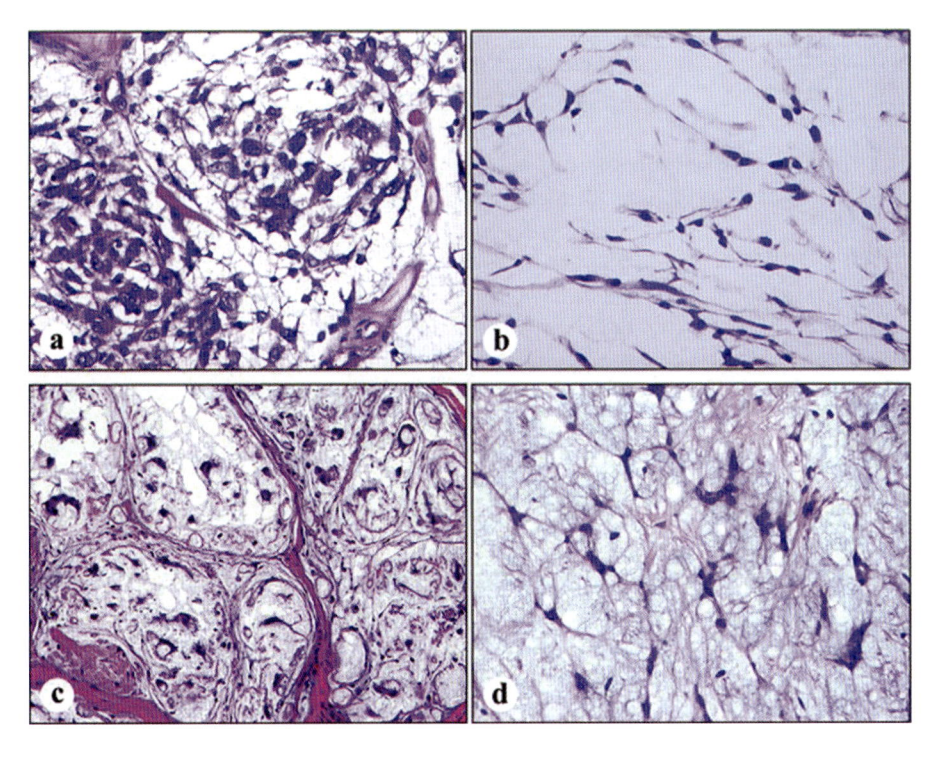

Figure 10.8 Comparison of neurothekeoma and nerve sheath myxoma. (**a**) Neurothekeoma shows confluent nodules. (**b**) Highly myxoid examples typically contain single cell cords. (**c,d**) Nerve sheath myxoma contains well-defined nodules with the tumor cells often forming multinuclear syncytia, and intracellular vacuoles are often present.

pictures of neurothekeoma (Figure 10.8a,b) and nerve sheath myxoma (Figure 10.8c,d). Neurothekeoma is negative for GFAP and S100 protein, in contrast to nerve sheath myxoma. This finding also helps to differentiate neurothekeoma from melanocytic lesions.

The multinodularity and the cytologic features resemble those of PFHT, but the latter lacks the myxoid component. The latter is also based in the subcutis, typically forming a poorly delineated mass with multiple "tentacles" in the fat and containing a more prominent fascicular fibroblastic component. Nevertheless, there may be some overlap between neurothekeoma and PFHT.[10] The epithelioid variant of fibrous histiocytoma is typically a polypoid lesion with epithelioid fibroblasts set in a diffuse, not micronodular, manner.

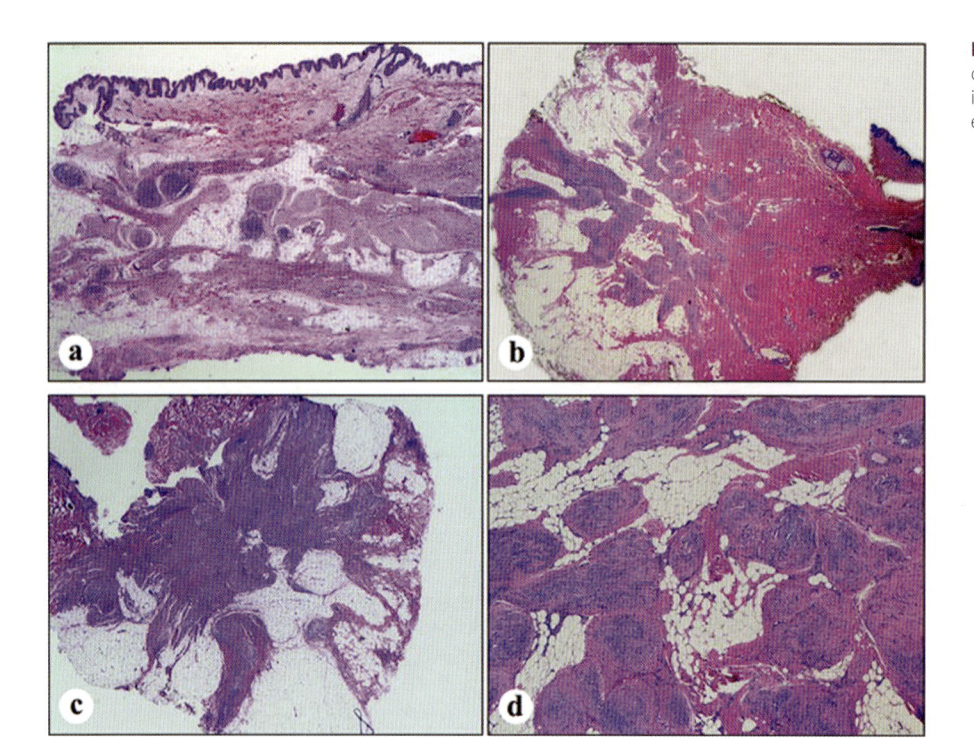

Figure 10.9 Plexiform fibrohistiocytic tumor is composed of multiple tumor nodules that center in the subcutis, often spreading as tentacle-like extensions.

Reticulohistiocytoma (solitary epithelioid histiocytoma) is a true histiocytic proliferation composed of CD163-positive epithelioid histiocytes.

Plexiform fibrohistiocytic tumor

This histologically distinctive multinodular subcutaneous tumor, typical in children and young adults, combines histiocyte-rich nodules and fibromatosis-like streaks in a plexiform pattern. This tumor was first described by Enzinger and Zhang.[11]

Clinical features

Over one half of all cases have been reported in children from infancy onward, and two-thirds occur in patients younger than the age of 20 years, with a mild female predominance. The tumor most commonly presents in the subcutis of the upper extremities (i.e., shoulder, forearm) and less often in the lower extremities, trunk, and head and neck. The tumor size varies from a small nodule of <1 cm to rare examples of >5 cm.[11–16]

The original series reported local recurrences in 37.5% of patients and lymph node metastases in 6% of cases, but there were no distant metastases. One subsequent series showed a lower rate of local recurrence (12.5%), lymph node metastases in one patient, and pulmonary metastases in another patient who was alive with tumors.[8]

Pathology

The tumor is typically centered in the subcutis but often extends into the lower dermis. It is poorly circumscribed and usually measures 1 cm to 3 cm in diameter; tumors are rarely larger than 5 cm.

Histologically, a typical tumor has a multinodular "plexiform" growth pattern and tentacle-like extensions radiating into the subcutaneous fat (Figure 10.9). The nodules and streaks of tumor are confluent or separated by dermal collagen and/or subcutaneous fat. Some nodules are surrounded by a dense band of fibrosis.

The tumor foci are composed of spindle cells with varying numbers of osteoclastic giant cells, sometimes seen in the center of the nodules, sometimes evenly dispersed. Some elements of the tumor are composed of spindle cells only, whereas others are rich in osteoclastic giant cells (Figure 10.10). The nodules often contain foci of hemosiderin and focal lymphocyte infiltration. Many examples have streaks of spindle cells in a collagenous background resembling fibromatosis. Focal cytologic atypia can be present, but mitotic activity usually does not exceed 3 mitoses per 10 HPFs, with atypical mitoses sometimes present.

Immunohistochemically the tumor cells are positive for SMA and include CD68-positive elements, whereas this tumor is negative for keratin cocktail AE1/AE3, CD34, desmin, HMB45, and S100 protein.[16]

Differential diagnosis

Tumors that can be confused with PFHT include giant cell tumor of soft parts, cellular neurothekeoma, and lipofibromatosis. Giant cell tumor of soft parts often has osteoclastic giant cells sprinkled throughout the tumor. It forms one nodule or multiple closely apposed nodules without complex tentacle-like extensions, sometimes with focal

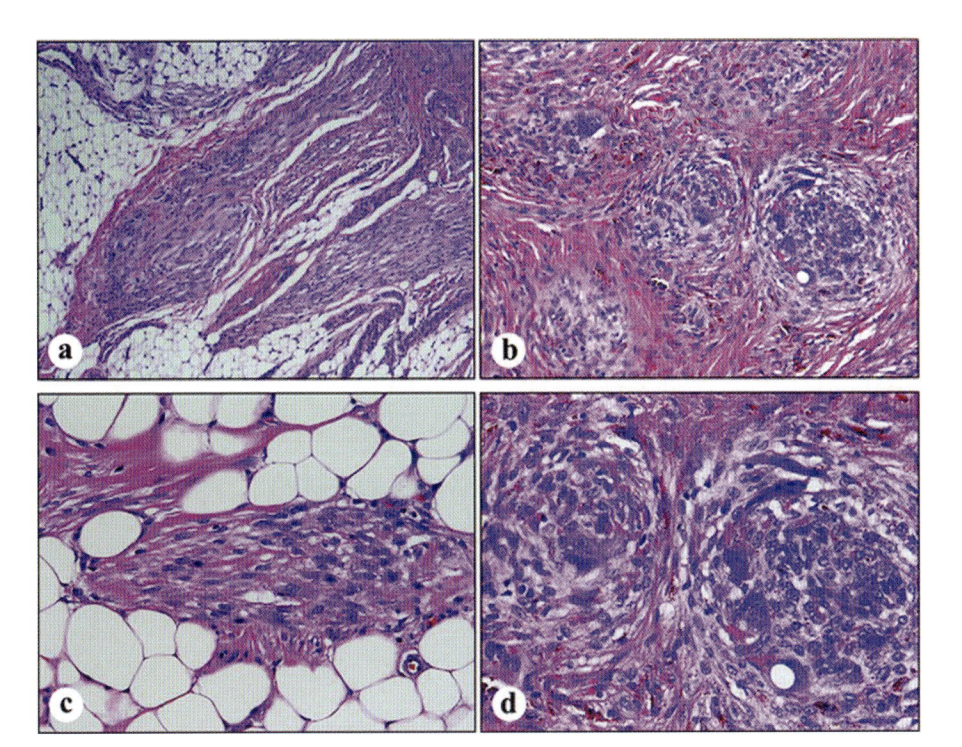

Figure 10.10 The individual tumor elements in plexiform fibrohistiocytic tumor are seen as fascicular formations and nodules. The cellular component includes myofibroblasts, osteoclastic giant cells are often present in the nodules.

cartilaginous or metaplastic osseous component. Cellular neurothekeoma has a micronodular pattern, with nodules composed of uniform ovoid cells arranged in concentric whorls around small vessels, in a predominantly dermal location. Lipofibromatosis can simulate PFHT by its tentacle-like units of fibrous proliferation among subcutaneous fatty elements, but this condition does not show the oval cell component and lacks osteoclasts.

Genetics

Two cytogenetic studies show different changes. One study showed a simple karyotype with a translocation t(4;15)(q21;q15).[17] Another study reported complex changes, including deletion of 4q25q31,[18] suggesting that involvement of a gene in 4q21–q25 is a common denominator.

Angiomatoid fibrous histiocytoma

AFH is a histologically distinctive childhood neoplasm composed of small round or spindled cells with common desmin positivity and relatively indolent behavior, rarely with lymph node metastasis. Histogenesis is unknown.

Clinical features

This tumor, originally described by Enzinger in 1979 as an angiomatoid malignant fibrous histiocytoma (MFH),[19] has more recently been renamed AFH because of its very uncommon malignant behavior.[20]

The tumor typically presents in superficial soft tissues of children and young adults. The median age in the three largest recent series has varied from 12 to 18 years; 80% of patients are under 20 years of age, and only 6% are over 40 years of age. Some series have shown a mild female predominance of 55%.[19–23] Age, sex, and anatomic distribution in the AFIP experience are shown in Figure 10.11.

The tumor occurs in the subcutis of the extremities, trunk, or less commonly the head and neck, often as a painless nodule clinically thought to be a cyst or a benign tumor. As many as one half of the tumors occur in major lymph node regions, such as the axilla, inguinal region, antecubital fossa, and head and neck. More recently examples have been reported in the lung, mediastinum, retroperitoneum, ovary, and vulva.[24,25] Some patients have larger tumors, and a few have constitutional symptoms, such as anemia, weight loss, and polyclonal gammopathy.[19]

Recurrence is rare following a local excision (<10%), and factors associated with increased risk of recurrence are infiltrative border and deep fascial, periosteal, or head and neck location.[20] Atypia and mitotic activity do not indicate an increased risk of recurrence, however. Lymph node metastases have been reported in 1% of cases; in more recent series, distant metastases have occurred, but are exceedingly rare.[20,21] Rare fatal cases continue to be reported, however, including one of a 9-year-old girl whose foot tumor caused massive metastases to liver and lung and killed the patient in 14 months.[26] The reported metastatic tumors have had some unusual features, such as deep intramuscular location or large tumor size (>10 cm).[19,23] Some of these cases might now be reclassified as other sarcomas. Local excision with negative margins is considered an optimal treatment, and adjuvant treatment is usually not advocated.

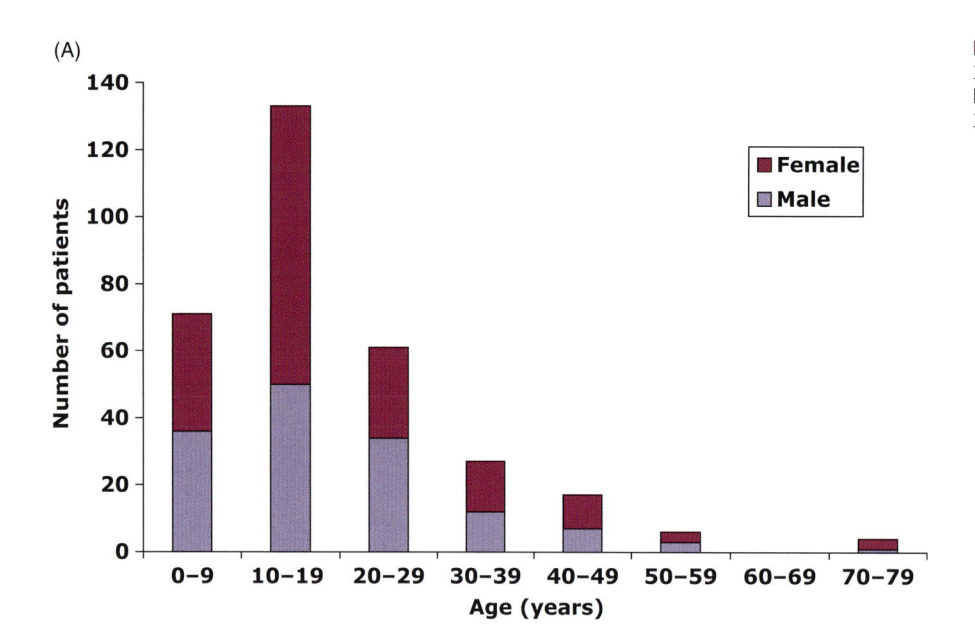

Figure 10.11 (A) Age and sex distribution of 320 patients with angiomatoid fibrous histiocytomas. (B) Anatomic distribution of 320 angiomatoid fibrous histiocytomas.

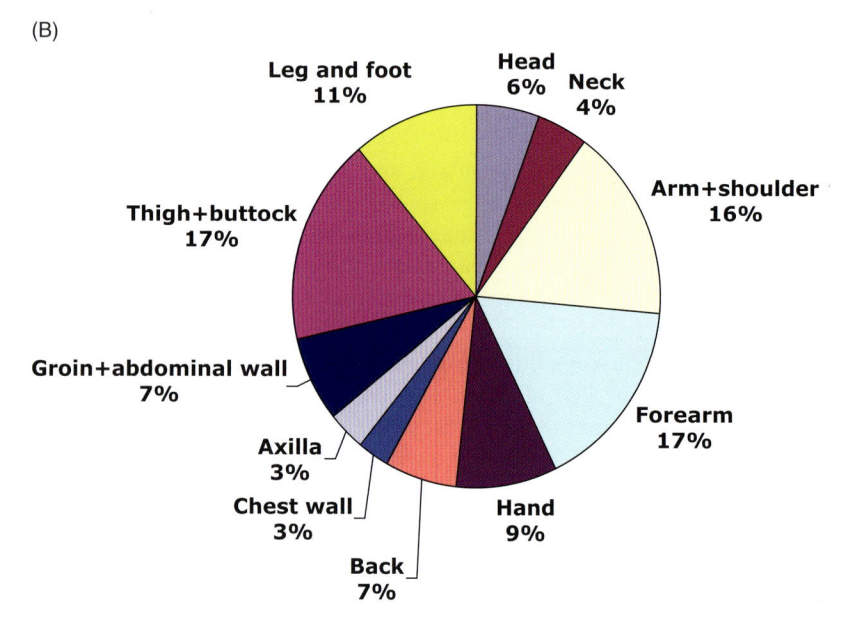

Pathology

Tumors usually measure 1 cm to 3 cm in diameter and rarely exceed 5 cm. Grossly, tumors vary from a tan homogeneous nodule to a hemorrhagic, cystic mass (Figure 10.12).

At low magnification, the tumor is typically surrounded by a peripheral band of dense fibrosis further encircled by a variably thick cuff of lymphoplasmacytic infiltration. The latter often contains germinal centers, which can lead to the incorrect assumption of a lymph node metastasis (Figure 10.13). No definite nodal elements such as sinuses occur in primary AFH, however. One exceptional case, accompanied by extensive lymphadenopathy simulating Castleman's disease, has been reported.[27]

Some examples contain multiple hemorrhagic cysts directly lined by tumor cells (Figure 10.14). Extensive cystic change in a thick-walled collagenous cyst with or without hemorrhage is seen in some cases, and in such examples the tumor cells are seen as small foci or narrow patches in the cyst wall (Figure 10.14b). Many examples lack hemorrhagic cysts, having the cells arranged in diffuse sheets, in a micronodular "cannon-ball" pattern or a vague storiform arrangement. The individual tumor nodules are often surrounded by lymphoplasmacytic infiltration (Figure 10.15). Rare variants have reticular or myxoid patterns (usually focally), and these can simulate extraskeletal myxoid chondrosarcoma.[28,29]

Cytologically, the tumors are equally often composed of spindle, oval to round cells, or an admixture thereof, and the uniform oval cells can resemble the appearance of Ewing sarcoma (Figure 10.16a). The tumor cells have poorly delineated borders with a moderate amount of pale eosinophilic cytoplasm. The nuclei are usually uniform with a pale or vesicular chromatin and a small nucleolus, but some tumors contain moderate atypia (Figure 10.16b–d), sometimes with nuclear pseudoinclusions. Mitotic rate is usually low, but some cases have 5 or more mitoses per 10 HPFs; this is not an alarming feature and has no adverse significance.

Degenerative changes, such as intratumoral hemorrhage and hemosiderin or hematin pigment deposition, are common in cystic variants (Figure 10.17). Rare findings include fascicular growth, xanthoma cells, hyaline eosinophilic globules (resembling those in Kaposi's sarcoma), and osteoclastic giant cells lining hemorrhagic cysts (Figure 10.18).

Immunohistochemistry

The tumor cells are usually positive for vimentin and calponin and variably for CD68, CD99, and desmin in 50% of cases.[22,30,31] They are occasionally positive for true smooth muscle markers, such as h-caldesmon and SMAs, but are negative for the skeletal muscle-specific transcriptional regulators MyoD1 and myogenin. The desmin-positive round tumor cells associated with lymphoid tissue have some analogy with the mold cells of lymphoid tissue, but a histogenetic link has not been proved.[22]

The tumor cells are negative for endothelial cell markers and CD34, keratins, S100 protein, and CD21 and CD35, markers for dendritic reticulum cells.[22] Some examples are positive for epithelial membrane antigen (EMA).[32]

Differential diagnosis

Lymph node metastases of melanoma, primary tumors of lymphoid tissue, such as the rare dendritic reticulum cell sarcoma, should be considered. AFH does not have a sinus system, however, thus differing from a lymph node; usually metastatic melanoma has greater cytologic atypia and is recognizable by immunohistochemical markers. The hemorrhagic examples of AFH differ from true vascular tumors by the lack of vasoformation by tumor cells, which is practically always evident in true endothelial neoplasms. The hemorrhagic, pleomorphic, and highly malignant giant cell MFHs (Chapter 13)

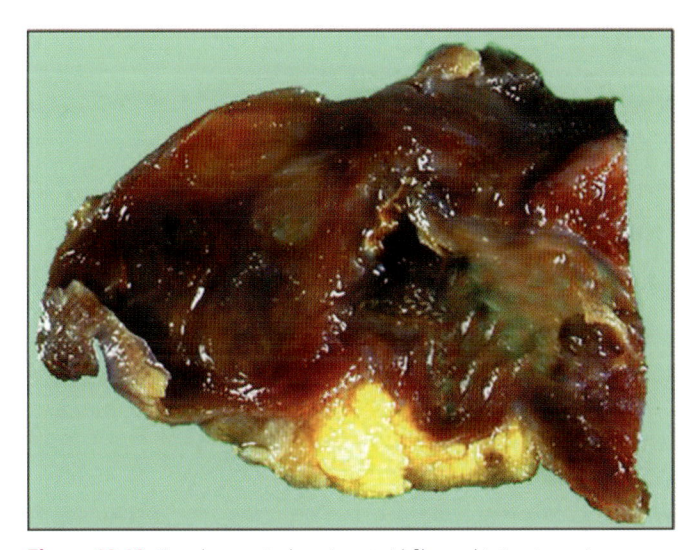

Figure 10.12 Grossly, a typical angiomatoid fibrous histiocytoma is multinodular and shows brown discoloration from hemosiderin content and focal cystic spaces.

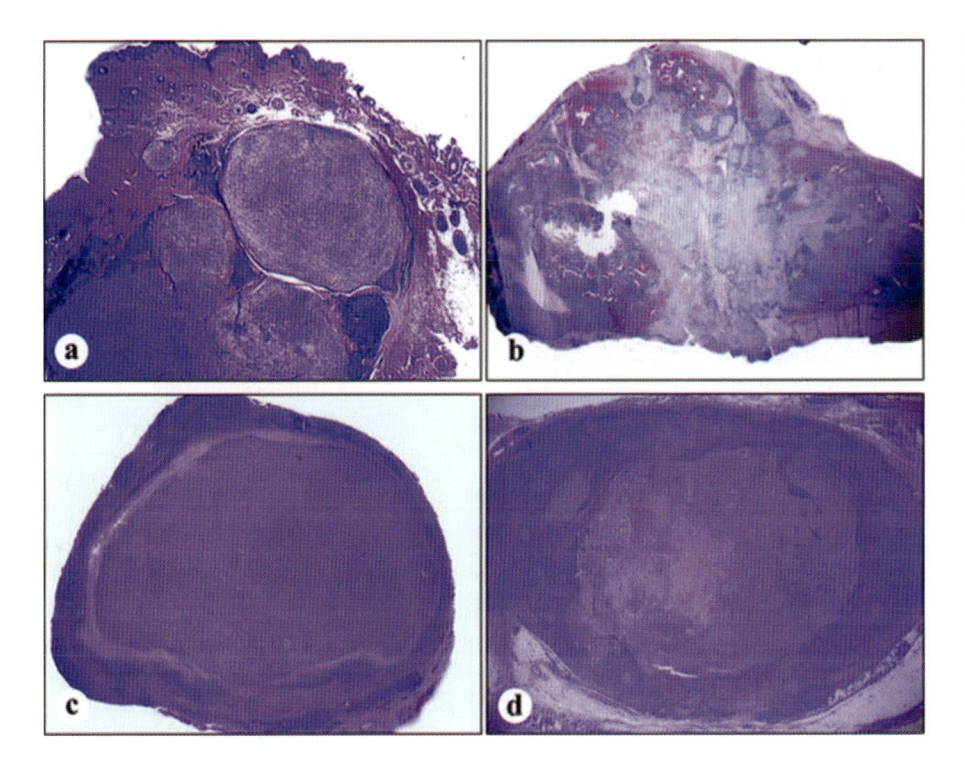

Figure 10.13 Low-magnification images of solid examples of angiomatoid fibrous histiocytomas. (**a**) The tumor can be multinodular or (**b**) composed of a single nodule. (**c**) Peripheral lymphoid tissue and a collagenous rim often surround the cellular component. (**d**) Cases with a prominent lymphoid element can resemble lymph node metastases.

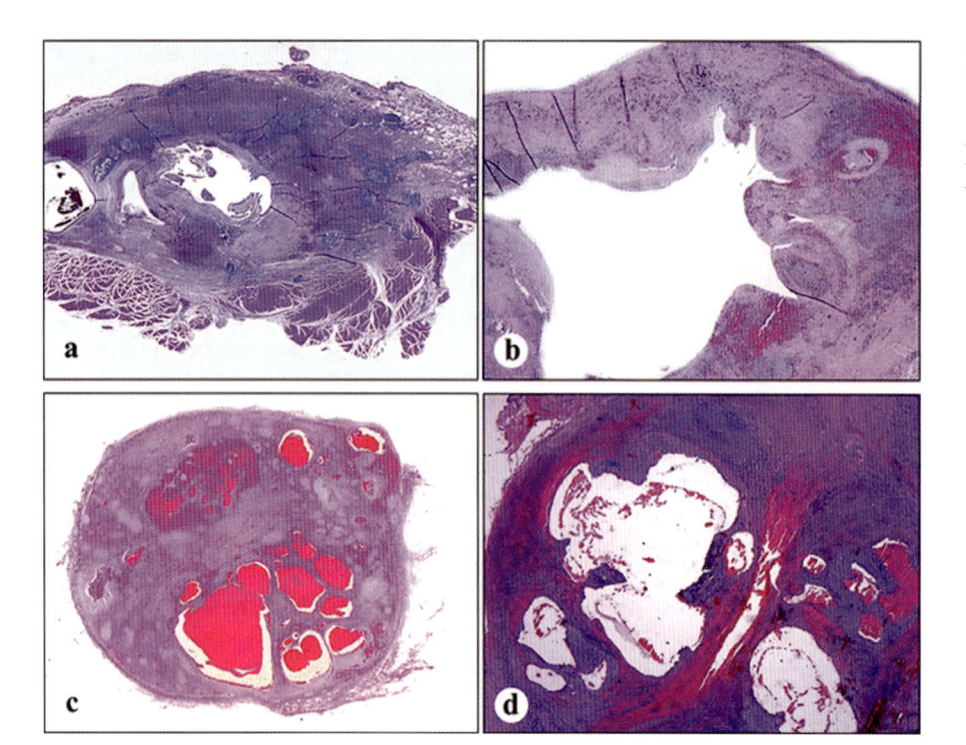

Chapter 10: Childhood fibroblastic and myofibroblastic proliferations of variable biologic potential

Figure 10.14 Low-magnification images of cystic examples of angiomatoid fibrous histiocytoma. (**a**) Peripheral lymphoid elements and fibrosis are prominent in some cases. (**b**) Some examples are dominated by a cystic space lined only focally by tumor elements. (**c,d**) The cystic spaces are variably filled by blood.

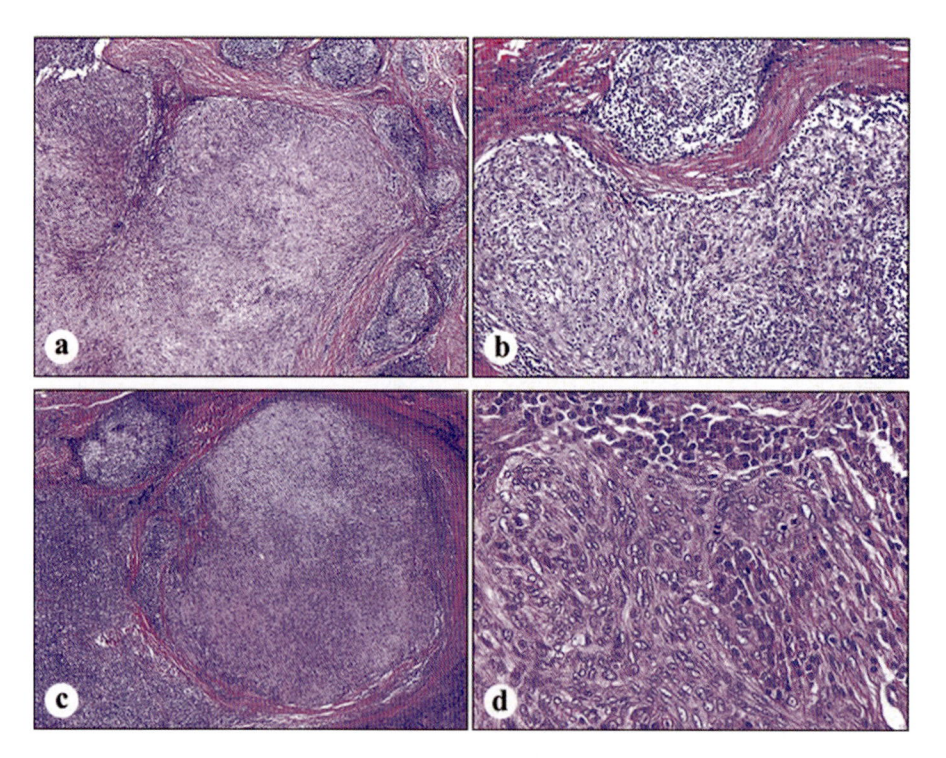

Figure 10.15 The tumor nodules are typically sharply demarcated from the surrounding lymphoid tissue and fibrous rim. (**d**) Some nodules are surrounded by large numbers of plasma cells.

in older patients should not be confused with AFH, which has a more uniform cellular composition with less atypia.[33]

The desmin-positive examples of AFH should not be confused with childhood rhabdomyosarcoma (RMS). The clinical presentation in superficial soft tissues and lack of histologic and immunohistochemical rhabdomyoblastic differentiation are the most significant differences. Similarly, the CD99 positivity should not lead to confusion with Ewing sarcoma; the latter is typically composed of densely packed small round cells with a smaller amount of cytoplasm.

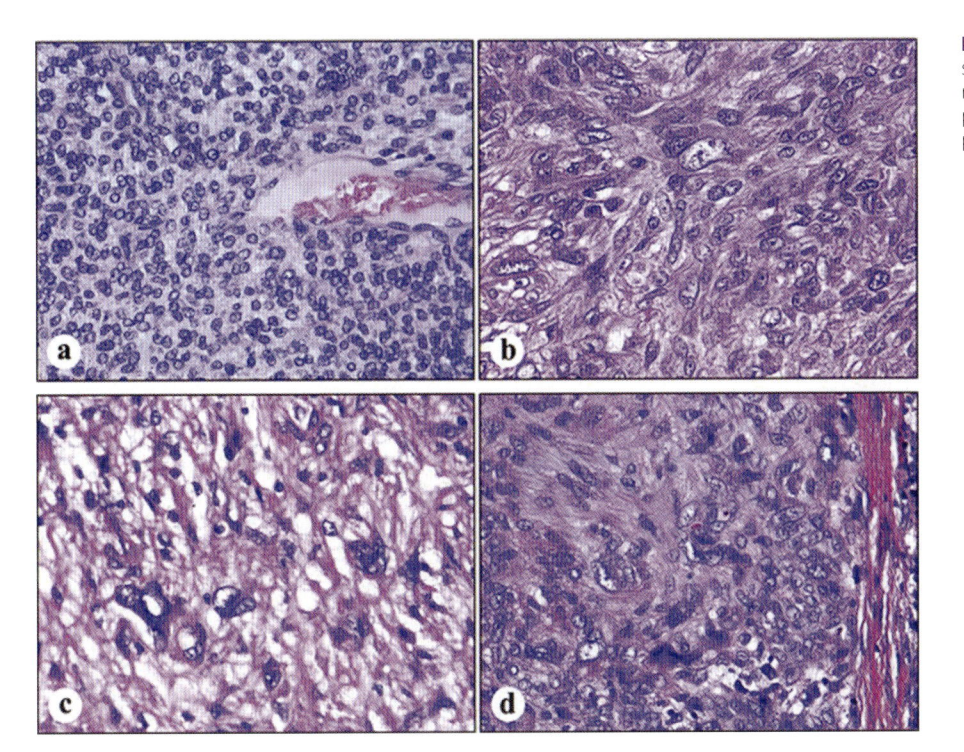

Figure 10.16 The tumor cells are seen as diffuse sheets and vary in appearance. (**a**) Example with round cell, Ewing-like morphology. (**b**) Spindled pattern with focal atypia. (**c,d**) Variable nuclear pleomorphism can be present.

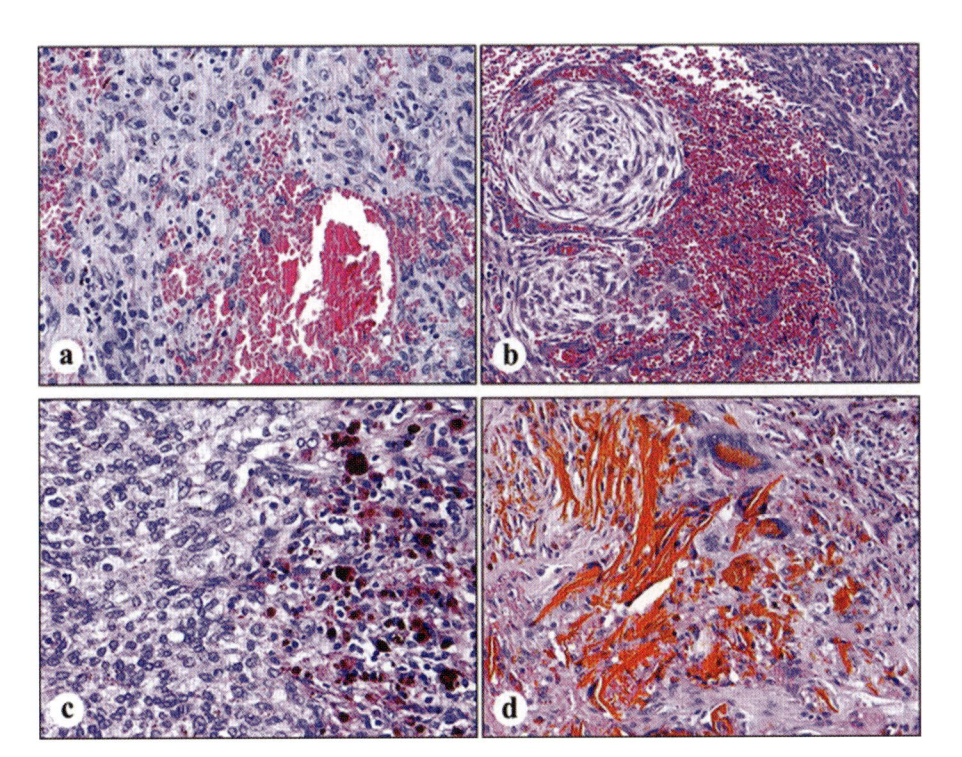

Figure 10.17 Hemorrhagic features in angiomatoid fibrous histiocytoma. (**a,b**) Blood-filled spaces between tumor cells. (**c,d**) Brown hemosiderin and yellow hematin pigment with multinucleated giant cells. The latter are non-neoplastic components.

Genetics

The most common recurrent gene fusion is EWSR1-CREB1. The corresponding translocation is t(2;22)(q34;q12), which also occurs in clear cell sarcoma of the gastrointestinal tract.[34] A rare fusion in AFH is EWSR1-ATF1, corresponding to t(12;22) – this fusion translocation also occurs in clear cell sarcoma of soft parts.[35] The FUS-ATF1 fusion corresponding to t(12;16)(q13;p11) has been reported in a few cases.[36,37] A FISH-based study of 14 cases showed that all cases had EWSR1 rearrangement, 13 of them with CREB1, and the remaining one with ATF1; in that study, no FUS rearrangements were detected. The common presence of EWSR1

287

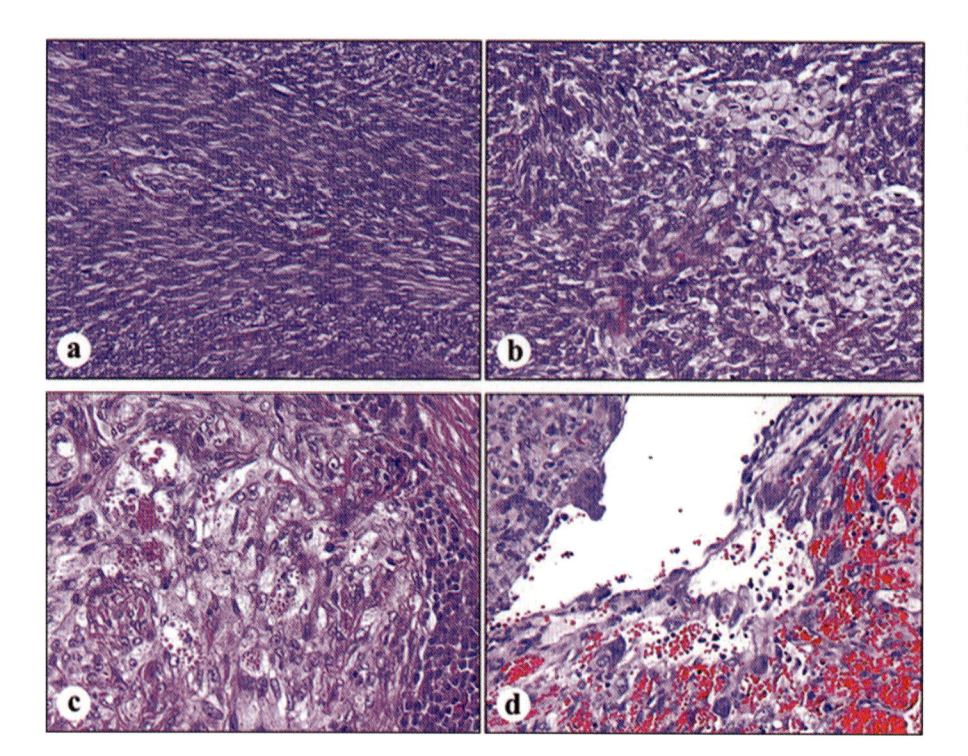

Figure 10.18 Histologic features in angiomatoid fibrous histiocytoma. (**a**) Fascicular pattern with spindle cells. (**b**) Collection of xanthoma cells. (**c**) Intracytoplasmic hyaline globules. (**d**) Osteoclastic giant cells lining a blood-containing space.

rearrangement in AFH should not lead to confusion with Ewing sarcoma.

Inflammatory myofibroblastic tumor

IMT is a distinctive childhood spindle cell tumor featuring spindle cells with abundant cytoplasm in an inflammatory lymphoplasmacytic background, often with immunohistochemical anaplastic lymphoma kinase (ALK) positivity. The presence of clonal chromosomal changes, especially ALK-activating translocations, indicates a neoplastic nature. Neither immunohistochemical ALK positivity nor ALK gene rearrangements have been required for definition in most studies, which has introduced a significant heterogeneity among the tumors reported as IMTs.

IMTs usually occur in children and young adults, but some recently reported variants often occur in older adults. The most common sites are abdominal cavity, lung, and urinary bladder, whereas location in peripheral soft tissues is rare.[38–40]

Terminology

The designation *inflammatory pseudotumor* was previously widely used for IMTs at different sites, but it has also been applied to unrelated inflammatory (infectious and noninfectious) tumefactions, such as mycobacterial pseudotumor, and therefore cannot be considered synonymous with IMT. Inflammatory pseudotumors of the lymph nodes, spleen, and liver are ALK negative and represent entities other than IMT, including inflammatory reactions, Epstein–Barr virus (EBV)-associated myofibroblastic proliferations, and dendritic reticulum cell sarcoma.[41–43]

In the urinary bladder, previous designations for IMT include (inflammatory) pseudosarcomatous fibromyxoid tumor[44] and pseudosarcomatous myofibroblastic tumor/proliferation.[45–47] Some lesions considered postoperative spindle cell nodules may also belong to this entity.[48] Some (but not all) pulmonary tumors previously called plasma cell granuloma, inflammatory pseudotumor, postinflammatory tumor, xanthomatous pseudotumor, and inflammatory myofibrohistiocytic tumor are now classified as IMTs.[49,50]

Inflammatory fibrosarcoma, a highly cellular abdominal myofibroblastic lesion found in children, is now considered as part of the spectrum of IMT.[51,52] In the author's experience, this group represents the most prototypic group of IMTs, with a high frequency of ALK immunoreactivity and ALK gene rearrangements. Many visceral IMTs have been designated as leiomyosarcomas or stromal tumors in the past, especially when located in the gastrointestinal tract and urinary bladder.

Clinical features

The characteristics of IMTs vary by site and are therefore summarized here separately for abdominal, urinary bladder, pulmonary, and other selected sites. Although the initial series emphasized their occurrence in children, many recent series have included substantial portions of adult or even elderly patients, with no clear gender predilection. Many series, particularly early studies, used designations other than IMT, so that their comparability to present-day IMTs is somewhat uncertain.

Abdominal IMTs occur mainly in children and young adults. They can cause intestinal obstruction and are often multiple and commonly large, >10 cm. They can involve omental and peritoneal soft tissues and less commonly any part of the gastrointestinal tract proper, and such tumors can clinically simulate a gastrointestinal stromal tumor (GIST). A minority of patients (15–30%) experiences constitutional symptoms that disappear after tumor removal, but can reappear with tumor recurrence. These symptoms include fever, malaise, and weight loss, and children also can experience growth retardation. Pathologic laboratory findings include anemia (normochromic or hypochromic, microcytic), leukocytosis, thrombocytosis, elevated erythrocyte sedimentation rate, and polyclonal hypergammaglobulinemia.[40–42] Abdominal IMTs recur in 25% of cases,[38,39,51] and pulmonary and brain metastases developed in 10% of patients in tumors reported as inflammatory fibrosarcomas.[52] In one pediatric series, most IMTs were abdominal, with one patient with abdominal IMT (6%) dying of disease, and local recurrences occurred in three additional patients (18%). This study also demonstrated association with predisposing conditions such as Fanconi anemia and the Fas-ligand deficiency type of immunodeficiency[53]

Pulmonary IMTs have occurred in ages ranging from 11 to 78 years (median, 36 years).[38,50] In one series, these tumor sizes varied from 3 cm to 15 cm.[50] Metastases to the mediastinum and chest wall have been reported in typical and atypical cases.[50,54]

Urinary bladder IMTs have been reported in a wide age range, including children, but the mean age in the two largest series has been 47 to 54 years.[48,55] These tumors usually present by gross hematuria, and occasional association with (sarcomatoid) transitional cell carcinoma has been observed.[55] In the two largest series, mean/median tumor size was 4 cm to 5 cm, and 10% to 30% of tumors recurred.[48,55] One series reported metastases in a malignant example.[55] In these series, ALK positivity was observed in 46% and 57% of cases.

Uterine IMTs have occurred from childhood to perimenopausal age, and most examples have been intracavitary ALK-positive tumors, varying from 1 cm to 12 cm in size. The behavior of the reported examples was nonaggressive.[56]

Laryngeal IMTs (no ALK studies reported) occurred predominantly in older adults as small pedunculated masses in vocal cords, consisted of SMA-positive myofibroblasts, and had a benign clinical course.[57]

Cardiac and central nervous system IMTs have occurred mostly in adult patients, but no convincing ALK positivity was present in these tumors, suggesting that they differ from typical IMTs. The cardiac examples were polypoid intracavitary tumors, some of which caused fatal embolic complications (e.g., brain, coronary artery), but otherwise these tumors followed a benign course.[58] CNS examples were mostly dura-based tumors, and late recurrences at 7 to 15 years occurred in two of nine patients, but with no distant metastases.[59]

Peripheral soft tissue IMTs are rare. In a recent series that included atypical and aggressive examples, 8% of IMTs occurred in the head and neck, 5% in the extremities, and none in the trunk wall. No patient developed metastases.[54] In one series predating ALK studies, such tumors occurred in adults of 28 to 83 years (median age, 54 years), and no recurrences or metastases were reported, although mortality was reported, related to comorbidities.[60] In one case, IMT of the forearm converted into a full-blown sarcoma after multiple recurrences; no ALK studies were available on this case.[61] At the author's institution, there have been sporadic soft tissue IMTs in the head and neck of children but no definitively ALK-positive IMTs in the extremities or trunk wall; the author therefore views the reported peripheral examples, especially those in older adults, with some skepticism.

In general, the prognosis of IMT is good, but well-documented examples that metastasized and caused patient death have occurred. Because of the paucity of malignant examples, no formula can be developed to separate tumors with benign and malignant behaviors. Surgery is the main treatment modality, and nonmutilating surgery is generally advocated. The ALK inhibitor crizotinib has emerged as a new targeted treatment and has shown potential for induction of remissions.[62,63]

Pathology

Grossly, the abdominal and pelvic lesions range widely in size, from 1 cm to 20 cm (median size is close to 10 cm). Bladder lesions are typically smaller, 5 cm on average. Multiple tumor nodules are common with lobular, multinodular, and sometimes whorled texture on sectioning. The tumor tissue is gray-white or tan-pink. The cut surface is usually rubbery, but can be myxoid in some areas (Figure 10.19).

Histologically, IMT is composed of a mixture of spindle or polygonal cells, (myo)fibroblasts, plasma cells, and lymphocytes in varying proportions (Figure 10.20). Some lesions contain eosinophils and neutrophils. The spindle cells

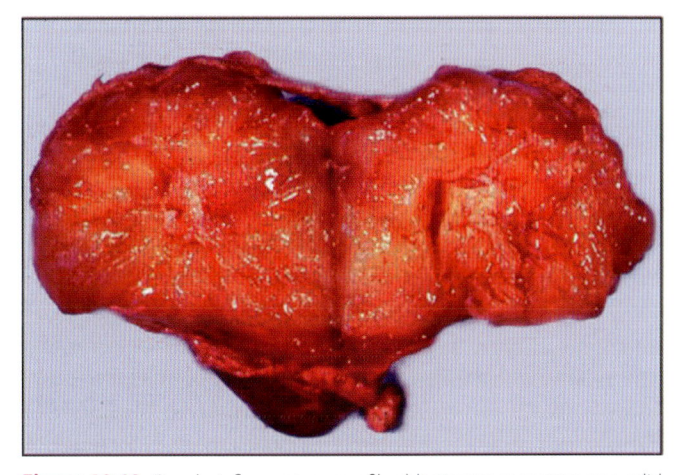

Figure 10.19 Grossly, inflammatory myofibroblastic tumor appears as a solid, homogeneous, fleshy mass measuring approximately 5 cm.

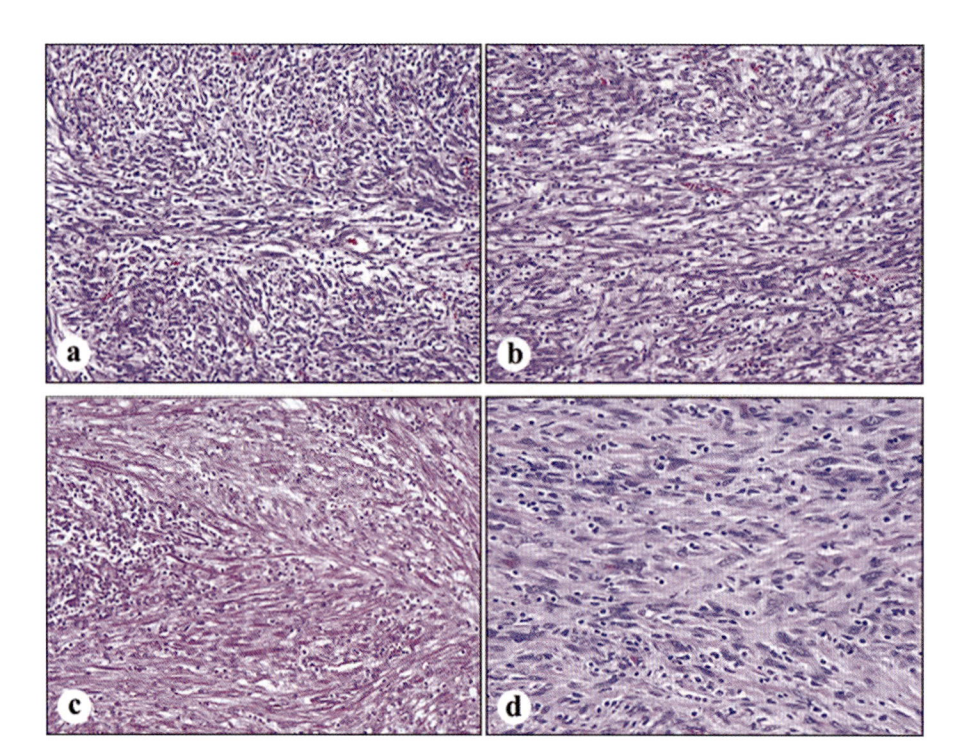

Figure 10.20 Low-magnification images of inflammatory myofibroblastic tumor reveal an admixture of spindle cells and lymphoplasmacytic cellular infiltration with varying degrees of fibrosis.

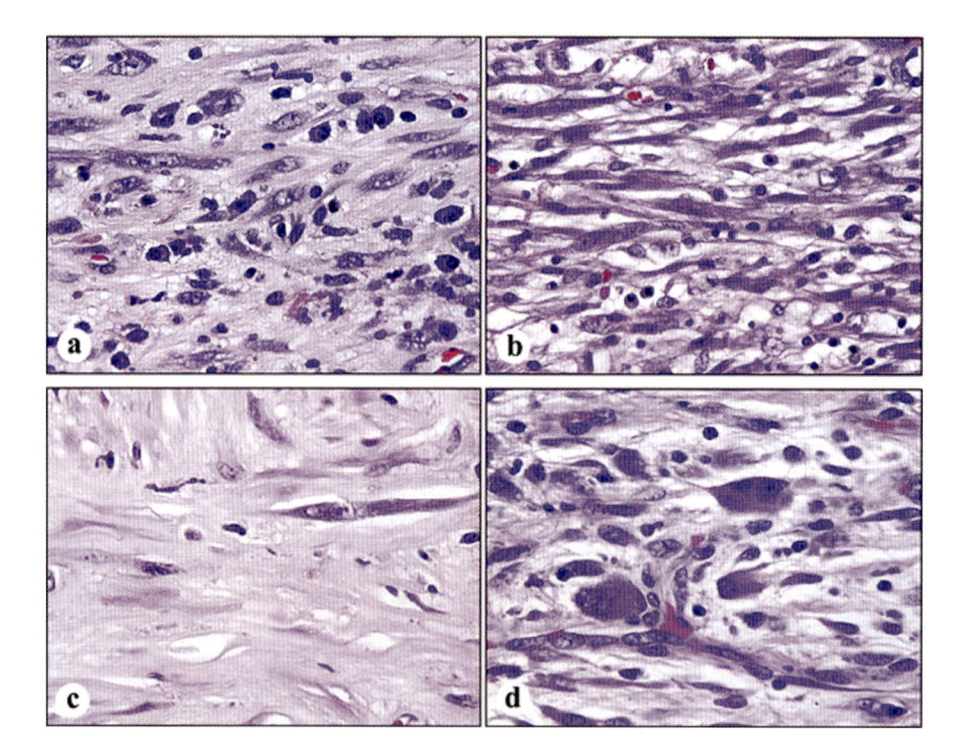

Figure 10.21 High magnification of IMT. (**a,b**) Alternating lymphoplasmacytic infiltration and elongated spindle cells. (**c**) Prominent fibrosclerotic matrix is present in some cases. (**d**) The tumor cells may also have an epithelioid appearance with abundant, amphophilic cytoplasm (**b**).

sometimes show a dense storiform or fascicular arrangement with relatively few inflammatory cells in between, or there can be extensive inflammatory infiltration, myxoid change, and fibrosis with only a scant neoplastic component. Calcifications have been reported but are rare.

Cytologically, the spindle cells have typically oval to elongated nuclei, variably prominent, eosinophilic nucleoli, and amphophilic, often abundant cytoplasm, creating some resemblance to skeletal muscle cells or ganglion cells (Figure 10.21). There is a histologic spectrum that includes relatively bland and paucicellular sclerotic lesions at one end, and more highly cellular and atypical lesions at the other. Mitotic frequency varies but is generally low, and necrosis is rare. Vascular invasion can occur, but this has no adverse prognostic significance.[51]

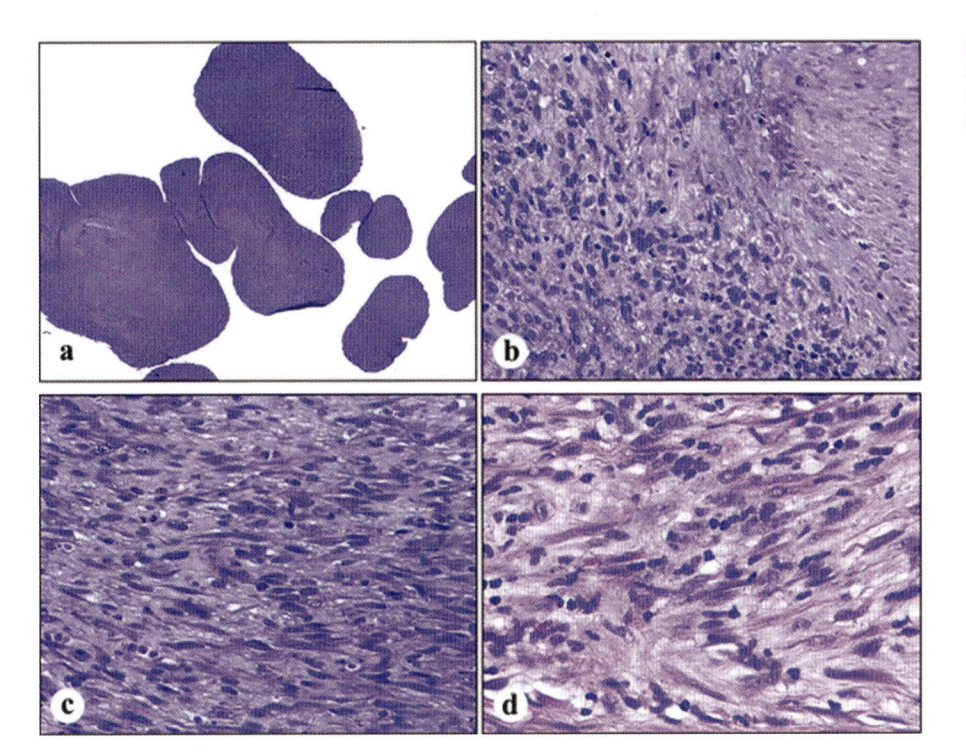

Figure 10.22 Cardiac IMT is composed of multiple papillary projections. The lesional cells are uniform myofibroblasts, and the larger papillae show central necrosis.

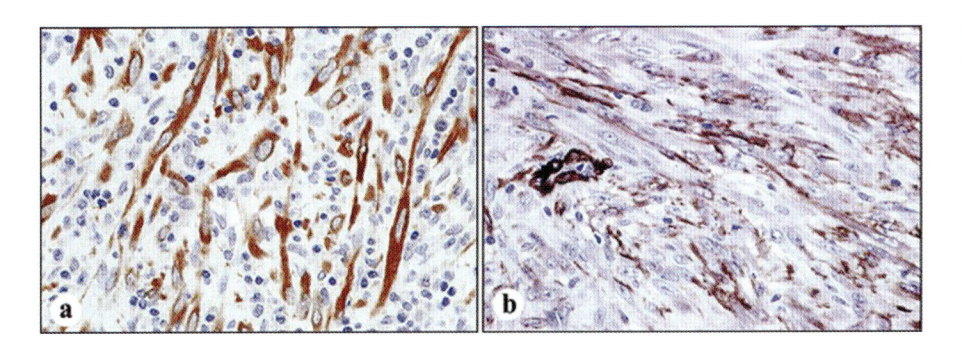

Figure 10.23 Immunohistochemically, the tumor in IMT cells shows strong cytoplasmic positivity for anaplastic lymphoma kinase (**a**), and many cells are also SMA positive (**b**).

A rare and clinical aggressive variant mostly reported in the abdominal cavity, often in older patients, is the epithelioid variant, also designated as epithelioid myofibroblastic sarcoma. This variant is composed of predominantly epithelioid, relatively uniform cells, which typically contain perinuclear ALK positivity (Figure 10.21).[64,65]

Cardiac inflammatory myofibroblastic tumors can form papillary endoluminal masses and are composed of a uniform myofibroblastic population sprinkled with lymphocytes (Figure 10.22).

Immunohistochemistry and other gene expression studies

The ample cytoplasmic myofibroblasts often show cytoplasmic positivity for ALK (CD246) (Figure 10.23a), although the percentage of positive cases varies. Some translocation variants show distinctive ALK localization, such as RANBP1-ALK fusion positive cases, which show a perinuclear zone of ALK positivity.[65] Immunohistochemical evaluation of ALK expression is useful in the differential diagnosis, because other inflammatory pseudotumors are usually ALK negative, although positivity has been reported in RMS and leiomyosarcoma, indicating that an immunohistochemical panel addressing these different diagnostic possibilities is necessary.[66,67]

IMTs are also positive for vimentin, and variably for muscle actin (HHF-35), alpha SMA (Figure 10.23b), and occasionally focally for desmin, although cases with primitive non-myoid phenotypes have also been reported.[68] Keratin positivity for AE1/AE3 cocktail and CAM5.2 antibodies is common, especially in the abdominal and urinary bladder tumors,[48,55] but seems to be less common in pulmonary lesions. The tumor cells are negative for myoglobin, MyoD1/myogenin, S100 protein, and KIT(CD117).[39,52]

PCR-based studies on abdominal lesions have been negative for EBV and cytomegalovirus (CMV) viral DNA.[52] Although human herpesvirus 8 (HHV8) was initially reported

Chapter 10: Childhood fibroblastic and myofibroblastic proliferations of variable biologic potential

Table 10.1 Reported partner genes involved in the ALK-gene fusion translocations in inflammatory myofibroblastic tumor[a]

Fusion partners	Expected translocation	Tumor location	Age	Sex	Ref.
TPM3-ALK	t(1;2)(q21;p23)[b]	Abdomen	23 y	Female	75
TPM4-ALK	t(2;19)(p23;p13)	Abdomen	1 y	Male	75
CLTC-ALK	t(2;17)(p23;q11)	Neck	3 y	Female	76
CARS-ALK	t(2;11)(p23;p15)	Cervical paraspinal soft tissue	10 y	Male	77
ATIC-ALK	t(2;2)(p23;q35)	Urinary bladder	46 y	Male	78
DCTN1-ALK	t(2;2)(p13;p23)	Neck	7 y	Female	79
EML-ALK	t(2;2)(p21;p23)	Lung	67 y	Male	80
PPFIBP1-ALK	t(2;12)(p23;p12)	Lung	34 y	Female	81
RANBP2-ALK	t(2;2)(p23;q12)	Base of mesentery	7 mos	Male	82
SEC31L1-ALK	t(2;4)(p23;q21)	Omentum	23 y	Male	83

[a] Only example cases are shown.
[b] Translocation has also been observed cytogenetically.

in IMT, more recent studies have shown that these tumors are HHV8 negative, both by immunohistochemistry and PCR.[69,70]

Differential diagnosis

The childhood lesions can resemble embryonal RMS, both clinically and histologically, especially when located in the bladder. The extensive inflammatory background differs from RMS, however. The large cytoplasmic cells are always negative for myoglobin, and MyoD1/myogenin, and IMT usually shows extensive SMA reactivity not typically seen in RMS.

IMT should be also separated as specific and nonspecific inflammatory pseudotumors, well-differentiated or dedifferentiated liposarcoma with inflammatory component, and from MFH. Increased numbers of IgG4-plasma cells reaching >10% of the total IgG4-positive cells have been noted in IMTs. Such a finding should not lead one to overlook the diagnostic features of IMT in favor of IgG4 disease.[71]

In the urinary bladder, sarcomatoid carcinoma can be a difficult differential diagnosis; however, the latter tumors typically have greater atypia and the tendency for more confluent sheet formation by the tumor cells. Searching for concomitant differentiated transitional cell carcinoma could yield a diagnostic clue.

Leiomyosarcomas typically have eosinophilic, fibrillary cytoplasm not seen in IMT. Compared with IMT, GISTs have a more homogeneous cellular composition, almost exclusively present in adults more than 40 years of age, and are almost always KIT positive. It seems likely that some early childhood tumors reported as GISTs included abdominal IMTs. Adult spindle cell sarcomas with prominent lymphoplasmacytic reaction usually represent other entities, such as dedifferentiated liposarcomas or myxofibrosarcomas.

Genetics

Cytogenetic studies on pulmonary[72] and abdominal IMTs[73,74] revealed clonal chromosomal changes involving 2p23. Subsequently, several fusion translocations were found that involved the ALK gene at 2p23 and several alternative fusion partners (Table 10.1).[75–83] Many of these translocations have been detected molecular genetically (RACE PCR technique) without prior cytogenetic evidence and can be diagnosed with appropriate FISH probes. Translocations involving tropomyosin 3 (TPM3) or tropomyosin 4 (TPM4) and ALK genes (also found in anaplastic large cell lymphoma) have been demonstrated to constitutionally activate ALK-tyrosine kinase in IMT.[75] Based on experience of urinary bladder tumors, ALK rearrangements are detected in 67% of cases, almost all of them being ALK positive by immunohistochemistry, whereas reactive myofibroblastic proliferations, leiomyosarcomas, RMSs, and sarcomatoid carcinomas are negative.[84] ROS1-involving gene rearrangements have also been detected in some inflammatory myofibroblastic tumors, and these could be treatable with crizotinib, similar to ALK-translocated tumors.[85]

Infantile fibrosarcoma (congenital fibrosarcoma)

This tumor typically occurs in infants and has been proved as a specific molecular entity with a typical gene fusion; the clinical features are also characteristic.

Clinical features

This rare tumor usually arises during the first year of life, and one half of the tumors are congenital; occurrence beyond 5 years of age is exceptional. The tumor has a 3:2 male predominance and mainly affects the distal parts of the upper

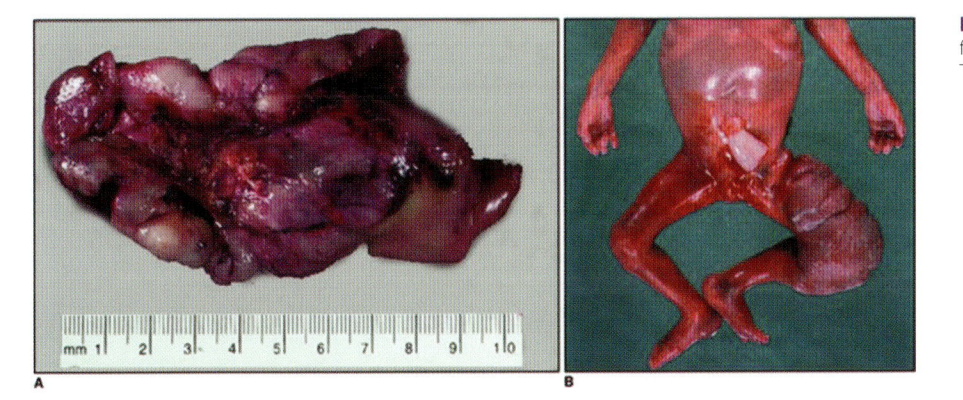

Figure 10.24 Two gross examples of infantile fibrosarcoma: (**A**) An example in the neck. (**B**) Tumor in the lower extremity of a stillborn infant.

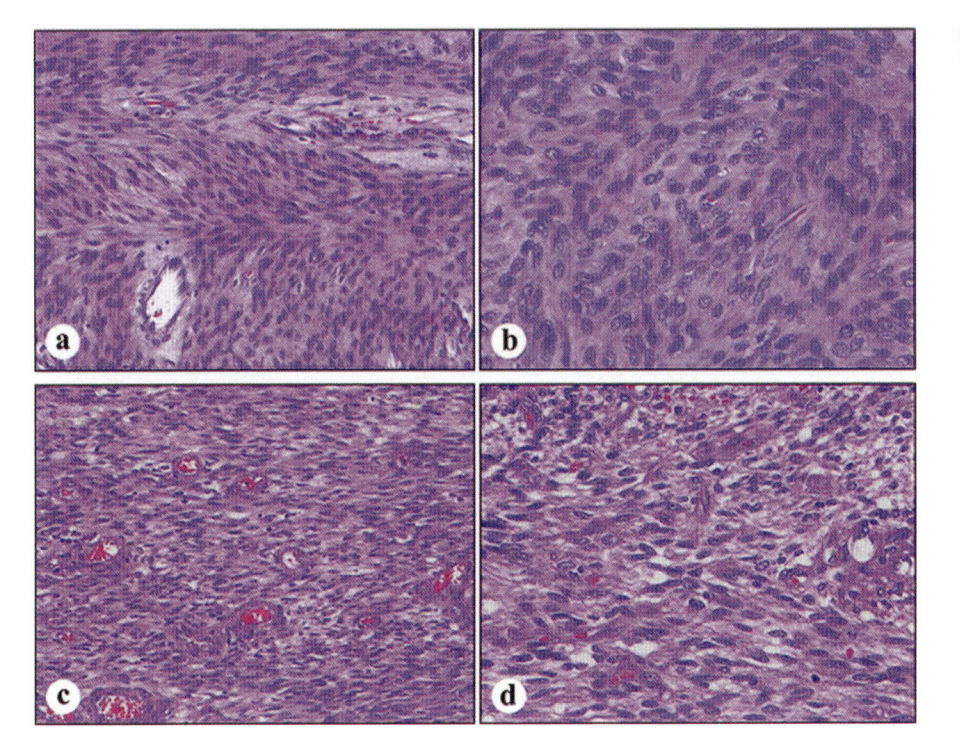

Figure 10.25 Well-differentiated, collagen-forming examples of infantile fibrosarcoma.

and lower extremities: hands, forearms, ankles, and feet. It is typically rapidly growing and poorly circumscribed and can infiltrate both superficial and deep soft tissues and reach a considerable size. Because this tumor has a relatively good prognosis despite the ominous clinical features, conservative primary treatment is now advocated as much as possible, although ideally a complete excision should be performed.[86–92] One study did not demonstrate an adverse effect of positive surgical margins, so that limb-sparing surgery may be appropriate.[93] Chemotherapy can obviate amputations in those cases in which amputation would be otherwise required for complete excision.[94] Occurrence in visceral locations has been reported in lung, heart, and intestines.[95,96]

Infantile fibrosarcoma is prone to local recurrences, but lung metastases are rare. The series of Chung and Enzinger showed a 5-year survival rate of 84%,[87] with a few fatalities caused by lung or brain metastases. Subsequent series have shown similar good survival results, but larger head and neck and to some degree centrally located truncal or proximal extremity tumors are associated with some mortality.[89–91] Compared with fibrosarcomas of older children, the prognosis is more favorable.[88] Nonaka and Sun[97] reported liver and adrenal metastases in a fetus with a large congenital fibrosarcoma involving the back and chest wall. Fetal mortality can occur following intrauterine tumor-associated hemorrhage.[98]

Pathology

On gross inspection, infantile fibrosarcoma varies from a small nodule to a bulky tumor, with a mean size of 4 cm in the largest series.[86–88] On sectioning, the tumor is pale tan or fleshy and varies from firm to soft (Figure 10.24). Larger tumors often have areas of hemorrhage and necrosis.

Histologically, infantile fibrosarcoma varies from well-differentiated collagen-forming fibrosarcomatous pattern to a highly cellular, noncollagenous spindle to oval cell tumor arranged in fascicles or diffuse sheets (Figures 10.25 and 10.26).

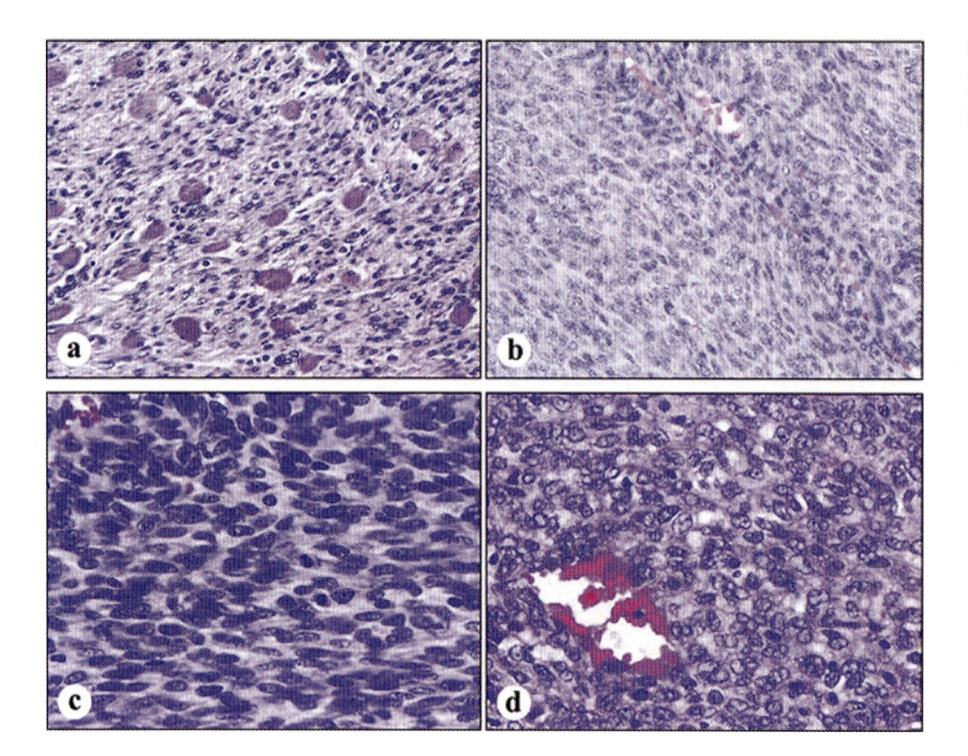

Figure 10.26 Highly cellular noncollagenous examples of infantile fibrosarcoma. (**a**) Skeletal muscle infiltration is present. Note mitotic activity in (**c**), and tendency to oval cell morphology in (**d**).

Mitotic activity is often high. MFH-like pleomorphic transformation can occur and has been shown to carry a poor prognosis.[99]

Immunohistochemically, infantile fibrosarcoma is positive for vimentin and variably positive for actins. Negative results have been obtained on desmin, keratins, EMA, and S100 protein.[90] Actin positivity correlates with myofibroblastic fibroblastic differentiation detected on ultrastructural studies.[90] Large studies on genetically verified cases are not available.

Differential diagnosis

Spindle cell and poorly differentiated embryonal RMS have to be ruled out, especially in cases involving the genital region, pelvis, and abdomen. Immunohistochemistry is required for this in every case, and a wide panel of markers, including desmin and skeletal muscle transcriptional regulators (MyoD1, myogenin), should be evaluated. Intratumoral hemorrhage, seen in some cases, should not be mistaken for a vascular tumor.

Dermatofibrosarcoma protuberans and giant cell fibroblastoma can be recognized by their variably storiform architecture, giant cells, superficial and subcutaneous fat involvement, and consistent immunohistochemical positivity for CD34.

Genetics

Trisomy of chromosome 11, along with variable trisomies of chromosomes 17 and 20 and occasionally of 8 and 10, occur as nonrandom cytogenetic changes with possible differential diagnostic value.[100-103]

A reciprocal translocation t(12;15)(p13;q25) resulting in the fusion of ETV6 and NTRK3 genes has been found in most cases.[104] The same translocation occurs in the cellular (but not the typical) variant of mesoblastic nephroma.[105,106] The translocation can be evaluated by RT-PCR using frozen or formalin-fixed and paraffin-embedded tissue.[107-110] This gene fusion has not been detected in other spindle cell tumors of childhood. The same gene fusion has been reported in secretory carcinoma of the breast, however.[111]

NTRK3 gene encodes for a tyrosine kinase receptor TRKC. According to two studies, this protein is similarly present in infantile fibrosarcoma and other spindle cell tumors of childhood, indicating that TRKC is not a tumor-specific immunohistochemical marker.[108,109]

References

Neurothekeoma

1. Gallager RL, Helwig EB. Neurothekeoma a benign cutaneous tumor of neural origin. *Am J Clin Pathol* 1980;74:759–764.
2. Laskin WB, Fetsch JF, Miettinen M. The neurothekeoma: immunohistochemical analysis distinguishes the true nerve sheath myxoma from its mimics. *Hum Pathol* 2000;31:1230–1241.
3. Barnhill RL, Dickersin GR, Nickeleit V, et al. Studies on the cellular origin of neurothekeoma: clinical, light microscopic, immunohistochemical and ultrastructural observations. *J Am Acad Dermatol* 1991;25:80–88.

4. Argenyi ZB, Le Boit PE, Santa Cruz D, Swanson PE, Kuzner H. Nerve sheath myxoma (neurothekeoma) of the skin: light microscopic and immunohistochemical reappraisal of the cellular variant. *J Cutan Pathol* 1993;20:294–303.

5. Fetsch JF, Laskin WB, Hallman JR, Lupton GP, Miettinen M. Neurothekeoma: an analysis of 178 tumors with detailed immunohistochemical data and long-term patient follow-up information. *Am J Surg Pathol* 2007;31:1103–1114.

6. Hornick JL, Fletcher CDM. Cellular neurothekeoma: detailed characterization in a series of 133 cases. *Am J Surg Pathol* 2007;31:329–340.

7. Page RN, King R, Mihm MC Jr, Googe PB. Microphthalmia transcription factor and NKI/C3 expression in cellular neurothekeoma. *Mod Pathol* 2004;17:230–234.

8. Suarez A, High WA. Immunohistochemical analysis of KBA.62 in 18 neurothekeomas: a potential marker for differentiating neurothekeoma, but a marker that may lead to confusion with melanocytic tumors. *J Cutan Pathol* 2014;41:36–41.

9. Sheth S, Li X, Binder S, Dry SM. Differential gene expression profiles of neurothekeomas and nerve sheath myxomas by microarray analysis. *Mod Pathol* 2011;24:343–354.

10. Jaffer S, Ambrosini-Spaltro A, Mancini AM, Eusebi V, Rosai J. Neurothekeoma and plexiform fibrohistiocytic tumor: mere histologic resemblance or histogenetic relationship? *Am J Surg Pathol* 2009;33:905–913.

Plexiform fibrohistiocytic tumor

11. Enzinger FM, Zhang R. Plexiform fibrohistiocytic tumor presenting in children and young adults: an analysis of 65 cases. *Am J Surg Pathol* 1988;12:818–826.

12. Hollowood K, Holley MP, Fletcher CDM. Plexiform fibrohistiocytic tumour: clinicopathological, immunohistochemical and ultrastructural analysis in favour of a myofibroblastic lesion. *Histopathology* 1991;19:503–513.

13. Zelger B, Weinlich G, Steiner H, Zelger BG, Egarter-Vigl E. Dermal and subcutaneous variants of plexiform fibrohistiocytic tumor. *Am J Surg Pathol* 1997;21:235–241.

14. Remstein ED, Arndt CAS, Nascimento AG. Plexiform fibrohistiocytic tumor: clinicopathologic analysis of 22 cases. *Am J Surg Pathol* 1999;23:662–670.

15. Leclerc S, Hamel-Teillac D, Oger P, Brousse N, Fraitag S. Plexiform fibrohistiocytic tumor: three unusual cases occurring in infancy. *J Cutan Pathol* 2005;32:572–576.

16. Moosavi C, Jha P, Fanburg-Smith JC. An update on plexiform fibrohistiocytic tumor and addition of 66 new cases from the Armed Forces Institute of Pathology, in honor of Franz M. Enzinger, MD. *Ann Diagn Pathol* 2007;11:313–319.

17. Redlich GC, Montgomery KD, Allgood GA, Joste NE. Plexiform fibrohistiocytic tumor with a clonal cytogenetic anomaly. *Cancer Genet Cytogenet* 1999;108:141–143.

18. Smith S, Fletcher CD, Smith MA, Gusterson BA. Cytogenetic analysis of a plexiform fibrohistiocytic tumor. *Cancer Genet Cytogenet* 1990;48:31–34.

Angiomatoid fibrous histiocytoma

19. Enzinger FM. Angiomatoid malignant fibrous histiocytoma: a distinct fibrohistiocytic tumor of children and young adults simulating a vascular neoplasm. *Cancer* 1979;44:2147–2157.

20. Costa MJ, Weiss SW. Angiomatoid malignant fibrous histiocytoma: a follow-up study of 108 cases with evaluation of possible histologic predictors of outcome. *Am J Surg Pathol* 1990;14:1126–1132.

21. Leu HJ, Makek M. Angiomatoid malignant fibrous histiocytoma. *Virchows Arch A Pathol Anat Histopathol* 1982;395:99–107.

22. Fanburg-Smith JF, Miettinen M. Angiomatoid "malignant" fibrous histiocytoma: a clinicopathologic study of 158 cases and further exploration of the myoid phenotype. *Hum Pathol* 1999;30:1336–1343.

23. Pettinato G, Manivel JC, De Rosa G, Petrella G, Jaszcz W. Angiomatoid malignant fibrous histiocytoma: cytologic, immunohistochemical, ultrastructural, and flow cytometric study of 20 cases. *Mod Pathol* 1990;3:479–487.

24. Chen G, Folpe AL, Colby TV, et al. Angiomatoid fibrous histiocytoma: unusual sites and unusual morphology. *Mod Pathol* 2011;24:1560–1570.

25. Thway K, Nicholson AG, Wallace WA, et al. Endobronchial pulmonary angiomatoid fibrous histiocytoma: two cases with EWSR1-CREB1 and EWSR1-ATF1 fusions. *Am J Surg Pathol* 2012;36:883–888.

26. Chow LT, Allen PW, Kumta SM, et al. Angiomatoid malignant fibrous histiocytoma: report of an unusual case with highly aggressive clinical course. *J Foot Ankle Surg* 1998;37:235–238.

27. Seo IS, Frizerra G, Coates TD, Mirkin LD, Cohen MD. Angiomatoid malignant fibrous histiocytoma with extensive lymphadenopathy simulating Castleman's disease. *Pediatr Pathol* 1986;6:233–247.

28. Moura RD, Wang X, Lonzo ML, et al. Reticular angiomatoid "malignant" fibrous histiocytoma–a case report with cytogenetics and molecular genetic analyses. *Hum Pathol* 2011;42:1359–1363.

29. Schaefer IM, Fletcher CD. Myxoid variant of so-called angiomatoid "malignant fibrous histiocytoma": clinicopathologic characterization in a series of 21 cases. *Am J Surg Pathol* 2014;38:816–823.

30. Fletcher CDM. Angiomatoid "malignant fibrous histiocytoma": an immunohistochemical study indicative of myoid differentiation. *Hum Pathol* 1991;22:563–568.

31. Smith MEF, Costa MJ, Weiss SW. Evaluation of CD68 and other histiocytic antigens in angiomatoid malignant fibrous histiocytoma. *Am J Surg Pathol* 1991;15:757–763.

32. Hasegawa T, Seki K, Ono K, Hirohashi S. Angiomatoid (malignant) fibrous histiocytoma: a peculiar low-grade tumor showing immunophenotypic heterogeneity and ultrastructural variations. *Pathol Int* 2000;50:731–738.

33. Costa MJ, McGlothlen L, Pierce M, Munn R, Vogt PJ. Angiomatoid features in fibrohistiocytic sarcomas. Immunohistochemical, ultrastructural, and clinical distinction from vascular neoplasms. *Arch Pathol Lab Med* 1995;119:1065–1071.

34. Antonescu CR, Dal Cin P, Nafa K, et al. EWSR1-CREB1 is the predominant gene fusion in angiomatoid fibrous histiocytoma. *Genes Chromosomes Cancer* 2007;46:1051–1060.

35. Hallor KH, Mertens F, Jin Y, et al. Fusion of the EWSR1 and ATF1 genes

without expression of the MITF-M transcript in angiomatoid fibrous histiocytoma. *Genes Chromosomes Cancer* 2005;44:97–102.

36. Hallor KH, Micci F, Meis-Kindblom J, et al. Fusion genes in angiomatoid fibrous histiocytoma. *Cancer Lett* 2007;251:158–163.

37. Waters BL, Panagopoulos I, Allen EF. Genetic characterization of angiomatoid fibrous histiocytoma identifies fusion of the FUS and ATF-1 genes induced by a chromosomal translocation involving bands 12q13 and 16p11. *Cancer Genet Cytogenet* 2000;121:109–116.

Inflammatory myofibroblastic tumor

38. Coffin CM, Watterson J, Priest JR, Dehner LP. Extrapulmonary inflammatory myofibroblastic tumor (inflammatory pseudotumor): a clinicopathologic and immunohistochemical study of 84 cases. *Am J Surg Pathol* 1995;19:859–872.

39. Coffin CM, Dehner LP, Meis-Kindblom JM. Inflammatory myofibroblastic tumor, inflammatory fibrosarcoma, and related lesions: an historical review with differential diagnostic considerations. *Semin Diagn Pathol* 1998;15:102–110.

40. Coffin CM, Humphrey PA, Dehner LP. Extrapulmonary inflammatory myofibroblastic tumor: a clinical and pathological survey. *Semin Diagn Pathol* 1998;15:85–101.

41. Neuhauser TS, Derringer GA, Thompson LD, et al. Splenic inflammatory myofibroblastic tumor (inflammatory pseudotumor): a clinicopathologic and immunophenotypic study of 12 cases. *Arch Pathol Lab Med* 2001;125:379–385.

42. Kutok JL, Pinkus GS, Dorfman DM, Fletcher CD. Inflammatory pseudotumor of lymph node and spleen: an entity biologically distinct from inflammatory myofibroblastic tumor. *Hum Pathol* 2001;32:1382–1387.

43. Chan JKC. Inflammatory pseudotumor: a family of lesions of diverse nature and etiologies. *Adv Anat Pathol* 1996;3:156–171.

44. Ro JY, el-Naggar AK, Amin MB, et al. Pseudosarcomatous fibromyxoid tumor of the urinary bladder and prostate: immunohistochemical, ultrastructural and DNA flow cytometric analyses of nine cases. *Hum Pathol* 1993;24:1203–1210.

45. Albores-Saavedra J, Manivel JC, Essenfeld H, et al. Pseudosarcomatous myofibroblastic proliferations in the urinary bladder of children. *Cancer* 1990;66:1234–1241.

46. Hojo H, Newton WA Jr, Hamoudi AB, et al. Pseudosarcomatous myofibroblastic tumor of the urinary bladder in children: a study of 11 cases with review of the literature. An intergroup rhabdomyosarcoma study. *Am J Surg Pathol* 1995;19:1224–1236.

47. Harik LR, Merino C, Coindre JM, et al. Pseudosarcomatous myofibroblastic proliferations of the bladder: a clinicopathologic study of 42 cases. *Am J Surg Pathol* 2006;30:787–794.

48. Proppe KH, Scully RE, Rosai J. Postoperative spindle cell nodules of genitourinary tract resembling sarcomas. *Am J Surg Pathol* 1984;8:101–108.

49. Pettinato G, Manivel JC, De Rosa N, Dehner LP. Inflammatory myofibroblastic tumor (plasma cell granuloma): clinicopathologic study of 20 cases with immunohistochemical and ultrastructural observations. *Am J Clin Pathol* 1990;94:538–546.

50. Yousem SA, Shaw H, Cieply K. Involvement of 2p23 in pulmonary inflammatory pseudotumors. *Hum Pathol* 2001;32:428–433.

51. Meis JM, Enzinger FM. Inflammatory fibrosarcoma of the mesentery and retroperitoneum: a tumor closely simulating inflammatory pseudotumor. *Am J Surg Pathol* 1991;15:1146–1156.

52. Meis-Kindblom JM, Kjellstrom C, Kindblom LG. Inflammatory fibrosarcoma: update, reappraisal, and perspective on its place in the spectrum of inflammatory myofibroblastic tumors. *Semin Diagn Pathol* 1998;15:133–143.

53. Alaggio R, Cecchetto G, Bisogno G, et al. Inflammatory myofibroblastic tumors in childhood: a report from the Italian Cooperative Group studies. *Cancer* 2010;116:216–226.

54. Coffin CM, Hornick JL, Fletcher CDM. Inflammatory myofibroblastic tumor: comparison of clinicopathologic, histologic, and immunohistochemical features including ALK-expression in atypical and aggressive cases. *Am J Surg Pathol* 2007;31:509–520.

55. Montgomery EA, Shuster DD, Burkart AL, et al. Inflammatory myofibroblastic tumors of the urinary tract: a clinicopathologic study of 46 cases, including a malignant example inflammatory fibrosarcoma and a subset associated with high-grade urothelial carcinoma. *Am J Surg Pathol* 2006;30;1502–1512.

56. Rabban JT, Zaloudek CJ, Shekitka KM, Tavassoli FA. Inflammatory myofibroblastic tumor of the uterus: a clinicopathologic study of 6 cases emphasizing distinction from aggressive mesenchymal tumors. *Am J Surg Pathol* 2005;29:1348–1355.

57. Wenig BM, Devaney K, Bisceglia M. Inflammatory myofibroblastic tumor of the larynx: a clinicopathologic study of eight cases simulating a malignant spindle cell neoplasm. *Cancer* 1995;76:2217–2229.

58. Burke A, Li L, Kling E, et al. Cardiac inflammatory myofibroblastic tumor: a "benign" neoplasm that may result in syncope, myocardial infarction, and sudden death. *Am J Surg Pathol* 2007;31:1115–1122.

59. Jeon YK, Chang KH, Suh YL, Jung HW, Park SH. Inflammatory myofibroblastic tumor of the central nervous system: clinicopathologic analysis of 10 cases. *J Neuropathol Exp Neurol* 2005;64:254–259.

60. Ramachandra S, Hollowood K, Bisceglia M, Fletcher CDM. Inflammatory pseudotumor of soft tissues: a clinicopathological and immunohistochemical analysis of 18 cases. *Histopathology* 1995;27:313–323.

61. Donner LR, Trompler RA, White RR. Progression of inflammatory myofibroblastic tumor (inflammatory pseudotumor) of soft tissue into sarcoma after several recurrences. *Hum Pathol* 1996;27:1095–1098.

62. Butrynski JE, D'Adamo DR, Hornick JL, et al. Crizotinib in ALK-rearranged inflammatory myofibroblastic tumor. *N Engl J Med* 2010;363:1727–1733.

63. Mossé YP, Lim MS, Voss SD, et al. Safety and activity of crizotinib for paediatric patients with refractory solid tumours or anaplastic large-cell lymphoma: a Children's Oncology Group phase 1 consortium study. *Lancet Oncol* 2013;14:472–480.

64. Chen ST, Lee JC. An inflammatory myofibroblastic tumor in liver with

ALK and RANBP2 gene rearrangement: combination of distinct morphologic, immunohistochemical, and genetic features. *Hum Pathol* 2008;39:1854–1858.
65. Mariño-Enríquez A, Wang WL, Roy A, et al. Epithelioid inflammatory myofibroblastic sarcoma: an aggressive intra-abdominal variant of inflammatory myofibroblastic tumor with nuclear membrane or perinuclear ALK. *Am J Surg Pathol* 2011;35:135–144.
66. Chan JKC, Cheuk W, Shimizu M. Anaplastic lymphoma kinase expression in inflammatory pseudotumors. *Am J Surg Pathol* 2001;25:761–768.
67. Cook JR, Dehner LP, Collins MH, et al. Anaplastic lymphoma kinase (ALK) expression in the inflammatory myofibroblastic tumor: a comparative immunohistochemical study. *Am J Surg Pathol* 2001;25:1364–1371.
68. Hisaoka M, Hussong JW, Brown M, et al. Comparison of DNA ploidy, histologic and immunohistochemical findings with clinical outcome in inflammatory myofibroblastic tumors. *Mod Pathol* 1999;12:279–286.
69. Yamamoto H, Kohashi K, Oda Y, et al. Absence of human herpesvirus-8 and Epstein–Barr virus in inflammatory myofibroblastic tumor with anaplastic large cell lymphoma kinase fusion gene. *Pathol Int* 2006;56:584–590.
70. Tavora F, Shilo K, Ozbudak IH, et al. Absence of human herpesvirus-8 in pulmonary inflammatory myofibroblastic tumor: immunohistochemical and molecular analysis of 20 cases. *Mod Pathol* 2007;20:995–999.
71. Saab ST, Hornick JL, Fletcher CD, Olson SJ, Coffin CM. IgG4 plasma cells in inflammatory myofibroblastic tumor: inflammatory marker or pathogenic link? *Mod Pathol* 2011;24:606–612.
72. Snyder CS, Dell-Aquila M, Haghighi P, et al. Clonal changes in inflammatory pseudotumor of the lung. *Cancer* 1995;76:1545–1549.
73. Su LD, Atayde-Perez A, Sheldon S, Fletcher JA, Weiss SW. Inflammatory myofibroblastic tumor: cytogenetic evidence supporting clonal origin. *Mod Pathol* 1998;11:364–368.
74. Griffin CA, Hawkins AL, Dvorak C, et al. Recurrent involvement of 2p23 in inflammatory myofibroblastic tumors. *Cancer Res* 1999;59:2776–2780.
75. Lawrence B, Perez-Atayde A, Hibbard MK, et al. TPM3-ALK and TPM4-ALK oncogenes in inflammatory myofibroblastic tumors. *Am J Pathol* 2000;157:377–384.
76. Bridge JA, Kanamori M, Ma Z, et al. Fusion of the ALK gene to the clathrin heavy chain gene, CLTC, in inflammatory myofibroblastic tumor. *Am J Pathol* 2001;159:411–415.
77. Debelenko LV, Arthur DC, Pack SD, et al. Identification of CARS-ALK fusion in primary and metastatic lesions of an inflammatory myofibroblastic tumor. *Lab Invest* 2003;83:1255–1265.
78. Debiec-Rychter M, Marynen P, Hagemeijer A, Pauwels P. ALK-ATIC fusion in urinary bladder inflammatory myofibroblastic tumor. *Genes Chromosomes Cancer* 2003;38:187–190.
79. Wang X, Krishnan C, Nguyen EP, et al. Fusion of dynactin 1 to anaplastic lymphoma kinase in inflammatory myofibroblastic tumor. *Hum Pathol* 2012;43:2047–2052.
80. Sokai A, Enaka M, Sokai R, et al. Pulmonary inflammatory myofibroblastic tumor harboring EML4-ALK fusion gene. *Jpn J Clin Oncol* 2014;44:93–96.
81. Takeuchi K, Soda M, Togashi Y, et al. Pulmonary inflammatory myofibroblastic tumor expressing a novel fusion, PPFIBP1-ALK: reappraisal of anti-ALK immunohistochemistry as a tool for novel ALK fusion identification. *Clin Cancer Res* 2011;17:3341–3348.
82. Ma Z, Hill DA, Collins MH, et al. Fusion of ALK to the RAN-binding protein 2 (RANBP2) gene in inflammatory myofibroblastic tumor. *Genes Chromosomes Cancer* 2003;37:98–105.
83. Panagopoulos I, Nilsson T, Domanski HA, et al. Fusion of the SEC31L1 and ALK genes in an inflammatory myofibroblastic tumor. *Int J Cancer* 2006;118:1181–1186.
84. Sukov WR, Cheville JC, Carlson AW, et al. utility of ALK-1 protein expression and ALK rearrangements in distinguishing inflammatory myofibrolastic tumor from malignant spindle cell lesions of the urinary bladder. *Mod Pathol* 2007;20:592–603.
85. Davies KD, Doebele RC. Molecular pathways: ROS1 fusion proteins in cancer. *Clin Cancer Res* 2013;19:4040–4045.

Infantile fibrosarcoma

86. Balsaver AM, Butler JJ, Martin RG. Congenital fibrosarcoma. *Cancer* 1967;20:1607–1616.
87. Chung EB, Enzinger FM. Infantile fibrosarcoma. *Cancer* 1976;38:729–739.
88. Soule EH, Pritchard DJ. Fibrosarcoma in infants and and children: a review of 110 cases. *Cancer* 1977;40:1711–1721.
89. Iwasaki H, Enjoji M. Infantile and adult fibrosarcomas of the soft tissues. *Acta Pathol Jpn* 1979;29:377–388.
90. Coffin CM, Jaszcz W, O'Shea PA, Dehner LP. So-called congenital-infantile fibrosarcoma: does it exist and what is it? *Pediatr Pathol* 1994;14:133–150.
91. Kodet R, Stejskal J, Pilat D, et al. Congenital-infantile fibrosarcoma: a clinicopathological study of five patients entered on the Prague children's tumor registry. *Pathol Res Pract* 1996;192:845–853.
92. Cofer BR, Vescio PJ, Wiener ES. Infantile fibrosarcoma: complete excision is the appropriate treatment. *Ann Surg Oncol* 1996;3:159–161.
93. Sulkowski JP, Raval MV, Browne M. Margin status and multimodal therapy in infantile fibrosarcoma. *Pediatr Surg Int* 2013;29:771–776.
94. Kynaston JA, Malcolm AJ, Craft AW, et al. Chemotherapy in the management of infantile fibrosarcoma. *Med Pediatr Oncol* 1993;21:488–493.
95. Steelman C, Katzenstein H, Parham D, et al. Unusual presentation of congenital infantile fibrosarcoma in seven infants with molecular-genetic analysis. *Fetal Pediatr Pathol* 2011;30:329–337.
96. Buccoliero AM, Castiglione F, Rossi Degl'Innocenti D, et al. Congenital/infantile fibrosarcoma of the colon: morphologic, immunohistochemical, molecular, and ultrastructural features of a relatively rare tumor in an extraordinary localization. *J Pediatr Hematol Oncol* 2008;30:723–727.
97. Nonaka D, Sun CCJ. Congenital fibrosarcoma with metastasis in a fetus. *Pediatr Dev Pathol* 2004;7:187–191.

98. Dumont C, Monforte M, Flandrin A, et al. Prenatal management of congenital infantile fibrosarcoma: unexpected outcome. *Ultrasound Obstet Gynecol* 2011;37:733–735.

99. Salloum E, Caillaud JM, Flamant F, Landman J, Lemerle J. Poor prognosis infantile fibrosarcoma with pathologic features of malignant fibrous histiocytoma after local recurrence. *Med Pediatr Oncol* 1990;18:295–298.

100. Adam LR, Davison EV, Malcolm AJ, Pearson AD, Craft AW. Cytogenetic analysis of a congenital fibrosarcoma. *Cancer Genet Cytogenet* 1991;52:37–41.

101. Dal Cin P, Brock P, Casteels-Van Daele M, et al. Cytogenetic characterization of congenital or infantile fibrosarcoma. *Eur J Pediatr* 1991;150:579–581.

102. Sankary S, Dickman PS, Wiener E, et al. Consistent numerical chromosome aberrations in congenital fibrosarcoma. *Cancer Genet Cytogenet* 1993;65:152–156.

103. Bernstein R, Zeltzer PM, Lin F, Carpenter PM. Trisomy 11 and other nonrandom trisomies in congenital fibrosarcoma. *Cancer Genet Cytogenet* 1994;78:82–86.

104. Knezevich SR, McTadden DE, Tao W, Lim JF, Sorensen PH. A novel ETV6-NTRK3 gene fusion in congenital fibrosarcoma. *Nat Genet* 1998;18:184–187.

105. Knezevich SR, Garnett MJ, Pysher TJ, et al. ETV6-NTRK3 gene fusions and trisomy 11 establish a histogenetic link between mesoblastic nephroma and congenital fibrosarcoma. *Cancer Res* 1998;58:5046–5048.

106. Rubin BP, Chen CJ, Morgan TW, et al. Congenital mesoblastic nephroma t(12;15) is associated with ETV6-NTRK3 gene fusion: cytogenetic and molecular relationship to congenital (infantile) fibrosarcoma. *Am J Pathol* 1998;153:1451–1458.

107. Argani P, Fritsch M, Kadkol SS, et al. Detection of the ETV6-NTRK3 chimeric RNA of infantile fibrosarcoma/cellular congenital mesoblastic nephroma in paraffin-embedded tissue: application to challenging pediatric renal stromal tumors. *Mod Pathol* 2000;13:29–36.

108. Bourgeois JM, Knezevich SR, Mathers JA, Sorensen PH. Molecular detection of the ETV6-NTRK3 gene fusion differentiates congenital fibrosarcoma from other childhood spindle cell tumors. *Am J Surg Pathol* 2000;24:937–946.

109. Dubus P, Coindre JM, Groppi A, et al. The detection of the Tel-TrkC chimeric transcripts is more specific than TrkC immunoreactivity for the diagnosis of congenital fibrosarcoma. *J Pathol* 2001;193:88–94.

110. Sheng WQ, Hisaoka M, Okamoto S, et al. Congenital-infantile fibrosarcoma: a clinicopathologic study of 10 cases and molecular detection of the ETV6-NTRK3 fusion transcripts using paraffin-embedded tissues. *Am J Clin Pathol* 2001;115:348–355.

111. Tognon C, Knezevich SR, Huntsman D, et al. Expression of the ETV6-NTRK3 gene fusion as a primary event in human secretory breast carcinoma. *Cancer Cell* 2002;2:367–376.

Chapter 11

Myxomas and ossifying fibromyxoid tumor

Markku Miettinen
National Cancer Institute, National Institutes of Health

Myxomas encompass a heterogeneous group of clinicopathologic entities, common to which is prominent content of extracellular mucin. Soft tissue myxomas are fibroblastic-myofibroblastic neoplasms, or in some cases possibly non-neoplastic proliferations.

Cardiac myxoma is thought to originate from primitive, multipotential endocardial cells. In addition to myxomas, ossifying fibromyxoid tumor, a mesenchymal tumor of uncertain histogenesis with potentially malignant variants, is discussed in this chapter.

Common to all myxomas is the production of a mucoid intercellular substance by the fibroblastic or myofibroblastic tumor cells. Although the chemical composition of the mucins is known to vary, its analysis is not of great practical significance. The constituents include glycoproteins and proteoglycans that include neutral and acidic carbohydrate components, such as hyaluronic acid.

The apparent character of the mucoid material on hematoxylin and eosin (HE) stain depends not only on its composition but also on fixation and staining. The mucinophilic (metachromatic) quality of HE stains varies. Mucin also becomes progressively extracted into the fixative over time, and therefore specimens kept longer in the fixative typically lose mucins and the stain is less metachromatic.

Intramuscular myxoma
Clinical features

This relatively rare tumor typically presents in middle-aged adults between 40 and 70 years, with a 2:1 female predominance (Figure 11.1). Nearly one-half of the lesions occur in the thigh, with the arm and buttocks being other common locations. A wide variety of extremity locations and the chest wall are among other possible sites (Figure 11.2). The tumor grows very slowly, and the size varies from a small nodule <1 cm to a sizeable mass exceeding 10 cm to 20 cm. The latter is especially true with tumors located in the thigh or buttock. Recurrence is rare after simple excision, and there is no risk of metastasis.[1–4]

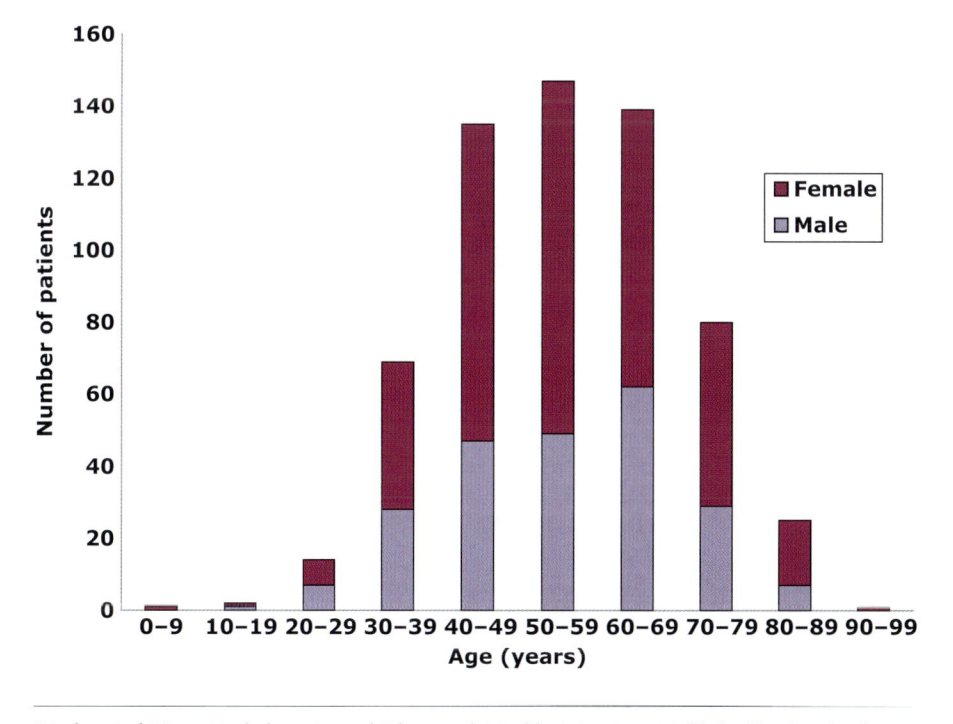

Figure 11.1 Age and sex distribution and anatomic location of 650 intramuscular myxomas.

Modern Soft Tissue Pathology, Second Edition, ed. Markku Miettinen. Published by Cambridge University Press. © Cambridge University Press 2016.

Specific preoperative diagnosis by radiologic studies (MRI) or biopsy can allow for more conservative surgery. The typical radiologic features include a peritumoral fat-like zone consisting of atrophic muscle, and water-like character (high signal in T2-weighted MRI) in the tumor, gadolinium enhancement in MR in 50% of cases, and often a cyst-like low attenuation in CT scans.[5–6]

Occasional (5–10%) patients have multiple intramuscular myxomas together with (polyostotic) fibrous dysplasia of bone, including lesions in adjacent bones. This constellation, first reported in the German literature in the 1920s, is presently known as *McCune–Albright syndrome* (especially when occurring with cutaneous hyperpigmentation and endocrine abnormalities), or as *Mazabraud syndrome*. The syndrome is not inheritable and might result from a mutation during early fetal development causing mosaicism.[7–11] Other minor radiologic bony abnormalities also can be present in sporadic myxomas, without a well-developed syndrome.[3]

Pathology

Grossly, the tumor appears as a demarcated mass, although along the periphery it often infiltrates between the skeletal muscle fibers. On sectioning it is gelatinous, gray-white, and glistening. Macroscopic cysts are sometimes present, especially in larger tumors (Figure 11.3).

Microscopically, intramuscular myxoma typically infiltrates between edematous skeletal muscle fibers in its periphery (Figure 11.4). The tumor is paucicellular and hypovascular, but contains a number of capillaries with small lumina (Figure 11.5). The inconspicuous fibroblasts and myofibroblasts are embedded in a myxoid, focally collagenous-fibrous matrix containing acid mucopolysaccharides, especially hyaluronic acid,[2] and the stroma can have a vacuolar appearance. The tumor cells contain small nuclei with dense chromatin and elongated but often weakly staining cytoplasm. Variants with a higher cellularity also occur (Figure 11.6); these may have some potential for recurrence, but none for metastasis.[12,13] Mitotic activity is inconspicuous. In some cases, the apparently higher cellularity is contributed by infiltration by numerous histiocytes. The tumor cells are immunohistochemically positive for vimentin and often focally smooth muscle actin and CD34, but are negative for S100 protein and desmin.[3–4]

Differential diagnosis

Myxofibrosarcoma/myxoid MFH generally shows a higher cellularity and more prominent vascularity, along with at least focal nuclear atypia (pleomorphism). Mitotic activity, including atypical mitoses, is generally also evident.

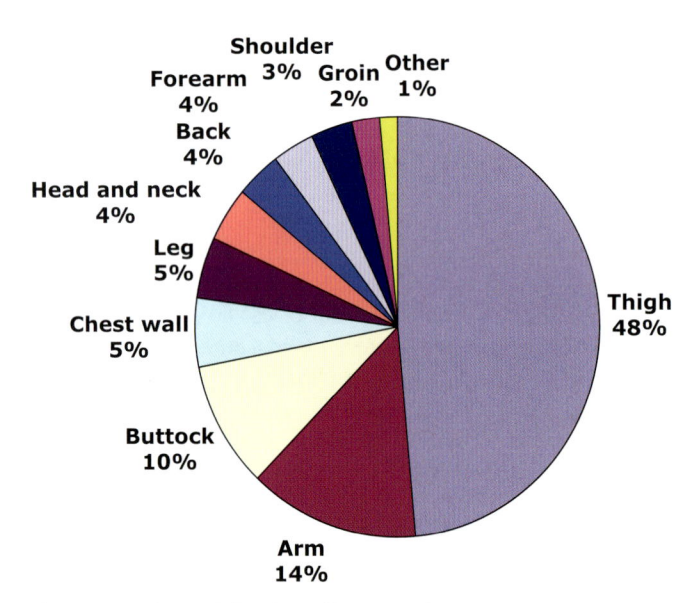

Figure 11.2 Anatomic locations of intramuscular myxoma.

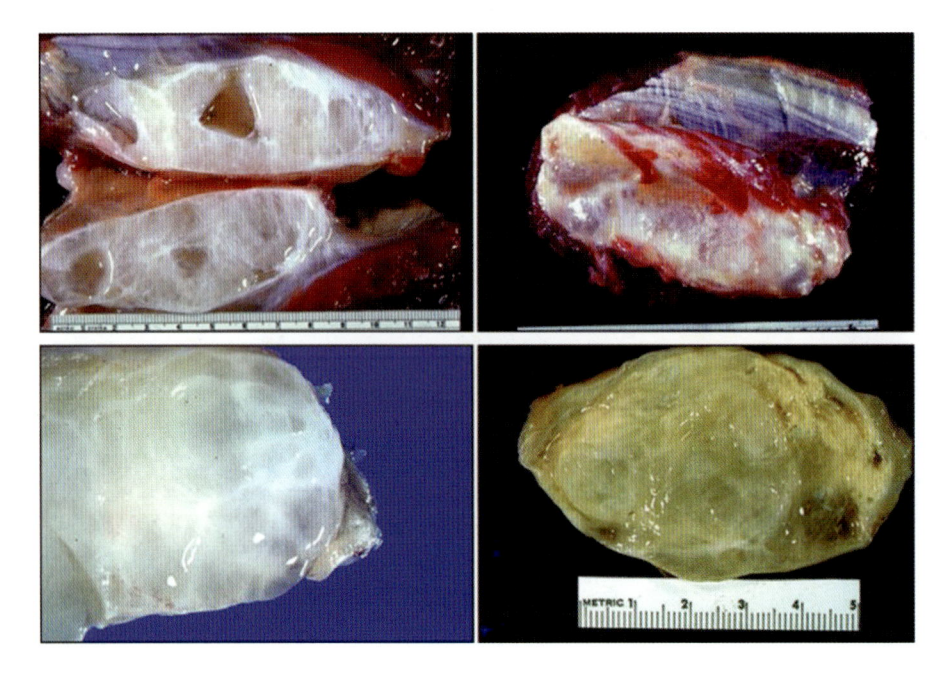

- Thigh 48%
- Arm 14%
- Buttock 10%
- Chest wall 5%
- Leg 5%
- Head and neck 4%
- Back 4%
- Forearm 4%
- Shoulder 3%
- Groin 2%
- Other 1%

Figure 11.3 Gross appearance of four intramuscular myxomas. The lesions show vague lobulation and typically have a translucent, gray-white to pale tan glistening cut surface. Larger tumors often contain mucus-filled cysts.

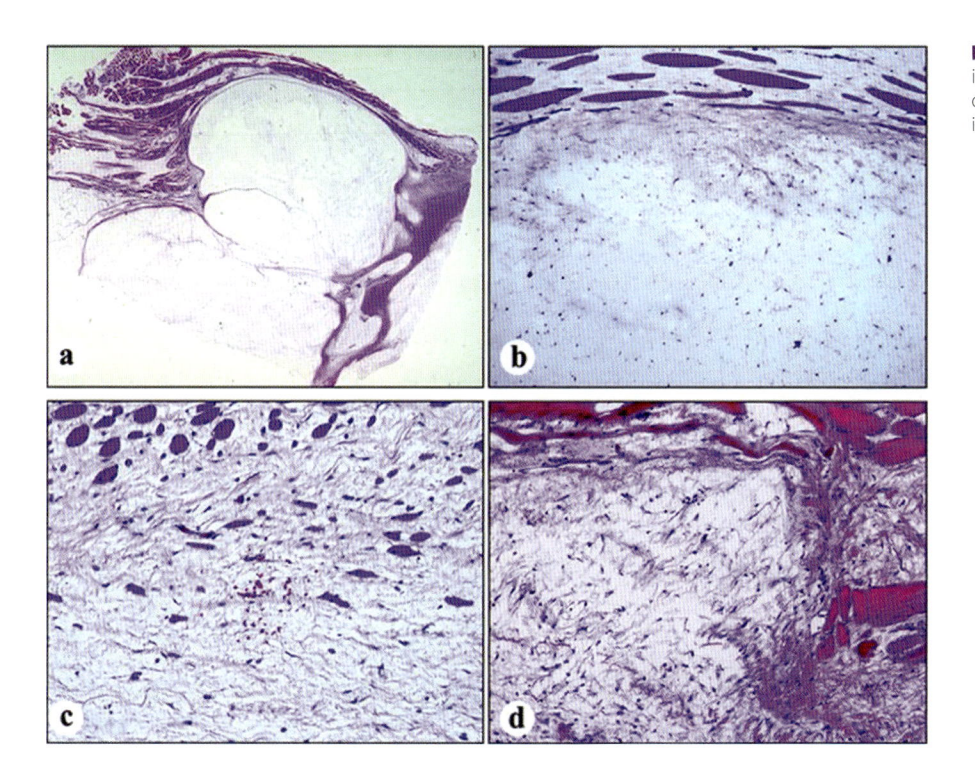

Chapter 11: Myxomas and ossifying fibromyxoid tumor

Figure 11.4 Low-magnification images of intramuscular myxoma show an apparently circumscribed mass, but at the margin, there is infiltration into the skeletal muscle.

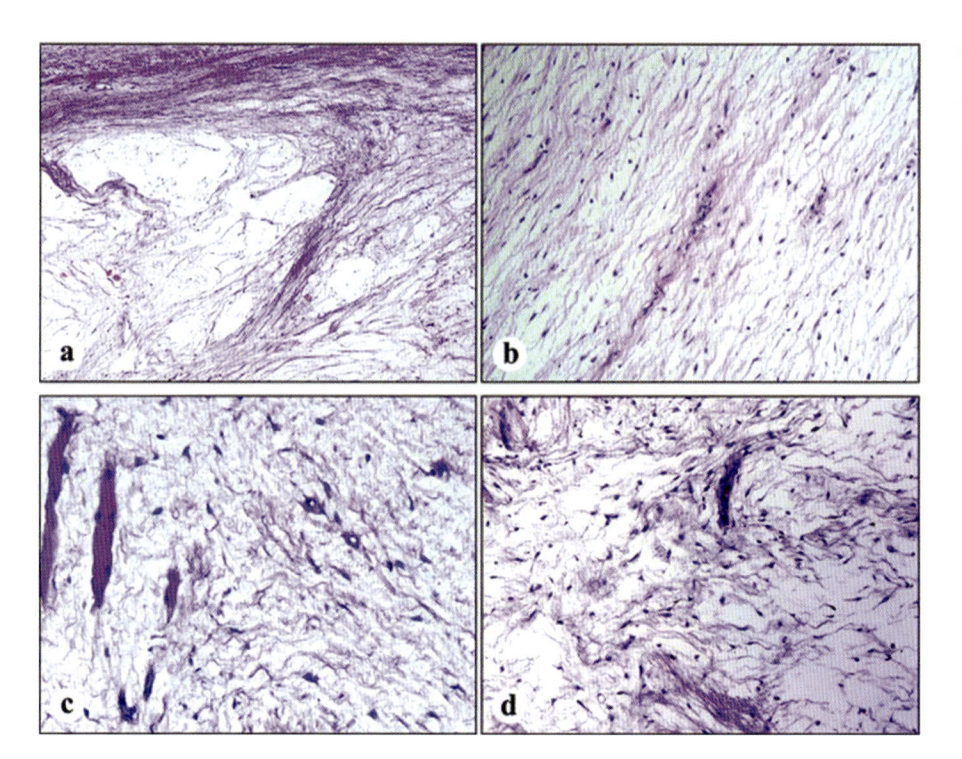

Figure 11.5 High magnification of intramuscular myxoma shows a paucicellular process with a low content of capillaries. (**a**) Microscopic cysts are present. (**b,d**) A minor collagenous component and delicate vessels are present. (**c**) Skeletal muscle infiltration.

Myxoid liposarcoma is also hypervascular, with a network of delicate capillaries surrounded by evenly spaced oval cells in a greater density than the cells in myxoma, with varying numbers of fat cells.

If the tumor is not muscle associated, then the generic designation of *myxoma* would be appropriate. Distinction from juxtaarticular myxoma in the absence of clinicoradiologic correlation can be difficult.

Genetics

Activating somatic mutations in the Gs alpha gene, encoding a signal transduction protein that regulates the cellular cyclic AMP level, have been demonstrated in intramuscular myxomas. In one study, these mutations (R261C or R261H involving exon 8) were seen in all three myxomas with fibrous dysplasia, and in two of three myxomas without fibrous

301

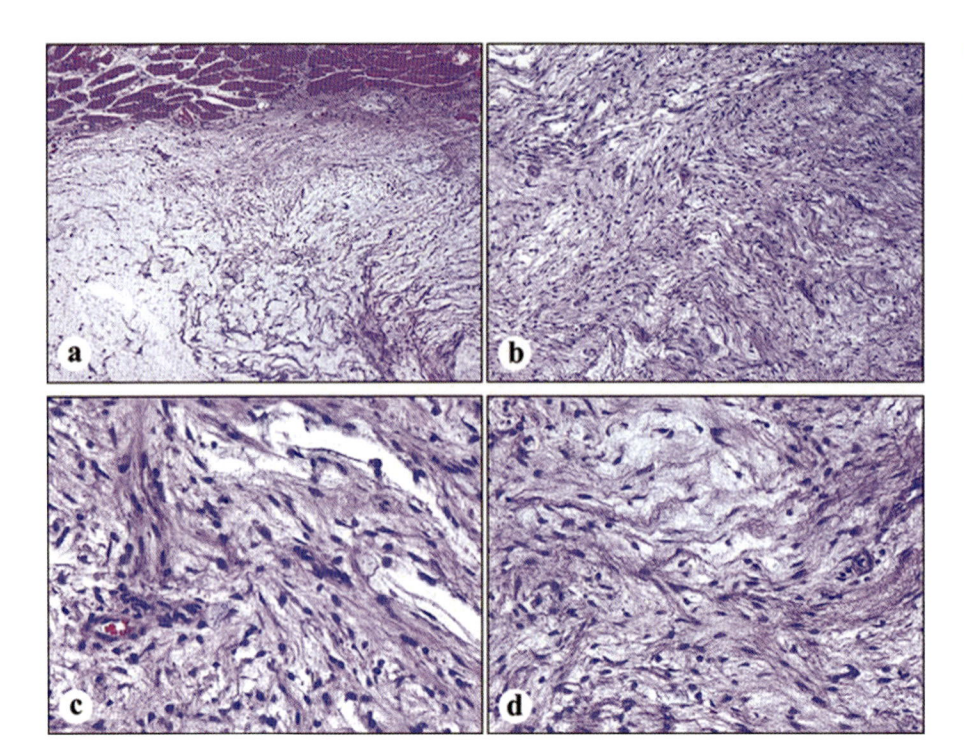

Figure 11.6 Cellular intramuscular myxoma shows denser cellularity, but no nuclear atypia is present.

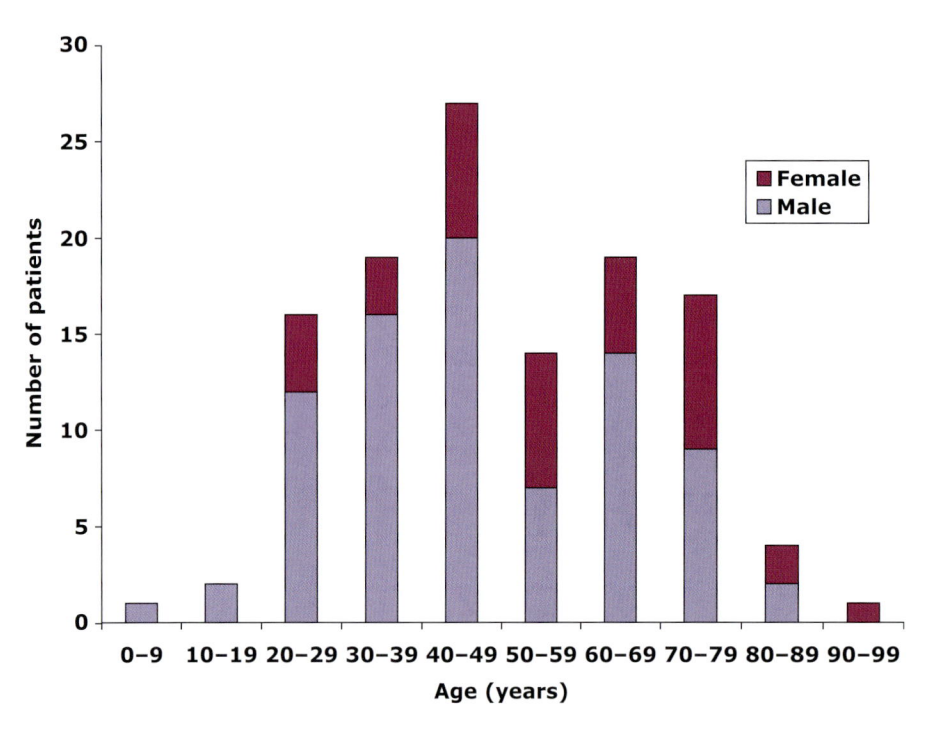

Figure 11.7 Age and sex distribution of 130 juxtaarticular myxomas.

dysplasia.[14] These mutations were similar to those seen in all six cases of fibrous dysplasia of bone, as previously reported in this disease,[15,16] indicating that these diverse pathologic processes could have the same molecular pathogenesis.

Juxtaarticular myxoma

Originally reported by Meis and Enzinger,[17] the nature of this knee- or other joint-associated myxoma is unclear. Although coexistence with internal knee derangements, such as a meniscal tear, and histologic resemblance to a ganglion cyst have been found in some cases, common potential for recurrence may suggest a neoplasm.

Clinical features

This form of myxoma occurs in young to older adults, but 60% of cases are seen between the ages of 20 and 50 years, with a 3:1 male predominance (Figure 11.7). Location is around large joints, mostly on the lateral or medial side of the knee

(80–90%) and involving the subcutis, joint capsule, and rarely, skeletal muscle. Rare examples occur around the shoulder, elbow, hip, or ankle (Figure 11.8). Most lesions are relatively small, <5 cm, with occasional ones being >10 cm.[17,18]

Clinical manifestations include swelling or pain, and some examples are detected incidentally. Recurrence seems to be relatively common and was seen in 34% of cases in the original series, still the largest to date. Most recurrences occurred within 2 years, but some were seen 10 to 15 years after the first surgery.[19] Radiologic studies commonly show joint pathology, such as a meniscal tear, raising the possibility that some examples of this tumor are reactive lesions related to joint pathology and extrusion of mucinous synovial fluid into surrounding soft tissue.[19,20]

Pathology

Grossly, the lesion is gelatinous, gray-white to pale tan. Microscopically it is paucicellular, with fine alternating streaks of collagen and myxoid areas, and commonly has ganglion-like cysts with a collagenous, acellular lining. There is a resemblance to intramuscular myxoma; however, juxtaarticular myxoma can have a more developed vascular pattern and show a higher cellularity than intramuscular myxoma (Figures 11.9 and 11.10). Some examples have focally hypercellular areas containing multinucleated histiocytoid cells.

Genetics

Several chromosomal changes were reported in one case of juxtaarticular myxoma. They involved an inversion in chromosome 2, a translocation t(8;22)(q11–12;q12–13), and trisomy of chromosome 7.[19] Gs alpha mutations typically seen in intramuscular myxoma were not found in the five cases examined in one study.[20]

Ganglion cyst
Clinical features

A *ganglion cyst* is a common, presumably non-neoplastic formation related to tenosynovial, synovial, or perisynovial tissues. These cysts most commonly occur in the dorsal wrist and knee region, but a wide variety of other locations are also possible, including the foot, shoulder, and hip. Some ganglion cysts have

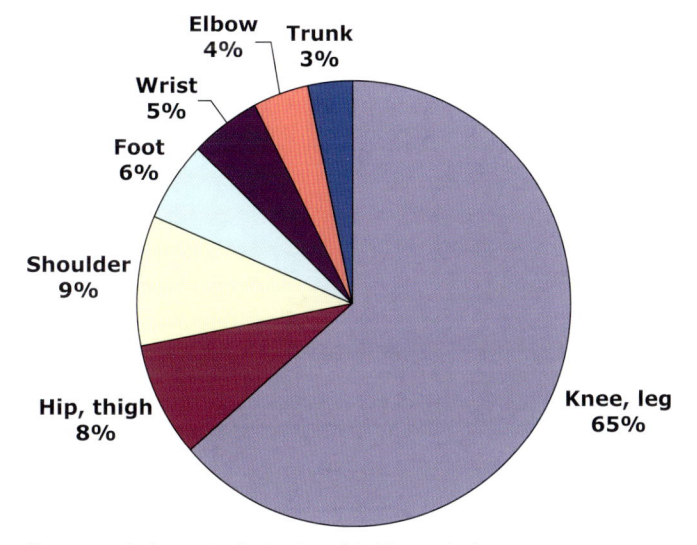

Figure 11.8 Anatomic distribution of 118 juxtaarticular myxomas.

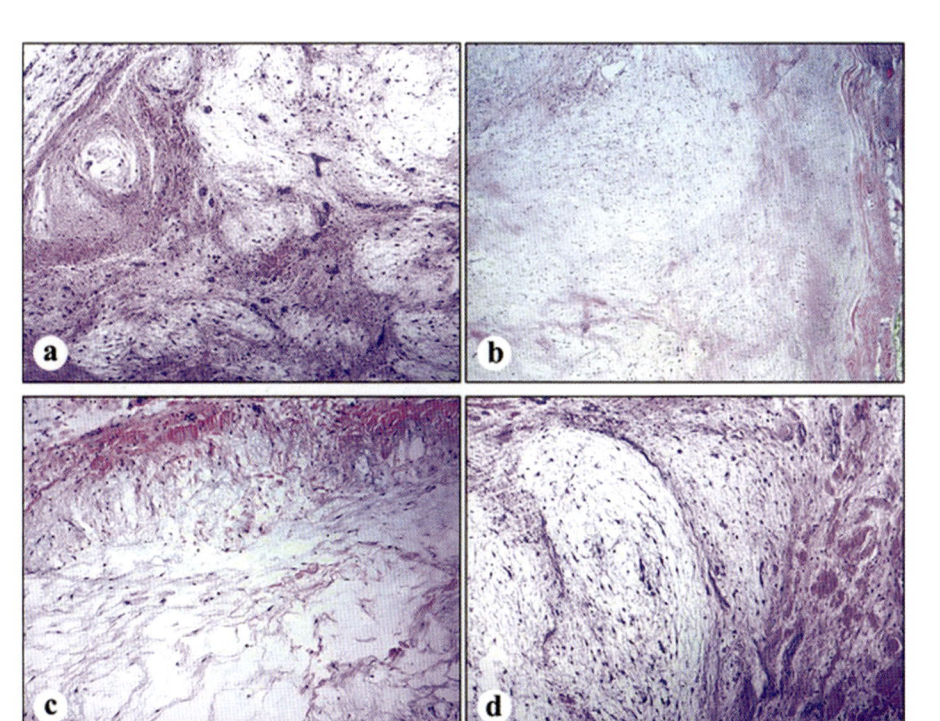

Figure 11.9 At low magnification, juxtaarticular myxoma shows variable nodularity and intermingling with collagenous elements, such as fasciae or ligaments.

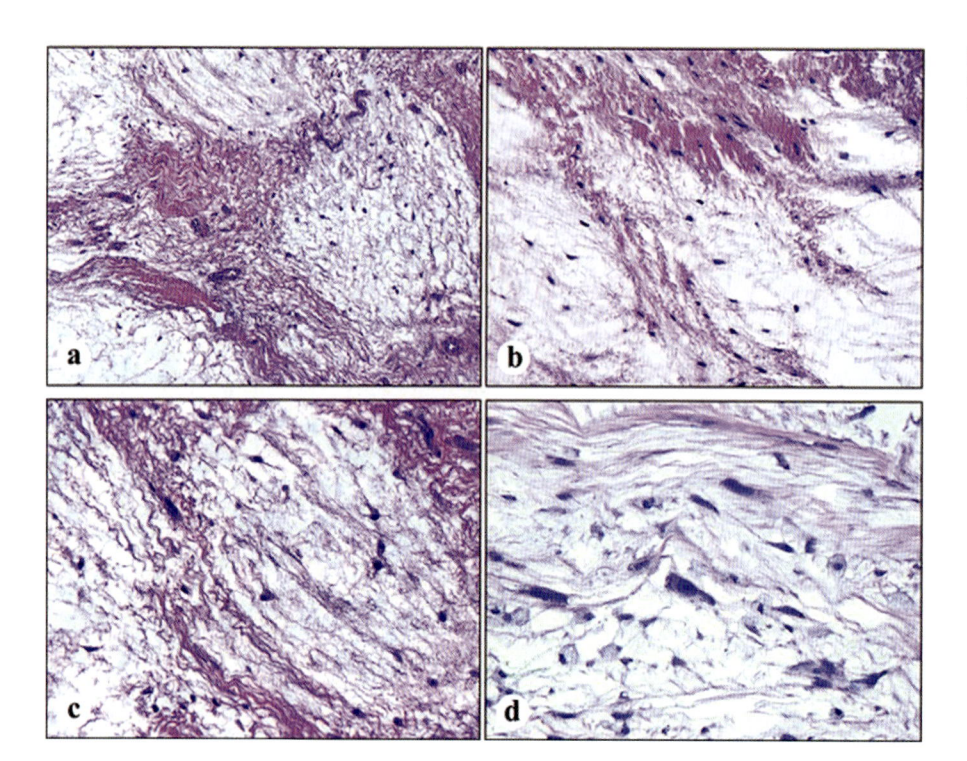

Figure 11.10 Paucicellular appearance, lack of atypia, and alternating myxoid and collagenous areas are typical of juxtaarticular myxoma.

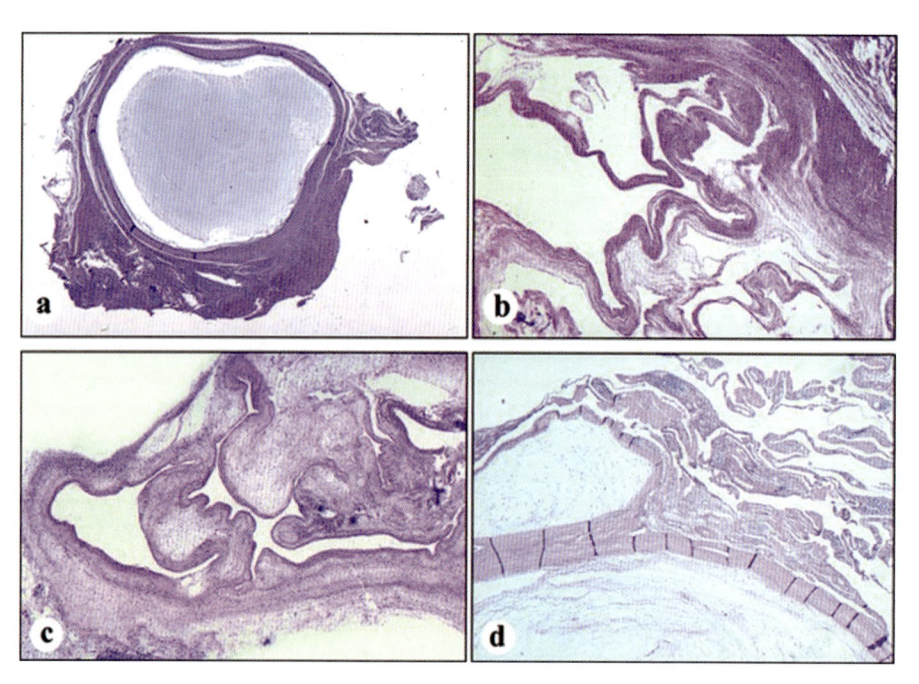

Figure 11.11 Ganglion cysts can be uni- or multilocular, variably containing mucous material. (**d**) Synovial elements (which can be present) are shown adjacent to the cyst.

both soft tissue and osseous components, and others are intraosseous, often in the lateral malleolus of the tibia and the small bones of the wrist, associated with articular damage and degenerative joint disease.[21] Some ganglion cysts involve nerves and have been designated intraneural ganglions or ganglions of a nerve.

Pathology

The pathogenesis of ganglion cysts is not clear, but it could involve cystic degeneration of ligamentous or meniscal tissue.

In some cases, ganglion cysts originate from synovial tissue, and synovial herniations (e.g., Baker's cyst of the knee) can create ganglion-like cysts.

Histologically, a ganglion cyst is lined by a dense collagenous capsule-like lining, and there are no distinct lining cells. The uni- or multilocular cysts can appear empty or filled with mucinous material rich in hyaluronic-acid-containing mucopolysaccharides (Figure 11.11).

Myxoid material accompanying benign fibroblastic proliferation is sometimes present around the cyst, and spillage of

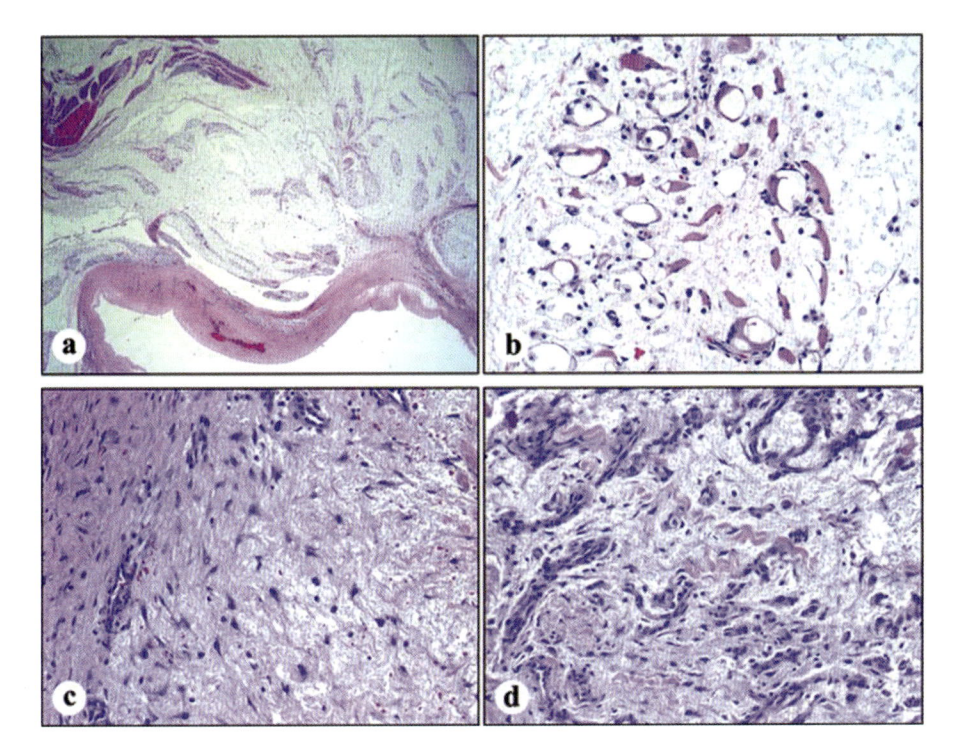

Figure 11.12 (**a**) Ganglion cyst surrounded by mucoid material spreading into soft tissue. (**b**) The mucoid material around the ganglion infiltrates skeletal muscle. (**c,d**) The mucoid material around the ganglion in a vascular background.

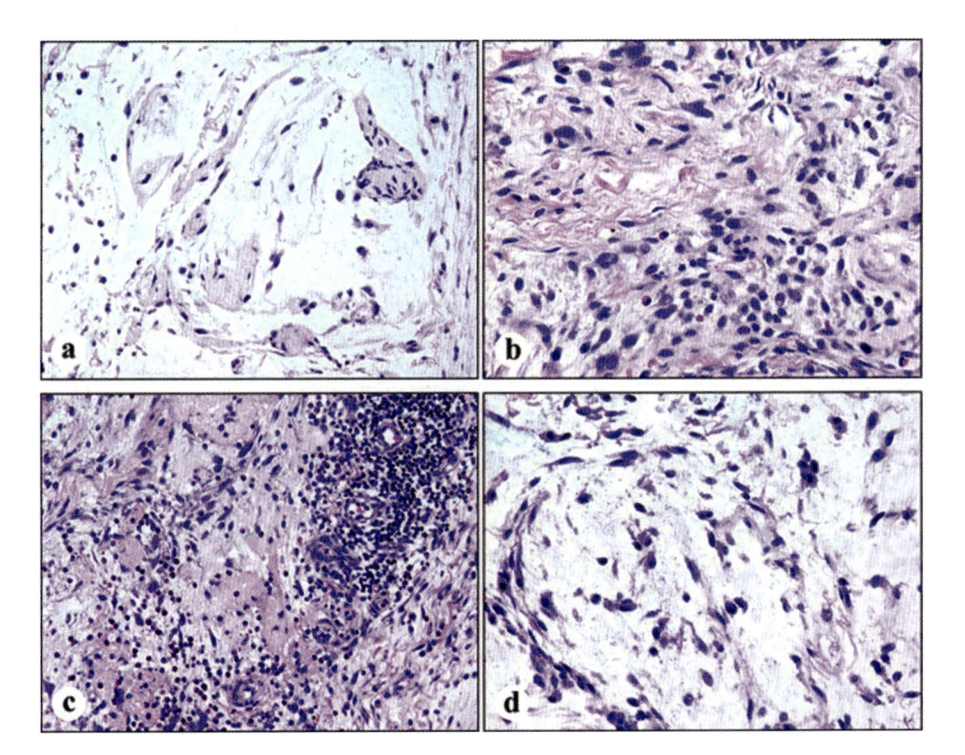

Figure 11.13 Different appearances of a cellular reaction around a ganglion cyst, probably related to extruded mucin. (**a,b**) Fibroblastic reaction. (**c**) Lymphoid and eosinophilic inflammatory reaction. (**d**) Myxoma-like focus around a ganglion cyst.

myxoid substance into the surrounding connective tissue can trigger reactive myofibroblastic or vascular proliferation (Figures 11.12 and 11.13). This can cause differential diagnostic concerns, especially with myxoma variants and even with low-grade myxofibrosarcoma. Differentiating features include lack of nuclear atypia, low mitotic activity, and lack of atypical mitoses.

Cutaneous myxoid cyst (digital mucoid cyst, ganglion of the distal interphalangeal joint)

Originally reported by Johnson *et al.*, this condition occurs in adults of all ages and equally in men and women.[22] Clinically it appears as a small, usually <1 cm cutaneous nodule in the dorsal aspect of any of the fingers, usually around the distal

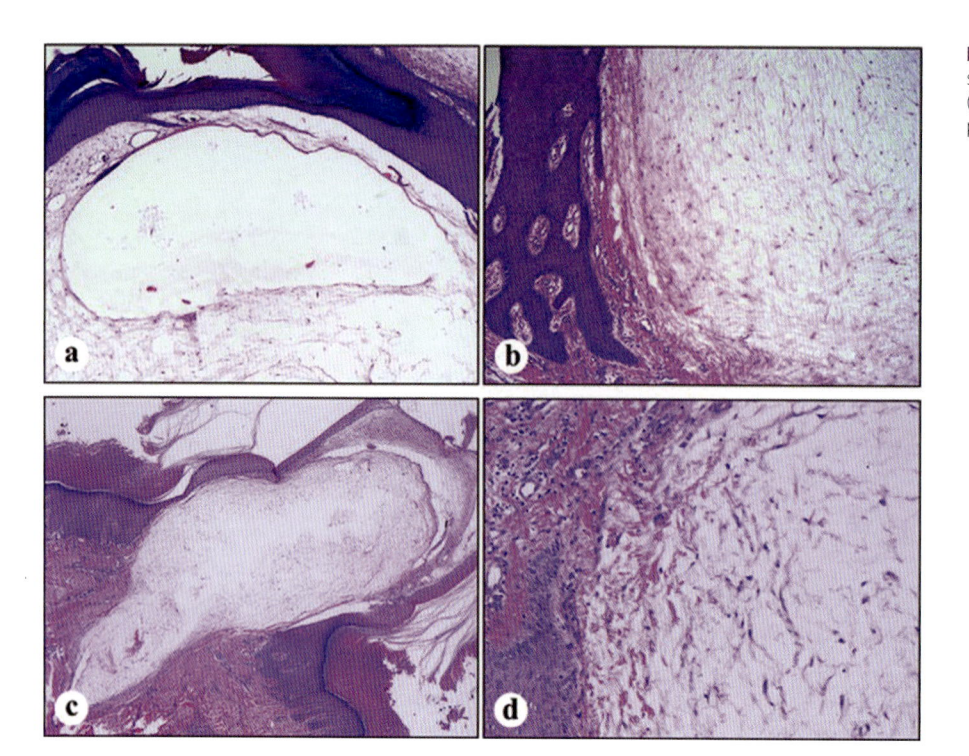

Figure 11.14 A digital mucous cyst forms a cystic space, often surrounded by loose myxoid matrix. (**c**) The lesion perforates to skin surface. (**d**) Note paucicellular appearance.

interphalangeal joint. Occurrence in the dorsal distal toe is also possible. Larger cysts can cause functional impairment.[23] The lesion typically has a translucent content consistent with a cyst, and sometimes communicates with the joint. It can be considered synonymous with a ganglion of an interphalangeal joint. Local excision can result in recurrence, considering the communication of the cyst with the joint. Corticosteroid injection into the cyst has also been effective in some cases.[22] Alternatively, ligation of the communication between the cyst and synovia has been used successfully.[24]

Histologically, there is a cystic cavity containing mucoid fluid, surrounded by a myxoid tissue zone containing scattered fibroblasts (Figure 11.14). This is probably a ganglion-related process that includes reaction to synovial fluid mucin extruded into periarticular soft tissue. It is unknown whether the periungual lesions included in the earlier series represent the same entity.

Superficial angiomyxoma (cutaneous myxoma)

Both *superficial angiomyxoma* (SAM) and *cutaneous myxoma* have been used to describe the same tumor, but the former designation seems to be the prevailing term. Cutaneous myxomas were described by Carney in connection with a syndrome now known as the *Carney complex*. This syndrome includes cutaneous and cardiac myxomas, spotty pigmentation of the skin, and endocrine overactivity (see more detailed discussion in Clinical features and genetics of Carney complex).[25–27] Earlier acronym designations for what appears to be the same syndrome are NAME (nevi, atrial myxoma, myxoid neurofibromata, and ephelides), and LAMB (lentigines, atrial myxoma, mucocutaneous myxoma, and blue nevi).[28–30] A series of similar sporadic tumors without syndrome association were first reported by Allen and subsequently by others as superficial angiomyxomas (SAMs).[31]

Clinical features

Sporadic, nonsyndromic SAMs form most of these tumors, and in fact, syndromic tumors were absent in the largest published series. Sporadic SAMs are almost always solitary and occur in a wide age range from early childhood to old age, with median/mean ages in different series varying between 39 and 46 years, with the peak incidence in the fourth or fifth decade. There is a 1.5:1 male predominance (Figure 11.15). Examples that arise in connection with Carney complex occur in younger patients (median age in the largest series, 17 years). There is no clear sex predilection in the familial cases. The presence of multiple SAMs in a young patient should raise a suspicion of Carney complex, especially to avoid potentially lethal complications of cardiac myxoma.

The most common locations for sporadic SAMs are the trunk, head and neck, and thigh (Figure 11.16). One series reported tumors in female external genitalia and scrotum.[32] Syndrome-associated SAMs can be multiple and have a predilection for head and neck sites such as eyelids and the external ear, and to nipples. Meticulous clinical examination of Carney complex patients often detects very small SAMs of only a few millimeters, especially those located in the facial area.

Clinically, SAM appears as a slowly growing polypoid or dome-shaped cutaneous lesion or a subcutaneous nodule, sometimes of several years' duration. This tumor can clinically

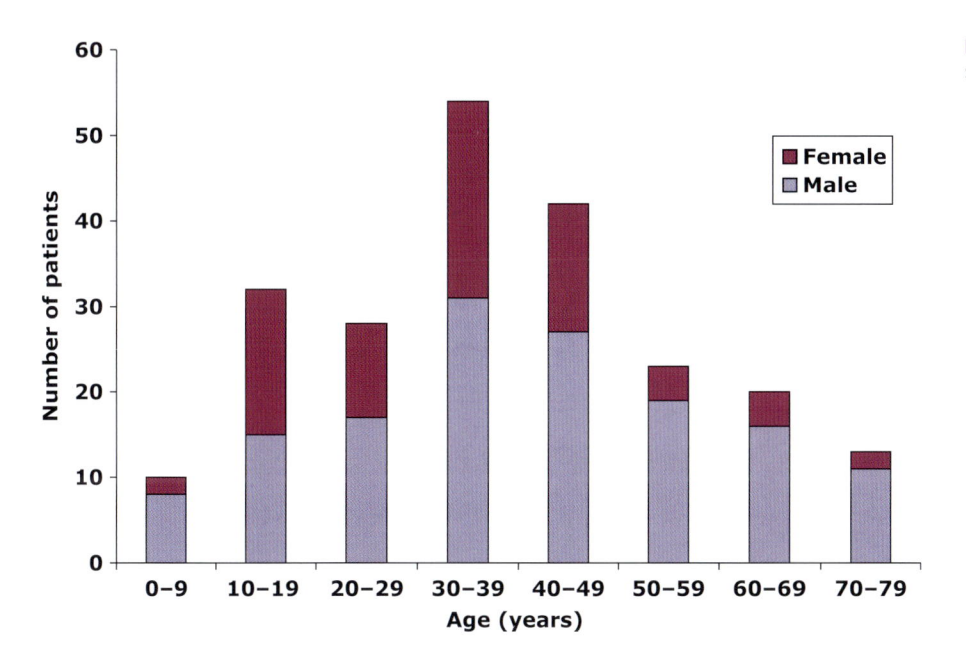

Figure 11.15 Age and sex distribution of superficial angiomyxoma.

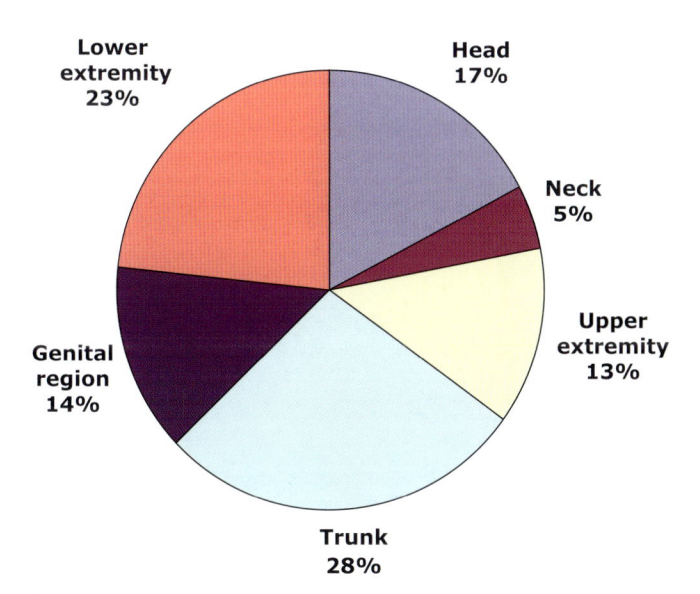

Figure 11.16 Anatomic distribution of superficial angiomyxoma.

resemble a cyst, skin tag, or neurofibroma. There is a significant, 20% to 30% tendency for local nondestructive recurrence for both sporadic and syndrome-associated SAMs.[31–33]

Pathology

The lesion usually measures between 1 cm and 5 cm. Incipient SAMs can be very small, but occasional examples have been >10 cm. Grossly, the tumor is soft, sometimes cystic, and has a gray-white, gelatinous surface on sectioning. Microscopically, the tumor usually involves both the dermis and subcutis (Figure 11.17). Larger lesions are composed of multiple lobules separated by collagenous septa. In many cases, low magnification reveals cleavage spaces and pools of mucin. Cystic skin adnexal or squamous elements, sometimes with benign epithelial proliferation, are entrapped in 20% to 25% of cases (Figure 11.18).

The tumor is paucicellular, with a loose, jelly-like, myxoid background and prominent capillaries surrounded by lymphocytes, plasma cells, and neutrophils, and in some cases, eosinophils and mast cells.

Cytologically, there is a mixture of small inconspicuous cells and larger irregularly shaped, mildly atypical cells surrounding prominent capillaries (Figure 11.19). Mitoses are scant. Variably observed features include cleavage spaces between microlobules, fascicular arrangement of spindled tumor cells, and vascular thrombosis (Figure 11.20). Some cases contain loosely arranged sheets of spindle cells with myofibroblastic features. Mitotic activity is inconspicuous.

Immunohistochemically, the tumor cells are variably, usually focally positive for muscle actins, CD34, and occasionally and focally positive for S100 protein; cells are negative for AE1/AE3 keratin cocktail, desmin, and estrogen receptor.[32]

Differential diagnosis

Designations that have partial overlap with SAM include cutaneous focal mucinosis, trichogenic myxoma, and trichodiscoma. *Cutaneous focal mucinosis* is a focal dermal collection of mucinous material in the context of low vascularity. *Trichodiscoma* contains proliferating hair shaft epithelia surrounded by poorly vascularized myxoid tissue. *A digital mucous cyst* is a small, cystic myxoid lesion in the fingers and toes related to a ganglion cyst. It has limited cellularity and lacks the prominent vascular component seen in SAM. *Superficial acral fibromyxoma* is a distinctive lesion often seen in the nail bed area. This tumor is less myxoid, lacking mucinous pools, and is usually more cellular than is SAM.

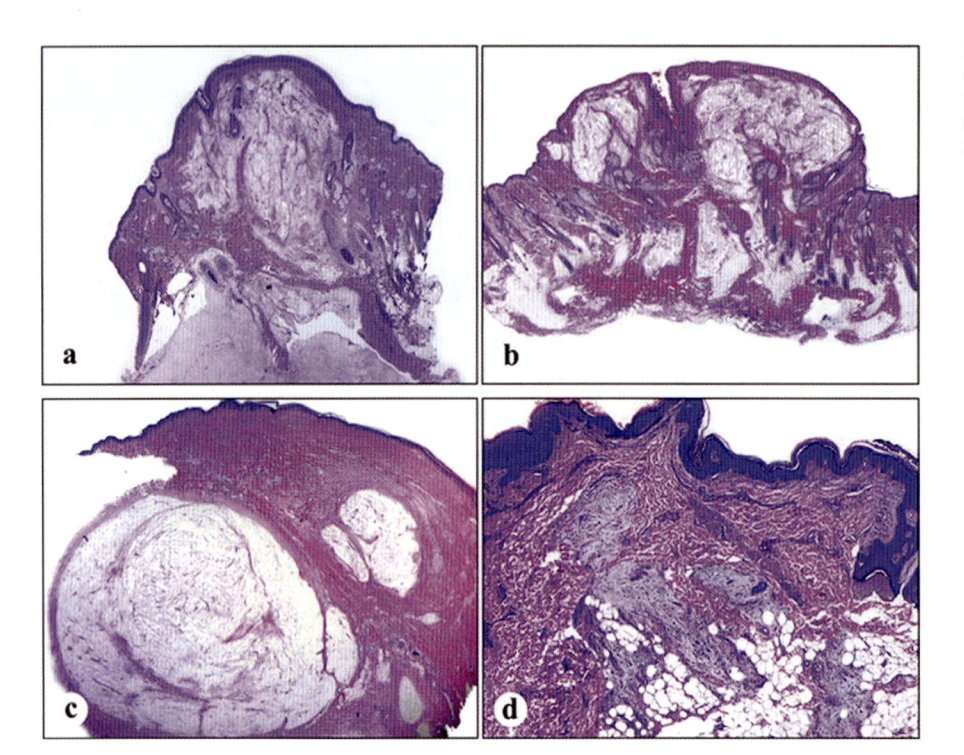

Chapter 11: Myxomas and ossifying fibromyxoid tumor

Figure 11.17 A superficial angiomyxoma can form a polypoid or a dome-shaped lesion. It contains multiple, variably coalescent mucoid nodules that can involve both the dermis and subcutis.

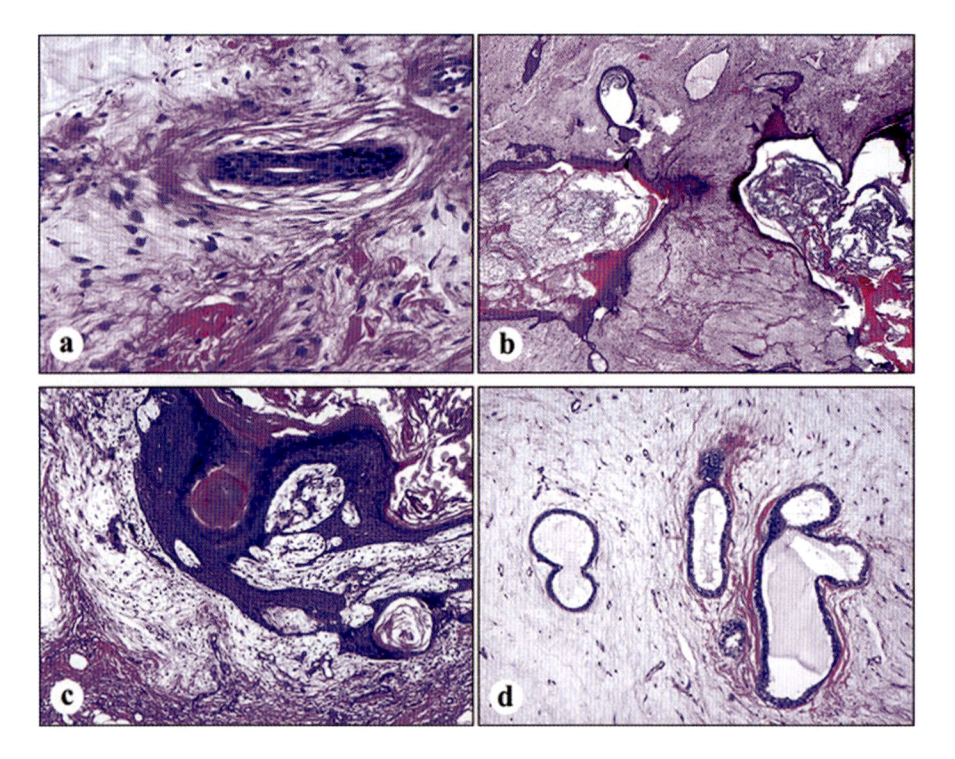

Figure 11.18 (**a–c**) Entrapped adnexa are often seen within the lesions of superficial angiomyxoma. (**d**) Angiomyxoma infiltrating around breast ducts. This superficial angiomyxoma is composed of coalescent myxoid lobules. Entrapped skin adnexa are present in this lesion (**a**). Higher magnification shows focal nuclear atypia and perivascular lymphocytes (**c**).

Extracellular escape of epithelial mucin in mucocele, usually in the minor salivary glands, especially in the oral region and lip, can sometimes resemble SAM histologically. Vascularity and mucous pools are often prominent, but a spindle cell component is inconspicuous. Instead, there are inflammatory cells, especially neutrophils and histiocytes. The presence of dilated, mucin-filled epithelial ducts is diagnostic of a mucocele (Figure 11.21).

Clinical features and genetics of Carney complex

Carney complex is inherited as an autosomal dominant trait in 70% of cases, the rest appearing sporadically, without a familial history.

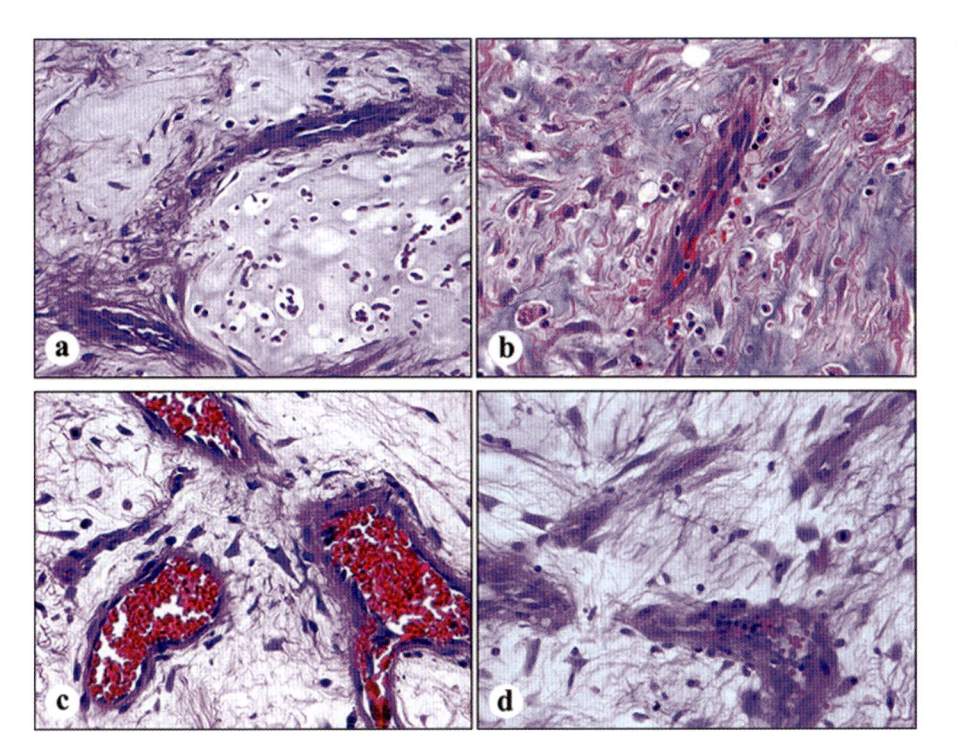

Chapter 11: Myxomas and ossifying fibromyxoid tumor

Figure 11.19 High vascularity is a typical feature. (**b**) Some vessels are surrounded by lymphoplasmacytic infiltration, and neutrophilic infiltration is also present. Note the relatively pale staining nuclei with small nucleoli.

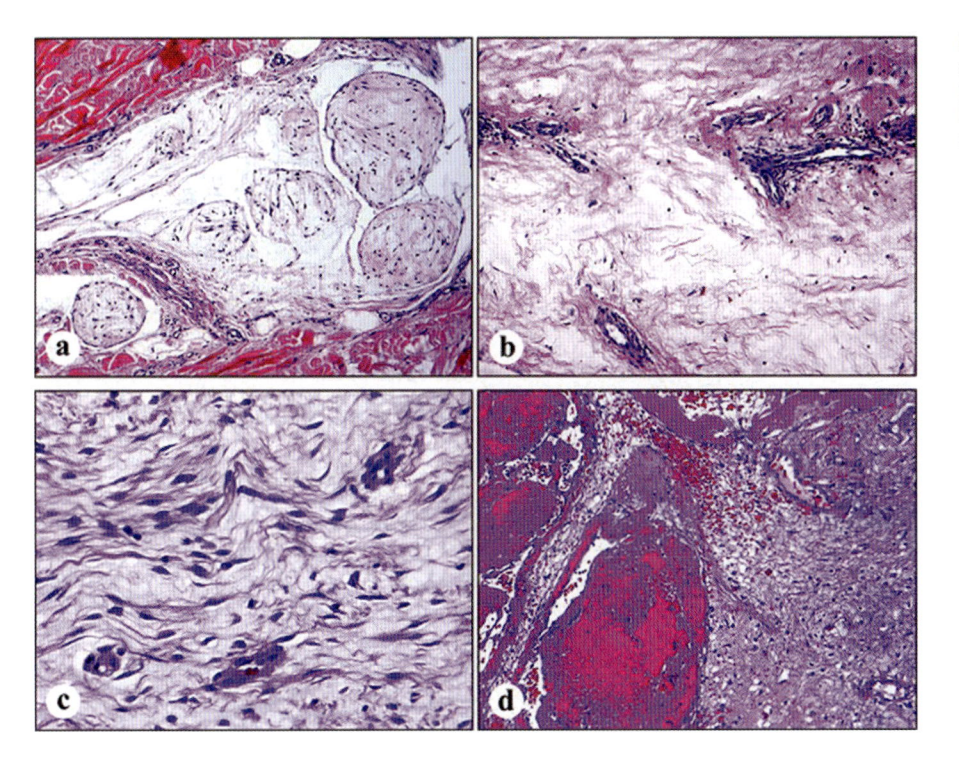

Figure 11.20 Common features in superficial angiomyxoma. (**a**) Cleft-like spaces within tumor lobules. (**b**) Perivascular lymphoid infiltration. Focal fascicular growth. Vascular thrombosis in dilated lesional vessels.

In addition to SAM (often in the head), the cutaneous manifestations include spotty pigmentation, especially periorally, around the lateral canthi of the eyes, and in the external genitalia. The typical endocrine manifestations include testicular calcifying Sertoli cell tumor (very common), and non-ACTH-dependent Cushing's syndrome with primary pigmented nodular adrenocortical disease (30%). Some patients have a pituitary adenoma with acromegaly or psammomatous melanotic schwannoma.[27] The potential occurrence of cardiac myxoma is significant for the possibility of serious complications, such as tumor embolization and sudden death.

Involvement of the PRKAR1A locus at 17p22–24 has been detected in members of Carney complex families and in some nonfamilial cases. PRKAR1A encodes the type 1

309

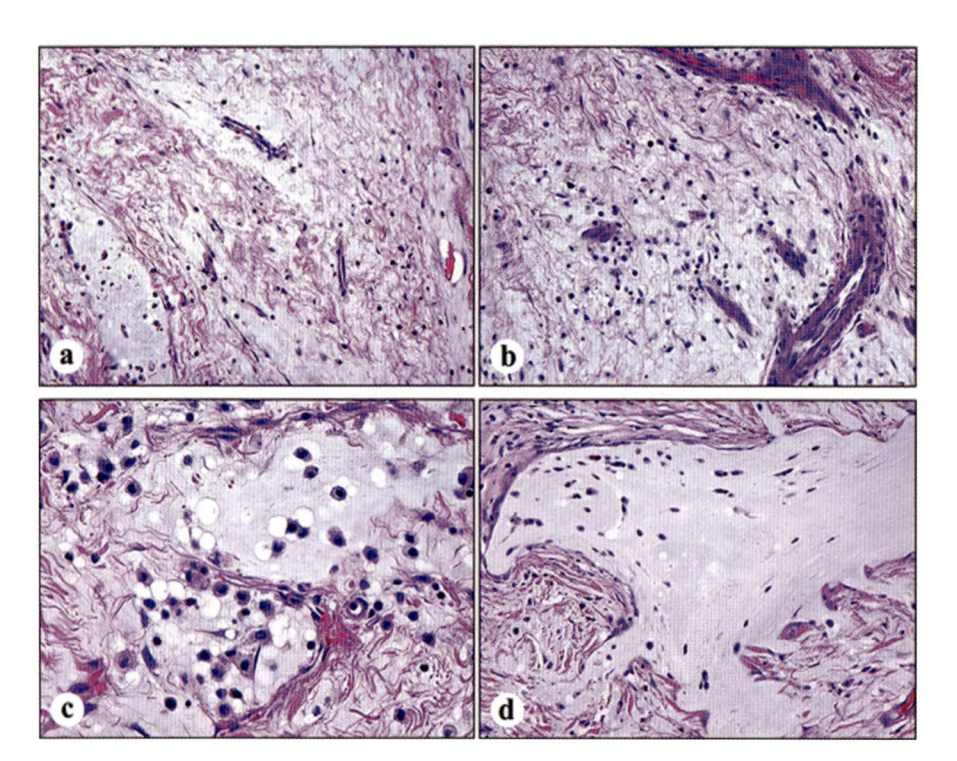

Figure 11.21 Extraductal escape of mucin in a mucocele can resemble superficial angiomyxoma in its vascular prominence and abundant mucoid matrix. The spindle cell component is very scant, however, and inflammatory cells, especially neutrophils and histiocytes, are present. This example is from a thyroglossal duct cyst, a rare site for mucocele. (**d**) Presence of a dilated, mucin-filled epithelial duct is diagnostic for a mucocele.

regulatory subunit of protein kinase A (PKA), a putative tumor-suppressor gene. Constitutional frameshift and point mutations resulting in stop codons have been identified in many patients, resulting in lower levels (haploinsufficiency) of this protein, and LOH has also been detected, suggesting a tumor-suppressor gene-like biallelic inactivation.[34–36] In a mouse model, however, experimental haploinsufficiency of this protein did not produce disease, indicating that additional genetic changes might be required.[37] There are no studies to document somatic PRKAR1A inactivation in non-syndromic SAM. In some involved familes, linkage to 2p16 chromosomal region (CNC2 locus) has been detected, but no specific gene involvement has yet been defined in this region.[36]

Focal cutaneous mucinosis

This entity, described by Johnson and Helwig, refers to focal mucinous expansion of the dermis without features typical of other myxomas. Because its description preceded that of cutaneous myxoma/SAM, early reports might have included some cases currently diagnosed as myxomas. Few reports exist on this entity, limiting the amount of definitive information.[38,39]

Clinical features

Focal cutaneous mucinosis forms a solitary, flesh-colored white papule or small cutaneous nodule usually <1 cm. It has a predilection for young adults, with male predominance, but also occurs in children. The anatomic location is broad. Simple excision seems to be curative.

Pathology

There is a dome-shaped or shallow nodular cutaneous elevation by a myxoid lesion that variably penetrates the dermis. The epidermis is typically unremarkable. The lesion is paucicellular, containing scattered oval to spindled fibroblasts embedded in basophilic mucin that contains dispersed, randomly oriented collagen fibers (Figure 11.22). Its paucicellular nature, lack of a vascular component, and absence of multinodularity separate focal mucinosis from SAM.

The lesional cells have been found to be variably positive for FXIIIa, but they are negative for desmin, smooth muscle actin, and S100 protein. Scattered CD34-positive cells are seen within the lesions.[39]

Superficial acral fibromyxoma
Clinical features

This designation, that of *superficial acral fibromyxoma* (SAF), refers to a relatively uncommon benign fibromyxoid neoplasm that typically occurs in the distal extremities. This tumor was originally reported by Fetsch *et al.*[40] The tumor often involves the nail bed region, but other segments of the fingers can also be involved, and occasional lesions have occurred in the palms and heels. The tumor occurs in a wide age range, with the median approximately 40 years, and has a significant male predominance (2:1). Most tumors are relatively small (<1–2 cm), and a long history of local lesions is common. Benign behavior has been documented, with a relatively low (10–20%) rate of recurrences.[40–42] No instances of malignant behavior have been documented. Periungual myxoid nodules designated as "cellular digital fibromas"[43] in actuality probably represent this entity.

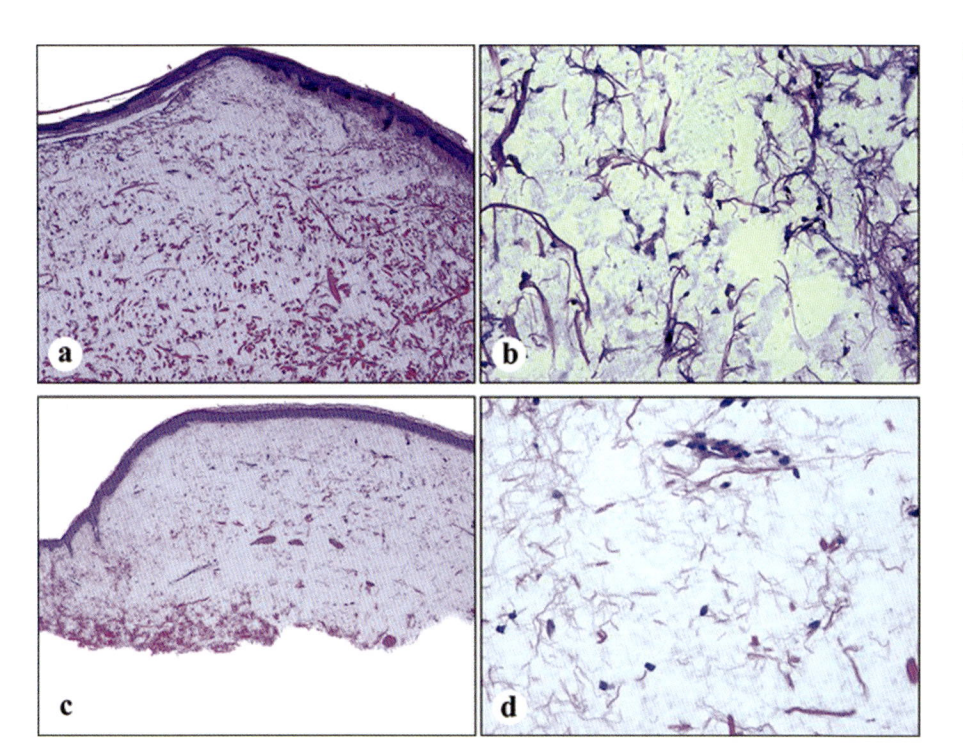

Chapter 11: Myxomas and ossifying fibromyxoid tumor

Figure 11.22 Two examples of focal cutaneous mucinosis. The lesion forms a dome-shaped or shallow dermal elevation, permeates through the dermis, and is paucicellular and hypovascular, containing scattered fibroblasts intermingling with collagen fibers in a myxoid matrix.

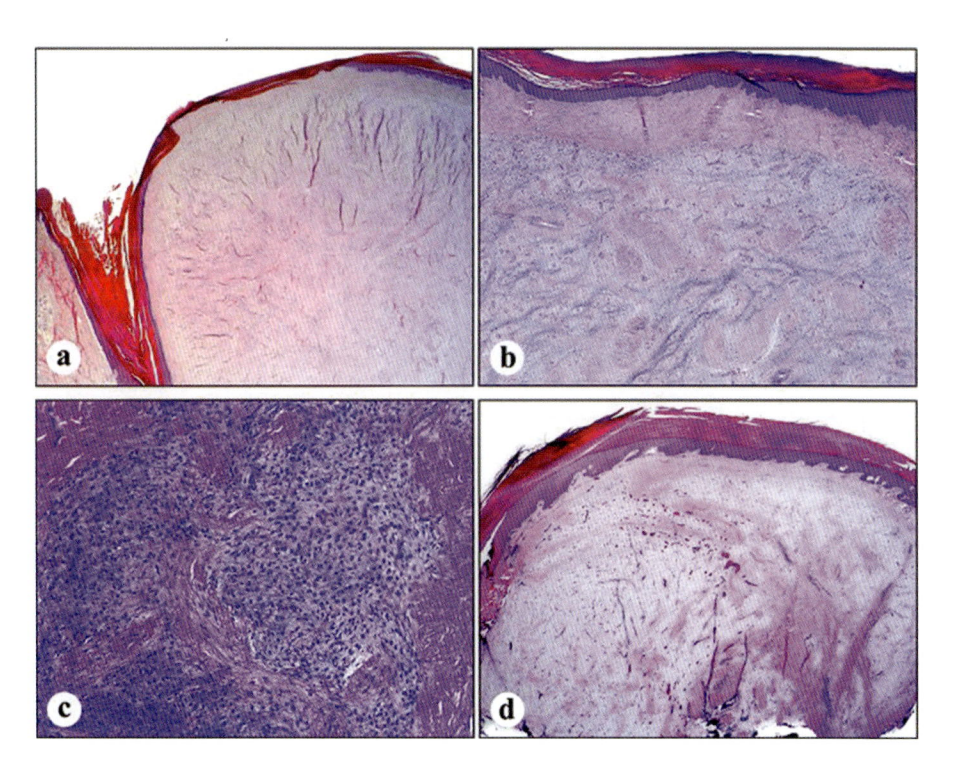

Figure 11.23 A superficial acral fibromyxoma often forms a polypoid nodule and contains microscopic lobules of tumor that spread among dense collagenous elements.

Pathology

Grossly, SAF is soft to firm and on sectioning has a gray-white, sometimes slightly gelatinous surface. The lesion often forms a protuberant mass, often near the nail bed area (Figure 11.23). The overlying nail bed epithelium often has a papillary configuration, and the tumor extends to the epithelial border (Figures 11.24 and 11.25). It is composed of hypervascular, myxoid spindle cell proliferation, often with alternating fibrous areas. Scattered mast cells are often present. The myxoid nature of the matrix varies from focal to extensive

311

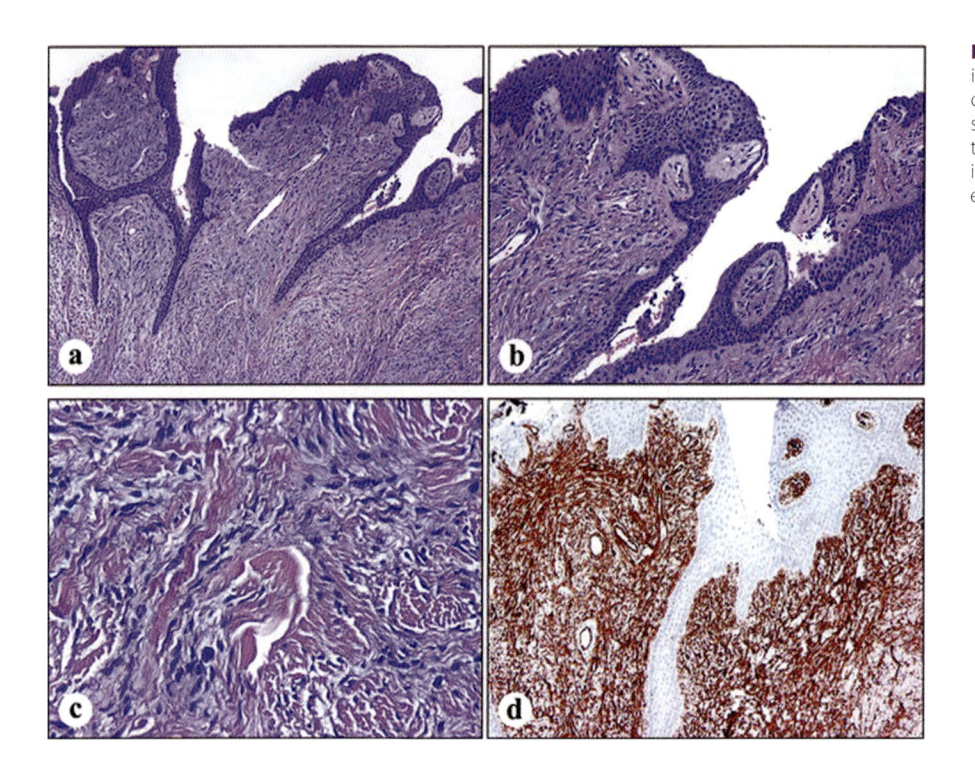

Chapter 11: Myxomas and ossifying fibromyxoid tumor

Figure 11.24 A superficial acral fibromyxoma involving the nail bed often has a papillary configuration. (**c**) High magnification shows bland spindle cells and occasional mast cells. (**d**) The tumor is strongly positive for CD34, and this immunostain shows tumor extension to the epithelial border.

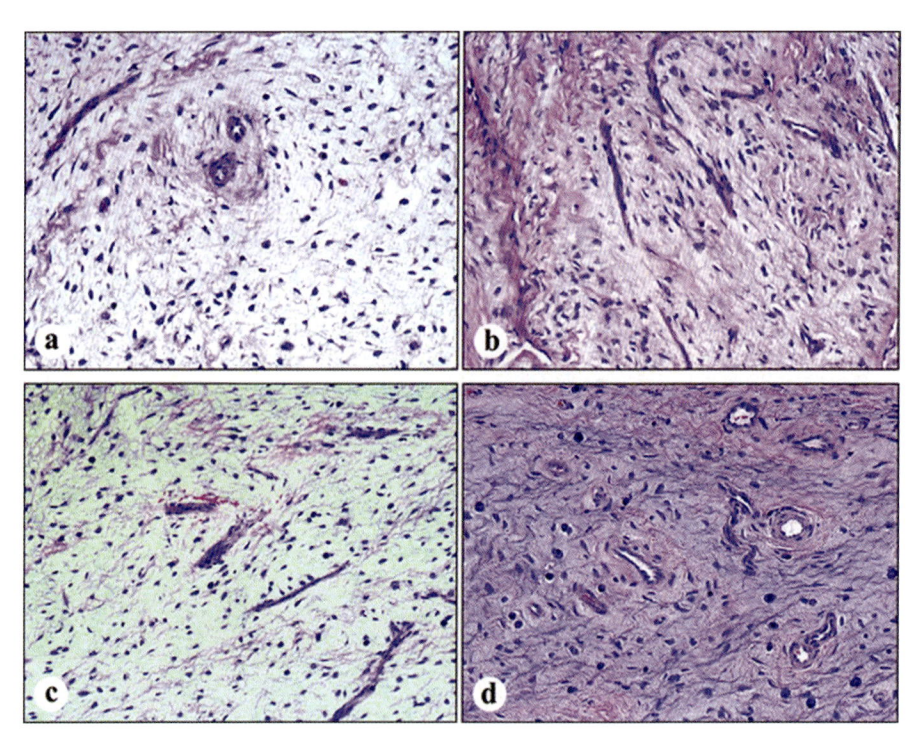

Figure 11.25 Higher magnification of a superficial acral fibromyxoma shows bland spindle cells and prominent capillaries. (**d**) The presence of numerous mast cells is a common finding.

(Figure 11.25). Nuclear atypia and mitotic activity are exceptional, but mildly atypical variants have been identified as part of the spectrum.

The tumor cells are usually (>90% of cases) positive for CD34 (Figure 11.24d) and often for EMA. They are negative for actins, desmin, keratins, and S100 protein.[40] There are no data on genetics.

Differential diagnosis

CD34 positivity should not lead to confusion with dermatofibrosarcoma protuberans. SAF is composed of spindle cells without any distinct tendency to a storiform pattern of subcutaneous fat infiltration. Cellular digital fibroma is probably a form of SAF, whereas conventional digital fibromas may be

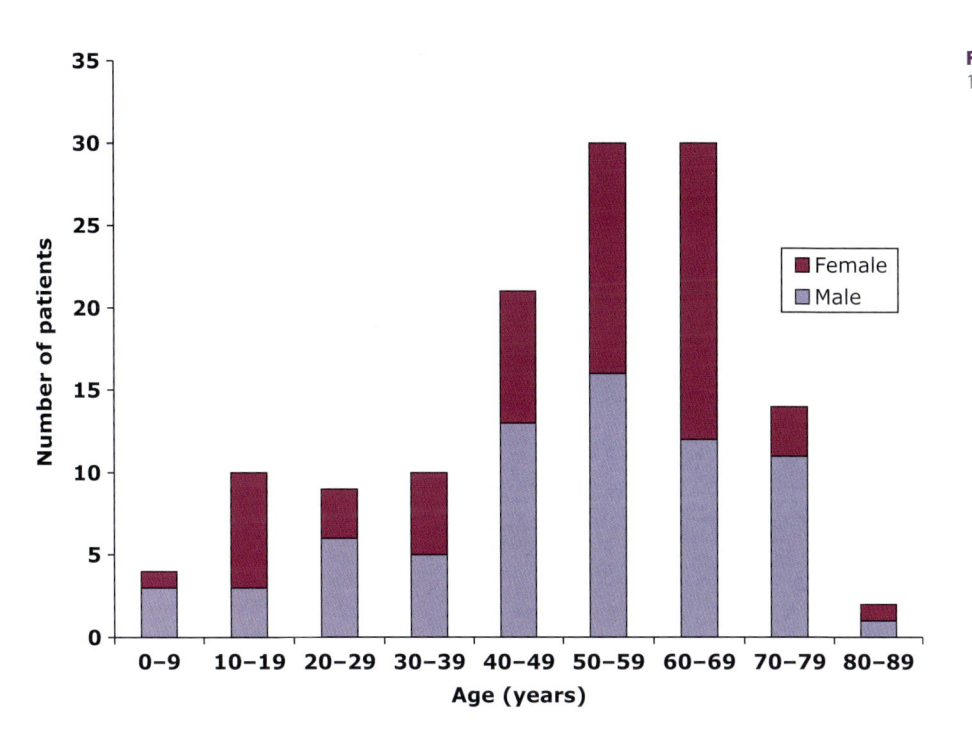

Figure 11.26 Age and sex distribution of 130 patients with cardiac myxoma.

different, as they are reported to be CD34 negative.[43] SAF may be related to onychomatricoma, a small subungual, CD34-positive mesenchymal proliferation.[44]

Cardiac myxoma

Cardiac myxoma, the most common tumor of the heart, is now considered a neoplasm, although in the past it was often considered a peculiar form of organized intracardiac thrombus.[45–50]

Clinical features

Cardiac myxoma is a relatively rare, unique, site-specific myxoma. It is the most common cardiac tumor, constituting >50% of all heart tumors. Cardiac myxoma occurs in all ages, from early childhood to old age, but it is most common in middle age and has an overall 2:1 female predominance (Figure 11.26). Most cases are sporadic, but 7% occur on a familial basis with Carney complex (i.e., endocrine overactivity, cutaneous pigmentation, SAMs, and PRKAR1A mutations; see earlier discussion of Carney complex). Syndrome-associated myxomas present at a younger age, even in childhood, are located at unusual places in the heart and can be multiple.[46,51,52]

The most common clinical presentation of left atrial myxoma is mitral valve obstruction. Such myxomas can also send tumor emboli into systemic circulation, leading into stroke, myocardial infarction, or other peripheral embolic complications. Right atrial myxoma can cause tricuspid valvular obstruction and embolize into lungs. Myxoma often simulates other heart conditions, especially mitral valve disease and bacterial endocarditis, but the latter can also complicate a myxoma.[46]

The etiology of cardiac myxoma is unknown. In one report, myxomas were found to be immunohistochemically positive for herpes simplex antigen in 12 of 17 cases, and 8 of 12 cases contained herpes simplex DNA, in contrast to control endocardial tissue. The pathogenetic association of this finding is not known.[53]

Cardiac myxoma can be most easily detected by echocardiography. Cardiac catheterization also has been used, although tumor manipulation carries a risk of myxoma embolization.[54] Prolapse of left atrial myxoma into the left ventricle is common. Prior to modern studies, myxoma was sometimes diagnosed based on position-dependent cardiac murmur. Some myxomas cause constitutional symptoms, such as fever, weight loss, and polyclonal marrow plasmacytosis; these symptoms might be caused by interleukin-6 secretion by the myxoma.[55]

Cardiac myxoma typically occurs in the left atrium (70%), and less commonly in the right atrium (20%) or in the ventricles (<5%). Rare myxomas arise from the mitral valve leaflets. Atrial myxomas usually originate at the foramen ovale, where they are often attached by a narrow pedicle,[45–48] Surgical excision is via open-heart or endoscopic surgery, and recurrances are rare except in familial cases, in which multifocal origin may be the cause for an apparent recurrence.[56] Brain metastases of cardiac myxoma have been reported,[57] but it is uncertain whether such a tumor was fully comparable with other myxomas.

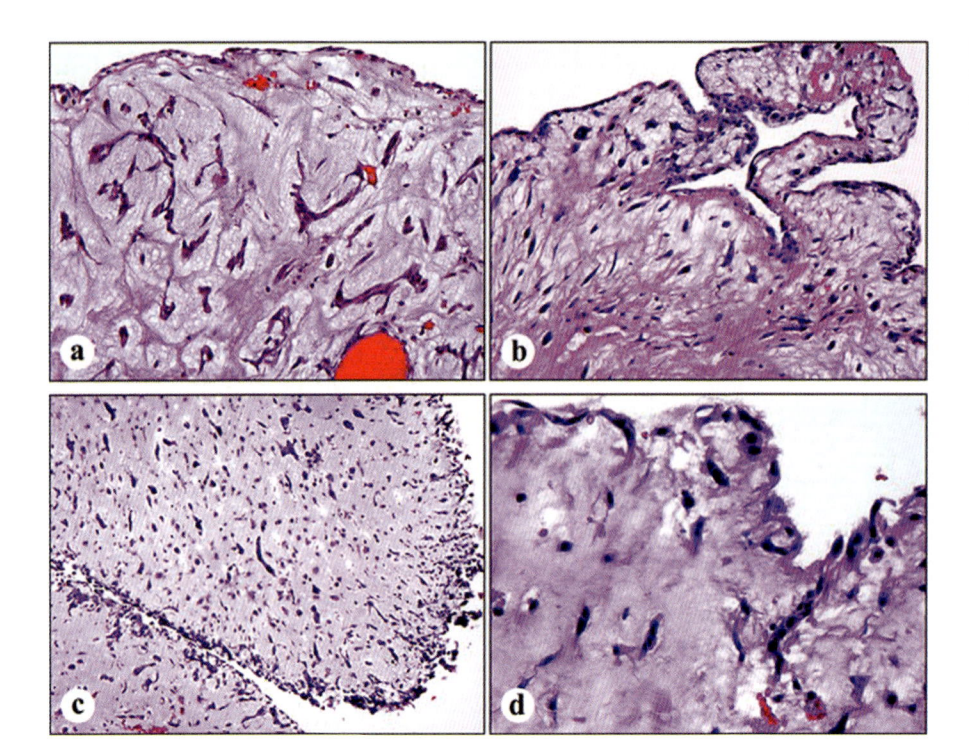

Figure 11.27 (**a**) Smooth luminal surface in cardiac myxoma. (**b–d**) Papillary formations on the surface of cardiac myxoma. These can be associated with risk for embolization.

Pathology

Grossly, cardiac myxomas form soft, gelatinous masses varying in size from a minimal and incidental (autopsy) finding to 10 cm in diameter (average, 5 cm). The myxoma is often attached to the atrial wall with a narrow, variably long pedicle that is recognizable when elements of endocardium or atrial myocardium are included in the specimen. The contour of the tumor can be smooth, or irregular and papilliform. The latter tumors have been associated with a greater risk of embolization.[45] The tumor surface contains a thrombus in nearly one-half of all cases, and this might have been a reason for the now-historical belief that cardiac myxoma is a form of thrombus.

Histologically, cardiac myxoma is relatively paucicellular, showing myxoma cells in an abundant mucoid stroma in various formations. These include myxoma cells surrounding tumor leaflets in a manner resembling the synovial lining, myxoma cells around capillaries as concentric ring-like clusters, and complex trabecular formations by myxoma cells (Figures 11.27 to 11.29) Myxoma cells do not form blood vessels, although they often associate with vessels, forming perivascular sheath-like structures. Mitoses are uncommon, but occasional regular and even atypical mitoses are seen in more cellular areas (Figure 11.28d). Pronounced atypia and mitotic activity should lead to a diagnosis of myxofibrosarcoma, however. These tumors are rare, and their relationship with cardiac myxoma is uncertain.

Degenerative and reactive changes, such as fibrosis, hemosiderin deposition, and clusters of plasma cells and lymphocytes are common (Figure 11.30). Calcification, metaplastic bone, and extramedullary hematopoiesis are uncommon findings. Calcification is more common in right atrial myxomas.

In rare examples (3%), glandular epithelial differentiation is present in addition to the usual myxoma elements. The epithelium shows mucinous differentiation, which can raise a differential diagnostic possibility of metastatic carcinoma (Figure 11.31). The presence of glandular structures in the context of myxoma strongly supports a primary tumor. The glandular elements have been thought to arise from the foregut rests occasionally encountered in myxoma.[58–60] In rare cases, thymic rests are encountered, and these have been offered as an explanation for the occurrence of thymoma within a myxoma.[61]

Immunohistochemically, the myxoma cells are positive for vimentin, CD34, CD31, and usually for calretinin and thrombomodulin (CD141). They have been reported as variably positive for desmin, smooth muscle actin, and S100 protein, and are negative for keratins and KIT.[59,62,63] The epithelial elements are positive for keratins 7 and 20, EMA (MUC1), and CEA.[58–60] Focal neuroendocrine differentiation has also been detected in these elements.[64] The MUC family proteins MUC1, MUC2, and MUC5A have been found in cardiac myxoma.[65]

Ultrastructural studies have shown prominent intermediate filaments (vimentin) but no cell-type-specific features.[66,67]

Genetics

Recurrent chromosomal changes have been reported, including rearrangements of 12p1 and 17p1, in addition to other,

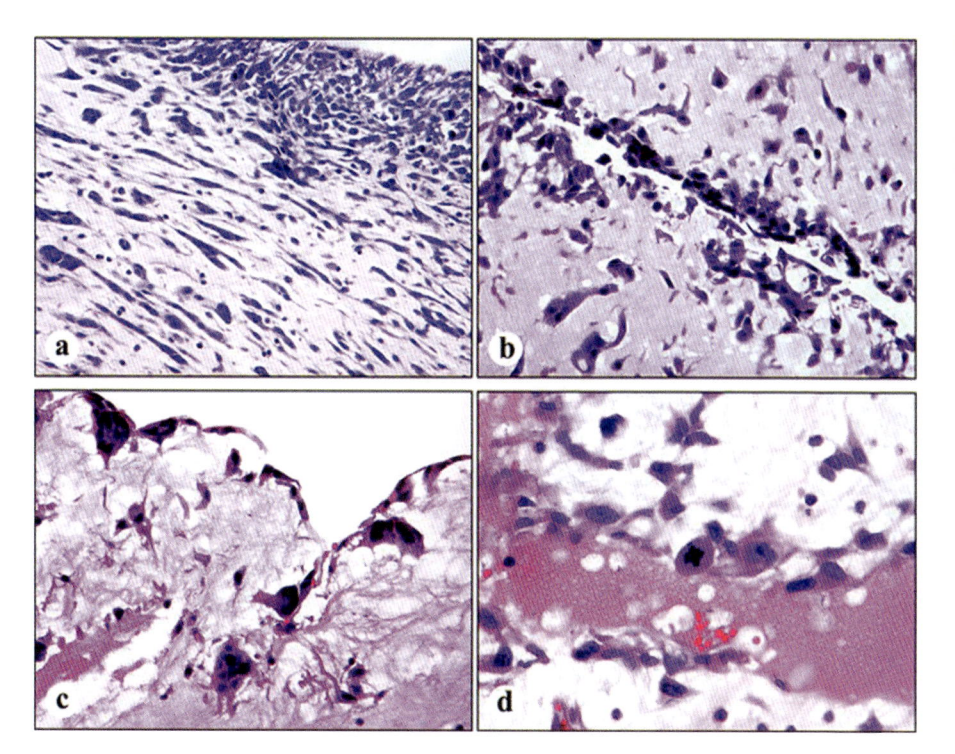

Figure 11.28 Cellular elements on the luminal surface of a cardiac myxoma. (**a**) Synovia-like cellular zone on the surface. (**b**) Moderate cellularity in papillary leaflets. (**c**) Multinucleated myxoma cells. (**d**) Rare example of an atypical mitosis in an otherwise typical myxoma.

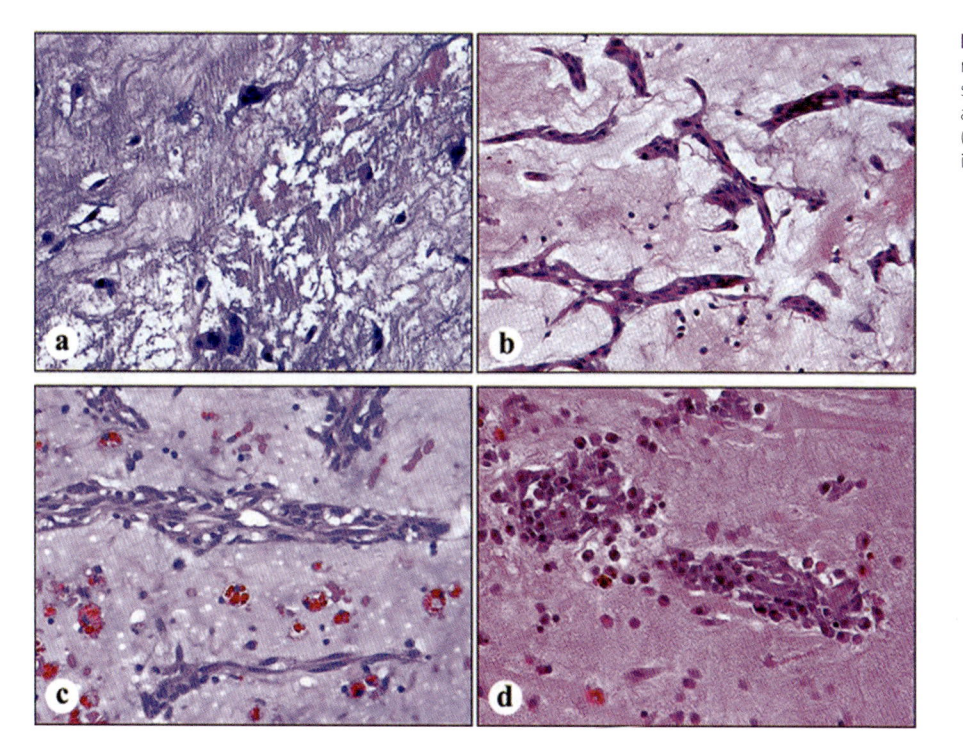

Figure 11.29 Cellular detail inside cardiac myxoma. (**a**) Randomly oriented myxoma cells seen singly. (**b**) Myxoma cells forming trabecular arrangements. (**c**) Growth as perivascular sheaths. (**d**) Perivascular sheaths with lymphoplasmacytic infiltration.

heterogeneous changes.[68,69] Addition of (1)(q32) has been reported in one case. PRKAR1A gene (at 17q24) mutations occur in Carney complex myxomas, but not in the sporadic ones.[37] However, a recent study detected PRKAR1A abnormalities in >30% of apparently sporadic myxomas, based on loss of the corresponding protein by immunohistochemistry.[70]

Odontogenic myxoma (myxoma of the jaw)

This myxoma is primarily a bone tumor, but it is included here because it can also involve the surrounding soft tissues. This rare tumor usually occurs in children or young adults between the second and fifth decades; many series show a female

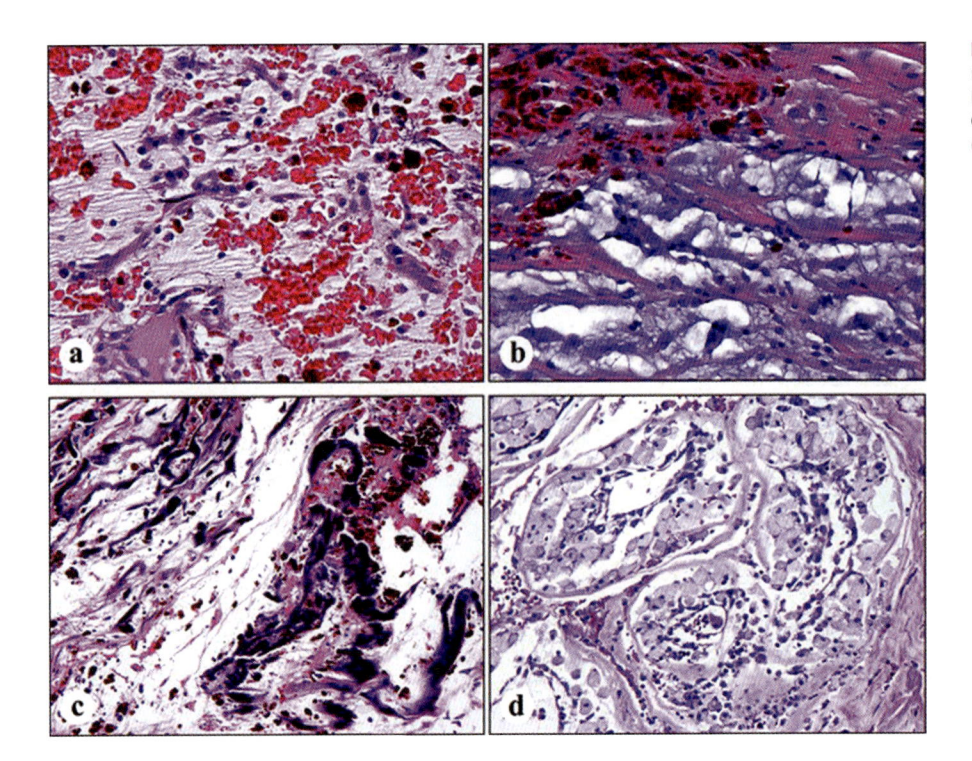

Figure 11.30 Regressive features in cardiac myxoma. (**a**) Intratumoral hemorrhage. (**b**) Hemosiderin deposition and fibrosis. (**c**) Gamna–Gandy bodies in a degenerative paucicellular area. (**d**) Xanthoma cells surrounding cellular elements.

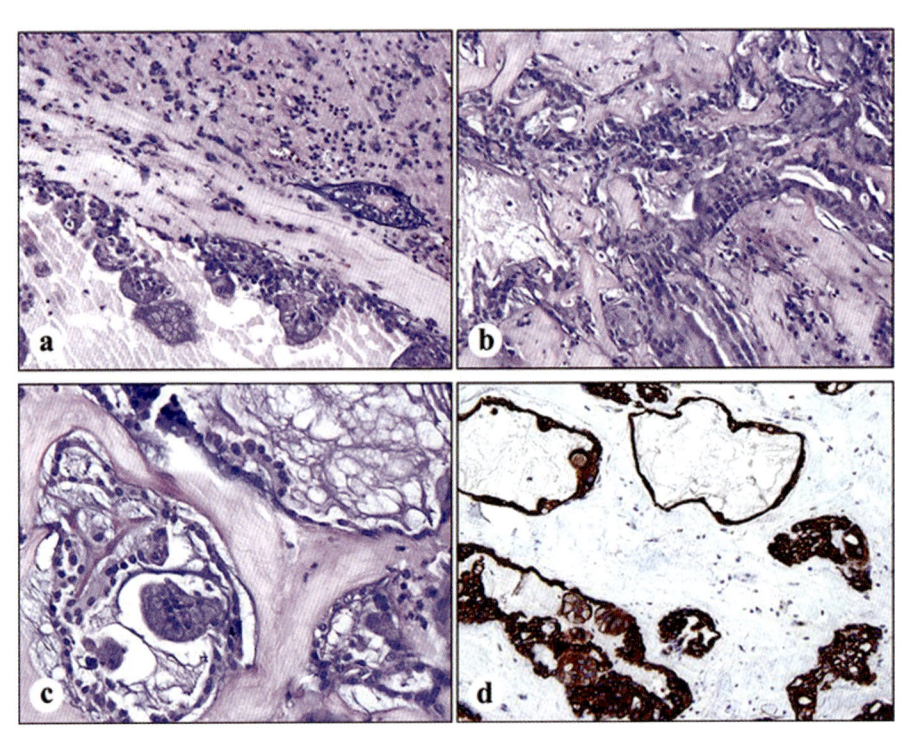

Figure 11.31 Mucinous epithelial differentiation in cardiac myxoma. (**d**) The epithelial cells are keratin positive. Note: this unusual finding should not be confused with a metastatic carcinoma.

predominance. This myxoma is thought to have an odontogenic mesenchymal origin, because the tumor most commonly occurs in the tooth-bearing areas of the jaws, especially in the premolar and molar regions. Occurrence in the maxilla might be more common than in the mandible. In the former location, myxoma of the jaw has the potential to involve the maxillary sinus. The tumor is often radiologically poorly defined. Small lesions can be removed by curettage, whereas larger and clinically more aggressive ones are often treated with en bloc resection. Incomplete resection is often followed by a local recurrence, but there is no metastatic potential.[71–74]

On gross inspection, jaw myxoma has a mucoid appearance on sectioning. Histologically, it is composed of spindled or scattered stellate cells embedded in an abundant myxoid or fibromyxoid matrix. Mitotic activity and nuclear pleomorphism are scant (Figure 11.32).

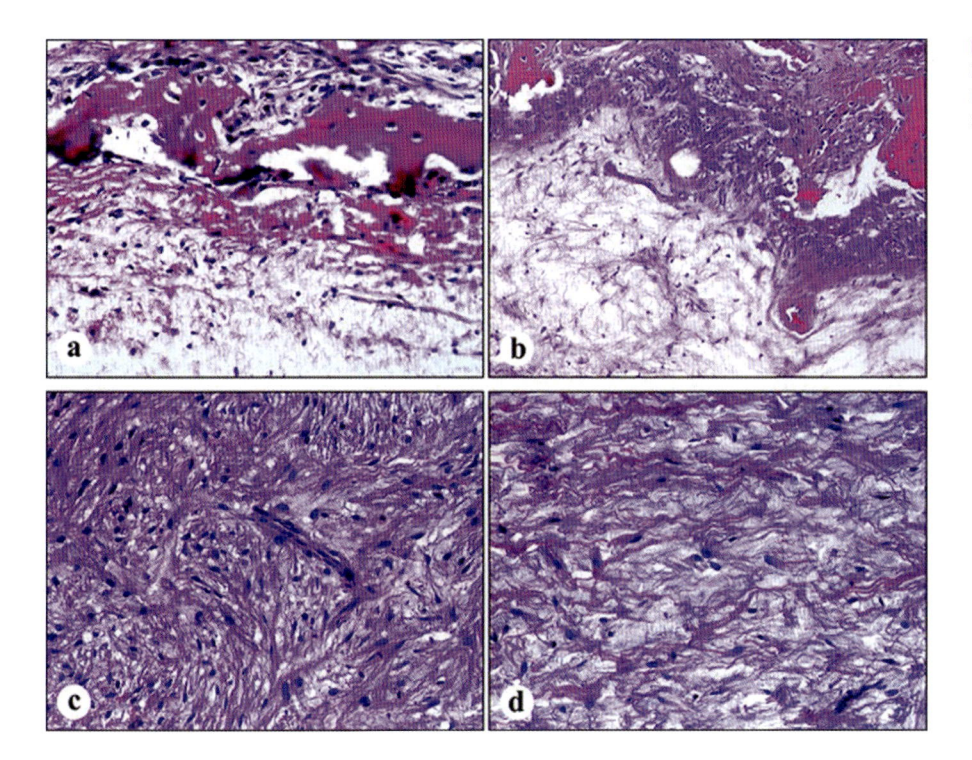

Figure 11.32 (**a,b**) Odontogenic myxoma involving jaw bone. (**c,d**) The collagen content and myxoid quality can vary. Note uniform cells without significant atypia.

Immunohistochemical analysis shows that odontogenic myxomas are positive for vimentin and often focally for smooth muscle actin, whereas they are negative for desmin, S100 protein, and keratins (except possible epithelial nests).[74]

Dental papillae, follicles, and pulp of a developing tooth should not be confused with myxoma of the jaw. Whenever developing dental tissues are seen in isolation, this error can occur; therefore, small biopsies should always be interpreted with clinicoradiological correlation.[75,76]

Ossifying fibromyxoid tumor

Originally reported by Enzinger et al., *ossifying fibromyxoid tumor* (OFT) is a neoplasm with generally low biologic potential and uncertain histogenesis.[77–85] There is some uncertainty about malignant potential, and the spectrum of this entity recently might have been unduly expanded by the addition of malignant variants.[77]

The histogenesis of OFT remains uncertain, and neither nerve sheath nor cartilaginous differentiation of the main cellular component has been proved, despite some immunophenotypic similarity, especially the S100 protein positivity.

Clinical features

OFT principally occurs in adults of all ages and occasionally has been reported in children. The median age in the largest series was 51 years with an age range from 21 to 80 years, and there is a 1.5:1 male predominance (Figure 11.33). The tumor occurs in a wide variety of locations and primarily involves the subcutis, and skeletal muscle in the head and neck only. This tumor is most common in the lower extremities and occurs with a lower, equal frequency in trunk wall, upper extremities, and head and neck (Figure 11.34). Many patients have a long history of tumor, 20 years or more, with a slowly growing, painless mass, usually in the subcutis. There is a moderate, 20% to 30% recurrence rate after local excision, and most recurrences are late, not infrequently 10 or more years after the first surgery. Development of metastases is exceptional if it occurs, and therefore complete excision and clinical follow-up is probably optimal.

Pathology

This tumor forms an oval-to-round, well-circumscribed mass, with the size varying from <1 cm to >10 cm (median size, 3 cm). In 80% of cases, the tumor is surrounded by a partial bony shell that can occasionally make up as much as one-half of the tumor volume (Figure 11.35).

Histologically, the tumor typically contains peripheral bone spicules that sometimes extend into the center. There is variably developed lobulation (Figure 11.36). A fibrous pseudocapsule typically surrounds the tumor, but small satellite nodules are sometimes present within or around the pseudocapsule.

The tumor cells often form diffuse arrangements within moderately collagenous, focally myxoid matrix. Some tumors contain cords or trabecular arrangements of cells (Figure 11.37).

Cytologically, the cells have epithelioid morphology and contain round, relatively uniform nuclei with small nucleoli,

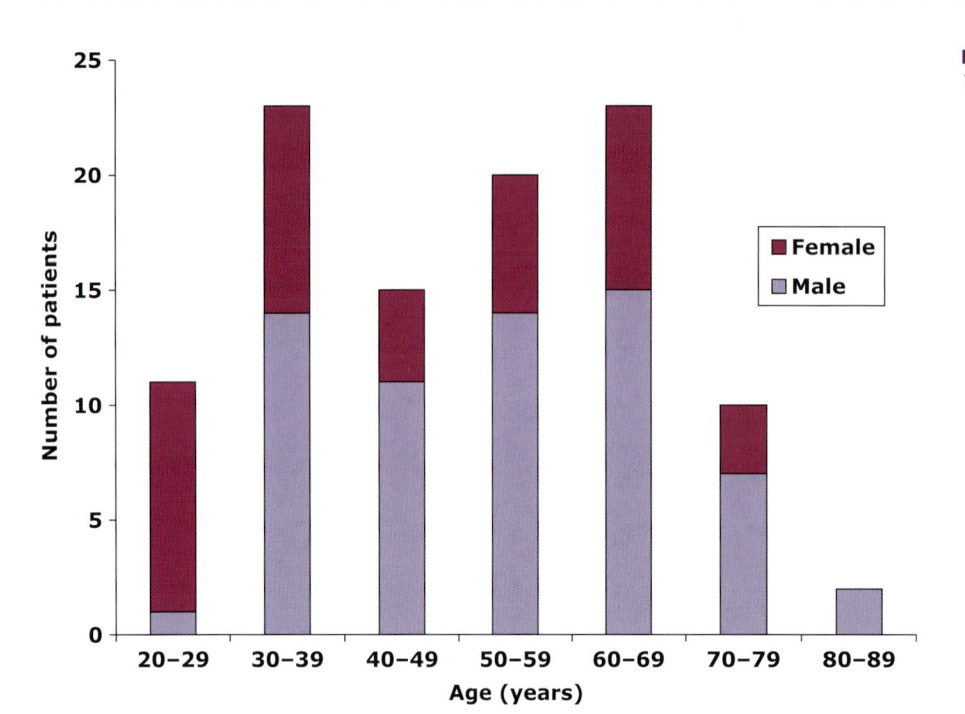

Chapter 11: Myxomas and ossifying fibromyxoid tumor

Figure 11.33 Age and sex distribution of 104 patients with ossifying fibromyxoid tumor.

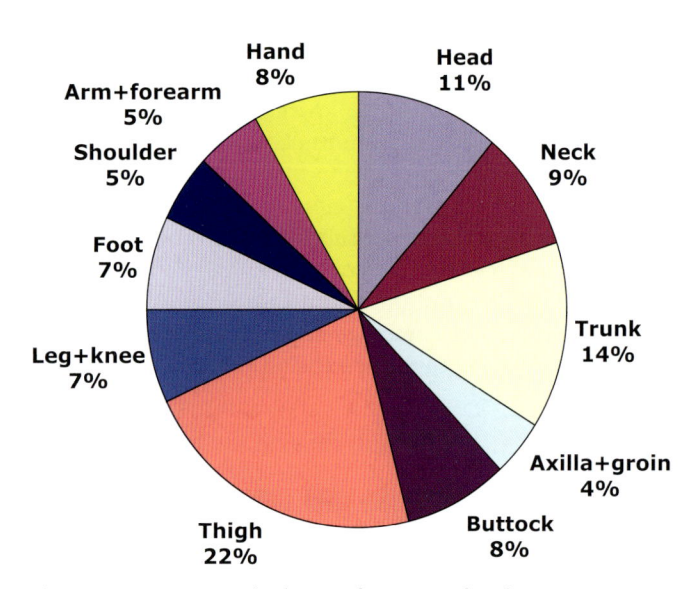

Figure 11.34 Anatomic distribution of 104 cases of ossifying fibromyxoid tumor.

but focal atypia can be present (Figure 11.38). The cytoplasm is variably eosinophilic and is sometimes surrounded by a shrinkage space. Mitotic activity varies from 0 to >10 mitoses per 10 HPFs. Tumors with mitotic rate >2 per 10 HPFs have an increased risk of recurrence; however, there seems to be no correlation between prognosis and tumor size, presence of satellite nodules, or coagulation necrosis.[78]

Atypical variants with metastatic risk have been reported. Such tumors differ from classic OFT, however, by their often overt sarcomatous features, deep location, and low or no expression of S100 protein.[86,87] Because some of these tumors have not contained conventional OFT components, the relationship of these tumors with OFTs is difficult to confirm. It is certainly more important to identify these highly malignant tumors as sarcomas rather than OFT variants.

A typical immunohistochemical finding is expression of vimentin and S100 protein (in nearly 100% of cases), and CD10, at least focally. Cases focally positive for keratin cocktail, desmin, or GFAP are in the minority.

Ultrastructural findings of interest include partial basement membranes and complex cell processes, but these findings have not helped to resolve the histogenesis of OFT.[77,80]

Cytogenetic data from two typical cases showed a simple karyotype, with a loss of chromosome 6 and an unbalanced t(6;14) translocation in one case[88] and a t(6;12)(p21;q24) translocation in another case.[89] Molecular genetic studies have shown PHF1 gene (at 6p21) rearrangements to be the typical recurrent changes in OFT. The detected gene fusions include EP400-PHF1 (most common), MEAF6-PHF1, and EPC1-PHF1.[89–92] In addition, ZC3HB7-BCOR fusion was reported in one case.[92] In one study, PHF1 fusion was detected only in one of six malignant cases,[90] whereas another study demonstrated PHF1 fusions equally in benign and malignant cases.[92] This leaves the question open as to whether some tumors reported as malignant OFTs are part of the same entity as the typical ones.

Differential diagnosis

Highly cellular and mitotically active tumors with atypia in the author's experience almost invariably represent other entities, with focal metaplastic bone formation and corded pattern. These include especially low-grade fibromyxoid sarcomas and

318

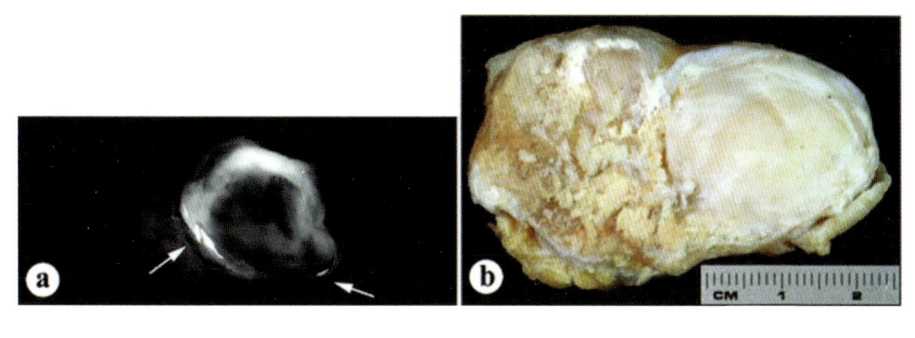

Chapter 11: Myxomas and ossifying fibromyxoid tumor

Figure 11.35 (**a**) X-ray image of an ossifying fibromyxoid tumor reveals streaks of calcifications in the tumor periphery (arrows). (**b**) Gross specimen shows whitish peripheral calcifications in the tumor periphery in the upper aspect. Calcifications extend inside the tumor on the left side, whereas the right side is composed of cellular elements typical of OFT.

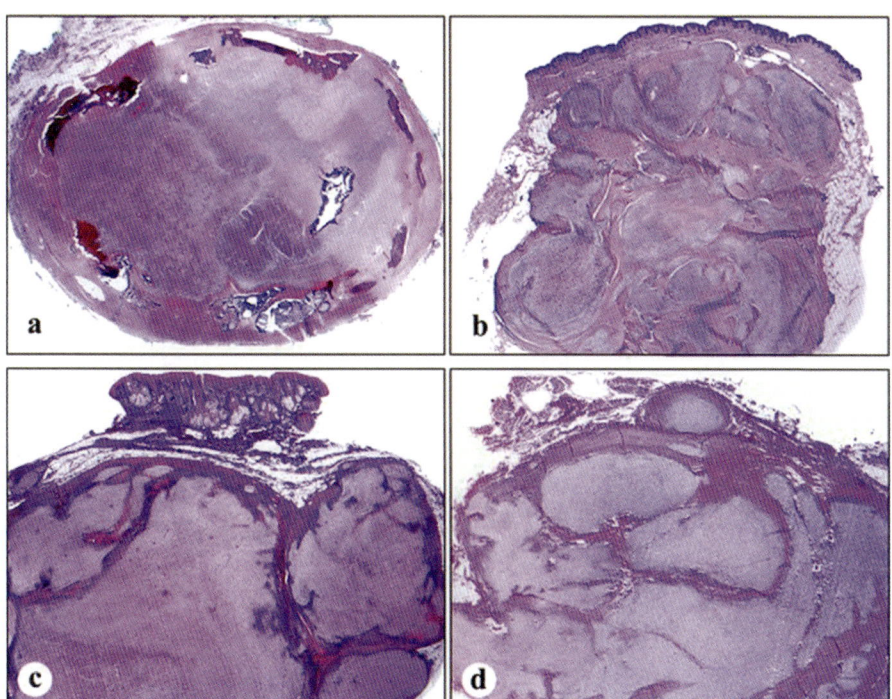

Figure 11.36 Low-magnification images of ossifying fibromyxoid tumor. (**a**) An incomplete bony shell surrounds the tumor. (**b**) The tumor involves the subcutis and shows lobulation. (**c,d**) Two examples with distinct lobulation and fibrous septa.

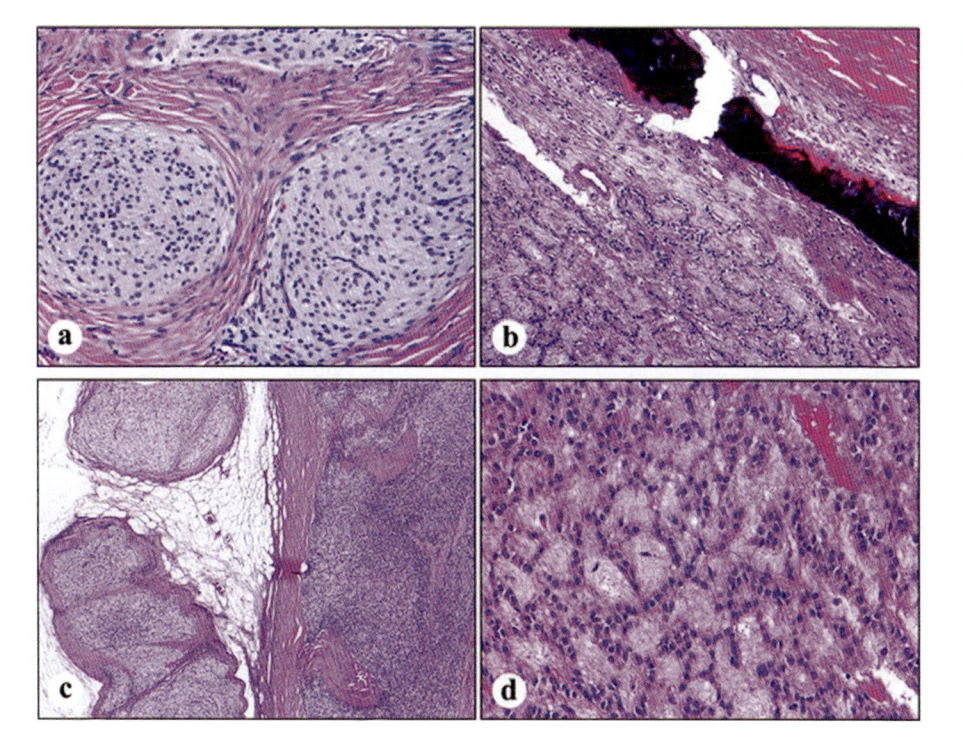

Figure 11.37 (**a**) Lobules of ossifying fibromyxoid tumor with delicate capillaries in the lobules. (**b**) A bony spicule in the tumor periphery. (**c**) Satellite nodules extending into surrounding fat. (**d**) Distinct trabecular arrangement is seen in some cases.

319

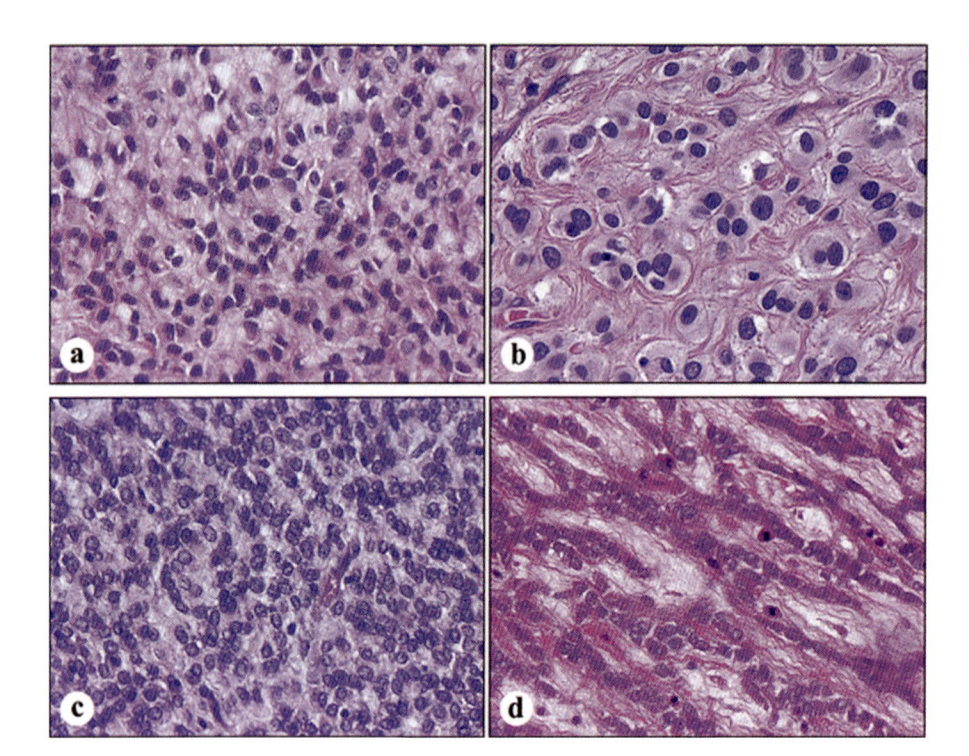

Figure 11.38 (**a**) Uniform cytologic appearance in ossifying fibromyxoid tumor. (**b**) Mild cytologic atypia and epithelioid cytomorphology. (**c**) A highly cellular example. (**d**) A tumor with significant mitotic activity (two mitotic figures seen to the right of center).

extraskeletal osteosarcoma variants. Many tumors reported in the literature as malignant OFTs have been large, deep intramuscular tumors. The author and colleagues have not been able to verify such tumors within the spectrum of OFT, although more studies incorporating genetic analysis are necessary for complete understanding. Tumors involving bones should not be classified as OFTs.

Tumors with neoplastic epithelial elements or significant keratin positivity are more consistent with mixed tumor (myoepithelioma variants). Some smooth muscle actin-positive myofibroblastic neoplasms have a resemblance to nonossifying variants of OFT. Their immunohistochemical profile, including negativity for S100 protein, helps to separate these from OFT.

References

Intramuscular myxoma

1. Enzinger FM. Intramuscular myxoma. *Am J Clin Pathol* 1965;43:104–110.
2. Kindblom LG, Stener B, Angervall L. Intramuscular myxoma. *Cancer* 1974;34:1737–1744.
3. Miettinen M, Hockerstedt K, Reitamo J, Totterman S. Intramuscular myxoma: a clinicopathological study of 23 cases. *Am J Clin Pathol* 1985;84:265–272.
4. Hashimoto H, Tsuneyoshi M, Daimaru Y, Enjoji M, Shinohara N. Intramuscular myxoma: a clinicopathologic, immunohistochemical, and electron microscopic study. *Cancer* 1986;58:740–747.
5. Bancroft LW, Kransdorf MJ, Menke DM, O'Connor MI, Foster WC. Intramuscular myxoma: characteristic MR imaging features. *AJR Am J Roentgenol* 2002;178:1255–1259.
6. Murphey MD, McRae GA, Fanburg-Smith JC, et al. Imaging of soft-tissue myxoma with emphasis on CT and MR and comparison of radiologic and pathologic findings. *Radiology* 2002;225:215–224.
7. Feldman F. Tuberous sclerosis, neurofibromatosis, and fibrous dysplasia. In *Diagnosis of Bone and Joint Diseases*, Vol 5, 5th edn. Resnick D (ed.) Philadelphia: W.B. Saunders Company; 1998.
8. Wirth WA, Leavitt D, Enzinger FM. Multiple intramuscular myxomas: another extraskeletal manifestation of fibrous dysplasia. *Cancer* 1971;27:1167–1173.
9. Ireland DC, Soule EH, Ivins JC. Myxoma of the somatic soft tissues: a report of 58 patients, 3 with multiple tumors and fibrous dysplasia of the bone. *Mayo Clin Proc* 1973;48:401–410.
10. Szendroi M, Rahoty P, Antal I, Kiss J. Fibrous dysplasia associated with intramuscular myxoma (Mazabraud's syndrome): a long-term follow-up of three cases. *J Cancer Res Clin Oncol* 1998;124:401–406.
11. Faivre L, Nivelon-Chevallier A, Kottler ML, et al. Mazabraud syndrome in two patients: clinical overlap with McCune-Albright syndrome. *Am J Med Genet* 2001;99:132–136.
12. Nielsen GP, O'Connell JX, Rosenberg AE. Intramuscular myxoma: a clinicopathologic study of 51 cases with emphasis on hypercellular and hypervascular variants. *Am J Surg Pathol* 1998;22:1222–1227.
13. van Roggen JF, McMenamin ME, Fletcher CD. Cellular myxoma of soft tissue: a clinicopathologic study of 38 cases confirming indolent behaviour. *Histopathology* 2001;39:287–297.
14. Okamoto S, Hisaoka M, Ushijima M, et al. Activating Gs mutation in intramuscular myxomas with and

without fibrous dysplasia of bone. *Virchows Arch* 2000;437:133–137.
15. Malchoff CD, Reardon G, Macgillivray DC, et al. An unusual presentation of McCune–Albright syndrome confirmed by an activating mutation of the Gs alpha-subunit from a bone lesion. *J Clin Endocrinol Metab* 1994;78:803–806.
16. Shenker A, Weinstein LS, Sweet DE, Spiegel AM. An activating Gs alpha mutation is present in fibrous dysplasia of bone in the McCune-Albright syndrome. *J Clin Endocrinol Metab* 1994;79:750–755.

Juxtaarticular myxoma, ganglion cyst, and cutaneous myxoid cyst
17. Meis JM, Enzinger FM. Juxtaarticular myxoma: a clinical and pathologic study of 65 cases. *Hum Pathol* 1992;23:639–646.
18. Minkoff J, Stecker S, Irizarry J, Whiteman M, Woodhouse S. Juxtaarticular myxoma: a rare cause of painful restricted motion of the knee. *Arthroscopy* 2003;19: E6–E13.
19. Sciot R, Dal Cin P, Samson I, Van Den Berghe H, van Damme B. Clonal chromosomal changes in juxtaarticular myxoma. *Virchows Arch* 1999;434:177–180.
20. Okamoto S, Hisaoka M, Meis-Kindblom JM, Kindblom LG, Hashimoto H. Juxtaarticular myxoma and intramuscular myxoma are two distinct entities: activating Gs alpha mutation at Arg 201 codon does not occur in juxtaarticular myxoma. *Virchows Arch* 2002;440:12–15.
21. Soren A. Pathogenesis, clinic and treatment of ganglion. *Arch Orthop Trauma Surg* 1982;99:247–252.
22. Johnson WC, Graham JH, Helwig EB. Cutaneous myxoid cyst: a clinicopathological and histochemical study. *JAMA* 1965;191:109–116.
23. Armijo M. Mucoid cysts of the fingers: differential diagnosis, ultrastructure, and surgical treatment. *J Dermatol Surg Oncol* 1981;7:317–322.
24. De Berker D, Lawrence C. Ganglion of the distal interphalangeal joint (myxoid cyst). *Arch Dermatol* 2001;137:607–610.

Superficial angiomyxoma (cutaneous myxoma) and focal cutaneous mucinosis
25. Carney JA, Headington JT, Su WP. Cutaneous myxomas: a major component of myxomas, spotty pigmentation, and endocrine overactivity. *Arch Dermatol* 1986;122:790–798.
26. Carney JA. Carney complex: the complex of myxomas, spotty pigmentation, endocrine overactivity, and schwannomas. *Semin Dermatol* 1995;14:90–98.
27. Boikos SA, Stratakis CA. Carney complex: the first 20 years. *Curr Opin Oncol* 2007;19:24–29.
28. Atherton DJ, Pitcher DW, Wells RS, MacDonald DM. A syndrome of various cutaneous pigmented lesions, myxoid neurofibromata, and atrial myxoma: the NAME syndrome. *Br J Dermatol* 1980;103:421–429.
29. Koopman RJJ, Happle R. Autosomal dominant transmission of the NAME syndrome (nevi, atrial myxoma, mucinosis of the skin and endocrine overactivity). *Hum Genet* 1991;86:300–304.
30. Rhodes AR, Silverman RA, Harrist TJ, Perez-Atayde AR. Mucocutaneous lentigines, cardiomucocutaneous myxomas, and multiple blue nevi: the "LAMB" syndrome. *J Am Acad Dermatol* 1984;10:72–82.
31. Allen PW, Dymock RB, MacCormac LB. Superficial angiomyxomas with and without epithelial components: report of 30 tumors in 28 patients. *Am J Surg Pathol* 1988;12:519–530.
32. Fetsch JF, Laskin WB, Tavassoli FA. Superficial angiomyxoma (cutaneous myxoma): a clinicopathologic study of 17 cases arising in the genital region. *Int J Gynecol Pathol* 1997;16:325–334.
33. Calonje E, Guerin D, McCormick D, Fletcher CD. Superficial angiomyxoma: clinicopathologic analysis of a series of distinctive but poorly recognized cutaneous tumors with tendency for recurrence. *Am J Surg Pathol* 1999;23:910–917.
34. Goldstein MM, Casey M, Carney JA, Basson CT. Molecular genetic diagnosis of the familial myxoma syndrome (Carney complex). *Am J Med Genet* 1999;86:62–65.
35. Kirschner LS, Carney JA, Pack SD, et al. Mutations of the gene encoding the protein kinase A type I-alpha regulatory subunit in patients with the Carney complex. *Nat Genet* 2000;26:89–92.
36. Bertherat J, Horvath A, Groussin L, et al. Mutations in regulatory subunit type 1A of cyclic adenosine 5′-monophosphate-dependent protein kinase (PRKAR1A): phenotype analysis in 353 patients and 80 different genotypes. *J Clin Endocrinol Metab* 2009;94:2085–2091.
37. Griffin KJ, Kirschner LS, Matyakhina L, et al. A transgenic mouse bearing an antisense construct of regulatory type subunit type 1A of protein kinase A develops endocrine and other tumours: comparison with Carney complex and other PRKAR1A induced lesions. *J Med Genet* 2004;41:924–931.
38. Johnson WC, Helwig EB. Focal cutaneous mucinosis. *Arch Dermatol* 1966;93:13–20.
39. Wilk M, Schmoeckel C. Cutaneous focal mucinosis: a histopathological and immunohistochemical analysis of 11 cases. *J Cutan Pathol* 1994;21:446–452.

Superficial acral fibromyxoma
40. Fetsch JF, Laskin WB, Miettinen M. Superficial acral fibromyxoma: a clinicopathologic and immunohistochemical analysis of 37 cases of a distinctive soft tissue tumor with a predilection for the fingers and toes. *Hum Pathol* 2001;32:704–714.
41. Al-Daraji W, Miettinen M. Superficial acral fibromyxoma: analysis of 27 cases, including four in the heel. *J Cutan Pathol* 2008;35:1020–1026.
42. Hollmann TJ, Bovée JV, Fletcher CD. Digital fibromyxoma (superficial acral fibromyxoma): a detailed characterization of 124 cases. *Am J Surg Pathol* 2012;36:789–798.
43. McNiff JM, Subtil A, Cowper SE, Lazova R, Glusac EJ. Cellular digital fibromas: distinctive CD34-positive lesions that may mimic dermatofibrosarcoma protuberans. *J Cutan Pathol* 2005;32:413–418.
44. Perrin C, Baran R, Balaguer T, et al. Onychomatricoma: new clinical and histological features: review of 19 tumors. *Am J Dermatopathol* 2010;32:1–8.

Cardiac myxoma
45. Burke AP, Virmani R. Cardiac myxoma: a clinicopathologic study. *Am J Clin Pathol* 1993;100:671–680.
46. Reynen K. Cardiac myxomas. *N Engl J Med* 1995;333:1610–1617.
47. Pucci A, Gagliardotto P, Zanini C, et al. Histopathologic and clinical characterization of cardiac myxoma:

48. Pinede L, Duhaut P, Loire R. Clinical presentation of left atrial cardiac myxoma: a series of 112 consecutive cases. *Medicine (Baltimore)* 2001;80:159–172.

49. Keeling IM, Oberwalder P, Anelli-Monti M, et al. Cardiac myxomas: 24 years of experience in 49 patients. *Eur J Cardiothorac Surg* 2002;22:971–977.

50. Wang JG, Li YJ, Liu H, et al. Clinicopathologic analysis of cardiac myxomas: seven years' experience with 61 patients. *J Thorac Dis* 2012;4:272–283.

51. Edwards A, Bermudez C, Plwonka G, et al. Carney's syndrome: complex myxomas. Report of four cases and review of the literature. *Cardiovasc Surg* 2002;10:264–275.

52. Mabuchi T, Shimizu M, Ino H, et al. PRKAR1A gene mutation in patients with cardiac myxoma. *Int J Cardiol* 2005;102:273–277.

53. Li Y, Pan Z, Ji Y, et al. Herpes simplex virus type 1 infection associated with atrial myxoma. *Am J Pathol* 2003;163:2407–2412.

54. Grebenc ML, Rosado-de-Christensen ML, Green CE, Burke AP, Galvin JR. Cardiac myxoma: imaging features in 83 patients. *Radiographics* 2002;22:673–689.

55. Jourdan M, Bataille R, Seguin J, et al. Constitutive production of interleukin-6 and immunologic features in cardiac myxomas. *Arthritis Rheum* 1990;33:398–402.

56. Deshpande RP, Casselman F, Bakir I, et al. Endoscopic cardiac tumor resection. *Ann Thorac Surg* 2007;83:2142–2146.

57. Altundag MB, Ertas G, Ucer AR, et al. Brain metastasis of cardiac myxoma: case report and review of the literature. *J Neurooncol* 2005;75:181–184.

58. Goldman BI, Frydman C, Harpaz N, Ryan SF, Loiterman D. Glandular cardiac myxomas: histologic, immunohistochemical, and ultrastructural evidence of epithelial differentiation. *Cancer* 1987;59:1767–1775.

59. Johansson L. Histogenesis of cardiac myxomas: an immunohistochemical study of 19 cases, including one with glandular structures, and review of the literature. *Arch Pathol Lab Med* 1989;113:735–741.

60. Abenoza P, Sibley RK. Cardiac myxoma with glandlike structures: an immunohistochemical study. *Arch Pathol Lab Med* 1986;110:736–739.

61. Miller DV, Tazelaar HD, Handy JR, Young DA, Hernandez JC. Thymoma arising within cardiac myxoma. *Am J Surg Pathol* 2005;29:1208–1213.

62. Terraciano LM, Mhawech P, Suess K, et al. Calretinin as a marker for cardiac myxoma: diagnostic and histogenetic considerations. *Am J Clin Pathol* 2000;114:754–759.

63. Acebo E, Val-Bernal JF, Gomez-Roman JJ. Thrombomodulin, calretinin and c-kit (CD117) expression in cardiac myxoma. *Histol Histopathol* 2001;16:1031–1036.

64. Pucci A, Bartoloni G, Tessitore E, Carney JA, Papotti M. Cytokeratin profile and neuroendocrine cells in the glandular component of cardiac myxoma. *Virchows Arch* 2003;443:618–624.

65. Chu PH, Jung SM, Yet TS, Lin HC, Chu JJ. MUC1, MUC2, and MUC5AC expressions in cardiac myxoma. *Virchows Arch* 2005;446:52–55.

66. Feldman PS, Horvath E, Kovacs K. An ultrastructural study of seven cardiac myxomas. *Cancer* 1979;40:2216–2232.

67. Govoni E, Severi B, Cenacchi G, et al. Ultrastructural and immunohistochemical contribution to the histogenesis of human cardiac myxomas. *Ultrastruct Pathol* 1988;12:221–233.

68. Dijkhuizen T, de Jong B, Meuzelaar JJ, Molenaar WM, Van Den Berg E. No cytogenetic evidence for involvement of gene(s) at 2p16 in sporadic cardiac myxomas: cytogenetic changes in ten sporadic cardiac myxomas. *Cancer Genet Cytogenet* 2001;126:162–165.

69. Guardiola T, Horton E, Lopez-Camarillo L, et al. Cardiac myxoma: a cytogenetic study of two cases. *Cancer Genet Cytogenet* 2004;148:145–147.

70. Maleszewski JJ, Larsen BT, Kip NS, et al. PRKAR1A in the development of cardiac myxoma: a study of 110 cases including isolated and syndromic tumors. *Am J Surg Pathol* 2014;38:1079–1087.

Odontogenic myxoma (myxoma of the jaw)

71. Ghosh BC, Huvos AG, Gerold FP, Miller TR. Myxoma of the jaw bones. *Cancer* 1973;31:237–240.

72. Barker BF. Odontogenic myxoma. *Semin Diagn Pathol* 1999;16:297–301.

73. Simon ENM, Merkx AW, Vuhahula E, Ngassapa N, Stoelinga PJW. Odontogenic myxoma: a clinicopathologic study of 33 cases. *Int J Oral Maxillofac Surg* 2004;33:333–337.

74. Li TJ, Sun LS, Luo HY. Odontogenic myxoma: a clinicopathologic study of 25 cases. *Arch Pathol Lab Med* 2006;130:1799–1806.

75. Suarez PA, Batsakis JG, El-Naggar AK. Don't confuse dental soft tissues with odontogenic tumors. *Ann Otol Rhinol Laryngol* 1996;105:490–494.

76. Fellegara G, Mody K, Kuhn E, Rosai J. Normal dental papilla simulating odontogenic myxoma. *Int J Surg Pathol* 2007;15:282–285.

Ossifying fibromyxoid tumor

77. Enzinger FM, Weiss SW, Liang CY. Ossifying fibromyxoid tumor of soft parts: a clinicopathologic analysis of 59 cases. *Am J Surg Pathol* 1989;13:817–827.

78. Miettinen M, Finnell V, Fetsch JF. Ossifying fibromyxoid tumor of soft parts: a clinicopathological and immunohistochemical study of 104 cases with long-term follow-up and a critical review of literature. *Am J Surg Pathol* 2008;32:996–1006.

79. Schofield JB, Krausz T, Stamp GW, et al. Ossifying fibromyxoid tumour of soft parts: immunohistochemical and ultrastructural analysis. *Histopathology* 1993;22:101–112.

80. Miettinen M. Ossifying fibromyxoid tumor of soft parts: additional observations of a distinctive soft tissue tumor. *Am J Clin Pathol* 1991;95:142–149.

81. Donner LR. Ossifying fibromyxoid tumor of soft parts: evidence supporting schwann cell origin. *Hum Pathol* 1992;23:200–202.

82. Zamecnik M, Michal M, Simpson RH, et al. Ossifying fibromyxoid tumor of soft parts: a report of 17 cases with emphasis on unusual histologic features. *Ann Diagn Pathol* 1997;1:73–81.

83. Williams SB, Ellis GL, Meis JM, et al. Ossifying fibromyxoid tumour (of soft parts) of the head and neck: a clinicopathological and immunohistochemical study of nine cases. *J Laryngol Otol* 1993;107:75–80.

84. Hanski V, Lewicki Z. New observations on three cases of ossifying fibromyxoid tumor of soft parts. *Pol J Pathol* 1994;45:231–238.

85. Holck S, Pederson JG, Ackermann T, et al. Ossifying fibromyxoid tumour of soft parts, with focus on unusual clinicopathological features. *Histopathology* 2003;42:599–604.

86. Kilpatrick SE, Ward WG, Mozes M. Atypical and malignant variants of ossifying fibromyxoid tumor: clinicopathologic analysis of six cases. *Am J Surg Pathol* 1995;19:1039–1046.

87. Folpe AL, Weiss SW. Ossifying fibromyxoid tumor of soft parts: a clinicopathologic study of 70 cases with emphasis on atypical and malignant variants. *Am J Surg Pathol* 2003;27:421–431.

88. Sovani V, Velagaleti GVN, Filipowicz E, Gatalica Z, Knisely AS. Ossifying fibromyxoid tumor of soft parts: report of a case with novel cytogenetic findings. *Cancer Genet Cytogenet* 2001;127:1–6.

89. Gebre-Medhin S, Nord KH, Möller E, et al. Recurrent rearrangement of the PHF1 gene in ossifying fibromyxoid tumors. *Am J Pathol* 2012;181:1069–1077.

90. Endo M, Kohashi K, Yamamoto H, et al. Ossifying fibromyxoid tumor presenting EP400-PHF1 fusion gene. *Hum Pathol* 2013;44(11):2603–2608.

91. Graham RP, Weiss SW, Sukov WR, et al. PHF1 rearrangements in ossifying fibromyxoid tumors of soft parts: a fluorescence in situ hybridization study of 41 cases with emphasis on the malignant variant. *Am J Surg Pathol* 2013;37:1751–1755.

92. Antonescu CR, Sung YS, Chen CL, et al. Novel ZC3H7B-BCOR, MEAF6-PHF1, and EPC1-PHF1 fusions in ossifying fibromyxoid tumors: molecular characterization shows genetic overlap with endometrial stromal sarcoma. *Genes Chromosomes Cancer* 2014;53:183–193.

Chapter 12

Solitary fibrous tumor, hemangiopericytoma, and related tumors

Markku Miettinen
National Cancer Institute, National Institutes of Health

Solitary fibrous tumor (SFT) varies from clinically benign to uncertain biologic potential and overt sarcoma, although the last is a rare occurrence. Giant cell angiofibroma is a morphologic variant of SFT. The World Health Organization (WHO) classification essentially merges SFT and hemangiopericytoma (HPC). Sinonasal HPC is a myopericytic or smooth muscle tumor and is discussed in Chapter 23.

Solitary fibrous tumor

SFT is a generally CD34-positive fibroblastic spindle cell neoplasm that can present in a wide variety of soft tissue and visceral locations, with a spectrum that ranges from benign to malignant.

History and terminology

Originally known as a neoplasm of the pleura[1,2] and mediastinum,[3] SFT was historically designated as (benign or malignant) fibrous mesothelioma. Its nonepithelial and nonmesothelial nature and lack of association with asbestosis prompted its renaming as a localized fibrous tumor, and more recently as SFT. Similar tumors were subsequently reported on peritoneal serous surfaces,[4,5] peripheral soft tissues,[6] and in a wide variety of visceral sites in different organ systems.[7]

Although most SFTs are benign, some recur locally and a few behave as soft tissue sarcomas and metastasize, indicating that this tumor has a clinical spectrum from benign to fully malignant. A mitotic rate of >4 per 10 HPFs, tumor necrosis, and atypia have been identified as risk factors for malignant behavior in pleural examples,[1] and similar criteria have been used in other sites. Previous designations for extrapleural SFTs have been deep fibrous histiocytoma, HPC, fibromatosis, and fibrosarcoma.

SFT of the pleura occurs in adults of all ages, but is rare before the age of 30. Nearly one half of these tumors are detected in the sixth and seventh decades (Figure 12.1A). SFTs of extrathoracic locations occur at all ages, and patients are of a younger age at presentation, probably because many of these tumors are detected earlier, perhaps because of their more accessible location (Figure 12.1B).

Clinical features at serosal and visceral sites

Pleural-thoracic SFTs constitute 30% of all SFTs, and 67% of patients are asymptomatic (Figure 12.2). Symptoms and signs include chest pain, pleural effusion, and rarely, clubbing of fingernails and hypoglycemia.[1] The latter is associated with large SFTs and resolves after tumor removal. The hypoglycemia has been attributed to secretion of insulin-like growth factors I and II by the tumor cells and has also been observed in extrapleural SFTs.[8,9] Historical reports of retroperitoneal fibrosarcomas with hypoglycemia have probably concerned SFTs.

Two-thirds of pleural SFTs occur in the visceral pleura, where the tumor is often attached to the lung by a narrow pedicle, and one-third in the parietal pleura, where the tumors are often larger, with a broad-based attachments from <1 cm to >30 cm. Larger and more aggressive examples can also involve adjacent soft tissues in the mediastinum, chest wall, and diaphragm, and can also extend into the abdomen. Clinical risk factors for pleural SFTs include large tumor size and sessile (as opposed to pedunculated) tumor.

The anatomic distribution of all SFTs is shown in Figure 12.2. Intra-abdominal SFTs in the peritoneum, retroperitoneum, and pelvis collectively constitute the largest site-related group in most series of extrapleural SFTS.[4,5,10–14] They are often large symptomatic masses, often >10 cm to 20 cm, and the tumors vary from paucicellular and clinically benign to sarcomas with metastatic potential. Occurrence in the liver, mostly intraparenchymally, has been reported, and such tumors have varied from 2 cm to 20 cm in greatest dimension, with the mitotic rate not exceeding 4 per 10 HPFs.[15]

SFTs in the orbit and sinonasal tract are usually relatively small, 1 cm to 3 cm, and most have been clinically benign with reported recurrences.[16–20] A review of 42 published orbital examples showed a 20% recurrence rate and one example with malignant transformation.[18]

>In the oral cavity, SFTs most commonly occur beneath the buccal mucosa (63%), tongue (10%), and lower lip (7%). These tumors are relatively small (<1–4 cm) and are usually clinically benign.[21]

Meningeal SFTs clinically resemble meningiomas and occur in the dura of the brain, mostly supratentorially, and

---

Modern Soft Tissue Pathology, Second Edition, ed. Markku Miettinen. Published by Cambridge University Press. © Cambridge University Press 2016.

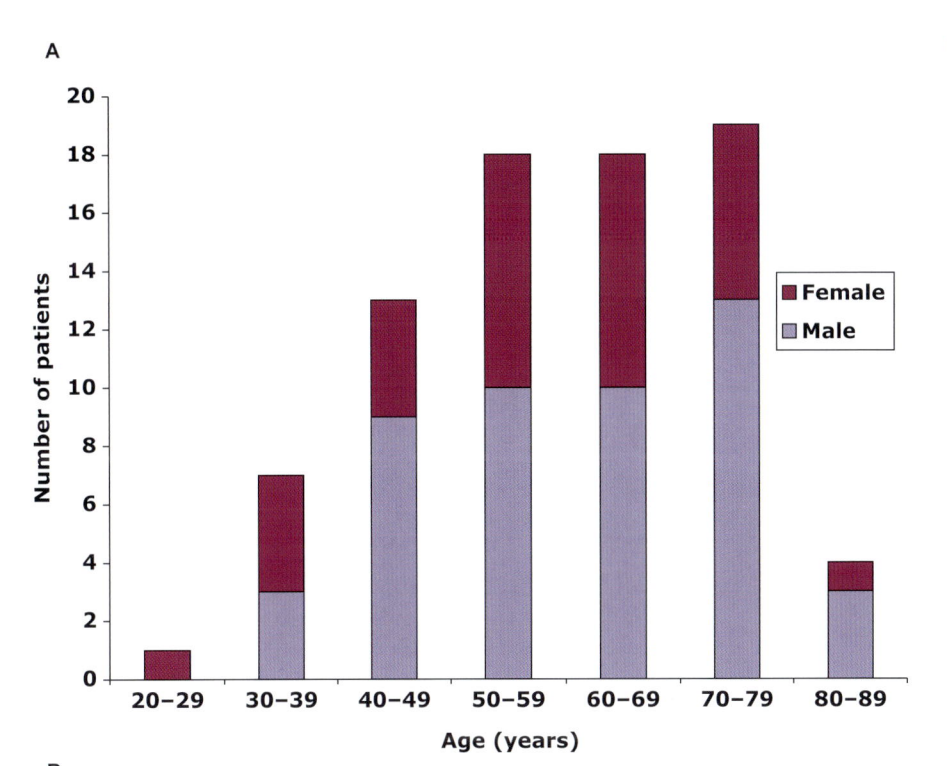

Chapter 12: Solitary fibrous tumor, hemangiopericytoma, and related tumors

Figure 12.1 (**A**) Age and sex distribution of 80 intrathoracic solitary fibrous tumors. (**B**) Age and sex distribution of 210 extrathoracic solitary fibrous tumors.

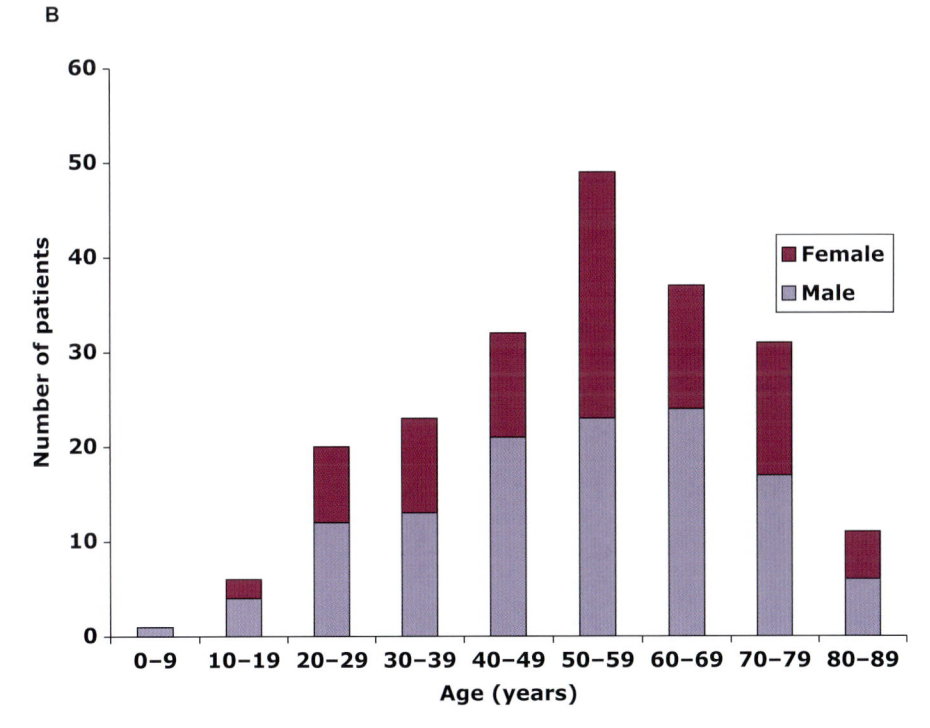

in the spinal canal. Recognition is based on histologic similarity to other SFTs and immunoreactivity for CD34 and none for EMA. Published series have shown atypical examples, rare recurrences, mostly related to primarily incomplete resection, but no extracranial metastases.[22,23] It is likely that tumors previously reported as meningeal SFTs have had lower cellularity and lower mitotic rates (lower-grade tumors), whereas tumors reported as HPCs have been more highly cellular and mitotically active (higher-grade tumors) in the spectrum of one family of tumors (see genetics).

In the urogenital tract, SFTs have been reported in the bladder, prostate, seminal vesicle, and kidney. In the last organ, these tumors are often >10 cm, with nonmalignant histologic features. Histologically malignant features are sometimes present, but these do not always correlate with a higher biologic potential.[24–26] Some SFTs counted as urogenital primaries could represent large pelvic SFTs extending into urogenital organs.

Clinical features and tumor behavior of SFTs in peripheral soft tissues

In the external soft tissues, SFTs occur in the head and neck and elsewhere (e.g., extremities and trunk wall) with approximately equal frequency, and such tumors have represented 0% to 67% in the series of nonpleural SFTs (the rest being internal head and neck and abdominal tumors). Cutaneous examples usually also involve the subcutis. Deep intramuscular location is also possible.[10–14,27–29]

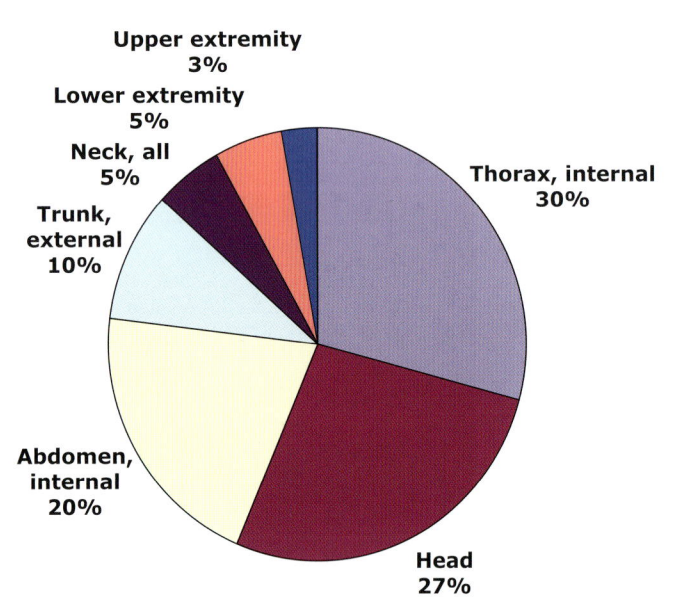

Figure 12.2 Anatomic distribution of 298 solitary fibrous tumors of all sites.

The sizes of soft tissue SFTs vary from <1 cm to >10 cm. Larger examples are typically seen in body cavities. Malignant and metastatic examples have been variably represented and defined, but local recurrences have been more common than distant metastases. Current information is insufficient for formulating site-, tumor-size-, or mitosis-rate-specific criteria for risk of clinically malignant behavior. The threshold 4 mitoses per 10 HPFs, as originally identified by Enzinger and Smith for soft tissue HPC, is commonly used for SFTs.[30] SFTs with atypia, mitotic activity, or size >5 cm should be approached with caution, and such lesions should be considered as having at least an uncertain (if not definite) malignant potential until more precise criteria to define tumor behavior become available.[31] Complete excision and long-term follow-up is generally indicated.

Pathology

SFTs vary greatly in size from 1 cm to more than 30 cm, and the largest examples typically occur in the body cavities. Grossly, the SFT is typically well circumscribed, but unencapsulated. Large tumors can contain cysts. On sectioning, the tumor is most commonly gray-white and varies from tan-yellow to pink-red. Some pleural and peritoneal examples are attached with a narrow pedicle to the outer surfaces of the lung, liver, or intestines. The tissue is homogeneously firm and gray-white or pale tan on sectioning and can show trabeculation (Figure 12.3). Malignant SFTs can appear softer and contain areas of gross necrosis.

Microscopically typical are relatively uniform spindled cells set in variably collagenous, cellular, and, in some cases, myxoid matrix. An HPC-like vascular pattern is seen at least focally in

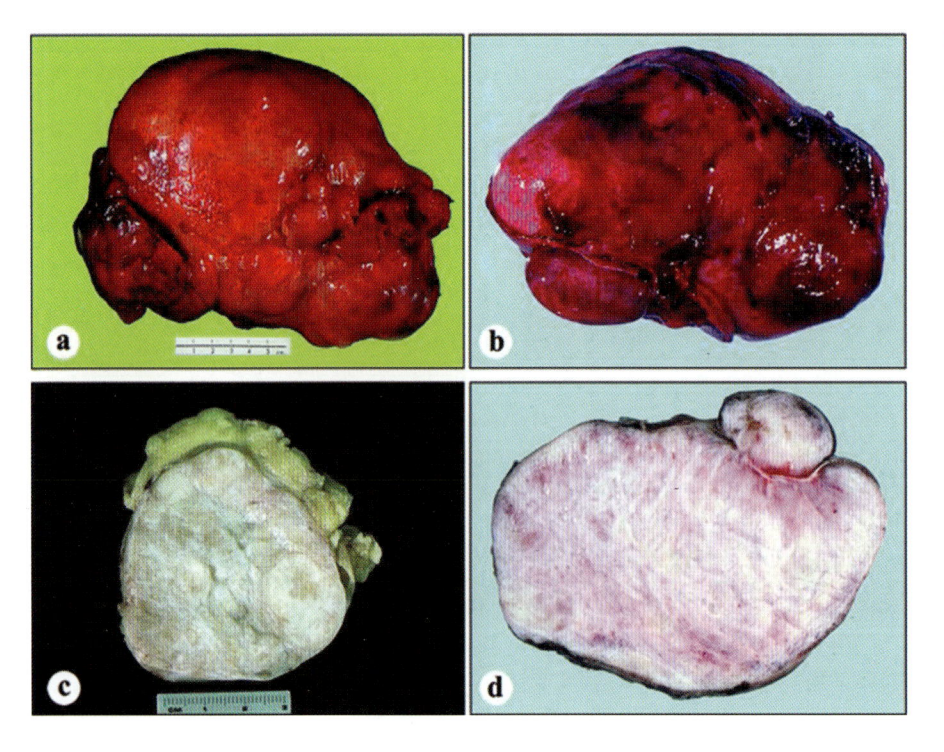

Figure 12.3 (a,b) Two examples of pleural solitary fibrous tumors composed of multiple nodules. (c) An inguinal formalin-fixed example shows a homogeneous grayish-white surface on sectioning. (d) This unfixed pleural solitary fibrous tumor has a trabeculated cut surface. (Same tumor as in b.)

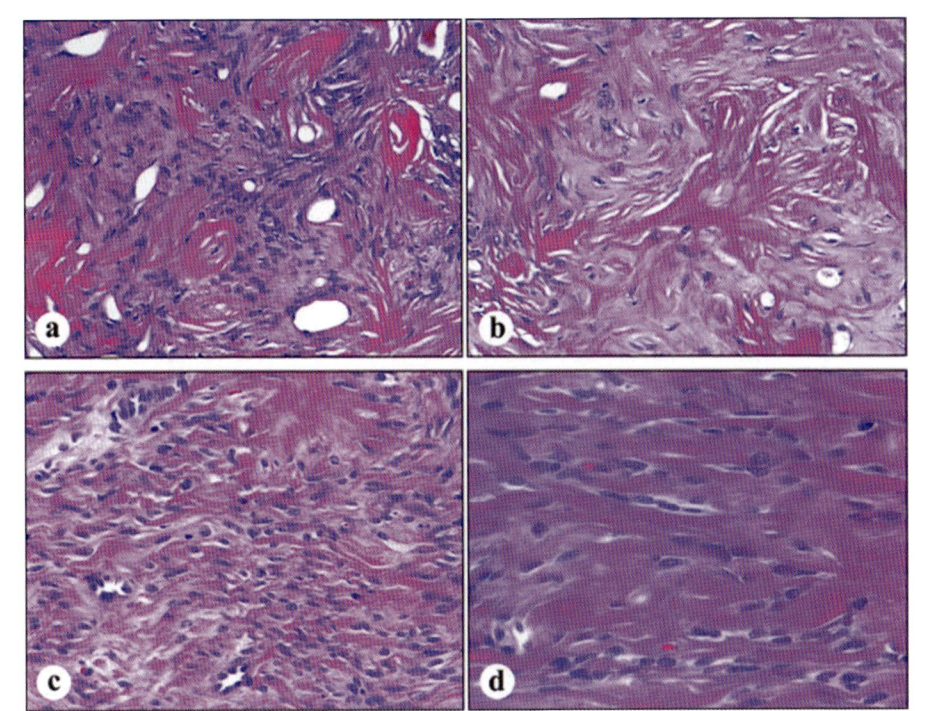

Figure 12.4 Examples of solitary fibrous tumor with overall paucicellular appearance. (a) Prominent vessels with hyalinized walls. (b) Extensive stromal sclerosis. (c) Moderate cellularity and prominent collagenous background. (d) Trabecular growth pattern and prominent collagenous matrix.

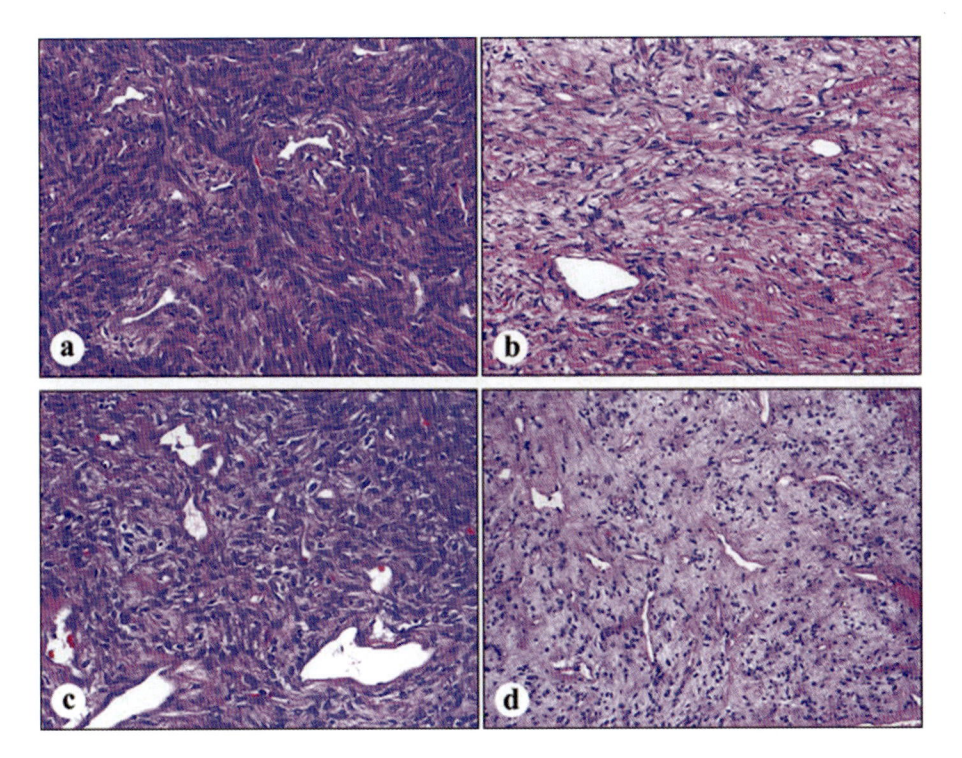

Figure 12.5 Variable cellularity and vascular patterns in solitary fibrous tumor. (d) In some cases, the collagenous matrix is myxoid.

most cases. The tumor cells form random or sometimes trabecular arrangements (Figures 12.4 to 12.6). Some variants are composed of slightly epithelioid cells in corded or reticular patterns. The tumor cells have indistinct cytoplasm and oval nuclei, usually with inconspicuous nucleoli. Mitotic activity is low (<2–3 mitoses per 10 HPFs) in most cases, but some tumors show overtly sarcomatous features with a high mitotic activity (Figure 12.7). Transition from histologically benign to malignant areas also can be observed. In some cases, a mitotically active focus is present in a tumor with conventional features (Figure 12.8). Because follow-up information about such cases is limited, their malignant potential is uncertain at the present. High-grade transformation (dedifferentiation) can occur and may also include osteosarcomatous differentiation.[32]

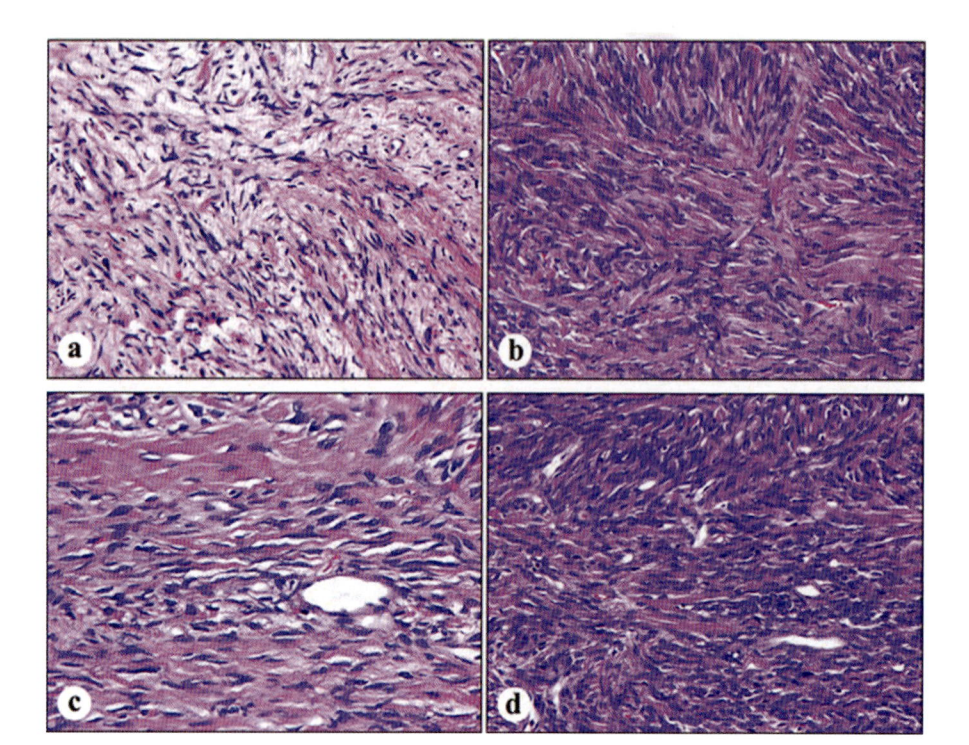

Figure 12.6 Moderately cellular solitary fibrous tumors with inconspicuous to distinct vascular pattern.

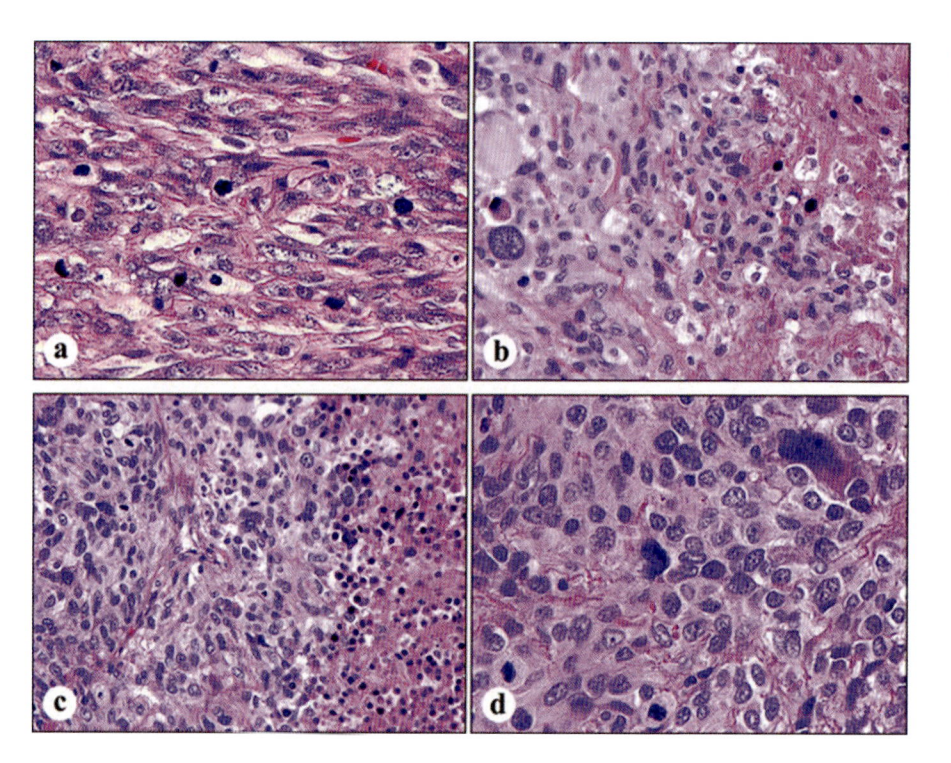

Figure 12.7 Overtly malignant solitary fibrous tumor, which was CD34 positive and showed conventional collagen-rich and mitotically active areas. (**a**) At least six mitotic figures are seen in this small area. (**b–d**) Note spindled to epithelioid cytology, zones of coagulative necrosis, general nuclear enlargement, and focal nuclear pleomorphism.

Immunohistochemistry

SFT cells are positive for vimentin, CD34, CD99, and BCL2, and generally negative for keratins and desmin.[33–35] Nuclear immunoreactivity for STAT6 is typical,[36–38] although not totally specific, as other tumors such as dedifferentiated liposarcoma can also be positive.[38,39]

Focal immunoreactivity for muscle actins does occur, but these tumors are negative for desmin, S100 protein, and EMA. Although most SFTs are negative for keratins, some examples, especially malignant ones, have shown keratin-positive cells.[7,17,28,31–33] Nuclear positivity for beta-catenin, known for desmoid tumor, is reportedly common in SFT

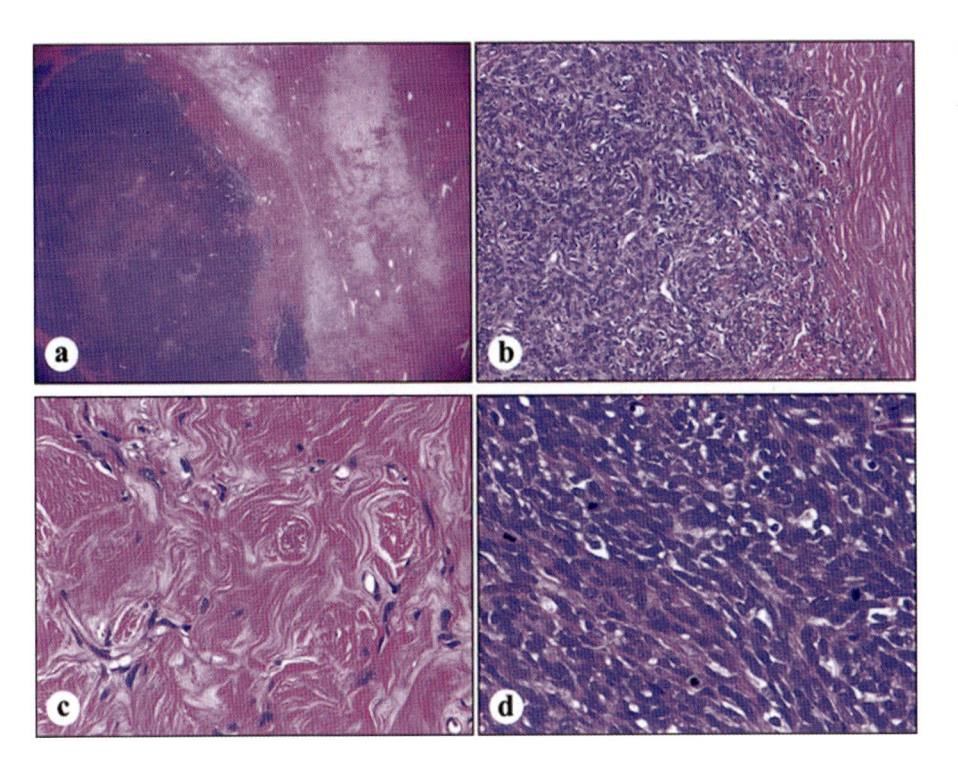

Figure 12.8 (**a**) Solitary fibrous tumor with a 1.5 cm hypercellular and mitotically active focus, judged to be of uncertain biologic potential. (**b**) The hypercellular area has a well-developed vascular pattern. (**c**) Most of the tumor consists of a collagen-rich paucicellular component. (**d**) The hypercellular focus contains numerous mitoses.

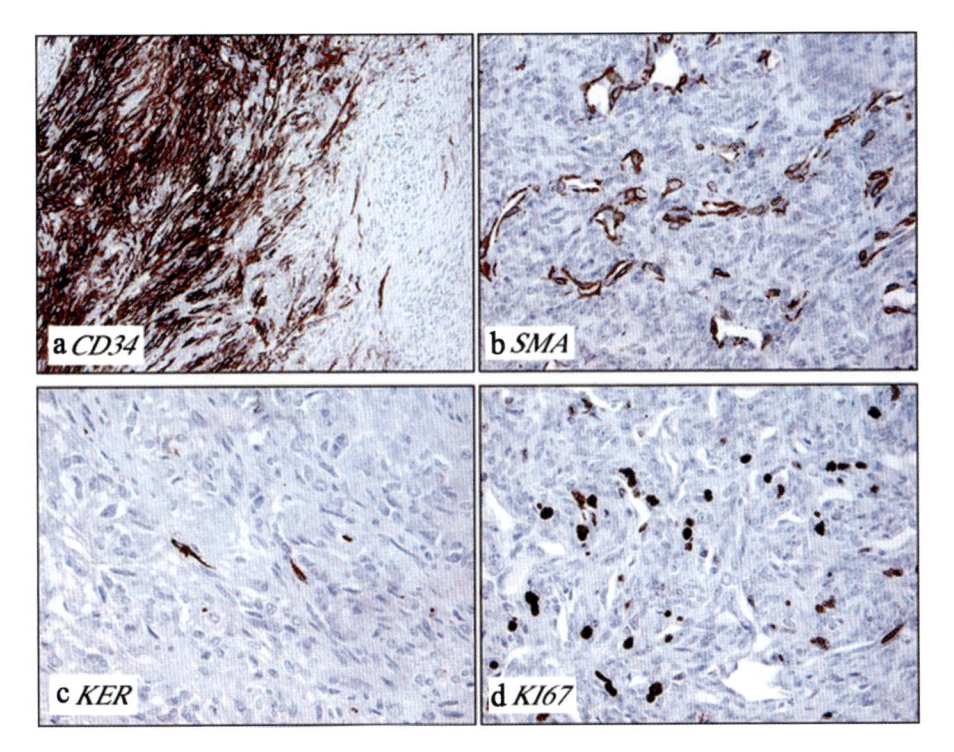

Figure 12.9 Immunohistochemical features of the hypercellular solitary fibrous tumor seen in Figure 12.8. (**a**) CD34 positivity, although typically strong and uniform, can be reduced in atypical and malignant variants. (**b**) Smooth muscle actin positivity is limited to normal pericytes. (**c**) Isolated keratin-positive cells can be seen in SFT. (**d**) Ki67-positive nuclei constitute approximately 5% of all tumor cell nuclei in this case, a higher index than is typically seen in SFT.

(40%).[40] The number of Ki67-positive nuclei is very low in conventional examples (1–2%), but is often elevated >5% in atypical and malignant variants. Definitive cut-off values for diagnostic use have not yet been developed, however (Figure 12.9).

Differential diagnosis

Lack of S100 protein reactivity separates this tumor from nerve sheath tumors. Dedifferentiated liposarcoma can have overlapping morphology with SFTs. Observation of atypical lipomatous component and demonstration of MDM2-positivity by immunohistochemistry or MDM2 gene amplification supports dedifferentiated liposarcoma.

Monophasic spindle cell synovial sarcoma is typically more homogeneous and contains more highly cellular spindle cell proliferation. It almost invariably shows immunoreactivity for keratins, EMA, or both, and almost never displays CD34 positivity.

Giant cell angiofibroma of the orbit and peripheral soft tissues is a variant of SFT; both are CD34-positive fibroblastic neoplasms (see Giant cell angiofibroma).

Genetics

NAB2-STAT6 gene fusion is the recurrent genetic change present in most SFTs of soft tissue. This fusion also occurs in both meningeal HPCs and SFTs, indicating that they belong to the spectrum of one family of tumors. The gene fusion induces BCL2 expression, shifting the balance of cell apoptosis to survival, and activates early growth response 1 (EGR1) signaling pathways.[36,41,42]

There may be phenotype-genotype correlation among the NAB2-STAT6 fusion variants. One study found NAB2 exon 4–STAT6 exon 2/3 fusion to correlate with generally benign pleural SFTs, whereas NAB2 exon 6–STAT exon 16/17 fusion was associated with highly cellular and deep HPC-like tumors.[43] However, it is likely that the accumulation of other, translocation-independent secondary changes also plays a role in the tumor progression.

Heterogeneous results have been obtained on cytogenetic analysis of isolated cases and have not been of diagnostic significance. None of those translocations involved the loci of NAB2 or STAT6 (both at 12q13). The NAB2-STAT6 fusion is very difficult to find by cytogenetics, as this translocation is an intrachromosomal inversion involving adjacent chromosomal loci. Comparative genomic hybridization studies have shown a few changes in small tumors, including losses in 13q, whereas the larger and histologically malignant tumors have shown numerous gains, including those in chromosome 8, as seen in some other soft tissue tumors.[44] Losses of 13q were also observed in another CGH study.[45]

Giant cell angiofibroma variant of solitary fibrous tumor

This tumor, originally reported by Dei Tos et al. in the orbit, especially the lacrimal gland region, is a morphologic and clinicopathologic variant of SFT.[46] Convincing arguments have been presented for its being a variant of SFT, especially because tumors with giant cell angiofibroma features often contain SFT-like components.[47]

Clinical features

Reported tumors have occurred in adult patients with a wide age range, but these tumors tend to occur at a younger age than do SFTs (median age, 46 years). The aggregate data show equal gender distribution, although orbital tumors were reported more often in men and peripheral examples in women. After the initial report, giant cell angiofibroma has been documented in other soft tissue locations, including head and neck, back, retroperitoneum, hip, and vulva. More often, the peripheral examples have been subcutaneous rather than deep. The orbital tumors are usually small (1–3 cm), whereas the peripheral tumors have varied from 1 cm to 11 cm. The tumor has a benign behavior, based on limited follow-up, although nondestructive recurrences were reported in the orbital tumors.[46–48]

Pathology

Histologically the tumor is highly vascular, often with a HPC-like vascular pattern with hyalinized vessel walls. The main tumor cell population is composed of uniform oval or spindled cells in a collagenous stroma, with little or no mitotic activity. Distinctive is the presence of scattered multinucleated giant cells, which often line pseudovascular cystic spaces along with mononuclear tumor cells (Figure 12.10).

The tumor cells are strongly positive for CD34 and CD99, and negative for desmin, similar to SFT and HPC. Although the original report found the tumor to be negative for smooth muscle actin and S100 protein, a later series reports some cases as positive for these markers.[47]

Hemangiopericytoma

HPC of soft tissues and SFT have histologic and immunophenotypic similarities (CD34 positive), and these entities are considered closely related in the 2010 WHO classification of soft tissue tumors.[49] NAB2-STAT6 gene fusion has been identified in both, although fusion types may at least partially segregate SFTs and HPCs (see genetics of SFT). For this reason we maintain some separation between SFTs and HPCs, while acknowledging that they belong to the same family of tumors.

Whereas CD34-positive tumors with dominating fibrous components and a spindle cell pattern are typically classified as SFTs, whether they have an HPC-like vascular pattern or not, highly cellular collagen-poor tumors composed of oval cells with an HPC-like vascular pattern have been classified as HPCs, as long as other diagnoses are ruled out. Many sarcomas, especially synovial sarcoma, mesenchymal chondrosarcoma, some liposarcomas, and certain carcinomas can have an HPC-like staghorn vascular pattern.[30,50,51] Therefore, these tumors must be ruled out before making a diagnosis of malignant HPC.

Despite the name, there is no evidence that HPC is a pericytic tumor. Although older cell culture and ultrastructural observations by Stout and Murray seemed to support a pericytic derivation,[52] this has not been confirmed by immunohistochemistry, because almost all HPCs lack the actins expressed in pericytes.[53,54] Two of four cases studied by Stout and Murray were tumors in small children, whose HPC-like

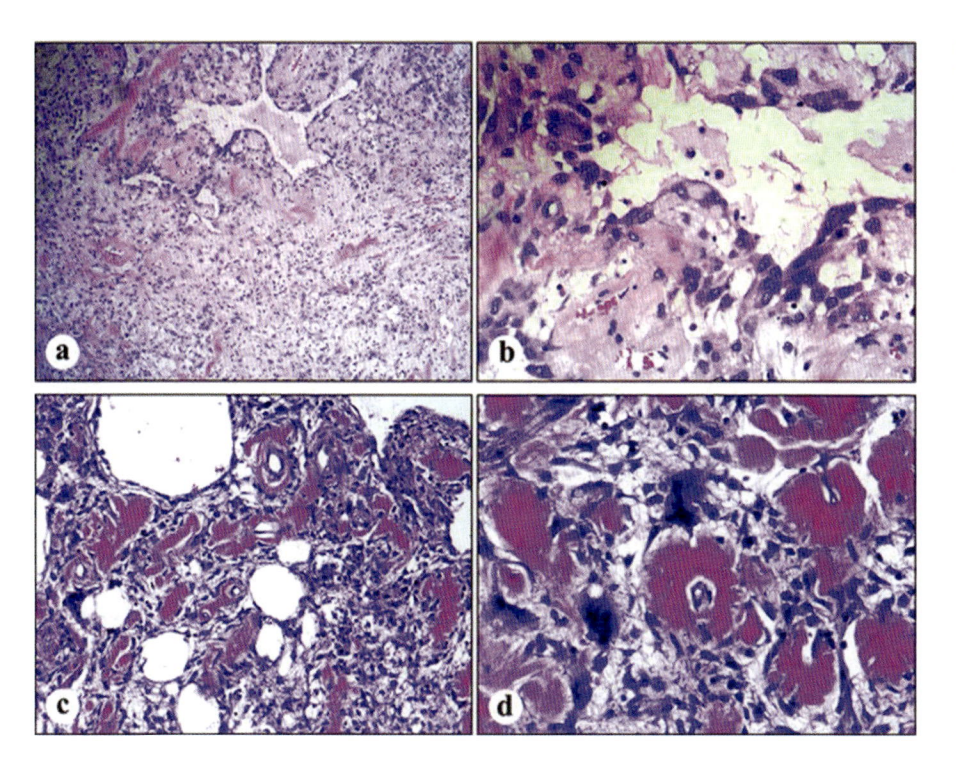

Figure 12.10 (**a,b**) Giant cell angiofibroma tends to form pseudovascular spaces lined by tumor cells, including multinucleated giant cells. (**c,d**) Another case also shows a microcystic appearance with hyalinized vessel walls and dispersed giant cells.

tumors have mostly been considered infantile myofibromas (Chapter 9), true myopericytic tumors, based on more recent studies.[55]

Sinonasal HPC is a peculiar, site-specific tumor that differs from ordinary HPC clinically, histologically, and immunohistochemically by its expression of actins.[56] This tumor is discussed in Chapter 23.

It seems that in the central nervous system (meninges) the designation HPC has been used for more highly cellular higher-grade tumors and SFT for lower-grade tumors, with the latter group seeming to have a better prognosis.[57–59]

Meningeal SFT/HPCs, previously called *angioblastic meningiomas*, have been well studied. They can occur in any part of the central nervous system, but are usually supratentorial and located at the meninges. They are difficult to control surgically and tend to recur repeatedly; some ultimately metastasize, often after long intervals. Distant metastases develop, especially to bones and to some degree to liver, lungs, and soft tissue, among others. The metastatic rate in the largest reported series varied from 14% to 30%, and long-term cure was achieved in only 25% of patients. Patients with tumor necrosis and whose tumors have a mitotic rate >5 per 10 HPFs fare worse than others.[57–59]

In contrast to their meningeal counterparts, most HPCs in peripheral soft tissues are clinically indolent. A higher mortality rate is associated with internal tumors, such as those in the pelvis and retroperitoneum. In two recent series from cancer hospitals, patients with HPC had a 71% to 86% 5-year actuarial survival.[60,61] In one series, overtreatment was also considered a potential problem in view of generally good prognosis.[61] Metastases (mostly to lungs or bones) rarely develop, sometimes long (10–15 years or more) after the primary surgery and sometimes following histologic sarcomatous transformation. Older series with substantially higher mortality numbers may have included sarcomas with an HPC-like pattern, especially synovial sarcomas.

In the largest series of peripheral HPCs by Enzinger and Smith (who already had the awareness to exclude other sarcomas with an HPC-pattern, e.g., synovial sarcoma), the recurrence rate was 17% and 12 of 93 patients (13%) died of metastatic disease, a figure comparable to those in modern series.[30] One family with three family members with HPC was reported but no candidate genes were identified.[62]

Pathology

Grossly, HPCs are circumscribed oval tumors surrounded by a fibrous pseudocapsule. On sectioning they can appear spongy because of vascular slits and vary from yellowish tan to reddish. Yellowish-green zones of necrosis are sometimes present. Examples in peripheral soft tissues vary from small subcutaneous nodules of 1 cm to 3 cm, to large, deep intramuscular and retroperitoneal masses that are commonly >10 cm and can reach a size of 20 cm.

Histologically, HPC has a prominent vascular pattern with gaping vessels lined by a single layer of normal-appearing or attenuated endothelia (Figure 12.11). The gaping or rounded profiles of vessels are sometimes surrounded by zones of perivascular hyalinization. Continuous intercellular network is typically highlighted with a reticulin stain, used in the classic definition of HPC.

Cytologically, the tumor cells are typically uniform, oval to slightly spindled, with round or oval nuclei and poorly

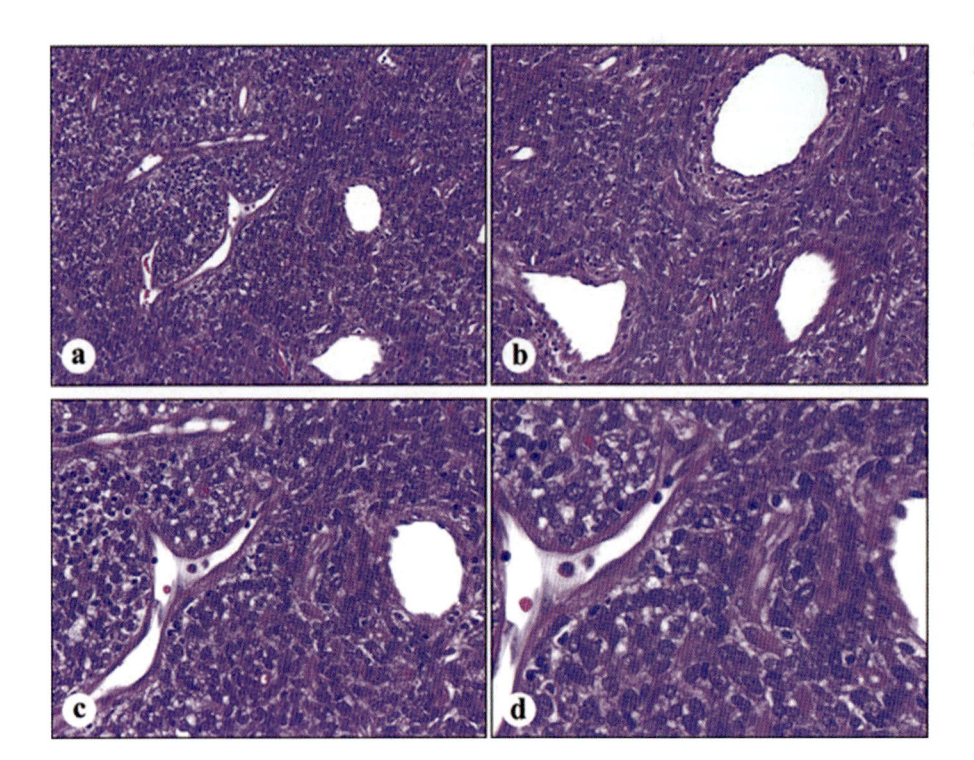

Figure 12.11 Prototypic hemangiopericytoma with a prominent, focally staghorn-like vascular profile. The tumor cells are oval and uniform, and collagenous matrix is less conspicuous than in solitary fibrous tumor.

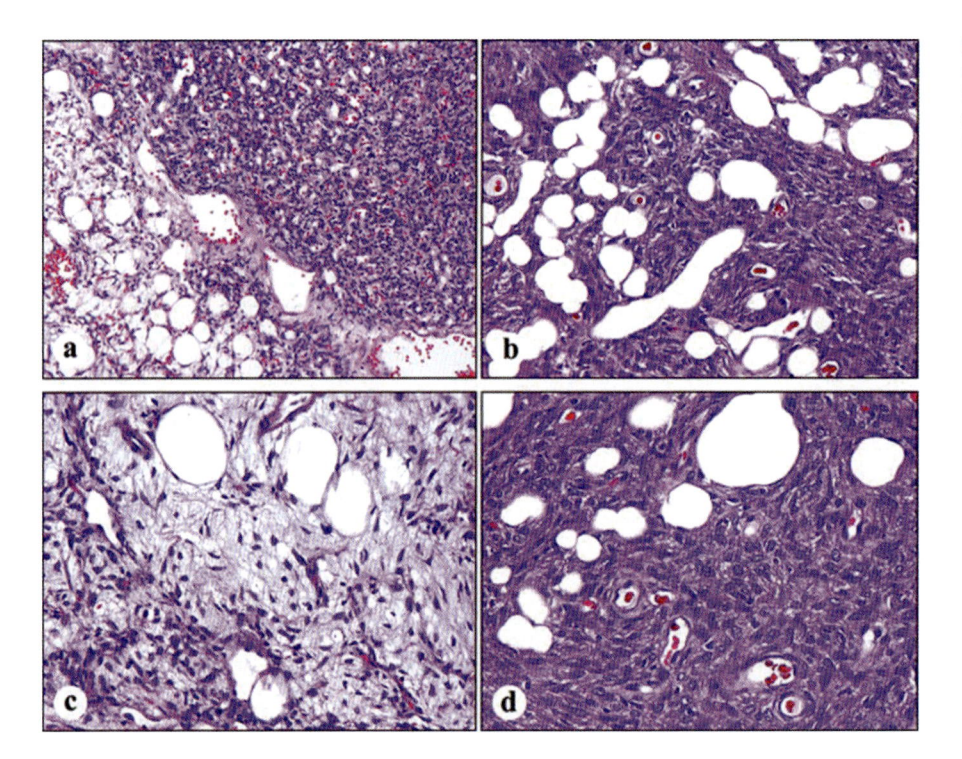

Figure 12.12 Lipomatous hemangiopericytoma. (**a**) Sharply demarcated areas of typical and fat-infiltrated hemangiopericytoma. (**b–d**) Fatty elements in typical and myxoid hemangiopericytoma.

visualized cell borders. The mitotic rate is generally low (up to 3 per 10 HPFs). A mitotic rate of more than 4 per 10 HPFs, and the presence of coagulative necrosis, nuclear pleomorphism, and areas of spindle cell sarcomatous transformation indicate malignant potential.

Histologic variant: lipomatous hemangiopericytoma

A peculiar variant with a nontypical mature fatty component in a histologically typical HPC has been termed *lipomatous HPC*. This component can be focal or dispersed (Figure 12.12).

Clinicopathologic features of lipomatous HPC are similar to those of the ordinary variant,[63-65] and thus far, no malignant examples have been reported. A variety of sites have been involved, including head and neck, extremities, abdomen-pelvis-retroperitoneum, and trunk wall. In one series, all extremity-based tumors were in a deep location.[64] This variant can also be viewed as a variant of SFT (preferred by some authors).[65]

Immunohistochemistry and ultrastructure

Similar to SFTs, HPCs also show nuclear STAT6 and cytoplasmic CD34 expression. The tumor cells are mesenchymal cells that do not have the phenotypic features of mature pericytes and are negative for smooth muscle actin. In general, immunohistochemical features are similar to those of SFT. On comparison with SFTs, meningeal HPCs may have more variable, sometimes only patchy CD34 expression.[66]

Earlier ultrastructural studies suggested pericytic-like differentiation based on cytoplasmic processes, basal-lamina material, and intimate relationship with complex capillaries.[63] In another study, the tumor cells were not considered comparable with adult but rather with immature pericytes.[67]

Differential diagnosis

Other tumors can have a HPC-like histologic pattern. Notable sarcomas with such features are monophasic and poorly differentiated synovial sarcomas, mesenchymal chondrosarcomas, and certain liposarcomas. Epithelial tumors that can have an HPC-like pattern include variants of thymomas and poorly differentiated thyroid carcinomas. These diagnoses should be ruled out before a diagnosis of thymic or thyroid HPC is made. Separation from SFTs is based on the prominent vascular pattern, relative paucity of interstitial collagen, and more ovoid versus spindle cell pattern in HPCs, but this distinction is arbitrary.

Genetics

The NAB2-STAT6 fusion has been discussed in the genetics of SFTs. Most reported genotypes of HPCs have had simple diploid or near-diploid karyotypes. A recurrent t(12;19)(q13;q13.3) translocation has been reported in retroperitoneal[68] and meningeal HPCs.[69] In addition, 12q13 and 19q13 have been separately involved in several complex rearrangements.[69] 12q13 harbors the region containing both NAB2 and STAT6 genes involved in the typical SFT/HPC translocation (see genetics of SFT). Meningeal HPCs lack neurofibromatosis type 2 gene mutations, as often seen in meningiomas, supporting a different pathogenesis.[70]

References

Solitary fibrous tumor

1. England DM, Hochholzer L, McCarthy MJ. Localized benign and malignant fibrous tumors of the pleura: a clinicopathologic review of 223 cases. *Am J Surg Pathol* 1989;13:640–658.
2. Briselli M, Mark EJ, Dickersin GR. Solitary fibrous tumors of the pleura: eight new cases and review of 360 cases in the literature. *Cancer* 1981;47:2678–2689.
3. Witkin GB, Rosai J. Solitary fibrous tumor of the mediastinum: a report of 14 cases. *Am J Surg Pathol* 1989;13:547–557.
4. el-Naggar AK, Ro JY, Ayala AG, Ward R, Ordonez NG. Localized fibrous tumor of the serosal cavities: immunohistochemical, electron-microscopic, and flow-cytometric DNA study. *Am J Clin Pathol* 1989;92:561–565.
5. Young RH, Clement PB, McCaughey WT. Solitary fibrous tumors ("fibrous mesotheliomas") of the peritoneum: a report on three cases and a review of literature. *Arch Pathol Lab Med* 1990;114:493–495.
6. Suster S, Nascimento AG, Miettinen M, Sickel JZ, Moran CA. Solitary fibrous tumor of soft tissue: a clinicopathologic and immunohistochemical study of 12 cases. *Am J Surg Pathol* 1995;19:1257–1266.
7. Nascimento AG. Solitary fibrous tumor: a ubiquitous neoplasm of mesenchymal differentiation. *Adv Anat Pathol* 1996;3:388–395.
8. Strom EH, Skjorten F, Aarseth LB, Haug E. Solitary fibrous tumor of the pleura: an immunohistochemical, electron microscopic and tissue culture study of a tumor producing insulin-like growth factor I in a patient with hypoglycemia. *Pathol Res Pract* 1991;187:109–113.
9. Fukasawa Y, Takada A, Tateno M, *et al.* Solitary fibrous tumor of the pleura causing recurrent hypoglycemia by secretion of insulin-like growth factor II. *Pathol Int* 1998;48:47–52.
10. Fukunaga M, Naganuma H, Nikaido T, Harada T, Ushigome S. Extrapleural solitary fibrous tumor: a report of seven cases. *Mod Pathol* 1997;10:443–450.
11. Vallat-Decouvelaere AV, Dry SM, Fletcher CD. Atypical and malignant solitary fibrous tumors in extrathoracic locations: evidence of their comparability to intra-thoracic tumors. *Am J Surg Pathol* 1998;22:1501–1511.
12. Brunnemann RB, Ro JY, Ordonez NG, *et al.* Extrapleural solitary fibrous tumor: a clinicopathologic study of 24 cases. *Mod Pathol* 1999;12:1034–1042.
13. Hasegawa T, Matsuno Y, Shimoda T, *et al.* Extrathoracic solitary fibrous tumors: their histological variability and potentially aggressive behavior. *Hum Pathol* 1999;30:1464–1473.
14. Morimitsu Y, Nakajima M, Hisaoka M, Hashimoto H. Extrapleural solitary fibrous tumor: clinicopathologic study of 17 cases and molecular analysis of the p53 pathway. *APMIS* 2000;108:617–625.
15. Moran CA, Ishak KG, Goodman ZD. Solitary fibrous tumor of the liver: a clinicopathologic and immunohistochemical study of nine cases. *Ann Diagn Pathol* 1998;2:19–24.
16. Dorfman DM, To K, Dickersin GR, Rosenberg AE, Pilch BZ. Solitary fibrous tumor of the orbit. *Am J Surg Pathol* 1994;18:281–287.
17. Westra WH, Gerald WL, Rosai J. Solitary fibrous tumor: consistent CD34 immunoreactivity and occurrence in

18. Bernardini FP, de Conciliis C, Schneider S, Kersten RC, Kulwin DR. Solitary fibrous tumor of the orbit: is it rare? Report of a case series and review of literature. *Ophthalmology* 2003;110:1442–1448.

19. Zukerberg LR, Rosenberg AE, Randolph G, Pilch BZ, Goodman ML. Solitary fibrous tumor of the nasal cavity and paranasal sinuses. *Am J Surg Pathol* 1991;15:126–130.

20. Witkin GB, Rosai J. Solitary fibrous tumor of the upper respiratory tract: a report of six cases. *Am J Surg Pathol* 1991;15:842–848.

21. Alawi F, Stratton D, Freedman PD. Solitary fibrous tumor of oral soft tissues: a clinicopathologic and immunohistochemical study of 16 cases. *Am J Surg Pathol* 2001;25:900–910.

22. Carneiro SS, Scheithauer BW, Nascimento AG, Hirose T, Davis DH. Solitary fibrous tumor of the meninges: a lesion distinctive from fibrous meningioma: a clinicopathologic and immunohistochemical study. *Am J Clin Pathol* 1996;106:217–224.

23. Tihan T, Viglione M, Rosenblum MK, Olivi A, Burger PC. Solitary fibrous tumor in the central nervous system: a clinicopathologic review of 18 cases and comparison to meningeal hemangiopericytomas. *Arch Pathol Lab Med* 2003;127:432–439.

24. Bainbridge TC, Singh RR, Mentzel T, Katenkamp D. Solitary fibrous tumor of urinary bladder: report of two cases. *Hum Pathol* 1997;28:1204–1206.

25. Westra WH, Grenko RT, Epstein J. Solitary fibrous tumor of the lower urogenital tract: a report of five cases involving the seminal vesicles, urinary bladder, and prostate. *Hum Pathol* 2000;31:63–68.

26. Wang J, Arder DA, Frankel K, Weiss LM. Large solitary fibrous tumor of the kidney: report of two cases and review of the literature. *Am J Surg Pathol* 2001;25:1194–1199.

27. Nielsen GP, O'Connell JX, Dickersin GR, Rosenberg AE. Solitary fibrous tumor of soft tissue: a report of 15 cases, including 5 malignant examples with light microscopic, immunohistochemical and ultrastructural data. *Mod Pathol* 1997;10:1028–1037.

28. Hanau CA, Miettinen M. Solitary fibrous tumor: histological and immunohistochemical spectrum of benign and malignant variants presenting at different sites. *Hum Pathol* 1995;26:440–449.

29. Erdag G, Qureshi HS, Petterson JW, Wick MR. Solitary fibrous tumor of the skin: a clinicopathologic study of 10 cases and review of the literature. *J Cutan Pathol* 2007;34:844–850.

30. Enzinger FM, Smith BH. Hemangiopericytoma: an analysis of 106 cases. *Hum Pathol* 1976;7:61–82.

31. Demicco EG, Park MS, Araujo DM, et al. Solitary fibrous tumor: a clinicopathological study of 110 cases and proposed risk assessment model. *Mod Pathol* 2012;25:1298–1306.

32. Thway K, Hayes A, Ieremia E, Fisher C. Heterologous osteosarcomatous and rhabdomyosarcomatous elements in dedifferentiated solitary fibrous tumor: further support for the concept of dedifferentiation in solitary fibrous tumor. *Ann Diagn Pathol* 2013;17:457–463.

33. van de Rijn M, Lombard CM, Rouse RV. Expression of CD34 by solitary fibrous tumors of the pleura, mediastinum, and lung. *Am J Surg Pathol* 1994;18:814–820.

34. Flint A, Weiss SW. CD-34 and keratin expression distinguishes solitary fibrous tumor (fibrous mesothelioma) of pleura from desmoplastic mesothelioma. *Hum Pathol* 1995;26:428–431.

35. Chilosi M, Facchetti F, Dei Tos AP, et al. bcl-2 expression in pleural and extrapleural solitary fibrous tumours. *J Pathol* 1997;181:362–367.

36. Schweizer L, Koelsche C, Sahm F, et al. Meningeal hemangiopericytoma and solitary fibrous tumors carry the NAB2-STAT6 fusion and can be diagnosed by nuclear expression of STAT6 protein. *Acta Neuropathol* 2013;125:651–658.

37. Doyle LA, Vivero M, Fletcher CD, Mertens F, Hornick JL. Nuclear expression of STAT6 distinguishes solitary fibrous tumor from histologic mimics. *Mod Pathol* 2014;27:390–395.

38. Yoshida A, Tsuta K, Ohno M, et al. STAT6 immunohistochemistry is helpful in the diagnosis of solitary fibrous tumors. *Am J Surg Pathol* 2014;38:552–559.

39. Doyle LA, Tao D, Mariño-Enríquez A. STAT6 is amplified in a subset of dedifferentiated liposarcoma. *Mod Pathol* 2014;27:1231–1237.

40. Ng TL, Gown AM, Barry TS, et al. Nuclear beta-catenin in mesenchymal tumors. *Mod Pathol* 2005;18:68–74.

41. Robinson DR, Wu YM, Kalyana-Sundaram S, et al. Identification of recurrent NAB2-STAT6 gene fusions in solitary fibrous tumor by integrative sequencing. *Nat Genet* 2013;45:180–185.

42. Mohajeri A, Tayebwa J, Collin A, et al. Comprehensive genetic analysis identifies a pathognomonic NAB2/STAT6 fusion gene, nonrandom secondary genomic imbalances, and a characteristic gene expression profile in solitary fibrous tumor. *Genes Chromosomes Cancer* 2013;52:873–886.

43. Barthelmeß S, Geddert H, Boltze C, et al. Solitary fibrous tumors/hemangiopericytomas with different variants of the NAB2-STAT6 gene fusion are characterized by specific histomorphology and distinct clinicopathological features. *Am J Pathol* 2014;184:1209–1218.

44. Miettinen MM, el-Rifai W, Sarlomo-Rikala M, Anderson LC, Knuutila S. Tumor size-related DNA copy number changes occur in solitary fibrous tumors but not in hemangiopericytomas. *Mod Pathol* 1997;10:1194–1200.

45. Ness GO, Lybaeck H, Ames J, Rodahl F. Chromosomal imbalances in a recurrent solitary fibrous tumor of the orbit. *Cancer Genet Cytogenet* 2005;162:38–44.

Giant cell angiofibroma

46. Dei Tos A, Seregard S, Calonje E, Chan JKC, Fletcher CDM. Giant cell angiofibroma: a distinctive orbital tumor in adults. *Am J Surg Pathol* 1995;19:1286–1293.

47. Guillou L, Gebhard S, Coindre JM. Orbital and extraorbital giant cell angiofibroma: a giant cell-rich variant of solitary fibrous tumor? Clinicopathologic, immunohistochemical, and ultrastructural analysis of a series in favor of a unifying concept. *Am J Surg Pathol* 2000;24:971–979.

48. Thomas R, Banerjee SS, Eyden BP, et al. A study of four cases of giant cell angiofibroma with documentation of some unusual features. *Histopathology* 2001;39:390–396.

Hemangiopericytoma

49. Fletcher CDM, Bridge JA, Lee JC. Extrapleural solitary fibrous tumour. In *Pathology and Genetics: Tumours of Soft Tissue and Bone*, Fletcher CDM, Bridge JA, Hogendoorn PCW, Mertens F (eds.) Lyon: WHO; 2013:80–82.

50. Tsuneyoshi M, Daimaru Y, Enjoji M. Malignant hemangiopericytoma and other sarcomas with hemangiopericytoma-like pattern. *Pathol Res Pract* 1984;178:446–453.

51. Nappi O, Ritter JH, Pettinato G, Wick MR. Hemangiopericytoma: histopathological pattern or clinicopathologic entity? *Semin Diagn Pathol* 1995;12:221–232.

52. Stout AP, Murray MR. Hemangiopericytoma: a vascular tumor featuring Zimmerman's pericytes. *Ann Surg* 1942;116:26–31.

53. Miettinen M. Antibody specific to muscle actins in the differential diagnosis and classification of soft tissue tumors. *Am J Pathol* 1988;130:205–210.

54. Porter PL, Bigler SA, McNutt M, Gown AM. The immunophenotype of hemangiopericytomas and glomus tumors, with special reference to muscle protein expression: an immunohistochemical study and review of the literature. *Mod Pathol* 1991;4:46–52.

55. Mentzel T, Calonje E, Nascimento AG, Fletcher CD. Infantile hemangiopericytoma versus infantile myofibromatosis: study of a series suggesting a continuous spectrum of infantile myofibroblastic lesions. *Am J Surg Pathol* 1994;18:922–930.

56. Thompson LD, Miettinen M, Wenig BM. Sinonasal type hemangiopericytoma: a clinicopathologic and immunophenotypic analysis of 104 cases showing perivascular myoid differentiation. *Am J Surg Pathol* 2003;27:737–749.

57. Jaaskelainen J, Servo A, Haltia M, Wahlstrom T, Valtonen S. Intracranial hemangiopericytoma: radiology, surgery, radiotherapy, and outcome in 21 patients. *Surg Neurol* 1985;23:227–236.

58. Guthrie BL, Ebersold MJ, Scheithauer BW, Shaw EG. Meningeal hemangiopericytoma: histopathological features, treatment, and long-term follow-up of 44 cases. *Neurosurgery* 1989;25:514–522.

59. Mena H, Ribas JL, Pezeshkpour GH, Cowan DN, Parisi JE. Hemangiopericytoma of the central nervous system: a review of 94 cases. *Hum Pathol* 1991;22:84–91.

60. Spitz FR, Bouvet M, Pisters PW, Pollock RE, Feig BW. Hemangiopericytoma: a 20-year single-institution experience. *Ann Surg Oncol* 1998;5:350–355.

61. Espat NJ, Lewis JJ, Leung D, et al. Conventional hemangiopericytoma: modern analysis of outcome. *Cancer* 2002;95:1746–1751.

62. Plukker JT, Koops HS, Molenaar I, et al. Malignant hemangiopericytoma in three kindred members of one family. *Cancer* 1988;61:841–844.

63. Nielsen GP, Dickersin GR, Provenzal JM, Rosenberg AE. Lipomatous hemangiopericytoma: a histologic, ultrastructural and immunohistochemical study of a unique variant of hemangiopericytoma. *Am J Surg Pathol* 1995;19:748–756.

64. Folpe AL, Devaney K, Weiss SW. Lipomatous hemangiopericytoma: a rare variant of hemangiopericytoma that may be confused with liposarcoma. *Am J Surg Pathol* 1999; 23:1201–1207.

65. Guillou L, Gebhard S, Coindre JM. Lipomatous hemangiopericytoma: a fat-containing variant of solitary fibrous tumor? Clinicopathologic, immunohistochemical, and ultrastructural analysis of a series in favor of a unifying concept. *Hum Pathol* 2000;31:1108–1115.

66. Perry A, Scheithauer BW, Nascimento AG. The immunophenotypic spectrum of meningeal hemangiopericytoma: a comparison with fibrous meningioma and solitary fibrous tumor of meninges. *Am J Surg Pathol* 1997;21:1354–1360.

67. Dardick I, Hammar SP, Scheithauer BW. Ultrastructural spectrum of hemangiopericytoma: a comparative study of fetal, adult, and neoplastic pericytes. *Ultrastruct Pathol* 1989; 13:111–154.

68. Streekantaiah C, Bridge JA, Rao UN, Neff JR, Sandberg AA. Clonal chromosomal abnormalities in hemangiopericytoma. *Cancer Genet Cytogenet* 1991;54:173–181.

69. Henn W, Wullich B, Thönnes M, et al. Recurrent t(12;19)(q13;q13.3) in intracranial and extracranial hemangiopericytoma. *Cancer Genet Cytogenet* 1993;71:151–154.

70. Joseph JT, Lisle DK, Jacoby LB, et al. NF2 gene analysis distinguishes hemangiopericytoma from meningiomas. *Am J Pathol* 1995;147:1450–1455.

Chapter 13
Fibroblastic and myofibroblastic neoplasms with malignant potential

Markku Miettinen
National Cancer Institute, National Institutes of Health

Fibroblastic and myofibroblastic neoplasms with malignant potential span a wide clinicopathological spectrum. Some of the entities listed here are nonmetastasizing and have recurrence potential only, whereas others have full metastatic potential. Solitary fibrous tumors are discussed in Chapter 12.

Dermatofibrosarcoma protuberans (DFSP) and its pigmented variant (*Bednar tumor*) are clinicopathologically distinctive CD34-positive fibroblastic tumors. Although typically they behave indolently if completely excised, they are capable of transformation to a more aggressive form that can metastasize. *Giant cell fibroblastoma* (GCF) is the juvenile variant of DFSP, with a lower biologic potential.

Low-grade fibromyxoid sarcoma (LGFMS) is a histologically distinctive tumor that can metastasize despite its bland appearance. This tumor is also capable of progressing to a more aggressive form.

Sclerosing epithelioid fibrosarcoma (SEF) is the designation for a clinically heterogeneous group of tumors with histologically distinctive features, and some tumors in this category represent progressive forms of LGFMS. Inflammatory fibrosarcoma belongs to the category of inflammatory myofibroblastic tumor and is discussed with fibroblastic tumors of children in Chapter 10.

Adult fibrosarcoma is the designation for a nonpleomorphic fibroblastic malignancy. This diagnosis is one of exclusion and is now rarely made because many tumors historically classified as adult fibrosarcomas are now diagnosed as monophasic synovial sarcoma, LGFMS, desmoid fibromatosis, solitary fibrous tumor, and even benign conditions such as florid nodular fasciitis.

Acral myxoinflammatory fibroblastic sarcoma (*inflammatory myxohyaline tumor of distal extremities*) is a newly described fibroblastic lesion of low malignant potential. Although this tumor commonly recurs, metastases are exceptional if they occur.

Undifferentiated pleomorphic sarcoma (UPS), also known as *malignant fibrous histiocytoma* (MFH), is a common pleomorphic sarcoma. MFH is handled by individual subtype, because each of them has a different histology also known as *pleomorphic* and terminologic and conceptual evolution. The tumor originally described as angiomatoid MFH, however, is now classified as *angiomatoid fibrous histiocytoma*. This specific childhood tumor is discussed in Chapter 10.

Myxofibrosarcoma (MFS), synonymous with myxoid MFH, is the most common MFH variant. It is understood as a fibroblastic and myofibroblastic tumor diagnosed by exclusion.

Immunohistochemistry is often necessary to rule out pleomorphic tumors with specific differentiation, such as carcinoma, melanoma, leiomyosarcoma, and others. The frequency of UPS is being debated, as some investigators maintain that this tumor is very rare after other sarcomas are excluded.

Inflammatory MFH is a rare variant that could be related to dedifferentiated liposarcoma. Giant cell MFH is a high-grade sarcoma with osteoclastic giant cells, and giant cell tumor of soft parts is its low-grade analog.

Pleomorphic hyalinizing angiectatic tumor is a clinicopathologically distinctive low-grade tumor with a predilection for the ankle and foot.

Dermatofibrosarcoma protuberans

DFSP is a relatively common superficial low-grade sarcoma mainly involving the skin and subcutis, with a predilection for truncal sites. Its variant, with scattered melanin-pigmented cells, is referred to as a Bednar tumor. GCF is the juvenile variant of DFSP, as supported by their occasional coexistence and immunohistochemical and genetic similarity. Positivity for CD34 and a specific t(17;22) translocation with COL1A-PDGFB gene fusion are important diagnostic features that can help to distinguish DFSP from other superficial sarcomas.

Clinical features

DFSP occurs in a wide age range, but is particularly common in young adults, with a significant male predominance. It also presents in older people and in children, including congenital occurrence (Figure 13.1). Many patients have a long history of tumor before surgery, indicating that true age of onset is much earlier than clinically detected.[1] Occurrence of multicentric lesions has been reported in association with adenosine deaminase immunorediciency syndrome.[2]

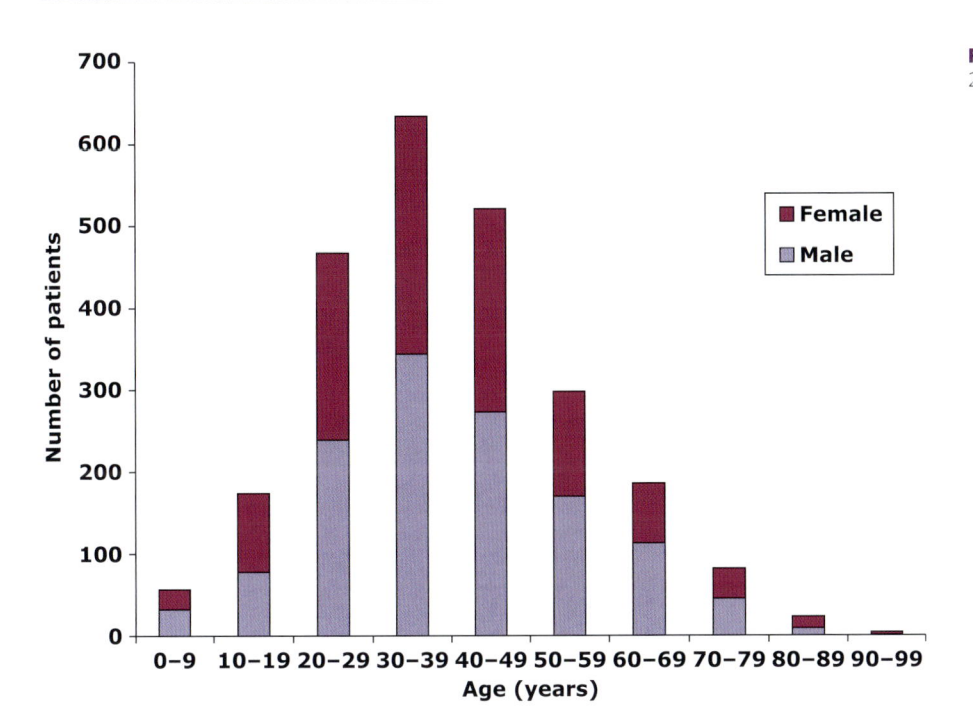

Figure 13.1 Age and sex distribution of 2466 cases of dermatofibrosarcoma protuberans.

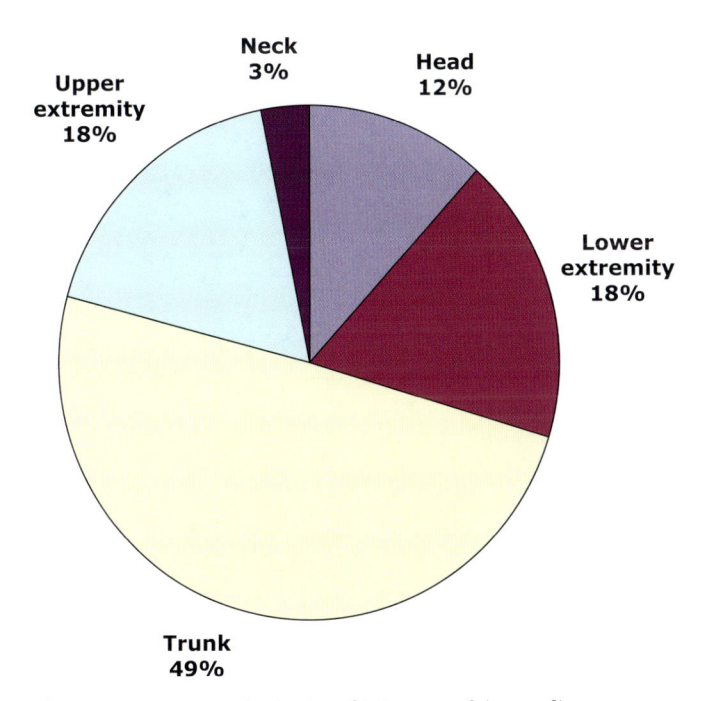

Figure 13.2 Anatomic distribution of 2494 cases of dermatofibrosarcoma protuberans.

There is a strong predilection for the trunk wall, with equal distribution in the back, chest wall, abdominal wall, and, rarely, involvement in the external genitalia.[1-7] In the upper extremities, DFSP has a strong predilection for proximal sites: shoulder and arm. In the lower extremities, however, the distribution is more even, and occurrence in the foot is not rare. In the head and neck, the scalp, forehead, and cheek are the most common sites (Figure 13.2).

An early lesion can present as a plaque-like cutaneous induration clinically resembling localized scleroderma, or as an apparently sharply demarcated cutaneous or subcutaneous nodule. More advanced tumors form multiple nodules protruding from the skin surface; these can develop into multiple nodules that can become fungating, ulcerated masses. Typically the tumor grows slowly, but rapid growth is sometimes observed during pregnancy[1,8] and with advanced tumors.

DFSP behaves as a low-grade sarcoma, with a strong tendency to recur locally unless completely excised, and recurrences often develop in 1 to 2 years, probably based on insidious growth of residual tumor. The prognosis after a complete excision is excellent, however. Typical DFSP rarely develops lymph node metastases, and distant metastases are essentially restricted to tumors with fibrosarcomatous transformation.[1-7,9-14]

Complete removal generally requires a wide excision, because the lesion typically infiltrates well beyond its grossly visible margins into the subcutaneous fat. Some authors have advocated Mohs micrographic surgery with incremental excision until normal tissue is obtained, as documented by repeated frozen sections. This might ensure complete excision during one procedure and optimize the margins, compared with the commonly used 2-cm to 3-cm gross margins in conventional surgery that can vary from excessive to insufficient if microscopically positive for tumor.[15] One meta-analysis examining data on 489 cases treated by wide local excision found a recurrence rate of 20%, whereas only 1 of 64 cases treated by Mohs micrographic surgery recurred (1.6%).[15]

Fibrosarcomatous (or rarely pleomorphic, MFH-like) transformation occurs more often at first presentation than in a recurrence, and such patients have a higher median age. Such a morphologic disease progression carries some risk

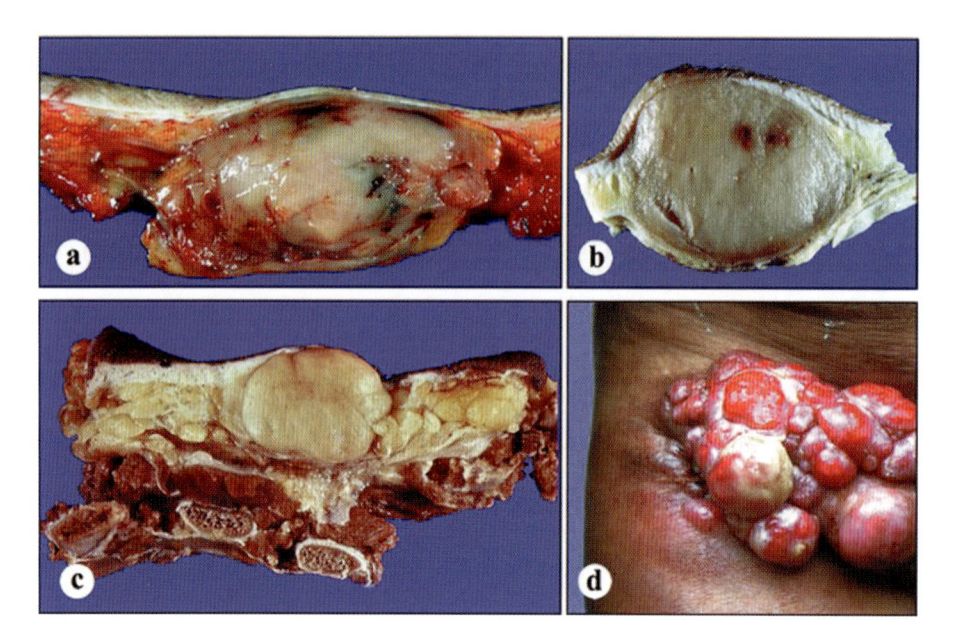

Figure 13.3 Gross appearances of dermatofibrosarcoma protuberans. (**a**) Excision specimen showing ample lateral margins, but tumor extending into the deep margin. (**b**) An apparently well-circumscribed tumor. (**c**) Wide excision specimen of a fibrosarcomatous dermatofibrosarcoma protuberans extending close to the skeletal muscle spans the entire thoracic wall. (**d**) An advanced dermatofibrosarcoma protuberans forming a fungating mass in the right groin. Note tumor extension beneath the intact skin.

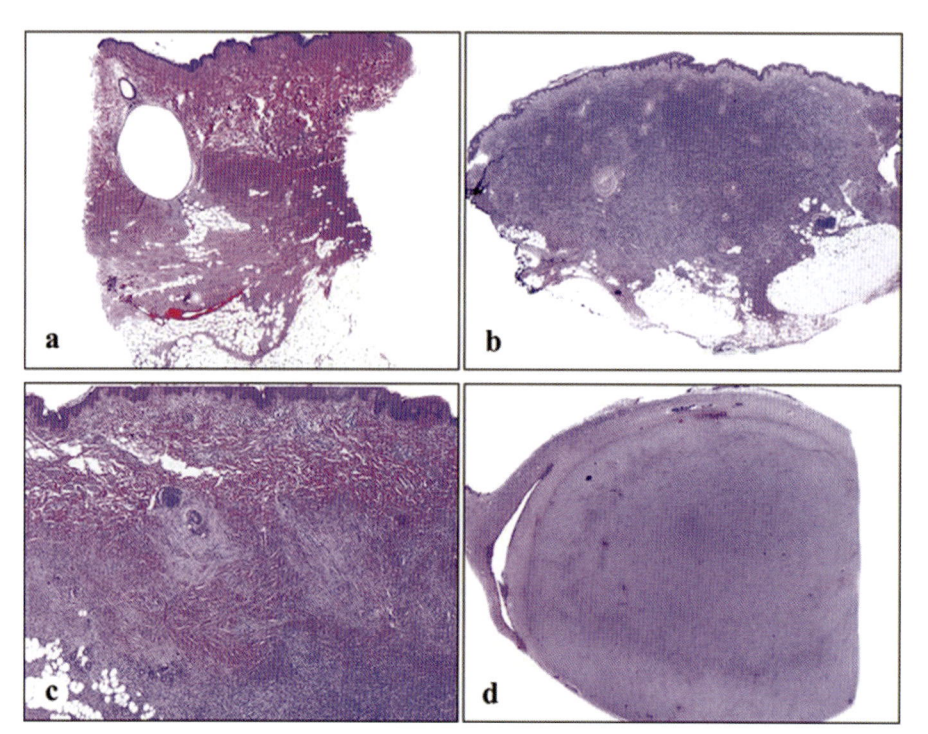

Figure 13.4 Profiles of four cases of dermatofibrosarcoma protuberans. (**a**) Plaque-like dermal lesion with subcutaneous extension. (**b**) Example involving the dermis and subcutis. (**c**) Sclerosing superficial and cellular deeper components. (**d**) An apparently sharply circumscribed subcutaneous example.

(0–10%) for distant metastasis (e.g., lungs, bone, soft tissues) and fatal outcome, although the incidence is difficult to estimate accurately because of variability in diagnostic criteria and patient populations, with the series from cancer hospitals containing more advanced tumors than those from pathology consultations.[1,3,7,9–14] Tyrosine kinase inhibitor imatinib may be applicable in fibrosarcomatous DFSP (see p. 345).

Pathology of typical DFSP

Grossly, an early DFSP might form a dense, dermal fibrous plaque or a solitary, circumscribed-appearing cutaneous-subcutaneous nodule, which is gray-white on sectioning. More advanced lesions are composed of multiple nodules or an exophytic, polypoid-to-fungating multinodular lesion (Figure 13.3). Pure subcutaneous location without a dermal component is possible.[16] Genetic analysis of subcutaneous fibrosarcomas has shown that this group also includes cases of DFSP-derived fibrosarcomas.[17]

Histologically, an early DFSP is plaque-like, whereas more developed lesions are nodular (Figure 13.4). The tumor typically diffusely infiltrates the dermis and subcutaneous fat (Figure 13.5) and also between adnexa. Some examples (approximately 10%) are well circumscribed and

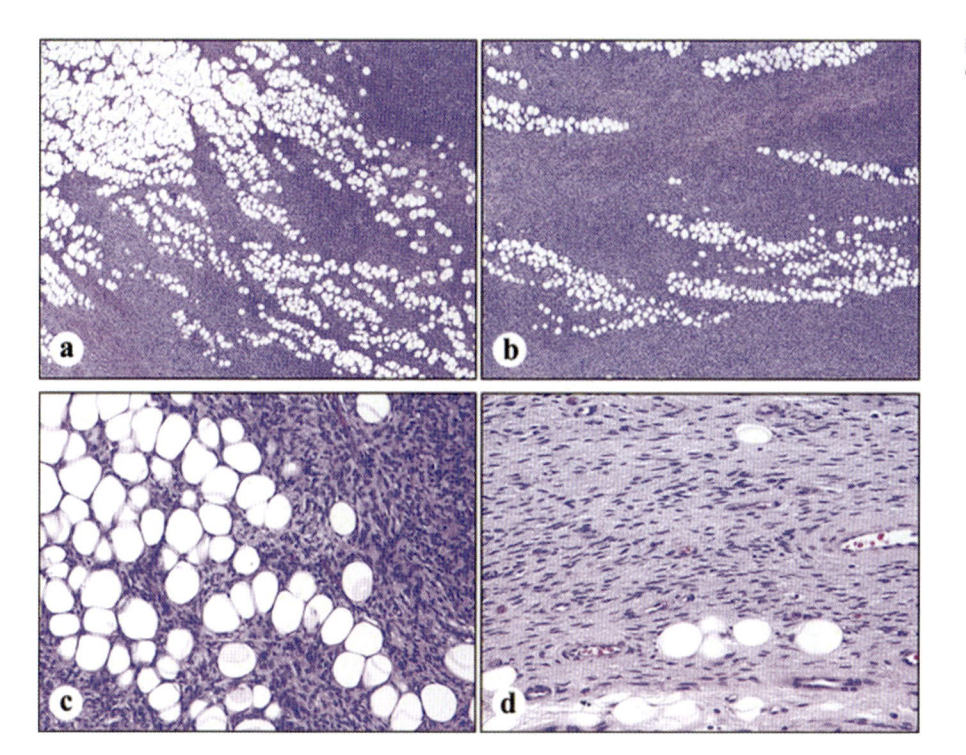

Figure 13.5 Subcutaneous fat infiltration in dermatofibrosarcoma protuberans.

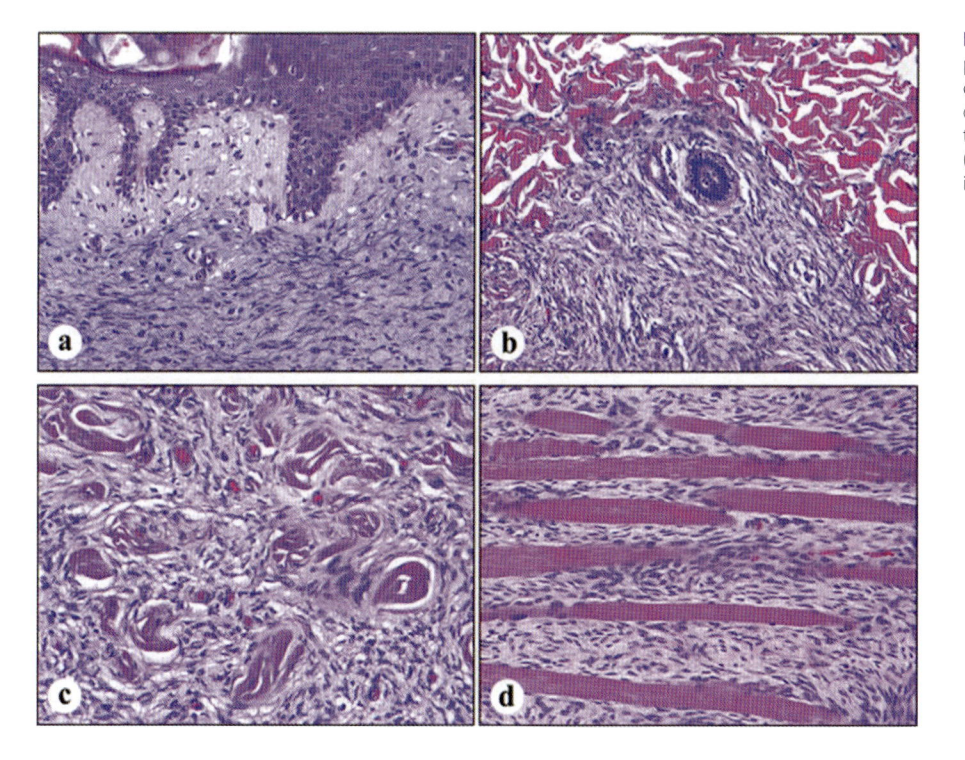

Figure 13.6 Borders of dermatofibrosarcoma protuberans. (**a**) Myxoid dermal surface component. (**b**) A smooth border to collagenized dermis. (**c**) Dermal collagen trapping resembling that more often seen in dermatofibroma. (**d**) Skeletal muscle infiltration is seen, especially in the head and neck.

even appear encapsulated, being surrounded by a dense, collagenous zone. The tumor can infiltrate the whole thickness of dermis but often leaves a hypocellular zone just below the epidermis. In the dermis, the tumor is sometimes better demarcated than dermatofibroma and can also show collagen trapping (Figure 13.6). Superficial portions of the tumor and the plaque-like lesions are often less cellular and more collagenous and can have a deceptively benign appearance on biopsy. Epidermal hyperplasia, usually seen in benign fibrous histiocytoma (BFH), is not present.

DFSP is typically composed of uniform, mildly atypical, oval fibroblasts often arranged in a storiform, cartwheel-like pattern. Sometimes the whorled pattern is very inconspicuous,

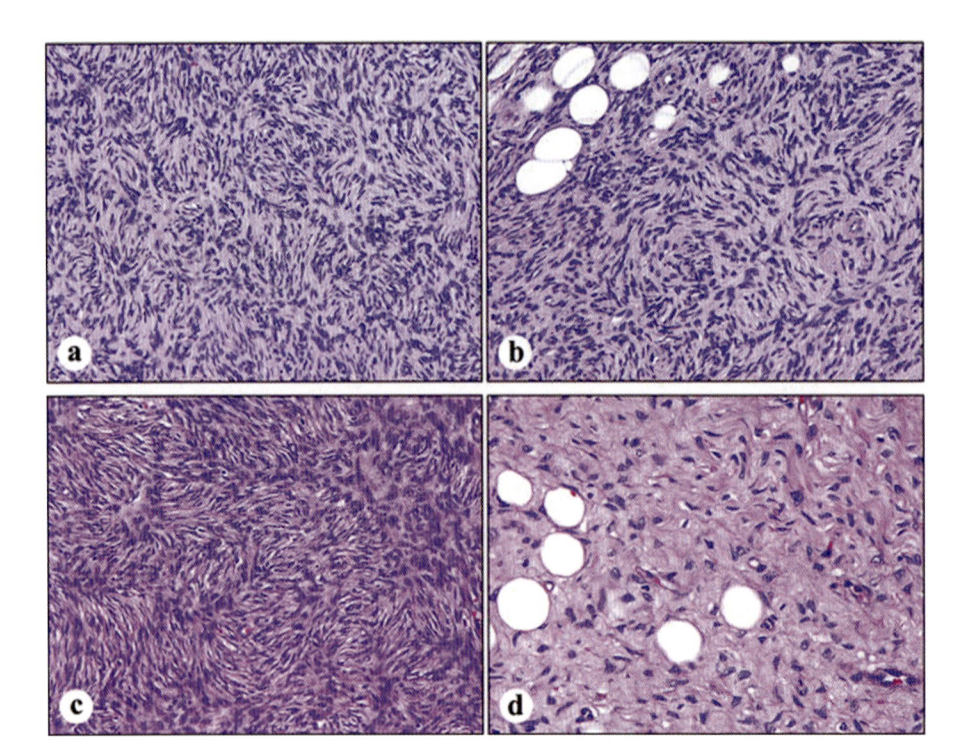

Chapter 13: Fibroblastic and myofibroblastic neoplasms with malignant potential

Figure 13.7 (**a–c**) Storiform and cartwheel patterns in dermatofibrosarcoma protuberans. (**d**) A more collagenous example.

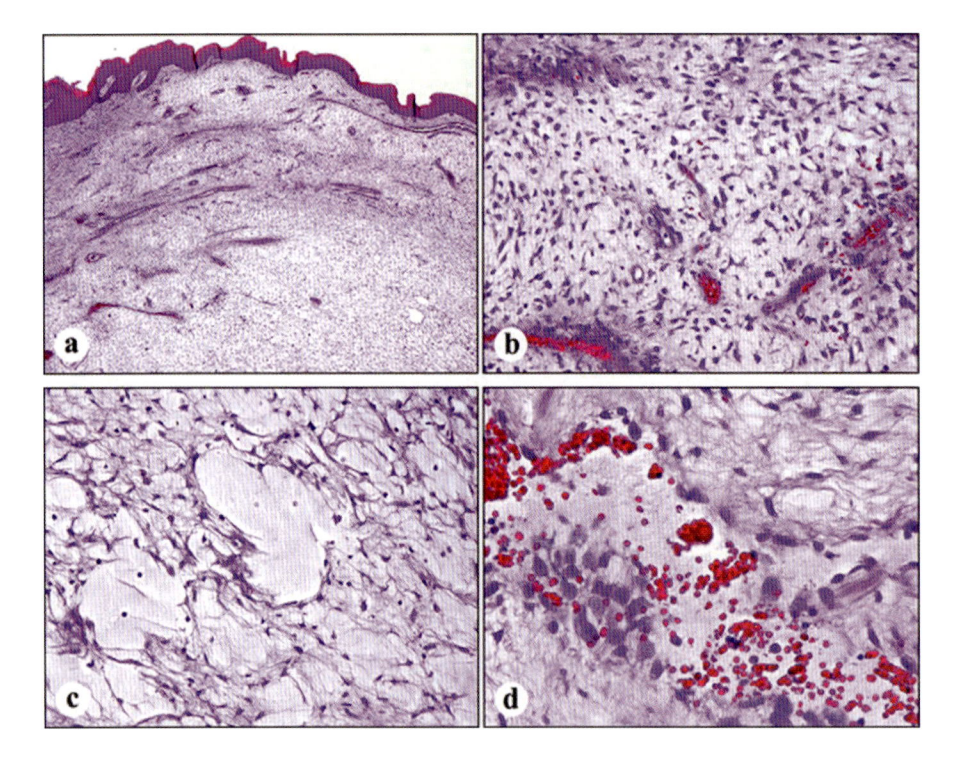

Figure 13.8 Myxoid dermatofibrosarcoma protuberans. (**a,b**) Prominent blood vessels. Focal mucinous pools. Pseudovascular space lined by mildly atypical tumor cells.

however, and the cells are more randomly or longitudinally oriented (Figure 13.7).

Significant myxoid matrix is seen in some cases, and some tumors contain stellate tumor cells as a major component in a predominantly myxoid matrix (Figure 13.8). These variants show no clinical difference from typical examples.[18,19]

Peculiar vascular myointimal proliferation occurs in some DFSPs with or without fibrosarcomatous transformation. This component is non-neoplastic and is analogous to the myofibroblastic component in myofibroma (Figure 13.9). It has a low mitotic activity and is SMA positive and CD34 and desmin negative.[20–22]

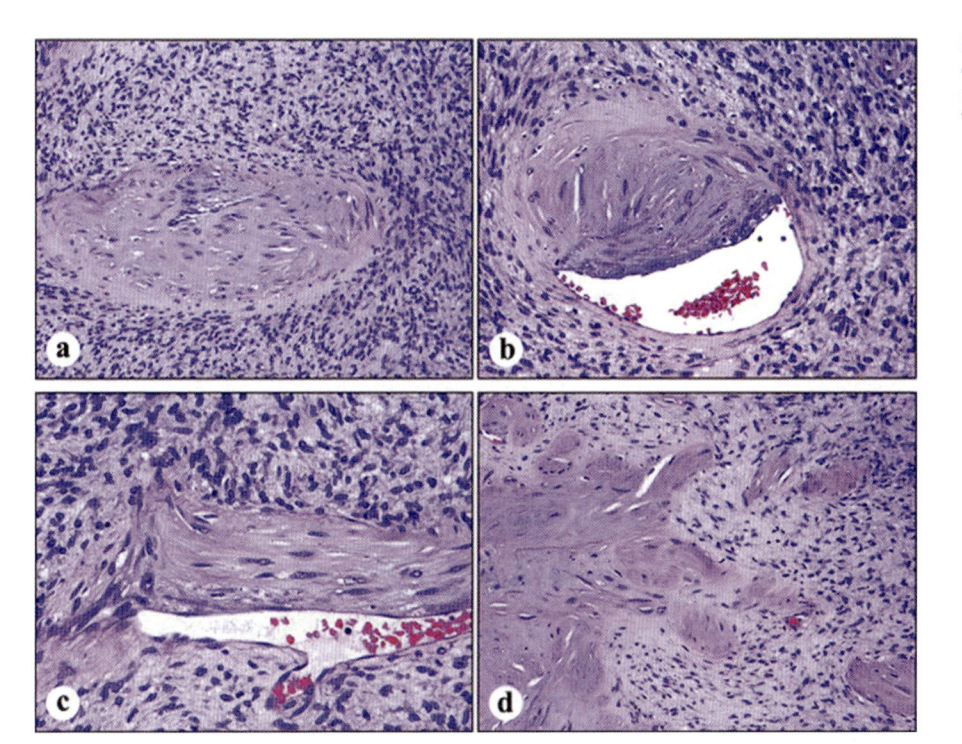

Chapter 13: Fibroblastic and myofibroblastic neoplasms with malignant potential

Figure 13.9 Myoid nodules in dermatofibrosarcoma protuberans. Note that these myointimal myofibroblastic nodules are associated with vessel walls.

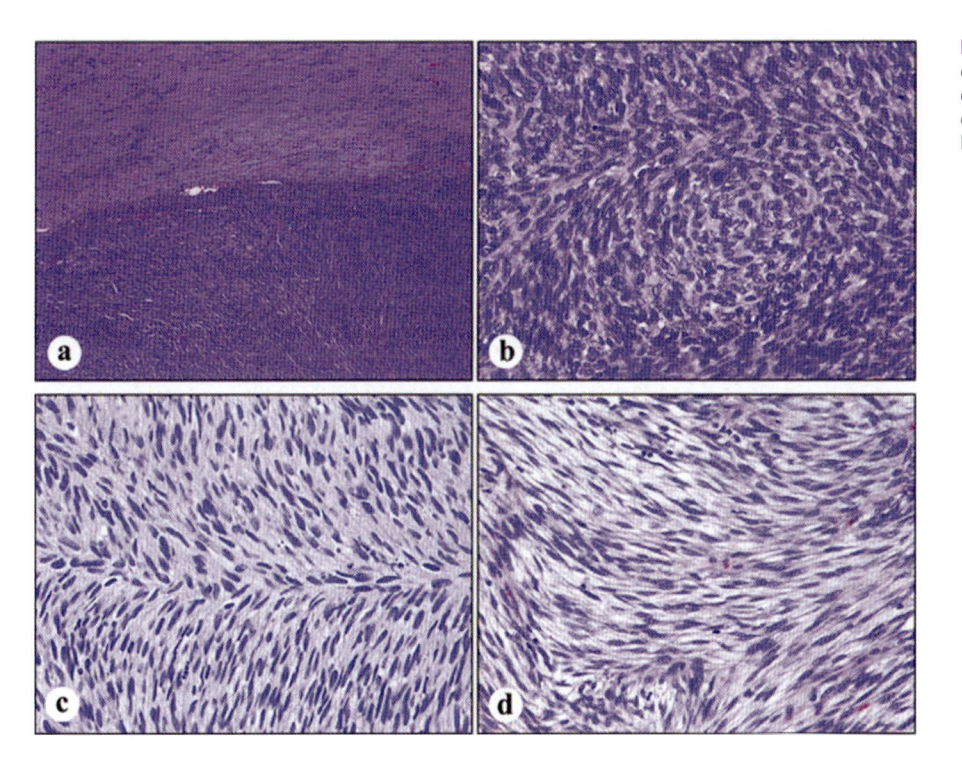

Figure 13.10 Fibrosarcomatous transformation of dermatofibrosarcoma protuberans. (**a**) Sharp demarcation between sclerosing and sarcomatous components. (**b–d**) Examples of fascicular and herringbone patterns.

Pathology of fibrosarcomatous transformation

Fibrosarcomatous (FS) transformation of DFSP (DFSP-FS) is the designation for DFSP that has attained a fascicular, no longer storiform, and more highly cellular appearance, often with a herringbone pattern, with the fascicles typically crossing at sharp angles. In some cases, the FS component is sharply demarcated from the conventional areas (Figure 13.10). It often (although not always) has a higher mitotic activity, often >10 mitoses per 10 HPFs and apparently higher cellularity than a typical DFSP, and can contain necrosis. This component usually dominates, with conventional areas sometimes limited to a small crescent in the tumor periphery. The FS component is usually low or intermediate grade and rarely high grade. Clinical implications vary, but metastatic risk is no higher than <5% to 10%. Occasionally, a pleomorphic MFH-like pattern develops in DFSP, and these tumors have had an aggressive clinical course.[23,24]

Pigmented DFSP (Bednar tumor)

Pigmented DFSP is a rare variant (3% of all DFSPs) containing scattered, melanin-pigmented cells. Whether these cells are neoplastic or derived from surrounding skin is unclear. This tumor, originally described as a *pigmented neurofibroma*, has a predilection for African Americans. Pigmented DFSP can also be seen in GCF. It does not differ clinically from ordinary DFSP, and similar to the usual variant, it can recur locally and undergo FS transformation.[25–27]

The pigmented cells are often concentrated in the middle of the tumor (Figure 13.11). Their number varies from a few to numerous, and they are often clustered or arranged in linear streaks. The pigmented cells vary from oval nondendritic to complex multipolar, elongated, tapered forms (Figure 13.12). The pigmented cells have melanocytic differentiation and accordingly have melanosomes ultrastructurally, and are variably positive for melanocytic markers such as S100 protein, tyrosinase, and melan A, in contrast to the major tumor cell population, which is negative for these markers.

Juvenile variant of DFSP: giant cell fibroblastoma

GCF is the juvenile variant of DFSP, although the typical DFSP also occurs in children.[28–31] Although GCF usually manifests in the first decade, it also can occur in older adults, with 15% occurring in adults >40 years in the largest series.[31] This tumor is predominantly located in the trunk or proximal lower extremities. Similar to typical DFSP, there is a >2:1 male predominance. Its relationship with DFSP is supported by coexistence of these lesions in 15% of cases, often with abrupt transition (Figure 13.13) and the similar COL1A-PDGF gene fusion. The tumor can recur, but FS transformation has been reported only in tumors with a DFSP component;[31] pigmented components have also been observed.[32]

GCF forms a painless, sometimes focally cystic mass of usually <5 cm. Histologically, the tumor typically lacks the organized, storiform pattern of DFSP, but the typical DFSP

Figure 13.11 Pigmented dermatofibrosarcoma protuberans. Center of the polypoid tumor is colored by melanin pigment.

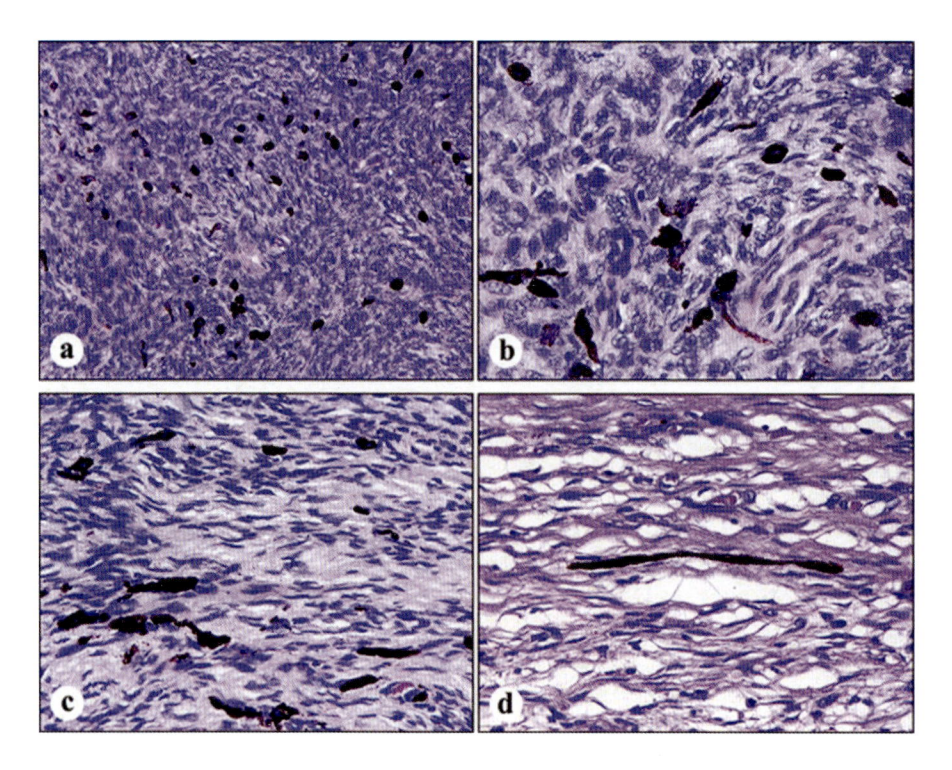

Figure 13.12 The pigmented cells in pigmented dermatofibrosarcoma protuberans often form clusters and streaks and vary from oval to bipolar and elongated.

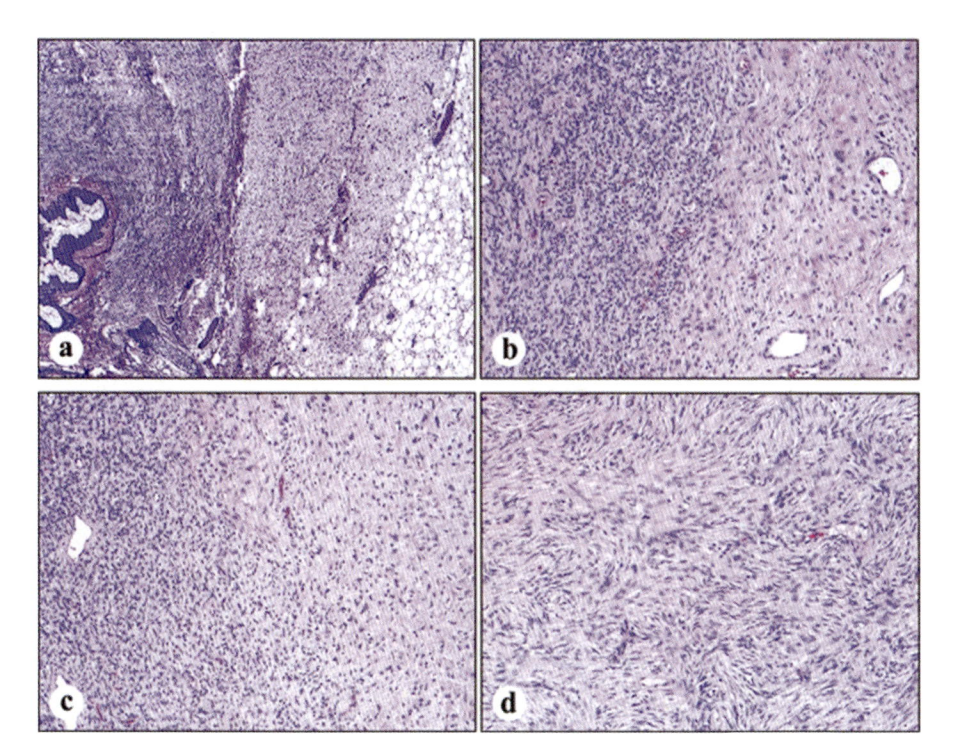

Chapter 13: Fibroblastic and myofibroblastic neoplasms with malignant potential

Figure 13.13 Giant cell fibroblastomas often form sharply demarcated areas when occurring together with conventional dermatofibrosarcoma protuberans. (**a–c**) Right side of the figures contain giant cell fibroblastoma elements. (**d**) Higher magnification of a typical dermatofibrosarcoma protuberans area seen with a giant cell fibroblastoma.

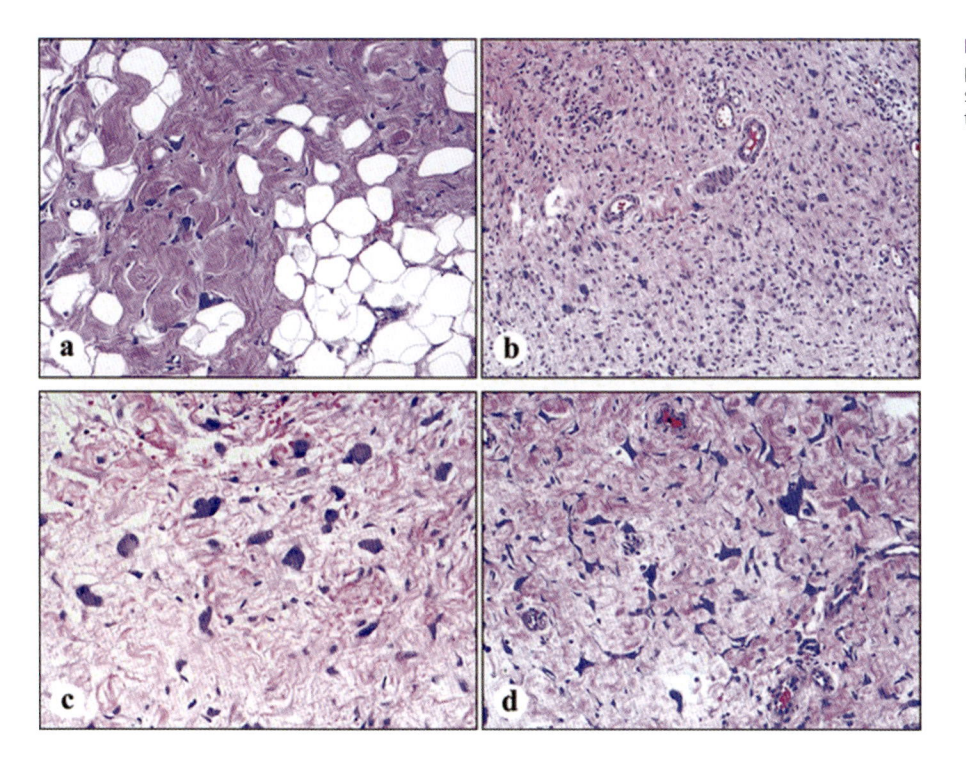

Figure 13.14 Giant cell fibroblastoma often has a prominent collagenous matrix containing scattered mononuclear and multinucleated tumor cells.

component is sometimes focally present. The cellular density is only moderate and lower than in typical DFSP, and the collagen content is often higher, giving the lesion a fibrosing appearance (Figure 13.14). Distinctive is the presence of large, hyperchromatic mono- or multinucleated tumor cells, often lining slit-like pseudovascular spaces containing slightly mucinous material that is surrounded by a collagenous background (Figures 13.15 and 13.16). There is not much mitotic activity. On immunohistochemical analysis, this tumor is positive for CD34, similar to typical DFSP.[31]

Immunohistochemistry of DFSP

DFSP cells in typical and variant tumors are positive for vimentin and CD34. The latter marker is also useful in

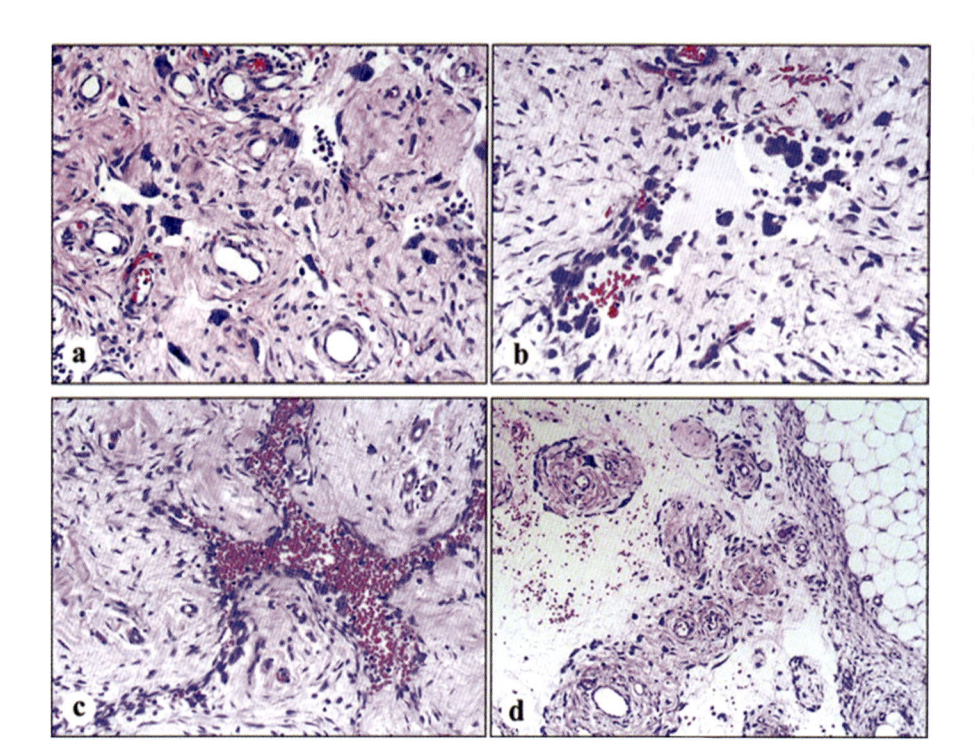

Figure 13.15 Pseudovascular spaces are typical of giant cell fibroblastoma and are often lined by tumor cells, including giant cell forms. Occasionally these spaces contain erythrocytes. (**d**) This example contains large pseudovascular spaces containing mucinous material.

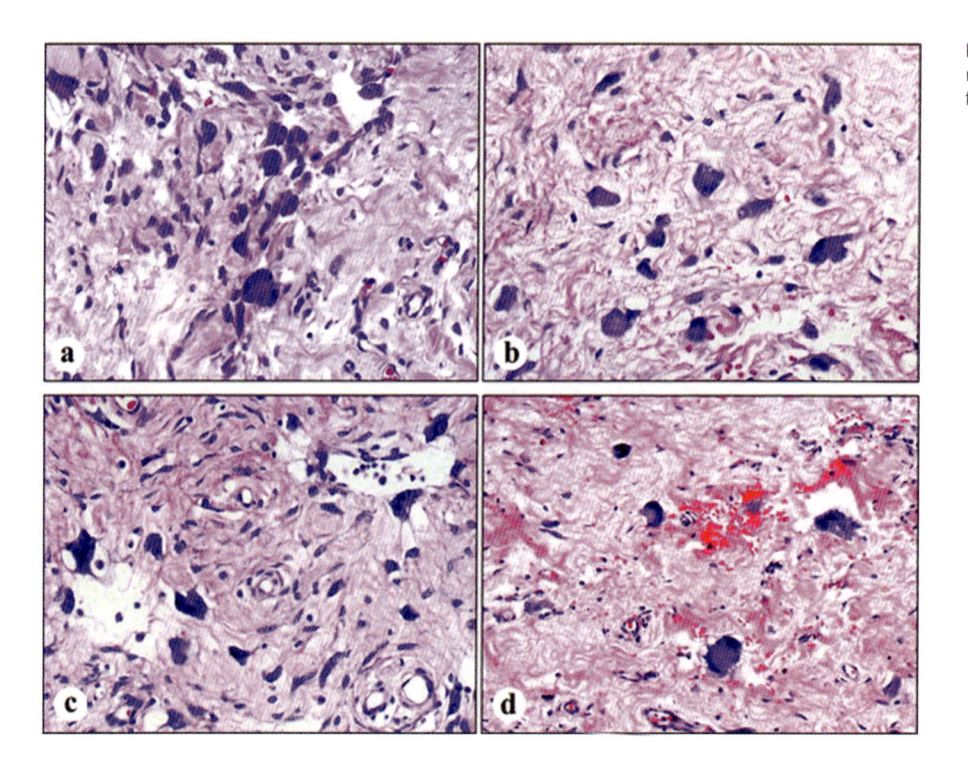

Figure 13.16 High-magnification appearance of multinucleated giant tumor cells in giant cell fibroblastoma.

evaluating the surgical margins and documenting residual disease (Figure 13.17). The tumor cells are negative for S100 protein, actins, desmin, keratins, and EMA.[33–37] Therefore, the tumor cell differentiation resembles that of CD34-positive dermal periadnexal fibroblasts. Although earlier ultrastructural studies suggested nerve sheath differentiation, immunohistochemical studies do not support this. The DFSP with FS transformation often shows more variable and less consistent CD34 positivity.[36,37] Another marker typically seen in DFSP, but not in BFH, is low-affinity nerve growth factor receptor, p75.[38] In contrast to BFH, DFSP does not have a tenascin-positive dermoepidermal zone.[39] Apolipoprotein D (APO D) has been found

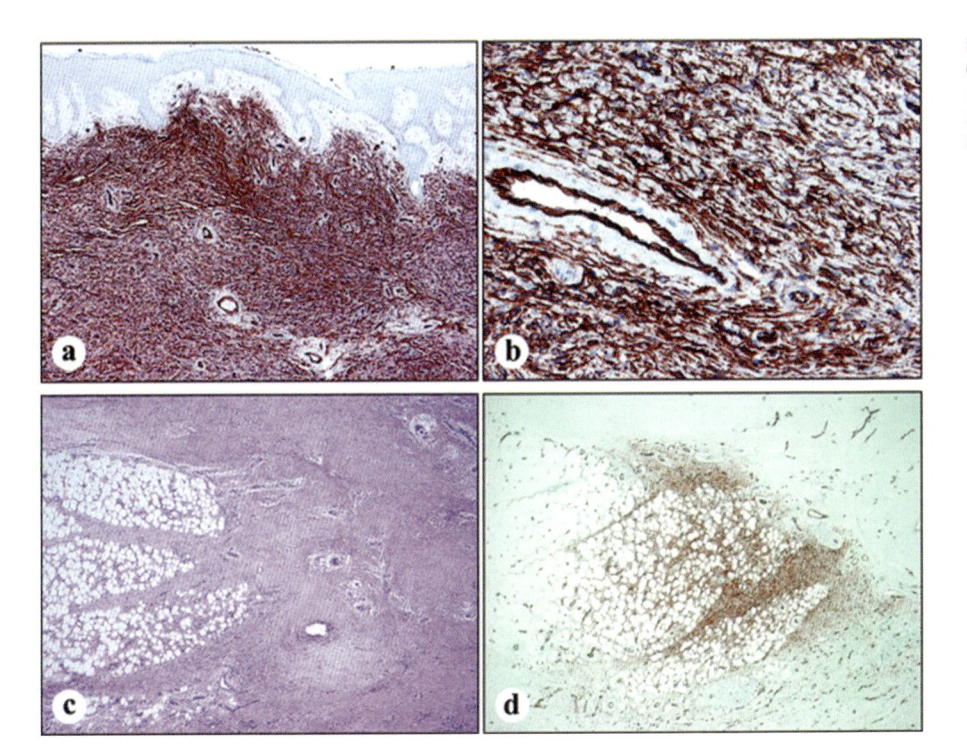

Figure 13.17 Strong CD34 positivity is typical of dermatofibrosarcoma protuberans. (**a**) Note a narrow uninvolved rim below the epidermis. (**c,d**) Minimal residual tumor amidst scar in an excision specimen is highlighted as a CD34-positive focus.

to be highly expressed in DFSP and its variants but absent in BFH and its variants, suggesting that this marker could be useful in differential diagnosis.[40]

Differential diagnosis

The lack of epidermal hyperplasia, relative cellular homogeneity, lesser amount of collagenous matrix, and diffuse subcutaneous infiltration separate DFSP from benign and cellular fibrous histiocytoma (dermatofibroma). Also helpful is the CD34 positivity of DFSP; BFHs are negative or have only a focally, usually peripherally, positive spindle cell component. DFSP and diffuse neurofibroma share diffuse subcutis infiltration; the latter is distinctly S100 protein positive, often with Meissnerian-like tactile corpuscles and sometimes with nerves with neurofibromatous transformation. CD34 positivity of the fibroblastic component should not lead a pathologist to mistake neurofibroma for DFSP.

Plaque-like CD34-positive dermal fibroma (medallion-like dermal dendrocytic hamartoma) is an upper dermal spindle cell proliferation that differs from DFSP based on its circumscription and also by lack of DFSP gene rearrangements (COL1A1-PDGF).[41]

True pigmented neurofibroma usually shows distinct foci of cells with melanocytic differentiation, and the tumor cells are S100 protein positive.[42]

Genetics

Near-diploid karyotypes with ring chromosomes containing low-level amplifications of material from chromosomes 17 and 22 are typical cytogenetic findings in DFSP.[43–45] The translocation t(17;22)(q22;q13), creating a fusion of the collagen type I gene α (COLIA) in chromosome 17q and platelet-derived growth factor β (PDGFB) in chromosome 22q, has been shown as a consistent event in DFSP and GCF.[46] Amplification of the DFSP translocation elements can often be detected, especially in the FS transformation. The fusion transcript has a transforming effect *in vitro* and is likely a key pathogenetic event, leading to autocrine activation of the PDGFB receptor.[47,48] Demonstration of the COL1A-PDGFB fusion transcript by RT-PCR or FISH is a diagnostic test for problem cases.[49,50] An unrelated translocation t(5;8), involving the CSPG2 and PTK2B genes, has been reported in one case of DFSP.[51]

New pathogenesis-based treatment of DFSP

The fusion translocation in DFSP involving PDGFB gene activates an autocrine loop in which the enhanced growth factor stimulus renders a constitutively activated status to the corresponding receptor, PDGFRB.[46,48,52–54] Because imatinib mesylate (Gleevec/Glivec, Novartis Pharma, New Hanover, NJ), a drug more commonly used in the treatment of chronic myelogenous leukemia and gastrointestinal stromal tumor, also inhibits PDGF receptors, there seems to be a scientific basis for its application to DFSP. Imatinib inhibited the growth of fibroblasts transformed with COL1A1/PDGFB sequences[52] and cultured DFSP cells[53] *in vitro*, induced tumor cell apoptosis, and reversed some of the effects of transformation. In some cases, dramatic clinical responses have been obtained in metastatic DFSPs with FS transformation. In one patient with unresectable metastatic DFSP, 4-month treatment by imatinib shrank the tumor by 75%, allowing surgical resection, whereupon no viable tumor was detected.[54]

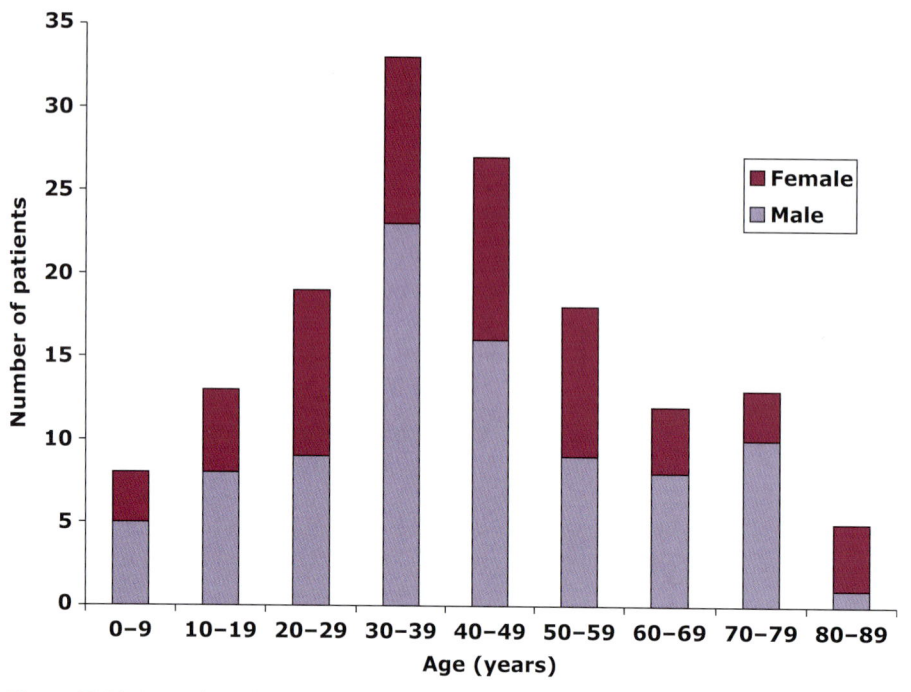

Figure 13.18 Age and sex distribution of 148 patients with low-grade fibromyxoid sarcomas.

Fibrosarcomatous metastatic and unresectable DFSP may be responsive to imatinib, but in some cases the responses have been short-lived.[55,56]

Low-grade fibromyxoid sarcoma
Clinical features

In 1987, Evans described this clinicopathologically distinctive low-grade sarcoma in three patients and again in 1993 in a larger series.[57] The tumor occurs predominantly in young and middle-aged adults between the ages of 25 and 45 years and also in children (Figure 13.18). Male predominance has been reported in some series. The tumor is usually intramuscular and is often relatively large, but some examples have been smaller, including those that have involved superficial soft tissues. The most common sites of presentation are the thigh and buttocks, inguinal area, shoulder region, and chest wall (Figure 13.19). LGFMS has also been reported in the retroperitoneum, mesentery, and small intestine.[57–65]

The clinical course is typically slow. Local recurrences can occur over a long time span, up to 50 years. Lung metastases developed in two early series in 7 of 12 and 1 of 11 cases, often after a long period of repeated local recurrences.[57,58] The author has seen cases in which pulmonary metastases developed before the detection of a large, occult primary soft tissue tumor involving the buttocks or abdominal soft tissues. In many cases, the patients live for a long time, even after developing pulmonary metastases, undergoing multiple thoracotomies for recurrent pulmonary metastases.

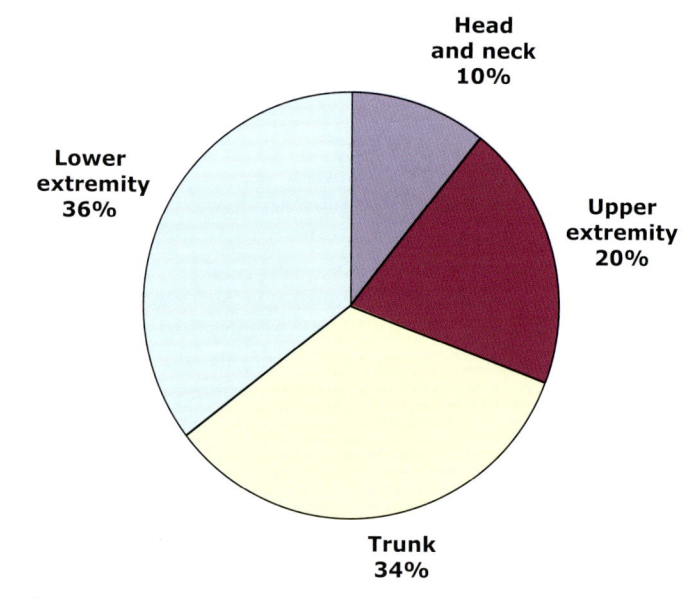

Figure 13.19 Anatomic distribution of 143 low-grade fibromyxoid sarcomas.

Pathology

The tumor size usually is in the range of 3 cm to 10 cm, but can be larger than 15 cm. Grossly, the tumor is firm and rubbery, but it can have a mildly mucoid appearance on sectioning. Although this tumor can appear well circumscribed, typically it infiltrates skeletal muscle microscopically (Figure 13.20).

Histologically, the tumor is often multinodular. The nodules typically show a myxoid character, although between them there are dense fibrous areas. The cellular nodules are

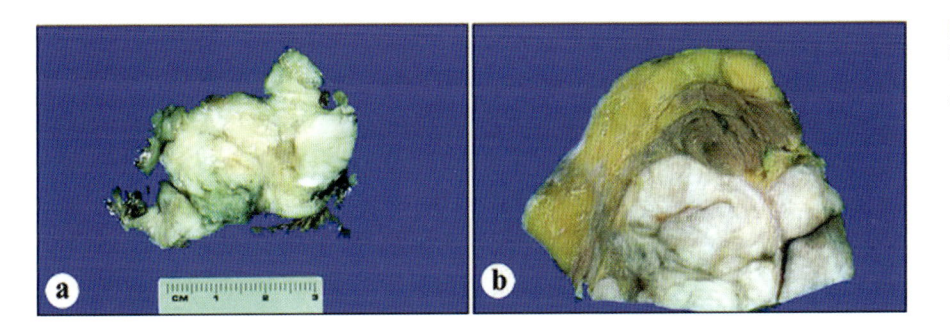

Figure 13.20 Two gross examples of low-grade fibromyxoid sarcoma. Gray-white cut surface with a mucoid appearance. This large example forms a deep intramuscular mass with a vague lobulation.

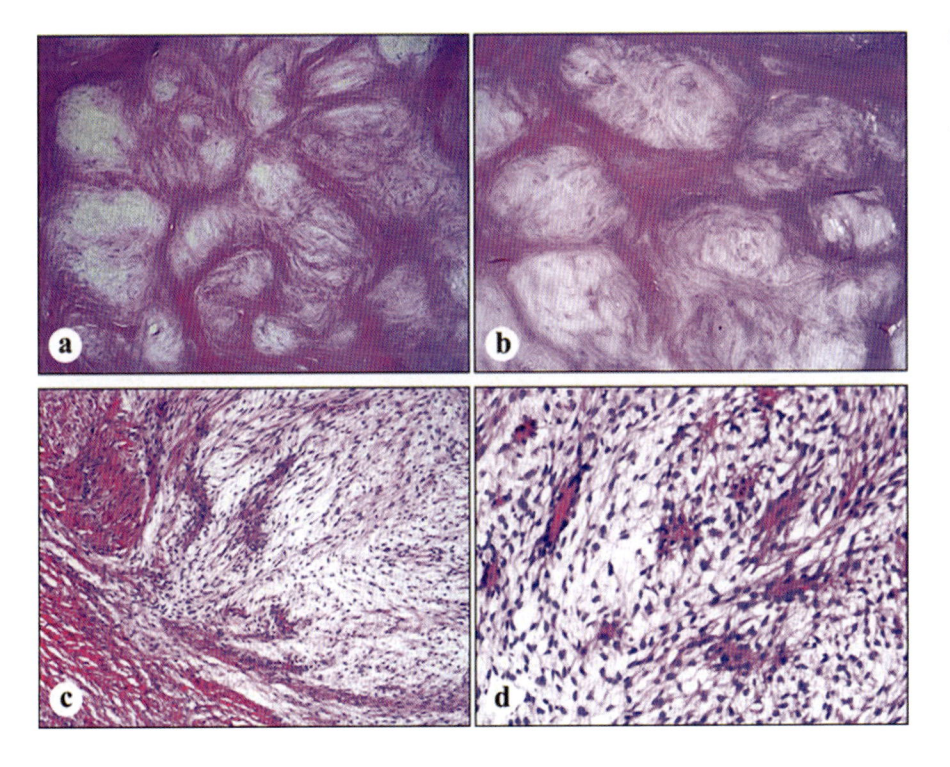

Figure 13.21 (**a,b**) Low-grade fibromyxoid sarcoma is often composed of myxoid nodules surrounded by fibrous areas. (**c,d**) The myxoid nodules show moderate cellularity and have variably prominent vessels.

hypervascular and contain mildly atypical fibroblasts in a fibromyxoid matrix (Figure 13.21). Some tumors are more evenly collagenous and can have fibromatosis or a neurofibroma-like appearance when wavy nuclei and collagen fibers are present (Figure 13.22). Some tumors contain perivascular cellular foci (Figure 13.23). In many areas, the tumor has a swirling or storiform appearance (Figure 13.24). Mitotic activity is low, but focal necrosis is sometimes present.

In some cases, the tumor undergoes a transformation to a higher histologic grade, with sheets or streaks of ovoid cells, and this appearance can closely resemble SEF (Figure 13.25). A more cellular, epithelioid appearance is often seen in the pulmonary metastases.[57,59]

Hyalinizing spindle cell tumor with giant rosettes is a histologic variant of LGFMS.[59,65,66] The reported cases predominantly occurred in young adults (mean age, 38 years) intramuscularly in the proximal extremities, showing similar clinicopathologic features as the main variant. This tumor contains peculiar giant rosettes formed by epithelioid fibroblasts with a core of dense, hyalinized collagen (Figures 13.26 and 13.27) in addition to features seen in a typical LGFMS.

Immunohistochemistry

LGFMS is typically positive for MUC4, a typical and diagnostic finding, and this marker is absent in most other spindle cells neoplasms.[67] Focal positivity for EMA is also common. SMA and CD34 positivity are occasional findings. These tumors are negative for desmin, S100 protein, and keratins.[59,65]

Differential diagnosis

Prior to its description, many examples of LGFMSs were undoubtedly diagnosed as benign tumors, with diagnoses ranging from fibroma or neurofibroma variants to fibromatosis and deep fibrous histiocytoma. A higher cellularity,

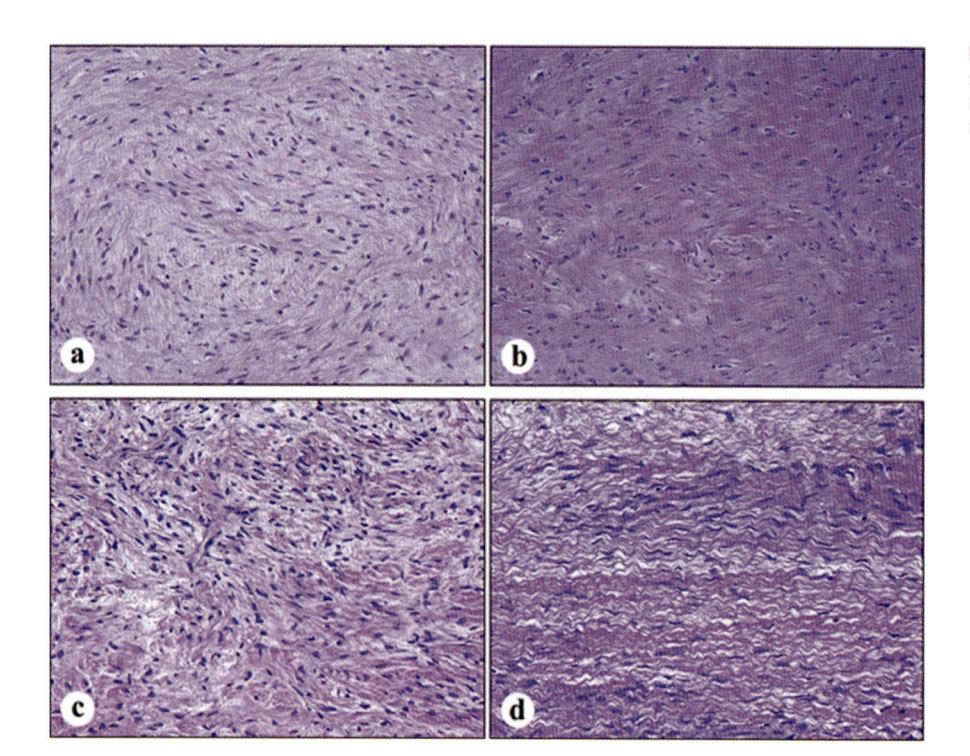

Figure 13.22 Examples of low-grade fibromyxoid sarcoma dominated by fibrous matrix. (**d**) Wavy nuclei and collagen fibers resemble features of neurofibroma.

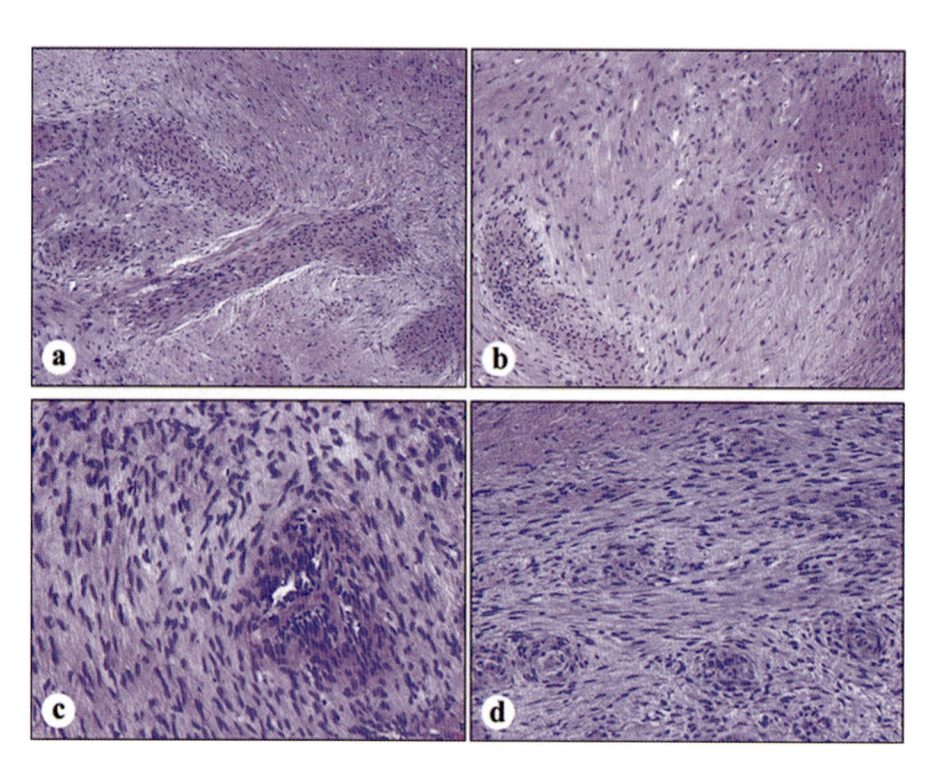

Figure 13.23 Perivascular hypercellularity characterizes some cases of low-grade fibromyxoid sarcoma.

multinodular, fibromyxoid character should allow one to distinguish LGFMS from benign fibroblastic tumors.

Low-grade MFS (low-grade myxoid MFH) differs from LGFMS by a more developed vascular pattern, more myxoid matrix, presence of greater nuclear atypia, and tendency for perivascular hypercellular zones.[68] Desmoid is less cellular and shows spindled cells with less atypia in a more uniformly collagenous matrix.

Genetics

The t(7;16)(q34;p11) translocation is a typical solitary cytogenetic finding in both the classic and the hyaline rosette variants.[65,69] The corresponding molecular genetic event is FUS-CREB3L2 gene (previously called BBF2H7) fusion.[69] An alternative gene fusion, FUS-CREB3L1, has been reported in a smaller number of cases. The latter change has not been

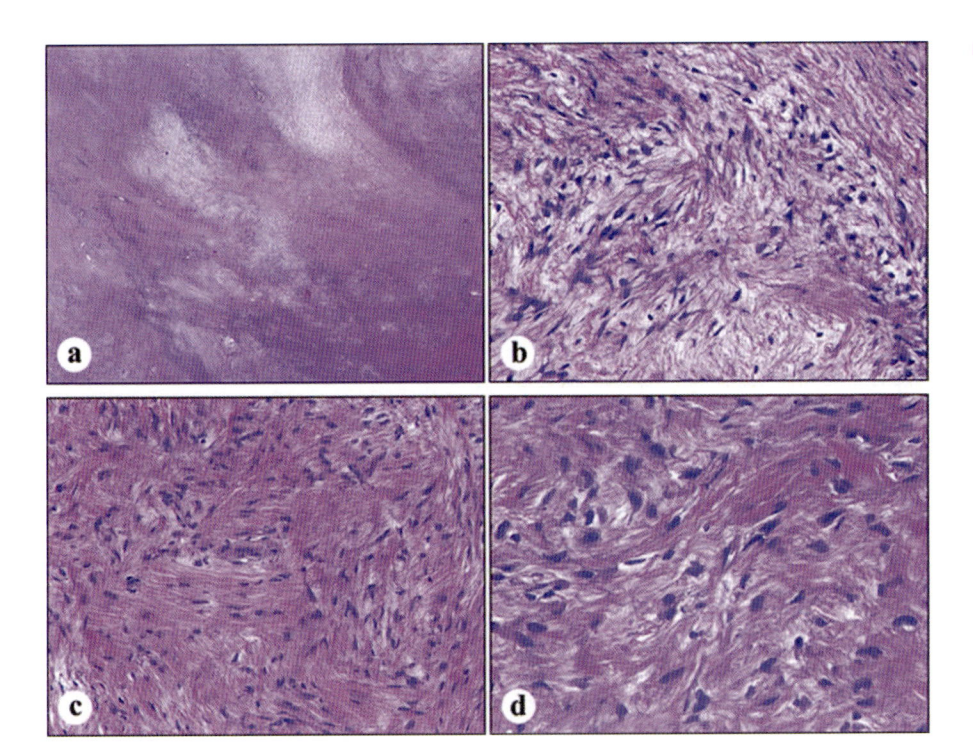

Figure 13.24 The tumor cells are sometimes arranged in a storiform manner. Fibroblasts show minimal atypia.

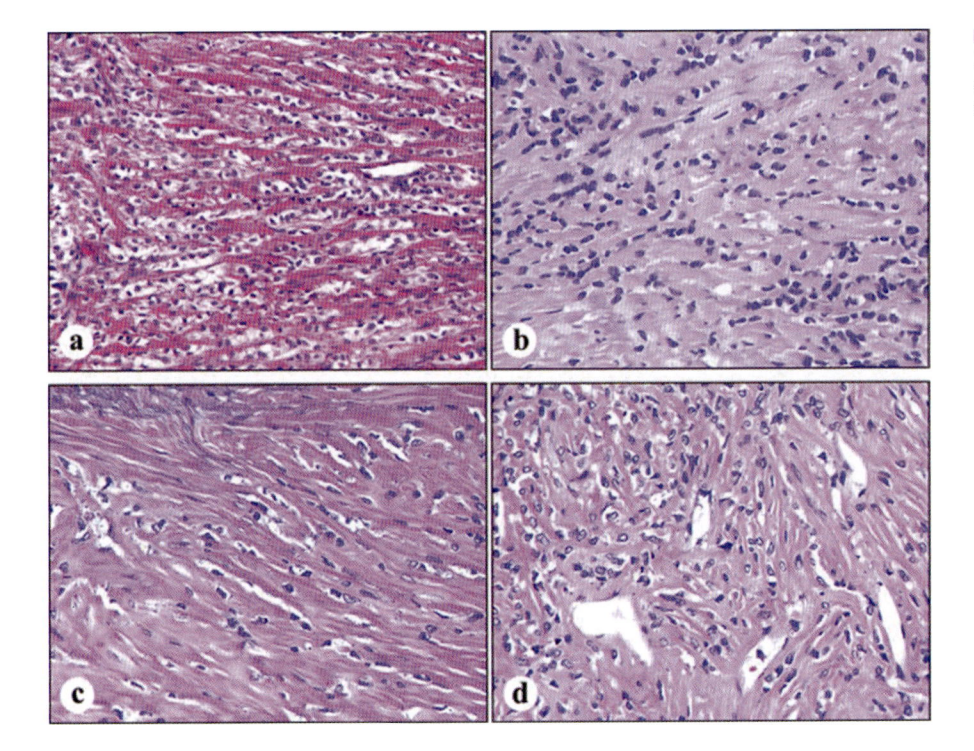

Figure 13.25 More highly cellular examples of low-grade fibromyxoid sarcoma showing streaks and clusters of tumor cells, with resemblance to the pattern of sclerosing epithelioid fibrosarcoma.

observed at the cytogenetic level, but the expected translocation would be t(11;16)(p11;p11).[70,71] CREB3L2 and CREB3L1 are closely related transcription factors that shuttle between intracellular membranes and nucleus and are activated by intramembrane proteolysis.[72]

PCR-based fusion transcript assay[73] and interphase FISH for FUS gene rearrangement can be used in the differential diagnosis of LGFMS and other fibroblastic tumors, but the same gene is also rearranged in myxoid liposarcoma translocation.[74]

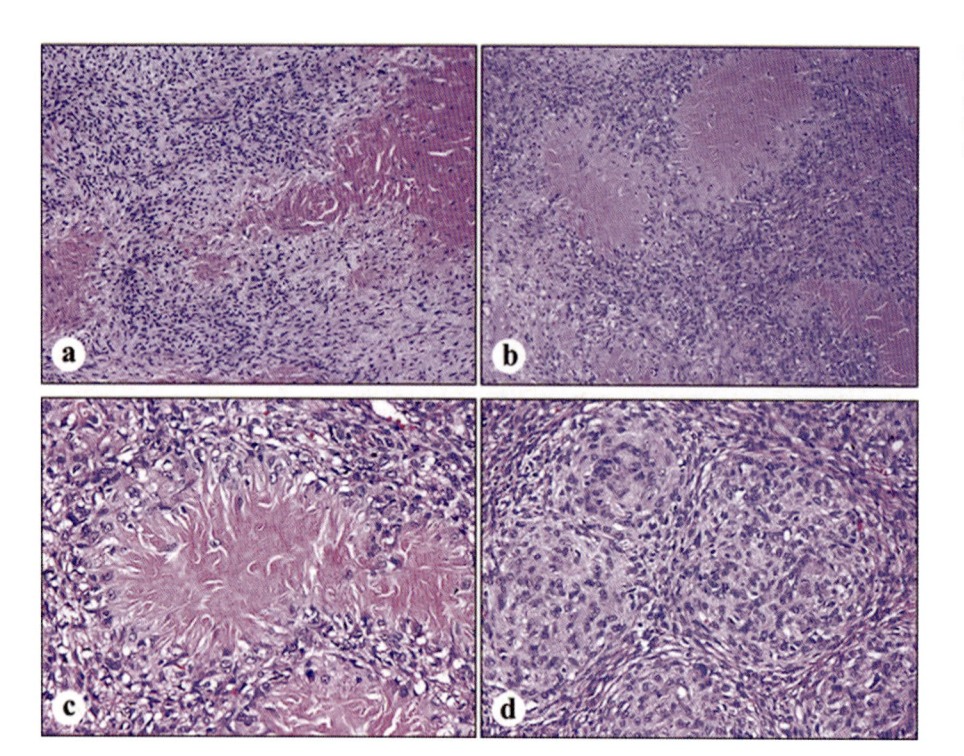

Figure 13.26 Low-grade fibromyxoid sarcoma variant with giant hyaline rosettes shows a collagen core in the rosettes, which are surrounded by epithelioid fibroblasts. (**d**) Dense clusters of epithelioid fibroblasts are present in some cases.

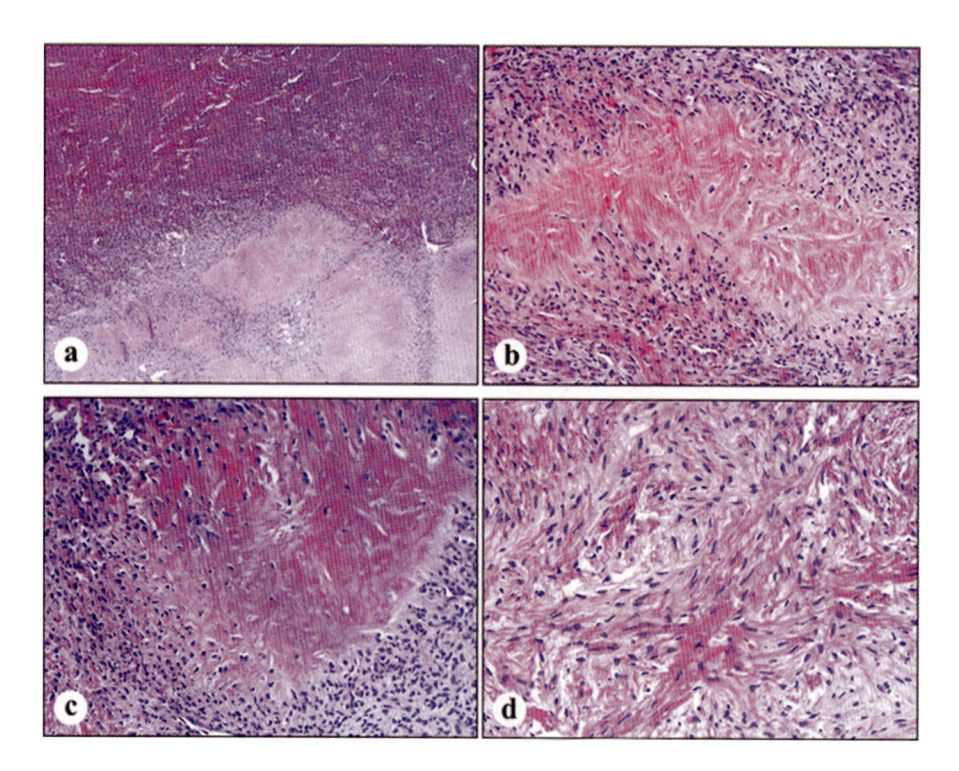

Figure 13.27 Giant hyaline rosettes of low-grade fibromyxoid sarcoma surrounded by fibroblasts. (**d**) Areas similar to usual variant of low-grade fibromyxoid sarcoma.

Sclerosing epithelioid fibrosarcoma

This histologic variant of fibrosarcoma was reported by Meis-Kindblom et al.[75] and subsequently by others.[76,77] Overlapping morphology, immunohistochemical expression of MUC4 and presence of FUS-CREB3L2/1 in many cases indicate a close relationship with low-grade fibromyxoid tumor, so that many cases of SEF can be viewed as progression forms of LGFMS. However, there seem to be a component in SEF that is independent of LGFMS, as the immunophenotypic and genetic markers for LGFMS are absent in some cases.

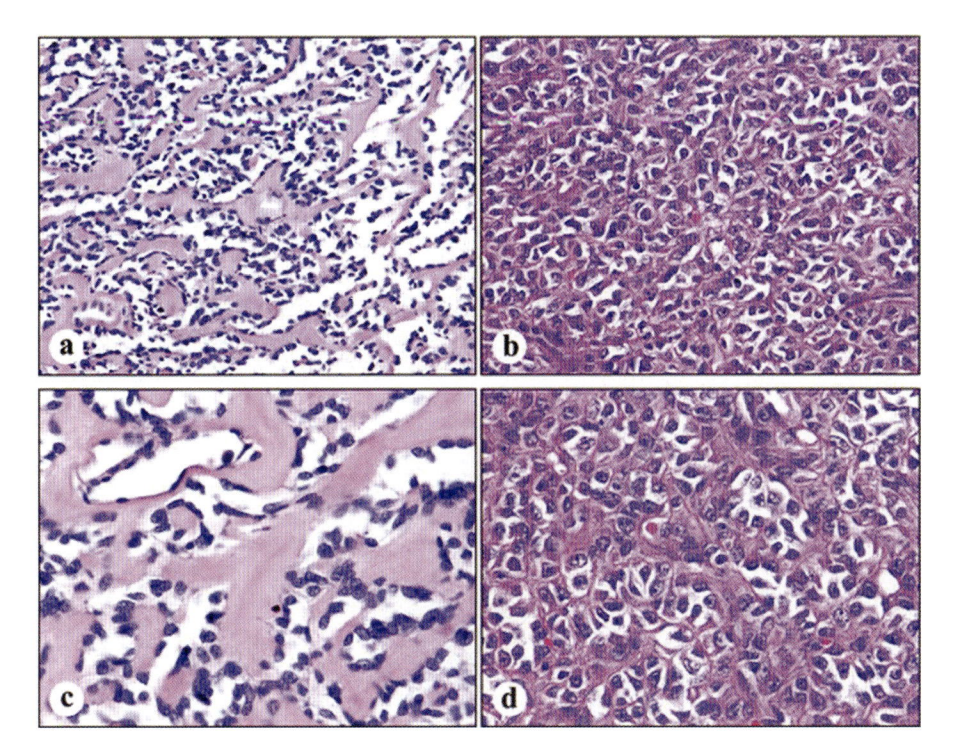

Figure 13.28 Sclerosing epithelioid fibrosarcoma shows cords of slightly epithelioid tumor cells with collagenous cores. Sclerosing rhabdomyosarcoma has to be ruled out by immunohistochemical analysis before this diagnosis is made.

Clinical features

The reported SEFs have occurred in a wide age range (14–87 years), with a mean age between 40 and 45 years and no predilection for either sex. Most SEFs have been deep soft tissue masses in the proximal parts of the extremities, trunk, and head and neck region. Nearly one-half of them have been 10 cm or larger.[75–77]

The tumors have behaved as fully malignant sarcomas, causing death in more than one-half of all patients. The most common metastatic sites are the lungs and bones. In some cases, the metastases have developed in less than 2 years, whereas in other cases they have developed 10 or more years after resection of the primary lesion.

Pathology

Grossly, SEF forms a large oval or discoid mass measuring 2 cm to 15 cm. The consistency is fleshy to hard on sectioning, often with necrosis.

A microscopically distinctive feature is the presence of oval to epithelioid cells with a clear cytoplasm, in a dense collagenous matrix (Figure 13.28). Mitotic activity can vary from 0 to 15 mitoses per 10 HPFs. The tumor can have areas of conventional fascicular fibrosarcoma and LGFMS elements may coexist, supporting the view that many SEFs are high-grade transformation from LGFMSs.[75] This is also supported by the presence of FUS-CREB3L2 gene fusion typical of LGFMS in some tumors with SEF morphology.[62]

Immunohistochemical features are also similar to LGFMS so that most cases are positive for MUC4[78] and some for for EMA. Keratins have also been dectected with AE1/AE3 and CAM5.2 antibodies.[75]

Differential diagnosis

Immunohistochemical studies for desmin and myogenin are necessary to rule out sclerosing spindle cell rhabdomyosarcoma, because this tumor can contain areas histologically resembling SEF. Sclerosing metaplastic carcinoma or malignant mixed tumor should be ruled out by immunohistochemistry for epithelial markers (extensive keratin positivity) and by clinicopathologic correlation. Although solitary fibrous tumors can have a somewhat similar histologic pattern, they differ from SEF by CD34 positivity.

Genetics

CREB3L2 fusion transcripts typical of LGFMS have been detected in some histologically defined SEFs, suggesting that some of these tumors might represent a progressive variant of LGFMS.[62] Amplification of 12q13 and 12q15 sequences has been reported in one case.[79] EWSR1-CREB3L1 fusions have also been detected as a recurrent change.[80] Based on one FISH study, FUS gene rearrangement does not seem to be common in SEF.[81]

Adult fibrosarcoma and myofibroblastic sarcoma

Adult fibrosarcoma is a collagen-forming, nonpleomorphic, spindle cell sarcoma that cannot be classified as belonging to any other category (i.e., malignant peripheral nerve sheath tumor, synovial sarcoma, solitary fibrous tumor, LGFMS,

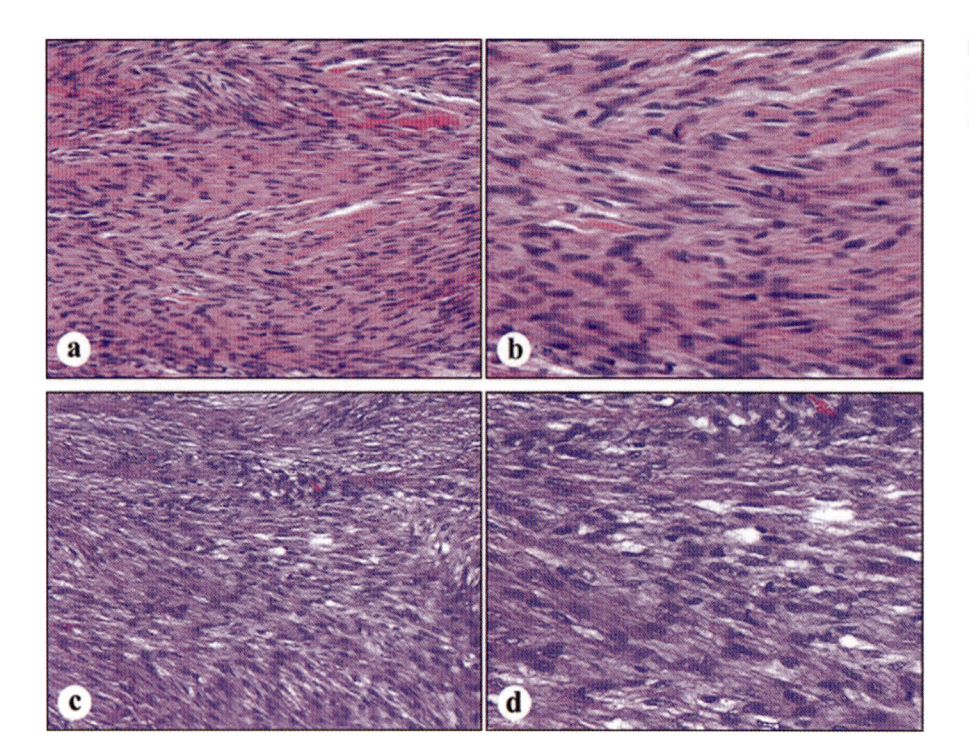

Figure 13.29 Adult fibrosarcoma is a highly cellular spindle cell proliferation with a fascicular pattern and a mildly collagenous matrix. The nuclei are enlarged but uniform.

cellular desmoid, or DMFS-FS transformation). Nowadays it is a rare diagnosis and one of exclusion. It is also conceivable that many superficial fibrosarcomas are DFSP-derived fibrosarcomas, based on demonstration of the COLIA1-PDGF fusion transcript in many such tumors.[17] Most fibroblastic sarcomas have at least some pleomorphic and myxoid features, and consequently they are currently usually classified as MFH or MFS.

Because published series on adult fibrosarcomas predate the present classification and emergence of diagnoses such as LGFMS, it is difficult to project the old data into the current classification.[82] Current data also are insufficient regarding age, sex, and anatomic site distribution. Older series undoubtedly include tumors that are now classified under the previously mentioned tumor entities. A recent reclassification study found that only 16% of tumors previously classified as fibrosarcoma satisfied the stringent inclusion criteria, while excluding histologic mimics.[83]

Most tumors reported as adult fibrosarcomas have occurred in deep soft tissues of the extremities or trunk in middle-aged adults.[82,83] A typical histologic finding is a fascicular spindle cell sarcoma in a variably collagenous matrix (Figures 13.29 and 13.30).

Myofibroblastic sarcoma (myofibrosarcoma) is a designation given to fibrosarcoma with myofibroblastic differentiation, as defined by histologic appearance: spindled cells with amphophilic abundant "myoid" cytoplasm and smooth muscle actin positivity (Figure 13.31). Electron microscopy shows myofibroblastic features with partial basement membranes, focal clusters of actin filaments, and attachment plaques at the cell membrane. Tumors reported by this name have mostly been low-grade sarcomas in adult patients, presenting in a wide variety of locations.[84,85] Many MFS and MFHs contain focal SMA positivity, indicating focal myofibroblastic differentiation and therefore potentially qualifying as myofibroblastic sarcomas. It is unknown whether myofibroblastic differentiation in this context has specific clinical significance.

Acral myxoinflammatory fibroblastic sarcoma (inflammatory myxohyaline tumor of distal extremities)

Clinical features

Three larger series have described this tumor under these two different names.[86–88] These reports collectively detailed similar clinically and histologically distinctive features in 95 examples. This tumor was formerly classified as a reactive process, variously termed *exuberant proliferative synovitis*, *atypical fibrohistiocytic neoplasm*, or a *low-grade myxoid MFH*.

The tumor has a predilection for adults, with the median age varying from 44 to 53 years, with equal occurrence in men and women (Figure 13.32). However, occurrence has also been reported in children.[89] There is a strong predilection for the distal extremities: fingers and hands, 56%; toes and feet, 17%; ankles/lower legs, 14%; and wrists/forearms, 11%. Rare examples occur in the forearm and arm. Some patients have a long history of a mass, but the median duration of symptoms is 1 year. The tumor has a predilection for the dorsal aspect of the hands and feet, and it often forms an ill-defined raised mass that is much wider than it is deep.[88]

Chapter 13: Fibroblastic and myofibroblastic neoplasms with malignant potential

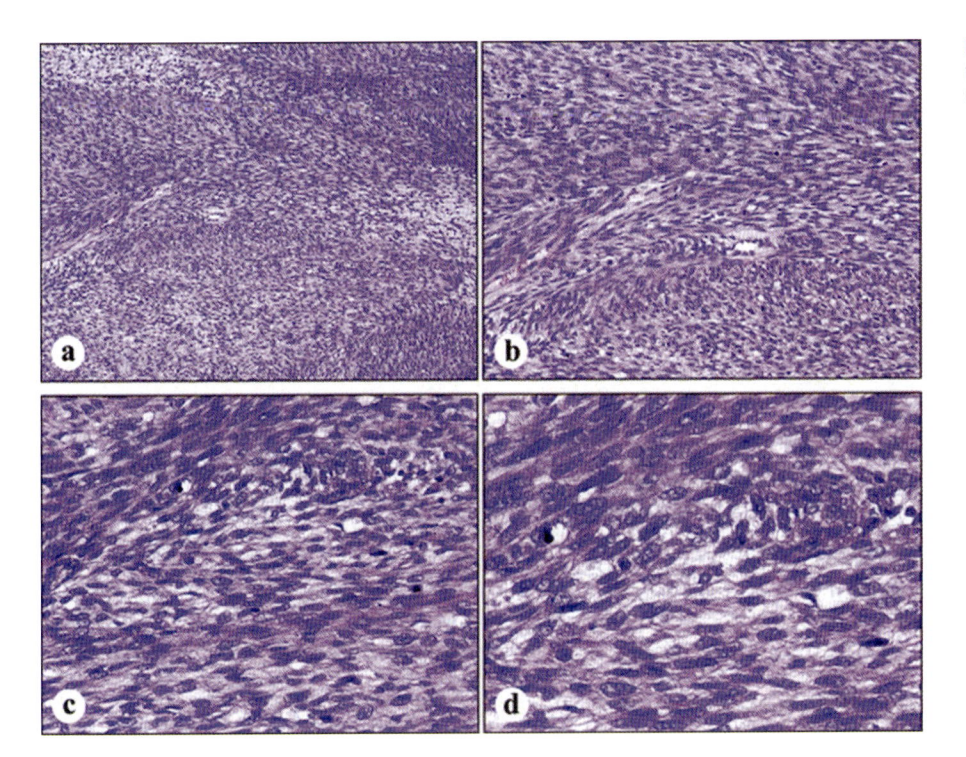

Figure 13.30 Another example of adult fibrosarcoma contains uniform spindle cells in scant collagenous matrix.

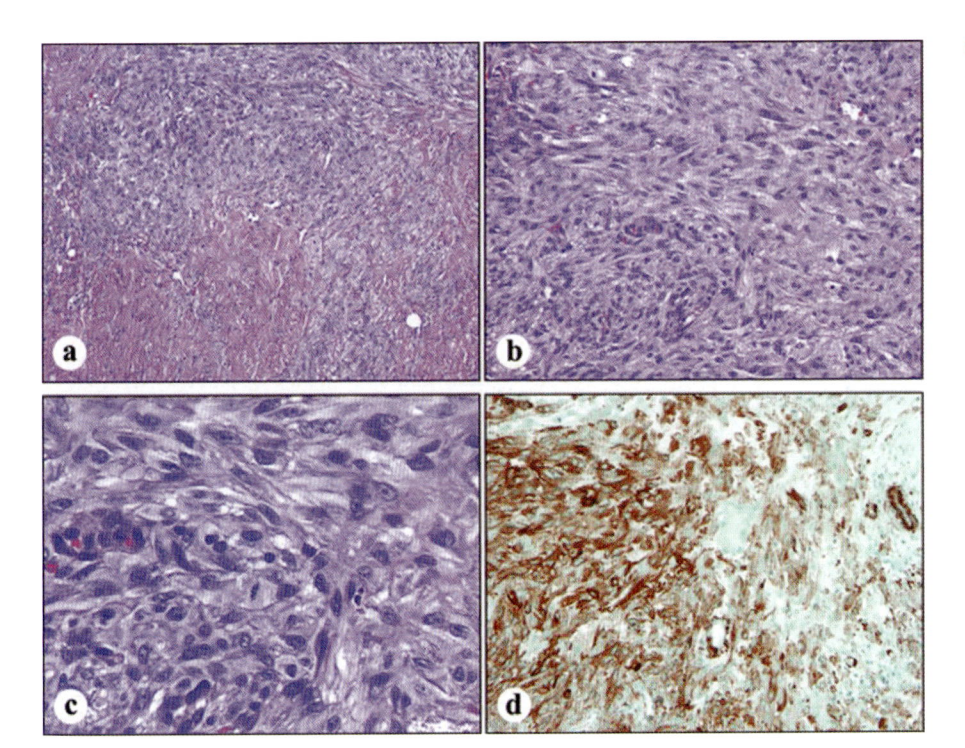

Figure 13.31 Myofibroblastic sarcoma (low-grade) is composed of mildly atypical, plump myofibroblasts with mitotic activity. The tumor involves fascia and is smooth muscle actin positive.

Tumor behavior varies, with common local recurrences, and some patients have multiple recurrences over several years. The metastatic potential of this tumor is questionable, however. One of the series had no metastases,[87] and the other had documented lymph node metastasis in one patient and a histologically undocumented pulmonary lesion in another.[86] The third series had regional lymph node metastasis in one case, but no distant metastases, despite inclusion of atypical and more cellular variants.[88]

The tumor should be excised completely, preferably in a function-preserving manner, although ray or finger amputation may be necessary in some cases.

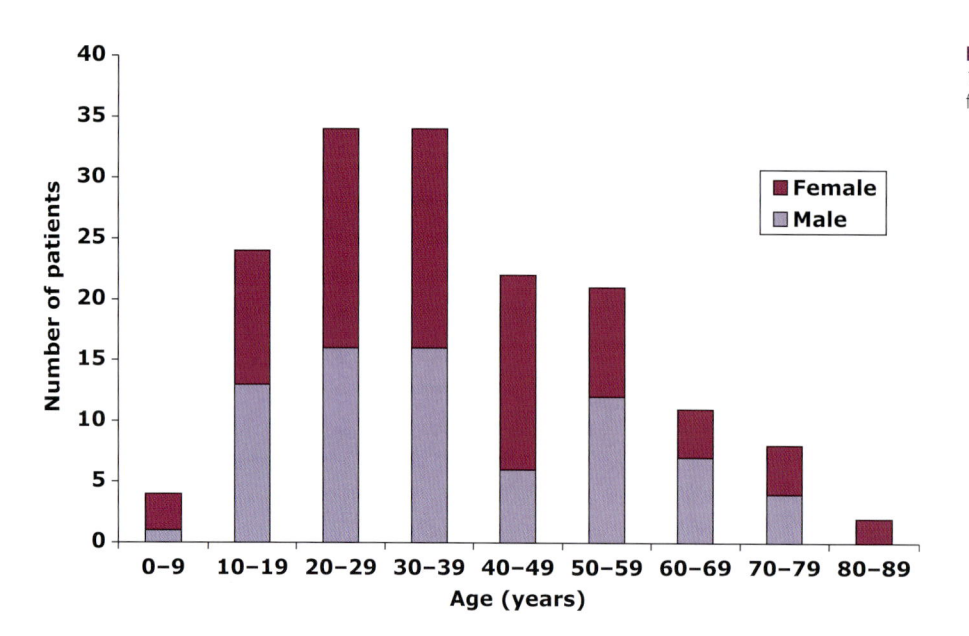

Figure 13.32 Age and sex distribution of 126 patients with acral myxoinflammatory fibroblastic sarcoma.

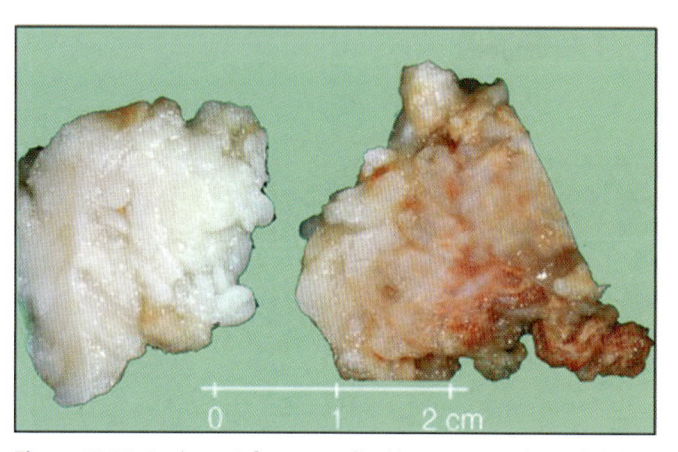

Figure 13.33 Acral myxoinflammatory fibroblastic sarcoma shows slightly mucoid appearance. Surface is gray-white to brownish on sectioning.

Pathology

The tumor typically measures 1 cm to 5 cm, with the median size approximately 3 cm. Grossly, the tumor is white, multinodular, often poorly circumscribed, and sometimes gelatinous (Figure 13.33). It primarily involves the subcutis, often synovia or tenosynovia, and occasionally skeletal muscle, tendons, or aponeuroses.

Histologically, the tumor is distinctive for a multinodular pattern on low magnification (Figure 13.34). There is variably prominent mixed inflammatory infiltrate composed of lymphocytes, plasma cells, neutrophils, and eosinophils in a variably myxoid and collagenous background. Germinal centers can also be present, especially in the periphery.[88]

The neoplastic cells often have an epithelioid appearance, although a spindle cell component can also be present (Figure 13.35). A distinctive feature is the occurrence of large, atypical, ganglion-like cells with prominent, eosinophilic nucleoli that can resemble Hodgkin cells or viral inclusion cells, although these cells typically have more cytoplasm than do Hodgkin cells (Figure 13.36). Multivacuolated polygonal fibroblasts with intracellular mucins can resemble lipoblasts and are often referred to as *pseudolipoblasts* (Figure 13.37). There is scant mitotic activity and no atypical mitoses.

Immunohistochemical features are nonspecific. High numbers of histiocytes can be demonstrated with stains for CD68 and CD163, but the large epithelioid cells are negative, especially for the more histiocyte-specific CD163. In 10% to 15% of cases, large atypical cells are focally keratin positive. Studies for CD15 and CD30 that may be prompted by Hodgkin-like appearance of atypical cells, and microorganisms (bacteria by Gram and acid-fast stains, and Gomori's methenamine silver [GMS] for fungi) are all negative. The tumor cells are also negative for desmin, SMA, CD34, and S100 protein. PCR-based assays for cytomegalovirus and Epstein–Barr virus have been negative.[86] The pathogenesis of this tumor is unknown, but seems to include a prominent lymphohistiocytic host response. Perhaps this tumor is an example of a strong immune response keeping a low-grade mesenchymal tumor under host control.

Differential diagnosis

This tumor can be confused with proliferative synovitis because of prominent inflammatory infiltration, and potentially with epithelioid sarcoma because of occurrence in the distal extremities; the presence of occasionally reported keratin-positive cells could also lead to such confusion. The distinctive Reed–Sternberg-like cells of the inflammatory myxohyaline tumor do not occur, in the mentioned conditions considered in the differential diagnosis.

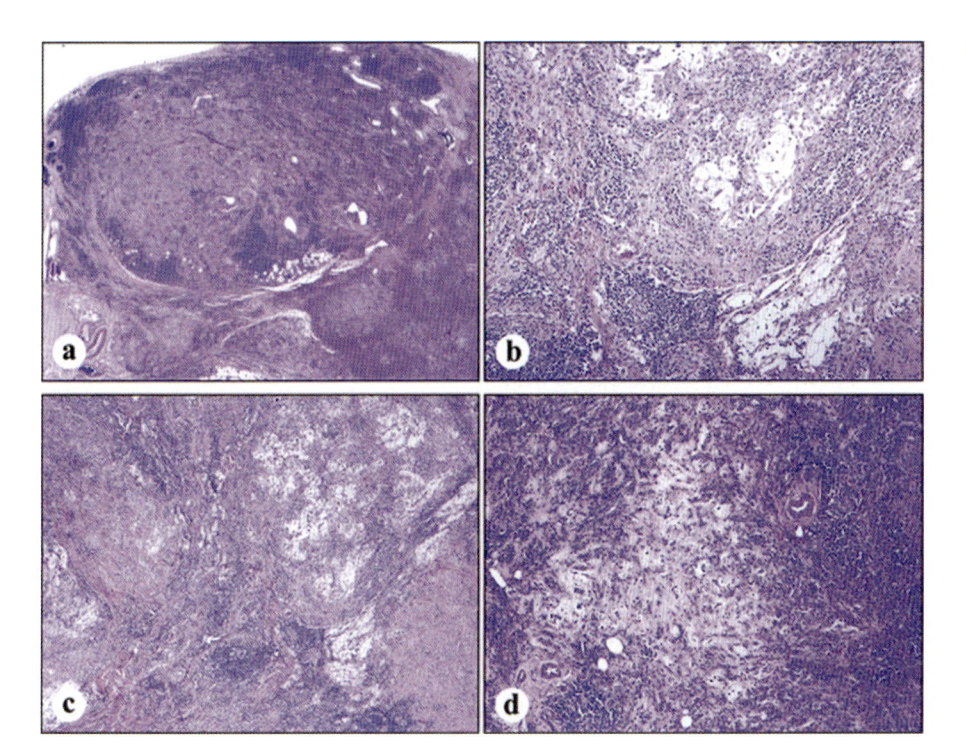

Figure 13.34 Low magnification of acral myxoinflammatory fibroblastic sarcoma shows multinodularity, myxoid zones, and patchy lymphoid infiltration.

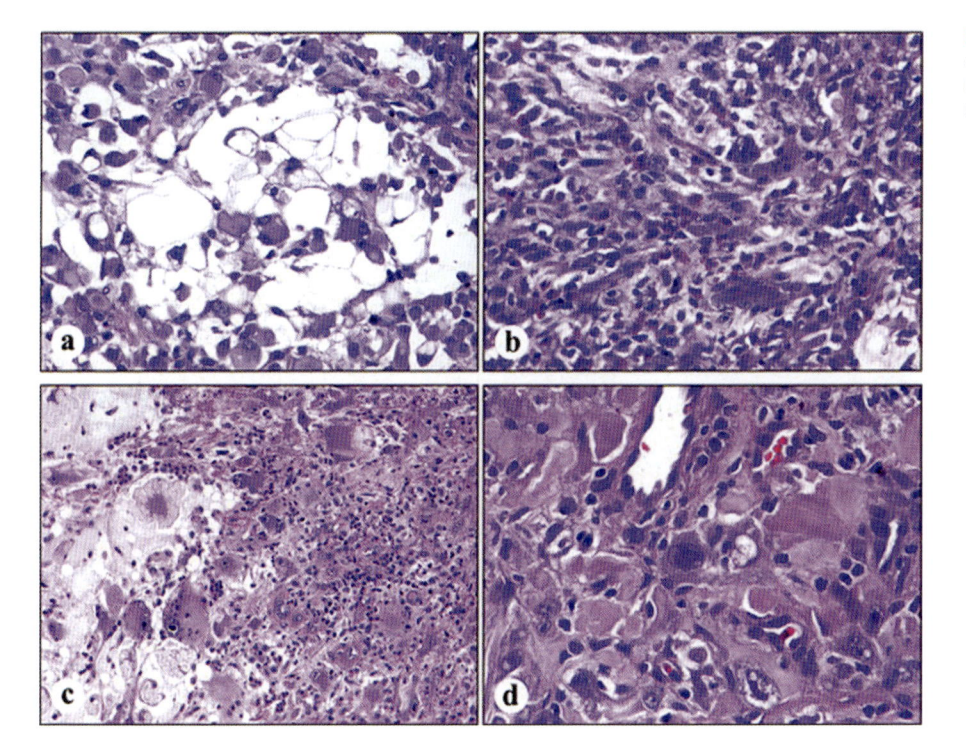

Figure 13.35 Epithelioid and myxoid, spindle cell, myxoinflammatory, and Reed–Sternberg-cell-like features in acral myxoinflammatory fibroblastic sarcoma.

Myxoid MFH contains a greater number of atypical cells, often in solid sheets. The author's opinion is that tumors containing solid sheets of atypical cells and atypical mitotic figures should be classified as myxoid MFHs rather than as inflammatory myxohyaline tumors.

Genetics

Heterogeneous changes have been reported in three cases. One of these had a reciprocal translocation t(1;10)(p22;q24) with loss of chromosomes 3 and 13.[90] Another case showed

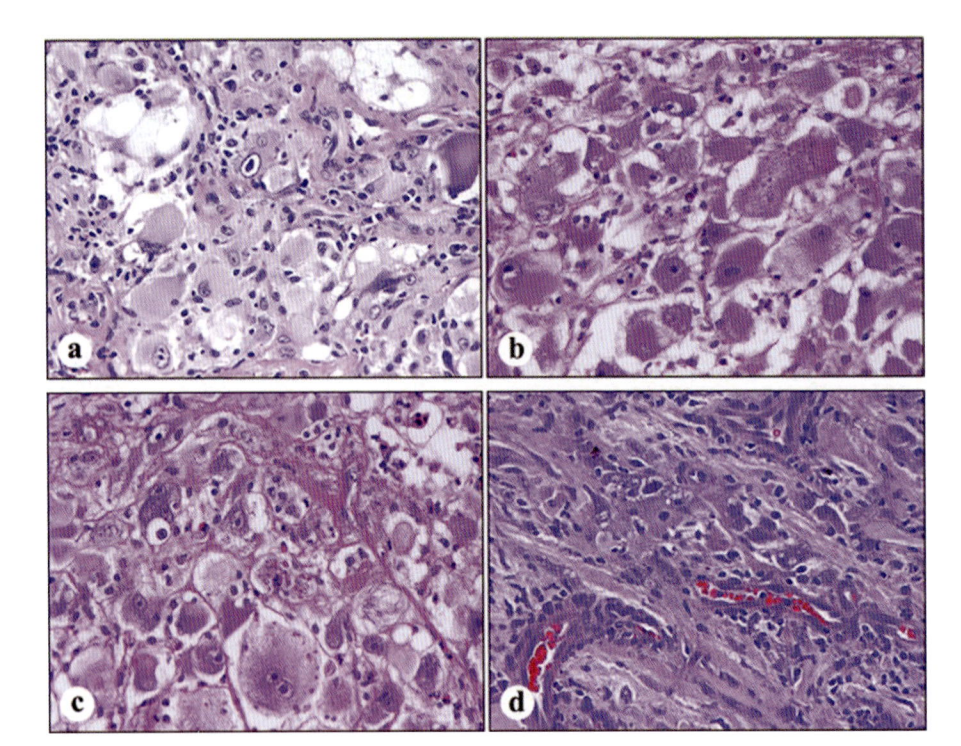

Figure 13.36 High magnification on Reed–Sternberg-like cells with large nuclei and prominent nucleoli in acral myxoinflammatory fibroblastic sarcoma.

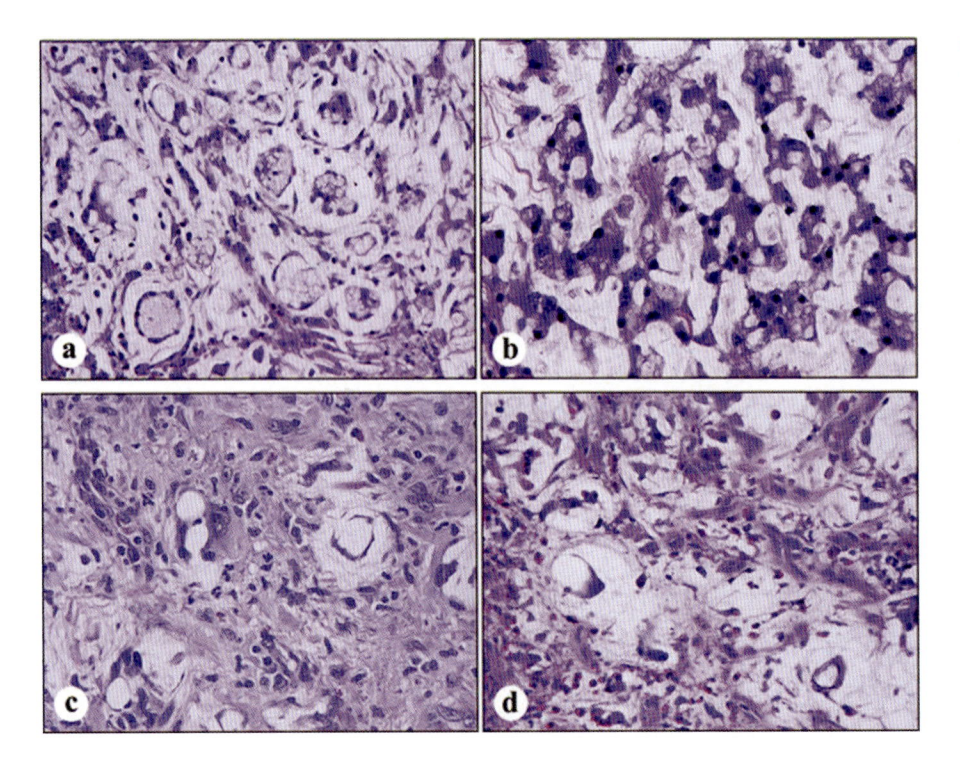

Figure 13.37 Intracytoplasmic vacuoles give a pseudolipoblastic appearance to some tumor cells in acral myxoinflammatory fibroblastic sarcoma. (**d**) Numerous eosinophils are present in some examples.

supernumerary ring chromosomes and a derivative chromosome 13.[91] A third case showed a reciprocal t(1;6)(q31;p21.3) translocation as the only cytogenetic change.[92] Molecular genetic analysis has shown amplification of 3p11–12 corresponding to VGLL3 gene amplification and TGFBR3 and MGEA5 gene rearrangements, although the number of reported cases remains small[93,94]

Myxofibrosarcoma (myxoid malignant fibrous histiocytoma)

In 1977, Angervall *et al.* reported myxofibrosarcoma (MFS)[95] and Weiss and Enzinger reported myxoid MFH[96] in two independent series. The former series included nonpleomorphic low-grade and pleomorphic higher-grade tumors,

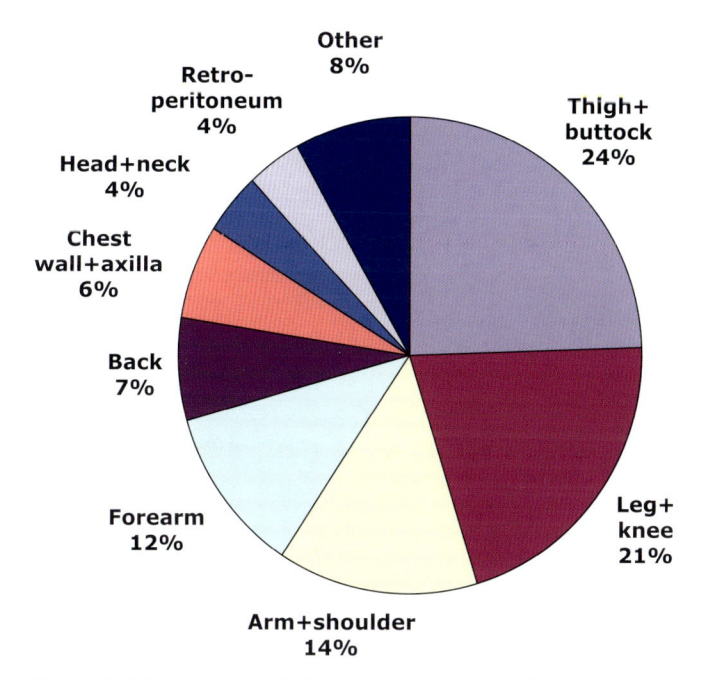

Figure 13.38 Age and sex distribution of 1141 patients with myxofibrosarcoma.

Location in the subcutis (60–80%) is more common than in deep soft tissues. The subcutaneous examples vary from small, 1- to 2-cm nodules to larger (>5 cm) masses, whereas most deep, intramuscular examples are large, often approaching or exceeding 10 cm in diameter.

Tumor behavior depends on malignancy grade, tumor size, depth, and local control. Whereas smaller subcutaneous examples essentially have recurrence potential only, larger, higher-grade, and deeper examples have a variable metastatic potential. In the largest series to date, patients with grade 1 tumors with no pleomorphism were found to have no metastatic potential and had a 90% 5-year overall survival rate, whereas the 5-year survival rate was 60% for those with grade 4 tumors.[99] In series containing tumors of all grades, the metastatic rate has been 20% to 25%.[96–98] In one series containing only low-grade tumors, however, the metastatic rate was 17%.[100] Local recurrence within a year of primary surgery is an adverse prognostic sign.[99] Distant metastases develop in the lungs, and rarely in bones. Lymph node metastases also occur. Based on the French Federation grading system (see Chapter 1), a majority of MFSs fall into the intermediate grade, approximately 20% to 30% into the high grade, and 10% to 20% into the low grade.

Pathology

Grossly, MFS forms a grayish, variably lobulated mass containing multiple nodules of mucoid and translucent appearance with solid nonmyxoid components; varying necrosis is present in larger examples (Figure 13.40).

Microscopic appearance varies by tumor grade and includes several patterns. Low-grade examples show a homogeneous architecture with uniformly myxoid matrix and distinctive, well-developed, somewhat coarsely built capillaries often forming curving (curvilinear) profiles. Nuclear atypia is mostly subtle, but scattered, more atypical forms are typically present, aiding in the recognition of this tumor. In some cases, the tumor cells contain large cytoplasmic vacuoles filled with alcian-blue-positive acid mucopolysaccharides. These cells are often designated as "pseudolipoblasts" and should not be considered evidence of fatty differentiation (Figures 13.41–13.43). Some tumors have an epithelioid clustered or trabecular pattern, mimicking chordoma or carcinoma (Figure 13.44). Higher-grade lesions can have solid pleomorphic MFH-like components (Figure 13.45). Mitotic activity varies widely and often exceeds 10 mitoses per 10 HPFs. Atypical mitoses are common in non-low-grade examples, and tumor necrosis is variably present.

In the author's experience, MFS shows nonspecific features on immunohistochemical analysis. These tumors are typically positive for vimentin and negative for desmin, keratins, and S100 protein. Positivity for CD34 is relatively common, and variable (usually focal) alpha SMA positivity can be observed.

Electron microscopic studies have shown mesenchymal cells with primitive undifferentiated and fibroblastic

whereas the latter group had pleomorphic features similar to those in other MFHs. The designations of MFS and myxoid MFH have been considered essentially synonymous in the 2002 WHO classification of soft tissue tumors.[97] That classification does not require any minimal amount of myxoid matrix for MFS designation, whereas myxoid MFH was defined to have at least 25% of myxoid matrix.[96] Other studies have also accepted a lesser amount of myxoid matrix, such as "at least 10% very prominent myxoid matrix" for the definition of MFS.[98] The lesser requirement of myxoid matrix has undoubtedly resulted in reclassification of many previous pleomorphic MFHs as MFSs. MFS and pleomorphic MFH form a histologic continuum without a sharp dividing line, but are discussed here separately.

Clinical features

MFS is one of the most common, if not *the* most common sarcoma in adults. It predominantly occurs in older age groups and occasionally in young adults, but almost never in children. The age range in the major series has been 21 to 91 years, with median ages 61 to 66 years, and the peak incidence is in the seventh decade. MFS has a predilection for the lower extremities (50–60%), and 20% to 35% occur in the upper extremities, 10% to 15% in the trunk wall, and <5% in the head and neck. In the extremities, proximal location is more common than distal Figure 13.38.[95–100] MFSs previously reported in the abdominal cavity probably contained large numbers of dedifferentiated liposarcomas, since these tumors can have morphology virtually indistinguishable from MFS.[101,102] Age and sex distribution of MFS are shown in Figure 13.39.

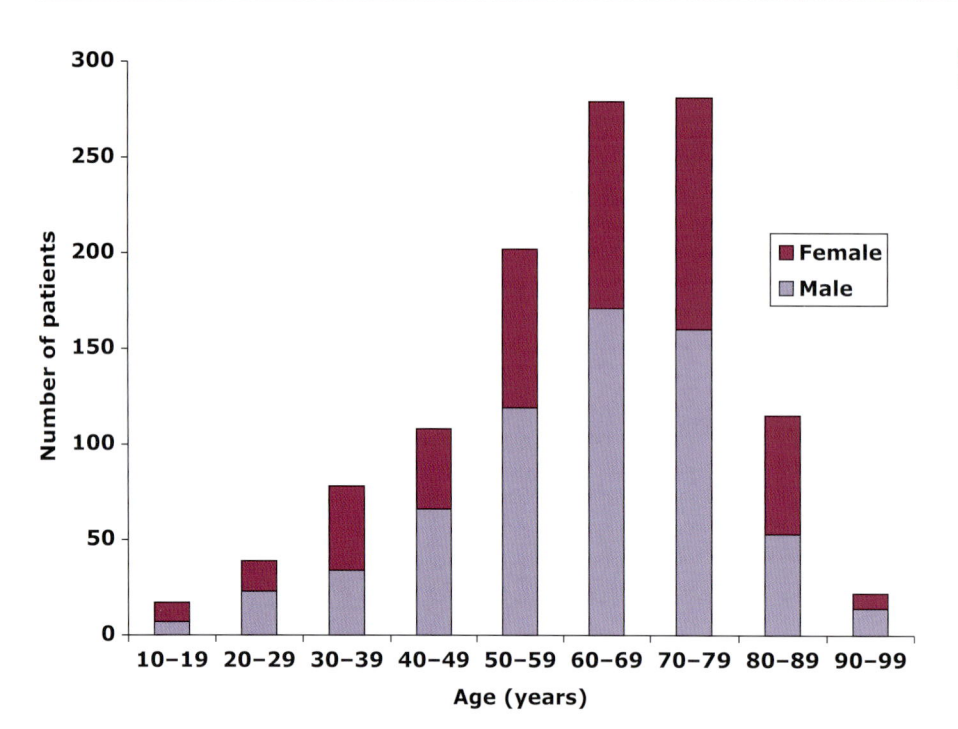

Chapter 13: Fibroblastic and myofibroblastic neoplasms with malignant potential

Figure 13.39 The eight most common sites for myxofibrosarcoma.

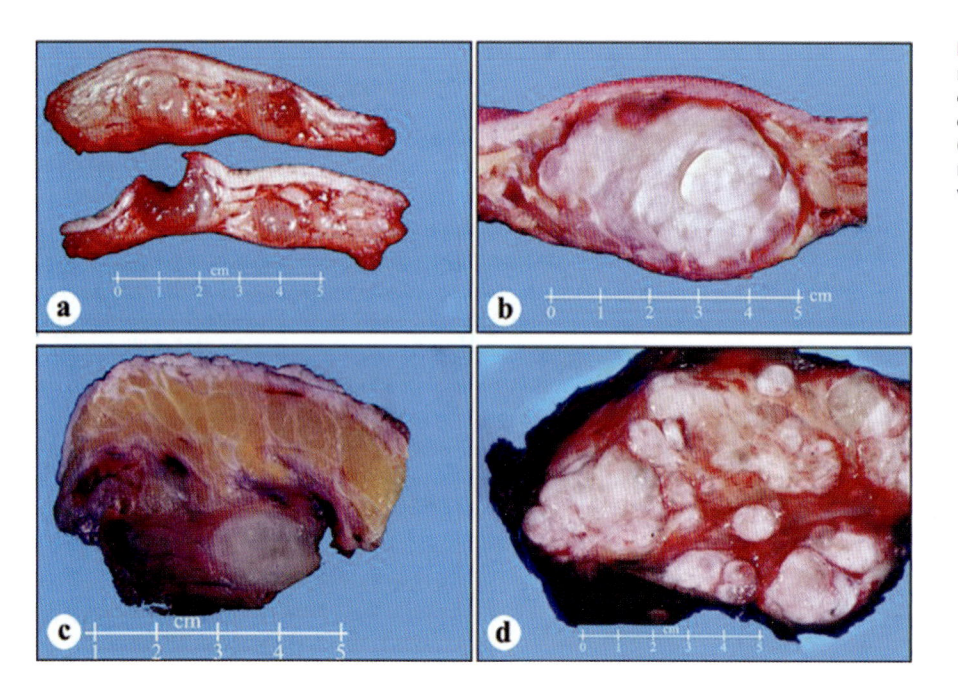

Figure 13.40 Gross appearances of myxofibrosarcoma. (**a**) Multinodular subcutaneous example with a gelatinous cut surface. (**b**) A less conspicuously myxoid subcutaneous example. (**c**) Intramuscular recurrence. (**d**) A large, multinodular intramuscular myxofibrosarcoma with focal mucoid appearance.

characteristics.[103,104] In one quantitative study, undifferentiated and histiocyte-like cells dominated over differentiated fibroblasts, and a minor myofibroblastic component was also present.[105]

Differential diagnosis

Myxoid liposarcoma has a more delicate capillary pattern and is composed of cytologically uniform oval cells, with variable, although sometimes limited, lipomatous differentiation. Cellular myxoma lacks cytologic atypia and has a less prominent and less complex vascular pattern. LGFMS typically shows alternating myxoid and dense fibrous areas, and nuclear atypia in this tumor is at best subtle. Myxoid DFSP shows a more homogeneous appearance, with the typical storiform and fat-infiltrating elements at least focally present in most cases. Dedifferentiated liposarcoma can be histologically very similar to MFS, and in the absence of detectable lipomatous

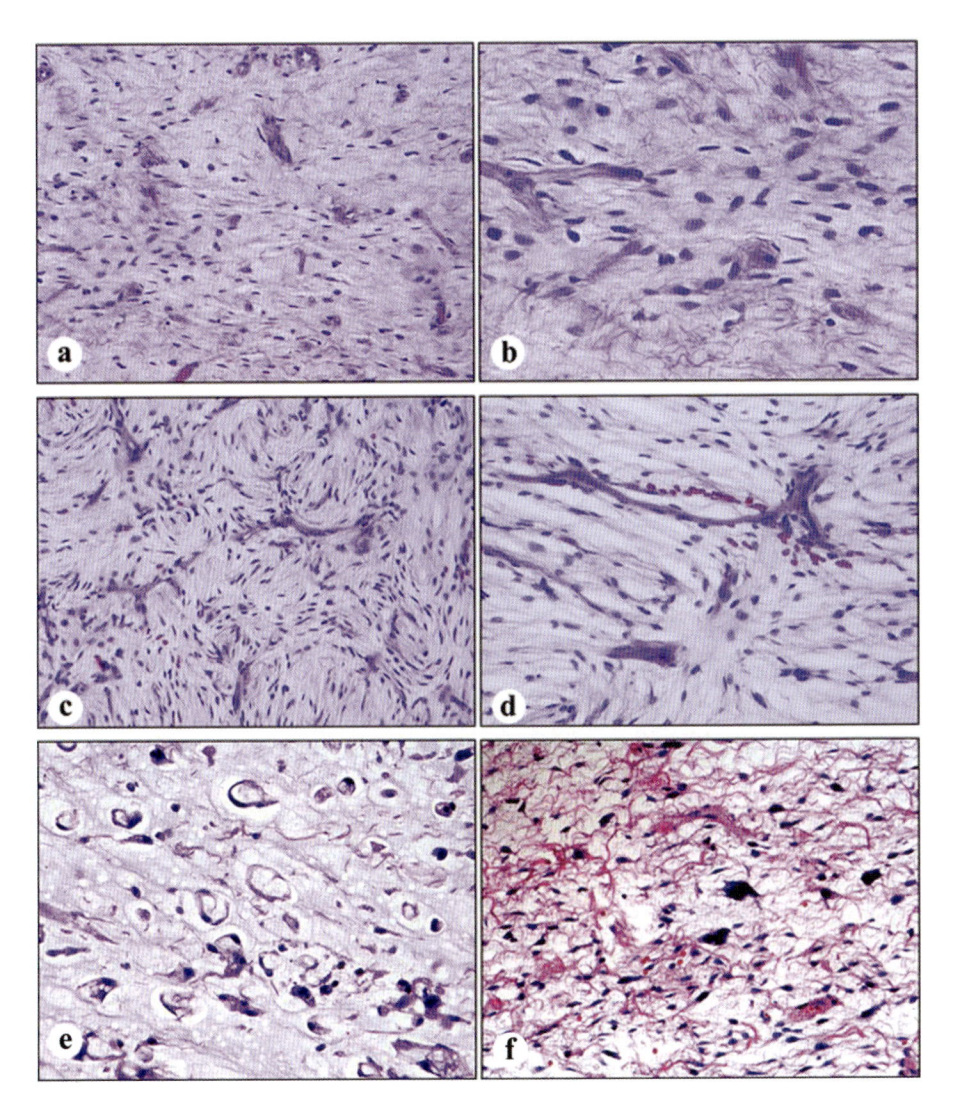

Figure 13.41 (**a–d**) Low-grade myxofibrosarcoma. Note moderate cellularity, mild atypia, prominent vascular pattern, and variably storiform achitecture. (**e**) Numerous pseudolipoblasts can occur, especially in the low-grade myxofibrosarcomas. (**f**) Focal pleomorphism is present in this tumor with a myxocollagenous matrix.

component, it can be impossible to differentiate this tumor from MFS. Clinicoradiologic correlation and extensive histologic sampling can be helpful in revealing a lipomatous component. Differentiation from pleomorphic MFH is based on the presence or absence of myxoid elements, and these categories form a biologic continuum.

Genetics

A relatively simple karyotype with t(2;15)(p23;q21.2) and interstitial deletion of 7q was reported in one low-grade MFS.[106] In general, typical of this tumor type are complex, nonreproducible karyotypes independent of histologic grade. The complexity often increases in tumor recurrences and is more pronounced in high-grade tumors.[107] One comparative genomic hybridization study showed gains at 7p21 and 12q15 most frequently, and most common losses at 13q14.3–q34.[108] Compared with myxoid liposarcoma, the changes are more numerous,[108] and FUS gene rearrangements are absent.[74]

Undifferentiated pleomorphic sarcoma (UPS, pleomorphic MFH)

UPS is an undifferentiated sarcoma diagnosed by exclusion. There are no specific markers to positively identify UPS, but use of an immunohistochemical panel is highly recommended to exclude sarcomas with specific differentiation, pleomorphic carcinomas, and melanomas.

Evolution of the MFH concept

It has long been agreed that UPS pleomorphic MFH is not a histiocytic neoplasm but a fibroblastic or undifferentiated one, because the tumor cells do not express cluster antigens specific to histiomonocytic lineage.[109] Electron microscopic studies have agreed on fibroblastic differentiation and similarities with fibrosarcomas, at variance with earlier studies that suggested differentiation toward fibroblasts and histiocytes with inter-conversions between these cell types.[110,111]

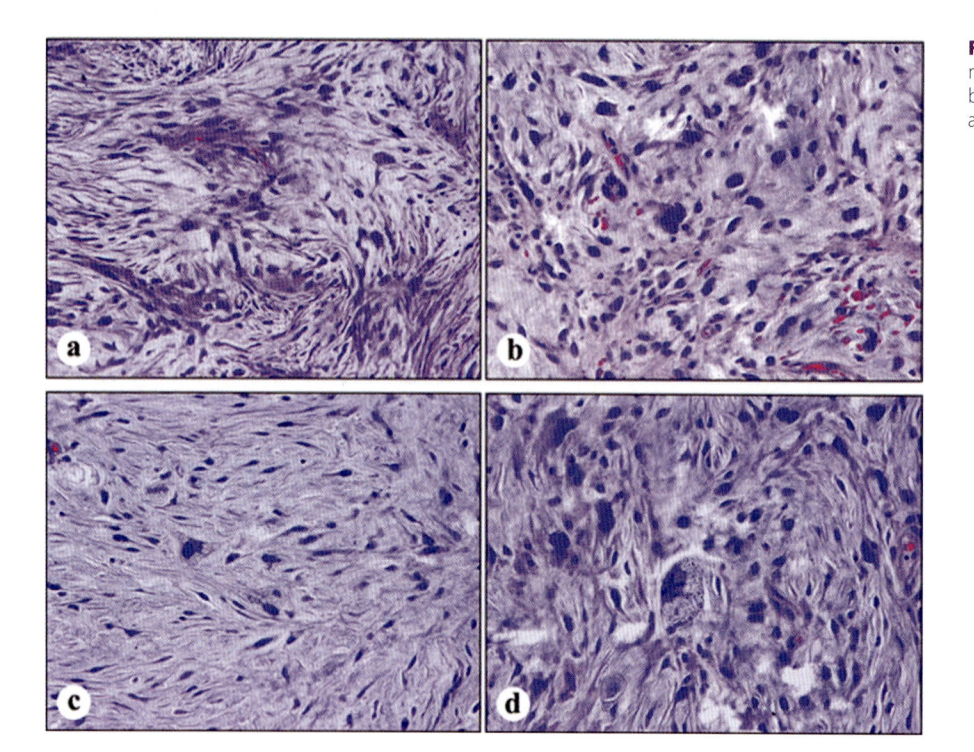

Figure 13.42 (**a–c**) Low-grade myxofibrosarcoma with focal to moderate atypia, but low mitotic activity (grade 2 of 4). (**d**) An atypical mitosis is present.

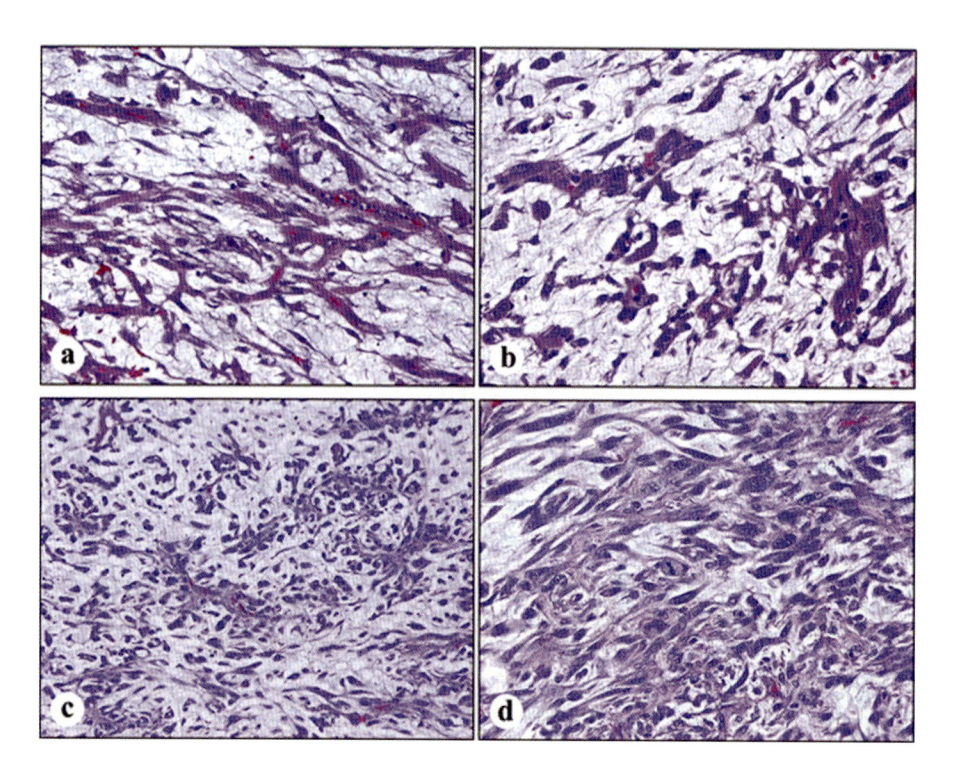

Figure 13.43 Intermediate-grade myxofibrosarcoma with moderate atypia, focal fascicular structure, and moderate mitotic activity (grade 2 of 3).

Immunohistochemical positivity for several histiocytic markers such as alpha-antitrypsin[112] and later CD68[113] seemed to support the existence of histiocytic differentiation in this tumor. Both of these markers, however, were shown to be non-lineage-specific, with a wide expression in different types of tumors of various lineages.[114,115] One monoclonal antibody study concluded that MFH shared phenotypic features with perivascular fibroblast-like mesenchymal cells.[116]

Understanding of the nonhistiocytic nature of pleomorphic MFH led to the WHO classification of soft tissue tumors of 2002 to favor the term undifferentiated pleomorphic sarcoma (UPS),[117] however, the term UPS, being well-known, might have clinical advantages.

Chapter 13: Fibroblastic and myofibroblastic neoplasms with malignant potential

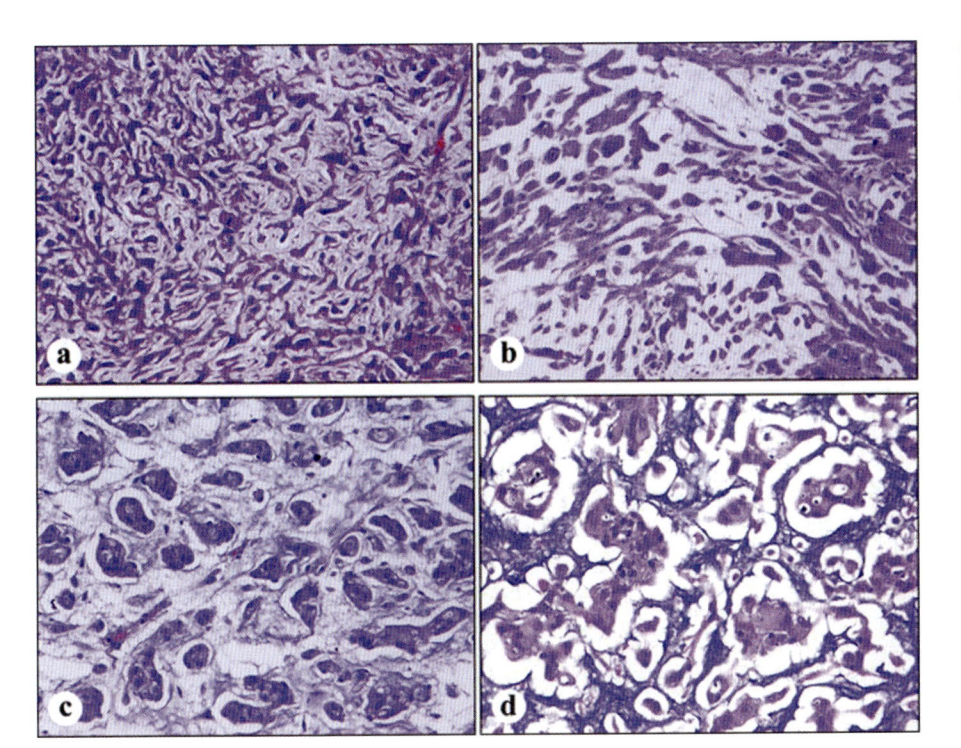

Figure 13.44 Myxofibrosarcoma with epithelioid cords and clusters resembling chordoma or other epithelial tumors. Intermediate grade (grade 2 of 3).

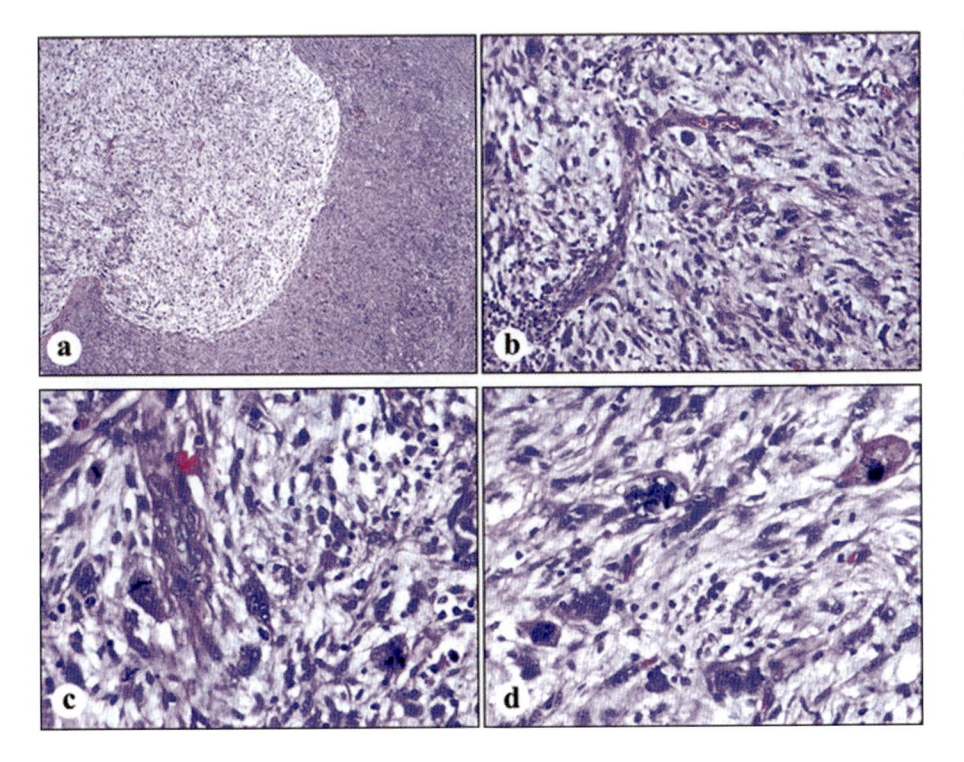

Figure 13.45 High-grade myxofibrosarcoma with marked pleomorphism and mitotic activity. (**a**) Solid areas are present. (**b,c**) Note prominent vessels. (**d**) Numerous atypical mitoses are present (grade 3–4). The presence of necrosis decides whether this tumor is high grade in the French Federation of Cancer Centers Grading System.

There is some uncertainty on the definition and frequency of pleomorphic MFH. Fletcher has expressed the view that pleomorphic MFH is a rare sarcoma type, suggesting that most tumors previously diagnosed as such can be reassigned to other diagnostic categories, leaving only 10% to 15% of tumors originally considered pleomorphic MFHs in this category.[118] In another study, Fletcher et al. reclassified only 12% of 100 sarcomas originally classified as MFH (different types) as pleomorphic undifferentiated sarcomas. The largest exclusion groups were MFS (29%), high-grade leiomyosarcoma (20%), myofibroblastic sarcoma (11%), and high-grade myogenic sarcoma (9%).[119] The definition of some of those diagnostic groups is also problematic. According to the author's experience, MFH remains a common sarcoma in older adults, even

361

after careful exclusion of other specific diagnostic categories, and we do not disqualify focally SMA-positive tumors from the UPS group.

The view that UPS might represent a final pathway in tumor progression of multiple specific tumor types, such as liposarcoma, seems quite plausible, as suggested by Brooks.[120] In many cases, however, the original differentiation of the tumor cannot be determined, leaving the tumor in the undifferentiated category.

Clinical features of UPS (pleomorphic MFH)

A population-based study from Sweden determined the annual incidence of MFH as 0.42 in 100, but this number also includes MFS.[121] UPS (Pleomorphic MFH) usually occurs in older adults, although an age range of 2 to 92 years has been collectively observed in the largest series, with very rare occurrence of this tumor during the first two decades, and rarity even before the age of 40 years. More than one-half of cases occur between the ages of 50 and 69 years. Most series have shown mild to moderate (55–60%) male predominance.[121–128]

UPS has a predilection for the extremities, being twice as common in the lower than the upper extremities. The most common sites are thigh and buttocks, arm and shoulder, leg, and forearm, and small numbers of cases present in the hands and feet. Frequency in the truncal locations is difficult to estimate, because older series contained many retroperitoneal examples that are now more often classified as dedifferentiated liposarcomas. In one series, 6% of cases occurred in the head and neck.[127] Most examples are deep, with intramuscular involvement. Organ-based location is very rare, and in these sites it is important to exclude sarcomatoid carcinoma.

Tumor size varies widely, ranging from <2 cm (rare) to >20 cm, and median tumor sizes in the largest series have been 6 cm to 8 cm. UPS is usually intermediate or high grade, with pulmonary metastases developing in up to 30% to 40% of cases. Other metastatic sites include liver, bone, and lymph node. Recurrence is at least equally common, and its occurrence within 1 year of primary surgery has been found to be an unfavorable prognostic sign.

Clinicopathologic series have demonstrated that superficially located and completely excised tumors, those <5 cm and not of high-grade histology, are associated with a better prognosis.[120–128] Some studies have also suggested that tumors that express a variety of myoid markers have a worse prognosis,[119,129,130] whereas one study did not observe this association.[131]

Pathology

UPS is often a large tumor, commonly measuring 10 cm to 15 cm in diameter, frequently involves both the muscles and the subcutis, and can extend into and ulcerate the overlying skin. Most tumors appear grossly well circumscribed, although histologically they typically infiltrate skeletal muscle at the periphery. On sectioning, UPS is often multilobulated and relatively soft; larger tumors usually have central necrosis. The color varies from gray-white to tan on sectioning.

Histologically, UPS can show random, irregular fascicular, or rarely a storiform pattern. It is composed of spindle-shaped, pleomorphic, occasionally polygonal epithelioid cells. Multinucleated tumor giant cells can be present. A collagenous matrix is variably present. Mitotic activity is usually easily identified and commonly exceeds 10 to 20 mitoses per 10 HPFs. Atypical mitoses are common (Figures 13.46–13.48).

Immunohistochemically, the tumor cells are positive for vimentin and typically negative for CD34, desmin, EMA, keratins, SMA, and S100 protein. The author, however, allows focal CD34, and SMA and even desmin expression that in the context probably identified fibroblastic and myofibroblastic differentiation. Nevertheless, desmin-positive tumors should be evaluated for MyoD1 or myogenin, and skeletal muscle transcription factors to rule out rhabdomyosarcoma. Focal keratin positivity can occur[132–134] and is probably comparable to keratin expression in transformed fibroblasts.[135,136]

Biology of UPS

Cytokine production of UPS cells has been demonstrated in some cases and could explain the high content of intratumoral histiocytes, the morphologic pattern of inflammatory UPS, and the constitutional symptoms observed in some patients. Indeed, the leukemoid reaction detected in some patients has been attributed to cytokines, such as interleukin-6 (IL-6) secreted by UPS cells,[137,138] and the production of multiple hematopoietic growth factors, such as granulocyte colony stimulating factor.[139]

Electron microscopy

Some studies emphasize fibroblastic and myofibroblastic features in the tumor cells. These typically have a prominent rough endoplasmic reticulum, often with dilated profiles, and can have scattered bundles of actin filaments.[110,111] Cells with truly histiocytic differentiation are tumor-infiltrating histiocytes.

Genetics

Complex cytogenetic changes are typical, but some recurrent changes have been found.[140–143] In one study, 19p13 aberration with the addition of new material from an unknown source correlated with an increased risk for recurrence.[143]

Some recurrent changes have also been identified in the comparative genomic hybridization (CGH) studies. Various studies highlight different genetic changes, however, and the CGH results are not reproducible from one study to another,

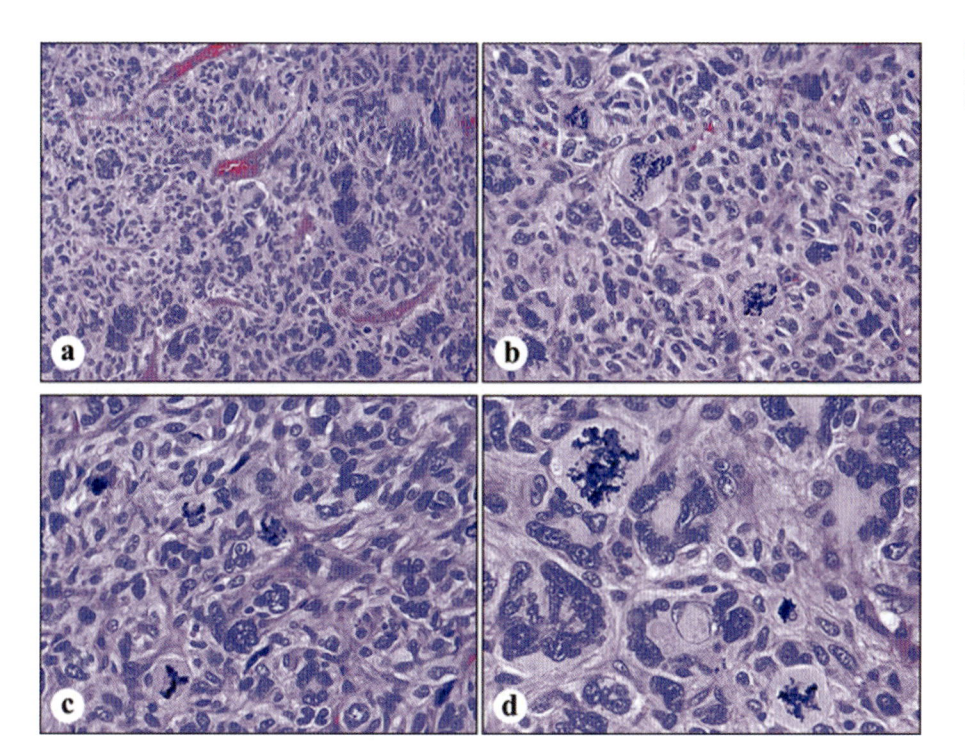

Figure 13.46 Undifferentiated pleomorphic sarcoma with severe nuclear atypia and numerous atypical mitoses.

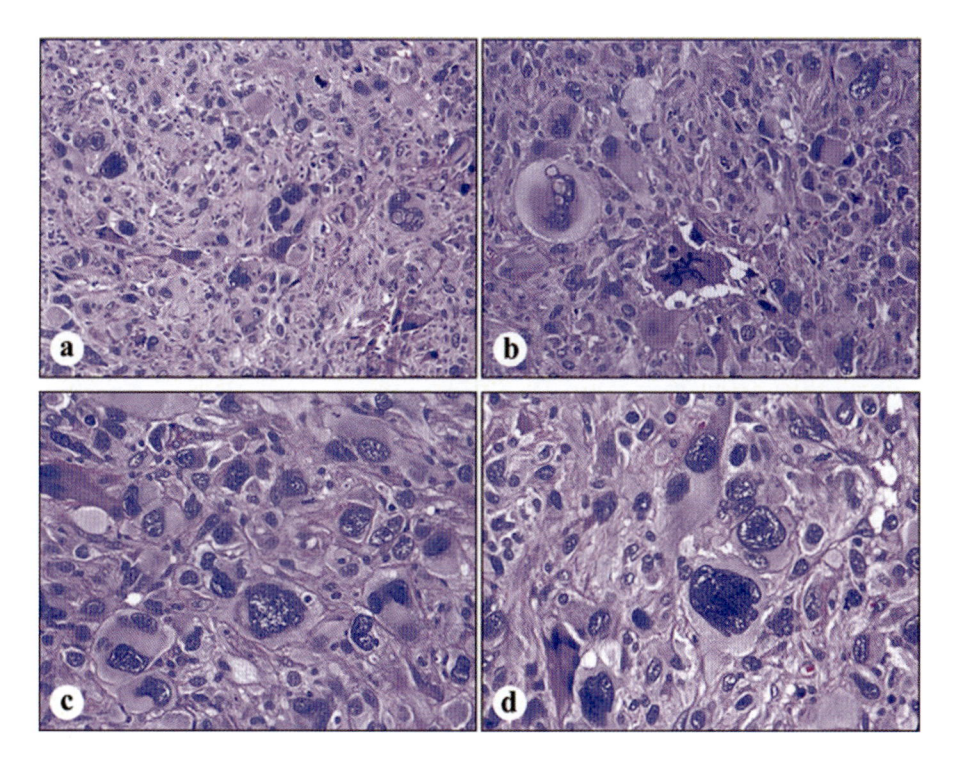

Figure 13.47 Undifferentiated pleomorphic sarcoma with epithelioid cytology.

perhaps because of differences in tumor material and techniques.[144–146] In one series, CGH studies identified gains in 7q32 as a possible prognostically adverse sign,[144] and another series found, among others, amplifications of 12q12–15, similar to those seen in many liposarcomas.[145] Yet another study identified the loss of chromosome 13 as the most common chromosomal imbalance.[146] Losses in 13q14–q21 were found in nearly 80% of tumors in another study, and the RB1 gene was included in the area with losses.[147] A new gene, named MASL1 (for *M*FH-*a*mplified *s*equences with *l*eucine-rich tandem repeats), was identified in the amplification region of 8p23.[148,149] A multimethod study using classic cytogenetics, CGH, and southern analysis found losses in 9p, including p16 INK4.[150] Gain of 12q13–21 and amplification of the MDM2

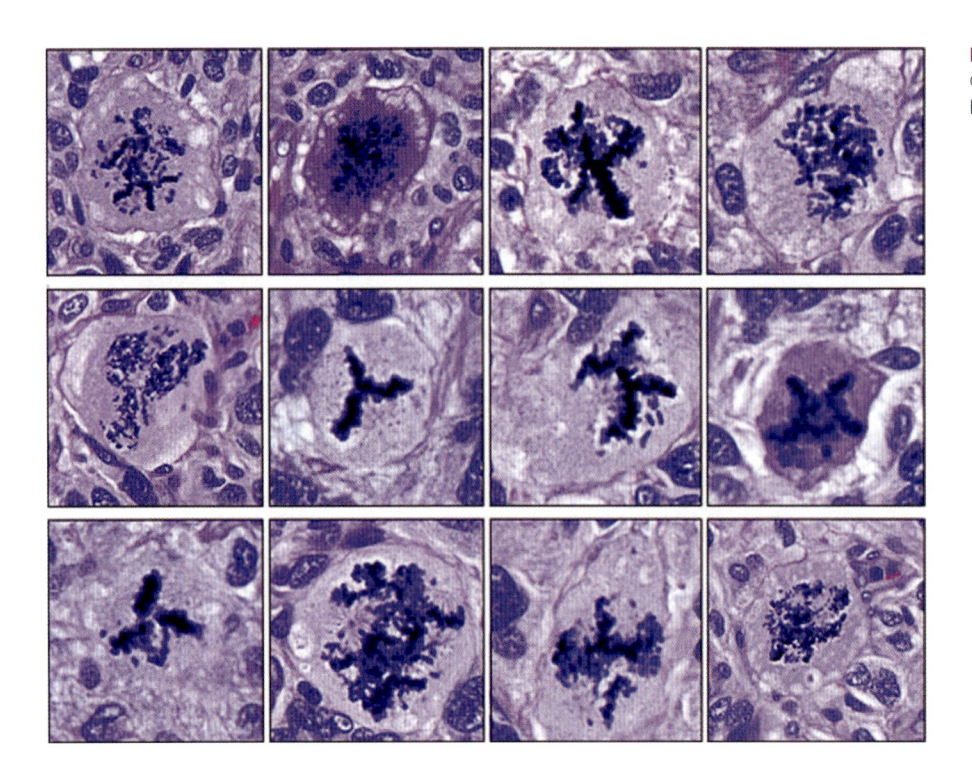

Figure 13.48 Examples of atypical mitoses with different configurations in undifferentiated pleomorphic sarcoma.

gene underscore genetic similarities and perhaps overlap with dedifferentiated liposarcoma in some cases.[145,151,152] Mutations in the beta-catenin gene[153] have also been reported.

Differential diagnosis

UPS pleomorphic MFH is a diagnosis of exclusion, and currently no specific markers are available to document this diagnosis.

Dedifferentiated liposarcoma should be ruled out with thorough gross examination and extensive sampling, especially in cases of retroperitoneal tumors. Radiologic studies (e.g., MRI) can be helpful in pinpointing a lipomatous component.

Differentiated myogenic sarcomas, sarcomatoid carcinoma, melanoma, and lymphoma need to be ruled out immunohistochemically before a diagnosis of UPS is made. Suggested work-up for a pleomorphic sarcomatous neoplasm is shown in Table 13.1.

Although UPS has been reported at almost any site, this diagnosis should be made with caution in parenchymal organs, for example in the kidney, where sarcomatoid carcinomas can have a pleomorphic, UPS-like appearance.

Extensive SMA or desmin positivity should lead to a classification of myogenic sarcoma, and further markers for smooth (heavy caldesmon) and skeletal muscle (myogenin, MyoD1) should be examined. If neither is found, then a designation of unclassified myogenic sarcoma is appropriate. The author allows limited SMA and desmin immunoreactivity in UPS and does not classify such tumors as myogenic sarcomas, but rather considers these features as expression of focal myofibroblastic differentiation, compatible with the diagnosis of UPS.

Atypical fibroxanthoma

Atypical fibroxanthoma (AFX) refers to a dermal analog of UPS pleomorphic MFH, and deeper undifferentiated tumors involving the subcutis should be classified as superficial MFHs. Pleomorphic malignancies of specific types should be ruled out, especially sarcomatoid spindle cell carcinoma of the skin, and cutaneous melanoma.

Clinical features

AFX typically occurs in elderly people: in the largest series, 75% of patients were 60 years or older. There is a predilection for Caucasians and other people with fair skin.[154–157] The causal role of ultraviolet and other radiation is suggested by several factors. AFX predominantly occurs on sun-exposed sites, and the incidence is higher in the southern latitudes of the United States, indicating a role of UV light carcinogenesis. Occurrence after therapeutic radiation has also been reported.[154,156] Impaired immunosurveillance could also be a factor, especially in post-transplant patients.[158]

AFX has a strong predilection for the head, especially the face around the nose, cheeks, and earlobes. It also occurs in the scalp, especially on bald areas. Occurrence in the skin of the trunk and extremities is uncommon.[154–157]

Despite an often highly malignant histologic appearance, tumor behavior is indolent if the tumor is adequately excised. Recurrence rate in the largest series was 10%.[154] If AFX is defined narrowly (not including tumors with deep extension), and carcinoma and melanoma are ruled out, then the metastatic rate of AFX is negligible.

Table 13.1 Suggested work up to rule out other diagnoses in pleomorphic malignancies suspected of MFH or atypical fibroxanthoma

Category to be excluded	Markers	Suggestions for use	Interpretation/further studies
Pleomorphic carcinoma	CK cocktail CK18	Always in visceral location, with discretion in peripheral sites	Extensive keratin positivity suggests carcinoma. Focal positivity requires clinical correlation, because it can also be seen in MFH. Some cases remain indeterminate
Sarcomatoid squamous cell carcinoma	CK5/6	Superficial tumors, or whenever squamous cell derivation is possible	Positivity (especially in superficial tumors) indicates squamous cell epithelial differentiation
Malignant melanoma	S100	Epithelioid and pleomorphic examples, others with discretion	S100 positivity suggests melanoma Additional melanoma markers are indicated.
	HMB45, melan A	With a high index of suspicion	Positivity strongly supports melanoma
Pleomorphic leiomyosarcoma	SMA, DES	In most cases, to rule out myogenic differentiation	Search for smooth muscle differentiation by additional sampling
	Heavy caldesmon	Further testing for smooth muscle differentiation	Positive results support the diagnosis of leiomyosarcoma
Rhabdomyosarcoma/ heterologous rhabdomyosarcomatous differentiation	Myogenin, MyoD1 Especially when SMA negative	Desmin-positive tumors, positivity supports rhabdomyosarcomatous differentiation. Clinical and morphologic correlation to determine whether this indicates heterologous rhabdomyosarcomatous differentiation (more common) or pleomorphic rhabdomyosarcoma (rare)	
Myogenic sarcoma, not further classified	SMA, DES	Most cases	Extensive positivity without smooth or skeletal muscle differentiation supports this designation
Lymphoma	CD45 (LCA)	With discretion	
Large cell anaplastic lymphoma	CD30	With high index of suspicion	

Metastatic AFX has been reported, however. In one study, locoregional metastases of eight AFXs of the head occurred in the parotid region, possibly involving the lymph nodes. In these cases, deep soft tissue and vascular invasion were noted in the primary tumor.[159] A more recent series also reported locoregional recurrences/metastases from as small as 1 cm AFX lesions.[160] Neither study employed rigorous immunohistochemical studies to rule out carcinoma and melanoma, however.

Pathology

AFX typically forms an elevated cutaneous nodule or ulcerated plaque usually measuring <1 cm (Figure 13.49). An epithelial collar often surrounds the exophytic tumor nodule (Figure 13.50). The tumor should be entirely confined to dermis, and examples with subcutaneous extension are better defined as superficial MFHs.

The histologic spectrum includes variants with conspicuous cellular pleomorphism and atypical mitoses, similar to

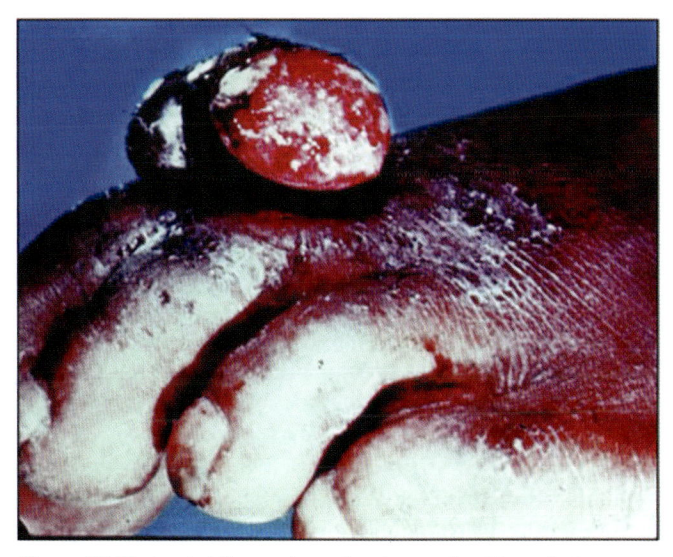

Figure 13.49 Atypical fibroxanthoma forming a polypoid exophytic mass on dorsal side of right big toe.

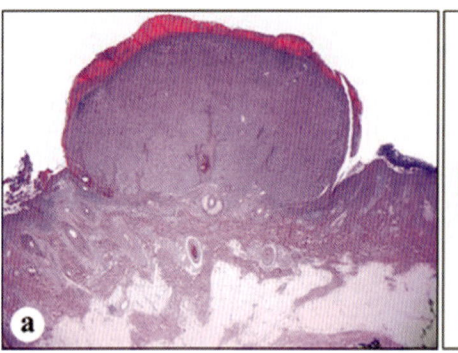

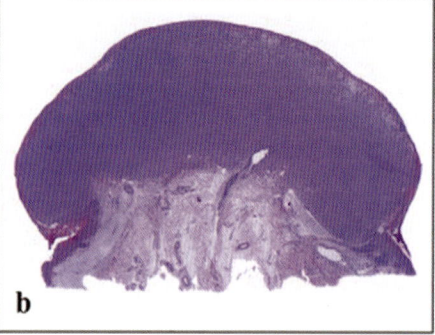

Figure 13.50 Low magnification of two atypical fibroxanthomas. Both examples form exophytic dermal masses with epithelial collars. Note confinement to the dermis.

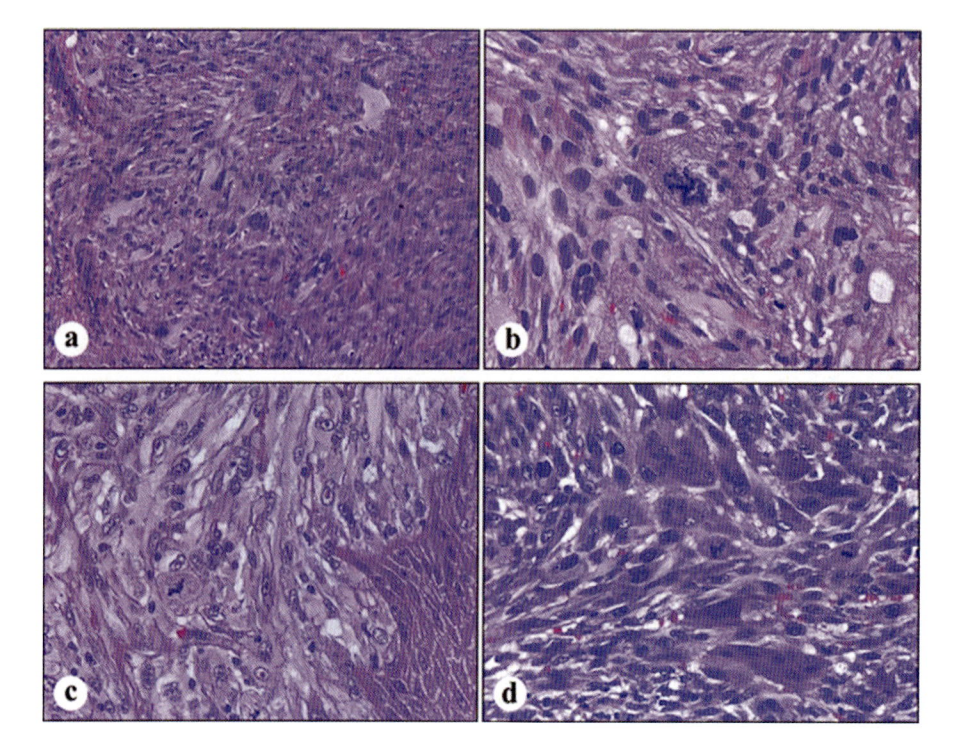

Figure 13.51 Marked nuclear pleomorphism in atypical fibroxanthoma. (**a**) Prominent vessels are present. (**b**) Tumor cells with spindled and epithelioid features and an atypical mitosis. (**c**) Elongated myofibroblastic tumor cells. (**d**) A spindled variant with pleomorphism and osteoclastic giant cells.

pleomorphic MFH. Variants composed of clear cells, nonpleomorphic spindle cells, and those containing osteoclast-like giant cells have been included under AFX (Figure 13.51).[154,161,162]

In general, AFX is defined as a tumor that lacks specific mesenchymal or epithelial differentiation and that is immunohistochemically negative for keratins, S100 protein, heavy caldesmon, and desmin (Figure 13.52).[163–166] Smooth muscle actin positivity is compatible with AFX in the absence of specific evidence for smooth muscle differentiation and reflects myofibroblastic differentiation. Expression of keratins 5/6, and p63 suggests sarcomatoid squamous cell carcinoma instead of AFX.[167]

AFX lesions have shown to have high TP53 expression by immunohistochemistry and TP53 mutations involving substitutions in cytosine residues,[168] similar to those previously seen in nonmelanoma skin tumors in xeroderma pigmentosum patients, indicating mutagenesis by UV radiation.[169]

Differential diagnosis

The two sides of this problem are differentiation from cutaneous tumors with specific differentiation and from MFH.

The diagnosis of AFX is one of exclusion, and similar work up is necessary as for MFH, with the addition of high-molecular-weight keratin immunostain to address sarcomatoid squamous carcinoma (Table 13.1).

Large cell anaplastic lymphoma can have pleomorphism, but usually shows a more uniform cellular population that is always CD30 positive and usually CD45 positive. Pleomorphic leiomyosarcomas are distinguished by the presence of more typical areas and expression of muscle actin, desmin, or both. Heavy caldesmon is another marker expressed in smooth muscle tumors, but not in AFX.[167] In small biopsies, without knowing lesion size, one has to be careful when using the designation of AFX; clinical correlation is needed to separate

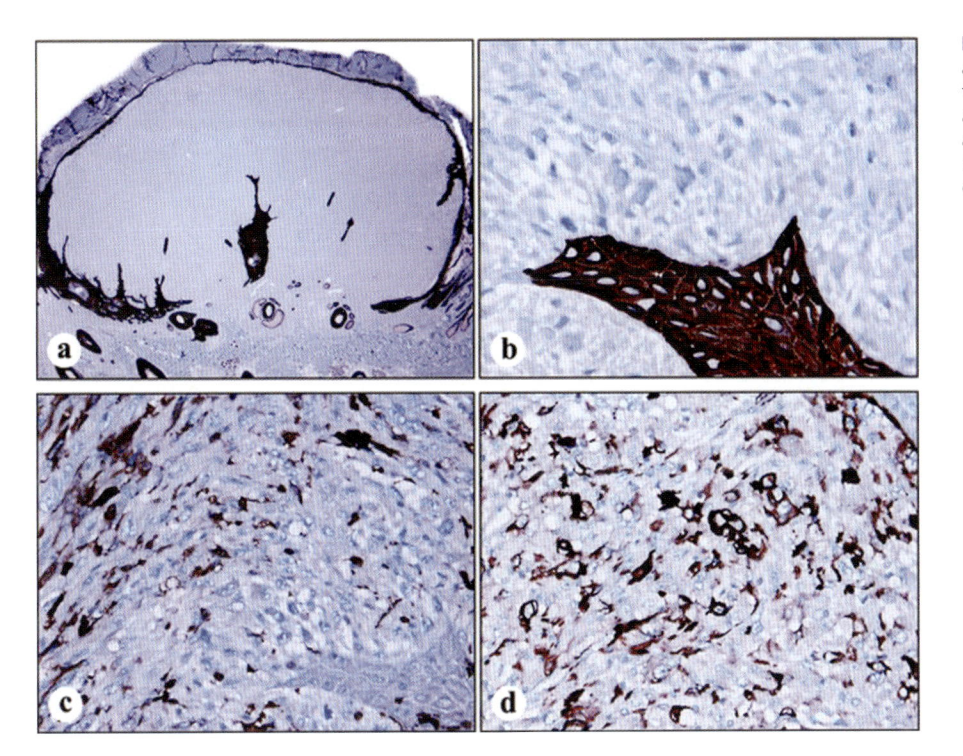

Figure 13.52 Immunohistochemical findings in atypical fibroxanthoma. Tumor cells are negative for keratin cocktail AE1/AE3, and CK5/6. Tumor cells are negative for S100 protein, but dendritic cells and epidermal Langerhans cells are positive. Moderate smooth muscle actin positivity is common in atypical fibroxanthoma.

this from (superficial) MFH. The latter diagnosis should be used in tumors involving the subcutis.

Inflammatory malignant fibrous histiocytoma

Originally reported by Kyriakos and Kempson in 1977, inflammatory MFH is a variant rich in neutrophils, histiocytes, or both.[170] The reported tumors occurred in older adults in both extremity-based and retroperitoneal locations. Cytokine production, possibly by tumor cells, probably plays a role in neutrophil recruitment and could also explain systemic symptoms, such as fever, weight loss, and systemic leukocytosis.[171] Inflammatory MFH is a rare variant with demographics similar to other MFHs. These tumors have been most commonly reported in the retroperitoneum. In fact, the presence of MDM2 gene amplification and morphological overlap support the assertion that inflammatory MFH largely represents a variant of dedifferentiated liposarcoma.

Histologically, inflammatory MFH has a prominent inflammatory infiltration containing neutrophils and variably xanthomatous histiocytes. Admixed within the inflammatory cells there are few atypical neoplastic cells that might have nuclear pleomorphism, multinucleation, and prominent nucleoli. Cytoplasmic engulfment of inflammatory cells can be seen in the atypical cells (the so-called emperipolesis phenomenon). These atypical cells are sometimes elusive, raising the differential diagnosis of an inflammatory process (Figure 13.53).

Differential diagnosis

The need to exclude other tumors applies to this variant as much as to other MFHs. The presence of concurrent lipomatous components, 12q13–15 amplification, and MDM2 and CDK4 gene amplification and immunohistochemical expression in most tumors previously classified as inflammatory MFHs suggests that inflammatory MFH largely represents a histologic variant of dedifferentiated liposarcoma.[172] The presence of a liposarcomatous component by radiologic studies (MRI) also supports the diagnosis of dedifferentiated liposarcoma.

Some carcinomas have prominent inflammatory infiltration, and keratin immunostaining is mandatory to rule out this possibility, especially in a visceral setting.

Lymphomas that can simulate MFH include rare variants of large cell anaplastic lymphoma and sarcomatoid variants of Hodgkin's disease. The diagnosis of the latter is facilitated by immunohistochemistry for leukocyte antigens CD15 and CD30, which are absent in the potentially lymphoma-like inflammatory variant of MFH.[173]

Giant cell tumors of soft parts

Giant cell tumors of soft tissue and giant cell MFH are a spectrum of giant-cell-rich tumors, the essential neoplastic component of which is probably fibroblastic or myofibroblastic. The benign end of their spectrum is now classified as *giant cell tumor of soft parts*, whereas the designation of *giant cell MFH* is reserved for overt sarcomas with marked nuclear atypia.

Salm and Sissons first described tumors that were histologically analogous to giant cell tumors of bone and lack an atypical mononuclear cell component.[174] Three subsequent series have reported similar tumors and emphasized that they should be separated from giant cell MFH because of their

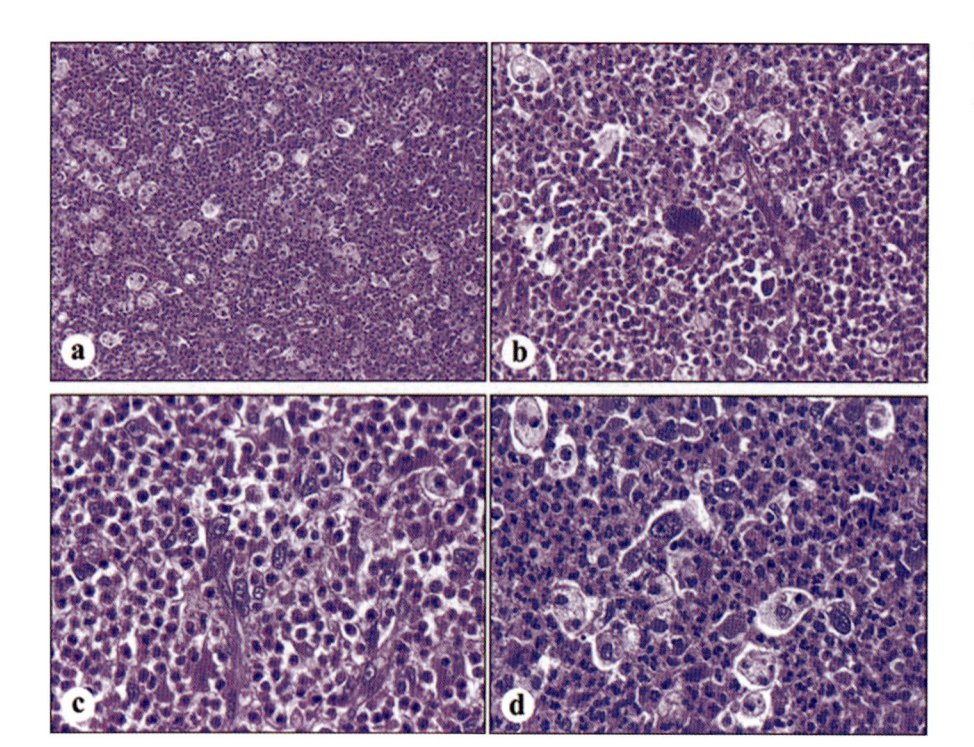

Figure 13.53 Inflammatory malignant fibrous histiocytoma with prominent neutrophil infiltration and scattered atypical cells.

limited malignant potential. Diagnostic criteria and opinions about whether these tumors should be considered nearly benign or of uncertain malignant potential vary, however.[175–177]

Clinical features

The tumor occurs in patients in a wide age range, including children. Collective data from the four series suggest a mild male predominance.[175–177] Most tumors occur in the proximal extremities, some in the trunk, and a few in intra-abdominal sites. Arm, thigh, knee, and leg are among the most common locations; examples have also been reported from the hands and feet. Approximately 60% of the tumors are subcutaneous or dermal, and 40% are deep to the superficial fascia. By definition, origin from bone has to be ruled out.

Provided that tumors with an atypical mononuclear cell population are excluded, the prognosis seems excellent, with potential for local recurrence after an incomplete excision. None of the series has had extensive long-term follow-up, however, and further studies are needed to confirm the biologic potential of these lesions. This applies especially to large and deep tumors.

Pathology

The tumor typically presents as a circumscribed nodule or larger mass measuring 1 cm to 10 cm. The median sizes in the four published series have been 2 cm to 4 cm. Grossly, the cut surface appears reddish or gray and varies from fleshy to rubbery.

Histologically typical is a multinodular pattern with fibrous septa dividing the tumor into cellular lobules, and a peripheral rim of metaplastic bone is sometimes present. The cellular areas are conspicuously rich in evenly distributed osteoclast-like giant cells. There is a mononuclear element, which can be composed of round, oval, or spindled cells. Atypia is limited in this component, and bizarre giant cells do not occur (Figure 13.54). The mitotic rate varies in a wide range, being on average 3 per 10 HPFs, but counts as high as >30 mitoses per 10 HPFs have been included under this definition. Necrosis is uncommon.

Opinions vary about whether these tumors should be classified as benign or potentially malignant; some investigators segregate a malignant group that seems to be comparable with giant cell MFH.[177]

Differential diagnosis

Plexiform fibrohistiocytic tumor is an architecturally distinctive giant-cell-rich soft tissue tumor, which is separate from the giant cell tumor of soft parts. This is a childhood tumor and is discussed in Chapter 10.

The tenosynovial giant cell tumor (giant cell tumor of tendon sheath) is a clinically and pathologically distinctive tumor that should be separated from this group. Tenosynovial giant cell tumors are almost always benign. They typically show an admixture of osteoclastic giant cells and small epithelioid neoplastic cells in a variably sclerosing stroma.

Giant cell malignant fibrous histiocytoma

This designation should be reserved for sarcomas with highly atypical pleomorphic or spindle cells admixed with osteoclast-like giant cells that do not contain malignant osteoid or specific differentiation, thus excluding giant cell tumors of soft parts

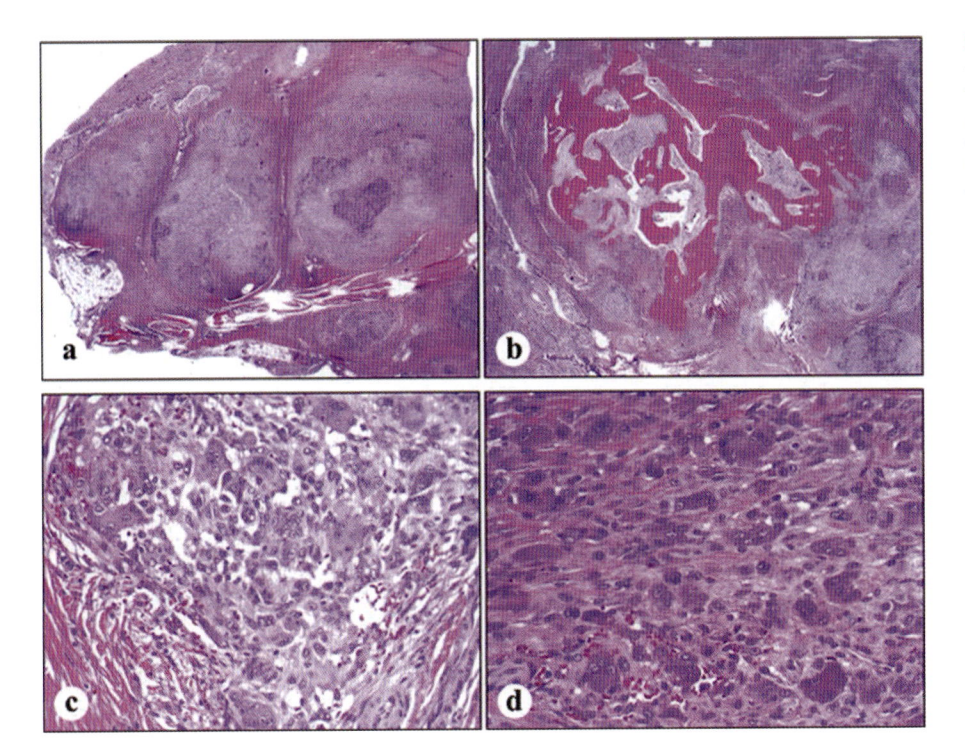

Figure 13.54 (**a**) Giant cell tumor of soft parts is composed of multiple nodules separated by fibrous bands. (**b**) Focal metaplastic bone may be present. (**c**) A tumor nodule rich in osteoclasts admixed with mononuclear cells. (**d**) The mononuclear tumor cells show limited nuclear atypia and mitotic activity.

(see above), extraskeletal osteosarcoma, other specific sarcomas, and carcinomas with osteoclastic giant cells. The last must be a serious consideration in visceral locations, such as thyroid and pancreas. In the author's experience, cases of giant cell MFH remain after these exclusions, although their number is lower than was the case prior to the availability of immunohistochemistry.

Clinical features

The tumor typically occurs in older adults in deep, intramuscular soft tissues of the extremities, and such tumors can reach a large size. However, smaller examples involving skin and subcutis also occur. The metastatic rate is high in deep tumors.[178,179]

Pathology

Grossly, giant cell MFH is typically lobulated by collagenous septa. On sectioning, the tumor is red-brown and often extensively hemorrhagic. Histologically, it is composed of cellular nodules separated by fibrous septa that also can contain tumor cells. The nodules are composed of an admixture of markedly atypical mononuclear cells, bizarre multinucleated tumor cells, and non-neoplastic osteoclast-like cells (Figures 13.55 and 13.56). Mitotic activity is often high, exceeding 10 to 20 mitoses per 10 HPFs, and atypical mitoses are often present. This and the common presence of necrosis typically results in a high-grade assignment.

Immunohistochemical studies for keratins, S100 protein, desmin, and smooth muscle actin should be performed to rule out melanoma, carcinoma, and leiomyosarcoma, as each of these tumor types can contain osteoclastic giant cells.[180]

Pleomorphic hyalinizing angiectatic tumor

Originally reported by Smith *et al.* in 1996, *pleomorphic hyalinizing angiectatic tumor* (PHAT) refers to a histologically distinctive tumor that is similar to pleomorphic MFH, but is of too low a biologic potential to warrant a separate designation.[181]

Clinical features

PHAT occurs at nearly all ages, usually in adults. Median age in the two published series is 54 years, with a mild overall female predominance. This tumor has a predilection for the foot, ankle, and leg. Nearly 80% occur in the lower extremity, and approximately 10% in the trunk and upper extremity (each). The tumor is subcutaneous in 90% of cases, but some reported examples have been intramuscular. Tumor size varies from <1 cm to 30 cm with a median size of 5.6 cm.[181–183] Metastases have not been reported. Evolution of PHAT into MFS have been reported in several cases, illustrating some malignant potential of this entity.[183–185]

In the author's experience, these tumors occur predominantly in the ankle and foot, and the size is usually <5 cm. There is a significant tendency for local recurrence after a simple excision, and therefore complete excision with negative margins is recommended. Recurrences can develop 10 or more years after primary surgery, necessitating long-term follow-up.

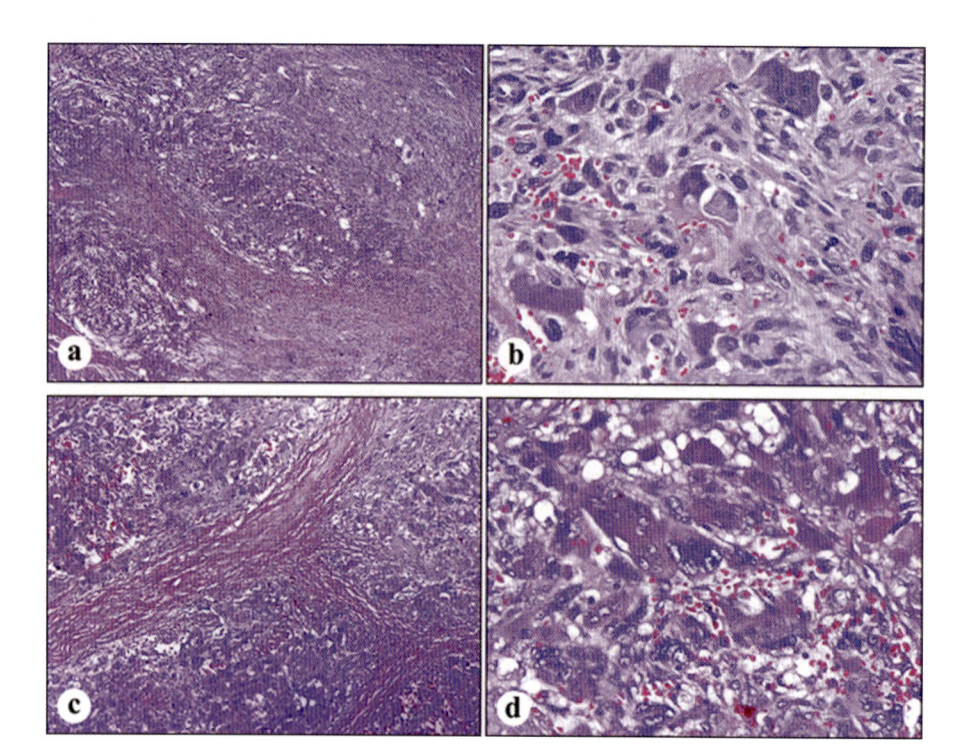

Figure 13.55 Giant cell MFH is also multinodular but contains highly atypical mononuclear cells and tumor giant cells between the osteoclasts.

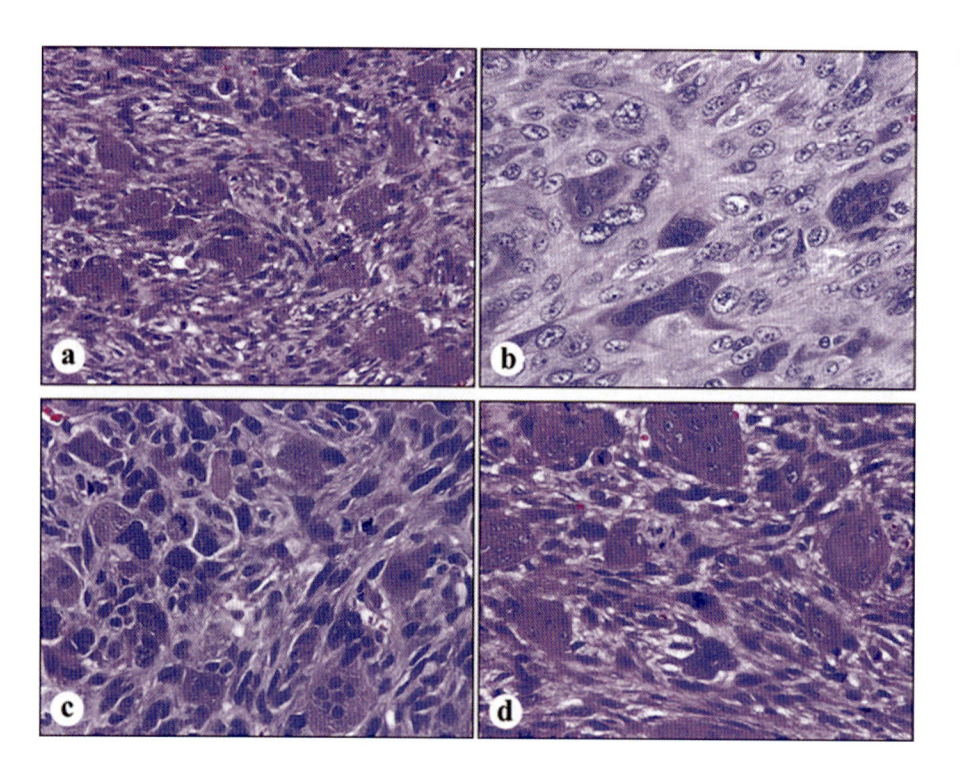

Figure 13.56 In giant cell MFH, the atypical cells can be mononuclear or multinucleated. Note that the atypical nuclei are much larger than the individual nuclei in the osteoclasts.

Pathology

Grossly, PHAT can be circumscribed or have diffuse borders. On sectioning, the tumor size varies from white to reddish-purple. Histologically distinctive is the presence of clusters of dilated blood vessels that often contain luminal thrombosis and fibrinous or hyaline material in the walls, resembling the vascular changes commonly seen in schwannoma (Figure 13.57). The cellular element often infiltrates subcutaneous fat and is spindled to epithelioid and pleomorphic. Focal cytoplasmic hemosiderin pigment is sometimes present, giving the cytoplasm a brownish discoloration. Moderate nuclear pleomorphism and intranuclear vacuolization (cytoplasmic inclusions) can occur (Figure 13.58). Mitotic rate is low, and atypical mitoses should not be accepted in

Chapter 13: Fibroblastic and myofibroblastic neoplasms with malignant potential

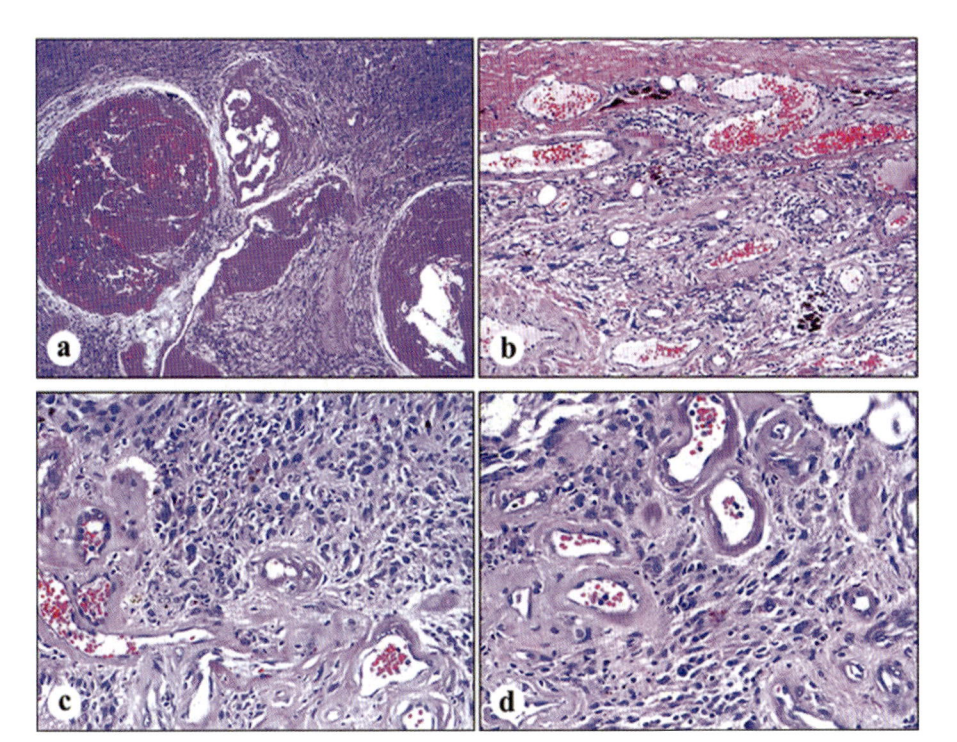

Figure 13.57 Pleomorphic hyaline angiectatic tumor contains dilated vascular channels with fibrinous walls. (**a**) Thrombosis may be present. (**b–d**) Between the vessels there are spindled or pleomorphic epitheloid cells, and focal hemosiderin is present.

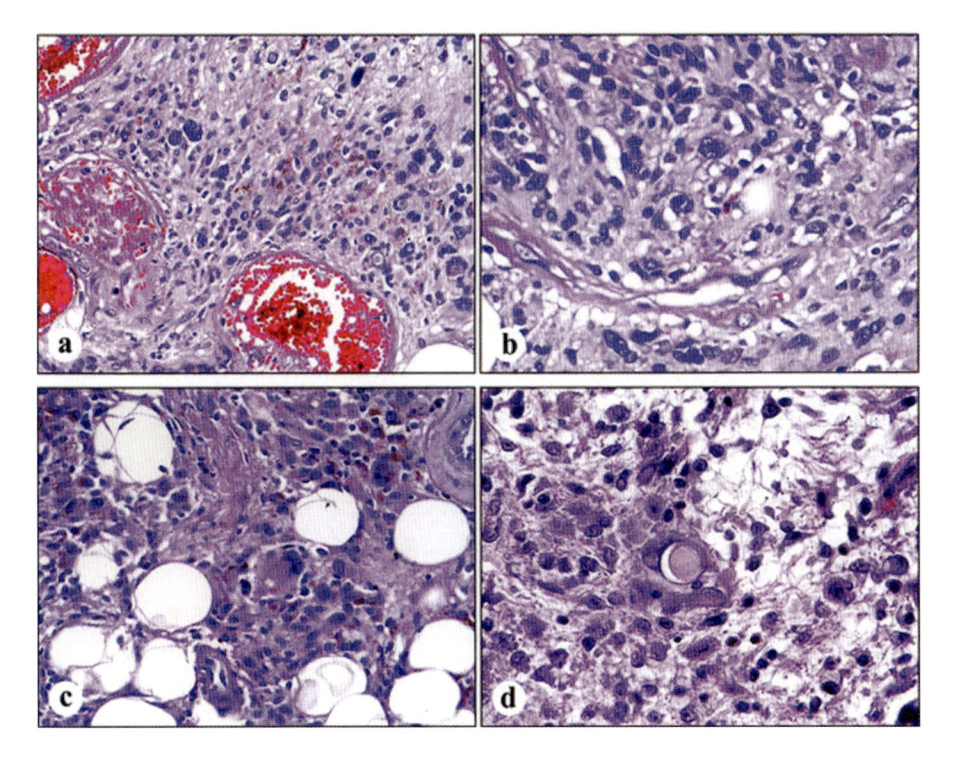

Figure 13.58 Cytologic features of pleomorphic hyaline angiectatic tumor. (**a,d**) Note nuclear variability, cytoplasmic hemosiderin, and intranuclear pseudoinclusions. (**c**) Tumor infiltrates fat.

this entity. A more recent series described spindle cell proliferations with similar vascular changes, suggesting that these could represent precursors of pleomorphic PHAT lesions.

Immunohistochemically, the tumor cells are often positive for CD34, but negative for S100 protein, SMA, and desmin.

Addressing the potential relationship of PHAT with myxoinflammatory fibroblastic sarcoma and hemosiderotic fibrohistiocytic lipomatous tumor, molecular genetic studies on PHAT have not found changes typical of those entities, such as TGFBR3 and MGEA5 gene rearrangements, or VGLL3 gene amplifications.[94,186]

Differential diagnosis

Pleomorphic neoplasms with PHAT-like vascular changes, but pleomorphic cellular elements with pronounced atypia and mitotic activity, especially when containing atypical mitoses, should be recognized as sarcomas (e.g., MFH) following immunohistochemical studies for exclusion of specific differentiation (i.e., keratins, S100 protein, SMA, desmin). Negativity for S100 protein helps to separate this tumor from schwannoma.

References

Dermatofibrosarcoma protuberans

1. Taylor HB, Helwig EB. Dermatofibrosarcoma protuberans: a study of 115 cases. *Cancer* 1962;15:717–725.
2. Kesserwan C, Sokolic R, Cowen EW, et al. Multicentric dermatofibrosarcoma protuberans in patients with adenosine deaminase-deficient severe combined immune deficiency. *J Allergy Clin Immunol* 2012;129:762.e1–769.e1.
3. McPeak CJ, Cruz T, Nicastri AD. Dermatofibrosarcoma protuberans: an analysis of 86 cases, five with metastasis. *Ann Surg* 1968;166:803–816.
4. Pappo AS, Rao BN, Cain A, Bodner S, Pratt CB. Dermatofibrosarcoma protuberans: the pediatric experience at St. Jude Children's Research Hospital. *Pediatr Hematol Oncol* 1997;14:563–568.
5. Terrier-Lacombe MJ, Guillou L, Maire G, et al. Dermatofibrosarcoma protuberans, giant cell fibroblastoma, and hybrid lesions in children: clinicopathologic comparative analysis of 28 cases with molecular data. A study from the French Federation of Cancer Centers Sarcoma Group. *Am J Surg Pathol* 2003;27:27–39.
6. Ghorbani RP, Malpica A, Ayala A. Dermatofibrosarcoma protuberans of the vulva: a clinicopathologic and immunohistochemical analysis of four cases, one with fibrosarcomatous change, and review of the literature. *Int J Gynecol Pathol* 1999;18:366–373.
7. Bowne WB, Antonescu CR, Leung DH, et al. Dermatofibrosarcoma protuberans: a clinicopathologic analysis of patients treated and followed at a single institution. *Cancer* 2000;88:2711–2720.
8. Parlette E, Smith KJ, Germain M, Rolfe A, Skelton H. Accelerated growth of dermatofibrosarcoma protuberans during pregnancy. *J Am Acad Dermatol* 1999;41:773–778.
9. Ratner D, Thomas CO, Johnson TM, et al. Mohs micrographic surgery for the treatment of dermatofibrosarcoma protuberans: results of a multi-institutional series with an analysis of the extent of microscopic spread. *J Am Acad Dermatol* 1997;37:600–613.
10. Wrotnowski U, Cooper PH, Shmookler BM. Fibrosarcomatous change in dermatofibrosarcoma protuberans. *Am J Surg Pathol* 1988;13:287–293.
11. Ding J, Hashimoto H, Enjoji M. Dermatofibrosarcoma protuberans with fibrosarcomatous areas: a clinicopathologic study of nine cases and comparison with allied tumors. *Cancer* 1989;64:721–729.
12. Connelly JH, Evans HL. Dermatofibrosarcoma protuberans: a clinicopathologic review with emphasis on fibrosarcomatous areas. *Am J Surg Pathol* 1992;16:921–925.
13. Mentzel T, Beham A, Katenkamp D, Dei Tos AP, Fletcher CDM. Fibrosarcomatous ("high-grade") dermatofibrosarcoma protuberans: clinicopathologic and immunohistochemical study of a series of 41 cases with emphasis on prognostic significance. *Am J Surg Pathol* 1998;22:576–587.
14. Goldblum JR, Reith JD, Weiss SW. Sarcomas arising in dermatofibrosarcoma protuberans: a reappraisal of biologic behavior in eighteen cases treated by wide local excision with extended clinical follow-up. *Am J Surg Pathol* 2000;24:1125–1130.
15. Gloster HM. Dermatofibrosarcoma protuberans. *J Am Acad Dermatol* 1996;35:355–374.
16. Diaz-Cascajo C, Weyers W, Rey-Lopez A, Borghi S. Deep dermatofibrosarcoma protuberans: a subcutaneous variant. *Histopathology* 1998;32:552–555.
17. Shen WQ, Hashimoto H, Okamoto S, et al. Expression of COLIAI-PDGFB fusion transcripts in superficial adult fibrosarcoma suggests close relationship to dermatofibrosarcoma protuberans. *J Pathol* 2001;194:88–94.
18. Frierson HF, Cooper PH. Myxoid variant of dermatofibrosarcoma protuberans. *Am J Surg Pathol* 1983;7:445–450.
19. Reimann JD, Fletcher CD. Myxoid dermatofibrosarcoma protuberans: a rare variant analyzed in a series of 23 cases. *Am J Surg Pathol* 2007;31:1371–1377.
20. Calonje E, Fletcher CDM. Myoid differentiation in dermatofibrosarcoma protuberans and its fibrosarcomatous variant: clinicopathologic analysis of 5 cases. *J Cutan Pathol* 1996;23:30–36.
21. Morimitsu Y, Hisaoka M, Okamoto S, Hashimoto H, Ushijima M. Dermatofibrosarcoma protuberans and its fibrosarcomatous variant with areas of myoid differentiation: a report of three cases. *Histopathology* 1998;32:547–551.
22. Sanz-Trelles A, Ayala-Carbonero A, Rodrigo-Fernandez I, Weil-Lara B. Leiomyomatous nodules and bundles of vascular origin in the fibrosarcomatous variant of dermatofibrosarcoma protuberans. *J Cutan Pathol* 1998;25:44–49.
23. O'Dowd J, Laidler P. Progression of dermatofibrosarcoma protuberans to malignant fibrous histiocytoma: report of a case with implications for tumor histogenesis. *Hum Pathol* 1988;19:368–370.
24. Swaby MG, Evans HL, Fletcher CD, et al. Dermatofibrosarcoma protuberans with unusual sarcomatous transformation: a series of 4 cases with molecular confirmation. *Am J Dermatopathol* 2011;33:354–360.
25. Dupree WB, Langloss JM, Weiss SW. Pigmented dermatofibrosarcoma protuberans (Bednar tumor): a pathologic, ultrastructural, and immunohistochemical study. *Am J Surg Pathol* 1985;9:630–639.
26. Fletcher CD, Theaker JM, Flanagan A, Krausz T. Pigmented dermatofibrosarcoma protuberans (Bednar tumour): melanocytic colonization or neuroectodermal differentiation? A clinicopathologic and

27. Ding JA, Hashimoto H, Sugimoto T, Tsuneyoshi M, Enjoji M. Bednar tumor (pigmented dermatofibrosarcoma protuberans): an analysis of six cases. *Acta Pathol Jpn* 1990;40:744–754.

28. Abdul-Karim FV, Evans HL, Silva EG. Giant cell fibroblastoma: a report of three cases. *Am J Clin Pathol* 1985;83:165–170.

29. Dymock RB, Allen PW, Stirling JW, Gilbert EF, Thornbery JM. Giant cell fibroblastoma: a distinctive, recurrent tumor of childhood. *Am J Surg Pathol* 1987;11:263–271.

30. Shmookler BM, Enzinger FM, Weiss SW. Giant cell fibroblastoma: a juvenile form of dermatofibrosarcoma protuberans. *Cancer* 1989;64:2154–2161.

31. Jha P, Moosavi C, Fanburg-Smith JC. Giant cell fibroblastoma: an update and addition of 86 new cases from the Armed Forces Institute of Pathology, in honor of Franz M. Enzinger. *Ann Diagn Pathol* 2007;11:81–88.

32. De Chadarevian JP, Coppola D, Billmire DF. Bednar tumor pattern in recurring giant cell fibroblastoma. *Am J Clin Pathol* 1993;100:164–166.

33. Aiba S, Tabata N, Ishii H, Ootani H, Tagami H. Dermatofibrosarcoma protuberans is a unique fibrohistiocytic tumour expressing CD34. *Br J Dermatol* 1992;127:79–84.

34. Altman DA, Nickoloff BJ, Fivenson DP. Differential expression of factor XIIIa and CD34 in cutaneous mesenchymal tumors. *J Cutan Pathol* 1993;20:154–158.

35. Kutzner H. Expression of the human progenitor cell antigen (CD34, HPCA1) distinguishes dermatofibrosarcoma protuberans from fibrous histiocytoma in formalin-fixed, paraffin-embedded tissue. *J Am Acad Dermatol* 1993;28:613–617.

36. Sato N, Kimura K, Tomita Y. Recurrent dermatofibrosarcoma protuberans with myxoid and fibrosarcomatous changes paralleled by loss of CD34 expression. *J Dermatol* 1995;22:665–672.

37. Goldblum JR, Tuthill RJ. CD34 and factor XIIIa immunoreactivity in dermatofibrosarcoma protuberans and dermatofibroma. *Am J Dermatopathol* 1997;19:147–153.

38. Fanburg-Smith JF, Miettinen M. Low-affinity nerve growth factor receptor (p75) in dermatofibrosarcoma protuberans and other non-neural tumors: a study of 1150 tumors and fetal and adult normal tissues. *Hum Pathol* 2001;32:976–983.

39. Kahn HJ, Fekete E, From L. Tenascin differentiates dermatofibroma from dermatofibrosarcoma protuberans: comparison with CD34 and factor XIIIa. *Hum Pathol* 2001;32:50–56.

40. West RB, Harvell J, Linn SC, et al. APO D in soft tissue tumors: a novel marker for dermatofibrosarcoma protuberans. *Am J Surg Pathol* 2004;28:1063–1069.

41. Kutzner H, Mentzel T, Palmedo G, et al. Plaque-like CD34-positive dermal fibroma ("medallion-like dermal dendrocyte hamartoma"): clinicopathologic, immunohistochemical, and molecular analysis of 5 cases emphasizing its distinction from superficial, plaque-like dermatofibrosarcoma protuberans. *Am J Surg Pathol* 2010;34:190–201.

42. Fetsch JF, Michal M, Miettinen M. Pigmented (melanotic) neurofibroma: a clinicopathologic and immunohistochemical analysis of 19 lesions from 17 patients. *Am J Surg Pathol* 2000;24:331–343.

43. Naeem R, Lux ML, Huang SF, et al. Ring chromosomes in dermatofibrosarcoma protuberans are composed of interspersed sequences from chromosomes 17 and 22. *Am J Pathol* 1995;147:1553–1558.

44. Pedeutour F, Simon MP, Minoletti F, et al. Ring 22 chromosomes in dermatofibrosarcoma protuberans are low-level amplifiers of chromosome 17 and 22 sequences. *Cancer Res* 1995;55:2400–2403.

45. Mandahl N, Limon J, Mertens F, Arheden K, Mitelman F. Ring marker containing 17q and chromosome 22 in a case of dermatofibrosarcoma protuberans. *Cancer Genet Cytogenet* 1996;89:88–91.

46. Simon MP, Pedeutour F, Sirvent N, et al. Deregulation of the platelet-derived growth factor B-chain via fusion with collagen gene COL1A1 in dermatofibrosarcoma protuberans and giant cell fibroblastoma. *Nat Genet* 1997;15:95–98.

47. Greco A, Fusetti L, Villa R, et al. Transforming activity of the chimeric sequence formed by the fusion of collagen gene COL1A1 and the platelet growth factor b-chain gene in dermatofibrosarcoma protuberans. *Oncogene* 1998;17:1313–1319.

48. Shimizu A, O'Brien KP, Sjoblom T, et al. The dermatofibrosarcoma protuberans-associated collagen type I alpha1/platelet-derived growth factor (PDGF) B-chain fusion gene generates a transforming protein that is processed to functional PDGF-BB. *Cancer Res* 1999;59:3719–3723.

49. Wang J, Hisaoka M, Shimajiri S, Morimitsu Y, Hashimoto H. Detection of COL1A1-PDGFB fusion transcripts in dermatofibrosarcoma protuberans by reverse transcription-polymerase chain reaction using archival formalin-fixed, paraffin-embedded tissues. *Diagn Mol Pathol* 1999;8:113–119.

50. Patel KU, Szabo SS, Hernandez VS, et al. Dermatofibrosarcoma protuberans COL1A1-PDGFB fusion is identified in virtually all dermatofibrosarcoma protuberans cases when investigated by newly developed multiplex reverse transcriptase polymerase chain reaction and in situ hybridization assays. *Hum Pathol* 2008;39:184–193.

51. Bianchini L, Maire G, Guillot B, et al. Complex t(5;8) involving the CSPG2 and PTK2B genes in a case of dermatofibrosarcoma protuberans without the COL1A1-PDGFB fusion. *Virchows Arch* 2008;452:689–696.

52. Greco A, Roccato E, Miranda C, et al. Growth-inhibitory effect of STI571 on cells transformed by the COL1A1/PDGFB rearrangement. *Int J Cancer* 2001;92:354–360.

53. Sjöblom T, Shimizu A, O'Brien KP, et al. Growth inhibition of dermatofibrosarcoma protuberans tumors by the platelet-derived growth factor receptor antagonist STI571 through induction of apoptosis. *Cancer Res* 2001;61:5778–5783.

54. Rubin BP, Schuetze SM, Eary JF, et al. Molecular targeting of platelet-derived growth factor B by imatinib mesylate in a patient with metastatic dermatofibrosarcoma protuberans. *J Clin Oncol* 2002;20:3586–3591.

55. Maki RG, Awan RA, Dixon RH, Jhanwar S, Antonescu CR. Differential sensitivity to imatinib of 2 patients with metastatic sarcoma arising from dermatofibrosarcoma

protuberans. *Int J Cancer* 2002;100:623–626.

56. Stacchiotti S, Pedeutour F, Negri T, *et al.* Dermatofibrosarcoma protuberans-derived fibrosarcoma: clinical history, biological profile and sensitivity to imatinib. *Int J Cancer* 2011;129:1761–1772.

Low-grade fibromyxoid sarcoma

57. Evans HL. Low-grade fibromyxoid sarcoma: a report of 12 cases. *Am J Surg Pathol* 1993;17:595–600.

58. Goodlad JR, Mentzel T, Fletcher CDM. Low-grade fibromyxoid sarcoma: clinicopathological analysis of eleven new cases in support of a distinct entity. *Histopathology* 1995;26:229–237.

59. Folpe AL, Lane KL, Paull G, Weiss SW. Low-grade fibromyxoid sarcoma and hyalinizing spindle cell tumor with giant rosettes: a clinicopathologic study of 73 cases supporting their identity and assessing the impact of high-grade areas. *Am J Surg Pathol* 2000;24:1353–1360.

60. Zamecnik M, Michal M. Low-grade fibromyxoid sarcoma: a report of eight cases with histologic, immunohistochemical, and ultrastructural study. *Ann Diagn Pathol* 2000;4:207–217.

61. Billings SD, Fanburg-Smith JC. Superficial low-grade fibromyxoid sarcoma (Evans tumor): a clinicopathologic analysis of 19 cases with a unique observation in the pediatric population. *Am J Surg Pathol* 2005;29:204–210.

62. Guillou L, Benhattar J, Gengler C, *et al.* Translocation-positive low-grade fibromyxoid sarcoma: clinicopathologic and molecular analysis of a series expanding the morphologic spectrum and suggesting a potential relationship with sclerosing epithelioid fibrosarcoma: a study from the French sarcoma group. *Am J Surg Pathol* 2007;31:1387–1402.

63. Evans HL. Low-grade fibromyxoid sarcoma: a clinicopathologic study of 33 cases with long-term follow-up. *Am J Surg Pathol* 2011;35:1450–1462.

64. Laurini JA, Zhang L, Goldblum JR, Montgomery E, Folpe AL. Low-grade fibromyxoid sarcoma of the small intestine: report of 4 cases with molecular cytogenetic confirmation. *Am J Surg Pathol* 2011;35:1069–1073.

65. Reid R, Chandu de Silva MV, Paterson L, Ryan E, Fisher C. Low-grade fibromyxoid sarcoma and hyalinizing spindle cell tumor with giant rosettes share a common t(7;16)(q34;p11) translocation. *Am J Surg Pathol* 2003;27:1229–1236.

66. Lane KL, Shannon RJ, Weiss SW. Hyalinizing spindle cell tumor with giant rosettes: a distinctive tumor closely resembling low-grade fibromyxoid sarcoma. *Am J Surg Pathol* 1997;21:1481–1488.

67. Doyle LA, Möller E, Dal Cin P, Fletcher CD, Mertens F, Hornick JL. MUC4 is a highly sensitive and specific marker for low-grade fibromyxoid sarcoma. *Am J Surg Pathol* 2011;35(5):733–741

68. Oda Y, Takahara T, Kawaguchi K, *et al.* Low-grade fibromyxoid sarcoma versus low-grade myxofibrosarcoma in the extremities and trunk: a comparison of clinicopathological and immunohistochemical features. *Histopathology* 2004;45:29–38.

69. Panagopoulos I, Storlazzi CT, Fletcher CD, *et al.* The chimeric FUS/CREB3L2 gene is specific for low-grade fibromyxoid sarcoma. *Genes Chromosomes Cancer* 2004;40:218–228.

70. Mertens F, Fletcher CDM, Antonescu CR, *et al.* Clinicopathologic and molecular genetic characterization of low-grade fibromyxoid sarcoma, and cloning of a novel FUS/CREB3L1 fusion gene. *Lab Invest* 2005;85:408–415.

71. Lau PP, Lui PC, Lau GT, *et al.* EWSR1-CREB3L1 gene fusion: a novel alternative molecular aberration of low-grade fibromyxoid sarcoma. *Am J Surg Pathol* 2013;37:734–738.

72. Panagopoulos I, Moller E, Dahlen A, *et al.* Characterization of the native CREB3L2 transcription factor and the FUS/CREB3L2 chimera. *Genes Chromosomes Cancer* 2007;46:181–191.

73. Matsuyama A, Hisaoka M, Shimajiri S, Hashimoto H. DNA-based polymerase chain reaction for detecting FUS-CREB3L2 in low-grade fibromyxoid sarcoma using formalin-fixed, paraffin-embedded tissue specimens. *Diagn Mol Pathol* 2008;17:237–240.

74. Downs-Kelly F, Goldblum JR, Patel RM, *et al.* The utility of fluorescence in-situ hybridization in the diagnosis of myxoid soft tissue neoplasms. *Am J Surg Pathol* 2008;32:8–13.

Other soft tissue fibrosarcomas

75. Meis-Kindblom JM, Kindblom LG, Enzinger FM. Sclerosing epithelioid fibrosarcoma: a variant of fibrosarcoma simulating carcinoma. *Am J Surg Pathol* 1995;19:979–993.

76. Eyden BP, Manson C, Banerjee SS, Roberts IS, Harris M. Sclerosing epithelioid fibrosarcoma: a study of five cases emphasizing diagnostic criteria. *Histopathology* 1998;33:354–360.

77. Antonescu C, Rosenblum MK, Pereira P, Nascimento AG, Woodruff JM. Sclerosing epithelioid fibrosarcoma: a study of 16 cases and confirmation of a clinicopathologic entity. *Am J Surg Pathol* 2001;25:699–709.

78. Doyle LA, Wang WL, Dal Cin P, *et al.* MUC4 is a sensitive and extremely useful marker for sclerosing epithelioid fibrosarcoma: association with FUS gene rearrangement. *Am J Surg Pathol* 2012;36:1444–1451.

79. Gisselsson D, Andreasson P, Meis-Kindblom JM, *et al.* Amplification of 12q13 and 12q15 sequences in a sclerosing epithelioid fibrosarcoma. *Cancer Genet Cytogenet* 1998;107:102–106.

80. Arbajian E, Puls F, Magnusson L, *et al.* Recurrent EWSR1-CREB3L1 gene fusions in sclerosing epithelioid fibrosarcoma. *Am J Surg Pathol* 2014;38:801–808.

81. Wang WL, Evans HL, Meis JM, *et al.* FUS rearrangements are rare in "pure" sclerosing epithelioid fibrosarcoma. *Mod Pathol* 2012;25:846–853.

82. Scott SM, Reiman HM, Pritchard DJ, Ilstrup DM. Soft tissue fibrosarcoma: a clinicopathologic study of 132 cases. *Cancer* 1989;64:925–931.

83. Bahrami A, Folpe AL. Adult-type fibrosarcoma: a reevaluation of 163 putative cases diagnosed at a single institution over a 48-year period. *Am J Surg Pathol* 2010;34:1504–1513.

84. Mentzel T, Dry S, Katenkamp D, Fletcher CDM. Low-grade myofibroblastic sarcoma: analysis of 18 cases in the spectrum of myofibroblastic tumors. *Am J Surg Pathol* 1998;22:1228–1238.

85. Montgomery EA, Goldblum JR, Fisher C. Myofibrosarcoma: a clinicopathologic study. *Am J Surg Pathol* 2001;25:219–228.

86. Meis-Kindblom JM, Kindblom LG. Acral myxoinflammatory fibroblastic sarcoma: a low grade tumor of the hands and feet. *Am J Surg Pathol* 1998;22:911–924.

87. Montgomery EA, Devaney KO, Giordano TJ, Weiss SW. Inflammatory myxohyaline tumor of distal extremities with virocyte or Reed-Sternberg-like cells: a distinctive lesion with features simulating inflammatory conditions, Hodgkin's disease, and various sarcomas. *Mod Pathol* 1998;11:384–391.

88. Laskin WB, Fetsch JF, Miettinen M. Myxoinflammatory fibroblastic sarcoma: a clinicopathologic analysis of 104 cases, with emphasis on predictors of outcome. *Am J Surg Pathol* 2014;38:1–12.

89. Weiss VL, Antonescu CR, Alaggio R, *et al*. Myxoinflammatory fibroblastic sarcoma in children and adolescents: clinicopathologic aspects of a rare neoplasm. *Pediatr Dev Pathol* 2013;16:425–431.

90. Lambert I, Debiec-Rychter M, Guelinckx P, Hagemeijer A, Sciot R. Acral myxoinflammatory fibroblastic sarcoma with unique clonal chromosomal changes. *Virchows Arch* 2001;438:509–512.

91. Mansoor A, Fidda N, Himoe E, *et al*. Myxoinflammatory fibroblastic sarcoma with complex supernumerary ring chromosomes composed of chromosome 3 segments. *Cancer Genet Cytogenet* 2004;142:61–65.

92. Ida CM, Rolig KA, Hulshizer RL, *et al*. Myxoinflammatory fibroblastic sarcoma showing t(2;6)(q31;p21.3) as a sole cytogenetic abnormality. *Cancer Genet Cytogenet* 2007;177:139–142.

93. Hallor KH, Sciot R, Staaf J, *et al*. Two genetic pathways, t(1;10) and amplification of 3p11-12, in myxoinflammatory fibroblastic sarcoma, haemosiderotic fibrolipomatous tumour, and morphologically similar lesions. *J Pathol* 2009;217:716–727.

94. Antonescu CR, Zhang L, Nielsen GP, *et al*. Consistent t(1;10) with rearrangements of TGFBR3 and MGEA5 in both myxoinflammatory fibroblastic sarcoma and hemosiderotic fibrolipomatous tumor. *Genes Chromosomes Cancer* 2011;50:757–764.

Myxofibrosarcoma and malignant fibrous histiocytoma

95. Angervall L, Kindblom LG, Merck C. Myxofibrosarcoma: a study of 30 cases. *Acta Pathol Microbiol Scand* 1977;85:127–140.

96. Weiss SW, Enzinger FM. Myxoid variant of malignant fibrous histiocytoma. *Cancer* 1977;39:1672–1685.

97. Mentzel T, Van Den Berg E, Molenaar WM. Myxofibrosarcoma. In *Pathology and Genetics of Tumours of Soft Tissue and Bone*. Fletcher CDM, Unni KK, Mertens F (eds.) Lyon: World Health Organization; 2002: 102–103.

98. Mentzel T, Calonje E, Wadden C, *et al*. Myxofibrosarcoma: clinicopathologic analysis of 75 cases with emphasis on the low-grade variant. *Am J Surg Pathol* 1996;20:391–405.

99. Merck C, Angervall L, Kindblom LG, Oden A. Myxofibrosarcoma: a malignant soft tissue tumor of fibroblastic-histiocytic origin. A clinicopathologic and prognostic study of 110 cases using a multivariate analysis. *APMIS* 1983;91(Suppl 282):1–40.

100. Huang HY, Lal P, Qin J, Brennan MF, Antonecu CR. Low-grade myxofibrosarcoma: a clinicopathologic analysis of 49 cases treated at a single institution with simultaneous assessment of the efficacy of 3-tier and 4-tier grading systems. *Hum Pathol* 2004;35:612–621.

101. Hisaoka M, Morimitsu Y, Hashimoto H, *et al*. Retroperitoneal liposarcoma with combined well-differentiated and myxoid malignant fibrous histiocytoma-like myxoid areas. *Am J Surg Pathol* 1999;23:1480–1492.

102. Coindre JM, Mariani O, Chibon F, *et al*. Most malignant fibrous histiocytomas developed in the retroperitoneum are dedifferentiated liposarcomas: a review of 25 cases initially diagnosed as malignant fibrous histiocytoma. *Mod Pathol* 2003;16:256–262.

103. Lagace R, Delage C, Seemayer TA. Myxoid variant of malignant fibrous histiocytoma: ultrastructural observations. *Cancer* 1979;43:526–534.

104. Kindblom LG, Merck C, Angervall L. The ultrastructure of myxofibrosarcoma: a study of 11 cases. *Virchows Arch Pathol Anat Histol* 1979;381:121–139.

105. Fukuda T, Tsuneyoshi M, Enjoji M. Malignant fibrous histiocytoma of soft parts: an ultrastructural quantitative study. *Ultrastruct Pathol* 1988;12:117–129.

106. Clawson K, Donner LR, Dobin SM. Translocation t(2;15)(p23;q21.2) and interstitial deletion of 7q in one case of low-grade myxofibrosarcoma. *Cancer Genet Cytogenet* 2001;127:140–142.

107. Willems SM, Debiec-Rychter M, Szuhai K, Hogendoorn PC, Sciot R. Local recurrence of myxofibrosarcoma is associated with increase in tumour grade and cytogenetic aberrations, suggesting a multistep tumour progression model. *Mod Pathol* 2006;19:407–416.

108. Ohquri T, Hisaoka M, Kawauchi S, *et al*. Cytogenetic analysis of myxoid liposarcoma and myxofibrosarcoma by array-based comparative genomic hybridization. *J Clin Pathol* 2006;59:978–983.

109. Wood GS, Beckstead JH, Turner RR, *et al*. Malignant fibrous histiocytoma tumor cells resemble fibroblasts. *Am J Surg Pathol* 1986;10:323–335.

110. Suh C, Ordonez NG, Mackay N. Malignant fibrous histiocytoma: an ultrastructural perspective. *Ultrastruct Pathol* 2001;24:243–250.

111. Erlandson RA, Antonescu CR. The rise and fall of malignant fibrous histiocytoma. *Ultrastruct Pathol* 2004;28:283–289.

112. Kindblom LG, Jacobsen GK, Jacobsen M. Immunohistochemical investigations of tumors of supposed fibroblastic-histiocytic origin. *Hum Pathol* 1982;13:834–840.

113. Binder SW, Said JW, Shintaku IP, Pinkus GS. A histiocyte-specific marker in the diagnosis of malignant fibrous histiocytoma: use of monoclonal antibody KP-1 (CD68). *Am J Clin Pathol* 1992;97:759–763.

114. Soini Y, Miettinen M. Alpha-1-antitrypsin and lysozyme: their limited significance in fibrohistiocytic tumors. *Am J Clin Pathol* 1989;91:515–521.

115. Weiss LM, Arber DA, Chang KL. CD68: a review. *Appl Immunohistochem* 1994;2:2–8.

116. Iwasaki H, Isayama T, Johzaki H, Kikuchi M. Malignant fibrous histiocytoma: evidence of perivascular mesenchymal cell origin: immunocytochemical studies with

monoclonal anti-MFH antibodies. *Am J Pathol* 1987;128:528–537.

117. Fletcher CDM, Van Den Berg E, Molenaar WM. Pleomorphic malignant fibrous histiocytoma/undifferentiated high grade pleomorphic sarcoma. In *Pathology and Genetics of Tumours of Soft Tissue and Bone*. Fletcher CDM, Unni KK, Mertens F (eds.) Lyon: World Health Organization; 2002: 120–122.

118. Fletcher CD. Pleomorphic malignant fibrous histiocytoma: fact or fiction? A critical reappraisal based on 159 tumors diagnosed as pleomorphic sarcoma. *Am J Surg Pathol* 1992;16:213–228.

119. Fletcher CDM, Gustafson P, Rydholm A, Willen H, Akerman M. Clinicopathologic re-evaluation of 100 malignant fibrous histiocytomas: prognostic relevance of subclassification. *J Clin Oncol* 2001;19:3045–3050.

120. Brooks JJ. The significance of double phenotypic patterns and markers in human sarcomas: a new model of mesenchymal differentiation. *Am J Pathol* 1986;125:113–123.

121. Rööser B, Willen H, Gustafson P, Alvegård TA, Rydholm A. Malignant fibrous histiocytoma of soft tissue: a population-based epidemiologic and prognostic study of 137 patients. *Cancer* 1991;67:499–505.

122. Weiss SW, Enzinger FM. Malignant fibrous histiocytoma: an analysis of 200 cases. *Cancer* 1978;41:2250–2266.

123. Enjoji M, Hashimoto H, Tsuneyoshi M, Iwasaki H. Malignant fibrous histiocytoma: a clinicopathologic study of 130 cases. *Acta Pathol Jpn* 1980;30:727–741.

124. Bertoni F, Capanna R, Biagini R, *et al*. Malignant fibrous histiocytoma of soft tissue: an analysis of 78 cases located and deeply seated in the extremities. *Cancer* 1985;56:356–367.

125. Rydholm A, Syk I. Malignant fibrous histiocytoma of soft tissue: correlation between clinical variables and histologic malignancy grade. *Cancer* 1986;57:2323–2324.

126. Pezzi CM, Rawlings MS, Esgro JJ, Pollock RE, Romsdahl MM. Prognostic factors in 227 patients with malignant fibrous histiocytoma. *Cancer* 1992;69:2098–2103.

127. LeDoussal V, Coindre JM, Leroux A, *et al*. Prognostic factors for patients with localized primary malignant fibrous histiocytoma: a multicenter study of 216 patients with multivariate analysis. *Cancer* 1996;77:1823–1830.

128. Salo JC, Lewis JJ, Woodruff JM, Leung DH, Brennan MF. Malignant fibrous histiocytoma of the extremity. *Cancer* 1999;85:1765–1772.

129. Deyrup AT, Haydon RC, Huo D, *et al*. Myoid differentiation and prognosis in adult pleomorphic sarcomas of the extremity: an analysis of 92 cases. *Cancer* 2003;98:805–813.

130. Massi D, Beltrami G, Capanna R, Franchi A. Histopathological re-classification of extremity pleomorphic soft tissue sarcoma has clinical relevance. *Eur J Surg Oncol* 2004;30:1131–1136.

131. Cipriani NA, Kurzawa P, Ahmad RA, *et al*. Prognostic value of myogenic differentiation in undifferentiated pleomorphic sarcomas of soft tissue. *Hum Pathol* 2014;45:1504–1508.

132. Miettinen M, Soini Y. Malignant fibrous histiocytoma: heterogeneous patterns of intermediate filament proteins by immunohistochemistry. *Arch Pathol Lab Med* 1989;113:1363–1366.

133. Litzky LA, Brooks JJ. Cytokeratin immunoreactivity in malignant fibrous histiocytoma and spindle cell tumors: comparison between frozen and paraffin-embedded tissues. *Mod Pathol* 1992;5:30–34.

134. Rosenberg AE, O'Connell JX, Dickersin GR, Bhan AK. Expression of epithelial markers in malignant fibrous histiocytoma of the musculoskeletal system: an immunohistochemical and electron microscopic study. *Hum Pathol* 1993;23:284–293.

135. von Koskull H, Virtanen I. Induction of cytokeratin expression in human mesenchymal cells. *J Cell Physiol* 1987;133:321–329.

136. Knapp AC, Franke WW. Spontaneous losses of control of cytokeratin gene expression in transformed, non-epithelial human cells occurring at different levels of regulation. *Cell* 1999;59:67–79.

137. Hamada T, Komiya S, Hiraoka K, *et al*. IL-6 in a pleomorphic type of malignant fibrous histiocytoma presenting with high fever. *Hum Pathol* 1998;29:758–761.

138. Reinecke P, Moll R, Hildebrandt B, *et al*. A novel human malignant fibrous histiocytoma cell line of heart (MFH-H) with secretion of hematopoietic growth factors. *Anticancer Res* 1999;19:1901–1907.

139. Mayumi E, Okuno T, Ogawa T, *et al*. Malignant fibrous histiocytoma of soft tissue producing granulocyte colony stimulating factor. *Intern Med* 2001;40:536–540.

140. Mandahl N, Heim S, Willen H, *et al*. Characteristic karyotypic anomalies identify subtypes of malignant fibrous histiocytoma. *Genes Chromosomes Cancer* 1989;1:9–14.

141. Szymanska J, Tarkkanen M, Wiklund T, *et al*. A cytogenetic study of malignant fibrous histiocytoma. *Cancer Genet Cytogenet* 1995;85:91–96.

142. Schmidt H, Korber S, Hinze R, *et al*. Cytogenetic characterization of ten malignant fibrous histiocytomas. *Cancer Genet Cytogenet* 1998;100:134–142.

143. Choong PF, Mandahl N, Mertens F, *et al*. 19p+ marker chromosome correlates with relapse in malignant fibrous histiocytoma. *Genes Chromosomes Cancer* 1996;16:88–93.

144. Larramendy ML, Tarkkanen M, Blomqvist C, *et al*. Comparative genomic hybridization of malignant fibrous histiocytoma reveals a novel prognostic marker. *Am J Pathol* 1997;151:1153–1161.

145. Hinze R, Schagdarsurengin U, Taubert H, *et al*. Assessment of genomic imbalances in malignant fibrous histiocytomas by comparative genomic hybridization. *Int J Mol Med* 1999;3:75–79.

146. Mairal A, Terrier P, Chibon F, *et al*. Loss of chromosome 13 is the most frequent genomic imbalance in malignant fibrous histiocytomas: a comparative genomic hybridization analysis of a series of 30 cases. *Cancer Genet Cytogenet* 1999;111:134–138.

147. Chibon F, Mairal A, Freneaux P, *et al*. The RB1 gene is the target of chromosome 13 deletions in malignant fibrous histiocytoma. *Cancer Res* 2000;60:6339–6345.

148. Sakabe T, Shinomiya T, Mori T, *et al*. Identification of a novel gene, MASL1, within an amplicon at 8p23.1 detected in malignant fibrous histiocytomas by

comparative genomic hybridization. *Cancer Res* 1999;59:511–515.
149. Weng WH, Weide J, Ahlen J, et al. Characterization of large chromosome markers in malignant fibrous histiocytoma by spectral karyotyping, comparative genomic hybridization (CGH), and array CGH. *Cancer Genet Cytogenet* 2004;150:27–32.
150. Simons A, Schepens M, Jeuken J, et al. Frequent loss of 9p21 (p16(INK4A)) and other genomic imbalances in human malignant fibrous histiocytoma. *Cancer Genet Cytogenet* 2000;118:89–98.
151. Reid AH, Tsai MM, Venzon DJ, et al. MDM2 amplification, P53 mutation, and accumulation of the P53 gene product in malignant fibrous histiocytoma. *Diagn Mol Pathol* 1996;5:65–73.
152. Molina P, Pellin A, Navarro S, et al. Analysis of p53 and mdm2 proteins in malignant fibrous histiocytoma in absence of gene alteration: prognostic significance. *Virchows Arch* 1999;435:596–605.
153. Iwao K, Miyoshi Y, Nawa G, et al. Frequent beta-catenin abnormalities in bone and soft tissue tumors. *Jpn J Cancer Res* 1999;90:205–209.

Atypical fibroxanthoma
154. Fretzin DF, Helwig EB. Atypical fibroxanthoma of the skin: a clinicopathologic study of 140 cases. *Cancer* 1973;31:1541–1552.
155. Dahl I. Atypical fibroxanthoma of the skin: a clinicopathological study of 57 cases. *Acta Pathol Microbiol Scand A* 1976;84:183–197.
156. Kuwano H, Hashimoto H, Enjoji M. Atypical fibroxanthoma distinguishable from spindle cell carcinoma in sarcoma-like skin lesions. *Cancer* 1985;55:172–180.
157. Mirza B, Weedon D. Atypical fibroxanthoma: a clinicopathological study of 89 cases. *Australas J Dermatol* 2005;46:235–238.
158. Hafner J, Kunzi W, Weinreich T. Malignant fibrous histiocytoma and atypical fibroxanthoma in renal transplant patients. *Dermatology* 1999;198:29–32.
159. Helwig EB, May D. Atypical fibroxanthoma of the skin with metastasis. *Cancer* 1986;57:368–376.

160. Cooper JZ, Newman SR, Scott GA, Brown MD. Metastasizing atypical fibroxanthoma (cutaneous malignant histiocytoma): report of five cases. *Dermatol Surg* 2005;31:221–225.
161. Calonje E, Wadden C, Wilson-Jones E, Fletcher CD. Spindle-cell non-pleomorphic atypical fibroxanthoma: analysis of a series and delineation of a distinctive variant. *Histopathology* 1993;22:247–254.
162. Tomaszewski MM, Lupton GP. Atypical fibroxanthoma: an unusual variant with osteoclast-like giant cells. *Am J Surg Pathol* 1997;21:213–221.
163. Longacre TA, Smoller BR, Rouse RV. Atypical fibroxanthoma: multiple immunohistologic profiles. *Am J Surg Pathol* 1993;17:1199–1209.
164. Gru AA, Santa Cruz DJ. Atypical fibroxanthoma: a selective review. *Semin Diagn Pathol* 2013;30:4–12
165. Kamino H, Salcedo E. Histopathologic and immunohistochemical diagnosis of benign and malignant fibrous and fibrohistiocytic tumors of the skin. *Dermatol Clin* 1999;17:487–505.
166. Sakamoto A, Oda Y, Yamamoto H, et al. Calponin and h-caldesmon expression in atypical fibroxanthoma and superficial leiomyosarcoma. *Virchows Arch* 2002;440:404–409.
167. Gleason BC, Calder KB, Cibull TL, et al. Utility of p63 in the differential diagnosis of atypical fibroxanthoma and spindle cell squamous cell carcinoma. *J Cutan Pathol* 2009;36:543–547.
168. Dei Tos AP, Maestro R, Doglioni C, et al. Ultraviolet-induced p53 mutations in atypical fibroxanthoma. *Am J Pathol* 1994;145:11–17.
169. Sato M, Nishigori C, Zghal M, Yagi T, Takebe H. Ultraviolet-specific mutations in p53 gene in skin tumors in xeroderma pigmentosum. *Cancer Res* 1993;53:2944–2946.

Inflammatory malignant fibrous histiocytoma
170. Kyriakos M, Kempson RL. Inflammatory fibrous histiocytoma: an aggressive and lethal lesion. *Cancer* 1976;37:1584–1606.
171. Melhem MF, Meisler AI, Saito R, et al. Cytokines in inflammatory malignant fibrous histiocytoma presenting with leukemoid reaction. *Blood* 1993;82:2038–2044.

172. Coindre JM, Hostein I, Maire G, et al. Inflammatory malignant fibrous histiocytomas and dedifferentiated liposarcomas: histologic review, genomic profile, and MDM2 and CDK4 status favour a single entity. *J Pathol* 2004;203:822–830.
173. Khalidi HS, Singleton TP, Weiss SW. Inflammatory malignant fibrous histiocytoma: distinction from Hodgkin's disease and non-Hodgkin's lymphoma by a panel of leukocyte markers. *Mod Pathol* 1997;10:438–442.

Giant cell MFH and giant cell tumor of soft parts
174. Salm R, Sissons HA. Giant cell tumours of soft tissues. *J Pathol* 1972;107:27–39.
175. Folpe AL, Morris RJ, Weiss SW. Soft tissue giant cell tumor of low malignant potential: a proposal for the reclassification of malignant giant cell tumor of soft parts. *Mod Pathol* 1999;12:894–902.
176. Oliveira AM, Dei Tos AP, Fletcher CDM, Nascimento AG. Primary giant cell tumor of soft tissues: a study of 22 cases. *Am J Surg Pathol* 2000;24:248–256.
177. O'Connell JX, Wehrli BM, Nielsen GP, Rosenberg AE. Giant cell tumors of soft tissue: a clinicopathologic study of 18 benign and malignant tumors. *Am J Surg Pathol* 2000;24:386–395.
178. Guccion JG, Enzinger FM. Malignant giant cell tumor of soft parts: an analysis of 32 cases. *Cancer* 1972;29:1518–1529.
179. Angervall L, Hagmar B, Kindblom LG, Merck C. Malignant giant cell tumor of soft tissues: a clinicopathologic, cytologic, ultrastructural, angiographic and microangiographic study. *Cancer* 1981;47:736–747.
180. Feng B, Rowe L, Zhang PJ, Khurana JS. Cutaneous sarcomatoid carcinoma with features of giant cell tumor of soft parts–a case report. *Am J Dermatopathol* 2008;30:395–397.

Pleomorphic hyalinizing angiectatic tumor
181. Smith MEF, Fisher C, Weiss SW. Pleomorphic hyalinizing angiectatic tumor of soft parts: a low-grade neoplasm resembling neurilemoma *Am J Surg Pathol* 1996;20:21–29.
182. Groisman GM, Bejar J, Amar M, Ben-Izhak O. Pleomorphic hyalinizing angiectatic tumor of soft parts:

immunohistochemical study including the expression of vascular endothelial growth factor. *Arch Pathol Lab* 2000;124:423–426.

183. Folpe AL, Weiss SW. Pleomorphic hyalinizing angiectatic tumor: analysis of 41 cases supporting evolution from a distinctive precursor lesion. *Am J Surg Pathol* 2004;28:1417–1425.

184. Kazakov DV, Pavlovsky M, Mukensnabl P, Michal M. Pleomorphic hyalinizing angiectatic tumor with a sarcomatous component recurring as high-grade myxofibrosarcoma. *Pathol Int* 2007;57:281–284.

185. Illueca C, Machado I, Cruz J, *et al.* Pleomorphic hyalinizing angiectatic tumor: a report of 3 new cases, 1 with sarcomatous myxofibrosarcoma component and another with unreported soft tissue palpebral location. *Appl Immunohistochem Mol Morphol* 2012;20:96–101.

186. Mohajeri A, Kindblom LG, Sumathi VP, *et al.* SNP array and FISH findings in two pleomorphic hyalinizing angiectatic tumors. *Cancer Genet* 2012;205:673–676.

Chapter 14

Lipoma variants and conditions simulating lipomatous tumors

Markku Miettinen
National Cancer Institute, National Institutes of Health

Ordinary lipoma and its variants contain mature white fat, sometimes with fibroblastic or other mesenchymal elements. The term *lipomatosis* covers a heterogeneous group of entities representing regional diffuse growth of mature fat that does not have known morphologically distinctive features, except for lipomatosis of the nerve, which is discussed in more detail later.

Some lesions designated as *lipomas*, such as synovial lipoma with villous pattern (lipoma arborescens) and lipoma of the hernia sac, are more likely reactive hyperplasias than true neoplasms. Furthermore, lumbosacral lipoma in children is associated with neural tube closure defects and could be a developmental anomaly.

Atypical lipomatous tumor is discussed in Chapter 15 with liposarcomas, because it is synonymous with well-differentiated liposarcoma.

Several special lipoma types are considered specific clinicopathologic entities, as listed previously. All of these, except angiolipoma and spindle cell lipoma, are quite rare. *Hibernoma* is a tumor resembling brown fat, and chondroid lipoma has some resemblance to it. *Lipoblastoma* is a clinically and genetically distinctive childhood lipomatous tumor. *Fibrohistiocytic lipoma* is a peculiar benign lipomatous tumor in the leg, having a mesenchymal fibrohistiocytic-like element.

Finally, fat atrophy, fat necrosis, sclerosing lipogranuloma silicone granuloma, and lymphedema in obese patients are discussed as examples of non-neoplastic lesions that can simulate lipomatous tumors. Angiomyolipoma is included among perivascular epithelioid cell tumors (PEComas), a group of tumors with HMB45-positive and smooth muscle actin-positive components (see Chapter 19).

Soft tissue lipomas
Clinical features

Lipomas are probably the most common soft tissue tumors in adults and have been estimated to constitute almost one-half of all of soft tissue tumors.[1] Hospital-based statistics underestimate the true incidence of lipomas, because many remain asymptomatic or patients having them never receive medical attention. Lipomas greatly outnumber liposarcomas, but the true ratio between these tumors is difficult to determine because lipomas are not included in tumor registries. In the author's experience at university hospitals with active musculoskeletal surgery, lipomas are only 10–20 times more common than liposarcomas, with relative over representation of the latter. Although malignant transformation of ordinary lipoma has not been thought to occur, many lipomas show involvement of the same chromosomal region, 12q13–15, as do atypical lipoma or well-differentiated liposarcoma,[2] suggesting that these groups could represent a biologic continuum.

Lipomas usually occur in adults older than 40 years and have a male predominance. They present in a wide variety of body sites, most commonly in the upper body, especially the back, shoulder, arm, forearm, and other extremity sites, especially the proximal ones.

Most soft tissue lipomas are subcutaneous, circumscribed, mobile tumors. According to a large series, 80% of them are smaller than 5 cm.[3] Multiple lipomas can occur, sometimes on a hereditary basis.[4] A minority of lipomas are intramuscular or located in intermuscular septa, located in body cavities or in visceral sites. Simple excision is the recommended treatment, but recurrence is possible, especially for the infiltrative variant of intramuscular lipoma.[5,6]

Pathology

On gross inspection and histologic analysis, lipomas resemble mature white adipose tissue. Although typically <5 cm, their largest diameter can reach 20 cm or more. Lipomas often form oval, flattened, or discoid subcutaneous masses, often with larger longitudinal and lateral, but smaller anteroposterior dimensions. They are often surrounded by a thin capsule-like membrane (*pseudocapsule*) and are sometimes lobulated. Intramuscular lipomas can be circumscribed or diffuse and tend to be larger than the subcutaneous ones. Fibrolipomas are firmer and can show grayish streaks of fibrous tissue. Some lipomas are mucoid (myxoid lipoma); this change is more common in spindle cell lipoma. On sectioning, the surface is typically golden yellow, especially in unfixed specimens; it can be paler after fixation (Figure 14.1).

Histologically, typical lipomas are surrounded by a thin collagenous pseudocapsule and composed of mature adipocytes indistinguishable from normal adipose tissue. Generally

Modern Soft Tissue Pathology, Second Edition, ed. Markku Miettinen. Published by Cambridge University Press. © Cambridge University Press 2016.

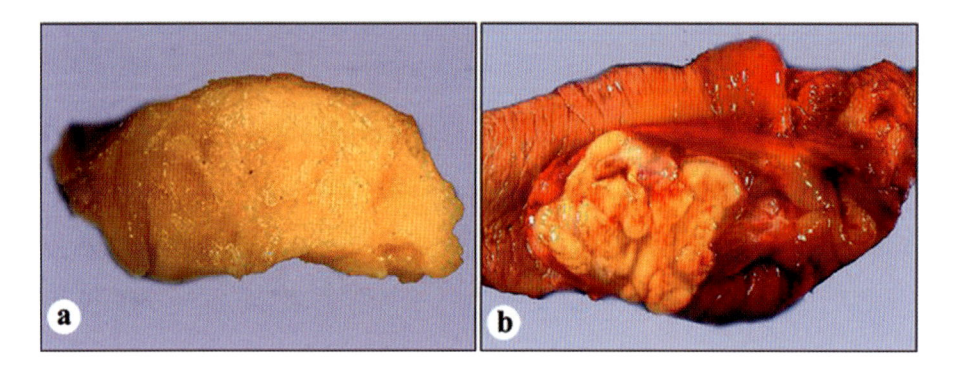

Figure 14.1 Gross appearance of lipoma. (**a**) Formalin-fixed chest wall lipoma is a demarcated mass with a yellow surface on sectioning. (**b**) Unfixed specimen of a submucous lipoma in the colon reveals a golden yellow surface on sectioning.

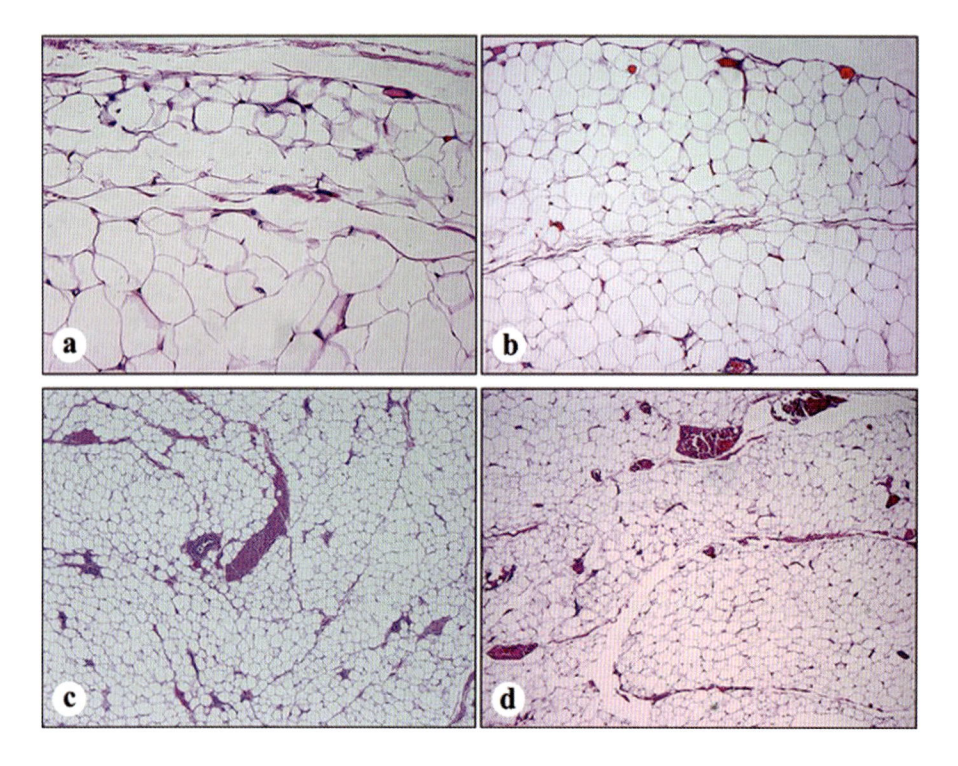

Figure 14.2 (**a,b**) Circumscription and a thin collagenous pseudocapsule is typical of lipoma. (**c,d**) Delicate collagenous septa often divide the lipoma into variably defined vague lobules. Note that there is only minor variation in adipocyte size, and the overall appearance is uniform.

only minor variation in adipocyte size occurs, and the overall appearance is usually "clean," being free of cellular infiltrates (Figure 14.2). Adipocyte nuclei are typically small and peripherally located, but because they are small, they are not seen in all cells in a given sectional plane. Tumors with significant nuclear atypia are classified as *atypical lipomatous tumors (ALTs)* or *well-differentiated liposarcomas* (notable exception: pleomorphic lipoma, a variant of spindle cell lipoma).

A common finding in lipoma, but not of any concern, is the occurrence of intranuclear vacuoles *(pseudoinclusions),* and focal collections of histiocytes (Figure 14.3a,b). Poor tissue sections, often resulting from inoptimal tissue processing, can be impossible to interpret (Figure 14.3c,d). Optimal tissue processing and sectioning can prevent diagnostic problems in lipomatous tumors. Slicing and overnight fixation is an easy way to facilitate preparation of high-quality tissue sections that are usually straightforward to interpret.

Intramuscular lipomas can involve skeletal muscle in a checkerboard pattern, with alternating fat and skeletal muscle cells (Figure 14.4), but they can also be well demarcated, similar to subcutaneous lipomas. Entrapped skeletal muscle giant cells *(sarcoplasmic giant cells)* should not be interpreted as atypical elements (Figure 14.4c,d).

The variants with prominent, paucicellular fibrous septa can be classified as fibrolipomas (Figure 14.5a,b). Examples with myxoid stroma (myxoid lipoma, myxolipoma; Figure 14.5c,d) should not be confused with myxoid liposarcoma. The lack of a prominent plexiform capillary pattern and lipoblasts separate them from the latter.

Differential diagnosis

In some instances it is difficult to determine whether well-differentiated fat represents lipoma or normal fat. Lipoma of the hernia sac more commonly contains reactive fat than a fatty neoplasm, but there is always the possibility that the lipomatous tumor caused the hernia. Demarcation, pseudoencapsulation, and lack of well-defined lobulation, as seen in normal fat, are features of lipoma. Recently, fat herniations in the orbit have been reported that contained normal

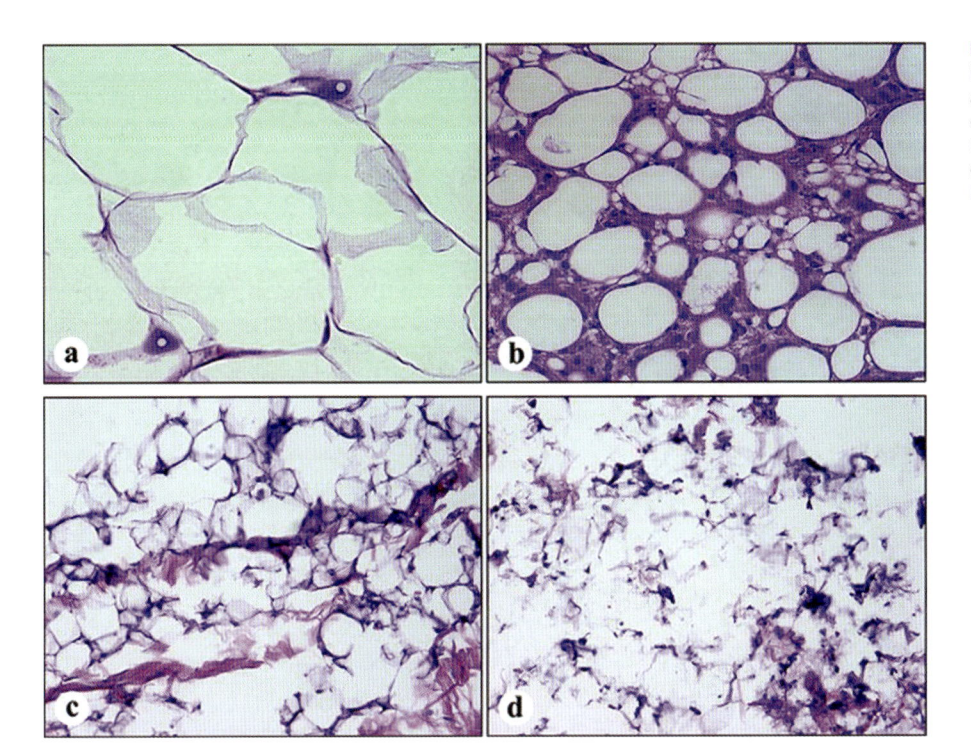

Figure 14.3 (**a**) Intranuclear vacuoles ("Lochkern") in adipocytes are common in lipomas and are not a sign of atypia. (**b**) Focal lipogranulomatous change should not be confused with atypia. (**c,d**) Nonoptimal tissue processing or sectioning can make observation of the cellular detail of a lipoma difficult or impossible.

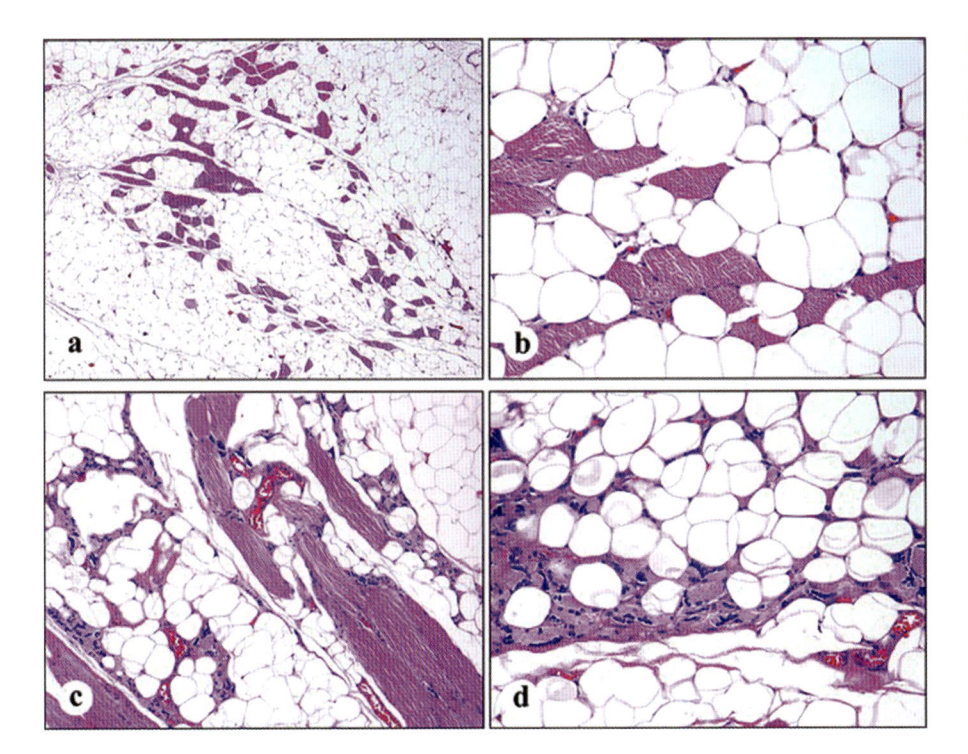

Figure 14.4 (**a,b**) Infiltrative type of intramuscular lipoma dissects between skeletal muscle fibers. (**c,d**) Reactive changes, including muscle giant cells, are often present in the skeletal muscle and should not be interpreted as atypia.

displaced fat; the differential diagnosis of this condition from a lipoma requires clinicoradiologic correlation.[7]

Large lipomas, especially the deep and intra-abdominal ones, should be sampled generously (at least one section per each cm of tumor diameter) to evaluate atypia comprehensively, because such atypia can be focal. Significant nuclear atypia in deep tumors confers the diagnosis of ALT/well- differentiated liposarcoma. The focal nature of atypia limits the value of small-needle biopsies to differentiate lipoma from atypical lipoma/well-differentiated liposarcoma.

Muscular dystrophy or atrophy triggered by an adjacent tumor can cause focal or massive regional fatty replacement in the involved muscles, potentially simulating a lipomatous process. An intramuscular hemangioma can have a significant

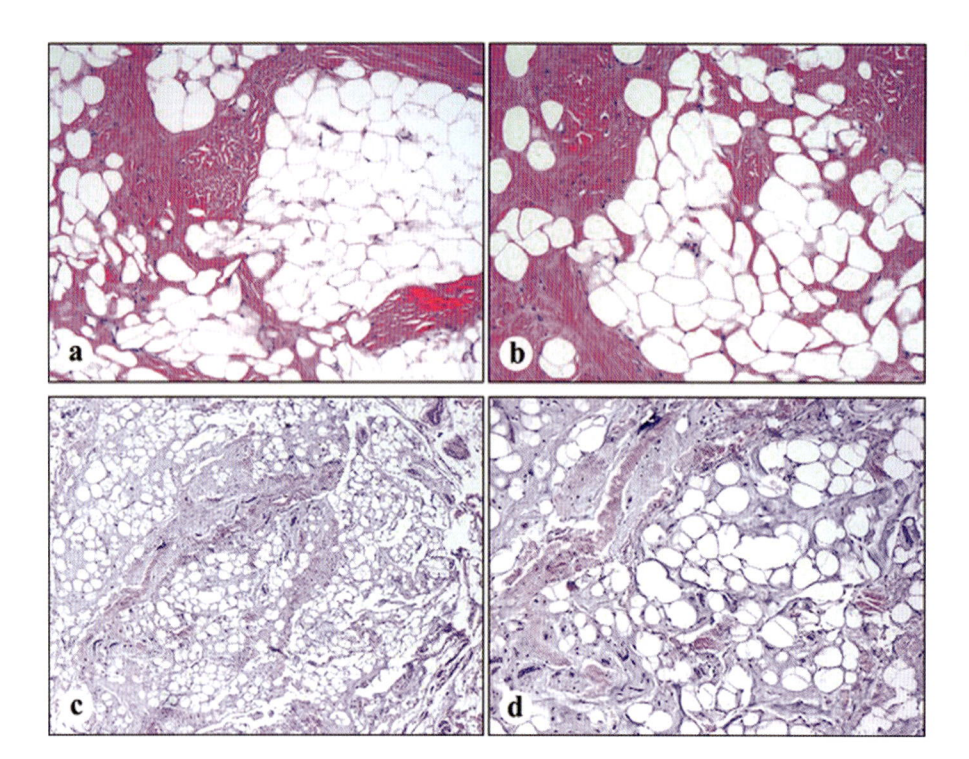

Figure 14.5 (**a,b**) Fibrolipoma has prominent, paucicellular fibrous septa. (**c,d**) Myxoid lipoma contains myxoid matrix between the relatively uniform adipocytes.

lipomatous component; in this tumor, the presence of the hemangiomatous component traditionally outweighs the presence of fat (although historically the designation of *infiltrative angiolipoma* has been used for these fat-containing hemangiomas).

Genetics

Ordinary lipomas show clonal chromosomal aberrations in nearly 80% of cases, and such changes seem to be more common in lipomas of older patients and those of larger size or located in the extremities, as opposed to lesions of the trunk wall.[8] Clonal chromosomal changes also support the idea that lipomas are clonal neoplasms and not fat hyperplasias.

The chromosome numbers in lipoma are usually diploid or near diploid and the karyotypes relatively simple, often characterized by one aberration only. Rare examples have shown polyploidy or complex changes, however.[2,9]

Balanced translocations are typical aberrations, and they most commonly involve chromosomal regions 12q13–15 (60%–70%), 13q12–22 (20%), 3q28, and 6p. Interstitial deletions often involve 13q13.[2,8–13] These data, apparently mostly based on ordinary soft tissue lipomas, do not necessarily apply to organ-based lipomas and special subtypes of ordinary lipomas.

The most common balanced translocation in lipomas is t(3;12)(q28;q14). The chromosome 12 breakpoint 12q13–15 involves the HMGA2 gene, whereas the 3q breakpoint involves the LPP ("LIM domain containing lipoma preferred partner") gene.[2] Although ring chromosomes derived from amplified 12q13–15 sequences (including genes such as MDM2)[14] usually occur in ALTs, they have also been reported in occasional conventional lipomas.[6] CGH studies, however, have suggested that lipomas and ALTs segregate well, with the latter having gains only in chromosome 12q.[15]

The HMGA2 gene at 12q15 encodes for a nuclear nonhistone protein, high-mobility group protein IC (formerly HMGIC) based, named for its high mobility in polyacrylamide gel electrophoresis. This DNA-binding protein participates in the transcriptional regulation as a cofactor (architectural transcription factor, regulating DNA configuration) and is commonly rearranged and overexpressed in different types of tumors in addition to lipoma, for example, pleomorphic adenoma of the salivary gland and uterine leiomyoma.[16]

Aberrations in 13q12 have been located in a gene named lipoma HMGA2/HMGIC fusion partner (LHFP). This is also a DNA-binding protein and a probable transcriptional regulator.[17]

Lipoma with metaplastic cartilage and bone

Rare examples of lipomas contain foci of mature hyaline cartilage, metaplastic bone, or both (Figure 14.6). Such tumors have sometimes been designated as chondrolipomas or osteolipomas.[18–21] The occurrence of mature hyaline cartilage differs from the appearance of chondroid lipoma, where the nonlipomatous component does not have the appearance of differentiated hyaline cartilage.

Lipomas with metaplastic bone and cartilage occur in a wide age range, from childhood to old age, and in locations such as the hand, trunk wall, breast, pharynx, and neck. Occurrence of osteocartilaginous differentiation in lipoma

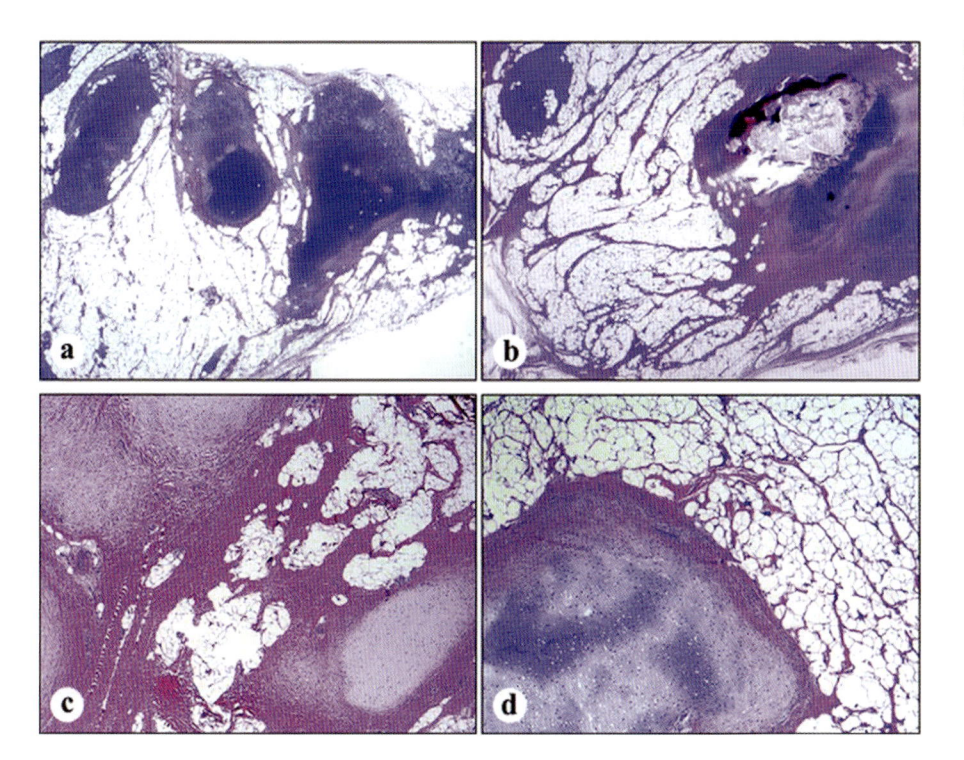

Chapter 14: Lipoma variants and conditions simulating lipomatous tumors

Figure 14.6 Lipoma with chondro-osseous metaplasia. Note cartilaginous foci in (**a–d**) and focal osseous metaplasia in (**b**). Paucicellular fibrous septa are also present.

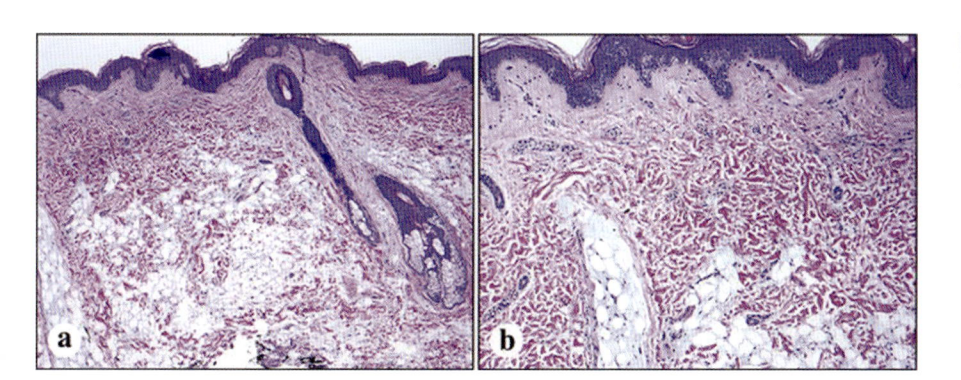

Figure 14.7 Nevus lipomatosus superficialis is histologically seen as an abnormal fatty element in the dermis. Note the mature appearance of the fat.

could be a reflection of the presence of multipotential stem-cell-like components in some lipomas.[22]

Nevus lipomatosus superficialis

Nevus lipomatosus superficialis is the term for the occurrence of linear streaks of lipomatous elements in the skin (Figure 14.7). These have a predilection for the proximal thigh, buttocks, and lower trunk. Clinically there are multiple papules (classic type) or a solitary papule; the latter variant is more common.[23,24] Histologically there are dermal islands of mature fat that dissect the various dermal and adnexal elements.

Lipoma arborescens (synovial lipomatosis)

Lipoma arborescens is a rare, histologically distinctive synovial fatty mass with a villiform appearance (Figure 14.8).

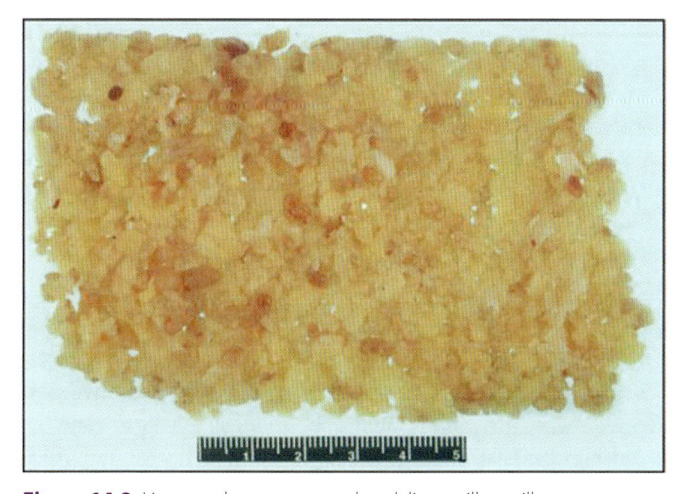

Figure 14.8 Lipoma arborescens reveals a delicate villopapillary structure on gross examination.

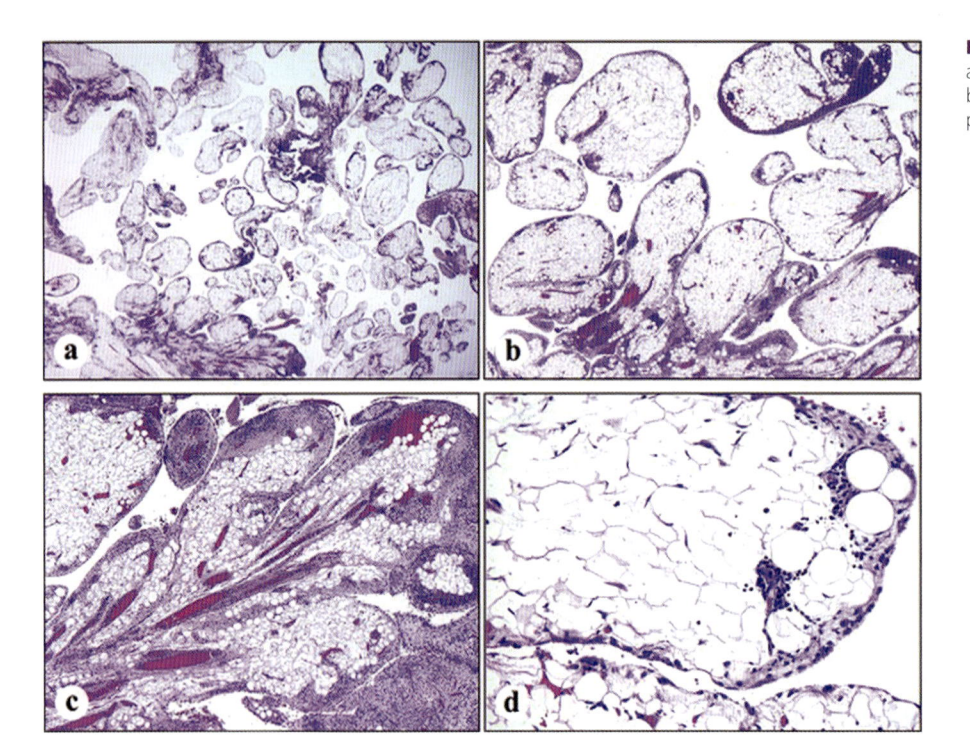

Figure 14.9 Complex villous structures of lipoma aborescens are covered by synovia. Hyperemic blood vessels and focal lymphoid infiltration are present.

Hoffa disease (sometimes incorrectly used as a synonym for lipoma arborescens) is usually understood as a chronic impingement of subpatellar fat pad into the articular cavity, resulting in degenerative changes such as fat necrosis and fibrosis.

Clinical features

This rare fatty lesion occurs predominantly in young adults without predilection for either sex, but can also occur in children and older adults. It usually involves the synovia of the knee, and less commonly other joints, such as the hip, elbow, ankle, or wrist. Clinically it manifests as painless joint swelling with effusion; bilateral knee involvement has been reported. Because this lipoma is often associated with joint pathology (e.g., osteoarthrosis, internal derangement), some have suggested that it is a reactive process triggered by joint pathology, although the opposite has also been proposed; namely, that lipoma arborescens is the cause of joint swelling, pain, and evolving osteoarthrosis. Synovectomy (arthroscopic or open) has been used as a treatment of lesions independent of joint pathology, although correction of the underlying condition (meniscal or ligament tear) is important when these secondary lesions are present.[25,26]

Pathology

Grossly, lipoma arborescens is a villiform proliferation of synovia, with the villous stroma containing the fatty element (Figure 14.8). Histologically it shows villous synovial hyperplasia with the cores markedly dilated by the content of mature adipose tissue without atypia. Focal lymphoid infiltration is sometimes present (Figure 14.9). There are no scientific studies addressing whether lipoma arborescens is a (polyclonal) fat hyperplasia or a clonal neoplasm.

Lumbosacral and spinal cord-associated lipoma

This lipoma group is clinically significant because of its nearly consistent association with spinal cord closure defects (spina bifida) and other anomalies. These lipomas are usually diagnosed in young children (<2 years of age) and only occasionally in adults. They variably involve the subcutis, extending to the distal spinal canal, or involve intraspinal spaces such as the spinal cord or conus medullaris only. There is a high association with spinal cord attachment, potentially causing stretching and damage of the spinal cord. Such attachment, commonly referred to as spinal cord tethering, can lead to neurologic deficits such as dysfunction of the autonomic nerves, especially those of the urinary bladder.[27–29] It is possible that these "lipomas" actually represent developmental anomalies or malformations rather than true fatty neoplasms.

Visceral lipomas

A wide variety of organ-based locations is possible, including the intracranial space, the heart, and the respiratory and gastrointestinal tracts (especially the colon). The diagnosis of mediastinal and intra-abdominal lipoma should be made with great caution, because well-differentiated liposarcomas are

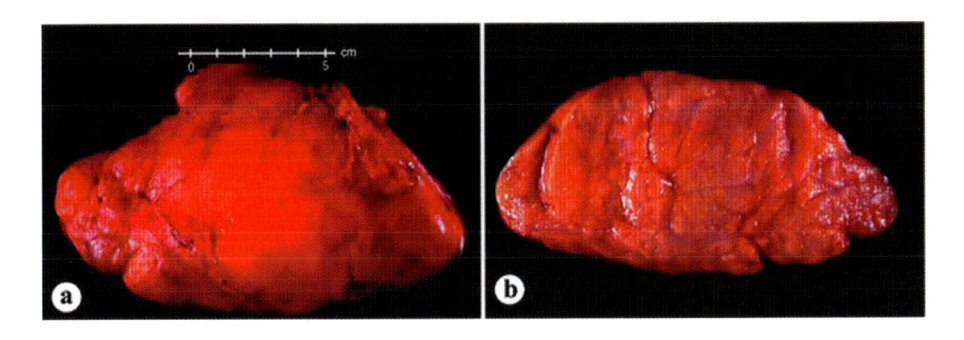

Figure 14.10 (**a**) Thymolipoma is grossly a well-delineated mass. (**b**) On sectioning the tissue is yellowish and lobulated, with grayish streaks representing the remaining thymic elements.

relatively more common in these locations. The histologic features of these visceral lipomas do not significantly differ from those of soft tissue lipomas.

Central nervous system lipomas are usually diagnosed in children and occur in the midline, including the corpus callosum and spinal canal. They can be asymptomatic or associated with various congenital malformations and epilepsy.[30,31] It is unknown whether these lipomas are true neoplasms or congenital malformations.

Lipomas and lipoma-like lesions of the heart include lipomatous hypertrophy of the interatrial septum and rare valvular lipomas. The former, although usually asymptomatic, can be associated with arrhythmias and can cause sudden death.[32]

Respiratory tract lipomas usually occur in older adults. Most of these lipomas are relatively small (1–3 cm), obstructive, endobronchial submucosal tumors arising from the proximal parts of the bronchial tree, but some occur in the peripheral lung, possibly arising from terminal branches of bronchi. Clinical suspicion of malignancy can lead to unnecessarily radical surgery if the lesion has not been preoperatively diagnosed.[33,34]

Gastrointestinal lipomas are usually small submucosal nodules, and most of them occur in the colon, although any part of the gastrointestinal tract can be involved. They are usually incidental findings during endoscopy, radiologic studies, or specimen examination, but larger lipomas can cause intestinal obstruction by intussusception, or they can ulcerate and cause bleeding.[35] Histologically they often have a prominent vascular component and have therefore sometimes been called "angiolipomas."

Lipoma of the hernia sac is a common designation for fibrolipomatous tissue fragments obtained from hernia surgery. This lipoma contains mature adipose tissue with reactive changes, such as myxoid stroma and mild myofibroblastic proliferation, and is probably a non-neoplastic process.

Thymolipoma

This is a specific anterior mediastinal tumor, histologically combining mature fat with epithelial elements of the thymus. It occurs in a wide age range (mean, 33 years) and equally in men and women, occasionally associated with myasthenia gravis. Tumor size varies from a few centimeters to > 30 cm. Gross appearance often differs from lipoma by the retention of thymic lobulation. Otherwise, cases with limited thymic elements resemble conventional lipomas, but more prominent thymic elements can give the tumor a gray-white appearance (Figure 14.10).

Histologically, there is a dominant fatty component and an interspersed thymic epithelial component varying from narrow, somewhat inconspicuous cords of epithelial elements to more complete representation of thymic elements with Hassal's corpuscles and lymphoid tissue (Figure 14.11).[36]

Lipomatosis of the nerve (fibrolipomatous hamartoma of nerve, neural fibrolipoma)

Lipomatosis of the nerve (also known under the designations cited here) is a rare tumor-like formation that usually occurs in the fingers and hands and rarely in the foot and elsewhere. It is not known whether this lesion is a malformation, a hyperplasia, or a true neoplasm.

Clinical features

This very rare condition occurs mostly in children and young adults and equally in men and women. In many cases the lesion is congenital. The median age in Armed Forces Institute of Pathology (AFIP) case files is 25 years. Clinically, there is an ill-defined, sometimes painful, often sausage-shaped tumor-like lesion usually developing on the volar aspects of the fingers and hands, usually around the median nerve (Figure 14.12); however, other nerves of the hand can also be involved (the ulnar nerve), and a similar process can involve the foot. Some patients have macrodactyly. Because radical excision leads to nerve sacrifice and neural deficits, this treatment is not advocated. Rather, this condition should be diagnosed by biopsy and treated by nerve decompression when necessary.[37,38]

Pathology

Grossly, the lesion represents a sausage-shaped or fusiform fibrofatty mass involving a nerve, usually the median nerve. It contains branches of the nerve separated by fibrofatty expansion of the epineurium with features of fibrolipoma. The nerve

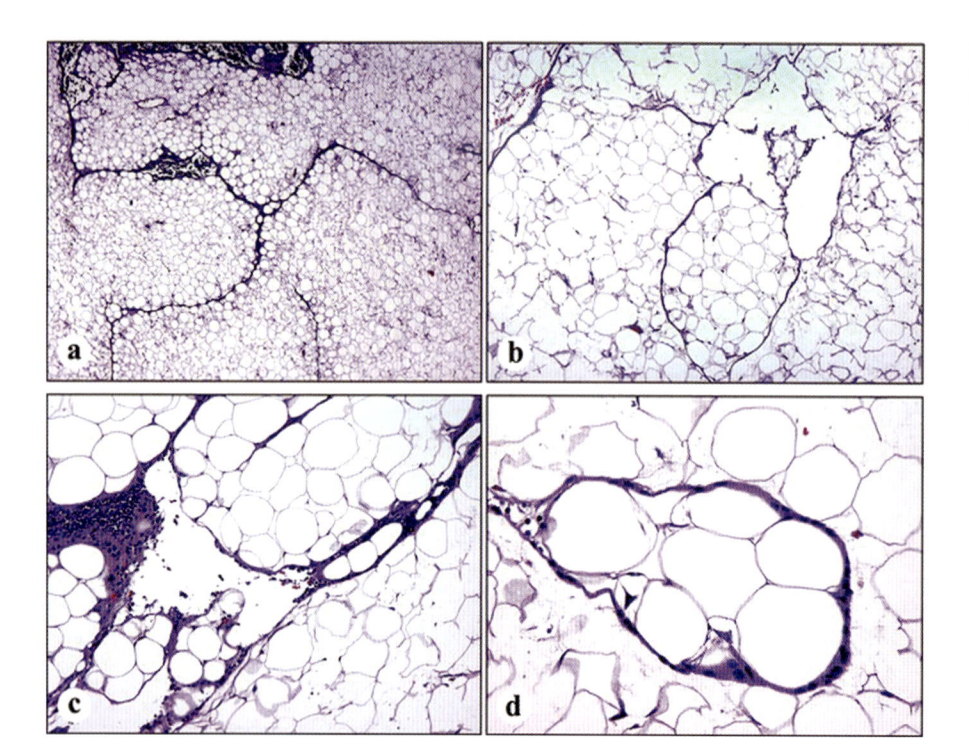

Chapter 14: Lipoma variants and conditions simulating lipomatous tumors

Figure 14.11 Thymolipoma contains mature fat with uniform fat cells and elements of thymic epithelium as narrow streaks, cystic structures, and pseudoglandular formations.

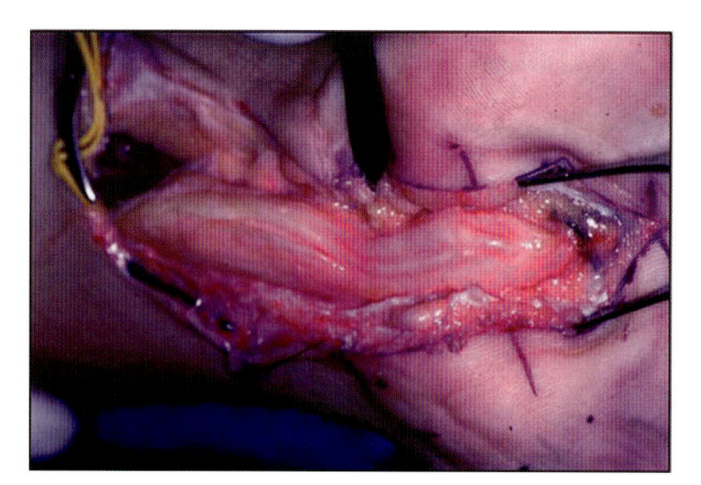

Figure 14.12 Neural fibrolipoma of the hand forms a sausage-shaped, elongated mass following the course of the median nerve.

branches, usually seen as cross-sections or sometimes longitudinally cut, vary from nearly normal to severely altered. They are often surrounded by epineurial fibrosis. Inside the nerve branches there is often endoneurial fibrosis. In some cases, onion bulb-like perineurial cell proliferation is present around the axons, reminiscent of intraneural perineurioma (Figure 14.13).

Lipomatoses
Clinical features
The *lipomatoses* are a clinicopathologically and etiologically diverse group of diffuse regional accumulations of adipose tissue without a well-defined tumor mass (Table 14.1). Clinically, these range from congenital malformation syndromes (encephalocraniocutaneous lipomatosis),[39–41] to disorders of fat metabolism (steroid lipomatosis), to tumor-like conditions (diffuse lipomatosis of extremities) and those related to mitochondrial myopathies (diffuse symmetric lipomatosis). Most cases are sporadic, but some diffuse symmetric lipomatosis cases are hereditary.[42–49]

Patients with diffuse symmetric lipomatosis have a diffuse fat collection in the anterior neck, shoulders, and upper body.[42–44] Many patients also have a peripheral neuropathy and myopathy, and a mutation in mitochondrial DNA is probably part of the pathogenesis.[44] The nature of the fat collection in the neck area resembles some forms of lipodystrophy associated with highly active antiretroviral therapy (HAART) used in the treatment of HIV-AIDS that could also be a mitochondriopathy by pathogenesis.[45]

Pelvic lipomatosis has a predilection for African American men, and the lipomatous masses can compromise the ureters, large vessels, or intestines.[46] There is also an association with glandular cystitis, and by virtue of the latter, a risk of adenocarcinoma of the urinary bladder exists.[47]

Pathology
Although different forms of lipomatosis are often clinically distinctive, their pathologic features are largely unknown because the lipomatoses are rarely, if ever, seen as surgical specimens. No diagnostic histologic changes have been described.

Chapter 14: Lipoma variants and conditions simulating lipomatous tumors

Table 14.1 Summary of the most important clinical lipomatosis syndromes

Condition	Description/management of lipomas
Diffuse lipomatosis	Lipomatous growth involving various tissue planes in an anatomic region, especially an extremity
Multiple symmetric lipomatosis (Madelung's disease, Launois–Bensaude syndrome)	Prominent, symmetric fat collection involving the anterior neck, upper trunk, and arm. Some cases are hereditary. Also reported in association with alcoholism. Connection with mitochondrial myopathy. Debulking to prevent neurovascular and respiratory compromise when necessary
Pelvic lipomatosis	Diffuse deposition of fat between bladder, rectum, and large vessels. Can cause urinary or colorectal obstruction. Predilection for black men. Associated with glandular metaplasia and adenocarcinoma of urinary bladder. Also an association with achondroplasia
Encephalocraniocutaneous lipomatosis	Congenital malformation syndrome with facial and ipsilateral oculocerebral malformations, hydrocephalus, seizures, and mental retardation. Multiple lipomas involving the skin of face and scalp, and intracranial space (meninges, cranial fossae). Lipomas can have fibrolipomatous features and are not the principal cause of morbidity
Spinal epidural lipomatosis	Can be associated with corticosteroid use or be idiopathic. Spinal decompression is sometimes needed
Steroid lipomatosis	Designation of lipomatous masses caused by excessive corticosteroid stimulation. Can be endogenous (Cushing's syndrome) or iatrogenic. A variety of sites can be involved. Predilection for the head and neck

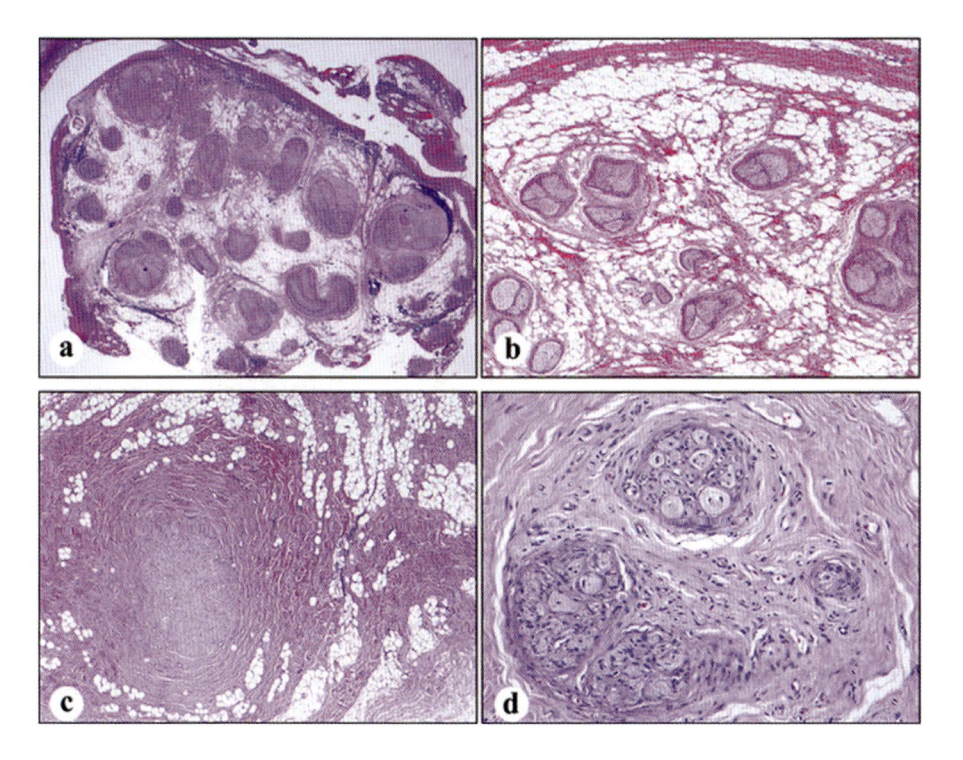

Figure 14.13 (**a,b**) Neural fibrolipoma is confined by a fibrous pseudocapsule and is composed of expansion of epineural fat, splaying the nerve fascicles apart. (**c,d**) The nerve fascicles are surrounded by fibrosis and can contain mild perineurial cell proliferation.

Genetics

Mutations in mitochondrial DNA, such as those in codon 8344 encoding the transfer RNA gene for lysine, have been detected in some patients with multiple symmetric lipomatosis.[44] Experimental truncation of the HMGA2 (previously HMG1C) gene in mice was found to lead to a condition comparable to human pelvic lipomatosis.[50] The HMGA2 gene is also known to be involved in typical and atypical lipomas and well-differentiated liposarcomas.

Figure 14.14 (**A**) Age and sex distribution of 1038 patients with angiolipoma. (**B**) Anatomic distribution of 950 angiolipomas.

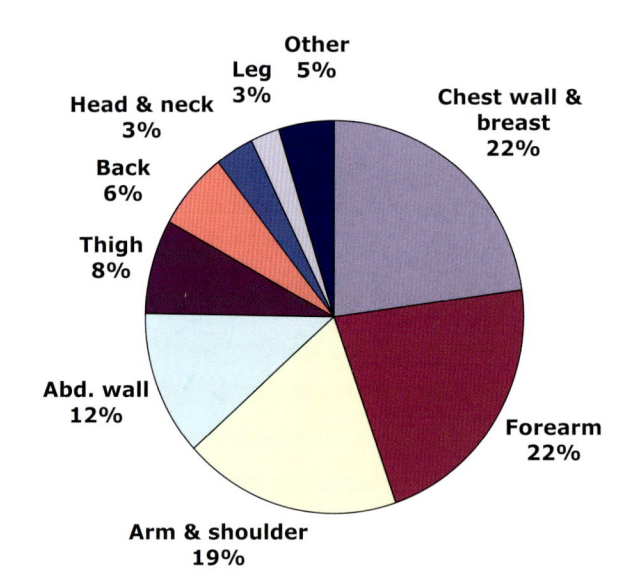

Angiolipoma

Angiolipoma is the designation for a lipoma variant that contains variable numbers of small-caliber vessels, often as streaks, with focal fibrin thrombi.[51] The term *angiolipoma*, especially in the form of *infiltrative angiolipoma* and *spinal angiolipoma*, has also been used for an unrelated lesion, intramuscular hemangioma, which often combines vascular and mature lipomatous elements.[52] Angiomyolipoma is an unrelated tumor that is discussed with the PEComas (Chapter 19).

Clinical features

An angiolipoma is a relatively common tumor that presents as a small, circumscribed, often painful subcutaneous nodule, typically in young to middle-aged adults, with a marked male predominance (Figure 14.14A). The most common locations are the breast and chest wall, forearm, arm, abdominal wall, thigh, and back. Distal extremity and head and neck locations are rare (Figure 14.14B). Multiple angiolipomas are seen in at least 10% of cases; some have occurred on a familial basis.[53] An exceptional case has been reported in a lymph node.[54] There is no significant tendency for recurrence.

Pathology

Grossly, angiolipomas are small, ovoid, circumscribed, pseudocapsulated lipomatous nodules usually measuring 1 cm to 2 cm in diameter. They can appear firmer than ordinary

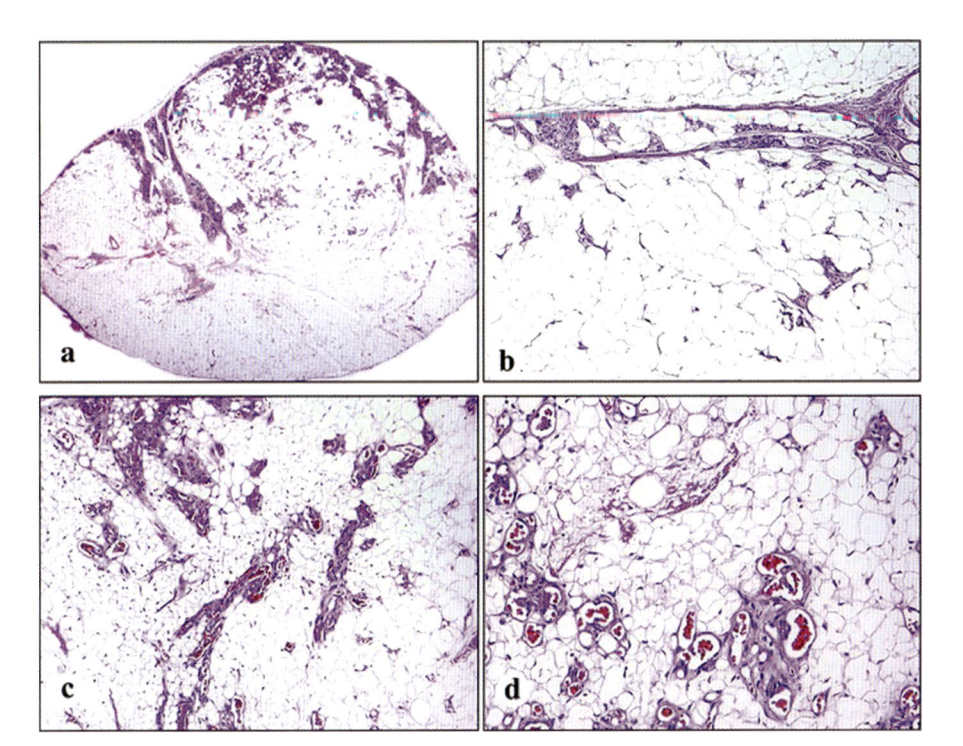

Chapter 14: Lipoma variants and conditions simulating lipomatous tumors

Figure 14.15 (**a**) Angiolipoma forms a sharply demarcated nodule containing capillaries, often as pericapsular foci or streaks streaming into the middle of the nodule. (**b**) The fatty component is mature, with only minor variation in adipocyte size. In this field, the capillaries are grouped around fibrous septa. (**c**) Typical streaks and clusters of capillaries in angiolipoma.

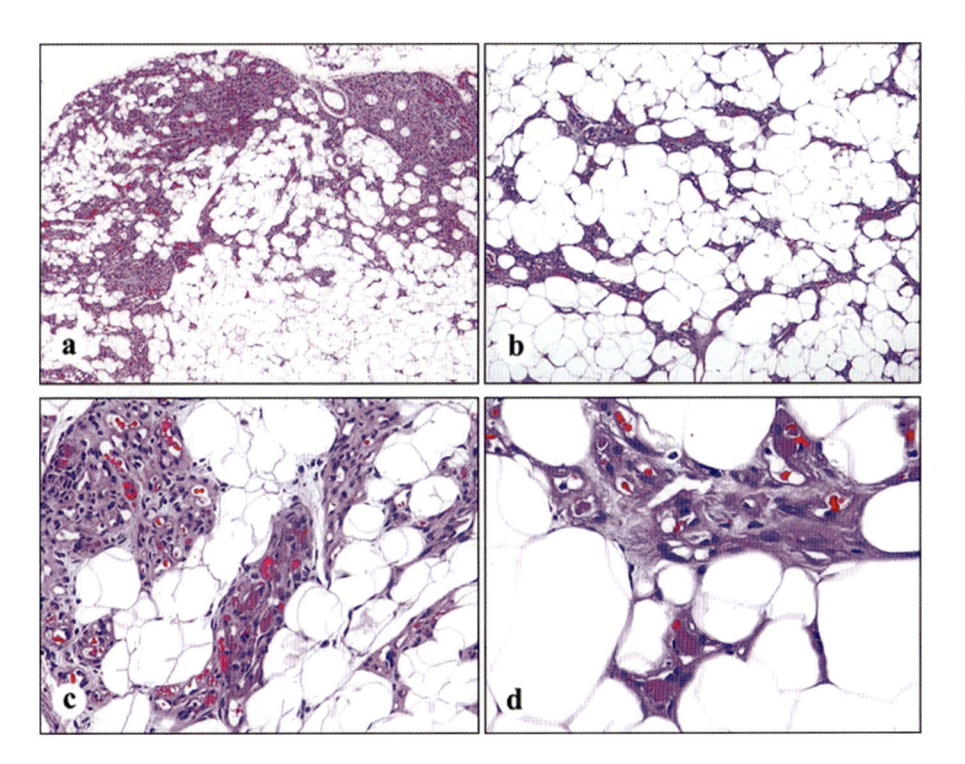

Figure 14.16 Angiolipoma capillaries are often hyperemic, and some contain fibrin thrombi, especially seen in (**c**) and (**d**).

lipomas and the cut surface can appear red owing to the capillary content.

Histologically, angiolipomas combine mature white fat with clusters of thick-walled capillaries, often located as streaks that radiate inward from the periphery of the tumor. The vessels are often congested, filled with erythrocytes, and some contain eosinophilic fibrin platelet microthrombi, a typical feature of this entity (Figures 14.15 and 14.16). Vessel number ranges from only a few to numerous.

Cellular angiolipoma refers to a form with an extensive vascular component and an inconspicuous fatty element, usually represented just by few fat cells in the periphery.[55] The cellular variant therefore resembles hemangioma on gross and histologic inspection, especially cellular capillary hemangioma,

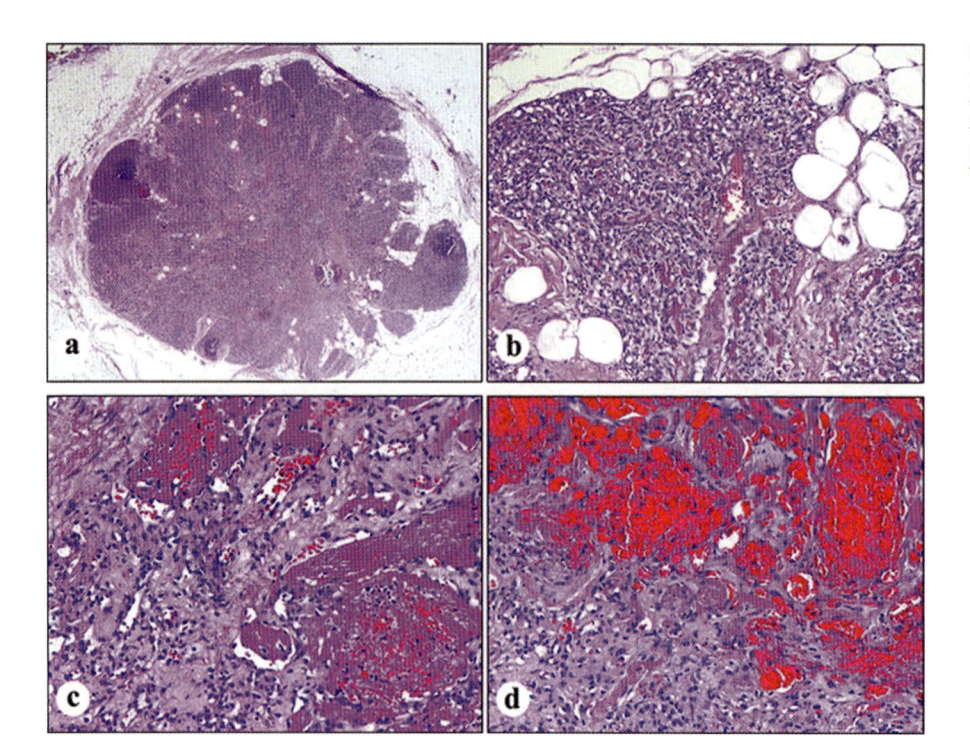

Figure 14.17 (**a,b**) Cellular angiolipoma is predominantly composed of lobules of capillaries, creating a hemangioma-like appearance, and only a minor fatty element is seen, often in the periphery of the tumor. (**c,d**) Extensive fibrin thrombi are typical.

spindle cell hemangioma, or even Kaposi's sarcoma (Figure 14.17). Although cellular angiolipoma can have focal mitotic activity, sharp demarcation, packeting of the vessels as lobules, and slit-like vascular spaces, the lack of endothelial atypia and multilayering, as well as the presence of distinctive fibrin thrombi help to separate it from borderline and malignant vascular tumors.[55]

Genetics

No cytogenetic alterations were reported in a study of 20 angiolipomas, suggesting a genetic difference from ordinary lipomas.[56] The apparent lack of genetic changes could suggest that these tumors have subtle submicroscopic chromosomal changes, such as gene mutations.

Myelolipoma

Myelolipoma is a rare internal tumor-like process usually occurring in the adrenal cortex and containing components of bone marrow: fat and hematopoietic elements. It is unknown whether myelolipoma is a peculiar tumor-like hyperplasia or a true neoplasm.

Clinical features

Myelolipoma occurs in older adults and has no connection with hematologic disease. It usually involves the adrenal cortex, but occasionally presents in the extra-adrenal soft tissues in the retroperitoneum, especially in the pelvis, or rarely in the mediastinum. Most myelolipomas are incidental radiologic or surgical findings during unrelated procedures, and if preoperatively diagnosed by radiology (fat-containing, small, circumscribed adrenal masses), they need not be surgically treated. Larger tumors or those that have ruptured can become symptomatic by abdominal pain or organ compression, and in these cases, simple excision is expected to be curative.

Myelolipoma is sometimes associated with obesity and hypercortisolism or hyperaldosteronism, either endogenous or iatrogenic, suggesting that corticosteroid stimulation might be a pathogenetic factor. Some patients have had bilateral adrenal myelolipomas or multiple extra-adrenal myelolipomas.[57–63]

Pathology

Grossly, myelolipoma forms a demarcated adrenal, retroperitoneal, or mediastinal mass. On sectioning it reveals lipoma-like mature adipose tissue and reddish components corresponding to hematopoietic bone marrow (Figure 14.18). Histologically, it is composed of areas of mature fat and cellular red bone marrow, especially containing maturing erythroid cells and megakaryocytes, and to lesser degree, myeloid elements (Figure 14.19). Lymphoid foci can be present and are prominent in some cases.

Myelolipoma should be separated from extramedullary hematopoiesis associated with myeloid neoplasia. Bone marrow studies and detection of dysplastic hematopoietic components help to differentiate these conditions. Extramedullary hematopoiesis also tends to be multifocal and is often associated with splenomegaly.

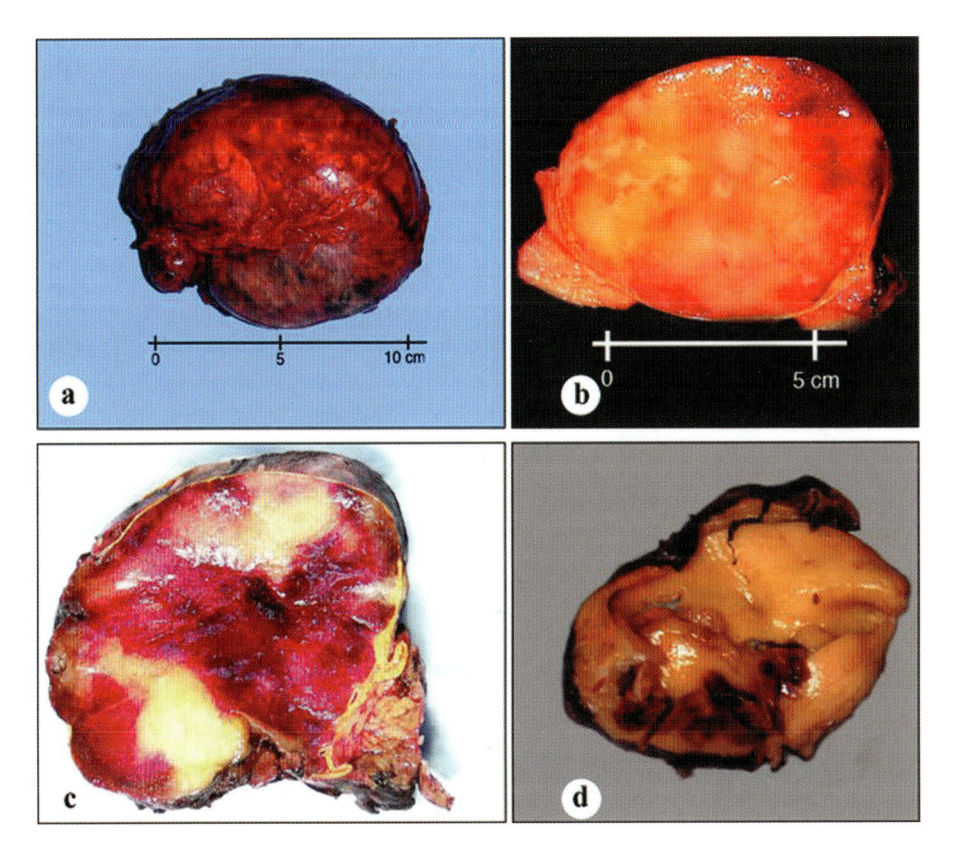

Chapter 14: Lipoma variants and conditions simulating lipomatous tumors

Figure 14.18 (**a–c**) Myelolipoma of the adrenal forms a sharply demarcated mass. Cut surfaces of unfixed specimens reveal yellowish fatty areas and red hematopoietic components, and a peripheral rim of adrenal tissue. (**d**) This retroperitoneal myelolipoma contains a major fatty component and a minor, red-brown hematopoietic component (fixed specimen).

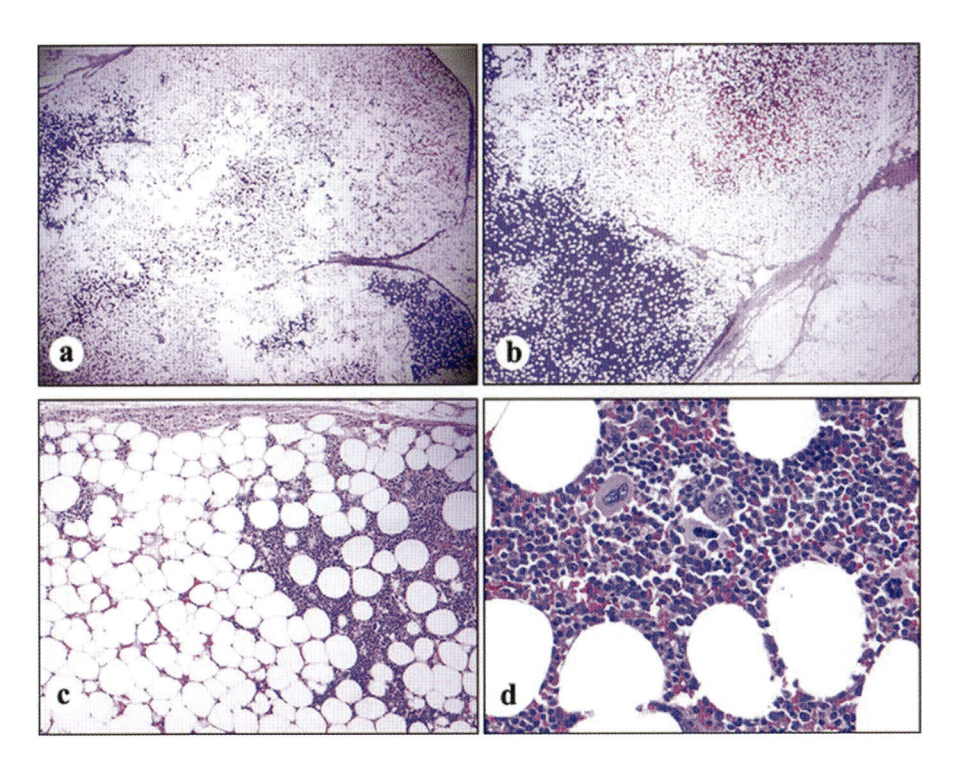

Figure 14.19 Histologically, myelolipoma contains mature fat and hematopoietic elements, and when seen together, the appearance resembles that of bone marrow.

Spindle cell and pleomorphic lipoma

First described by Enzinger and Harvey in 1975,[64] *spindle cell lipoma* is a relatively uncommon benign lipomatous tumor that shows a variable, nonlipogenic, spindle cell component. Pleomorphic lipoma, described by Shmookler and Enzinger in 1981, is its histologic variant, with floret-like adipocytes with nuclear pleomorphism.[65]

391

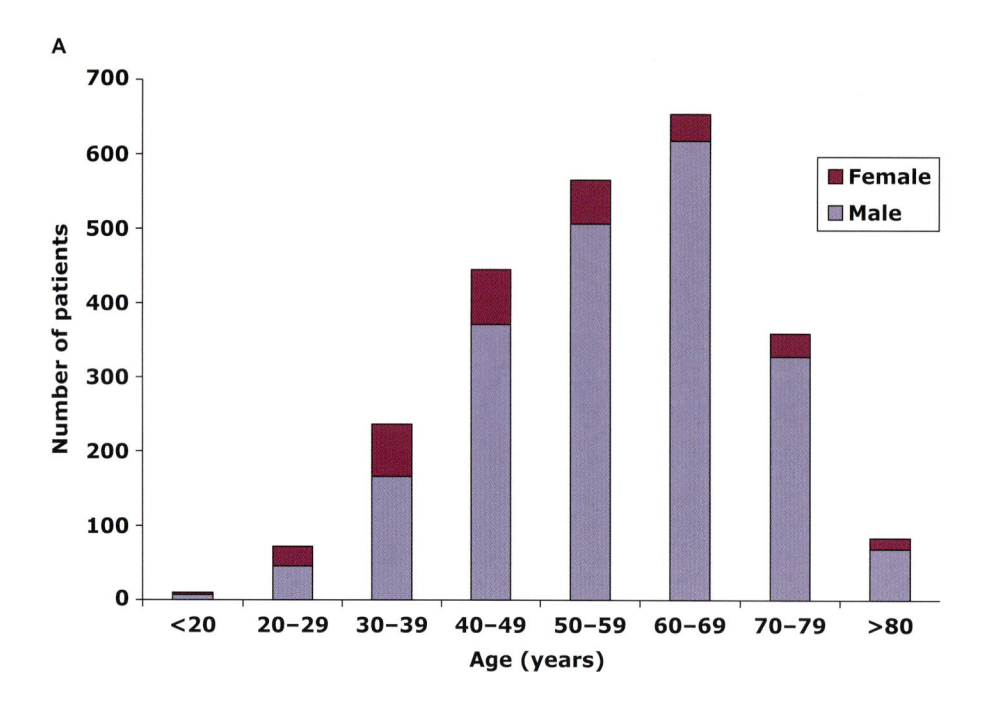

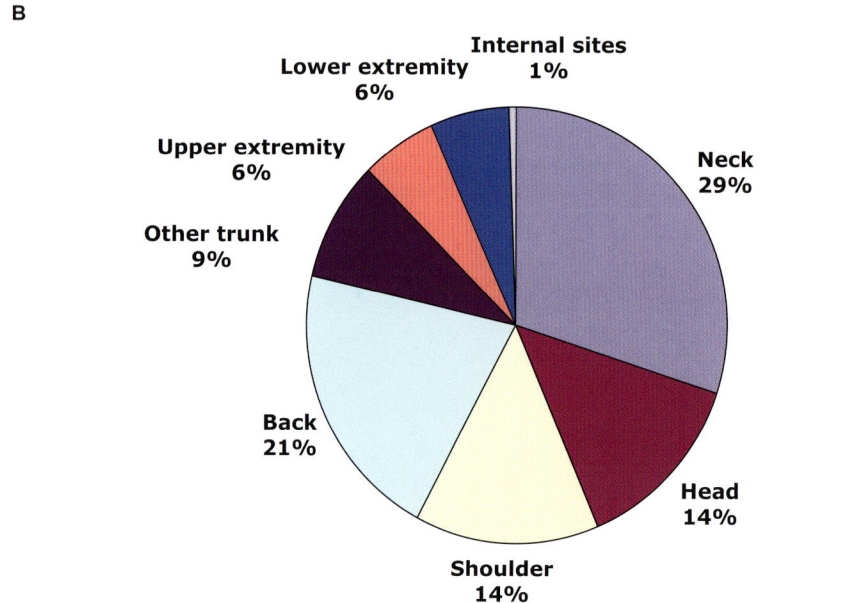

Chapter 14: Lipoma variants and conditions simulating lipomatous tumors

Figure 14.20 (**A**) Age and sex distribution of 2427 patients with spindle cell lipoma. (**B**) Anatomic distribution of 2477 patients with spindle cell lipoma.

Clinical features

Both spindle cell and pleomorphic lipomas (Figures 14.20A and 14.21A) typically present in older men (85–90%) with a median age of more than 55 years.[64–69] Patients often have a long history of the lesions, and over 80% occur in the subcutis of the posterior neck, back, and shoulder area (Figures 14.20B and 14.21B). In an AFIP study, nearly 50% of oral and maxillofacial lipomas were spindle cell or pleomorphic lipomas, and these tumors were a predominant lipoma type in the lip and parotid region.[69] On occasion, spindle cell and pleomorphic lipomas present elsewhere in the trunk or in the extremities, orbit, and skin, but very rarely if ever in deep soft tissues. It is uncertain whether lipomas with focal spindle cell components that sometimes present in other sites, such as distal extremities, are fully comparable to spindle cell lipomas.

Spindle cell lipomas, although usually relatively small (2–5 cm), can reach a size of >10 cm. Some patients have multiple lesions, and familial occurrence has been reported, mostly in men.[70] Spindle cell and pleomorphic lipomas have a benign behavior, and simple local excision is considered sufficient.

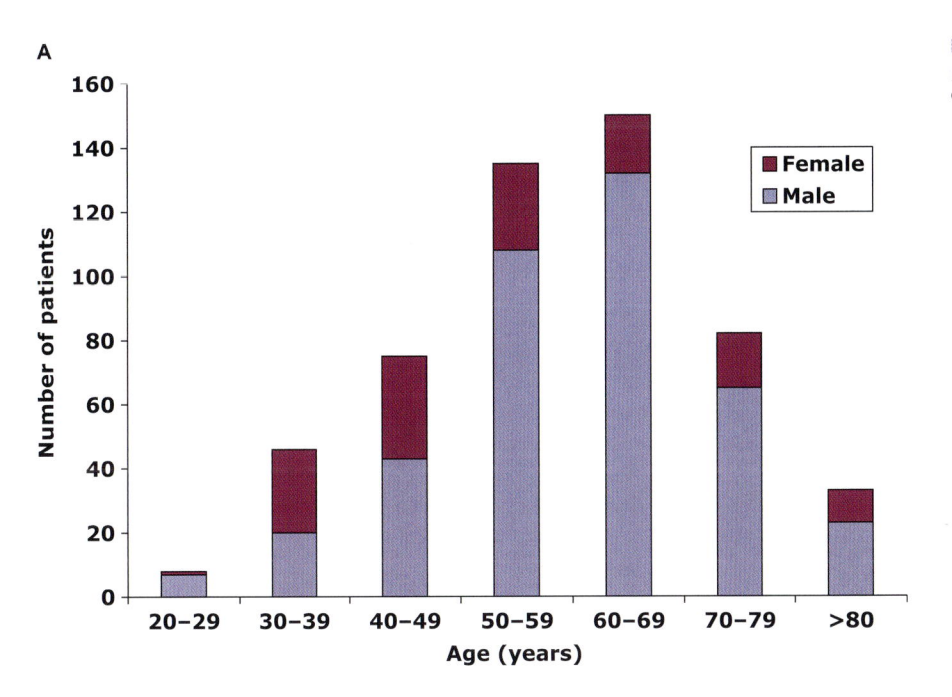

Figure 14.21 (**A**) Age and sex distribution of 534 pleomorphic cell lipomas. (**B**) Anatomic distribution of 537 pleomorphic lipomas.

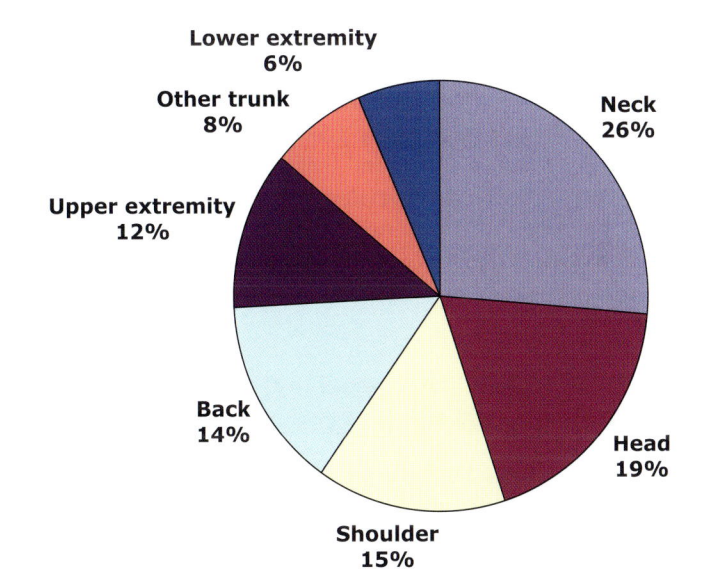

Pathology

Grossly, spindle cell lipoma forms an oval or discoid, yellowish to grayish-white mass, depending on the relative extent of the fatty and spindle cell components (Figure 14.22). The tumor has a firmer texture than that of ordinary lipoma.

Histologically, spindle cell lipoma is composed of mature fat and bland spindled mesenchymal cells that can present as small clusters between the fat cells or dominate the tumor. The spectrum of spindle cell lipomas varies from tumors that resemble ordinary lipomas, but have narrow streaks of spindle cells, to tumors that are mostly composed of spindle cells with just a few fat cells (Figure 14.23). Mast cells are often scattered between the spindle cells, and lymphocytes and plasma cells can occur, especially in the pleomorphic lipoma. Coarse "rope-like" collagen bands are common amid the cellular elements. Unusual features include nuclear palisading and perivascular hyalinization (Figure 14.24). Some spindle cell lipomas have myxoid stroma, which can be a dominant feature (Figure 14.25). Mitoses are exceptional. Some spindle cell lipomas contain cleavage spaces resembling vascular slits (Figure 14.26); this has been called *pseudoangiomatoid variant*.[71] The plexiform variant contains multiple, separate nodules in fibrous stroma.[72] Cytologically the spindle cells are uniform, have often pointed ends, and can appear slightly hyperchromatic (Figure 14.27).

Pleomorphic lipoma is a variant of spindle cell lipoma. In addition to the features mentioned here, this tumor contains multinucleated floret-like giant cells, so named because of their radially arranged nuclei, like the petals of flowers (Figure 14.28). Some of these tumors have prominent nuclear atypia with hyperchromatism and even occasional atypical mitoses. In such instances, the border between pleomorphic lipoma and atypical lipoma is arbitrary. Clinical features and behavior are similar to spindle cell lipoma.

Immunohistochemically, the spindle cells in both spindle cell and pleomorphic lipomas are strongly positive for CD34 (Figure 14.29), but negative for S100 protein and smooth muscle actin.[73–75] Some cases (10–20%) can be desmin positive, a finding that should not lead to a misdiagnosis of a smooth muscle tumor.[76] Immunohistochemical loss of retinoblastoma protein 1 (RB1) is characteristic of spindle cell and pleomorphic lipoma and is also observed in cellular angiofibroma and mammary myofibroblastoma.[77]

Differential diagnosis

Uniformity of spindle cells, mature collagen fibers, and the absence of lipoblasts separate spindle cell and pleomorphic lipoma from liposarcoma. The subcutaneous location also differs from that of liposarcoma.

Genetics

Chromosome 16q losses with partial monosomy are typical of spindle cell and pleomorphic lipoma and differ from the changes seen in other lipomas. The involved genes have not been specifically identified. Recurrent involvement of chromosome 13q has also been reported; however, spindle cell lipomas characteristically lack the 12q13–15 alterations that are common in conventional and atypical lipomas.[78,79] Atypical spindle cell lipomas are genetically more related to ordinary spindle cell lipomas, and they lack MDM2 amplification typical of atypical lipomatous tumors, but may instead have polysomy of chromosome 12 loci, which results in increased MDM2 (as well as centromeric control) signals.[80]

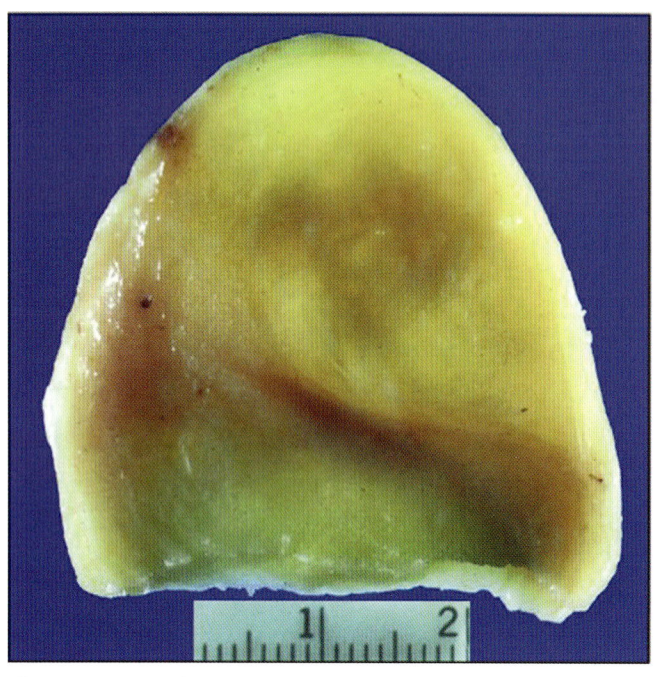

Figure 14.22 Grossly, a spindle cell lipoma is often paler than an ordinary lipoma and contains varying amounts of gray-white fibrous components.

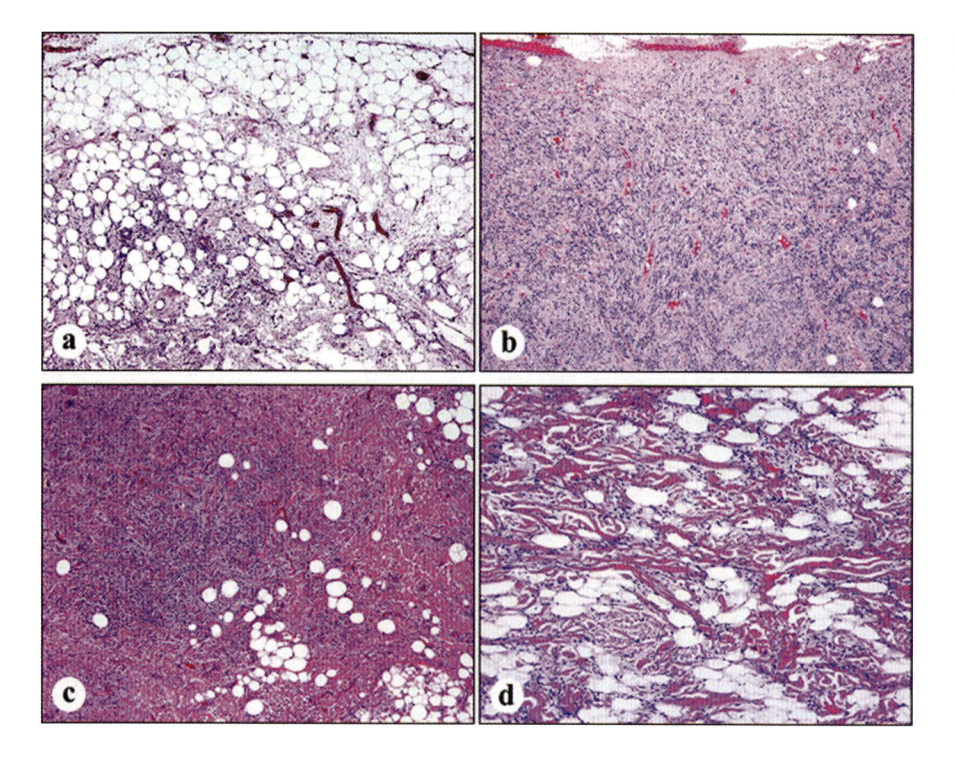

Figure 14.23 Spindle cell lipomas can contain mature fat, spindle cells, and coarse collagen fibers in varying proportions.

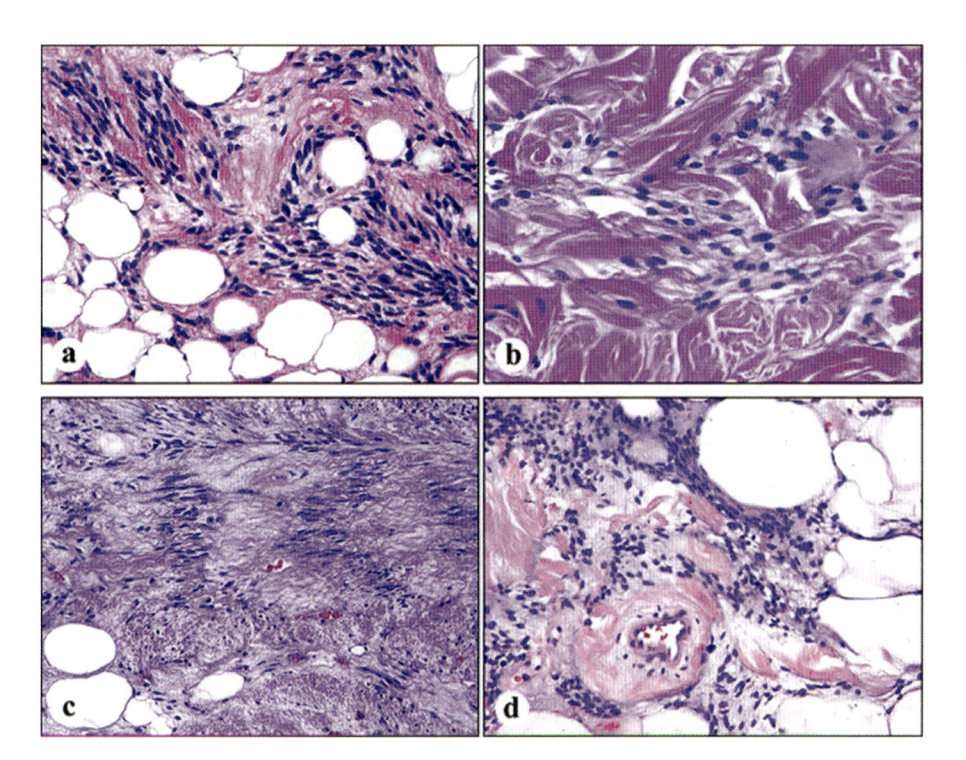

Figure 14.24 (**a,b**) In spindle cell lipoma, the uniform spindle cells can be admixed with fat or lie in a dense collagenous matrix. (**c**) Distinct nuclear palisading is an uncommon finding. (**d**) Perivascular hyalinization is sometimes present.

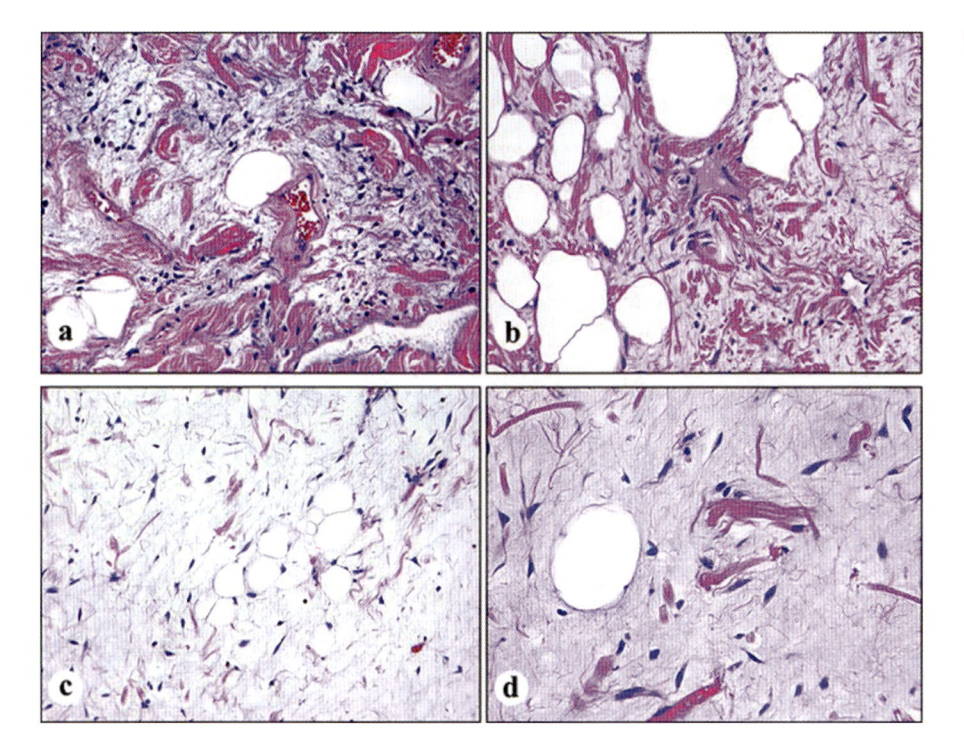

Figure 14.25 Spindle cell lipoma with prominent myxoid matrix.

Sclerotic lipoma

This unusual fibroma-like lipoma usually forms a small nodule in one of the fingers, or occasionally in the hands, toes, or scalp. Sclerotic lipomas are slightly more common in men and occur in a wide age range, with a median age around 40 years. The tumor measures from a few millimeters to 2 cm and has no tendency to recur after a simple excision.[81,82]

Grossly, sclerotic lipomas are egg-shaped rubbery masses with gray-white to pink surface on sectioning. Histologic

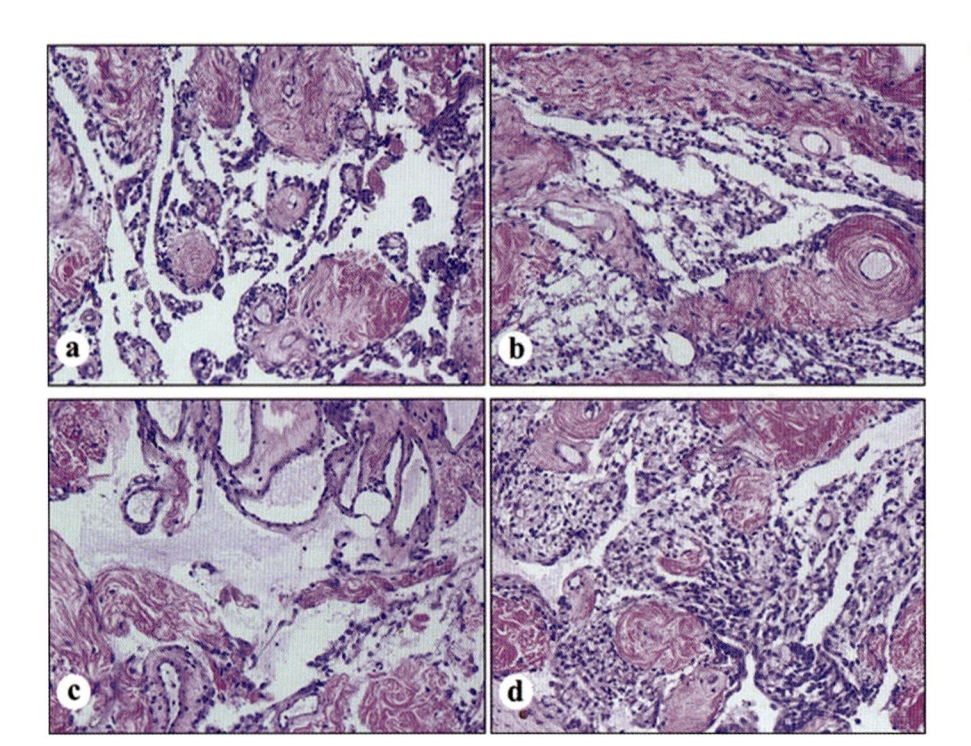

Figure 14.26 Pseudoangiomatous pattern in spindle cell lipoma is created by stromal mucinous degeneration.

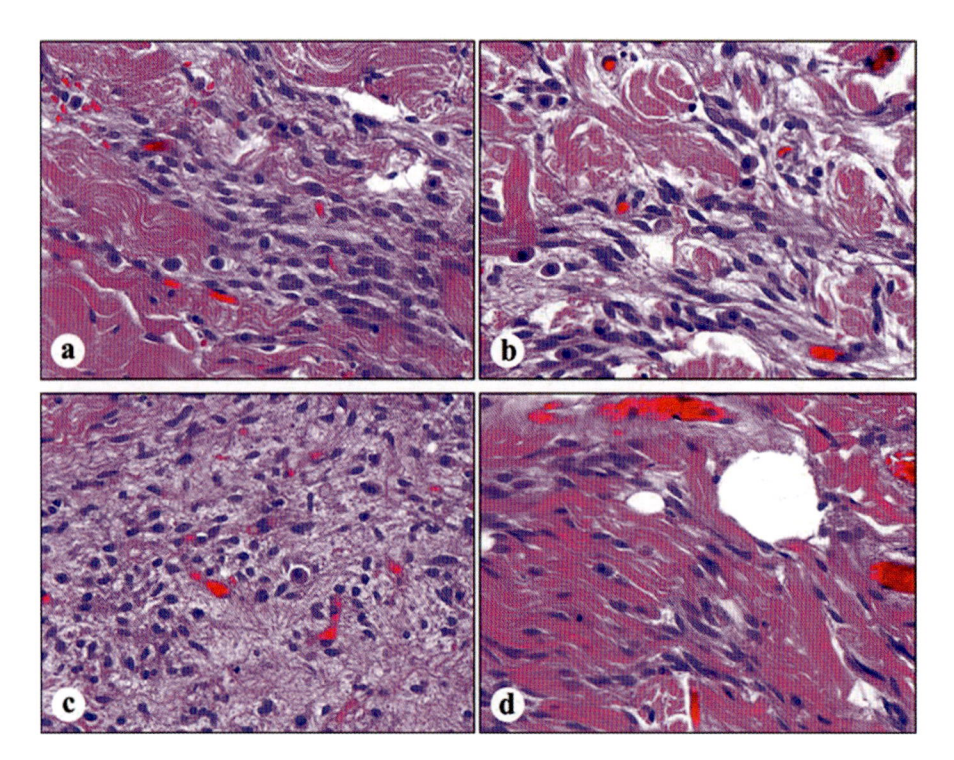

Figure 14.27 On cytological analysis, the spindle cell lipoma cells are uniform with pointed ends; scattered mast cells are often present.

analysis shows a dominant fibrous or fibromyxoid matrix that contains scattered mature fat cells (Figure 14.30). The fibrous component can show a lamellar pattern with cleavages, resembling the features of sclerotic fibroma. No evidence for association with Cowden syndrome was obtained in one of the published series, as would be expected for true sclerotic fibromas.[82] The fatty component varies from scattered cells to nearly one-half of the mass. The spindle cells are often focally positive for CD34 and S100 protein.

Hibernoma

A *hibernoma* is a rare tumor that mimics the differentiation of brown fat, which is normally involved in metabolic

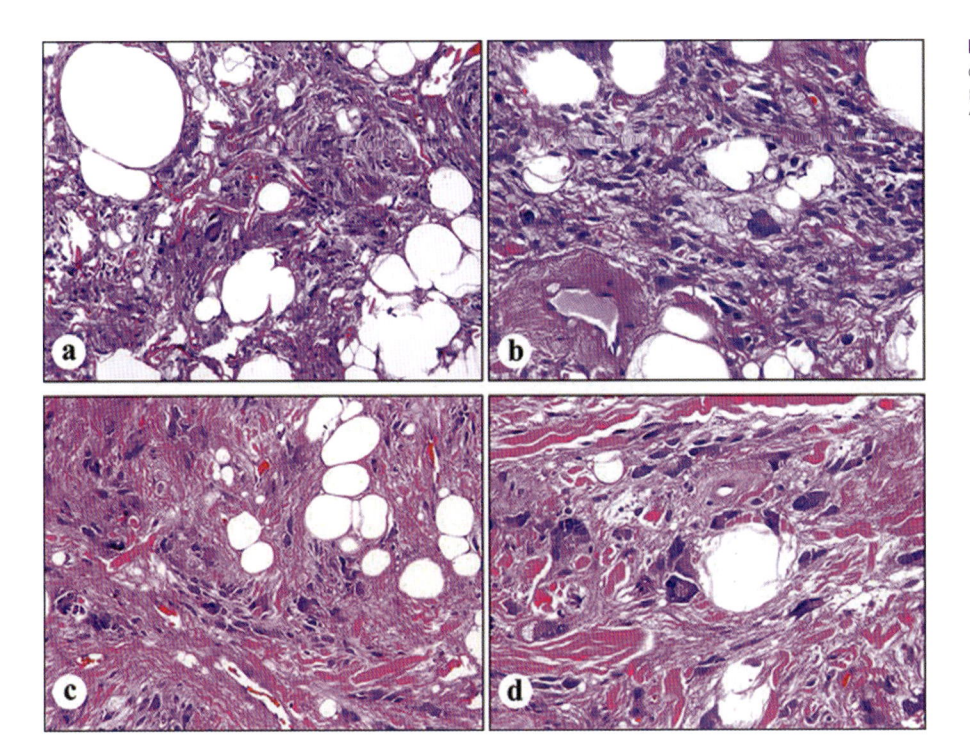

Chapter 14: Lipoma variants and conditions simulating lipomatous tumors

Figure 14.28 Pleomorphic lipoma is a spindle cell lipoma variant with multinucleated giant cells, the nuclei of which are often arranged in a radial, "floral" pattern.

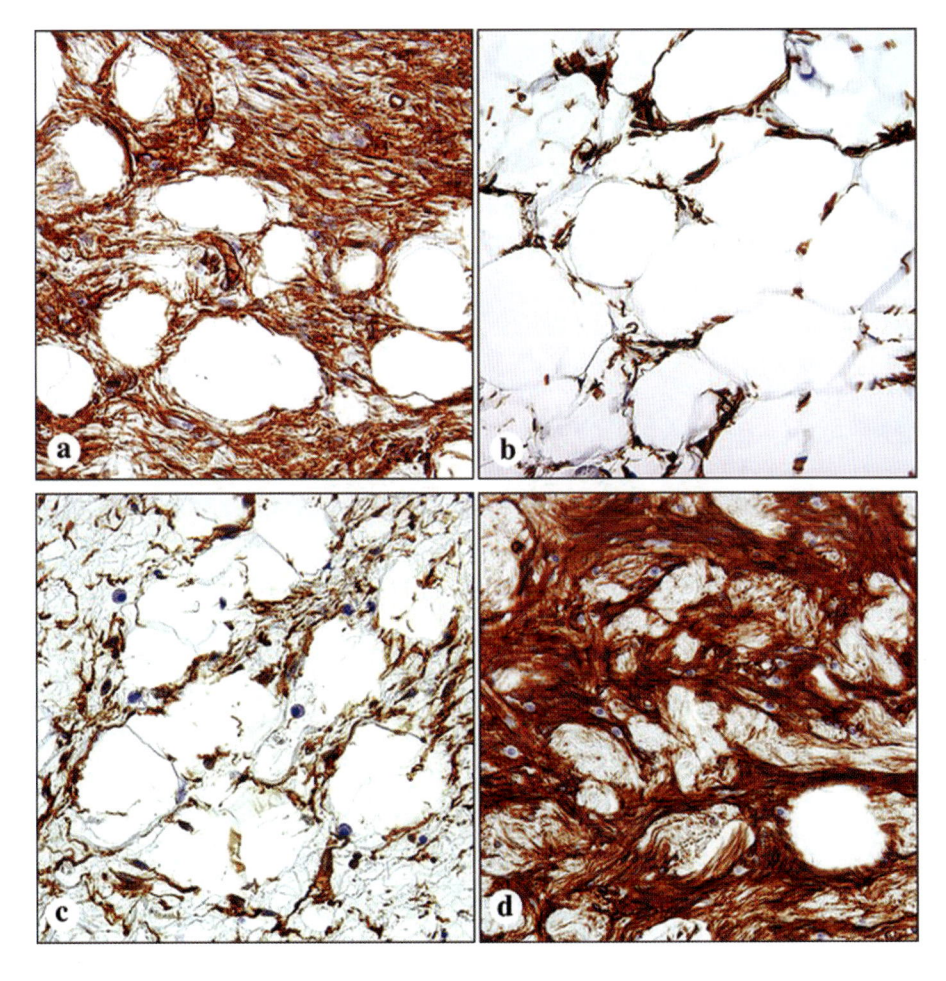

Figure 14.29 Immunohistochemical analysis shows spindle cell lipoma to be positive for CD34. Various constellations of fatty versus spindle cell components create different staining patterns.

397

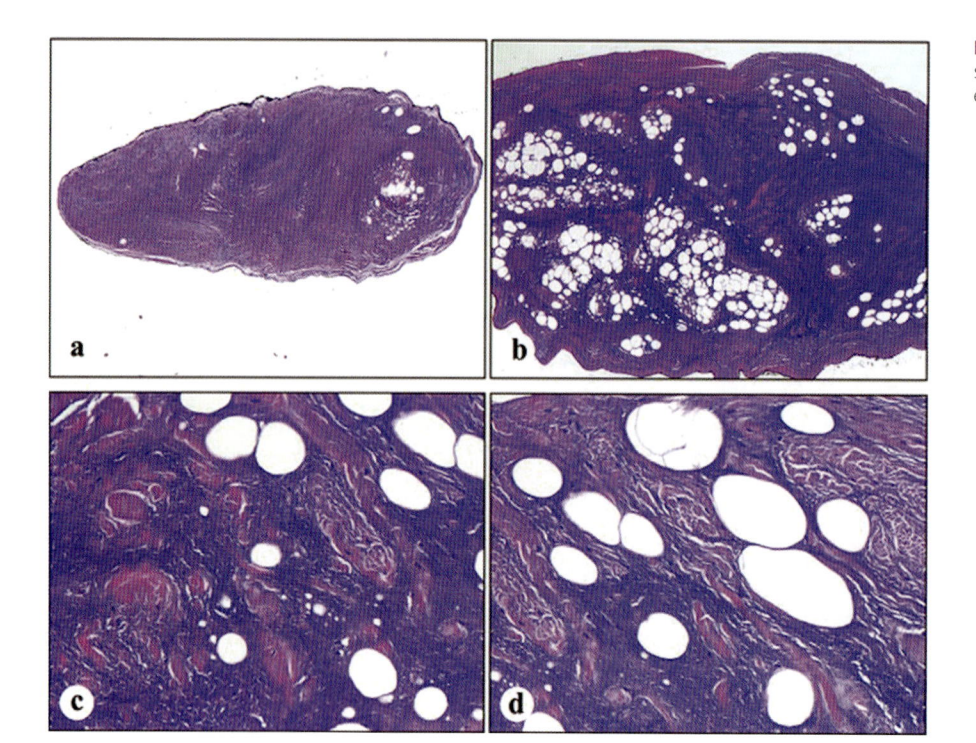

Chapter 14: Lipoma variants and conditions simulating lipomatous tumors

Figure 14.30 Sclerotic lipoma of the finger is a small, demarcated nodule with mature fatty elements in fibrosclerosing or myxoid stroma.

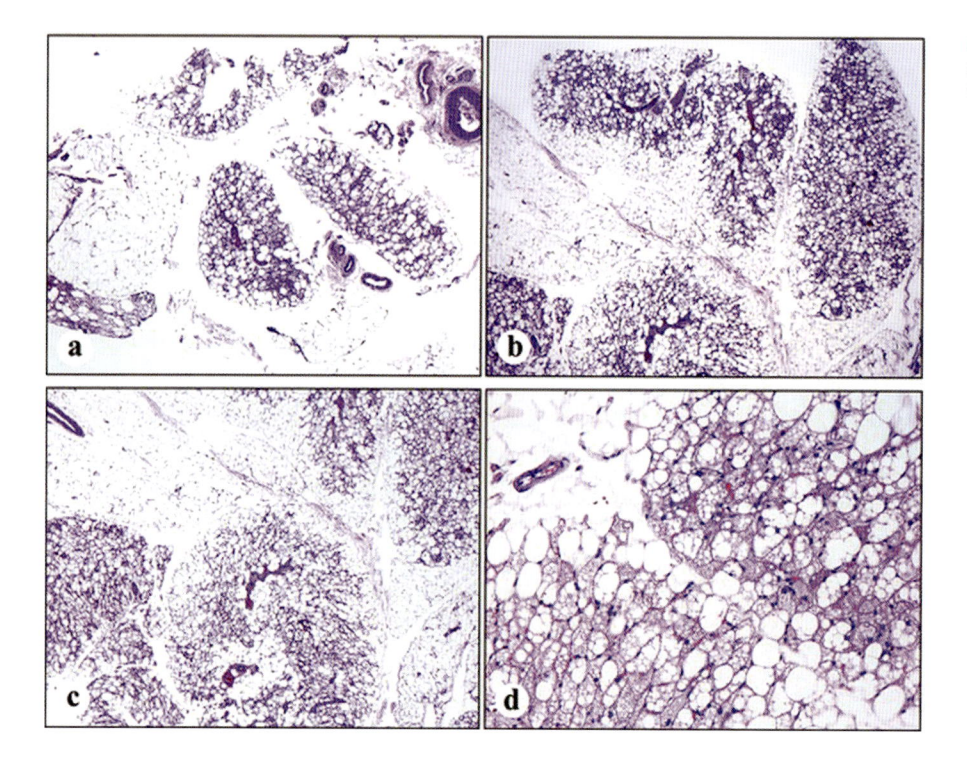

Figure 14.31 Histologically, brown fat is often seen as small nodules amid white fat, especially in the neck and axilla.

thermogenesis in hibernating animals by regulating the uncoupling of mitochondrial oxidative phosphorylation. Brown fat is normally seen in the posterior neck and axilla, for example, around the cervical and axillary lymph nodes (Figure 14.31). It is more commonly encountered in children and young adults.[83,84]

Clinical features

Based on AFIP data on 170 cases, hibernoma predominantly occurs in young adults: 61% in the third and fourth decades (Figure 14.32A). The youngest patients have been children 13 to 15 years old (5%). Rarely, hibernomas are seen in persons

Figure 14.32 (**A**) Age and sex distribution of 281 patients with hibernomas. (**B**) Site distribution of 257 hibernomas.

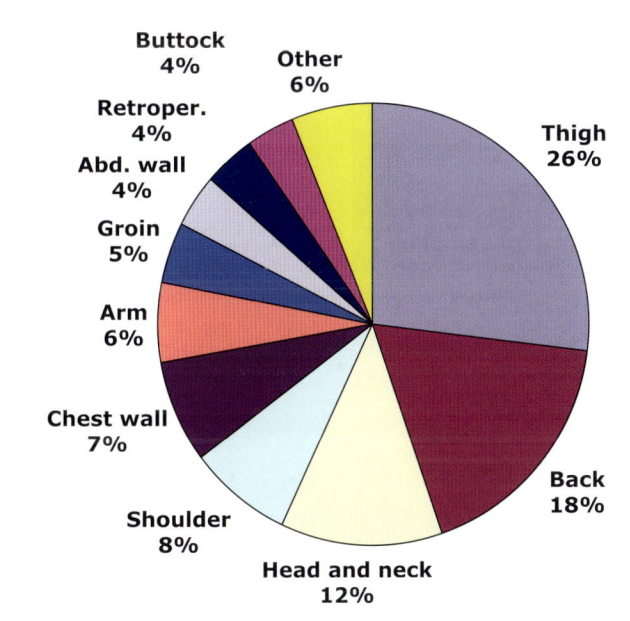

older than 60 years (7%). These tumors most commonly occur in the upper trunk (40% from neck to chest wall).[85] One-third of them present in the deep thigh and inguinal region (Figure 14.32B). This tumor can be subcutaneous or intramuscular. Its behavior does not differ from that of ordinary lipoma, and recurrences are rare.[83–85]

Pathology

Grossly, hibernomas that show brown fat cells amid white fat cannot be distinguished from ordinary lipomas. Those tumors that have extensive brown fat differentiation are yellow-brown to brown. Otherwise they resemble lipomas and usually measure between 3 cm and 10 cm in diameter. These tumors usually appear well demarcated (Figure 14.33).

Histologically, hibernomas vary, being composed of multivacuolated eosinophilic or pale cells (typical variant, >80%), and some tumors have a dominant white fat component (lipoma-like variant, 10% of all hibernomas). Rare variants include myxoid and spindle cell hibernomas (<10% of all cases).

Hibernomas often show lobulation by thin fibrous streaks or mildly dilated medium-sized vessels that can be quite prominent in this tumor (Figure 14.34). The eosinophilic cell subtype of the typical variant showing variably eosinophilic, multivacuolated cytoplasm in >50% of tumor cells is not the

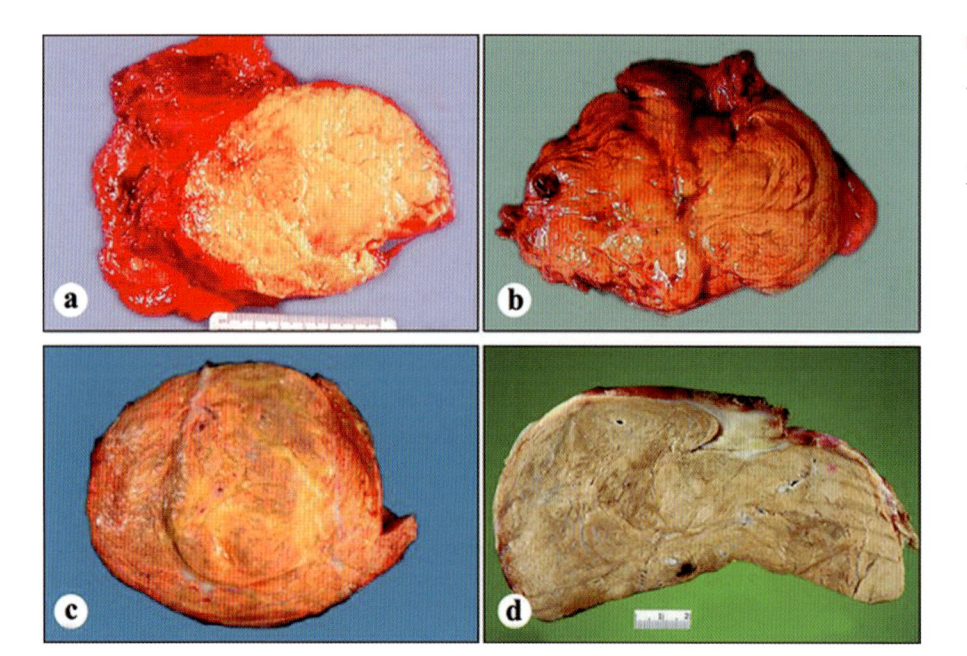

Chapter 14: Lipoma variants and conditions simulating lipomatous tumors

Figure 14.33 The gross appearance of hibernoma varies. (**a**) An example dominated by white fat does not differ from ordinary lipoma. (**b–d**) Examples rich in hibernoma cells have a brownish appearance, although this is more prominent in fresh unfixed specimens (**b**) than those fixed in formalin (**c,d**).

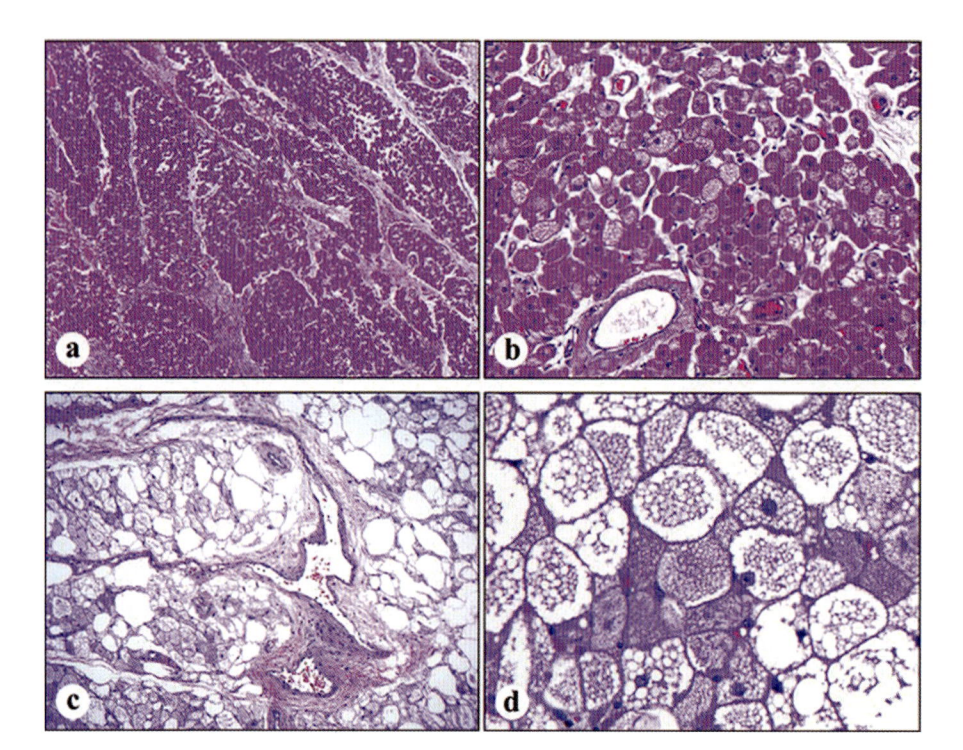

Figure 14.34 (**a–b**) Hibernomas composed of eosinophilic cells. (**c**) Prominent blood vessels are a typical finding in this tumor. (**d**) Cytologically, the hibernoma cells often contain a centrally placed nucleus with a prominent nucleolus.

most common (only 15% of all hibernomas), but it is the most distinctive and diagnostically unmistakable subtype (Figure 14.34). This subtype occurs more commonly in the upper extremities and upper body.

The pale cell subtype of the typical variant is the most common (>40%), showing multivacuolated hibernoma cells with pale staining cytoplasm (Figure 14.35). This subtype has a predilection for the thigh.

White fat is commonly present as a component (Figure 14.35), and in rare cases (lipoma-like subtype), hibernoma cells are seen among a dominant white fat component. A myxoid variant with prominent myxoid stroma does occur, but is rare (Figure 14.36a,b).[85,86] Another rare variant contains a spindle cell component not unlike the one seen in spindle cell lipoma (Figure 14.36c,d); this component is also CD34 positive.[85]

Cytologically, hibernoma cells are multivacuolated and have small, often centrally placed, nuclei; they also can have prominent nucleoli (Figures 14.34d, 14.35c,d, and 14.36b). Mitoses are exceptional.

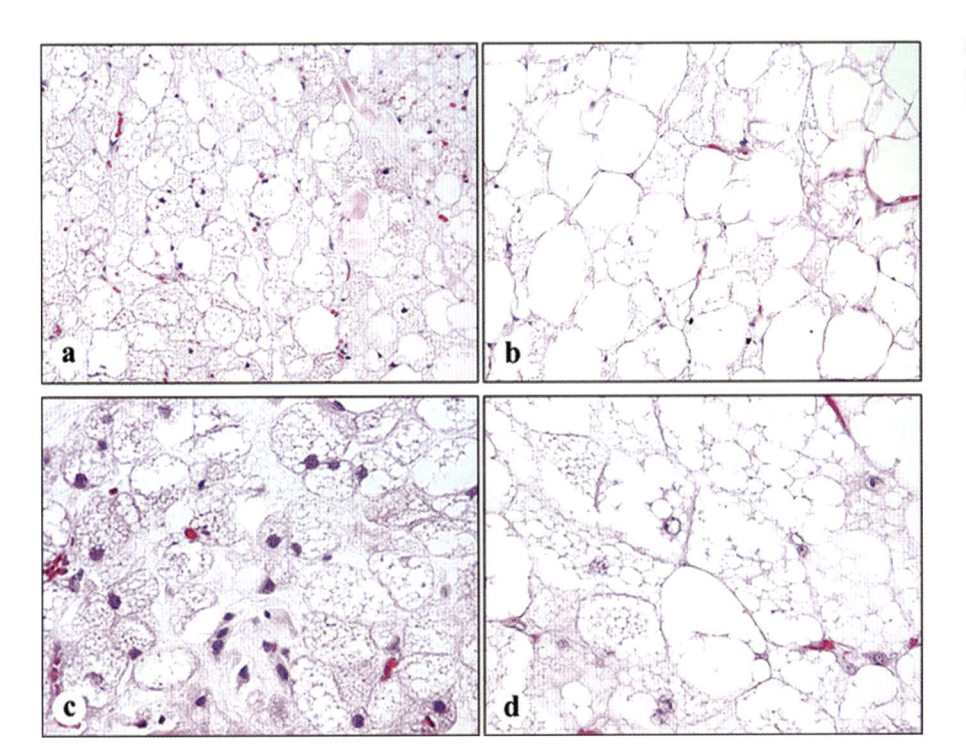

Chapter 14: Lipoma variants and conditions simulating lipomatous tumors

Figure 14.35 Examples of hibernoma composed of pale cells. Univacuolar white fat cells are also present in (**a**) and (**b**).

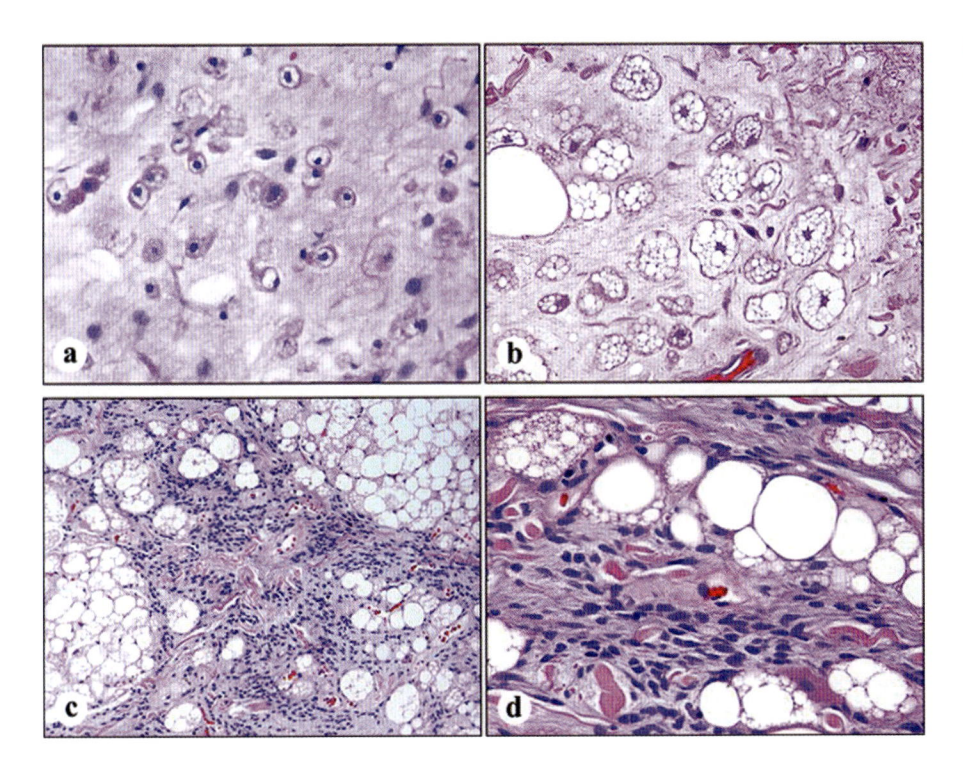

Figure 14.36 (**a,b**) Hibernoma with prominent stromal myxoid change. Note the multivacuolated hibernoma cells with centrally placed nuclei. (**c,d**) A spindle cell component is a rare finding in hibernoma, and this variant shows a resemblance to spindle cell lipoma.

Immunohistochemically, hibernoma cells are variably, but often strongly S100 protein positive. Expression of uncoupling protein, typical of brown fat, has been shown in hibernoma cells by immunohistochemistry.[87]

Differential diagnosis

Myxoid liposarcoma and rarely well-differentiated liposarcoma can contain hibernoma-like multivacuolated cells.

Attention should be paid to the diagnostic features of these liposarcoma types, especially the presence of an atypical adipocytic component with enlarged and hyperchromatic adipocyte nuclei.

Genetics

Rearrangement of 11q13, occurring in somewhat heterogeneous, sometimes complex translocations involving multiple chromosomes, appears to be the most common recurrent cytogenetic change.[88–91] Those cytogenetic changes are associated with deletions of multiple endocrine neoplasia I (MEN1) and AIP genes (the latter encoding aryl hydrocarbon receptor interacting protein).[92]

Chondroid lipoma
Clinical features

This uncommon lipoma subtype, described by Meis and Enzinger,[93] is a peculiar, benign tumor that combines cartilage-like myxoid and lipomatous elements. Prior to its description, chondroid lipomas were sometimes classified as myxoid or round cell liposarcomas. The tumor chiefly presents in young adults, with a female predominance (80%). The most common locations are the shoulder, arm, and thigh.[93,94] It has been also reported in the oral cavity.[69] The tumor can be subcutaneous or intramuscular. The lesion size averages from 3 cm to 4 cm, but it can exceed 10 cm (largest reported case, 11 cm).[93] The behavior is benign, but the author has seen occasional local recurrences.

Pathology

Grossly, the tumor is typically well-circumscribed, golden yellow, and slightly firmer than lipoma, often with visible lobulation and gray-white fibrous septa (Figure 14.37). Microscopically it consists of foci or intermingled white fat cells and distinctive, chondroid lipoma cells in nests, sheets, or cords. These cells are relatively small, often with multivacuolated, bubbly cytoplasm resembling hibernoma cells, and they can sit in a lacunar space surrounded by matrix varying from basophilic myxoid to fibrin-like or deeply eosinophilic, showing some resemblance to cartilage (Figure 14.38).

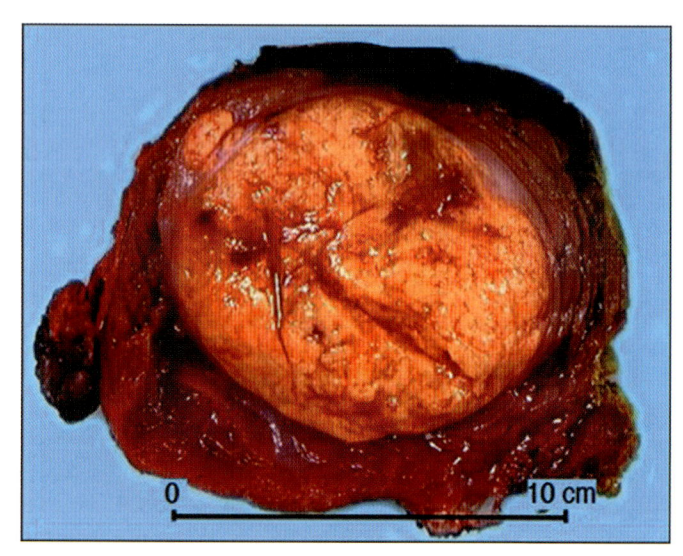

Figure 14.37 This large chondroid lipoma forms a well-demarcated intramuscular mass with a golden yellow-brownish cut surface. (Courtesy of Dr. Uma Rao, Pittsburgh, PA.)

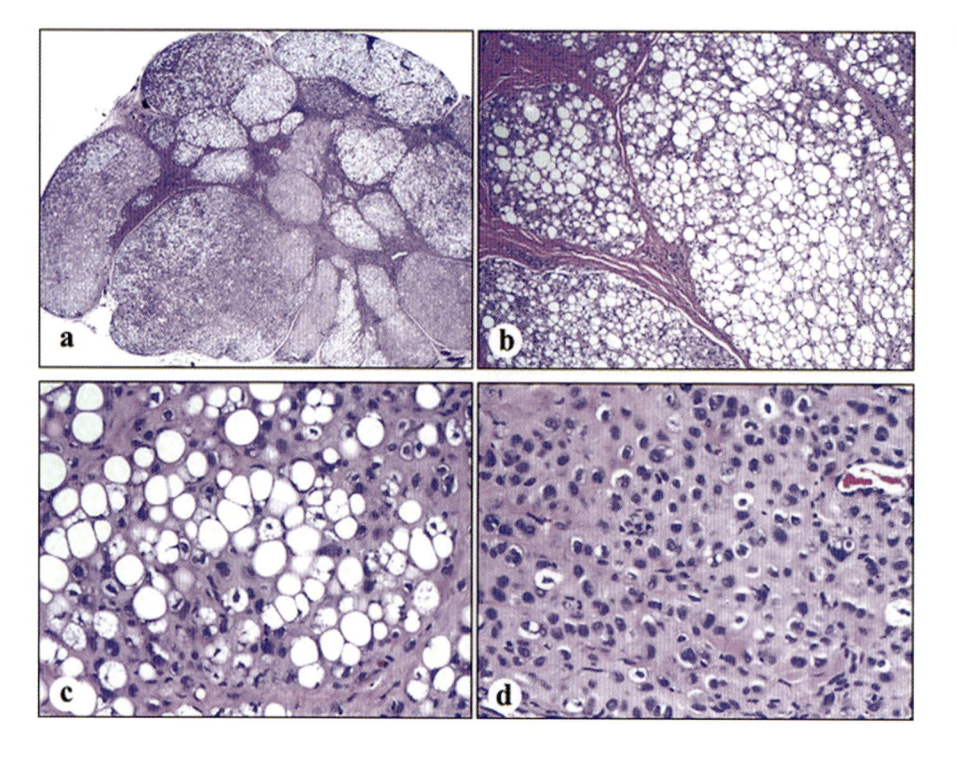

Figure 14.38 (**a**) Chondroid lipoma reveals microscopic lobulation at a low magnification. (**b**) The lobules can contain a significant amount of white fat. (**c,d**) A distinctive feature is the presence of multivacuolated cells resembling those in hibernoma and eosinophilic fibrous matrix.

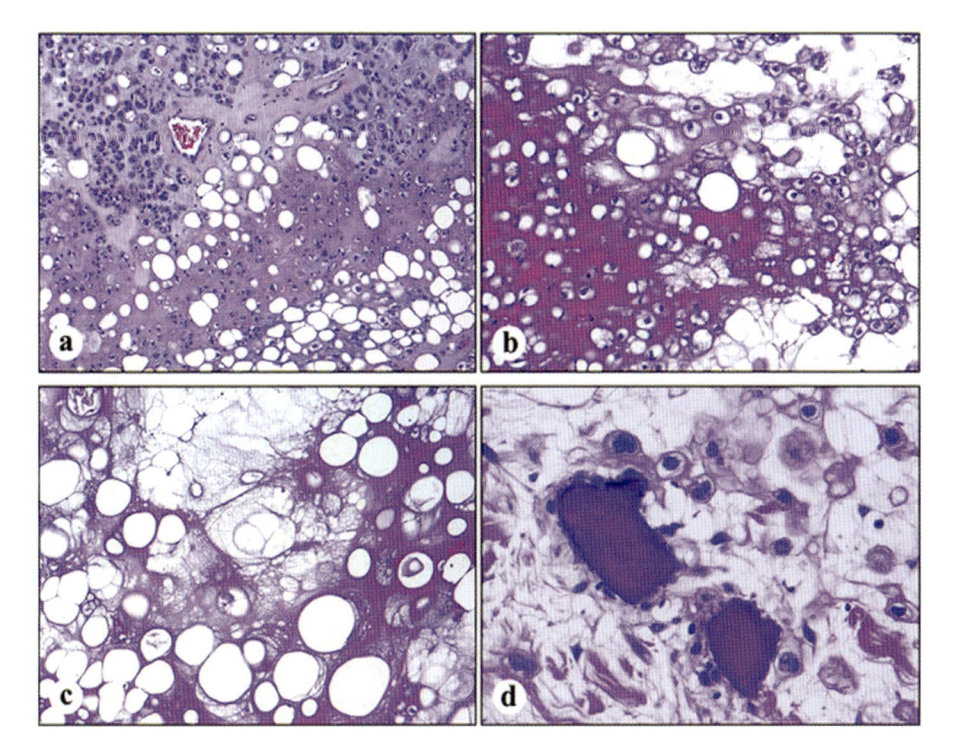

Figure 14.39 (**a**) Chondroid lipoma matrix varies from fibrous, fibrin-like, and myxoid. (**b**) This example shows prominent fibrinoid matrix. (**c**) Necrotic areas devoid of viable cells can be present. (**d**) Rarely, chondroid lipoma contains dystrophic calcifications.

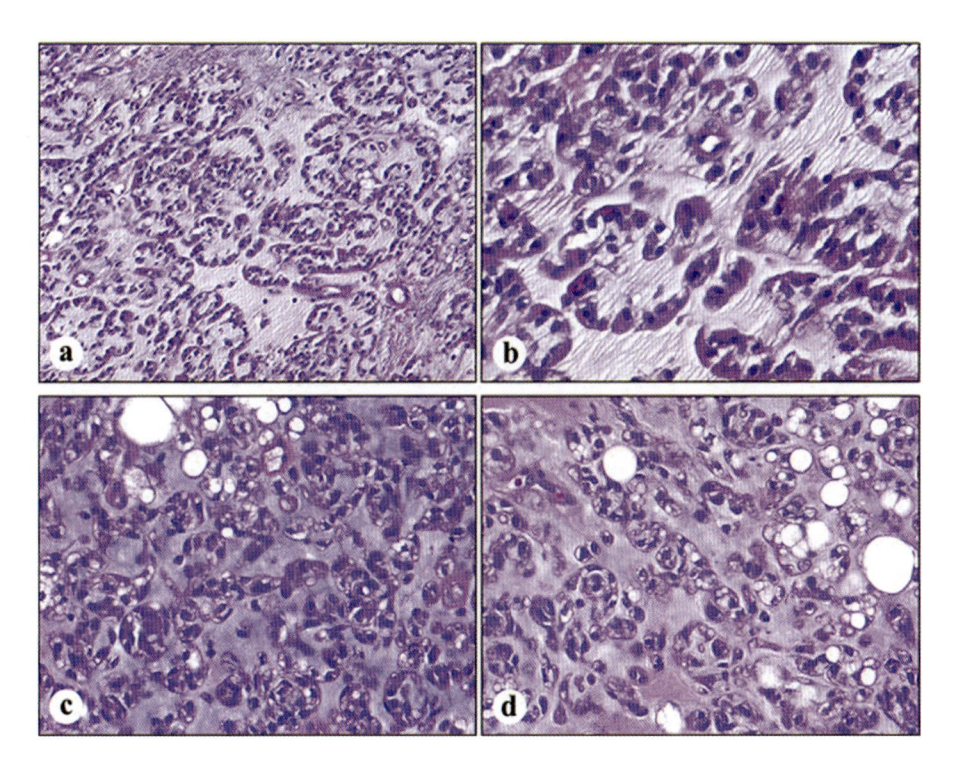

Figure 14.40 Chondroid lipoma with prominent myxoid matrix. (**a,b**) The cells form trabecular arrangements resembling those of extraskeletal myxoid chondrosarcoma. (**c,d**) Cluster and trabecular in myxoid matrix. Typical multivacuolated chondroid lipoma cells are also evident.

Well-differentiated cartilage is not present, however. Chondroid lipoma cells typically have complex nuclear outlines, including curved, C-shaped forms. Mitotic activity is difficult to detect if found at all. A focal to extensive deeply eosinophilic fibrinous matrix is often present, and fat necrosis and focal calcification are occasionally present (Figure 14.39). Some tumors have a focal or widespread corded pattern resembling the appearance of extraskeletal myxoid chondrosarcoma (Figure 14.40).

The tumor cells contain glycogen and are periodic acid–Schiff (PAS) positive. The myxoid matrix is alcian blue positive at an acidic pH similar to that of a chondroid matrix. Immunohistochemically the tumor cells are variably positive for S100 protein and CD68, and they can also be focally keratin cocktail positive. Collagen IV immunoreactivity around tumor cells reflects the presence of basement membranes.

Electron microscopic findings typically include prominent pinocytic vesicles and numerous cytoplasmic lipid vacuoles, interpreted to support white fat differentiation.[78]

Differential diagnosis

Lipomas with sharply demarcated foci of cartilaginous differentiation in the middle of ordinary fat are not chondroid lipomas, but are classified as *lipomas with chondroid metaplasia (chondrolipoma)*. Some atypical lipomatous tumors with enlarged and hyperchromatic nuclei have chondroid lipoma-like elements. Although they can represent variants of chondroid lipoma, it is important to recognize the atypia that could possibly signify higher biologic potential. Mixed tumors with a fatty component should not be confused with chondroid lipoma. Glandular epithelial differentiation is generally present, and expression of epithelial-myoepithelial markers (i.e., keratins, GFAP) is helpful.

Genetics

Balanced translocation t(11;16)(q13;p12–13) has been described in three cases, suggesting that this aberration is characteristic of chondroid lipoma. The breakpoint in chromosome 11 is similar to that of hibernoma, whereas the breakpoint in chromosome 16 differs from that of myxoid liposarcoma.[95–97] The reported cytogenetic changes correspond to C11orf95-MKL2 gene fusion.[98]

Myolipoma

Myolipoma is the designation for an uncommon, fat-dominated tumor containing an interspersed, usually minor smooth muscle element.[99] Myolipomas present in adult patients, with female predominance, and occur in a variety of locations including the axilla, other peripheral soft tissues, and retroperitoneum.

Grossly, the periphery of the tumor is fatty. Histologically, myolipomas contain mature fatty elements admixed with the foci of mature smooth muscle cells. The smooth muscle component is positive for smooth muscle actin and desmin, and often also for estrogen and progesterone receptors.[100]

Tumors with predominant smooth muscle elements with focal fat are better designated as *lipoleiomyomas*. Many myolipomas reported in the uterus and retroperitoneum might be more appropriately classified as lipoleiomyomas.

Hemosiderotic fibrolipomatous tumor

This condition, originally reported by Marshall-Taylor and Fanburg-Smith,[101] occurs mainly in middle-aged patients, with a female predominance, but it has also been reported in children. Most examples have occurred in the dorsal aspect of the foot/ankle, and a few in the hand and cheek.

Grossly, this tumor resembles a lipoma, but contains foci of yellow-brown hemosiderin pigment. Tumor size varies in a broad range from 1 cm to >10 cm. Histologically it is characterized by fat cells with minimal size variation and no atypia, and quilt-like, periseptal, periadipocytic, and perivascular foci of spindle cells, histiocytes, mast cells, and coarsely granulated iron pigment (Figure 14.41). The spindle cells resemble those of spindle cell lipoma and are CD34 positive; calponin

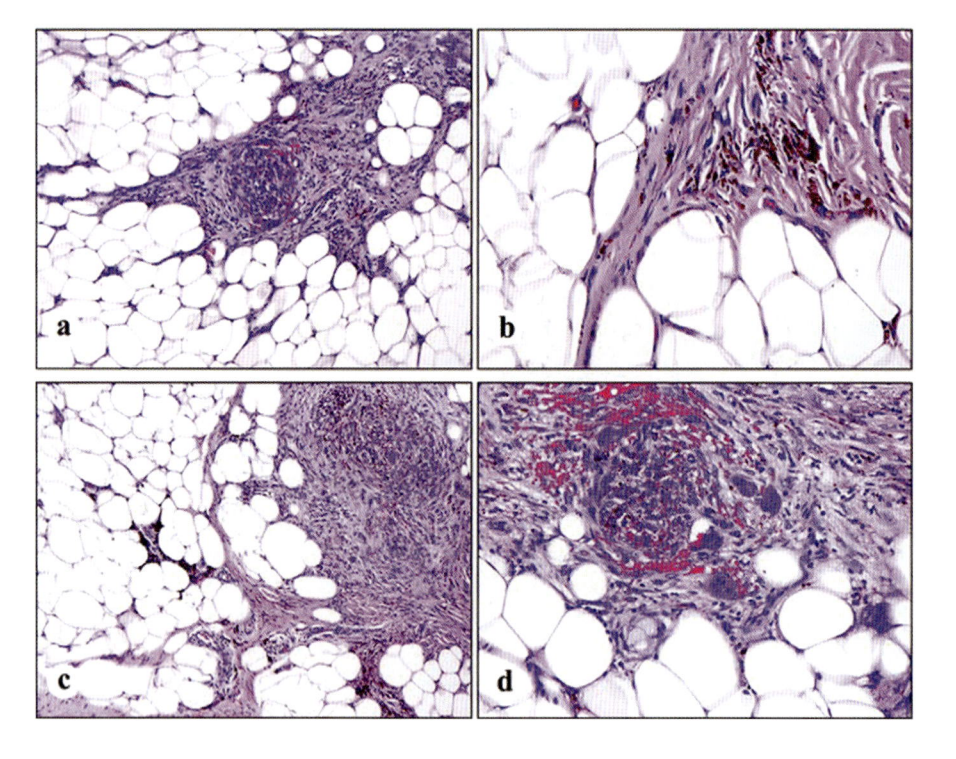

Figure 14.41 Fibrohistiocytic lipoma contains streaks of spindle cells in a fibrous stroma with frequent hemosiderin deposits; scattered osteoclast-like giant cells are also present.

Figure 14.42 (**A**) Age distribution of lipoblastoma. (**B**) Site distribution of lipoblastoma.

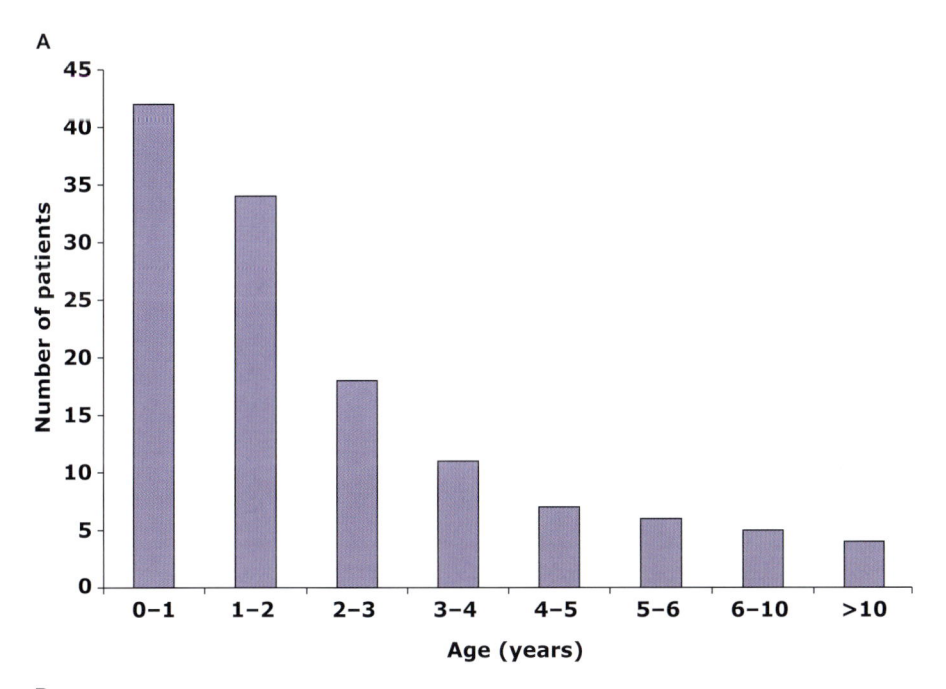

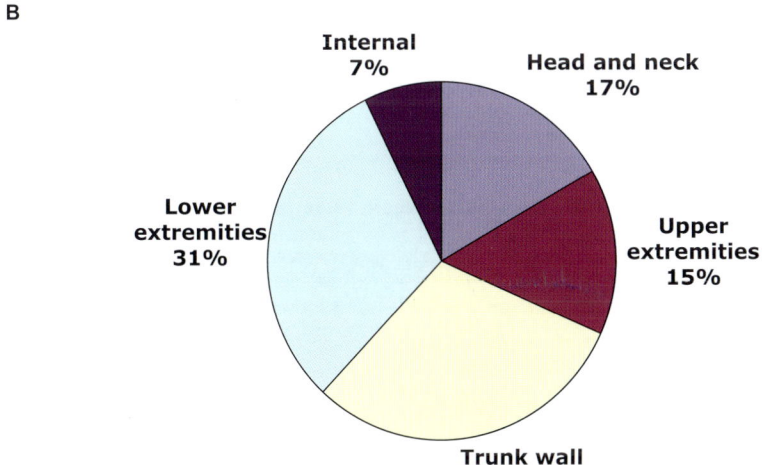

positivity suggesting myofibroblastic differentiation has also been reported. Local recurrence is common, but long-term follow-up in the initial series did not identify further morbidity. The spindle cell component is negative for desmin, smooth muscle actin, and S100 protein.[101,102] However, transformation into a high-grade rapidly metastasizing sarcoma fatal in 2 years from onset was reported in one case.[103]

There is some overlap with myxoinflammatory fibroblastic sarcoma histologically, with some cases showing features of both. Indeed, these tumors may share the same genetic change, the translocation t(1;10) with TGFBR3 and MGEA5 gene rearrangements.[104–106]

Lipoblastoma

The first English language description of lipoblastoma was by Vellios *et al.* in 1958 under the name "lipoblastomatosis."[107] Currently, lipoblastoma is a designation for specific, rare, grossly well-defined benign fatty tumors occurring in young children. The term *lipoblastomatosis* has been used for similar, diffusely infiltrative lesions.

Clinical features

Lipoblastoma is a tumor of childhood, and 75% of patients are under 3 years of age (Figure 14.42A). There is a nearly 2:1 male predominance in most published series. It typically presents in the extremities, trunk, head and neck, and sometimes in body cavities, specifically the retroperitoneum and mediastinum (Figure 14.42B). Most examples are subcutaneous, but intramuscular locations are also possible. The tumor is benign but can recur locally, especially if diffuse. Because, otherwise, long-term prognosis is excellent, complete but conservative excision is advocated.[107–113]

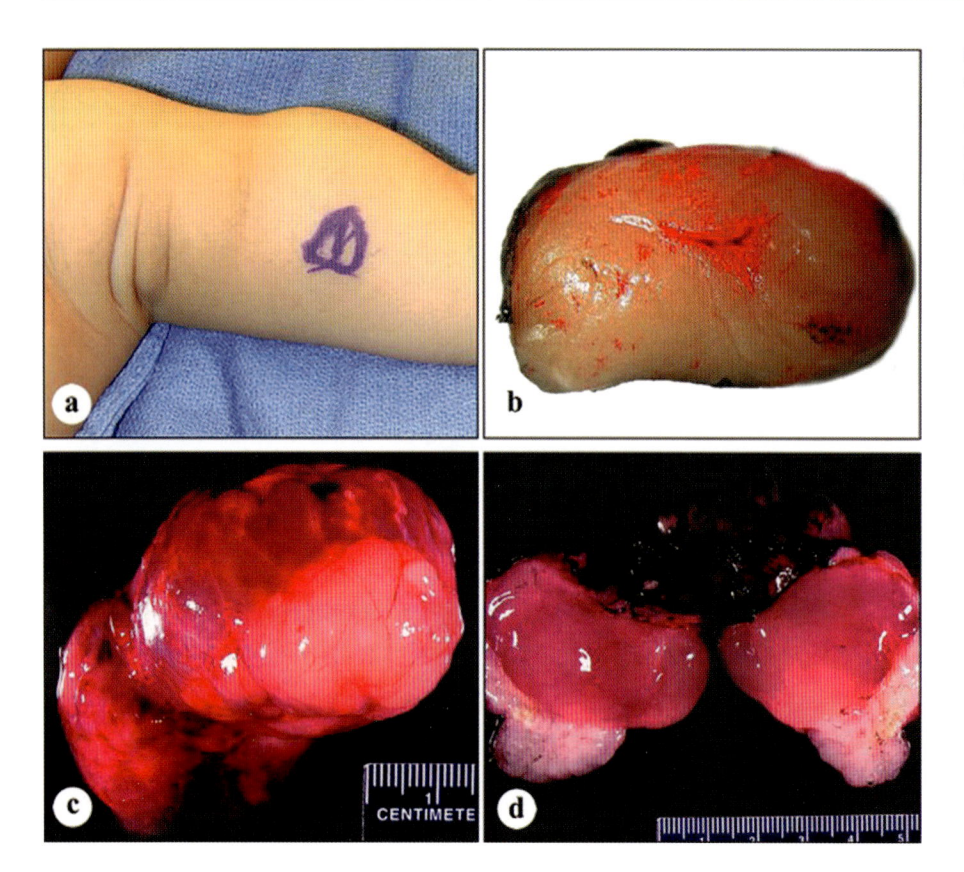

Chapter 14: Lipoma variants and conditions simulating lipomatous tumors

Figure 14.43 (**a**) Lipoblastoma in the proximal arm of a 1-year-old. (**b**) A retroperitoneal lipoblastoma is a pale yellowish mass. (**c,d**) Lipoblastoma of the neck forms a demarcated mass that has a gelatinous, pinkish-tan surface on sectioning.

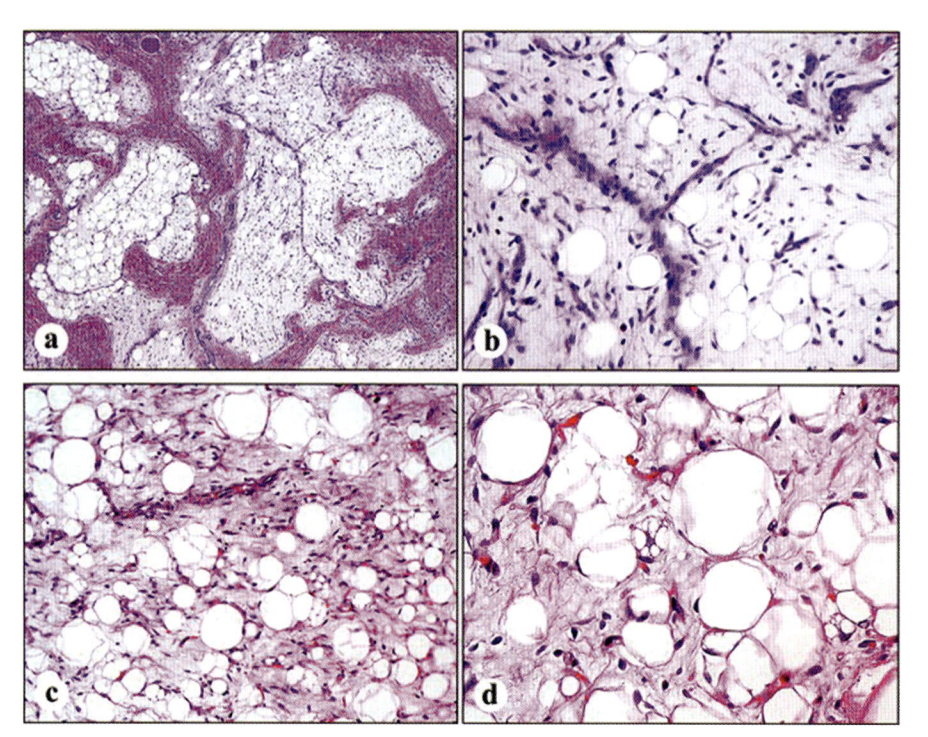

Figure 14.44 (**a**) Lipoblastoma is composed of multiple lobules separated by variably developed fibrous septa. (**b,c**) Prominent capillaries and focally myxoid matrix are typical features. (**d**) Multivacuolated cells resembling those in hibernoma are sometimes present.

Pathology

Grossly, lipoblastoma is yellow or yellowish gray, often lobulated, and shows a lipoma-like or myxoid appearance on sectioning (Figure 14.43). Most examples measure <5 cm, but a lipoblastoma >20 cm has been reported in the retroperitoneum.[112]

Histologically, lipoblastoma and lipoblastomatosis are typically lobulated by connective tissue septa (Figure 14.44).

406

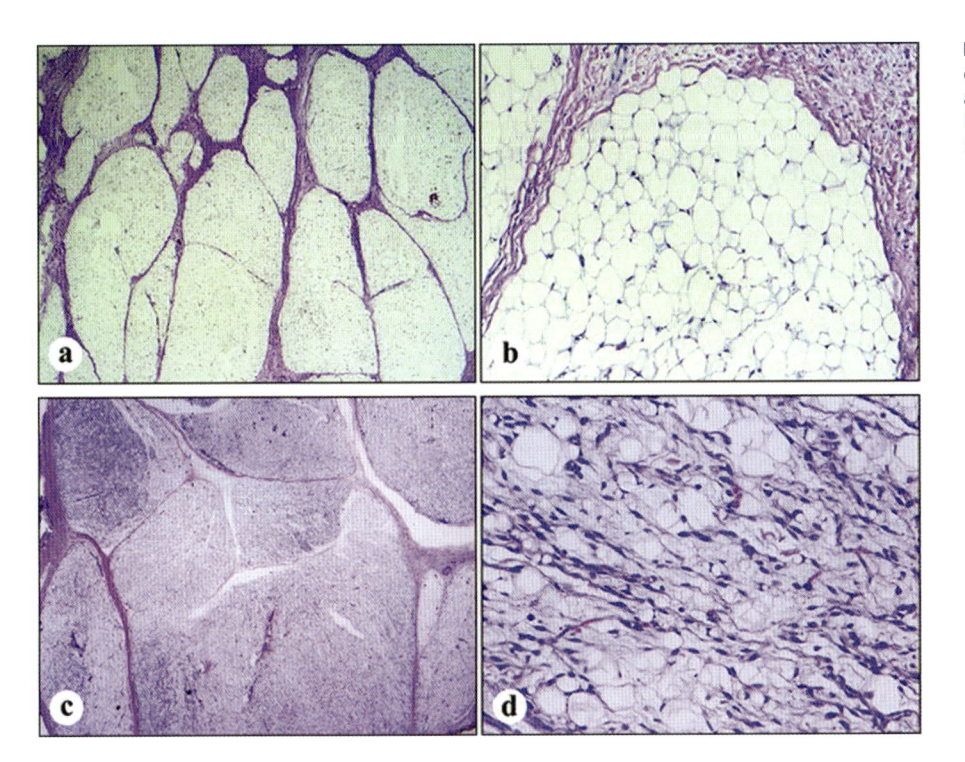

Figure 14.45 (**a,b**) Variant of lipoblastoma dominated by mature fat. Note the fibrous septa and focal myxoid stroma. (**c,d**) Lobulated lipoblastoma with a spindle cell component and a myxoid stroma.

The lobules are composed of highly vascular fat with variably myxoid stroma resembling embryonic fat. Foci of multivacuolated, hibernoma-like cells can occur. Some lipoblastomas are composed of nearly mature fat with focal myxoid change, resembling fibrolipoma (Figure 14.45a,b), and others have a dominant spindle cell component (Figure 14.45c,d), and this can be desmin positive.[113] Mitotic activity is inconspicuous. Lipoma-like features have also been reported in recurrent lipoblastoma, suggesting that the lesions undergo maturation over time.

Distinction from myxoid liposarcoma is aided by the lobular and more organized nature, lack of perivascular hypercellularity, and more commonly subcutaneous location of lipoblastoma. Lipofibromatosis contains cellular fibrous septa with fibromatosis-like appearance.

Genetics

Genetic changes described in lipoblastoma typically include structural rearrangements, especially translocations and inversions involving the region of chromosome 8q11–14.[114–124] PLAG1 is the target gene in this region, and its overexpression has been demonstrated as a probable pathogenetic mechanism. Reported gene rearrangements include PLAG1-RAD51L1 gene fusion.[125] Alternately, some lipoblastomas may harbor HMGA2 gene rearrangements with PLAG1 alterations.[126]

Fat atrophy (lipodystrophy)

Atrophy of adipose tissue can result from metabolic factors that range from localized to systemic. Corticosteroid injection can induce localized or even systemic fat atrophy,[127] whereas the etiology of systemic fat atrophy includes malnutrition, tumor cachexia, and highly active antiretroviral therapy (HAART) used for the treatment of HIV infection. The peripheral lipodystrophy and central body fat accumulation typically occurring during long-term HAART has been attributed to the mitochondrial toxicity caused by antiretroviral drugs, such as protease and reverse transcriptase inhibitors.[128] Systemic fat atrophy associated with HAART is also linked to metabolic syndrome and its cardiovascular complications.[129]

Clinically, localized fat atrophy causes an area of skin depression that can on inspection resemble a hypertrophic scar. Histologically, it has a pseudohypercellular appearance because of shrunken, partially fat-depleted fatty lobules containing a prominent capillary pattern. There is no adipocytic atypia (Figure 14.46).

Fat necrosis and lipogranuloma

The response of adipose tissue to accidental or surgical trauma, inflammation, or enzymatic damage is fat necrosis and histiocytic infiltration (lipogranuloma formation). It is commonly seen at surgical sites and apparently sometimes as small subcutaneous or intra-abdominal nodules.

Histologically, nonenzymatic fat necrosis is characterized by the absence of fat cell nuclei. The amount of reactive, predominantly histiocytic cellular component varies depending on factors such as local circulation. Fat necrosis caused by tissue ischemia (e.g., by torsion of a

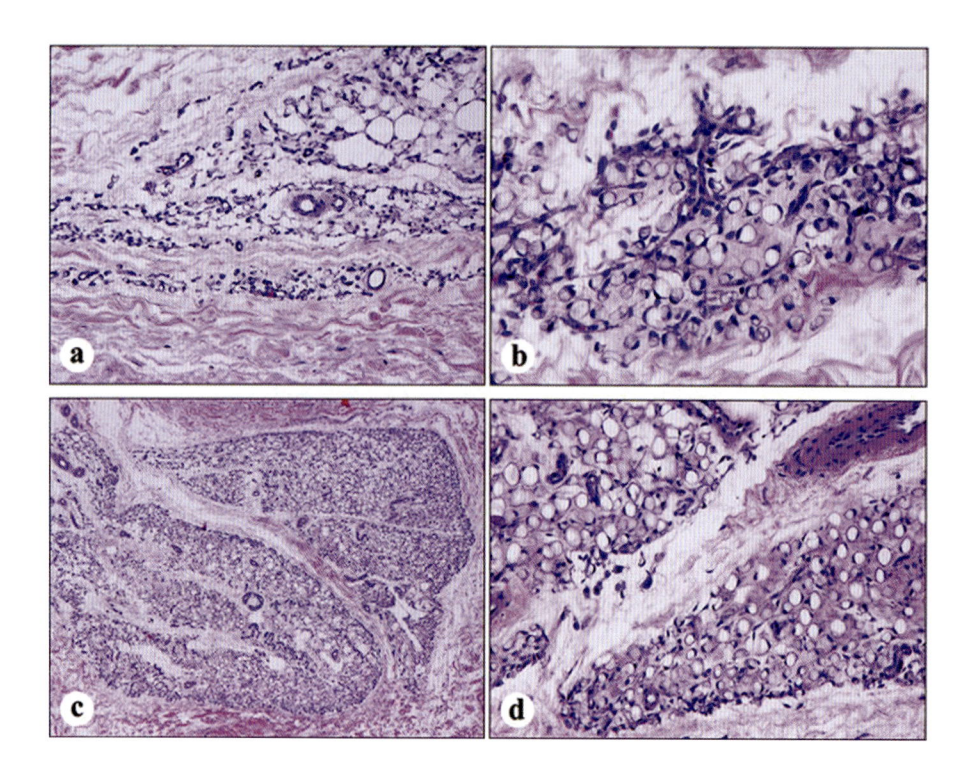

Chapter 14: Lipoma variants and conditions simulating lipomatous tumors

Figure 14.46 Fat atrophy shows streaks and clusters of fat from shrunken fat lobules, often grouped around blood vessels.

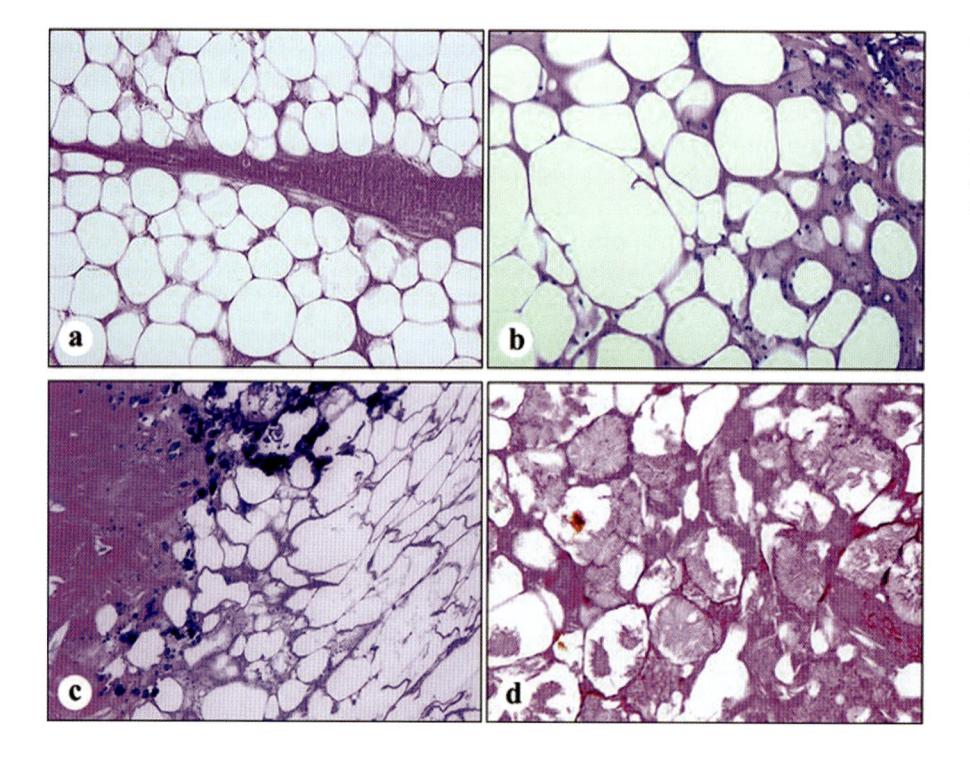

Figure 14.47 Different appearances of fat necrosis. (**a**) Example with no cellular reaction, as seen in recent fat necrosis or in poorly vascularized infarcted tissue. (**b**) Fat necrosis with histiocytic infiltration. (**c**) Dystrophic calcification in fat necrosis. (**d**) Enzymatic fat necrosis with fat saponification in massive soft tissue infection in necrotizing fasciitis.

pedunculated lipoma or a fatty mesenteric appendage [appendix epiploica]) shows scant if any cellular infiltration, with the lesion often undergoing eventual calcification.

Fat necrosis with histiocytic infiltration often contains xanthoma cells and mononuclear and multinuclear histiocytes, the latter of which should not be mistaken for atypical cells (Figure 14.47a–c). The presence of adipocytic

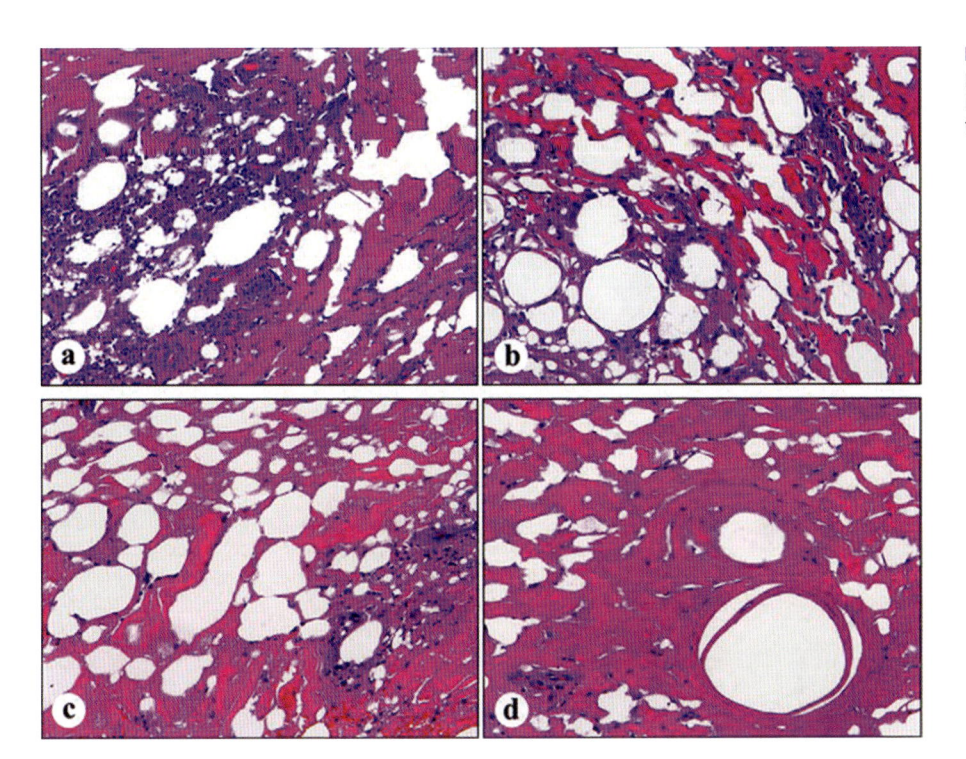

Figure 14.48 Sclerosing lipogranuloma of the penis. There are rounded or oval spaces created by lipid material, surrounded by fibrous reaction and focal lymphoid infiltration.

atypia has to be ruled out, however, because lipomatous neoplasms often contain focal fat necrosis and histiocytic infiltration.

Enzymatic fat necrosis usually occurs around the pancreas, generally related to pancreatitis or pancreatic trauma. This type of fat necrosis can also occur in sites of massive bacterial infection, such as necrotizing fasciitis (Figure 14.47d)

Paraffinoma and sclerosing lipogranuloma

This condition is a reaction to a lipid substance, which can be exogenous, paraffin oil related, or injected for cosmetic purposes or body part augmentation (paraffinoma). In some cases, it can be endogenous. The lesions most commonly occur in the external genitalia and breast. The lipid material is found in extracellular spaces surrounded by sclerotic collagen, often having lymphoplasmacytic reaction and fat necrosis (Figure 14.48). Exogenous paraffin hydrocarbons can be demonstrated by infrared spectrophotometry in most cases. Systemic and tissue eosinophilia can accompany endogenous lipogranulomas.[130,131]

Silicone granuloma

Silicone seepage from silicone-containing tissue implants creates vacuolization that histologically resembles a fatty lesion. Most commonly this is seen in the breast implant capsules, a dense collagenous zone formed around the implant. The silicone material from breast implants can also reach axillary lymph nodes.

Histologically, silicone material in a lymph node can simulate a lipogranuloma, but it is apparent that the vacuolar material in the lymph node is not fat but related to silicone (Figure 14.49a–c). Around silicone breast implants, there is typically a dense fibrous, nearly acellular, capsule-like reactive zone, often comprising histiocytes containing silicone material (Figure 14.49d). Less commonly, silicone becomes liberated as wear particles from silicone bone implants, especially when used as prosthetic replacement for small wrist bones. In such cases, silicone particles are seen engulfed in histiocytes in adjacent synovial tissue.

The extracellular silicone material becomes more easily visible as refractile material when lowering the microscope condenser. When present in vacuolated nodal histiocytes, it can be less easy to detect. Histiocytic giant cells are present in some cases. If specific verification is desired, silicone can be detected by X-ray microanalysis or laser Raman microprobe analysis from lesional cells or tissue sections.[132–134]

Massive localized lymphedema in the obese patient

Clinical features

This tumor-like condition usually occurs in middle-aged, seriously to extremely obese patients who weigh >150 kg to 200 kg or more. The condition is more common in women (Figure 14.50). Clinically there is a large, pendulous tumor-like

409

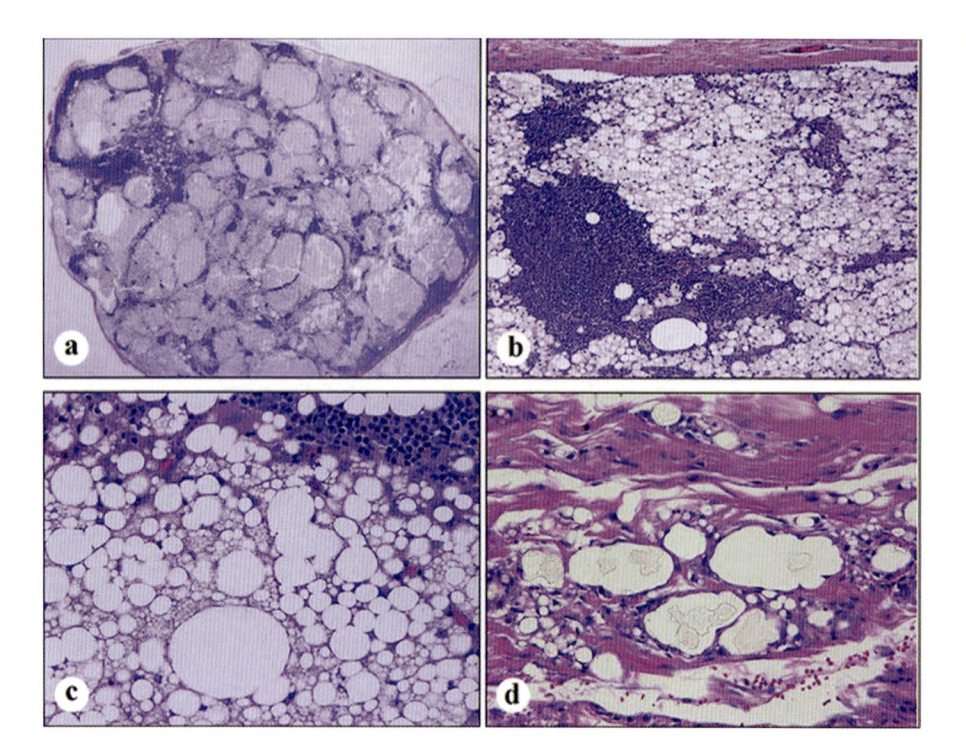

Figure 14.49 Silicone deposition in an axillary lymph node from a patient with a breast implant. (**a**) The lymph node is nearly replaced by the silicone material. (**b,c**) Silicone material creates an impression of fat; in these sections foreign material is not readily apparent. (**d**) In a breast implant capsule, refractile silicone material is seen in empty spaces after the microscope aperture is closed.

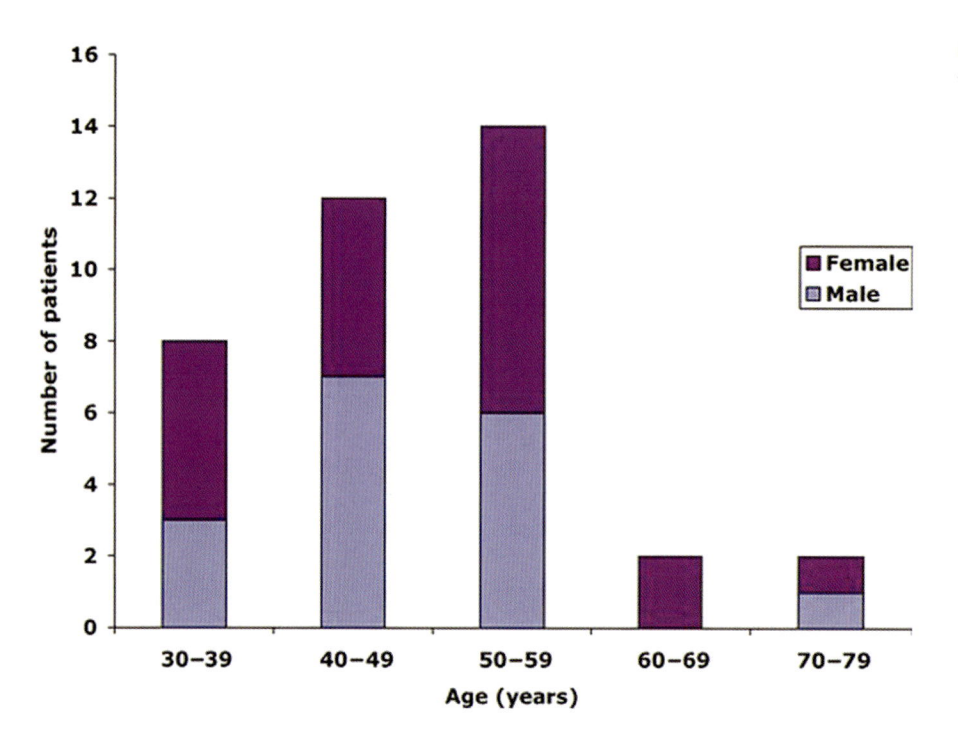

Figure 14.50 Age and sex distribution of 40 obese patients with massive localized lymphedema.

panniculus that forms on the medial proximal thigh, lower abdominal wall, or less commonly, the scrotum or proximal arm (Figure 14.51). In some cases, the condition has been precipitated by lymphadenectomy for carcinoma, varicose vein stripping, or trauma.[135–139] In some patients, hypothyroidism has been an associated factor.[136]

The tumor-like fat pannus commonly weighs 5 kg to 10 kg or more and measures 20 cm to 30 cm or more longitudinally and in width, and up to 5 cm or more in depth. Bilaterality sometimes occurs. Clinically, such a lesion is usually appropriately identified as a massively enlarged fat pannus, but in some cases it may raise a suspicion of tumor. Regrowth is possible after excision.

Development of cutaneous angiosarcoma has been reported in one case.[140] This may suggest that massive localized lymphedema also has some risk for angiosarcoma, as is known for hereditary and acquired lymphedema.

Pathology

The process involves the skin and subcutis, which are usually included in the excision specimen. Grossly, the overlying skin is markedly thickened and indurated, with an orange peel or cobblestone-like appearance. The subcutaneous fat is greatly expanded. On sectioning, the adipose tissue is edematous and punctuated by thickened and edematous fibrous septa.

Marked fibrous thickening of the dermis is histologically typical, often with dilated lymphatics and lymphangiectasia in some cases. The subcutaneous fat is lobulated by thickened, edematous, paucicellular fibrous septa containing reactive fibroblasts and myofibroblasts. The fat is also edematous and can contain foci of fat necrosis and mild diffuse lymphohistiocytic infiltration, but there is no adipocytic atypia. A zone of neovascular capillaries often borders the fatty lobules (Figure 14.52). These features, plus the large specimen consisting of elements of skin and subcutis, are clues to this diagnosis.

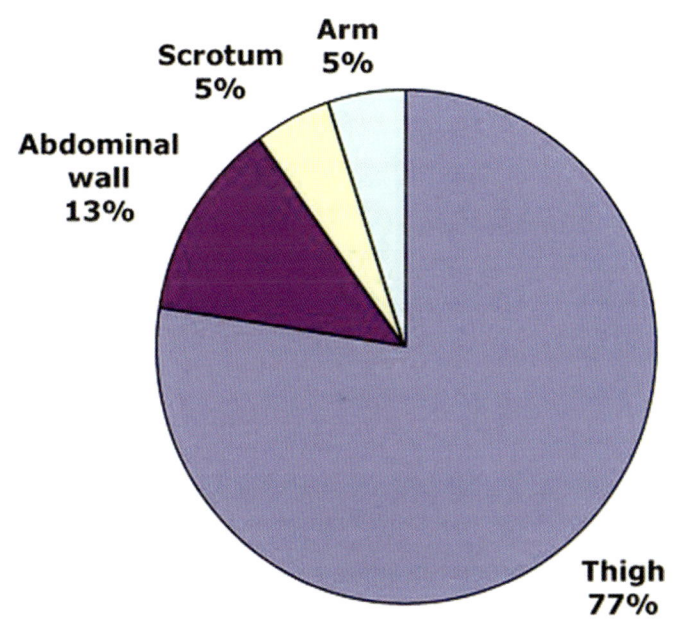

Figure 14.51 Anatomic distribution of 40 cases of massive localized lymphedema in the obese.

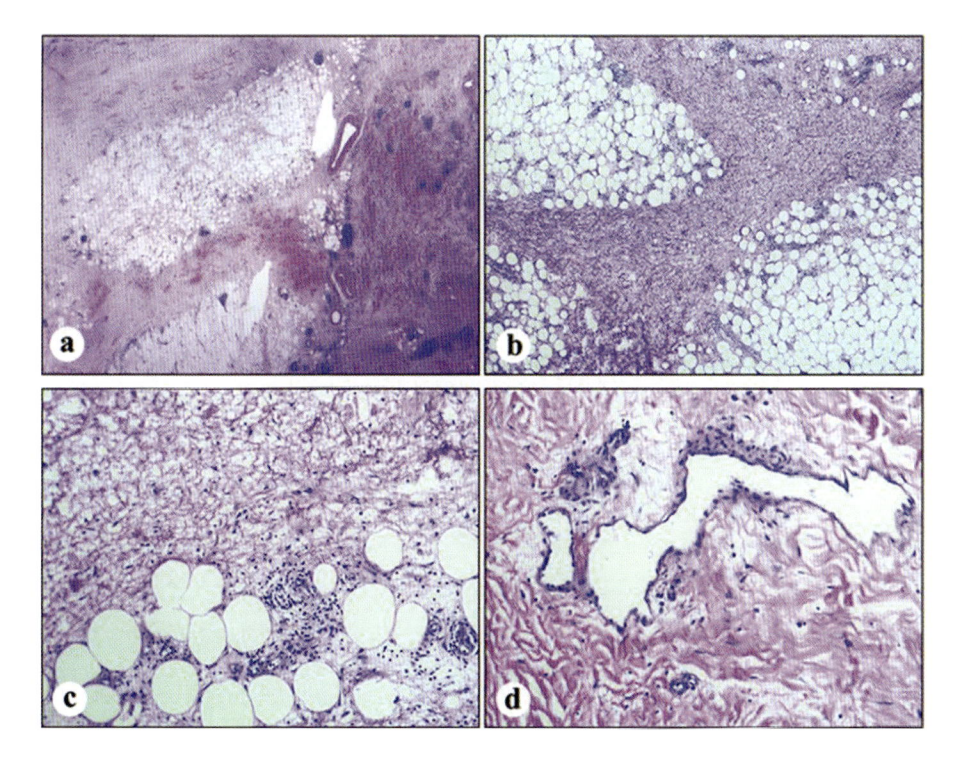

Figure 14.52 Massive localized lymphedema in the obese. Note edematous expansion of fat septa, showing fibrosis and focal lymphoid infiltration. (**b**) Neovascular capillaries are present in the periphery of fat lobules. (**d**) Dilated lymphatics in the overlying dermis.

References

Lipoma

1. Myhre-Jensen O. A consecutive 7-year series of 1431 benign soft tissue tumours: clinicopathologic data. Comparison with sarcomas. *Acta Orthop Scand* 1981;52:287–293.
2. Sandberg AA. Updates on the cytogenetics and molecular genetics of bone and soft tissue tumors: lipoma. *Cancer Genet Cytogenet* 2004;150:93–115.
3. Rydholm A, Berg NO. Size, site and clinical incidence of lipoma: factors in the differential diagnosis of lipoma and sarcoma. *Acta Orthop Scand* 1983;54:929–934.
4. Shanks JA, Paranchych W, Tuba J. Familial multiple lipomatosis. *Can Med Assoc J* 1957;77:881–884.
5. Bjerregaard P, Hagen K, Daugaard S, Kofoed H. Intramuscular lipoma of the lower limb: long-term follow-up after local resection. *J Bone Joint Surg [Br]* 1989;71:812–815.
6. Fletcher CD, Martin-Bates E. Intramuscular and intermuscular lipoma: neglected diagnoses. *Histopathology* 1988;12:275–287.
7. Schmack I, Patel RM, Folpe AL, et al. Subconjunctival herniated orbital fat: a benign adipocytic lesion that may mimic pleomorphic lipoma and atypical lipomatous tumor. *Am J Surg Pathol* 2007;31:193–198.
8. Willen H, Akerman M, Dal Cin P, et al. Comparison of chromosomal patterns with clinical features in 165 lipomas: a report of the CHAMP study group. *Cancer Genet Cytogenet* 1998;102:46–49.
9. Sreekantaiah C, Leong SPL, Karakousis CP, et al. Cytogenetic profile of 109 lipomas. *Cancer Res* 1991;51:422–433.
10. Mrozek K, Karakousis CP, Bloomfield CD. Chromosome 12 breakpoints are cytogenetically different in benign and malignant lipogenic tumors: localization of breakpoints in lipoma to 12q15 and in myxoid liposarcoma to 12q13.3. *Cancer Res* 1993;53:1670–1675.
11. Sait SNJ, Dal Cin P, Sandberg AA, et al. Involvement of 6p in benign lipomas: a new cytogenetic entity. *Cancer Genet Cytogenet* 1989;37:281–283.
12. Mandahl N, Heim S, Arheden K, et al. Four cytogenetic subgroups can be identified in lipomas. *Hum Genet* 1988;79:203–208.
13. Fletcher CDM, Akerman M, Dal Cin P, et al. Correlation between clinicopathological features and karyotype in lipomatous tumors: a report of 178 cases from the chromosomes and morphology (CHAMP) collaborative study group. *Am J Pathol* 1996;148:623–630.
14. Nilbert M, Rydholm A, Willen H, Mitelman F, Mandahl N. MDM2 gene amplification correlates with ring chromosomes in soft tissue tumors. *Genes Chromosomes Cancer* 1994;9:261–265.
15. Szymanska J, Virolainen M, Tarkkanen M, et al. Overrepresentation of 1q21–23 and 12q13–21 in lipoma-like liposarcomas but not in benign lipomas: a comparative genomic hybridization study. *Cancer Genet Cytogenet* 1997;99:14–18.
16. Tallini G, Dal Cin P. HMGI(Y) and HMGI-C dysregulation: a common occurrence in human tumors. *Adv Anat Pathol* 1999;6:237–246.
17. Petit MM, Schoenmakers EF, Huysmans C, et al. LHFP, a novel translocation partner gene of HMGIC in a lipoma, is a member of a new family of LHFP-like genes. *Genomics* 1999;57:438–441.

Variants of ordinary lipoma

18. Marsh WL, Lucas JG, Olsen J. Chondrolipoma of breast. *Arch Pathol Lab Med* 1989;113:369–371.
19. Nwaorgy OK, Akang EE, Ahmad BM, Nwachokor FN, Olu-Eddo AN. Pharyngeal lipoma with cartilaginous metaplasia (chondrolipoma): a case report and literature review. *J Laryngol Otol* 1997;111:656–658.
20. Hopkins JDF, Rayan GM. Osteolipoma of the hand: a case report. *J Okla State Med Assoc* 1999;92:535–537.
21. Rau T, Soeder S, Olk A, Aigner T. Parosteal lipoma of the thigh with cartilaginous and osseous differentiation: an osteochondrolipoma. *Ann Diagn Pathol* 2006;10:279–282.
22. Lin TM, Chang HW, Wang KH, et al. Isolation and identification of mesenchymal stem cells from human lipoma tissue. *Biochem Biophys Res Commun* 2007;361:883–889.
23. Mehregan AH, Tavafoghi V, Ghandchi A. Nevus lipomatosus cutaneus superficialis (Hoffmann-Zurhelle). *J Cutan Pathol* 1975;2:307–313.
24. Wilson-Jones EW, Marks R, Pongsehirun D. Naevus superficialis lipomatosus: a clinicopathological report of twenty cases. *Br J Dermatol* 1975;93:121–133.
25. Hallel T, Lew S, Kfar-Saba I, Bansal M. Villous lipomatous proliferation of the synovial membrane (lipoma arborescens). *J Bone Joint Surg* 1988;70A:264–270.
26. Yildiz G, Deveci MS, Ozcan A, et al. Lipoma arborescens (diffuse articular lipomatosis). *J South Orthoped Assoc* 2003;12:163–166.
27. Lassman LP, James CCM. Lumbosacral lipomas: critical survey of 26 cases submitted to laminectomy. *J Neurol Neurosurg Psychiatry* 1967;30:174–181.
28. Kieck CF, De Villiers JC. Subcutaneous lumbosacral lipomas. *S Afr Med J* 1975;49:1563–1566.
29. Arai H, Sato K, Okuda O, et al. Surgical experience of 120 patients with lumbosacral lipomas. *Acta Neurochir* 2001;143:857–864.
30. Maiuri F, Cirillo S, Simonetti L, De Simone MR, Gangemi M. Intracranial lipomas: diagnostic and therapeutic considerations. *Neurosurg Sci* 1988;32:161–167.
31. Donati F, Vassella F, Kaiser G, Blumberg A. Intracranial lipomas. *Neuropediatrics* 1992;23:32–38.
32. O'Connor S, Recavarren R, Nichols LC, Parvani AV. Lipomatous hypertrophy of the interatrial septum: an overview. *Arch Pathol Lab Med* 2006;130:397–399.
33. Moran CA, Suster S, Koss MN. Endobronchial lipomas: a clinicopathologic study of four cases. *Mod Pathol* 1994;7:212–214.
34. Muraoka M, Oka T, Akamine S, et al. Endobronchial lipoma: review of 64 cases reported in Japan. *Chest* 2003;123:293–296.
35. Rogy MA, Mirza D, Berlakovich G, Winkelbauer F, Rauhs R. Submucous large-bowel lipomas, presentation and management: an 18-year study. *Eur J Surg* 1991;157:51–55.
36. Moran CA, Rosado-de-Christenson M, Suster S. Thymolipoma: clinicopathologic review of 33 cases. *Mod Pathol* 1995;8:741–744.

Lipomatoses

37. Silverman TA, Enzinger FM. Fibrolipomatous hamartoma of the nerve: a clinicopathologic analysis of 26 cases. *Am J Surg Pathol* 1985;9:7–14.
38. Brodwater BK, Major NM, Goldner R, Layfield LJ. Macrodystrophia lipomatosa with associated fibrolipomatous hamartoma of the median nerve. *Pediatr Surg Int* 2000;16:216–218.
39. Haberland C, Perou M. Encephalocraniocutaneous lipomatosis: a new example of ectomesodermal dysgenesis. *Arch Neurol* 1970;22:144–155.
40. Sanchez NP, Rhodes AR, Mandell F, Mihm MC. Encephalocraniocutaneous lipomatosis: a new neurocutaneous syndrome. *Br J Dermatol* 1981;104:89–96.
41. Legius E, Wu R, Eyssen M, *et al* Encephalocranio cutaneous lipomatosis with a mutation in the NF1 gene. *J Med Genet* 1995;32:316–319.
42. Enzi G. Multiple symmetric lipomatosis: updated clinical report. *Medicine* 1984;63:56–64.
43. Smith PD, Stedelmann WK, Wassermann RJ, Kearney RE. Benign symmetric lipomatosis (Madelung's disease). *Ann Plast Surg* 1998;41:671–673.
44. Munoz-Malaga A, Bautista J, Salazar JA, *et al*. Lipomatosis, proximal myopathy, and the mitochondrial 8344 mutation: a lipid storage myopathy? *Muscle Nerve* 2000;23:538–542.
45. Behrens GM, Stoll M, Schmidt RE. Lipodystrophy syndrome in HIV-infection: what is it, what causes it and how it can be managed. *Drug Saf* 2000;23:57–76.
46. Klein FA, Smith MJ, Kasenetz I. Pelvic lipomatosis: 35-year experience. *J Urol* 1988;139:998–1001.
47. Sozen S, Gurocak S, Uzum N, *et al*. The importance of re-evaluation of patients with cystitis glandularis associated with pelvic lipomatosis: a case report. *Urol Oncol* 2004;22:428–430.
48. Stern JD, Quint DJ, Swaesey TA, Hoff JT. Spinal epidural lipomatosis: two new idiopathic cases and a review of the literature. *J Spinal Disord* 1994;7:343–349.
49. Carlsen A, Thomsen M. Different clinical types of lipomatosis: a case report. *Scand J Plast Reconstr Surg* 1978;12:75–79.
50. Battista S, Fidanza V, Fedele M, *et al*. The expression of truncated HMGI-C gene induces giantism associated with lipomatosis. *Cancer Res* 1999;59:4793–4797.

Angiolipoma

51. Howard WR, Helwig EB. Angiolipoma. *Arch Dermatol* 1960;82:924–931.
52. Lin JJ, Lin F. Two entities in angiolipoma: a study of 459 cases of lipoma with review of literature on infiltrating angiolipoma. *Cancer* 1974;34:720–727.
53. Namba M, Kohda M, Mimura S, Nakagawa S, Ueki H. Angiolipoma in brothers. *J Dermatol* 1977;4:255–257.
54. Kazakov DV, Hes O, Hora M, Sima R, Michal M. Primary intranodal angiolipoma. *Int J Surg Pathol* 2005;13:99–101.
55. Hunt SJ, Santa-Cruz DJ, Barr RJ. Cellular angiolipoma. *Am J Surg Pathol* 1990;14:75–81.
56. Sciot R, Akerman M, Dal Cin P, *et al*. Cytogenetic analysis of subcutaneous angiolipoma: further evidence supporting its difference from ordinary pure lipomas: a report of the CHAMP study group. *Am J Surg Pathol* 1997;21:441–444.

Myelolipoma

57. Han M, Burnett AL, Fishman EK, Marshall FF. The natural history and treatment of adrenal myelolipoma. *J Urol* 1997;157:1213–1216.
58. Noble MJ, Montague DK, Levin HS. Myelolipoma: an unusual surgical lesion of the adrenal gland. *Cancer* 1982;49:952–958.
59. Fowler MR, Williams RM, Alba JM, Burd CR. Extra-adrenal myelolipomas compared with extramedullary hematopoietic tumors: a case of presacral myelolipoma. *Am J Surg Pathol* 1983;6:363–374.
60. Hunter SB, Schemankewitz EH, Patterson C, Varma VA. Extraadrenal myelolipoma. A report of two cases. *Am J Clin Pathol* 1992;97:402–404.
61. Sanders R, Bissada N, Cutty N, Gordon B. Clinical spectrum of adrenal myelolipoma: analysis of 8 tumors in 7 patients. *J Urol* 1995;153:1791–1793.
62. Shapiro JL, Goldblum JR, Bobrow DA, Ratliff NB. Giant bilateral extra-adrenal myelolipoma. *Arch Pathol Lab Med* 1995;119:283–285.
63. Allison KH, Mann GN, Norwood TH, Rubin BP. An unusual case of multiple giant myelolipomas: clinical and pathogenetic implications. *Endocr Pathol* 2003;14:93–100.

Spindle cell, pleomorphic, and sclerotic lipoma

64. Enzinger FM, Harvey DA. Spindle cell lipoma. *Cancer* 1975;36:1852–1859.
65. Shmookler BM, Enzinger FM. Pleomorphic lipoma: a benign tumor simulating liposarcoma. A clinicopathologic analysis of 48 cases. *Cancer* 1981;47:126–133.
66. Angervall L, Dahl I, Kindblom LG, Save-Soderbergh J. Spindle cell lipoma. *Acta Pathol Microbiol Scand A* 1976;84:477–487.
67. Fletcher CD, Martin-Bates E. Spindle cell lipoma: a clinicopathologic study with some original observations. *Histopathology* 1987;11:803–817.
68. Azzopardi JG, Iocco J, Salm R. Pleomorphic lipoma: a tumor simulating liposarcoma. *Histopathology* 1983;7:511–523.
69. Furlong MA, Fanburg-Smith JC, Childers EL. Lipoma of the oral and maxillofacial region: site and subclassification of 125 cases. *Oral Surg Oral Med Oral Pathol Oral Radiol Endod* 2004;98:441–450.
70. Fanburg-Smith JF, Miettinen M, Weiss SW. Multiple spindle cell lipomas. *Am J Surg Pathol* 1998;22:40–48.
71. Hawley ICV, Krausz T, Evans DJ, Fletcher CD. Spindle cell lipoma: a pseudoangiomatous variant. *Histopathology* 1994;24:565–569.
72. Zelger BW, Zelger BG, Plorer A, Steiner H, Fritsch PO. Dermal spindle cell lipoma: plexiform and nodular variants. *Histopathology* 1995;27:533–540.
73. Templeton SF, Solomon AR Jr. Spindle cell lipoma is strongly CD34-positive: an immunohistochemical study. *J Cutan Pathol* 1996;23:546–550.
74. Suster S, Fisher C. Immunoreactivity for the human hematopoietic progenitor cell antigen (CD34) in lipomatous tumors. *Am J Surg Pathol* 1997;21:195–200.
75. Beham A, Schmid C, Hodl S, Fletcher CD. Spindle cell and pleomorphic lipoma: an immunohistochemical study

and histogenetic analysis. *J Pathol* 1989;158:219–222.

76. Tardio JC, Aramburu JA, Santonja C. Desmin expression in spindle cell lipomas: a potential diagnostic pitfall. *Virchows Arch* 2004;445:354–358.

77. Chen BJ, Mariño-Enríquez A, Fletcher CD, Hornick JL. Loss of retinoblastoma protein expression in spindle cell/pleomorphic lipomas and cytogenetically related tumors: an immunohistochemical study with diagnostic implications. *Am J Surg Pathol* 2012;36:1119–1128.

78. Mandahl N, Mertens F, Willen H, *et al.* A new cytogenetic subgroup in lipomas: loss of chromosome 16 material in spindle cell and pleomorphic lipomas. *J Cancer Res Clin Oncol* 1994;120:707–711.

79. Dal Cin P, Sciot R, Polito P, *et al.* Lesions of 13q may occur independently of deletion of 16q in spindle cell/pleomorphic lipomas. *Histopathology* 1997;31:222–225.

80. Creytens D, van Gorp J, Savola S, *et al.* Atypical spindle cell lipoma: a clinicopathologic, immunohistochemical, and molecular study emphasizing its relationship to classical spindle cell lipoma. *Virchows Arch* 2014;465:97–108.

81. Zelger BG, Zelger B, Steiner H, Rütten A. Sclerotic lipomas: lipomas simulating sclerotic fibroma. *Histopathology* 1997;31:174–181.

82. Laskin WB, Fetsch JF, Miettinen M. Sclerotic (fibromalike) lipoma: a distinctive lipoma variant with a predilection for the distal extremities. *Am J Dermatopathol* 2006;28:308–316.

Hibernoma

83. Kindblom LG, Angervall L, Stener B, Wickbom I. Intramuscular and intermuscular lipomas and hibernomas: a clinical roentgenologic, histologic and prognostic study of 46 cases. *Cancer* 1974;33:754–762.

84. Gaffney EF, Hargreaves HK, Semple E, Vellios F. Hibernoma: distinctive light and electron microscopic features and relationship to brown adipose tissue. *Hum Pathol* 1983;14:677–687.

85. Furlong MA, Fanburg-Smith JC, Miettinen M. Hibernoma. *Am J Surg Pathol* 2001;25:809–814.

86. Crieac LR, Dekmezian RH, Ayala A. Characterization of the myxoid variant of hibernoma. *Ann Diagn Pathol* 2006;10:104–106.

87. Zancanaro C, Pelosi G, Accordini C, *et al.* Immunohistochemical identification of the uncoupling protein in human hibernoma. *Biol Cell* 1994;80:75–78.

88. Meloni AM, Spanier SS, Bush CH, Stone JF, Sandberg AA. Involvement of 10q22 and 11q13 in hibernoma. *Cancer Genet Cytogenet* 1994;72:59–64.

89. Mertens F, Rydholm A, Brosjo O, *et al.* Hibernomas are characterized by rearrangements of chromosome bands 11q13–21. *Int J Cancer* 1994;58:503–505.

90. Mrozek K, Karakousis CP, Bloomfield CD. Band 11q13 is nonrandomly rearranged in hibernomas. *Genes Chromosomes Cancer* 1994;9:145–147.

91. Maire G, Forus A, Foa C, *et al.* 11q13 alterations in two cases of hibernoma: large heterozygous deletions and rearrangement breakpoints near GARP in 11q13.5. *Genes Chromosomes Cancer* 2003;37:389–395.

92. Nord KH, Magnusson L, Isaksson M, *et al.* Concomitant deletions of tumor suppressor genes MEN1 and AIP are essential for the pathogenesis of the brown fat tumor hibernoma. *Proc Natl Acad Sci USA*. 2010;107:21122–21127.

Chondroid lipoma, myolipoma, and hemosiderotic fibrohistiocytic lipomatous lesion

93. Meis JM, Enzinger FM. Chondroid lipoma: a unique tumor simulating liposarcoma and myxoid chondrosarcoma. *Am J Surg Pathol* 1993;17:1103–1112.

94. Nielsen GP, O'Connell JX, Dickersin GR, Rosenberg AE. Chondroid lipoma. *Am J Surg Pathol* 1995;19:1272–1276.

95. Thomson TA, Horsman D, Bainbridge TC. Cytogenetic and cytologic features of chondroid lipoma of soft tissue. *Mod Pathol* 1999;12:88–91.

96. Gisselsson D, Domanski HA, Hoglund M, *et al.* Unique cytological features and chromosome aberrations in chondroid lipoma: a case report based on fine-needle aspiration cytology, histopathology, electron microscopy, chromosome banding, and molecular cytogenetics. *Am J Surg Pathol* 1999;23:1300–1304.

97. Ballaux F, Debiec-Rychter M, De Wever I, Sciot R. Chondroid lipoma is characterized by t(11;16)(q13;p12–13). *Virchows Arch* 2004;444:208–210.

98. Huang D, Sumegi J, Dal Cin P, *et al.* C11orf95-MKL2 is the resulting fusion oncogene of t(11;16)(q13;p13) in chondroid lipoma. *Genes Chromosomes Cancer* 2010;49:810–818.

99. Meis JM, Enzinger FM. Myolipoma of soft tissue. *Am J Surg Pathol* 1991;15:121–125.

100. Ben-Izhak O, Elmalach I, Kerner H, Best LA. Pericardial myolipoma: a tumour presenting as a mediastinal mass and containing estrogen receptors. *Histopathology* 1996;29:184–186.

101. Marshall-Taylor C, Fanburg-Smith JC. Hemosiderotic fibrohistiocytic lesion: 10 cases of a previously undescribed fatty lesion of the foot/ankle. *Mod Pathol* 2000;13:1192–1199.

102. Browne TJ, Fletcher CD. Haemosiderotic fibrolipomatous tumour (so-called haemosiderotic fibrohistiocytic lipomatous tumour): analysis of 13 new cases in support of a distinct entity. *Histopathology* 2006;48:453–461.

103. Solomon DA, Antonescu CR, Link TM, *et al.* Hemosiderotic fibrolipomatous tumor, not an entirely benign entity. *Am J Surg Pathol* 2013;37:1627–1630.

104. Wettach GR, Boyd LJ, Lawce HJ, Magenis RE, Mansoor A. Cytogenetic analysis of a hemosiderotic fibrolipomatous tumor. *Cancer Genet Cytogenet* 2008;182:140–143.

105. Elco CP, Mariño-Enríquez A, Abraham JA, Dal Cin P, Hornick JL. Hybrid myxoinflammatory fibroblastic sarcoma/hemosiderotic fibrolipomatous tumor: report of a case providing further evidence for a pathogenetic link. *Am J Surg Pathol* 2010;34:1723–1727.

106. Antonescu CR, Zhang L, Nielsen GP, *et al.* Consistent t(1;10) with rearrangements of TGFBR3 and MGEA5 in both myxoinflammatory fibroblastic sarcoma and hemosiderotic fibrolipomatous tumor. *Genes Chromosomes Cancer* 2011;50:757–764.

Lipoblastoma

107. Vellios F, Baez J, Shumacker HB. Lipoblastomatosis: a tumor of fetal fat different from hibernoma; report of a case, with observations on the embryogenesis of human adipose tissue. *Am J Pathol* 1958;34:1149–1159.

108. Chung EB, Enzinger FM. Benign lipoblastomatosis: an analysis of 35 cases. *Cancer* 1973;32:482–492.
109. Mahour GH, Bryan BJ, Isaacs H. Lipoblastoma and lipoblastomatosis: a report of six cases. *Surgery* 1988; 104:577–579.
110. Mentzel T, Calonje E, Fletcher CD. Lipoblastoma and lipoblastomatosis: a clinicopathological study of 14 cases. *Histopathology* 1993;23:527–533.
111. Coffin CM. Lipoblastoma: an embryonal tumor of soft tissue related to organogenesis. *Semin Diagn Pathol* 1994;11:98–103.
112. Collins MH, Chatten J. Lipoblastoma/lipoblastomatosis: a clinicopathologic study of 25 tumors. *Am J Surg Pathol* 1997;21:1131–1137.
113. Chun YS, Kim WK, Park KW, Lee SC, Jung SE. Lipoblastoma. *J Pediatr Surg* 2001;36:905–907.
114. Sciot R, De Wever I, Debiec-Rychter M. Lipoblastoma in a 23-year-old male: distinction from atypical lipomatous tumor using cytogenetic and fluorescence in-situ hybridization analysis. *Virchows Arch* 2003;442:468–471.
115. Miller GC, Yanchar NL, Magee JF, Blair GK. Lipoblastoma and liposarcoma in children: an analysis of 9 cases and review of the literature. *Can J Surg* 1998;41:455–458.
116. Hicks J, Dilley A, Patel D, et al. Lipoblastoma and lipoblastomatosis in infancy and childhood: histopathologic, ultrastructural, and cytogenetic features. *Ultrastruct Pathol* 2001;25:321–333.
117. Jung SM, Chang PY, Luo CC, et al. Lipoblastoma/lipoblastomatosis: a clinicopathologic study of 16 cases in Taiwan. *Pediatr Surg Int* 2005;21:809–812.
118. Ohjimi Y, Iwasaki H, Kaneko Y, et al. A case of lipoblastoma with t(3;8)(q12; q11.2). *Cancer Genet Cytogenet* 1992;62:103–105.
119. Fletcher JA, Kozakewich HP, Schoenberg ML, Morton CC. Cytogenetic findings in pediatric adipose tissue tumors: consistent rearrangement of chromosome 8 in lipoblastoma. *Genes Chromosomes Cancer* 1993;6:24–29.
120. Dal Cin P, Sciot R, DeWever I, van Damme B, Van Den Berghe H. New discriminative chromosomal marker in adipose tissue tumors: the chromosome 8q11–q13 region in lipoblastoma. *Cancer Genet Cytogenet* 1994;78:232–235.
121. Astrom A, DíAmore ES, Sainati L, et al. Evidence of involvement of the PLAG1 gene in lipoblastomas. *Int J Oncol* 2000;16:1107–1110.
122. Hibbard MK, Kozakewich H, Dal Cin P, et al. PLAG1 fusion oncogenes in lipoblastoma. *Cancer Res* 2000;60:4869–4872.
123. Gisselsson D, Hibbard MK, Dal Cin P, et al. PLAG1 alterations in lipoblastoma: involvement in varied mesenchymal cell types and evidence for alternative oncogenic mechanisms. *Am J Pathol* 2001;159:955–962.
124. Morerio C, Rapelola A, Rosanda C, et al. PLAG1-HAS2 fusion in lipoblastoma with masked 8q intrachromosomal rearrangement. *Cancer Genet Cytogenet* 2005;156:183–184.
125. Deen M, Ebrahim S, Schloff D, Mohamed AN. A novel PLAG1-RAD51L1 gene fusion resulting from a t(8;14)(q12;q24) in a case of lipoblastoma. *Cancer Genet* 2013;206:233–237.
126. Pedeutour F, Deville A, Steyaert H, et al. Rearrangement of HMGA2 in a case of infantile lipoblastoma without Plag1 alteration. *Pediatr Blood Cancer* 2012;58:798–800.

Lipoatrophy/lipodystrophy, lipogranuloma, and silicone

127. Bauerschmitz J, Bork K. Multifocal disseminated lipoatrophy secondary to intravenous corticosteroid administration in a patient with adrenal insufficiency. *J Am Acad Dermatol* 2002;46:S130–S132.
128. McGomsey GA, Walker UA. Role of mitochondria in HIV lipoatrophy: insight into pathogenesis and potential therapies. *Mitochondrion* 2004;4:111–118.
129. Estrada V, Martinez-Larrad MT, Gonzalez-Sanchez JL, et al. Lipodystrophy and metabolic syndrome in HIV-infected patients treated with antiretroviral therapy. *Metabolism* 2006;55:940–945.
130. Oertel YC, Johnson FB. Sclerosing lipogranuloma of male genitalia: review of 23 cases. *Arch Pathol Lab Med* 1977;101:321–326.
131. Matsuda T, Shichiri Y, Hida S, et al. Eosinophilic sclerosing lipogranuloma of the male genitalia not caused by exogenous lipids. *J Urol* 1988;140:1021–1024.
132. Greene WB, Raso DS, Walsh LG, Harley RA, Silver RM. Electron probe microanalysis of silicon and the role of macrophage in proximal (capsule) and distant sites in augmentation mammoplasty patients. *Plast Reconstr Surg* 1995;95:513–519.
133. Luke JL, Kalasinsky VF, Turnicky RP, et al. Pathological and biophysical findings associated with silicone breast implants: a study of capsular tissues from 86 cases. *Plast Reconstr Surg* 1997;100:1558–1565.
134. Centeno JA, Mullick FG, Panos RG, Miller FW, Valenzuela-Espinoza A. Laser-Raman microprobe identification of inclusions in capsules associated with silicone gel breast implants. *Mod Pathol* 1999;12:714–721.

Massive localized lymphedema

135. Farshid G, Weiss SW. Massive localized lymphedema in the morbidly obese: a histologically distinct reactive lesion simulating liposarcoma. *Am J Surg Pathol* 1998;22:1277–1283.
136. Wu D, Gibbs J, Cotral D, Intengan M, Brooks JJ. Massive localized lymphedema. *Hum Pathol* 2000;31:1162–1168.
137. Oswald TM, Lineweaver W. Limited segmental resection on symptomatic lower extremity lymphodystrophic tissue in high-risk patients. *South Med J* 2003;96:689–691.
138. Goshtasby P, Dawson J, Agarwal N. Pseudosarcoma: massive localized lymphedema in the obese. *Obes Surg* 2006;16:88–93.
139. Manduch M, Oliveira AM, Nascimento AG, Folpe AL. Massive localized lymphedema: a clinicopathologic study of 22 cases and review of the literature. *J Clin Pathol* 2009;62:808–811.
140. Shon W, Ida CM, Boland-Froemming JM, Rose PS, Folpe A. Cutaneous angiosarcoma arising in massive localized lymphedema of the morbidly obese: a report of five cases and review of the literature. *J Cutan Pathol* 2011;38:560–564.

Chapter 15
Atypical lipomatous tumors and liposarcomas

Markku Miettinen
National Cancer Institute, National Institutes of Health

Atypical lipomatous tumor (ALT)/well-differentiated liposarcoma (WDLS) and three other clinicopathologically or genetically distinct types of liposarcomas, sarcomas with fatty differentiation, are discussed in this chapter. Table 15.1 reviews their clinicopathologic features. The present classification of liposarcoma has evolved, especially from work by Enterline et al.,[1] Enzinger and Winslow,[2] and Evans.[3-4]

ALT, synonymous with WDLS, is usually recognized by the presence of atypical fat cells in the background of a well-differentiated lipomatous tumor, but there are many variants that are described in detail below. This tumor has no metastatic potential.

Dedifferentiated liposarcoma (DDLS) is a transformed variant of well-differentiated liposarcoma, representing its histologic and biologic disease progression. This type is most common in the retroperitoneum. It has a wide variation in morphology with potential heterologous rhabdomyosarcomatous, leiomyosarcomatous, and osteosarcomatous components. This type has metastatic potential.

Myxoid liposarcoma is a clinicopathologically and genetically distinct type characterized by its common occurrence in young adults, its location in the thigh, and the presence of a t(12;16) translocation. Round cell liposarcoma is a high-grade variant of myxoid liposarcoma (MLS).

Pleomorphic liposarcoma is a histologically well-defined high-grade liposarcoma. Its genetics and relationship with the other liposarcoma types are incompletely understood, although it may have overlap with DDLS being its component in some cases.

No significant etiologic clues are available for liposarcoma; however, liposarcomas seem to be distinctly uncommon among postirradiation sarcomas. Studies on liposarcomas have led to the discovery of many genes important in the transcriptional regulation and metabolism of adipose tissue.

Atypical lipomatous tumor/ well-differentiated liposarcoma

ALT and WDLS are synonymous and refer to well-differentiated ALTs that have capacity for local recurrence only and no metastatic potential. Many authors prefer to use the term ALT (or atypical lipoma) for subcutaneous (superficial) lesions and the term WDLS for intramuscular and body

Table 15.1 Overview of liposarcoma types and their clinicopathologic features

Type	Estimated frequency (%)	Age at presentation	Typical sites	Behavior	Genetics
ALT/WDLS	>50–60	Middle-aged to old	Retroperitoneum, extremities, trunk wall	Local recurrence and risk of dedifferentiation	12q13–15 MDM2 amplification
Dedifferentiated liposarcoma	15–20	Middle-aged to old	Retroperitoneum, extremities, trunk wall	Risk for metastasis, especially with high-grade dedifferentiation	12q13–15 MDM2 amplification
Myxoid liposarcoma	20–25	Adult, often <40 years, rare in childhood	Thigh, other extremity sites, very rare in retroperitoneum	Recurrence common. Metastatic rate 30–40% in long-term follow-up	t(12;16) with DDIT3-FUS gene fusion
Pleomorphic liposarcoma	<5	Old adults	Extremities, trunk wall	High risk for recurrence and metastasis	Complex, poorly understood

Modern Soft Tissue Pathology, Second Edition, ed. Markku Miettinen. Published by Cambridge University Press. © Cambridge University Press 2016.

Chapter 15: Atypical lipomatous tumors and liposarcomas

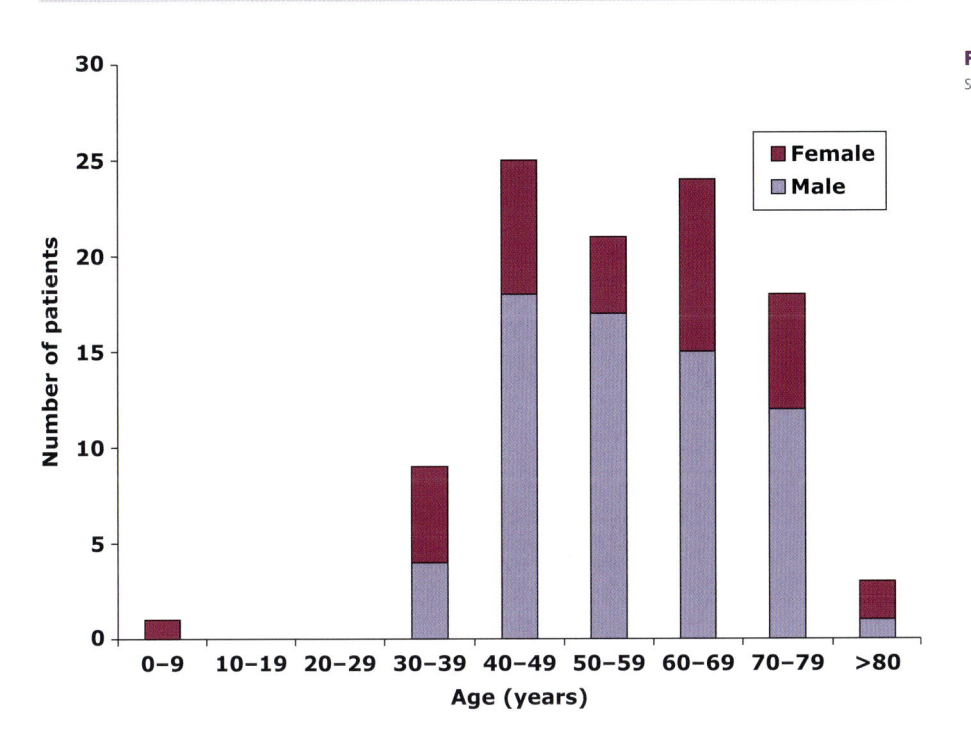

Figure 15.1 Age and sex distribution of subcutaneous ALT/WDLS in 101 civilian patients.

cavity (deep) lesions, as the latter has a greater potential for disease progression. ALT/WDLS has a wide histological spectrum and can at times simulate other types of liposarcomas. Another important problem is how to distinguish ALT/WDLS from ordinary lipoma and its variants.

Clinical features

Superficial (subcutaneous) ALT/WDLS, or simply ALT or atypical lipoma, occurs predominantly in middle-aged and older adults from the fourth decade. AFIP statistics show a significant, 3:1 male predominance (Figure 15.1). It presents in a wide variety of extremity and trunk wall locations. The most common sites are the back, thigh, buttocks, and shoulder. Occurrence in distal extremities is rare (Figure 15.2). Tumor size varies from a small nodule to a mass >10 cm, but an average size for subcutaneous tumors is 2 cm to 5 cm. Although many follow-up studies indicate a low recurrence rate,[4–8] one small study had a 50% rate of recurrence (5/10).[9] Transformation into DDLS is rare but has been reported.[10] Considering its generally indolent behavior and accessibility for monitoring for recurrences, subcutaneous ALTs can be treated by simple local excision, combined with follow-up.

Deep ALT/WDLS also occurs almost exclusively in adults, usually in older age groups in the fifth to the eighth decades, and there seems to be a male predilection, as observed for superficial tumors. Only 5% of cases occur in patients younger than 40 years (Figure 15.3).[5–16] Rare occurrences in children have been reported, especially with Li–Fraumeni syndrome.[17] Extremity-based intramuscular ALT/WDLS often forms large tumors (>10 cm), and has a predilection for the thigh and the shoulder girdle.[5,7,8] The

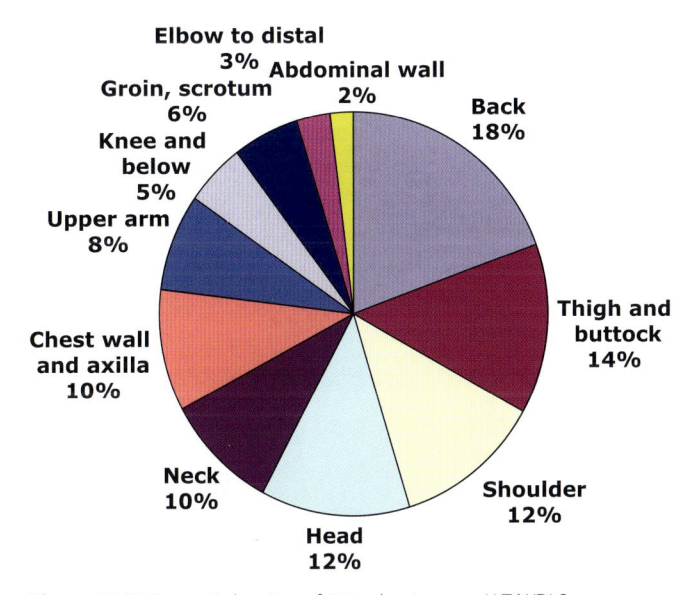

Figure 15.2 Anatomic location of 196 subcutaneous ALT/WDLS.

retroperitoneum is probably the most common site, and scrotum and trunk wall are additional common locations (Figure 15.4). Wide local excision is the preferred treatment. Recurrences are common, especially in body cavity tumors because of the difficulty of radical excision, but metastases are rare, even after dedifferentiation. Only a few patients succumb to locally aggressive recurrences without metastases.[5–9,11–13]

The retroperitoneal ALT/WDLS often reaches a large size, >10 cm to 20 cm or more and up to several kilograms. Tumors can approach or involve kidney, adrenal, and

Chapter 15: Atypical lipomatous tumors and liposarcomas

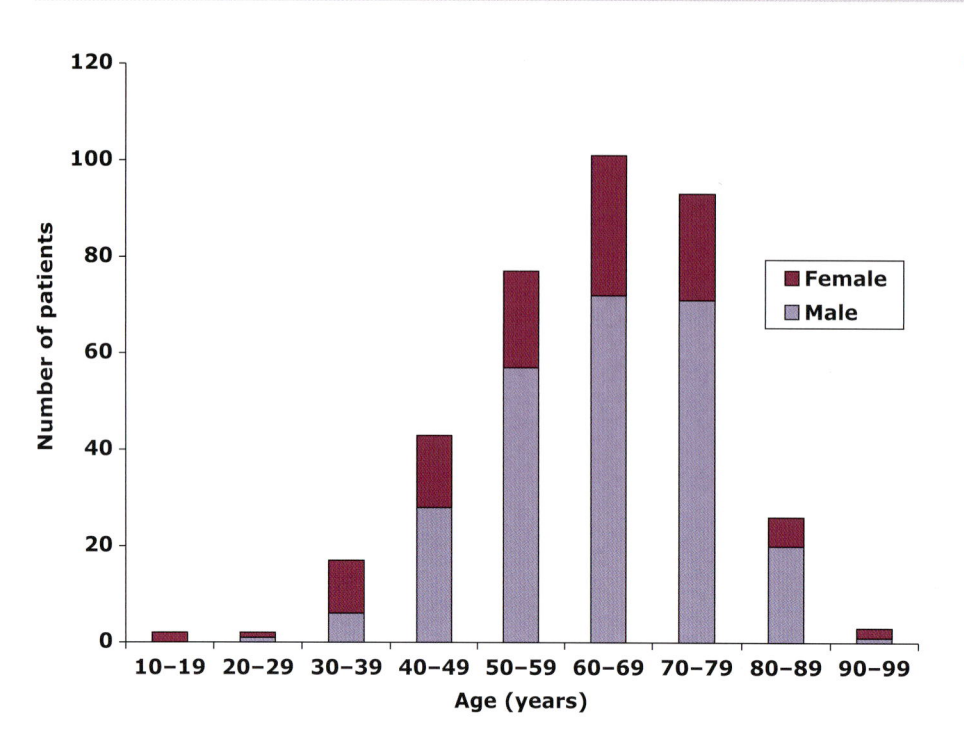

Figure 15.3 Age and sex distribution of 364 deep-seated ALT/WDLS.

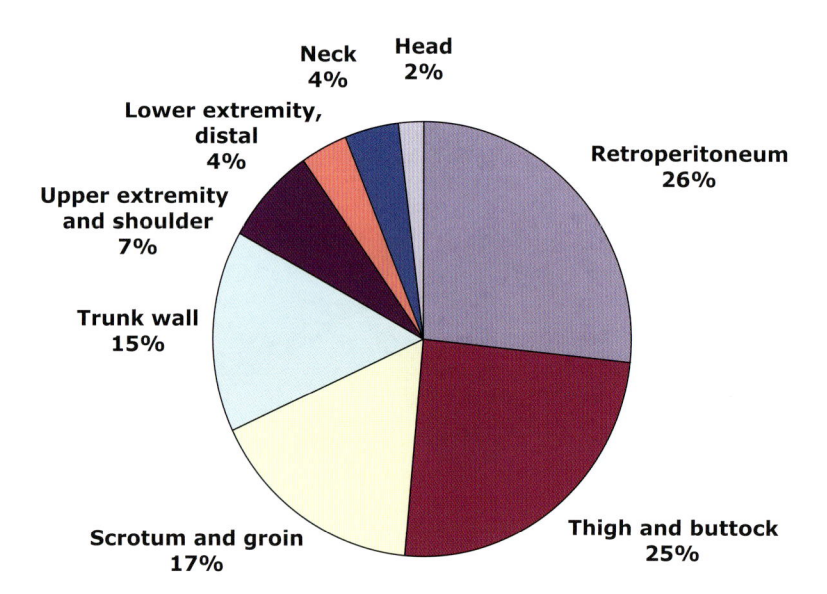

Figure 15.4 Anatomic location of 366 ALT/WDLS.

mesenteries, often necessitating nephrectomy or intestinal resections. Two published series found 14% mortality rate related to retroperitoneal disease.[12,13] This figure, however, underestimates the long-term mortality of retroperitoneal WDLSs that is increased by frequent dedifferentiation with increased potential for metastases.

Most scrotal/paratesticular liposarcomas are ALT/WDLS, and these tumors variably involve paratesticular soft tissue, spermatic cord, and sometimes the testicular adnexa. Tumor size often exceeds 10 cm. There is a moderate tendency toward dedifferentiation, but there are no metastases or mortality without dedifferentiation.[14] Inguinal tumors are often extensions from scrotal or retroperitoneal disease, and clinical and radiologic studies to delineate the extent of tumor and rule out an intra-abdominal component are warranted.

Mediastinal ALT/WDLS constitutes most liposarcomas of this site. These tumors are often large when detected and occur in the anterior and posterior mediastinum. Some anterior mediastinal examples arise in the thymus and have been designated thymoliposarcomas.[15] In one study, there was 30% tumor-related mortality rate from incontrollable recurrences, but no metastases.[16]

Deep head and neck ALT/WDLS in the larynx and hypopharynx can cause a medical emergency by respiratory compromise. They have a high (80%) tendency to recur after local

Chapter 15: Atypical lipomatous tumors and liposarcomas

excision, but according to the largest study to date, all ten tumors were ultimately controlled by radical surgery, such as laryngectomy, sometimes supplemented by local radiation, and no metastases developed.[18] Esophageal and hypopharyngeal examples can take the form of a giant fibrovascular polyp and it seems likely that many of the giant fibrovascular polyps in these locations in fact are ALT/WDLS.[19–21]

Pathology

On gross inspection, ALT/WDLS often has a paler yellow color than lipoma and contains more prominent, grossly visible fibrous septal elements, giving the tumor a firmer texture than that observed in typical lipomas. Larger retroperitoneal tumors are often grossly heterogeneous, containing soft, lipoma-like, yellow or gray, fibrous, rubbery, or gelatinous areas (Figure 15.5).

Three main histologic patterns are recognized in ALT/WDLS: lipoma-like, sclerosing, and inflammatory (lymphocyte-rich). The patterns sometimes coexist in the same tumor, especially in the larger retroperitoneal liposarcomas. Common to all subtypes is low to moderate cellularity, a variable atypical adipocytic component, and a virtual lack of mitotic activity. By these features, WDLS is a grade 1 sarcoma in the French Federation of Cancer Centers' classification (Chapter 1).

Lipoma-like ALT/WDLS

A lipoma-like pattern dominates in subcutaneous deep extremity lesions and is usually at least focally present in body cavity tumors. Low magnification (4–10×) typically reveals adipocyte size variation (Figure 15.6). Cellular infiltration by focal clusters or macrophages or lymphocytes is also common, giving ALT/WDLS a more cellular appearance than that seen in ordinary lipomas. The atypical adipocytes show a wide variety of appearances, common to which is nuclear elongation, irregular enlargement, and hyperchromasia (Figure 15.7). Most of these are univacuolar, but multivacuolated forms (lipoblasts) can also be present, but are not required for diagnosis. Expanded fibrous septa containing mildly atypical spindled cells are common, and in some examples, the atypia is essentially restricted to these septal elements (Figure 15.8). Larger tumors can have degenerative changes, such as fat necrosis and dystrophic calcification.

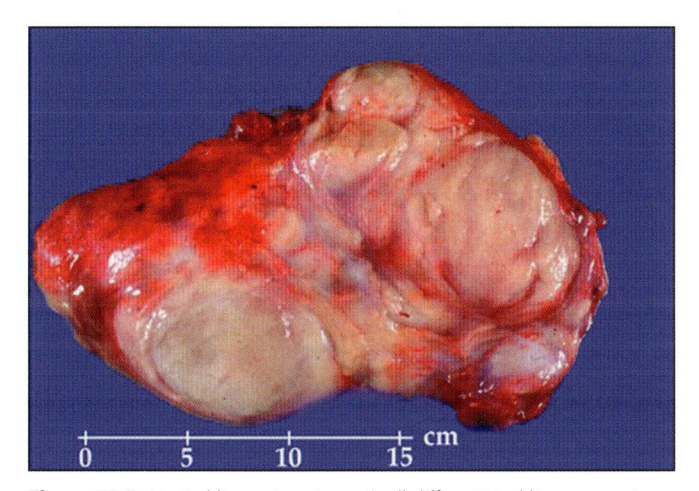

Figure 15.5 Atypical lipomatous tumor/well-differentiated liposarcoma is typically grossly paler than typical lipoma and contains grayish zones representing expanded fibrous septa.

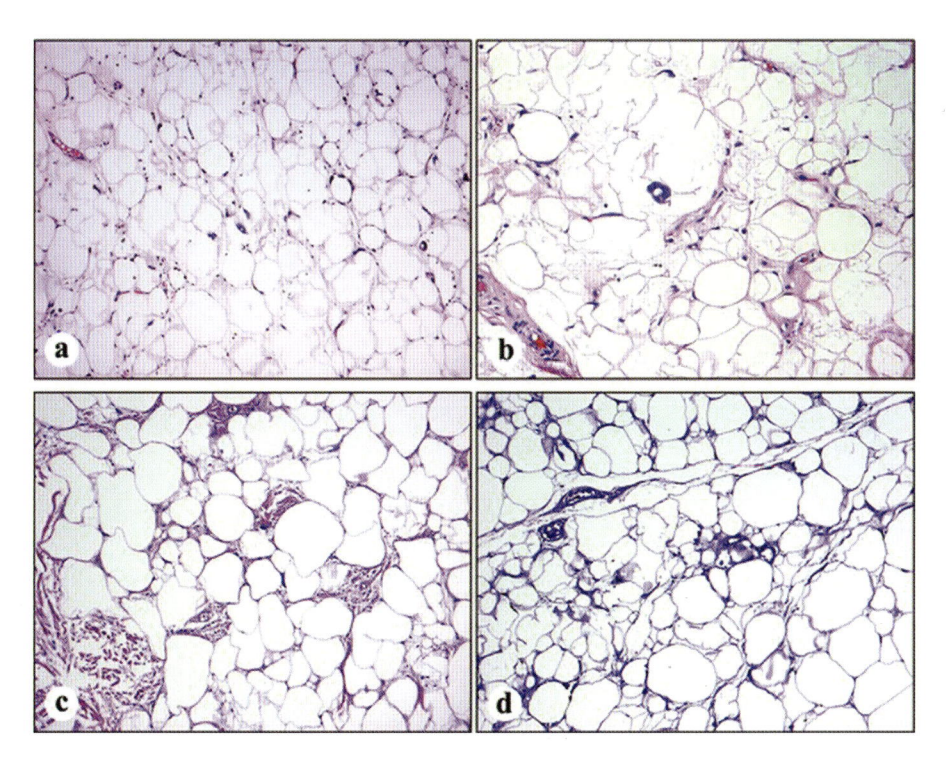

Figure 15.6 ALT/lipoma-like WDLS shows significant variation in adipocyte size, and focal nuclear atypia is present. Note also increased cellularity in **a**, representing histiocytic infiltration.

419

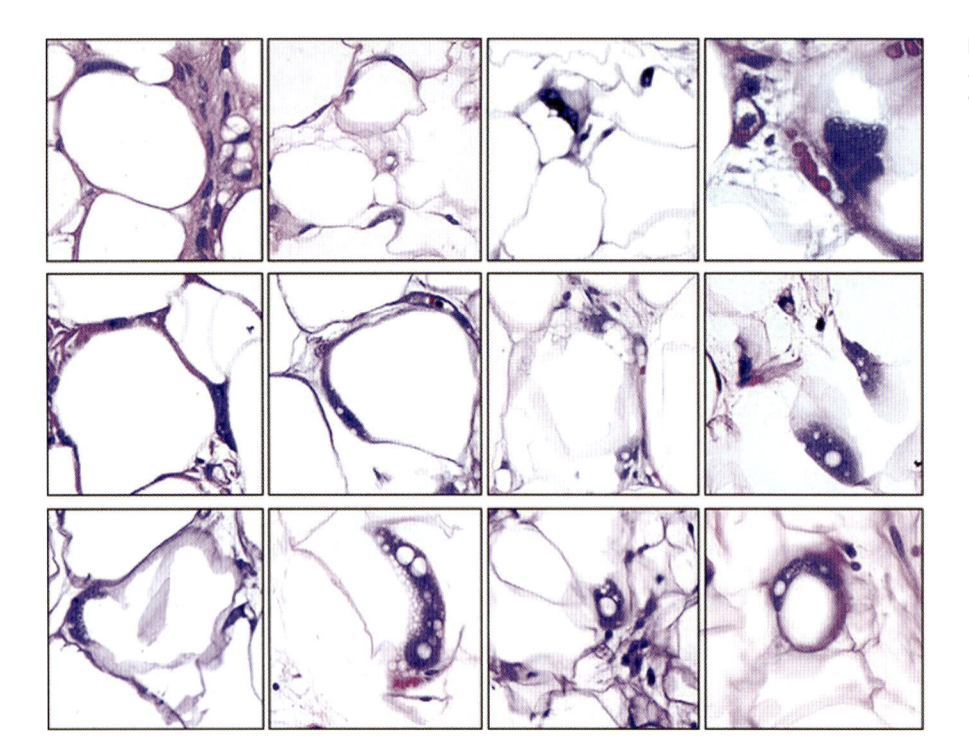

Figure 15.7 Examples of atypical adipocytes in ALT/WDLS. Note enlarged nuclei, hyperchromasia, and intranuclear vacuoles in the atypical cells.

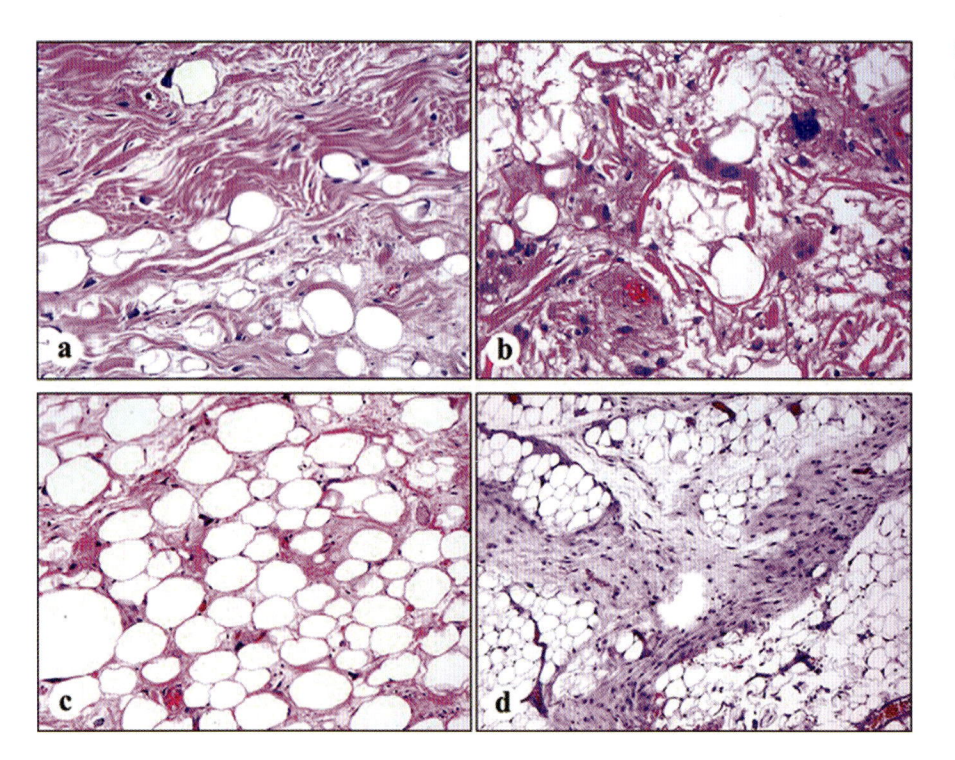

Figure 15.8 Lipoma-like ALT/WDLS with stromal sclerosis and expanded, hypercellular fibrous septa.

Sclerosing ALT/WDLS

Sclerosing ALT/WDLS is typically seen in the scrotum and inguinal region as a dominant component and often focally in retroperitoneal tumors. This variant appears on gross inspection as grayish, firm, and rubbery, but a yellow fatty component is usually noted in other areas of tumor.[2]

Microscopically, this variant contains scant atypical to mildly pleomorphic cells embedded in a dense collagenous, often focally myxoid matrix. Multinucleated forms with floral-type nuclear arrangements are common. Fatty differentiation can be scant (Figure 15.9). In the context of a paucicellular tumor with a low mitotic rate (1–3 mitoses per 10 HPFs),

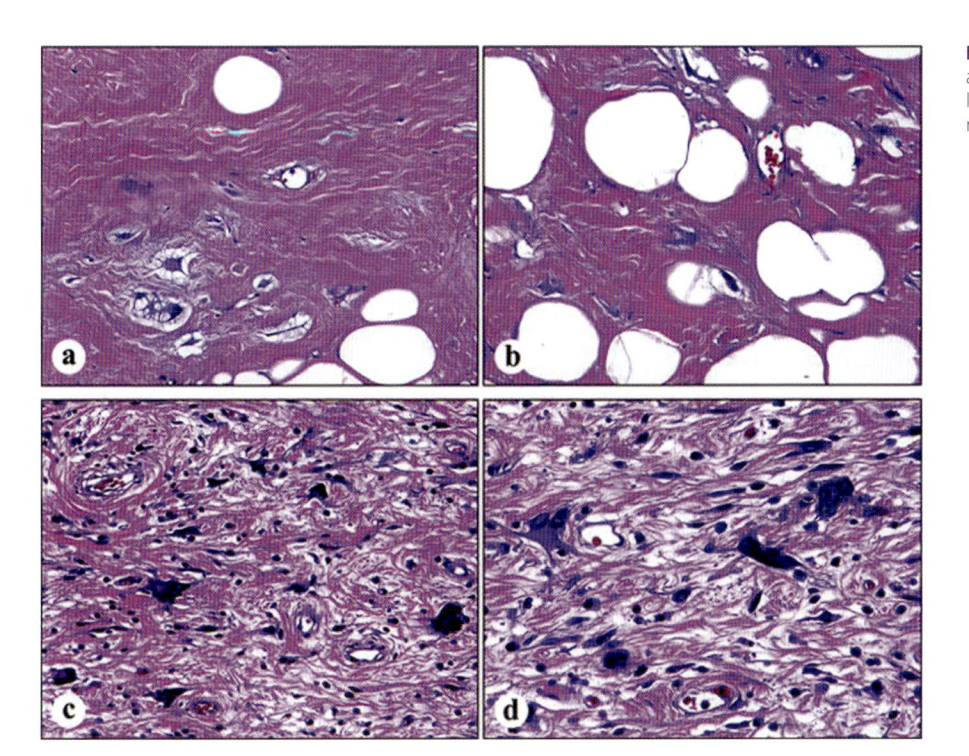

Figure 15.9 Sclerosing ALT/WDLS. Although adipocytes are scant or absent, this pattern with low cellularity and abundant collagenous matrix is not considered dedifferentiation.

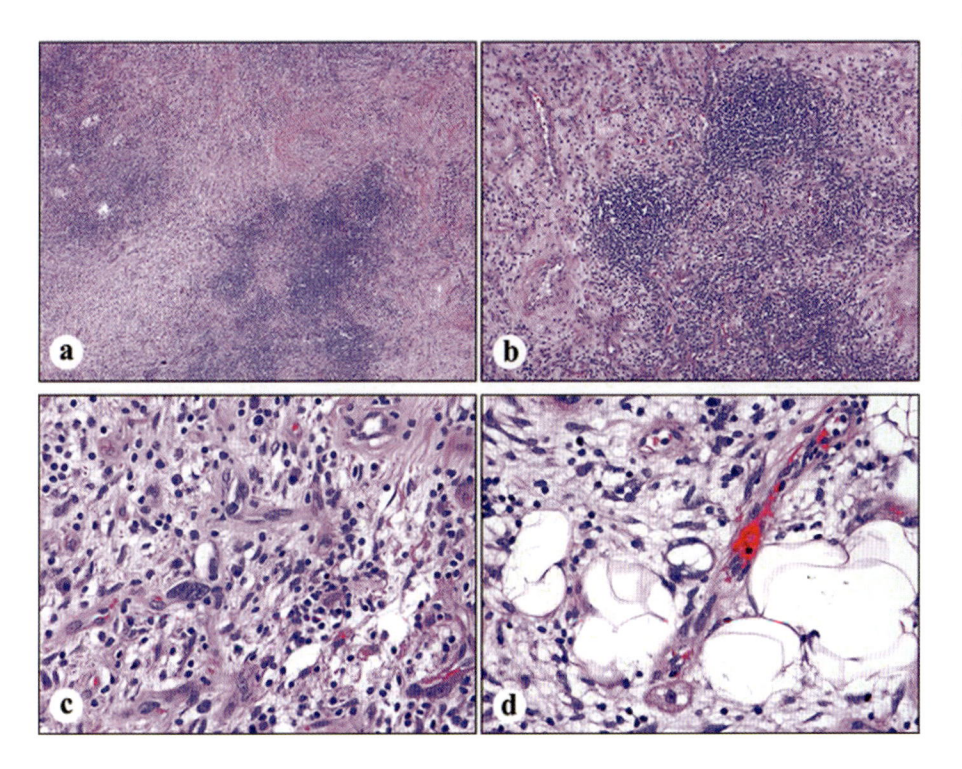

Figure 15.10 Inflammatory ALT/WDLS contains lymphoplasmacytic infiltration and fibrosclerosing stroma. Foci of atypical lipomatous components are present.

a sclerosing liposarcoma pattern is not to be confused with dedifferentiation.

Inflammatory (lymphocyte-rich) ALT/WDLS

The inflammatory variant of WDLS is most often seen in the retroperitoneum, but can also occur in peripheral soft tissues, such as the head and neck, trunk wall, and thigh.[22,23] This variant contains a prominent lymphoid infiltration, sometimes with germinal center formation, and in some cases this infiltration is grossly visible as discrete nodules. The tumor can have a myxoid-sclerosing stroma and contain abundant plasma cells, and should not be confused with an inflammatory process, such as fibrosclerosing disease or Rosai–Dorfman disease. Lipomatous atypia, the presence of lipoblasts, or both in the fatty components allows for the identification of this as a WDLS (Figure 15.10). The nonlipomatous atypical cells in

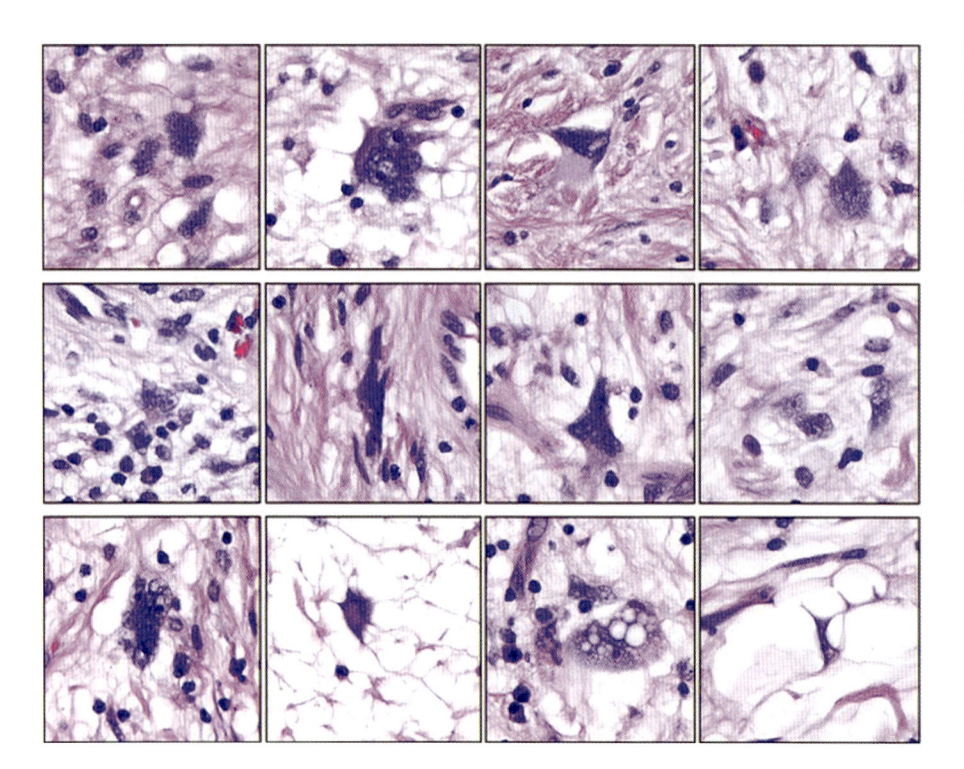

Figure 15.11 Examples of atypical nonlipogenic cells and atypical adipocytes in ALT/WDLS with inflammatory features. Among the atypical cells, nonlipogenic ones predominate, including multinucleated forms and those resembling Hodgkin cells. The two lower right panels show atypical multivacuolated adipocytes.

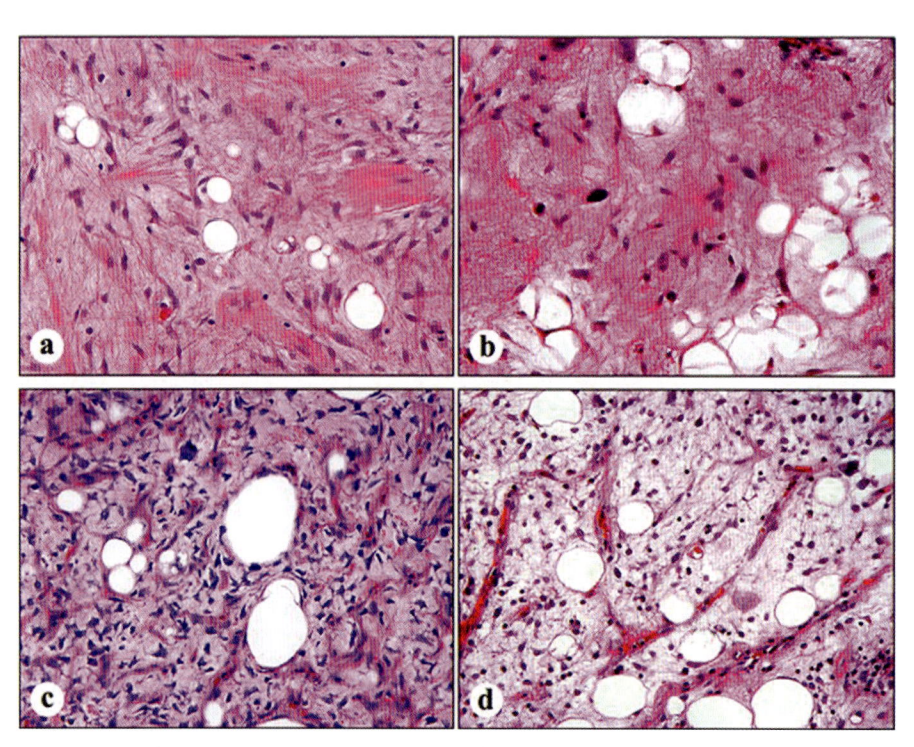

Figure 15.12 Spectrum of ALT/WDLS of the retroperitoneum with myxoid features. Note the focal presence of multivacuolated lipoblasts (**a,c**), and prominent capillaries with some resemblance to myxoid liposarcoma (**c,d**).

ALT/WDLS have many variations, including binucleated cells with large nucleoli resembling Hodgkin cells (Figure 15.11).

The lymphoid component has been found to contain B-cell nodules[22] and diffuse T-cell infiltration, and T-cell gene rearrangement studies have shown no evidence of clonality, supporting the fact that the lymphoid element is reactive, probably a host reaction to the liposarcoma element.[23]

Myxoid components in ALT/WDLS

Significant myxoid component is often present, especially in the retroperitoneal variant of WDLS. This can create a resemblance to MLS, but the presence of scattered atypical adipocytes is diagnostic of WDLS (Figure 15.12). At present, the difference between the myxoid WDLS form and MLS is

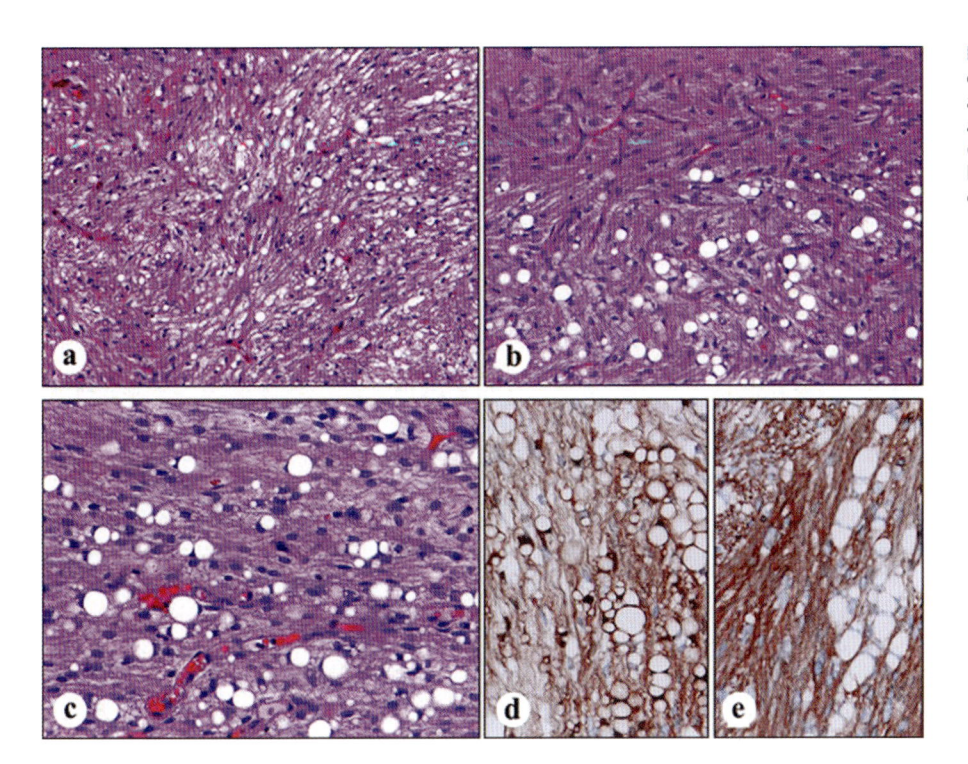

Figure 15.13 (a–c) Spindle cell liposarcoma is composed of mildly atypical spindled cells admixed with atypical, mainly univacuolar adipocytes. Note a prominent capillary pattern. (d) Significant S100 protein positivity is present in both adipocytes and the nonlipogenic spindle cells. (e) The spindle cells are positive for CD34.

well understood; the latter type contains delicate capillaries and evenly spaced small oval cells and is very rare in the retroperitoneum. In the past, however, myxoid WDLS in the retroperitoneum was sometimes classified as MLS. Myxoid ALT/WDLS, even with no adipocytes in some areas, is not to be classified as dedifferentiation when maintaining a low cellularity and low mitotic rate (only occasional mitoses present).

Spindle cell liposarcoma

Spindle cell liposarcoma is a designation that has been applied variably in the literature. The concept is here aligned with ALT. Earlier reports of spindle cell liposarcoma mostly included superficial nonmetastasizing tumors, but rare metastases were reported for retroperitoneal tumors, probably explained by overlap with forms of DDLSs.[24] Newer studies incorporating morphology and genetics indicate that a majority of tumors considered spindle cell liposarcomas are atypical variants of spindle cell lipoma, based on genetic studies. Some cases have MDM2 amplification and thus are probably related to ALT/WDLS.[25,26]

Fibrosarcoma-like lipomatous neoplasm is a term coined by Deyrup et al. for tumors that are in the spectrum of spindle cell liposarcoma.[26] This definition excluded cases related to spindle cell lipoma (based on losses in 13q) and also those that are related to ALT/WDLS or DDLS. These tumors occurred in various locations with some predilection to thigh and inguinal region. Histologically typical of spindle cell liposarcoma/fibrosarcoma-like lipomatous neoplasm is the presence of mildly atypical spindle cell proliferation containing scattered atypical adipocytes, with prominent perinuclear vacuoles. These cells have been characterized as "ice cream cone lipoblasts."[26] Myxoid matrix is variably present (Figure 15.13).

Immunohistochemistry of ALT/WDLS

Like normal fat cells, atypical adipocytes are variably S100 protein positive, and lipoblasts are often highlighted. CD34 positivity is common in the spindle cells and fat cells in well-differentiated liposarcomas (Figure 15.13e).[27] Desmin and even focal keratin positivity also can occur. Nuclear MDM2 positivity in ALT/WDLS can help to separate them from ordinary lipomas, but one has to consider that not all ALTs are MDM2 positive and also some nonlipomatous cells (histiocytes) can be positive.[28–31] Li–Fraumeni-syndrome-associated ALTs are specifically MDM2 negative, but instead express p53.[17] Nuclear cyclin-dependent kinase 4 (CDK4) positivity also reflecting an amplified and overexpressed gene has been applied as an alternative marker to MDM2.[28–31] Another potential but less-documented marker is nuclear p16 positivity.[32,34]

Genetics

The most important genetic change described in ALT/WDLS is the gene amplification 12q13–15. The giant marker chromosomes and ring chromosomes seen in cytogenetic studies are derived from this chromosomal region.[35–39] The most important genes located in this area are MDM2, and CDK4. Other genes include Gli-1 and tetraspanin 31/SAS (sarcoma amplified sequences).

Table 15.2 Recommendations for use of MDM2 amplification assay for evaluation of lipomatous tumors

1. Lipomatous tumors with equivocal/uncertain atypia
2. Recurrent tumors diagnosed as lipomas
3. Deep (intramuscular) extremity-based lipomas without atypia if >15 cm
4. Retroperitoneal and mediastinal (body-cavity-based) lipomatous tumors without cytologic atypia

Adapted from Zhang et al.[42]

MDM2 amplification tested by FISH is generally considered superior to immunohistochemical testing. However, the sensitivity of the MDM2 amplification test depends on the set threshold for percentage of amplification-positive cells (often 10%), so that negative results may not rule out ALT/WDLS. The other reason for false-negative results on MDM2 amplification is that this test does not detect ALTs arising via alternative mechanisms, such as p53 aberrations, as seen in Li–Fraumeni syndrome.[17] When using MDM2 immunohistochemistry as a screening test, one should note that nuclear MDM2 positivity occurs in nonlipomatous cells, such as histiocytes, so that attention to the character of positive cells is necessary.[31,40–42] The recommended diagnostic approach for evaluation of possible ALT by MDM2-FISH has been summarized in Table 15.2.

Differential diagnosis

Conditions that can simulate ALT/WDLS include fat necrosis, reactive fatty changes in extremely obese persons, and categories of benign lipoma, especially spindle cell, pleomorphic, and chondroid lipoma. Lipogranuloma and silicone deposition can resemble WDLS, at least superficially. Mild nuclear atypia and multinucleation can occur in retroperitoneal fat around tumors such as kidney cancers.[43] Clinicoradiologic correlation for a lipomatous mass and MDM2 amplification studies can be helpful in problem cases. However, MDM2 amplification is not a prerequisite for ALT/WDLS, as other mechanisms may be responsible in some cases.

Sclerosing extramedullary myeloid tumor with atypical megakaryocytes in a fibrosclerotic background also simulates liposarcoma; these tumors occur in connection with chronic myeloproliferative disorders. This distinction is aided by documentation of the large atypical cells as megakaryocytes (i.e., FVIIIRAg, CD61 positive), and observation of other immature hematopoietic components, especially eosinophilic myelocytes.[44]

Deep lipomatous tumors with limited atypia (especially those in the retroperitoneum) are often problematic, and they should be sampled extensively. Even in the absence of significant atypia, small biopsies of large retroperitoneal tumors should be diagnosed as well-differentiated lipomatous tumors, with the statement added that the possibility of WDLS cannot be definitely ruled out.

Dedifferentiated liposarcoma

DDLS refers to the presence of a highly cellular and mitotically active nonlipogenic component devoid of intermingling fat cells in conjunction with well-differentiated liposarcoma. An area of $1\,cm^2$ in diameter or a low-power field have been suggested for a minimum criterion.[12,45] Dedifferentiation is a gradual event, so that there are cases in which its definition is arbitrary. Evans also includes a mitotic rate threshold, a minimum of 5 mitoses per 10 HPFs for a definition of DDLS.[4]

Radiologic studies, specifically MRI, are often helpful in pinpointing a lipomatous component in an undifferentiated sarcoma and suggesting the diagnosis of DDLS when sampling is limited.[46] This scenario especially applies to needle biopsies and is common in retroperitoneal tumors. Thorough sampling and special attention to possible lipomatous tumor elements in an undifferentiated sarcoma can help to identify DDLS. For example, studies on retroperitoneal pleomorphic sarcomas originally diagnosed as malignant fibrous histiocytoma (MFH) have found a high frequency of DDLS on careful reclassification. In those studies, retroperitoneal MFH sometimes recurred as WDLS, supporting the idea that many of these tumors in fact are variants of DDLS.[47–49]

Clinical features

DDLS predominantly occurs in older adults 50 years of age or older, with a significant male predominance (Figure 15.14). Nearly one-half of these cases occur in the retroperitoneum. Other relatively common locations are thigh, scrotum, and groin area (Figure 15.15).[14,45–55]

Dedifferentiation is more often already present in the primary tumor, but in a minority of cases it is seen in a recurrence. Retroperitoneal DDLSs often form large (>20 cm) complex masses that involve neighboring organs, such as the kidneys and intestines, requiring resection of those organs when curative surgery is attempted. Multiple noncontiguous masses are sometimes present. In the abdomen, DDLS can clinically simulate other tumors, especially gastrointestinal stromal tumor (GIST).

Based on follow-up studies, DDLSs have a strong tendency to recur locally, and in the abdominal and thoracic cavity these recurrences often become uncontrollable over time and can cause death, even without metastases. Tumor-related mortality has varied from 30% to 50% for body cavity sites.[48–56] Pulmonary or liver metastases develop mostly in high-grade examples. DDLS patients with lung metastases often also have extrapulmonary metastases and clinically progressive disease.[57,58] In one study, median survival was only 5 months after development of pulmonary metastases.[57] In two series of scrotal DDLS, 11–20% of patients died of metastatic disease.[14,45] In peripheral soft tissue, including the extremities, trunk wall, and neck, the prognosis is better. In one study, 4 of 27 peripheral DDLSs recurred and 3 of 27 (11%)

Figure 15.14 Age and sex distribution of 226 patients with dedifferentiated liposarcoma.

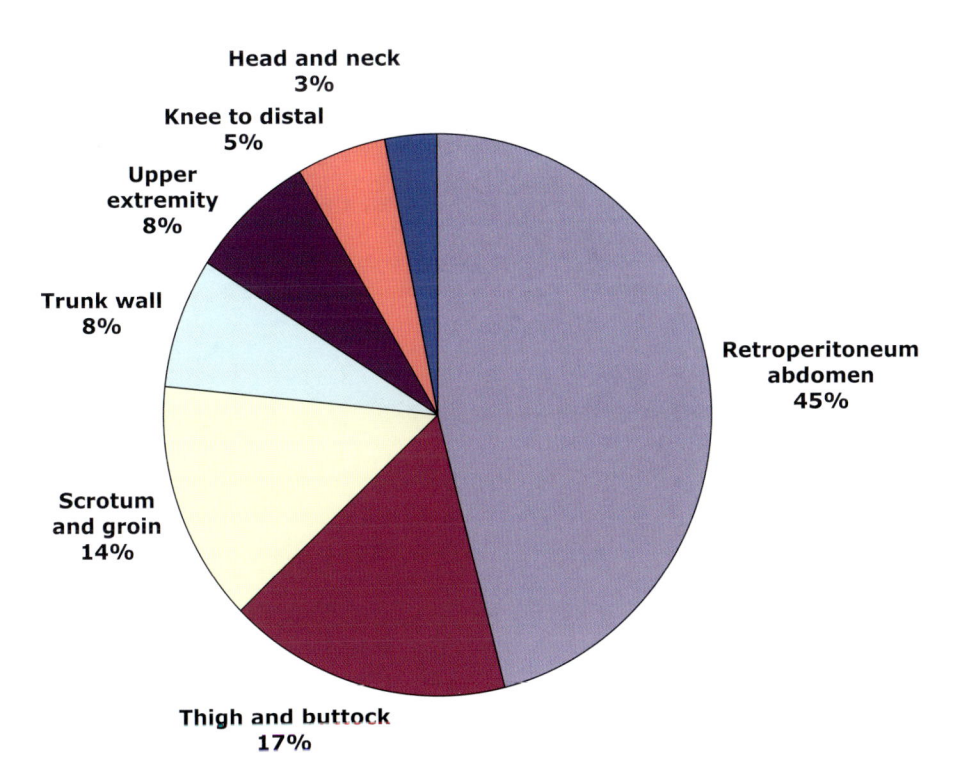

Figure 15.15 Anatomic distribution of 229 dedifferentiated liposarcomas.

metastasized.[45] Grade, multifocality, and resectability are prognostic factors.[47]

Pathology

Grossly, the dedifferentiated components are often distinctive. Areas of dedifferentiation often appear as sharply demarcated, soft white nodules flanked by the differentiated lipomatous component (Figure 15.16). The low-grade dedifferentiated components can be firm and white to tan, whereas the high-grade components are often variegated by areas of gross necrosis.

Histologically, the dedifferentiated components vary from a minor to an overwhelmingly dominant component, and they can have an abrupt or gradual transition from the well-differentiated areas (Figure 15.17). Their histologic

Chapter 15: Atypical lipomatous tumors and liposarcomas

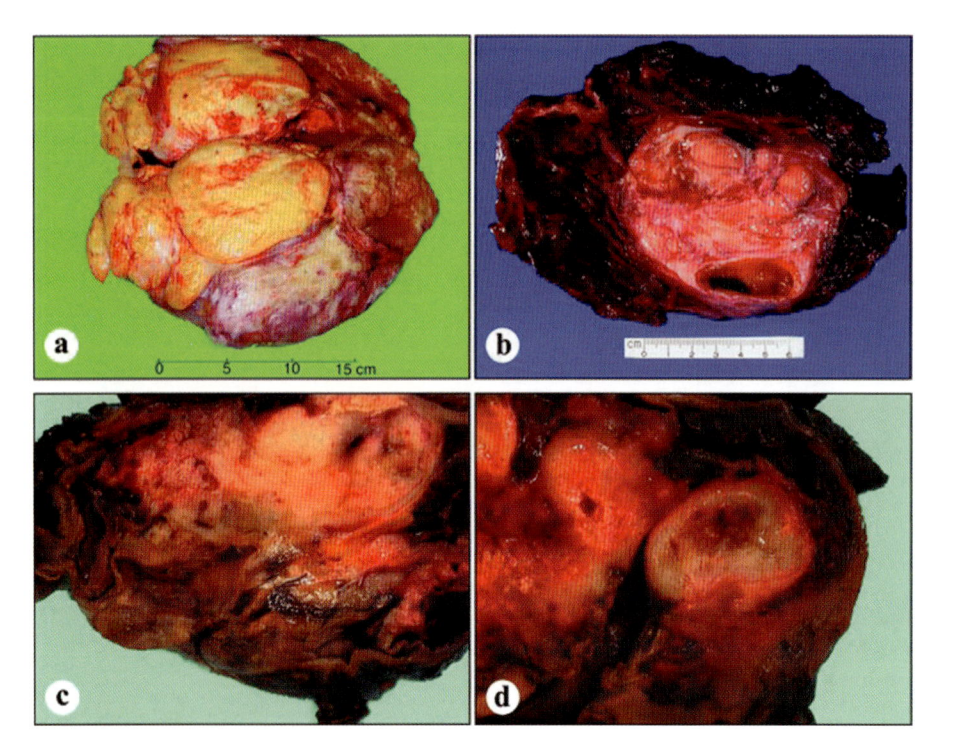

Figure 15.16 Four gross examples of dedifferentiated liposarcoma. The lipomatous component is seen as a yellow portion in the center (**a**) or periphery of the lesion (**c**), but it can also be difficult to notice on gross examination (**b,d**). The dedifferentiated components vary from grayish to tan, and all cases show areas of yellow-green necrosis.

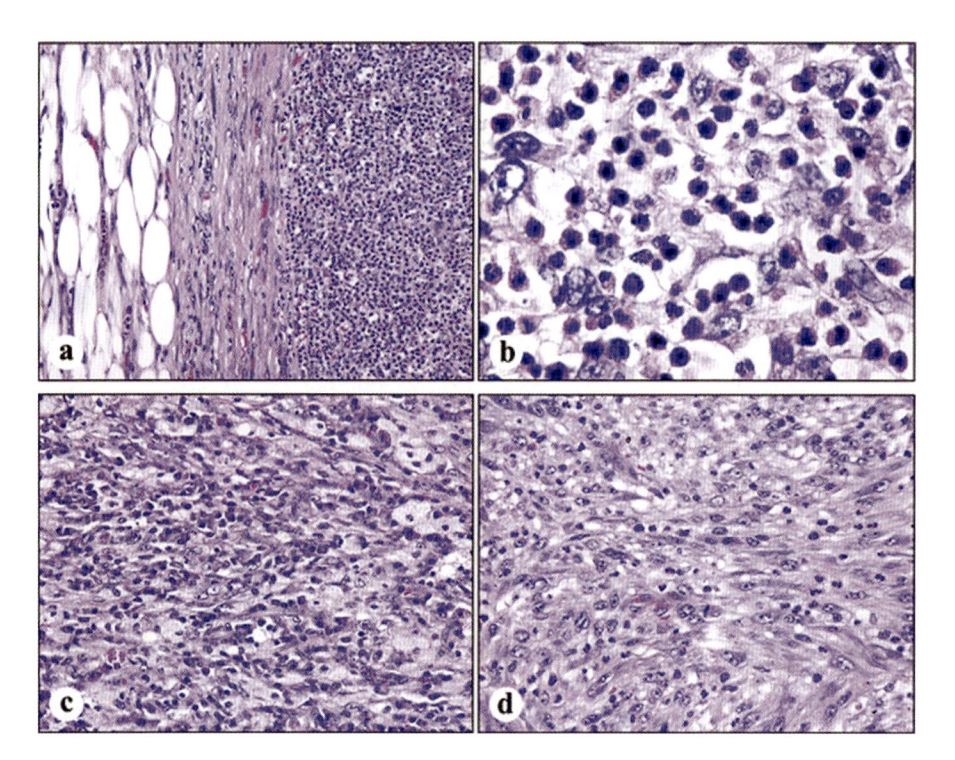

Figure 15.17 Example of dedifferentiated liposarcoma resembling inflammatory malignant fibrous histiocytoma. (**a**) Low magnification shows sharp demarcation between the components. (**b**) Scattered atypical cells are present among neutrophils. (**c**) Abundant xanthoma cells and plasma cells obscure the tumor cells. (**d**) Spindle cell sarcomatous pattern resembling myofibroblastic sarcoma.

appearances vary greatly, including low- and high-grade variants of different types. These include patterns resembling myxofibrosarcoma, solitary fibrous tumor/hemangiopericytoma, fibromatosis, fibrosarcoma, and GIST (Figure 15.18). Inflammatory and pleomorphic MFH patterns also occur (Figure 15.17). Low- or intermediate-grade tumors with relatively low mitotic count are more common than high-grade ones. Pleomorphic liposarcoma pattern can occur in DDLS and has been referred to as "homologous dedifferentiation."[59]

Meningothelial-like whorls forming concentric perivascular proliferations of slender spindle cells with minor atypia occur in rare examples of DDLS, especially in the retroperitoneum (Figures 15.19 and 15.20) This pattern is often

Chapter 15: Atypical lipomatous tumors and liposarcomas

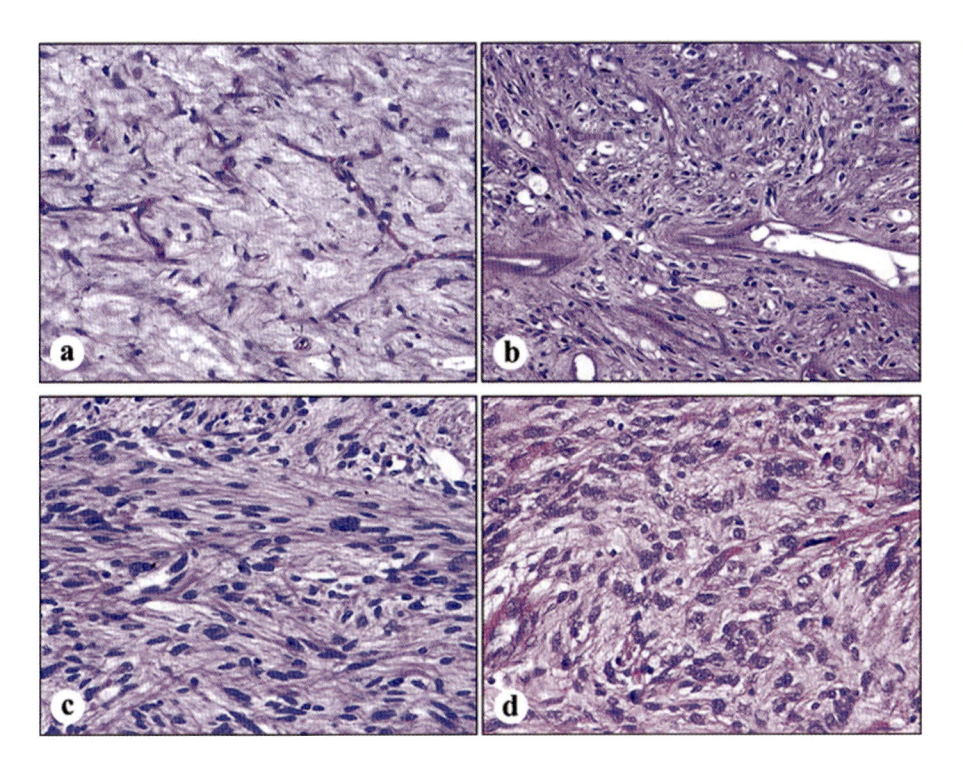

Figure 15.18 Four examples of dedifferentiated liposarcoma simulating different sarcoma types: (**a**) Myxofibrosarcoma. (**b**) Solitary fibrous tumor. (**c**) Fibrosarcoma. (**d**) Gastrointestinal stromal tumor.

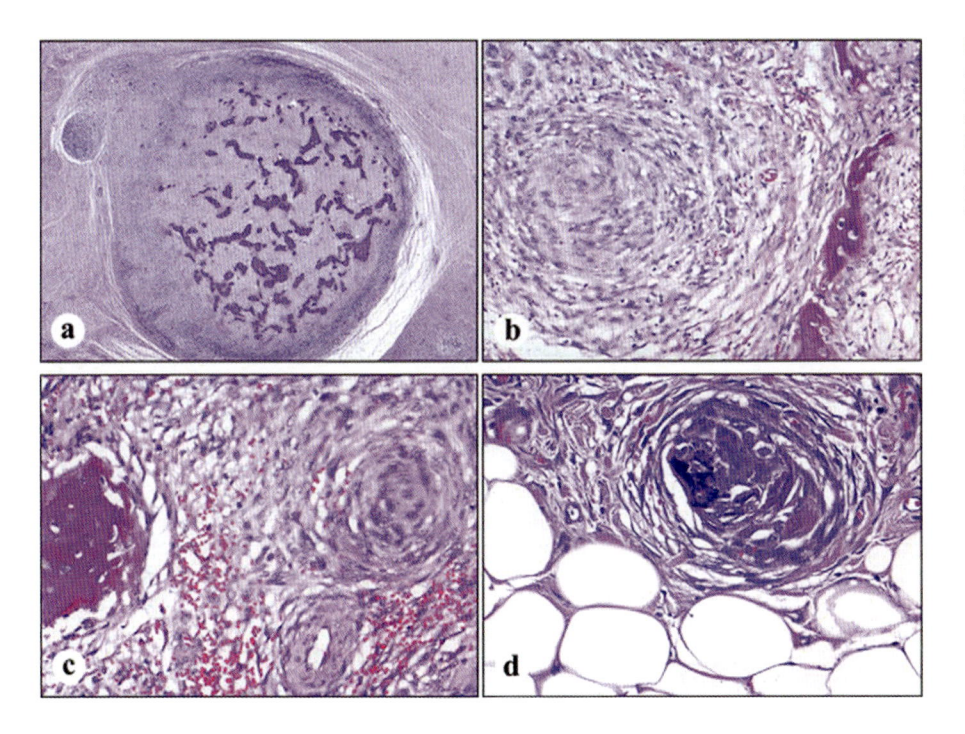

Figure 15.19 Dedifferentiated liposarcoma with concentric perivascular spindle cells resembling meningothelial whorls. (**a**) Such nodules are often associated with osseous differentiation. (**b**) Meningothelial nodule at 10 o'clock on side of an area with bone formation. (**c,d**). The meningothelial nodule is intimately associated with bone formation.

accompanied by metaplastic bone and displays myofibroblastic, not true meningothelial, differentiation on immunohistochemical and ultrastructural studies. Local recurrences are common, and metastases occur in one-third of patients in long-term follow-up. Some patients live a long time even after pulmonary metastases, but long-term prognosis is in the range of DDLS in general.[53–56]

Heterologous elements, especially rhabdomyosarcomatous, leiomyosarcomatous, osteosarcomatous, and chondrosarcomatous elements occur in some DDLSs. In some cases, this is histologically obvious, but more often rhabdomyosarcomatous differentiation is a subtle change documented by immunohistochemistry only (Figures 15.21–15.23).[60–64] Some DDLS with desmin positivity lacks the morphologic

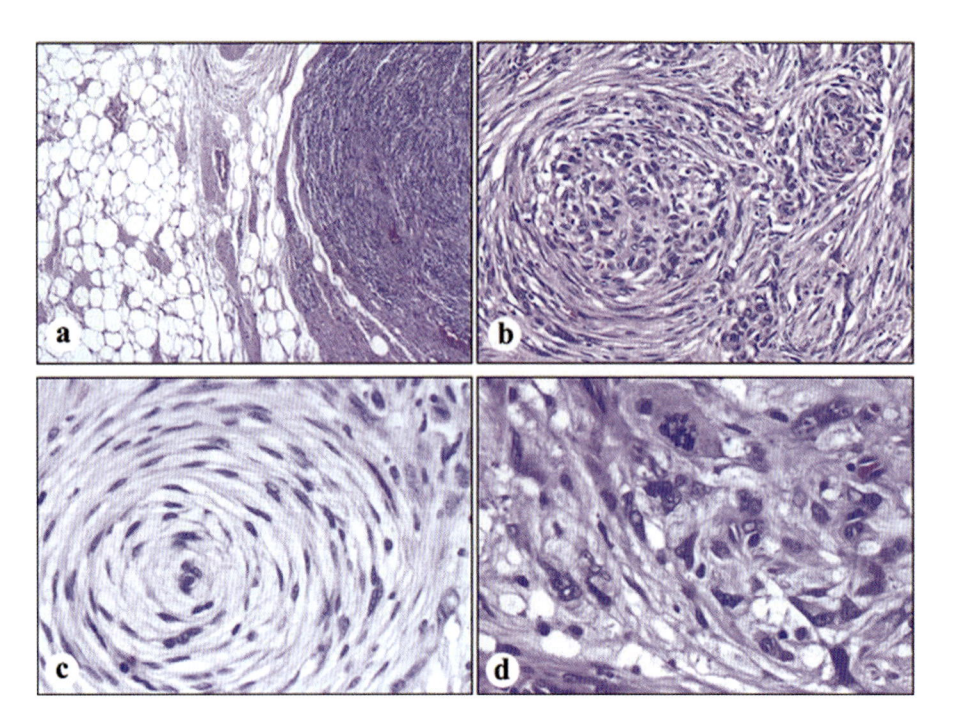

Figure 15.20 Dedifferentiated liposarcoma with meningothelial-like whorls. (**a**) Abrupt demarcation between the well-differentiated and dedifferentiated components. (**b**) In some cases, meningothelial whorls are composed of more highly cellular oval cells. (**c**) Delicate spindle cells forming a distinct whorl composed on concentric lamellae of spindle cells. (**d**) An atypical mitosis in a dedifferentiated liposarcoma with meningothelial whorls.

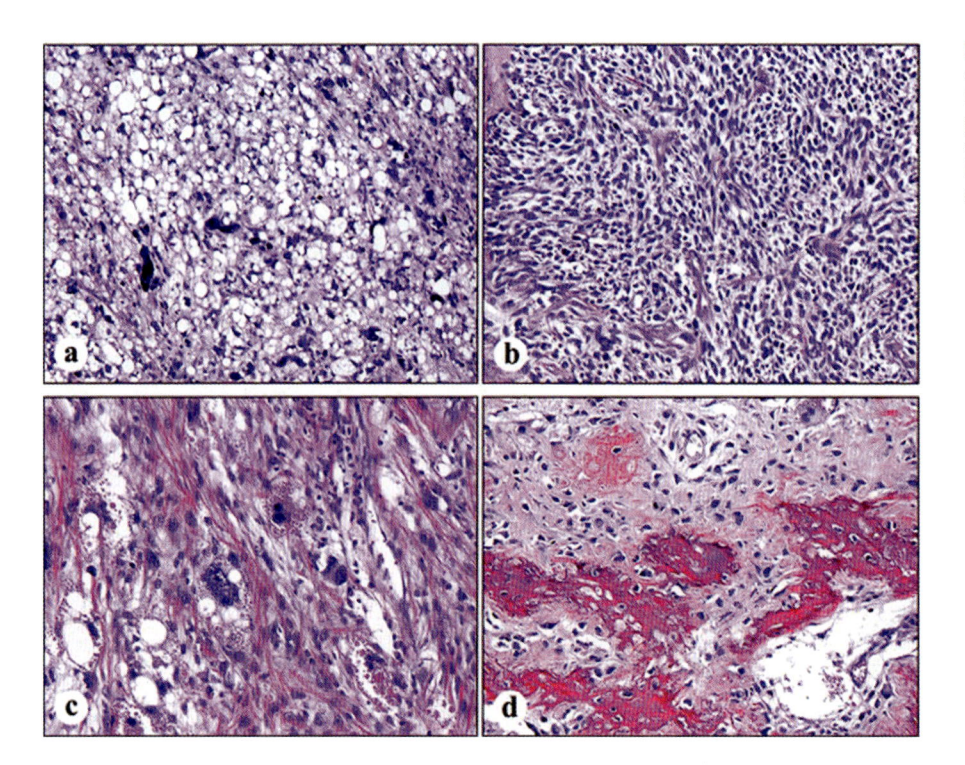

Figure 15.21 Different patterns in dedifferentiated liposarcoma. (**a**) A pleomorphic liposarcomatous component. (**b**) A mildly pleomorphic spindle cell sarcoma. (**c**) Pleomorphic component with large cells containing eosinophilic hyaline globules. (**d**) Osteosarcomatous differentiation.

and immunohistochemical (i.e., heavy caldesmon, myogenin/MyoD1) evidence for specific smooth or skeletal muscle differentiation.

Skeletal muscle differentiation in DDLS does not seem to influence the prognosis, because such tumors tend to have a metastatic potential as low as other DDLSs, especially compared with retroperitoneal leiomyosarcomas, based on one study.[64] This is also consistent with the finding of skeletal muscle differentiation in some cases of ALT/WDLS.[4] According to the French Federation's grading system (Chapter 1), DDLS is assigned 3 points for differentiation, resulting in an intermediate grade at the minimum. Tumors with combinations of necrosis and mitotic rate ≥ 10 per 10 HPFs would thereby be assigned a high grade.

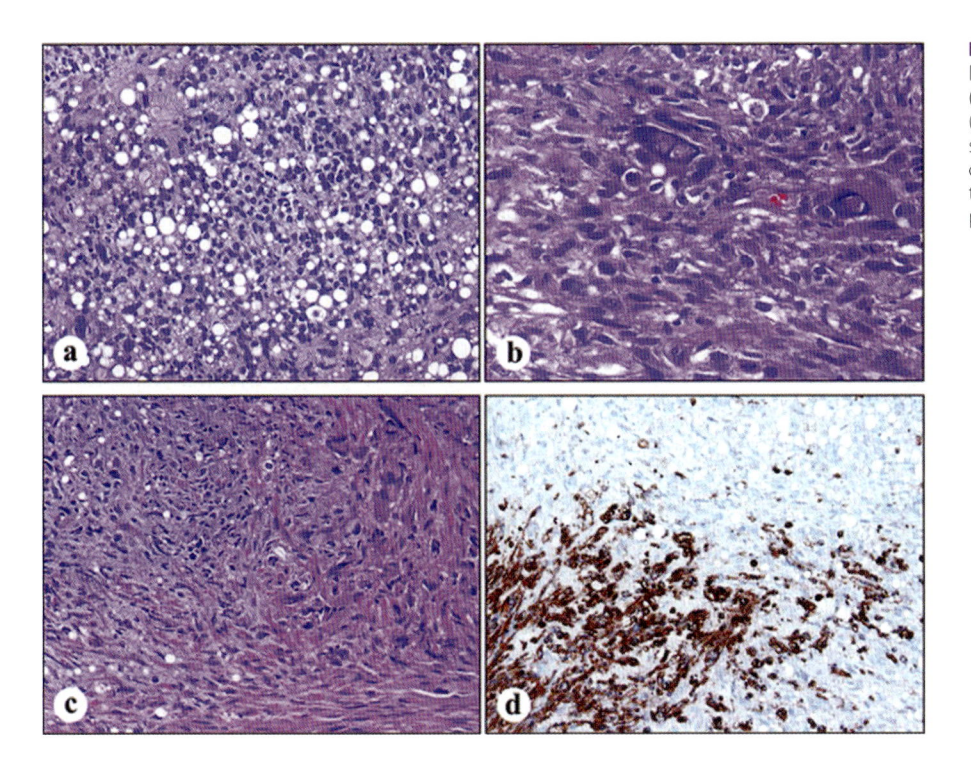

Figure 15.22 Dedifferentiated liposarcoma with leiomyosarcomatous differentiation (**a**) Pleomorphic liposarcoma component. (**b**) A pleomorphic MFH-like component without specific differentiation. (**c**) A leiomyosarcomatous component is distinguished by its eosinophilic tinctorial quality. (**d**) This component is strongly positive for heavy caldesmon.

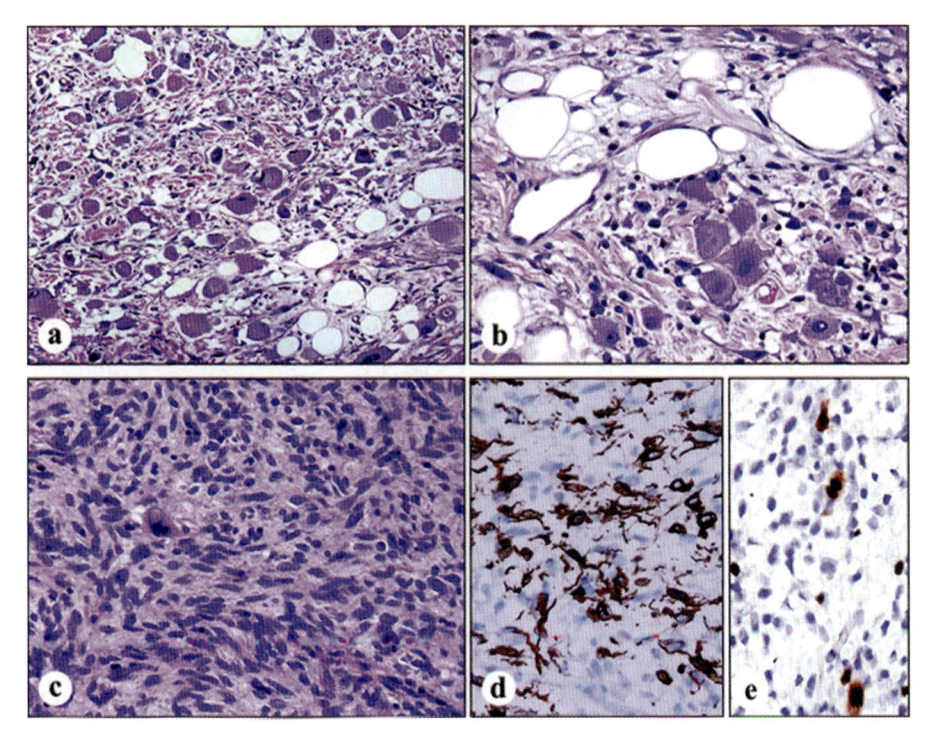

Figure 15.23 Two examples of dedifferentiated liposarcoma with rhabdomyosarcomatous differentiation. (**a,b**) An example with well-differentiated rhabdomyoblasts admixed with a lipomatous component. (**c**) A histologically less obvious example with scattered large rhabdomyoblasts. (**d**) The number of desmin-positive cells is larger than expected based on HE stain. (**e**) Myogenin-positive nuclei confirm skeletal muscle differentiation.

Immunohistochemistry

Nuclear MDM2 expression is typical of DDLS and appears in varying proportions of tumor cell nuclei (10–50% or more). In addition, DDLSs are often positive for CD34. Tumors with heterologous smooth or skeletal muscle differentiation are positive for desmin and those with skeletal muscle differentiation, contain MyoD1 and myogenin-positive nuclei in varying numbers.

Differential diagnosis

DDLS can simulate many other sarcoma types and contain differentiation typical of tumors such as extraskeletal osteosarcoma and leiomyosarcoma. Very likely many tumors previously considered "malignant mesenchymomas" are DDLSs. Undifferentiated pleomorphic sarcoma/MFH-like histological pattern is common, and its association with DDLS is signaled by MDM2 immunoreactivity or gene

Figure 15.24 Age and sex distribution of 970 patients with myxoid liposarcoma.

amplification.[65] Fibrosarcoma-like examples can contain lymphoid reaction and should not be confused with inflammatory myofibroblastic tumors.[66] Such tumors typically occur in older adults. Immunohistochemical detection of KIT or DOG1/Ano1 supports the diagnosis of GISTs over DDLS.

Genetics and cell biology

MDM2 amplification is the typical finding in DDLS.[31] Giant marker chromosomes with amplification of 12q segments also characterized a cell line established from a retroperitoneal DDLS.[67] Loss of retinoblastoma protein (RB1 gene product) through allelic losses and promoter methylation also seems to be common in DDLS.[68] However, pathogenesis alternative to MDM2 gene amplification is possible, and therefore this finding is not absolutely required if other features are diagnostic.[69,70]

Expression profiling studies have also found overexpression of the proteins corresponding to the 12q13–15 amplicon, especially MDM2, CDK4, and GLI1.[6,56] One study found a >100-fold overexpression of DNA-topoisomerase II alpha (TOP2A) gene against adipose tissue (located at 17q21).[6]

Myxoid liposarcoma (including round cell liposarcoma)

Clinical features

MLS is the second most common liposarcoma type after WDLS/ALT and constitutes 20–25% of all liposarcomas. It occurs most commonly in young and middle-aged adults (Figure 15.24). The median age was 44 years, and there was no sex predilection in the largest modern series.[71] MLS is occasionally seen in children from 10 to 16 years of age and represents most of the pediatric liposarcomas.[71–74] Cellular myxolipomatous tumors in the first decade are usually lipoblastomas, however.

MLS usually occurs intramuscularly in the deep soft tissues of the extremities. Nearly one-half of cases are located in the thigh, where the tumor often reaches a size of >15 cm. Another common site is the knee region, especially the popliteal fossa, but MLS almost never occurs primarily in the retroperitoneum (Figure 15.25). Most intra-abdominal

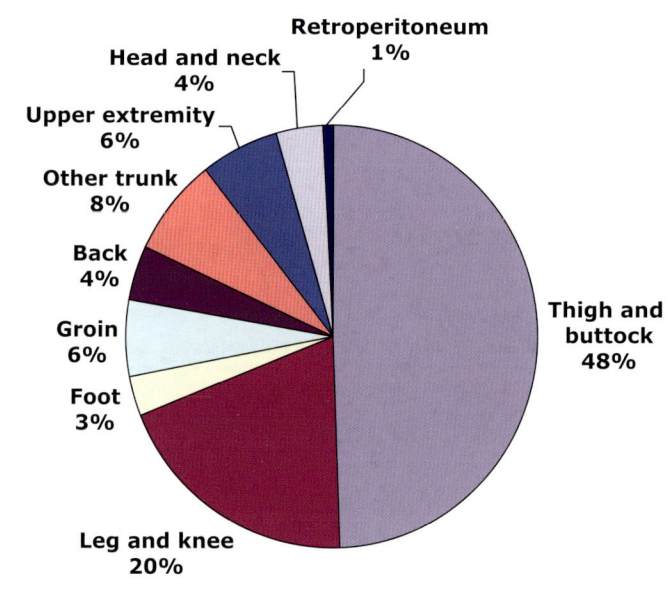

Figure 15.25 Anatomic distribution of 973 cases of myxoid liposarcoma.

liposarcomas with grossly myxoid features are not MLSs but rather WDLSs with a myxoid component. Some MLSs are primarily subcutaneous, and distal examples (e.g., in the foot) often appear superficial, but the true level of involvement is difficult to ascertain in those sites without a radiologic correlation.

MLS has a high recurrence rate if it is incompletely excised. Approximately 20% to 40% of patients develop metastases in long-term follow-up.[71,75] MLS has a peculiar tendency to metastasize to peripheral soft tissues, which raises the question of multiple primary tumors versus metastatic disease. Genetic comparison has established single clonal origin in such cases, confirming metastases rather than multiple independent tumors.[76] Lung and liver metastases occur in a few cases. Two large studies found metastases in 30% to 35% of patients.[71,77] The author has seen one patient who died of massive pericardial sac metastasis 2 years after the excision of MLS of the thigh. Bone metastases also can occur.

Tumors with a round cell component show a significantly higher risk for metastasis. Significance of the round cell component has been shown at the levels of 25%[71] and another at 5% in different studies.[75] Spontaneous tumor necrosis has also been identified as an unfavorable prognostic factor and there seems to be worse prognosis by increased proportion of round cell areas.[71] One study found infiltrative margins to be a significant adverse prognostic factor.[78]

The optimal treatment is wide excision. If the tumor cannot be excised widely because of its proximity to neurovascular bundles, postoperative radiation or chemotherapy may prevent recurrence or metastasis.[79] Although one study found a median survival of 35 months following distant metastases, excision of metastases from soft tissue and lung has been found to be advantageous.[77]

Gross pathology

Grossly, MLS is sometimes detected as a small tumor <5 cm in diameter, especially when located in the upper extremity. When located in the deep thigh muscles, these tumors typically reach a size of 15 cm or larger. On gross inspection, MLS is typically gelatinous, jelly-like, and soft and friable on sectioning. The cut surface is glistening and tan, often with a red tinge from its high vascularity (Figure 15.26). Large tumors can have significant spontaneous necrosis.

Histologic features

Histologically, a typical MLS is composed of evenly dispersed, relatively small oval or plump cells with scant cytoplasm in myxoid matrix with variable numbers of fat cells (Figure 15.27). There is a background with prominent, arborizing, thin-walled, fine capillary vessels, often seen in a "chicken-footprint" configuration (Figure 15.28). In some cases, there is abundant well-differentiated fat, especially in the periphery of the tumor, and this focally resembles

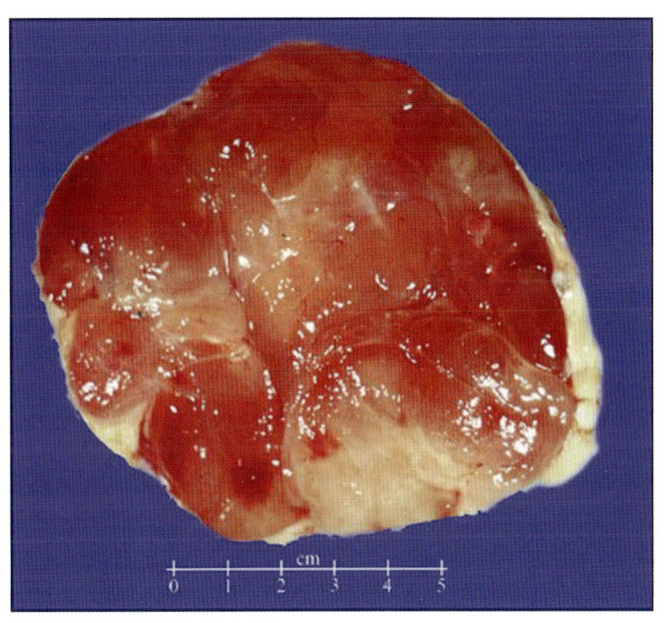

Figure 15.26 Grossly, myxoid liposarcoma has a gelatinous yellowish appearance with red-tinged foci.

WDLS or even lipoma. Small numbers of signet ring cells and multivacuolated lipoblasts are usually present, but are not required for diagnosis. Prominent cystic change reminiscent of histologic patterns of lymphangioma or pulmonary edema is a common feature (Figure 15.29). Some tumors contain multivacuolated, hibernoma-like fat cells (Figure 15.30). Chondroid differentiation has been described.[80]

Evolution into dense, hypercellular areas (round cell liposarcoma pattern) can be seen focally in an otherwise typical MLS, or rarely as the only pattern in a tumor (pure round cell liposarcoma). These tumors show corded arrangements or solid sheets of round cells with only focal vacuolization as evidence of adipocytic differentiation (Figure 15.31). The hypercellular components can have a hemangiopericytoma-like vascular pattern, and highly cellular cases can resemble lymphoma or other round cell tumors (Figure 15.32).

MLS following preoperative radiation typically has a nonspecific histologic appearance because the tumor cells and the prominent vasculature are replaced by nearly acellular hyalinized tissue with only scattered fat cells and mucous pools (Figure 15.33).[81]

Immunohistochemistry

S100 protein positivity, if found in round cell liposarcoma, may help to identify lipoblasts. In contrast to WDLSs, myxoid and round cell variants are usually negative for CD34, except for the endothelia.

According to one study, upregulation of CDK4 or CDK6 is very common (85%), and one-third of tumors have loss of retinoblastoma locus/product; overexpression of MDM2 gene is also common.[82] Expression of p27 (Kip1)[83] and p14 (ARF)[84] proteins has been found to be a favorable prognostic factor.

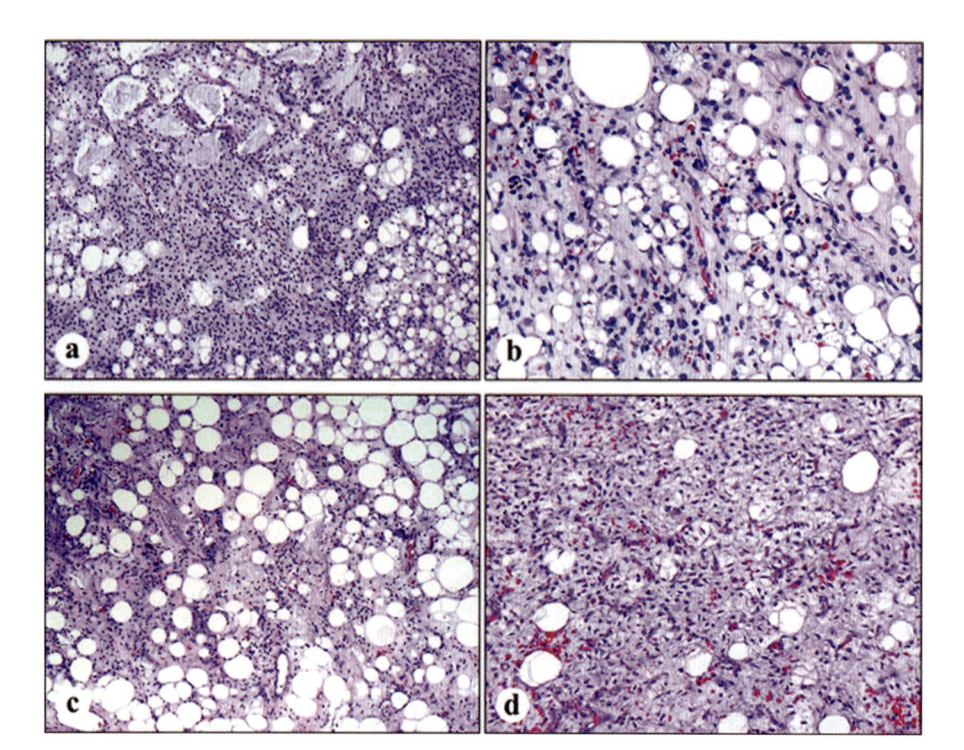

Chapter 15: Atypical lipomatous tumors and liposarcomas

Figure 15.27 Low magnification of myxoid liposarcoma showing variable numbers of adipocytes and a prominent capillary pattern.

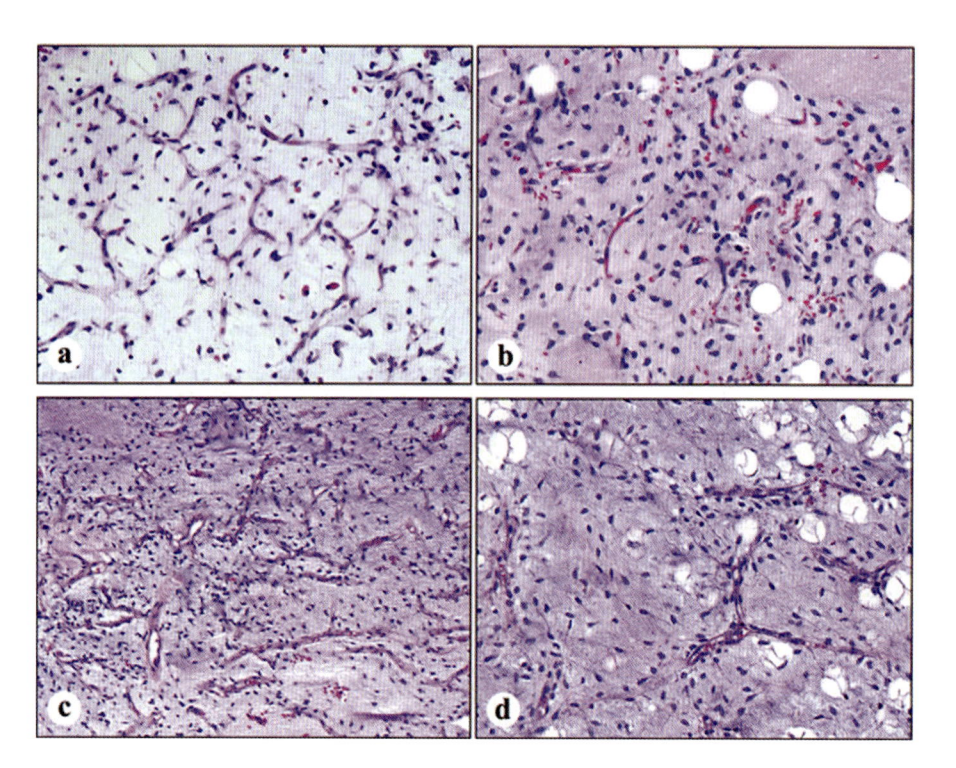

Figure 15.28 Myxoid liposarcoma with relatively low cellularity shows evenly dispersed, uniform oval tumor cells and a prominent, branching capillary pattern with variable numbers of nonatypical adipocytes.

Differential diagnosis

Intramuscular myxoma, lipoblastoma, WDLS with myxoid stroma, and myxofibrosarcoma (myxoid MFH), and myxoid spindle cell lipoma are the most important differential diagnoses. Intramuscular myxoma is a paucicellular tumor having a scant vascular pattern with relatively inconspicuous and small vessel profiles.

Lipoblastoma shows distinct lobulation and typically has greater maturation. It usually occurs during the first decade of life. Myxoid MFH and myxofibrosarcoma have a prominent

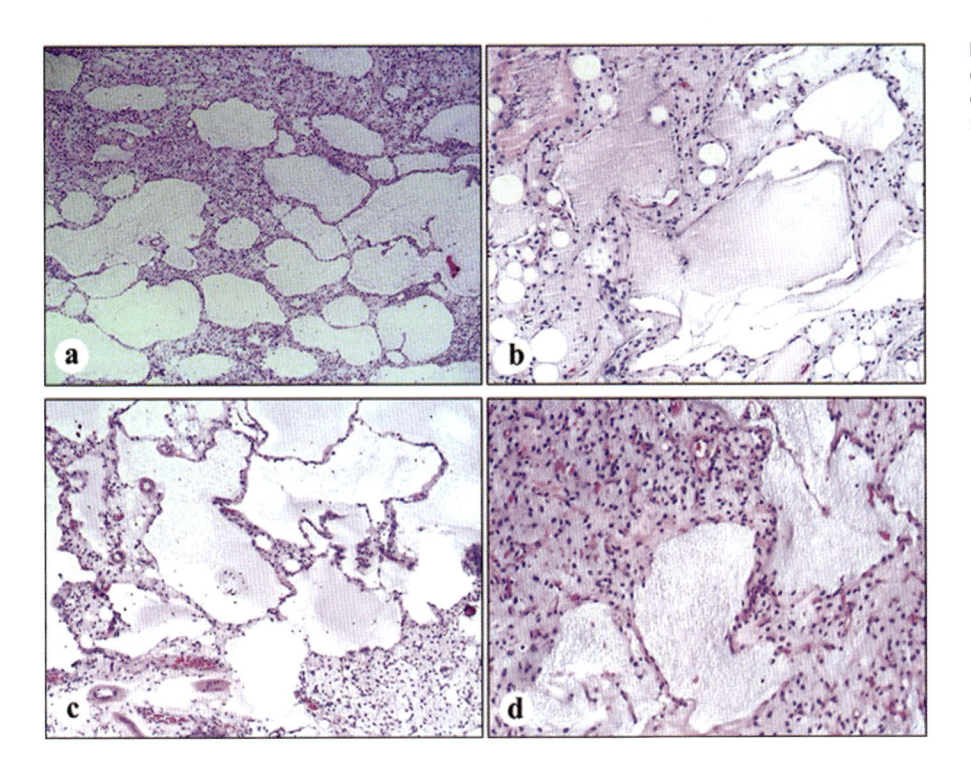

Chapter 15: Atypical lipomatous tumors and liposarcomas

Figure 15.29 Lymphangioma-like or pulmonary edema-like pattern in myxoid liposarcoma containing acellular spaces filled with mucoid material.

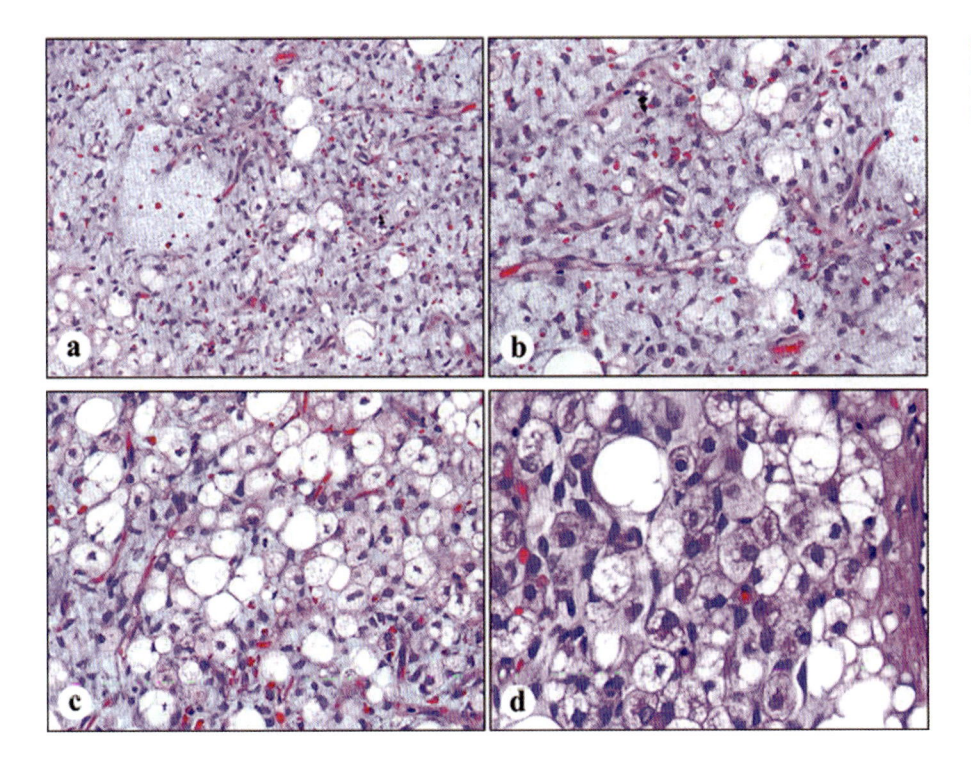

Figure 15.30 Hibernoma-like pattern in myxoid liposarcoma. Note multivacuolated hibernoma-like cells. White fat-like cells and a prominent, branching capillary pattern are also present.

vascular pattern, but the vessels are often thick walled and appear coarse; these tumors tend to have focal fibrous matrix and variable cellular pleomorphism. Extraskeletal myxoid chondrosarcoma is similarly myxoid, but the tumor cells have a rim of eosinophilic cytoplasm arranged in cords or rounded clusters in a hypovascular background. Myxoid spindle cell lipoma shows lesser vascularity, common infiltration of mast cells, and is CD34 positive, in contrast to MLS.

Round cell liposarcoma can resemble lymphoma or another round cell tumor, especially in a small biopsy without fatty differentiation. Immunohistochemical studies usually easily resolve these problems.

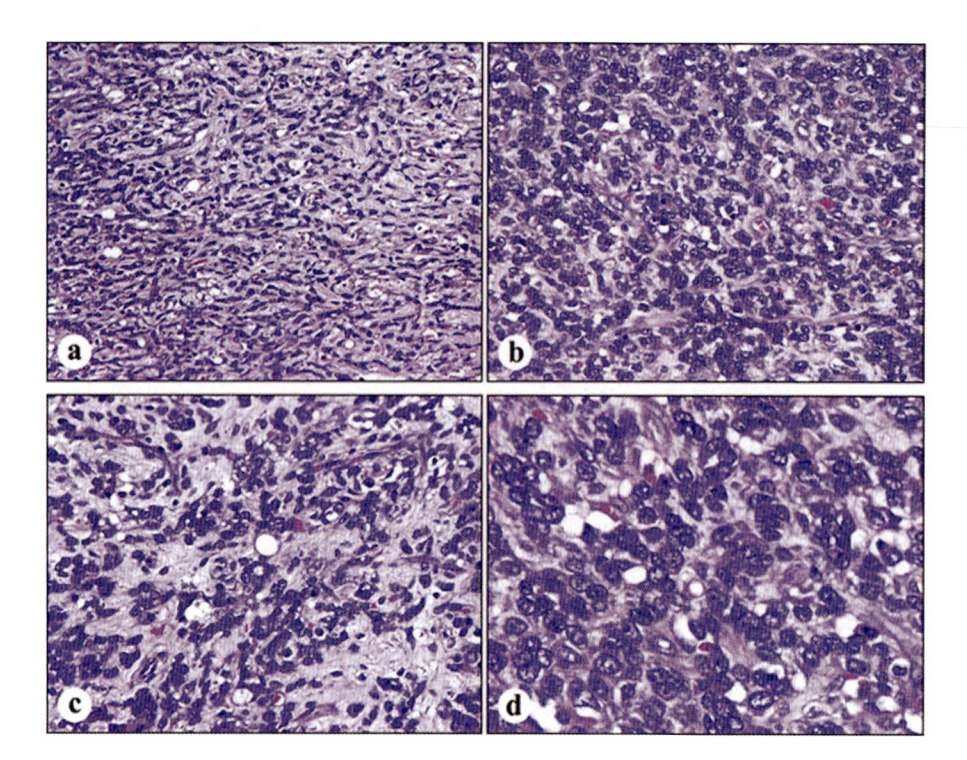

Figure 15.31 Myxoid liposarcoma with a highly cellular round cell liposarcoma pattern containing few cells with adipocytic differentiation. This tumor is typical of low-grade liposarcoma. (**a**) A trabecular pattern. (**b–d**) Dense round cell population and a scant adipocytic component can create an illusion of lymphoma.

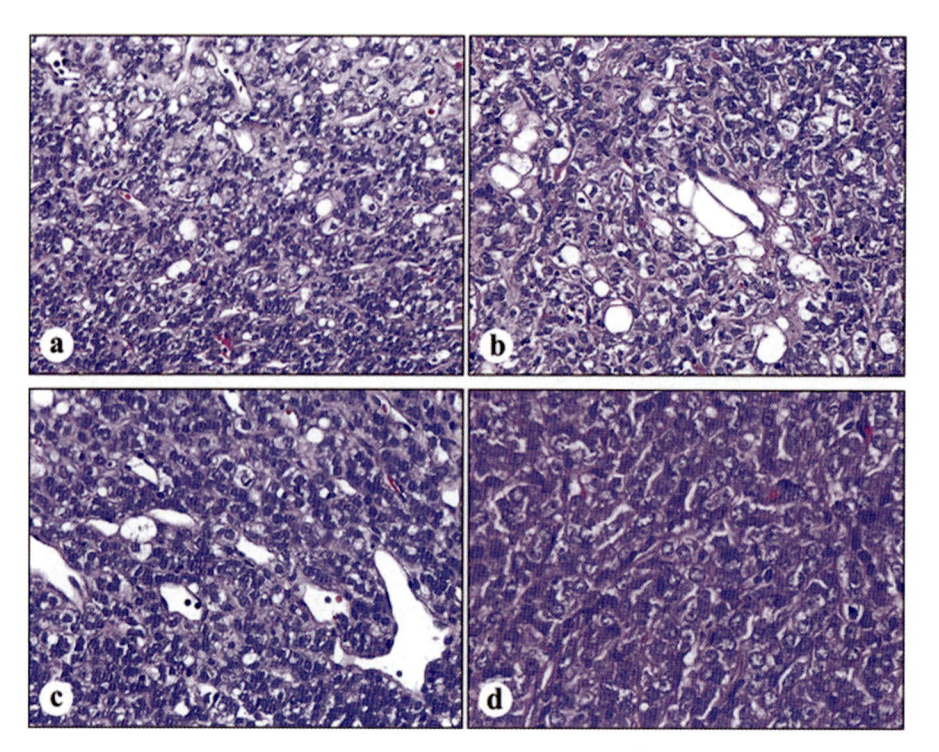

Figure 15.32 Myxoid liposarcoma with a round-cell liposarcoma pattern can show variable collagenous matrix (**a,b**), a hemangiopericytoma-like vascular pattern (**c**), and a solid cellular pattern resembling round cell sarcomas or lymphoma (**d**).

Genetics

The most common cytogenetic change in MLS is translocation t(12;16)(q13;p11).[85–90] The involved gene in the chromosome 12q13 is named DDIT3 (formerly CHOP); this gene encodes for a DNA-binding transcriptional regulator protein. The involved gene in the short arm of chromosome 16p11 is named FUS (formerly translocated in liposarcoma [TLS]), and it encodes for an RNA-binding protein. The FUS gene is also involved in translocations in low-grade fibromyxoid sarcoma and acute myeloid leukemia.[90] Variant translocations in MLS include one involving the Ewing sarcoma gene (EWS) in 22q12, another gene encoding an RNA-binding protein, instead of FUS.[91–93]

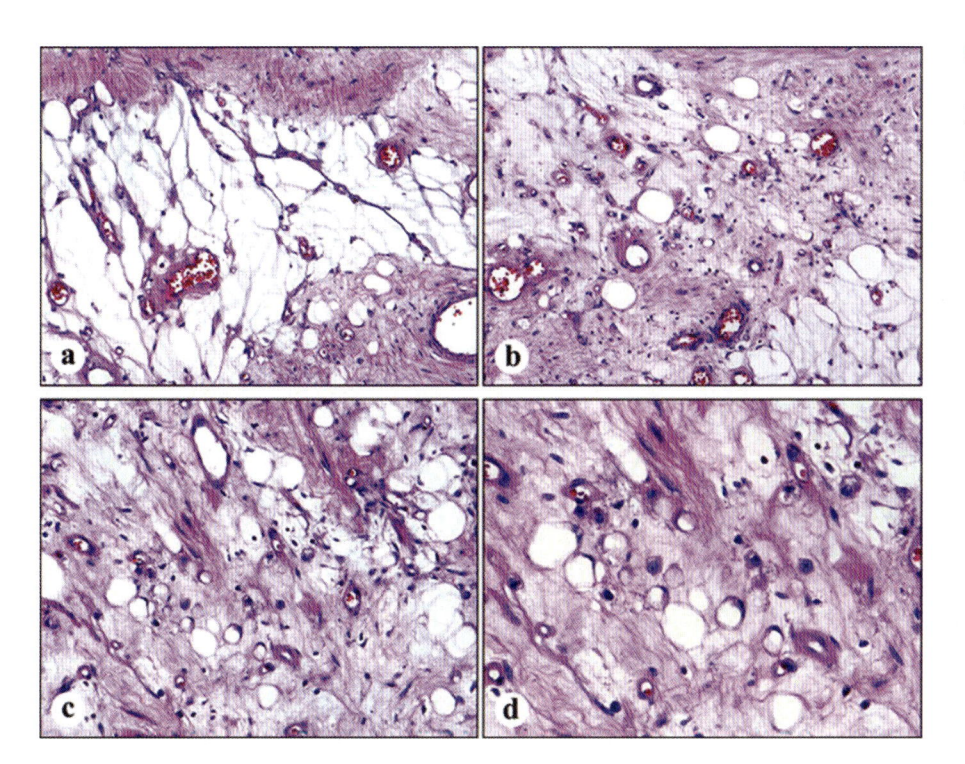

Figure 15.33 Myxoid liposarcoma following radiation therapy typically shows hyalinized vessels and low cellularity with scattered fat cells. These features are no longer diagnostic of myxoid liposarcoma, although the pretreatment biopsy showed typical features.

These translocations lead to the formation of fusion transcripts that can be detected by RT-PCR of FISH.[94] Production of FUS-DDIT3 oncoprotein has also been detected on the t(12;16) translocation. The fusion protein of the translocation appears to be pathogenetically significant, and it has been shown to prevent normal adipocytic differentiation by abnormal expression of the DOL54 gene, normally only transiently expressed during adipocyte differentiation.[95,96] FUS-DDIT3 fusion protein and normal DDIT3 have been found to induce lipomatous differentiation and liposarcomatous phenotype in human fibrosarcoma cells.[97]

FUS-DDIT3 fusion transcript assay can be performed on fresh or fixed tissues, but the heterogeneity of the breakpoints necessitates the use of multiple primer systems.[94,98,99] One study detected the fusion transcript in the peripheral blood, indicating the presence of circulating tumor cells.[100] FISH studies for FUS or DDIT3 gene rearrangements are more practical tests for genetic confirmation of MLS diagnosis. However, it has to be taken into account that FUS gene rearrangements also occur in other tumors, such as low-grade fibromyxoid sarcoma.[101,102]

Pleomorphic liposarcoma

Clinical features

Pleomorphic liposarcoma is a rare tumor, constituting <5% of all liposarcomas. It usually occurs in older adults (Figure 15.34) in intramuscular extremity locations and occasionally in the retroperitoneum (Figure 15.35).[103–109] Rare examples of cutaneous pleomorphic liposarcomas (with occasional examples of cutaneous liposarcomas of other types) have been described. In contrast to other pleomorphic liposarcomas, the cutaneous examples are indolent tumors if completely excised, but local recurrence is possible.[110,111] Wide surgical excision is the standard treatment for limited tumors, and adjuvant therapy is a consideration.

Metastatic and mortality rates in major series have varied in the range of 32% to 45% and 30% to 50%, respectively. Tumors <5 cm and those located superficially and in the extremities seem to have a better prognosis.[103–109]

Pathology

On gross inspection, pleomorphic liposarcomas are often yellow, but tumors with limited fatty differentiation can be gray-white, without specific gross evidence for lipomatous differentiation. These tumors are typically large intramuscular masses (Figure 15.36). They are classified as intermediate or high grade in the French Federation of Cancer Centers' grading system, depending on the mitotic rate (often high, >20 per 10 HPFs), and necrosis (common).

Histologic variants include pleomorphic MFH-like pleomorphic tumors, which show focal evidence of lipomatous differentiation in atypical multinucleated tumor cells (Figure 15.37). Some cases show large, bizarre lipoblasts (Figure 15.38), and the number of lipoblasts is typically higher in this variant compared with WDLS. Some authors have included tumors with extensive undifferentiated nonlipogenic components in this category. Fat stains are not useful in the differential diagnosis, because fat is present in other tumors, for example in MFH (undifferentiated sarcoma) and in nonsarcomatous tumors, such as renal cell carcinoma.

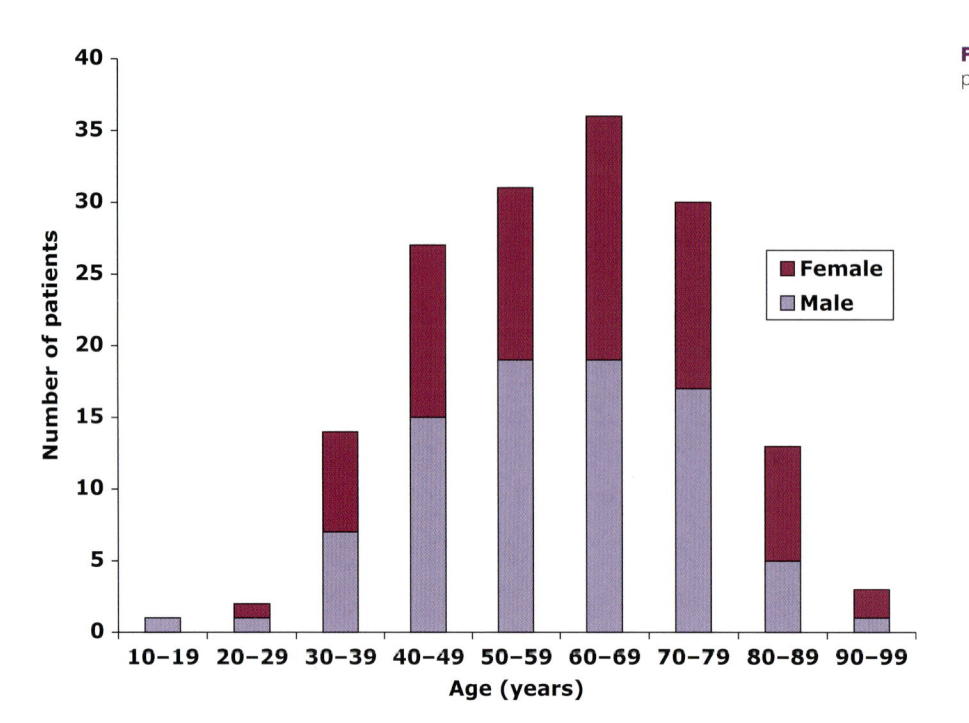

Figure 15.34 Age and sex distribution of 157 pleomorphic liposarcomas in five series.

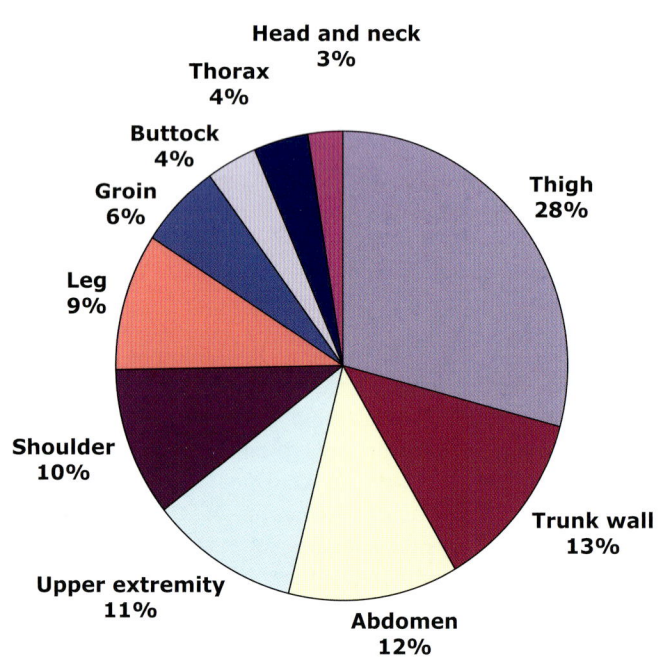

Figure 15.35 Anatomic distribution of 157 pleomorphic liposarcomas in five series.

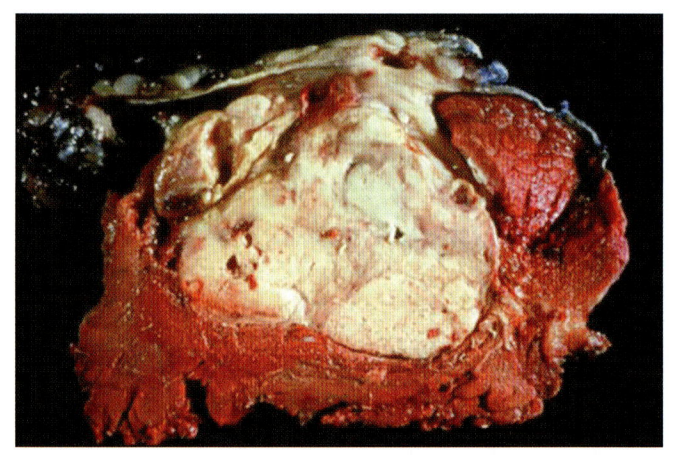

Figure 15.36 Grossly, an intramuscular pleomorphic liposarcoma appears as a solid yellowish mass.

The epithelioid variant is composed of solid sheets of epithelioid-appearing cells with areas of obvious fatty differentiation.[108,109] The epithelioid variant should not be confused with metastatic carcinoma, such as adrenocortical or renal clear cell carcinoma (Figure 15.39).

Immunohistochemical analysis shows that the lipogenic tumor cells can be positive for S100 protein, but the reaction varies, and the lack of it does not rule out adipose tissue differentiation. Those pleomorphic liposarcomas that contain keratin-positive cells (in the author's experience, especially keratins 7, 8, and 18) should not be confused with carcinomas.[108,109] Smooth muscle actin, desmin, and EMA are also detectable.[104] One study found melan A (A103) immunoreactivity in four of six epithelioid variants, creating further resemblance to adrenal cortical carcinoma, but inhibin was not detected in any cases.[109]

Chapter 15: Atypical lipomatous tumors and liposarcomas

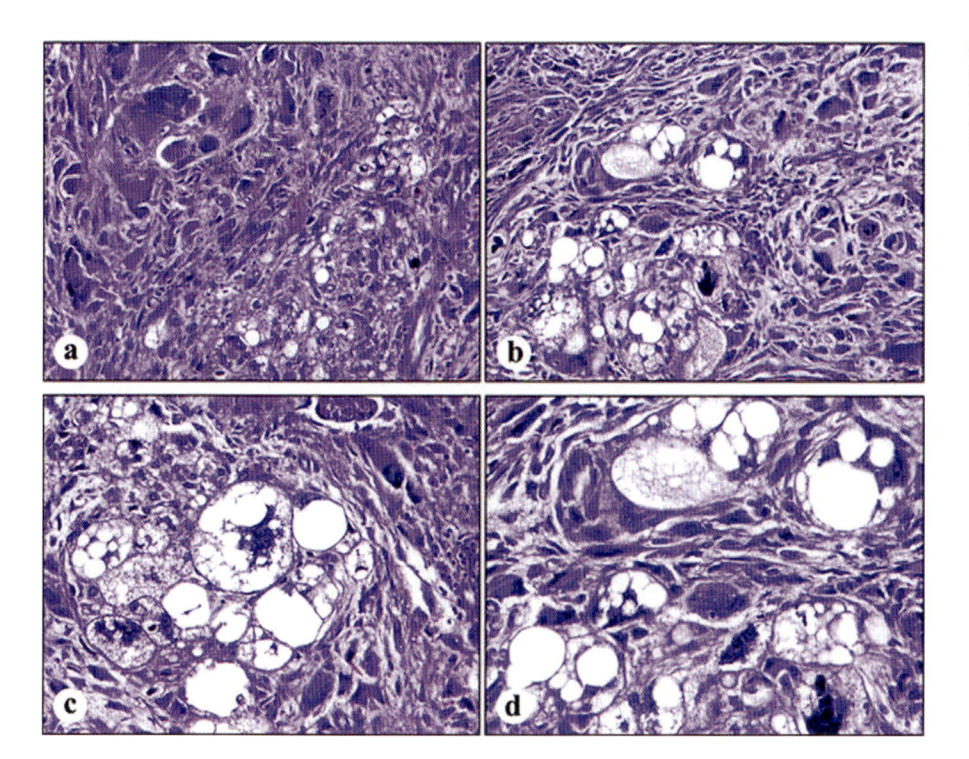

Figure 15.37 The most common variant of pleomorphic liposarcomas contains multivacuolated, highly atypical tumor cells and undifferentiated large cells similar to those seen in MFH.

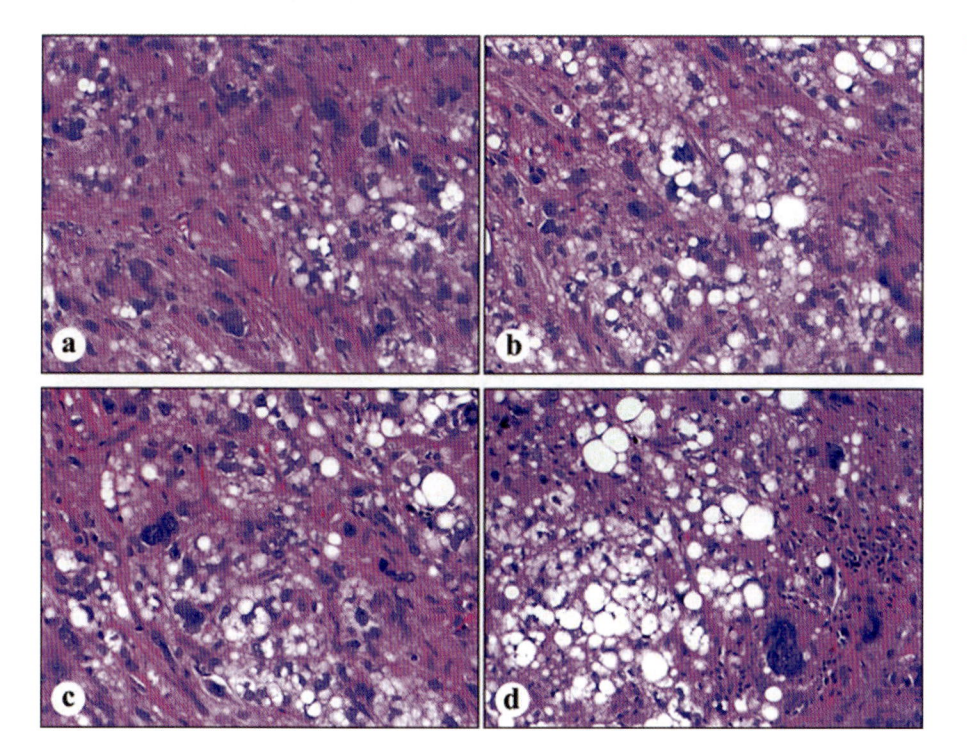

Figure 15.38 Pleomorphic liposarcoma with markedly atypical nuclei and high cellularity.

Genetics

Data on genetics are scant. The cytogenetic profile is complex, with numerous heterogeneous structural aberrations and polyploidy in some cases.[93] Two comparative genomic hybridization studies did not find 12q gains in pleomorphic liposarcoma,[112,113] and in fact, tumors containing MDM2-amplification are probably better classified as DDLSs. In one study, gains in chromosomes 1, 5p, 19p, 19q, and 20q were

437

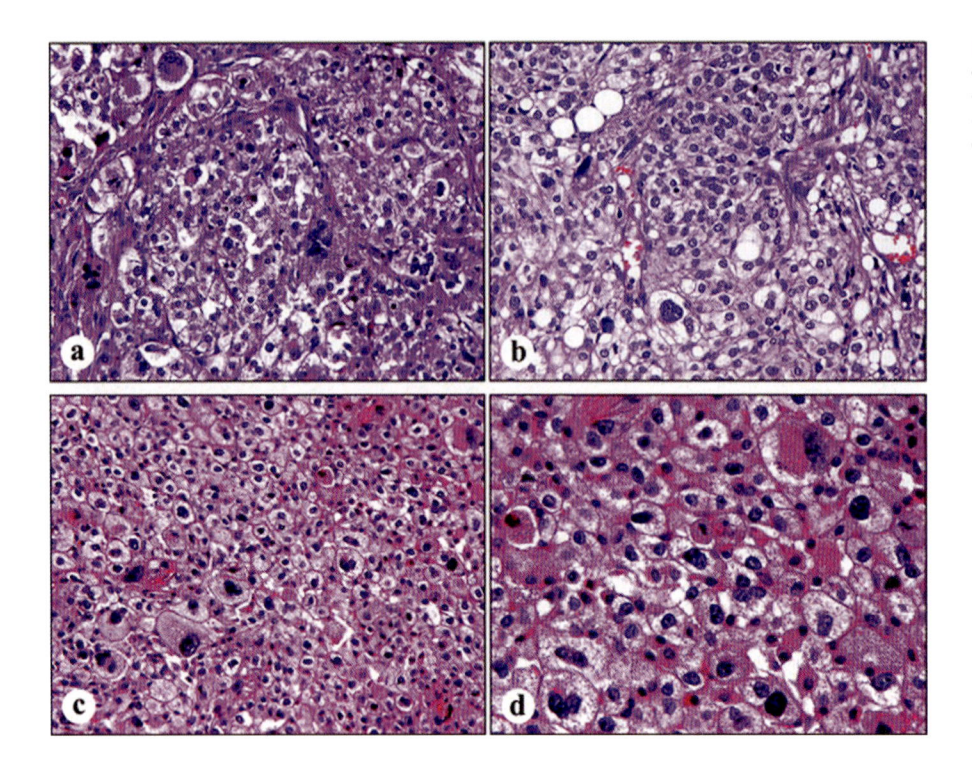

Figure 15.39 Pleomorphic liposarcoma with epithelioid features. Note distinct cell borders creating a resemblance to adrenal cortical or renal cell carcinoma. Adipocytic differentiation is diagnostic.

found, similar to myxoid MFHs studied for comparison.[113] The presence of t(12;16) translocation in some pleomorphic liposarcomas seems to support their origin from myxoid/round cell liposarcoma.[114]

References

Liposarcoma (general)

1. Enterline HT, Culberson JD, Rochlin DB, Brady LW. Liposarcoma: a clinical and pathological study of 53 cases. *Cancer* 1960;13:932–950.
2. Enzinger FM, Winslow DJ. Liposarcoma: a study of 103 cases. *Virchows Arch Anat Pathol A* 1962;335:367–388.
3. Evans HL. Liposarcoma: a study of 55 cases with a reassessment of its classification. *Am J Surg Pathol* 1979;3:507–523.
4. Evans HL. Atypical lipomatous tumor, its variants, and its combined forms: a study of 61 cases, with a minimum follow-up of 10 years. *Am J Surg Pathol* 2007;31:1–14.

Atypical lipomatous tumor/ well-differentiated liposarcoma

5. Azumi N, Curtis J, Kempson RL. Atypical and malignant neoplasms showing lipomatous differentiation: a study of 111 cases. *Am J Surg Pathol* 1987;11:161–183.
6. Allen PW, Strungs I, MacCormac LB. Atypical subcutaneous fatty tumors: a review of 37 referred cases. *Pathology* 1988;30:123–135.
7. Evans HL, Soule EH, Winkelmann RK. Atypical lipoma, atypical intramuscular lipoma, and well-differentiated retroperitoneal liposarcoma: a reappraisal of 30 cases formerly classified as well-differentiated liposarcoma. *Cancer* 1979;43:574–584.
8. Evans HL. Liposarcomas and atypical lipomatous tumors: a study of 66 cases followed for a minimum of 10 years. *Surg Pathol* 1988;1:41–54.
9. Kindblom LG, Angervall L, Fassina AS. Atypical lipoma. *Acta Pathol Microbiol Immunol Scand A* 1982;90:27–36.
10. Yoshikawa H, Ueda T, Mori S, et al. Dedifferentiated liposarcoma of the subcutis. *Am J Surg Pathol* 1996;20:1525–1530.
11. Kindblom LG, Angervall L, Svendsen P. Liposarcoma: a clinicopathologic, radiographic, and prognostic study. *Acta Pathol Microbiol Scand Suppl* 1992;253:1–71.
12. Weiss SW, Rao VK. Well-differentiated liposarcoma (atypical lipoma) of deep soft tissue of the extremities, retroperitoneum, and miscellaneous sites: a follow-up study of 92 cases with analysis of the incidence of "dedifferentiation." *Am J Surg Pathol* 1992;167:1051–1058.
13. Lucas DR, Nascimento AG, Sanjay BK, Rock MG. Well-differentiated liposarcoma: The Mayo Clinic experience with 58 cases. *Am J Clin Pathol* 1994;102:677–683.
14. Montgomery EA, Fisher C. Paratesticular liposarcoma: a clinicopathologic study. *Am J Surg Pathol* 2003;27:40–47.
15. Klimstra DS, Moran CA, Perino G, Koss MN, Rosai J. Liposarcoma of the anterior mediastinum and thymus: a clinicopathologic study of 28 cases. *Am J Surg Pathol* 1995;19:782–791.
16. Hahn HP, Fletcher CD. Primary mediastinal liposarcoma: clinicopathologic analysis of 24 cases. *Am J Surg Pathol* 2007;31:1868–1874.
17. Debelenko LV, Perez-Atayde AR, Dubois SG, et al. p53+/mdm2- atypical lipomatous tumor/well-differentiated liposarcoma in young children: an early expression of Li-Fraumeni syndrome. *Pediatr Dev Pathol* 2010;13:218–224.

18. Wenig BM, Weiss SW, Gnepp DR. Laryngeal and hypopharyngeal liposarcoma: a clinicopathologic study of 10 cases with a comparison to soft-tissue counterparts. *Am J Surg Pathol* 1990;14:134–141.

19. McQueen C, Montgomery E, Dufour B, Olney MS, Illei PB. Giant hypopharyngeal atypical lipomatous tumor. *Adv Anat Pathol* 2010;17(1):38–41.

20. Jakowski JD, Wakely PE Jr. Rhabdomyomatous well-differentiated liposarcoma arising in giant fibrovascular polyp of the esophagus. *Ann Diagn Pathol* 2009;13(4):263–268.

21. Boni A, Lisovsky M, Dal Cin P, Rosenberg AE, Srivastava A. Atypical lipomatous tumor mimicking giant fibrovascular polyp of the esophagus: report of a case and a critical review of literature. *Hum Pathol* 2013;44:1165–1170.

22. Argani P, Facchetti F, Inghirami G, Rosai J. Lymphocyte-rich well-differentiated liposarcoma: report of nine cases. *Am J Surg Pathol* 1997;21:884–895.

23. Kraus MD, Guillou L, Fletcher CD. Well-differentiated inflammatory liposarcoma: an uncommon and easily overlooked variant of a common sarcoma. *Am J Surg Pathol* 1997;21:518–527.

24. Dei Tos AP, Mentzel T, Newman PL, Fletcher CDM. Spindle cell liposarcoma, a hitherto unrecognized variant of liposarcoma. Analysis of six cases. *Am J Surg Pathol* 1994;18:913–921.

25. Mentzel T, Palmedo G, Kuhnen C. Well-differentiated spindle cell liposarcoma ("atypical spindle cell lipomatous tumor") does not belong to the spectrum of atypical lipomatous tumor but has a close relationship to spindle cell lipoma: clinicopathologic, immunohistochemical, and molecular analysis of six cases. *Mod Pathol* 2010;23:729–736.

26. Deyrup AT, Chibon F, Guillou L, *et al*. Fibrosarcoma-like lipomatous neoplasm: a reappraisal of so-called spindle cell liposarcoma defining a unique lipomatous tumor unrelated to other liposarcomas. *Am J Surg Pathol* 2013;37:1373–1378.

27. Suster S. Fisher C. Immunoreactivity for the human hematopoietic progenitor cell antigen (CD34) in lipomatous tumors. *Am J Surg Pathol* 1997;21:195–200.

28. Dei Tos AP, Doglioni C, Piccinin S, *et al*. Coordinated expression and amplification of the MDM2, CDK4, and HMGI-C genes in atypical lipomatous tumors. *J Pathol* 2000;190:531–536.

29. Pilotti S, Della Torre G, Mezzelani A, *et al*. The expression of MDM2/CDK4 gene product in the differential diagnosis of well-differentiated liposarcoma and large deep-seated lipoma. *Br J Cancer* 2000;82:1271–1275.

30. Binh MB, Sastre-Garau X, Guillou L, *et al*. MDM2 and CDK4 immunostainings are useful adjuncts in diagnosing well-differentiated and dedifferentiated liposarcoma subtypes: a comparative analysis of 559 soft tissue neoplasms with genetic data. *Am J Surg Pathol* 2005;29:1340–1347.

31. Coindre JM, Pédeutour F, Aurias A. Well-differentiated and dedifferentiated liposarcomas. *Virchows Arch* 2010;456:167–179.

32. He M, Aisner S, Benevenia J, *et al*. p16 immunohistochemistry as an alternative marker to distinguish atypical lipomatous tumor from deep-seated lipoma. *Appl Immunohistochem Mol Morphol* 2009;17:51–56.

33. Thway K, Flora R, Shah C, Olmos D, Fisher C. Diagnostic utility of p16, CDK4, and MDM2 as an immunohistochemical panel in distinguishing well-differentiated and dedifferentiated liposarcomas from other adipocytic tumors. *Am J Surg Pathol* 2012;36:462–469.

34. Gonzalez RS, McClain CM, Chamberlain BK, Coffin CM, Cates JM. Cyclin-dependent kinase inhibitor 2A (p16) distinguishes well-differentiated liposarcoma from lipoma. *Histopathology* 2013;62(7):1109–1111.

35. Stephenson CF, Berger CS, Leong SP, Davis JR, Sandberg AA. Analysis of a giant marker chromosome in well-differentiated liposarcoma using cytogenetics and fluorescence in situ hybridization. *Cancer Genet Cytogenet* 1992;15:134–138.

36. Dal Cin P, Kools P, Sciot R, *et al*. Cytogenetic and fluorescence in situ hybridization investigation of ring chromosomes characterizing a specific pathologic subgroup of adipose tissue tumors. *Cancer Genet Cytogenet* 1993;68:85–90.

37. Pedeutour F, Suikerbuijk RF, Forus A, *et al*. Complex composition and co-amplification of SAS and MDM2 in ring and giant rod marker chromosomes in well-differentiated liposarcomas. *Genes Chromosomes Cancer* 1994;10:85–94.

38. Szymanska J, Virolainen M, Tarkkanen M, *et al*. Overrepresentation of 1q21–23 and 12q13–21 in lipoma-like liposarcomas but not in benign lipomas: a comparative genomic hybridization study. *Cancer Genet Cytogenet* 1997;99:14–18.

39. Rosai J, Akerman M, Dal Cin P, *et al*. Combined morphologic and karyotypic study of 59 atypical lipomatous tumors: evaluation of their relationship and differential diagnosis with other adipose tissue tumors: a report of the CHAMP study group. *Am J Surg Pathol* 1996;20:1182–1189.

40. Sirvent N, Coindre JM, Maire G, *et al*. Detection of MDM2-CDK4 amplification by fluorescence in situ hybridization in 200 paraffin-embedded tumor samples: utility in diagnosing adipocytic lesions and comparison with immunohistochemistry and real-time PCR. *Am J Surg Pathol* 2007;31(10):1476–1489.

41. Weaver J, Downs-Kelly E, Goldblum JR, *et al*. Fluorescence in situ hybridization for MDM2 gene amplification as a diagnostic tool in lipomatous neoplasms. *Mod Pathol* 2008;21:943–949.

42. Zhang H, Erickson-Johnson M, Wang X, *et al*. Molecular testing for lipomatous tumors: critical analysis and test recommendations based on the analysis of 405 extremity-based tumors. *Am J Surg Pathol* 2010;34:1304–1311.

43. Tanas MR, Sthapanachai C, Nonaka D, *et al*. Pseudosarcomatous fibroblastic/myofibroblastic proliferation in perinephric adipose tissue adjacent to renal cell carcinoma: a lesion mimicking well-differentiated liposarcoma. *Mod Pathol* 2009;22:1196–1200.

44. Remstein E, Kurtin PJ, Nascimento AG. Sclerosing extramedullary hematopoietic tumor in chronic myeloproliferative disorders. *Am J Surg Pathol* 2000;24:51–55.

Dedifferentiated liposarcoma

45. Henricks W, Chu YC, Goldblum JR, Weiss SW. Dedifferentiated

liposarcoma: a clinicopathological analysis of 155 cases with a proposal for an expanded definition of dedifferentiation. *Am J Surg Pathol* 1997;21:271–281.

46. Kransdorf MJ, Meis JM, Jelinek JS. Dedifferentiated liposarcoma of the extremities: imaging finding in four patients. *AJR Am J Roentgenol* 1993;161:127–130.

47. Keung EZ, Hornick JL, Bertagnolli MM, Baldini EH, Raut CP. Predictors of outcomes in patients with primary retroperitoneal dedifferentiated liposarcoma undergoing surgery. *J Am Coll Surg* 2014;218:206–217.

48. Hisaoka M, Morimitsu Y, Hashimoto H, et al. Retroperitoneal liposarcoma with combined well-differentiated and myxoid malignant fibrous histiocytoma-like myxoid areas. *Am J Surg Pathol* 1999;23:1480–1492.

49. Coindre JM, Hostein I, Maire G, et al. Inflammatory malignant fibrous histiocytomas and dedifferentiated liposarcomas: histologic review, genetic profile, and MDM2 and CDK4 status favor a single entity. *J Pathol* 2004;203:822–830.

50. Coindre JM, Mariani O, Chibon F, et al. Most malignant fibrous histiocytomas developed in the retroperitoneum are dedifferentiated liposarcomas: a review of 25 cases initially diagnosed as malignant fibrous histiocytomas. *Mod Pathol* 2003;16:256–262.

51. McCormick D, Mentzel T, Beham A, Fletcher CDM. Dedifferentiated liposarcoma: clinicopathologic analysis of 32 cases suggesting a better prognostic subgroup among pleomorphic sarcomas. *Am J Surg Pathol* 1994;18:1213–1223.

52. Elgar F, Goldblum JR. Well-differentiated liposarcoma of the retroperitoneum: a clinicopathologic analysis of 20 cases, with particular attention to the extent of low-grade dedifferentiation. *Mod Pathol* 1997;10:113–120.

53. Hasegawa T, Seki K, Hasegawa F, et al. Dedifferentiated liposarcoma of retroperitoneum and mesentery: varied growth patterns and histological grades. A clinicopathologic study of 32 cases. *Hum Pathol* 2000;31:717–727.

54. Nascimento AG, Kurtin PJ, Guillou L, Fletcher CD. Dedifferentiated liposarcoma: a report of nine cases with a peculiar neural-like whorling pattern associated with metaplastic bone formation. *Am J Surg Pathol* 1998;22:945–955.

55. Fanburg-Smith J, Miettinen M. Liposarcoma with meningothelial whorls: an analysis of 17 cases of a distinctive histologic pattern associated with dedifferentiated liposarcoma. *Histopathology* 1998;33:414–424.

56. Thway K, Robertson D, Thway Y, Fisher C. Dedifferentiated liposarcoma with meningothelial-like whorls, metaplastic bone formation, and CDK4, MDM2, and p16 expression: a morphologic and immunohistochemical study. *Am J Surg Pathol* 2011;35:356–363.

57. Nicolas M, Moran CA, Suster S. Pulmonary metastasis from liposarcoma: a clinicopathologic and immunohistochemical study of 24 cases. *Am J Clin Pathol* 2005;123:265–275.

58. Huang HY, Brennan MF, Singer S, Antonescu CR. Distant metastases in retroperitoneal dedifferentiated liposarcoma is rare and rapidly fatal: a clinicopathologic study with emphasis on the low-grade myxofibrosarcomatous pattern as an early sign of differentiation. *Mod Pathol* 2005;18:976–984.

59. Mariño-Enríquez A, Fletcher CD, Dal Cin P, Hornick JL. Dedifferentiated liposarcoma with "homologous" lipoblastic (pleomorphic liposarcoma-like) differentiation: clinicopathologic and molecular analysis of a series suggesting revised diagnostic criteria. *Am J Surg Pathol* 2010;34:1122–1131.

60. Tallini G, Erlandson RA, Brennan MF, Woodruff JM. Divergent myosarcomatous differentiation in retroperitoneal liposarcoma. *Am J Surg Pathol* 1993;17:546–556.

61. Evans HL, Khurana KK, Kemp BL, Ayala AG. Heterologous elements in the dedifferentiated component of dedifferentiated liposarcoma. *Am J Surg Pathol* 1994;18:1150–1157.

62. Suster S, Wong TY, Moran CA. Sarcomas with combined features of liposarcoma and leiomyosarcoma: study of two cases of an unusual soft tissue tumor showing dual lineage differentiation. *Am J Surg Pathol* 1993;17:905–911.

63. Folpe AL Wiess SW. Lipoleiomyosarcoma (well-differentiated liposarcoma with leiomyosarcomatous differentiation): a clinicopathologic study of nine cases including one with dedifferentiation. *Am J Surg Pathol* 2002;26:742–749.

64. Binh MBN, Guillou L, Hostein I, et al. Dedifferentiated liposarcomas with divergent myosarcomatous differentiation developed in the internal trunk: a study of 27 cases and comparison to conventional dedifferentiated liposarcomas and leiomyosarcomas. *Am J Surg Pathol* 2007;31:1557–1566.

65. Chung L, Lau SK, Jiang Z, et al. Overlapping features between dedifferentiated liposarcoma and undifferentiated high-grade pleomorphic sarcoma. *Am J Surg Pathol* 2009;33:1594–1600.

66. Lucas DR, Shukla A, Thomas DG, et al. Dedifferentiated liposarcoma with inflammatory myofibroblastic tumor-like features. *Am J Surg Pathol* 2010;34:844–851.

67. Nishio J, Iwasaki H, Ishiguro M, et al. Establishment of a novel human dedifferentiated liposarcoma cell line, FU-DDLS-1: conventional and molecular cytogenetic characterization. *Int J Oncol* 2003;22:535–542.

68. Takahira T, Oda Y, Tamiya S, et al. Alterations of the RB1 gene in dedifferentiated liposarcoma. *Mod Pathol* 2005;18:1461–1470.

69. Rieker RJ, Weitz J, Lehner B, et al. Genomic profiling reveals subsets of dedifferentiated liposarcoma to follow separate molecular pathways. *Virchows Arch* 2010;456:277–285.

70. Sadri N, Surrey LF, Fraker DL, Zhang PJ. Retroperitoneal dedifferentiated liposarcoma lacking MDM2 amplification in a patient with a germ line CHEK2 mutation. *Virchows Arch* 2014;464:505–509.

Myxoid and round cell liposarcoma

71. Kilpatrick SE, Doyon J, Choong PF, Sim FH, Nascimento AG. The clinicopathologic spectrum of myxoid and round cell liposarcoma: a study of 95 cases. *Cancer* 1996;77:1450–1458.

72. Shmookler BM, Enzinger FM. Liposarcomas occurring in children: an

73. La Quaglia MP, Spiro SA, Ghavimi F, et al. Liposarcoma in patients younger than or equal to 22 years of age. *Cancer* 1993;72:3114–3119.
74. Alaggio R, Coffin CM, Weiss SW, et al. Liposarcomas in young patients: a study of 82 cases occurring in patients younger than 22 years of age. *Am J Surg Pathol* 2009;33:645–658.
75. Smith TA, Easley KA, Goldblum JR. Myxoid/round cell liposarcoma of the extremities: a clinicopathologic study of 29 cases with particular attention to extent of round cell liposarcoma. *Am J Surg Pathol* 1996;20:171–180.
76. Antonescu CR, Elahi A, Healey JH, et al. Monoclonality of multifocal myxoid liposarcoma: confirmation by analysis of TLS-CHOP or EWS-CHOP rearrangements. *Clin Cancer Res* 2000;6:2788–2793.
77. Spillane AJ, Fisher C, Thomas JM. Myxoid liposarcoma: the frequency and the natural history of nonpulmonary soft tissue metastases. *Ann Surg Oncol* 1999;6:389–394.
78. Fukuda T, Oshiro Y, Yamamoto I, Tsuneyoshi M. Long-term follow-up of pure myxoid liposarcomas with special reference to local recurrence and progression to round cell lesions. *Pathol Int* 1999;49:710–715.
79. Fritchie KJ, Goldblum JR, Tubbs RR, et al. The expanded histologic spectrum of myxoid liposarcoma with an emphasis on newly described patterns: implications for diagnosis on small biopsy specimens. *Am J Clin Pathol* 2012;137:229–239.
80. Siebert JD, Williams RP, Pulitzer DR. Myxoid liposarcoma with cartilaginous differentiation. *Mod Pathol* 1996;9:249–252.
81. Engstrom K, Bergh P, Cederlund CG, et al. Irradiation of myxoid/round cell liposarcoma induces volume reduction and lipoma-like morphology. *Acta Oncol* 2007;46:838–845.
82. Dei Tos AP, Piccinin S, Doglioni C, et al. Molecular aberrations of the G1-S checkpoint in myxoid and round cell liposarcoma. *Am J Pathol* 1997;151:1531–1539.
83. Oliveira AM, Nascimento AG, Okuno SH, Lloyd RV. P27 (kip1) protein expression correlates with survival in myxoid and round-cell liposarcoma. *J Clin Oncol* 2000;18:2888–2893.
84. Oda Y, Yamamoto H, Takahira T, et al. Frequent alteration of p16(INK4a)/p14 (ARF) and p53 pathways in the round cell component of myxoid/round cell liposarcoma: p53 gene alterations and reduced p14 (ARF) expression both correlate with poor prognosis. *J Pathol* 2005;207:410–421.
85. Crozat A, Aman P, Mandahl N, Ron D. Fusion of CHOP to a novel RNA-binding protein in human myxoid liposarcoma. *Nature* 1993;363:640–644.
86. Panagopoulos I, Mandahl N, Ron D, et al. Characterization of the CHOP breakpoints and fusion transcripts in myxoid liposarcomas with the 12;16 translocation. *Cancer Res* 1994;54:6500–6503.
87. Gibas Z, Miettinen M, Limon J, et al. Cytogenetic and immunohistochemical profile of myxoid liposarcoma. *Am J Clin Pathol* 1995;103:20–26.
88. Knight JC, Renwick PJ, Dal Cin PD, Van Den Berghe H, Fletcher CD. Translocation t(12;16)(q13;p11) in myxoid liposarcoma and round cell liposarcoma: molecular and cytogenetic analysis. *Cancer Res* 1995;55:24–27.
89. Tallini G, Akerman M, Dal Cin P, et al. Combined morphologic and karyotypic study of 28 myxoid liposarcomas: implications for a revised morphological typing (a report from the CHAMP-group). *Am J Surg Pathol* 1996;20:1047–1055.
90. Panagopoulos I, Mandahl N, Mitelman F, Aman P. Two distinct FUS breakpoint clusters in myxoid liposarcoma and acute myeloid leukemia with the translocations t(12;16) and t(16;21). *Oncogene* 1995;11:1133–1137.
91. Panagopoulos I, Hoglund M, Mertens F, et al. Fusion of the EWS and CHOP genes in myxoid liposarcoma. *Oncogene* 1996;12:489–494.
92. Dal Cin P, Sciot R, Panagopoulos I, et al. Additional evidence of a variant translocation t(12;22) with EWS/CHOP fusion in myxoid liposarcoma: clinicopathological features. *J Pathol* 1997;182:437–441.
93. Panagopoulos I, Lassen C, Isaksson M, et al. Characteristic sequence motifs at the breakpoints of the hybrid genes FUS/CHOP, EWS/CHOP and FUS/ERG in myxoid liposarcoma and acute myeloid leukemia. *Oncogene* 1997;15:1357–1362.
94. Kuroda M, Ishida T, Horiuchi H, et al. Chimeric TLS/FUS-CHOP gene expression and the heterogeneity of its junction in human myxoid and round cell liposarcoma. *Am J Pathol* 1995;147:1221–1227.
95. Kuroda M, Wang X, Sok J, et al. Induction of a secreted protein by the myxoid liposarcoma oncogene. *Proc Natl Acad Sci USA* 1999;96:5025–5030.
96. Adelmant G, Gilbert JD, Freytag SO. Human translocation liposarcoma-CCAAT/enhancer binding protein (C/EBP) homologous protein (TLS/CHOP) oncoprotein prevents adipocyte differentiation by directly interfering with C/EBP beta function. *J Biol Chem* 1998;273:15574–15581.
97. Engstrom K, Willen H, Kabjorn-Gustafsson C, et al. The myxoid/round cell liposarcoma fusion oncogene FUS-DDIT3 and the normal DDIT3 induce a liposarcoma phenotype in transfected human fibrosarcoma cells. *Am J Pathol* 2006;168:1642–1653.
98. Hisaoka M, Tsuji S, Morimitsu Y, et al. Detection of TLS/FUS-CHOP fusion transcript in myxoid and round cell liposarcomas by nested reverse transcription-polymerase chain reaction using archival paraffin-embedded tissues. *Diagn Mol Pathol* 1998;7:96–101.
99. Willeke F, Ridder R, Mechtersheimer G, et al. Analysis of FUS-CHOP fusion transcripts in different types of soft tissue liposarcoma and their diagnostic implications. *Clin Cancer Res* 1998;4:1779–1784.
100. Panagopoulos I, Aman P, Mertens F, et al. Genomic PCR detects tumor cells in peripheral blood from patients with myxoid liposarcoma. *Genes Chromosomes Cancer* 1996;17:102–107.
101. Narendra S, Valente A, Tull J, Zhang S. DDIT3 gene break-apart as a molecular marker for diagnosis of myxoid liposarcoma–assay validation and clinical experience. *Diagn Mol Pathol* 2011;20:218–224.
102. Powers MP, Wang WL, Hernandez VS, et al. Detection of myxoid liposarcoma-associated FUS-DDIT3 rearrangement variants including a newly identified breakpoint using an optimized RT-PCR assay. *Mod Pathol* 2010;23:1307–1315.

Pleomorphic liposarcoma

103. Downes KA, Goldblum JR, Montgomery EA, Fisher C. Pleomorphic liposarcoma: a clinicopathologic analysis of 19 cases. *Mod Pathol* 2001;14:179–184.

104. Gebhard S, Coindre JM, Michels JJ, et al. Pleomorphic liposarcoma: clinicopathologic, immunohistochemical, and follow-up analysis of 63 cases: a study from the French Federation of Cancer Centers Sarcoma Group. *Am J Surg Pathol* 2002;26:601–616.

105. Hornick JL, Bosenberg MW, Mentzel T, et al. Pleomorphic liposarcoma: clinicopathologic analysis of 57 cases. *Am J Surg Pathol* 2004;28:1257–1267.

106. Wang L, Ren W, Zhou X, Sheng W, Wang J. Pleomorphic liposarcoma: a clinicopathological, immunohistochemical and molecular cytogenetic study of 32 additional cases. *Pathol Int* 2013;63:523–531.

107. Ghadimi MP, Liu P, Peng T, et al. Pleomorphic liposarcoma: clinical observations and molecular variables. *Cancer* 2011;117:5359–5369.

108. Miettinen M, Enzinger FM. Epithelioid variant of pleomorphic liposarcoma: a study of 12 cases of a distinctive variant of high-grade liposarcoma. *Mod Pathol* 1999;12:722–728.

109. Huang HY, Antonescu CR. Epithelioid variant of pleomorphic liposarcoma: a comparative immunohistochemical and ultrastructural analysis of six cases with emphasis on overlapping features with epithelioid malignancies. *Ultrastruct Pathol* 2002;26:299–308.

110. Dei Tos AP, Mentzel T, Fletcher CD. Primary liposarcoma of the skin: a rare neoplasm with unusual high-grade features. *Am J Dermatopathol* 1998;20:332–338.

111. Gardner JM, Dandekar M, Thomas D, et al. Cutaneous and subcutaneous pleomorphic liposarcoma: a clinicopathologic study of 29 cases with evaluation of MDM2 gene amplification in 26. *Am J Surg Pathol* 2012;36:1047–1051.

112. Idbaih A, Coindre JM, Derre M, et al. Myxoid malignant fibrous histiocytoma and pleomorphic liposarcoma share very similar genomic imbalance. *Lab Invest* 2005;85:176–181.

113. Rieker RJ, Joos S, Bartsch C, et al. Distinct chromosomal imbalances in pleomorphic and high-grade dedifferentiated liposarcomas. *Int J Cancer* 2002;99:68–73.

114. De Cecco L, Gariboldi M, Reid JF, et al. Gene expression profile identifies a rare epithelioid variant case of pleomorphic liposarcoma carrying a FUS-CHOP transcript. *Histopathology* 2005;46:334–341.

Chapter 16

Smooth muscle tumors

Markku Miettinen
National Cancer Institute, National Institutes of Health

Smooth muscle tumors almost invariably arise from tissues with normal smooth muscle components, consistent with their probable origin from the stem-cell-like, replication-capable pool of local smooth muscle cells. This pool can be differently represented in various smooth muscle elements, reflecting variance in the incidence of smooth muscle tumors at different body sites. The nearly ubiquitous presence of vascular smooth muscle explains the origin of smooth muscle tumors at a wide variety of sites.

Benign smooth muscle tumors, leiomyomas, and their malignant counterparts leiomyosarcomas, are histologically separated by mitotic count and atypia. Supplemental parameters, such as the presence of coagulative necrosis, are sometimes used. Despite the fact that benign leiomyomas and leiomyosarcomas usually segregate well by morphology, there are instances in which the designation of "smooth muscle tumor of uncertain malignant potential" is prudent, to denote a borderline morphology with behavior too difficult to predict. In general, smooth muscle tumors of peripheral soft tissues in the extremities, trunk, and retroperitoneum should be considered at least potentially malignant if both atypia and any significant (more than occasional) mitotic activity are present.

Estrogen- and progesterone-sensitive benign and malignant smooth muscle tumors occur in the uterus, pelvis, and abdominal cavity. Their malignancy criteria differ from those of peripheral soft tissue tumors, especially in that more mitotic activity is allowed if atypia is not present. These tumors are discussed separately.

Leiomyomas of peripheral soft tissues include those arising from the dermal pilar smooth muscle (piloleiomyomas) and angioleiomyomas (vascular leiomyomas) that chiefly present in the subcutis and arise from vascular smooth muscle. Organ-based benign leiomyomas can arise in the respiratory and genitourinary tracts. In the gastrointestinal tract, true leiomyomas are rare and occur mainly in the esophagus, colon, and rectum. They are much less common than gastrointestinal stromal tumors (GISTs), which constitute most gastrointestinal mesenchymal tumors (see Chapter 17).

Soft tissue leiomyosarcomas are histologically relatively similar regardless of the site of origin, but their clinicopathologic features differ, especially in that the rate of malignancy increases by the depth of the tumor's location. Three important subcategories are the peripheral types arising from arrector pili, from vascular structures, and in internal organs (visceral leiomyosarcomas).

Lymphangioleiomyoma of retroperitoneal lymph nodes and pulmonary lymphangioleiomyomatosis are related to angiomyolipoma and belong to the group of perivascular epithelioid cell tumors (PEComas, Chapter 19). The Epstein–Barr virus (EBV)-associated smooth muscle tumor/leiomyosarcomas (LMSs) in immunosuppressed patients is a specific clinicopathologic group and is discussed separately.

Smooth muscle hamartoma of skin

The name refers to a rare cutaneous lesion featuring a localized patch or linear streak of hyperplasia of arrector pili smooth muscle. The designation *smooth muscle hamartoma* has been sporadically used on other types of lesions, including hamartomas of breast and scrotal smooth muscle hyperplasia (see later discussion).

Clinical features

This rare cutaneous lesion is often congenital or presents shortly after birth and has mild male predominance; however, cases have also been reported in older children or adults and are then designated as *acquired smooth muscle hamartoma*. The anatomic location is predominantly in the lower trunk or the proximal parts of extremities: lower back, buttock, thigh, and arm.

The clinical appearance is that of a hyperpigmented or flesh-colored plaque or patch measuring up to several centimeters in diameter. The lesion can be hairy, clinically resembling a large pigmented nevus or even a cutaneous mastocytoma because it can elicit a pseudo-Darier sign (elevation after touching the lesion), possibly because of the degranulation of resident mast cells.[1–6] Some of the reported children had linear lesions with follicular papules spanning a large area in the trunk or extremities.[3,7,8] Familial presentation has also been reported.[9] Preoperative clinical diagnosis can allow more conservative treatment, unlike congenital pigmented nevi, which require excision because of risk of melanoma.

Modern Soft Tissue Pathology, Second Edition, ed. Markku Miettinen. Published by Cambridge University Press. © Cambridge University Press 2016.

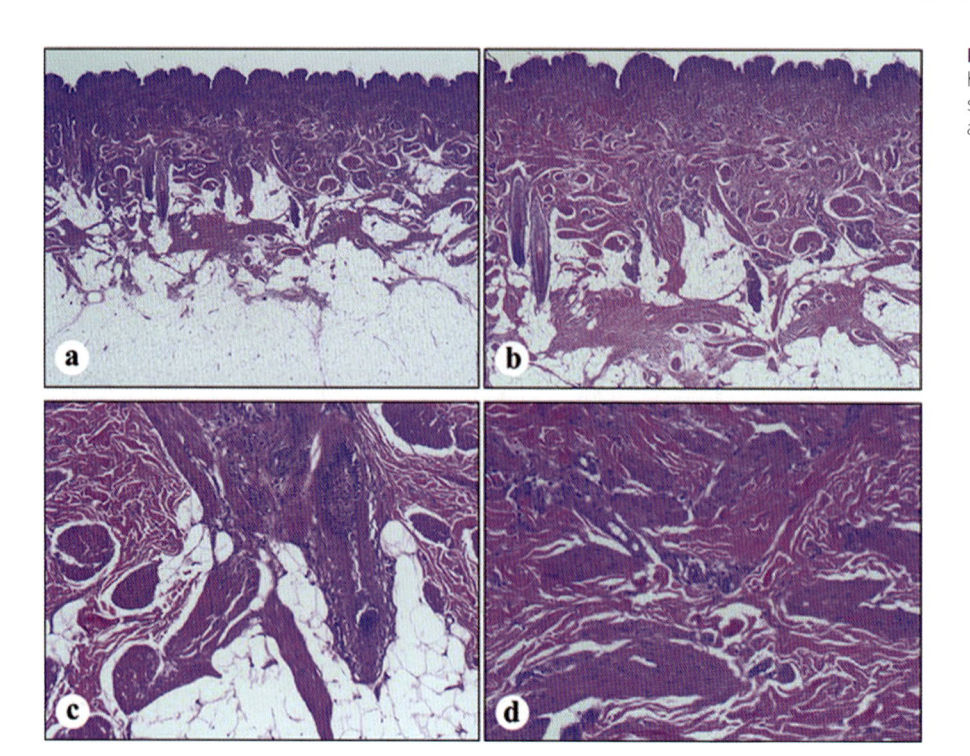

Figure 16.1 Congenital smooth muscle hamartoma contains linearly arranged arrector pili smooth muscle elements, sometimes grouped around hair follicles.

Pathology

Histologically, smooth muscle hamartoma forms a plaque-like lesion containing linear dermal bundles of well-differentiated arrector pili smooth muscle without atypia or mitotic activity. The lesion also can involve superficial subcutis, where smooth muscle elements are often grouped around hair follicles (Figure 16.1). Some smooth muscle hamartomas overlap with Becker's nevus/melanosis. Unlike congenital nevus, smooth muscle hamartoma does not contain a melanocytic component, but has basal cell hyperpigmentation.

Smooth muscle (myoid) hamartomas have also been reported in the breast.[10] They contain a constellation of fat, fibrous tissue, and disorganized elements of smooth muscle without atypia or mitotic activity (Figure 16.2).

Scrotal smooth muscle hyperplasia

A peculiar mass-forming smooth muscle hyperplasia of the testicular adnexa can clinically simulate a tumor.[11] Such lesions have been reported in middle-aged to older men. The size of the lesion varies widely (<1–>5 cm), and the smooth muscle hyperplasia occurs around the spermatic cord or associated with the tunica albuginea periductally, perivascularly, or interstitially. The proliferation differs from true leiomyoma by being less compact and more orderly and "organoid" in nature. Some lesions previously designated as smooth muscle hamartomas of the scrotum could also represent this hyperplastic process. This hyperplasia is clinically indolent, with no tendency to recur.

Angiomyomatous hamartoma of the lymph nodes

Clinical features

This designation, coined by Chan *et al.*, refers to a benign angioleiomyoma-like vascular smooth muscle proliferation involving a lymph node hilus.[12] The lesion forms a mass, which almost always involves an inguinal lymph node, but it has also been reported in a cervical lymph node.[13] This lesion occurs in a wide age range, with an apparent male predominance.

Pathology

On gross inspection, firm white tissue replaces much of the involved lymph node. The reported lesions have measured 1 cm to 3.5 cm (median, 2 cm). The tumor involves principally the hilus, but also can replace the nodal cortex. Histologically, the proliferation resembles angioleiomyoma, but has additional fibrous stroma and occasionally contains hemangioma-like elements (Figure 16.3). There is neither smooth muscle atypia nor mitotic activity, so that the lesion is unlikely to be confused with lymph node metastasis of a LMS (which is extremely rare in all circumstances).

Pilar leiomyoma (piloleiomyoma)

Clinical features

This rare skin tumor, derived from the arrector pili smooth muscle, usually occurs as multiple linear or clustered small cutaneous papules and less commonly as a single nodule that

Chapter 16: Smooth muscle tumors

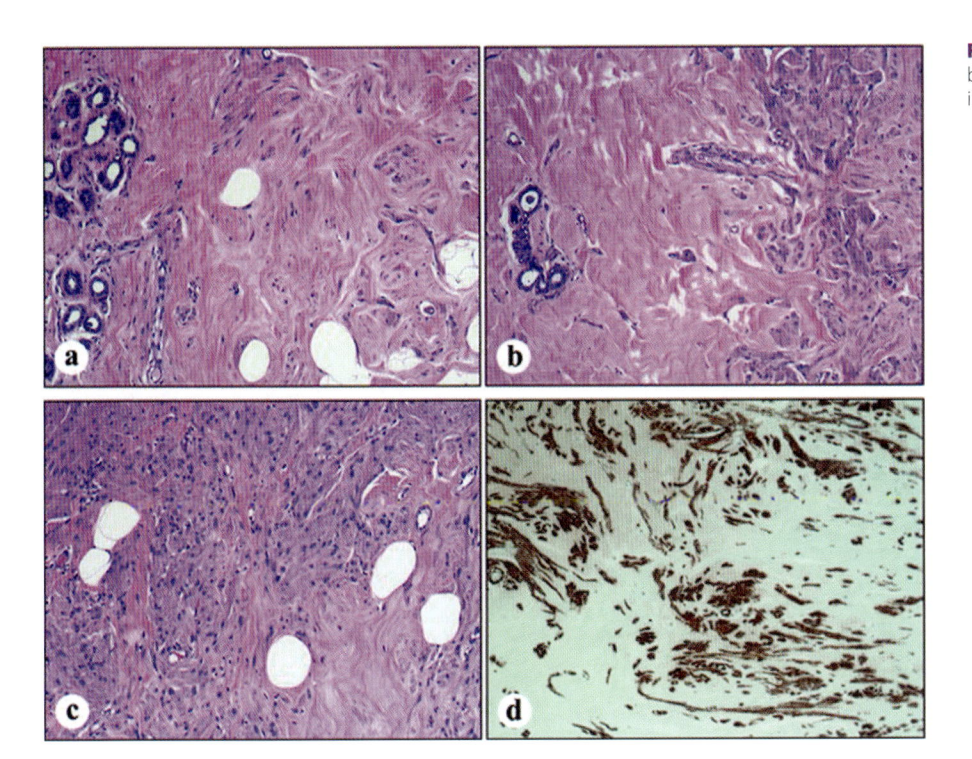

Figure 16.2 Smooth muscle hamartoma of the breast contains benign smooth muscle elements as irregular clusters that are SMA positive.

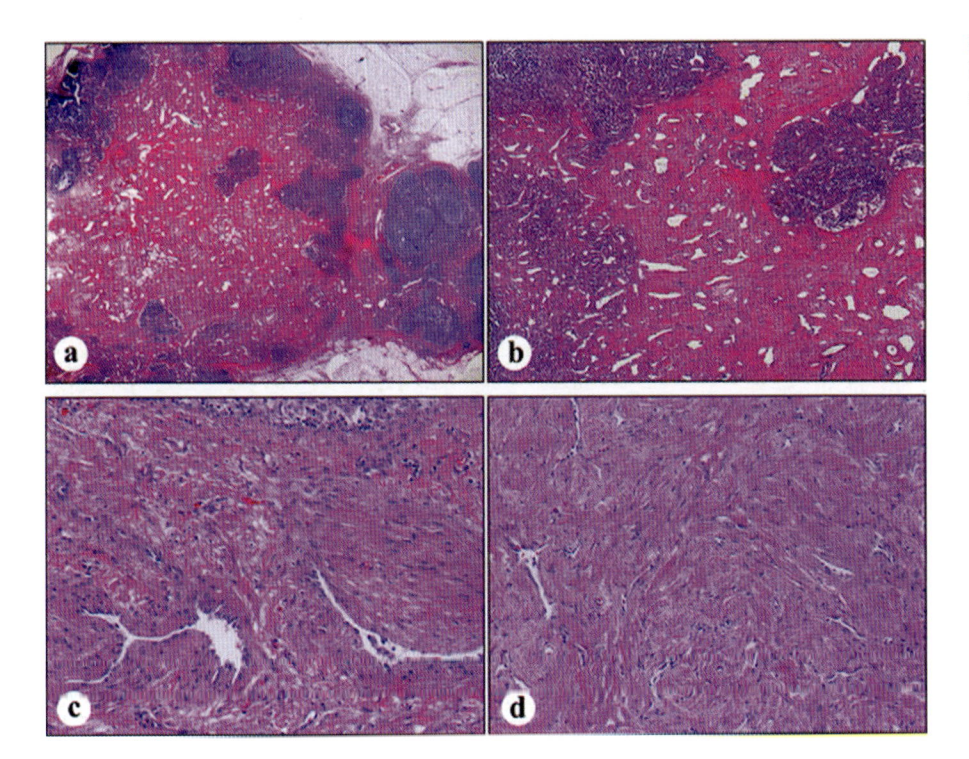

Figure 16.3 Angiomyomatous hamartoma of the lymph nodes contains angioleiomyoma-like areas sharply demarcated from the lymphoid tissue.

can vary from a few millimeters to 2 cm in diameter. Tumors presenting as single lesions are generally larger than the multiple piloleiomyomas. Pilar leiomyomas most commonly occur in young adults, usually on the extensor surfaces of legs and arms. The lesions are typically chronically painful, causing a burning or pinching sensation, and similar pain can be elicited by emotions.[14–16]

Hereditary leiomyomatosis and renal cell cancer

Hereditary leiomyomatosis and renal cell cancer (HLRCC) is a familial tumor syndrome with an autosomal dominant inheritance. It includes multiple cutaneous and early onset uterine leiomyomas, and clinically aggressive papillary type II renal carcinoma. The syndrome is caused by loss of function

445

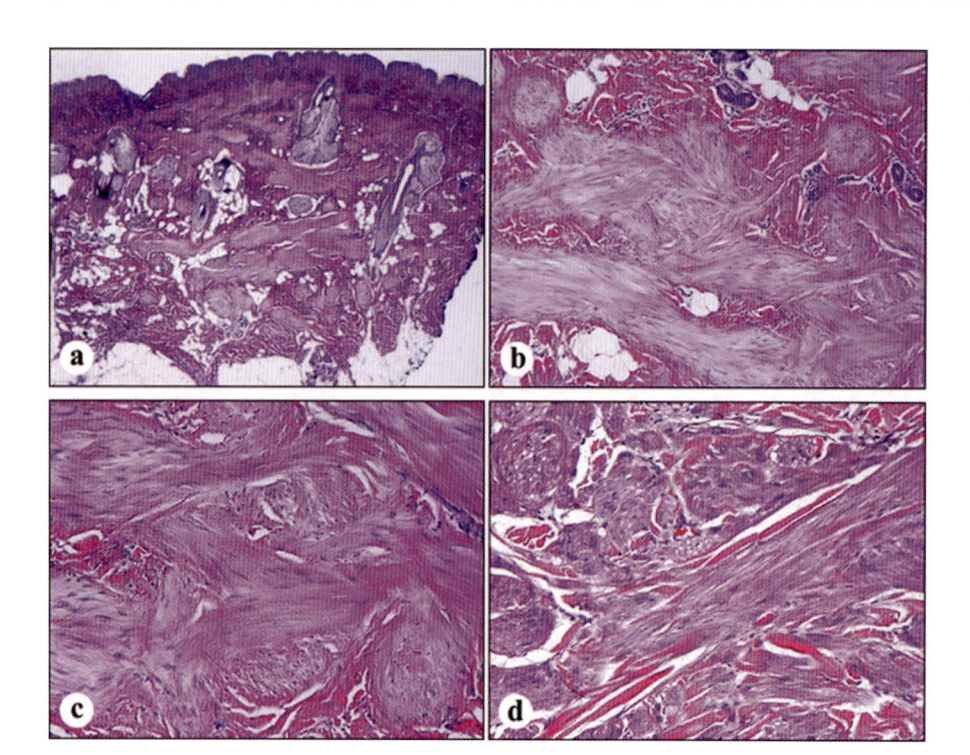

Figure 16.4 Pilar leiomyoma is composed of clusters of well-differentiated smooth muscle cells, the configuration of which resembles arrector pili muscles.

germline mutations of the fumarate hydratase (fumarase) gene located in chromosome 1q[17] in combination with somatic allelic losses. It was originally reported from northern Europe[17] and subsequently from the United States.[18,19] Lifetime renal carcinoma risk for mutation carriers has been estimated as 15%.[20] The early clinical descriptions from the 1960s and 1970s termed the syndrome *familial uterocutaneous leiomyomatosis,*[21] cutaneous leiomyomata with uterine leiomyomata,[22] and *Reed's syndrome*.

Pathology

Pilar leiomyomas are composed of radially or haphazardly arranged, poorly circumscribed clusters of mature smooth muscle cells that can resemble the configuration of expanded arrector pili muscles, separated from each other by abundant collagenous stroma (Figure 16.4). Small lesions can be concentrated around adnexa. Focal nuclear atypia does occur, but with no mitotic activity. A less organized architecture and presentation in adults separates this tumor from a smooth muscle hamartoma.

Angioleiomyoma (angiomyoma, vascular leiomyoma)

Angioleiomyoma is a benign, well-circumscribed smooth muscle tumor arising from the walls of small veins, usually in peripheral subcutaneous tissue.[23–26]

Clinical features

Angioleiomyoma occurs in all ages, but is more common in middle-aged to older adults, with an overall mild male predominance (Figure 16.5). In the largest series, there was female predominance in the lower extremities and male predominance in the upper extremities.[23] In the Armed Forces Institute of Pathology (AFIP) civilian cases, there is a 2:1 male predominance in upper extremity lesions, but even sex distribution in lower extremity lesions.

The lesion typically appears as a small, solitary, often painful subcutaneous nodule, usually in the extremities. In AFIP civilian cases, 60% of angioleiomyomas occurred in the lower extremity below the thigh (Figure 16.6). In the upper extremity, distal sites predominate. Occurrence in the trunk wall or head and neck is rare. Angioleiomyomas can occur in various locations in the oral cavity and perioral region, especially the lip and palate.[26] Simple excision is curative.

Pathology

Grossly and histologically, angioleiomyoma is a sharply circumscribed, whitish rubbery nodule that usually measures 1 cm to 2 cm (Figure 16.7). The lesions that have calcification can appear gritty on sectioning. Three histologic variants have been recognized. The most common (67%) is the capillary or solid type, with large numbers of narrow vascular slits surrounded by well-differentiated smooth muscle cells (Figure 16.8a). The vessels are almost always small veins, but on rare occasions, arterial involvement has been noted. Foci of dystrophic calcification and fat cells are sometimes present, and the tumors of older persons tend to be less cellular and more often calcified, perhaps a sign of tumor regression.

The venous type (23%) has gaping venous lumina surrounded by a thick smooth muscle layer that merges with adjacent vessel walls. The cavernous type (11%) has wide

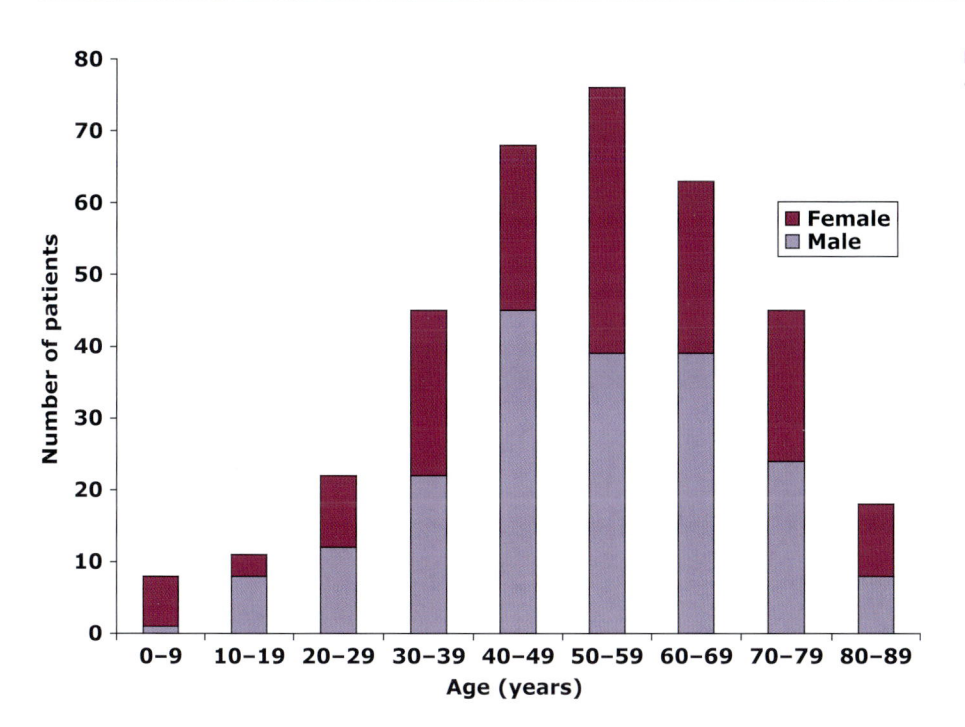

Figure 16.5 Age and sex distribution of angioleiomyomas in 356 civilian patients.

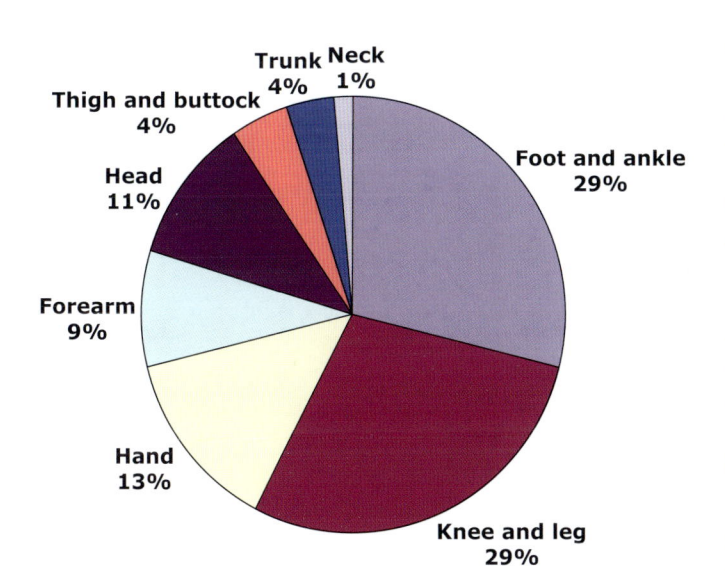

Figure 16.6 Anatomic distribution of 361 angioleiomyomas.

vascular lumina resembling cavernous hemangioma, but the septal elements are composed of smooth muscle cells (Figure 16.8b–d). Unlike the solid variant, the cavernous version has a significant male predominance and occurs more often in the upper extremities. The venous variant has a mild male predominance.[23]

Histologic variations in angioleiomyoma include lipomatous metaplasia (Figure 16.9a,b), paucicellular examples with calcification (Figure 16.9c), and focal nuclear atypia (Figure 16.9d). Such atypia without mitotic activity does not seem to have any significance. Angioleiomyomas only rarely recur, according to the largest series.[23] Lipomatous angioleiomyomas are sometimes erroneously considered "angiomyolipomas." These tumors are not related to PEComas, however, and do not have an HMB45-positive component.[27]

Over one-half of all angioleiomyomas are painful, and this has been attributed to the presence of nerve fibers within the external aspect of the tumor. Some have suggested that only the painful tumors contain nerves.[28,29]

Immunohistochemically, all variants of angioleiomyomas are strongly positive for SMA and heavy caldesmon, whereas desmin positivity is variable: 75% in solid, >50% in venous, and <20% in cavernous variants.[30]

Genetics

Heterogeneous clonal chromosomal changes have been reported in each of the three published cytogenetic reports on angioleiomyomas.[31–33] They include interstitial deletions of 6p21-23 and 21q21, monosomy of chromosome 13, and translocation X;10(q22;q23.2). A comparative genomic hybridization study showed copy number changes in 8 of 13 cases and recurrent losses were found in chromosome 22q11.[34]

Leiomyomas of the male external genitalia

Scrotal leiomyomas can arise from smooth muscle elements of the dartos muscle or those in and around the spermatic cord, and a few of these are skin based. The internal examples can reach a larger size (Figure 16.10).[35] Nuclear atypia in a scrotal leiomyoma should lead to an intense search for mitoses. Even if mitotic activity is not found, one should approach these lesions with great caution. Although such

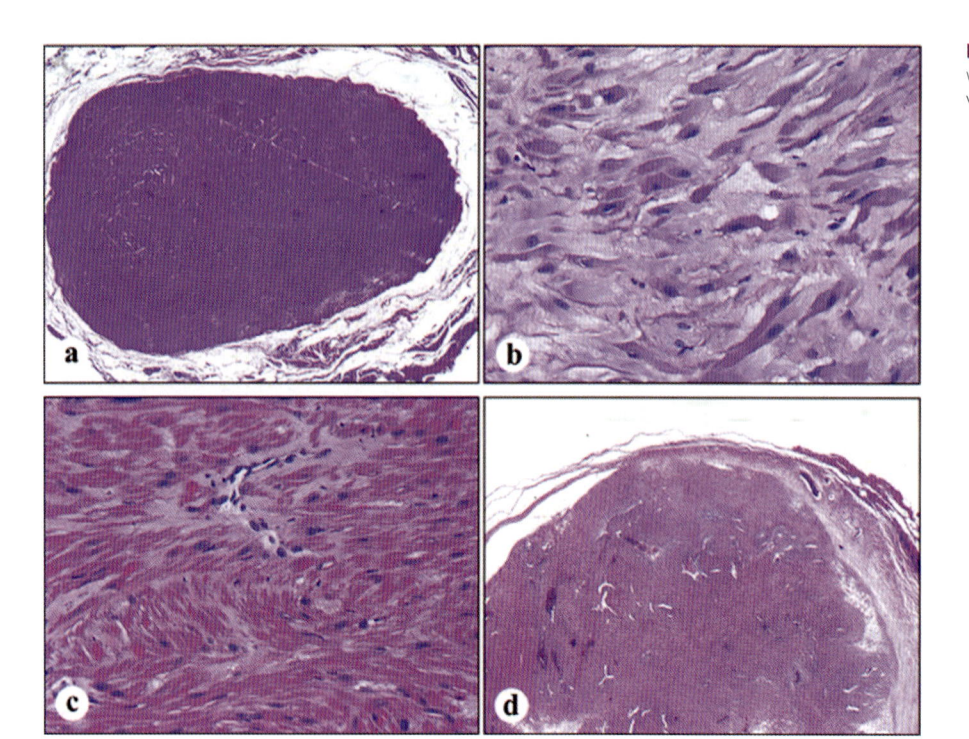

Figure 16.7 Angioleiomyoma forms a well-circumscribed nodule composed of mature vascular smooth muscle cells.

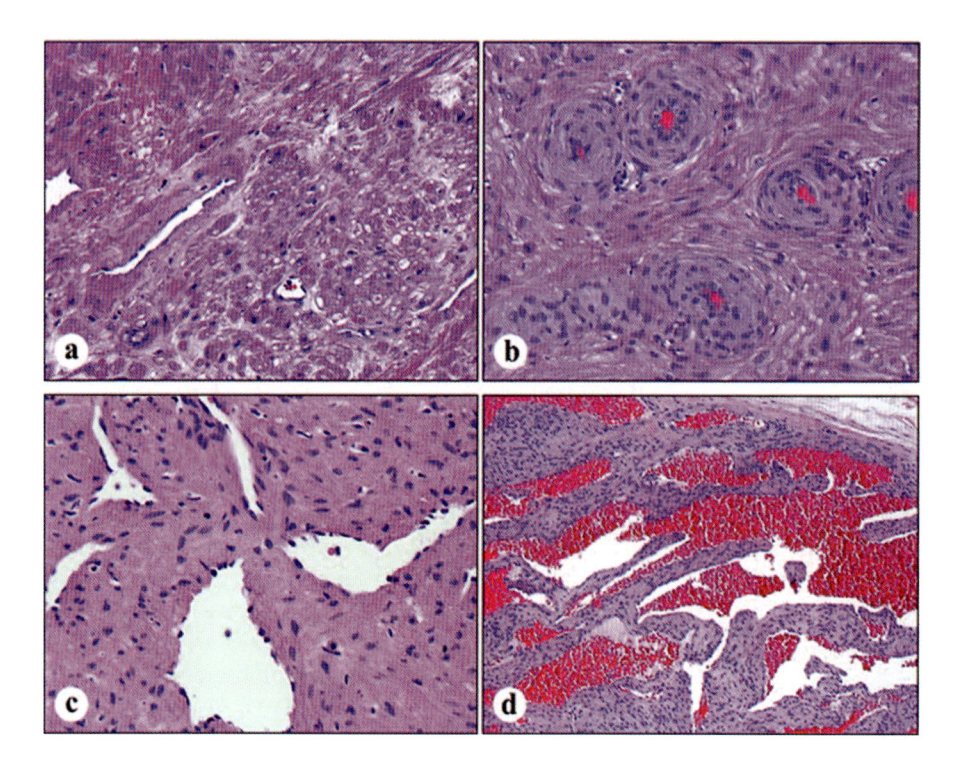

Figure 16.8 Histologic spectrum of angioleiomyoma. (**a**) Solid pattern. (**b,c**) Venous variant composed of vein-like vessels surrounded by smooth muscle proliferation. (**d**) Cavernous hemangioma-like variant.

tumors have been named "bizarre" or "symplastic leiomyomas,"[36] insufficient follow-up information exists to confirm their indolent nature. Practically any mitotic activity in scrotal smooth muscle tumors signifies the risk of malignant behavior, and therefore the designation of LMS would be more appropriate. Cutaneous leiomyomas also rarely occur in penile skin.

Gastrointestinal leiomyomas

Two types of leiomyomas occur in the gastrointestinal tract: intramural and muscularis mucosae leiomyomas. The former occur throughout the gastrointestinal tract, but are most common in the esophagus. The latter are usually seen in the colon and rectum, and rarely in the small intestine. All of these

Chapter 16: Smooth muscle tumors

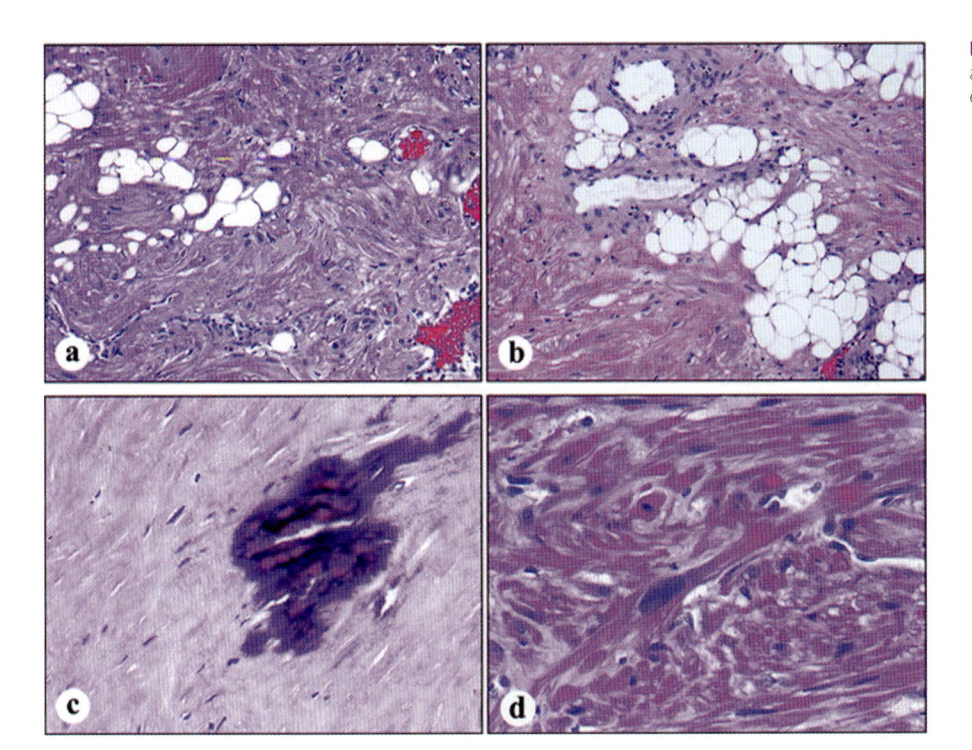

Figure 16.9 (**a,b**) Lipomatous component in angioleiomyoma. (**c**) A calcified paucicellular example. (**d**) Nuclear atypia in angioleiomyoma.

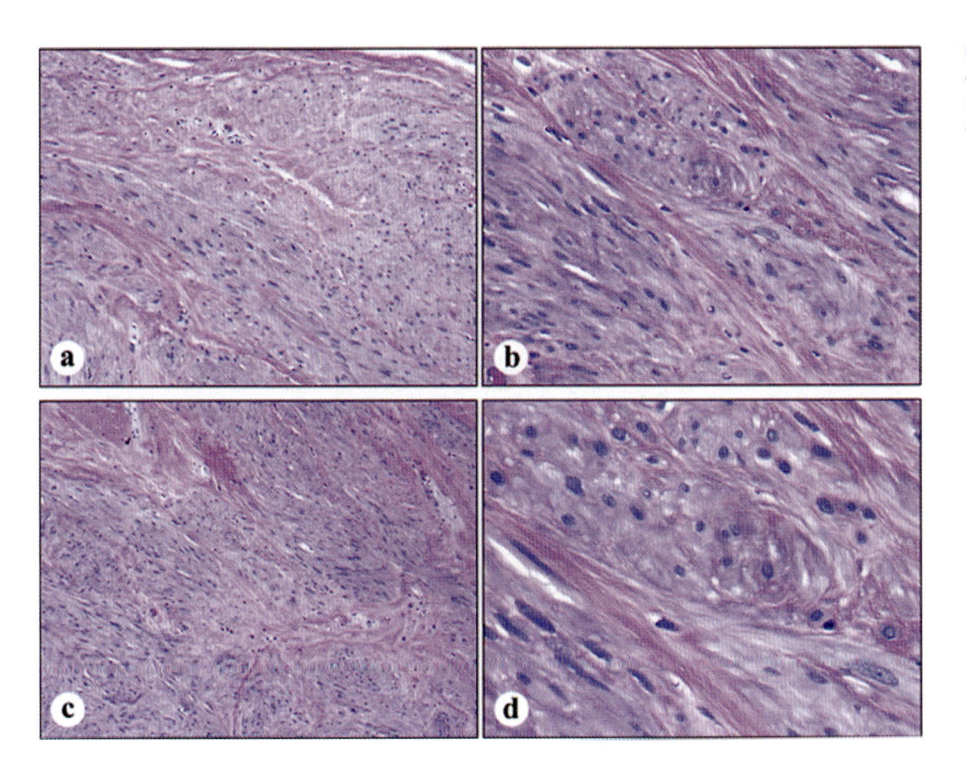

Figure 16.10 A solitary scrotal leiomyoma is composed of solid sheets of well-differentiated smooth muscle cells with no atypia and no mitotic activity.

tumors are composed of mature smooth muscle cells with the corresponding immunophenotype. They are KIT negative and should be separated from GISTs. In addition to these leiomyomas, estrogen receptor/progesterone receptor (ER/PR)-positive leiomyomas in women can involve intestines secondarily; these types of tumors are discussed separately with abdominal leiomyomas.

Intramural leiomyoma and leiomyomatosis
Clinical features

Intramural leiomyomas are most common in the esophagus, where they are at least three times more common than GISTs, and occur equally in men and women in a broad age range, with a peak incidence in the fourth decade (Figure 16.11).

449

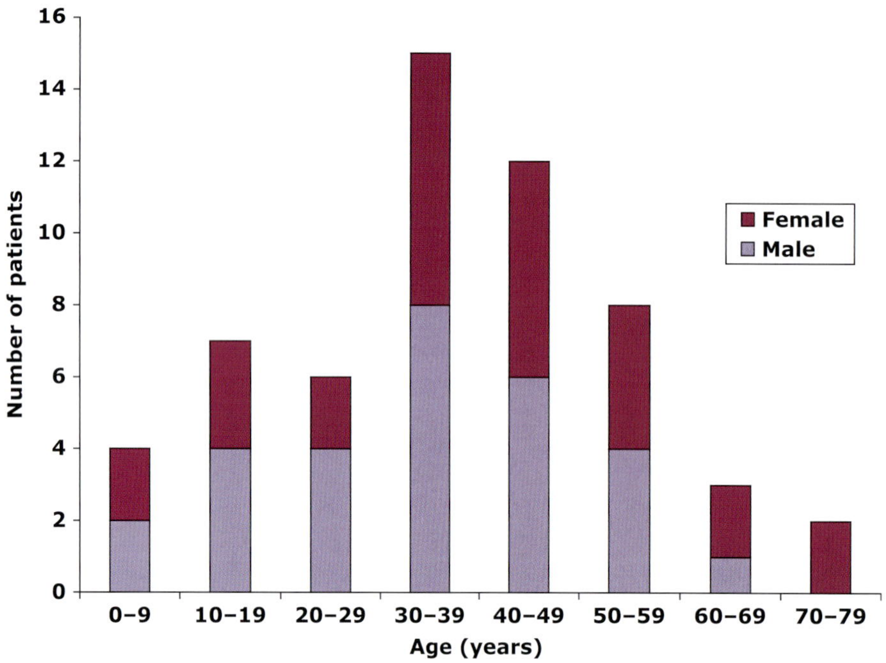

Figure 16.11 Age and sex distribution of 58 civilian patients with esophageal leiomyoma.

There is a predilection for the lower third of the esophagus. Rarely, esophageal leiomyomas occur in the middle portion, and very rarely in the upper third. Dysphagia is a typical complaint, but one-half of these tumors are asymptomatic and detected incidentally by endoscope or radiologic studies. Size varies in a wide range. Because of their intramural location they are not easily accessible by conventional endoscopic biopsy, and when surgery is necessary, they can almost always be excised extramucosally. Recurrence is rare after simple enucleation.[37–39]

Esophageal leiomyomatosis is a rare familial condition with an extensive longitudinal involvement of a leiomyoma. Most patients are young, and some are children, and there is a connection with Alport syndrome, a usually X-linked recessive syndrome having manifestations more common in men. This syndrome includes glomerular basement membrane disease, eye lens abnormalities, and sensorineural hearing loss, based on mutations in genes encoding basement membrane collagens.[40–44]

Intramural leiomyomas occur rarely in the stomach or any segment of the intestines, but they are much rarer than GISTs in all intestinal locations. They vary from asymptomatic mural nodules to ulcerated tumors manifested by gastrointestinal bleeding.

Pathology

Grossly, intramural leiomyoma forms a circumscribed white rubbery mass that usually measures 3 cm to 5 cm, but is sometimes detected as a microscopic focus. Intramural leiomyomas can occasionally reach a large size (>500 g and bulge into the mediastinum. Some esophageal leiomyomas form a circumferential mass, and others have a large longitudinal dimension, presenting as sausage-shaped masses along the muscle layer (Figure 16.12). Microscopically, intramural leiomyomas are paucicellular and show sparsely distributed small nuclei, while the cell borders are histologically indistinct (Figure 16.13a,b). Focal nuclear atypia can be present, but mitotic activity is rare. Eosinophilic cytoplasmic inclusions are common (Figure 16.13c). There can be eosinophilic granulocyte infiltration within the tumor, and the presence of numerous mast cells is common. Larger leiomyomas in particular can have extensive calcification. True leiomyomas can be distinguished from GISTs by their overall eosinophilic appearance and low cellularity.

Immunohistochemical analysis demonstrates that gastrointestinal intramural leiomyomas are positive for SMA, heavy caldesmon, and desmin, with the last marker highlighting the eosinophilic cytoplasmic inclusions (Figure 16.13d). They are usually negative for vimentin, and uniformly negative for CD34 and KIT, but numerous KIT-positive mast cells (Figure 16.13e) and occasional Cajal cells are sometimes present. The presence of these KIT-positive elements should not be taken as evidence of a KIT-positive GIST. Desmin expression is also in sharp contrast with GIST.

Genetics

Deletions in the genes encoding the basement membrane collagen type IV alpha 5 and alpha 6 have been reported in both sporadic esophageal leiomyoma and leiomyomatosis in connection with Alport syndrome.[40–44] Similar genetic changes suggest that sporadic leiomyoma and leiomyomatosis could be pathogenetically related, and the basement membrane collagen abnormality could be related to their pathogenesis.

Chapter 16: Smooth muscle tumors

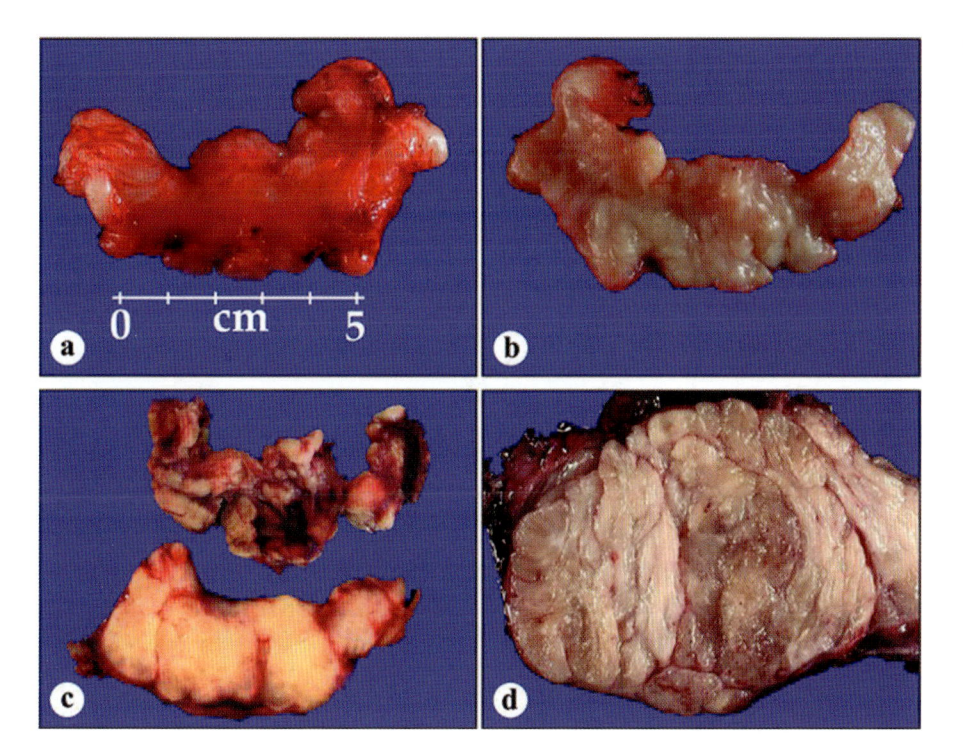

Figure 16.12 Gross appearance of esophageal leiomyoma. The tumors can have complex and convoluted outlines or be oval masses. The cut surface is grayish tan.

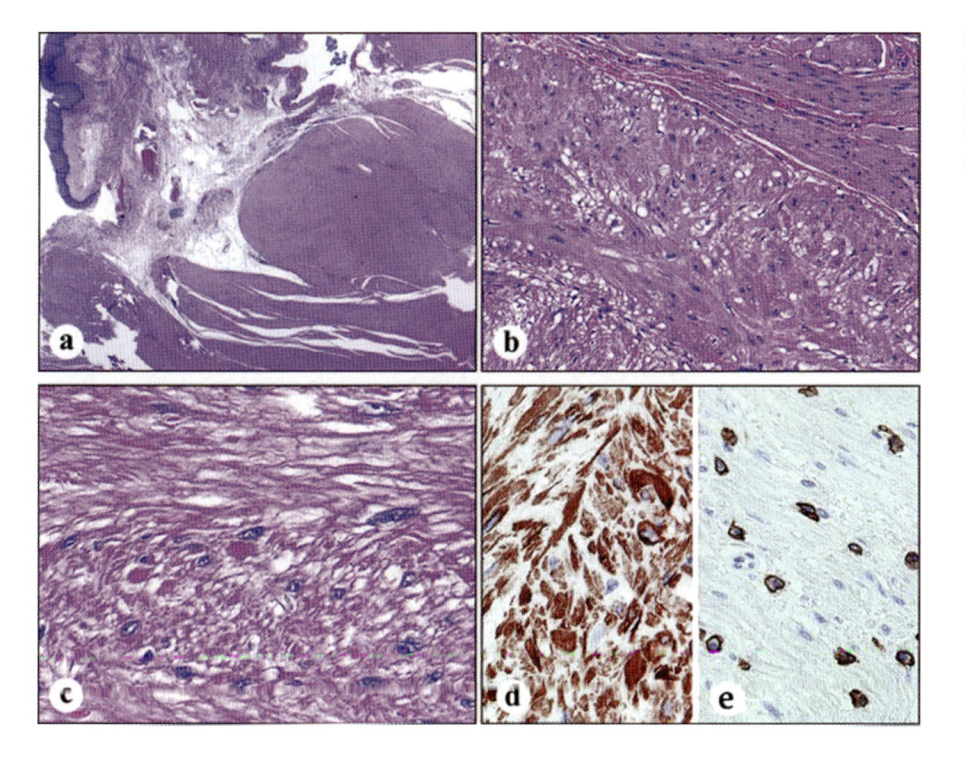

Figure 16.13 Esophageal leiomyoma is composed of well-differentiated smooth muscle cells. The tumor cells are negative for KIT, in contrast to gastrointestinal stromal tumor, but a high number of KIT-positive mast cells are typically present.

Esophageal leiomyomas commonly show gains in chromosome 5 in comparative genomic hybridization.[45]

Leiomyoma of muscularis mucosae
Clinical features
These tumors appear as small pedunculated polyps that almost always occur in the colon and rectum and are usually incidental findings during colonoscopy in older adults (50–80 years). There is a male predilection, and no morbidity is associated with this lesion.[46] Rarely, similar tumors occur in the small intestine, but apparently never in the stomach. Because the lesion is limited to the mucosa, it can be pulled out endoscopically from the retractable submucosa.

451

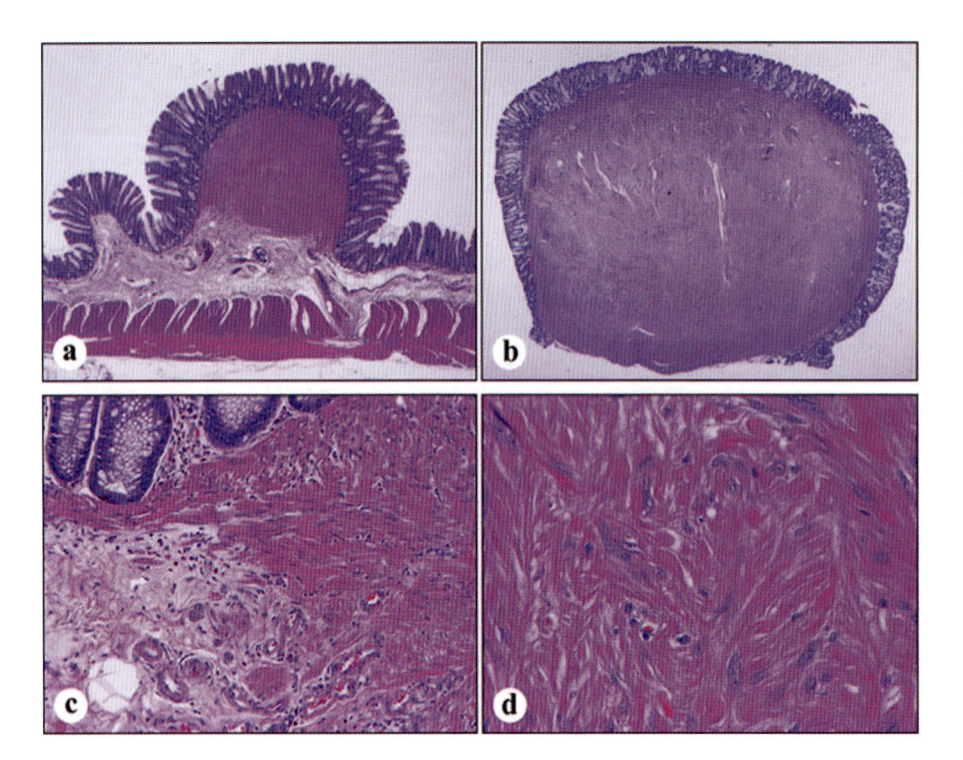

Figure 16.14 Leiomyoma of the muscularis mucosae. (**a**) This unusual specimen contained an incidental leiomyoma in a rectal resection for a mucinous carcinoma. (**b**) Appearance of a typical endoscopically excised lesion. (**c**) The tumor is composed of well-differentiated smooth muscle cells that merge with the muscularis mucosae. (**d**) Eosinophilic cytoplasmic globules are present (they are desmin positive).

Pathology

Grossly, the lesions form sessile polyps and usually measure <1 cm (average, 4 mm); only rarely do they reach 2 cm. The lesion is firm, grayish-yellow, and almost always covered by an intact mucosa. Histologic analysis shows it to be composed of well-differentiated smooth muscle cells similar to those in the muscularis mucosae, where the lesions merge and apparently arise from (Figure 16.14). Focal nuclear atypia is rarely present. Immunohistochemical features are similar to intramural leiomyomas.[46]

Other visceral leiomyomas (excluding estrogen-receptor/progesterone-receptor-positive tumors)

Leiomyomas arising from either proprietary smooth muscle components or vascular smooth muscle cells can occur in a wide variety of organ locations, including the oral cavity, and the respiratory and genitourinary tracts.

Sinonasal leiomyomas occur in the nasal cavity and less often in the sinuses and can originate from the vascular smooth muscle elements of the turbinates.[47] In the lung, leiomyomas can occur as symptomatic, obstructive intraluminal tumors, or sometimes as asymptomatic pulmonary nodules.[48] Many of these tumors seem to arise from bronchial smooth muscle elements. Benign metastasizing leiomyoma is an example of an ER-positive tumor; it is discussed under Estrogen-receptor-positive smooth muscle tumors.

In the genitourinary tract, leiomyomas are most common in the urinary bladder.[49] In this location, leiomyomas occur in all ages, and two-thirds are seen in women. Leiomyomas of the urinary bladder vary from small mucosal nodules to large mural tumors. Rarely, leiomyomas occur in the prostate or urethra; however, there are no large studies of leiomyomas in these locations.

Soft tissue and visceral leiomyosarcomas excluding uterine leiomyosarcomas

LMSs (excluding those of the uterus and female genital regions) are defined as smooth muscle tumors with nuclear atypia in combination with any level of mitotic activity. These tumors occur in almost all soft tissue and organ locations, probably because they can arise from the ubiquitous vascular smooth muscle elements. Other possible origins are pilar smooth muscle in the skin and parenchymal smooth muscle elements in various organs. In general, LMSs are aggressive tumors with a high rate of lung metastasis, especially in high-grade tumors.

Uterine and other ER/PR-positive LMSs are discussed separately. Most tumors formerly considered gastrointestinal LMSs are now classified as GISTs. These tumors are discussed in Chapter 17.

Pathology of visceral leiomyosarcomas

Grossly, LMSs vary according to tumor differentiation. The low-grade tumors resemble leiomyomas and tend to be hard masses with a whitish, whorled surface on sectioning. High-grade tumors resemble other high-grade sarcomas and are typically composed of homogeneous, soft, gray-tan tissue. Many large high-grade LMSs show extensive central necrosis,

Chapter 16: Smooth muscle tumors

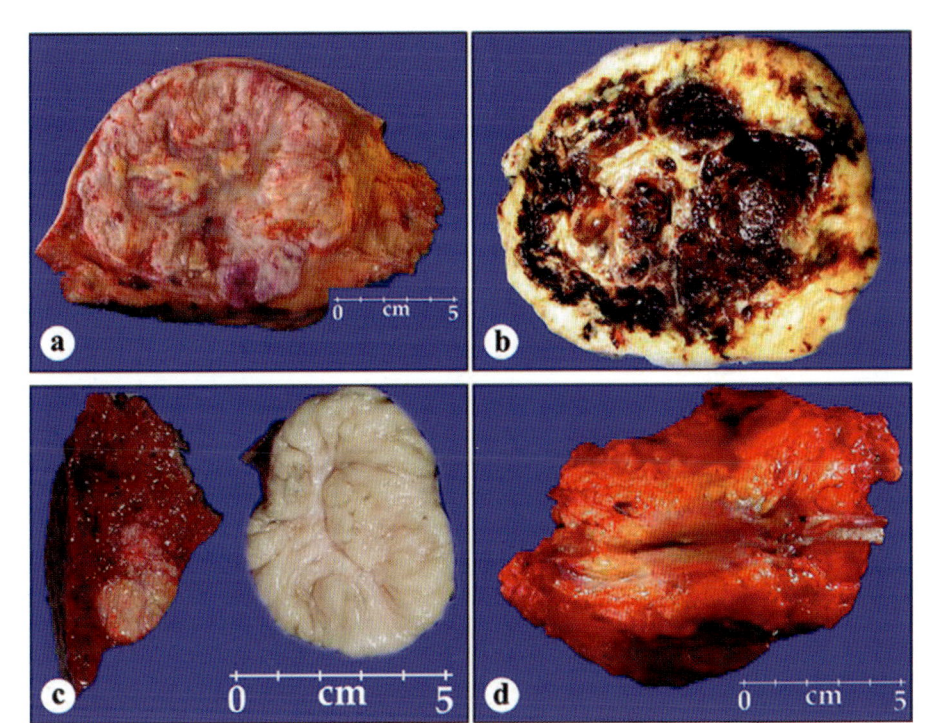

Figure 16.15 Gross appearances of soft tissue leiomyosarcoma (LMS). (**a**) This subcutaneous high-grade LMS shows a variegated cross section with foci of yellowish necrosis. (**b**) A large retroperitoneal LMS with extensive central hemorrhagic necrosis. (**c**) A 5-cm retroperitoneal LMS shows a homogeneous, grayish cut surface and liver metastasis at presentation. (**d**) A vascular LMS forms a whitish plaque-like lesion originating from a venous segment.

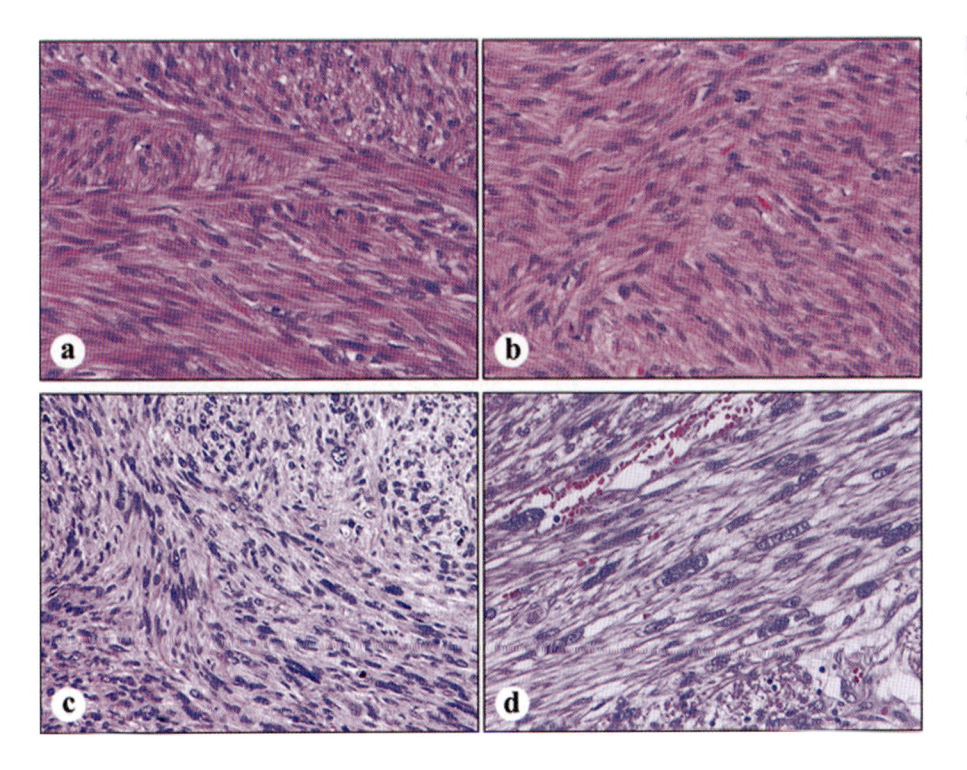

Figure 16.16 Examples of well-differentiated LMS. Note intersecting fascicles in (**a–c**), and blunt-ended, "cigar-shaped" nuclei. Note cytoplasmic eosinophilic floccular material in (**d**), representing contraction bands.

which appears as a yellowish central area with a peripheral rim of viable tumor tissue (Figure 16.15).

Microscopically, LMSs are almost always composed of spindled cells, which commonly show longitudinally oriented intersecting fascicles. Cytologically, the tumor cells typically have blunt-ended "cigar-shaped" nuclei, and the cytoplasm is variably eosinophilic, sometimes distinctly clumped, resembling the contraction bands commonly seen in histologically processed benign and malignant smooth muscle (Figure 16.16).[50] Some tumors have myxoid stroma, with the tumor cells arranged in a "basket-weave" pattern. Focal pleomorphism is common (Figure 16.17), and some cases show extensive pleomorphism, resembling malignant fibrous histiocytoma (MFH; Figure 16.18). In such tumors, the nature of the tumor as a poorly differentiated LMS might become apparent only after extensive sampling. These tumors occur especially in the retroperitoneum. The designation

453

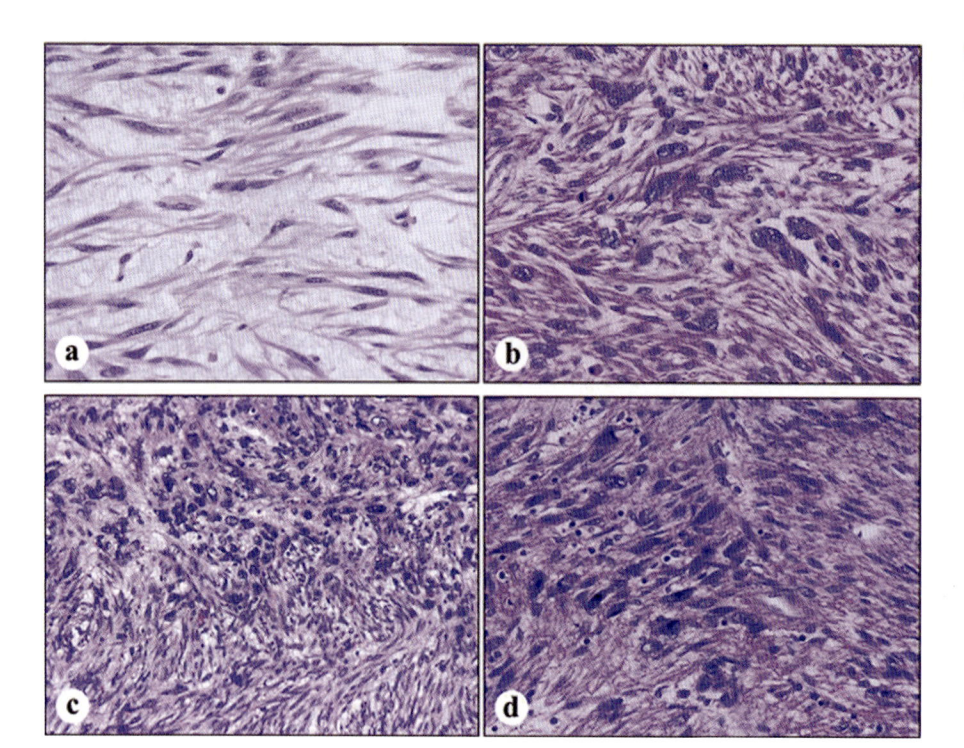

Figure 16.17 (**a,b**) Myxoid and focally myxoid LMS with mucoid matrix. (**c,d**) Moderate nuclear pleomorphism in LMS.

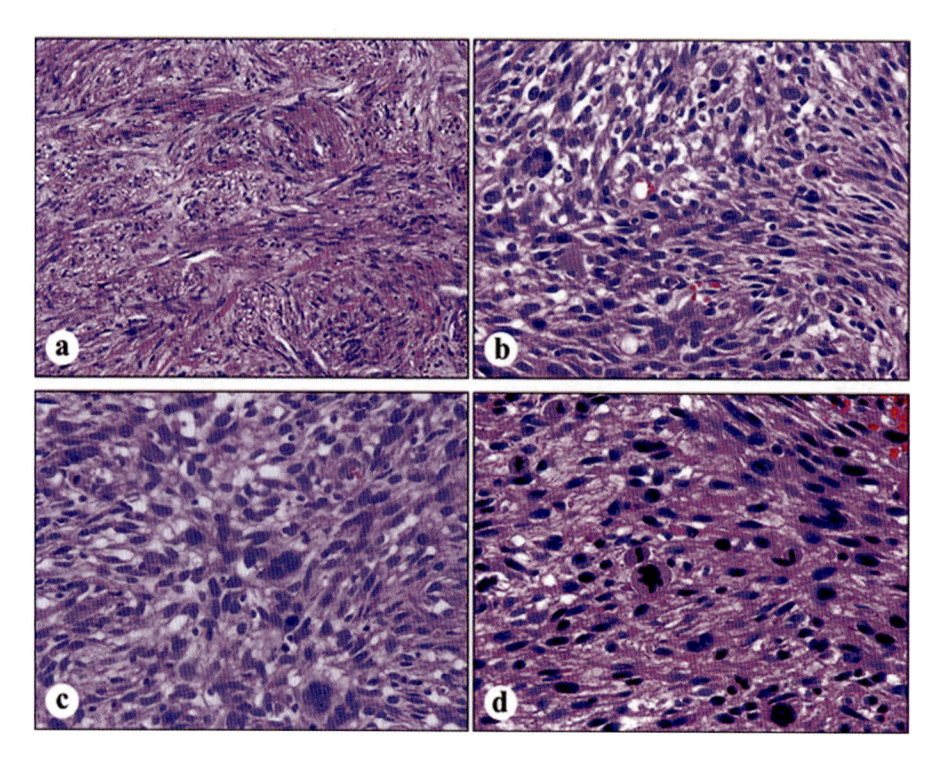

Figure 16.18 Examples of severely pleomorphic LMS. (**a**) Conventional area similar to one present in each tumor. (**b–d**) Examples of pleomorphic areas, including atypical mitoses. These areas are not diagnostic of LMS unless differentiated areas are present.

"dedifferentiated leiomyosarcoma" has been applied with these pleomorphic tumors.[51] Prominent inflammatory infiltration is present in rare examples (Figure 16.19).

Immunohistochemistry

LMSs are nearly always positive for alpha SMA and muscle actins (HHF-35), which can be considered part of the definition. Desmin positivity varies and is overall 70% to 80%, with poorly differentiated tumors being less consistently, often only focally, positive (Figure 16.20).[52,53]

Heavy caldesmon is a smooth-muscle-specific cytoskeletal component that is typically although not invariably present in LMS.[54,55] CD34 is expressed in a minority (<30%) of typical retroperitoneal LMSs, in addition to its common expression in GIST.[56,57] LMSs are generally negative for KIT (CD117), in contrast to GISTs.[57]

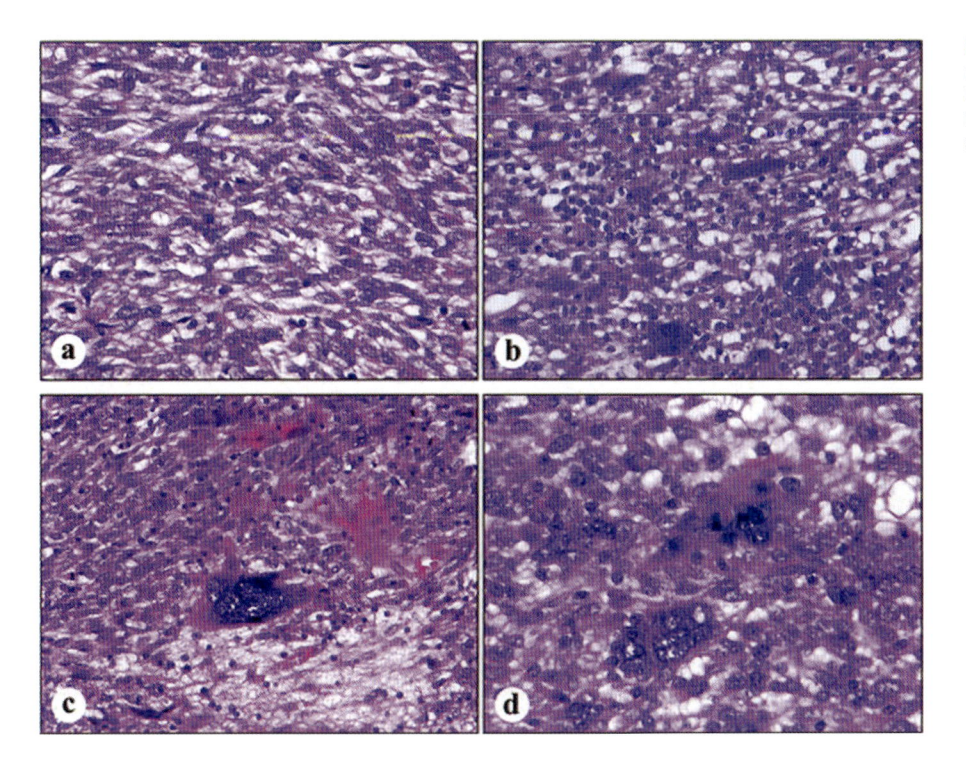

Figure 16.19 An example of LMS with prominent inflammatory component. (**a**) Conventional differentiated component in the tumor. (**b–d**) Note variable lymphoid infiltration and significant nuclear pleomorphism.

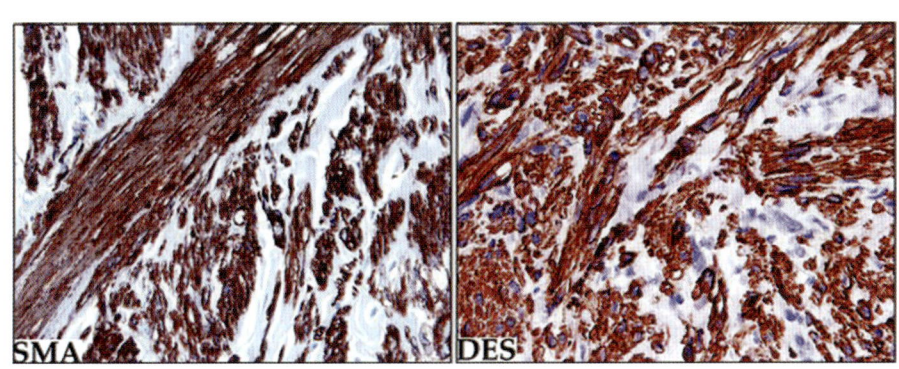

Figure 16.20 Immunohistochemical features of LMS. This cutaneous example shows strong positivity for both SMA and desmin.

Epithelial markers, including keratins of simple epithelia and epithelial membrane antigen (EMA), are expressed in approximately 10% to 30% of LMSs. The expression of keratins can be especially prominent in uterine and cutaneous LMS, but also can be seen in retroperitoneal and vascular LMSs.[58,59] The keratin polypeptides expressed in LMSs, in the author's experience, especially include keratins 8 and 18. Most LMSs are positive for vimentin, but some well-differentiated tumors are negative, similar to many normal smooth muscle cells.

LMSs have been reported to have p53 mutations with varying frequency.[60–62] One study reported 10% to 20% frequency and showed more common loss of the retinoblastoma gene protein 1 (RB1) and overexpression of cyclin D1, suggesting that RB1-cyclin D1 pathway might be more important in their deranged cell cycle control.[63]

Ultrastructure

Electron microscopic findings of typical smooth muscle tumors include abundant cytoplasmic actin filaments with dense bodies and variable intermediate filaments, actin filaments as attachment plaques at the cell membrane, pinocytic vesicles, and basement membranes surrounding the cells. These features are often incompletely developed in LMSs. The actin filaments are usually abundant, but basement membranes are often incomplete.

Genetics

Cytogenetically, LMSs show complex karyotypes with heterogeneous changes without any characteristic diagnostic aberrations. Irregular chromosome numbers (aneuploidy and polyploidy) are common.[64–66] Comparative genomic hybridization studies have also shown multiple, complex, heterogeneous changes.[67–69]

Cutaneous LMS

Cutaneous, purely dermal LMSs occur predominantly in older adults with a male predominance in some series[68,69] and mild

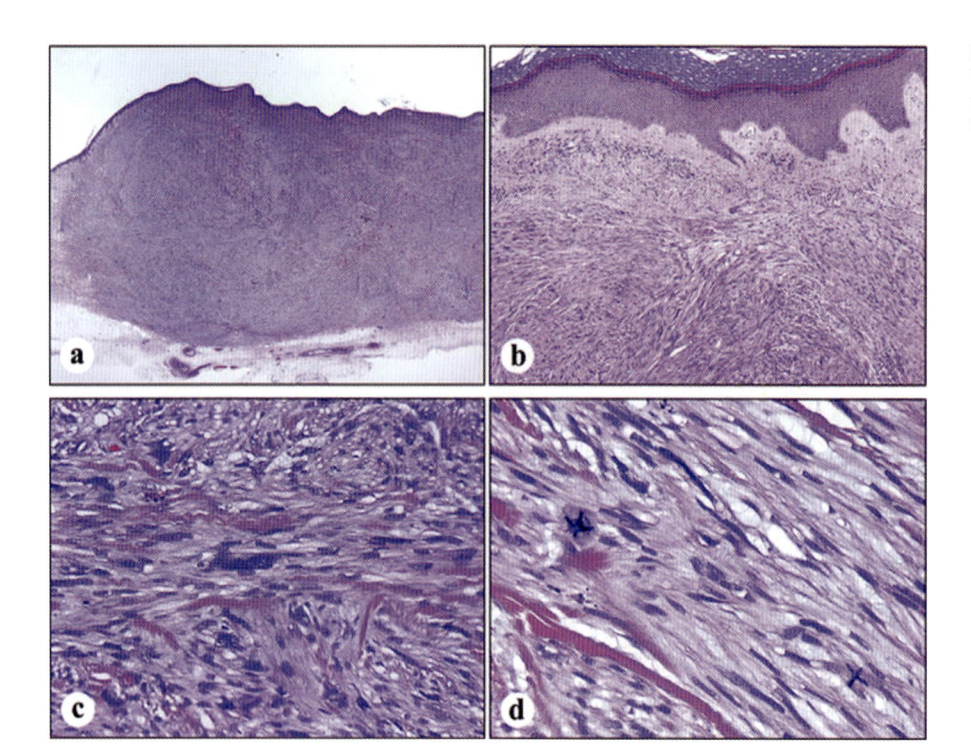

Figure 16.21 Cutaneous involvement of LMS. The tumor extends close to the epidermis in both examples. Significant cytologic atypia and atypical mitoses are present.

female predominance in others.[70] Although some series show a predilection for the extremities (especially the thigh),[68,71,72] others show common occurrence in exposed sites in the head and neck,[70] which suggests the role of ultraviolet carcinogenesis.

Cutaneous LMSs present as relatively small solitary plaque-like or elevated skin nodules that can be ulcerated and resemble common skin cancers. Purely cutaneous LMSs are usually <2 cm, but some studies have included tumors >5 cm, which probably has allowed some subcutaneous extension. Histologically, these tumors are often poorly circumscribed, intermingling with dermal collagen, which suggests origin from pilar smooth muscle elements (Figure 16.21). Recurrence is common if treated by enucleation only, but otherwise purely cutaneous LMSs have an excellent prognosis, independent of histology. Although most series have shown no metastases,[68,70,71,73] occasional metastases have been reported.[69] Complete excision with negative margins and follow-up are the optimal treatments. A proposal has been made to designate purely cutaneous LMSs as "atypical intradermal smooth muscle neoplasms" based on their apparent lack of metastatic potential (provided that a narrow definition for "dermal" is being applied). Nevertheless, other recent studies have noted existing low metastatic potential for purely cutaneous LMSs so that categoric exclusion of these tumors from malignancy may be premature.[74,75]

Subcutaneous and intramuscular LMS

The annual population incidence of noncutaneous soft tissue LMS in the extremities and trunk wall was estimated as 1.3 per 1 million inhabitants in one study from Sweden.[76] Soft tissue LMSs occur in a wide age range, with a mild male predominance and predilection for older adults and extremity locations, especially the thigh.[77] Origin from small vessels can sometimes be demonstrated, and some such tumors have features comparable to malignant counterparts of angioleiomyomas.[72,78,79] Metastasis from a previously diagnosed or an occult internal LMS (often retroperitoneal) must be excluded, especially with multiple subcutaneous LMSs.[72]

Subcutaneous LMS often forms an ovoid mass with a median size of 4 cm, but it can be >10 cm (Figure 16.15a). Subcutaneous LMSs differ from the cutaneous variety by their moderate potential to metastasize. A larger series showed predilection for older adults with no gender predilection, and a metastatic rate of 20%.[73] Other studies have shown metastatic rates from 0%[77] to 33%,[68] or even higher metastatic rates, such as 8 of 12 in one study.[69]

Deep intramuscular LMS are usually diagnosed as advanced large tumors, and this is probably a significant factor in their poorer prognosis. They most commonly occur in the thigh and buttocks, and in the largest series showed a high metastatic rate, approaching 60%,[73] with higher (≥90%) metastatic rates in a smaller series.[77] Deep vascular LMSs in the extremities have a poor prognosis.[80]

Retroperitoneal LMS

The retroperitoneum is one of the most common locations for soft tissue LMS. These tumors occur chiefly in older middle age with a female predominance (Figure 16.22). Tumor size is usually >10 cm at the time of diagnosis, and prognosis is poor, with tumor-related mortality rates of 80% and metastases frequently to lungs and liver, and less commonly to peripheral

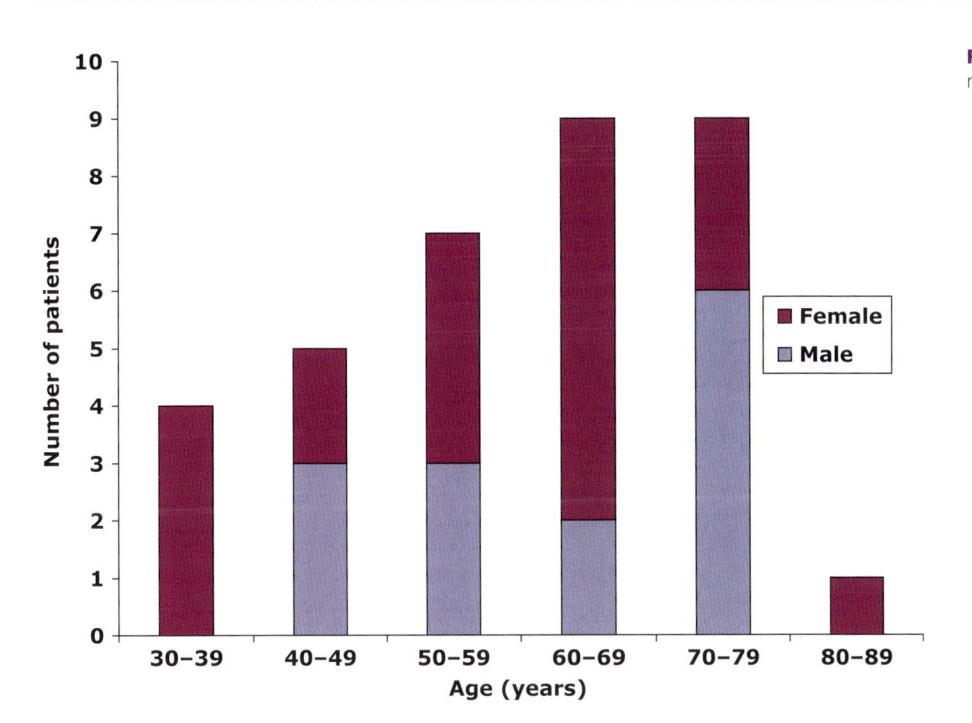

Figure 16.22 Age and sex distribution of 38 reclassified retroperitoneal LMSs.

soft tissues and bone. Nearly uniformly poor prognosis limits the opportunity to analyze prognostic factors. Patients with smaller tumors and those with lower mitotic rates (and lower grades) might have longer survival rates, however.[81]

GIST (usually KIT, CD117 positive; see Chapter 17) is a serious differential diagnosis. In our experience, true LMSs and GISTs are nearly equally represented among cases historically considered "retroperitoneal LMS." This also complicates the interpretation of older series. Uterine LMSs not infrequently metastasize to the retroperitoneum.

Gastrointestinal leiomyosarcoma

True LMSs with complete phenotypic properties of smooth muscle cells (SMA, heavy caldesmon, and desmin positive, KIT negative) are very rare in the gastrointestinal tract, because most tumors previously classified as gastrointestinal LMSs are now classified as GISTs based on their KIT positivity and general negativity for desmin.

Gastrointestinal LMS is extremely rare in the esophagus and stomach but occurs in all intestines and seems to be relatively most common in the colon and rectum, mainly affecting older and middle-aged adults. These tumors commonly cause gastrointestinal bleeding, obstruction, or both. Many well-documented LMSs of the gastrointestinal tract are polypoid intraluminal tumors, and such tumors in the colorectum might be derived from the muscularis mucosae structures. Most primary LMSs of the gastrointestinal tract are clinically and histologically high-grade tumors, but there have been long-term survivors among patients with relatively small polypoid intraluminal tumors. Intestinal LMSs seem to have a better prognosis than GISTs of similar size and mitotic rates.[82,83] Metastatic LMSs can occur in the gastrointestinal tract either as polypoid, often multiple, intraluminal tumors, or as external masses attached to serosae. Clinical correlation therefore is required to rule out a tumor of metastatic origin.

Leiomyosarcoma of the vena cava and other veins

Vena cava LMS has a marked female predominance, with various series containing 60% to 83% of female patients. The age range in those series is 15 to 85 years (median ages, 54–57 years). The middle segment of vena cava between the renal and hepatic veins is the area most commonly involved (43–65%), followed by the lower segment between the iliac and the renal veins (20–30%), and the upper segment (5–19%). The location and mode of vascular involvement determine the clinical features and prognosis. Over one-half of these tumors have intraluminal growth, which is an unfavorable prognostic factor, especially in the upper segment, often resulting in obstruction of the vena cava and hepatic vein causing Budd–Chiari syndrome. Extension into the right side of the heart can also complicate this tumor (Figure 16.23). Five-year survivals in surgical series with long-term follow-up vary from 49% to 76%, and are close to 0% in inoperable cases. Tumor size <10 cm and negative resection margins, and possibly low histologic grade are favorable prognostic factors. Rare occurrences in the iliac and renal veins have also been reported.[84–87]

Venous origin is detected in some peripheral LMSs, sometimes during surgery, and thorough gross and microscopic examination reveals vascular origin in a small number of additional cases (Figure 16.24). Overall, the true frequency of vascular origin of LMS of soft tissues is probably underestimated. In the author's experience, the most commonly involved vessels are the saphenous vein and small unnamed peripheral veins.

Chapter 16: Smooth muscle tumors

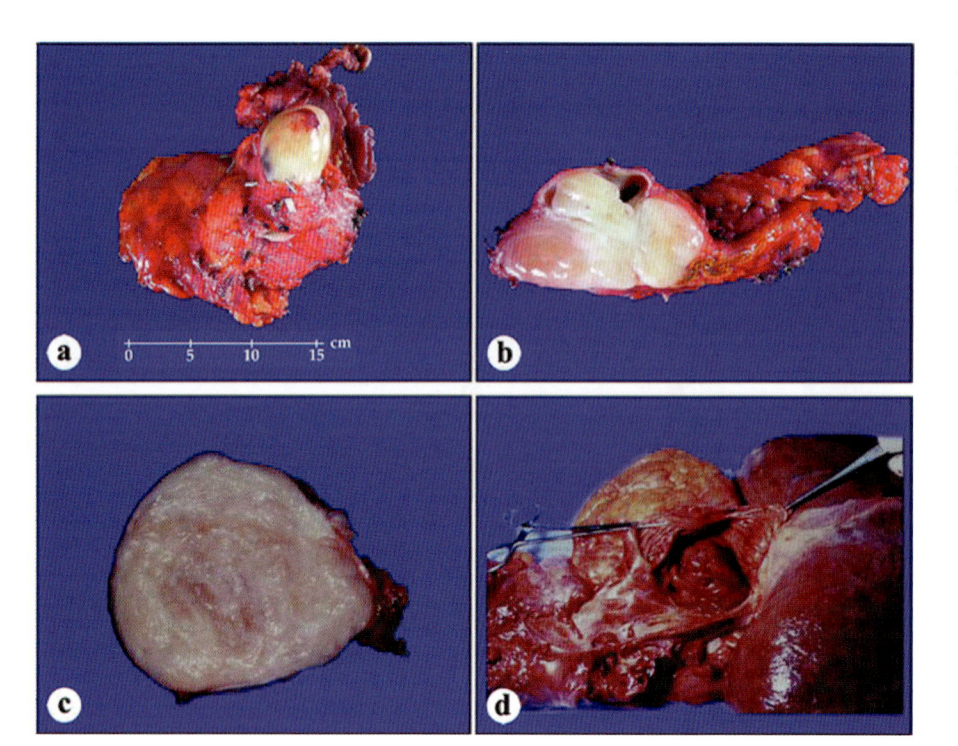

Figure 16.23 Gross appearances of LMS of the vena cava. (**a,b**) Tumor with intraluminal involvement. (**c**) Example with prominent extraluminal mass showing a pale off-white cut surface and a weak trabecular pattern. (**d**) Tumor of upper segment of vena cava extending into the right side of heart.

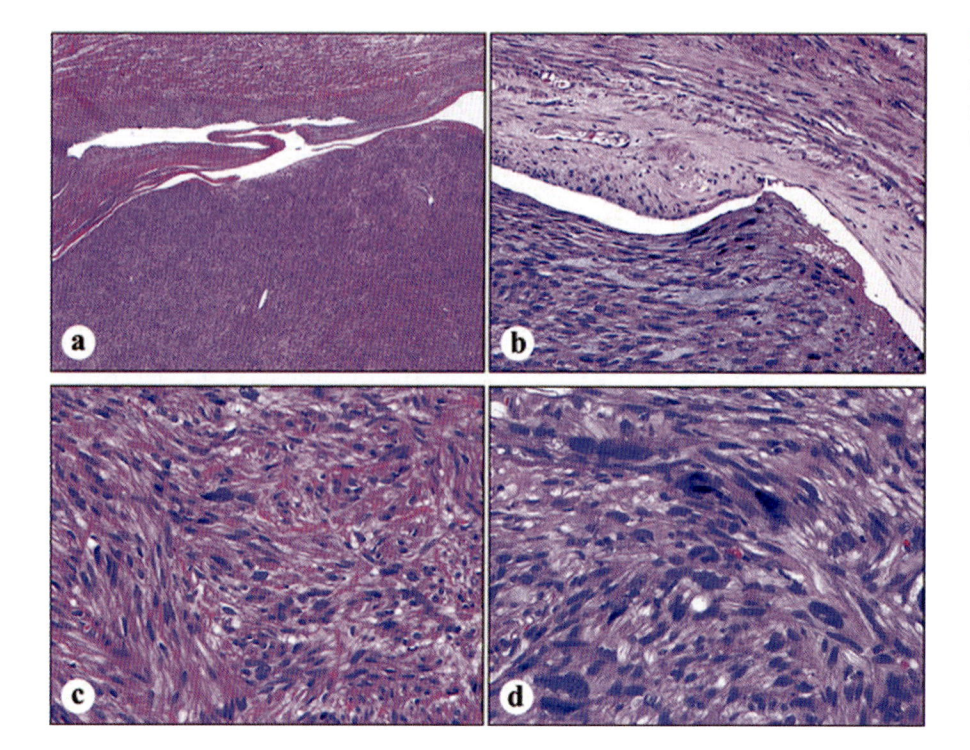

Figure 16.24 Vascular LMS arising from a vein wall. (**a,b**) Low magnification shows tumor in the vein lumen. (**c,d**) High magnification shows features typical of LMS with variable nuclear pleomorphism.

Other organ-based leiomyosarcomas

Less common sites for LMSs reported in clinicopathologic series are the sinonasal tract,[88] oral cavity,[89] lungs,[90] urinary bladder,[91] prostate,[92] penis,[93] and testicular adenxa,[94] among others. All of these tumors occur predominantly in older patients. Prognosis varies, and some grade and tumor size dependency of survival has been observed, for example with paratesticular LMS. Superficial and smaller LMSs in the penis have a better prognosis than deep tumors.[93]

Leiomyosarcoma of children

LMSs very rarely occur in children; among those that do, the most common are LMSs in immunosuppressed patients (see discussion later). According to the largest series, LMSs in immunocompetent children present at a variety of sites, and most are low-grade tumors with a good prognosis.[95,96] According to the author's experience, most tumors originally considered intra-abdominal LMSs in children are actually inflammatory myofibroblastic tumors (synonymous to inflammatory fibrosarcoma).

Estrogen-receptor-positive smooth muscle tumors in women

Leiomyomas in abdominal soft tissues

Clinical features

This is a newly discovered category of tumors that occur exclusively in women, typically in early middle age, and show a histologic spectrum similar to that of uterine leiomyomas.[97,98] In fact, they can be considered extrauterine examples of uterine-type, ER/PR-positive leiomyomas. The most common locations include various pelvic sites and the retroperitoneum. In the author's experience, similar tumors also occur elsewhere in the abdominal cavity attached to the intestines, omentum, abdominal wall, and inguinal region.[99] In the latter location, leiomyomas could originate from the round ligament.

Abdominal leiomyomas vary from small nodules <1 cm to larger masses; the latter is especially true for retroperitoneal leiomyomas. Prognosis of these tumors is excellent, as long as mitotic activity is low (≤3 mitoses per 50 HPFs).[98] Tumor behavior has not been verified for those examples with higher mitotic rates, which would still qualify as benign in the uterus. Therefore, such tumors with nonatypical features with higher mitotic rates must be considered of uncertain biologic potential until more experience with them has been obtained.

Pathology

Abdominal and especially retroperitoneal leiomyomas, are often large, >10 cm to 15 cm, and can weigh more than 1 kg. Grossly they are firm, gray-white masses, often with a whorled appearance similar to uterine leiomyomas, and cysts can be present on sectioning (Figure 16.25). Similar tumors in the omentum and around the intestines are typically smaller.

Histologically they are generally paucicellular to moderately cellular and have a spectrum similar to that of uterine leiomyomas. Myxoid change, hyalinization, and trabecular patterns are part of their spectrum (Figure 16.26). Examples with fat infiltration are called *lipoleiomyomas* (Figure 16.27). All these patterns are also seen in uterine leiomyomas.

Similar to uterine leiomyomas, abdominal leiomyomas are immunohistochemically positive for smooth muscle actin, desmin, heavy caldesmon, and ER/PRs (Figure 16.28). Examples

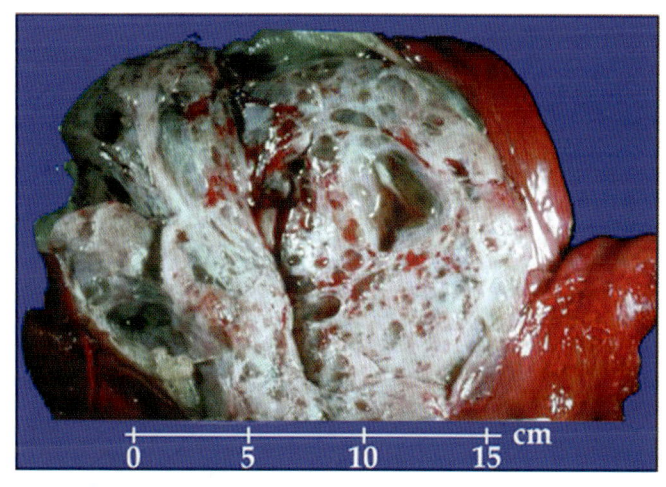

Figure 16.25 Gross appearance of a retroperitoneal leiomyoma. Note multiple cysts and gray but otherwise homogeneous texture.

with significant mitotic activity (>5 per 50 HPFs) should be classified as at least having uncertain malignant potential, even if they are hormone-receptor-positive, as data on such extrauterine tumors are limited

Leiomyomas of the female external genitalia and uterus

Leiomyoma of external genitalia

Leiomyomas of the vulva are typically solitary, well-circumscribed nodules, and it is likely that in these locations they are related to the locally abundant, subcutaneous band-like smooth muscle elements. In this location, leiomyomas typically arise from the hormonally responsive ER-positive smooth muscle in the labia majora, where they appear as painless masses.[35,100,101] Some leiomyomas of this site are of vascular origin (angioleiomyomas). Vulvar leiomyomas are mostly hormonally responsive tumors, and those cases with ER positivity can be allowed some mitotic activity, up to 5 per 10 HPFs; in fact only one in five tumors having mitotic activity of >5 per 10 HPFs recurred in one study. Pregnancy-associated smooth muscle tumors of the external genitalia tend to have a low mitotic activity and a benign course.[99] Another series showed local recurrences but no metastases in tumors diagnosed as leiomyomas based on criteria for uterine tumors.[100] Older series of external genital leiomyomas predating newer entities, such as angiomyofibroblastoma and cellular angiofibroma, have probably included such tumors as "leiomyomas."

Uterine leiomyoma

Uterine leiomyomas are extremely common tumors and are present in a majority of women undergoing hysterectomy. They vary from minute (<1 mm) nodules to large masses, and are often multiple, extend intraluminally or outward and cause varying symptoms such as abnormal uterine bleeding, a feeling of pelvic pressure, and a palpable mass. They can be

Chapter 16: Smooth muscle tumors

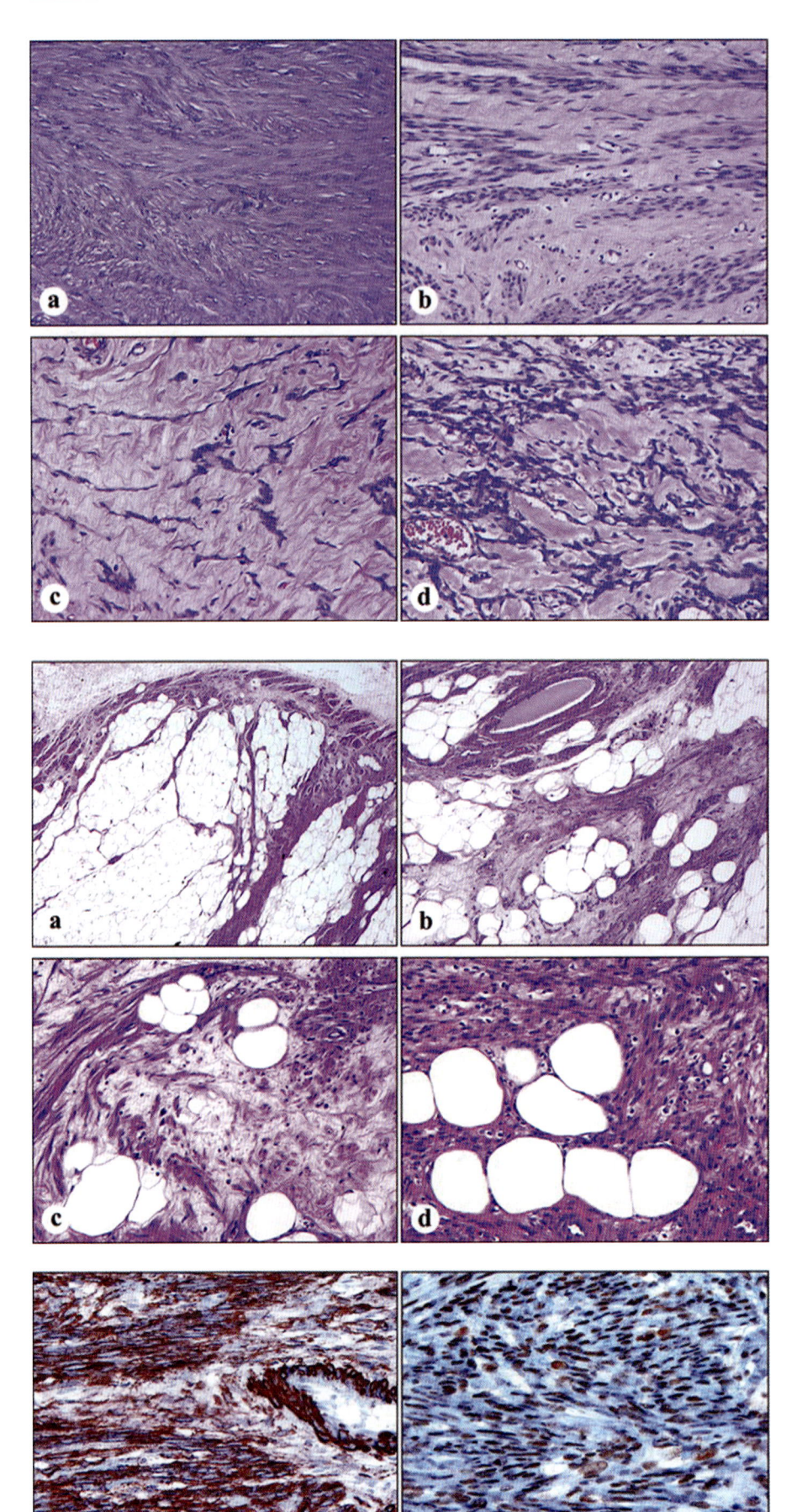

Figure 16.26 Histologic variation of retroperitoneal uterine-type leiomyomas. (**a**) A solid variant with a diffuse pattern. (**b**) This leiomyoma shows an "organoid" compartmental pattern. (**c**) A trabecular pattern is seen in this retroperitoneal leiomyoma. (**d**) The tumor cells are positive for ER.

Figure 16.27 Lipoleiomyoma shows foci of mature fat amid well-differentiated smooth muscle cells.

Figure 16.28 Retroperitoneal as well as female genital leiomyomas in women are positive for SMA and ER.

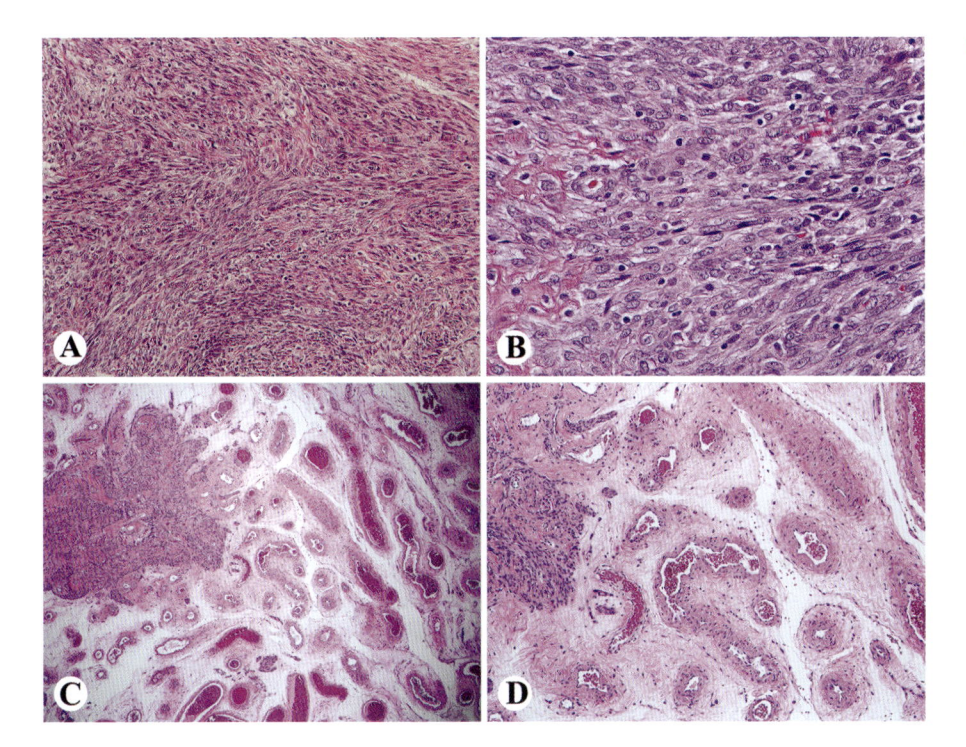

Figure 16.29 (**A,B**) Cellular leiomyoma contains sheets of spindle cells without significant nuclear atypia with inconspicuous matrix. (**C,D**) Cotyledonoid leiomyoma contains prominent engorged vessels and cleavage spaces.

submucous, intramural, and subserosal. Especially the latter ones are often amenable to uterus-sparing myomectomy usually practiced in the fertile age groups. Grossly diffuse leiomyomatous overgrowth of the uterus typically represents adenomyosis.

A typical leiomyoma bulges outward from a freshly cut surface, is grossly white, firm, and has a whorled surface on cross sectioning. Necrotic examples can show red-brown or greenish-yellowish surfaces on sectioning. Cellular and atypical leiomyomas are typically softer and may be yellow and lack trabeculation, and cytoledonoid leiomyoma has a "granular" appearance due to extensive clefting inside the tumor.

Histologically, uterine leiomyomas typically show bundles of smooth muscle interspersed with collagenous bands, but variants contain macro- and microtrabecular arrangements and fatty differentiation, as shown for abdominal extrauterine ER-positive leiomyomas (Figures 16.26 and 16.27). Cellular leiomyomas contain sheets of nonatypical smooth muscle cells with very little collagenous matrix often limited to narrow perivascular zones (Figure 16.29A,B). These tumors may contain considerable mitotic activity, but numerous mitoses may also be present in conventional noncellular leiomyomas.[102,103]

Mitotic activity up to 10–15 per 10 HPFs without atypia is not considered indicative of malignancy. However, there is some uncertainty about leiomyomas with mitotic rate >15 per 10 HPFs, even without atypia, because of insufficient follow-up information on such tumors.[104–106]

Cotyledonoid (dissecting or grape-like) leiomyomas contain clefts with very loose myxoid matrix and prominent engorged blood vessels (Figure 16.29C,D). Atypia, significant mitotic activity or intravascular spread is not present. They typically extend externally to the pelvic cavity and due to their clefting and vascular engorgement, can grossly resemble placental tissue.[107]

Atypical leiomyomas can contain moderate to severe nuclear atypia, but have low mitotic activity (<10 per 10 HPFs without necrosis, <5 per 10 HPFs if necrosis present). This group is pathogenetically heterogeneous, including tumors that by prominent cytologic atypia, even atypical mitotic figures, resemble LMS (Figure 16.30A,B). However, they are excluded from LMSs based on low mitotic activity and often by their small size and location, especially when confined in the hysterectomy specimen.[104–106] Diagnostic criteria for atypical leiomyomas are summarized in Table 16.1.

A significant subgroup of atypical leiomyoma is fumarate-hydratase (fumarase)-deficient leiomyoma associated with HLRCC syndrome. These tumors typically occur in young women and are enriched in myomectomy specimens. Their histology can be distinctive, including ribbon-like arrangements of nuclei, multinucleated and hyperchromatic cells, sometimes with prominent eosinophilic nuclei, and occurrence of cytoplasmic eosinophilic globules (Figure 16.30C,D).[107,108] Mitotic activity is typically low, and so HLRCC-associated, fumarase-deficient leiomyomas may have been classified as "symplastic leiomyomas," a designation previously used of atypical smooth muscle tumors with low mitotic activity. They typically show loss of fumarase expression in the tumor cells, while expression is retained in vascular and other non-neoplastic elements (Figure 16.31).[109,110]

Chapter 16: Smooth muscle tumors

Table 16.1 Assessment of malignancy of conventional uterine smooth muscle tumors

(A) Tumors with no more than insignificant (mild) nuclear atypia

	Mitotic rate per 10 HPFs	Diagnostic conclusion
No coagulation necrosis present	0–9	Leiomyoma
	≥10–19	Leiomyoma with increased mitotic activity[a]
	≥20	Leiomyoma with increased mitotic activity, but experience is limited[a]
Coagulation necrosis present	<10	Smooth muscle tumor of low malignant potential, but experience is limited
	≥10	Leiomyosarcoma (high risk of recurrence and metastasis)

(B) Tumors with significant (moderate or severe) nuclear atypia

	Mitotic rate per 10 HPFs	Diagnostic conclusion
No coagulation necrosis present	0–9	Atypical leiomyoma with low risk of recurrence[b]
	≥10	Leiomyosarcoma: Tumors with 10–19 mitoses with only focal atypia can be designated as atypical leiomyomas, but experience is limited by rarity of such cases
Coagulation necrosis present	Any	Leiomyosarcoma

The criteria follow those of Bell et al.,[104,106] with the exceptions noted below.
[a] Some experts prefer to designate those tumors with >15 mitoses per 10 HPFs as "SMT of uncertain malignant potential (UMP)."
[b] Some experts regard uterine SMTs with significant atypia, no coagulation necrosis, and 5–9 mitoses per 10 HPFs as not less than an "SMT of UMP." Necrosis is specifically qualified as coagulative tumor necrosis, abruptly transitioning from tumor cells without an intervening hyalinized zone.

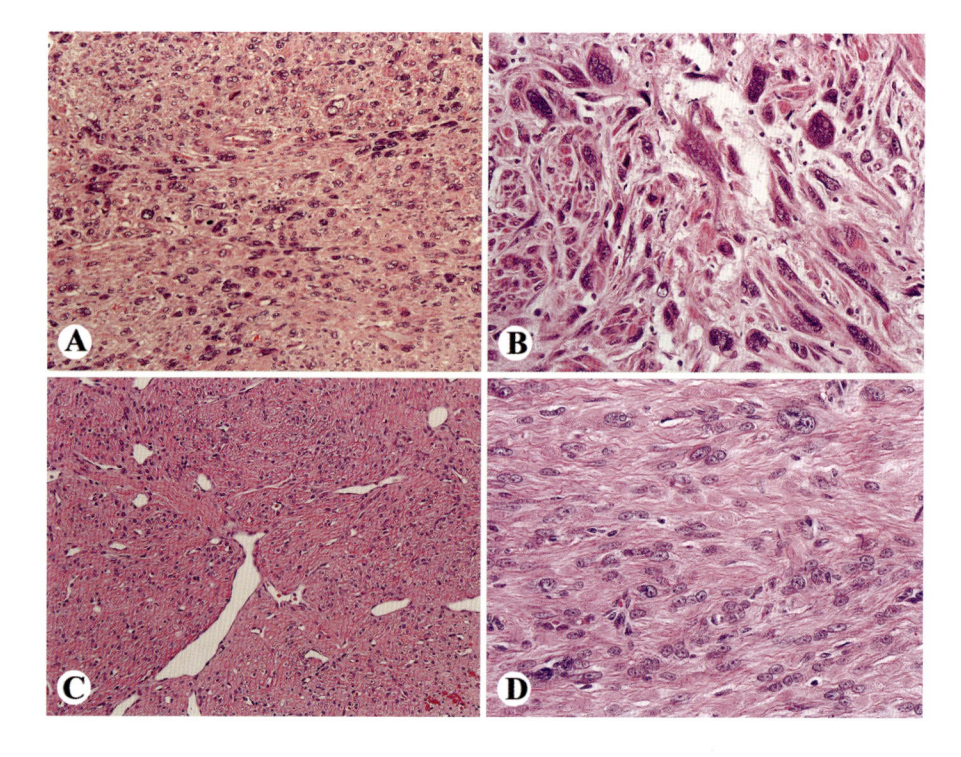

Figure 16.30 (**A**) Atypical uterine leiomyoma with moderate nuclear atypia, but very low mitotic activity. (**B**) Atypical leiomyoma with severe atypia and (**C**) hereditary leiomyomatosis and kidney cancer syndrome-associated uterine leiomyomas often contain prominent staghorn vessels and have less collagenous matrix than usual in leiomyomas. (**D**) At high magnification, they tend to show ribbon-like nuclear arrangements, focal moderate nuclear atypia, and eosinophilic cytoplasmic inclusion bodies (just above the middle center).

Genetics

Genetically, uterine leiomyomas are heterogeneous. The most common known mutations seem to be gain-of-function mutations in the MED12 gene enclding mediator complex member #12 of chromatin regulating proteins.[111,112] Some examples contain mutations in nuclear high mobility group protein HMGA2 (formerly HGMC1).[112,113] HLRCC-associated fumarase-deficient leiomyomas are associated with loss-of-function germline mutations in the fumarate hydratase gene combined with somatic losses according to the classic tumor-suppressor gene mechanism.[18]

Epithelioid smooth muscle tumors of the gynecologic tract

These smooth muscle tumors usually occur in the uterus, but sometimes also in the external genitalia. Histologically, they are composed of polygonal cells with abundant eosinophilic cytoplasm (Figure 16.32). Tumors with low mitotic activity (<5 per 10 HPFs) are classified as benign, because follow-up studies of such tumors show benign behavior.[114] Tumors with higher mitotic counts have to be approached with caution, as they may already have malignant potential. Tumors with >10 mitoses per 10 HPFs can be considered malignant. Epithelioid smooth muscle tumors are consistently positive for SMA and variably for heavy caldesmon and desmin, ER, and PR.

Some tumors formerly classified as epithelioid leiomyomas, especially those with clear cell features, are variably HMB45 positive, and are now classified as PEComas (Chapter 19). Most abdominal tumors previously considered epithelioid leiomyomas or LMSs are KIT positive and are now classified as GISTs (Chapter 17).

Myxoid smooth muscle tumors of the uterus

This small group is heterogeneous. It contains tumors that have a smooth muscle phenotype, and those that are negative for smooth muscle markers and can be characterized as myxomas or myxofibrosarcomas, when they have mitotic activity.[115] As previously pointed out, tumors with mitotic rates as low as 1 to 2 per 10 HPFs have metastatic potential.[116]

Intravenous leiomyomatosis
Clinical features

This rare condition occurs in women, primarily in the uterus. It is defined as nodular intravascular growth of histologically benign smooth muscle cells, extending beyond a uterine leiomyoma confined in a vessel or heart chamber outside an organ location, and coexisting with leiomyoma. An apparent

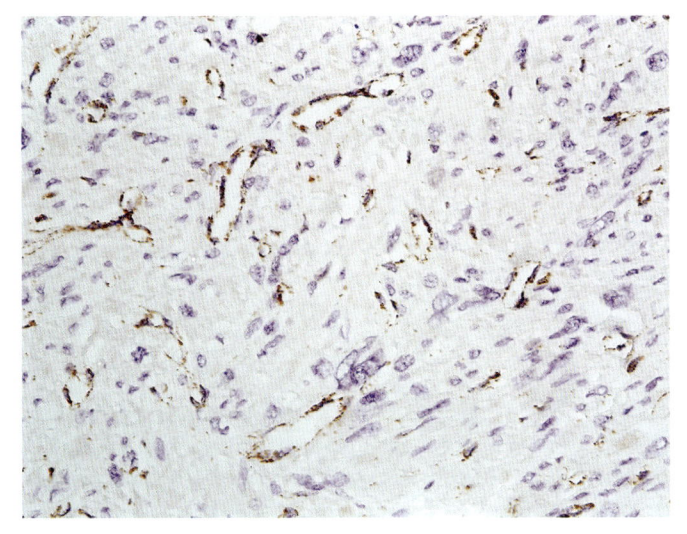

Figure 16.31 Immunohistochemical loss of fumarate hydratase is a characteristic feature of HLRCC-associated uterine leiomyomas. Note that fumarase expression is retained in non-neoplastic elements, such as blood vessels.

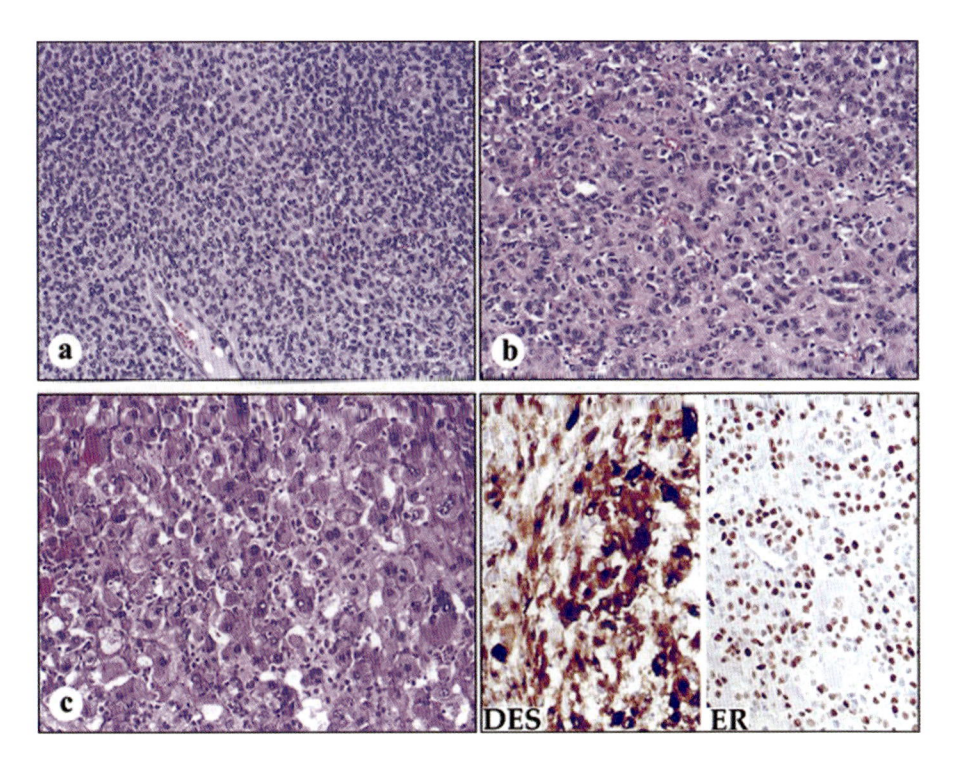

Figure 16.32 Epithelioid leiomyoma is composed of relatively uniform epithelioid cells with abundant cytoplasm and inconspicuous mitotic activity (**a,b**). (**c**) Multinucleated tumor cells can be present. Immunohistochemical positivity for desmin and estrogen receptor is common.

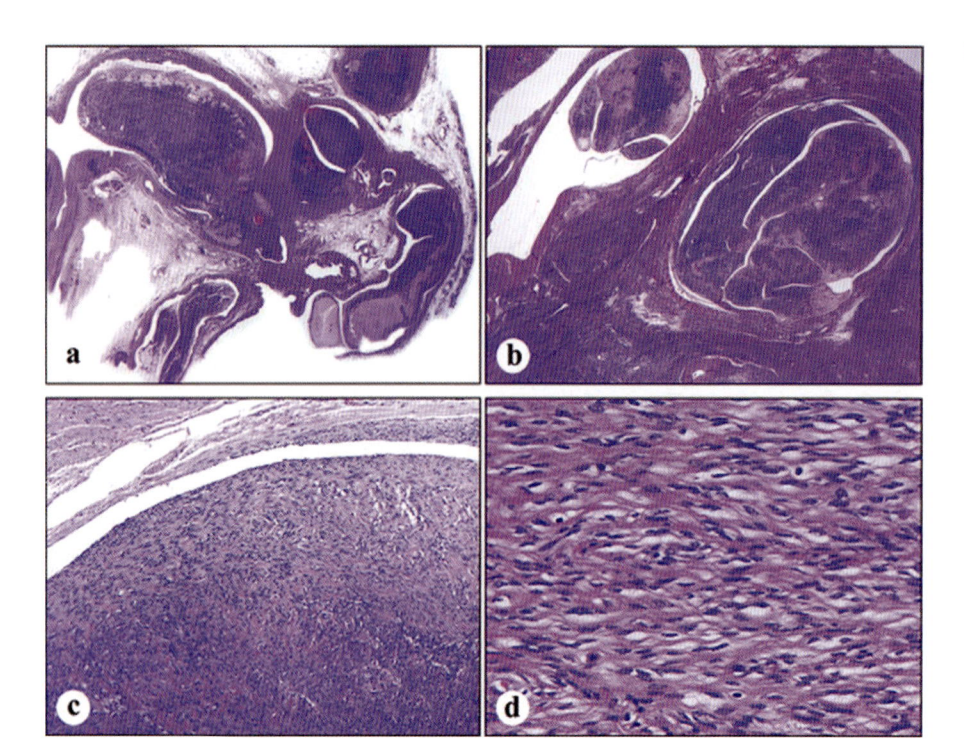

Figure 16.33 (a–c) Intravenous leiomyomatosis of the uterus shows an intravenous plug of histologically benign leiomyoma. (d) Cytologic appearance is similar to uterine leiomyoma.

intravascular smooth muscle protrusion within a uterine leiomyoma does not constitute intravenous leiomyomatosis, and an organ-based tumor other than one in the heart chamber is considered metastasis.[106] Intravenous leiomyomatosis usually occurs in women between the ages of 35 and 50, and occasionally at an older age.[117–121] The symptoms are similar to those seen with uterine leiomyoma. The tumor presents as a pelvic mass and usually (in 80% of cases) extends outside the uterus into the pelvic veins. It extends into the vena cava in 30% of cases and can extend as an intravenous column up to the right side of the heart and rarely into the pulmonary arteries. Cardiac involvement is often surgically treatable, and patients can have a surprisingly good prognosis, but this tumor can prove fatal in some cases. Rarely, the cardiac involvement has been the first sign of the tumor.

The treatment consists of total abdominal hysterectomy and removal of the extrauterine intravenous extension whenever possible. Overall prognosis is good, but recurrent tumor develops in 30% of cases and can involve the pelvic veins and the right side of the heart. Rare pulmonary metastases, with an apparently indolent course, have been reported. These appear to be comparable to "benign metastasizing leiomyoma." Surgery is advocated in recurrences, and tamoxifen can be useful.

Pathology
Grossly, intravenous leiomyomatosis typically forms a large uterine mass (mean size 6–7 cm, up to 20 cm) containing convoluted, coiled, or worm-like intravenous plugs. The mass sometimes involves the entire uterus. Extension into pelvic veins is often noted by the surgeon.

The intravascular plugs have varied histologic appearances in the spectrum of uterine leiomyomas, but they rarely contain mitotic activity (Figure 16.33). The patterns include hyaline change and hydropic degeneration.[117–121] In some cases, the smooth muscle proliferation also extends to the walls of veins. Morphologic variants of uterine leiomyoma can be represented by myxoid, epithelioid, bizarre (atypia without mitoses), and clear cell changes. Lipoleiomyoma-like lesions containing mature fat have also been reported.[120] Mitotic activity is usually low (<1 per 10 HPFs, occasionally up to 4 per 10 HPFs).

On immunohistochemical analysis, intravascular leiomyomatosis lesions are essentially similar to uterine leiomyomas. They are positive for desmin, SMA, and ER/PRs.

Intravenous leiomyomatosis differs from LMS by lack of atypia; it also differs from endometrial stromal sarcoma because the latter has a round cell pattern. Immunohistochemically, stromal sarcoma is positive for CD10, but it can have variable desmin and SMA positivity; indeed, there is some overlap between smooth muscle and stromal tumors. The myometrial vascular invasion of endometrial stromal sarcoma is usually not evident on gross inspection.

Benign metastasizing leiomyoma
This designation is generally used for pulmonary nodules in women who have had previous surgery for histologically nonmalignant uterine smooth muscle tumors. These are composed of ER/PR-positive well-differentiated smooth muscle cells with limited, if any, atypia and low mitotic activity, similar to the phenotype of nonmalignant uterine smooth muscle tumors. The nodules can appear long (>10 years) after hysterectomy.

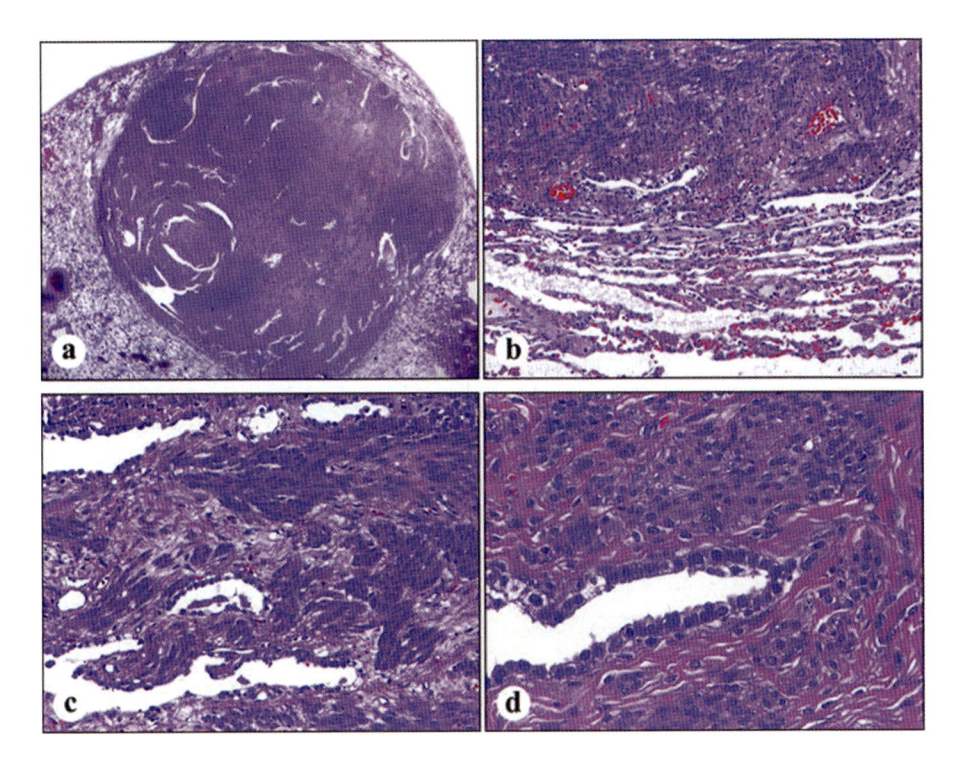

Figure 16.34 (**a**) Benign metastasizing leiomyoma forms a rounded intraparenchymal pulmonary nodule. (**b**) Entrapped alveoli and peripheral airways are seen in tumor periphery. (**c,d**) The tumor is composed of bundles of uniform smooth muscle cells in a collagenous matrix. Entrapped alveolar elements are present.

The pulmonary lesions can multiply and recur locally, but generally the behavior is indolent.[122–125] Some authors, however, consider benign metastasizing leiomyoma to be a slowly progressive variant of LMS.[113]

Such pulmonary nodules are relatively small (1–2 cm), and the smooth muscle often intermingles with mildly proliferating pulmonary epithelial elements. Histologic features are those of a cellular uterine leiomyoma (Figure 16.34). Immunohistochemically, the lesions are positive for desmin, SMA, heavy caldesmon, and ER/PRs.[124] These leiomyomas have typical cytogenetic changes: terminal deletions of chromosomes 19q and 22q.[126]

Occasionally, this designation has also been applied to similar tumors in pelvic and even peripheral soft tissues.[127] This condition must be separated from true metastases of LMS and PEComa. The latter includes lymphangiomyoma and lymphangioleiomyomatosis of the lung, which also combines uterine/pelvic and pulmonary involvement. PEComas are characterized by clear cell morphology and coexpression of smooth muscle markers and HMB45 (Chapter 19).

Peritoneal leiomyomatosis (leiomyomatosis peritonealis disseminata)

Clinical features

This rare condition refers to the development of innumerable small smooth muscle nodules on peritoneal surfaces in a manner that clinically and grossly simulates peritoneal carcinomatosis. The condition occurs almost exclusively in women of childbearing age, and only occasionally in postmenopausal women. Hormonal stimulation by pregnancy or exogenous estrogen intake is thought to be a triggering mechanism in most cases. In one case, a granulosa cell tumor provided estrogenic stimulus.[128–131]

In most cases, the condition is asymptomatic and diagnosed incidentally at cesarean section or surgery for uterine leiomyomas, which usually coexist with the lesion. Concurrent endometriosis also seems relatively common. Treatment is usually conservative and includes antiestrogen or gonadotropin agonists. Prognosis is excellent even if the lesions are not completely excised.

Rare examples of malignant transformation to LMS have been reported, but in such cases, the identity of the lesion with peritoneal leiomyomatosis is unclear.[132]

Pathology

Typical on gross inspection is the presence of variable numbers, often innumerable, small and white rubbery-to-firm nodules in the omentum and serosal surfaces of intestines, inner abdominal wall, and pelvic organs. These nodules vary from microscopic (<1 mm) to rarely >1 cm. Sometimes the nodules form coalescent small masses (Figure 16.35); in some cases, peritoneal leiomyomatosis coexists with larger abdominal leiomyomas.

Histologically, the leiomyomatosis nodules are composed of well-differentiated smooth muscle cells with no atypia and low mitotic activity, in the range of uterine leiomyomas, but usually fewer than 2 mitoses per 10 HPFs. Decidual stromal components can coexist in the lesions in pregnant patients. This condition probably represents the multipotentiality of the peritoneal mesenchyma and its capability to develop into smooth muscle metaplasia. In this sense, peritoneal

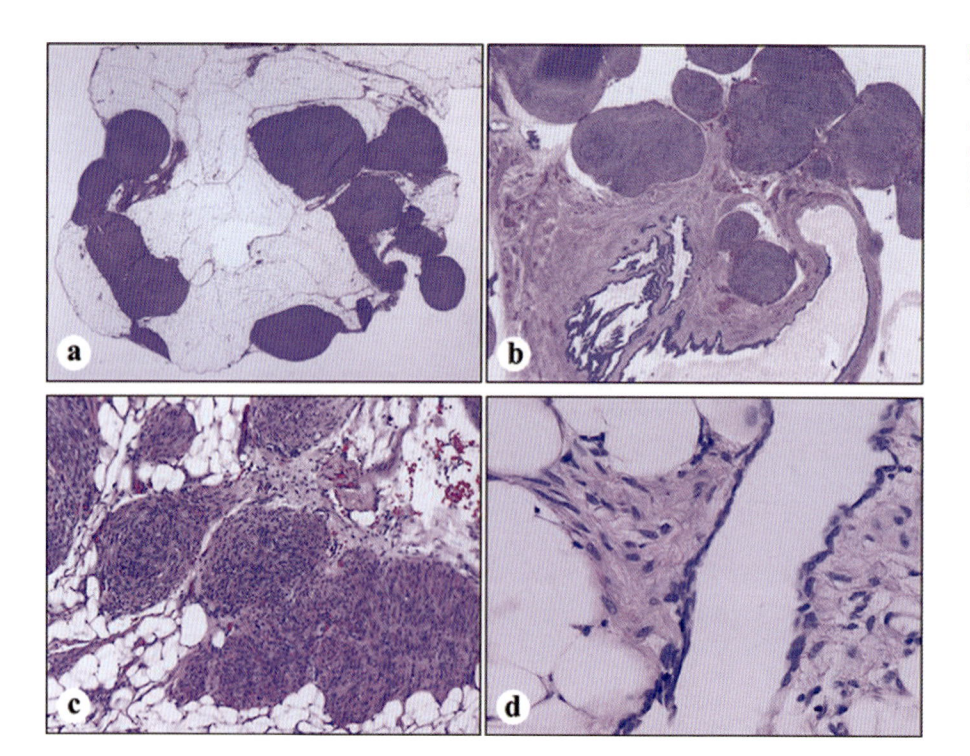

Figure 16.35 Leiomyomatosis peritonealis disseminata forms multiple, sometimes innumerable smooth muscle nodules. (**a**) Involvement in omentum. (**b**) Parametrium. (**c**) Cytologic detail shows resemblance to uterine leiomyoma. (**d**) An incipient minute lesion just beneath the peritoneal mesothelium.

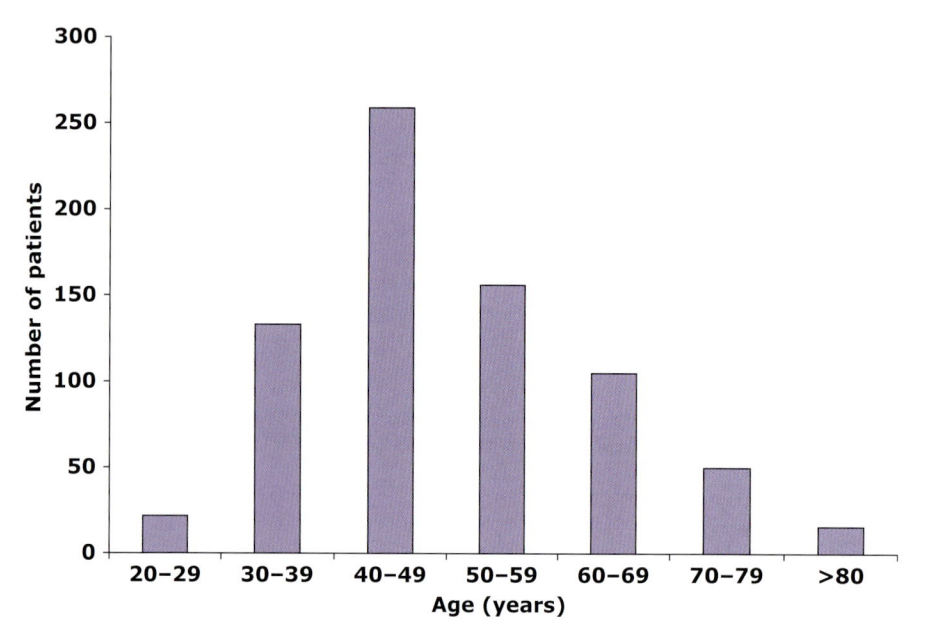

Figure 16.36 Age distribution of 743 patients with uterine LMSs.

leiomyomatosis can be analogous to endometriosis, and these lesions in fact can coexist.[131]

Immunohistochemically, the nodules show the phenotype of hormonally sensitive gynecologic smooth muscle cells and are positive for SMA, desmin, and ER/PRs, and are negative for KIT.

Differential diagnosis

This condition must be separated from disseminated LMS and GIST. Lack of cytologic atypia, low mitotic activity, and absence of a dominant tumor differ from the entities discussed previously. GISTs are generally positive for KIT (CD117) and are negative for ER; PR positivity, however, does occur. In some cases, it can be difficult to determine whether larger pelvic and abdominal lesions represent concurrent leiomyomas or larger examples of leiomyomatosis lesions.

Uterine leiomyosarcoma

Uterine LMS probably constitutes the single largest site-specific group of LMSs. Uterine LMSs occur from the third decade into old age, but are most common in the perimenopausal age group in the fifth decade (Figure 16.36). Most LMSs

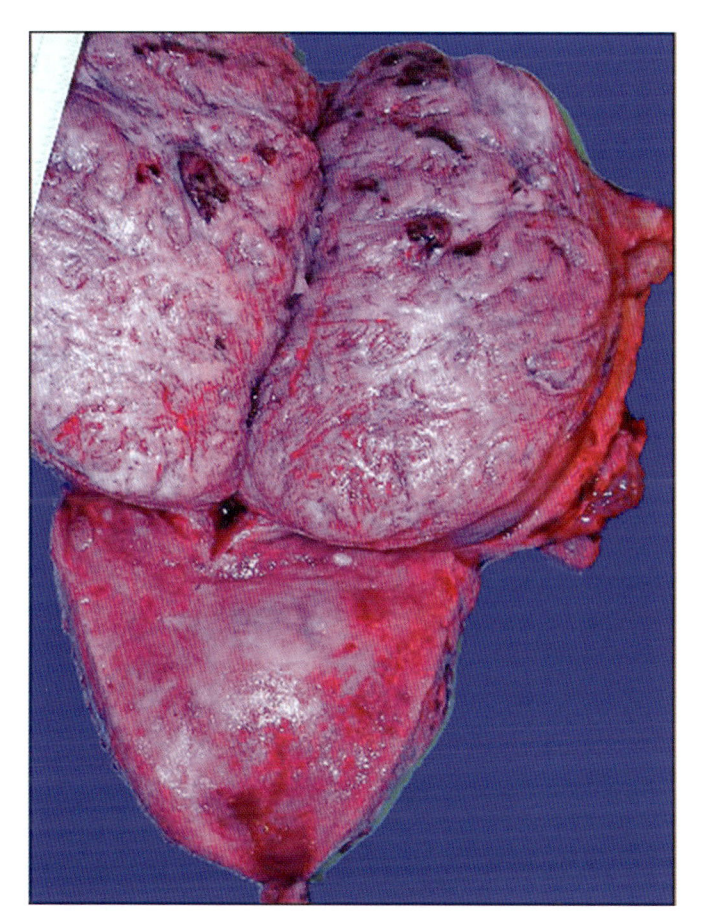

Figure 16.37 Gross appearance of a large uterine LMS.

diagnostic criteria with indeterminate features are best classified as smooth muscle tumors of uncertain (or low) malignant potential (Table 16.1).

Immunohistochemically, uterine LMSs are typically positive for SMA, h-caldesmon, and desmin, but the latter can be absent in poorly differentiated or dedifferentiated tumors, especially when metastatic. A majority of uterine LMSs are ER-positive, based on two studies.[139,140]

Diagnostic criteria

The diagnostic criteria are summarized in Table 16.1. Uterine smooth muscle tumors are classified as LMSs if they have coagulative necrosis or at least moderate atypia and have mitoses >5 per 10 HPF. Cellular leiomyomas without significant atypia can have mitoses at least up to 10 per 10 HPFs; however, some authors accept mitotic counts as high as 15–20 per 10 HPFs.

Tumors with no or minimal atypia are allowed 15 (possibly even more) mitoses per 10 HPF, and they are classified as cellular leiomyomas with mitotic activity. Tumors with significant atypia are classified as LMS if 10 or more mitoses per 10 HPFs are present; below that they can be classified as bizarre symplastic leiomyomas, although such tumors usually have fewer than 2 to 3 mitoses per 10 HPFs. Coagulative tumor necrosis is a warning sign and should lead to a search for other signs of malignancy, such as mitotic activity and moderate to severe atypia. The presence of coagulation necrosis has been suggested to upgrade a tumor to low-grade malignant, even if the mitotic count is <10 per 10 HPFs, and atypia is limited. The classification of myxoid and epithelioid tumors as LMS has been suggested with mitotic counts as low as 0 to 2 mitoses per 10 HPFs.[104,106,141]

Epstein–Barr-virus-associated leiomyosarcoma/smooth muscle tumor
Clinical features

This rare tumor occurs in immunosuppressed patients, and EBV is considered to be the etiologic factor.[142] The clinical background is AIDS,[143–147] immunosuppression following a solid organ transplantation,[147–150] or, rarely, corticosteroid treatment[147] or congenital immunodeficiency syndrome.[151] Because of uncertainty as to whether these tumors are all malignant, EBV-associated smooth muscle tumor has been used as an alternative term.[150] Most of the reported AIDS- and congenital-immunodeficiency-related EBV-associated LMSs have occurred in children and young adults, whereas the post-transplantation-related cases have occurred mostly in adults. These LMSs are often multiple and involve both peripheral and visceral sites, including the gastrointestinal tract, urinary bladder, thyroid, and dura.[142–151] The multiple tumors seem to be independent clones and not metastases from a single clone. The EBV2 serotype is specifically associated with these tumors.[146]

are large tumors (>10 cm; Figure 16.37).[133–137] Indicators such as age <50 years, stage 1, smaller tumor size, and lower grade were favorable prognostically in the largest series. In this series, the disease-specific 10-year survival rate was approximately 60% for patients whose tumors were <5 cm, whereas this rate dipped to 30% for patients whose tumors were >5 cm.[125] Premenopausal patients and those whose tumors are confined to a single nodule or have a low mitotic count have a better prognosis, with the overall 5-year survival rate being 60% to 70%.[133–137]

With multiple smooth muscle tumors present, one should pay close attention to nodules that are large and grossly different from LMS (softer and more necrotic), because such tumors have a greater likelihood of representing sarcoma. Uterine LMSs are often less circumscribed than leiomyomas, but some are well-defined and can simulate a leiomyoma.

Significant mitotic activity >10 per 10 HPFs combined with significant nuclear atypia and coagulative necrosis are features typically seen in LMS, and by these definitions, uterine LMSs are generally high grade. One large retrospective study focusing on tumors diagnosed as low-grade LMS concluded that this is a heterogeneous category of tumors with low biologic potential more aligned with atypical smooth muscle tumors than true LMS.[138] Tumors that fall between the

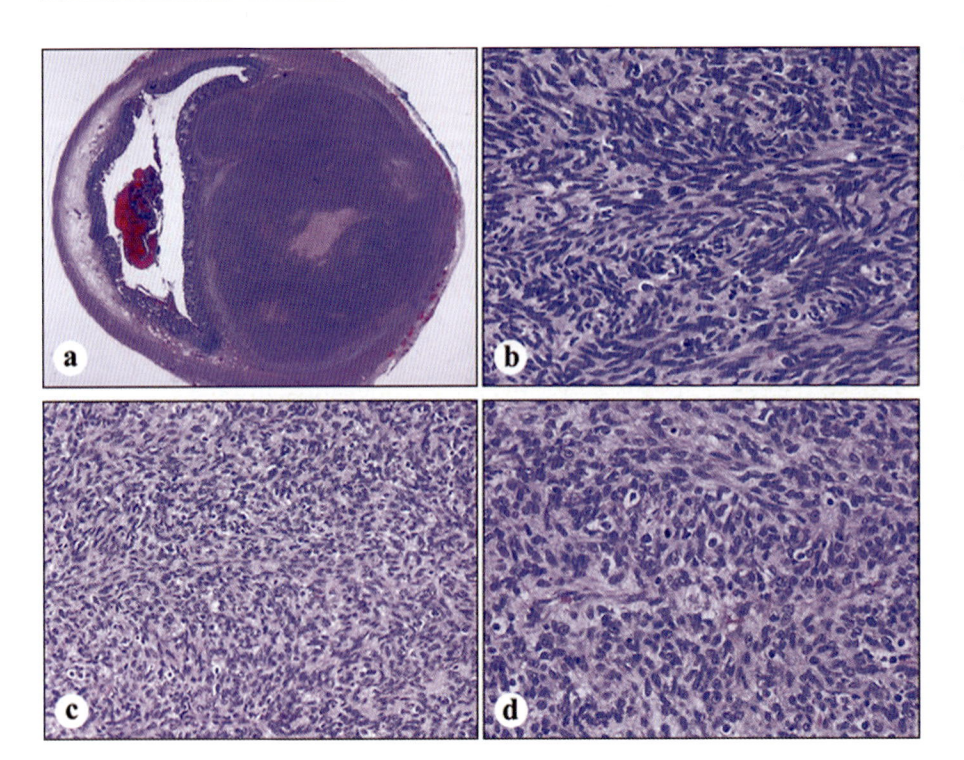

Figure 16.38 Two examples of EBV-associated LMSs, both in AIDS patients. (**a**) The tumor involves the appendix as an eccentric mass. (**b**) It has a spindle cell pattern, but is less differentiated than a typical LMS. (**c,d**) An oval cell pattern in a groin tumor.

In post-transplantation patients, the LMS has been found to originate either from the donor cells when in the graft or from recipient tissue when arising outside of the graft.[148,149] High numbers of EBV copies have been reported in the tumor tissue but not in the normal tissue or in control LMSs in nonimmunosuppressed patients, suggesting the etiologic role of EBV in LMSs of immunosuppressed patients.[152,153] Many immunosuppression-associated LMSs are low-grade tumors, and antiviral therapy has proved useful in some cases. The prognosis greatly depends on the underlying disease, and metastases and tumor-related deaths are distinctly rare. In the largest series, only 1 of 19 patients died of disease, although persistent disease was observed in many.[147]

Pathology

Histologically, EBV-associated LMS has a spectrum from well-differentiated to a less-differentiated form composed of oval cells, with a smaller amount of eosinophilic cytoplasm. Both atypia and mitotic activity are low. The tumor cells are typically positive for alpha SMA and show a low frequency of desmin positivity (Figure 16.38). *In situ* hybridization for EBV-encoded RNA (EBER) typically shows strong nuclear positivity (Figure 16.39).

Smooth muscle tumors of uncertain malignant potential (STUMP)

This category does not refer to a specific biologic or clinicopathologic entity, but rather to a diagnostic situation in which either the quantity of material is insufficient or diagnostic

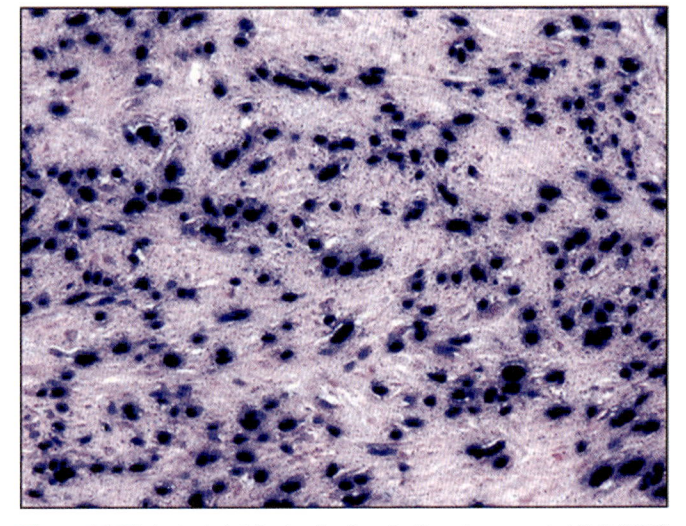

Figure 16.39 *In situ* hybridization for Epstein–Barr-virus-encoded RNA (EBER) shows strong nuclear positivity (blue chromogen).

parameters, such as tumor subclassification, mitotic rate, or necrosis are ambiguous for determination of the biologic potential. The term can be applied to soft tissue as well as visceral and uterine smooth muscle tumors.

The diagnosis of smooth muscle tumors of uncertain malignant potential (STUMP) can be applied to small biopsy specimens, which are insufficient to reveal potential atypia or mitotic activity, especially when the tumor is large. STUMP is also a prudent diagnosis for an atypical smooth muscle tumor of soft tissue with no mitotic activity despite prolonged search, or in some cases with a necrotic smooth muscle tumor that lacks apparent atypia. In some cases, the STUMP designation reflects limited information on tumors at specific sites. Clinicoradiologic correlation can be useful, especially in cases with small biopsy specimens. Additional sampling, or outright complete excision and follow-up are generally advisable for these tumors

References

Smooth muscle hamartomas

1. Urbanek RW, Johnson WC. Smooth muscle hamartomas associated with Becker's nevus. *Arch Dermatol* 1978;114:104–106.
2. Berger TG, Levin MW. Congenital smooth muscle hamartoma. *J Am Acad Dermatol* 1984;11:709–712.
3. Slifman NR, Harrist TJ, Rhodes AR. Congenital arrector pili hamartoma: a case report and review of the spectrum of Becker's melanosis and pilar smooth-muscle hamartoma. *Arch Dermatol* 1985;121:1034–1037.
4. Johnson MD, Jacobs AH. Congenital smooth muscle hamartoma: a report of six cases and review of the literature. *Arch Dermatol* 1989;125:820–822.
5. Zvulunov A, Rotem A, Merlob P, Metzker A. Congenital smooth muscle hamartoma: prevalence, clinical findings, and follow-up in 15 patients. *Am J Dis Child* 1990;144:782–784.
6. Gagne EJ, Su WP. Congenital smooth muscle hamartoma of the skin. *Pediatr Dermatol* 1993;10:142–145.
7. Grau-Massanes M, Raimer S, Colome-Grimmer M, Yen A, Sanchez RL. Congenital smooth muscle hamartoma presenting as a linear atrophic plaque: case report and review of the literature. *Pediatr Dermatol* 1996;13:222–225.
8. Jang HS, Kim MB, Oh CK, Kwon KS, Chung TA. Linear congenital smooth muscle hamartoma with follicular spotted appearance. *Br J Dermatol* 2000;142:138–142.
9. Gualandri L, Cambiaghi S, Ermacora E, et al. Multiple familial smooth muscle hamartomas. *Pediatr Dermatol* 2001;18:17–20.
10. Daroca PJ Jr, Reed RJ, Love GL, Kraus SD. Myoid hamartomas of the breast. *Hum Pathol* 1985;16:212–219.
11. Barton JH, Davis CJ Jr, Sesterhenn IA, Mostofi FK. Smooth muscle hyperplasia of the testicular adnexa clinically mimicking neoplasia: clinicopathologic study of sixteen cases. *Am J Surg Pathol* 1999;23:903–909.
12. Chan JKC, Frizzera G, Fletcher CDM, Rosai J. Primary vascular tumors of lymph nodes other than Kaposi's sarcoma. *Am J Surg Pathol* 1992;16:335–350.
13. Laeng RH, Hotz MA, Borisch B. Angiomyomatous hamartoma of a cervical lymph node combined with hemangiomatosis and vascular transformation of sinuses. *Histopathology* 1996;29:80–84.

Piloleiomyoma

14. Fisher WC, Helwig EB. Leiomyomas of the skin. *Arch Dermatol* 1963;88:78–88.
15. Peters CW, Hanke CW, Reed JC. Nevus leiomyomatosus systematicus. *Cutis* 1981;27:484–486.
16. Raj S, Calonje E, Kraus M, et al. Cutaneous pilar leiomyoma: clinicopathologic analysis of 53 lesions in 45 patients. *Am J Dermatopathol* 1997;19:2–9.
17. Kiuru M, Launonen V, Hietala M, et al. Familial cutaneous leiomyomatosis is a two-hit condition associated with renal cell cancer of characteristic histology. *Am J Pathol* 2001;159:825–829.
18. Alam NA, Rowan AJ, Wortham NC, et al. Genetic and functional analyses of FH mutations in multiple cutaneous and uterine leiomyomatosis, hereditary leiomyomatosis and renal cancer, and fumarate hydratase deficiency. *Hum Mol Genet* 2003;159:825–829.
19. Toro RJ, Nickerson ML, Wei MH, et al. Mutations in the fumarate hydratase gene cause hereditary leiomyomatosis and renal cell cancer in families in North America. *Am J Hum Genet* 2003;73:95–106.
20. Menko FH, Maher ER, Schmidt LS, et al. Hereditary leiomyomatosis and renal cell cancer (HLRCC): renal cancer risk, surveillance and treatment. *Fam Cancer* 2014;13:637–644.
21. Knoth W, Knoth-Born RC. Familial uterocutaneous leiomyomatosis. *Z Haut Geschlechtskr* 1964;37:191–206.
22. Reed WB, Walker R, Horowitz R. Cutaneous leiomyomata with uterine leiomyomata. *Acta Derm Venereol* 1973;53:318–320.

Angioleiomyoma

23. Hachisuga T, Hashimoto H, Enjoji M. Angioleiomyoma: a clinicopathologic reappraisal of 562 cases. *Cancer* 1984;54:126–130.
24. Katenkamp D, Kohmehl H, Lengbein L. Angiomyome: Eine pathologish-anatomische Analyse von 229 Fallen. *Zentralbl Allg Pathol* 1988;134:423–433.
25. Calle SC, Eaton RG, Littler JW. Vascular leiomyomas in the hand. *J Hand Surg [Am]* 1994;19:281–286.
26. Brooks JK, Nikitakis NG, Goodman NJ, Levy BA. Clinicopathologic characterization of oral angioleiomyomas. *Oral Surg Oral Med Oral Pathol Oral Radiol Endod* 2002;94:221–227.
27. Beer TW. Cutaneous angiomyolipomas are HMB45 negative, not associated with tuberous sclerosis, and should be considered as angioleiomyomas with fat. *Am J Dermatopathol* 2005;27:418–421.
28. Fox SB, Heryet A, Khong TY. Angioleiomyomas: an immunohistochemical study. *Histopathology* 1990;16:495–496.
29. Hasegawa T, Seki K, Yang P, Hirose T, Hizawa K. Mechanism of pain and cytoskeletal properties in angioleiomyomas: an immunohistochemical study. *Pathol Int* 1994;44:66–72.
30. Matsuyama A, Hisaoka M, Hashimoto H. Angioleiomyoma: a clinicopathologic and immunohistochemical reappraisal with special reference to the correlation with myopericytoma. *Hum Pathol* 2007;38:645–651.
31. Heim S, Mandahl N, Kristofferson U, et al. Structural chromosome aberrations in a case of angioleiomyoma. *Cancer Genet Cytogenet* 1986;20:325–330.
32. Nilbert M, Mandahl N, Heim S, et al. Cytogenetic abnormalities in angioleiomyoma. *Cancer Genet Cytogenet* 1989;37:61–64.
33. Sonobe H, Ohtsuki Y, Mizobuchi H, Toda M, Shimizu K. An angiomyoma with t(X;10)(q22;q23.2). *Cancer Genet Cytogenet* 1996;90:54–56.
34. Nishio J, Iwasaki H, Ohjimi Y, et al. Chromosomal imbalances in

angioleiomyomas by comparative genomic hybridization. *Int J Mol Med* 2004;13:13–16.

Leiomyomas of male genitalia

35. Newman PL, Fletcher CDM. Smooth muscle tumours of the external genitalia: clinicopathologic analysis of a series. *Histopathology* 1991;18:523–529.
36. Slone S, O'Connor D. Scrotal leiomyomas with bizarre nuclei: a report of three cases. *Mod Pathol* 1998;11:282–287.

Gastrointestinal leiomyomas

37. Seremetis MG, Lyons WS, de Guzman VC, Peabody JW. Leiomyomata of the esophagus: an analysis of 838 cases. *Cancer* 1976;38:2166–2177.
38. Miettinen M, Sarlomo-Rikala M, Sobin LH, Lasota, J. Esophageal stromal tumors: a clinicopathologic, immunohistochemical and molecular genetic study of seventeen cases and comparison with esophageal leiomyomas and leiomyosarcomas. *Am J Surg Pathol* 2000;23:121–132.
39. Hatch GF 3rd, Wertheimer-Hatch L, Hatch KF, et al. Tumors of the esophagus. *World J Surg* 2000;24:401–411.
40. Lonsdale RN, Roberts PF, Vaughan R, Thiru S. Familial oesophageal leiomyomatosis and nephropathy. *Histopathology* 1993;20:127–133.
41. Heidet L, Boye E, Cai Y, et al. Somatic deletion of the 5' ends of both the COL4A5 and COL4A6 genes in a sporadic leiomyoma of the esophagus. *Am J Pathol* 1998;152:673–678.
42. Ueki Y, Naito I, Oohashi T, et al. Topoisomerase I and II consensus sequences in a 17-kb deletion junction of the COL4A5 and COL4A6 genes and immunohistochemical analysis of esophageal leiomyomatosis associated with Alport syndrome. *Am J Hum Genet* 1998;62:253–261.
43. Sado Y, Kagawa M, Naito I, et al. Organization and expression of basement membrane collagen IV genes and their roles in human disorders. *J Biochem* 1998;123:767–776.
44. Segal Y, Peissel B, Renieri A, et al. LINE-1 elements at the sites of molecular rearrangements in Alport syndrome-diffuse leiomyomatosis. *Am J Hum Genet* 1999;64:62–69.
45. Sarlomo-Rikala M, El-Rifai W, Andersson L, Miettinen M, Knuutila S. Different patterns of DNA copy number changes in gastrointestinal stromal tumors, leiomyomas and schwannomas. *Hum Pathol* 1998;29:476–481.
46. Miettinen M, Sarlomo-Rikala M, Sobin LH. Mesenchymal tumors of muscularis mucosae of colon and rectum are benign leiomyomas that should be separated from gastrointestinal stromal tumors: a clinicopathologic and immunohistochemical study of 88 cases. *Mod Pathol* 2001;14:950–956.

Leiomyomas of other visceral nongynecologic sites

47. Huang HY, Antonescu CR. Sinonasal smooth muscle tumors: a clinicopathologic and immunohistochemical analysis of 12 cases with emphasis on the low-grade end of the spectrum. *Arch Pathol* 2003;127:297–304.
48. Yellin A, Rosenman Y, Lieberman Y. Review of smooth muscle tumours of the lower respiratory tract. *Br J Dis Chest* 1984;78:337–351.
49. Martin SA, Sears DL, Sebo TJ, Lohse CM, Cheville JC. Smooth muscle neoplasms of the urinary bladder: a clinicopathologic comparison of leiomyoma and leiomyosarcoma. *Am J Surg Pathol* 2002;26:292–300.

Pathology of leiomyosarcoma

50. Venance SL, Burns KL, Veinot JP, Walley VM. Contraction bands in visceral and vascular smooth muscle. *Hum Pathol* 1996;27:1035–1041.
51. Chen E, O'Connell F, Fletcher CD. Dedifferentiated leiomyosarcoma: clinicopathological analysis of 18 cases. *Histopathology* 2011;59:1135–1143.
52. Schürch W, Skalli O, Seemayer TA, Gabbiani G. Intermediate filament proteins and actin isoforms as markers for soft tissue tumor differentiation and origin: I. Smooth muscle tumors. *Am J Pathol* 1987;128:91–103.
53. Azumi N, Ben-Ezra J, Battofora H. Immunophenotypic diagnosis of leiomyosarcomas and rhabdomyosarcomas with monoclonal antibodies to muscle-specific actin and desmin in formalin-fixed tissue. *Mod Pathol* 1988;1:469–474.
54. Watanabe K, Kusakabe T, Hoshi N, Saito A, Suzuki T. h-Caldesmon in leiomyosarcoma and tumors with smooth muscle cell-like differentiation: its specific expression in the smooth muscle cell tumor. *Hum Pathol* 1999;30:392–396.
55. Miettinen M, Sarlomo-Rikala M, Kovatich AJ, Lasota J. Calponin and h-caldesmon in soft tissue tumors: consistent h-caldesmon immunoreactivity in gastrointestinal stromal tumors indicates traits of smooth muscle differentiation. *Mod Pathol* 1999;12:756–762.
56. Rizeq MN, van de Rijn M, Hendrickson MR, Rouse RV. A comparative immunohistochemical study of uterine smooth muscle neoplasms with emphasis on the epithelioid variant. *Hum Pathol* 1994;25:671–677.
57. Miettinen M, Sobin LH, Sarlomo-Rikala M. Immunohistochemical spectrum of gastrointestinal stromal tumors at different sites and their differential diagnosis with other tumors with a special reference to CD117 (KIT). *Mod Pathol* 2000;13:1134–1142.
58. Miettinen M. Immunoreactivity for cytokeratin and epithelial membrane antigen in leiomyosarcoma. *Arch Pathol Lab Med* 1988;112:637–640.
59. Iwata J, Fletcher CD. Immunohistochemical detection of cytokeratin and epithelial membrane antigen in leiomyosarcoma: a systematic study of 100 cases. *Pathol Int* 2000;50:7–14.
60. de Vos S, Wilczynski SP, Fleischhacker M, Koeffler P. p53 alterations in uterine leiomyosarcomas versus leiomyomas. *Gynecol Oncol* 1994;54:205–208.
61. Patterson H, Gill S, Fisher C, et al. Abnormalities of the p53, MDM2 and DCC genes in human leiomyosarcomas. *Br J Cancer* 1994;69:1052–1058.
62. Wurl P, Taubert H, Bache M, et al. Frequent occurrence of p53 mutations in rhabdomyosarcoma and leiomyosarcoma, but not in fibrosarcoma and malignant neural tumors. *Int J Cancer* 1996;69:317–323.
63. Dei Tos AP, Maestro R, Doglioni C, et al. Tumor suppressor genes and related molecules in leiomyosarcoma. *Am J Pathol* 1996;148:1037–1045.
64. Sreekantaiah C, Davis JR, Sandberg AA. Chromosomal abnormalities in leiomyosarcomas. *Am J Pathol* 1993;142:293–305.

65. Mandahl N, Fletcher CD, Dal Cin P, et al. Comparative cytogenetic study of spindle cell and pleomorphic leiomyosarcomas of soft tissues: a report from the CHAMP study group. *Cancer Genet Cytogenet* 2000;116:66–73.

66. Sandberg AA. Updates on the cytogenetics and molecular genetics of bone and soft tissue tumors: leiomyosarcoma. *Cancer Genet Cytogenet* 2005;161:1–19.

67. El-Rifai W, Sarlomo-Rikala M, Knuutila S, Miettinen M. DNA copy number changes in development and progression in leiomyosarcomas of soft tissues. *Am J Pathol* 1998;153:985–990.

Leiomyosarcoma of soft tissues

68. Fields JP, Helwig EB. Leiomyosarcoma of the skin and subcutaneous tissue. *Cancer* 1981;47:156–169.

69. Bernstein SC, Roenigk RK. Leiomyosarcoma of the skin: treatment of 34 cases. *Dermatol Surg* 1996;22:631–635.

70. Kaddu S, Beham A, Cerroni L, et al. Cutaneous leiomyosarcoma. *Am J Surg Pathol* 1997;21:979–987.

71. Jensen ML, Jensen OM, Michalski W, Nielsen OS, Keller J. Intradermal and subcutaneous leiomyosarcoma: a clinicopathological and immunohistochemical study of 41 cases. *J Cutan Pathol* 1996;23:458–463.

72. Dahl I, Angervall L. Cutaneous and subcutaneous leiomyosarcoma: a clinicopathologic study of 47 patients. *Pathol Eur* 1974;9:307–315.

73. Svarvar K, Böhling T, Berlin Ö, et al. Clinical course of nonvisceral soft tissue leiomyosarcoma in 225 patients from the Scandinavian sarcoma group. *Cancer* 2007;109:282–291.

74. Kraft S, Fletcher CD. Atypical intradermal smooth muscle neoplasms: clinicopathologic analysis of 84 cases and a reappraisal of cutaneous "leiomyosarcoma." *Am J Surg Pathol* 2011;35:599–607.

75. Winchester DS, Hocker TL, Brewer JD, et al. Leiomyosarcoma of the skin: clinical, histopathologic, and prognostic factors that influence outcomes. *J Am Acad Dermatol* 2014;71:919–925.

76. Gustafson P, Willen H, Baldetrop B, et al. Soft tissue leiomyosarcoma: a population-based epidemiologic and prognostic study of 48 patients, including cellular DNA content. *Cancer* 1992;70:114–119.

77. Hashimoto H, Daimaru Y, Tsuneyoshi M, Enjoji M. Leiomyosarcoma of the external soft tissues: a clinicopathologic, immunohistochemical and electron microscopic study. *Cancer* 1986;57:2077–2088.

78. Farshid G, Pradham M, Goldblum J, Weiss SW. Leiomyosarcoma of somatic soft tissues: a tumor of vascular origin with multivariate analysis of outcome in 42 cases. *Am J Surg Pathol* 2002;26:14–24.

79. Varela-Duran J, Oliva H, Rosai J. Vascular leiomyosarcoma: the malignant counterpart of vascular leiomyoma. *Cancer* 1979;44:1684–1691.

80. Berlin O, Stener B, Kindblom LG, Angervall L. Leiomyosarcoma of venous origin in the extremities: a correlated clinical, roentgenologic, and morphologic study with diagnostic surgical implications. *Cancer* 1984;54:2147–2159.

81. Rajani B, Smith TA, Reith JD, Goldblum JR. Retroperitoneal leiomyosarcomas unassociated with the gastrointestinal tract: a clinicopathologic analysis of 17 cases. *Mod Pathol* 1999;12:21–28.

82. Miettinen M, Sarlomo-Rikala M, Sobin LH, Lasota J. Gastrointestinal stromal tumors and leiomyosarcomas in the colon: a clinicopathologic, immunohistochemical, and molecular genetic study of 44 cases. *Am J Surg Pathol* 2000;24:1339–1352.

83. Miettinen M, Furlong M, Sarlomo-Rikala M, et al. Gastrointestinal stromal tumors, intramural leiomyomas, and leiomyosarcomas in the rectum and anus: a clinicopathologic, immunohistochemical, and molecular genetic study of 144 cases. *Am J Surg Pathol* 2001;25:1121–1133.

84. Mingoli A, Cavallaro A, Sapienza P, et al. International registry of inferior vena cava leiomyosarcoma: analysis of a world series on 218 patients. *Anticancer Res* 1996;16:3201–3205.

85. Hines OJ, Nelson S, Quinones-Baldrich WJ, Eilber FR. Leiomyosarcoma of the inferior vena cava: prognosis and comparison with leiomyosarcoma of other anatomic sites. *Cancer* 1999;85:1077–1083.

86. Hollenbeck ST, Grobmyer SR, Kent KC, Brennan MF. Surgical treatment and outcomes in patients with primary inferior vena cava leiomyosarcoma. *J Am Coll Surg* 2003;197:575–579.

87. Ito H, Hornick JL, Bertagnolli MM, et al. Leiomyosarcoma of the inferior vena cava: survival after aggressive management. *Ann Surg Oncol* 2007;14:3534–3541.

88. Kuruvilla A, Wenig BM, Humphrey DM, Heffner DK. Leiomyosarcoma of the sinonasal tract: a clinicopathologic study of nine cases. *Arch Otolaryngol Head Neck Surg* 1990;116:1278–1286.

89. Dry SM, Jorgensen JL, Fletcher CD. Leiomyosarcomas of the oral cavity: an unusual topographic subset easily mistaken for nonmesenchymal tumours. *Histopathology* 2000;36:210–220.

90. Moran CA, Suster S, Abbondanzo SL, Koss MN. Primary leiomyosarcomas of the lung: a clinicopathologic and immunohistochemical study of 18 cases. *Mod Pathol* 1997;10:121–128.

91. Mills SE, Bova GS, Wick MR, Young RH. Leiomyosarcoma of the urinary bladder: a clinicopathologic and immunohistochemical study of 15 cases. *Am J Surg Pathol* 1989;13:480–489.

92. Cheville JC, Dundore PA, Nascimento AG, et al. Leiomyosarcoma of the prostate: report of 23 cases *Cancer* 1995;76:1422–1427.

93. Fetsch JF, Davis CJ Jr, Miettinen M, Sesterhenn IA. Leiomyosarcoma of the penis: a clinicopathologic study of 14 cases with review of the literature and discussion of the differential diagnosis. *Am J Surg Pathol* 2004;28:115–125.

94. Fisher C, Goldblum JR, Epstein JI, Montgomery E. Leiomyosarcoma of the paratesticular region: a clinicopathologic study. *Am J Surg Pathol* 2001;25:1143–1149.

95. Swanson PE, Wick MR, Dehner LP. Leiomyosarcoma of somatic soft tissues in childhood: an immunohistochemical analysis of six cases with ultrastructural correlation. *Hum Pathol* 1991;22:569–577.

96. de Saint Aubain Somerhausen N, Fletcher CD. Leiomyosarcoma of soft tissues in children: clinicopathologic analysis of 20 cases. *Am J Surg Pathol* 1999;23:755–763.

Estrogen-positive smooth muscle tumors in women

97. Billings SD, Folpe Al, Weiss SW. Do leiomyomas of deep soft tissue exist?: an analysis of highly differentiated smooth muscle tumors of deep soft tissue supporting two distinct subtypes. *Am J Surg Pathol* 2001;25:1134–1142.

98. Paal E, Miettinen M. Retroperitoneal leiomyomas: a clinicopathologic and immunohistochemical study of 56 cases with a comparison to retroperitoneal leiomyosarcomas. *Am J Surg Pathol* 2001;25:1355–1363.

99. Patil DT, Laskin WB, Fetsch JF, Miettinen M. Inguinal smooth muscle tumors in women: a dichotomous group consisting of Mullerian-type leiomyomas and soft tissue leiomyosarcomas, an analysis of 55 cases. *Am J Surg Pathol* 2011;35:315–324.

100. Tavassoli FA, Norris HJ. Smooth muscle tumors of the vulva. *Obstet Gynecol* 1979;53:213–217.

101. Nielsen GP, Rosenberg AE, Koerner FC, Young RH, Scully RE. Smooth-muscle tumors of the vulva: a clinicopathological study of 25 cases and review of the literature. *Am J Surg Pathol* 1996;20:779–793.

102. O'Connor DM, Norris HJ. Mitotically active leiomyomas of the uterus. *Hum Pathol* 1990;21:223–227.

103. Oliva E, Young RH, Clement PB, Bhan AK, Scully RE. Cellular benign mesenchymal tumors of the uterus: a comparative morphologic and immunohistochemical analysis of 33 highly cellular leiomyomas and six endometrial stromal nodules, two frequently confused tumors. *Am J Surg Pathol* 1995;19:757–768.

104. Bell SW, Kempson RL, Hendrickson MR. Problematic uterine smooth muscle neoplasms: a clinicopathologic study of 213 cases. *Am J Surg Pathol* 1994;18:535–558.

105. Ly A, Mills AM, McKenney JK, *et al.* Atypical leiomyomas of the uterus: a clinicopathologic study of 51 cases. *Am J Surg Pathol* 2013;37:643–649.

106. Kempson RL, Hendrickson MR. Smooth muscle, endometrial stromal, and mixed mullerian tumors of the uterus. *Mod Pathol* 2000;13:328–342.

107. Roth LM, Reed RJ, Sternberg WH. Cotyledonoid dissecting leiomyoma of the uterus: the Sternberg tumor. *Am J Surg Pathol* 1996;20:1455–1461.

108. Sanz-Ortega J, Vocke C, Stratton P, Linehan WM, Merino MJ. Morphologic and molecular characteristics of uterine leiomyomas in hereditary leiomyomatosis and renal cancer (HLRCC) syndrome. *Am J Surg Pathol* 2013;37:74–80.

109. Reyes C, Karamurzin Y, Frizzell N, *et al.* Uterine smooth muscle tumors with features suggesting fumarate hydratase aberration: detailed morphologic analysis and correlation with S-(2-succino)-cysteine immunohistochemistry. *Mod Pathol* 2014;27:1020–1027.

110. Llamas-Velasco M, Requena L, Kutzner H, *et al.* Fumarate hydratase immunohistochemical staining may help to identify patients with multiple cutaneous and uterine leiomyomatosis (MCUL) and hereditary leiomyomatosis and renal cell cancer (HLRCC) syndrome. *J Cutan Pathol* 2014;41:859–865.

111. Mäkinen N, Mehine M, Tolvanen J, *et al.* MED12, the mediator complex subunit 12 gene, is mutated at high frequency in uterine leiomyomas. *Science* 2011;334:252–255.

112. McGuire MM, Yatsenko A, Hoffner L, *et al.* Whole exome sequencing in a random sample of North American women with leiomyomas identifies MED12 mutations in majority of uterine leiomyomas. *PLoS One* 2012;7:e33251.

113. Bertsch E, Qiang W, Zhang Q, *et al.* MED12 and HMGA2 mutations: two independent genetic events in uterine leiomyoma and leiomyosarcoma. *Mod Pathol* 2014;27:1144–1153.

114. Prayson RA, Goldblum JR, Hart WR. Epithelioid smooth-muscle tumors of the uterus: a clinicopathologic study of 18 patients. *Am J Surg Pathol* 1997;21:383–391.

115. Anandan V, Moosavi C, Miettinen M. Clinicopathologic characterization of myxoid mesenchymal neoplasms of the uterus. *Mod Pathol* 2008;21(Suppl 1):191A–195A. Abstract.

116. King ME, Dickersin GR, Scully RE. Myxoid leiomyosarcoma of the uterus. *Am J Surg Pathol* 1981;6:589–598.

Intravenous and peritoneal leiomyomatosis and benign metastasizing leiomyoma

117. Clement PB. Intravenous leiomyomatosis of the uterus. *Pathol Annu* 1988;23(Part 2):153–183.

118. Norris HJ, Parley T. Mesenchymal tumors of the uterus. V. Intravenous leiomyomatosis: a clinical and pathological study of 14 cases. *Cancer* 1975;36:2164–2178.

119. Clement PB, Young RH, Scully RE. Intravenous leiomyomatosis of the uterus: a clinicopathological analysis of 16 cases with unusual histologic features. *Am J Surg Pathol* 1988;12:932–945.

120. Mulvany NJ, Slavin JL, Ostor AG, Fortune DW. Intravenous leiomyomatosis of the uterus: a clinicopathologic study of 22 cases. *Int J Gynecol Pathol* 1994;13:1–9.

121. Brescia RJ, Tazelaar HD, Hobbs HJ, Miller AW. Intravascular leiomyomatosis: a report of two cases. *Hum Pathol* 1989;20:252–256.

122. Wolff M, Silva F, Kaye G. Pulmonary metastases (with admixed epithelial elements) from smooth muscle neoplasms: report of nine cases, including three in men. *Am J Surg Pathol* 1979;3:325–342.

123. Esteban JM, Allen WM, Schaerf RH. Benign metastasizing leiomyoma of the uterus: histologic and immunohistochemical characterization of primary and metastatic lesions. *Arch Pathol Lab Med* 1999;123:960–962.

124. Jautzke G, Müller-Ruchholtz E, Thalmann U. Immunohistochemical detection of estrogen and progesterone receptors in multiple and well-differentiated leiomyomatous lung tumors in women with uterine leiomyomas (so-called benign metastasizing leiomyomas): a report on 5 cases. *Pathol Res Pract* 1996;192:215–223.

125. Kayser K, Zink S, Scheider T, *et al.* Benign metastasizing leiomyoma of the uterus: documentation of clinical, immunohistochemical, and lectin histochemical data on ten cases. *Virchows Arch* 2000;437:284–292.

126. Nucci MR, Drapkin R, Dal Cin P, Fletcher CDM, Fletcher JA. Distinctive cytogenetic profile in benign metastasizing leiomyoma: pathologic

implications. *Am J Surg Pathol* 2007;31:737–743.

127. Alessi G, Lemmerling M, Vereecken L, De Waele L. Benign metastasizing leiomyoma to skull base and spine: report of two cases. *Clin Neurol Neurosurg* 2003;105:170–174.

128. Tavassoli FA, Norris HJ. Peritoneal leiomyomatosis (leiomyomatosis peritonealis disseminata): a clinicopathologic study of 20 cases with ultrastructural observations. *Int J Gynecol Pathol* 1982;1:59–74.

129. Valente PT. Leiomyomatosis peritonealis disseminata: a report of two cases and review of the literature. *Arch Pathol Lab Med* 1984;108:669–672.

130. Hardman WJ, Majmudar B. Leiomyomatosis peritonealis disseminata: clinicopathologic analysis of five cases. *South Med J* 1986;89:291–294.

131. Zotalis G, Nayar R, Hicks DG. Leiomyomatosis peritonealis disseminata, endometriosis and multicystic mesothelioma: an unusual association. *Int J Gynecol Pathol* 1998;17:178–182.

132. Bekkers RL, Willemsen WN, Shijf CP, *et al.* Leiomyomatosis peritonealis disseminata: does malignant transformation occur? A literature review. *Gynecol Oncol* 1999;75:158–163.

Uterine leiomyosarcoma

133. Van Dinh T, Woodruff JD. Leiomyosarcoma of the uterus. *Am J Obstet Gynecol* 1982;144:817–823.

134. Larson B, Silfversward C, Nilsson B, Petterson F. Prognostic factors in uterine leiomyosarcoma: a clinical and histopathological study of 143 cases. The Radiumhemmet series 1936–1981. *Acta Oncol* 1990;29:185–191.

135. Jones MW, Norris HJ. Clinicopathologic study of 28 uterine leiomyosarcomas with metastasis. *Int J Gynecol Pathol* 1995;14:243–249.

136. Mayerhofer K, Obermair A, Windbichler G, *et al.* Leiomyosarcoma of the uterus: a clinicopathologic multicenter study of 71 cases. *Gynecol Oncol* 1999;74:196–201.

137. Giuntoli RL, Metzinger DS, DiMarco CS, *et al.* Retrospective review of 208 patients with leiomyosarcoma of the uterus: prognostic indicators, surgical management, and adjuvant therapy. *Gynecol Oncol* 2003;89:460–469.

138. Veras E, Zivanovic O, Jacks L, *et al.* "Low-grade leiomyosarcoma" and late-recurring smooth muscle tumors of the uterus: heterogeneous collection of frequently misdiagnosed tumors associated with an overall favorable prognosis relative to conventional uterine leiomyosarcomas. *Am J Surg Pathol* 2011;35:1626–1637.

139. Rao U, Finkestein SD, Jones MW. Comparative immunohistochemical and molecular analysis of uterine and extrauterine leiomyosarcomas. *Mod Pathol* 1999;12:1001–1009.

140. Bodner K, Bodner-Adler B, Kimberger O, *et al.* Estrogen and progesterone receptor expression in patients with uterine leiomyosarcoma and correlation with different clinicopathological parameters. *Anticancer Res* 2003;23:729–732.

141. Miettinen M, Fetsch JF. Evaluation of biologic potential of smooth muscle tumors. *Histopathology* 2006;48:97–105.

EBV-associated leiomyosarcoma

142. McClain KL, Leach CT, Jenson HB, *et al.* Association of Epstein–Barr virus with leiomyosarcomas in children with AIDS. *N Engl J Med* 1995;332:12–18.

143. Ross JS, Del Rosario A, Bui HX, Sonbati H, Solis O. Primary hepatic leiomyosarcoma in a child with the acquired immunodeficiency syndrome. *Hum Pathol* 1992;23:69–72.

144. Orlow SJ, Kamino H, Lawrence RL. Multiple subcutaneous leiomyosarcomas in an adolescent with AIDS. *Am J Pediatr Hematol Oncol* 1992;14:265–268.

145. van Hoeven KH, Factor SM, Kress Y, Woodruff JM. Visceral myogenic tumors: a manifestation of HIV infection in children. *Am J Surg Pathol* 1993;17:1176–1181.

146. Litofsky NS, Pihan G, Corvi F, Smith TW. Intracranial leiomyosarcoma: a neuro-oncological consequence of acquired immunodeficiency syndrome. *J Neurooncol* 1998;40:179–183.

147. Timmons CF, Dawson DB, Richards CS, Andrews WS, Katz JA. Epstein-Barr virus-associated leiomyosarcomas in liver transplantation recipients: origin from either donor or recipient tissue. *Cancer* 1995;76:1481–1489.

148. Somers GR, Tesoriero AA, Hartland E, *et al.* Multiple leiomyosarcomas of both donor and recipient origin arising in heart-lung transplant patients. *Am J Surg Pathol* 1998;22:1423–1428.

149. Rogatsch H, Bonatti H, Menet A, *et al.* Epstein–Barr virus-associated multicentric leiomyosarcoma in an adult patient after heart transplantation: case report and review of the literature. *Am J Surg Pathol* 2000;24:614–621.

150. Deyrup AT, Lee VK, Hill CE, *et al.* Epstein-Barr virus-associated smooth muscle tumors are distinctive mesenchymal tumors reflecting multiple infection events: a clinicopathologic and molecular analysis of 29 tumors from 19 patients. *Am J Surg Pathol* 2006;30:75–82.

151. Tulbah A, Al-Dayel F, Fawaz I, Rosai J. Epstein–Barr virus-associated leiomyosarcoma of the thyroid in a child with congenital immunodeficiency: a case report. *Am J Surg Pathol* 1998;23:473–476.

152. Boman F, Gultekin H, Dickman PS. Latent Epstein–Barr virus infection demonstrated in low-grade leiomyosarcomas of adults with acquired immunodeficiency syndrome, but not in adjacent Kaposi's lesion or smooth muscle tumors in immunocompetent patients. *Arch Pathol Lab Med* 1997;121:834–838.

153. Hill MA, Araya JC, Eckert MW, *et al.* Tumor-specific Epstein–Barr virus infection is not associated with leiomyosarcoma in human immunodeficiency virus negative individuals. *Cancer* 1997;80:204–210.

Chapter 17

Gastrointestinal stromal tumor (GIST)

Markku Miettinen
National Cancer Institute, National Institutes of Health

Gastrointestinal stromal tumor (GIST) is defined as a specific mesenchymal tumor of the gastrointestinal (GI) tract that has a characteristic histologic spectrum, is generally KIT (CD117) positive, and is driven by oncogenic KIT or platelet-derived growth factor receptor alpha (PDGFRA) mutations in 85% of cases, with recently discovered alternative pathogenesis in a subset of cases. The definition of GIST excludes GI smooth muscle and nerve sheath tumors. Availability of KIT tyrosine kinase inhibitor drug therapy (imatinib mesylate and second-generation inhibitors) has greatly magnified the interest in GIST and vastly increased the importance of its specific diagnosis.

GISTs occur along the entire length of the GI tract, from the esophagus to the anus, and similar tumors can occur anywhere in the abdomen, usually representing metastases or tumors detached from their GI tract site of origin. GISTs have a wide clinical spectrum, ranging from benign, small incidentally detected nodules to massive, frankly malignant tumors that can fill the entire abdomen.

This chapter summarizes the clinical, pathologic, and molecular genetic features of GISTs. Although all GISTs share many features, tumors at each site have distinctive clinicopathologic and prognostic differences and are therefore reviewed site by site. Small intestinal GISTs in particular show malignant behavior more frequently than gastric tumors do. In addition to KIT/PDGFRA mutations (85%), GISTs can be driven by succinate dehydrogenase SDH deficiency (4%), neurofibromatosis 1 (1%), and BRAF mutation (<1%), and these categories are discussed separately.

Tumor size, mitotic rate per 50 HPFs, tumor margin status, and pathogenetic category should be included in the pathology report if possible.

History and terminology

Specific recognition of GIST as a biologic entity was based on studies by Hirota et al., who first described KIT expression and KIT-activated mutations as a pathogenesis of GIST.[1] Earlier observations that most GI mesenchymal tumors lacked the histologic, electron microscopic, and immunohistochemical features of smooth muscle or nerve sheath tumors led to the introduction of a histogenetically neutral and subsequently widely used term "stromal tumor."[2] A wide array of diagnoses such as gastrointestinal smooth muscle tumor, leiomyoma, leiomyoblastoma, leiomyosarcoma, neurofibroma, or schwannoma were used for GISTs in earlier literature. Because most GI mesenchymal tumors are actually GISTs, older series of GI mesenchymal tumors (irrespective of the original designation) largely represent GISTs. Tumors previously classified as GI autonomic nerve tumors (GANTs), also represent GISTs, because they are KIT positive and have GIST-specific mutations.[3]

Epidemiology

Population-based epidemiologic studies from Sweden, Iceland, and Norway have estimated GIST incidence between 11 and 14.5 per million.[4–6] Assuming the same incidence, these numbers would translate to 3300 to 4350 cases annually in the United States. Benign examples clearly outnumber malignant ones. The common occurrence of minimal incidental GISTs indicates an even higher incidence than that cited previously. For example, a Japanese study reported 50 microscopic GISTs in 35 of 100 thoroughly sampled gastrectomy specimens.[7] In a study from the United States, such tumors were found in 10% of gastroesophageal junction carcinoma specimens.[8] In another study from Germany, small GISTs (1–4 mm) were grossly identified in 22.5% of consecutive autopsies.[9] These results, together with a much lower population incidence of GISTs, indicate that only a few of the minute GISTs progress into a clinically significant tumor.

GISTs typically occur in patients older than 50 years of age and are rare in patients younger than 40 years of age. The median ages in the largest GIST series of different locations range between 58 and 65 years. Some series, especially those of malignant GISTs, have shown a mild male predominance (Figure 17.1). GISTs are very rare in children; tumors earlier classified as intestinal smooth muscle tumors in children <5 years are usually inflammatory myofibroblastic tumors or other entities unrelated to GIST.[10]

Locations

GISTs occur most commonly in the stomach (60%), followed by the small intestine (30%), duodenum (5%), colon and

Modern Soft Tissue Pathology, Second Edition, ed. Markku Miettinen. Published by Cambridge University Press. © Cambridge University Press 2016.

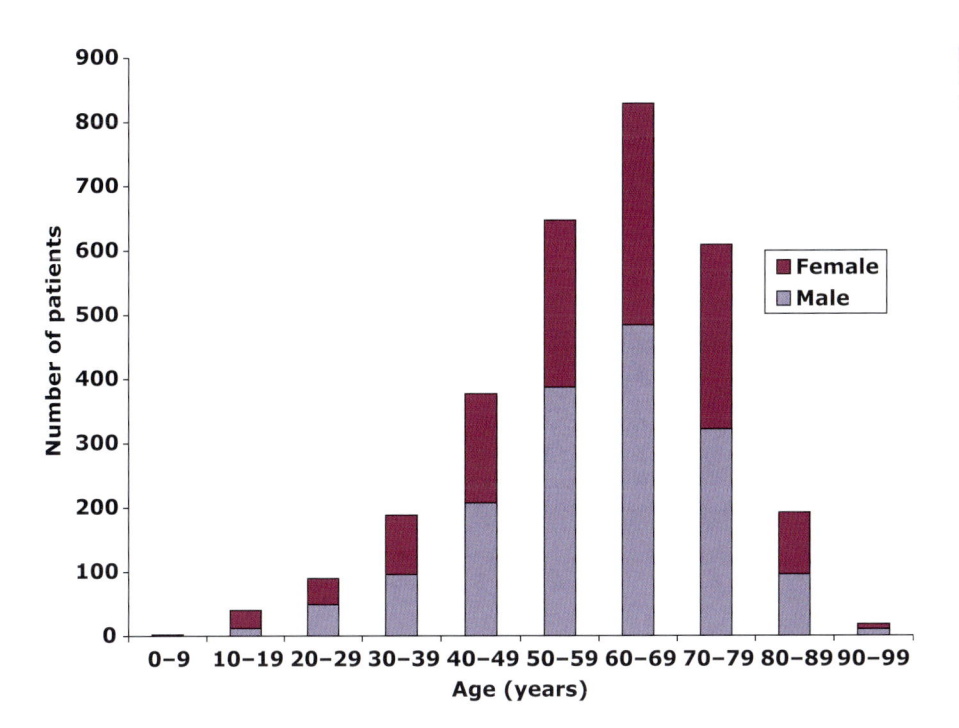

Figure 17.1 Age and sex distribution of 2992 GISTs of the stomach, small intestine, and rectum.

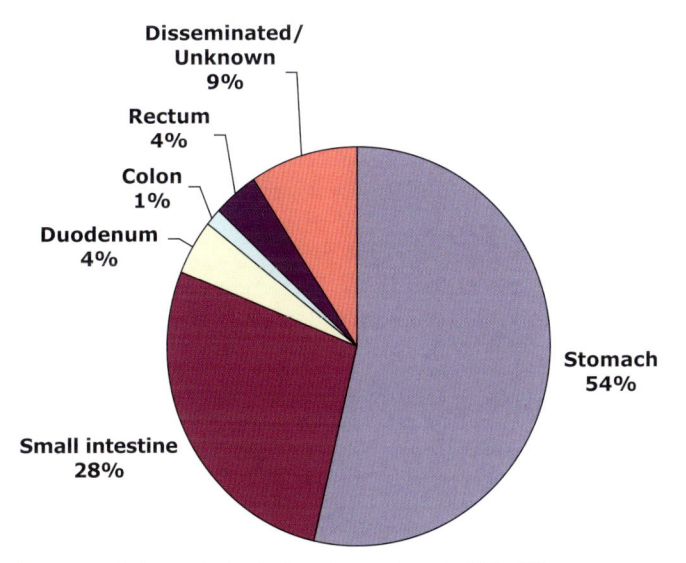

Figure 17.2 Anatomic distribution of approximately 3400 GISTs.

rectum (<5%), and esophagus (<1%). Up to 10% of GISTs are primarily disseminated in the omentum, mesentery, and retroperitoneum, but solitary tumors in these locations may be localized GISTs detached from their gastrointestinal organ of origin and may behave similarly to corresponding primary GISTs (Figure 17.2). In all, approximately 30% of all GISTs (excluding the apparently common minimal tumors) are clinically malignant.[10,11]

Histogenesis and pathogenesis

A stem-cell-like subset of interstitial cells of Cajal, the intermediaries between the GI autonomic nervous system and the smooth muscle cells, are the ancestry for GISTs.[1,12] These cells, first discovered by Santiago Ramon y Cajal in the late 1800s by histochemical stains, are difficult to visualize with hematoxylin and eosin (HE) staining under normal circumstances, but are readily detected by KIT immunostaining as slender, elongated spindle cells in and around the myenteric plexus, and in some locations, inside the muscularis propria. The development and function of Cajal cells are KIT dependent, and these cells modulate autonomous nerve function and regulate autonomous GI peristalsis.[13,14] Cajal cells or their precursors can differentiate into smooth muscle, especially if KIT signaling is disrupted, indicating their multipotential and stem-cell-like qualities.[15] Shared embryonic myosin expression in Cajal cells and GISTs[16] and the coexpression of KIT and CD34 in explanted murine Cajal cells[17] have also been interpreted to support the relationship of GIST and Cajal cells.

A key element in the pathogenesis of GISTs is mutational activation of KIT or PDGFRA receptor tyrosine kinases.[1,18,19] These mutations lead to constitutively activated status of the corresponding KIT and PDGFRA proteins, and activation of the signaling pathways that ultimately leads to cellular proliferation and decrease in apoptosis. Mouse models support a central pathogenetic role for these mutations: introduced constitutional KIT mutations similar to those in human GISTs produce a hereditary GIST syndrome.[20–22]

Clinical features and prognosis

Small GISTs are often incidentally detected as gastric and small intestinal serosal or mural masses during surgery, endoscopy, or radiologic studies for other conditions, such as surgery for GI cancers.

Patients having symptomatic GISTs of the esophagus typically present with dysphagia, or occasionally these GISTs

Table 17.1 Rates of metastases or tumor-related death in GISTs of stomach, duodenum, jejunum and ileum, and rectum by mitotic rate and tumor size

Tumor parameters			Patients with progressive disease during long-term follow-up and characterization of the risk for metastasis (%)			
Group	Size	Mitotic rate	Gastric GISTs	Jejunal and ileal GISTs	Duodenal GISTs	Rectal GISTs
1	≤2 cm	≤5 per 50 HPFs	0 none	0 none	0 none	0 none
2	>2≤5 cm	≤5 per 50 HPFs	1.9 very low	4.3 low	8.3 low	8.5% low
3a	>5≤10 cm	≤5 per 50 HPFs	3.6 low	24 moderate		
3b	>10 cm	≤5 per 50 HPFs	12 moderate	52 high	34 high[b]	57[a] high[b]
4	≤2 cm	>5 per 50 HPFs	0[a]	50[a]	[c]	54 high
5	>2≤5 cm	>5 per 50 HPFs	16 moderate	73 high	50 high	52 high
6a	>5≤10 cm	>5 per 50 HPFs	55 high	85 high		
6b	>10 cm	>5 per 50 HPFs	86 high	90 high	86 high[b]	71 high[b]

Based on previously published long-term follow-up studies on 1055 gastric, 629 small intestinal, and 111 rectal GISTs.
[a] Denotes tumor categories with very small numbers of cases.
[b] Groups 3a and 3b or 6a and 6b combined in duodenal and rectal GISTs because of small numbers of cases.
[c] No tumors of such category were included in the study. Note that small intestinal and other intestinal GISTs show a markedly worse prognosis in many mitosis and size categories than gastric GISTs.

present as mediastinal tumors connected to the esophagus. Gastric GISTs usually give vague symptoms that lead to their gastroscopic detection, and those that ulcerate often present with acute or chronic upper GI bleeding, or anemia caused by insidious chronic bleeding. Intestinal GISTs can cause vague abdominal symptoms, bleeding, anemia, or an acute abdomen due to obstruction or tumor rupture. Preoperative specific diagnosis can be successful by endoscopic biopsy, but those GISTs that are not accessible by conventional endoscopic biopsy can be reached by endoscopic ultrasonography or CT-guided biopsy.

Malignant GISTs (20–30% of all GISTs) typically metastasize in any part of the abdominal cavity, often forming multiple, sometimes confluent, masses that can fill the abdomen. Gastric GISTs in particular can involve the splenic and hepatic regions. GISTs of all sites also commonly metastasize to the liver, whereas lung metastases are distinctly rare, in contrast to many other sarcoma types. Bone and soft tissue metastases occur rarely. The most common bone metastases are in the axial skeleton, and soft tissue metastases have a predilection for the abdominal wall, with rare involvement of peripheral soft tissues.

The time course of metastatic development varies. More aggressive tumors, recognizable by high mitotic activity, often metastasize in 1 to 2 years from apparently complete excision, whereas some GISTs (tumors with lower mitotic counts) metastasize 5 to 10 years after surgery, or occasionally much later, indicating the importance of long-term follow-up.

The major histologic prognostic factors for GIST are mitotic activity and tumor size. In contrast to many soft tissue sarcomas, mitotic rates of >5 per 50 HPFs (corresponding to an average of 1 mitotic figure per 10 HPFs) imply metastatic risk. Tumor size is another well-studied prognostic parameter. Prognosis by tumor size and mitotic rate categories based on follow-up studies of >2000 GISTs is summarized in Table 17.1.[10]

Esophageal GISTs

GISTs are very rare in the esophagus and comprise <1% of all GISTs. They usually occur in the lower third of the esophagus. The relatively small number of reported examples has mostly included advanced malignant tumors and occasional small, incidentally detected examples. Larger esophageal GISTs can extend into the mediastinum, and clinically and radiologically simulate a primary mediastinal tumor. The small number of reported cases does not allow formulation of prognostic factors, although small tumor size and low mitotic rate have been favorable features.[23]

Most esophageal GISTs are spindle cell tumors, and many are sarcomatous examples with significant mitotic activity. Rare examples have epithelioid morphology (Figure 17.3). On immunohistochemical analysis, esophageal GISTs appear consistently positive for KIT and CD34, but usually negative for SMA and desmin.

Gastric GISTs in adults

This is the largest and most complex group of GISTs (nearly 60% of all GISTs). These GISTs encompass a broad clinical spectrum from minimal incidental tumors to sarcomas, and there is also great histologic variation, sometimes making the recognition of GISTs challenging.[24,25] Gastric GISTs include higher numbers of indolent tumors than intestinal GISTs. They are also molecular genetically more heterogeneous, especially because they include a significant subgroup of PDGFRA-mutant and SDH-deficient tumors in children and young adults (separate text below).

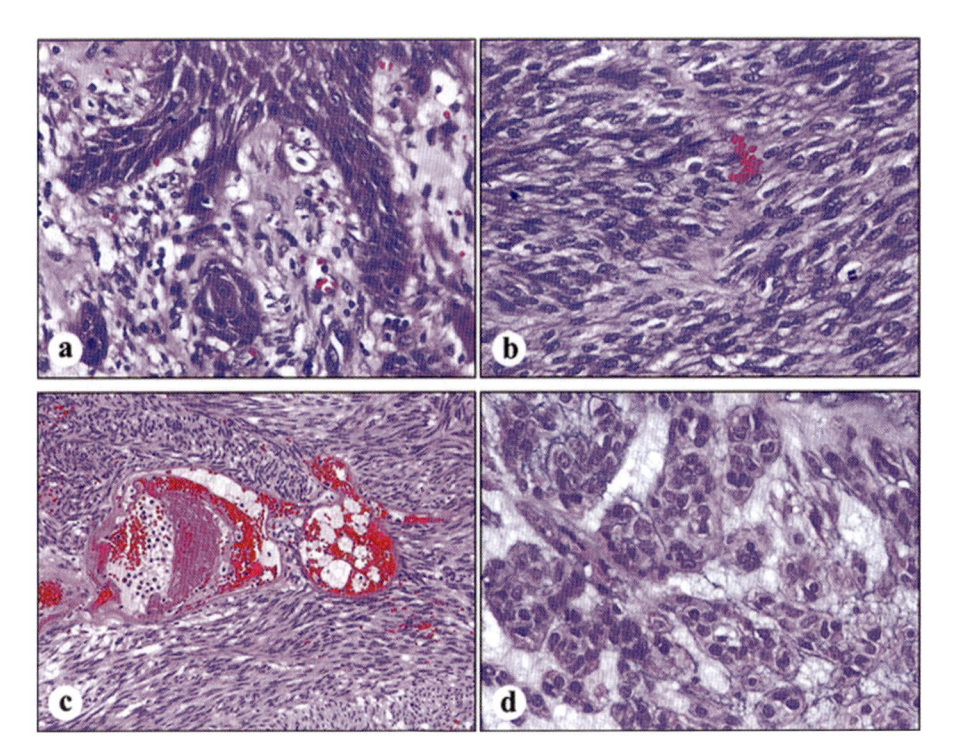

Chapter 17: Gastrointestinal stromal tumor (GIST)

Figure 17.3 Histologic spectrum of esophageal GISTs. (**a**) Tumor extending below squamous mucosa and eliciting epithelial hyperplasia. (**b**) A spindle cell tumor with significant mitotic activity (sarcomatous GIST). (**c**) A spindle cell GIST with dilated and thrombosed vessels resembling similar findings seen in schwannoma. (**d**) A malignant epithelioid GIST of the esophagus.

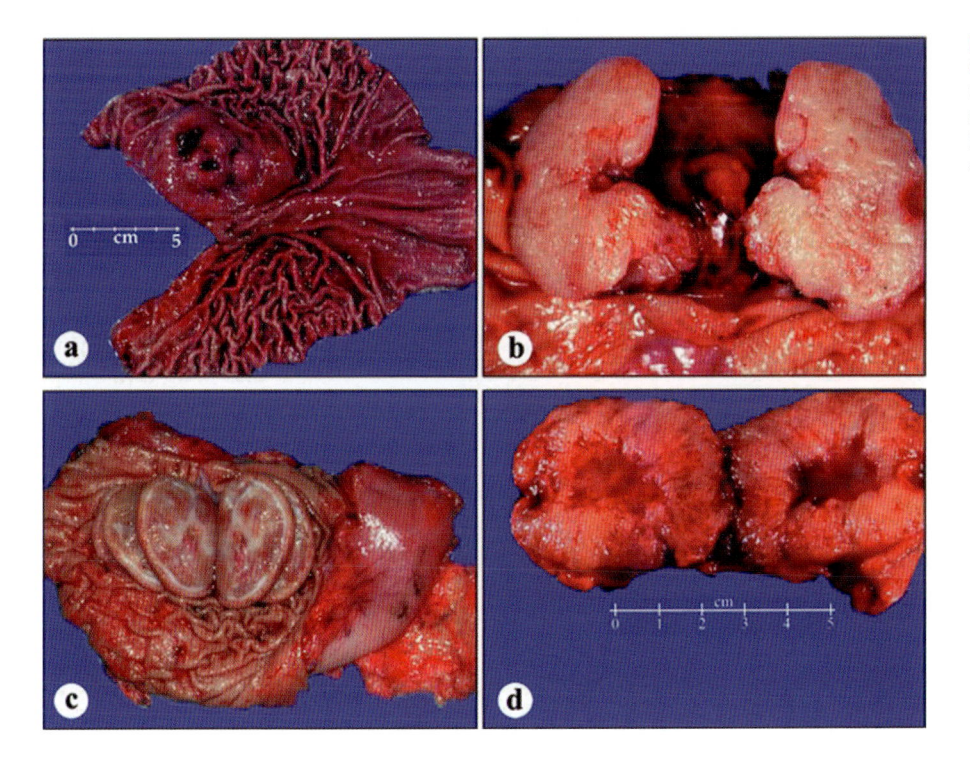

Figure 17.4 Gross appearances of gastric GISTs. (**a**) An internally bulging GIST with mucosal ulceration. (**b**) Homogeneous pale tan surface in a bisected GIST. (**c**) Gastric GIST forming a sessile intraluminal mass. (**d**) A tumor with central cystic cavity, a common finding in larger GISTs.

Grossly, gastric GISTs vary; they can be firm, gray-white, mural or serosal nodules in the stomach or intestines, but they can also form intraluminal polypoid masses. Larger tumors often show both intra- and extraluminal components and show structural heterogeneity, including hemorrhagic and cystic changes (Figure 17.4). Many larger gastric GISTs are extensively cystic, and in some examples, only a narrow rim of tumor remains in the cyst wall. Tumors that have reached a moderate size (<5 cm) commonly contain a central cyst, and occurrence of numerous microscopic cysts is a common finding. Some gastric GISTs are attached to the gastric outer wall with a narrow pedicle, and others are

477

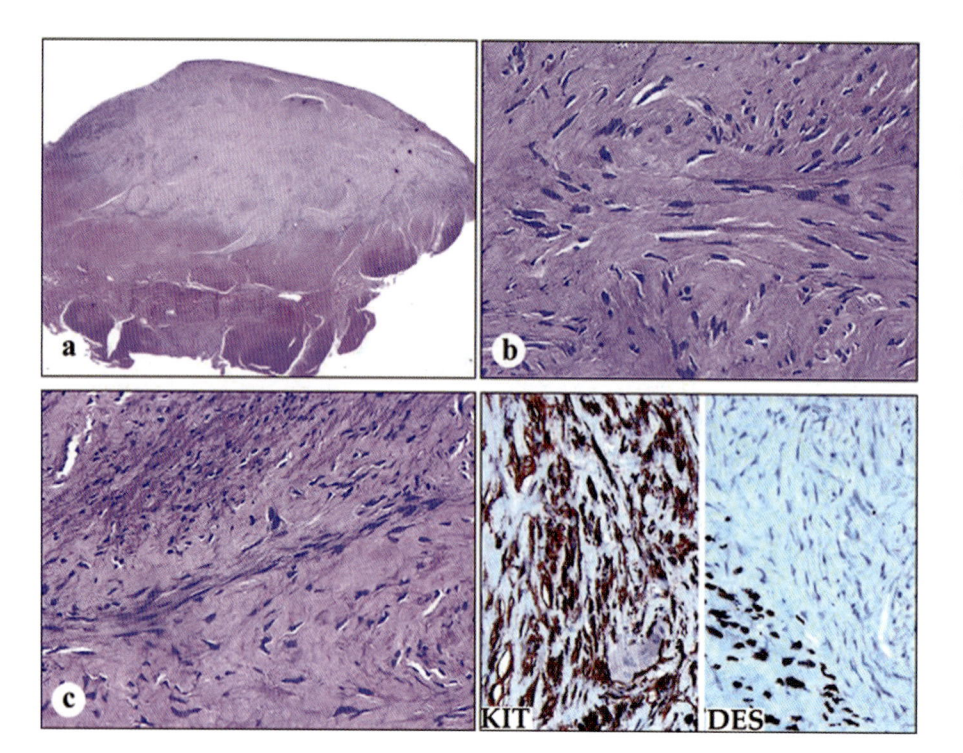

Figure 17.5 Small (<2 cm) gastric GIST (sclerosing spindle cell type) forming a serosal nodule. (**a,b**) The tumor is composed of bland spindle cells that intermingle with smooth muscle cells. (**c**) The tumor is KIT positive, and desmin-positive elements represent normal smooth muscle.

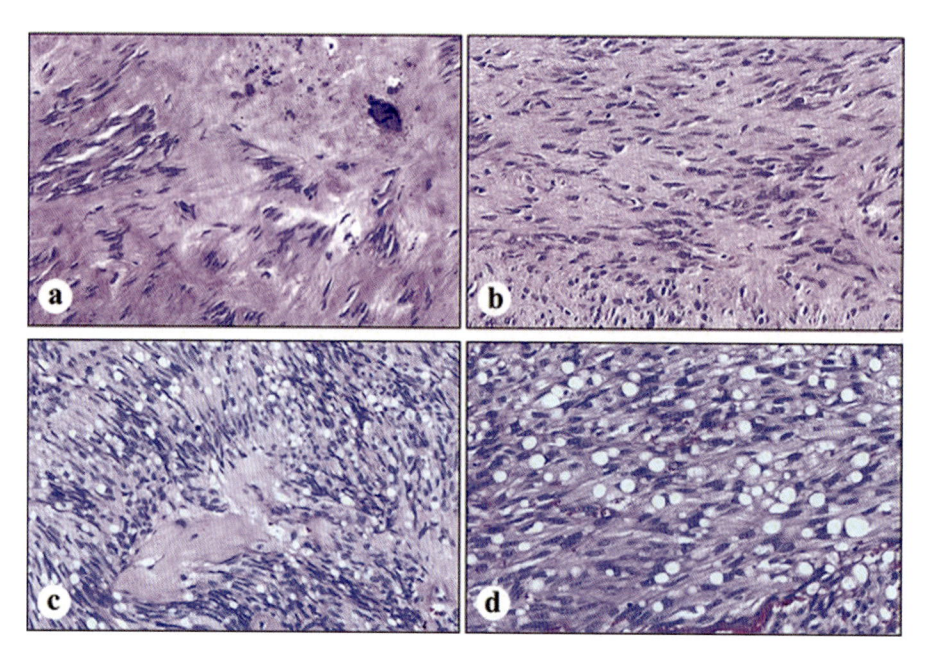

Figure 17.6 (**a,b**) Sclerosing spindle cell GIST is paucicellular and contains calcification. (**c,d**) Spindle cell GIST of palisaded-vacuolated type.

embedded in the major or minor omentum, clinically and pathologically simulating an omental primary tumor. The pedicle of such GISTs typically contains elements of tumor infiltrating the gastric wall, so that a patch of gastric wall must be removed to ensure a complete excision. Otherwise, smaller gastric GISTs are typically managed by wedge resections, whereas larger tumors require partial or occasionally total gastrectomies or more extensive surgery, including additional splenectomy or partial colectomy, hepatic, or pancreatic resection, depending on the tumor involvement.

Histologic variants of gastric gastrointestinal stromal tumors

Histologically and clinicopathologically, gastric GISTs include several distinct (although to some degree overlapping) groups. Recognition of these types helps to identify GISTs specifically on histologic grounds.[25]

Sclerosing spindle cell GISTs (Figures 17.5 and 17.6) are often small tumors. They are paucicellular, contain high amounts of collagenous matrix, and often have dystrophic calcification. Mitotic rate is usually low, and the prognosis is good.

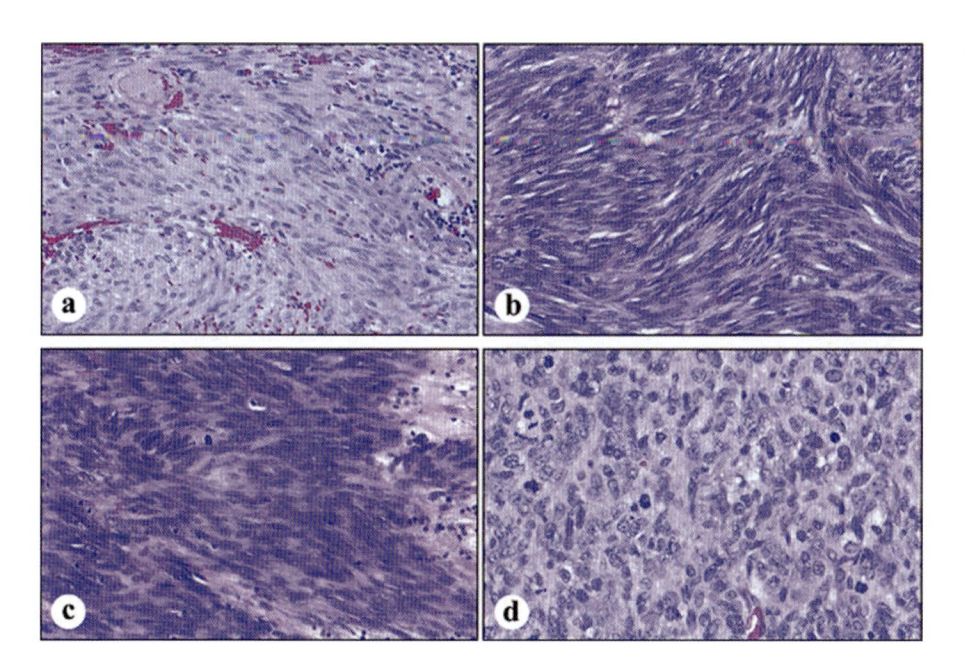

Figure 17.7 (**a**) Gastric hypercellular spindle cell GIST. (**b–d**) Variants of sarcomatous spindle cell GISTs with significant mitotic activity.

The palisaded-vacuolated type (Figure 17.6c,d) is perhaps the most common histologic variant of gastric GISTs. These tumors can reach a significant size, but they usually have a low mitotic rate and a good prognosis. Histologically typical is a spindle cell pattern with rhythmic nuclear palisading, somewhat resembling that of schwannomas, and variably prominent perinuclear vacuolization.

Hypercellular spindle cell GISTs (Figure 17.7a) show solid sheets of spindle cells with a relatively low mitotic activity and atypia. Neither nuclear palisading nor perinuclear vacuolization are prominent. These tumors often have an elevated mitotic rate, and there is some tumor-related mortality.

Sarcomatous spindle cell GISTs are tumors with a significant mitotic rate (often over 20 per 50 HPFs) and nuclear atypia, generally manifesting as an increased nucleocytoplasmic ratio. Tumor necrosis is common. This group is characterized clinically by a high tumor-related mortality rate (Figure 17.7b–d).

Epithelioid gastric GISTs (30–40% of all gastric GISTs) range from mitotically inactive and generally clinically indolent tumors to occasional sarcomas. SDH-deficient GISTs are also usually epithelioid, and their distinctive features are discussed below.

Their variants include sclerosing tumors with content of collagenous matrix (Figure 17.8a) and more cellular examples, often composed of poorly cohesive tumor cells (Figure 17.8b). Examples of progressive higher cellularity include hypercellular, slightly mitotically active tumors and those that have sarcomatous features with marked mitotic activity (Figure 17.8c). Occasionally, gastric epithelioid GISTs have a paraganglioma or carcinoid-like compartmental pattern. Common to many epithelioid gastric GISTs is the presence of multinucleation and relatively common focal nuclear atypia, including pleomorphic forms (Figure 17.8d).

Duodenal and small intestinal gastrointestinal stromal tumors

This is the second largest group of GISTs, constituting 35% of all GISTs (approximately 5% of all GISTs are duodenal, and 30% occur in the jejunum or ileum). Like gastric GISTs, these tumors have a broad clinical spectrum. As a group, however, they are more commonly malignant than are gastric GISTs, with a 40% overall tumor-related mortality.[26,27]

Small tumors (<2 cm) are often incidental findings during another surgery or radiologic study. Larger tumors can cause symptoms such as acute or chronic GI bleeding, or an acute abdomen by obstruction or perforation. Duodenal GISTs, especially those involving the second portion, often secondarily involve the pancreas. Such tumors can require pancreaticoduodenectomy for curative resection. Otherwise, intestinal GISTs are typically managed by segmental (sleeve) resections, or rarely by a smaller wedge resection in the case of very small tumors.

Grossly, small GISTs often form externally bulging nodules, and larger examples can form a dumbbell-shaped mass with internal and external components or a pedunculated external mass (Figure 17.9). Many larger tumors form complex cystic masses, and some examples can cause diverticular-like outpouching that can be suspected clinically as being a tumor originating from a Meckel's diverticulum. Well-documented cases originating from a pre-existing diverticulum have not been reported, and in the author's experience, the diverticulum is a secondary formation generated by the tumor.

Histologically, duodenal and small intestinal GISTs are typically spindle cell tumors. Approximately 40% of them contain distinctive rounded or elongated, eosinophilic,

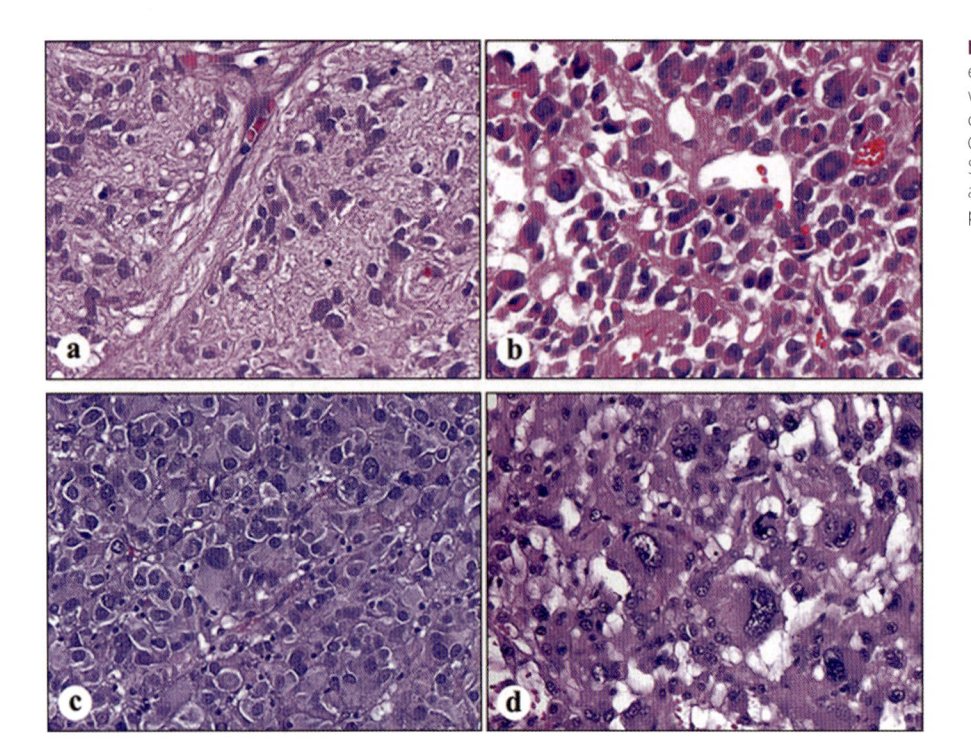

Figure 17.8 Histologic variation of gastric epithelioid GISTs. (**a**) Epithelioid sclerosing GIST with a relatively low cellularity and a prominent collagenous matrix. (**b**) Epithelioid dyscohesive GIST with scant extracellular matrix. (**c**) Sarcomatous epithelioid GIST with a high mitotic activity. (**d**) Epithelioid GIST with a prominent pleomorphism, but inconspicuous mitotic activity.

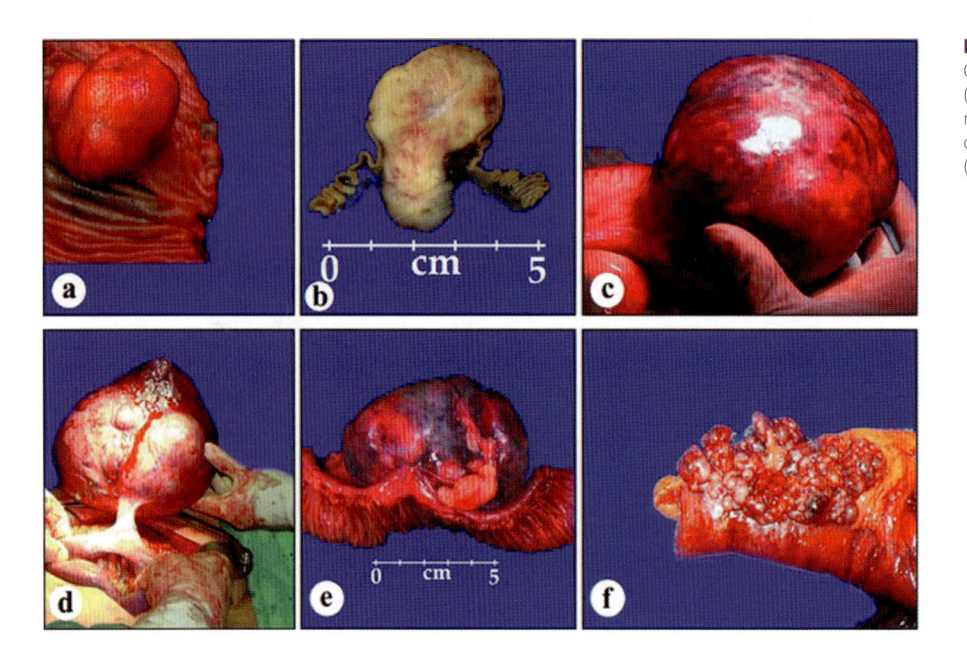

Figure 17.9 Gross appearances of small intestinal GISTs. (**a**) Tumor forming an endoluminal mass. (**b**) A dumbbell-shaped example containing a major endoluminal and a minor external component. (**c–e**) GISTs forming external masses. (**f**) Example with multiple nodules on the serosa.

PAS-positive, extracellular collagen globules, termed *skeinoid fibers*,[28] based on their ultrastructural yarn-like appearance (Figure 17.10a,b). The occurrence of anuclear zones resembling neuropil is common, and indeed, these foci represented complex cytoplasmic processes (Figure 17.10c). Highly mitotically active tumors may have spindle cell or epithelioid appearance, with the latter pattern probably representing morphologic transformation from spindle cell tumors. Significant pleomorphism is rare, but can occur (Figure 17.10d).

Gastrointestinal stromal tumors of the colon and appendix

Primary colonic and appendiceal GISTs are rare (1% of all GISTs), but secondary involvement of the colonic serosal surface is quite common in malignant GISTs of any other origin. Clinically and histologically, colonic GISTs are similar to small intestinal GISTs. The sigmoid colon seems to be the most common site. The occurrence of skeinoid fibers is not uncommon, especially in those examples with scant mitotic activity.

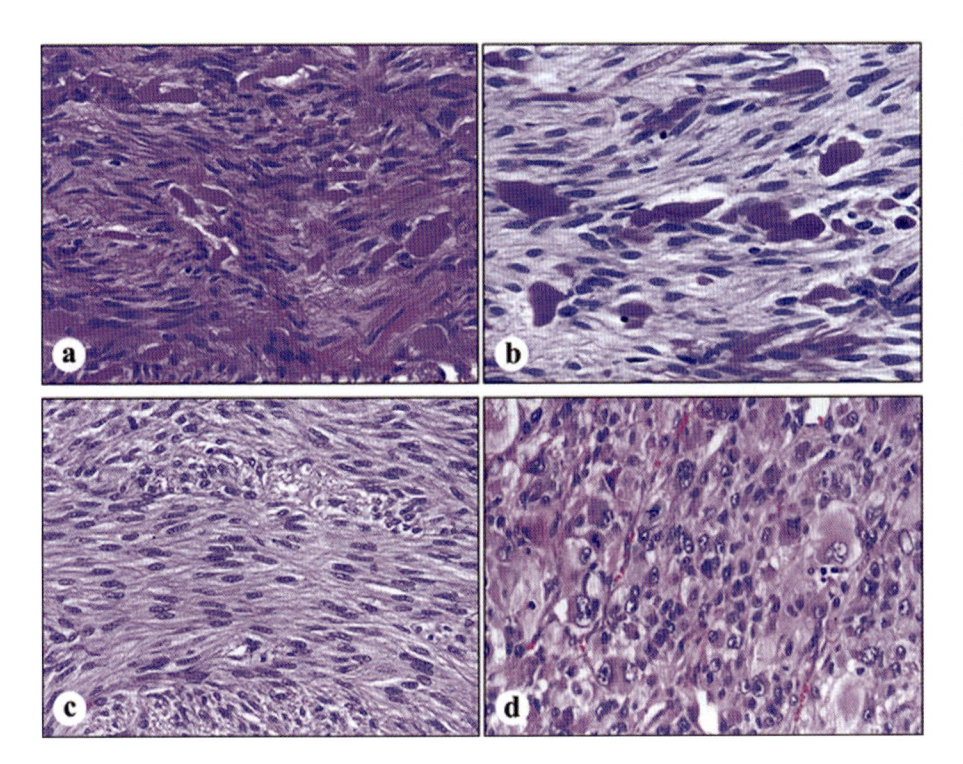

Figure 17.10 Histologic variation of small intestinal GIST. (**a**) This GIST is composed of tapered spindled cells, and there are numerous extracellular collagen globules (skeinoid fibers). (**b**) The skeinoid fibers are PAS positive. (**c**) This cellular spindle cell GIST contains prominent cell processes seen as anucleated areas that can resemble neuropil of neuroblastoma (also composed of prominent cell processes). (**d**) Pleomorphism is rarely conspicuous in a small intestinal GIST.

Positivity for KIT is a consistent feature for both colonic and appendiceal GISTs, but expression of CD34 and SMA varies.[29,30]

Fewer than ten cases of appendiceal GISTs have been reported. These are typically small, incidental, clinically indolent tumors sometimes discovered with appendicitis. Some examples form pedunculated external masses, and others are intramural thickenings. Histologic features are similar to intestinal GISTs, including the presence of skeinoid fibers in some cases.

Gastrointestinal stromal tumor of the rectum

This is a small subgroup constituting no more than 4% to 5% of all GISTs. Clinicopathologically, the group is heterogeneous, containing small tumors incidentally palpated during prostate or pelvic examination. Not all of these small tumors are indolent, however, in contrast to the very small tumors of stomach and small intestine. The rectum is the most common site for small GISTs with significant mitotic activity, and patients with these tumors also have significant risk of tumor progression. Many rectal GISTs are relatively large and often clinically malignant tumors that commonly infiltrate adjacent organs. In men, rectal GISTs can be attached to the posterior aspect of the prostate, and sometimes they are initially suspected as prostatic stromal tumors or GISTs of the prostate.[31]

GISTs of the rectum often develop pelvic recurrences or metastases, and liver metastases are common. Bone metastases may be more common in this subgroup of GISTs.

Grossly, small rectal GISTs can appear as intraluminal polyps, despite their origin from the muscularis propria region. Larger tumors form solid intramural masses that are often centrally cystic.

Histologically, most rectal GISTs are spindle cell tumors, and very few epithelioid examples have been reported. Nuclear palisading is common, but perinuclear vacuolization is generally not prominent. These GISTs typically infiltrate between normal smooth muscle elements (Figure 17.11). Mitotic rate varies, and with size this is a good prognosticator (Table 17.1). In the rectum, the relative number of small tumors with mitotic activity is the highest, and such tumors have a substantial metastatic rate.

The main differential diagnoses of rectal GIST are true smooth muscle tumors. These tumors range from small, muscularis mucosae leiomyomas to larger intramural smooth muscle tumors. The latter tumors include examples with no mitotic activity that can be confidently considered benign leiomyomas. Also included here are mitotically active tumors with variable atypia, and these must be considered leiomyosarcomas. Immunohistochemical analysis makes the distinction of GIST and smooth muscle tumors straightforward, these tumor types having opposite marker profiles, GISTs being positive for KIT and CD34, and smooth muscle tumors for SMA and desmin.

In women, uterine-type leiomyomas can be attached to the rectum or present as perirectal masses. These tumors are negative for KIT, have smooth muscle markers, and are also positive for estrogen receptors.

Extragastrointestinal, gastrointestinal stromal tumors, and gastrointestinal stromal tumors disseminated at presentation

GISTs occurring outside the tubular GI tract are sometimes called extragastrointestinal GISTs (EGISTs). Most GISTs

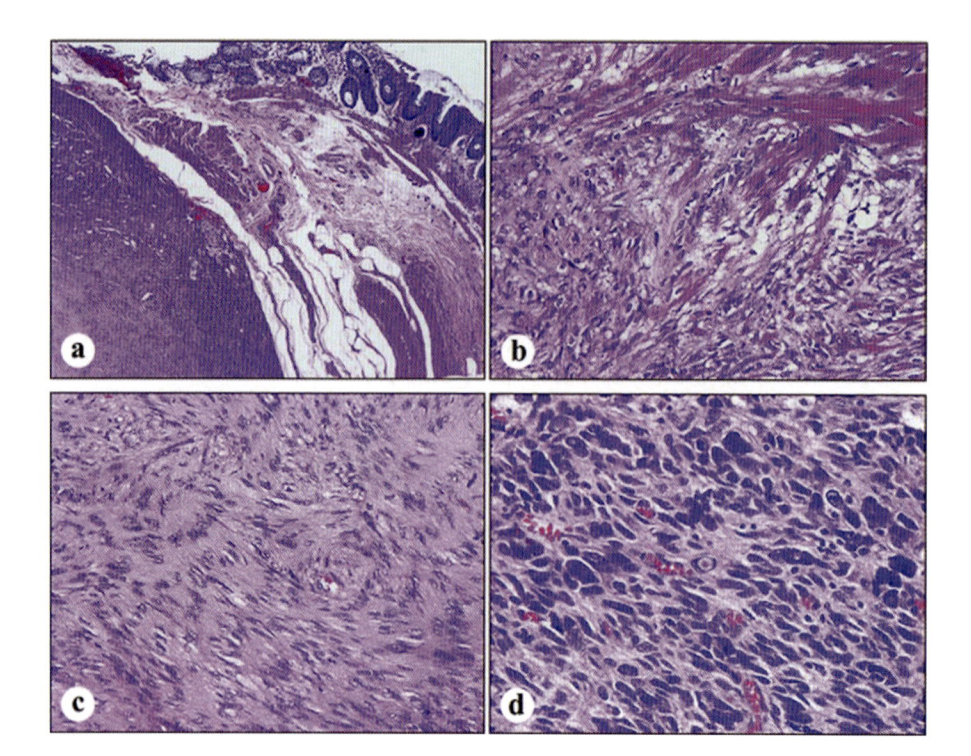

Figure 17.11 Histologic spectrum of rectal GIST. (**a**) A small tumor that involves the rectal muscularis propria underneath a normal mucosa. (**b**) The tumor intermingles between the smooth muscle cells of muscularis propria that appear to be more eosinophilic elements. (**c**) Nuclear palisading in a rectal GIST. (**d**) Pleomorphism is a rare finding.

seen in the omentum, mesenteries, or in the retroperitoneum are metastases from GI primaries or direct extensions from GI primary tumors. Solitary primary EGISTs have been reported, although truly primary EGISTs are very rare or their existence is doubtful, and follow-up data on such cases are scant.[32–34]

Omental GISTs are often large tumors, 10 cm to 20 cm in diameter. They are often cystic hemorrhagic masses, and many are attached to the stomach, suggesting that in fact these tumors are gastric GISTs forming external masses.

Omental GISTs often have an epithelioid cytology resembling gastric GISTs. Some of these tumors also have PDGFRA mutations, again similar to gastric GISTs. It seems likely that omental GISTs are of gastric origin, probably representing tumors that during their evolution have been detached from the stomach and become secondarily attached to the omentum. Tumors in the major omentum might have arisen from the anterior gastric wall, whereas those involving the lesser omental sack are probably of posterior gastric wall origin.

Morphologically, mesenteric GISTs resemble small intestinal GISTs, and some have skeinoid fibers typical of intestinal GISTs. GIST metastases from GI primaries often involve various regions of the abdominal cavity and retroperitoneum. In the author's experience, most tumors previously considered abdominal leiomyosarcomas are actually GISTs. In the retroperitoneum, tumors previously classified as leiomyosarcomas seem to contain approximately equal numbers of GISTs and true smooth muscle tumors.

Succinate-dehydrogenase-deficient GISTs

SDH-deficient GISTs typically occur in children and young adults and comprise the vast majority of GISTs in patients <30 years of age. In these tumors the neoplastic cells have lost function of the mitochondrial SDH complex. SDH-deficiency in the tumors is similar to that observed in paragangliomas, and silencing of one subunit gene impairs formation of a functional complex.[35,36]

SDH-deficient GISTs occur exclusively in the stomach and comprise 7.5% of all gastric GISTs. Age of onset is usually from the second decade on, but sometimes in the later first decade (8–10 years of age). Below the age of 16 years, SDH-deficient GISTs occur almost exclusively in females, but gender difference diminishes with age. Some patients (10–20% of SDH-deficient GISTs) have other tumor manifestations of SDH-deficiency syndrome: pulmonary chondromas and paragangliomas, or, rarely, renal cancers.[35–37]

The designation Carney triad had been used for the occurrence of GIST with pulmonary chondroma, paraganglioma, or both, without SDH mutations.[38] Occurrence of GIST with paraganglioma in connection with SDH germline mutations defines Carney–Stratakis syndrome.[39] Both of these syndromes involve SDH-deficient GISTs, but only a minority of all SDH-deficient GISTs fulfills the criteria for these syndromes. In many cases the different manifestations of those syndromes occur years or even decades apart, which makes it difficult to clinically define these syndromes.

SDH-deficient GISTs are in many ways clinically and pathologically distinctive. Tumors are often multifocal, and local gastric wall recurrences are common and metastases develop in at least 20% of patients, sometimes 20 years or longer after the primary tumor. The clinical course is often slow and the patient can live for a long time even with liver

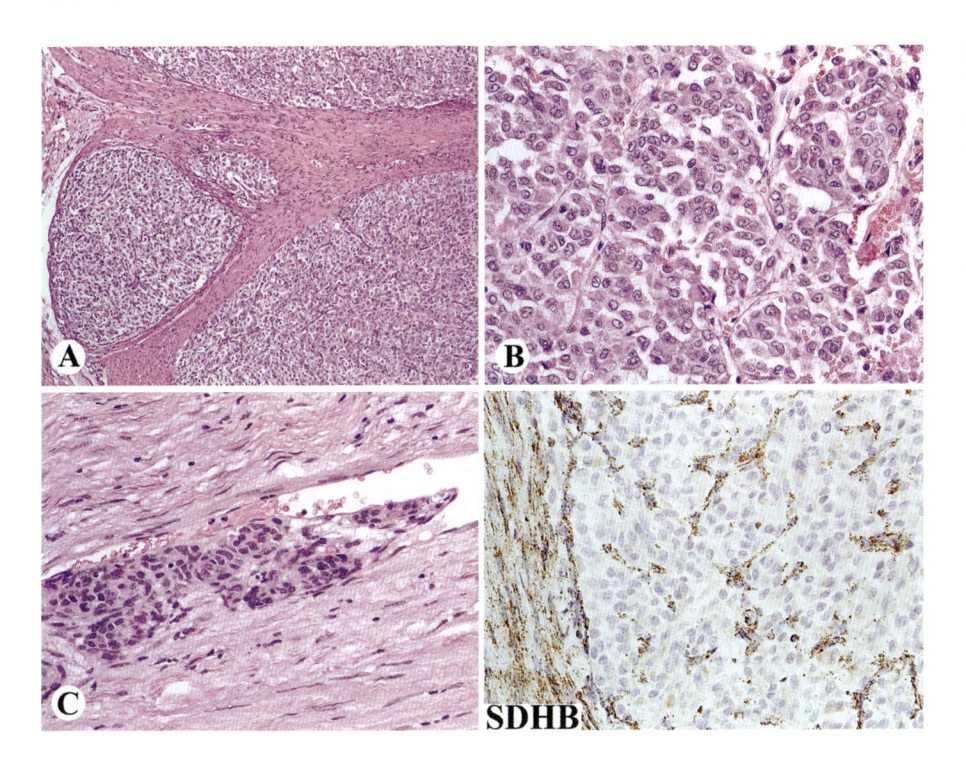

Figure 17.12 (**A**) Multinodular involvement in the gastric wall is typical of succinate-dehydrogenase-deficient GISTs. (**B**) These tumors typically have epithelioid histology, sometimes with a nested pattern. (**C**) Vascular invasion is common in the SDH-deficient GIST. Diagnostic of SDH-deficient GISTs is immunohistochemial loss of SDHB, while this protein is retained in non-neoplastic components such as gastric smooth muscle and vascular elements.

metastases, in contrast to KIT-mutant GISTs that are usually rapidly progressive when metastatic.[37]

Histologically typical features include multinodular involvement of the gastric wall spaced by strands of smooth muscle, and epithelioid cytology, sometimes with significant pleomorphism. A characteristic feature of SDH-deficient GISTs is lymphovascular invasion, and lymph node metastases are detected in 10–20% of patients (Figure 17.12). Mitotic rate varies, but it does not have the predictive value in this group as that seen in KIT-mutant GISTs. Thus tumors with low mitotic rates can metastasize and tumors with 20 mitoses per 10 HPFs can be indolent after complete excision.[37]

The most practical confirmatory test is SDH subunit B (SDHB) immunohistochemistry, which demonstrates loss in tumor cells and retained expression in non-neoplastic cells.[36,37] SDHA is lost in cases that are associated with SDHA mutations.[40–42] Otherwise SDH-deficient GISTs have KIT expression similar to other GISTs and most are also CD34 positive. However, these GISTs are usually SMA negative. A characteristic feature is IGF1R expression, which reflects activation of the IGF1R signaling in this group of GISTs, and this signal pathway is a potential treatment target.[43–45]

Genetically SDH-deficient GISTs are characterized as KIT/PDGFRA wild-type tumors (no mutations in these genes).[35] Approximately 50–60% of patients have SDH-subunit mutations, typically germline. Most common are SDHA mutations (30% of all SDH-deficient GISTs), followed by SDHB (10–15%) and SDHC and SDHD (10% together).[42,46] Less than a half of the cases show SDH wild-type sequences and are associated with epigenetic silencing of the SDH complex. Current data indicates that pathogenesis in these cases is indeed specific epigenetic SDHC silencing, which disables the entire SDH complex.[47] In addition, SDH-deficient GISTs show abnormal genomic methylation in global methylation assays.[48] Genotype-phenotype correlation is still incompletely developed, but it seems that the epigenetic silencing group manifests earlier and SDHA mutants later, usually in young adults.

Gastrointestinal stromal tumors in neurofibromatosis 1 patients

Neurofibromatosis 1 (NF1) syndrome-associated GISTs comprise 4% to 6% of all intestinal GISTs (Figure 17.13).[27] NF1 patients have a markedly increased incidence of GIST, and in these patients, GISTs occur at a lower median age than do sporadic tumors. One autopsy study showed small GISTs in one-third of NF1 patients.[49] NF1 patients undergoing abdominal surgery are often diagnosed with incidental GISTs. In fact, GISTs in these patients often occur as multiple separate nodules. The tumors are usually located in the small intestine and duodenum and only rarely in the stomach and colon. Most GISTs in NF1 patients are indolent, but some of these patients develop malignant GISTs.[49,50] The relative frequency of malignancy seems greater in duodenal than in small intestinal GISTs.[50]

Histologically, NF1-associated GISTs are usually spindle cell tumors. They resemble sporadic intestinal tumors by a common content of skeinoid fibers. Cajal cell hyperplasia, possibly a GIST precursor, is common (Figure 17.14). Incidental neurofibromas are also sometimes present. Although NF1-associated GISTs are KIT positive (as are other intestinal GISTs), most investigators have found them to have wild-type KIT and PDGFRA sequences, indicating a different pathogenesis.[49–51] The genesis of NF1-associated GISTs might be related to the pathologically activated Ras pathway, related to the loss of the NF1 tumor-suppressor protein function.[52]

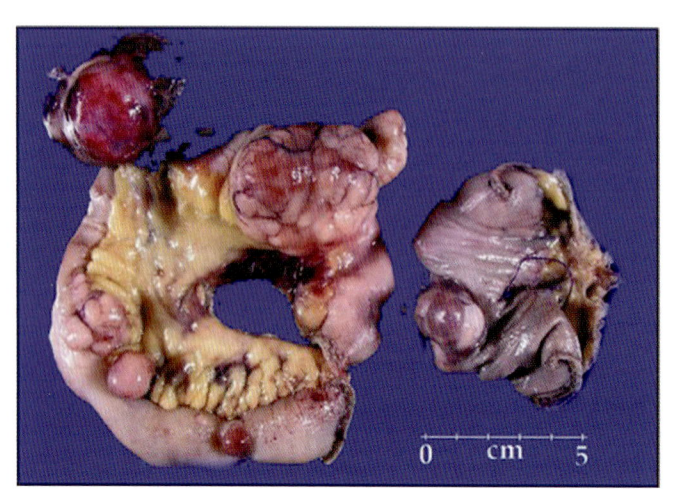

Figure 17.13 Multiple GISTs were detected in the small intestine of this patient with NF type 1. Several GIST nodules are seen in the larger intestinal segment, and one tumor is present in a separate noncontiguous segment.

Immunohistochemistry of gastrointestinal stromal tumors

The most important markers in the evaluation of a potential GIST are KIT (CD117), Ano1-DOG1, CD34, alpha SMA, desmin, and S100 protein. Heavy caldesmon is a smooth muscle marker commonly expressed in GISTs, and other markers are sometimes necessary in making the distinction between GISTs and other entities. Immunohistochemical testing for SDH subunits SDHB and SDHA is important in KIT/PDGFRA wild-type cases and is discussed earlier.

KIT (CD117 antigen) is typically strongly expressed in GISTs (Figure 17.15a), but the patterns of immunoreactivity vary. Some cases show Golgi zone accentuation, and in a small number of cases, a Golgi zone dot-like pattern is predominant (Figure 17.15b,c). KIT immunoreactivity is often more focal and weaker in epithelioid variants (Figure 17.15e). Very rarely, such tumors are entirely negative for KIT (Figure 17.15f). The latter scenario usually occurs in gastric epithelioid GISTs, and <5% of gastric GISTs are KIT negative.

Because KIT is relatively restricted to GIST among sarcomas and other tumors, it is very useful in the diagnosis of GIST.[53,54] Despite occasional statements to the contrary, high KIT expression in GIST is not a consequence of mutation.

Anoctamin-1 (Ano1), also known as DOG1 from "discovered in GIST" is consistently expressed in GIST and nearly specific among mesenchymal tumors (Figure 17.16). Ano1 is a chloride channel protein selectively expressed in Cajal cells and GISTs and only sporadically in non-GIST mesenchymal tumors: some smooth muscle tumors and occasional synovial

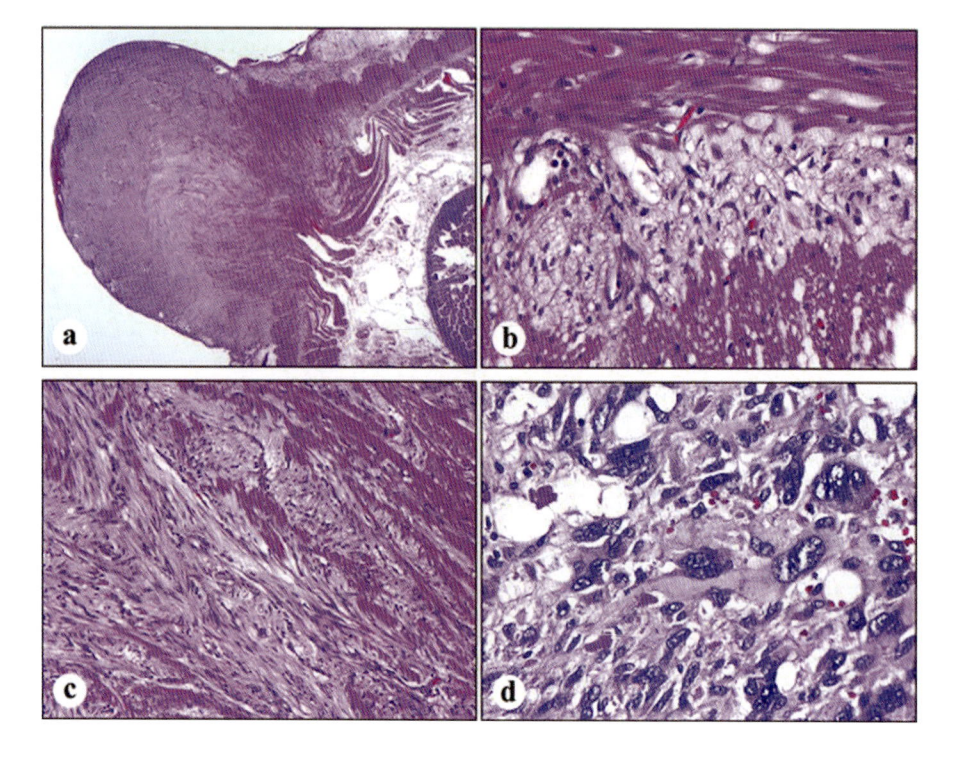

Figure 17.14 (**a**) Histologically, a small jejunal GIST in an NF1 patient forms an externally bulging mass. (**b**) Diffuse Cajal cell hyperplasia is present in the myenteric plexus, seen as a pale staining cellular zone between the smooth muscle layers. (**c,d**) Spindle cell GIST infiltrating between muscularis propria elements.

Chapter 17: Gastrointestinal stromal tumor (GIST)

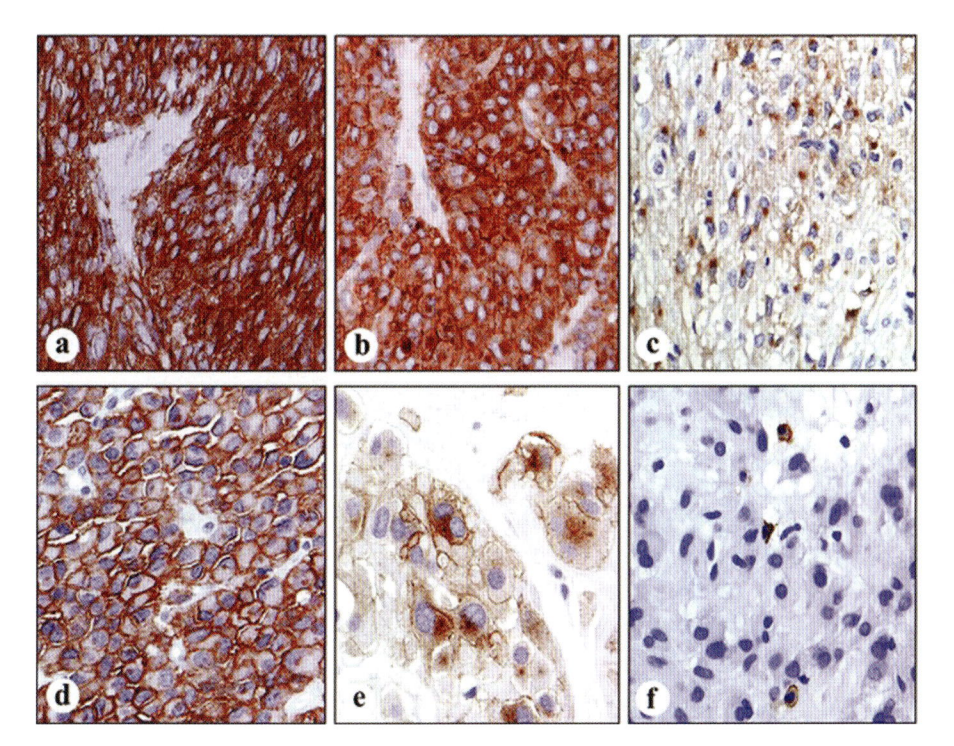

Figure 17.15 Patterns of immunohistochemical KIT positivity in GISTs. (**a**) Strong apparently pancytoplasmic positivity is the most common pattern. (**b**) Some GISTs show perinuclear (Golgi zone) accentuation of KIT positivity. (c) A few GISTs have a predominantly perinuclear Golgi zone distribution. (**d**) Membrane staining is seen in some, especially epithelioid GISTs. (**e**) Some epithelioid GISTs show spotty KIT positivity. (**f**) Few GISTs (usually with epithelioid morphology) are entirely KIT negative; only mast cells are positive.

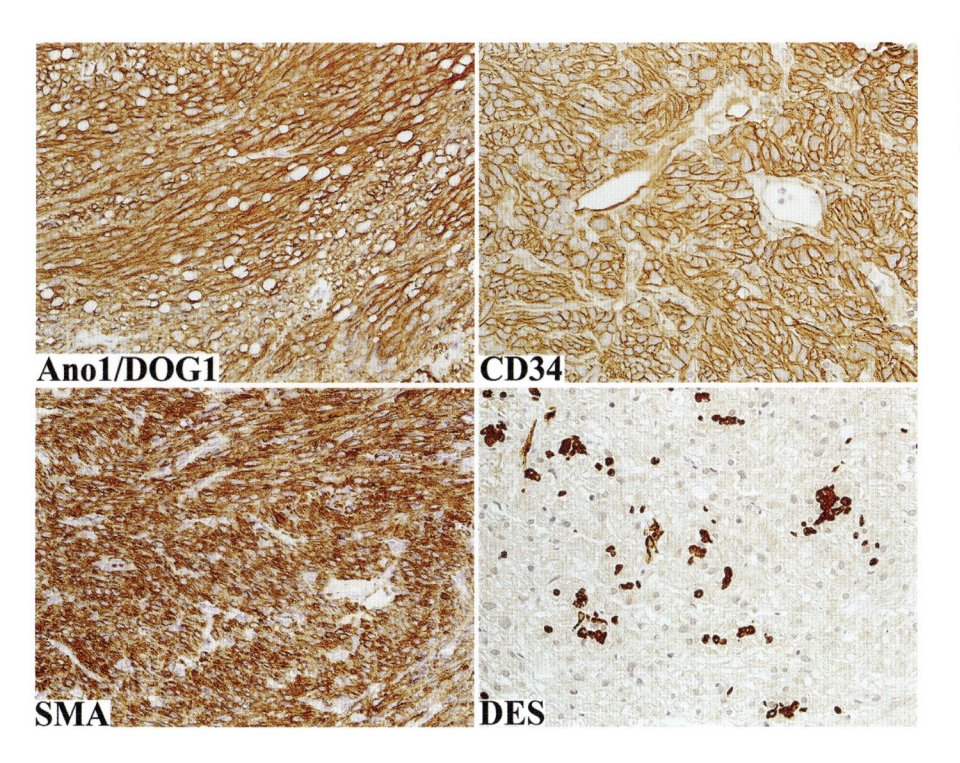

Figure 17.16 The vast majority of GISTs are positive for Ano1-DOG1 and a majority also for CD34. Smooth muscle actin positivity occurs in 30% of GISTs. Some GISTs, especially epithelioid examples, tumor cells are focally desmin positive.

sarcomas. However, it is commonly expressed in carcinomas, especially squamous ones.[55,56]

PDGFRA immunohistochemistry is not useful in the diagnosis of GISTs as it is rather widespread in different types of tumors. Alternative markers include protein kinase C theta, a downstream effector in the KIT signaling pathway, but specificity is low.[57]

CD34 is expressed in most spindle GISTs. In epithelioid GISTs, a distinctive membrane staining is often observed (Figure 17.16). However, this marker is not specific for GISTs and is expressed in potential GIST mimics such as solitary fibrous tumor and dedifferentiated liposarcoma. SMA positivity is present in approximately 20% of gastric and 35% of small intestinal GISTs. The positivity varies from focal to extensive

Table 17.2 Summary of the clinical significance of the most important KIT and PDGFRA mutations in GIST

Mutation type	Overall frequency in GISTs (%)	Comments
KIT mutations (all)	70–75	
Exon 11 deletions/deletion-insertions	45	Gastric GISTs with exon 11 deletions are more aggressive than those with exon 11 substitutions (gastric GISTs)
Exon 11 substitutions	10–15	Usually involve codons 557, 559, 560, and 576 Tumors more indolent than KIT-deletion mutants (gastric GISTs)
Exon 11 duplications	5	Predilection for gastric GISTs Tendency toward favorable prognosis in gastric GISTs
Exon 9 duplication	5	Usually AlaTyr 502–503 duplication. Predilection to intestinal GISTs, exceptional in gastric GISTs Reduced imatinib sensitivity, dose escalation recommended
Exon 13 substitutions	1	Rare, limited experience
Exon 17 substitutions	1	Rare, limited experience
PDGFRA	10–15	Essentially restricted to gastric GISTs, and often occur in GISTs with epithelioid morphology Also reported in duodenum
Exon 12 deletions and substitutions	<5	
Exon 14 substitutions	1	
Exon 18 substitutions and deletions	10	Majority of PDGFRA mutations Most common variant D842V mutants are imatinib resistant.
No KIT or PDGFRA mutation	15–20	Typical finding in GISTs in neurofibromatosis 1 patients and children

(Figure 17.16). SMA-positive gastric and small intestinal GISTs may be prognostically more favorable than the negative tumors.[25,27]

Desmin expression is relatively rare and is usually focal. In most GISTs, desmin-positive elements are entrapped smooth muscle cells (Figure 17.16). Approximately 5% of gastric GISTs, especially the epithelioid ones, contain scattered desmin-positive cells.[25] Heavy-caldesmon positivity is common in GISTs, being present in 50% to 60% of cases.[58] This marker is therefore not applicable in the differential diagnosis between GISTs and true smooth muscle tumors.

Keratin positivity (specifically keratin 18, and to lesser degree, keratin 8) occurs in a minority of GISTs; however, GISTs are generally negative when tested with AE1/AE3 keratin antibodies. Various neural markers, such as neurofilament 68 have also been reported, but their significance is uncertain.

KIT and PDGFRA mutations

Somatic KIT mutations occur in 60% to 70% of gastric GISTs, and a small number of families with constitutional mutations have been reported. KIT mutations usually occur in exon 11 and most commonly represent in-frame deletions/deletion-insertions, single nucleotide substitutions (point mutations), and duplications. Most of these mutations occur in the 5′ region of exon 11, involving codons 550–560. These mutations are believed to disrupt the anti-dimerization motif of GIST causing spontaneous ligand-independ KIT activation (phosphorylation), one of the key driving forces of KIT mutant GIST. Tumors with the latter two types of mutations have a statistically better prognosis than those with deletions, especially in the stomach (Table 17.2).[18,19]

PDGFRA mutations are essentially limited to gastric GISTs. They often occur in epithelioid GISTs and usually involve exon 18. The most common mutation is D842V substitution. Occasional epithelioid gastric GISTs have PDGFRA exon 12 or 14 mutations (in frame deletions or substitutions). Most patients with GISTs with PDGRA mutations have a favorable course.[18,19]

Secondary KIT mutations mainly in kinase domains during imatinib mesylate therapy cause drug resistance and are a substantial problem limiting long-term success of this treatment. In some cases, D842V PDGFRA mutation has been reported as a secondary mutation following imatinib therapy.[19,59]

Familial gastrointestinal stromal tumor syndrome

Inheritable GIST-specific heterozygous germline KIT or PDGFRA mutation (the latter being described in one family)

results in a familial GIST syndrome that manifests with multiple GISTs and is inherited in an autosomal dominant pattern. About 20 such families have been reported from Japan, North America, and Europe.[18,19,60–65] The familial GIST syndrome has been reproduced in three mouse models, where introduced germline mutations KIT V558D del and a mutation corresponding to human V652K, and D820Y caused multiple GISTs with a high penetrance in transgenic mice.[20–22] In some families, additional manifestations of KIT activation have included mastocytoma, cutaneous hyperpigmentation, and esophageal dysmotility.

The tumors in familial GIST syndrome usually manifest in middle age and rarely in early life. Both gastric and intestinal GISTs appear, and diffuse Cajal cell hyperplasia is a typical finding. In some segments of the intestine, this can evolve into diffuse, longitudinally extensive GIST involvement. The clinical course is often long, but ultimately many patients develop malignant GISTs and die of the tumors. Experience with tyrosine kinase inhibitor treatment is limited. Tumor multiplicity and diffuse nature often makes curative surgery impossible.

SDH-deficient GISTs are also forms of familial GISTs, and these were discussed above.

Genetic changes other than KIT and PDGFRA mutations

Very rare GISTs (<1–2%) have BRAF mutations, typically the V600E mutant, which is also common in malignant melanoma, colon carcinoma, and many other tumors. BRAF V600E mutant GISTs typically occur in the small intestine.[66] One small series suggests that these tumors are relatively indolent.[67] The significance of detection of BRAF mutation is that BRAF kinase inhibitor might be clinically useful, and indeed in one case activity was reported, but ultimately complicated with other mutations in PIK3CA and CDKN2A genes.[68]

As KIT mutations already seem to occur in the smallest of GISTs, which usually never develop into clinical disease, additional genetic changes must occur before a clinically manifest GIST develops. These changes are currently less well understood. The present understanding of the genetic changes in GISTs is based mainly on data from cytogenetics and comparative genomic hybridization. Common cytogenetic changes include losses of chromosomes 1p, 9p, 14, 15, and 22,[69] and comparative genomic hybridization studies have shown corresponding losses. The number of losses is greater, and chromosomal gains are more common in malignant GISTs.[70,71]

Differential diagnosis of GIST

True smooth muscle tumors in the GI tract include leiomyomas and leiomyosarcomas. The former most commonly occur in the esophagus as intramural tumors, and in the muscularis mucosae of the colon and rectum. They are paucicellular spindle cell tumors with histologic resemblance to normal smooth muscle. True leiomyosarcomas also resemble differentiated smooth muscle cells and are positive for smooth muscle actin and desmin. They are very rare in the stomach and small intestine, but seem to be relatively more common in the colon and rectum.

Schwannomas are rare in the GI tract, where they are most common in stomach, followed by the colon, and very rare in the small intestine. They are often surrounded by a discontinuous lymphoid cuff and composed of spindle cells with vague nuclear palisading. They are strongly S100 and usually GFAP positive and are negative for KIT, SMA, and desmin, and usually are also negative for CD34.

Intra-abdominal desmoid fibromatosis forms mesenteric masses that can also involve the gastric and intestinal walls. These tumors are usually more collagenous and less cellular, and they are negative for KIT.

Inflammatory myofibroblastic tumor occurs especially in children. This tumor can form a GIST-like transmural mass in the stomach or intestines, although it is more often omental or mesenteric. The spindled tumor cells are typically admixed with lymphocytes and plasma cells and are typically positive for ALK and variably positive for actins, but they are negative for KIT.

Dedifferentiated liposarcoma can histologically mimic GIST. However, these tumors typically occur in the retroperitoneum and are immunohistochemically negative for KIT usually showing nuclear MDM2 positivity.

There is a small, somewhat problematic group of tumors that is in the histologic range, but does not express KIT. These are typically high-grade, undifferentiated mesenchymal tumors that are also negative for muscle markers and CD34.

Nonepithelial tumors that are often KIT positive, but must be separated from GISTs include metastatic melanoma, small cell carcinoma (pulmonary), angiosarcoma, and Ewing sarcoma.

Comments on treatment

Complete surgical excision is sufficient for smaller low-risk tumors. Because GISTs less than 5 cm (in stomach ≤10 cm) with low mitotic rates have a generally benign behavior with low/very low metastatic rates (<4%), these tumors could be managed by resection alone. In an NCI-sponsored trial, however, imatinib mesylate has been used for adjuvant treatment on a trial basis in tumors larger than 3 cm.[72]

Metastatic and unresectable GISTs are now routinely treated with KIT/PDGFRA tyrosine kinase inhibitors, of which the first-line agent is imatinib mesylate. Practice for adjuvant treatment for high-risk and possibly other GISTs is being formulated in clinical trials. Most GISTs are primarily imatinib sensitive, although some mutants are primarily resistant, especially PDGFRA D842V mutants and some exon 13 and exon 17 mutants. Patients with exon 9 mutant GISTs may benefit from higher imatinib doses. Because resistance to imatinib often occurs in 1 to 2 years, second-line inhibitors, such as sunitinib malate and others, are used. Several other experimental targeted modalities are being tested, including abolition of KIT, heat shock protein inhibitors, and mammalian target of rapamycin (mTOR) inhibitors, such as everolimus and sirolimus.[59,72]

Treatment of SDH-deficient GISTs is evolving. These tumors do not generally respond well to tyrosine kinase inhibitors so that alternative treatment modalities have been sought. These include inhibitors of IGF1R signaling, such as vandetanib, which is now in trial phase.[72]

Treatment can induce morphological changes. Most commonly imatinib-treated GISTs with a good response become paucicellular with abundant myxoid matrix.[73] In some cases, therapy-resistant tumors undergo dedifferentiation, which can take a form of KIT-negative rhabdomyosarcomatous or spindle cell sarcoma.[74,75] In some cases, such dedifferentiation has also occurred without kinase inhibitor therapy, so that it may be a rare event in natural history of GIST.[75]

References

1. Hirota S, Isozaki K, Moriyama Y, et al. Gain-of-function mutations of ckit in human gastrointestinal stromal tumors. Science 1998;279:577–580.
2. Mazur M, Clark HB. Gastric stromal tumors: reappraisal of histogenesis. Am J Surg Pathol 1983;7:507–519.
3. Lee JR, Joshi V, Griffin JW Jr, Lasota J, Miettinen M. Gastrointestinal autonomic nerve tumor: immunohistochemical and molecular identity with gastrointestinal stromal tumor. Am J Surg Pathol 2001;25:979–987.
4. Nilsson B, Bumming P, Meis-Kindblom JM, et al. Gastrointestinal stromal tumors: the incidence, prevalence, clinical course, and prognostication in the preimatinib mesylate era – a population-based study in western Sweden. Cancer 2005;103:821–829.
5. Tryggvason G, Gislason HG, Magnusson MK, Jonasson JG. Gastrointestinal stromal tumors in Iceland, 1990–2003: the Icelandic GIST study, a population-based incidence and pathologic risk stratification study. Int J Cancer 2005;117:289–293.
6. Steigen SE, Eide TJ. Trends in incidence and survival of mesenchymal neoplasm of the digestive tract within a defined population of northern Norway. APMIS 2006;114:192–200.
7. Kawanowa K, Sakuma Y, Sakurai S, et al. High incidence of microscopic gastrointestinal stromal tumors in the stomach. Hum Pathol 2006;37:1527–1535.
8. Abraham SC, Krasinskas AM, Hofstetter WL, Swisher SG, Wu TT. "Seedling" mesenchymal tumors (gastrointestinal stromal tumors and leiomyomas) are common incidental tumors of the gastroesophageal junction. Am J Surg Pathol 2007;31:1629–1635.
9. Agaimy A, Wunsch PH, Hofstaedter F, et al. Minute gastric sclerosing stromal tumors (GIST tumorlets) are common in adults and frequently show KIT mutations. Am J Surg Pathol 2007;31:113–120.
10. Miettinen M, Lasota J. Gastrointestinal stromal tumors: pathology and prognosis at different sites. Semin Diagn Pathol 2006;23:70–83.
11. DeMatteo RP, Lewis JJ, Leung D, et al. Two hundred gastrointestinal stromal tumors: recurrence patterns and prognostic factors for survival. Ann Surg 2000;231:51–58.
12. Kindblom LG, Remotti HE, Aldenborg F, Meis-Kindblom JM. Gastrointestinal pacemaker cell tumor (GIPACT). Gastrointestinal stromal tumors show phenotypic characteristics of the interstitial cells of Cajal. Am J Pathol 1998;153:1259–1269.
13. Maeda H, Yamagata A, Nishikawa S, et al. Requirement of c-kit for development of intestinal pacemaker system. Development 1992;116:369–375.
14. Huizinga JD, Thuneberg L, Klüppel M, et al. W/kit gene required for interstitial cells of Cajal and for intestinal pacemaker activity. Nature 1993;373:347–349.
15. Torihashi S, Nishi K, Tokutomi Y, et al. Blockade of kit signaling induces transdifferentiation of interstitial cells of Cajal to a smooth muscle phenotype. Gastroenterology 1999;117:140–148.
16. Sakurai S, Fukusawa T, Chong JM, Tanaka A, Fukuyama M. Embryonic form of smooth muscle myosin heavy chain (SEmb/MCH-B) in gastrointestinal stromal tumor and interstitial cells of Cajal. Am J Pathol 1999;154:23–28.
17. Robinson TL, Sircar K, Hewlett BR, et al. Gastrointestinal stromal tumors may originate from a subset of CD34-positive interstitial cells of Cajal. Am J Pathol 2000;156:1157–1163.
18. Corless CL, Barnett CM, Heinrich MC. Gastrointestinal stromal tumours: origin and molecular oncology. Nat Rev Cancer 2011;11:865–878.
19. Lasota J, Miettinen M. Clinical significance of KIT and PDGFRA mutations in gastrointestinal stromal tumours. Histopathology 2008;53:245–266.
20. Sommer G, Agosti V, Ehlers I, et al. Gastrointestinal stromal tumors in a mouse model by targeted mutation of the Kit receptor tyrosine kinase. Proc Natl Acad Sci USA 2003;100:6706–6711.
21. Rubin BP, Antonescu CR, Scott-Browne JP, et al. A knock-in mouse model of gastrointestinal stromal tumor harboring kit K641E. Cancer Res 2005;65:6631–6639.
22. Nakai N, Ishikawa T, Nishitani A, et al. A mouse model of a human multiple GIST family with KIT-Asp820Tyr mutation generated by a knock-in strategy. J Pathol 2008;214:302–311.
23. Miettinen M, Sarlomo-Rikala M, Sobin LH, Lasota J. Esophageal stromal tumors: a clinicopathologic, immunohistochemical and molecular genetic study of seventeen cases and comparison with esophageal leiomyomas and leiomyosarcomas. Am J Surg Pathol 2000;23:121–132.
24. Wong NACS, Young R, Malcomson RDG, et al. Prognostic indicators for gastrointestinal stromal tumours: a clinicopathological and immunohistochemical study of 108 resected cases of the stomach. Histopathology 2003;43:118–126.
25. Miettinen M, Sobin LH, Lasota J. Gastrointestinal stromal tumors of the stomach: a clinicopathologic, immunohistochemical, and molecular genetic study of 1765 cases

with long-term follow-up. *Am J Surg Pathol* 2005;29:52–68.

26. Miettinen M, Kopczynski J, Maklouf HR, et al. Gastrointestinal stromal tumors, intramural leiomyomas and leiomyosarcomas in the duodenum: a clinicopathologic, immunohistochemical and molecular genetic study of 167 cases. *Am J Surg Pathol* 2003;27:625–641.

27. Miettinen M, Makhlouf HR, Sobin LH, Lasota J. Gastrointestinal stromal tumors of the jejunum and ileum: a clinicopathologic, immunohistochemical and molecular genetic study of 906 cases prior to imatinib with long-term follow-up. *Am J Surg Pathol* 2006;30:477–489.

28. Min K-W. Small intestinal stromal tumors with skeinoid fibers: clinicopathological, immunohistochemical, and ultra-structural investigations. *Am J Surg Pathol* 1992;16:145–155.

29. Miettinen M, Sarlomo-Rikala M, Sobin LH, Lasota, J. Gastrointestinal stromal tumors and leiomyosarcomas in the colon: a clinicopathologic, immunohistochemical and molecular genetic study of 44 cases. *Am J Surg Pathol* 2000;24:1339–1352.

30. Miettinen M, Sobin LH. Gastrointestinal stromal tumors in the appendix: a clinicopathologic and immunohistochemical study of four cases. *Am J Surg Pathol* 2001;25:1433–1437.

31. Miettinen M, Furlong M, Sarlomo-Rikala M, et al. Gastrointestinal stromal tumors, intramural leiomyomas, and leiomyosarcomas in the rectum and anus: a clinicopathologic, immunohistochemical, and molecular genetic study of 144 cases. *Am J Surg Pathol* 2001;25:1121–1133.

32. Miettinen M, Sobin LH, Lasota J. Gastrointestinal stromal tumors presenting as omental masses: a clinicopathologic analysis of 95 cases. *Am J Surg Pathol* 2009;33:1267–1275.

33. Reith JD, Goldblum JR, Lyles RH, Weiss SW. Extragastrointestinal (soft tissue) stromal tumors: an analysis of 48 cases with emphasis on histologic predictors of outcome. *Mod Pathol* 2000;13:577–585.

34. Agaimy A, Wünsch PH. Gastrointestinal stromal tumours: a regular origin in the muscularis propria, but an extremely diverse gross presentation: a review of 200 cases to critically re-evaluate the concept of so-called extra-gastrointestinal stromal tumours. *Langenbecks Arch Surg* 2006;391:322–329.

35. Janeway KA, Kim SY, Lodish M, et al. Defects in succinate dehydrogenase in gastrointestinal stromal tumors lacking KIT and PDGFRA mutations. *Proc Natl Acad Sci USA* 2011;108:314–318.

36. Gill AJ. Succinate dehydrogenase (SDH) and mitochondrial driven neoplasia. *Pathology* 2012;44:285–292.

37. Miettinen M, Wang ZF, Sarlomo-Rikala M, et al. Succinate dehydrogenase-deficient GISTs: a clinicopathologic, immunohistochemical, and molecular genetic study of 66 gastric GISTs with predilection to young age. *Am J Surg Pathol* 2011;35:1712–1721.

38. Carney JA. Gastric stromal sarcoma, pulmonary chondroma, and extra-adrenal paraganglioma (Carney triad): natural history, adrenocortical component, and possible familial occurrence. *Mayo Clin Proc* 1999;74:543–552.

39. Carney JA, Stratakis CA. Familial paraganglioma and gastric stromal sarcoma: a new syndrome distinct from the Carney triad. *Am J Med Genet* 2002;108:132–139.

40. Italiano A, Chen CL, Sung YS, et al. SDHA loss of function mutations in a subset of young adult wild-type gastrointestinal stromal tumors. *BMC Cancer* 2012;12:408.

41. Wagner AJ, Remillard SP, Zhang YX, et al. Loss of expression of SDHA predicts SDHA mutations in gastrointestinal stromal tumors. *Mod Pathol* 2013;26:289–294.

42. Miettinen M, Killian JK, Wang ZF, et al. Immunohistochemical loss of succinate dehydrogenase subunit A (SDHA) in gastrointestinal stromal tumors (GISTs) signals SDHA germline mutation. *Am J Surg Pathol* 2013;37:234–240.

43. Tarn C, Rink L, Merkel E, et al. Insulin-like growth factor 1 receptor is a potential therapeutic target for gastrointestinal stromal tumors. *Proc Natl Acad Sci USA* 2008;105:8387–8392.

44. Pantaleo MA, Astolfi A, Di Battista M, et al. Insulin-like growth factor 1 receptor expression in wild-type GISTs: a potential novel therapeutic target. *Int J Cancer* 2009;125:2991–2994.

45. Lasota J, Wang Z, Kim SY, Helman L, Miettinen M. Expression of the receptor for type i insulin-like growth factor (IGF1R) in gastrointestinal stromal tumors: an immunohistochemical study of 1078 cases with diagnostic and therapeutic implications. *Am J Surg Pathol* 2013;37:114–119.

46. Pantaleo MA, Astolfi A, Urbini M, et al. Analysis of all subunits, SDHA, SDHB, SDHC, SDHD, of the succinate dehydrogenase complex in KIT/PDGFRA wild-type GIST. *Eur J Hum Genet* 2014;22:32–39.

47. Killian JK, Miettinen M, Walker RL, et al. Recurrent epimutation of SDHC in gastrointestinal stromal tumors. *Sci Transl Med* 2014;6:268ra177.

48. Killian JK, Kim SY, Miettinen M, et al. Succinate dehydrogenase mutation underlies global epigenomic divergence in gastrointestinal stromal tumor. *Cancer Discov* 2013;3:648–657.

49. Andersson J, Sihto H, Meis-Kindblom JM, et al. NF1-associated gastrointestinal stromal tumors have unique clinical, phenotypic, and genotypic characteristics. *Am J Surg Pathol* 2005;29:1170–1176.

50. Miettinen M, Fetsch JF, Sobin LH, Lasota J. Gastrointestinal stromal tumors in patients with neurofibromatosis. 1: A clinicopathologic study of 45 patients with long-term follow-up. *Am J Surg Pathol* 2006;30:90–96.

51. Kinoshita K, Hirota S, Isozaki K, et al. Absence of c-kit gene mutations in gastrointestinal stromal tumours from neurofibromatosis type 1 patients. *J Pathol* 2004;202:80–85.

52. Maertens O, Prenen H, Debiec-Rychter M, et al. Molecular pathogenesis of multiple gastrointestinal stromal tumors in NF1 patients. *Hum Mol Genet* 2006;15:1015–1023.

53. Sarlomo-Rikala M, Kovatich A, Barusevicius A, Miettinen M. CD117: a sensitive marker for gastrointestinal stromal tumors that is more specific than CD34. *Mod Pathol* 1998;11:728–734.

54. Hornick JL, Fletcher CD. The significance of KIT (CD117) in gastrointestinal stromal tumors. *Int J Surg Pathol* 2004;12:93–97.

55. Espinosa I, Lee CH, Kim MK, et al. A novel monoclonal antibody against DOG1 is a sensitive and specific marker for gastrointestinal stromal tumors. *Am J Surg Pathol* 2008;32(2):210–218.

56. Miettinen M, Wang ZF, Lasota J. DOG1 antibody in the differential diagnosis of gastrointestinal stromal tumors: a study of 1840 cases. *Am J Surg Pathol* 2009;33:1401–1408.

57. Ou WB, Zhu MJ, Demetri GD, Fletcher CD, Fletcher JA. Protein kinase C-theta regulates KIT expression and proliferation in gastrointestinal stromal tumors. *Oncogene* 2008;27:5624–5634.

58. Miettinen M, Sarlomo-Rikala M, Kovatich AJ, Lasota J. Calponin and h-caldesmon in soft tissue tumors: consistent h-caldesmon immunoreactivity in gastrointestinal stromal tumors indicates traits of smooth muscle differentiation. *Mod Pathol* 1999;12:1109–1118.

59. Casali PG, Fumagalli E, Gronchi A. Adjuvant therapy of gastrointestinal stromal tumors (GISTs). *Curr Treat Options Oncol* 2012;13:277–284.

60. Nishida T, Hirota S, Taniguchi M, et al. Familial gastrointestinal stromal tumours with germline mutation of the KIT gene. *Nat Genet* 1998;19:323–324.

61. Isozaki K, Terris B, Belghiti J, et al. Germline-activating mutation in the kinase domain of KIT gene in familial gastrointestinal stromal tumors. *Am J Pathol* 2000;157:1581–1585.

62. Beghini A, Tibiletti MG, Roversi G, et al. Germline mutation in the juxtamembrane domain of the KIT gene in a family with gastrointestinal stromal tumors and urticaria pigmentosa. *Cancer* 2001;92:657–662.

63. Maeyama H, Hidaka E, Ota H, et al. Familial gastrointestinal stromal tumor with hyperpigmentation: association with a germline mutation of the c-kit gene. *Gastroenterology* 2001;120:210–215.

64. Chompret A, Kannengiesser C, Barrois M, et al. PDGFRA germline mutation in a family with multiple cases of gastrointestinal stromal tumor. *Gastroenterology* 2004;126:318–321.

65. Robson ME, Glogowski E, Sommer G, et al. Pleomorphic characteristics of a germ-line KIT mutation in a large kindred with gastrointestinal stromal tumors, hyperpigmentation, and dysphagia. *Clin Cancer Res* 2004;10:1250–1254.

66. Hostein I, Faur N, Primois C, et al. BRAF mutation status in gastrointestinal stromal tumors. *Am J Clin Pathol* 2010;133:141–148.

67. Rossi S, Gasparotto D, Miceli R, et al. KIT, PDGFRA, and BRAF mutational spectrum impacts on the natural history of imatinib-naive localized GIST: a population-based study. *Am J Surg Pathol* 2015;39:922–930.

68. Falchook GS, Trent JC, Heinrich MC, et al. BRAF mutant gastrointestinal stromal tumor: first report of regression with BRAF inhibitor dabrafenib (GSK2118436) and whole exomic sequencing for analysis of acquired resistance. *Oncotarget* 2013;4:310–315.

69. Gunawan B, Bergmann F, Hoer J, et al. Biological and clinical significance of cytogenetic abnormalities in low-risk and high-risk gastrointestinal stromal tumors. *Hum Pathol* 2002;33:316–321.

70. El-Rifai W, Sarlomo-Rikala M, Miettinen M, Knuutila S, Andersson LCA. DNA copy number losses in chromosome 14: an early change in gastrointestinal stromal tumors. *Cancer Res* 1996;56:3230–3233.

71. El-Rifai W, Sarlomo-Rikala M, Andersson L, Knuutila S, Miettinen M. Prognostic significance of DNA copy number changes in benign and malignant GISTs. *Cancer Res* 2000;60:3899–3903.

72. van Mehren M. The role of adjuvant and neoadjuvant therapy for gastrointestinal stromal tumors. *Curr Opin Oncol* 2008;20:428–432.

73. Sciot R, Debiec-Rychter M. GIST under imatinib therapy. *Semin Diagn Pathol* 2006;23:84–90.

74. Liegl B, Hornick JL, Antonescu CR, Corless CL, Fletcher CD. Rhabdomyosarcomatous differentiation in gastrointestinal stromal tumors after tyrosine kinase inhibitor therapy: a novel form of tumor progression. *Am J Surg Pathol* 2009;33:218–226.

75. Antonescu CR, Romeo S, Zhang L, et al. Dedifferentiation in gastrointestinal stromal tumor to an anaplastic KIT-negative phenotype: a diagnostic pitfall: morphologic and molecular characterization of 8 cases occurring either de novo or after imatinib therapy. *Am J Surg Pathol* 2013;37:385–392.

Chapter 18

Stromal tumors and tumor-like lesions of the female genital tract

Markku Miettinen
National Cancer Institute, National Institutes of Health

This chapter is devoted to specific mesenchymal tumors of the vulva, vagina, and adjacent soft tissues, and common to most of them is their origin in the hormonally responsive stromal elements, as illustrated by their general estrogen and progesterone receptor positivity. A wide range of other mesenchymal tumors also occurs in this region, but these tumors are not related to the specific stromal tissues and more commonly occur elsewhere in the body. These tumors are discussed in the appropriate chapters. Smooth muscle tumors of the uterus and other female genitalia are included in Chapter 16.

Although most entities in this chapter are specific to women, cellular angiofibroma occurs in women and men. Tumors reported as aggressive angiomyxomas in male patients belong mostly to other entities, especially cellular angiofibroma.

Sarcomas that are usually primary tumors in the uterus, such as endometrial stromal sarcoma (ESS) and malignant mixed müllerian tumor (carcinosarcoma), sometimes arise outside of the genitalia in the pelvis or other abdominal locations. These tumors can also metastasize almost anywhere in the abdominal cavity and sometimes outside it, especially the lungs. Among such tumors is also granulosa cell tumor of the ovary, which can recur after long periods and sometimes simulate an intra-abdominal mesenchymal tumor. The specific recognition of genital stromal tumors is important because some of them respond to hormonal or antihormonal treatment.

Mesodermal stromal polyp (pseudosarcomatous stromal polyp)

These polyps vary from those with bland cytology to those that are highly cellular and even pseudosarcomatous, with atypia and mitotic activity. Nevertheless, all of these behave indolently based on available follow-up information.

Clinical features

These polyps occur in the mucosa of the vagina, vulva, or rarely, the cervix (Figure 18.1). They occur in women of all ages (Figure 18.2), although some published reports have emphasized occurrence in premenopausal women, with a median age between 35 and 45 years in different series. One half of the patients have been asymptomatic, and the remaining patients have had local discomfort or bleeding. A significant number of reported patients have been pregnant, and some have had two or more polyps. The behavior has been benign in published studies, but rare recurrences have been reported.[1-7] Regression of polyps in infant patients have been noted, warranting an extremely conservative management, as long as botryoid-embryonal rhabdomyosarcoma has been ruled out.[1] Because there is some evidence that tumors with more atypical features recur more often,[6] polyps with significant atypia should be excised with normal tissue margins. Additional long-term follow-up studies are necessary to rule out a tumor cohort with more adverse behavior.

Pathology

Grossly, the polyp can be smooth-surfaced and pedunculated, or composed of multiple finger-like polypoid

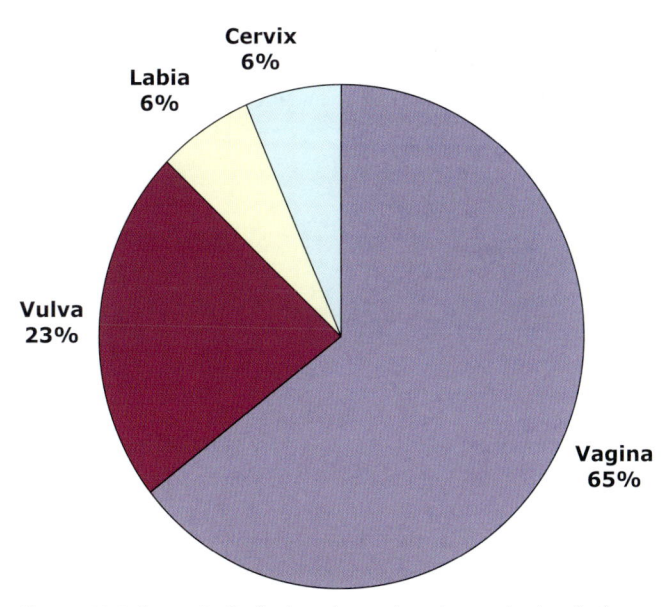

Figure 18.1 Anatomic distribution of mesodermal stromal polyps in the genital tract.

Modern Soft Tissue Pathology, Second Edition, ed. Markku Miettinen. Published by Cambridge University Press. © Cambridge University Press 2016.

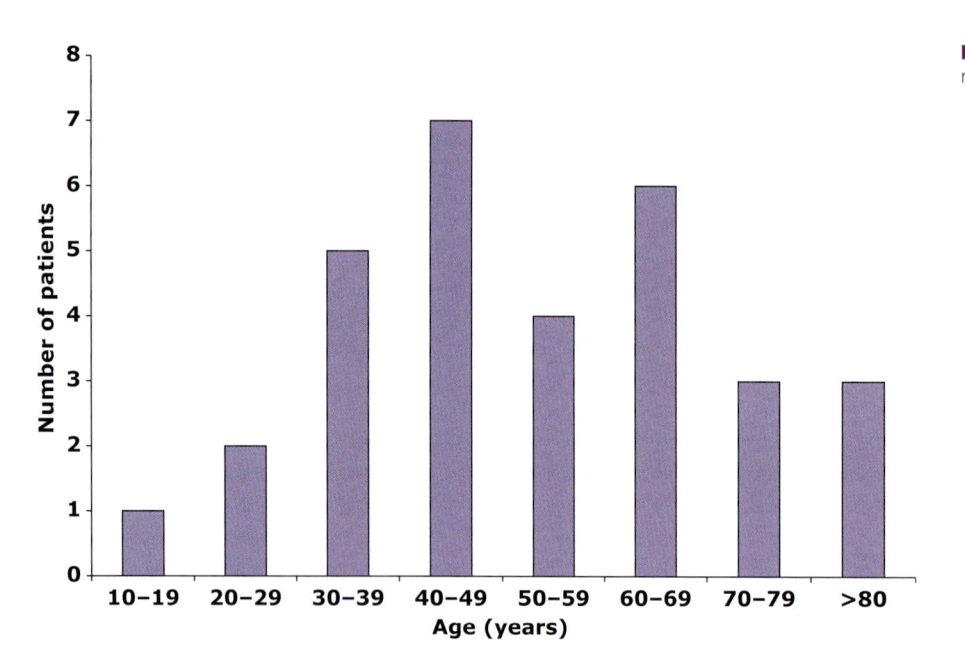

Figure 18.2 Age distribution of 31 women with mesodermal stromal polyps of the genital tract.

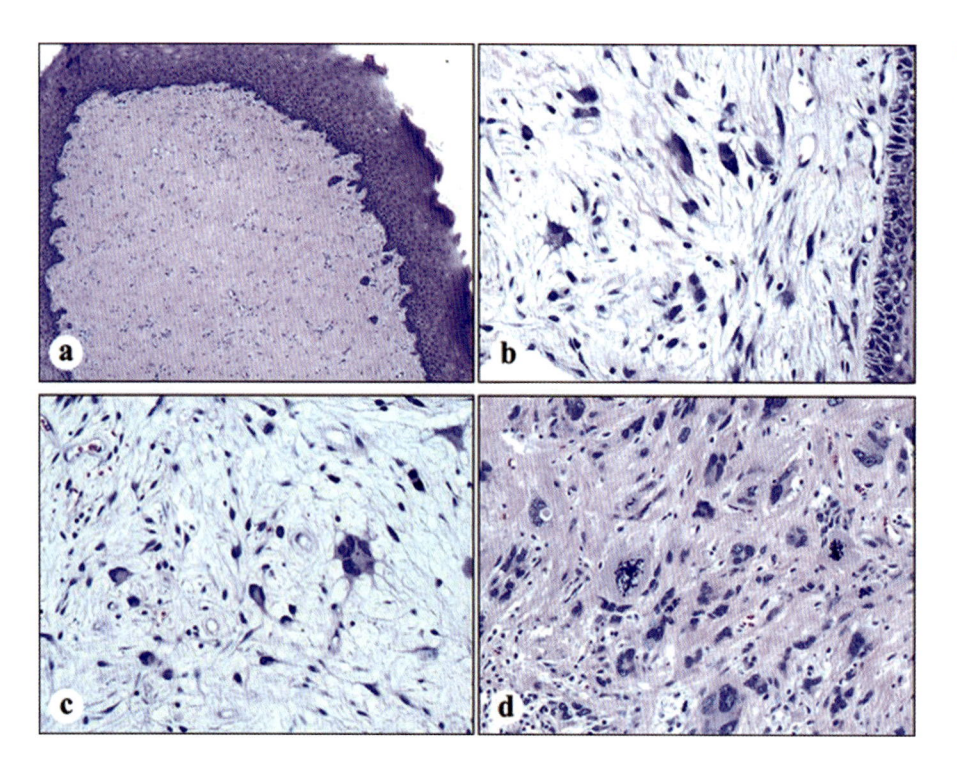

Figure 18.3 Histologic appearances of mesodermal stromal polyps. (**a–c**) Lesion with limited cellularity with numerous multinucleated giant cells and limited atypia. (**d**) A highly cellular example with prominent nuclear atypia and mitotic activity with atypical mitoses.

protrusions, potentially simulating the appearance of botryoid rhabdomyosarcoma.

Histologically, the lesions are heterogeneous, generally occupying the entire subepithelial zone. They can be bland and edematous, as seen in the lesions of infants, or composed of stellate, sometimes multinucleated fibroblasts (Figure 18.3a–c). Some lesions are highly cellular and atypical, having frequent bizarre nuclei, high mitotic activity (over 10 per 10 HPFs) and even atypical mitoses (Figure 18.3d). This variant has been referred to as *pseudosarcomatous stromal polyp* and is histologically analogous to atypical fibroxanthoma of the skin. Such atypical polyps are associated with pregnancy in particular.

Immunohistochemically typical is expression of desmin and estrogen and progesterone receptors, whereas the tumor cells are generally negative for actins, S100 protein, histiocytic markers, and keratins.[5–8]

Differential diagnosis

The desmin-positive mesodermal stromal polyps should not be confused with botryoid rhabdomyosarcoma, which usually occurs in small children, is more highly cellular at the stromoepithelial interface, and contains differentiated rhabdomyoblasts, at least focally. Fortunately, the rarely reported

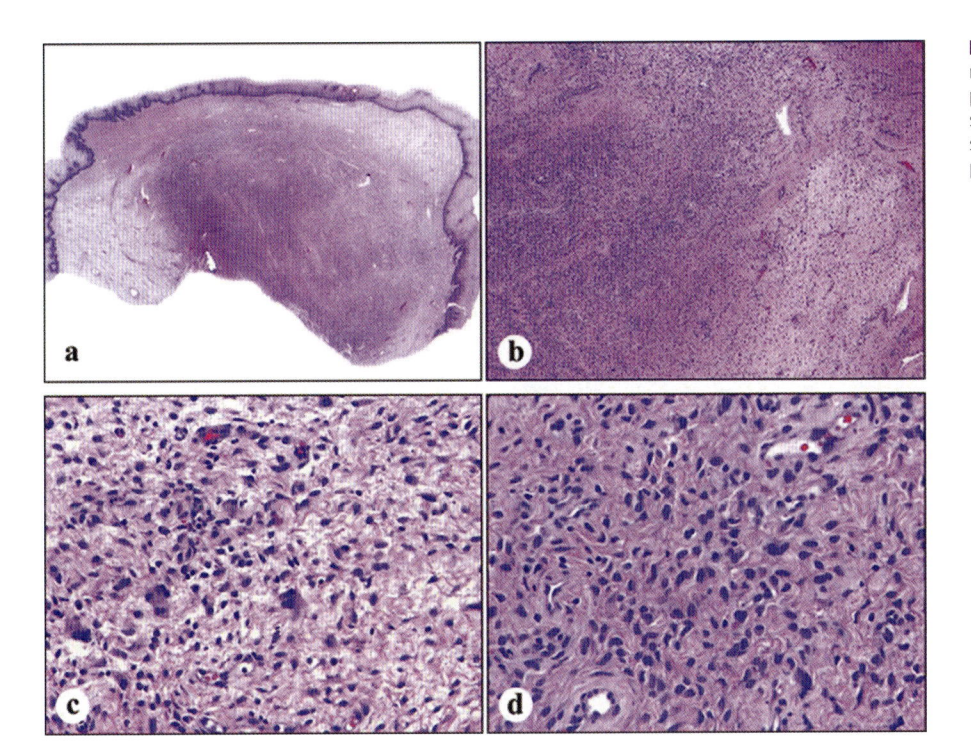

Figure 18.4 Superficial cervicovaginal myofibroblastoma. (**a**) The tumor forms a sessile polypoid lesion in the cervix. (**b**) Low magnification shows moderate cellularity. (**c,d**) The tumor cells show limited atypia, and multinucleated cells are present. (**d**) Numerous mast cells are also present.

polyps in infants have been paucicellular and histologically do not pose a differential diagnostic problem.[1] Genital rhabdomyoma occurs as a polypoid lesion in adult women. These tumors show well-differentiated skeletal muscle cells without mitotic activity.

Superficial cervicovaginal myofibroblastoma forming a broad-based dome-shaped lesion has been previously included among mesodermal stromal polyps, but is now considered to be a separate entity and is discussed later in this chapter.

Pedunculated fibroepithelial polyps of the skin of the labia majora and elsewhere in the genital region represent a different entity, comparable to other fibroepithelial polyps of the skin. These polyps generally have a paucicellular stroma with a fatty component.

Some mucosal melanomas and carcinomas (e.g., metastases from renal cell carcinoma) can form polypoid lesions; exclusion of these entities and establishing the diagnosis of a pseudosarcomatous polyp might require immunohistochemical studies.

Superficial cervicovaginal myofibroblastoma

This lesion has clinicopathologic features distinct from the mesodermal stromal polyp, but such cases have apparently been included in previous series of the latter.

Clinical features

The tumor occurs in older patients than the mesodermal stromal polyp, and 10 of 14 patients in the seminal series were more than 50 years old. The mass is located in the vagina or less commonly in the cervix and usually measures 1 cm to 2 cm, occasionally more than 5 cm. Some cases have been associated with tamoxifen use. Clinical behavior has been benign in all cases, although late recurrences can occur.[9–11]

Pathology

Grossly typical is a broad-based, dome-shaped elevated lesion below the mucosa, which can be hyperplastic or, less commonly, ulcerated. On sectioning, the tumor is glistening, mucoid or fleshy, and varies from white to pink.

Histologically, the lesion is often located beneath a zone of normal superficial stroma. It is composed of alternating cellular collagenous and myxoid areas, sometimes with a lace-like pattern with prominent capillaries (Figure 18.4). The tumor cells are oval, uniform, and have no distinct cell borders. Small amounts of eosinophilic cytoplasm can be seen as bipolar processes, especially in vimentin or desmin immunostains. The nuclei are oval, with an even chromatin pattern and a small nucleolus. Mitotic activity is scant (0–2 per 50 HPFs). Mast cells are often present.

Immunohistochemical analysis shows that the tumor cells are positive for vimentin, desmin, and estrogen and progesterone receptors (Figure 18.5). They are also variably positive for CD34 and occasionally for muscle actins, but are negative for S100 protein, epithelial membrane antibody (EMA), and keratins.

Genital (vaginal) rhabdomyoma

This rare, site-specific benign tumor usually occurs in the vagina. It differs from adult rhabdomyoma, which occurs in the head and neck of adults and contains complex-shaped, typically cross-sectional profiles of skeletal muscle cells. Genital rhabdomyoma has some similarities to fetal rhabdomyoma,

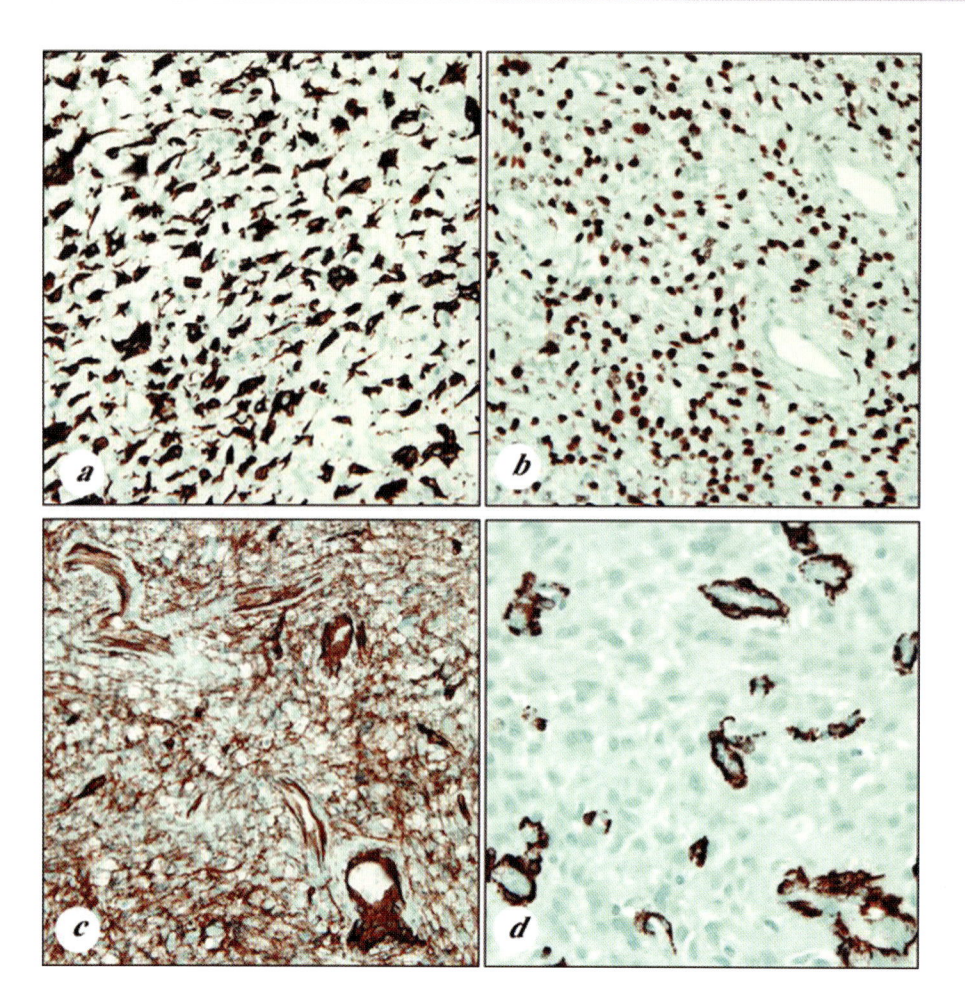

Figure 18.5 Immunohistochemically superficial cervicovaginal myofibroblastoma cells are strongly positive for (**a**) desmin, (**b**) estrogen receptor, (**c**) CD34. (**d**) The tumor cells are negative for alpha smooth muscle actin.

a highly cellular tumor usually seen in the head and neck of children and composed of less-differentiated, elongated rhabdomyoblasts (Chapter 13).

Clinical features

This very rare, histologically distinctive polypoid tumor typically occurs in women of early middle age; fewer than 25 cases have been reported.[12–15] It presents as a small, broad-based mucosal polyp usually in the vagina and occasionally in the vulva or cervix. Clinical symptoms are similar to those of mesodermal stromal polyps, and behavior is benign following a simple excision, although local recurrence developed in one case 4 years after the primary surgery.[15] Apparently similar skeletal muscle differentiation has been reported in the wall of ovarian serous cystadenoma.[16]

Pathology

Grossly, genital rhabdomyoma typically forms a small lobulated polyp measuring 1 cm to 2 cm. Histologically it is lined by normal or hyperplastic squamous mucosa. The stroma, especially superficially, contains lobules or random clusters of well-differentiated but immature striated muscle cells with eosinophilic cytoplasm and appearances of maturing rhabdomyoblasts in a collagenous background (Figure 18.6). Some of these are oval, and others are elongated, probably depending on the plane of section; some have cytoplasmic cross-striations. A prominent stromal vascular pattern is often present. Some of these polyps also contain nonspecific stromal elements without skeletal muscle cells. Mature histology and lack of atypia and mitotic activity separate this tumor from embryonal (botryoid) rhabdomyosarcoma, which in this region usually occurs in small children.

Immunohistochemically, the main lesional cells are positive for desmin and myoglobin; data on hormone receptors, MyoD1, and myogenin are not available. Electron microscopic studies have shown features of differentiated striated muscle cells with well-aligned sarcomeres.

Angiomyofibroblastoma

Clinical features

This histologically distinctive tumor forms a relatively small (usually 2–3 cm), well-circumscribed, asymptomatic mass in the vulva or labia majora, clinically simulating a Bartholin's cyst.[17–23] Rarely, it occurs in the perineum and inguinal area.

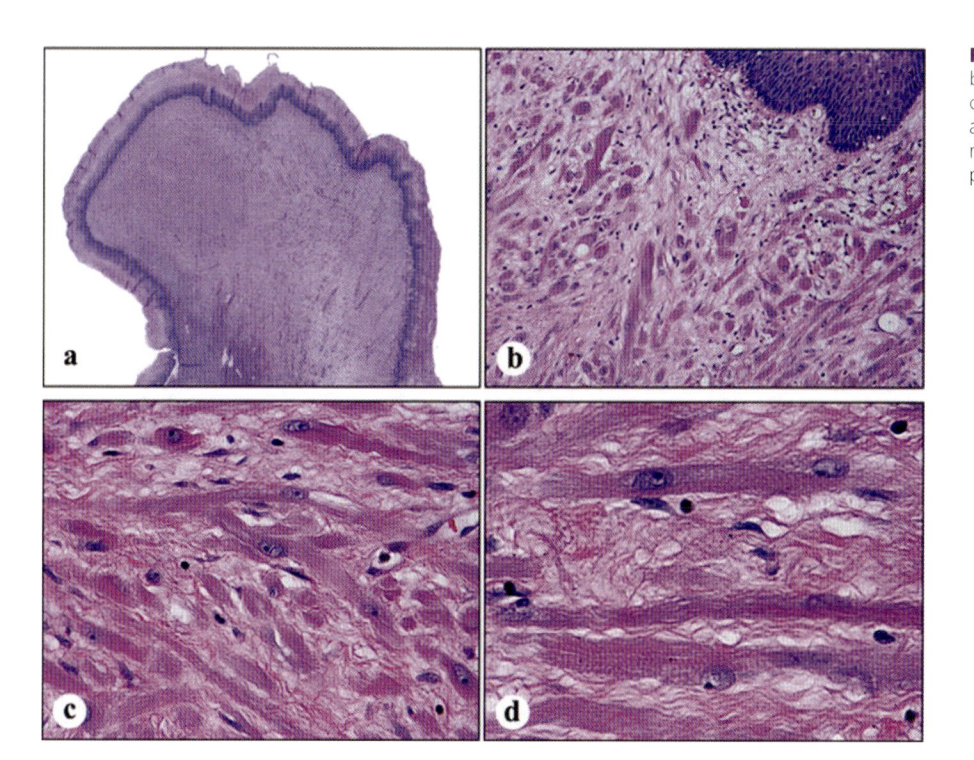

Figure 18.6 Vaginal rhabdomyoma forms a broad-based vaginal polyp with moderate cellularity. Below the epithelium there are partially aligned well-differentiated but immature skeletal muscle cells without atypia. Cross-striations are present.

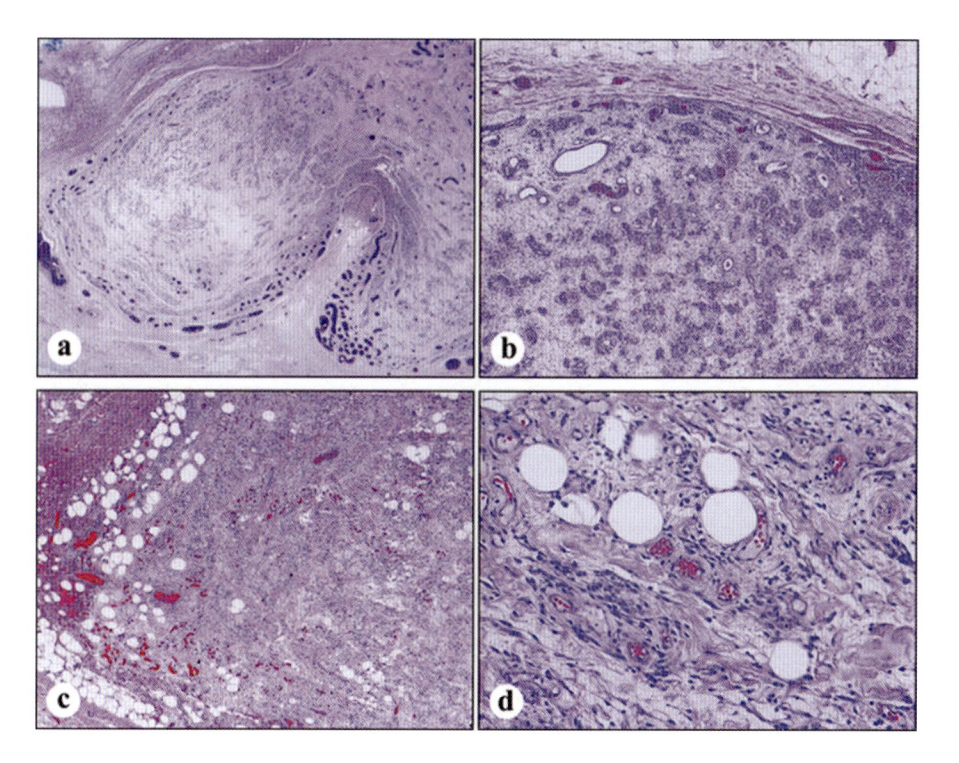

Figure 18.7 (**a,b**) Angiomyofibroblastoma of the vulva forms a sharply demarcated nodule. (**c,d**) This tumor has a fatty component and shows cytologic resemblance to spindle cell lipoma.

The tumor occurs mostly in peri- and postmenopausal women, with the median age in the largest series being 46 years.[21] Simple excision is usually curative, and there is no tendency to recur. A case interpreted as sarcomatous transformation of a recurrent angiomyofibroblastoma was reported in an 80-year-old patient,[24] but the identity of this tumor with typical angiomyofibroblastoma is subject to interpretation.

Pathology

Grossly, the tumor forms a circumscribed, solid, tan nodule measuring 1 cm to 5 cm that is sometimes mucoid on sectioning. Histologically, the tumor is composed of alternating cellular and myxoid areas, and some lesions have an admixed mature lipomatous component (Figure 18.7).

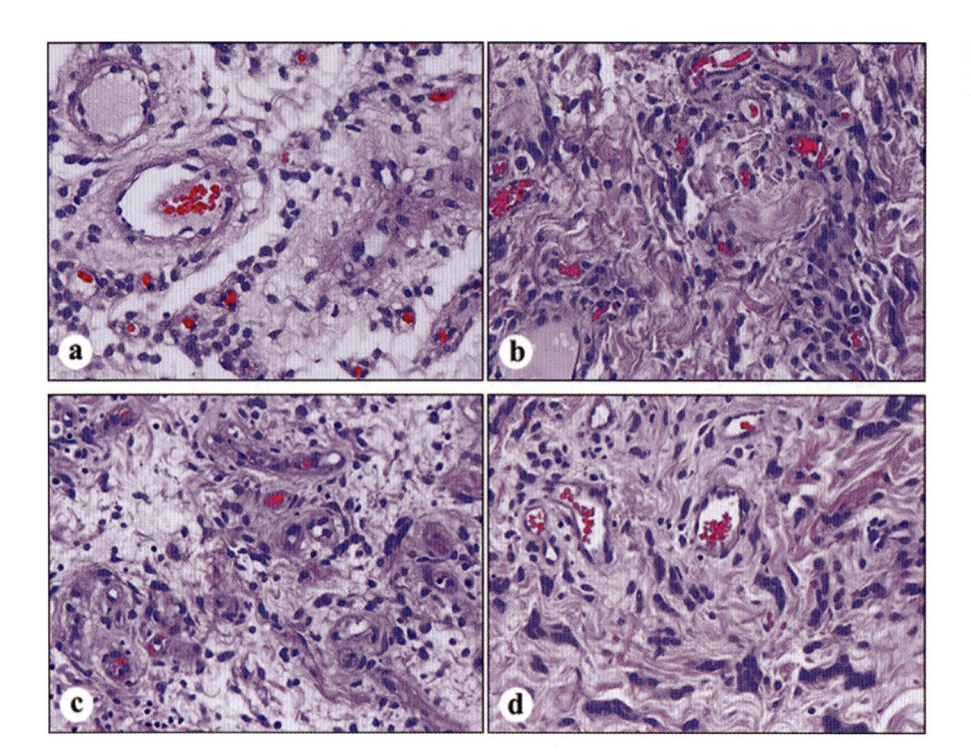

Figure 18.8 Angiomyofibroblastoma with an epithelioid cellular component, which has a tendency to form perivascular arrangements or cords and clusters.

The tumor cells are often grouped as trabeculae, cords, and cohesive nests around prominent, dilated thin-walled or sclerosing capillaries. The lesional cells vary from spindled to epithelioid and can form clusters (Figure 18.8). Some tumor cells have abundant cytoplasm and an eccentric nucleolus, imparting a plasmacytoid appearance. There is only minimal atypia, and the mitotic activity is low.

Immunohistochemically, the tumor cells are positive for vimentin and usually but variably for desmin, but rarely for smooth muscle actin or CD34. They are also positive for estrogen and progesterone receptors. The tumor cells are negative for S100 protein and keratins.[21–23,25] Ultrastructurally, the tumor cells have myofibroblastic features. The genetics are not well understood, but these tumors seem to lack the FOXO1 deletions present in cellular angiofibromas.[26]

Angiomyofibroblastoma-like tumors also occur in men (see next section). These tumors are identical to those reported as cellular angiofibroma.[27,28]

Cellular angiofibroma and male angiomyofibroblastoma-like tumor

These two designations pertain to histologically similar, if not identical, tumors. The former has been described in women and the latter in men in the external genital and inguinal regions. Most tumors reported as aggressive angiomyxomas (AAMs) in men probably belong to this category.

Clinical features

Initially reported in the vulval area of middle-aged women by Nucci et al.,[27] and older men in the scrotal area by Laskin et al.,[28] a series of 51 cases by Iwasa and Fletcher[29] reported similar numbers of tumors in men and women. Based on these studies, the tumors in women are typically smaller (<1 cm to >5 cm, median <3 cm) and have a predilection to the vulva and labia, where this tumor clinically simulates a Bartholin gland cyst. Rare occurrence in the vagina has also been reported.

In contrast, the tumors in men are larger (median size 6 cm, maximum 25 cm in the retroperitoneum) and have a predilection for the scrotal and inguinal region, where they clinically simulate a hernia. Smaller tumors form circumscribed nodules in the vulval region or scrotum. Occurrence in the anogenital region (i.e., anus, perineum) has been reported in both sexes. Long-term follow-up studies on male tumors have revealed one late recurrence after 13 years,[28,29] but the series on female tumors did not report recurrences during a relatively limited follow-up, despite positive margins in some cases.[30] Examples with sarcomatous evolution have been reported. By morphology, these include areas of atypical lipomatous tumor, pleomorphic liposarcoma, and pleomorphic sarcoma, not otherwise specified. Relatively short follow-up did not reveal metastases so that the biologic potential of this group remains somewhat uncertain.[31]

Pathology

Grossly, the tumors are usually circumscribed, lobulated, soft to rubbery, to slightly mucoid masses that often show a reddish surface because of prominent engorged vessels (Figure 18.9). Histologically, they resemble angiomyofibroblastoma, but have more prominent large vessels with frequent perivascular hyalinization and alternating less and more cellular areas, with tapered, uniform spindle cells separated by thin collagen fibers and collagenized areas. Myxoid stroma can also be present (Figures 18.10 and 18.11). The tumor cells generally have a more spindled morphology than angiomyofibroblastomas, with focal epithelioid change in some cases. Moderate mitotic activity was reported in some of the tumors from women (even occasionally exceeding 10 per 10 HPFs), whereas those tumors from men are mitotically less active.

Tumors from both men and women can have a minor fatty component, with some cases showing resemblance to cellular spindle cell lipoma.

Immunohistochemically, the tumor cells are often (>50%) positive for CD34, variably positive for muscle actins and desmin, and negative for S100 protein. Estrogen and progesterone receptor positivity occurs in one half of the tumors in women, but it has also been reported in some tumors found in men.[29]

Differential diagnosis

Cellular angiofibroma differs from angiomyofibroblastoma by a spindled versus epithelioid cellular component without clustered or corded patterns, and by its less common desmin positivity. AAM contains evenly spaced, delicate, oval to spindled cells with generally less prominent vascular pattern, and it is almost always desmin and estrogen receptor positive. Hemangiopericytoma and solitary fibrous tumor have a variable staghorn vascular pattern and densely collagenous stroma. Nerve sheath tumors show variable nuclear palisading (schwannoma) and have prominent S100 protein positivity, which is absent in cellular angiofibroma.

Genetics

Loss of the 13q14 chromosomal region, also detected in spindle cell lipoma, has been observed in two cases.[32,33] Monoallelic loss of the retinoblastoma gene RB1 is associated with this loss.[34]

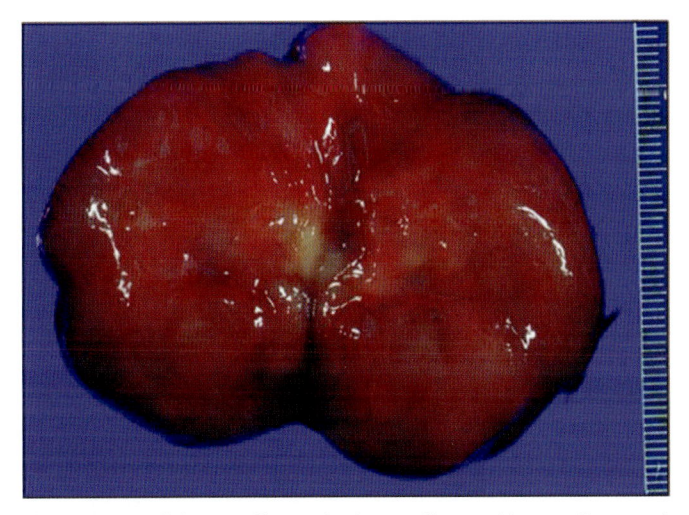

Figure 18.9 Cellular angiofibroma (male angiofibromyoblastoma-like tumor) from the lower pelvis of a man shows a glistening, mucoid, and hemorrhagic surface on sectioning.

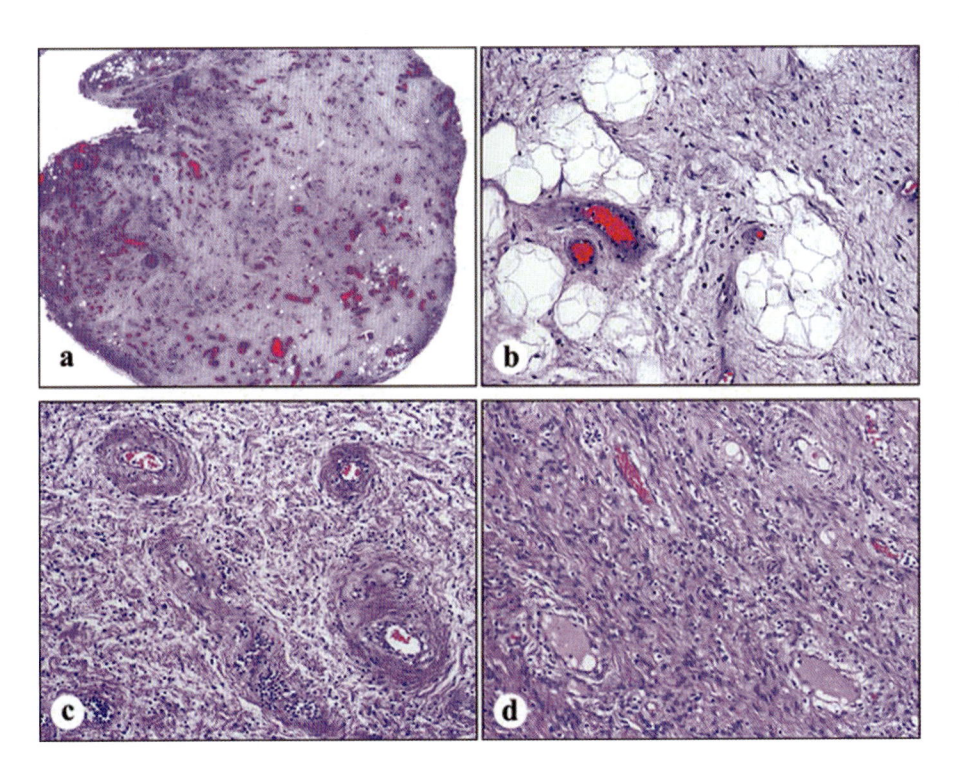

Figure 18.10 (**a**) Cellular angiofibroma forms a demarcated nodule containing large numbers of engorged vessels. (**b**) The tumor shows a fatty component. (**c,d**) The overall appearance shows some resemblance to neurofibroma, and prominent blood vessels are present.

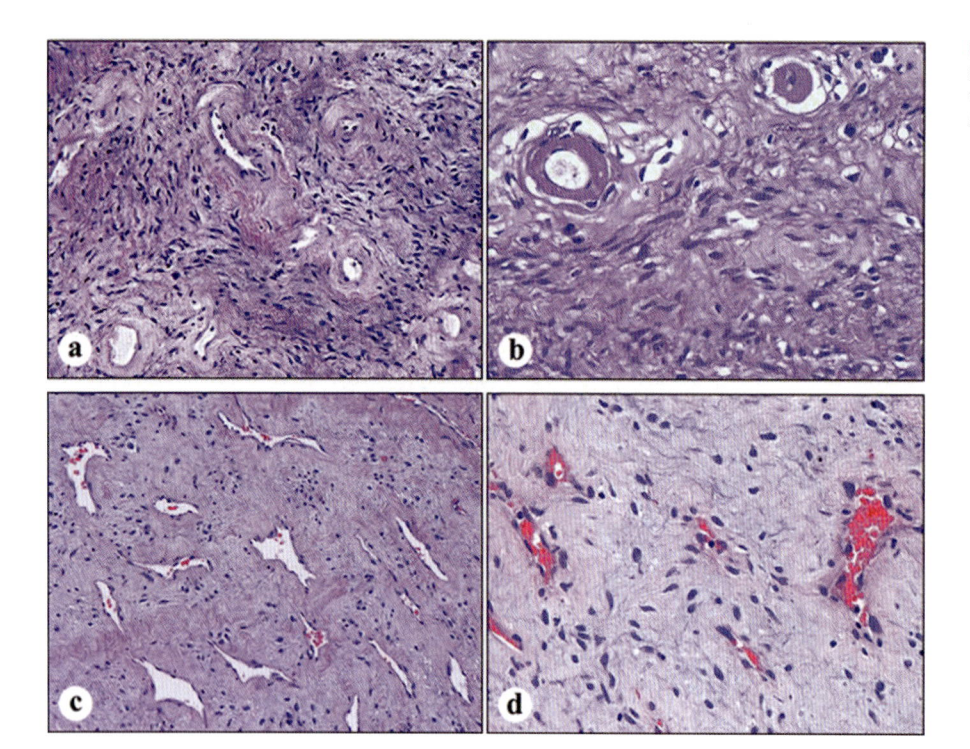

Figure 18.11 Four different examples of cellular angiofibroma show prominent blood vessels, variably myxoid stroma with uniform spindled cells, and scattered mast cells.

Aggressive angiomyxoma

Clinical features

Originally reported in 1983 by Steeper and Rosai, AAM is a specific fibromyxoid tumor of the hormonally sensitive gynecologic stromal tissues.[35] It presents exclusively in adult women from the third decade on, usually less than 50 years of age, however. This indolent tumor seems to persist into old age, however, unless it is surgically excised, and therefore it can be encountered in patients until old age (Figure 18.12). The tumor originates in the perineopelvic region, usually forming a large pelvic mass often extending into the external genital area. It can thereby clinically simulate an inguinal hernia or a Bartholin's cyst.[35–41]

AAM is usually large, more than 10 cm in diameter and sometimes much larger, filling the whole pelvic cavity. It typically has infiltrative margins at the bordering fat. The tumor has a moderate tendency to recur, often a long time after the primary surgery (in 30–40% of cases), but it does not metastasize. Given the problems of radical resection in a woman of childbearing age, incomplete excision is acceptable in some cases, supplemented by treatments such as gonadotropin releasing hormone agonist.[42] In some cases, surgery has been performed followed by selective embolization of the tumor, and in some cases, incomplete excision has been supplemented by postoperative radiation.[43]

Tumors reported as AAMs in men in the scrotum, pelvis, and inguinal region[44–46] seem to be mostly different, representing other entities, such as cellular angiofibroma (see previous section). These tumors have occurred in men between the ages of 18 and 70 years, with a median of 52 years, older than the median age of AAM. More than one half of the tumors have been located in the scrotum, and others in the spermatic cord, perineum, or groin. Some of them were reportedly actin positive, but desmin negative, differing from the typical AAM. Occasionally in literature AAM is applied as a diagnostic term for tumors in nongynecological tract locations; these tumors represent unrelated entities, such as other fibromyxoid or lipomatous neoplasms.

Pathology

On gross inspection, AAM can appear as an apparently circumscribed or complex, multilobular solid mass sometimes showing finger-like projections representing tumor infiltration into soft tissues surrounding the vagina and rectum. On sectioning, AAM is gray-tan and rubbery or rubbery-myxoid, with a glistening, sometimes slightly mucoid surface (Figure 18.13).

Microscopically, the tumor has an infiltrative margin to the surrounding fat and Bartholin's glands, with nerve entrapment occurring frequently. It is relatively paucicellular and highly vascular, with dilated, often hyperemic thick-walled vessels. Recurrent tumors are sometimes more cellular than the primary ones. Small clusters or fascicles of well-differentiated smooth muscle cells that radiate outward from the vessels often surround the dilated blood vessels. Between tumor cells there is a background of edematous, fine collagen fibers (Figures 18.14 and 18.15).

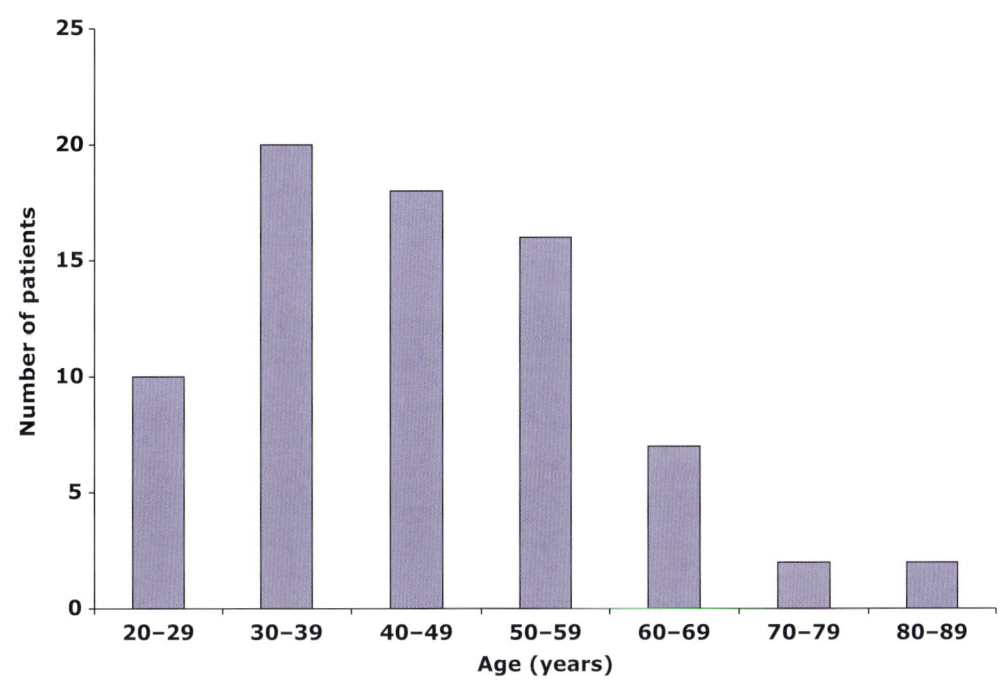

Chapter 18: Stromal tumors and tumor-like lesions of the female genital tract

Figure 18.12 Age distribution of 75 women with aggressive angiomyoma.

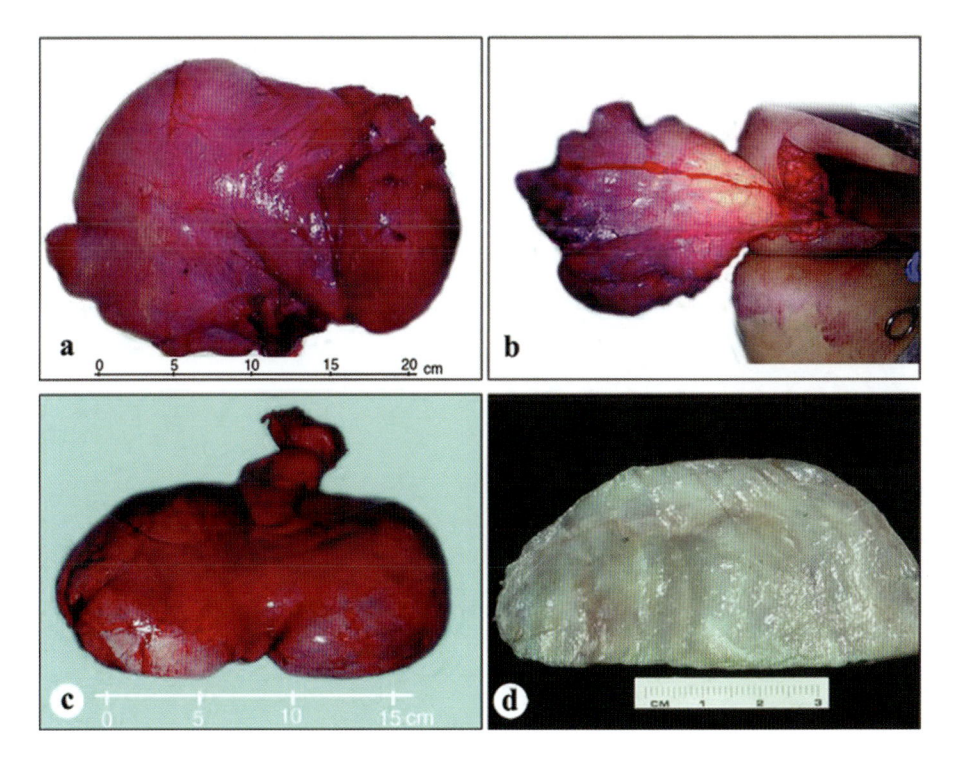

Figure 18.13 Gross appearances of aggressive angiomyxoma. (**a**) A large, apparently circumscribed mass. (**b**) A large example removed from the right pelvis through a perineal incision. (**c**) A large aggressive angiomyxoma with a smooth surface. (**d**) Glistening, pale gray cut surface of a formalin-fixed tumor.

The tumor cells are relatively small, uniform oval to spindle, often with bipolar cytoplasmic processes. The nuclei are oval and uniform, with a delicate nucleolus. Mitotic activity is scant (0–4 per 50 HPFs).

Immunohistochemical analysis shows the tumor cells to be typically positive for desmin, and estrogen and progesterone receptors, variably positive for muscle actins and CD34, and negative for S100 protein.[37,38,41,47] The percentage of MIB1-positive proliferative cells is low, less than 1%.[37,41]

Ultrastructural studies have shown myofibroblastic differentiation, with focal densities of actin filaments and partial basal laminas.[47,48]

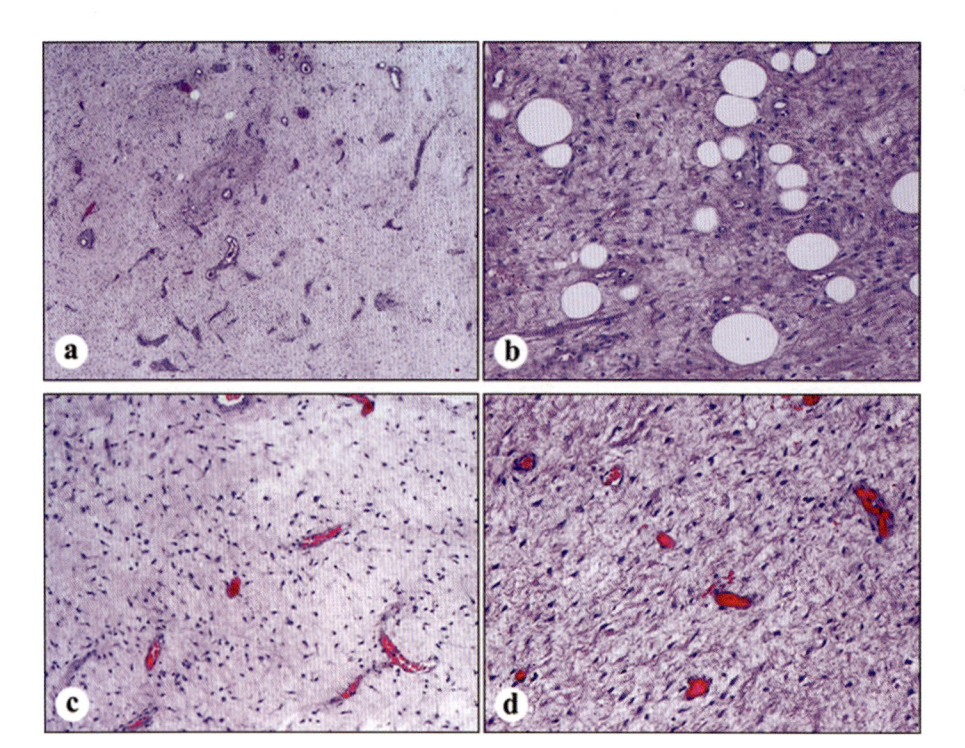

Chapter 18: Stromal tumors and tumor-like lesions of the female genital tract

Figure 18.14 Aggressive angiomyxoma shows moderate cellularity and high capillary density. The tumor can infiltrate adipose tissue.

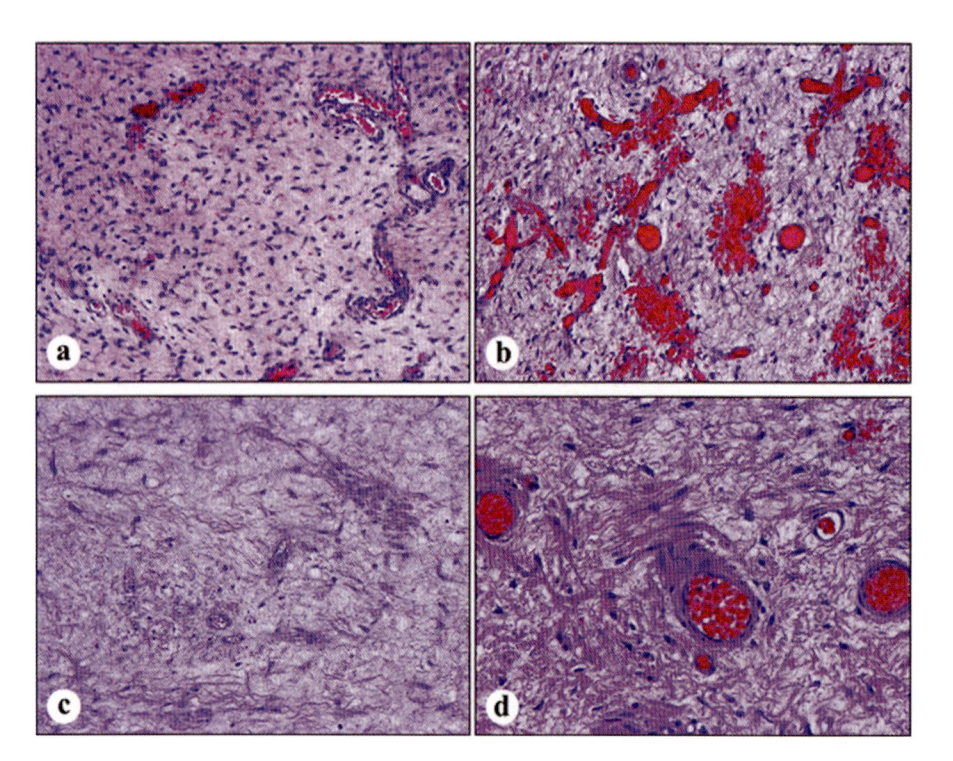

Figure 18.15 (**a,b**) Hyperemic vessels and uniform cellular element. (**c,d**) Perivascular foci of smooth muscle cells is a typical finding.

Differential diagnosis

Angiomyofibroblastoma is typically a well-circumscribed, relatively small (<5 cm) tumor in the labial region. Histologically different from angiomyoma are delicate capillaries and a corded perivascular pattern of tumor cells that are often epithelioid.

Superficial angiomyxoma (Chapter 11) can present in the external genital region. It is typically a multilobulated cutaneous or subcutaneous nodule, has a loose texture, is relatively paucicellular with abundant acid mucopolysaccharide matrix, and commonly contains perivascular lymphoid infiltration. This tumor is estrogen and progesterone receptor negative.

Intramuscular myxoma (Chapter 11) can occur in the buttocks and inguinal region, but is rarely seen in the genital region. It is a paucicellular tumor with a sparse vascularity and abundant acid mucopolysaccharides.

Myxoid liposarcoma and myxoid malignant fibrous histiocytoma (MFH) contain a prominent arborizing vascular pattern; the former contains uniform evenly distributed cells with possible lipoblasts, whereas the latter typically has a pleomorphic component and perivascular zones of higher cellularity.

Myxoid leiomyomas usually have bundles of smooth muscle cells at least focally, and they are typically positive for both desmin and smooth muscle actin.

Developmental asymmetry of the vulva and vulval lymphedema[49] can create a tumor-like bulge into the labia, and if this is biopsied, a slightly myxoid stroma could lead to an incorrect diagnosis of AAM.

Genetics

Cytogenetic studies have shown rearrangement of 12q15 commonly involved in various translocations.[50–54] The high-mobility group AT-hook 2 (HMGA2) gene (previously designated as HMGIC) located in this region has been shown to be involved and aberrantly expressed in the tumor cells;[51–55] normally the expression of this DNA architectural organizer protein is restricted to fetal tissues.[52] HMGA2 rearrangements seem to be relatively specific for AAM, but have also been found in some genital leiomyomas.[53] One AAM was reported to have t(5;8)(p15;q22) translocation as the sole chromosomal change.[55] AAM also expresses HMGA2 protein, but the marker is not specific for this tumor.[57]

Endometrial stromal sarcoma (stromatosis, intralymphangial stromal myosis)

This generally low-grade tumor is composed of cells closely resembling those of the proliferative endometrium; it has a slow clinical course. Its older synonyms reflect the relatively indolent clinical course and commonly prominent intralymphatic invasion. Common abdominal metastases and potential primary origin outside the uterus, possibly from endometriosis, make this tumor important for soft tissue pathology. Primary extrauterine tumors have been referred to as "endometrioid stromal sarcoma."

Clinical features

ESS typically occurs in adult women younger than 60 years old, the median ages in the largest series being around 45 years.[58–60] Most frequently, uterine tumors present with abnormal bleeding. Although most tumors are endometrial primaries and involve the uterine corpus only, some extend into the cervix, parametria, and elsewhere in the abdominal cavity, even occasionally into the great vessels and right side of the heart.[61] Extrauterine origin apparently from endometrosis explains occurrence in the abdomen in the absence of concurrent or previous uterine tumor. Most common locations are ovaries, intestinal wall, abdomen/peritoneum, and pelvic soft tissues. Behavior is rather indolent with frequent local recurrences, but dedifferentiated high-grade examples are aggressive.[62–64]

Abdominal metastases often occur in the soft tissues of the pelvis, the external aspect of the intestines, or on the peritoneal surfaces, and sometimes metastases extend into the buttocks and proximal thigh. Pulmonary metastases (more often multiple than solitary) also can develop, and this occurrence is not incompatible with long survival.

In one study, the interval from primary surgery to lung metastasis varied from 2.5 to 20 years (mean, 10 years).[65] Tumors confined to the uterus have better long-term prognosis than those that have spread or originated outside of it; however, one-third of patients with localized disease develop recurrences. The 10-year actuarial survival is stage dependent and varies from 75% to 90%.[60] ESS typically has a long course of disease, and late recurrences 10 to 30 years after the first surgery can occur. This tumor is hormonally dependent, and treatment with progesterone can stabilize disease.

Pathology

Grossly, ESS is often distinctive by involvement of multiple vessels in a worm-like or complex sausage-like pattern, which can be seen in the primary and recurrent extrauterine ESS and their abdominal metastases.

Histologically, the low-grade tumor is highly distinctive and generally retains the typical features in its soft tissue primaries or metastases (Figures 18.16 and 18.17). When involving the uterus, this tumor typically extends to the myometrium as tongues and involves lymphatics in the uterine wall; hence the previous designation *intralymphangial stromal myosis*. It is composed of uniform, round, oval, or sometimes spindled cells with little variation. The tumor cells are grouped around prominent, small-caliber blood vessels that can resemble the endometrial spiral arterioles or the vessels of hemangiopericytoma, whenever gaping or staghorn-shaped vessels are present.[58,59] Some tumors have hyalinized areas, and sex cord-like or glandular epithelial differentiation has been noted on rare occasions. Endometrial type glands can occur in these tumors, whether they are primary in the uterus or peritoneum.[60,61] The tumor usually has a low to moderate mitotic activity (up to 10 per 10 HPFs), but cases with higher mitotic counts (usually no more than 15 mitoses per 10 HPFs) may belong to the same biologic group, as long as they have the same overall morphology.[59,60] Current classification recognizes high-grade variants. These usually have high mitotic activity: >15–20 per 10 HPFs and are clinically aggressive. They include examples with YWHAE-FAM22 fusion translocation.[66]

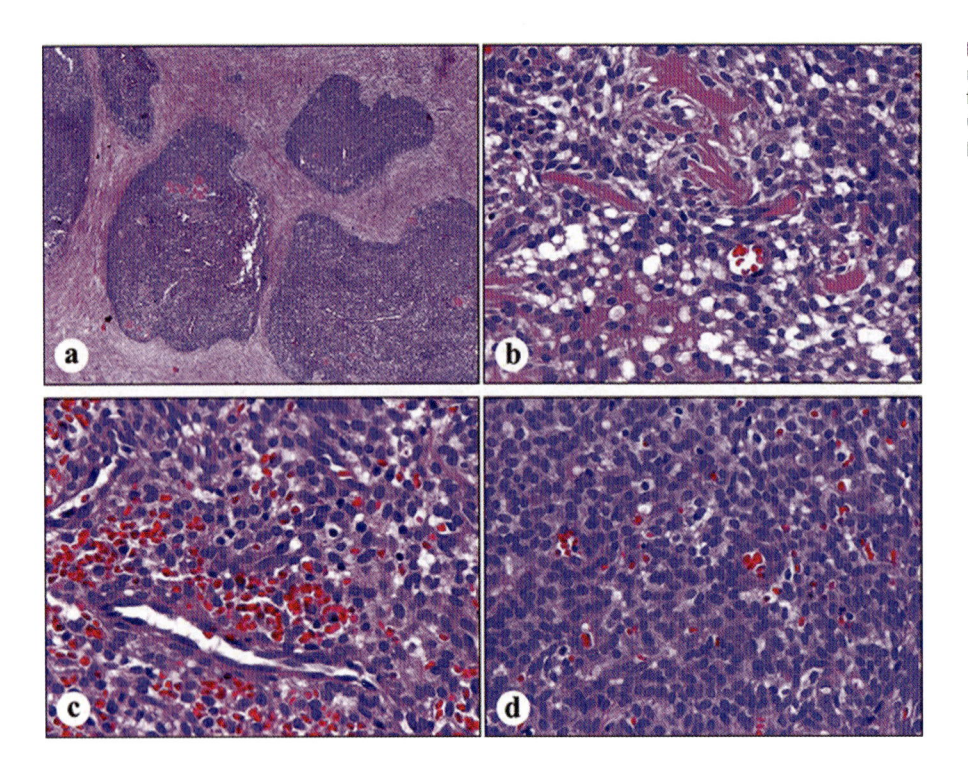

Figure 18.16 Endometrial stromal sarcoma metastatic to the pelvis forms large cellular nests in fibrous stroma. The tumor cells are round and uniform and often arranged in perivascular patterns.

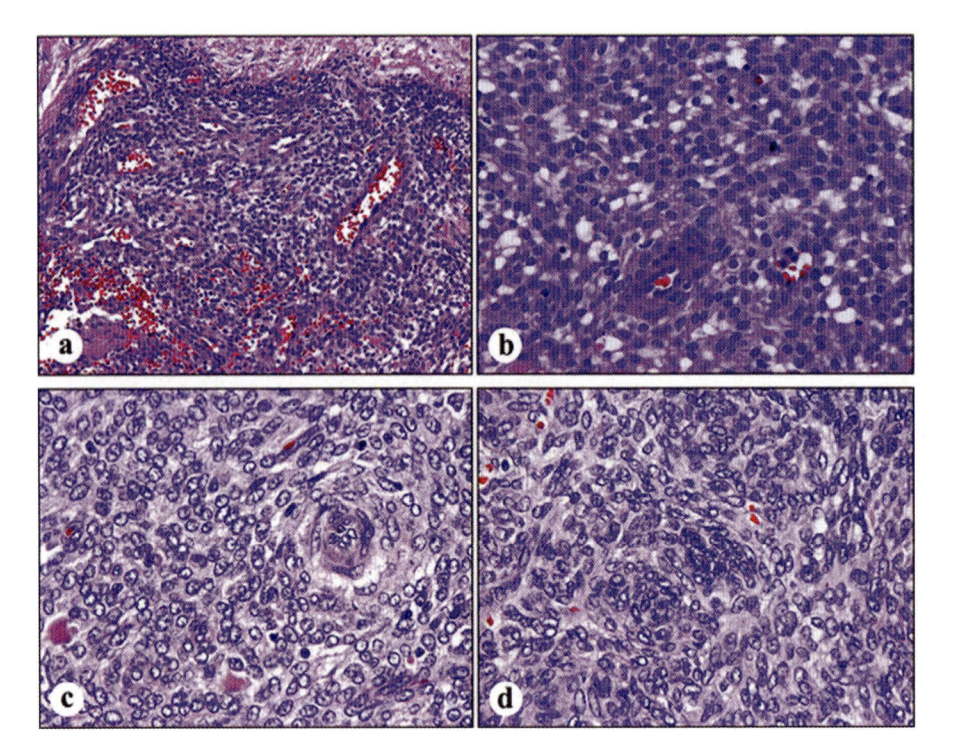

Figure 18.17 Higher magnification of endometrial stromal sarcoma reveals a tumor with prominent capillary pattern. The tumor cells often form perivascular whorls.

Immunohistochemically, the tumor cells are strongly positive for estrogen and progesterone receptors (Figure 18.18).[67] They are also positive for CD10 and can be variably positive for desmin, muscle actins, and keratins, but are usually negative for EMA.[68–71] Contrasted with smooth muscle tumors, the lesional cells are generally negative for h-caldesmon, which can be of differential diagnostic aid for separating these tumors.[69–71] High-grade variants are distinct from low-grade tumors in their cyclin D1 overexpression.[66]

Differential diagnosis

Endometrial stromal nodule is a rare, histologically similar lesion, which is noninfiltrative with a pushing border and

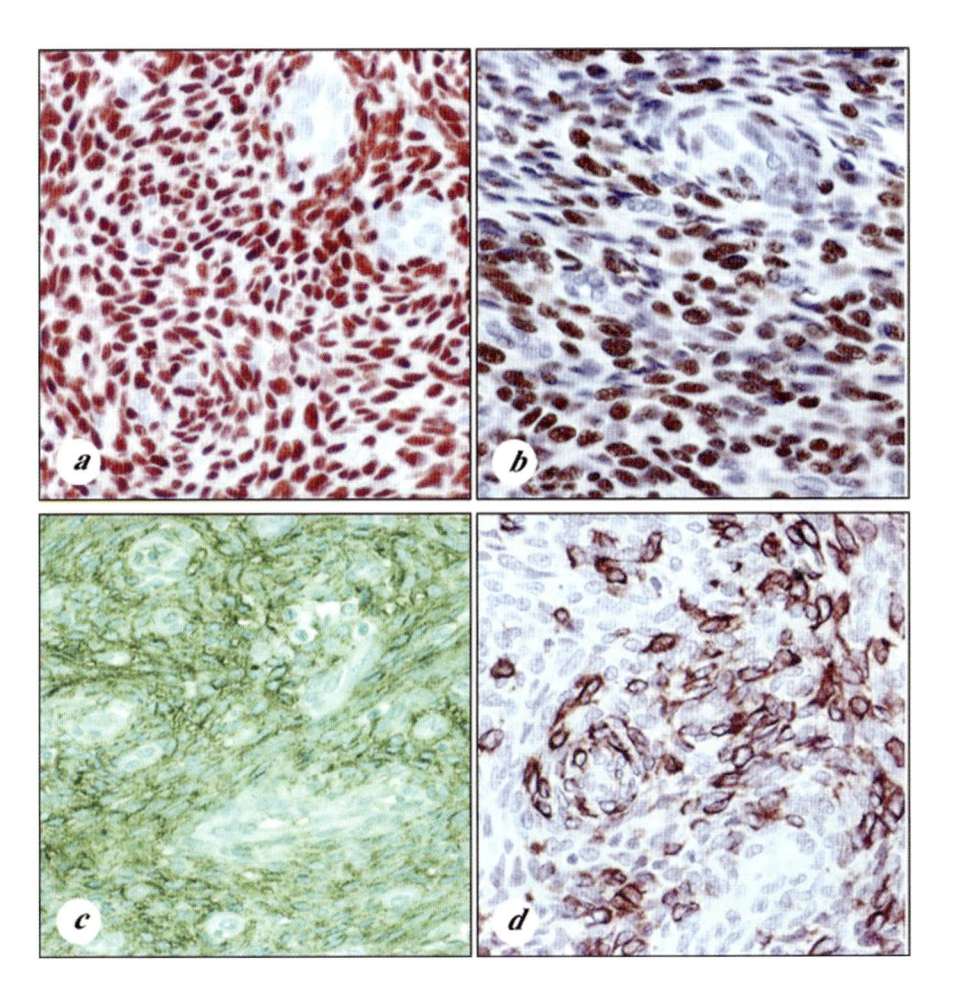

Figure 18.18 Immunohistochemically, endometrial stromal sarcoma is positive for (**a**) estrogen and (**b**) progesterone receptors, and (**c**) CD10. (**d**) Focal smooth muscle actin positivity is present.

can occur in the myometrium or be limited to the endometrium.[58]

Immunohistochemistry for epithelial, lymphoid, and myeloid markers can be necessary to separate low-grade stromal sarcoma from metastatic carcinoma (especially lobular carcinoma of the breast) and hematopoietic tumors, lymphoma, and extramedullary myeloid tumor.

Genetics

The most common gene fusion translocation is JAZF1-SUZ12/JJAZ1.[72,73] Additional fusion translocations are EPC1-PHF1, JAZF1-PHF1, MEAF6-PHF1, ZC3H7B-BCOR, and MBTD1-CXorf67.[74–77] The YWHAE-FAM22 gene fusion occurs in high-grade ESSs.[78]

Carcinosarcoma (malignant mixed müllerian tumor, malignant mixed mesodermal tumor)

This tumor is now considered an aggressive form of endometrial carcinoma with sarcomatous transformation, and therefore other synonyms include carcinoma with heterologous mesenchymal elements (metaplastic carcinoma). The likeness to carcinomas is expressed in the treatment of this tumor, which can have nodal dissemination and therefore requires nodal evaluation/evacuation procedures.

Clinical features

This relatively rare tumor, composing less than 5% of malignant uterine tumors, usually occurs in older women between 60 and 80 years and older. Patients typically present with postmenopausal bleeding. Tumors involving the uterus frequently can be detected by endometrial curettage and variably by endometrial cytologic sampling because of their intracavitary involvement. In some cases, this tumor has arisen in patients who had received radiation, suggesting radiation-related etiology. Estrogen replacement therapy could also be a predisposing factor.[79–85]

The tumor most commonly originates in the endometrium, but primary occurrence in the vagina, cervix, fallopian tube, ovary, and peritoneal surfaces is also possible, with origin in the last location most likely being explained by endometriosis. Primary abdominal tumors can present in the pelvic peritoneum, exterior surface of the intestines, adjacent to the liver, omentum, or retroperitoneum. Abdominal metastases can involve the same sites, and pulmonary metastases are common. Most tumors are highly aggressive, and an extended survival is rare. Exceptional, small polypoid, anatomically limited lesions can be more favorable.[79–85] Because this tumor is related more to carcinoma than a true sarcoma, it often involves lymph nodes, and therefore lymph node sampling and lymphadenectomy are often performed for staging and therapeutic purposes.[86]

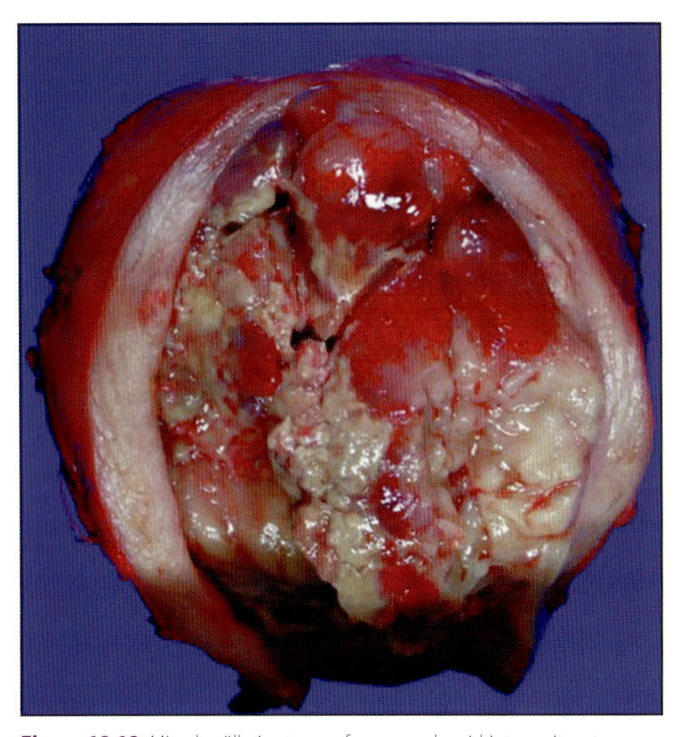

Figure 18.19 Mixed müllerian tumor forms a polypoid intracavitary tumor that fills the endometrial cavity.

Pathology

Grossly, a uterine carcinosarcoma typically forms a polypoid, intracavitary mass with variable, often extensive involvement of the myometrium. It can reach a large size, involve a large volume of the uterus, and also extend outside. On sectioning, the tumor is soft, gray-tan, and often variegated by extensive hemorrhage and necrosis (Figure 18.19). Some tumors have firmer, osteocartilaginous components.

By definition, the tumor has microscopic carcinomatous and sarcomatous components (Figure 18.20). The former can form sharply demarcated clusters, often with a complex cribriform pattern, having endometrioid, clear-cell, serous, and often focally squamoid features. Poorly differentiated examples and metastases can have an entirely sarcomatous appearance (Figure 18.20d), although the sarcomatous component can display subtle epithelial differentiation, as detected by immunohistochemistry.

The sarcomatous component can be monomorphous and undifferentiated (homologous carcinosarcoma) or it can be composed of multiple mesenchymal components (heterologous carcinosarcoma). Cytologically the sarcomatous component varies from round to spindle shaped, sometimes with moderate pleomorphism. All components are typically high grade, and the mitotic rate is high, especially in the epithelial component.

The heterologous sarcomatous components often include embryonal rhabdomyosarcoma-like elements, with varying numbers of differentiating rhabdomyoblasts (Figures 18.20 and 18.21). Cartilaginous, osseous, and lipomatous differentiation also can be present. Some tumors have PAS-positive

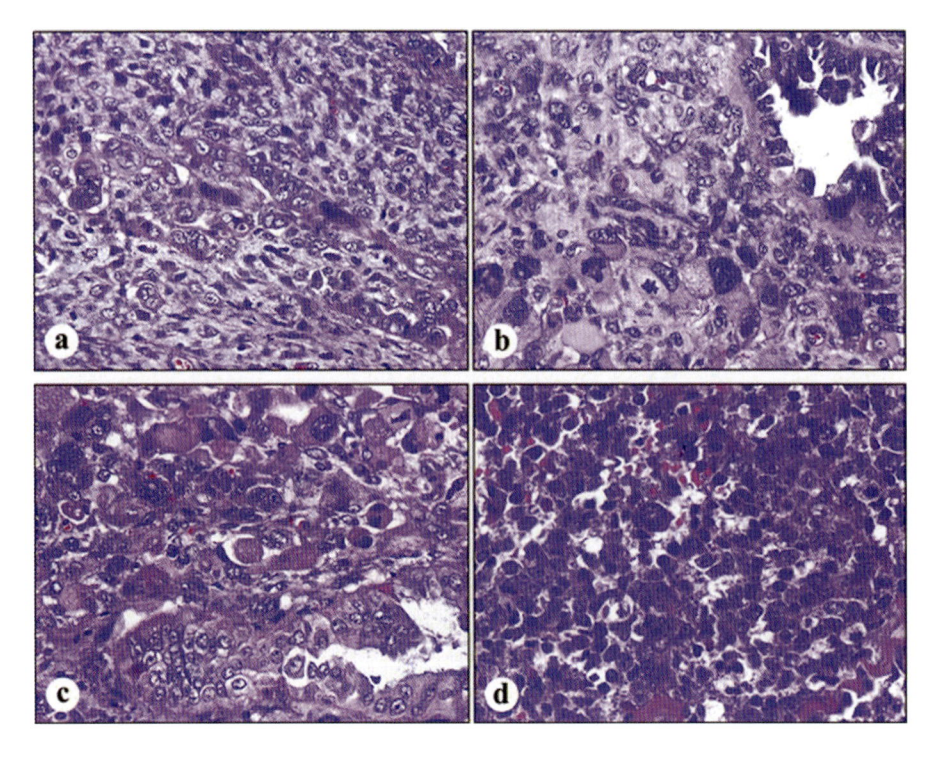

Figure 18.20 (**a,b**) Mixed müllerian tumor contains atypical epithelial elements and a primitive-appearing sarcomatous component. (**c**) The large cells with eosinophilic cytoplasm are rhabdomyoblasts. (**d**) A more primitive appearance is seen in an inguinal metastasis of the same tumor.

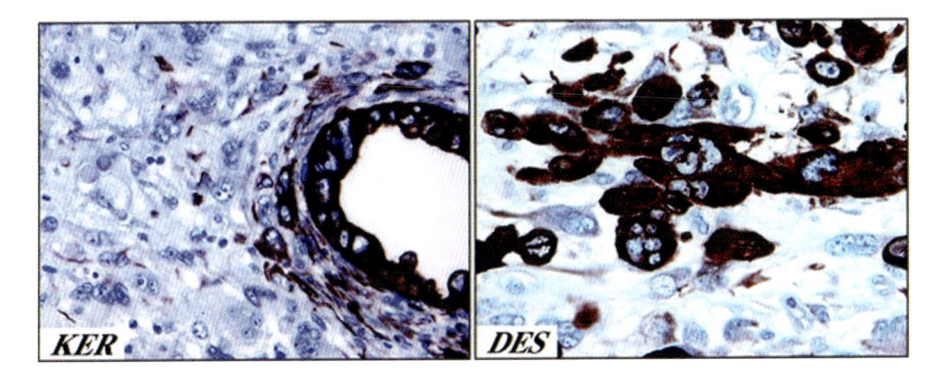

Figure 18.21 The epithelial elements and some sarcomatoid components are keratin positive. The rhabdomyoblastic component is strongly desmin positive, and cross-striations are visible in some cells.

hyaline globules or droplets resembling those seen in some germ cell tumors.

Metastases can be composed of purely epithelial, sarcomatous, or mixed elements. The nature and extent of heterologous differentiation does not seem to have prognostic significance.

Immunohistochemical studies have shown keratin expression in both epithelial and sarcomatous components, and this has been grounds for considering these tumors to be sarcomatoid carcinomas.[87–89] Desmin (Figure 18.21), myoglobin, and MyoD1/myogenin positivity highlight heterologous skeletal muscle differentiation, and S100 protein positivity that of cartilage. The epithelial component has been found variably but inconsistently positive for estrogen and progesterone receptors, and the sarcomatous component is only rarely positive. Extensive epitheliomesenchymal transition and associated markers characterize uterine heterologous carcinosarcomas[90]

Genetics and histogenesis

Complex cytogenetic changes have been detected, including inversion of 12p[91] and a translocation t(8;22)(q24;q12);[92] however, no recurrent translocations or mutations have been reported. Similar molecular genetic changes, namely identical p53 mutations[93] and similar patterns of allelic losses,[94] have been found in the carcinomatous and sarcomatous components. This confirms the common clonal origin of both components and supports the concept that this tumor is indeed a metaplastic carcinoma or sarcomatoid carcinoma. This is also supported by the differentiation of carcinomatous elements into skeletal muscle in cell culture.[95]

Related terms and tumors

Adenosarcoma (müllerian adenosarcoma) refers to a neoplasm that typically shows a low-grade sarcomatous stroma and a benign-appearing epithelial component. It usually occurs in the endometrium or cervix, but sarcomatous components of this tumor can disseminate as intra-abdominal metastases.[79–82]

Adenofibroma has a benign-appearing epithelium and stroma and most commonly occurs in the endometrium, cervix, and ovary.[57]

Adult granulosa cell tumor

The significance of this tumor in nongynecologic pathology is its common occurrence as an intra-abdominal metastasis and occasional presentation as a primary nonovarian tumor, often long after initial surgery. The rare juvenile variant and granulosa cell tumors of the testis are clinicopathologically different tumors and are not discussed here.

Clinical features

This tumor, classified among the sex cord stromal tumors, almost always presents as an ovarian primary tumor in middle-aged and older women and very rarely in children. A few potentially primary tumors elsewhere in the abdomen (pelvic wall) have also been reported. Patients with this tumor often present with postmenopausal bleeding, because granulosa cell tumors typically have estrogenic activity and reactivate the endometrium; some examples have no hormonal activity, and few have androgenic activity.[96–101]

Granulosa cell tumors commonly recur in the abdominal cavity, often after a long asymptomatic period of 10 to 20 years or even more. In such a situation the history of a primary tumor is often unknown, and the diagnosis can be challenging.

Pathology

Adult granulosa cell tumors of the ovary are typically relatively large, often exceeding 10 cm. They are often grossly cystic and hemorrhagic, and the tumor tissue is typically gray-tan or yellowish (Figure 18.22). Microscopically, the histologic patterns include microfollicular and follicular ones somewhat simulating the normal Graaffian follicles. Many tumors have microscopic rosettes containing eosinophilic material referred to as *Call-Exner bodies*, and some have a trabecular pattern with or without intervening fibrous tissues. A pattern with curved trabeculae is commonly referred to as a "watered-silk" pattern. Some tumors have a diffuse growth with no structural differentiation (Figure 18.23). Common to all variants is dense cellularity and uniform round to oval cells with scant cytoplasm. Varying numbers of the nuclei have deep grooves, imparting a "coffee bean" shape. Focal pleomorphism occurs, but rarely. Mitotic activity varies but is usually less than 5 to 7 per 10 HPFs.

The abdominal recurrences usually have a diffuse or partly trabecular pattern, but sarcomatous transformation[102] and hemangiopericytoma-like appearance[103] can occur, increasing the risk of misdiagnosis (Figure 18.23).

An immunohistochemically typical and diagnostically useful feature is cytoplasmic positivity for alpha-inhibin (Figure 18.24). This distinguishes granulosa cell tumors from carcinomas, which are negative, whereas fibrothecomas are also alpha-inhibin positive.[103–106] The tumor cells are also positive for vimentin and are variably but less strikingly keratin 8 or keratin 18 positive (often with paranuclear dot-like pattern) than ovarian carcinomas in general.[107,108] They are consistently more calretinin positive than other sex cord stromal tumors,[109] and are S100 protein positive in 50% of the cases, but are EMA negative, in contrast to most ovarian and abdominal carcinomas.[108]

Genetics

Recurrent chromosomal changes, including trisomy 12[110] and trisomy 14,[111–113] have been found in granulosa cell tumors. Nearly consistent FOXL2 mutation (402 C>G, Cys134Trp) characterizes most adult granulosa cell tumors, and similar mutations occur in a subset of fibrothecomas.[114] These mutations are likely cause for proliferation although the mechanism is not completely understood. FOXL2 expression is also detectable in most granulosa cell tumors, as well as fibrothecomas and some other sex cord stromal tumors.[115]

Endometriosis

Endometriosis refers to the ectopic occurrence of endometrial tissue outside the uterus. The occurrence of endometrial tissue in the myometrium is termed *adenomyosis*. The importance of endometriosis for soft tissue pathology is its occurrence in peripheral soft tissues, especially the abdominal wall and inguinal region. This section focuses on abdominal wall endometriosis.

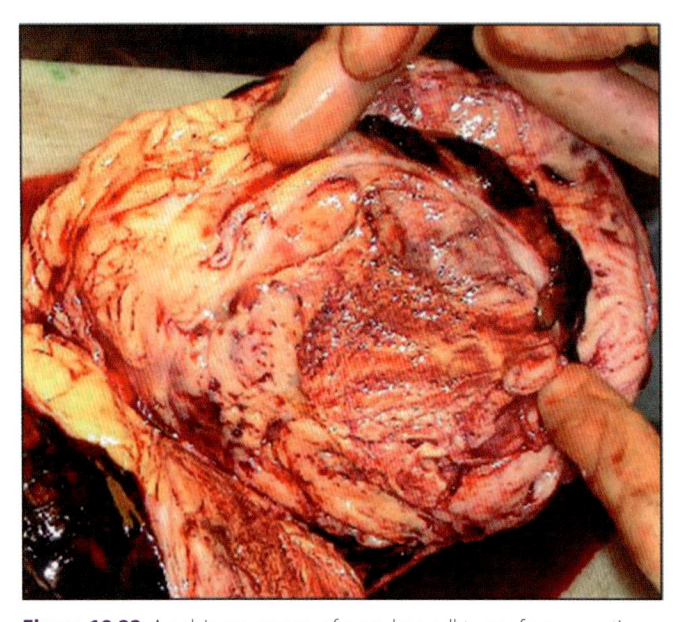

Figure 18.22 A pelvic recurrence of granulosa cell tumor forms a cystic, partly necrotic mass.

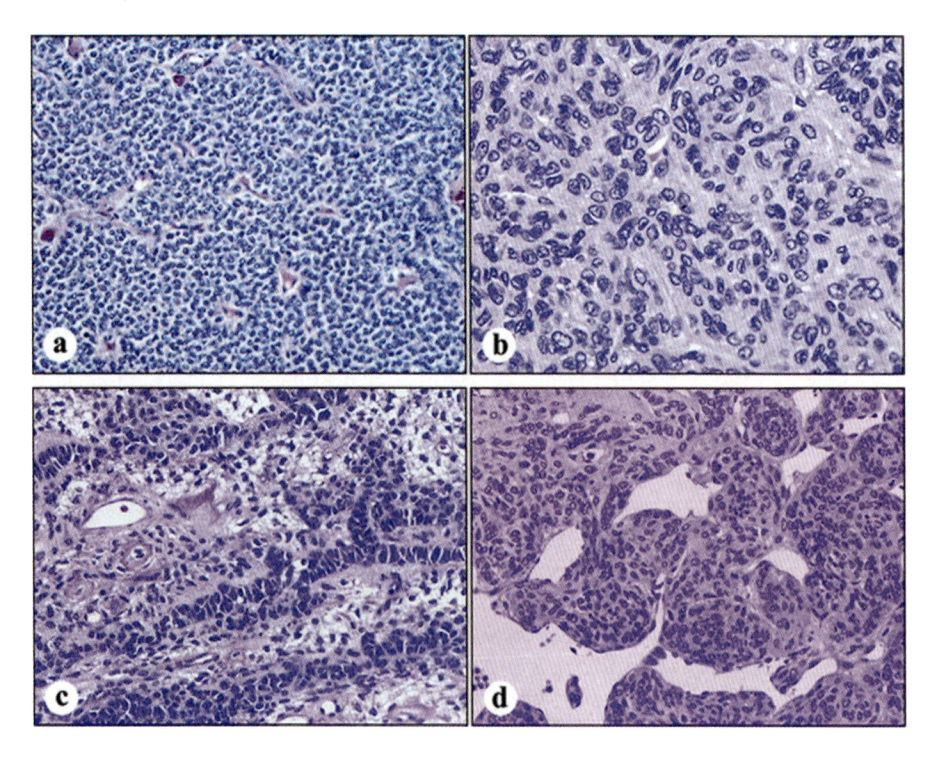

Figure 18.23 (**a**) Histologically, granulosa cell tumor is composed of relatively uniform round cells. (**b**) The tumor cells can have a diffuse or (**c**) trabecular arrangement. (**d**) A metastatic granulose cell tumor shows a hemangiopericytoma-like pattern.

Clinical features

Endometriosis itself is very common, but its occurrence in the abdominal wall is less common and is reported in less than 5% of all patients with endometriosis. Abdominal wall endometriosis usually occurs in the scar following a cesarean section, with an estimated frequency of 1% to 2%. Less frequently it occurs after hysterectomy or other surgery. In rare instances, it has occurred in needle tracts or spontaneously without previous surgery. In the latter case, the most common site is the umbilical region. In the abdominal wall it typically appears as a painful mass, and in some cases, the pain is related to the menstrual cycle. Occurrence in the inguinal region and proximal thigh is also possible.[116–121]

Malignancy is an uncommon complication of endometriosis, and both carcinomas (mainly endometrioid and clear cell) and müllerian sarcomas (carcinosarcomas) have been reported.[122–125] Therefore, endometriosis must be considered as the origin in the case of a primary abdominal wall or peritoneal carcinoma.

Pathology

Surgical specimens on abdominal wall endometriosis can reveal cystic spaces, reflecting dilated endometrial glands. Histologically, the presence of endometrial glandular tissue together with stroma is diagnostic, but a diagnosis can be made based on one element only, especially the stromal one.[113] The appearance of the epithelium varies from examples closely simulating a normal proliferative endometrium to those with small foci of epithelial tissue with scant stroma, and vice versa. Well-developed secretory phase change can occur (Figure 18.25). Decidual change is typically prominent in postpartum lesions (Figure 18.26). Florid decidual reaction can simulate a neoplasia, unless properly recognized.[126] In some cases, hyperplasia and atypia in endometriosis can foretell the development of carcinoma.[127]

The immunohistochemical profile can assist in specific diagnosis. The glandular and epithelial element can be highlighted with antibodies to keratins, whereas stroma is vimentin positive. Both epithelial and stromal components are positive for estrogen and progesterone receptors, and stroma for CD10 and WT1. These immunostains can help to highlight minute endometriosis nests on peritoneal surfaces and other abdominal sites.

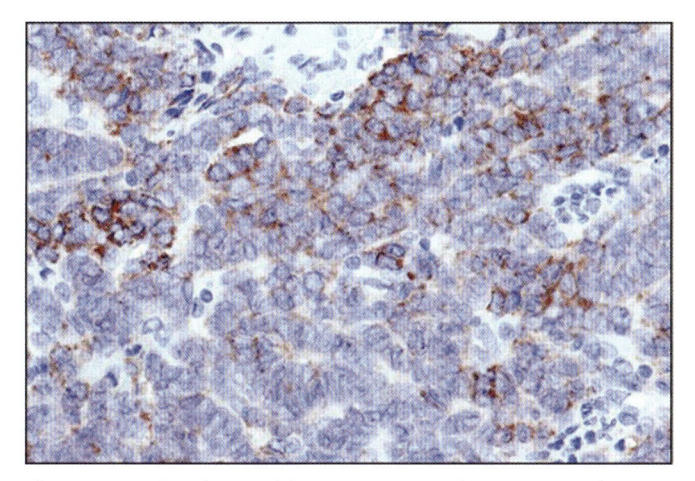

Figure 18.24 Cytoplasmic inhibin positivity in a pelvic metastasis of a granulosa cell tumor.

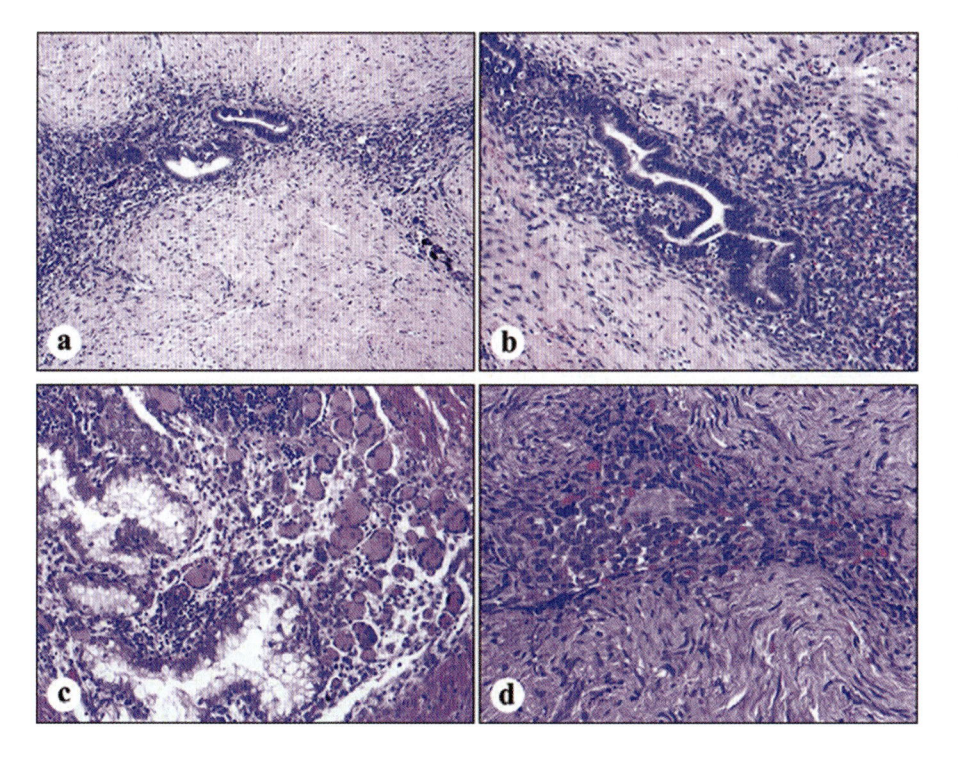

Figure 18.25 (**a,b**) Abdominal wall endometriosis with glandular and stromal elements. The epithelium shows proliferative phase features. (**c**) Abdominal wall endometriosis involving skeletal muscle. Note the secretory phase features in the epithelium. (**d**) Endometriosis nest composed of stromal cells only.

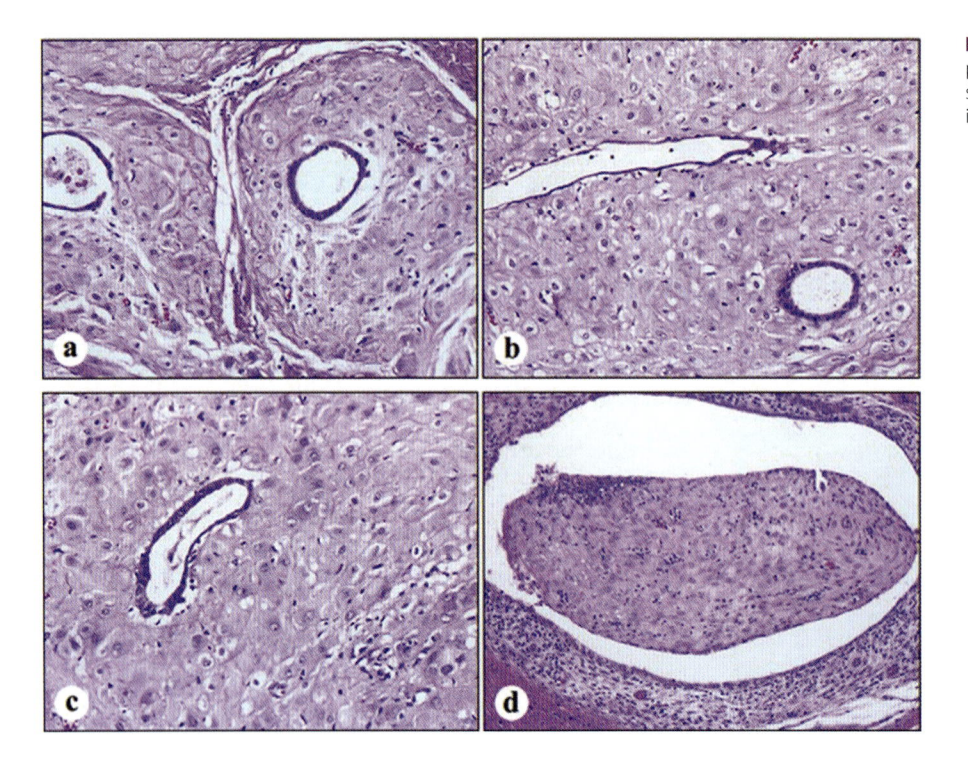

Figure 18.26 Abdominal endometriosis in a pregnant patient shows a prominent decidual stromal change. The epithelial element is inconspicuous and atrophic.

References

Mesodermal stromal polyp and related lesions

1. Norris HJ, Taylor HB. Polyps of the vagina: a benign lesion resembling sarcoma botryoides. *Cancer* 1966;19:227–232.
2. Chirayil SJ, Tobon H. Polyps of the vagina: a clinicopathologic study of 18 cases. *Cancer* 1981;47:2904–2907.
3. Miettinen M, Wahlström T, Vesterinen E, Saksela E. Vaginal polyps with pseudosarcomatous features: a clinicopathologic study of seven cases. *Cancer* 1983;51:1148–1151.
4. Östor AG, Fortune DW, Riley CB. Fibroepithelial polyps with atypical stromal cells (pseudosarcoma botryoides) of vulva and vagina: a report of 13 cases. *Int J Gynecol Pathol* 1988;7:351–360.
5. Mucitelli DR, Charles EZ, Kraus FT. Vulvovaginal polyps: histologic appearance, ultrastructure, immunocytochemical characteristics, and clinicopathological correlations. *Int J Gynecol Pathol* 1990;9:20–40.
6. Nucci MR, Young RH, Fletcher CDM. Cellular pseudosarcomatous fibroepithelial stromal polyps of the lower female genital tract: an underrecognized lesion often misdiagnosed as sarcoma. *Am J Surg Pathol* 2000;24:231–240.
7. Rollason TP, Byrne P, Williams A. Immunohistochemical and electron microscopic findings in benign fibroepithelial vaginal polyps. *J Clin Pathol* 1990;43:224–229.
8. Hartmann CA, Sperling M, Stein H. So-called fibroepithelial polyps of the vagina exhibiting an unusual but uniform antigen profile characterized by expression of desmin and steroid hormone receptors but no muscle-specific actin or macrophage markers. *Am J Clin Pathol* 1990;93:604–608.
9. Laskin WB, Fetsch JF, Tavassoli FA. Superficial cervicovaginal myofibroblastoma: fourteen cases of a distinctive mesenchymal tumor arising from the specialized subepithelial stroma of the lower female genital tract. *Hum Pathol* 2001;32:715–725.
10. Stewart CJ, Amanuel B, Brennan BA, et al. Superficial cervicovaginal myofibroblastoma: a report of five cases. *Pathology* 2005;37:144–148.
11. Ganesan R, McCluggage WG, Hirschowitz L, Rollason TP. Superficial myofibroblastoma of the lower female genital tract: report of a series including tumours with a vulval location. *Histopathology* 2005;46:137–143.

Genital rhabdomyoma

12. Gold JH, Bossen EH. Benign vaginal rhabdomyoma: a light and electron microscopic study. *Cancer* 1976;37:2283–2294.
13. Chabrel CM, Beilby JOW. Vaginal rhabdomyoma. *Histopathology* 1980;4:645–651.
14. Iversen UM. Two cases of benign vaginal rhabdomyoma. *APMIS* 1996;104:575–578.
15. Losi L, Choreutaki T, Nascetti D, Eusebi V. Recurrence in a case of rhabdomyoma of the vagina. *Pathologica* 1995;87:704–708.
16. Huang TY, Chen JT, Ho WL. Ovarian serous cystadenoma with mural nodules of genital rhabdomyoma. *Hum Pathol* 2005;36:433–435.

Angiomyofibroblastoma

17. Fletcher CDM, Tsang WYW, Fisher C, Lee KC, Chan JKC. Angiomyofibroblastoma of the vulva: a benign neoplasm distinct from aggressive angiomyxoma. *Am J Surg Pathol* 1992;16:373–382.
18. Hisaoka M, Kouho H, Aoki T, Daimaru Y, Hashimoto H. Angiomyofibroblastoma of the vulva: a clinicopathologic study of seven cases. *Pathol Int* 1995;45:487–492.

19. Nielsen GP, Rosenberg AE, Young RH, et al. Angiomyofibroblastoma of the vulva and vagina. *Mod Pathol* 1996;9:284–291.

20. Fukunaga M, Nomura K, Matsumoto K, et al. Vulval angiomyofibroblastoma: clinicopathologic analysis of six cases. *Am J Clin Pathol* 1997;107:45–51.

21. Laskin WB, Fetsch JF, Tavassoli FA. Angiomyofibroblastoma of the female genital tract: analysis of 17 cases including a lipomatous variant. *Hum Pathol* 1997;28:1046–1055.

22. Ockner DM, Sayadi H, Swanson PE, Ritter JH, Wick MR. Genital angiomyofibroblastoma: comparison with aggressive angiomyxoma and other myxoid neoplasms of skin and soft tissue. *Am J Clin Pathol* 1997;107:36–44.

23. Vasquez MD, Ro JY, Park YW, et al. Angiomyofibroblastoma: a clinicopathologic study of eight cases and review of the literature. *Int J Surg Pathol* 1999;7:161–169.

24. Nielsen GP, Young RH, Dickersin GR, Rosenberg AE. Angiomyofibroblastoma of the vulva with sarcomatous transformation ("angiomyofibrosarcoma"). *Am J Surg Pathol* 1997;21:1104–1108.

25. Bigotti G, Coli A, Gasbarri A, et al. Angiomyofibroblastoma and aggressive angiomyxoma: two benign mesenchymal neoplasms of the female genital tract. An immunohistochemical study. *Pathol Res Pract* 1999;195:39–44.

26. Magro G, Righi A, Caltabiano R, Casorzo L, Michal M. Vulvovaginal angiomyofibroblastomas: morphologic, immunohistochemical, and fluorescence in situ hybridization analysis for deletion of 13q14 region. *Hum Pathol* 2014;45:1647–1655.

Cellular angiofibroma

27. Nucci MR, Granter SR, Fletcher CDM. Cellular angiofibroma: a benign neoplasm distinct from angiomyofibroblastoma and spindle cell lipoma. *Am J Surg Pathol* 1997;21:636–644.

28. Laskin WB, Fetsch JF, Mostofi FK. Angiomyofibroblastoma-like tumor of the male genital tract: analysis of 11 cases with comparison to female angiomyofibroblastoma and spindle cell lipoma. *Am J Surg Pathol* 1998;22:6–16.

29. Iwasa Y, Fletcher CD. Cellular angiofibroma: clinicopathologic and immunohistochemical analysis of 51 cases. *Am J Surg Pathol* 2004;28:1426–1435.

30. McCluggage WG, Ganesan R, Hirschowitz L, Rollason TP. Cellular angiofibroma and related fibromatous lesions of the vulva: report of a series of cases with a morphological spectrum wider than previously described. *Histopathology* 2004;45:360–368.

31. Chen E, Fletcher CD. Cellular angiofibroma with atypia or sarcomatous transformation: clinicopathologic analysis of 13 cases. *Am J Surg Pathol* 2010;34(5):707–714.

32. Maggiani F, Debiec-Rychter M, Vanbockrijck M, Sciot R. Cellular angiofibroma: another mesenchymal tumor with 13q14 involvement, suggesting a link with spindle cell lipoma and (extra)mammary myofibroblastoma. *Histopathology* 2007;51:410–412.

33. Hameed M, Clarke K, Amer HZ, Mahmet K, Aisner S. Cellular angiofibroma is genetically similar to spindle cell lipoma: a case report. *Cancer Genet Cytogenet* 2007;177:131–134.

34. Flucke U, van Krieken JH, Mentzel T. Cellular angiofibroma: analysis of 25 cases emphasizing its relationship to spindle cell lipoma and mammary-type myofibroblastoma. *Mod Pathol* 2011;24:82–89.

Aggressive angiomyxoma

35. Steeper TA, Rosai J. Aggressive angiomyxoma of the female pelvis and perineum: report of nine cases of a distinctive type of gynecologic soft-tissue neoplasm. *Am J Surg Pathol* 1983;7:463–475.

36. Begin LR, Clement PB, Kirk ME, et al. Aggressive angiomyxoma of pelvic soft parts: a clinicopathologic study of nine cases. *Hum Pathol* 1985;16:621–628.

37. Fetsch JF, Laskin WB, Lefkowitz M, Kindblom LG, Meis-Kindblom JM. Aggressive angiomyxoma: a clinicopathologic study of 29 female patients. *Cancer* 1996;78:79–90.

38. Granter SR, Nucci MR, Fletcher CDM. Aggressive angiomyxoma: reappraisal of its relationship to angiomyofibroblastoma in a series of 16 cases. *Histopathology* 1997;30:3–10.

39. Magtibay PM, Salmon Z, Keeney GL, Podratz KC. Aggressive angiomyxoma of the female pelvis and perineum: a case series. *Int J Gynecol Cancer* 2006;16:396–401.

40. Fine BA, Munoz AK, Litz CE, Gershenson DM. Primary medical management of recurrent aggressive angiomyxoma with a gonadotropin-releasing hormone agonist. *Gynecol Oncol* 2001;81:120–122.

41. van Roggen JF, van Unnik JA, Briare-de Bruijn IH, Hogendoorn PC. Aggressive angiomyxoma: a clinicopathologic and immunohistochemical study of 11 cases with long-term follow-up. *Virchows Arch* 2005;446:157–163.

42. McCluggage WG, Jamieson T, Dobbs SP, Grey A. Aggressive angiomyxoma of the vulva: dramatic response to gonadotropin releasing hormone agonist therapy. *Gynecol Oncol* 2006;100:623–625.

43. Han-Geurts IJ, van Geel AN, van Dorn L, et al. Aggressive angiomyxoma: multimodality treatments can avoid mutilating surgery. *Eur J Surg Oncol* 2006;32:1217–1221.

44. Tsang WYW, Chan JKC, Lee KC, Fisher C, Fletcher CDM. Aggressive angiomyxoma: a report of four cases occurring in men. *Am J Surg Pathol* 1992;16:1059–1065.

45. Clatch RJ, Drake WK, Conzalez JG. Aggressive angiomyxoma in men: a report of two cases associated with inguinal hernias. *Arch Pathol Lab Med* 1993;117:911–913.

46. Iezzoni JC, Fechner RE, Wong LS, Rosai J. Aggressive angiomyxoma in males: a report of four cases. *Am J Clin Pathol* 1995;104:391–396.

47. Skalova A, Michal M, Husek K, Zamecnik M, Leivo I. Aggressive angiomyxoma of the pelvicoperineal region: immunohistochemical and ultrastructural study of seven cases. *Am J Dermatopathol* 1993;15:446–451.

48. Martinez MA, Ballestin C, Carabias E, Gonzalez LC. Aggressive angiomyxoma: an ultrastructural study of four cases. *Ultrastruct Pathol* 2003;27:227–233.

49. Vang R, Connelly JH, Hammill HA, Shannon RL. Vulvar hypertrophy with lymphedema: a mimicker of aggressive

angiomyxoma. *Arch Lab Pathol Med* 2000;124:1697–1699.

50. Kazmierczak B, Wanschura S, Meyer-Bolte K, et al. Cytogenetic and molecular analysis of an aggressive angiomyxoma. *Am J Pathol* 1995;147:580–585.

51. Nucci MR, Weremowicz S, Neskey DM, et al. Chromosomal translocation t(8;12) induces aberrant HMGIC expression in aggressive angiomyxoma of the vulva. *Genes Chromosomes Cancer* 2001;32:172–176.

52. Hess JL. Chromosomal translocations in benign tumors: the HMGI proteins. *Am J Clin Pathol* 1998;109:251–261.

53. Micci F, Panagopoulos I, Bjerkehagen B, Heim S. Deregulation of HMGA2 in an aggressive angiomyxoma with t(11;12)(q23;q15). *Virchows Arch* 2006;448:838–842.

54. Rawlinson NJ, West WW, Nelson M, Bridge JA. Aggressive angiomyxoma with t(12;21) and HMGA2 rearrangement: report of a case and review of the literature. *Cancer Genet Cytogenet* 2008;181:119–124.

55. Medeiros F, Ericson-Johnson MR, Keeney GL, et al. Frequency and characterization of HMGA2 and HMGA1 rearrangements in mesenchymal tumors of the lower genital tract. *Genes Chromosomes Cancer* 2007;46:981–990.

56. Tsuji T, Yoshinaga M, Inomoto Y, Taguchi S, Douchi T. Aggressive angiomyxoma of the vulva with a sole t(5;8)(p15;q22) chromosome change. *Int J Gynecol Pathol* 2007;26:494–496.

57. McCluggage WG, Connolly L, McBride HA. HMGA2 is a sensitive but not specific immunohistochemical marker of vulvovaginal aggressive angiomyxoma. *Am J Surg Pathol* 2010;34:1037–1042.

Endometrial stromal sarcoma

58. Norris HJ, Taylor HB. Mesenchymal tumors of the uterus: I. A clinical and pathological study of 53 endometrial stromal tumors. *Cancer* 1966;19:755–766.

59. Evans HL. Endometrial stromal sarcoma and poorly differentiated endometrial sarcoma. *Cancer* 1982;50:2170–2182.

60. Chang KL, Crabtree GS, Lim-Tan SK, Kempson RL, Hendrickson MR. Primary uterine stromal neoplasms: a study of 117 cases. *Am J Surg Pathol* 1990;14:415–438.

61. Tabata T, Takeshima N, Hirai Y, Hasumi K. Low-grade endometrial stromal sarcoma with cardiovascular involvement: a report of three cases. *Gynecol Oncol* 1999;75:495–498.

62. Chang KL, Crabtree GS, Lim-Tan SK, Kempson RL, Hendrickson MR. Primary extrauterine endometrial stromal neoplasms: a clinicopathologic study of 20 cases and a review of the literature. *Int J Gynecol Pathol* 1993;12:282–296.

63. Fukunaga M, Ishihara A, Ushigome S. Extrauterine low-grade endometrial stromal sarcoma: report of three cases. *Pathol Int* 1998;48:297–302.

64. Masand RP, Euscher ED, Deavers MT, Malpica A. Endometrioid stromal sarcoma: a clinicopathologic study of 63 cases. *Am J Surg Pathol* 2013;37:1635–1647.

65. Abrams J, Talcott J, Corson JM. Pulmonary metastases in patients with low-grade endometrial stromal sarcoma: clinicopathologic findings with immunohistochemical characterization. *Am J Surg Pathol* 1989;13:133–140.

66. Sciallis AP, Bedroske PP, Schoolmeester JK, et al. High-grade endometrial stromal sarcomas: a clinicopathologic study of a group of tumors with heterogenous morphologic and genetic features. *Am J Surg Pathol* 2014;38:1161–1172.

67. Navarro D, Cabrera JJ, Leon L, et al. Endometrial stromal sarcoma expression of estrogen receptors, progesterone receptors and estrogen-induced srp27 (24K) suggests hormone responsiveness. *J Steroid Biochem Mol Biol* 1992;41:589–596.

68. Farhood AI, Abrams J. Immunohistochemistry of endometrial stromal sarcoma. *Hum Pathol* 1991;22:224–230.

69. Chu PG, Arber DA, Weiss LM, Chang KL. Utility of CD10 in distinguishing between endometrial stromal sarcoma and uterine smooth muscle tumors: an immunohistochemical comparison of 34 cases. *Mod Pathol* 2001;14:465–471.

70. Rush DS, Tan J, Baergen RN, Soslow RA. h-Caldesmon, a novel smooth muscle-specific antibody, distinguishes between cellular leiomyoma and endometrial stromal sarcoma. *Am J Surg Pathol* 2001;25:253–258.

71. Nucci MR, O'Connell JT, Huettner PC, et al. h-Caldesmon expression effectively distinguishes endometrial stromal tumors from uterine smooth muscle tumors. *Am J Surg Pathol* 2001;25:455–463.

72. Koontz JI, Soreng AL, Nucci M, et al. Frequent fusion of the JAZF1 and JJAZ1 genes in endometrial stromal tumors. *Proc Natl Acad Sci USA* 2001;98:6348–6353.

73. Micci F, Walter CU, Texeira MR, et al. Cytogenetic and molecular genetic analyses of endometrial stromal sarcoma: nonrandom involvement of chromosome arms 6p and 7p and confirmation of JAZF1/JJAZ1 gene fusion in t(7;17). *Cancer Genet Cytogenet* 2003;144:119–124.

74. Micci F, Panagopoulos I, Bjerkehagen B, Heim S. Consistent rearrangement of chromosomal band 6p21 with generation of fusion genes JAZF1/PHF1 and EPC1/PHF1 in endometrial stromal sarcoma. *Cancer Res* 2006;66:107–112.

75. Panagopoulos I, Micci F, Thorsen J, et al. Novel fusion of MYST/Esa1-associated factor 6 and PHF1 in endometrial stromal sarcoma. *PLoS One* 2012;7:e39354.

76. Panagopoulos I, Thorsen J, Gorunova L, et al. Fusion of the ZC3H7B and BCOR genes in endometrial stromal sarcomas carrying an X;22-translocation. *Genes Chromosomes Cancer* 2013;52:610–618.

77. Dewaele B, Przybyl J, Quattrone A, et al. Identification of a novel, recurrent MBTD1-CXorf67 fusion in low-grade endometrial stromal sarcoma. *Int J Cancer* 2014;134:1112–1122.

78. Lee CH, Ou WB, Mariño-Enriquez A, et al. 14-3-3 fusion oncogenes in high-grade endometrial stromal sarcoma. *Proc Natl Acad Sci USA* 2012;109:929–934.

Malignant müllerian mixed tumor (heterologous carcinosarcoma)

79. Silverberg SG. Mixed müllerian tumors. *Curr Top Pathol* 1992;85:35–56.

80. Kempson RL, Hendrickson MR. Smooth muscle, endometrial stromal, and mixed Müllerian tumors of the uterus. *Mod Pathol* 2000;13:328–342.

81. Silverberg SG, Major FJ, Blessing JA, et al. Carcinosarcoma (malignant mixed mesodermal tumor) of the uterus: a gynecologic oncology group pathologic study of 203 cases. *Int J Gynecol Pathol* 1990;9:1–19.

82. Colombi RP. Sarcomatoid carcinomas of the female genital tract (malignant mixed müllerian tumors). *Semin Diagn Pathol* 1993;10:169–175.

83. Pfeiffer P, Hardt-Madsen M, Rex S, Holund B, Bertelsen K. Malignant mixed müllerian tumors of the ovary: report of 13 cases. *Acta Obstet Gynecol Scand* 1991;70:79–83.

84. Garamvoelgyi E, Guillou L, Gebhard S, et al. Primary malignant mixed Müllerian tumor (metaplastic carcinoma) of the female peritoneum: a clinical, pathologic, and immunohistochemical study of three cases and review of the literature. *Cancer* 1994;74:854–863.

85. Rose PG, Rodriguez M, Abdul-Karim FW. Malignant mixed müllerian tumor of the female peritoneum: treatment and outcome of three cases. *Gynecol Oncol* 1997;65:523–525.

86. Inthasorn P, Carter J, Valmadre S, et al. Analysis of clinicopathologic factors in malignant mixed Müllerian tumors of uterine corpus. *Int J Gynecol Cancer* 2002;12:348–353.

87. Costa MJ, Khan R, Judd R. Carcinosarcoma (malignant mixed müllerian [mesodermal] tumor) of the uterus and ovary: correlation of clinical, pathologic, and immunohistochemical features in 29 cases. *Arch Pathol Lab Med* 1991;115:583–590.

88. Meis JM, Lawrence WD. The immunohistochemical profile of malignant mixed müllerian tumor: overlap with endometrial adenocarcinoma. *Am J Clin Pathol* 1990;94:1–7.

89. George E, Manivel JC, Dehner LP, Wick MR. Malignant mixed müllerian tumors: an immunohistochemical study of 47 cases, with histogenetic considerations and clinical correlation. *Hum Pathol* 1991;22:215–223.

90. Mirantes C, Espinosa I, Ferrer I. Epithelial-to-mesenchymal transition and stem cells in endometrial cancer. *Hum Pathol* 2013;44:1973–1981.

91. Streekantaiah C, Rao UN, Sandberg AA. Complex karyotypic aberrations, including i(12p), in malignant mixed müllerian tumor of uterus. *Cancer Genet Cytogenet* 1992;60:78–81.

92. Streekantaiah C, Kwark E, Chuang LT, Ladanyi M. Cytogenetic and molecular characterization of a malignant mixed müllerian tumor of the uterus with a t(8;22)(q24.1;q12). *Cancer Genet Cytogenet* 1999;115:73–76.

93. Kounelis S, Jones MW, Papadaki H, et al. Carcinosarcomas (malignant mixed müllerian tumors) of the female genital tract: comparative molecular analysis of epithelial and mesenchymal components. *Hum Pathol* 1998;29:82–87.

94. Abeln EC, Smit VT, Wessels JW, et al. Molecular genetic evidence for the conversion hypothesis of the origin of malignant mixed müllerian tumours. *J Pathol* 1997;183:424–431.

95. Eimoto M, Iwasaki H, Kikuchi M, Shirakawa K. Characteristics of cloned cells of mixed müllerian tumor of the human uterus: carcinoma cells showing myogenic differentiation in vitro. *Cancer* 1993;71:3065–3075.

Granulosa cell tumor

96. Ayhan A, Tuncer ZS, Tuncer R, et al. Granulosa cell tumor of the ovary: a clinicopathological evaluation of 60 cases. *Eur J Gynaecol Oncol* 1994;15:320–324.

97. Miller BE, Barron BA, Wan JY, et al. Prognostic factors in adult granulosa cell tumor of the ovary. *Cancer* 1997;79:1951–1955.

98. Cronje HS, Niemand I, Bam RH, Woodruff JD. Review of the granulosa-theca cell tumors from the Emil Novak ovarian tumor registry. *Am J Obstet Gynecol* 1999;180:323–327.

99. Robinson JB, Im DD, Logan L, McGuire WP, Rosenshein NB. Extraovarian granulosa cell tumor. *Gynecol Oncol* 1999;74:123–127.

100. Stuart GC, Dawson LM. Update on granulos cell tumours of the ovary. *Curr Opin Obstet Gynecol* 2003;15:33–37.

101. Sehouli J, Drescher FS, Mustea A, et al. Granulosa cell tumor of the ovary: 10 years follow-up data of 65 patients. *Anticancer Res* 2004;24:1223–1229.

102. Susil BJ, Sumithran E. Sarcomatous change in granulosa cell tumor. *Hum Pathol* 1987;18:397–399.

103. Flemming P, Wellmann A, Maschek H, Lang H, Georgii A. Monoclonal antibodies against inhibin represent key markers of adult granulosa cell tumors of the ovary even in their metastases: a report of three cases with late metastases, being previously interpreted as hemangiopericytoma. *Am J Surg Pathol* 1995;19:927–933.

104. Rishi M, Howard LN, Bratthauer GL, Tavassoli FA. Use of monoclonal antibody against human inhibin as a marker for sex cord-stromal tumors of the ovary. *Am J Surg Pathol* 1997;21:583–589.

105. McCluggage WG, Maxwell P, Sloan JM. Immunohistochemical staining of ovarian granulosa cell tumors with monoclonal antibody against inhibin. *Hum Pathol* 1997;28:1034–1038.

106. Hildebrandt RH, Rouse RV, Longacre TA. Value of inhibin in the identification of granulosa cell tumors of the ovary. *Hum Pathol* 1997;28:1387–1395.

107. Otis CN, Powell JL, Barbuto D, Carcangiu ML. Intermediate filamentous proteins in adult granulosa cell tumors: an immunohistochemical study of 25 cases. *Am J Surg Pathol* 1992;16:962–968.

108. Costa MJ, DeRose PB, Roth LM, et al. Immunohistochemical phenotype of ovarian granulosa cell tumors: absence of epithelial membrane antigen has diagnostic value. *Hum Pathol* 1994;25:60–66.

109. Movahedi-Lankarani S, Kurman RJ. Calretinin, a more sensitive but less specific marker than alpha-inhibin for ovarian sex-cord stromal neoplasms: an immunohistochemical study of 215 cases. *Am J Surg Pathol* 2002;26:1477–1483.

110. Fletcher JA, Gibas Z, Donovan K, et al. Ovarian granulosa-stromal cell tumors are characterized by trisomy 12. *Am J Pathol* 1991;138:515–520.

111. Gorski GK, McMorrow LE, Blumstein L, Faasse D, Donaldson MH. Trisomy 14 in two cases of granulosa cell tumor of the ovary. *Cancer Genet Cytogenet* 1992;60:202–205.

112. Lindgren V, Waggoner S, Rotmensch J. Monosomy 22 in two ovarian granulosa cell tumors. *Cancer Genet Cytogenet* 1996;89:93–97.

113. Van Den Berghe I, Dal Cin P, De Groef K, Michielssen P, Van Der Berghe H. Monosomy 22 and trisomy 14 may be early events in the tumorigenesis of adult granulosa cell tumor. *Cancer Genet Cytogenet* 1999;112:46–48.

114. Shah SP, Köbel M, Senz J, et al. Mutation of FOXL2 in granulosa-cell tumors of the ovary. *N Engl J Med* 2009;360:2719–2729.

115. Al-Agha OM, Huwait HF, Chow C, et al. FOXL2 is a sensitive and specific marker for sex cord-stromal tumors of the ovary. *Am J Surg Pathol* 2011;35:484–494.

Endometriosis

116. Clement PB. Pathology of endometriosis. *Pathol Annu* 1990;25 (Part 1):245–295.

117. Clement PB. The pathology of endometriosis: a survey of the many faces of a common disease emphasizing diagnostic pitfalls and unusual newly appreciated aspects. *Adv Anat Pathol* 2007;14:241–260.

118. Singh KK, Lessells AM, Adam DJ, et al. Presentation of endometriosis to general surgeons: a 10-year experience. *Br J Surg* 1995;82:1349–1351.

119. Blanco RG, Periphivel VS, Shah AK, et al. Abdominal wall endometriomas. *Am J Surg* 2003;185:596–598.

120. Zhao X, Lang J, Leng J, et al. Abdominal wall endometriosis. *Int J Gynecol Obstet* 2005;90:218–222.

121. Horton JD, DeZee KJ, Ahnfeldt EP, Wagner M. Abdominal wall endometriosis: a surgeon's perspective and review of 445 cases. *Am J Surg* 2008;196:207–212.

122. Brooks JJ, Wheeler JE. Malignancy arising in extragonadal endometriosis: a case report and summary of the world literature. *Cancer* 1977;40:3065–3073.

123. Stern RC, Dash R, Bentley RC, et al. Malignancy in endometriosis: frequency and comparison of ovarian and extraovarian types. *Int J Gynecol Pathol* 2001;20:133–139.

124. Leng J, Lang J, Guo L, Li H, Liu Z. Carcinosarcoma arising from atypical endometriosis in a cesarean section. *Int J Gynecol Cancer* 2006;16:432–435.

125. Shalin SC, Haws AL, Carter DG, Zarrin-Khameh N. Clear cell adenocarcinoma arising from endometriosis in abdominal wall cesarean section scar: a case report and review of the literature. *J Cutan Pathol* 2012;39:1035–1041.

126. Begin LR. Florid soft tissue decidual reaction: a potential mimic of neoplasia. *Am J Surg Pathol* 1997;21:348–353.

127. Seidman JD. Prognostic importance of hyperplasia and atypia in endometriosis. *Int J Gynecol Pathol* 1996;15:1–9.

Chapter 19
Angiomyolipoma and other perivascular epithelioid cell tumors (PEComas)

Markku Miettinen
National Cancer Institute, National Institutes of Health

Perivascular epithelioid cell tumors (PEComas) are a group of smooth-muscle-related mesenchymal tumors that also contain melanocytic antigens, especially HMB45, and variably melan A.[1]

Angiomyolipoma (AML) occurs primarily in the kidney but is rare in the liver or retroperitoneum (Figure 19.1). It can be associated with tuberous sclerosis.[2]

Other PEComas include retroperitoneal lymphangiomyoma, pulmonary lymphangiomyomatosis, and clear cell "sugar" tumor. Other PEComas (PEComa, not otherwise specified) usually occur in the abdomen, especially in the uterus and elsewhere in the abdominal cavity, and rarely, in peripheral soft tissues. Cardiac rhabdomyoma is discussed in this chapter, although it is not a PEComa. It is associated with the tuberous sclerosis complex (TSC), however. Some PEComas contain losses in the tuberous sclerosis genes 1 and 2 (TSC1/2), whereas others have overexpression of TFE3 gene as the possible alternative mechanism.

Angiomyolipoma
Clinical features

AML of the kidney usually presents in adults and rarely in children, with an overall median age ~50 years. There is a significant (3:1 or higher) female predominance (Figure 19.2). Retroperitoneal AML has similar demographics, and in fact some of such tumors are extensions of a renal primary tumor. Approximately 20% of renal AMLs occur in patients with TSC. In this context, these AMLs occur in patients 20 years younger than the sporadic cases; these tumors are often multiple and sometimes bilateral. Most childhood AMLs are seen in TSC patients.[2]

TSC is an inheritable disease with an autosomal dominant pattern of inheritance, but more than one-half of these patients have new mutations. Manifestations also include subependymal astrocytic nodules, pulmonary lymphangioleiomyomatosis, facial cutaneous nodules (named angiofibromas), and cardiac rhabdomyoma. Epilepsy and mild cognitive impairment are common.

In the kidney, AMLs range from minute incidental nodules to large tumors >10 cm to 20 cm involving much of the kidney, with an average size of 5 cm in recent major series. Larger tumors can cause hematuria or abdominal pain. Vessel rupture in large tumors can cause a life-threatening intra-abdominal hemorrhage.[2–6]

Because small renal AMLs are indolent tumors with minimal risk of bleeding or malignant evolution, preoperative diagnosis by combined radiologic and cytologic studies can allow for more conservative management options. These include follow-up care only, selective embolization of tumor vessels, or a conservative kidney-sparing resection. This applies especially to TSC patients, who may have multiple and bilateral tumors with radical surgery compromising renal function.[3–5]

Hepatic AML 8% of all AMLs. The largest series show a significant female predominance and 2–10% association with

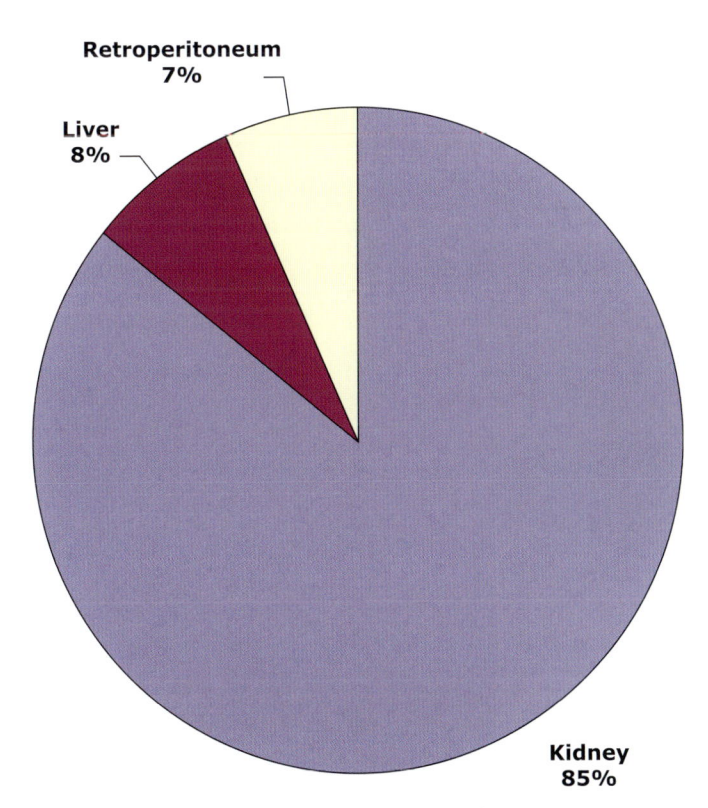

Figure 19.1 Anatomic distribution of 984 angiomyolipomas.
- Kidney 85%
- Liver 8%
- Retroperitoneum 7%

Modern Soft Tissue Pathology, Second Edition, ed. Markku Miettinen. Published by Cambridge University Press. © Cambridge University Press 2016.

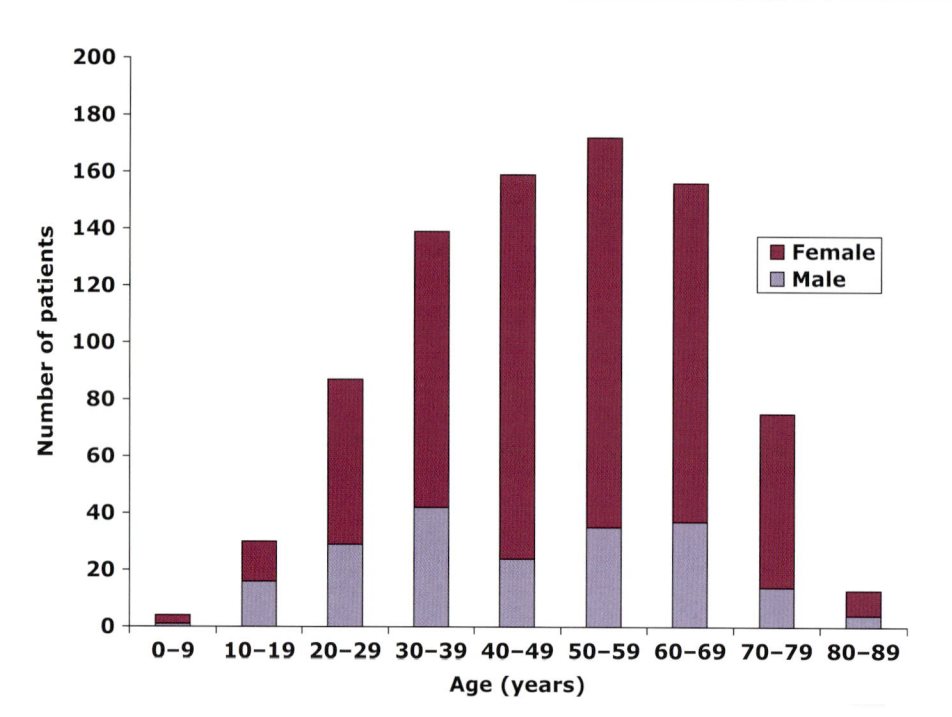

Figure 19.2 Age and sex distribution of 835 patients with renal angiomyolipomas.

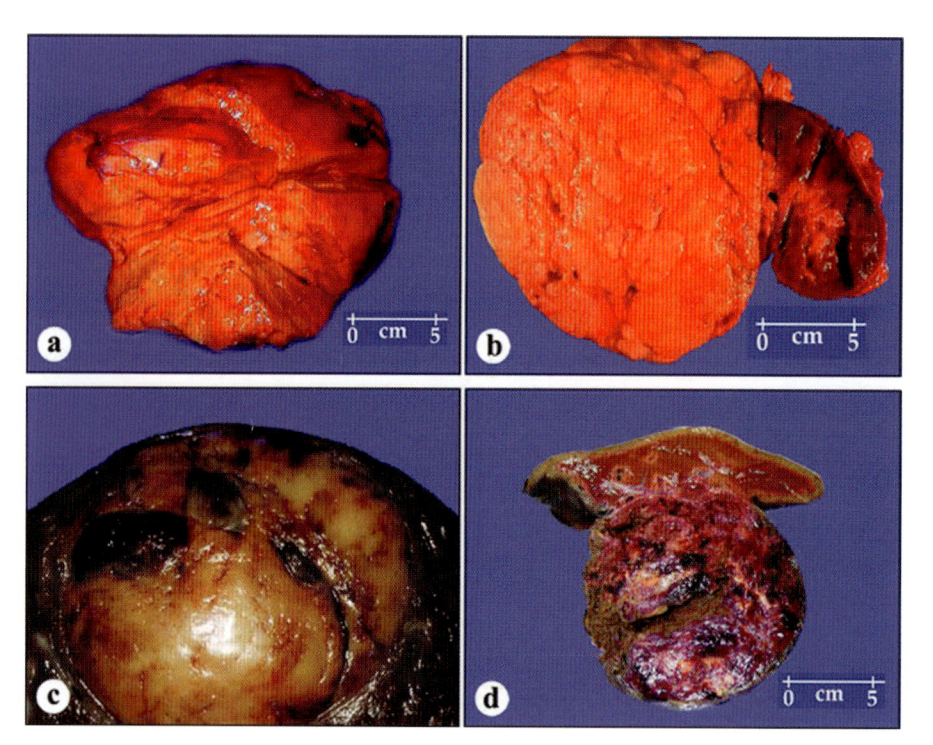

Figure 19.3 Gross appearance of angiomyolipoma. (**a,b**) Renal tumors with a prominent, yellowish lipomatous component. (**c**) Hepatic angiomyolipoma showing a yellowish surface. (**d**) An epithelioid hepatic angiomyolipoma forming a pedunculated external mass. The sectioned surface is reddish with focal yellow fat.

tuberous sclerosis.[7,8] The reported cases have varied from incidentally detected tumors <2 cm to large masses that can be >20 cm.[7–9]

Behavior of the usual AML is benign, but at least three cases of sarcomatous transformation of conventional sporadic AMLs without tuberous sclerosis have been reported. In each case, the sarcomatoid component was epithelioid, resembling a similar focal component in the primary tumor. Liver, lung, and bone metastases developed, and one patient died of the tumor.[10–12] Epithelioid AML in general carries some risk for malignant behavior, but risk estimates vary widely (5–50%).[13–15]

Pathology

The gross appearance of renal AML varies according to the proportions of the components. Predominantly fatty tumors

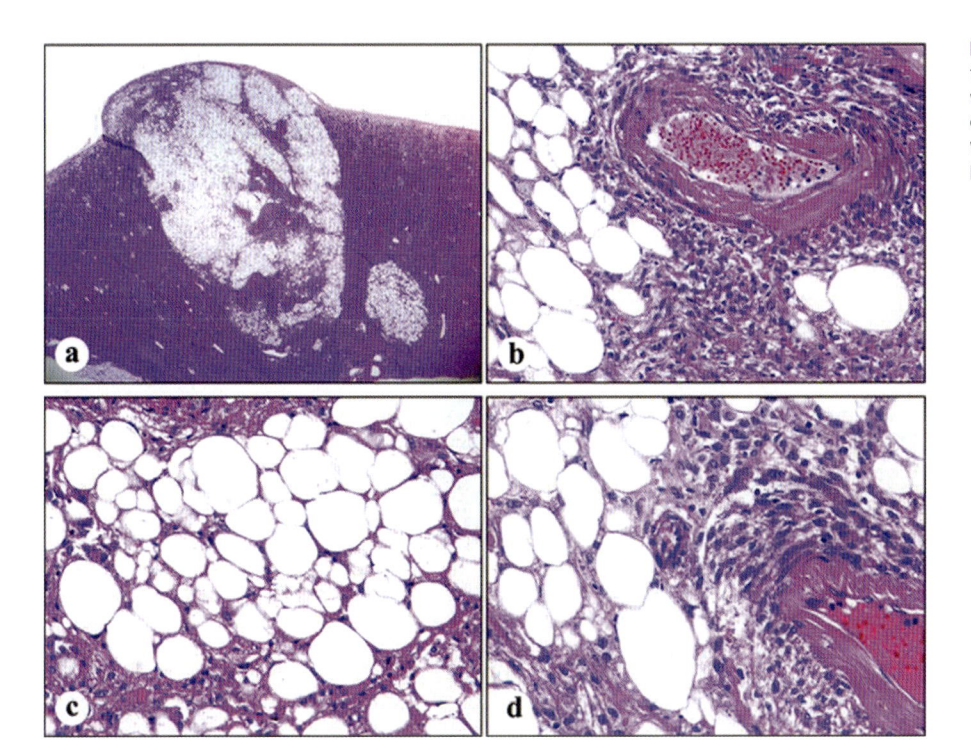

Figure 19.4 (**a**) Renal angiomyolipoma in a tuberous sclerosis patient forms a cortical nodule, with a smaller adjacent nodule. (**b–d**) The typical components: mature fat, vessels with hyalinized walls, and smooth-muscle-like cells, often forming perivascular arrangements.

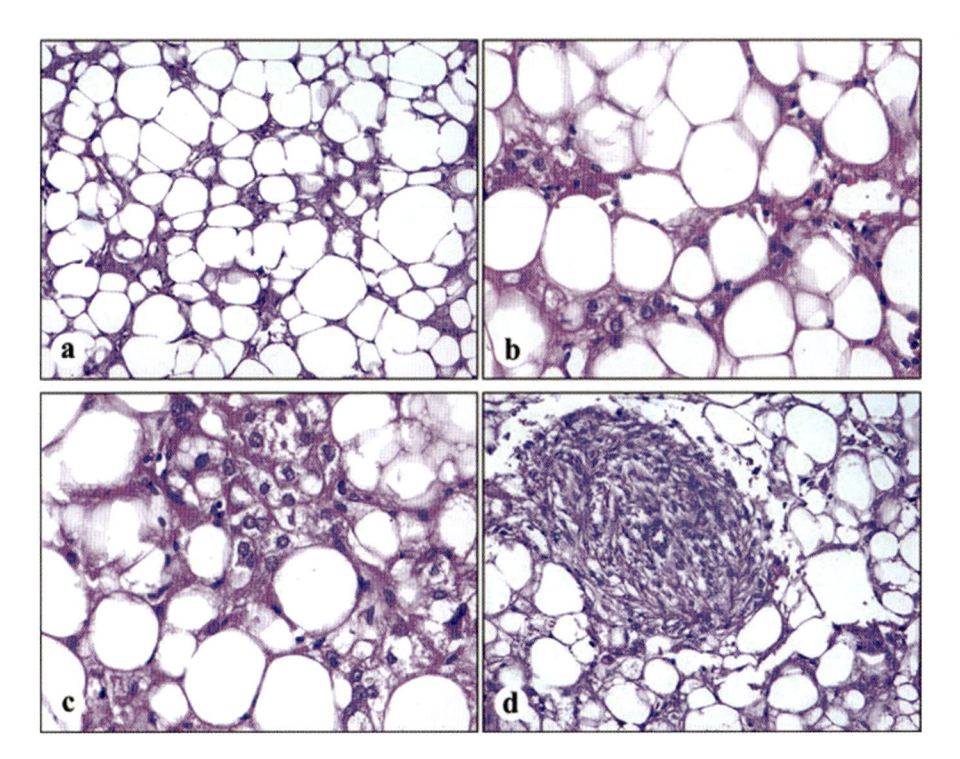

Figure 19.5 Lipomatous component in angiomyolipoma can be dominant in areas containing variable numbers of scattered or clustered smooth-muscle-like cells.

are pale yellow, whereas tumors with an extensive smooth muscle component are brownish (Figure 19.3). Hepatic AML can form an intraparenchymal or occasionally a pedunculated exophytic mass.

Histologically, mature adipose tissue, smooth muscle, and vascular components occur in various proportions (Figure 19.4). Some tumors are predominantly composed of mature fat that lacks atypia, but adipocyte size can vary greatly. These tumors, especially the extrarenal type in the abdomen, are easily misdiagnosed as liposarcomas if the PEComa component is not detected (Figure 19.5). The fatty component can also contain multivacuolated, lipoblast-like cells. These tumors show only inconspicuous clusters of immature smooth muscle cells typically

515

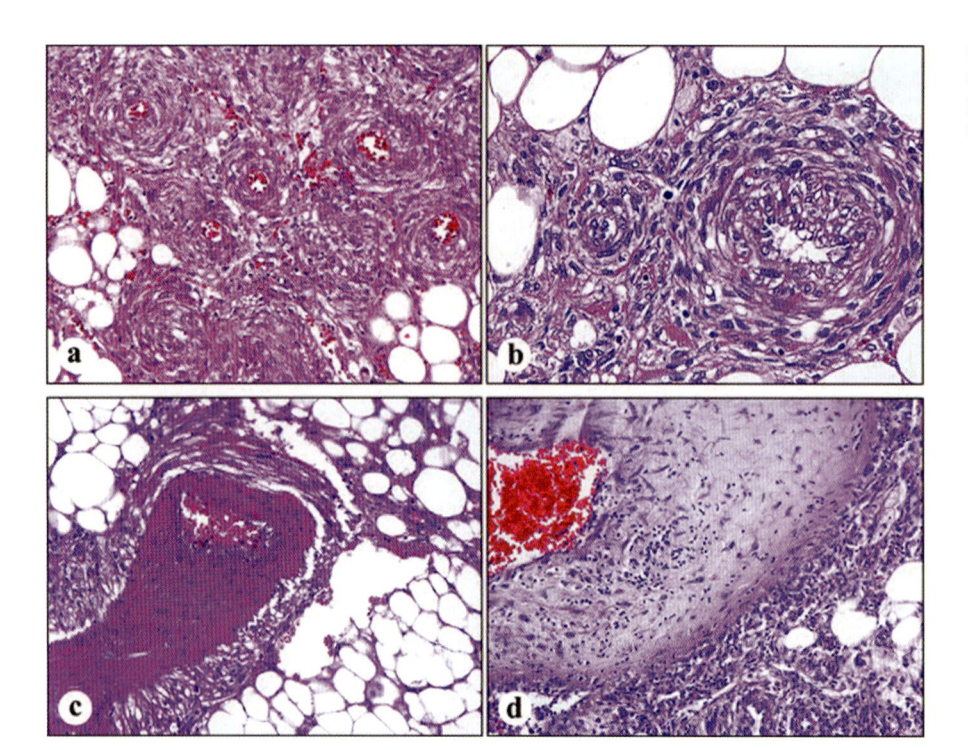

Figure 19.6 Vascular and perivascular features in angiomyolipoma. (**a–c**) Tumor cells forming perivascular whorls. (**d**) Vascular myointimal proliferation in an angiomyolipoma vessel.

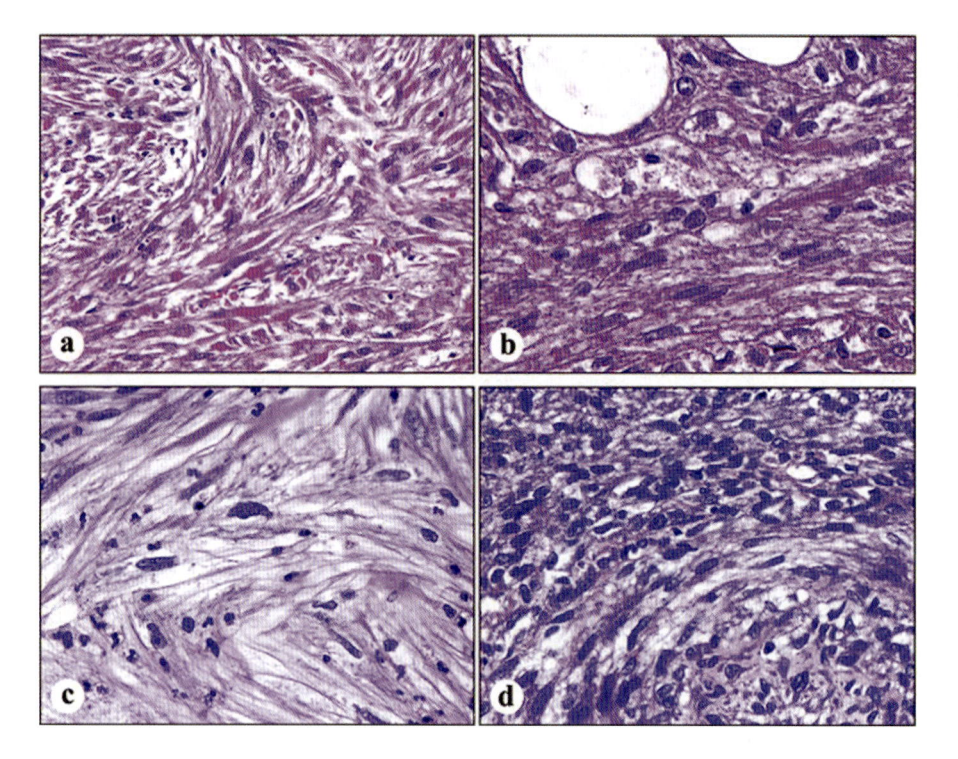

Figure 19.7 Smooth-muscle-like spindle cell element in typical angiomyolipoma. This element shows a variably eosinophilic, often partly cleared cytoplasm.

composed of hollow-appearing, spindled, or polygonal smooth muscle cells that can be present in clusters around the blood vessels or form inconspicuous foci within the fatty component.[16]

The vascular component is composed of blood vessels, often with a thick smooth muscle layer sometimes surrounded by periadventitial clusters of the specific spindle cell or epithelioid smooth-muscle-like PEComa component (Figure 19.6).

Massive intratumoral hemorrhage, however, can disrupt the architecture and make the diagnosis difficult. In addition to the above features, hepatic AML often shows extramedullary hematopoiesis (usually focal, rarely extensive).[7,9]

A smooth muscle component is predominant in some cases,[17] presenting either as a spindled or an epithelioid component. This component can form solid sheets of fascicles, and tumor necrosis can be present. The smooth-muscle-like

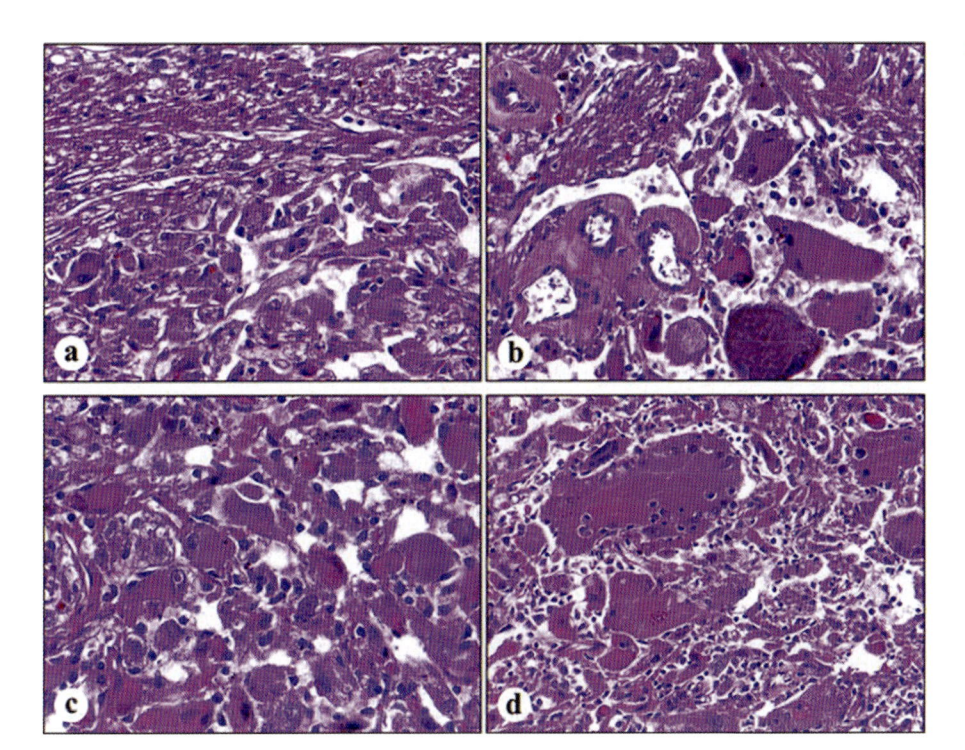

Figure 19.8 Epithelioid angiomyolipoma shows a sharp demarcation between the typical spindle cell and the epithelioid element. (**b–d**) The epithelioid cells have abundant cytoplasm, and some cells are multinucleated and may resemble ganglion cells or histiocytes.

component often shows a cleared or hollow-appearing, "moth-eaten" cytoplasm, typical of this tumor (Figure 19.7).

Approximately 5–10% of AMLs, sporadic or TSC-associated, have a prominent epithelioid component. This component can be seen as solid sheets or interspersed with sclerosing stroma. The epithelioid cells can be small or large, and pleomorphism and mitotic activity are sometimes present.[18–22] Some epithelioid cells have eccentric and multiple nuclei, and others resemble ganglion cells (Figure 19.8). Malignant behavior has been reported in epithelioid AMLs. Adverse histologic factors include carcinoma-like solid growth, vascular invasion, and possibly tumor necrosis.[23] Some studies indicate frequent malignant potential (49%),[14] but others have found only rare malignant evolution (5%);[15] the higher estimate may be related to selection bias in consultation material.

An unusual form of AML has been reported, with cysts lined by hobnail epithelial cells probably derived from renal collecting ducts. These cysts can be visible on gross inspection.[24,25]

Occasionally, AML involves lymph nodes; this does not indicate malignant behavior, but multifocality rather than true metastasis. Vascular invasion into the renal vein or vena cava has been rarely noted and has no adverse prognostic significance.[1] Histologic features associated with tuberous sclerosis include epithelial cysts, epithelioid histology, and multiple satellite tumors.[26]

Immunohistochemistry

The spindle cell smooth muscle component of AML is positive for SMA and variably for desmin. HMB45 positivity is present, especially in epithelioid cells and in a subset of spindle cells.[27–31] Melan A[29] and microphthalmia transcription factor immunoreactivity are variably present.[32] A few tumors show focal cytoplasmic and nuclear S100 protein positivity,[30] but they are negative for CD34. Estrogen and progesterone receptor alpha immunoreactivity has been detected in 25% of sporadic cases and probably more often in TSC-associated AMLs. Estrogen receptor beta is uniformly expressed, however, and androgen receptors also in many cases.[33] One study suggested that loss of calponin h1 correlates with aggressive behavior in AML.[34]

Differential diagnosis

Retroperitoneal AML with dominant fatty component must be differentiated from well-differentiated lipoma-like liposarcoma.[16] Lack of adipocytic atypia and the presence of HMB45 and melan A-positive smooth muscle elements are diagnostic. The epithelioid variants should not be confused with renal cell carcinoma. Immunoreactivity for keratins, including keratin 18, and EMA supports the diagnosis of renal carcinoma. Most nasal and cutaneous tumors reported as AMLs represent angioleiomyomas with a lipomatous component and not AMLs.

Genetics and biology

Although previously characterized as hamartomas, AML and related tumors are more recently understood as clonal neoplasms. Involvement of the tuberous sclerosis complex 2 (TSC2) tumor-suppressor gene in chromosome band 16p13 is the best-known underlying genetic change in AML.[35] Allelic losses at 16p13 have been found in both sporadic AMLs and those associated with TSC, and TSC patients have loss of another copy of TSC2 protein (tuberin) because of a loss-of-function germline mutation in this gene.[35,36] Lack of hamartin

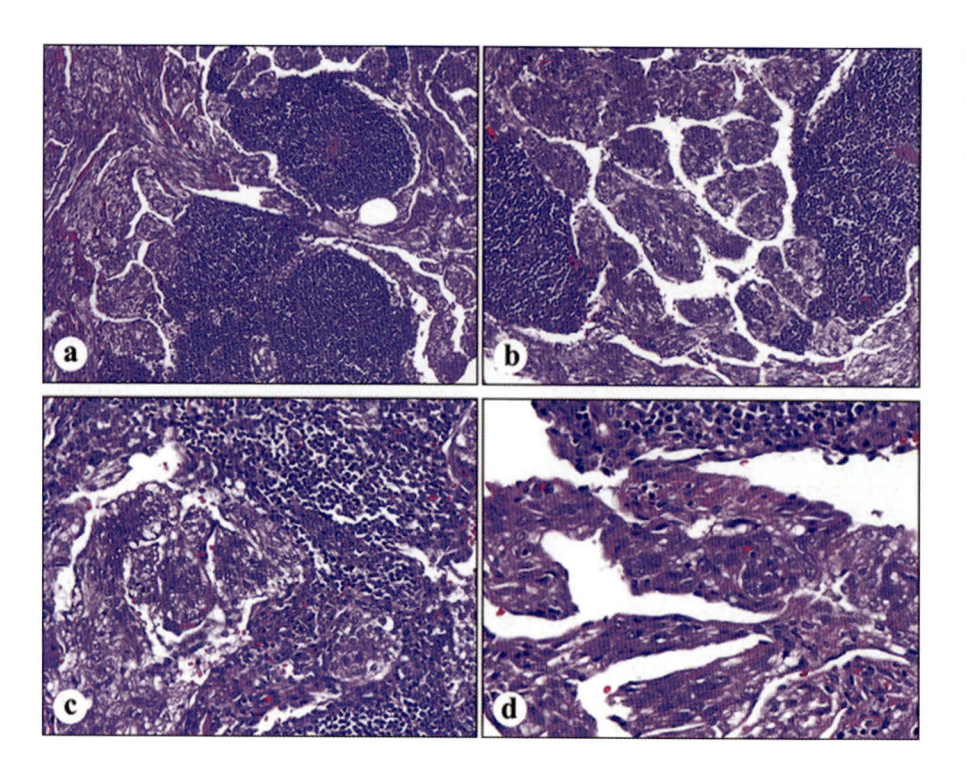

Figure 19.9 Lymphangiomyoma in a lymph node shows intermingling of smooth-muscle-like elements and remains of lymphoid tissue. The smooth-muscle-like elements fill the walls of nodal sinuses.

expression has also been reported, consistent with the concept that AML development can also be caused by lack of TSC1 protein function.[37] Loss of TSC2 leads to activation of the mTOR pathway in sporadic AMLs and most extrarenal PEComas, and therefore rapamycin-like inhibitors of this pathway could be candidates for targeted therapy.[38] Clinical trials on one of the inhibitors, sirolimus or temsirolimus, have shown some effectiveness, although the responses have been short.[39]

Comparative genomic hybridization studies have shown recurrent losses in 5q33–q34, suggesting clonality and the possible involvement of a tumor-suppressor gene in this region.[40] Clonal nature is also supported by human androgen receptor assay (HUMARA) analysis.[41]

Lymphangiomyoma and lymphangiomyomatosis

The two manifestations of lymphangiomyoma and lymphangiomyomatosis (LAM) – mass-like involvement of lymph nodes (lymphangiomyoma) and diffuse involvement of lungs by HMB45-positive smooth-muscle-like cells (lymphangiomyomatosis) – are within the spectrum of one disease.

Clinical features

This disease occurs exclusively in women, usually younger than 50 years; median ages in the clinical series have varied from 40 to 45 years. There is a high association with TSC, and 30% of patients with this complex have been estimated to develop LAM, in addition to having AML.[42,43]

Nodal involvement of LAM occurs in the upper retroperitoneum, pelvis, and posterior mediastinum. It can be associated with chylous ascites or chylothorax as a result of lymphatic obstruction by nodal involvement of LAM. Although initially presenting with apparently local disease, most if not all patients ultimately develop pulmonary disease.[44]

Pulmonary LAM develops insidiously and usually is manifested by dyspnea or spontaneous pneumothorax. Decrease of pulmonary function follows over long periods; however, the natural course of disease is often slow, with survivals reported at 5 years as 91%, at 10 years as 79%, and at 15 years as 71%.[45] Unilateral or bilateral lung transplantation has been successfully performed on many patients, but the disease can recur in the transplant.[45]

Pathology

LAM, whose nodal involvement was originally termed *lymphangiopericytoma*, can form solitary or multiple masses that usually measure <5 cm but can be >10 cm.[44,46] Gross chylous cysts can be present. Nodal LAM typically partially replaces the architecture, leaving small foci of residual lymphoid tissue between the fascicular proliferation of spindled cells, with characteristics of the smooth muscle component of AML and hollow cytoplasm. These fascicles are located in expanded walls of lymphatics lined by small endothelial cells and forming sinusoidal spaces (Figure 19.9). Neither atypia nor mitotic activity is present.

Early pulmonary involvement is not grossly distinctive. When fully developed, the involvement is grossly remarkable as cystic lung texture resembling severe emphysema (Figure 19.10). The emphysematous change has been attributed to activation of the matrix metalloproteinase system in this disease.[47,48]

Histologically, the lung lesions are seen as microscopic nodules in the walls of the peripheral airways involving the interstitial lymphatics. Microscopically, foci of lymphangiomyoma are seen perivascularly, peribronchially, and interstitially, often associated with emphysematous cyst formation.[47,48] The smooth muscle cells resemble those of AML and often have a "hollow" vacuolated, clear cell appearance (Figure 19.11). Immunohistochemical analysis shows that the lesional cells are also similar to the smooth muscle components of AML, being positive for SMA, desmin, and usually HMB45.[24,48] Mutations of the TSC2 gene have been reported in pulmonary lymphangiomyomatosis, similar to other AML family tumors.[49]

Pulmonary clear cell "sugar" tumor

This very rare tumor, originally reported by Liebow and Castleman,[50] usually presents as a solitary pulmonary lesion; however, association with systemic pulmonary lymphangiomyomatosis and multifocal pneumocyte hyperplasia has been reported.[51] One similar tumor has also been reported in the trachea.[52] The tumors are usually incidental radiologic findings manifesting as spherical "coin" lesions in the lung periphery, and 75% occur in women, usually between the ages of 30 and 65 years, but cases have been reported in children. The behavior is usually benign,[50,53–55] but a fatal course with late hepatic metastases was reported in one case. The primary tumor in this patient featured necrosis.[56]

The pulmonary clear cell tumor forms a circumscribed mass <1 cm to 3 cm (rarely >5 cm), with a pushing border. Histologically, the tumor often infiltrates between small airway epithelia, especially on its periphery, and infiltrative margin is not a sign of malignancy for this tumor. There is a prominent vascular pattern that can have a hemangiopericytoma-like character, and tumor cells are sometimes seen around blood vessels, similar to the appearance in AML (Figure 19.12). The tumor cells are principally epithelioid, with abundant floccular, eosinophilic or clear, glycogen-rich cytoplasm (PAS-positive, diastase-sensitive). Variable features include focal calcifications and stromal sclerosis (Figure 19.12d). Mitotic activity is usually undetectable, but mild focal atypia can be present.

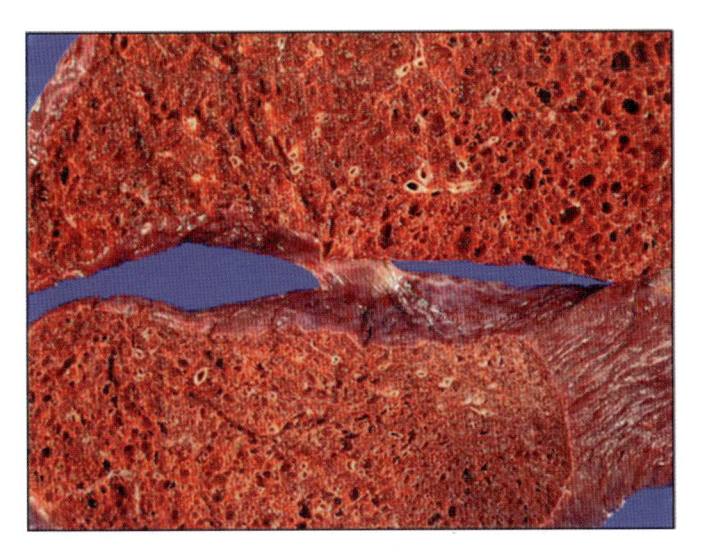

Figure 19.10 Lung involved by advanced lymphangiomyomatosis shows cystic texture resembling severe emphysema.

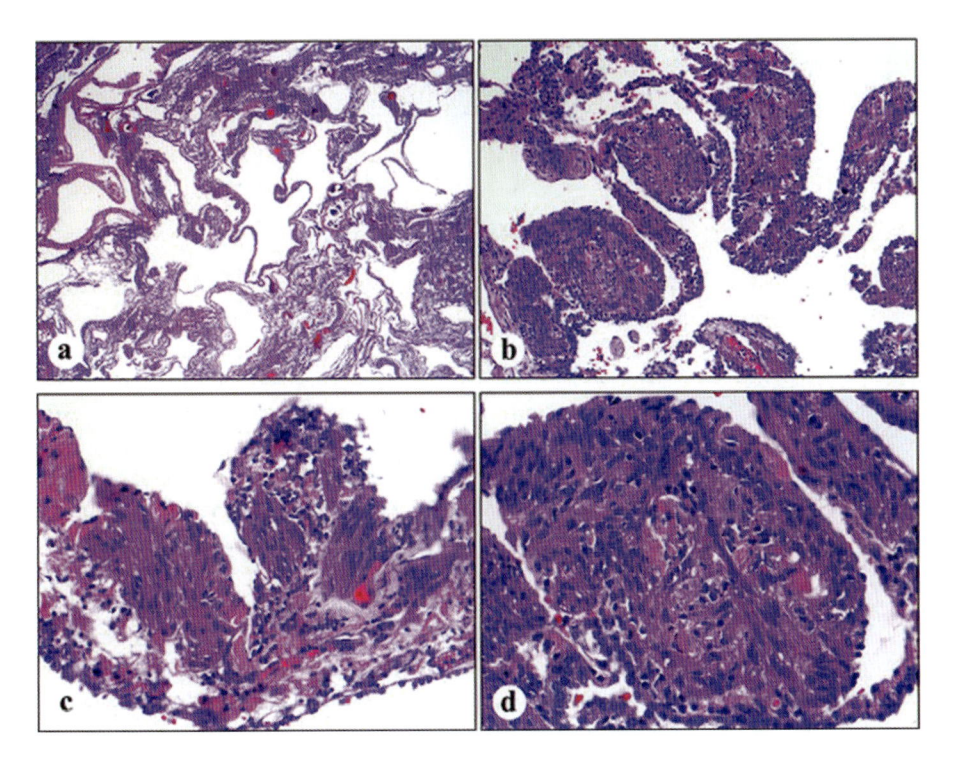

Figure 19.11 (**a**) Low magnification of pulmonary lymphangiomyomatosis is remarkable for cystically dilated airways. (**b–d**) Alveolar walls contain clusters and streaks of smooth-muscle-like cells.

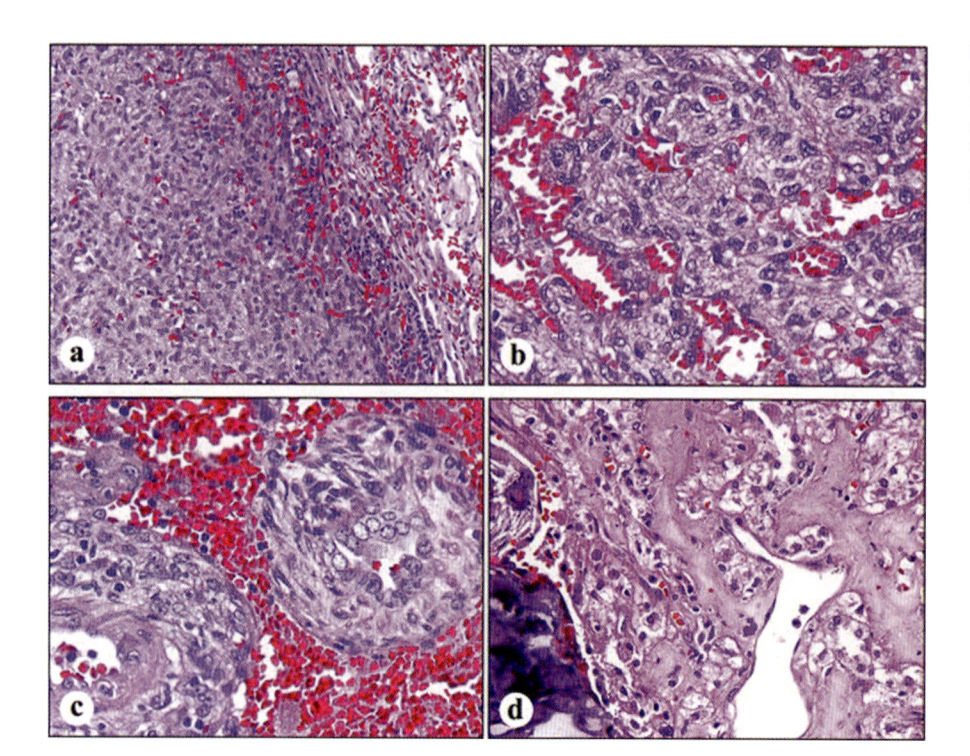

Figure 19.12 Pulmonary clear cell "sugar" tumor. (**a**) Sharp demarcation to lung tissue. (**b**) Tumor cells are epithelioid and have eosinophilic to clear cytoplasm. (**c**) Tumor cells clustered around blood vessels. (**d**) Focal calcification and stromal sclerosis can be present. Note the clear cytoplasmic tumor cells.

Immunohistochemical features are similar to those of epithelioid AMLs, with HMB45 positivity and variable, usually limited (if any) S100 protein positivity and no expression of keratins or EMA.[53,57,58] These features also help to distinguish clear cell tumor from melanoma and metastatic clear cell carcinoma. CD34 and collagen IV expression were detected in both examined cases in one study.[58] Ultrastructural studies have revealed the presence of melanosomes, as well as abundant glycogen particles, especially within lysosomes.[53,54,58]

Other PEComas

These tumors belong to the AML group because of their dual expression of smooth muscle markers and HMB45. They are composed of elements similar to those of the smooth-muscle-like component in AML, often also showing AML-like vascular changes while lacking a fatty component.

Renal capsule

Renal capsular tumors with HMB45, SMA, and desmin positivity (*renal capsular leiomyomas* or *capsulomas*) were first identified by Bonsib.[59] The reported tumors were mostly small, 2 mm to 2.4 cm, but the largest was 5.5 cm. Three of the four patients were women, and all were adults aged 50 to 70 years. None of the patients had tuberous sclerosis, but two had a concomitant renal cell carcinoma.[59]

Grossly, renal capsular PEComas abut kidney tissue and appear as gray-white, homogeneous masses (Figure 19.13). Histologically, they are spindle cell tumors with a resemblance to the spindled component of AML, with low mitotic activity.

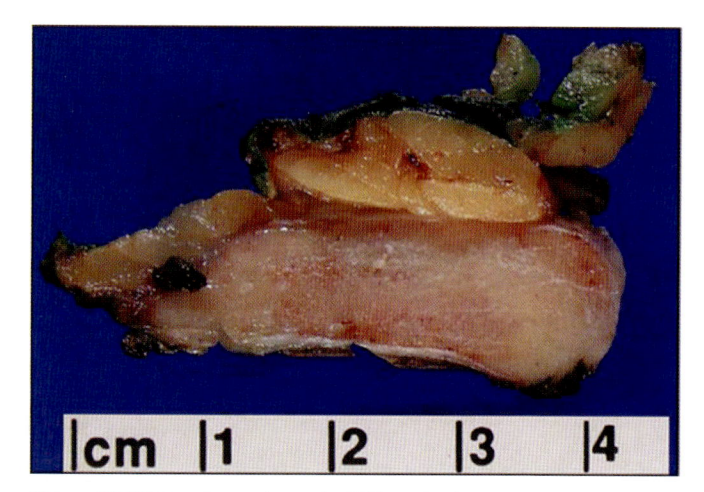

Figure 19.13 Grossly, renal capsular PEComa is a homogeneous pale-tan mass.

The uterus and other abdominal sites

Based on the number of reported cases, the uterus is the most common site for PEComas outside of the kidney and lungs. These tumors occur in adults, often peri- or postmenopausally. The tumors vary from small nodules of <2 cm to large masses measuring >10 cm, although most are relatively small and apparently clinically indolent. In our experience, PEComas constitute 20% of tumors previously uniformly designated as "epithelioid leiomyomas" (excluding tumors now classified as GISTs). Clues for PEComa in the uterus especially include the combination of epithelioid and clear cell morphology when seen in mesenchymal tumors.

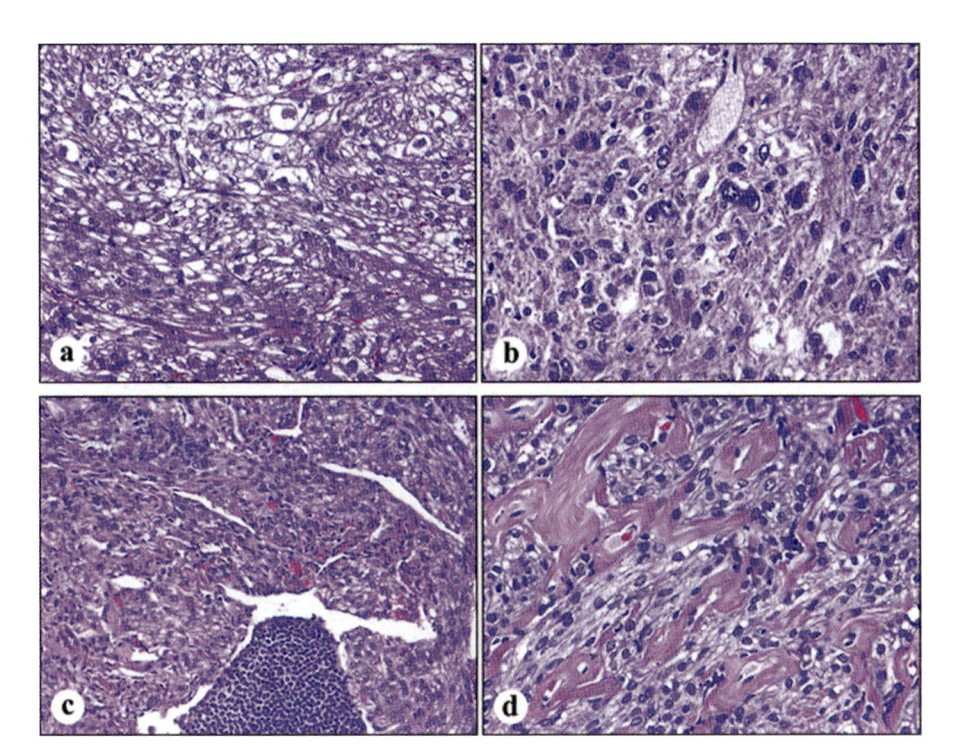

Figure 19.14 Two examples of PEComas of the abdominal cavity. (**a,b**) A mesenteric example with spindled cells and partly clear cytoplasm. Focal nuclear atypia is present. (**c,d**) A pelvic PEComa that also involves a lymph node. The tumor is composed of spindled cells with mildly eosinophilic to clear cytoplasm and vessel walls with sclerosis.

There is a debate over whether such HMB45-positive uterine tumors are to be considered just phenotypic variants of uterine smooth muscle tumors or an independent tumor type (PEComa).[60,61] Relatively high frequency of HMB45 positivity has also been found in uterine leiomyosarcomas. Malignant uterine PEComas have been reported with a low frequency.[60–64]

PEComas also occur in various locations in the abdominal cavity, such as the gastrointestinal tract, retroperitoneum and mesenteries. In the GI tract, PEComas most commonly present in the colon and small intestine, but can occur in other locations, such as the gallbladder. In one series, 40% of tumors had malignant behavior. Factors predicting malignancy included marked nuclear atypia, diffuse pleomorphism, and mitotic rate >2 per 10 HPFs.[65] Occurrence in the omentum, pancreas,[66,67] pelvis,[68] urinary bladder,[69] and ligamentum teres of the liver[70] has also been reported. Folpe et al. reported seven PEComas in the falciform ligament of the liver in young adults of 3–29 years (six female and one male).[71] None of these patients died of the disease, although follow-up was short. A radiologically suspected lung metastasis was detected in one patient, however.[71] In the author's experience, all tumors previously classified as leiomyosarcomas of the falciform ligament are PEComas showing a histologic spectrum that variably resembles the smooth muscle component of AML (Figure 19.14).

Behavior of PEComas

The number of cases with long-term follow-up is very limited; however, most PEComas follow a benign course. Mitotic activity >1 per 50 HPFs, tumor size >5 cm, and coagulation necrosis have been identified as factors statistically associated with malignant tumor behavior.[63] One fatal abdominal PEComa disseminated into the liver; that tumor had only "rare mitotic figures."[70] Because of limited data on tumor behavior, complete excision and follow-up are indicated in all cases.

Pathology of abdominal PEComas

Histologic features of uterine, other female genital-related, and other intra-abdominal PEComas generally fall within the same spectrum. Common themes include epithelioid cells with variably nested patterns showing clear and eosinophilic cytoplasm. Malignant examples typically show pleomorphism and mitotic activity (Figure 19.15). Approximately 10–15% of uterine epithelioid smooth muscle tumors contain HMB45-positive cells, thus qualifying as PEComas.

Most abdominal PEComas are composed of vague nests of epithelioid cells with variably clear cytoplasm, resembling those in pulmonary clear cell tumors, but some have spindle cell morphology resembling the spindle cell component of AML. A prominent vascular pattern with perivascular hyalinization is often present.[72] Lymph nodes can be involved, similar to AML (Figure 19.14). Most examples have a low mitotic activity, but some have mitotic activity >10 per 10 HPFs and coagulation necrosis; cases with such features can be labeled unequivocally malignant. Rarely, soft tissue PEComas can resemble pleomorphic sarcomas histologically.

Immunohistochemically, abdominal and uterine PEComas are typically positive for SMA, desmin, and variably for HMB45, other melanocyte markers such as melan A and PNL2, and in some cases, for estrogen and progesterone receptors. TFE3 positivity occurs in a subset of PEComas and may

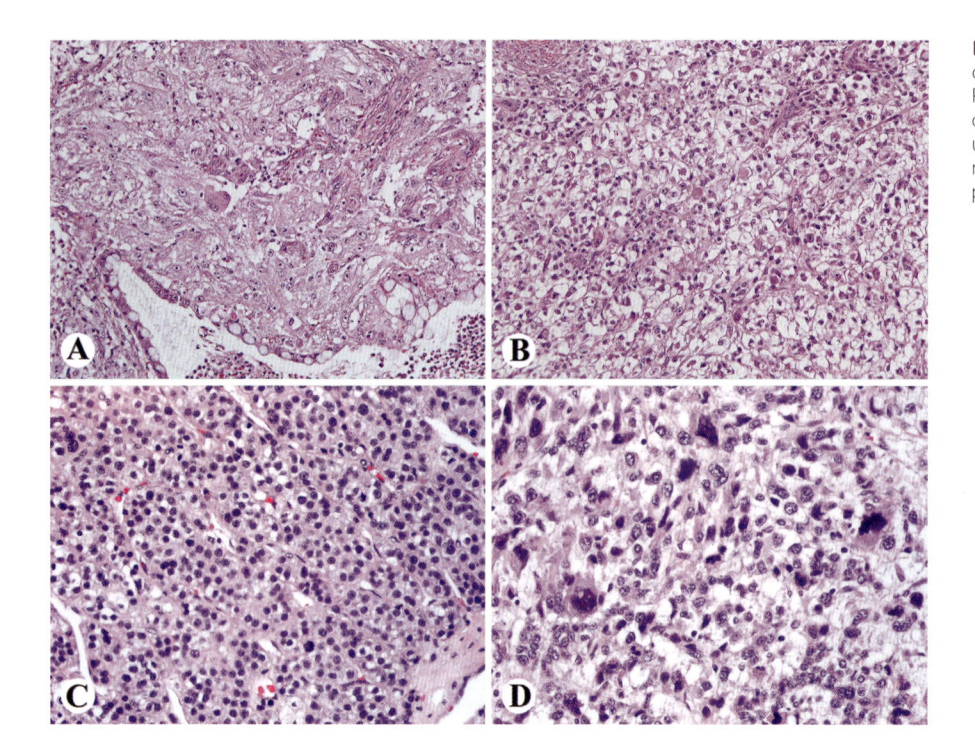

Chapter 19: Angiomyolipoma and other perivascular epithelioid cell tumors (PEComas)

Figure 19.15 (**A**) Colonic PEComa is composed of epithelioid cells with pale cytoplasm. (**B**). Vulvar PEComa shows epithelioid cells with variably clear cell and eosinophilic cytoplasm. (**C,D**) Malignant uterine PEComa that metastasized is composed of nested epithelioid cells and there is focal pleomorphism.

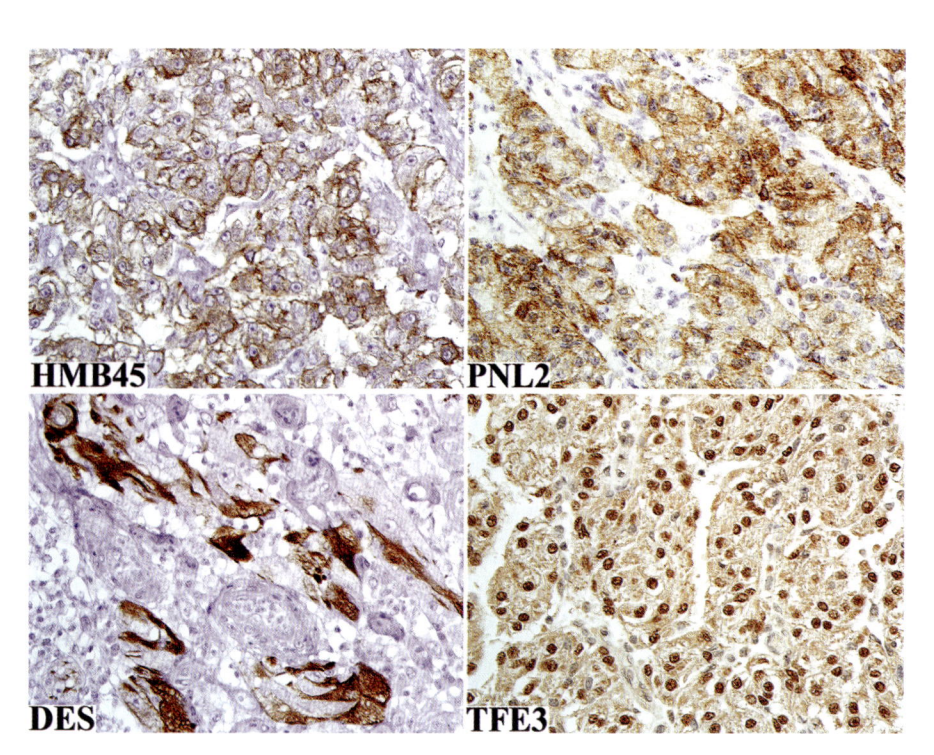

Figure 19.16 Immunohistochemical features of PEComa. The tumor cells show extensive positivity for HMB45 and PNL2-positive cells. There is focal positivity for desmin. Nuclear TFE3 immunoreactivity is seen in a subset of PEComas.

be associated with TFE3 gene rearrangements (Figure 19.16). Potential expression of KIT should not lead to confusion with gastrointestinal stromal tumor.

Soft tissue and cutaneous PEComas

Although PEComas of peripheral soft tissues are rare, cutaneous PEComas have been reported.[73,74] These tumors occurred in the skin and subcutis of adult female patients with a median age of 42 years, and measured 0.5 cm to 3 cm. They had nonmalignant histologic features, with limited atypia and mitotic activity and uneventful short-term follow-up. One PEComa was reported in the breast parenchyma,[75] and another in deep soft tissues of the thigh. The latter tumor had low mitotic activity.[76]

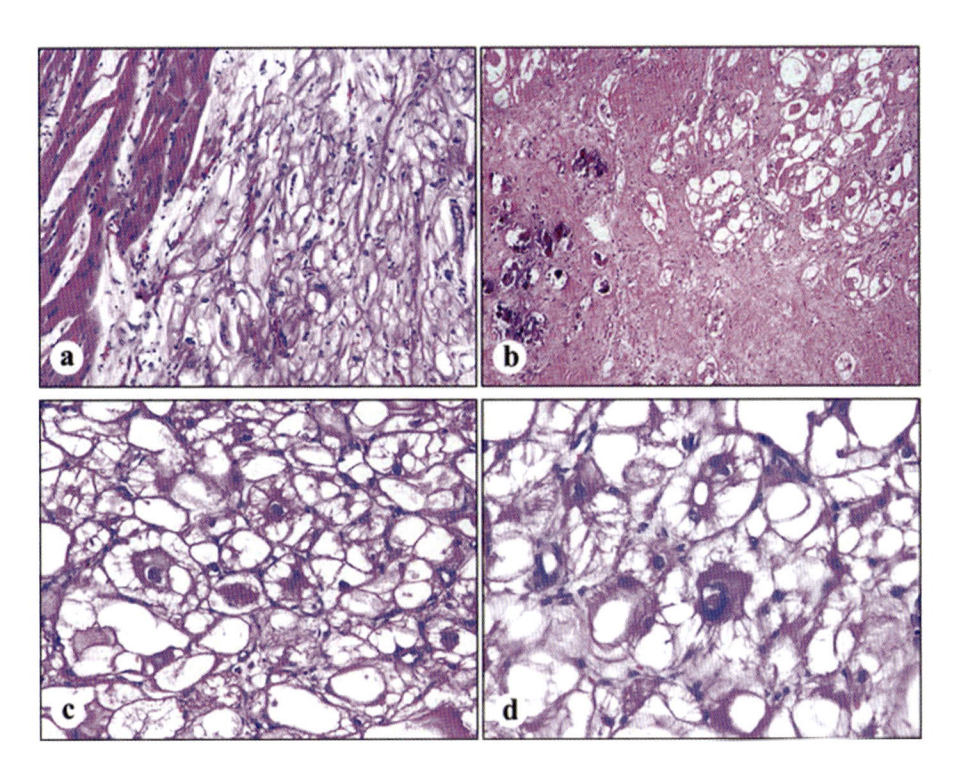

Figure 19.17 (**a**) Cardiac rhabdomyoma forms a mass that is sharply demarcated from the myocardium. (**b**) The tumor cells have a clear cytoplasm. (**c,d**) Cells with multipolar eosinophilic cytoplasm in a spiderweb pattern are present.

Genetics of PEComas

Genetic loss-of-function alterations in the TSC1/2 similar to those in AML have also been reported in other PEComas.[77] Comparative genomic hybridization studies on PEComas (including renal AMLs) have also found nearly consistent 16p losses (including TSC2 locus), along with other recurrent losses in chromosomes 1p, 17p, 18p, 19, and recurrent gains in chromosomes 2q, 3q, 5, 12q, and X. The patterns were mostly similar in renal and extrarenal tumors.[78] Alternatively, PEComas can have TFE3 gene rearrangements and overexpression, which typically manifests by nuclear TFE3-immunolabeling.[79] It also appears that the TFE3 gene-rearranged PEComas lack TSC gene alterations.[80]

Differential diagnosis of PEComas

PEComas prompt a broad differential diagnosis that depends on the site. The most important entities involved are metastatic melanoma, alveolar soft part sarcoma, leiomyosarcoma, and GIST. In contrast to melanomas, PEComas are usually negative for S100 protein and typically are (variably) positive for SMA and desmin. When showing discohesive patterns, PEComa can be distinguished from alveolar soft part sarcoma based on a solid epithelioid or spindle cell component and positivity for SMA; notably desmin can be present in both. Expression of HMB45, melan A, and microphthalmia transcription factor distinguish PEComa from smooth muscle tumors. Potentially KIT-positive abdominal PEComas are distinguished from GIST by their epithelioid, often granular or clear cytoplasm and variable presence of HMB45, melan A, and desmin. TFE gene-rearranged t(6;x) renal carcinomas can overlap with PEComas by their expression of melanocytic markers. However, these tumors typically express PAX2 and PAX8 and are negative for MITF, in contrast to true PEComas.[81]

Cardiac rhabdomyoma

These tumors have a significant male predominance and are typically diagnosed in infants during the first year of life and rarely later during the first decade. Some cases occur in connection with tuberous sclerosis or with valvular and ventricular malformations, but most are sporadic. Multiplicity is common regardless of associated disease.[82,83] Although historically most cases have been autopsy findings, many have recently been diagnosed prenatally, allowing an early intervention and possible surgical cure. When associated with tuberous sclerosis, cardiac rhabdomyomas have a tendency to regress over time, as demonstrated by echocardiography.[84]

Histologically, cardiac rhabdomyoma forms a well-demarcated mass, but multiple tumors are present in more than one-half of all patients. The tumor is composed of clear cells, some of which have eosinophilic cytoplasm, sometimes radiating to the cell periphery like a spiderweb ("spider cells"; Figure 19.17). Ultrastructural studies have demonstrated skeletal muscle differentiation,[81] and immunohistochemically typical is positivity for myoglobin and desmin.[83] Recently, HMB45 positivity has been detected, suggesting that cardiac rhabdomyomas might belong to the family of PEComas.[85] Evidence for this is yet incomplete, however, and the author and colleagues have not been able to confirm this after studying a small number of cases.

References

Angiomyolipoma and abdominal PEComas

1. Bonetti F, Pea M, Martignoni G, et al. Clear cell "sugar" tumor of the lung is a lesion strictly related to angiomyolipoma: the concept of a family of lesions characterized by the presence of perivascular epithelioid cell (PEC). Pathology 1994;26:230–236.

2. Eble JN. Angiomyolipoma of the kidney. Semin Diagn Pathol 1998;15:21–40.

3. Tong YC, Chieng PU, Tsai TC, Lin SN. Renal angiomyolipoma: report of 24 cases. Br J Urol 1990;66:585–589.

4. Kennelly MJ, Grossman HB, Cho KJ. Outcome analysis of 42 cases of renal angiomyolipoma. J Urol 1994;152:1988–1991.

5. Chen SS, Lin AT, Chen KK, Chang LS. Renal angiomyolipoma: experience of 20 years in Taiwan. Eur Urol 1997;32:175–178.

6. L'Hostis H, Deminiere C, Ferriere JM, Coindre JM. Renal angiomyolipoma: a clinicopathologic, immunohistochemical, and follow-up study of 46 cases. Am J Surg Pathol 1999;23:1011–1020.

7. Tsui WM, Colombari R, Portmann BC, et al. Hepatic angiomyolipoma: a clinicopathologic study of 30 cases and delineation of unusual morphologic variants. Am J Surg Pathol 1999;23:34–48.

8. Nonomura A, Enomoto Y, Takeda M, et al. Angiomyolipoma of the liver: a reappraisal of morphological features and delineation of new characteristic histological features from the clinicopathological findings of 55 tumours in 47 patients. Histopathology 2012;61:863–880.

9. Goodman ZD, Ishak KG. Angiomyolipomas of the liver. Am J Surg Pathol 1984;8:745–750.

10. Ferry JA, Malt RA, Young RH. Renal angiomyolipoma with sarcomatous transformation and pulmonary metastases. Am J Surg Pathol 1991;15:1083–1088.

11. Cibas ES, Goss CA, Kulke MH, Demetri GD, Fletcher CDM. Malignant epithelioid angiomyolipoma ("sarcoma ex angiomyolipoma") of the kidney: a case report and review of the literature. Am J Surg Pathol 2001;25:121–126.

12. Martignoni G, Pea M, Rigaud G, et al. Renal angiomyolipoma with epithelioid sarcomatous transformation and metastases: demonstration of the same genetic defects in the primary and metastatic lesions. Am J Surg Pathol 2000;24:889–894.

13. Dalle I, Sciot R, de Vos R, et al. Malignant angiomyolipoma of the liver: a hitherto unreported variant. Histopathology 2001;36:443–450.

14. Nese N, Martignoni G, Fletcher CD, et al. Pure epithelioid PEComas (so-called epithelioid angiomyolipoma) of the kidney: a clinicopathologic study of 41 cases: detailed assessment of morphology and risk stratification. Am J Surg Pathol 2011;35:161–176.

15. He W, Cheville JC, Sadow PM, et al. Epithelioid angiomyolipoma of the kidney: pathological features and clinical outcome in a series of consecutively resected tumors. Mod Pathol 2013;26:1355–1364.

16. Hruban RH, Bhagavan BS, Epstein JH. Massive retroperitoneal angiomyolipoma: a lesion that may be confused with well-differentiated liposarcoma. Am J Clin Pathol 1989;92:805–808.

17. Nonomura A, Minato H, Kurumaya H. Angiomyolipoma predominantly composed of smooth muscle cells: problems in histological diagnosis. Histopathology 1998;33:20–27.

18. Mai KT, Perkins DG, Collins JP. Epithelioid cell variant of renal angiomyolipoma. Histopathology 1996;28:277–280.

19. Eble JN, Amin MB, Young RH. Epithelioid angiomyolipoma of the kidney: a report of five cases with a prominent and diagnostically confusing epithelioid smooth muscle component. Am J Surg Pathol 1997;21:1123–1130.

20. Martignoni G, Pea M, Bonetti F, et al. Carcinomalike monotypic epithelioid angiomyolipoma in patients without evidence of tuberous sclerosis: a clinicopathologic and genetic study. Am J Surg Pathol 1998;22:663–672.

21. Pea M, Bonetti F, Martignoni G, et al. Apparent renal cell carcinomas in tuberous sclerosis are heterogeneous: the identification of malignant epithelioid angiomyolipomas. Am J Surg Pathol 1998;22:180–187.

22. Delgado R., de Leon Bojorge B, Albores-Saavedra J. Atypical angiomyolipoma of the kidney: a distinct morphologic variant that is easily confused with a variety of malignant neoplasms. Cancer 1998;83:1581–1592.

23. Lau SK, Marchevsky A, McKenna RJ, Luhringer DJ. Malignant monotypic epithelioid angiomyolipoma of the retroperitoneum. Int J Surg Pathol 2003;11:223–228.

24. Fine SW, Reuter VE, Epstein JL, Argani P. Angiomyolipoma with epithelial cysts (AMLEC): a distinct cystic variant of angiomyolipoma. Am J Surg Pathol 2006;30:593–599.

25. Davis CJ, Barton JH, Sesterhenn IA. Cystic angiomyolipoma of the kidney: a clinicopathologic description of 11 cases. Mod Pathol 2006;19:669–674.

26. Aydin H, Magi-Galluzzi C, Lane BR, et al. Renal angiomyolipoma: clinicopathologic study of 194 cases with emphasis on the epithelioid histology and tuberous sclerosis association. Am J Surg Pathol 2009;33:289–297.

27. Pea M, Bonetti F, Zamboni G, et al. Melanocyte-marker HMB-45 is regularly expressed in angiomyolipoma of the kidney. Pathology 1991;23:185–188.

28. Chan JK, Tsang WY, Pau MY, et al. Lymphangiomyomatosis and angiomyolipoma: closely related entities characterized by hamartomatous proliferation of HMB-45-positive smooth muscle cells. Histopathology 1993;22:445–455.

29. Fetsch PA, Fetsch JF, Marincola FM, et al. Comparison of melanoma antigen recognized by T-cell (MART-1) to HMB-45: additional evidence to support a common lineage for angiomyolipoma, lymphangiomyomatosis and clear cell sugar tumor. Mod Pathol 1998;11:699–703.

30. Bernard M, Lajoie G. Angiomyolipoma: immunohistochemical and ultrastructural study of 14 cases. Ultrastruct Pathol 2001;25:21–29.

31. Stone CH, Lee MW, Amin MB, et al. Renal angiomyolipoma: further immunophenotypic characterization of an expanding morphologic spectrum. Arch Lab Pathol Med 2001;125:751–758.

32. Zavala-Pompa A, Folpe AL, Jimenez RE, et al. Immunohistochemical study

of microphthalmia transcription factor and tyrosinase in angiomyolipoma of the kidney, renal cell carcinoma, and renal and retroperitoneal sarcomas. *Am J Surg Pathol* 2001;25:65–70.

33. Boorijan SA, Sheinin Y, Crispen PL, et al. Hormone receptor expression in renal angiomyolipoma: clinicopathologic correlation. *Urology* 2008;72:927–943.

34. Islam AHMM, Ehara T, Kato H, Hayma M, Nishizawa O. Loss of calponin h1 in renal angiomyolipoma correlates with aggressive behavior. *Urology* 2004;64:468–473.

35. Green AJ, Smith M, Yater RRW. Loss of heterozygosity on chromosome 16p13.3 in hamartomas from tuberous sclerosis patients. *Nat Genet* 1994;6:193–196.

36. Henske EP, Neumann HP, Scheithauer BW, et al. Loss of heterozygosity in the tuberous sclerosis (TSC2) region of chromosome band 16p13 occurs in sporadic as well as TSC-associated angiomyolipomas. *Genes Chromosomes Cancer* 1995;13:295–298.

37. Plank TL, Loggindou H, Klein-Szanto A, Henske EP. The expression of hamartin, the product of the TSC1 gene, in normal human tissues and in TSC1- and TSC2-linked angiomyolipomas. *Mod Pathol* 1999;12:539–545.

38. Kenerson H, Folpe AL, Takayama TK, Yeung RS. Activation of the mTOR pathway in sporadic angiomyolipomas and other perivascular epithelioid cell neoplasms. *Hum Pathol* 2007;38:1361–1371.

39. Benson C, Vitfell-Rasmussen J, Maruzzo M, et al. A retrospective study of patients with malignant PEComa receiving treatment with sirolimus or temsirolimus: the Royal Marsden Hospital experience. *Anticancer Res* 2014;34:3663–3668.

40. Kattar MM, Grignon DJ, Eble JN, et al. Chromosomal analysis of renal angiomyolipoma by comparative genomic hybridization: evidence for clonal origin. *Hum Pathol* 1999;30:295–299.

41. Green AJ, Sepp T, Yates JRW. Clonality of tuberous sclerosis hamartomas shown by non-random X-chromosome inactivation. *Hum Genet* 1996;97:240–243.

Lymphangiomyoma, lymphangiomyomatosis, and clear cell tumor of lung

42. McCormack FX. Lymphangioleiomyomatosis: a clinical update. *Chest* 2008;133:507–516.

43. Costello LC, Hartman TE, Ryu JH. High frequency of pulmonary lymphangioleiomyomatosis in women with tuberous sclerosis complex. *Mayo Clin Proc* 2000;75:591–594.

44. Matsui K, Tatsuguchi A, Valencia J, et al. Extrapulmonary lymphangioleiomyomatosis (LAM): clinicopathologic features in 22 cases. *Hum Pathol* 2000;31:1242–1248.

45. Urban T, Lazor R, Lacronique J, et al. Pulmonary lymphangioleiomyomatosis: a study of 69 patients. *Medicine (Baltimore)* 1999;78:321–337.

46. Cornog JL, Enterline HT. Lymphangiomyoma, a benign lesion of chyliferous lymphatics synonymous with lymphangiopericytoma. *Cancer* 1966;19:1909–1930.

47. Corrin B, Liebow AA, Friedman PJ. Pulmonary lymphangiomyomatosis. *Am J Pathol* 1975;79:348–382.

48. Zhang X, Travis WD. Pulmonary lymphangioleiomyomatosis. *Arch Pathol Lab Med* 2010;134:1823–1828.

49. Carsillo T, Astrinidis A, Henske EP. Mutations in the tuberous sclerosis complex gene TSC2 are a cause of sporadic pulmonary lymphangioleiomyomatosis. *Proc Natl Acad Sci USA* 2000;97:6085–6090.

Pulmonary clear cell "sugar" tumor

50. Liebow AA, Castleman B. Benign clear cell "sugar" tumor of the lung. *Yale J Biol Med* 1971;43:213–222.

51. Flieder DB, Travis WD. Clear cell "sugar" tumor of the lung: association with lymphangioleiomyomatosis and multifocal micronodular pneumocyte hyperplasia in a patient with tuberous sclerosis. *Am J Surg Pathol* 1997;21:1242–1247.

52. Kung M, Landa JF, Lubin J. Benign clear cell tumor ("sugar tumor") of the trachea. *Cancer* 1984;54:517–519.

53. Gaffey MJ, Mills SE, Askin FB, et al. Clear cell tumor of the lung: a clinicopathologic, immunohistochemical, and ultrastructural study of eight cases. *Am J Surg Pathol* 1990;14:248–259.

54. Andrion A, Mazzucco G, Gugliotta P, Monga G. Benign clear cell tumor ("sugar tumor") of the lung: a light microscopical, histochemical, and ultrastructural study with review of the literature. *Cancer* 1985;56:2657–2663.

55. Dail DH. Benign clear cell ("sugar") tumor of lung. *Arch Pathol Lab Med* 1989;113:573–574.

56. Sale GF, Kulander BG. "Benign" clear-cell tumor (sugar tumor) of the lung with hepatic metastases ten years after resection of pulmonary primary tumor. *Arch Pathol Lab Med* 1988;112:1177–1178.

57. Gal AA, Koss MN, Hochholzer L, Chejfec G. An immunohistochemical study of benign clear cell ("sugar") tumor of the lung. *Arch Pathol Lab Med* 1991;115:1034–1038.

58. Lantuejoul S, Isaac S, Pinel N, et al. Clear cell tumor of the lung: an immunohistochemical and ultrastructural study supporting a pericytic differentiation. *Mod Pathol* 1997;10:1001–1008.

Abdominal and uterine PEComas

59. Bonsib SM. HMB-45 reactivity in renal leiomyomas and leiomyosarcomas. *Mod Pathol* 1996;9:664–669.

60. Vang R, Kempson RL. Perivascular epithelioid cell tumor ("PEComa") of the uterus: a subset of HMB-45 positive epithelioid mesenchymal neoplasms with uncertain relationship with pure smooth muscle tumors. *Am J Surg Pathol* 2002;26:1–13.

61. Silva EG, Deavers MT, Bodurka DC, Malpica A. Uterine epithelioid leiomyosarcomas with clear cells: reactivity with HMB45 and the concept of PEComa. *Am J Surg Pathol* 2004;28:244–249.

62. Fukunaga M. Perivascular epithelioid cell tumor of the uterus: report of four cases. *Int J Gynecol Pathol* 2005;24:341–346.

63. Folpe AL, Mentzel T, Lehr HA, et al. Perivascular epithelioid cell neoplasms of soft tissue and gynecological origin: a clinicopathologic study of 26 cases and review of the literature. *Am J Surg Pathol* 2005;29:1558–1575.

64. Schoolmeester JK, Howitt BE, Hirsch MS, et al. Perivascular epithelioid cell neoplasm (PEComa) of the gynecologic tract: clinicopathologic and immunohistochemical characterization

of 16 cases. *Am J Surg Pathol* 2014;38:176–188.

65. Doyle LA, Hornick JL, Fletcher CD. PEComa of the gastrointestinal tract: clinicopathologic study of 35 cases with evaluation of prognostic parameters. *Am J Surg Pathol* 2013;37:1769–1782.

66. Zamboni G, Pea M, Martignoni G, et al. Clear cell "sugar" tumor of the pancreas: a novel member of the family of lesions characterized by the presence of perivascular epithelioid cells. *Am J Surg Pathol* 1996;20:722–730.

67. Ramuz O, Lelong B, Giovannini M, et al. "Sugar" tumor of the pancreas: a rare entity that is diagnosable on preoperative fine-needle biopsies. *Virchows Arch* 2005;446:555–559.

68. Bonetti F, Martignoni G, Colato C, et al. Abdominopelvic sarcoma of perivascular epithelioid cells: report of four cases in young women, one with tuberous sclerosis. *Mod Pathol* 2001;14:563–568.

69. Williamson SR, Bunde PJ, Montironi R, et al. Malignant perivascular epithelioid cell neoplasm (PEComa) of the urinary bladder with TFE3 gene rearrangement: clinicopathologic, immunohistochemical, and molecular features. *Am J Surg Pathol* 2013;37:1619–1626.

70. Tanaka Y, Ijiri R, Kato K, et al. HMB45/melan-A and smooth muscle actin-positive clear-cell epithelioid tumor arising in the ligamentum teres hepatis: additional example of clear cell "sugar" tumors. *Am J Surg Pathol* 2000;24:1295–1299.

71. Folpe AL, Goodman ZD, Ishak KG, et al. Myomelanocytic tumor of falciform ligament/ligamentum teres: a novel member of the perivascular epithelioid clear cell family of tumors with a predilection for children and young adults. *Am J Surg Pathol* 2000;24:1239–1246.

72. Hornick JL, Fletcher CD. Sclerosing PEComa: clinicopathologic analysis of a distinctive variant with a predilection for the retroperitoneum. *Am J Surg Pathol* 2008;32(4):493–501.

PEComa in peripheral soft tissues

73. Mentzel T, Reisshauer S, Rütten A, et al. Cutaneous clear cell myomelanocytic tumour: a new member of the growing family of perivascular epithelioid cell tumours (PEComas). Clinicopathological and immunohistochemical analysis of seven cases. *Histopathology* 2005;46:498–504.

74. Liegl B, Hornick JL, Fletcher CD. Primary cutaneous PEComa: distinctive clear cell lesions of skin. *Am J Surg Pathol* 2008;32:608–614.

75. Govender D, Sabaratnam RM, Essa AS. Clear cell "sugar" tumor of the breast: another extrapulmonary site and review of the literature. *Am J Surg Pathol* 2002;26:670–675.

76. Folpe AL, McKenney JK, Li Z, Smith SJ, Weiss SW. Clear cell myomelanocytic tumor of the thigh: report of a unique case. *Am J Surg Pathol* 2002;26:809–812.

77. Pan CC, Chung MY, Ng KF, et al. Constant allelic alteration on chromosome 16p (TSC2 gene) in perivascular epithelioid cell tumour (PEComa) with angiomyolipoma. *J Pathol* 2008;214:387–393.

78. Pan CC, Jong YJ, Chai CY, Huang SH, Chen YJ. Comparative genomic hybridization study of perivascular epithelioid cell tumor: molecular genetic evidence of perivascular epithelioid cell tumor as a distinctive neoplasm. *Hum Pathol* 2006;37:606–612.

79. Argani P, Aulmann S, Illei PB, et al. A distinctive subset of PEComas harbors TFE3 gene fusions. *Am J Surg Pathol* 2010;34:1395–1406.

80. Malinowska I, Kwiatkowski DJ, Weiss S, et al. Perivascular epithelioid cell tumors (PEComas) harboring TFE3 gene rearrangements lack the TSC2 alterations characteristic of conventional PEComas: further evidence for a biological distinction. *Am J Surg Pathol* 2012;36:783–784.

81. Argani P, Hicks J, De Marzo AM, et al. Xp11 translocation renal cell carcinoma (RCC): extended immunohistochemical profile emphasizing novel RCC markers. *Am J Surg Pathol* 2010;34:1295–1303.

Cardiac rhabdomyoma

82. Fenoglio JJ, McAllister HA, Ferrans VJ. Cardiac rhabdomyoma: a clinicopathologic and electron microscopic study. *Am J Cardiol* 1976;38:241–248.

83. Burke AP, Virmani R. Cardiac rhabdomyoma: a clinicopathologic study. *Mod Pathol* 1991;4:70–74.

84. Dimario F, Diana D, Leopold H, Chameides L. Evolution of cardiac rhabdomyoma in tuberous sclerosis complex. *Clin Pediatr* 1996;35:615–619.

85. Weeks DA, Chase DR, Malott RL, et al. HMB staining in angiomyolipoma, cardiac rhabdomyoma, other mesenchymal processes, and tuberous sclerosis-associated brain lesions. *Int J Surg Pathol* 1994;1:191–197.

Chapter 20

Rhabdomyomas and rhabdomyosarcomas

Erin R. Rudzinski
Seattle Children's Hospital

Vinay Prasad
Ohio State University

Kadria Sayed
Arkansas Children's Hospital

Faridali Ramji
University of Oklahoma

David M. Parham
University of Southern California Keck School of Medicine

Tumors composed of skeletal muscle are the most common soft tissue sarcomas in the pediatric age group, but make up only a small percentage of sarcomas in adults. In this chapter we discuss benign and malignant soft tissue tumors that have skeletal muscle differentiation as a principal component (Table 20.1). Tumors may also display heterologous skeletal muscle differentiation (Table 20.2). For this reason, when one encounters rhabdomyosarcomatous differentiation, these diagnoses must be excluded, especially in adults.

Unlike tumors of adipose tissue, blood vessels, or fibrous tissue, malignant tumors with skeletal muscle differentiation are more common than benign ones. A prevalent theory is that an undifferentiated mesenchymal stem cell becomes malignant and then differentiates along several tissue pathways, giving rise to sarcomas having various histologic characteristics. Pre-existing skeletal muscle is not a prerequisite for the development of malignant rhabdomyoblastic tumors, as evidenced by rhabdomyosarcomas (RMSs) arising in sites such as the urinary bladder. However, evidence also exists for the origination of some RMSs from differentiated myocytes.[1]

The reasons for the predominance of RMS in childhood and their rarity in adults remain a mystery. During fetal development, normal myogenesis is completed early in the second trimester. The association of RMS with younger age groups suggests that primitive myogenic stem cells are perturbed by proliferation factors, genetic associations, and tissue differentiation processes rather than external carcinogenic factors.[1,2] The paucity of recurrent point mutations typical of adult malignancies further supports this hypothesis.

Rhabdomyoma

Benign soft tissue tumors that primarily differentiate into skeletal muscle are designated as rhabdomyomas. They are

Table 20.1 WHO classification of skeletal muscle tumors

Benign
    Rhabdomyoma
        Adult type
        Fetal type
        Genital type

Malignant
    Embryonal rhabdomyosarcoma
    Alveolar rhabdomyosarcoma
    Spindle cell/sclerosing rhabdomyosarcoma
    Pleomorphic rhabdomyosarcoma

Table 20.2 Tumors other than rhabdomyoma and rhabdomyosarcoma that may contain focal skeletal muscle differentiation

Benign	Malignant
Mature teratoma	Medullomyoblastoma
Benign triton tumor	Malignant peripheral nerve sheath tumors
Thymoma	Wilms' tumor (nephroblastoma)
Complex congenital melanocytic nevi (neurocristic hamartoma)	Mixed müllerian tumor
	Carcinosarcoma
	Immature and malignant teratoma
	Pleuropulmonary blastoma
	Hepatoblastoma
	Ectomesenchymoma
	Malignant mesenchymoma
	Osteosarcoma
	Dedifferentiated sarcomas: chondrosarcoma, liposarcoma
	Neuroendocrine carcinomas
	Melanoma
	Sertoli–Leydig cell tumor

Modern Soft Tissue Pathology, Second Edition, ed. Markku Miettinen. Published by Cambridge University Press. © Cambridge University Press 2016.

rare neoplasms that account for 2% or fewer of all tumors showing primary skeletal muscle differentiation.[3] Rhabdomyomas are broadly classified into cardiac and noncardiac subtypes. Cardiac rhabdomyomas have a strong association with tuberous sclerosis and are discussed in Chapter 19.

Noncardiac rhabdomyomas include fetal, adult, and genital rhabdomyomas. The last type is discussed in Chapter 18.

Fetal rhabdomyoma

First described by Pendle in 1897,[4] fetal rhabdomyomas are rare benign mesenchymal tumors with immature skeletal muscle differentiation and a predilection for the soft tissues of the head and neck.[5]

Clinical features

Fetal rhabdomyomas usually occur in children younger than 3 years of age; but can be seen in older children, adults,[6] and even the elderly.[7] They occur more frequently in boys. Their age range overlaps with RMS and distinction between these tumors can be a source of clinical and histologic confusion. These tumors occur sporadically in children without tuberous sclerosis or other genetic conditions or they can be associated with the basal cell nevus syndrome (Gorlin–Goltz syndrome).[7]

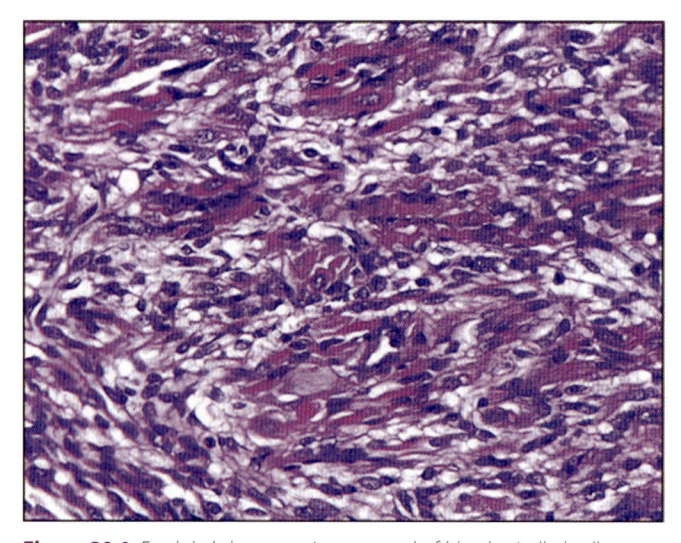

Figure 20.1 Fetal rhabdomyoma is composed of bland spindled cells, some of which show skeletal muscle differentiation.

Fetal rhabdomyomas often occur in the head and neck, but RMS vastly outnumbers them by at least 50 to 1.[4] Postauricular soft tissue shows a particular propensity for these lesions. Other sites of presentation are possible, although subcutaneous location is the rule, and deep examples are rare.

Fetal rhabdomyomas can be cured with a local excision, although progression to RMS,[8] and recurrent disease have been reported.[5]

Histologic features

Fetal rhabdomyomas (classic, immature) contain arrays of spindle cells and myoid tubules, embedded in a myxoid background (Figure 20.1). Less-differentiated cells separate these arrays into bundles, creating a distinctive biphasic pattern. Some tumor cells contain cross-striations, and some tumors contain bundles of eosinophilic strap cells with little or no stroma. Significant nuclear pleomorphism is absent, and mitotic activity is low, usually ≤1 per 10 high-power fields (HPFs).

Intermediate forms of rhabdomyomas, also called juvenile rhabdomyomas, show an appearance that is "intermediate" between fetal and adult rhabdomyoma with loss of immature cells (Figure 20.2).[9] Tumor cells possess features of both myofibroblasts and rhabdomyoblasts.

Differential diagnosis

Differential diagnosis of a fetal rhabdomyoma includes spindle cell and embryonal RMS. The presence of the cambium layer, necrosis, significant hypercellularity, and nuclear or mitotic atypia favors a malignant diagnosis. Circumscription and a low mitotic rate (<5 per 10 HPFs) indicates rhabdomyoma. Chemotherapy-treated RMSs also may be included in the differential diagnosis; however, clinical history and review of the prior material should exclude this possibility.

Immunohistochemistry

Fetal rhabdomyomas express markers of differentiated skeletal muscle, such as desmin, myoglobin, and muscle-specific actin. These tumors do not express keratin, epithelial membrane antigen (EMA), or histiocytic markers such as CD68.[6,7] The latter marker is important, because histiocytic tumors at times strongly resemble differentiated myoblastic lesions.

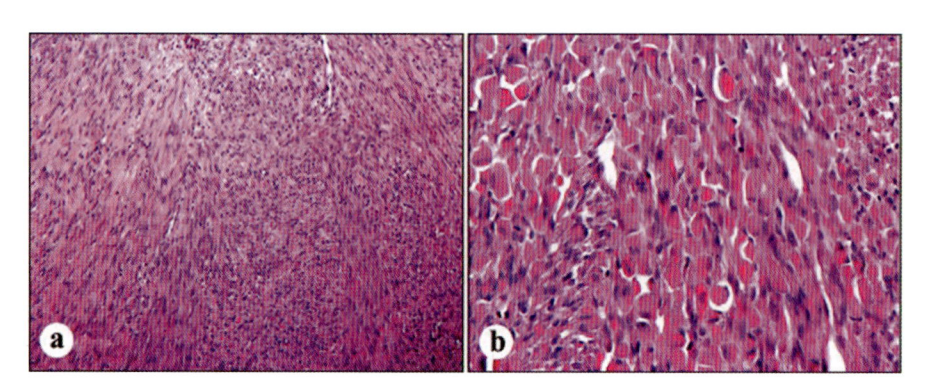

Figure 20.2 So-called juvenile rhabdomyoma. (**a**) A tumor of the posterior auricular soft tissues of a young child consists of interweaving fascicles of spindle cells. (**b**) At higher power, the lesion contains mature-appearing myocytes with abundant eosinophilic cytoplasm.

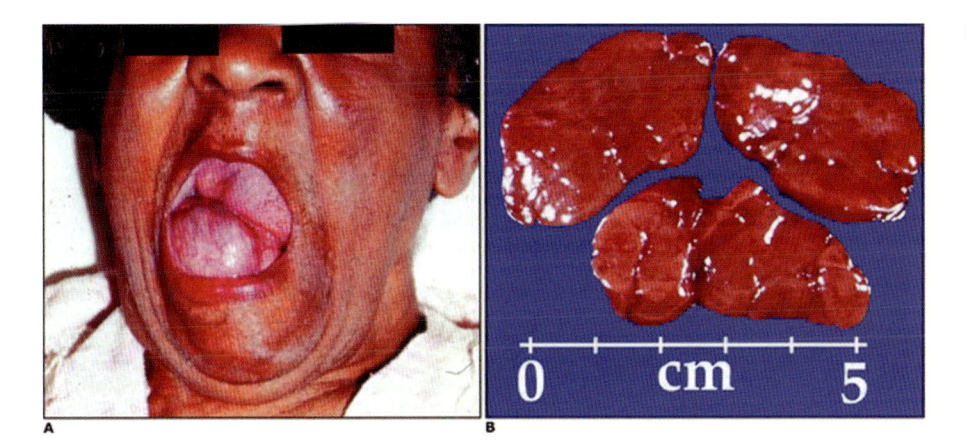

Figure 20.3 (**A**) An adult rhabdomyoma in the floor of the mouth forms a rounded sublingual elevation. (**B**) Grossly, the tumor, removed in several pieces, is brown and lobulated.

Genetics

Multiple cases of fetal rhabdomyoma have been reported in patients with the nevoid basal cell carcinoma syndrome, which is caused by mutations in the tumor-suppressor gene *PTCH1*. *PTCH1* encodes a negative regulator of the hedgehog signaling pathway. In syndromic cases, loss of function mutations in *PTCH1* may fully inactivate PTCH function, thereby activating hedgehog signaling. The hedgehog pathway is activated by various mechanisms, including *PTCH1* frameshift mutations and homozygous deletions, in nonsyndromic fetal rhabdomyoma.[10,11]

Adult rhabdomyoma

Adult rhabdomyomas are rare benign neoplasms predominantly composed of mature, differentiated skeletal muscle cells and typically arising after adolescence.

Clinical features

Adult rhabdomyomas usually present in the oral cavity or the superficial soft tissues of the head and neck of adults, more often in men (Figure 20.3).[12,13] The mean patient age is 50 years, but pediatric cases have been reported.[14] Origin of this tumor from the branchial musculature (third and fourth branchial arches) has been suggested, owing to its unique restriction to the head and neck; up to 20% of cases are multifocal.[1] Although the tumor is benign, local recurrences occur, sometimes years after primary surgery.

Pathologic features

Adult rhabdomyomas form brown, lobulated masses (Figure 20.3). Histologically, they comprise of large round and polygonal cells with abundant clear to eosinophilic cytoplasm and centrally or peripherally located nuclei (Figure 20.4).[13,15] The cells are highlighted by periodic-acid–Schiff stain (PAS), reflecting their high glycogen content. Cytoplasmic cross-striations can sometimes be identified.

Ultrastructurally, adult rhabdomyomas have features of disorganized skeletal muscle cells. Thick and thin filaments are randomly dispersed with pleomorphic mitochondria, and the tumor cells are invested by a continuous basal lamina. Some tumors contain cytoplasmic inclusions similar to those of nemaline myopathy, corresponding to hypertrophic Z bands by electron microscopy.

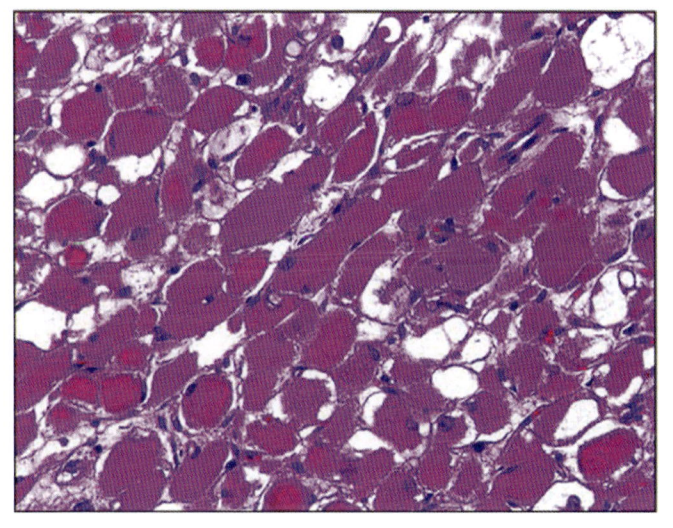

Figure 20.4 Histologically, this adult rhabdomyoma contains closely apposed, mature rhabdomyoblasts with abundant eosinophilic cytoplasm and small nuclei.

Immunohistochemically, adult rhabdomyomas are positive for smooth muscle actin, desmin, and myoglobin, and variably positive for vimentin and alpha smooth muscle actin. Neural markers such as S100 protein and CD57 may also be positive.[13,15]

Differential diagnosis

Because of their eosinophilic cytoplasm, the differential diagnosis of adult rhabdomyoma includes granular cell tumor and hibernoma. Ultrastructurally, rhabdomyomas contain thick and thin filaments, whereas granular cell tumors contain intracellular autophagocytic granules, an extracellular basal lamina. Immunohistochemically they are S100 protein positive. Hibernomas contain lipid droplets and usually have obvious fatty differentiation. Immunohistochemistry usually suffices for ancillary diagnosis, because, with rare exceptions, neither granular cell tumors nor lipomas express muscle markers.

An unusual B-cell lymphoma with collections of peculiar, large, polygonal eosinophilic histiocytes, crystal-storing

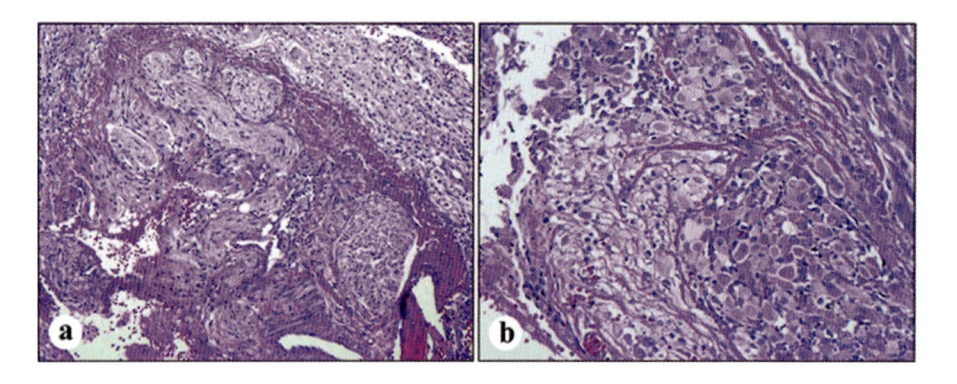

Figure 20.5 Neuromuscular choristoma. (**a**) The nasal mucosa in a child with a growing nasal mass contains hypertrophic, proliferating nerve fibers adjacent to chronic inflammation. (**b**) In other areas, the tumor contains well-differentiated muscle fibers arising from the peripheral nerve, adjacent to a rhabdomyomatous proliferation of myocytes.

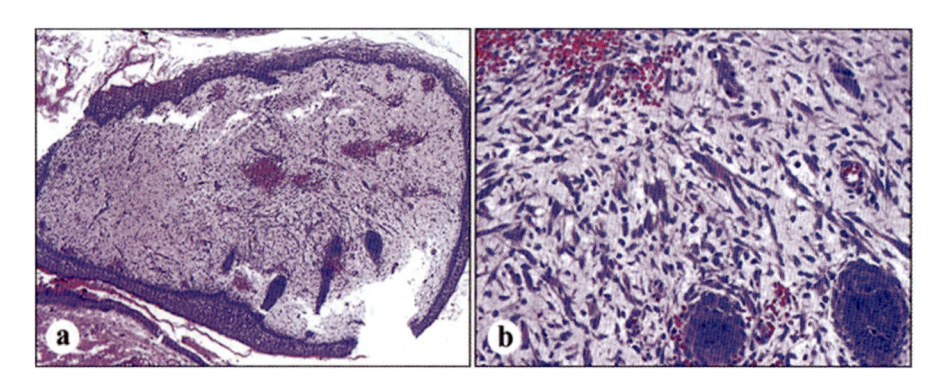

Figure 20.6 Rhabdomyomatous mesenchymal hamartoma. (**a**) This polypoid lesion arose from the region of a first branchial cleft defect and contains a loose mesenchymal core lined by epidermis. (**b**) At higher power, benign cells have features of well-differentiated striated muscle cells.

histiocytosis, also enters into the differential diagnosis of adult rhabdomyoma.[16] Electron microscopy demonstrates cytoplasmic rhomboid crystals that impart the eosinophilia noted on light microscopy. Immunohistochemical negativity for muscle markers further separates these lesions from rhabdomyomas.

Genetics

Cytogenetic characteristics of adult rhabdomyoma are not well known. In a case of a recurrent parapharyngeal rhabdomyoma, Gibas and Miettinen demonstrated clonal structural abnormalities affecting 60% of metaphases and reciprocal translocations involving chromosomes 15 and 17.[17]

Genital rhabdomyoma

Genital rhabdomyomas are benign myogenic tumors that usually arise from the superficial soft tissues of the female genital tract. They probably arise from myogenic stem cells located in the subepithelial stroma. Genital rhabdomyomas usually occur in the vagina or cervix of postpubescent, reproductive-age women.[18,19] Some lesions occur during pregnancy, possibly because of hormonal influence. Genital rhabdomyoma has also been reported in male genitalia.[20] These lesions have a benign course and are cured by local excision.

Genital rhabdomyomas have some similarity to fetal rhabdomyomas and do not contain significant nuclear pleomorphism. Usually exophytic, they contain muscle cells showing focal cross-striations and ultrastructurally haphazardly arranged myofilaments.[21,22] This tumor is further discussed and illustrated in Chapter 18.

Miscellaneous benign myogenous tumors

Rare benign tumors and tumor-like lesions displaying skeletal muscle differentiation include neuromuscular choristoma, rhabdomyomatous mesenchymal hamartoma, and focal myositis.

Neuromuscular choristoma (synonym: *benign triton tumor*, *neuromuscular hamartoma*) is a tumor composed of skeletal muscle and mature neural elements, occurring in infancy and the first decade of life. In most cases these lesions closely follow the course of nerves or a neural plexus. Most descriptions comprise isolated case reports of head and neck lesions, which highlights their rarity.[23–26] These multinodular tumors consist of a disordered bundles of nerve fibers and mature muscle bundles with focal chronic inflammatory infiltrates (Figure 20.5a). Careful examination discloses the presence of myocytes within the perineurial and endoneurial sheaths of the enclosed peripheral nerves (Figure 20.5b). These lesions can be quite extensive and, similar to hamartomas, do not show independent growth. Some lesions have a rhabdomyomatous component or, rarely, a smooth muscle component.[27]

A *rhabdomyomatous mesenchymal hamartoma* is another rare tumor with myogeneous features. Published descriptions are primarily isolated case reports of head and neck lesions[28–33] sometimes associated with other congenital defects.[34,35] An extremely rare entity, this tumor predominantly affects the skin of the face of newborns and young children. Clinically they are ill-defined, firm, and sometimes polypoid, resembling a skin tag.

Histologically, rhabdomyomatous mesenchymal hamartomas contain immature skeletal muscle and varying amounts of fibrous tissue, vessels, and mature adipose tissue (Figure 20.6).

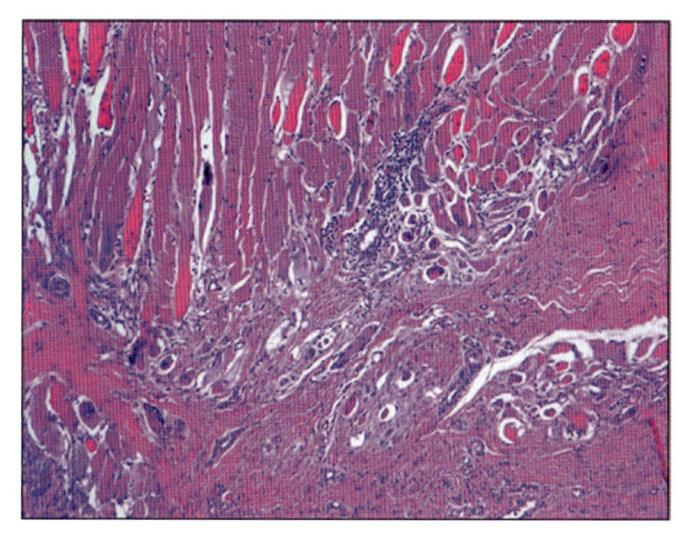

Figure 20.7 Focal myositis. This fibroinflammatory lesion arose from the perioral tissues of a child and presented as a soft tissue mass. It is composed of mature skeletal muscle containing an infiltrate of mononuclear cells that permeates the muscle bundle and surrounds individual fibers. An extensive infectious and autoimmune workup failed to reveal an etiology for this lesion.

They may strongly resemble fibrous hamartoma of infancy, with skeletal muscle instead of a myofibromatous component. The skeletal muscle cells resemble myotubes and contain cross-striations.

Focal myositis

This tumor-like lesion arises in relation to skeletal muscle in a variety of locations. Although a neoplasm might be suspected clinically, histologic examination discloses only degenerating and regenerating skeletal muscle containing infiltrates of lymphocytes and macrophages (Figure 20.7).[36–38] The etiology of these lesions is uncertain, but based on enzyme histochemical studies, a denervating process has been suspected.[39] Focal myositis does not usually progress into polymyositis, but in some cases a relationship has been suggested.[40]

Rhabdomyosarcoma

Most commonly arising in soft tissue, RMS is characterized by a tendency to exhibit the histologic and molecular features of skeletal myogenesis. RMS usually arises in children, although some affect adults. It is now recognized as a class of lesions with differing clinical, cytologic, architectural, and molecular genetic features, all sharing potential for rhabdopoiesis.

Because of their distinctive myogenesis histology and relatively high frequency among childhood cancers, RMSs were well described at the dawn of the histopathologic era of the nineteenth century. The tongue was the site of the initial case description, in 1854.[41] In 1894, Berard first described the features of embryonal RMS,[42] but it was not until 1956 that Theriault and Riopelle described alveolar RMS.[43] In 1958, Horn and Enterline[44] established a formal classification that included alveolar, embryonal, botryoid, and pleomorphic variants. The International Classification, published in 1995, provided a prognostically relevant classification system,[45,46] but this system has recently been superseded by fusion status as the main prognostic marker for RMS.[47] Currently, the WHO classifies RMS into embryonal, alveolar, spindle cell/sclerosing and pleomorphic subtypes.[48]

Introduction and clinical features: general

RMSs are the most common malignant soft tissue tumors in children, but they can affect patients of all ages. The clinical presentation varies by age and histologic subtype (Figure 20.8). Patients with RMS usually present with a painless mass and rarely have systemic complaints, although hypercalcemia has been reported with osteolytic lesions.[49] Symptoms also sometimes relate to affected organs: urinary retention occurs with lesions of the prostate, bile retention follows lesions involving the gallbladder, and diplopia results from orbital tumors. RMS more commonly occurs in males and Caucasians.[50] Children with RMS may have associated genetic conditions such as Beckwith–Weidemann syndrome, neurofibromatosis type 1 (NF1), or the basal cell nevus syndrome.

RMS is generally defined as sarcoma with pure skeletal muscle differentiation, although only a minority of these tumors occurs in extremity skeletal muscle.[46] RMS has histologic features that can be strikingly similar to embryonic and early fetal skeletal muscle. In development, immature mesenchymal cells that resemble fibroblasts develop increased amounts of cytoplasm. These *myoblasts* become mitotically active and then fuse to form myotubes. As the myotubes acquire increased cytoplasm, they become aggregated into densities that outline primordial muscles (Figure 20.9a,b). This aggregation is often a feature of RMS (Figure 20.9c). With increasing differentiation, multinucleated myotubes elongate as their nuclei migrate to the periphery, forming definitive myofibers. With differentiation, the primitive myofibers acquire progressively greater amounts of eosinophilic cytoplasm, coincident with translation and organization of myogenic proteins. Terminally differentiated cells can sometimes be seen in RMS, particularly after chemotherapy (Figure 20.9d).[51] The presence of terminally differentiated cells after therapy does not appear to affect outcome and is currently not an indication for radical surgery or additional aggressive therapy.[52]

Currently, most pediatric cases in developed countries are treated on multi-institutional protocols such as those of the Children's Oncology Group (COG) in North America or the European Pediatric Soft Tissue Sarcoma Group. Treatment of RMS depends on fusion status, location, stage, and other factors, and is generally subdivided into low-, intermediate-, and high-risk groups.[53]

Pathologic features: general

Grossly, RMS can be a myxoid, gray-white mass or a fleshy "lymphoma-like" tumor. These features depend on the relative content of cells and stroma, as well as histologic subtype. RMS may appear encapsulated, but microscopic examination

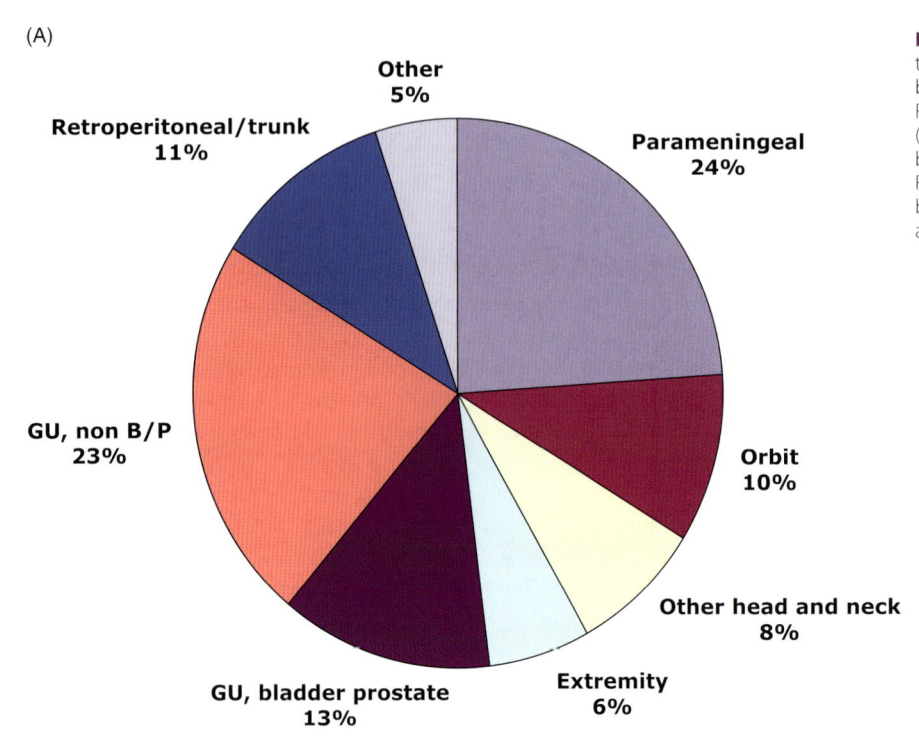

Figure 20.8 Age and anatomic distribution of RMS treated by the Intergroup Rhabdomyosarcoma Study between 1987 and 1999. There were 1279 embryonal RMS patients (72%) and 488 alveolar RMS patients (28%). (**A**) Embryonal RMS by site. (**B**) Embryonal RMS by age in years. (**C**) Alveolar RMSs by site. (**D**) Alveolar RMSs by age in years. (GU = genitourinary; B/P = bladder/prostate. Data courtesy of Dr. James Anderson and the Children's Oncology Group.)

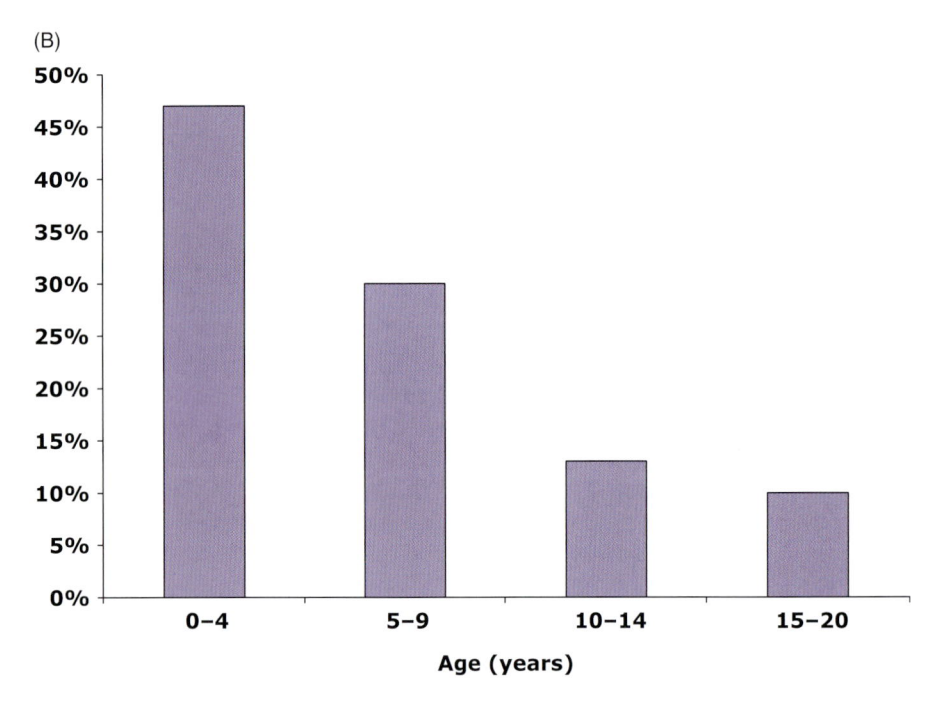

typically reveals poor delineation or capsular invasion. Other lesions are locally invasive, particularly lesions located in the head and neck, and the degree of permeation into adjacent tissues can be extensive. With successful multiagent therapy, the masses typically shrink and become progressively more fibrotic. Examination of inked margins is imperative with resections, because positive margins imply greater risk of local recurrence and require additional therapy. It is important to examine all lymph nodes, particularly with alveolar lesions.

Microscopically, RMS exhibit skeletal muscle differentiation in a strikingly wide histologic spectrum, from primitive undifferentiated cells to rhabdomyoblasts, myotubules, multinucleated cells, and fully differentiated muscle cells with tandem nuclei. Because of limited differentiation, RMS diagnosis generally requires documentation of skeletal muscle differentiation by immunohistochemistry or electron microscopy. Genetic findings can assist in the diagnosis of undifferentiated lesions or limited samples.

Figure 20.8 (cont.)

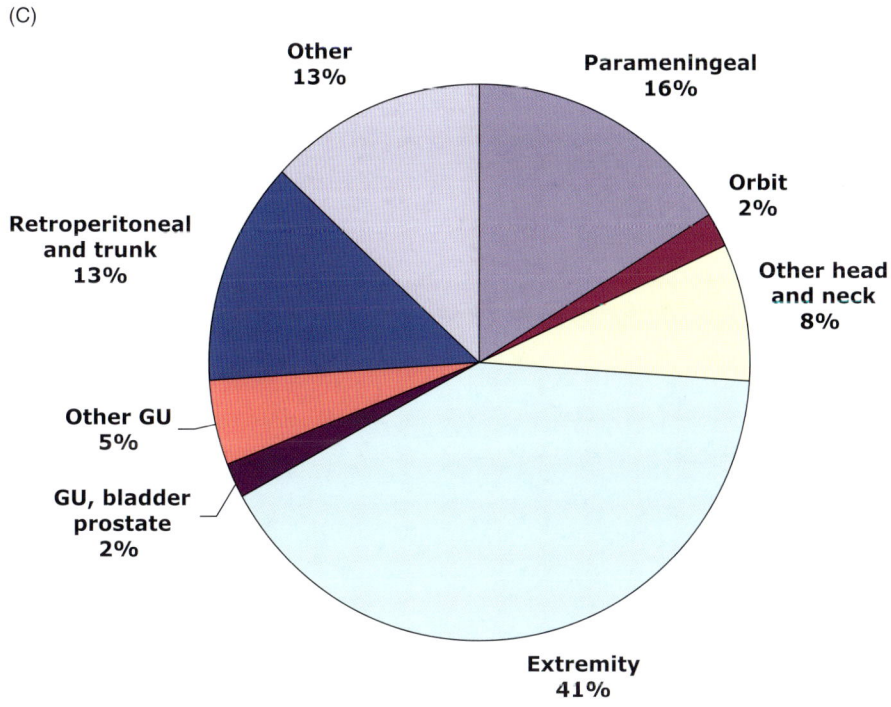

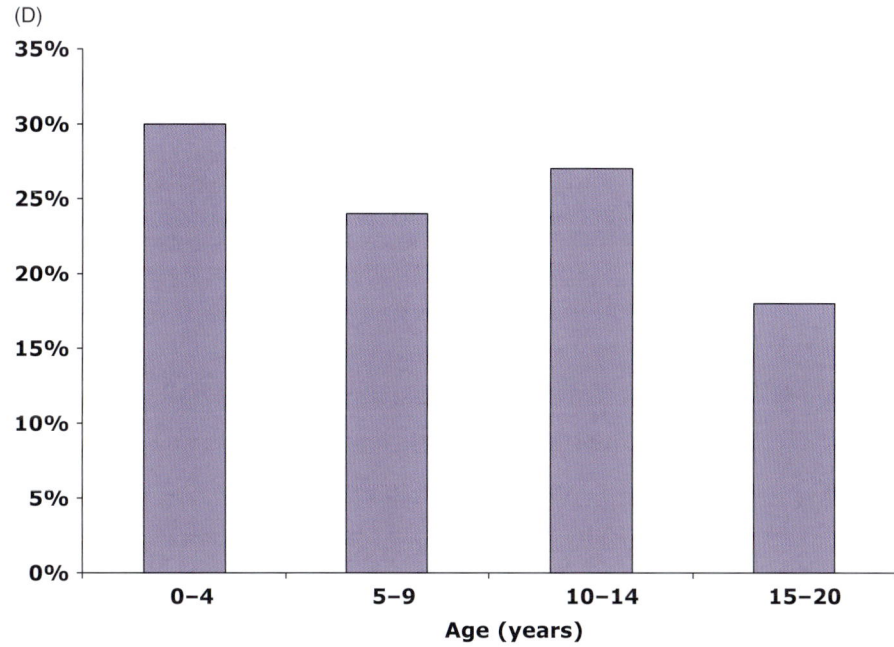

(C)
- Other 13%
- Parameningeal 16%
- Orbit 2%
- Other head and neck 8%
- Extremity 41%
- GU, bladder prostate 2%
- Other GU 5%
- Retroperitoneal and trunk 13%

(D) Age (years): 0–4: 30%, 5–9: 24%, 10–14: 27%, 15–20: 18%

Embryonal rhabdomyosarcoma

Embryonal RMS is a primitive, malignant soft tissue tumor with phenotypic features of embryonic skeletal muscle.

Clinical features

Embryonal RMS accounted for over one-half of all RMSs.[54] Usually embryonal RMS occurs in prepubertal children who present with a genitourinary (Figure 20.10), abdominal, or head and neck mass (Figure 20.11a,b,c). Occurrence in older adults has rarely been reported.[55]

Several patterns of embryonal RMS occur in the pediatric population. Perhaps the most easily recognizable, botryoid RMS is defined by its occurrence adjacent to an epithelial surface; it typically arises in the urinary bladder, vagina, and bile ducts. Less frequently the botyroid subtype is seen in the pharynx, gingivae, cheeks, conjunctiva, ear, and uterus. Historically, the botyroid subtype was associated with an excellent

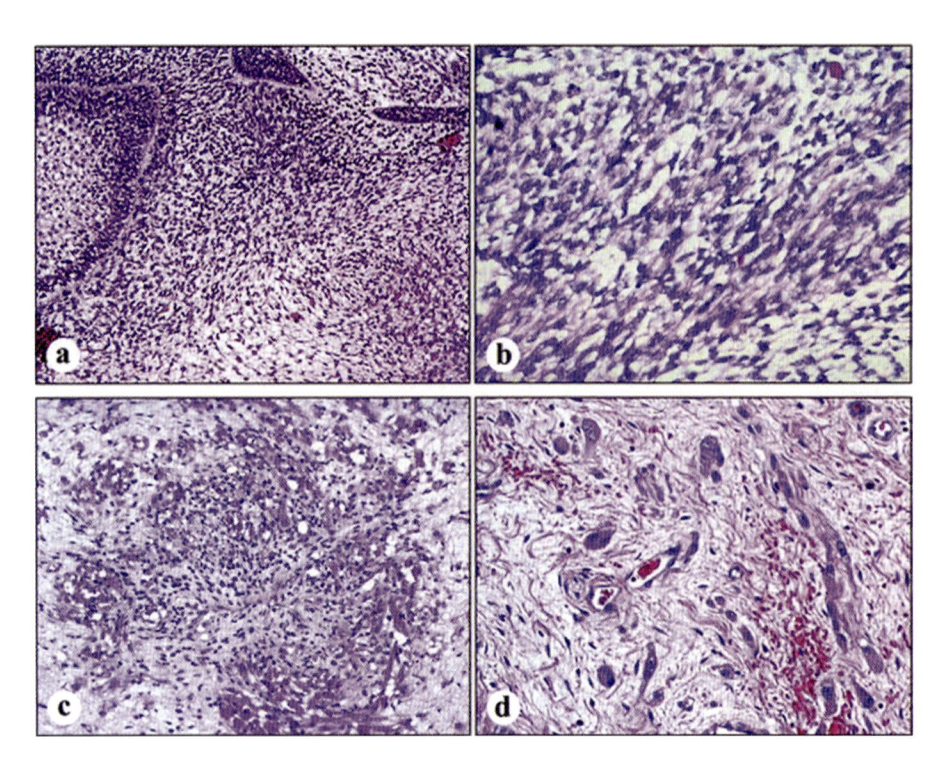

Figure 20.9 Embryonic myogenesis compared with rhabdomyosarcoma (RMS). (**a**) Embryonic myogenesis. Primitive mesenchyme adjacent to cartilage forms dense cellular aggregates. (**b**) This aggregate shows early myotube formation, with the acquisition of elongated eosinophilic cytoplasm. (**c**) Embryonal RMS. Differentiating neoplastic rhabdomyoblasts aggregate into a discrete cluster, similar to embryonic myogenesis. (**d**) Treated RMS. Residual tumor, removed after combination chemotherapy, consists of scattered, terminally differentiated rhabdomyoblasts within a loose collagenous stroma.

prognosis, with a 95% 5-year survival rate.[46] Because the clinical outcome of RMSs is highly site dependent, however, the better prognosis of botryoid tumors could relate to their occurrence in favorable risk sites.[53,56] Botyroid RMS is therefore no longer separated into a distinct subtype but instead is included within embryonal RMS.

Histologic features

Embryonal RMS usually occurs in young children and displays features of embryonic myogenesis. It is composed of haphazardly arranged cells, often having a myxoid stromal background (Figure 20.12). At lower magnification, embryonal RMS typically displays alternating areas of increased and decreased cellular density, imparting a "loose and dense" architecture. This feature recapitulates the focal cellular aggregates noted in embryonic muscle. The loose areas contain a mucoid, intercellular ground substance composed of hyaluronidates and chondroitin, staining with alcian blue, but not with mucicarmine. The amount of matrix production varies considerably from tumor to tumor and ranges from loose, hypocellular, myxoid stroma to denser, immature mesenchyme to mature collagen.

The constituent cells of typical embryonal RMS show wide variation in differentiation, some being primitive and stellate, others having moderate eosinophilic cytoplasm, and still others showing terminal differentiation with abundant eccentric eosinophilic cytoplasm and cross-striations. Careful observation generally discloses rhabdomyoblasts with features of embryonic myocytes. A good hematoxylin and eosin (HE) stain should reveal varying numbers of cells with increased, brightly eosinophilic cytoplasm; these tumoral rhabdomyoblasts are the key to histologic diagnosis. The eosinophilic cytoplasm of rhabdomyoblasts often shows some degree of graininess or striation, and their nuclei should appear viable to distinguish them from apoptotic cells with cytoplasmic eosinophilia. Multinucleate tumor giant cells can recapitulate the fused myotubes of myogenesis (Figure 20.12c).

Rhabdomyoblasts can assume a variety of shapes and appearances that have been likened to a collection of things as diverse as tadpoles, razor straps, spiders, tennis racquets, and broken straws. Potential mimics of rhabdomyoblasts include macrophages, squamous cells, rhabdoid cells, and crystal-storing cells. In case of doubt, immunohistochemical confirmation is advisable. Entrapped native myocytes have the same cytologic features, and immunostains decorate primitive cells as well as differentiated ones.

Embryonal RMSs vary from hypercellular (Figure 20.12a) to hypocellular (Figure 20.12b). Cellularity is sometimes enhanced around blood vessels. A tumor can have any combination of these stromal or cellular features. Hypocellular examples might initially lead one to question whether a tumor was present. Hypercellular examples can consist entirely of dense areas, particularly on limited biopsies, and thus invite confusion with alveolar RMS and other primitive round cell tumors.[57] The cytologic features of the primitive cells assume particular importance in densely cellular samples. Densely cellular embryonal RMS contains cells that are stellate or spindly (Figure 20.12d), in contradistinction to the uniform, lymphocyte-like round cells of "solid" alveolar RMS. Shifts in the histologic criteria of RMS have further confounded this distinction, and a recent study resulted in the reclassification of up to one-third of alveolar RMS as embryonal RMS. In over half of these cases, a dense pattern of embryonal RMS was the source of diagnostic difficulty. Molecular studies for the presence of a FOXO1 rearrangement should be performed, as, with rare exceptions, embryonal RMS, is fusion negative.

The botryoid subtype of embryonal RMS, also called *sarcoma botryoides*, arises adjacent to epithelial surfaces and is characterized by a subepithelial concentration of tumor cells. This so-called cambium layer resembles the zone of cellular condensation seen in the branches of plants. Grossly, these tumors are polypoid, similar to a bunch of grapes (Greek, "botryos"; Figure 20.13). On limited needle biopsies, this pattern might not be apparent. Botryoid RMS can be quite gelatinous, with a hypocellular myxoid stroma, or the cambium layer can be dense. In larger lesions, dense cellularity is not uncommon in the deeper portions. Histologically, they show features of embryonal RMS in the polypoid elements (Figure 20.14).

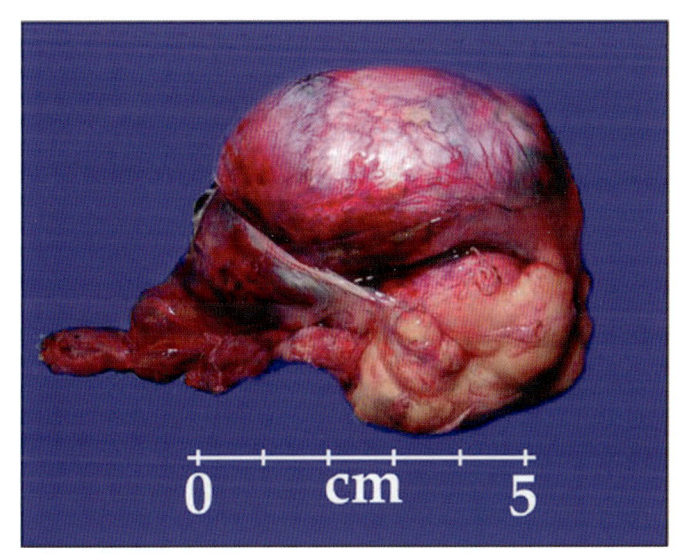

Figure 20.10 Gross appearance of a bosselated paratesticular embryonal RMS shows a yellow-tan tumor arising adjacent to a testicle. (Courtesy of Mr. Armando Jackson and Dr. Cyril D'Cruz.)

Genetics of embryonal rhabdomyosarcoma

Embryonal RMS exhibits no single specific structural cytogenetic features, but instead shows a variety of chromosomal gains and losses.[54,58] A profound loss of heterozygosity occurs on chromosome 11 (11p15.5) in embryonal RMS. The locus causing the Beckwith–Wiedemann cancer predisposition syndrome is also localized to this region. This locus is hypothesized to carry an imprinted tumor-suppressor gene. In syndromic cases, loss of the active maternal allele and retention of the inactive paternal allele leads to loss of expression of tumor-suppressor genes.[59] Candidate genes include *CDKN1C*, *H19*, and *HOTS*.

Mutational analyses reveal a variety of point mutations in ERMS, including predominantly *RAS* family mutations (12%), with a lower frequency of *FGFR4*, *PIK3CA*, *CTNNB1*, *BRAF*, and *PTPN11* mutations. Gene fusions are described in a small subset of embryonal RMS, including one case with a *PAX3-NCOA2* and one with an *EWSR1-DUX4* rearrangement. Embryonal RMSs nearly always lack evidence of a *PAX/FOXO1* gene fusion, although the authors have seen rare exceptions, particularly for embryonal RMS occurring in the extremities. There are no reported genetic features that distinguish the dense or botryoid patterns of RMS from typical embryonal RMS.

Differential diagnosis

Differential diagnosis of embryonal RMS includes malignant peripheral nerve sheath tumors, myofibroblastic tumors, or other myxoid lesions. Botryoid tumors of the urinary bladder should not be confused with reactive spindle cell nodules or inflammatory myofibroblastic tumors (see Chapter 10), or with inflammatory processes such as eosinophilic cystitis.[60]

Spindle cell/sclerosing rhabdomyosarcoma

Spindle cell RMS is composed of tight bundles of spindle cells, resembling smooth muscle and fibrous neoplasms. In some cases, the stroma may show the extensive hyalinization characteristic of the sclerosing variant of RMS.

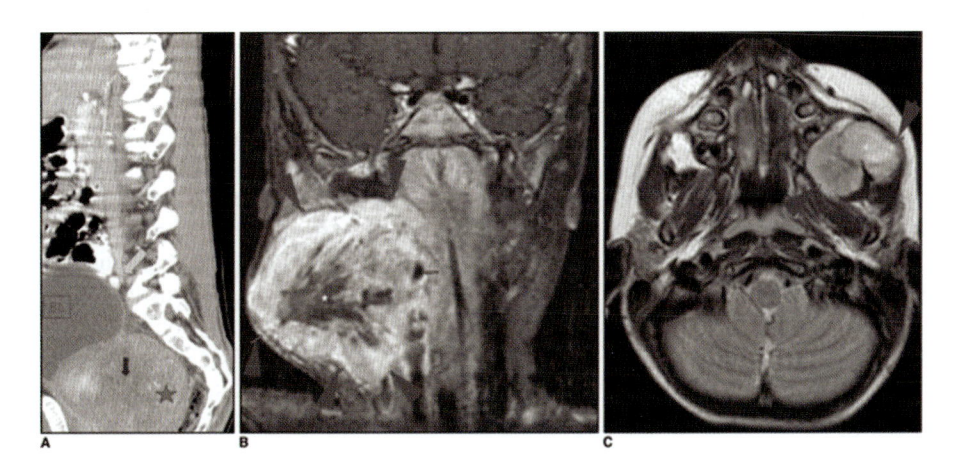

Figure 20.11 Radiologic appearances of embryonal RMS. (**A**) Prostatic RMS. A large mass pushes the base of the bladder cranially. The mass is solid with enhancement (star), but there is a central area of decreased attenuation suggestive of some necrosis (black arrow). Note the deformed bladder (BL) and the slightly dilated ureter (gray arrow). (**B**) Parapharyngeal RMS, coronal MRI, T1-weighted with fat suppression. A mass fills the right parapharyngeal space and shows brisk enhancement (arrowheads) and central necrosis (star). Notice the flow void in the carotid artery (small arrow), which is displaced and partially surrounded by the mass. (**C**) RMS, axial MRI, T2-weighted. A solid mass (arrowheads) is demonstrated on the left side in the masticator space of the face. The mass is well defined.

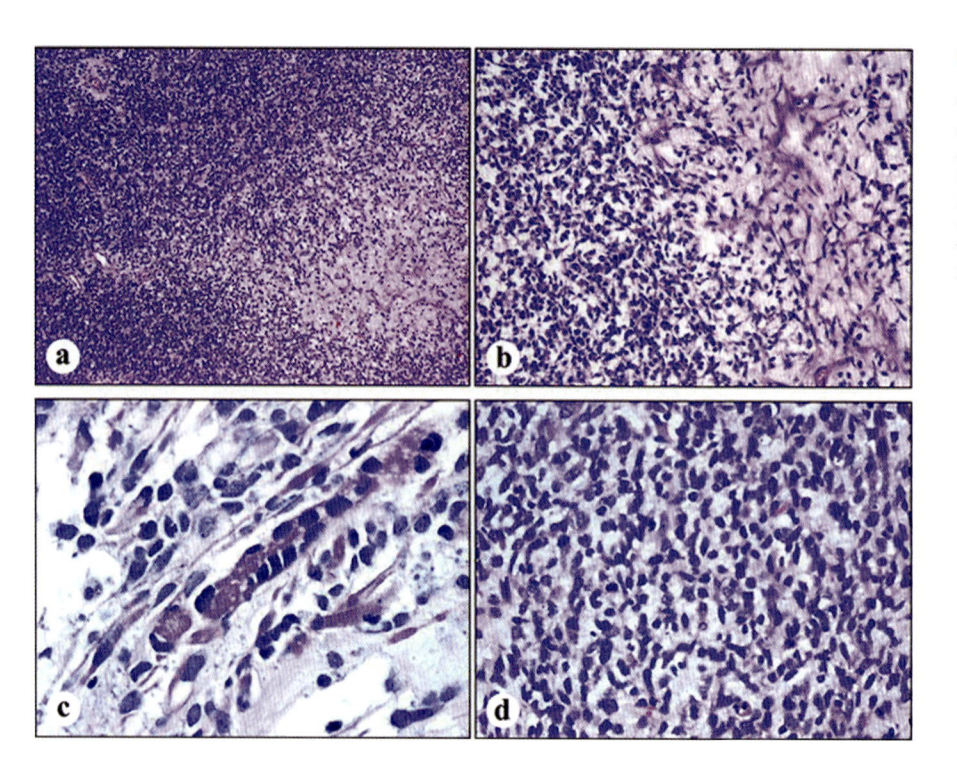

Figure 20.12 Embryonal RMS. (**a**) This example has a predominantly dense architecture, with tightly aggregated cells. (**b**) Compact, dense areas alternate with hypocellular foci containing abundant myxoid stroma. (**c**) A neoplastic myotube has elongated cell boundaries and contains multiple peripheral nuclei in a tandem array. (**d**) This dense aggregate contains primitive cells with short spindly contours and elongated nuclei. Some cells contain brightly eosinophilic cytoplasm.

Figure 20.13 Botryoid RMS expands and fills the common bile duct in a "bunch of grapes" fashion.

Clinical features

In children, spindle cell RMS has a strong predilection for the paratesticular region.[61] Spindle cell RMS also occurs in the head and neck and the extremities; the head and neck is the most common site in adults. In both adults and children, sclerosing lesions are more common in the extremities, followed by the head and neck.

In the classic paratesticular spindle cell RMS, lesions are detected early so that children have a good prognosis with metastasis occurring in a minority of cases. Pelvic, inguinal, and para-aortic lymph nodes are the most common sites of metastatic disease in paratesticular tumors. Approximately 95% of affected patients are alive at 5 years.[12] Conversely, in adults, spindle cell/sclerosing RMS is an aggressive tumor with a recurrence and metastasis rate of 40–50%. Unlike pure embryonal RMS, spindle cell RMS appears in older children, in whom it may behave more like an adult tumor.

Histologic features

Spindle cell RMS mostly contains cells with elongated borders and a relatively dense, collagenous background stroma (Figure 20.15). The amount of spindle cell component required to place a tumor in this category is not well defined, but studies suggest this pattern should occupy at least 80% of the tumor. Spindle cell RMS simulates leiomyosarcomas or fibrosarcomas, yet it should have definitive skeletal muscle rather than smooth muscle differentiation, as revealed by immunohistochemistry (Figure 20.16). Cells with the properties of smooth muscle can also be present, however. Spindle cell RMS can also contain abundant or scant collagen between tumor cells.[61]

Sclerosing RMS was first described in 2000 by Mentzel and Katencamp,[62] who reported a group of three adults with a lesion characterized by an abundant hyalinizing matrix surrounding and entrapping tumor cells (Figure 20.17). These lesions could also contain foci with a chondroid or osteoid appearance. In 2002, Folpe et al. described four adults with the same features and coined the term "sclerosing RMS."[63] Later, Chiles et al. described 13 childhood cases.[64] Recent studies

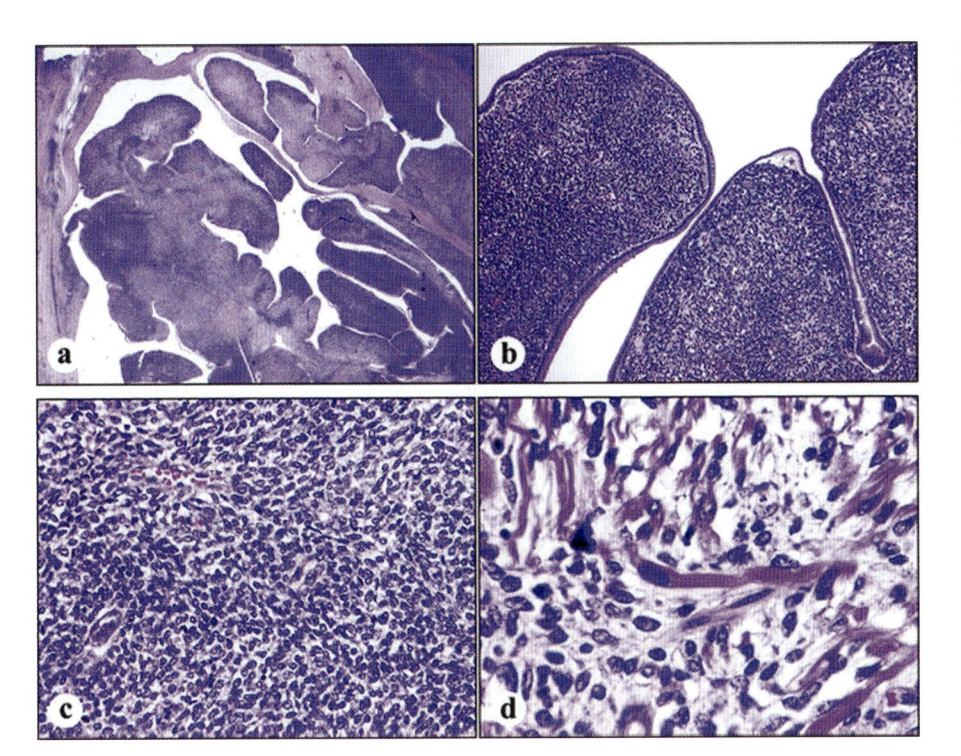

Figure 20.14 Histologically, this biliary duct RMS shows multiple polypoid protrusions lined by epithelium. Many tumor cells are primitive, but differentiated rhabdomyoblasts are present, some with cross-striations.

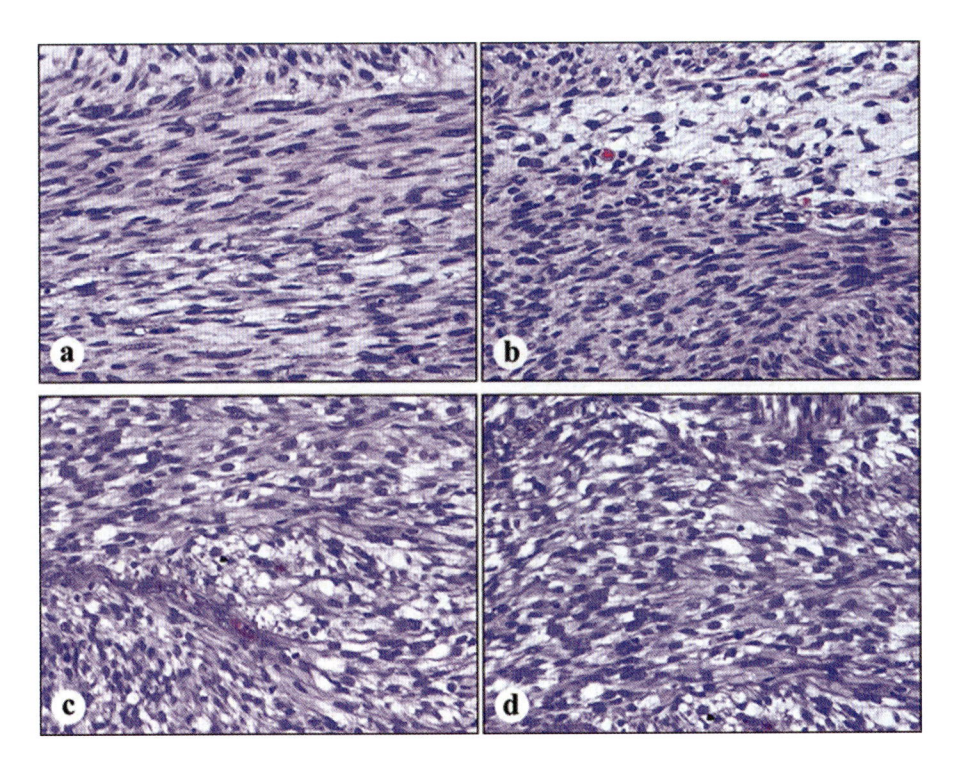

Figure 20.15 Spindle cell RMS. The lesion is composed of closely packed fascicles of spindle cells, similar to leiomyosarcoma but showing striated muscle differentiation.

suggest there may be a continuum between spindle cell RMS and sclerosing RMS. Tumor cells may be round to spindled and cling to the thickened septae creating a pseudovascular or microalveolar appearance. In contrast to alveolar RMS, islands of tumor cells floating within the alveolar space are not seen in the sclerosing pattern. Sclerosing RMS demonstrates variable expression of myogenin but strong expression of MyoD. The weak myogenin expression and fusion negativity distinguish it from alveolar RMS, which it can strongly resemble.

The relationship of spindle cell/sclerosing RMS to embryonal RMS is unclear in the pediatric population, and often embryonal RMS shows foci of spindle cell or sclerosing patterns. This is particularly true in sites where a primary resection is common, such as in the retroperitoneum or paratesticular region.

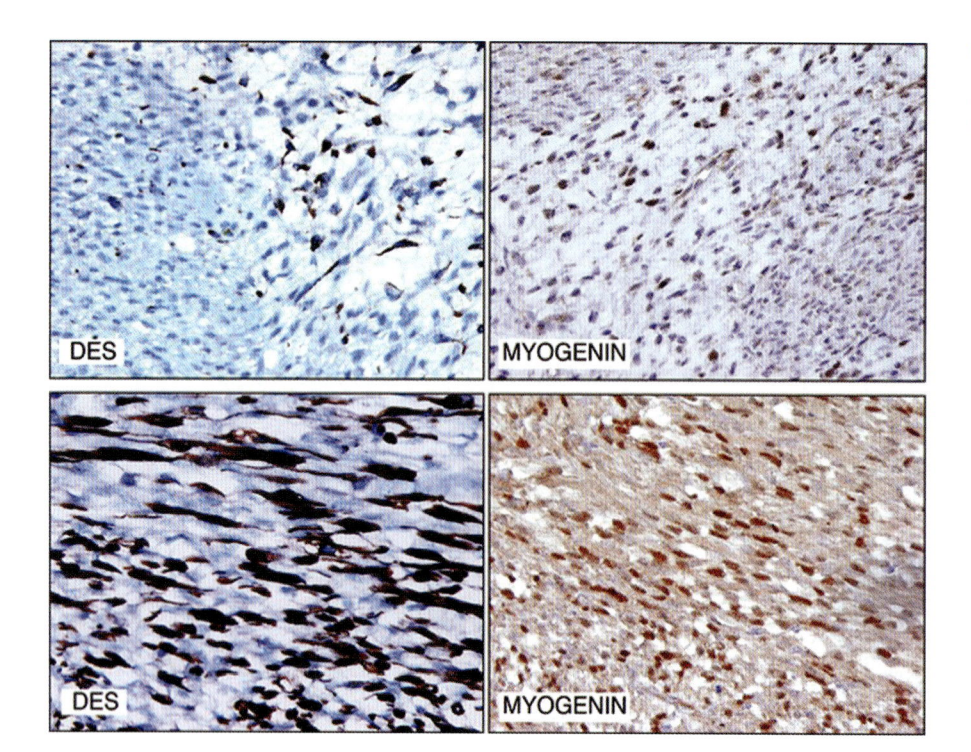

Figure 20.16 Immunohistochemically, spindle cell RMS contains variable desmin positivity, and nuclear positivity for myogenin is present in a portion of the tumor cells.

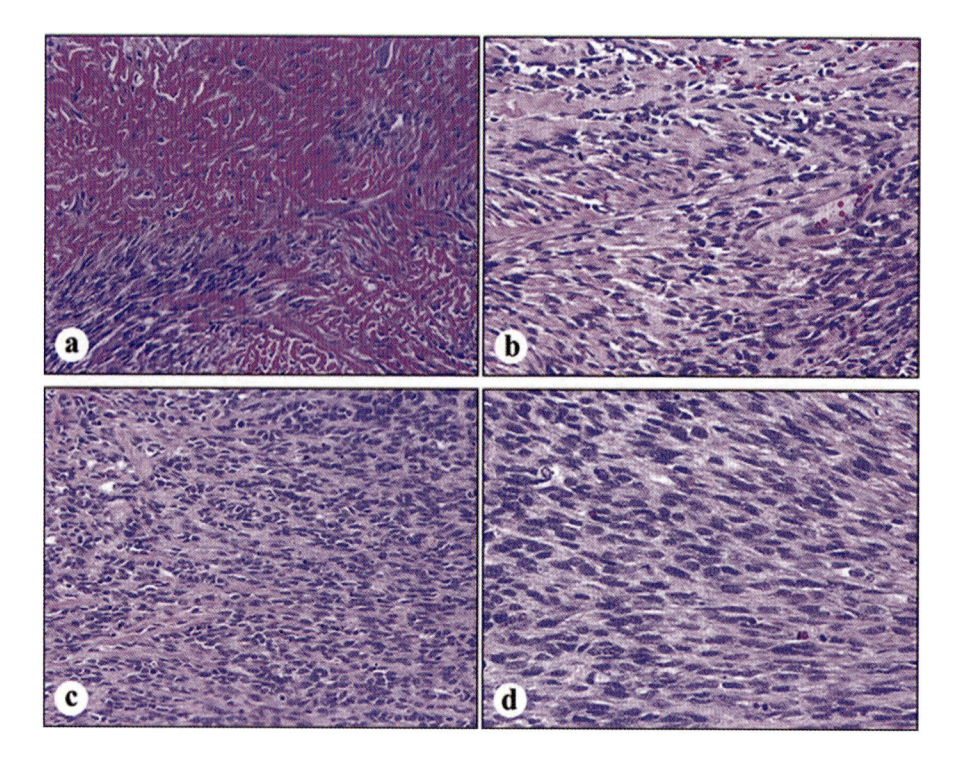

Figure 20.17 Sclerosing RMS. (**a,b**) A densely sclerotic stroma encases primitive cells, creating an appearance of fibrosarcoma. (**c,d**) Highly cellular areas are also present, but there is no histologic evidence for skeletal muscle differentiation.

Genetics of spindle cell/sclerosing rhabdomyosarcoma

Cytogenetic studies in these tumors are limited, but recent studies have shown transactivating mutations in *MyoD1* in up to 40% of adult spindle cell RMS. Aneuploidy with whole chromosome gains and nonrecurrent structural changes is described. Among them are *MDM2* amplification and a *PIK3CA* point mutation reported in one case of sclerosing RMS.[65] *NCOA2* gene rearrangements were detected in three spindle cell RMS of infants.[66] Spindle cell/sclerosing RMS virtually always lacks evidence of a *FOXO1* rearrangement.[67–69]

Differential diagnosis

Spindle cell tumors can be confused with leiomyosarcomas, fibrosarcomas, monophasic synovial sarcomas, and inflammatory (myofibroblastic) tumors. Sclerosing RMS should be distinguished from sclerosing epithelioid fibrosarcoma, infiltrating carcinoma, osteosarcoma, and angiosarcoma.

Alveolar rhabdomyosarcoma

Alveolar RMS contains a uniform population of primitive cells with round, even nuclear margins. It might contain fibrovascular septa (typical pattern) or patternless sheets of cells with little stroma (solid pattern). Most lesions contain a *PAX/FOXO1* fusion gene. Historical synonyms include monomorphous round cell RMS and alveolar sarcoma.

Clinical features

Alveolar RMS more often present in the extremities or the head and neck region (sinonasal region and parameningeal regions) in adolescents and young adults (see Figure 20.8C and D), but all ages and sites can be affected. Of note is that visceral organs are rarely the site of origin and that a sizeable percentage arises from the sinonasal region (Figure 20.18). Whatever the origin, they are aggressive neoplasms that can respond to initial chemotherapy, but recur with a vengeance.[53] Alveolar RMSs often metastasize to regional lymph nodes, where they usually maintain the same architectural pattern as seen in the primary tumors. Rare cases simulate hematopoietic neoplasms (Figure 20.19) and have no apparent primary site.[49,70]

Histologic features

Alveolar RMS is classically defined as a tumor composed of neoplastic cells that appear to line hollow spaces. *Alveolus* is derived from Latin, meaning "trough, hollow sac, or cavity." Tumors with this architecture typically contain fibrous septa lined by tumor cells, forming the alveolar space. The attached tumor cells adhere to the septa as a single row, creating a picket-fence-like arrangement. The septa encompass a central cavity that often contains a "floating cluster" of cells surrounded by a cleft of discohesive tumor cells (Figure 20.20). Some tumors simply present a nesting pattern outlined by septa.

Tsokos and coworkers[71] expanded alveolar RMS to include "solid alveolar" types (Figure 20.21). These possess typical cytologic features, with even, round nuclear margins reminiscent of lymphoma or Ewing sarcoma. The cytologic features of both solid alveolar RMS and typical alveolar RMS are those of a "small round cell tumor" with homogeneous, dark chromatin with or without nucleoli. Mitotic activity is often frequent. The cytologic distinctions between embryonal and alveolar RMS have been tested by morphometric analysis, which confirms the clinical utility of separating round cell tumors from spindle cell lesions.[72] Solid variant tumors possess the cytologic features of alveolar RMSs but lack septa or discohesion. They form hypercellular, patternless sheets punctuated by occasional small vessels.[71] Cytogenetic studies and clinical features of "solid alveolar" RMS support their inclusion with typical alveolar RMS.[73] Some alveolar RMSs contain rhabdomyoblasts with eccentric eosinophilic cytoplasm and prominent tumor giant cells with brightly eosinophilic cytoplasm (Figure 20.22a). Rare cases also contain areas with an epithelioid or rhabdoid appearance (Figure 20.22b). Readily identifiable rhabdomyoblasts are often scanty or absent, so that supplemental studies to document rhabdomyoblastic differentiation are usually required.

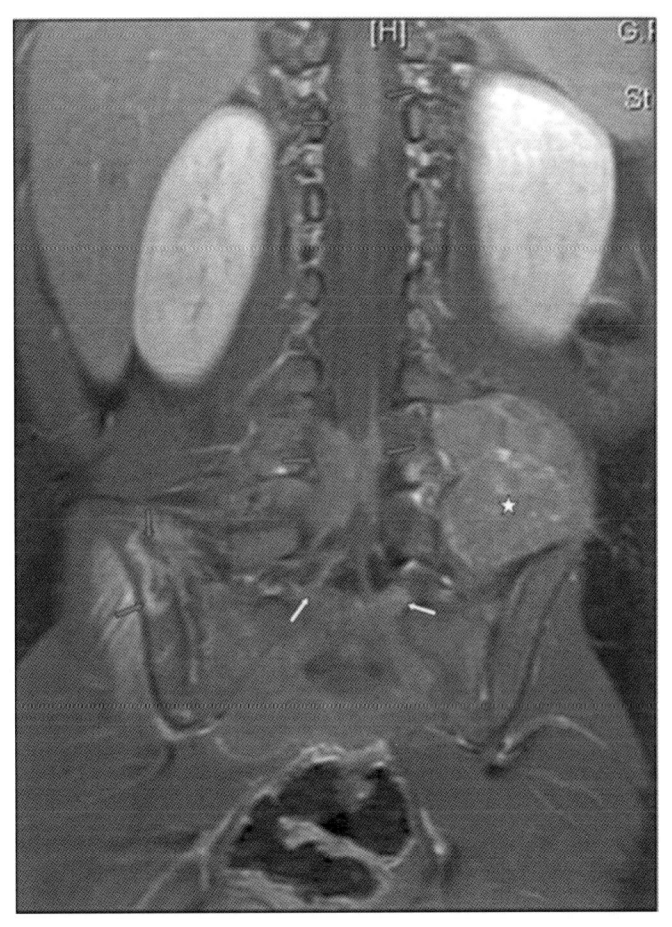

Figure 20.18 Paraspinal alveolar RMS, sagittal postgadolinium coronal MRI, T1-weighted, with fat suppression. A mass arises in the soft tissues of the back (star) and shows intraspinal metastatic spread in separate locations, superiorly and inferiorly (arrows). The mass extends along the nerve root in the subpedicular recess, which is filled by the tumor. The tumor has also metastasized to the bone marrow (arrows) of the right iliac wing.

Genetics of alveolar rhabdomyosarcoma

Alveolar RMS is associated with two recurring cytogenetic translocations in approximately 80% of cases.[59] This has led some investigators[74] to consider these translocations to be diagnostic of the alveolar type of RMS. The more common

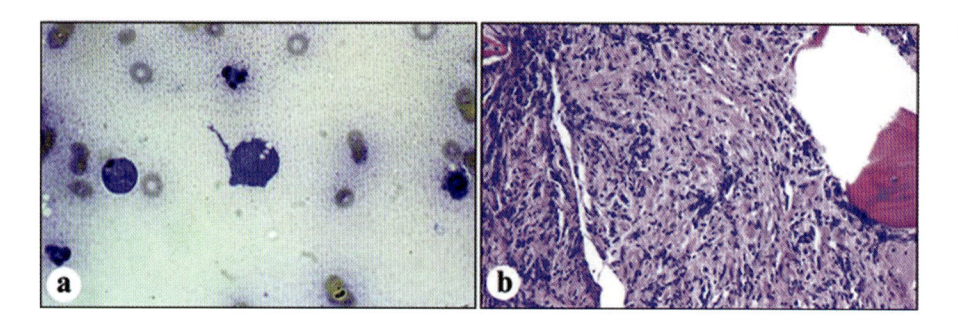

Figure 20.19 Leukemoid RMS. (**a**) A Wright stain of peripheral blood reveals a tumor cell with round nucleus and cytoplasmic vacuoles, resembling lymphoblastic leukemia. (**b**) After a course of chemotherapy, rhabdomyoblastic cells become apparent in bone marrow biopsy.

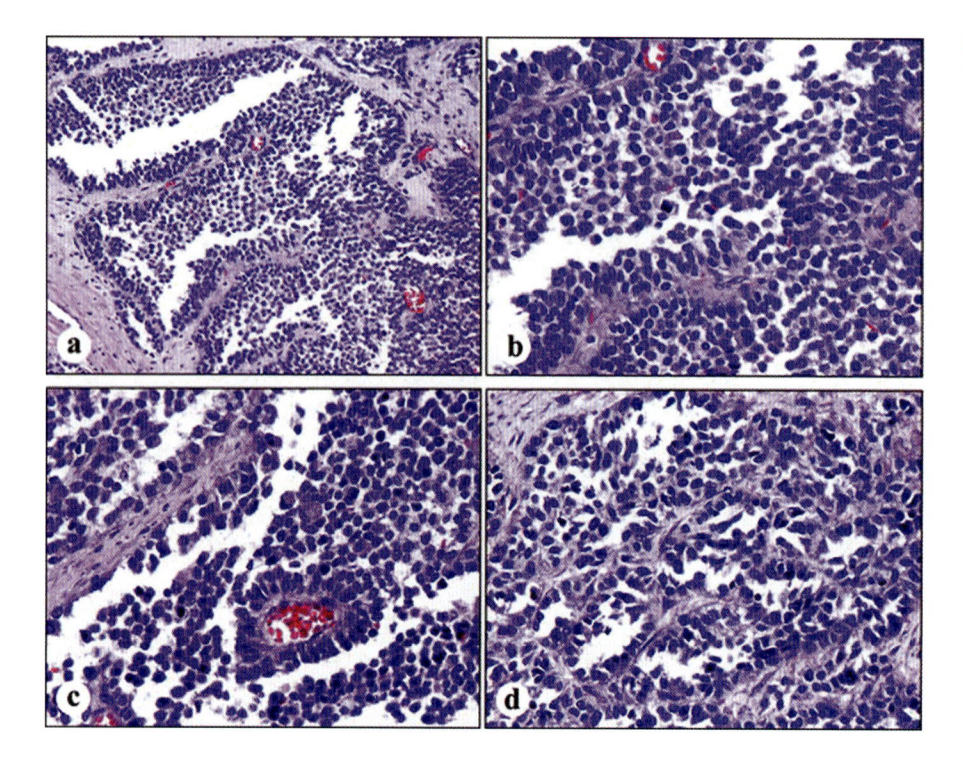

Figure 20.20 Alveolar RMS shows discohesive tumor cells with prominent fibrovascular septa.

translocation, t(2;13)(q35;q14), occurs in about 60–70% of alveolar RMS, whereas a t(1;13)(p36;q14) translocation is less common and is seen in about 12–20%. These translocations might be part of a complex karyotype or partially demonstrated (Figure 20.23). Both of these translocations fuse *FOXO1* (formerly *FKHR*, or forkhead receptor gene) on 13q14 with one of the *PAX* genes (*PAX3* on 2q35; *PAX7* on 1p36; Figure 20.24). The fusion product activates aberrant DNA transcription, leading to tumorigenesis.

PAX fusion transcripts can be identified by reverse transcriptase polymerase chain reaction (RT-PCR) of tumor samples and cell lines, and rearrangement caused by the translocation can be identified by fluorescence *in situ* hybridization (FISH) testing. *PAX7-FOXO1* translocations have a high frequency of subsequent amplification, which is rarely seen in the *PAX3-FOXO1* translocation. This amplification may obscure detection of the t(1;13) translocation by routine karyotyping, but it can serve as a marker of *PAX7* involvement when using *FOXO1* break-apart FISH to determine fusion status. Tumors with t(1;13) may possess a different clinical phenotype than the more common t(2;13), although this data is not conclusive and currently patients are treated the same.[75]

A subset of alveolar RMS does not have cytogenetic or PCR evidence of the characteristic translocations. In a minority of cases, low-level transcription of the fusion gene or alternative fusion partners such as *PAX3-FOXO4 (AFX)*, *PAX3-NCOA1* or *NCOA2* or *FOXO1-FGFR1* may explain this finding.[76] Approximately 10–20% of confirmed alveolar RMS lacks any evidence of a gene fusion, however. This fusion-negative alveolar RMS appears to show no identifiable genetic differences from embryonal RMS.[77] Recent studies confirm that fusion-negative alveolar RMS is clinically and biologically indistinguishable from embryonal RMS, and fusion status is the more important biological risk factor for poor outcome in children with RMS.[78] Based on this data, fusion status is required to determine risk stratification and treatment for alveolar RMSs. This testing has utility in the diagnosis of small samples of tissue and distinguishing between dense embryonal RMS or

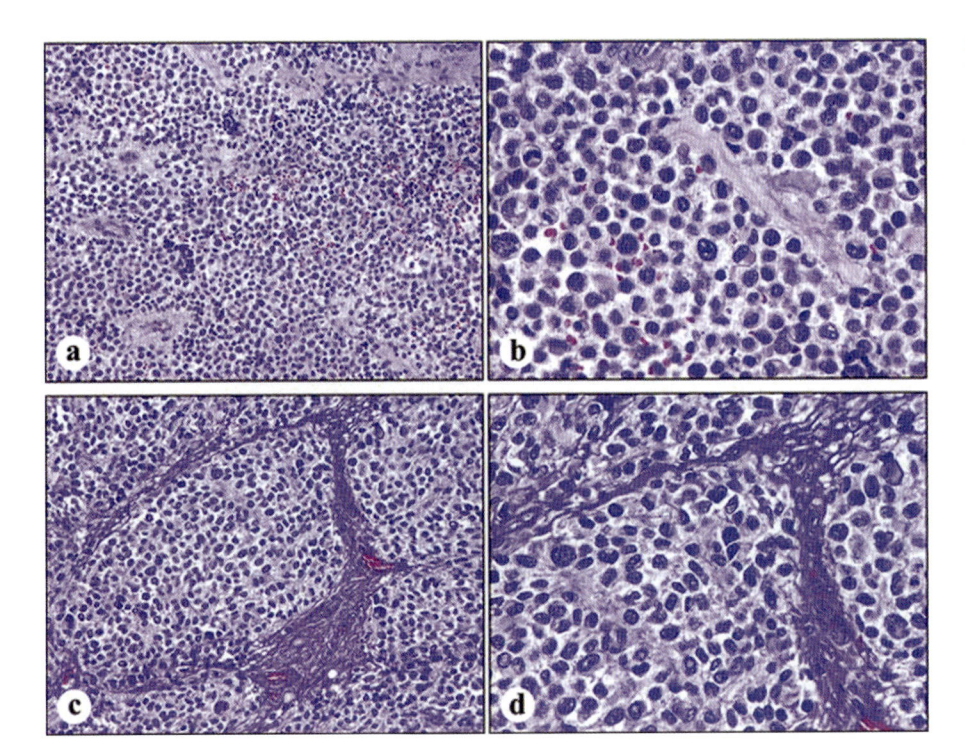

Figure 20.21 Alveolar RMS, solid variant. The tumor contains densely apposed primitive cells with monomorphous round nuclei. Scattered tumor cells contain abundant eosinophilic cytoplasm. Fibrous septa and clefting are not apparent.

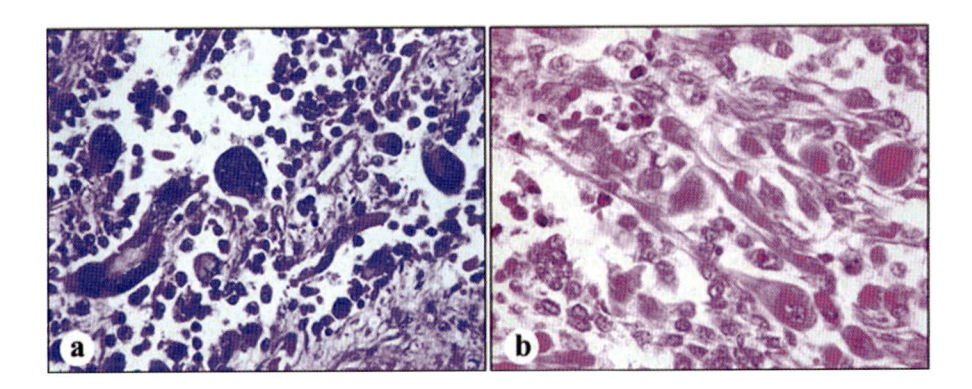

Figure 20.22 (**a**) Pleomorphic giant cells are present in some alveolar RMSs. (**b**) Rhabdoid features are a rare finding in alveolar RMS.

solid alveolar RMS. Fusion testing is usually not required for embryonal or spindle cell/sclerosing RMS, as *FOXO1* fusion transcripts are exceedingly rare in these subtypes. However, gene fusions have been documented in rare cases of embryonal RMS, particularly those occurring in the extremities or other unfavorable sites.

Differential diagnosis

The differential diagnosis of RMS depends on the age of the patient and the tumor subtype. In alveolar RMS of children, the differential diagnosis can include any and all of the "small blue cell tumors," including hematopoietic malignancies (lymphoma or leukemia), the Ewing family of tumors, poorly differentiated or undifferentiated neuroblastoma, rhabdoid tumors (especially in patients <1 year of age), and desmoplastic small round cell tumor. In adults, the differential diagnosis includes hematopoietic neoplasms, Ewing tumors, small cell carcinomas, peripheral neuroectodermal tumors, and desmoplastic small round cell tumors.

Pleomorphic rhabdomyosarcoma

Pleomorphic RMS is a high-grade soft tissue sarcoma that tends to arise in the extremities of adults. It has cytologic and pathologic features typical of undifferentiated pleomorphic sarcomas (malignant fibrous histiocytomas; MFHs), but by definition it contains variable definitive foci of rhabdomyogenic differentiation. The identification of myogenesis usually requires immunohistochemical or ultrastructural analysis.

Clinical features

In one recent large study,[79] pleomorphic RMS occurred in the lower extremities, abdomen/retroperitoneum, chest/abdominal wall, spermatic cord/testes, upper extremity, mouth, and

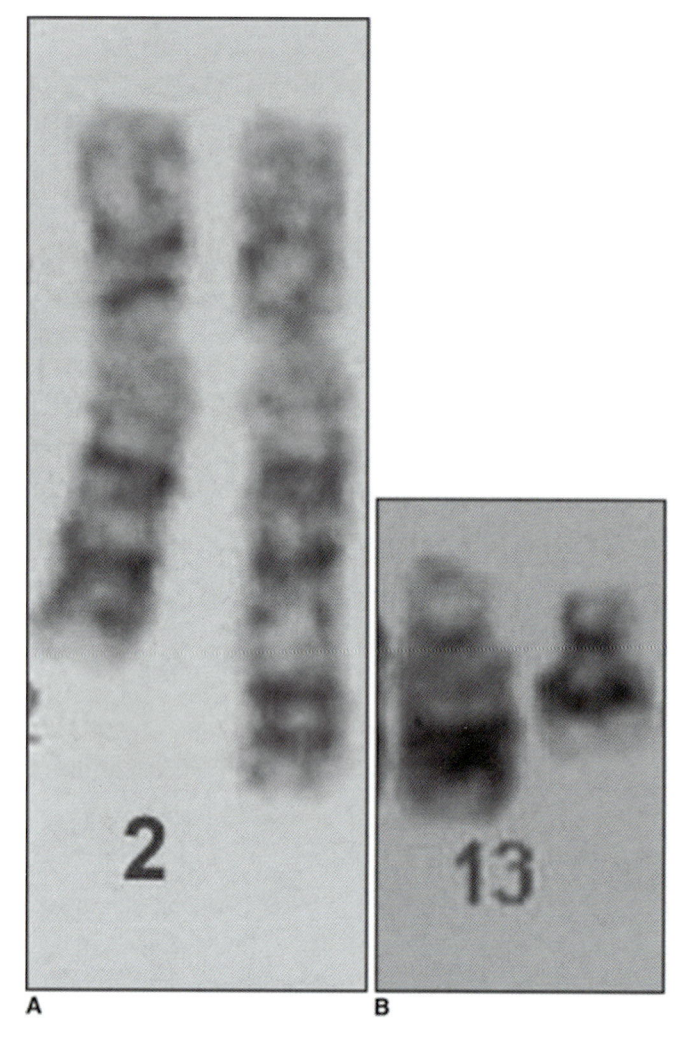

Figure 20.23 Karyotypic features of alveolar RMS. The t(2;13)(q34;q14) is found in most alveolar RMS. Note that a large portion of chromosome 13 is attached to the q arm of the chromosome 2 on the right, whereas the chromosome 13 on the right contains a shortened q arm, with a small portion of chromosome 2 attached to the distal end.

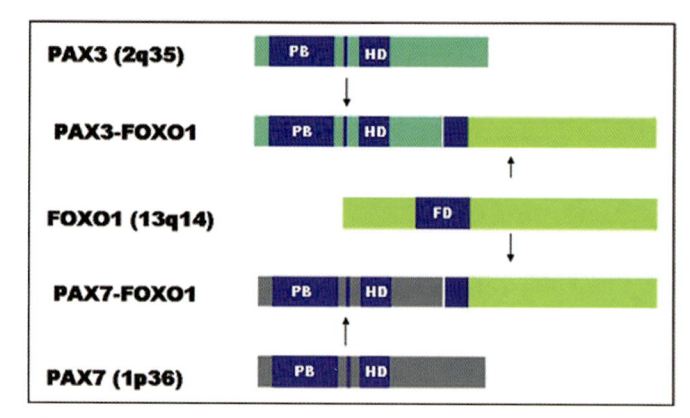

Figure 20.24 Molecular features of alveolar RMS. In the t(2;13) translocation (second from the top), the carboxyl terminal of the FOXO1 gene on chromosome 13 fuses with the amino terminal of the PAX3 gene on chromosome 2. Alternatively, the FOXO1 gene can fuse with the PAX7 gene on chromosome 1 in the t(1;13) (second from the bottom). Both fusions retain the paired box (PB) and homeodomain (HD) of the PAX genes, but part of the forkhead domain (FK) in the FOXO1 gene is lost.

orbit, in decreasing order of frequency. Ages ranged from 21 to 81 years, with a median of 54 years. Clinical outcome was poor, and 70% of patients died from the tumor after a mean survival time of 20 months.

Pathologic features

Pleomorphic RMSs are composed of pleomorphic spindle cells and tumor giant cells, arranged in a storiform or fascicular pattern. Nuclei are large and pleomorphic and the cytoplasm is variably eosinophilic (Figure 20.25). Diagnostic considerations include liposarcoma, leiomyosarcoma, and undifferentiated pleomorphic sarcoma, formerly MFH. The presence of tumor cells with cytoplasmic rhabdomyoblastic differentiation is key to diagnosis. The diagnosis is usually based on immunohistochemical and ultrastructural verification of skeletal muscle differentiation in a high-grade, pleomorphic sarcoma (Figure 20.26).[80,81] Other tumors that can have rhabdomyoblastic components must be considered (Table 20.2). Diagnosis can be problematic, because myogenin and MyoD studies can be negative and desmin can be non-specific.[81]

Pleomorphic RMSs are rare tumors that have been defined[46] as occurring exclusively in the adult population, most often on the trunk and extremities; however, this exclusive distribution solely in adults has been challenged by descriptions of pediatric tumors.[79] Similar lesions have rarely occurred as IRS and COG cases, but they have operationally been called *anaplastic spindle cell RMSs*. Because there are no established histologic criteria to differentiate pleomorphic RMS from anaplastic spindle cell RMS, the authors' opinion is that these two lesions might have some overlap.

Genetics of pleomorphic rhabdomyosarcoma

Few molecular cytogenetic studies have been performed for pleomorphic RMS. One recent comparative genomic hybridization study, using a small number of well-characterized cases, found regions of genetic gain and loss in multiple chromosomes.[82] Most contained areas of genomic amplification distinct from those associated with alveolar RMS. Not surprisingly, the genetic signature of this pleomorphic RMS appeared to be akin to pleomorphic undifferentiated sarcoma (MFH) and osteosarcoma. Interestingly, one pleomorphic RMS contained a *PAX3/FOXO1* fusion gene; the authors have recently seen a similar case in their own material (unpublished data).

Differential diagnosis

Pleomorphic RMS should not be confused with undifferentiated pleomorphic sarcoma (MFH) and various dedifferentiated sarcomas.

Chapter 20: Rhabdomyomas and rhabdomyosarcomas

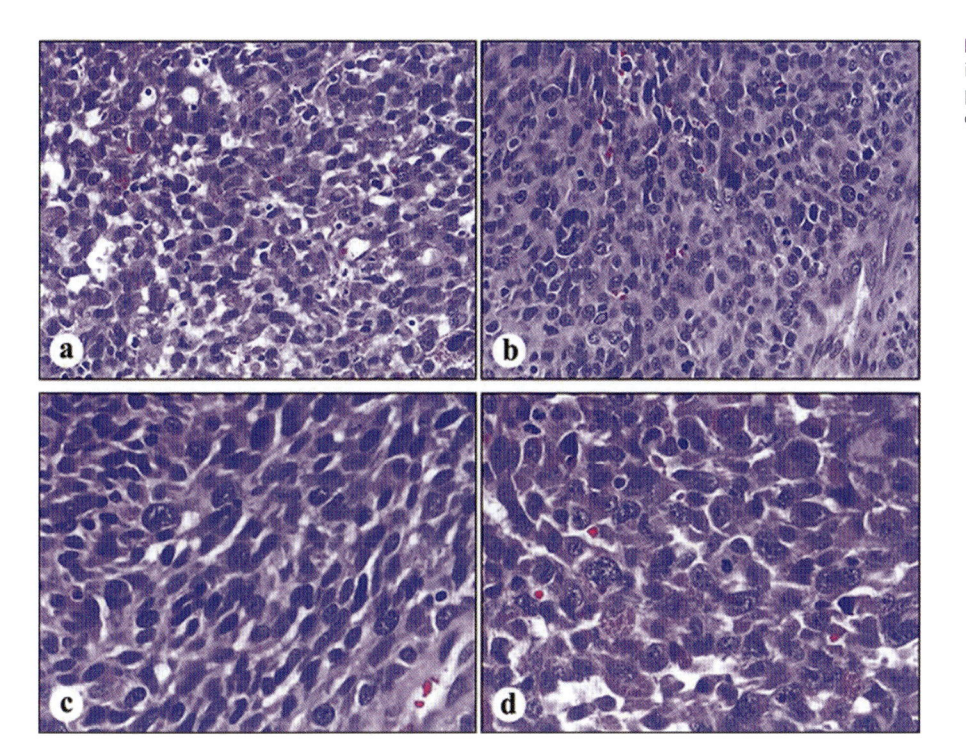

Figure 20.25 Retroperitoneal pleomorphic RMS in an adult patient. The lesion is composed of pleomorphic cells with variably eosinophilic cytoplasm.

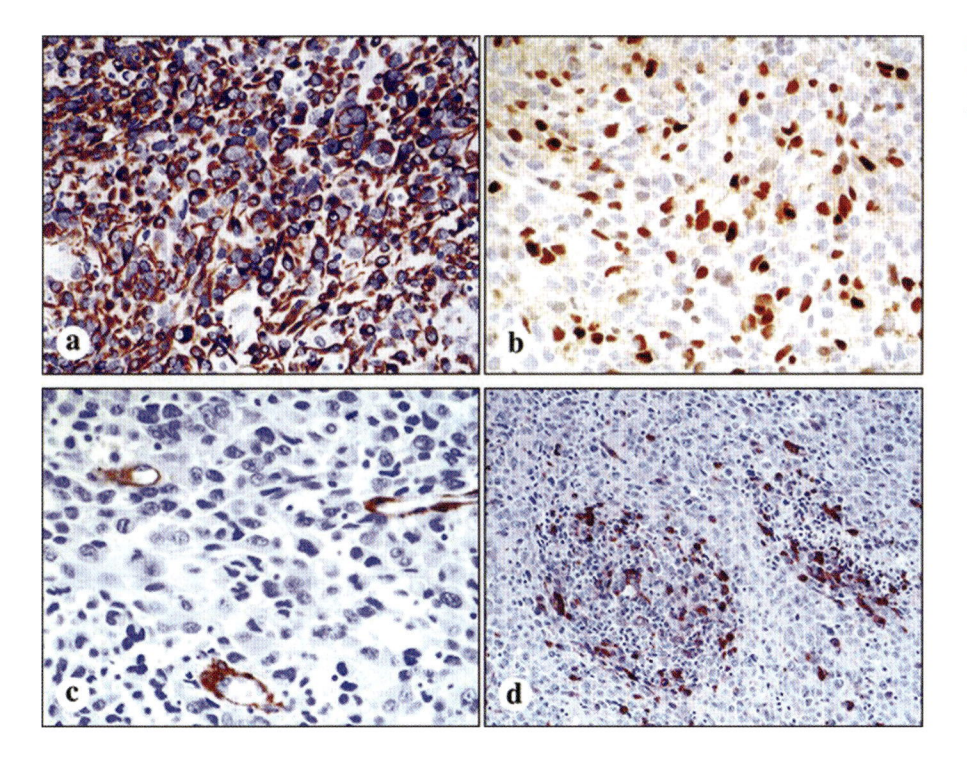

Figure 20.26 (**a**) Immunohistochemically, pleomorphic RMS is strongly positive for desmin. (**b**) It contains myogenin-positive nuclei. (**c,d**) The tumor is negative for smooth muscle actin and S100 protein.

Rhabdomyosarcoma with anaplasia

In children, up to 13% of RMS shows anaplasia. Anaplasia in this context is defined by enlarged hyperchromatic nuclei that are three times larger than adjacent tumor nuclei at their narrowest diameter, similar to the criteria for Wilms' tumors (Figure 20.27). Atypical mitotic figures were not required for the diagnosis of anaplasia in Kodet's study,[83] but were an important histologic feature. These features should be apparent at low power (10×) to avoid confusion with "nuclear unrest," which is characterized by mild hyperchromasia and nuclear atypia. Cells with nuclear unrest alone do not show 3X enlargement, or contain bizarre mitoses, or affect outcome to the same degree as anaplastic ones. *Diffuse anaplasia* is defined as the presence of clonal aggregates of anaplastic cells, whereas

543

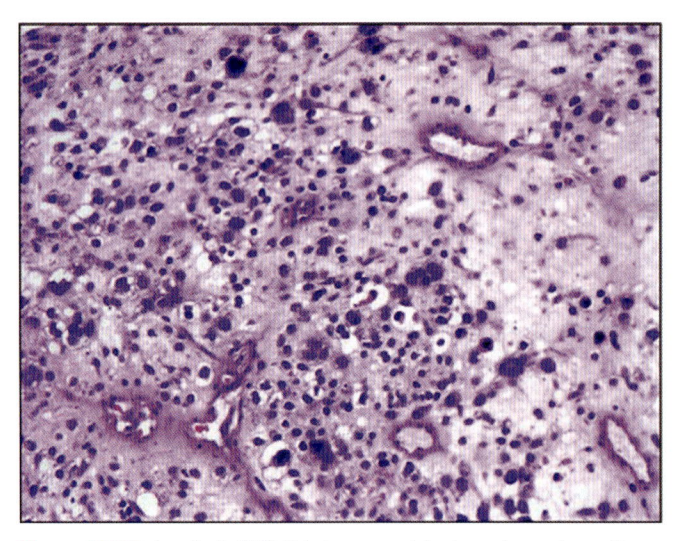

Figure 20.27 Anaplastic RMS. This tumor contains large, hyperchromatic, pleomorphic nuclei that overtly contrast with the small primitive cells.

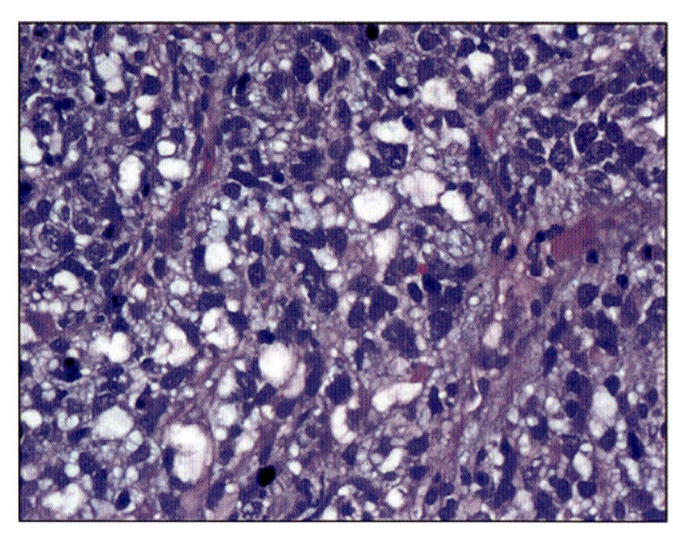

Figure 20.28 Clear cell RMS. The tumor cells contain amounts of cytoplasmic vacuoles, resembling lipoblasts.

focal anaplasia refers to the occurrence of scattered, widely separated anaplastic cells.

The presence of anaplasia indicates genomic instability, and it occurs in all forms of RMS.[83] Anaplasia suggests a worse prognosis if it is diffuse rather than focal, the major criterion for diffuse anaplasia being the presence of sheets of anaplastic cells rather than isolated clusters.[83] Genetic analysis indicates that anaplastic embryonal RMS frequently exhibits gene amplification, distinguishing it from typical embryonal RMS and aligning it with alveolar RMS.[60] Similarly, it exhibits more aggressive clinical behavior,[83] although this feature is currently not used to determine risk groups.

Unusual rhabdomyosarcoma variants
Mixed alveolar and embryonal rhabdomyosarcoma

Some RMS appears to possess features of both alveolar and embryonal RMS and resemble nested or collision tumors. All subtypes of alveolar and embryonal RMS have been seen. While most mixed alveolar and embryonal RMS is fusion negative, some cases have been positive for a PAX/FOXO1 rearrangement. Regardless of fusion status, both components appear to be the same genetically, despite their differing histologic appearance. No mixed RMS has shown differing fusion status between the two components, although a single report demonstrated *MYCN* amplification in the alveolar component without amplification of the embryonal component. Tumors should be treated based on their fusion status rather than the predominant histologic component.

Clear cell rhabdomyosarcoma

Clear cell RMS[84–86] is characterized by cells with vacuolated cytoplasm, resembling lipoblasts (Figure 20.28). By electron microscopy, these tumor cells contain abundant glycogen, similar to Ewing sarcoma cells. Reports of alveolar architecture and nodal metastasis suggest that these lesions are alveolar RMS, but genetic data are lacking for confirmation. Reports of this lesion are very sparse. In the authors' experience, clear cell foci occur relatively commonly in otherwise typical RMS, but tumors exclusively composed of clear cells are rare.

Lipid-rich RMS appears similar to the clear cell variety, but on electron microscopy the vacuolated cells demonstrate lipid droplets instead of glycogen. This phenotype of RMS might contain varying amounts of both substances, analogous to clear cell renal carcinoma, however.[87,88] Potential for diagnostic confusion with entities such as liposarcoma should be considered.

Rhabdomyoma-like rhabdomyosarcoma

This RMS is difficult to diagnose, but fortunately occurs rarely. Kodet *et al.*[8] described a tumor of the tongue with features that are consistent with the diagnosis of fetal cellular rhabdomyoma in an 18-month-old infant. This lesion ultimately metastasized. Rhabdomyoma-like RMS can be difficult to distinguish from cellular rhabdomyoma, which should not contain more than 15 mitoses per 50 HPFs. Other worrisome features that favor malignancy include atypical mitoses, anaplastic cells, and a cambium layer.[6]

Rhabdomyoma-like RMS contains well-differentiated myoblasts with prominent cross-striations (Figure 20.29), and primitive cells can be sparse. A layered biphasic appearance can create a similarity to infantile rhabdomyoma. Capsular penetration and invasion of adjacent tissue favor malignancy.

Epithelioid/rhabdoid rhabdomyosarcoma

In 1991, Kodet *et al.* described an unusual set of RMS tumors that exhibited globular inclusions resembling rhabdoid tumor (Figure 20.30).[89] Since then there have been additional reports

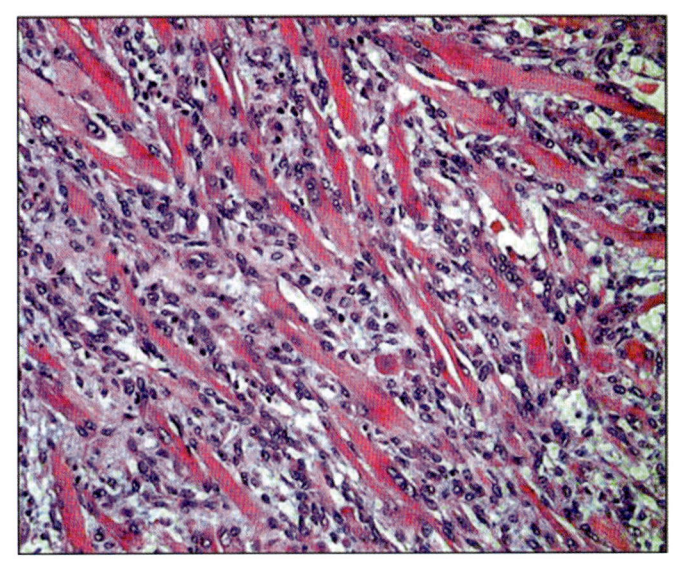

Figure 20.29 Rhabdomyoma-like RMS. The tumor contains bundles of primitive cells intersected by maturing myofibers, similar to infantile rhabdomyoma.

Figure 20.30 Rhabdoid RMS. Occasionally RMS contains prominent hyaline cytoplasmic inclusions formed from whorled intermediate filaments, similar to rhabdoid tumors of the kidney and soft tissues.

of RMS with rhabdoid features, including a primary cerebral tumor.[90] In 2011, Jo et al. described epithelioid RMS,[91] a subset of which contained rhabdoid-like cytoplasmic inclusions, as an aggressive neoplasm affecting older patients. In 2014, Zin et al. described a series of epithelioid RMS occurring in the pediatric population.[92] Evidence that these may represent a variant pattern of embryonal RMS includes lack of fusion transcripts, weak staining for myogenin, and an overall favorable clinical course.

In the pediatric population, the differential diagnosis of epithelioid/rhabdoid RMS includes malignant rhabdoid tumor. Diagnostic confusion can be compounded by desmin and actin positivity in true rhabdoid tumors, which show polyphenotypia with staining for cytokeratin, vimentin, and EMA. Lack of INI1 immunostaining characterizes true rhabdoid tumor, whereas RMS is INI1 positive.[93] Similarly, RMS generally does not show the alterations in the *hSNF5/INI1* gene that are typical of rhabdoid tumor.[94] Both RMS and rhabdoid tumors show aberrations of chromosome 11p15,[95,96] and both can contain mutations of *STIM1* (alias *GOK*), a tumor-suppressor gene located in that region.[97] Of particular concern is a report indicating that RMS could show alterations in *INI1*,[98] suggesting that the absence of INI1 staining or mutations of the gene might not be infallible in separating these two lesions. Fortunately, myogenin does not appear to stain rhabdoid tumor nuclei,[99] but the immunophenotypic promiscuity of these lesions is a source of concern. The diagnosis of rhabdoid rhabdomyosarcoma does not portend the dire clinical behavior of rhabdoid tumors.[89]

In adults, the differential diagnosis of epithelioid/rhabdoid RMS includes epithelial malignancies and melanoma. Immunohistochemistry aids in this distinction, as desmin expression is virtually nonexistent in epithelial tumors. Although focal desmin expression may be seen in melanomas, epithelioid RMS shows strong and diffuse desmin expression.

Ectomesenchymoma

This designation refers to a tumor that combines RMS with neoplastic neuroblasts or neurons, with or without ganglion cells. This is an exceedingly rare tumor and occurs most commonly in young male patients. Grossly, the lesion is well circumscribed.[100] Histologically, this tumor can have areas that appear similar to embryonal, spindle cell, or alveolar RMS and with variable, but often scarce, mature neurons or neuroblasts. The authors have seen one case of ectomesenchymoma with alveolar RMS that harbored a *FOXO1* gene rearrangement.

Electron microscopy can be used to confirm neural and myogenic differentiation in ectomesenchymoma (Figure 20.31). Immunohistochemistry highlights neural differentiation in at least some neoplastic cells (by definition) with positive staining for neuron-specific enolase, S100 protein, synaptophysin, neurofilament, glial fibrillary acidic protein, and protein gene product (PGP) 9.5. The predominant rhabdomyosarcomatous components stain similarly to RMS. Rare cases stain with keratin (AE-1/AE-3).[100] However, Bahrami et al.[101] demonstrated at least focal staining for neural markers, including synaptophysin and chromogranin, in 25–35% of alveolar RMS. Additionally, CD56 is a nonspecific marker expressed in the majority of alveolar RMS. Positive immunostaining for neural markers therefore does not constitute a diagnosis of ectomesenchymoma, unless light microscopic or ultrastructural evidence of neuronal differentiation is also apparent.

Rarely, RMS metastasizes as an ectomesenchymoma.[102] Because a small neural component can be easily missed, this

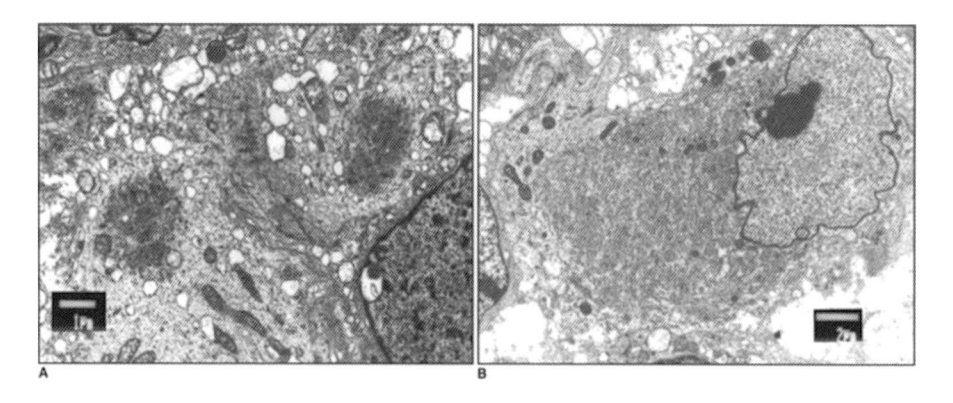

Figure 20.31 Electron microscopic features of RMS. (**A**) Tumor cells with filamentous cytoplasmic inclusions containing arrays of thick and thin filaments and Z band densities. (**B**) Rhabdomyoblast with cytoplasmic arrays of myofilaments. Z bands are not apparent, but the filaments have the diameter of myosin and actin.

is a tumor that may be under-recognized. Clinically, ectomesenchymoma can be considered a histologic variant of RMS and not a "unique" tumor type; however, biologically it might represent a form of PNET (see Chapter 31). Ectomesenchymomas are currently treated with IRSG protocols, as they behave similarly to RMS.

Immunohistochemistry of rhabdomyosarcoma
Diagnostic markers for rhabdomyosarcoma

Muscle actins, desmin, and myogenic regulatory proteins are the usual diagnostic markers of RMS. RMS typically contains muscle-specific actin, which also detects smooth muscle and cardiac muscle actin.[103,104] Smooth muscle actin is more specific for smooth muscle tumors, but occasional RMS (10%)[105] and rhabdomyoma[6] can be positive.

Because its expresssion occurs in desmoplastic small round cell tumors, various sarcomas, and giant cell tumors of the tendon sheath, the diagnosis of RMS cannot be based on desmin positivity alone.[106,107] Desmin is an otherwise reliable marker of myogenic differentiation in RMS and it may be positive in tumors that are negative for muscle actins.

Myoglobin is specific for skeletal muscle differentiation, but it suffers from lack of sensitivity.[108] It can be useful in the evaluation of post-treatment specimens with atypical cytologic features. In this situation, reactive myofibroblasts, which are actin positive and occasionally desmin positive, might suggest a recurrent or persistent tumor. Positivity for myoglobin is then useful to confirm terminal skeletal muscle differentiation, but does not exclude regenerative skeletal muscle. Myogenic cells with voluminous eosinophilic cytoplasm should be myoglobin positive; negativity in this instance may indicate an alternative diagnosis. Be aware that myoglobin blush occurs as a result of prefixation leakage of adjacent skeletal muscle into nonmyogenous tumors.[109]

Myogenic transcription factors such as MyoD (myf3) and myogenin (myf4) interact with the promoter regions of other myogenic genes (e.g., desmin, myosin, and creatine kinase), inducing their transcription and leading to skeletal muscle differentiation. The MyoD family of proteins initiates determination of a myogenesis and terminal differentiation.[110]

These genes are usually silent in differentiated muscle cells, but stay active in RMS. The relationships that are involved in this group of transcription factors are complex, but they are specific and sensitive markers of skeletal muscle differentiation.[111]

MyoD immunostains are best performed on freshly cut sections, because antigen reactivity diminishes with slides exposed to room temperature with storage.[112] Caution must be taken to consider only nuclear staining as truly positive since many other lesions show nonspecific cytoplasmic positivity.[111] Myogenin stains are not susceptible to cytoplasmic staining, and the commercial antibodies yield robust staining, making this the stain of choice. The combination of both stains is more sensitive, particularly for the diagnosis of sclerosing RMS.[113]

A word of caution about the diagnosis of RMS: positive MyoD or myogenin stains indicate the presence of myogenesis, which can occur as a secondary phenomenon in lesions such as Wilms' tumor, pleuropulmonary blastoma, and carcinosarcoma. Myogenin staining also occurs in a nonmyogenic but pleuripotent tumor, melanotic neuroectodermal tumor.[114]

The strong, homogeneous myogenin immunostaining of alveolar RMS offers a potent means of diagnosis in round cell tumors, and a fairly reliable but not absolute means of separating it from embryonal RMS, which generally shows a patchy, heterogeneous pattern (Figure 20.32).[115,116] The methylation status of the gene regulatory CpG sites in the *MyoD1* gene promoter region also correlates with tumor type. Most alveolar RMSs have unmethylated CpG sites associated with the *MyoD1* gene promoter, whereas the majority (90%) of embryonal RMS is partially methylated (similar to fetal muscle).[117]

Surrogate markers of fusion status in rhabdomyosarcoma

Immunohistochemical antibodies may also act as surrogate markers of fusion status in RMS. As noted above, strong, homogeneous myogenin immunostaining characterizes alveolar RMS. Nearly all fusion-positive RMS shows at least 50% nuclear immunostaining for myogenin. Additional antibodies aid in the determination of fusion status. Fusion-negative tumors overexpress HMGA2, EGFR, and fibrillin-2, while fusion-positive tumors preferentially express AP2beta, NOS-1, and P-cadherin. Panels combining these antibodies are up to 96% sensitive and 92% specific for predicting fusion status in RMS. Such panels may also offer a method to detect fusion variant tumors not involving rearrangement of *FOXO1* gene that could otherwise be missed by conventional molecular genetic methods.[118]

Prognosis and clinical course of rhabdomyosarcoma

Fusion status recently surpassed histologic subtype as the defining pathologic risk factor in RMS. Failure-free survival of fusion-positive RMS is worse than that of fusion-negative RMS, irrespective of histologic classification. The prognostic significance of the fusion type is less clear. Studies suggest PAX7-FOXO1 confers a more favorable prognosis than PAX3-FOXO1 in patients with high-risk, metastatic disease. For intermediate-risk disease, Skapek *et al.* demonstrated no significant difference in failure-free survival for patients with PAX3 versus PAX7-FOXO1-positive tumors.[47] However, PAX7-FOXO1-positive patients had an 87% overall survival compared with a 64% overall survival for PAX3-FOXO1-positive patients. This suggests that patients with PAX7-FOXO1-positive RMS have a similar relapse rate to patients with PAX3-FOXO1-positive RMS, but have a greater chance of salvage with additional treatment. Adult RMS tumors are so small in number that adequate comparison with childhood RMS is not possible; however, the experience with pleomorphic RMS indicates that adult lesions have an unfavorable outcome.[119]

In children, survival also depends largely on the stage and the extent of resection. Overall, the 5-year survival rate in children with localized RMS is >70%.[120] The wide array of primary sites complicates management, however, as some structures show repeatable patterns of invasiveness, spread, and treatment response. Two different staging systems are used: the clinicopathologic group (Table 20.3) and the TNM staging system (Table 20.4).[46] Grouping is based on pathologic findings, whereas stage is based purely on clinical studies. Both stratifications are used to separate patients into low-risk, intermediate-risk, and high-risk subtypes for therapy (Table 20.5).

In the most recent clinical trials, which reflect risk assignment based on histology rather than fusion status, low-risk RMS had an overall survival of 85–95%. The lowest-risk subset comprised those with nonmetastatic embryonal RMS at favorable sites, totally resected tumors at unfavorable sites, and Group I to III orbital disease; overall survival reached 98%. Overall survival for intermediate risk is 60–80%, while for high-risk RMS the overall survival is approximately 15%.

RMS requires multimodality therapy, necessitating input from a variety of specialists, including oncologists, radiation oncologists, surgeons, radiologists, and pathologists.[120] This complicated therapy can cause severe toxicity and long-term problems, so that tailoring to individual patients and dose reductions in treatment have been recurrent protocol aims. Complete excision of the tumor is a primary treatment goal. Patients with incompletely excised tumors must receive radiation therapy and more potent systemic chemotherapy. Combination chemotherapy (i.e., vincristine, dactinomycin, and

Table 20.3 Clinical group staging for rhabdomyosarcoma

Clinical group	Disease extent/surgical results
I	Localized tumor, confined to site of origin, completely resected
II	Localized tumor, infiltrating beyond the site of resection A. Localized tumor, gross total resection, but with residual microscopic disease B. Locally extensive tumor (spread to regional lymph nodes), completely resected C. Locally extensive tumor (spread to regional lymph nodes), gross total resection, but microscopic residual disease after biopsy
III	A. Localized or locally extensive, gross residual only B. Localized or locally extensive, gross residual and major resection (>50% debulking)
IV	Any size primary tumor with or without regional lymph node involvement, with distant metastases, without respect to surgical approach to primary tumor

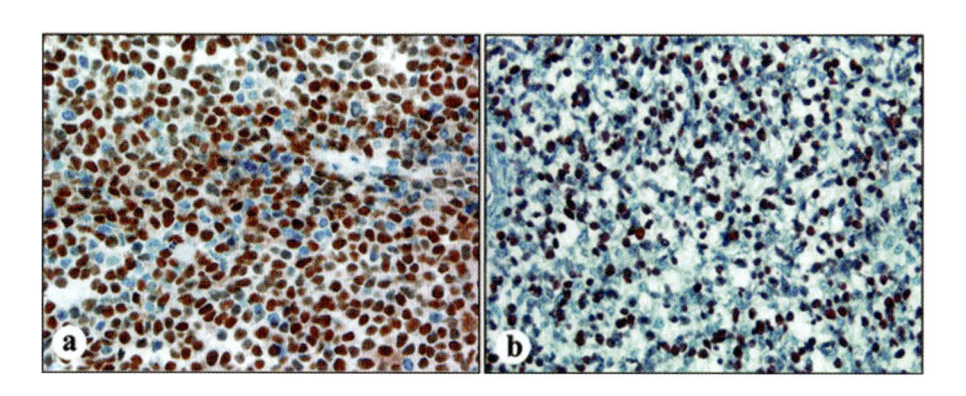

Figure 20.32 Myogenin staining in RMS. (**a**) Alveolar RMS typically shows strong, diffuse nuclear staining. (**b**) Embryonal RMS typically shows a patchy, heterogeneous staining pattern.

Table 20.4 TNM pretreatment (IRS-IV) staging of childhood rhabdomyosarcoma

Stage	Sites	T	T size	Node status	Metastases
1	Favorable site (orbit; head and neck, excluding parameningeal; and genitourinary, nonbladder/prostate)	T1 or T2	Any	N0, N1, or NX	M0
2	Unfavorable site (any site not listed above)	T1 or T2	<5 cm	N0 or NX	M0
3	Unfavorable site	T1 or T2	<5 cm	N1	M0
			>5 cm	N0, N1, or NX	
4	Any	T1 or T2	Any	N0, N1, or NX	M1

Key: T = tumor; N = node; M = metastasis; T1 = tumor confined to site of origin; T2 = locally invasive tumor; N0 = lymph nodes negative for tumor; N1 = lymph nodes positive for metastatic tumor; NX = indeterminate nodal status; M0 = no distant metastasis; M1 = distant metastasis present.

Table 20.5 Updated risk groups for rhabdomyosarcoma, children's oncology group

Risk group	Stage/group	Fusion status
Low	Stage 1, Group III (orbit)	Negative
	Stage 1, Group I-II	
	Stage 2, Group I-II	
Intermediate, subset 1	Stage 1, Group III (non-orbit)	Negative
	Stage 3, Group I-II	
Intermediate, subset 2	Stage 2-3, Group III	Negative
	Stage 1-3, Group I-III	Positive
Intermediate, subset 3	Stage 4, Group IV	Negative
High	Stage 4, Group IV	Positive

either cyclophosphamide [VAC] or ifosfamide [VAI]) has long been the backbone of therapy.

High-risk patients include those with metastatic or recurrent disease, for whom the long-term outlook is poor. These groups have been the focus of the most experimental portions of therapy arms. Even within these groups, however, one sees differences in the outcome of patients having tumors with differing histologies. It is hoped that newer therapies might be devised based on the heterogeneity of the biologic features of this group.

A percentage of RMS cells undergoes differentiation (Figure 20.9d) and have decreased mitotic activity after systemic therapy. In these patients, the determination of whether a differentiated tumor is present, whether it is capable of further growth and patient injury, and whether it requires further therapeutic strategy can be a problem. Cytodifferentiation of tumor cells has been demonstrated *in vivo* and *in vitro*, with a reduction in metastatic behavior.[52,121,122] This differentiation includes increased amounts of eosinophilic cytoplasm, often with more obvious cross-striations or the formation of myotubules. Coffin and colleagues[51] have suggested that alveolar RMS shows differentiation less commonly than do embryonal tumors and that this differentiation might relate to improved survival. Proliferation markers, such as MIB1, could be useful in clinical decision-making and determining prognosis, but this is not yet clear. In some cases with differentiation, expression of myogenin and MyoD1 markedly decreases in post-therapy differentiated tumor cells, perhaps owing to the interplay between MyoD1 protein and cell cycle regulation.[123] Recent studies indicate that occasional residual differentiated rhabdomyoblasts do not affect outcome or justify further radical surgery.[52]

Other genetic changes of potential clinical utility

Most patients with RMS have no underlying risk factors; however, patients with Li–Fraumeni syndrome and a constitutional allelic deletion of p53 have an increased risk of developing RMS. Germline mutations of p53 are more likely in young patients with RMS,[124] and an increased incidence of anaplastic RMS was recently described in p53 germline carriers.[125] About 50% of all RMS has mutations of p53. The mutant form of p53 has a longer half-life and might disregulate the cell cycle. Wild-type p53 inhibits transcription of IGF2,[126] which might explain why some cases with mutant p53 overexpress this gene.

Flow cytometric and image analysis studies suggest an association of ploidy and histologic subtype. Embryonal histology correlates with hyperdiploidy, whereas alveolar subtype correlates with near-tetraploidy; both subtypes can be diploid. No correlation with patient survival was demonstrated in one study,[127] but in another study, diploid and near-tetraploidy had a 5-year survival rate of approximately 30%, whereas hyperdiploidy had a survival rate of 73% at 5 years.[128]

Acknowledgments

The authors acknowledge their keen appreciation and gratitude for the services of Ms. Rossana Desrochers, who ably assisted in the editing of this chapter.

References

1. Rubin BP, Nishijo K, Chen HI, et al. Evidence for an unanticipated relationship between undifferentiated pleomorphic sarcoma and embryonal rhabdomyosarcoma. *Cancer Cell* 2011;19:177–191.
2. Shern JF, Chen L, Chmielecki J, et al. Comprehensive genomic analysis of rhabdomyosarcoma reveals a landscape of alterations affecting a common genetic axis in fusion-positive and fusion-negative tumors. *Cancer Discov* 2014;4:216–231.
3. Enzinger FM, Weiss SW. Rhabdomyoma and rhabdomyosarcoma. In *Soft Tissue Tumors*, 4th edn. St. Louis: Mosby; 2001: 771.
4. Pendle F. Uber ein congenitales Rhabdomyom der Zunge. *Ztschn Neilkund* 1897;18:457–468.
5. Dehner LP, Enzinger FM, Font RL. Fetal rhabdomyoma: an analysis of nine cases. *Cancer* 1972;30:160–166.
6. Kapadia SB, Meis JM, Frisman DM, Ellis GL, Heffner DK. Fetal rhabdomyoma of the head and neck: a clinicopathologic and immunophenotypic study of 24 cases. *Hum Pathol* 1993;24:754–765.
7. O'Shea P. *Pediatric Soft Tissue Tumors: A Clinical, Pathologic, and Therapeutic Approach*. Baltimore: Williams & Wilkins; 1997: 214.
8. Kodet R, Fajstavr J, Kabelka Z, et al. Is fetal cellular rhabdomyoma an entity or a differentiated rhabdomyosarcoma?: a study of patients with rhabdomyoma of the tongue and sarcoma of the tongue enrolled in the intergroup rhabdomyosarcoma studies I, II, and III. *Cancer* 1991;67:2907–2913.
9. Crotty PL, Nakhleh RE, Dehner LP. Juvenile rhabdomyoma: an intermediate form of skeletal muscle tumor in children. *Arch Pathol Lab Med* 1993;117:43–47.
10. Tostar U, Malm CJ, Meis-Kindblom JM, et al. Deregulation of the hedgehog signalling pathway: a possible role for the PTCH and SUFU genes in human rhabdomyoma and rhabdomyosarcoma development. *J Pathol* 2006;208:17–25.
11. Hettmer S, Teot LA, van Hummelen P, et al. Mutations in Hedgehog pathway genes in fetal rhabdomyomas. *J Pathol* 2013;231:44–52.
12. Agamanolis DP, Dasu S, Krill CE, Jr. Tumors of skeletal muscle. *Hum Pathol* 1986;17:778–795.
13. Kapadia SB, Meis JM, Frisman DM, et al. Adult rhabdomyoma of the head and neck: a clinicopathologic and immunophenotypic study. *Hum Pathol* 1993;24:608 617.
14. Solomon MP, Tolete-Velcek F. Lingual rhabdomyoma (adult variant) in a child. *J Pediatr Surg* 1979;14:91–94.
15. Hansen T, Katenkamp D. Rhabdomyoma of the head and neck: morphology and differential diagnosis. *Virchows Arch* 2005;447:849–854.
16. Friedman MT, Molho L, Valderrama E, Kahn LB. Crystal-storing histiocytosis associated with a lymphoplasmacytic neoplasm mimicking adult rhabdomyoma: a case report and review of the literature. *Arch Pathol Lab Med* 1996;120:1133–1136.
17. Gibas Z, Miettinen M. Recurrent parapharyngeal rhabdomyoma: evidence of neoplastic nature of the tumor from cytogenetic study. *Am J Surg Pathol* 1992;16:721–728.
18. Willis J, Abdul-Karim FW, di Sant'Agnese PA. Extracardiac rhabdomyomas. *Semin Diagn Pathol* 1994;11:15–25.
19. Konrad EA, Meister P, Hubner G. Extracardiac rhabdomyoma: report of different types with light microscopic and ultrastructural studies. *Cancer* 1982;49:898–907.
20. Tanda F, Rocca PC, Bosincu L, et al. Rhabdomyoma of the tunica vaginalis of the testis: a histologic, immunohistochemical, and ultrastructural study. *Mod Pathol* 1997;10:608–611.
21. Hanski W, Hagel-Lewicka E, Daniszewski K. Rhabdomyomas of female genital tract: report on two cases. *Zentralbl Pathol* 1991;137:439–442.
22. Ostor AG, Fortune DW, Riley CB. Fibroepithelial polyps with atypical stromal cells (pseudosarcoma botryoides) of vulva and vagina: a report of 13 cases. *Int J Gynecol Pathol* 1988;7:351–360.
23. Kawamoto S, Matsuda H, Ueki K, Okada Y, Kim P. Neuromuscular choristoma of the oculomotor nerve: case report. *Neurosurgery* 2007;60:E777–E778; discussion E778.
24. Tobias S, Kim CH, Sade B, Staugaitis SM, Lee JH. Neuromuscular hamartoma of the trigeminal nerve in an adult. *Acta Neurochir* 2006;148:83–87; discussion 87.
25. Uysal A, Sungur N, Kocer U, et al. Neuromuscular hamartoma of the occipital nerve: clinical report. *J Craniofacial Surg* 2005;16:740–742.
26. Tiffee JC, Barnes EL. Neuromuscular hamartomas of the head and neck. *Arch Otolaryngol Head Neck Surg* 1998;124:212–216.
27. Van Dorpe J, Sciot R, De Vos R, et al. Neuromuscular choristoma (hamartoma) with smooth and striated muscle component: case report with immunohistochemical and ultrastructural analysis. *Am J Surg Pathol* 1997;21:1090–1095.
28. Mavrikakis I, White VA, Heran M, Rootman J. Orbital mesenchymal hamartoma with rhabdomyomatous features. *Br J Ophthalmol* 2007;91:692–693.
29. De la Sotta P, Salomone C, Gonzalez S. Rhabdomyomatous (mesenchymal) hamartoma of the tongue: report of a case. *J Oral Pathol Med* 2007;36:58–59.
30. Ortak T, Orbay H, Unlu E, et al. Rhabdomyomatous mesenchymal hamartoma. *J Craniofacial Surg* 2005;16:1135–1137.
31. Magro G, Di Benedetto A, Sanges G, Scalisi F, Alaggio R. Rhabdomyomatous mesenchymal hamartoma of oral cavity: an unusual location for such a rare lesion. *Virchows Arch* 2005;446:346–347.
32. Rosenberg AS, Kirk J, Morgan MB. Rhabdomyomatous mesenchymal hamartoma: an unusual dermal entity with a report of two cases and a review of the literature. *J Cutan Pathol* 2002;29:238–243.
33. Read RW, Burnstine M, Rowland JM, Zamir E, Rao NA. Rhabdomyomatous mesenchymal hamartoma of the eyelid: report of a case and literature review. *Ophthalmology* 2001;108:798–804.
34. Takeyama J, Hayashi T, Sanada T, et al. Rhabdomyomatous mesenchymal hamartoma associated with nasofrontal meningocele and dermoid cyst. *J Cutan Pathol* 2005;32:310–313.
35. Adam MP, Abramowsky CR, Brady AN, Coleman K, Todd NW.

Rhabdomyomatous hamartomata of the pharyngeal region with bilateral microtia and aural atresia: a new association? Birth defects research Part A. *Clin Mol Teratol* 2007;79:242–248.

36. Heffner RR, Jr., Armbrustmacher VW, Earle KM. Focal myositis. *Cancer* 1977;40:301–306.

37. Colding-Jorgensen E, Laursen H, Lauritzen M. Focal myositis of the thigh: report of two cases. *Acta Neurol Scand* 1993;88:289–292.

38. Ellis GL, Brannon RB. Focal myositis of the perioral musculature. *Oral Surg Oral Med Oral Pathol* 1979;48:337–341.

39. Heffner RR, Jr., Barron SA. Denervating changes in focal myositis, a benign inflammatory pseudotumor. *Arch Pathol Lab Med* 1980;104:261–264.

40. Flaisler F, Blin D, Asencio G, Lopez FM, Combe B. Focal myositis: a localized form of polymyositis? *J Rheumatol* 1993;20:1414–1416.

41. Weber C. Anatomishche Untersuchung einer hypertrophische Zunge nebst Bemerkunger uber die Neubildung quergestreifter Muskelfasern. *Virchows Arch* 1854;7:115.

42. Berard M. Tumeur embryonnaire du muscle striae. *Lyon Med* 1894:52.

43. Riopelle JL, Theriault JP. [An unknown type of soft part sarcoma: alveolar rhabdomyosarcoma]. *Ann Anat Pathol (Paris)* 1956;1:88–111.

44. Horn RC, Jr., Enterline HT. Rhabdomyosarcoma: a clinicopathological study and classification of 39 cases. *Cancer* 1958;11:181–199.

45. Newton WA, Jr., Gehan EA, Webber BL, *et al*. Classification of rhabdomyosarcomas and related sarcomas: pathologic aspects and proposal for a new classification – an Intergroup Rhabdomyosarcoma Study. *Cancer* 1995;76:1073–1085.

46. Qualman SJ, Coffin CM, Newton WA, *et al*. Intergroup Rhabdomyosarcoma Study: update for pathologists. *Pediatr Dev Pathol* 1998;1:550–561.

47. Skapek SX, Anderson J, Barr FG, *et al*. PAX-FOXO1 fusion status drives unfavorable outcome for children with rhabdomyosarcoma: a children's oncology group report. *Pediatr Blood Cancer* 2013;60:1411–1417.

48. Fletcher CDM, Bridge JA, Hogendoorn PCW, Mertens F. *WHO Classification of Tumours of Soft Tissue and Bone*. Lyon: International Agency for Research of Cancer; 2013.

49. Locatelli F, Tonani P, Porta F, *et al*. Rhabdomyosarcoma with primary osteolytic lesions simulating non-Hodgkin's lymphoma. *Pediatr Hematol Oncol* 1991;8:159–164.

50. Arndt CA, Crist WM. Common musculoskeletal tumors of childhood and adolescence. *N Engl J Med* 1999;341:342–352.

51. Coffin CM, Rulon J, Smith L, Bruggers C, White FV. Pathologic features of rhabdomyosarcoma before and after treatment: a clinicopathologic and immunohistochemical analysis. *Mod Pathol* 1997;10:1175–1187.

52. Arndt CA, Hammond S, Rodeberg D, Qualman S. Significance of persistent mature rhabdomyoblasts in bladder/prostate rhabdomyosarcoma: results from IRS IV. *J Pediatr Hematol Oncol* 2006;28:563–567.

53. Meza JL, Anderson J, Pappo AS, Meyer WH. Analysis of prognostic factors in patients with nonmetastatic rhabdomyosarcoma treated on intergroup rhabdomyosarcoma studies III and IV: the Children's Oncology Group. *J Clin Oncol* 2006;24:3844–3851.

54. Parham DM, Barr FG. Classification of rhabdomyosarcoma and its molecular basis. *Adv Anat Pathol* 2013;20:387–397.

55. Scott RS, Jagirdar J. Right atrial botryoid rhabdomyosarcoma in an adult patient with recurrent pleomorphic rhabdomyosarcomas following doxorubicin therapy. *Ann Diagn Pathol* 2007;11:274–276.

56. Smith LM, Anderson JR, Qualman SJ, *et al*. Which patients with microscopic disease and rhabdomyosarcoma experience relapse after therapy?: a report from the soft tissue sarcoma committee of the children's oncology group. *J Clin Oncol* 2001;19:4058–4064.

57. Rudzinski ER, Teot LA, Anderson JR, *et al*. Dense pattern of embryonal rhabdomyosarcoma, a lesion easily confused with alveolar rhabdomyosarcoma: a report from the Soft Tissue Sarcoma Committee of the Children's Oncology Group. *Am J Clin Pathol* 2013;140:82–90.

58. Bridge JA, Liu J, Weibolt V, *et al*. Novel genomic imbalances in embryonal rhabdomyosarcoma revealed by comparative genomic hybridization and fluorescence in situ hybridization: an intergroup rhabdomyosarcoma study. *Genes Chromosomes Cancer* 2000;27:337–344.

59. Merlino G, Helman LJ. Rhabdomyosarcoma: working out the pathways. *Oncogene* 1999;18:5340–5348.

60. Bridge JA, Liu J, Qualman SJ, *et al*. Genomic gains and losses are similar in genetic and histologic subsets of rhabdomyosarcoma, whereas amplification predominates in embryonal with anaplasia and alveolar subtypes. *Genes Chromosomes Cancer* 2002;33:310–321.

61. Leuschner I, Newton WA, Jr., Schmidt D, *et al*. Spindle cell variants of embryonal rhabdomyosarcoma in the paratesticular region. A report of the Intergroup Rhabdomyosarcoma Study. *Am J Surg Pathol* 1993;17:221–230.

62. Mentzel T, Katenkamp D. Sclerosing, pseudovascular rhabdomyosarcoma in adults: clinicopathological and immunohistochemical analysis of three cases. *Virchows Arch* 2000;436:305–311.

63. Folpe AL, McKenney JK, Bridge JA, Weiss SW. Sclerosing rhabdomyosarcoma in adults: report of four cases of a hyalinizing, matrix-rich variant of rhabdomyosarcoma that may be confused with osteosarcoma, chondrosarcoma, or angiosarcoma. *Am J Surg Pathol* 2002;26:1175–1183.

64. Chiles MC, Parham DM, Qualman SJ, *et al*. Sclerosing rhabdomyosarcomas in children and adolescents: a clinicopathologic review of 13 cases from the Intergroup Rhabdomyosarcoma Study Group and Children's Oncology Group. *Pediatr Dev Pathol* 2004;7:583–594.

65. Kikuchi K, Wettach GR, Ryan CW, *et al*. MDM2 amplification and PI3KCA mutation in a case of sclerosing rhabdomyosarcoma. *Sarcoma* 2013;2013:520858.

66. Mosquera JM, Sboner A, Zhang L, *et al*. Recurrent NCOA2 gene rearrangements in congenital/infantile spindle cell rhabdomyosarcoma. *Genes Chromosomes Cancer* 2013;52:538–550.

67. Parham DM, Ellison DA. Rhabdomyosarcomas in adults and children: an update. *Arch Pathol Lab Med* 2006;130:1454–1465.

68. Kuhnen C, Herter P, Leuschner I, et al. Sclerosing pseudovascular rhabdomyosarcoma-immunohistochemical, ultrastructural, and genetic findings indicating a distinct subtype of rhabdomyosarcoma. *Virchows Arch* 2006;449:572–578.

69. Croes R, Debiec-Rychter M, Cokelaere K, et al. Adult sclerosing rhabdomyosarcoma: cytogenetic link with embryonal rhabdomyosarcoma. *Virchows Arch* 2005;446:64–67.

70. Morandi S, Manna A, Sabattini E, Porcellini A. Rhabdomyosarcoma presenting as acute leukemia. *J Pediatr Hematol Oncol* 1996; 18:305–307.

71. Tsokos M, Webber BL, Parham DM, et al. Rhabdomyosarcoma: a new classification scheme related to prognosis. *Arch Pathol Lab Med* 1992;116:847–855.

72. Kazanowska B, Jelen M, Reich A, Tarnawski W, Chybicka A. The role of nuclear morphometry in prediction of prognosis for rhabdomyosarcoma in children. *Histopathology* 2004;45:352–359.

73. Parham DM, Shapiro DN, Downing JR, Webber BL, Douglass EC. Solid alveolar rhabdomyosarcomas with the t(2;13): report of two cases with diagnostic implications. *Am J Surg Pathol* 1994;18:474–478.

74. Scrable H, Witte D, Shimada H, et al. Molecular differential pathology of rhabdomyosarcoma. *Genes Chromosomes Cancer* 1989;1:23–35.

75. Kelly KM, Womer RB, Sorensen PH, Xiong QB, Barr FG. Common and variant gene fusions predict distinct clinical phenotypes in rhabdomyosarcoma. *J Clin Oncol* 1997;15:1831–1836.

76. Barr FG, Qualman SJ, Macris MH, et al. Genetic heterogeneity in the alveolar rhabdomyosarcoma subset without typical gene fusions. *Cancer Res* 2002;62:4704–4710.

77. Davicioni E, Finckenstein FG, Shahbazian V, et al. Identification of a PAX-FKHR gene expression signature that defines molecular classes and determines the prognosis of alveolar rhabdomyosarcomas. *Cancer Res* 2006;66:6936–6946.

78. Missiaglia E, Williamson D, Chisholm J, et al. PAX3/FOXO1 fusion gene status is the key prognostic molecular marker in rhabdomyosarcoma and significantly improves current risk stratification. *J Clin Oncol* 2012;30:1670–1677.

79. Furlong MA, Fanburg-Smith JC. Pleomorphic rhabdomyosarcoma in children: four cases in the pediatric age group. *Ann Diagn Pathol* 2001;5:199–206.

80. Gaffney EF, Dervan PA, Fletcher CD. Pleomorphic rhabdomyosarcoma in adulthood: analysis of 11 cases with definition of diagnostic criteria. *Am J Surg Pathol* 1993;17:601–609.

81. Wesche WA, Fletcher CD, Dias P, Houghton PJ, Parham DM. Immunohistochemistry of MyoD1 in adult pleomorphic soft tissue sarcomas. *Am J Surg Pathol* 1995;19:261–269.

82. Gordon A, McManus A, Anderson J, et al. Chromosomal imbalances in pleomorphic rhabdomyosarcomas and identification of the alveolar rhabdomyosarcoma-associated PAX3-FOXO1A fusion gene in one case. *Cancer Genet Cytogenet* 2003;140:73–77.

83. Kodet R, Newton WA, Jr., Hamoudi AB, et al. Childhood rhabdomyosarcoma with anaplastic (pleomorphic) features: a report of the Intergroup Rhabdomyosarcoma Study. *Am J Surg Pathol* 1993;17:443–453.

84. Govender D, Chetty R. Clear cell (glycogen-rich) rhabdomyosarcoma presenting as cervical lymphadenopathy. *ORL* 1999;61:52–54.

85. Boman F, Champigneulle J, Schmitt C, et al. Clear cell rhabdomyosarcoma. *Pediatr Pathol Lab Med* 1996;16:951–959.

86. Chan JK, Ng HK, Wan KY, et al. Clear cell rhabdomyosarcoma of the nasal cavity and paranasal sinuses. *Histopathology* 1989;14:391–399.

87. Quincey C, Banerjee SS, Eyden BP, Vasudev KS. Lipid rich rhabdomyosarcoma. *J Clin Pathol* 1994;47:280–282.

88. Zuppan CW, Mierau GW, Weeks DA. Lipid-rich rhabdomyosarcoma: a potential source of diagnostic confusion. *Ultrastruct Pathol* 1991;15:353–359.

89. Kodet R, Newton WA, Jr., Hamoudi AB, Asmar L. Rhabdomyosarcomas with intermediate-filament inclusions and features of rhabdoid tumors: light microscopic and immunohistochemical study. *Am J Surg Pathol* 1991;15:257–267.

90. Caputo V, Repetti ML, Grimoldi N, et al. Cerebral rhabdomyosarcoma with rhabdoid tumor-like features. *J Neurooncol* 1997;32:81–86.

91. Jo VY, Marino-Enriquez ML, Fletcher CD. Epithelioid rhabdomyosarcoma: clinicopathologic analysis of 16 cases of a morphologically distinct variant of rhabdomyosarcoma. *AM J Surg Pathol* 2011;35:1523–1530.

92. Zin A, Bertorelle R, Dall'Igna P, et al. Epithelioid rhabdomyosarcoma: a clinicopathologic and molecular study. *AM J Surg Pathol* 2014;38:273–278.

93. Hoot AC, Russo P, Judkins AR, Perlman EJ, Biegel JA. Immunohistochemical analysis of hSNF5/INI1 distinguishes renal and extra-renal malignant rhabdoid tumors from other pediatric soft tissue tumors. *Am J Surg Pathol* 2004;28:1485–1491.

94. Uno K, Takita J, Yokomori K, et al. Aberrations of the hSNF5/INI1 gene are restricted to malignant rhabdoid tumors or atypical teratoid/rhabdoid tumors in pediatric solid tumors. *Genes Chromosomes Cancer* 2002;34:33–41.

95. Newsham I, Daub D, Besnard-Guerin C, Cavenee W. Molecular sublocalization and characterization of the 11;22 translocation breakpoint in a malignant rhabdoid tumor. *Genomics* 1994;19:433–440.

96. Karnes PS, Tran TN, Cui MY, et al. Establishment of a rhabdoid tumor cell line with a specific chromosomal abnormality, 46,XY,t(11;22)(p15.5;q11.23). *Cancer Genet Cytogenet* 1991;56:31–38.

97. Sabbioni S, Veronese A, Trubia M, et al. Exon structure and promoter identification of STIM1 (alias GOK), a human gene causing growth arrest of the human tumor cell lines G401 and RD. *Cytogenet Cell Genet* 1999;86:214–218.

98. DeCristofaro MF, Betz BL, Wang W, Weissman BE. Alteration of hSNF5/INI1/BAF47 detected in rhabdoid cell lines and primary rhabdomyosarcomas but not Wilms' tumors. *Oncogene* 1999;18:7559–7565.

99. Kumar S, Perlman E, Harris CA, Raffeld M, Tsokos M. Myogenin is a specific marker for rhabdomyosarcoma: an immunohistochemical study in paraffin-embedded tissues. *Mod Pathol* 2000;13:988–993.

100. Mouton SC, Rosenberg HS, Cohen MC, et al. Malignant ectomesenchymoma in childhood. *Pediatr Pathol Lab Med* 1996;16:607–624.

101. Bahrami A, Gown AM, Baird GS, Hicks MJ, Folpe AL. Aberrant expression of epithelial and neuroendocrine markers in alveolar rhabdomyosarcoma: a potentially serious diagnostic pitfall. *Mod Pathol* 2008;21:795–806.

102. Edwards V, Tse G, Doucet J, Pearl R, Phillips MJ. Rhabdomyosarcoma metastasizing as a malignant ectomesenchymoma. *Ultrastruct Pathol* 1999;23:267–273.

103. Brooks J. Immunohistochemistry in the differential diagnosis of soft tissue tumors. In *Soft Tissue Tumors*, Weiss SW, Brooks JSJ (eds.) No 38 in the USCAP Monographs in Pathology series, Baltimore: Williams & Wilkins; 1996: 65.

104. Azumi N, Ben-Ezra J, Battifora H. Immunophenotypic diagnosis of leiomyosarcomas and rhabdomyosarcomas with monoclonal antibodies to muscle-specific actin and desmin in formalin-fixed tissue. *Mod Pathol* 1988;1:469–474.

105. Parham DM, Reynolds AB, Webber BL. Use of monoclonal antibody 1H1, anticortactin, to distinguish normal and neoplastic smooth muscle cells: comparison with anti-alpha-smooth muscle actin and antimuscle-specific actin. *Hum Pathol* 1995;26:776–783.

106. Parham DM, Dias P, Kelly DR, Rutledge JC, Houghton P. Desmin positivity in primitive neuroectodermal tumors of childhood. *Am J Surg Pathol* 1992;16:483–492.

107. Somerhausen NS, Fletcher CD. Diffuse-type giant cell tumor: clinicopathologic and immunohistochemical analysis of 50 cases with extraarticular disease. *Am J Surg Pathol* 2000;24:479–492.

108. Leader M, Patel J, Collins M, Henry K. Myoglobin: an evaluation of its role as a marker of rhabdomyosarcomas. *Br J Cancer* 1989;59:106–109.

109. Eusebi V, Bondi A, Rosai J. Immunohistochemical localization of myoglobin in nonmuscular cells. *Am J Surg Pathol* 1984;8:51–55.

110. Dias P, Dilling M, Houghton P. The molecular basis of skeletal muscle differentiation. *Semin Diagn Pathol* 1994;11:3–14.

111. Wang NP, Marx J, McNutt MA, Rutledge JC, Gown AM. Expression of myogenic regulatory proteins (myogenin and MyoD1) in small blue round cell tumors of childhood. *Am J Pathol* 1995;147:1799–1810.

112. Mukunyadzi P. DPHPea: comparison of MyoD1 immunostaining of pediatric tumors using frozen or paraffin-embedded sections. *Appl Immunohistochem Mol Morphol* 1999;7:260.

113. Mentzel T, Kuhnen C. Spindle cell rhabdomyosarcoma in adults: clinicopathological and immunohistochemical analysis of seven new cases. *Virchows Arch* 2006;449:554–560.

114. Ellison DA, Adada B, Qualman SJ, Parham DM. Melanotic neuroectodermal tumor of infancy: report of a case with myogenic differentiation. *Pediatr Dev Pathol* 2007;10:157–160.

115. Dias P, Chen B, Dilday B, et al. Strong immunostaining for myogenin in rhabdomyosarcoma is significantly associated with tumors of the alveolar subclass. *Am J Pathol* 2000;156:399–408.

116. Morotti RA, Nicol KK, Parham DM, et al. An immunohistochemical algorithm to facilitate diagnosis and subtyping of rhabdomyosarcoma: the Children's Oncology Group experience. *Am J Surg Pathol* 2006;30:962–968.

117. Chen B, Dias P, Jenkins JJ, 3rd, Savell VH, Parham DM. Methylation alterations of the MyoD1 upstream region are predictive of subclassification of human rhabdomyosarcomas. *Am J Pathol* 1998;152:1071–1079.

118. Rudzinski ER, Anderson JR, Lyden ER, et al. Myogenin, AP2beta, NOS-1, and HMGA2 are surrogate markers of fusion status in rhabdomyosarcoma: a report from the soft tissue sarcoma committee of the children's oncology group. *Am J Surg Pathol* 2014;38:654–659.

119. Furlong MA, Mentzel T, Fanburg-Smith JC. Pleomorphic rhabdomyosarcoma in adults: a clinicopathologic study of 38 cases with emphasis on morphologic variants and recent skeletal muscle-specific markers. *Mod Pathol* 2001;14:595–603.

120. Walterhouse D, Watson A. Optimal management strategies for rhabdomyosarcoma in children. *Paediatr Drugs* 2007;9:391–400.

121. d'Amore ES, Tollot M, Stracca-Pansa V, et al. Therapy associated differentiation in rhabdomyosarcomas. *Mod Pathol* 1994;7:69–75.

122. Smith LM, Anderson JR, Coffin CM. Cytodifferentiation and clinical outcome after chemotherapy and radiation therapy for rhabdomyosarcoma (RMS). *Med Pediatr Oncol* 2002;38:398–404.

123. Gu W, Schneider JW, Condorelli G, et al. Interaction of myogenic factors and the retinoblastoma protein mediates muscle cell commitment and differentiation. *Cell* 1993;72:309–324.

124. Diller L, Sexsmith E, Gottlieb A, Li FP, Malkin D. Germline p53 mutations are frequently detected in young children with rhabdomyosarcoma. *J Clin Invest* 1995;95:1606–1611.

125. Hettmer S, Archer NM, Somers GR, et al. Anaplastic rhabdomyosarcoma in TP53 germline mutation carriers. *Cancer* 2014;120:1068–1075.

126. Zhang L, Zhan Q, Zhan S, et al. p53 regulates human insulin-like growth factor II gene expression through active P4 promoter in rhabdomyosarcoma cells. *DNA Cell Biol* 1998;17:125–131.

127. Kilpatrick SE, Teot LA, Geisinger KR, et al. Relationship of DNA ploidy to histology and prognosis in rhabdomyosarcoma: comparison of flow cytometry and image analysis. *Cancer* 1994;74:3227–3233.

128. De Zen L, Sommaggio A, d'Amore ES, et al. Clinical relevance of DNA ploidy and proliferative activity in childhood rhabdomyosarcoma: a retrospective analysis of patients enrolled onto the Italian Cooperative Rhabdomyosarcoma Study RMS88. *J Clin Oncol* 1997;15:1198–1205.

Chapter 21

Hemangiomas, lymphangiomas, and reactive vascular proliferations

Markku Miettinen
National Cancer Institute, National Institutes of Health

Benign vascular tumors and tumor-like conditions include a wide variety of clinicopathologic entities, among them different types of hemangiomas, vascular malformations, telangiectasiae, hemangioma-like reactive vascular proliferations, and lymphangiomas. Also included in this discussion is kaposiform hemangioendothelioma (KH), a hemangioma variant without metastatic potential (although one having potential for fatal complications through thrombocytopenia). The discussion of specific tumor entities is preceded by a synopsis of endothelia and angiogenesis given for biologic background information relevant to vascular tumors.

The genetic background of inheritable hemangiomas and vascular malformations is becoming known from the hereditary syndromes, but the somatic genetics of hemangiomas is still poorly understood.

Borderline (hemangioendotheliomas) and malignant vascular tumors (angiosarcoma) are discussed in Chapter 22. Hemangiopericytoma is considered related to solitary fibrous tumors, and these tumors are discussed together (Chapter 12). Vascular smooth muscle tumors, vascular leiomyoma, and leiomyosarcoma, are discussed with other smooth muscle tumors (Chapter 16), and lymphangiomyoma is included among PEComas (Chapter 19). Glomus tumor and angiomyopericytoma have significant vascular components while also containing smooth-muscle-related elements (Chapter 23).

Biology of endothelia and angiogenesis

Endothelial cells are mesenchymal cells that not only form a protective epithelial-like inner lining in the lumen of the blood vessels and lymphatics but also have several other vasoregulatory functions. These include regulation of vascular permeability and caliber, transport of metabolites (e.g., cholesterol), and regulation of leukocyte adhesion, trafficking, and hemostasis. External signals from growth factors/cytokines secreted by a variety of cells modulate the differentiation, antigen expression, and functional state of endothelial cells. Hormonal activity has also been demonstrated, since endothelial cells secrete vasoregulatory peptides such as endothelin. The endothelial cells of an average human body have been estimated to line an area of 6 m^2 to 7 m^2 and weigh 1 kg in aggregate.[1,2]

The dynamic growth of endothelial cells and capillaries is important for optimal oxygen supply of tissues, inflammatory response, and wound healing, and it is one of the tissue responses to ischemia. Regulators of the endothelial proliferation include angiopoietins and several families of vascular endothelial growth factors (VEGF): VEGF, VEGF-B, VEGF-C, and VEGF-D. The most widely expressed in normal tissues and cancer cells are the growth factors of the VEGF family, but growth factors of other families have also been detected in cancer cells. Some types of endothelia are under the influence of fibroblast growth factors and have corresponding receptors.

Endothelial cells have several types of vascular endothelial growth factor and related receptors. They include receptor tyrosine kinases VEGFR1, VEGFR2, VEGFR3, and TIE1, which carry the external growth factor signal to the nucleus through a complex phosphorylation cascade referred to as the *signal transduction pathway*. Interaction with the extracellular matrix through a complex set of receptors, such as the integrins, also regulates endothelial proliferation and migration. Specific knowledge of the role of their functional alterations in vascular neoplasia is still limited, however.[3–6]

Angiogenesis, the formation of new blood vessels, is a requirement for any solid tumor to grow beyond the size of a few millimeters. This process also occurs during wound healing, but in normal tissues only rarely, for example in the cyclic growth of the endometrium. During angiogenesis, the new capillaries arise by sprouting from the existing ones. Understanding of the molecular mechanism of angiogenesis and discovery of antiangiogenic substances such as angiostatin and endostatin has therefore been of great interest as a potential cancer treatment. Thalidomide, a drug originally used for its antiemetic and sedative effects during pregnancy, has been shown to be antiangiogenic. This effect is also thought to be related to its teratogenic complications, such as the truncation-type extremity malformations (*phocomelia*). Promotion of angiogenesis, on the other hand, could be utilized to generate new blood vessels to replace the defunct ones.[3–6]

Vasculogenesis is the process of formation of the blood vessels during embryonic life. The early blood vessels come from putative mesodermal angioblasts surrounding early hematopoietic elements, but the source of the angioblasts and the process of

Modern Soft Tissue Pathology, Second Edition, ed. Markku Miettinen. Published by Cambridge University Press. © Cambridge University Press 2016.

very early vasculogenesis is not completely understood. The cardiovascular system develops further by branching out from the heart. There is some evidence that during one's lifetime, the endothelia of blood vessels are replenished by circulating bone marrow stem cells. This would also explain how some endothelial neoplasms could be of marrow stem cell origin, and why some hemangiomas and angiosarcomas are multifocal.

Classification of hemangiomas

Hemangiomas are among the most common benign soft tissue tumors, together with lipomas. They typically form well-differentiated blood vessels containing both endothelial and pericytic elements and having a limited proliferative capacity. They can occur in any age, but are particularly common in children. Hemangiomas can be classified by different parameters, generating overlapping categories. They can be defined by: (1) vessel type (capillary, cavernous, venous), (2) location (cutaneous, intramuscular, synovial, among others), (3) characteristic cell type (epithelioid, spindle cell, hobnail), (4) neoplastic status (true neoplasia, vascular malformation, telangiectasia, reactive hyperplasia), and (5) by patient age (juvenile, senile). The specific types of hemangioma discussed herein have been variously defined by these parameters, but are considered distinct clinicopathologic entities.

Morphologically, hemangiomas are traditionally classified into capillary, cavernous, or venous types according to the prevalent histologic pattern, and these descriptors are used for large numbers of hemangiomas that do not fall into the categories described here. Many hemangiomas also combine different patterns. In addition, many also contain lymphatic vessels, so that the description "mixed angioma" characterizes them more accurately than the term hemangioma.

By location, hemangiomas can be classified as cutaneous, subcutaneous, synovial, and intramuscular. In this chapter, only intramuscular hemangioma is discussed as a separate entity. Many specific hemangioma types (hobnail and glomerular) are cutaneous tumors, however. The location-based categories obviously overlap with those defined by vessel type.

Epithelioid, spindle cell, and hobnail hemangiomas are distinct clinicopathologic entities. Epithelioid hemangioma is a vascular proliferation arising in the background of an inflammatory response of unknown origin. This condition is different from Kimura disease, which is a lymphoid proliferation with a reactive, nonepithelioid, vascular component (Chapter 35). Spindle cell hemangioma (SCH) is a distinct type with strong acral predilection. In general, typical of all hemangiomas is composition of well-differentiated blood vessels featuring both endothelial cells and pericytes.[7–11]

Hemangioma versus vascular malformation versus telangiectasia

Classification by neoplastic status is problematic and somewhat controversial. In clinical pediatrics and pediatric pathology in particular, there is a tendency to classify many lesions traditionally considered hemangiomas as vascular malformations, although juvenile hemangioma and lobular capillary hemangioma are examples of retention of the term *hemangioma*.[12–14]

Features considered clinically distinctive for vascular malformations are the congenital nature of the lesion (although some hemangiomas are also congenital), lack of a proliferative endothelial component, growth proportional to the growth of the child, and lack of spontaneous regression. Some authors consider most if not all intramuscular hemangiomas and lymphangiomas to be vascular malformations. The distinction of vascular malformation versus hemangioma by the previous criteria requires strong clinical correlation and cannot generally be made by morphology.[12–14] Therefore, the term *hemangioma* is used for all lesions here. There is probably more agreement among specialists that large vessel-type hemangiomas (*venous hemangiomas*) are mostly vascular malformations rather than true neoplasms.

In contrast, hemangiomas are usually acquired, contain a proliferative endothelial component, and grow disproportionally to the affected body part, and some of them regress. The last is especially true for hemangiomas of infancy. In some cases, the determination of whether a vascular lesion is a telangiectasia, malformation, or true hemangioma is arbitrary, and this determination is often based on conventions. Hard evidence of the clonal neoplastic nature of hemangiomas and other benign vascular tumors is currently scant.

Telangiectasia implies a lesion with dilated vascular lumina without vascular proliferation. Typical examples of telangiectasiae are common cutaneous telangiectasia and the facial ones (e.g., port-wine stain, nevus flammeus) in Sturge–Weber syndrome, and venous lake.[9]

Infantile hemangioma and related entities

The main group of these lesions is infantile (juvenile capillary) hemangioma, which develops during infancy, but regresses over time. There is a small group of congenital hemangiomas that differ from this type (rapidly involuting congenital hemangioma and noninvoluting congenital hemangioma); they are discussed briefly at the end of this section.

Clinical features

This common hemangioma has been estimated to be present in 10% of infants. It is present in 2% of newborns, more often in premature infants. It usually appears during the first year of life, but can persist in older children despite involution. There is a predilection for female children (Figure 21.1) and Caucasians, and the preferred location is the head and neck (Figure 21.2). These hemangiomas usually involve the skin and can extend to the subcutis, but they can also present in deeper locations (e.g., orbit and parotid gland, among others).[15–17]

Juvenile hemangiomas grow rapidly during the first postnatal weeks, after which a slow regression follows over several

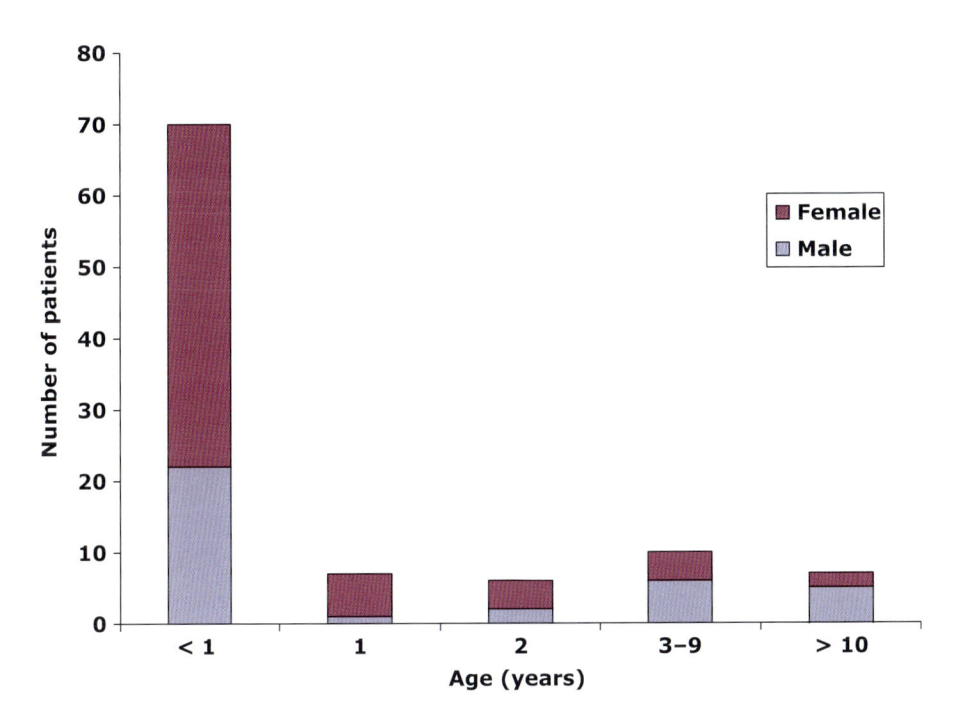

Figure 21.1 Age and sex distribution of 100 patients with juvenile hemangiomas.

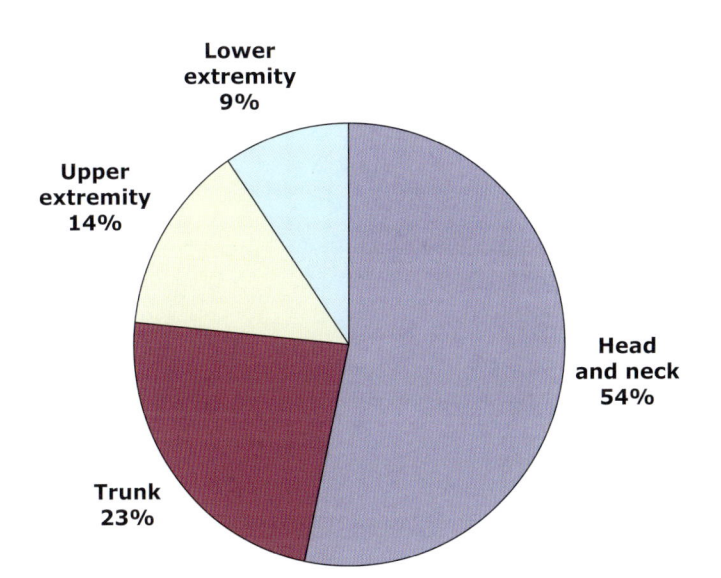

Figure 21.2 Anatomic distribution of 100 juvenile hemangiomas.

years. Small hemangiomas often need no treatment, but those that are troublesome can be surgically excised. Extensive hemangiomas can cause high-output cardiac failure.[18] Nonsurgical treatment options for large or inaccessible tumors include corticosteroids and interferon alpha-2, both of which promote hemangioma regression.[16–18] Oral propranolol (beta-blocker) has been more recently used as nonsurgical therapy.[19]

Pathology

Grossly, juvenile capillary hemangiomas often form elevated, bright-red, nodular cutaneous lesions, and therefore they are referred to as *strawberry nevi*. Histologically, these hemangiomas are typically lobulated, but in some instances they form homogeneous contiguous masses. They entrap surrounding elements such as skin adnexa, subcutaneous fat, and salivary gland tissue. The degree of vasoformation varies; there can be densely cellular areas having closely packed capillaries, with inconspicuous vascular lumina often containing a single erythrocyte. Mitotic activity is often present in the endothelial cells. These hemangiomas were historically called "infantile hemangioendotheliomas," a term that should no longer be used because of potential confusion with borderline malignant vascular tumors. More well-defined and wider vascular lumina also occur, sometimes in combination with solid areas (Figures 21.3 and 21.4). A prominent pericytic, actin-positive component typically accompanies the endothelial cells. Infantile hemangiomas contain numerous mast cells and infiltrating histiomonocytic cells.[16,20]

Immunohistochemically, the endothelial cells of juvenile capillary hemangiomas are uniquely positive for glucose transporter type I (GLUT1), similar to the capillaries of placental villi, and for Lewis Y blood group antigen (CD174). This phenotypic similarity to placental capillaries has been suggested to reflect that fetal autotransplantation of placental endothelia gives rise to these hemangiomas.[20,21] The capillary endothelia are positive for endothelial markers, and there is a substantial pericytic smooth-muscle actin-positive component.[20]

Electron microscopy has revealed peculiar lamellar crystalline endothelial cell cytoplasmic inclusions that measure 0.5 μm to 2 μm and have a periodic substructure.[20,23]

One study showed a somatic mutation in VEGFR2 in one of 15 juvenile hemangiomas, suggesting that pathologic activation of the VEGF signaling pathway could be responsible for the pathogenesis of some of these hemangiomas.[24]

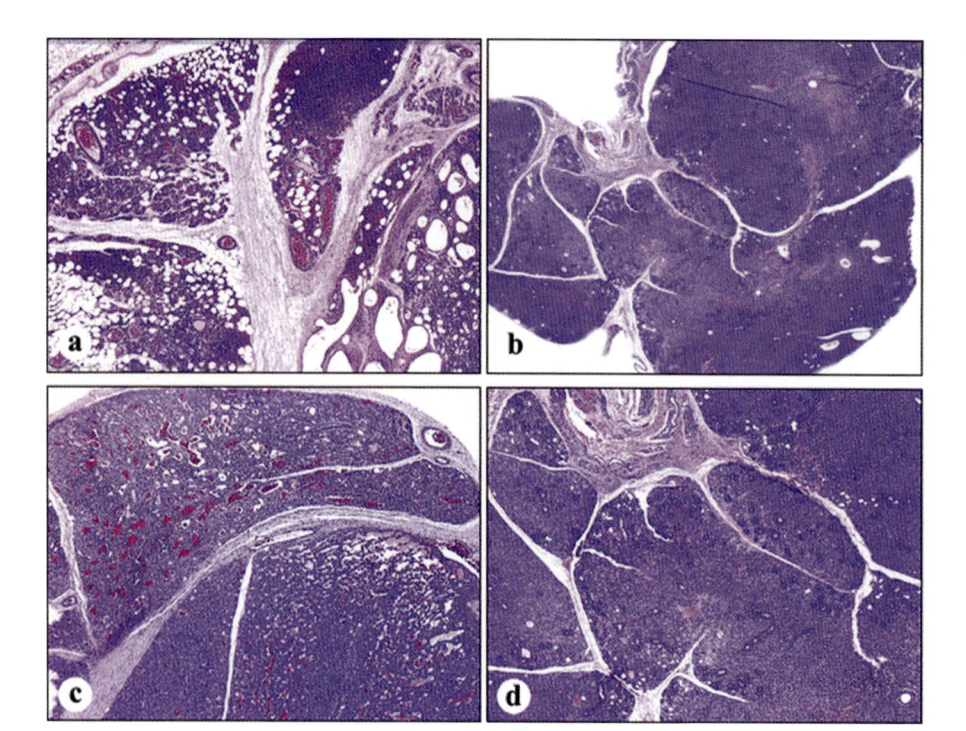

Figure 21.3 A juvenile capillary hemangioma is composed of lobules of capillary hemangioma-like units that can infiltrate fat.

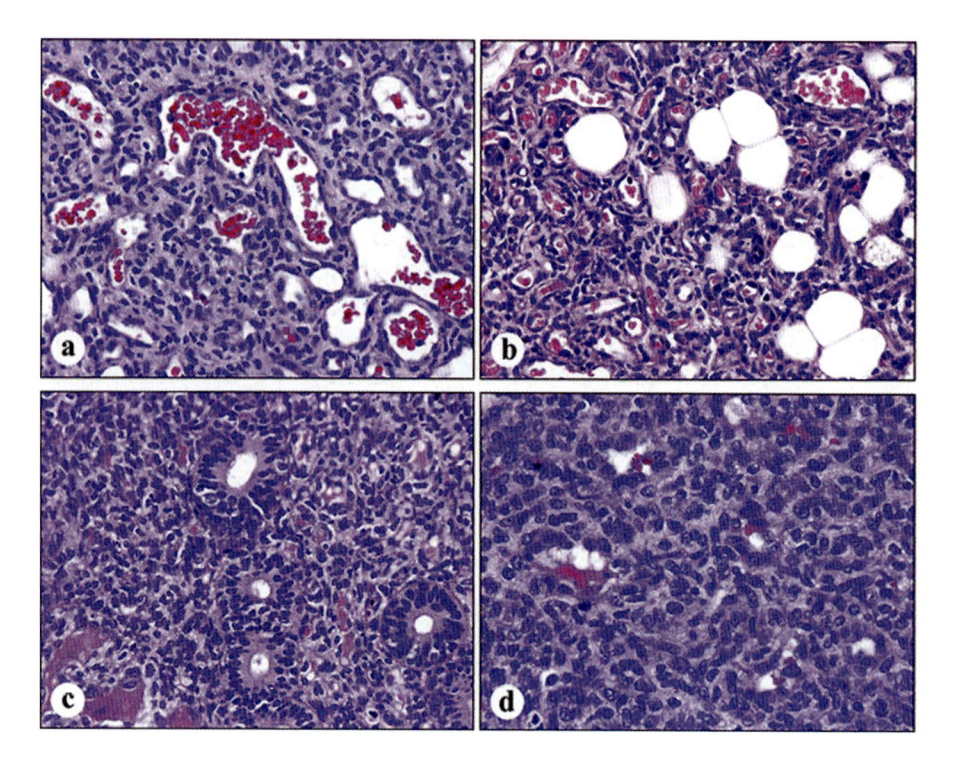

Figure 21.4 A juvenile capillary hemangioma shows a spectrum of well to poorly defined vasoformation. These tumors infiltrate fat and existing glandular structures, such as lacrimal glands.

Related terms and entities

The terms *rapidly involuting congenital hemangioma, noninvoluting congenital hemangioma*,[25] and *congenital nonprogressive hemangioma*[26] refer to uncommon congenital hemangiomas that mainly occur in the head and neck, have no gender bias, are quite large (often >5 cm), and sometimes are associated with mild hemangioma-related platelet trapping without full-blown disseminated intravascular coagulation (Kasabach–Merritt syndrome).

Histologically, they do not have unifying morphology but show a spectrum of changes. Many are lobulated, and some resemble lobular capillary hemangioma, tufted angioma, or KH. These hemangiomas are presently better recognized as clinical rather than distinct histopathologic entities; however,

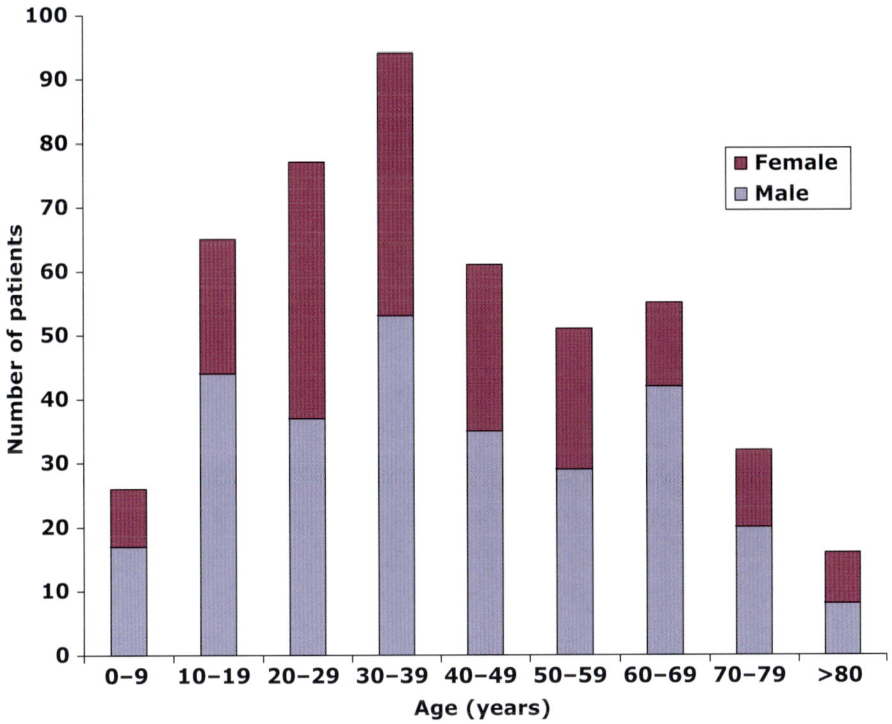

Figure 21.5 Age and sex distribution of 477 lobular capillary hemangiomas.

Lobular capillary hemangioma (pyogenic granuloma)

Clinical features

This lesion occurs in a wide age range, but is common in children and young adults, with a male predominance, especially in children (Figure 21.5). One large series in children also noted a male predominance.[27] Occurrence during pregnancy is common, especially known in gingival mucosa (granuloma gravidarum). A large series of oral examples noted 8-month average preoperative duration.[28]

This clinically distinctive form of capillary hemangioma often presents as a small (<1 cm) purplish red, rapidly growing polypoid protrusion on the skin or mucosal surfaces, especially in the head and neck, most often in the lip or gingival mucosa.[27,28] A large series has also been reported in the nasal cavity.[29] The involved cutaneous or mucosal surface is often ulcerated, resulting in a secondary inflammatory response. Cutaneous examples, also designated as lobular capillary hemangiomas, have been suggested as the underlying lesion in any pyogenic granuloma.[30] A subcutaneous location is also possible although relatively rare.[31] Behavior is benign, but local recurrences can occur.[30] A large series of pediatric examples reported no recurrences,[27] although the recurrence rate in an oral series including all ages was 16%.[28] These hemangiomas in children differ from the juvenile versions, they seem to differ from juvenile capillary hemangiomas by their lack of GLUT1 and Lewis Y blood group antigen expression.[26]

having almost no tendency for spontaneous regression.[27] Multiplicity[27] or satellite nodules have been occasionally reported.[32]

Pathology

Grossly, the superficial examples often present as dome-shaped reddish masses, often covered by a crusted ulceration. Microscopically, the lesion consists of proliferative capillaries arranged in lobules; mitotic activity can be present. A collar-like epithelial lining often surrounds cutaneous and mucosal lesions (Figure 21.6). Variably developed lobulation is typically evident, and there are sometimes larger central vessels called *feeder vessels* (Figure 21.7). The endothelial cells are typically slightly enlarged, forming vessels with variably developed lumina. Mitotic activity can be present, and occasionally there is mild endothelial atypia.

The endothelia are positive for CD31, CD34, and von Willebrand factor (VWF). The pericytic population is typically well-developed as a sign of relative vascular maturity and can be highlighted by immunohistochemistry for actins.[33]

Intravascular pyogenic granuloma refers to a histologically similar lesion that is located inside the lumen of a small vein (Figure 21.8). This tumor typically occurs in the head and neck area or in the upper extremity, with a predilection for young adults, and has a benign course.[34,35]

Florid lobular capillary hemangiomas with focally solid, nonlobular histology can occur especially in the oral and nasal mucosas, and these lesions should not be misdiagnosed as hemangioendotheliomas or angiosarcomas. The lack of significant endothelial tufting or atypia can be of help in this

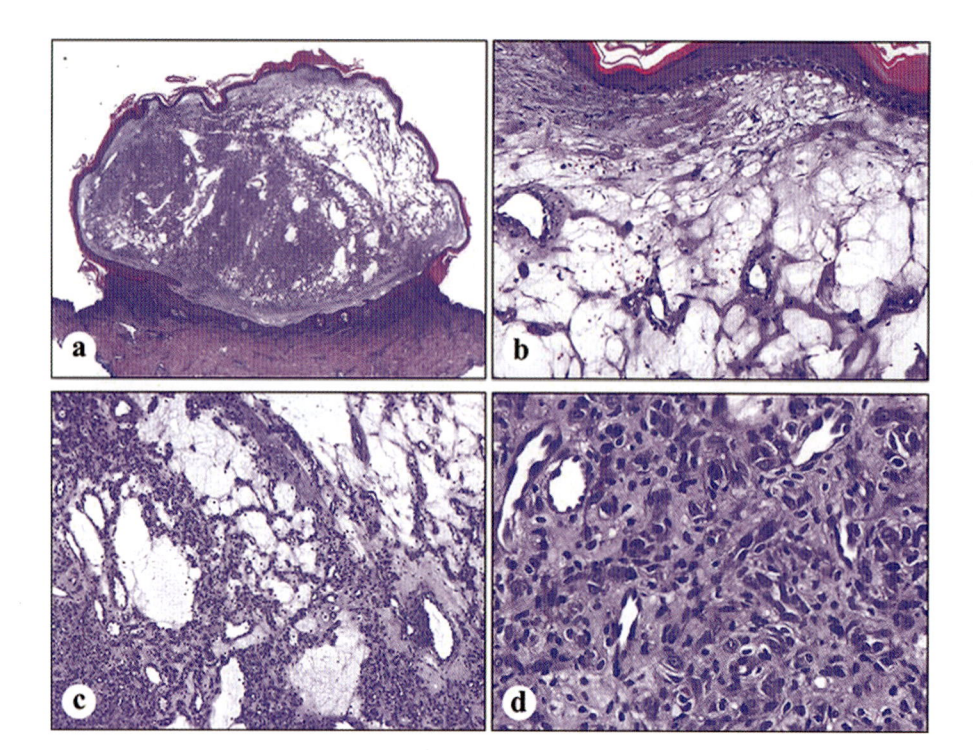

Figure 21.6 (**a**) A superficial lobular capillary hemangioma (pyogenic granuloma) forms an exophytic cutaneous nodule lined by an epidermal collar. (**b,c**) Variably mature capillaries are present in an edematous stroma. (**d**) The capillaries vary from closed to open, and there is no endothelial cell atypia.

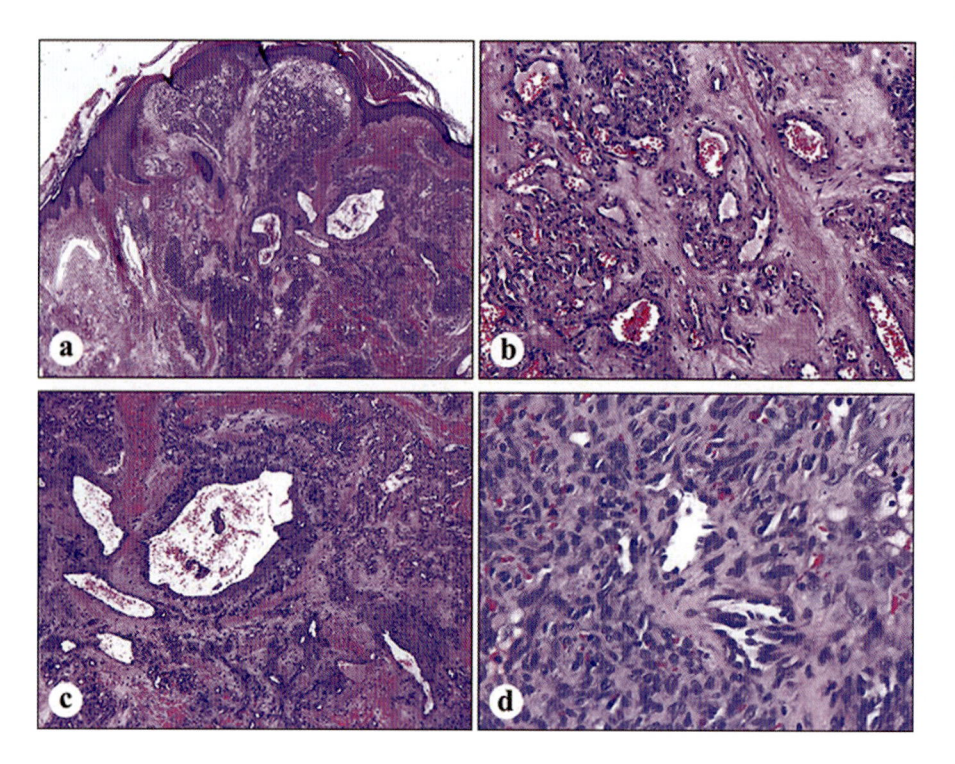

Figure 21.7 (**a–d**). Another example of a dermal lobular capillary hemangioma. Note lobulation and larger vessels (**c**) that are feeder vessels.

distinction.[36] Recognition of the actin-positive pericytic component is also useful. Focal granulocytic infiltration is common, especially adjacent to ulceration, but the presence of neutrophilic microabscesses should lead to suspicion of bacillary angiomatosis, especially in immunocompromised patients. Whether lobular capillary hemangioma is a reactive lesion or a true neoplasm is unknown. A deletion in q21.2q22.12 has been reported as the sole cytogenetic abnormality in a nasal example.[37]

Cavernous hemangioma

These common hemangiomas occur with no age and sex predilection in a wide variety of locations, especially in the

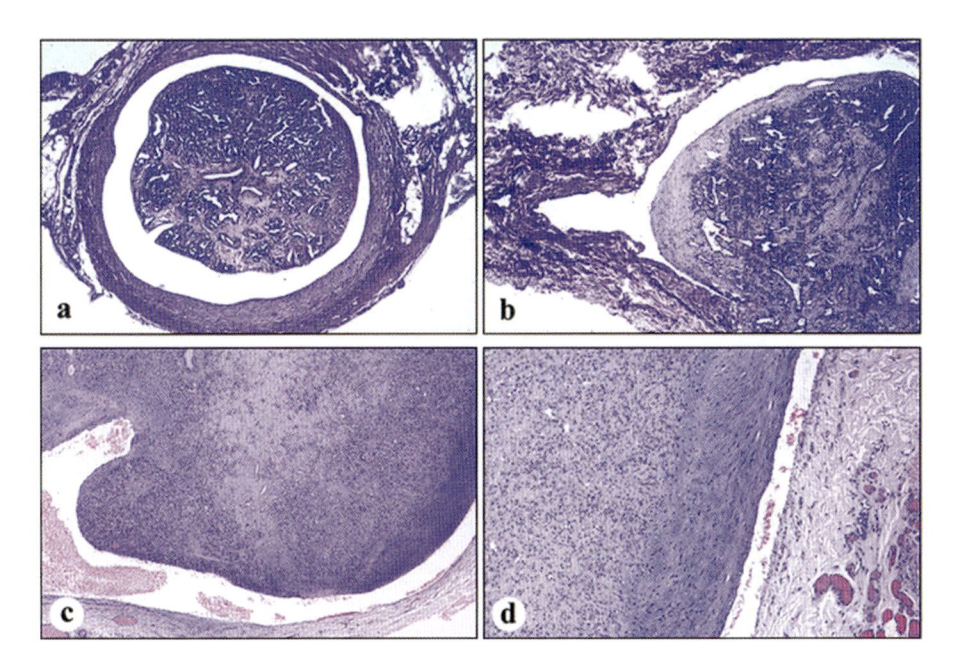

Figure 21.8 Four examples of intravascular pyogenic granuloma.

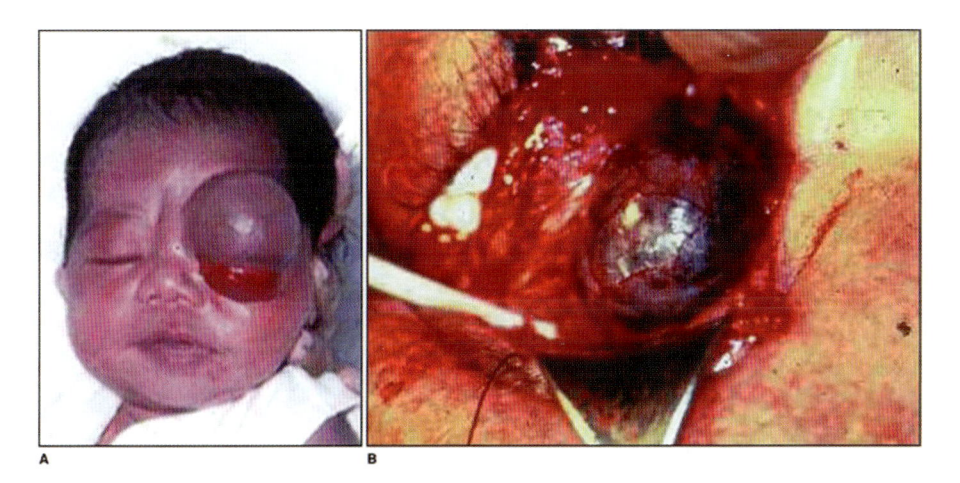

Figure 21.9 Clinical and operative appearances of a cavernous hemangioma of the orbit. The exposed tumor appears as a red, spherical, demarcated mass.

upper body and often in the orbit (Figure 21.9). The lesions can involve the skin, subcutis, deep soft tissues, and bone. A common site of presentation is the liver, where the tumor can be a small incidental nodule or a large space-occupying lesion. Cavernous hemangiomas of the skin do not typically regress as the juvenile capillary ones do.

Grossly, the larger lesions have a spongy appearance because of the large vascular spaces. Histologically, the lesions are composed of a conglomerate of widely open vascular lumina lined by relatively inconspicuous endothelial cells and separated by thick nearly acellular fibrous septa (Figure 21.10). Some tumors combine the features of cavernous and capillary hemangiomas.

Sinusoidal hemangioma is the designation given to a variant of cavernous hemangioma that occurs in adults and forms sinusoidal spaces lined by delicate septa.[38]

Angiokeratoma

Angiokeratoma refers to a group of clinicopathologically heterogeneous but morphologically rather similar superficial cutaneous vascular tumors that are generally thought to be telangiectasiae and not true hemangiomas. Their clinical significance varies from clinical melanoma simulators to bleeding angiomas and sentinel lesions of the hereditary Fabry disease.[39–41]

Clinical features

The four generally recognized forms of angiokeratoma are Mibelli angiokeratoma, Fordyce angiokeratoma, diffuse angiokeratoma (usually associated with the hereditary Fabry disease), and angiokeratomas not belonging to these groups; the last group is the largest and includes cases with single or

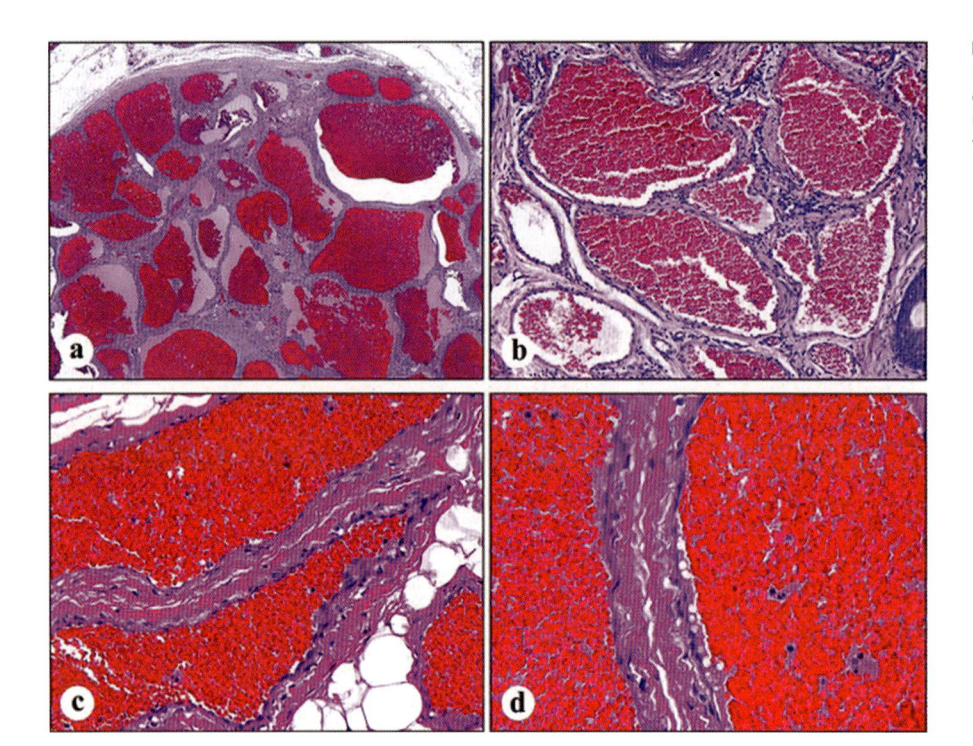

Figure 21.10 (**a,b**) Histologically, a cavernous hemangioma forms a well-delineated mass composed of blood-filled dilated vascular spaces. (**c,d**) Note nearly acellular fibrous septa dividing the tumor into lobules.

multiple lesions. Many of the forms were originally described more than a century ago, but the modern descriptions are by Imperial and Helwig, who also listed "circumscribed angiokeratoma," a congenital lesion, as a separate group that is here merged with angiokeratoma, not otherwise specified.[39–41]

Mibelli angiokeratoma occurs in the fingers and hands of children and young adults, with a female predominance. These warty, elevated, often multiple lesions typically occur over bony prominences in the hands and feet.[39]

Fordyce angiokeratoma forms multiple reddish papules, mainly on the scrotal skin, sometimes also involving adjacent sites, and mostly occurring in adults of varying ages. Some patients have a varicocele, which has led to the belief that angiokeratomas are superficial venous varicosities.[40]

Diffuse angiokeratomas in Fabry disease, also designated as *angiokeratoma corporis diffusum*, occur as numerous small purplish papules between the lower trunk and the knees, usually with onset at young adulthood.

Fabry disease is a hereditary, X-linked recessive lysosomal storage disease that affects only boys, although female subjects are carriers with possible mild disease. This syndrome is caused by germline mutations (missense mutations, deletions) in the gene encoding alpha-galactosidase A at Xq21–22, leading to an enzyme defect and accumulation of glycosphingolipids in multiple organs. Significant long-term clinical manifestations include brain damage and renal failure.[41]

If angiokeratoma can be clinically recognized, no specific therapy is usually necessary. Some lesions are removed for diagnostic purposes, especially those that are blackish nodules and clinically simulate malignant melanoma.

Pathology

Histologically, angiokeratomas typically appear as elevated cutaneous lesions surrounded by a hyperplastic epidermal collar containing several dilated, typically markedly hyperemic capillaries. These vascular elements are lined by small, relatively inconspicuous endothelial cells (Figure 21.11).

Angiokeratoma in Fabry disease is often less keratotic but otherwise histologically similar to the nonsyndromic type. In angiokeratomas in Fabry disease, however, there are ultrastructurally distinctive multilamellar, lysosomal myelinoid inclusions.

Verrucous hemangioma

Clinical features

This rare cutaneous hemangioma occurs mainly in children and young adults and is generally present at birth. There is no gender predilection, and the most commonly involved sites are the foot and ankle. The lesion can enlarge over several years to form a cluster of irregularly shaped purple nodules that later become hyperkeratotic. Some patients have multiple lesions arranged in a linear pattern. Because this angioma extends upward to the epidermis, it can bleed after any minor trauma. The lesional size varies from a small to a large plaque-like lesion of several centimeters in diameter. Recurrence is common after an incomplete excision.[42–46]

Pathology

Histologically, there is verrucous papillary epidermal hyperplasia and dilated capillaries; cavernous, thick-walled,

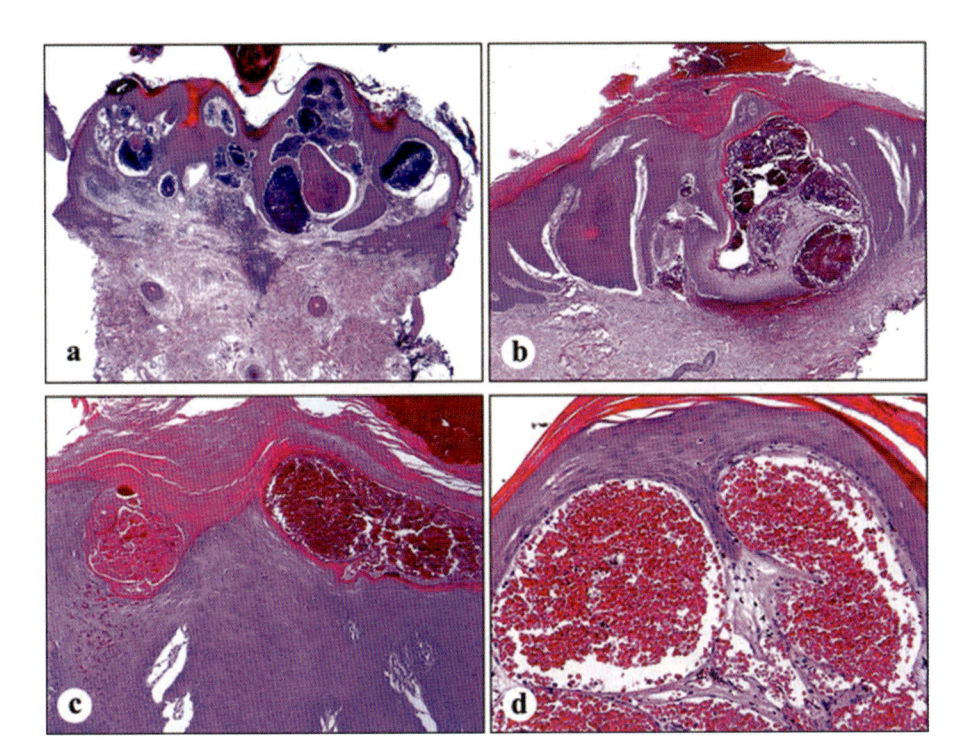

Figure 21.11 An angiokeratoma forms a superficial vascular lesion showing dilated capillaries immediately below the epidermis. (**d**) The endothelial cells are small and few in number.

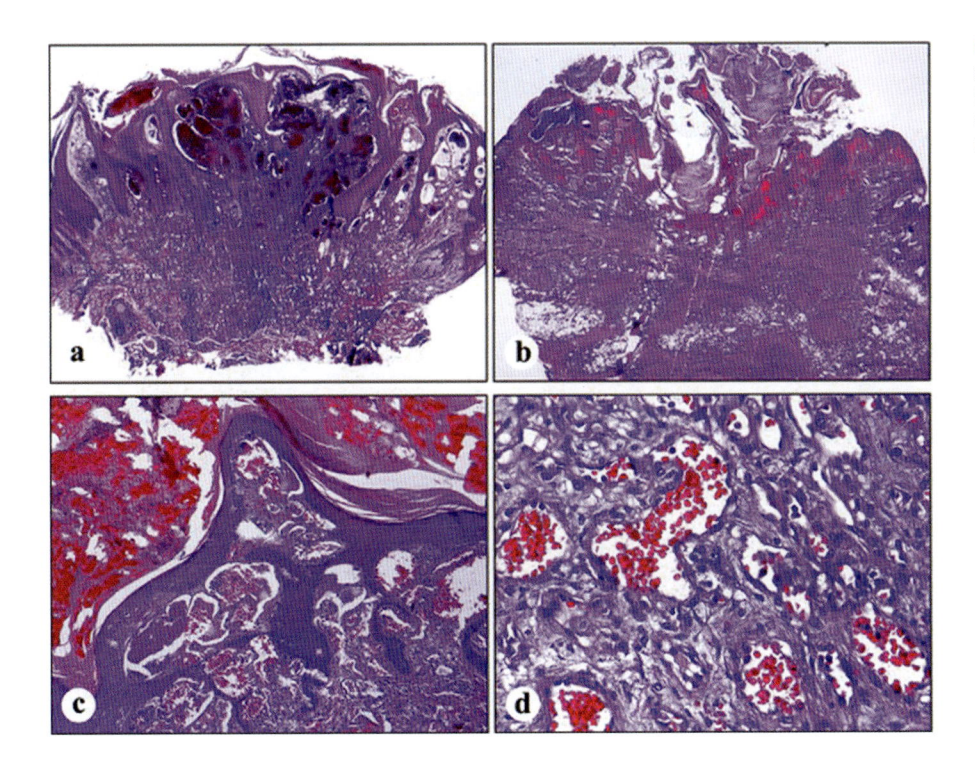

Figure 21.12 A verrucous hemangioma is covered by hyperkeratotic epidermis with wart-like folding. (**b**) Lobules of capillaries extend into the deep dermis. (**d**) Endothelial cells are small and regular.

blood-filled vascular spaces can extend from just below the epidermis into the deep dermis and sometimes into the subcutis. A deeper, capillary-like element with smaller lumina is also present, in contrast to angiokeratoma. The endothelial lining forms an inconspicuous monolayer without atypia (Figure 21.12).

Arteriovenous hemangioma of the skin (acral arteriovenous tumor, cirsoid aneurysm)

This distinctive lesion, originally reported by Carapeto et al. in 1977, refers to a hemangioma containing a cluster of large-profile vessels with muscular walls, with features of arteries

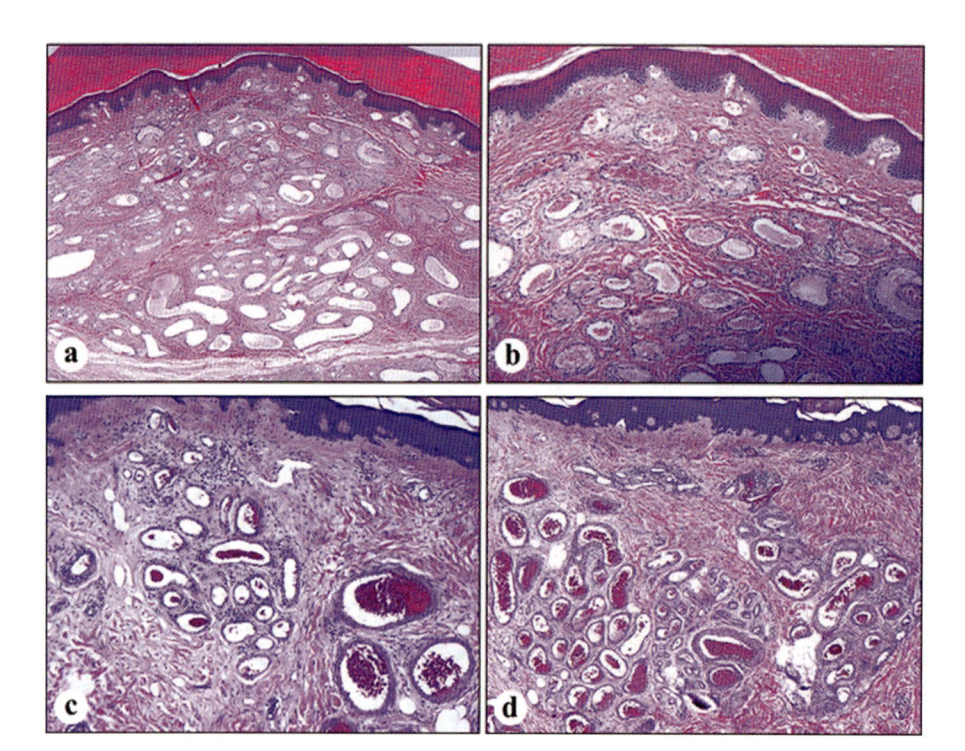

Figure 21.13 An arteriovenous hemangioma contains defined clusters of thick-walled vessels with wide lumina.

and veins. It occurs predominantly in middle-aged and older adults, with a male predominance, but can occur at all ages. The typical locations are the head and neck, especially the orofacial region, and the extremities; occurrence in the trunk is rare. There is no clinical evidence for arteriovenous shunting.[47–49]

Clinically, cutaneous arteriovenous hemangioma forms an asymptomatic, small (<1 cm), red, raised papule of plaque, often with a history of several years. There is no tendency to recur after a simple excision.

Histologically, the formation consists of a well-circumscribed dermal cluster of profiles of muscular veins and arteries (Figure 21.13). Some lesion vessels can undergo thrombosis. A superficial component consisting of capillaries is sometimes present.

Venous hemangioma

This designation refers to a hemangioma that is composed of dilated vein-like vascular units with incompletely developed, discontinuous smooth muscle layers, which distinguish it from a cavernous hemangioma. The lumina are typically lined by endothelial cells with small nuclei (Figure 21.14). These lesions are often considered vascular malformations.

Venous hemangiomas often occur intramuscularly,[50] but they can also be present in superficial soft tissues, such as the breast.[51] Occurrence of venous hemangioma of the foot (pedal hemangioma) has been reported in connection with Turner syndrome.[52]

Intramuscular hemangioma
Clinical features

This rare presentation of hemangioma occurs primarily in young adults <20 years, equally in men and women, some being congenital (Figure 21.15). The most commonly involved sites are the muscles of the thigh, followed by the upper extremity and chest wall (Figure 21.16). The size of the lesion varies and can be large, >10 cm to 15 cm.

Intramuscular hemangiomas have a high recurrence rate, because complete excision of these poorly marginated tumors can be difficult.[53,54]

Pathology

Grossly, intramuscular hemangiomas vary. The small-vessel examples are often pale, nonhemorrhagic nodules (Figure 21.17), whereas the large-vessel angiomas form blood-colored, often poorly delineated masses. Histologically, the intramuscular hemangiomas are composed of well-differentiated vessels and can show capillary, cavernous, or venous patterns in various combinations. The capillary variants are the most common (Figure 21.18). Some lesions contain a fatty component, and such tumors have been earlier referred to as deep or infiltrative angiolipomas.

Angiomatosis

Angiomatosis is defined as a hemangioma involving multiple contiguous tissue planes. These lesions are often diagnosed in

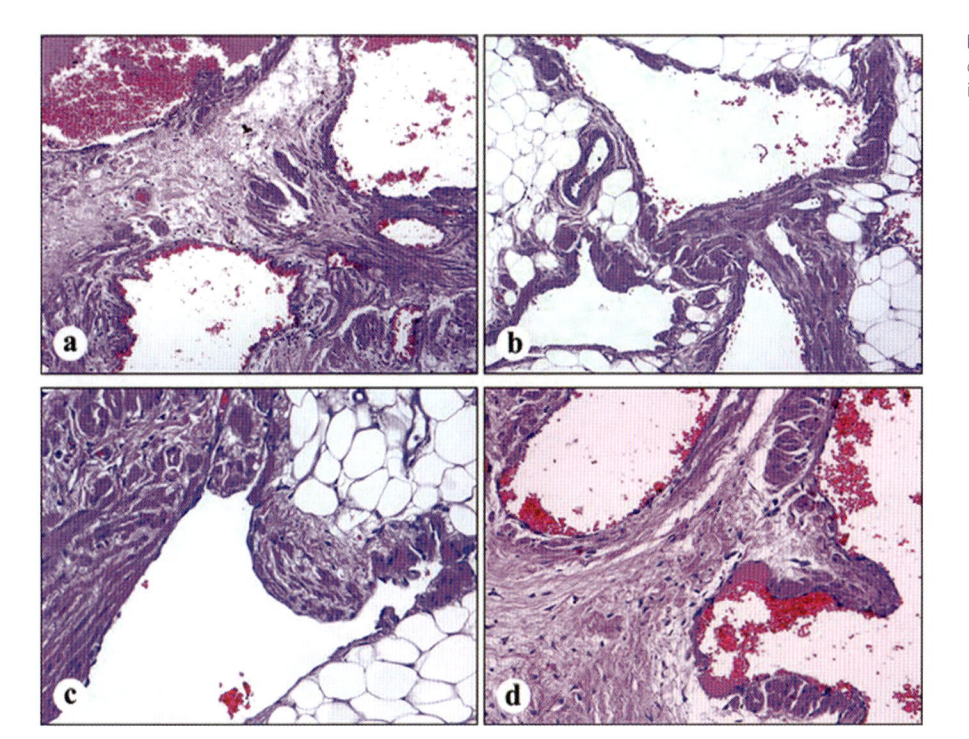

Figure 21.14 A venous hemangioma is composed of vein-like vascular units with an incompletely developed smooth muscle layer.

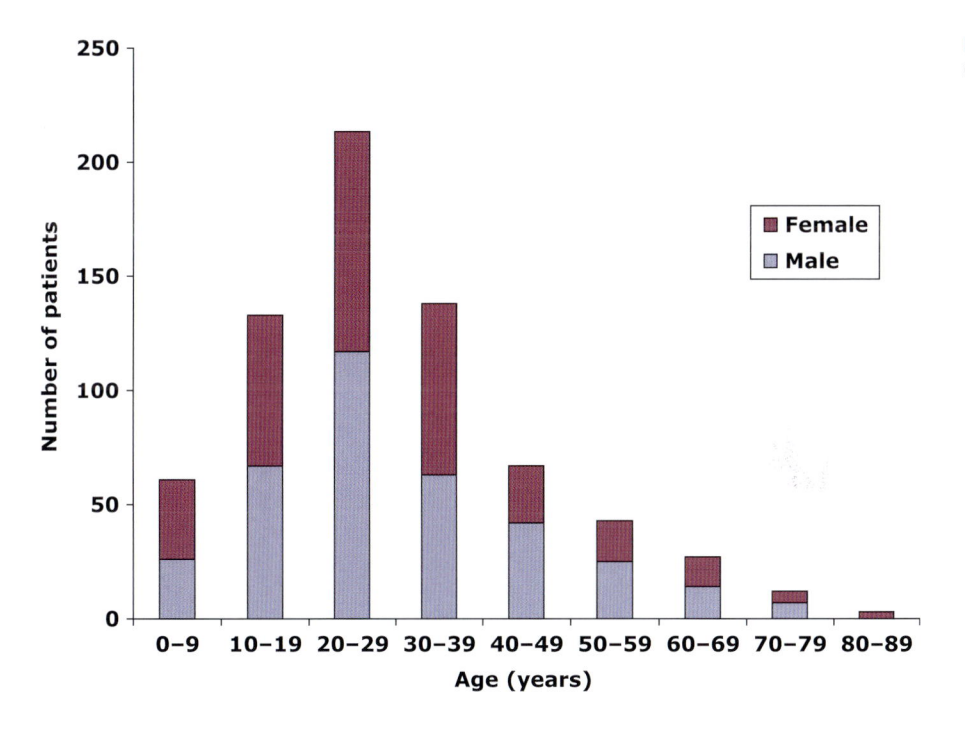

Figure 21.15 Age and sex distribution of 697 patients with intramuscular hemangiomas.

early childhood, and more than one half of the tumors are diagnosed by the age of 20 years, with a female predilection (1.5:1). The condition is often brought to attention by pain and swelling of the involved limb or region. Notably, skin extension is rare. The most common region of presentation is the lower extremity, with the foot, ankle, and leg being the most common sites of involvement, but occurrence in the trunk wall is also possible. The diagnosis is based on clinical and radiologic evidence of an extensive lesion with histologic features of a hemangioma in a biopsy. Although the lesion might recur, malignant transformation is not a feature of angiomatosis. Conservative management is therefore possible, in view of the difficulty of complete removal without structural sacrifice.[55,56]

Histologically, the angiomatosis can show combinations of capillary, cavernous, venous, and even SCH. The presence of increased numbers of capillaries in the vein walls is a

Chapter 21: Hemangiomas, lymphangiomas, and reactive vascular proliferations

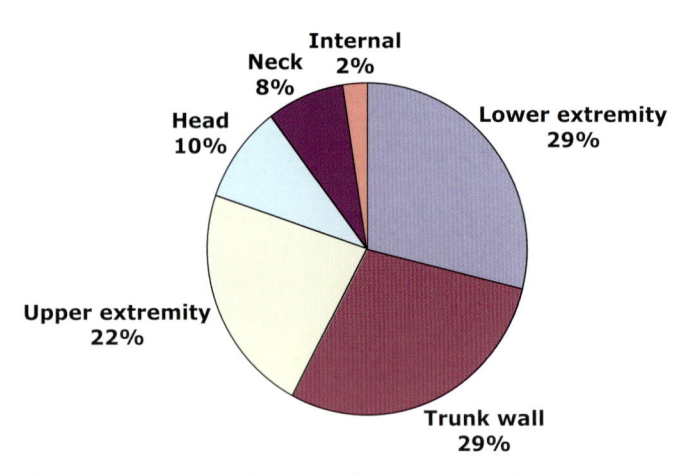

Figure 21.16 Anatomic distribution of 677 intramuscular hemangiomas.

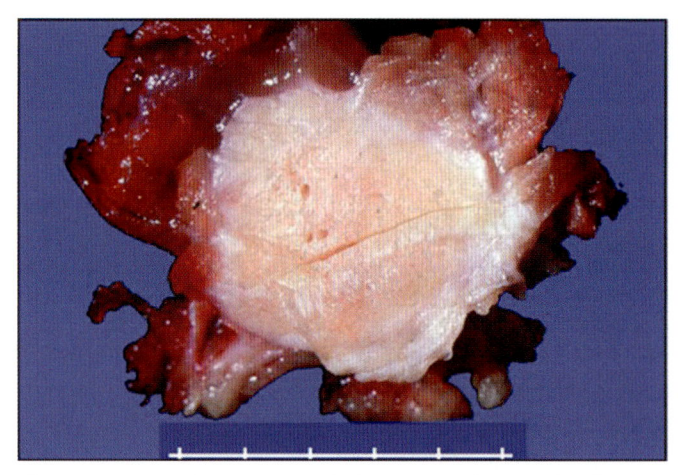

Figure 21.17 Grossly, an intramuscular hemangioma forms a pale yellow-gray mass. Small cysts on the sectional surface are dilated blood vessels.

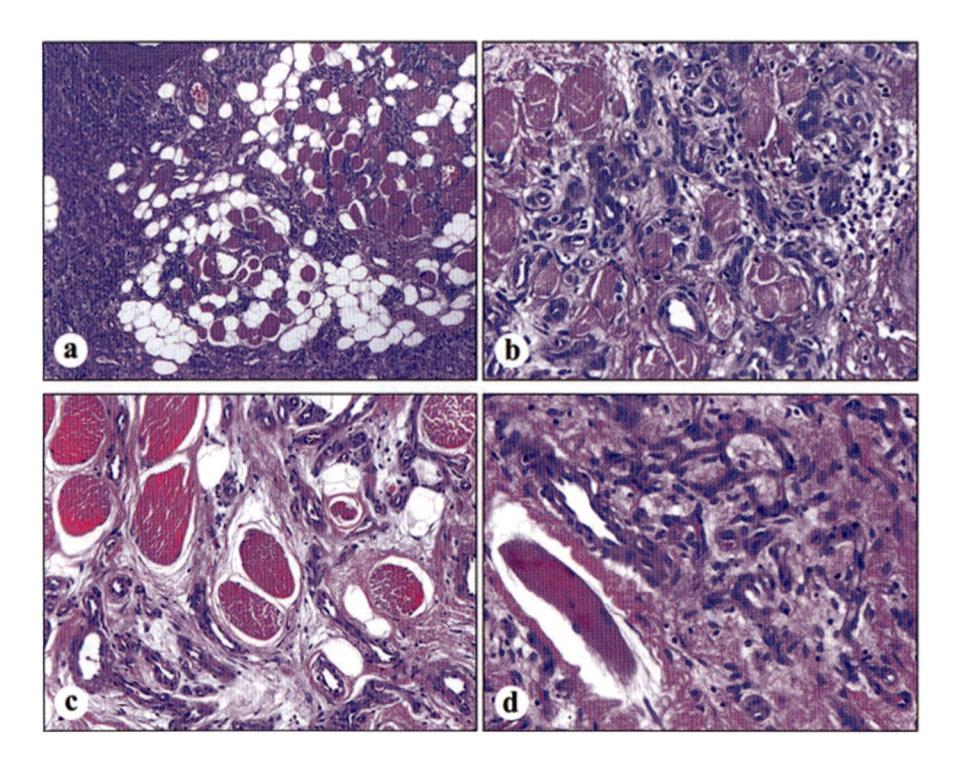

Figure 21.18 Intramuscular hemangiomas are most often capillary hemangiomas. They infiltrate between skeletal muscle fibers, and fat can be present.

common feature of this condition, apparently representing budding of capillaries from the walls of small veins (Figure 21.19).

Glomeruloid hemangioma

This designation refers to a rare form of dermal hemangioma that forms intravascular papillary proliferations, with the intravascular containment creating an appearance that resembles renal glomeruli.

Clinical features

This cutaneous vascular proliferation oocurs mainly in older adults, often in the head and neck region. It can occur alone or be a manifestation of the rare POEMS syndrome. The key pathogenetic factor in POEMS syndrome (p = polyneuropathy, o = organomegaly, e = endocrinopathy, m = monoclonal gammopathy, s = skin lesions) is probably the underlying plasma cell neoplasia, which is seen in all patients with this syndrome. The organomegaly entails hepato- and splenomegaly, and the endocrinopathy includes hypothyroidism, adrenal insufficiency, and amenorrhea. The plasma cell neoplasia is the key element, and usually has a lambda light chain; sclerotic bone lesions are often present. The skin lesions are glomeruloid hemangiomas. Lymphadenopathy, glomerular kidney disease, and pulmonary hypertension are other common features of the syndrome.[57–59] In POEMS syndrome, the lymph nodes have histologic features identical with multicentric

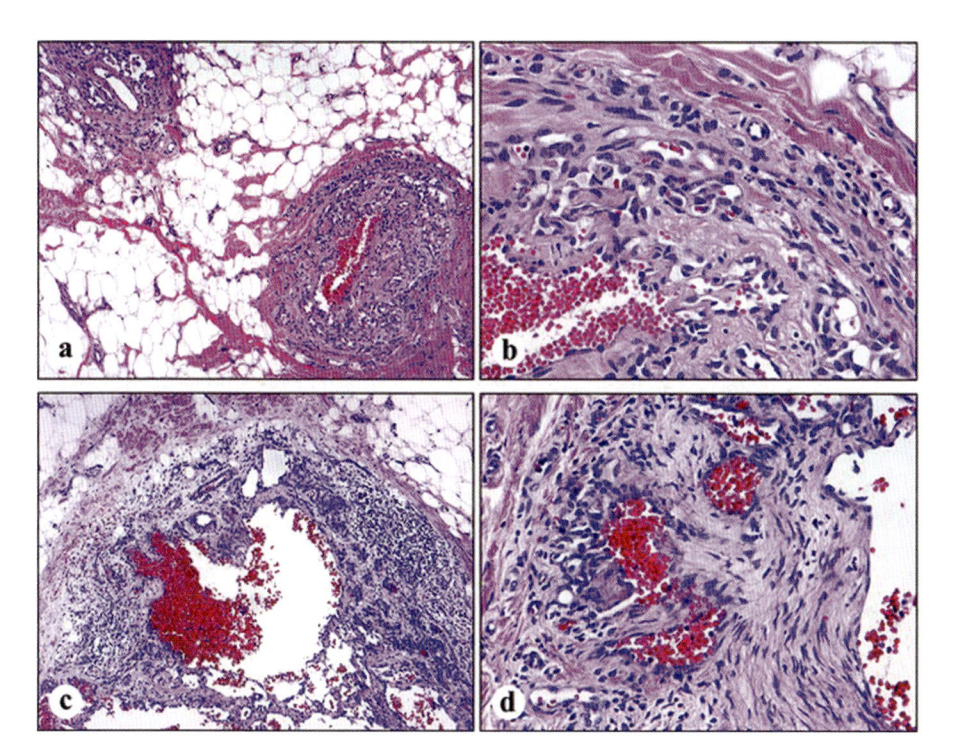

Figure 21.19 Angiomatosis lesions containing clusters of vessels surrounded by hemangiomatous proliferation. The endothelial element is prominent but nonatypical.

Castleman's disease, in which human herpesvirus 8 has been detected.[60] Interestingly, this virus has been also detected in Kaposi's sarcoma and multiple myeloma.

Recently, many examples of glomeruloid hemangioma have been reported without POEMS syndrome.[61] The designation *papillary hemangioma* has also been used for such sporadic tumors.[62] All examples of glomeruloid hemangioma in the author's experience have been nonsyndromic, presenting in older adults in the head and neck, or trunk. Nevertheless, it is wise to evaluate patients with a glomeruloid hemangioma for monoclonal serum immunoglobulins, because glomeruloid hemangioma can be a presenting sign for POEMS syndrome.

Pathology

The skin lesions form small red-to-purple papules or nodules measuring 1 mm to 10 mm. Histologically, they consist of dilated blood vessels containing multiple profiles of tortuous capillaries, resembling the appearance of a renal glomerulus. Endothelial cells are swollen, and some contain eosinophilic hyaline globules. These globules can represent engulfed degraded erythrocytes (Figures 21.20 and 21.21).

Hobnail hemangioma (targetoid hemosiderotic hemangioma)

This cutaneous hemangioma was originally named for its often target-like clinical appearance: there is a rounded erythematous area, which especially in an early lesion has a darker, mildly elevated central zone. Because the clinical gross appearance is sometimes only part of its clinical spectrum, the designation of *hobnail hemangioma*, referring to its characteristic histologic appearance, has been applied to this subtype of capillary hemangioma.[63-66]

Clinical features

These tumors occur from childhood to old age, with a predilection for young adults; the median age in the three largest series varied between 28 and 32 years, with an apparent male predominance in two series. This hemangioma occurs in a wide variety of locations. It seems to be most common in the lower extremities (nearly 50% of all cases), with a strong predilection for the thigh and buttocks. In the upper extremities (22%), there is a predilection for proximal sites, the shoulder and forearm. The trunk wall, especially the back, is often involved (25%), whereas only one series reported cases in the head and neck (4% of all reported cases). The lesions are typically small and superficial, usually measuring <1 cm. They have a uniformly benign behavior with no tendency for recurrence, and therefore simple excision is sufficient.[63-65]

Pathology

These hemangiomas are histologically distinctive for their mildly dilated, irregularly shaped, superficial dermal vascular channels lined by variably luminally protruding "hobnail" endothelial cells, which can have mild nuclear hyperchromasia, but lack a multilayered pattern and significant nuclear atypia (Figure 21.22). Focal small intravascular protrusions resembling those in papillary intralymphatic angioendothelioma (Dabska tumor) can occur, but are less prominent than those seen in the latter entity. The deeper dermal component of this

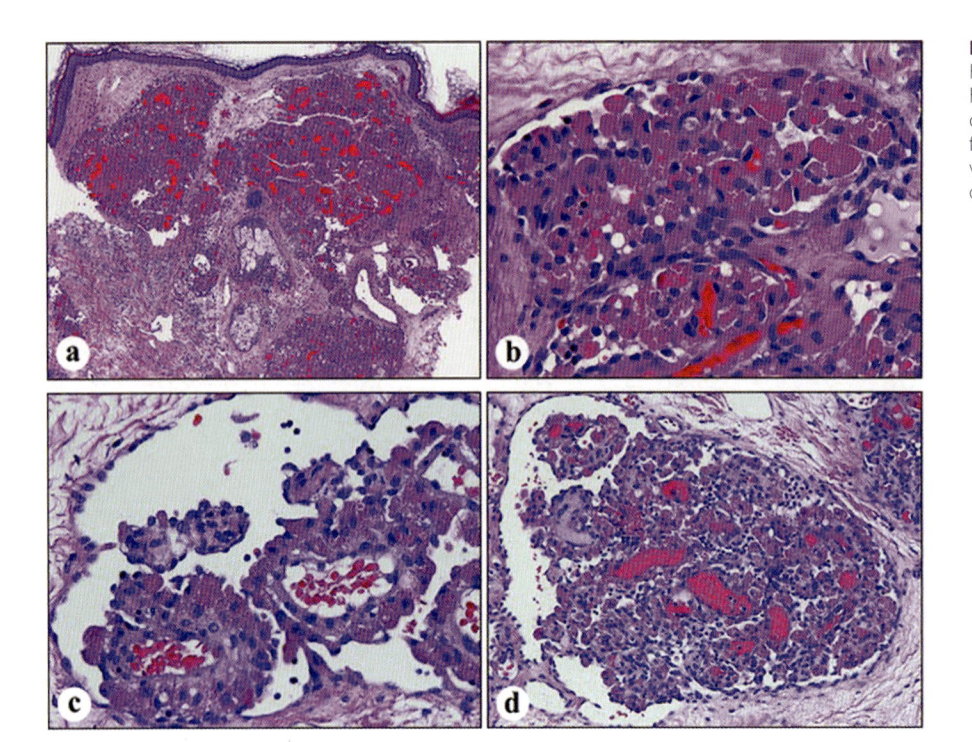

Figure 21.20 Example of a glomeruloid hemangioma. (**a**) A lobulated dermal tumor with a hemorrhagic appearance. (**b**) The vascular lumina contain prominent clusters of endothelial cells filled with eosinophilic globules. (**c,d**) The dilated vessels with angiomatoid proliferation have a configuration reminiscent of a renal glomerulus.

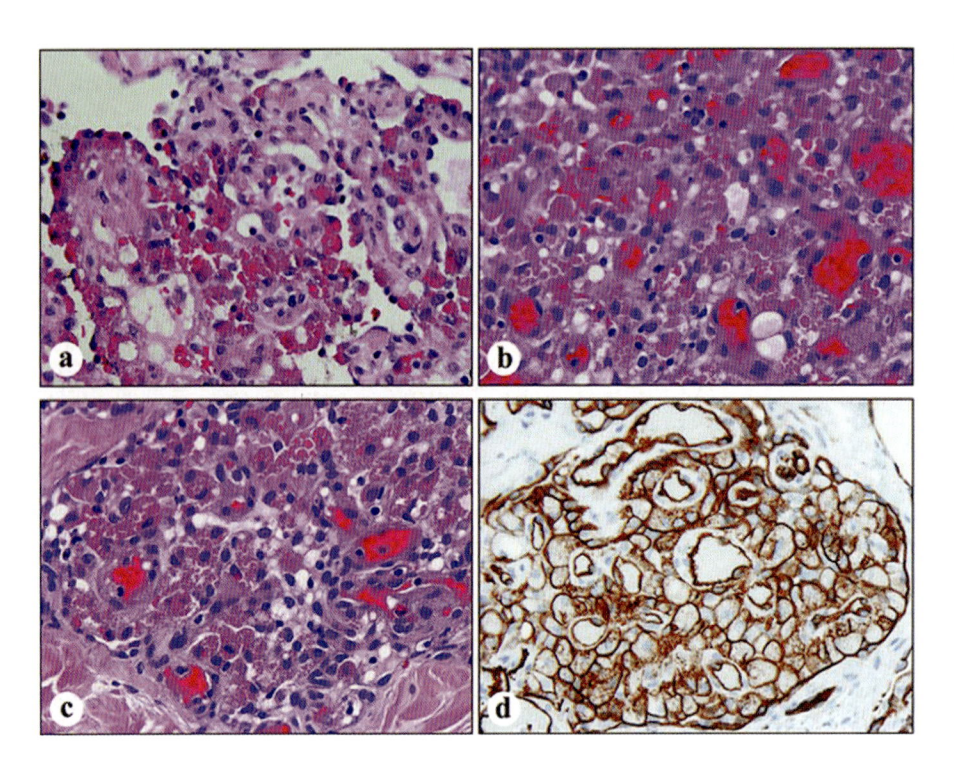

Figure 21.21 Another example of glomeruloid hemangioma with numerous eosinophilic hyaline globules. (**d**) The endothelial cells are CD31 positive.

hemangioma typically has slit-like lumina. The perivascular stroma can contain hemosiderin and scattered lymphocytes.

Hobnail hemangioma somewhat resembles retiform hemangioendothelioma, from which it differs by lack of complex anastomosing vascular channels that show a higher cellularity and greater nuclear atypia.

Immunohistochemically typical is expression of VEGFR3 and podoplanin, indicating a lymphatic vessel-like phenotype.[65,66]

Acquired tufted angioma (progressive angioma, angioblastoma of Nakagawa)

Clinical features

This rare form of cellular capillary hemangioma occurs most commonly in children, but has also been reported in older adults. The lesion often develops in early childhood, with some

Chapter 21: Hemangiomas, lymphangiomas, and reactive vascular proliferations

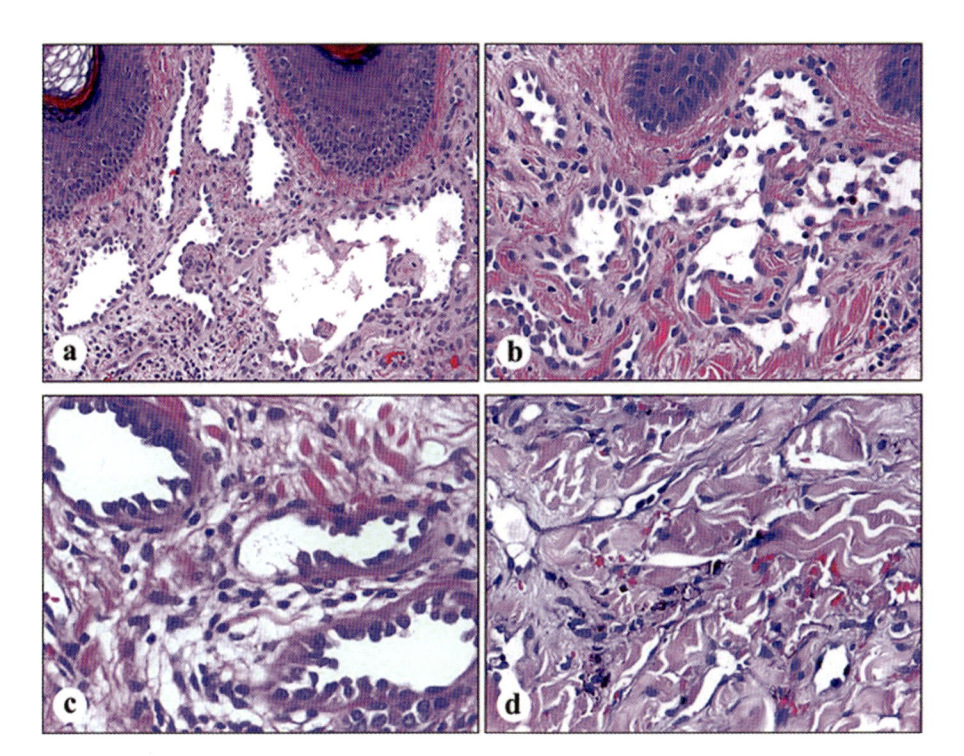

Figure 21.22 A hobnail hemangioma is superficially located in the dermis and contains dilated capillaries with prominent, inward protruding "hobnail" endothelia. (**d**) Stromal hemosiderin deposition can be present.

lesions being congenital, and it often forms a progressively enlarging cutaneous macule over years. It therefore has also been referred to as *progressive angioma*.

It most commonly occurs in the head and neck, and upper trunk wall, but it has also been reported in the extremities and genital region.[67–70] Some tufted angiomas have been associated with severe platelet consumption (Kasabach–Merritt syndrome), which is more commonly seen with KH.[71]

Pathology

Histologically, the lesion is composed of multiple, sometimes band-like, dermal nodules with an overall appearance reminiscent of that of a cellular lobular capillary hemangioma. The multiple, scattered, spherical dermal nodules have been described as having a "cannonball pattern" (Figure 21.23). The sharply demarcated nature of these elements is often a result of their intravascular location. Indeed, a narrow vascular luminal space is often seen around the cellular clusters. The capillaries often have inconspicuous, closed lumina, similar to cellular (juvenile) capillary hemangioma. There is a single endothelial cell layer with no atypia, and a pericytic SMA-smooth-muscle-actin positive element is prominent and sometimes gives the lesion a spindled appearance.

Kaposiform hemangioendothelioma

Although this tumor was originally described as hemangioma with Kaposi's sarcoma-like features[72] or Kaposi-like infantile hemangioendothelioma,[73] Zukerberg *et al.* subsequently renamed this tumor *kaposiform hemangioendothelioma*

(KH).[74] KH is considered closely related to acquired tufted angioma, perhaps the latter representing its deep and more extensive form.[75] Nevertheless, small cutaneous lesions can still be designated as acquired tufted angiomas, as discussed earlier.

Clinical features

This tumor occurs predominantly in young children (median age 2 years) and occasionally in young adults, without a clear sex predilection. KH might be more common in peripheral soft tissues, although initial cases were disproprotionately retroperitoneal.[74–76] The tumor can involve both superficial and deep soft tissues. Clinically, extremity-based lesions typically appear as blue-red patches that vary from 2 cm to several centimeters. A distinctive complication is platelet trapping, and large tumors can lead to potentially fatal consumption coagulopathy (Kasabach–Merritt syndrome). According to two studies, this syndrome is nearly always associated with KH, although milder forms of platelet trapping occur in other hemangiomas.[77,78] KH can spread as microscopic nodules to surrounding tissues and involve regional lymph nodes, but true metastases have not been reported.[74–76,79] While the superficial lesions are generally indolent, large retroperitoneal and multiorgan lesions are associated with neonatal mortality due to Kasabach–Merritt syndrome.[80]

Pathology

Grossly, the excised lesions are typically grayish white. Histologically, KH forms a lobulated mass divided by fibrous septa. In some cases, the lesional cells form uniform, Kaposi's-sarcoma-like vascular slits (Figure 21.24). In some cases, there

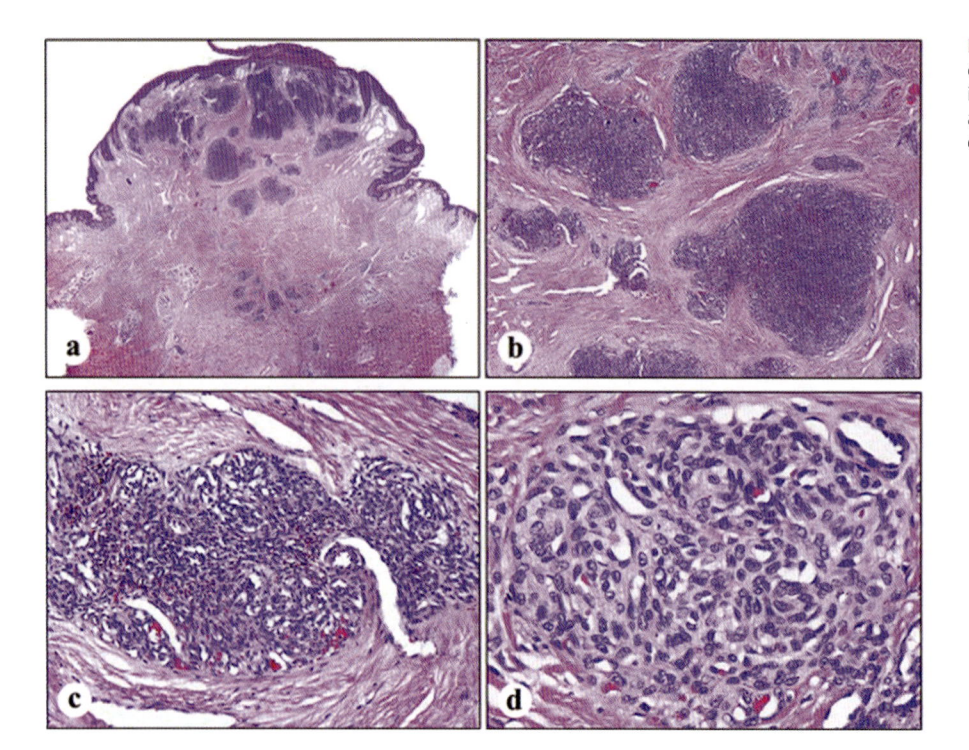

Figure 21.23 An acquired tufted angioma contains multiple lobules of capillary hemangioma in the dermis. The lobules are well-demarcated and are composed of slit-like capillaries lined by oval endothelial cells.

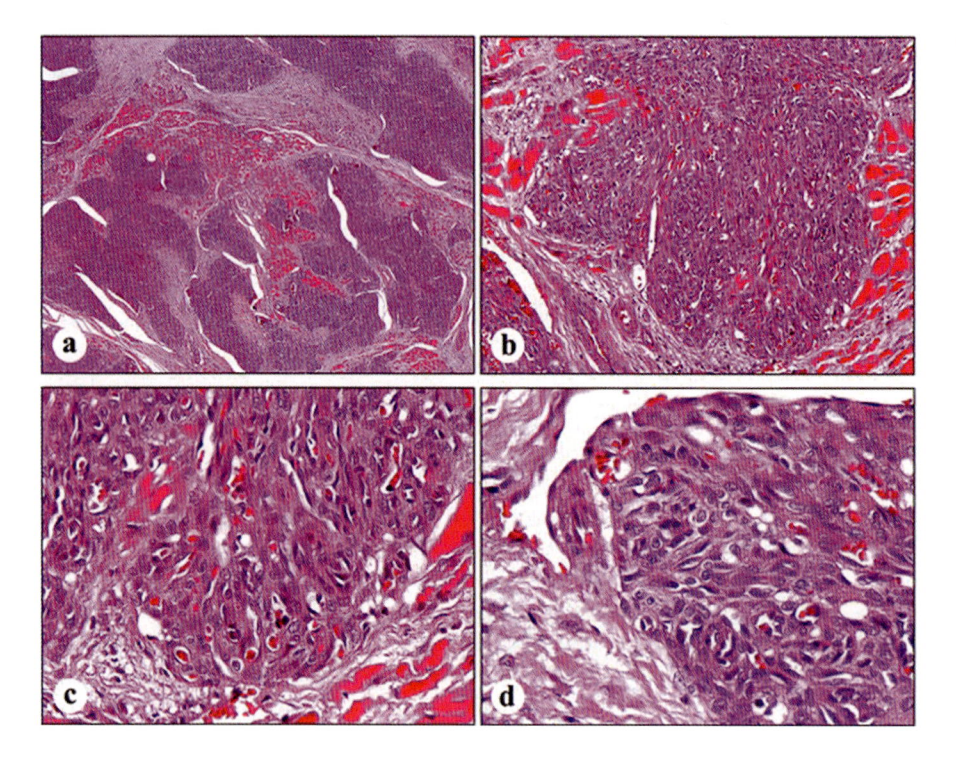

Figure 21.24 A kaposiform hemangioendothelioma inside of skeletal muscle. Note uniform vessels with narrow slits. The tumor cells have a spindled appearance.

are cellular pericyte-containing whorls and areas of hemorrhage and thrombosis among the spindled elements (Figure 21.25). The spindled endothelial proliferation has limited atypia and low mitotic activity. Hyaline globules similar to those in Kaposi's sarcoma can be present. Lymphangiomatous components have been noted in some examples.

Immunohistochemically typical is positivity for podoplanin and PROX1, indicative of lymphatic endothelial phenotype, and negative results for GLUT1, in contrast to juvenile capillary hemangioma.[75,81,82] Actin-positive pericytes are highlighted by actin immunostains, especially in the whorled (glomerular-like) areas. Various numbers of platelet thrombi are highlighted by immunostain for CD61.[75]

Chapter 21: Hemangiomas, lymphangiomas, and reactive vascular proliferations

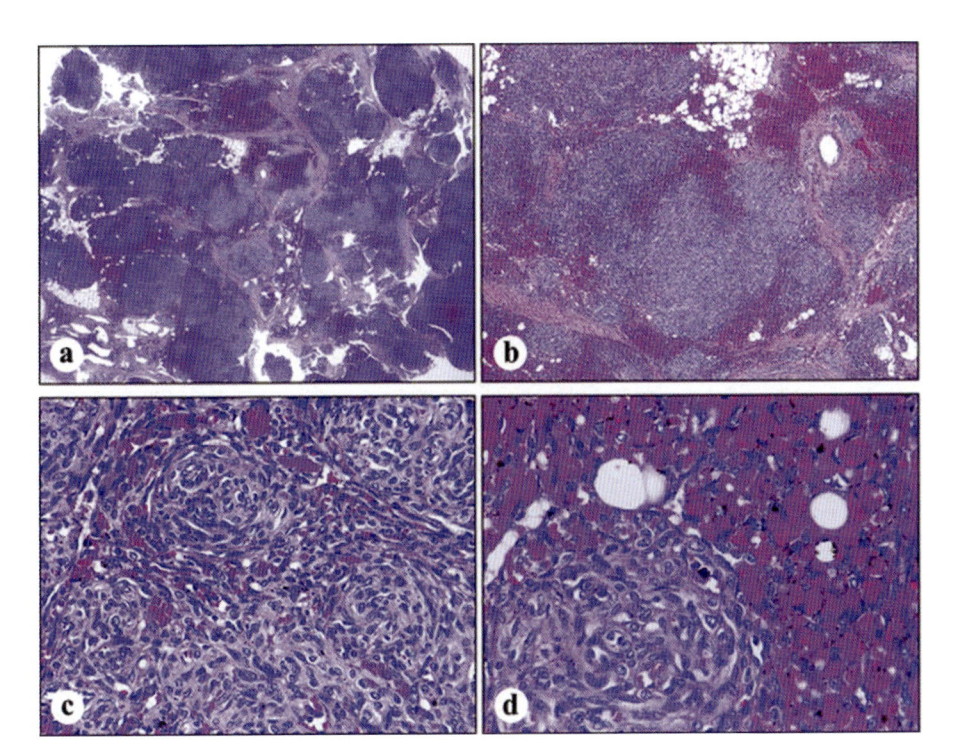

Figure 21.25 (**a,b**) A retroperitoneal example of kaposiform hemangioendothelioma. Note lobular structure. (**c,d**) The tumor is highly cellular and contains whorl-like arrangements of spindled cells. Large hemorrhagic foci are also present.

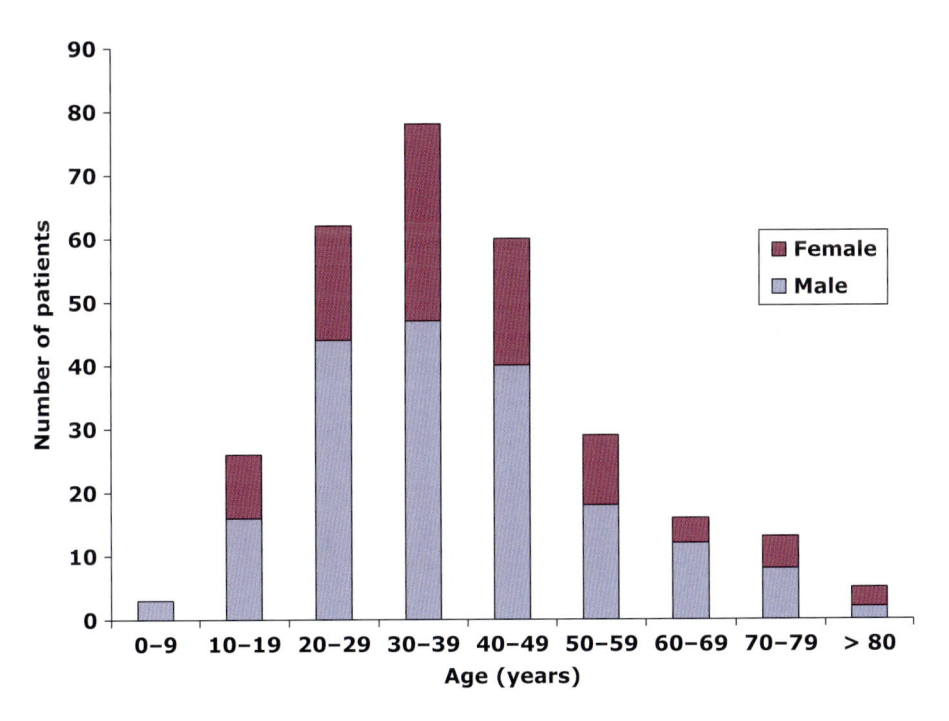

Figure 21.26 Age and sex distribution of 302 patients with epithelioid hemangiomas.

Epithelioid hemangioma (angiolymphoid hyperplasia with eosinophilia)

This diagnostic category, containing benign vascular proliferation with distinctively epithelioid endothelial cells, has evolved over the past 40 years as a clinicopathologic entity encompassing many tumors previously designated as histiocytoid hemangiomas. Although epithelioid hemangioma (EH) was earlier considered to be synonymous with Kimura disease, these entities are now considered unrelated.[83–85]

Clinical features

EH occurs in a wide range of ages but has a predilection for young adults, with a male predominance (Figure 21.26). It most commonly occurs in the head and neck, where

569

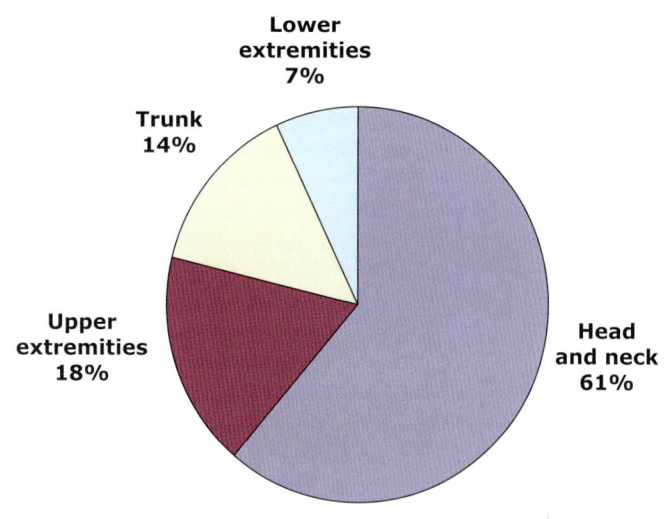

Figure 21.27 Anatomic distribution of 303 epithelioid hemangiomas.

preferential sites include the temporal region, forehead, and the postauricular scalp.[86–88] In these sites, EH often involves branches of the external carotid artery. In the extremities, distal sites are often affected (Figure 21.27). Bone involvement is well known, and in some cases, these lesions also involve adjacent soft tissues.[89] Rare sites include the penis[90] and the lung.[91] The author and colleagues have seen one case with metachronous soft tissue and pulmonary involvement.

EH can form solitary or multiple cutaneous papules or present as a subcutaneous mass measuring <1 cm to 8 cm (Figure 21.28). A mean duration of >1 year has been reported. There is a 30% rate of local recurrence, but truly aggressive behavior does not occur.[87] Rare lymph node involvement has occurred, without this indicating an adverse prognosis. It is unknown whether EH is a reactive or neoplastic condition; however, the etiologic role of arterial trauma has been suspected, considering common involvement of superficial arterial segments that often have reactive changes that can be shown histologically.[88]

Pathology

Histologically, EH forms a nodule often containing lymphoid infiltration, sometimes with germinal centers, and eosinophils in the periphery or throughout the nodule. More than one half of all cases, especially the subcutaneous ones, demonstrate involvement of an artery, where epithelioid endothelial cells can proliferate in the luminal side or in the artery wall (Figure 21.29). EH contains large numbers of capillaries, some of which are lined by epithelioid endothelial cells with abundant eosinophilic cytoplasm and low nucleocytoplasmic ratio (Figure 21.30). Endothelial proliferation can include intraluminal clustering of the epithelioid endothelia, and apparently solid epithelioid cell proliferation occurs in cases designated as exuberant EHs (Figure 21.31). Even in such cases, mitotic activity is low (1–2 per 10 HPFs at the most), and there is no atypia.

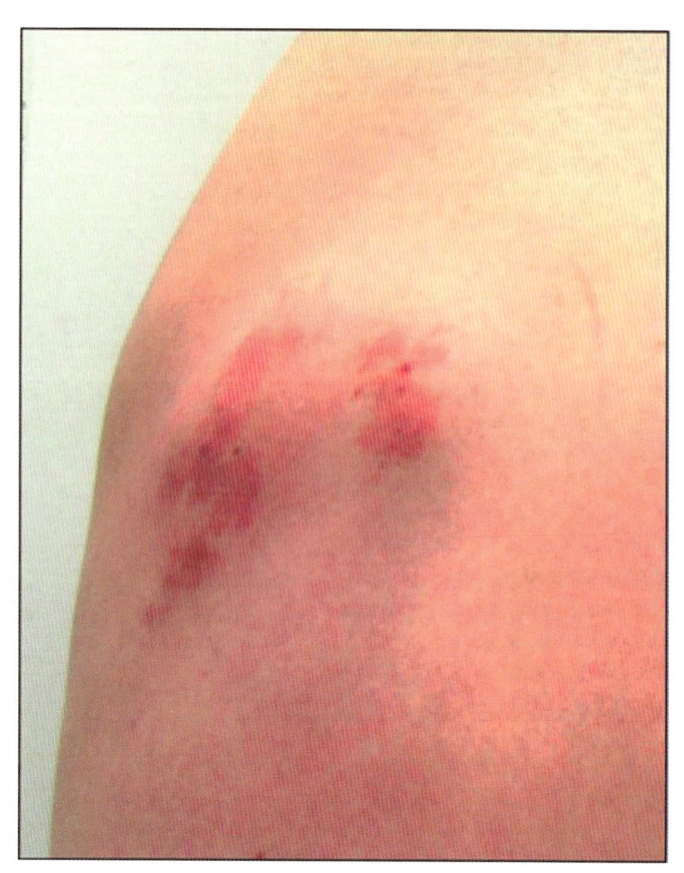

Figure 21.28 A cutaneous and subcutaneous epithelioid hemangioma appears as a mildly elevated nodule with a focally ecchymotic surface.

Immunohistochemically, the capillaries lined by epithelioid endothelial cells are surrounded by well-developed actin-positive pericytes, and the vascular lumina show positivity for vWF (Figure 21.32). The endothelial cells are also usually positive for CD31 and CD34.

Epithelioid hemangiomas of deeper soft tissue and bone (noncutaneous, non-head and neck examples) often contain FOS-LMNA gene fusions and atypical examples may harbor ZEP36-FOSB gene fusions.[92]

Differential diagnosis

The epithelioid endothelial morphology, lack of prominent lymphoid component, and fibrosis separate this entity from Kimura disease. The latter does not contain a distinctive epithelioid vascular component.[93–95]

The so-called epithelioid hemangioma (histiocytoid hemangioma) of the heart is a mesothelial rather than an endothelial proliferation.[96]

Epithelioid hemangioendothelioma does not contain well-defined vasoformative elements, but is instead composed of cords of epithelioid cells, often in a myxohyaline stroma (Chapter 22).

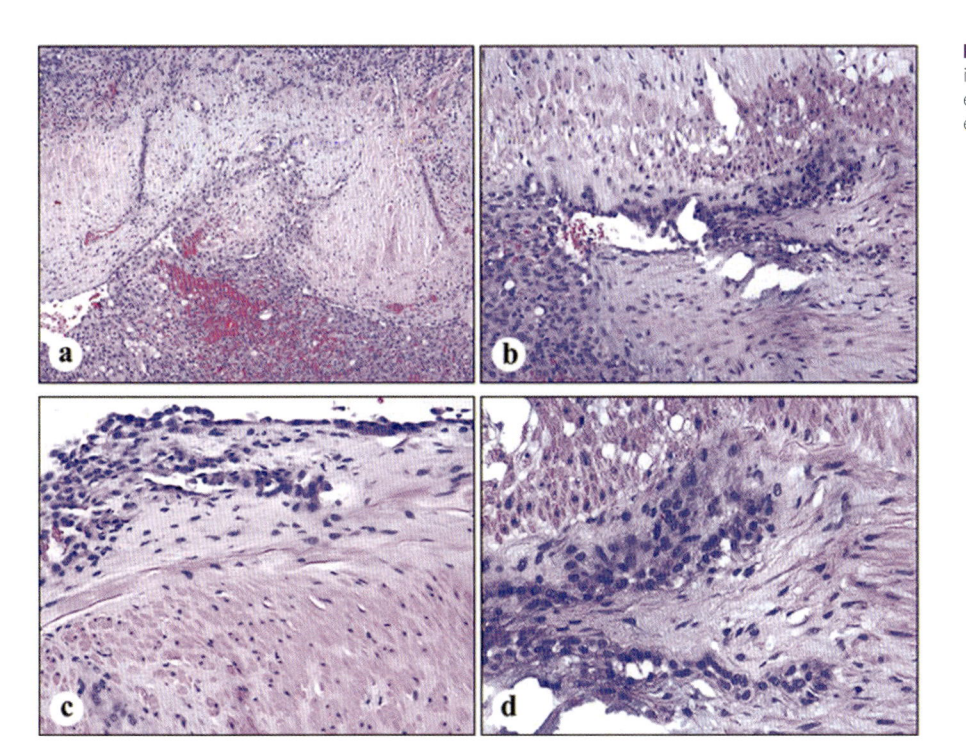

Chapter 21: Hemangiomas, lymphangiomas, and reactive vascular proliferations

Figure 21.29 An epithelioid hemangioma involving an artery wall. Note epithelioid endothelial proliferation on the luminal side extending inside the muscular layer.

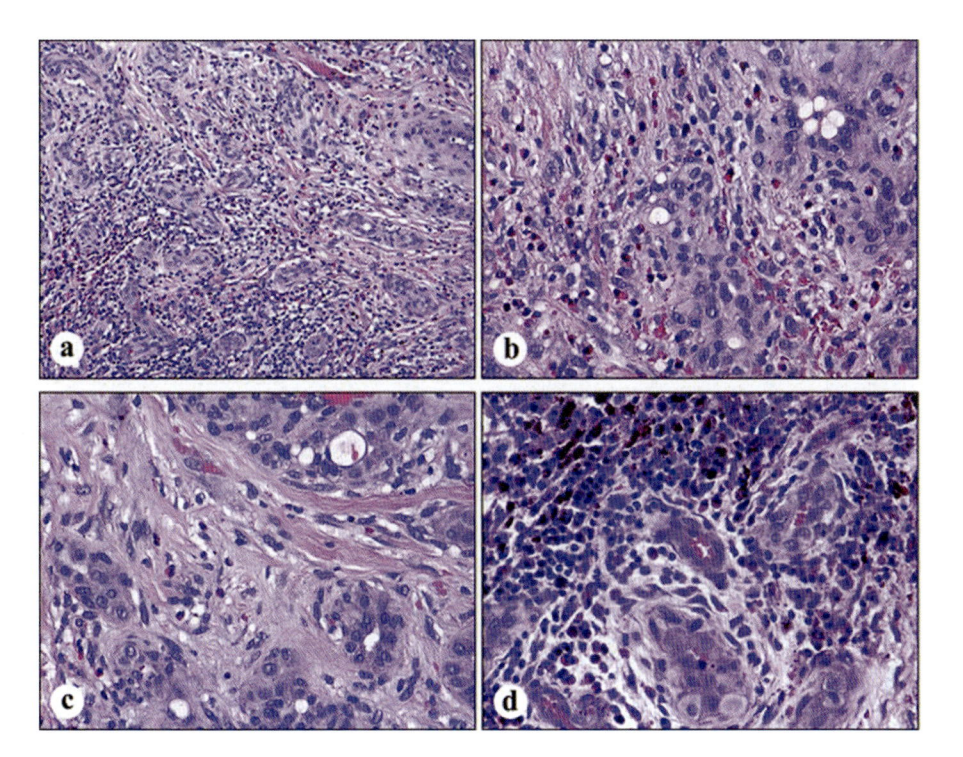

Figure 21.30 An epithelioid hemangioma with variable infiltration of eosinophils, lymphocytes, plasma cells, and clusters of epithelioid endothelial cells.

Spindle cell hemangioma

SCH, formerly known as spindle cell hemangioendothelioma and originally thought to have malignant potential, is now classified among hemangiomas.[97,98]

Clinical features

Spindle cell hemangioma occurs in all ages, but has some predilection for young adults, with a female predominance (Figure 21.33). It most commonly presents in the distal parts of the extremities (i.e., the hands and feet) but also occurs infrequently in various proximal locations (Figure 21.34). It usually develops as a slowly growing, single purplish nodule or multiple lesions <1 cm in diameter, with cutaneous or subcutaneous involvement. Rare lesions are intramuscular. Many patients have a long history, and multiple lesions occur in more than one half of patients.[97–101] Although at earlier times

571

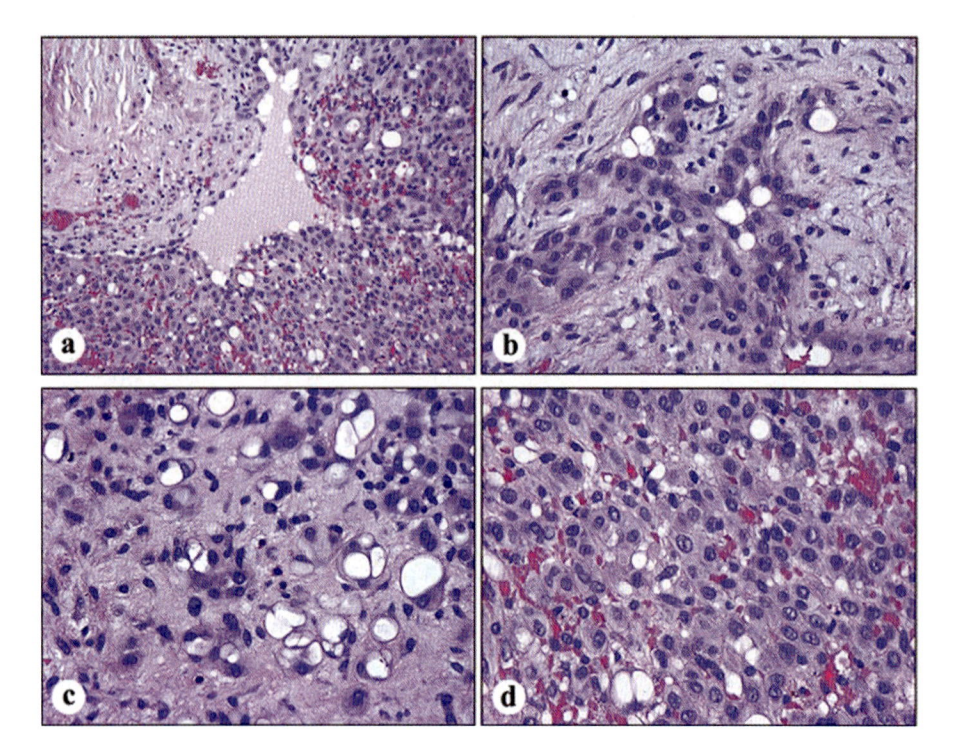

Figure 21.31 Florid epithelioid endothelial cell proliferation in an epithelioid hemangioma. (**c**) Focal vacuolization is present. (**d**) Note that the epithelioid endothelial cells have a voluminous cytoplasm and regular nuclei with small nucleoli.

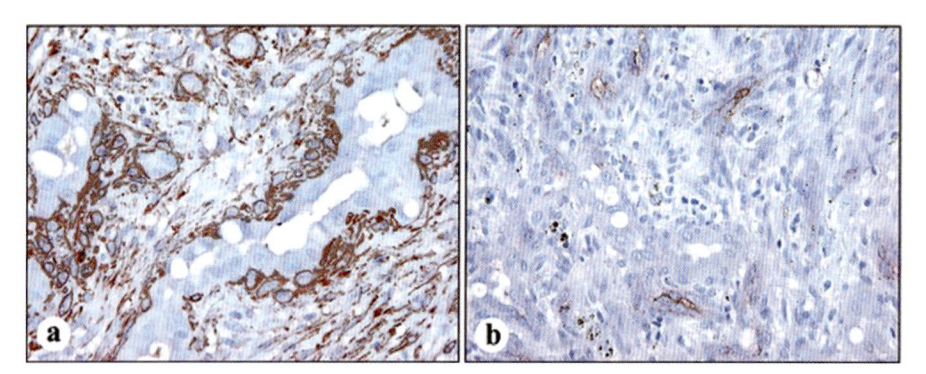

Figure 21.32 (**a**) Smooth-muscle actin-positive pericytes surround the vessels lined by epithelioid endothelial cells in epithelioid hemangioma. (**b**) The vascular lumina show von Willebrand factor positivity.

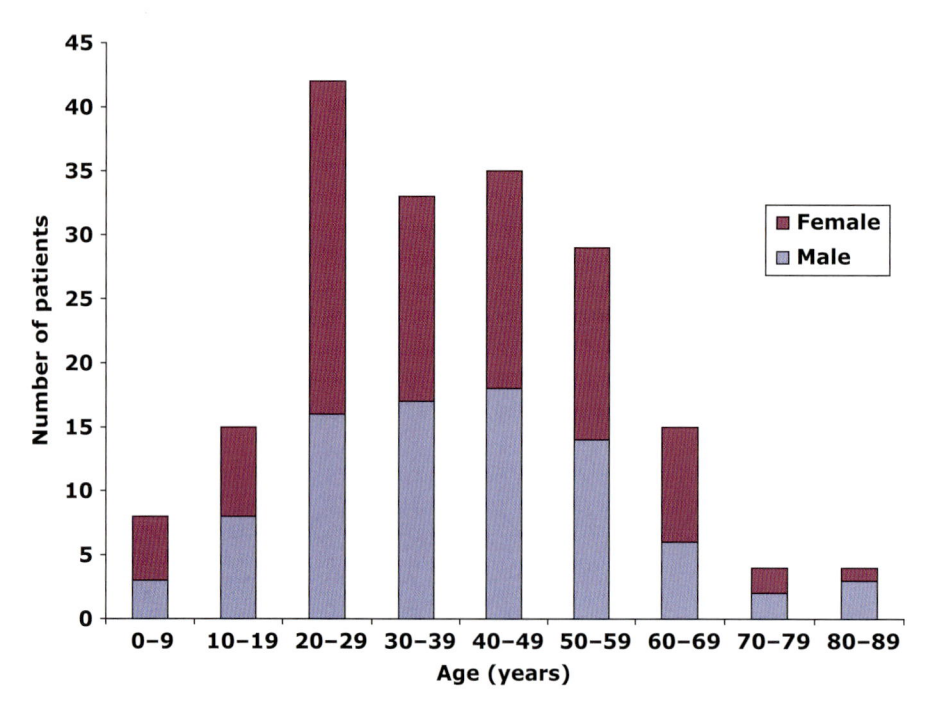

Figure 21.33 Age and sex distribution of 187 patients with spindle cell hemangiomas.

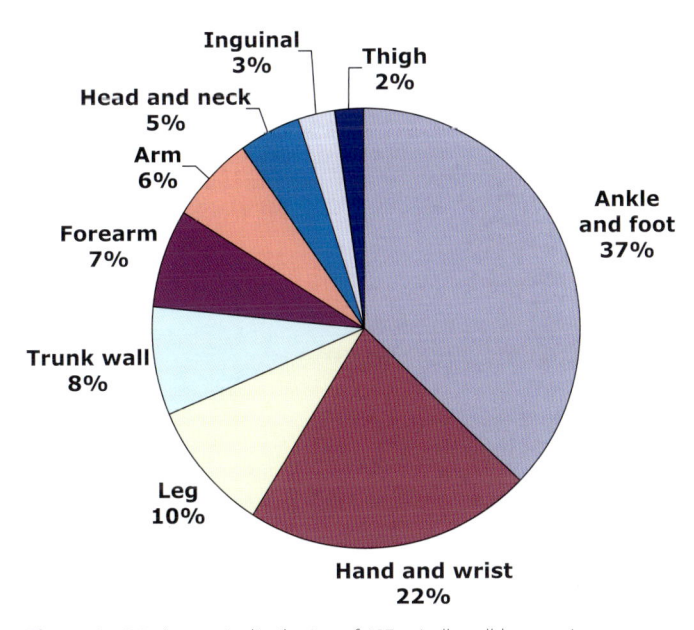

Chapter 21: Hemangiomas, lymphangiomas, and reactive vascular proliferations

Figure 21.34 Anatomic distribution of 187 spindle cell hemangiomas.

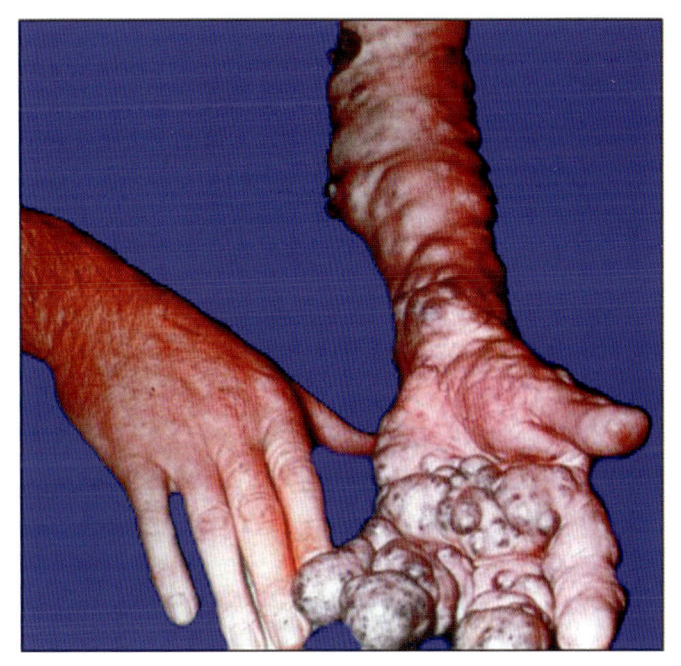

Figure 21.35 Extensive nodules of spindle cell hemangioma in the hand and forearm of a patient with Maffucci syndrome.

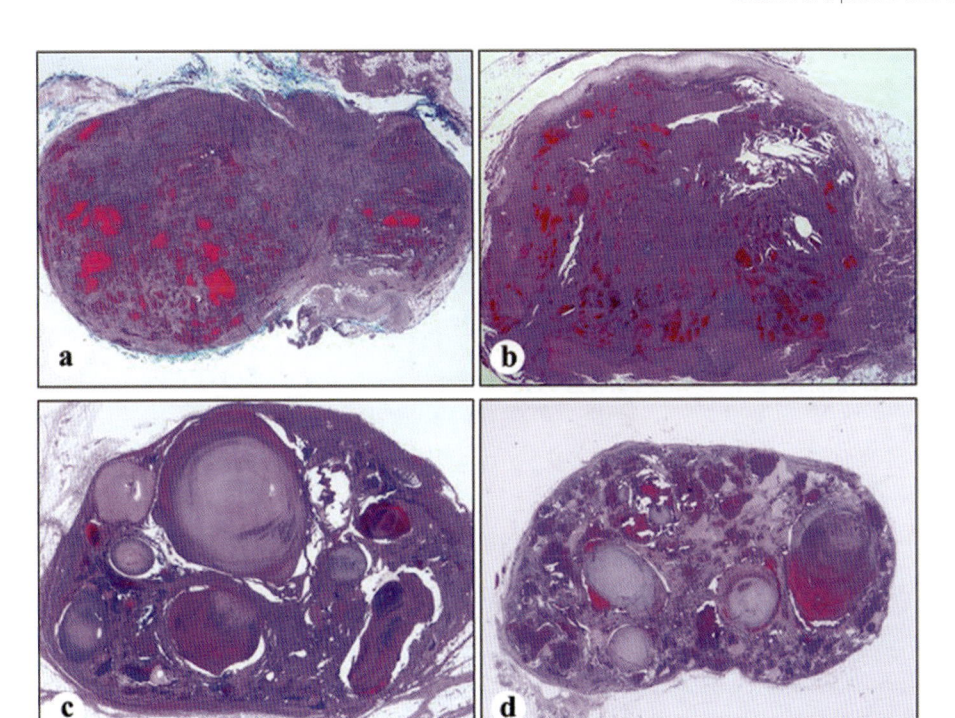

Figure 21.36 Spindle cell hemangioma typically forms a 1-cm to 2-cm demarcated nodule. Some of the vascular lumina show cavernous dilatation and contain organized and hyalinized thrombi called phleboliths.

considered cavernous hemangiomas, multiple vascular tumors in patients with Maffucci syndrome are SCHs.[102] This syndrome also includes cartilaginous bone tumors.[103] In some patients, multiple SCHs form extensive tuberosity on the involved sites (Figure 21.35).

SCH has a >50% local recurrence rate, but it does not metastasize. New lesions often develop in the same region, over several years, indicating a chronic, multifocal regional process. The rare existence of hypercellular and atypical variants might have earned this lesion its previously applied borderline designation.

Pathology

SCH typically forms a 1-cm to 2-cm demarcated nodule in the subcutaneous fat (Figure 21.36). It resembles a cavernous hemangioma, with dilated vascular channels, but has greater cellularity in the septa that typically contain uniform

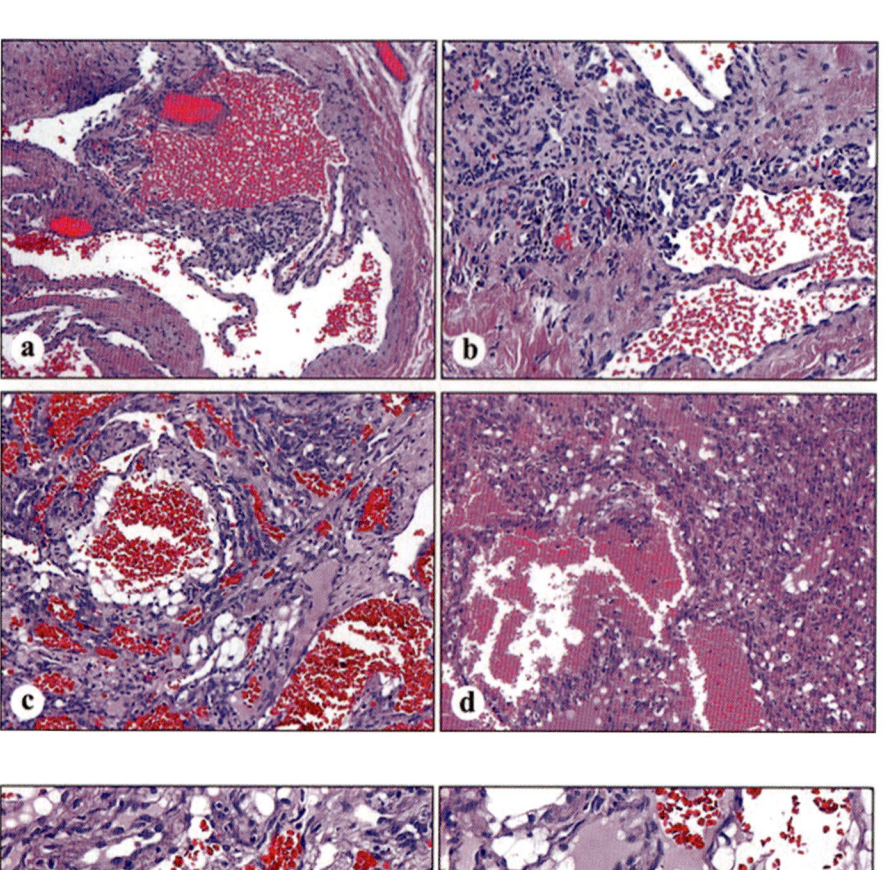

Figure 21.37 The cavernous vascular lumina in spindle cell hemangioma contain variable numbers of spindle cells. Note that the septa are more cellular than those in cavernous hemangioma.

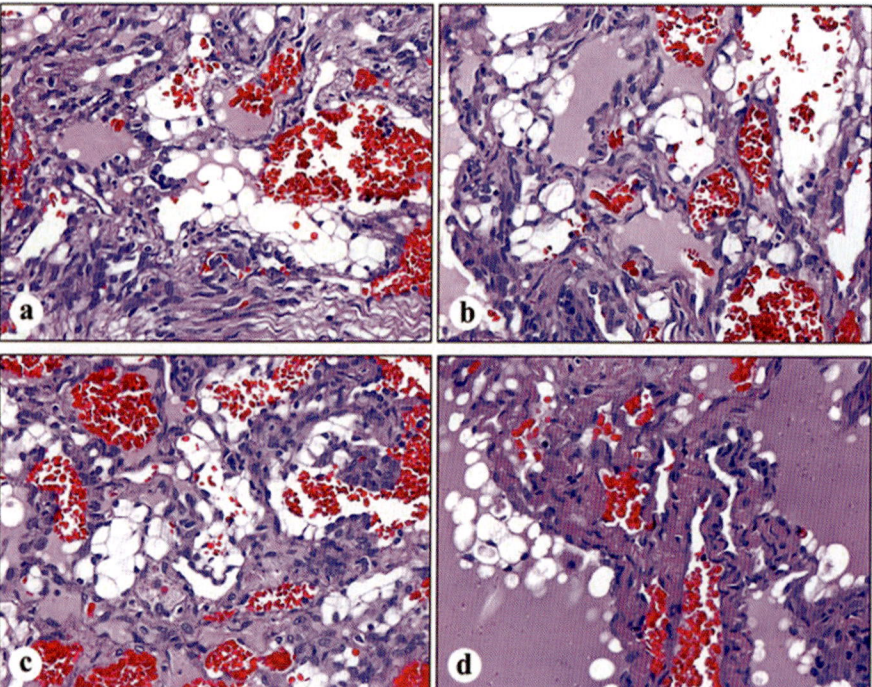

Figure 21.38 Endothelial cell vacuolization is a typical focal finding in spindle cell hemangioma.

spindled cells with tapered ends in a visibly collagenous background (Figure 21.37). Some lesions are entirely intravascular, and a rim of smooth muscle (probably derived from wall of a vein) often partly surrounds the lesion. The cavernous spaces can contain calcified thrombi referred to as *phleboliths*. The endothelial cells often have focal epithelioid features and show focal cytoplasmic vacuolization (Figure 21.38). In some cases, the spindle cells form more solid areas with inconspicuous vasoformation (Figure 21.39). Mitotic activity and atypia are low.

Immunohistochemically, the slit-lining cells and the epithelioid cells have an endothelial phenotype and are positive for vWF, UEA1, and CD31, whereas the septal spindle cells have smooth muscle or pericytic features, shown by their positivity for smooth muscle actin and focally for desmin. All components have been found to be negative for S100 protein.[99]

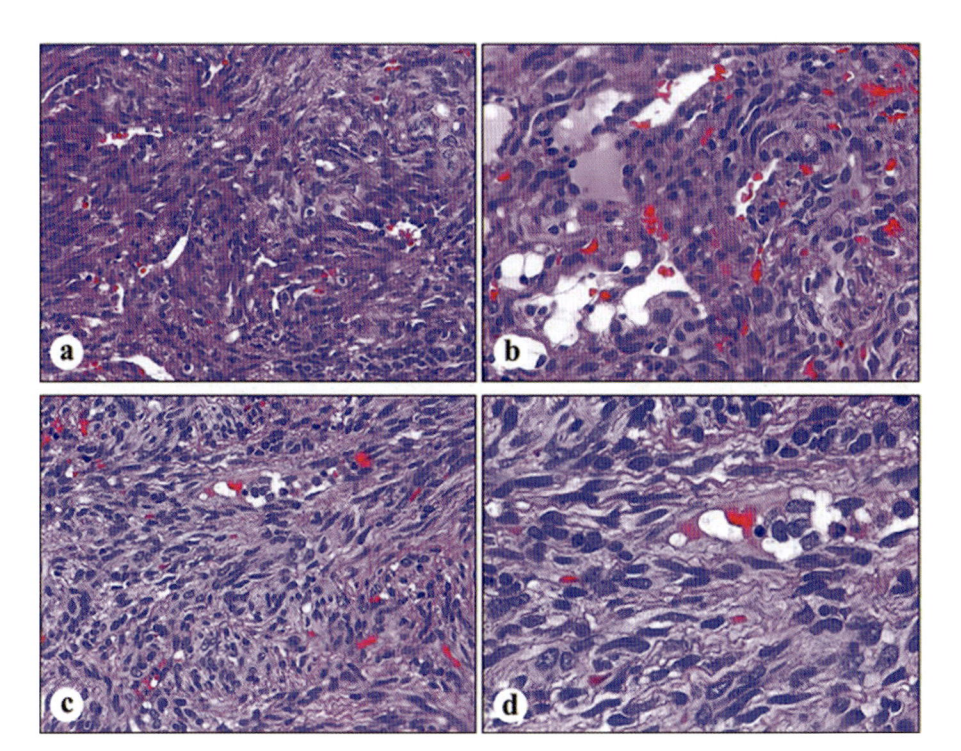

Figure 21.39 Focal solid spindle cell areas can be present in spindle cell hemangioma.

Somatic mosaic mutations in IDH1 and IDH2 genes characterize SCHs arising in Maffucci syndrome and are the probable pathogenesis. Somatic mosaicism also explains regionally limited occurrence of tumors.[104] IDH1 or IDH2 mutations outside of Maffucci syndrome have also been found in 71% of SCHs, but such mutations were absent in other types of hemangiomas and vascular malformations. Small number cases with germline testing ruled out a germline mutation, but somatic mosaicism could not be ruled out.[105]

Other hemangiomas

There are several cutaneous capillary hemangioma types that vary from common to those reported in small numbers. A brief discussion of three examples follows: cherry hemangioma, microvenular hemangioma, and acquired elastotic hemangioma.

Cherry hemangioma

Cherry hemangioma is a common capillary hemangioma that often occurs in older persons, also referred to as a *senile hemangioma*. This hemangioma forms a superficial, well-demarcated dermal nodule containing dilated capillaries with slightly prominent endothelial cells (Figure 21.40).[10,11]

Microvenular hemangioma

This cutaneous hemangioma is a rarely diagnosed capillary hemangioma type, occurring mainly in young to middle-aged adults of either sex. Clinically, it manifests as a small reddish nodule, especially in the forearm and other extremity locations.

Histologically, the lesion is composed of venule-like vascular elements with slit-like lumina that often run parallel to the skin surface dissecting the dermal collagen (Figure 21.41). The endothelial cells form a nonatypical monolayer, and a pericytic layer is evident; however, vascular smooth muscle elements are not present. Only a small number of these hemangiomas have been reported since the original description; however, these angiomas are probably under-reported. Clinical behavior is indolent, and simple excision is sufficient.[106,107]

Acquired elastotic hemangioma

This capillary hemangioma variant occurs in sun-exposed skin in the setting of extensive elastotic damage. It forms a 2-cm to 5-cm vascular lesion composed of dilated capillaries within the elastosis, often parallel to the epidermis (Figure 21.42). A more lobular histologic pattern is also possible. Most reported cases have occurred in older women in the forearm, with one lesion reported in the neck.[108]

Telangiectasiae

These conditions involve dilatation of vessels instead of tumor formation, although clinically they appear hemangiomatous. The main distinction from an angioma is the lack of increase in numbers of vessels; instead, merely dilatation of existing vessels is present. Two examples are presented here: venous lake and port-wine stain (nevus flammeus).

Port-wine stain (nevus flammeus)

This cutaneous vascular abnormality is considered a telangiectasia and not a true hemangioma. Clinically

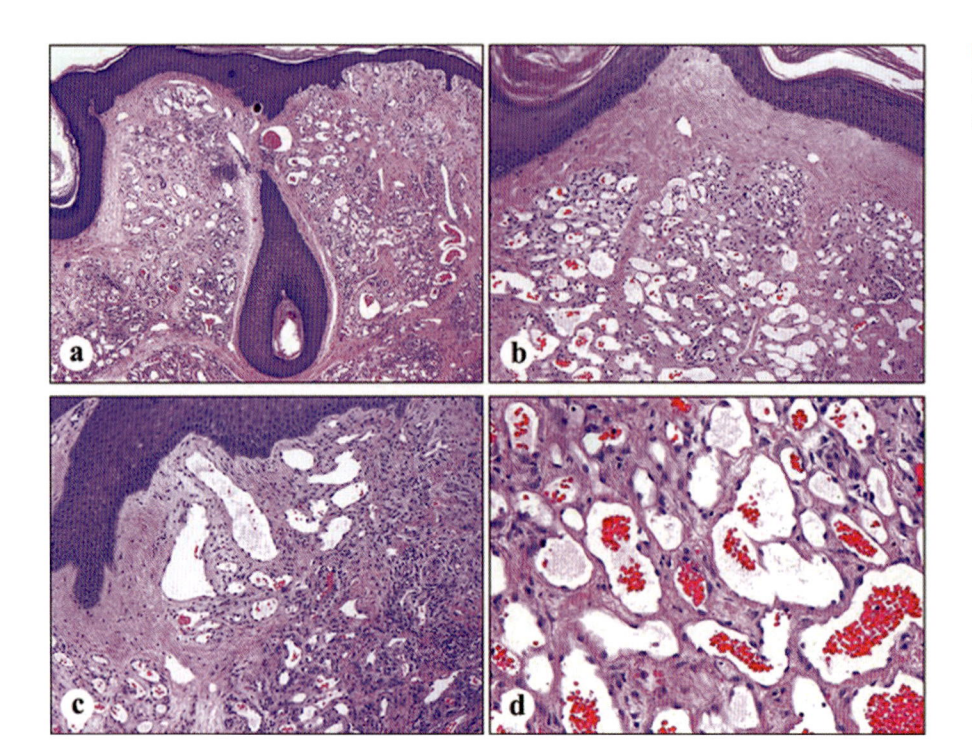

Figure 21.40 A cherry hemangioma is a superficial capillary hemangioma composed of vague lobules of capillaries with mildly protuberant endothelial cells.

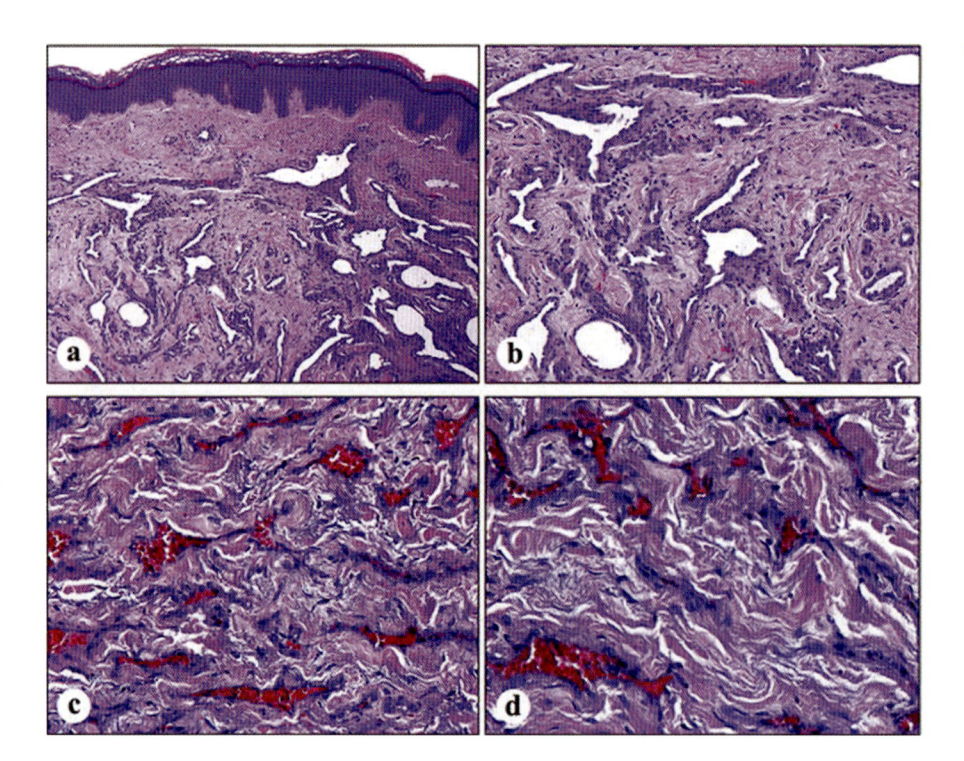

Figure 21.41 A microvenular hemangioma contains slit-like and occasionally gaping dermal capillaries that dissect collagen and often run parallel to the epidermis.

there is a bright to dark red stained area that varies in size. This finding can be sporadic or syndrome associated and can encompass one half of the face (i.e., in Sturge–Weber syndrome) or a portion of the lower extremity (i.e., in Klippel–Trenaunay syndrome).[109–111] The pathogenesis is unclear, but related to Sturge–Weber syndrome, localized lack of vasomotoric tone has been suggested to lead to inappropriate, constitutive vascular dilatation.[112]

Histologically, port-wine stains show multiple dilated hyperemic capillaries or venules, with no proliferative neoplastic component (Figure 21.43a,b).

Venous lake

This designation refers to a superficial dermal venectasia, dilatation of a single vein causing a bluish to blackish protuberance of the skin that can clinically resemble malignant melanoma. The condition occurs in older adults, usually in the head and neck region.[113]

Histologically, there is a dilated, engorged vein lined by regular endothelial cells (Figure 21.43c,d). Thrombosis is possible.

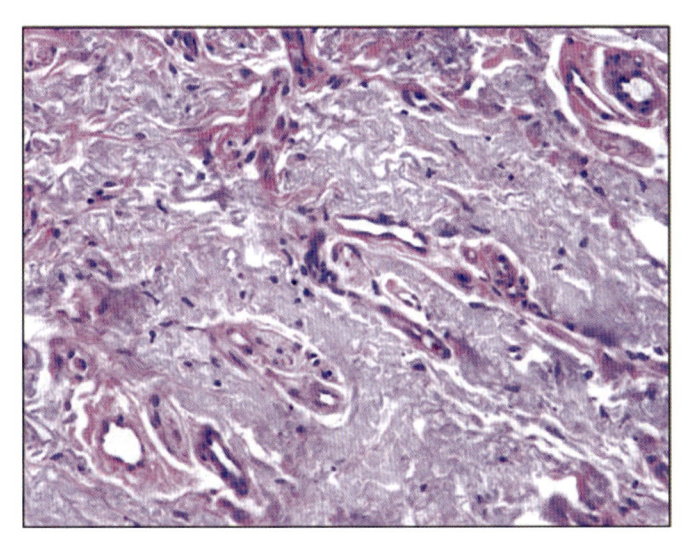

Figure 21.42 Acquired elastotic hemangioma contains scattered dilated capillaries in a markedly elastotic stroma.

Papillary endothelial hyperplasia (Masson tumor)

Clinical features

This benign condition is significant because it can be confused with angiosarcoma, although it represents an organizing thrombus with florid endothelial hyperplasia. A Masson tumor usually forms a small (1–2 cm), circumscribed, superficial nodule in the peripheral parts of extremities (especially the fingers and hand) or head (Figure 21.44). In some cases, papillary endothelial hyperplasia arises in a pre-existing hemangioma. Focal papillary endothelial hyperplasia is a relatively common finding in surgically excised hemorrhoids.[114–117]

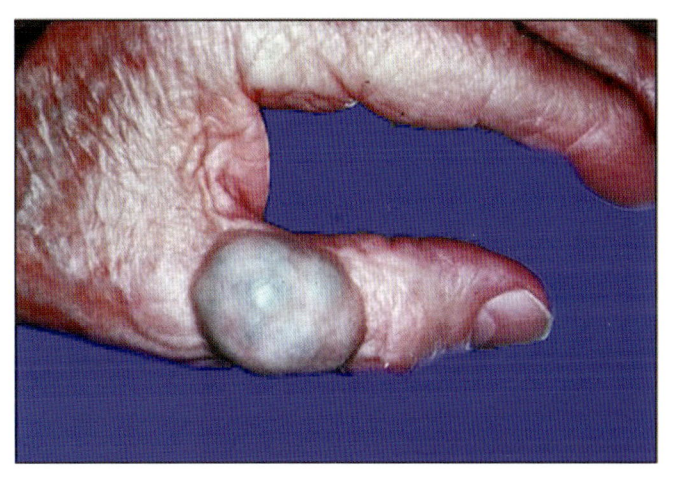

Figure 21.44 Papillary endothelial hyperplasia forms a dome-shaped bluish protuberance on the dorsal side of the thumb.

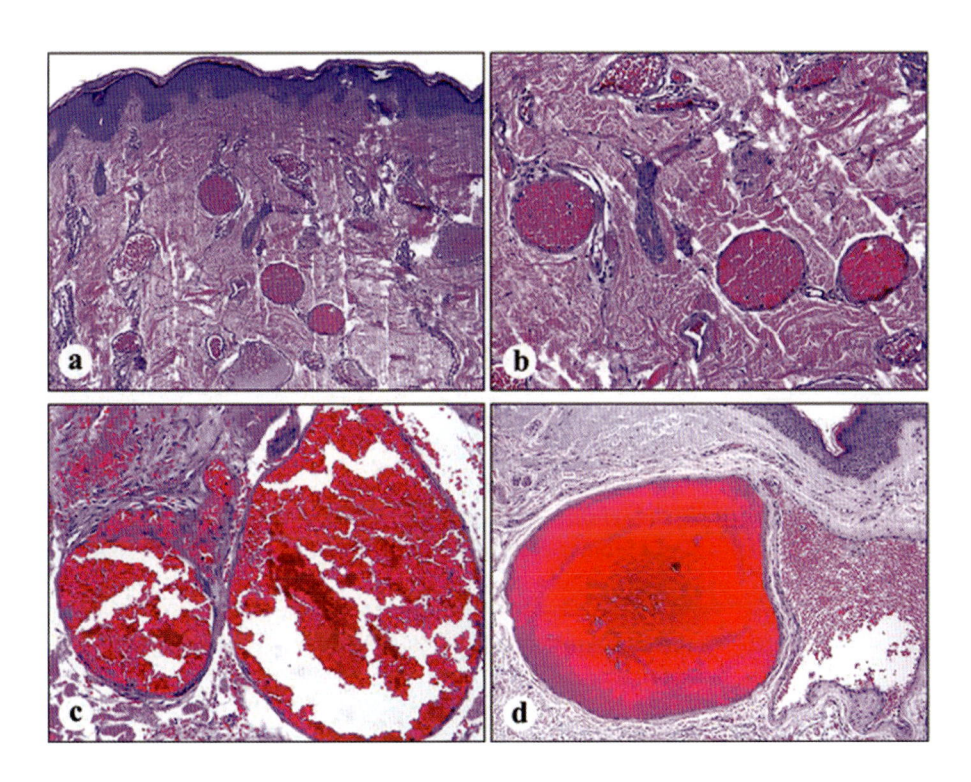

Figure 21.43 (a,b) Nevus flammeus (port-wine stain) in a telangiectasia containing dilated, engorged dermal capillaries. Fibrous stroma is a typical finding. (c,d) Venous lake is a superficial telangiectatic small vein.

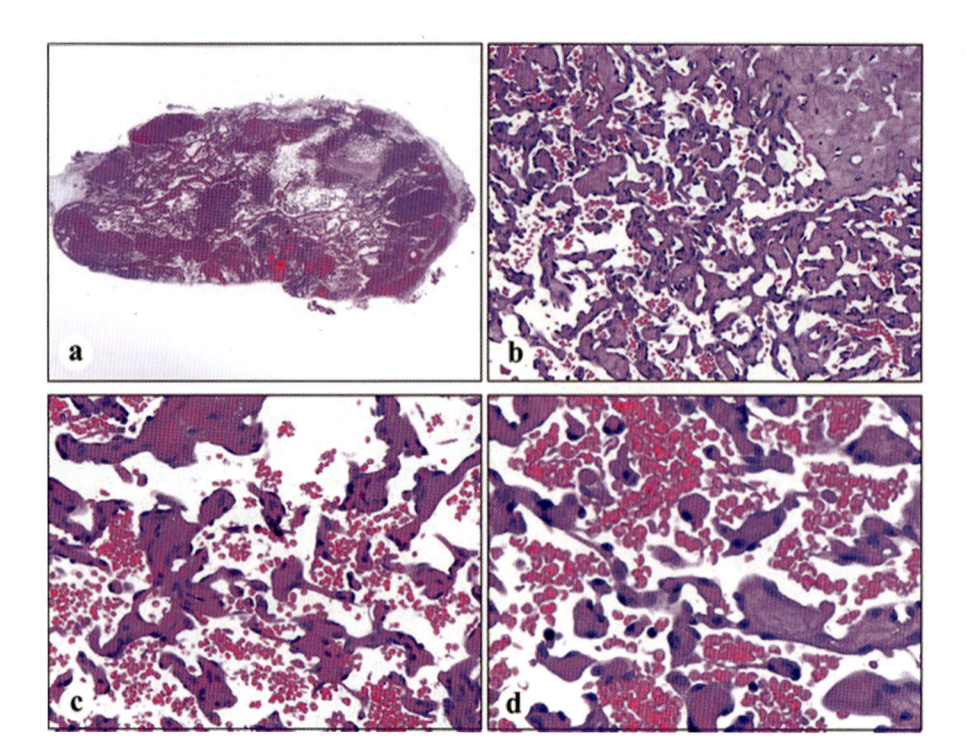

Figure 21.45 (**a**) Papillary endothelial hyperplasia forms a sharply demarcated nodule. (**b–d**) The lesion contains multiple papillae with fibrous cores lined by nonatypical endothelial cells with relatively small nuclei.

Pathology

The intravenous location and coexistence with thrombosis point to an origin from an organizing vascular thrombus; although rarely, similar changes are seen outside of vessels, and rarely in an organizing hematoma. Some lesions have elements of a cavernous hemangioma, indicating a benign vascular tumor as the origin. Histologically typical are numerous intravascular papillary projections, with fibrous stroma and a lining of hyperplastic endothelia, giving the lesion a placental-like villous appearance (Figure 21.45).

The features that differentiate this condition from an angiosarcoma are its typical intravascular location, one-layered endothelial lining, limited endothelial atypia, lack of solid growth and necrosis, and slight if any mitotic activity.

Reactive vascular proliferations

Reactive localized vascular proliferations of infectious origin are bacillary angiomatosis and verruga peruana. In these conditions, the bacterial infection triggers a hemangioma-like local endothelial cell proliferation. Recognition of these conditions is important for specific treatment. Other reactive vascular proliferations include acroangiodermatitis, a condition related to venous stasis, and vascular proliferation associated with intestinal intussusception.

Bacillary angiomatosis

Clinical features

This rare, reactive hemangioma-like vascular proliferation is a manifestation of an opportunistic infection caused by Gram-negative, rickettsia-like bacteria of the genus *Bartonella* (formerly named *Rochalimae*). Two species, *B. henselae* and *B. quintana*, are equally involved, with the spectrum of clinical features somewhat different with each. Most cases occur in AIDS patients or persons with other immunosuppressive conditions.[118–123] Only rarely are immunocompetent persons affected, and in these circumstances the cutaneous lesions tend to be fewer in number. More recently, effective antiretroviral treatment and infection control measures in AIDS patients have lowered the incidence of bacillary angiomatosis.

The vascular lesions occur primarily in the skin, and the lesions are usually multiple. Deep soft tissues also can be involved, and dissemination into the lymph nodes, mucosal surfaces, liver, and spleen is possible.[120]

Because bacillary angiomatosis is treatable with antibiotics (e.g., erythromycin and others), its specific recognition is important, especially because an untreated infection can disseminate into internal organs and become life-threatening. The nodules undergo complete regression with antibiotic therapy.

Pathology

Histologically, the cutaneous lesion forms nodules resembling florid lobular capillary hemangioma (pyogenic granuloma). It consists of lobular vascular proliferation, and the endothelia often show mild epithelioid change and sometimes focal nuclear atypia. Stromal neutrophilic microabscesses are a characteristic finding, and their presence in a lobular capillary hemangioma-like process without ulceration should raise suspicions of bacillary angiomatosis (Figure 21.46).

Chapter 21: Hemangiomas, lymphangiomas, and reactive vascular proliferations

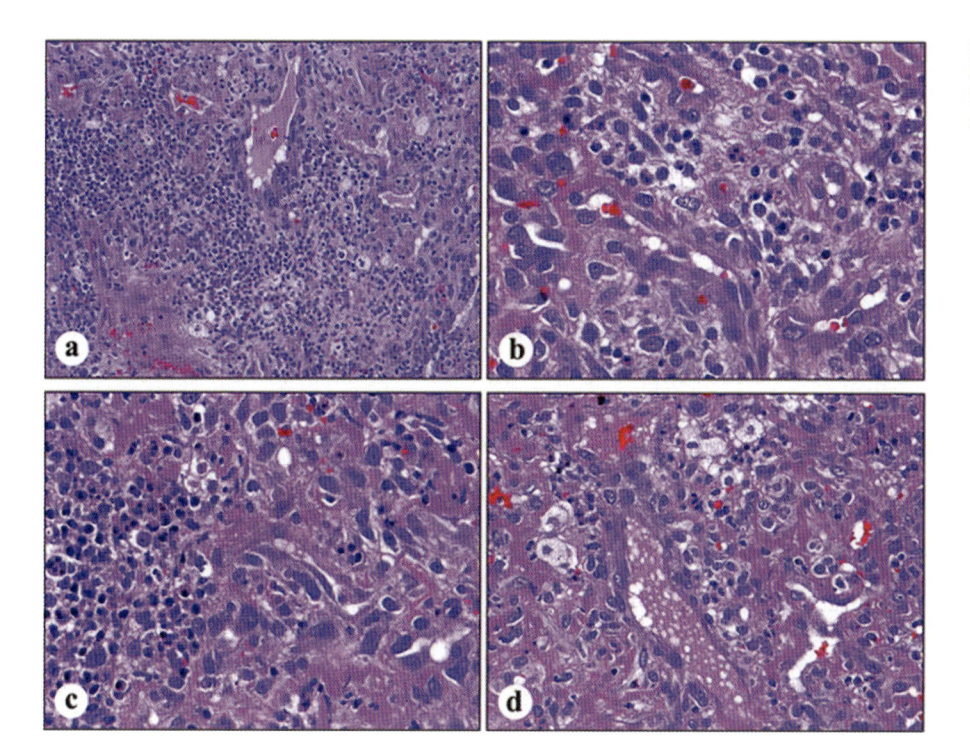

Figure 21.46 The overall appearance of a bacillary angiomatosis lesion resembles that of a lobular capillary hemangioma, but with numerous neutrophilic microabscesses.

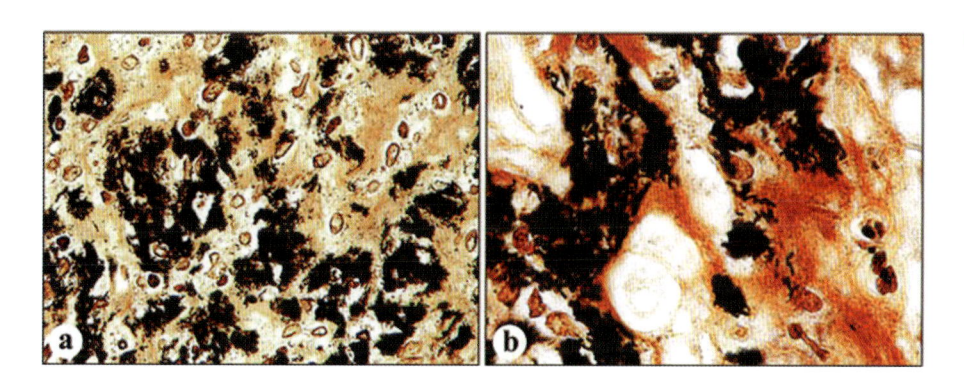

Figure 21.47 Warthin–Starry stain demonstrates numerous clusters of bacilli. Their morphology is evident at a higher magnification.

The microabscesses and the adjacent amorphous eosinophilic extracellular material contain rod-like *Bartonella* spp. bacteria of 3 μm, which can be highlighted with Warthin–Starry or Giemsa staining (Figure 21.47). Immunohistochemical identification with antibodies to *Bartonella* can also be useful.[124]

Verruga peruana (acute bartonellosis, Carrion's disease, Oroya fever)

Clinical features

This febrile infectious disease is geographically restricted to South America and occurs endemically in certain regions of Peru. A large epidemic occurred in the last century in the region of Oroya, and smaller epidemics have been detected since then. The disease is caused by Gram-negative bacteria *Bartonella bacilliformis*. The skin lesions appear during the course of the systemic disease, often as verruca-like exophytic growths.[125,126]

Pathology

The highly cellular and vascular skin lesions resemble a vascular neoplasm because of florid endothelial proliferation and mild endothelial atypia. There is a histologic resemblance to bacillary angiomatosis. The causative bacteria can be seen in the cytoplasm of endothelial cells as so-called Rocha–Lima inclusions, and they are best demonstrated with a Giemsa stain. The *Bartonella* bacteria have been shown to secrete angiogenic factors that are the likely source for the pathogenesis of the vascular proliferation.[127]

Reactive angioendotheliomatosis

Multiple reddish cutaneous plaques histologically representing focal reactive vascular proliferation, sometimes with papillary intraluminal endothelial hyperplasia and fibrin microthrombi, can occur in connection with systemic infections, such as endocarditis, cardiac valvular disease,

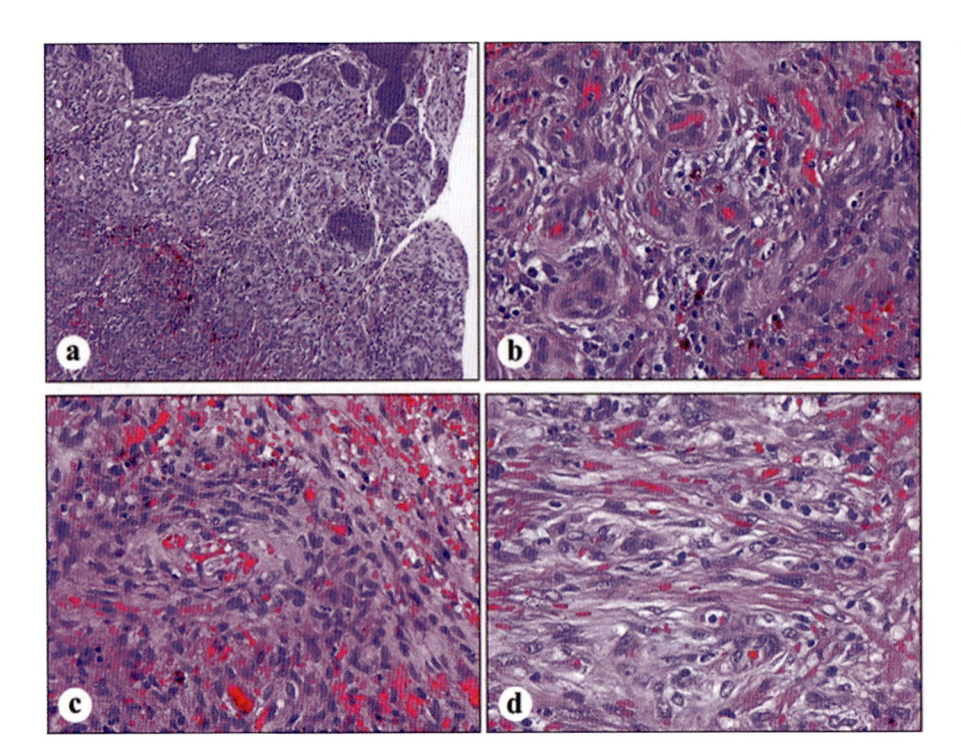

Figure 21.48 Acroangiodermatitis lesion shows hyperplasia in the overlying epidermis. The lesion contains lobules of capillaries that often contain thick walls. (**c**) Pericytic proliferation or (**d**) a spindled myofibroblastic component is sometimes present.

or chronic renal disease. Pathogenesis is likely related to angiogenic factors elicited by the infection, cytokine reaction to it, or vascular occlusion. Small lesional size, circumscribed nature, and the lack of true endothelial proliferation and atypia separate this lesion from vascular neoplasia.[128,129]

Acroangiodermatitis (pseudo-Kaposi's sarcoma)

This reactive pseudoneoplastic condition occurs in connection with venous stasis related to idiopathic chronic venous insufficiency or one associated with Klippel–Trenaunay syndrome, primary lymphedema, congenital arteriovenous malformations, or iatrogenic arteriovenous shunts. It has also been reported in amputation stumps and in association with coagulopathies.

Acroangiodermatitis usually manifests on the extensor surfaces of the feet and legs below the knee and includes chronic leg ulcers, often bilaterally. It can clinically simulate Kaposi's sarcoma by the presence of multiple elevated violaceous plaques.[128,130]

Histologically, it shows a vascular proliferation with a reactive appearance composed of lobules of thick-walled capillaries surrounded by fibrosis and variable hemosiderin deposition. Pericytic and stromal myofibroblastic proliferation can create an appearance of a spindle cell component. The lesions are typically associated with ulceration and reactive epidermal hyperplasia (Figure 21.48). CD34 positivity is limited to endothelial cells, and there is no CD34-positive spindle cell component, in contrast with Kaposi's sarcoma.

Florid vascular proliferation related to intestinal intussusception

Exuberant vascular proliferation can occur in connection with intestinal intussusception. The patients reported in one study included men and women of different ages (36–82 years). Two patients were HIV-positive young adults.[131]

Histologically, there is florid capillary proliferation, often in a zonal organized pattern (Figure 21.49). This proliferation can permeate throughout the intestinal wall and extend into the mesenteric fat. It can be admixed with inflammatory cells. The neovascular capillaries are immature but contain pericytes. They lack cytologic atypia, but there can be mitotic activity.

Superficial vascular pseudoaneurysm

This reactive myointimal and vascular wall proliferation occurs mainly in the superficial branches of the external carotid artery in the temporal region or forehead, locations where EH also commonly occurs. The involved patients have mostly been young men, and some patients have had a history of prior trauma.[132,133]

Histologically, there is a dilated arterial segment containing intramural proliferation of myointimal myofibroblasts, often with myxoid stroma (Figure 21.50). The involved vascular segment can undergo thrombosis. This process differs from EH, being centered in the vessel wall and not containing epithelioid endothelial cell proliferation. The elastic lamina is typically disrupted. Immunohistochemical analysis shows the myointimal component to be smooth muscle actin positive (Figure 21.51).

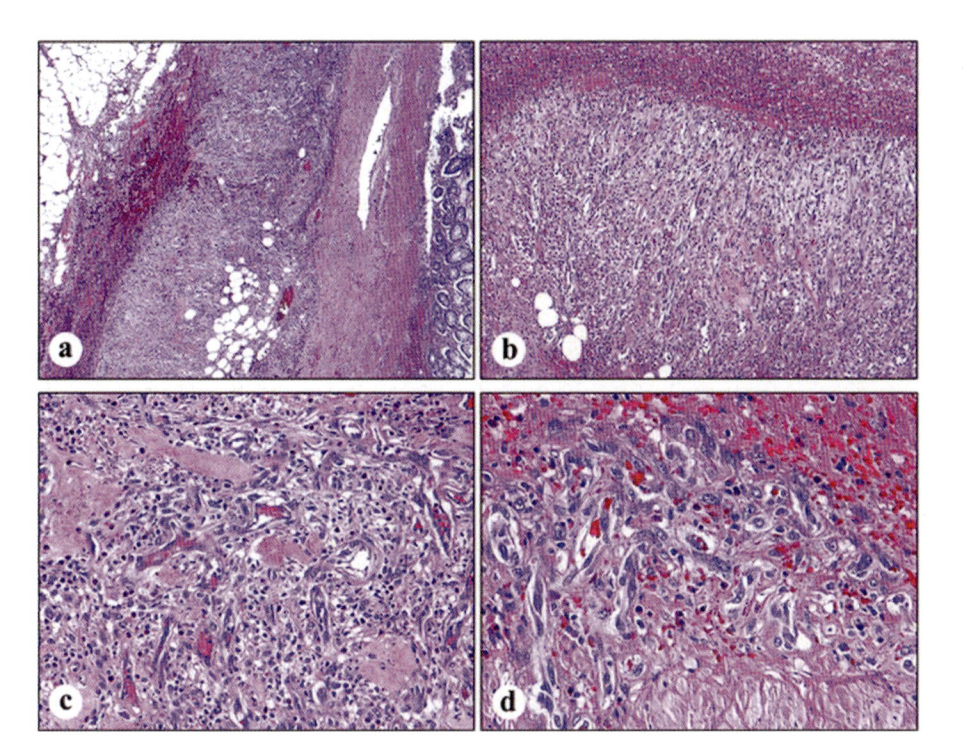

Figure 21.49 Vascular proliferation associated with an intestinal intussusception. Note zones of neovascular capillaries that permeate the muscular wall.

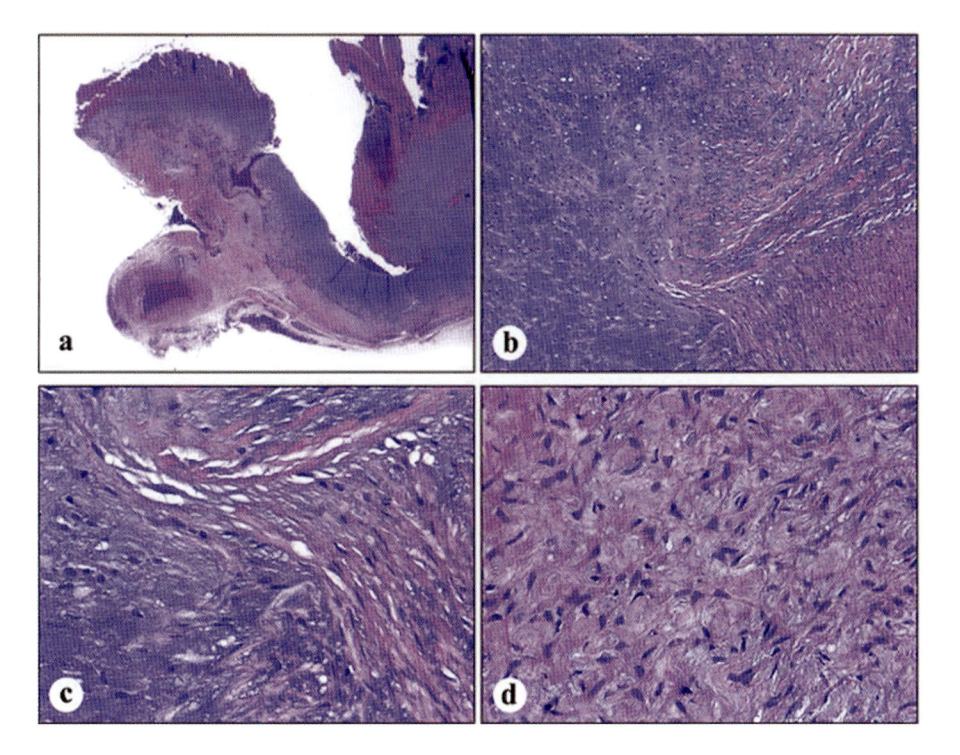

Figure 21.50 (**a**) Vascular pseudoaneurysm appears as a thickening of a small artery containing basophilic material. (**c**) The basophilic appearance results from myointimal proliferation within a myxoid matrix. (**d**) Stellate myointimal cells are uniform.

Lymphangioma and related tumors

Lymphangioma, lymphangiomatosis, and lymphangioendothelioma (acquired progressive lymphangioma) are discussed in this section.

Lymphangiomyoma(tosis) is a smooth-muscle-like proliferation related to the smooth muscle elements in the walls of the lymphatic vessels. The tumor is now classified as a perivascular epithelioid cell tumor (PEComa, Chapter 19). These tumors are unique for their coexpression of HMB45, a melanoma marker, and smooth muscle markers.

Lymphangioma

Lymphangioma is a benign vascular tumor composed of lymphatic-like vessels. This designation encompasses a range

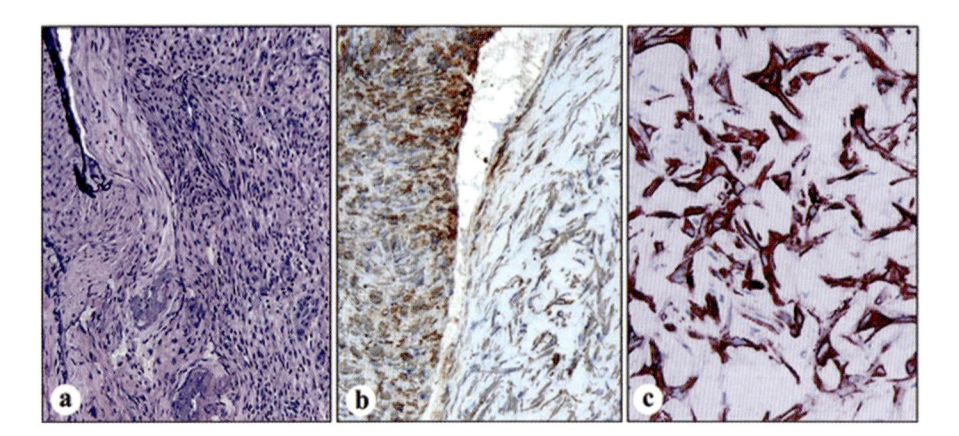

Figure 21.51 (a) Disrupted elastic lamina of vascular pseudoaneurysm as seen in elastin stain. (b,c) Smooth-muscle actin-positive myofibroblasts proliferate in the damaged arterial wall.

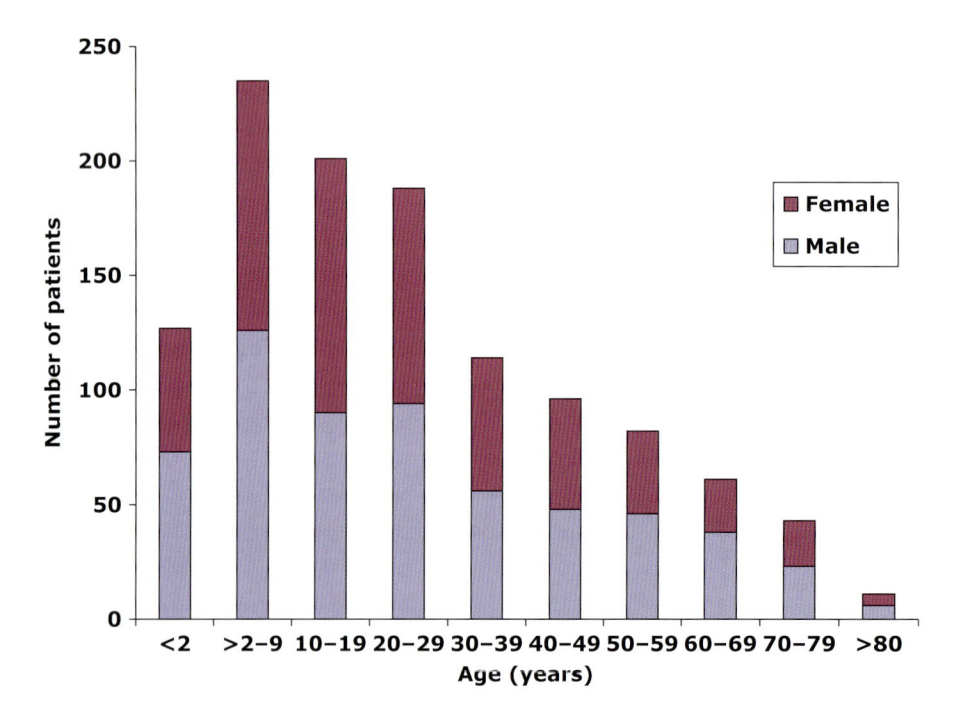

Figure 21.52 Age and sex distribution of 1177 lymphangiomas.

of lymphatic vascular proliferations in the skin, other superficial locations, and deep locations, including the body cavities. The pathogenesis of lymphangioma in early childhood is probably related to focal malformation, and therefore many lymphangiomas can be characterized as vascular malformations. Some adult-onset lymphangiomas could be related to disruption of lymphatic flow by trauma, and therefore some of these lymphangiomas might be lymphangiectasiae.

Clinical features

Lymphangiomas most commonly occur in children, but they are seen with a lower frequency in adults of various ages, without any clear gender bias (Figure 21.52). Cutaneous examples have the highest relative frequency in adults. The most common locations are the head and neck, and the abdomen, whereas occurrence in the extremities is less common (Figure 21.53a). In the head and neck, the most commonly involved areas are the neck and eye/orbital region (Figure 21.53b).[134–136]

The most common lymphangiomas in the neck (cystic hygromas) occur from infancy on. Such cystic lymphangiomas seem to have a nonrandom association with Turner syndrome.[137]

Intra-abdominal or intrathoracic lymphangiomas form cystic masses, and these lesions can be extensive. They occur in all ages. Gastrointestinal tract lymphangiomas vary from endoscopically visible small mucosal masses to large transmural tumors that more often primarily involve the mesenteries and sometimes several contiguous organs.[136,138]

Cutaneous lymphangiomas are often referred to as *lymphangioma circumscriptum*. They include minute superficial variants extending close to the epidermis. There are also deeper dermal types forming larger masses, often with subcutaneous extension.

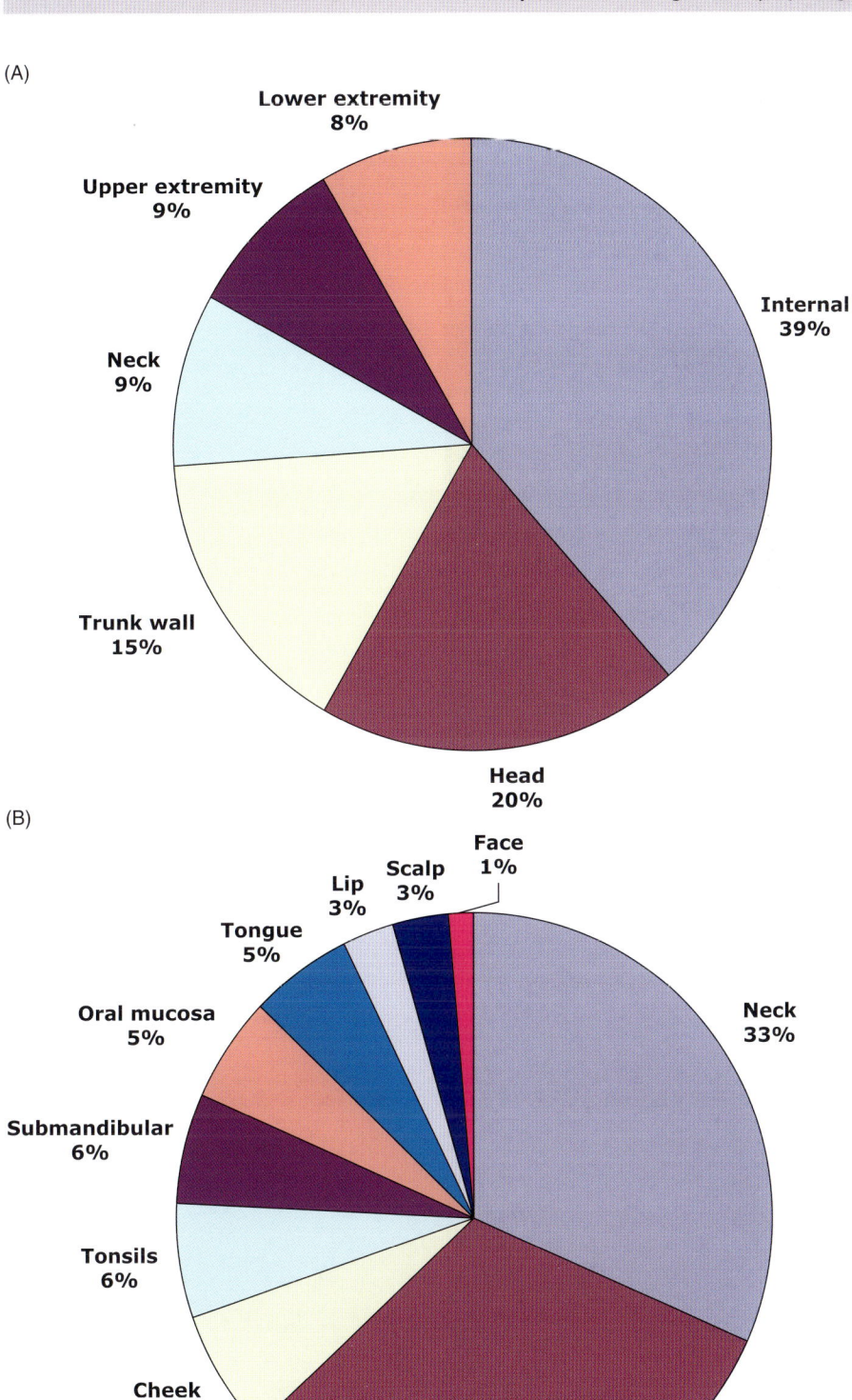

Chapter 21: Hemangiomas, lymphangiomas, and reactive vascular proliferations

Figure 21.53 (**a**) Anatomic distribution of 1159 lymphangiomas. (**b**) Anatomic distribution of 457 lymphangiomas in the head and neck.

(A)
- Lower extremity 8%
- Upper extremity 9%
- Neck 9%
- Trunk wall 15%
- Head 20%
- Internal 39%

(B)
- Face 1%
- Scalp 3%
- Lip 3%
- Tongue 5%
- Oral mucosa 5%
- Submandibular 6%
- Tonsils 6%
- Cheek 7%
- Eye+orbit 31%
- Neck 33%

583

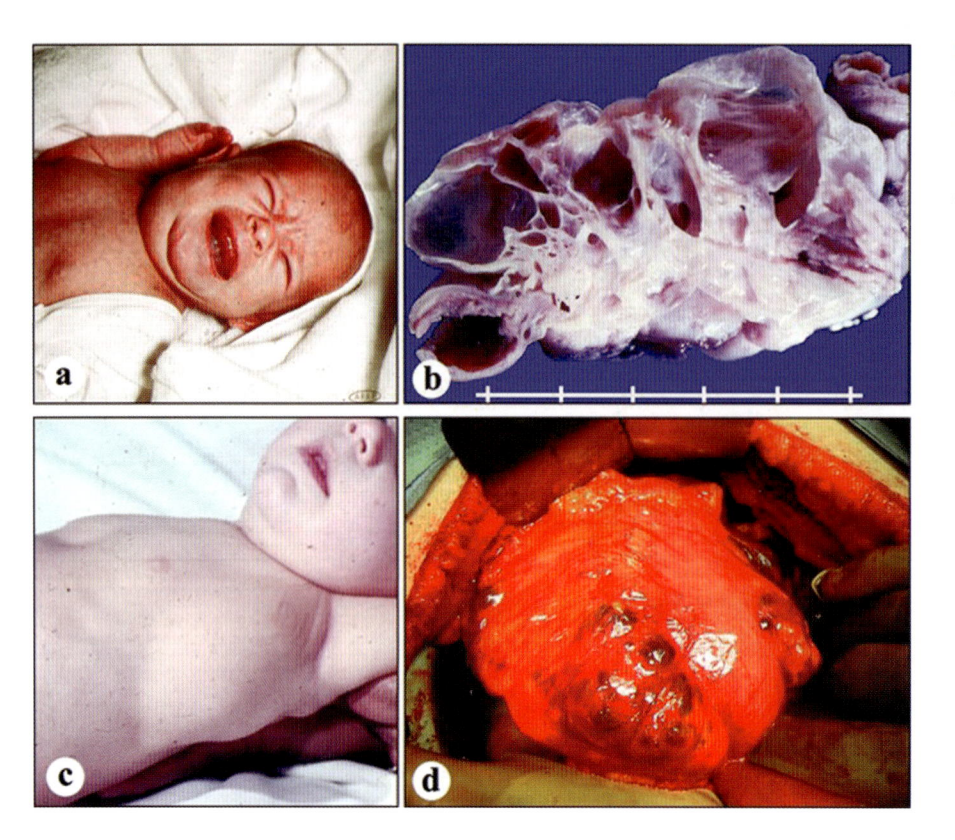

Figure 21.54 Clinical and gross appearances of lymphangiomas. (**a**) A cystic lymphangioma in the right cheek and neck. (**b**) Cross-section reveals multiple thin-walled cystic cavities. (**c**) Lymphangioma in the upper lateral chest wall. (**d**) Abdominal lymphangioma *in situ*. The cysts are visible as cobblestone-like structures.

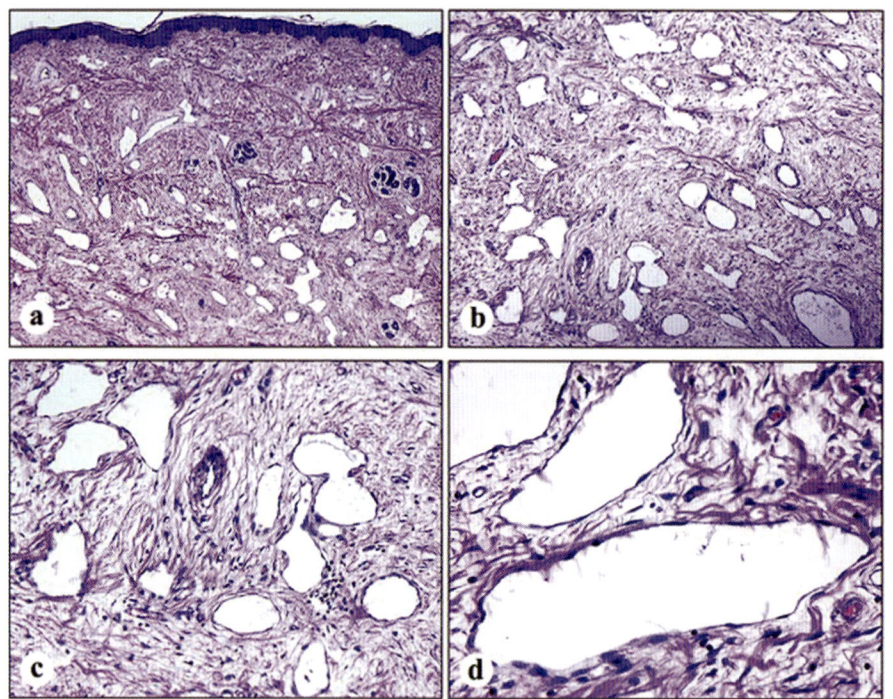

Figure 21.55 A dermal lymphangioma containing lymphatic channels lined by attenuated endothelia.

Pathology

Grossly, lymphangiomas are often unimpressive because the lymphatic vessels collapse into a slit-like form in the surgical specimen. *In situ* they have a spongy appearance, however, and many are cystic on sectioning (Figure 21.54). Superficial examples can show dilated lymphatics as blister-like cutaneous formations.

Histologically, there are two kinds of lymphangiomas. The superficial types are composed of scattered small-caliber lymphatic vessels lined by attenuated lymphatic endothelia (Figure 21.55). Clusters of lymphocytes can be present around or inside the lumina.

Deep lymphangiomas are composed of varyingly dilated lymphatic channels filled with clear or proteinaceous fluid and

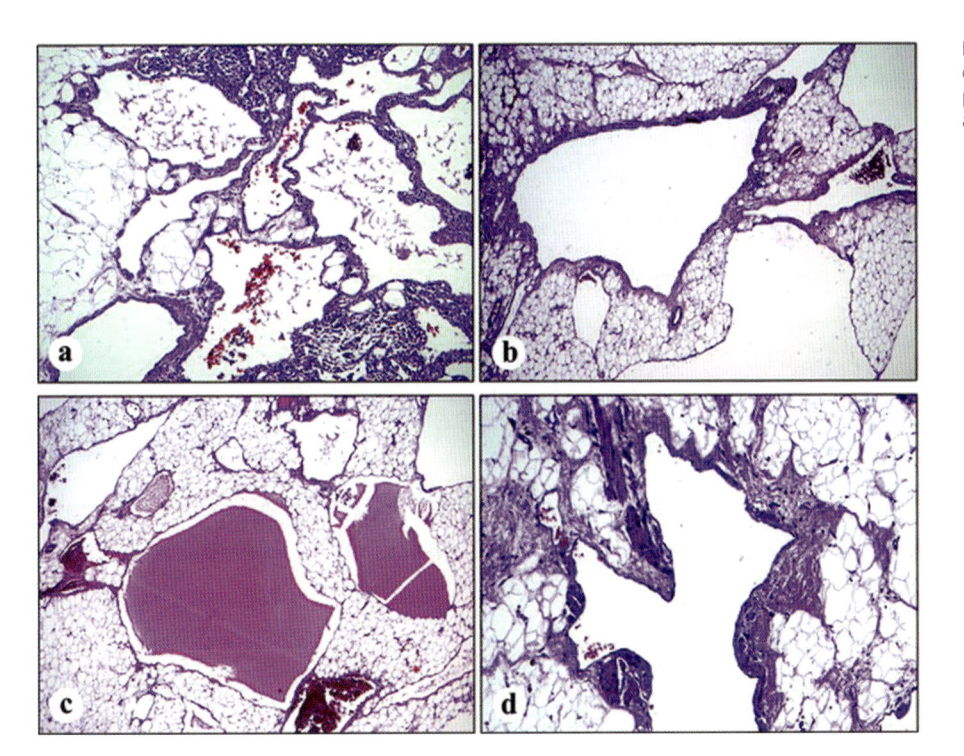

Figure 21.56 A deep lymphangioma composed of cavernous lumina, some of which contain proteinaceous fluid. (**d**) Smooth muscle elements are present in some vessel walls.

sometimes containing lymphocytes. The vessels have thin walls and flaccid contours, but often contain vascular smooth muscle elements. Lymphoid aggregates or more completely developed lymph nodes are often seen in the septa (Figure 21.56).

Immunohistochemically distinctive for lymphatic vessels is the expression of three vascular markers: VEGFR 3, PROX1, and podoplanin (D2–40). Neither of these is entirely specific for lymphatic vessels.[139,140] Positivity for CD31 and CD34 is often weaker than that seen in blood vessel endothelia.

Differential diagnosis

Several unrelated tumors can simulate lymphangioma morphologically. These include cystic mesothelial proliferations in the serous cavities, and myxoid liposarcoma.

In the abdomen and spleen, cystic mesothelial proliferations can closely simulate lymphangiomas. A more prominent epithelioid lining cell element is usually visible, however, and if there is any doubt, the immunohistochemistry for epithelial/mesothelial and vascular markers can clarify the situation.

Myxoid liposarcoma often contains "lymphangioma-like" spaces filled by proteinaceous fluid. A prominent plexiform capillary pattern, lack of true lining cells in the cystic spaces, and the usually focal presence of fatty differentiation help to distinguish this entity from lymphangioma, however.

Lymphangiomyoma(tosis) is a smooth muscle proliferation related to the smooth muscle elements in the walls of lymphatic vessels. The tumor is unique for its coexpression of HMB45, a melanoma marker, and smooth muscle markers and is therefore included among PEComas (Chapter 19).

Benign lymphangioendothelioma (acquired progressive lymphangioma)

Originally reported as acquired progressive lymphangioma by Watanabe et al. in 1983,[141] this lesion was renamed *benign lymphangioendothelioma* by Wilson Jones et al. in 1990.[142] This process can be considered a large, superficially spreading example of lymphangioma.

Clinical features

This rare cutaneous tumor occurs in a wide age range from early childhood to old age; some lesions have been apparently congenital. There is no predilection for either sex. The lesion usually appears as a discolored, reddish-to-purplish macule or plaque, and rarely an elevated lump that often enlarges over several years, sometimes exceeding 10 cm in diameter; hence the original name *progressive lymphangioma*. The largest series reported seven of ten lesions to measure 2 cm or less. Multiple lesions have been reported in some patients.

Based on four published series, the most commonly involved cutaneous sites include the upper extremity, especially the shoulder and forearm (7/19), head and neck (6/19), and lower extremity and trunk (each: 3/19). In the first series, malignancies (i.e., hepatoma, pulmonary small cell cancer) preceded the skin lesion by 0.5 to 2 years. The tumor can persist or recur after an incomplete excision, but there is no malignant potential. Although postoperative follow-up data in the published series are scant, a preoperative history of several years in many patients supports a benign process.[141–145]

A clinical syndrome with multifocal lymphangioendotheliomas (lymphangioendotheliomatosis) with thrombocytopenia has been reported in children. These patients had innumerable

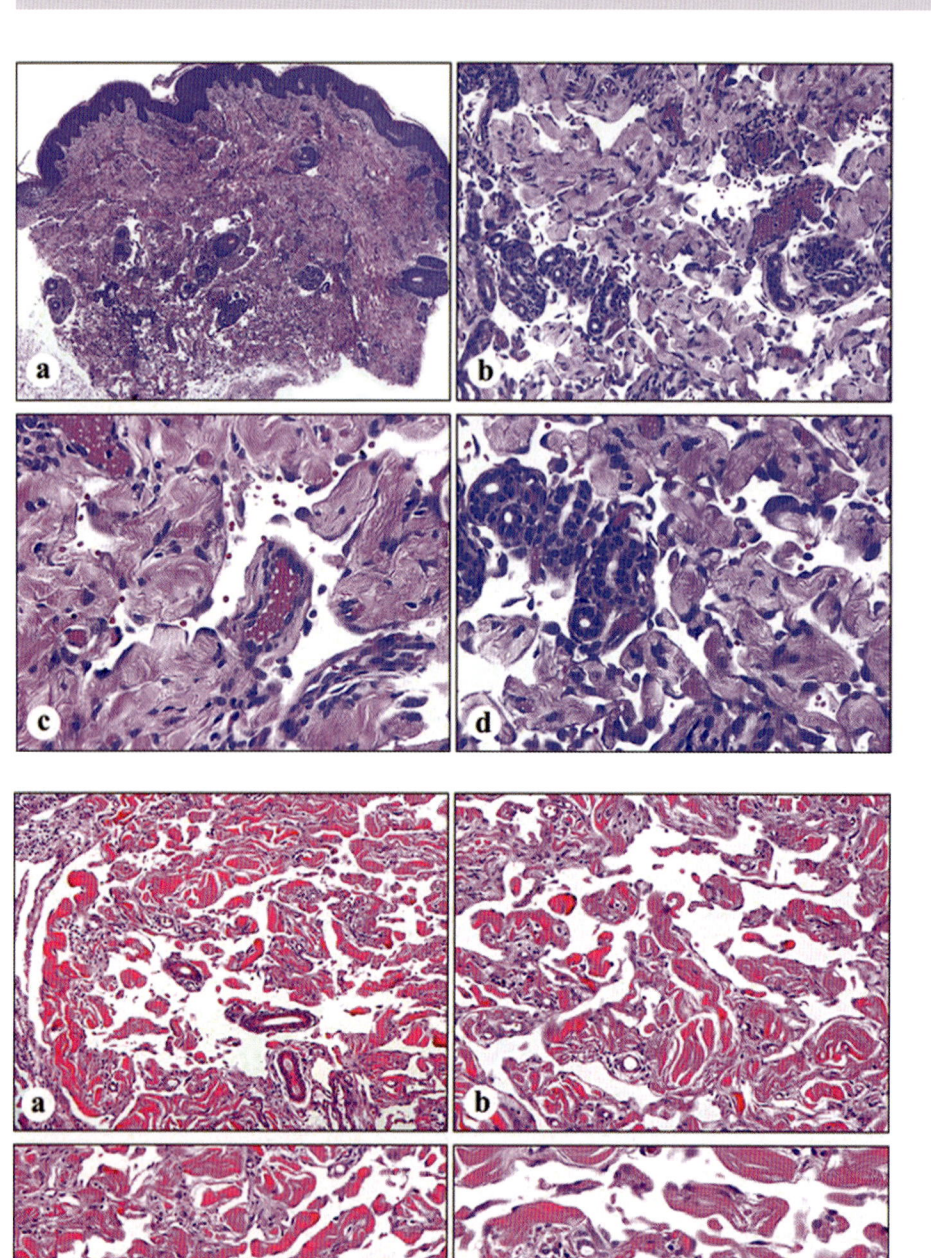

Chapter 21: Hemangiomas, lymphangiomas, and reactive vascular proliferations

Figure 21.57 Example of a lymphangioendothelioma. This is a small dermal lesion. Note anastomosing lymphatic vascular channels lined by focally prominent but nonatypical endothelial cells, with focal papillary intraluminal proliferation.

Figure 21.58 Another example of a lymphangioendothelioma containing anastomosing lymphatic channels. Papillary intraluminal proliferation is evident in this case.

congenital cutaneous and multiple gastrointestinal lesions with severe gastrointestinal bleeding.[146] Histologic features differ from adult lymphangioendothelioma and seem to resemble those of papillary intralymphatic angioendothelioma.

Pathology

No gross descriptions have been reported. Histologically, the lesions involve the dermis, and the larger examples usually extend into the subcutis. Secondary ulceration on traumatized areas is possible; otherwise the epidermis is normal. There is a permeative, anastomosing network of lymphatic-like vascular channels. These channels tend to run horizontally and are more open in the upper dermis and more slit-like or closed in the lower dermis. The permeative nature resembles that of angiosarcoma (Figures 21.57 and 21.58). Cytologically, however, the lining cells typically form a single layer of attenuated endothelial cells without mitotic activity, although the cellularity is generally greater than that in the lymphatic vessels. A focal hobnail appearance and papillary intraluminal proliferation resembling papillary endothelial hyperplasia are relatively common, and some vascular channels contain focal smooth muscle elements. The vascular channels generally appear empty, without much proteinaceous fluid, but they can contain erythrocytes. Scattered lymphoid cells are often present around the vascular channels, but are generally not prominent, and hemosiderin deposition occurs infrequently. The lining endothelia have been reported to be positive for CD31, CD34, and FVIIIRAg, but no distinctive immunohistochemical features have been reported so far.

Differential diagnosis

The presence of perivascular inflammation, extravasated erythrocytes, and often multiplicity characterize the lesions of Kaposi's saroma, the lymphangioma-like variant of which can otherwise resemble benign lymphangioendothelioma. HHV8 positivity is the best discriminatory marker to be positive in Kaposi's sarcoma. Benign lymphangioendothelioma lacks the endothelial multilayering, nuclear atypia, and frequent mitotic activity that are typically seen in angiosarcoma.

Hobnail hemangioma typically consists of separate, nonanastomosing vascular lumina with a conspicuous, intraluminally protuberant, hobnail appearance. Lymphangiomatosis is histologically very similar to benign lymphangioendothelioma and most significantly differs from it by greater extent of the lesion with prominent subcutaneous and common intramuscular involvement; visceral cases are also classified as lymphangiomatosis.

Lymphangiomatosis

This designation refers to a diffuse lymphangioma involving an anatomic region. The involvement varies from one limited to an extremity to extensive visceral involvement, especially the thorax. There seems to be male predominance in the reported series, but no familial tendency has been reported.[147,148]

Extremity-based examples manifest in early childhood, often at birth. There is chronic swelling of a leg with marked subcutaneous expansion (Figure 21.59) and often brownish

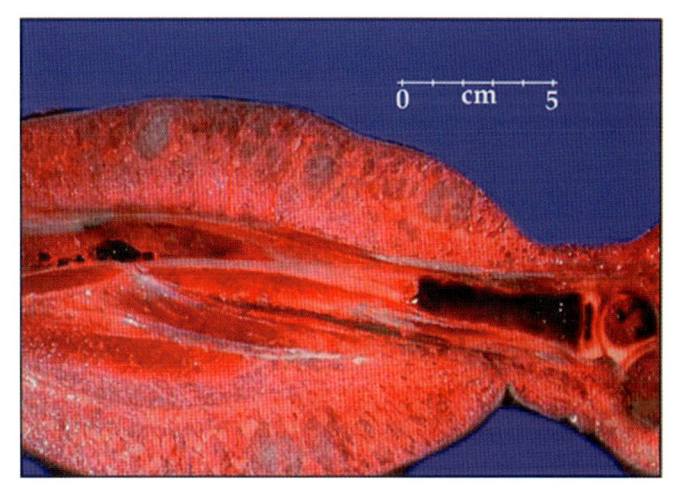

Figure 21.59 Lymphangiomatosis involving the leg forms a massive subcutaneous swelling. The cystic appearance results from the presence of multiple dilated lymphatic channels.

Table 21.1 Summary of selected syndromes affecting the peripheral vascular system and their genetic basis[a]

Disease	Mutated gene	Locus	Comment
Primary congenital lymphedema (Milroy disease)	VEGFR3	5q35.3	Mutations that inactivate the VEGFR3 tyrosine kinase
Maffucci syndrome	Unknown	–	Multiple spindle cell hemangiomas, cartilaginous bone tumors
Klippel–Trenaunay syndrome	RASA1	5q13.3s	Large cutaneous telangiectasiae, vascular malformations/angiodysplasia in the lower extremity
Blue rubber bleb nevus syndrome	Unknown	–	Cutaneous, gastrointestinal, pulmonary, and central nervous system hemangiomas. Gastrointestinal bleeding and chronic anemia. Vascular obliteration by hemangioma thrombosis
Cowden syndrome	PTEN	10q23	Hemangiomas, hemihypertrophy, fibromas
Venous malformations	TIE2	9p21	Multiple cutaneous and mucosal venous malformations
Hereditary hemorrhagic telangiectasia			Vascular telangiectasias in various organs. Bleeding, especially gastrointestinal
Osler (Osler–Rendu–Weber)			
Types 1–3	Endoglin ALK1 ?	9q34.1 12q11–q14 5q31.5–32	
Sturge–Weber	?		Port-wine stain (facial telangiectasiae), epilepsy, mental retardation, hemangiomas of the leptomeninges, adjacent to cerebral cortex

[a] Data based on references 149–153.

cutaneous discoloration. Histologically, the process shows contiguous subcutaneous spaces lined by attenuated endothelia, similar to features of lymphangioendothelioma.[147]

Lymphangiomatosis involving the thorax also manifests in children, either in infancy or around the age of 10 years. There is diffuse involvement of the lungs and pleura, including the parietal one, and often chylothorax. Such involvement has a high complication rate and mortality. Three of four patients reported in one series died within 3 years. The histologic appearance is that of an anastomosing network of lymphatic channels with attenuated endothelia.[148]

Syndromes featuring hemangiomas and vascular malformations: genetic background

Numerous malformation syndromes of early childhood and some syndromes manifesting in adulthood have a component of telangiectasia, vascular malformation, or hemangioma. Syndromes related to the cerebrovascular system are excluded.

The most important syndromes featuring vascular lesions and their genetic backgrounds are listed in Table 21.1.

References

Biology of endothelia and angiogenesis

1. Cines DB, Pollak ES, Buck CA, et al. Endothelial cells in physiology and in the pathophysiology of vascular disorders. Blood 1998;91:3527–3561.
2. Cotran RS, Mayadas-Norton T. Endothelial adhesion molecules in health and disease. Pathol Biol (Paris) 1998;46:164–170.
3. Parikh AA, Ellis LM. The vascular endothelial growth factor family and its receptors. Hematol Oncol Clin North Am 2004;18:951–971.
4. Carmeliet P. Mechanisms of angiogenesis and arteriogenesis. Nat Med 2000;6:389–395.
5. Folkman J. Angiogenesis and angiogenesis inhibition: an overview. EXS 1997;79:1–8.
6. Ferrara N, Alitalo K. Clinical applications of angiogenic growth factors and their inhibitors. Nat Med 1999;5:1359–1364.

Classification of vascular tumors

7. Kempson RL, Fletcher CDM, Evans HL, Hendrickson MR, Sibley RK. Vascular tumors. In Atlas of Soft Tissue Tumors. Washington, DC: AFIP; 2001: 307–370.
8. Marler JJ, Mulliken JV, Vascular anomalies: classification, diagnosis, and natural history. Facial Plast Surg Clin North Am 2001;9:495–504.
9. Requena L, Sangueza OP. Cutaneous vascular anomalies. Part 1. Hamartomas, malformations, and dilatation of preexisting vessels. J Am Acad Dermatol 1997;37:523–549.
10. Requena L, Sangueza OP. Cutaneous vascular proliferations. Part II. Hyperplasias and benign vascular neoplasms. J Am Acad Dermatol 1997;37:887–920.
11. Hunt SJ, Santa Cruz DJ. Vascular tumors of the skin: a selective review. Semin Diagn Pathol 2004;21:166–218.
12. Mulliken JB, Glowacki J. Hemangiomas and vascular malformations in infants and children: a classification based on endothelial characteristics. Plast Reconstr Surg 1982;69:412–422.
13. Enjolras O, Mulliken JB. Vascular tumors and vascular malformations: new issues. Adv Dermatol 1997;13:375–423.
14. Cohen MM. Vascular update: morphogenesis, tumors, malformations, and molecular dimensions. Am J Med Genet 2006;140A:2013–2038.

Juvenile capillary hemangioma

15. Coffin CM, Dehner LP. Vascular tumors in children and adolescents: a clinicopathologic study of 228 tumors in 222 patients. Pathol Annu 1993;1:97–120.
16. Takahashi K, Mulliken JB, Kozakewich HP, et al. Cellular markers that distinguish the phases of hemangioma during infancy and childhood. J Clin Invest 1994;93:2357–2364.
17. Smolinski KN, Yan AC. Hemangiomas of infancy: clinical and biological characteristics. Clin Pediatr 2005;44:747–766.
18. Enjolras O, Riche MC, Merland JJ, Escande JP. Management of alarming hemangiomas of infancy: a review of 25 cases. Pediatrics 1990;85:491–498.
19. Chen TS, Eichenfield LF, Friedlander SF. Infantile hemangiomas: an update on pathogenesis and therapy. Pediatrics. 2013;131:99–108.
20. Gonzalez-Crussi F, Reyes-Mugica M: Cellular hemangiomas of infancy ("hemangioendotheliomas"): light microscopic, immunohistochemical, and ultrastructural observations. Am J Surg Pathol 1991;15.769–778.
21. North PE, Waner M, Mizeracki A, Mihm M, Jr. GLUT1: a newly discovered immunohistochemical marker for juvenile hemangiomas. Hum Pathol 2000;31:11–22.
22. North PE, Waner M, Mizeracki A, Mrak RE. A unique microvascular phenotype shared by juvenile hemangiomas and human placenta. Arch Dermatol 2001;137:559–570.
23. Kumakiri M, Muramoto F, Tsukinaga I, et al. Crystalline lamellae in the endothelial cells of a type of hemangioma characterized by the proliferation of immature endothelial cells and pericytes-angioblastoma. J Am Acad Dermatol 1983;8:68–75.
24. Walter JW, North PE, Waner M, et al. Somatic mutation of vascular endothelial growth factor receptors in juvenile hemangioma. Genes Chromosomes Cancer 2002;33:295–303.
25. Berenguer B, Mulliken JB, Enjolras O, et al. Rapidly involuting congenital hemangioma: clinical and histopathologic features. Pediatr Dev Pathol 2003;6:495–510.
26. North PE, Waner M, James CA, et al. Congenital nonprogressive hemangioma: a distinct clinicopathologic entity unlike infantile hemangioma. Arch Pathol 2001;137:1607–1620.

Pyogenic granuloma (lobular capillary hemangioma) and cavernous hemangioma

27. Patrice SJ, Wiss K, Mulliken J. Pyogenic granuloma (lobular capillary hemangioma) pathologic study of 178 cases. Pediatr Dermatol 1991;8:267–276.

28. Bhaskar SN, Jacoway JR. Pyogenic granuloma – clinical features, incidence, histology, and result of treatment: report of 242 cases. *J Oral Surg* 1966;24:391–398.
29. Puxeddu R, Berlucchi M, Ledda GP, et al. Lobular capillary hemangioma of the nasal cavity: a retrospective study of 40 patients. *Am J Rhinol* 2006;20:480–484.
30. Mills SE, Cooper PH, Fechner RE. Lobular capillary hemangioma: the underlying lesion of pyogenic granuloma. *Am J Surg Pathol* 1980;4:471–479.
31. Cooper PH, Mills SE. Subcutaneous granuloma pyogenicum: lobular capillary hemangioma. *Arch Dermatol* 1982;118:30–33.
32. Warner J, Wilson-Jones E. Pyogenic granuloma with multiple satellites: a report of 11 cases. *Br J Dermatol* 1968;80:218–227.
33. Toida M, Hasegawa T, Watanabe F, et al. Lobular capillary hemangioma of the oral mucosa: clinicopathological study of 43 cases with a special reference to immunohistochemical characterization of the vascular elements. *Pathol Int* 2003;53:1–7.
34. Cooper PH, McAllister HA, Helwig EB. Intravenous pyogenic granuloma: a study of 18 cases. *Am J Surg Pathol* 1979;3:221–228.
35. Ulbright TM, Santa-Cruz DJ. Intravenous pyogenic granuloma. *Cancer* 1980;45:1646–1652.
36. Kapadia SB, Heffner DK. Pitfalls in the histopathological diagnosis of pyogenic granuloma. *Eur Arch Otolaryngol* 1992;249:195–200.
37. Truss L, Dobin SM, Donner LR. Deletion (21)(q21.2q22.12) as the sole cytogenetic abnormality in lobular capillary hemangioma of the nasal cavity. *Cancer Genet Cytogenet* 2006;170:69–70.
38. Calonje E, Fletcher CDM. Sinusoidal hemangioma: a distinctive benign vascular neoplasm within the group of cavernous hemangiomas. *Am J Surg Pathol* 1991;14:1130–1135.

Angiokeratoma and verrucous hemangioma

39. Imperial R, Helwig EB. Angiokeratoma: a clinicopathologic study. *Arch Dermatol* 1967;95:166–175.
40. Imperial R, Helwig EB. Angiokeratoma of the scrotum (Fordyce type). *J Urol* 1967;98:379–387.
41. Schiller PI, Itin PH. Angiokeratomas: an update. *Dermatology* 1996;193:275–282.
42. Imperial R, Helwig EB. Verrucous hemangioma: a clinicopathological study of 21 cases. *Arch Dermatol* 1967;96:247–253.
43. Cruces MJ, De la Torre C. Multiple eruptive verrucous hemangiomas: a variant of multiple hemangiomatosis. *Dermatologica* 1985;171:106–111.
44. Puig L, Llistosella E, Moreno A, de Moragas JM. Verrucous hemangioma. *J Dermatol Surg Oncol* 1987;13:1089–1092.
45. Yang CH, Ohara K. Successful surgical treatment of verrucous hemangioma: a combined approach. *Dermatol Surg* 2002;28:913–919.
46. Tennant LB, Mulliken JB, Perez-Atayde AR, Kozakewich HP. Verrucous hemangioma revisited. *Pediatr Dermatol* 2006;23:208–215.

Arteriovenous hemangioma (acral arteriovenous tumor) and venous hemangioma

47. Girard C, Graham J, Johnson WC. Arteriovenous hemangioma (arteriovenous shunt): a clinicopathological and histochemical study. *J Cutan Pathol* 1974;1:73–87.
48. Carapeto FJ, Garcia-Perez A, Winkelmann RK. Acral arteriovenous tumor. *Acta Dermatovenereol* 1977;157:155–158.
49. Connelly MG, Winkelmann RK. Acral arteriovenous tumor: a clinicopathologic review. *Am J Surg Pathol* 1985;9:15–21.
50. Light RA. Venous hemangioma of skeletal muscle: case report. *Ann Surg* 1943;118:465–468.
51. Rosen PP, Jezefzyk MA, Boram LH. Vascular tumors of the breast, IV. The venous hemangioma. *Am J Surg Pathol* 1985;9:659–665.
52. Weiss SW. Pedal hemangioma (venous malformation) occurring in Turner's syndrome: an additional manifestation of the syndrome. *Hum Pathol* 1988;19:1015–1018.

Intramuscular hemangioma and angiomatosis

53. Allen PW, Enzinger FM. Hemangiomas of skeletal muscle: an analysis of 89 cases. *Cancer* 1972;29:8–23.
54. Beham A, Fletcher CDM. Intramuscular angioma: a clinicopathological analysis of 74 cases. *Histopathology* 1991;18:53–59.
55. Howat AJ, Campbell PE. Angiomatosis: a vascular malformation of infancy and childhood. *Pathology* 1987;19:377–382.
56. Rao VK, Weiss SW. Angiomatosis of soft tissue: an analysis of the histological features and clinical outcome in 51 cases. *Am J Surg Pathol* 1992;16:764–771.

Glomeruloid hemangioma

57. Soubrier MJ, Dubost JJ, Sauvezie, BJ. POEMS syndrome: a study of 25 cases and a review of the literature. French study Group on POEMS syndrome. *Am J Med* 1994;97:543–553.
58. Chan JK, Fletcher CD, Hicklin GA, Rosai J. Glomeruloid hemangioma: a distinctive cutaneous lesion of multicentric Castleman's disease associated with POEMS syndrome. *Am J Surg Pathol* 1990;14:1036–1046.
59. Kanitakis J, Roger H. Soubrier M. Cutaneous angiomas in POEMS syndrome: an ultrastructural and immunohistochemical study. *Arch Dermatol* 1988;124:695–698.
60. Belec L, Mohamed AS, Authier FJ, et al. Human herpesvirus 8 infection in patients with POEMS syndrome-associated multicentric Castleman's disease. *Blood* 1999;93:3643–3653.
61. Forman SB, Tyler WB, Ferringer TC, Elston DM. Glomeruloid hemangioma without POEMS syndome: series of three cases. *J Cutan Pathol* 2007;34:956–957.
62. Suurmeijer AJ, Fletcher CD. Papillary hemangioma: a distinctive cutaneous haemangioma of the head and neck area containing eosinophilic hyaline globules. *Histopathology* 2007;51:638–648.

Hobnail hemangioma (targetoid hemosiderotic hemangioma)

63. Santa-Cruz DJ, Aronberg J. Targetoid hemosiderotic hemangioma. *J Am Acad Dermatol* 1988;19:550–558.
64. Guillou L, Calonje E, Speight P, Rosai J, Fletcher CD. Hobnail hemangioma: a pseudomalignant vascular lesion with a reappraisal of targetoid hemosiderotic hemangioma. *Am J Surg Pathol* 1999;23:97–105.
65. Mentzel T, Partanen T, Kutzner H. Hobnail hemangioma ("targetoid

hemosiderotic hemangioma"): clinicopathologic and immunohistochemical analysis of 62 cases. *J Cutan Pathol* 1999;26:279–286.

66. Franke FE, Steger K, Marks A, Kutzner H, Mentzel T. Hobnail hemangiomas (targetoid hemosiderotic hemangiomas) are true lymphangiomas. *J Cutan Pathol* 2004;31:362–367.

Acquired tufted angioma (angioblastoma of Nakagawa)

67. Wilson-Jones E, Orkin M. Tufted angioma (angioblastoma): a benign progressive angioma not to be confused with Kaposi's sarcoma or low-grade angiosarcoma. *J Am Acad Dermatol* 1989;20:214–225.

68. Alessi E, Bertani E, Sala F. Acquired tufted angioma. *Am J Dermatopathol* 1986;8:426–429.

69. Padilla RS, Orkin M, Rosai J. Acquired "tufted" angioma (progressive capillary hemangioma): a distinctive clinicopathologic entity related to lobular capillary hemangioma. *Am J Dermatopathol* 1987;9:292–300.

70. Cho KH, Kim SH, Park KC, et al. Angioblastoma (Nakagawa): is it the same as tufted angioma? *Clin Exp Dermatol* 1991;16:110–113.

71. Seo SK, Suh JC, Na GY, Kim IS, Sohn KR. Kasabach–Merritt syndrome: identification of platelet trapping in a tufted angioma by immunohistochemistry technique using monoclonal antibody to CD61. *Pediatr Dermatol* 1999;16:392–394.

Kaposiform hemangioendothelioma

72. Niedt GW, Greco MA, Wieczorek R, Blanc WA, Knowles DM. Hemangioma with Kaposi's sarcoma-like features: report of 2 cases. *Pediatr Pathol* 1989;9:567–575.

73. Tsang WYW, Chan JKC. Kaposi-like infantile hemangioendothelioma: a distinctive vascular neoplasm of the retroperitoneum. *Am J Surg Pathol* 1991;15:982–989.

74. Zukerberg LR, Nickoloff BJ, Weiss SW. Kaposiform hemangioendothelioma of infancy and childhood: an aggressive neoplasm associated with Kasabach-Merritt syndrome and lymphangiomatosis. *Am J Surg Pathol* 1993;17:321–328.

75. Lyons LL, North PE, Mac-Moune LF, et al. Kaposiform hemangioendothelioma: a study of 33 cases emphasizing its pathologic, immunophenotypic, and biologic uniqueness from juvenile hemangioma. *Am J Surg Pathol* 2004;28:559–568.

76. Mentzel T, Mazzoleni G, Deios A, Fletcher CD. Kaposi-form hemangioendothelioma in adults: clinicopathologic and immunohistochemical analysis of three cases. *Am J Clin Pathol* 1997;108:450–455.

77. Deraedt K, Vander Poorten V, van Geet C, et al. Multifocal Kaposiform hemangioendothelioma. *Virchows Arch* 2006;448:843–846.

78. Sarkar M, Mulliken JB, Kozakewich HP, Robertson RL, Burrows PE. Thrombocytopenic coagulopathy (Kasabach-Merritt phenomenon) is associated with Kaposiform hemangioendothelioma and not with common infantile hemangioma. *Plast Reconstr Surg* 1997;100:1377–1386.

79. Enroljas O, Wassef M, Mazoyer E, et al. Infants with Kasabach-Merritt syndrome do not have "true" hemangiomas. *J Pediatr* 1997;30:631–640.

80. Nakaya T, Morita K, Kurata A, et al. Multifocal kaposiform hemangioendothelioma in multiple visceral organs: an autopsy of 9-day-old female baby. *Hum Pathol* 2014;45:1773–1777.

81. Debelenko LV, Perez-Atayde AR, Mulliken JB, et al. D2-40 immunohistochemical analysis of pediatric vascular tumors reveals positivity in Kaposiform hemangioendothelioma. *Mod Pathol* 2005;18:1454–1460.

82. Le Huu AR, Jokinen CH, Rubin BP, et al. Expression of prox1, lymphatic endothelial nuclear transcription factor, in Kaposiform hemangioendothelioma and tufted angioma. *Am J Surg Pathol* 2010;34:1563–1573.

Epithelioid hemangioma

83. Allen PW, Ramakrishna B, MacCormac LB. The histiocytoid hemangiomas and other controversies. *Pathol Annu* 1992;27(Part 1):51–87.

84. Tsang WYW, Chan JKC. The family of epithelioid vascular tumors. *Histol Histopathol* 1993;8:187–212.

85. Rosai J. Angiolymphoid hyperplasia with eosinophilia of the skin: its nosological position in the spectrum of histiocytoid hemangioma. *Am J Dermatopathol* 1992;4:175–184.

86. Castro C, Winkelmann RK. Angiolymphoid hyperplasia with eosinophilia in the skin. *Cancer* 1974;34:1696–1705.

87. Olsen TG, Helwig EB. Angiolymphoid hyperplasia with eosinophilia: a clinicopathologic study of 116 patients. *J Am Acad Dermatol* 1985;12:781–796.

88. Fetsch JF, Weiss SW. Observations concerning the pathogenesis of epithelioid hemangioma (angiolymphoid hyperplasia). *Mod Pathol* 1991;4:449–455.

89. O'Connell JX, Kattapuram SV, Mankin HJ, Bhan AK, Rosenberg AE. Epithelioid hemangioma of the bone: a tumor often mistaken for low-grade angiosarcoma or malignant hemangioendothelioma. *Am J Surg Pathol* 1993;17:610–617.

90. Fetsch JF, Sesterhenn IA, Miettinen M, Davis CJ. Epithelioid hemangioma of the penis: a clinicopathologic and immunohistochemical analysis of 19 cases, with special reference to exuberant examples often confused with epithelioid hemangioendothelioma and epithelioid angiosarcoma. *Am J Surg Pathol* 2004;28:523–533.

91. Moran CA, Suster S. Angiolymphoid hyperplasia with eosinophilia (epithelioid hemangioma) of the lung: a clinicopathologic and immunohistochemical study. *Am J Clin Pathol* 2005;123:762–765.

92. Huang SC, Zhang L, Sung YS, et al. Frequent FOS gene rearrangements in epithelioid hemangioma: a molecular study of 58 cases with morphologic reappraisal. *Am J Surg Pathol* 2015;39:1313–1321.

93. Kung ITM, Gibson JB, Bannatyne PM. Kimura's disease: a clinicopathologic study of 21 cases and its distinction from angiolymphoid hyperplasia with eosinophilia. *Pathology* 1984;16:39–44.

94. Urabe A, Tsuneyoshi M, Enjoji M. Epithelioid hemangioma versus Kimura's disease: a comparative clinicopathologic study. *Am J Surg Pathol* 1987;11:758–766.

95. Motoi M, Wahid S, Horie Y, Akagi T. Kimura's disease: clinical, histological and immunohistochemical studies. *Acta Med Okayama* 1992;46:449–455.

96. Luthringer DJ, Virmani R, Weiss SW, Rosai J. A distinctive cardiovascular lesion resembling histiocytoid (epithelioid) hemangioma: evidence suggesting mesothelial participation. *Am J Surg Pathol* 1990;14:993–1000.

Spindle cell hemangioma (hemangioendothelioma)

97. Weiss SW, Enzinger FM. Spindle cell hemangioendothelioma: a low-grade angiosarcoma resembling a cavernous hemangioma and Kaposi's sarcoma. *Am J Surg Pathol* 1986;10:521–530.

98. Perkins P, Weiss SW. Spindle cell hemangioendothelioma: an analysis of 78 cases with reassessment of its pathogenesis and biologic behavior. *Am J Surg Pathol* 1996;20:1196–1204.

99. Scott GA, Rosai J. Spindle cell hemangioendothelioma: report of seven additional cases of a recently described vascular neoplasm. *Am J Dermatopathol* 1988;10:281–288.

100. Ding J, Hashimoto H, Imayama S, Tsuneyoshi M, Enjoji M. Spindle cell haemangioendothelioma: probably a benign vascular lesion not a low-grade angiosarcoma. A clinicopathological, ultrastructural and immunohistochemical study. *Virchows Arch A Pathol Anat Histopathol* 1992;420:77–85.

101. Fukunaga M, Ushigome S, Nikaido T, Ishikawa E, Nakamori K. Spindle cell hemangioendothelioma: an immunohistochemical and flow cytometric study of six cases. *Pathol Int* 1995;45:589–595.

102. Fanburg JC, Meis-Kindblom JM, Rosenberg AE. Multiple enchondromas associated with spindle cell hemangioendotheliomas: an overlooked variant of Maffucci's syndrome. *Am J Surg Pathol* 1995;19:1029–1038.

103. Lewis RJ, Ketcham AS. Mafucci's syndrome: functional and neoplastic significance. *J Bone Joint Surg* 1973;55A:1465–1479.

104. Pansuriya TC, van Eijk R, d'Adamo P, et al. Somatic mosaic IDH1 and IDH2 mutations are associated with enchondroma and spindle cell hemangioma in Ollier disease and Maffucci syndrome. *Nat Genet* 2011;43(12):1256–1261.

105. Kurek KC, Pansuriya TC, van Ruler MA, et al. R132C IDH1 mutations are found in spindle cell hemangiomas and not in other vascular tumors or malformations. *Am J Pathol* 2013;182:1494–1500.

Other hemangiomas and telangiectasiae

106. Hunt SJ, Santa Cruz DJ, Barr RJ. Microvenular hemangioma. *J Cutan Pathol* 1991;18:235–240.

107. Aloi F, Tomasini C, Pippione M. Microvenular hemangioma. *Am J Dermatopathol* 1993;15:534–538.

108. Requena L, Kutzner H, Mentzel T. Acquired elastotic hemangioma: a clinicopathologic variant of hemangioma. *J Am Acad Dermatol* 2002;47:371–376.

109. Baselga E. Sturge-Weber syndrome. *Semin Cutan Med Surg* 2004;23:87–98.

110. Di Rocco C, Tamburrini G. Sturge–Weber syndrome. *Childs Nerv Syst* 2006;22:909–921.

111. Timur AA, Driscoll DJ, Wang Q. Biomedicine and diseases: the Klippel–Trenaunay syndrome, vascular anomalies and vascular morphogenesis. *Cell Mol Life Sci* 2005;62:1434–1447.

112. Rosen S, Smoller BR. Port-wine stains: a new hypothesis. *J Am Acad Dermatol* 1987;17:164–166.

113. Bean WB, Walsh JR. Venous lakes. *AMA Arch Derm* 1956;74:459–463.

Papillary endothelial hyperplasia

114. Kuo TT, Salyers CP, Rosai J. Masson's "vegetant intravascular hemangioendothelioma": a lesion often mistaken for angiosarcoma. A study of seventeen cases located in the skin and soft tissues. *Cancer* 1976;38:1227–1236.

115. Clearkin KP, Enzinger FM. Intravascular papillary endothelial hyperplasia. *Arch Pathol Lab Med* 1976;100:441–444.

116. Hashimoto H, Daimaru Y, Enjoji M. Intravascular papillary endothelial hyperplasia: a clinicopathologic study of 91 cases. *Am J Dermatopathol* 1983;5:539–546.

117. Amerigo J, Berry CL. Intravascular papillary endothelial hyperplasia in the skin and subcutaneous tissue. *Virchows Arch A Pathol Anat Histopathol* 1980;387;81–90.

Bacillary angiomatosis and verruca peruana

118. Stoler MH, Bonfiglio TA, Steigbigel RT, Pereira M. An atypical subcutaneous infection associated with acquired immunodeficiency syndrome. *Am J Clin Pathol* 1983;80:714–718.

119. LeBoit PE, Berger TG, Egbert BM, et al. Bacillary angiomatosis: the histopathology and differential diagnosis of a pseudoneoplastic infection in patients with human immunodeficiency virus disease. *Am J Surg Pathol* 1989;13:909–920.

120. Schinella RA, Greco MA. Bacillary angiomatosis presenting as a soft-tissue tumor without skin involvement. *Hum Pathol* 1990;21:567–569.

121. Cockerell CJ, Tierno PM, Friedman-Kien AE, Kim KS. Clinical, histologic, microbiologic, and biochemical characterization of the causative agent of bacillary (epithelioid) angiomatosis: a rickettsial illness with features of bartonellosis. *J Invest Dermatol* 1991;97:812–817.

122. Tsang WY, Chan JK. Bacillary angiomatosis: a "new" disease with a broadening clinicopathologic spectrum. *Histol Histopathol* 1992;7:143–152.

123. Koehler JE, Sanchez MA, Garrido CS, et al. Molecular epidemiology of bartonella infections in patients with bacillary angiomatosis-peliosis. *N Engl J Med* 1997;337:1876–1883.

124. Reed JA, Brigati DJ, Flynn SD, et al. Immunocytochemical identification of Rochalimaea henselae in bacillary (epithelioid) angiomatosis, parenchymal bacillary peliosis, and persistent fever with bacteremia. *Am J Surg Pathol* 1992;16:650–657.

125. Arias-Stella J, Lieberman PH, Erlandson RA, Arias-Stella J Jr. Histology, immunohistochemistry, and ultrastructure of the verruga in Carrion's disease. *Am J Surg Pathol* 1986;10:595–610.

126. Arias-Stella J, Lieberman PH, Garcia-Caceres U, et al. Verruga peruana mimicking malignant neoplasms. *Am J Dermatopathol* 1987;9:279–291.

127. Garcia FU, Wojta J, Broadley KN, Davidson JM, Hoover RL. Bartonella bacilliformis stimulates endothelial cells in vitro and is angiogenic in vivo. *Am J Pathol* 1990;136:1125–1135.

Other reactive vascular proliferations

128. Rongioletti F, Rebora A. Cutaneous reactive angiomatosis: patterns and classification of reactive vascular proliferation. *J Am Acad Dermatol* 2003;49:887–896.

129. McMenamin ME, Fletcher CD. Reactive angioendotheliomatosis: a study of

15 cases demonstrating a wide clinicopathologic spectrum. *Am J Surg Pathol* 2002;26:685–697.

130. Heller M, Karen JK, Fangman W. Acroangiodermatitis. *Dermatol Online J* 2007;13:2.

131. Bavikatty NR, Goldblum JR, Abdul-Karim FW, Nielsen SL, Greenson JK. Florid vascular proliferation of the colon related to intussusception and mucosal prolapse: potential diagnostic confusion with angiosarcoma. *Mod Pathol* 2001;14:1114–1118.

132. Vadlamudi G, Schinella R. Traumatic pseudoaneurysm: a possibly early lesion in the spectrum of epithelioid hemangioma/angiolymphoid hyperplasia with eosinophilia. *Am J Dermatopathol* 1998;20:113–117.

133. Burke AP, Jarvelainen H, Kolodgie FD, et al. Superficial pseudoaneurysms: clinicopathologic aspects and involvement of extracellular matrix proteoglycans. *Mod Pathol* 2004;17:482–488.

Lymphangioma and related tumors

134. Radhakrishnan K, Rockson SG. The clinical spectrum of lymphatic disease. *Ann NY Acad Sci* 2008;1131:155–184.

135. Flanagan BP, Helwig EB. Cutaneous lymphangioma. *Arch Dermatol* 1977;113:24–30.

136. Allen JG, Riall TS, Cameron JL, et al. Abdominal lymphangiomas in adults. *J Gastrointest Surg* 2006;10:746–751.

137. Byrne J, Blanc WA, Warburton D, Wigger J. The significance of cystic hygroma in fetuses. *Hum Pathol* 1984;15:61–67.

138. Kim KM, Choi KY, Lee A, Kim BK. Lymphangioma of large intestine: report of ten cases with endoscopic and pathologic correlation. *Gastrointest Endosc* 2000;52:255–259.

139. Partanen TA, Alitalo K, Miettinen M. Lack of lymphatic vascular specificity of vascular endothelial growth factor receptor 3 in 185 vascular tumors. *Cancer* 1999;86:2406–2412.

140. Kahn HJ, Bailey D, Marks A. Monoclonal antibody D2-40, a new marker of lymphatic endothelium, reacts with Kaposi sarcoma and a subset of angiosarcomas. *Mod Pathol* 2002;15:434–440.

141. Watanabe M, Kishiyama K Ohkawara A. Acquired progressive lymphangioma. *J Am Acad Dermatol* 1983;8:663–667.

142. Wilson Jones E, Winkelmann RK, Zachary CB, Reda AM. Benign lymphangioendothelioma. *J Am Acad Dermatol* 1990;23:229–235.

143. Tadaki T, Aiba S, Masu S, et al. Acquired progressive lymphangioma as a flat erythematous patch on the abdominal wall of a child. *Arch Dermatol* 1988;124:699–701.

144. Ramani P, Shah A. Lymphangiomatosis: histologic and immunohistochemical analysis of four cases. *Am J Surg Pathol* 1993;17:329–335.

145. North PE, Kahn T, Kordisco MR, et al. Multifocal lymphangioendotheliomatosis with thrombocytopenia: a newly recognized clinicopathological entity. *Arch Dermatol* 2004;140:599–606.

146. Guillou L, Fletcher CDM. Benign lymphangioendothelioma (acquired progressive lymphangioma): a lesion not to be confused with well-differentiated angiosarcoma and patch stage Kaposi's sarcoma. *Am J Surg Pathol* 2000;24:1047–1057.

147. Ramani P, Shah A. Lymphangiomatosis: histologic and immunohistochemical analysis of four cases. *Am J Surg Pathol* 1993;17:329–335.

148. Gomez CS, Calonje E, Ferrar DW, Browse NL, Fletcher CDM. Lymphangiomatosis of the limbs: clinicopathologic analysis of a series with a good prognosis. *Am J Surg Pathol* 1995;19:125–133.

Hemangioma-lymphangioma syndromes and their genetics

149. Wang QK. Update on the molecular genetics of vascular anomalies. *Lymphat Res Biol* 2005;3:226–233.

150. Brouillard P, Vikkula M. Genetic causes of vascular malformations. *Hum Mol Genet* 2007;16:R140–R149.

151. Fernandes S, Silva A, Coelho A, Campos M, Pontes F. Blue rubber bleb nevus: case report and literature review. *Eur J Gastroenterol Hepatol* 1999;11:455–457.

152. You CK, Rees J, Gillis DA, Steeves J. Klippel-Trenaunay syndrome: a review. *Can J Surg* 1983;26:399–403.

153. Paller A. The Sturge-Weber syndrome. *Pediatr Dermatol* 1987;4:300–304.

Chapter 22
Hemangioendotheliomas, angiosarcomas, and Kaposi's sarcoma

Markku Miettinen
National Cancer Institute, National Institutes of Health

This chapter includes borderline to low-grade malignant vascular tumors (hemangioendotheliomas) and malignant vascular tumors (angiosarcoma and Kaposi's sarcoma).

Hemangioendothelioma is the designation for a group of borderline or outright malignant vascular tumors. They are capable of aggressive local growth, recurrence, or both, and some, especially epithelioid hemangioendothelioma, can also metastasize.

Angiosarcomas (ASs) are malignant endothelial neoplasms. They are very rare compared with hemangiomas and comprise <1% of all vascular tumors and only 1% to 2% of all sarcomas. ASs are essentially composed of malignant endothelial cells representing endothelial cell neoplasms. They include rare low-grade variants but most are the high-grade variety. The features that aid in recognition of an AS are disorganized and anastomosing vascular channels, permeative growth versus lobulation of the vascular units, and multilayered, atypical endothelia, often with solid areas, and a general lack of pericytes of the vessel wall. Lymphangiosarcomas are not recognized separately. ASs are frequently multifocal, making it sometimes difficult to determine the primary site. Because of this multifocal tendency, ASs are excluded from sarcoma staging systems.

Radiation and environmental carcinogens play a causal role, the latter especially in hepatic ASs. By their location, the endothelial cells are directly exposed to circulating carcinogens. The genetic mechanisms related to the pathogenesis of ASs are much less understood than those for many other sarcomas, however.

Kaposi's sarcoma is etiologically related to human herpesvirus 8 (HHV8), which is one of the strongest connections between a virus and human mesenchymal neoplasia. Several clinicopathologic forms are distinguished: classic, epidemic, endemic, and transplantation associated.

Papillary intralymphatic angioendothelioma (Dabska tumor)

Clinical features

Dabska described this tumor in children.[1] Similar cases reported subsequently seem to have a predilection for children and young adults.[2] The lesions occur in the skin and subcutis in soft tissues, in a wide variety of anatomic regions. Spleen involvement is also possible. Although the initial reports suggested metastatic potential (mainly lymph node metastases, but also distant metastases in one case),[1] the largest series to date from the Armed Forces Institute of Pathology (AFIP) found favorable outcomes in all cases.[3] Complete, preferably wide excision seems to be the optimal treatment.

Pathology

Grossly, the tumor tissue is pale tan to yellowish, and dilated vascular lumina can give the cut surface a focally cystic appearance (Figure 22.1). The size of the lesion generally varies from a small cutaneous nodule <1 cm to a subcutaneous mass of 5 cm.

Histologically, the tumor is composed of anastomosing, dilated vascular channels that often have papillary intravascular proliferations, with radially arranged endothelial cells surrounding a fibrous core in a "match-stick pattern" (Figure 22.2). Lymphangioma-like components with vascular lumina containing proteinaceous fluid and lymphocytes are seen in some tumors, suggesting a relationship with lymphangioma (Figure 22.3). Occurrence with a lymphangioma circumscriptum has also been reported.[4] Consistent positivity for vascular endothelial growth factor receptor 3 (VEGFR3) (Figure 22.4)

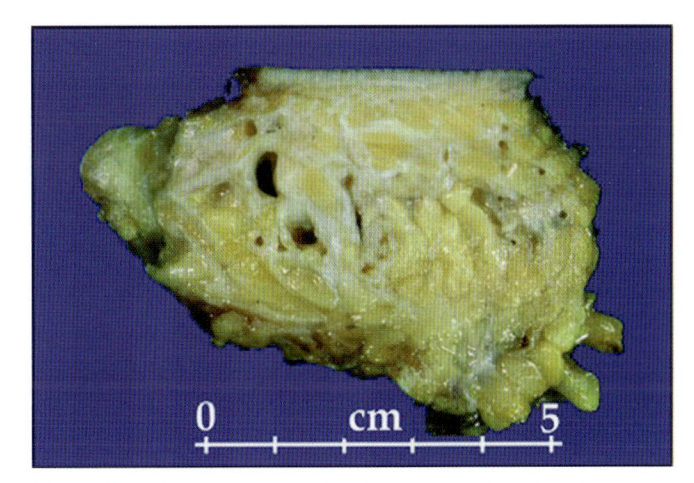

Figure 22.1 Grossly, papillary intralymphatic angioendothelioma (PILA) appears as multiple subcutaneous cystic spaces representing dilated (lymphatic) vascular spaces surrounded by whitish streaks of tumor tissue.

Modern Soft Tissue Pathology, Second Edition, ed. Markku Miettinen. Published by Cambridge University Press. © Cambridge University Press 2016.

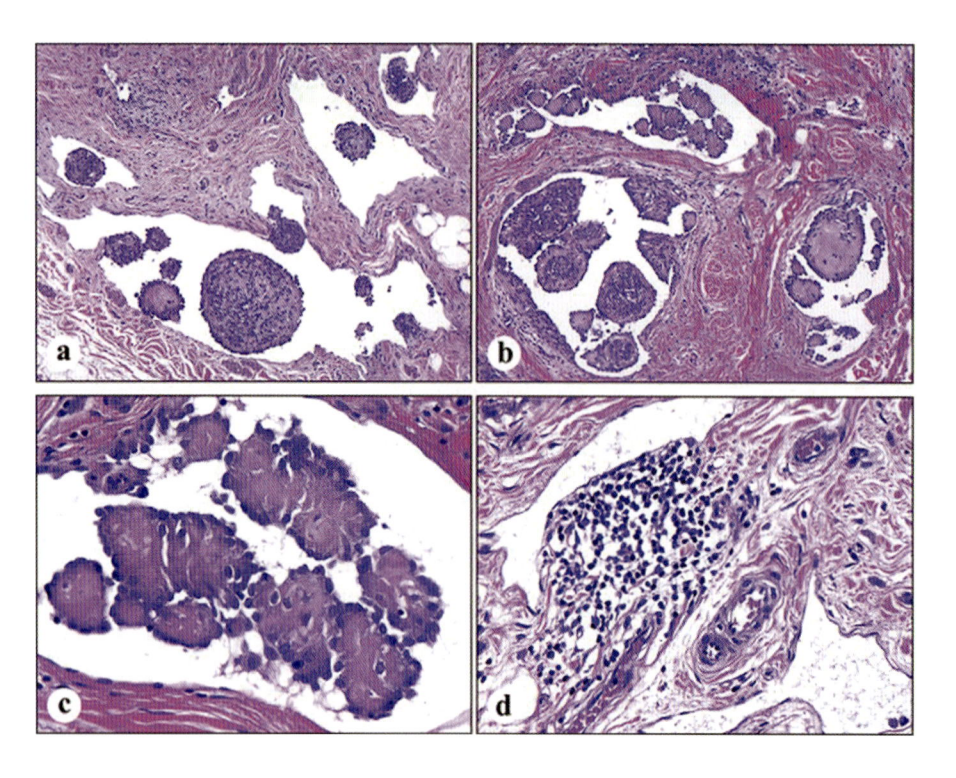

Figure 22.2 (**a,b**) Histologically typical of PILA are dilated lymphatic vascular lumina containing papillary endothelial proliferations. (**c**) The papillae have fibrous cores and are typically lined by cytologically uniform, slightly enlarged endothelial cells. (**d**) Focal lymphoid infiltration is often seen around lymphatic type vessels.

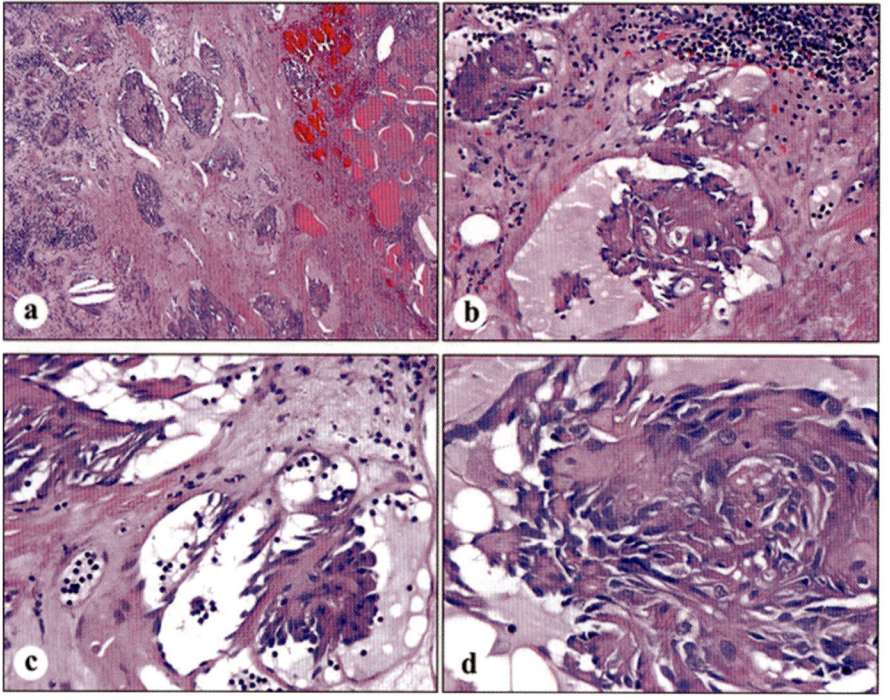

Figure 22.3 Another example of a PILA involving the spleen. (**a**) Low magnification shows components of PILA (left) together with lymphangioma component. (**b,c**) The dilated lymphatic vascular lumina contain papillary intravascular proliferation, proteinaceous fluid, and scattered lymphocytes. (**d**) The papillary projections are lined by endothelial cells with no significant atypia or mitotic activity.

also supports lymphatic vascular origin, although this receptor is not entirely specific for lymphatic vessels.[3]

Retiform hemangioendothelioma
Clinical features
This cutaneous tumor was originally classified as a low-grade AS,[5] but more recently it has been included among borderline malignant tumors.[6] It often occurs in peripheral superficial locations, such as the hands and feet, mainly in young adults of either sex. Local recurrence is common, but metastasis is unusual and seems limited to lymph nodes. In one report, multiple synchronous tumors were described in one patient.[7]

Pathology
The tumor is usually relatively small, measuring 1 cm to 2 cm. It is typically centered in the mid-dermis, but can extend to the subcutis. Histologically typical are gaping "retiform"

Chapter 22: Hemangioendotheliomas, angiosarcomas, and Kaposi's sarcoma

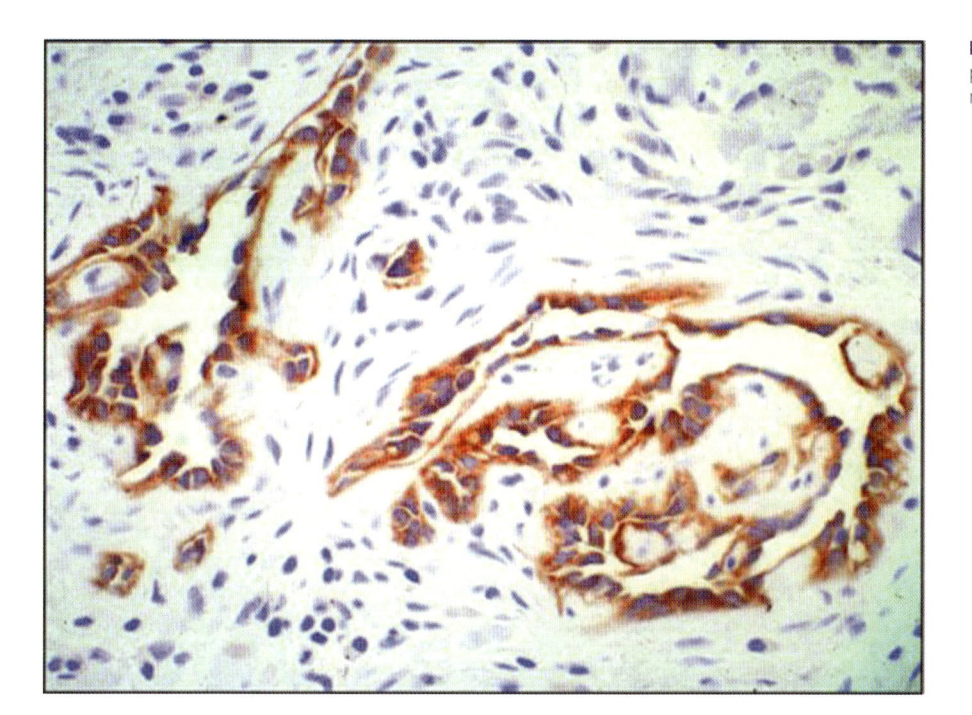

Figure 22.4 The endothelial cells in PILA are positive for vascular endothelial growth factor receptor 3 (VEGFR3)

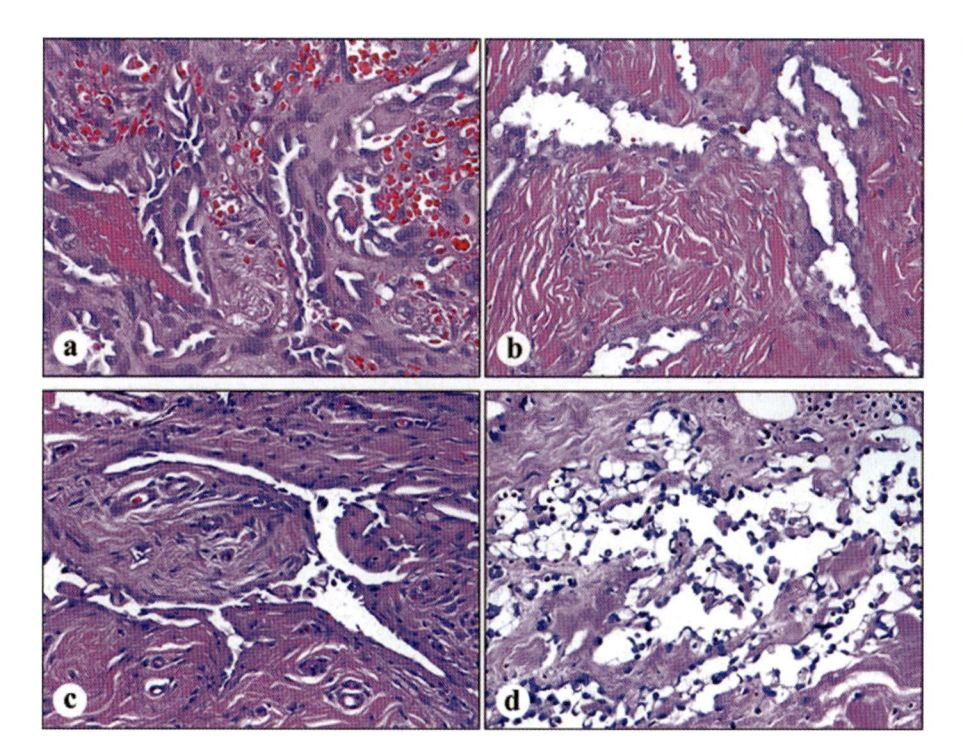

Figure 22.5 A retiform hemangioendothelioma forms multiple dilated vascular lumina that can anastomose. They are lined by endothelia protruding into the lumina, showing hobnail features.

vessels resembling rete testis structures. The endothelium is tall, with protruding cytoplasm referred to as "hobnail cells," somewhat similar, but more proliferative than those found in hobnail hemangioma (Figure 22.5). Focal solid areas are often present, but atypia and mitotic activity are limited. One study showed these hemangioendotheliomas almost consistently podoplanin and VEGFR3 negative.[8] In this respect, they seem to differ from papillary intralymphatic angioendotheliomas.

Epithelioid hemangioendothelioma
Clinical features

Epithelioid hemangioendothelioma (EHE) is a low-grade malignant vascular endothelial tumor, originally reported by Weiss and Enzinger in 1982.[9,10] It occurs in adults of all ages, and rarely in children, most commonly in the liver, soft tissue, lung, and bones.[9–14]

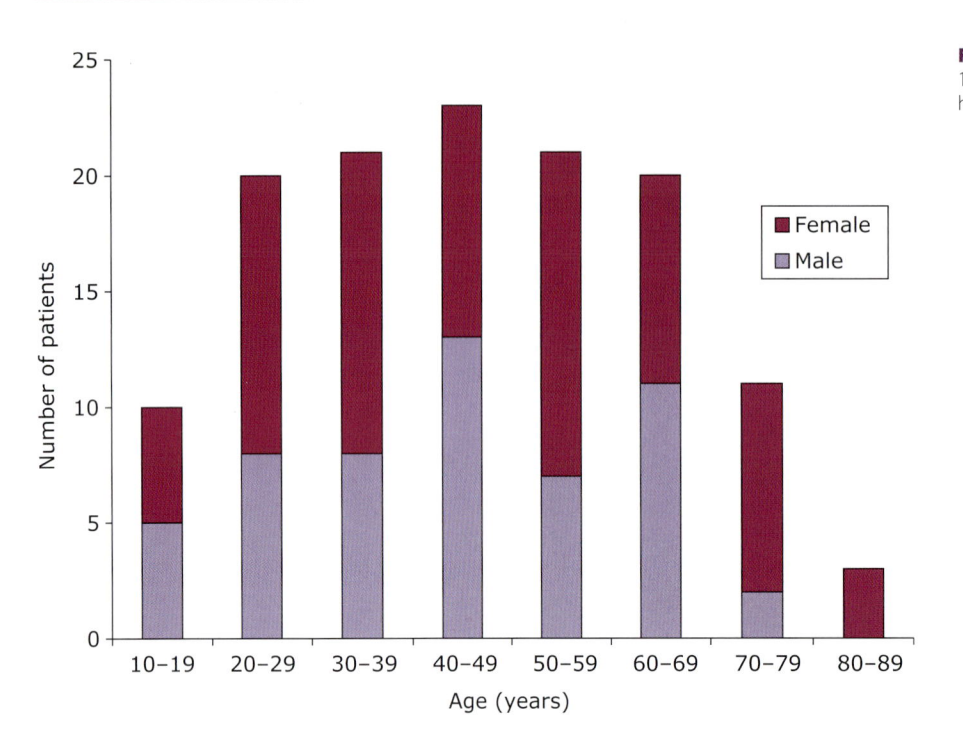

Figure 22.6 Age and sex distribution of 129 patients with epithelioid hemangioendothelioma of soft tissues.

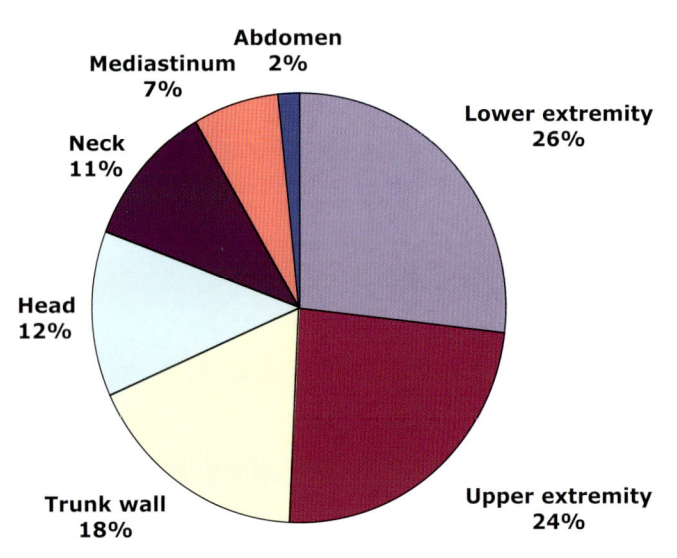

Figure 22.7 Anatomic distribution of 130 cases of epithelioid hemangioendothelioma of soft tissues.

In soft tissues, EHE occurs from the second decade on, with a moderate female predominance (Figure 22.6). These tumors most commonly present in the subcutis or deep soft tissues and less commonly in the skin in a variety of anatomic sites (Figure 22.7). The tumor has an unpredictable behavior, and 20% to 30% of patients with peripheral soft tissue tumors develop metastases in the liver, bones, or lungs, often long after the primary surgery. Regional lymph node metastases also occur.

Hepatic EHE, previously often confused with sclerosing cholangiocarcinoma, occurs in a wide age range from the second decade into old age, with a 3:2 female predominance. These patients usually have multiple tumors that frequently involve most of the liver, and <20% of patients have a solitary tumor. There is a 43% tumor-related mortality, but liver transplantation can be curative, although recurrence in the transplant does occur.[13,14] A large series documented 43% 5-year survival and 25% 10-year survival rates, with tumor-related mortality continuing over 25 years of follow-up.[14]

Pulmonary EHE (previously known as "intravascular bronchioloalveolar tumor") has a predilection for young adult females. One-half of the patients are <40 years of age, and 80% are women. This tumor typically forms multiple nodules, often bilaterally.[15,16] Some pulmonary EHEs have been proved to be pulmonary metastases of peripheral primary tumors.[17,18] Search for an occult soft tissue EHE is therefore warranted in pulmonary lesions, especially when they are multifocal. The same might be true for hepatic and other internal organ EHEs. Many patients with pulmonary EHEs survive for a long time despite multiple tumors. Pleural examples have been reported, and some of them clinically and grossly can simulate a mesothelioma.[19]

Pathology

Grossly, EHE varies, but is often pale gray, despite its angiomatous nature (Figure 22.8). Some EHEs are grossly distinctive, involving a vein filled by a hemorrhagic tumor thrombus. Tumor size varies from a small superficial nodule <1 cm to >10 cm. Median size in one series was 2.5 cm.[11] Hepatic examples typically are ill-defined, with infiltrative margins, and many involve much of the liver. Pulmonary EHEs form demarcated nodules of varying sizes, usually <1 cm to 2 cm.

Microscopically, EHE consists of small clusters, cords, or more solid nests of epithelioid to slightly spindled cells, with variably eosinophilic cytoplasm in a myxoid or collagenous matrix. The cellularity varies from paucicellular to focally highly cellular. Intraluminal growth in small- to medium-sized vessels is often observed (Figure 22.9). These vessels are usually veins, but small arteries can also be involved. The tumor cells often have intracytoplasmic vacuoles that are primitive vascular lumina; some of these can contain erythrocytes (Figure 22.10). In rare cases, a concomitant spindle cell component resembling spindle cell hemangioma suggests that some EHEs could originate from a spindle cell hemangioma.

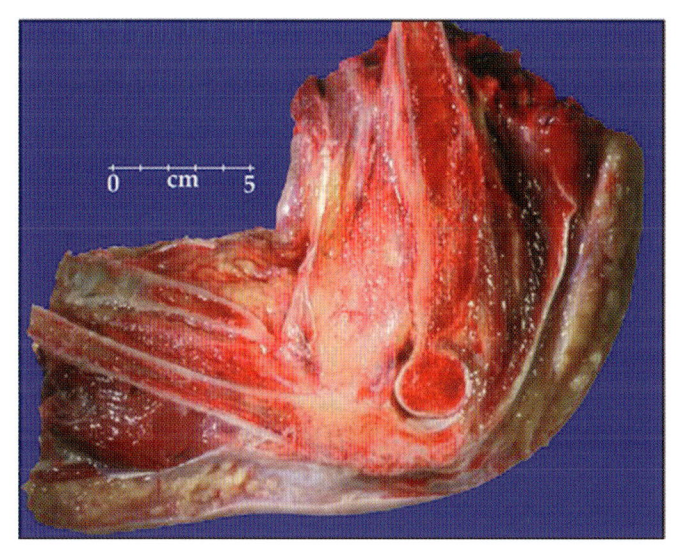

Figure 22.8 A large, deep, epithelioid hemangioendothelioma involving the antecubital fossa forms a solid, reddish-yellow fusiform mass.

In the liver, the tumor can grow diffusely between focally preserved trabeculae of parenchymal cells. Cytologically, the tumor cells have small- to medium-sized pale nuclei, mildly to moderately prominent nucleoli, and low mitotic activity (Figure 22.11). Areas of extensive necrosis or collagenous necrobiosis are often present. Pronounced atypia and brisk mitotic activity are not features of this lesion, and if they are present, epithelioid AS is usually a more appropriate diagnosis.

Pulmonary EHE typically fills alveolar spaces, often in a polypoid manner in the tumor's periphery, and involves vascular lumina; hence the previous designation "intravascular bronchioloalveolar tumor" (Figure 22.12). The tumor nodules also can contain extensive central necrosis. Overall, they often have a deciduoid appearance (Figure 22.13).

Histologic features that have been found to correlate with better outcome in soft tissue tumors were low cellularity and low proliferative rate, lack of striking atypia, and a low mitotic rate (2 per 10 HPFs or less), whereas tumor size was not.[11] A large follow-up study stratified soft tissue EHEs into low-risk tumors based on mitotic activity <3 per 50 HPFs and tumor size; patients alive after follow-up had a median tumor size of 1.3 cm, whereas those who died of the tumor had a median tumor size of 3.5 cm. In that study, the degree of atypia was not a significant factor.[20] Among hepatic tumors, those with low cellularity had a better outcome with a 17% mortality rate, whereas those with high cellularity had a mortality rate of 84%. Lack of necrosis was also a favorable sign.[14]

Immunohistochemically, the lesional cells show membrane staining for CD31, nuclear staining for ERG, and usually at least focal granular cytoplasmic positivity for von Willebrand

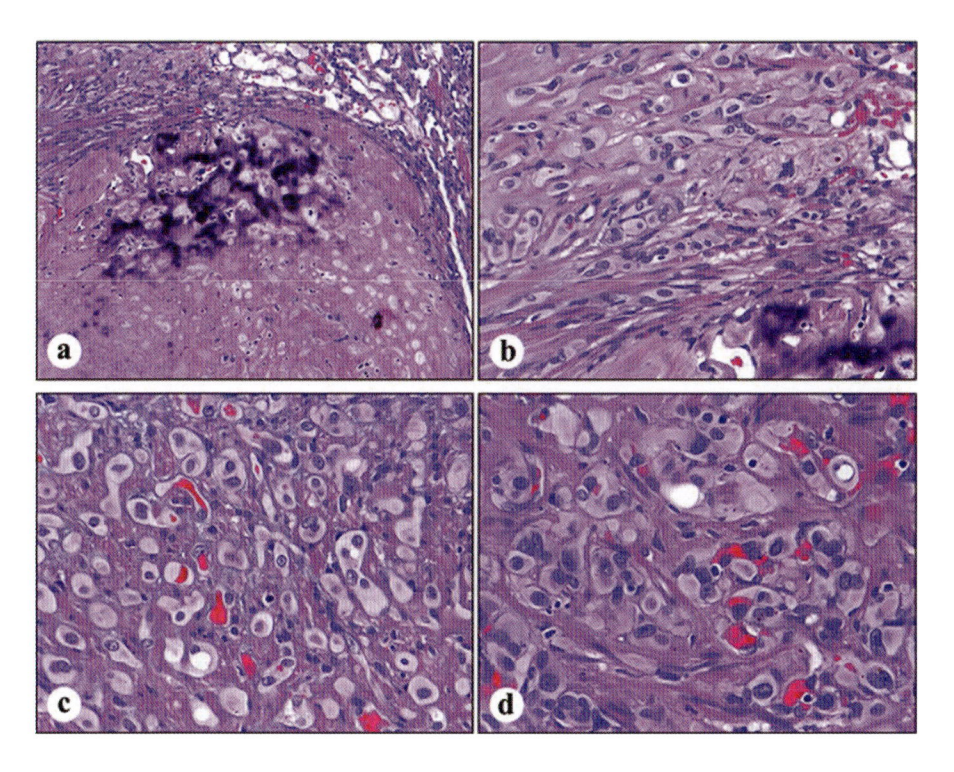

Figure 22.9 (**a**) An epithelioid hemangioendothelioma forms a sharply demarcated nodule because of intraluminal involvement of a vein. Focal calcification is present. (**b**) The tumor extends outside the vein lumen into soft tissues and is composed of cords of pale staining cells. (**c,d**) Some of the cellular cords often contain cytoplasmic vacuoles and erythrocytes.

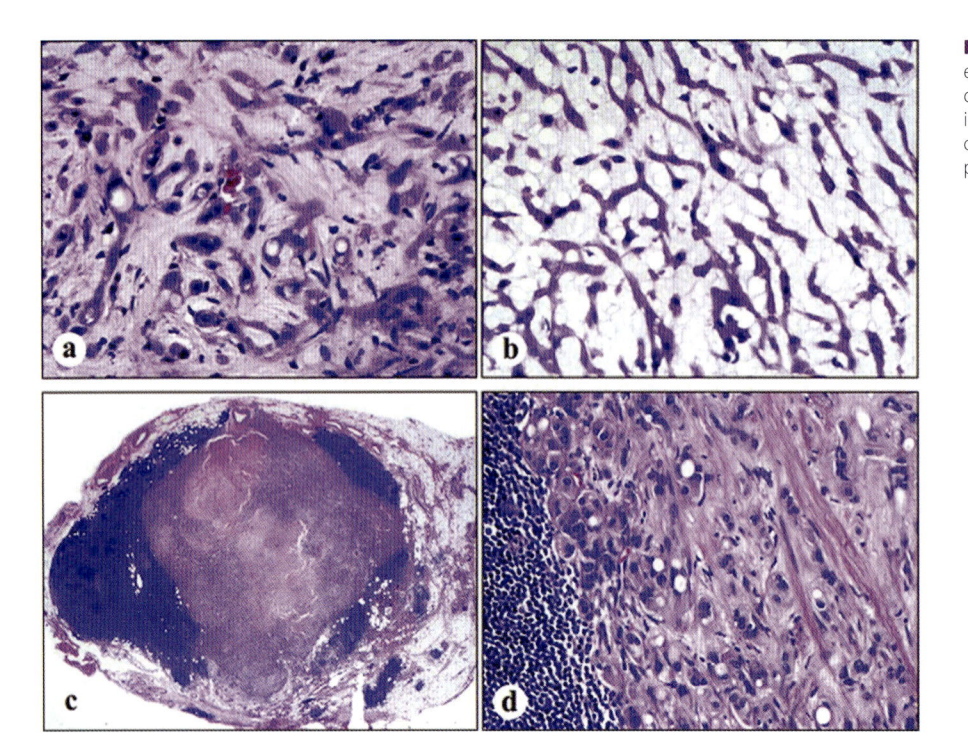

Figure 22.10 (**a,b**) The cellular cords of epithelioid hemangioendothelioma infiltrate the collagenous or myxoid matrix. (**c,d**) The tumor can involve regional lymph nodes, in which similar corded patterns and intracytoplasmic vacuoles are present.

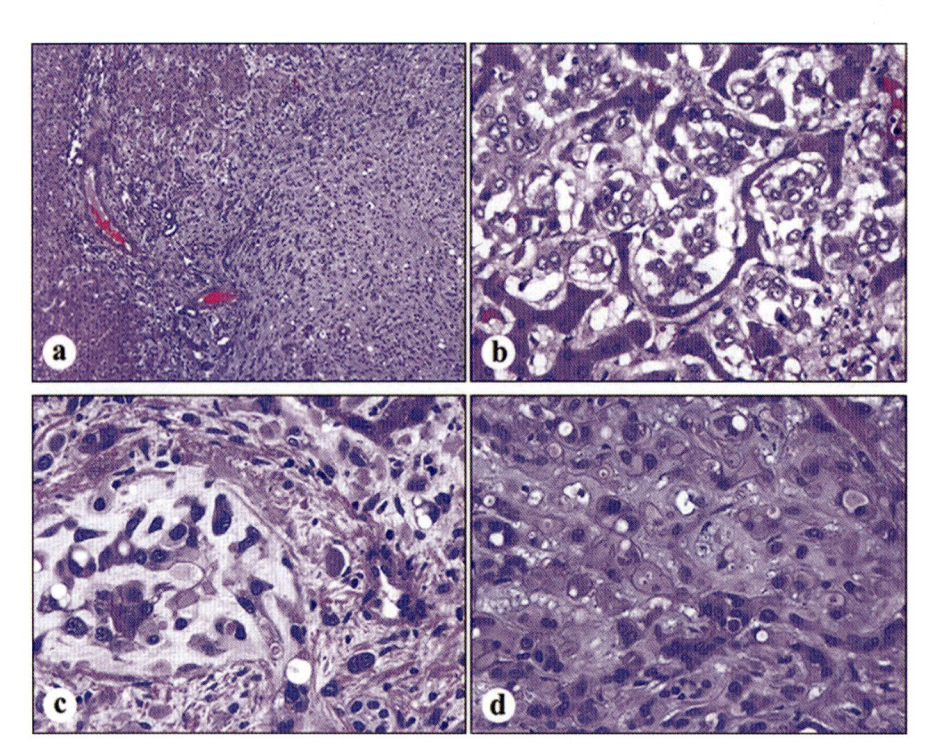

Figure 22.11 (**a**) Hepatic epithelioid hemangioendotheliomas can form distinctive nodules. The basophilic appearance comes from myxoid stroma. (**b**) Some examples demonstrate diffuse infiltration between the trabeculae of hepatocytes. (**c**) A sharply demarcated infiltrate is intravascular involvement. (**d**) The tumors have abundant eosinophilic cytoplasm and form cords showing focal cytoplasmic vacuolization, similar to soft tissue EHE.

factor, often in intracellular lumina. Approximately 50% of cases are positive for CD34. Keratins are commonly present (Figure 22.14). K18 seems to be consistently expressed in 100% of cases, K7 and K8 in 25% of cases, but K19 in none of the cases; antibodies to the K19 or AE1 monoclonal antibody are therefore more useful in the differential diagnosis of EHE and carcinoma. Infrequent AE1/AE3 keratin cocktail positivity probably reflects K7 positivity. Epithelial membrane antigen (EMA) is rarely expressed, usually by weak luminal staining in the intracytoplasmic vacuoles at the most.[21]

Genetics

WWTR-CAMTA1 gene fusion is a typical genetic event and a diagnostic marker.[22,23] This corresponds to the t(1;3)

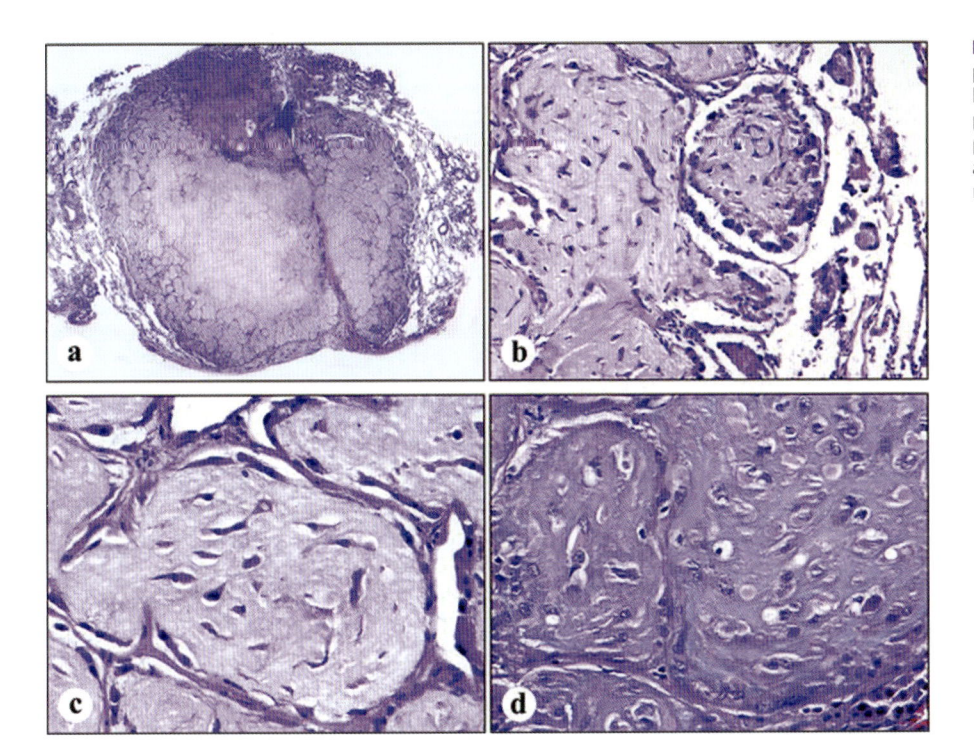

Figure 22.12 (**a**) An apparently demarcated pulmonary nodule of an epithelioid hemangioendothelioma. (**b**) The tumor forms polypoid protrusions into alveolar spaces in its periphery. (**c,d**) The tumor fills the alveolar lumina and has a variably myxoid or myxocollagenous matrix containing corded or scattered tumor cells.

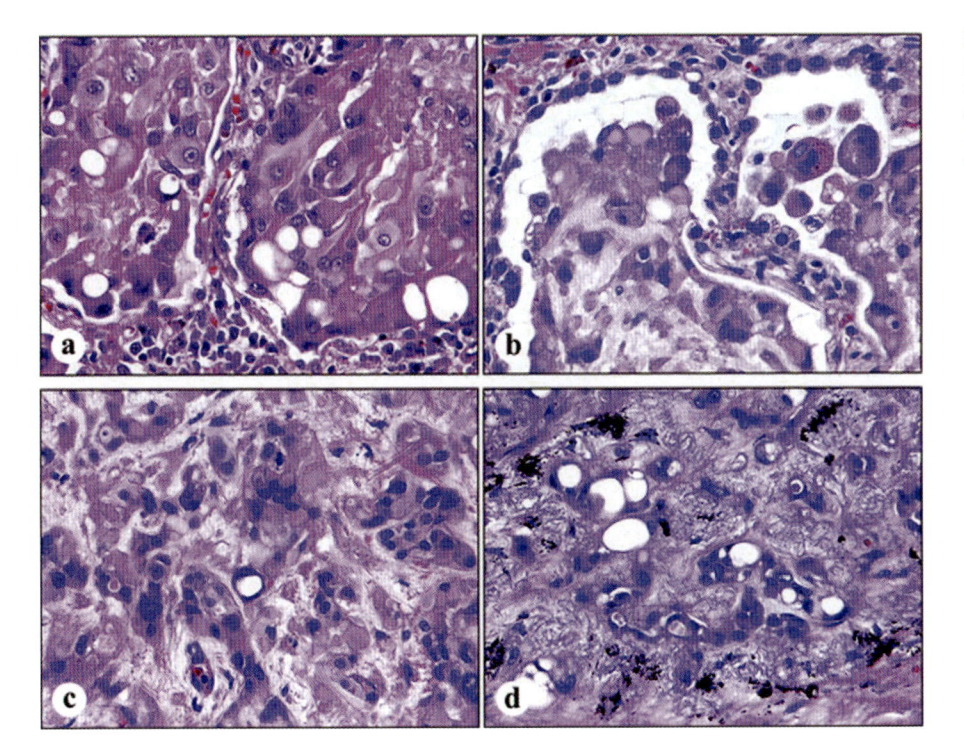

Figure 22.13 (**a,b**) The intra-alveolar papillary protrusions of epithelioid hemangioendothelioma can also consist of more solid sheets of epithelioid cells having eosinophilic cytoplasm with vacuoles in some cells. (**c,d**) The corded patterns of epithelioid cells with vacuoles, typical of EHE.

(p36;q25) translocation previously reported in EHE.[23] EHE has also been reported with YAP1-TFE3 gene fusion, (estimate: <10% of all cases) and these tumors are immunohistochemically TFE3 positive.[25,26] Such an EHE variant may be morphologically distinctive showing sheets of epithelioid cells with some tendency for vasoformation (Figure 22.15). However, TFE3 immunopositivity also occurs in other EHEs, so that genetic analysis is necessary for the diagnosis of TFE-translocated hemangioendothelioma.[26]

Epithelioid sarcoma-like (pseudomyogenic) hemangioendothelioma

This tumor was originally described as epithelioid sarcoma-like hemangioendothelioma and subsequently under the name

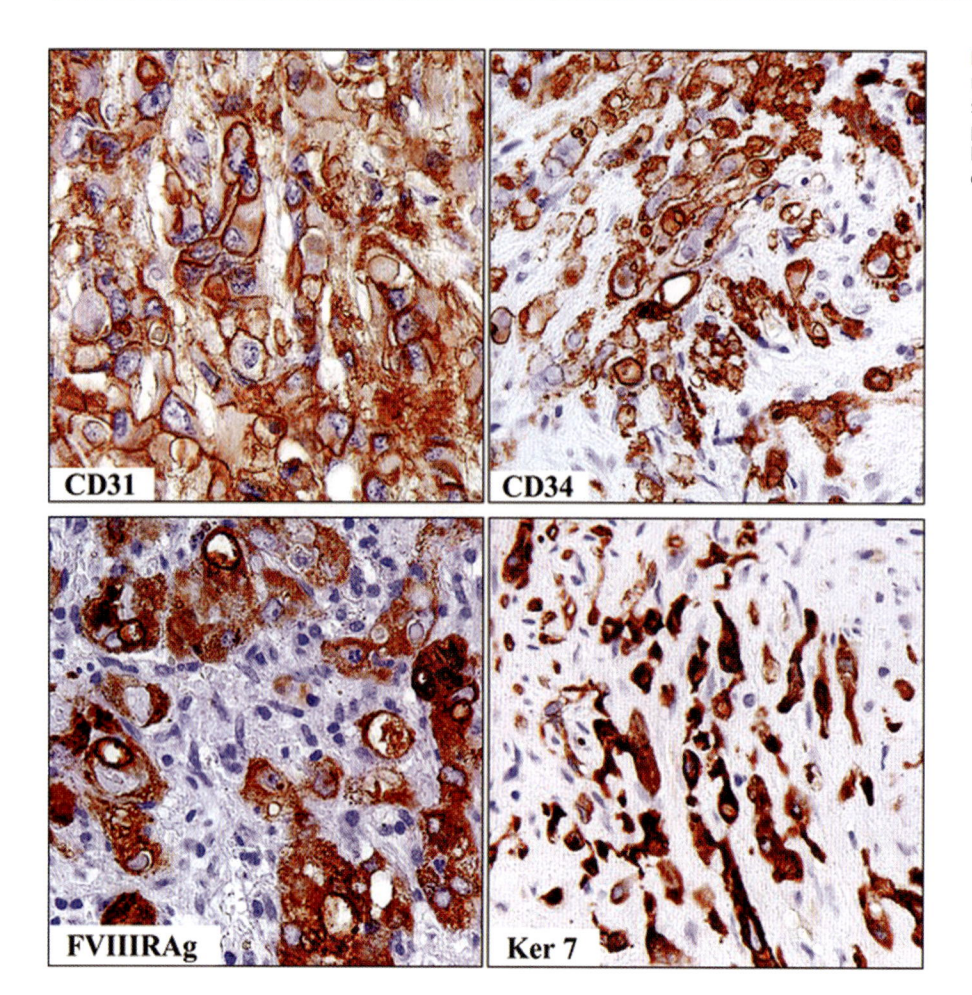

Figure 22.14 Immunohistochemically, the neoplastic epithelioid endothelial cells in EHE show membrane staining for CD31, are often positive for CD34, and usually positive for Factor-VIII-related antigen. Approximately 30% of cases are positive for keratin 7.

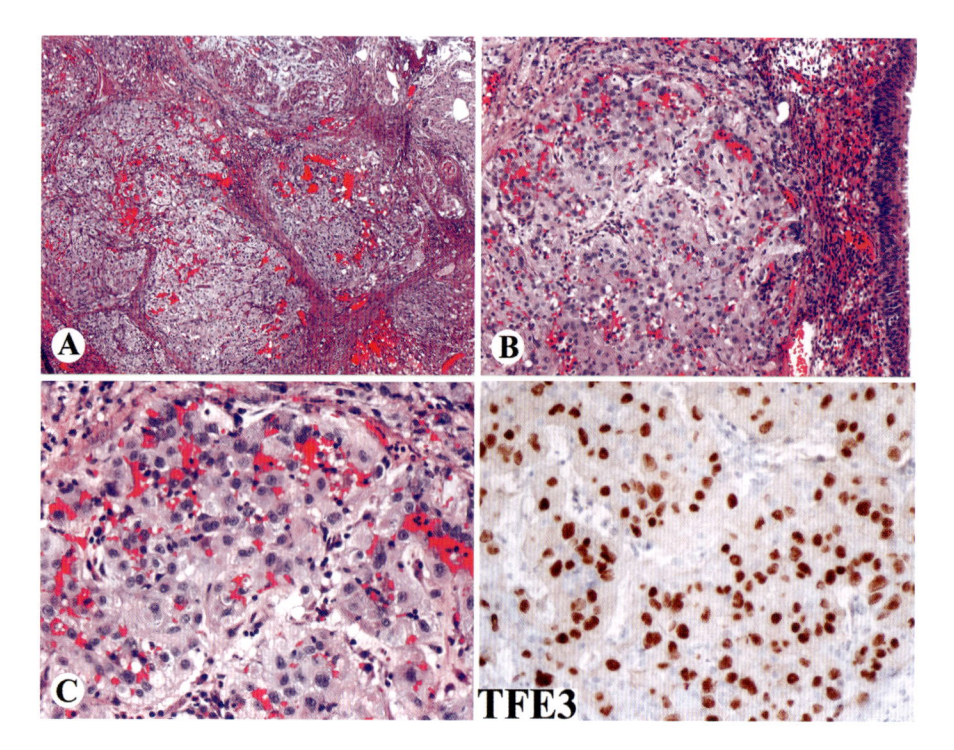

Figure 22.15 TFE-translocated variant of epithelioid hemangioendothelioma. (**A**) Tumor is multinodular. (**B,C**) It shows sheets of mildly atypical epithelioid cells with some tendency for vasoformation. Tumor cells are TFE3 positive. However, genetic studies are required to confirm translocation.

pseudomyogenic hemangioendothelioma. It shows spindled to epithelioid morphology without vasoformation and shares antigens of epithelia (keratins) and endothelia (CD31, ERG).

Clinical features

The tumor has predilection for young adults and occurs in superficial and deep soft tissues, sometimes multifocally. Bone involvement (metastases) can be seen as a part of multifocality. Despite multifocality, the tumor is believed to be generally indolent. However, long-term follow-up data are scant so that prognosis remains uncertain.[27–29]

Pathology

The tumor forms grayish nodules, which may contain necrosis. Microscopically it may be composed of irregular fascicles or solid sheets of epithelioid cells or spindled cells in a mildly collagenous matrix. The spindled cells contain eosinophilic cytoplasm and may resemble rhabdomyoblasts. Cytoplasmic vacuolization may be present, but is less prominent than usually seen in EHE. Nuclear atypia is mild and mitotic activity is low, usually <2 per 10 HPFs (Figure 22.16).

Immunohistochemically, most tumors are extensively positive for keratins (AE1/AE3). Although most cases are positive for ERG and FLI1, not all cases are CD31 positive. The tumor cells are typically negative for CD34, in contrast to epithelioid sarcoma. It also differs from epithelioid sarcoma in retaining INI1 expression (Figure 22.16). A balanced t(7;19)(q22;q13) translocation was reported in one case and a gene rearrangement involving this region in another case, although no gene fusions have been specified so far.[30]

Angiosarcoma

AS is the designation for a sarcoma that shows endothelial cell differentiation and vasoformation to various degrees. The diagnosis, in general, should be supported by immunohistochemical demonstration of endothelial-specific markers that allow separation from other malignant tumors. Several hemangioma variants that were historically sometimes included in AS are now recognized as benign entities (Chapter 21); therefore, the overall frequency of AS is lower than previously, and in our estimation, is now no more than 1% to 2% of all sarcomas. An epidemiologic study of >25 000 sarcomas reported a 4.1% frequency for AS, but reclassification was not included.[31] Clinical and gross features of clinicopathologic subsets (Table 22.1) are discussed separately, and their common histologic and immunohistochemical features are discussed thereafter. It has been noted that patient median age, gender ratio, and survival vary in different clinicopathologic subgroups of AS; the overall survival rate was 31%. Paclitaxel and doxorubicin were among the active chemotherapy agents used.[32] In one institutional series of 82 cases, there was a 56% male predominance and median patient age of 65 years (range, 22–91 years). The clinicopathologic subgroups included cutaneous (39%), deep soft tissue (27%), radiation- or lymphedema-related (12%), breast parenchymal (8.5%), bone (10%), and other (3.5%).[33] AS often metastasizes to the lungs,[34] and sometimes to brain, bone, and bone marrow, and abdominal cavity and pleura. Some ASs extensively involve serous cavities without distant spread. One large series reported a 43% 5-year overall survival rate.[35] Age/sex (Figure 22.17) and anatomic distributions of ASs are wide (Figure 22.18).

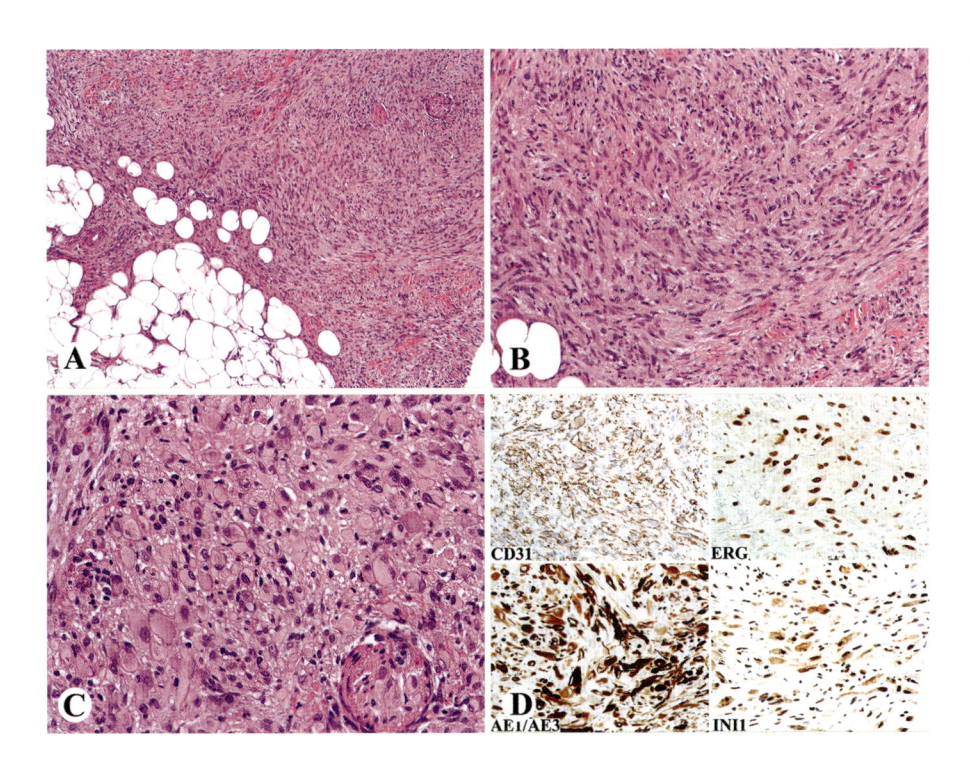

Figure 22.16 (A) Epithelioid sarcoma-like hemangioendothelioma is composed of spindled cells with prominently eosinophilic cytoplasm with overall resemblance to rhabdomyoblasts. (B) These tumors can also have epithelioid cytology. (C) The spindled cells are here set in a collagenous matrix. (D) Immunohistochemically these tumors are typically positive for CD31 (variably), ERG, keratins, and they retain INI1, in contrast to epithelioid sarcoma.

Chapter 22: Hemangioendotheliomas, angiosarcomas, and Kaposi's sarcoma

Table 22.1 Clinical associations and subgroups of angiosarcoma

Radiation-associated AS
    In breast or chest wall following radiation for breast carcinoma
    In abdominal wall following radiation for bladder or uterine cancer
    In peritoneal cavity following radiation for lymphoma or other cancer
    Thorium oxide (thorotrast) in hepatic angiosarcoma
    UV radiation carcinogenesis in angiosarcoma of scalp?

Occupational exposure to chemical carcinogens
    Vinyl chloride in hepatic angiosarcoma

Lymphedema
    Postmastectomy
    Hereditary primary lymphedema
    Filiaria-associated lymphedema

With other tumors
    Benign and malignant peripheral nerve sheath tumors
    Germ cell tumors, as somatic evolution
    Ovarian cystadenocarcinoma (serous, mucinous)

With foreign bodies
    Retained shrapnel, bullets
    Retained surgical sponges
    Dacron vascular graft
    Long-standing gouty tophus

Site-related
    Hepatic and splenic, mammary parenchymal
    Cardiac

Pediatric

Cutaneous angiosarcoma of the scalp, face, and other locations

Occurrence in the scalp and face is probably the most common setting for AS.[36–43] These ASs mainly occur in older patients, with the peak incidence between 70 and 79 years, with a marked male predominance. The tumor typically forms violaceous nodules or plaques, sometimes with multiple lesions (Figure 22.19). In the face, the upper and middle parts are more commonly involved than the lower parts.

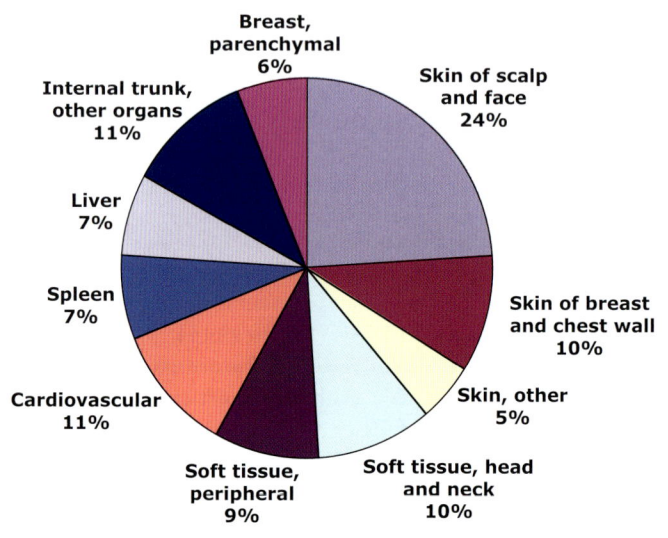

Figure 22.18 Anatomic distribution of 100 ASs. Bone tumors are excluded.

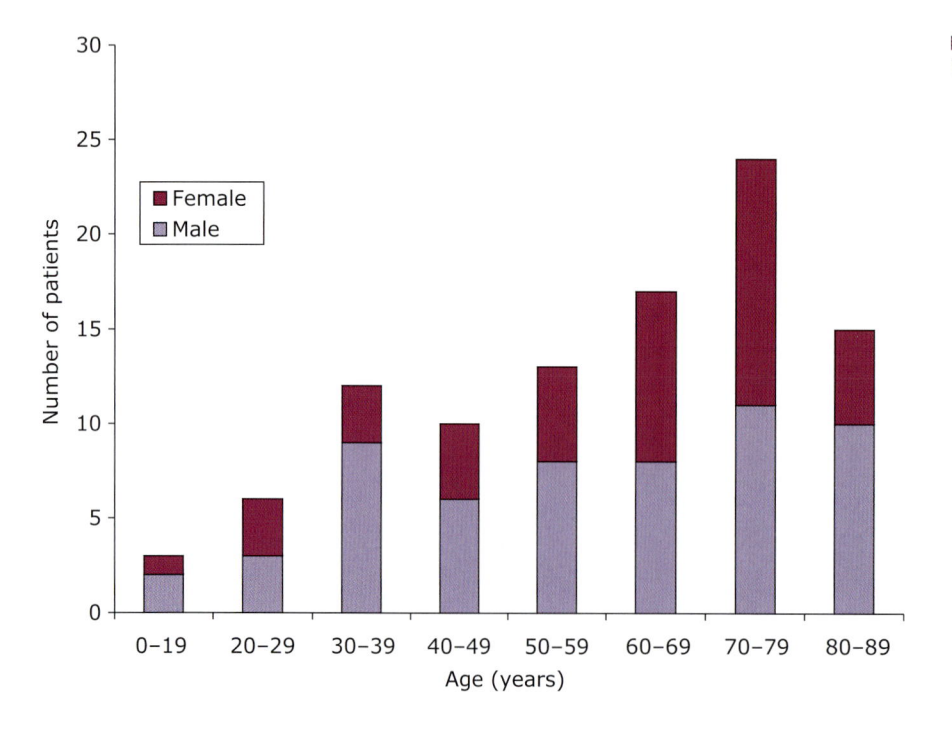

Figure 22.17 Age and sex distribution of 100 patients with angiosarcoma (AS).

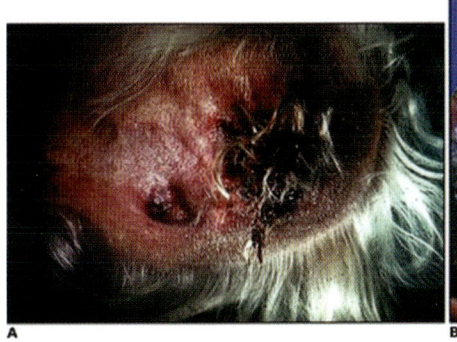

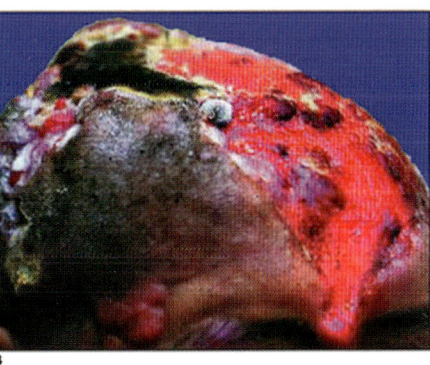

Figure 22.19 AS of the scalp usually forms multiple purplish, sometimes encrusted, nodules. Complete excision is difficult despite extensive surgery, often resulting in troublesome multinodular local recurrences.

Complete excision is difficult to achieve due to indistinct margins and occurrence of satellite lesions. Both local recurrences and distant metastases are common, and the overall prognosis is poor, 50% of patients being dead of disease in 15 months and only 12% surviving >5 years, according to one large study.[39] Patients with small and solitary lesions could have a better prognosis. One study identified tumor size <5 cm and lymphoid infiltration as prognostically favorable signs.[36]

Cutaneous AS also occurs in a wide variety of other sites, especially the lower extremities. In these locations, the demographics are more variable, including more frequent occurrence in women and younger patients.

Angiosarcoma and atypical vascular lesions following irradiation

Postradiation AS is on the rise, and this is the largest subgroup of ASs in some centers. The most common scenario is occurrence in the breast or chest wall following surgery and postoperative radiation for breast carcinoma.[44-50] The latency period for AS after radiation is shorter than for other post-irradiation sarcomas (mean: 5-7 years), but this may be related to earlier detection of these typically skin-involving tumors. Most of these tumors are high grade, and prognosis is generally poor, with an estimated 3-year survival rate of only 20%.

Atypical vascular lesions that could be precursors to AS of breast skin are increasingly recognized after radiotherapy. They can occur after a shorter latency following irradiation (2-5 years). Prognosis seems to be good, but because recurrences and evolution into AS are possible, complete excision and follow-up are necessary. The presence of multiple non-biopsied lesions requires special caution.[51-54]

Histologically, the postradiation atypical vascular lesions vary, showing lymphangioma- or hemangioma-like appearances with anastomosing patterns. Mild endothelial atypia can be present, but the cellularity is limited, and mitotic activity and endothelial multilayering are not evident in the lesion, thus falling short of AS (Figure 22.20).

External beam radiation for other cancers, most often cervical, endometrial, or bladder carcinoma, also can be complicated by AS of the lower abdominal wall, gastrointestinal tract, or peritoneum.[55-57] MYC-gene amplification is a distinctive feature predominantly observed in postradiation AS.

AS can also follow radiation for benign conditions, such as hemangioma. In such circumstances, the interval from radiation to tumor formation has usually been longer, as was the case with an atom bomb survivor, who supposedly developed an AS in a pre-existing osseous hemangioma >50 years later.[58] Some ASs, however, have developed spontaneously in hemangiomas/vascular malformations.[59] These findings indicate that different events could trigger an AS in a benign vascular tumor.

Angiosarcoma arising in lymphedematous extremities

This rare condition is also known as the Stewart–Treves syndrome, after the authors of the first report.[60] It is a rare complication of lymphedema, which historically was usually associated with radical breast cancer surgery. The incidence appears to be decreasing owing to less radical practices in breast cancer surgery. Most of these tumors arise in the upper arm of middle-aged or older women with long-standing lymphedema, on average 10 years after mastectomy and lymphadenectomy (Figure 22.21), but in some cases, the interval has been >30 years.[60-65] Radiation has been thought to have a contributory role in some cases, especially in causing blockage of axillary lymphatics. Occurrence in congenital (Milroy disease) and filarial infection-associated lymphedema has been reported, but rarely, suggesting that the lymphedema itself is a key factor.[61,62] Rare cases have also been reported in massive localized lymphedema in obese patients.[65] The pathogenesis has been suggested to be related to disrupted lymphatic circulation interfering with local immunosurveillance.[66] ASs arising in lymphedematous extremities are highly malignant, and in the absence of radical surgery tend to spread distally and proximally into the chest wall. Most of these patients have died in 2 to 3 years with metastases. Although earlier studies found amputation to improve the prognosis,[61,62] one study found similar results with wide excision,[63] whereas others suggest that major ablative surgery such as forequarter amputation might offer the best chance of cure.[64,65] The overall survival is poor, and long-term survival is observed in only 10% to 15% of patients.[61,62]

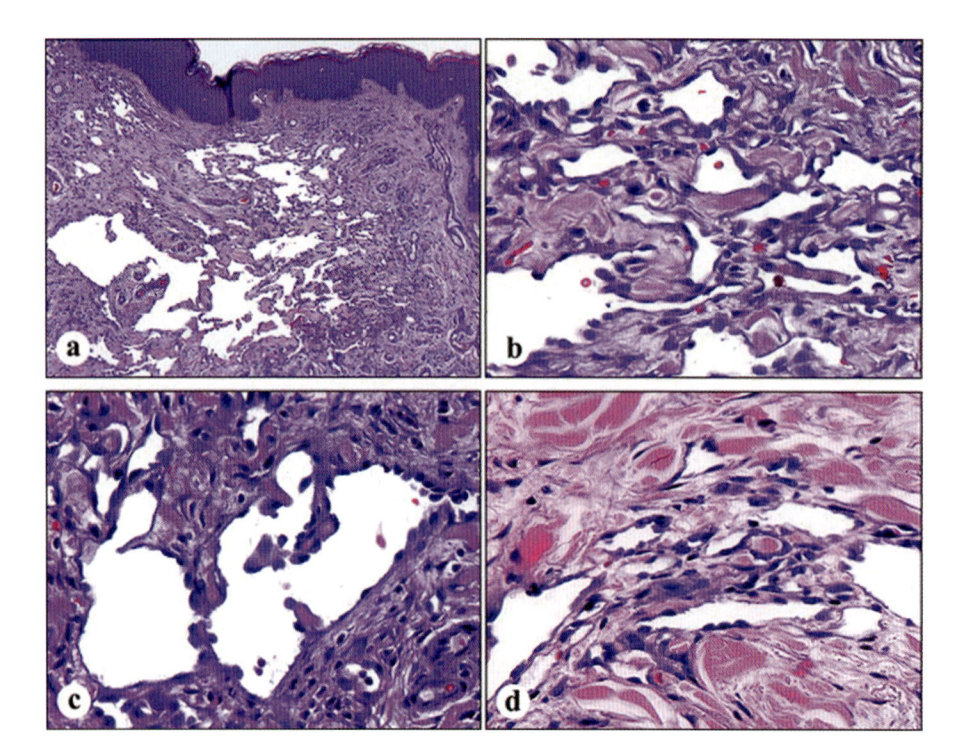

Figure 22.20 Atypical vascular lesions in the breast skin following radiation. (**a**) The lesion shows multiple anastomosing vascular channels. Note subepithelial fibrosis in the radiated skin. (**b,c**) Mild endothelial atypia and tufting are present. (**d**) Another example showing dermal fibrosclerosis following radiation and irregular vascular channels lined by mildly atypical endothelial cells.

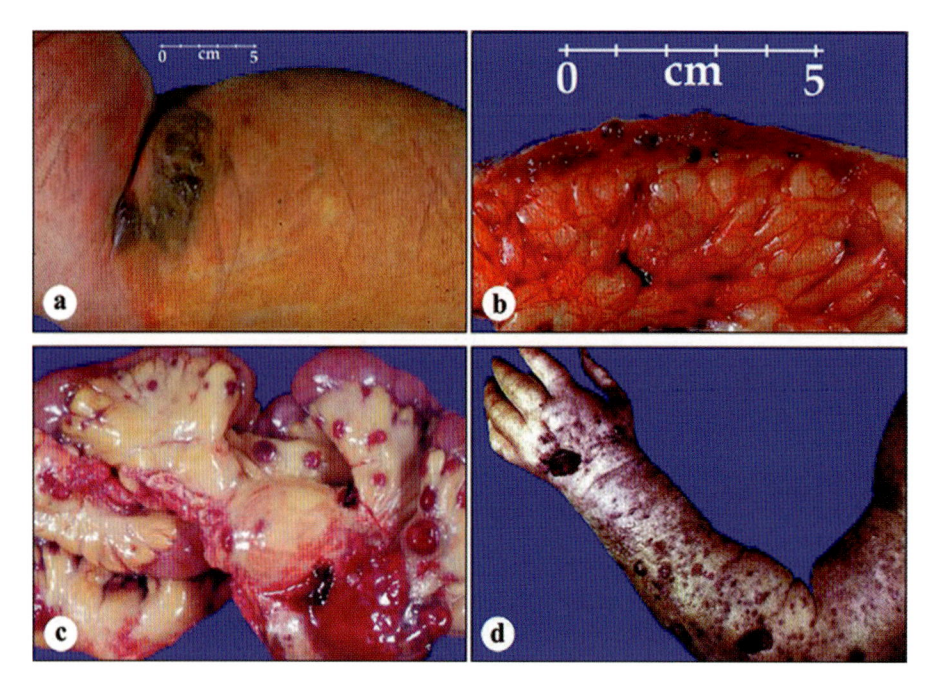

Figure 22.21 (**a,b**) AS arising in a lymphedematous arm following mastectomy forms multiple purplish cutaneous nodules. The nodules involve the skin and superficial subcutis. (**c**) Metastatic postmastectomy AS involving the mesenteries. (**d**) Multiple cutaneous purplish nodules of AS involving a lymphedematous arm.

Grossly, these tumors form multiple purplish cutaneous plaques or nodules. Histologically, they show a spectrum from well-differentiated vasoformative tumors to solid sheets of undifferentiated tumor cells. Tumors usually involve the skin and subcutis, but deep soft tissues can also be involved.

Angiosarcoma of the breast parenchyma

These ASs occur in a wide age range (second to eighth decade), but they have a predilection for young women, with a peak incidence between 30 and 50 years. They present as breast masses of 4 cm to 5 cm in median size, often with purplish skin discoloration, and usually require at least subcutaneous mastectomy for complete excision. Despite radical surgery, metastases often develop, most commonly in the lungs, liver, bones, soft tissues, and skin, including the contralateral breast (Figure 22.22).[67–72] While one series showed survival advantage for low-grade tumors,[69] another demonstrated no such advantage.[70]

Grossly, mammary ASs are typically hemorrhagic intraparenchymal masses that can simulate a hematoma on gross inspection. Histologically, these tumors vary from well to poorly differentiated and infiltrate interstitially in the stroma and diffusely in the fat. The well-differentiated components can have a deceptively benign appearance (Figure 22.23).

The occurrence of a wide variety of benign hemangiomas in the breast should be recognized in the differential diagnosis.[73] Of these, perilobular hemangioma is common, occurring in 12% of women in an autopsy study.[74] These tumors can be separated from AS based on their small size (usually <2 cm), lobulation, lack of anastomosing vascular channels, and inconspicuous endothelial cell nuclei.

Angiosarcoma of deep soft tissues

This is a heterogeneous group of tumors, constituting 20% to 25% of all ASs. According to a series of 80 cases, they occur in a wide age range, from childhood to old age, with the peak incidence in the seventh decade and a moderate (5:3) male predominance. The patients include those with syndromes such as Klippel–Trenaunay and Maffucci syndrome, suggesting that AS might originate in a pre-existing vascular tumor or malformation. This group also includes tumors arising from radiation therapy.[75] The most common locations are the extremities, especially the thigh (54%); trunk, especially the retroperitoneum (35%); and head and neck (11%). More than one-half of the patients develop distant metastases and die of the tumor. The prognosis is poorer with older patients and those with intra-abdominal tumors. Overall, >50% of patients in the largest series died of the tumor in a median time of 11 months.[75]

Soft tissue ASs often form large hemorrhagic masses that can grossly and even microscopically simulate a hematoma (Figure 22.24). Some examples become manifest by serious local thrombotic complications. Histology varies, but epithelioid morphology is common.[75,76]

A small number of ASs have been reported as arising in old arteriovenous fistulas in renal failure patients who underwent kidney transplantation. Many have been of the epithelioid variety.[77–80] These could be examples of immunosuppression-associated malignancies. Indeed, AS has also been reported in other sites in chronically immunosuppressed renal transplantation patients.[81]

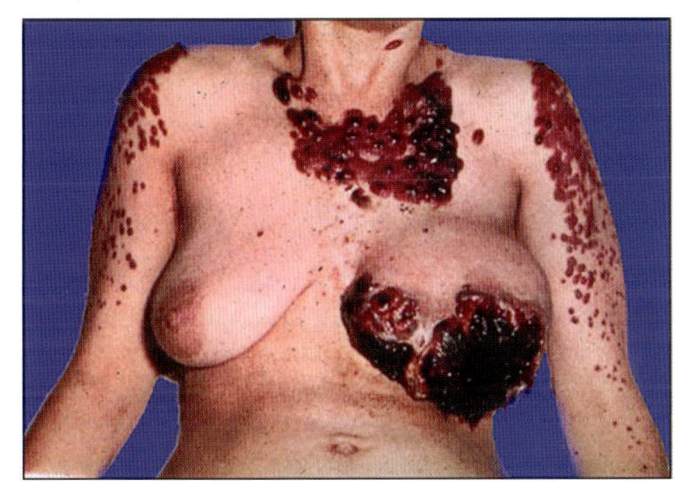

Figure 22.22 An advanced AS of the breast parenchyma forms a large ulcerated tumor, and numerous cutaneous satellite lesions are present in the neck and arms.

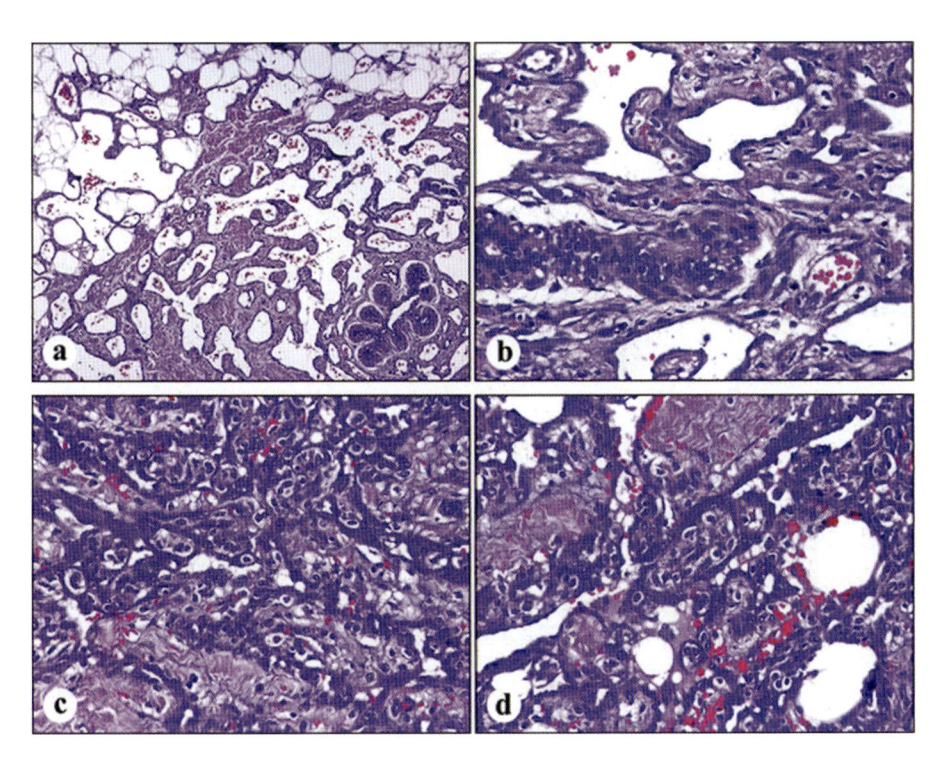

Figure 22.23 (**a**) AS of the breast grows permeatively in the breast parenchyma and subcutaneous fat. (**b**) The tumor contains well-differentiated vascular lumina with limited endothelial atypia. (**c,d**) More atypical endothelial cells establish the diagnosis of AS.

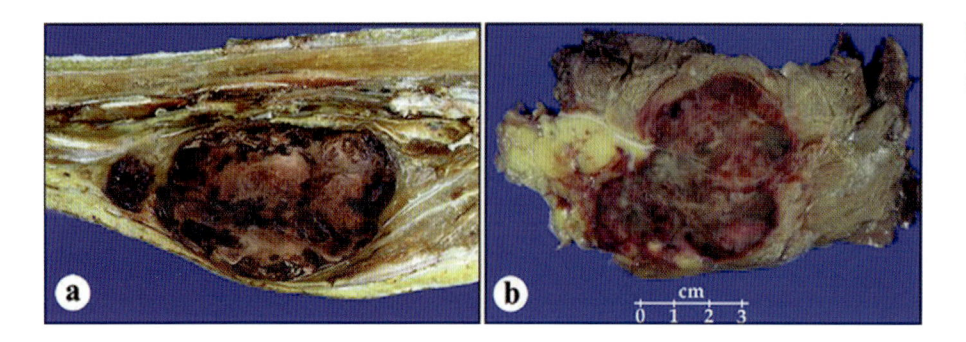

Figure 22.24 Deep soft tissue AS forms reddish hemorrhagic masses discolored by yellowish-gray thrombosis and necrosis.

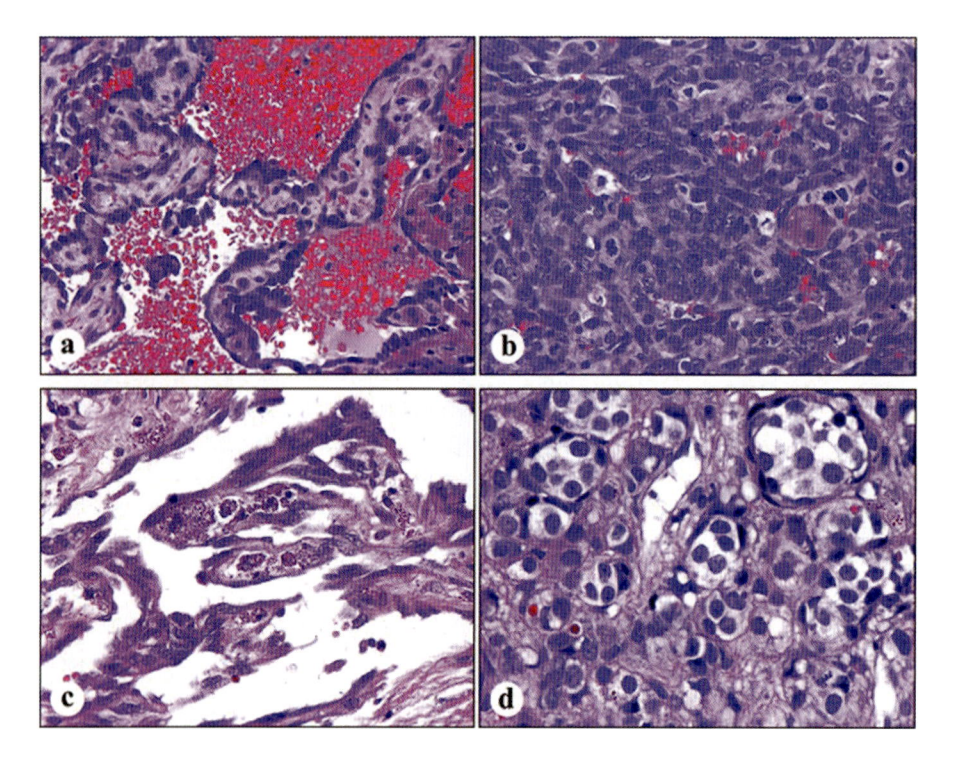

Figure 22.25 (**a,b**) Hepatic AS with vasoformative and solid areas. (**b**) An eosinophilic cluster is composed of residual hepatocytes. (**c**) Gaping vascular lumina lined by atypical spindled endothelial cells. The stromal granular material is thorotrast, an obsolete radioactive contrast medium linked to ASs. (**d**) A variant of AS composed of small clusters of endothelial cells. Note erythrocytes in vacuoles in some of the clusters.

Hepatic, splenic, and other internal angiosarcomas

Hepatic ASs are remarkable for several known occupational and iatrogenic causal connections; many of these tumors also involve the spleen and vice versa. Because the circulating carcinogens pass through the liver and could be metabolically activated there, the hepatic sinusoidal endothelia are exposed to carcinogens more than any other endothelia. The carcinogens specifically associated with hepatic AS are vinyl chloride (chemical used in plastic manufacturing), thorium oxide (radioactive compound thorotrast, used as radiologic contrast medium until the 1940s), and arsenic compounds that were used as pesticides.[82–86] The rare occurrence of hepatic ASs in children has also been reported. These tumors occur mostly in small children (mean age of 4 years). They tend to have a spindle cell pattern with eosinophilic globules, and the prognosis is poor.[87] Histologic features of hepatic ASs vary from well to poorly differentiated (Figure 22.25).

Splenic ASs vary from well-differentiated vasoformative to spindle cell and anaplastic tumors that can be difficult to recognize as ASs (Figure 22.26). They occur in adult patients, sometimes together with hepatic involvement, and most are highly malignant.[88–90] ASs from other locations can also metastasize to the spleen.

The splenic hemangiomas and littoral cell angiomas can be distinguished from AS based on their relative circumscription, organized vascular pattern, and lack of endothelial proliferation (multilayering) and atypia.

Angiosarcoma in the cardiovascular system

Several ASs have been reported in the heart and some as having arisen from arteries.[91–94] AS is the most common sarcoma in the heart, where it usually occurs in the right atrium. Possible presentations include hemorrhagic pericardial tamponation, arrhythmias, and sudden death.[91,92] Cardiac ASs occur in a wide age range with a male predominance, and they have also been reported in young children, although most pediatric patients have been 15 years or older.[91] The reported arterial intimal ASs have arisen from the aorta or

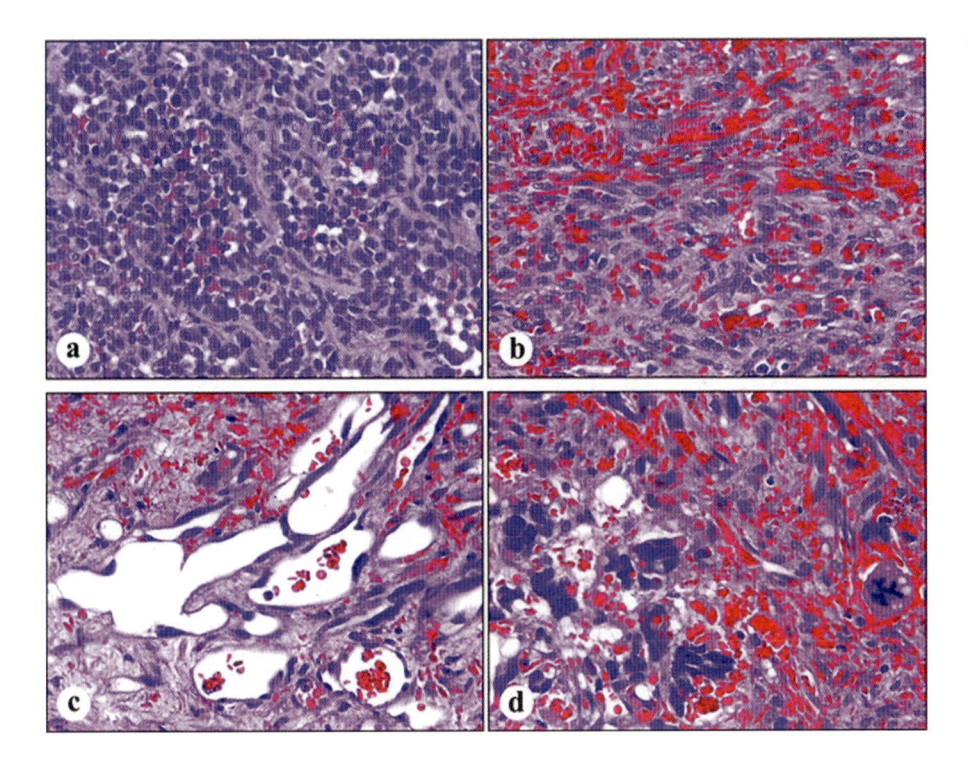

Figure 22.26 Examples of splenic AS. (**a**) Vasoformative pattern. (**b**) Spindle cell pattern without obvious vascular lumina. (**c,d**) An example containing differentiated vascular lumina and pleomorphic elements. Note that some of the multinucleated tumor cells contain erythrocytes, possibly reflecting attempts of primitive vasoformation.

carotid artery in older adults as stenosing vascular lesions, and all have been of the epithelioid variety.[94]

Angiosarcoma of rare sites

AS has been reported as a primary tumor at almost any site, including the brain;[95] thyroid;[96] oral cavity;[97,98] adrenal glands;[99] the gastrointestinal,[100,101] urogenital, and gynecologic tracts (i.e., ovary, uterus); the lung; and in the pleural and peritoneal surfaces. In the latter two sites, AS can clinically, grossly, and even histologically simulate a diffuse epithelial mesothelioma. Some of the gastrointestinal and serosal examples have occurred in patients who have had previous radiation treatment.[55,102–105] Metastatic AS often occurs in the previously mentioned sites, so that clinical correlation is required to determine the primary site.

Angiosarcoma arising in nerve sheath tumors

A small number of benign schwannomas and more cases of malignant peripheral nerve sheath tumors (MPNST) with AS components have been reported. The nature of this apparent nonrandom association is not clear, but malignant transformation of the proliferative angiomatous component (especially in MPNST), or divergent differentiation of the neuroectodermal elements are possibilities.

In benign schwannoma, the reported AS has usually been an epithelioid variant, and there has been a predilection for older men and a neck location associated with the vagus nerve. Some examples have involved the thigh, and occurrence in a deep plexiform schwannoma has been reported once.[106–110] None of the patients has had neurofibromatosis 1 (NF1). Based on a small number of cases, the prognosis seems better than that of AS with MPNST.

Most ASs in MPNSTs have occurred in NF1 patients, many of which have been young adults or children. The locations have varied, including thigh/sciatic nerve, forearm, mediastinum, and neck.[107,111–118] AS is usually a focal finding when it occurs in MPNST. Prognosis is poor, but it is difficult to determine to what extent this prognosis in MPNST-associated tumors is specifically related to the AS. In some cases, however, the distant metastases have been composed of AS only.[111,115]

Angiosarcoma arising in germ cell tumors and ovarian carcinomas

After rhabdomyosarcoma, AS seems to be the second most common sarcoma arising from a germ cell tumor by somatic evolution. These germ cell tumors have been located in the mediastinum and ovary, and in the ovary these tumors have included mature cystic teratomas.[119–122] In one series of ovarian AS, two of seven examples arose from a mature cystic teratoma.[122] Single cases of AS arising in different ovarian non-germ-cell tumors, mucinous[123] and serous cystadenocarcinoma,[124] have also been reported.

Angiosarcoma arising around a foreign body

AS developing around a long-standing foreign body, such as a bullet, shrapnel, or an accidentally retained surgical sponge, has been reported in a small number of cases. The latency period can be very long, exceeding 50 years in some cases.

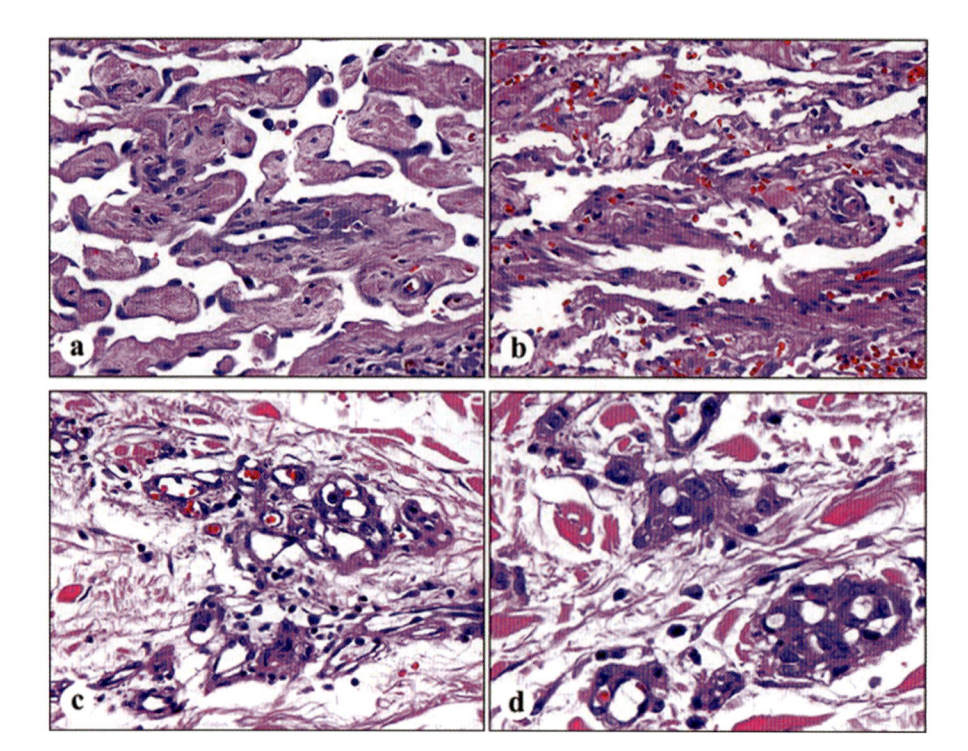

Figure 22.27 (a,b) Well-differentiated appearances of cutaneous AS. Note anastomosing vascular channels lined by cytologically atypical endothelial cells. (c,d) These well-formed clusters of vessels in AS contain markedly atypical endothelial cells apparent at a higher magnification.

The chronic carcinogenic nature of long-standing metal implants has been suspected. ASs related to retained surgical sponges have developed in the capsule of the reactive mass. A few ASs have been reported developing around surgically implanted material: a Dacron vascular graft or orthopedic hardware (fixation plate).[125–128] In one case, AS reportedly developed in a long-standing gouty tophus.[129] Foreign-body-associated ASs are almost invariably high-grade tumors with poor prognosis.

Angiosarcoma in children

Small numbers of ASs have been reported in children.[130,131] AS in children is rare, and many such childhood tumors earlier considered ASs are now classified as hemangioma and hemangioendothelioma variants.

Childhood AS includes a significant number of visceral tumors (e.g., heart, liver), and those arising in an MPNST. Rare examples present in the skin, mostly in the extremities, with a predilection for the foot. Predisposing or associated factors are xeroderma pigmentosum, radiation for hemangioma, and Aicardi syndrome. Although many tumors behave aggressively, there have been long-term survivors of cutaneous ASs.[132–134]

Histopathology of angiosarcoma

The histologic spectrum of AS includes several overlapping patterns that can occur in one tumor: well-differentiated vasoformative, poorly differentiated solid, spindle cell, and epithelioid.

Well-differentiated AS dissects between collagen fibers and fat and often occurs in the skin, breast, and soft tissue, including ASs arising in lymphedema. Anastomosing vascular channels in AS have some resemblance to lymphangioendothelioma (Chapter 21); however, nuclear atypia with nuclear irregularity, enlargement, and hyperchromasia are present (Figure 22.27). Some scalp ASs contain prominent lymphoid infiltration (Figure 22.28).

Spindle cell AS can occur in the skin, and it is particularly common in the liver and spleen. This pattern shows some resemblance to Kaposi's sarcoma. Definitive vasoformation by spindle cells is evident, although solid areas are often present.

Epithelioid AS shows large neoplastic cells with abundant eosinophilic cytoplasm and large nuclei with prominent nucleoli. The degree of vasoformation varies from distinct lumen-forming vascular units to solid sheets of epithelioid cells (Figures 22.29 and 22.30). These tumors are especially common in deep soft tissues.

A solid, poorly differentiated pattern that resembles a carcinoma, melanoma, or lymphoma can be seen in any poorly differentiated AS (Figure 22.31).

Most ASs are high-grade tumors, and some tumors previously considered low-grade ASs have been renamed borderline malignant tumors (hemangioendotheliomas).

Differential diagnosis

Features helpful in distinguishing AS from benign vascular tumors are shown in Table 22.2. ASs tend to have a randomly oriented (never entirely lobulated) vascular pattern infiltrating between connective or adipose tissue elements. The vascular channels often also have irregularly curving contours. The presence of cytologic atypia in the endothelial

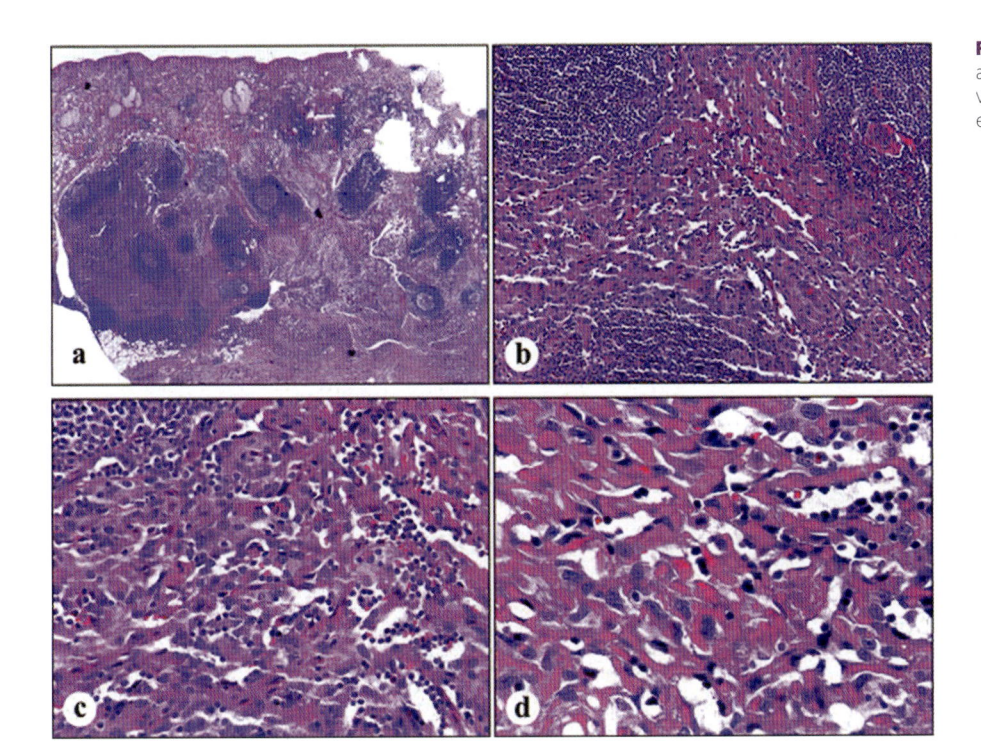

Figure 22.28 Example of a cutaneous AS with a prominent lymphoid infiltration. Note slit-like vascular lumina lined by moderately atypical endothelial cells.

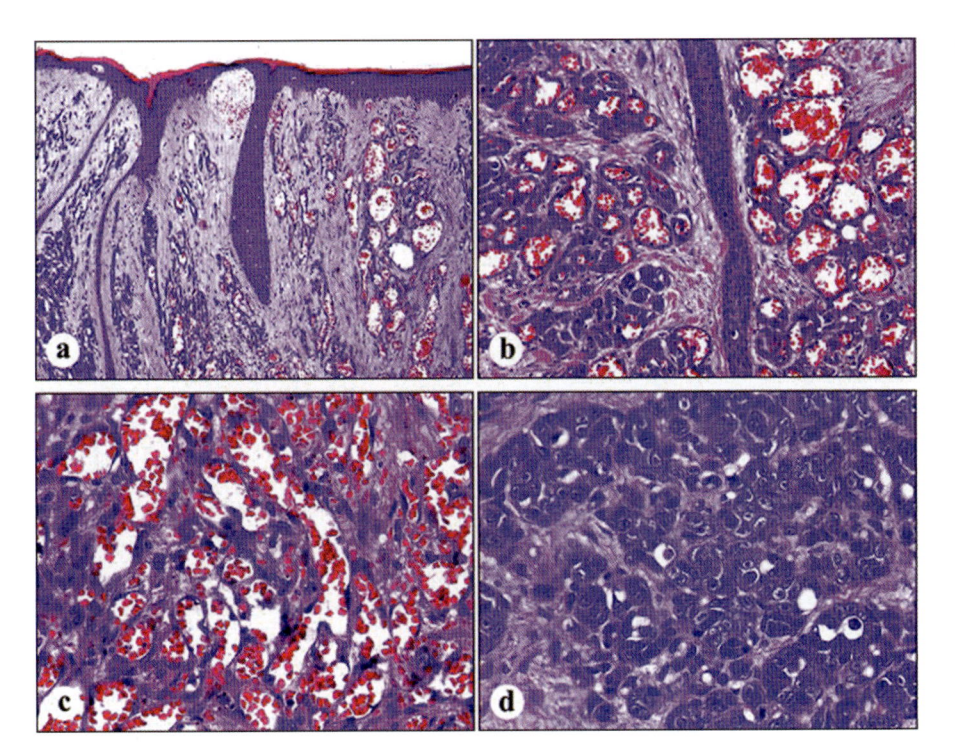

Figure 22.29 Cutaneous AS with epithelioid cytologic features.

cells, tendency to at least focally multilayered growth, and lack of actin-positive pericytes are additional features typical of AS. Mitotic activity and atypical mitoses are often present, but regular mitoses also can occur in benign proliferating endothelia. High-grade tumors often have significant hemorrhage and necrosis, sometimes obscuring their neoplastic nature.

The clinical context is very helpful in separating Kaposi's sarcoma from spindle cell AS. In problematic cases, immunostain for HHV8 is useful – it is positive in Kaposi's sarcoma and negative in AS (as well as hemangioma).

Many other malignant tumors (e.g., renal carcinoma, various undifferentiated carcinomas, epithelioid sarcoma,

Chapter 22: Hemangioendotheliomas, angiosarcomas, and Kaposi's sarcoma

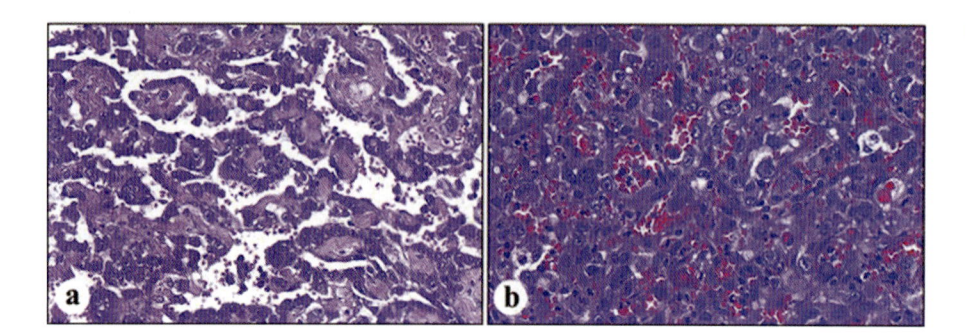

Figure 22.30 Anastomosing vascular channels and solid sheets in epithelioid AS involving deep soft tissue.

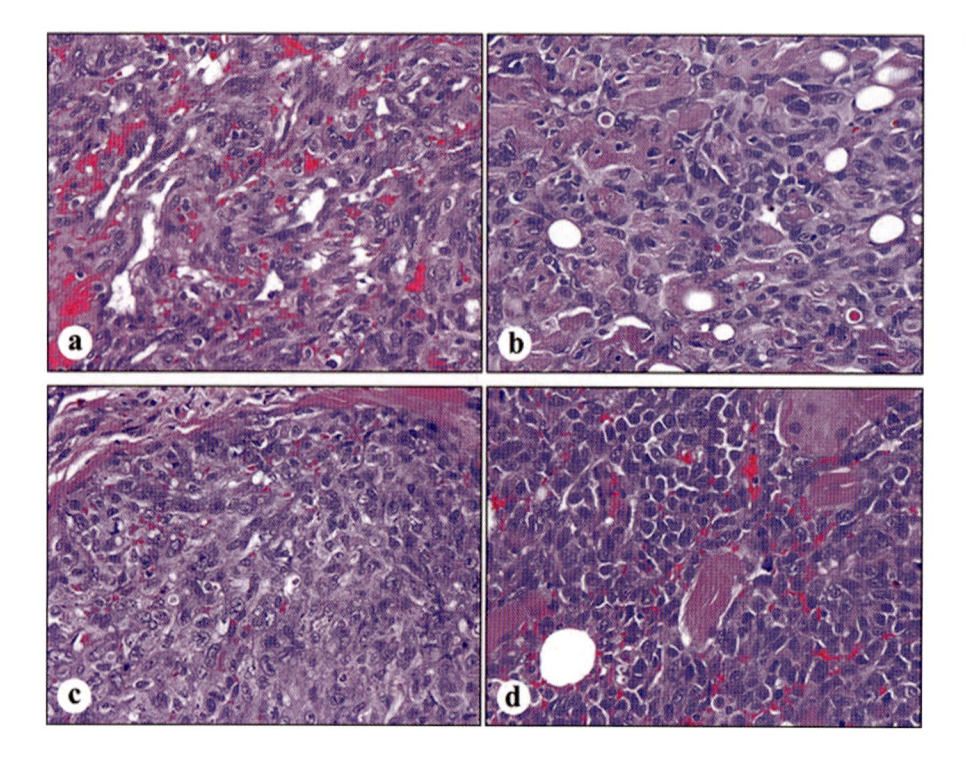

Figure 22.31 Poorly differentiated appearances of cutaneous AS. (**a**) Some luminal differentiation is present. (**b,c**) The tumor cells form solid sheets, with areas of little evidence of vasoformation. (**d**) Deep soft tissue involvement of an angiosarcoma arising in postmastectomy lymphedema in skeletal muscle.

melanoma) can be highly vascular and hemorrhagic. These tumors should not be mistaken for AS, which almost always shows true vasoformation by tumor cells, at least focally; immunohistochemistry for CD31, keratins, EMA, and S100 protein are useful.

Immunohistochemistry

ERG and CD31 are the most reliable markers, seen with a membrane-staining pattern in almost all ASs (Figure 22.32). The positivity of platelets and histiocytes (also the membrane pattern) should not be misconstrued as evidence for endothelial differentiation.[135–137] Previous studies showing a lower percentage of CD31 expression in AS, such as 33% in one study,[75] probably reflect insufficient antigen retrieval modalities available in older studies.

CD34 is expressed in approximately 50% of ASs somewhat unpredictably; both well and poorly differentiated tumors are variably positive.[136–138]

Von Willebrand factor (Factor VIII-related antigen) can be demonstrated in well-differentiated ASs, but usually only focally (if at all) in less-differentiated tumors. Positivity typically appears as granular cytoplasmic staining.

ERG transcription factor is typically expressed in benign and malignant endothelia and is retained in ASs, making ERG an excellent phenotypic marker. However, one has to consider that ERG is also expressed in up to 50% of prostate carcinomas and frequently in epithelioid sarcomas.[139]

Expression of thrombomodulin (CD141) has been suggested as useful for AS.[140] This marker is shared by endothelia and mesothelia, but according to the author's experience, expression is variable and inconsistent (only 30% of ASs are positive). VEGFR3 can be demonstrated in approximately 50% to 80% of ASs, less commonly so in the epithelioid variants.[141,142] Another endothelial marker normally expressed in lymphatics, podoplanin, is also present in subsets of ASs, but experience is limited.[143,144]

Table 22.2 Features useful in the differential diagnosis of angiosarcoma and benign vascular tumors

Feature	Angiosarcoma	Hemangioma
Lobulation of vascular units	Usually absent	Present in many
Contour of vessels	Often irregular, angulated	Typically smooth
Endothelial proliferation	Multilayering common	Multilayering rare
Mitoses	Variable, atypical mitoses can be present	Regular mitoses only, mitoses infrequent
Endothelial atypia	Large nuclei and nucleoli often present	Minimal if present. Small nucleoli
Pericytic component	Usually absent. SMA-positive component absent	Usually present. SMA-positive component present

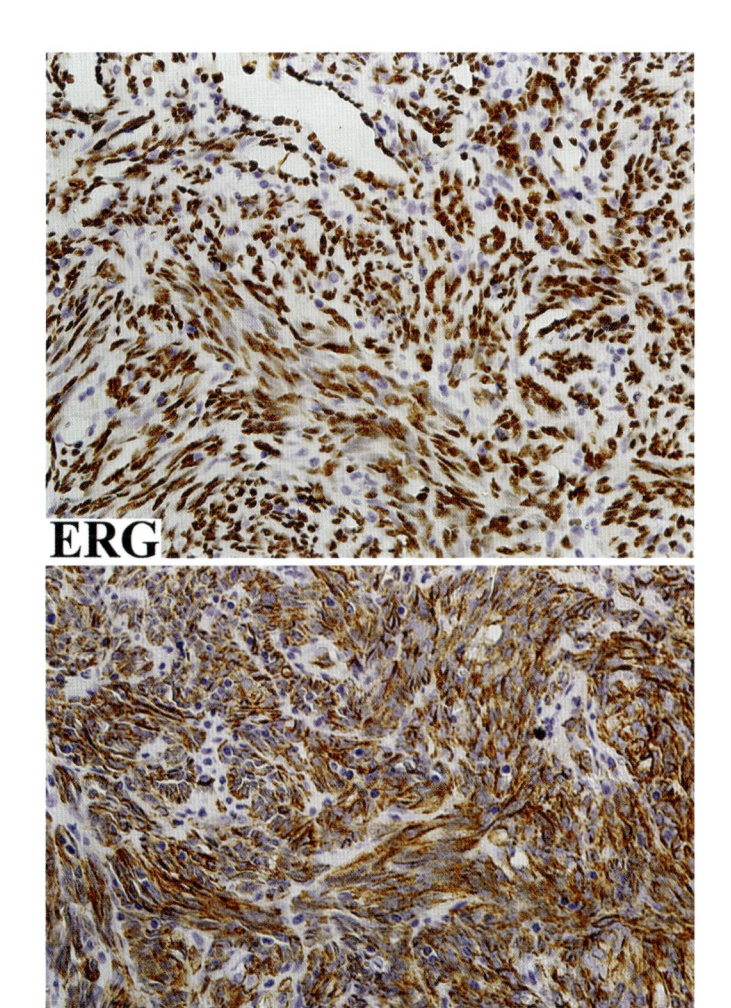

Figure 22.32 Nuclear ERG and membranous CD31 positivity are observed in nearly all ASs.

The transcriptional regulator gene FLI1, involved in the most common variant of Ewing sarcoma translocation, is constitutionally expressed in endothelial cells and has been suggested as an auxiliary marker for malignant vascular tumors; its expression seems to be conserved in AS.[145]

Keratins, especially K18, and to lesser degree K7 and K8, are present in 20% to 50% of ASs, more often in the epithelioid ones. EMA can also be present, making it challenging in some cases to differentiate between AS and carcinoma, and emphasizing the interpretation of all clinicopathologic data together. The expression of K7 and K18 in AS could reflect the phenotypic features of normal endothelia, which also express these keratins, whereas the presence of K8 more likely reflects neo-expression in transformed endothelia.[146,147]

Approximately 50% to 60% of ASs are positive for CD117 (KIT), similar to fetal endothelial cells, consistent with oncofetal expression of this marker, but KIT mutations in exons 11 and 17 have not been reported.[148] Despite KIT expression, there is no evidence that KIT tyrosine kinase inhibitors are effective in AS.

Lack of a well-defined, actin-positive pericytic layer could be helpful in differentiating AS from benign vascular proliferations, which tend to have preserved actin-positive pericytes.

MYC-gene amplification mostly seen in postradiation AS can be detected with FISH-based gene amplification studies.[149–152] Immunohistochemical detection closely parallels that amplification so that MYC immunohistochemistry is also useful. Although MYC gene amplification and immunoreactivity are characteristic of postradiation AS, they can be also found in other ASs.[153] These tests are also potentially useful in distinguishing atypical vascular proliferation from postradiation angiosarcoma, as the latter only tends to be positive.

Genetics of angiosarcoma

Mutations involving genes regulating angiogenesis have been detected in a subset of ASs. Gain-of-function mutations in proangiogenic KDR (VEGFR2) were detected in 10% of cases and were restricted to ASs of the breast, including patients with and without history of radiation.[154] One series demonstrated activating mutations in the PLCG1 signal transducer (9%), and isolated mutations in the receptor tyrosine kinase gene FLT4 downstream kinase PIK3CA, and various RAS-genes.[143g] Loss-of-function mutations of PTPRB (negative regulator of angiogenesis) were seen in 26% of ASs with apparent biallelic inactivation.[155] Cytogenetic studies on cutaneous, postmastectomy, deep soft tissue ASs, have shown complex changes. Observed recurrent changes include losses of chromosome 22 and 7pter-p15, and gains of 5pter-p11 and 8p12-qter, and 20pter-q12.[156–160]

Kaposi's sarcoma

Kaposi's sarcoma was named after an Austro-Hungarian dermatologist Moritz Kaposi, who reported the first examples of this tumor.[161] Kaposi's sarcoma is a primarily cutaneous malignant vascular tumor, but it can also involve the lymph nodes and almost any internal organ. It occurs in four clinicopathologically distinctive forms: classic, epidemic associated with AIDS, endemic in Africa, and iatrogenic, immunosuppression-associated.[162–164] The histologic features of the different forms are essentially identical. Oncogenic HHV8 infection of endothelial cells, often in the setting of immunosuppression, is thought to be the main pathogenetic factor.

Etiology and pathogenesis

The uniformly strong association with Kaposi's sarcoma and HHV8 (also known as Kaposi's sarcoma-associated herpes virus) indicates that this virus is the key etiologic factor for Kaposi's sarcoma; this is one of the best-documented connections between a virus and human mesenchymal neoplasia.[165–167]

HHV8 is a gamma herpes virus that can be demonstrated in all forms of Kaposi's sarcoma, either by PCR-based tests or by immunohistochemistry by detection of latent nuclear antigen (LANA1/LNA1). Transmission by bodily secretions (especially saliva) and parenteral transmission are the most important routes of HHV8 infection, and sexual transmission is considered less significant. Immunosuppression is an important factor promoting the formation of Kaposi's sarcoma in HHV8-infected AIDS and post-transplantation patients.[165–167]

In addition, HHV8 virus has also been implicated in certain body-cavity-based B-cell lymphomas, phenotypically similar lymphomas elsewhere, and multicentric Castleman's disease with plasmacytoid lymphoma.[168]

The mechanism of HHV8 tumorigenesis probably relates to interference of the viral proteins, such as cyclin and BCL2 analogs, with the host tumor-suppressor pathways and natural apoptosis. Endothelial growth-promoting cytokine/growth factor analogs are also produced by the tumor. Viral interferon regulatory factor in turn causes overexpression of the endogenous c-myc (onco)gene.[163] Multiple Kaposi's sarcoma lesions seem to be polyclonally infected by the virus, indicating multiple independent primaries.[169]

Classic chronic Kaposi's sarcoma

Classic (chronic) Kaposi's sarcoma presents in older patients, usually >60 years, with a marked (>10:1) male predominance. The populations most commonly affected are those of Ashkenazi-Jewish or Mediterranean origin; the latter population also appears to have a higher seroprevalence of HHV8 infection than control populations. In the classic form, single or multiple, purple to reddish skin lesions develop predominantly in the distal parts of lower extremities, sometimes in the hands and occasionally in other acral sites, such as the penile skin. Significant association with other malignancies, especially lymphomas and leukemias, has been demonstrated.

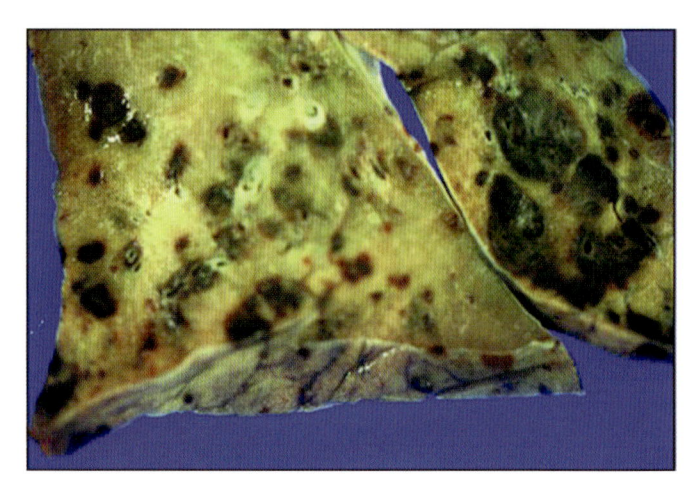

Figure 22.33 Multiple purplish pulmonary infiltrates of Kaposi's sarcoma. Pneumonic consolidation is also present.

Chronic Kaposi's sarcoma is usually clinically indolent, although the tumors are typically multifocal and recur locally.[162–164,170,171] Local excisions and radiation therapy in some cases usually keep the tumor under a satisfactory control.[163]

Epidemic Kaposi's sarcoma

Epidemic Kaposi's sarcoma is associated with AIDS and globally is probably the most common form. It occurs especially in male homosexual AIDS patients and is rare in those who have contracted AIDS by transfusion or intravenous drug abuse. Its incidence in AIDS patients has been markedly decreasing after highly effective antiretroviral therapy in AIDS.[163,172–175]

AIDS-associated Kaposi's sarcoma can involve a wide variety of sites in addition to the skin, especially oral and gastrointestinal mucosa, and lymph nodes. In some cases, internal organs, such as lungs (Figure 22.33), intestines, abdominal cavity, bones, and deep soft tissues, are discovered in autopsy studies.[174,175] Disseminated Kaposi's sarcoma historically occurred in 10% to 20% of patients with AIDS-associated Kaposi's sarcoma, but is now less common. The overall survival of AIDS-Kaposi's sarcoma patients was only 17 months in the early 1990s before modern HIV treatment. Although one would expect the opposite, one study found the patch and plaque stage of Kaposi's sarcoma in AIDS to be associated with opportunistic infections and a worse prognosis than the nodular form.[173] One possible explanation is that poor immunologic response favors the formation of such plaque and patch lesions, as opposed to nodular skin lesions.

Endemic Kaposi's sarcoma

Endemic Kaposi's sarcoma occurs in sub-Saharan equatorial Africa, where it is one of the most common cancer types. It involves both children and adults and ranges from an indolent peripheral skin disease to one with aggressive local infiltration of soft tissue and bones, and multiple internal organs. The aggressive form often occurs in young children in the first decade and typically involves lymph nodes in a generalized manner (lymphadenopathic Kaposi's sarcoma).[176] Systemic chemotherapy and radiation have been the most important forms of treatment.[177,178] The reason for the severity in many childhood cases is unknown, but immunosuppression caused by malnutrition and malarial infection could be a factor.

Iatrogenic immunosuppression-associated Kaposi's sarcoma

Iatrogenic immunosuppression-associated Kaposi's sarcoma mainly occurs in kidney transplant patients, and less commonly in other transplant patients who receive antirejection immunosuppressive therapy. In this setting, Kaposi's sarcoma develops on average 16 months after the transplantation, and has a lesser male predominance than other forms of Kaposi's sarcoma. In transplant patients, Kaposi's sarcoma probably represents reactivation of existing low-level HHV8 infection triggered by immunosuppression. This is supported by a markedly higher incidence of Kaposi's sarcoma in patients from ethnic groups or areas with a higher incidence of Kaposi's sarcoma and higher seroprevalence of HHV8. Titration of the immunosuppression to the lowest possible level is one of the main treatment strategies. Besides the skin, this form of Kaposi's sarcoma can involve mucosal surfaces and internal organs.[179–181] More recently, antiherpes agents have also been used.[181]

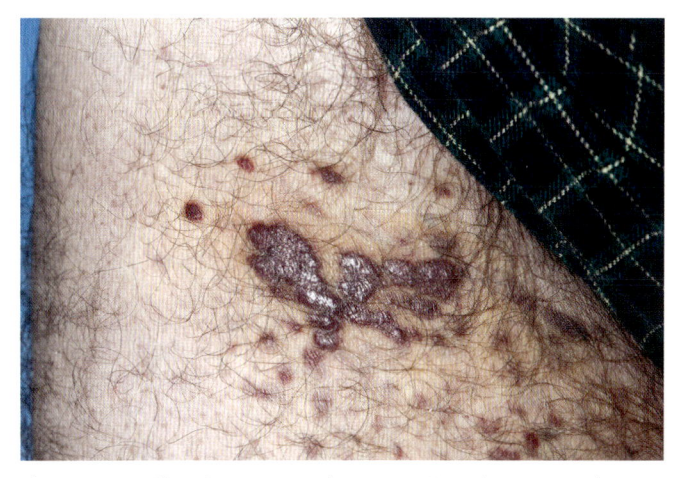

Figure 22.34 Clinical appearance of cutaneous Kaposi's sarcoma with a combination of patch, plaque, and early nodular lesions. Courtesy of Dr. Edward Cowen, Bethesda, Maryland.

Pathology

The skin lesions are thought to develop through several stages: patch, plaque-like, and nodular. The early patch-stage lesion is grossly a macular, nonpalpable lesion. The earliest patch lesions are often difficult to diagnose histologically, because they mainly consist of dilated, irregularly shaped vascular channels, perivascular lymphocytes, and plasma cells. There is capillary neovascularization arising from dilated vessels, however, mainly involving the upper or entire reticular dermis (Figure 22.34). The presence of small numbers of atypical spindle cells around the neovascularization is diagnostically

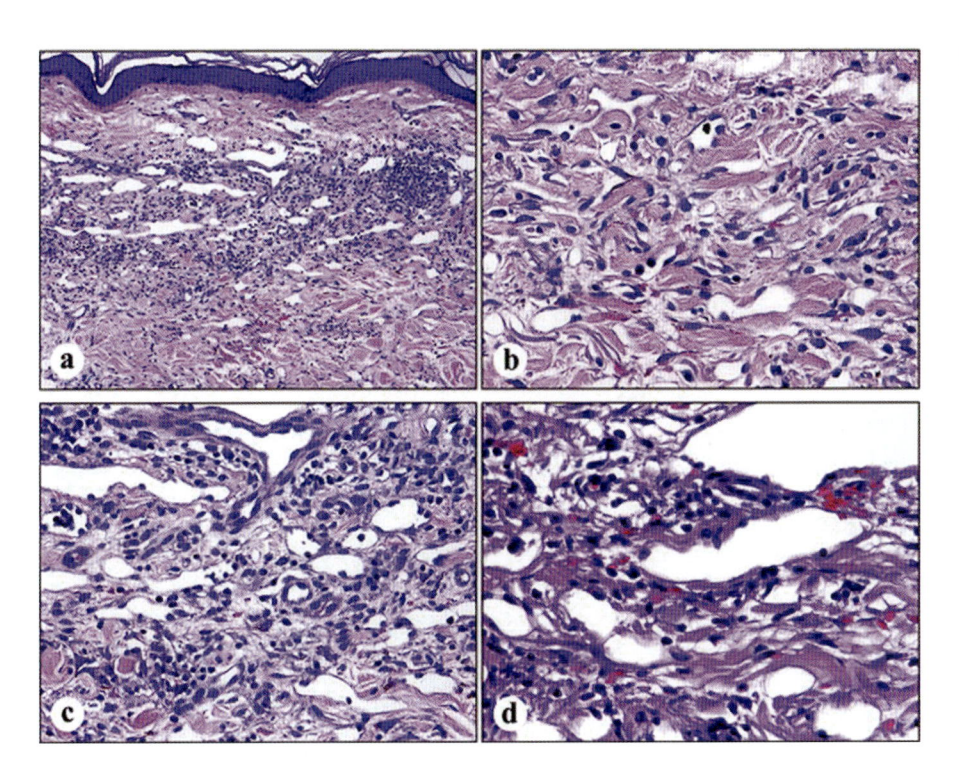

Figure 22.35 An example of a patch-stage Kaposi's sarcoma with numerous dilated lymphatic vessels. Note prominent lymphoplasmacytic infiltration. The spindle cell component is scant.

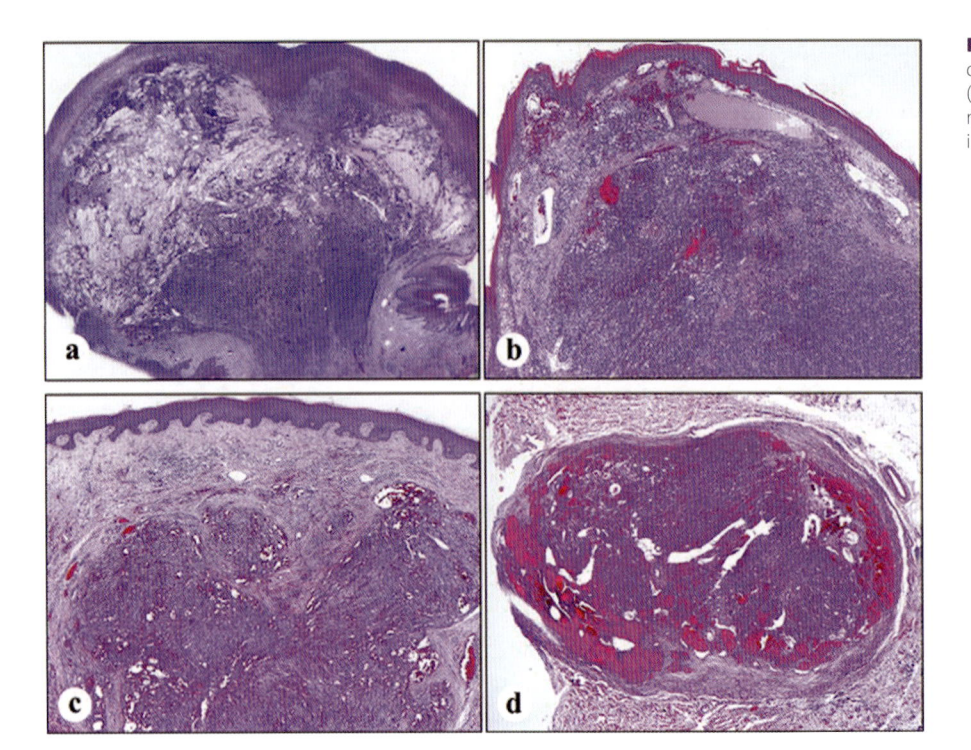

Figure 22.36 Low-magnification examples of Kaposi's sarcoma involving the skin. (**a,b**) Exophytic nodular lesions. (**c**) A plaque-like mildly elevated lesion. (**d**) A deeper dermal infiltrate.

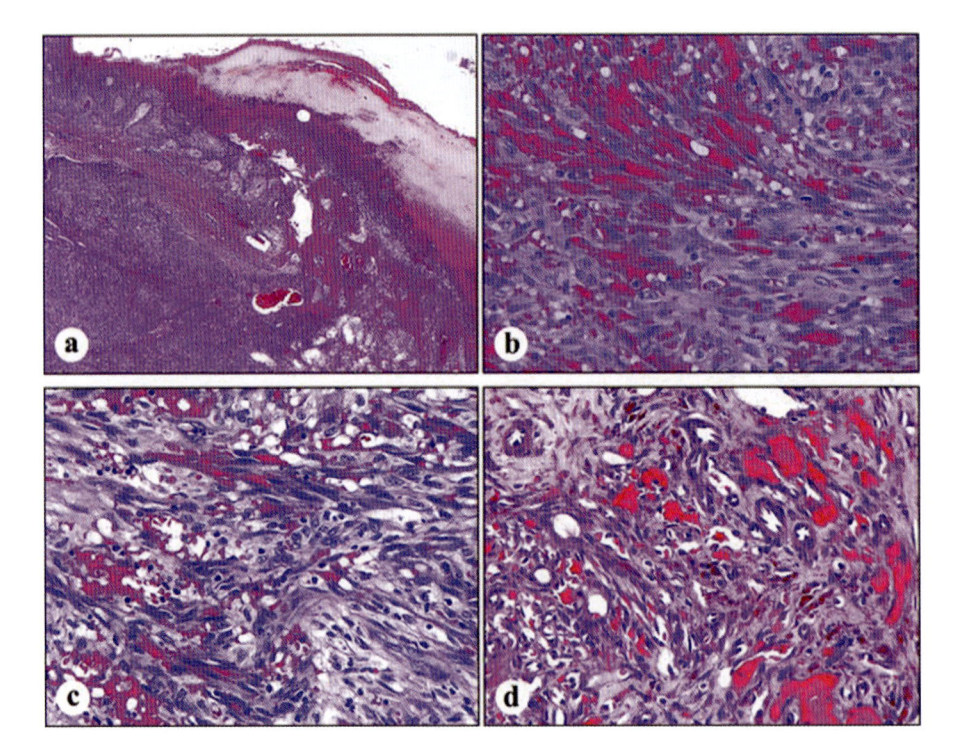

Figure 22.37 (**a**) Ulcerated and crusted cutaneous lesion of Kaposi's sarcoma. (**b–d**) Typical histologic appearances with spindle cells with focal vacuolization and variably forming vascular slits.

helpful.[182–183] The lymphangioma-like Kaposi's sarcoma essentially corresponds to this lesional stage, containing permeative infiltration of lymphatic-like vessels, scant, if any, spindle cell component, and focal lymphoplasmacytic infiltration (Figure 22.35). Immunohistochemistry for HHV8 is helpful in these cases. The plaque-stage lesion is the intermediate stage between the patch and the nodular lesion, or it can be understood as an early nodular lesion with lesser protuberance.

The nodular lesion is the typical manifestation of classic Kaposi's sarcoma. It forms a purplish cutaneous nodule

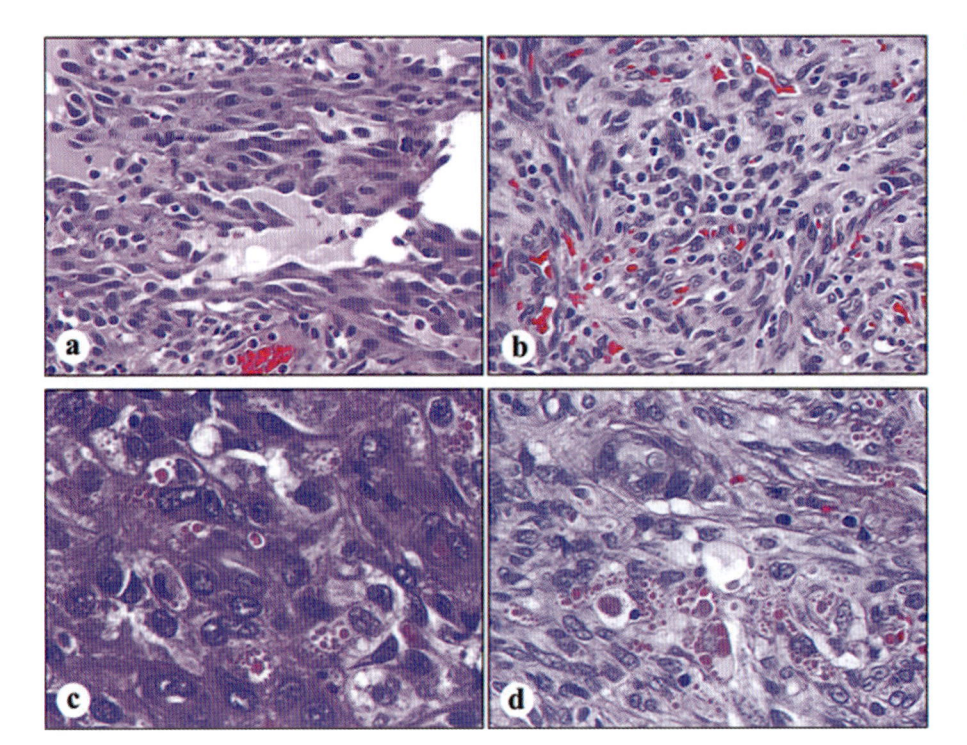

Figure 22.38 Typical histologic features of Kaposi's sarcoma. (**a**) Intravascular protrusion of tumor cells. (**b**) Cluster of lymphocytes and plasma cells. (**c,d**) Cytoplasmic eosinophilic hyaline globules.

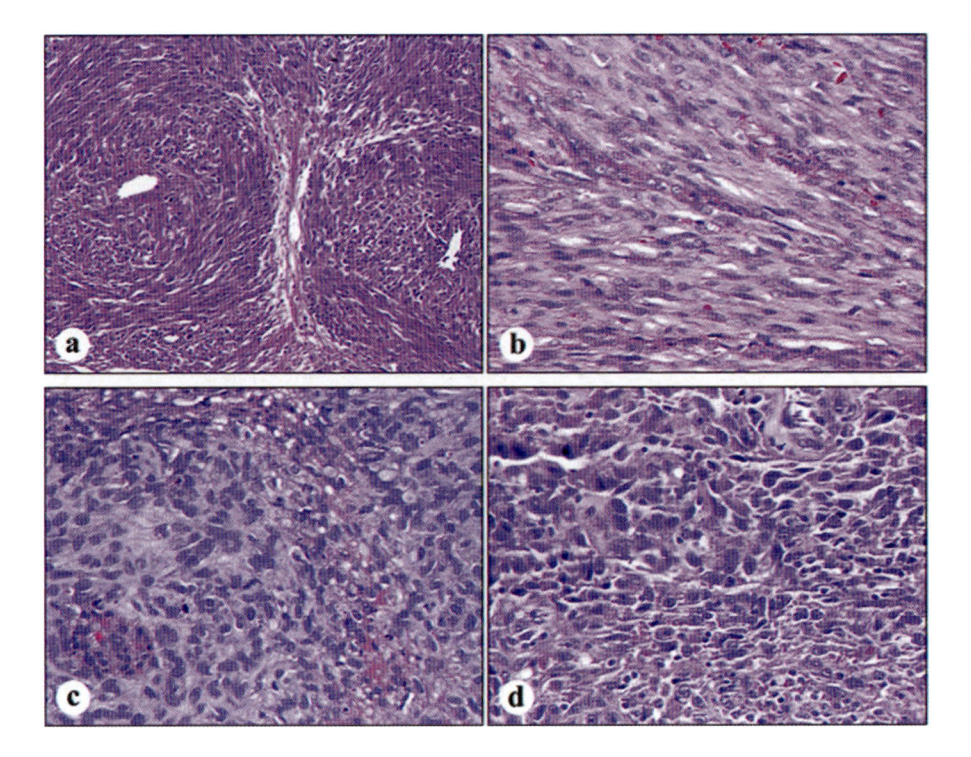

Figure 22.39 Sarcomatoid features in Kaposi's sarcoma. (**a**) Sheets of spindle cells forming perivascular collars. (**b**) Solid areas with fascicles of spindle cells can resemble the appearance of leiomyosarcoma. (**c,d**) Solid sheets of tumor cells with mild pleomorphism.

overlaid by a normal epidermis, or an exophytic growth surrounded by an epithelial collar, resembling a lobular capillary hemangioma (Figure 22.36). The lesion is composed of bundles or sheets of mildly atypical, uniform spindle cells forming cleft-like spaces often containing erythrocytes (Figure 22.37). Typical are cytoplasmic pink hyaline globules seen in at least some cells in most lesions, although a small number of globules can also be seen in the earlier stages of the lesion. Other helpful hints are spindle cells protruding into vascular lumina and patchy lymphoplasmacytic infiltration

615

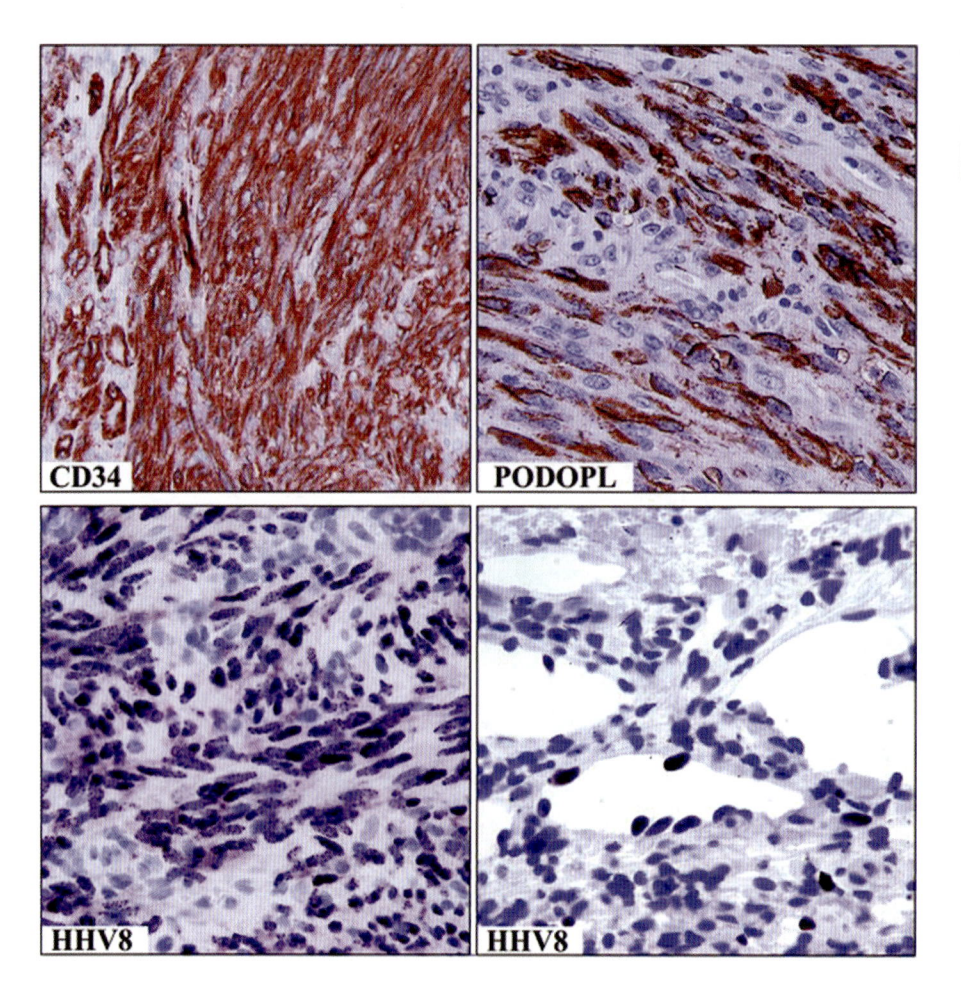

Figure 22.40 Typical Kaposi's sarcoma is positive for CD34 and podoplanin. There is nuclear positivity for HHV8 (red chromogen). Note that an early lesion, a lymphangioma-like patch stage of Kaposi's sarcoma, also shows HHV8-positive nuclei.

(Figure 22.38). These eosinophilic hyaline globules can be highlighted with periodic-acid–Schiff (PAS) stain and appear to represent the partially digested intralysosomal remains of erythrocytes.[182–184] A plaque-like lesion is a combination of this pattern, containing dilated vascular channels and spindle cell elements.

Lymph node involvement can be focal and seen only in the subcapsular area, but when extensive this may cause nodal effacement. Its cellular composition is similar to that of nodular cutaneous Kaposi's sarcoma.

Advanced Kaposi's sarcoma with extensive organ involvement can simulate other sarcomas, especially AS, and even solid spindle cell tumors in particular sites, such as gastrointestinal stromal tumors in the gastrointestinal tract (Figure 22.39).

Immunohistochemistry

The spindle cells in Kaposi's sarcoma lesions are neoplastic endothelial cells positive for CD31 and CD34;[138,185] they are only variably positive or negative for von Willebrand factor. Nuclear positivity for HHV8 LNA-1 is an important confirmatory test, being positive in virtually all cases (Figure 22.40).[186,187] Negative immunostaining results for S100 protein, actins, desmin, keratins, and EMA help to separate Kaposi's sarcoma from smooth muscle, myofibroblastic, and epithelial tumors.

Kaposi's sarcoma cells have many lymphatic endothelial markers, indicating a lymphatic endothelial phenotype and possible lymphatic endothelial origin. These markers include enzyme histochemical positivity for 5′-nucleotidase[188] and immunohistochemical positivity for VEGFR3[189] and podoplanin (D2-40).[144,190]

Differential diagnosis of Kaposi's sarcoma

The clinical setting and generally less-developed vasoformation separate Kaposi's sarcoma from AS, but the morphology of advanced and systemic lesions can approach AS. Although early reports to the contrary exist,[191] ASs of liver,[192] soft tissue, and skin[193] have been found to be negative for HHV8 by PCR and immunohistochemistry.[194]

Vascular transformation of lymph node sinuses is a condition associated with metastatic deposits and lymphatic obstruction. The hemorrhagic spindle cell proliferation involves the whole node, and solid spindle cell sheets are not present.[195–197]

Cellular hemangioma shows distinct luminal differentiation in the vascular proliferation, in contrast with nodal Kaposi's sarcoma.[197] In problematic cases, HHV8 immunohistochemistry is helpful.

References

Papillary intralymphatic angioendothelioma (malignant endovascular papillary angioendothelioma) and retiform hemangioendothelioma

1. Dabska M. Malignant endovascular papillary angioendothelioma of the skin in childhood: clinicopathologic study of six cases. *Cancer* 1969;24:503–510.
2. Schwartz RA, Dabski C, Dabska M. The Dabska tumor: a thirty-year retrospective. *Dermatology* 2000;201:1–5.
3. Fanburg-Smith JC, Michal M, Partanen TA, Alitalo K, Miettinen M. Papillary intralymphatic angioendothelioma (PILA): a report of twelve cases of a distinctive vascular tumor with phenotypic features of lymphatic vessels. *Am J Surg Pathol* 1999;23:1004–1010.
4. Emanuel PO, Lin R, Silver L, et al. Dabska tumor arising in lymphangioma circumscriptum. *J Cutan Pathol* 2008;35:65–69.
5. Calonje E, Fletcher CD, Wilson-Jones E, Rosai J. Retiform hemangioendothelioma: a distinctive form of low-grade angiosarcoma delineated in a series of 15 cases. *Am J Surg Pathol* 1994;18:115–125.
6. Kempson RL, Fletcher CDM, Evans HL, Hendrickson MR, Sibley RK. Vascular tumors. In *Atlas of Soft Tissue Tumors*. Washington, DC: AFIP; 2001: 307–370.
7. Duke D, Dvorak AM, Harris TJ, Cohen LM. Multiple retiform hemangioendotheliomas: a low grade angiosarcoma. *Am J Dermatopathol* 1996;18:606–610.
8. Parsons A, Sheehan DJ, Sangueza OP. Retiform hemangioendotheliomas usually do not express D2–40 and VEGFR-3. *Am J Dermatopathol* 2008;30:31–33.

Epithelioid hemangioendothelioma

9. Weiss SW, Enzinger FM. Epithelioid hemangioendothelioma: a vascular tumor often mistaken for a carcinoma. *Cancer* 1982;50:970–981.
10. Weiss SW, Ishak KG, Dail DH, Sweet DE, Enzinger FM. Epithelioid hemangioendothelioma and related lesions. *Semin Diagn Pathol* 1986;3:259–287.
11. Mentzel T, Beham A, Calonje E, Katenkamp D, Fletcher CD. Epithelioid hemangioendothelioma of skin and soft tissues: clinicopathologic and immunohistochemical study of 30 cases. *Am J Surg Pathol* 1997;21:363–374.
12. Ishak KG, Sesterhenn IA, Goodman ZD, Rabin L, Stromeyer FW. Epithelioid hemangioendothelioma of the liver: a clinicopathologic and follow-up study of 32 cases. *Hum Pathol* 1984;15:839–852.
13. Kelleher MB, Iwatsuki S, Sheahan DG. Epithelioid hemangioendothelioma of liver: clinicopathological correlations of 10 cases treated by orthotopic liver transplantation. *Am J Surg Pathol* 1989;13:999–1008.
14. Makhlouf HR, Ishak KG, Goodman ZD. Epithelioid hemangioendothelioma of the liver: a clinicopathologic study of 137 cases. *Cancer* 1999;85:562–582.
15. Dail DH, Liebow AA. Gmelich JT. Intravascular bronchioloalveolar tumor of lung: an analysis of twenty cases of a peculiar sclerosing endothelial tumor. *Cancer* 1983;51:452–464.
16. Bagan P, Hassan M, Le Pimpec Barthes F, et al. Prognostic factors and surgical indications of pulmonary epithelioid hemangioendothelioma: a review of the literature. *Ann Thorac Surg* 2006;82:2010–2013.
17. Verbeken E, Beyls J, Moerman P, et al. Lung metastasis of malignant epithelioid hemangioendothelioma mimicking a primary intravascular bronchoalveolar tumor: a histologic, ultrastructural, and immunohistochemical study. *Cancer* 1985;55:1741–1746.
18. Sortini A, Santini M, Carcoforo P, et al. Primary lung epithelioid hemangioendothelioma with multiple bilateral metachronous localizations: case report and review. *Int Surg* 2000;85:336–338.
19. Attanoos RL, Suvarna SK, Rhead E, et al. Malignant vascular tumours of the pleura in "asbestos" workers and endothelial differentiation in malignant mesothelioma. *Thorax* 2000;55:860–863.
20. Deyrup AT, Tighiouart M, Montag AG, Weiss SW. Epithelioid hemangioendothelioma of soft tissue: a proposal for risk stratification based on 49 cases. *Am J Surg Pathol* 2008;32:924–927.
21. Miettinen M, Fetsch JF. Distribution of keratins in normal endothelial cells and in a spectrum of vascular tumors: implications in tumor diagnosis. *Hum Pathol* 2000;31:1062–1067.
22. Errani C, Zhang L, Sung YS, et al. A novel WWTR1-CAMTA1 gene fusion is a consistent abnormality in epithelioid hemangioendothelioma of different anatomic sites. *Genes Chromosomes Cancer* 2011;50:644–653.
23. Tanas MR, Sboner A, Oliveira AM, et al. Identification of a disease-defining gene fusion epithelioid hemangioendothelioma. *Sci Transl Med* 2011;3:98ra82.
24. Mendlick MR, Nelson M, Pickering D, et al. Translocation t(1;3)(p36.3;q25) is a nonrandom aberration in epithelioid hemangioendothelioma. *Am J Surg Pathol* 2001;25:684–687.
25. Antonescu CR, Le Loarer F, Mosquera JM, et al. Novel YAP1-TFE3 fusion defines a distinct subset of epithelioid hemangioendothelioma. *Genes Chromosomes Cancer* 2013;52:775–784.
26. Flucke U, Vogels RJ, de Saint Aubain Somerhausen N, et al. Epithelioid hemangioendothelioma: clinicopathologic, immunhistochemical, and molecular genetic analysis of 39 cases. *Diagn Pathol* 2014;9:131.

Epithelioid sarcoma-like hemangioendothelioma

27. Billings SD, Folpe AL, Weiss SW. Epithelioid sarcoma-like hemangioendothelioma. *Am J Surg Pathol* 2003;27:48–57.
28. Hornick JL, Fletcher CD. Pseudomyogenic hemangioendothelioma: a distinctive, often multicentric tumor with indolent behavior. *Am J Surg Pathol* 2011;35:190–201.
29. Amary MF, O'Donnell P, Berisha F, et al. Pseudomyogenic (epithelioid sarcoma-like) hemangioendothelioma: characterization of five cases. *Skeletal Radiol* 2013;42:947–957.
30. Trombetta D, Magnusson L, von Steyern FV, et al. Translocation t(7;19)(q22;q13): a recurrent chromosome aberration in pseudomyogenic hemangioendothelioma? *Cancer Genet* 2011;204:211–215.

Angiosarcoma: general

31. Toro JR, Travis LB, Wu HJ, et al. Incidence patterns of soft tissue sarcomas, regardless of primary site, in the surveillance, epidemiology, and end results program, 1978–2001: an analysis of 26,758 cases. *Int J Cancer* 2006;119:2922–2930.

32. Fury MG, Antonescu CR, Van Zee KJ, Brennan MF, Maki RG. A 14-year retrospective review of angiosarcoma: clinical characteristics, prognostic factors, and treatment outcomes with surgery and chemotherapy. *Cancer J* 2005;11:241–247.

33. Abraham JA, Hornicek EJ, Kaufman AM, et al. Treatment and outcome of 82 patients with angiosarcoma. *Ann Surg Oncol* 2007;14:1953–1967.

34. Bocklage T, Leslie KO, Yousem S, Colby T. Extracutaneous angiosarcoma metastatic to lungs: clinical and pathologic features of twenty-one cases. *Mod Pathol* 2001;14:1216–1225.

35. Fayette J, Martin E, Piperno-Neumann S, et al. Angiosarcoma: a heterogeneous group of sarcomas with specific behavior depending on primary site: a retrospective study of 161 cases. *Ann Oncol* 2007;18:2030–2036.

Cutaneous angiosarcoma

36. Hodgkinson DJ, Soule EH, Woods JE. Cutaneous angiosarcoma of the head and neck. *Cancer* 1979;44:1106–1113.

37. Maddox JC, Evans HL. Angiosarcoma of skin and soft tissue: a study of forty-four cases. *Cancer* 1981;51:1907–1921.

38. Pawlik TM, Paulino AF, McGinn CJ, et al. Cutaneous angiosarcoma of the scalp: a multidisciplinary approach. *Cancer* 2003;98:1716–1726.

39. Morgan MB, Swann M, Somach S, Eng W, Smoller B. Cutaneous angiosarcoma: a case series with prognostic correlation. *J Am Acad Dermatol* 2004;50:867–874.

40. Deyrup AT, McKenney JK, Tighiouart M, Folpe AL, Weiss SW. Sporadic cutaneous angiosarcomas: a proposal for risk stratification based on 89 cases. *Am J Surg Pathol* 2008;32:72–77.

41. Guadagnolo BA, Zagars GK, Araujo D, et al. Outcomes after definitive treatment for cutaneous angiosarcoma of the face and scalp. *Head Neck* 2011;33:661–667.

42. Lindet C, Neuville A, Penel N, et al. Localised angiosarcomas: the identification of prognostic factors and analysis of treatment impact. A retrospective analysis from the French Sarcoma Group (GSF/GETO). *Eur J Cancer* 2013;49:369–376.

43. Perez MC, Padhya TA, Messina JL, et al. Cutaneous angiosarcoma: a single-institution experience. *Ann Surg Oncol* 2013;20:3391–3397.

Radiation-related angiosarcoma

44. Otis CN, Perschel R, McKhann C, Merino MJ, Duray PH. The rapid onset of cutaneous angiosarcoma after radiotherapy for breast cancer. *Cancer* 1986;57:2130–2134.

45. Monroe AT, Feigenberg SJ, Mendenhall NP. Angiosarcoma after breast-conserving therapy. *Cancer* 2003;97:1832–1840.

46. Billings SD, McKenney JK, Folpe AL, Hardacre MC, Weiss SW. Cutaneous angiosarcoma following breast-conserving surgery and radiation: an analysis of 27 cases. *Am J Surg Pathol* 2004;28:781–788.

47. Fodor J, Orosz Z, Szabo E, et al. Angiosarcoma after conservation treatment for breast carcinoma: our experience and review of the literature. *J Am Acad Dermatol* 2006;54:499–504.

48. Morgan EA, Kozono DE, Wang Q, et al. Cutaneous radiation-associated angiosarcoma of the breast: poor prognosis in a rare secondary malignancy. *Ann Surg Oncol* 2012;19:3801–3808.

49. Seinen JM, Styring E, Verstappen V, et al. Radiation-associated angiosarcoma after breast cancer: high recurrence rate and poor survival despite surgical treatment with R0 resection. *Ann Surg Oncol* 2012;19:2700–2706.

50. D'Angelo SP, Antonescu CR, Kuk D, et al. High-risk features in radiation-associated breast angiosarcomas. *Br J Cancer* 2013;109:2340–2346.

51. Fineberg S, Rosen PP. Cutaneous angiosarcoma and atypical vascular lesions of the skin and breast after radiation therapy for breast carcinoma. *Am J Clin Pathol* 1994;102:757–763.

52. Sener SF, Milos S, Feldman JL, et al. The spectrum of vascular lesions in the mammary skin, including angiosarcoma, after breast conservation treatment for breast cancer. *J Am Coll Surg* 2001;193:22–28.

53. Brenn T, Fletcher CD. Radiation-associated cutaneous atypical vascular lesions and angiosarcoma: clinicopathological analysis of 42 cases. *Am J Surg Pathol* 2005;29:983–996.

54. Patton KT, Deyrup AT, Weiss SW. Atypical vascular lesions after surgery and radiation of the breast: a clinicopathologic study of 32 cases analyzing histologic heterogeneity and association with angiosarcoma. *Am J Surg Pathol* 2008;32:943–950.

55. Wolov RB, Sato N, Azumi N, Lack EE. Intra-abdominal "angiosarcomatosis": report of two cases after pelvic irradiation. *Cancer* 1991;67:2275–2279.

56. Suzuki F, Saito A, Ishi K, et al. Intra-abdominal angiosarcomatosis after radiotherapy. *J Gastroenterol Hepatol* 1999;14:289–292.

57. Aitola P, Poutiainen A, Nordback I. Small-bowel angiosarcoma after pelvic irradiation: a report of two cases. *Int J Colorect Dis* 1999;14:308–310.

58. Yamamoto T, Iwasaki Y, Kurosaka M, Minami R. Angiosarcoma arising from skeletal hemangiomatosis in an atomic bomb survivor. *J Clin Pathol* 2001;54:716–717.

59. Rossi S, Fletcher CD. Angiosarcoma arising in hemangioma/vascular malformation: report of four cases and review of the literature. *Am J Surg Pathol* 2002;26:1319–1329.

Angiosarcoma arising in lymphedema

60. Stewart FW, Treves N. Lymphangiosarcoma in postmastectomy lymphedema. *Cancer* 1948;1:1674–1678.

61. Woodward AH, Ivins JC, Soule EH. Lymphangiosarcoma arising in chronic lymphedematous extremities. *Cancer* 1972;30:562–572.

62. Sordillo P, Chapman R, Hajdu SI, Magill GB, Golbey RB. Lymphangiosarcoma. *Cancer* 1981;48:1674–1679.

63. Grobmyer SJ, Daly JM, Glotzback RE, Grobmyer AJ 3rd. Role of surgery in the management of postmastectomy extremity angiosarcoma (Stewart–Treves syndrome). *J Surg Oncol* 2000;73:182–188.

64. Roy P, Clark MA, Thomas JM. Stewart–Treves syndrome: treatment and outcome in six patients from a single centre. *Eur J Surg Oncol* 2004;30:982–986.
65. Shon W, Ida CM, Boland-Froemming JM, Rose PS, Folpe A. Cutaneous angiosarcoma arising in massive localized lymphedema of the morbidly obese: a report of five cases and review of the literature. *J Cutan Pathol* 2011;38:560–564.
66. Schreiber H, Barry FM, Russell WC, et al. Stewart–Treves syndrome: a lethal complication of postmastectomy lymphedema and regional immune deficiency. *Arch Surg* 1979;114:82–85.

Angiosarcoma in breast parenchyma
67. Steingaszner LC, Enzinger FM, Taylor HB. Hemangiosarcoma of the breast. *Cancer* 1965;18:352–361.
68. Merino MJ, Berman M, Carter D. Angiosarcoma of the breast. *Am J Surg Pathol* 1983;7:53–60.
69. Rosen PP, Kimmel M, Ernsberger D. Mammary angiosarcoma: the prognostic significance of tumor differentiation. *Cancer* 1988;62:2145–2151.
70. Nascimento AF, Raut CP, Fletcher CD. Primary angiosarcoma of the breast: clinicopathologic analysis of 49 cases, suggesting that grade is not prognostic. *Am J Surg Pathol* 2008;32:1896–1904.
71. Sher T, Hennessy BT, Valero V, et al. Primary angiosarcomas of the breast. *Cancer* 2007;110:173–178.
72. Hodgson NC, Bowen-Wells C, Moffat F, Franceschi D, Avisar E. Angiosarcoma of the breast: a review of 70 cases. *Am J Clin Oncol* 2007;30:570–573.
73. Jozefczyk MA, Rosen PP. Vascular tumors of the breast. II. Perilobular hemangiomas and hemangiomas. *Am J Surg Pathol* 1985;9:491–503.
74. Lesueur GC, Brown RW, Bhathal PS. Incidence of perilobular hemangioma in the female breast. *Arch Pathol Lab Med* 1983;107:308–310.

Deep soft tissue angiosarcoma
75. Meis-Kindblom JM, Kindblom LG. Angiosarcoma of soft tissue: a study of 80 cases. *Am J Surg Pathol* 1998;22:683–697.
76. Fletcher CD, Beham A, Bekir S, Clarke AM, Marley NJ. Epithelioid angiosarcoma of deep soft tissue: a distinctive tumor readily mistaken for an epithelial neoplasm. *Am J Surg Pathol* 1991;15:915–924.
77. Byers RJ, McMahon RF, Freemont AJ, Parrott NR, Newstead CG. Epithelioid angiosarcoma arising in an arteriovenous fistula. *Histopathology* 1992;21:87–89.
78. Keane MM, Carney DN. Angiosarcoma arising from a defunctionalized arteriovenous fistula. *J Urol* 1993;149:364–365.
79. Wehrli BM, Janzen DL, Shokeir O, et al. Epithelioid angiosarcoma arising in a surgically constructed arteriovenous fistula: a rare complication of chronic immunosuppression in the setting of renal transplantation. *Am J Surg Pathol* 1998;22:1154–1159.
80. Faraq R, Schulak JA, Abdul-Karim FW, Wasman JK. Angiosarcoma arising in an arteriovenous fistula site in a renal transplant patient: a case report and literature review. *Clin Nephrol* 2005;63:408–412.
81. Ahmed I, Hamacher KL. Angiosarcoma in a chronically immunosuppressed renal transplant recipient: report of a case and review of the literature. *Am J Dermatopathol* 2002;24:330–335.

Angiosarcomas in other locations
82. Neshiwat LF, Friedland ML, Suhorr-Lesnick B, et al. Hepatic angiosarcoma. *Am J Med* 1992;93:219–222.
83. Popper H, Thomas LB, Telles NC. Development of hepatic angiosarcoma in man induced by vinyl chloride, thorotrast and arsenic. *Am J Pathol* 1978;92:349–376.
84. Alrenga DP. Primary angiosarcoma of the liver. *Int Surg* 1975;60:198–203.
85. Kojiro M, Nakashima T, Ito Y. Thorium dioxide-related angiosarcoma of the liver: pathomorphologic study of 29 autopsy cases. *Arch Pathol Lab Med* 1985;109:853–857.
86. Lander JJ, Stanley RJ, Sumner HW. Angiosarcoma of the liver associated with Fowler's solution (potassium arsenate). *Gastroenterology* 1975;68:1582–1586.
87. Selby DM, Stocker JT, Ishak KG. Angiosarcoma of the liver in childhood: a clinicopathologic and follow-up study of 10 cases. *Pediatr Pathol* 1992;12:485–498.
88. Falk S, Krishnan J, Meis JM. Primary angiosarcoma of the spleen: a clinicopathologic study of 40 cases. *Am J Surg Pathol* 1993;17:959–970.
89. Neuhauser T, Derringer G, Thompson LDR, et al. Splenic angiosarcoma: a clinicopathologic and immunohistochemical study of 27 cases. *Mod Pathol* 2000;13:978–987.
90. Mikami T, Saegusa M, Akino F, et al. A Kaposi-like variant of splenic angiosarcoma lacking association with human herpesvirus 8. *Arch Pathol Lab Med* 2002;126:191–194.
91. Burke AP, Cowan D, Virmani R. Primary sarcoma of the heart. *Cancer* 1992;69:387–395.
92. Murinello A, Mendonca P, Abreu A, et al. Cardiac angiosarcoma: a review. *Rev Port Cardiol* 2007;28:577–584.
93. Booth AM, LeGallo RD, Stoler MH, Waldron PE, Cerilli LA. Pediatric angiosarcoma of the heart: a unique presentation and metastatic pattern. *Pediatr Dev Pathol* 2001;4:490–495.
94. Hottenrott G, Mentzel T, Peters A, Schröder A, Katenkamp D. Intravascular ("intimal") epithelioid angiosarcoma: clinicopathological and immunohistochemical analysis of three cases. *Virchows Arch* 1999;435:473–478.
95. Mena H, Ribas JL, Enzinger FM, Parisi JE. Primary angiosarcoma of the central nervous system: study of eight cases and review of the literature. *J Neurosurg* 1991;75:73–76.
96. Eusebi V, Carcangiu ML, Dina R, Rosai J. Keratin-positive epithelioid angiosarcoma of thyroid: a report of four cases. *Am J Surg Pathol* 1990;14:737–747.
97. Fanburg-Smith JC, Furlong MA, Childers EL. Oral and salivary gland angiosarcoma: a clinicopathologic study of 29 cases. *Mod Pathol* 2003;16:263–271.
98. Favia G, Lo Muzio L, Serpico R, Maiorano E. Angiosarcoma of the head and neck with intra-oral presentation: a clinicopathological study of four cases. *Oral Oncol* 2002;38:757–762.
99. Wenig BM, Abbondanzo SL, Heffess CS. Epithelioid angiosarcoma of the adrenal glands: a clinicopathologic study of nine cases with a discussion of the implications of finding

"epithelial-specific" markers. *Am J Surg Pathol* 1994;18:62–73.

100. Taxy JB, Battifora H. Angiosarcoma of the gastrointestinal tract: a report of three cases. *Cancer* 1988;62:210–216.

101. Allison KH, Yoder BJ, Bronner MP, Goldblum JR, Rubin BP. Angiosarcoma involving the gastrointestinal tract: a series of primary and metastatic cases. *Am J Surg Pathol* 2004;28:298–307.

102. McCaughey WTE, Dardick I, Barr R. Angiosarcoma of serous membranes. *Arch Pathol Lab Med* 1983;107:304–307.

103. Lin BT, Colby T, Gown AM, et al. Malignant vascular tumors of the serous membranes mimicking mesothelioma: a report of 14 cases. *Am J Surg Pathol* 1996;20:1431–1439.

104. Zhang PJ, Livolsi VA, Brooks JJ. Malignant epithelioid vascular tumors of the pleura: report of a series and literature review. *Hum Pathol* 2000;31:29–34.

105. Del Frate C, Mortele K, Zanardi R, et al. Pseudomesotheliomatous angiosarcoma of the chest wall and pleura. *J Thorac Imaging* 2003;18:200–203.

Angiosarcoma arising in nerve sheath tumors

106. Trassard M, Le Doussal V, Bui BN, Coindre JM. Angiosarcoma arising in a solitary schwannoma (neurilemmoma) of the sciatic nerve. *Am J Surg Pathol* 1996;20:1412–1417.

107. Mentzel T, Katenkamp D. Intraneural angiosarcoma and angiosarcoma arising in benign and malignant nerve sheath tumours: clinicopathological and immunohistochemical analysis of four cases. *Histopathology* 1999;35:114–120.

108. Rückert RI, Fleige B, Rogalla P, Woodruff JM. Schwannoma with angiosarcoma: report of a case and comparison with other type of nerve sheath tumors with angiosarcoma. *Cancer* 2000;89:1577–1585.

109. McMenamin ME, Fletcher CD. Expanding the spectrum of malignant change in schwannomas: epithelioid malignant change, epithelioid malignant peripheral nerve sheath tumor, and epithelioid angiosarcoma. A study of 17 cases. *Am J Surg Pathol* 2001;25:13–25.

110. Lee FY, Wen MC, Wang J. Epithelioid angiosarcoma arising in a deep-seated plexiform schwannoma: a case report and literature review. *Hum Pathol* 2007;38:1096–1101.

111. Macaulay RA. Neurofibrosarcoma of the radial nerve in von Recklinghausen's disease with metastatic angiosarcoma. *J Neurol Neurosurg Psychiatry* 1978;41:474–478.

112. Millstein DI, Tang CK, Campvell EW Jr. Angiosarcoma developing in a patient with neurofibromatosis (von Recklinghausen's disease). *Cancer* 1981;47:950–954.

113. Ducatman BS, Scheithauer BW. Malignant peripheral nerve sheath tumors with divergent differentiation. *Cancer* 1984;54:1049–1057.

114. Riccardi VM, Wheeler TM, Pickard LR, King B. The pathophysiology of neurofibromatosis: II. Angiosarcoma as a complication. *Cancer Genet Cytogenet* 1984;12:275–280.

115. Lederman SM, Martin EC, Laffey KT, Lefkowitch JH. Hepatic neurofibromatosis, malignant schwannoma, and angiosarcoma in von Recklihause's disease. *Gastroenterology* 1987;92:234–239.

116. Brown RW, Tornos C, Evans HL. Angiosarcoma arising from malignant schwannoma in a patient with neurofibromatosis. *Cancer* 1992;70:1141–1144.

117. Morphopoulos GD, Banerjee SS, Ali HH, et al. Malignant peripheral nerve sheath tumour with vascular differentiation: a report of four cases. *Histopathology* 1996;28:401–410.

118. Elli M, Can B, Ceyhan M, et al. Intrathoracic malignant peripheral nerve sheath tumor with angiosarcoma in a child with NF1. *Tumori* 2007;93:641–644.

Angiosarcoma in germ cell and ovarian tumors

119. Malagon HD, Valdez AM, Moran CA, Suster S. Germ cell tumors with sarcomatous components: a clinicopathologic and immunohistochemical study of 46 cases. *Am J Surg Pathol* 2007;31:1356–1362.

120. den Bakker MA, Ansink AC, Ewing-Graham PC. "Cutaneous-type" angiosarcoma arising in a mature cystic teratoma of the ovary. *J Clin Pathol* 2006;59:658–660.

121. Sahoo S, Ryan CW, Recant WM, Yang XJ. Angiosarcoma masquerading as embryonal carcinoma in the metastasis from a mature testicular teratoma. *Arch Pathol Lab Med* 2003;127:360–363.

122. Nielsen GP, Young RH, Prat J, Scully RR. Primary angiosarcoma of the ovary: a report of seven cases and review of the literature. *Int J Gynecol Pathol* 1997;16:378–382.

123. Jylling AM, Jörgensen L, Holund B. Mucinous cystadenocarcinoma in combination with hemangiosarcoma in the ovary. *Pathol Oncol Res* 1999;5:318–319.

124. Pillay K, Essa AS, Chetty R. Borderline serous cystadenocarcinoma with coexistent angiosarcoma: an unusual form of ovarian carcinoma. *Int J Surg Pathol* 2001;9:317–321.

Angiosarcoma associated with foreign bodies

125. Jennings TA, Peterson L, Axiotis CA, et al. Angiosarcoma associated with foreign body material: a report of three cases. *Cancer* 1988;62:2436–2444.

126. Ben-Izhak O, Kerner H, Brenner B, Lichtig C. Angiosarcoma of the colon developing in a capsule of a foreign body: report of a case with associated hemorrhagic diathesis. *Am J Clin Pathol* 1992;97:416–420.

127. Ben-Izhak O, Vlodavsky E, Ofer A, et al. Epithelioid angiosarcoma associated with a Dacron vascular graft. *Am J Surg Pathol* 1999;23:1418–1422.

128. Cokelaere K, Vanvuchelen J, Michielsen P, Sciot R. Epithelioid angiosarcoma of the splenic capsule: report of a case reiterating the concept of inert foreign body tumorigenesis. *Virchows Arch* 2001;438:398–403.

129. Folpe Al, Johnston CA, Weiss SW. Cutaneous angiosarcoma arising in a gouty tophus: report of a unique case and review of foreign material-associated angiosarcomas. *Am J Dermatopathol* 2000;22:418–421.

Pediatric angiosarcoma

130. Lezama-del Valle P, Gerald WL, Tsai J, Meyers P, La Quaglia MP. Malignant vascular tumors in young patients. *Cancer* 1998;83:1634–1639.

131. Ferrari A, Casanova M, Bisogno G, et al. Malignant vascular tumors in children and adolescents: a report from the Italian and German Soft Tissue

Sarcoma Cooperative Group. *Med Pediatr Oncol* 2002;39:109–114.

132. Leake J, Sheehan MP, Rampling D, Ramani P, Atherton DJ. Angiosarcoma complicating xeroderma pigmentosum. *Histopathology* 1992;21:179–181.

133. Deyrup AT, Miettinen M, North PE, et al. Pediatric cutaneous angiosarcomas: a clinicopathologic study of 10 cases. *Am J Surg Pathol* 2011;35:70–75.

134. Deyrup AT, Miettinen M, North PE, et al. Angiosarcomas arising in the viscera and soft tissue of children and young adults: a clinicopathologic study of 15 cases. *Am J Surg Pathol* 2009;33:264–269.

Immunohistochemistry

135. Kuzu I, Bicknell R, Harris AL, et al. Heterogeneity of vascular endothelial cells with relevance to diagnosis of vascular tumours. *J Clin Pathol* 1992;45:143–148.

136. Miettinen M, Lindenmayer AE, Chaubal A. Endothelial cell markers CD31, CD34, and BNH9 antibody to H- and Y-antigens: evaluation of their specificity and sensitivity in the diagnosis of vascular tumors and comparison with von Willebrand's factor. *Mod Pathol* 1994;7:82–90.

137. DeYoung BR, Swanson PE, Angenyi ZB, et al. CD31 immunoreactivity in mesenchymal neoplasms of the skin and subcutis: report of 145 cases and review of putative immunohistologic markers of endothelial cell differentiation. *J Cutan Pathol* 1995;22:215–222.

138. Traweek ST, Kandalaft P, Mehta P, Battifora H. The human progenitor cell antigen (CD34) in vascular neoplasia. *Am J Clin Pathol* 1991;96:25–31.

139. Miettinen M, Wang ZF, Paetau A, et al. ERG transcription factor as an immunohistochemical marker for vascular endothelial tumors and prostatic carcinoma. *Am J Surg Pathol* 2011;35:432–441.

140. Appleton MA, Attanoos RL, Jasani B. Thrombomodulin as a marker of vascular and lymphatic tumours. *Histopathology* 1996;29:153–157.

141. Partanen TA, Alitalo K, Miettinen M. Lack of lymphatic vascular specificity of vascular endothelial growth factor receptor 3 in 185 vascular tumors. *Cancer* 1999;86:2406–2412.

142. Itakura E, Yamamoto H, Oda Y, Tsuneyoshi M. Detection and characterization of vascular endothelial growth factors and their receptors in a series of angiosarcomas. *J Surg Oncol* 2008;97:74–81.

143. Breiteneder-Geleff S, Soleiman A, Kowalski H, et al. Angiosarcomas express mixed endothelial phenotypes of blood and lymphatic capillaries: podoplanin as a specific marker for lymphatic endothelium. *Am J Pathol* 1999;154:385–394.

144. Kahn HJ, Bailey D, Marks A. Monoclonal antibody D2-40, a new marker of lymphatic endothelium, reacts with Kaposi sarcoma and a subset of angiosarcomas. *Mod Pathol* 2002;15:434–440.

145. Folpe AL, Chand EM, Goldblum JR, Weiss SW. Expression of Fli-1, a nuclear transcription factor, distinguishes vascular neoplasms from potential mimics. *Am J Surg Pathol* 20001;25:1061–1066.

146. Miettinen M, Fetsch JF. Distribution of keratins in normal endothelial cells and in a spectrum of vascular tumors: implications in tumor diagnosis. *Hum Pathol* 2000;31:1062–1067.

147. Al-Abbadi MA, Almasri NM, Al-Quran S, Wilkinson EJ. Cytokeratin and epithelial membrane antigen expression in angiosarcomas: an Immunohistochemical study of 33 cases. *Arch Pathol Lab Med* 2007;131:288–292.

148. Miettinen M, Lasota J. KIT expression in angiosarcomas and in fetal endothelial cells: lack of c-kit mutations in exons 11 and 17 in angiosarcoma. *Mod Pathol* 2000;13:536–541.

149. Manner J, Radlwimmer B, Hohenberger P, et al. MYC high level gene amplification is a distinctive feature of angiosarcomas after irradiation or chronic lymphedema. *Am J Pathol* 2010;176:34–39.

150. Guo T, Zhang L, Chang NE, et al. Consistent MYC and FLT4 gene amplification in radiation-induced angiosarcoma but not in other radiation-associated atypical vascular lesions. *Genes Chromosomes Cancer* 2011;50:25–33.

151. Mentzel T, Schildhaus HU, Palmedo G, Büttner R, Kutzner H. Postradiation cutaneous angiosarcoma after treatment of breast carcinoma is characterized by MYC amplification in contrast to atypical vascular lesions after radiotherapy and control cases: clinicopathological, immunohistochemical and molecular analysis of 66 cases. *Mod Pathol* 2012;25:75–85.

152. Fernandez AP, Sun Y, Tubbs RR, Goldblum JR, Billings SD. FISH for MYC amplification and anti-MYC immunohistochemistry: useful diagnostic tools in the assessment of secondary angiosarcoma and atypical vascular proliferations. *J Cutan Pathol* 2012;39:234–242.

153. Shon W, Sukov WR, Jenkins SM, Folpe AL. MYC amplification and overexpression in primary cutaneous angiosarcoma: a fluorescence in-situ hybridization and immunohistochemical study. *Mod Pathol* 2014;27:509–515.

154. Antonescu CR, Yoshida A, Guo T, et al. KDR activating mutations in human angiosarcomas are sensitive to specific kinase inhibitors. *Cancer Res* 2009;69:7175–7179.

155. Behjati S, Tarpey PS, Sheldon H, et al. Recurrent PTPRB and PLCG1 mutations in angiosarcoma. *Nat Genet* 2014;46:376–379.

156. Kindblom LG, Stenman G, Angervall L. Morphological and cytogenetic studies of angiosarcoma in Stewart-Treves syndrome. *Virchows Arch A Pathol Anat Histopathol* 1991;419:439–445.

157. Gill-Benso R, Lopez-Gines C, Soriano P, et al. Cytogenetic study of angiosarcoma of the breast. *Genes Chromosomes Cancer* 1994;10:210–212.

158. Schuborg C, Mertens F, Rydholm A, et al. Cytogenetic analysis of four angiosarcomas from deep and superficial soft tissue. *Cancer Genet Cytogenet* 1998;100:52–56.

159. Wong KF, So CC, Wong N, et al. Sinonasal angiosarcoma with marrow involvement at presentation mimicking malignant lymphoma: cytogenetic analysis using multiple techniques. *Cancer Genet Cytogenet* 2001;129:64–68.

160. Zu Y, Pele MA, Yan Z, et al. Chromosomal abnormalities and p53 gene mutation in a cardiac angiosarcoma. *Appl Immunohistochem Mol Morphol* 2001;9:24–28.

Kaposi's sarcoma

161. Oriel JD. Moritz Kaposi (1837–1902). *Int J STD AIDS* 1997;8:715–717.

162. Tappero JW, Conant MA, Wolfe SF, Berger TG. Kaposi's sarcoma: epidemiology, pathogenesis, histology, clinical spectrum, staging criteria and therapy. *J Am Acad Dermatol* 1993;28:371–395.

163. Uldrick TS, Whitby D. Update on KSHV epidemiology, Kaposi Sarcoma pathogenesis, and treatment of Kaposi Sarcoma. *Cancer Lett* 2011;305:150–162.

164. Ganem D. KSHV and the pathogenesis of Kaposi sarcoma: listening to human biology and medicine. *J Clin Invest* 2010;120:939–949.

165. Chang Y, Cesarman E, Pessin MS, *et al.* Identification of herpes virus-like DNA sequences in AIDS-associated Kaposi's sarcoma. *Science* 1994;266:1865–1869.

166. Moore PJ, Chang Y. Detection of Herpes virus-like DNA sequences in Kaposi's sarcoma patients with and those without HIV-infection. *N Engl J Med* 1995;332:1181–1185.

167. Ambroziak JA, Blackbourn DJ, Herndier BG, *et al.* Herpes-like sequences in HIV-infected and uninfected Kaposi's sarcoma patients. *Science* 1995;268:582–583.

168. Du MQ, Bacon CM, Isaacson PG. Kaposi sarcoma-associated herpesvirus/human herpesvirus 8 and lymphoproliferative disorders. *J Clin Pathol* 2007;60:1350–1357.

169. Duprez R, Lacoste V, Briere J, *et al.* Evidence for a multifocal origin of multicentric advanced lesions of Kaposi sarcoma. *J Natl Cancer Inst* 2007;99:1086–1094.

170. Iscovich J, Boffetta P, Franceschi S, Azizi E, Sarid R. Classic Kaposi sarcoma: epidemiology and risk factors. *Cancer* 2000;88:500–517.

171. Hiatt KM, Nelson AM, Lichy J, Fanburg-Smith JC. Classic Kaposi sarcoma in the United States over the last two decades: a clinicopathologic and molecular study of 438 non-HIV-related Kaposi sarcoma patients with comparison to HIV-related Kaposi sarcoma. *Mod Pathol* 2008;21:572–582.

172. Gottlieb GJ, Ackerman AB. Kaposi's sarcoma: an extensively disseminated form in young homosexual men. *Hum Pathol* 1982;13:882–892.

173. Niedt GW, Myskowski PL, Urmacher C, *et al.* Histologic predictors of survival in acquired immunodeficiency syndrome-associated Kaposi's sarcoma. *Hum Pathol* 1992;23:1419–1426.

174. Moskowitz LB, Hensley GT, Gould EW, Weiss SD. Frequency and anatomic distribution of lymphadenopathic Kaposi's sarcoma in the acquired immunodeficiency syndrome. *Hum Pathol* 1985;16:447–456.

175. Ioachim HL, Adsay V, Giancotti FR, Dorsett B, Melamed J. Kaposi's sarcoma of internal organs: a multiparameter study of 86 cases. *Cancer* 1995;75:1376–1385.

176. O'Connell KM. Kaposi's sarcoma: histopathological study of 159 cases from Malawi. *J Clin Pathol* 1977;30:687–695.

177. Slavin G, Cameron HM, Forbes C, Mitchell RM. Kaposi's sarcoma in East African children: a report of 51 cases. *J Pathol* 1970;100:198–199.

178. Taylor JF, Templeton AC, Vogel CL, Ziegler JL, Kyalwazi SK. Kaposi's sarcoma in Uganda: a clinicopathological study. *Int J Cancer* 1971;8:122–135.

179. Penn I. Kaposi's sarcoma in transplant recipients. *Transplantation* 1997;64:669–673.

180. Farge D, Lebbe C, Marjanovic Z, *et al.* Human herpes virus-8 and other risk factors for Kaposi's sarcoma in kidney transplant recipients. *Transplantation* 1999;67:1236–1242.

181. Lebbé C, Legendre C, Francès C. Kaposi sarcoma in transplantation. *Transplant Rev (Orlando)* 2008;22:252–261.

182. Ackerman AB. Subtle clues to diagnosis by conventional microscopy: the patch stage of Kaposi's sarcoma. *Am J Dermatopathol* 1979;1:165–172.

183. Radu O, Pantanowitz L. Kaposi sarcoma. *Arch Pathol Lab Med* 2013;137:289–294.

184. Kao G, Johnson FB, Sulica VI. The nature of hyaline (eosinophilic) globules and vascular slits in Kaposi's sarcoma. *Am J Dermatopathol* 1990;12:256–267.

185. Nickoloff BJ. The human progenitor cell antigen (CD34) is localized on endothelial cells, dermal dendritic cells, and perifollicular cells in formalin-fixed normal skin, and on proliferatin endothelial cells and stromal spindle-shaped cells in Kaposi's sarcoma. *Arch Dermatol* 1991;127:523–529.

186. Cheuk W, Wong KO, Wong CS, *et al.* Immunostaining for human herpesvirus 8 latent nuclear antigen-1 helps distinguish Kaposi sarcoma from its mimickers. *Am J Clin Pathol* 2004;121:335–342.

187. Patel RM, Goldblum JR, Hsi ED. Immunohistochemical detection of human herpes virus-8 latent nuclear antigen-1 is useful in the diagnosis of Kaposi sarcoma. *Mod Pathol* 2004;17:456–460.

188. Beckstead JH, Wood GS, Fletcher V. Evidence for the origin of Kaposi's sarcoma from lymphatic endothelium. *Am J Pathol* 1985;119;294–300.

189. Jussila L, Valtola R, Partanen TA, *et al.* Lymphatic endothelium and Kaposi's sarcoma spindle cells detected by antibodies against the vascular endothelial growth factor receptor-3. *Cancer Res* 1998;58:1599–1604.

190. Weninger WA, Partanen TA, Breiteneder-Geleff S, *et al.* Expression of vascular endothelial growth factor receptor-3 and podoplanin suggests a lymphatic endothelial cell origin of Kaposi's sarcoma tumor cells. *Lab Invest* 1999;79:243–251.

191. McDonagh DP, Liu J, Gaffey MJ, *et al.* Detection of Kaposi's sarcoma-associated herpesvirus-like DNA sequence in angiosarcoma. *Am J Pathol* 1996;149:1363–1368.

192. Ishak KG, Bijwaard KE, Makhlouf HR, *et al.* Absence of human herpesvirus 8 DNA sequences in vascular tumors of the liver. *Liver* 1998;18:124–127.

193. Lasota J, Miettinen M. Absence of Kaposi's sarcoma-associated virus (human herpesvirus-8) sequences in angiosarcoma. *Virchows Arch* 1999;434:51–56.

194. Schmid H, Zietz C. Human herpesvirus 8 and angiosarcoma: analysis of 40 cases and review of the literature. *Pathology* 2005;37:284–287.

195. Haferkamp O, Rosenau W, Lennert K. Vascular transformation of lymph node

sinuses due to venous obstruction. *Arch Pathol* 1971;92:81–83.

196. Cook PD, Czerniak B, Chan JK, *et al.* Nodular spindle-cell vascular transformation of lymph nodes: a benign process occurring predominantly in retroperitoneal lymph nodes draining carcinomas that can simulate Kaposi's sarcoma or metastatic tumor. *Am J Surg Pathol* 1995;19:1010–1020.

197. Chan JK, Frizzera G, Fletcher CD, Rosai J. Primary vascular tumors of the lymph nodes other than Kaposi's sarcoma: analysis of 39 cases and delineation of two new entities. *Am J Surg Pathol* 1992;16:335–358.

Chapter 23
Glomus tumor, glomangiopericytoma, myopericytoma, and juxtaglomerular tumor

Markku Miettinen
National Cancer Institute, National Institutes of Health

This chapter covers glomus tumor and related entities: glomangiopericytoma (sinonasal hemangiopericytoma) and myopericytoma. These tumors are sometimes classified as *perivascular cell tumors*. The term *glomus tumor* has historically also been applied to jugulotympanic and some other paragangliomas (glomus jugulare, glomus tympanicum); however, *paraganglioma* is the proper designation for these tumors. Hemangiopericytoma of soft tissues, which is essentially synonymous to solitary fibrous tumor, is discussed in Chapter 12. Myofibromas, although sometimes likened to myopericytomas, are discussed in Chapter 9. Juxtaglomerular cell tumor is included because of its histologic resemblance to glomus tumor.

Glomus tumor

A glomus tumor shows mesenchymal differentiation similar to the specialized smooth muscle cells of the glomus bodies that regulate peripheral blood flow. Most glomus tumors are benign, but rare atypical and malignant examples exist.

Glomus bodies

Normal glomus bodies are present in the distal extremities, such as fingers, and other acral locations, but they are detected infrequently because of their small size. Perhaps the most common location to encounter normal glomus bodies is the coccygeal area, where small clusters of glomus cells are located ventral to the tip of coccyx.[1] In one study, these bodies were detected in 18 of 37 coccygectomy specimens,[2] and another study revealed 2 glomus bodies in 2 of 382 excisions of pilonidal sinus (0.5%).[3] The author and colleagues have also encountered glomus bodies in a wall of a tailgut cyst (Figure 23.1), and such an instance has been reported at least once.[4]

The glomus cell clusters in the coccygeal region can be present in an area several millimeters in diameter, and in some cases the distinction from a small glomus tumor can become arbitrary, especially if larger solid clusters of glomus cells are present.

Clinical features

Glomus tumors are more commonly diagnosed in young adults, but they occur in a wide age range (Figure 23.2). Glomus tumors arise in various superficial soft tissue locations, but most distinctive is their occurrence in the nail bed area, consistent with the concept that they are related to the perivascular glomus cells that regulate blood flow in distal extremities. Subungual lesions have a female predominance.[5] Glomus tumors also can occur in more proximal parts of the extremities and sometimes in the trunk wall, but they are rare in the head and neck (Figure 23.3A,B).

Subungual glomus tumors are typically small, only a few millimeters in diameter, and typically painful, as are glomus tumors of other soft tissue sites.[5] These glomus tumors have an association with neurofibromatosis type 1 (NF1).[6,7]

The stomach is the origin for nearly all gastrointestinal glomus tumors, but rare examples have occurred in the colon.[8,9] According to our experience, these tumors occur in

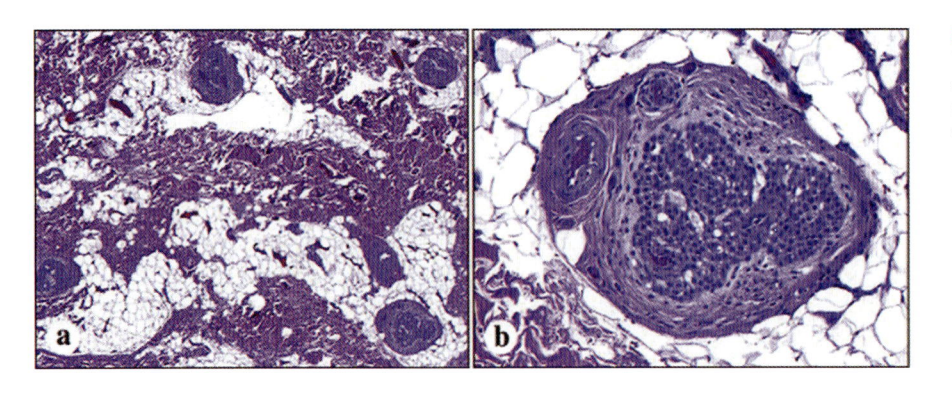

Figure 23.1 Glomus bodies in a tailgut cyst wall in the coccygeal area form small cell nests scattered in a fibrous stroma. The nests are surrounded by a fibrous band and are composed of uniform epithelioid cells.

Modern Soft Tissue Pathology, Second Edition, ed. Markku Miettinen. Published by Cambridge University Press. © Cambridge University Press 2016.

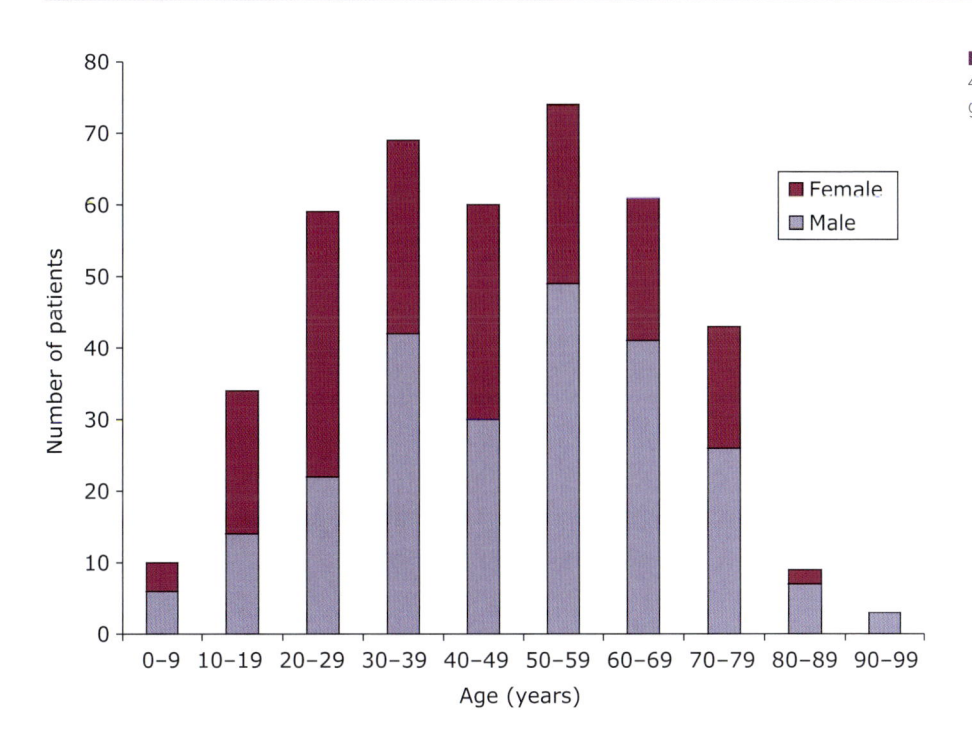

Figure 23.2 Age and sex distribution of 432 civilian patients with glomus tumors. Visceral glomus tumors are excluded.

adult patients in a wide age range, with a significant female predominance. Tumor size varies from 1 cm to 7 cm, and mitotic activity is low, usually <1 per 50 HPFs and generally <5 per 50 HPFs in the mitotically active cases. Nearly all follow a benign clinical course, despite many tumors having apparent vascular invasion and focal cytologic atypia. One malignant gastric glomus tumor metastasized to the liver, however; this tumor was large, 6.5 cm, and had only 1 mitosis per 50 HPFs. In the author's experience, malignant behavior is a rare and unpredictable occurrence in a gastric glomus tumor. Multiple apparently intravascular tumors have also been reported.[10]

Pulmonary glomus tumors are usually small, 1-cm to 3-cm nodules, and they can form intrabronchial lesions, clinically simulating a carcinoid tumor. There is a predilection for young to middle-aged adults with an apparent male predominance. A small number of mediastinal examples have also been reported. Malignant glomus tumors have occurred, both in the lungs and the mediastinum, and these have been large tumors usually (>5 cm), with mitotic activity often approaching 10 mitoses per 10 HPFs.[11] The possibility of a metastasis from another source must be considered in a malignant pulmonary glomus tumor. Contiguous lymph node involvement has been reported in a pulmonary glomus tumor with low mitotic activity.[12]

The kidney is a rare site for a glomus tumor. One series of three cases included men aged from 36 to 81 years with tumors of 3 cm to 7 cm in diameter. All tumors were located beneath the kidney capsule, abutting it in some cases, and all patients were alive and well after a relatively short follow-up. Histologically, these tumors showed a spectrum of glomus tumor with solid, glomangioma-like and glomangiomyoma-like features.[13] Occurrence in the renal pelvis has also been reported, where a glomus tumor can cause ureteropelvic obstruction and hydronephrosis, and simulate a renal pelvic carcinoma on clinical and gross inspection.[14]

Benign glomus tumors can recur, and rare malignant variants occur. They most commonly metastasize to the lungs and sometimes to the abdominal cavity, or into the intestines or mesenteries.

Pathology

Most glomus tumors are circumscribed ovoid or round superficial yellowish to tan-red nodules that measure a few millimeters in diameter. Some reach a larger size, usually <2 cm. Most glomus tumors are located in the dermis or subcutis, and rare examples are intramuscular. The tumor can be solid or mucoid on sectioning, and some examples are cystic and others surrounded by a fibrous capsule. Some examples appear hemorrhagic (Figure 23.4).

Histologically, the peripheral glomus tumors of skin and soft tissues form variably circumscribed nodules composed of clusters or solid sheets of glomus cells. Some lesions are surrounded by collagenous capsule-like bands (Figure 23.5). The cellular nests can be set in a myxoid or collagenous matrix, with some examples having an angiomatoid appearance (Figure 23.6). Cytologically, the solid examples show uniform, round or slightly polygonal epithelioid-like cells with sharp cellular borders and eosinophilic, even oncocytic cytoplasm. The nuclei are small and uniform (Figure 23.7). Some examples contain glomus cells lining cavernously dilated vascular spaces. These variants have also been called glomangiomas (Figure 23.8). Rare examples are intramuscular with diffuse growth, and intravascular growth is also observed (Figure 23.9). Some glomus tumors are dispersed as small clusters in fibrous stroma, and many have areas of hemorrhage with focal hemosiderin.

Glomangiomyoma is the designation for a tumor that combines the features of a glomus tumor and an angioleiomyoma: glomus cells lining vascular spaces interspersed by mature vascular smooth muscle cells (Figure 23.10).

Chapter 23: Glomus tumor, glomangiopericytoma, myopericytoma, and juxtaglomerular tumor

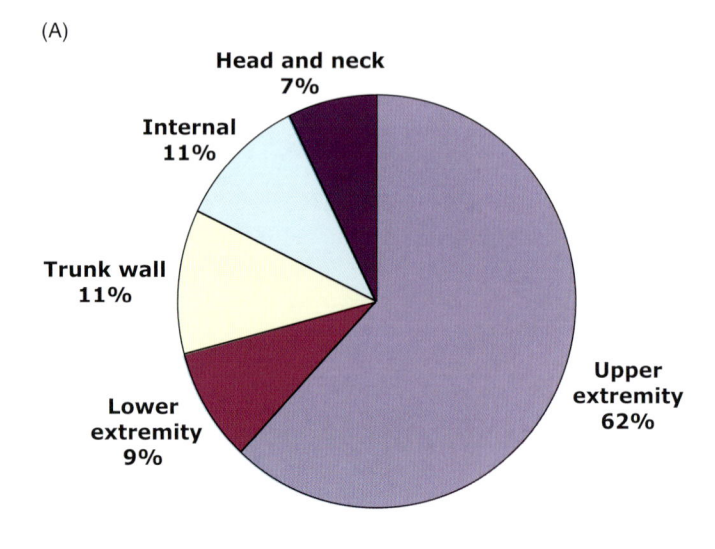

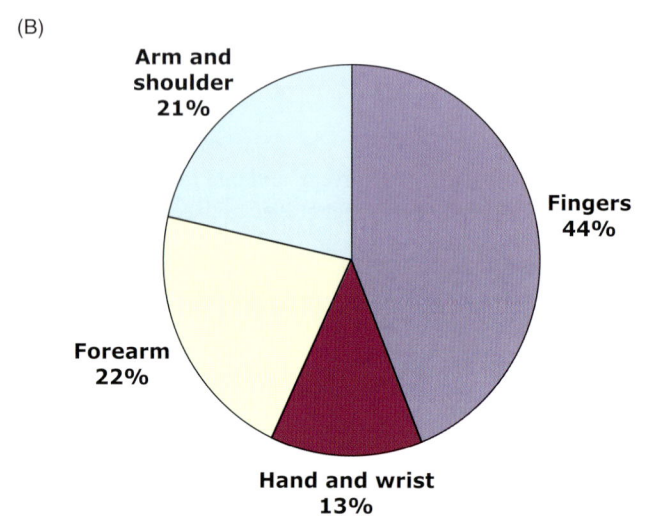

Figure 23.3 (A) Location of 359 glomus tumors. (B) Location of 221 glomus tumors in the upper extremities.

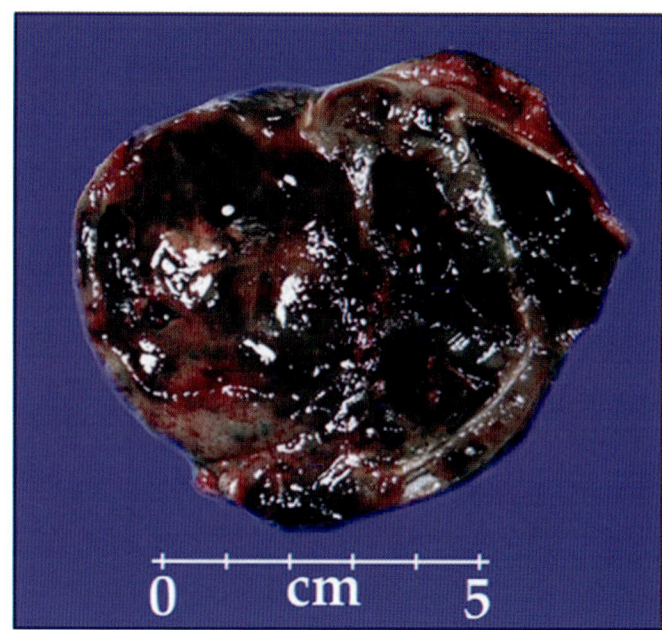

Figure 23.4 Grossly, some glomus tumors form hemorrhagic-cystic masses.

Gastric glomus tumors are usually relatively small (2–4 cm), but they occasionally exceed 5 cm. They are often surrounded by a fibrous capsule-like band and usually have a significant solid component, often with a hemangiopericytoma-like pattern, but are otherwise similar to their peripheral counterparts. Despite the presence of apparent vascular invasion (intravascular growth) in some cases, most tumors behave in a benign manner. It is uncertain at present which criteria might predict malignant behavior of a gastric glomus tumor, but the presence of atypia or increased mitotic rate (>5 per 50 HPFs) must be viewed as worrisome, similar to the criteria employed for the gastrointestinal stromal tumors.

Atypical and malignant glomus tumors

Cytologically atypical, mitotically active, or overtly sarcomatous glomus tumors are rare, probably <1% of all glomus tumors. The recognition of such sarcomatous tumors as glomus tumors must be based on the recognition of foci of conventional glomus tumor, usually seen in the periphery. The proposed definitions, based on a recent large series, are:[15]

(1) Glomus tumor with nuclear atypia (symplastic glomus tumor) usually forms a small nodule with marked focal atypia, but low mitotic rate. Tumors with these features do not have an increased risk of recurrence (Figure 23.11a,b).

(2) Glomus tumors with risk for malignant behavior (malignant glomus tumor, glomangiosarcoma) are those that are intramuscular and >2 cm, or tumors with atypical mitoses, or marked atypia with regular mitoses (>5 per 50 HPFs). Of such tumors, 38% were found to metastasize.

Malignant glomus tumors might have a round cell (glomus cell) or spindle cell morphology (Figures 23.11c,d and 23.12). Vascular invasion can also occur; however, this feature is also seen in benign glomus tumors. Consistent with expectation, a 2.8-cm glomus tumor in the thumb of a 48-year-old woman with 12 mitoses per 10 HPFs metastasized into the lungs 8 months after surgery.[16] Glomus tumors with a superficial location but mitotic activity >5 per 50 HPFs, and all large and deep tumors must be considered as having an uncertain malignant potential. Infiltrative glomus tumors in deep locations often recur locally, but they do not seem to metastasize.[17]

Immunohistochemistry and ultrastructure of glomus tumors

Glomus tumor cells are positive for vimentin and SMA, and 20% to 30% are focally or extensively positive for CD34. Laminin- and collagen IV-positive basement membranes

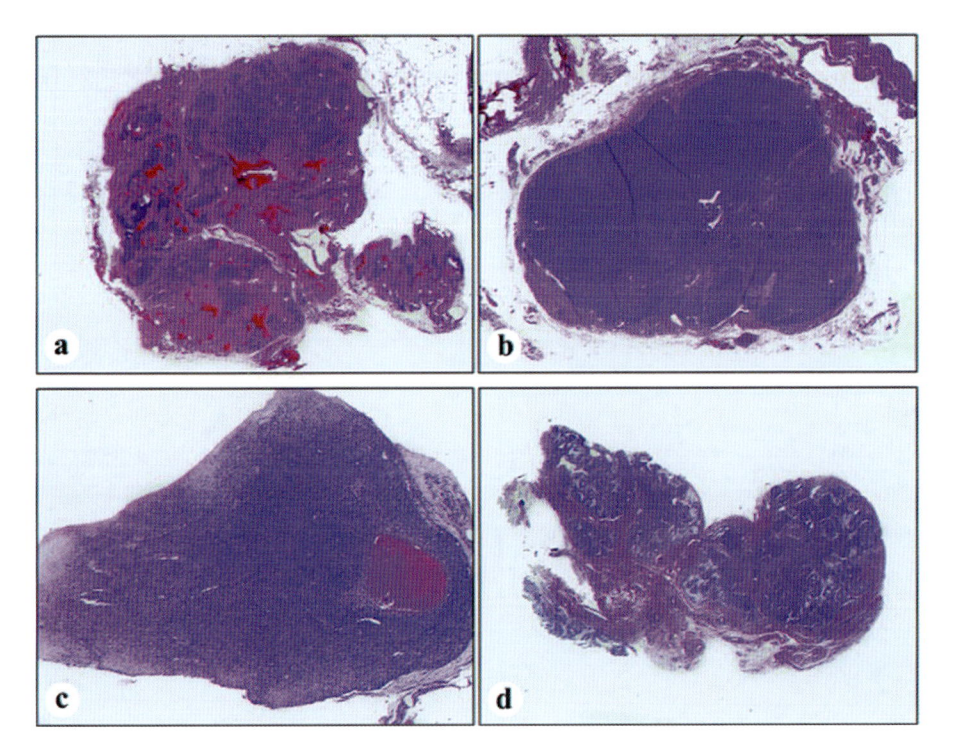

Figure 23.5 Low-magnification appearance of glomus tumor. (**a,d**) Examples composed of nests of glomus cells in a fibrous matrix. (**b,c**) Examples composed of solid sheets of glomus cells with occasional dilated vessels.

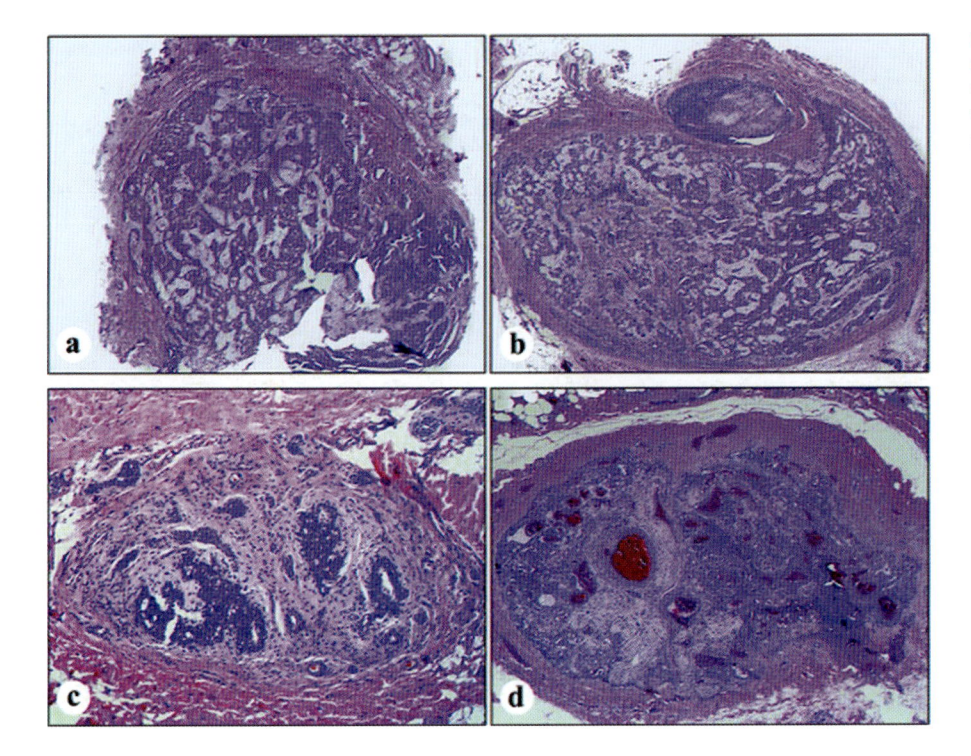

Figure 23.6 Examples of glomus tumors. (**a,b**) Nested pattern with myxoid stroma. (**c**) A small glomus tumor with fibrous stroma. (**d**) A well-demarcated, possibly intravascular glomus tumor with dilated vascular lumina.

typically surround the tumor cells in a net-like pattern (Figure 23.13). Heavy caldesmon and calponin expression varies. All glomus tumors are negative for desmin, keratins, and S100 protein.[18,19] Glomus tumors are negative for chromogranin, but focal synaptophysin positivity can occur, especially in the gastric examples.[9] S100 protein immunohistochemistry often reveals minimal nerve twigs within the tumor.

Electron microscopic studies have documented smooth muscle features in the glomus cells: prominent actin bundles, pinocytic vesicles, and basement membranes.[5,20] Genetic changes in sporadic glomus tumors are unknown; those in

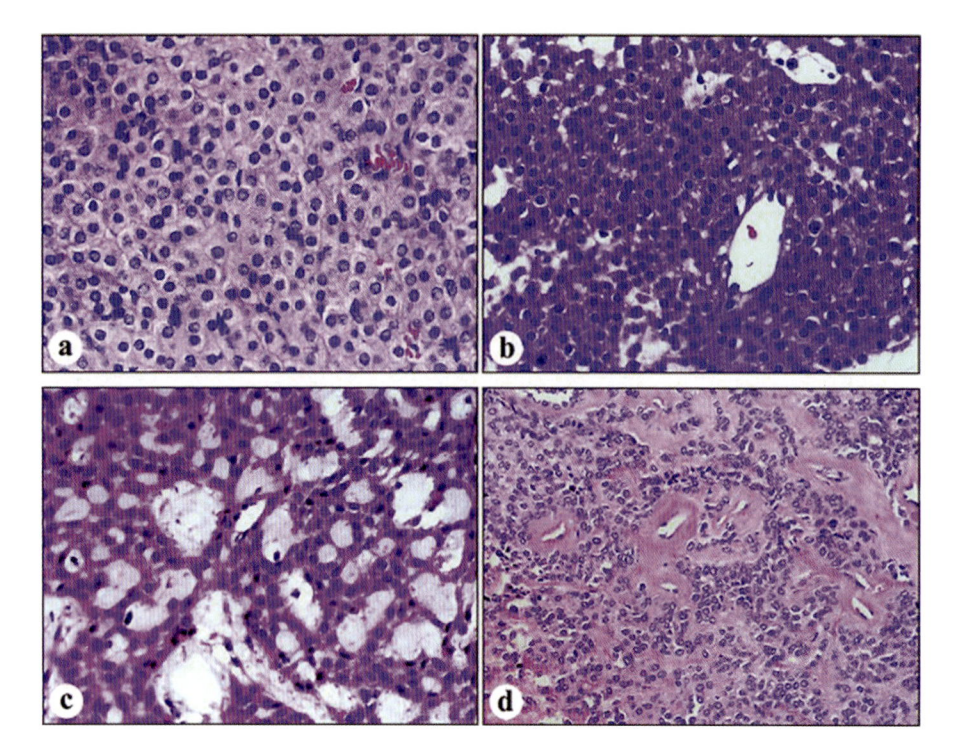

Figure 23.7 (a–c) Cytologic features of glomus tumor. The tumor cells are uniform, with relatively small nucleoli and variably eosinophilic, abundant cytoplasm. (d) Perivascular sclerosis and interstitial fibrosis can be present.

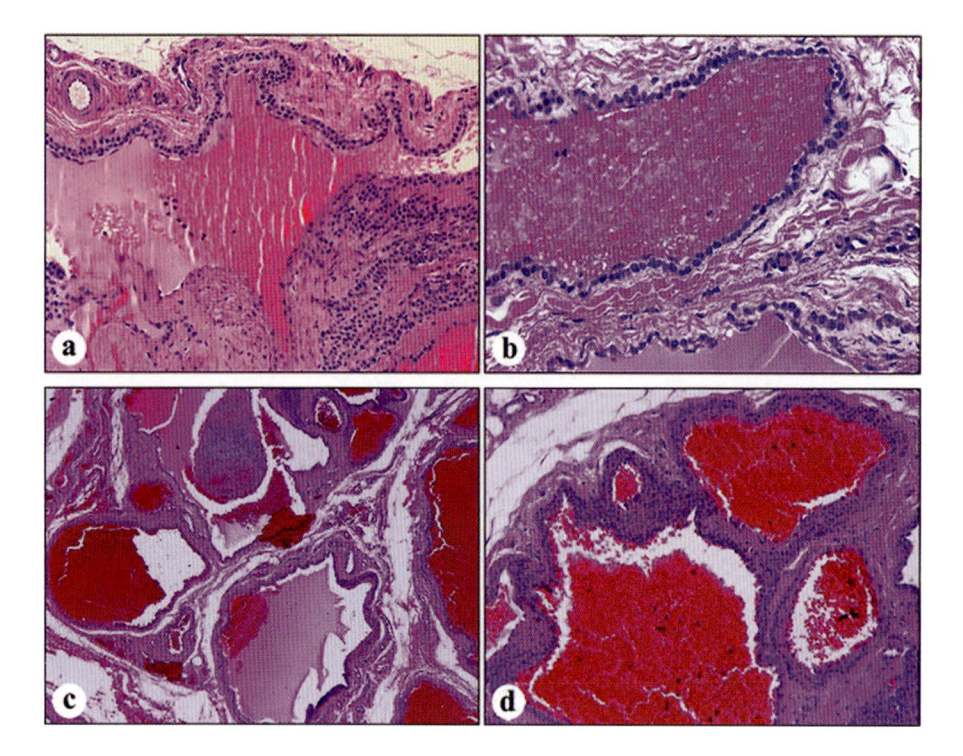

Figure 23.8 A glomangioma variant of a glomus tumor with markedly dilated vascular lumina lined by glomus cells.

familial glomus tumors (glomuvenous malformations) are discussed later in this chapter.

Differential diagnosis

Glomus tumors can resemble certain skin adnexal tumors, especially the more solid variants of hidradenoma (eccrine acrospiroma). Presence of ductal structures, at least focally clearer cytoplasm, and keratin expression help to identify the latter. Superficial variants of the Ewing family of tumors can be distinguished by their much higher cellularity and scant cytoplasm.

In the lung and gastrointestinal tract, a glomus tumor can be easily confused with a carcinoid or low-grade neuroendocrine carcinoma. A variably developed nested pattern is typical

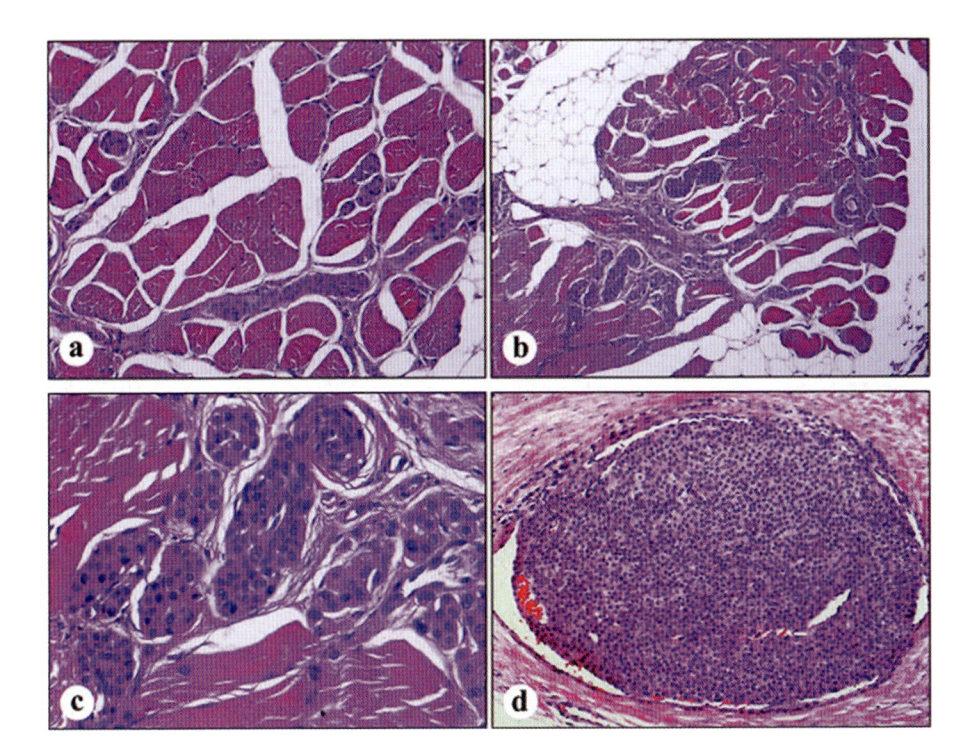

Chapter 23: Glomus tumor, glomangiopericytoma, myopericytoma, and juxtaglomerular tumor

Figure 23.9 (**a–c**) A rare example of an intramuscular glomus tumor infiltrating between skeletal muscle fibers. (**d**) An intravascular glomus tumor.

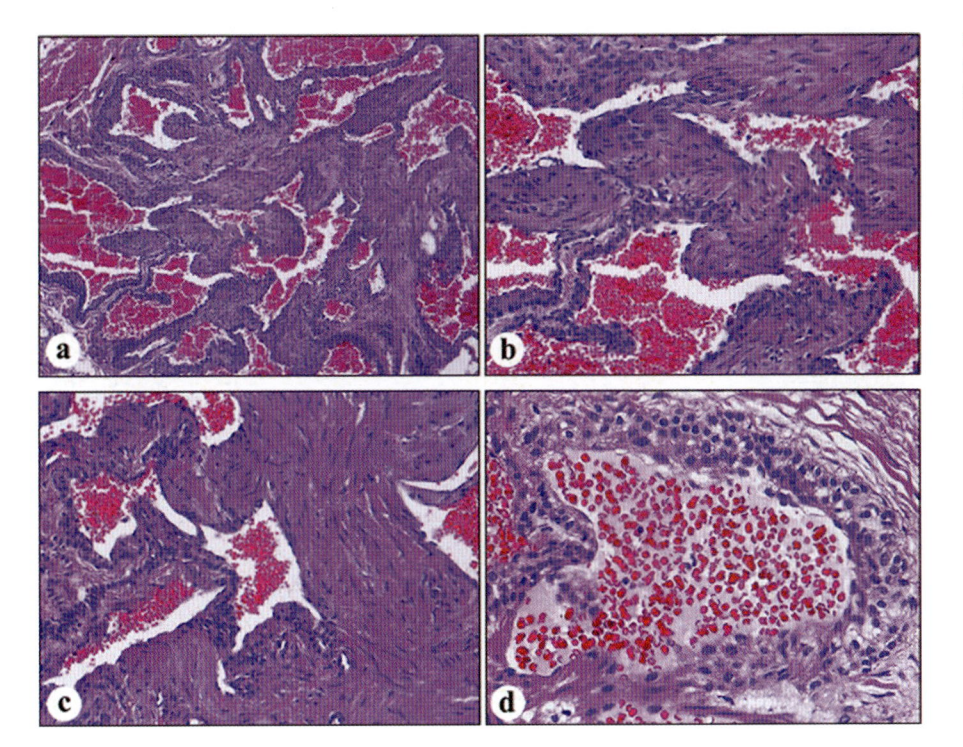

Figure 23.10 Glomangiomyoma is the designation for a glomus tumor combining glomus cells and vascular smooth muscle elements.

of neuroendocrine tumors, whereas glomus tumor cells are usually in diffuse sheets or large nests or compartments. Immunohistochemistry offers the easiest resolution in problem cases. Positivity for keratins, chromogranin, and synaptophysin with actin negativity supports an epithelial neuroendocrine neoplasm. Some glomus tumors can be focally positive for synaptophysin, but are uniformly negative for chromogranin and keratin, and positive for SMA.

A hemangiopericytoma is composed of less-differentiated cells with scant cytoplasm. This tumor almost uniformly lacks markers for smooth muscle cells, in contrast to a glomus tumor.

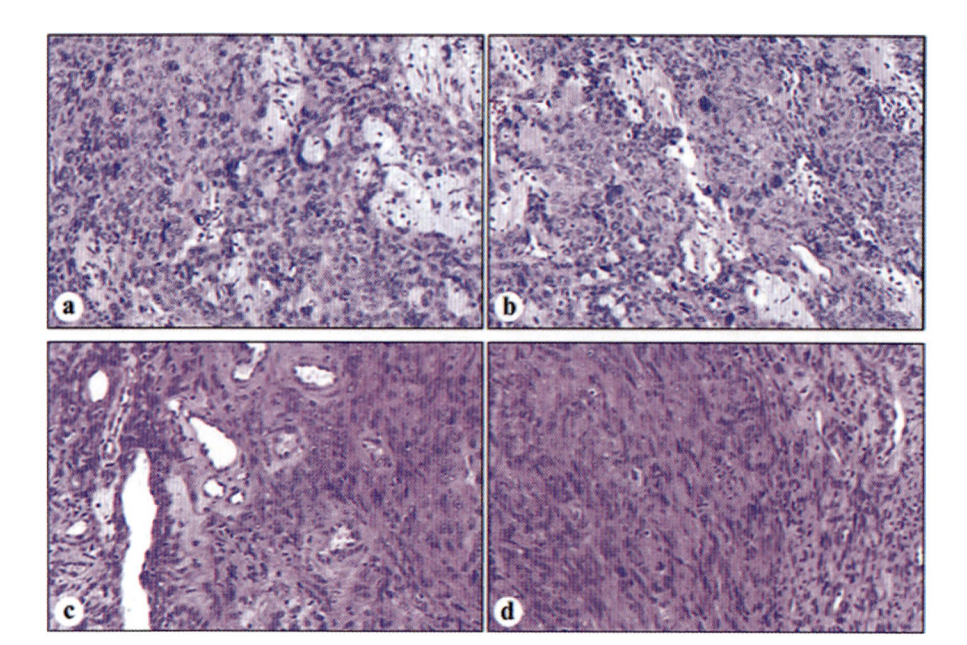

Figure 23.11 (a,b) Glomus tumor with cytologic atypia. (c,d) Glomangiosarcoma with spindle cell proliferation, which is mitotically active. Typical glomus tumor elements were seen elsewhere in the tumor.

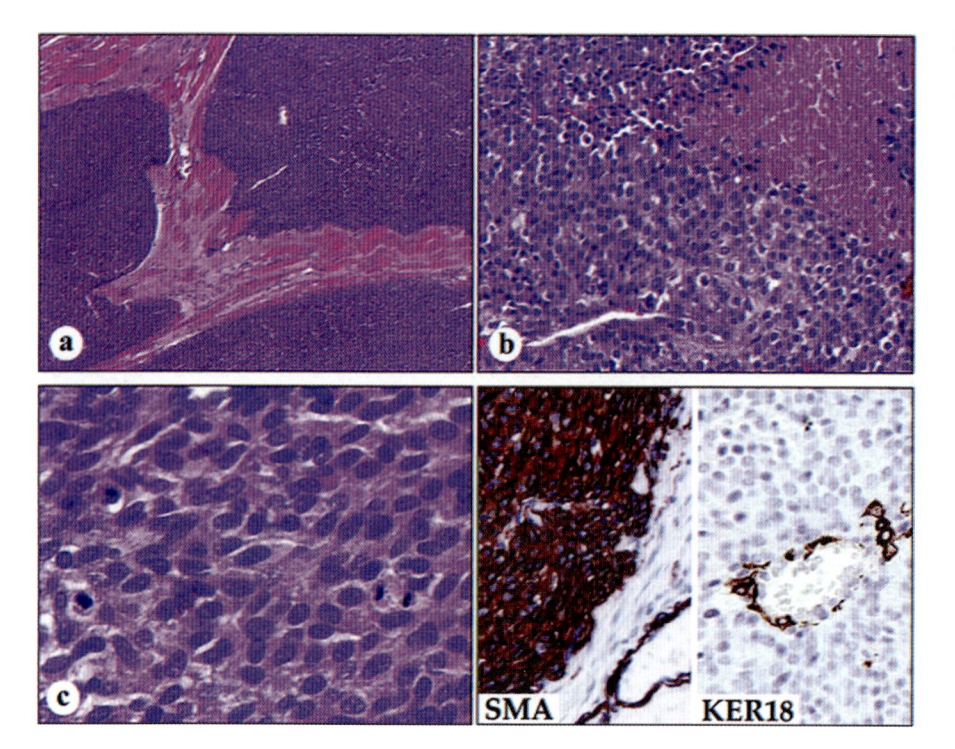

Figure 23.12 (a) A malignant glomus tumor in the esophagus shows a plexiform growth pattern between normal smooth muscle elements; this pattern is common in gastrointestinal glomus tumors in general. (b) The tumor forms diffuse sheets of epithelioid eosinophilic cells with foci of coagulative necrosis. (c) Significant mitotic activity (6 per 10 HPFs) was present. (d) Immunohistochemical features: positivity for SMA and negativity for keratin 18 help to separate this tumor from a neuroendocrine carcinoma.

Genetics of glomus tumor

Glomuvenous malformations are relatively rare compared with venous malformations, and they often occur on a familial basis. The lesions are often present at birth, expand during childhood, and are usually evident by the age of 20 years. Clinically, they often appear as cobblestone-like pink to purplish blue skin lesions in the extremities, and the subcutis can also be involved. Like ordinary glomus tumors, the lesions are often painful.[21]

Affected persons with familial glomuvenous malformations have nonsense germline mutations in the gene encoding for glomulin, located at 1p21–22. In limited studies of lesional tissue, a second-hit nonsense mutation involving the glomulin gene seems to be present, leading to complete inactivation of the glomulin gene and lack of glomulin protein. Glomulin alterations in this disease therefore follow the principle of biallelic inactivation according to

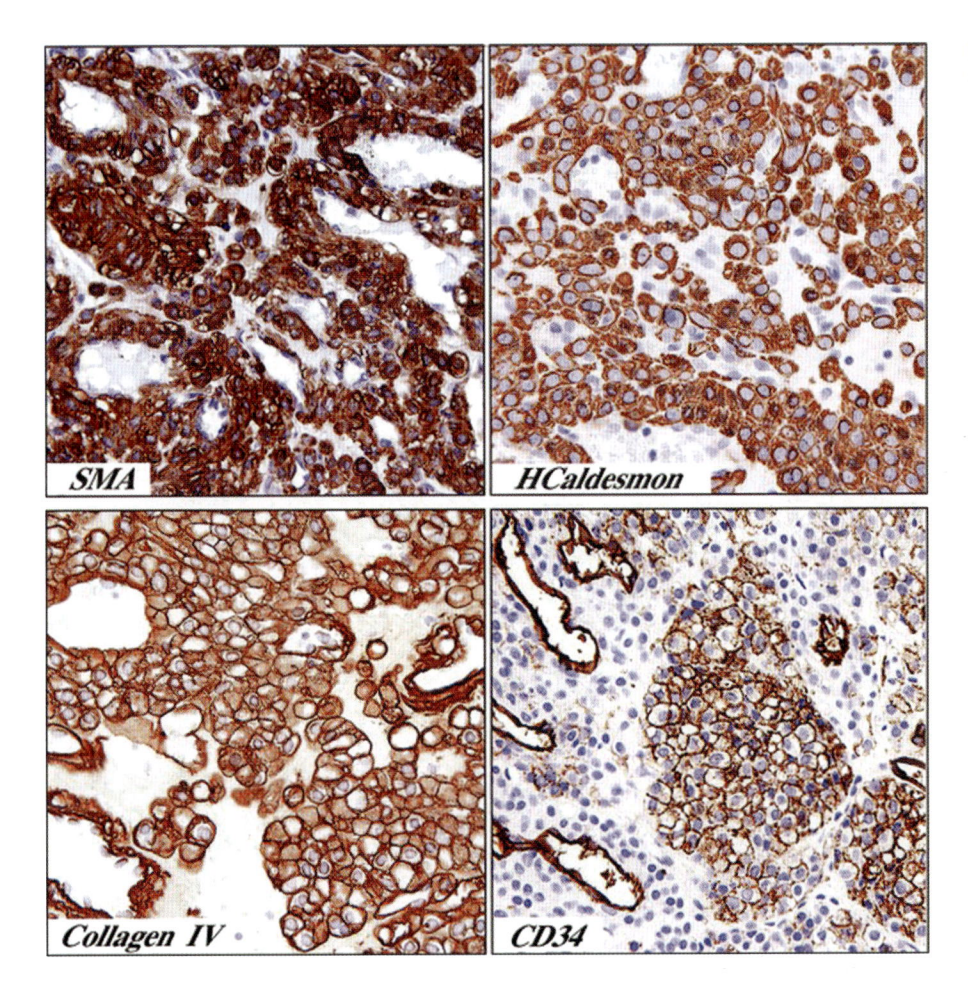

Figure 23.13 Immunohistochemical features of glomus tumor. There is strong positivity for alpha SMA and heavy caldesmon, membrane positivity for collagen IV, and focal positivity for CD34.

the principle for classic tumor-suppressor genes. The function of glomulin is not known, but it is ubiquitously expressed.[22,23]

Gene fusions involving MIR143-NOTCH genes have been detected in both benign and malignant glomus tumors.[24] Neurofibromatosis 1-associated glomus tumors have been shown to contain biallelic NF1 gene inactivation, the same mechanism observed in neurofibromas of NF1 patients.[25]

Glomangiopericytoma

Clinical features

Glomangiopericytoma (sinonasal-type hemangiopericytoma [SNH]) is a rare tumor. It occurs in a wide age range, most commonly in the sixth and seventh decades, and is slightly more common in women. SNH can involve the nasal cavity only, paranasal sinuses, or both. Clinically it can resemble a nasal inflammatory polyp. It has a moderate tendency for recurrence (20–25%), often after a long interval, but it does not metastasize.[26–29] Conventional hemangiopericytoma/solitary fibrous tumors can also occur in the sinonasal area.

Pathology

Grossly, the tumor is a polypoid submucosal lesion of <1 cm to 8 cm, with a median size of 3 cm. Histologically, the tumor is composed of syncytial sheets of uniform spindled or epithelioid cells in fascicular, diffuse, or whorled patterns. It is highly vascularized, but the vessels have narrow lumina, sometimes having a staghorn-like profile of vascular lumina, as typically observed in hemangiopericytoma/solitary fibrous tumor (Figures 23.14 and 23.15). Cytoplasmic retraction often gives tumor cells a clear cell appearance. Mitotic rate is low, but focal atypia and multinucleated tumor cells can be present.

Immunohistochemically, SNH is positive for SMA (Figure 23.15d), vimentin, and laminin, but is negative for desmin and keratins. Focal CD34 positivity occurs infrequently (<10%). In contrast, peripheral hemangiopericytoma is CD34 positive.

The typical genetic change in glomangiopericytoma is a beta-catenin gene (CTNNB1) mutation. Immunohistochemically this is reflected by strong nuclear beta-catenin positivity, which can also be a diagnostic marker.[30]

Differential diagnosis

Solitary fibrous tumor/hemangiopericytoma is a highly vascular spindle cell tumor usually containing gaping vessels, often

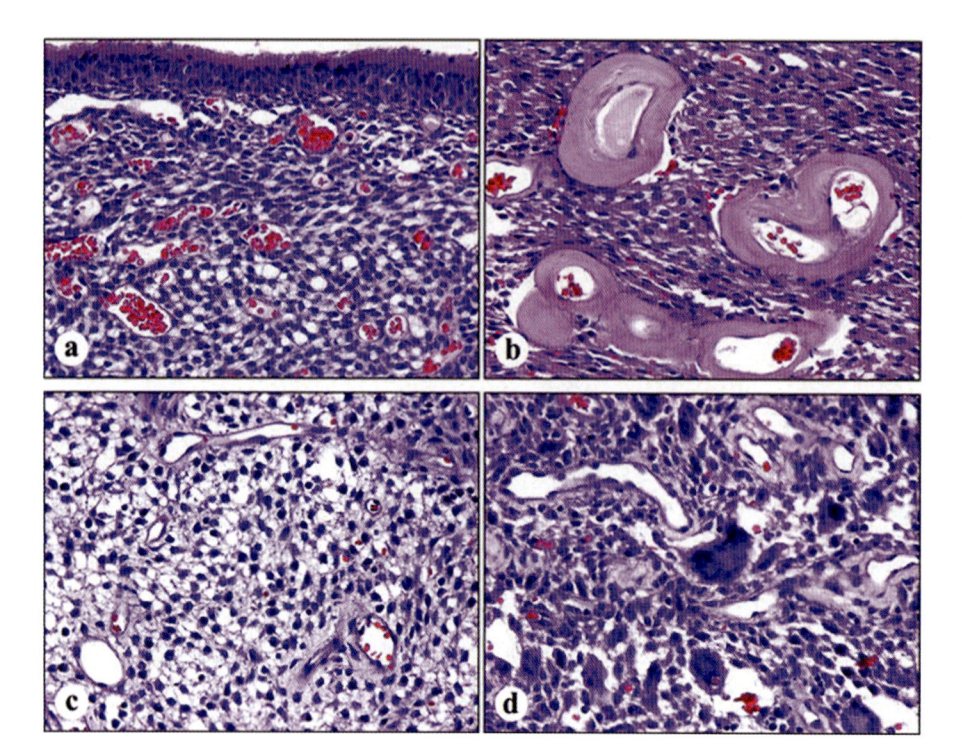

Figure 23.14 (a) Glomangiopericytoma Sinonasal hemangiopericytoma beneath intact respiratory mucosa. Note hyperemic vessels and uniform cytology, with uniform spindled to epithelioid tumor cells. (b) Perivascular hyalinization is a common feature. (c) This example is composed of uniform epithelioid cells in an edematous matrix. (d) Multinucleated tumor giant cells are a rare finding.

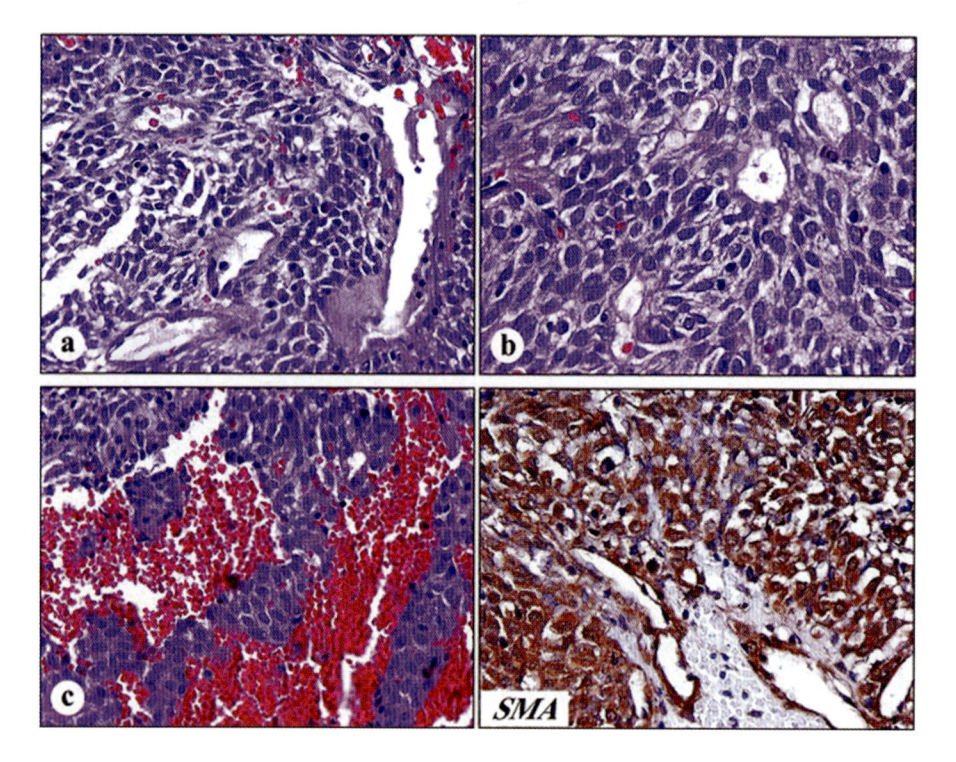

Figure 23.15 (a,b) Glomangiopericytoma Sinonasal hemangiopericytoma composed of epithelioid cells with a prominent hemangiopericytoma-like vascular pattern. (c) Hemorrhagic stroma can give the impression of a vascular tumor. (d) Strong positivity for alpha SMA is a typical finding.

with staghorn- or antler-like vessel profiles. This tumor is almost uniformly CD34 positive, and basement membrane proteins are usually limited to the perivascular areas.

Myopericytoma

This tumor is understood to be a hybrid of a glomus tumor and a vascular leiomyoma. Examples that correspond to myofibroma (with myofibroblastic and vascular hemangiopericytoma-like components) should be considered myofibromas (Chapter 9).

Clinical features

Myopericytoma occurs in a wide age range from the second decade to old age, but there is predilection for young and middle-aged adults, with a 2:1 male predominance. The most

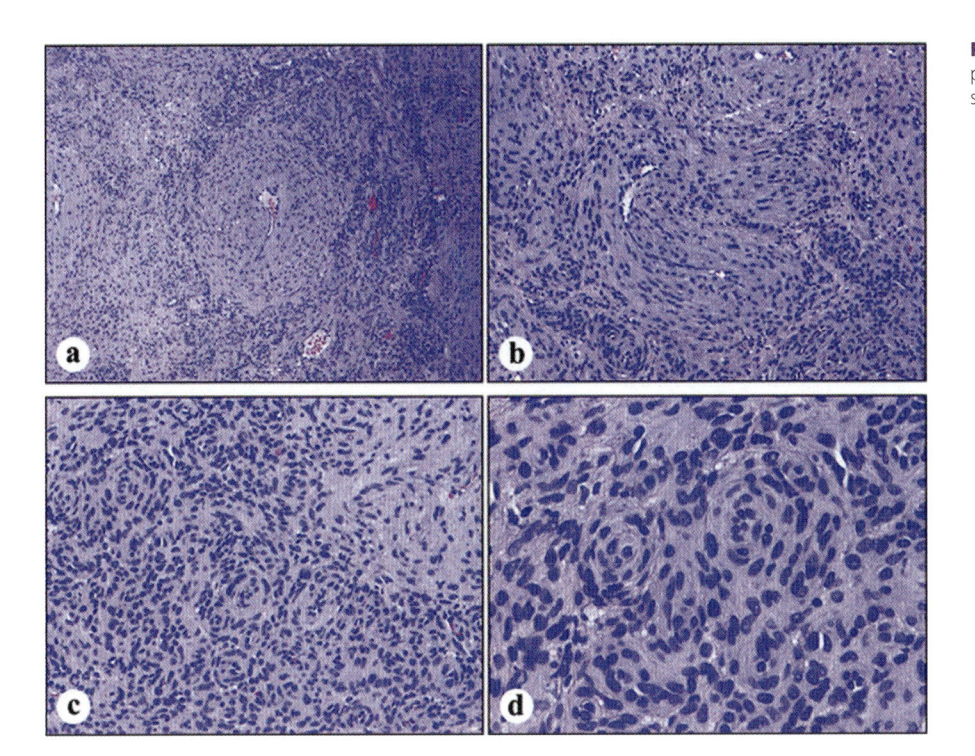

Figure 23.16 Myopericytoma containing prominent blood vessels and uniform ovoid-to-spindled cells arranged around the vessels.

common locations are the leg/knee, forearm/hand, and most cases have been reported in the extremities, more frequently distally. Rare examples have been reported in the head and neck, and trunk wall. The tumor usually forms a small subcutaneous nodule that can be painful. The duration of the lesion varies and can exceed 10 years. Local excision is sufficient, but local recurrence is possible.[31–33] Malignant variants reported in one series metastasized in 4 of 5 cases, and 3 of 5 patients died within one year.[34]

Pathology

A myopericytoma typically forms a well-demarcated subcutaneous nodule measuring 1 cm to 2 cm. The tumor has a prominent vascular pattern, often containing round vascular profiles with patent lumina (Figures 23.16 and 23.17). Surrounding the vessels are spindled or epithelioid (glomus-tumor-like) cells, often in a concentric pattern. Mitotic activity and atypia are limited in the conventional variant. Malignant examples with high mitotic activity have been reported, however.[34]

Immunohistochemically, myopericytomas are variably positive for alpha SMA (Figure 23.17d) and CD34, but they are negative for desmin, keratins, and S100 protein. Caldesmon positivity is detectable in some cases.

Juxtaglomerular cell tumor
Clinical features

The rare renal juxtaglomerular cell tumor was first reported in the late 1960s with to date just over 100 cases described.[35,36] This tumor occurs predominantly in young adults and with a 2:1 female predilection. It also can occur in children and older adults.[37,38] Clinically the tumor is typically associated with hypertension and hypokalemia, which resolve after tumor removal. Renin secretion by tumor cells causes secondary hyperaldosteronism leading to hypokalemia. Most tumors are clinically benign, but a case with metastases has been reported.[39] In most cases, nephron-sparing partial nephrectomy is the treatment of choice.

Pathology

Juxtaglomerular cell tumor of the kidney forms a sharply demarcated nodule that often measures up to 2–3 cm, but in some cases tumor has been >5 cm. The tumor is located in kidney parenchyma, usually in the cortex, but sometimes around the pelvis. Grossly the tumor is typically yellowish to pale brown and focal hemorrhage or fibrous septa can be present. A fibrous capsule typically surrounds the tumor.[38,40]

Histologically, the tumor often shows features of hemangiopericytoma by its content of staghorn vessels, and it is often composed of sheets or nests of epithelioid cells resembling glomus tumor cells (Figure 23.18). In some cases, proliferation of intervening non-neoplastic epithelioid elements creates a papillary architecture. The cell borders are usually not distinct. Significant focal nuclear atypia can be present, but mitotic activity is difficult to find. Occurrence of numerous mast cells is typical, and some cases have prominent lymphoid reaction. Some cases show loose myxoid areas with tumor cells grouped around blood vessels. Necrosis is rare.[35,36,40–42]

Immunohistochemically typical is universal positivity for CD34 with focal SMA and weak KIT expression (Figure 23.18). KIT also highlights numerous mast cells.

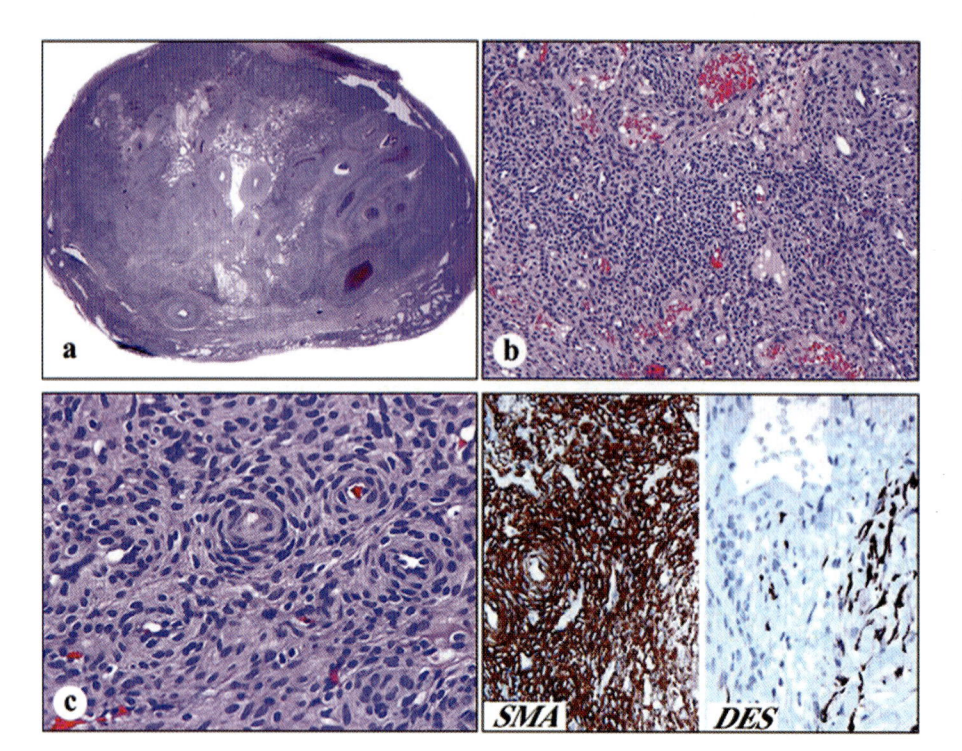

Chapter 23: Glomus tumor, glomangiopericytoma, myopericytoma, and juxtaglomerular tumor

Figure 23.17 (**a**) A myopericytoma forms a sharply demarcated mass with prominent blood vessels. (**b**) Some areas of the tumor resemble a glomus tumor. (**c**) Other areas show spindle cells concentrically arranged around blood vessels. (**d**) Strong positivity for alpha smooth muscle actin is typical.

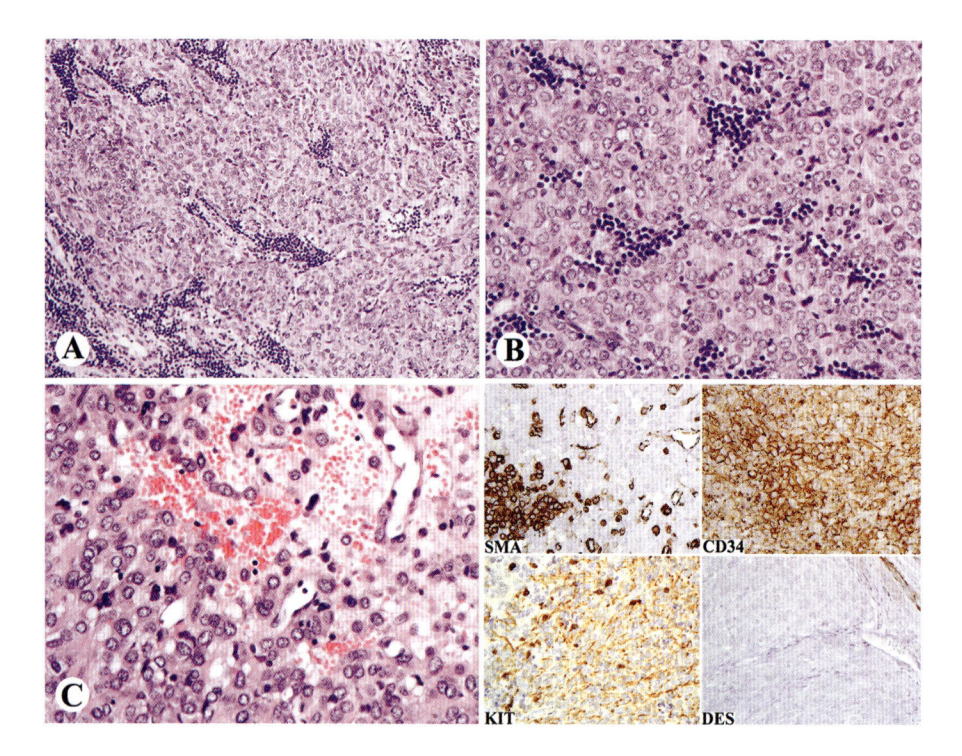

Figure 23.18 (**A–C**) Histological features of juxtaglomerular cell tumor. The tumor is composed of nests or solid sheets of ovoid to epithelioid cells and may contain abundant lymphoid reaction. Immunohistochemically the tumor cells are typically strongly positive for CD34 and focally positive for SMA and sometimes weakly for KIT, while being negative for desmin.

Renin expression has been demonstrated by immunohistochemistry, but has lesser practical significance. The tumor is typically negative for desmin and keratins.[41,42] Ultrastructurally typical are rhomboid crystals and round membrane-bound granules with renin-containing structures.[43,44] Loss of part of choromosome 9 may be a recurrent genetic change.[45]

634

References

Glomus tumor

1. Albreacht S, Zbieranowski I. Incidental glomus coccygeum: when a normal structure looks like a tumor. *Am J Surg Pathol* 1990;14:922–924.
2. Gatalica Z, Wang L, Lucio ET, Miettinen M. Glomus coccygeum in surgical pathology specimens: small troublemaker. *Arch Pathol Lab Med* 1999;123:905–908.
3. Santos LD, Chow C, Kennerson AR. Glomus coccygeum may mimic glomus tumour. *Pathology* 2002;34:339–343.
4. McDermott NC, Newman J. Tailgut cyst (retrorectal cystic hamartoma) with prominent glomus bodies. *Histopathology* 1991;18:265–266.
5. Tsuneyoshi M, Enjoji M. Glomus tumor: a clinicopathologic and electron microscopic study. *Cancer* 1982;50:1601–1607.
6. Sawada S, Honda M, Kamide R, Niimura M. Three cases of subungual glomus tumors with von Recklinghausen neurofibromatosis. *J Am Acad Dermatol* 1995;32(2 Pt 1):277–278.
7. Stewart DR, Sloan JL, Yao L, et al. Diagnosis, management, and complications of glomus tumours of the digits in neurofibromatosis type 1. *J Med Genet* 2010;47:525–532.
8. Appelman HD, Helwig EB. Glomus tumor of the stomach. *Cancer* 1969;23:203–213.
9. Miettinen M, Paal E, Lasota J, Sobin LH. Gastrointestinal glomus tumors: a clinicopathological, immunohistochemical, and molecular genetic study of 32 cases. *Am J Surg Pathol* 2002;26:301–311.
10. Haque S, Modlin IM, West AB. Multiple glomus tumors of the stomach with intravascular spread. *Am J Surg Pathol* 1992;16:291–299.
11. Gaertner EM, Steinberg DM, Huber M, et al. Pulmonary and mediastinal glomus tumors: report of five cases including pulmonary glomangiosarcoma. A clinicopathologic study with literature review. *Am J Surg Pathol* 2000;24:1105–1114.
12. Zhang Y, England DM. Primary pulmonary glomus tumor with contiguous spread to a peribronchial lymph node. *Ann Diagn Pathol* 2003;7:245–248.
13. Al-Ahmadie HA, Yilmaz A, Olgac S, Reuter VE. Glomus tumor in the kidney: a report of 3 cases involving renal parenchyma and review of the literature. *Am J Surg Pathol* 2007;31:585–591.
14. Herawi M, Parwani AV, Eldow D, Smolev JK, Epstein JI. Glomus tumor of renal pelvis: a case report and review of the literature. *Hum Pathol* 2005;36:299–302.
15. Folpe AL, Fanburg-Smith JC, Miettinen M, Weiss SW. Atypical and malignant glomus tumors: analysis of 52 cases with a proposal for the reclassification of glomus tumors. *Am J Surg Pathol* 2001;25:1–12.
16. Khoury T, Balos L, McGrath B, et al. Malignant glomus tumor: a case report and review of the literature, focusing on its clinicopathologic features and immunohistochemical profile. *Am J Dermatopathol* 2005;27:428–431.
17. Gould EW, Manivel JC, Albores-Saavedra J. Locally infiltrative glomus tumors and glomangiosarcomas: a clinical, ultrastructural, and immunohistochemical study. *Cancer* 1990;65:310–318.
18. Miettinen M, Lehto VP, Virtanen I. Glomus tumor cells: evaluation of smooth muscle and endothelial cell properties. *Virchows Arch B Cell Pathol* 1983;43:139–149.
19. Porter PL, Bigler SA, McNutt M, Gown AM. The immunophenotype of hemangiopericytomas and glomus tumors, with special reference to muscle protein expression: an immunohistochemical study and review of the literature. *Mod Pathol* 1991;4:46–52.
20. Murad TJ, Von Hasam Z, Murthy MSN. The ultrastructure of hemangiopericytoma and glomus tumor. *Cancer* 1968;22:1239–1249.
21. Boon LM, Mulliken JB, Enjolras O, Vikkula M. Glomuvenous malformation (glomangioma) and venous malformation: distinct clinicopathologic and genetic entities. *Arch Dermatol* 2004;140:971–976.
22. Brouillard P, Boon LM, Mulliken JM, et al. Mutations in a novel factor "glomulin" are responsible for glomuvenous malformations ("glomangiomas"). *Am J Hum Genet* 2002;70:866–874.
23. Brouillard P, Ghassibe M, Penington A, et al. Four common glomulin mutations cause two thirds of glomuvenous malformations ("familial glomangiomas"): evidence for a founder effect. *J Med Genet* 2005;42:e13.
24. Mosquera JM, Sboner A, Zhang L, et al. Novel MIR143-NOTCH fusions in benign and malignant glomus tumors. *Genes Chromosomes Cancer* 2013;52:1075–1087.
25. Brems H, Park C, Maertens O, et al. Glomus tumors in neurofibromatosis type 1: genetic, functional, and clinical evidence of a novel association. *Cancer Res* 2009;69:7393–7401.

Glomangiopericytoma (sinonasal hemangiopericytoma)

26. Compagno J, Hyams VJ. Hemangiopericytoma-like intranasal tumors: a clinicopathologic study of 23 cases. *Am J Clin Pathol* 1976;66:672–683.
27. Eichhorn JH, Dickersin GR, Bhan AK, Goodman ML. Sinonasal hemangiopericytoma: a reassessment with electron microscopy, immunohistochemistry, and long-term follow-up. *Am J Surg Pathol* 1990;14:856–866.
28. Thompson LD, Miettinen M, Wenig BM. Sinonasal-type hemangiopericytoma: a clinicopathologic and immunophenotypic analysis of 104 cases showing perivascular myoid differentiation. *Am J Surg Pathol* 2003;27:737–749.
29. Li XQ, Hisaoka M, Morio T, Hashimoto H. Intranasal pericytic tumors (glomus tumor and sinonasal hemangiopericytoma-like tumor): report of two cases and review of the literature. *Pathol Int* 2003;53:303–308.
30. Lasota J, Felisiak-Golabek A, Thompson LD, Wang ZF, Miettinen M. Nuclear expression and gain-of-function β-catenin mutation in glomangiopericytoma (sinonasal-type hemangiopericytoma): insight into pathogenesis and a diagnostic marker. *Mod Pathol* 2015;28:715–720.

Myopericytoma

31. Granter SR, Badizadegan K, Fletcher CD. Myofibromatosis in adults, glomangiopericytoma, and myopericytoma: a spectrum of tumors with perivascular myoid differentiation. *Am J Surg Pathol* 1998;22:513–525.

32. Mentzel T, Dei Tos AP, Sapi Z, Kutzner H. Myopericytoma of the skin and soft tissues: clinicopathologic and immunohistochemical study of 54 cases. *Am J Surg Pathol* 2006;30:104–113.

33. Matsuyama A, Hisaoka M, Hashimoto H. Angioleiomyoma: a clinicopathologic and immunohistochemical reappraisal with special reference to the correlation with myopericytoma. *Hum Pathol* 2007;38:645–651.

34. McMenamin ME, Fletcher CD. Malignant myopericytoma: expanding the spectrum of tumours with myopericytic differentiation. *Histopathology* 2002;41:450–460.

Juxtaglomerular cell tumor

35. Robertson PW, Klidjian A, Harding LK, et al. Hypertension due to a renin-secreting renal tumour. *Am J Med* 1967;43:963–976.

36. Kihara I, Kitamura S, Hoshino T, Seida H, Watanabe T. A hitherto unreported vascular tumor of the kidney: a proposal of "juxtaglomerular cell tumor." *Acta Pathol Jpn* 1968;18:197–206.

37. Haab F, Duclos JM, Guyenne T, Plouin PF, Corvol P. Renin secreting tumors: diagnosis, conservative surgical approach and long-term results. *J Urol* 1995;153:1781–1784.

38. McVicar M, Carman C, Chandra M, et al. Hypertension secondary to renin-secreting juxtaglomerular cell tumor: case report and review of 38 cases. *Pediatr Nephrol* 1993;7:404–412.

39. Duan X, Bruneval P, Hammadeh R, et al. Metastatic juxtaglomerular cell tumor in a 52-year-old man. *Am J Surg Pathol* 2004;28:1098–1102.

40. Martin SA, Mynderse LA, Lager DJ, Cheville JC. Juxtaglomerular cell tumor: a clinicopathologic study of four cases and review of the literature. *Am J Clin Pathol* 2001;116:854–863.

41. Kim HJ, Kim CH, Choi YJ, et al. Juxtaglomerular cell tumor of kidney with CD34 and CD117 immunoreactivity: report of 5 cases. *Arch Pathol Lab Med* 2006;130:707–711.

42. Kuroda N, Maris S, Monzon FA, et al. Juxtaglomerular cell tumor: a morphological, immunohistochemical and genetic study of six cases. *Hum Pathol* 2013;44:47–54.

43. Kodet R, Taylor M, Váchalová H, Pýcha K. Juxtaglomerular cell tumor. An immunohistochemical, electron-microscopic, and in situ hybridization study. *Am J Surg Pathol* 1994;18:837–842.

44. Hasegawa A. Juxtaglomerular cells tumor of the kidney: a case report with electron microscopic and flow cytometric investigation. *Ultrastruct Pathol* 1997;21:201–208.

45. Capovilla M, Couturier J, Molinié V, et al. Loss of chromosomes 9 and 11 may be recurrent chromosome imbalances in juxtaglomerular cell tumors. *Hum Pathol* 2008;39:459–462.

Chapter 24

Nerve sheath tumors

Markku Miettinen
National Cancer Institute, National Institutes of Health

The benign nerve sheath tumors discussed in this chapter include variants of neurofibroma, schwannoma, and granular cell tumors, and the rare perineurial cell tumors (perineuriomas) and nerve sheath myxomas. *Malignant peripheral nerve sheath tumor* (MPNST) is the term for malignant tumors arising in a neurofibroma. Nerve sheath tumors occur in both peripheral and cranial nerves. Their classification is discussed following the principles in the Armed Forces Institute of Pathology (AFIP) atlas of tumors of the peripheral nervous system.[1]

Alterations in the neurofibromatosis type 1 (NF1) gene leading to the absence or loss of function of the NF1 protein (neurofibromin) are the basis of the NF1 syndrome and associated neurofibromas. There is also evidence for somatic NF1 gene mutations being responsible for sporadic neurofibromas. Analogous to this, changes in the neurofibromatosis type 2 (NF2) gene that lead to the loss of function or absence of NF2 protein (merlin) are the pathogenesis of the NF2 syndrome. This syndrome entails bilateral vestibular schwannomas, meningiomas, and certain gliomas. Corresponding somatic genetic changes are thought to be behind the pathogenesis of sporadic schwannomas.

True neural tumors, such as paragangliomas and related tumors, and neuroectodermal tumors, such as malignant melanoma and clear cell sarcoma, are discussed in Chapters 25 and 26. Gastrointestinal autonomic nerve tumor (GANT) is now classified among the gastrointestinal stromal tumors (GISTs) as the ultrastructural variant of a GIST (Chapter 17). Although GANTs and GISTs are not nerve sheath tumors, they have a weak association with NF1 syndrome.

Neuroblastoma, a primitive childhood tumor with sympathetic nerve differentiation, peripheral primitive neuroectodermal tumor (PNET) and peripheral neuroepithelioma of the Ewing sarcoma family are discussed with small round cell tumors (Chapter 31).

Cellular components of nerve sheaths and their significance for tumorigenesis

The peripheral nerves contain long axons that originate from the spinal ganglia. The larger nerves form nerve trunks, which are composed of multiple nerve fascicles. Each fascicle is surrounded by the *perineurium*. The fascicles contain numerous axons immediately surrounded by Schwann cells. The nerve fascicles are separated by the *endoneurium*, a fibrous matrix containing fibroblasts. The nerve trunks are surrounded by the epineurium. Small nerves seen in skin and peripheral soft tissues represent nearly bare nerve fascicles that are separated from the surrounding tissue by the perineurium and an inconspicuous epineurium.

Autonomic nerves of visceral organs originate from the visceral ganglions. In the gastrointestinal tract, there are two neural plexuses with ganglia: the submucous (Auerbach's) and myenteric plexus (Meissner's) between the muscle layers. Both of these contain ganglion cells and nerves. In addition, numerous small nerves dispersed between the smooth muscle cells can be highlighted, for example with S100 protein immunostaining.

All components of the nerve sheath have been implicated in the pathogenesis of nerve sheath tumors. Schwann cells are spindled cells with elongated, complex, intertwining cell processes and well-developed basement membranes (basal or external laminas). The myelinated Schwann cells provide the myelin for the myelinated nerves, and the nonmyelinated cells surround nonmyelinated nerves. They are immunohistochemically recognized as S100-protein-positive cells surrounded by laminin and collagen type-IV-positive sheaths. They are the principal cells of schwannomas and are an essential cellular component in neurofibromas.

The endoneurium inside the nerve fascicles contains fibroblasts, many of which belong to the specific subset of CD34-positive fibroblasts. They participate in the genesis of neurofibromas with Schwann cells.

Perineurial cells are usually seen as a multilamellar sheath of spindled, elongated cells around the nerve fascicles. They represent a peripheral continuation of the pia-arachnoid membrane of the central nervous system and are identified as slender EMA-positive and S100-protein-negative cells that are surrounded by basement membranes. Perineuriomas are rare tumors that show perineurial cell differentiation.

The epineurium does not contain cell types specific for nerves. It consists of loose connective tissue containing

---

Modern Soft Tissue Pathology, Second Edition, ed. Markku Miettinen. Published by Cambridge University Press. © Cambridge University Press 2016.

fibroblasts, mast cells, histiocytes, lymphatics, and blood vessels. These components are the likely origin for nonschwannian, nonperineurial mesenchymal tumors of the peripheral nerves. All of these conditions are very rare and include hemangiomas, lipomas, and ganglion cysts of the nerve.

Neuromuscular hamartoma (neuromuscular choristoma, benign triton tumor)

This very rare lesion contains an intimate admixture of peripheral nerves and skeletal muscle cells, typically involving a nerve. Most cases have been reported in infants, with some of the masses being present at birth. Occurrence in older children and young adults has also been reported. The usual clinical finding is a mass of 2 cm to 4 cm that can involve a major nerve or nerve plexus, which can cause neurologic signs.[2-6] The involved sites especially include the head and neck, and the upper trunk.

The gross appearance is one of a rubbery to firm, sometimes encapsulated mass. Histologically typical is the intermingling of peripheral nerves and clusters of skeletal muscle fibers (Figure 24.1). Although the components are usually readily apparent in hematoxylin-eosin (HE) staining, the neural component can be highlighted as paler staining areas and the muscle as red foci in Masson trichrome staining (Figure 24.1d). The nerves can also be visualized as S100 protein positive and the skeletal muscle as desmin positive.

Occurrence with desmoid tumor has been reported in some cases.[4,5] The possible association with beta-catenin signaling is unclear. Possible histogenetic explanations for neuromuscular hamartoma include incorporation of skeletal muscle inside the nerve sheaths during development or abnormal differentiation of neuroectodermal elements. The designation *benign Triton tumor* has been applied to a neurofibroma with skeletal muscle differentiation, a very rare event.[7]

Morton neuroma (localized interdigital neuritis, intermetatarsal compression neuritis)

This term refers to a reactive condition involving one of the digital nerves around the distal metatarsal. Similar lesions occur in the hand, but rarely. The term *localized interdigital neuritis* is used in the AFIP fascicle on peripheral nerve tumors.[1]

The lesion most commonly occurs in middle-aged women and typically presents with a swelling and paroxysmal pain between the distal ends of the second and third, or third and fourth metatarsal bones. It is thought to result from minor chronic repetitive trauma combined with local ischemia, probably related to mechanical factors such as compressing shoes. Surgical excision of the fibrous nodule is sometimes necessary for pain relief.[8,9] Conservative alternatives beyond correction of footwear include corticosteroid, alcohol, or phenol injections.[10]

Histologically, the nodule (usually <1 cm) consists of fibrofatty tissue containing enlarged nerves expanded by marked epineurial and sometimes endoneurial fibrosis and myxoid stromal change (Figure 24.2). The number of axons and Schwann cells can be reduced, reflecting nerve damage. Fibrous thickening of the blood vessel walls is common, and vascular thrombosis can be present.[11-13]

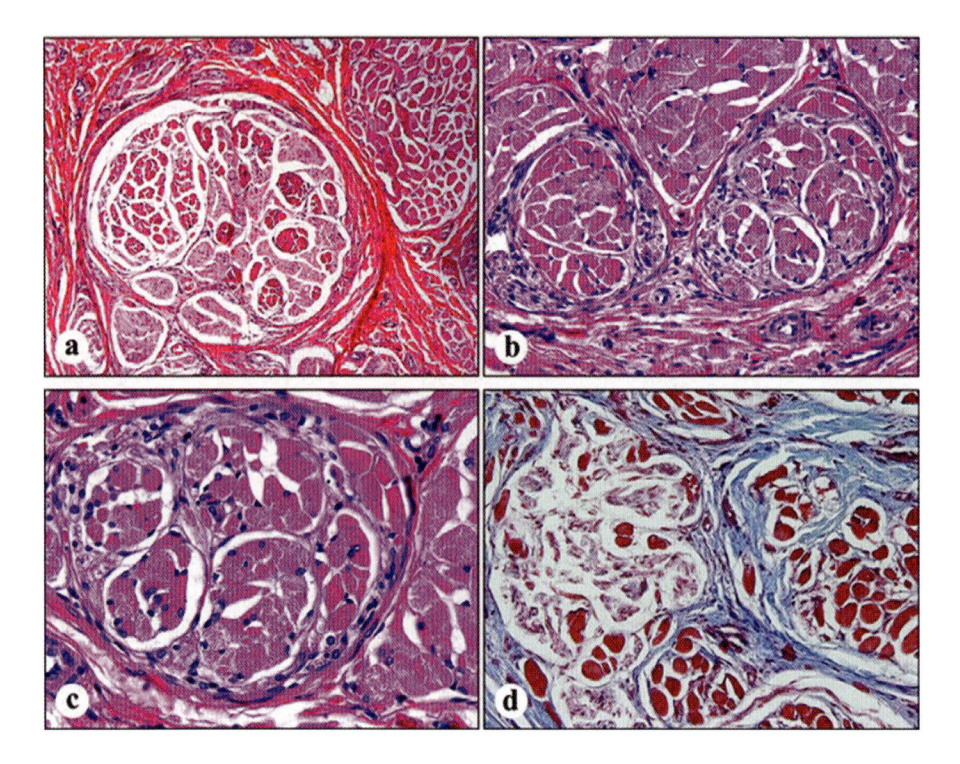

Figure 24.1 (a–c) Neuromuscular hamartoma consists of enlarged nerves that contain clusters of skeletal muscle fibers. (d) In trichrome staining, the neural elements are pale, and the muscle is red.

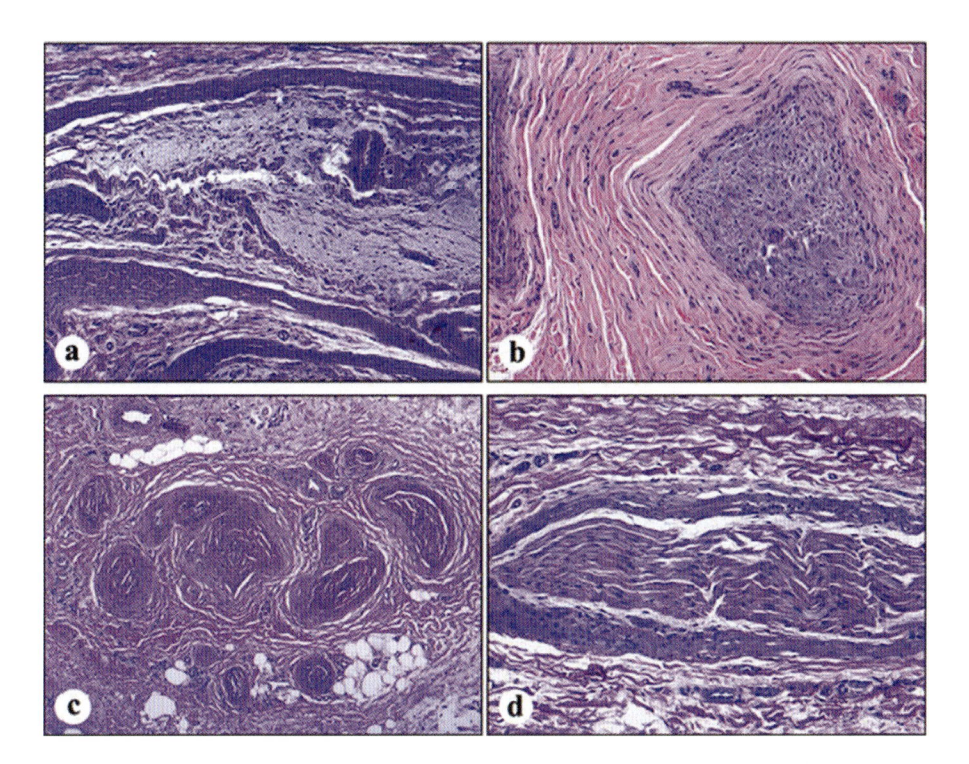

Figure 24.2 (a–c) Morton neuroma contains nerves with variable epineurial fibrosis. The nerve shows extensive stromal edema. (d) The nerve has mild myxoid change and prominent epineurial fibrosis.

Traumatic neuroma (amputation neuroma)

Traumatic neuroma represents an attempted, but insufficient reparative process to a totally or partially severed nerve, thus failing to re-establish the axonal connections, resulting in disorganized, tangled axonal proliferation. It can follow an amputation (hence the older term, *amputation neuroma*), other surgery, or nerve damage following other injuries.

Clinically, traumatic neuroma can appear as a small, painful nodule in the amputation stump or site of trauma or previous surgery, most commonly in the extremities. In some cases, the neuroma forms an asymptomatic nodule that is suspected of being a local tumor recurrence.[14] Small traumatic neuromas are the most common form, and these are typically incidental findings in re-excision specimens. Traumatic neuromas can be seen around the cystic or biliary ducts after cholecystectomy or other biliary surgery. Although these neuromas are usually asymptomatic incidental findings, they can compress bile ducts and cause biliary obstruction.[15,16]

Histologically, a traumatic neuroma consists of tangled, disorganized nerve fascicles containing axons and Schwann cells surrounded by perineurial cells, thus containing the complete set of cell types in a nerve. An amputation neuroma typically forms a sharply demarcated nodule that is encased by the epineurium (Figure 24.3), whereas the incidental traumatic neuromas form poorly delineated masses by irregularly criss crossing proliferating small nerves, often in the middle of collagenous scar tissue (Figure 24.4). These neural and schwannian components are immunohistochemically positive for neurofilaments and S100 protein, and are surrounded by EMA-positive perineurial cells.

Mucosal neuromas and ganglioneuromatosis

These lesions usually occur in connection with multiple endocrine neoplasia type 2b (MEN2b), with medullary thyroid carcinoma and pheochromocytoma in various parts of the gastrointestinal tract. The cause of this syndrome is an activating germline mutation in the gene encoding the RET receptor tyrosine kinase.[17–19]

Mucosal neuromas are visible in the lips and tongue as micronodular elevations or overall enlargement of the lips, and they are important clinical signs for MEN2b syndrome. Such lesions are seen only rarely as surgical specimens. Histologically, they contain profiles of hyperplastic nerves that can occur singly or in clusters, and these can have a thickened perineurium and prominent, hyperemic capillaries (Figure 24.5). A prominent EMA-positive perineurial cell layer has been suggested as a distinctive immunohistochemical finding.[20]

In some cases, mucosal biopsies of the gastrointestinal tract have been helpful in the diagnosis of MEN2b syndrome. The gastrointestinal lesions in MEN2b can involve diffusely the submucous and myenteric plexuses in a band-like manner, and typically they have both a neural and ganglionic component representing ganglioneuromatosis.[17–19] Solitary ganglioneuromas do not have any significant association with MEN syndrome, and they can occur in colonic adenomas.[20,21]

A mucosal ganglioneuroma can form plaque-like superficial lesions or polyps. The cellular components include ganglion cells and spindled Schwann cells. In addition, mixed inflammatory cells are sometimes present, and some cases contain prominent eosinophils (Figure 24.6).

Chapter 24: Nerve sheath tumors

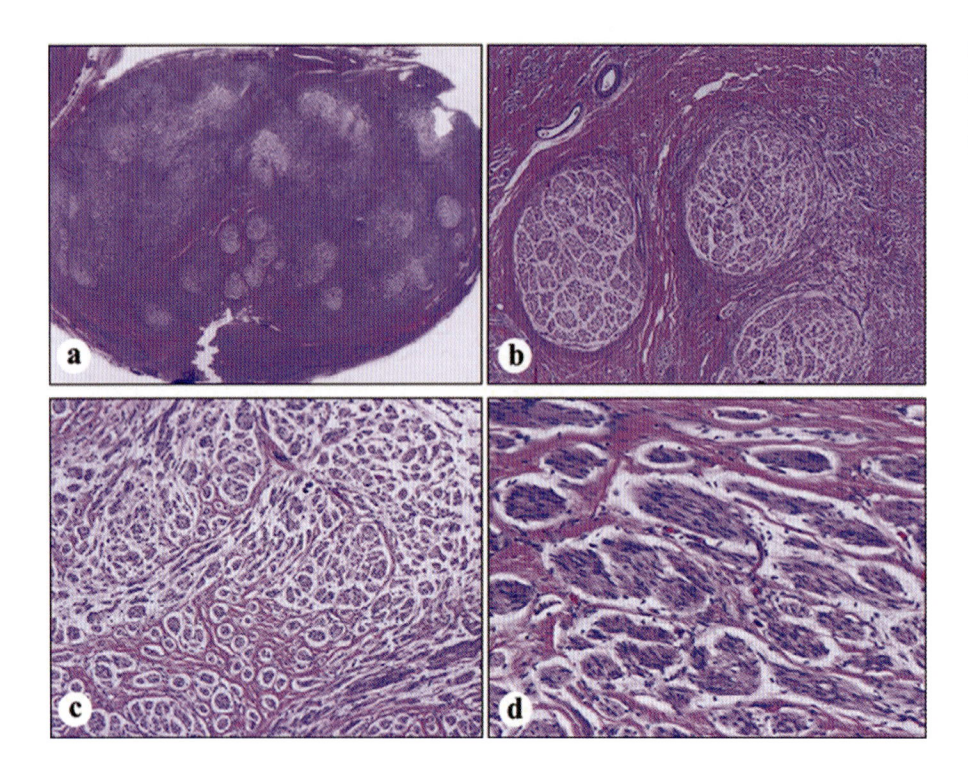

Figure 24.3 (a,b) Traumatic neuroma of the amputation neuroma type shows a nodule composed of regenerating nerves and marked epineurial fibrosis. (c,d) Irregular neuroaxonal bundles are present.

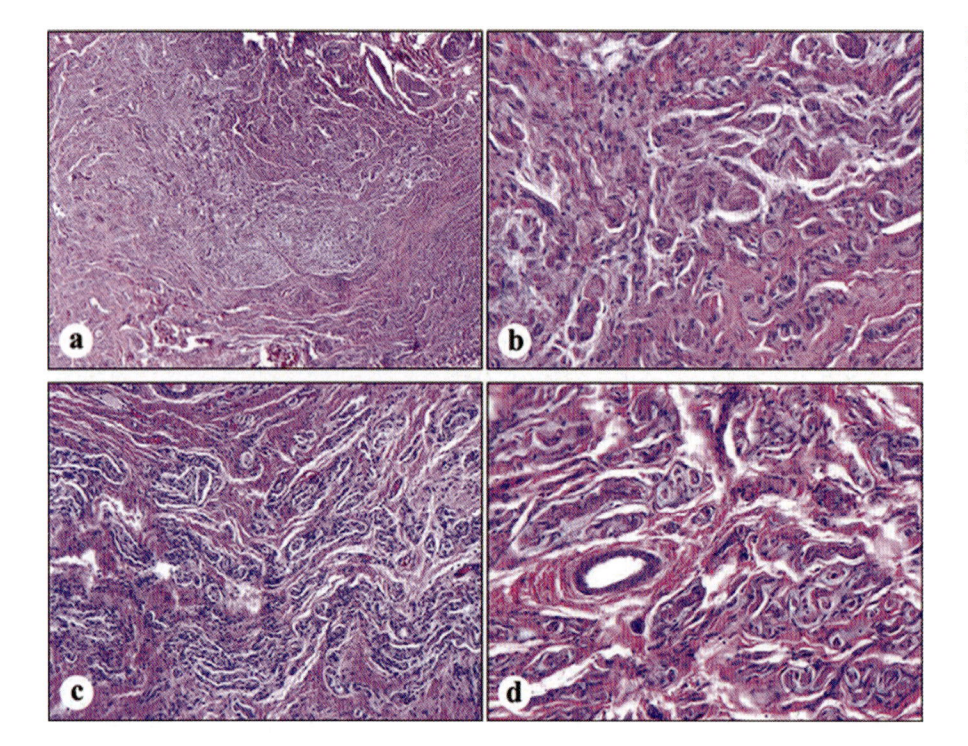

Figure 24.4 (a,b) Most frequently, traumatic neuromas are microscopic findings in surgical excisions, especially re-excisions. They are composed of irregularly arranged partially regenerated nerves that form microscopic tangled masses. (c,d) Myxoid matrix is visible in the nerves.

Pacinian neuroma (pacinian neurofibroma)

This designation refers to an uncommon hyperplasia of the *pacinian corpuscles*, the specialized nerve endings functioning as pressure receptors. The lesion (<1 cm) typically occurs in the fingers or hands of middle-aged adults and often causes chronic pain. It does not recur following simple excision. Some examples have developed after a local trauma.[22–24] Occurrence in the foot with symptoms resembling Morton neuroma[24] and occasional association with NF1[25] have been reported.

Chapter 24: Nerve sheath tumors

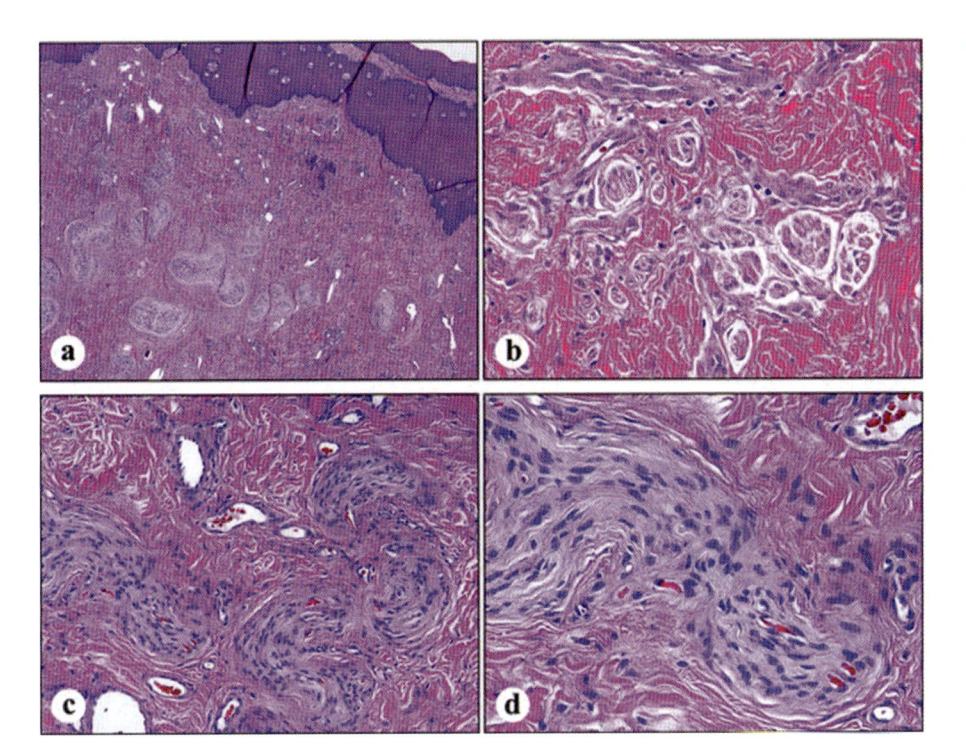

Figure 24.5 (**a**) Mucosal neuroma of the lip is remarkable for pale appearing clusters of prominent nerves. (**b**) Some of these nerves resemble the constellation more commonly seen in traumatic neuroma. (**c,d**) The nerves are mildly hypercellular and often contain prominent hyperemic intraneural capillaries.

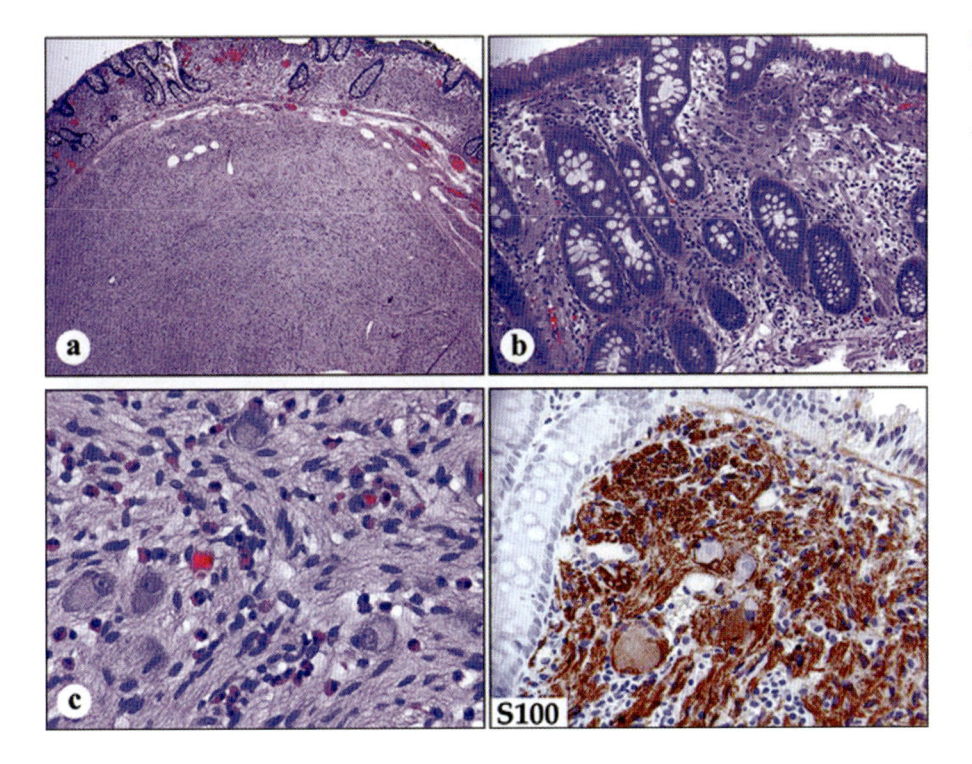

Figure 24.6 Mucosal ganglioneuroma of the colon. (**a**) A polypoid example. (**b**) A plaque-like lesion with mucosal expansion. (**c**) An admixture of spindled Schwann cells and ganglion cells is typical. (**d**) The lesional elements are S100 protein positive.

Microscopically, a pacinian neuroma is composed of multiple clustered pacinian corpuscles (Figure 24.7), or occasionally of a single enlarged one. The lamellar units are EMA-positive perineurial cells, and the center contains an S100-protein-positive schwannian core and a small axon, similar to a normal pacinian corpuscle.[22,23] A tumor earlier reported as a pacinian neurofibroma[26] is different, however: its histologic appearance resembles nerve sheath myxoma.

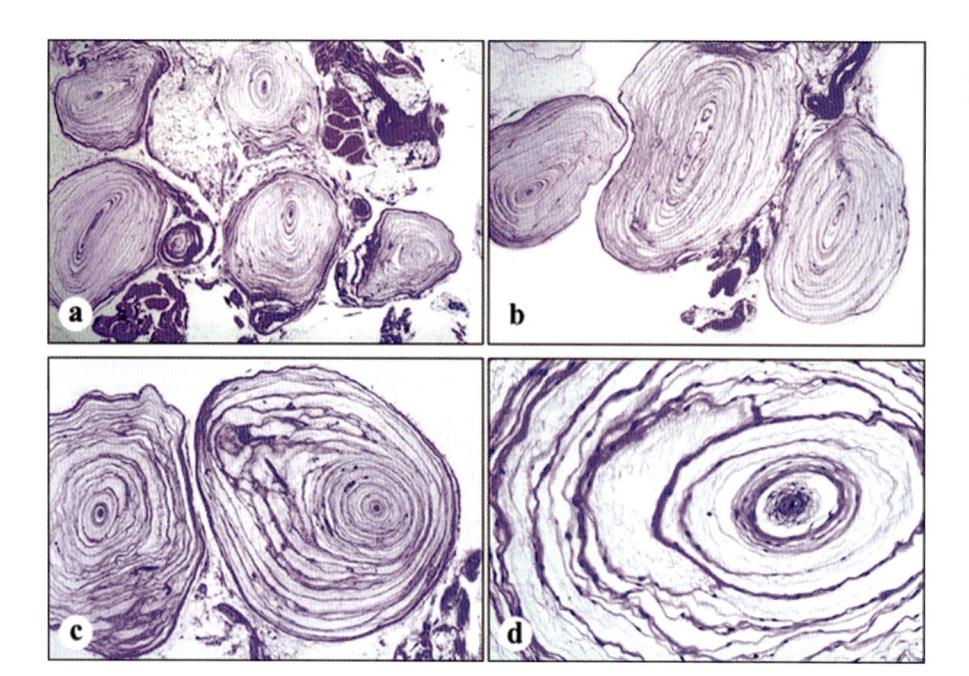

Figure 24.7 Pacinian neuroma is composed of an aggregate of well-differentiated, enlarged pacinian corpuscles that have a distinct lamellar structure.

Palisaded encapsulated neuroma (solitary circumscribed neuroma)

This designation applies to small, circumscribed benign Schwann cell tumors that have a strong predilection for facial skin.[27] Because palisading and true encapsulation are not consistent features, the name *solitary circumscribed neuroma* has been suggested as being perhaps a more accurate term.[28]

Clinical features

This small, benign, solitary cutaneous tumor most commonly occurs in young to middle-aged adults, affecting men and women equally, based on published series, but showing a marked male predominance in AFIP series (Figure 24.8). Most of the lesions occur in the head, most commonly in the cheek, nose, and face (Figure 24.9).[27–30] Ocurrence has been also reported in the oral mucosa (mostly palate and gingiva)[31] and glans penis.[32] The tumor presents as a small, solitary, flesh-colored, dome-shaped papule or nodule clinically resembling a nevus or basal cell carcinoma. Concomitant acne-like changes have been reported, leading to speculation about a reactive origin, and the same has been suspected based on the presence of numerous lesional axons.[33] The tumor has a benign behavior following simple excision. There is no association with NF1 or NF2, or MEN (mucosal neuromas are associated with the latter).

Pathology

On sectioning, the tumor is a small, whitish nodule usually measuring 2 mm to 5 mm. Histologically, it usually forms an oval nodule or cluster of nodules limited to the dermis,

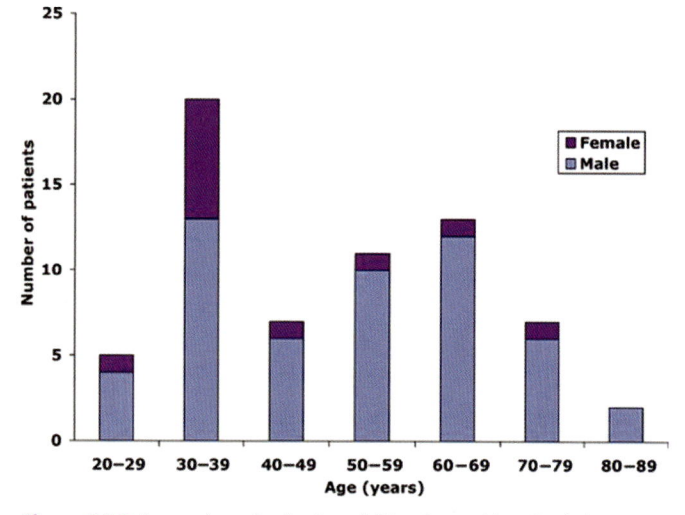

Figure 24.8 Age and sex distribution of 67 patients with palisaded encapsulated neuromas. The male dominance is probably explained by the nature of a largely military patient cohort.

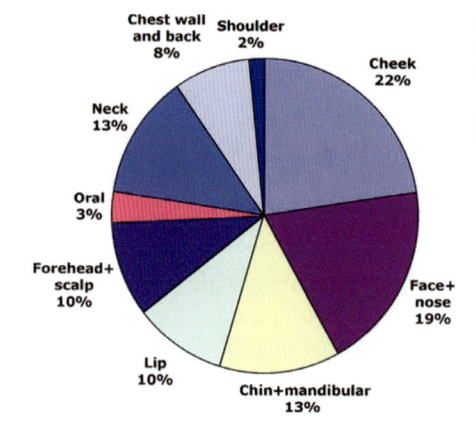

Figure 24.9 Anatomic distribution of 67 palisaded encapsulated neuromas. Note strong predilection for the head and neck.

occasionally extending to the subcutis, and variably surrounded by a collagenous capsule-like rim (Figure 24.10). Occasional examples are funnel-shaped, narrowing to the deeper dermis, and some lesions contain multiple nodules with a plexiform appearance.[33] The cellular nodules contain spindled Schwann cells often arranged in fascicles, sometimes separated by narrow clefts. Focal nuclear palisading is often present, although typical Verocay bodies are rare (Figure 24.11). An adjacent peripheral nerve can be identified in many cases.

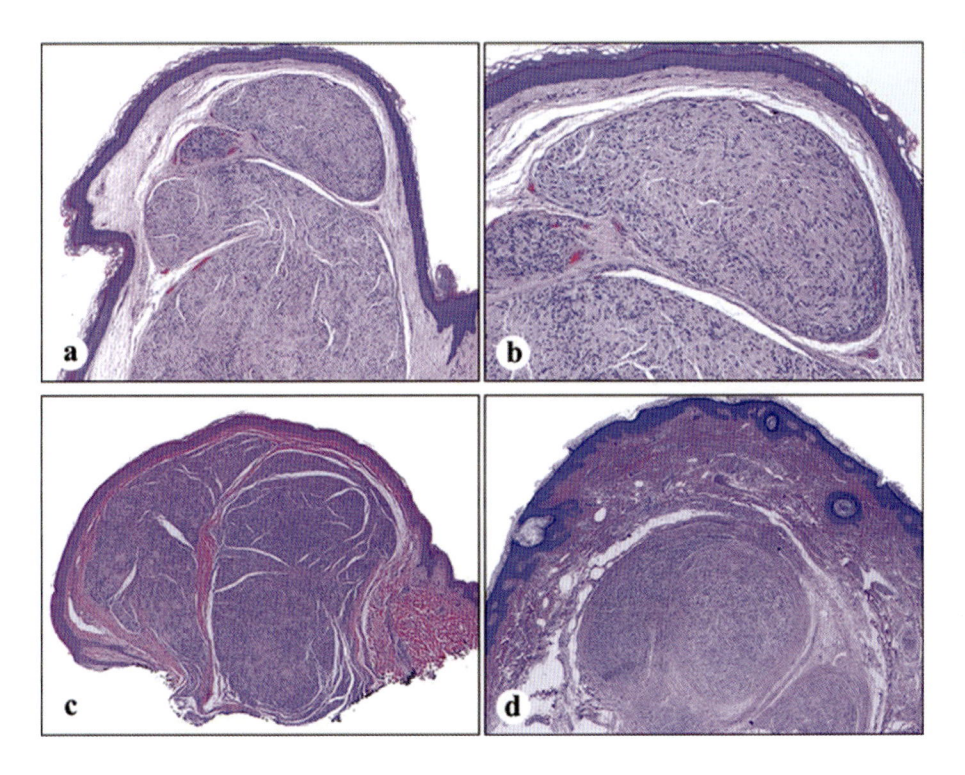

Figure 24.10 Palisaded encapsulated neuroma typically forms a dome-shaped or sessile polypoid lesion underneath an intact epidermis. The tumor can be lobulated, creating a microplexiform pattern.

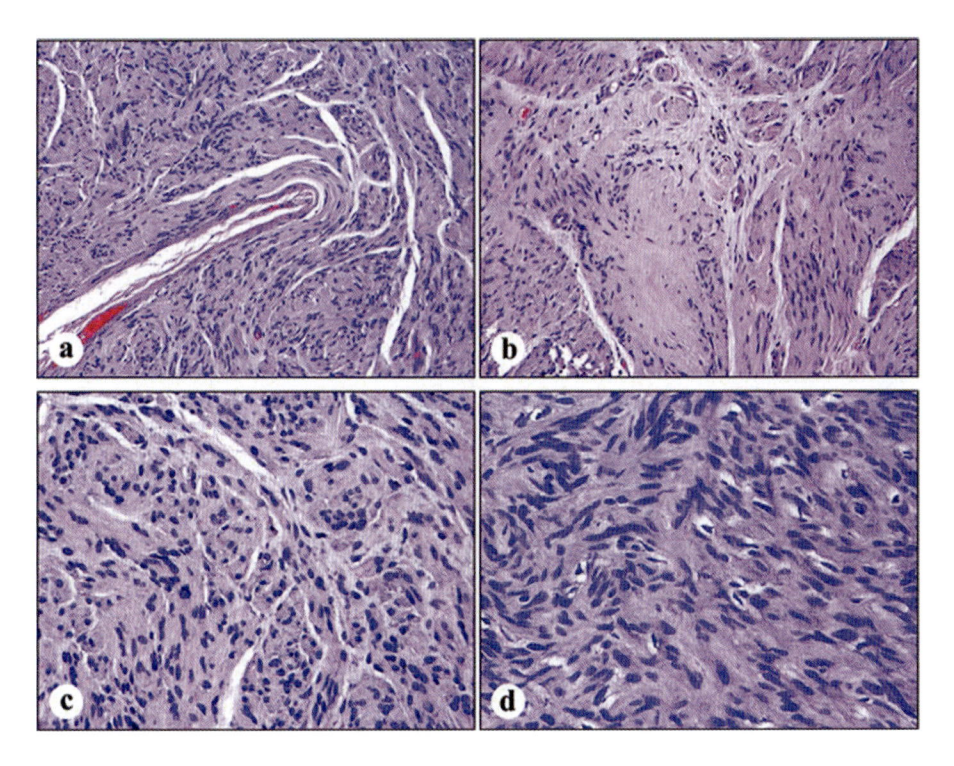

Figure 24.11 High magnification of palisaded encapsulated neuroma shows tapered spindled cells with vague palisading and occasional Verocay-body-like areas.

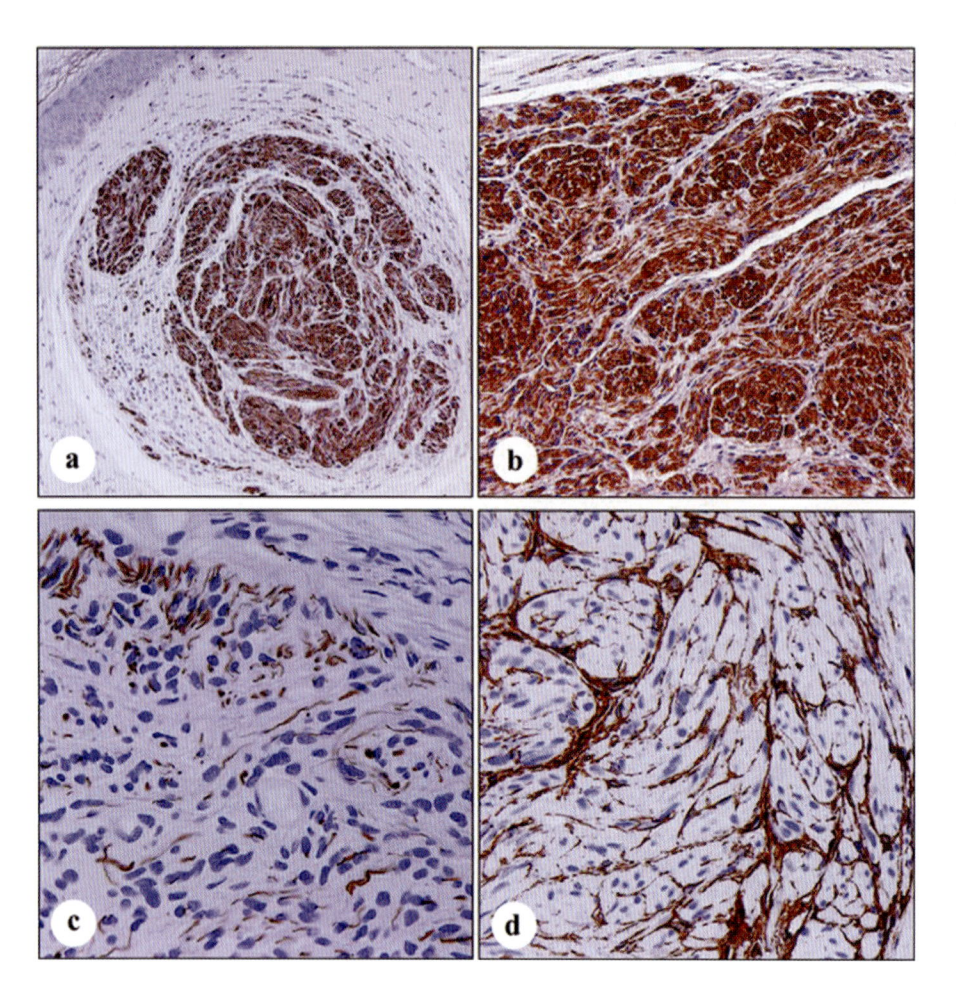

Figure 24.12 Immunohistochemical features of palisaded encapsulated neuroma. (**a,b**) Strong immunoreactivity for S100 protein is a universal feature. (**c**) Numerous neurofilament-positive axons are typically present, especially at the lesional periphery. (**d**) The neural microcompartments are often delineated by CD34-positive fibroblasts.

Immunohistochemically, the lesional cells are strongly positive for S100 protein, collagen IV, and for CD57. Variable numbers of entrapped axons can be (at least focally) demonstrated by neurofilament immunostaining, and EMA-positive perineurial cells are seen in the perineurium of nerves and also at the periphery of the lesion. All tested examples have been negative for glial fibrillary acid protein (GFAP).[34,35] CD34 fibroblasts can be present in a pattern commonly seen in neurofibroma (Figure 24.12).

Neurofibroma and neurofibromatosis

Neurofibroma[1] is the designation for a group of common, closely related benign nerve sheath tumors that according to present evidence have a similar molecular pathogenesis. It seems likely that loss-of-function alterations in the NF1 gene play a role in both NF1-associated and sporadic neurofibromas. Neurofibromas are composed of a dual population of Schwann cells and fibroblasts, and entrapped axons are often present in intraneural lesions.

Neurofibromas occur in people of all ages, but they are most commonly diagnosed in young adults. They can present in superficial or deep soft tissues, and each subtype can present in a wide variety of anatomic locations. The most common form is solitary cutaneous neurofibroma. Other forms discussed here are intraneural neurofibroma, which is called plexiform when forming a complex tortuous mass. *Cellular*, *diffuse*, and *pigmented* neurofibromas are designations for histologically distinctive less common subtypes.[1]

Intraneural, plexiform, cutaneous, and diffuse neurofibromas are associated with NF1 and also can occur in children, although they rarely appear before the age of 5 years. Intraneural neurofibromas can cause neurologic symptoms when involving major nerves. Diffuse neurofibromas, including pigmented variants, can be massive and usually occur in NF1 patients.[1,36–39]

Cutaneous neurofibroma

Cutaneous neurofibroma occurs in subjects of all ages (Figure 24.13). It more commonly forms a small solitary skin nodule, but in patients with NF1, innumerable lesions can occur. A neurofibroma can present as a dome-shaped or polypoid elevation of the skin and grossly can resemble a fibroepithelial polyp. It usually measures from a few millimeters to <2 cm. All anatomic regions can be involved, but occurrence in the trunk and head seems to be more common (Figure 24.14).

Grossly, a neurofibroma is rubbery, gray, and glistening on sectioning. Histologically, a cutaneous neurofibroma shows a mixture of elongated spindled Schwann cells and fibroblasts in a background of wavy collagenous fibers. The degree of cellularity varies, and the lesion can extend into the subcutaneous fat (Figure 24.15). The Schwann cells have small oval to elongated irregular nuclei with a dense chromatin structure. Varying numbers of mast cells are commonly present. There is no visible association with nerves, unless an intraneural neurofibroma component is present, and the origin is apparently in microscopic nerves or nerve endings. Entrapped skin adnexal structures are not uncommon, and such epithelial elements are no longer considered heterologous epithelial differentiation.

A peculiar histologic variant of cutaneous neurofibroma contains large Schwann cells with dendritic processes (Figure 24.16), surrounded by smaller cells.[40] Both components are S100 protein positive (Figure 24.16d). These tumors are usually sporadic and solitary.

Cellular neurofibroma

This rare variant usually presents in the skin and subcutis, where it forms a solid, rubbery, white 2-cm to 4-cm nodule.

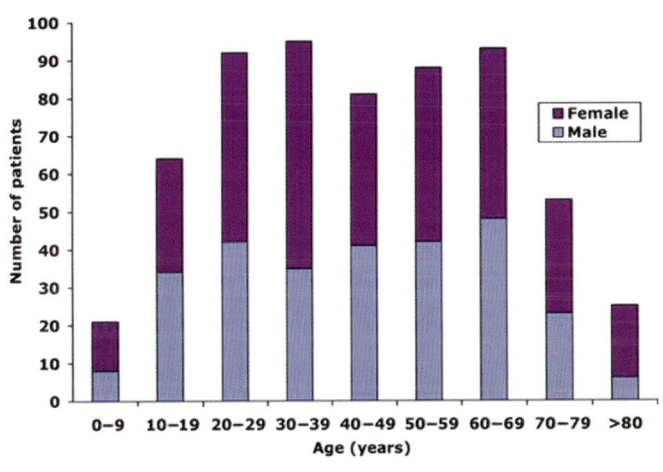

Figure 24.13 Age and sex distribution of 612 civilian patients with cutaneous neurofibromas. Note occurrence in all ages with a mild female predominance.

Figure 24.14 Wide anatomic distribution of cutaneous neurofibromas in 595 civilian patients.

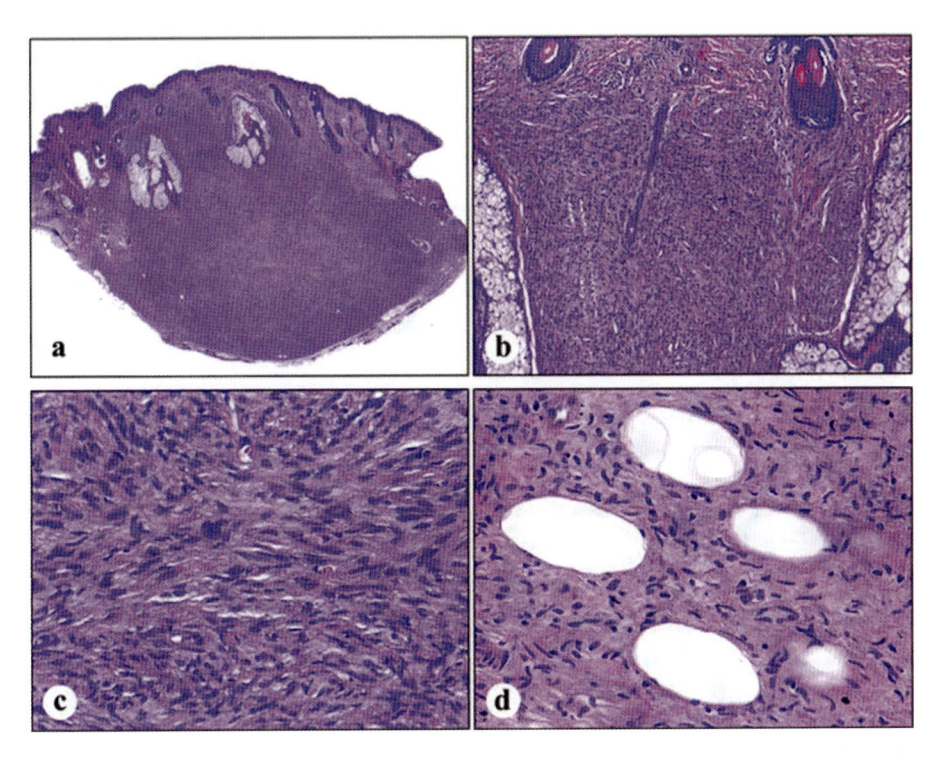

Figure 24.15 (a,b) Cutaneous neurofibroma forms an ill-defined nodule containing entrapped skin adnexa. The tumor is composed of bland spindle cells that show vague palisading or are randomly arranged. (c) A more cellular example. (d) Extension into subcutaneous fat is common.

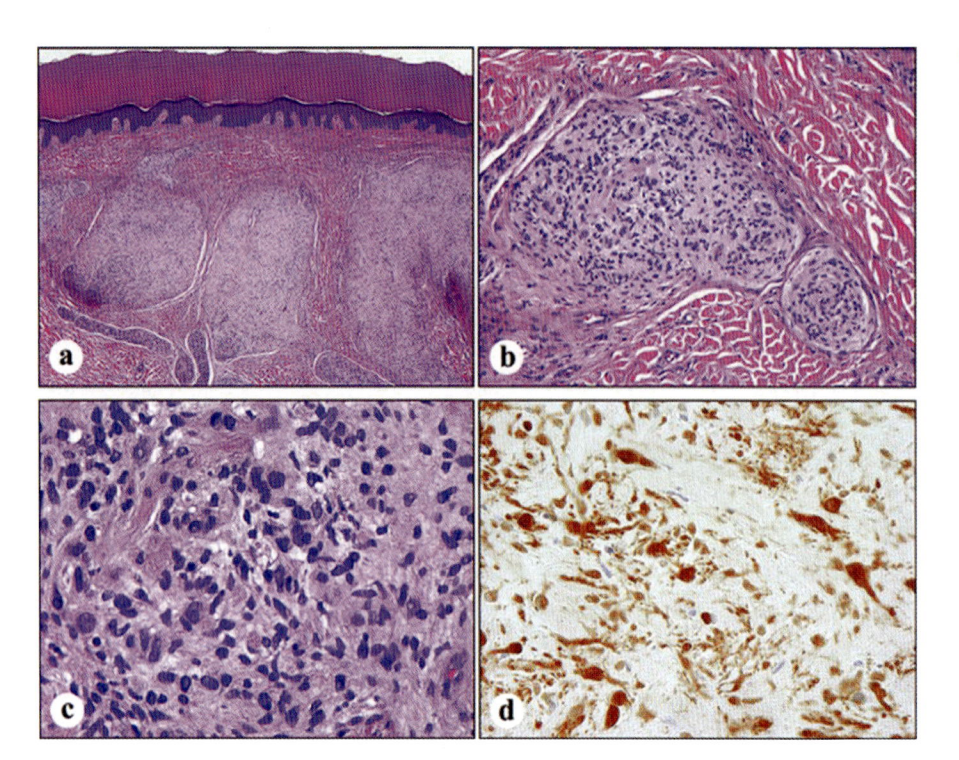

Figure 24.16 (a–c) An unusual variant of cutaneous neurofibroma containing large cells with dendritic processes. These lesions often have a multinodular "microplexiform" pattern. (d) The dendritic processes are highlighted with an S100 protein immunostain.

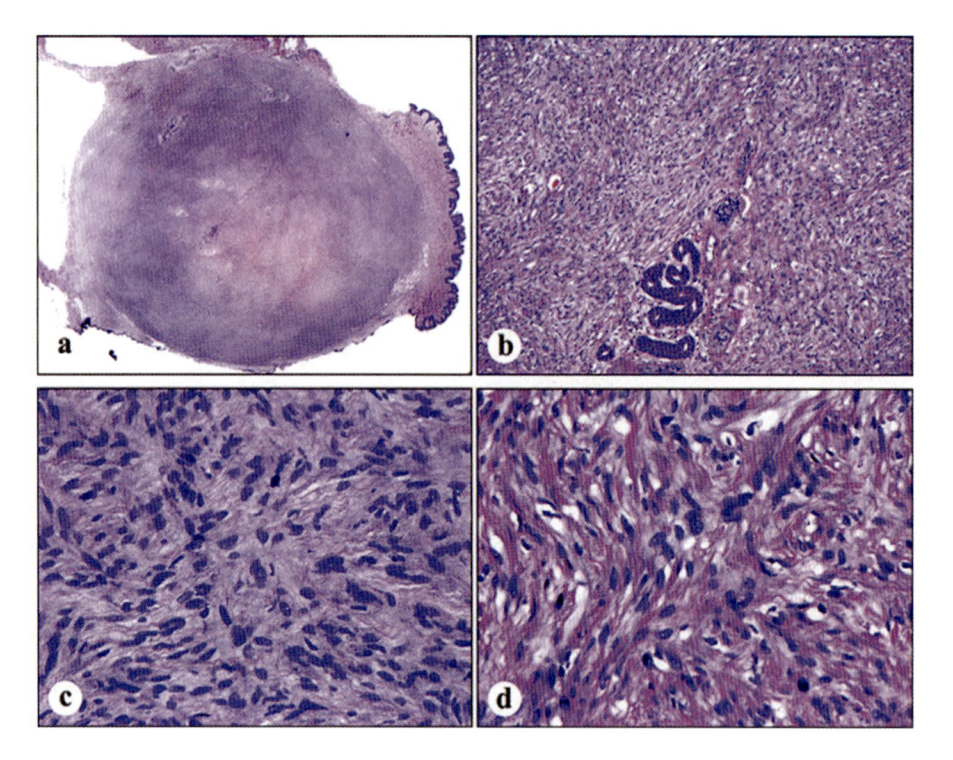

Figure 24.17 Cellular neurofibroma forms a subcutaneous mass that infiltrates between skin adnexa. It is composed of uniform, mildly plump, spindled cells in a variably collagenous matrix.

The tumor is more highly cellular than a typical neurofibroma and often lacks the typical wavy appearance of collagen. Because of its high cellularity, this neurofibroma can be easily confused with low-grade sarcomas, especially dermatofibrosarcoma protuberans (DFSP). Focal atypia can also be present, but mitotic activity is very low (Figure 24.17). Immunohistochemical demonstration of the S100 protein and scattered CD34 fibroblasts is the key to the diagnosis of cellular neurofibroma (Figure 24.18).

Intraneural neurofibroma

Intraneural neurofibroma is confined to the epineurium and is therefore truly encapsulated. It can involve superficial small nerves, deep somatic and visceral nerves, or major nerve trunks. Grossly, solitary intraneural neurofibromas are typically fusiform, well-defined lesions that are confined within a nerve, and therefore the lesions involving larger nerves are sharply circumscribed. Microscopic examples involving smaller cutaneous nerves blend with dense dermal collagen. On sectioning they are gray-white and vary from mucoid to rubbery or gelatinous (Figure 24.19).

Histologically, an intraneural neurofibroma is encased by the epineurium (Figure 24.20) and is therefore well demarcated. It consists of elongated spindle cells diffusely dispersed or forming cable-like structures with a variably collagenous or myxoid matrix (Figure 24.21). The Schwann cells can be spindled or rounded; the latter superficially resemble lymphocytes. Focal atypia in the schwannian elements with enlarged, hyperchromatic nuclei is common and has no specific significance (Figure 24.22), unless accompanied by marked increase in cellularity and mitotic activity.[41]

As with cutaneous neurofibromas, there is a mixture of S100-protein-positive Schwann cells and CD34-positive fibroblasts, the latter often seen in a net-like pattern. Residual axons can often be demonstrated with immunostaining for neurofilament proteins in intraneural neurofibromas.

Plexiform neurofibroma

Plexiform neurofibroma is a gross pathologic or low-magnification histologic distinction. It represents a complex aggregate of intraneural neurofibromas, that is, transformation of multiple adjacent nerves into complex tortuous masses, or a rope-like diffuse thickening of a large nerve trunk, especially when involving the sciatic nerve (Figure 24.23). Some tumors reach a large size, up to several kilograms.

The clinical significance of this variant lies in its consistent association with NF1 syndrome, and that it has some, although low, potential for malignant transformation to MPNST. Plexiform neurofibroma typically has an early onset, with a peak incidence in the first two decades (Figure 24.24). Locations vary, but there is a predilection for the head and neck (Figure 24.25). Histologically, plexiform neurofibroma is similar to the usual type of intraneural neurofibroma, except that it is composed of multiple juxtaposed nerve fascicles transformed by the neurofibroma (Figure 24.26).

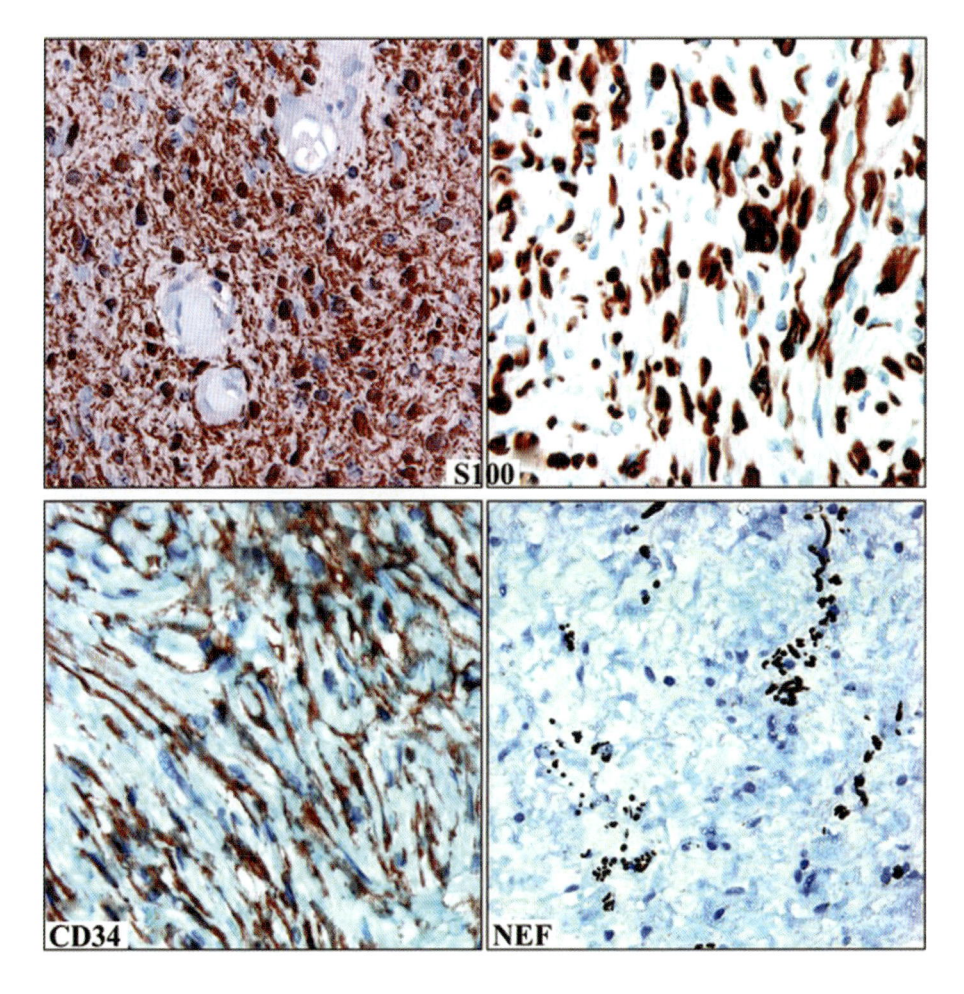

Figure 24.18 Immunohistochemical features of cellular neurofibroma. The Schwann cells are S100 protein positive showing both nuclear and cytoplasmic positivity. There is a substantial CD34-positive fibroblastic component. Neurofilament-positive axons are often present.

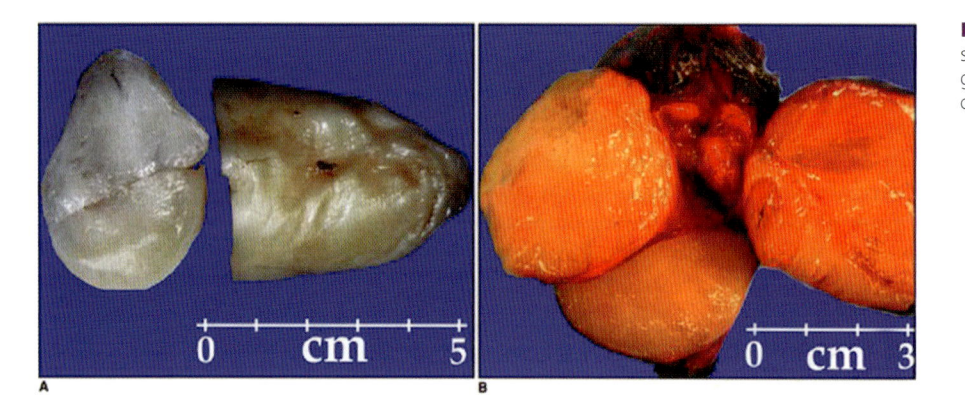

Figure 24.19 Intraneural neurofibroma forms a sharply demarcated nodule that in a fixed state is grayish, but in the fresh state can have a yellowish color resembling a lipoma.

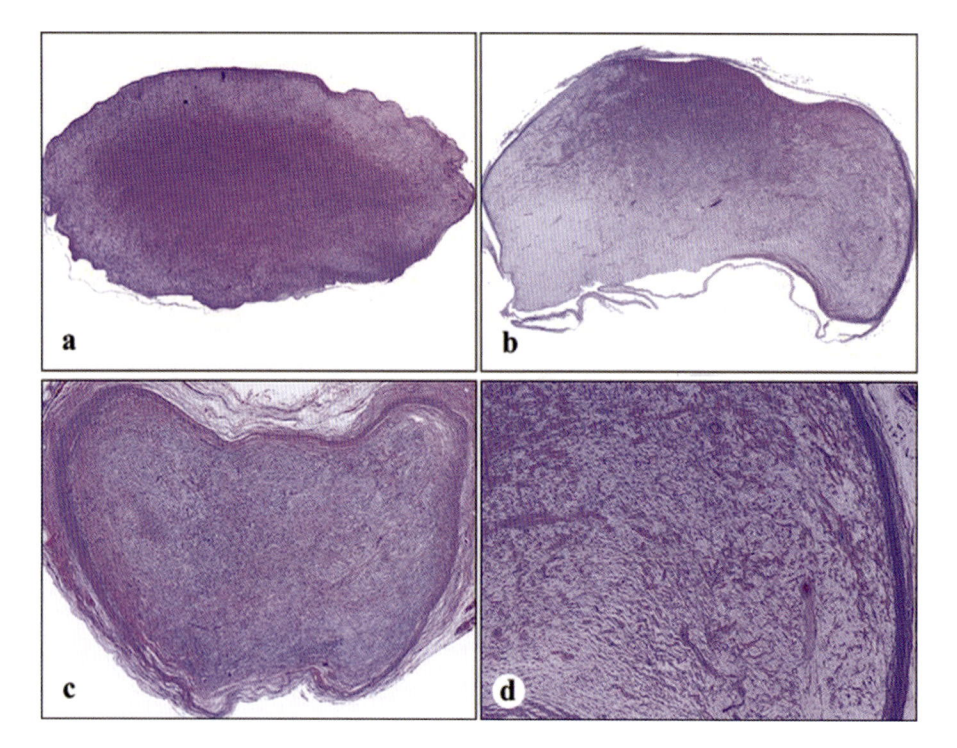

Figure 24.20 Low-magnification examples of intraneural neurofibroma are encapsulated masses with variably myxoid matrix. The tumor is encased by the epineurium.

Diffuse neurofibroma

This subtype occurs in patients with the NF1 syndrome, beginning in the latter part of the first decade. A wide variety of soft tissues throughout the body can be involved, but most commonly the lesions involve skin and subcutis. Malignant transformation of diffuse neurofibroma is very rare.

Grossly, the lesions vary from plaque-like to large contiguous masses that can reach disfiguring proportions. On sectioning, they are textureless, gray-white, and vary from soft, slightly mucoid to firm and rubbery. The excised tumor from locations such as the eyelids is often relatively small, but in the trunk and extremities, the lesion often measures 5 cm or more, and can reach massive proportions in NF1 patients. The borders are indistinct, and the subcutis is primarily involved.

Histologically, diffuse neurofibroma resembles cutaneous neurofibroma, but it has prominent, diffuse fat infiltration (Figure 24.27). In some cases, the abundance of fat masks the neurofibromatous component and simulates a fibrolipoma (Figure 24.28). A common histologic feature is the presence of Wagner–Meissner-like tactile corpuscles.[42] Some lesions are more highly cellular and some may contain elements of intraneural/plexiform neurofibroma. The cellular examples can be distinguished from MPNSTs based on very low mitotic activity and relative lack of atypia.

Pigmented neurofibroma

Pigmented neurofibroma is a rare variant of diffuse neurofibroma with focal melanocytic differentiation (pigmented or melanocytic neurofibroma). These tumors predominantly occur in African Americans in a wide age range. The most common locations are head and neck, or buttocks and leg. Similar to other diffuse neurofibromas, association with NF1 is common. Behavior is similar to diffuse neurofibroma.[43]

Chapter 24: Nerve sheath tumors

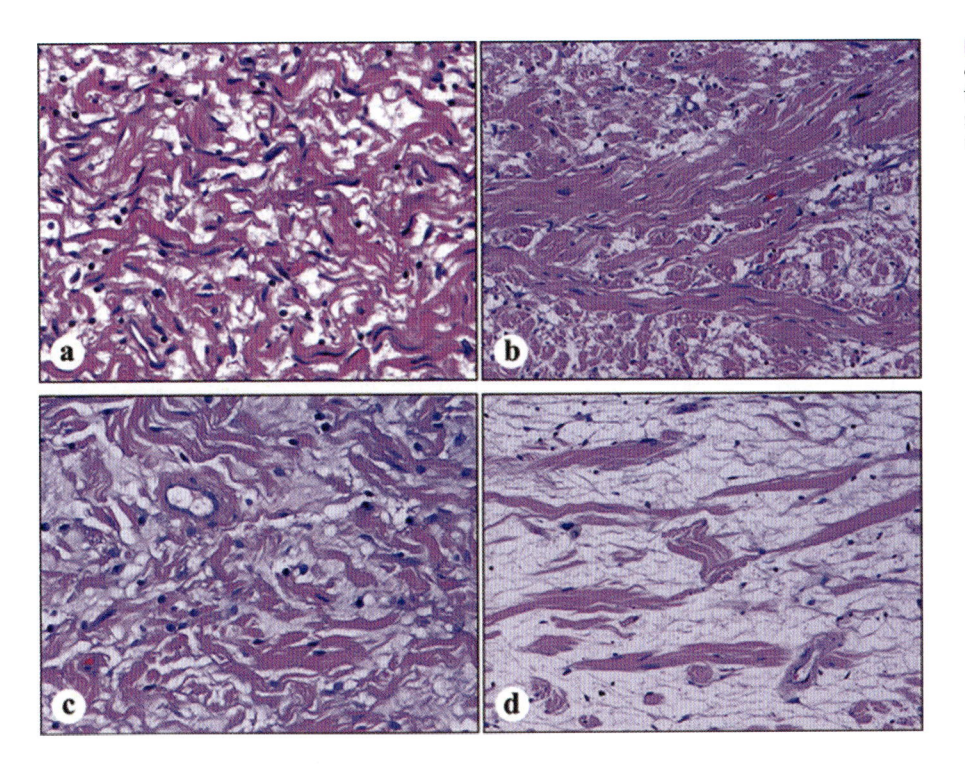

Figure 24.21 Intraneural neurofibroma shows cells with tapering, "wavy" nuclei, and collagen fibers that vary from diffusely arranged to those seen in cable-like arrangements. The extracellular matrix varies from collagenous to myxoid.

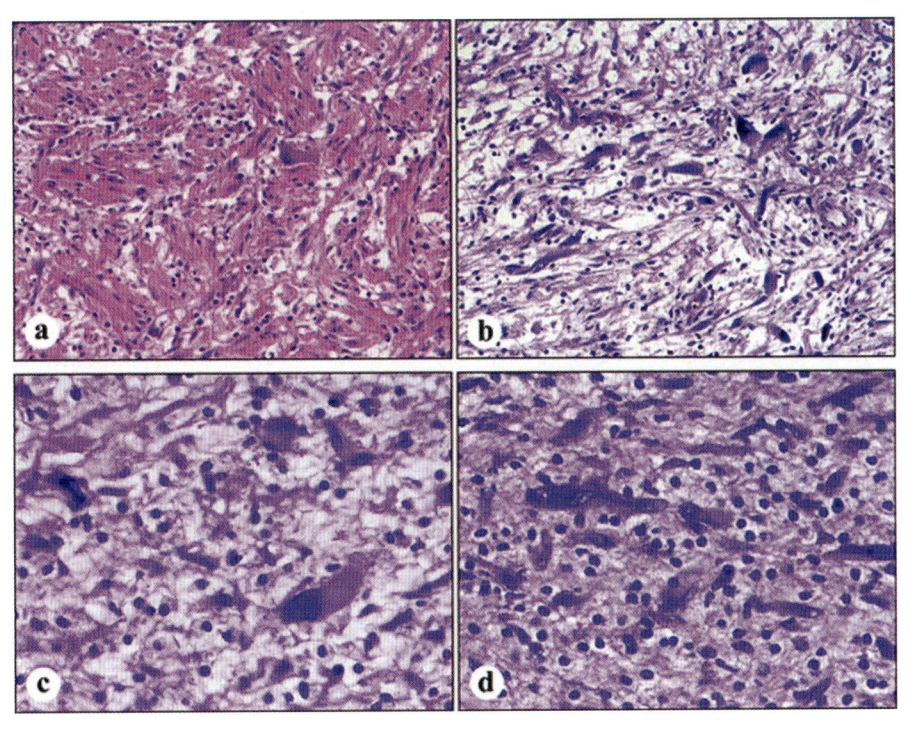

Figure 24.22 Nuclear atypia in intraneural neurofibroma. The atypia varies from focal to more extensive, but overall cellularity is moderate at most, and there is no mitotic activity.

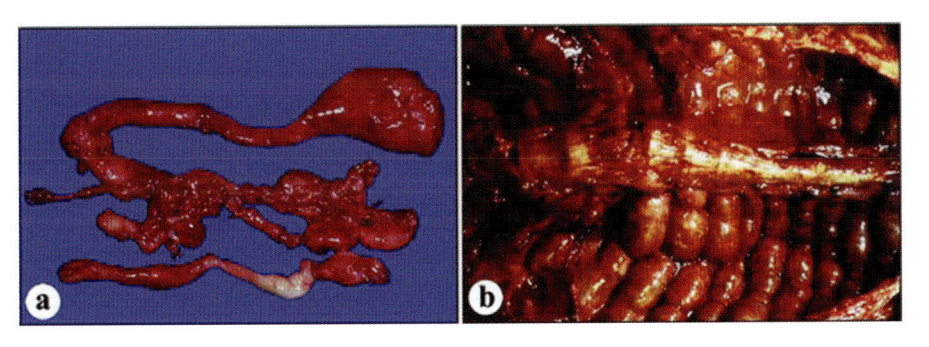

Figure 24.23 (a) A grossly plexiform neurofibroma forms a tortuous cable-like structure composed of multiple adjacent nerve fascicles variably thickened by the neurofibroma formation. (b) Plexiform neurofibroma extensively involving multiple intercostal nerves at autopsy. Note also scoliosis, a common finding in neurofibromatosis 1 (NF1) patients.

649

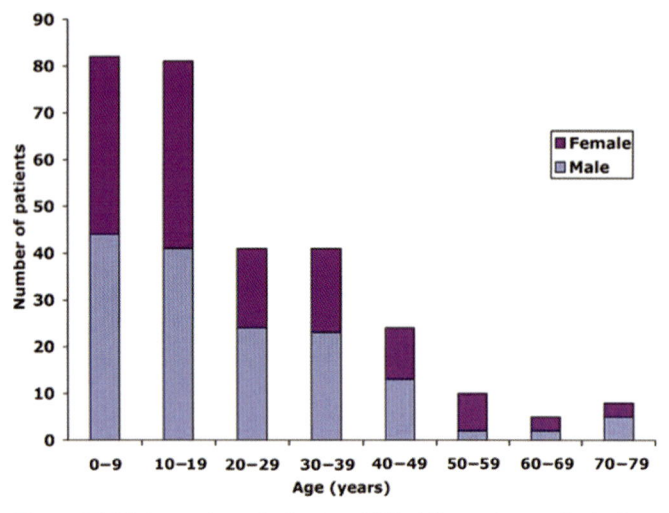

Figure 24.24 Age and sex distribution of 292 civilian patients with plexiform neurofibromas. Note predilection for children.

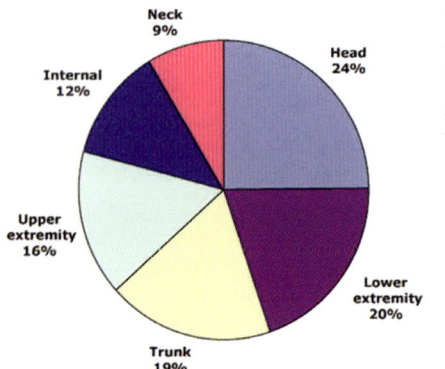

Figure 24.25 Wide anatomic distribution of 278 plexiform neurofibromas. Note common occurrence in the head and neck.

The pigmented neurofibroma most commonly involves the subcutis and skin, but occasionally extends to skeletal muscle. Its features are generally similar to diffuse neurofibroma. In addition, there are scattered or larger clusters of dendritic pigmented cells (Figure 24.29). These cells have melanocytic markers HMB45, tyrosinase, and melan A, and like the non-pigmented cells, they are also S100 protein positive.[43]

Critical review of historic reports on pigmented neurofibroma reveals that many older cases are more likely pigmented dermatofibrosarcomas (DFSPs, Bednar tumors). Bednar's series on what are now classified as pigmented DFSPs was originally published as pigmented neurofibromas.

Immunohistochemistry and ultrastructure of neurofibroma

Neurofibromas are typically composed of a dual population of wavy Schwann cells and fibroblasts. The former typically have elongated S100-protein-positive cytoplasmic processes and nuclei.[44] The Wagner–Meissner-like bodies are typically strongly S100 protein positive. There is a variably prominent CD34-positive fibroblastic population.[45,46] These cells also have elongated cytoplasmic processes and are typically intermingled with the schwannian component, often in a net-like pattern. CD57 (Leu7) also marks the schwannian component.[47] Neurofilament-protein-positive residual axons can be demonstrated, especially in intraneural neurofibromas, but axons are not always present and are not required for the diagnosis. EMA-positive perineurial cells can be present, usually in the periphery of intraneural neurofibromas. Basement membrane proteins laminin and collagen IV surround the

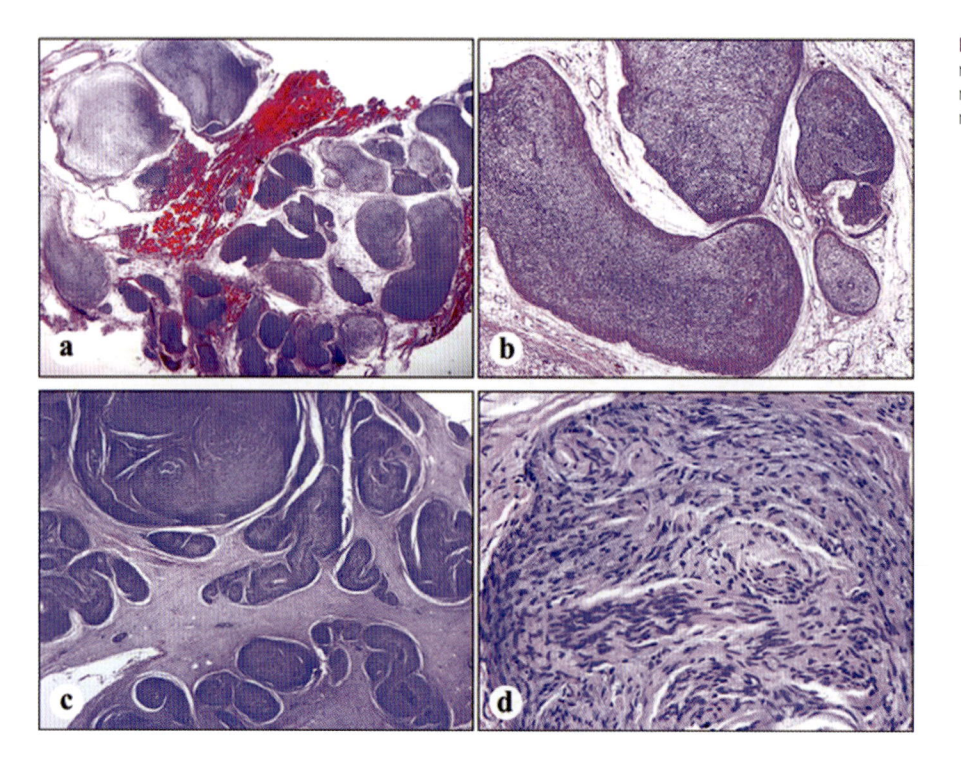

Figure 24.26 Low magnification of plexiform neurofibromas shows masses composed of multiple adjacent nerve fascicles involved with the neurofibroma.

Chapter 24: Nerve sheath tumors

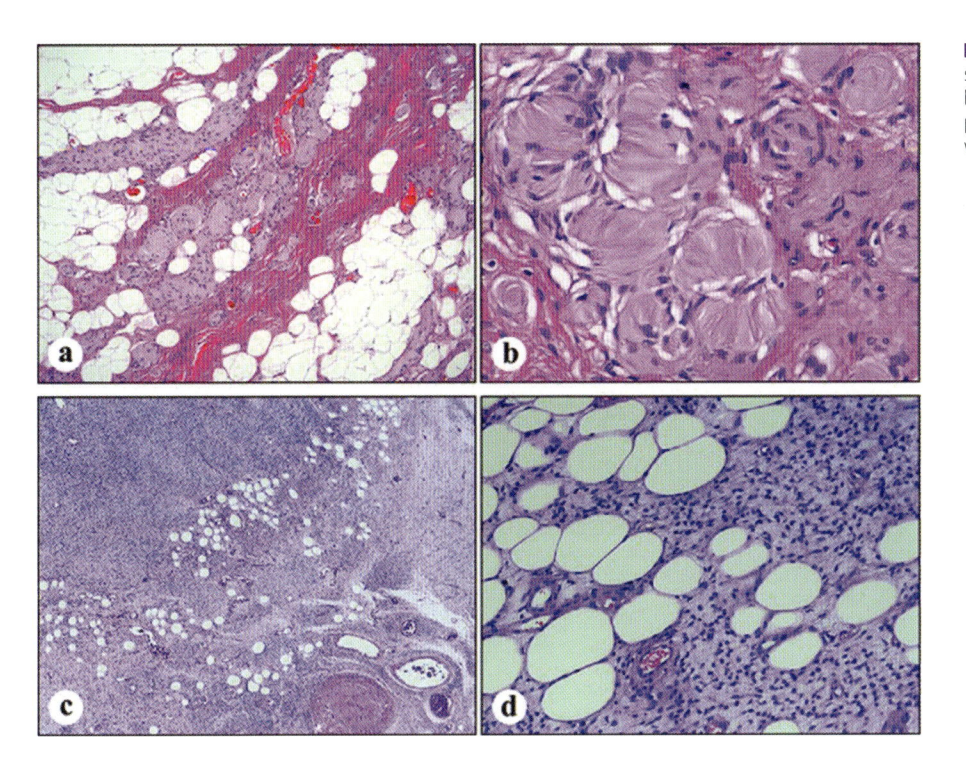

Figure 24.27 (**a**) Diffuse neurofibroma involves subcutaneous fat. (**b**) Numerous Meissner-like bodies is a typical finding. (**c,d**) Some cases are predominantly composed of oval or spindled cells with diffuse subcutaneous fat infiltration.

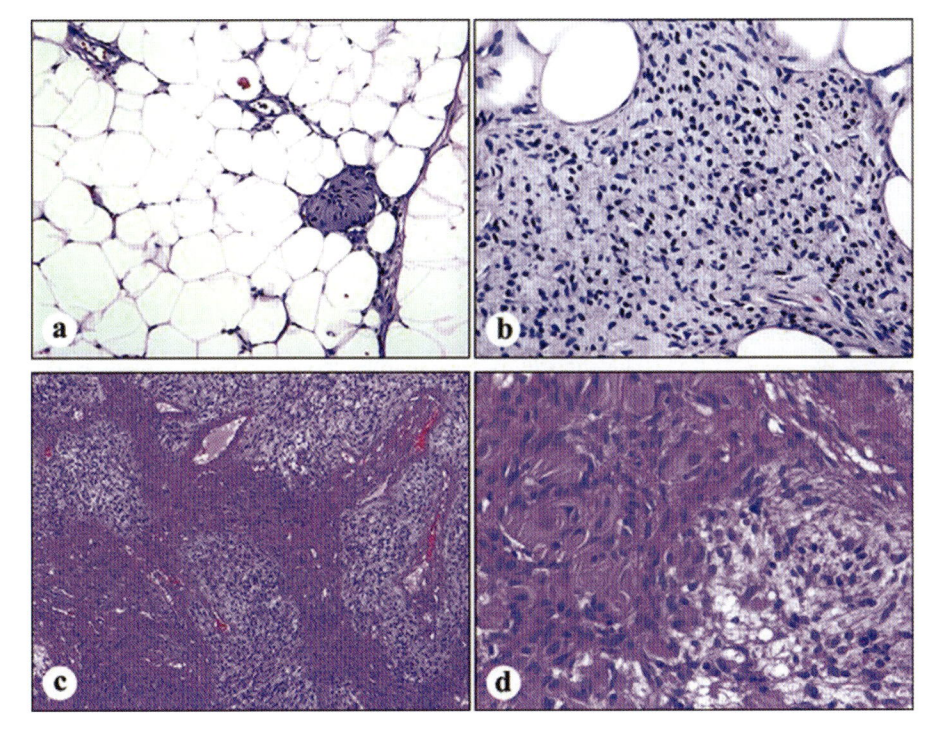

Figure 24.28 Variable cellularity in diffuse neurofibroma. (**a**) An example with limited numbers of neurofibroma elements in the fat should not be confused with a fatty neoplasm. (**b**) Uniform oval cells in a myxoid matrix. (**c,d**) Alternating collagenous and myxoid areas containing spindle cells and Meissner bodies.

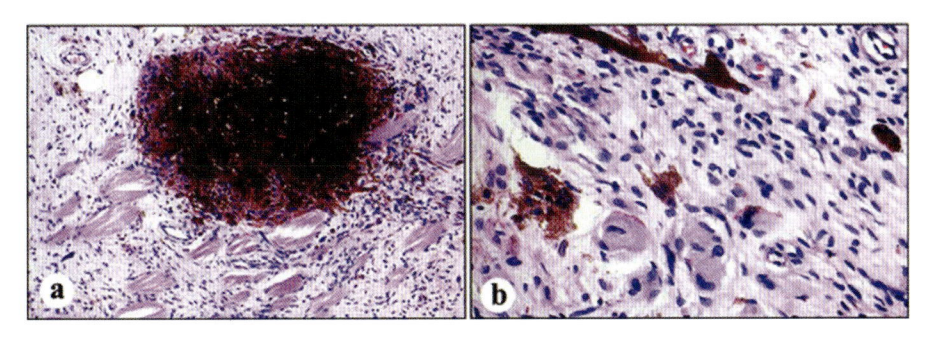

Figure 24.29 Pigmented diffuse neurofibroma contains melanin-pigmented cells as dense clusters or scattered between spindle cell elements and Meissner-like bodies.

Schwann cells, but positivity in neurofibroma is usually less prominent than that in schwannoma.[48]

It has been suggested that a low Ki67 analog score and negativity for p53 are features that distinguish neurofibroma from MPNSTs.[49,50] It has not been convincingly demonstrated, however, that these features are useful in borderline tumors, and therefore information from these markers should be evaluated with caution.

Ultrastructural studies have also shown the dual composition of neurofibromas of basement-membrane-positive Schwann cells and negative fibroblasts.[51]

Transformation and differential diagnosis of neurofibroma

Transformation

The NF1 syndrome carries a risk of malignant transformation of neurofibroma to MPNST, which has been estimated to occur in 2% to 4% of patients during their lifetime.[52] If the development of a central nervous system tumor is taken into account, the frequency of development of a malignancy in NF1 patients is approximately 5%. The risk seems to be greatest for intraneural neurofibromas, especially the plexiform types.

The transformation of neurofibroma to MPNST is a biologic continuum, and therefore the line between neurofibroma and MPNST is not always clear, and there are no absolute criteria for this distinction. Atypia by itself is not indicative of malignancy, because it is commonly seen in many variants of neurofibromas. Indicative of malignancy are atypia with high cellularity and significantly elevated mitotic rate (more than 1–2 mitoses found in a small area). Which size of foci with increased mitotic rate indicates malignancy is not clear. Clinical significance of early malignant transformation of neurofibroma and what constitutes it should be addressed in future clinicopathologic studies. Tumors with histologically suspicious features should be excised completely and preferably widely to prevent local disease progression. Radical treatment such as amputation is probably not warranted for such borderline or low-grade tumors involving major nerves.

Differential diagnosis

Differentiation of neurofibroma from schwannoma is sometimes problematic (Table 24.1). Most tumors that are difficult in this respect turn out to be neurofibromas. As a general rule, neurofibromas can have schwannoma-like areas, but generally, schwannomas do not have neurofibroma-like areas. Content of residual axons and presence of a dual schwannian S100-protein-positive and fibroblastic CD34-positive cell populations are features of neurofibroma, often useful in the differential diagnosis.

Although the typical cutaneous, intraneural, and plexiform neurofibromas are histologically highly distinctive, the diffuse and cellular types can be easily confused with other tumors.

The pattern of diffuse subcutaneous infiltration of diffuse neurofibroma resembles that of DFSP, but the tumor cells of neurofibroma less frequently have a storiform pattern. In neurofibroma, immunohistochemical studies reveal the dual population of S100-protein-positive Schwann cells and CD34-positive fibroblasts (occasionally, the latter is absent). DFSP lesions are typically negative for S100 protein and positive for CD34. Some variants (Bednar tumor), however, might contain isolated S100-protein-positive tumor cells.

Cellular neurofibroma is sometimes confused with low-grade sarcoma, especially DFSP and low-grade fibromyxoid sarcoma. An immunohistochemical demonstration of a dual population of S100-protein- and CD34-positive cells is one of the most helpful features for identifying these neurofibromas. Myofibroblastic and smooth muscle tumors are usually structurally more homogeneous, and their principal components are myofibroblasts and smooth muscle cells that are positive for muscle markers.

The specific myxoid neoplasms of the vulvovaginal region (Chapter 18) have prominent vascular patterns and are negative for S100 protein.

Table 24.1 Differential diagnosis between intraneural neurofibroma and schwannoma

Differential feature	Neurofibroma	Schwannoma
Relationship to nerve of origin	Fusiform mass involving the whole nerve	Forms an eccentric mass around the nerve
Circumscription	Encased in a nerve	Usually encapsulated
Nuclear palisading	Absent	Often present
Myxoid stroma	Often present	Rarely present
Vascular dilatation	Not prominent	Often present
Xanthoma cells	Usually absent	Often present
S100 protein in cellular areas	Variable numbers of cells positive	All cells positive
CD34	Positive cells in the cellular areas	Cellular Schwann cell areas negative
Neurofilament	Residual axons positive	Positive elements usually absent

Neurofibromatosis 1 tumor syndrome

NF1 (von Recklinghausen's disease) is caused by loss-of-function mutations and deletions of the NF1 gene. It is one of the most common autosomal dominant disorders, with a birth incidence of 1:3000. Approximately one-half of the NF1 cases result from new germline mutations in the NF1 gene, and the other half are inherited.[53–55] The diagnostic criteria of NF1, based on a NIH-sponsored consensus conference, are following:[53]

(1) Two or more neurofibromas of any type, or one of the plexiform type
(2) Six or more lightly pigmented macules (café au lait spots): >5 mm in prepubertal patients, >15 mm in postpubertal patients
(3) Freckling in axilla or inguinal regions
(4) Optic glioma
(5) Two or more iris hamartomas (Lisch nodules)
(6) Osseous lesions: skeletal dysplasia, cortical thinning of a long bone
(7) A first-degree relative with NF1.

The presence of two or more of these clinical criteria defines the NF1. Many other tumors occur with increased frequency in NF1 patients. Such tumors especially include GISTs, certain gliomas, duodenal carcinoid with psammoma bodies, and adrenal pheochromocytoma. The severity of NF1 manifestations in different patients varies in a very wide range. Clinical manifestations are shown in Figures 24.30 and 24.31.

NF1 gene alterations

The NF1 gene is located pericentromerically at chromosome 17q11.2. The gene encodes for a protein named *neurofibromin*, which is nearly ubiquitously expressed and is thought to have an inhibitory role in the RAS signal transduction pathway. The NF1 gene is large, and the distribution and patterns of germline mutations and other changes in the NF1 gene are as yet incompletely characterized.[54,55]

Tumorigenesis in NF1 has been assumed to follow a two-hit hypothesis for recessive tumor-suppressor genes. Neurofibroma lesions presumably arise when both alleles are inactivated, by a mutation or an allelic loss.[56–58] In neurofibroma, the Schwann cells, but not the fibroblasts, seem to have the NF1 gene alterations.[59]

The large size of the NF1 gene (60 exons, spanning 350 kilobases of genomic DNA) and the existence of homologous pseudogene sequences elsewhere in the genome make the comprehensive analysis of mutations tedious and difficult. Nevertheless, exhaustive analysis of NF1 sequences has allowed a high frequency of mutation detection.[60–62]

Subgroups of neurofibromatosis

Segmental neurofibromatosis refers to a NF1 clinical subtype that involves a body segment containing neurofibromas and harboring a systemic NF1 mutation involving that whole segment. This form is thought to be the result of an early embryonic NF1 gene mutation that involves a segment of the body during development. The person is therefore said to have a mosaicism in relation to the configuration of the NF1 gene.[63–66]

Gastrointestinal neurofibromatosis is essentially a historical designation that refers to the occurrence of different tumors in the gastrointestinal tract in NF1 patients. In many cases, this designation probably referred to multiple GISTs occurring in NF1 patients, or in some cases, familial GIST syndrome with multiple or diffuse GIST formations in the gastrointestinal tract. As such, this designation is less useful than the tumor-type-specific diagnoses. Plexiform neurofibromas can also involve the gastrointestinal and urogenital tracts in NF1 patients on rare occasions.

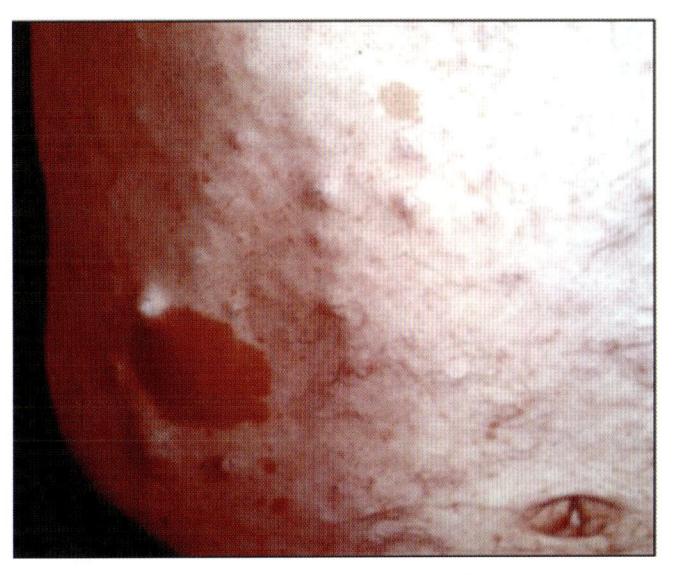

Figure 24.30 Café au lait spots in the abdominal wall skin in a neurofibromatosis type 1 patient.

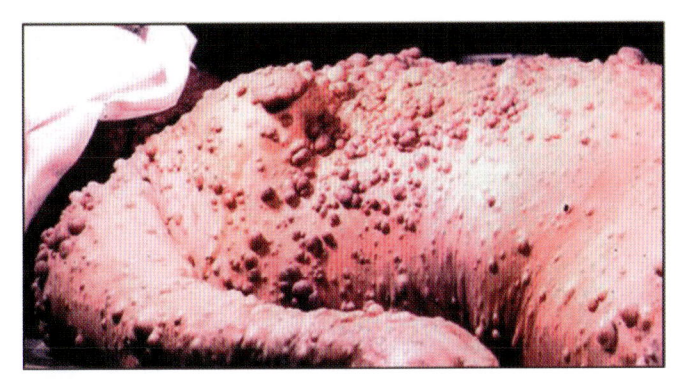

Figure 24.31 Extensive cutaneous neurofibromas in an autopsy of a patient with NF1.

Schwannoma (neurilemmoma) and neurofibromatosis 2

A *schwannoma* is a common peripheral nerve sheath tumor that occurs in a wide variety of locations. Certain schwannomas, especially the bilateral vestibular ones, are associated with NF2 syndrome and hereditary NF2 gene mutations. Somatic alterations in this gene seem to be key events for the tumorigenesis of sporadic schwannoma. The rare syndrome of multiple peripheral schwannomas without vestibular tumors is called *schwannomatosis*.

This text discusses classical schwannoma and its variants and reviews the NF2 syndrome separately. Gastrointestinal tract schwannomas differ from the peripheral variety and are discussed separately. The rare psammomatous melanotic schwannoma is not associated with NF2.

Conventional schwannoma

Clinical features

Peripheral soft tissue schwannomas occur in patients of all ages, although they are rare in children. Most commonly they occur in middle age, without a clear sex predilection. Schwannomas usually present in superficial soft tissues of the extremities, perhaps relatively most often in the head and neck region and the distal parts of the extremities. In peripheral soft tissues they vary from small nodules to tumors of 5 cm, but are rarely much larger than that. Peripheral soft tissue schwannomas usually present as asymptomatic nodules, but some patients experience nerve-related symptoms such as a mild twinge of pain if the tumor is touched. Most schwannomas appear to originate in sensory nerves. A significant group of schwannomas occur in the thoracic cavity, with a strong predilection for the posterior mediastinum around the spinal column at the spinal nerve roots. In this location, schwannomas often reach a larger size than in the periphery.

The retroperitoneal space in the abdominal cavity is a relatively common site for these tumors. The presacral pelvic cavity is the most common retroperitoneal location, but the upper abdominal cavity, especially the kidney region, also can be involved. Retroperitoneal schwannoma patients have a median age of 48 years, with mild female predominance, 60:40 (Figure 24.32). There seems to be a bimodal age distribution, with two peaks of incidence in both sexes at the fourth and seventh decades. The median size for retroperitoneal schwannomas in our series is 9 cm. Smaller retroperitoneal schwannomas are typically incidental findings during pelvic examination, palpation, or radiologic studies, whereas large schwannomas are often symptomatic, causing organ compression manifesting as hydronephrosis, constipation, and dysuria. Typical schwannomas can also involve visceral sites primarily, such as the kidney.[28,67,68]

Schwannomas of cranial nerves present in the head and neck area and intracranially, especially those of the eighth cranial nerve, especially its portion of the vestibular nerve, where a bilateral schwannoma is diagnostic of the NF2 tumor syndrome.

Pathology: gross

Schwannomas typically arise in the peripheral aspect of the nerve and are encapsulated by a fibrous band. Although the normal nerve can usually be surgically identified in the periphery of the tumor, it is rarely present in the surgical specimen, because nerve sacrifice is avoided. On sectioning, schwannomas are firm, elastic, and can be slightly mucoid. The color varies from gray-tan to yellowish, and many have yellow foci or are homogeneously yellow. Hemorrhagic foci are common, and cystic change can be seen, especially in larger tumors, such as those in the retroperitoneum (Figure 24.33). Some schwannomas, especially larger examples in the body cavities, are extensively cystic, with only a peripheral rim of preserved tumor tissue.

The schwannomas in superficial soft tissue are usually relatively small, measuring 1 cm to 5 cm, whereas mediastinal

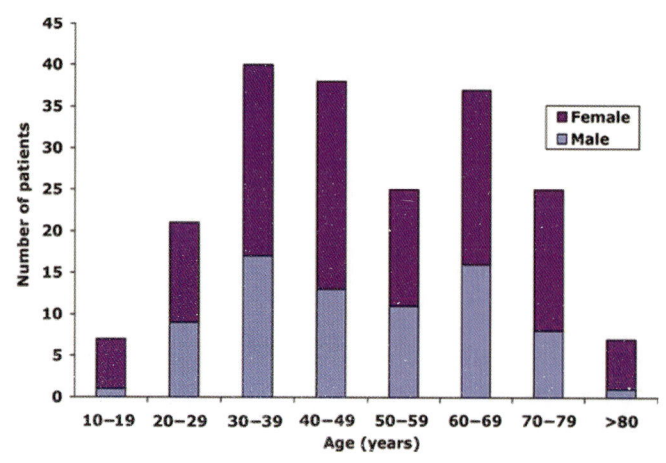

Figure 24.32 Age and sex distribution of 100 patients with retroperitoneal schwannomas.

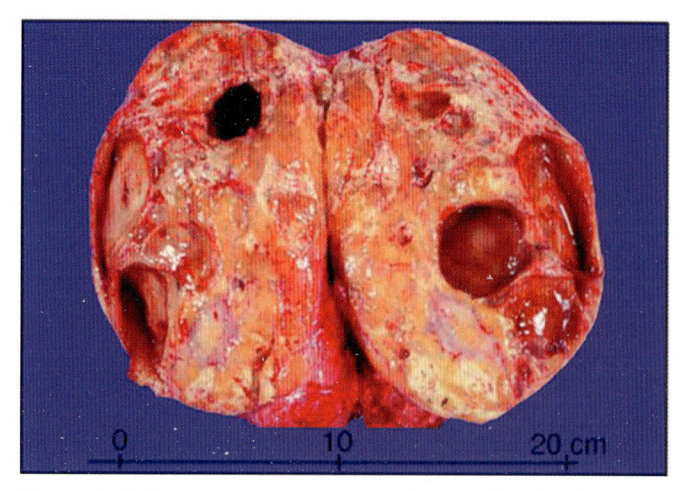

Figure 24.33 Retroperitoneal schwannoma shows a yellowish surface on sectioning, and cysts are present.

and retroperitoneal tumors often reach a large size, the latter often >10 cm.

Pathology: microscopic

Histologically, schwannomas are spindle cell neoplasms that typically have two components: compact spindle cell Schwann cell components (*Antoni A areas*), and a loosely textured, histiocyte-rich component (*Antoni B areas*). The schwannian component has a syncytial pattern and at least focal nuclear palisading, with an adjacent light, microscopically amorphous-appearing pool of cellular processes (Verocay bodies). Some tumors show a palisading pattern throughout, and some are essentially composed of cellular palisading areas only (Figure 24.34a,b), whereas some tumors have focal palisading only. The nuclei vary from oval and blunt-ended to elongated, and focal atypia can occur. The tumor cells can have nuclear vacuolization. Tumors with prominent benign atypia have been referred to as *ancient schwannomas*.[69] The presence of slight mitotic activity in an otherwise typical encapsulated schwannoma is acceptable. The histiocyte-rich component contains sheets of xanthoma cells sprinkled with lymphocytes, and these areas correspond to the deeply yellow foci seen grossly. Many schwannomas have thick-walled vessels with fibrinoid and hyaline changes in the vessel walls (Figure 24.34c,d).

Cystic change can be extensive, transforming the tumor into a cyst that contains residual schwannoma only as a narrow rim in the cyst walls (Figure 24.35). Necrosis can occur, possibly representing ischemic infarction of an encapsulated tumor under pressure, but this is not an alarming feature in a cytologically regular schwannoma.

Immunohistochemistry of schwannoma

Schwannoma cells are strongly and uniformly positive for S100 protein with both nuclear and cytoplasmic staining. Scattered EMA-positive perineurial cells are often seen in the capsular area, whereas the tumor cells are negative. There are CD34-positive fibroblasts, especially adjacent to the capsule at the tumor periphery, perivascularly associated with the degenerative areas, but the cellular areas are CD34 negative. Schwannomas are consistently positive for vimentin, and about one-half of them have GFAP-positive Schwann cells.[70-72] Keratin positivity (detected with AE1 antibody or cocktails) is quite common and probably results from cross-reaction of these antibodies with GFAP.[72] Schwannoma cells can be positive for CD68 because of their high lysosomal content.[73] The strong immunoreactivity for laminin and collagen type IV reflects the presence of abundant basement membrane material.[74]

Ultrastructure of schwannoma

The cellular areas of schwannoma are composed of Schwann cells with prominent, continuous basal laminas, and the complex cellular processes are also surrounded by basal laminas. Many tumors have extracellular foci of banded collagen, the so-called long-spacing collagen.[51,75,76]

Cellular schwannoma

Highly cellular schwannomas with a solid cellular pattern without typical Verocay bodies have been designated as *cellular schwannomas*. The greatest significance of this variant is its recognition as a benign nerve sheath tumor and its separation

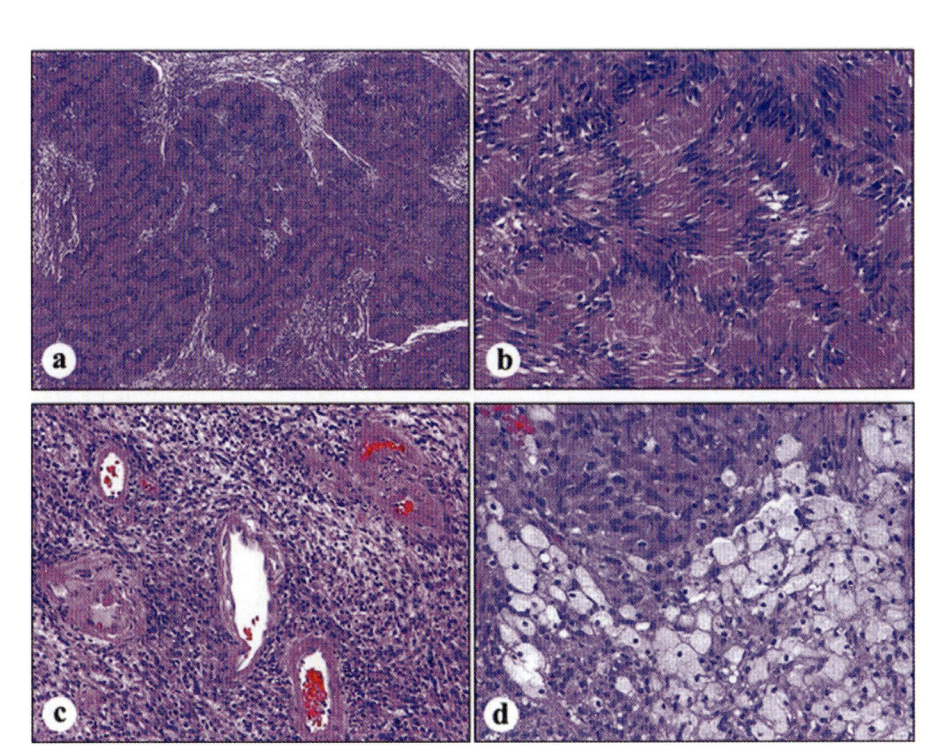

Figure 24.34 (a,b) Prominent nuclear palisading in schwannoma. (c,d) Hyalinized dilated blood vessels and foci of xanthoma cells are typical features.

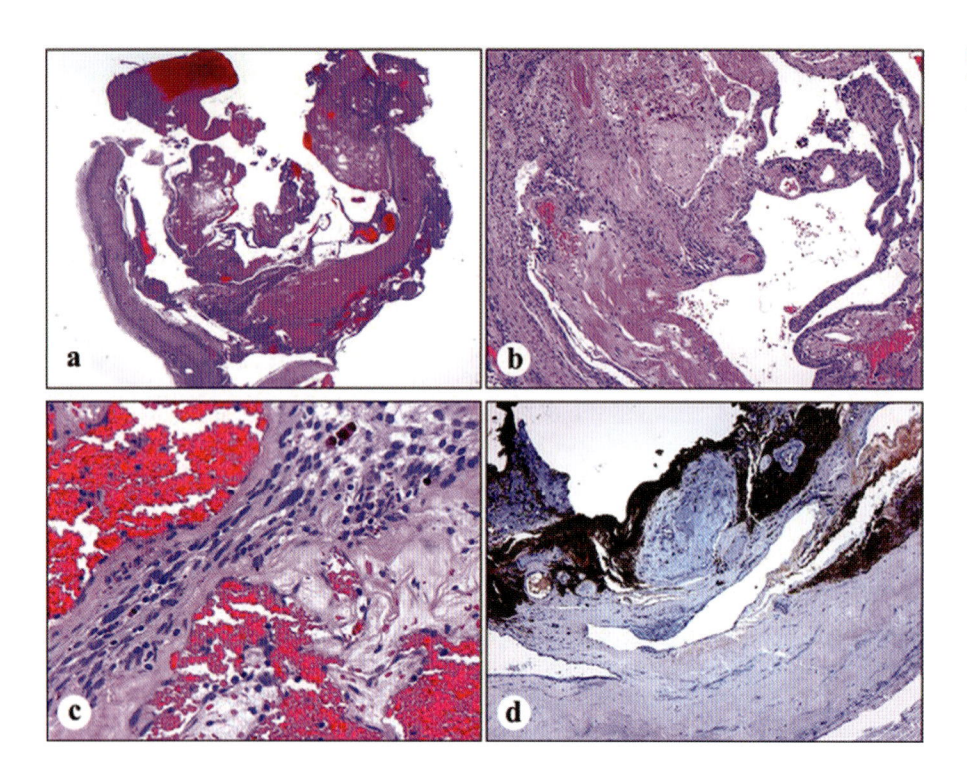

Figure 24.35 Cystic schwannoma. In the walls of the cystic tumor, scant S100-protein-positive Schwann cell elements are present.

from MPNSTs; apparently many of these tumors were diagnosed previously as malignant schwannomas.

Clinical features

According to the three largest clinicopathologic series,[77–80] cellular schwannomas occur in a wide age range, with a predominance in young and middle-aged adults, with the peak incidence in the fourth decade and median and mean ages between 40 and 54 years. The two largest series report a significant female predominance of 63% to 72%.[78,79] Most cellular schwannomas occur in the retroperitoneum and posterior mediastinum, although a few are seen in the head and neck, and the extremities. Tumors adjacent to the spine or ribs can cause bone erosion. Although their behavior is benign, one series reported a local recurrence rate of 23%.[80]

Pathology

Grossly, cellular schwannomas are well demarcated and usually encapsulated spherical or ovoid masses, similar to conventional schwannomas. They tend to be grayish-white on sectioning and more homogeneous than ordinary schwannomas, but yellow histiocytic foci can be present. The tumor size varies over a wide range (1–20 cm), with the median size being 6 cm in the largest series.

Histologically, tumors are composed of sheets and fascicles of spindle cells, which are more slender than smooth muscle cells and sometimes form perivascular whorls (Figure 24.36). Many tumors have foci of loose texture with xanthomatous histiocytes. Lymphoid clusters can be seen in the capsular region. Mitotic activity is present, but is usually limited to 4 per 10 HPFs; however, foci of higher mitotic activity can be present. Necrosis occurs rarely, and when present is limited to small foci. Pleomorphism is not a feature of cellular schwannoma.

Immunohistochemically, cellular schwannomas are similar to the ordinary variety, with a strong S100 protein positivity. This is one distinguishing feature from MPNSTs, which are usually much less S100 protein positive than cellular schwannomas.

Differential diagnosis

Lack of atypia or significant necrosis and the presence of elements seen in the context of ordinary schwannoma (i.e., vascular changes, xanthoma cells), and uniform S100 protein positivity support the diagnosis of cellular schwannoma over MPNST. Proliferation marker studies showed that the Ki67 index was 6% to 8% in both nonrecurrent and recurrent tumors, and this parameter was not helpful to predict recurrence. Focal p53 positivity was also common, despite the benign nature of these tumors.

Other tumors with a variable nuclear palisading pattern include GISTs, which are KIT positive (Chapter 17), and monophasic synovial sarcomas (Chapter 27). Neither of these is an encapsulated tumor.

Plexiform schwannoma

This schwannoma variant can be considered to include two distinct clinicopathologic entities: superficial (cutaneous) plexiform schwannoma, and deep plexiform schwannoma. Cellular plexiform schwannoma is a rare childhood tumor that is discussed separately.

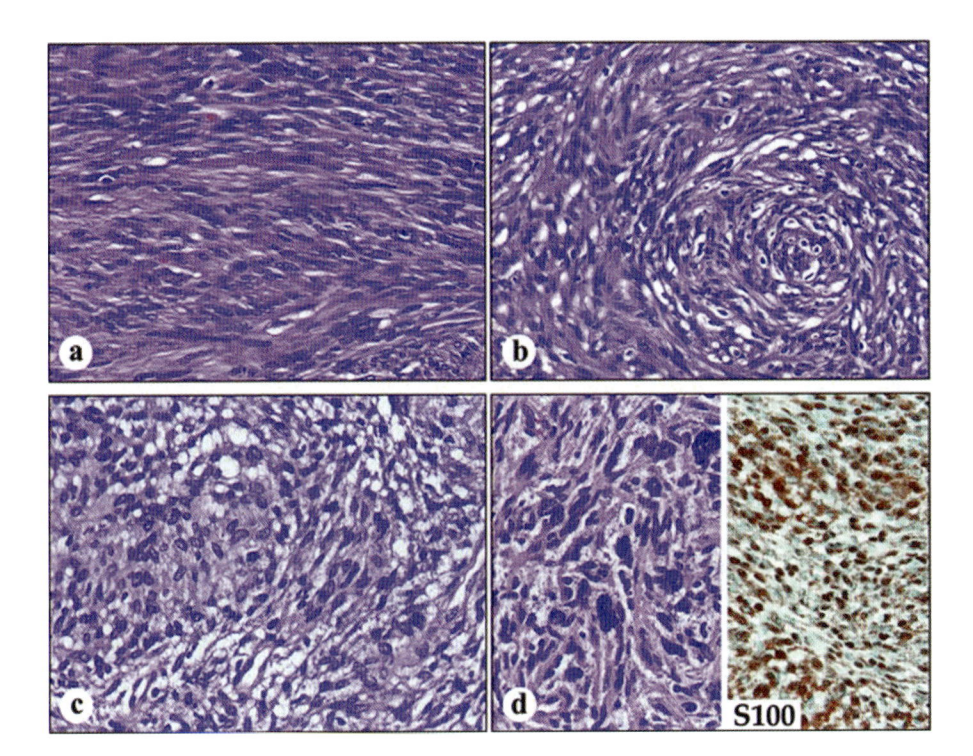

Figure 24.36 (a) Cellular schwannoma contains irregular fascicles of spindle cells with vague palisading. (b) Perivascular whorling can be present. (c,d) An epithelioid example with focal atypia is positive for S100 protein.

Clinical features

Superficial plexiform schwannoma is a relatively rare, multinodular schwannoma variant that usually presents as a small solitary cutaneous or subcutaneous nodule, most commonly in the extremities. There is a predilection for young adults.[81–86] Rare association with NF2 syndrome has been reported.[87] Local recurrences have occurred in some cases, but otherwise the behavior is benign.

Deep plexiform schwannoma is a rare tumor that can occur in the peripheral soft tissues intramuscularly, and in visceral sites, especially in the gastrointestinal tract.[85]

Pathology

Grossly, the lesion is multinodular and gray or yellowish-gray. Histologically, it consists of multiple round to oval nodules composed of spindled Schwann cells. Nuclear palisading and Verocay bodies are usually present, but xanthoma cells and lymphocytic infiltration are uncommon (Figure 24.37). Diagnosis is supported by identification of a purely S100-protein-positive Schwann cell population. Some plexiform neurofibromas can have areas resembling schwannoma, but these tumors can be diagnosed as neurofibroma by the typical areas elsewhere in the tumor and by the presence of a prominent CD34-positive cellular component.

Cellular plexiform schwannoma

This tumor was originally described as "plexiform MPNST," but its behavior was subsequently shown to be benign. The reported tumors mainly occurred in peripheral soft tissues in young children and measured 1.5 cm to 8 cm. Histologically, these tumors were composed of bundles of spindle cells, giving them a microscopically plexiform architecture. The cellularity was high, and the mitotic rate varied from 1 to 18 per 10 HPFs; however, only one tumor was fatal in the original series; subsequent series showed a uniform benign course.[88,89]

Epithelioid schwannoma

This rare variant of schwannoma with epithelioid cellular morphology has been reported in the peripheral soft tissues and in the neck of the urinary bladder in adult patients. The author and colleagues have seen similar tumors in the peripheral soft tissues and in the submucosa of the colon. These tumors are typically small (<1–2 cm) and have a benign course.[90]

Immunohistochemical identification of S100-protein-positive pure schwannian element is a key feature, and lack of atypia and significant mitotic activity help to distinguish this tumor from malignant epithelioid MPNST and melanoma; the latter can be identified based on expression of melanoma-specific markers (Figure 24.38). When the tumor is intestinal, the lack of neuroendocrine markers and keratins help to separate it from carcinoid.

Schwannoma with rosettes and neuroblastoma-like features

Schwannomas composed of densely packed oval cells with rosette-like structures have been called *neuroblastoma-like schwannomas*.[91] The reported three tumors occurred in young or middle-aged adults. Such tumors show high cellularity,

Chapter 24: Nerve sheath tumors

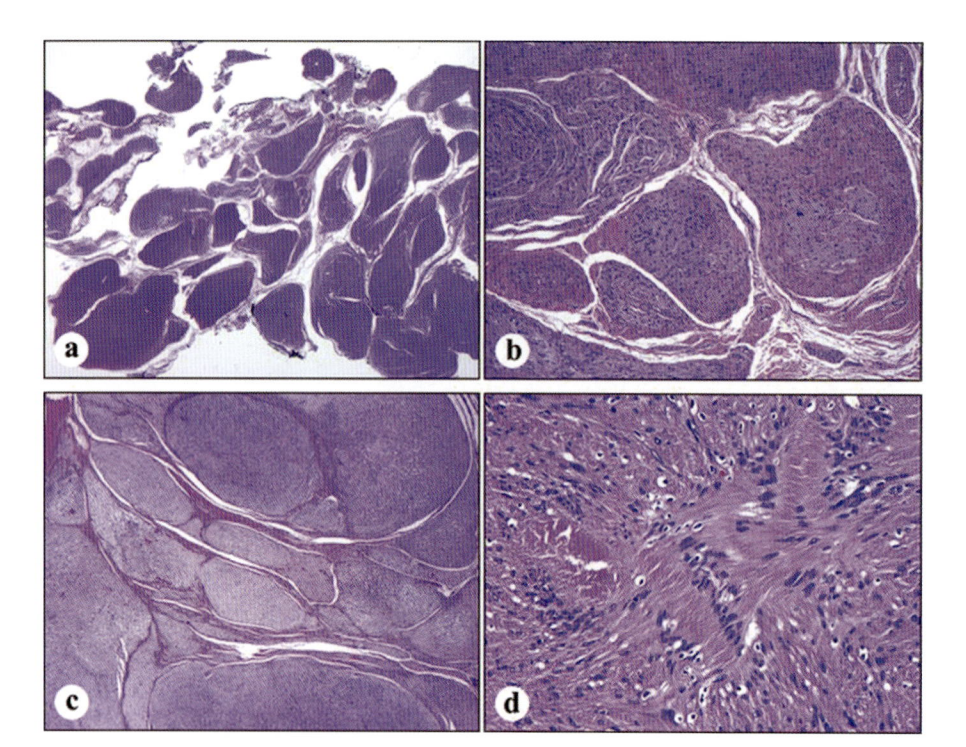

Figure 24.37 Plexiform schwannoma is composed of multiple nodules of Schwann cell proliferation.

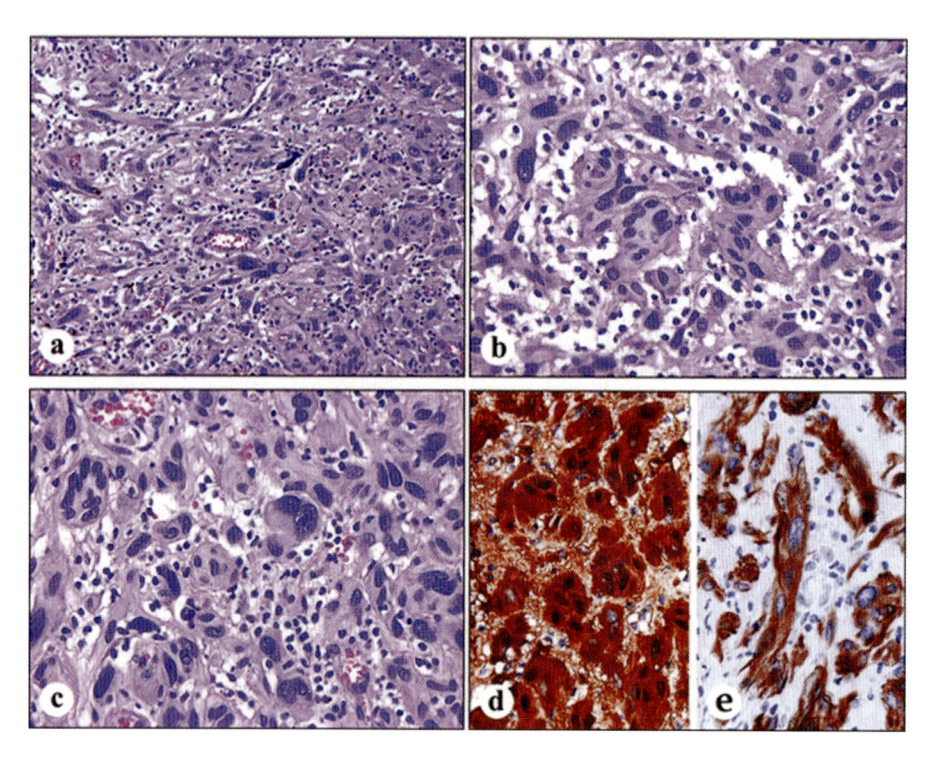

Figure 24.38 Schwannoma with epithelioid features. (d,e) The tumor cells are positive for S100 protein and GFAP.

areas of tumor cells with oval to round cell morphology, and peculiar rosettes surrounding amorphous material, somewhat resembling Homer Wright rosettes of neuroblastoma (Figure 24.39). The immunophenotypical features are those of schwannoma, with uniform and strong S100 protein positivity and negativity for the true neural markers. The tumor appears benign, although follow-up of the reported cases has been short.

Sarcoma arising in schwannoma

Malignant transformation of an ordinary schwannoma is an extremely rare event, but isolated cases and small series have been reported. As defined by Woodruff et al., such a diagnosis requires documented origin of the malignant component from a pre-existing schwannoma, proof of malignant behavior of the transformed component, and exclusion of the possibility

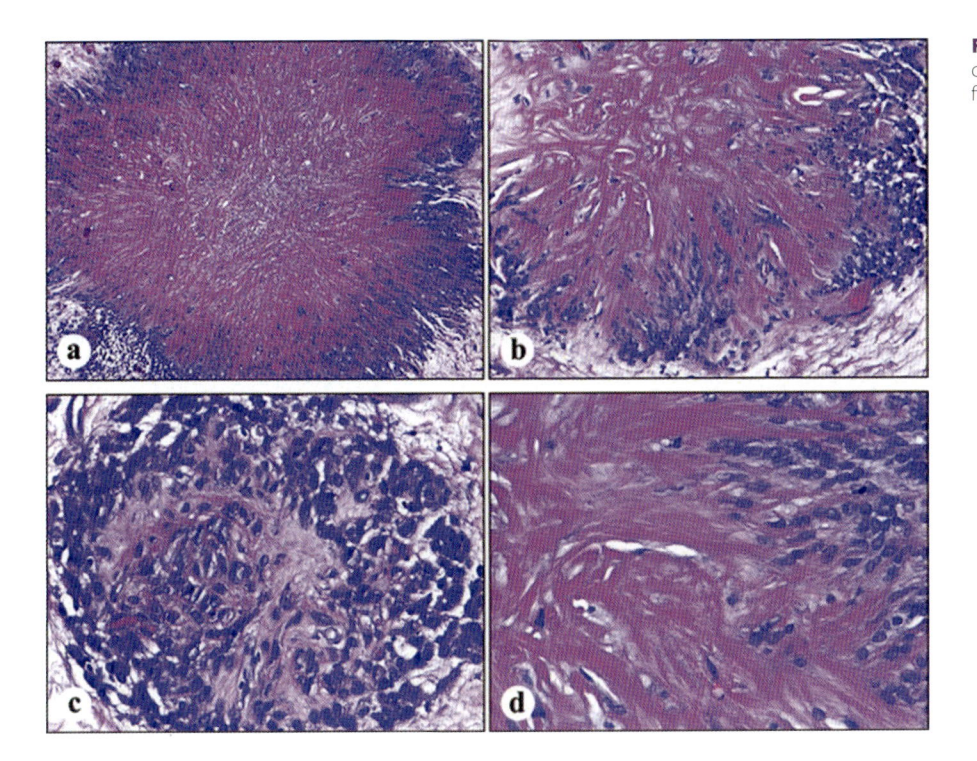

Figure 24.39 Schwannoma with hyaline rosettes composed of rounded tumor cells surrounding fibrous matrix ("neuroblastoma-like schwannoma").

that the malignant component is a metastasis from another source.[91]

According to a review of well-documented cases, schwannomas with malignant transformation occur in middle-aged patients and have a high tumor-related mortality, frequently with liver and lung metastases. Histologically, they are high-grade tumors, usually with an epithelioid morphology. Keratin positivity is common, and S100 protein positivity is patchy rather than diffuse.[92]

Angiosarcoma can arise in schwannoma. The reported tumors have often occurred in the neck region in older men, and many of them originated from the vagus nerve. Some clinically simulated a carotid body tumor, and most were histologically epithelioid angiosarcomas.[92–95]

Neurofibromatosis 2 and neurofibromatosis 2 gene alterations

NF2 is an autosomal dominant disorder caused by a germline mutation in the NF2 gene, located in the pericentromeric region of chromosome arm 22q11.2. NF2 syndrome is much less common than NF1, with the birth incidence of 1:40 000.[96,97] Similar to NF1, one-half of all NF2 cases result from new mutations.[98,99] NF2 predominantly involves the central nervous system, and bilateral vestibular schwannomas are typical and diagnostic; NF2 has been synonymously called *central neurofibromatosis* or *bilateral vestibular neurofibromatosis*. Multiple meningiomas and gliomas, especially spinal types, and meningoangiomatosis are also manifestations of NF2 syndrome.

Diagnosis of NF2 syndrome

Bilateral vestibular schwannomas are diagnostic of NF2. According to the NIH consensus conference, additional diagnostic criteria of NF2 are the following:[53]

(1) Family history of NF2 and occurrence of unilateral vestibular schwannoma <30 years
(2) Family history of NF2 and two of the following in any combination: glioma, meningioma, schwannoma, juvenile cortical cataract (posterior eye lens opacities).

Probable diagnosis (requiring further evaluations) is based on both of the these findings without family history of NF2. Multiple meningiomas either with unilateral vestibular schwannoma <30 years of age or any of the signs of (2) is suspicious for NF2 and requires further evaluation. Demonstration of a germline mutation in the NF2 gene from peripheral blood can confirm the diagnosis.

Pathogenesis of NF2 syndrome and NF2-associated tumors

The key event in pathogenesis of both NF2-associated and sporadic schwannomas (and many types of meningiomas) seems to be loss of function of the NF2 encoded protein merlin (schwannomin), following a two-hit mechanism. This occurs by a combination of inactivating NF2 mutations and allelic losses, analogous to the changes in the NF1 gene in NF1 syndrome.[98,99] Experimental expression of the mutant NF2 protein causes Schwann cell hyperplasia and tumors, supporting the direct role of abrogation of this protein.[100] There seems to be a molecular correlation with the severity of disease. Truncation of NF2 (premature stop codons) results in a more severe phenotype than missense mutations.[101,102] In addition to mutation, calcium-activated proteolytic degradation of NF2 protein might be an alternative disease mechanism in some cases.[103]

Schwannomatosis

Schwannomatosis is the designation for a very rare condition, which manifests as multiple peripheral schwannomas. These tumors have been reported in superficial soft tissues and in the spinal region and occasionally in the vestibular nerve. The patients do not have bilateral vestibular tumors and lack germline mutations of NF2 gene, in contrast with NF2 syndrome.[104–108] In general, schwannomatosis involves somatic biallelic activation of the NF2 gene associated with loss-of-function germline mutations in other genes nearby in chromosome 22q, especially SMARCB1.[109–113] Germline mutations in SMARCB1 gene related to partial loss of function occur in some patients. Loss-of-function germline mutations in another gene located at 22q11, LZTR1, has been reported in a majority of schwannomatosis cases that lack SMARCB1 mutations.[114]

Hybrid nerve sheath tumors

Some nerve sheath tumors contain several cellular components and are considered to represent hybrid tumors. These include hybrid schwannoma-neurofibroma tumors, which histologically and immunohistochemically have features of both. These tumors are variably associated with neurofibromatoses (especially NF2), or schwannomatosis.[115,116] Hybrid schwannoma-perineuriomas[117,118] or less often, neurofibroma-perineuriomas[119] have also been reported, generally unassociated with neurofibromatoses. They contain a distinct non-overlapping perineurial cell component, in addition to schwannian and fibroblastic components.

Gastrointestinal schwannoma

This rare tumor differs sufficiently from conventional schwannoma to be placed in a separate clinicopathologic category, and its pathogenesis could also be different from that of schwannoma. Otherwise visceral schwannomas are rare.

Clinical features

Schwannomas in the gastrointestinal tract most commonly occur in the stomach (60–70%). The next most common site is the colon, whereas only isolated cases have been reported in the esophagus and small intestine.[120–124] Gastrointestinal schwannomas are much rarer than GISTs, occurring in a ratio of approximately 1 schwannoma to 50 GISTs.

Gastrointestinal schwannomas occur in a wide age range, but most commonly in older adults, similar to GISTs. There is no clear sex predilection. Many patients present with gastrointestinal bleeding, but some tumors are incidental findings during endoscopy or imaging studies for cancer surveillance.

Follow-up studies indicate a uniformly benign clinical course, and there is no association with NF1 or NF2.[120–124]

Pathology

Grossly, the gastrointestinal schwannomas are typically ovoid, yellowish nodules that appear sharply demarcated, although unencapsulated. The tumor usually measures 2 cm to 4 cm, and occasionally >10 cm. In the stomach, schwannomas usually present as circumscribed intramural tumors (Figure 24.40), whereas many colonic ones appear as polypoid intraluminal masses, although with transmural involvement.

Histologically, the most common variant of gastrointestinal schwannomas is a spindle cell tumor at least focally surrounded by a lymphoid cuff, often with germinal centers that can extend into the peripheral portion of the tumor (Figure 24.41a). The tumor cells are often arranged in fascicles in a microtrabecular pattern between thick vascular septa, and focal nuclear atypia and mild lymphoplasmacytic infiltration may be present throughout the tumor. Mitotic activity almost never exceeds 5 per 50 HPFs. Vascular hyaline changes, nuclear palisading, and xanthoma cells are not prominent features, in contrast to peripheral schwannomas (Figure 24.41b–d).

A rare variant of gastrointestinal schwannoma has signet-ring cell-like morphology, and this should not be confused with signet ring cell carcinoma on biopsy.[125] This variant has also been reported as "microcystic schwannoma" and may also occur outside the gastrointestinal tract.[126]

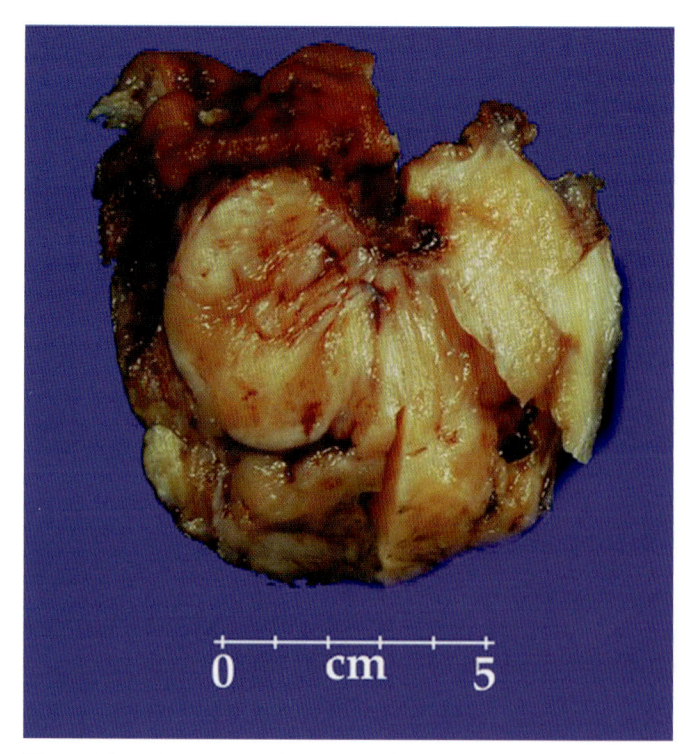

Figure 24.40 Gastrointestinal schwannoma forms a fleshy, pale tan mass underneath intact mucosa (on top left).

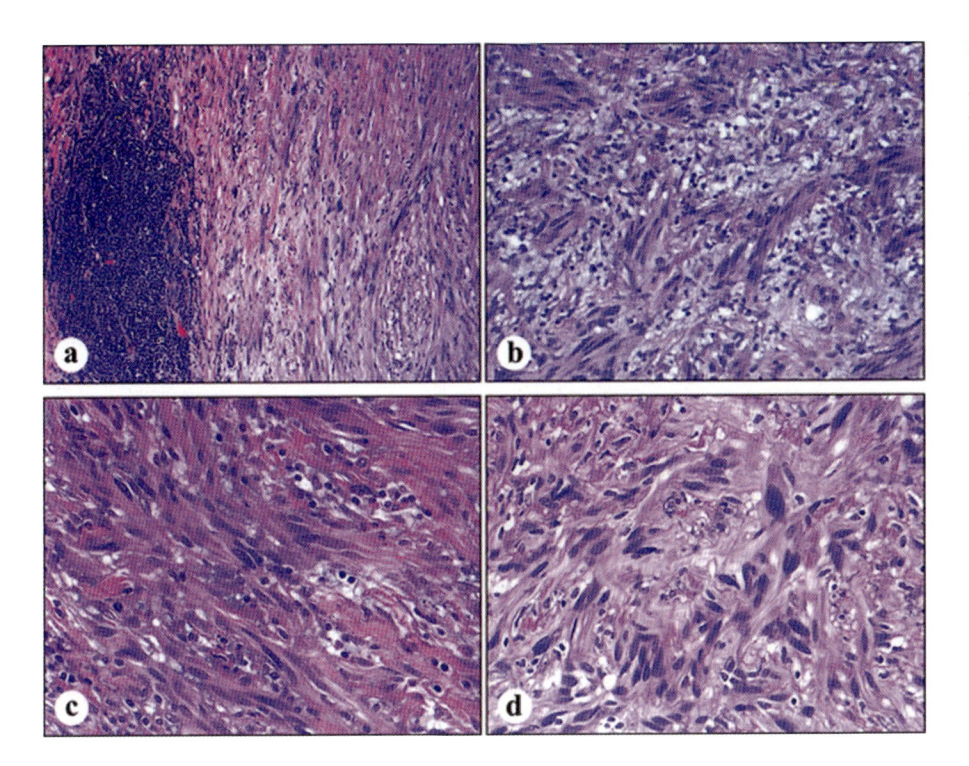

Figure 24.41 (a) Peritumoral lymphoid infiltrate is typical of gastrointestinal schwannoma. (b,c) The tumor cells are often arranged in bundles set in a fibromyxoid matrix. (d) Focal nuclear atypia is often present.

Plexiform and epithelioid schwannomas, similar to those seen in peripheral soft tissues, can also occur in the gastrointestinal tract. The former present as multinodular masses and can reach a considerable size. The latter can form small intraluminal polyps, especially in the colon.[123,127]

Immunohistochemical analysis shows tumor cells that are typically strongly positive for S100 protein and usually for GFAP and NGFR/p75. Most tumors are negative for CD34, and these tumors are uniformly negative for KIT (CD117).

Genetically, gastrointestinal schwannomas differ from the peripheral variety by the lack of NF2 gene alterations. Instead, many of these tumors show allelic losses in NF1, indicating that the gastrointestinal schwannomas might be genetically related more to neurofibromas than schwannomas.[128]

Nerve sheath tumors (indeterminate whether schwannomas or neurofibromas) can also present as small mucosal polyps, usually in the colon. These polyps typically show interstitial Schwann cell proliferation between the crypt elements, and vague nuclear palisading can be present. The spindle cell element is positive for S100 protein, and a CD34-positive fibroblastic element can also be present. This lesion is KIT negative, however, which helps to rule out a GIST (Figure 24.42).

Differential diagnosis

GISTs are KIT-positive gastrointestinal mesenchymal tumors, and a minority of them (10–15%) is also S100 protein positive. Many GISTs show prominent nuclear palisading, which should not lead to confusion with schwannoma. True neurofibromas occur in the gastrointestinal tract in two different clinical contexts: plexiform and diffuse neurofibromas in young NF1 patients, and sporadic small polypoid mucosal neurofibromas in adults. Most, if not all, highly malignant, strongly S100-protein-positive spindle cell and epithelioid tumors of the intestines are metastatic melanomas; some of them are polypoid. Their diagnosis may be aided by positivity for melanoma-specific markers and clinical history.

Melanotic schwannoma

This very rare Schwann cell tumor combines the features of schwannoma (schwannian differentiation) and melanoma (presence of melanosomes and melanin pigmentation). Of note is its association with the Carney complex. Psammoma bodies are often, but not always, present ("psammomatous melanotic schwannoma").

Clinical features

In up to half of the cases, this tumor occurs as a part of an inheritable syndrome, named the *Carney complex* based on its first description by Carney in 1985. The other manifestations of the Carney complex include spotty cutaneous hyperpigmentation, cutaneous myxoma (superficial angiomyxoma), cardiac myxoma, and endocrine overactivity (e.g., Cushing's syndrome, acromegaly, calcifying Sertoli cell tumor). According to Carney, 14% of patients with Carney complex have a (psammomatous) melanotic schwannoma. There is no association with neurofibromatoses.[129]

Patients with this tumor usually present at a relatively young age, between 20 and 40 years. Patients with the Carney complex are almost 10 years younger (average 23 years) than those without it (average 33 years). The lesions often occur near the midline in the skin, subcutis, or deep soft tissues or

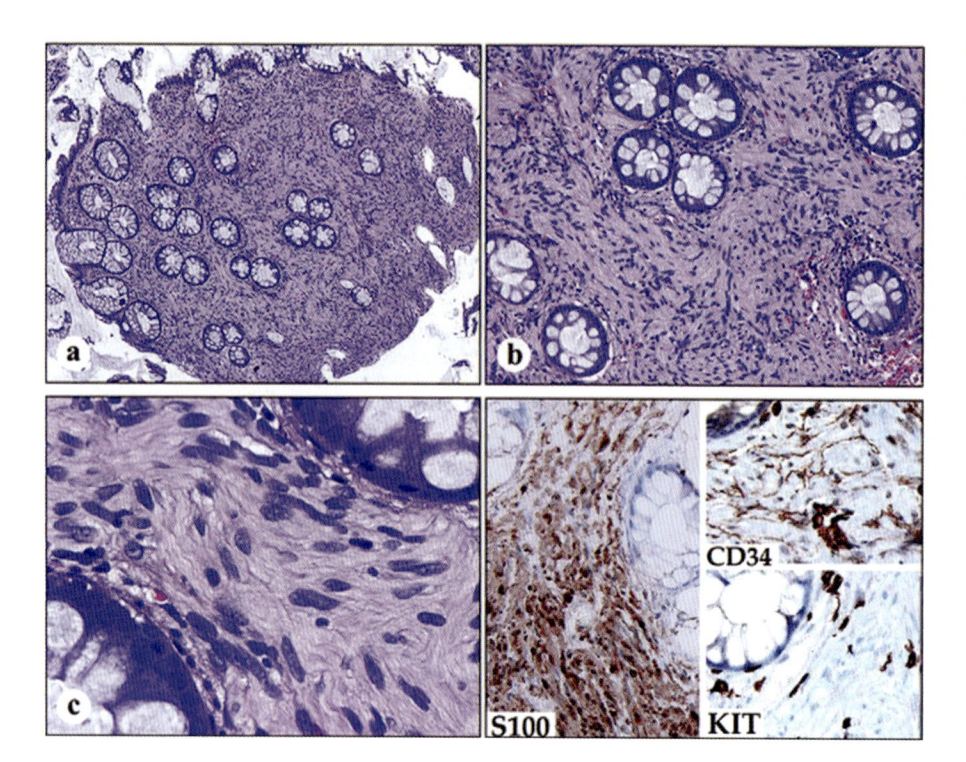

Figure 24.42 Mucosal nerve sheath tumor (neurofibroma) of the colon shows bland spindle cell proliferation expanding the mucosal stroma. A typical immunohistochemical finding is strong positivity for S100 protein with variable CD34 reactivity and negativity for KIT, except that mast cells are positive.

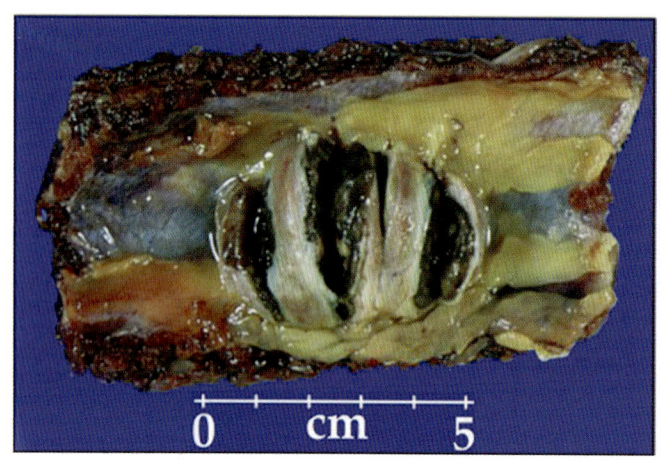

Figure 24.43 Psammomatous melanotic schwannoma forms an intercostal mass bulging into the pleural cavity. Note a thick fibrous capsule and brown-pigmented cut surface.

viscera. The three most common locations are the spinal nerve roots (back), and the soft tissues of the trunk and of the extremities.[129–131] Destructive bone involvement is also possible, and isolated tumors have occurred in the heart, liver, and bronchus. Some patients have multiple tumors, a feature potentially difficult to distinguish from metastatic disease. Primary bone tumors of this type have been reported.[132]

Melanotic schwannomas vary in their behavior. The Carney series included mostly clinically benign cases, with 3 of 31 patients dying of metastatic disease within 2 to 7 years.[129] However, another series showed 44% frequency of metastases.[131] Complete excision with tumor-free margins is thought to be the optimal treatment. Metastases develop in the lungs, pleura, liver, and spleen.

Pathology

Melanotic schwannoma ranges from a small nodule <1 cm to a bulky tumor >20 cm; most examples are >5 cm. Gross inspection shows that melanotic schwannomas are less often encapsulated than typical schwannomas, but some are surrounded by a thick capsule. Like conventional schwannomas, the tumors are often fusiform, sausage-like, or dumbbell-shaped. On sectioning, they typically have a gray-to-black surface with paler foci from variable melanin pigmentation (Figure 24.43).

Histologically, the tumor can have a solid, fascicular, or whorled pattern (Figure 24.44). It is composed of spindled to epithelioid cells with moderate to abundant cytoplasm, and delicate spindled cells with dendritic processes also occur. Schwannoma-like nuclear palisading is not typical, but can be present. The occurrence of spherical, PAS-positive multilaminar psammomatous calcifications in varying numbers is unique to this tumor, although these are not present in all cases. In some degenerated examples, psammoma bodies and fibrosis can be the dominating elements, and fat-like vacuolization of neoplastic Schwann cells is common (Figure 24.45). Pigmentation varies from zones of dark melanin-laden histiocytes to fine, cytoplasmic punctate melanin.

Cytologically, most tumor cells have limited atypia, with mildly atypical nuclei having a small nucleolus. The spindle

Chapter 24: Nerve sheath tumors

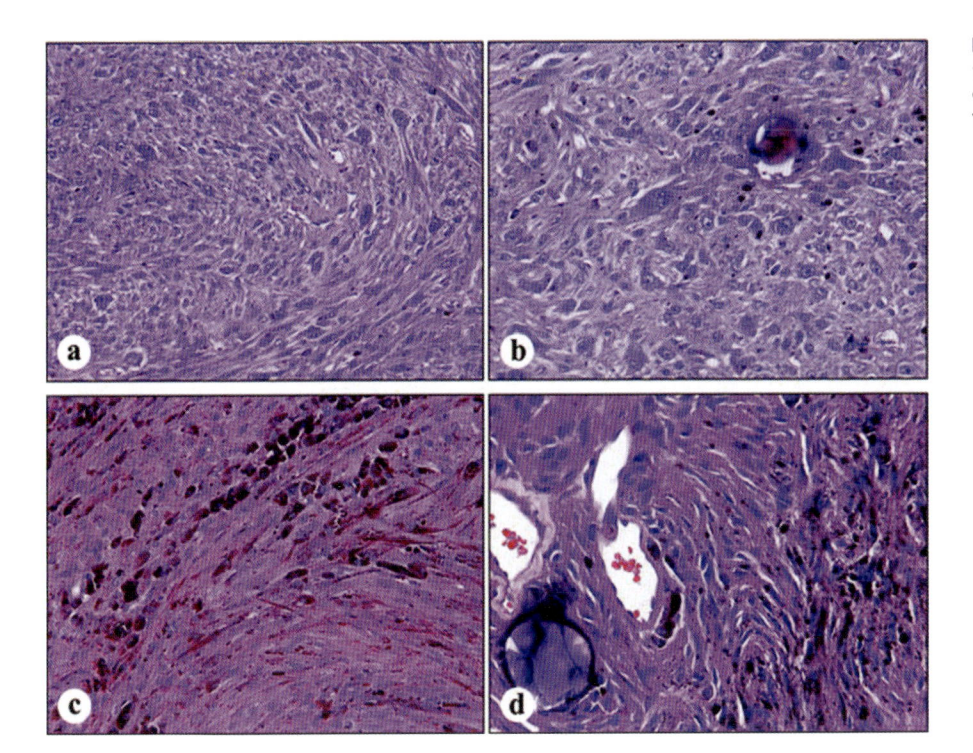

Figure 24.44 Psammomatous melanotic schwannoma consists of solid sheets of spindled to epithelioid cells with focal psammoma bodies and varying amounts of melanin pigment.

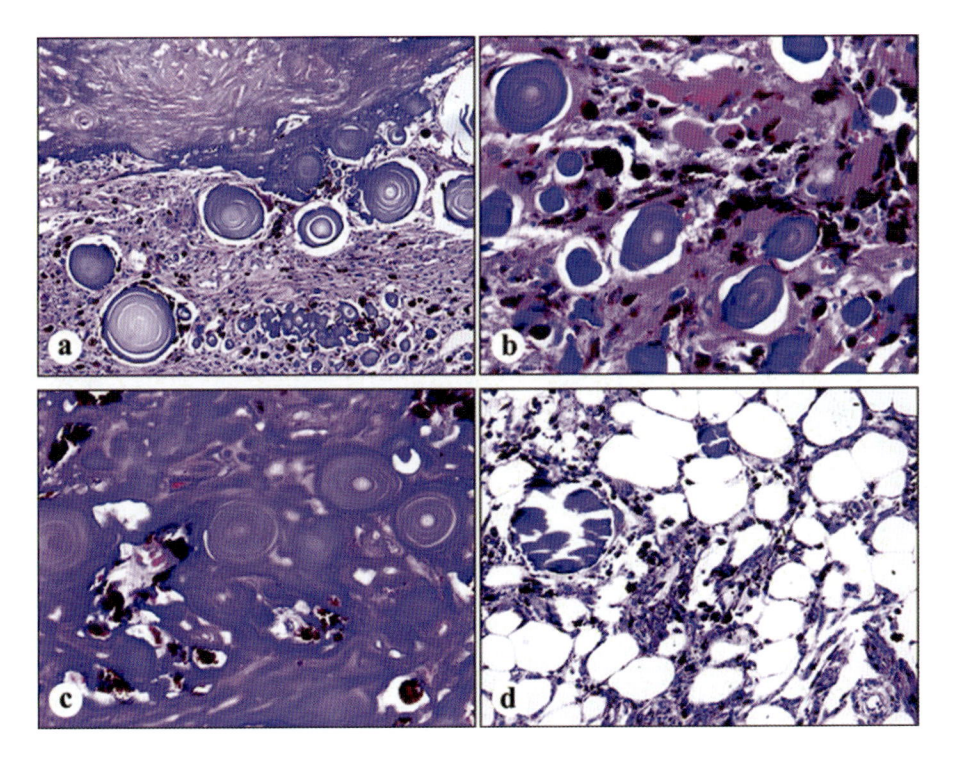

Figure 24.45 Extensive degenerative features in psammomatous melanotic schwannoma with numerous psammoma bodies, extensive hyalinization, and fat-like vacuolization. The presence of melanin pigment is distinctive.

cells can have a dendritic morphology. Nuclear pseudoinclusions, similar to those seen in other Schwann cell tumors, are sometimes present. Some tumor cells have large, sometimes multiple, nuclei with prominent nucleoli and even macronucleoli. Increased nuclear atypia with prominent nucleoli and tumor necrosis are features seen in malignant variants (Figure 24.46). A mitotic rate of 2 per 10 HPFs has been found to be associated with higher risk of malignant behavior.[131]

Ultrastructurally, the tumors combine schwannian features, such as complex cell processes and well-developed basement membranes, with the presence of melanosomes. Specific genetic information is not available for this tumor.

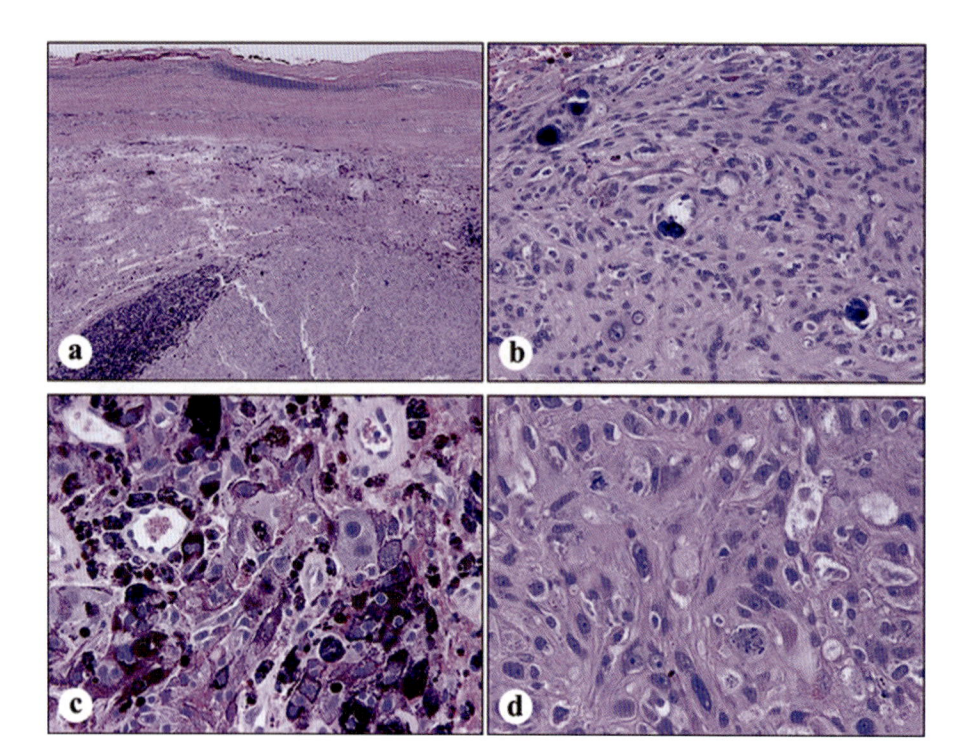

Figure 24.46 Histologically malignant features in psammomatous melanotic schwannoma with high cellularity and nuclear atypia, including prominent nucleoli.

Immunohistochemistry

The pigmented melanotic schwannomas are typically positive for both S100 protein and HMB45, and focally for melan A. Like ordinary schwannomas, melanotic schwannomas are positive for basement membrane components laminin and collagen type IV. The tumor cells are negative for keratins and EMA. Loss of PRKAR1A expression can be an immunohistochemical marker for association with Carney complex.[131]

Differential diagnosis

Tumors that resemble psammomatous melanotic schwannoma include metastatic melanoma, clear cell sarcoma, and pigmented neurofibroma. The presence of dendritic, bipolar cytologic features is more typical of malignant pigmented schwannoma than melanoma. Both psammoma bodies and a fatty component are specific for melanocytic schwannoma and do not occur in other melanocytic tumors. Clear cell sarcoma typically shows a compartmental pattern, and pigmented neurofibroma almost always shows areas typical of neurofibroma elsewhere in the tumor.

Benign epithelioid peripheral nerve sheath tumor

This designation is based on the finding that many epithelioid peripheral nerve sheath tumors are indeterminate as to whether they are schwannomas or neurofibromas, although some show features of the latter. The epithelioid peripheral nerve sheath tumor primarily occurs in the skin and superficial soft tissues. It occurs in a wide age range with a predilection for young adults. Most examples are located in the extremities, where the tumor usually forms a small (1–2 cm) circumscribed dermal or subcutaneous nodule. Rarely does the size exceed 5 cm. The clinical course is indolent following a local excision.

Histologically, this tumor is typically composed of rounded clusters of epithelioid Schwann cells with abundant eosinophilic cytoplasm (Figure 24.47). Sclerosing or myxoid stroma can be present, and mitotic activity almost never exceeds 5 per 50 HPFs. Focal nuclear atypia can be present. If significant atypia and higher mitotic activity are present, the diagnosis of malignant epithelioid peripheral nerve sheath tumor is appropriate. Nevomelanocytic tumors (mostly benign) have to be considered in the differential diagnosis, and the immunohistochemistry for melanoma/nevus markers such as HMB45, melan A, and tyrosinase is negative. Mixed tumor of the skin is in the differential diagnosis morphologically and because of its S100 protein positivity. The identification of mixed tumor is based on epithelial and sometimes glandular elements, keratin expression, and possibly expression of other myoepithelial markers (e.g., calponin, actins).

Molecular genetic studies found losses in both NF1 and NF2 genes and were not able to pin down their nature as unequivocal neurofibromas or schwannomas.[133]

These nerve sheath tumors differ from superficial epithelioid MPNSTs by low mitotic activity and limited nuclear atypia. Atypical examples with borderline features should be excised with negative margins, complemented with clinical follow-up.

Chapter 24: Nerve sheath tumors

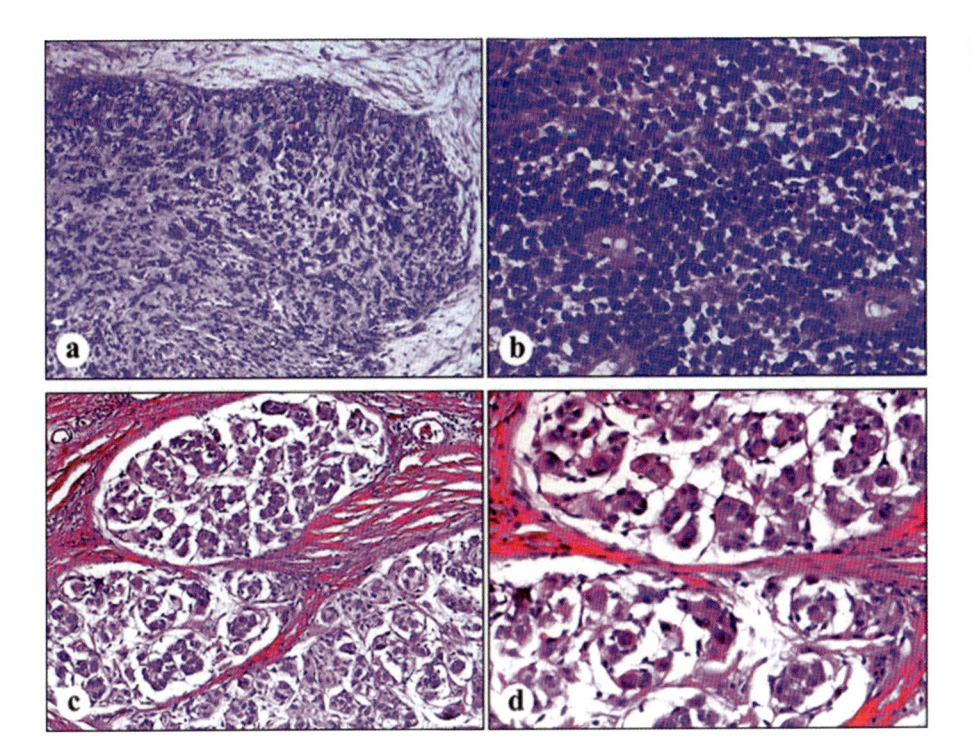

Figure 24.47 Spectrum of epithelioid peripheral nerve sheath tumor.

Nerve sheath myxoma (previously myxoid neurothekeoma)

The nerve sheath myxoma was originally introduced by Harkin and Reed in 1969 in the AFIP atlas of peripheral nerve sheath tumors.[134] The newest AFIP series conclusively defined *nerve sheath myxoma* as a rare subset of true Schwann cell tumors among tumors previously uniformly classified as "neurothekeomas."[135,136] In 1980, Gallager and Helwig introduced the term *neurothekeoma*,[137] but this and a subsequent series included fibrohistiocytic "neurothekeomas" with some nerve sheath myxomas. Neurothekeomas do not have schwannian features and can be best classified with fibroblastic or myofibroblastic neoplasms. These tumors also differ clinicopathologically from true nerve sheath myxomas in that they occur at a young age, often in the head and neck, and rarely recur. They are discussed in Chapter 10 and also here for differential diagnosis.

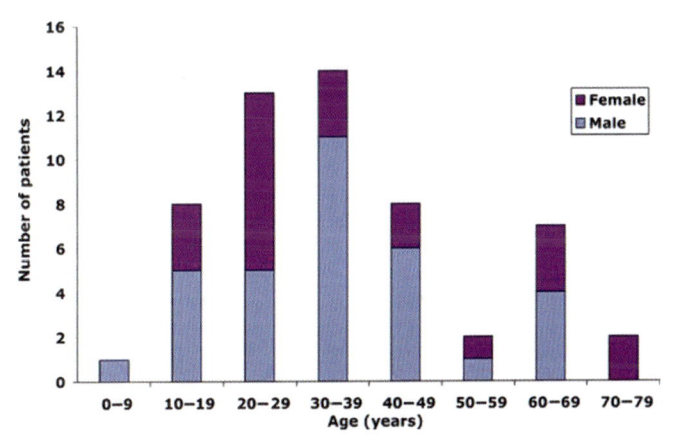

Figure 24.48 Age and sex distribution of 56 patients with nerve sheath myxoma.

Clinical features

Nerve sheath myxoma occurs in a wide age range, but it has a predilection for young adults, with a mild male predominance (Figure 24.48). It typically presents as a small (usually <2 cm) superficial painless nodule, and rare examples form polypoid cutaneous lesions. The tumor has a predilection for the distal extremities, with one-third of all tumors occurring in the fingers (Figure 24.49). Other common locations are the leg and knee region, whereas this tumor is rare in the trunk wall or the head and neck, in contrast to neurothekeoma. The tumor has a high rate of local recurrence following an intralesional

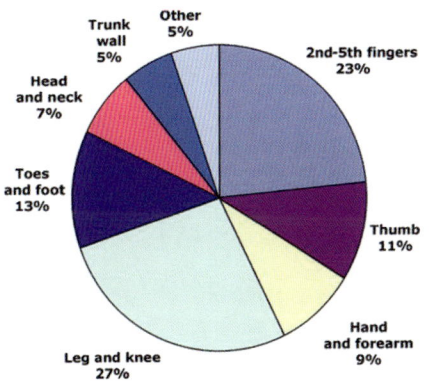

Figure 24.49 Anatomic distribution of 57 cases of nerve sheath myxoma. Note predilection for distal extremity sites.

665

excision (50%), but because the recurrences are nondestructive and asymptomatic they are often managed rather conservatively.[136–140] Occurrence in the tongue has also been reported.[141]

Pathology

This dermal or subcutaneous tumor typically appears as a multinodular mucoid mass that usually measures 0.5 cm to 2 cm and is almost never >4 cm. Histologically, the tumor is conspicuously lobulated by distinct collagenous septa and is often surrounded by a thick, collagenous, capsule-like fibrous band (Figure 24.50). Some tumors have smaller satellite nodules around the main mass, sometimes resulting in a plexiform pattern. The individual nodules are paucicellular, with a prominent myxoid variably metachromatically staining matrix.

The nodules contain cords, bundles, or syncytial clusters of small spindled or mildly epithelioid cells, occasionally forming a net-like pattern (Figure 24.51). Blood vessels are sometimes prominent, with some cases showing vessels with aneurysmal dilatation. Peritumoral lymphoid clusters are occasionally present.

The tumor cells often form large intracytoplasmic vacuoles with a signet ring cell appearance. Some cases have more compact foci, forming Verocay bodies, suggesting a relation to schwannoma (Figure 24.51d). The tumor cells typically have small hypochromatic nuclei, frequently with intranuclear vacuoles. Mitotic activity is inconspicuous, rarely exceeding 3 mitoses in 25 wide HPFs.

Immunohistochemically, the tumor cells are positive for S100 protein, with strong cytoplasmic and nuclear staining. They are also positive for GFAP, CD57, NGFR (p75), collagen IV, and vimentin, consistent with a schwannian phenotype. A peripheral zone of EMA-positive cells and scattered CD34-positive fibroblasts might be present, probably reflecting the participation of perineurial cells and neural fibroblasts, respectively (Figure 24.52). Nerve sheath myxomas are negative for HMB45, SMA, and desmin.

Differential diagnosis

Neurothekeoma is a non-neural tumor that often contains compartments of epithelioid cells with only limited, if any, myxoid matrix. Very myxoid neurothekeomas exist, however. The tumor cells do not usually form trabecular arrangements, and cytoplasmic vacuoles typical of nerve sheath myxoma are absent. S100 protein positivity is limited to scattered dendritic cells.

Superficial angiomyxoma (cutaneous myxoma) has an extensively mucoid matrix with scattered, spindled, focally atypical tumor cells that lack the organization typical of nerve sheath myxoma. S100 protein expression is limited, if it occurs.

Perineurioma: a tumor of perineurial cells

Perineurial cells form the outer cellular lining around the peripheral nerve fascicles and small nerves. These cells represent the peripheral continuation of the meningeal cells in the pia-arachnoid membranes. The normal perineurial cells are typically elongated mesenchymal cells that are often stacked in a lamellar manner. Their typical ultrastructural features include elongated cell processes, basement membranes, and pinocytic vesicles. Immunohistochemically, the perineurial

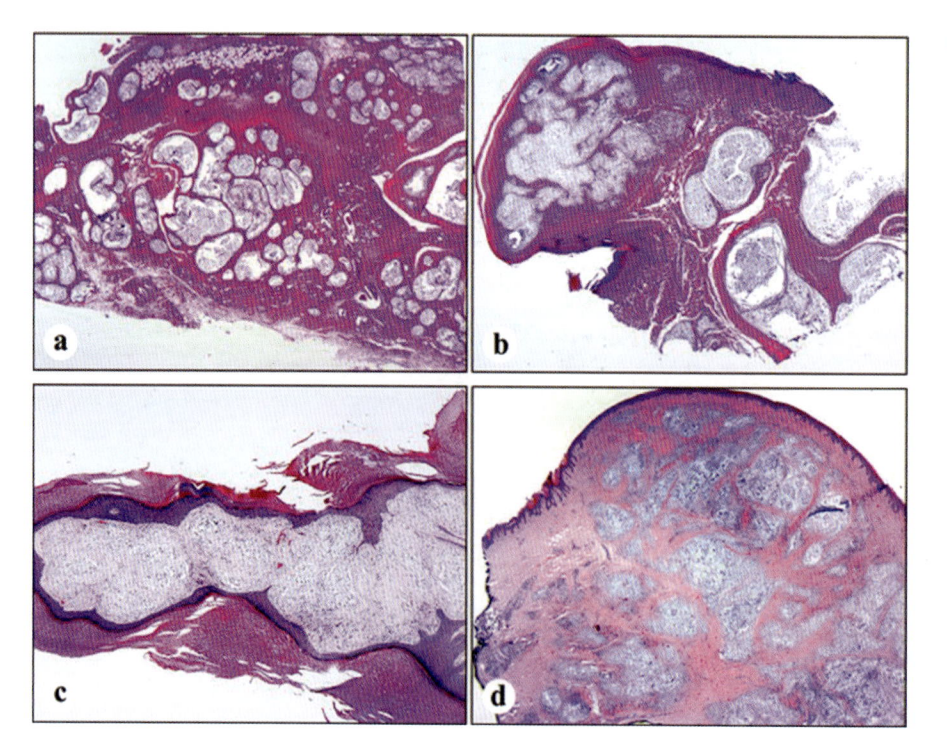

Figure 24.50 Examples of low-magnification configuration of nerve sheath myxoma. (**a**) A polypoid example. (**b–d**) Multinodular tumors with fibrous septa.

Chapter 24: Nerve sheath tumors

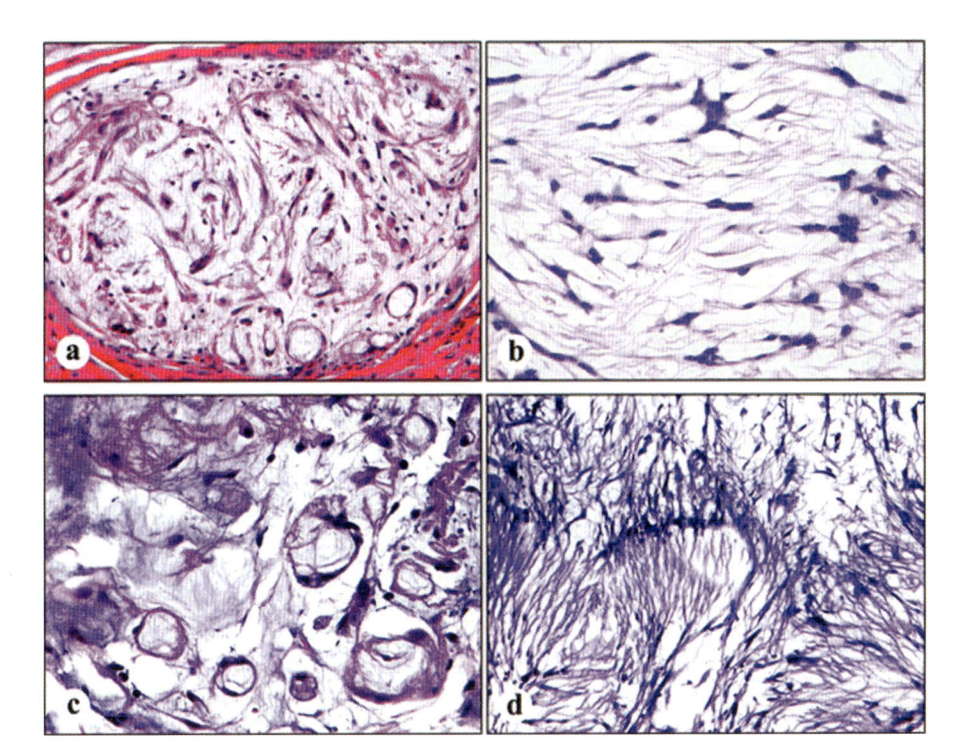

Figure 24.51 Cellular detail of nerve sheath myxoma. (a) Tumor cell lobule is composed of cords on tumor cells in myxoid matrix. (b) A typical corded pattern with syncytial clusters of tumor cells. (c) Signet-ring cell-like intracytoplasmic vacuoles. (d) Focal nuclear palisading can be present.

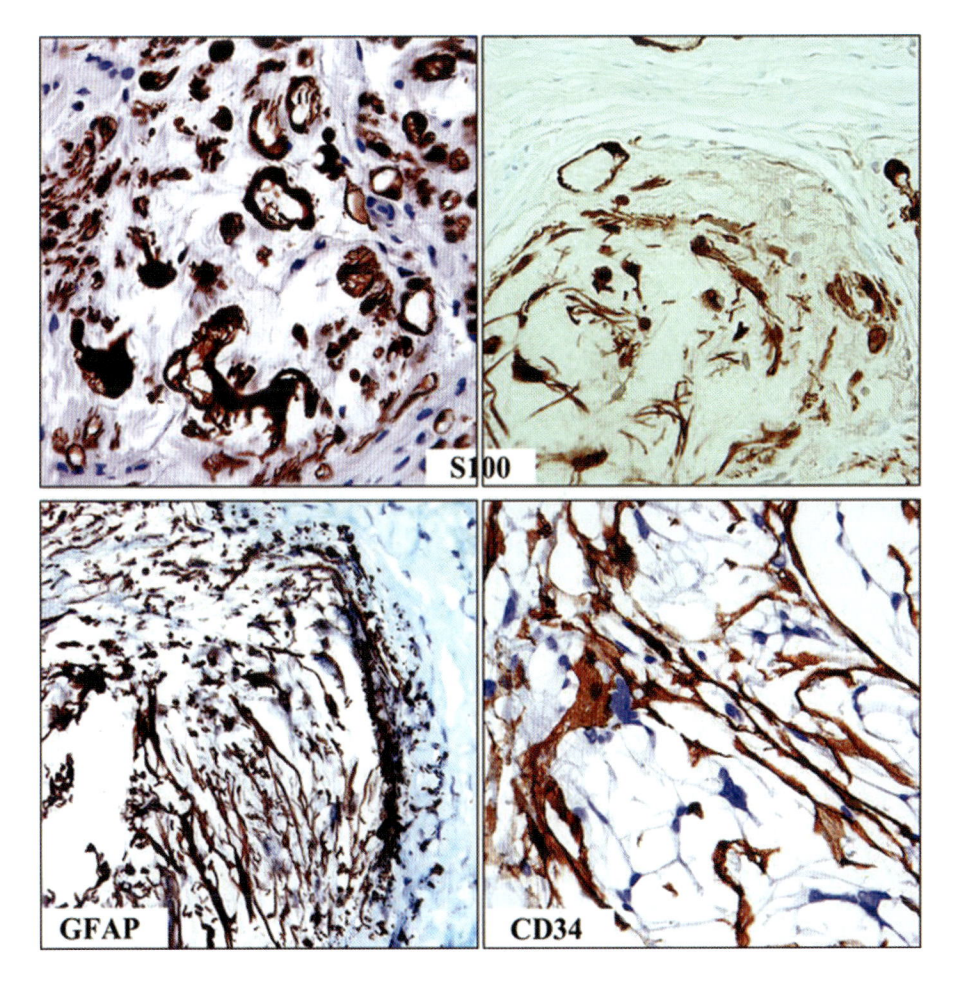

Figure 24.52 Immunohistochemical features of nerve sheath myxoma. S100 protien and GFAP positivity are nearly uniform features, and CD34-positive fibroblasts are often present.

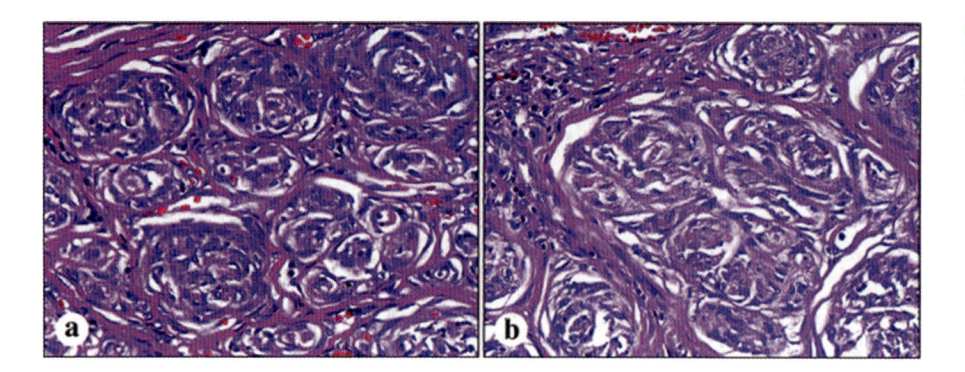

Figure 24.53 An intraneural perineurioma is composed of nerve fascicles containing extensive perineurial cell proliferation spaced by fibrous matrix.

cells are positive for vimentin and EMA and are negative for S100 protein.[142-144]

Perineuriomas (perineurial cell tumors) are very rare nerve sheath tumors that display pure perineurial cell differentiation without Schwann cell components. The group is clinically and histologically heterogeneous and contains two main types: intraneural and soft tissue perineurioma. Although many of these tumors are histologically very distinctive, the diagnosis of soft tissue perineuriomas is usually based on immunohistochemical demonstration of EMA positivity, ultrastructural analysis, or both. Because immunohistochemical positivity for EMA is not specific for perineurial cells, any tumors not confirmed with the known variants of perineurioma should not be diagnosed solely based on EMA positivity; the identification of unusual variants of soft tissue perineurioma also requires ultrastructural documentation.

Under electron microscopy, perineurial cells and perineuriomas show elongated cells with partial basal lamina and prominent pinocytic vesicles. The rough endoplasmic reticulum is less prominent than in fibroblasts.[145-147]

Intraneural perineurioma

Previously, this very rare tumor was referred to variously as localized hypertrophic neuropathy, hypertrophic mononeuropathy, localized hypertrophic neurofibrosis, hypertrophic interstitial neuritis, and intraneural neurofibroma.[148-150] The tumor occurs in children and young adults, with a mild female predominance. It usually involves major nerves in the extremities, such as the brachial plexus, the major nerves of the upper extremities, and the sciatic, femoral, and peroneal nerves. In these locations, the tumor causes motor or, less often, sensory neurologic deficits, often with a long history. Because this condition is benign, conservation of the involved major nerve is generally advocated, and treatment should be limited to biopsy only, unless the function has already been lost. Rare occurrence has been reported in the oral region (e.g., the mandible and tongue).

An intraneural perineurioma forms a fusiform symmetric expansion of the involved nerve that usually measures 2 cm to 10 cm (median, 2-3 cm), but longitudinal involvement can be extensive, up to 30 cm. Histologically, the tumor consists of innumerable expanded nerve fascicles surrounded by concentric, onion-bulb-like spindled perineurial cell proliferation. A central degenerating axon and small amounts of Schwann cells remain in the core of each concentric unit (Figure 24.53).

Immunohistochemically the tumor cells are positive for EMA and negative for S100 protein and Sox10. Small numbers of residual Schwann cells are S100 protein positive, and a neurofilament-positive central axon can usually be demonstrated.

The presence of clonal chromosomal changes, namely, the loss of chromosome 22,[150] supports the neoplastic nature of this lesion, which earlier was considered a hyperplastic process.

Sclerosing perineurioma

This is a characteristic variant of soft tissue perineurioma and is probably the most common clinicopathologic variant. Based on a series of 19 cases from the AFIP, this tumor typically occurs as a circumscribed, small (usually <2 cm), painless nodule in the fingers and hands. Subsequently the author and colleagues have encountered sporadic cases in the trunk wall (back) and head (earlobe). The tumor has a strong predilection for young adults in the third and fourth decades, with a 2:1 male predominance. Single cases have been seen in children and in older adults. There is often a long history of a tumor that remains stationary or slowly growing over the observation period of up to 40 years. The tumor is benign and does not have a significant tendency to recur.[151]

Grossly, the tumor forms a well-circumscribed nodule. On sectioning, it is off-white, with a rubbery and occasionally cartilage-like consistency. The lesion usually measures <1 cm to >3 cm (median, 1.5 cm). Histologically, a sclerosing perineurioma is a densely collagenous mass. It contains clusters, cords, or concentric arrangements of small, cuboidal, epithelioid-appearing tumor cells in a densely collagenous background (Figures 24.54 and 24.55), sometimes clustered around small nerves or vessels.

Immunohistochemically, the tumor cells are positive for EMA and basement membrane proteins laminin and collagen IV. They are negative for S100 protein and CD34, and only occasionally positive for keratins and variably (focally) for SMA.[151] Positivity for GLUT1 is also typical, consistent with a perineurial phenotype.[152]

Chapter 24: Nerve sheath tumors

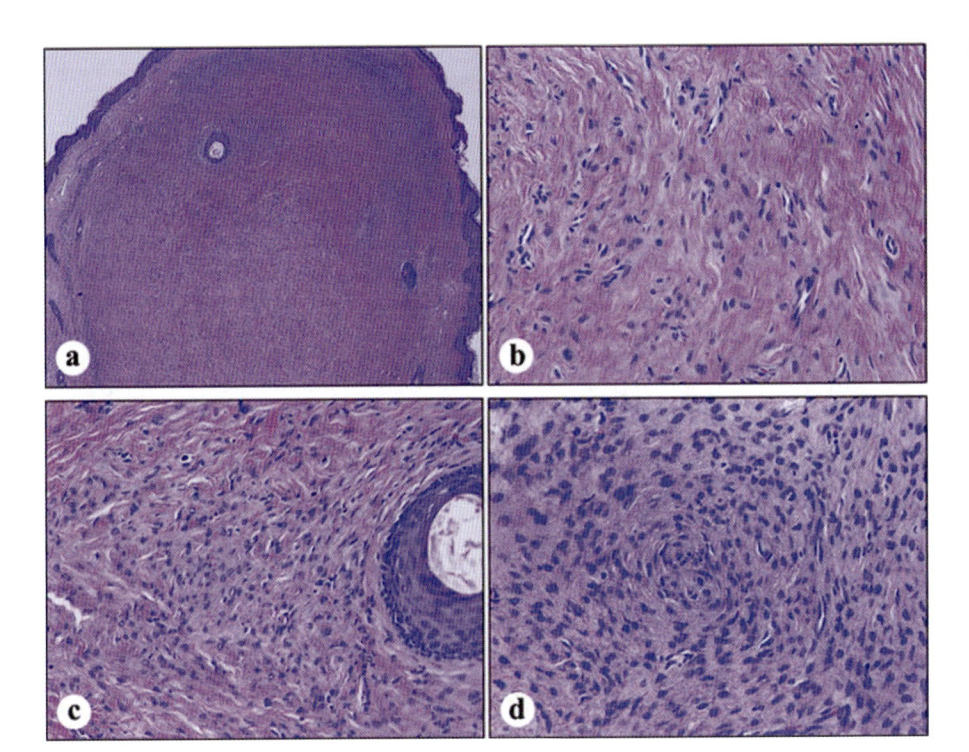

Figure 24.54 Sclerosing perineurioma of the earlobe contains scattered spindled to oval cells focally forming concentric whorls around nerves.

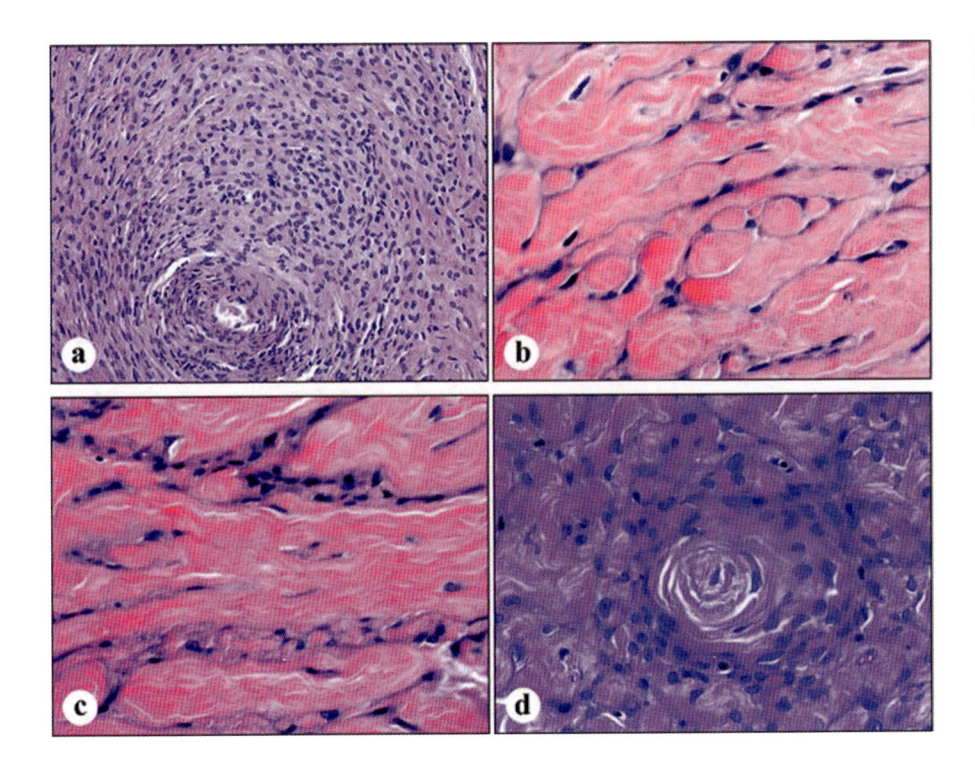

Figure 24.55 Sclerosing perineurioma of the finger. Tumor cells often form corded patterns in a collagenous stroma. (**d**) Note concentric perineurial cells around a small nerve.

A cryptic NF2 gene deletion has been reported in one case of sclerosing perineurioma.[153] Missense NF2 mutations and allelic losses of the NF2 gene seem to occur in sclerosing perineuriomas, suggesting a two-hit involvement of the NF2 gene in a manner similar to that found in meningiomas and schwannomas.[154] Cytogenetic studies have shown diploid or near-diploid karyotypes, with a t(2;10)(p23;q24) translocation in one case and deletions involving 22q in another case, among others.[155]

Sclerosing perineurioma should be distinguished from sclerosing glomus tumor and mixed tumor of the skin. The former is a smooth-muscle-related tumor strongly positive for SMA and having vascular association, and the latter usually shows evidence of epithelial differentiation and is keratin

669

positive. The corded pattern also sometimes resembles an epithelioid hemangioendothelioma, although perineurioma lacks cytoplasmic vacuolization and endothelial markers.

Retiform (reticular) perineurioma

This distinctive, very rare variant was first described by Ushigome et al.[156] The subsequently reported 11 tumors have occurred in young to middle-aged patients, with 7:4 female predominance. There seems to be a predilection for the upper extremity (more commonly distal), with single cases being reported in the groin and oral cavity. Tumor size varies from 1.5 cm to 10 cm (median, 3 cm). The clinical course has been indolent, although there are no long-term follow-up data.[157–159]

The slender tumor cells with small nuclei form a net-like pattern and are set in a myxoid or occasionally denser collagenous matrix, and mitotic activity is not detectable (Figure 24.56). Immunohistochemical EMA and GLUT1 positivity are typical findings.

Other soft tissue perineuriomas

The most authentic examples include peripheral soft tissue tumors that have some resemblance to meningiomas, showing perivascular whorling and concentric lamellae of slender spindle cells (Figure 24.57).[160–163]

Other tumors reported as soft tissue perineuriomas are a heterogeneous group. The reported soft tissue perineuriomas have presented in a wide variety of body sites in adult patients. A large series reported perineuriomas in almost all ages except in early childhood, and in a wide variety of anatomic locations.[164] Composition of slender spindle cells has been a common theme. Some series report high frequency of CD34 positivity.[164] Soft tissue perineuriomas not belonging to the

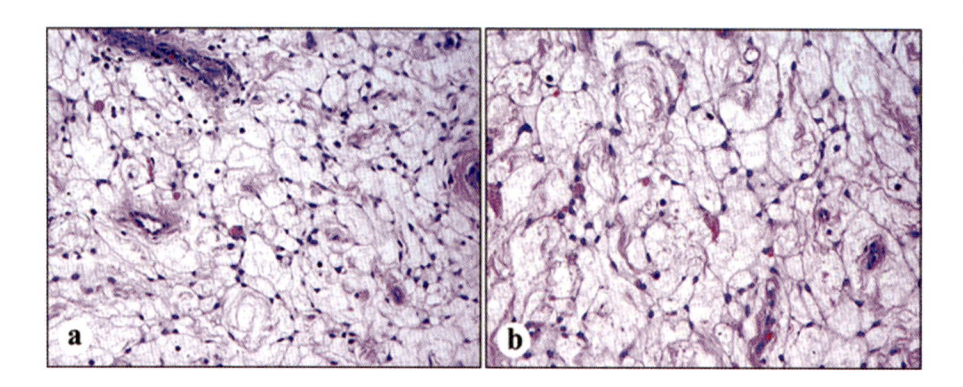

Figure 24.56 Retiform perineurioma. Tumor cells form net-like patterns in a fibromyxoid matrix.

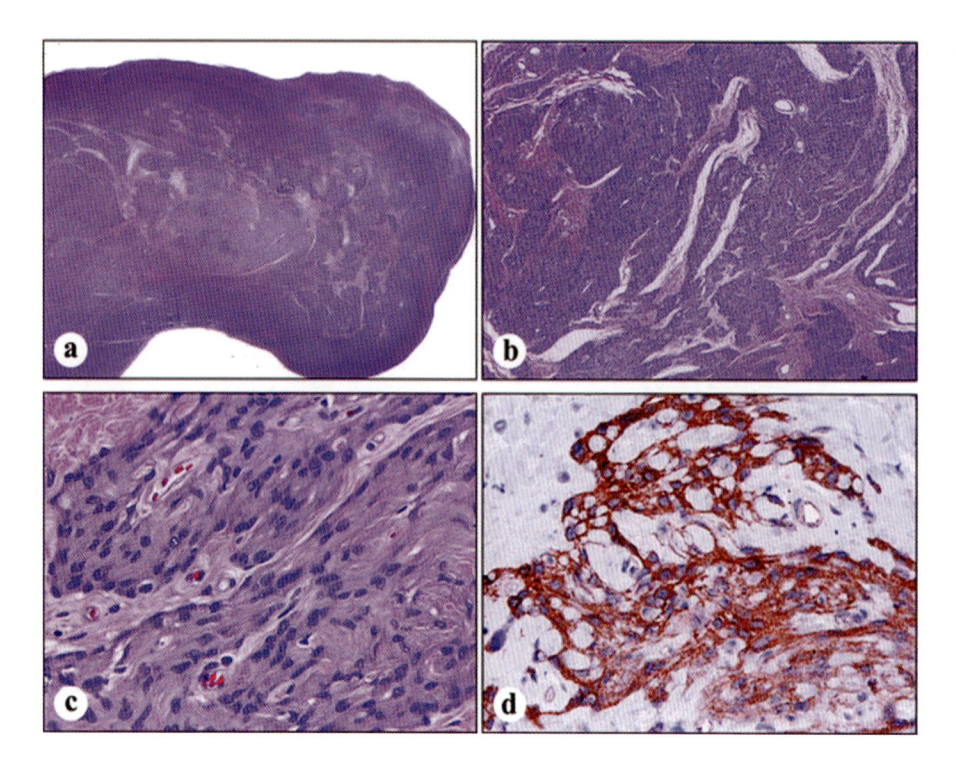

Figure 24.57 Soft tissue perineurioma. The tumor is composed of islands of spindled to epithelioid cells that are strongly positive for EMA.

previously discussed histologic groups need critical re-examination.

In general, perineuriomas are distinctive in being positive for EMA (Figure 24.58) and basement membrane proteins, collagen IV, and laminin, whereas they are negative for S100 protein and keratins. GLUT1 and claudin are also expressed.[165] The diagnostic value of these newer markers is somewhat open, however, because of limited experience and lack of distinct perineurial cell specificity. The latter is especially true for malignant tumors.

Ultrastructural features of perineurioma include features of perineurial cells such as elongated cytoplasmic processes, basement membranes/external lamina, and pinocytic vesicles.[146,147]

At least some soft tissue perineuriomas have losses in chromosome 22,[162] but NF2 gene mutations have not been found to be a consistent feature in tumors other than sclerosing perineuriomas.[154]

Differential diagnosis of perineurioma

Over-reliance on EMA immunoreactivity as a diagnostic marker can cause substantial overdiagnosis of perineurioma. Mesenchymal spindle cell tumors that are also EMA positive in particular include low-grade fibromyxoid sarcoma and superficial acral fibromyxoma of the fingers and toes. DFSP consists of slender spindle cells that are typically EMA negative and CD34 positive.

Malignant perineurioma

Although perineuriomas are generally benign tumors, occasional malignant examples have been reported.[166] Conceptually these tumors can be alternatively considered MPNSTs with perineurial cell differentiation.

The seven cases described by Hirose et al. occurred in a wide range of age and location.[166] The tumors varied from a small nodule in various truncal, head, or extremity locations to a 30-cm retroperitoneal tumor.[166] Perivascular whorling is a typical histologic feature, and mitotic activity and grade ranged from low to high. A concentric lamellar pattern of slender spindle cells was the most conspicuous ultrastructural finding, and EMA positivity and S100 protein negativity were immunohistochemical findings. Although most tumors were high grade histologically, they had a better prognosis than MPNSTs in general. Others have reported rare similar tumors.[167]

One tumor considered to be malignant perineurioma had myxofibrosarcoma myxoid MFH-like features and was ascertained by EMA positivity.[168]

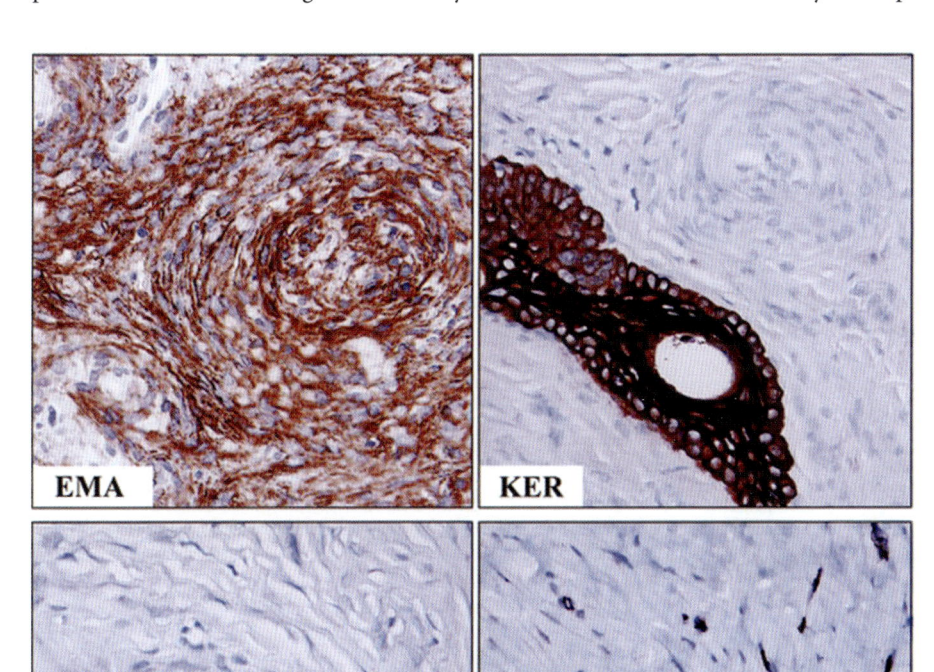

Figure 24.58 Immunohistochemical positivity for EMA is typical of perineurioma, whereas the tumor cells are typically negative for keratins, S100 protein, and CD34.

Differential diagnosis

The identification of malignant perineurioma is based on compatible morphology and EMA positivity, ideally supported by electron microscopy. Many unrelated malignant spindle cell tumors can be focally EMA positive, and they should not be confused with perineurioma. Some dedifferentiated liposarcomas have meningothelial-like whorls; these tumors can be properly identified by their atypical lipomatous component and common bony metaplasia, which are not features of perineurioma.

Granular cell tumor

Granular cell tumor (GCT), historically called "granular cell myoblastoma," is an S100 protein-positive Schwann-cell-related neoplasm usually presenting in superficial soft tissues and in a wide variety of organ-based sites, especially in the respiratory and gastrointestinal tracts. (Rare malignant variants are presented in a separate section.) Gingival GCT in infants (congenital epulis) is a separate entity clinically and histogenetically unrelated to ordinary GCT and is discussed separately in the section following malignant GCT.

Clinical features

GCT occurs in all age groups, although very rarely under the age of 5 years and uncommonly after the eighth decade. There is some predilection for young adults and a 5:4 female predominance (Figure 24.59). Other studies have noted a relative predilection for African Americans.[169,170]

GCT occurs in a wide variety of sites, of which the peripheral soft tissues are collectively the most common (Figure 24.60). In these sites, GCT usually involves the skin or subcutis and is rarely intramuscular. GCTs of the breast can simulate breast carcinoma both clinically and radiologically.[171] In the vulva, pseudoepitheliomatous hyperplasia commonly accompanies a GCT, and this should not be confused with squamous cell carcinoma.[172] The same is true for other mucosal sites.

In the oral cavity, GCT has a predilection for the tongue, especially its anterior and dorsal surfaces, and most lesions are small.[169,173] Internal organ sites include the gastrointestinal tract, especially the esophagus and the colon, where GCT can be a small incidental nodule or form an obstructing mass (especially in the esophagus).[174] Laryngeal examples often involve vocal cords, causing hoarseness and can clinically (and histologically, by pseudoepitheliomatous hyperplasia) simulate a squamous carcinoma.[175] In the lung, GCT presents incidentally as often as it forms an obstructive lesion. Most tumors are small, approximately 1 cm, and involve the bronchial wall.[176] Extrahepatic biliary tract and pituitary neurohypophysis are additional visceral sites.[177]

Multiplicity has been estimated to occur in 5% of patients. These lesions can occur in peripheral soft tissues, internal sites, or both. Multiplicity should not be taken as evidence of malignancy.[169,170] In one patient, 52 separate GCTs were reported in a colectomy specimen.[178]

Familial occurrence has been reported, but there is no clear evidence for a syndrome. In one case, multiple GCTs occurred in a mother and son. Other examples include a mother with multiple tumors and a daughter with a solitary tumor. In a third family, two brothers were involved, one with multiple tumors.[179]

Pathology

Most CCTs are 1 cm to 2 cm in diameter, and occasional tumors reach the size of 5 cm or more. Grossly, the tumor typically forms an oval nodule that varies from sharply circumscribed to ill-defined, having a variably yellow or grayish surface on sectioning (Figure 24.61).

Microscopically, GCTs can form solid compact sheets, variably defined compartments, or clusters diffusely dispersed in collagenous matrix (Figure 24.62). In the tongue, the tumor cells can form a checkerboard pattern with the entrapped skeletal muscle cells.

Pseudoepitheliomatous hyperplasia is common at mucosal surfaces, especially the tongue, larynx, and vulva, but this is

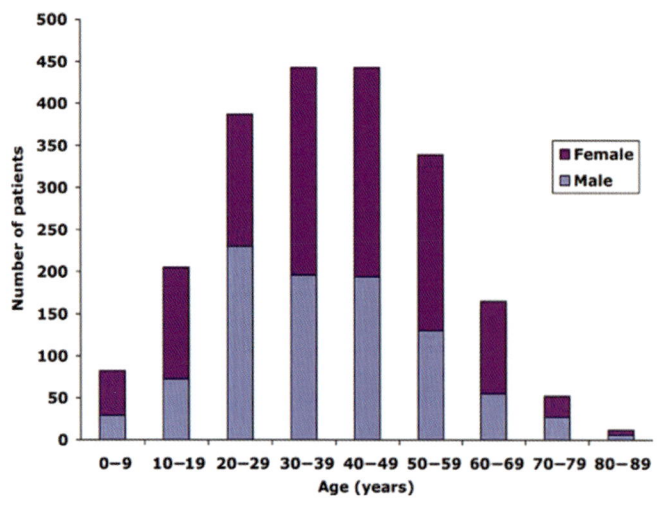

Figure 24.59 Age and sex distribution of 2128 patients with granular cell tumor (GCT).

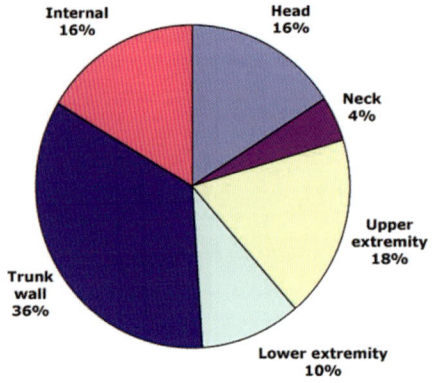

Figure 24.60 Anatomic distribution of 2164 GCTs.

less common in peripheral GCTs. Especially when superficial elements of GCT are scant, there is a danger of confusing this reaction with squamous cell carcinoma. Nuclear atypia is mild (Figure 24.63). In some cases, the granular cells involve or surround small nerves, consistent with origin from Schwann cell elements (Figure 24.64).

The tumor cells vary from ellipsoid to slightly elongated and can have well-defined cell borders or present in a syncytial pattern. The cytoplasm is abundant and granular and can contain larger granular condensations, representing giant lysosomal aggregates. The nuclei vary from small and hyperchromatic with a dense chromatin to medium-sized with open chromatin and distinct nucleoli. Nuclear pseudoinclusions and focal atypia can occur, but mitoses are rare (Figure 24.65). The granular cytoplasm is variably PAS positive.

Immunohistochemistry and ultrastructure

Immunohistochemically, a GCT is consistently positive for S100 protein (Figure 24.66) and Sox10 and negative for muscle cell and epithelial markers.[180–182] Because of the high lysosomal content, the tumor cells are also positive for CD68. The intermediate filaments are composed of vimentin only, and basement membrane proteins can often be demonstrated around cell clusters.[183] GCTs of the peripheral soft tissues and hepatobiliary tract are alpha-inhibin positive.[177,184] In the author's experience, GCTs are the only peripheral nerve sheath tumors to be inhibin positive, since neurofibromas and schwannomas are negative.

Ultrastructurally, the granular cytoplasm contains numerous phagolysosomes (autophagic vacuoles), and some cells also contain large, boat-shaped cytoplasmic crystals referred to as *angulated bodies*.[185]

Differential diagnosis

True GCTs are thought to be Schwann cell related and consistently S100 protein positive, which separates them from most other granular cell lesions, including those of histiocytic, fibroblastic, or smooth muscle derivation. Some smooth muscle tumors have a granular cell appearance; their diagnosis

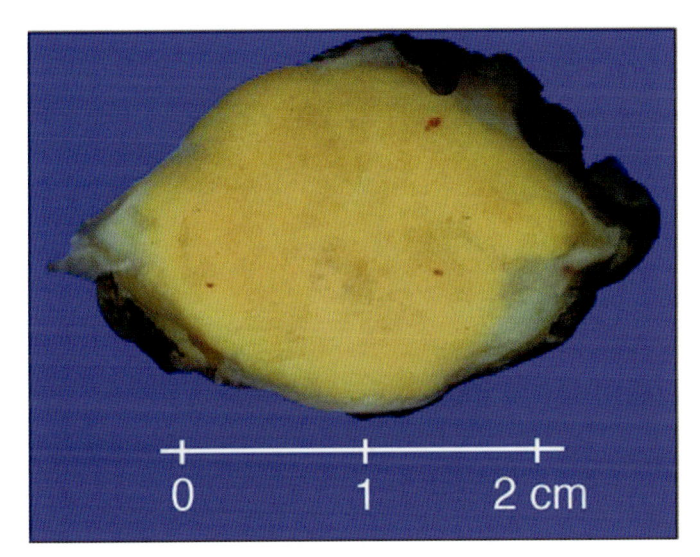

Figure 24.61 Grossly, a GCT forms a demarcated yellowish mass.

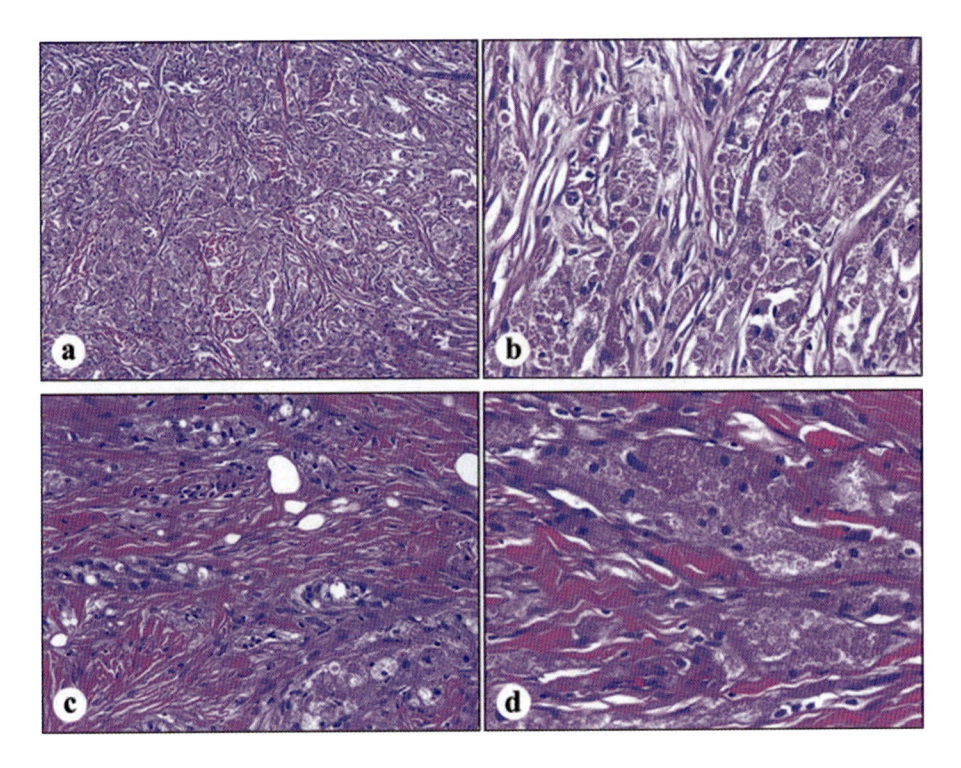

Figure 24.62 Histologically, GCTs can be composed of solid sheets or clusters of polygonal epithelioid cells with granular cytoplasm.

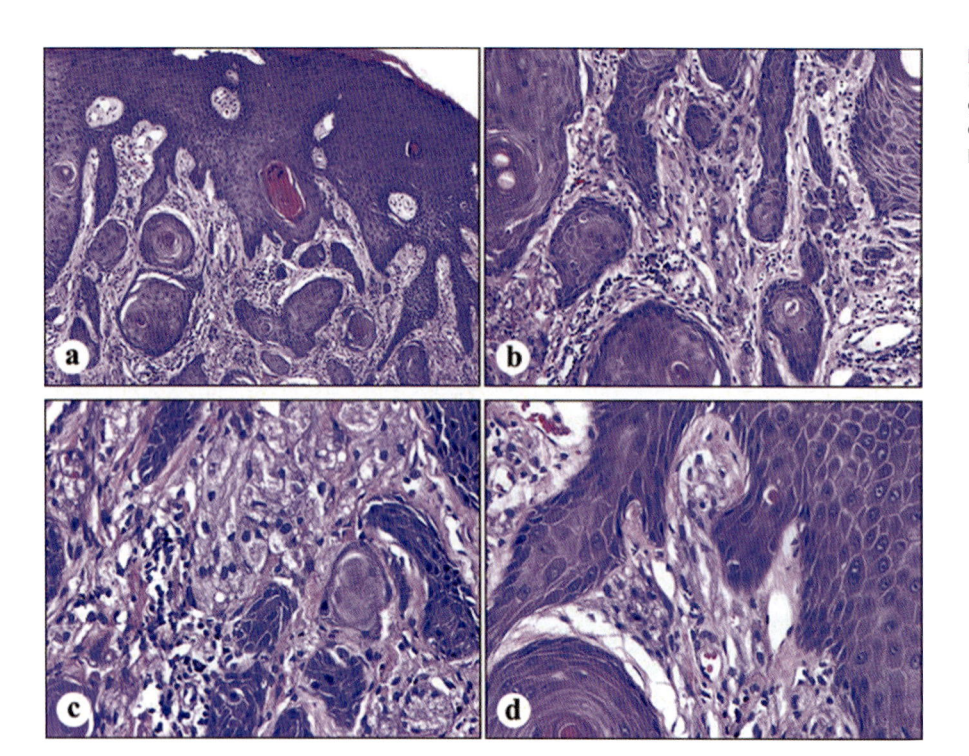

Figure 24.63 Pseudoepitheliomatous hyperplasia is a common finding on mucosal granular cell tumors. Higher magnification demonstrates nests of granular cells adjacent to proliferating squamous epithelia.

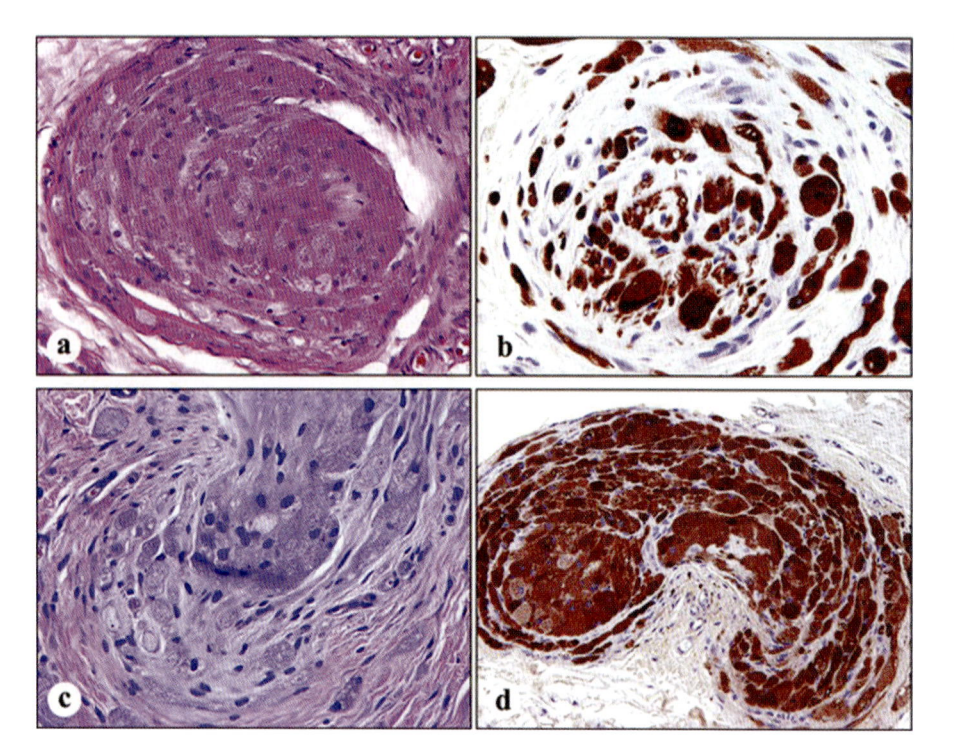

Figure 24.64 Neural involvement of granular cell tumor is highlighted by S100 protein immunostain.

is based on other features of smooth muscle tumors and immunohistochemical demonstration of actins, desmin, or both.[186]

Granular cell histiocytic reaction usually contains more conventional histiocytes, and multinucleated forms and xanthoma cells are often present, distinct from GCT. Epithelial tumors, such as jaw ameloblastomas and carcinomas with granular cell features, should not be confused with true GCTs – they are keratin-positive epithelial tumors. Granular cell fibrohistiocytic neoplasms (Chapter 7) are uncommon neoplasms with strikingly granular cytoplasm. These tumors often form small well-demarcated nodules, but their distinction from GCT is ultimately by immunohistochemistry: they are S100 protein negative.

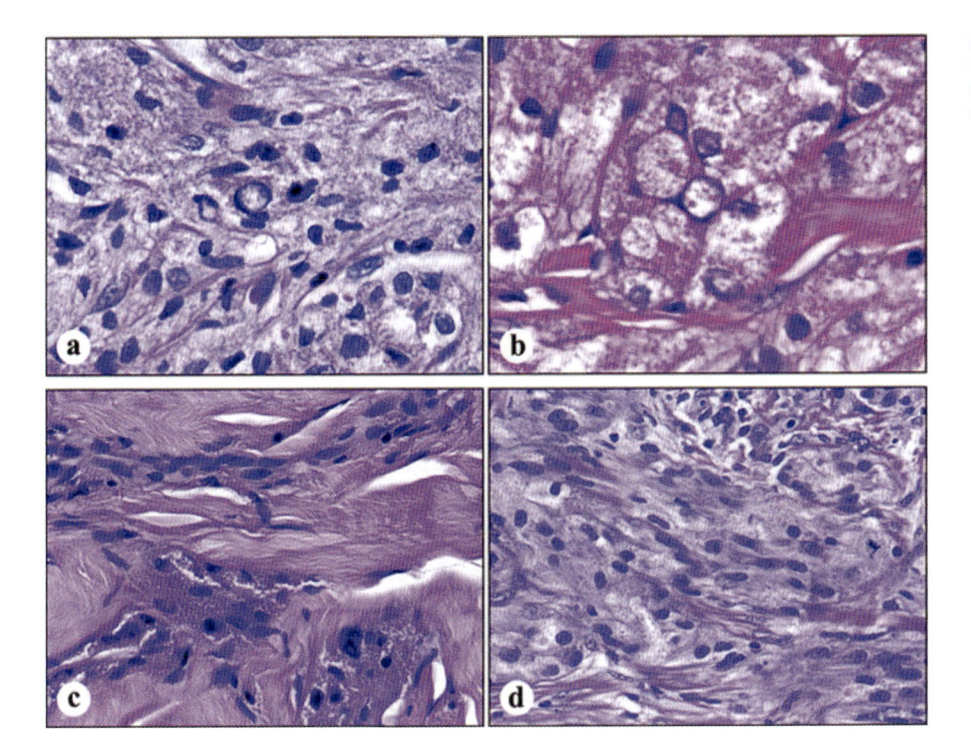

Figure 24.65 Cellular features of granular cell tumor. (a,b) Intranuclear inclusions and otherwise uniform cytologic features. (c) Focal nuclear atypia. (d) Spindling of tumor cells.

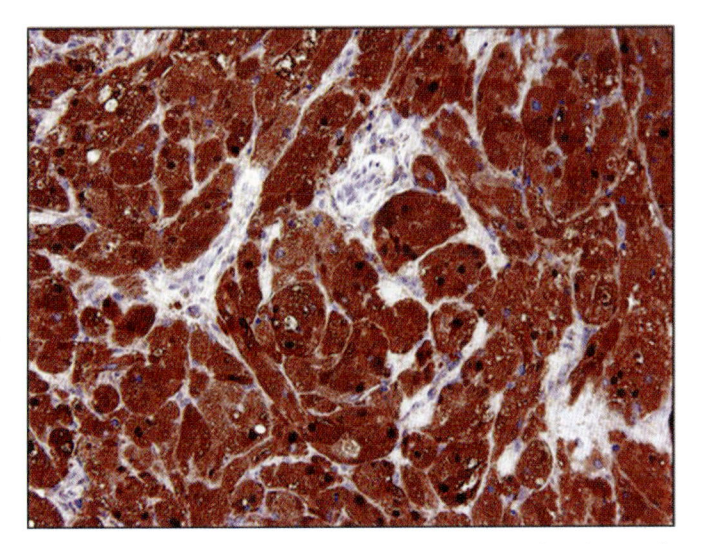

Figure 24.66 Strong S100 protein immunoreactivity is a uniform feature of GCTs.

Figure 24.67 Grossly, a malignant GCT forms a demarcated pale grayish mass. Malignant examples are often >5 cm.

Malignant granular cell tumor

Malignant behavior of GCT is rare (<1%), and the event of clinical malignancy might be impossible to predict histologically. Compared with their benign counterparts, these tumors tend to be larger and occur in older patients, >50 years of age (Figure 24.67). Malignant GCTs often recur locally with multiple skin satellite nodules. Metastases most commonly develop in the lymph nodes, lungs, and bones, and rarely in the intestines, liver, or brain. Tumor-related deaths occurred in 11 of 28 patients, with a median time of 3 years (range, <1–9 years).[187]

According to the largest published series, the following features are associated with an increased metastatic risk: mitotic rate of 2 or more per 10 HPFs; significant nuclear atypia with high nucleocytoplasmic ratio and prominent nucleoli, cell spindling, tumor necrosis; and Ki67 index >10% (Figure 24.68). It was found that tumors with three or more of these features had a high frequency of malignant behavior, with one-half of the patients developing distant metastases. Malignant GCTs are often larger, and in the largest series, 6 of 11 metastasizing tumors were >5 cm, although one 1-cm tumor of the groin was reported to have metastasized in the bone and brain, and was fatal in 4 years.[187] Others have

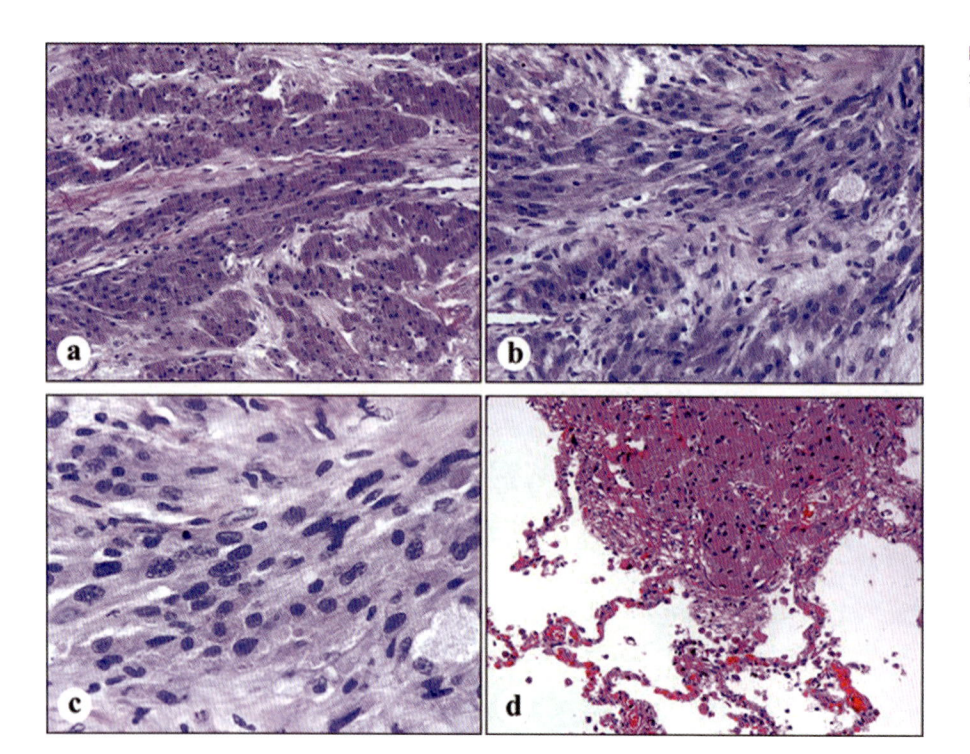

Figure 24.68 Malignant GCT shows fascicles of spindle cells with nuclear atypia. (**d**) Lung metastasis.

confirmed high Ki67 index and p53 positivity to be findings associated with malignant GCT.[188]

Gingival granular cell tumor of the newborn

Gingival GCT is not a nerve sheath tumor, but has fibroblastic or pericytic origin. It is a clinicopathologically distinctive tumor of infants. Even before the availability of electron microscopy and immunohistochemistry, Custer and Fust correctly pointed out its histogenetic unrelatedness to ordinary GCTs.[189]

Clinical features

This rare congenital tumor has also been referred to as "gingival epulis of newborn." It predominantly (90%) occurs in female infants and typically presents as a polypoid pale to pinkish mass bulging from the anterior alveolar ridge, more commonly in the maxilla than the mandible. It can reach considerable size and interfere with feeding. The average size is 2.0 cm to 2.5 cm, but the largest examples have been >5 cm. Multiple tumors occur in 10% of cases, and there is no racial predilection. Conservative excision, sparing the unerupted teeth, is sufficient, because incompletely removed tumors do not have a tendency to recur, and the tumor can involute over time.[190,191]

Disseminated congenital GCT has been reported once, in a stillborn fetus. In addition to the gingiva, it involved most of the major internal organs, the pituitary, and the leptomeninges, and similar to the gingival tumors, it was S100 protein negative.[192] This occurrence raises the possibility of some form of (lysosomal) storage disease as an alternative diagnosis.

Pathology

Gingival GCT forms a sessile polypoid lesion that is often focally ulcerated (Figure 24.69). The lesion does not elicit the epithelial hyperplasia often seen with ordinary GCTs. Histologically, this tumor is composed of sheets of ovoid or rounded granular cells arranged between prominent networks of branching capillaries that vary from slit-like to those with gaping lumina or a staghorn configuration (Figure 24.70). The collagenous matrix is typically inconspicuous, but can be visible in some cases, and the granular cells typically have well-defined cell borders. The nuclei are small and centrally located with open chromatin, and there is no mitotic activity or atypia. Odontogenic epithelial inclusions are a rare finding.

Similar to ordinary GCTs, gingival GCTs in newborn are PAS positive, but in contrast to those GCTs of soft tissues, the gingival variety are negative for S100 protein and Sox10 (Figure 24.71). They are vimentin positive and are thought to be of fibroblastic, pericytic, or primitive mesenchymal cell origin, based on ultrastructural and immunohistochemical features.[191–196]

Malignant peripheral nerve sheath tumor

This designation has replaced the older terms of *neurofibrosarcoma* and *malignant schwannoma*. When applied strictly, the diagnosis should be used only for malignant tumors that originate from a neurofibroma or nerve. A sarcoma in a patient with NF1 has a good likelihood for being an MPNST, and approximately 5% of patients with NF1 will eventually be diagnosed with a MPNST.

Chapter 24: Nerve sheath tumors

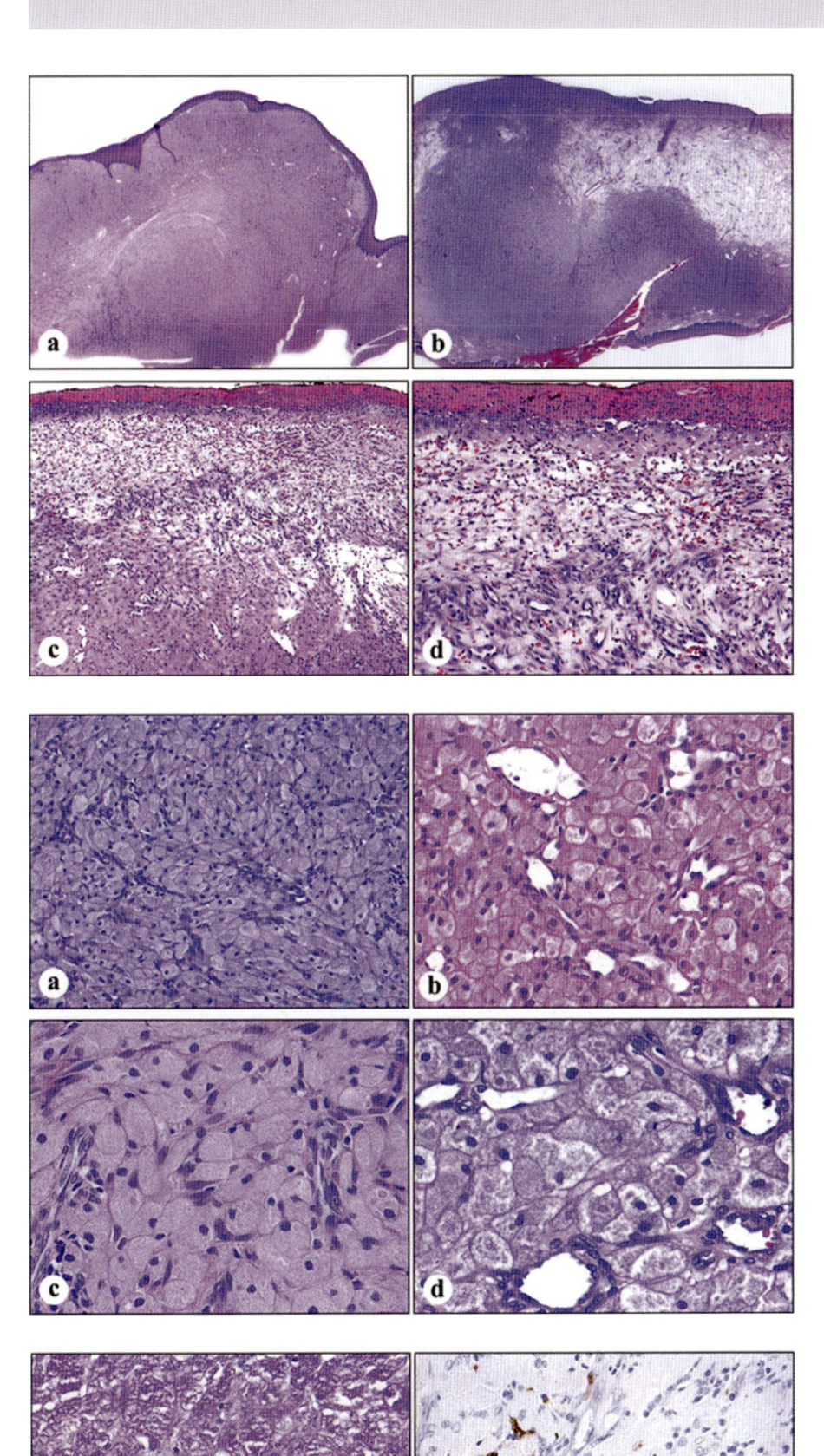

Figure 24.69 Low magnification appearance of gingival GCT of infants. (a,b) Sessile polypoid lesions. (c,d) There is mucosal ulceration with a superficial zone of nonspecific granulation tissue.

Figure 24.70 Prominent vascular capillary network and uniform granular cells are typical features of gingival granular cell tumor.

Figure 24.71 Gingival GCT shows cytoplasmic PAS positivity, similar to the ordinary GCTs. The granular cells are S100 protein negative, but scattered dendritic cells are positive.

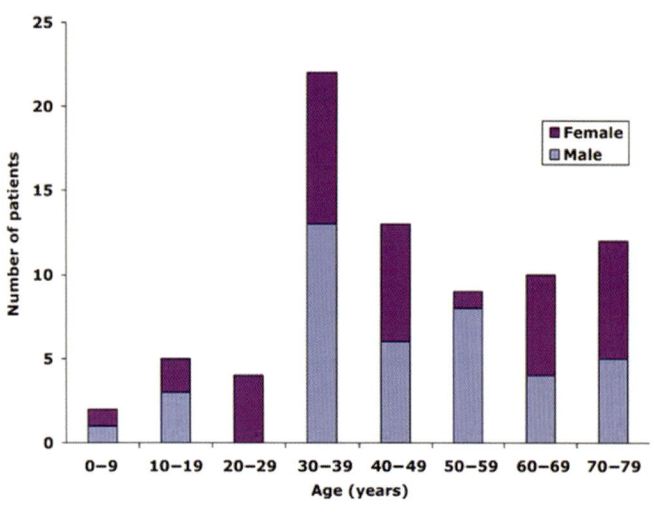

Figure 24.72 Age and sex distribution of 65 patients with malignant peripheral nerve sheath tumors (MPNSTs).

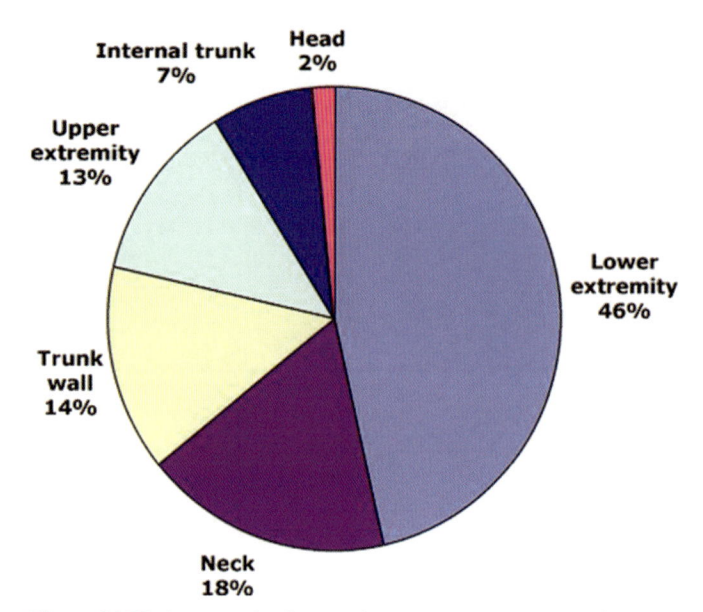

Figure 24.73 Anatomic distribution of 65 MPNSTs arising in neurofibroma.

Some tumors can be identified as probable examples of MPNST by their histologic similarity to the definitive examples. This includes epithelioid MPNST in particular, a histologic subgroup, which is discussed separately. S100 protein positivity is inconsistent in MPNST. Loss of histone trimethylation is emerging as a new potential masters for sporadic MPNST.[197]

Clinical features

MPNST occurs in all ages from early childhood, but it is most common in early middle age, with the median ages in the largest series varying between 35 and 45 years. AFIP experience is shown in Figure 24.72. Most series show a mild female predominance. This is especially true for patients with NF1, who also have a younger age at presentation.[198–208] A small percentage of MPNSTs arise on the site of previous irradiation (postirradiation sarcoma), and radiation can accelerate the development of MPNST in NF1 patients.[209,210]

The tumor most commonly presents as a mass, often with pain, and some patients experience focal neurologic deficits or tingling. The most common sites are the proximal parts of the lower and upper extremities, paraspinal region of the trunk, and neck (Figure 24.73). Five-year survival rates vary from 15% to 40%, and patients with NF1 have a poorer prognosis. Other adverse factors are tumor size >5 cm, mitotic rate over 20 per 10 HPFs, central location, and incomplete resection. The local recurrence rate is high, and metastases develop in more than one-half of patients. Metastases occur most often in the lungs, bones, pleura, and liver, and spread by the meningeal route is also possible.

Pathology

Grossly, the tumor typically forms an oval or fusiform mass arising from a nerve, often involved by an intraneural, often plexiform neurofibroma (Figure 24.74). The tumor often measures >10 cm, sometimes >25 cm. It typically shows a variegated surface on sectioning, with areas of hemorrhage and necrosis. Elements of the pre-existing neurofibroma can be visible as focal rubbery or fibromyxoid areas either in the periphery of the tumor or in the ends of a fusiform tumor.

Identification of elements of pre-existing neurofibroma helps in the diagnosis of those MPNSTs that do not arise from a major nerve; such elements and the MPNST often show a sharp demarcation (Figure 24.75).

Histologically, most MPNSTs are high-grade tumors with a high mitotic rate and frequent necrosis. The most common histologic patterns include a high-grade fibrosarcomatous type composed of densely packed sheets of plump but relatively uniform spindle or oval cells. A herringbone pattern also can be present (Figure 24.76). Other patterns include myxoid, pleomorphic, round cell, Ewing-like, and synovial-sarcoma-like (Figure 24.77). Many tumors have geographic necrosis (Figure 24.78) and complex-to-glomeruloid vascular proliferation (Figure 24.79) resembling those commonly seen in glioblastoma multiforme. There can also be a prominent vascular pattern slightly reminiscent of that of hemangiopericytoma. The tumor cells often have spindled, bent nuclei with diffuse hyperchromasia (the pathology of epithelioid MPNST is discussed separately later).

Tumors originally reported in children as plexiform MPNSTs that had a good prognosis[211] have been reinterpreted as cellular plexiform schwannomas.[212]

Immunohistochemistry

MPNSTs often show nonspecific or less commonly schwannian immunohistochemical features. Only a third of them are extensively S100 protein positive, whereas many are focally positive (Figure 24.80); in many cases, the S100-protein-

Chapter 24: Nerve sheath tumors

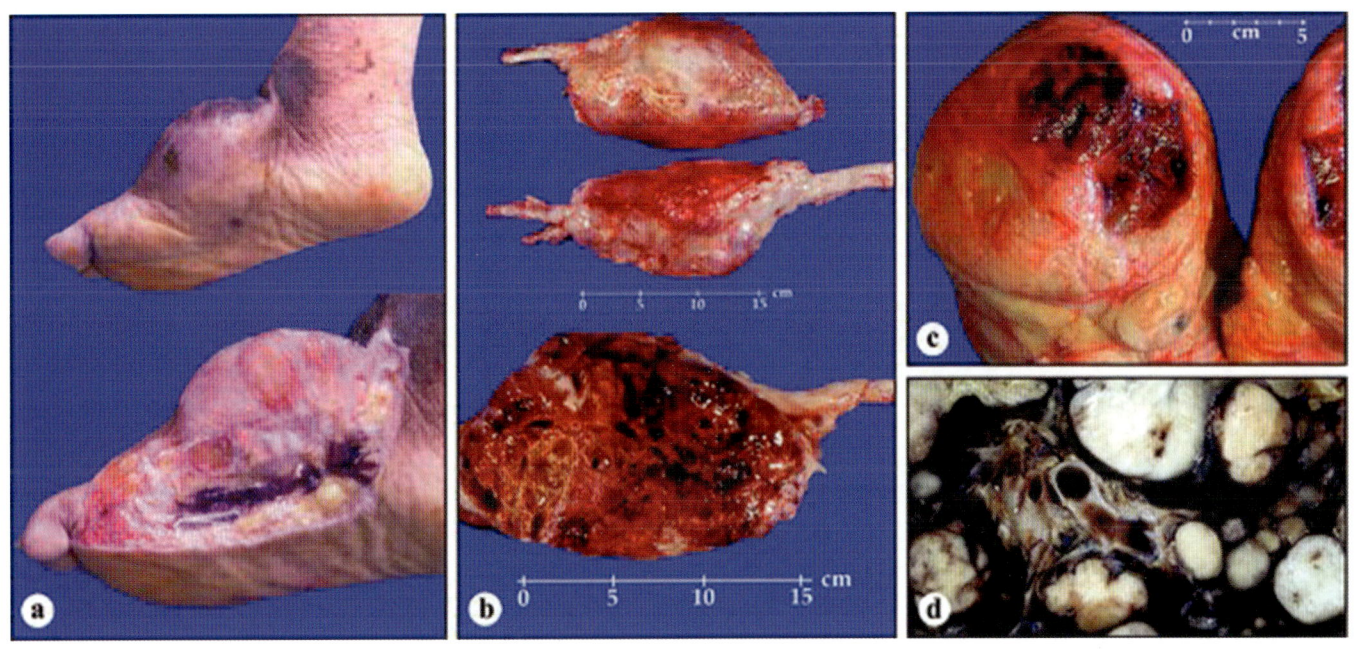

Figure 24.74 Gross appearances of MPNSTs. (**a**) Superficial example on the dorsal aspect of foot. (**b**) Sciatic nerve MPNST arising in a plexiform neurofibroma shows a hemorrhagic necrotic surface on sectioning. Note that the ends of the tumor contain elements of plexiform neurofibroma. (**c**) Another example arising from plexiform neurofibroma shows hemorrhagic necrotic tumor, and there is a pale separate nodule (center, bottom) representing the plexiform neurofibroma component. (**d**) Lung metastases of MPNST.

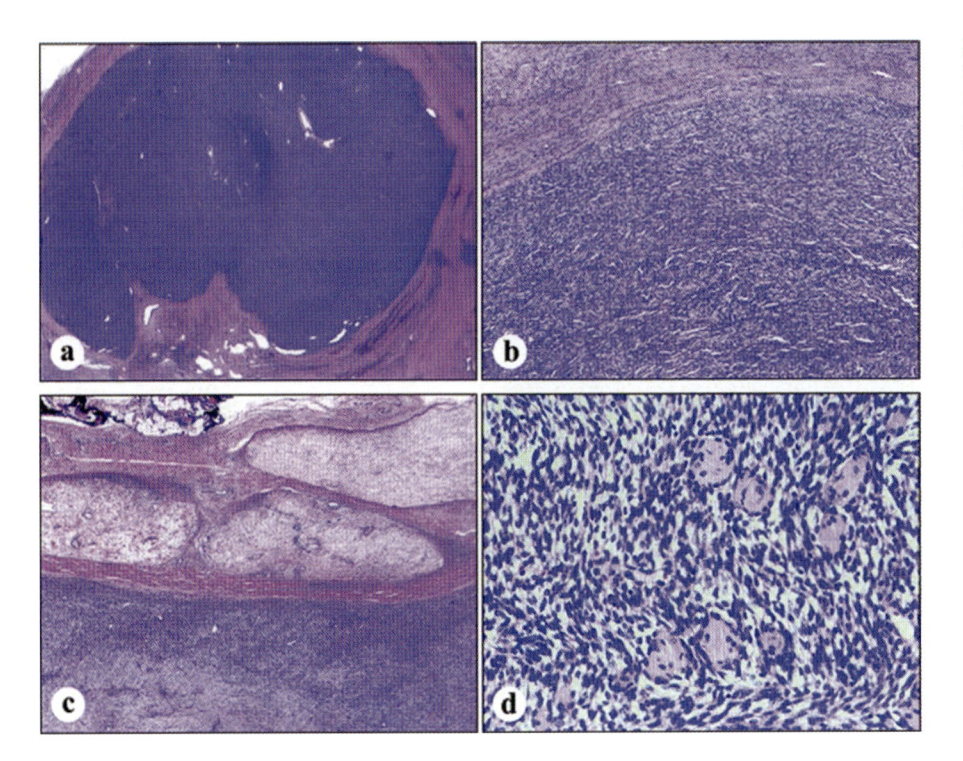

Figure 24.75 MPNSTs arising in a neurofibroma. (**a**) The malignant component forms a highly cellular, demarcated area. (**b**) MPNST and neurofibroma components are also well demarcated in this case. (**c**) Plexiform neurofibroma components are seen in the periphery of this MPNST. (**d**) Unusual example of an MPNST arising in a diffuse neurofibroma. Note remnants of Meissnerian bodies.

positive elements are intermingled remnants of residual nerve or neurofibroma.[213,214] In contrast to neurofibroma, CD34 expression is uncommon. In the author's experience, focal keratin K8 and K18 positivity is quite common, but K7 and K19 are absent (Figures 24.80 and 24.81). The presence of desmin and actins usually reflects the presence of a heterologous rhabdomyoblastic component (triton tumor, see discussion later in this chapter). Considering that synovial sarcomas can also be S100 protein positive, immunohistochemical results must be interpreted with caution.

679

Chapter 24: Nerve sheath tumors

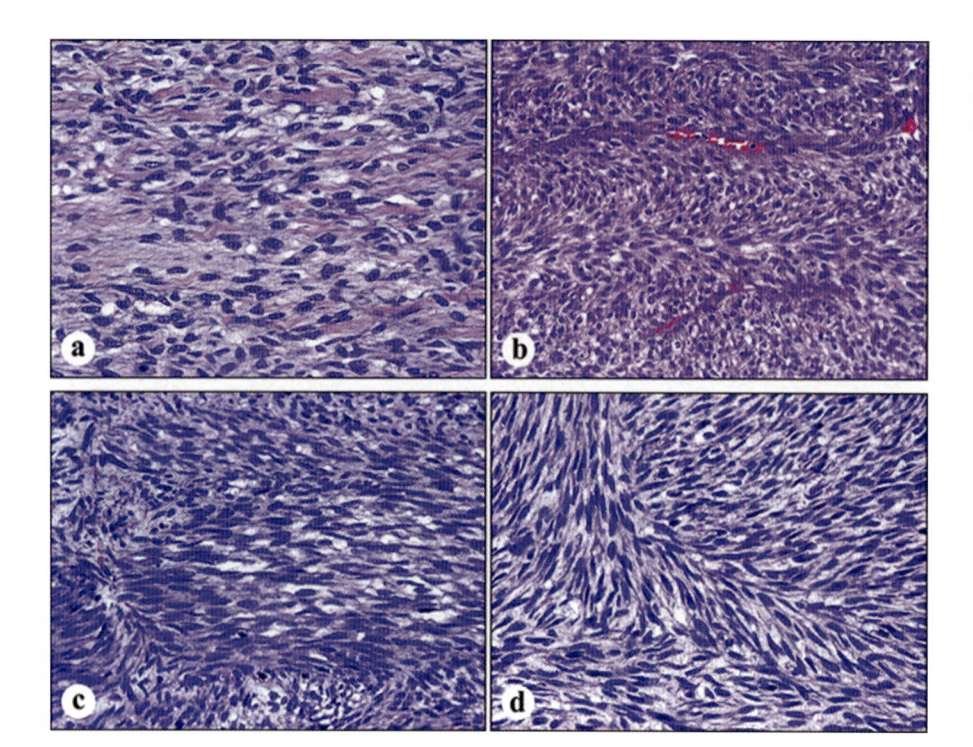

Figure 24.76 Fibrosarcoma-like components are common in MPNSTs. (**a**) Moderate interstitial collagen is sometimes present. (**b–d**) Fascicular and herringbone-like arrangements are often present.

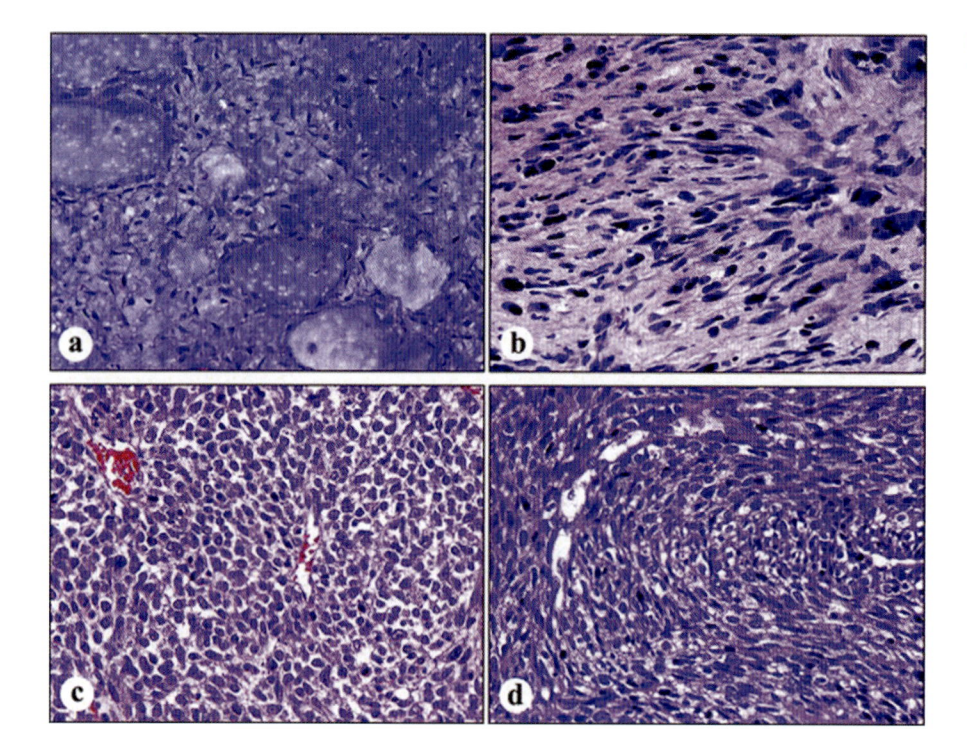

Figure 24.77 Morphologic variation in MPNST. (**a**) Myxoid features resembling those of extraskeletal myxoid chondrosarcoma. (**b**) Pleomorphism. (**c**) Example composed of uniform oval cells resembling a small round cell tumor. (**d**) A spindle cell example resembling synovial sarcoma.

Ultrastructure

Some MPNSTs show schwannian differentiation, such as complex cell processes and variably developed basal laminas. Many of these tumors have no specific ultrastructural features, however.[215] The role of electron microscopy in defining a poorly differentiated sarcoma as being an MPNST is dubious.

Differential diagnosis

A synovial sarcoma that arises close to nerves, contains benign neural proliferation, or has nuclear palisading should not be confused with MPNST. These tumors, when of a monophasic spindle cell type, can often be identified by scattered EMA- and keratin-positive cells. In problematic cases, demonstration of

Chapter 24: Nerve sheath tumors

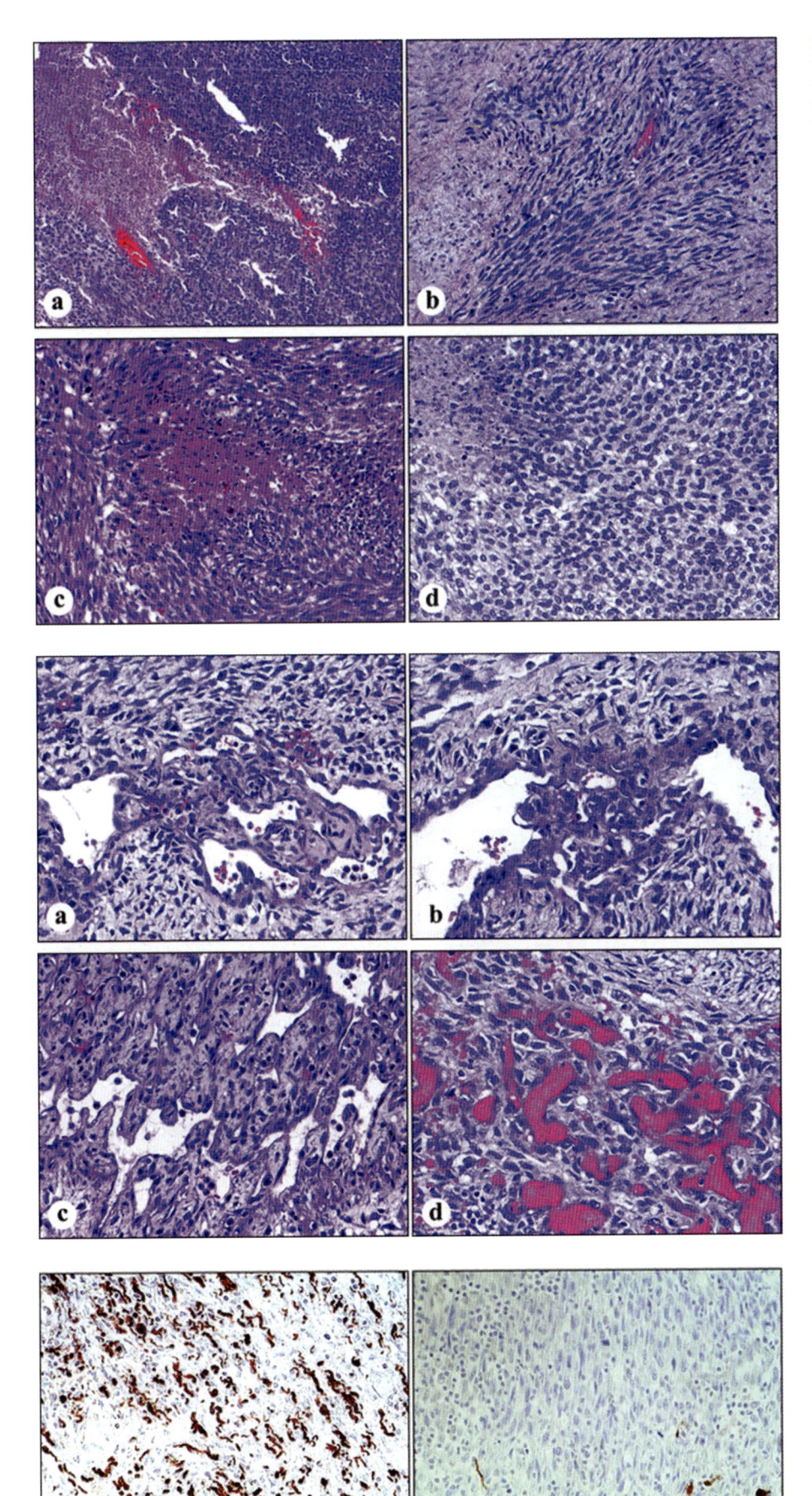

Figure 24.78 Geographic necrosis resembling that of glioblastoma multiforme is a common finding in MPNST. In some cases, the necrosis is extensive, leaving only perivascular collars of viable tumor cells.

Figure 24.79 Complex vascular proliferation in MPNST. (**a–c**) Glomeruloid or hemangiopericytoma-like vascular proliferation. (**d**) Atypical endothelial proliferation with an angiosarcoma-like focus.

Figure 24.80 (**a**) S100-protein-positive cells in MPNST often are elements of pre-existing neurofibroma. (**b**) In this example, the only S100-protein-positive cells are dendritic antigen-presenting cells.

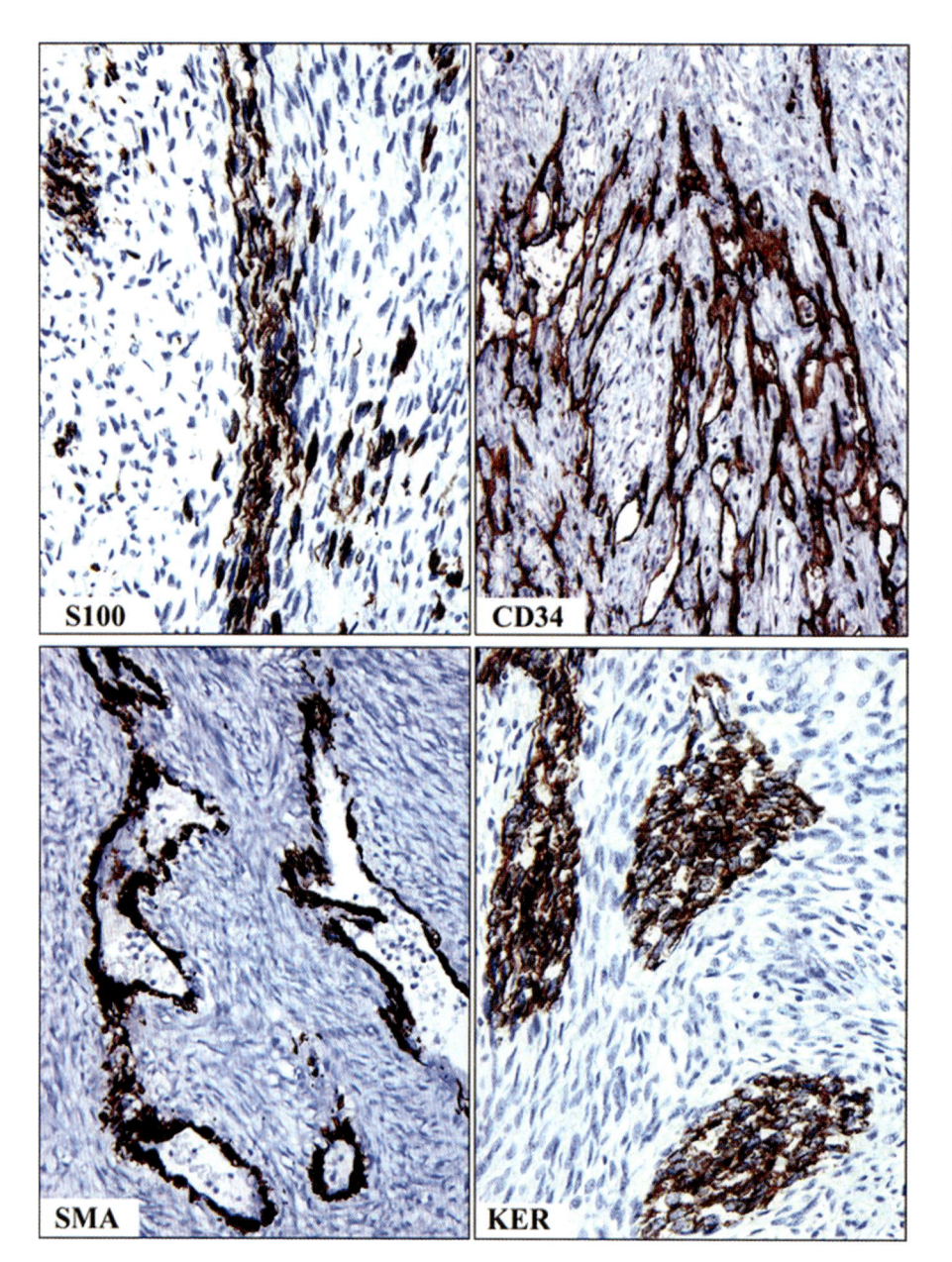

Figure 24.81 Immunohistochemical findings in MPNST. (S100) S100-protein-positive components are often elements of pre-existing neurofibroma. (CD34) Prominent vascular proliferation is highlighted with endothelial markers. (SMA) SMA is essentially limited to pericytes. (KER) Keratin-positive elements can be present, with or without glands. The corresponding HE is shown in Figure 24.83c.

the SYT-SSX fusion transcript is helpful. Such synovial sarcomas could account for some tumors reported as SYT-SSX-positive MPNSTs. The histologic distinction between MPNST and fibrosarcoma is more difficult, and the identification of MPNST in this context usually requires the proper context (i.e., NF1, nerve, or neurofibroma origin).

The diagnosis of cellular schwannoma should be seriously considered in well-differentiated tumors with sparse mitotic activity and extensive S100 protein positivity. These tumors are encapsulated and often contain xanthoma cells.

Metastatic (or primary large) malignant melanomas should be ruled out in the case of a strongly S100-protein-positive, spindled or epithelioid malignant tumor, especially when seen in a lymph node. Additional melanoma markers and clinical correlation (information on previous melanoma) are helpful.

In the gastrointestinal tract, practically all NF1-associated sarcomas are GISTs; the author's review has not identified any bona fide gastrointestinal MPNSTs in the AFIP files.

Genetics of MPNST

Cytogenetic studies have shown complex clonal abnormalities in most cases.[216–218] Rare cases have had single abnormalities, such as +7 and −22; neither is diagnostically specific.[218] Homozygous deletion of the INK4A/CDKN2A gene at 9p, encoding for cyclin-dependent kinase inhibitor cell-cycle regulators such as p16 and p19, has been shown to be typical of MPNST, but not of neurofibromas. Mutations or methylation of this gene was not observed in two studies.[219,220]

Loss of expression of cell-cycle regulator p27 (kip) also seems to be common in MPNST, as opposed to neurofibroma.[221]

Malignant peripheral nerve sheath tumor with divergent differentiation

The most common divergent differentiation (heterologous element) of MPNSTs is rhabdomyosarcomatous differentiation (malignant triton tumor). A malignant glandular MPNST with epithelial elements is very rare. Additional heterologous elements include cartilaginous and osteosarcomatous ones, which may occur together with the other heterologous components. Heterologous elements more commonly occur in tumors of NF1 patients and can be associated with more malignant tumor behavior.

An angiosarcoma originating in an MPNST has been reported occasionally. This event might represent an extreme level of tumor angiogenesis with malignant transformation.

Malignant triton tumor

Triton tumor is the designation for an MPNST with rhabdomyosarcomatous differentiation. The name is derived from the triton salamander, in which the development of both myoid and neural elements from a transplanted sciatic nerve were experimentally demonstrated.[222]

Clinical features

Triton tumors are clinically similar to other MPNSTs, except that they are more often NF1 associated (60–70%). They also occur at a younger age (mean, 33 years) when associated with NF1, whereas the patients without NF1 have been slightly older (mean, 42 years). The tumors present in a wide variety of locations, with the head and neck, or trunk being the most common. The clinicopathologic series show a highly malignant behavior, with the 5-year survival rates being 10% to 20%.[222–225]

Pathology

Characteristic of this tumor is focal rhabdomyosarcomatous differentiation occurring in a background of a high-grade spindle cell or pleomorphic MPNST (Figure 24.82). Well-differentiated rhabdoblastic components are scattered between the spindle cells or seen in clusters. They are positive for desmin, myoglobin, and MyoD1. S100 protein positivity varies, and remnants of neurofibroma can often be demonstrated. Electron microscopy is also useful in verifying the skeletal muscle differentiation.

Glandular malignant peripheral nerve sheath tumors

This variant of MPNST, with heterologous epithelial differentiation, is extremely rare. Many of the reported cases have been associated with NF1 syndrome. The tumors have occurred in children or young adults in a wide variety of locations, and most have shown malignant behavior.[226,227]

Histologically remarkable is the occurrence of glandular elements in the background of MPNST, generally with a spindle cell sarcomatous pattern (Figure 24.83). Other heterologous elements, such as cartilage and bone, can also occur in glandular MPNST.

Differential diagnosis

Glandular differentiation of MPNST must be separated from entrapped sweat glands in nerve sheath tumors; some earlier reports on "glandular schwannomas" might have included such tumors.

Pseudoglandular spaces can occur in MPNSTs; these are not lined by true epithelial cells, but are related to vascular

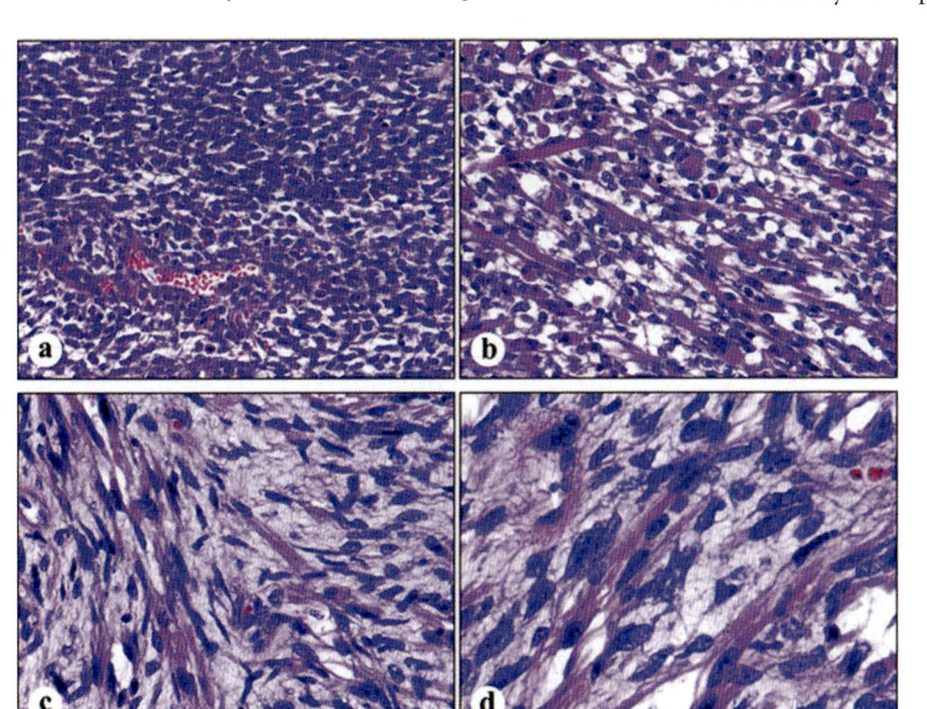

Figure 24.82 MPNST with rhabdomyosarcomatous differentiation (malignant triton tumor). (**a**) Overall view showing a highly cellular spindle cell tumor with prominent vessels. (**b–d**) Varying numbers of differentiated rhabdomyoblasts are present. (**d**) Cross-striations are focally visible.

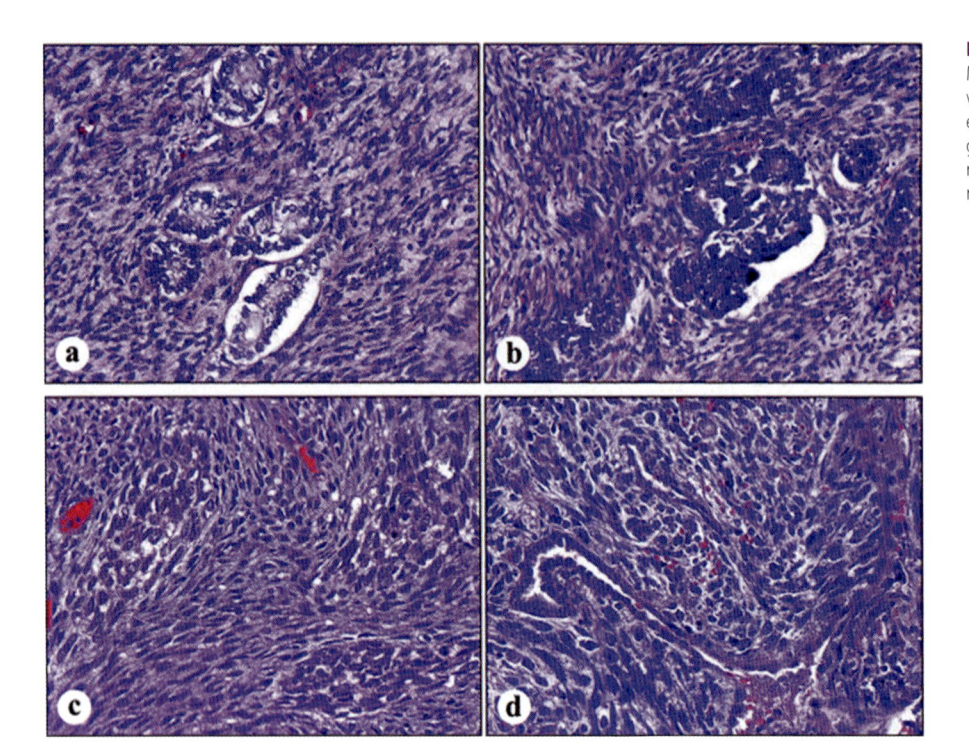

Figure 24.83 (**a,b**) Glandular elements in MPNST. (**c**) Epithelial differentiation without well-developed glands. (**d**) Entrapped epithelial elements in MPNST should not be confused with glandular differentiation, which in this field represents terminal airway epithelia in this lung metastasis of MPNST.

elements. Those biphasic synovial sarcomas that are S100 protein positive should not be confused with malignant glandular MPNST. In synovial sarcoma, the epithelial elements are prominently K7 positive, whereas those in glandular MPNST tend to be negative. K20 can be present in both. In problematic cases, distinction from synovial sarcoma is aided by SYT-SSX gene rearrangement studies, which are negative in glandular MPNST.

Epithelioid malignant peripheral nerve sheath tumor

This histologically distinctive, rare variant of MPNST often arises in a nerve and has a predilection for young adults; in contrast to the other types of MPNST; however, association with NF1 syndrome is weak.

Clinical features

Based on four clinicopathologic series, epithelioid MPNST occurs from childhood to old age, but has a predilection for young adults, with the median ages in the three series being between 29 and 39 years. The combined data do not suggest any sex predilection.[228–231]

Sixty percent of cases in the three earlier series occurred in a nerve, superficial nerves, the sciatic nerve, and the brachial plexus being the most common locations. The most commonly involved sites are the thigh and the upper extremities, with most tumors being proximally located and only a few occurring in the hands and feet. Despite frequent nerve involvement, neurologic symptoms are rare. Only occasional cases have been associated with NF1.

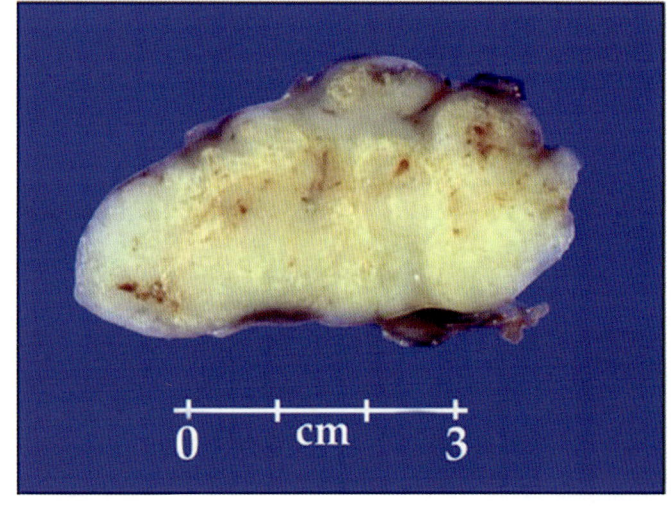

Figure 24.84 Grossly, an epithelioid MPNST forms a demarcated whitish-yellow nodule.

An earlier series showed a significant risk for metastasis to the lungs, pleura, and liver, but superficial tumors had a better prognosis.[230] A more recent series had only 15% metastatic rate. However, this series contained more superficial and small tumors, including tumors with low mitotic rates of 1 per 10 HPFs, and had a relative short follow-up.[231]

Pathology

Grossly, the tumor forms a white nodule or mass that usually varies from 1 cm to 15 cm (median, 4–5 cm); the superficial tumors are smaller than deep ones (Figure 24.84).

Histologically, the tumor is typically composed of multiple nodules of polygonal epithelioid cells separated by fibrovascular septa. In some cases, the tumor cells spread in the enlarged nerve fascicles. The tumor cells often form trabeculae or cords separated by variably myxoid stroma, but they can also form solid areas (Figure 24.85). Cytologically, the tumor cells have large nuclei with prominent nucleoli and a moderate amount of eosinophilic or amphophilic cytoplasm (Figure 24.86). Focal fat-like vacuolization and rhabdoid tumor-like cytoplasmic eosinophilic inclusions can be present.

Immunohistochemically, these tumors are usually strongly positive for S100 protein (Figure 24.86d) and NSE, but negative for HMB45 and tyrosinase; the latter findings are useful in distinguishing these tumors from metastatic melanoma.

Figure 24.85 (a) Epithelioid MPNST forms a relatively small well-demarcated nodule. (b) Epithelioid cytology and prominent mitotic activity are evident. (c,d) This example is composed of sheets of relatively uniform epithelioid cells.

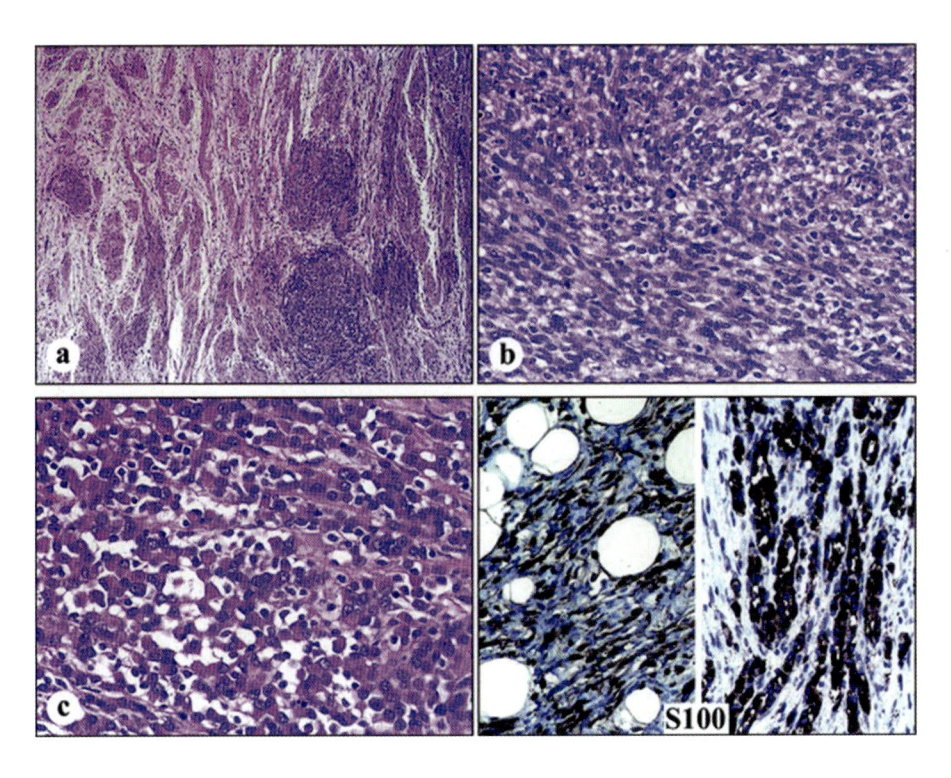

Figure 24.86 Malignant epithelioid MPNST. (a) The malignant components and pre-existing neurofibroma intermingle. (b) The malignant component is highly cellular with mitotic activity. (c) Some tumor areas show eosinophilic cytoplasm with rhabdoid features. (d) Both the pre-existing neurofibroma components and the epithelioid elements are strongly S100 protein positive.

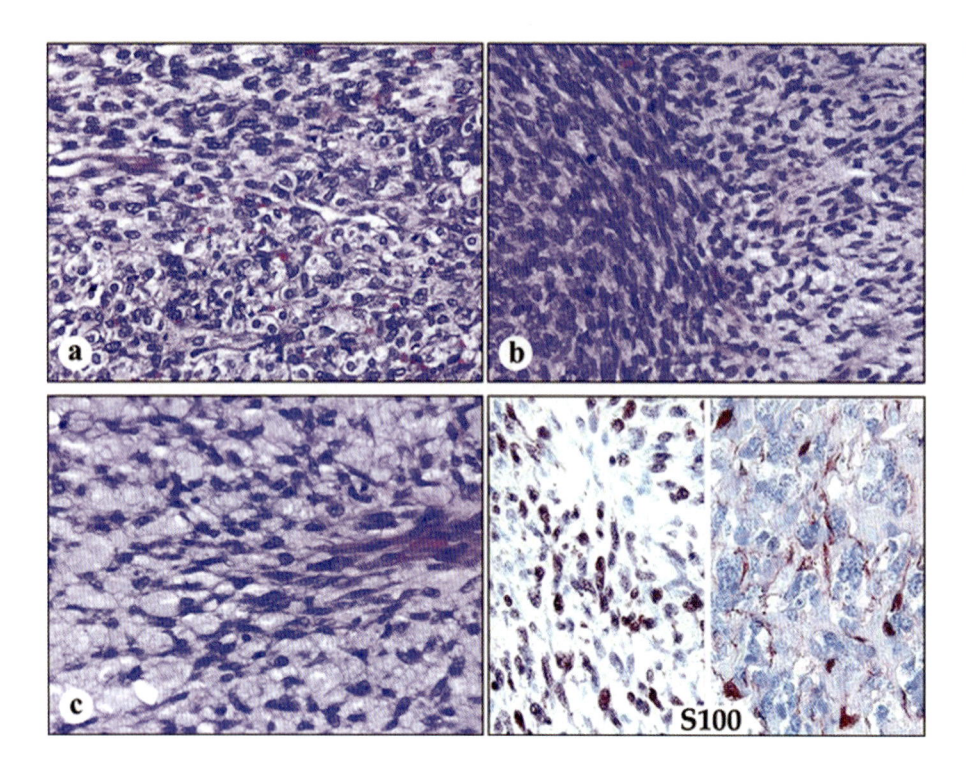

Figure 24.87 MPNSTs arising in a pheochromocytoma. (**a**) There is a sharp demarcation between the pheochromocytoma element (lower right half) and the adjacent spindle cell sarcoma. (**b,c**) The MPNST component shows a fibrosarcoma-like pattern with myxoid matrix. The MPNST element is S100 protein positive, whereas only the sustentacular cells are positive in the pheochromocytoma component.

The tumors have been reported as negative for keratins. Loss of INI1 has been reported in two-thirds of cases.[231]

Malignant peripheral nerve sheath tumors arising from ganglioneuromas and pheochromocytomas

A small number of MPNSTs have been reported as having arisen from a ganglioneuroma, ganglioneuroblastoma, or pheochromocytoma, mostly in the adrenal glands.[232–236]

These tumors have mostly been large, occurred in young adults, and had an unfavorable course. Histologically, they have been high-grade spindle cell sarcomas, and the specific diagnosis has been based on the immunohistochemical demonstration of S100 protein and identification of a juxtaposed pheochromocytoma or ganglioneuroma element (Figure 24.87). It is possible that some of these tumors represent a malignant transformation of the S100-protein-positive sustentacular cell element in pheochromocytomas.

References

General

1. Antonescu CR, Scheithauer BW, Woodruff JM. *Tumors of the Peripheral Nervous System. AFIP Atlas of Tumor Pathology*, Fourth series, fascicle 19. Silver Spring, Maryland: ARP Press, 2013.

Neuromuscular hamartoma

2. Louhimo I, Rapola J. Intraneural muscular hamartoma: a report of two cases in small children. *J Pediatr Surg* 1972;7:696–699.
3. Markel SF, Enzinger FM. Neuromuscular hamartoma: a benign "triton tumor" composed of mature neural and striated muscle elements. *Cancer* 1982;49:140–144.
4. Bonneau R, Brochu P. Neuromuscular choristoma: a clinicopathologic study of two cases. *Am J Surg Pathol* 1983;7:521–528.
5. Mitchell A, Scheithauer BW, Ostertag H, Sepehrnia A, Sav A. Neuromuscular choristoma. *Am J Clin Pathol* 1995;103:460–465.
6. Tiffee JC, Barnes EL. Neuromuscular hamartomas of the head and neck. *Arch Otolaryngol Head Neck Surg* 1998;124:212–216.
7. Azzopardi JG, Eusebi V, Tison V, Betts BM. Neurofibroma with rhabdomyomatous differentiation: benign "Triton" tumor of the vagina. *Histopathology* 1983;7:561–572.

Morton neuroma and traumatic neuroma

8. Wu KK. Morton neuroma and metatarsalgia. *Curr Opin Rheumatol* 2000;12:131–142.
9. Bennett GL, Graham CE, Mauldin DM. Morton's interdigital neuroma: a comprehensive treatment protocol. *Foot Ankle Int* 1995;16:760–763.
10. Saygi B, Yildirim Y, Saygi EK, Kara H, Esemenli T. Morton neuroma: comparative results of two conservative methods. *Foot Ankle Int* 2005;26:556–559.
11. Reed RJ, Bliss BO. Morton's neuroma: regressive and productive intermetatarsal elastofibrositis. *Arch Pathol* 1973;95:123–129.
12. Lassmann G, Lassmann H, Stockinger L. Morton's metatarsalgia: light and electron microscopic observations and their relation to entrapment neuropathies. *Virchows Arch A Pathol Anat Histopathol* 1976;370:307–321.

13. Bourke G, Owen J, Machet D. Histological comparison of the third interdigital nerve in patients with Morton's metatarsalgia and control patients. *Aust N Z J Surg* 1994;64:421–424.

14. Vernadakis AJ, Koch H, Mackinnon SE. Management of neuromas. *Clin Plast Surg* 2003;30:247–268.

15. Larson DM, Storsteen KA. Traumatic neuroma of the bile ducts with intrahepatic extension causing obstructive jaundice. *Hum Pathol* 1984;15:287–289.

16. van Gulik TM, Brummelkamp WH, Lygidakis NJ. Traumatic neuroma giving rise to biliary obstruction after reconstructive surgery for iatrogenic lesions of the biliary tract: a report of three cases. *Hepatogastroenterology* 1989;36:255–257.

Mucosal neuroma and ganglioneuroma(tosis)

17. Carney JA, Sizemore GW, Hayles AB. Multiple endocrine neoplasia type 2b. *Pathobiol Annu* 1978;8:105–153.

18. Giangarella J, Jagirdar J, Adelman H, Budzilovich G, Greco MA. Mucosal neuromas and plexiform neurofibromas: an immunocytochemical study. *Pediatr Pathol* 1993;13:281–288.

19. d'Amore ESG, Manivel JC, Pettinato G, Niehans GA, Snover DC. Intestinal ganglioneuromatosis: mucosal and transmural types. A clinicopathologic and immunohistochemical study of six cases. *Hum Pathol* 1991;22:276–286.

20. Shekitka KM, Sobin LH. Ganglioneuromas of the gastrointestinal tract: relation to von Recklinghausen's disease and other multiple tumor syndromes. *Am J Surg Pathol* 1994;18:250–257.

21. Weidner N, Flanders DJ, Mitros FA. Mucosal ganglioneuromatosis associated with multiple colonic polyps. *Am J Surg Pathol* 1984;8:779–786.

Pacinian neuroma

22. Fletcher CD, Theaker JM. Digital pacinian neuroma: a distinctive hyperplastic lesion. *Histopathology* 1989;15:249–256.

23. Reznik M, Thiry A, Fridman V. Painful hyperplasia and hypertrophy of pacinian corpuscles in the hand: report of two cases with immunohistochemical and ultrastructural studies, and a review of the literature. *Am J Dermatopathol* 1998;20:203–207.

24. Satge D, Nabhan J, Nandigou Y, et al. A pacinian hyperplasia of the foot. *Foot Ankle Int* 2001;22:342–344.

25. Yan S, Horangic NJ, Harris BT. Hypertrophy of Pacinian corpuscles in a young patient with neurofibromatosis. *Am J Dermatopathol* 2006;28:202–204.

26. MacDonald DM, Wilson-Jones E. Pacinian neurofibroma. *Histopathology* 1977;1:247–255.

Palisaded encapsulated neuroma

27. Reed RJ, Fine RM, Meltzer HD. Palisaded, encapsulated neuromas of the skin. *Arch Dermatol* 1972;106:865–870.

28. Fletcher CD. Solitary circumscribed neuroma (so-called palisaded encapsulated neuroma): a clinicopathologic and immunohistochemical study. *Am Surg Pathol* 1989;13:574–580.

29. Dover JS, From L, Lewis A. Palisaded encapsulated neuromas: a clinicopathologic study. *Arch Dermatol* 1989;125:386–389.

30. Dakin MC, Leppard B, Theaker JM. The palisaded encapsulated neuroma (solitary circumscribed neuroma). *Histopathology* 1992;20:405–410.

31. Koutlas IG, Scheithauer BW. Palisaded encapsulated ("solitary circumscribed") neuroma of the oral cavity: a review of 55 cases. *Head Neck Pathol* 2010;4:15–26.

32. Navarro M, Vilata H, Requena C, Aliaga A. Palisaded encapsulated neuroma (solitary circumscribed neuroma) of the glans penis. *Br J Dermatol* 2000;142:1061–1062.

33. Argenyi ZB, Cooper PH, Santa Cruz D. Plexiform and other unusual variants of palisaded encapsulated neuroma. *J Cutan Pathol* 1993;20:34–39.

34. Albrecht S, Kahn HJ, From L. Palisaded encapsulated neuroma: an immunohistochemical study. *Mod Pathol* 1989;2:403–406.

35. Argenyi ZB. Immunohistochemical characterization of palisaded, encapsulated neuroma. *J Cutan Pathol* 1990;17:329–335.

Neurofibroma: clinicopathologic and histopathologic studies

36. Reed ML, Jacoby RA. Cutaneous neuroanatomy and neuropathology. *Am J Dermatopathol* 1983;5:335–362.

37. Fletcher CD. Peripheral nerve sheath tumors: a clinicopathologic update. *Pathol Ann* 1990;25(Part 1):53–74.

38. Megahed M. Histopathological variants of neurofibroma: a study of 114 lesions. *Am J Dermatopathol* 1994;16:486–495.

39. Requena L, Sangueza OP. Benign neoplasms with neural differentiation: a review. *Am J Dermatopathol* 1995;17:75–96.

40. Michal M, Fanburg-Smith JC, Mentzel T, et al. Dendritic cell neurofibroma with pseudorosettes: a report of 18 cases of a distinct and hitherto unrecognized neurofibroma variant. *Am J Surg Pathol* 2001;25:587–594.

41. Lin BT, Weiss LM, Medeiros LJ. Neurofibroma and cellular neurofibroma with atypia: a report of 14 tumors. *Am J Surg Pathol* 1997;21:1443–1449.

42. Kaiserling E, Geerts ML. Tumour of Wagner–Meissner touch corpuscles: a Wagner–Meissner neurilemmoma. *Virchows Arch A Pathol Anat Histopathol* 1986;409:241–250.

43. Fetsch JF, Michal M, Miettinen M. Pigmented (melanotic) neurofibroma: a clinicopathologic and immunohistochemical analysis of 19 lesions from 17 patients. *Am J Surg Pathol* 2000;24:331–343.

Neurofibroma: immunohistochemistry and ultrastructure

44. Weiss SW, Langloss JM, Enzinger FM. Value of S100-protein in the diagnosis of soft tissue tumors with particular reference to benign and malignant Schwann cell tumors. *Lab Invest* 1983;49:299–308.

45. Weiss SW, Nickoloff BJ. CD34 is expressed by a distinctive cell population in peripheral nerve, nerve sheath tumors and related lesions. *Am J Surg Pathol* 1993;17:1039–1045.

46. Chaubal A, Paetau A, Zoltick P, Miettinen M. CD34 immunoreactivity in nervous system tumors. *Acta Neuropathol* 1994;88:454–458.

47. Perentes E, Rubinstein LJ. Immunohistochemical recognition of human nerve sheath tumors by anti-Leu 7 antibodies (HNK-1) monoclonal antibody. *Acta Neuropathol* 1986;69:227–233.

48. Chanoki M, Ishii M, Fukai K, et al. Immunohistochemical localization of type I, III, IV, V and VI collagens and laminin in neurofibroma and neurofibrosarcoma. *Am J Dermatopathol* 1991;13:365–373.

49. Kindblom LG, Ahlden M, Meis-Kindblom JM, Stenman G. Immunohistochemical and molecular analysis of p53, MDM2, proliferating cell nuclear antigen and Ki-67 in benign and malignant nerve sheath tumors. *Virchows Arch* 1995;427:19–26.

50. Halling KC, Scheithauer BW, Halling AC, et al. p53 expression in neurofibroma and malignant peripheral nerve sheath tumor: an immunohistochemical study of sporadic and NF1-associated tumors. *Am J Clin Pathol* 1996;106:282–288.

51. Erlandson RA, Woodruff JM. Peripheral nerve sheath tumors: an electron microscopic study of 43 cases. *Cancer* 1982;49:273–287.

Genetics of neurofibroma and neurofibromatosis 1

52. Sorensen SA, Mulvihill JJ, Nielsen A. Long-term follow-up of von Recklinghausen neurofibromatosis: survival and malignant neoplasms. *N Engl J Med* 1986;314:1010–1015.

53. Mulvihill JJ, Parry DM, Sherman JL, et al. NIH conference: neurofibromatosis 1 (Recklinghausen disease) and neurofibromatosis 2 (bilateral acoustic neurofibromatosis). An update. *Ann Intern Med* 1990;113:39–52.

54. Gutmann DH, Aylsworth A, Carey JC, et al. The diagnostic evaluation and multidisciplinary management of neurofibromatosis 1 and neurofibromatosis 2. *JAMA* 1997;278:51–57.

55. Shen MH, Harper PS, Upadhyay M. Molecular genetics of neurofibromatosis type 1 (NF1). *J Med Genet* 1996;22:2–17.

56. Rasmussen SA, Friedman JM. NF1 gene and neurofibromatosis 1. *Am J Epidemiol* 2000;151:33–40.

57. Serra E, Puig S, Otero D, et al. Confirmation of a double-hit model for the NF1 gene in benign neurofibromas. *J Med Genet* 1997;61:512–519.

58. Sawada S, Florell S, Purandare SM, et al. Identification of NF1 mutations in both alleles of a dermal neurofibroma. *Nat Genet* 1996;14:110–112.

59. Kluwe L, Friedrich R, Mautner VF. Loss of NF1 allele in schwann cells but not in fibroblasts derived from an NF1-associated neurofibroma. *Genes Chromosomes Cancer* 1999;24:283–285.

60. Eisenbarth I, Beyer K, Krone W, Assum G. Toward a survey of somatic mutation of the NF1 gene in benign neurofibromas of patients with neurofibromatosis type 1. *Am J Hum Genet* 2000;66:393–401.

61. John AM, Ruggieri M, Ferner R, Upadhyaya M. A search for evidence of somatic mutations in the NF1 gene. *J Med Genet* 2000;37:44–49.

62. Messiaen LM, Callens T, Mortier G, et al. Exhaustive mutation analysis of the NF1 gene allows identification of 95% of mutations and reveals a high frequency of unusual splicing defects. *Hum Mutat* 2000;15:541–555.

63. Jung EG. Segmental neurofibromatosis. *Neurofibromatosis* 1988;1:306–311.

64. Listernick R, Mancini AJ, Charrow J. Segmental neurofibromatosis in childhood. *Am J Med Genet A* 2003;121A:132–135.

65. Tinschert S, Naumann I, Stegmann E, et al. Segmental neurofibromatosis is caused by somatic mutation of the neurofibromatosis type 1 (NF1) gene. *Eur J Hum Genet* 2000;8:455–459.

66. Colman SD, Rasmussen SA, Ho VT, Abernathy CR, Wallace MR. Somatic mosaicism in a patient with neurofibromatosis type 1. *Am J Hum Genet* 1996;58:484–490.

Classic schwannoma (neurilemmoma): clinicopathologic studies and immunohistochemistry

67. Das Gupta TK, Brasfield RD, Strong EW, Hajdu SI. Benign solitary schwannomas (neurilemomas). *Cancer* 1969;24:355–366.

68. Alvarado-Cabrero I, Folpe AL, Srigley JR, et al. Intararenal schwannoma: a report of four cases including three cellular variants. *Mod Pathol* 2000;13:851–856.

69. Dahl I. Ancient neurilemmoma (schwannoma). *Acta Pathol Microbiol Scand A* 1977;85:812–818.

70. Gould VE, Moll R, Moll I, Lee I, Schwechheimer K. The intermediate filament complement of the spectrum of nerve sheath neoplasms. *Lab Invest* 1986;55:463–474.

71. Kawahara E, Oda Y, Ooi A, et al. Expression of glial fibrillary acidic protein (GFAP) in peripheral nerve sheath tumors: a comparative study of immunoreactivity of GFAP, vimentin, S100-protein and neurofilament in 38 schwannomas and 18 neurofibromas. *Am J Surg Pathol* 1988;12:115–120.

72. Kaiserling E, Xiao JC, Ruck P, Horny HP. Aberrant expression of macrophage-associated antigens (CD68 and Ki-M1P) by Schwann cells in reactive and neoplastic neural tissue. *Mod Pathol* 1994;6:463–468.

73. Gray MH, Rosenberg AE, Dickersin GR, Bhan AK. Glial fibrillary acidic protein and keratin expression by benign and malignant nerve sheath tumors. *Hum Pathol* 1989;20:1089–1096.

74. Oda Y, Kawahara E, Minamoto T, Tsuenyoshi M. Immunohistochemical studies on the tissue localization of collagen types I, III, IV, V and VI in schwannomas. *Virchows Arch B Cell Pathol Incl Mol Pathol* 1988;56:153–163.

75. Waggener JD. Ultrastructure of benign peripheral nerve sheath tumors. *Cancer* 1966;19:699–709.

76. Dickersin GR. The electron microscopic spectrum of nerve sheath neoplasms. *Ultrastruct Pathol* 1987;11:103–146.

Variants of schwannoma and malignant transformation

77. Woodruff JM, Godwin TA, Erlandson RA, Susin M, Martini N. Cellular schwannoma: a variety of schwannoma sometimes mistaken for a malignant tumor. *Am J Surg Pathol* 1981;5:733–744.

78. Lodding P, Kindblom LG, Angervall L, Stenman G. Cellular schwannoma: a clinicopathologic study of 29 cases. *Virchows Arch A Pathol Anat Histopathol* 1990;416:237–244.

79. White W, Shiu MH, Rosenblum MK, Erlandson RA, Woodruff JM. Cellular schwannoma: a clinicopathologic study of 57 patients and 58 tumors. *Cancer* 1990;66:1266–1275.

80. Casadei GR, Scheihauer BW, Hirose T, et al. Cellular schwannoma: a clinicopathologic, DNA flow-cytometric and proliferation marker study of 71 cases. *Cancer* 1995;75:1109–1119.

81. Woodruff JM, Marshall ML, Goodwin TA, et al. Plexiform (multinodular) schwannoma: a tumor simulating plexiform neurofibroma. *Am J Surg Pathol* 1983;7:691–697.

82. Fletcher CDM, Davies SE. Benign plexiform (multinodular) schwannoma: a rare tumor unassociated with neurofibromatosis. *Histopathology* 1986;10:971–980.

83. Iwashita T, Enjoji M. Plexiform neurilemmoma: a clinicopathological and immunohistochemical analysis of 23 tumors from 20 patients. *Virchows Arch* 1987;411:305–309.

84. Kao GR, Laskin WB, Olsen TG. Solitary cutaneous plexiform neurilemmoma (schwannoma): a clinicopathologic, immunohistochemical, and ultrastructural study of 11 cases. *Mod Pathol* 1989;2:20–26.

85. Hirose T, Scheithauer BW, Sano T. Giant plexiform schwannoma: a report of two cases with soft tissue and visceral involvement. *Mod Pathol* 1997;10:1075–1081.

86. Agaram NP, Prakash S, Antonescy CR. Deep-seated plexiform schwannoma. *Am J Surg Pathol* 2005;29:1042–1048.

87. Ishida T, Kuroda M, Motoi T, et al. Phenotypic diversity of neurofibromatosis 2: association with plexiform schwannoma. *Histopathology* 1998;32:264–270.

88. Woodruff JM, Scheithauer BW, Kurtakaya-Yapicier O, et al. Congenital and childhood plexiform (multinodular) schwannoma: a troublesome mimic of malignant peripheral nerve sheath tumor. *Am J Surg Pathol* 2003;27:1321–1329.

89. Kindblom LG, Meis-Kindblom JM, Havel G, Busch C. Benign epithelioid schwannoma. *Am J Surg Pathol* 1998;22:762–770.

Epithelioid schwannoma

90. Goldblum JR, Beals TF, Weiss SW. Neuroblastoma-like neurilemoma. *Am J Surg Pathol* 1994;18:266–273.

91. Woodruff JM, Selig AM, Crowley K, Allen RW. Schwannoma (neurilemmoma) with malignant transformation: a rare, distinctive peripheral nerve tumor. *Am J Surg Pathol* 1994;18:882–895.

92. Joste NE, Racz MI, Montgomery KD, et al. Clonal chromosome abnormalities in a plexiform schwannoma. *Cancer Genet Cytogenet* 2004;150:73–77.

93. Mentzel T, Katenkamp D. Intraneural angiosarcoma and angiosarcoma arising in benign and malignant peripheral nerve sheath tumours: clinicopathological and immunohistochemical analysis of four cases. *Histopathology* 1999;35:114–120.

94. Ruckert RI, Fleige B, Rogalla P, Woodruff JM. Schwannoma with angiosarcoma: report of a case and comparison with other types of nerve sheath tumors with angiosarcoma. *Cancer* 2000;89:1577–1585.

95. McMenamin ME, Fletcher CD. Expanding the spectrum of malignant change in schwannomas: epithelioid malignant change, epithelioid malignant peripheral nerve sheath tumor, and epithelioid angiosarcoma. A study of 17 cases. *Am J Surg Pathol* 2001;25:13–25.

Neurofibromatosis 2 and genetics of schwannoma

96. Rouleau GA, Merel P, Lutchman M, et al. Alteration in a new gene encoding a putative membrane-organizing protein causes neurofibromatosis type 2. *Nature* 1993;363:515–521.

97. Trofatter JA, MacCollin MM, Rutter JL, et al. A novel moesin-, ezrin-, radixin-like gene is a candidate for the neurofibromatosis 2 tumor suppressor. *Cell* 1993;72:791–800.

98. Louis DN, Ramesh V, Gusella JF. Neuropathology and molecular genetics of neurofibromatosis 2 and related tumors. *Brain Pathol* 1995;5:163–172.

99. Evans DG, Sainio M, Baser ME. Neurofibromatosis type 2. *J Med Genet* 2000;37:897–904.

100. Giovannini M, Robanus-Maandag E, Niwa-Kawakita M, et al. Schwann cell hyperplasia and tumors in transgenic mice expressing a naturally occurring mutant NF2 protein. *Genes Dev* 1999;15:978–986.

101. Ruttledge MH, Andermann AA, Phelan CM, et al. Type of mutation in the neurofibromatosis 2 gene frequently determines severity of disease. *Am J Hum Genet* 1996;59:331–342.

102. Parry DM, McCollin MM, Kaiser-Kupfer MI, et al. Germ-line mutations in the neurofibromatosis 2 gene: correlations with disease severity and retinal abnormalities. *Am J Hum Genet* 1996;59:529–539.

103. Kimura Y, Saya H, Nakao M. Calpain-dependent proteolysis of NF2 protein: involvement in schwannomas and meningiomas. *Neuropathology* 2000;20:153–160.

Schwannomatosis

104. Purcell SM, Dixon SL. Schwannomatosis: an unusual variant of neurofibromatosis or a distinct clinical entity. *Arch Dermatol* 1989;125:390–393.

105. MacCollin M, Woodfin W, Kronn D, Short MP. Schwannomatosis: a clinical and pathologic study. *Neurology* 1996;46:1072–1079.

106. Evans DG, Mason S, Huson SM, et al. Spinal and cutaneous schwannomatosis is a variant form of type 2 neurofibromatosis: a clinical and molecular study. *J Neurol Neurosurg Psychiatry* 1997;62:361–366.

107. Jacoby LB, Jones D, Davis K, et al. Molecular analysis of NF2 tumor-suppressor gene in schwannomatosis. *Am J Hum Genet* 1997;61:1293–1302.

108. Seppälä MT, Sainio MA, Haltia MJ, et al. Multiple schwannomas: schwannomatosis or neurofibromatosis type 2? *J Neurosurg* 1998;89:36–41.

109. Hulsebos TJ, Plomp AS, Wolterman RA, et al. Germline mutation of INI1/SMARCB1 in familial schwannomatosis. *Am J Hum Genet* 2007;80:805–810.

110. Sestini R, Bacci C, Provenzano A, Genuardi M, Papi L. Evidence of a four-hit mechanism involving SMARCB1 and NF2 in schwannomatosis-associated

111. Boyd C, Smith MJ, Kluwe L, et al. Alterations in the SMARCB1 (INI1) tumor suppressor gene in familial schwannomatosis. *Clin Genet* 2008;74:358–366.

112. Swensen JJ, Keyser J, Coffin CM, et al. Familial occurrence of schwannomas and malignant rhabdoid tumour associated with a duplication in SMARCB1. *J Med Genet* 2009;46:68–72.

113. Rousseau G, Noguchi T, Bourdon V, Sobol H, Olschwang S. SMARCB1/INI1 germline mutations contribute to 10% of sporadic schwannomatosis. *BMC Neurol* 2011;11:9.

114. Piotrowski A, Xie J, Lu YF, et al. Germline loss-of-function mutations in LZTR1 predispose to an inherited disorder of multiple schwannomas. *Nat Genet* 2014;46:182–187.

Hybrid nerve sheath tumors

115. Feany MB, Anthony DC, Fletcher CD. Nerve sheath tumours with hybrid features of neurofibroma and schwannoma: a conceptual challenge. *Histopathology* 1998;32:405–410.

116. Harder A, Wesemann M, Hagel C, et al. Hybrid neurofibroma/schwannoma is overrepresented among schwannomatosis and neurofibromatosis patients. *Am J Surg Pathol* 2012;36:702–709.

117. Michal M, Kazakov DV, Belousova I, et al. A benign neoplasm with histopathological features of both schwannoma and retiform perineurioma (benign schwannoma-perineurioma): a report of six cases of a distinctive soft tissue tumor with a predilection for the fingers. *Virchows Arch* 2004;445:347–353.

118. Hornick JL, Bundock EA, Fletcher CD. Hybrid schwannoma/perineurioma: clinicopathologic analysis of 42 distinctive benign nerve sheath tumors. *Am J Surg Pathol* 2009;33:1554–1561.

119. Kazakov DV, Pitha J, Sima R, et al. Hybrid peripheral nerve sheath tumors: schwannoma-perineurioma and neurofibroma-perineurioma. A report of three cases in extradigital locations. *Ann Diagn Pathol* 2005;9:16–23.

Gastrointestinal schwannoma

120. Daimaru Y, Kido H, Hashimoto H, Enjoji M. Benign schwannoma of the gastrointestinal tract: a clinicopathologic and immunohistochemical study. *Hum Pathol* 1988;19:257–264.

121. Sarlomo-Rikala M, Miettinen M. Gastric schwannoma: a clinicopathologic analysis of six cases. *Histopathology* 1995;27:355–360.

122. Prevot S, Bienvenu L, Vaillant JC, de Saint-Maur PP. Benign schwannoma of the digestive tract: a clinicopathologic and immunohistochemical study of five cases, including a case of esophageal tumor. *Am J Surg Pathol* 1999;23:431–436.

123. Miettinen M, Shekitka KM, Sobin LH. Schwannomas in the colon and rectum: clinicopathologic and immunohistochemical study of 20 cases. *Am J Surg Pathol* 2001;25:846–855.

124. Voltaggio L, Murray R, Lasota J, Miettinen M. Gastric schwannoma: a clinicopathologic study of 51 cases and critical review of the literature. *Hum Pathol* 2012;43:650–659.

125. Tozbikian G, Shen R, Suster S. Signet ring cell gastric schwannoma: report of a new distinctive morphological variant. *Ann Diagn Pathol* 2008;12:146–152.

126. Liegl B, Bennett MW, Fletcher CD. Microcystic/reticular schwannoma: a distinct variant with predilection for visceral locations. *Am J Surg Pathol* 2008;32:1080–1087.

127. Lewin MR, Dilworth HP, Abu Alfa AK, Epstein JI, Montgomery E. Mucosal benign epithelioid nerve sheath tumors. *Am J Surg Pathol* 2005;29:1310–1315.

128. Lasota J, Wasag B, Dansonka-Mieszkzkowska A, et al. Evaluation of NF2 and NF1 tumor suppressor genes in distinctive gastrointestinal nerve sheath tumors traditionally diagnosed as benign schwannomas: a study of 20 cases. *Lab Invest* 2003;83:1361–1371.

Melanotic schwannoma

129. Carney JA. Psammomatous melanotic schwannoma: a distinctive, heritable tumor with special associations, including cardiac myxoma and the Cushing syndrome. *Am J Surg Pathol* 1990;14:206–222.

130. Thornton CM, Handley J, Bingham EA, Toner PG, Walsh MY. Psammomatous melanotic schwannoma arising in the dermis in a patient with Carney's complex. *Histopathology* 1992;20:71–73.

131. Torres-Mora J, Dry S, Li X, et al. Malignant melanotic schwannian tumor: a clinicopathologic, immunohistochemical, and gene expression profiling study of 40 cases, with a proposal for the reclassification of "melanotic schwannoma". *Am J Surg Pathol* 2014;38:94–105.

132. Myers JL, Bernreuter W, Dunham W. Melanotic schwannoma of bone: clinicopathologic, immunohistochemical and ultrastructural features of a rare primary bone tumor. *Am J Clin Pathol* 1990;93:424–429.

Nerve sheath myxoma and other nerve sheath tumors

133. Laskin WB, Fetsch JF, Lasota J, Miettinen M. Benign epithelioid peripheral nerve sheath tumors of the soft tissues: clinicopathologic spectrum of 33 cases. *Am J Surg Pathol* 2005;29:39–51.

134. Harkin JC, Reed RJ. Myxoma of nerve sheath: tumors of the peripheral nervous system. In *Atlas of Tumor Pathology*, 2nd series, fascicle 3. Washington, DC: AFIP; 1969: 60–65.

135. Laskin WB, Fetsch JF, Miettinen M. The "neurothekeoma". Immunohistochemical analysis distinguishes the true nerve sheath myxoma from its mimics. *Hum Pathol* 2000;31:1230–1241.

136. Fetsch JF, Laskin WB, Miettinen M. Nerve sheath myxoma: a clinicopathological and immunohistochemical analysis of 57 morphologically distinctive, S100 protein- and GFAP-positive, myxoid peripheral nerve sheath tumors with a predilection for the extremities and a high recurrence rate. *Am J Surg Pathol* 2005;29:1615–1624.

137. Gallager RL, Helwig EB. Neurothekeoma: a benign cutaneous tumor of neural origin. *Am J Clin Pathol* 1980;74:759–764.

138. Holden CA, Wilson-Jones E, MacDonald DM. Cutaneous lobular neuromyxoma. *Br J Dermatol* 1982;106:211–215.

139. Angervall L, Kindblom LG, Haglid K. Dermal nerve sheath myxoma: a light and electron microscopic,

histochemical and immunohistochemical study. *Cancer* 1984;53:1752–1759.

140. Pulitzer DR, Reed RJ. Nerve sheath myxoma (perineurial myxoma). *Am J Dermatopathol* 1985;7:409–421.

141. Mincer HH, Spears KD. Nerve sheath myxoma in the tongue. *Oral Surg Oral Med Oral Pathol* 1974;37:428–430.

Perineurioma

142. Perentes E, Nakagawa Y, Ross GW, Stanton C, Rubinstein LJ. Expression of epithelial membrane antigen in perineurial cells and their derivatives: an immunohistochemical study with multiple markers. *Acta Neuropathol* 1987;75:160–165.

143. Ariza A, Bilbao JM, Rosai J. Immunohistochemical detection of epithelial membrane antigen in normal perineurial cells and perineurioma. *Am J Surg Pathol* 1988;12:678–683.

144. Theaker JM, Fletcher CD. Epithelial membrane antigen expression by the perineurial cell: further studies of peripheral nerve lesions. *Histopathology* 1989;14:581–591.

145. Macarenko RS, Oliveira AM. A distinctive and underrecognized peripheral nerve sheath neoplasm. *Arch Pathol Lab Med* 2007;131:625–636.

146. Lazarus SS, Trombetta LD. Ultrastructural identification of a benign perineurial cell tumor. *Cancer* 1978;41:1823–1829.

147. Erlandson RA. The enigmatic perineurial cell and its participation in tumors and tumorlike entities. *Ultrastruct Pathol* 1991;15:335–351.

148. Bilbao JM, Khoury NJS, Hudson AR, Briggs SJ. Perineurioma (localized hypertrophic neuropathy). *Arch Pathol Lab Med* 1984;108:557–560.

149. Boyanton BL, Jones JK, Shenaq SM, Hicks MJ, Bhattacharjee MB. Intraneural perineurioma: a systematic review with illustrative cases. *Arch Pathol Lab Med* 2007;131:1382–1392.

150. Emory TS, Scheithauer BW, Hirose T, et al. Intraneural perineurioma: a clonal neoplasm associated with abnormalities of chromosome 22. *Am J Clin Pathol* 1995;103:696–704.

151. Fetsch JF, Miettinen M. Sclerosing perineurioma: a clinicopathologic study of 19 cases of a distinctive soft tissue lesion with a predilection for the fingers and palms of young adults. *Am J Surg Pathol* 1997;21:1433–1442.

152. Yamaguchi U, Hasegawa T, Hirose T, et al. Sclerosing perineurioma: a clinicopathologic study of five cases and diagnostic utility of immunohistochemical staining for GLUT1. *Virchows Arch* 2003;443:159–163.

153. Sciot R, Dal Cin P. Cutaneous sclerosing perineurioma with cryptic NF2 gene deletion. *Am J Surg Pathol* 1999;23:849–853.

154. Lasota J, Fetsch JF, Wozniak A, et al. The neurofibromatosis type 2 gene is mutated in perineurial cell tumors: a molecular genetic study of eight cases. *Am J Pathol* 2001;158:1223–1229.

155. Brock JE, Perez-Atayde AR, Kozakewich PW, et al. Cytogenetic aberrations in perineurioma: variation with subtype. *Am J Surg Pathol* 2005;29:1164–1169.

156. Ushigome S, Takakuwa T, Hyuga M, Tadokoro M, Shinagawa T. Perineurial cell tumor and the significance of the perineurial cells in neurofibroma. *Acta Pathol Jpn* 1986;36:973–987.

157. Michal M. Extraneural retiform perineuriomas: a report of four cases. *Pathol Res Pract* 1999;195:759–763.

158. Graadt van Roggen JF, McMenamin ME, Belchis DA, et al. Reticular perineurioma: a distinctive variant of soft tissue perineurioma. *Am J Surg Pathol* 2001;25:485–493.

159. Mentzel T, Kutzner H. Reticular and plexiform perineurioma: clinicopathological and immunohistochemical analysis of two cases and review of perineurial neoplasms of skin and soft tissues. *Virchows Arch* 2005;447:677–682.

160. Tsang WY, Chan JKC, Chow LTC, Tse CCH. Perineurioma: an uncommon soft tissue neoplasm distinct from localized hypertrophic neuropathy and neurofibroma. *Am J Surg Pathol* 1992;16:756–763.

161. Mentzel T, Dei Tos AP, Fletcher CDM. Perineurioma (storiform perineurial fibroma): clinicopathological analysis of four cases. *Histopathology* 1994;25:261–267.

162. Giannini C, Scheithauer BW, Jenkins RB, et al. Soft tissue perineurioma: evidence for an abnormality of chromosome 22, criteria for diagnosis, and review of the literature. *Am J Surg Pathol* 1997;21:164–173.

163. Rankine AJ, Filion PR, Plattewn MA, Spagnolo DV. Perineurioma: a clinicopathologic study of eight cases. *Pathology* 2004;36:309–315.

164. Hornick JL, Fletcher CDM. Soft tissue perineurioma: clinicopathologic analysis of 81 cases including those with atypical histologic features. *Am J Surg Pathol* 2005;29:845–858.

165. Folpe AL, Billings SD, McKenney JK, et al. Expression of claudin-1, a recently described tight junction-associated protein, distinguishes soft tissue perineurioma from potential mimics. *Am J Surg Pathol* 2002;26:1620–1626.

166. Hirose T, Scheithauer BW, Sano T. Perineurial malignant peripheral nerve sheath tumor (MPNST): a clinicopathologic, immunohistochemical, and ultrastructural study of seven cases. *Am J Surg Pathol* 1998;22:1368–1378.

167. Zamecnik M, Michal M. Malignant peripheral nerve sheath tumor with perineurial cell differentiation (malignant perineurioma). *Pathol Int* 1999;49:69–73.

168. Rosenberg AS, Langee CL, Stevens GL, Morgan MB. Malignant peripheral nerve sheath tumor with perineurial cell differentiation: malignant perineurioma. *J Cutan Pathol* 2002;29:362–367.

Granular cell tumor of soft tissues

169. Lack EE, Worsham GF, Callihan MD, et al. Granular cell tumor: a clinicopathologic study of 110 patients. *J Surg Oncol* 1980;13:301–316.

170. Ordonez NG. Granular cell tumor: a review and update. *Adv Anat Pathol* 1999;6:186–203.

171. Damiani S, Koerner FC, Dickersin GR, Cook MG, Eusebi V. Granular cell tumour of the breast. *Virchows Arch A Pathol Anat Histopathol* 1992;420:219–226.

172. Wolber RA, Talerman A, Wilinson EJ, Clement PB. Vulvar granular cell tumors with pseudocarcinomatous hyperplasia: a comparative analysis with well-differentiated squamous carcinoma. *Int J Gynecol Pathol* 1991;10:59–66.

173. Chaudhry AP, Jacobs MS, SunderRaj M, et al. A clinicopathologic study of

50 adult oral granular cell tumors. *J Oral Med* 1984;39:97–103.

174. Johnston J, Helwig EB. Granular cell tumors of the gastrointestinal tract and perianal region: a study of 74 cases. *Dig Dis Sci* 1981;26:807–816.

175. Compagno J, Hyams VJ, Sainte-Marie P. Benign granular cell tumors of the larynx: a review of 36 cases with clinicopathologic data. *Ann Otol Rhinol Laryngol* 1975;84:308–314.

176. Deavers M, Guinee D, Koss MN, Travis WD. Granular cell tumors of the lung: clinicopathologic study of 20 cases. *Am J Surg Pathol* 1995;19:627–635.

177. Murakata LA, Ishak KG. Expression of inhibin-alpha by granular cell tumors of the gallbladder and extrahepatic bile ducts. *Am J Surg Pathol* 2001;25:1200–1203.

178. Melo CR, Melo IS, Schmitt FC, Fagundes R, Amendola D. Multicentric granular cell tumor of the colon: report of a patient with 52 tumors. *Am J Gastroenterol* 1993;88:1785–1787.

179. Rifkin RH, Blocker SH, Palmer JO, Ternberg JL. Multiple granular cell tumors. A familial occurrence in children. *Arch Surg* 1986;121:945–947.

180. Nakazato Y, Ishizeki J, Takahashi K, Yamaguchi H. Immunohistochemical localization of S100-protein in granular cell myoblastoma. *Cancer* 1982;49:1624–1628.

181. Mazur MT, Schultz JJ, Myers JL. Granular cell tumor: immunohistochemical analysis of 21 benign tumors and one malignant tumor. *Arch Pathol Lab Med* 1990;114:692–696.

182. Filie AC, Lage JM, Azumi N. Immunoreactivity of S100-protein, alpha-1-antitrypsin, and CD68 in adult and congenital granular cell tumors. *Mod Pathol* 1996;9:888–892.

183. Miettinen M, Lehtonen E, Lehtola H, et al. Histogenesis of granular cell tumour: an immunohistochemical and ultrastructural study. *J Pathol* 1984;142:221–229.

184. Le BH, Boyer PJ, Lewis JE, Kapadia SB. Granular cell tumor: immunohistochemical assessment of inhibin-α, protein gene product 9.5, S100 protein, CD68, and Ki67 proliferative index with clinical correlation. *Arch Pathol Lab Med* 2004;128:771–775.

185. Carstens PH, Yacoub O. Importance of the angulate bodies in the diagnosis of granular cell tumors (schwannomas). *Ultrastruct Pathol* 1993;17:271–278.

186. Mentzel T, Wadden C, Fletcher CD. Granular cell change in smooth muscle tumours of skin and soft tissue. *Histopathology* 1994;24:223–231.

187. Fanburg-Smith JC, Meis-Kindblom JM, Fante R, Kindblom LG. Malignant granular cell tumor of soft tissue: diagnostic criteria and clinicopathologic correlation. *Am J Surg Pathol* 1998;22:779–794.

188. Cruz-Mojarrieta J, Navarro S, Gomez-Cabrera E, et al. Malignant granular cell tumor of soft tissues: a study of two new cases. *Int J Surg Pathol* 2001;9:255–259.

Gingival granular cell tumor

189. Custer RP, Fust JA. Congenital epulis. *Am J Clin Pathol* 1952;2:1044–1053.

190. Lack EE, Worsham GF, Callihan MD, Crawford BE, Vawter GF. Gingival granular cell tumors of the newborn (congenital "epulis"): a clinical and pathologic study of 21 patients. *Am J Surg Pathol* 1981;5:37–46.

191. Childers EL, Fanburg-Smith JC. Congenital epulis of the newborn: 10 new cases of a rare oral tumor. *Ann Diagn Pathol* 2011;15:157–161.

192. Park SH, Kim TJ, Chi JG. Congenital granular cell tumor with systemic involvement: immunohistochemical and ultrastructural study. *Arch Pathol Lab Med* 1991;115:934–938.

193. Rohrer MD, Young SK. Congenital epulis (gingival granular cell tumor): ultrastructural evidence of origin from pericytes. *Oral Surg Oral Med Oral Pathol* 1982;53:56–63.

194. Zarbo RJ, Lloyd RV, Beals TF, McClatchey KD. Congenital gingival granular cell tumor with smooth muscle cytodifferentiation. *Oral Surg Oral Med Oral Pathol* 1983;56:512–520.

195. Takahashi H, Fujita S, Satoh H, Okade H. Immunohistochemical study of congenital gingival granular cell tumor (congenital epulis). *J Oral Pathol Med* 1990;19:492–496.

196. Tucker MC, Rusnock EJ, Azumi N, Hoy GR, Lack EE. Gingival granular cell tumors of the newborn: an ultrastructural and immunohistochemical study. *Arch Pathol Lab Med* 1990;114:895–898.

197. Schaefer IM, Fletcher CD, Hornicle JL. Loss of H3K27 trimethylation distinguishes malignant peripheral nerve sheath tumor from histologic mimics. *Mod Path* 2016;29:4–13.

Malignant peripheral nerve sheath tumor

198. Ghosh BC, Ghosh L, Huvos AG, Fortner JG. Malignant schwannoma: a clinicopathologic study. *Cancer* 1973;31:184–190.

199. Guccion JG, Enzinger FM. Malignant schwannoma associated with von Recklinghausen's neurofibromatosis. *Virchows Arch A Pathol Anat Histol* 1979;383:43–57.

200. Tsuneyoshi M, Enjoji M. Primary malignant peripheral nerve tumors (malignant schwannomas): a clinicopathologic and electron microscopic study. *Acta Pathol Jpn* 1979;29:363–375.

201. Matsunou H, Shimoda T, Kakimoto S, et al. Histopathologic and immunohistochemical study of malignant tumors of peripheral nerve sheaths (malignant schwannoma). *Cancer* 1985;56:2269–2279.

202. Ducatman BS, Scheithauer BW, Piepgras DW, Reiman HM, Ilstrup DM. Malignant peripheral nerve sheath tumors: a clinicopathologic study of 120 cases. *Cancer* 1986;57:2006–2021.

203. Vauthey JN, Woodruff JM, Brennan MF. Extremity malignant peripheral nerve sheath tumors (neurogenic sarcomas): a 10-year experience. *Ann Surg Oncol* 1995;2:126–131.

204. Hruban RH, Shiu MH, Senie RT, Woodruff JM. Malignant peripheral nerve sheath tumors of the buttock and lower extremity: a study of 43 cases. *Cancer* 1990;66:1253–1265.

205. Ducatman BS, Scheithauer BW, Piepgras DG, Reiman HM. Malignant peripheral nerve sheath tumors in childhood. *J Neurooncol* 1984;2:241–248.

206. Meis JM, Enzinger FM, Martz KL, Neal JA. Malignant peripheral nerve sheath tumors (malignant schwannomas) in children. *Am J Surg Pathol* 1992;16:694–707.

207. Ramanathan RC, Thomas JM. Malignant peripheral nerve sheath tumors associated with von Recklinghausen's neurofibromatosis. *Eur J Surg Oncol* 1999;25:190–193.

208. Kourea HP, Bilsky MH, Leung DHY, Lewis JJ, Woodruff JM. Subdiaphragmatic and intrathoracic paraspinal malignant peripheral nerve sheath tumors: a clinicopathologic study of 25 patients and 26 tumors. *Cancer* 1998;82:2191–2203.

209. Foley KM, Woodruff JM, Ellis FT, Posner JB. Radiation-induced malignant and atypical peripheral nerve sheath tumors. *Ann Neurol* 1980;7:311–318.

210. Ducatman BS, Scheithauer BW. Postirradiation neurofibrosarcoma. *Cancer* 1983;51:1028–1033.

211. Meis-Kindblom JM, Enzinger FM. Plexiform malignant peripheral nerve sheath tumor of infancy and childhood. *Am J Surg Pathol* 1994;18:479–485.

212. Woodruff JM, Erlandson RA, Scheithauer BW. Nerve sheath tumors: letter to the editor. *Am J Surg Pathol* 1995;19:608–609.

213. Daimaru Y, Hashimoto H, Enjoji M. Malignant peripheral nerve sheath tumors (malignant schwannomas): an immunohistochemical study of 29 cases. *Am J Surg Pathol* 1985;9:434–444.

214. Wick MR, Swanson PE, Scheithauer BW, Manivel JC. Malignant peripheral nerve sheath tumors: an immunohistochemical study of 62 cases. *Am J Clin Pathol* 1987;87:425–433.

215. Herrera GA, de Moraes PH. Neurogenic sarcoma in patients with neurofibromatosis (von Recklinghausen's disease): light, electron microscopy and immunohistochemistry study. *Virchows Arch A Pathol Anat Histopathol* 1984;403:361–376.

216. Jhanwar SC, Chen Q, Li FP, Brennan MF, Woodruff JM. Cytogenetic analysis of soft tissue sarcomas: recurrent chromosome abnormalities in malignant peripheral nerve sheath tumors (MPNST). *Cancer Genet Cytogenet* 1994;78:138–144.

217. Plaat BEC, Molenaar WM, Mastik MF, et al. Computer-assisted cytogenetic analysis of 51 malignant peripheral-nerve-sheath-tumors: sporadic versus neurofibromatosis-type-1 associated malignant schwannomas. *Int J Cancer* 1999;83:171–178.

218. Mertens F, Dal Cin P, de Wever I, et al. Cytogenetic characterization of peripheral nerve sheath tumours: a report of the CHAMP study group. *J Pathol* 2000;190:31–38.

219. Kourea HP, Orlow I, Scheithauer BW, Cordon-Cardo C, Woodruff JM. Deletions of the INK4A gene occur in malignant peripheral nerve sheath tumors but not in neurofibromas. *Am J Pathol* 1999;155:1855–1860.

220. Nielsen GP, Stemmer-Rachamimov AO, Ino Y, et al. Malignant transformation of neurofibromas in neurofibromatosis 1 is associated with CDKN2A/p16 inactivation. *Am J Pathol* 1999;155:1879–1884.

221. Kourea HP, Cordon-Cardo C, Dudas M, Leung D, Woodruff JM. Expression of p27 (kip) and other cell cycle regulators in malignant peripheral nerve sheath tumors and neurofibromas: the emerging role of p27 (kip) in malignant transformation of neurofibroma. *Am J Pathol* 1999;155:1885–1891.

Malignant peripheral nerve sheath tumors with divergent differentiation

222. Brooks JSJ, Freeman M, Enterline HT. Malignant "Triton" tumors: natural history and immunohistochemistry of nine new cases with literature review. *Cancer* 1985;55:2543–2549.

223. Woodruff JM, Chernik NL, Smith MC, Millett WB, Foote FW. Peripheral nerve tumors with rhabdomyosarcomatous differentiation (malignant "triton" tumors). *Cancer* 1973;32:426–439.

224. Daimaru Y, Hashimoto H, Enjoji M. Malignant "triton" tumors: a clinicopathologic and immunohistochemical study of nine cases. *Hum Pathol* 1984;15:768–778.

225. Ducatman BS, Scheithauer BW. Malignant peripheral nerve sheath tumors with divergent differentiation. *Cancer* 1984;54:1049–1057.

226. Woodruff JM. Peripheral nerve sheath tumors showing glandular differentiation (glandular schwannomas). *Cancer* 1976;37:2399–2413.

227. Christensen WN, Strong EW, Bains MS, Woodruff JM. Neuroendocrine differentiation in glandular peripheral nerve sheath tumor: pathologic distinction from the biphasic synovial sarcoma with glands. *Am J Surg Pathol* 1988;12:417–426.

Epithelioid malignant peripheral nerve sheath tumor

228. Lodding P, Kindblom LG, Angervall L. Epithelioid malignant schwannoma: a study of 14 cases. *Virchows Arch A Pathol Anat Histopathol* 1986;409:433–451.

229. DiCarlo EF, Woodruff JM, Bansal M, Erlandson RA. The purely epithelioid malignant peripheral nerve sheath tumor. *Am J Surg Pathol* 1986;10:478–490.

230. Laskin WB, Weiss SW, Bratthauer GL. Epithelioid variant of malignant peripheral nerve sheath tumor (malignant epithelioid schwannoma). *Am J Surg Pathol* 1991;15:1136–1145.

231. Jo VY, Fletcher CD. Epithelioid malignant peripheral nerve sheath tumor: clinicopathologic analysis of 53 cases. *Am J Surg Pathol* 2015;39:673–682.

Malignant peripheral nerve sheath tumors arising from ganglioneuroma and pheochromocytoma

232. Ricci A, Parham DM, Woodruff JM, et al. Malignant peripheral nerve sheath tumors arising from ganglioneuromas. *Am J Surg Pathol* 1984;8:19–29.

233. Fletcher CD, Fernando IN, Braimbridge MV, McKee PH, Lyall JR. Malignant nerve sheath tumor arising in a ganglioneuroma. *Histopathology* 1988;12:445–448.

234. Damiani S, Manetto V, Carrillo G, et al. Malignant peripheral nerve sheath tumor arising in a de novo ganglioneuroma. *Tumori* 1991;77:90–93.

235. Min KW, Clemens A, Bell J, Dick H. Malignant peripheral nerve sheath tumor and pheochromocytoma: a composite tumor of the adrenal. *Arch Pathol Lab Med* 1988;112:266–270.

236. Miettinen M, Saari A. Pheochromocytoma combined with malignant schwannoma: unusual neoplasm of the adrenal medulla. *Ultrastruct Pathol* 1988;12:513–527.

Chapter 25

Neuroectodermal tumors: melanocytic, glial, and meningeal neoplasms

Markku Miettinen
National Cancer Institute, National Institutes of Health

Although most neuroectodermal tumors are not soft tissue sarcomas, they are important in the differential diagnosis of primary soft tissue tumors. Neuroectodermal tumors discussed here include soft tissue nevi, metastatic melanoma, and clear cell sarcoma. Cutaneous nevi and ordinary cutaneous melanoma are not discussed here, but the reader is referred to other dermatopathology texts such as the Armed Forces Institute of Pathology (AFIP) atlas of melanocytic tumors of the skin.[1] Pigmented neuroectodermal tumor of infancy, extra-central nervous system gliomas, and meningothelial neoplasms relevant to the soft tissues are also discussed. Primitive neuroectodermal tumors belong to the Ewing sarcoma family of tumors and are discussed in Chapter 31.

Cellular blue nevus

The *cellular blue nevus* (CBN) is a relatively uncommon melanocytic neoplasm that can form a soft tissue mass. It should be distinguished from malignant melanoma and soft tissue sarcoma, especially clear cell sarcoma and malignant peripheral nerve sheath tumor (MPNST). Atypical and malignant forms occur, however, and the latter especially must be distinguished from the indolent variants.

Clinical features

CBN typically forms a bluish-black cutaneous nodule of 1 cm to 2 cm that can be larger, but rarely >5 cm; on occasion the nodule is ulcerated. This tumor often occurs in young adults, occasionally on a congenital basis, and less often in older adults. The average age in the largest series was 33 years, and there was a >2:1 female predominance.[2] Other series have shown an even gender distribution, also with a median age in the early thirties.[1,3]

CBN most often occurs in the buttocks and lower back, in the sacrococcygeal region. Less common sites include the scalp and dorsum of the foot and hand. Among rare sites are the proximal upper extremities, trunk wall (chest), gynecologic tract (e.g., the vagina and cervix), spermatic cord region, and breast.[1–3] Ordinary CBN are benign, and complete conservative excision is sufficient.

Rare atypical and malignant variants have been reported, and these tumors behave in a manner similar to aggressive melanomas, with a high metastatic rate and tumor-related mortality.[4,5] Atypical examples without overtly malignant features (mildly increased mitotic activity) have behaved in a benign manner and therefore can be treated similarly to ordinary CBN, although complete excision with negative margins and follow-up are advisable. The overtly malignant examples should be treated in a manner similar to melanoma, with complete wide excision and attention to lymph node status (i.e., with sentinel biopsies or sampling of suspicious nodes).

Pathology

Typical CBN involves the dermis and subcutis, often forming a circumscribed subcutaneous mass. The deeper element is sometimes connected to the superficial component, creating a dumbbell-shaped lesion. A typical example is composed of rounded nodules surrounded by a fascicular spindle cell component. Around the nodules there is often a loose or hemorrhagic matrix containing deeply pigmented melanophages. A superficial ordinary blue nevus component is often detectable (Figure 25.1). The spindle cells are uniform, having tapered nuclei with open chromatin and small nucleoli (Figure 25.2). Multinucleated giant cells with a wreath-like nuclear arrangement are sometimes present. Mitotic activity is inconspicuous, generally ≤1 per 10 HPFs. Lymph node involvement (often focal and capsular) has been detected in some CBNs, and this does not seem to have any adverse significance. Melanin-poor variants have been reported as part of the spectrum.[6]

Immunohistochemical studies typically show a full complement of melanocytic markers: S100 protein, HMB45, melan A, tyrosinase, and microphthalmia transcription factor.

Atypical and malignant variants and differential diagnosis

Variants with increased mitotic rate (1 to 2 per 10 HPFs) and focal atypia, but without other worrisome features are designated as atypical CBNs. Examples with significant nuclear atypia, prominent nucleoli, elevated mitotic activity >2 per 10 HPFs often including atypical mitoses, infiltrative growth,

Modern Soft Tissue Pathology, Second Edition, ed. Markku Miettinen. Published by Cambridge University Press. © Cambridge University Press 2016.

Chapter 25: Neuroectodermal tumors

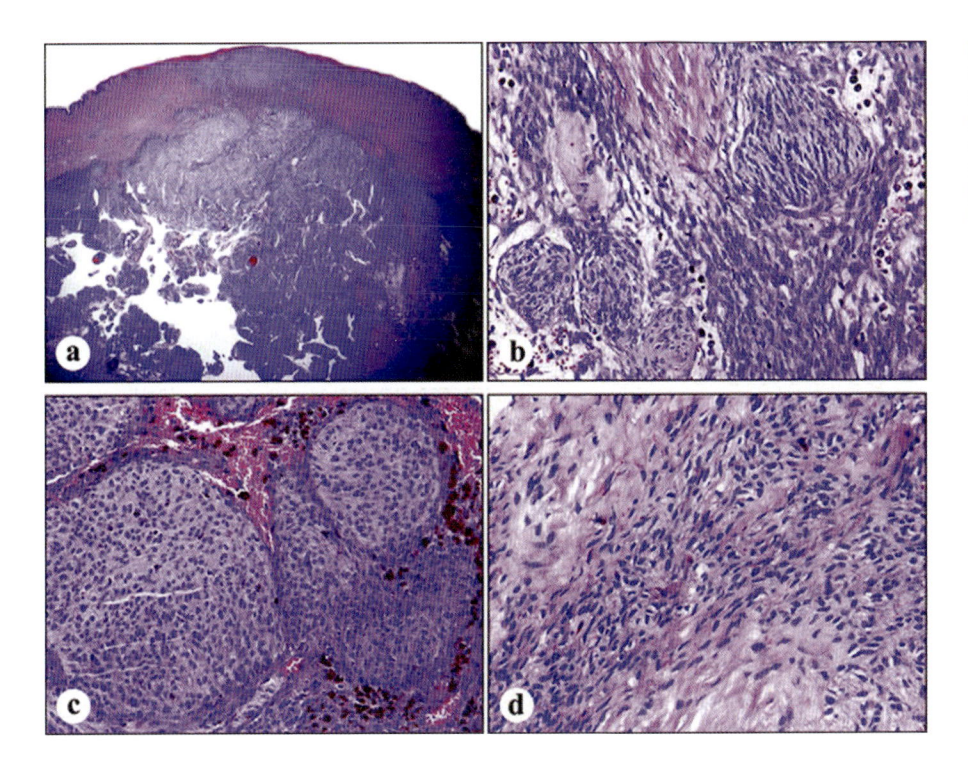

Figure 25.1 (a–c) Cellular blue nevus (CBN) forms a cutaneous and subcutaneous nodule underneath an intact epidermis. The tumor is composed of multiple round nodules surrounded by fascicles of spindle cells, and between the nodules there are clusters of melanophages in a loose or hemorrhagic stroma. (d) A conventional dermal blue nevus component is often present.

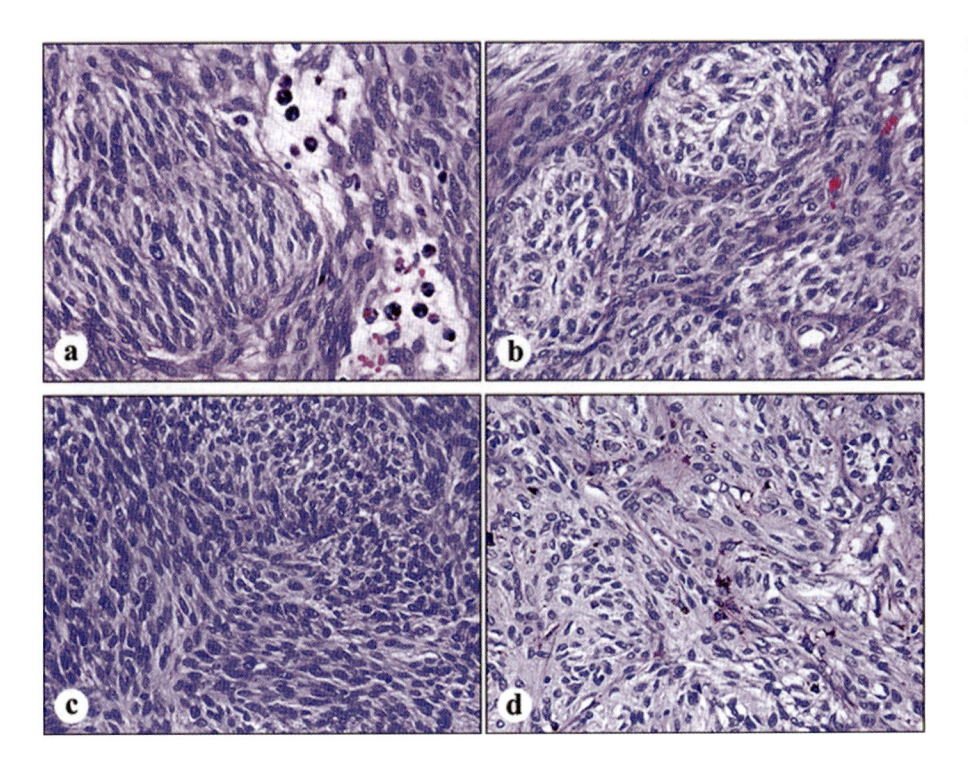

Figure 25.2 (a–c) Cytologic detail of a CBN. The cells have oval, uniform nuclei with small nucleoli. (d) Blue-nevus-like components show similar cytology.

and in some cases, tumor necrosis, are designated as malignant CBNs (Figure 25.3).[4–7] Reliable application of these criteria is difficult, however, because interobservational agreement among experts is far from complete.[8] A practical solution to these problems of definition is to treat worrisome examples by complete excision with documented negative margins, combined with a long-term follow-up.

Metastatic melanoma

Metastatic malignant melanoma is a common problem in pathology. Melanoma metastases can present at almost any surgical site, sometimes long after the primary diagnosis. This makes it often difficult to connect the metastasis with the primary tumor. The wide variety of morphologic

695

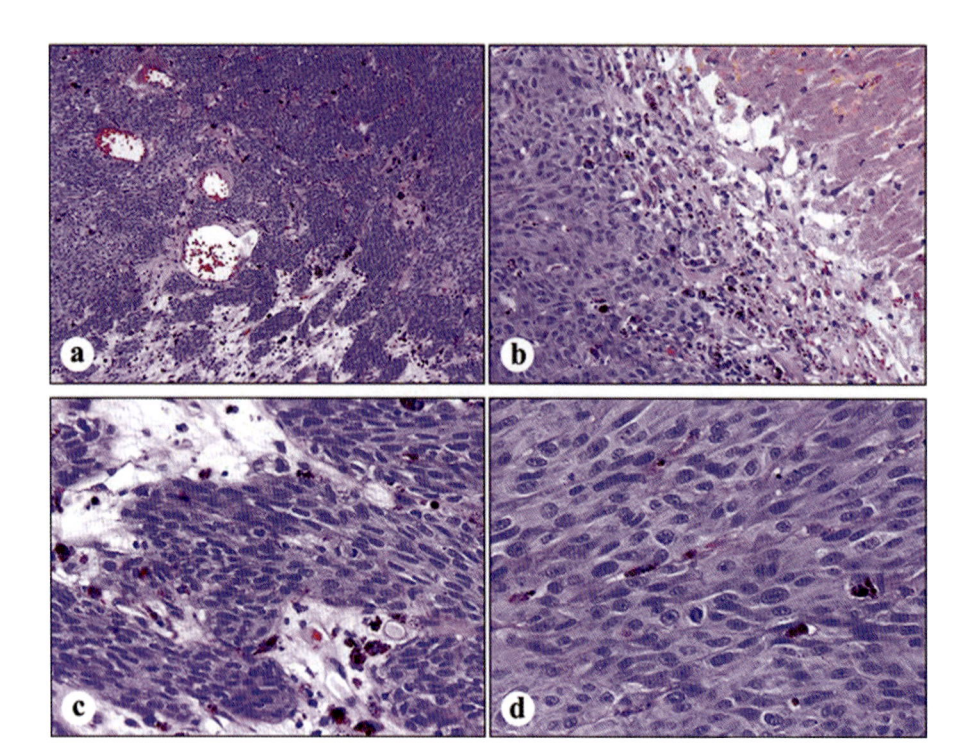

Figure 25.3 Malignant CBN. (**a**) Architectural appearance shows overall resemblance to ordinary CBN, but compartment formation is less distinct. (**b**) Tumor coagulation necrosis is present. (**c**) Mitotic activity is easily detectable. (**d**) Most cells have prominent, eosinophilic nucleoli.

appearances of malignant melanoma contributes to the problem.[9] In the author's experience, this spectrum is even wider in metastatic melanomas. Metastatic melanoma must always be considered in the differential diagnosis of poorly differentiated tumors at virtually all sites. In one study, positron emission tomography (PET) imaging detected soft tissue metastases of melanoma in nearly 10% of patients.[10]

Clinical features

Melanoma metastases occur equally in men and women of all ages, but are uncommon in children and in persons having dark skin color. Metastases can develop at varying amounts of time after the primary tumor, and a long delay, even 10 to 15 or more years, is not uncommon.

These metastases are common in regional lymph nodes, and the skin and subcutis, but they can also occur at unexpected locations. Axillary and inguinal lymph node metastasis can develop into large conglomerates with extensive or predominant extranodal involvement, further introducing the clinical picture of a soft tissue sarcoma. Cerebral, gastrointestinal, osseous, and pulmonary metastases are relatively common, whereas other parenchymal metastases, involving the liver, pancreas, and ovaries, are encountered more rarely as surgical specimens and sometimes mimic primary tumors of those sites.

Some metastatic melanomas have no known primary tumor. The search for occult cutaneous, ocular, or mucosal melanoma is warranted in search of the primary tumor. Explanations offered for lack of a primary tumor after a thorough search include a previously undiagnosed melanoma (e.g., a discarded "nevus" with no histologic examination), regression of the cutaneous primary melanoma, and origin from nodal melanocytic rests.

Paradoxically, large primary cutaneous melanomas can be difficult to identify as such, and they can be confused with soft tissue sarcomas such as clear cell sarcoma or MPNST, if junctional involvement is not detected (Figure 25.4).

Pathology

Grossly, melanoma metastases form fleshy nodules and masses, which vary from black pigmented to white nonpigmented. Many are tan to brown, reflecting mild melanin production. In some cases, the entire lesion is pigmented, whereas in others, small pigmented areas are present in an otherwise pale tumor.

The histologic variation in metastatic melanoma is extensive, and resemblance to a wide variety of malignant tumors is possible, including various carcinomas, sarcomas (e.g., MPNSTs[11,12]), and lymphomas. An epithelioid pattern is common, and there is often resemblance to an endocrine or other tumor with organoid pattern, such as an adrenal cortical carcinoma, alveolar soft part sarcoma, or paraganglioma (Figure 25.5).

Perivascular sparing and trabecular endocrine carcinoma-like architectural arrangements are common. A diffuse spindle cell pattern can resemble that of an MPNST, and rhabdoid cytology can invoke similarity to rhabdoid sarcomas: rhabdoid tumor and epithelioid sarcoma with rhabdoid cytology (Figure 25.6). In rare cases, fat-like vacuolization and lipoblast-like cells can give an impression of liposarcoma, whereas an alveolar discohesive pattern can resemble that seen

Chapter 25: Neuroectodermal tumors

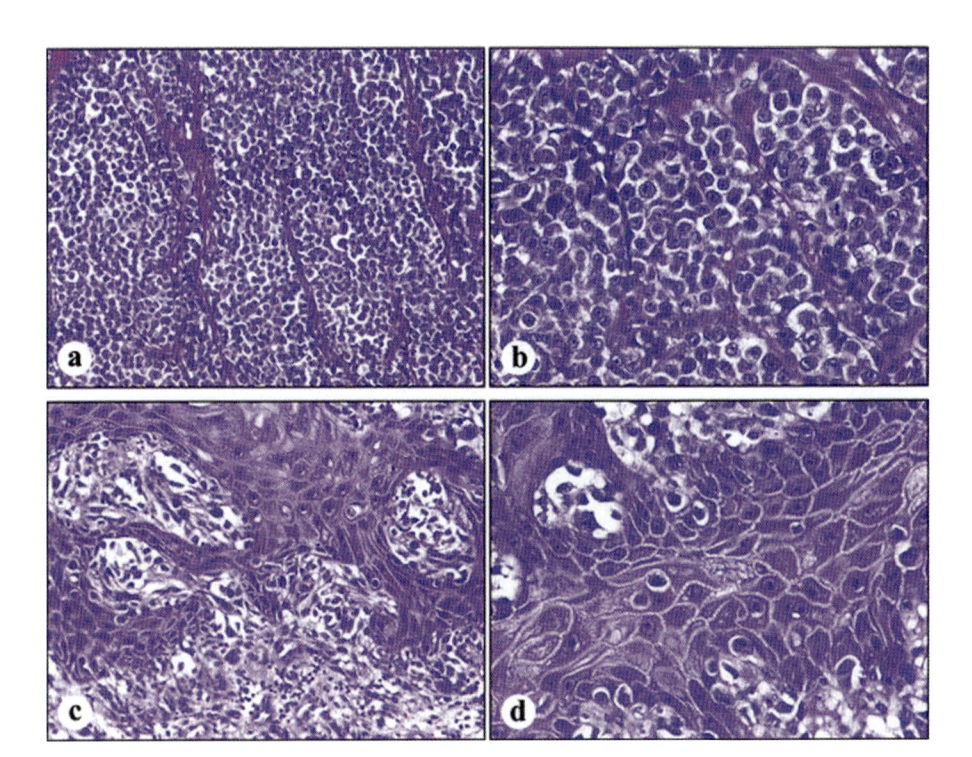

Figure 25.4 (a,b) A large cutaneous melanoma can resemble clear cell sarcoma or epithelioid sarcoma. (c,d) Detection of junctional activity helps to identify this tumor as a primary cutaneous melanoma.

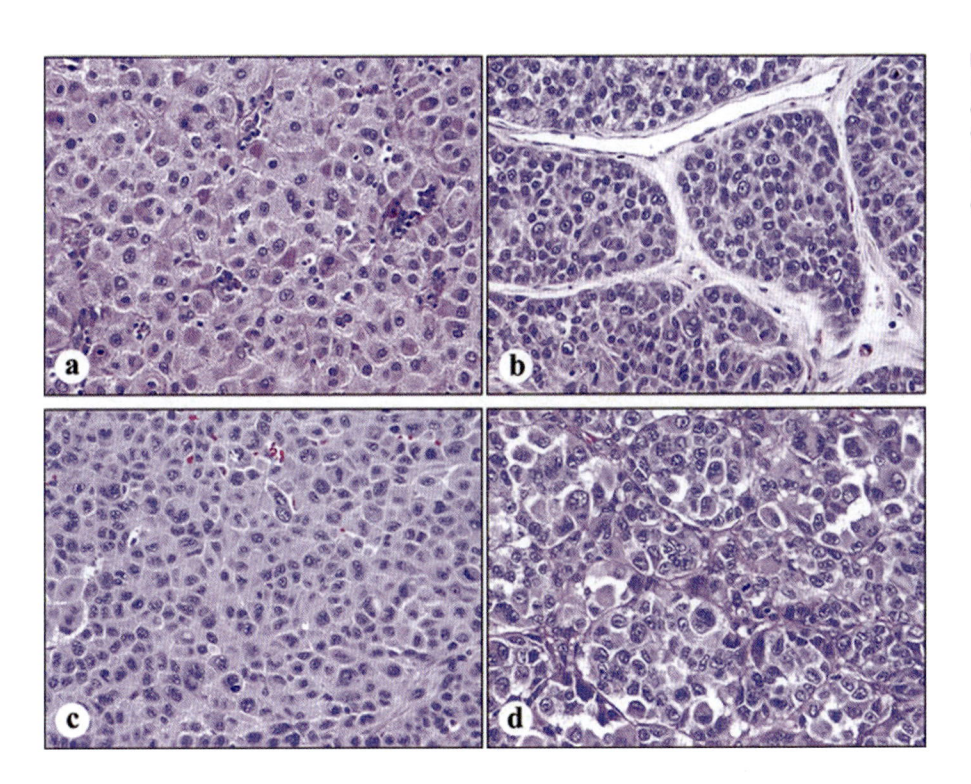

Figure 25.5 Four metastatic melanomas showing histologic patterns resembling the following endocrine-like tumors or carcinomas. (a) Adrenal cortical carcinoma. (b) Alveolar soft part sarcoma. (c) Poorly differentiated carcinoma. (d) Resemblance to paraganglioma by a compartmental pattern.

in alveolar rhabdomyosarcoma. Neuropil-like material can give the impression of neuroepithelial differentiation (Figure 25.7). Occasionally, osteoclastic giant cells can mask the presence of a melanoma population and give the impression of a giant cell tumor of soft parts or giant cell malignant fibrous histiocytoma (MFH; Figure 25.8). Marked pleomorphism is uncommon, but when it occurs, it resembles pleomorphic sarcoma or lymphoma (Figure 25.9). By relative lack of cell cohesion and cellular morphology, metastatic melanoma can also resemble various hematopoietic tumors, such as a diffuse lymphoma, histiocytic neoplasm, Hodgkin's disease, and plasmacytoma (Figure 25.10).

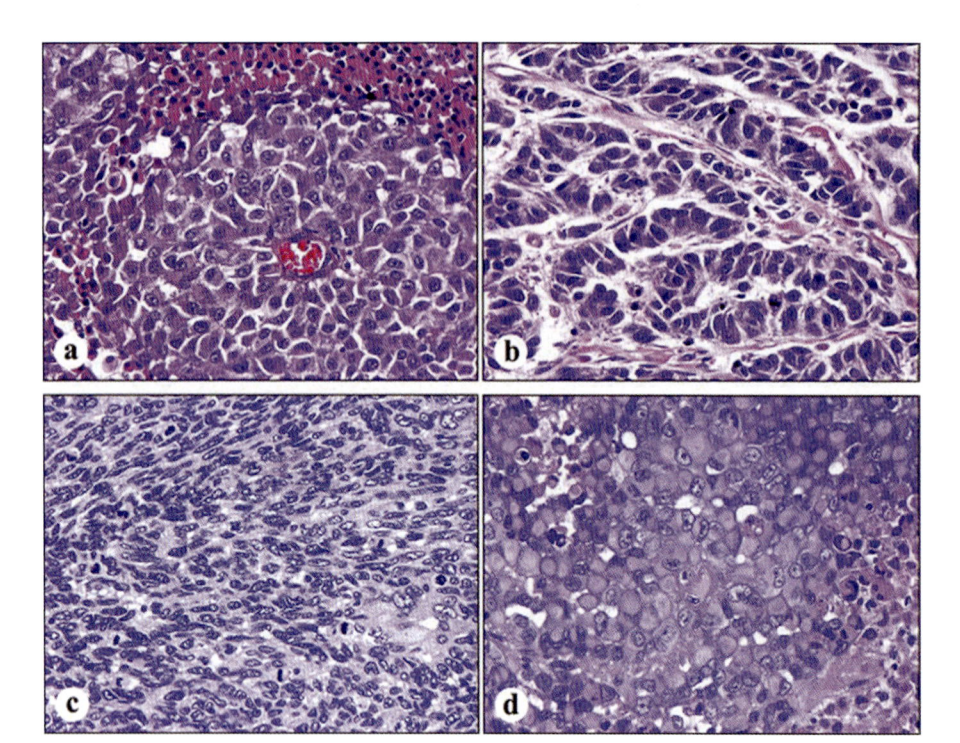

Figure 25.6 (**a**) Perivascular sparing pattern in metastatic melanoma. (**b**) An endocrine carcinoma-like ribbon pattern. (**c**) Spindle cell sarcomatous appearance resembling a high-grade MPNST. (**d**) Rhabdoid cells showing cytologic resemblance to rhabdoid tumor or epithelioid sarcoma cells.

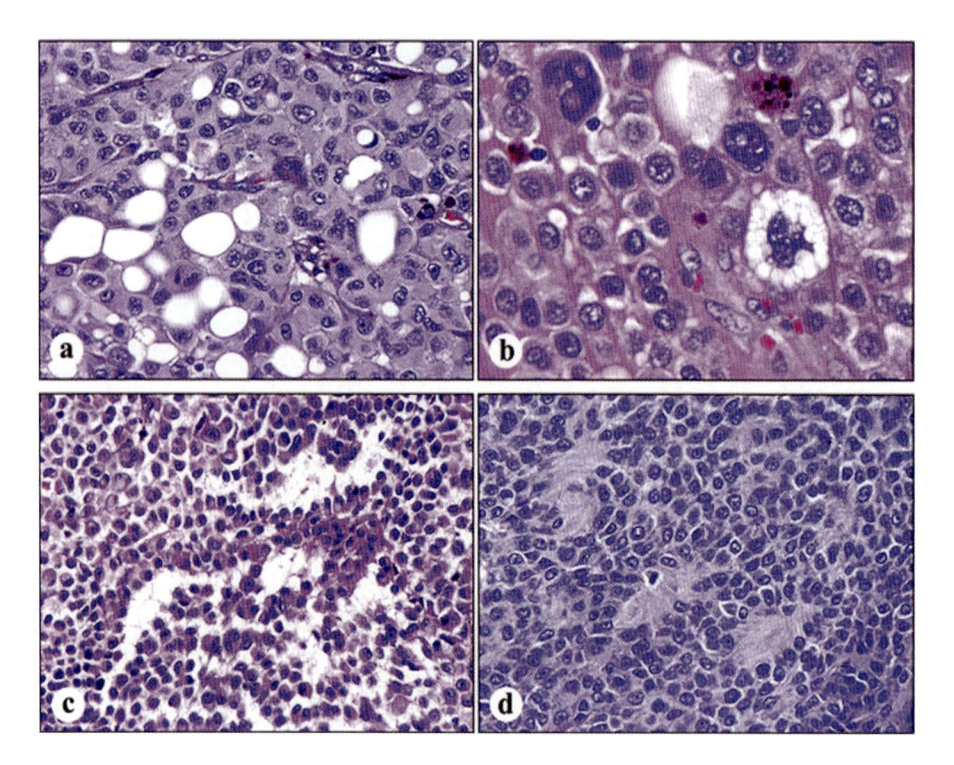

Figure 25.7 (**a,b**) Metastatic melanoma showing vacuolization and lipoblast-like cells resembling liposarcoma. (**c**) Alveolar pattern resembling alveolar rhabdomyosarcoma. (**d**) Neuropil-like material in metastatic melanoma.

Eosinophilic coagulation necrosis is common and can be extensive, leaving only preserved tumor cells as narrow perivascular collars. Glomeruloid vascular proliferation, similar to that seen in MPNST, glioblastoma, and neuroendocrine tumors, can occur.[13] The pigmentation seen in tumor cells or streaks of melanophages is helpful, if present.

Cytologically, metastatic melanoma often has well-defined cell borders. The cytoplasm varies from pale, variably basophilic to deeply eosinophilic. The nuclei typically have complex outlines with grooves and cleavages sometimes resulting in nuclear pseudoinclusions (cytoplasmic invaginations), similar to those seen in Schwann cell tumors. The nucleoli are often, but not always, prominent.

Chapter 25: Neuroectodermal tumors

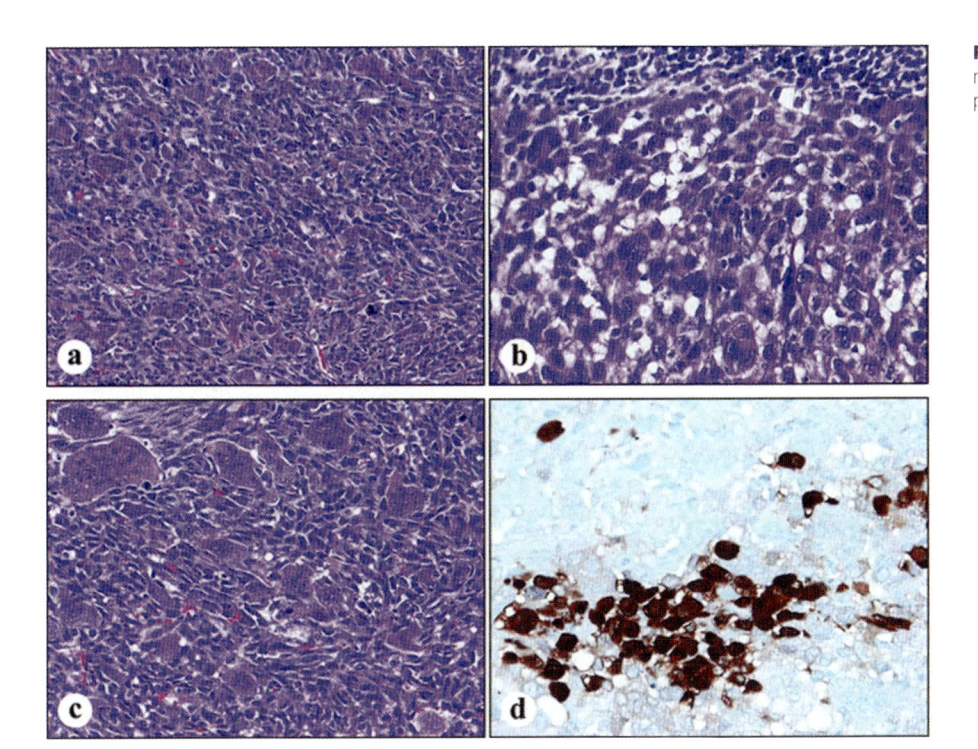

Figure 25.8 (a–c) Osteoclast-like giant cells in metastatic melanoma. (d) Melanoma cell population is S100 protein positive.

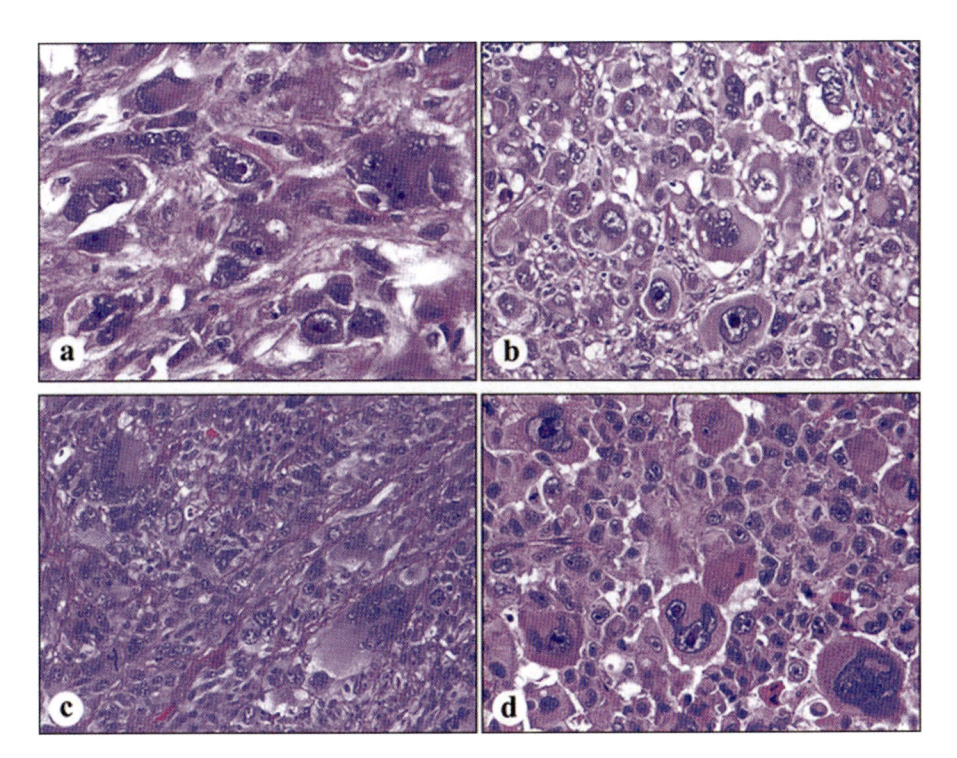

Figure 25.9 Pleomorphism in metastatic melanoma simulating MFH or a pleomorphic lymphoma.

Immunohistochemistry

Nearly all metastatic melanomas are strongly and globally S100 protein positive, but a small percentage of them show reduced or no S100 protein expression. Only 3% to 4% of metastatic melanomas are S100 protein negative (Figure 25.11). Fortunately, such cases show variable positivity for other melanoma markers, so that when there is a significant suspicion of metastatic melanoma from the histologic appearance or clinical history, other melanoma markers should be tested in S100-protein-negative tumors.[6,7] Sox10 is an excellent alternative marker for melanoma, as well as other melanocytic, Schwann cell, and myoepithelial tumors.[14,15]

699

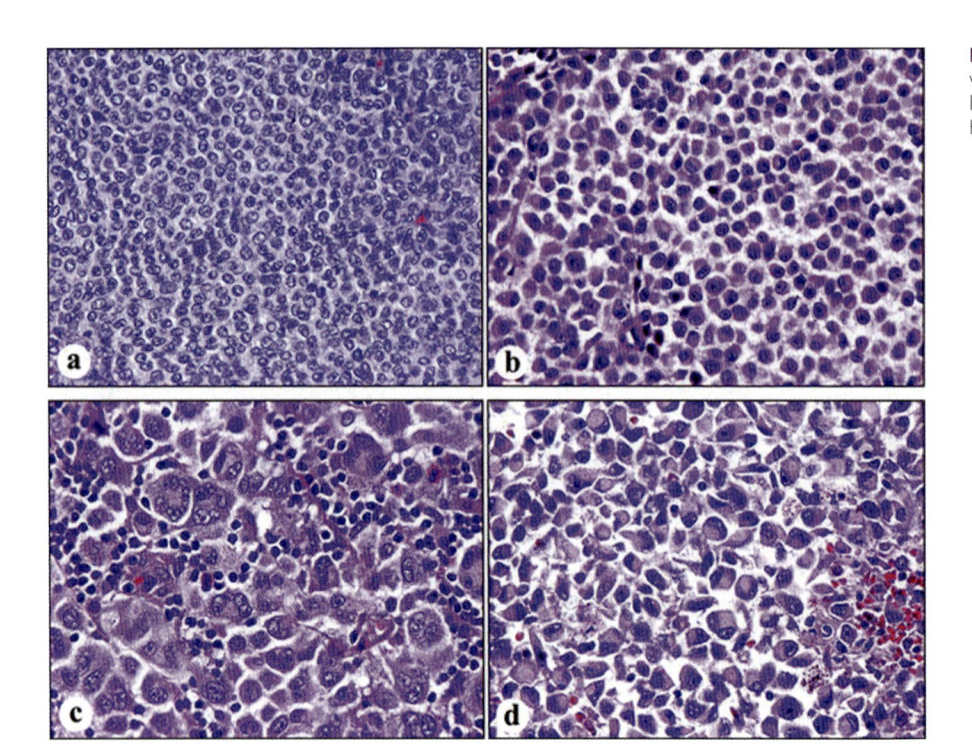

Figure 25.10 Metastatic melanoma simulating various hematopoietic tumors. (**a**) Diffuse large cell lymphoma. (**b**) Histiocytic sarcoma. (**c**) Pleomorphic Hodgkin's disease. (**d**) Plasmacytoma.

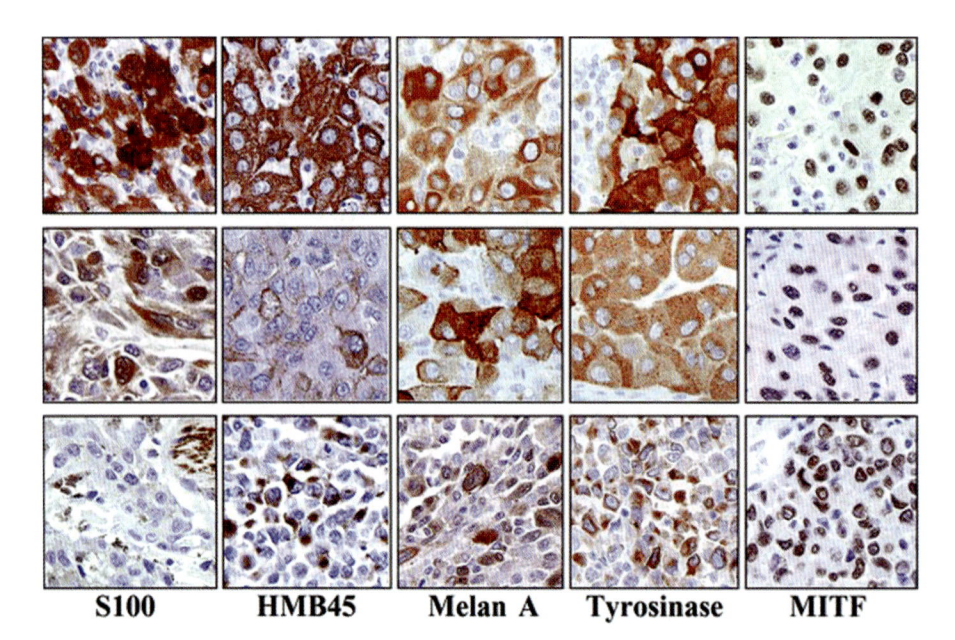

Figure 25.11 Examples of immunostains for five melanoma markers on three metastatic melanomas. The rows represent results of one case, and the columns represent different immunostains, as specified at the bottom. Note that S100 protein positivity is seen in both the cytoplasm and nuclei, and rare cases are negative. Positivity for HMB45, melan A, and tyrosinase is often granular. In some tumors, positivity appears as cytoplasmic dots. Nuclear microphthalmia transcription factor positivity is seen in most metastatic melanomas.

Of other newer markers, KBA62 is less specific and PNL2 less sensitive, resembling HMB45 in performance.

HMB45-positive cells are present in 70% to 80% of cases, but often only focally. Melan A and tyrosinase are present in 75% of cases. Microphthalmia transcription factor is present in most metastatic melanomas, although this marker is typically absent in desmoplastic melanomas, as are HMB45, melan A, and tyrosinase.[16,17]

In some metastases (20%), especially sarcomatoid spindle cell and rhabdoid ones, S100 protein is the only positive marker (Figure 25.12). In some cases, foci of more differentiated positive components selectively positive for melanocyte-specific markers are present (Figure 25.13). These markers should be used for S100-protein-negative tumors suspected of melanoma, because some of these tumors are positive for additional melanoma markers.

Keratin positivity, especially keratin 18, is seen in 20% of metastatic melanomas, but there is no keratin 19, which can be helpful in the differential diagnosis between melanoma and metastatic carcinoma. The presence of keratins 8 and 18 in

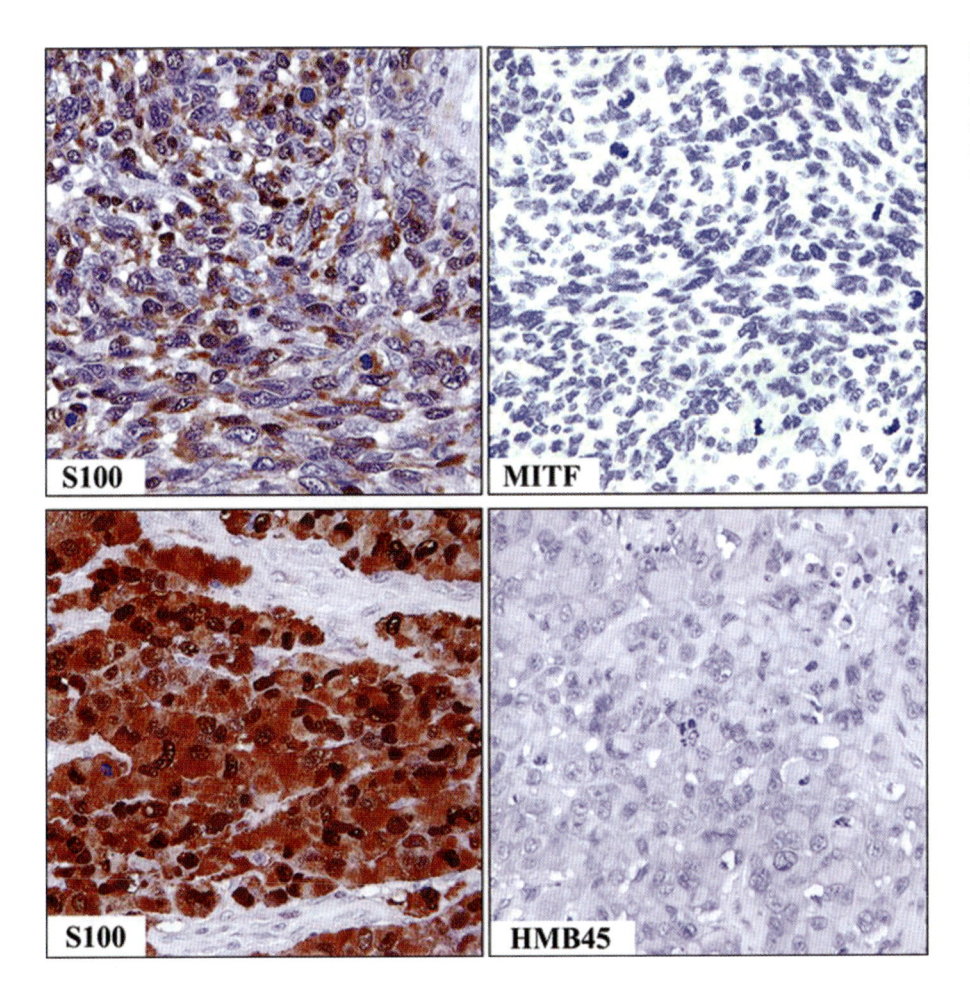

Figure 25.12 Two examples of metastatic melanomas that were negative for all markers other than S100 protein. The upper panel depicts a sarcomatous spindle cell melanoma. The lower panel shows a melanoma with cytoplasmic rhabdoid inclusions.

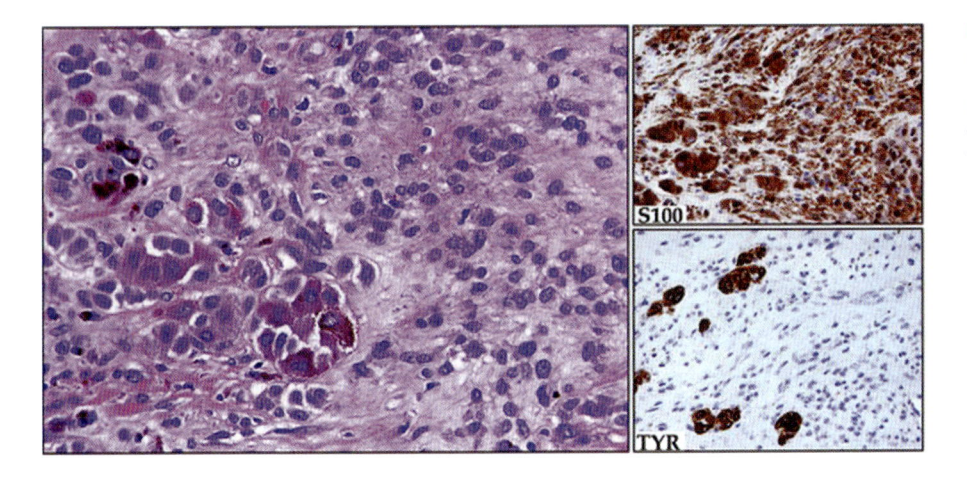

Figure 25.13 Examples of different marker positivity in various components of metastatic melanoma. All tumor cells are positive for S100 protein in this melanoma, combining spindle cell and epithelioid components. (TYR) Only the epithelioid clusters are positive for tyrosinase.

metastatic melanoma has also been confirmed by Western blotting.[18] Similar to many carcinomas, melanoma metastases are often positive for E-cadherin.

The common expression of histiocytic markers in melanomas, such as alpha-1-antitrypsin, alpha-1-antichymotrypsin, and CD68, should not lead to confusion with true histiocytic tumors. Microphthalmia transcription factor also can be expressed in mononuclear, as well as multinucleated histiocytes.

Differential diagnosis

When evaluating a strongly S100-protein-positive malignant soft tissue tumor, malignant melanoma must be a serious consideration. In fact, alternative diagnoses are far less common (e.g., clear cell sarcoma, MPNST, or rare S100-protein-positive carcinomas, such as myoepithelial ones).

Metastatic melanoma should be ruled out before a superficial epithelioid MPNST is diagnosed. In fact, the latter diagnosis should essentially be restricted to cases in which elements of pre-existing neurofibroma are seen adjacent to the malignant component.

Clear cell sarcoma is a clinically and histologically distinctive related tumor, which typically has an organoid pattern divided by fibrous septa. In problematic cases, demonstration of EWSR1 gene rearrangement by fluorescence *in situ* hybridization (FISH) or polymerase chain reaction (PCR) can support clear cell sarcoma.

Psammomatous melanotic schwannoma and certain variants of MPNSTs can closely simulate melanoma phenotypically by their S100 protein positivity. The presence of psammoma bodies is typical of the former, however, and the latter do not express melanocytic markers such as HMB45, melan A, and tyrosinase. The clinical context can also be helpful.

Epithelioid sarcoma has typically large cells with abundant eosinophilic cytoplasm. In contrast to melanoma, S100 protein positivity is rare, and keratin positivity is nearly uniform.

Myoepithelial carcinoma (malignant mixed tumor) usually has a differentiated epithelial component that is keratin positive. Although prominent S100 protein positivity is a shared feature, the presence of melanoma-specific markers is unique to malignant melanoma.

Interdigitating reticulum cell sarcoma is sometimes considered for S100-protein-positive nodal malignancies, especially when the tumor cells intermingle with lymphocytes. Compared with melanoma, this tumor is extremely rare. Its identification can be aided by the demonstration of leukocyte-specific antigens and by clinical correlation.

Genetics

Malignant melanomas typically show complex cytogenetic changes, and no specific diagnostic features have emerged. They do not have the t(12;22) translocation seen in clear cell sarcoma. Melanomas often have mutations in the BRAF protein kinase gene.[19,20] BRAF V600E mutation is the most common, and there is an antibody to detect mutant protein. However, diagnostically, BRAF mutation detection may have limitations, because some MPNSTs and many other tumors of diverse histogenesis, can have the same mutation.[19–21]

Desmoplastic melanoma

Desmoplastic melanoma, originally delineated by Conley *et al.*, is a rare melanoma subtype that typically occurs in the dermis and subcutis and contains dense fibrous stroma.[22] Its variant, neurotropic melanoma, described by Reed and Leonard, has a propensity to infiltrate along small nerves.[22] Approximately one-third of desmoplastic melanomas do not contain an atypical junctional component, and they appear as primary soft tissue tumors. Most desmoplastic melanomas are amelanotic and therefore subject to diagnostic difficulty and underdiagnosis.

Clinical features

Desmoplastic melanoma generally occurs at an older age than conventional melanoma. The median patient age in the largest series has been approximately 60 years. Occasional cases have been reported in children, especially in connection with xeroderma pigmentosum. Most series have shown approximately 2:1 male predominance.

Most desmoplastic melanomas occur in the head and neck, especially face and cheek. Nearly any anatomic site can be involved, however, with a predilection for sun-exposed areas in the extremities and upper trunk.[22–33] Presentation in the oral mucosa has been reported rarely.[34]

Clinically, the lesion often develops from a superficial lentigo-maligna-type melanoma, which sometimes precedes the desmoplastic melanoma by 10 to 20 years. In up to one-third of cases, no preceding or simultaneous melanocytic lesion is present, and the desmoplastic melanoma is considered to have arisen *de novo*. The lesion typically forms a shallow tumor-like elevation below intact skin. Its generally nonpigmented nature does not give any clinical indication of melanoma, and therefore the diagnosis is often greatly delayed.

Earlier reports showed a recurrence rate >50%, which is down to 10% to 15% with wide excision. Lymph node metastasis occurs, but is less common than in conventional melanoma. Nevertheless, sentinel node biopsy has been advocated, and in one study, this yielded positive nodes in 4 of 33 patients without clinical evidence for metastasis.[35] Distant metastasis to the lungs, liver, and central nervous system can develop. Although earlier series showed a grim prognosis, probably because many tumors were diagnosed at an advanced stage, the latest series show 5-year survival rates as high as 70% to 75% and demonstrate better survival than for conventional melanomas of similar thickness. Female patients might have better survival rates.[29,30,33]

Pathology

On sectioning, the lesion appears as a dense, plaque-like dermal-subcutaneous expansion, which is usually classified as being Clark level IV or V. The reported tumors have usually been 2 mm to 10 mm thick, and laterally measured 1 cm to 3 cm, but superficial and early lesions are being recognized. Despite their small overall dimensions (5 mm maximum diameter), microscopic deep extension around adnexa and nerves is common.[36]

Histologically, desmoplastic melanoma typically forms a dermal thickening extending to the subcutis as fibrous streaks. The lesion is often deceptively bland, and it can resemble a scar or an inflammatory process, with the tumor cells dispersed in a fibrous desmoplastic stroma (Figure 25.14). The tumor cells often form curved, narrow fascicles, bundles, or a sheet-like

Chapter 25: Neuroectodermal tumors

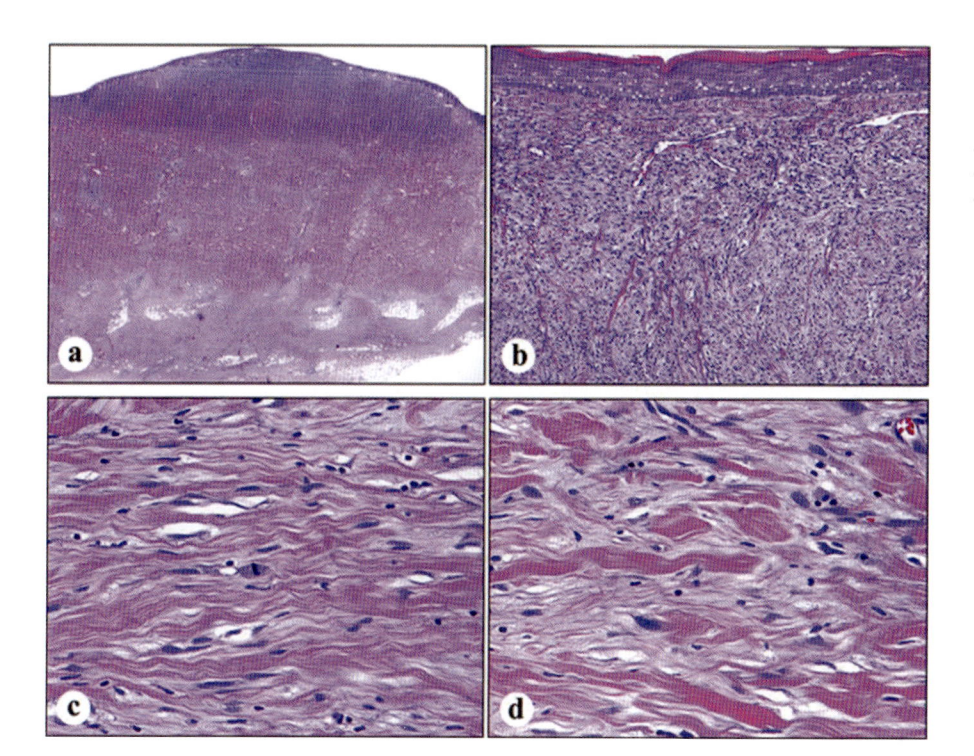

Figure 25.14 (a) Desmoplastic melanoma forms a plaque-like thickening of the subcutis. (b) In the superficial dermis, there is a more cellular zone of spindle cell proliferation. (c,d) The deeper dermis contains a relatively paucicellular, collagen-rich component that includes atypical cells with mitotic activity.

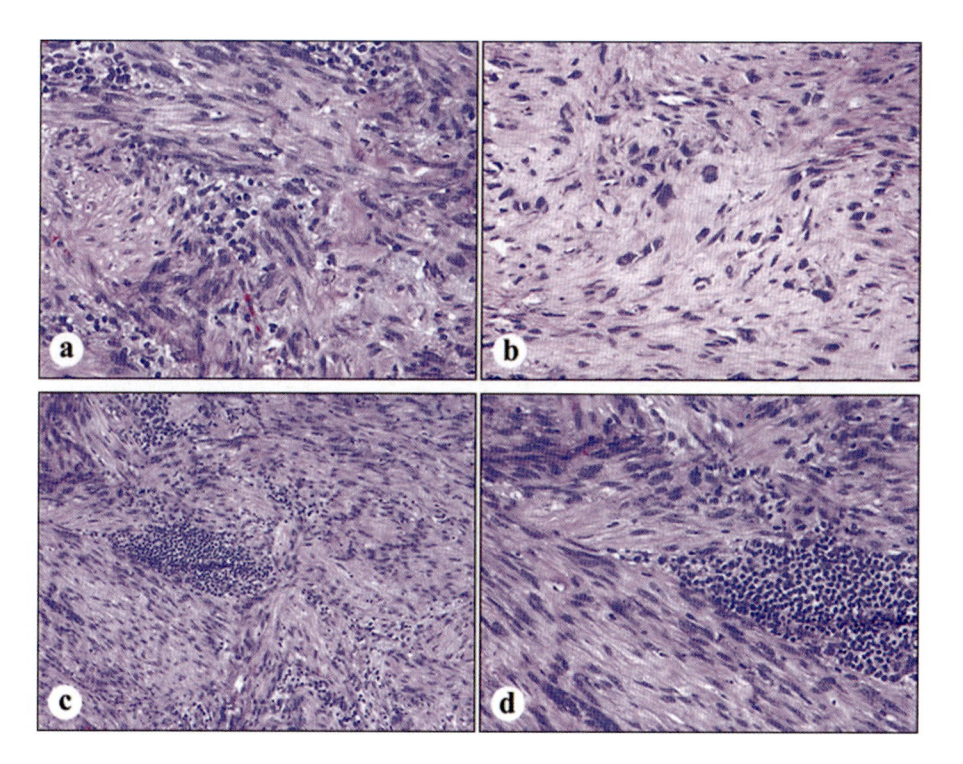

Figure 25.15 Typical features of desmoplastic melanoma include irregular fascicles of spindled cells, variable collagen deposition, and patchy lymphoplasmacytic infiltration.

pattern similar to that of conventional spindle cell melanoma. In some cases, the stromal collagen has a keloid-like appearance. Lymphoplasmacytic infiltration is typical and can be extensive enough to mask the neoplastic population.

Cytologically, at least some tumor cells show mild to moderate atypia, but many lesions show at least some markedly atypical cells with hyperchromatic, irregularly shaped nuclei (Figure 25.15). The mitotic rate varies widely, and atypical mitoses may be present.

Unfavorable histologic features are increasing tumor thickness, subcutaneous extension, and an accompanying conventional melanoma component.[32] In one study, the presence of stromal mucin was found to be an additional unfavorable feature thought to promote tumor invasion.[29]

Immunohistochemistry

The tumor cells of desmoplastic melanoma are almost always strongly positive for S100 protein, Sox10, and vimentin (Figure 25.16). Most series have found these tumors uniformly or mostly negative for HMB45,[17,37] whereas others have found a minority of cases to be positive.[28,38] Tyrosinase and melan A are typically absent, but the results of microphthalmia transcription factor expression vary from 5% to 50% in two series; this is seen in the epithelioid components, when present.[17,39] In the author's experience, the presence of melanoma markers other than S100 protein could reflect combined-type histology with conventional melanoma components being present.

The tumor cells are also more consistently positive for the low-affinity nerve growth factor receptor p75, which is not true of conventional melanoma.[40] Keratin positivity is also less common than in conventional melanoma. The significant smooth-muscle actin-positive spindle cell component has been demonstrated to represent an S100-protein-negative reactive myofibroblastic population on double staining.[37] Desmoplastic melanoma has a high expression of clusterin, compared with conventional melanoma. This has been demonstrated both by gene expression arrays and immunohistochemistry.[41]

Differential diagnosis

A high index of suspicion should be maintained, and scarring, plaque-like dermal lesions with atypia should be tested for S100 protein to rule out desmoplastic melanoma. Benign nerve sheath tumors are typically more organized and cytologically bland. It can be difficult to distinguish desmoplastic melanoma from a superficial MPNST; actually, most tumors with this differential diagnostic dilemma are desmoplastic melanomas. Only in the presence of a pre-existing neurofibroma does the author side with superficial MPNST.

Genetics

Allelic losses of the NF1 gene have been reported as being common in desmoplastic melanoma.[42]

Clear cell sarcoma of tendons and aponeuroses

This histologically distinctive rare tumor, named by Enzinger in 1965 and usually simply designated as clear cell sarcoma, typically presents in the distal extremity sites in young adults.[43] It constitutes <1% of all sarcomas. Although clear cell sarcoma shares many histologic similarities with melanoma, this tumor is genetically different from melanoma by its unique t(12;22)

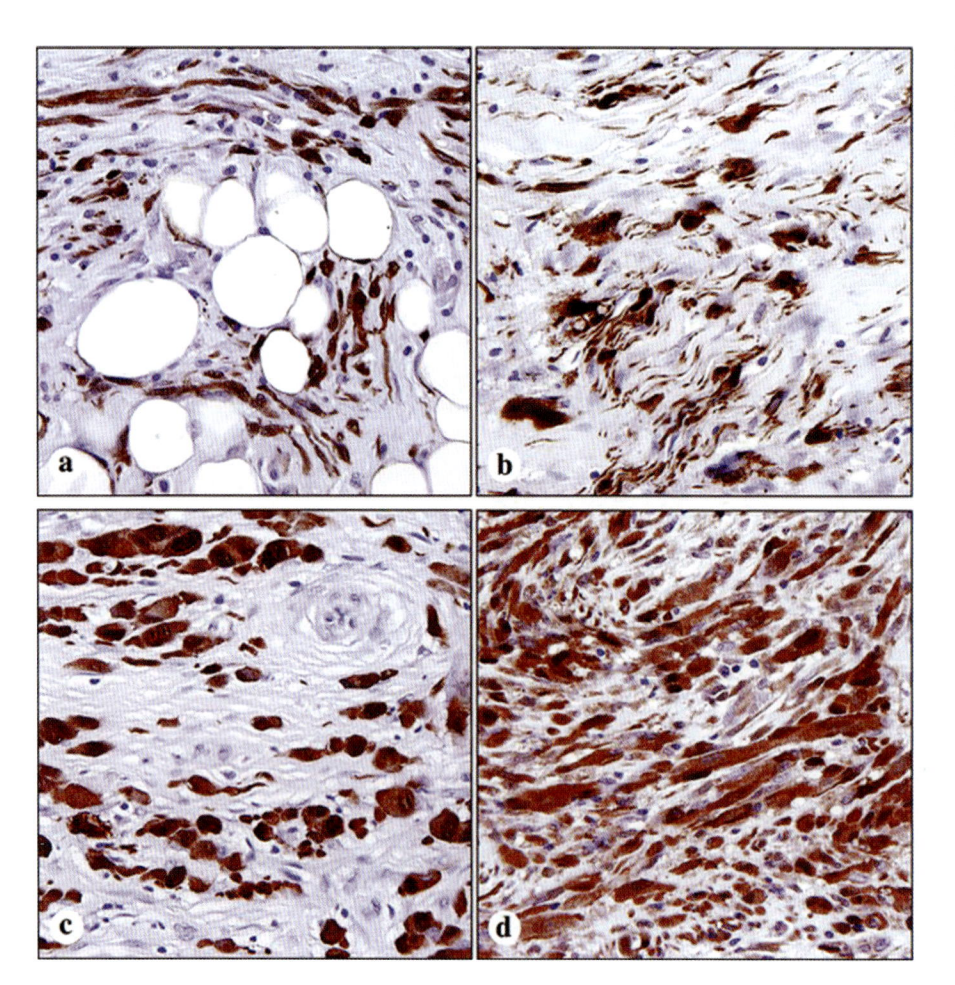

Figure 25.16 S100 protein immunostain highlights the neoplastic component of desmoplastic melanoma involving sclerosing subcutaneous fat and dermis. (**b**) Note elongated cells, and the vague resemblance to neurofibroma.

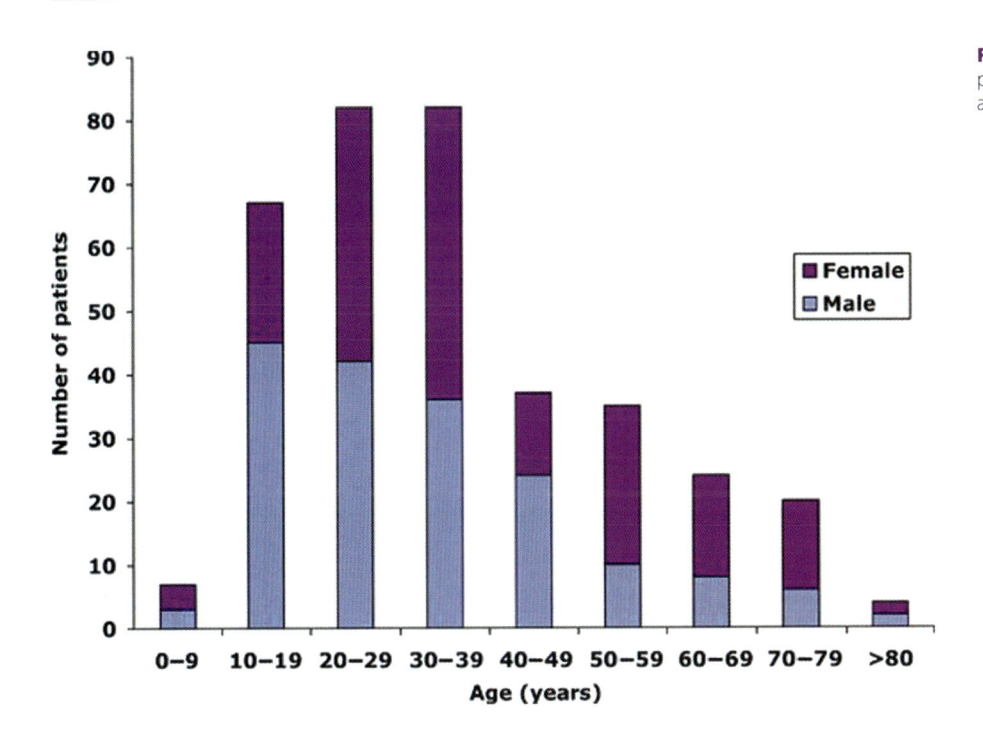

Figure 25.17 Age and sex distribution of 368 patients with clear cell sarcoma of tendons and aponeuroses.

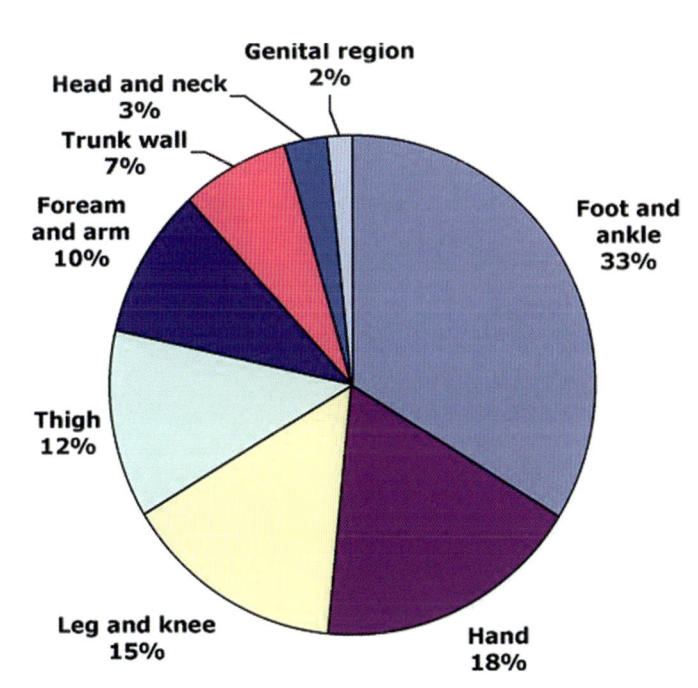

Figure 25.18 Anatomic distribution of 368 cases of clear cell sarcoma of tendons and aponeuroses.

translocation, involving EWSR1-ATF1 fusion. Therefore, the term clear cell sarcoma is preferable to *malignant melanoma of soft parts*.[44–46] Clear cell sarcoma also occurs in the gastrointestinal tract, and in this location, the tumor can be histologically, immunohistochemically, and genetically somewhat different, and it is discussed separately. The designation *clear cell sarcoma of the kidney* refers to an unrelated childhood tumor.

Clinical features

Clear cell sarcoma of tendons and aponeuroses occurs predominantly in young adults of ages between the second and fourth decades (Figure 25.17). Most major series show a mild (≤3:2) female predominance. One-third of all tumors occur in the foot and ankle, followed by the hand, leg, and knee. Occurrence in the proximal extremities, trunk, and the head and neck is less common (Figure 25.18).[47–55] Well-documented cases positive for the clear cell sarcoma translocation or fusion transcript have been reported in the kidney.[56]

Clear cell sarcoma most commonly forms a 2-cm to 4-cm dome-shaped, elevated mass, and in some cases, there is a long history of a slowly growing tumor, sometimes 5 to 10 years or more. In the thigh and trunk wall, the tumor is often larger and can exceed 10 cm, with highly malignant behavior. Recurrences are common. Metastases develop to lymph nodes, liver, bones, and lungs, and the more recent major series report 5-year survivals of 40% to 70%.[52–55]

Pathology of clear cell sarcomas

Peripheral clear cell sarcomas typically occur as a small, elevated, and palpable nodule around the tendons of the foot and ankle, usually measuring 1 cm to 3 cm. The proximally located tumors tend to be larger and usually measure 5 cm to 10 cm. Grossly, the mass is ill defined and can merge with tendons. On sectioning, the tumor varies from white to gray. Larger tumors sometimes contain necrosis.

Histologically typical of this tumor is division into compartments separated by dense collagenous septa (Figure 25.19). Oval cellular compartments, fascicles of tumor cells, or alveolar

Chapter 25: Neuroectodermal tumors

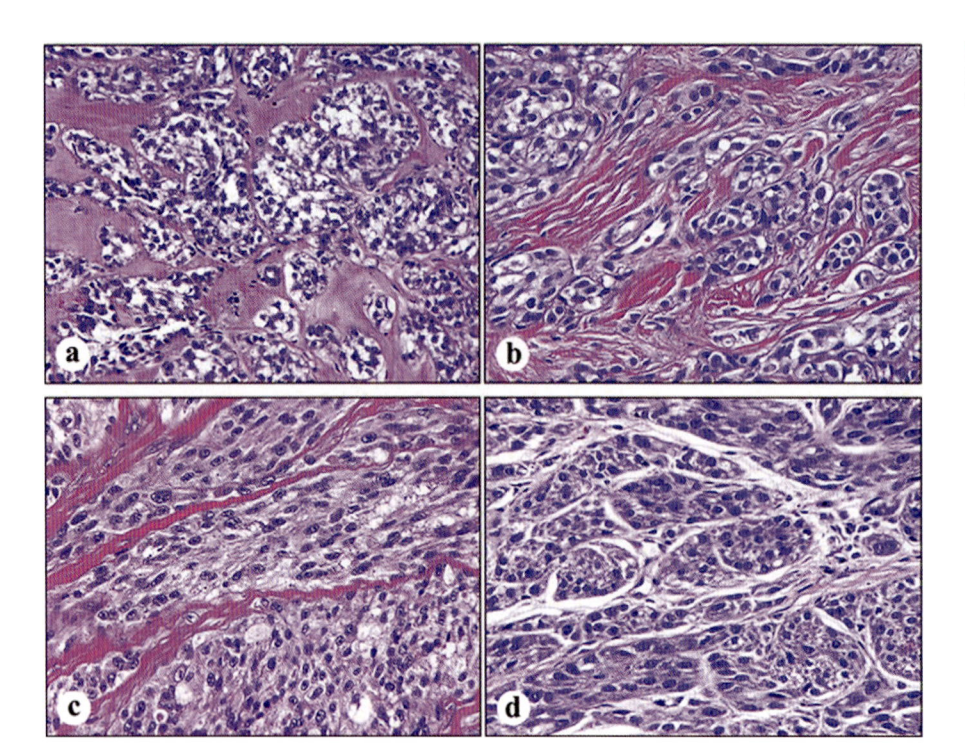

Figure 25.19 Clear cell sarcoma typically forms packets of tumor cells surrounded by variable amounts of sclerosing collagen or collagen fibers.

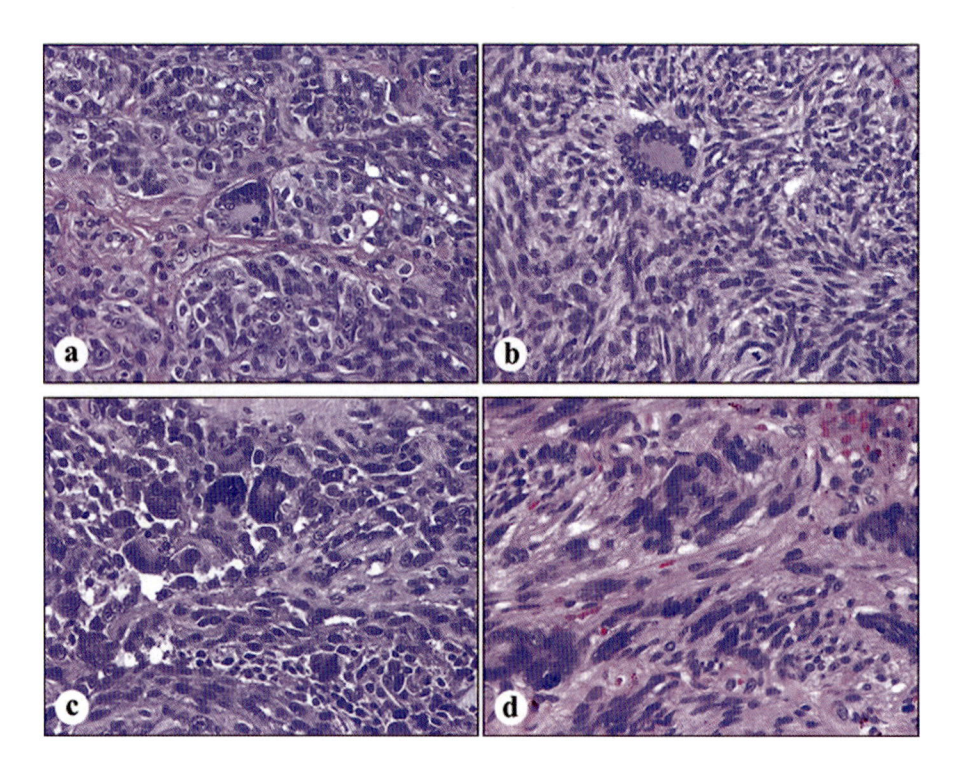

Figure 25.20 (a–c) Multinucleated tumor giant cells with nuclei arranged in a horseshoe or wreath-like pattern are often present in clear cell sarcoma. (d) Pleomorphism may occur focally.

patterns can be seen, although some tumors, especially recurrent and metastatic ones, may show a lesser degree of organization. The tumor cells are spindled or polygonal, and have a moderate amount of cytoplasm that varies from slightly clear to mildly or moderately eosinophilic. Some tumor cells are multinucleated, with a Touton cell- or wreath-like nuclear arrangement (Figure 25.20). The nuclei are round, uniform, and often have large nucleoli that vary from amphophilic to deeply eosinophilic (Figure 25.21). Nuclear pseudoinclusions, similar to those seen in melanoma, can be present. The clarity of the cytoplasm is contributed by prominent glycogen content (PAS-positive, diastase-sensitive). Melanin stains are positive in at least one-half of all cases. The mitotic rate varies from 1 to >10 per 10 HPFs; however, recurrent and metastatic tumors

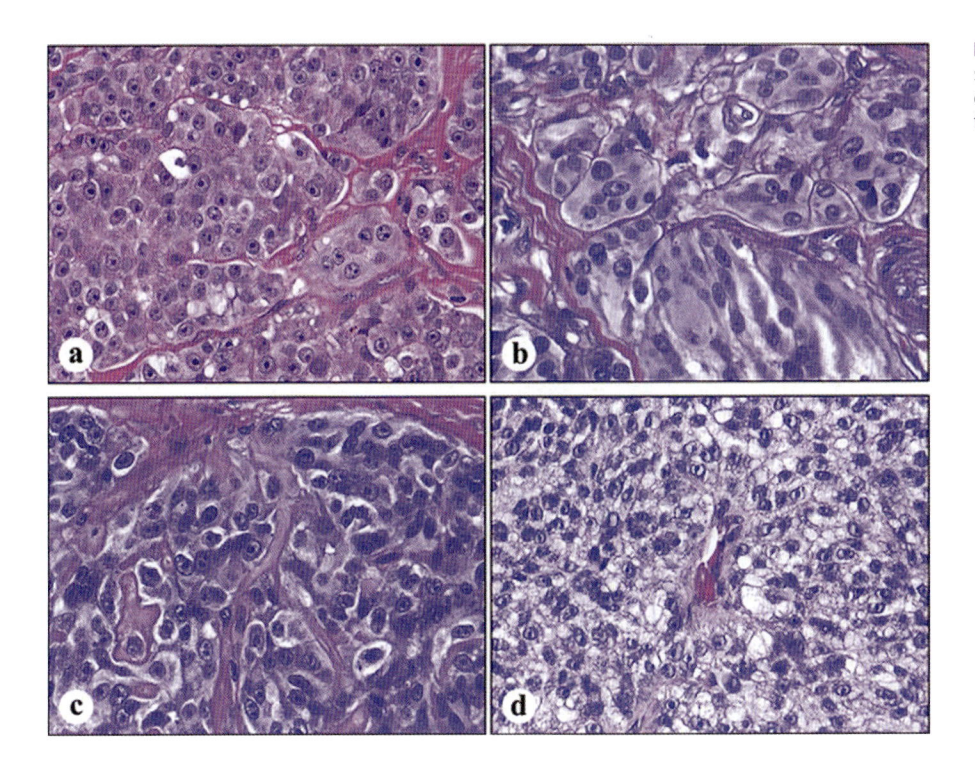

Figure 25.21 Cytologic detail of clear cell sarcoma. Note large nuclei with prominent variably eosinophilic nuclei. The quality of cytoplasm varies from mildly basophilic to eosinophilic and clear.

can show greater atypia and less-specific histologic features. Melanin pigmentation occurs in some cases, and melanosomes have been demonstrated by electron microscopy.

Adverse histologic prognostic factors are tumor size >5 cm, presence of tumor necrosis, and a high proliferation index.[48–55]

Immunohistochemistry

Immunohistochemically, clear cell sarcoma of peripheral soft tissues is similar to and virtually indistinguishable from malignant melanoma (Figure 25.22). The tumor cells are positive for S100 protein, Sox10, HMB45, melan A, neuron-specific enolase (NSE), CD57, and vimentin,[57,58] and in the author's experience, often (but variably) positive for tyrosinase, microphthalmia transcription factor, and CD117 (KIT). The level of positivity for S100 protein and HMB45 varies from case to case. The tumor cells are usually negative for keratins, EMA, muscle actins, and desmin, although in one series, focal keratin positivity was reported in one-third of genetically confirmed cases.[55]

Differential diagnosis

Primary or metastatic malignant melanoma should be ruled out. Typical tumor location, architectural and histologic features with compartmentalization and fibrous septa, lack of cutaneous involvement, and history of melanoma are helpful.

CBN usually presents as a well-circumscribed cutaneous or subcutaneous nodule. It is almost always at least focally pigmented with melanophages, and there are more spindled cellular components, typically with smaller nuclei and less prominent nucleoli. Malignant CBN can be difficult to distinguish from clear cell sarcoma, however, and genetic testing is sometimes necessary.

Skin adnexal and metastatic clear cell tumors can simulate clear cell sarcoma, but they show histologic signs or markers for epithelial differentiation and usually have a more prominent clear cell pattern than is seen in an average clear cell sarcoma.

Cytogenetic or molecular genetic verification is highly desirable, especially in tumors that present at unusual sites.

Genetics

The typical diagnostic genetic change in clear cell sarcoma is t(12;22)(q13;q12) translocation,[44,45,59–61] leading to fusion of genes activating transcription factor 1 (ATF1) in 12q13 and EWSR1 (Ewing sarcoma) in the 22q12.[62–64] The fusion protein leads to constitutive activation of ATF1 and has been shown to be important for tumor cell viability.[65] A limited number of cytogenetic changes, diploid pattern on flow cytometry, lack of microsatellite instability, and paucity of LOHs suggest a lack of complex genetic changes in most cases of clear cell sarcoma.[66] The clear cell sarcoma translocation has not been found in malignant melanomas.[67,68]

The fusion transcript can be demonstrated by RT-PCR for a diagnostic test, because this translocation does not occur in malignant melanoma;[62–64,69,70] however, one should be aware that the same translocation has been reported in angiomatoid fibrous histiocytoma.[71] FISH analysis of the EWSR1 gene rearrangement can also be useful to distinguish clear cell sarcoma from melanoma.[72] In contrast to melanoma, clear cell sarcoma also seems to lack BRAF and NRAS mutations.[73,74]

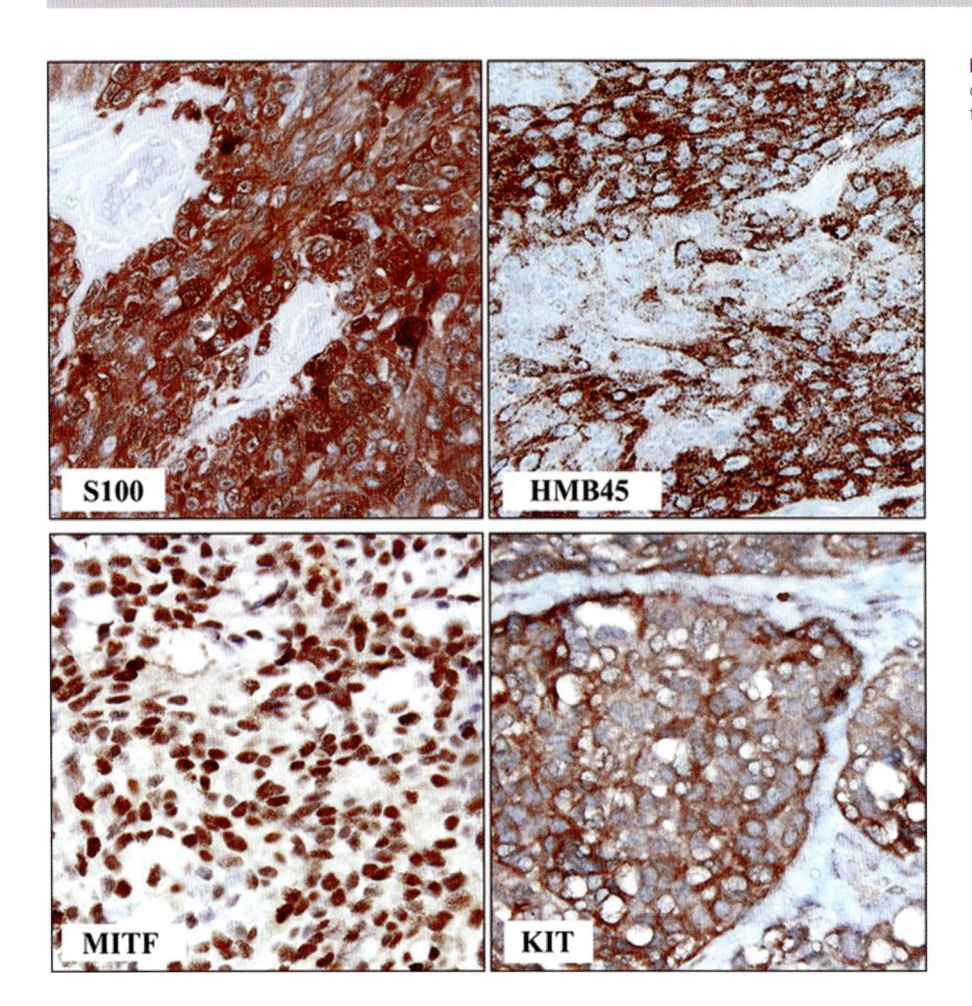

Figure 25.22 Immunohistochemical findings in clear cell sarcoma. Melanocytic markers are typically expressed, whereas KIT positivity varies.

Even if EWSR1-CREB1 gene fusions are better known in gastrointestinal clear cell sarcoma, they have also been reported in clear cell sarcoma of soft tissue.[75]

Gastrointestinal clear cell sarcoma (malignant gastrointestinal neuroectodermal tumor)

The most common sites for this tumor are the small intestine and stomach, and a small number of cases have been reported in the colon. The age of presentation varies, but most patients are between 30 and 50 years. Occurrence in the elderly is also possible. Clinical presentation can greatly simulate that of a gastrointestinal stromal tumor (GIST). Metastases often develop in regional lymph nodes, unlike in GIST, and in the liver.[76–81] The recently introduced alternative term reflects differences from clear cell sarcoma of soft tissue.[81]

The tumors usually form a mural mass of 3 cm to 6 cm that is often concentrated on the luminal aspect below the mucosa, and the mucosa can be ulcerated. On sectioning, the tumor varies from grayish to pink tan. Pigmentation is typically absent.

Histologically, there is a multinodular to diffuse proliferation of oval to round cells, and numerous osteoclastic giant cells may be present. Packeting by fibrous septa, as typically seen in peripheral clear cell sarcomas, is not a typical feature of gastrointestinal clear cell sarcoma (Figure 25.23).

Immunohistochemically, gastrointestinal clear cell sarcoma is positive for S100 protein and Sox10, but in contrast to the peripheral examples, it is generally negative for other melanocytic markers: HMB45, melan A, tyrosinase, and microphthalmia transcription factor. Positivity for NSE, synaptophysin, and CD56 seems to be consistent based on a small number of examined cases. The examined tumors have been negative for KIT, CD34, chromogranin A, keratins, and muscle markers.[81]

Most gastrointestinal clear cell sarcomas contain EWSR1 gene rearrangement by FISH. The t(12;22) translocation typical of peripheral clear cell sarcoma[77–80] is present in half or more cases. EWS-CREB1 fusion often occurs in this group (frequency in two series 23% and 100%).[80,81] The expected translocation would be t(2;22)(q34;q12).

Melanotic neuroectodermal tumor of infancy (melanotic progonoma, retinal anlage tumor)

Clinical features

This rare pediatric tumor usually occurs in infants during the first six months of life and rarely in children over the age of 1 year; published series have shown a male predominance. The tumor presents as a rapidly growing mass in the

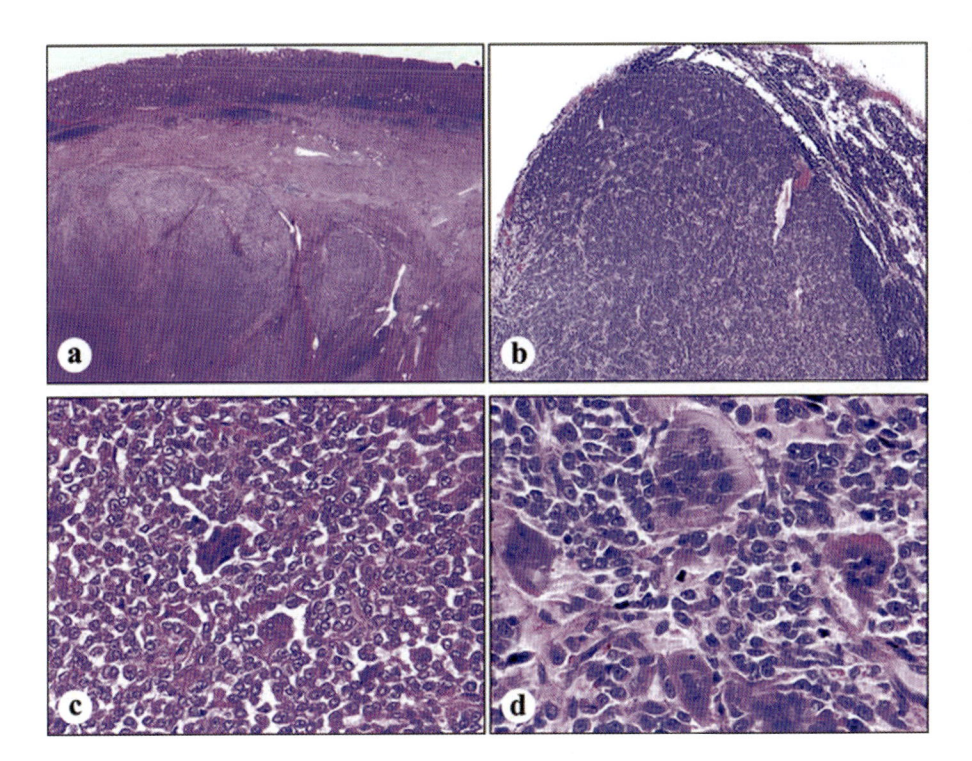

Figure 25.23 Gastrointestinal clear cell sarcoma. (a) A multinodular tumor involving the gastric wall. (b) Lymph node involvement is a relatively common feature. (c,d) Osteoclastic giant cells are scattered among tumor cells that have varying eosinophilic cytoplasm and prominent nucleoli.

maxilla or, less commonly, in the mandible, where it commonly measures 1 cm to 2 cm in diameter.[82–85] Rare examples have occurred intracranially in the dura or brain[86,87] and in the epididymis of male infants, where the tumor has reached a size of 2 cm to 3 cm and typically resulted in orchiectomy.[88–90] One bone tumor was reported in the femur;[83] some reports of tumors in peripheral soft tissues have been disputed. It remains to be seen whether the epididymal examples are truly identical to the typical examples in the head.

In some cases, serum metanephrine levels have been elevated. Despite its histologically immature appearance, this tumor is usually benign; however, local recurrences, lymph node metastases and even distant metastasis with fatal outcome have been reported. Based on a review of literature, the rate of malignancy has been estimated as 3% to 4%.[82]

Pathology

Histologic analysis shows that the tumor is composed of cellular nests spaced by fibrous septa with prominent, thick-walled capillaries. The cellular nests contain small, undifferentiated round cells in the center, and larger, somewhat epithelioid, finely pigmented cells in the periphery. These components are sometimes separated by clefts, and the proportions of the components can vary (Figure 25.24). Foci of eosinophilic rhabdomyoblasts also can be present. Glial (astrocytic) differentiation occurs in a few cases (Figure 25.25). No clear criteria exist for prediction of malignancy.

This tumor must be differentiated from malignant teratomas with pigmented neuroectodermal components and from pigmented melanotic schwannomas.

Immunohistochemically, the small cells are positive for neuroendocrine and neural markers such as NSE, synaptophysin, microtubule-associated protein 2 (MAP 2), occasionally for chromogranin, and variably for neurofilaments. The large epithelioid cells are variably positive for neuroendocrine markers, but are consistently positive for keratins and HMB45, similar to retinal pigment epithelium, which this tumor cell appears to simulate. The large cells also can be focally positive for GFAP. In contrast to melanoma and related tumors, the tumor cells are negative for S100 protein. Desmin is present in the rhabdomyoblastic components.[85,91]

Histogenesis

Similarities with the fetal pineal gland have been noted and interpreted to suggest pineal-like differentiation in this tumor[92] that ultrastructurally combines neuroblastic and melanocytic differentiation.[93]

Molecular genetics

Studies of three typical maxillary tumors revealed a lack of MYCN amplification and no losses in chromosome 1p. The lack of these alterations, commonly seen in neuroblastoma, suggests that there is no relationship between these tumors. Lack of t(11;12) translocations, similar to those in Ewing sarcoma and small round cell desmoplastic tumor, have also been noted.[94]

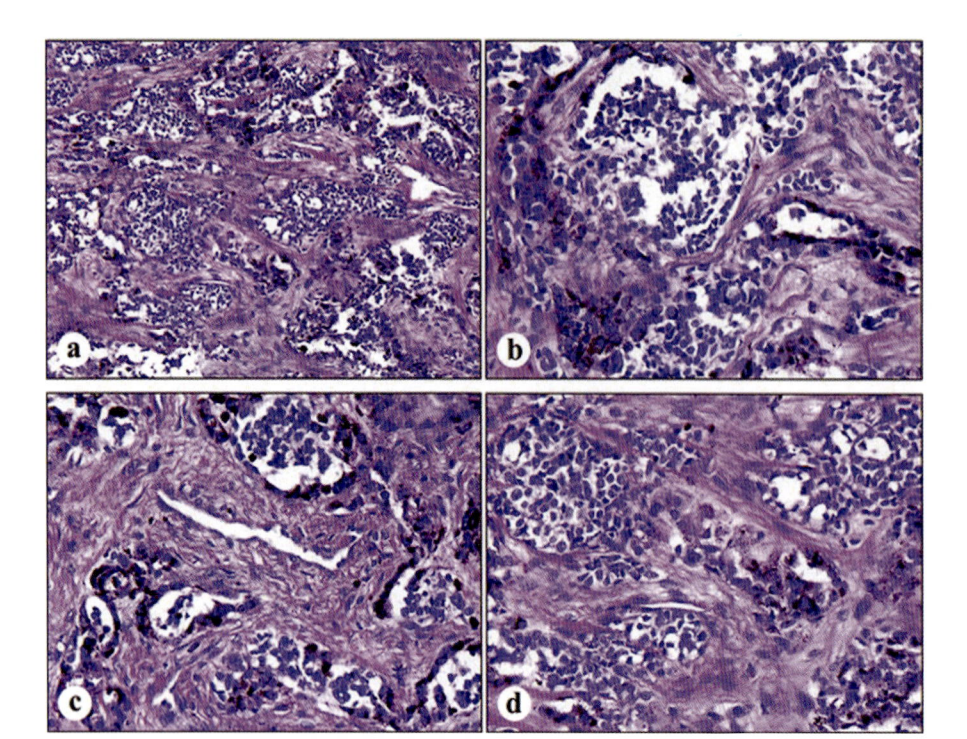

Figure 25.24 Pigmented melanotic neuroectodermal tumor of infancy is composed of nests of small round cell tumor cells. A large epithelioid cell component with melanin pigmentation is present, often in the periphery of the cellular nests.

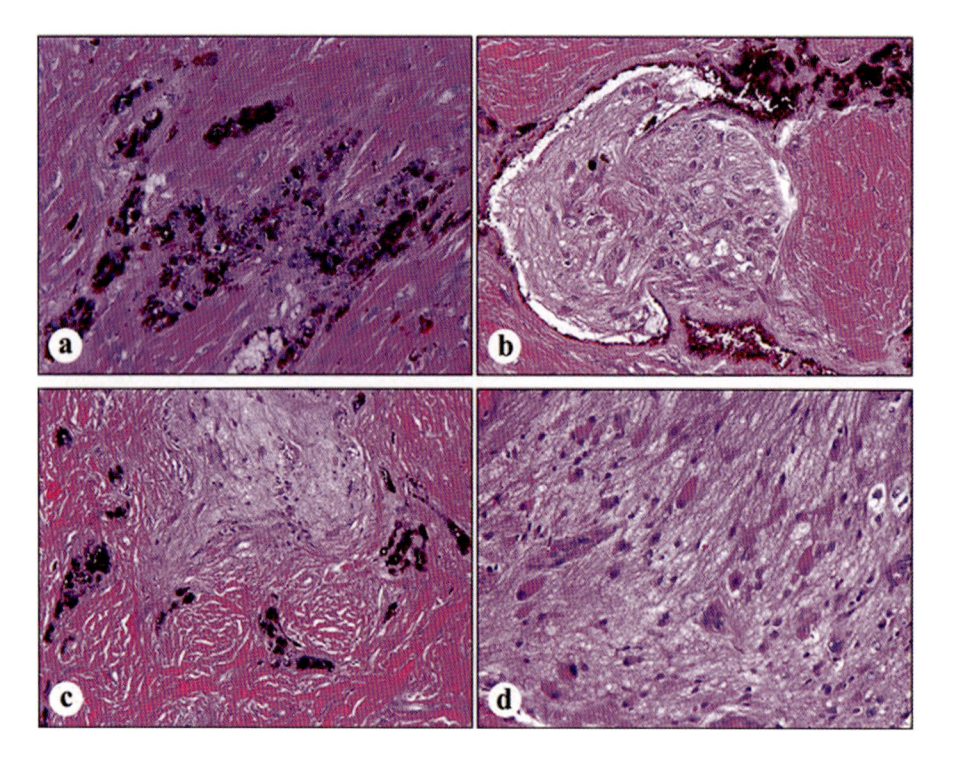

Figure 25.25 An unusual example of pigmented melanotic neuroectodermal tumor of infancy showing glial differentiation. The glial component contains reactive type astrocytes with voluminous cytoplasm.

Ectopic glial tissue: nasal glioma and soft tissue gliomatosis

Ectopic glial tissue most commonly occurs in the nasal region (*glioma nasi*) in small children. The lesion can be located inside the nose or in the nasal skin or subcutis, and probably is congenital. Nasal glioma has been viewed as a disconnected (sequestered) encephalocele, a congenital malformation. True encephaloceles, lesions connected with the brain, also occur. The possible communication with the liquor space must be considered in the surgical treatment.[95-97]

Rare examples of cutaneous and subcutaneous glial lesions, termed *soft tissue gliomatosis*, have been reported in non-nasal sites. Most reported cases have occurred as solitary masses in the chest wall of infants and children younger than 2 years.[98,99]

Occasional examples in the scalp have been considered to be sequestered encephaloceles, and occurrence in the sacrococcygeal region has been explained as a manifestation of sacrococcygeal teratoma with glial elements. Peritoneal gliomatosis represents multifocal peritoneal growth of a ruptured ovarian teratoma with glial elements. All reported cases of peripheral soft tissue gliomatosis have been clinically benign.[98,99]

Histologically, the ectopic glial elements are seen as pale staining, small, irregularly shaped nests surrounded by fibrous tissue (Figure 25.26). There is no atypia or mitotic activity. The glial elements are positive for GFAP and S100 protein.

Myxopapillary ependymoma
Clinical features

More commonly known as a central nervous system tumor that is virtually specific to the terminal part of the spinal cord, the filum terminale and cauda equina, this form of ependymoma rarely occurs as a primary soft tissue tumor in the subcutis of sacrococcygeal region, usually around the tip of the coccyx often in young adults. These tumors are not connected with the spinal cord.

Myxopapillary ependymomas have to be considered at least potentially malignant, because 15% of them eventually develop distant metastases, most commonly to the lungs. In some cases, the metastasis has occurred 20 years after the primary tumor.[100-103]

Pathology

Grossly, myxopapillary ependymomas are sharply circumscribed, often dome-shaped, elevated, cutaneous or subcutaneous masses of 1 cm to >5 cm. They appear friable and gelatinous on sectioning. Histologically, a myxopapillary ependymoma is composed of epithelioid papillary projections with a central vascular core (Figure 25.27). The papillary units are surrounded by cuboidal epithelioid cells arranged in one or more layers and can contain fibrous or myxoid matrix and be dispersed in myxoid matrix. The tumor cells are strongly positive for GFAP (Figure 25.28).

Minute ependymal rests also occur in the same locations; these lesions can be microscopic incidental findings and measure only a few millimeters; they are benign and could represent precursors of myxopapillary ependymoma.[104]

Genetics

Clonal chromosomal changes identified in these tumors include telomeric fusions and chromosome instability.[105]

Abdominal ependymomas

Ependymomas other than the myxopapillary ones rarely present in the abdominal cavity. These tumors occur almost exclusively in female patients, with a predilection for children and young adults. Most of them are pelvic tumors originating from the ovary or the adjacent ligaments. Origin from an ovarian or sacrococcygeal teratoma is a leading histogenetic consideration, but a metaplastic origin has also been suggested. These tumors often metastasize into the abdominal cavity,

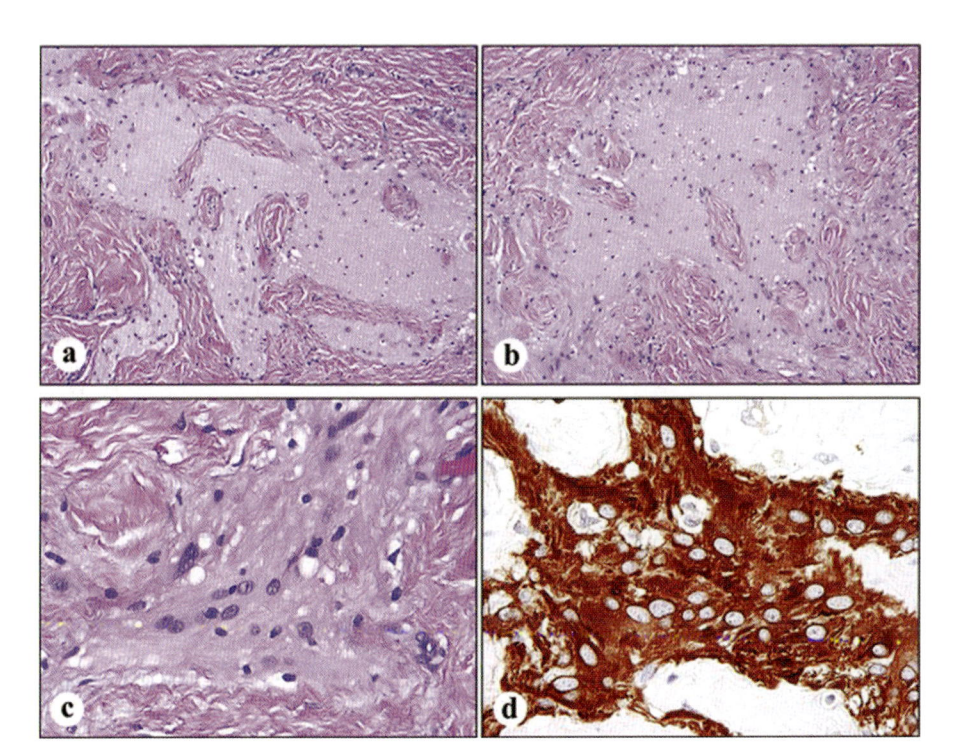

Figure 25.26 Soft tissue gliomatosis is composed of streaks of glial cells in a fibrous background. The glial element is GFAP positive.

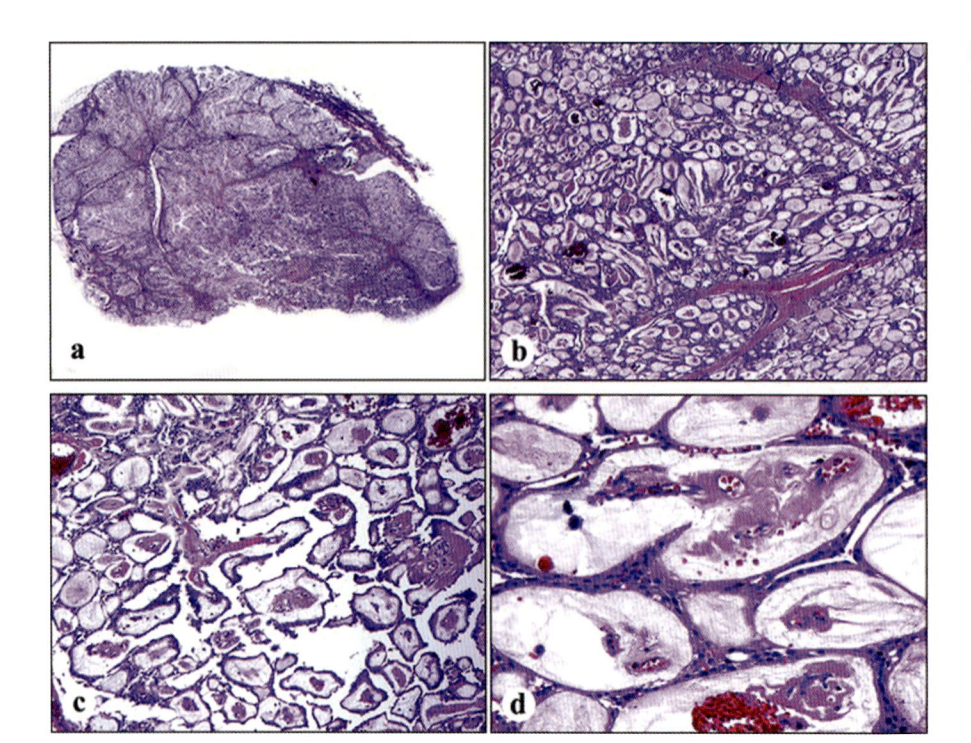

Figure 25.27 Myxopapillary ependymoma forms a solid mass composed of pseudoglandular and papillary structures. The architecture is partly papillary, partly pseudoglandular.

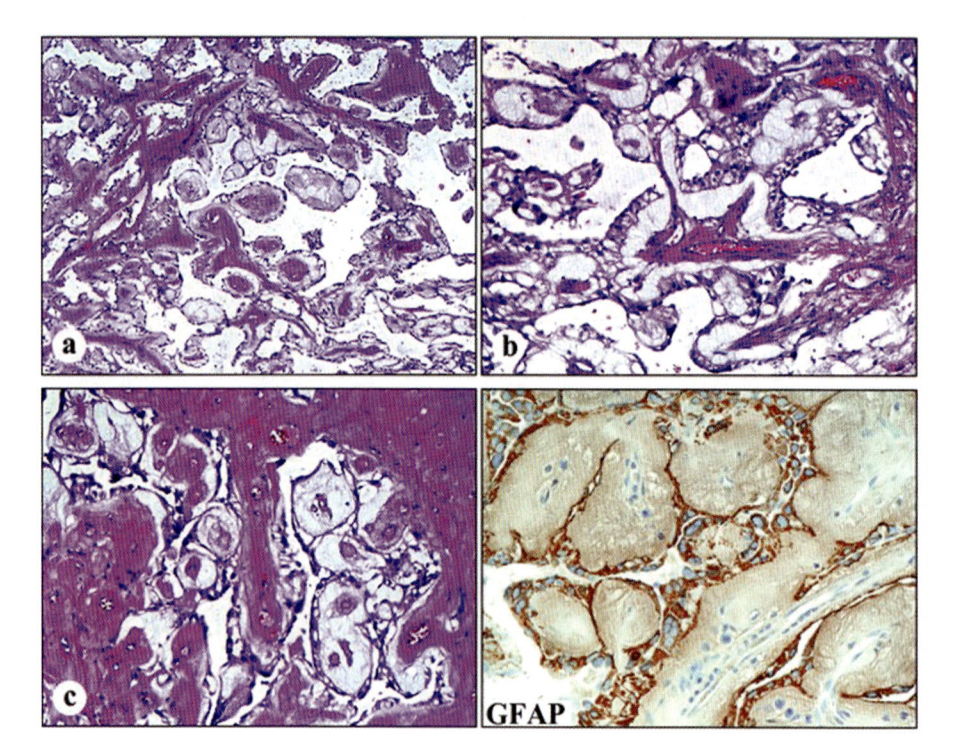

Figure 25.28 The papillae have a vascular core surrounded by edematous stroma and are lined by cuboidal ependymal-like cells that are positive for GFAP.

including the omentum, but generally they exhibit low-grade behavior.[106–109]

Pelvic ependymoma often forms a large cystic mass >10 cm. Histologically, there is great variability, including trabecular, cystic, and solid architectural patterns. Ependymal rosettes can be present (Figure 25.29). The mitotic rate is usually low, but in some cases it has exceeded 5 to 10 per 50 HPFs.

Immunohistochemically, these ependymomas are consistently GFAP positive. In contrast to central nervous system

Chapter 25: Neuroectodermal tumors

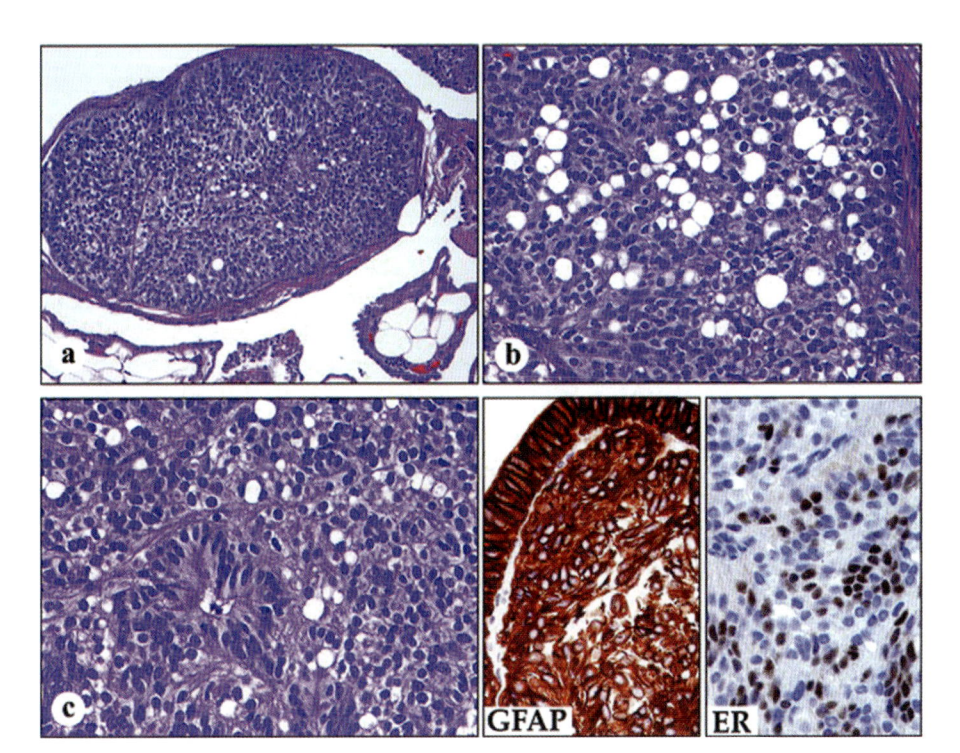

Figure 25.29 (a) An ependymoma presenting as multiple omental nodules in a young woman. (b) The tumor infiltrates the omental fat and is composed of uniform epithelioid cells. (c) Ependymal luminal differentiation is present. The tumor cells are positive for GFAP and partly positive for estrogen receptor.

ependymomas, they are positive for keratins (K18 and often K7), and estrogen receptors. Focal expression of GFAP in ovarian serous carcinoma should not lead to a misclassification as an ependymoma; such carcinomas typically have high-grade histology, including high mitotic activity and nuclear atypia.[109]

Peritoneal metastases from ventriculoperitoneal shunts can develop in central nervous system ependymomas, and the reported tumors have occurred in young children.[110] A few mediastinal ependymomas have been reported, including paravertebral examples.[111,112]

Meningioma outside the central nervous system

Meningiomas and meningioma-like nodules outside the central nervous system (ectopic meningiomas) mainly present in five types of locations and clinical settings: (1) scalp (often congenital); (2) paraspinal soft tissue; (3) sinonasal, middle ear, and orbital, often as extensions from a central nervous system meningioma; (4) lung and pleura; (5) in the soft tissues, as distant metastases of meningioma. Common to all of them is a histologic and immunophenotypic similarity to meningioma of the central nervous system. In the first three scenarios, direct extension from an intracranial meningioma must be ruled out. The anatomic distribution of meningiomas outside the central nervous system is shown in Figure 25.30. Clinicoradiologic correlation is needed to rule out extracranial extension of a central nervous system

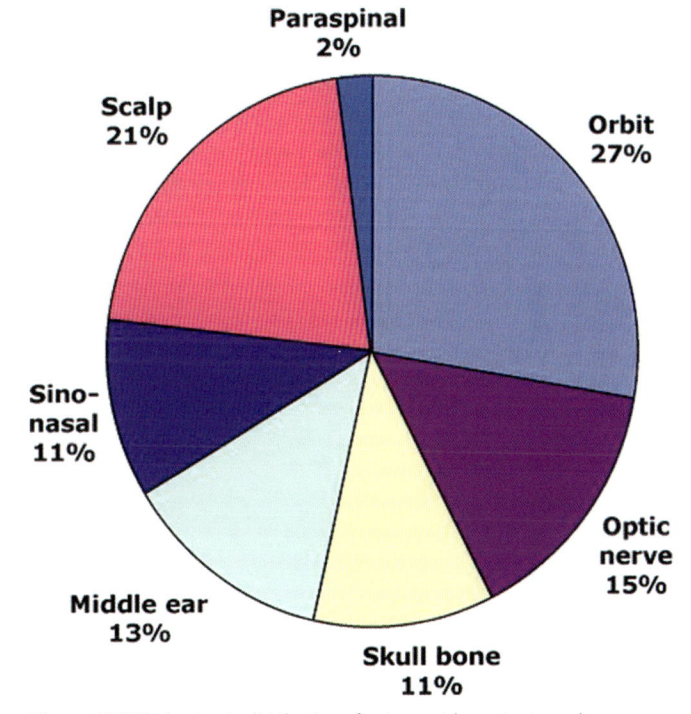

Figure 25.30 Anatomic distribution of extracranial meningioma; however, most orbital, optic nerve, skull bone, and middle ear examples are of central nervous system origin. Some sinonasal meningiomas are also extensions of a brain tumor, whereas many examples in the scalp and paraspinal region are legitimate examples of ectopic meningiomas.

meningioma. In a large series 44% of the lesions were located in the scalp, 26% in temporal bone, and 24% in the sinonasal tract.[113]

Ectopic meningiothelial elements of the scalp (meningioma of the scalp, meningothelial hamartoma, meningothelial nodule, hamartoma with ectopic meningothelial elements, heterotopic neural nodule)

Terminology and synonyms

The variegated terminology of the heading for this section reflects different thoughts about the nature of meningothelial proliferations in the scalp. Such meningothelial proliferations have sometimes been divided into meningothelial hamartomas, congenital formations that usually have more subtle meningothelial elements, and cutaneous meningiomas that contain more extensive, solid meningothelial proliferations.[114] This subdivision is subjective, however, because the groups overlap and probably represent a continuum. The designation *cutaneous meningioma* is largely a misnomer, because most such lesions primarily involve the subcutis and not the skin. Simpler, unifying terminology that groups all ectopic meningothelial elements of the scalp is usually used.

Clinical features

Ectopic meningothelial elements of the scalp usually occur in children and young adults, and most of them are thought to be congenital (Figure 25.31). Clinicoradiologic correlation is needed to rule out a calvarial defect and connection with the dural space, to avoid complications from penetrating the intracranial space.[115] These meningeal proliferations in infants typically occur in the midline, usually in the parietal or occipital scalp, often with a hairless patch surrounded by focal hypertrichosis.[115,116] Similar nodules have also been reported in young and older adults in the forehead, mastoid, and zygomatic areas.[114,117] The lesion size generally varies from 1 cm to 3 cm in diameter. Connection with intracranial space and content of glial elements are grounds for the designation of meningocele or meningoencephalocele. It also has been postulated that many meningothelial hamartomas in fact are rudimentary meningoceles that have lost their dural connection.[114,116] Meningioma of the scalp in a child with NF1[118] and in two siblings has been reported.[119]

The pathogenesis of the more solid ectopic meningiomas in the scalp of older adults is uncertain. These nodules more closely resemble intracranial meningiomas, and indeed, this possibility should be ruled out by clinicoradiologic correlation. It is possible that they have arisen from dormant, clinically undetected meningothelial rests or hamartomas that have proliferated to form a solid nodule.

Rarely, an intracranial meningioma directly involves the overlying soft tissues of the scalp. In such cases, the tumor has often persisted for a long time in the skull bone, ultimately extending into overlying soft tissue. Clinical correlation is needed to rule out an extension of an intracranial meningioma.

Pathology

Histologically, an ectopic meningothelial nodule/meningothelial hamartoma contains scattered, microscopic nests and cords of meningothelial cells in a fibrous, hypervascular

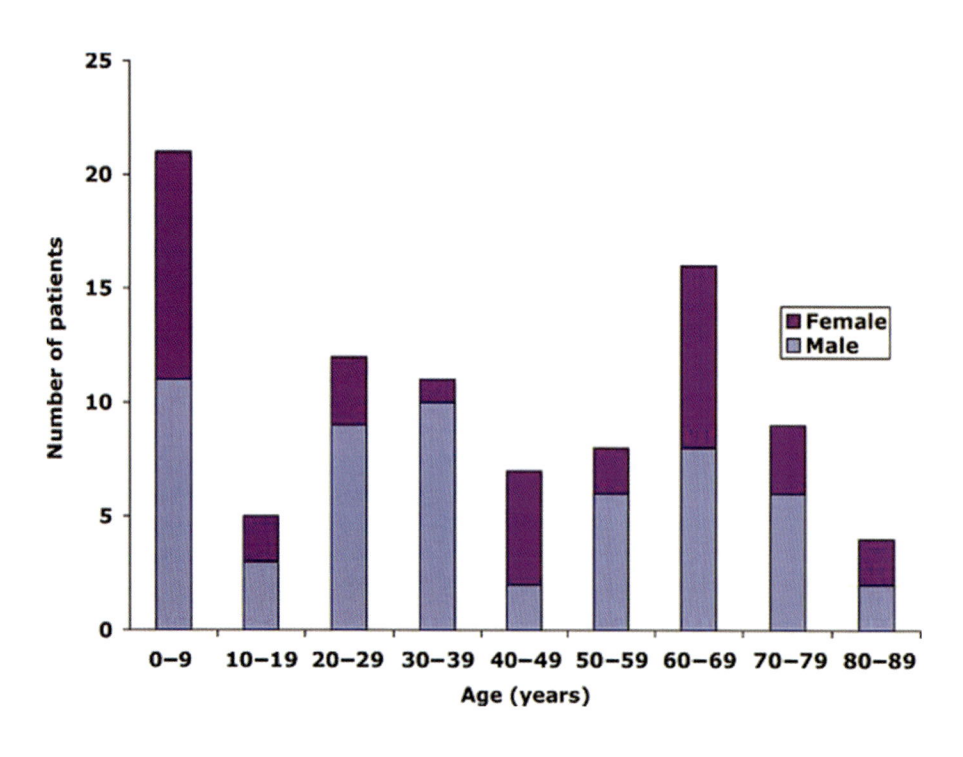

Figure 25.31 Age and sex distribution of 93 patients with extracranial meningioma. Tumors involving the head (i.e., orbit, sinonasal tract, middle ear) are excluded, and only tumors in the scalp and paraspinal soft tissues are included.

stroma, often with prominent cleft-like spaces, and there are usually no well-formed meningothelial whorls (Figure 25.32). Such a constellation can resemble a vascular tumor, even an angiosarcoma[120] or tumors such as lymphangioendothelioma (see Chapter 21). Examples extending to the intradural space can be considered meningoceles, and some of them also include glial elements, warranting a designation of meningoencephalocele. Lesions sometimes designated as cutaneous meningiomas form more solid nests and sheets of cells with meningothelial whorls (Figure 25.33).

Immunohistochemically, these meningothelial nests and ectopic meningiomas show features similar to meningioma with positivity for vimentin and EMA, often for keratin 18, and are negative for vascular markers such as CD31.

Figure 25.32 Ectopic meningioma in the scalp. (**a,b**) Meningothelial clusters are set in a fibrous stroma. (**c,d**) The presence of stromal hemorrhage or vascular-like slits can create a false impression of a vascular tumor.

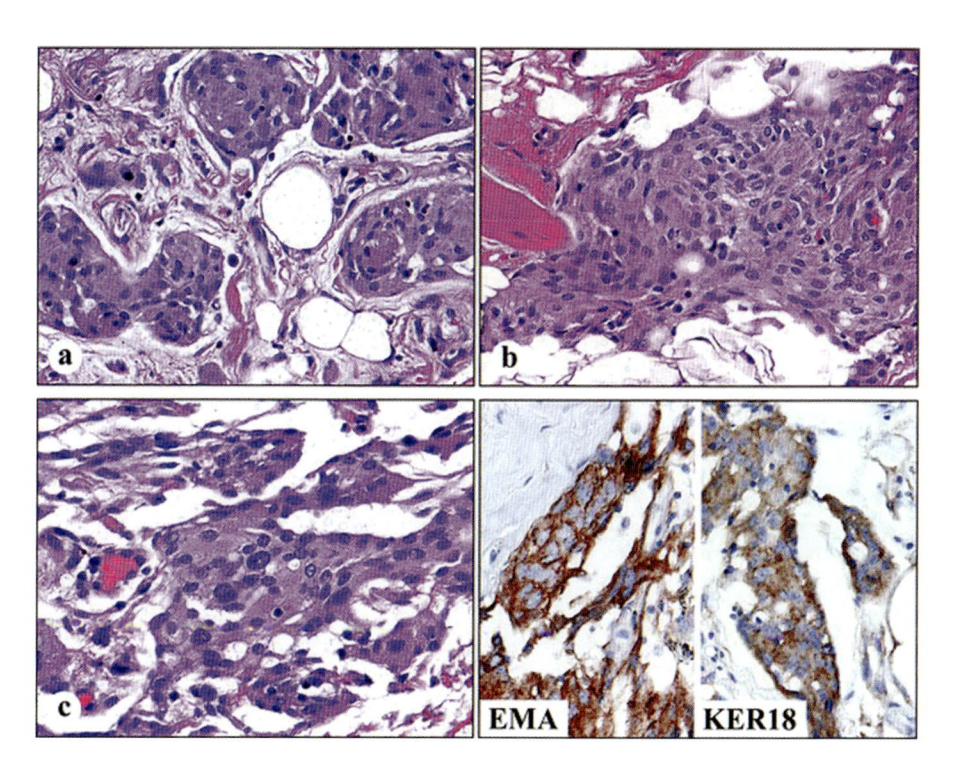

Figure 25.33 Ectopic meningioma in the scalp forming more solid clusters of epithelioid meningioma elements. These are immunohistochemically positive for EMA and keratin 18.

Ectopic meningioma in the paraspinal tissues

Only a small number of ectopic meningothelial proliferations have been reported in the back, in the mid-line overlying the vertebral column. All examples have occurred in children, often on a congenital basis. Grossly, meningiomas in the back are typically pedunculated lesions of 2 cm to 3 cm, resembling fibroepithelial polyps, and some of them have been connected with the spinal dura. Histologically, the meningothelial elements often form subtle aggregates in fibrous stroma, but psammoma bodies occur.[114,116,121,122] Paraspinal meningiomas can also occur internally in the pelvis or the abdominal cavity. These tumors form solid masses that can histologically resemble ordinary meningothelial meningioma (Figure 25.34a–c).

Sinonasal, external ear, and orbital meningiomas

Meningiomas can involve the sinonasal tract and external ear primarily or secondarily, and clinicoradiologic correlation is required to determine whether such a tumor is an isolated process or an extension of a primary central nervous system tumor. Penetration of intracranial meningioma into the sinonasal system from the bottom of the frontal or middle cranial fossa occurs through the lamina cribrosa or the bony sinus walls, whereas middle and external ear meningiomas usually have migrated through the conduits of vestibulocochlear nerve and normal porous structures of the ear.[123,124]

An orbital meningioma can be an extension from a central nervous system meningioma of the skull base, such as one in the sphenoid wing, along the optic nerve canal. In some cases this tumor originates from meningothelial elements along the optic nerve.

In the largest series of sinonasal meningiomas, 6 of 30 tumors (20%) were known to be connected to a central nervous system meningioma.[124] Pathogenesis of the isolated sinonasal and middle ear meningiomas is uncertain, but origin from arachnoid cells associated with major cranial nerves or arachnoid rests are possible explanations.

Primary sinonasal meningiomas are rare and occur from the teenage years to old age (median age approximately 45 years), with a possible mild female predominance. They most often involve the nasal cavity and paranasal sinuses, but can involve any of the sinuses, most often the maxillary and adjoining sinuses. In the largest series, the tumor size varied from 1 cm to 8 cm, and the median size was 3.5 cm. Most examples were infiltrative, with only few being circumscribed masses.[124]

Although most sinonasal meningiomas are benign, they can cause potentially fatal complications by local invasion, similar to intracranial meningiomas. In the largest series, 3 of 28 patients died of septic complications of an inoperable sinonasal meningioma that extended into the skull base.[124]

Most sinonasal meningiomas are of the meningothelial type, consisting of whorls of syncytially arranged meningothelial cells, often with psammoma bodies, but a fibroblastic meningioma appearance with a spindle cell pattern is also possible (Figure 25.34d). Metaplastic meningiomas with lipomatous components, and atypical meningiomas also can occur in this location. On immunohistochemical analysis, ectopic meningiomas are usually at least focally positive for EMA and generally negative for the keratin cocktail, chromogranin,

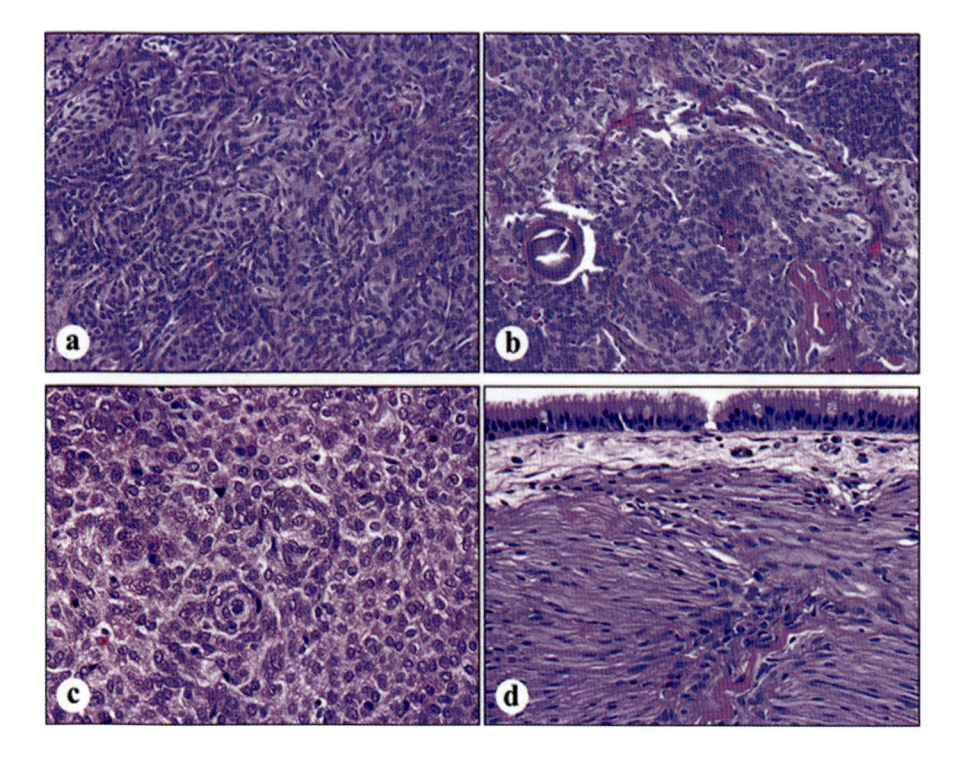

Figure 25.34 (a–c) Solid ectopic meningiomas in a paraspinal location resemble intracranial meningiomas. (a) Transitional meningioma. (b,c) Meningothelial meningiomas with psammoma bodies and meningothelial whorls. (d) A fibroblastic meningioma involving the paranasal sinus seen underneath the respiratory mucosa. This tumor originated in the skull base.

synaptophysin, S100 protein, and CD34, helping to separate them from carcinoma, melanoma, paraganglioma, solitary fibrous tumor or olfactory neuroblastoma. Positivity for keratin 18 occurs, similar to other meningiomas, however.

Middle ear and orbital meningiomas represent a wide spectrum of meningiomas, including meningothelial, transitional, and fibroblastic types.

Pulmonary meningothelial nodules

Originally considered minute pulmonary chemodectomas (paragangliomas), these minute lesions were found ultrastructurally to differ from paragangliomas and show meningioma-like features.[125] Subsequent immunohistochemical studies confirmed meningothelial differentiation, leading to the present terminology.[126]

Pulmonary meningothelial nodules are small (<1–3 mm), whitish to tan-yellow nodules in the visceral pleura or lung parenchyma. Such nodules are usually incidental findings, and multiple lesions are relatively common. In one study from the United States, meningothelial nodules were found in 6 of 356 autopsies, a nearly 2% frequency.[126] In another study from Japan, such nodules were present in 20 of 214 specimens (nearly 10%) of pulmonary adenocarcinoma resections.[127] Both studies showed female predominance and occurrence primarily in older adults. Careful examination of any pulmonary surgery specimen can reveal meningothelial nodules, more often at the pleural surface.

In rare instances, symptomatic bilateral diffuse interstitial pulmonary involvement has been reported. This can cause restrictive pulmonary disease with dyspnea and shortness of breath. Of five reported patients with such lesions, four were female and one was male (age range, 54–75 years).[128] Microscopically, these pulmonary nodules are composed of clusters of ovoid cells resembling meningothelial cells (Figure 25.35). Immunohistochemically, they are positive for EMA and vimentin, similar to meningioma. In one study, human androgen receptor assay (HUMARA) analysis showed clonality in approximately one-half of the nodules, and multiple nodules from one patient appeared to represent multiple independent clones.[127]

Metastatic meningioma

Soft tissue metastases of meningioma (excluding noncontinuous growth from an intracranial tumor) are very rare. The author and colleagues have seen only a few examples, such as metastatic central nervous system meningioma in the chest wall that had nonmalignant histologic features and had infiltrated the skeletal muscle. The designation of *benign metastasizing meningioma* might be appropriate for such an occasion. One report described an isolated chest wall metastasis from a recurrent intracranial meningioma that clinically simulated a soft tissue sarcoma.[129] There is also a report of accidental implantation of histologically malignant meningioma into an abdominal wall tissue donor site during surgery for recurrent orbital malignant meningioma, presumably by a surgical instrument contaminated by autologous tumor cells.[130] In one patient, soft tissue metastasis of meningioma developed at the site of a pin used for intraoperative skull fixation.[131]

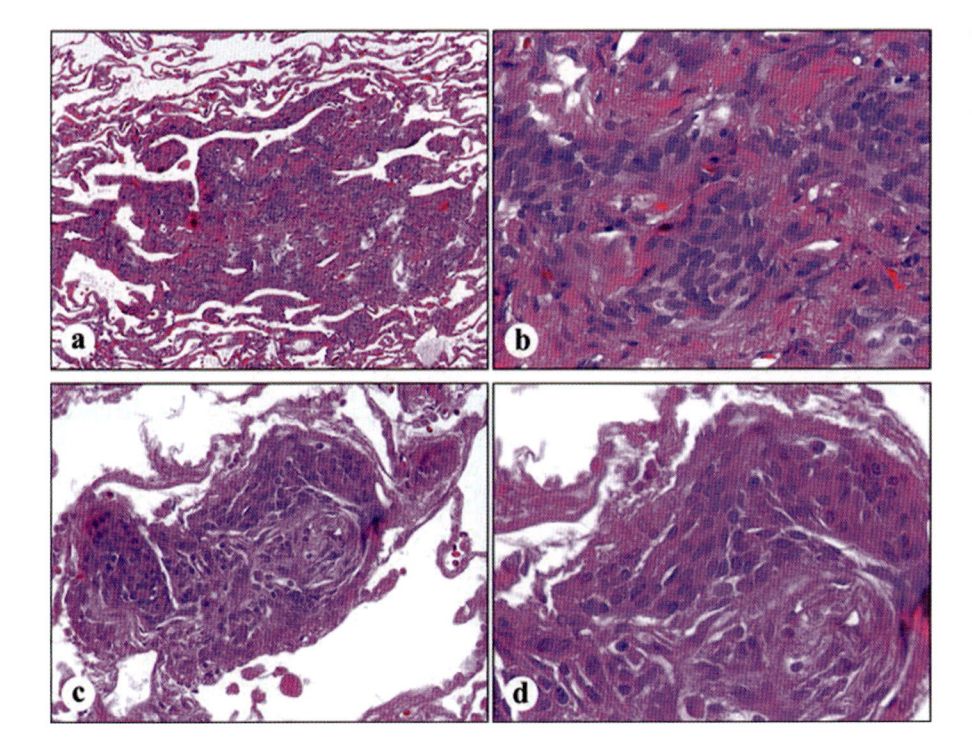

Figure 25.35 Minute pulmonary meningothelial nodules. Two examples of intraparenchymal nodules. Cytologic features correspond to those of meningothelial cells.

Table 25.1 Immunohistochemical distinction of ectopic meningioma from other tumors

	EMA	S100	SOX10	Chromogranin/ synaptophysin	Keratin cocktail	Keratin 18
Meningioma	+ (variable)	–/+ rarely positive	–	–	–/+ rarely positive	+ variable
Carcinoma	++	–/+	–/(+)	–/+[a]	+	++
Melanoma	–	+	+	–	–	–/+ occasionally positive
Paraganglioma	–	–/+[b]	=	+	–	–
Olfactory neuroblastoma	–	–	–	+	–	–

[a] Positive in examples with neuroendocrine differentiation.
[b] Minor elements, sustentacular cells positive.

Differential diagnosis of ectopic meningioma

Meningothelial whorls, sometimes with metaplastic bone, occur in dedifferentiated liposarcomas, usually in the retroperitoneum. Close observation of adjacent fatty components provides a diagnostic clue (see Chapter 15). Some cutaneous/soft tissue tumors, such as neurothekeoma, commonly seen in the head and neck (see Chapter 10), also can create a resemblance to a meningothelial nodule by the presence of multiple round compartments. In the sinonasal tract, external ear, and orbit, a meningioma must be differentiated from other tumors with a nesting pattern, especially poorly differentiated sinonasal carcinoma, sinonasal melanoma, paraganglioma, and olfactory neuroblastoma. Immunohistochemistry provides a distinction in each case (Table 25.1).

References

Cellular blue nevus

1. Elder DE, Murphy GF. *Melanocytic Tumors of the Skin*, third series, fascicle 2. Washington, DC: Armed Forces Institute of Pathology, 1996.
2. Rodriguez H, Ackerman LV. Cellular blue nevus. *Cancer* 1968;21:393–405.
3. Temple-Camp CR, Saxe N, King H. Benign and malignant cellular blue nevus: a clinicopathologic study of 30 cases. *Am J Dermatopathol* 1988;10:289–296.
4. Connelly J, Smith JL. Malignant blue nevus. *Cancer* 1991;67:2653–2657.
5. Granter SR, McKee PH, Calonje E, Mihm MC, Busam K. Melanoma associated with blue nevus and melanoma mimicking cellular blue nevus: a clinicopathologic study of 10 cases of the spectrum of so-called malignant blue nevus. *Am J Surg Pathol* 2001;25:316–323.
6. Zembowicz A, Granter SR, McKee PH, Mihm MC. Amelanotic cellular blue nevus: a hypopigmented variant of the cellular blue nevus: clinicopathologic analysis of 20 cases. *Am J Surg Pathol* 2002;26:1493–1500.
7. Tran TA, Carlson JA, Basaca PC, Mihm MC. Cellular blue nevus with atypia (atypical cellular blue nevus): a clinicopathologic study of nine cases. *J Cutan Pathol* 1998;25:252–258.
8. Barnhill RL, Argenyi Z, Berwick M, et al. Atypical cellular blue nevi (cellular blue nevi with atypical features): lack of consensus for diagnosis and distinction from cellular blue nevi and malignant melanoma ("malignant blue nevus"). *Am J Surg Pathol* 2008;32:36–44.

Metastatic melanoma and related problems

9. Nakhleh RE, Wick MR, Rocamora A, Swanson PE, Dehner LP. Morphologic diversity in malignant melanomas. *Am J Clin Pathol* 1990;93:731–740.
10. Nguyen NC, Chaar BT, Osman MM. Prevalence and patterns of soft tissue metastasis: detection with true whole-body F-18 FDG PET/CT. *BMC Med Imaging* 2007;7:8.
11. Lodding P, Kindblom LG, Angervall L. Metastases of malignant melanoma simulating soft tissue sarcoma: a clinicopathological, light and electron microscopic and immunohistochemical study of 21 cases. *Virchows Arch Pathol Anat Histopathol A* 1990;417:377–388.
12. King R, Busam K, Rosai J. Metastatic malignant melanoma resembling malignant peripheral nerve sheath tumor: report of 16 cases. *Am J Surg Pathol* 1999;23:1499–1505.
13. Gaudin P, Rosai J. Florid vascular proliferation associated with neural and neuroendocrine neoplasms: a diagnostic clue and potential pitfall. *Am J Surg Pathol* 1995;19:642–652.
14. Nonaka D, Chriboga L, Rubin BP. Sox10: a pan-schwannian and melanocytic marker. *Am J Surg Pathol* 2008;32:1291–1298.
15. Miettinen M, McCue PA, Sarlomo-Rikala M, et al. Sox10: a marker for not only schwannian and melanocytic tumors but also myoepithelial tumors of soft tissue: a systematic analysis of 5134 tumors. *Am J Surg Pathol* 2015;39:826–835.
16. Kaufmann O, Koch S, Burghardt J, Audring H, Dietel M. Tyrosinase, melan-A, KBA62 as markers for the immunohistochemical identification of metastatic amelanotic melanomas on paraffin sections. *Mod Pathol* 1998;11:740–746.
17. Miettinen M, Fernandez M, Franssila KO, et al. Microphthalmia transcription factor in the immunohistochemical diagnosis of metastatic melanoma: comparison with four other melanoma markers. *Am J Surg Pathol* 2001;25:205–211.

18. Zarbo RJ, Gown AM, Nagle RB, Visscher DW, Crissman JD. Anomalous cytokeratin expression in malignant melanoma: one- and two-dimensional western blot analysis and immunohistochemical survey of 100 melanomas. *Mod Pathol* 1990;3:494–501.

19. Busam KJ. Molecular pathology of melanocytic tumors. *Semin Diagn Pathol* 2013;30:362–374.

20. Marin C, Beauchet A, Capper D, *et al.* Detection of BRAF p.V600E mutations in melanoma by immunohistochemistry has a good interobserver reproducibility. *Arch Pathol Lab Med* 2014;138:71–75.

21. Serrano C, Simonetti S, Hernández-Losa J, *et al.* BRAF V600E and KRAS G12S mutations in peripheral nerve sheath tumours. *Histopathology* 2013;62:499–504.

Desmoplastic melanoma

22. Conley J, Lattes R, Orr W. Desmoplastic malignant melanoma (a rare variant of spindle cell melanoma). *Cancer* 1971;28:914–936.

23. Reed RJ, Leonard DD. Neurotropic melanoma: a variant of desmoplastic melanoma. *Am J Surg Pathol* 1979;3:301–311.

24. Reiman HM, Goellner JR, Woods JE, Mixter RC. Desmoplastic melanoma of the head and neck. *Cancer* 1987;60:2269–2274.

25. Egbert B, Kempson R, Sagebiel R. Desmoplastic malignant melanoma: a clinicohistopathologic study of 25 cases. *Cancer* 1988;62:2033–2041.

26. Jain S, Allen PW. Desmoplastic malignant melanoma and its variants: a study of 45 cases. *Am J Surg Pathol* 1989;13:358–373.

27. Smithers BM, McLeod GR, Little JH. Desmoplastic, neural transforming and neurotropic melanoma: a review of 45 cases. *Aust N Z J Surg* 1990;60:967–972.

28. Carlson JA, Dickersin GR, Sober AJ, Barnhill RL. Desmoplastic neurotropic melanoma: a clinicopathologic analysis of 28 cases. *Cancer* 1995;75:478–494.

29. Skelton HG, Smith KJ, Laskin WB, *et al.* Desmoplastic malignant melanoma. *J Am Acad Dermatol* 1995;32:717–725.

30. Tsao H, Sober AJ, Barnhill RL. Desmoplastic neurotropic melanoma. *Semin Cutan Med Surg* 1997;16:131–136.

31. Quinn MJ, Crotty KA, Thompson JF, *et al.* Desmoplastic and desmoplastic neurotropic melanoma: experience with 280 patients. *Cancer* 1998;83:1128–1135.

32. Busam KJ, Mujumdar U, Hummer AJ, *et al.* Cutaneous desmoplastic melanoma: reappraisal of morphologic heterogeneity and prognostic factors. *Am J Surg Pathol* 2004;28:1518–1525.

33. Chen LL, Jaimes N, Barker CA, Busam KJ, Marghoob AA. Desmoplastic melanoma: a review. *J Am Acad Dermatol* 2013;68:825–833.

34. Kilpatrick SE, White WL, Browne JD. Desmoplastic malignant melanoma of the oral mucosa: an underrecognized diagnostic pitfall. *Cancer* 1996;78:383–389.

35. Su LD, Fullen DR, Lowe L, *et al.* Desmoplastic and neurotropic melanoma: analysis of 33 patients with lymphatic mapping and sentinel lymph node biopsy. *Cancer* 2004;100:598–604.

36. Wharton JM, Carlson JA, Mihm MC. Desmoplastic malignant melanoma: diagnosis of early clinical lesions. *Hum Pathol* 1999;30:537–542.

37. Longacre TA, Egbert BM, Rouse RV. Desmoplastic and spindle-cell malignant melanoma: an immunohistochemical study. *Am J Surg Pathol* 1996;20:1489–1500.

38. Anstey A, Cerio R, Ramnarain N, *et al.* Desmoplastic malignant melanoma: an immunocytochemical study of 25 cases. *Am J Dermatopathol* 1994;16:14–22.

39. Koch MB, Shih IM, Weiss SW, Folpe AL. Microphthalmia transcription factor and melanoma cell adhesion molecule expression distinguish desmoplastic/spindle cell melanoma from morphologic mimics. *Am J Surg Pathol* 2001;25:58–64.

40. Kanik AB, Yaar M, Bhawan J. p75 nerve growth factor receptor staining helps identify desmoplastic and neurotropic melanoma. *J Cutan Pathol* 1996;23:205–210.

41. Busam KJ, Zhao H, Coit DG, *et al.* Distinction of desmoplastic melanoma from non-desmoplastic melanoma by gene expression profiling. *J Invest Dermatol* 2005;124:412–418.

42. Gutzmer R, Herbst RA, Mommert S, *et al.* Allelic loss at the neurofibromatosis type 1 (NF1) gene locus is frequent in desmoplastic neurotropic melanoma. *Hum Genet* 2000;107:357–361.

Clear cell sarcoma

43. Enzinger FM. Clear cell sarcoma of tendons and aponeuroses: an analysis of 21 cases. *Cancer* 1965;18:1163–1174.

44. Peulve P, Michot C, Vannier JP, Tron P, Hemet J. Clear cell sarcoma with t(12;22)(q13–14;q12). *Genes Chromosomes Cancer* 1991;3:400–402.

45. Sandberg AA, Bridge JA. Updates on the cytogenetic and molecular genetics of bone and soft tissue tumors: clear cell sarcoma (malignant melanoma of soft parts). *Cancer Genet Cytogenet* 2001;130:1–7.

46. Chung EM, Enzinger FM. Malignant melanoma of soft parts: reassessment of clear cell sarcoma. *Am J Surg Pathol* 1983;7:405–413.

47. Sara AS, Evans HL, Benjamin RS. Malignant melanoma of soft parts (clear cell sarcoma): a study of 17 cases, with emphasis on prognostic factors. *Cancer* 1990;65:367–374.

48. Lucas DR, Nascimento AG, Sim FH. Clear cell sarcoma of soft tissues: Mayo Clinic experience with 35 cases. *Am J Surg Pathol* 1992;16:1197–1204.

49. Montgomery EA, Meis JM, Ramos AG, Martz KL. Clear cell sarcoma of tendons and aponeuroses: a clinicopathologic study of 58 cases with analysis of prognostic factors. *Int J Surg Pathol* 1993;1:89–100.

50. Deenik W, Mooi WJ, Rutgers EJ, *et al.* Clear cell sarcoma (malignant melanoma) of soft parts: a clinicopathologic study of 30 cases. *Cancer* 1999;86:969–975.

51. Ferrari A, Casanova M, Bisogno G, *et al.* Clear cell sarcoma of tendons and aponeuroses in pediatric patients: report from the Italian and German Soft Tissue Sarcoma Cooperative Group. *Cancer* 2002;984:3269–3276.

52. Coindre JM, Hostein I, Terrier P, *et al.* Diagnosis of clear cell sarcoma by real-time reverse transcriptase-polymerase chain reaction analysis of paraffin-embedded tissues: clinicopathologic and molecular analysis of 44 patients from the French sarcoma group. *Cancer* 2006;107:1055–1064.

53. Kawai A, Hosono A, Nakayama R, *et al.* Clear cell sarcoma of tendons and aponeuroses: a study of 75 patients. *Cancer* 2007;109:109–116.

54. Hisaoka M, Ishda T, Kuo TT, et al. Clear cell sarcoma of soft tissue: a clinicopathological, immunohistochemical, and molecular analysis of 33 cases. *Am J Surg Pathol* 2008;32:452–460.

55. Bianchi G, Charoenlap C, Cocchi S, et al. Clear cell sarcoma of soft tissue: a retrospective review and analysis of 31 cases treated at Instituto Ortopedico Rizzoli. *Eur J Surg Oncol* 2014;40:505–510.

56. Rubin BP, Fletcher JA, Renshaw AA. Clear cell sarcoma of soft parts: report of a case primary in the kidney with cytogenetic confirmation. *Am J Surg Pathol* 1999;23:589–594.

57. Swanson PE, Wick MR. Clear cell sarcoma: an immunohistochemical analysis of six cases and comparison with other epithelioid neoplasms of soft tissue. *Arch Pathol Lab Med* 1989;113:55–60.

58. Graadt van Roggen JF, Mooi WJ, Hogendoorn PC. Clear cell sarcoma of tendons and aponeuroses (malignant melanoma of soft parts) and cutaneous melanoma: exploring the histogenetic relationship between these two clinicopathological entities. *J Pathol* 1998;186:3–7.

59. Stenman G, Kindblom LG, Angervall L. Reciprocal translocation t(12;22)(q13;q13) in clear cell sarcoma of tendons and aponeuroses. *Genes Chromosomes Cancer* 1992;4:122–127.

60. Reeves BR, Fletcher CD, Gusterson BA. Translocation t(12;22)(q13;q13) is a nonrandom rearrangement in clear cell sarcoma. *Cancer Genet Cytogenet* 1992;64:101–103.

61. Mrozek K, Karakousis CP, Perez-Mesa C, Bloomfield CD: Translocation t(12;22)(q13;q12.2-12.3) in a clear cell sarcoma of tendons and aponeuroses. *Genes Chromosomes Cancer* 1993;6:249–252.

62. Fujimura Y, Ohno T, Siddique H, et al. The EWS-ATF-1 gene involved in malignant melanoma of soft parts with t(12;22) chromosome translocation encodes a constitutive transcriptional activator. *Oncogene* 1996;4:159–167.

63. Pellin A, Monteagudo C, Lopez-Gines C, et al. New type of chimeric fusion product between the EWS and ATFI genes in clear cell sarcoma (malignant melanoma of soft parts). *Genes Chromosomes Cancer* 1998;23:358–360.

64. Sonobe H, Taguchi T, Shimizu K, et al. Further characterization of the human clear cell sarcoma cell line HS-MM demonstrating a specific t(12;22)(q13;q12) translocation and hybrid EWS/ATF-1 transcript. *J Pathol* 1999;187:594–597.

65. Bosilevac JM, Olsen RJ, Bridge JA, Hinrichs SH. Tumor cell viability in clear cell sarcoma requires DNA binding activity of the EWS/ATF1 fusion protein. *J Biol Chem* 1999;274:34811–34818.

66. Aue G, Hedges LK, Schwartz HS, et al. Clear cell sarcoma or malignant melanoma of soft parts: molecular analysis of microsatellite instability with clinical correlation. *Cancer Genet Cytogenet* 1998;105:24–28.

67. Segal NH, Pavlidis P, Noble WS, et al. Classification of clear cell sarcoma as a subtype of melanoma by genomic profiling. *J Clin Oncol* 2003;21:1775–1781.

68. Langezaal SM, Graadt van Roggen JF, Cleton-Jansen AM, Baelde JJ, Hogendoorn PC. Malignant melanoma is genetically distinct from clear cell sarcoma of tendons and aponeuroses (malignant melanoma of soft parts). *Br J Cancer* 2001;84:535–538.

69. Antonescu CR, Tschermyavsky SJ, Woodruff JM, et al. Molecular diagnosis of clear cell sarcoma: detection of EWS-ATF1 and MITF-M transcripts and histopathological and ultrastructural analysis of 12 cases. *J Mol Diagn* 2002;4:44–52.

70. Panagopoulos I, Mertens F, Debiec-Rychter M, et al. Molecular genetic characterization of the EWS/ATF1 fusion gene in cell sarcoma of tendons and aponeuroses. *Int J Cancer* 2002;99:560–567.

71. Hallor KH, Mertens F, Jin Y, et al. Fusion of the EWSR1 and ATF1 genes without expression of the MITF-M transcript in angiomatoid fibrous histiocytoma. *Genes Chromosomes Cancer* 2005;44:97–102.

72. Patel RM, Downs-Kelly F, Weiss SW, et al. Dual-color, break-apart fluorescence in situ hybridization for EWS gene rearrangement distinguishes clear cell sarcoma of soft tissue from malignant melanoma. *Mod Pathol* 2005;18:1585–1590.

73. Panagopoulos I, Mertens F, Isaksson M, Mandahl N. Absence of mutations of the BRAF gene in malignant melanoma of soft parts (clear cell sarcoma of tendons and aponeuroses). *Cancer Genet Cytogenet* 2005;156:74–76.

74. Yang L, Chen Y, Cui T, et al. Identification of biomarkers to distinguish clear cell sarcoma from malignant melanoma. *Hum Pathol* 2012;43:1463–1470.

75. Wang WL, Mayordomo E, Zhang W, et al. Detection and characterization of EWSR1/ATF1 and EWSR1/CREB1 chimeric transcripts in clear cell sarcoma (melanoma of soft parts). *Mod Pathol* 2009;22:1201–1209.

Gastrointestinal clear cell sarcoma

76. Ekfors TO, Kujari H, Isomaki M. Clear cell sarcoma of tendons and aponeuroses (malignant melanoma of soft parts) in the duodenum: the first visceral case. *Histopathology* 1993;22:255–259.

77. Donner LR, Trompler RA, Dobin S. Clear cell sarcoma of the ileum: the crucial role of cytogenetics for the diagnosis. *Am J Surg Pathol* 1998;22:121–124.

78. Pauwels P, Debiec-Rychter M, Sciot R, et al. Clear cell sarcoma of the stomach. *Histopathology* 2002;41:526–530.

79. Zambrano E, Reyes-Mugica M, Franchi A, Rosai J. An osteoclast-rich tumor of the gastrointestinal tract with features resembling clear cell sarcoma of soft parts: reports of 6 cases of a GIST simulator. *Int J Surg Pathol* 2003;11:75–81.

80. Antonescu CR, Nafa K, Segal NH, Dal Cin P, Ladanyi M. EWS-CREB1: a recurrent variant fusion in clear cell sarcoma. Association with gastrointestinal location and absence of melanocytic differentiation. *Clin Cancer Res* 2006;12:5356–5362.

81. Stockman DL, Miettinen M, Suster S, et al. Malignant gastrointestinal neuroectodermal tumor: clinicopathologic, immunohistochemical, ultrastructural, and molecular analysis of 16 cases with a reappraisal of clear cell sarcoma-like tumors of the gastrointestinal tract. *Am J Surg Pathol* 2012;36:857–868.

Pigmented neuroectodermal tumor of infancy

82. Cutler LS, Chaudry AP, Topazian R. Melanotic neuroectodermal tumor of infancy: an ultrastructural study,

literature review, and re-evaluation. *Cancer* 1981;48:257–270.

83. Johnson RE, Scheithauer BW, Dahlin DC. Melanotic neuroectodermal tumor of infancy: a review of seven cases. *Cancer* 1983;52:661–666.

84. Pettinato G, Manivel JC, D'Amore ES, Jaszcz W, Gorlin RL. Melanotic neuroectodermal tumor of infancy: a reexamination of a histogenetic problem based on immunohistochemical, flow cytometric, and ultrastructural study of 10 cases. *Am J Surg Pathol* 1991;15:233–245.

85. Kapadia SB, Frisman DM, Hitchcock CL, Ellis GL, Popek EJ. Melanotic neuroectodermal tumor of infancy: clinicopathological, immunohistochemical, and flow cytometric study. *Am J Surg Pathol* 1993;17:566–573.

86. Pierre-Kahn A, Cinalli G, Lellouch-Tubiana A, et al. Melanotic neuroectodermal tumor of the skull and meninges in infancy. *Pediatr Neurosurg* 1992;18:6–15.

87. Yu JS, Moore MR, Kupsky WJ, Scott RM. Intracranial melanotic neuroectodermal tumor of infancy: two case reports. *Surg Neurol* 1992;37:123–129.

88. Ricketts RR, Majmuddar B. Epididymal melanotic neuroectodermal tumor of infancy. *Hum Pathol* 1985;16:416–420.

89. Calabrese F, Danieli D, Valente M. Melanotic neuroectodermal tumor of the epididymis in infancy: case report and review of the literature. *Urology* 1995;46:415–418.

90. Kobayashi T, Kunimi K, Imao T, et al. Melanotic neuroectodermal tumor of infancy in the epididymis: case report and literature review. *Urol Int* 1996;57:262–265.

91. Raju U, Zarbo RJ, Regezi JA, Krutchkoff D, Perrin E. Melanotic neuroectodermal tumor of infancy: intermediate filaments, neuroendocrine and melanoma-associated antigen profiles. *Appl Immunohistochem* 1993;1:69–76.

92. Dooling EC, Chi JG, Gilles FH. Melanotic neuroectodermal tumor of infancy: its histological similarities to fetal pineal gland. *Cancer* 1977;39:1535–1541.

93. Navas-Palacios JJ. Malignant melanotic neuroectodermal tumor: light and electron microscopic study. *Cancer* 1980;46:529–536.

94. Khoddami M, Squire J, Zielenska M, Thorner P. Melanotic neuroectodermal tumor of infancy: a molecular genetic study. *Pediatr Dev Pathol* 1998; 1:295–299.

Glial hamartoma and soft tissue gliomatosis

95. Hirsh LF, Stool SE, Langfitt RF, Schut L. Nasal glioma. *J Neurosurg* 1977;46:85–91.

96. Azumi N, Matsuno T, Tateyama M, Inoue K. So-called nasal glioma. *Acta Pathol Jpn* 1984;34:215–220.

97. Fletcher CD, Carpenter G, McKee PH. Nasal glioma: a rarity. *Am J Dermatopathol* 1986;8:341–346.

98. Shepherd NA, Coates PJ, Brown AA. Soft tissue gliomatosis – heterotopic glial tissue in the subcutis: a case report. *Histopathology* 1987;11:655–660.

99. McDermott MB, Glasner SD, Nielsen PL, Dehner LP. Soft tissue gliomatosis: morphologic unity and histogenetic diversity. *Am J Surg Pathol* 1996;20:148–155.

Ependymoma

100. Wolff M, Santiago H, Duby MM. Delayed distant metastasis from a subcutaneous sacrococcygeal myxopapillary ependymoma: case report, with tissue culture, ultrastructural observations, and review of the literature. *Cancer* 1972;30:1046–1067.

101. Helwig EB, Stern JB. Subcutaneous sacrococcygeal myxopapillary ependymoma: a clinicopathologic study of 32 cases. *Am J Clin Pathol* 1984;81:156–161.

102. Kindblom LG, Lodding P, Hagmar B, Stenman G. Metastasizing myxopapillary ependymoma of the sacrococcygeal region: a clinicopathologic, light- and electron microscopic, immunohistochemical, tissue culture, and cytogenetic analysis of a case. *Acta Pathol Microbiol Immunol Scand A* 1986;94:79–90.

103. Sonneland PR, Scheithauer BW, Onofrio BM. Myxopapillary ependymoma: a clinicopathologic and immunocytochemical study of 77 cases. *Cancer* 1985;56:883–893.

104. Pulitzer DR, Martin PC, Collins PC, Ralph DR. Subcutaneous sacrococcygeal ("myxopapillary") ependymal rests. *Am J Surg Pathol* 1988;12:672–677.

105. Sawyer JR, Miller JP, Ellison DA. Clonal telomeric fusions and chromosome instability in a subcutaneous sacrococcygeal myxopapillary ependymoma. *Cancer Genet Cytogenet* 1998;100:169–175.

106. Bell DA, Woodruff JM, Scully RE. Ependymoma of the broad ligament. *Am J Surg Pathol* 1984;8:203–209.

107. Duggan MA, Hugh J, Nation JG, Robertson DI, Stuart GCE. Ependymoma of the uterosacral ligament. *Cancer* 1989;64:2565–2571.

108. Kleinman GM, Young RH, Scully RE. Ependymoma of the ovary: report of three cases. *Hum Pathol* 1984;15:632–638.

109. Idowu MO, Rosenblum MK, Wei XJ, Edgar MA, Soslow RA. Ependymomas of the central nervous system and adult extra-axial ependymomas are morphologically and immunohistochemically distinct: a comparative study with assessment of ovarian carcinomas for expression of glial fibrillary protein. *Am J Surg Pathol* 2008;32:710–718.

110. Newton HB, Henson J, Walker RW. Extraneural metastases in ependymoma. *J Neurooncol* 1992;14:135–142.

111. Wilson RW, Moran CA. Primary ependymoma of the mediastinum: a clinicopathologic study of three cases. *Ann Diagn Pathol* 1998;2:293–300.

112. Mogler C, Kohlhof P, Penzel R, et al. A primary malignant ependymoma of the abdominal cavity: a case report and review of the literature. *Virchows Arch* 2009;454:475–478.

Meningioma outside the central nervous system

113. Rushing EJ, Bouffard JP, McCall S, et al. Primary extracranial meningiomas: an analysis of 146 cases. *Head Neck Pathol* 2009;3:116–130.

114. Lopez DA, Silvers DN, Helwig EB. Cutaneous meningiomas: a clinicopathologic study. *Cancer* 1974;34:728–744.

115. Rogers GF, Mulliken JB, Kozakewich HPW. Heterotopic neural nodules of the scalp. *Plast Reconstr Surg* 2005;115:376–382.

116. Sibley DA, Cooper PH. Rudimentary meningocele: a variant of "primary cutaneous meningioma." *J Cutan Pathol* 1989;16:72–80.

117. Argenyi ZB. Cutaneous neural heterotopias and related tumors relevant for the dermatopathologist. *Semin Diagn Pathol* 1996;13:60–71.

118. Argenyi ZB, Thieberg MD, Hayes CM, Whitaker DC. Primary cutaneous meningioma associated with von Recklinghausen's disease. *J Cutan Pathol* 1994;21:549–556.

119. Miyamoto T, Mihara M, Hagari Y, Shimao S. Primary cutaneous meningioma on the scalp: report of two siblings. *J Dermatol* 1995;22:611–619.

120. Suster S, Rosai J. Hamartoma of the scalp with meningothelial elements: a distinctive benign soft tissue lesion that may simulate angiosarcoma. *Am J Surg Pathol* 1990;14:1–11.

121. Theaker JM, Fletcher CD, Tudway AJ. Cutaneous heterotopic meningeal nodules. *Histopathology* 1990;16:475–479.

122. Gelli MC, Pasquinelli G, Martinelli G, Gardini G. Cutaneous meningioma: histochemical, immunohistochemical and ultrastructural investigation. *Histopathology* 1993;23:576–578.

123. Perzin KH, Pushparaj N. Nonepithelial tumors of the nasal cavity, paranasal sinuses, and nasopharynx: a clinicopathologic study. XIII. Meningiomas. *Cancer* 1984;54:1860–1869.

124. Thompson LD, Guyre K. Extracranial sinonasal tract meningiomas: a clinicopathologic study of 30 cases with a review of the literature. *Am J Surg Pathol* 2000;24:640–650.

125. Kuhn C, Askin FB. The fine structure of so-called minute pulmonary chemodectomas. *Hum Pathol* 1975;6:681–691.

126. Gaffey MJ, Mills SE, Askin FB. Minute pulmonary meningothelial-like nodules: a clinicopathologic study of the so-called minute pulmonary chemodectoma. *Am J Surg Pathol* 1988;12:167–175.

127. Niho S, Yokose T, Nishiwaki Y, Mukai K. Immunohistochemical and clonal analysis of minute pulmonary meningothelial-like nodules. *Hum Pathol* 1999;30:425–429.

128. Suster S, Moran C. Diffuse pulmonary meningotheliomatosis. *Am J Surg Pathol* 2007;31:624–631.

129. Williamson BE, Stanton CA, Levine EA. Chest wall metastasis from recurrent meningioma. *Am Surg* 2001;67:966–968.

130. Sadahira Y, Sugihara K, Manabe T. Iatrogenic implantation of malignant meningioma to the abdominal wall. *Virchows Arch* 2001;438:316–318.

131. Ozer E, Kalemei O, Acar UD, Canda S. Pin site metastasis of meningioma. *Br J Neurosurg* 2007;21:524–527.

Chapter 26

Paragangliomas

Markku Miettinen
National Cancer Institute, National Institutes of Health

Paragangliomas include pheochromocytomas and other paraganglionic cell tumors that are neural autonomic nervous-system-related tumors showing neural differentiation. Gastrointestinal autonomic nerve tumor (GANT) is in fact gastrointestinal stromal tumor (GIST) and is discussed in Chapter 17. Neuroblastoma, a primitive childhood tumor with sympathetic nerve differentiation, and ganglioneuroma, the most differentiated member of the same family, are included in Chapter 31.

Overview of the paraganglia

The *paraganglia* are collections of specialized neural cells that serve neurosecretory and neuroreceptive functions at various sites, such as oxygen sensing in the carotid bodies. The paraganglia include the adrenal medulla, and small paraganglia along the para-aortic sympathetic nervous chains around the aorta in the abdomen and thorax (sympathetic paraganglia), and associated with the parasympathetic nervous system, especially in the neck, linked with the carotid arteries (carotid body) and the vagus nerve (vagal paraganglia). Organ-based paraganglia include the adrenal medulla and small clusters of paraganglionic cells that occur in the urinary bladder and the mesenteries, and around the gallbladder, heart, and lungs, among others. Adrenal medulla and some other paraganglia secrete catecholamines that regulate the heart and the cardiovascular system.[1]

Normal paraganglia are composed of neural cells, the chief cells that are typically clustered as spherical collections ("cell balls," originally described as *Zellballen* in German texts). These structures are sometimes seen in surgical specimens, especially from the abdomen (Figure 26.1).

The paraganglioma chief cells are surrounded by Schwann-cell-like elongated cells, the so-called sustentacular cells. The chief cells are positive for neural markers, including neuron-specific enolase (NSE), chromogranin, synaptophysin, and neurofilament protein, whereas the spindled Schwann-cell-like sustentacular cells are positive for S100 protein and sometimes for GFAP.

Overview of paragangliomas

Paragangliomas can be grouped in many ways by several parameters into variably distinctive clinicopathologic groups (Table 26.1). The most common are adrenal medullary paragangliomas (pheochromocytomas), which constitute over one-half of all paragangliomas. They are usually hormonally active, with catecholamine production. Extra-adrenal paragangliomas compose the other half of all paragangliomas and occur in several sites, most commonly in the retroperitoneum and neck (Figure 26.2). Their classification was developed by Glenner and Grimley.[2] Extra-adrenal retroperitoneal paragangliomas, those arising in the posterior mediastinum adjacent to the sympathetic chain (aortosympathetic paragangliomas) and urinary bladder paragangliomas are closely related to the adrenal types, because they also often produce catecholamines.

Head and neck paragangliomas (those of the carotid body, glomus jugulare, and vagal body) related to the parasympathetic

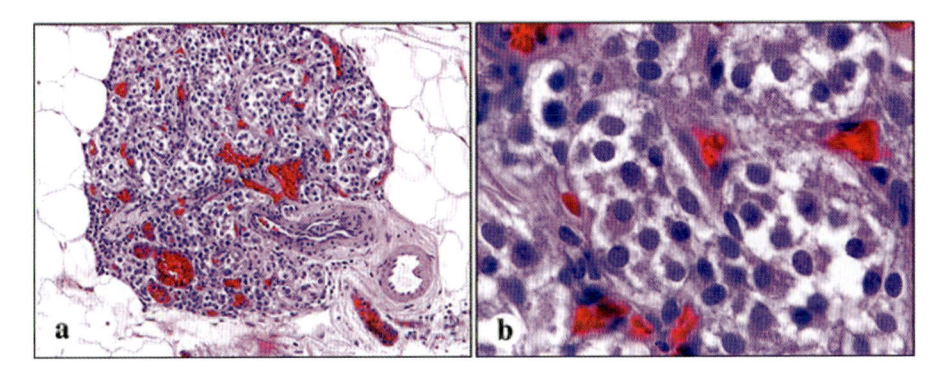

Figure 26.1 A small visceral paraganglion located in the fat around the gallbladder. Note an organoid pattern, similar to the one commonly seen in paragangliomas.

Modern Soft Tissue Pathology, Second Edition, ed. Markku Miettinen. Published by Cambridge University Press. © Cambridge University Press 2016.

Table 26.1 Classification of paragangliomas by different sets of parameters

Grimley–Glenner classification of extra-adrenal paragangliomas

Sympathicoadrenal (retroperitoneal)
Branchiomeric (neck)
Aorticopulmonary
Visceral

Location and paraganglionic origin

Adrenal
Extra-adrenal (abdominal)
  Retroperitoneal
  Mesenteric
Urinary bladder
Mediastinal
  Aorticopulmonary
  Posterior (paraspinal)
Head and neck (mostly parasympathetic)
  Carotid body
  Vagal
  Jugulotympanic
  Spinal
  Duodenal (gangliocytic)

Autonomic nervous system type

Sympathicoadrenal
Parasympathetic

Syndrome status

Sporadic
Syndromic (germline mutation syndromes)

Clinical

Benign
Malignant

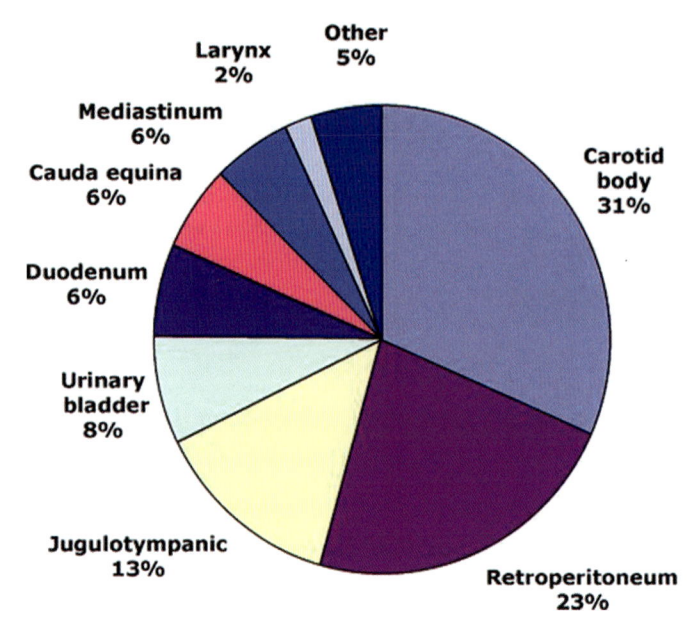

Figure 26.2 Anatomic distribution of extra-adrenal paragangliomas.

- Carotid body 31%
- Retroperitoneum 23%
- Jugulotympanic 13%
- Urinary bladder 8%
- Duodenum 6%
- Cauda equina 6%
- Mediastinum 6%
- Larynx 2%
- Other 5%

nervous system are generally hormonally inactive. A *duodenal gangliocytic paraganglioma* is a unique tumor that combines elements of a paraganglioma with epithelial, ganglionic, and schwannian components; paraganglioma of cauda equina is related to this type. Most paragangliomas are well-differentiated tumors that show a close homology to the corresponding normal paraganglia, the adrenal medulla, extra-adrenal paraganglia, or the paraganglia of the head and neck. Most are clinically benign, but prediction of future behavior is notoriously difficult with this group of tumors.

Paraganglioma syndromes

Paragangliomas are associated with several syndromes, each of which is characterized by specific germline mutations that are autosomally dominantly inherited (Table 26.2). It is difficult to estimate the percentage of syndromic cases, but the total percentage may exceed 40%, especially if syndromic is defined by the presence of germline mutation and genetic testing for germline mutations in each gene-involved syndrome is performed. In general, syndromic paragangliomas occur at an earlier age and are often multicentric. Genetic testing for germline mutations is therefore becoming a new standard of care for patients with paragangliomas to detect an occult hereditary syndrome, because this can help to determine inheritability for genetic counseling, anticipate possible multicentricity, and determine the prognosis in some cases.[3–5]

Paraganglioma syndromes (PGL) associated with losses of the succinate dehydrogenase (SDH) complex

These PGLs have been summarized in Table 26.2. Paraganglioma is the most common tumor associated with losses in the SDH complex, collectively often designated as SDH-deficient tumors. Other tumors in this group include subsets of GISTs of the stomach (Chapter 17), rare renal cell carcinomas, and occasional pituitary adenomas.[6] Up to 30–40% of extra-adrenal paragangliomas are SDH-deficient, whereas only 2% of adrenal pheochromocytomas belong to this group.

SDH (respiratory complex II) is a heteropolymeric enzyme complex located in the inner mitochondrial membrane. It functions in the tricarboxylic acid cycle converting succinate to fumarate, and in the electron transfer of the respiratory chain. All SDH subunit genes are encoded by the chromosomal DNA.[7] Inactivation of the SDH-complex in paragangliomas is typically associated with loss-of-function germline mutations in one of the SDH-subunit genes, combined with a somatic loss of the other allele in tumor cells according to the

Table 26.2 Summary of the paraganglioma syndromes (PGLs) and other most important mutation syndromes that have paraganglioma as a component

Syndrome	Locus and gene name	OMIM number[a]	Type of mutation	Types of paraganglioma
PGL1	11q23 SDHD	#168000	Inactivating	Mostly head and neck
PGL2	11q13.1 SDHAF2	#601650	Inactivating	Head and neck
PGL3	1q23.3 SDHC	#605373	Inactivating	Head and neck and thorax
PGL4	1p36.1-p35 SDHB	#115310	Inactivating	Various
PGL5	5p15 SDHA	#614165	Inactivating	Very rare (new)
MEN2A, MEN2B	10q11.2 RET	#171400, #162300	Activating	Mostly adrenal
Von Hippel–Lindau	3p26-p25 VHL	#193300	Inactivating	Various
Neurofibromatosis 1	17q11.2 NF1	+162200	Inactivating	Mostly adrenal

[a] OMIM = Online Mendelian Inheritance in Man, a database created, maintained, and made available online by Johns Hopkins University, in Baltimore, MD, at National Center for Biotechnology Information (NCBI) website. Description of documentation level for the syndromes: # phenotype and molecular basis known; + gene with known sequence and phenotype.

classic tumor-suppressor gene model.[8–11] Epigenetic silencing may also play a role, as high levels of genomic methylation are seen in SDH-associated paragangliomas.[12]

Any SDH gene mutation causes loss of the SDH-complex function. This is practically detectable by immunohistochemical loss of SDHB in the tumor cells.[13,14] Immunoreactivity for SDHA is lost in those rare tumors that are associated with SDHA losses.[15] The loss of SDH-complex activates pseudohypoxia signaling via HIF1A and HIF2A. This in turn leads to paraganglionic hyperplasia and neoplasia.[16] Paragangliomas observed at high altitude may also be associated with SDH mutations.[17]

The disease penetrance for paraganglioma development in various SDH-syndromes varies, and is often lower than observed with oncogene mutation syndromes. Therefore, SDH-syndromes have been viewed as tumor susceptibility syndromes rather than tumor syndromes. The inheritance pattern of SDHD mutations suggests maternal imprinting, i.e., epigenetic inactivation of the maternal allele. This means that the syndrome is inherited almost exclusively from mutation-carrier fathers.[18] Imprinting might therefore lead to SDHD inactivation without two genetic hits, differing from the situation usually seen in tumor-suppressor gene syndromes.

Histologically SDH-deficient paragangliomas are not distinctive. Their known clinical associations include predilection to extra-adrenal locations for all mutation types. SDHB mutant tumors have a higher rate of malignancy than other paragangliomas. SDHC mutant tumors may have predilection for thorax and SDHD mutant tumors in the head and neck.[3,4] In addition, mutations in SDHAF define a syndrome previously designated as PGL2 that is associated with head and neck paragangliomas.[19]

Multiple endocrine neoplasia 2A and 2B

Multiple endocrine neoplasia (MEN) types 2A and 2B feature usually bilateral adrenal pheochromocytoma, medullary thyroid carcinoma, and in addition, mucosal neuromas in MEN2B. Patients with these syndromes have receptor tyrosine kinase (RET)-activating mutations that drive neural and neuroendocrine tumorigenesis in this disease. Some sporadic nonsyndromic tumors have similar somatic mutations.[20–22] MEN2-syndrome-associated pheochromocytoma does not show any significant malignant potential.[23]

Von Hippel–Lindau disease

Patients with von Hippel–Lindau disease type 2 (VHL-2), characterized by loss-of-function, missense type mutations in the VHL gene, are prone to adrenosympathetic paragangliomas, especially adrenal pheochromocytomas. Other tumors in this syndrome include renal clear cell carcinoma, cerebellar hemangioblastoma, and pancreatic neuroendcrine tumors, and these tumors occur in various frequencies in different forms of VHL-2 disease.[24–27] The pathomechanism of pheochromocytoma genesis is related to VHL protein loss of function, activating pseudohypoxia signaling via overexpression of hypoxia inducible factors (HIFs). This in turn leads to altered expression of genes regulating angiogenesis, mitogenesis, and apoptosis. Sporadic pheochromocytomas do not generally have somatic VHL gene mutations, however. VHL-associated pheochromocytomas are often multiple, but clinical malignancy is uncommon.[27,28]

Neurofibromatosis 1

Neurofibromatosis 1 (NF1) tumor syndrome, with the loss of function of the NF1 tumor-suppressor gene, is associated with pheochromocytoma, among many other tumors and neurofibromas. NF1-related pheochromocytomas often have somatic LOH in the NF1 gene, suggesting biallelic inactivation according to the classic tumor-suppressor gene model. The phenotypic and prognostic profile of NF1-associated pheochromocytomas does not seem to differ from that of sporadic tumors and most of these tumors are benign.[27,29–31]

In addition, frequent somatic mutations in the NF1 gene have been reported in pheochromocytomas unassociated with NF1 syndrome.[32]

Newer mutation syndromes associated with paraganglioma-pheochromocytoma

Other tumor-suppressor genes with loss-of-function mutations (mostly germline) involved in pheochromocytoma/paraganglioma pathogenesis include EGLN1/PHD2 and EGLN2/PHD1 (HIF prolyl hydroxylases),[32,33] and EPAS1 (HIF2A).[34] In the former syndromes, loss of the prolyl hydroxylase proteins upholds the levels of HIFs and promotes pseudohypoxia signaling, whereas gain-of-function mutations in EPAS1 mutation syndrome are causative. Other genes associated with paraganglioma are TMEM127, a negative regulator of the mTOR pathway[35] and KIF1B, a proapoptotic factor for neural crest cells.[36] MAX is a MYC-associated X factor mutated in 1% of paragangliomas.[37,38] Because information on these paraganglioma syndromes is new, their frequency and clinical spectrum is incompletely understood.

Pheochromocytoma (paraganglioma) of the adrenal medulla

Paragangliomas of adrenal medullary origin are called *pheochromocytomas*. The adrenal pheochromocytomas are catecholamine-secreting tumors, and they are almost ten times more common than the closely related extra-adrenal retroperitoneal paragangliomas.

Clinical features

Although many paragangliomas are associated with various germline syndromes, adrenal pheochromocytomas are less often associated with known syndromes.[39] Their population incidence, based on studies from several countries, varies between two and eight per million. The tumor is most common in middle age, but occurs from childhood to old age without a clear predilection for either sex. Occurrence at a young age and multiplicity is typically associated with hereditary syndromes, which together could account for 20% to 25% of all cases.[40,41] The sporadic tumors are often discovered based on hormonal symptoms, such as sustained or paroxysmal hypertension, or paroxysms of sweating, headaches, or attacks of anxiety. Some are incidentally discovered in asymptomatic patients. In clinically suspected cases, the diagnosis is aided by measurement of urine metanephrines, the most relevant catecholamine metabolites in this context. Computerized tomography and scintigraphy with radioiodine-labeled metaiodobenzylguanidine, a catecholamine analog taken up by the tumors, are used to localize the suspected tumors in patients with elevated catecholamine secretion. Prior to modern treatment, many patients died of stroke or a hypertensive crisis attributed to catecholamine release from the tumor.[42]

Malignancy is rare in adrenal pheochromocytomas (2–5%), but it is difficult to predict histologically. Metastases develop in the liver, lungs, and bones, or sometimes in lymph nodes; however, some patients may survive a long time with metastases.

Pathology

Grossly, pheochromocytomas are usually circumscribed nodular tumors. They vary from small (1–2 cm) adrenal medullary nodules to bulky masses that can exceed 20 cm in diameter and 2 kg in weight. Residual normal adrenal cortex is usually found after a careful search as a yellow rim around the tumor. On sectioning, the tumors are soft and often hemorrhagic. The color is usually homogeneously brown to dark brown, but is occasionally pale with brown foci.[43] After the tumor is kept a few days in formalin, the fixative typically turns brown, a feature quite unique for pheochromocytoma. A color change by an oxidative compound in a chromaffin reaction led to the historical designation of pheochromocytoma as a "chromaffin paraglioma."

Histologically, pheochromocytomas vary (Figure 26.3). They typically have an "organoid" structure, being composed of cells arranged in compartments separated by thin vascular septa. Some of the compartments are round (referred to as *Zellballen*), and others are irregularly shaped, trabecular, or poorly developed, with the tumor showing a more solid pattern. The cell size varies from small to large, and shapes vary from round to epithelioid, and are occasionally spindled. The cytoplasm varies from eosinophilic to basophilic; the latter variants often have a granular cytoplasm. Nuclear pleomorphism is not uncommon, and some tumors have nuclear pseudoinclusions, cytoplasmic invaginations (Figure 26.4). Eosinophilic cytoplasmic globules occur in some cases, and a few have lipofuscin (neuromelanin) pigment.[44] Extensive hemorrhage or stromal fibrosis is sometimes present.

Pheochromocytoma-ganglioneuroma composite tumors are rare. There are isolated reports of spindle cell sarcomas, usually MPNST, arising in pheochromocytomas.[1] These tumors are discussed and illustrated in Chapter 24.

Pathology correlated with genetic syndromes

Detailed analysis of pheochromocytomas in patients having MEN2 syndromes has shown multicentric adrenal involvement, a high rate of bilaterality, and concomitant adrenal medullary hyperplasia.[45] In VHL-disease-associated pheochromocytomas, a thick fibrous capsule, myxoid and hyaline stromal change, and lack of adrenal medullary hyperplasia and intracytoplasmic hyaline globules have been reported as typical features.[46]

Prediction of malignancy

Histologic assessment of malignancy is notoriously difficult in paragangliomas. The *pheochromocytoma of adrenal scaled*

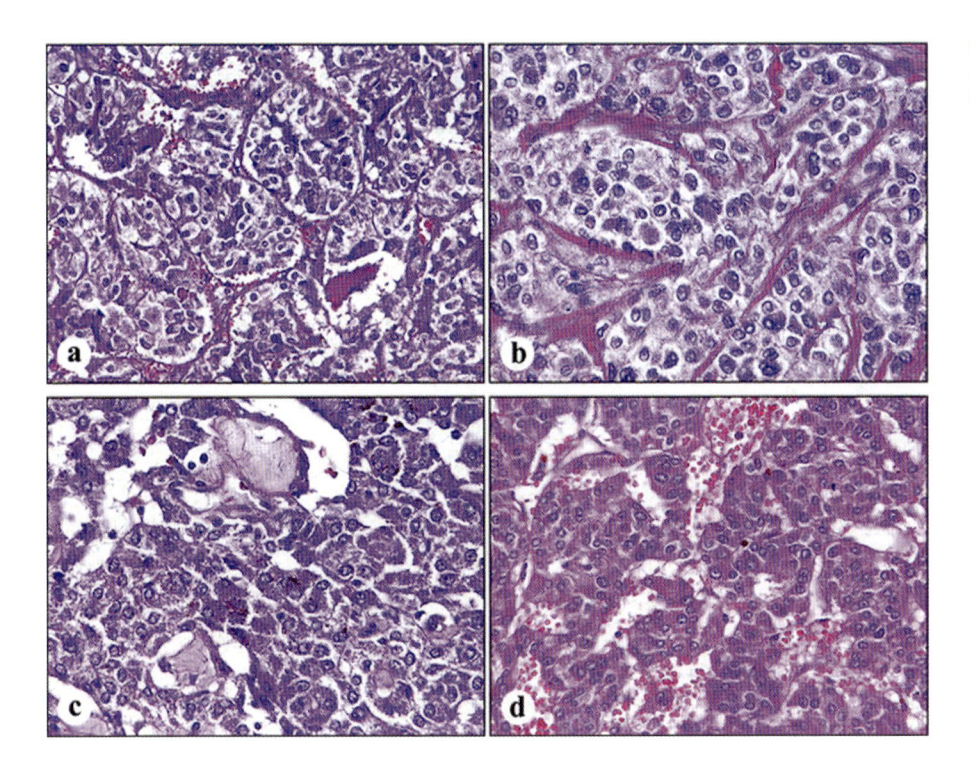

Figure 26.3 Architectural patterns and tinctorial quality of adrenal pheochromocytomas. The organoid pattern is variably developed.

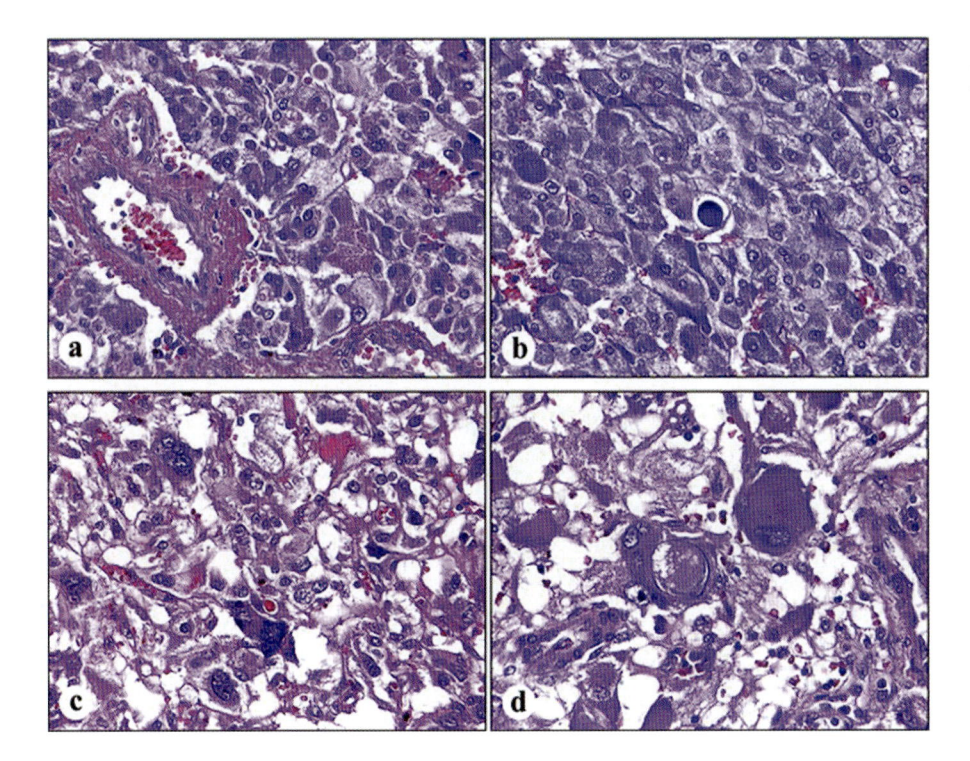

Figure 26.4 Architectural and cytologic features of pheochromocytomas. (**a**) Prominent blood vessels. (**b**) Hyaline globule in a tumor with a basophilic cytoplasm. (**c**) Focal nuclear atypia. (**d**) Focal nuclear atypia and intranuclear inclusion.

score (PASS), developed by Thompson by comparing the features of clinically benign and malignant cases, identified mitotic activity >3 per 10 HPFs, atypical mitoses, tumor cell spindling, cellular monotony, and tumor extension into fat as being among those features specifically seen in malignant examples.[47]

One factor that other studies identified more frequently in malignant tumors is coarse nodularity; vascular invasion and increased mitotic activity were not reliable indicators.[43] In another study, the presence of more than one of the following features was predictive of malignancy: coarse nodularity, tumor necrosis, and absence of hyaline globules.[44]

Immunohistochemistry

Immunohistochemically, the chief cells of pheochromocytoma are consistently positive for the neuroendocrine markers NSE, chromogranin, and synaptophysin. Neurofilaments of 68 kD can be typically demonstrated in frozen sections, but only inconsistently in paraffin-embedded tissue. All components are usually negative for keratins.

The neuropeptides leu- and met-encephalin, somatostatin, and pancreatic polypeptide have been found in most tumors, and other neuropeptides (e.g., ACTH, calcitonin, bombesin) in a few tumors.[48,49] Reduced expression of neuropeptides has been observed in malignant variants.[50] One study identified increased tenascin expression in malignant pheochromocytomas, but the number of observations was small.[51]

A Schwann-cell-like S100-protein-positive spindled sustentacular cell component is variably present around the chief cell compartments.[52,53] This cell population is often less prominent in pheochromocytoma than in carotid body paraganglioma. In contrast to head and neck paragangliomas, the adrenal ones seem to lack GFAP in the sustentacular cells.[54] The sustentacular cell component can be reduced or absent in malignant tumors, but its presence seems to vary from one tumor to another. Pheochromocytomas in MEN patients have been reported to have a greater number of S100-protein-positive sustentacular cells than those associated with NF1 syndrome.[53]

A proliferative index greater than 2% or 2.5% by MIB immunostaining has been suggested to be indicative of malignancy.[54,56] This criterion was not very sensitive for identifying malignant examples, however, because only one-half of the malignant cases had values exceeding the threshold of the two different studies mentioned here.[46,47] Because the MIB indexes of benign and malignant pheochromocytomas overlap, and the differences in the index between benign and malignant pheochromocytomas are often small, the MIB1 index is difficult to utilize in a practical setting.

Ultrastructure

Pheochromocytoma cells are rich in dense core granules of 150 nm to 250 nm. Among them there are granules filled with an electron-dense substance and those with an empty, halo-like space. The morphologic correlation with secretion type seems to be incomplete.[43]

Retroperitoneal extra-adrenal paragangliomas

Pheochromocytoma-like paragangliomas can occur para-aortally in the retroperitoneum and mediastinum, but these are much rarer than those occurring in the adrenals.[57,58] In the retroperitoneum, some of these tumors arise from the aortic paraganglia (organ of Zuckerkandl). Location at the mesenteric arteries is also possible.

Retroperitoneal paragangliomas typically occur in young adults (Figure 26.5). More of these tumors are malignant than are the adrenal paragangliomas; in nonselected series the frequency of malignancy is 15% to 20%.[57] Histologically predictive features are the same as found for adrenal paragangliomas: confluent tumor necrosis, extensive soft tissue or vascular invasion, and lack of hyaline globules. Malignant variants can have low mitotic activity, but the architectural patterns do not seem to have discriminatory value.[44]

The gross, histologic, and immunohistochemical features vary. The tumors often form encapsulated masses, which on

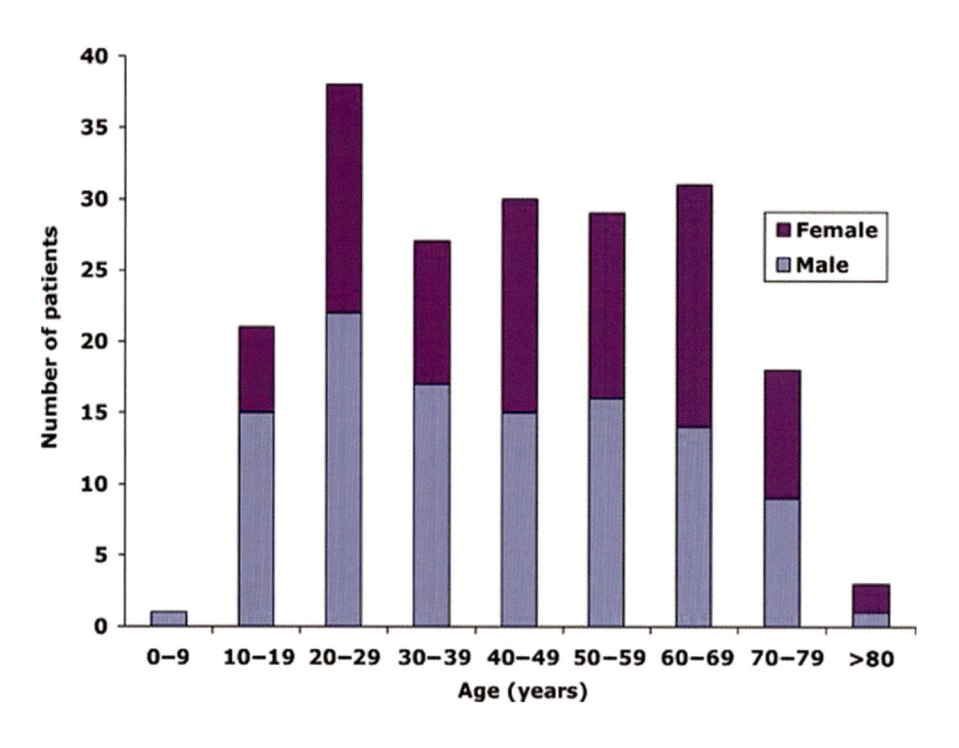

Figure 26.5 Age and sex distribution of 200 patients with retroperitoneal paraganglioma.

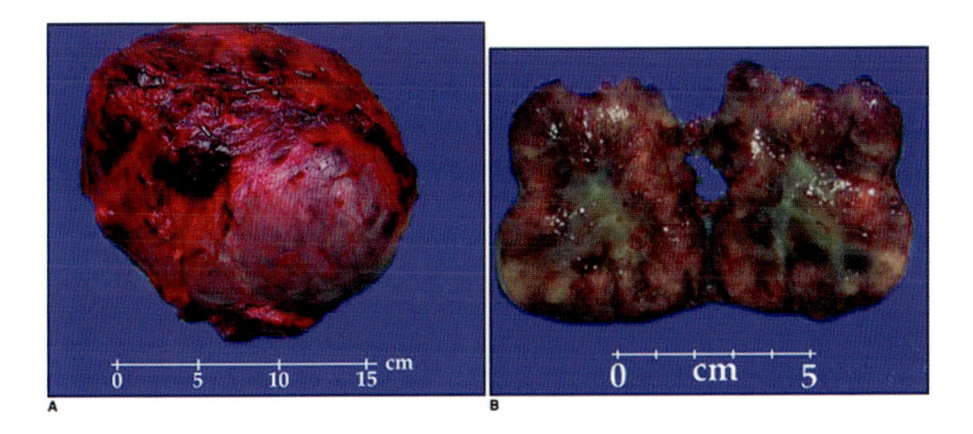

Figure 26.6 Gross appearances of retroperitoneal paraganglioma. (**A**) External view of a large tumor shows a globular, apparently encapsulated mass. (**B**) On sectioning, another example shows a brownish surface and a central area with myxoid hyaline change.

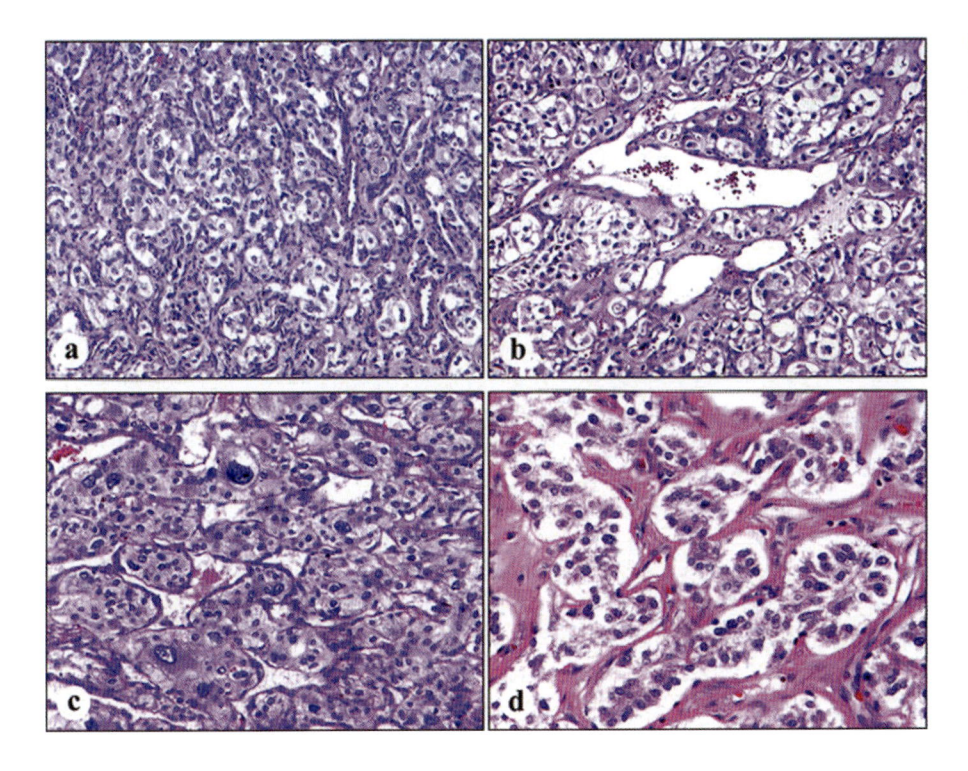

Figure 26.7 Typical organoid patterns at low magnification in retroperitoneal paragangliomas. Note prominent vessels, focal atypia, and variable fibrous stroma.

sectioning are often brown or red (Figure 26.6). The compartmental pattern is variably developed (Figure 26.7) and includes morphologic variations such as pseudoglomerular, trabecular, pseudopapillary, and spindle cell patterns (Figure 26.8). Many examples show histologic and cytologic features, such as cytoplasmic basophilia, closely mirroring those of adrenal pheochromocytoma (Figure 26.9). Pigmented variants occur, and variants with extensive spindling and atypia are among those having increased risk for malignant behavior (Figure 26.10). Prominent nuclear atypia is sometimes present, but is not related to malignancy (Figure 26.11). The pigment is usually lipofuscin, and the pigmented tumors have been negative for S100 protein and HMB45.[58] Similar to adrenal tumors, some retroperitoneal paragangliomas have combined features of pheochromocytomas and ganglioneuromas (Figure 26.12).[1]

Immunohistochemically typical is expression of chromogranin and synaptophysin and NSE, and a minor S100-protein-positive sustentacular cell component is variably present (Figure 26.13). Paragangliomas are negative for keratins, however.

Paraganglioma (pheochromocytoma) of the urinary bladder

These paragangliomas have the character of sympathetic nervous system tumors with hormonal activity, and histologically they resemble the spectrum of pheochromocytoma. They can be associated with paraganglioma syndromes, and these patients can have multiple paragangliomas. Their origin is probably from normal paraganglia in the bladder.

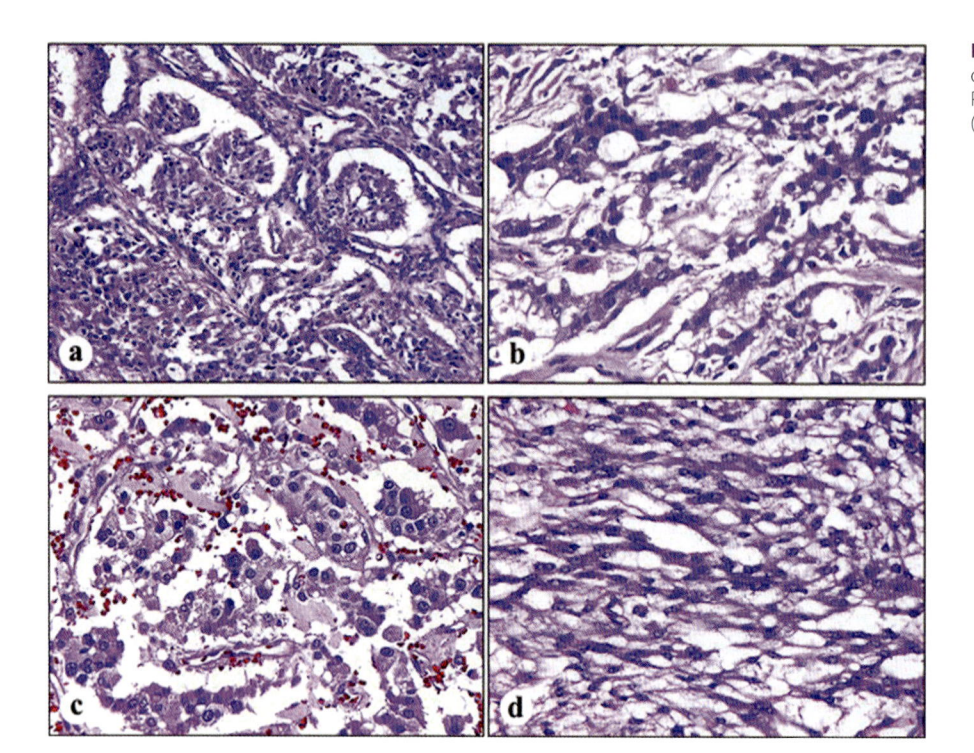

Figure 26.8 Variations in the organoid patterns of retroperitoneal paraganglioma. (**a**) Pseudoglomeruloid appearance. (**b**) Trabecular. (**c**) Pseudopapillary. (**d**) Spindle cell pattern.

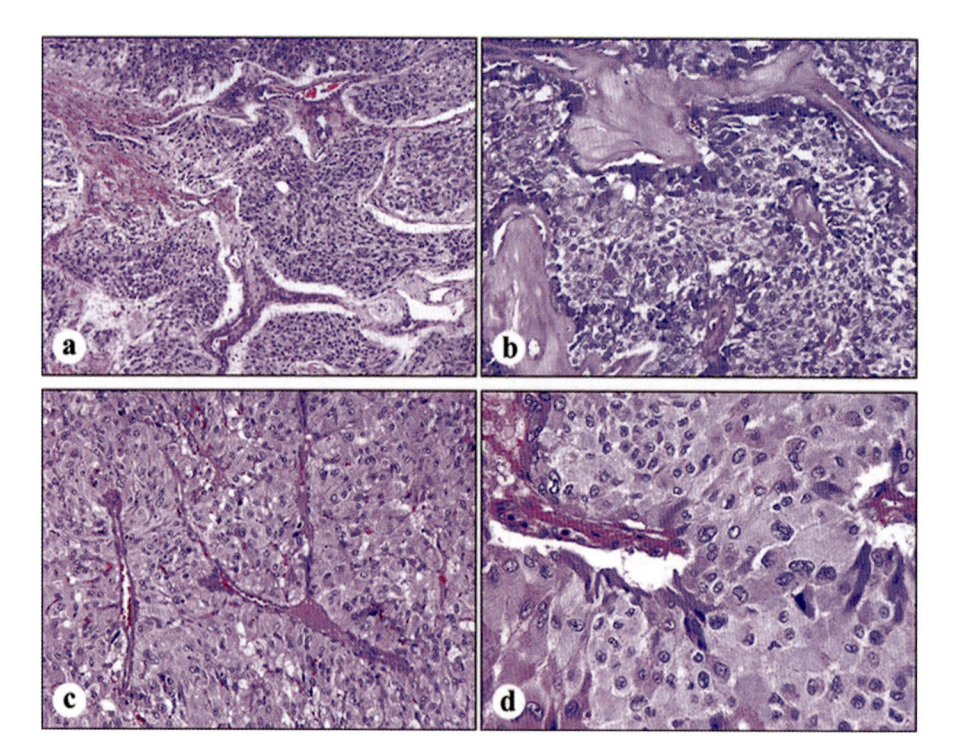

Figure 26.9 Pheochromocytoma-like features in retroperitoneal paraganglioma. Large, sometimes less-defined compartments separated by vessels, sometimes with hyalinized walls are observed.

This finding was demonstrated in more than one-half of the bladders examined in one autopsy study.[59]

Clinical features

Paragangliomas of the urinary bladder are rare, and fewer than 100 cases have been reported. These tumors occur in a wide age range, with a female predominance in some series.[60–62] Patients with these tumors can experience hematuria, or adrenosympathetic hormonal symptoms, or both. In some patients, micturition triggers a paroxysmal attack by catecholamine secretion; however, patients with small tumors can be asymptomatic. The most common location is near the

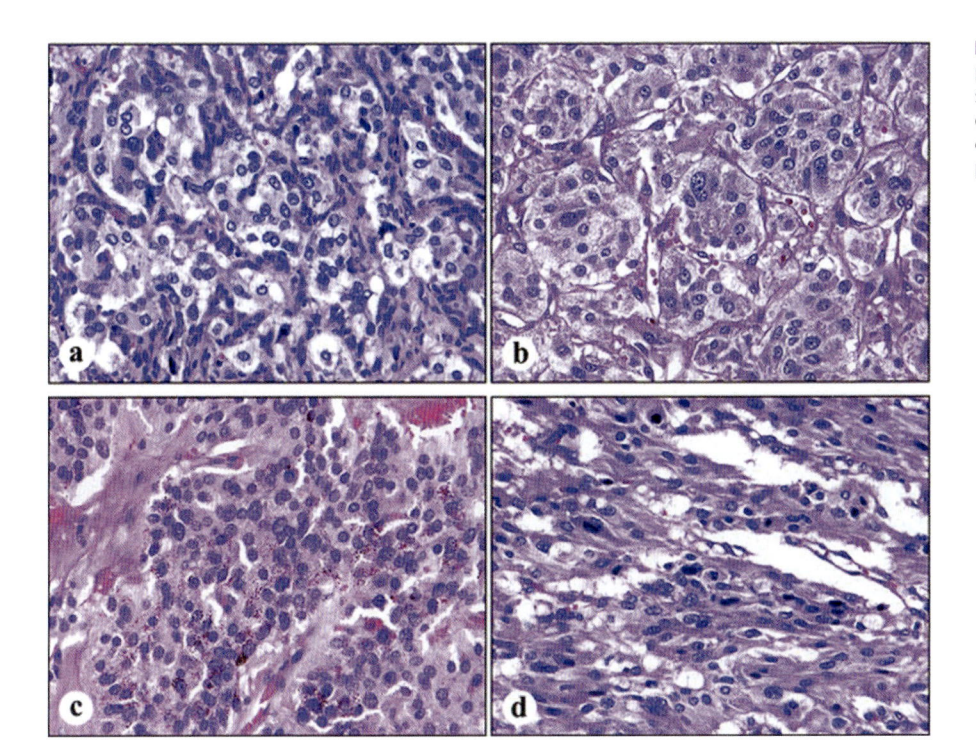

Figure 26.10 Variations of retroperitoneal paraganglioma. (**a**) Organoid pattern with fibrous stroma. (**b**) Organoid pattern with dilated capillaries. (**c**) Pigmented paraganglioma containing lipofuscin. (**d**) Spindle cell paraganglioma with nuclear atypia.

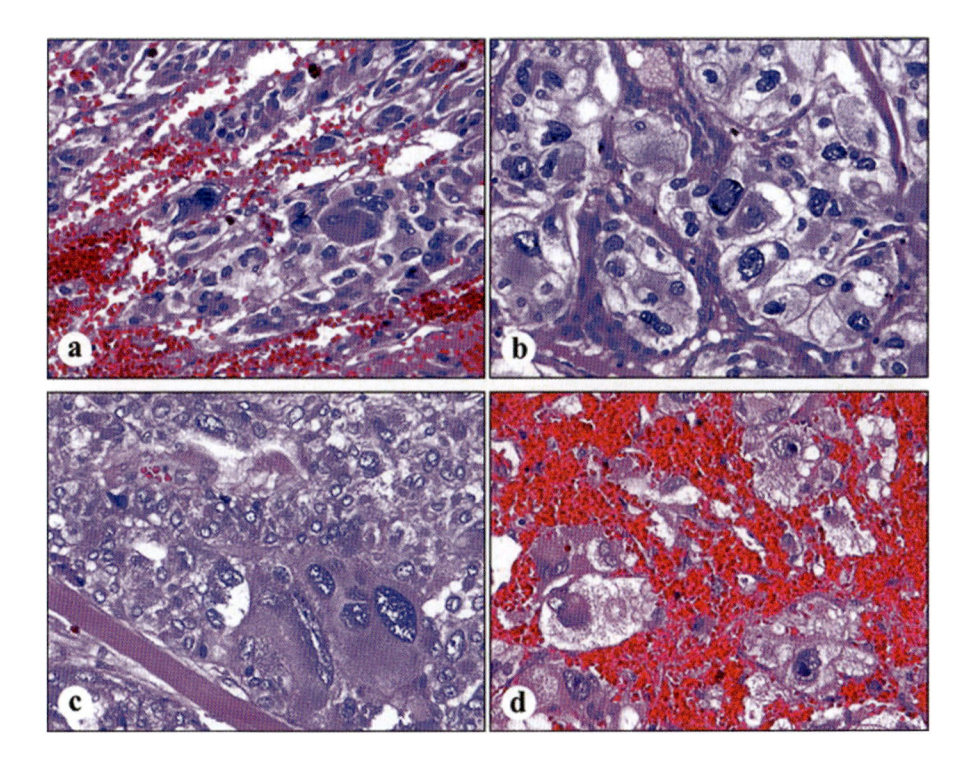

Figure 26.11 Nuclear atypia in retroperitoneal paraganglioma. This feature alone would not define malignancy.

trigonum, often close to the ureteral orifices. A smaller number of paragangliomas occurs in the lateral wall, dome, anterior wall, and elsewhere.[60]

Most tumors can be treated by transurethral resection, but larger and more deeply penetrating tumors might require partial cystectomy. The frequency of malignancy has been 10% to 15%. Malignant behavior is linked to large tumor size and extravesical extension. Late recurrences and metastases can occur, indicating the necessity of long-term follow-up.[61,62]

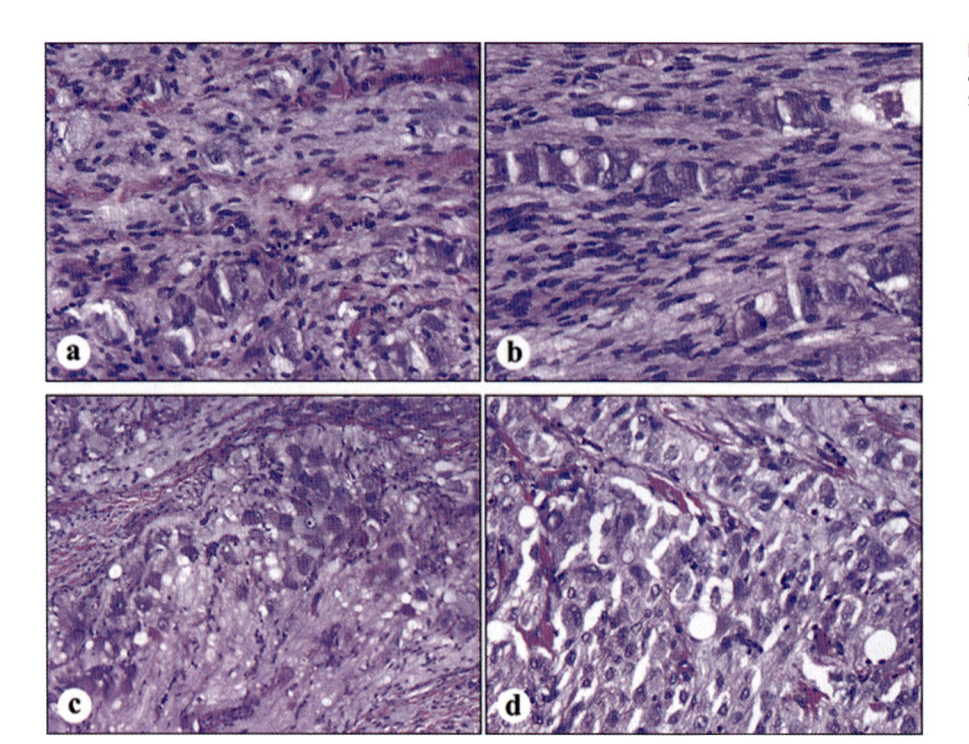

Figure 26.12 Combined pheochromocytoma and ganglioneuroma. Note spindle cell schwannian stroma and ganglion cells.

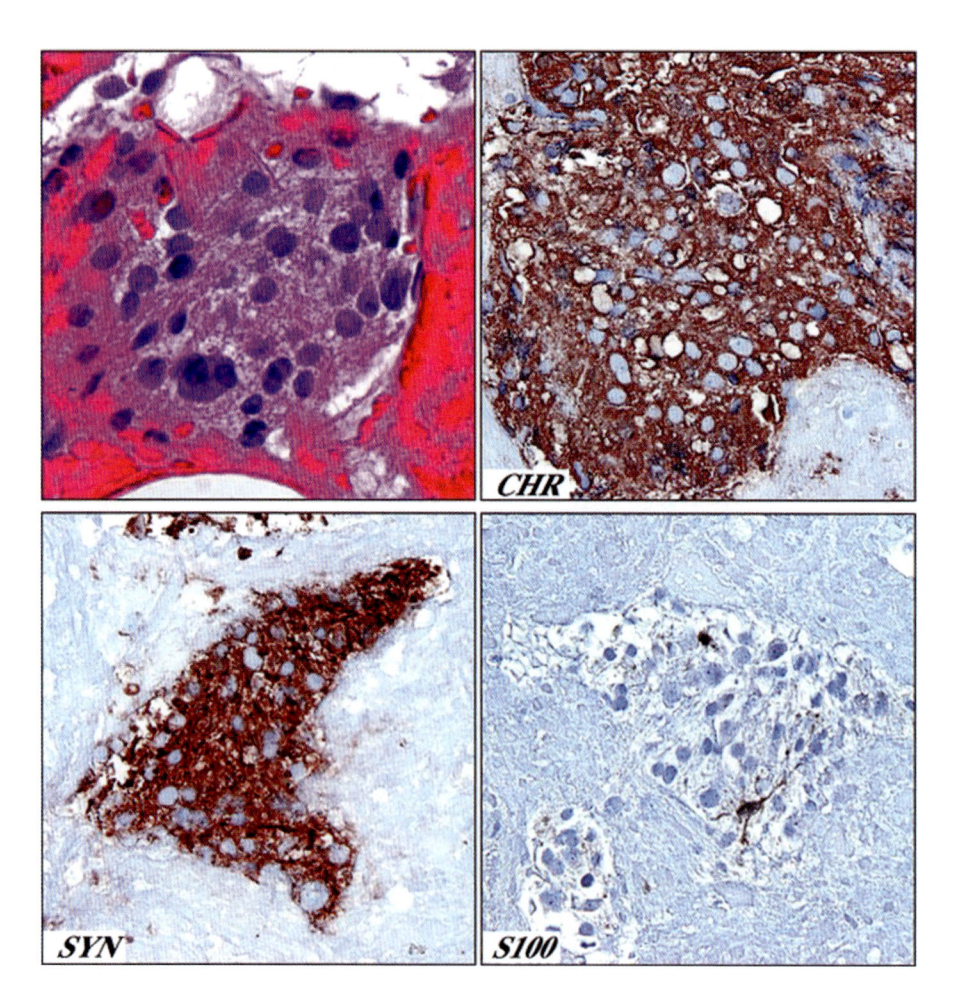

Figure 26.13 Typical immunohistochemical features of a cytologic specimen of a retroperitoneal paraganglioma. The tumor cells are strongly positive for both chromogranin and synaptophysin, and a small number of S100-protein-positive sustentacular cells are present.

Pathology

The tumor often forms a polypoid intraluminal mass. It is typically overlaid by attenuated or sometimes denuded epithelium (Figure 26.14). Histologically, the compartmental pattern is often less developed than in a typical pheochromocytoma. Instead, there is a vague macrotrabecular pattern, with the tumor cells growing as sheets between vascular septa. Cytologically, the tumor cells often have cytoplasmic basophilia, similar to that commonly seen in pheochromocytoma. A few cases have a well-developed compartmental pattern. Focal nuclear atypia is not uncommon, but typical cases do not reveal mitotic activity (Figure 26.15).

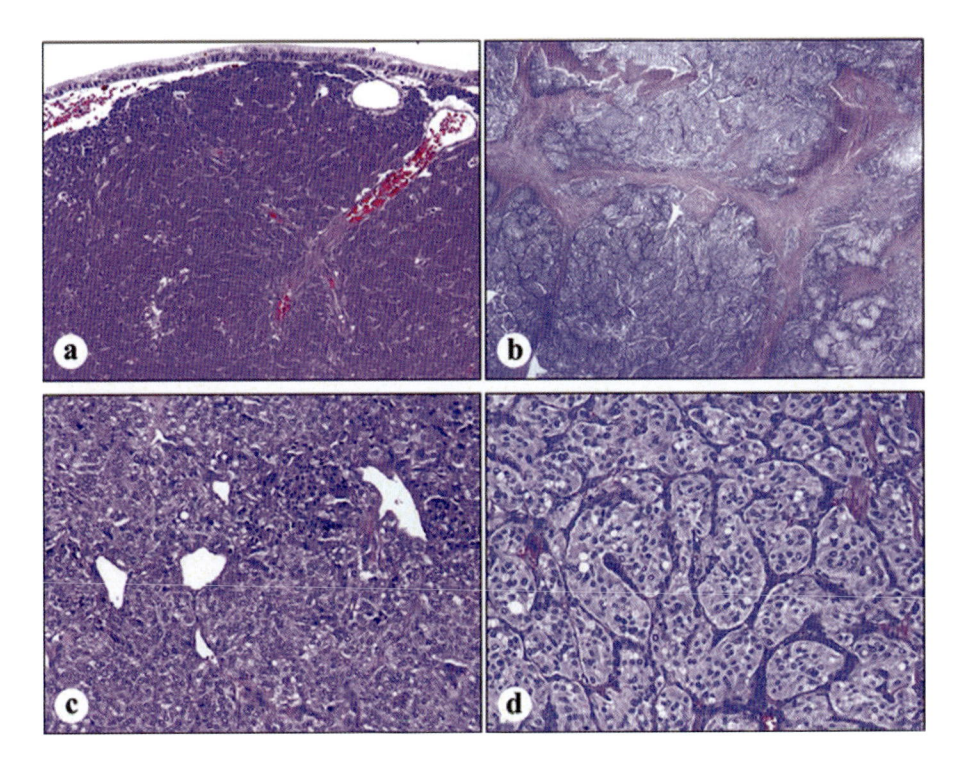

Figure 26.14 A urinary bladder paraganglioma with various architectural patterns. (**a**) A pheochromocytoma-like tumor extending immediately beneath the urothelial surface. (**b**) A tumor infiltrating the bladder wall as large islands consisting of conglomerated compartments. (**c**) A hemangiopericytoma-like pattern. (**d**) A well-developed organoid pattern, a less common occurrence in paragangliomas in the bladder.

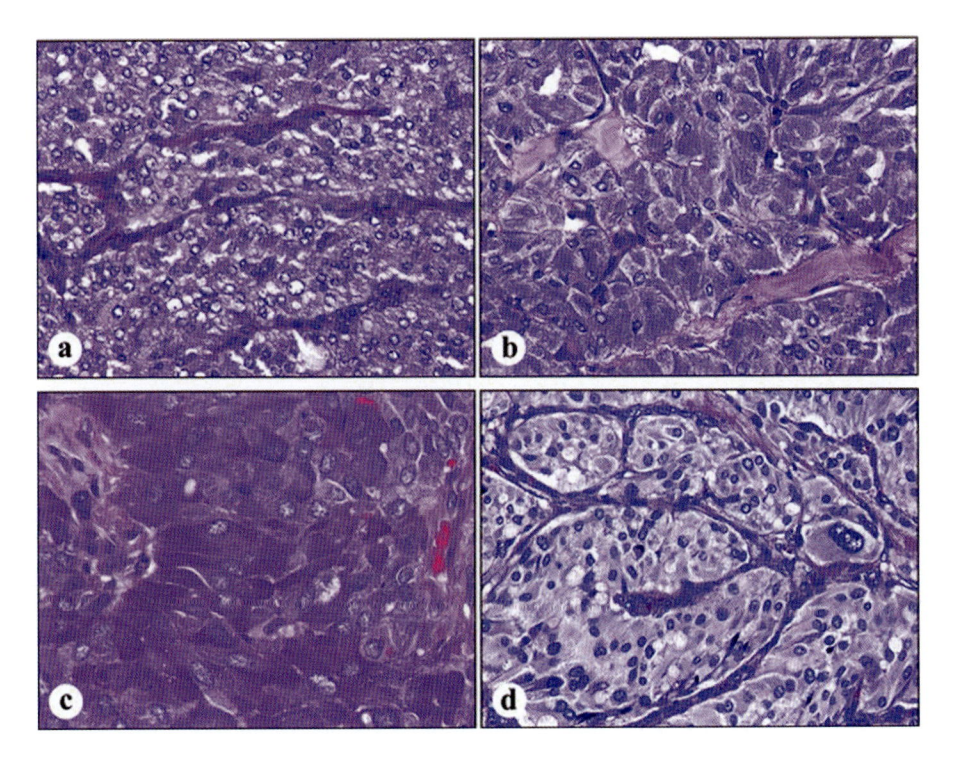

Figure 26.15 Cytoarchitectural features of urinary bladder paragangliomas. (**a**) Trabecular pattern by elongated compartments between thin vascular septa. (**b**) A cytologic appearance with basophilia similar to that commonly seen in pheochromocytoma. (**c**) Variably deeply eosinophilic to amphophilic cytoplasm. (**d**) An organoid example with tumor cells containing pale eosinophilic cytoplasm and focal nuclear atypia.

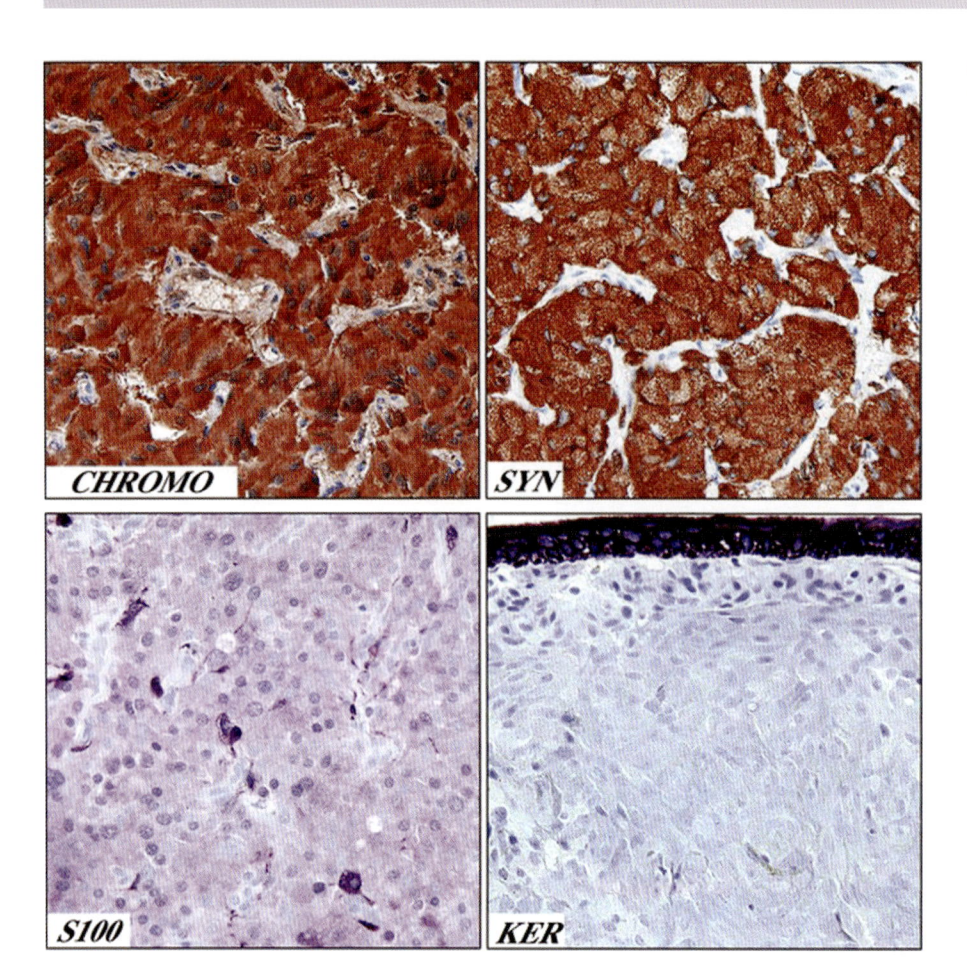

Figure 26.16 Immunohistochemical features of urinary bladder paraganglioma. There is uniform and strong positivity for chromogranin and synaptophysin, whereas the number of S100-protein-positive sustentacular cells is small. The tumor cells are negative for keratins, in contrast with urothelial carcinomas.

The diagnosis is aided by an index of suspicion and immunohistochemical stains for chromogranin and synaptophysin, which are positive. The S100-protein-positive sustentacular cell element is typically poorly represented, and some chief cells have nuclear S100 protein positivity. Unlike bladder carcinomas, paragangliomas are keratin negative (Figure 26.16). Nearly 30% of cases are SDH-deficient (SDHB negative by immunohistochemistry).[63] For small paraganglionic lesions in the bladder, the differential diagnosis of hyperplastic paraganglion versus small paraganglioma can be difficult and ultimately arbitrary.

Thoracic paragangliomas

A small number of paragangliomas present in the mediastinum: either adjacent to the aorta, in the heart, or along the paravertebral sympathetic chain in the costovertebral sulcus.[64–68]

Aorticopulmonary paragangliomas occur along the aortic arch, and by their anatomic location, they usually form anterior or superior mediastinal masses in young to middle-aged adults. Some aorticopulmonary paragangliomas associated with the posterior aspect of the aortic arch also can be characterized as posterior mediastinal tumors. These paragangliomas mostly present as mass lesions and do not secrete catecholamines. Larger tumors can cause superior vena cava syndrome. In a survey of 41 cases, 7 patients developed metastases, 2 died of local complications of inoperable tumor, and 2 died perioperatively.[64]

Cardiac paragangliomas have a predilection for young adults. They are typically 5-cm to 7-cm masses that often secrete catecholamines and are associated with related symptoms, including hypertension and headaches. They are usually located in the posterior wall of the left atrium, and surgical excision is difficult, sometimes requiring coronary artery and atrial wall reconstruction.[66]

Paravertebral paragangliomas are thought to arise from paraganglia along the sympathetic chain located paravertebrally in the posterior mediastinum. They have a predilection for young adults and most frequently occur in the midthoracic region, at the level of the fifth to the seventh ribs. One-half of patients have symptoms associated with catecholamine secretion.[67] Histologically, some of these paragangliomas have pheochromocytoma-like features, with basophilic granular cytoplasm. Of the SDH-deficient syndromes, SDHC-mutant cases may be overrepresented among thoracic paragangliomas.[69]

Paraganglioma of the neck

Carotid body paragangliomas constitute most of the paragangliomas in the neck. Smaller groups include vagal paraganglioma and rare paragangliomas occurring in the larynx, presumably originating from laryngeal paraganglia. Small numbers of paragangliomas have also been reported in the thyroid.[1] In general, these tumors resemble carotid body paragangliomas, both on histological and immunohistochemical studies.

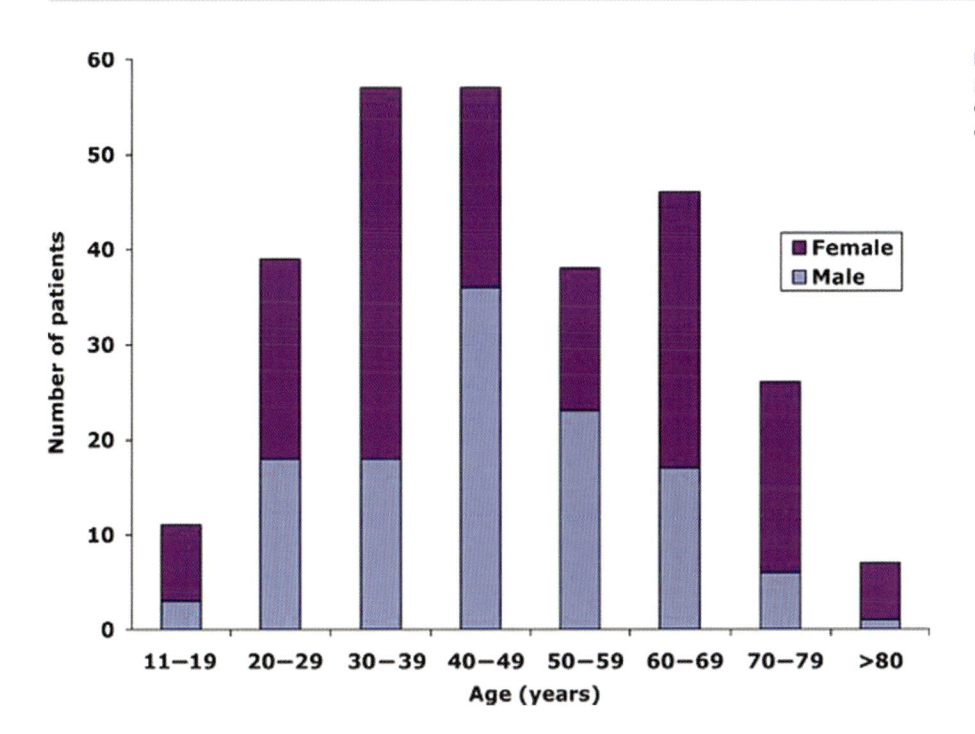

Figure 26.17 Age and sex distribution of paragangliomas in the neck. This group is mainly composed of carotid body tumors, with a minor component of vagal paragangliomas.

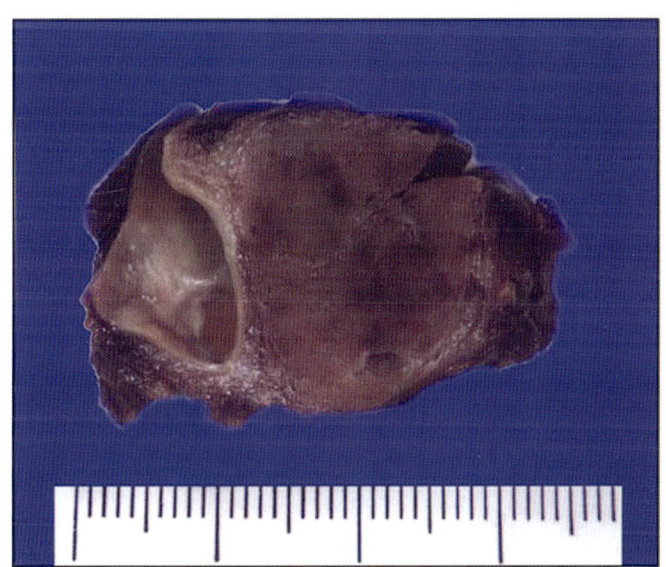

Figure 26.18 Grossly, a carotid body paraganglioma demonstrates close association with the carotid artery. This fixed specimen shows a brown surface on sectioning.

The carotid body paraganglia are located in the carotid bifurcation and act as chemoreceptors for oxygen tension. Chronic hypoxia, such as living at high altitudes or chronic obstructive pulmonary disease, causes hyperplasia of the chief cells, and permanent exposure to high altitude can also predispose to a carotid body tumor. In Peru, the incidence of carotid body tumors has been estimated to be ten times higher in the high altitudes than at sea level.[70–72]

Carotid body paragangliomas

Carotid body paragangliomas generally occur in middle-aged patients and equally in men and women (Figure 26.17). These tumors grow slowly, with an estimated doubling time of 4.2 years in one study.[73] They are usually hormonally silent and become symptomatic by their space-occupying nature. Most tumors measure between 2 cm and 6 cm. They are usually benign, but based on larger series, metastases develop in 5% to 10% of cases, most commonly in bones, lungs, and liver.[74–77] Patients who develop distant metastases have only a 10% to 15% 5-year survival rate, whereas patients with regional (nodal) metastases have 75% 5-year survival rate.[78]

Smaller tumors and those loosely attached to the carotid artery can often be excised without difficulty, but surgery for larger tumors and those attached to the carotid artery often require complex vascular reconstructions. Preoperative embolization of the tumor can facilitate surgery on these highly vascular tumors. SDH-deficiency is common.

Vagal body paragangliomas

Vagal body paragangliomas occur in the head and neck, arising from the vagal body paraganglia almost anywhere in the course of the vagus nerve. They most commonly present in middle-aged patients, with a 2:1 female predominance. Vagal paragangliomas are usually benign, but can reach a size of 5 cm or larger. Tumors extending to the skull base are difficult to manage and can cause serious complications.[79,80] Occasional metastasizing[81] and rare catecholamine-secreting examples have been reported.[82]

Pathology of paragangliomas of the neck

Grossly, the carotid body and vagal body paragangliomas are typically fleshy, red to brownish, homogeneous oval or dumbbell-shaped pseudoencapsulated masses (Figure 26.18). In carotid body tumors, organoid clustering with compartments (*Zellballen*) is typically well developed, but the compartment sizes vary greatly (Figure 26.19). The tumors are highly vascular and can have foci with hemosiderin and fibrosis; they sometimes have a hemangioma or hemangiopericytoma-like overall pattern. Some tumors have wide fibrous septa, and in some cases, extensive sclerosis obscures the paraganglionic component.[83]

Cytologically the tumor cells have variably clear to basophilic to eosinophilic cytoplasm. The nuclei have usually finely distributed chromatin, but atypical, hyperchromatic, bizarre nuclei occur and do not indicate malignancy (Figure 26.20). Nuclear pseudoinclusions similar to those seen in adrenal tumors occur. Mitotic activity is usually inconspicuous. It is difficult to predict tumor behavior

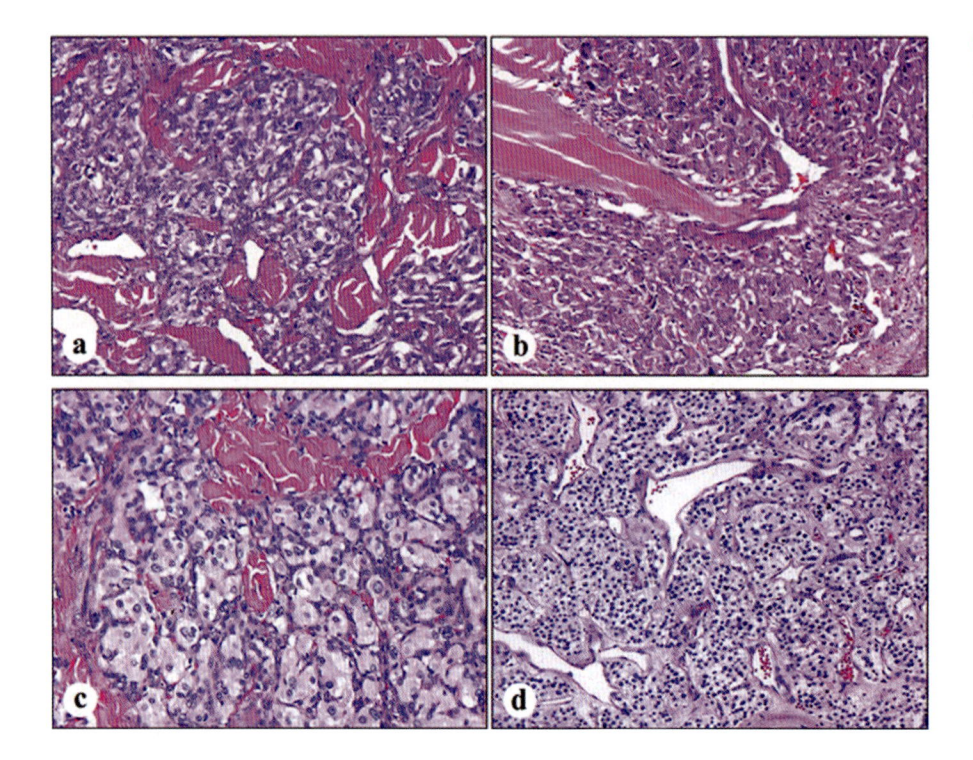

Figure 26.19 Architectural patterns in carotid body paragangliomas. Note the variably defined compartments, often interspersed by fibrous septa or blood vessels, some of which show staghorn morphology similar to the vessels in hemangiopericytoma.

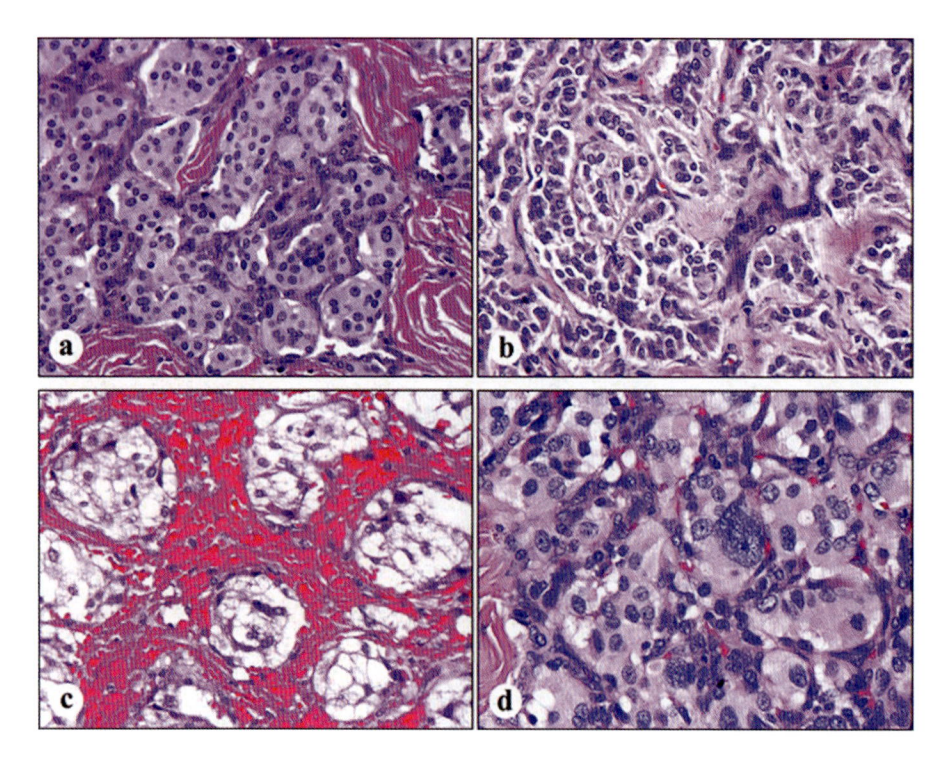

Figure 26.20 (a–c) A closer view of a carotid body paraganglioma, with tumor cells forming compartments or trabecular patterns. The cytoplasm varies from pale eosinophilic to clear. (d) Focal nuclear atypia is not uncommon.

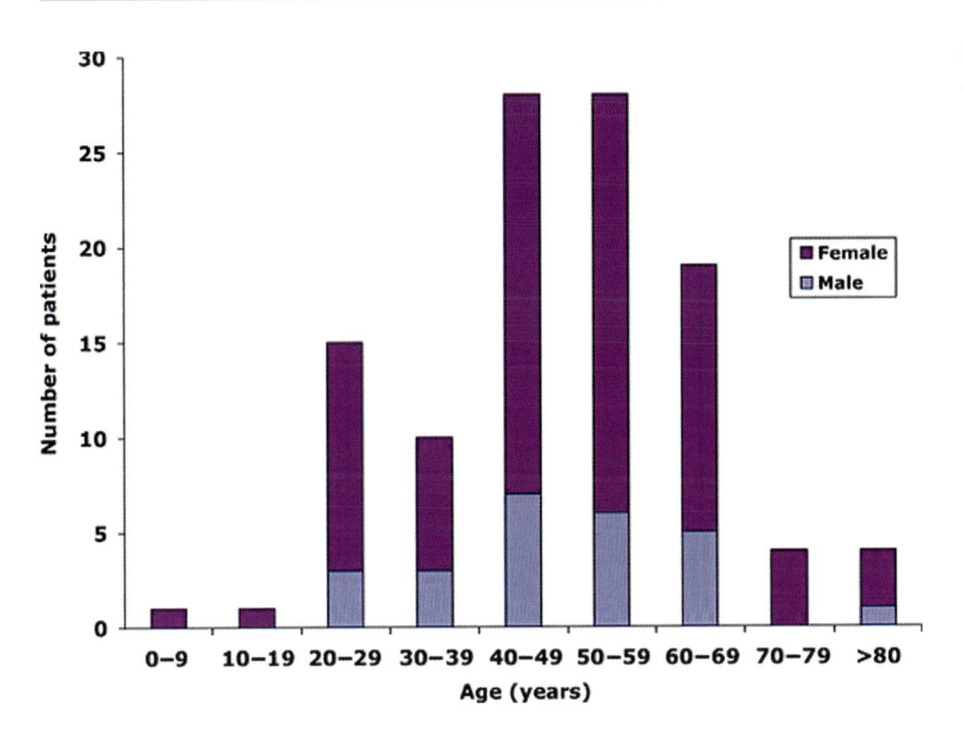

Figure 26.21 Age and sex distribution of jugulotympanic paraganglioma. Note the significant female predominance.

histologically, but vascular invasion, mitotic activity, and tumor necrosis have been suggested as features correlating to malignant behavior.[74–77]

Immunohistochemistry

All head and neck paragangliomas, similar to the adrenal tumors, are positive for the neuroendocrine markers NSE, chromogranin, and synaptophysin, and negative for keratins.[84,85] S100-protein- and GFAP-positive sustentacular cells can usually be identified, but this component might be less prevalent in malignant tumors.[86] SDHB loss is observed in 30% of cases, indicating pathogenesis related to SDH-gene mutations.

Differential diagnosis

The differential diagnosis is site dependent. Neuroendocrine carcinomas with an organoid appearance are composed of epithelial neuroendocrine cells that are positive for keratins and synaptophysin. Among these tumors, medullary carcinoma of the thyroid and its metastases often have a paraganglioma-like organoid pattern simulating paraganglioma. The most common middle ear tumor is middle ear adenoma, the primary carcinoid tumor in this location. In contrast to paraganglioma, this tumor is composed of ribbons of cells, which are epithelial neuroendocrine cells positive for keratins.

Meningioma and epithelioid GISTs of the stomach with organoid appearance and composed of rounded clusters of epithelioid tumor cells should not be confused with paraganglioma.

Alveolar soft part sarcoma, originally thought to be a paraganglioma variant, could structurally simulate the organoid or solid variants of paraganglioma. The cells are larger in this tumor, with ample eosinophilic cytoplasm, and neuroendocrine markers are absent. This tumor is discussed in Chapter 32.

Jugulotympanic paraganglioma

Jugulotympanic paragangliomas are still often referred to as "glomus tumors" or "glomus jugulare tumors" in the clinical literature. They typically occur in middle-aged to older patients, with a female predominance (Figure 26.21). They are usually confined to the middle ear and often cause hearing loss or tinnitus.[86]

Most of these tumors are small middle ear nodules (<1 cm) attached to the tympanic membrane, and they are easily excised. Larger masses invading the petrous bone are difficult to manage, and complications of the tumor excision have to be weighed against risks of a conservative approach or radiation therapy.[87] Radiosurgery (gamma knife) has emerged as a promising, effective, and well-tolerated method of treatment in more extensive tumors.[88]

Histologic specimens of jugulotympanic paragangliomas are often small. Although they can show features similar to other paragangliomas of the neck, many tumors are rich in fibrovascular stroma, with the paraganglioma cells spread as small perivascular cellular clusters (Figure 26.22).

Gangliocytic paraganglioma of duodenum

The gangliocytic paraganglioma is a rare neuroendocrine tumor almost specific to the duodenum, originally described as a duodenal ganglioneuroma.[89–93] Similar histologic features occur in paragangliomas of the cauda equina. The presence of

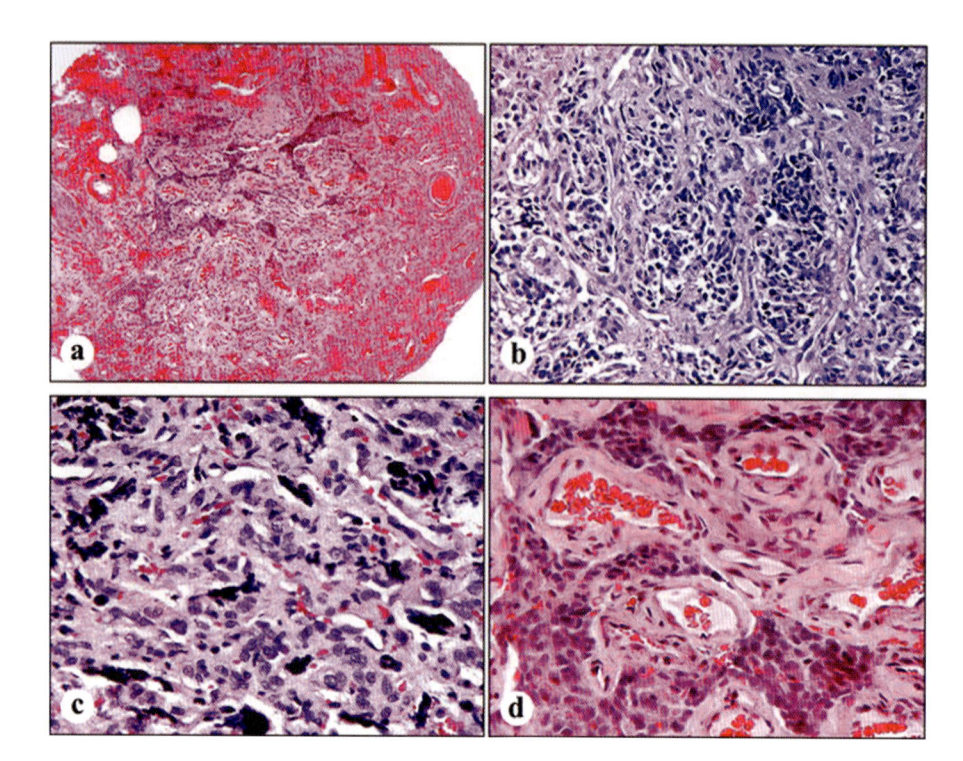

Figure 26.22 (a) Tympanic paragangliomas typically form small nodules in the middle ear. (b) The organoid pattern can be well developed. (c,d) More frequently, there is a vague organoid pattern with the tumor cells surrounding vessels as diffuse arrangements or clusters, with the latter resembling the appearance of a glomus tumor.

pancreatic-like ducts and expression of pancreatic polypeptide has led to the suggestion that this tumor is histogenetically related to ectopic pancreatic tissue.[91]

Clinical features

Gangliocytic paragangliomas occur in adults in a wide age range, with a median age approximately 50 to 55 years and a male predominance. Association with NF1 is rare. Most tumors occur in the second part of the duodenum, and a small number of cases have been reported in the third, first, and fourth parts of the duodenum and the pylorus. Some patients present with gastrointestinal bleeding, a few with biliary obstruction, and some tumors are incidental findings during surgery or autopsy. The behavior is almost invariably benign, but local recurrences and lymph node metastases have been reported on rare occasions.[94,95] Complete conservative excision and follow-up are generally considered sufficient.

Pathology

Grossly, the tumor forms a relatively small (1–3 cm) intraluminal pedunculated or sessile polyp, which is sometimes ulcerated; bigger examples (up to 10 cm) are on record. The tumor involves the submucosa and often infiltrates in the muscularis propria. Histologically this tumor often shows an organoid *Zellballen* or trabecular pattern. The tumor is composed of three components that are present in various proportions: epithelial cells often with spindled morphology, spindled Schwann-like cells resembling those seen in ganglioneuroma, and polygonal ganglion cell components. Some tumors contain psammoma bodies like those more commonly seen in duodenal carcinoids. The epithelial component is often prevalent and forms solid sheets or ribbons (Figure 26.23). Mitotic activity is low.

The epithelial elements are positive for keratins, which are rarely seen in other paragangliomas. The spindle cell component with a Schwann-cell-like appearance is S100 protein positive; S100-protein-positive sustentacular cells can also be seen around the epithelial elements.[91–93,96] The ganglion cells are scattered in and adjacent to the epithelial and schwannian elements, and they are positive for neural markers (e.g., NSE, synaptophysin, neurofilament proteins). Synaptophysin and chromogranin can also be demonstrated in the epithelial component. Like pancreatic neuroendocrine tumors, the epithelial component in gangliocytic paraganglioma can be neurofilament NF68 positive.[95] The neuropeptides and neurotransmitters often detected include pancreatic polypeptide, somatostatin, and serotonin; others have been found sporadically.

Paraganglioma of the cauda equina and spinal canal
Clinical features

This rare paraganglioma usually occurs in middle-aged adult patients and often causes lower back pain and sometimes sciatica symptoms: incontinence or paraplegia (cauda equina syndrome). The largest series reported an 18:13 male predominance.[81] The tumor is usually well circumscribed and limited to the filum terminale. It is almost always benign, but local complications, including paraplegia, are possible.[97,98] Paragangliomas can also occur in other regions of the spinal canal. In one study they occurred with decreasing frequency in the lumbar, cervical, and thoracic regions. Endocrine function is

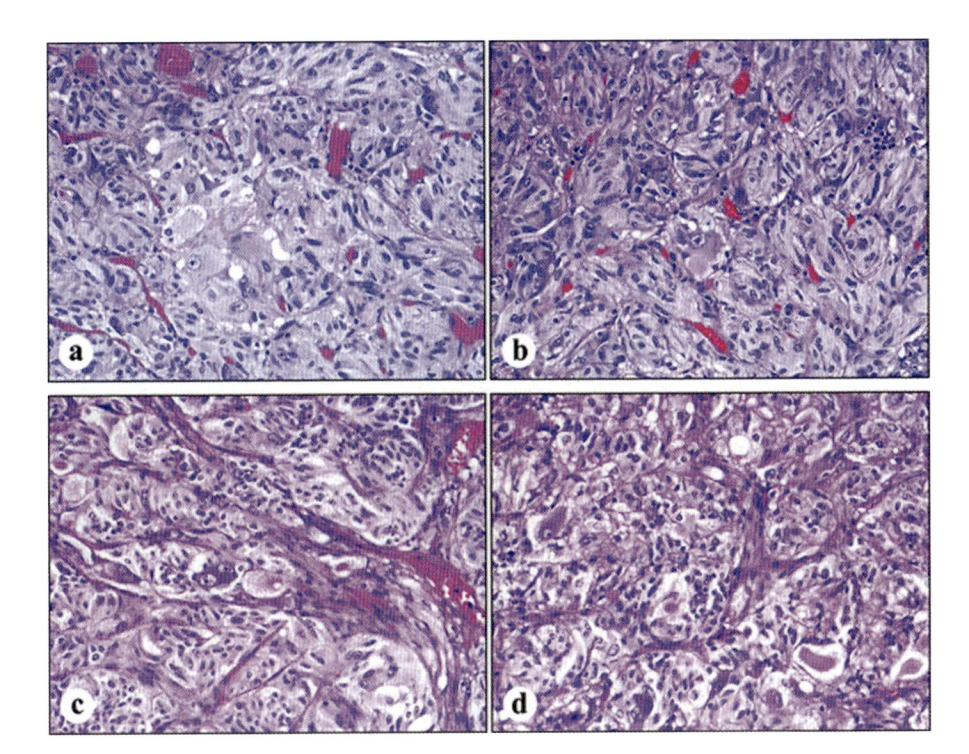

Figure 26.23 Duodenal gangliocytic paraganglioma. (**a,b**) Ganglion-like cells are present between the cellular nests. (**c,d**) A spindled schwannian stroma is variably present between the cellular nests.

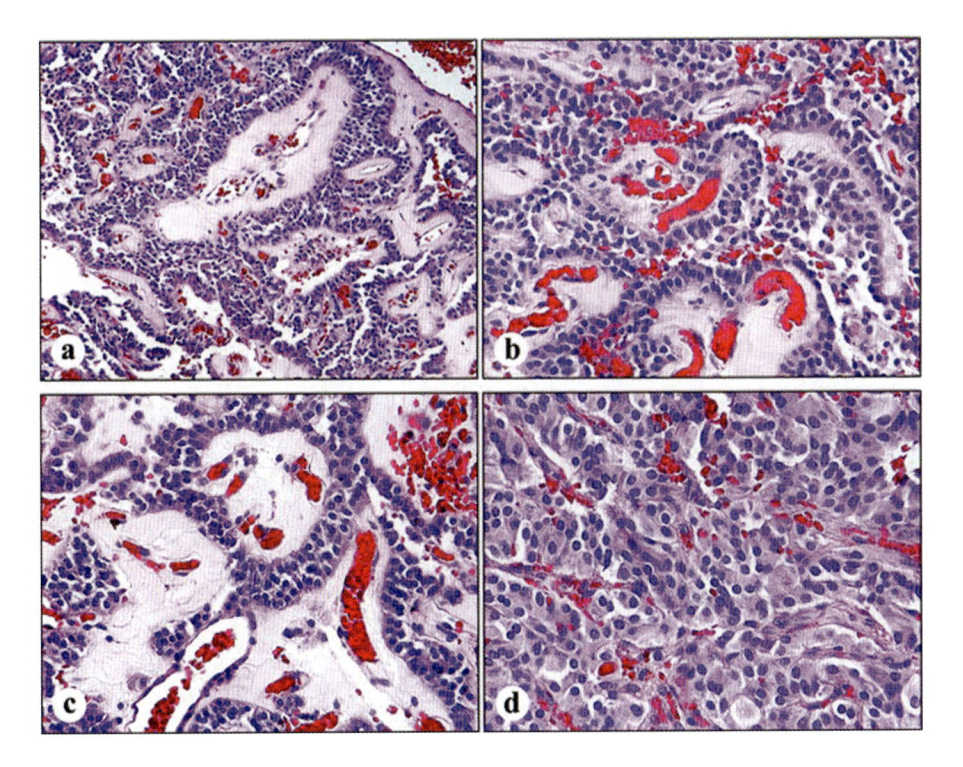

Figure 26.24 Paraganglioma of the cauda equina demonstrates perivascular rosette-like arrangements, and the edematous vessel wall creates some resemblance to a myxopapillary ependymoma.

usually not detectable in these tumors, and malignant behavior seems rare, but local recurrences can occur.[97]

Pathology

Grossly, the tumor is typically an encapsulated, intradural, extraspinal mass and measures 1.5 cm to 5 cm in maximum diameter. The histologic appearance varies. It can resemble that of a carotid body tumor, but in some cases, there is prominent perivascular rosette formation, with some resemblance to myxopapillary ependymoma (Figure 26.24) Nearly one-half of all cases have ganglion cell differentiation, and some cases resemble gangliocytic paraganglioma of the duodenum. Mitotic activity is usually low.

Immunohistochemically, the tumor cells are positive for the neuroendocrine markers (e.g., NSE, chromogranin, synaptophysin, and neurofilament proteins), and S100-protein- and GFAP-positive sustentacular cells are present in varying numbers. 5-Hydroxytryptamine and somatostatin are often present.[97] In the author's experience, the tumor is often keratin positive, similar to duodenal gangliocytic paragangliomas, but different from most other paragangliomas.

Differential diagnosis

The paraganglioma of the cauda equina must be separated from ependymoma, which in this location is usually of myxopapillary and less commonly of a solid type. Ependymomas lack the organoid structure and are globally positive for GFAP; they do not show a biphasic structure with chief cell and sustentacular cell components, and there are no ganglion cells.

References

General

1. Lack EE. *Tumors of the Adrenal Glands and Extra-Adrenal Paraganglia*, Fourth series, Fascicle 8. Washington, DC: Armed Forces Institute of Pathology, 2008.
2. Glenner GG, Grimley PM. *Tumors of the Extra-Adrenal Paraganglion System (Including Chemoreceptors)*. Armed Forces Institute of Pathology Atlas of Tumor Pathology, Second series, Fascicle 9. Washington, DC: Armed Forces Institute of Pathology, 1974.

Paraganglioma syndromes

3. Gimenez-Roqueplo AP, Dahia PL, Robledo M. An update on the genetics of paraganglioma, pheochromocytoma, and associated hereditary syndromes. *Horm Metab Res* 2012;44:328–333.
4. King KS, Pacak K. Familial pheochromocytomas and paragangliomas. *Mol Cell Endocrinol* 2014;386:92–100.
5. Welander J, Andreasson A, Juhlin CC, et al. Rare germline mutations identified by targeted next-generation sequencing of susceptibility genes in pheochromocytoma and paraganglioma. *J Clin Endocrinol Metab* 2014;99:E1352–E1360.
6. Gill AJ. Succinate dehydrogenase (SDH) and mitochondrial driven neoplasia. *Pathology* 2012;44:285–292.
7. Hoekstra AS, Bayley JP. The role of complex II in disease. *Biochim Biophys Acta* 2013;1827:543–551.
8. Baysal BE, Ferrell RE, Willet-Brozick JE, et al. Mutations in SDHD, a mitochondrial complex II gene, in hereditary paraganglioma. *Science* 2000;287:848–851.
9. Niemann S, Muller U. Mutations in SDHC cause autosomal dominant paraganglioma, type 4. *Nat Genet* 2000;26:268–270.
10. Astuti D, Latif F, Dallol A, et al. Gene mutations in the succinate dehydrogenase subunit SDHB cause susceptibility to familial pheochromocytoma and to familial paraganglioma. *Am J Hum Genet* 2001;69:49–54.
11. Burnichon N, Brière JJ, Libé R, et al. SDHA is a tumor suppressor gene causing paraganglioma. *Hum Mol Genet* 2010;19:3011–3020.
12. Letouzé E, Martinelli C, Loriot C, et al. SDH mutations establish a hypermethylator phenotype in paraganglioma. *Cancer Cell* 2013;23:739–752.
13. van Nederveen FH, Gaal J, Favier J, et al. An immunohistochemical procedure to detect patients with paraganglioma and phaeochromocytoma with germline SDHB, SDHC, or SDHD gene mutations: a retrospective and prospective analysis. *Lancet Oncol* 2009;10:764–771.
14. Gill AJ, Benn DE, Chou A, et al. Immunohistochemistry for SDHB triages genetic testing of SDHB, SDHC, and SDHD in paraganglioma-pheochromocytoma syndromes. *Hum Pathol* 2010;41:805–814.
15. Korpershoek E, Favier J, Gaal J, et al. SDHA immunohistochemistry detects germline SDHA gene mutations in apparently sporadic paragangliomas and pheochromocytomas. *J Clin Endocrinol Metab* 2011;96:E1472–E1476.
16. Jochmanová I, Zelinka T, Widimský J Jr, Pacak K. HIF signaling pathway in pheochromocytoma and other neuroendocrine tumors. *Physiol Res* 2014;63(Suppl 2):S251–S262.
17. Cerecer-Gil NY, Figuera LE, Llamas FJ, et al. Mutation of SDHB is a cause of hypoxia-related high-altitude paraganglioma. *Clin Cancer Res* 2010;16:4148–4154.
18. Baysal BE, McKay SE, Kim YJ, et al. Genomic imprinting at a boundary element flanking the SDHD locus. *Hum Mol Genet* 2011;20:4452–4461.
19. Kunst HP, Rutten MH, de Mönnink JP, et al. SDHAF2 (PGL2-SDH5) and hereditary head and neck paraganglioma. *Clin Cancer Res* 2011;17(2):247–254.
20. Mulligan LM, Kwok JB, Healey CS, et al. Germline mutations of the RET proto-oncogene in multiple endocrine neoplasia type 2A. *Nature* 1993;363:458–460.
21. Hofstra RM, Landsvater RM, Ceccherini I, et al. A mutation in the RET proto-oncogene associated with multiple endocrine neoplasia type 2B and sporadic medullary thyroid carcinoma. *Nature* 1994;367:375–376.
22. Komminoth P, Kunz E, Hiort O, et al. Detection of RET protooncogene point mutations in paraffin-embedded pheochromocytoma specimens in non-radioactive single-strand conformational polymorphism and direct sequencing. *Am J Pathol* 1994;145:922–929.
23. Eng C. RET proto-oncogene in the development of human cancer. *J Clin Oncol* 1999;17:380–393.
24. Hes F, Zewald R, Peeters T, et al. Genotype-phenotype correlations in families with deletions in the von Hippel–Lindau (VHL) gene. *Hum Genet* 2000;106:425–431.
25. Kaelin WG. Von Hippel–Lindau disease. *Annu Rev Pathol Mech Dis* 2007;2:145–173.
26. Chou A, Toon C, Pickett J, Gill AJ. von Hippel–Lindau syndrome. *Front Horm Res* 2013;41:30–49.
27. Opocher G, Conton P, Sciavi F, Macino B, Mantero F. Pheochromocytoma in von Hippel-Lindau disease and neurofibromatosis type 1. *Fam Cancer* 2005;4:13–16.
28. Bar M, Friedman E, Jakobovitz O, et al. Sporadic pheochromocytomas are rarely associated with germline

mutations in the von Hippel–Lindau and RET genes. *Clin Endocrinol* 1997;47:707–712.

29. Bausch B, Borozdin W, Mautner VF, et al. Germline NF1 mutational spectra and loss of heterozygosity analyses in patients with pheochromocytoma and neurofibromatosis type 1. *J Clin Endocrinol Metab* 2007;92:2784–2792.

30. Bausch B, Borozdin W, Neumann HP, European-American Pheochromocytoma Study Group. Clinical and genetic characteristics of patients with neurofibromatosis type 1 and pheochromocytoma. *N Engl J Med* 2006;354:2729–2731.

31. Burnichon N, Buffet A, Parfait B, et al. Somatic NF1 inactivation is a frequent event in sporadic pheochromocytoma. *Hum Mol Genet* 2012;21:5397–5405.

32. Ladroue C, Carcenac R, Leporrier M, et al. PHD2 mutation and congenital erythrocytosis with paraganglioma. *N Engl J Med* 2008;359(25):2685–2692.

33. Yang C, Zhuang Z, Fliedner SM, et al. Germ-line *PHD1* and *PHD2* mutations detected in patients with pheochromocytoma/paraganglioma-polycythemia. *J Mol Med (Berl)* 2015;93:93–104.

34. Lorenzo FR, Yang C, Ng Tang Fui M, et al. A novel EPAS1/HIF2A germline mutation in a congenital polycythemia with paraganglioma. *J Mol Med (Berl)* 2013;91:507–512.

35. Qin Y, Yao L, King EE, et al. Germline mutations in TMEM127 confer susceptibility to pheochromocytoma. *Nat Genet* 2010;42:229–233.

36. Yeh IT, Lenci RE, Qin Y, et al. A germline mutation of the KIF1B beta gene on 1p36 in a family with neural and nonneural tumors. *Hum Genet* 2008;124:279–285.

37. Comino-Méndez I, Gracia-Aznárez FJ, Schiavi F, et al. Exome sequencing identifies MAX mutations as a cause of hereditary pheochromocytoma. *Nat Genet* 2011;43:663–667.

38. Burnichon N, Cascón A, Schiavi F, et al. MAX mutations cause hereditary and sporadic pheochromocytoma and paraganglioma. *Clin Cancer Res* 2012;18:2828–2837.

Pheochromocytoma

39. Buffet A, Venisse A, Nau V, et al. A decade (2001–2010) of genetic testing for pheochromocytoma and paraganglioma. *Horm Metab Res* 2012;44:359–366.

40. Tischler AS. Pheochromocytoma and extra-adrenal paraganglioma: updates. *Arch Pathol Lab Med* 2008;132:1272–1284.

41. Lendrers JWM, Eisenhofer G, Mannelli M, Pacak K. Phaeochromocytoma. *Lancet* 2005;366:665–675.

42. Melicow MM. One hundred cases of pheochromocytoma (107 tumors) at the Columbia-Presbyterian Medical Center, 1926–1976. *Cancer* 1977;40:1987–2004.

43. Medeiros LJ, Wolf BC, Balogh K, Federman M. Adrenal pheochromocytoma: a clinicopathologic review of 60 cases. *Hum Pathol* 1985;16:580–589.

44. Linnoila RI, Keiser HR, Steinberg SM, Lack EE. Histopathology of benign versus malignant sympathoadrenal paragangliomas: clinicopathologic study of 120 cases including unusual histologic features. *Hum Pathol* 1990;21:1168–1180.

45. Webb TA, Sheps SG, Carney JA. Differences between sporadic pheochromocytoma and pheochromocytoma in multiple endocrine neoplasia, type 2. *Am J Surg Pathol* 1980;4:121–126.

46. Koch CA, Mauro D, Walther MM, et al. Pheochromocytoma in von Hippel–Lindau disease: distinct histopathologic phenotype compared to pheochromocytoma in multiple endocrine neoplasia type 2. *Endocr Pathol* 2002;13:17–27.

47. Thompson LDR. Pheochromocytoma of the adrenal gland scaled score (PASS) to separate benign from malignant neoplasms. *Am J Surg Pathol* 2002;26:551–566.

48. Hassoun J, Monges G, Giraud B, et al. Immunohistochemical study of pheochromocytomas: an investigation of methionine-enkephalin, vasoactive intestinal polypeptide, somatostatin, corticotropin, β-endorphin, and calcitonin in 16 tumors. *Am J Pathol* 1984;114:56–63.

49. Hacker GW, Bishop AE, Terenghi G, et al. Multiple peptide production and presence of general neuroendocrine markers detected in 12 cases of human phaeochromocytoma and in mammalian adrenal glands. *Virchows Arch A Pathol Anat Histopathol* 1988;412:399–411.

50. Linnoila RI, Lack EE, Steinberg SM, Keiser HR. Decreased expression of neuropeptides in malignant paragangliomas: an immunohistochemical study. *Hum Pathol* 1988;19:41–50.

51. Salmenkivi K, Haglund C, Arola J, Heikkilä P. Increased expression of tenascin in pheochromocytomas correlates with malignancy. *Am J Surg Pathol* 2001;25:1419–1423.

52. Lloyd RV, Blaivas M, Wilson BS. Distribution of chromogranin and S100 protein in normal and abnormal adrenal medullary tissues. *Arch Pathol Lab Med* 1985;109:633–635.

53. Achilles E, Padberg BC, Holl K, Klöppel G, Schröder S. Immunocytochemistry of paragangliomas: value of staining for S-100 protein and glial fibrillary acid protein in diagnosis and prognosis. *Histopathology* 1991;18:453–458.

54. Nagura S, Katoh R, Kawaoi A, et al. Immunohistochemical estimations of growth activity to predict biological behavior of pheochromocytomas. *Mod Pathol* 1999;12:1107–1111.

55. Van Der Harst E, Bruining HA, Jaap Bonjer H, et al. Proliferative index in pheochromocytomas: does it predict the occurrence of metastases? *J Pathol* 2000;191:175–180.

56. Brown HM, Komorowski RA, Wilson SD, Demeure MJ, Zhy Y. Predicting metastases of pheochromocytomas using DNA flow cytometry and immunohistochemical markers of cell proliferation. *Cancer* 1999;86:1583–1589.

Retroperitoneal and urinary bladder paraganglioma

57. Lack EE, Cubilla AL, Woodruff JM, Lieberman PH. Extra-adrenal paragangliomas of the retroperitoneum: a clinicopathologic study of 12 tumors. *Am J Surg Pathol* 1980;4:109–120.

58. Moran CA, Albores-Saavedra J, Wenig BM, Mena H. Pigmented extraadrenal paragangliomas: a clinicopathologic and immunohistochemical study of five cases. *Cancer* 1997;79:398–402.

59. Honma K. Paraganglia of the urinary bladder: an autopsy study. *Zentralbl Pathol* 1994;139:465–469.
60. Leestma JE, Price EB. Paraganglioma of the urinary bladder. *Cancer* 1971;28:1063–1073.
61. Grignon DJ, Ro JY, Mackay B, et al. Paraganglioma of the urinary bladder: immunohistochemical, ultrastructural, and DNA flow cytometric studies. *Hum Pathol* 1991;22:1162–1169.
62. Cheng L, Leibovich BC, Cheville JC, et al. Paraganglioma of the urinary bladder: can biologic potential be predicted? *Cancer* 2000;88:844–852.
63. Mason EF, Sadow PM, Wagner AJ, et al. Identification of succinate dehydrogenase-deficient bladder paragangliomas. *Am J Surg Pathol* 2013;37:1612–1618.

Thoracic paraganglioma

64. Olson JL, Salyer WR. Mediastinal paragangliomas (aortic body tumor): a report of four cases and a review of literature. *Cancer* 1978;41P:2405–2412.
65. Lack EE, Stillinger RA, Colvin DB, Groves RM, Burnette DG. Aorticopulmonary paraganglioma: report of a case with ultrastructural study and review of the literature. *Cancer* 1979;43:269–278.
66. Johnson TL, Lloyd RV, Shapiro B, et al. Cardiac paragangliomas: a clinicopathologic and immunohistochemical study of four cases. *Am J Surg Pathol* 1985;9:827–833.
67. Gallivan MVE, Chun B, Rowden G, Lack EE. Intrathoracic paravertebral paraganglioma. *Arch Pathol Lab Med* 1980;104:46–51.
68. Moran CA, Suster S, Fishback N, Koss MN. Mediastinal paragangliomas: a clinicopathologic and immunohistochemical study of 16 cases. *Cancer* 1993;72:2358–2364.
69. Else T, Marvin ML, Everett JN, et al. The clinical phenotype of SDHC-associated hereditary paraganglioma syndrome (PGL3). *J Clin Endocrinol Metab* 2014;99:E1482–E1486.

Paraganglioma of the head and neck

70. Saldana MJ, Salem LE, Travezan R. High altitude hypoxia and chemodectomas. *Hum Pathol* 1973;4:251–263.
71. Arias-Stella J, Valcarcel J. Chief cell hyperplasia in the human carotid body at high altitudes: physiologic and pathologic significance. *Hum Pathol* 1976;7:361–373.
72. Rodriguez-Cuevas S, Lopez-Garza J, Labastida-Almendaro S. Carotid body tumors in inhabitants of altitudes higher than 2000 meters above sea level. *Head Neck* 1998;20:374–378.
73. Jansen JC, Van Den Berg R, Kuiper A, et al. Estimation of growth rate in patients with head and neck paragangliomas influences the treatment proposal. *Cancer* 2000;88:2811–2816.
74. Shamblin WR, ReMine WH, Sheps SG, Harrison EG. Carotid body tumor (chemodectoma): clinicopathologic analysis of 90 cases. *Am J Surg* 1971;122:732–739.
75. Lack EE, Cubilla AL, Woodruff JM. Paragangliomas of the head and neck region: a pathologic study of tumors from 71 patients. *Hum Pathol* 1979;10:191–218.
76. Hodge KM, Byers RM, Peters LJ. Paragangliomas of the head and neck. *Arch Otolaryngol Head Neck Surg* 1988;114:872–877.
77. Nora JD, Hallett JW, Jr, O'Brien PC, et al. Surgical resection of carotid body tumors: long-term survival, recurrence and metastasis. *Mayo Clin Proc* 1988;63:348–352.
78. Lee JH, Barich F, Karnell LH, et al. National cancer database report on malignant paragangliomas of the head and neck. *Cancer* 2002;94:730–737.
79. Netterville JL, Jackson CG, Miller FR, Wanamaker JR, Glasscock ME. Vagal paraganglioma: a review of 46 patients treated during a 20-year period. *Arch Otolaryngol Head Neck Surg* 1998;124:1133–1140.
80. Miller RB, Boon MS, Atkins JP, Lowry LD. Vagal paraganglioma: the Jefferson experience. *Otolaryngol Head Neck Surg* 2000;122:482–487.
81. Heinrich MC, Harris AE, Bell WR. Metastatic intravagal paraganglioma: case report and review of the literature. *Am J Med* 1985;78:1017–1024.
82. Tannir NM, Cortas N, Allam C. A functioning catecholamine-secreting vagal body tumor: a case report and review of literature. *Cancer* 1983;52:932–935.
83. Plaza JA, Wakely PE, Moran C, Fletcher CDM, Suster S. Sclerosing paraganglioma: report of 19 cases of an unusual variant of neuroendocrine tumor that may be mistaken for an aggressive malignant neoplasm. *Am J Surg Pathol* 2006;30:7–12.
84. Hamid Q, Varndell IM, Ibrahim NB, Mingazzini P, Polak JM. Extraadrenal paragangliomas: an immunocytochemical and ultrastructural report. *Cancer* 1987;60:1776–1781.
85. Kliewer KE, Wen DR, Cancilla PA, Cochran AJ. Paragangliomas: assessment of prognosis by histologic, immunohistochemical, and ultrastructural techniques. *Hum Pathol* 1989;20:29–39.
86. Brown JS. Glomus jugulare tumors revisited: a ten-year statistical follow-up of 231 cases. *Laryngoscope* 1985;95:284–288.
87. Van Der Mey AG, Frijns JH, Cornelisse CJ, et al. Does intervention improve the natural course of glomus tumors?: a series of 108 patients seen in a 32-year period. *Ann Otol Rhinol Laryngol* 1992;101:635–642.
88. Lim M, Bower R, Nangiana JS, Adler JR, Chang SD. Radiosurgery for glomus jugulare tumors. *Technol Cancer Res Treat* 2007; 6:419–423.

Gangliocytic paraganglioma

89. Kepes JJ, Zacharias DL. Gangliocytic paraganglioma of the duodenum: a report of two cases with light and electron microscopic examination. *Cancer* 1971;27:61–70.
90. Reed RJ, Daroca PJ Jr, Harkin JC. Gangliocytic paraganglioma. *Am J Surg Pathol* 1977;1:207–216.
91. Perrone T, Sibley RK, Rosai J. Duodenal gangliocytic paraganglioma: an immunohistochemical and ultrastructural study and hypothesis concerning its origin. *Am J Surg Pathol* 1985;7:31–41.
92. Scheithauer BW, Nora FE, Lechago J, et al. Duodenal gangliocytic paraganglioma: clinicopathologic and immunohistochemical study of 11 cases. *Am J Clin Pathol* 1986;86:559–565.

93. Burke AP, Helwig EB. Gangliocytic paraganglioma. *Am J Clin Pathol* 1989;92:1–9.
94. Inai K, Kobuke T, Yonehara S, Tokuoka S. Duodenal gangliocytic paraganglioma with lymph node metastasis in a 17-year-old boy. *Cancer* 1989;63:2540–2545.
95. Dookhan DB, Miettinen M, Finkel G, Gibas Z. Recurrent duodenal gangliocytic paraganglioma with lymph node metastases. *Histopathology* 1993;22:399–401.
96. Hamid QA, Bishop AE, Rode J, et al. Duodenal gangliocytic paragangliomas: a study of 10 cases with immunocytochemical neuroendocrine markers. *Hum Pathol* 1986;17:1151–1157.

Cauda equina and spinal paraganglioma

97. Sonneland PRL, Scheithauer BW, Lechago J, Crawford BG, Onofrio BM. Paraganglioma of the cauda equina region: clinicopathologic study of 31 cases with special reference to immunocytology and ultrastructure. *Cancer* 1986;58:1720–1735.
98. Moran CA, Rush W, Mena H. Primary spinal paragangliomas: a clinicopathological and immunohistochemical study of 30 cases. *Histopathology* 1997;31:167–171.

Chapter 27
Primary soft tissue tumors with epithelial differentiation

Markku Miettinen
National Cancer Institute, National Institutes of Health

Epithelial differentiation is often easily detected histologically and is almost always highlighted with immunohistochemical markers (i.e., keratins, epithelial membrane antigen [EMA], and others). Mesothelial proliferations and tumors are discussed in Chapter 28, and Merkel cell carcinoma and metastatic carcinomas in Chapter 29.

Some other nonepithelial cells and tumors also can express keratins and other epithelial markers. Examples of normal varyingly keratin-positive mesenchymal cells include endothelial and smooth muscle cells and some myofibroblasts. Many nonepithelial tumors, such as melanomas and many sarcomas, occasionally are keratin positive. These are discussed in the appropriate chapters.

Synovial sarcoma and epithelioid sarcoma are epithelial tumors of soft tissue of unknown histogenesis. These tumors are not thought to be closely related.

Mixed tumors and related skin adnexal tumors, and chordoma, a tumor mimicking the features of notochord cells, and adamantinoma, a leg tumor, are also reviewed here.

Synovial sarcoma

A synovial sarcoma is a specific soft tissue sarcoma with dual epithelial and mesenchymal differentiation. The tumor is not related to synovial lining cells, and its association with synovia, if any, is only incidental. The name is based on historical reasons, because the epithelial component of the biphasic tumors was once thought to replicate synovial slits.[1,2] The t(X;18) translocation with SS18-SSX gene fusion is a useful, and based on present information, specific diagnostic marker.

Demographics of synovial sarcoma

Synovial sarcomas are relatively common sarcomas, constituting from 7% to 10% of all soft tissue sarcomas. This tumor typically presents in young adults, with a median age of approximately 30 years and a mild male predominance. It can occur in children from the first decade on, but very rarely before the age of 5 years.[3-4] It is rarely seen in older adults in the seventh and eighth decades (Figure 27.1).

Synovial sarcomas primarily occur in deep, intramuscular soft tissues and have a strong predilection for the extremities (>70%). The most common sites are the thigh, knee region, ankles and feet, hands, and the other parts of the upper extremities. Occurrence in the proximal parts of extremities is slightly more common than in the distal areas (Figure 27.2).

Figure 27.1 Age and sex distribution of 2130 patients with synovial sarcomas.

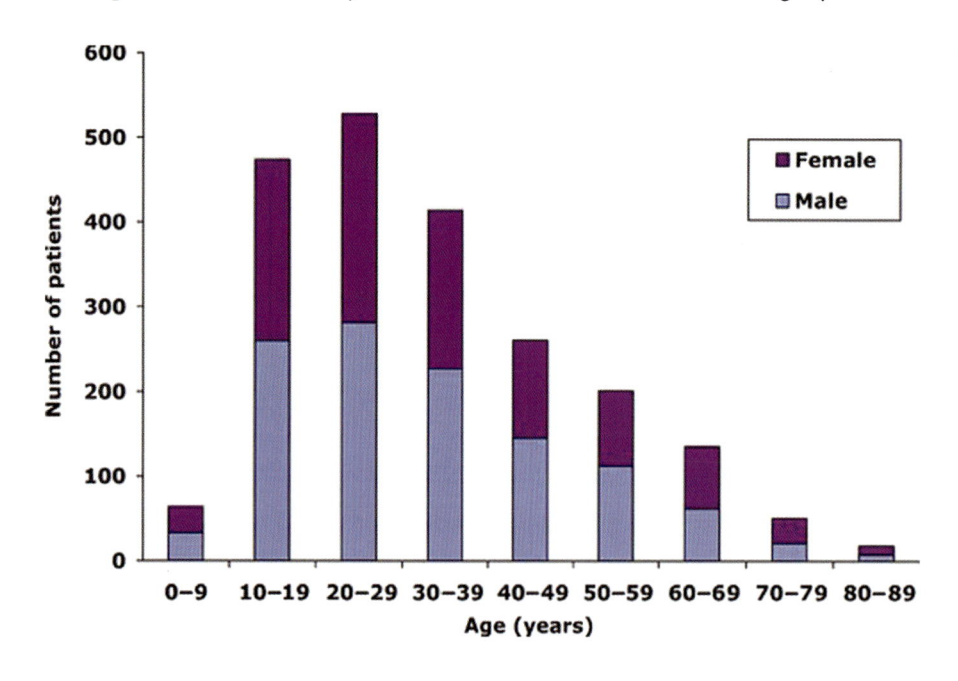

Modern Soft Tissue Pathology, Second Edition, ed. Markku Miettinen. Published by Cambridge University Press. © Cambridge University Press 2016.

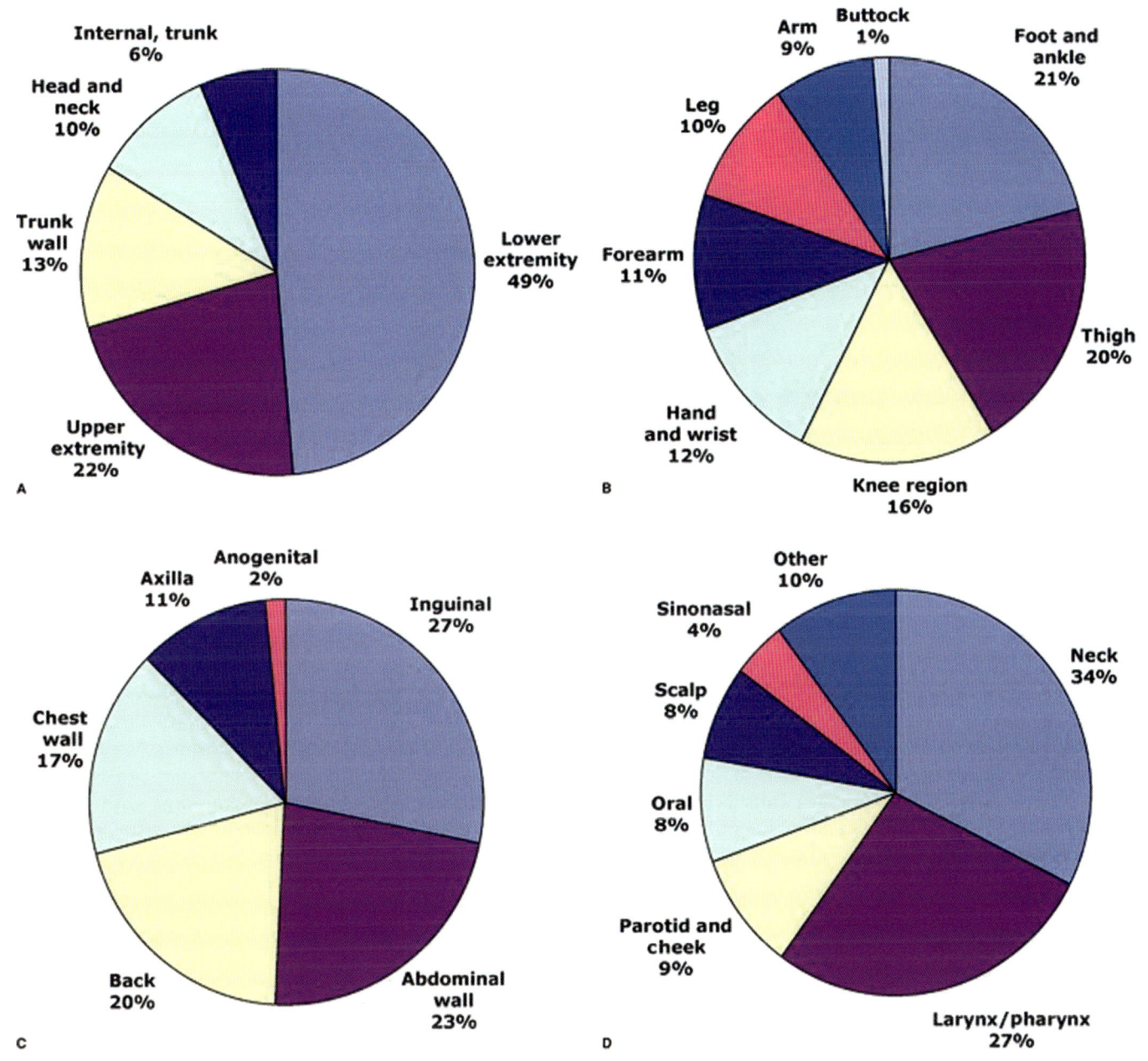

Figure 27.2 Anatomic distribution of 2130 cases of synovial sarcoma. (**A**) Overall distribution. (**B**) Distribution in the extremities. (**C**) Distribution in the trunk wall. (**D**) Distribution in the head and neck.

Direct synovial involvement is exceptional, mostly related to the bulge of a periarticular tumor into the synovial space.

In the trunk wall, the most common locations are the inguinal region and the abdominal wall.[5] In the neck, the hypopharynx is the most common location.[6,7] In the head, the temporal scalp and orofacial region are most commonly involved.[8,9] (Figure 27.2).

In the body cavities, rare examples have been reported in the retroperitoneum[10] and mediastinum, both anterior and posterior.[11–13]

Clinical features

Synovial sarcomas of soft tissues are often painful, but otherwise they are not clinically distinctive from other soft tissue sarcomas. Small tumors of the hand and wrist can be clinically mistaken for ganglion cysts or other benign processes. In some cases, slow tumor growth has been documented, with a 10- to 20-year history before the diagnosis.

Local recurrence is rare after adequate wide excision, but is common if the tumor is simply shelled out. Lymph node metastases, rare in other sarcomas, sometimes develop long

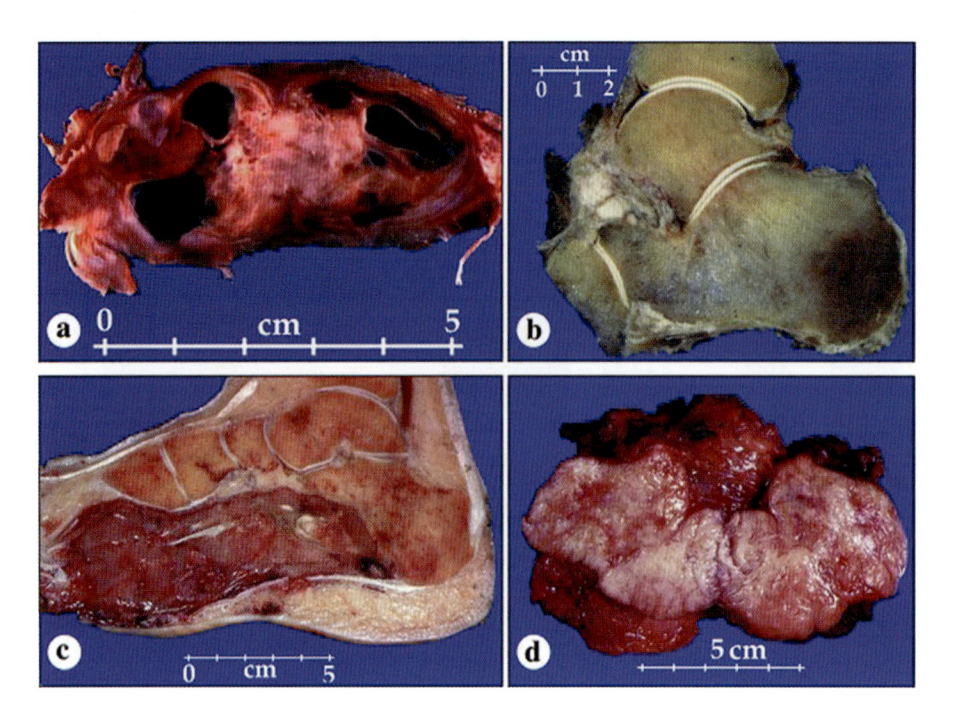

Figure 27.3 Gross appearances of synovial sarcoma. (**a**) A cystic example. (**b**) A small binodular tumor just above the distal end of calcaneus. (**c**) A large example involving much of the plantar foot. (**d**) An intramuscular complex-shaped synovial sarcoma.

after the primary surgery. Distant metastases are diagnosed in 40% to 50% of patients. The most common metastatic site is the lung, and massive pleuropulmonary metastasis often extending into chest wall is the leading cause of death. Liver and brain metastases also occur.

The 5-year survival rate for tumors other than the poorly differentiated types is 60% to 70%. Poorly differentiated tumors have a markedly worse 5-year survival rate, only 20% to 30%. Clinically favorable prognostic factors include age <20 years, tumor size <5 cm, and a peripheral tumor location.[14–19] Late metastases 5–15 years after primary presentation are relatively common.[20]

Synovial sarcoma at visceral sites

The possibility of metastasis from a soft tissue primary tumor must be excluded before one accepts the diagnosis of a primary organ-based synovial sarcoma. The most commonly involved visceral site is the lung, and some examples have involved the pleural cavity.[13,14,21–23] Pleuropulmonary synovial sarcomas occur in older patients than the peripheral tumors, with a median age of 47 years. The tumors vary from nodules <3 cm to large masses >10 cm; most are >5 cm. The disease-specific survival rate is lower (around 30%) than for soft tissue synovial sarcomas.[12]

In the cardiovascular system, intravascular presentation in the femoral vein[24,25] and occurrence in the atrium of the heart[26] has been reported. In one series, 2 of 27 cardiac sarcomas were synovial sarcomas.[27]

A small number of synovial sarcomas have been reported in the gastrointestinal tract, especially the esophagus, stomach, and (exceptionally) in the duodenum.[28–30] Esophageal synovial sarcomas can form a polypoid intraluminal mass. In the stomach, a synovial sarcoma often forms a small mucosal or submucosal plaque, or a cup-shaped lesion of 1 cm to 2 cm. Large transmural examples also occur. Small tumors and those without a poorly differentiated, mitotically highly active component have a better prognosis.[29]

In the urogenital tract, some renal tumors originally classified as embryonal sarcomas of the kidney have recently been reclassified as primary renal synovial sarcomas, based on compatible histology and demonstration of the SYT-SSX gene rearrangements. These tumors were large, histologically monophasic or poorly differentiated, and often contained entrapped cystic renal tubular elements.[31] Isolated cases have also been reported in the prostate, but it is not certain if these tumors were truly intraparenchymal.[32,33]

Pathology

Grossly, synovial sarcoma is rarely diagnostic. Tumor size varies broadly. Peripheral tumors can be 1 cm or smaller, although these tumors more commonly reach the size of at least 3 cm to 10 cm in more proximal locations and up to 10 cm to 20 cm in the deep thigh.

On sectioning, the tumor is soft, gray-white, and resembles the flesh of a fish with often slightly mucoid character. Some examples are extensively cystic, lined by a narrow rim of tumor tissue (Figure 27.3). Such tumors were historically sometimes assumed to have originated in the bursa. Gross calcification and ossification can occur, whereas necrosis is essentially restricted to high-grade examples.

Synovial sarcoma has a wide histologic spectrum that includes three variants that sometimes coexist in one tumor. A monophasic spindle cell pattern is the most common, followed by biphasic synovial sarcoma with epithelial, usually glandular elements, and spindle cells. Poorly differentiated synovial sarcomas refers to tumors with undifferentiated,

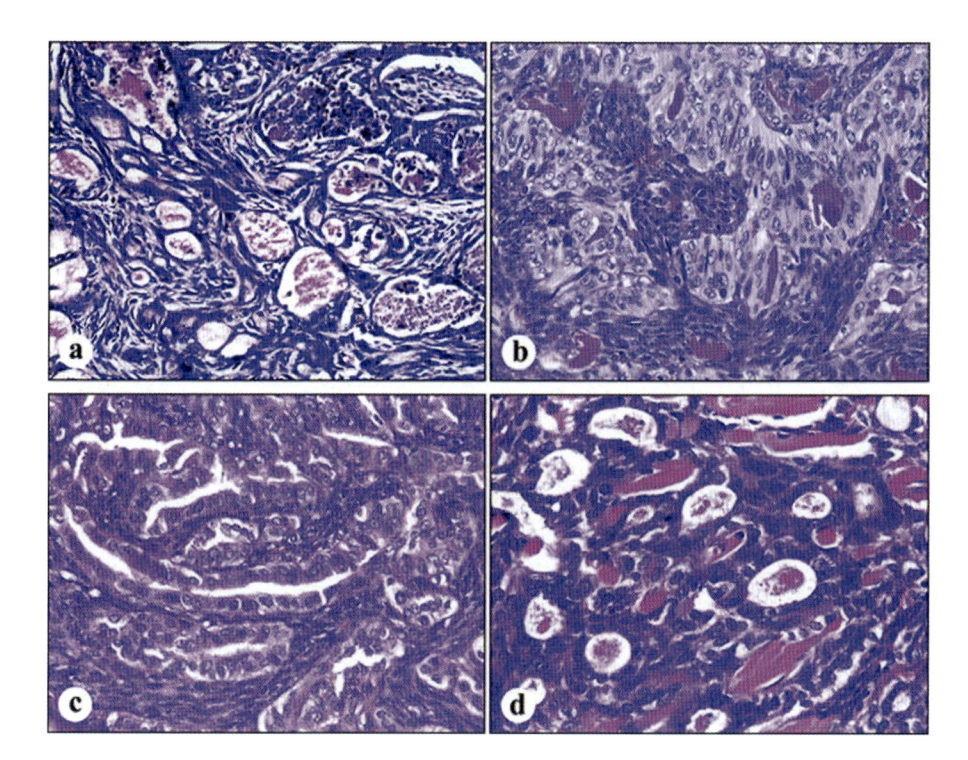

Figure 27.4 Histologic spectrum of biphasic synovial sarcoma with different histologic patterns of epithelial differentiation. (**a**) Glandular epithelia with luminal secretion. (**b**) Pale staining glandular epithelia with eosinophilic luminal content. (**c**) Glands with columnar epithelium. (**d**) An example with the glandular epithelia as the dominant component.

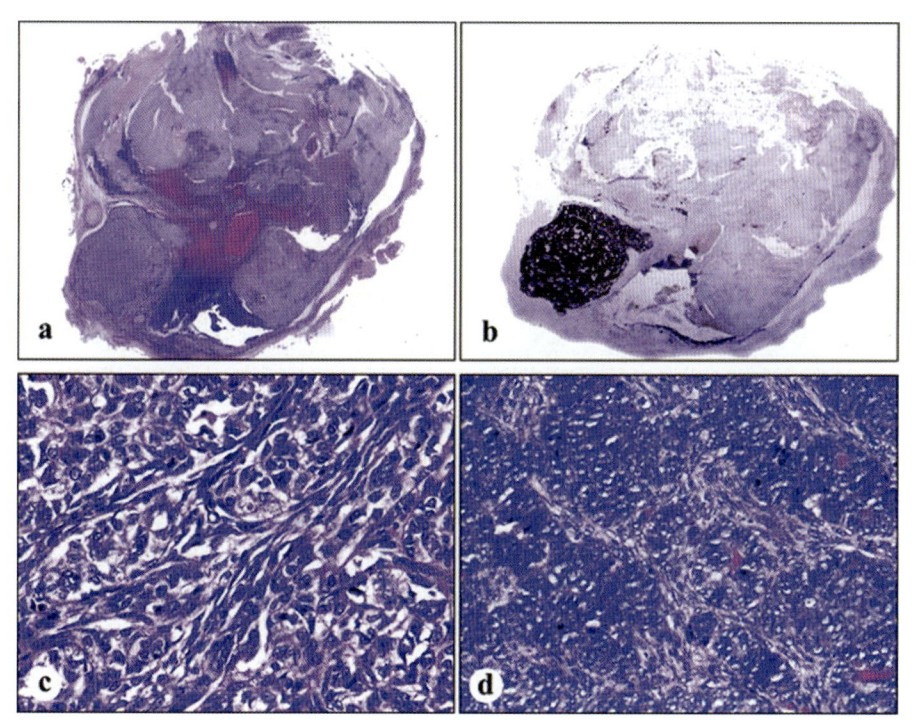

Figure 27.5 (**a,b**) Synovial sarcoma with a small, distinct biphasic, keratin-positive component. (**c,d**) Examples of biphasic tumors with atypical epithelial element.

high-grade appearance with mitotic rate >15 per 10 HPFs. These tumors can also contain a more differentiated biphasic or monophasic component.

Pathology of the biphasic type

Biphasic synovial sarcoma (20–30%) is the most distinctive, although not the most common variant. It contains a combination of epithelial and spindle cell elements. The former often appear as well-formed glandular epithelial structures or solid epithelial sheets surrounded by a basement membrane. The lumen may contain an inspissated secretion that is periodic-acid–Schiff (PAS) positive. The appearance of the glandular epithelium varies from cuboidal to tall columnar (Figure 27.4). Occasionally there is squamous differentiation with keratinization.[34] The proportion of the epithelial component varies widely. Some tumors are monophasic-like or poorly differentiated, containing a focal biphasic component or poorly formed glands with highly atypical epithelium (Figure 27.5).

Chapter 27: Primary soft tissue tumors with epithelial differentiation

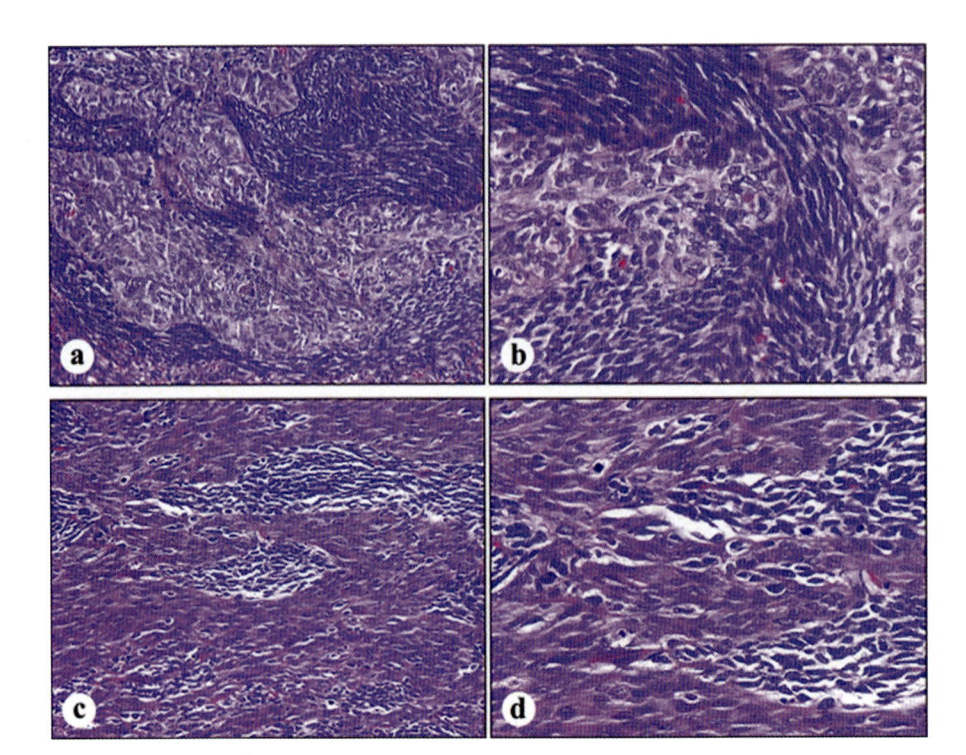

Figure 27.6 (a,b) Biphasic synovial sarcoma with no distinct luminal differentiation. (c,d) The eosinophilic component is the epithelial one that forms solid sheets.

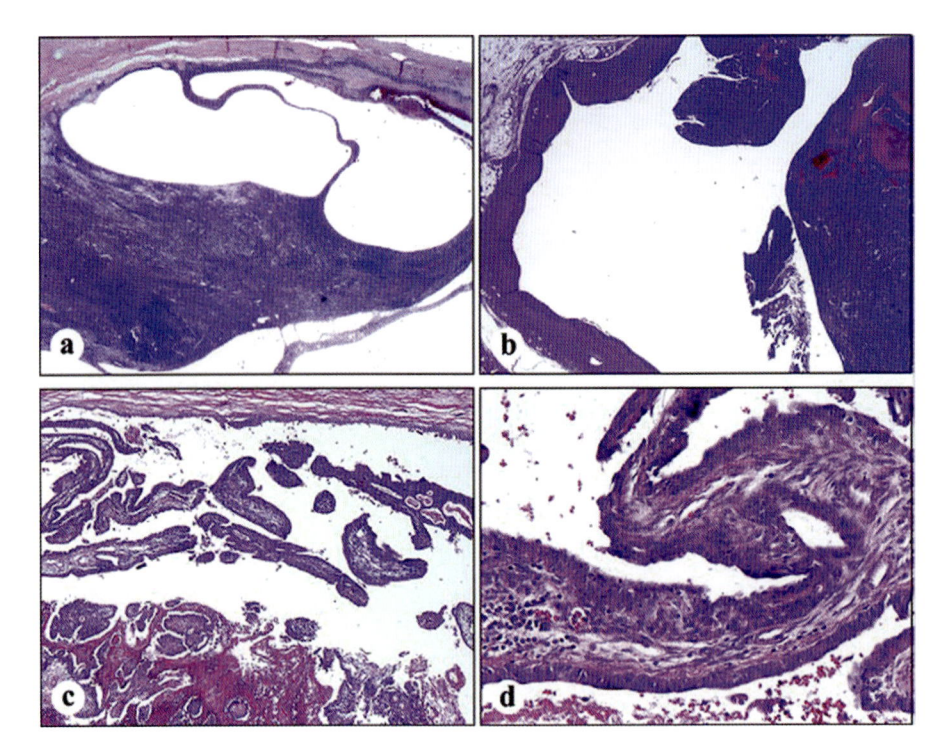

Figure 27.7 Cystic biphasic synovial sarcoma. The cysts are lined by glandular epithelia.

Some biphasic tumors contain an extensive epithelial element visible as paler or more eosinophilic zones, but glandular differentiation is rudimentary or absent (Figure 27.6). The spindle cell component is composed of uniform, mildly hyperchromatic spindle cells, similar to those in the monophasic variants. The cystic elements may be lined by glandular epithelium (Figure 27.7).

Pathology of the monophasic spindle cell type

A great many spindle cell sarcomas earlier classified as fibrosarcomas are actually monophasic synovial sarcomas, as defined in the early 1980s.[35–37] The monophasic spindle cell variant (50–60%) is composed of sheets and fascicles of relatively uniform spindle cells, which are densely packed or

Chapter 27: Primary soft tissue tumors with epithelial differentiation

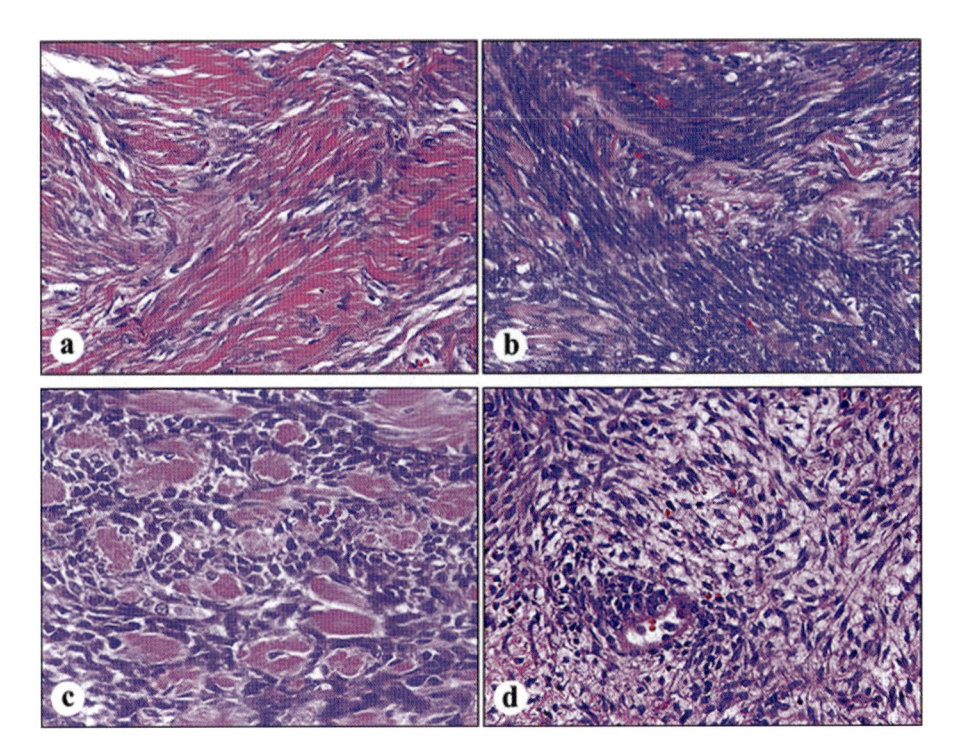

Figure 27.8 Histologic spectrum of monophasic synovial sarcoma. (a–c) Variable patterns of alternating cellular and collagenous areas. (d) A variant with myxoid stroma.

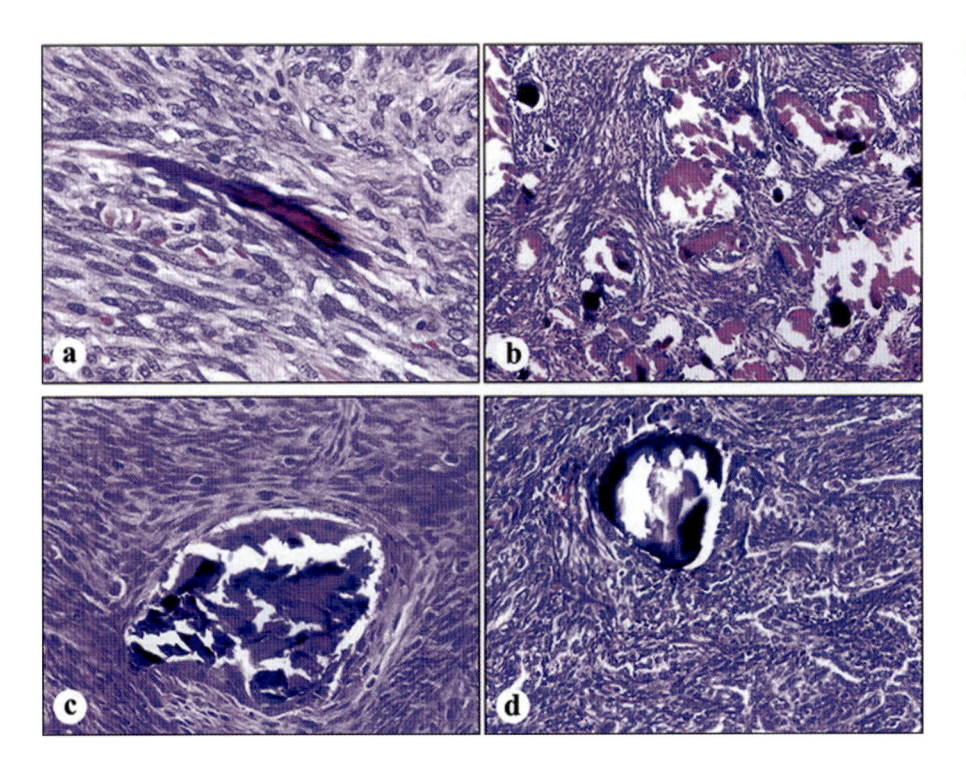

Figure 27.9 (a,b) Monophasic synovial sarcomas with calcification. (c,d) Biphasic synovial sarcomas with intraluminal calcification.

arranged in a matrix that varies from myxoid to densely collagenous, sometimes in alternating patterns within one tumor. The lack of pleomorphic nuclei is typical of synovial sarcoma (Figure 27.8). Focal calcifications are common in the fibrous matrix, and extensive calcification[38] and ossification[39] may also occur (Figure 27.9). Cellular clusters with vague epithelioid appearance can be seen among the spindle cells, and mast cell infiltration is typical. A hemangiopericytoma-like pattern is quite common. Some tumors have schwannoma-like nuclear palisading. Cytologically, the nuclei are elongated with pointed ends. The cytoplasm is basophilic, with inconspicuous cell borders.

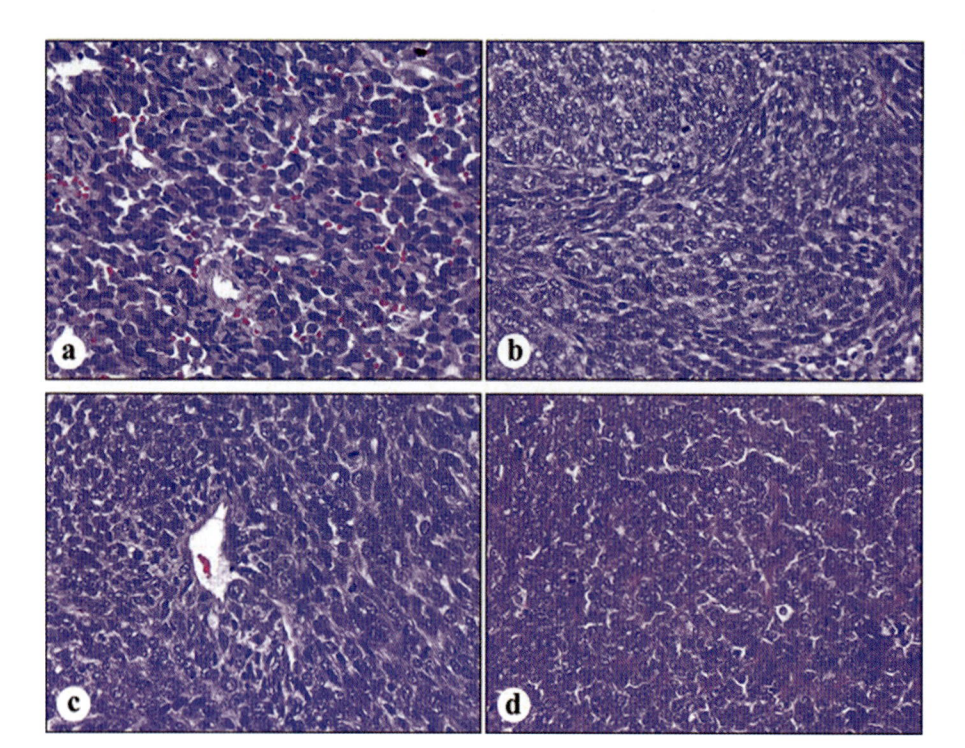

Figure 27.10 Variants of poorly differentiated synovial sarcoma. (**a**) Example with rhabdoid cytology. (**b**) High-grade fibrosarcoma or MPNST-like pattern. (**c**) Hemangiopericytoma-like pattern. (**d**) Ewing–PNET-like round cell pattern.

Pathology of poorly differentiated synovial sarcoma

Poorly differentiated synovial sarcoma (10–15% of all cases) refers to a histologic pattern that has features of neither the biphasic nor the monophasic type and has a high mitotic rate (>15 mitoses per 10 HPFs). This pattern often occurs in metastatic synovial sarcoma. Recognition of these tumors as synovial sarcomas often requires molecular studies, unless differentiated components are present in other areas. Extensive sampling is recommended, because this can reveal the diagnosis more easily than any special studies, especially if glands are present.[40]

A high-grade fibrosarcoma-like pattern with fascicular appearance and a round cell pattern simulating small round cell tumors (e.g., extraskeletal Ewing's sarcoma or peripheral neuroepithelioma) are the most common variants. Some cases contain rhabdoid cells with perinuclear cytoplasmic inclusions of intermediate filaments. These tumors also sometimes show a hemangiopericytoma-like vascular pattern with sheets of oval tumor cells between the vessels (Figure 27.10).

Hybrid synovial sarcoma and extraskeletal myxoid chondrosarcoma

A unique case of hybrid tumor with histological features and gene fusions of well-differentiated synovial sarcoma and extraskeletal myxoid chondrosarcoma has been reported.[41]

Prognostic pathologic factors

Reported favorable prognostic factors are tumor size <5 cm, a peripheral location, young age (<20 years), high mast cell content (>20 per 10 HPFs), and extensive calcifications. Unfavorable features are mitotic rate >10–15 per 10 HPFs, the presence of a rhabdoid cellular component, tumor necrosis, poorly differentiated histology, high stage, and incomplete primary excision.[42]

The prognostic significance of monophasic versus biphasic type and SS18-SSX fusion type is still controversial. Some studies suggested SSX2 fusion to be prognostically favorable,[43,44] but other studies have not found significant differences in the prognosis of SSX1 versus SSX2 fusion variants.[19]

Immunohistochemistry

Immunohistochemical analysis is essential for the diagnosis of monophasic and poorly differentiated variants, although the latter might show limited positivity for epithelial markers.

The epithelial cells in biphasic tumors and scattered cells (or nests) and small clusters of spindle cells are positive for keratins and EMA.[2,45–49]

The epithelial cells in biphasic synovial sarcoma are positive for most keratin cocktails, such as the AE1/AE3 (Figure 27.11). These cells contain simple epithelial keratins (i.e., K7, K8, K18, and K19) and additionally high-molecular-weight (HMW) keratins of complex epithelia K14 and K17, and sporadically K13, K16, and K20. Examples with squamous differentiation might show K10 in the keratinizing epithelia.[49]

The monophasic tumors contain scattered cells that are positive for keratins K7, K8, K18, and K19, whereas other keratins are rare in the monophasic tumors. EMA reactivity is seen in patches, and the number of positive cells can be greater than that for keratins (Figure 27.12).

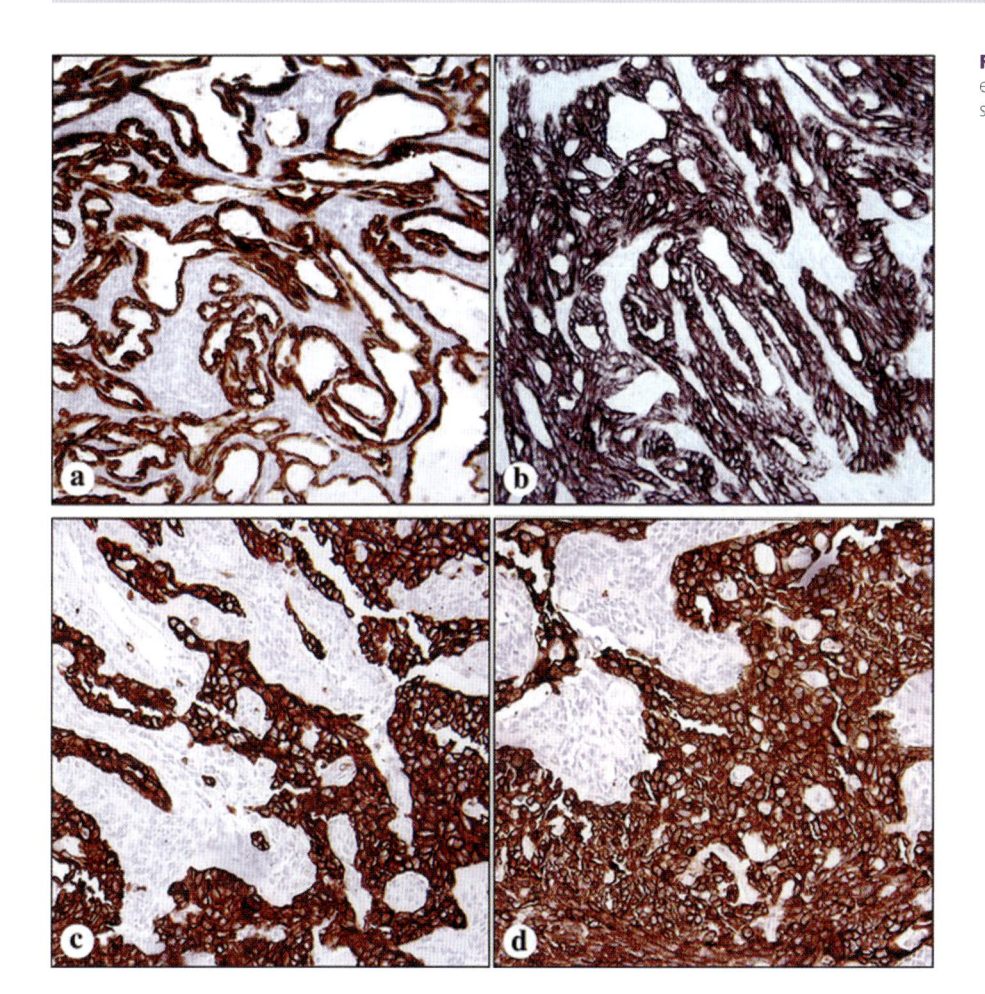

Chapter 27: Primary soft tissue tumors with epithelial differentiation

Figure 27.11 Keratin positivity highlights the epithelial components in biphasic synovial sarcoma.

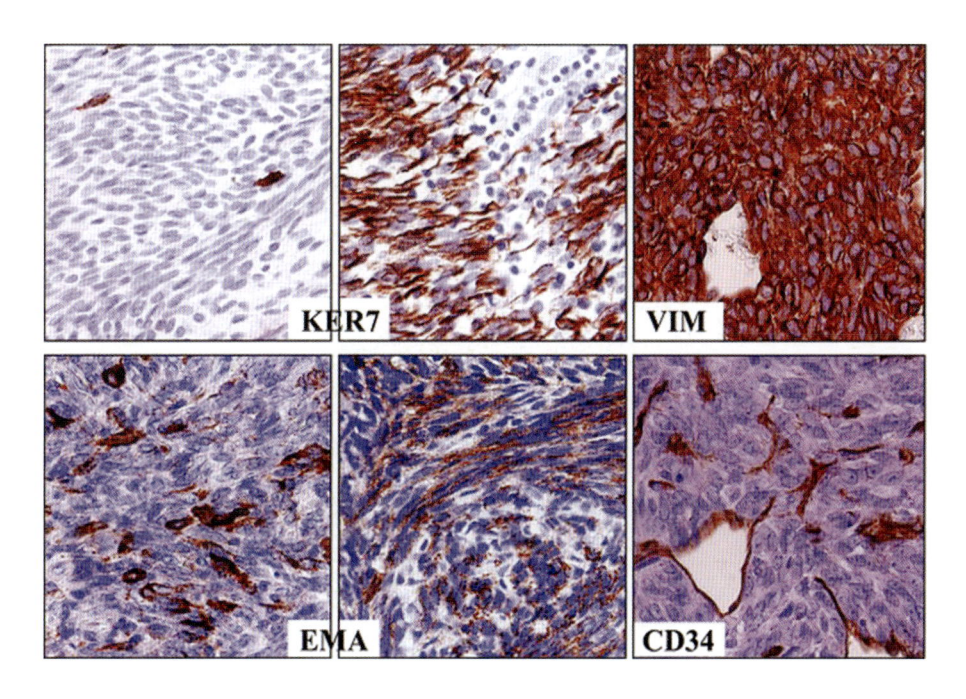

Figure 27.12 Immunohistochemistry of monophasic synovial sarcoma. There is variable K7 positivity, although the tumor cells often show greater numbers of EMA-positive cells. All cells are vimentin positive, whereas CD34 immunostain highlights the vascular endothelia only.

751

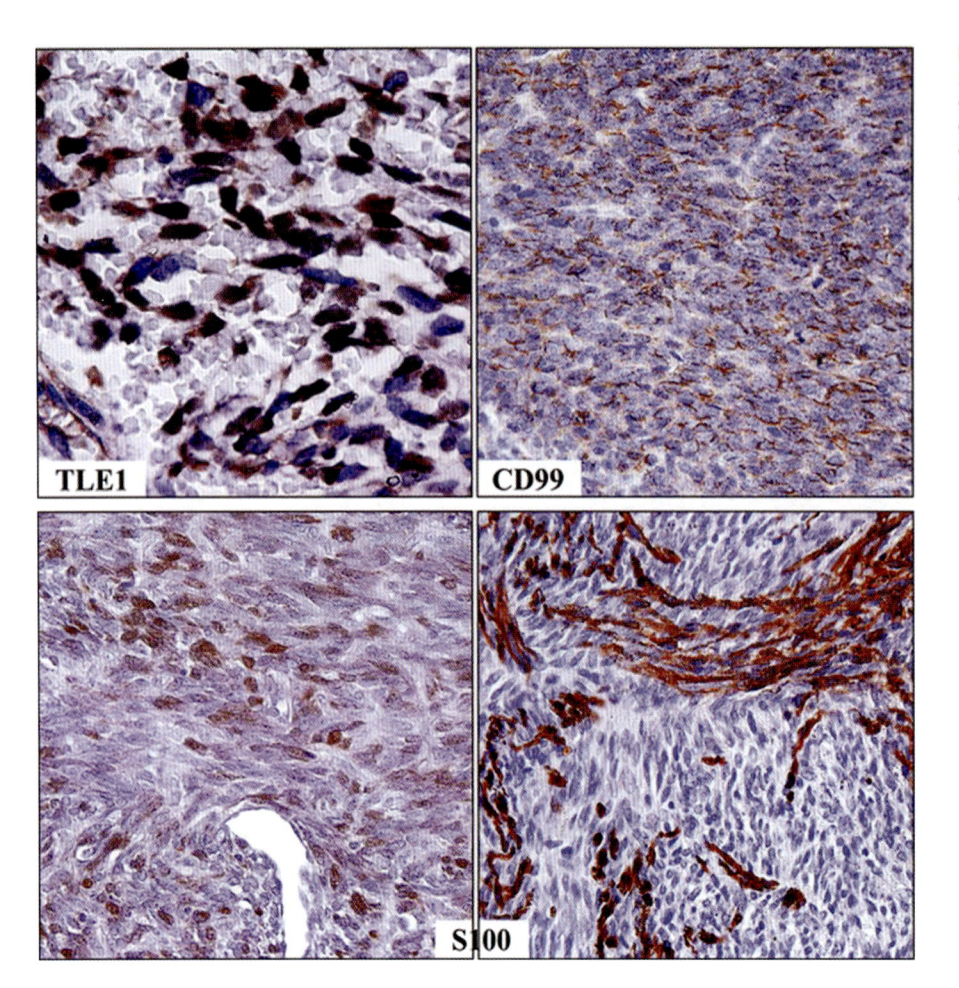

Figure 27.13 Synovial sarcoma cells typically show nuclear positivity for TLE1. CD99 positivity is common, and S100 protein positivity is seen in 30% of cases. In some monophasic synovial sarcomas, especially small tumors, there is S100-protein-positive neural proliferation; however, the tumor cells are negative in this case.

E-cadherin and N-cadherin are variably expressed, especially in the epithelial component; the former is usually more prevalent.[50,51]

Some markers often used in the positive identification of mesothelioma (*mesothelial markers*) are expressed in synovial sarcoma. In the biphasic tumors, the glandular epithelial cells are HBME1 positive, whereas the monophasic and poorly differentiated tumors are typically negative. The epithelial cells can also be positive for calretinin, which is more commonly expressed in the spindle cell component, monophasic tumors, and poorly differentiated components. Approximately 70% of synovial sarcomas have calretinin-positive cells; this should not lead to its being confused with mesothelioma. Of the other mesothelioma markers, thrombomodulin (CD141) is rarely expressed in synovial sarcoma, whereas WT1 is not. Ber-EP4 positivity occurs in the epithelial components.[52]

Transducin-like enhancer of split 1 (TLE1), a nuclear transcription factor in the WNT signaling pathway, is emerging as a new marker. TLE1 positivity appears in the nuclei of most synovial sarcoma cells of all types, and only nuclear positivity is significant (Figure 27.13). This marker is not totally specific, because solitary fibrous tumors and malignant peripheral nerve sheath tumors (MPNSTs) can also be positive.[53]

S100 protein positivity is seen in 30% of synovial sarcomas, sometimes as extensive reactivity in most of the tumor cells.[54] Positivity can occur in all types of synovial sarcoma, although it is more common in biphasic than in monophasic tumors (Figure 27.13). In addition, small hyperplastic nerve twigs are highlighted as being S100 protein positive, especially in small monophasic synovial sarcomas (Figure 27.13).

Rarely, biphasic tumors have a focally CD34-positive spindle cell component, which is practically uniformly negative for SMA and desmin.

Vimentin is expressed both in spindle cell and epithelial components, the latter often showing a basal cytoplasmic pattern. Common CD99 positivity should not lead to confusion with Ewing sarcoma (Figure 27.13).[55] The spindle cells of synovial sarcoma are typically positive for BCL2, whereas the epithelial component is usually negative.[56] CD56 positivity is a typical but nonspecific finding.

Some poorly differentiated synovial sarcomas show limited, if any, keratin and EMA positivity, and immunohistochemistry can be inconclusive. Recognition of the poorly differentiated variant often requires cytogenetic or molecular diagnosis, unless better differentiated areas or typical patterns of keratin or EMA expression are present.

Synovial sarcoma is a unique spindle cell sarcoma to express the NYESO cancer testis antigen, and this may have diagnostic utility.[57] Partial loss of nuclear INI1 (SMARCB1 gene product) has also been suggested as a distinctive feature of synovial sarcoma.[58,59]

Genetics

Cytogenetically, synovial sarcomas typically show a reciprocal t(X;18)(p11.2;q11.2) translocation that creates a large derivative X where a major portion of the long arm of chromosome 18 is translocated to the short arm of the X chromosome. This distinctive chromosomal change appears to be specific for synovial sarcoma and has not been documented in other sarcomas or tumors.[60,61]

The genes involved in the synovial sarcoma translocation are SS18 (previously known as SYT) in chromosome 18 and one of the SSX genes in the X chromosome: SSX1, SSX2,[62-66] or very rarely, SSX4.[67]

The synovial sarcoma translocation creates a gene fusion, with a resulting fusion transcript (chimeric transcript) and fusion protein. Both SS18 and SSX gene products are expressed in the nuclei. The SS18 gene apparently is a transcriptional activator, and SSX genes are transcriptional repressors. The altered regulatory function of these genes could be the key event in the pathogenesis of synovial sarcoma.[66] The SSX genes are normally expressed in the testis and also have been detected in many cancers, representing one of the so-called testis cancer antigens.[68]

The translocation can be diagnosed by fluorescent *in situ* hybridization (FISH)[69-75] or by reverse transcription polymerase chain reaction (RT-PCR).[43,44,76-81] FISH studies have expectedly shown that both the epithelial and spindle cells in biphasic tumors carry the translocation.[74] Determination of whether SSX1 or SSX2 is involved is also possible by FISH, using appropriate probes.[75] The commercially available SS18 break-apart probe is a convenient tool for FISH diagnosis of synovial sarcoma.

It has been shown that many biphasic tumors have SSX1 gene involvement, whereas the monophasic variety has either SSX1 or SSX2 with equal frequency.[43] The tumors with SSX1 fusion have been shown to have a higher proliferative activity.[80]

Moderately large numbers of other sarcomas have been evaluated for the SS18-SSX fusions, and all have been found to be negative.[81,82] Occasional opposite results have been presented, including the apparent presence of SS18-SSX fusions in neurofibroma and MPNSTs.[83] These results have been viewed with great skepticism, however, because of the uniform lack of t(X;18) translocation in the cytogenetic studies of large numbers of peripheral nerve sheath tumors.[82] Potential reasons for the differences include different diagnostic criteria for tumors and false-positive results in the translocation assay.

Additional secondary genetic changes often produce complex karyotypes, and they probably represent genetic disease progression, because such changes are more common in recurrent tumors.[84] Comparative genomic hybridization has shown no DNA copy number changes in 50% of cases, apparently representing the cases with balanced translocation only. In the remaining 50% of cases, the most common additional DNA copy number changes have been gains of 8q and 12q and losses of 13q21–31 and 3p.[85]

Differential diagnosis

Monophasic synovial sarcoma often resembles fibrosarcoma, and actually many deep spindle cell sarcomas previously diagnosed as fibrosarcomas are actually synovial sarcomas. Tumors with palisades and S100 protein positivity could resemble cellular schwannoma; however, monophasic synovial sarcoma is more highly cellular and more homogeneous in composition than schwannoma.

Small synovial sarcomas in the peripheral extremities often elicit a significant neural proliferation inside the tumor and should not be misinterpreted to be neurofibromas or MPNSTs. The cellular components between the neuroma-like elements are similar to monophasic (or rarely biphasic) synovial sarcoma.

The biphasic tumors are sometimes suspected of being teratomas or mixed tumors, especially when they occur in the lower abdominal wall or inguinal region. Teratomas do not generally occur in these soft tissue locations in adults, and mixed tumors do not have a true biphasic pattern, as is seen in synovial sarcoma, but are composed of keratin-positive epithelial cells with a variable morphology and having a possible chondroid, keratin-negative component.

The poorly differentiated tumors might simulate the Ewing family of tumors both histologically and immunohistochemically (being CD99 positive). If no differentiated components are present, the patterns of keratins and EMA could be helpful, synovial sarcomas being more often K7 and EMA positive. In keratin-negative cases, molecular diagnosis of specific translocations (SS18-SSX versus EWS rearrangements) is often necessary for the definitive diagnosis.

Intrathoracic synovial sarcomas should be separated from sarcomatoid carcinoma and mesothelioma. The lack of a pleomorphic component and limited keratin positivity in the spindle cell component support the diagnosis of synovial sarcoma, but molecular genetic studies could be necessary for definitive diagnosis.

Epithelioid sarcoma

Named and described in detail by Enzinger in 1970, an *epithelioid sarcoma* is a histologically distinctive, rare soft tissue tumor with true epithelial differentiation, predominantly occurring in the distal extremities of young adults; the tumor is notorious for its potential resemblance to necrotizing granuloma and carcinoma.[86] The normal cell counterpart of epithelioid sarcomas is unknown. Epithelioid sarcoma is a rare tumor and constitutes no more than 1% to 2% of all sarcomas.

Variants composed of large cells, commonly with rhabdoid epithelioid cytology and often presenting in proximal sites

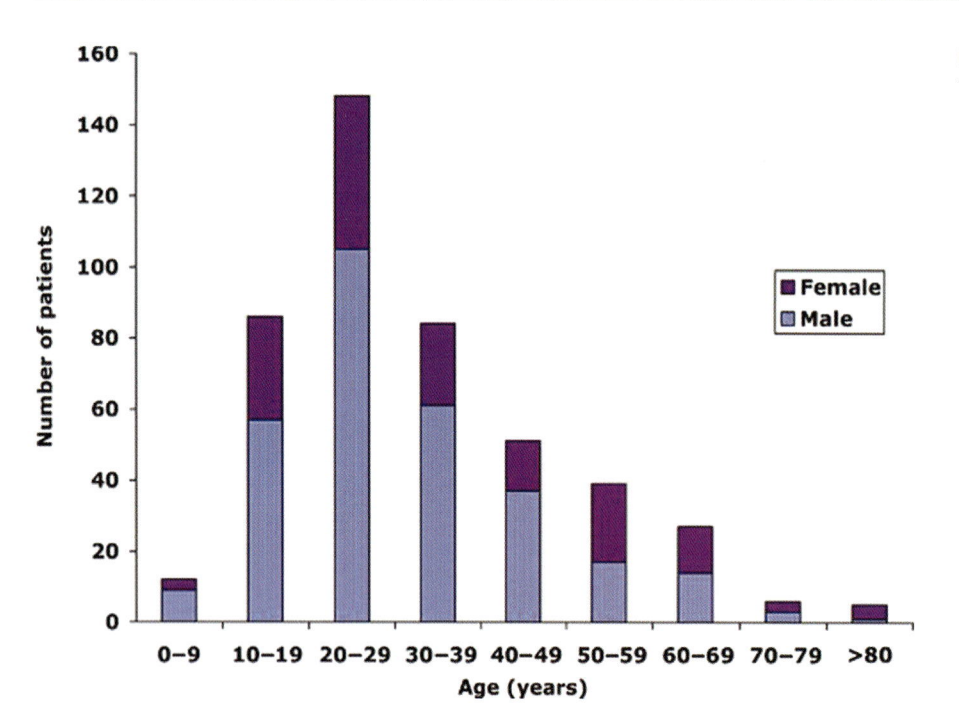

Figure 27.14 Age and sex distribution of 458 patients with epithelioid sarcoma.

("proximal type of epithelioid sarcoma"),[87] differ from the typical variant and is discussed separately.

Clinical features

Epithelioid sarcomas have a predilection for young men, with the peak incidence in the third decade. According to Armed Forces Institute of Pathology (AFIP) statistics, 69% of all cases occur between the ages of 10 and 39 years, and the median age is 28 years (Figure 27.14). Epithelioid sarcoma also occurs in younger children and older adults with a lower frequency, but it is exceptionally rare before the age of 5 years and after the age of 70 years.[86,88–107] If only civilian patients are included, the male-to-female ratio is still 1.8:1 in the AFIP series. Some series have shown a similar male predominance.[91–95]

The upper extremities is the most commonly involved region (57%), and there is a predilection for the fingers and hands (Figure 27.15). In the fingers and hand, the volar side is more commonly involved, whereas in the forearm, the extensor surface is more often affected.[86] The lower extremities are involved in 28% of patients, and the foot, ankle, and pretibial region are most commonly involved. Perineal, vulvar, and penile skin are involved in 7% of cases. Occurrence in the trunk wall is rare, and the involved sites include the chest wall and axillary region. In the head, the scalp is most commonly involved, but this is also a common site for metastatic epithelioid sarcoma. Reported rare sites include the oral cavity[108] and intra-articular extremity locations,[109] while presentation in the neck seems to be very rare.

Epithelioid sarcoma can be clinically elusive, often leading to diagnostic delay. In the fingers and hands, the tumor can appear as a nonspecific swelling suggesting a harmless reactive condition. In some cases, the tumor forms a cutaneous ulcer with raised margins. In the largest series, the mean duration of symptoms was 2 years, with some patients having a 25-year history prior to diagnosis.[89] A more recent series still found a 1.5-year duration of symptoms prior to diagnosis.[94] A history of significant prior trauma, such as bone fracture or muscle tear at the site where the tumor developed, does not seem to be uncommon.[89]

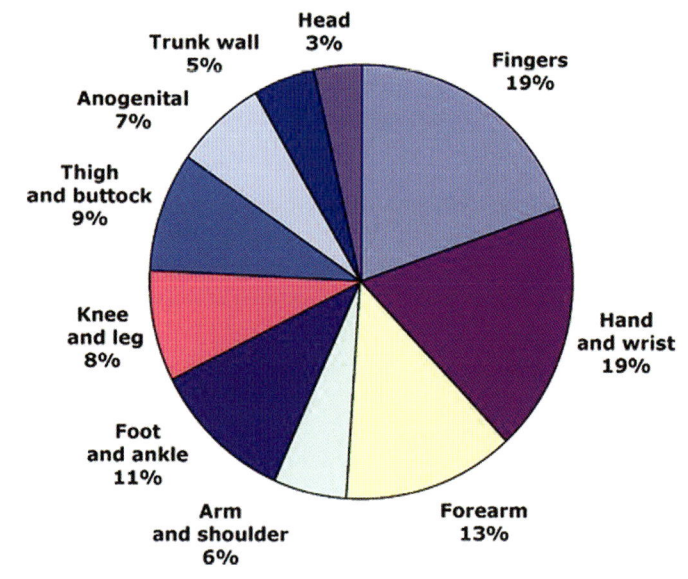

Figure 27.15 Anatomic distribution of 458 epithelioid sarcomas.

The local recurrence rate in the older series was very high (77%), but has been lower (35–69%, average: 42%) in more recent tertiary care series, probably because of more aggressive

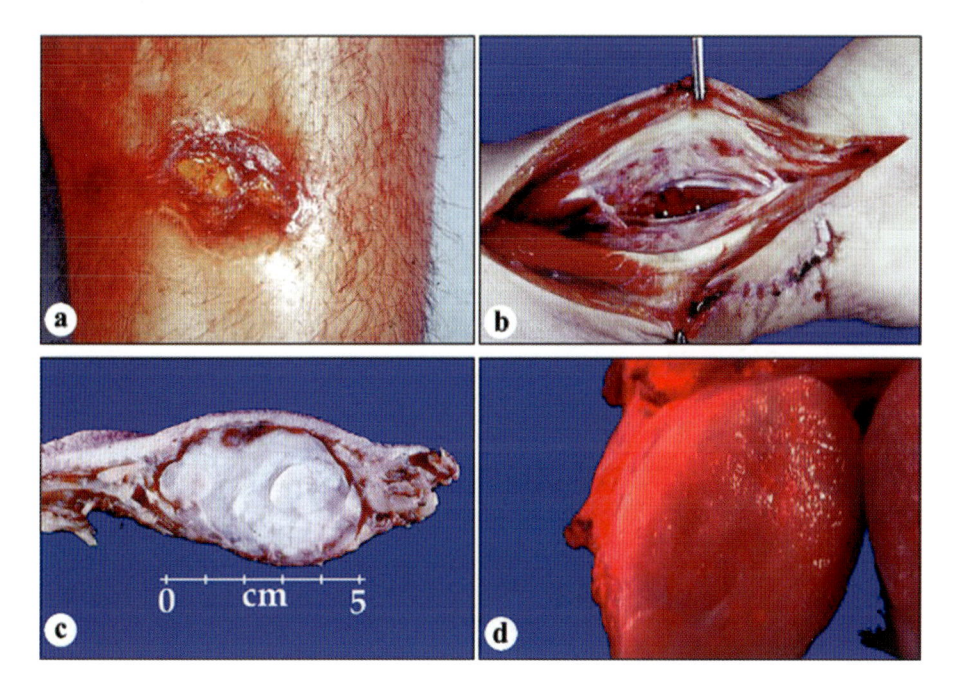

Figure 27.16 (a) Epithelioid sarcoma sometimes forms an ulcerated lesion, as seen in this tumor involving pretibial soft tissue. (b) Epithelioid sarcoma forming a whitish deep mass. (c) Subcutaneous nodule. (d) On sectioning, an unfixed epithelioid sarcoma shows a homogeneous pale tan surface.

treatment.[90,93–95] Regional lymph node and distant metastases occur equally often in 30% to 50% of patients. Distant metastatic sites include the lung and pleura, the soft tissue of the scalp, and, rarely, the brain.[89,93,95] Although the 5-year survival rate in the largest series was 75%, the 10-year survival rate dropped to 50% as distant metastases continued to develop.[89] More recent series have reported 60% to 66% actuarial 5-year survival rates,[94–96] but in a recent large study, the median survival after distant metastasis was only 8 months.[95] Wide excision is generally necessary for local control, and in some cases, this requires ray amputation of a digit, hand, or foot. Limb-sparing surgery supplemented with postoperative radiation therapy was reported to give results equal to radical surgery in one series, however.[96] Systemic chemotherapy has been used for metastatic tumors, but experience is limited because of the rarity of this tumor.[94]

Pathology

Grossly, an epithelioid sarcoma can form a sharply circumscribed nodule, usually 2 cm to 5 cm, but the extent of recurrent tumor is often impossible to determine on gross examination. The tumor nodules typically reveal a gray-white surface on sectioning, and necrosis can be identifiable (Figure 27.16).

Histologically, these lesions are composed of nodules and clusters of deeply eosinophilic epithelioid or spindled cells. The cellular areas can form a central geographic necrosis "garland pattern" or invade as narrow streaks in a dense, fibrous stroma (Figure 27.17) that may contain metaplastic bone. The tumor cells have complex nuclear outlines and delicate nucleoli with little nuclear pleomorphism. The epithelioid sarcoma cells typically have an abundant, strongly eosinophilic cytoplasm (Figure 27.18).

In some cases, hemorrhage develops in large epithelial nests (Figure 27.19a), and this occurrence should not be confused with angiosarcoma.[98] A rare variant shows a prominent spindle cell component simulating fibrous histiocytoma (Figure 27.19b–d), but infiltration by the epithelioid tumor cells occurs among the major spindle cell components.[99]

Immunohistochemistry

Epithelioid sarcoma cells are almost always strongly positive for keratins, EMA, and vimentin.[100–106] Among the keratin polypeptides present are keratins 8, 18, and usually 19, typically with most of the tumor cells being strongly positive (Figure 27.20). Keratin 7 is focally present in 20% of cases, keratins 5, 6, and 14 focally in rare cases, and K20 almost never.[106] Multidirectional differentiation, including expression of neurofilament proteins, has been reported.[107,110] One half of all cases are positive for CD34 with a membrane pattern, and approximately one-third react with muscle actins (HHF-35).[105,111] Immunoreactivity for S100 protein and desmin is very rare. In contrast with many other epithelial tumors, epithelioid sarcomas have been reported negative for E-cadherin, but positive for vascular cadherin (V-cadherin).[112] Immunohistochemical lack of the nuclear SMARCB1 (INI1) gene product is typical of all types of epithelioid sarcoma, and this is a useful, although not totally specific, diagnostic test.[113] Among others, renal and extrarenal rhabdoid tumors, some extraskeletal myxoid chondrosarcomas, mixed tumors/myoepitheliomas, and rare carcinomas can have a similar loss of INI1.[114,115] Common nuclear ERG-immunoreactivity has been reported with an antibody recognizing the aminoterminus,[116] but not with an antibody recognizing the carboxyterminus.[117] ERG positivity should not lead to confusion with vascular tumors.

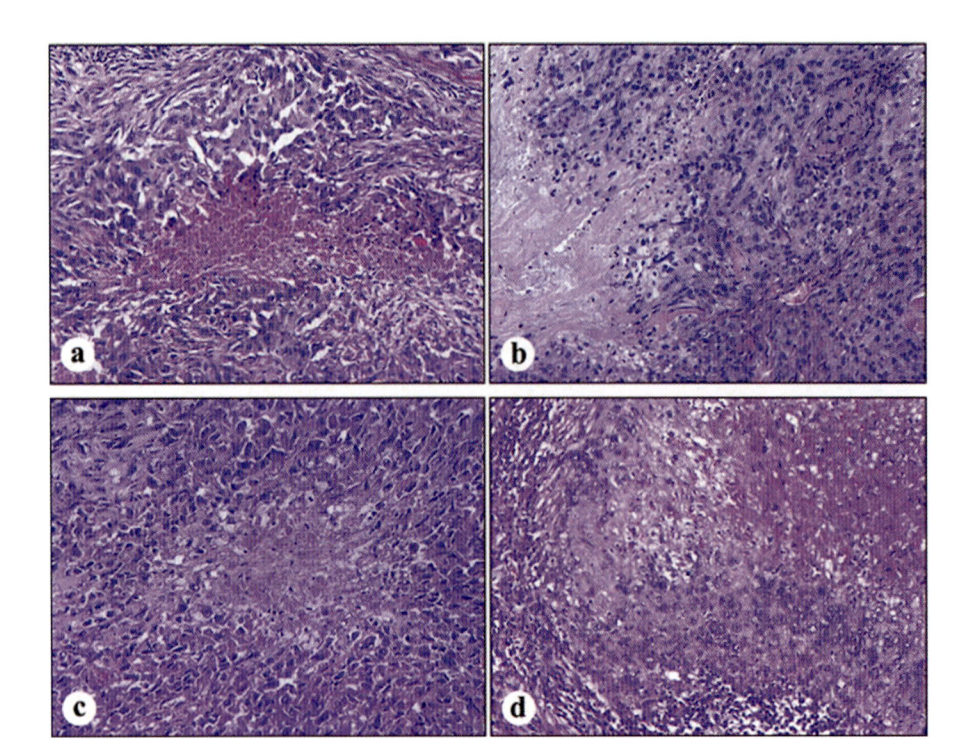

Figure 27.17 Low magnification of an epithelioid sarcoma, highlighting different patterns of epithelial element with central necrosis.

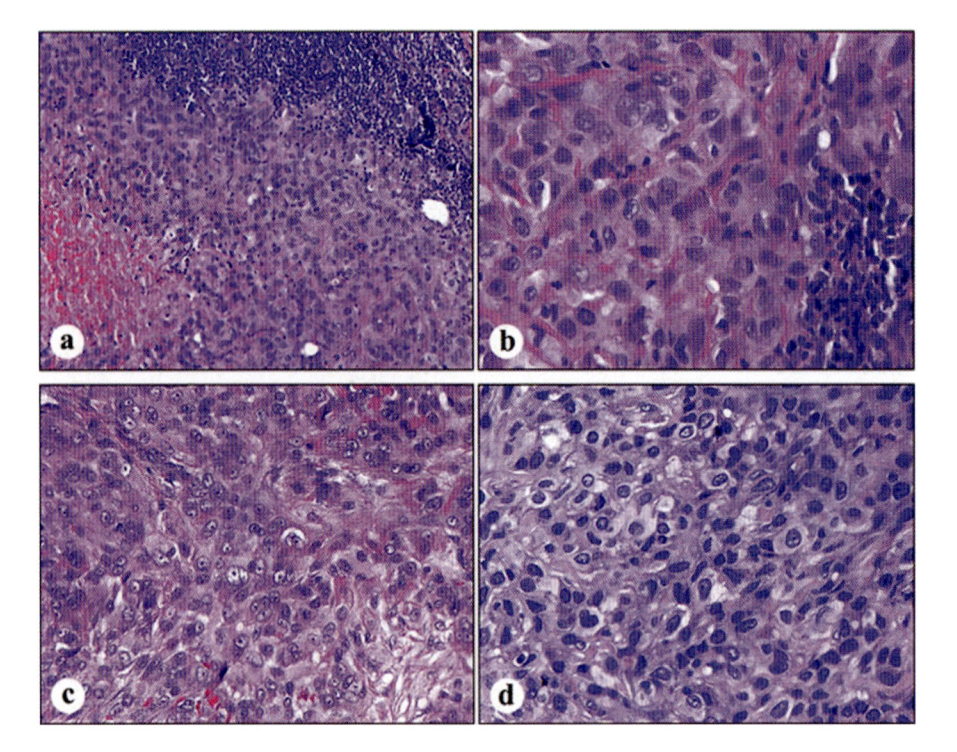

Figure 27.18 High magnification of epithelioid sarcoma shows eosinophilic cells with abundant cytoplasm and relatively uniform nuclei.

Differential diagnosis

The rarity of epithelioid sarcoma and its variable histologic patterns can make this diagnosis difficult, resulting in diagnostic delay. Indeed, epithelioid sarcomas are among the most common soft tissue tumors associated with claims of delayed pathologic diagnosis.

The differential diagnostic problems range from proper recognition of an epithelioid sarcoma as a malignant process to separating it from benign conditions, such as necrobiotic granuloma and benign lesions with osseous metaplasia. Malignancies that can resemble epithelioid sarcoma include squamous cell, skin adnexal and metastatic carcinoma, and melanoma and clear cell sarcoma.

Chapter 27: Primary soft tissue tumors with epithelial differentiation

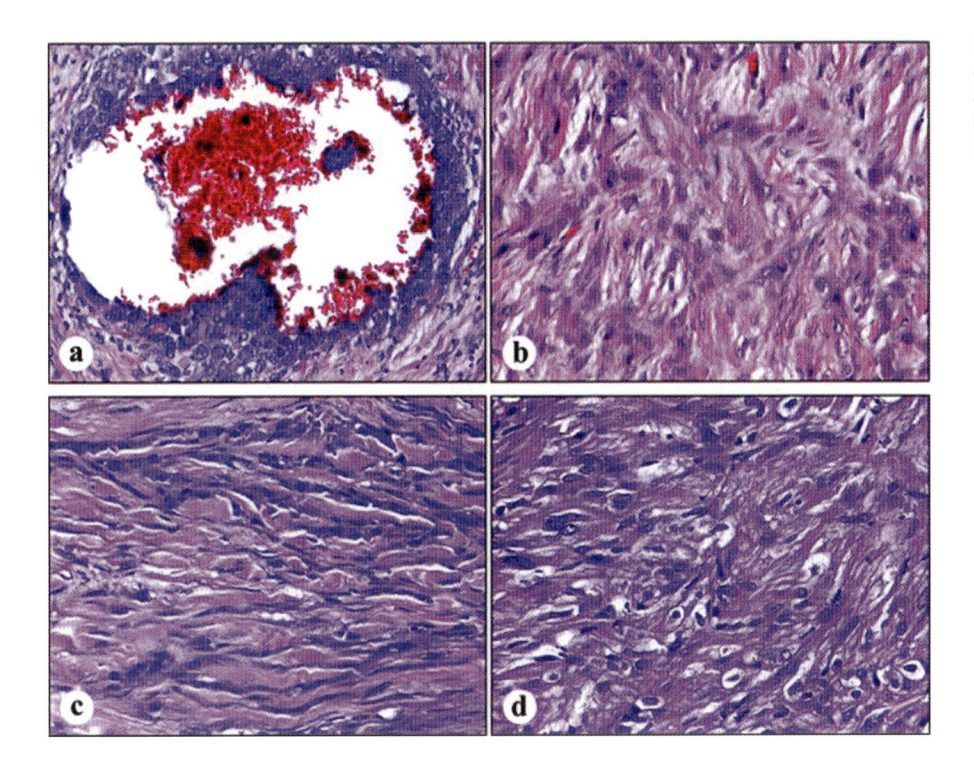

Figure 27.19 (a) Intranodular hemorrhage in epithelioid sarcomas can create the appearance of a vascular tumor. (b–d) An epithelioid sarcoma growing as narrow streaks in a fibrous stroma can simulate variants of fibrous histiocytoma.

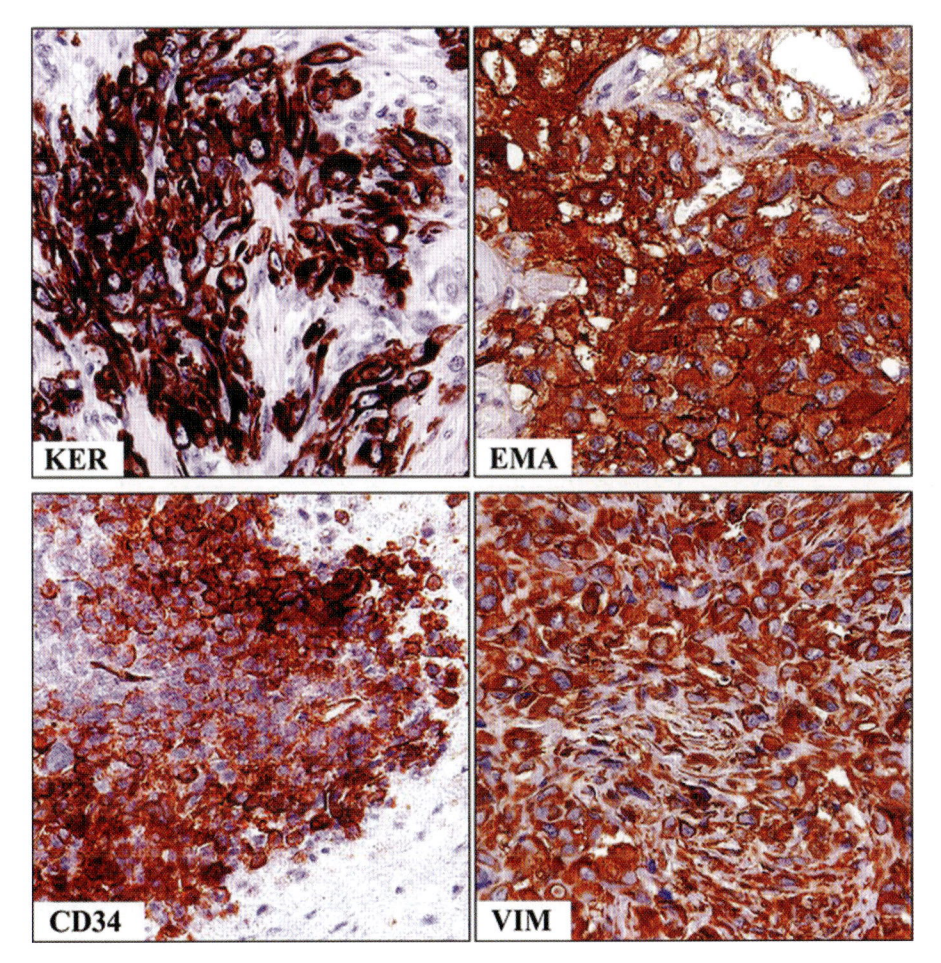

Figure 27.20 Immunohistochemical features of an epithelioid sarcoma. The tumor cells are positive for keratin, EMA, CD34, and vimentin.

The presence of reactive osteoid spicules should not lead to confusion with panniculitis or myositis ossificans. The distinctive multinodular growth, containing eosinophilic keratin-positive and often CD34-positive tumor cells, should allow identification of the epithelioid sarcoma elements.

Necrobiotic granulomas are lined by smaller histiocytic cells that lack the prominently eosinophilic, keratin-positive cytoplasm. The lining cell population is also heterogeneous, containing lymphoid cells admixed with histiocytes that express CD45 (LCA), and histiocytic markers such as CD68 and CD163.[118]

Squamous cell carcinomas often reveal squamous pearls on closer inspection and can contain a malignant or premalignant epidermal element. They are more often positive for keratins 5 and 6, and p63 than are epithelioid sarcomas, and rarely if ever express CD34.[106,119] Metastatic carcinomas, especially from the lung and kidney, can sometimes simulate epithelioid sarcoma when showing nodules with central necrosis. The tumor cells typically display greater cytologic atypia than is observed in epithelioid sarcoma, however. Multiple lesions at different sites should also raise the possibility of a metastatic tumor.

Melanoma and clear cell sarcoma cells can have prominently eosinophilic cytoplasm, but these tumors do not usually show the centrally necrotic nodules. They are easily separated from epithelioid sarcomas by their immunohistochemical positivity for melanoma markers, especially S100 protein and HMB45. At metastatic sites, such as the brain, the geographic necrosis of epithelioid sarcoma should not be confused with glioblastoma multiforme.[120]

Genetics

Cytogenetic studies report heterogeneous changes, and no consistent alterations have been observed in the different studies. However, molecular genetic studies on different subtypes of epithelioid sarcoma have revealed allelic loss in chromosome 22, including the SMARCB1 (INI1) locus. Concurrent loss-of-function mutation of this gene leads to homozygous loss of expression of the protein, which is a practical immunohistochemical diagnostic marker.[121-123]

Comparative genomic hybridization has shown gains in 11q13 as the most common recurrent changes, possibly including amplification of the cyclin D1 gene.[121] FISH studies have not revealed cyclin D amplification, but this protein is consistently expressed.[122] A well-characterized epithelioid sarcoma cell line was reported with hyperdiploid karyotype and several complex translocations.[123]

Proximal-type epithelioid sarcoma

These epithelioid sarcomas, which are also characterized by losses of INI1/SMRCB1, are typically composed of larger epithelioid cells, often with rhabdoid cytoplasmic intermediate filament inclusions.[87,124] This variant, first reported by Guillou et al., typically occurs in the proximal extremities in deep locations (e.g., proximal thigh, buttocks) and truncal locations (e.g., the anogenital region, trunk wall, and axilla).[87] The tumors vary from small (1–3 cm) to large masses (>10 cm). There is a >50% rate of tumor-related mortality.[87,124-126]

Histologic appearance varies, but most examples show solid sheets of epithelioid cells, and necrotizing granuloma-like patterns, as seen in typical epithelioid sarcoma, are only rarely present. Mitotic activity is usually higher than in typical epithelioid sarcomas, often exceeding 10 per 10 HPFs. Rhabdoid cytoplasmic inclusions are a typical feature seen in all cases, although usually only in a few tumor cells (Figure 27.21). Immunohistochemical features for keratin, vimentin, EMA, and CD34 expression are similar to those found in typical epithelioid sarcomas, but in addition, desmin has been reported in one half of the cases.[87]

Differential diagnosis from extrarenal rhabdoid tumor is problematic. Rhabdoid tumors typically occur at a very young age and are often composed of more primitive and rounded cells with a lesser amount of cytoplasm.

Mixed tumor and myoepithelioma

Mixed tumors of the skin and soft tissues are analogous to corresponding tumors commonly seen in salivary glands. They are synonymous with *chondroid syringomas*, a name used particularly in older dermatopathology literature. These tumors have variable epithelial and myoepithelial components and typically have metaplastic cartilage-like elements. In the skin and subcutis, mixed tumors most likely originate in the sweat glands or their ducts; however, it is unclear how the occurrence of similar tumors in intramuscular soft tissues can be explained.

In the current dermatopathology literature, *mixed tumor* has been the preferred term, although myoepitheliomas have also been reported. In the literature on soft tissue, the term *myoepithelioma* has been applied especially to tumors with no glandular elements. It seems clear that these groups represent a continuum, and therefore they are combined here. Rare malignant variants with metastases occur.

Clinical features

Cutaneous and soft tissue examples have occurred in a wide age range, from 2 to 93 years; the median ages in various series range from 30 to 60 years, with apparent male predominance.

Whereas mixed tumor of the skin has a strong predilection for the skin of the head (e.g., face, nose, forehead, upper lip), the reported soft tissue examples have occurred in a broader range of sites without significant regional predilection, including presentation in distal and proximal extremities, trunk, and the head and neck.[127-133] Most examples involve the dermis, subcutis, or both, and some extend into the fascia or skeletal muscle. Tumors considered to be cutaneous myoepitheliomas have also been identified, and these seem to occur outside of the head and neck, in contrast to cutaneous mixed tumors.[134,135] The tumor size in the cutaneous examples varies

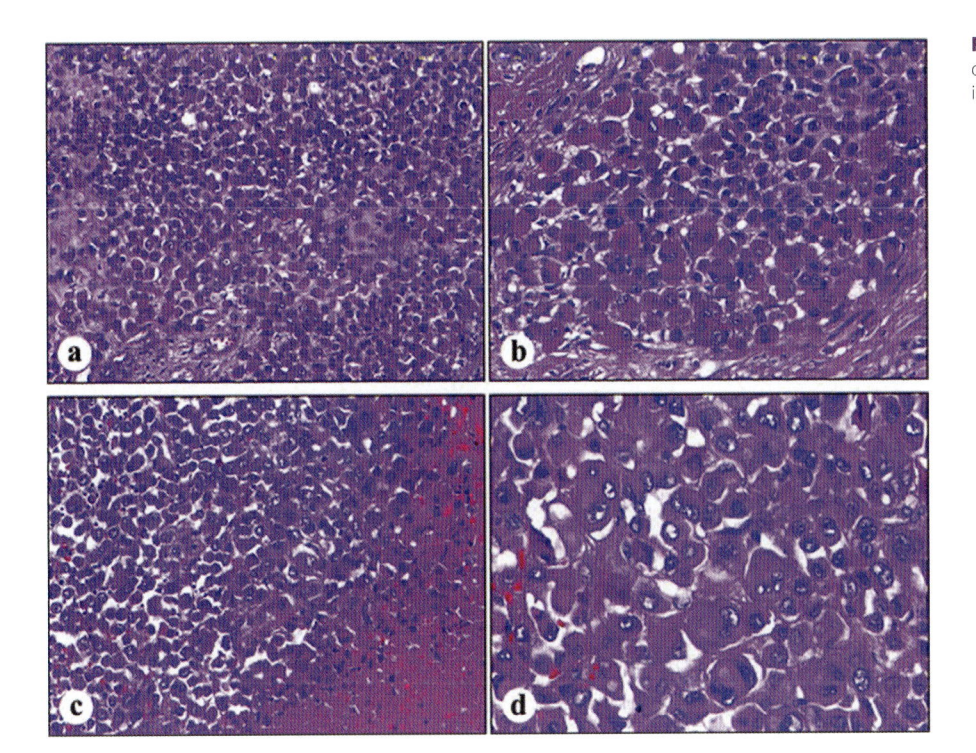

Figure 27.21 An epithelioid sarcoma composed of large cells with rhabdoid-type cytoplasmic inclusions.

(~1 cm), whereas the soft tissue examples have been larger, on average 2 cm to 3 cm, and rare examples are >5 cm. Local recurrence in soft tissue examples has been reported with a frequency of approximately 20%, and distant metastases with fatal outcome have also been reported.[133,136] The behavior of this tumor can be difficult to predict histologically, and the tumor should be completely excised, at least when atypical features are present. Occurrence in deep periosteal location and bone have also been reported. Malignant examples are discussed in a separate section.

Pathology

Grossly, mixed tumors form circumscribed and often lobulated masses, which can be mucoid or firm. The reported tumors have measured 1 cm to 17 cm, but most have been <3 cm.[127–134]

Microscopically, the tumors can be recognized by their similarity to the analogous tumors of the salivary glands and skin. The composition varies from case to case, but most examples show epithelioid tumor cells forming cords, nests, and occasional tubular epithelial structures in a loose myxoid or dense hyaline matrix, and true cartilage can also be present (Figure 27.22). Some variants are composed of solid sheets of epithelioid cells, and there are examples composed of cords of cells in a myxoid matrix. The cytoplasm is usually eosinophilic, but some tumors have clear cell features, and rhabdoid cytology is a feature of some examples (Figure 27.23).[137] Mitotic activity is typically low, and nuclear atypia is limited. Tumors with significant atypia and mitotic activity must be considered as being of worrisome or malignant potential (see discussion later on malignant mixed tumor). Earlier authors have separated cutaneous mixed tumors into apocrine and eccrine types. This distinction is based on epithelial stratification: tumors with two epithelial layers in the ductal epithelia have been classified as *apocrine*, and those with the ducts lined by a single epithelial layer as *eccrine*.[138] This distinction does not seem to have any more practical significance, however.

Immunohistochemistry

Mixed tumors are typically positive for keratins, and many tumors also show significant positivity for S100 protein. Expression of GFAP and SMA, calponin, and other determinants of myoepithelial cells varies.[130,133,135] EMA is usually restricted to the glandular luminal aspect (Figure 27.24).

The simple epithelial keratins (i.e., K7, K8, K18, K19) are usually expressed, whereas the myoepithelial keratins K5 and K6 are only variably present, and K20 is essentially absent. Keratin cocktails, such as AE1/AE3, typically show extensive positivity.

Differential diagnosis

Although some histologic features can resemble those of extraskeletal myxoid chondrosarcoma, the latter lacks epithelial differentiation and almost uniformly epithelial markers. Parachordoma might be a related tumor, but is discussed here separately. Parachordomas tend to show less complex

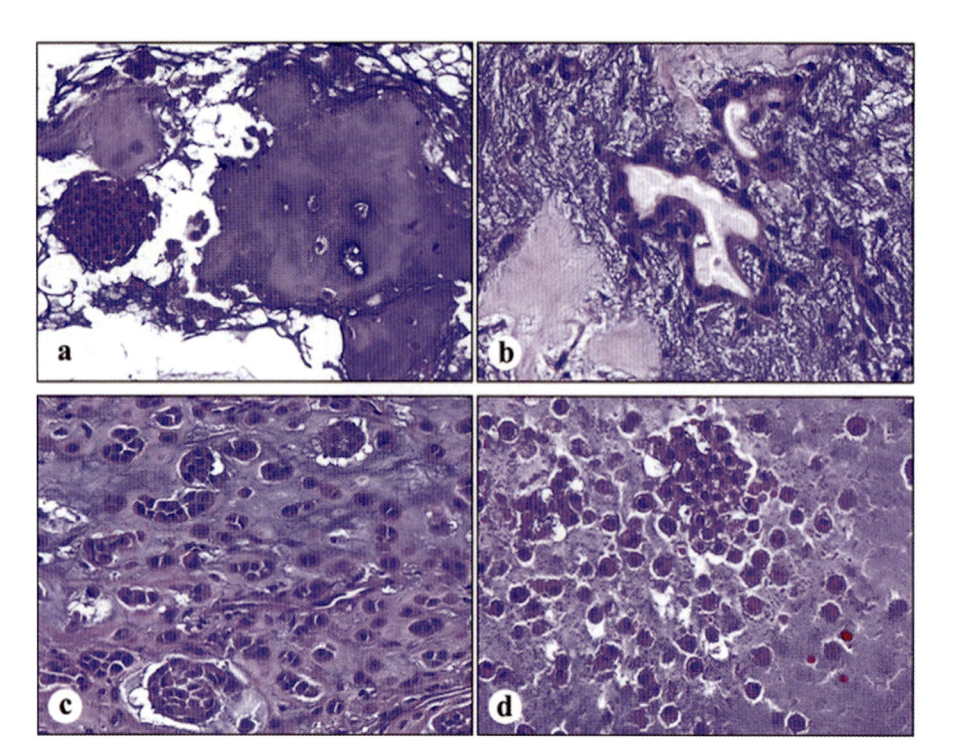

Figure 27.22 (a) Mixed tumor containing epithelial clusters and cartilage-like matrix. (b) Epithelial glandular elements in myxoid matrix. (c,d) Clusters or scattered epithelioid cells in myxoid matrix.

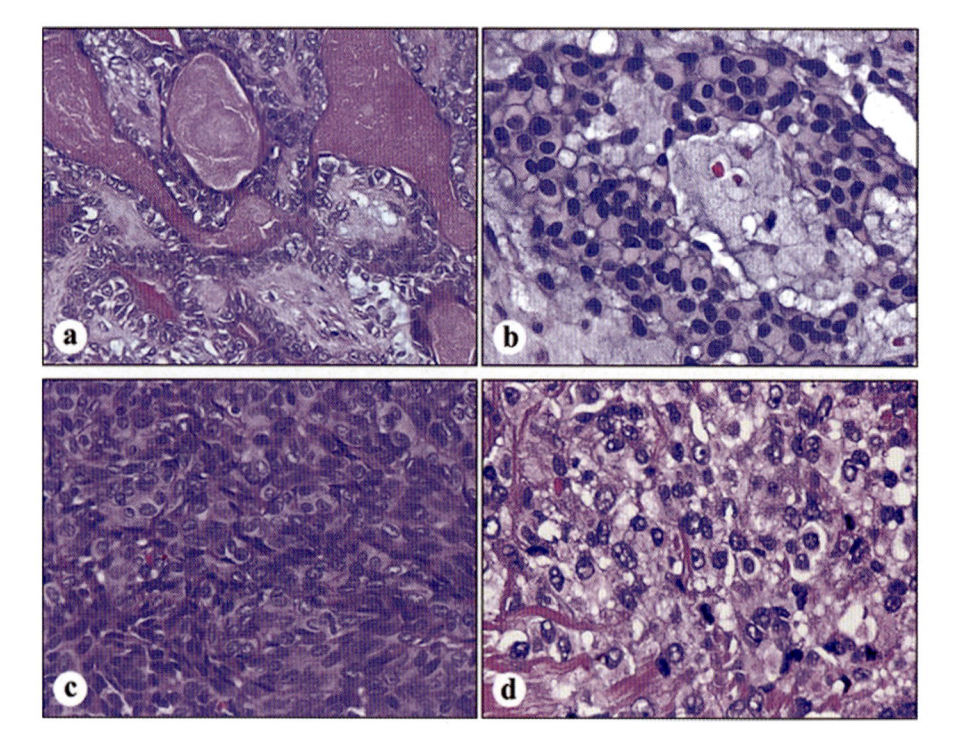

Figure 27.23 Variations in the epithelial component in a mixed tumor. (a) Squamoid appearance. (b) Tumor with rhabdoid cytology. (c) A spindled pattern that can resemble synovial sarcoma. (d) Example with moderate cytologic atypia.

epithelial differentiation, with lack of K19, but complete comparison of these tumor types is not available.

The diagnoses of metastatic carcinoma and synovial sarcoma also should be considered in the case of deep soft tissue tumors with epithelial differentiation.

Genetics

EWSR1 gene rearrangements, including gene fusion EWSR1-PBX1,[139] EWSR1-POU5F1,[140] and EWSR1-ATF1,[141] and FUS-POU5F1[142] have been reported in soft tissue

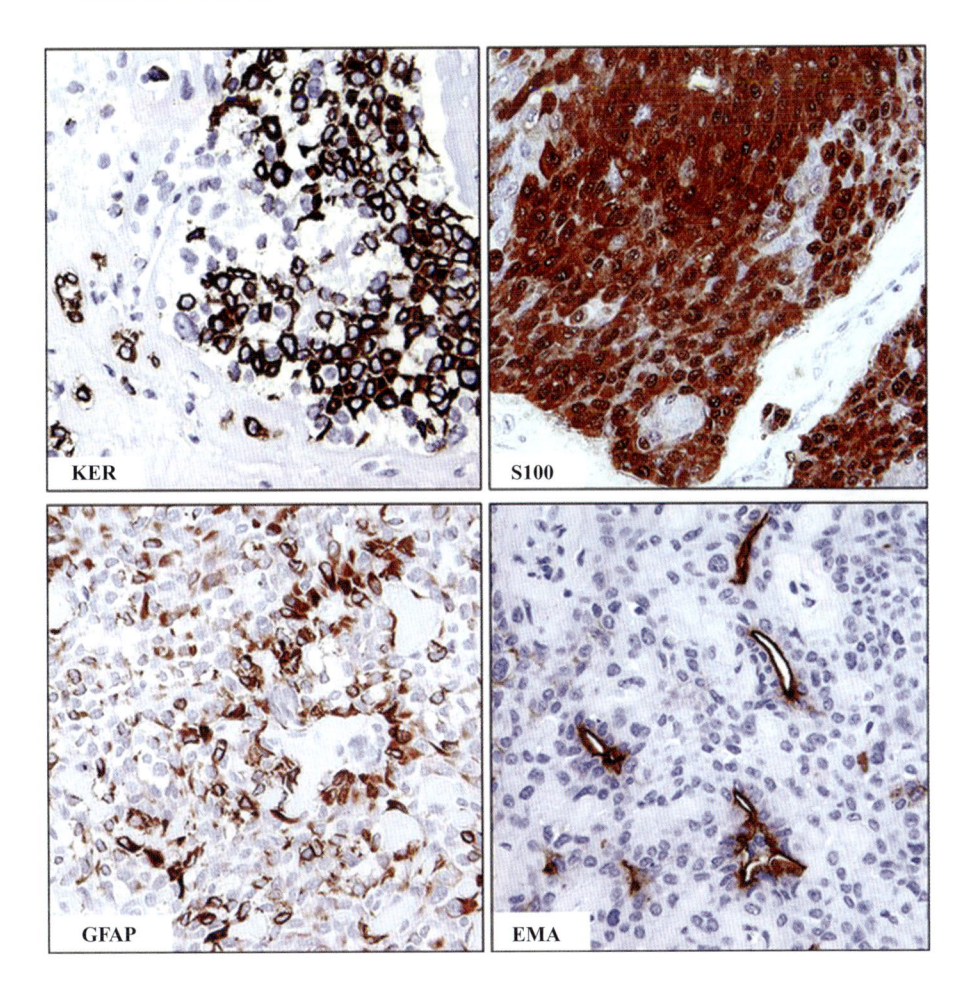

Figure 27.24 The epithelial elements in mixed tumors are usually variably keratin and S100 protein positive, and GFAP-positive cells are a common finding, reflecting myoepithelial differentiation. EMA positivity is usually restricted to luminal aspect of ducts.

myoepitheliomas. Translocations and other gene rearrangements of the PLAG1 gene occur in mixed-tumor-like morphology with epithelial elements, in tumors analogous to salivary gland mixed tumors.[143]

Parachordoma

Originally named by Dabska,[144] a *parachordoma* is rare soft tissue tumor, and very few series about it have been published. The histogenesis of this tumor is unresolved, but there are similarities with deep mixed tumors of soft tissues, and for now it is appropriate to consider this tumor as a morphologic variant of mixed tumor/myoepithelioma.

Clinical features

This tumor occurs in patients of all ages, from 4 to 62 years with predilection for young adults; the collective median age of the three largest series is 30 years.[144–146] There is no predilection for either sex. The tumor typically presents in deep soft tissue of the extremities. The clinical course seems favorable, although data on long-term follow-up are few.

Pathology

Grossly, parachordomas form a 1-cm to 3-cm circumscribed whitish nodule in the subcutis or deep tissue. Histologically, the tumor is typically multinodular. The nodules are composed of cords of epithelioid cells separated from a myxoid stroma (Figure 27.25). Parachordomas resemble chordomas in these features. Immunohistochemically, the tumor cells are positive for keratins and S100 protein. However, a small number of soft tissue tumors quite similar to chordoma and expressing nuclear brachyury have been reported and some of these tumors have been considered peripheral analogs to chordoma.[147–149]

Differential diagnosis

Most parachordomas differ from chordoma histologically and immunohistochemically with a less complex keratin pattern (no K19). They also differ from extraskeletal myxoid chondrosarcoma, which is typically negative for epithelial markers. Otherwise, immunohistochemical parameters place parachordoma tumor within the spectrum of mixed tumor/myoepithelioma.

Malignant mixed tumor (myoepithelioma)

Examples of mixed tumors having malignant clinical behavior, including distant metastases and death from the tumor, have been reported as isolated cases and part of some clinicopathologic series. Such tumors have usually occurred in older

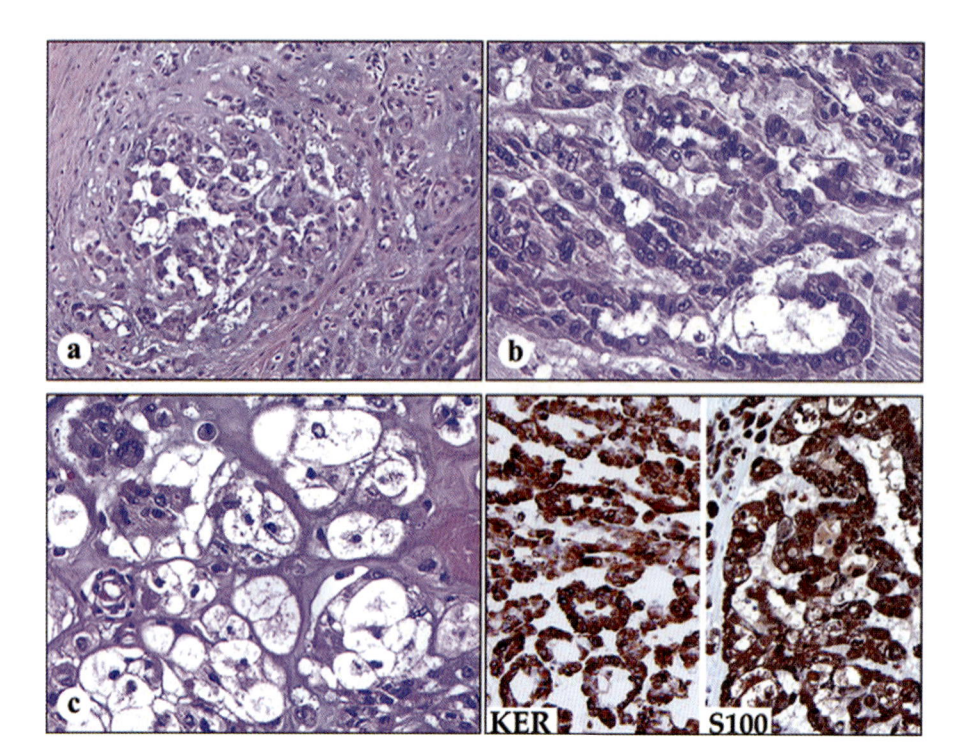

Figure 27.25 A parachordoma is composed of multiple lobules of tumor cells arranged in trabeculae separated by myxoid stroma.

patients and have often been fairly large (≥5 cm); distant metastasis has been preceded by local recurrence and regional nodal metastasis. In some cases, the metastases have developed years after the primary surgery.[133,136] Examples of aggressive malignant myoepitheliomas have also been reported in children. Some of these tumors have had fairly undifferentiated morphology differing from typical myoepitheliomas.[150]

Histologically, the malignant mixed tumors have an increased mitotic activity and cellular pleomorphism, but the published data do not allow delineation of distinct criteria for malignancy. It is best to consider tumors with significant atypia, mitotic activity, or invasive behavior to have at least the potential for malignant behavior until better diagnostic criteria are developed. Large tumors should be extensively sampled. In these cases, complete excision with negative margins and long-term follow-up are necessary.

Hidradenoma (eccrine acrospiroma)

This relatively common skin adnexal tumor occurs in a broad age range, with a predilection for young adults. A wide variety of cutaneous locations can be involved, and this tumor is rarely painful. There is a small (10%) potential for local recurrence, which is usually related to incomplete excision.[151,152]

Pathology

The size of the lesion is usually small, often approximately 1 cm, but exceptional examples that are >10 cm have been reported. Typical histologic features include a well-circumscribed nodule with cystic and solid epithelial elements, often surrounded by dense, hyaline collagenous matrix. The epithelial element can show focal squamoid or clear cell cytology (Figure 27.26). The clear cell change might correlate with cytoplasmic glycogen content. Atypia, mitotic activity, and invasive epithelial clusters around the capsule are not features of this tumor, and their presence suggests a more aggressive neoplasm, such as a variant of hidradenocarcinoma.

Immunohistochemically typical is positivity for both low-molecular-weight (LMW) and HMW keratins, EMA, and focally and variably for S100 protein in the basaloid cells facing the stroma.[153,154]

Eccrine spiradenoma

Described in detail in 1956 in the classic AFIP series by Kersting and Helwig,[155] this skin adnexal tumor of sweat gland origin is histologically distinctive, with rare malignant variants also described.

Clinical features

This not-so-uncommon skin adnexal tumor has a predilection for young adults, with almost 90% of the patients being younger than 40 years old, and there is a male predominance. Occurrence in infancy has also been reported.[156] Nearly any skin site can be involved, but the favored sites include the ventral side of the upper trunk and the head (i.e., the face/forehead).[155,157]

Clinically, there is a small, usually 1-cm to 2-cm dermal or subcutaneous nodule that can be skin colored, or blue or reddish. The nodule is painful in one half of the cases, and many patients have a long, often more than 5-year history of a local, slowly enlarging nodule. The behavior of

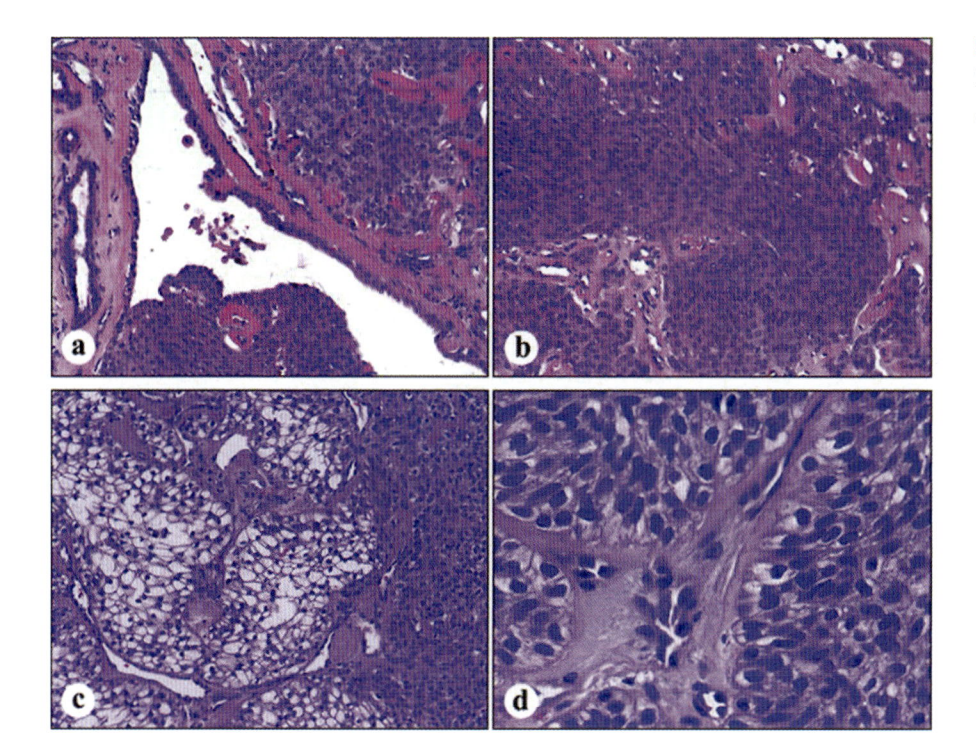

Figure 27.26 Eccrine acrospiroma contains cystic ductal elements and a solid epithelial component. The epithelia can show clear cell change and uniform cytology.

ordinary eccrine spiradenoma is benign, although recurrence (regrowth) is possible after an incomplete excision.

Rare malignant variants have a metastatic potential, and some cases have been fatal. In the malignant examples, marked enlargement of a long-standing tumor has been sometimes observed. Metastases have developed in 20% to 25% of patients into the lymph nodes, lungs, liver, and brain. Some patients with malignant examples have suffered from cylindromatosis syndrome, which carries an increased risk of malignancy.[158–162]

Pathology

Benign eccrine spiradenoma is usually a 1-cm to 2-cm, non-ulcerated cutaneous lesion, sometimes <0.5 cm, but almost never >5 cm. It appears as a sharply demarcated nodule that has a grayish or pinkish cut surface. Histologically typical is a multinodular pattern, although small lesions can be composed of a single nodule. The tumor and the individual nodules are surrounded by a thin capsule-like streak of connective tissue. The nodules are composed of varyingly defined microscopic lobules that in their periphery typically contain darker cells having small nuclei with dense chromatin and invisible nucleoli. Dilated lymphatic-like vascular spaces can be present (Figure 27.27).[163] Some cells are typically SMA and S100 protein positive, indicating myoepithelial differentiation. Most of the cells are paler, with larger vesicular nuclei and delicate nucleoli. In the center of the lobules, a ductal structure can be distinguished, but luminal differentiation is clearly highlighted by immunostains with EMA or carcinoembryonic antigen (CEA). Mitotic activity is inconspicuous or absent.

Keratin positivity is detected in most of the cells (K18, K19, K5, K6, especially in the central lobules).[154] A comparative study concluded that the keratin positivity was similar to that in the transitional portion of eccrine glands between the secretory elements and coiled ducts.[164] Immunohistochemical studies have shown large numbers of S100 protein-positive Langerhans cells within the epithelial lobules, whereas S100 protein positivity in the epithelial cells varies.[165]

Rare malignant variants have contained a small element of benign eccrine spiradenoma with a carcinomatous component having marked atypia and mitotic activity.[158–161] A sarcomatoid carcinoma (carcinosarcoma) component, with heterologous rhabdomyosarcomatous or osteosarcomatous components, has been reported.[162]

Cylindroma

This skin adnexal tumor usually originates in the scalp and adjacent regions, especially the forehead. More often than not it forms a solitary nodule.[166] In familial cylindromatosis (e.g., Brooke–Spiegler syndrome), multiple to innumerable cylindromas develop and grow into large confluent masses sometimes referred to as *turban tumors*. Other skin adnexal tumors, specifically eccrine spiradenomas and trichoepitheliomas, are also part of the syndrome. These patients carry a loss-of-function germline mutation in the CYLD gene (at 16q12.1). Loss of heterozygosity (LOH) at this locus is observed in cylindromas, indicating a type of mechanism similar to that of tumor-suppressor genes.[167–169]

Histologically, there is a sharply demarcated nodule typically surrounded by a fibrous capsule. This nodule is subdivided into multiple, irregularly shaped micronodules that are

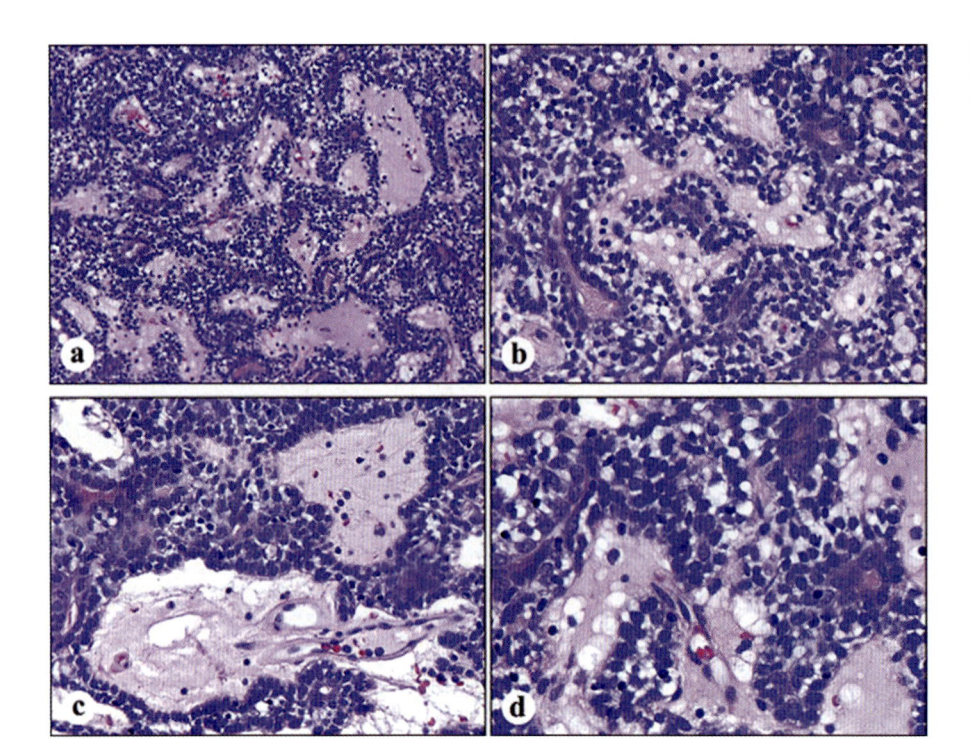

Chapter 27: Primary soft tissue tumors with epithelial differentiation

Figure 27.27 An eccrine spiradenoma contains epithelial cells and septa with perivascular edema. Note also the scattered lymphoid cells.

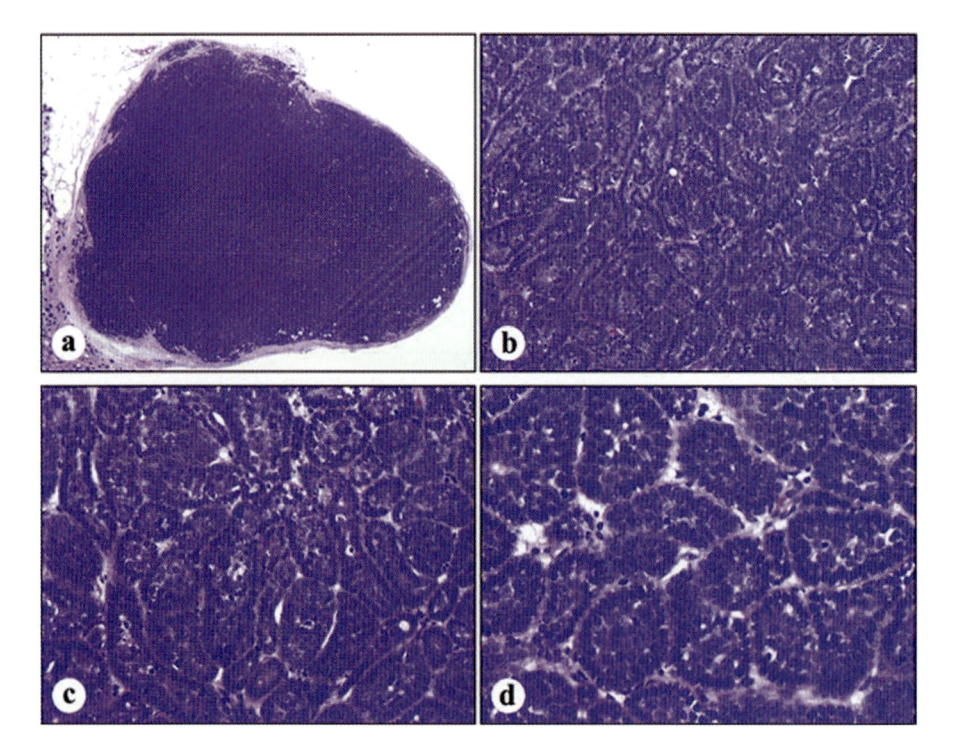

Figure 27.28 Cylindromas form a sharply demarcated nodule that consists of microscopic lobules surrounded by basement membranes.

surrounded by an eosinophilic hyaline zone composed of basement membrane proteins. All cells in the nodule are similar, showing nuclei of moderate size often containing delicate nucleoli. Cytoplasm is relatively scant. A smaller number of lymphoid cells are sometimes present (Figure 27.28).

Aggressive digital papillary adenocarcinoma

Tumors of this entity were previously stratified into two variants: *digital papillary adenomas* and *malignant digital papillary adenocarcinomas*.[170] More recent series have concluded that

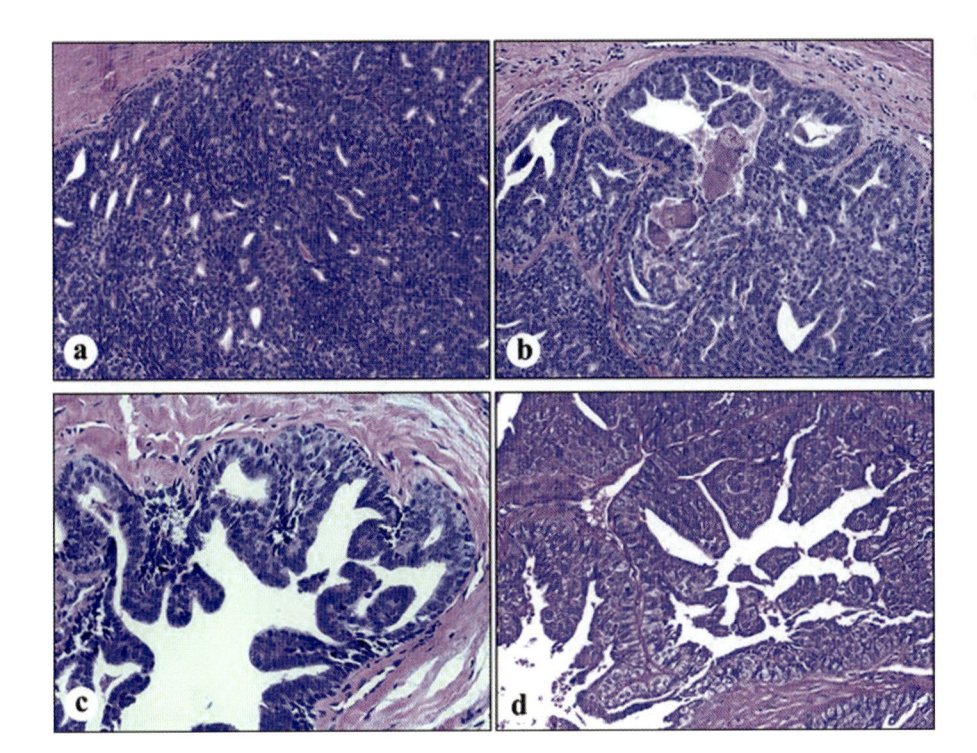

Figure 27.29 Examples of aggressive digital papillary adenocarcinoma. Note papillary tufting, cribriform, and focally solid areas.

this separation is not practical, because morphology (degree of atypia and mitotic activity) is not predictive of tumor behavior.[171,172]

Clinical features

This rare skin adnexal carcinoma occurs in adults of all ages, with a median age of approximately 50 years. There is a significant male predominance. These tumors occur exclusively in the distal extremities. Nearly 80% present in the fingers, usually on the volar surface of the distal phalanx. In the foot, the usual location is in the toe, and rarely in the sole of the foot. A few cases have been reported in the palm of the hand.[170] Local recurrence is common after nonradical surgery. Metastases to the regional lymph nodes and lungs can develop. In the largest series of 67 cases, 6 patients developed pulmonary metastases, and 3 patients (4%) had progressive disease or died of metastatic disease.[171,172]

Pathology

The tumor forms a solid and cystic mass of <1 cm to 4 cm (mean size, 1.7 cm), and is composed of epithelial nests in a dense collagenous stroma (Figure 27.29). The epithelial nests vary from cystic types with various papillary intracystic elements to solid types that often contain central comedonecrosis (Figure 27.30). The mitotic rate varies from 0 to 60 per 10 HPFs.

Differential diagnosis

Metastatic disease in the lymph nodes should be separated from other epithelial tumors, especially synovial sarcoma.

In aggressive digital papillary adenocarcinoma, there is only one component, the epithelial one, in contrast with the biphasic population in synovial sarcoma.

Chordoma

A *chordoma* is primarily a bone tumor with epithelial differentiation, probably arising in notochordal rests. Its importance in soft tissue pathology lies in the fact that chordomas often form a significant soft tissue mass at the site of origin, most commonly in the sacrococcygeal region. Chordomas can also metastasize to distant soft tissue sites.[173,174]

Clinical features

Chordomas occur predominantly in older adults but occasionally are seen in children, in this case almost exclusively occurring in the axial skeleton in the spine and skull base, most commonly in the sacrococcygeal region. It can also bulge into body cavities (e.g., the mediastinum or retroperitoneum), skin, and the subcutis as the soft tissue component of a primary spinal neoplasm.[173–175] The sacral tumors frequently bulge outward, forming a subcutaneous mass. Chordomas can metastasize to the skin and deeper soft tissues.

Pathology

Grossly, a chordoma typically forms a pale multinodular mass that often has a mucoid appearance on sectioning (Figure 27.31). Histologically distinctive is a multinodular pattern of growth, similar to cartilaginous tumors. The tumor cells can form cords in a myxoid matrix or sometimes as solid sheets of epithelioid cells (Figures 27.32 and 27.33). The nuclei

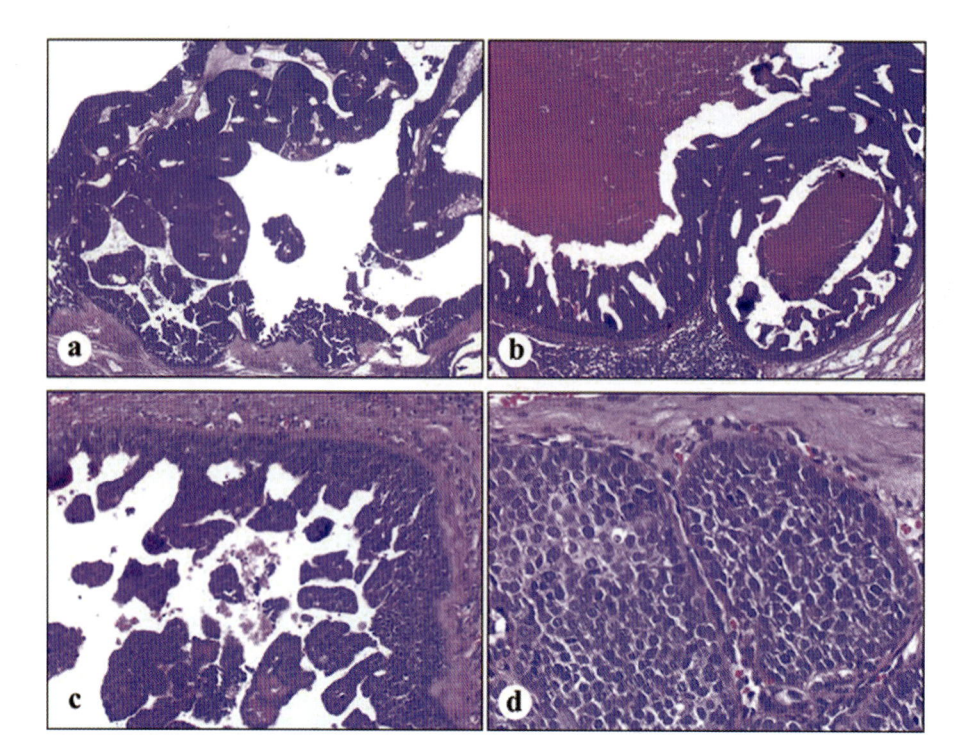

Figure 27.30 Lymph node metastasis of aggressive digital papillary adenocarcinoma contains papillary and solid elements that sometimes contain central comedonecrosis.

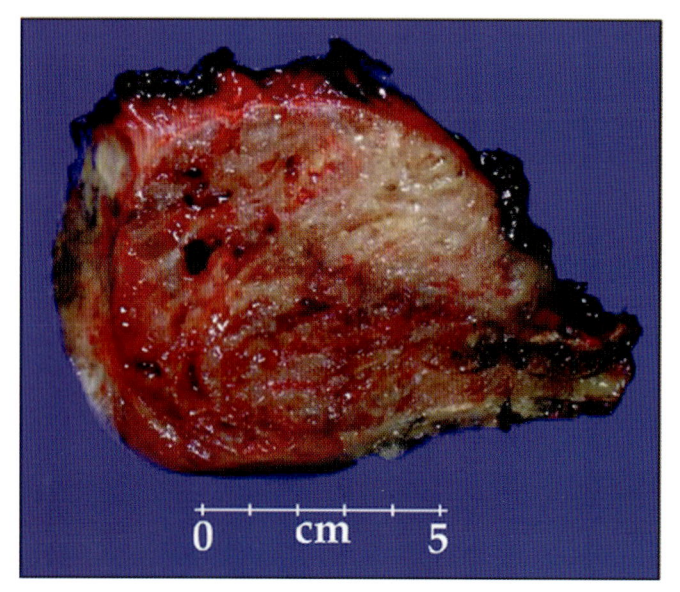

Figure 27.31 Grossly, a chordoma forms a solid mass with a mucoid texture on sectioning.

are usually small, but some chordomas have focal nuclear pleomorphism. Fibrosarcoma- or MFH-like transformation can occur.

Immunohistochemically, chordoma cells are typically positive for the simple epithelial keratins K8, K18, and K19 (Figure 27.33), whereas they are usually negative for K7 (although focal positivity is possible). Most cases are also positive for EMA. Vimentin positivity is consistent, but the chordoma cells are only variably S100 protein positive.[176] Brachyury transcription factor is consistently detectable in nuclei, although it can be absent in fully sarcomatoid examples.[147,149,177] One has to consider that brachyury is not totally specific for chordoma as it can also be expressed in nuclei of germ cell tumors, small cell carcinomas, and, rarely other carcinomas, sarcomas, and melanomas.[149] Brachyury gene (T-gene) polymorphisms and duplications are associated with development of chordoma.[178,179]

Differential diagnosis

Architectural differences, especially a trabecular epithelioid pattern, and the presence of epithelial markers, keratins, and EMA, separate chordomas from conventional and extraskeletal myxoid chondrosarcoma. Chordoma metastasis can be separated from mixed tumors of the skin and soft tissues by their structural heterogeneity and the presence of tubular epithelial structures and cartilage-like differentiation in the latter.

Branchial anlage mixed tumor (ectopic hamartomatous thymoma)

This peculiar and rare epithelial and spindle cell tumor, originally described by Rosai and colleagues as "ectopic hamartomatous thymoma," is a clinicopathologically distinctive entity.[180] Prior to its description, it was variably considered a variant of mixed tumor or squamous cell carcinoma. Based on its typical location and histology, this tumor has been suggested to represent a branchial anlage mixed tumor with a prominent myoepithelial component, possibly originating from residual elements of the cervical sinus of His that failed to regress. Arguments against its thymic derivation include lack of evidence of thymic components, sparse if any lymphoid

Chapter 27: Primary soft tissue tumors with epithelial differentiation

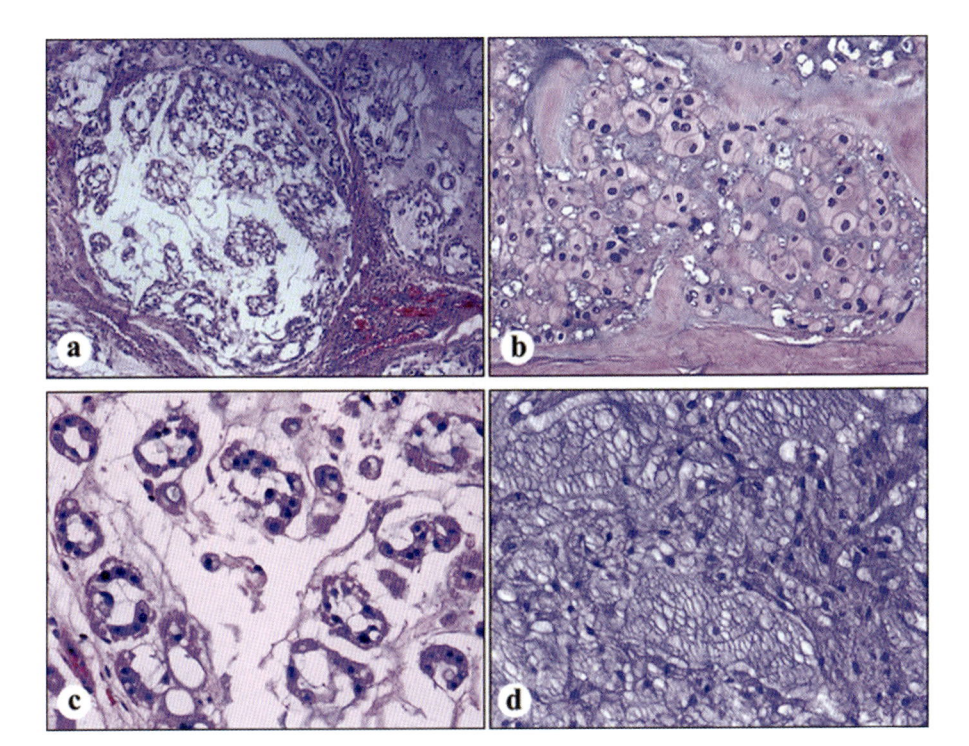

Figure 27.32 Chordoma. (**a**) Typical lobulation. (**b**) Eosinophilic epithelial cells forming discohesive clusters in mucoid stroma. (**c**) An example with pseudoglandular arrangement of tumor cells. (**d**) A trabecular pattern.

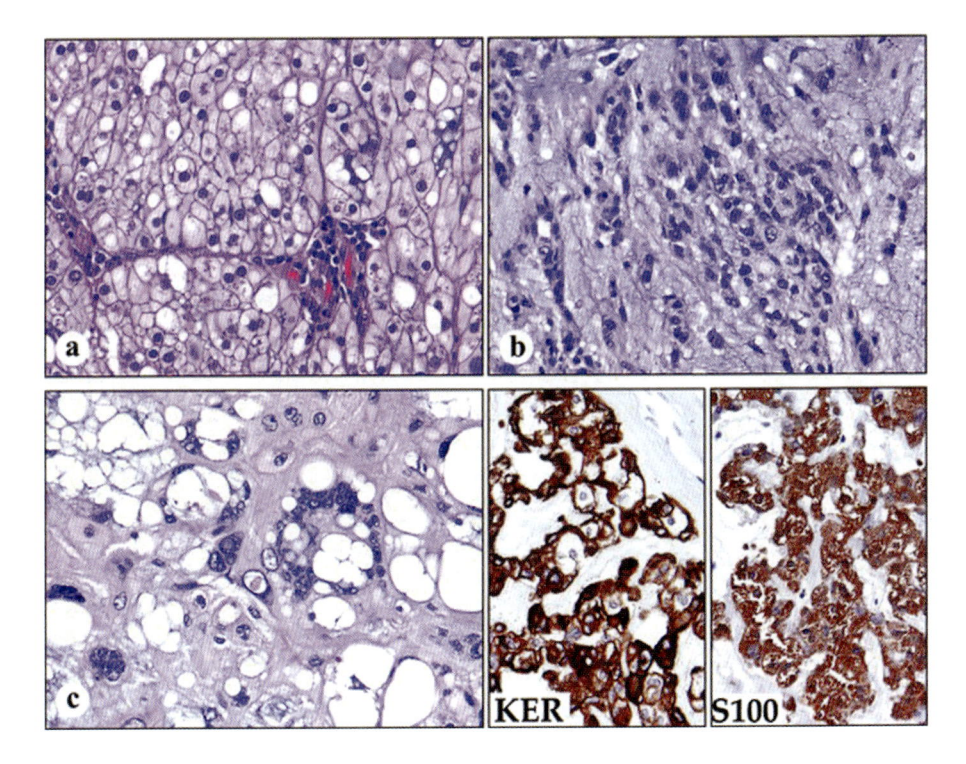

Figure 27.33 (**a**) A chordoma forming solid sheets of epithelial cells. (**b**) An example with an atypical spindle cell component. (**c**) A chordoma with nuclear atypia and vacuolization resembling that of a liposarcoma. The tumor cells are strongly positive for keratins, and in this case, also for S100 protein.

component, lack of lobulation typical of many thymic neoplasms, and lack of reported cases in the thymus.[181]

Clinical features

The tumor specifically occurs in the lower end of the sternocleidomastoid muscle, just above the mid- or medial clavicle. It predominantly occurs in young to middle-aged men (median age, 40 years) and is rare (<10%) in women. The tumor presents as a painless, often long-standing subcutaneous mass that usually measures <5 cm. According to follow-up information (some spanning >20 years), the tumor is benign, but local recurrence is possible, generally representing regrowth of residual tumor after a biopsy or incomplete excision.[180–185]

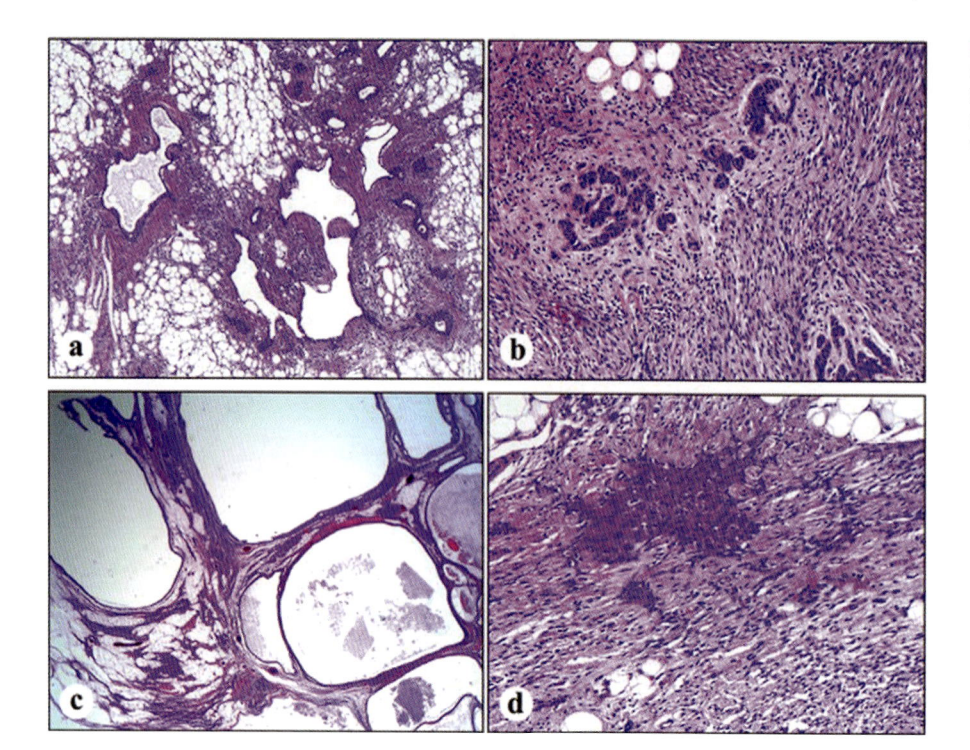

Figure 27.34 Branchial anlage mixed tumor (ectopic hamartomatous thymoma). Low magnification reveals epithelial cysts, and higher magnification shows epithelial foci in the midst of uniform spindle cells.

Pathology

Grossly, the tumor is well circumscribed but unencapsulated, and based on collective reports, measures from 1.5 cm to 19 cm (median size, 4 cm). The tissue has a firm or rubbery consistence and is mottled gray-white to tan with yellow foci on sectioning. Fluid-containing cysts might be recognizable to the naked eye.

Histologically, the lesion exhibits a biphasic pattern with epithelial and spindle cell elements. The former contains cleft-like and slit-like epithelial structures lined variably by cuboidal, glandular, or keratinizing squamous epithelia; occasionally the last element predominates. The spindle cell component has a diffuse or vague fascicular pattern and usually forms a majority element (Figure 27.34). It is cytologically bland, with elongated nuclei having pointed or blunt ends and delicate nucleoli. The cytoplasm varies from pale to mildly eosinophilic, and cell borders are vague with variable presence of extracellular collagen. Mitotic activity is low, only rarely exceeding 5 per 50 HPFs; it is rarely seen in epithelial elements. Foci of adipose tissue are almost uniformly present (Figure 27.35). In some cases, the cords and clusters of epithelial cells are disorganized, suggesting carcinomatous transformation, but mitotic activity is scant, and follow-up studies have not demonstrated malignant behavior.

Immunohistochemistry and ultrastructure

Both the epithelial and spindle cell components are positive for keratins (Figure 27.36) and variably for EMA, and contain both LMW, simple epithelial keratins (K7, K8, K18, K19), and medium MW keratins of myoepithelial cells (K5, K6, K13, and K14), although K10 and K16 are limited to keratinizing squamous elements and K20 to occasional glandular cells. Some glandular cells can be KIT positive. CEA and Ber-EP4 positivity is present in the glandular and focally in the squamous epithelia, and CEA is also focally present in the spindled component.

The spindle cell component is variably positive for other myoepithelial antigens such as CD10, calponin, and SMA. Focal positivity for CD34 can be present in a subset of spindled and glandular components. Both epithelial and spindled elements can show limited calretinin positivity. All elements are negative for desmin, S100 protein, and TTF1.[181]

Electron microscopic studies have shown tonofilaments and desmosomes in the spindle cell component.[180,185] No genetic studies are available to date.

Differential diagnosis

Mixed tumor of skin adnexa and salivary glands, synovial sarcoma, teratoma, and glandular schwannoma are examples of other biphasic tumors that should not be confused with branchial anlage mixed tumor. Lack of evidence for chondroid or nerve sheath differentiation and the differentiating nature of the epithelial elements are distinguishing features. When sampled alone, the spindle cell component should not be confused with mesenchymal spindle cell neoplasms; its bland cytology differs from that of spindle cell carcinoma, and keratin positivity and myoepithelial markers, in addition to the specific location of the tumor, are unique to branchial anlage mixed tumors.

Chapter 27: Primary soft tissue tumors with epithelial differentiation

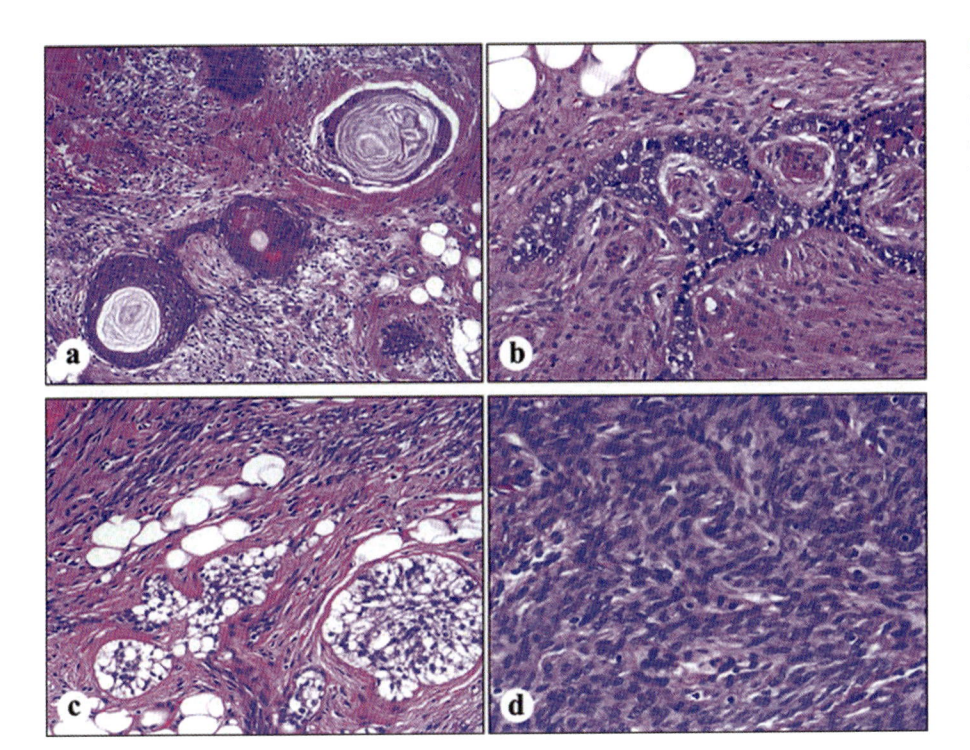

Figure 27.35 Features of branchial anlage mixed tumor. (**a**) Squamous cysts. (**b**) Complex epithelial clusters surrounded by spindle cell element. (**c**) Epithelial nests with clear cells, spindle cell component, and focal fat. (**d**) The spindle cell component shows uniform cytology.

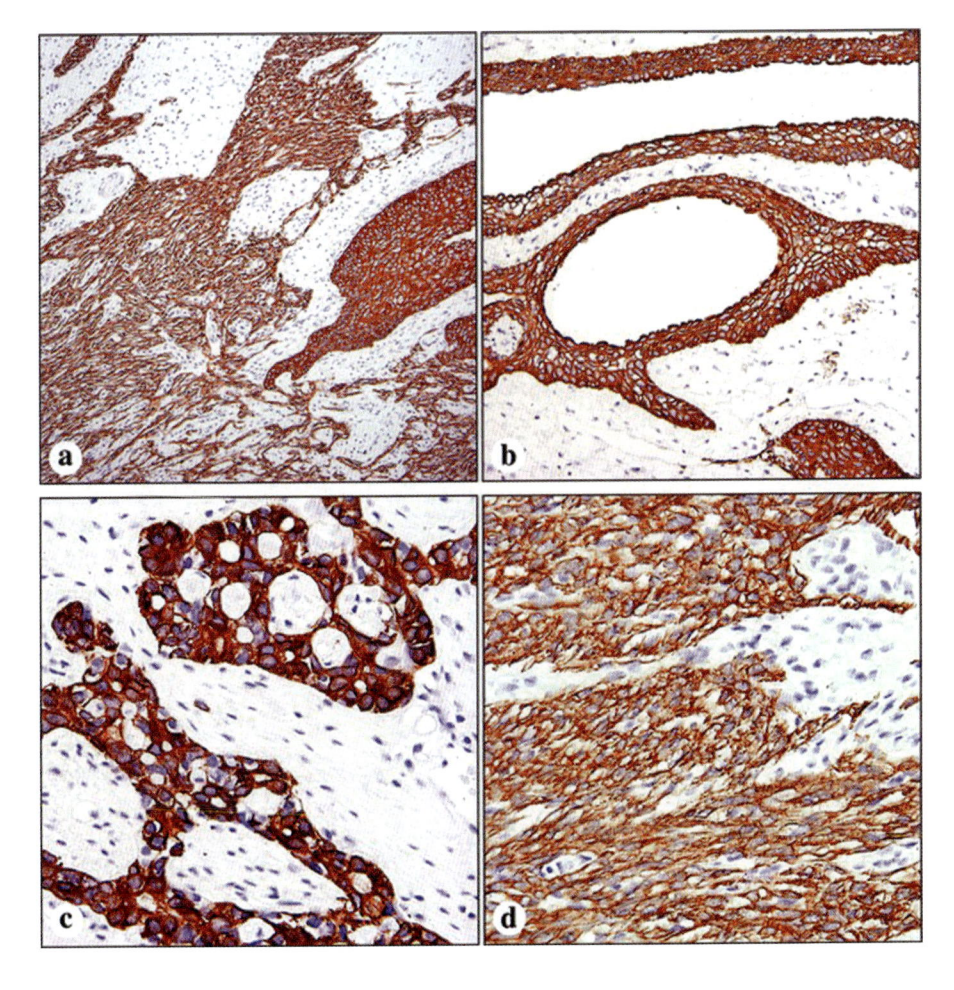

Figure 27.36 Keratin positivity is present in both the glandular and spindle cell elements of a branchial anlage mixed tumor.

769

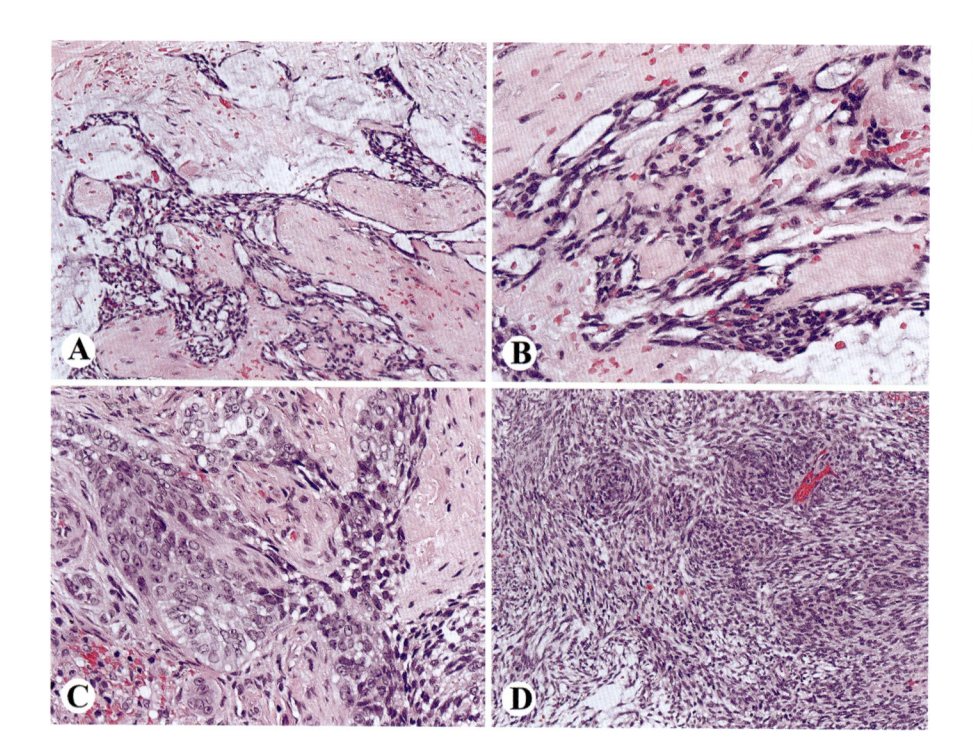

Figure 27.37 (A–C) Adamantinoma contains epithelial cells arranged in corded, pseudoglandular or squamoid epithelial formations. (D) Rare examples contain solid sheets of spindled cells.

Adamantinoma

Adamantinoma is a rare epithelial neoplasm usually involving the tibia bone. However, it can form a pretibial soft tissue mass, possibly independent of a tibial lesion and is therefore included in this chapter. Especially in older literature, the term adamantinoma was also used for clinicopathologically distinct, but histogenetically related tumors. The name adamantinoma has been adopted for this bone/soft tissue tumor from the old name of ameloblastoma of the jaw, and this name has also been used for craniopharyngeoma of the sella.

Clinical features

Adamantinoma usually occurs in young adults and most commonly arises in the tibial diaphysis or metaphysis. It can also involve pretibial soft tissue simulating a soft tissue tumor and possibly even arise primarily in the pretibial soft tissue. Rare osseous sites include fibula and ulna.[186–188] The tumor is low-grade malignant, but has the capability to metastasize to lymph nodes and distant sites, especially the lung. Largest clinicopathologic series have reported pulmonary metastases in 10–15% of patients.[186] Limb-sparing interstitial resection of the involved bone with surrounding soft tissue is the usual treatment. Bone involvement has to be ruled out for all apparent pretibial soft tissue lesions using advanced radiology and biopsy if indicated.

Pathology

Histologically adamantinoma is usually composed of clusters, streaks, and pseudoglandular or pseudoangiomatoid formations of basaloid epithelial cells. Some variants contain focal or occasionally dominant spindle cell components (Figure 27.37). However, immunohistochemically, all lesional epithelioid cells are positive of keratin cocktail AE1/AE3, keratins 5/6, EMA, and show nuclear labeling for p63.[189–191] Rare dedifferentiation with loss of epithelial markers has been reported.[192] The variants with a dominant spindle cell pattern have to be separated from spindle cell sarcomas such as synovial sarcoma. In adamantinoma, all neoplastic cells are positive for keratins and p63, in contrast with monophasic synovial sarcoma where only a minor subset of tumor cells are keratin positive and generally p63 negative.

References

Synovial sarcoma: general reviews and clinicopathologic series

1. Fisher C. Synovial sarcoma. *Ann Diagn Pathol* 1998;2:401–421.
2. Miettinen M, Virtanen I. Synovial sarcoma: a misnomer. *Am J Pathol* 1984;117:18–25.
3. Schmidt D, Thum P, Harms D, Treuner J. Synovial sarcoma in children and adolescents: a report from the Kiel Pediatric Tumor Registry. *Cancer* 1991;67:1667–1672.
4. Pappo AS, Fontanesi J, Luo X, et al. Synovial sarcoma in children and adolescents: The St. Jude Children's Research Hospital experience. *J Clin Oncol* 1994;12:2360–2366.
5. Fetsch JF, Meis JM. Synovial sarcoma of the abdominal wall. *Cancer* 1993;72:469–477.
6. Roth JA, Enzinger FM, Tannenbaum M. Synovial sarcoma of the neck: a follow-up study of 24 cases. *Cancer* 1975;35:1243–1253.

7. Dei Tos AP, Dal Cin P, Sciot R, et al. Synovial sarcoma of the larynx and hypopharynx. *Ann Otol Rhinol Laryngol* 1998;107:1080–1085.

8. Shmookler BM, Enzinger FM, Brannon RB. Orofacial synovial sarcoma: a clinicopathologic study of 11 new cases and review of the literature. *Cancer* 1982;50:269–276.

9. Al-Daraji W, Lasota J, Foss R, Miettinen M. Synovial sarcoma involving the head: analysis of 36 cases with predilection to the parotid and temporal regions. *Am J Surg Pathol* 2009;33:1494–1503.

10. Shmookler BM. Retroperitoneal synovial sarcoma: a report of four cases. *Am J Clin Pathol* 1982;77:686–691.

11. Witkin G, Miettinen M, Rosai J. A biphasic tumor of the mediastinum with features of synovial sarcoma: a report of four cases. *Am J Surg Pathol* 1989;13:490–499.

12. Begueret H, Galateau-Salle F, Guillou L, et al. Primary intrathoracic synovial sarcoma: a clinicopathologic study of 46 t(X;18)-positive cases from the French Sarcoma Group and the Mesopath Group. *Am J Surg Pathol* 2005;29:339–346.

13. Hartel PH, Fanburg-Smith JC, Frazier AA, et al. Primary pulmonary and mediastinal synovial sarcoma: a clinicopathologic study of 60 cases and comparison with five prior series. *Mod Pathol* 2007;20:760–769.

14. Singer S, Baldini EH, Demetri GC, Fletcher JA, Corson JM. Synovial sarcoma: prognostic significance of tumor size, margin of resection, and mitotic activity for survival. *J Clin Oncol* 1996;14:1201–1208.

15. Bergh P, Meis-Kindblom JM, Gherlinzoni F, et al. Synovial sarcoma: identification of low- and high-risk groups. *Cancer* 1999;85:2596–2607.

16. Machen SK, Fisher C, Gautam RS, Tubbs RR, Goldblum JR. Synovial sarcoma of the extremities: a clinicopathologic study of 34 cases, including semi-quantitative analysis of spindled, epithelial, and poorly differentiated areas. *Am J Surg Pathol* 1999;23:268–275.

17. Lewis JJ, Antonescu CR, Leung DH, et al. Synovial sarcoma: a multivariate analysis of prognostic factors in 112 patients with primary localized tumors of the extremity. *J Clin Oncol* 2000;18:2087–2094.

18. Spillane AJ, A'Hern R, Judson IR, Fisher C, Thomas JM. Synovial sarcoma: a clinicopathologic, staging, and prognostic assessment. *J Clin Oncol* 2000;18:3794–3800.

19. Trassard M, LeDoussal V, Hacene K, et al. Prognostic factors in localized primary synovial sarcoma: a multicenter study of 128 adult patients. *J Clin Oncol* 2001;19:525–534.

20. Krieg AH, Hefti F, Speth BM, et al. Synovial sarcomas usually metastasize after >5 years: a multicenter retrospective analysis with minimum follow-up of 10 years for survivors. *Ann Oncol* 2011;22:458–467.

21. Zeren H, Moran CA, Suster S, Fishback NF, Koss MN. Primary pulmonary sarcomas with features of monophasic synovial sarcoma: a clinicopathological, immunohistochemical and ultrastructural study of 25 cases. *Hum Pathol* 1995;26:474–480.

22. Kaplan MA, Goodman MD, Satish J, Bhagavan BS, Travis WD. Primary pulmonary sarcoma with morphologic features of monophasic synovial sarcoma and chromosome translocation t(X;18). *Am J Clin Pathol* 1996;105:195–199.

23. Hisaoka M, Hashimoto H, Iwamasa T, Ishikawa K, Aoki T. Primary synovial sarcoma of the lung: report of two cases confirmed by molecular detection of SYT-SSX fusion gene transcript. *Histopathology* 1999;34:205–210.

24. Miettinen M, Santavirta S, Slatis P. Intravascular synovial sarcoma. *Hum Pathol* 1987;18:1075–1077.

25. Robertson NJ, Halawa MH, Smith ME. Intravascular synovial sarcoma. *J Clin Pathol* 1998;51:172–173.

26. Karn CM, Socinski MA, Fletcher JA, Corson JM, Craighead JE. Cardiac synovial sarcoma with translocation (X;18) associated with asbestosis exposure. *Cancer* 1994;73:74–78.

27. Zhang PJ, Brooks JS, Goldblum JR, et al. Primary cardiac sarcomas: a clinicopathologic analysis of a series with follow-up information in 17 patients and emphasis on long-term survival. *Hum Pathol* 2008;39:1385–1395.

28. Billings SD, Meisner FL, Cummings OW, Tejada E. Synovial sarcoma of the upper digestive tract: a report of two cases with demonstration of the X;18 translocation by fluorescence in situ hybridization. *Mod Pathol* 2000;13:68–76.

29. Makhlouf HR, Ahrens W, Agarwal B, et al. Synovial sarcoma of the stomach: a clinicopathologic, immunohistochemical, and molecular genetic study of 10 cases. *Am J Surg Pathol* 2008;32:275–281.

30. Schreiber-Facklam H, Bode-Lesniewska B, Frigerio S, Flury R. Primary monophasic synovial sarcoma of the duodenum with SYT/SSX2 types of translocation. *Hum Pathol* 2007;38:946–949.

31. Argani P, Faria PA, Epstein JI, et al. Primary renal synovial sarcoma: molecular and morphologic delineation of an entity previously included among embryonal sarcomas of the kidney. *Am J Surg Pathol* 2000;24:1087–1096.

32. Iwasaki H, Ishiguro M, Ohjimi Y, et al. Synovial sarcoma of the prostate with t(X;18)(p11.2;q11.2). *Am J Surg Pathol* 1999;23:220–226.

33. Fritsch M, Epstein JI, Perlman EJ, Watts JC, Argani P. Molecularly confirmed primary prostatic synovial sarcoma. *Hum Pathol* 2000;31:246–250.

Synovial sarcoma: histologic variants and prognostic factors

34. Mirra JM, Wang S, Bhuta S. Synovial sarcoma with squamous differentiation of its mesenchymal glandular elements: a case report with light-microscopic, ultramicroscopic, and immunologic correlation. *Am J Surg Pathol* 1984;8:791–796.

35. Mackenzie DH. Monophasic synovial sarcoma: a histologic entity? *Histopathology* 1977;1:151–157.

36. Evans HL. Synovial sarcoma: a study of 23 biphasic and 17 probable monophasic examples. *Pathol Annu* 1980;15:309–313.

37. Krall RA, Kostianovsky M, Patchefsky AS. Synovial sarcoma: a clinical, pathological and ultrastructural study of 26 cases supporting the recognition of a monophasic variant. *Am J Surg Pathol* 1981;5:137–151.

38. Varela-Duran J, Enzinger FM. Calcifying synovial sarcoma. *Cancer* 1982;50:345–352.

39. Milchgrub S, Ghandur-Mnaymneh L, Dorfman HD, Albores-Saavedra J.

Synovial sarcoma with extensive osteoid and bone formation. *Am J Surg Pathol* 1993;17:357–363.

40. van de Rijn M, Barr FG, Xiong QB, et al. Poorly differentiated synovial sarcoma: an analysis of clinical, pathologic, and molecular genetic features. *Am J Surg Pathol* 1999;23:106–112.

41. Vergara-Lluri ME, Stohr BA, Puligandla B, Brenholz P, Horvai AE. A novel sarcoma with dual differentiation: clinicopathologic and molecular characterization of a combined synovial sarcoma and extraskeletal myxoid chondrosarcoma. *Am J Surg Pathol* 2012;36:1093–1098.

42. Oda Y, Hashimoto H, Tsuneyoshi M, Takeshita S. Survival in synovial sarcoma: a multivariate study of prognostic factors with special emphasis on the comparison between early death and long-term survival. *Am J Surg Pathol* 1993;17:35–44.

43. Kawai A, Woodruff J, Healey JH, et al. SYT-SSX gene fusion as a determinant of morphology and prognosis in synovial sarcoma. *N Engl J Med* 1998;338:153–160.

44. Ladanyi M, Antonescu CR, Leung DH, et al. Impact of SYT-SSX fusion type on the clinical behavior of synovial sarcoma: a multi-institutional retrospective study of 243 patients. *Cancer Res* 2002;62:135–140.

Synovial sarcoma: immunohistochemistry

45. Ordonez NG, Mahfouz SM, Mackay B. Synovial sarcoma: an immunohistochemical and ultrastructural study. *Hum Pathol* 1990;21:733–749.

46. Miettinen M. Keratin subsets in spindle cell sarcomas: keratins are widespread but synovial sarcoma contains a distinctive keratin polypeptide pattern and desmoplakins. *Am J Pathol* 1991;138:505–513.

47. Lopes JM, Bjerkehagen B, Holm R, et al. Immunohistochemical profile of synovial sarcoma with emphasis on the epithelial-type differentiation: a study of 49 primary tumors, recurrences and metastases. *Pathol Res Pract* 1994;190:168–177.

48. Folpe AL, Gown AM. Poorly differentiated synovial sarcoma: immunohistochemical distinction from primitive neuroectodermal tumors and high-grade malignant peripheral nerve sheath tumors. *Am J Surg Pathol* 1998;22:673–682.

49. Miettinen M, Limon J, Niezabitowski A, Lasota J. Patterns in keratin polypeptides in 110 biphasic, monophasic and poorly differentiated synovial sarcomas. *Virchows Arch* 2000;438:275–283.

50. Sato H, Hasegawa T, Abe Y, Sakai H, Hirohashi S. Expression of E-cadherin in bone and soft tissue sarcomas: a possible role in epithelial differentiation. *Hum Pathol* 1999;30:1344–1349.

51. Laskin WB, Miettinen M. Epithelial-type and neural-type cadherin expression in malignant noncarcinomatous neoplasms with epithelioid features that involve the soft tissues. *Arch Pathol Lab Med* 2002;126:425–431.

52. Miettinen M, Limon J, Niezabitowski A, Lasota J. Calretinin expression in synovial sarcoma: analysis of similarities and differences from malignant mesothelioma. *Am J Surg Pathol* 2001;25:610–617.

53. Terry J, Saito T, Subramanian S, et al. TLE1 as a diagnostic immunohistochemical marker for synovial sarcoma emerging from gene expression profiling studies. *Am J Surg Pathol* 2007;31:240–246.

54. Guillou L, Wadden C, Kraus MD, Dei Tos AP, Fletcher CDM. S-100 protein reactivity in synovial sarcomas: a potentially frequent diagnostic pitfall: immunohistochemical analysis of 100 cases. *Appl Immunohistochem* 1996;4:167–175.

55. Stevenson AJ, Chatten J, Bertoni F, Miettinen M. CD99 (p30/32 -MIC2) neuroectodermal/Ewing sarcoma antigen as an immunohistochemical marker: review of more than 600 tumors and the literature experience. *Appl Immunohistochem* 1994;2:231–240.

56. Suster S, Fisher C, Moran CA. Expression of bcl2 oncoprotein in benign and malignant spindle cell tumors of soft tissue, skin, serosal surfaces, and gastrointestinal tract. *Am J Surg Pathol* 1998;22:863–872.

57. Lai JP, Robbins PF, Raffeld M, et al. NY-ESO-1 expression in synovial sarcoma and other mesenchymal tumors: significance for NY-ESO-1-based targeted therapy and differential diagnosis. *Mod Pathol* 2012;25:854–858.

58. Mularz K, Harazin-Lechowska A, Ambicka A, et al. Specificity and sensitivity of INI-1 labeling in epithelioid sarcoma. Loss of INI1 expression as a frequent immunohistochemical event in synovial sarcoma. *Pol J Pathol* 2012;63:179–183.

59. Arnold MA, Arnold CA, Li G, et al. A unique pattern of INI1 immunohistochemistry distinguishes synovial sarcoma from its histologic mimics. *Hum Pathol* 2013;44:881–887.

Synovial sarcoma: genetics

60. Limon J, Mrozek K, Mandahl N, et al. Cytogenetics of synovial sarcoma: presentation of ten new cases and review of the literature. *Genes Chromosomes Cancer* 1991;3:338–345.

61. Dal Cin P, Rao U, Jani-Sait S, Karakousis C, Sandberg AA. Chromosomes in the diagnosis of soft tissue tumors: I. Synovial sarcoma. *Mod Pathol* 1992;5:57–62.

62. Clark J, Rocques PJ, Crew AJ, et al. Identification of novel genes, SYT and SSX, involved in the t(X;18)(p11.2; q11.2) translocation found in human synovial sarcoma. *Nat Genet* 1994;7:502–508.

63. de Leeuw B, Balemans M, Olde Weghuis D, Geurts van Kessel A. Identification of two alternative fusion genes, SYT-SSX1 and SYT-SSX2, in t(X;18)(p11.2;q11.2)-positive synovial sarcomas. *Hum Mol Genet* 1995;4:1097–1099.

64. Crew J, Clark J, Fisher C, et al. Fusion of SYT to two genes, SSX1 and SSX2, encoding proteins with homology to the Kruppel-associated box in human synovial sarcoma. *EMBO J* 1995;14: 2333–2340.

65. Renwick PJ, Reeves BR, Dal Cin P, et al. Two categories of synovial sarcoma defined by divergent chromosome translocation breakpoints in Xp11.2, with implications for the histologic sub-classification of synovial sarcoma. *Cytogenet Cell Genet* 1995;70:58–63.

66. dos Santos NR, de Bruijn DRH, van Kessel AG. Molecular mechanisms underlying human synovial sarcoma development. *Genes Chromosomes Cancer* 2001;30:1–14.

67. Skytting B, Nilsson G, Brodin B, et al. A novel fusion gene, SYT-SSX4, in synovial sarcoma. *J Natl Cancer Inst* 1999;91:974–975.
68. Gure AO, Türeci Ö, Sahin U, et al. SSX: a multigene family with several members transcribed in normal testis and human cancer. *Int J Cancer* 1997;72:965–971.
69. de Leeuw B, Suijkerbujik, RF, Olde Weghuis D, et al. Distinct Xp11.2 breakpoint regions in synovial sarcoma revealed by metaphase and interphase FISH: relationship to histologic subtypes. *Cancer Genet Cytogenet* 1994;73:89–94.
70. Poteat HT, Corson JM, Fletcher JA. Detection of chromosome 18 rearrangement in synovial sarcoma by fluorescence in situ hybridization. *Cancer Genet Cytogenet* 1995;84:76–81.
71. Nagao K, Ito H, Yoshida H. Chromosomal translocation t(X;18) in human synovial sarcomas analyzed by fluorescence in situ hybridization using paraffin-embedded tissue. *Am J Pathol* 1996;148:601–609.
72. Shipley J, Crew J, Birdsall S, et al. Interphase fluorescence in situ hybridization and reverse transcription polymerase chain reaction as a diagnostic aid for synovial sarcoma. *Am J Pathol* 1996;148:559–567.
73. Zilmer M. Use of nonbreakpoint DNA probes to detect the t(X;18) in interphase cells from synovial sarcoma: implications for detection of diagnostic tumor translocations. *Am J Pathol* 1998;152:1171–1177.
74. Birdsall S, Osin P, Ly YJ, Fisher C, Shipley J. Synovial sarcoma specific translocation associated with both epithelial and spindle cell components. *Int J Cancer* 1999;82:605–608.
75. Lu YJ, Birdsall S, Summersgill B, et al. Dual colour fluorescence in situ hybridization to paraffin-embedded samples to deduce the presence of the der(x)t(x;18)(p11.2;q11.2) and involvement of either the SSX1 or SSX2 gene: a diagnostic and prognostic aid for synovial sarcoma. *J Pathol* 1999;187:490–496.
76. Fligman I, Leonardo F, Jhanwar SC, et al. Molecular diagnosis of synovial sarcoma and characterization of a variant SYT-SSX2 fusion transcript. *Am J Pathol* 1995;147:1592–1599.
77. Argani P, Zakowski MF, Klimstra DS, Rosai J, Ladanyi M. Detection of the SYT-SSX chimeric RNA of synovial sarcoma in paraffin-embedded tissue and its application in problematic cases. *Mod Pathol* 1997;11:65–71.
78. Lasota J, Jasinski M, Debiec-Rychter M, Limon J, Miettinen M. Detection of the SYT-SSX fusion transcripts in formaldehyde-fixed, paraffin embedded tissue: a reverse transcription polymerase chain reaction amplification assay useful in the diagnosis of synovial sarcoma. *Mod Pathol* 1998;11:626–633.
79. Tsuji S, Hashimoto H. Detection of SYT-SSX fusion transcripts in synovial sarcoma by reverse transcription-polymerase chain reaction using archival paraffin-embedded tissues. *Am J Pathol* 1998;153:1807–1812.
80. Nilsson G, Skytting B, Xie Y, et al. The SYT-SSX1 variant of synovial sarcoma is associated with a high rate of tumor cell proliferation and poor clinical outcome. *Cancer Res* 1999;59:3180–3184.
81. van de Rijn M, Barr FG, Collins MH, Xiong QB, Fisher C. Absence of SYT-SSX fusion products in soft tissue tumors other than synovial sarcoma. *Am J Clin Pathol* 1999;112:43–49.
82. Ladanyi M, Woodruff JM, Scheithauer BW, et al. Letter to editor. Re: O'Sullivan MJ, Kyriakos M, Zhu X, Wick MR, Swanson PE, Dehner LP, Humphrey PA, Pfeifer JD: Malignant peripheral nerve sheath tumors with t(X;18): a pathologic and molecular genetic study. *Mod Pathol* 2000;13:1336–46. *Mod Pathol* 2001;14:733–737.
83. O'Sullivan MJ, Kyriakos M, Zhu X, et al. Malignant peripheral nerve sheath tumors with t(X;18): a pathologic and molecular genetic study. *Mod Pathol* 2000;13:1336–1346.
84. Mandahl N, Limon J, Mertens F, et al. Nonrandom secondary chromosome aberrations in synovial sarcomas with t(X;18). *Int J Oncol* 1995;7:495–499.
85. Szymanska J, Serra M, Skytting B, et al. Genetic imbalances in 67 synovial sarcomas evaluated by comparative genomic hybridization. *Genes Chromosomes Cancer* 1998;23:213–219.

Epithelioid sarcoma

86. Enzinger FM. Epithelioid sarcoma: a sarcoma simulating a granuloma or a carcinoma. *Cancer* 1970;26:1029–1041.
87. Guillou L, Wadden C, Coindre J-M, Krausz T, Fletcher CDM. "Proximal-type" epithelioid sarcoma, a distinctive aggressive neoplasm showing rhabdoid features: clinicopathological, immunohistochemical and ultrastructural study of a series. *Am J Surg Pathol* 1997;21:130–146.
88. Prat J, Woodruff JM, Marcove RC. Epithelioid sarcoma: an analysis of 22 cases indicating the prognostic significance of vascular invasion and regional lymph node metastasis. *Cancer* 1978;41:1472–1487.
89. Chase DR, Enzinger FM. Epithelioid sarcoma: diagnosis, prognostic indicators, and treatment. *Am J Surg Pathol* 1985;9:241–263.
90. Steinberg BD, Gelberman RH, Mankin HJ, Rosenberg AE. Epithelioid sarcoma in the upper extremity. *J Bone Joint Surg Am* 1992;74:28–35.
91. Evans HL, Baer SC. Epithelioid sarcoma: a clinicopathologic and prognostic study of 26 cases. *Semin Diagn Pathol* 1993;10:286–291.
92. Kodet R, Smelhaus V, Newton WA, et al. Epithelioid sarcoma in childhood: an immunohistochemical, electron microscopic, and clinicopathologic study of 11 cases under 15 years of age and review of the literature. *Pediatr Pathol* 1994;14:433–451.
93. Halling AC, Wollen PC, Pritchard DJ, Vlasak R, Nascimento AG. Epithelioid sarcoma: a clinicopathologic review of 55 cases. *Mayo Clin Proc* 1996;71:636–642.
94. Ross HM, Lewis JJ, Woodruff JM, Brennan MF. Epithelioid sarcoma: clinical behavior and prognostic factors of survival. *Ann Surg Oncol* 1997;4:491–495.
95. Spillane AJ, Thomas JM, Fisher C. Epithelioid sarcoma: the clinicopathological complexities of this rare soft tissue sarcoma. *Ann Surg Oncol* 2000;7:218–225.
96. Callister MD, Bello MT, Pisters PWT, et al. Epithelioid sarcoma: results of conservative surgery and radiotherapy. *Int J Radiat Oncol Biol Phys* 2001;51:384–391.
97. Casanova M, Ferrari A, Collini P, et al. Epithelioid sarcoma in children and adolescents: a report from the Italian Soft Tissue Sarcoma Committee. *Cancer* 2006;106:708–717.

98. von Hochstetter AR, Meyer VE, Grant JW, Honegger HP, Schreiber A. Epithelioid sarcoma mimicking angiosarcoma: the value of immunohistochemistry in the differential diagnosis. *Virchows Arch A Pathol Anat Histopathol* 1991;418:271–278.

99. Mirra JM, Kessler S, Bhuta S, Eckardt J. The fibroma-like variant of epithelioid sarcoma: a fibrohistiocytic/myoid cell lesion often confused with benign and malignant spindle cell tumors. *Cancer* 1992;69:1382–1395.

100. Chase DR, Weiss SW, Enzinger FM, Langloss JM. Keratin in epithelioid sarcoma: an immunohistochemical study. *Am J Surg Pathol* 1984;8:435–441.

101. Mukai M, Torikata C, Iri H, et al. Cellular differentiation of epithelioid sarcoma: an electron microscopic, enzyme histochemical, and immunohistochemical study. *Am J Pathol* 1985;119:44–56.

102. Daimaru Y, Hashimoto H, Tsuneoshi M, Enjoji M. Epithelial profile of epithelioid sarcoma: an immunohistochemical analysis of eight cases. *Cancer* 1987;59:131–141.

103. Manivel JC, Wick MR, Dehner LP, Sibley RK. Epithelioid sarcoma: an immunohistochemical study. *Am J Clin Pathol* 1987;87:319–326.

104. Meis JM, Mackay B, Ordonez NG. Epithelioid sarcoma: an immunohistochemical and ultrastructural study. *Surg Pathol* 1988;1:13–31.

105. Miettinen M, Fanburg-Smith JC, Virolainen M, Shmookler BM, Fetsch JF. Epithelioid sarcomas: an immunohistochemical analysis of 112 classical and variant cases and a discussion of the differential diagnosis. *Hum Pathol* 1999;30:934–942.

106. Laskin WB, Miettinen M. Epithelioid sarcoma: new insights based on extensive immunohistochemical analysis. *Arch Pathol Lab Med* 2003;127:1161–1168.

107. Gerharz CD, Moll R, Meister P, Knuth A, Gabbert H. Cytoskeletal heterogeneity of an epithelioid sarcoma with expression of vimentin, cytokeratins, and neurofilaments. *Am J Surg Pathol* 1990;14:274–283.

108. Hagström J, Mesimäki K, Apajalahti S, et al. A rare case of oral epithelioid sarcoma of the gingiva. *Oral Surg Oral Med Oral Pathol Oral Radiol Endod* 2011;111(4):e25–e28.

109. Kosemehmetoglu K, Kaygusuz G, Bahrami A, et al. Intra-articular epithelioid sarcoma showing mixed classic and proximal-type features: report of 2 cases, with immunohistochemical and molecular cytogenetic INI-1 study. *Am J Surg Pathol* 2011;14(6):891–897.

110. Gerharz CD, Moll R, Ramp U, Mellin W, Gabbert HE. Multidirectional differentiation in a newly established human epithelioid sarcoma cell line (GRU-1) with co-expression of vimentin, cytokeratin and neurofilament proteins. *Int J Cancer* 1990;45:143–152.

111. Sirgi KE, Wick MR, Swanson PE. B72.3 and CD34 immunoreactivity in malignant epithelioid soft tissue tumors: adjuncts in the recognition of endothelial neoplasms. *Am J Surg Pathol* 1993;17:179–185.

112. Smith MEF, Brown JI, Fisher C. Epithelioid sarcoma: presence of vascular-endothelial cadherin and lack of epithelial cadherin. *Histopathology* 1998;33:425–431.

113. Hornick JL, Dal Cin P, Fletcher CD. Loss of INI1 expression is characteristic of both conventional and proximal-type epithelioid sarcoma. *Am J Surg Pathol* 2009;33:542–550.

114. Hollmann TJ, Hornick JL. INI1-deficient tumors: diagnostic features and molecular genetics. *Am J Surg Pathol* 2011;35:e47–e63.

115. Agaimy A, Rau TT, Hartmann A, Stoehr R. SMARCB1 (INI1)-negative rhabdoid carcinomas of the gastrointestinal tract: clinicopathologic and molecular study of a highly aggressive variant with literature review. *Am J Surg Pathol* 2014;38:910–920.

116. Miettinen M, Wang Z, Sarlomo-Rikala M, et al. ERG expression in epithelioid sarcoma: a diagnostic pitfall. *Am J Surg Pathol* 2013;37:1580–1585.

117. Stockman DL, Hornick JL, Deavers MT, et al. ERG and FLI1 protein expression in epithelioid sarcoma. *Mod Pathol* 2014;27:496–501.

118. Wick MR, Manivel JC. Epithelioid sarcoma and isolated necrobiotic granuloma: a comparative immunocytochemical study. *J Cutan Pathol* 1986;13:253–260.

119. Lin L, Skacel M, Sigel JE, et al. Epithelioid sarcoma: an immunohistochemical analysis evaluating the utility of cytokeratin 5/6 in distinguishing superficial epithelioid sarcoma from spindle cell squamous cell carcinoma. *J Cutan Pathol* 2003;30:114–117.

120. Prayson RA, Chahlavi A. Metastatic epithelioid sarcoma to the brain: palisaded necrosis mimicking glioblastoma multiforme. *Ann Diagn Pathol* 2002;6:272–280.

Genetics of epithelioid sarcoma

121. Quezado MM, Middleton LP, Bryant B, et al. Allelic loss on chromosome 22q in epithelioid sarcomas. *Hum Pathol* 1998;29:604–608.

122. Sullivan LM, Folpe AL, Pawel BR, Judkins AR, Biegel JA. Epithelioid sarcoma is associated with a high percentage of SMARCB1 deletions. *Mod Pathol* 2013;26:385–392.

123. Le Loarer F, Zhang L, Fletcher CD, et al. Consistent SMARCB1 homozygous deletions in epithelioid sarcoma and in a subset of myoepithelial carcinomas can be reliably detected by FISH in archival material. *Genes Chromosomes Cancer* 2014;53:475–486.

Large cell (proximal-type) epithelioid sarcoma

124. Hasegawa T, Matsuno Y, Shimoda T, et al. Proximal-type epithelioid sarcoma: a clinicopathologic study of 20 cases. *Mod Pathol* 2001;14:655–663.

125. Modena P, Lualdi E, Facchninetti F, et al. SMARCB1/INI1 tumor suppressor gene is frequently inactivated in epithelioid sarcomas. *Cancer Res* 2005;65:4012–4019.

126. Rakheja D, Wilson KS, Meehan J, Schultz RA, Gomez AM. "Proximal-type" and classic epithelioid sarcomas represent a clinicopathologic continuum: a case report. *Pediatr Dev Pathol* 2005;8:105–114.

Mixed tumor/myoepithelioma and parachordoma

127. Hirsch P, Helwig EB. Chondroid syringoma: mixed tumor of skin, salivary gland type. *Arch Dermatol* 1961;84:835–847.

128. Stout AP, Gorman G. Mixed tumors of the skin of the salivary gland type. *Cancer* 1959;12:537–543.

129. Kunikane H, Ishikura H, Yamaguchi J, et al. Chondroid syringoma (mixed

130. Hassab-El-Naby HM, Tam S, White WL, Ackerman AB. Mixed tumors of the skin: a histological and immunohistochemical study. *Am J Dermatopathol* 1989;11:413–428.

131. Kilpatrick SE, Hitchcock MG, Kraus MD, Calonje E, Fletcher CD. Mixed tumors and myoepitheliomas of soft tissue: a clinicopathologic study of 19 cases with a unifying concept. *Am J Surg Pathol* 1997;21:13–22.

132. Michal M, Miettinen M. Myoepitheliomas of skin and soft tissues. *Virchows Arch* 1999;434:393–400.

133. Hornick JL, Fletcher CD. Myoepithelial tumors of soft tissue: a clinicopathologic and immunohistochemical study of 101 cases with evaluation of prognostic parameters. *Am J Surg Pathol* 2003;27:1183–1196.

134. Mentzel T, Requena L, Kaddu S, et al. Cutaneous myoepithelial neoplasms: clinicopathologic and immunohistochemical study of 20 cases suggesting a continuous spectrum ranging from benign mixed tumor of the skin to cutaneous myoepithelioma and myoepithelial carcinoma. *J Cutan Pathol* 2003;30:294–302.

135. Hornick JL, Fletcher CD. Cutaneous myoepithelioma: a clinicopathologic and immunohistochemical study of 14 cases. *Hum Pathol* 2004;35:14–24.

136. Ishimura E, Iwamoto H, Kobashi Y, Yamabe H, Ichijima K. Malignant chondroid syringoma: report of a case with widespread metastasis and review of pertinent literature. *Cancer* 1983;52:1966–1973.

137. Ferreiro JA, Nascimento AG. Hyaline-cell rich chondroid syringoma: a tumor mimicking malignancy. *Am J Surg Pathol* 1995;19:912–917.

138. Headington JT. Mixed tumor of the skin: eccrine and apocrine type. *Arch Dermatol* 1961;84:989–996.

139. Brandal P, Panagopoulos I, Bjerkehagen B, et al. Detection of a t(1;22)(q23;q12) translocation leading to an EWSR1-PBX1 fusion gene in myoepithelioma. *Genes Chromosomes Cancer* 2008;47:558–564.

140. Antonescu CR, Zhang L, Chang NE, et al. EWSR1-POU5F1 fusion in soft tissue myoepithelial tumors: a molecular analysis of sixty-six cases, including soft tissue, bone, and visceral lesions, showing common involvement of the EWSR1 gene. *Genes Chromosomes Cancer* 2010;49:1114–1124.

141. Flucke U, Mentzel T, Verdijk MA, et al. EWSR1-ATF1 chimeric transcript in a myoepithelial tumor of soft tissue: a case report. *Hum Pathol* 2012;43:764–768.

142. Puls F, Arbajian E, Magnusson L, et al. Myoepithelioma of bone with a novel FUS-POU5F1 fusion gene. *Histopathology* 2014;65:917–922.

143. Antonescu CR, Zhang L, Shao SY, et al. Frequent PLAG1 gene rearrangements in skin and soft tissue myoepithelioma with ductal differentiation. *Genes Chromosomes Cancer* 2013;52:675–682.

144. Dabska M. Parachordoma: a new clinical entity. *Cancer* 1977;40:1586–1592.

145. Fisher C, Miettinen M. Parachordoma: clinicopathologic and immunohistochemical study of four cases of an unusual soft tissue neoplasm. *Ann Diagn Pathol* 1997;1:3–10.

146. Folpe AL, Agoff SN, Willis J, Weiss SW. Parachordoma is immunohistochemically and cytogenetically distinct from axial chordoma and extraskeletal myxoid chondrosarcoma. *Am J Surg Pathol* 1999;23:1059–1067.

147. Tirabosco R, Mangham DC, Rosenberg AE, et al. Brachyury expression in extra-axial skeletal and soft tissue chordomas: a marker that distinguishes chordoma from mixed tumor/myoepithelioma/parachordoma in soft tissue. *Am J Surg Pathol* 2008;32:572–580.

148. Lauer SR, Edgar MA, Gardner JM, Sebastian A, Weiss SW. Soft tissue chordomas: a clinicopathologic analysis of 11 cases. *Am J Surg Pathol* 2013;37:719–726.

149. Miettinen M, Wang ZF, Lasota J, et al. Nuclear brachyury expression is consistent in chordoma, common in germ cell tumors and small cell carcinomas and rare in other carcinomas and sarcomas. An immunohistochemical study of 5229 cases. *Am J Surg Pathol* 2015;39:1305–1312.

150. Gleason BC, Fletcher CD. Myoepithelial carcinoma of soft tissue in children: an aggressive neoplasm analyzed in a series of 29 cases. *Am J Surg Pathol* 2007;31:1813–1824.

Other adnexal tumors

151. Johnson BL, Helwig EB. Eccrine acrospiroma: a clinicopathologic study. *Cancer* 1969;23:641–657.

152. Winkelmann RK, Wolff K. Solid-cystic hidradenoma. *Arch Dermatol* 1968;97:651–661.

153. Wiley EL, Milchgrub S, Freeman RG, Kim ES. Sweat gland adenomas: immunohistochemical study with emphasis on myoepithelial differentiation. *J Cutan Pathol* 1993;20:337–343.

154. Demirkesen C, Hoede N, Moll R. Epithelial markers and differentiation in adnexal neoplasms of the skin: an immunohistochemical study including individual cytokeratins. *J Cutan Pathol* 1995;22:518–535.

155. Kersting DW, Helwig EB. Eccrine spiradenoma. *AMA Arch Dermatol* 1956;73:199–227.

156. Kao GF, Laskin WB, Weiss SW. Eccrine spiradenoma occurring in infancy mimicking mesenchymal tumor. *J Cutan Pathol* 1990;17:214–219.

157. Mambo NC. Eccrine spiradenoma: clinical and pathologic study of 49 tumors. *J Cutan Pathol* 1983;10:312–320.

158. Evans HL, Daniel WP, Smith JL, Winkelmann RK. Carcinoma arising in eccrine spiradenoma. *Cancer* 1979;43:1881–1884.

159. Wick MR, Swanson PE, Kaye VN, Pittelkow MR. Sweat gland carcinoma ex eccrine spiradenoma. *Am J Dermatopathol* 1987;9:90–98.

160. Argenyi ZB, Nguyen AV, Balogh K, Sears JK, Whitaker DC. Malignant eccrine spiradenoma: a clinicopathologic study. *Am J Dermatopathol* 1992;14:381–390.

161. Granter SR, Seeger K, Calonje E, Busam K, McKee PH. Malignant eccrine spiradenoma (spiradenocarcinoma): a clinicopathologic study of 12 cases. *Am J Dermatopathol* 2000;22:97–103.

162. McKee PH, Fletcher CDM, Stavrinos P, Pambakian H. Carcinosarcoma arising in eccrine spiradenoma: a clinicopathologic and immunohistochemical study of two

163. Van Den Oord JJ, De Wolf-Peeters C. Perivascular spaces in eccrine spiradenoma: a clue to its histological diagnosis. *Am J Dermatopathol* 1995;17:266–270.

164. Watanabe S, Hirose M, Sato S, Takahashi H. Immunohistochemical analysis of cytokeratin expression in eccrine spiradenoma: similarities to the transitional portions between secretory segments and coiled ducts of eccrine glands. *Br J Dermatol* 1994;131:799–807.

165. Al-Nafussi A, Blessing K, Rahilly M. Non-epithelial cellular components in eccrine spiradenoma: a histological and immunohistochemical study of 20 cases. *Histopathology* 1991;18:155–160.

166. Crain RC, Helwig EB. Dermal cylindroma (dermal eccrine cylindroma). *Am J Clin Pathol* 1961;35:504–515.

167. Takahashi M, Rapley E, Biggs PJ, et al. Linkage and LOH studies in 19 cylindromatosis families show no evidence of genetic heterogeneity and define the CYLD locus on chromosome 16q12-q13. *Hum Genet* 2000;106:58–65.

168. Lee DA, Grossman ME, Scheinerman P, Celebi JT. Genetics of skin appendage neoplasms and related syndromes. *J Med Genet* 2005;42:811–819.

169. Massoumi R, Paus R. Cylindromatosis and the CYLD gene: new lessons on the molecular principles of epithelial growth control. *Bioessays* 2007;29:1203–1214.

170. Kao GF, Helwig EB, Graham JH. Aggressive digital papillary adenoma and adenocarcinoma: a clinicopathological study of 57 patients, with histochemical, immunopathological, and ultrastructural observations. *J Cutan Pathol* 1987;14:129–146.

171. Duke WH, Sherrod TT, Lupton GP. Aggressive digital papillary adenocarcinoma (aggressive digital papillary adenoma and adenocarcinoma revisited). *Am J Surg Pathol* 2000;24:775–784.

172. Suchak R, Wang WL, Prieto VG, et al. Cutaneous digital papillary adenocarcinoma: a clinicopathologic study of 31 cases of a rare neoplasm with new observations. *Am J Surg Pathol* 2012;36:1883–1891.

173. Chambers PW, Schwinn CP. Chordoma: a clinicopathologic study of metastasis. *Am J Clin Pathol* 1979;72:765–776.

174. Su WPD, Louback JB, Gagne EJ, Scheithauer BW. Chordoma cutis: a report of nineteen patients with cutaneous involvement of chordoma. *J Am Acad Dermatol* 1993;29:63–66.

175. Suster S, Moran C. Chordomas of the mediastinum: clinicopathologic, immunohistochemical, and ultrastructural study of six cases presenting as posterior mediastinal masses. *Hum Pathol* 1995;26:1354–1362.

176. O'Hara BJ, Paetau A, Miettinen M. Keratin subsets and monoclonal antibody HBME-1 in chordoma: immunohistochemical differential diagnosis between tumors simulating chordoma. *Hum Pathol* 1998;29:119–126.

177. Vujovic S, Henderson S, Presneau N, et al. Brachyury, a crucial regulator of notochordal development, is a novel biomarker for chordomas. *J Pathol* 2006;209:157–165.

178. Pillay N, Plagnol V, Tarpey PS, et al. A common single-nucleotide variant in T is strongly associated with chordoma. *Nat Genet* 2012;44:1185–1187.

179. Yang XR, Ng D, Alcorta DA, et al. T (brachyury) gene duplication confers major susceptibility to familial chordoma. *Nat Genet* 2009;41:1176–1178.

Ectopic hamartomatous thymoma

180. Rosai J, Limas C, Husband EM. Ectopic hamartomatous thymoma: a distinctive benign lesion of lower neck. *Am J Surg Pathol* 1984;8:501–513.

181. Fetsch JF, Laskin WB, Michal M, et al. Ectopic hamartomatous thymoma: a clinicopathologic and immunohistochemical analysis of 21 cases with data supporting reclassification as a branchial anlage mixed tumor. *Am J Surg Pathol* 2004;28:1360–1370.

182. Fetsch JF, Weiss SW. Ectopic hamartomatous thymoma: clinicopathologic, immunohistochemical, and histogenetic considerations in four new cases. *Hum Pathol* 1990;21:662–668.

183. Chan JK, Rosai J. Tumors of the neck showing thymic or related branchial pouch differentiation: a unifying concept. *Hum Pathol* 1991;22:349–367.

184. Michal M, Zamecnik M, Gogora M, Mukensnabl P, Neubauer L. Pitfalls in the diagnosis of ectopic hamartomatous thymoma. *Histopathology* 1996;29:549–555.

185. Fukunaga M. Ectopic hamartomatous thymoma: a case report with immunohistochemical and ultrastructural studies. *APMIS* 2002;110:565–570.

Adamantinoma

186. Keeney GL, Unni KK, Beabout JW, Pritchard DJ. Adamantinoma of long bones: a clinicopathologic study of 85 cases. *Cancer* 1989;64:730–737.

187. Jundt G, Remberger K, Roessner A, Schulz A, Bohndorf K. Adamantinoma of long bones: a histopathological and immunohistochemical study of 23 cases. *Pathol Res Pract* 1995;191:112–120.

188. Mills SE, Rosai J. Adamantinoma of the pretibial soft tissue: clinicopathologic features, differential diagnosis, and possible relationship to intraosseous disease. *Am J Clin Pathol* 1985;83:108–114.

189. Rosai J, Pinkus GS. Immunohistochemical demonstration of epithelial differentiation in adamantinoma of the tibia. *Am J Surg Pathol* 1982;6(5):427–434.

190. Benassi MS, Campanacci L, Gamberi G, et al. Cytokeratin expression and distribution in adamantinoma of the long bones and osteofibrous dysplasia of tibia and fibula: an immunohistochemical study correlated to histogenesis. *Histopathology* 1994;25(1):71–76.

191. Dickson BC, Gortzak Y, Bell RS, et al. p63 expression in adamantinoma. *Virchows Arch* 2011;459(1):109–113.

192. Izquierdo FM, Ramos LR, Sánchez-Herráez S, et al. Dedifferentiated classic adamantinoma of the tibia: a report of a case with eventual complete revertant mesenchymal phenotype. *Am J Surg Pathol* 2010;34:1388–1392.

Chapter 28

Malignant mesothelioma and other mesothelial proliferations

Markku Miettinen
National Cancer Institute, National Institutes of Health

Benign mesothelial tumors and malignant mesothelioma are important soft tissue tumors that can present at various sites in the body cavities, and the latter also presents as metastatic tumors in the peripheral soft tissues.

The tumor formerly called a *fibrous mesothelioma* or *localized fibrous tumor* is a mesenchymal, fibroblastic, nonmesothelial tumor and is now known as a *solitary fibrous tumor*. This tumor is unrelated to mesothelioma and is included in Chapter 12.

Adenomatoid tumor

An *adenomatoid tumor* is a benign mesothelial proliferation that is best known for its occurrence in the internal female and male genitalia: the uterus, fallopian tubes, and epididymis, but it can occur almost anywhere around the serosal surfaces, both in the abdominal and thoracic cavities. It is not uniformly agreed whether an adenomatoid tumor is a benign neoplasm or a form of localized mesothelial hyperplasia.

Clinical features

Adenomatoid tumors are usually detected in young to middle-aged adults, often incidentally during surgery or imaging studies for unrelated conditions. The most common sites in women are the fallopian tube and uterus, where a small 1-cm to 2-cm adenomatoid tumor is not an uncommon finding in a hysterectomy specimen.[1,2] In a unique case, an adenomatoid tumor involved the uterus diffusely in a renal transplant patient.[3] Involvement of the ovary has been rarely reported.[4]

In men, the epididymis is the most common site, but the spermatic cord or tunica albuginea can also be involved. In the latter case, recognition of a sharply demarcated lesion based on the tunica albuginea helps to distinguish an adenomatoid from a primary testicular tumor, which allows a local excision instead of an orchiectomy. Male adenomatoid tumors are usually detected as palpable 1-cm to 2-cm scrotal masses.[5–8]

Rarely, adenomatoid tumors have been associated with or adjacent to the peritoneal mesothelia in locations such as the adrenal gland, mesentery, and pancreas. All reported adrenal examples have been diagnosed in male patients and measured from 1 cm to 3.5 cm.[9–11] Occurrence in the thoracic cavity (i.e., the pericardium, mediastinum, and mediastinal lymph nodes) is rare. Most of these adenomatoid tumors have been incidental findings.[12–14] Local excision is curative.

Pathology

Uterine adenomatoid tumors typically occur near the serosa, often at the site of mesothelial invagination at the junction of the fallopian tube and fundus. They can be recognizable on gross inspection as distinct masses that are yellowish on sectioning, but some blend into the myometrial tissue.[1,2] Some examples are cystic, and a component resembling multicystic peritoneal mesothelioma is sometimes present.[4] Adenomatoid tumors of the epididymis typically form well-demarcated, unencapsulated masses that are gray-white to yellowish on sectioning, and cysts can be visible. Rarely, an adenomatoid tumor occurs in the tunica albuginea or inside it.[5,6]

Histologically, the lesional mesothelial cells form cords, and gland-like, and cystic structures in fibrous or smooth muscle stroma (Figure 28.1). The mesothelial appearance varies from flattened to cuboid. The cytoplasm is abundant and eosinophilic, and cytoplasmic vacuolization and lumen-like formations in the tubular epithelial formations are typical findings (Figure 28.2). Extensive degenerative changes, such as infarction and calcification, can mask adenomatoid tumors on histologic analysis and cause diagnostic difficulties.[15] Variants showing small mesothelial cysts surrounded by myometrium are sometimes overlooked on histologic studies, unless close attention is paid to the lining cells.

Immunohistochemically, the tumor cells are similar to mesothelial cells and are positive for keratin cocktails, usually keratins 5/6, and 7; calretinin; podoplanin; Wilms' tumor 1 (WT1); and thrombomodulin, but they are negative for the endothelial cell markers.[2,9,16]

Electron microscopic studies have demonstrated tall cytoplasmic microvillous extensions, similar to other differentiated mesothelial tumors.[2]

For the differential diagnosis from malignant mesothelioma, one must keep in mind that malignant mesotheliomas can have tubular differentiation with central vacuolization, resembling the appearance of an adenomatoid tumor, but the proliferation shows nuclear atypia, mitotic activity, and usually is extensive in scope with infiltrative growth, differing from the more limited nature of adenomatoid tumors.

Chapter 28: Malignant mesothelioma and other mesothelial proliferations

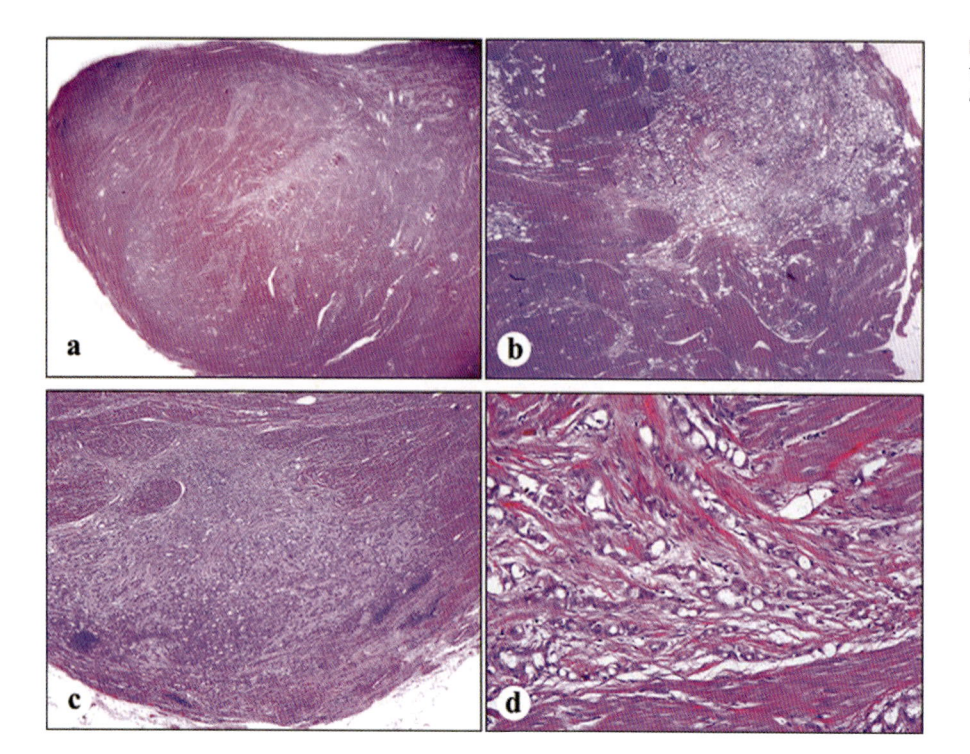

Figure 28.1 Four examples of adenomatoid tumor form demarcated nodules at the outer aspect of myometrium.

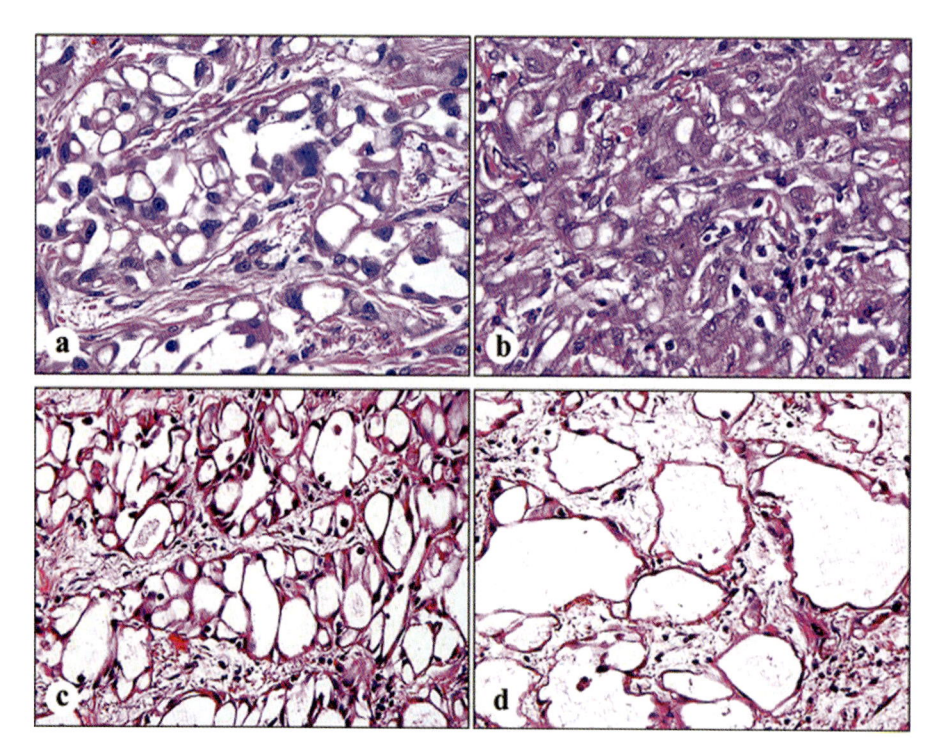

Figure 28.2 Multiple gland-like structures with large lumen-like vacuoles are typical of adenomatoid tumor.

Multicystic peritoneal mesothelioma (multilocular peritoneal inclusion cyst)

This histologically distinctive, rare, benign mesothelial proliferation is a heterogeneous group that includes cases that are most likely reactive mesothelial proliferations (reflecting the alternative diagnostic term). This entity also includes probable true neoplasms that have a greater proliferative capacity and can recur, however. Distinguishing one from the other, although difficult, should always be attempted.

Clinical features

Multicystic peritoneal mesothelioma typically occurs in young female patients, in the pelvic peritoneum, adjacent to the uterus and other pelvic organs, or in the upper peritoneum, including the omentum.[17–21] Many patients have a history of previous abdominal surgery. Very few cases have occurred in men. The median age in a large series was 38 years for women and 47 years for men. Approximately 10% of these tumors occurred in children.[20] Occurrence with endometriosis, leiomyomatosis peritonealis disseminata, and pseudomyxoma peritonei from an appendiceal mucinous tumor has been reported.[22–24] Lower abdominal pain is the most common presenting symptom. The prognosis has been generally excellent, especially in those cases in which complete excision of the tumor has been performed. Local recurrences can develop in up to one-half of all cases, however.[17–21] Large clinical series report long survivals with cytoreductive surgery and hyperthermic intraperitoneal chemotherapy (HIPEC).[21] A similar multicystic mesothelial lesion was reported in the pleura of a 37-year-old woman.[25]

Pathology

Grossly, the lesions form multiple and varyingly confluent cysts. The lesions contain cysts filled with clear or blood-tinged fluid. The cysts range from a few millimeters in diameter to several centimeters (Figure 28.3).

Histologically, those examples that consist of multiple cysts lined by a single layer of mesothelial cells and spaced by loose, reactive-appearing myofibroblastic proliferation and mild chronic inflammation have an appearance that suggests a reactive condition (Figure 28.4). In some cases, a higher amount of fibrous stroma is present (Figure 28.5). Those cases having multiple mesothelial cysts lined by cuboid to focally proliferative epithelium with intracystic papillary formations could represent neoplastic variants (Figure 28.6). Adenomatoid-tumor-like features can also be present. The cyst-lining epithelial components are positive for keratins, calretinin, and podoplanin. In some cases, immunohistochemical studies reveal concomitant endometriosis elements that are positive for estrogen and progesterone receptors, but the cyst epithelium is negative.

Differential diagnosis

Lymphangioma has a similar multicystic quality, but the lining cells are attenuated and express endothelial and not

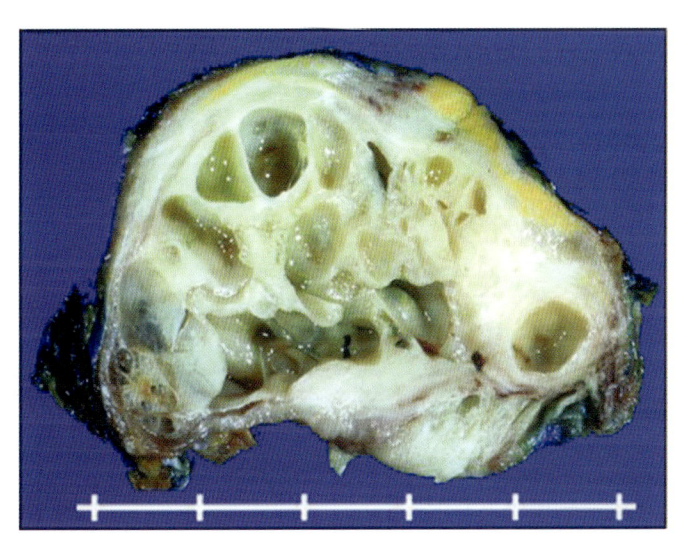

Figure 28.3 Grossly, multicystic peritoneal mesothelioma shows multiple cysts contained in a demarcated nodular tumor.

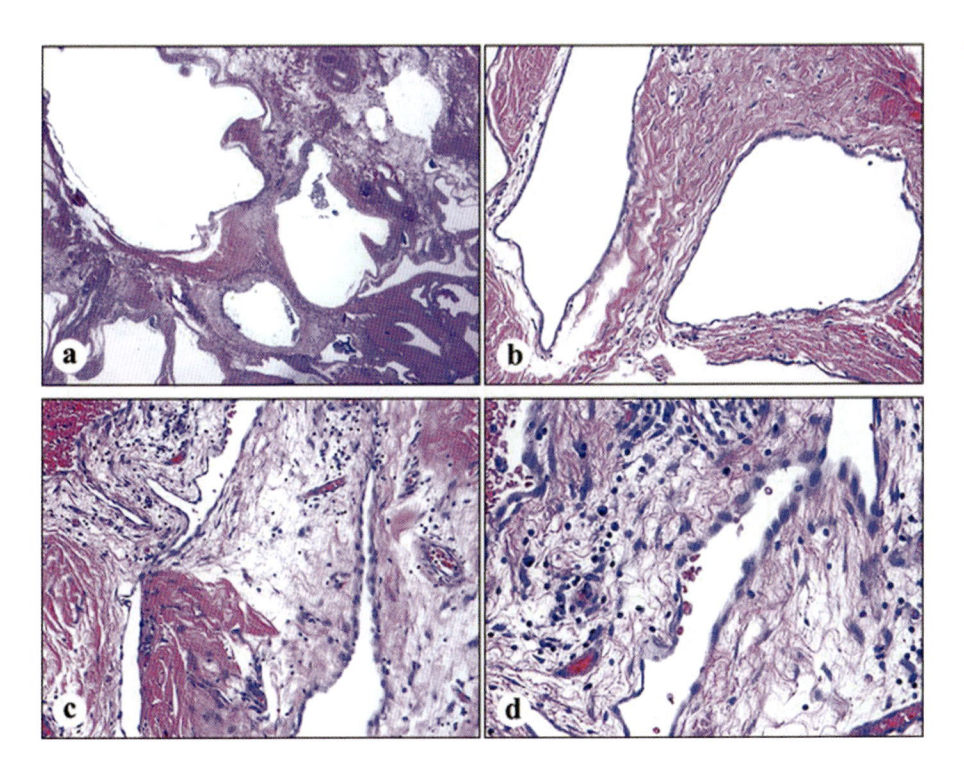

Figure 28.4 Multicystic peritoneal mesothelioma showing a multiloculated mesothelial cyst lined by a single mesothelial layer.

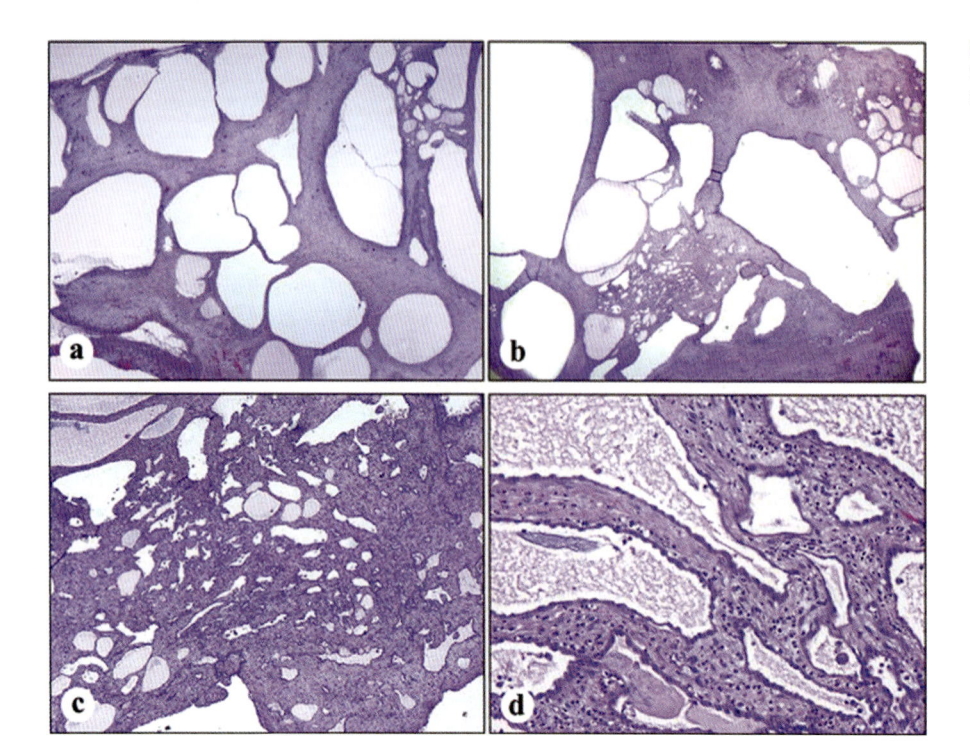

Figure 28.5 Multicystic peritoneal mesothelioma with a greater amount of fibrous stroma. Some cysts show hobnail morphology.

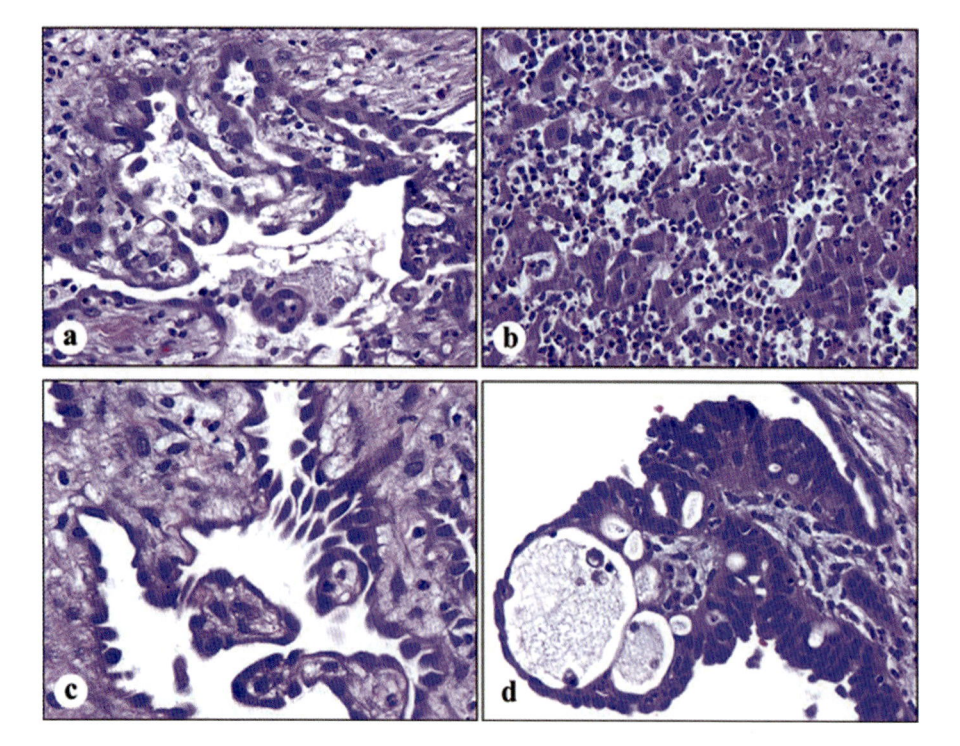

Figure 28.6 An example of multicystic peritoneal mesothelioma showing intracystic papillary proliferations. This tumor recurred after excision.

mesothelial markers, although they are positive for podoplanin, and some keratin expression (keratins K7 and K18) can occur. Endometriosis lesions rarely have a distinctive multicystic appearance, are estrogen receptor positive, and are often more heterogeneous, with hemorrhage, hemosiderin deposition, and fibrosis.

Well-differentiated papillary mesothelioma

Originally known to be in the peritoneal cavity, similar tumors have also been reported in the pleura. This form of mesothelioma could represent an early phase of mesothelial neoplasia. Although the prognosis has been favorable in many cases,

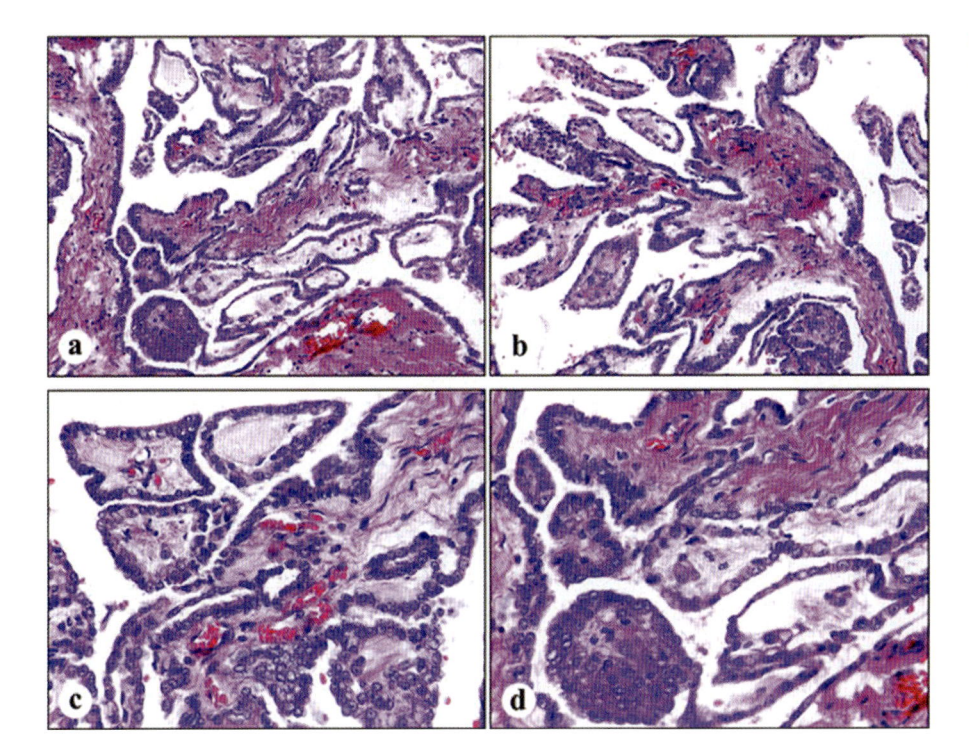

Figure 28.7 Well-differentiated papillary mesothelioma is composed of papillary proliferations lined by a nonatypical single mesothelial layer.

a small number of peritoneal cases and many pleural types have shown progression to clinically malignant mesothelioma, suggesting that some examples of this entity represent an early phase of malignant mesothelioma.

Clinical features

Well-differentiated papillary mesothelioma,[26] also reported as benign papillary mesothelioma,[27] has been reported mostly in the peritoneal cavity, including its extension, the tunica vaginalis.[26–33] These tumors predominantly occur in young and middle-aged women, with <20% having been diagnosed in men. In many patients, papillary mesothelioma is an incidental finding during surgery for unrelated conditions, but some patients have had ascites. No relationship to asbestos exposure has been established.

Most patients have multiple peritoneal nodules and some have mat-like peritoneal involvement, although a few have had solitary papillary lesions that usually measure 0.5 cm to 2 cm. In general, surgical excision has been considered sufficient, but adjuvant treatment has been used in patients with extensive peritoneal disease and chronic ascites.[29] Solitary lesions have an excellent prognosis, and in some cases they may be difficult to distinguish from florid papillary mesothelial hyperplasia. Multifocality and mat-like diffuse growth are worrisome signs that suggest progression to malignant mesothelioma. The reported cases have shown a 5% tumor-related mortality rate so that at least in earlier series there has been some overlap with malignant mesothelioma.[31]

Pleural examples typically manifest by unilateral pleural effusion and only rarely have been incidental findings. They occur with a more even male-to-female ratio at an older age than do the peritoneal examples, and asbestos exposure has been detected in one-half of the patients. Slow progression to diffuse malignant mesothelioma in up to 10% to 15% has been observed in some cases, although follow-up information is limited.[34]

Pathology

Histologically, papillary mesotheliomas are composed of well-differentiated papillary structures with paucicellular and often edematous stromal cores. The histologic appearance can remotely resemble early placental tissue. The papillae are lined by a single cell layer of well-differentiated cuboid mesothelial cells that have limited atypia and inconspicuous mitotic activity (Figure 28.7).[26–32] Superficial stromal invasion has been allowed in some studies for well-differentiated papillary mesothelioma. Tumors with invasive foci have a greater potential for local recurrence, but are not generally associated with progressive disease.[33] However, more extensive invasion should lead to diagnosis of malignant mesothelioma. Occurrence has been reported together with adenomatoid tumor and multicystic peritoneal mesothelioma.

Immunohistochemical features are similar to those of well-differentiated malignant mesothelioma, with consistent expression of keratins, including K5 and K6, calretinin, podoplanin, and WT1.

Because malignant mesothelioma can also contain well-differentiated papillary elements, the diagnosis of well-differentiated papillary mesothelioma requires clinicopathologic correlation and extensive sampling. This diagnosis therefore cannot be made in isolation based on histology, especially of a small biopsy specimen only.

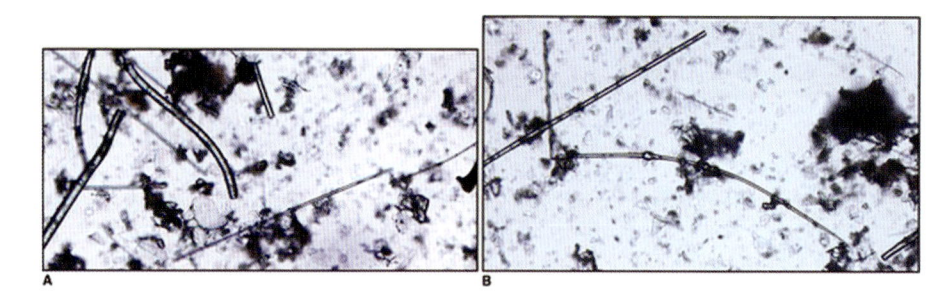

Figure 28.8 Asbestos fibers deposited with dust on a histologic slide. The thin straight fibers have the morphology of amphibole asbestos fibers.

Malignant mesothelioma and its variants

Malignant mesothelioma is an epithelial neoplasm of mesothelial origin. It most commonly occurs in the pleural cavity; approximately 20% occur in the peritoneum and a few tumors (<1%) in the pericardium and the tunica vaginalis testis, a continuation of the peritoneal surface. There is a strong etiologic association with asbestos exposure. Most malignant mesotheliomas are diffuse, with extensive involvement on serosal surfaces, but rare examples are solitary and are referred to as *localized malignant mesotheliomas*. Tumors formerly known as benign and malignant fibrous mesotheliomas have been reclassified as benign and malignant solitary fibrous tumors, which are unrelated fibroblastic neoplasms (see Chapter 12).

Epidemiology of mesothelioma

The incidence varies globally and regionally, often correlating with asbestos exposure. The highest national incidence has been recorded in Australia, Great Britain, and Belgium, where approximately 30 mesotheliomas occur per 1 million inhabitants. In most countries of Western and Northern Europe, there are approximately 15 mesotheliomas per 1 million inhabitants. Although the incidence was rising until the late 1980s, it has since then shown a tendency to decline.[35] In the United States, the age-adjusted incidence was around 20 per million in 2000, and a decline in overall incidence has been recorded from the late 1980s. The incidence is markedly higher in men, reflecting occupational exposure patterns. The latency of asbestos-related carcinogenesis is long (20–40 years or longer), and sustained exposure is more damaging than a transient one.[36] The connection of mesothelioma with industrial exposure is further confirmed by the finding that mesothelioma by autopsy data did not occur 100 years ago.[37]

Patient cohorts who have suffered from asbestosis and mesothelioma include workers in asbestos mines and manufacturing plants for asbestos products, shipbuilders, construction and demolition workers. In some instances, family members and general populations in mining towns have also been exposed. In addition, asbestos causes pulmonary fibrosis (asbestosis) and is strongly linked with pulmonary carcinomas and possibly with gastrointestinal carcinomas.

Mesothelioma and asbestos

Asbestos exposure is believed to be the single most important etiologic factor for malignant mesothelioma, both pleural and peritoneal. *Asbestos* refers to a group of naturally occurring, iron-containing complex magnesium silicate minerals, whose fiber-generating crystalline structure made them valuable as insulation material and fire retardants until limited by regulations based on recognized health hazards.

Asbestos fibers are prone to shed as microscopic needle-like particles (Figure 28.8) that stay in the air and are therefore inhaled. Those fibers >5 μm long and with a length-to-width ratio >5:1 in particular can enter the airways and accumulate in the peripheral lung. They then migrate into the pleura by means of macrophages and lymphatic circulation and act as carcinogens, causing asbestos-related pulmonary fibrosis (asbestosis) on their deposition into tissue.

There are six main types of asbestos fibers, which vary in their chemical and physical composition. Their overall usage, distribution, and potency for causing mesothelioma varies. The most widely used is chrysotile (a serpentine-type asbestos with curled fibers) that seems to have relatively low propensity for causing mesothelioma. A relatively rare type, crocidolite (an amphibole-type asbestos) with straight, needle-like fibers, has a high potency to cause mesothelioma and has been linked to significant local mesothelioma epidemics in Western Australia and South Africa, where the first connection was made between occupational asbestos exposure and mesothelioma. Amosite, another amphibole-type asbestos, is also strongly linked to mesothelioma.[36–41]

The biologic background for the carcinogenicity of asbestos fibers is not well understood. Hypotheses include formation of free oxygen radicals, and mechanical DNA damage caused by the needle-like fibers.

In the lung and pleura, asbestos fibers are surrounded by an iron-containing coating composed of cellular secretions and extracellular matrix components to form so-called ferruginous bodies (asbestos bodies). These structures contain an asbestos fiber in the center, and around the fibers there are brownish, rounded, cuff-like structures wrapped like pearls on a chain, although some appear linear and spear-like. Ferruginous bodies are highlighted with iron stains (Figure 28.9). Their presence in tissue sections reflects a high asbestos load in the lung. Ferruginous bodies are usually found in the peripheral lung underneath the pleura, but rarely in the mesothelioma

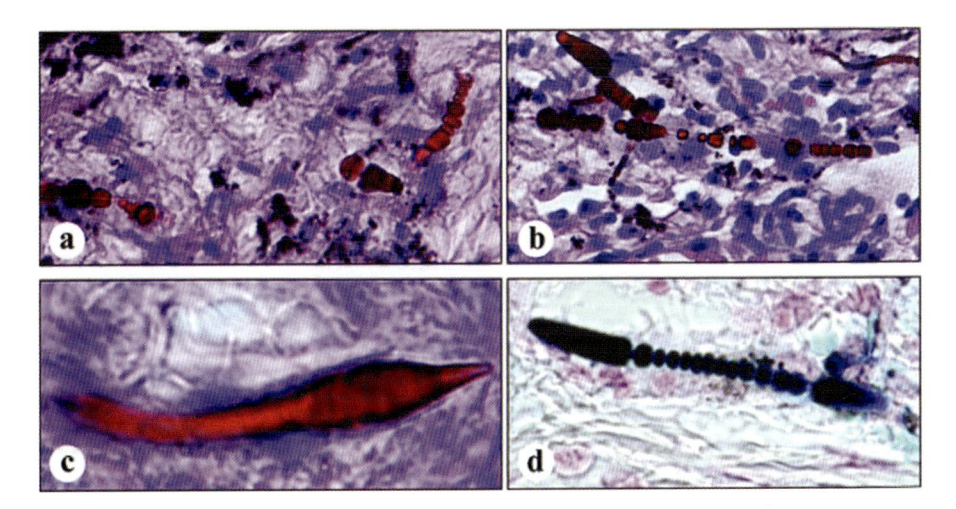

Figure 28.9 Ferruginous bodies contain a central core of an asbestos fiber that is surrounded by brownish material, often arranged as spherical pearl-like structures. Some asbestos bodies are linear rods, and others have dumbbell morphology. Ferruginous bodies are positive (blue) in an iron stain.

tissue. They can also be recovered from the sputum in heavily exposed patients.

Other possible causal factors

Simian virus 40 (SV40) DNA sequences have been demonstrated in mesothelioma tissue by polymerase chain reaction (PCR),[42] but some studies by similar methods have failed to find them and also found no serologic evidence for SV40 infection, casting doubt on the possible role of this virus.[43]

Radiation therapy for a cancer (e.g., lymphoma or breast cancer) has preceded pleural and rarely peritoneal mesothelioma. These mesotheliomas can occur 20 years or longer after the radiation therapy, although the average latency from literature observations has been <5 years.[44]

Clinical features of pleural mesothelioma

Pleural mesotheliomas constitute 70% to 80% of all mesotheliomas. They usually occur in middle-aged and older adults, with at least a 3:1 male predominance in most series. The median ages in major series have been between 54 and 59 years. Pleural mesothelioma rarely occurs in children and young adults. Dyspnea, chest pain, or both are among the most common symptoms, usually associated with extensive pleural effusion, and chest or shoulder pain is often a sign of chest wall infiltration.[45–49]

Clinicopathologically favorable prognostic factors are age <50 years, good performance status, less extensive disease, and epithelioid histology. The prognosis is generally poor, but pleurectomy or pleuropneumonectomy and chemotherapy can prolong life in some cases.[49] Newer immunotherapy approaches such as immunotoxin coupled with tumor antigens such as mesothelin are being investigated.[50] Local spread into the pleural cavity, pericardium, and chest wall is usually more significant for the cause of death than metastases. Metastases can develop in regional lymph nodes, bones, liver, and other organs. Very rare, truly localized forms of malignant mesothelioma have a better prognosis.[45–48]

Clinical features of peritoneal mesothelioma

Peritoneal mesotheliomas constitute 10% to 20% of all malignant mesotheliomas, and they also are often associated with asbestos. There is a lesser male predominance than for pleural mesothelioma. Clinical manifestations include abdominal pain, ascites, and weight loss. Prognosis varies by level of peritoneal involvement, but is generally better than for pleural mesothelioma, with a significant percentage of patients surviving >5 years. The treatment includes surgical debulking and HIPEC.[51–54]

Rare mesotheliomas in male patients involve the tunica vaginalis in the scrotum; this is an extension of peritoneal mesothelium. Patients with malignant mesothelioma in this region have a better prognosis, but the disease often spreads into the genitoinguinal region and the retroperitoneum.[55,56]

Gross features of malignant mesothelioma

The gross features vary widely, depending on the stage of the disease. The early lesions are inconspicuous on gross inspection and difficult to discern, especially when growing in a pleural fibrous plaque. The insidious nodular pleural densities gradually spread along the serous surface and form coalescing nodules that ultimately encase the lung in pleural mesothelioma (Figure 28.10), and various abdominal organs (e.g., the spleen) in peritoneal mesothelioma. Rare instances have been reported with malignant mesothelioma forming a localized mass (Figure 28.11) ≤10 cm in diameter, and patients with these tumors have a better prognosis.[57,58]

Histopathology of malignant mesothelioma

Histologically, malignant mesothelioma shows an essentially similar spectrum regardless of the site of origin. Most tumors have an epithelioid morphology. Better differentiated tumors show tubulopapillary and less differentiated solid epithelial patterns. The sarcomatoid mesotheliomas have spindle cell (or pleomorphic) patterns and include desmoplastic mesothelioma as a variant.

Tubulopapillary malignant mesothelioma

Tubular and papillary patterns are common in the better-differentiated variants of epithelial mesothelioma (Figure 28.12). The papillary elements can have an arching appearance, or when cross-sectioned, they can have a rosette-like appearance. Adenomatoid tumor-like morphology is also possible, but atypia and the extent of the lesion are greater (Figure 28.13). The malignant papillary tumors are microscopically invasive and have greater cytologic atypia than do the well-differentiated papillary mesotheliomas (Figure 28.14). A glycogen-rich clear cell pattern and mucinous stroma are also possible.

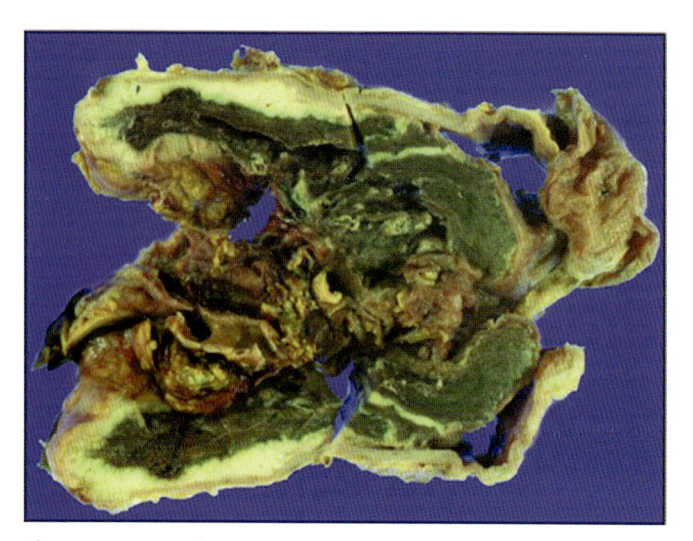

Figure 28.10 An advanced pleural mesothelioma surrounds the lung, forming a thick white covering.

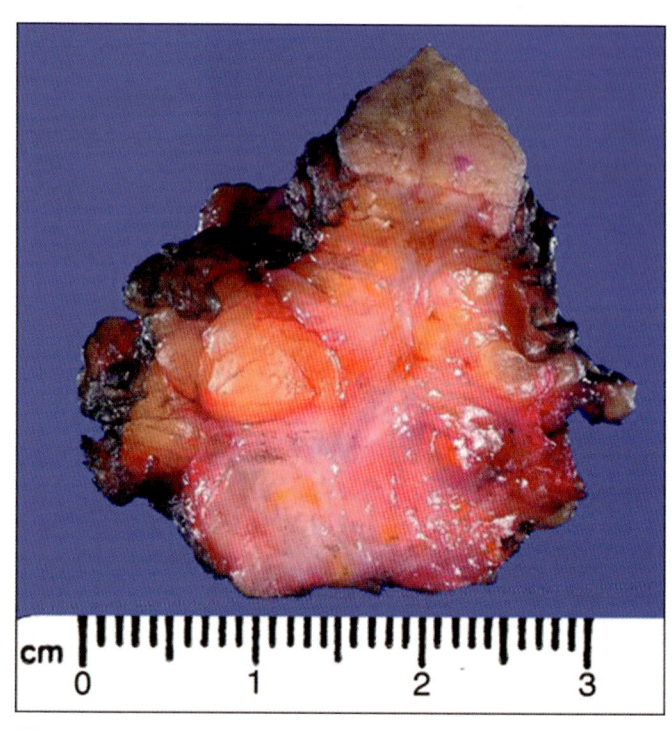

Figure 28.11 Localized malignant mesothelioma forms a whitish juxtaperitoneal mass in the pubic region.

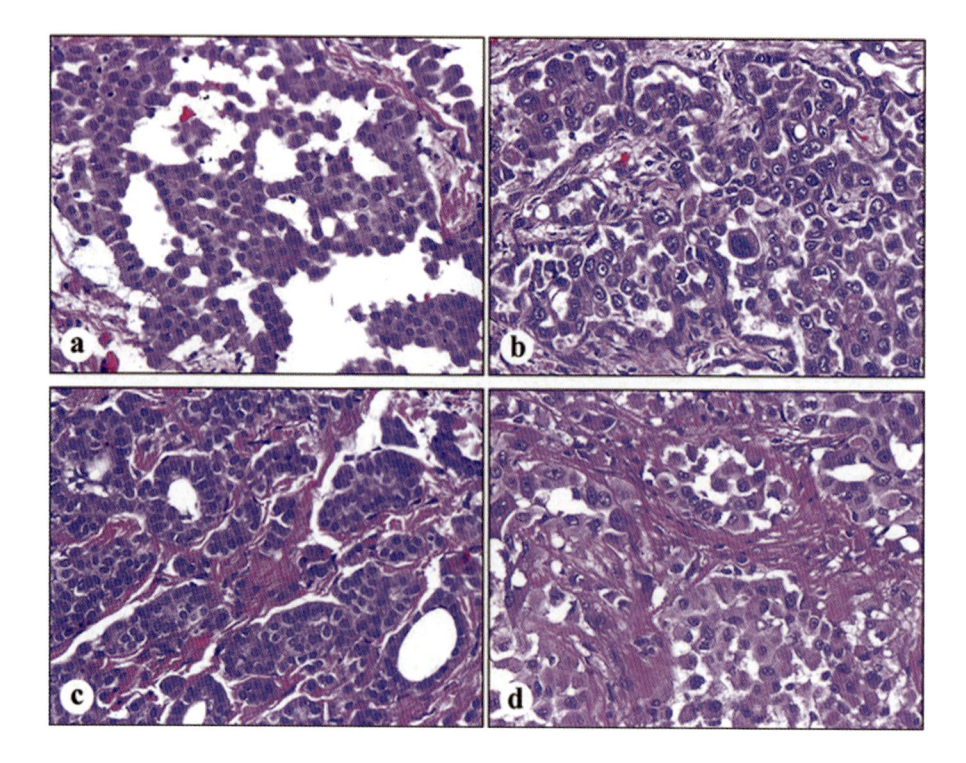

Figure 28.12 Variants of tubulopapillary malignant epithelial mesothelioma of the pleura. The degree of luminal differentiation varies from one case to another.

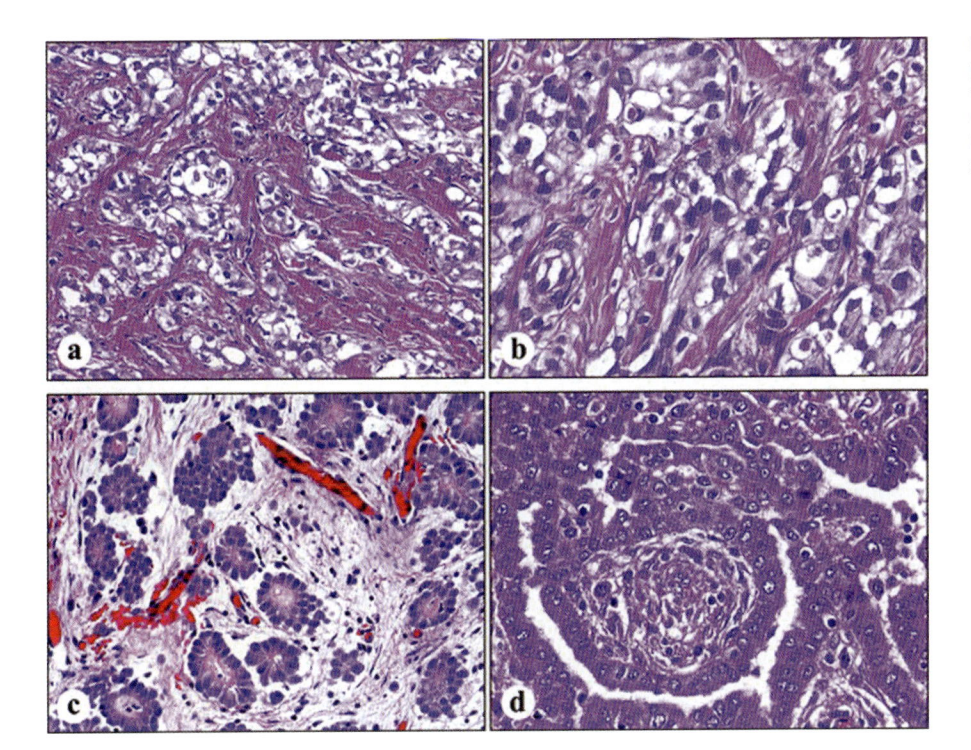

Chapter 28: Malignant mesothelioma and other mesothelial proliferations

Figure 28.13 (a,b) Adenomatoid-tumor-like pattern in a malignant epithelioid mesothelioma of the peritoneum. Note significant cytologic atypia. (c) Cross-sectioned papillary structures create a rosette-like impression. (d) Arching papillary structures in peritoneal mesothelioma.

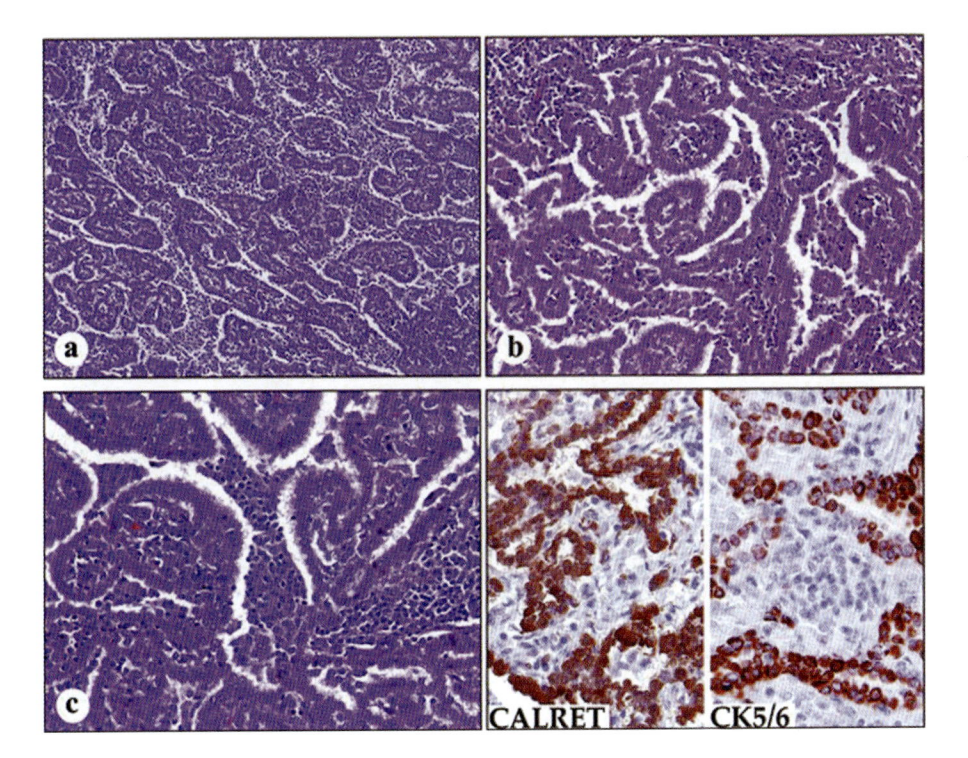

Figure 28.14 Well-differentiated malignant mesothelioma with a papillary pattern. This tumor had an infiltrative component. Note the prominent lymphoplasmacytic stroma. Immunohistochemical positivity for calretinin and keratins 5 and 6 is typical.

Solid epithelial mesothelioma

Many epithelioid mesotheliomas grow as solid sheets, with only rudimentary, if any, papillary or tubular differentiation, and these can be understood to be structurally less-differentiated variants, although they differ from sarcomatoid mesotheliomas by their prominently epithelioid morphology (Figure 28.15).

Deciduoid mesothelioma is the designation for a rare, solid variant of epithelial mesothelioma. This variant was originally reported in the peritoneum in a child,[59] and subsequently in female and male patients, both in the peritoneum and pleura, and with or without asbestos history.[60,61] Histologically, the deciduoid mesothelioma is composed of diffuse sheets of large epithelioid cells with

785

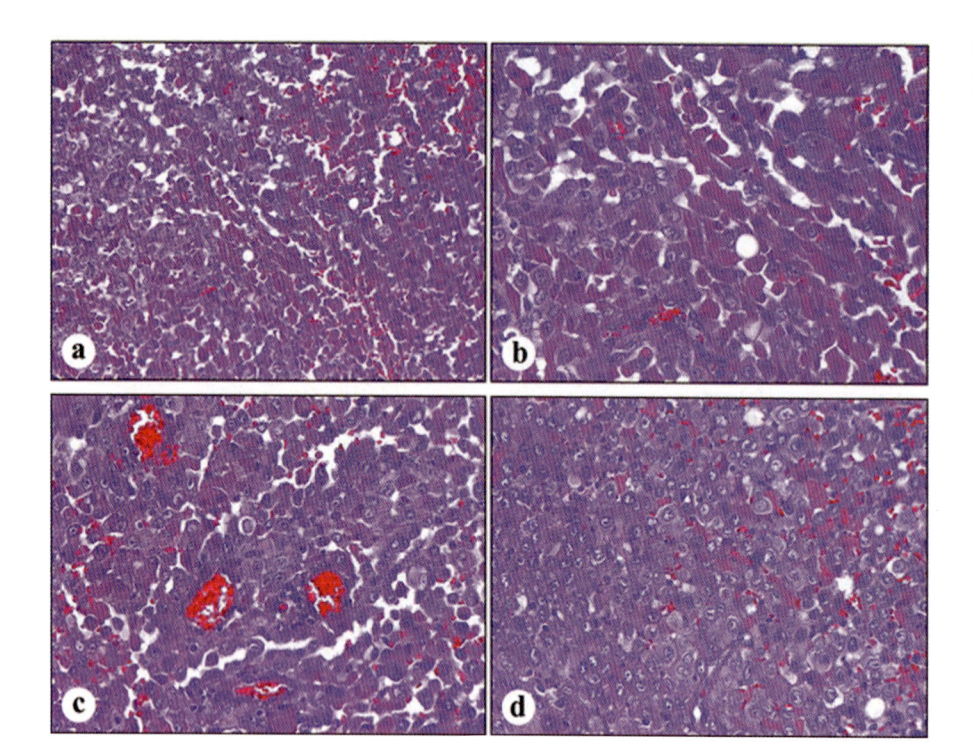

Figure 28.15 An epithelioid mesothelioma with poorly developed tubulopapillary differentiation. Large cytoplasmic vacuoles are present.

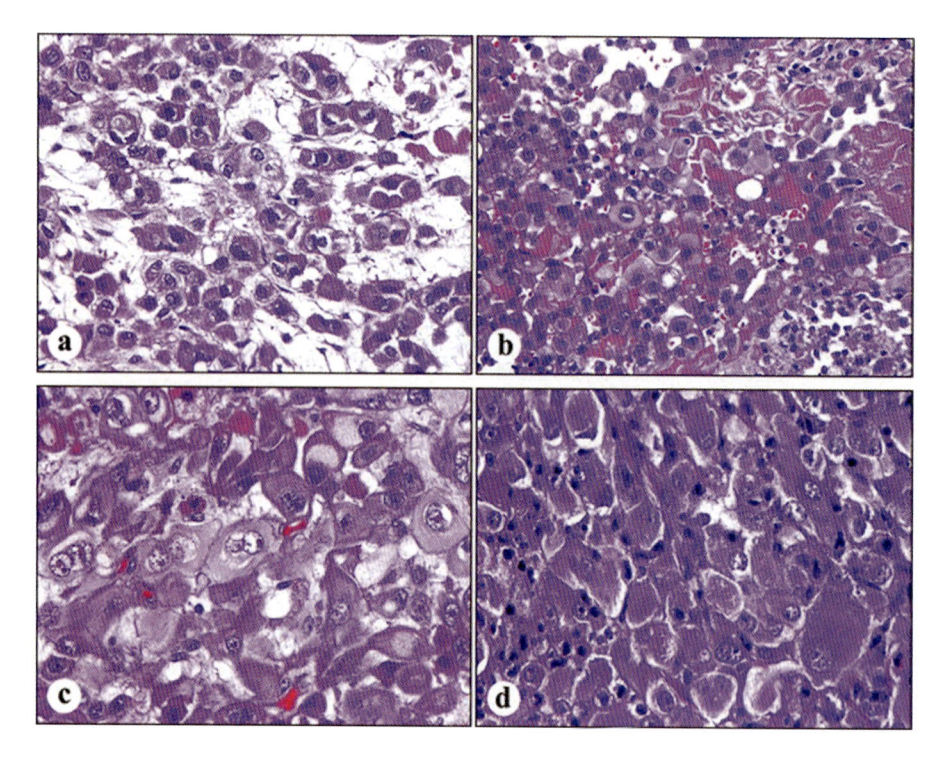

Figure 28.16 An epithelial malignant mesothelioma composed of solid sheets of large epithelioid cells with abundant cytoplasm is often called deciduoid mesothelioma.

pale-staining abundant cytoplasm, imparting a decidual-cell-like appearance (Figure 28.16).

Biphasic mesothelioma

Malignant mesotheliomas with both epithelial and sarcomatoid growth are called *biphasic mesotheliomas*. This pattern is quite rare and most likely represents a transitional form between epithelial and sarcomatoid mesothelioma.

Histologically, there are glandular or tubular elements interspersed with solid elements composed of oval or spindled cells. Immunohistochemically, the glandular element is strongly positive for keratins and calretinin, whereas the sarcomatous component is often weakly keratin positive and can be calretinin negative (Figure 28.17). In rare cases,

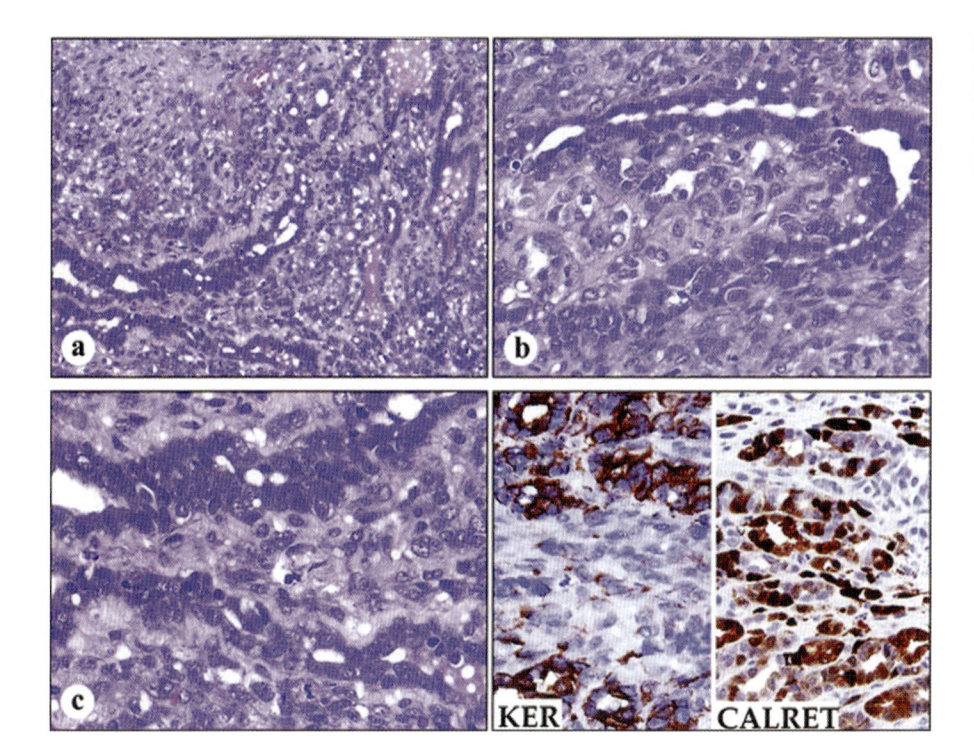

Figure 28.17 Biphasic malignant mesothelioma shows tubuloglandular elements and an intervening sarcomatoid component. The epithelial elements are positive for keratins and calretinin, whereas the sarcomatoid component is weakly keratin positive, and negative for calretinin.

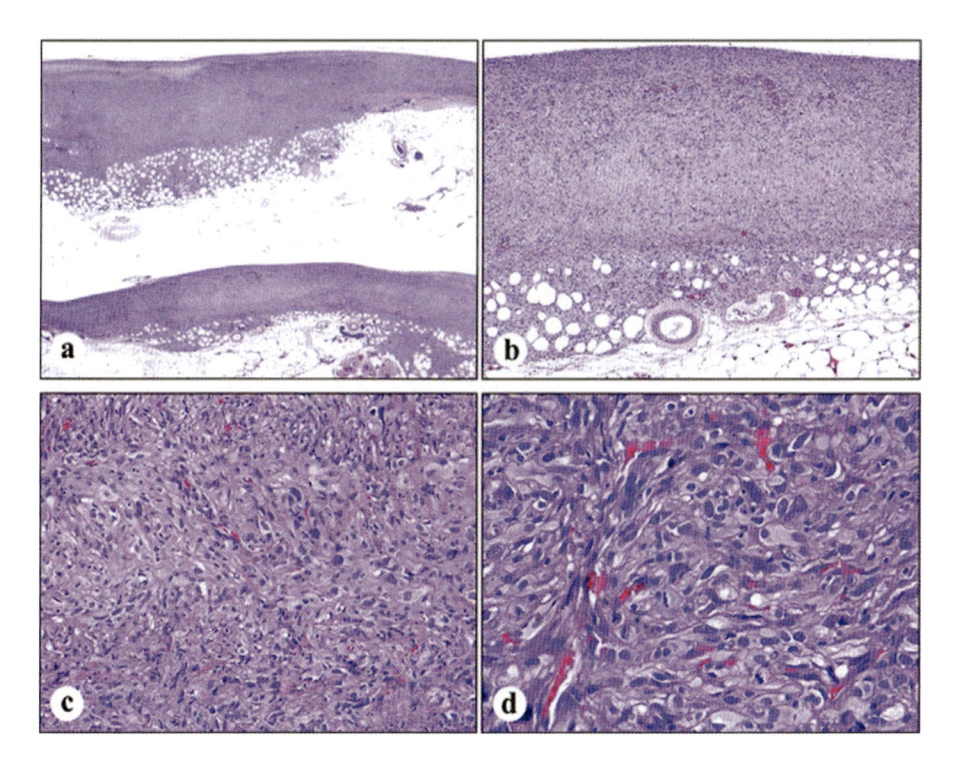

Figure 28.18 Sarcomatoid pleural mesothelioma grows diffusely in the pleura and infiltrates the adipose tissue.

mesothelioma can extensively infiltrate lungs, simulating interstitial disease.[62]

Sarcomatoid mesothelioma

Sarcomatoid mesotheliomas vary from spindle cell to pleomorphic, potentially simulating soft tissue sarcomas such as fibrosarcoma and undifferentiated sarcomas/malignant fibrous histiocytoma. Demonstration of strong keratin and focal calretinin expression is diagnostically helpful (Figure 28.18), whereas other markers typical for mesotheliomas are inconsistently expressed (see Immunohistochemistry section). Geographic necrosis is also typical of sarcomatoid mesothelioma (Figure 28.19). Because heterologous osteosarcomatous and

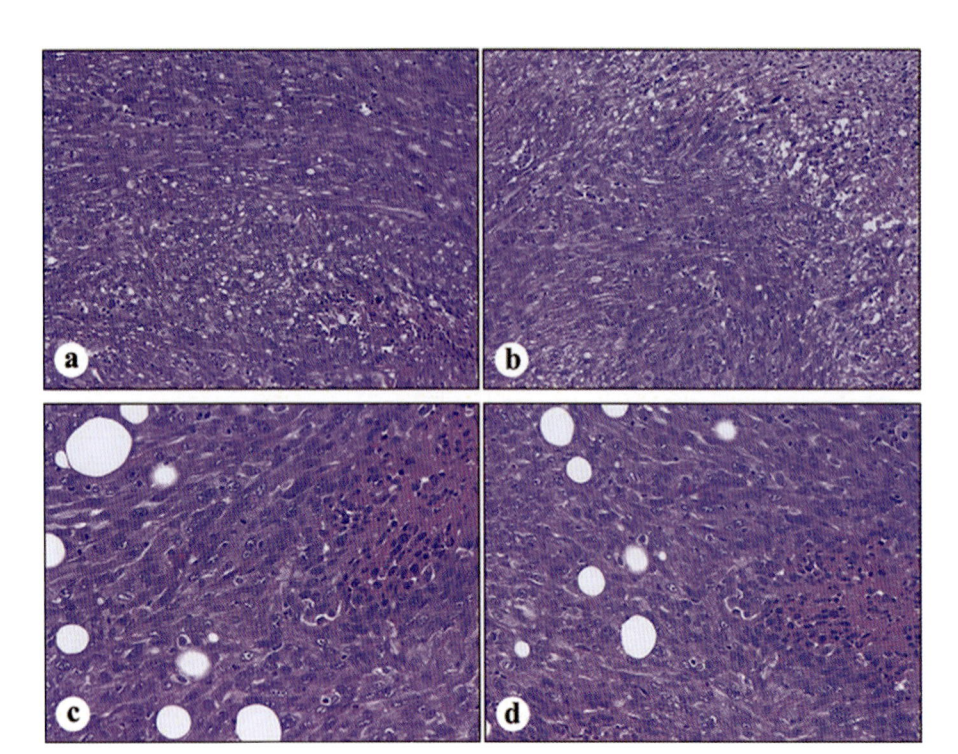

Figure 28.19 Sarcomatoid peritoneal mesothelioma with areas of geographic necrosis.

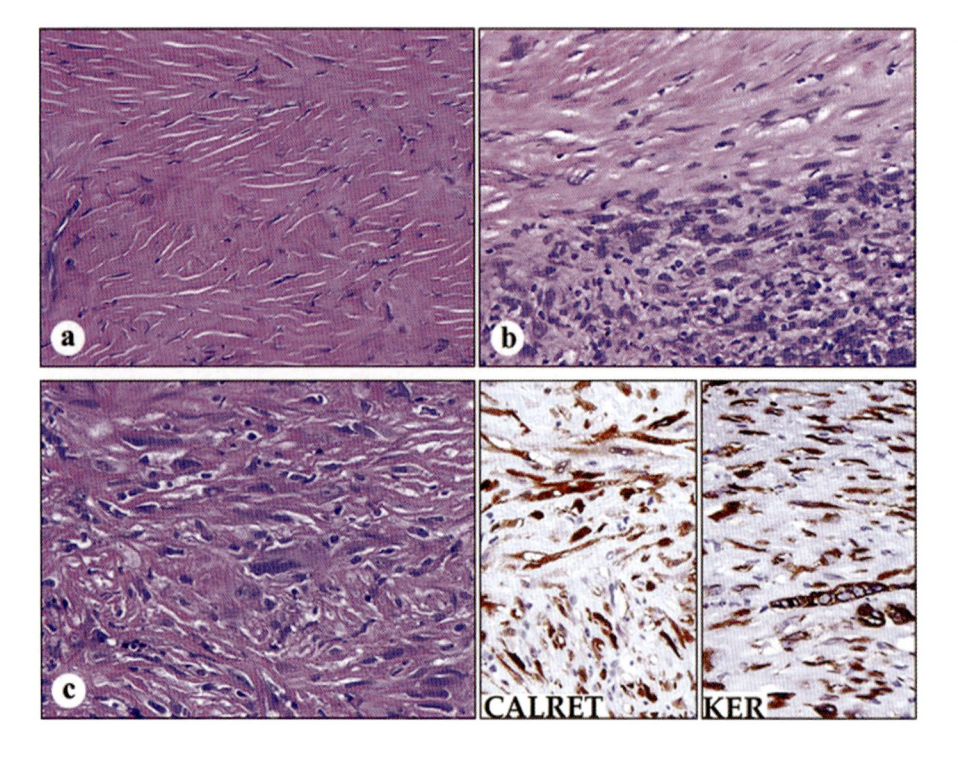

Figure 28.20 Desmoplastic mesothelioma is relatively paucicellular, but atypical cells are present in a dense collagenous matrix, and these elements are calretinin and keratin positive.

rhabdomyosarcomatous differentiation may occur, encountering these components in serous cavity locations should prompt a further search for mesothelial differentiation.[63]

Desmoplastic mesothelioma

This rare form usually is a variant of sarcomatoid mesothelioma, but some cases have biphasic histology, containing differentiated tubular (or solid) epithelial elements. It is characterized by a densely collagenous, desmoplastic stroma containing atypical spindle cells that can also be pleomorphic. In some cases, the desmoplastic appearance is related to involvement of a pleural plaque.[64]

Strong pan-keratin expression is typical, but calretinin positivity varies (Figure 28.20). Although the tumor can have paucicellular areas, it typically shows higher cellularity than

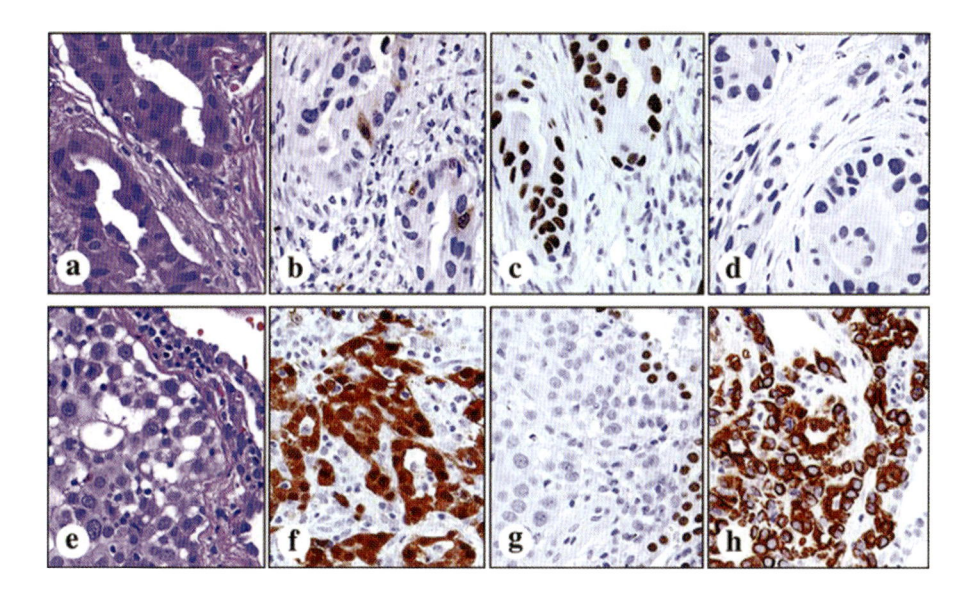

Figure 28.21 Differential diagnosis of mesothelioma and adenocarcinoma. (a) This pleural epithelial neoplasm is positive for TTF1, (c) and is essentially negative for calretinin and CK5/6 (b,d), indicating that this is a pulmonary adenocarcinoma. (e) A pulmonary solid epithelial infiltration is positive for calretinin (f) and keratins 5 and 6 (h), supporting the diagnosis of mesothelioma. (g) Tumor is nagative for TTF1, but positive pneumocytes are present.

ordinary pleural plaques. The diagnosis is aided by demonstration of an overtly sarcomatous hypercellular component and invasion into chest wall fat. The presence of tumor cell necrosis is supportive of malignancy, although this finding is not specific.[64,65] Small biopsy specimens of fibrosclerosing pleural and mediastinal infiltrations should be evaluated with caution, and clinicoradiologic correlation can be helpful.

Immunohistochemistry of malignant mesothelioma

Best characterized immunohistochemical markers for the positive identification of mesothelioma are calretinin, keratins 5 and 6, Wilms' tumor protein 1 (WT1), and mesothelin.[66–76] Calretinin and mesothelin are also expressed in some pulmonary and other carcinomas.[72] Podoplanin may also be useful, but is less specific (Table 28.2).[75,76] Markers variably expressed in adenocarcinoma, but usually not in mesothelioma include the epithelial antigens CEA, CD15, and epithelial cell adhesion molecule (EpCAM) recognized by Ber-EP4and MOC31 antibodies. Examples of the application of immunohistochemistry are shown in Figure 28.21.

Mesothelioma markers are only variably expressed in sarcomatoid mesothelioma. These tumors are variably keratin cocktail and keratin 7 positive but are usually negative for keratins 5 and 6. They often show at least focal calretinin positivity, and are often podoplanin positive (Figure 28.22).[77–79]

Ultrastructure of mesothelioma

Typical of mesothelioma are tall and thin microvilli on the cell surface (Figure 28.23). It has been suggested that only microvilli with length exceeding the width by a margin of 15:1 are diagnostic of mesothelioma.[80] Because the microvilli are often poorly developed in sarcomatoid mesotheliomas, electron microscopy is of limited value in their diagnosis.

Differential diagnosis of malignant mesothelioma

The most important histologic problems related to mesothelioma include the differential diagnosis from benign mesothelial proliferations and carcinomas, including adenocarcinoma (especially pulmonary); pleural and peritoneal carcinomas from other sources (e.g., primary peritoneal and metastatic serous carcinoma); and the differential diagnosis of desmoplastic mesothelioma and reactive proliferations, and sarcomatoid mesothelioma from sarcomas.

Mesothelioma versus reactive mesothelial proliferation

Fat invasion, when present, supports the diagnosis of malignant mesothelioma. The submesothelial cells are slender spindled to epithelioid cells typically seen on reactive peritoneal tissues and pleural plaques. They are keratin positive, which should be noted in the differential diagnosis of mesothelioma and reactive conditions. Therefore, demonstration of keratin-positive cells per se has no value in the diagnosis of mesothelioma versus pleuritis or pleural plaque (Figure 28.24).[81,82] Immunohistochemistry may assist distinction of benign versus malignant mesothelial cells in surgical and cytological specimens. Markers reported in benign, but generally not malignant mesothelia include desmin, whereas EMA, p53, and an oncofetal protein IMP3 are typically expressed in neoplastic, but not reactive mesothelia.[83,84]

In the hernia sac and sometimes also inside the cysts of multicystic peritoneal mesothelioma, nodular cellular proliferation can occur (Figure 28.25). This phenomenon was originally described as *nodular mesothelial hyperplasia,*[85] Subsequently, major histiocytic (CD68, CD163 positive) and minor mesothelial (keratin and calretinin positive) components were verified by immunohistochemistry.[86]

Papillary mesothelial hyperplasia can form a cluster composed of multiple adjacent papillary processes in a limited area lacking cytologic atypia (Figure 28.26). Such a hyperplasia

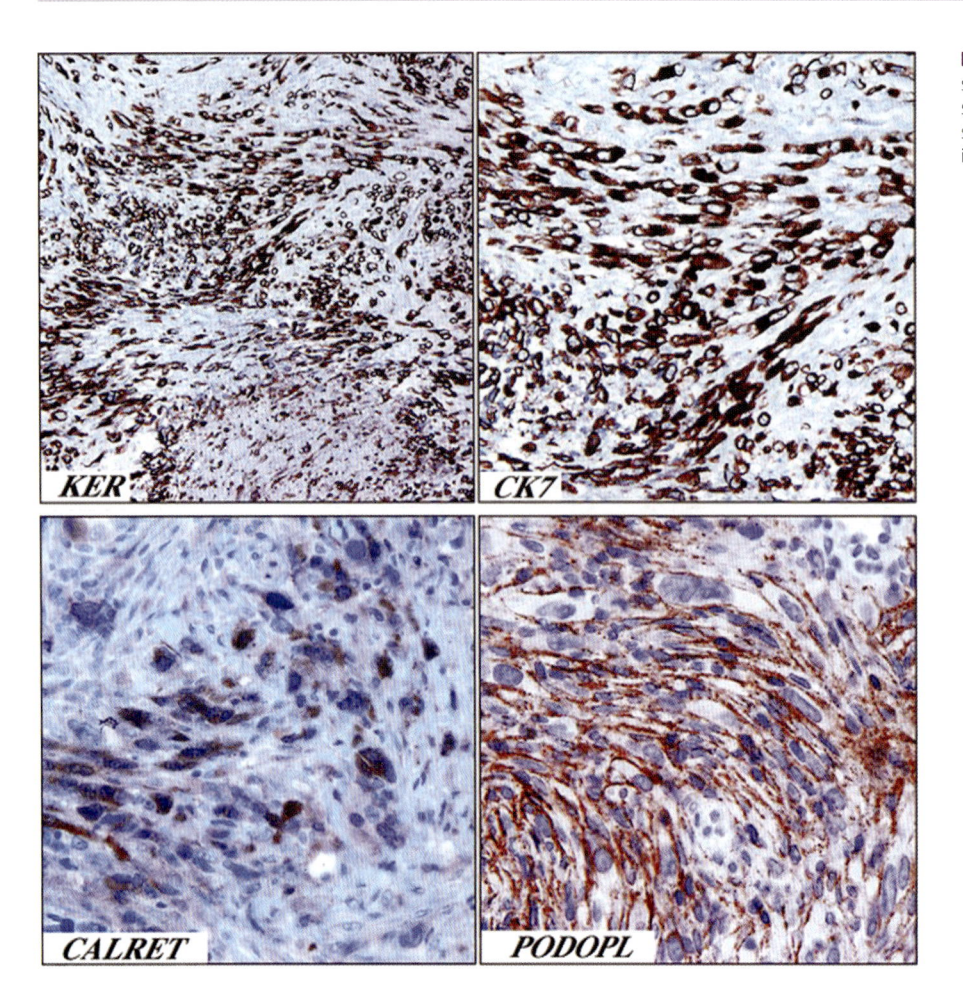

Figure 28.22 Immunohistochemically, sarcomatoid mesotheliomas usually show significant keratin positivity. Calretinin positivity is seen in only a portion of the tumor cells, but there is extensive podoplanin positivity.

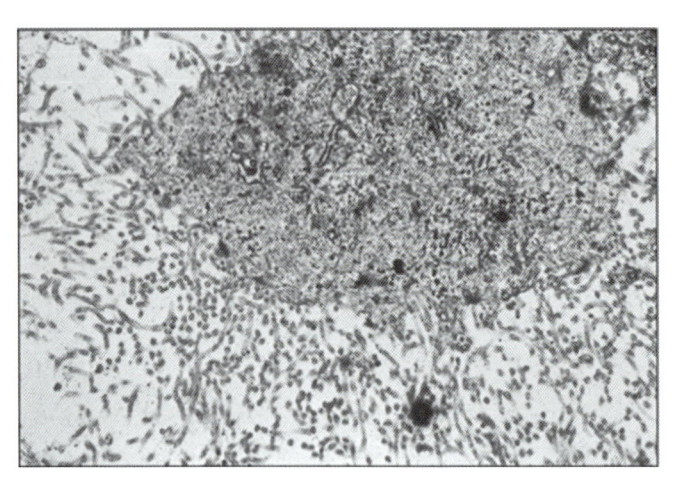

Figure 28.23 Typical findings of epithelial mesothelioma on electron microscopy are long, tall cytoplasmic microvillous extensions.

found adjacent to ovarian tumors should not lead to a diagnosis of peritoneal spread of a papillary carcinoma.[87]

Most problems are related to small specimens, such as needle biopsy, and the scant sampling is often a factor that limits definitive conclusions. The US and Canadian Mesothelioma Reference Panel has reviewed large numbers of problem cases and reported on the issues in differential diagnosis.[88] In some cases, it is difficult even for the most experienced pathologists to predict the biologic potential, and then the designation *atypical mesothelial hyperplasia* can be used. Clinicoradiologic correlation, additional sampling, and follow-up study are primary considerations in such cases. In some instances, malignant mesothelioma has developed years later, indicating that some atypical mesothelial proliferations can be premalignant. Caution is always needed in the assessment of biologic potential based on small samples of large lesions, because the power to make a definitive diagnosis of mesothelioma or to rule it out is limited.

Histologically, benign mesothelial cells have been reported in the lymph node sinuses of patients with pleural effusions, but no evidence of mesothelioma. Long-term follow-up study and careful clinical correlation are necessary to rule out the possibility of metastatic mesothelioma.[89,90]

The distinction of desmoplastic mesothelioma from a pleural plaque or other pleural fibrosclerosing process can be difficult, especially in a small biopsy specimen. The presence of atypical cells, a hypercellular area with atypical cells, and fat infiltration are signs of desmoplastic mesothelioma.[64,65,88]

Florid mesothelial proliferations in the tunica vaginalis associated with the hydrocele sack can raise the differential diagnosis of mesothelioma. The presence of tissue clefts with

Chapter 28: Malignant mesothelioma and other mesothelial proliferations

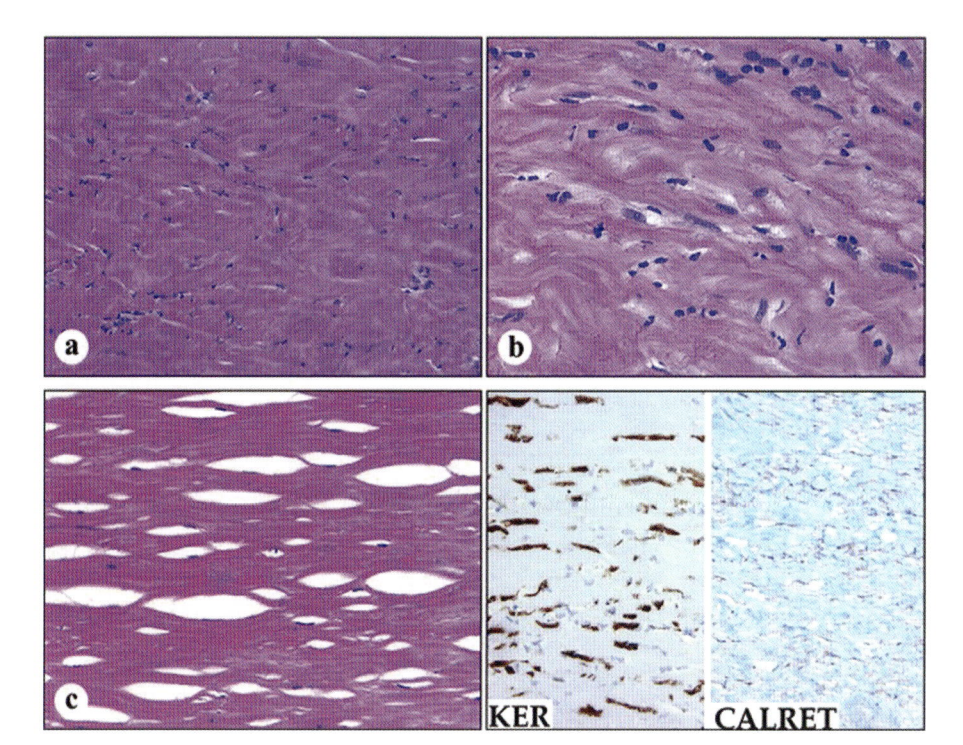

Figure 28.24 Pleural plaque is paucicellular, composed mainly of collagen. Immunohistochemistry demonstrates keratin-positive submesothelial cells, but these elements are negative for calretinin.

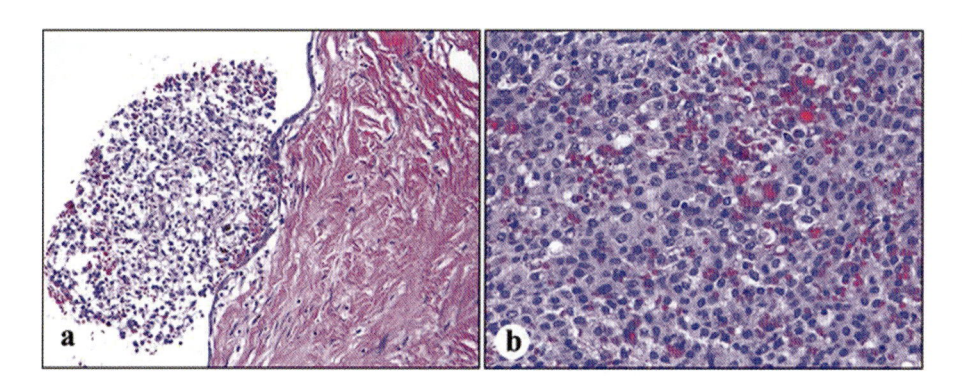

Figure 28.25 A cellular nodule inside the hernia sac is composed of histiocytes and mesothelial cells, representing a reactive process.

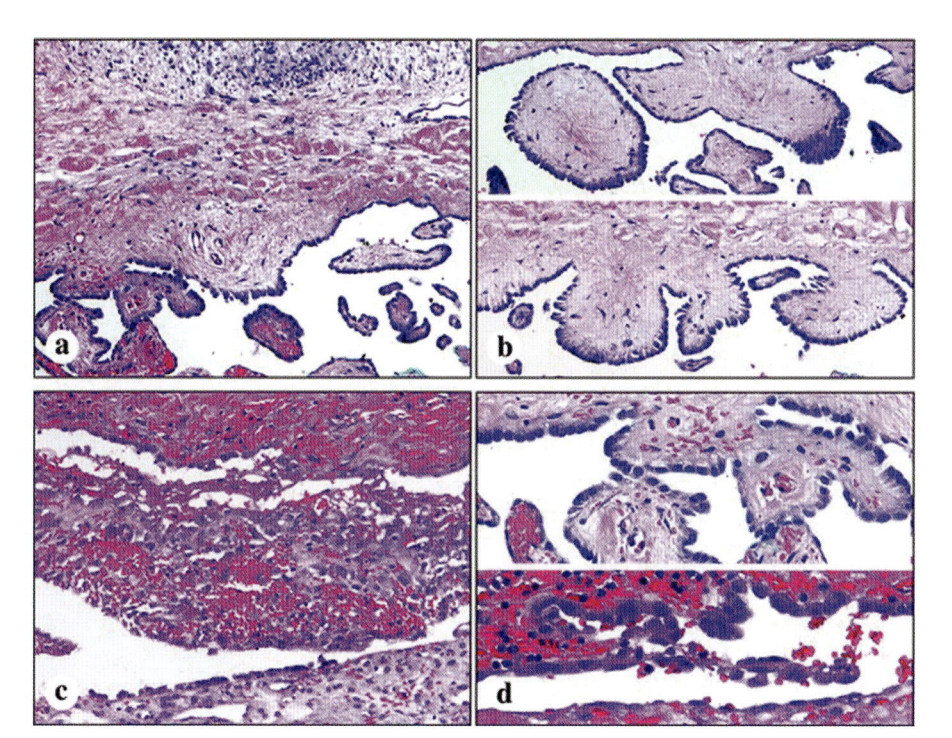

Figure 28.26 Papillary reactive peritoneal mesothelial proliferation in the gastric serosa adjacent to two examples of gastrointestinal stromal tumor.

Chapter 28: Malignant mesothelioma and other mesothelial proliferations

Table 28.1 Summary of immunohistochemical markers in the differential diagnosis of mesothelioma and adenocarcinoma

Typically positive in adenocarcinoma and negative in mesothelioma	Mesothelioma markers typically positive in mesothelioma and usually negative in adenocarcinoma	Other tumors positive for mesothelioma markers (shown on the same line)
CEA	Calretinin	Synovial sarcoma, small cell carcinoma, sex-cord-stromal tumors of ovary
CD15 (LeuM1)	Keratins 5/6	Squamous cell carcinoma, myoepithelial carcinoma
Ber-EP4	Podoplanin (D2–40)	Some angiosarcomas, seminoma
MOC31	Thrombomodulin (CD141)	Squamous cell and urothelial carcinomas
TTF1 (pulmonary adenocarcinoma)	WT1	Serous ovarian/peritoneal and some other müllerian duct carcinomas

Note that each marker can occasionally be positive in either tumor. Data from references 66–79.

Table 28.2 Other markers usually not expressed in mesothelioma but positive in certain types of carcinomas

p63	Especially in squamous cell carcinoma
Estrogen receptor	Many breast, ovarian, and endometrial carcinomas
Renal Ca marker	Renal cell carcinoma

Data from references 78, 79, and 94.

linear strands of isolated mesothelial proliferation spaced by solid areas of collagen indicate reactive proliferation.[91]

Mesothelioma versus primary peritoneal or metastatic serous papillary carcinoma

Serous carcinomas of ovarian or primary peritoneal origin can create extensive and diffuse peritoneal coating, similar to mesothelioma. The presence of numerous psammoma bodies is more typical of serous carcinoma (especially primary peritoneal), although small numbers of psammoma bodies also occur in mesothelioma (Figure 28.27).

The best immunohistochemically positive markers for serous carcinoma in this distinction are estrogen receptors, Ber-EP4, and MOC31. Calretinin and podoplanin are the best markers to support mesothelioma in this differential diagnosis; nuclear WT1 is expressed in both.[92]

Mesothelioma versus other carcinoma

Many immunohistochemical markers have been evaluated in the distinction of pulmonary adenocarcinoma and mesothelioma. These markers are summarized in Tables 28.1 and 28.2.

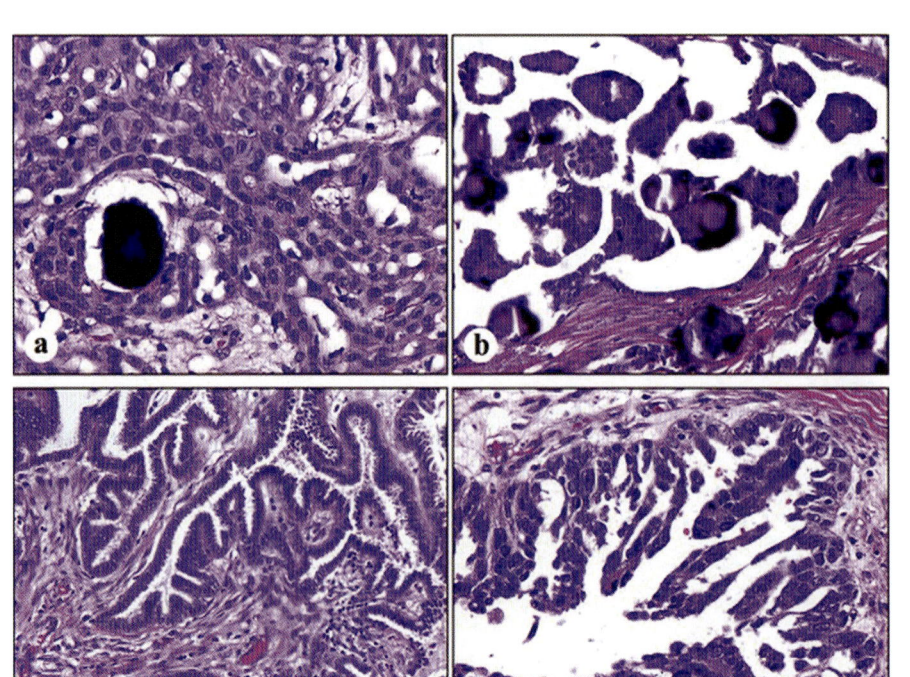

Figure 28.27 (a) Rarely, malignant mesothelioma contains psammoma bodies. (b–d) Ovarian or peritoneal serous carcinoma contains varying numbers of psammoma bodies and papillary epithelial projections. (c) A low-grade example. (d) A tumor with greater atypia and mitotic activity.

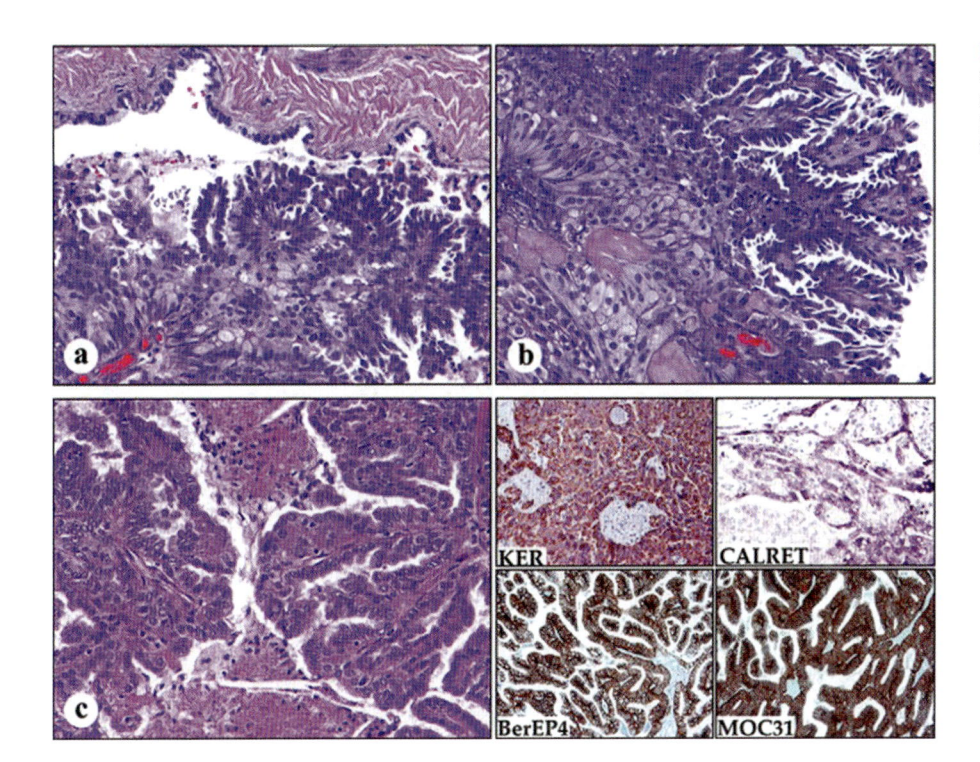

Figure 28.28 Serous carcinoma of the tunica vaginalis contains papillary projections lined by cuboid, columnar, or hobnail-like epithelium. Strong positivity for Ber-EP4 and MOC31 are features distinct from malignant mesothelioma.

The best positive markers for mesothelioma are calretinin, keratins 5 and 6, podoplanin, and WT1. Each of these can be present in some adenocarcinomas, however. The best positive markers for pulmonary adenocarcinoma are TTF1 (almost never present in mesothelioma), and for adenocarcinomas of any origin, antibodies to EpCAM (Ber-EP4 and MOC31).[73-76] Examples of the immunohistochemical evaluation of this problem are shown in Figure 28.21.

In distinguishing mesothelioma from other pulmonary carcinomas, consider the fact that other types of pulmonary carcinomas (including squamous cell carcinomas) have a higher frequency of expression for calretinin, keratins 5 and 6, and thrombomodulin.[72] One marker selectively expressed in squamous cell carcinoma versus mesothelioma is p63.[93]

In the peritoneal cavity, carcinomatosis (other than that related to serous carcinoma) can also clinically resemble diffuse mesothelioma. The histologic diagnosis is usually straightforward, however. An immunohistochemical panel using the more specific mesothelioma markers (e.g., calretinin, keratins 5 and 6, and WT1) and carcinoma markers such as EpCAM, are helpful in problem cases. Other "organ-specific" markers (e.g., estrogen receptor, CDX2) are useful on a case-by-case basis. The limitations of WT1 should be considered as carcinomas of müllerian origin are variably positive.

In the scrotum, mesothelioma should be separated from paratesticular serous papillary carcinoma.[94] These tumors show some resemblance to their female analogs. Histologically, they differ from mesothelioma by a complex, well-developed, branching papillary pattern of tall, columnar epithelial cells (Figure 28.28). Immunohistochemically they are positive for epithelial markers such as Ber-EP4 and MOC31, and focal calretinin positivity is possible.

Mesothelioma versus primary mesenchymal tumors

The most common mesenchymal tumor that creates confusion in the differential diagnosis of mesothelioma is the solitary fibrous tumor (see Chapter 12). These tumors occur both in the pleura and the peritoneal cavity. They can contain entrapped epithelial components (especially when involving the lung), but otherwise are composed of generally keratin-negative and CD34-positive mesenchymal cells. However, one should consider that malignant solitary fibrous tumors can be focally keratin positive.

Pleuropulmonary metastases of synovial sarcoma may be confused with mesothelioma unless appropriate clinical correlation is known. Their usually monophasic histology in metastases with possible round cell patterns is typical, and keratin positivity is often, although not invariably, detected. Calretinin expression is fairly common in synovial sarcoma and should not lead to confusion with mesothelioma. In problem cases, molecular genetic testing for synovial sarcoma gene fusions is helpful (see Chapter 27).

Mesothelioma versus vascular tumors

The rare epithelioid vascular tumors involving the pleura and peritoneum, especially diffuse angiosarcoma, can clinically and histologically simulate a diffuse mesothelioma. Their diagnosis is based on the immunohistochemistry of endothelial markers, CD31 consistently and CD34 variably, strong positivity for vimentin and limited or variable positivity for keratins.[95,96]

Calretinin positivity supports the diagnosis of mesothelioma as opposed to a vascular tumor. Vimentin expression does not reliably discriminate between mesothelioma and carcinoma, although the former is more commonly positive.

Genetics

Mesotheliomas commonly have losses in the NF2 gene, which may play a role in pathogenesis.[97] Recently, loss-of-function mutations in the BAP1 gene encoding BRACA1-associated protein have been identified as a pathogenetic factor.[98] Germ-line mutations of BAP1 predispose to mesothelioma and some other cancers, such as melanoma and renal carcinoma.[99] Immunohistochemical loss of BAP1 protein is a common, but not universal feature of malignant mesothelioma and a promising tool to screen for functional BAP gene aberrations (Figure 28.29).

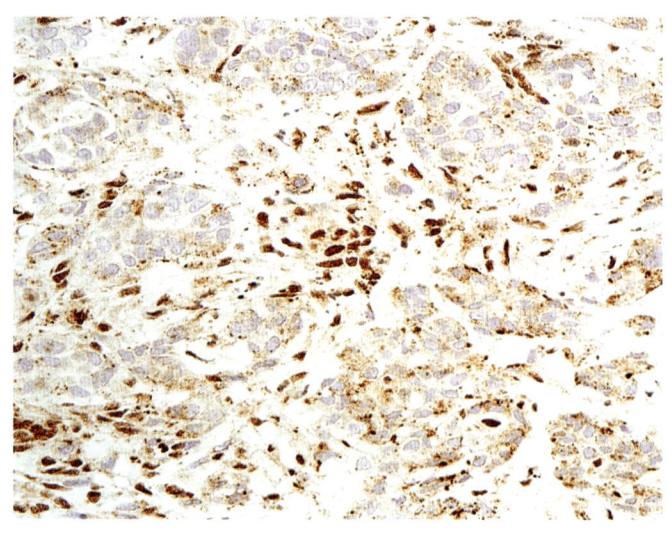

Figure 28.29 Immunohistochemical loss of nuclear BAP1 expression can give insight into the occurrence of BAP1 mutations.

References

Adenomatoid tumor

1. Huang CC, Chang DY, Chen CK, Chou YY, Huang SC. Adenomatoid tumor of the female genital tract. *Int J Gynecol Pathol* 1995;50:275–280.
2. Nogales FF, Isaac MA, Hardisson D, et al. Adenomatoid tumor of the uterus: an analysis of 60 cases. *Int J Gynecol Pathol* 2001;21:31–40.
3. Cheng CL, Wee A. Diffuse uterine adenomatoid tumor in an immunosuppressed renal transplant recipient. *Int J Gynecol Pathol* 2003;22:198–201.
4. Zamecnik M, Gomolcak P. Composite multicystic mesothelioma and adenomatoid tumor of the ovary: additional observation suggesting common histogenesis of both lesions. *Cesk Patol* 2000;38:160–162.
5. Broth G, Bullock WK, Morrow J. Epididymal tumors: 1. Report of 15 new cases including review of literature. 2: Histochemical study of the so-called adenomatoid tumor. *J Urol* 1968;100:530–536.
6. Yasuma T, Saito S. Adenomatoid tumor of the male genital tract and a pathological study of eight cases and review of the literature. *Acta Pathol Jpn* 1980;30:883–906.
7. Barry P, Chan KG, Hsu J, Quek ML. Adenomatoid tumor of the tunica albuginea. *Int J Urol* 2005;12:516–518.
8. Raaf HN, Grant LD, Santoscoy C, Levin HS, Abdul-Karin FW. Adenomatoid tumor of the adrenal gland: a report of four new cases and a review of the literature. *Mod Pathol* 1996;9:1046–1051.
9. Isotalo PA, Keeney GL, Sebo TJ, Riehle DL, Cheville JC. Adenomatoid tumor of the adrenal gland: a clinicopathologic study of five cases and review of the literature. *Am J Surg Pathol* 2003;27:969–977.
10. Craig JR, Hart WR. Extragenital adenomatoid tumor: evidence for the mesothelial theory of origin. *Cancer* 1979;43:1678–1681.
11. Overstreet K, Wixom C, Shabaik A, Bouvet M, Herndier B. Adenomatoid tumor of the pancreas: a case report with comparison of histology and aspiration cytology. *Mod Pathol* 2003;16:613–617.
12. Plaza J, Dominguez F, Suster S. Cystic adenomatoid tumor of the mediastinum. *Am J Surg Pathol* 2004;28:132–138.
13. Natarajan S, Luthringer DJ, Fishbein MC. Adenomatoid tumor of the heart: report of a case. *Am J Surg Pathol* 1997;21:1378–1380.
14. Isotalo PA, Nascimento AG, Trastek VF, Wold LE, Cheville JC. Extragenital adenomatoid tumor of a mediastinal lymph node. *Mayo Clin Proc* 2003;78:350–354.
15. Skinnider BF, Young RH. Infarcted adenomatoid tumor: a report of five cases of a facet of a benign neoplasm that may cause diagnostic difficulty. *Am J Surg Pathol* 2004;28:77–83.
16. Lehto VP, Miettinen M, Virtanen I. Adenomatoid tumor: immunohistochemical features suggesting a mesothelial origin. *Virchows Arch B Cell Pathol Incl Mol Pathol* 1983;42:153–159.

Multicystic peritoneal mesothelioma

17. Katsube Y, Mukai K, Silverberg SG. Cystic mesothelioma of the peritoneum: a report of five cases and review of literature. *Cancer* 1982;50:1615–1622.
18. Carpenter HA, Laskaster JR, Lee RA. Multilocular cysts of the peritoneum. *Mayo Clin Proc* 1982;57:634–638.
19. Schneider V, Partridge JR, Gutierrez F, et al. Benign cystic mesothelioma involving the female genital tract: report of four cases. *Am J Obstet Gynecol* 1983;145:355–359.
20. Weiss SW, Tavassoli FA: Multicystic mesothelioma: an analysis of pathological findings and biologic behavior in 37 cases. *Am J Surg Pathol* 1988;12:737–746.
21. Chua TC, Yan TD, Deraco M, et al. Multi-institutional experience of diffuse intra-abdominal multicystic peritoneal mesothelioma. *Br J Surg* 2011;98:60–64.
22. Groisman GM, Kerner H. Multicystic mesothelioma with endometriosis. *Acta Obstet Gynecol Scand* 1992;71:642–644.
23. Zotalis G, Nayar R, Hicks DG. Leiomyomatosis peritonealis disseminata, endometriosis, and multicystic peritoneal mesothelioma: an unusual association. *Int J Gynecol Pathol* 1998;17:178–182.

24. Kusuyama T, Fujita M. Appendiceal mucinous cystadenoma associated with pseudomyxoma peritonei and multicystic peritoneal mesothelioma: report of a case. *Surg Today* 1995;25:745–749.

25. Ball NJ, Urbanski SJ, Green FH, Kieser T. Pleural multicystic mesothelial proliferation: the so-called multicystic mesothelioma. *Am J Surg Pathol* 1990;14:375–378.

Well-differentiated papillary mesothelioma

26. Daya D, McCaughey WTE. Well-differentiated papillary mesothelioma of the peritoneum: a clinicopathologic study of 22 cases. *Cancer* 1990;65:292–296.

27. Goepel JR. Benign papillary mesothelioma of the peritoneum: a histological, histochemical and ultrastructural study of six cases. *Histopathology* 1981;5:21–30.

28. Goldblum J, Hart WR. Localized and diffuse mesotheliomas of the genital tract and peritoneum in women: a clinicopathologic study of nineteen true mesothelial neoplasms, other than adenomatoid tumors, multicystic mesotheliomas, and localized fibrous tumors. *Am J Surg Pathol* 1995;19:1124–1137.

29. Hoekstra AV, Riben MW, Frumovitz M, Liu J, Ramirez PT. Well-differentiated papillary mesothelioma of the peritoneum: a pathological analysis and review of the literature. *Gynecol Oncol* 2005;98:161–167.

30. Butnor KJ, Sporn TA, Hammar SP, Roggli VL. Well-differentiated papillary mesothelioma. *Am J Surg Pathol* 2001;25:1304–1309.

31. Malpica A, Sant'Ambrogio S, Deavers MT, Silva EG. Well-differentiated papillary mesothelioma of the female peritoneum: a clinicopathologic study of 26 cases. *Am J Surg Pathol* 2012;36:117–127.

32. Chen X, Sheng W, Wang J. Well-differentiated papillary mesothelioma: a clinicopathological and immunohistochemical study of 18 cases with additional observation. *Histopathology* 2013;62:805–813.

33. Cagle PT, Galateau-Sallé F, Hwang H, et al. Well-differentiated papillary mesothelioma with invasive foci. *Am J Surg Pathol* 2014;38:990–998.

34. Galateau-Salle F, Vignaud JM, Burke L, et al. Well-differentiated papillary mesothelioma of the pleura: a series of 24 cases. *Am J Surg Pathol* 2004;28:534–540.

Malignant mesothelioma

35. Bianchi C, Bianchi T. Malignant mesothelioma: global incidence and relationship with asbestos. *Industr Health* 2007;45:379–387.

36. Weill H, Hughes JM, Churg AM. Changing trends in US mesothelioma incidence. *Occup Environ Med* 2004;61:438–441.

37. Strauchen JA. Rarity of malignant mesothelioma prior to the widespread commercial introduction of asbestos: the Mount Sinai autopsy experience 1883–1910. *Am J Ind Med* 2011;54:467–469.

38. Wagner JC, Sleggs CA, Marchand P. Diffuse pleural mesothelioma and asbestos exposure in the northwestern Cape province. *Br J Industr Med* 1980;17:260–271.

39. Reid A, de Klerk NH, Magnani C, et al. Mesothelioma risk after 40 years since first exposure to asbestos: a pooled analysis. *Thorax* 2014;69:843–850.

40. McDonald JC, McDonald A. Mesothelioma and asbestos exposure. In *Malignant Mesothelioma*. Pass HI, Vogelzang NJ, Carbone M (eds.) Berlin-Heidelberg-New York: Springer; 2005.

41. Roggli VA, Sharma A, Butnor KJ, Sporn T, Vollmer RT. Malignant mesothelioma and occupational exposure to asbestos: a clinicopathologic correlation of 1445 cases. *Ultrastruct Pathol* 2002;26:55–65.

42. Carbone M, Pass HI, Rizzo P, et al. Simian virus 40-like DNA sequences human pleural mesothelioma. *Oncogene* 1994;9:1781–1790.

43. Strickler HD, Goedert JJ, Fleming M, et al. Simian virus 40 and pleural mesothelioma in humans. *Cancer Epidemiol Biomarkers Prev* 1996:5:473–475.

44. Cavazza A, Travis LB, Travis WD, et al. Post-irradiation malignant mesothelioma. *Cancer* 1996;77:1379–1385.

45. Adams VI, Unni KK, Muhn JR, et al. Diffuse malignant mesothelioma of pleura: diagnosis and survival in 92 cases. *Cancer* 1986;58:1540–1551.

46. Antman K, Shemin R, Ryan L, et al. Malignant mesothelioma: prognostic variables in a registry of 180 patients, the Dana-Farber Cancer Institute and Brigham and Women's Hospital experience over two decades, 1965–1985. *J Clin Oncol* 1988;6:147–153.

47. Ruffie P, Feld R, Minkin S, et al. Diffuse malignant mesothelioma of the pleura in Ontario and Quebec: a retrospective study of 332 patients. *J Clin Oncol* 1989;7:1157–1168.

48. Chahinian AP. Clinical presentation and natural history of mesothelioma: pleural and pericardial. In *Malignant Mesothelioma*. Pass HI, Vogelzang NJ, Carbone M (eds.) Berlin-Heidelberg-New York: Springer; 2005.

49. Sugarbaker DJ, Richards WG, Bueno R. Extrapleural pneumonectomy in the treatment of epithelioid malignant pleural mesothelioma: novel prognostic implications of combined N1 and N2 nodal involvement based on experience in 529 patients. *Ann Surg* 2014;260:577–580.

50. Pastan I, Hassan R. Discovery of mesothelin and exploiting it as a target for immunotherapy. *Cancer Res* 2014;74:2907–2912.

51. Kerrigan SAJ, Turnnir RT, Clement PB, Young RH, Churg A. Diffuse malignant epithelial mesotheliomas of the peritoneum in women: a clinicopathologic study of 25 cases. *Cancer* 2002;94:378–385.

52. Baker PM, Clement PB, Young RH. Malignant peritoneal mesothelioma in women: a study of 75 cases with emphasis on their morphologic spectrum and differential diagnosis. *Am J Clin Pathol* 2005;123:724–737.

53. Hassan R, Alexander R, Antman K, et al. Current treatment options and biology of peritoneal mesothelioma: meeting summary of the first NIH peritoneal mesothelioma conference. *Ann Oncol* 2006;17:1615–1619.

54. Schaub NP, Alimchandani M, Quezado M, et al. A novel nomogram for peritoneal mesothelioma predicts survival. *Ann Surg Oncol* 2013;20:555–561.

55. Jones MA, Young RH, Scully RA. Malignant mesothelioma of the tunica vaginalis: a clinicopathologic analysis of 11 cases with review of the literature. *Am J Surg Pathol* 1995;19:815–825.

56. Churg A. Paratesticular mesothelial proliferations. *Semin Diag Pathol* 2003;20:272–278.

57. Crotty TB, Myers JL, Katzenstein AL, et al. Localized malignant mesothelioma: a clinicopathologic and flow cytometric study. *Am J Surg Pathol* 1994;18:357–363.

58. Allen TG, Cagle PT, Churg AM, et al. Localized malignant mesothelioma. *Am J Surg Pathol* 2005;29:866–873.

59. Talerman A, Chilcote RR, Montero JR, Okagaki T. Diffuse malignant peritoneal mesothelioma in a 13-year-old girl. *Am J Surg Pathol* 1985;9:73–80.

60. Nascimento AG, Keeney GL, Fletcher CDM. Deciduoid peritoneal mesothelioma: an unusual phenotype affecting young females. *Am J Surg Pathol* 1994;18:439–445.

61. Shanks JH, Harris M, Banerjee SS, et al. Mesotheliomas with deciduoid morphology: a morphologic spectrum and a variant not confined to young females. *Am J Surg Pathol* 2000;24:285–294.

62. Klebe S, Mahar A, Henderson DW, Roggli VL. Malignant mesothelioma with heterologous elements: clinicopathologic correlation of 27 cases and literature review. *Mod Pathol* 2008;21:1084–1094.

63. Larsen BT, Klein JR, Hornychová H, et al. Diffuse intrapulmonary malignant mesothelioma masquerading as interstitial lung disease: a distinctive variant of mesothelioma. *Am J Surg Pathol* 2013;37:1555–1564.

64. Cantin R, Al-Jabi M, McCaughey WTE. Desmoplastic diffuse mesothelioma. *Am J Surg Pathol* 1982;6:215–222.

65. Mangano WE, Cagle PT, Churg A, Vollmer RT, Roggli VL. The diagnosis of desmoplastic malignant mesothelioma and its distinction from fibrous pleurisy: a histologic and immunohistochemical analysis of 31 cases including p53 immunostaining. *Am J Clin Pathol* 1998;110:191–199.

66. Moll R, Dhouailly D, Sun TT. Expression of keratin 5 as a distinctive feature of epithelial and biphasic mesotheliomas: an immunohistochemical study using monoclonal antibody AE14. *Virchows Arch B Cell Pathol Incl Mol Pathol* 1989;58:129–145.

67. Wick MR, Loy T, Mills SE, Legier JF, Manivel JC. Malignant epithelioid pleural mesothelioma versus peripheral pulmonary adenocarcinoma: a histochemical, ultrastructural, and immunohistologic study of 103 cases. *Hum Pathol* 1990;21:759–766.

68. Sheibani K, Shin SS, Kezirian J, Weiss LM. Ber-Ep4 antibody as a discriminant in the differential diagnosis of malignant mesothelioma versus adenocarcinoma. *Am J Surg Pathol* 1991;15:779–784.

69. Amin KM, Litzky LA, Smythe WR, et al. Wilms tumor 1 susceptibility (WT1) gene products are selectively expressed in malignant mesothelioma. *Am J Pathol* 1995;146:344–356.

70. Gotzos V, Vogt P, Celio MR. The calcium binding protein calretinin is a selective marker for malignant pleural mesotheliomas of the epithelial type. *Pathol Res Pract* 1996;192:137–147.

71. Dei Tos AP, Doglioni C. Calretinin: a novel tool for diagnostic immunohistochemistry. *Adv Anat Pathol* 1998;5:61–66.

72. Miettinen M, Sarlomo-Rikala M. Expression of calretinin, thrombomodulin, keratin 5, and mesothelin in lung carcinomas of different types: an immunohistochemical analysis of 596 tumors in comparison with epithelioid mesotheliomas of the pleura. *Am J Surg Pathol* 2003;27:150–158.

73. Suster S, Moran CA. Applications and limitations of immunohistochemistry in the diagnosis of malignant mesothelioma. *Adv Anat Pathol* 2006;13:316–329.

74. Marchevsky AM. Application of immunohistochemistry to the diagnosis of malignant mesothelioma. *Arch Pathol Lab Med* 2008;132:397–401.

75. Betta PG, Magnani C, Bensi T, Trincheri NF, Orecchia S. Immunohistochemistry and molecular diagnostics of pleural malignant mesothelioma. *Arch Pathol Lab Med* 2012;136:253–261.

76. Ordóñez NG. Application of immunohistochemistry in the diagnosis of epithelioid mesothelioma: a review and update. *Hum Pathol* 2013;44:1–19.

77. Lucas DR, Pass HI, Madan SK, et al. Sarcomatoid mesothelioma and its histological mimics: a comparative immunohistochemical study. *Histopathology* 2003;42:270–279.

78. Hinterberger M, Reineke T, Storz M, et al. D2-40 and calretinin: a tissue microarray analysis of 341 malignant mesotheliomas with emphasis on sarcomatoid differentiation. *Mod Pathol* 2007;20:248–255.

79. Padgett DM, Cathro HP, Wick MR, Mills SE. Podoplanin is a better immunohistochemical marker for sarcomatoid mesothelioma than calretinin. *Am J Surg Pathol* 2008;32:123–127.

80. Oury TD, Hammar SP, Roggli VL. Ultrastructural features of diffuse malignant mesotheliomas. *Hum Pathol* 1998;29:1382–1392.

81. Bolen JW, Hammar SP, McNutt MA. Reactive and neoplastic serosal tissue: a light-microscopic, ultrastructural, and immunocytochemical study. *Am J Surg Pathol* 1986;10:34–47.

82. Epstein JI, Budin RE. Keratin and epithelial membrane antigen immunoreactivity in nonneoplastic fibrous pleural lesions: implications for the diagnosis of desmoplastic mesothelioma. *Hum Pathol* 1986;17:514–519.

83. Attanoos RL, Griffin A, Gibbs AR. The use of immunohistochemistry in distinguishing reactive from neoplastic mesothelium: a novel use for desmin and comparative evaluation with epithelial membrane antigen, p53, platelet-derived growth factor-receptor, P-glycoprotein and Bcl-2. *Histopathology* 2003;43:231–238.

84. Shi M, Fraire AE, Chu P, et al. Oncofetal protein IMP3, a new diagnostic biomarker to distinguish malignant mesothelioma from reactive mesothelial proliferation. *Am J Surg Pathol* 2011;35:878–882.

85. Rosai J, Dehner LP. Nodular mesothelial hyperplasia in hernia sacs: a benign reactive condition simulating a neoplastic process. *Cancer* 1975;35:165–175.

86. Ordonez NG, Ro JY, Ayasla AG. Lesions described as nodular mesothelial hyperplasia are primarily composed of histiocytes. *Am J Surg Pathol* 1998;22:285–292.

87. Clement PB, Young RH. Florid mesothelial hyperplasia associated with ovarian tumors: a potential source of error in tumor diagnosis and staging. *Int J Gynecol Pathol* 1993;12:51–58.

88. Churg A, Colby TV, Cagle P, et al. The separation of benign and malignant mesothelial proliferations. *Am J Surg Pathol* 2000;24:1183–1200.

89. Brooks JSJ, Livolsi VA, Pietra GG. Mesothelial cell inclusions in

mediastinal lymph nodes mimicking metastatic carcinoma. *Am J Clin Pathol* 1990;93:741–748.

90. Argani P, Rosai J. Hyperplastic mesothelial cells in lymph nodes: report of six cases of a benign process that can simulate metastatic involvement by mesothelioma or carcinoma. *Hum Pathol* 1998;29:339–346.

91. Lee S, Illei PB, Han JS, Epstein JI. Florid mesothelial hyperplasia of the tunica vaginalis mimicking malignant mesothelioma: a clinicopathologic study of 12 cases. *Am J Surg Pathol* 2014;38:54–59.

92. Ordonez NG. Value of immunohistochemistry in distinguishing peritoneal mesothelioma from serous carcinoma of the ovary and peritoneum: a review and update. *Adv Anat Pathol* 2006;13:16–25.

93. Ordonez NG. The diagnostic utility of immunohistochemistry in distinguishing between epithelioid mesotheliomas and squamous carcinomas of the lung: a comparative study. *Mod Pathol* 2006;19:417–428.

94. Jones MA, Young RH, Srigley JR, Scully RE. Paratesticular serous papillary carcinoma: a report of six cases. *Am J Surg Pathol* 1995;19:1359–1365.

95. Lin BT, Colby T, Gown AM, *et al*. Malignant vascular tumors of the serous membranes mimicking mesothelioma: a report of 14 cases. *Am J Surg Pathol* 1996;20:1431–1439.

96. Zhang PJ, Livolsi VA, Brooks JJ. Malignant epithelioid vascular tumors of the pleura: report of a series and literature review. *Hum Pathol* 2000;31:29–34.

97. Sekido Y. Molecular pathogenesis of malignant mesothelioma. *Carcinogenesis* 2013;34:1413–1419.

98. Yoshikawa Y, Sato A, Tsujimura T, *et al*. Frequent inactivation of the BAP1 gene in epithelioid-type malignant mesothelioma. *Cancer Sci* 2012;103:868–874.

99. Cheung M, Talarchek J, Schindeler K, *et al*. Further evidence for germline BAP1 mutations predisposing to melanoma and malignant mesothelioma. *Cancer Genet* 2013;206:206–210.

Chapter 29

Merkel cell carcinoma and metastatic and sarcomatoid carcinomas involving soft tissue

Markku Miettinen
National Cancer Institute, National Institutes of Health

A *Merkel cell carcinoma* is a primary cutaneous neuroendocrine carcinoma that often involves soft tissue, especially the subcutis. Metastatic carcinomas are common diagnostic problems, especially in determining their primary origin. When sarcomatoid, carcinomas often simulate primary soft tissue sarcomas. The discussion of carcinomas from specific sources is therefore twofold in this chapter: (1) histologic features and specific markers to evaluate the site of origin, and (2) problems related to sarcomatoid carcinomas of a particular organ site or tumor type. This chapter includes the most common carcinomas, and selected practical, well-established, and important markers are emphasized. Clinical or radiologic correlation is always necessary to confirm the primary site for a metastasis from an unknown source. Carcinosarcoma of the gynecological tract is discussed in Chapter 18.

Merkel cell carcinoma (primary neuroendocrine carcinoma of the skin, trabecular carcinoma of the skin)

Merkel cell carcinoma is a distinctive, relatively rare primary cutaneous or subcutaneous high-grade neuroendocrine carcinoma, originally reported as *trabecular carcinoma of the skin* by Toker in 1972.[1] Its ultrastructural resemblance to Merkel cells, the cutaneous neuroendocrine cells that are scattered in the basal epidermis and hair shafts, led to its being named *Merkel cell carcinoma* during the early 1980s.[2] Positivity for keratin 20 also mirrors Merkel cells, but in neurofilament positivity it differs from normal Merkel cells.[3] Therefore, the histogenetic origin of Merkel cell carcinoma from Merkel cells is not uniformly agreed on, and some authors prefer the designation of *primary neuroendocrine carcinoma of the skin*. These designations seem synonymous, because there are no other well-recognized neuroendocrine carcinomas in the skin.

Clinical features

Merkel cell carcinoma typically occurs in older adults between the ages of 70 and 90 years and is rare before the age of 60 years. Based on Surveillance Epidemiology End Results (SEER) data, the incidence might have increased during the last decades, being 0.36 per 100 000 persons in the United States in 2000. This translates to >1000 new cases annually.[4] Most patients (76%) are 65 years or older, and there is a 2:1 male predominance and strong predilection (94%) for Caucasians with rare to very rare occurrence in Asians and Africans. In the Armed Forces Institute of Pathology (AFIP) files, the median age is 72 years, and there is a 2:1 male predominance. Occurrence before the age of 30 years is exceptionally rare (Figure 29.1).

Ultraviolet (UV)-radiation-based carcinogenesis is thought to be a significant causal factor. The tumor favors sun-exposed skin, and incidence in sun-exposed sites parallels the UVB index, similar to malignant melanoma.[4] Immunosuppressed patients, especially those having had solid organ transplantation[5] or those with HIV/AIDS, have an increased risk, with a younger age of presentation.[6] Association with Merkel cell polyomavirus may be causal and links this tumor with immunosuppression.[7]

Arsenic exposure is another rare causal factor, as reported in patients with arsenic poisoning related to environmental contamination.[8] Possible postradiation etiology has also been reported.[9]

Nearly half of Merkel cell carcinomas occur in the head and neck region, especially the face and eyelids, 40% in the extremities, and 10–15% in the trunk and external genitals.[4] Salivary glands are among the more common extracutaneous sites, and isolated cases have been reported in the oral mucosa. AFIP data show an essentially similar distribution (Figure 29.2).

Cases have been reported in inguinal and axillary lymph nodes without a primary tumor elsewhere and have been considered nodal primary tumors;[10] however, an alternative explanation is a regressed or occult primary tumor. Some studies report better prognosis for such cases with an unknown (or nodal) primary.[11] Metastases often develop in regional lymph nodes and sometimes systemically, occasionally long after the primary tumor. It has been estimated that 30% of the patients die of the tumor, although a 5-year disease-specific survival rate of 75% was found in one series.[12] Younger persons and those having distally located and small tumors (<2 cm) have a better prognosis.[13] Systemic metastases commonly develop in skin, and in some cases in the liver, lungs, and brain. A small number of patients with apparent spontaneous tumor regression have been reported.[14]

Modern Soft Tissue Pathology, Second Edition, ed. Markku Miettinen. Published by Cambridge University Press. © Cambridge University Press 2016.

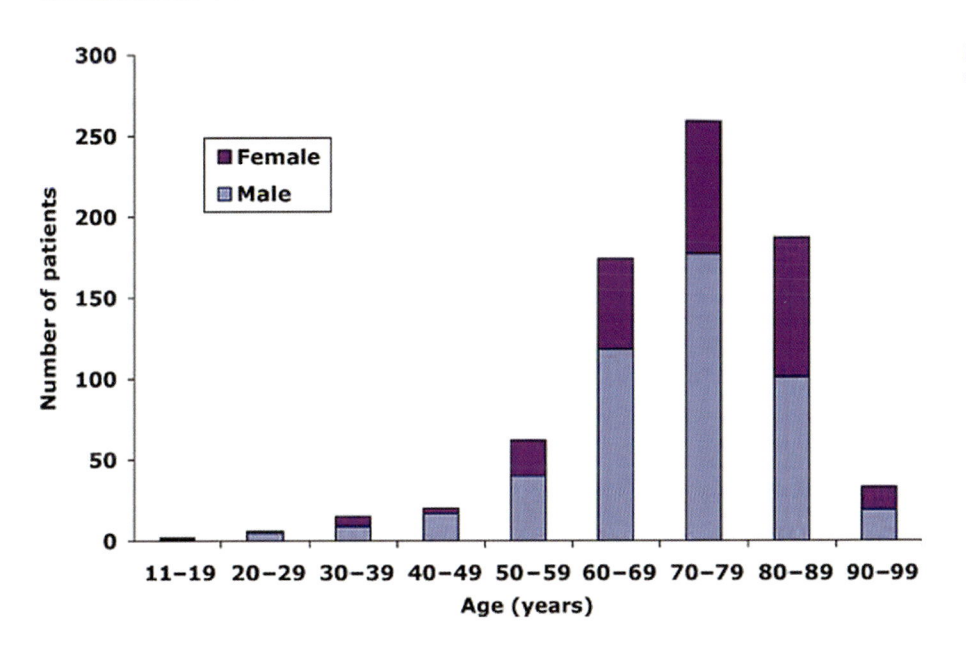

Figure 29.1 Age and sex distribution of 658 patients with Merkel cell carcinoma.

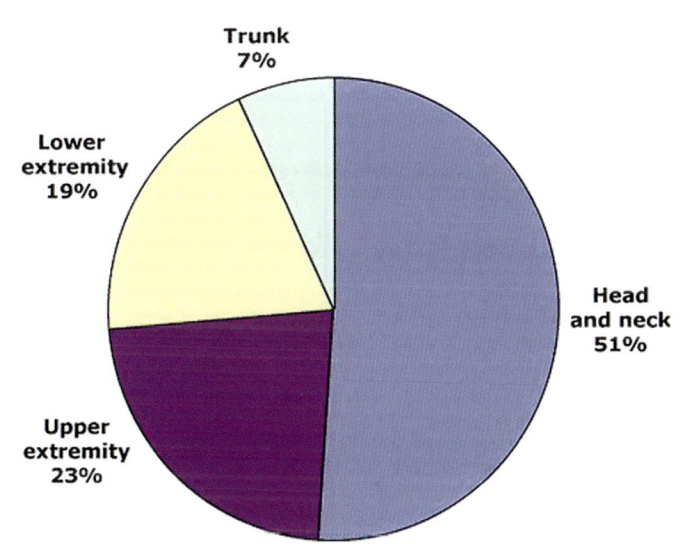

Figure 29.2 Anatomic distribution of 730 Merkel cell carcinomas.

Current treatment recommendations (2014) include preoperative biopsy, sentinel node biopsy, complete wide excision of primary tumor if possible, but tissue-saving surgery modalities in sensitive sites. Gross nodal disease or positive sentinel biopsies indicate nodal dissection or radiation, which is also applied postoperatively at the primary site.[15]

Pathology

Clinically and grossly, Merkel cell carcinoma can be difficult to distinguish from more common other skin cancers, such as basal or squamous cell carcinoma. It often forms a fleshy or purplish nodule that is sometimes ulcerated. Grossly, it typically measures 1 cm to 3 cm in diameter, but some are only a few millimeters, and others >5 cm to 10 cm, especially in the extremities and trunk. In a large series, the median diameter was 1.5 cm.[12] On sectioning, the tumor is reddish or white, and grossly it can resemble a lymphoma (Figure 29.3).

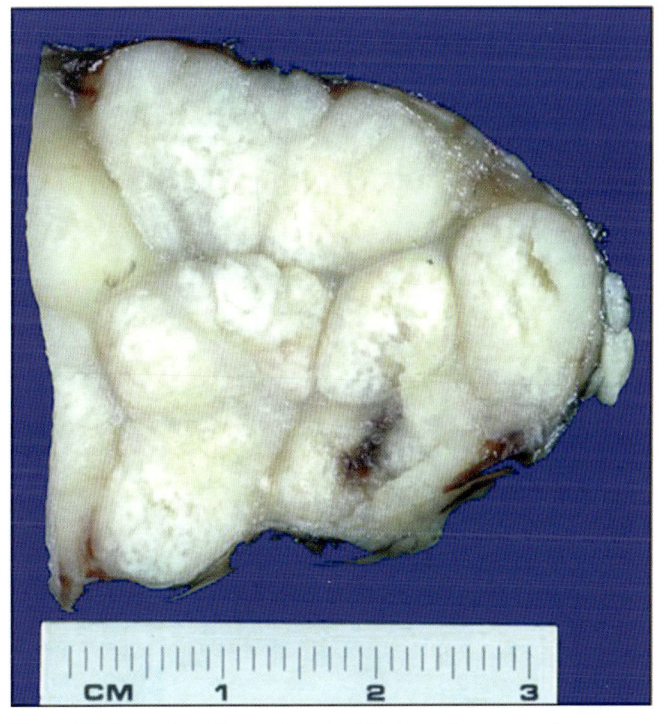

Figure 29.3 Grossly, axillary metastasis of Merkel cell carcinoma forms a multilobulated mass that is white on sectioning, but discolored by necrosis.

The tumor usually involves both the dermis and subcutaneous fat. Some cases also involve the epidermis in a pagetoid manner. Histologically, three different cell-type variants are recognized, all of which are composed of uniform cells without much nuclear variation. The most common is the *intermediate cell variant*, composed of medium-sized cells with round, vesicular nuclei (Figure 29.4). The *small cell variant*

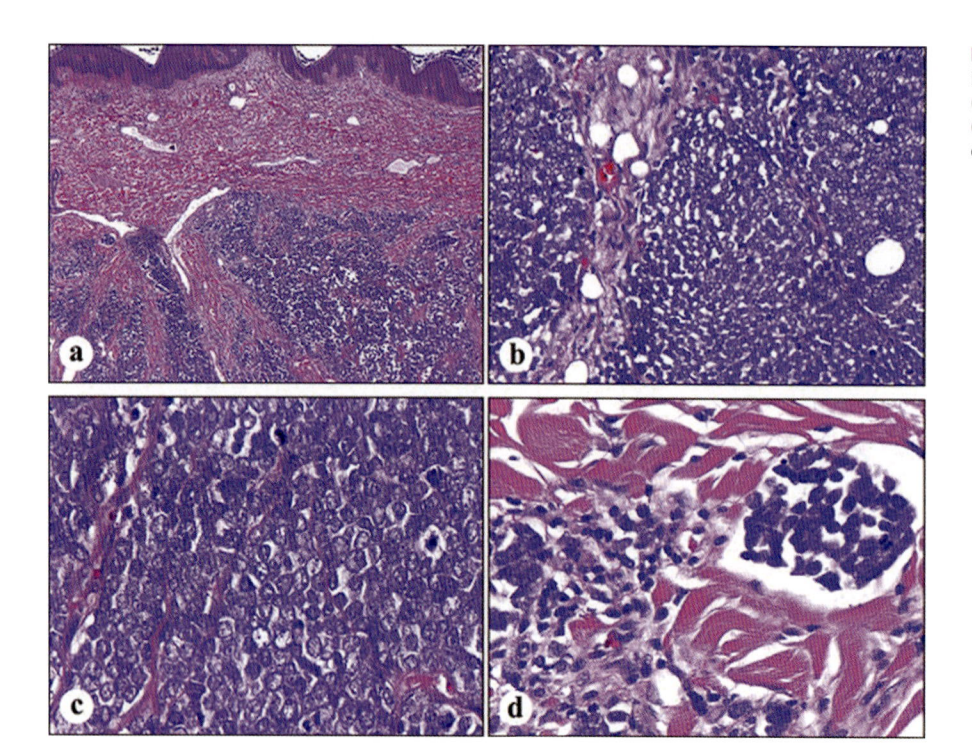

Figure 29.4 A typical Merkel cell carcinoma involving skin and extending to subcutaneous fat. (**c**) Note large, vesicular nuclei and mitotic activity. (**d**) Lymphatic invasion is not uncommon in Merkel cell carcinoma.

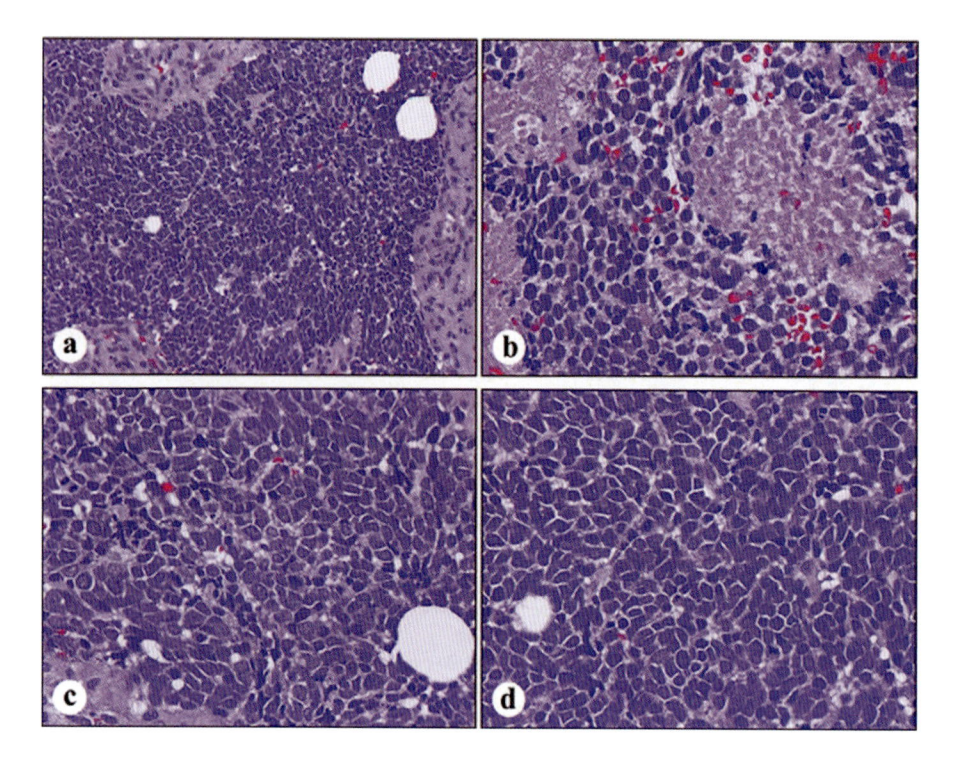

Figure 29.5 Merkel cell carcinoma resembling pulmonary small cell carcinoma. Note relatively small, uniform, often angulated nuclei. (**b**) Tumor necrosis is present. (**c,d**) The tumor infiltrates the subcutaneous fat.

resembles pulmonary small cell carcinoma, being composed of relatively small, hyperchromatic cells (Figure 29.5). Other features more typical of pulmonary tumors, such as the so-called Azzopardi phenomenon, or encrustation of basophilic material in vessel walls, can be present. The least common is the trabecular variant, which is composed of ribbons and cords, or medium-sized cells with a neuroendocrine appearance. Poorly cohesive examples can resemble a lymphoma (Figure 29.6). Rare examples contain neuroendocrine rosettes or spindle cells potentially resembling neuroblastoma (Figure 29.7).[16–18] Mitotic activity is typically high, and apoptotic bodies can resemble those commonly seen in Burkitt lymphoma. Lymphatic vascular invasion is common and an early event.[19]

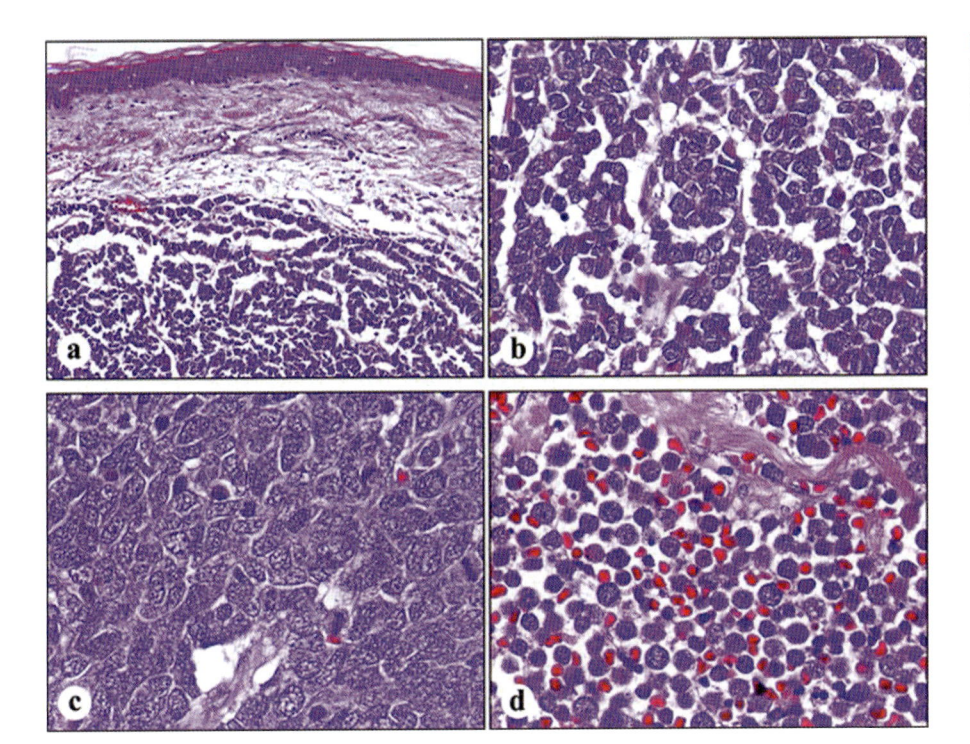

Figure 29.6 (a,b) Trabecular histological pattern in Merkel cell carcinoma. (c) A solid pattern. (d) Discohesive Merkel cell carcinoma resembling a lymphoma.

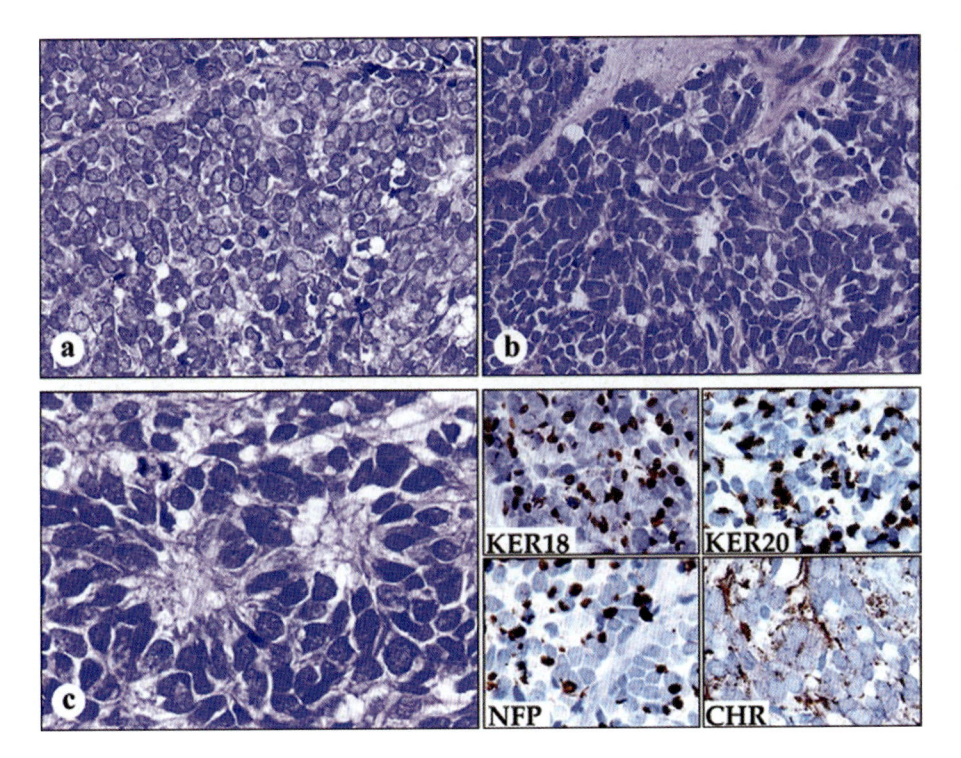

Figure 29.7 Merkel cell carcinoma resembling primitive neuroectodermal tumor of neuroblastoma. (a) The nuclei are relatively uniform and vesicular. (b,c) Rosette formation is present, and there are Homer–Wright-type rosettes with a central anuclear zone. Immunohistochemically this example is typical of Merkel cell carcinoma, with dot-like positivity for keratins 18 and 20 and neurofilament protein, as well as chromogranin positivity.

Some tumors contain a squamous cell carcinoma component, or have basal-cell carcinoma-like features.[13,16–18,20,21] Occurrence with actinic keratosis and a separate squamous cell carcinoma is also common.[20] Heterologous skeletal muscle differentiation has occurred in rare cases.[22–24]

Ultrastructurally typical is the presence of perinuclear intermediate filament whorls and scattered dense core neuroendocrine granules; detection of the latter might require more extensive search because of their low numbers.[2,3]

Polyomavirus in Merkel cell carcinoma

Viral sequences named Merkel cell polyomavirus (MCV) have been identified in Merkel cell carcinomas. The viral genome is integrated into the tumor genome in a clonal pattern.[7] Normal tissues and control skin samples contained these sequences in 8% to 16% of cases, suggesting that MCV might be a relatively common viral infection in humans.[25] There might be geographical differences, however, because Australian Merkel cell carcinoma patients have this virus less often than do European or North American Merkel cell carcinoma patients. Viral contribution to the origin of Merkel cell carcinoma could also explain the increased incidence in immunosuppressed patients.[26–28]

Immunohistochemistry

Immunohistochemically characteristic of this tumor is positivity for keratins 8 and 18, as detected with monospecific antibodies and suitable keratin cocktails (i.e., AE1/AE3), and nearly consistent, or at least focal, positivity for keratin 20. This tumor is usually negative for keratin 7, however. Many cases show punctate perinuclear positivity, whereas in some cases, the immunoreactivity is seen in the entire cytoplasm.[16,29] Positivity for neurofilaments (especially NF68 and NF160) is seen in a similar pattern in most cases, but NF200 is not present. Most tumors are positive for synaptophysin and neuron-specific enolase (NSE) and some for chromogranin (Figure 29.8). Various neuropeptides, such as pancreatic polypeptide and vasoactive intestinal peptide, have been detected in a few cases.[26] Negativity for thyroid transcription factor 1 (TTF1) helps to distinguish this tumor from metastatic pulmonary small cell carcinoma.[16,29] Merkel cell carcinoma is negative for S100 protein. One-half of all cases are positive for p63, and positivity for this marker has been found to be an adverse prognostic factor.[30] KIT (CD117) expression has been reported to be a common occurrence, although most cases show only weak or moderate positivity.[31]

Over half of Merkel cell carcinomas express PAX5 and TDT. Expression of these lymphoid markers should be recognized as a diagnostic pitfall in the differential between Merkel cell carcinoma and lymphomas.[32,33] In addition to these markers, some studies have found even immunoglobulin light chain antigens and gene rearrangements in Merkel cell carcinoma, possibly suggesting B-cell related ancestry in some cases.[34]

Differential diagnosis

Pulmonary small cell carcinoma tends to have a more prominent vascular pattern and often has more spindled and

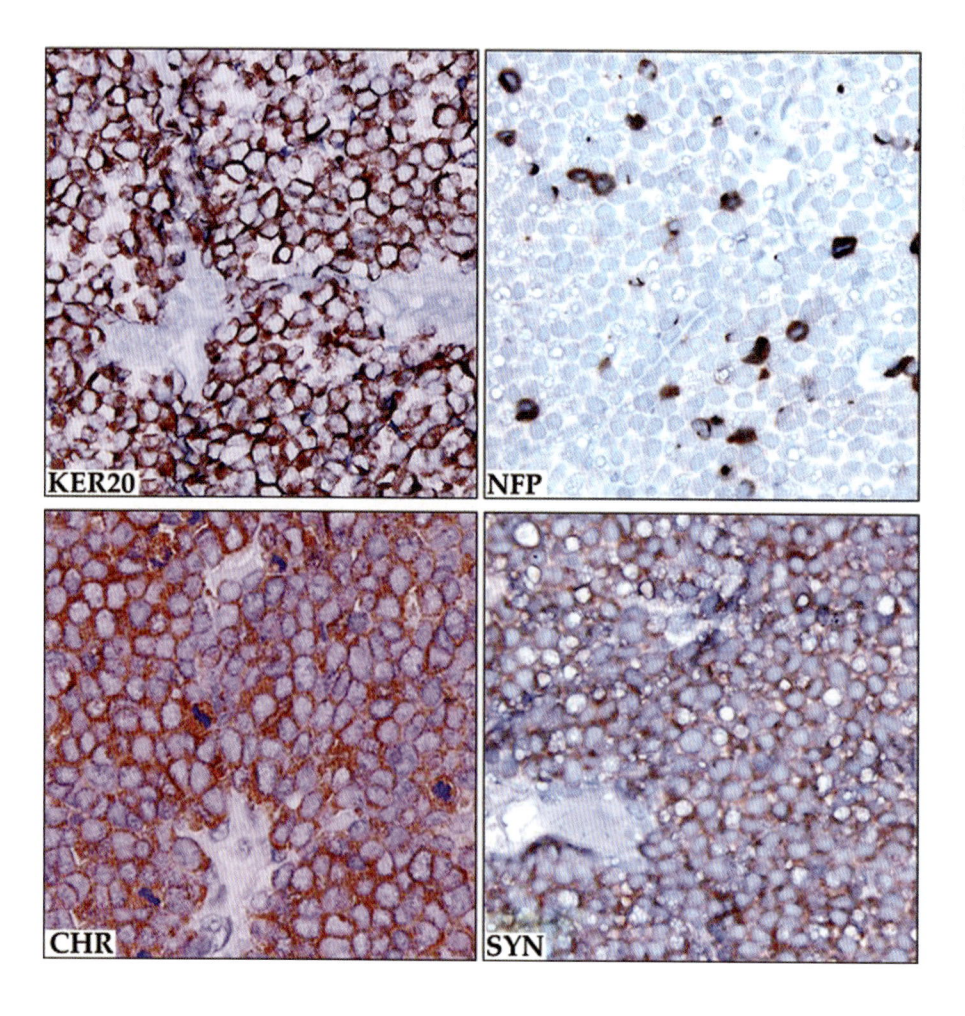

Figure 29.8 Immunohistochemical features of Merkel cell carcinoma. This example is positive for keratin 20 in a pancytoplasmic pattern, and scattered neurofilament-positive cells are present. The tumor shows strong positivity for chromogranin and is also positive for synaptophysin.

hyperchromatic nuclei. The common presence of TTF1 and the absence of K20 are features that distinguish it from Merkel cell carcinoma. Immunohistochemical evaluation for leukocyte markers CD20, CD30, and CD45 are helpful in separating Merkel cell carcinoma from cutaneous lymphomas, because Merkel cell carcinoma lacks these lymphoid markers, even though it may express lymphoid markers PAX5 and TDT.[16]

Genetics currently plays no role in the diagnosis, except for detection of polyoma virus DNA or associated antigens.[23,24]

Squamous cell carcinoma

In general, there are no specific markers to determine the primary origin of metastatic squamous cell carcinoma of unknown origin.[35] In most cases, this determination must be made by clinicoradiologic correlation. Studies of human papillomavirus 16 DNA in squamous cell carcinomas of various head and neck origins have suggested that the presence of the HPV16 genome is linked with oropharyngeal (tonsillar) origin. The presence of immunohistochemical p16 positivity is an indirect marker that correlates with HPV16 positivity and also points to an oropharyngeal origin for head and neck squamous cell carcinomas. In addition, this is observed with genital squamous cell carcinomas.[36]

Sarcomatoid squamous cell carcinomas can occur in nearly every site that ordinary squamous cell carcinomas occur. The most common sites, however, are the skin, oral cavity, larynx, and lung.[37–42]

Cutaneous sarcomatoid (spindle cell) carcinoma is usually from a squamous carcinoma and rarely a basal cell carcinoma origin. It forms a small nodule or a large tumor involving the subcutis. Some of these tumors occur in immunosuppressed patients.

Histologically, a sarcomatoid squamous cell carcinoma is often composed of spindled variably atypical cells resembling an atypical fibroxanthoma or fibrosarcoma (Figure 29.9). Some examples are composed of epithelioid cells interspersed with prominent mucoid stroma (Figure 29.10). Laryngeal sarcomatoid carcinomas can contain differentiated squamous and sarcomatoid elements in various proportions (Figure 29.11). Identification of keratin positivity (low- and high-molecular-weight [LMW, HMW] keratins) and a differentiated squamous cell carcinoma component is diagnostic. Antibodies that recognize keratins 5 and 6 have been found especially useful.[43,44] Positivity for p63 seems also to be typical.[45] Recently, antibodies to p40 have been used to the same purpose.[46]

Metastatic carcinoma of the breast

Soft tissue metastases of breast cancer are most common in the chest wall and breast, but they can occur in a wide variety of locations.

Metastatic ductal carcinomas have a broad spectrum of morphology. Their specific identification by histology only can be difficult. GATA3 has emerged the potentially most sensitive (>90%), although not specific, marker as it is expressed in urothelial carcinoma and some other carcinomas with a lower frequency.[47–49] Gross cystic disease fluid protein (GCDFP) and estrogen (ER) and progesterone (PR) receptors are other markers.[50–52] ER positivity is also frequently present in gynecologic carcinomas and rarely in some others,

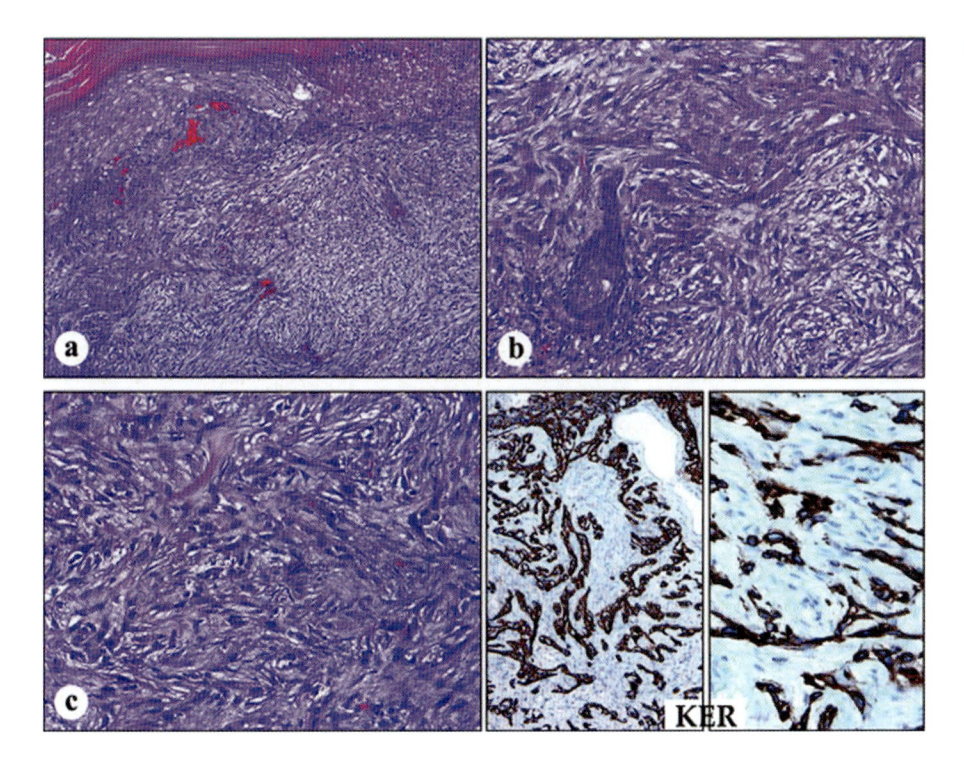

Figure 29.9 Sarcomatoid squamous cell carcinoma of the skin contains eosinophilic and pale spindled cells. The tumor shows streaks of keratin-positive cells corresponding to the streaks of eosinophilic cells.

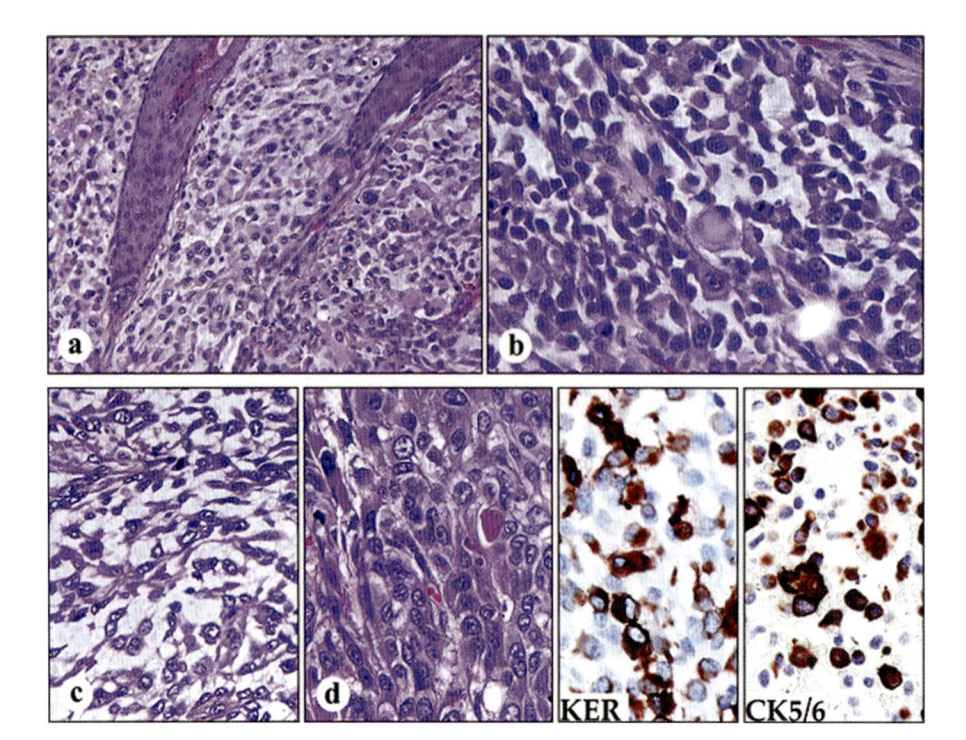

Figure 29.10 Another example of sarcomatoid squamous cell carcinoma of the skin. The tumor cells have epithelioid to spindled morphology and are set in a myxoid matrix. (**d**) A minor differentiated squamous cell carcinoma component is present. The tumor cells are positive for keratin cocktail and keratins 5 and 6.

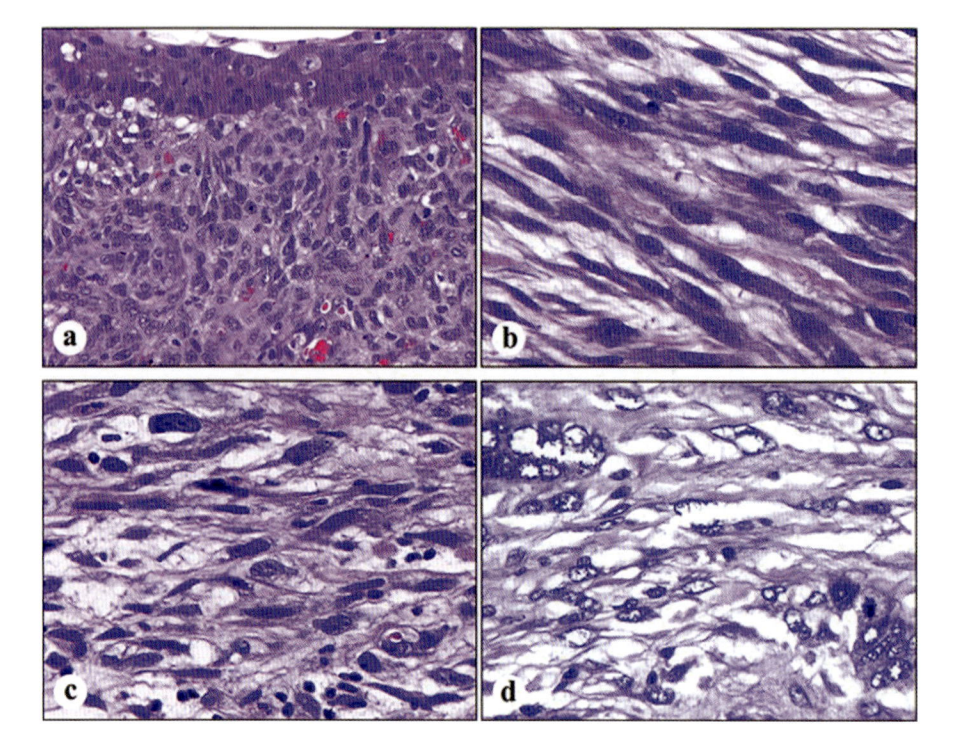

Figure 29.11 Sarcomatoid (spindle cell) carcinoma of the larynx (**a**) Solid sheets of epithelioid to spindled cells. (**b,c**) Spindle cells with variable atypia. (**d**) Spindle cells admixed with distinct epithelial clusters.

including occasional pulmonary, bladder, prostate, and even thyroid carcinoma.[53] Negative results for other site-specific markers, such as TTF1 and caudal type homeobox transcription factor 2 (CDX2), are diagnostic to rule out other cancers.

Metastatic lobular carcinoma can occur long after the primary tumor, sometimes >20 years. These tumors can involve soft tissues, especially skin and subcutis, and they have some tendency to diffuse peritoneal involvement. Bone is a common site.

Histologic features include cords of medium-sized cells in sclerosing stroma. Some examples show signet ring cell morphology. Immunohistochemically, these tumors are

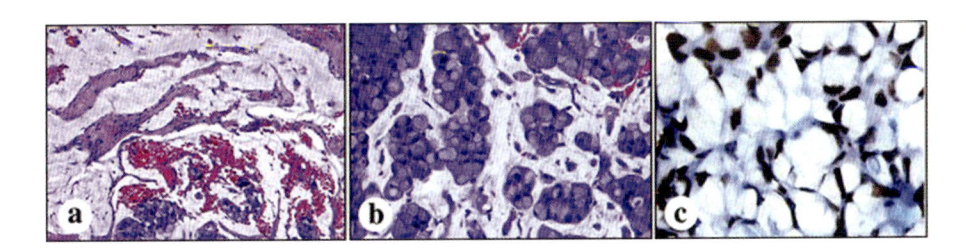

Figure 29.12 Metastatic lobular carcinoma of the breast with signet ring cell features. (c) The tumor cells are strongly positive for keratins and estrogen receptor (ER), supporting breast origin.

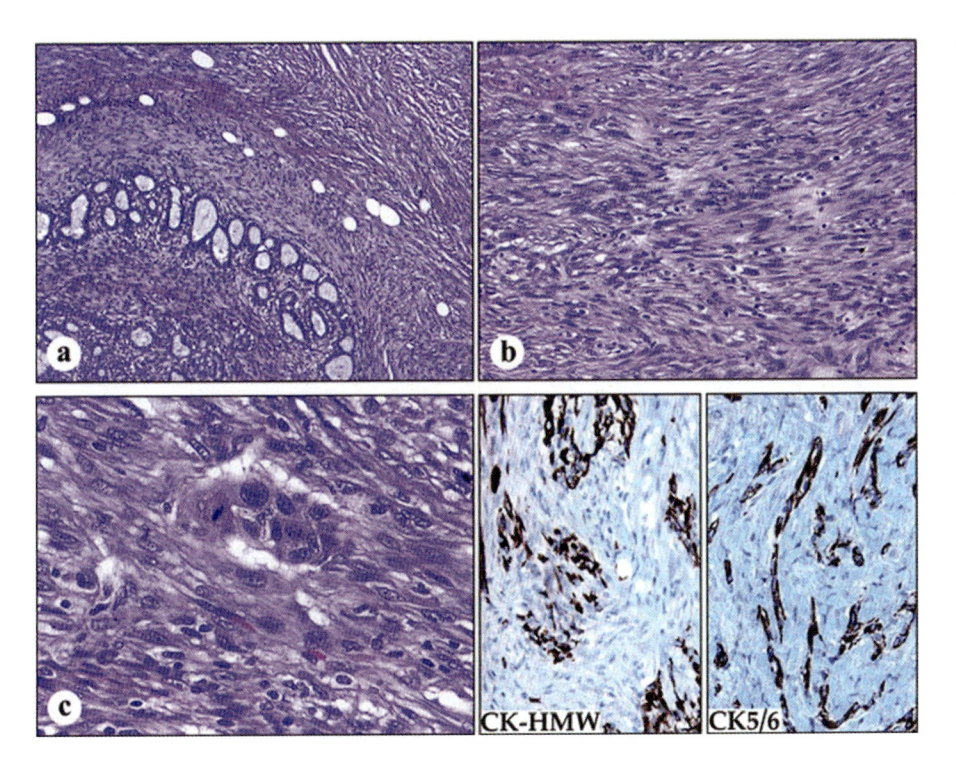

Figure 29.13 (a) Sarcomatoid carcinoma of the breast infiltrates between hyperplastic epithelial elements. (b,c) The spindle cell element contains an atypical epithelioid cell population, especially detectable at a high magnification. Streaks of sarcomatoid carcinoma cells are positive for HMW keratins, such as keratins 5 and 6.

typically positive for ERs, helping to differentiate them from signet ring cell carcinomas from other sources (Figure 29.12).

Sarcomatoid (spindle cell or metaplastic) carcinoma of the breast

Sarcomatoid carcinoma occurs in older women, with the mean ages in most series being approximately 65 years or older. The tumors vary from small to those >10 cm. They have a high rate of local recurrence, and high-grade variants have a substantial metastatic rate, although they do not usually involve the axillary lymph nodes.[54–61]

Histologically, these tumors are often difficult to identify as epithelial neoplasms unless an *in situ* or invasive ductal carcinoma component is present. Low-grade variants can resemble nodular fasciitis or fibromatosis, and high-grade variants have features simulating fibrosarcoma or malignant fibrous histiocytoma (MFH). Heterologous osteosarcomatous components can be present.

Immunohistochemistry is the key to the identification of sarcomatoid carcinoma. These tumors show streaks of keratin-positive cells, and positivity for HMW keratins, such as keratins 5 and 6, is typical (Figure 29.13). Keratin immunostaining is therefore mandatory in sarcomatoid tumors of the breast and is also recommended for low-grade spindle cell tumors. Other recommended markers are p63 or p40.

Metastatic pulmonary adenocarcinoma and large cell carcinoma

Metastatic pulmonary adenocarcinoma commonly involves the subcutis, and sometimes the deep soft tissues, especially in the chest wall region.

Immunohistochemical findings that are typically positive in these tumors include keratin 7 (in nearly all cases) and TTF1, and Napsin A can be demonstrated in 60% to 80% of cases of adenocarcinomas (Figure 29.14). This marker is superior to surfactant proteins.[62–64] TTF1 positivity in undifferentiated large cell carcinoma is less common and squamous carcinomas are almost always negative. A negative result for TTF1 does not rule out pulmonary origin, however, and in rare cases, adenocarcinomas of other than pulmonary (or thyroid) origin, for example, occasional uterine cervical adenocarcinomas, have been positive.[64]

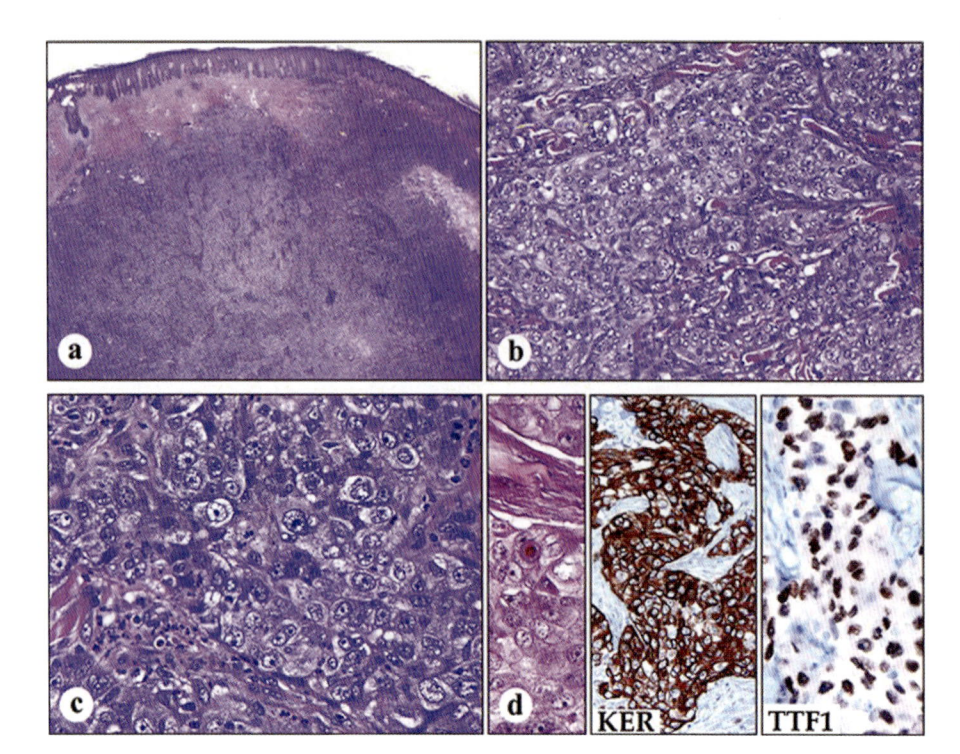

Figure 29.14 (a) Metastatic poorly differentiated pulmonary adenocarcinoma in the skin. (b,c) The tumor is composed of uniform large cells with prominent nucleoli. (d) Intraluminal mucicarmin positivity indicates adenocarcinoma type of differentiation. All tumor cells are positive for keratin cocktail, and there is strong positivity for TTF1.

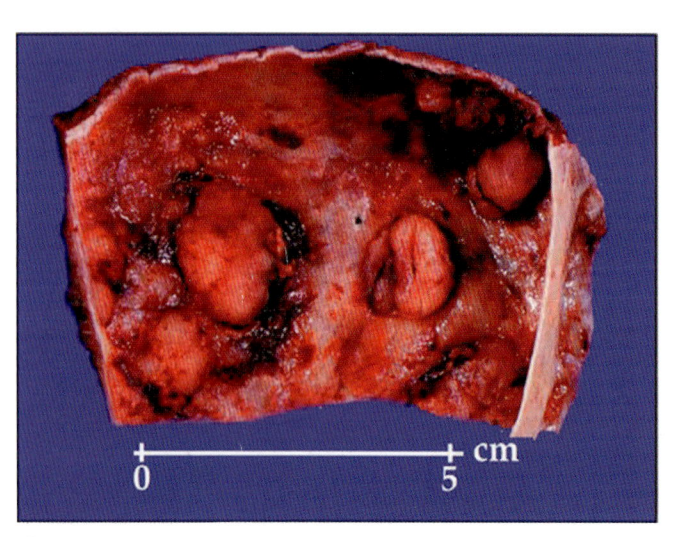

Figure 29.15 Metastatic pulmonary small cell carcinoma, forming an unusual intracranial mass involving the inner side of the dura and simulating a multifocal meningioma.

Metastatic pulmonary small cell carcinoma and other neuroendocrine carcinomas

This high-grade neuroendocrine carcinoma can be seen as a metastasis in the skin and subcutis of any location, and in such circumstances, the differential diagnosis from Merkel cell carcinoma may be problematic. Unusual soft tissue locations include subdural masses simulating meningioma (Figure 29.15).

Histologically, pulmonary small cell carcinoma usually shows more oval or spindled cell morphology, as opposed to the uniformly round and vesicular nuclei typically seen in Merkel cell carcinoma, and neural rosette-like formations can be present (Figure 29.16). Separation from small-cell carcinoma-like variants of Merkel cell carcinoma is more problematic and must be based on immunohistochemistry and clinical correlation. The most sensitive detection of keratins is by antibody to keratin 18.

Immunohistochemical features that strongly favor a pulmonary origin of small cell carcinoma is positivity for TTF1 and keratin 7 (the latter is only rarely positive in Merkel cell carcinoma). Expression of TTF1 has been reported in nonpulmonary small cell carcinomas other than Merkel cell carcinoma, however.[65,66] Pulmonary small cell carcinomas are almost always negative for keratin 20, in contrast to Merkel cell carcinoma. They are usually positive for NSE, synaptophysin, and CD56, and are variably positive for chromogranin (Figure 29.17). In one study, CD56 was found to be the most sensitive neuroendocrine marker.[67]

Carcinoid tumors of the lung are typically positive for TTF1, whereas intestinal carcinoids are negative for TTF1 and typically positive for CDX2, a transcription factor expressed in intestinal epithelia.[68]

Metastatic intra-abdominal neuroendocrine carcinomas can simulate desmoplastic small round cell tumors when forming desmoplastic stroma. These tumors are typically strongly positive for keratins and synaptophysin (Figure 29.18). They are negative for EWSR1 gene rearrangements and Wilms' tumor-suppressor gene (WT1), in contrast to desmoplastic small round cell tumor.

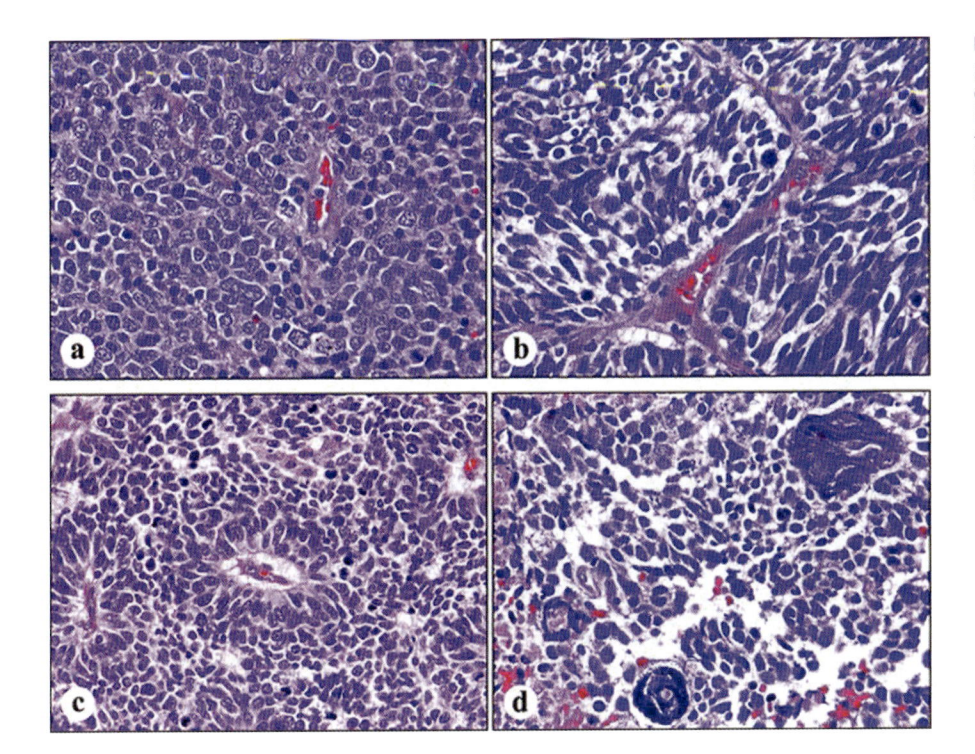

Figure 29.16 (a) Histologically, metastatic pulmonary small cell carcinoma can be composed of uniform relatively small cells. (b) Some examples have spindled morphology. (c) Perivascular rosette formation is an unusual feature. (d) Perivascular or periseptal calcification (Azzopardi phenomenon) is present in this example.

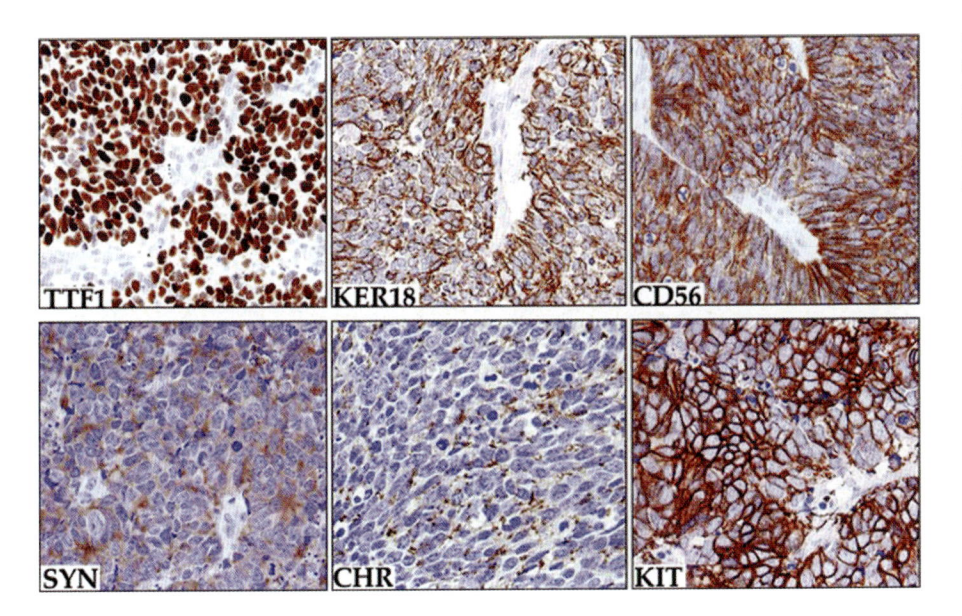

Figure 29.17 Typical immunohistochemical features of metastatic small cell carcinoma. The tumor cells are positive for TTF1, keratin 18, and CD56 in a membrane pattern. Synaptophysin positivity is typically stronger than chromogranin immunoreactivity. Approximately one-half of small cell carcinomas are KIT positive.

Sarcomatoid pulmonary carcinoma

Sarcomatoid pulmonary carcinomas can be composed of sarcomatoid elements only, or they can also contain a differentiated element of adeno-, squamous, or large cell carcinoma. As many tumors are large, prognosis is poor. However, by some studies the prognosis does not differ from differentiated carcinomas of comparable stages.[69–71] The sarcomatous elements can have spindle cell or pleomorphic morphology, and heterologous rhabdomyosarcomatous differentiation may also occur.[72] Examples with biphasic histology can also resemble biphasic malignant mesothelioma (Figure 29.19). The epithelial components are consistently and sarcomatous components variably keratin positive. TTF1 expression occurs in adenocarcinoma-related lesions.

Similar K-ras mutations in the carcinomatous and sarcomatous components of carcinosarcoma support the common clonal origin and evolution of the sarcoma from the carcinomatous element.[73]

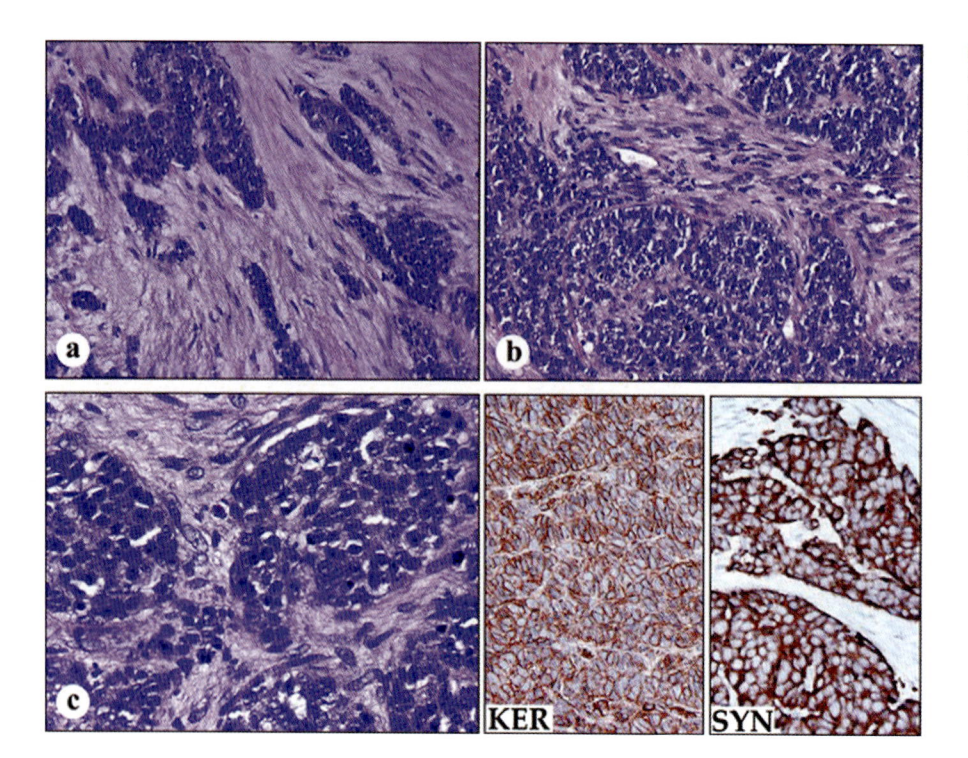

Figure 29.18 Metastatic neuroendocrine carcinoma in the abdomen resembling desmoplastic small round cell tumor. This tumor is strongly positive for keratins and synaptophysin, but negative for desmin and EWSR1 gene rearrangement.

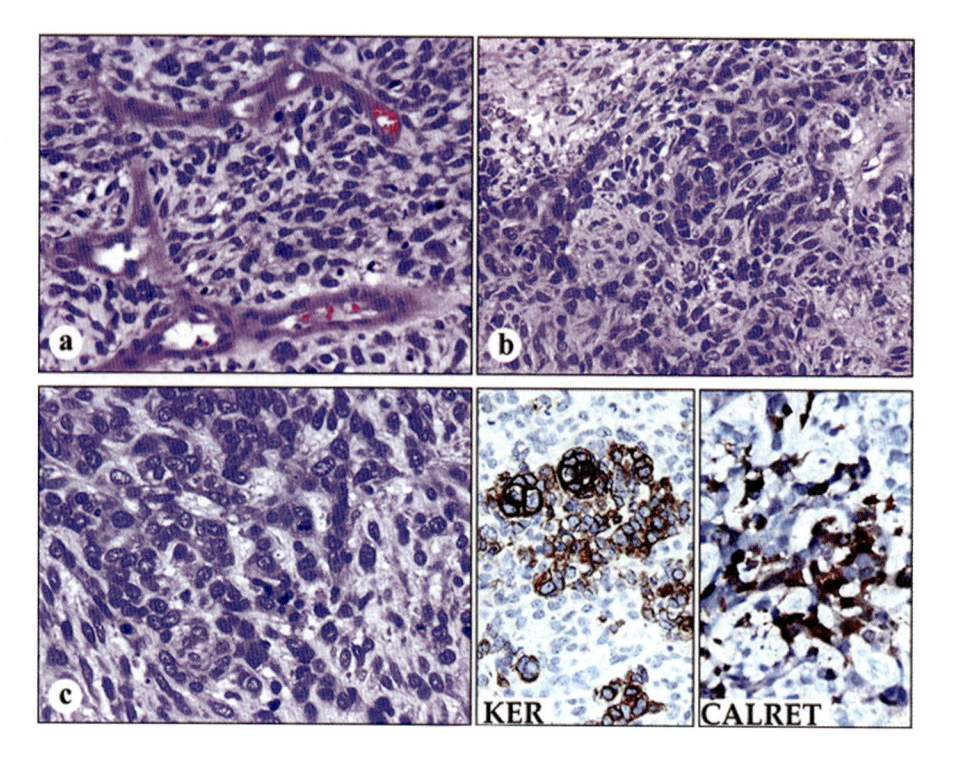

Figure 29.19 Intraparenchymal sarcomatoid carcinoma (carcinosarcoma of the lung). (**a**) Sarcomatoid element. (**b,c**) Epithelioid and sarcomatoid elements. The tumor shows keratin positivity in the epithelial components, and the epithelium is focally calretinin positive. These tumors can be difficult to distinguish from mesothelioma.

Metastatic gastrointestinal and pancreatic adenocarcinoma

The characteristic histologic appearance of colon carcinoma usually gives a strong clue to this site, although similar intestinal-type carcinomas can arise from the small intestine, ampulla, stomach, and, rarely, the nasal cavity. These tumors are typically positive for CDX2 (nuclear positivity) and keratin 20, and they are mostly negative for keratin 7.[74,75]

Gastric and gastroesophageal carcinomas (other than colon-carcinoma-like intestinal carcinomas) are heterogeneous both histologically and immunohistochemically. They are variably positive for CDX2 and keratin 20), and usually positive for keratin 7. CDX2 is also expressed in intestinal-type

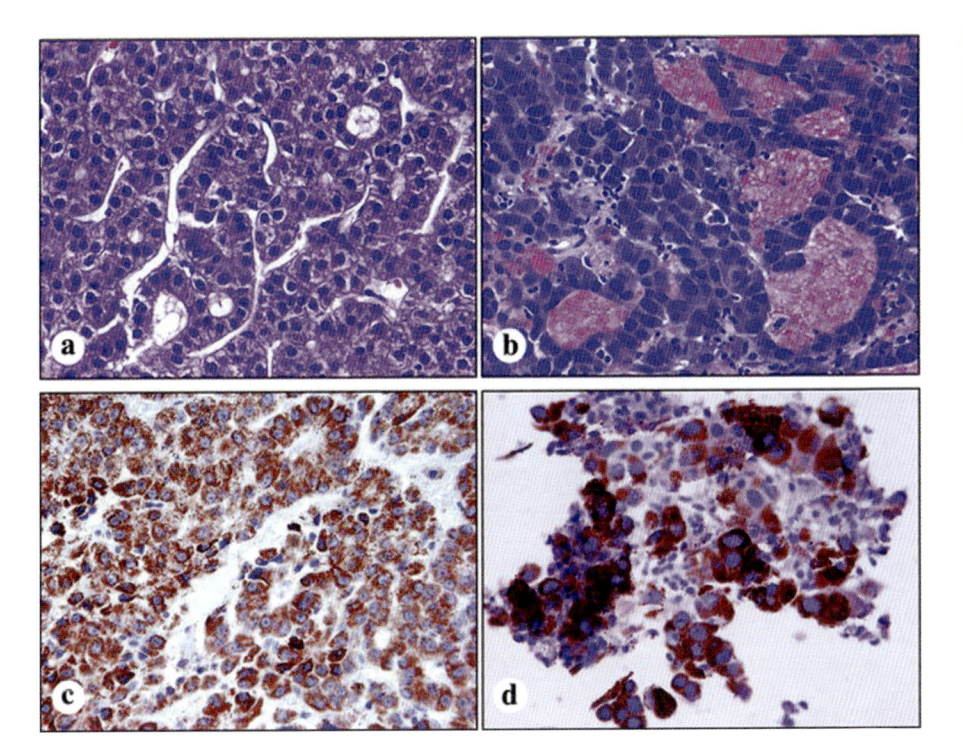

Figure 29.20 Metastatic hepatocellular carcinoma variably replicates hepatocyte-like morphology, and variably differentiated acini are present. The tumor cells are positive for HepPar1, a monoclonal antibody to a hepatocyte antigen.

carcinomas of other origins, such as the urinary bladder,[75] ovary (primary ovarian mucinous tumors),[76] cervix,[77] and nose.[78]

There are no specific markers for metastatic pancreatic adenocarcinomas. Prostate stem-cell antigen, fascin, and mesothelin are commonly expressed,[79] but their specificity is so far incompletely documented. Mesothelin is expressed in many other tumors, for example, ovarian serous carcinomas and mesotheliomas.

Hepatocellular carcinoma

Hepatocellular carcinoma can spread hematogenously into the peripheral soft tissues, bone, and skin. When sufficiently differentiated, these metastases can be specifically identified by their cytologic appearance, imitating hepatocytic differentiation: large cells with abundant eosinophilic cytoplasm, and acinar or pseudoglandular differentiation, sometimes with intraluminal bile formation.

Immunohistochemically, hepatocellular carcinomas are most consistently positive for keratins 8 and 18, and often for a hepatocyte antigen HepPar1 in a granular pattern (Figure 29.20). Although most types of adenocarcinomas are HepPar1 negative, commonly positive types include gastric and esophageal adenocarcinomas, and occasionally positive are pulmonary, pancreatic, and urinary bladder adenocarcinomas.[80,81] Differentiated hepatocellular carcinomas also can express CEA in the luminal aspect of canalicular differentiation, whereas others might show a membrane pattern.[82,83] CD10 is another canalicular marker, but it seems to be expressed in only 50% of hepatocellular carcinomas.[80] Arginase may be a more sensitive marker for hepatocellular carcinoma than the previous markers, although it has not been extensively studied in metastases.[84]

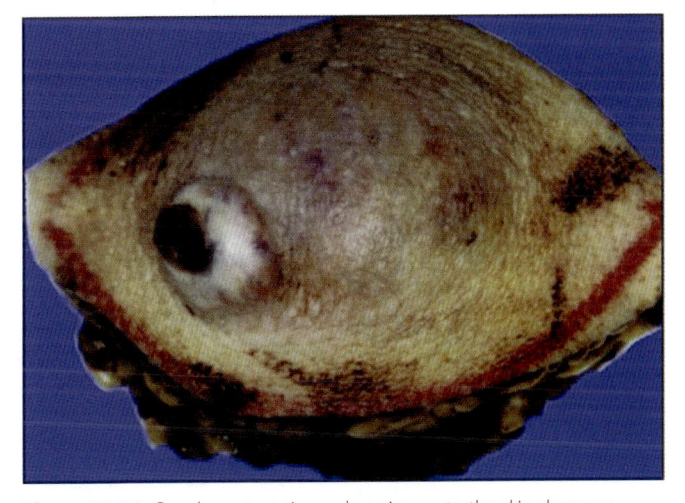

Figure 29.21 Grossly, metastatic renal carcinoma to the skin shows an elevated cutaneous lesion that contains a small umbilicated and ulcerated nodule. Courtesy of Dr. Maarit Sarlomo-Rikala, Helsinki, Finland.

Metastatic renal cell carcinoma

Renal cell carcinoma (clear cell and other variants) commonly metastasizes in the peripheral soft tissues in a wide variety of locations, and detection of metastasis before the primary tumor is not rare.[85] Metastases can occur long after primary surgery, and some patients enjoy long remission after metastasectomy.[86] The most commonly involved soft tissue sites are the female lower genital tract, scalp, and other cutaneous and subcutaneous sites (Figure 29.21), and occasionally deep soft tissues. Bones are also commonly involved. Because of their high vascularity, renal carcinoma metastases typically bleed heavily during surgery, and therefore this is often considered a clinical sign suggesting renal origin of a metastasis.

When the tumor has histologically clear cell features and a high vascularity, renal origin has to be seriously considered (Figure 29.22). Examples with more solid and eosinophilic cytoplasm can be more difficult to connect with a renal primary tumor, and examples with polygonal cytology and pleomorphism sometimes resemble epithelioid angiomyolipoma (Figure 29.23). Those soft tissue metastases that contain necrotizing-granuloma-like formations can be easily confused with primary tumors, especially epithelioid sarcomas (Figure 29.24).

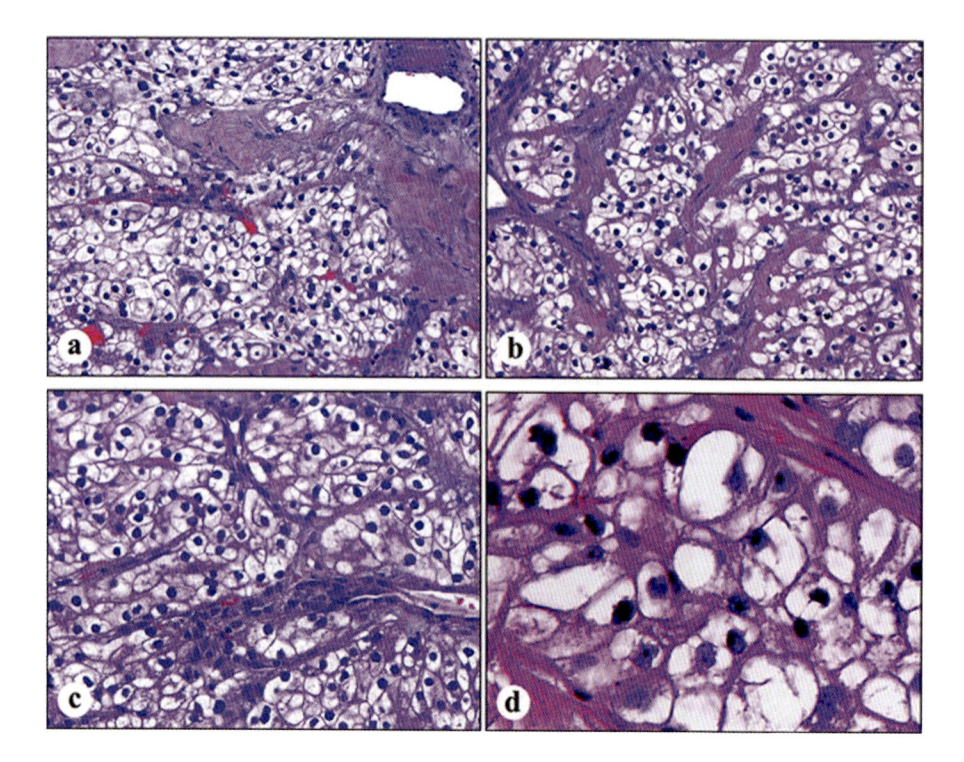

Figure 29.22 Metastatic clear cell carcinoma of the kidney with typical features.

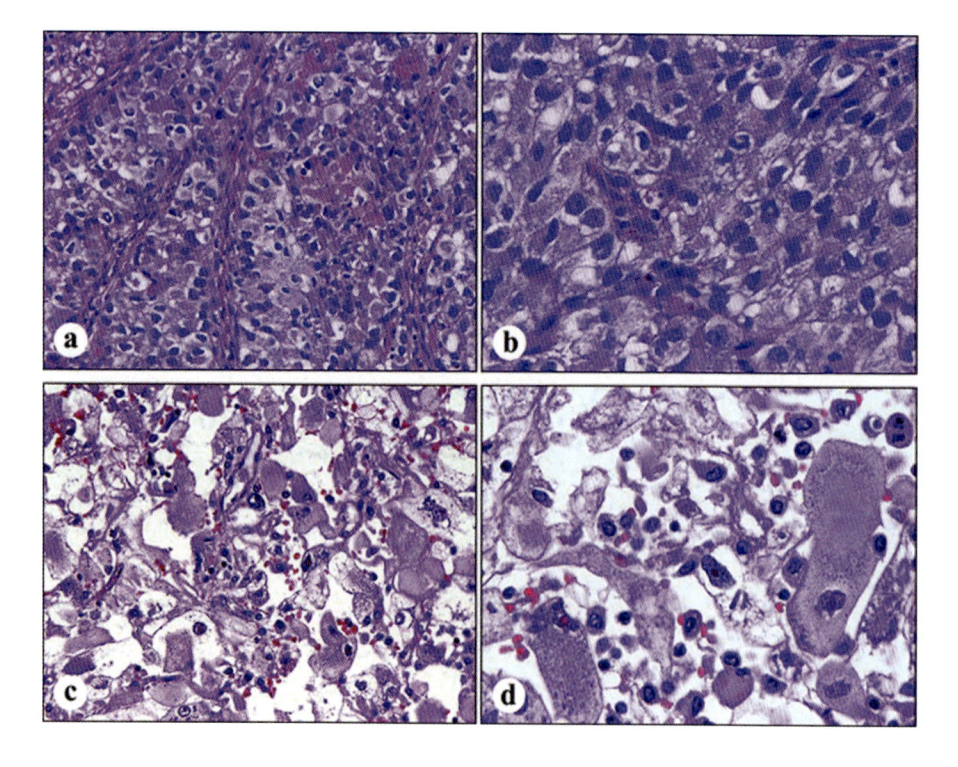

Figure 29.23 (a,b) Metastatic renal cell carcinoma with eosinophilic cytoplasm. (c,d) An unusual metastatic renal cell carcinoma in the abdomen simulating an epithelioid angiomyolipoma.

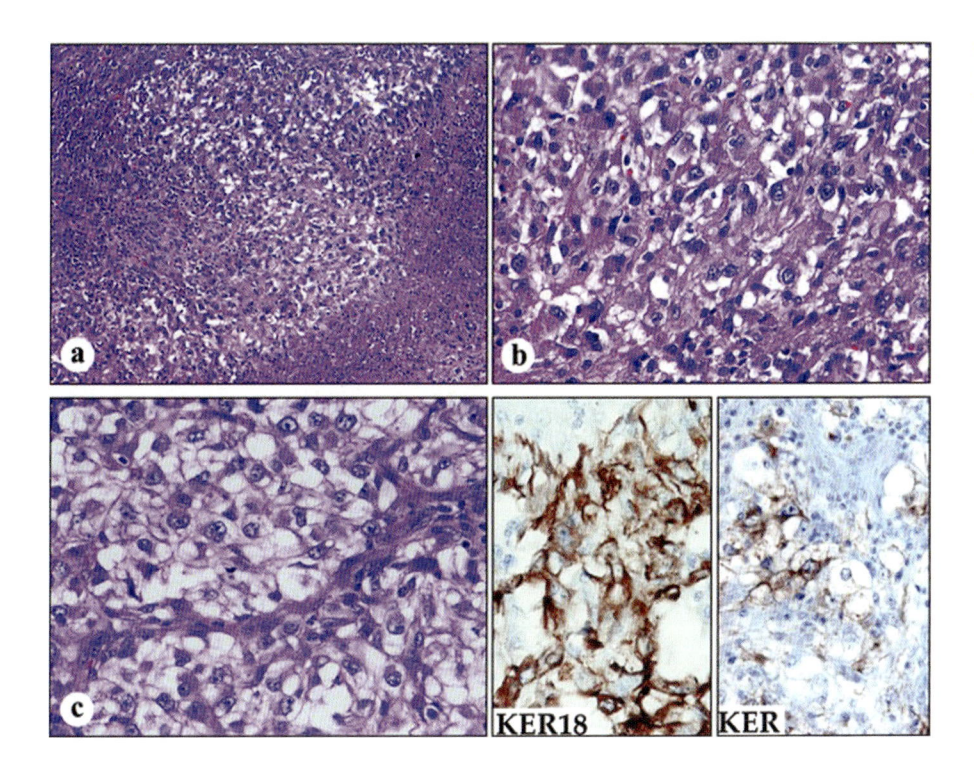

Figure 29.24 Metastatic renal cell carcinoma in the thigh with a necrotic granuloma-like formation simulating epithelioid sarcoma. Some areas of the tumor show clearer cytoplasm. The tumor cells are positive for keratin 18 and focally for keratin cocktail AE1/AE3.

Metastatic renal carcinomas are typically positive for keratins 8 and 18, EMA, variably for keratin 19, and rarely for keratin 7. Markers that can be helpful in differentiated metastases include CD10 and renal carcinoma marker (RCC); the latter marker recognizes an unknown protein (Figure 29.25). Negative results for either marker do not rule out renal cell carcinoma, however.[87–90] Especially, poorly differentiated renal carcinomas often lack RCC reactivity (a more specific marker), whereas CD10 (a less specific marker) is often conserved. A common feature for renal clear cell carcinoma is S100 protein positivity.[91] PAX2, PAX8, and human kidney injury molecule (hKIM-1) are potential new markers for renal carcinoma.[92–94] None of these is totally specific, but each is also present in ovarian clear cell carcinomas (Table 29.1).

Sarcomatoid renal cell carcinoma

Sarcomatoid morphology has been estimated to be present in 5% to 8% of renal carcinomas and these tumors are aggressive, with a high metastatic rate.[94–102] The frequency of sarcomatoid transformation can be higher in collecting duct and chromophobe carcinomas.[97,98]

Metastases with sarcomatoid features can be very difficult to link to the renal primary tumor and are more likely to be confused with sarcomas. It is prudent to clinicoradiologically rule out the possibility of occult sarcomatoid renal carcinoma before diagnosing a keratin-positive soft tissue or osseous MFH.

Extensive sampling is also recommended, because discovery of a differentiated component helps to establish the diagnosis of carcinoma.

The sarcomatoid component can be spindled or pleomorphic and intermixed with the differentiated element. Keratin positivity, especially for keratin 18, is typically present at least in a portion of tumor cells, and CD10 is also often conserved (Figure 29.26). RCC is not generally detectable in poorly differentiated examples, however. In one study, 94% of cases were positive for keratins and 50% of cases for EMA. PAX8 is detected in 70% of cases.[103] Muscle markers, especially muscle-specific actin can be expressed.[101] Heterologous liposarcomatous and osteosarcomatous components can be present,[104] sometimes intermingling with the differentiated carcinoma component, such as a chromophobe carcinoma (Figure 29.27). Extensive sampling of sarcomatoid renal tumors is therefore recommended to allow detection of minor differentiated components.

Urothelial (transitional cell) carcinoma

Urothelial tumors are most common in the urinary bladder, and less frequently they occur in the renal pelvis, ureter, and the urethra. Soft tissue metastases of urothelial carcinomas most commonly occur in the pelvis and abdominal wall. These are always high-grade tumors, and the typical expression patterns do not necessarily apply. Sarcomatoid components are not uncommon in high-grade examples.[105]

Keratins 7 and 20 are expressed in differentiated variants,[106] and keratins 5 and 6 are limited to tumors with squamous differentiation. Uroplakin 2 (UP II) is a differentiation marker thought to be specific for transitional cell carcinoma. In one study, UP II RNA was detected in 100% of bladder cancers and in 94% of nodal metastases, and it was a more

Chapter 29: Merkel cell carcinoma and metastatic and sarcomatoid carcinomas

Table 29.1 Nuclear transcription factor markers useful to determine origin for metastatic carcinomas.[47–49, 62–66, 74–78, 94, 113, 135–137]

Marker	Primary target carcinomas	Examples of additional positive tumor types (variable subsets, variably positive)
CDX2	Intestinal Colon, small intestine, stomach	Other carcinomas with intestinal differentiation (lung, ovary)
ERG	Prostate (40–50%)	Seems to be highly specific for prostate cancer Among carcinomas
GATA3	Breast, skin, and salivary duct Urothelial, choriocarcinoma	Kidney cancer, especially chromophobe mesothelioma, pancreatic cancer (minority of cases)
NKX3.1	Prostate (nearly 100%)	Breast cancer, lobular and ductal (10–30%)
PAX 8	Renal, thyroid, endometrial, cervical	Occasional thymic, pulmonary, gastric, and serous carcinoma of ovary, subsets of gastric and other cancers
SALL4	Seminoma Embryonal carcinoma Yolk sac tumor	Minor subsets of gastric, ovarian, urothelial carcinoma, and others
TTF1	Thyroid (differentiated) including medullary carcinoma Pulmonary small cell Pulmonary adenocarcinoma (60%)	Some small cell carcinomas of other organs

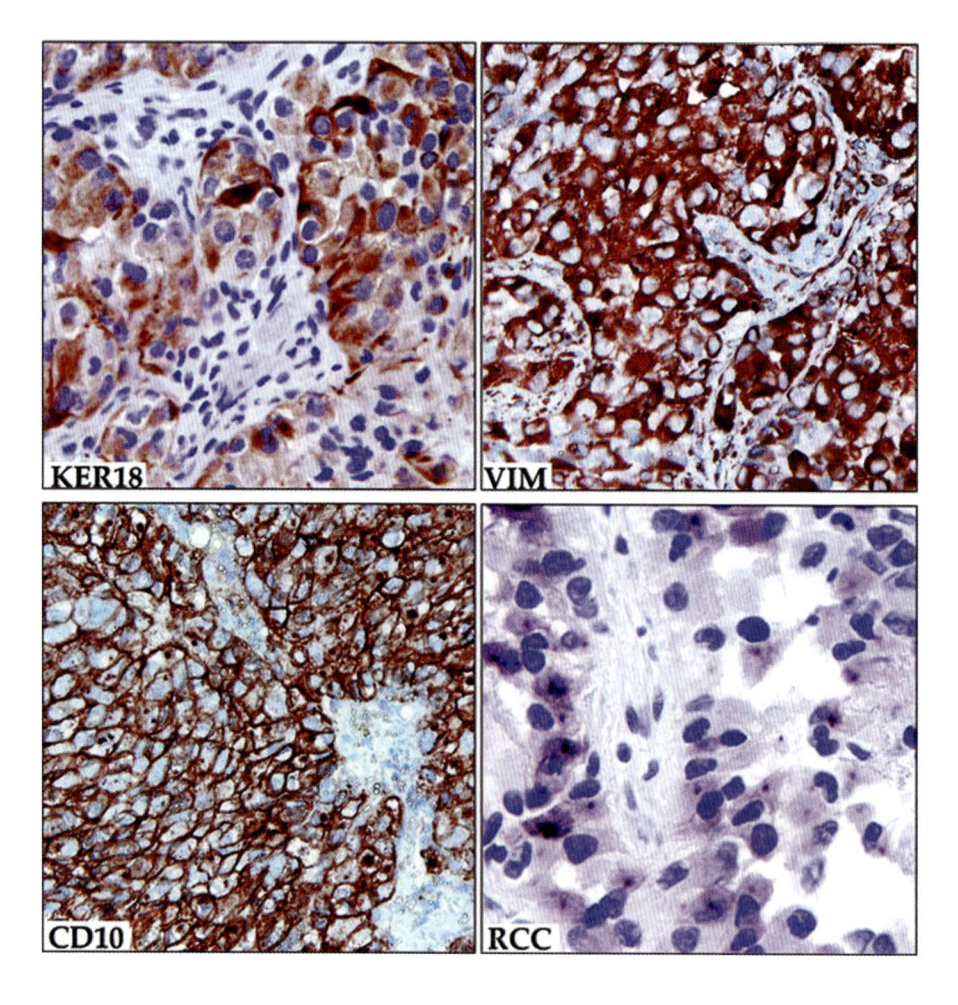

Figure 29.25 Immunohistochemical features typical of metastatic renal cell carcinoma. The tumor cells are almost always positive for keratin 18, and are often positive for vimentin and CD10. A tumor with poor luminal differentiation, as seen in this case, can contain dot-like cytoplasmic positivity for renal cell carcinoma antigen.

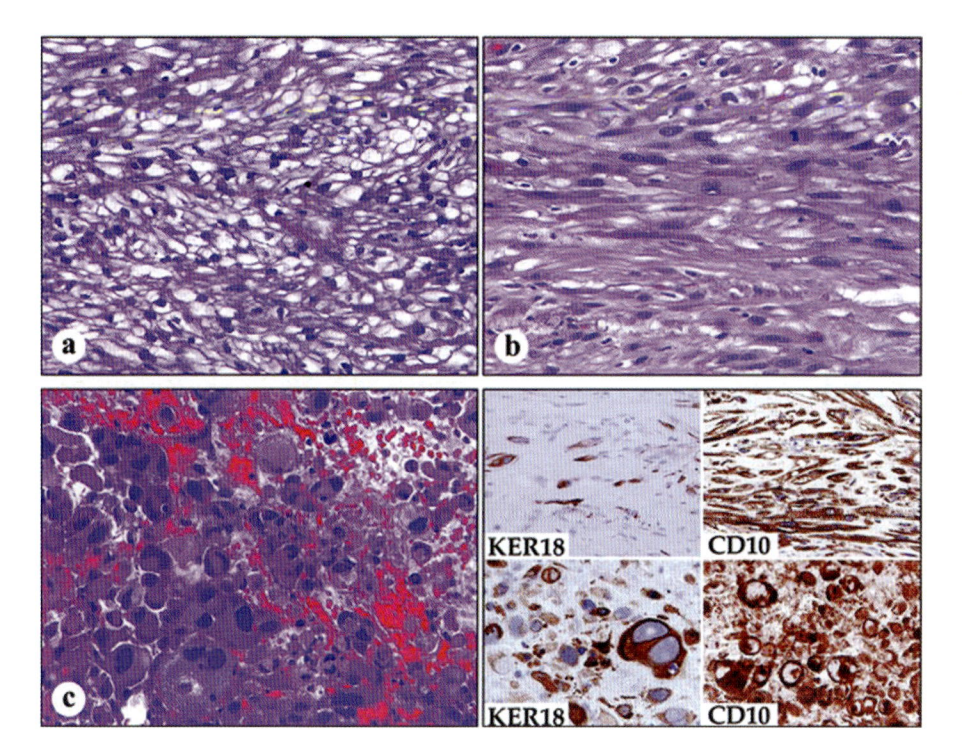

Figure 29.26 Examples of sarcomatoid renal carcinoma. (**a**) Spindle cells with a vague compartmental pattern and focal clear cell change. (**b**) Spindle cell carcinoma resembling a sarcoma. (**c**) A pleomorphic carcinoma that can resemble undifferentiated sarcoma. Immunohistochemical findings in cases shown in (**b**) and (**c**). Prominent positivity is present for both keratin 18 and CD10.

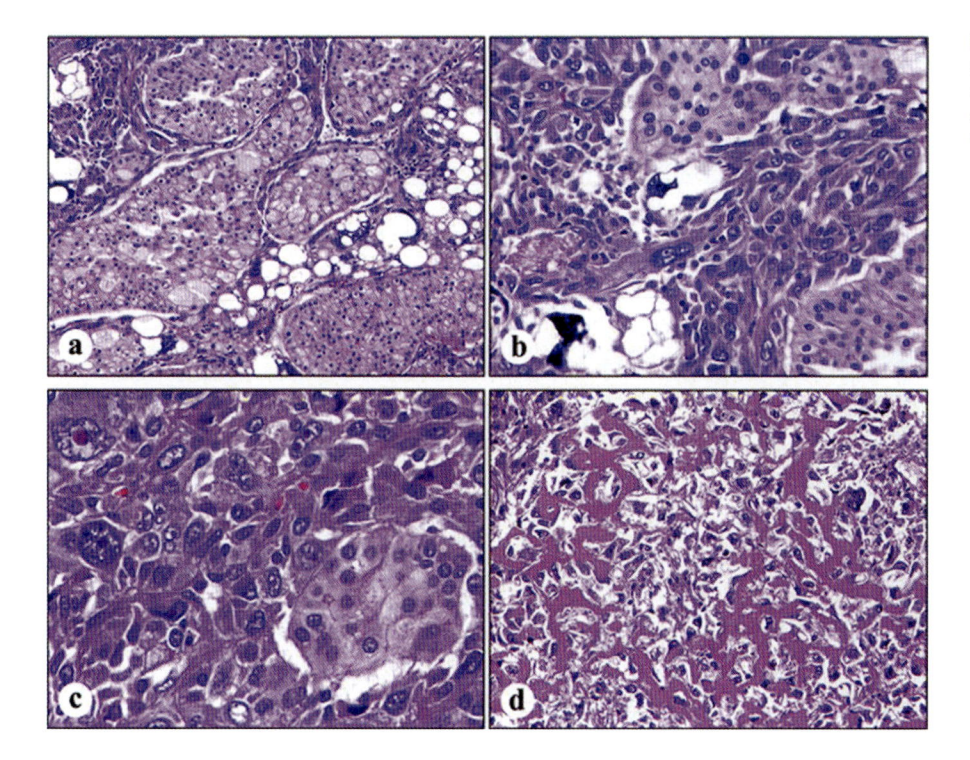

Figure 29.27 Sarcomatoid chromophobe renal carcinoma (eosinophilic variant). The chromophobe carcinoma and sarcomatoid elements are intermingled. There is liposarcoma (**a,b**) and osteosarcoma-like differentiation (**d**).

sensitive marker than keratin 20.[107] Immunohistochemical studies have shown UP expression to be reduced in muscle-invasive transitional cell carcinomas (52%), and detectable in 66% of lymph node metastases of urothelial carcinomas.[108] Therefore, their value seems only moderate in assessing the urothelial origin of distant metastases. GATA3 has emerged as a very sensitive new marker for urothelial carcinoma, and detects up to 30% of sarcomatoid urothelial carcinomas of the bladder, and PAX8 is rarely detected.[103]

Metastatic prostatic carcinoma

When extending into pelvic soft tissues by contiguous growth, or presenting as pelvic, cervical, or sometimes other lymph node metastases, prostatic carcinoma can be difficult to recognize specifically. Distant metastases can also occur in peripheral soft tissue, in some cases as a direct extension from bone metastases (e.g., from the ribs).

Histologically, many such metastases are high-grade tumors with a partially solid growth pattern, although variable acinar differentiation is often present (Figure 29.28). They are composed of large cells, often with prominent nucleoli, although this is not invariably the case. The tumor cells can be arranged in neuroendocrine-like rosette-forming patterns. Poorly differentiated examples can show rhabdoid or plasmacytoid features, yet often maintain expression of epithelial and prostatic markers (Figure 29.29).

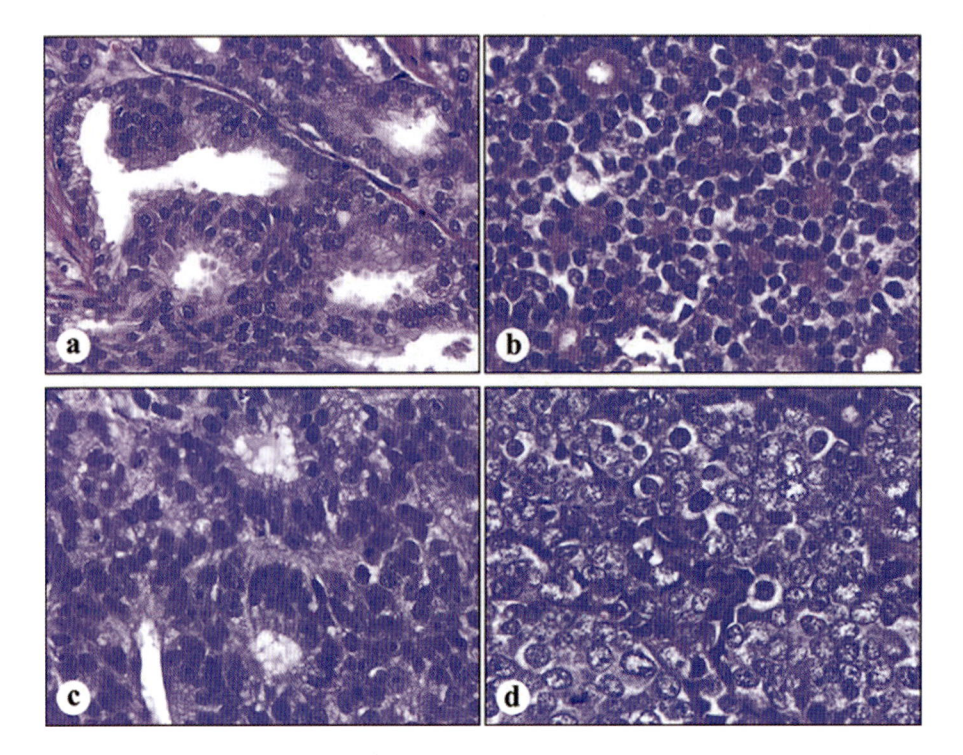

Figure 29.28 Metastatic prostate carcinoma. (**a**) Well-differentiated pelvic nodal metastases can resemble prostate epithelium. (**b**) Microacinar arrangement is common, and this case shows an eosinophilic perilumical appearance, a Paneth-cell-like feature. (**c**) Prominent microacini and large nuclei with prominent nucleoli are typical findings. (**d**) Example of a solid metastasis composed of large, uniform cells with very prominent nucleoli.

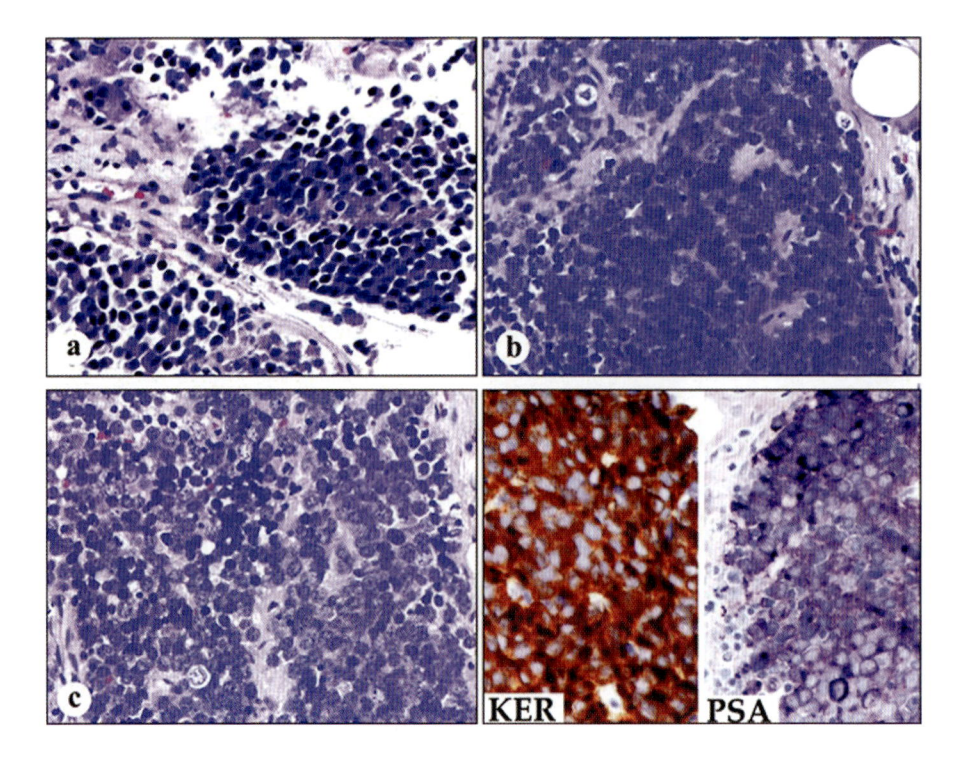

Figure 29.29 Metastatic prostate carcinoma in the pelvic soft tissues. The tumor cells contain large, round nuclei with variably prominent nucleoli. Positivity for keratin and prostate-specific antigen supports a prostatic origin for this tumor.

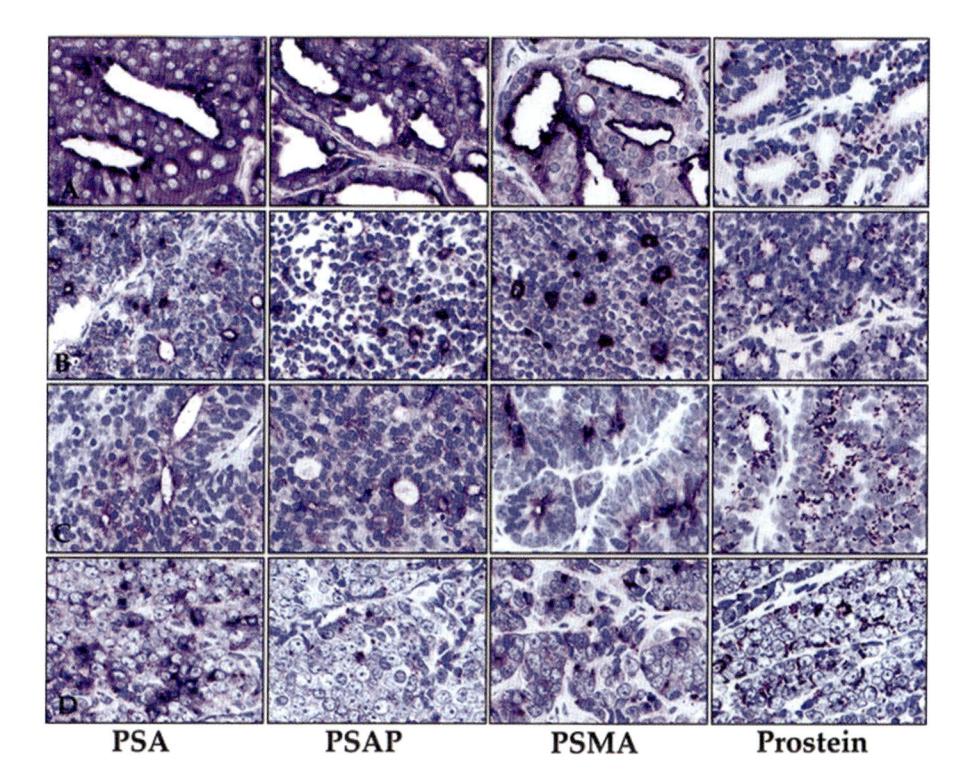

Figure 29.30 Immunohistochemical findings for four prostatic markers in four different metastatic prostate carcinomas. (The letters correspond to the cases in Figure 29.28.) Note that the staining patterns vary from cytoplasmic to luminal and Golgi-like. Stains courtesy of Dr. Isabell Sesterhenn, AFIP, Washington, DC.

Immunohistochemically, these tumors are typically positive for keratin cocktails, keratins 8 and 18, and usually negative for keratins 7 and 20. Positivity for either (usually focal) is detectable in a few cases.

Immunoreactivity for prostate-specific antigen (PSA) and prostate acid phosphatase (PAP) is very helpful in the diagnosis; however, PSA and PAP may also be expressed in other tumors, such as salivary gland and carcinoid tumors.[109,110] Other markers, such as androgen receptor,[106] prostate-specific membrane antigen (PSMA),[111] and prostein (p501S)[112] can also be useful, and use of all markers is indicated in problem cases. NKX3.1 transcription factor is a new sensitive marker that detects a great majority of prostate cancers, irrespective of grade. The only other tumor group with a significance percentage of positivity is breast cancer, especially the lobular form.[113] ERG is expressed in 30 to 40% of cases.

More differentiated carcinomas usually show stronger positivity, but poorly differentiated examples usually show at least focal staining. The pattern varies from cytoplasmic to luminal and Golgi-like; the latter two patterns are often seen with PSMA and prostein (Figure 29.30).

Metastatic adrenal cortical carcinoma

Adrenal cortical carcinoma is a rare tumor that can metastasize into the peripheral soft tissue. The author and colleagues have seen isolated cases in sites such as the neck, arm, and gastric mucosa. More frequently, these tumors form large retroperitoneal masses that can infiltrate well beyond the boundaries of adrenal glands and thereby simulate other neoplasms, including sarcomas, and also appear as abdominal and retroperitoneal metastases after primary surgery. Other tumors that these carcinomas can histologically resemble especially include renal carcinoma, alveolar soft part sarcoma, perivascular epithelioid cell tumors (PEComas), and pleomorphic liposarcoma.

Adrenal cortical carcinoma is often histologically distinctive. The tumor has a variably organoid, packet-like architecture. The tumor cells have abundant, variably eosinophilic to focally clear cytoplasm (Figures 29.31 and 29.32). The nuclei vary from uniformly enlarged to variable in size, sometimes with extensive pleomorphism.

Immunohistochemical markers that are especially valuable include melan A and inhibin. Adrenal carcinomas are typically positive for vimentin, whereas their keratin expression varies from none to moderate; rarely is there extensive keratin positivity.[114–117] Keratins that are more often expressed include keratins 8 and 18 (Figure 29.32). Synaptophysin positivity is common, but chromogranin is not expressed. In the author's experience of a limited number of cases, adrenal carcinomas are negative for TFE3. They are also negative for EMA, desmin, HMB45, and S100 protein.

Ovarian carcinomas

This is a complex group of tumors that frequently involve pelvic and peritoneal surfaces. Metastases involving other soft tissues are less common. Typical immunohistochemical findings include positivity for ER and CA125. Neither of these markers is specific, because ER positivity is also seen in breast and endometrial carcinomas, and CA125 is often (>50%) positive in pancreatobiliary carcinomas and sometimes (13–20%) in breast and lung carcinomas.[118,119]

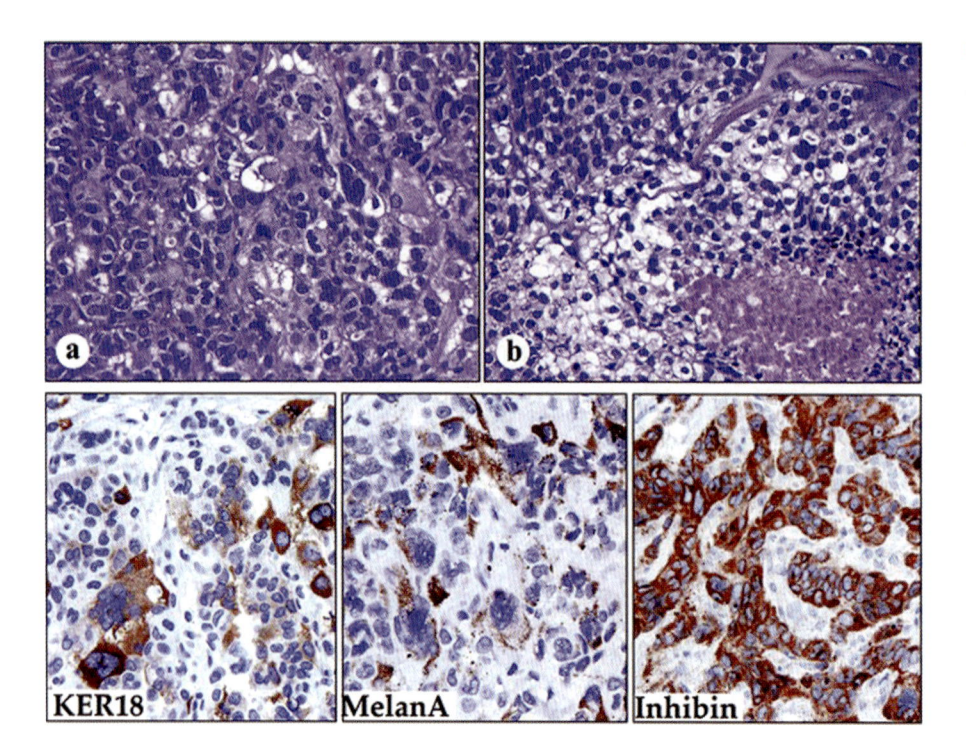

Figure 29.31 Metastatic adrenocortical carcinoma in the neck detected prior to adrenal tumor. The epithelioid cells show focal clear cell appearance. The tumor cells are focally positive for keratin 18. There is focal positivity for melan A and strong positivity for inhibin.

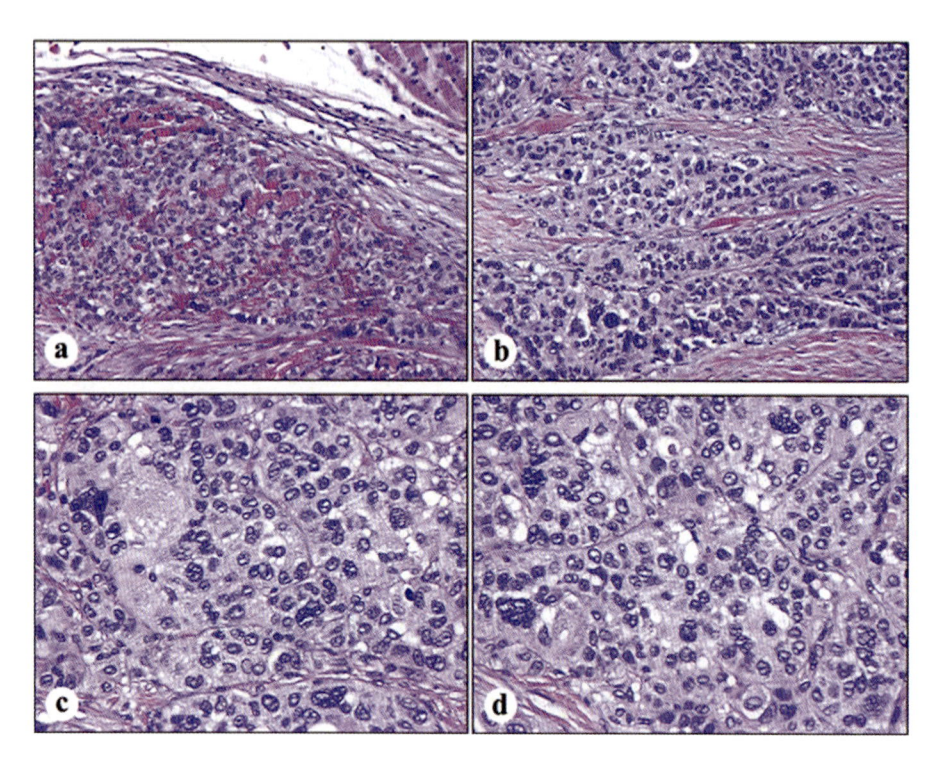

Figure 29.32 Another example of metastatic adrenocortical carcinoma involving the skeletal muscle of the arm. There is a nested pattern with focal cytoplasmic vacuolization.

Nuclear WT1 expression is typical of serous carcinomas, and it is usually absent from endometrioid, clear cell, and mucinous carcinomas.[120–123]

Thyroid carcinomas

The diagnosis of metastatic papillary and follicular carcinomas is usually straightforward, but when there is any doubt, immunohistochemical evaluation for TTF1 and thyroglobulin is useful.[124]

Anaplastic (undifferentiated) carcinoma poses a greater problem, because these tumors are morphologically nonspecific. They are usually limited to the neck, but they can infiltrate deeper tissues, including the upper airways. When of spindle cell type, these tumors simulate soft tissue sarcomas. Keratins can usually be demonstrated, especially keratins 8 and 18.

TTF1 is only rarely expressed in anaplastic carcinoma, however.[124,125]

Medullary thyroid carcinoma can develop soft tissue metastases, usually in the neck, chest wall, or back. These tumors can have spindle cell or epithelioid morphology, and they may have an organoid pattern similar to paraganglioma, with which they can be easily confused (Figure 29.33). Immunohistochemical studies verifying neuroendocrine differentiation (e.g., chromogranin, synaptophysin) and keratin, TTF1, and calcitonin positivity are diagnostic.[126] CEA is another marker typically expressed in medullary thyroid carcinoma.[127] However, calcitonin can be expressed in pulmonary carcinoid so that this antigen alone cannot be used to diagnose metastatic medullary carcinoma

Seminoma/dysgerminoma

The retroperitoneal lymph nodes are the most common site for metastatic seminoma, and in most cases, it is related to lymph node involvement. Some patients have a testicular seminoma, but many have a testicular scar thought to be a regressed primary tumor that is the probable source of the metastases. These tumors occur predominantly in middle-aged men (Figure 29.34).

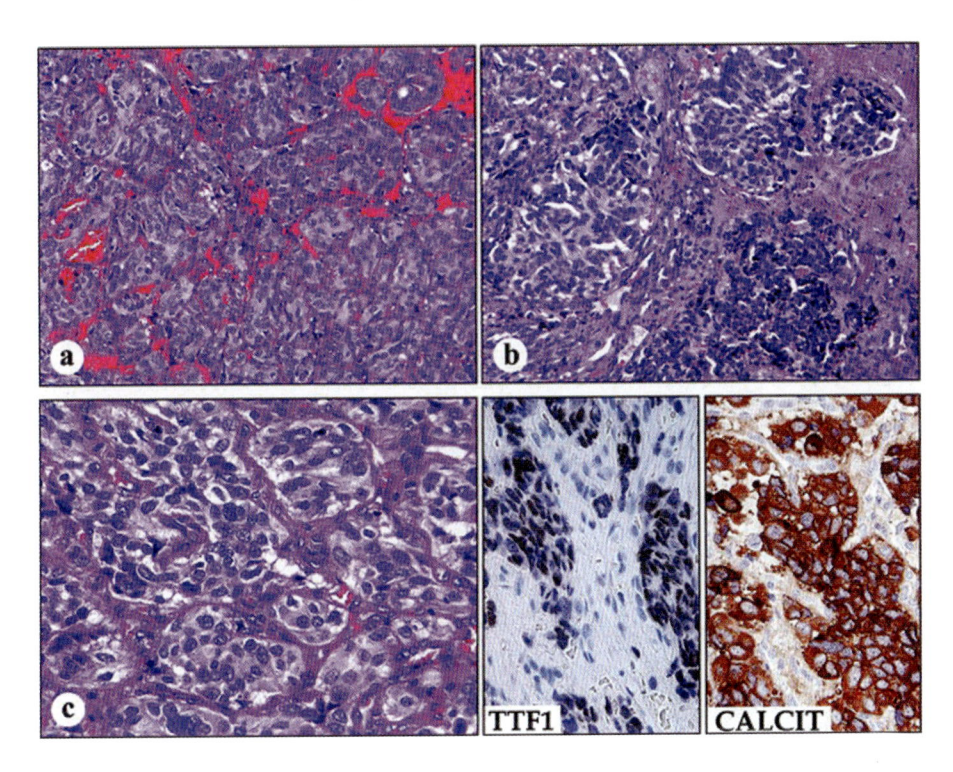

Figure 29.33 Metastatic thyroid medullary carcinoma in the back. The tumor forms a nested pattern resembling paraganglioma. Immunohistochemical positivity for TTF1 and calcitonin supports thyroid origin.

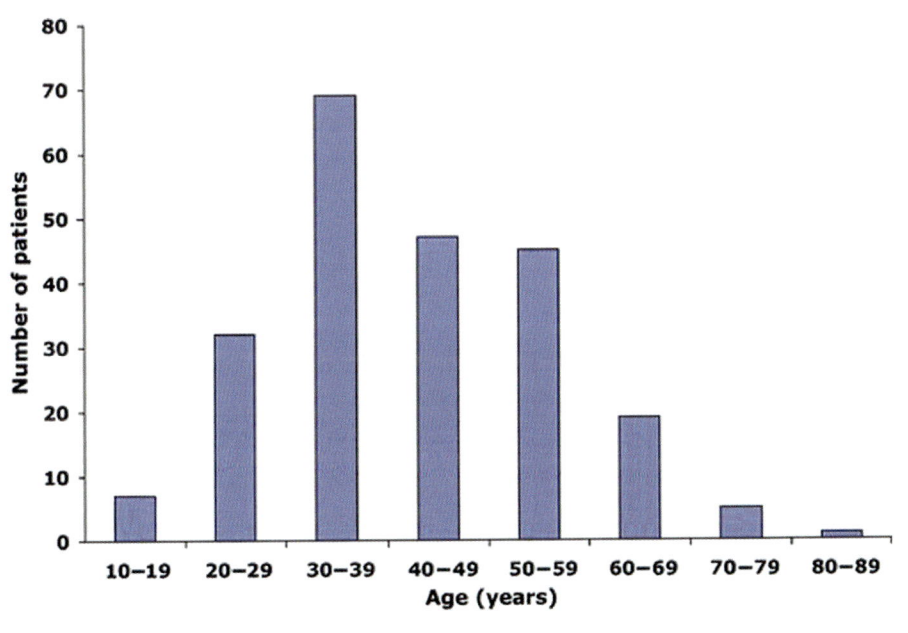

Figure 29.34 Age distribution of 230 men with metastatic retroperitoneal seminoma.

Table 29.2 Immunohistochemical markers useful in the determination of the primary site for a metastatic adenocarcinoma

Primary Site	GCDFP-15	TTF1	CDX2	K7	K20	ER	CA125	Mesothelin
Breast	65	0	0	91	0	60	13	3
Lung	0–5	75	0–2	98	9	3	20	24
Stomach	0–3	0–3	20	51	50	0–2	9	21
Pancreas	0–2	0–2	0–32	94	37	0	48	75
Colon	0–9	0	90	7	88	0–13	7	4
Prostate	0–15	0–11	0–4	12	0–21	11	0–2	0
Ovary, nonmucinous	0–6	0	0	91	0–19	34	85	95

Data adapted and condensed from Dennis et al.[52]
A single number is the median percentage value of positive cases in the studied series. A range is given with predominantly negative results to the highest reported percentage of positive cases.

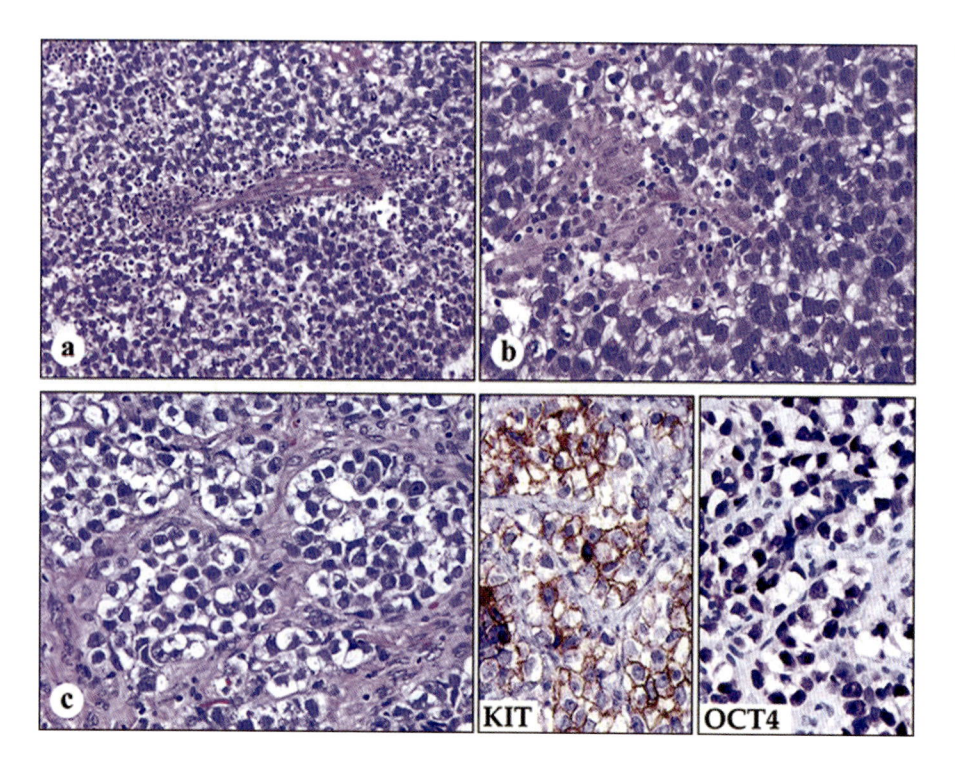

Figure 29.35 Seminoma/dysgerminoma is composed of large cells with clear cytoplasm, often interspersed with lymphocytes. (b) Granuloma-like collections of epithelioid histiocytes are present in some cases. The tumor cells show membrane positivity for KIT (CD117), and nuclear positivity for OCT4.

Metastatic seminoma from testicular origin can be difficult to identify. This is especially true for small biopsies from a retroperitoneum, where testicular seminomas often form nodal metastases. In women, a large ovarian dysgerminoma can be difficult to identify clinically and radiologically as being of ovarian origin; it can simulate many other tumors.

Histologically, seminoma, primary and metastatic, is typically composed of relatively large cells with medium-sized basophilic nuclei, prominent nucleoli, and variably clear cytoplasm. Lymphoid infiltration or a granulomatous response is often present (Figure 29.35).[128]

The best immunohistochemical markers to confirm metastatic seminoma are the nuclear transcription factors OCT4 and NANOG (stem cell pluripotency factors), which seem to be better markers than the previously used placental-like alkaline phosphatase (Figure 29.35).[129–134] Sall4 is also consistently present in seminoma, but in addition is expressed in most other germ cell tumors and also some non-germ-cell carcinomas.[135–138] KIT (CD117) is also expressed in seminomas, often in a distinct membrane pattern (Figure 29.35), and KIT mutations (exon 17, second tyrosine kinase domain) are common.[139,140]

Immunohistochemical markers in metastatic carcinoma of unknown origin: summary

Table 29.2 summarizes the most important and most practical cell-type markers in the evaluation of metastatic carcinoma of unknown origin. This is a simplified summary of the most common patterns of reactivity; however, in most sites, there are exceptional carcinoma types that differ from the main patterns.

Table 29.3 Simplified chart for the application of keratin antibodies in the determination of origin of metastatic carcinoma

	CK7	CK5/6	CK20	CK18
Breast	+	–/+	–/(+)	+
Lung	+	–/+	–/(+)	+
Stomach	+	–	–/(+)	+
Colon	–/(+)	–	+	+
Kidney	–/+	–	–	+
Urinary bladder	+	–	+	+
Prostate	–/(+)	–	–/(+)	+
Ovary	+	–/+	–/+	+

Key: –, usually negative; –/(+), rarely positive; –/+, variably negative or positive; +, usually positive.
Based on refs. 135 and 136.

Table 29.3 summarizes the application of single polypeptide-specific keratin antibodies. Again, among carcinomas of many organs there are subtypes having different keratin profiles. In general, keratin polypeptide profiles can help to narrow the list of possibilities for primary origin, but they are less powerful in pinpointing the organ site than the organ-specific markers.[141,142]

Ultimately, clinicoradiologic correlation is needed to determine the primary site. CT scans and PET scans can be useful for detecting the site of origin for a metastatic carcinoma of unknown source.[143]

Large-scale new-generation studies of gene expression have shown the potential of this analysis to determine the primary origin for metastatic carcinoma. It is almost certain that many new immunohistochemically applicable markers will be discovered based on these data.[144–146]

References

Merkel cell carcinoma

1. Toker C. Trabecular carcinoma of the skin. *Arch Dermatol* 1972;105:107–110.
2. Sibley RK, Dehner LP, Rosai J. Primary neuroendocrine (Merkel cell) carcinoma of the skin. I: A clinicopathologic and ultrastructural study. *Am J Surg Pathol* 1985;9:95–108.
3. Gould VE, Moll R, Moll I, Lee I, Franke WW. Neuroendocrine (Merkel) cells of the skin: hyperplasias, dysplasias and neoplasms. *Lab Invest* 1985;52:334–353.
4. Agelli M, Clegg LX. Epidemiology of primary Merkel cell carcinoma in the United States. *J Am Acad Dermatol* 2003;49:832–841.
5. Penn I, First MR. Merkel cell carcinoma in organ recipients: report of 41 cases. *Transplantation* 1999;68:1717–1721.
6. Engels EA, Frisch M, Goedert JJ, Biggar RJ, Miller RW. Merkel cell carcinoma and HIV infection. *Lancet* 2002;359:497–498.
7. Chang Y, Moore PS. Merkel cell carcinoma: a virus-induced human cancer. *Annu Rev Pathol* 2012;7:123–144.
8. Lien HC, Tsai TF, Lee YY, Hsiao CH. Merkel cell carcinoma and chronic arsenicism. *J Am Acad Dermatol* 1999;41:641–643.
9. Tuneu A, Pujol RM, Moreno A, Barnadas MA, De Moragas JM. Postirradiation Merkel cell carcinoma. *J Am Acad Dermatol* 1989;20:506–507.
10. Eusebi V, Capella C, Cossu A, Rosai J. Neuroendocrine carcinoma within lymph nodes in the absence of a primary tumor, with a special reference to Merkel cell carcinoma. *Am J Surg Pathol* 1992;16:658–666.
11. Chen KT, Papavasiliou P, Edwards K, et al. A better prognosis for Merkel cell carcinoma of unknown primary origin. *Am J Surg* 2013;206:752–757.
12. Allen PJ, Bowne WB, Jaques DP, et al. Merkel cell carcinoma: prognosis and treatment of patients from a single institution. *J Clin Oncol* 2005;23:2300–2309.
13. Skelton HG, Smith KJ, Hitchcock CL, et al. Merkel cell carcinoma: analysis of clinical, histologic, and immunohistologic features of 132 cases with relation to survival. *J Am Acad Dermatol* 1997;37:734–739.
14. Connelly TJ, Cribier B, Brown TJ, Yanguas I. Complete spontaneous regression of Merkel cell carcinoma: a review of 10 reported cases. *Dermatol Surg* 2000;26:853–856.
15. Bichakjian CK, Olencki T, Alam M, et al. Merkel cell carcinoma, version 1.2014. *J Natl Compr Canc Netw* 2014;12:410–424.
16. Pulitzer MP, Amin BD, Busam KJ. Merkel cell carcinoma: review. *Adv Anat Pathol* 2009;16:135–144.
17. Plaza JA, Suster S. The Toker tumor: spectrum of morphologic features in primary neuroendocrine carcinomas of the skin (Merkel cell carcinoma). *Ann Diagn Pathol* 2006;10:376–385.
18. Walsh NMG. Primary neuroendocrine (Merkel cell) carcinoma of the skin: morphologic diversity and implications thereof. *Hum Pathol* 2001;32:680–689.
19. Kukko HM, Koljonen VS, Tukiainen EJ, Haglund CH, Böhling TO. Vascular invasion is an early event in pathogenesis of Merkel cell carcinoma. *Mod Pathol* 2010;23:1151–1156.
20. Gomez LG, DiMaio S, Silva EG, Mackay B. Association between neuroendocrine (Merkel cell) carcinoma and squamous carcinoma of the skin. *Am J Surg Pathol* 1983;7:171–177.
21. Ball NJ, Tanhuanco-Kho G. Merkel cell carcinoma frequently shows histologic features of basal cell carcinoma: a study of 30 cases. *J Cutan Pathol* 2007;34:612–619.
22. Foschini MP, Eusebi V. Divergent differentiation in endocrine and nonendocrine tumors of the skin. *Semin Diagn Pathol* 2000;17:162–168.
23. Adhikari LA, McCalmont TH, Folpe AL. Merkel cell carcinoma with heterologous rhabdomyoblastic differentiation: the role of immunohistochemistry for Merkel cell

polyomavirus large T-antigen in confirmation. *J Cutan Pathol* 2012;39:47–51.

24. Martin B, Poblet E, Rios JJ, et al. Merkel cell carcinoma with divergent differentiation: histopathological and immunohistochemical study of 15 cases with PCR analysis for Merkel cell polyomavirus. *Histopathology* 2013;62:711–722.

25. Feng H, Shuda M, Chang Y, Moore PS. Clonal integration of a polyomavirus in human Merkel cell carcinoma. *Science* 2008;319:1096–1100.

26. Kassem A, Schöpflin A, Diaz C, et al. Frequent detection of Merkel cell polyomavirus in human Merkel cell carcinomas and identification of a unique deletion in the VP1 gene. *Cancer Res* 2008;68:5009–5013.

27. Becker JC, Houben R, Ugurel S, et al. MC polyomavirus is frequently present in Merkel cell carcinoma in European patients. *J Invest Dermatol* 2009;129:248–250.

28. Garneski KM, Warcola AH, Feng O, et al. Merkel cell carcinoma polyomavirus is more frequently present in North American than Australian Merkel cell carcinoma tumors. *J Invest Dermatol* 2009;129:246–248.

29. Byrd-Gloster AL, Khoor A, Glass LF, et al. Differential expression of thyroid transcription factor I in small cell lung carcinomas and Merkel cell tumor. *Hum Pathol* 2000;31:58–62.

30. Asioli S, Righi A, Volante M, Eusebi V, Bussolati G. p63 expression as a new prognostic marker in Merkel cell carcinoma. *Cancer* 2007;110:640–647.

31. Su LD, Fullen DR, Lowe L, et al. CD117 (KIT receptor) expression in Merkel cell carcinoma. *Am J Dermatopathol* 2002;24:289–293.

32. Sur M, AlArdati H, Ross C, Alowami S. TdT expression in Merkel cell carcinoma: potential diagnostic pitfall with blastic hematological malignancies and expanded immunohistochemical analysis. *Mod Pathol* 2007;20:1113–1120.

33. Dong HY, Liu W, Cohen P, Mahle CE, Zhang W. B-cell specific activation protein encoded by the PAX-5 gene is commonly expressed in merkel cell carcinoma and small cell carcinomas. *Am J Surg Pathol* 2005;29:687–692.

34. Zur Hausen A, Rennspiess D, Winnepenninckx V, Speel EJ, Kurz AK. Early B-cell differentiation in Merkel cell carcinomas: clues to cellular ancestry. *Cancer Res* 2013;73:4982–4987.

Squamous cell carcinoma

35. Pereira TC, Share SM, Magalhães AV, Silverman JF. Can we tell the site of origin of metastatic squamous cell carcinoma?: an immunohistochemical tissue microarray study of 194 cases. *Appl Immunohistochem Mol Morphol* 2011;19:10–14.

36. Begum S, Cao D, Gillison M, Zahurak M, Westra WH. Tissue distribution of human papillomavirus 16 DNA integration in patients with tonsillar carcinoma. *Clin Cancer Res* 2006;11:5694–5699.

37. Leventon GS, Evans HL. Sarcomatoid squamous cell carcinoma of the mucous membranes of the head and neck: a clinicopathologic study of 20 cases. *Cancer* 1981;48:994–1003.

38. Patel NK, McKee PH, Smith NP, Fletcher CD. Primary metaplastic carcinoma (carcinosarcoma) of the skin: a clinicopathologic study of four cases and review of the literature. *Am J Dermatopathol* 1997;19:363–372.

39. Takata T, Ito H, Ogawa I, et al. Spindle cell squamous carcinoma of the oral region: an immunohistochemical and ultrastructural study on the histogenesis and differential diagnosis with a clinicopathologic analysis of six cases. *Virchows Arch A Pathol Anat Histopathol* 1991;419:177–182.

40. Lewis JE, Olsen KD, Sebo TJ. Spindle cell carcinoma of the larynx: review of 26 cases including DNA content and immunohistochemistry. *Hum Pathol* 1997;28:664–673.

41. Thompson LD, Wieneke JA, Miettinen M, Heffner DK. Spindle cell (sarcomatoid) carcinomas of the larynx: a clinicopathologic study of 187 cases. *Am J Surg Pathol* 2002;26:153–170.

42. Wick MR, Ritter JH, Humphrey PA. Sarcomatoid carcinomas of the lung: a clinicopathologic review. *Am J Clin Pathol* 1997;108:40–53.

43. Sigel JE, Skacel M, Bergfeld WF, et al. The utility of cytokeratin 5/6 in the recognition of cutaneous spindle cell squamous cell carcinoma. *J Cutan Pathol* 2001;28:520–524.

44. Morgan MB, Purohit C, Anglin TR. Immunohistochemical distinction of cutaneous spindle cell carcinoma. *Am J Dermatopathol* 2008;30:228–232.

45. Dotto JE, Glusac EJ. p63 is useful marker for cutaneous spindle cell squamous cell carcinoma. *J Cutan Pathol* 2006;33:413–417.

46. Bishop JA, Montgomery EA, Westra WH. Use of p40 and p63 immunohistochemistry and human papillomavirus testing as ancillary tools for the recognition of head and neck sarcomatoid carcinoma and its distinction from benign and malignant mesenchymal processes. *Am J Surg Pathol* 2014;38:257–264.

Carcinoma of the breast

47. Liu H, Shi J, Wilkerson ML, Lin F. Immunohistochemical evaluation of GATA3 expression in tumors and normal tissues: a useful immunomarker for breast and urothelial carcinomas. *Am J Clin Pathol* 2012;138:57–64.

48. Cimino-Mathews A, Subhawong AP, Illei PB, et al. GATA3 expression in breast carcinoma: utility in triple-negative, sarcomatoid, and metastatic carcinomas. *Hum Pathol* 2013;44:1341–1349.

49. Miettinen M, McCue PA, Sarlomo-Rikala M, et al. GATA3: a multispecific but potentially useful marker in surgical pathology: a systematic analysis of 2500 epithelial and nonepithelial tumors. *Am J Surg Pathol* 2014;38:13–22.

50. Wick MR, Lillemoe TJ, Copland GT, et al. Gross cystic disease fluid protein-15 as a marker for breast cancer: immunohistochemical analysis of 690 human neoplasms and comparison with alpha-lactalbumin. *Hum Pathol* 1989;20:281–287.

51. Kaufmann O, Deidesheimer T, Meuhlenberg M, Deicke P, Dietel M. Immunohistochemical differentiation of breast carcinomas from metastatic adenocarcinomas of other common sites. *Histopathology* 1996;29:233–240.

52. Dennis JL, Hvidsten TR, Wit EC, et al. Markers of adenocarcinoma characteristic of the site of origin: development of a diagnostic algorithm. *Clin Cancer Res* 2005;11:3766–3772.

53. De Young BR, Wick MR. Immunohistologic evaluation of metastatic carcinomas of unknown origin: an algorithmic approach. *Semin Diagn Pathol* 2000;17:184–193.

54. Oberman HA. Metaplastic carcinoma of the breast: a clinicopathologic study of 29 cases. *Am J Surg Pathol* 1987;11:918–929.

55. Wargotz ES, Deos PH, Norris HJ. Metaplastic carcinomas of the breast. II: Spindle cell carcinoma. *Hum Pathol* 1989;20:732–740.

56. Wargotz ES, Norris HJ. Metaplastic carcinomas of the breast. I: Matrix-producing carcinoma. *Hum Pathol* 1989;20:628–635.

57. Foschini MP, Dina RE, Eusebi V. Sarcomatoid neoplasms of the breast: proposed definition for biphasic and monophasic sarcomatoid mammary carcinomas. *Semin Diagn Pathol* 1993;10:128–136.

58. Gobbi H, Simpson JF, Borowsky A, Jensen RA, Page DL. Metaplastic breast tumors with a dominant fibromatosis-like phenotype have a high risk of local recurrence. *Cancer* 1999;85:2170–2182.

59. Sneide N, Yaziji H, Mandavilli SR, et al. Low-grade (fibromatosis-like) spindle cell carcinoma of the breast. *Am J Surg Pathol* 2001;25:1009–1016.

60. Kurian KM, Al-Nafussi A. Sarcomatoid/metaplastic carcinoma of the breast: a clinicopathologic study of 12 cases. *Histopathology* 2002;40:58–64.

61. Carter MR, Hornick JL, Lester S, Fletcher CDM. Spindle cell (sarcomatoid) carcinoma of the breast: a clinicopathologic and immunohistochemical analysis of 29 cases. *Am J Surg Pathol* 2006;30:300–309.

Lung cancer

62. Reis-Filho JS, Carrilho C, Valenti C, et al. Is TTF1 a good immunohistochemical marker to distinguish primary from metastatic lung adenocarcinomas. *Pathol Res Pract* 2000;196:835–840.

63. Kaufmann O, Dietel M. Thyroid transcription factor 1 is a superior immunohistochemical marker for pulmonary adenocarcinoma and large cell carcinoma compared to surfactant proteins A and B. *Histopathology* 2000;36:8–16.

64. Ordóñez NG. Value of thyroid transcription factor-1 immunostaining in tumor diagnosis: a review and update. *Appl Immunohistochem Mol Morphol* 2012;20:429–444.

65. Agoff SN, Lamps LW, Philip AT, et al. Thyroid transcription factor-1 is expressed in extrapulmonary small cell carcinomas but not in other extrapulmonary neuroendocrine tumors. *Mod Pathol* 2000;13;238–242.

66. Cheuk W, Kwan MY, Suster S, Chank JK. Immunostaining for thyroid transcription factor 1 and cytokeratin 20 aids the distinction of small cell carcinoma from Merkel cell carcinoma, but not pulmonary from extrapulmonary small cell carcinoma. *Arch Pathol Lab Med* 2001;125:228–231.

67. Kaufmann O, Georgi T, Dietel M. Utility of 123C3 monoclonal antibody against CD56 (NCAM) for the diagnosis of small cell carcinomas on paraffin sections. *Hum Pathol* 1998;28:1373–1378.

68. Lin X, Saad RS, Lukasevic TM, Selverman JF, Liu Y. Diagnostic value of CDX-2 and TTF-1 expressions in separating metastatic neuroendocrine neoplasms of unknown origin. *Appl Immunohistochem Mol Morphol* 2007;15:407–414.

69. Nakajima M, Kasai T, Hashimoto H, Iwata Y, Manabe H. Sarcomatoid carcinoma of the lung: a clinicopathologic study of 37 cases. *Cancer* 1999;86:608–616.

70. Rossi G, Cavazza A, Sturm N, et al. Pulmonary carcinomas with pleomorphic, sarcomatoid, or sarcomatous elements: a clinicopathologic and immunohistochemical study of 75 cases. *Am J Surg Pathol* 2003;27:311–324.

71. Franks TJ, Galvin JR. Sarcomatoid carcinoma of the lung: histologic criteria and common lesions in the differential diagnosis. *Arch Pathol Lab Med* 2010;134(1):49–54.

72. Turk F, Yuncu G, Bir F, Ozturk G, Ekinci Y. Squamotous-type sarcomatoid carcinoma of the lung with rhabdomyosarcomatous components. *J Cancer Res Ther* 2012;8:148–150.

73. Pelosi G, Scarpa A, Manzotti M, et al. K-ras gene mutational analysis supports a monoclonal origin of biphasic pleomorphic carcinoma of the lung. *Mod Pathol* 2004;17:538–546.

Gastrointestinal, pancreatic, and hepatocellular carcinoma

74. Werling RW, Yaziji H, Bacchi CE, Gown AM. CDX2, a highly sensitive and specific marker of adenocarcinomas of intestinal origin: an immunohistochemical survey of 476 primary and metastatic carcinomas. *Am J Surg Pathol* 2003;27:303–310.

75. Moskaluk CA, Zhang H, Powell SM, et al. CDX2 protein expression in normal and malignant human tissues: an immunohistochemical survey using tissue microarrays. *Mod Pathol* 2003;16:913–919.

76. Vang R, Gown AM, Wu LS, et al. Immunohistochemical expression of CDX2 in primary ovarian mucinous tumors and metastatic mucinous carcinomas involving the ovary: comparison with CK20 and correlation with coordinate expression of CK7. *Mod Pathol* 2006;19:1421–1428.

77. McCluggage WG, Shar R, Connolly LE, McBride HA. Intestinal type cervical adenocarcinoma in situ and adenocarcinomas exhibit a partial enteric immunophenotype with consistent expression of CDX2. *Int J Gynecol Pathol* 2008;27:92–100.

78. Cathro HP, Mills SE. Immunophenotypic differences between intestinal-type and low-grade papillary sinonasal adenocarcinomas: an immunohistochemical study of 2 cases utilizing CDX2 and MUC2. *Am J Surg Pathol* 2004;28:1026–1032.

79. Swierczynski SL, Maitra A, Abraham SC, et al. Analysis of novel tumor markers in pancreatic and biliary carcinomas using tissue microarrays. *Hum Pathol* 2004;35:357–366.

80. Chu PG, Ishizawa S, Wu E, Weiss LM. Hepatocyte antigen as a marker for hepatocellular carcinoma: an immunohistochemical comparison to carcinoembryonic antigen, CD10, and alpha-fetoprotein. *Am J Surg Pathol* 2002;26:978–988.

81. Kakar S, Muir T, Murphy LM, Lloyd RV, Burgart LJ. Immunoreactivity of Hep Par 1 in hepatic and extrahepatic tumors and its correlation with albumin in situ hybridization in hepatocellular carcinoma. *Am J Clin Pathol* 2003;119:361–366.

82. Morrison C, Marsh W Jr, Frankel WL. A comparison of CD10 to pCEA, MOC-31, and hepatocyte for the distinction of malignant tumors in the liver. *Mod Pathol* 2002;15:1279–1287.

83. Varma V, Cohen C. Immunohistochemical and molecular markers in the diagnosis of hepatocellular carcinoma. *Adv Anat Pathol* 2004;11:239–249.

84. Yan BC, Gong C, Song J, et al. Arginase-1: a new immunohistochemical marker of hepatocytes and hepatocellular neoplasms. *Am J Surg Pathol* 2010;34:1147–1154.

Renal cell carcinoma

85. Togral G, Arıkan M, Gungor S. Rare skeletal muscle metastasis after radical nephrectomy for renal cell carcinoma: evaluation of two cases. *J Surg Case Rep* 2014;2014(10):rju101.

86. Kierney PC, van Heerden JA, Sgura JW, Weaver AL. Surgeon's role in the management of solitary renal cell carcinoma metastases occurring subsequent to initial curative nephrectomy: an institutional review. *Ann Surg Oncol* 1994;1:345–352.

87. McGregor DK, Khurana KK, Cao C, et al. Diagnosing primary and metastatic renal cell carcinoma: the use of the monoclonal antibody "renal carcinoma marker." *Am J Surg Pathol* 2001;25:1485–1492.

88. Avery AK, Beckstead J, Renshaw AA, Corless CL. Use of antibodies to RCC and CD10 in the differential diagnosis of renal neoplasms. *Am J Surg Pathol* 2000;24:203–210.

89. Pan CC, Chen PC, Ho DM. The diagnostic utility of MOC31, BerEp4, RCC marker and CD10 in the classification of renal carcinoma and renal oncocytoma: an immunohistochemical analysis of 328 cases. *Histopathology* 2004;45:452–459.

90. Bakshi N, Kunju LP, Giordano T, Shah RB. Expression of renal cell carcinoma antigen (RCC) in renal epithelial and nonrenal tumors: diagnostic implications. *Appl Immunohistochem Mol Morphol* 2007;15:310–315.

91. Takashi M, Haimoto H, Murase T, Mitsuya H, Kato K. An immunochemical and immunohistochemical study of S100 protein in renal cell carcinoma. *Cancer* 1988;61:889–895.

92. Gokden N, Gokden M, Phan DC, McKenney JK. The utility of PAX-2 in distinguishing metastatic clear cell renal cell carcinoma from its morphological mimics: an immunohistochemical study with comparison to renal cell carcinoma marker. *Am J Surg Pathol* 2008;32:1462–1467.

93. Sangoi AR, Karamchandani J, Kim J, Pai RK, McKenney JK. The use of immunohistochemistry in the diagnosis of metastatic clear cell renal cell carcinoma: a review of PAX-8, PAX-2, hKIM-1, RCCma, and CD10. *Adv Anat Pathol* 2010;17(6):377–393.

94. Laury AR, Perets R, Piao H, et al. A comprehensive analysis of PAX8 expression in human epithelial tumors. *Am J Surg Pathol* 2011;35:816–826.

95. Farrow GM, Harrison EG, Utz DC. Sarcomas and sarcomatoid and mixed malignant tumors of the kidney in adults. *Cancer* 1968;22:556–563.

96. Ro JY, Ayala AG, Sella A, Samuels ML, Sanson DA. Sarcomatoid renal cell carcinoma: a clinicopathologic study of 42 cases. *Cancer* 1987;59:516–526.

97. Baer SC, Ro JY, Ordonez NG, et al. Sarcomatoid collecting duct carcinoma: a clinicopathologic and immunohistochemical study of five cases. *Hum Pathol* 1993;24:1017–1022.

98. Akhtar M, Tulbah A, Kardar AH, Ali MA. Sarcomatoid renal cell carcinoma: the chromphobe connection. *Am J Surg Pathol* 1997;21:1188–1195.

99. da Peralta Venturina M, Moch H, Amin M, et al. Sarcomatoid differentiation in renal cell carcinoma: a study of 101 cases. *Am J Surg Pathol* 2001;25:275–284.

100. Cheville JC, Lohse CM, Zincke H, et al. Sarcomatoid renal cell carcinoma: an examination of underlying histologic subtype and an analysis of association with patient outcome. *Am J Surg Pathol* 2004;28:435–441.

101. DeLong W, Grignon DJ, Eberwein P, Shum DT, Wyatt JK. Sarcomatoid renal cell carcinoma: an immunohistochemical study of 18 cases. *Arch Pathol Lab Med* 1993;117:636–640.

102. Li L, Teichberg S, Steckel J, Chen QH. Sarcomatoid renal cell carcinoma with divergent sarcomatoid growth patterns: a case report and review of the literature. *Arch Pathol Lab Med* 2005;129:1057–1060.

103. Chang A, Brimo F, Montgomery EA, Epstein JI. Use of PAX8 and GATA3 in diagnosing sarcomatoid renal cell carcinoma and sarcomatoid urothelial carcinoma. *Hum Pathol* 2013;44(8):1563–1568.

104. Itoh T, Chikai K, Ota S, et al. Chromophobe renal cell carcinoma with osteosarcoma-like differentiation. *Am J Surg Pathol* 2002;26:1358–1362.

Urothelial and prostate carcinoma

105. Cheng L, Zhang S, Alexander R, et al. Sarcomatoid carcinoma of the urinary bladder: the final common pathway of urothelial carcinoma dedifferentiation. *Am J Surg Pathol* 2011;35:e34–e46.

106. Jiang J, Ulbright TM, Younger C, et al. Cytokeratin 7 and cytokeratin 20 in primary urinary bladder carcinoma and matched lymph node metastases. *Arch Pathol Lab Med* 2001;125:921–923.

107. Wu X, Kakehi Y, Zeng Y, et al. Uroplakin II as a promising marker for molecular diagnosis of nodal metastases from bladder cancer: comparison with cytokeratin 20. *J Urol* 2005;174:2138–2142.

108. Huang HY, Shariat SR, Sun TT, et al. Persistent uroplakin expression in advanced urothelial carcinomas: implications in urothelial tumor progression and clinical outcome. *Hum Pathol* 2007;38:1703–1713.

109. Fan CY, Wang J, Barnes EL. Expression of androgen receptor and prostatic specific markers in salivary duct carcinoma: an immunohistochemical analysis of 13 cases and review of the literature. *Am J Surg Pathol* 2000;24:579–586.

110. Hameed O, Humphrey PA. Immunohistochemistry in diagnostic surgical pathology of the prostate. *Semin Diagn Pathol* 2005;22:88–104.

111. Bostwick DG, Pacelli A, Blute M, Roche P, Murphy GP. Prostate-specific membrane antigen expression in prostatic intraepithelial neoplasia and adenocarcinoma: a study of 184 cases. *Cancer* 1998;82:2256–2261.

112. Sheridan T, Herawi M, Epstein JI, Illei PB. The role of P501S and PSA in the diagnosis of metastatic adenocarcinoma of the prostate. *Am J Surg Pathol* 2007;31:1351–1355.

113. Gurel B, Ali TZ, Montgomery EA, et al. NKX3.1 as a marker of prostatic origin in metastatic tumors. *Am J Surg Pathol* 2010;34:1097–1105.

Adrenal cortical carcinoma

114. Busam KJ, Iversen K, Coplan KA, et al. Immunoreactivity for A103, an antibody to melan-A (Mart-1), in adrenocortical and other steroid cell tumors. *Am J Surg Pathol* 1998;22:57–63.

115. Arola J, Liu J, Heikkila P, Voutilainen R, Kahri A. Expression of inhibin alpha in the human adrenal gland and adrenocortical tumors. *Endocr Res* 1998;24:865–867.

116. Renshaw AA, Granter SR. A comparison of A103 and inhibin reactivity in adrenal cortical tumors: distinction from hepatocellular carcinoma and renal tumors. *Mod Pathol* 1999;11:1160–1164.

117. Pan CC, Chen PC, Tsay SH, Ho DM. Differential immunoprofiles of hepatocellular carcinoma, renal cell carcinoma, and adrenocortical carcinoma: a systemic immunohistochemical survey using tissue array technique. *Appl Immunohistochem Mol Morphol* 2005;13:347–352.

Ovarian carcinoma

118. Loy TS, Quesenberry JT, Sharp SC. Distribution of CA125 in adenocarcinomas: an immunohistochemical study of 481 cases. *Am J Clin Pathol* 1992;98:175–179.

119. Langendijk DH, Mullink H, Van Diest PJ, Meijer GA, Meijer CJ. Tracing the origin of adenocarcinomas with unknown primary using immunohistochemistry: differential diagnosis between colonic and ovarian carcinomas as primary sites. *Hum Pathol* 1998;29:491–497.

120. Baker TM, Oliva E. Immunohistochemistry as a tool in the differential diagnosis of ovarian tumors: an update. *Int J Gynecol Pathol* 2005;24:39–55.

121. Soslow RA. Histologic subtypes of ovarian carcinoma: an overview. *Int J Gynecol Pathol* 2008;27:161–174.

122. Acs G, Pasha T, Zhang PJ. WT1 is differentially expressed in serous, endometrioid, clear cell, and mucinous carcinomas of the peritoneum, fallopian tube, ovary, and endometrium. *Int J Gynecol Pathol* 2004;23:110–118.

123. Tornos C, Soslow R, Chen S, et al. Expression of WT1, CA125, and GCDFP-15 as useful markers in the differential diagnosis of primary ovarian carcinomas versus metastatic breast cancer to the ovary. *Am J Surg Pathol* 2005;29:1482–1489.

Thyroid carcinoma

124. Rosai J. Immunohistochemical markers of thyroid tumors: significance and diagnostic applications. *Tumori* 2003;89:517–519.

125. Miettinen M, Franssila K. Variable expression of keratins and nearly uniform lack of thyroid transcription factor 1 in thyroid anaplastic carcinoma. *Hum Pathol* 2000;31:1139–1145.

126. Bussolati G, Papotti M, Pagani A. Diagnostic problems in medullary carcinoma of the thyroid. *Pathol Res Pract* 1995;191:332–344.

127. Hamada S, Hamada S. Localization of carcinoembryonic antigen in medullary thyroid carcinoma by immunofluorescent techniques. *Br J Cancer* 1977;36:572–576.

Germ cell tumors

128. Sung MT, MacLennan GT, Cheng L. Retroperitoneal seminoma in limited biopsies: morphologic criteria and immunohistochemical findings in 30 cases. *Am J Surg Pathol* 2006;30:766–773.

129. Jones TD, Ulbright TM, Ebje JN, Beldridge LA, Cheng L. OCT4 staining in testicular tumors: a sensitive and specific marker for seminoma and embryonal carcinoma. *Am J Surg Pathol* 2004;28:935–940.

130. Cheng L. Establishing germ cell origin for metastatic tumors using OCT4 immunohistochemistry. *Cancer* 2004;101:2006–2010.

131. Emerson RE, Ulbright TM. The use of immunohistochemistry in the differential diagnosis of tumors of the testis and paratestis. *Semin Diagn Pathol* 2005;22:33–50.

132. Hoei-Hansen CE, Almstrup K, Nielsen JE, et al. Stem cell pluripotency factor NANOG is expressed in human fetal gonocytes, testicular carcinoma in situ and germ cell tumors. *Histopathology* 2005;47;48–66.

133. Hart AH, Hartley L, Parker K, et al. The pluripotency homeobox gene is expressed in human germ cell tumors. *Cancer* 2005;104:2092–2098.

134. Santagata S, Ligon KL, Hornick JL. Embryonic stem cell transcription factor signatures in the diagnosis of primary and metastatic germ cell tumors. *Am J Surg Pathol* 2007;31:836–845.

135. Cao D, Li J, Guo CC, Allan RW, Humphrey PA. SALL4 is a novel diagnostic marker for testicular germ cell tumors. *Am J Surg Pathol* 2009;33:1065–1077.

136. Camparo P, Comperat EM. SALL4 is a useful marker in the diagnostic work-up of germ cell tumors in extra-testicular locations. *Virchows Arch* 2013;462:337–341.

137. Miettinen M, Wang Z, McCue PA, et al. SALL4 expression in germ cell and non-germ cell tumors: a systematic immunohistochemical study of 3215 cases. *Am J Surg Pathol* 2014;38:410–420.

138. Ulbright TM, Tickoo SK, Berney DM, Srigley JR; Members of the ISUP Immunohistochemistry in Diagnostic Urologic Pathology Group. Best practices recommendations in the application of immunohistochemistry in testicular tumors: report from the International Society of Urological Pathology consensus conference. *Am J Surg Pathol* 2014;38: e50–e59.

139. Biermann K, Klingmuller D, Koch A, et al. Diagnostic value of markers M2A, OCT3/4, AP-2 gamma, PLAP and c-KIT in the detection of extragonadal seminomas. *Histopathology* 2006;49:290–297.

140. Kemmer K, Corless CL, Fletcher JA, et al. KIT mutations are common in testicular seminoma. *Am J Pathol* 2004;164:305–313.

Carcinomas of unknown primary

141. Moll R. Cytokeratins in the histologic diagnosis of malignant tumors. *Int J Biol Markers* 1994;9:63–69.

142. Chu PG, Weiss LM. Keratin expression in human tissues and neoplasms. *Histopathology* 2002;40:403–439.

143. Paul SA, Stoeckli SJ, von Schulthess GK, Goerres GW. FDG PET and PET/CT

for the detection of the primary tumour in patients with cervical non-squamous cell carcinoma metastasis of an unknown primary. *Eur Arch Otorhinolaryngol* 2007;264:189–195.

144. Tothill RW, Kowalczyk A, Rischin D, *et al.* An expression-based site of origin diagnostic method designed for clinical application to cancer of unknown origin. *Cancer Res* 2005;65:4031–4040.

145. Horlings HM, van Laar RK, Kerst JM, *et al.* Gene expression profiling to identify the histogenetic origin of metastatic adenocarcinomas of unknown primary. *J Clin Oncol* 2008;26:4435–4441.

146. Varadhachary GR, Talantov D, Raber MN, *et al.* Molecular profiling of carcinoma of unknown primary and correlation with clinical evaluation. *J Clin Oncol* 2008;26:4442–4448.

Chapter 30
Cartilage- and bone-forming tumors

Julie C. Fanburg-Smith
Uniformed Services University of the Health Sciences (USUHS) Bethesda, Maryland

Mark D. Murphey
Armed Forces Institute of Pathology Washington, DC

This chapter presents cartilage- and bone-forming tumors and tumor-like lesions in soft tissue. Just like for intraosseous bone- and cartilage-forming tumors, preoperative radiologic correlation is advised for complete pathologic classification. Some of these bone-forming lesions are more common in soft tissue. Clinicopathologic features and genetic changes may vary between these counterpart intraosseous and extraskeletal tumors.

From a historical viewpoint, most of these lesions have been well defined for at least the last one to five decades. Regarding cartilage tumors, Lichtenstein and Goldman[1] were the first to describe *chondroma of soft parts* in the hands and feet in 1964; this tumor had been previously described by Jaffe[2] in 1958 in periarticular and intracapsular locations. *Synovial chondromatosis* was first identified by Ambroise Paré in *Monsters and Prodigies* in 1558 and Laennec in 1813;[3] in 1900 and depicted in the German literature by Reichel.[4] Although Stout and Verner[5] described extraskeletal chondrosarcomas in 1953, the most common variant, *extraskeletal myxoid chondrosarcoma* (EMC), was defined by Enzinger and Shiraki in 1972.[6] *Mesenchymal chondrosarcoma* was first classified by Lichtenstein and Bernstein in 1959.[7] Bone- or matrix-producing lesions also have been known for a long time. Although it has been known for more than a half-century that tumors can cause oncogenic osteomalacia,[8] the histologic appearance of *phosphaturic mesenchymal tumor of the mixed connective tissue type* was first reported by Weidner and Santa Cruz in 1987.[9] *Tumoral calcinosis* has been recognized since 1899 by Duret, yet it was first described using the terminology "tumoral calcinosis" in 1943 by Inclan.[10] Calcium pyrophosphate deposition disease (CPPD) was originally recognized in 1958,[11] but it was McCarty *et al.* who linked pseudogout to rhomboid crystals in 1962.[12] Furthermore, the term *tophaceous pseudogout* was first observed in the literature in 1982.[13] Tumors and lesions forming bone, including reactive and malignant types, have also been described for a half-century to a few centuries. The development of a mass after trauma has been known since the eighteenth century, and clinical descriptions as early as 1692 have been identified; however, *myositis ossificans* was first described as a clinicopathologic entity by Ackerman in 1958.[14] The first known description of *fibrodysplasia ossificans progressiva* was in 1962, by a French physician, Guy-Patin, in a letter to a colleague.[15] Finally, one of the earliest reports of *extraskeletal osteosarcoma* was that of Fine and Stout in 1956.[16]

Chondroma of soft parts (extraskeletal chondroma, soft part chondroma)

Clinical features

Soft tissue chondromas usually occur in adults between the ages of 30 and 60 years, with a male predominance.[17–19] They often present as small, painless solitary masses of the hands and feet (particularly the fingers and toes) most commonly the volar hands, proximal digits.[19] Intraosseous, periosteal, or juxtacortical chondromas are more common than soft tissue chondromas. Periosteal chondroma occurs in the second decade, typically as an incidental finding on radiographic studies. The metaphysis and diaphysis are most commonly involved. This tumor may be palpable on physical examination and painless. Chondromas also occur in para-articular or intracapsular[20] locations of large joints, usually the knee. The intracapsular or para-articular lesions are associated with pain and decreased motion.

Pulmonary chondroma can be associated with the Carney triad. This rare nonheritable syndrome usually presents in young women and is defined by the presence of at least two of the components: pulmonary chondroma, gastrointestinal stromal tumor, and functioning extra-adrenal paraganglioma.[21] The syndrome is discussed in detail in Chapter 17.

Extraskeletal chondroma can recur locally (10–30%);[17–19] there have been no reports of chondrosarcomatous transformation, as commonly proposed for long-bone chondrosarcoma arising from an intraosseous enchondroma. Complete local excision is the treatment of choice for all variants of chondroma.

Imaging studies

Radiographs or computed tomography (CT) of soft tissue chondroma often reveal focal calcifications with a characteristic ring-and-arc appearance typical of cartilage neoplasms. Areas of metaplastic ossification can also be apparent. Magnetic resonance (MR) imaging of these lesions usually reveals predominantly low signal intensity on T1-weighting and very high signal intensity on T2-weighting, reflecting the high water content of cartilaginous lesions.[22] Peripheral and septal contrast enhancement, common in cartilage lesions, is also common. Foci of calcification usually remain at low signal intensity on all pulse sequences (Figure 30.1).

Modern Soft Tissue Pathology, Second Edition, ed. Markku Miettinen. Published by Cambridge University Press. © Cambridge University Press 2016.

Chapter 30: Cartilage- and bone-forming tumors

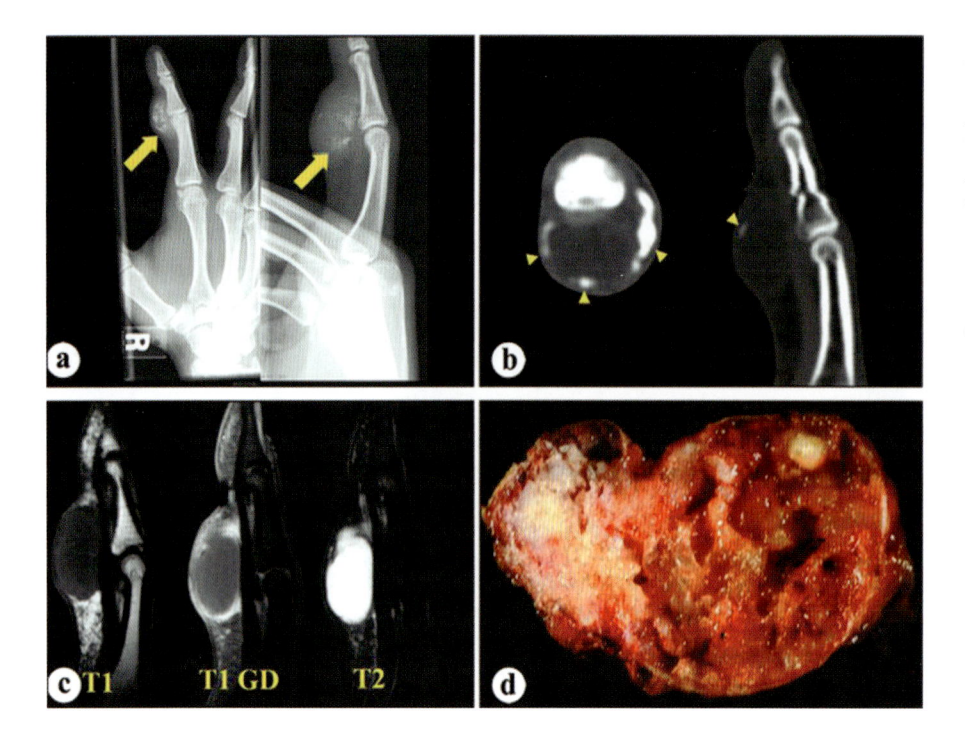

Figure 30.1 Chondroma of soft parts of the index finger (**a**) Radiographs show a calcified soft tissue mass *(arrow)* volar to the proximal interphalangeal joint (**b**) Axial (left image) and sagittal (right image) CT also reveal the calcified soft tissue mass *(arrowheads)*. (**c**) Sagittal T1-weighted (left image), T1-weighted postcontrast (middle image) and T2-weighted (left image) MR images show the lesion with low signal intensity on T1 and prominent high signal intensity on T2. There is peripheral enhancement on the postcontrast image. (**d**) Grossly, there is a gray-white, mucoid cartilaginous mass.

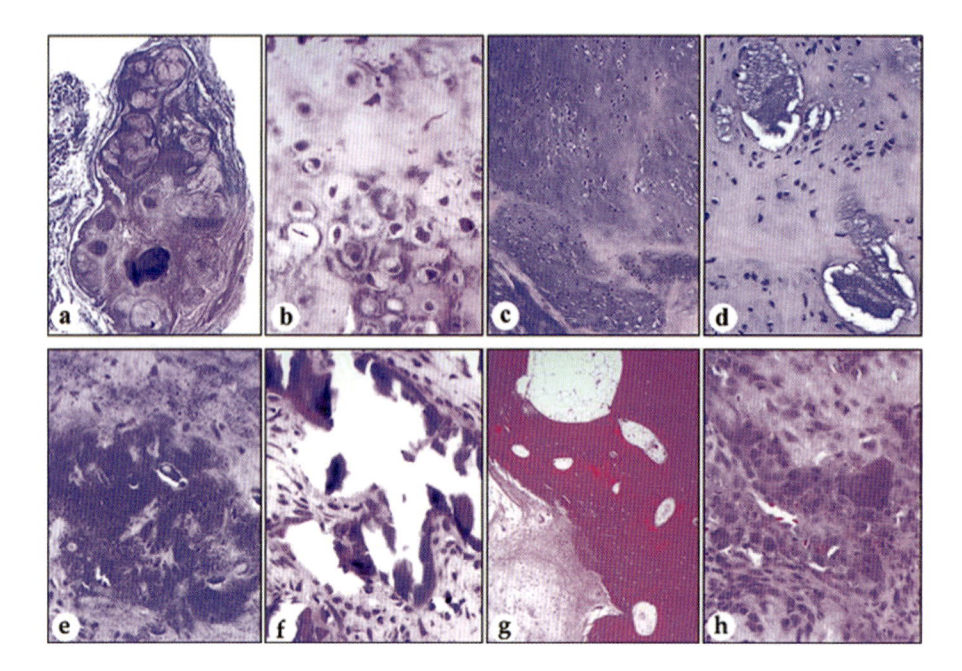

Figure 30.2 (**a–d**) By histology, soft tissue chondroma is composed of hyaline cartilage with variable myxoid change, hyalinization, finely stippled calcification (**e**), coarse calcification (**f**), and ossification, with production of fatty marrow (**g**). A subtype that represents about 10% to 15% of soft tissue chondromas includes histiocytoid chondrocytes and osteoclast-type giant cells (**h**).

Pathology

Grossly, a soft tissue chondroma is a small, solitary, well-circumscribed mass, usually <3 cm. It can be lobulated and associated with the tendon, tendon sheath, or joint capsule of the hands or feet. The intracapsular or para-articular variant averages 5 cm and is usually in an infrapatellar location. Both the soft tissue and the pulmonary variants can have extensive calcification. On sectioning, chondromas can be gray-white, firm and glistening, or gelatinous and myxoid (Figure 30.1).

By histology, soft tissue chondroma is usually a solitary mass containing hyaline cartilage and variable amounts of secondary ossification, fibrosis, giant cell or histiocytic reaction, and myxoid change (Figure 30.2).

Occasional findings include peripheral residual fat and bone with fatty marrow. One theory for the development of chondromas is that fat undergoes myxoid change, then chondromyxoid change, and finally chondroid changes, with different stages of cartilage development. These should not be called "osteochondromas," however, as metaphyseal skeletal osteochondroma is a lesion with a cortex-attached stalk and cartilage cap, with the interior of the lesion being continuous with bone marrow. Soft tissue chondroma, on the other hand, is a purely soft tissue mass. The hyaline cartilage of soft tissue chondroma can be predominantly myxoid; with the chondrocytes in the same stage of development, similar to the lobules of enchondroma in bone. Variants of extraskeletal chondroma include those with granuloma-like areas with epithelioid chondrocytes and multinucleated osteoclast type giant cells, constituting approximately 10% of cases (Figure 30.2), and those with immature chondroblasts and prominent myxoid change. Increased cellularity, slight pleomorphism, and binucleation are acceptable and do not indicate malignancy. These changes are particularly common in the myxoid subtype of chondroma. Imaging studies are often useful in confirming the lesional location, extent, and relationship to bone.[22]

By immunohistochemistry, extraskeletal chondromas are positive for S100 protein and vimentin. These tumors are also positive for SOX9 and ERG (nuclear positivity).[23]

The differential diagnosis of chondroma most importantly includes osseous chondrosarcoma extending or metastatic into soft tissue, and clinicoradiologic correlation may be required in this regard. In the chest wall, chondroid neoplasms are most often chondrosarcomas arising in the rib or sternum. Osseous chondrosarcomas are notoriously rare in fingers and hands, the favored locations for chondroma of soft parts. EMC shows cord-like arrangement of tumor cells, typically lacks hyaline cartilage, and occurs in proximal or axial locations. Synovial chondromatosis and its extra-articular variant, tenosynovial chondromatosis, form multiple nodules of hyaline cartilage, rather than a solitary mass and are associated with synovia.[24] Myxomas are much less cellular than myxoid chondromas and lack cartilage.

Genetics

Monosomy of chromosome 6 and rearrangement of chromosome 11q13 have been reported for chondroma, as well as 46, XY inv 12 (p12.3q15).[25,26] Additional findings include different clones with t(6;12)(q12;p11.2), t(3;7)(q13;p12), and der(2)t(2;18)(p11.2;q11.2). Changes detected in a periosteal and pulmonary chondroma include nonrandom 12q13-q15 rearrangements involving the HMGA2 gene.[27–29]

Synovial chondromatosis (synovial osteochondromatosis, synovial chondrometaplasia, synovial osteochondrosis, Reichel disease, and henderson-jones disease)

Clinical features

Primary synovial chondromatosis is a benign lesion that is composed of multiple lobules of hyaline cartilage, surrounded by synovium, and located in the subsynovial tissue of a joint, tendon sheath, or bursa. It occurs in the third to the fifth decades, with a mean age of approximately 40 years.[30–33] Twice as common in males as females, anatomic locations involve the knee in two-thirds of cases, followed by the hip, elbow, and other synovial joints. Hands and feet can be involved, and some of these are entirely extra-articular, termed *tenosynovial chondromatosis*.[24] Axial joints also can be involved, including the temporomandibular joint. Presenting symptoms include pain, swelling, palpable loose bodies, joint clicking and locking, and loss of joint movement. Osteoarthritis is a common complication.

Primary synovial chondromatosis has the potential for recurrence in approximately 15–20% of cases. There is an estimated 5% potential for malignant transformation to chondrosarcoma, typically after multiple recurrences. These examples have been typically defined by bone invasion without distant metastases.[33–35]

Treatment for primary synovial chondromatosis includes removal of the loose bodies by partial or sometimes total synovectomy by arthroscopic surgery – unless there is evidence for extra-articular disease. This decreases the moderate to high potential for local recurrence. In rare cases of malignant transformation, more aggressive therapy such as amputation might be necessary.[35]

Secondary synovial chondromatosis refers to deposition of dislodged articular cartilage fragments (loose bodies) into periarticular soft tissue as a result of trauma or arthritic joint abnormalities.[36] It most commonly involves the knee and hip and typically occurs in older patients.

Imaging studies

Radiologic studies reveal intra-articular calcifications (70–95% of cases) of similar size and shape, distributed throughout the joint, with typical ring-and-arc chondroid mineralization.[36,37] Extrinsic erosion of bone is observed in 20% to 50% of cases. CT demonstrates calcified intra-articular fragments and extrinsic bone erosion. MR images vary, depending on the degree of mineralization, with the most common pattern (77% of cases) portraying low to intermediate signal intensity on T1- and high signal on T2-weighted images (Figure 30.3). Secondary synovial chondromatosis can be identified radiologically by the underlying articular disease (typically osteoarthritis) and fewer chondral bodies of varying size and shape.[36,37]

Pathology

Grossly, multiple (often numerous, up to >1000) gray-white cobblestone-like cartilaginous nodules are seen in synovial-lined locations, both intra- and extra-articularly (Figure 30.4).[32] These are often similar in size and can be loose within the joint or become reattached and reabsorbed. They can also coalesce and form large calcified lesions (≤20 cm), termed *giant* or *massive* synovial osteochondromatosis. Synovial chondromatosis can invade bone. Extra-articular (tenosynovial chondromatosis) is

Chapter 30: Cartilage- and bone-forming tumors

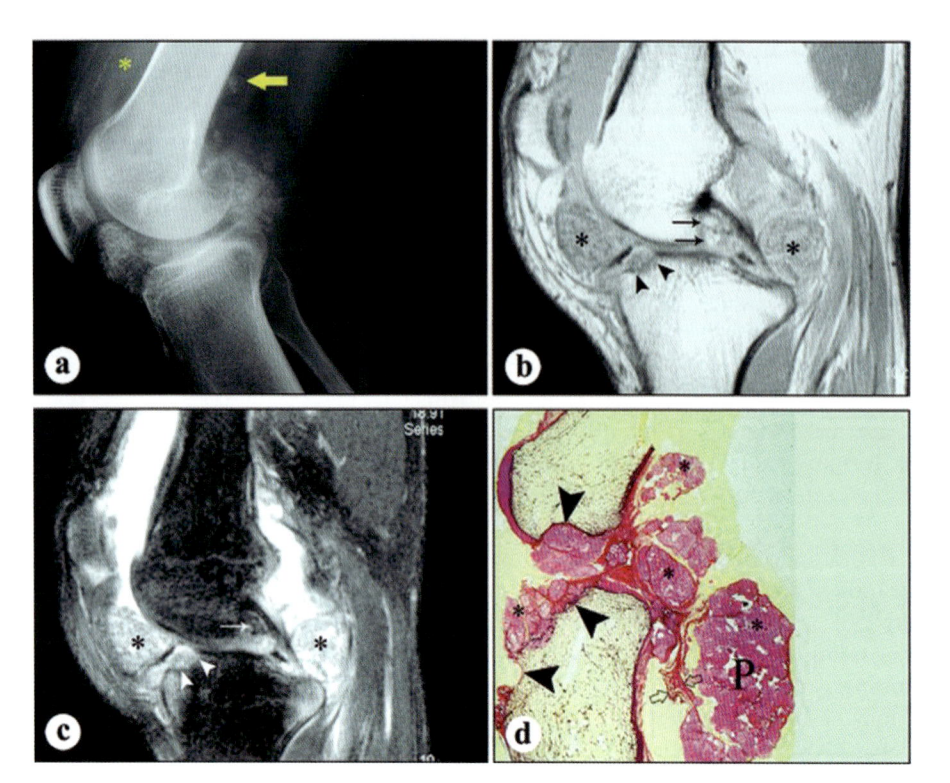

Figure 30.3 Synovial chondromatosis of the knee. (**a**) Lateral radiograph shows the joint effusion (*) and multiple intra-articular osteochondral bodies (arrow). (**b,c**) Sagittal proton density and T2-weighted (**c**) MR images reveal intra-articular chondral neoplasia (* and arrows) and extrinsic bone erosion (arrowheads). (**d**) This whole-mount histologic section demonstrates the multiple nodules of synovial chondromatosis (*), erosions of bone at arrowheads, and involvement of popliteal bursa by synovial chondromatosis (P, bursal synovium between arrows).

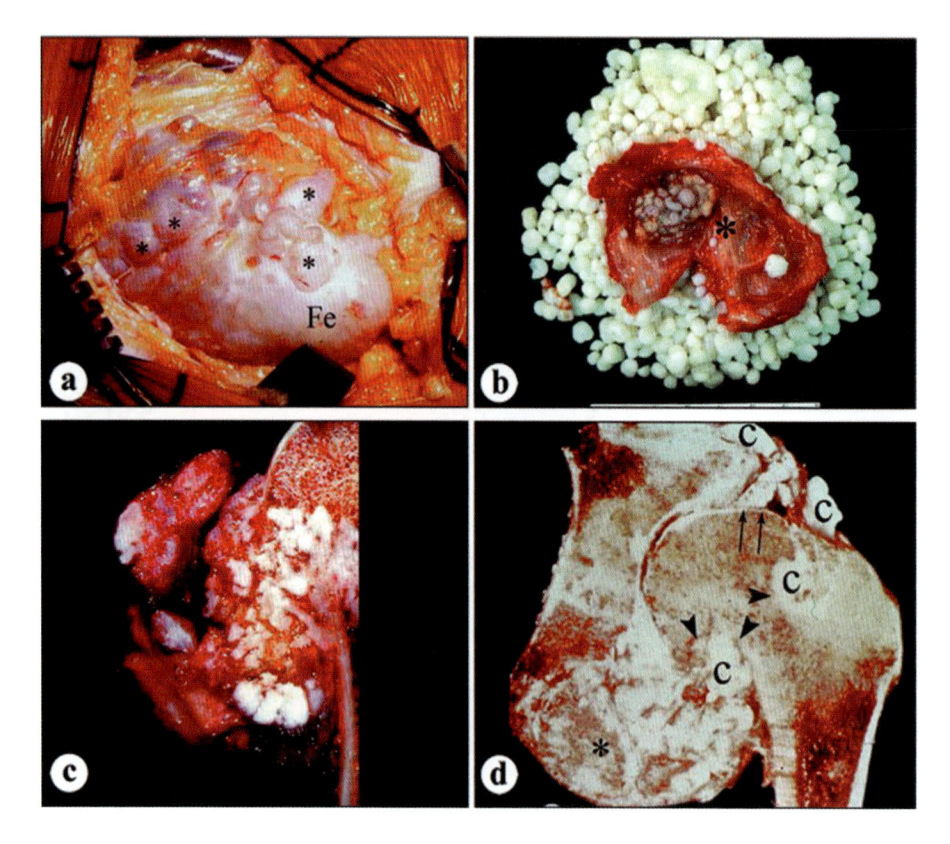

Figure 30.4 Surgical exposure of extensive intra-articular primary synovial chondromatosis (**a**, Fe = femur, * = nodules of primary synovial chondromatosis), intra- and extra-articular disease, multiple nodules (**b**), glistening gray gross appearance (**c**), encasement of bone and erosion into and involvement of bone (**d**, [c] cartilage, evidence for erosion at arrowheads).

subsynovial and extends around bursae or along tendon sheaths.[24]

By histology, primary synovial chondromatosis is composed of multiple uniform-sized lobules of hyaline cartilage surrounded by a synovial lining (Figure 30.5), within the synovium of the joint, bursae, or tendon sheath. The cellularity and pleomorphism in synovial chondromatosis can be striking, with double and multiple nuclei within individual

Chapter 30: Cartilage- and bone-forming tumors

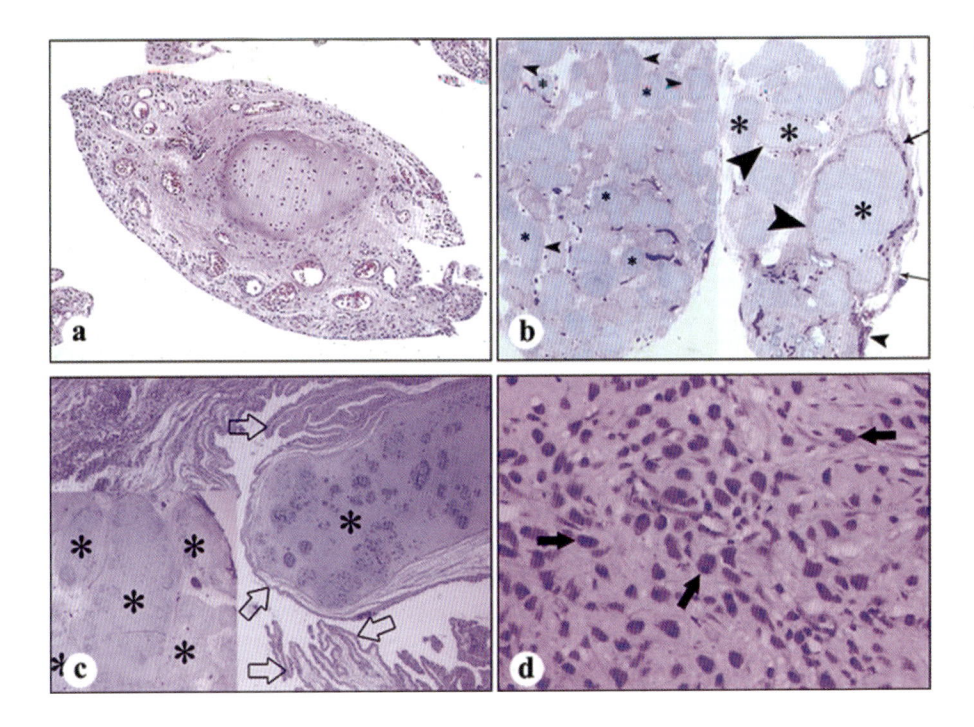

Figure 30.5 (**a**) One nodule of primary synovial chondromatosis embedded in synovium, depicting hyaline cartilage surrounded by synovial lining. (**b**) Typical appearance of multiple nodules of hyaline cartilage surrounded by synovial lining *(arrows and arrowheads demonstrating synovium;* is hyaline cartilage)*. (**c**) A closer demonstration of synovial lining *(arrows)*, synovium, and hyaline cartilage (*). (**d**) Atypia, including binucleation, myxoid change, and nuclear enlargement are acceptable findings in benign synovial chondromatosis.

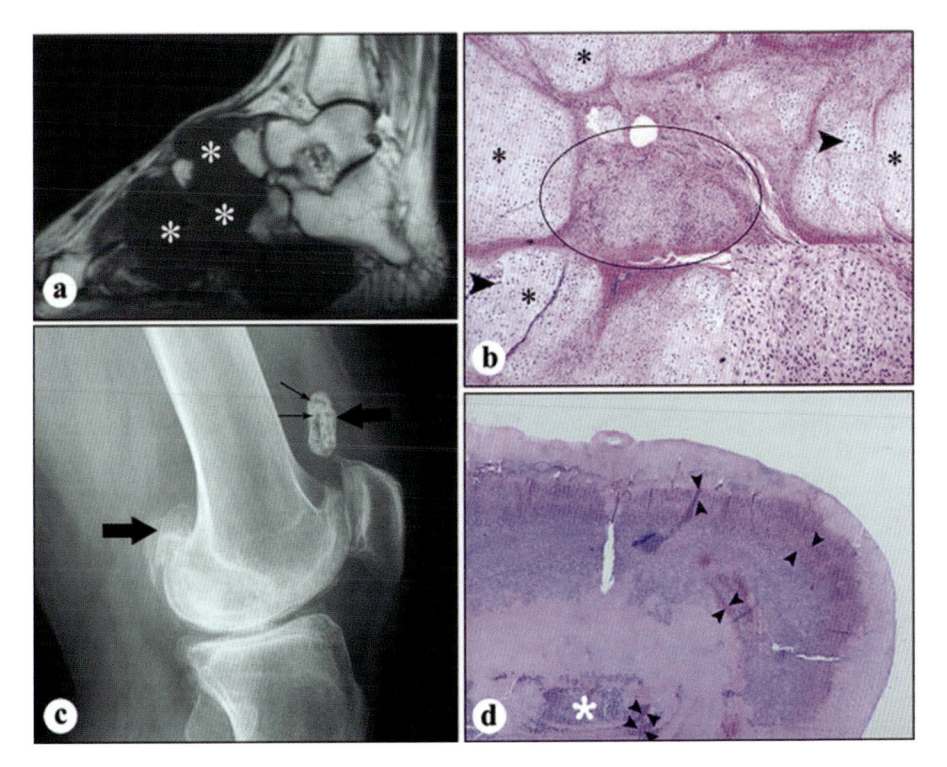

Figure 30.6 Malignant transformation of primary synovial chondromatosis. (**a**) Sagittal T1-weighted MR image shows a large midfoot mass (*) with marrow invasion. (**b**) Multiple nodules of synovial chondromatosis (*) with central cellular atypical nodule representing malignant transformation to chondrosarcoma *(circled malignant area)*, increased cellularity *(arrowheads)*. (**c**) Secondary synovial chondromatosis: lateral radiograph of the knee shows osteoarthritic changes with several intra-articular osteochondral fragments *(large arrows)*. (**d**) Several rings of calcification area seen in the anterior fragment *(arrowheads)*. Correlating histologic section demonstrating "rings of growth" of this osteochondral fragment.

chondrocyte lacunae, acceptable findings in this benign process, but mitotic activity and necrosis are rare. Calcification and disorganized ossification can occur but usually represent <10% of the lesion. This should not be called "osteochondromatosis" because this name creates confusion with exostosis and osteochondromas. In synovial chondromatosis, sheets of atypical chondrocytes, marked myxoid stromal change, necrosis, mitotic figures, and crowding and spindling of the nuclei at the periphery of the lobules or overtaking lobules (Figure 30.6) have been suggested to indicate local aggressiveness with potential for bone invasion.[33–35]

Secondary synovial chondromatosis has a central nidus of hypocellular, non-neoplastic cartilage, with characteristic concentric rings of growth. It usually lacks atypia and has fragments of articular cartilage in the synovium. The nodules vary in size and shape and are fewer than in primary synovial chondromatosis (Figure 30.6).

The differential diagnosis includes soft part chondroma, juxtacortical chondroma, intracortical chondroma, osteochondroma, and soft tissue extension of primary osseous chondrosarcoma. Soft tissue chondroma is unilocular and not surrounded by synovium. Imaging is often required for definitive classification of synovial chondromatosis and its distinction from the previously mentioned entities.

Genetics

Clonal chromosomal changes suggest that primary synovial chondromatosis is a true neoplastic lesion.[38] In addition, brothers with the same joint primary synovial chondromatosis, demonstrating abnormalities of C-Erb-B2, have been identified.[39] Observed chromosomal abnormalities include complex changes with multiple translocations and loss of chromosome copies. Common findings are the loss of band 10q26 and rearrangements of 1p13 and 12q13–15. The latter segment is also recurrently rearranged in other chondromatous tumors. Hedgehog signaling might play a role in synovial chondromatosis.[40]

Extraskeletal myxoid chondrosarcoma
Clinical features

Most patients with EMC are middle-aged adults, with a median age in the fifth decade and a slight male predominance.[6] This tumor can also occur in children, however.[41] A common location is the deep thigh. Other sites include the hand, retroperitoneum, and head and neck. Most examples present as painless masses.

Prognosis for EMC was originally thought to be favorable; however, late metastases are often detected during long-term follow-up studies.[42] High-grade variants and "dedifferentiation" into a primitive phenotype also occur.[43–45] In some series, the metastatic potential for high-grade tumors has been as high as 82%. Metastases are generally to the lungs, lymph nodes, or soft tissue. The estimated 5-, 10- and 15-year survival rates are 90%, 70%, and 60%, respectively.[45] High cellularity, pleomorphism, high mitotic activity, older patient age, larger tumor size, metastases, and proximal location are adverse prognostic factors.[45] Wide local excision, with consideration to adjuvant therapy in high-grade tumors, is the treatment of choice. A more recent study confirms a poor response to chemotherapy, and therefore aggressive early surgical control of localized disease is advised.[46] Newer therapeutic modalities include inhibitors of the mammalian target of rapamycin (mTOR).[47]

Imaging studies

Imaging of EMC is frequently nonspecific, showing a relatively large soft tissue mass. Foci of punctate calcification can be seen on radiographs and CT.[48] The CT attenuation is predominantly low. On MR imaging, low to intermediate signal intensity is seen on T1-weighting and high signal intensity on T2-weighting.[49] Unlike other cartilage neoplasms, areas of hemorrhage (high signal intensity on all pulse sequences or fluid levels) are common. Peripheral and septal enhancement with nodularity is common after intravenous contrast administration. Regional lymphadenopathy can also be depicted on CT or MR imaging (Figure 30.7).

Pathology

Most EMC are intramuscular, but 25% are located in the subcutis.[5] The median tumor size is approximately 11 cm (range 5–15 cm). Grossly, the tumors are gelatinous and often hemorrhagic; they can simulate hematomas.

By histology, the tumors are typically composed of multiple myxoid lobules, separated by fibrous septa (Figure 30.8). The tumor cells are arranged in cords or strands in a net-like or linear interconnected pattern, from the periphery into the center of the lobules, often with central clustered nests of four to ten cells. These central cell clusters can occasionally mimic glands (Figure 30.9). Intralesional hemorrhage is common (Figure 30.8). The cells are round to stellate with hyperchromatic, uniform, round nuclei, and scant to moderately abundant eosinophilic cytoplasm; the stellate cytoplasm from one cell to another is connected by strands like a "string of pearls" from the periphery (Figure 30.10). Mitotic activity is generally low, and the usual type is a low-grade tumor. Highly cellular, less myxoid areas can occur in higher-grade tumors, with epithelioid areas or areas resembling malignant fibrous histiocytoma (MFH).[44]

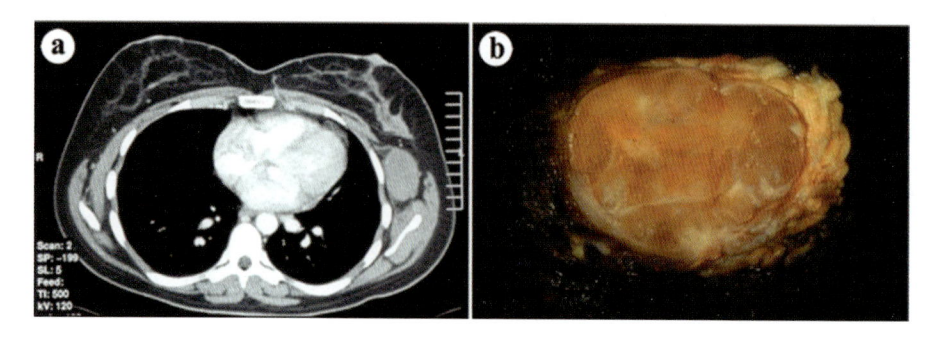

Figure 30.7 Extraskeletal myxoid chondrosarcoma: (**a**) CT of the chest shows a predominantly low attenuation soft tissue mass (arrows) in the chest wall. (**b**) Gross demonstrating myxoid and gelatinous appearance of this subcutaneous or intramuscular mass.

Chapter 30: Cartilage- and bone-forming tumors

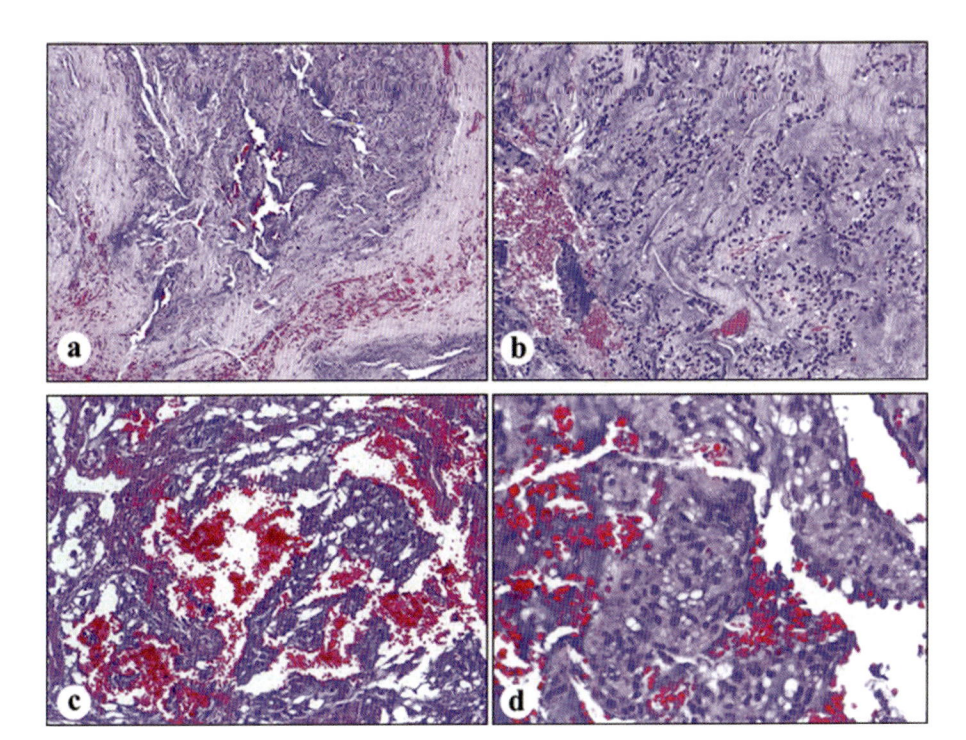

Figure 30.8 (**a**) An extraskeletal myxoid chondrosarcoma is composed histologically of myxoid lobules of spindled to epithelioid cells surrounded by fibrous septa. (**b–d**) Extravasation of erythrocytes can mimic hematoma.

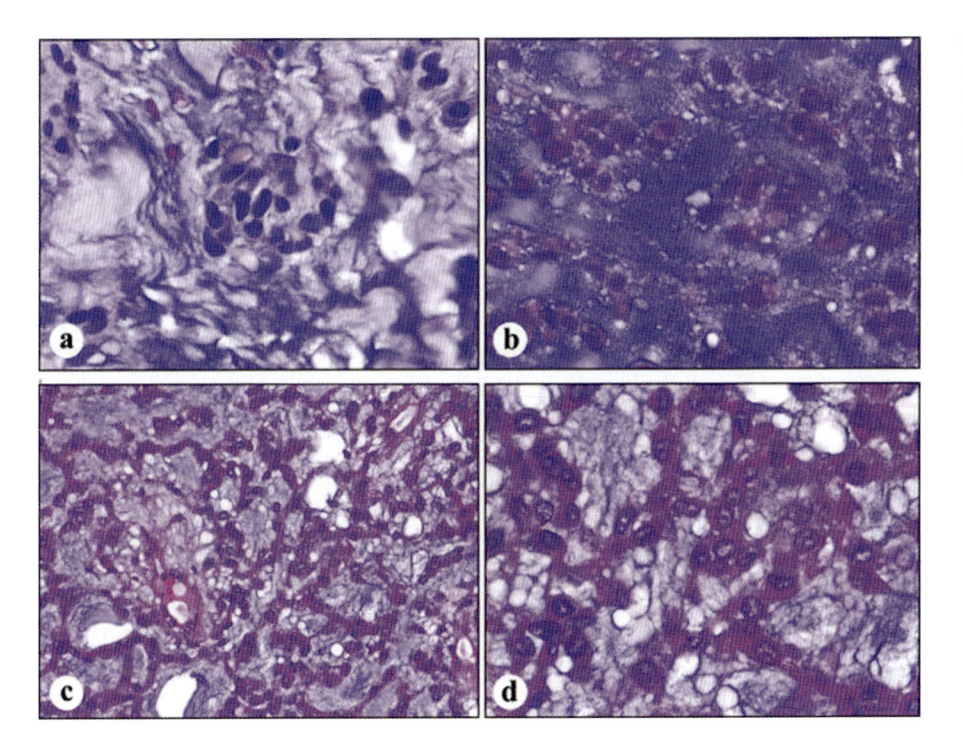

Figure 30.9 (**a,b**) The cells in extraskeletal myxoid chondrosarcoma, unlike intraosseous myxoid chondrosarcoma, have central four- to ten-cell clusters of medium-sized epithelioid cells. (**c,d**) Peripheral interconnection of spindled chondrocytes creates a fishnet-like appearance.

By immunohistochemistry, a subset, including examples with rhabdoid cells, shows loss of INI1,[50] but in contrast to ordinary chondrosarcoma, these tumors are only variably positive for S100 protein.[51,52] These tumors can be positive for epithelial membrane antigen (EMA),[51] but they are generally negative for keratins (except rare CAM5.2-positive examples), and glial fibrillary acid protein (GFAP) showing some immunophenotypic overlap with myoepithelioma.[51–53]

The differential diagnosis includes soft part chondroma, parosteal chondrosarcoma, metastatic chordoma, mixed tumors of the salivary or sweat glands or soft tissue myoepithelioma, myxoma, myxoid liposarcoma, myxoid MFH, myxoid malignant peripheral nerve sheath tumor (MPNST), ossifying fibromyxoid tumor (OFT), and myxopapillary ependymoma. The site and the presence of hyaline cartilage make the first two lesions easy to distinguish. Chordomas and mixed tumors are

831

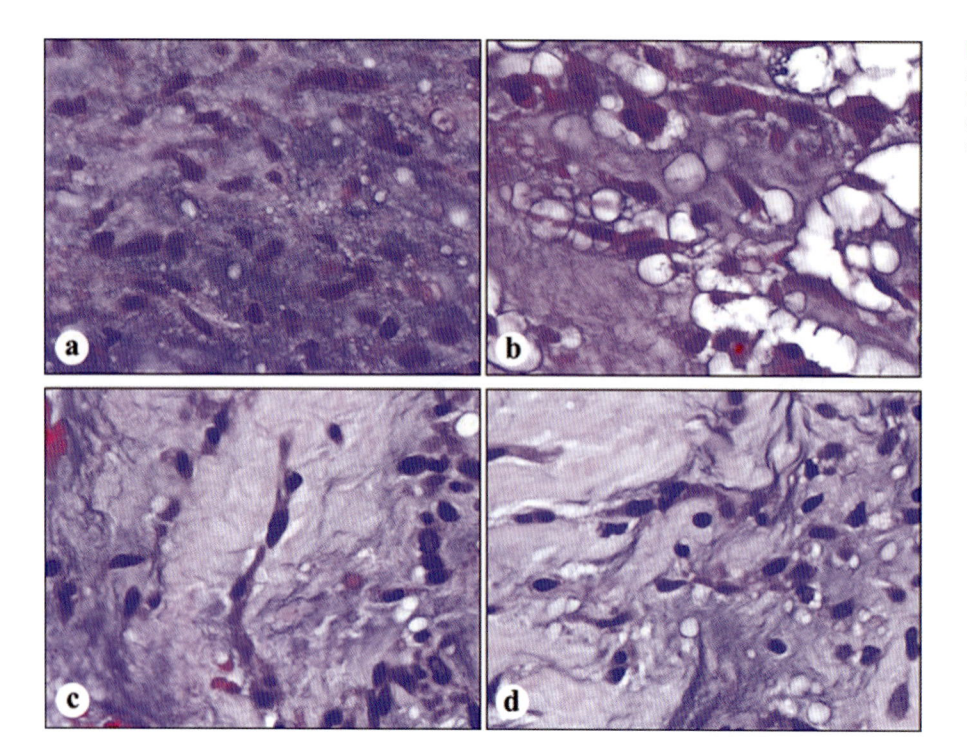

Figure 30.10 Extraskeletal myxoid chondrosarcoma spindled cells are stellate, and their processes like to "reach out and touch one another" like a string of pearls from the periphery of the lobules.

generally keratin or EMA positive and positive for SMA, as well as S100 protein.[52,53] Chordomas differ from EMC, by having more abundant cytoplasm in the large polygonal cells that form cords or sheets. OFT has a distinctive metaplastic osseous peripheral rim, more prominent vasculature, and more collagenous stroma, with larger epithelioid cells with distinctive cytoplasmic borders. Strong GFAP positivity distinguishes myxopapillary ependymoma from EMC.

Genetics

EMC is marked by a reciprocal translocation: t(9;22)(q22;q12) in 75% of cases.[54–56] This change corresponds to the EWSR1/NR4A3 gene fusion.[57,58] Although most osseous myxoid chondrosarcomas are pathologically and genetically different and do not have this translocation, there are rare bone tumors similar to EMC with this translocation.[59,60] A variant t(9;17)(q22;q11) translocation with a novel non-EWS fusion product = TCF12 or TAF15 (an EWSR1-related gene)/NR4A3 has also been identified in EMC.[61–63] This variant is associated with rhabdoid phenotype and high-grade features.[63] FISH detection of the EWSR1 gene rearrangement can be helpful[64] but the occurrence of EWSR1 gene rearrangements in different tumors has to be considered (see Chapter 5).

Extraskeletal mesenchymal chondrosarcoma
Clinical features

Mesenchymal chondrosarcoma (MCHSA) also occurs as a primary soft tissue tumor although it is 3–4 times more common in bone. It typically presents in teenagers and young adults, with a marked female predilection.[65–70] Head and neck, including orbit and meninges are the most common sites, and the thigh is the most common peripheral site. MCHSA is a high-grade tumor with a 10-year survival rate of only 25%. Metastases are generally to the lungs. The treatment of choice is surgery and adjuvant therapy. Newer chemotherapies include docetaxel and bevacizumab.[69,70]

Imaging studies

Imaging of extraskeletal MCHSA can reveal a nonspecific or calcified soft tissue mass on radiographs and CT (62% of cases by CT).[71,72] The calcification is typically mild, although it can be extensive (Figure 30.11a). In contradistinction to other chondroid neoplasms, the nonmineralized component does not show low attenuation on CT or markedly higher signal intensity on T2-weighted MR. This is likely a reflection of the higher degree of cellularity, with less myxoid matrix and lower water content. MR imaging usually reveals intermediate signal intensity on both T1-weighted and T2-weighted sequences. Intense enhancement is sometimes apparent on both CT and MR imaging following intravenous contrast administration, owing to the high vascularity (Figure 30.11).

Pathology

The size of MCHSA can range from 2.5 cm to >30 cm. Grossly, these tumors are circumscribed and fleshy, sometimes with visible chondroid foci (Figure 30.12).

By histology, MCHSA is characterized by primitive small round or slightly spindled cells, often with a hemangiopericytoma-like growth pattern (Figure 30.12), with

Chapter 30: Cartilage- and bone-forming tumors

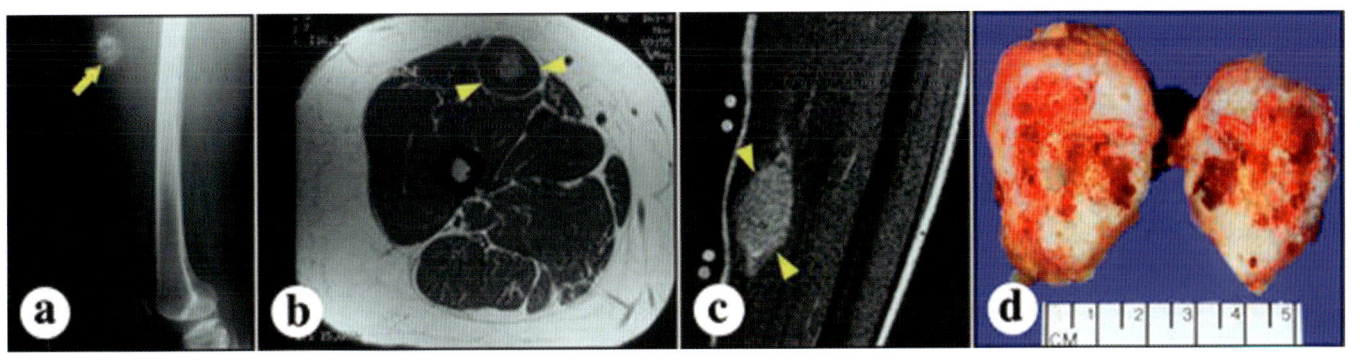

Figure 30.11 (**a**) Mesenchymal chondrosarcoma: lateral humeral radiograph shows a largely calcified soft tissue mass (arrow). (**b,c**) An example in the anterior thigh: axial T1-weighted and sagittal T2-weighted MR images reveal a soft tissue mass (arrowheads) with nonspecific signal intensity. (**d**) Mesenchymal chondrosarcoma of soft tissue grossly has glistening gray (representing hyaline cartilage) and white firm areas (representing round tumor cells).

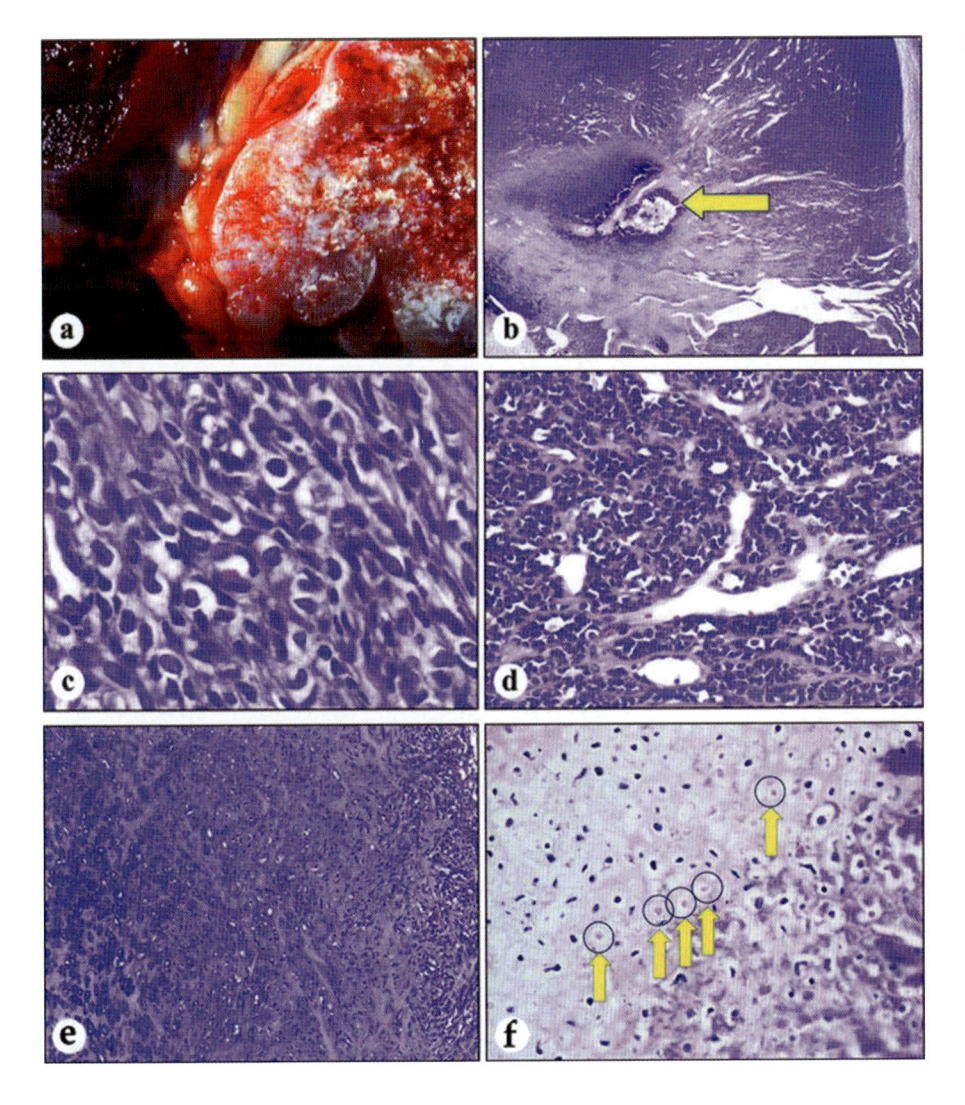

Figure 30.12 (**a**) Grossly, mesenchymal chondrosarcoma reveals glistening gray cartilage, solid tumor areas, and flecks of bone. (**b**) This tumor has a zone distribution of central hyaline cartilage, compared with peripheral round cells histologically. (**c,d**) The primitive round cells in mesenchymal chondrosarcoma demonstrate a hemangiopericytoma-like vascular pattern. (**e**) Fairly abrupt transition of the primitive cells to hyaline cartilage is typical of mesenchymal chondrosarcoma. (**f**) Cell death and endochondral ossification (arrows) and replacement by bone.

usually an abrupt transition to single or coalescing, centrally located lobules of well-differentiated hyaline cartilage. The cartilage shows zonal transition from small primitive chondrocytes to proliferating and hypertrophic chondrocytes that simulate endochondral bone formation in the center of the cartilage islands (Figure 30.12).

By immunohistochemistry, the chondrocytes are positive for S100 protein and most of the small round cells negative.

833

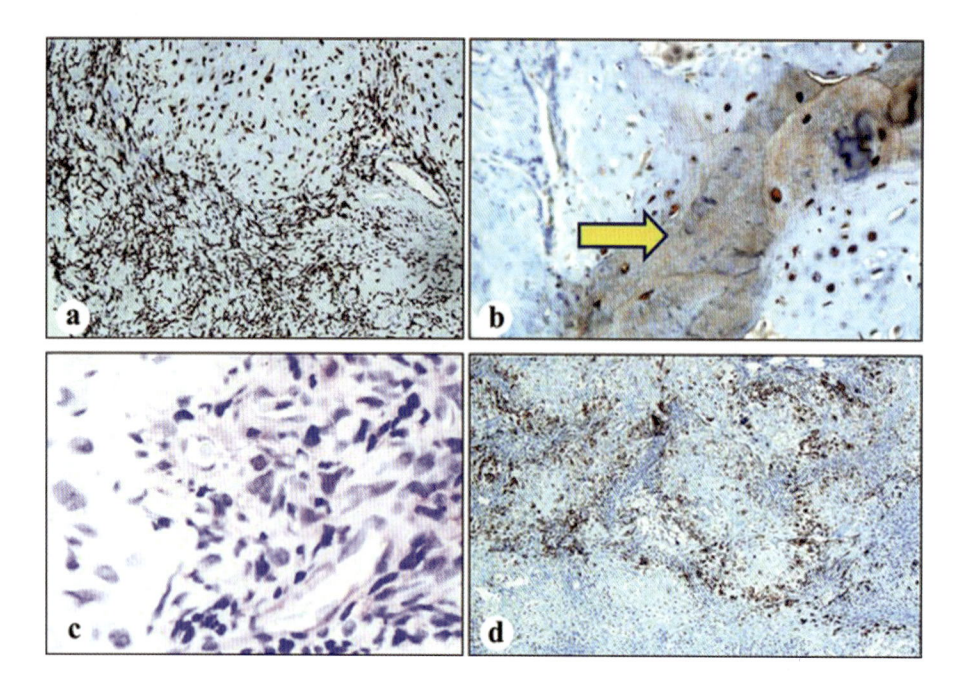

Figure 30.13 (**a**) Immunohistochemically typical of mesenchymal chondrosarcoma is SOX9 nuclear staining in both round cells and cartilage components. (**b**) Central cartilage island cells and matrix are positive for osteocalcin. (**c**) Focal nuclear beta-catenin immunoreactivity at the interface between round cells and hyaline cartilage. (**d**) Focal desmin staining in MCHSA, but negative for skeletal-muscle-specific markers. S100 protein is focally positive in round cells.

SOX9, the master regulator of chondrogenesis, is expressed in both round cells and hyaline cartilage (Figure 30.13).[73,74] Rare tumor cells are immunoreactive with desmin (Figure 30.13) and EMA, but keratins, SMA, myogenin, MyoD1, and synaptophysin, are negative. INI1 is retained in MCHSA.[73,74]

The differential diagnosis for MCHSA in soft tissue includes small round cell tumors such as Ewing sarcoma, small cell osteosarcoma, and rhabdomyosarcoma, and poorly differentiated synovial sarcoma. Attention to the unique constellation of cartilage islands and small round cells in a dense matrix allows identification of MCHSA. In minimal samples, molecular genetic studies may be necessary.

Genetics

HEY1-NCOA2 fusion appears to be the defining gene fusion in MCHSA present in most cases.[75] Previous cytogenetic studies have shown complex karyotypes with numerous chromosomal rearrangements, such as t(13;21)(q10;q10) and t(11;22)(q24;q12) in one case each.[76,77]

Phosphaturic mesenchymal tumor of mixed connective tissue variant (oncogenic osteomalacia)

Clinical features

Phosphaturic mesenchymal tumor (PMT) is a very rare tumor. It is clinically associated with *oncogenic osteomalacia*, a rare paraneoplastic syndrome resulting in phosphate wasting, profound weakness, and pathologic fractures. Chemically there is hypophosphatemia, normocalcemia, and increased levels of alkaline phosphatase. Overexpression of fibroblast growth factor-23 (FGF-23) by the tumor cells is the probable underlying cause implicated in renal tubular phosphate loss. Patient ages range from 9 to 80 years (median 53 years).[78,79] Recognition of PMT is critical, because tumor removal cures the oncogenic osteomalacia, with postoperative resolution of serum chemistries, yet often leaving residual radiologic evidence for a lesser degree of osteomalacia, especially after a long-standing history. Most of these tumors are benign, with a low potential for recurrence after surgical excision. Rare malignant examples with lung metastases have been reported.[79] Serum FGF-23 levels can be measured in patients to track their postoperative tumor-free status.[80] If surgical treatment is not possible, pharmacologic induction of hypoparathyroidism can partially resolve osteomalacia.[81] Most of these tumors are histologically characterized as *phosphaturic mesenchymal tumors of mixed connective tissue variant* (PMTMCT).[9,79]

Imaging studies

PMT has a nonspecific imaging (i.e., ultrasonography, CT, or MR) appearance. Contrast enhancement of these lesions is common, reflecting prominent vascularity. Radiographic findings of multiple insufficiency fractures in young patients with unexplained osteopenia and indistinct trabeculae (favoring osteomalacia as opposed to osteoporosis) are common. This constellation of findings and lack of other causes of osteomalacia or palpable soft tissue mass can require a diligent search for the offending lesion by whole-body CT, or MR imaging or nuclear medicine studies, including fluorine-18 fluoro-2-deoxy-D-glucose positron emission tomography (FDG-PET) or octreotide scan (Figure 30.14).[82]

Pathology

Grossly, tumors range from <2 cm to >10 cm (Figure 30.14). Histologically common is resemblance to hemangiopericytoma (Figure 30.15) with a mesenchymal milieu including myofibroblasts, fat, and grungy calcification (Figure 30.16), as well as osteoclast-type giant cells (Figure 30.17), with mimicry of other giant-cell-rich tumors. Bone and myxochondroid features are also present (Figure 30.18). Several other bone and soft tissue tumors, such as sclerosing osteosarcoma, giant cell tumor, fibrous dysplasia, sinonasal hemangiopericytoma, hemangiopericytoma, and others have been reported to cause oncogenic osteomalacia, but some of these might in fact have been examples of PMTMCT type.[79] These tumors usually have low cellularity, myxoid change, bland spindled cells, distinctive

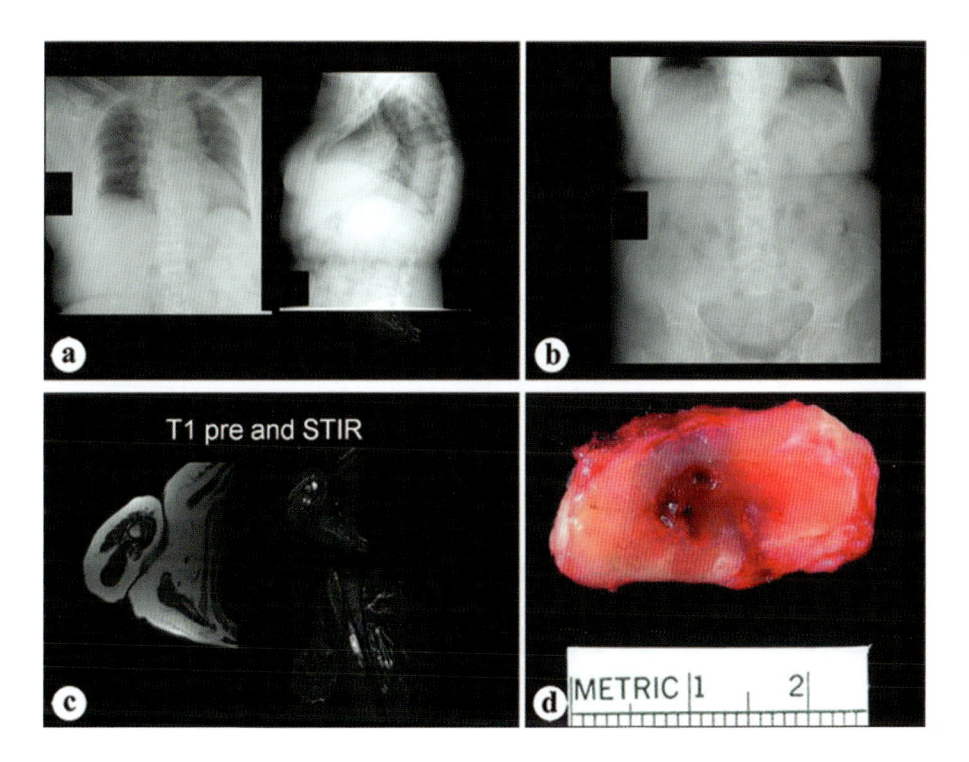

Figure 30.14 Phosphaturic mesenchymal tumor with associated osteomalacia. (**a,b**) Chest and abdominal radiographs show prominent osteopenia with insufficiency fractures involving the spine, ribs, and pubic ramus. (**c**) Multiple MR images demonstrate a nonspecific small soft tissue mass in the upper arm representing the causative neoplasm. (**d**) Gross image showing delineated process with solid, gelatinous (myxoid), calcified areas.

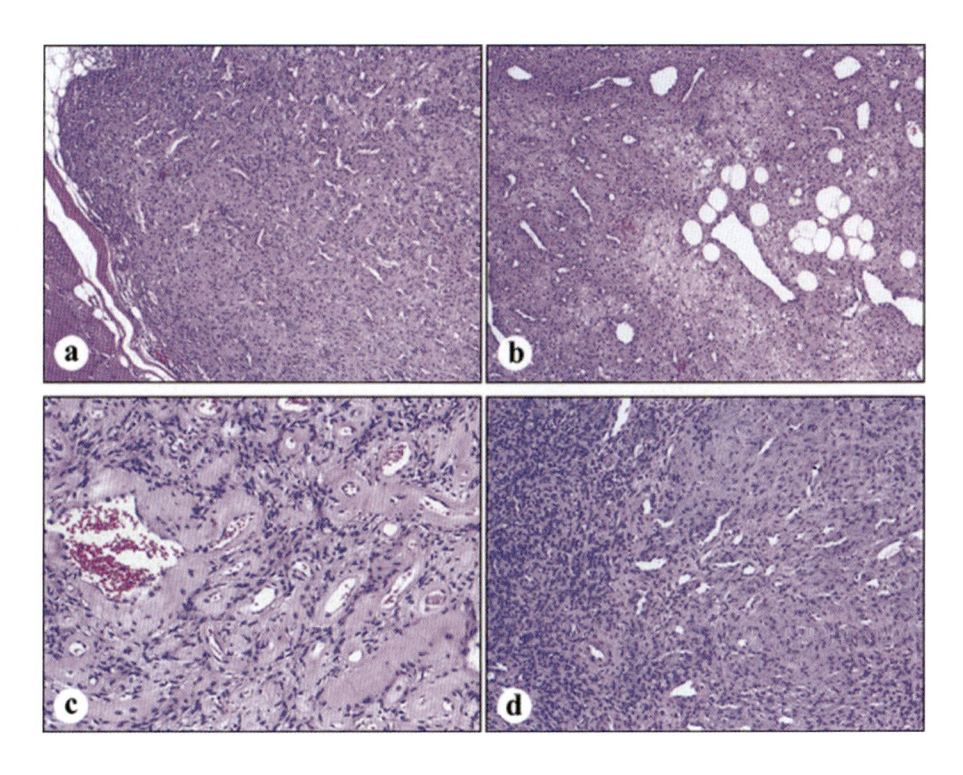

Figure 30.15 (**a,b**) By histology, phosphaturic mesenchymal tumor, mixed connective tissue type, shows hemangiopericytoma-like cells with occasional fat cells. (**c,d**) Perivascular hyalinization and staghorn vasculature are also typical.

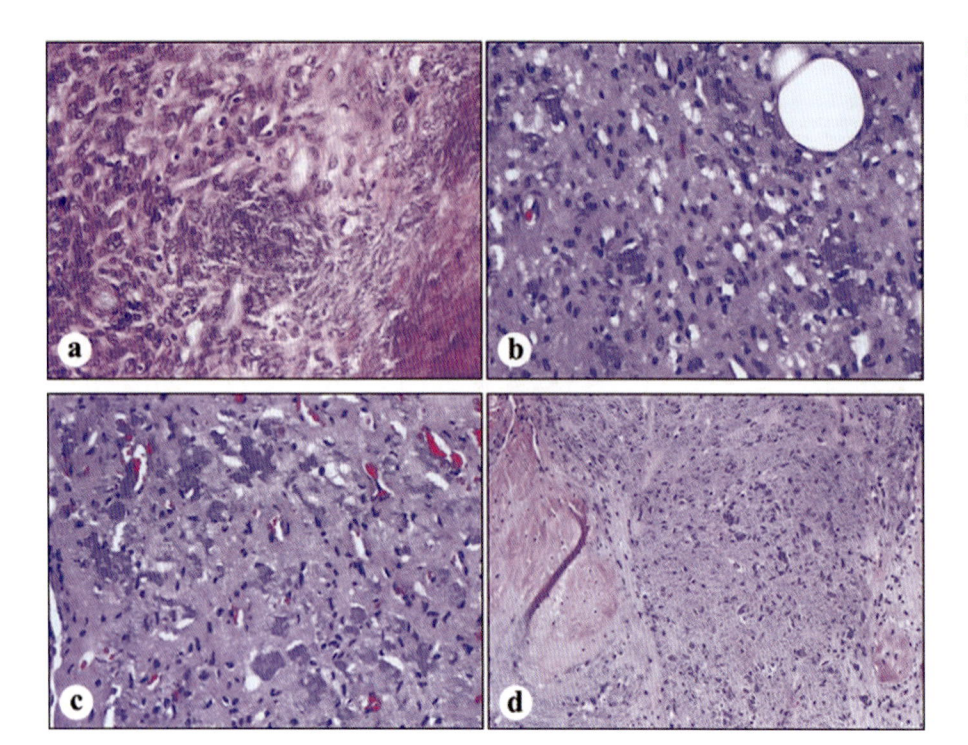

Figure 30.16 Additional features of phosphaturic mesenchymal tumor: (**a**) Grungy stippled calcification. (**b,c**) Multiple foci of calcification. (**d**) Myxoid change and hyalinization.

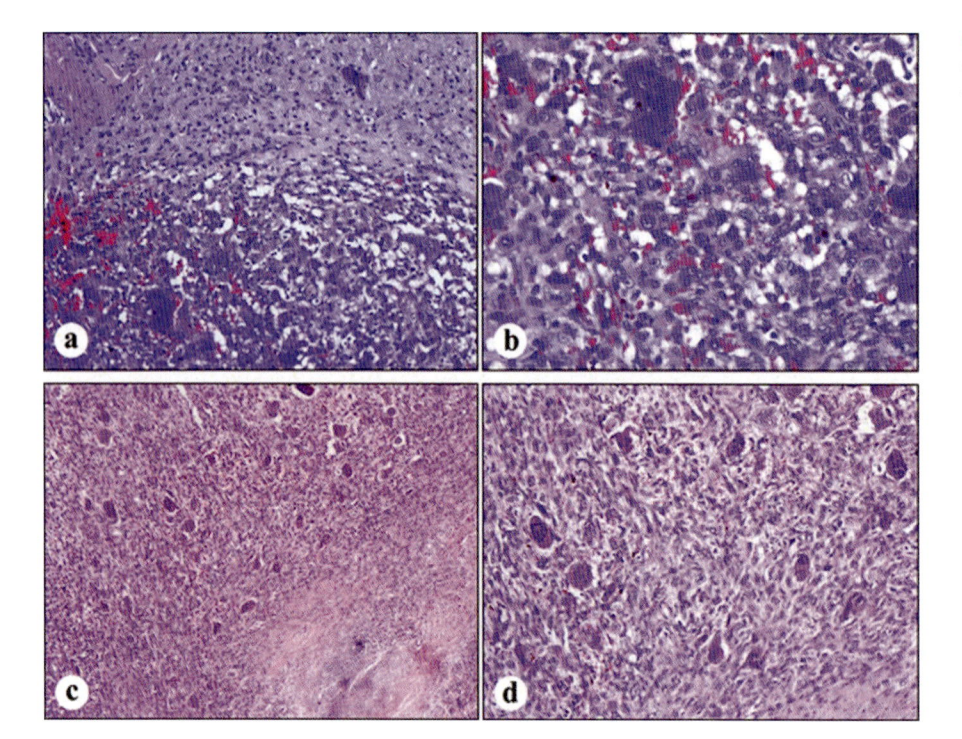

Figure 30.17 Numerous osteoclast-like giant cells can mimic giant cell tumors of soft tissues and bone.

"grungy" calcified matrix, fat, HPC-like vessels, microcysts, hemorrhage, osteoclasts, and an incomplete rim of intramembranous ossification. Four of these benign-appearing PMTs contained an osteoid-like matrix. Three other PMTs were hypercellular and cytologically atypical and were considered malignant.[79]

By immunohistochemistry, PMT is negative for CD34, desmin, S100 protein, and keratin, but can be positive for actin.[79] SSTR2A (somatostatin receptor 2A) is reported as a new immunohistochemical marker for PMT.[83]

The differential diagnosis includes solitary fibrous tumor/ hemangiopericytoma, other tumors that produce grungy

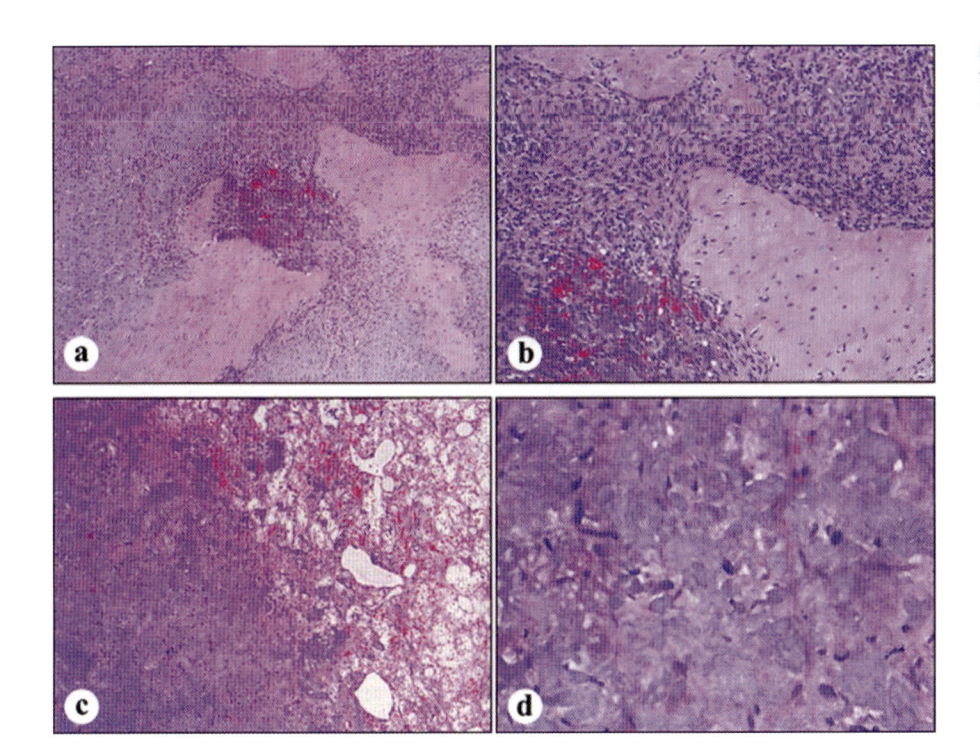

Figure 30.18 (**a,b**) Phosphaturic mesenchymal tumor can have islands of bone. (**c,d**) Myxochondroid appearance is seen in some cases.

calcification, such as soft tissue chondroma, and other osteoclast-type giant-cell-rich lesions. Most important, PMTs are negative for CD34, unlike the usual solitary fibrous tumor/hemangiopericytoma, with the latter lacking osteoclast-type giant cells or grungy calcification. History and morphology can separate the other tumors from PMT.

RT-PCR for FGF-23 confirms the role of this protein in PMT-associated oncogenic osteomalacia.[79] A novel FN1-FGFR1 fusion gene discovered by next-generation RNA sequencing and FISH analysis seems to be present in a majority of cases.[84]

Tumoral calcinosis
Clinical features

Tumoral calcinosis is the designation for an extraskeletal soft tissue hydroxyapatite calcification with a granulomatous response that develops in patients with secondary hyperparathyroidism or hypercalcemia, usually idiopathic or because of end-stage kidney disease. Clinical classification of these lesions into three types has been proposed: solitary lesion without hyperphosphatemia, multiple lesions associated with hyperphosphatemia, but not hypercalcemia on a familial basis, and lesions associated with hypercalcemia secondary to renal disease.[85,86] In addition, there is a variant occurring in peripheral extremities, termed "tumoral calcinosis-like lesion."[87] Historical terminology for tumoral calcinosis and related lesions includes calcifying collagenolysis, calcifying bursitis, tumoral lipocalcinosis, pseudotumoral calcinosis, and hip stone disease.

In the first clinical group, the solitary lesions occur in patients without hyperphosphatemia (*primary normophosphatemic*). These patients are often from tropical or subtropic areas, and the lesions can clinically simulate parasite granulomas.[85]

In the second group, multiple familial lesions usually occurred in African American adolescents with hyperphosphatemia (*primary hyperphosphatemic*), who have elevated serum 1, 25-dihydroxyvitamin D levels. This type is inherited as an autosomal recessive or dominant pattern.[86,88]

The third group (*secondary tumoral calcinosis*) is the most common. In this group, the lesions occurred with secondary hypercalcemia or hyperphosphatemia, often linked to end-stage chronic renal disease. These patients are older, do not have familial disease, and can have visceral involvement. Less frequently, tumoral calcinosis is associated with a hypercalcemia state, with massive bone destruction by metastases, hypervitaminosis D, or other causes.[85]

Tumoral calcinosis arises from an error in phosphorus metabolism that leads to extracellular deposition of calcium hydroxyapatite crystals.[88] These patients also have dental abnormalities, transient periosteal thickenings, and eye lesions.[89]

Tumoral calcinosis typically involves the periarticular areas of the shoulder, scapulae, hips, buttocks, and elbows, but is not intra-articular. Symptoms include a firm nontender mass, often with signs of inflammation. Sinus tracts with excretion of white creamy material sometimes occur. The natural history for tumoral calcinosis is one of progressive slow growth and disfigurement, with limitation of joint function.[85] Treatment is

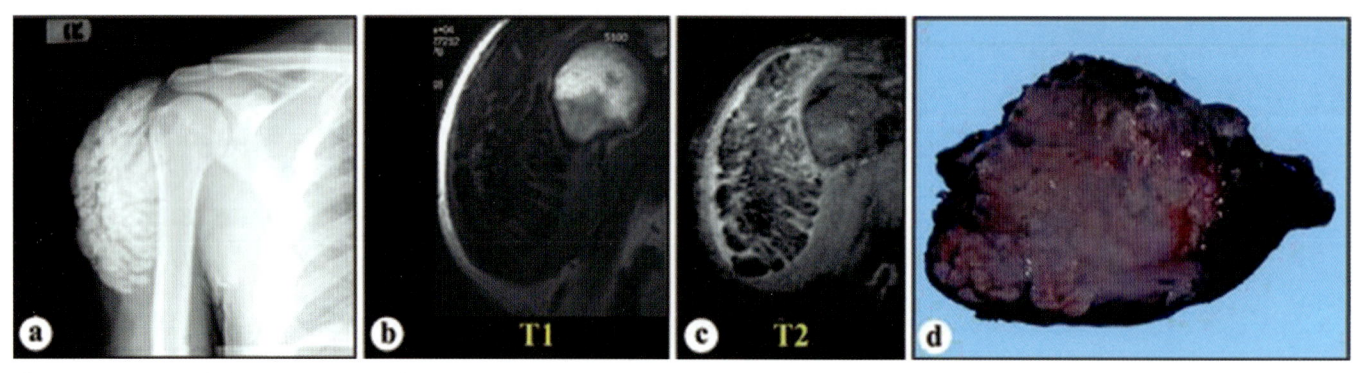

Figure 30.19 Tumoral calcinosis around the shoulder. (**a**) Radiograph shows a large calcified periarticular soft tissue mass with radiolucent septations. (**b**) Coronal T1-weighted image demonstrates peripheral and septal areas of high signal intensity. (**c**) T2-weighted MR image also reveals the large soft tissue mass with predominantly low signal intensity. (**d**) Grossly, tumoral calcinosis has a lobular glistening to fleck-like appearance.

difficult because local excision of a large mass is often impossible and results in rapid recurrence. The first group can receive surgical treatment alone.[90] There has been some success with surgical treatment in the second group as well.[90] Treatment of the third group is mainly medical, aimed at correction of the metabolic error by dietary restriction of phosphorus, use of phosphate binders to lower the serum phosphorus levels, a low-calcium dialysate for dialysis, renal transplantation, and parathyroidectomy.[91]

Tumoral calcinosis-like lesions of the distal extremities are sporadic, have a Caucasian female predominance, and occur in all ages.[87] The hands, fingers, and wrists are more common locations than the feet, toes, and ankles, and most are solitary, but can be multiple. Chief initial complaints included the presence of either a painful or asymptomatic mass and limitation of joint mobility.[87] Pertinent clinical associations included antecedent trauma, scleroderma, long-standing osteoarthritis, bony deformities (including infants with congenital hand malformations), and chronic renal failure.[87] Most patients have no evidence of recurrent disease, with only rare development of additional lesions after simple (local) excision. In some cases, acral lesions can be a sign of scleroderma.[87]

Imaging studies

Tumoral calcinosis reveals a large area of dense periarticular calcification on radiographs or CT.[92] Radiolucent septations are also frequently seen. MR imaging usually demonstrates low signal intensity on all pulse sequences. The calcium fluid level is also sometimes apparent on CT or MR imaging. A small peripheral rim and septa of higher signal intensity is common on T2-weighted images, and these areas might be enhanced following contrast administration. These lesions show intense radionuclide uptake on bone scintigraphy, reflecting the increased calcium and phosphate turnover. The radiographic picture, with fluid–fluid levels, is classic (Figure 30.19).

Pathology

Grossly, these lesions vary in size. The distal extremity lesions range from 0.3 cm to 4.5 cm, but the proximal lesions can reach as large as 30 cm in diameter. Grossly, the lesions are a single, firm, multiloculated, cystic mass with a rubbery texture that contains milky white to gritty yellow fluid (Figure 30.19). Cutaneous ulceration can develop over subcutaneous deposits.

By histology, tumoral calcinosis lesions are essentially similar, irrespective of pathogenesis showing amorphous multilocular calcifications surrounded by multinucleated histiocytes of foreign body and Langhans type (Figure 30.20).[93,94] Cysts lined by fibrous walls are also common, and osseous metaplasia can develop in long-standing lesions.[93–95] The distal lesions located in tenosynovial/fascial tissue more commonly contain small calcospherites as opposed to amorphous calcifications.[87] Like PMT (above), tumoral calcinosis can also have FGF-23 overexpression, yet with hyperphosphatemia.[96]

The differential diagnosis includes calcified soft tissue chondroma having granuloma-like features; the presence of hyaline cartilage (in chondroma) can aid in distinguishing the two lesions. Tophaceous pseudogout has calcium pyrophosphate rhomboid crystals.

Genetics

Germline mutations in three genes regulating calcium and phosphorus metabolism, GALNT3, FGF23, and KL (klotho), commonly occur in hyperphosphatemic familial tumoral calcinosis patients.[96–99] SAMD9 mutations have been detected in normophosphatemic patients[100] and GATA3 mutation in hypophosphatemic patients with familial tumoral calcinosis.[101]

Tophaceous pseudogout (focal tumoral calcium pyrophosphate dihydrate deposition or crystal deposition disease)

Clinical features

This rare condition can form a mass with abundant calcium pyrophosphate crystals. Focal tumoral calcium pyrophosphate dehydrate (tophaceous pseudogout, T-CPPD) is distinctive from polyarticular chondrocalcinosis and polyarticular calcium

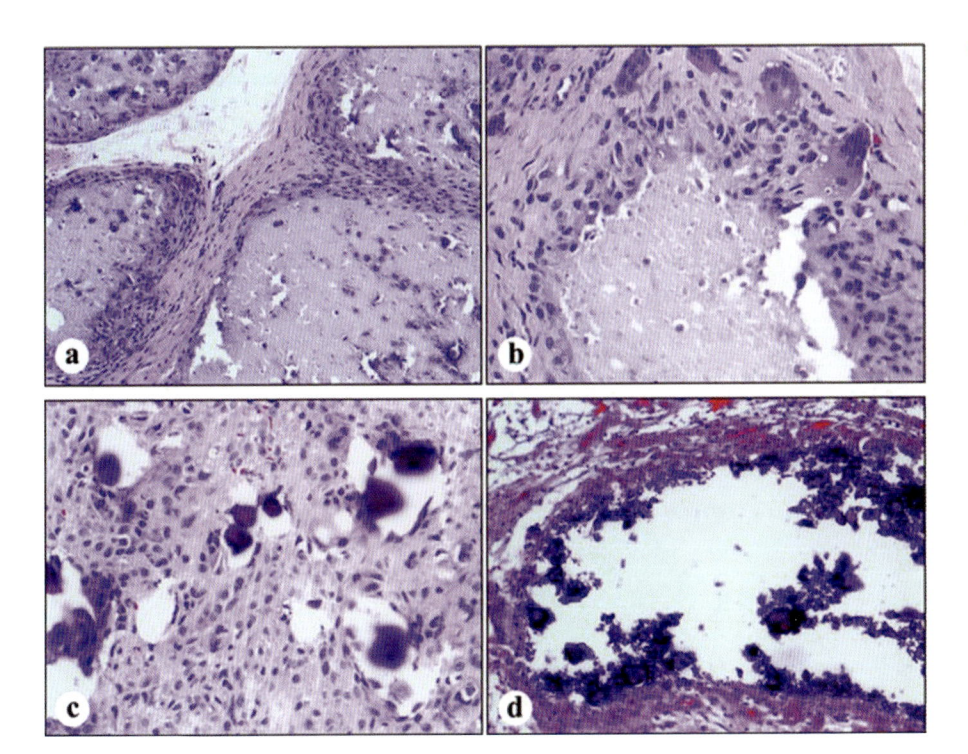

Figure 30.20 (**a,b**) Microscopically, tumoral calcinosis has central amorphous hydroxyapatite material surrounded by a granulomatous (histiocytoid) response, with multinucleated giant cells. (**c,d**) Sometimes the hydroxyapatite is psammoma-like or chunky, but does not demonstrate rhomboid crystals, as seen in calcium pyrophosphate. The histiocytic host response is absent.

pyrophosphate deposition disease (CPPD). It typically occurs in middle-aged and older adults, with a female predominance. Locations include the temporomandibular,[102] metacarpophalangeal, first metatarsophalangeal, paraspinal, and rarely large joints, in descending order of frequency.[103] Presenting symptoms include swelling, pain, or neurovascular compromise. The clinical, radiographic, and pathologic features can resemble chondrosarcoma, and it can destroy bone.[102,103] Although local recurrence can happen after complete or incomplete surgical excision, local excision is usually curative.

Imaging studies

Imaging of tophaceous pseudogout is nonspecific, with a peri-articular calcified soft tissue mass. Unlike tumoral calcinosis, however, these calcifications are punctuate and not as extensive. CT and MR imaging studies reveal a heterogeneous mass. There is usually low to intermediate signal intensity on all MR pulse sequences. Joint involvement and cartilage calcification (chondrocalcinosis) are typically absent, unlike CPPD arthropathy (Figure 30.21).

Pathology

Grossly, tophaceous pseudogout is a whitish-gray mass with a chalky appearance. Involvement of the adjacent bone from pressure erosion is common.[102] The gross features can mimic a benign or malignant chondroid neoplasm.

By histology, the crystals of tophaceous pseudogout are rhomboid (Figure 30.22) and stacked, and may be observed on H&E, although identification requires polarization and these crystals are weakly positively birefringent by polarization. Sometimes one has to search for a while to find crystals. These crystals may also be detected by alizarin red stain. A foreign body granulomatous response is often observed, but may be absent. Hyaline and fibrocartilage surround the crystals and often exhibit cellular atypia, mimicking benign and malignant chondroid neoplasms. Occasionally a florid myofibroblastic proliferation is observed.[104]

The differential diagnosis includes benign chondroid neoplasms. The chalky gross appearance, the presence of a granulomatous response, and the crystals separate this from chondroma of soft parts, although a case has been reported of soft tissue chondroma with calcium pyrophosphate.[105] PMT of the mixed connective tissue type can also have grungy calcification that mimics tumoral calcinosis or tophaceous pseudogout. Tumoral calcinosis is composed of hydroxyapatite crystals, rather than calcium pyrophosphate, and the patient has a different clinical history. Gouty lesions are surrounded by histiocytes and giant cells, with stellate, pointed crystals, rather than rhomboid, which are instead negatively birefringent on polarized microscopy (Figure 30.22).

Genetics

There is no known genetic association for tumoral calcinosis, although systemic calcium pyrophosphate deposition has been considered to be associated with mutations of the ANKH gene, located at 5p15.1.[106]

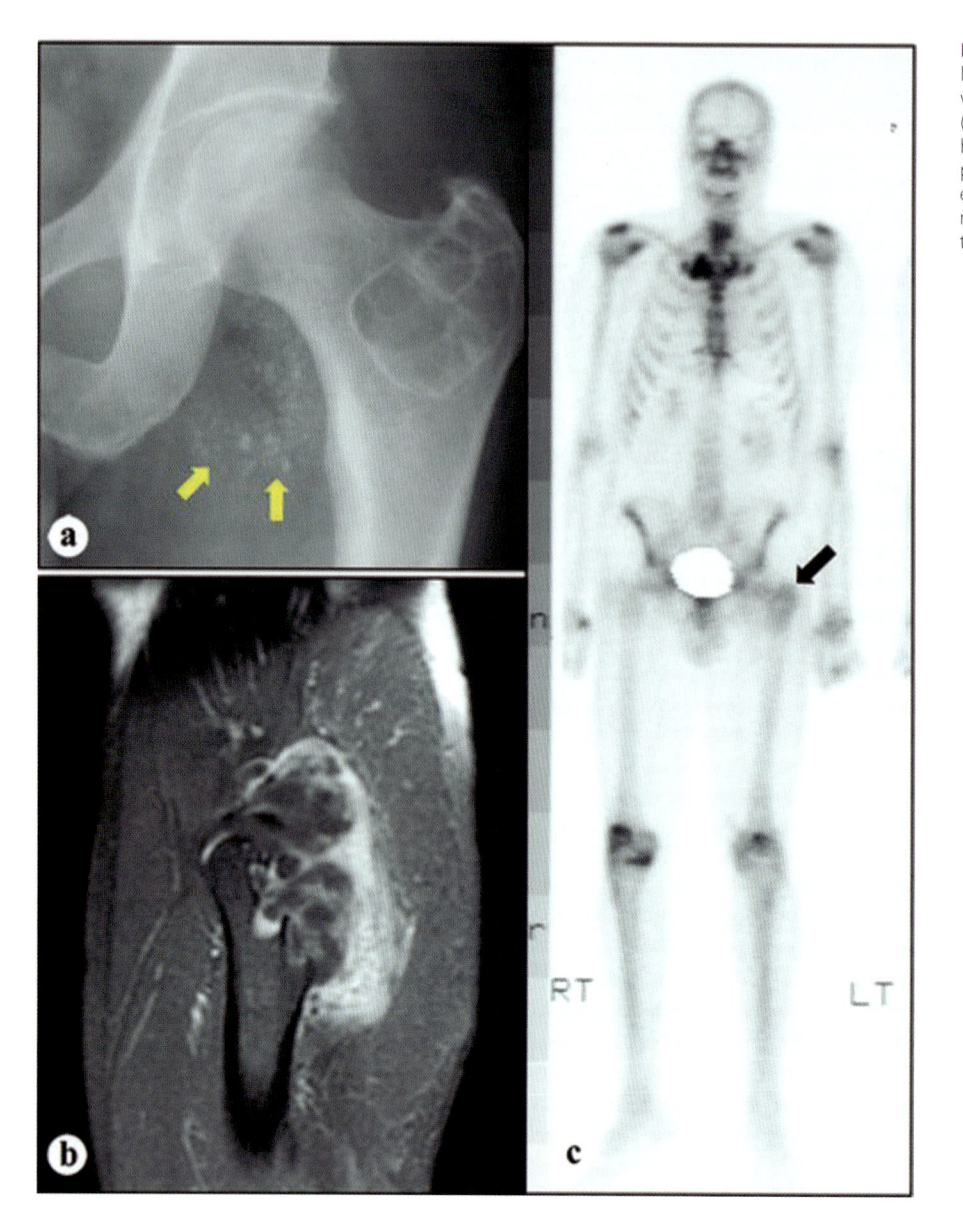

Figure 30.21 Tophaceous pseudogout about the left hip. (**a**) Hip radiograph shows a soft tissue mass with small punctate calcifications (arrows). (**b**) Sagittal T2-weighted MR image demonstrates a heterogeneous soft tissue mass with predominately low signal intensity and extrinsic erosion of the femur. (**c**) Bone scintigraphy reveals mild increased radionuclide uptake (arrow) about the left hip.

Calciphylaxis (systemic calcinosis, calcific uremic arteriolopathy, or metastatic calcification)

Clinical features

Calciphylaxis[107–110] (Figure 30.24) refers to small vessel calcification, occlusion, and resultant skin necrosis and subcutaneous acute panniculitis, and sepsis. It is usually associated with end-stage renal disease, other causes of secondary hyperparathyroidism, associated collagen vascular disease, or massive obesity. Calciphylaxis is a clinical emergency, most often developing in patients with end-stage renal disease. Painful, violaceous mottling or erythematous papules, plaques, or nodules develop in the distal lower extremities and rapidly progress. These patients typically develop symmetric superficial skin necrosis, leading to nonhealing ulcers.[107]

Cutaneous calcific uremic arteriolopathy is more frequent than systemic and clinically presents as a first phase of arteriolar lesion demonstrated by cutaneous hardening and erythema, followed by a second phase of tissue ischemic damage shown by ulcerations and scars. Calciphylaxis is associated with elevated parathyroid hormone levels and a dysregulation of the calcium/phosphate metabolism. Vascular calciphylaxis is a component of the widespread soft tissue calcification that occurs when the physiologic calcium phosphate solubility threshold is exceeded.[111] Systemic calcific uremic arteriolopathy is rare. Visceral organs can be involved, and there is high mortality rate, 60% to 80%,[107,108] usually from sepsis.

Chapter 30: Cartilage- and bone-forming tumors

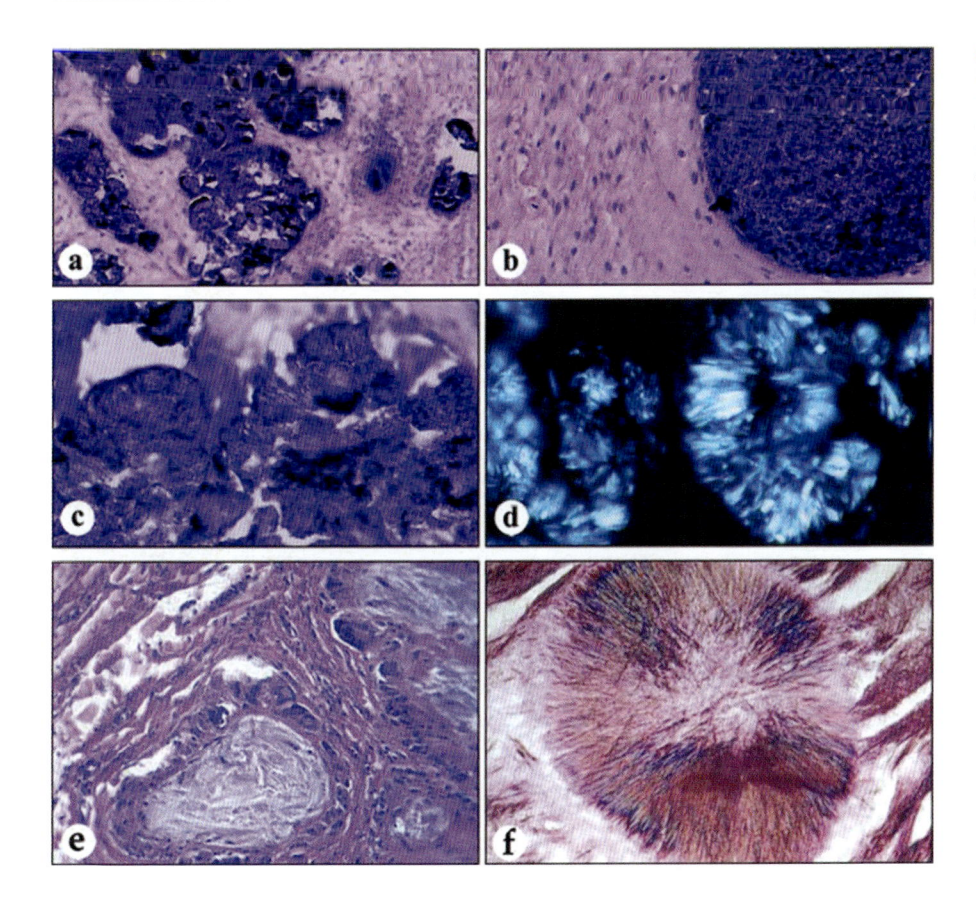

Figure 30.22 (**a,b**) Tophaceous pseudogout is histologically demonstrated by purple crystalline material that has spindled fibroblasts, but not much histiocytic or giant cell response. (**c**) Higher magnification demonstrates the rhomboid structure of the calcium pyrophosphate crystals. (**d**) The crystals are positively birefringent on polarized microscopy. (**e**) Gout, in comparison, has more pink material surrounded by a marked histiocytic and giant cell response. (**f**) The urate crystals are needle-shaped and negatively birefringent on polarized microscopy.

Imaging studies

The "pipe stem" calcifications of large vessels can be detectable radiographically.[92,112] This is also true for slowly developing metastatic calcifications involving larger vessels (Figure 30.23). Rapidly developing acute calcifications in small vessels are difficult to detect radiologically. Bone scintigraphy, however, can reveal areas of calcification with diffuse uptake of radionuclide in the soft tissue, as opposed to normal osseous activity, as diffuse blurry findings on a bone scan (Figure 30.23).

Pathology

Grossly, this condition involves soft tissue, especially diffuse calcification of the medial layer of small- to medium-sized vessels. By histology, small- to medium-sized vessels in the subcutis show medial calcification and endovascular fibrosis. Sometimes the calcifications can be observed occluding the lumen and in soft tissue adjacent to vessels. There is often associated acute panniculitis and dermal-epidermal necrosis, with apoptosis of dermal structures and sloughing (Figure 30.24). There is no known genetic background for this disorder.

Myositis ossificans

Clinical features and related lesions

Myositis ossificans is a reactive, potentially pseudosarcomatous fibro-osseous lesion in the skeletal muscle of the proximal extremities and trunk of young patients. It generally has a zonal pattern and matures over several weeks to form a peripheral rim of bone. This lesion typically occurs in children and young adults equally in both sexes.[113–116] The most common locations are the anterior muscles of the thigh or the buttocks. Its relationship with the so-called soft tissue aneurysmal bone cyst has been discussed, but is unclear.[115] Alternative or older terminology includes myositis ossificans circumscripta, pseudomalignant heterotopic ossification, pseudomalignant osseous tumor of soft tissue, or simply, heterotopic ossification.

Panniculis ossificans involves the subcutis of the upper extremities of mainly female patients and demonstrates a less prominent zoning phenomenon than myositis ossificans.[114] Fascial lesions are termed *fasciitis ossificans*. Small reactive heterotopic ossifications (sometimes named "osteomas"), other than myositis ossificans, can be seen in sites of trauma (e.g., tendon avulsion), as a post-traumatic ossifying lesion. These heterotopic calcifications are intramembranous and endochondral. They are metaplastic bone that lacks the zonation and clinicopathologic features of myositis ossificans, composed primarily of mature lamellar bone with a well-defined Haversian system with bone marrow, as well as myxoid vascular and fibrous connective tissue between the bony trabeculae. Sites include the abdomen, posterior tongue, and the thigh.[117]

Myositis ossificans is associated with trauma in 75% of cases. It is a rapidly growing painful mass that expands for

Chapter 30: Cartilage- and bone-forming tumors

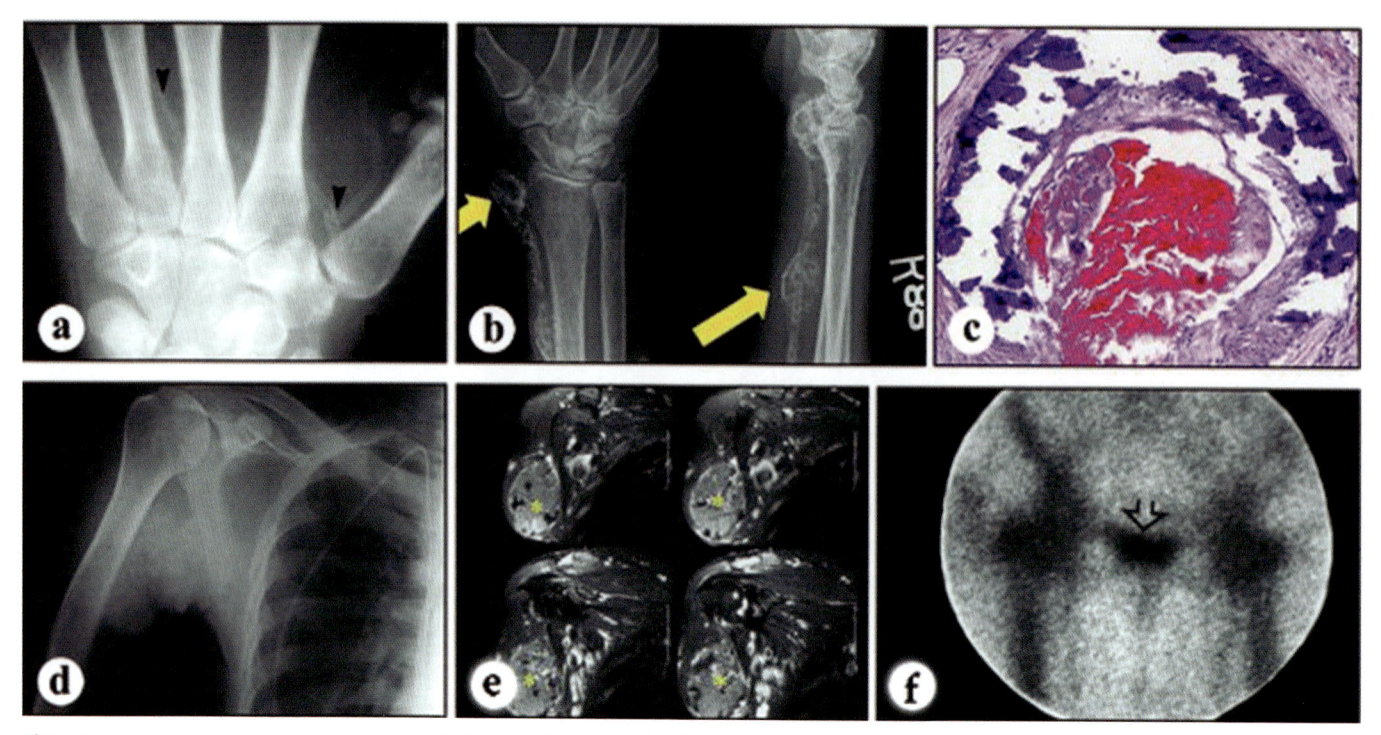

Figure 30.23 Long-standing renal disease and soft tissue calcifications. (**a,b**) Wrist radiographs show "pipestem" vascular calcifications (*arrowheads*) and shunt aneurysm formation (*arrows*). (**c**) Histologically there is medial calcification of larger vessels. (**d**) The radiograph shows a vaguely calcified soft tissue mass near the shoulder, similar to tumoral calcinosis; chest tubes are for hemodialysis access. (**e**) Multiple coronal T2-weighted MR images reveal a heterogeneous soft tissue mass. (**f**) Calciphylaxis instead contains calcifications of small vessels and tissues; bone scintigraphy shows extensive soft tissue radionuclide uptake causing poorly defined osseous structures (*open arrow*) because of microscopic soft tissue calcification.

6 to 8 weeks, often with prodromal viral-like symptoms. After active growth, the pain ceases, and the lesion matures over several months. It can shrink and undergoes regression in 35% of cases.[118]

Surgical excision is necessary only if the lesions are unusually large, painful, restrict motion, or if the diagnosis is uncertain.[119] Myositis ossificans is a benign reactive process; a rare case is reported to have undergone "malignant transformation."[116,120] Many pathologists view such cases as well-differentiated extraskeletal osteosarcoma from their inception.

Imaging and pathology

The radiographic images and gross pathology are characteristic (Figure 30.25).[121] A well-developed lesion 3 to 6 weeks after trauma has a mature eggshell-like rim of bone at the periphery, both radiographically and grossly, which helps to distinguish this lesion from an osteosarcoma. In contrast, an osteosarcoma has reversed zonation with central bone formation infiltrating into soft tissue. Myositis ossificans is usually seen over the shaft of the bone and not at the metaphyseal junction, as in osteosarcoma.[119] Myositis ossificans is also limited to soft tissue, without attachment to bone. The outer shell of bone and soft tissue location also distinguishes myositis ossificans from parosteal osteosarcoma, which arises from the surface of bone and has central maturation rather than peripheral bone maturation. Radiographs of myositis ossificans demonstrate the zonal phenomenon seen pathologically, with a peripheral rim of ossification representing more mature bone. This calcification can require 3 to 6 weeks to form (dependent on the patient age), however, and is optimally detected by CT. Lesions typically mature over the ensuing months, forming a complete outer cortex and inner trabecular bone with yellow marrow when mature. Bone scintigraphy reveals markedly increased radionuclide activity on all phases initially, with gradual reduction in the degree of uptake with maturation. MR imaging often simulates a more aggressive process because the peripheral rim of ossification is difficult to detect and characterize as calcification. The soft tissue mass is intermediate in signal intensity on T1-weighting, has high signal intensity on T2-weighting, and enhances with contrast. There is a prominent zone of irregular surrounding edema that gradually recedes toward the soft tissue mass with maturation. Mature myositis ossificans demonstrates internal yellow marrow signal characteristics on MR imaging.[121]

Grossly, the lesion is white and glistening or has a gritty cut surface, rarely larger than 6 cm. Spicules of trabeculae radiate inward toward a less mature zone.

By histology, zonal maturing of the bone from the center to the periphery is the pathognomonic feature (Figure 30.26). The center often has a cellular fibroblastic proliferation similar to nodular fasciitis, with abundant mitoses and osteoclast-type giant cells. Osteoid bone that has not yet undergone maturation to woven bone can be present in the center as well

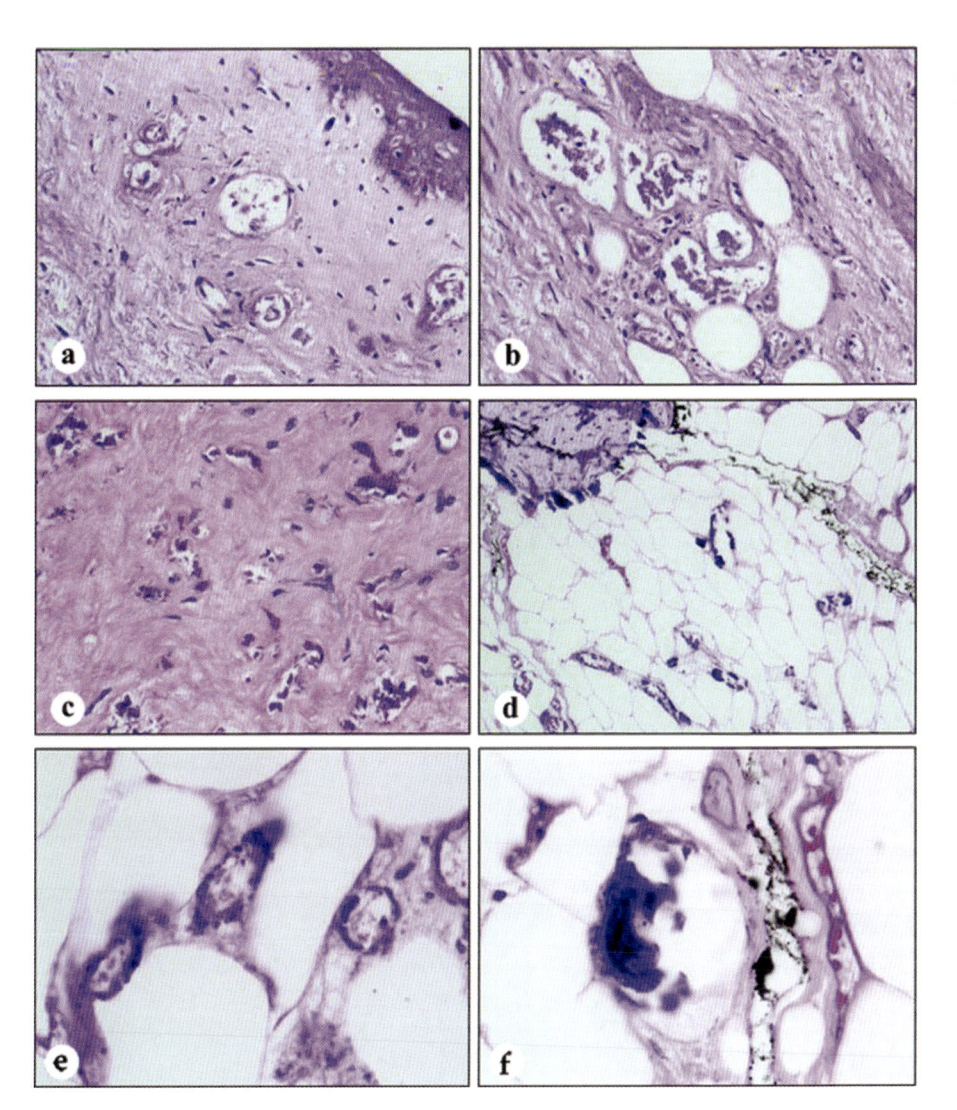

Figure 30.24 (**a,b**) Microscopically, calciphylaxis is often associated with dermal necrosis, including necrosis and sloughing of adnexal structures. (**c–f**) The calcifications can be located in soft tissue as well as in small vessels in the subcutis.

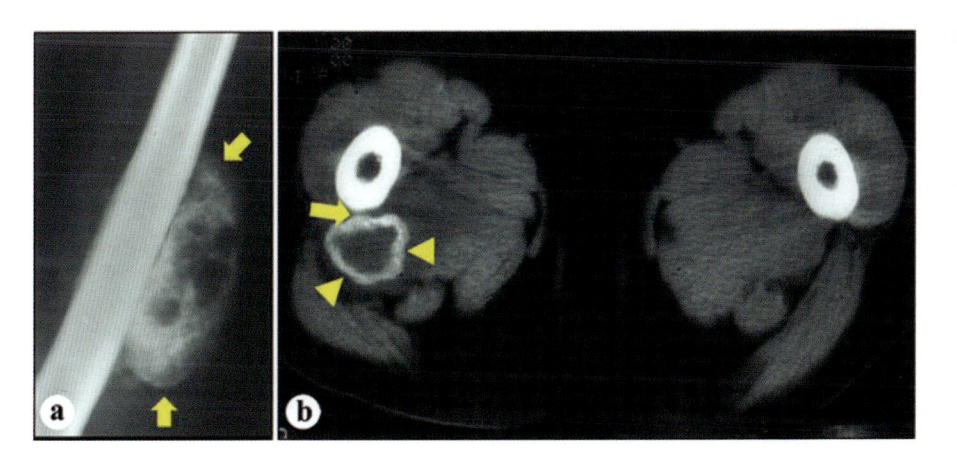

Figure 30.25 (**a**) Radiographic peripheral eggshell-like calcification distinguishes myositis ossificans lesions from osteosarcoma, which typically has central mature ossification and invasion of immature areas into soft tissue. (**b**) A CT scan of myositis ossificans of the right thigh reveals a lesion with a peripheral rim of calcification and separation of the mass *(arrow)* from the adjacent femoral cortex.

(Figure 30.26). This center initially is worrisome, suggesting osteosarcoma; waiting to rebiopsy over several weeks leads to the correct diagnosis, however. Cartilage also is sometimes present as part of the endochondral ossification (Figure 30.26). The rare occurrence of "malignant transformation" is hard to prove, but by radiologic imaging these entities look like myositis ossificans with a peripheral rim of bone, and histologically they have mature bone with fat or marrow and a "blowout" of one wall with true malignant neoplasm, producing malignant osteoid bone, namely, osteosarcoma (Figure 30.26).

Chapter 30: Cartilage- and bone-forming tumors

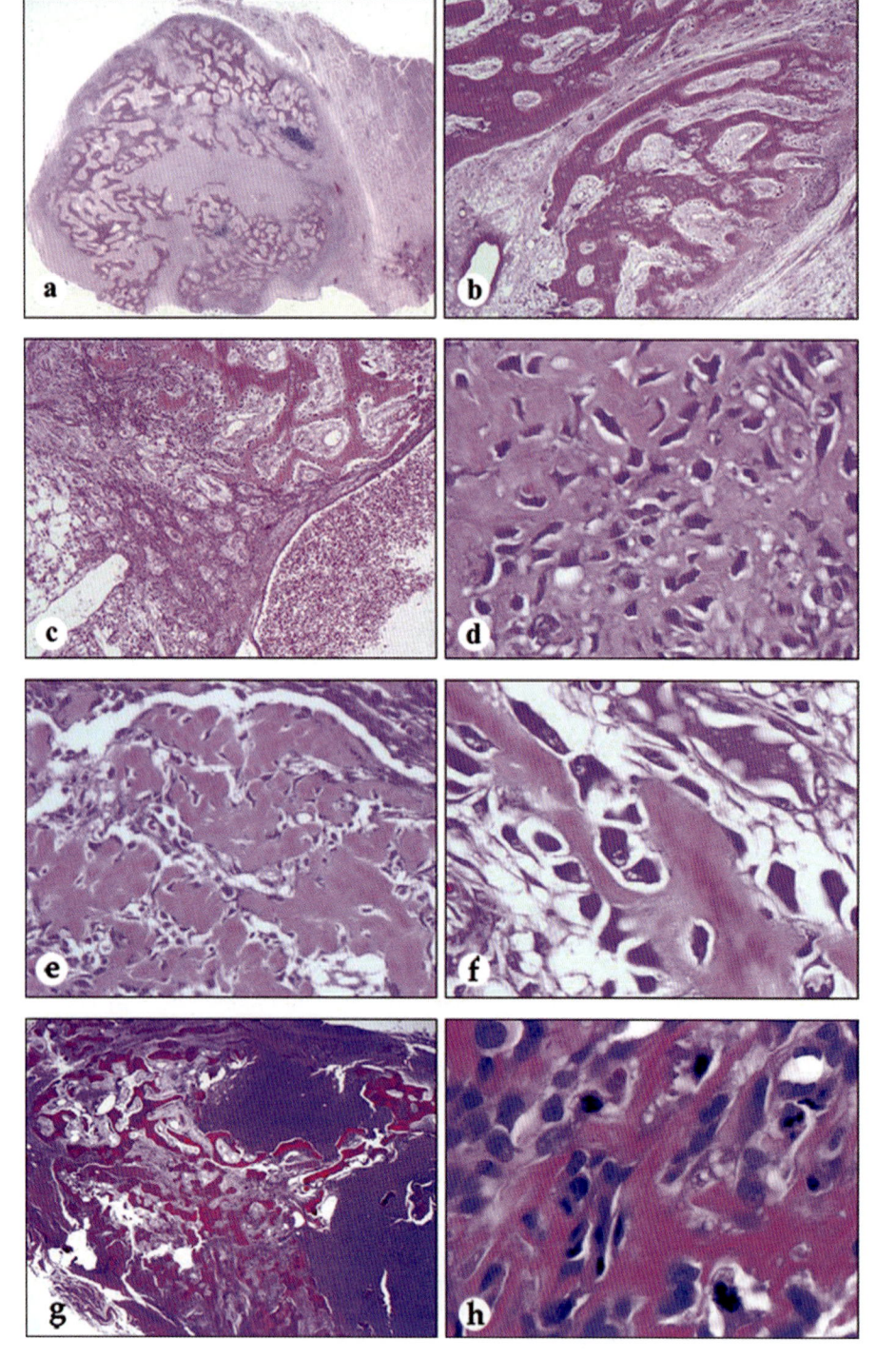

Figure 30.26 (**a,b**) Microscopically, myositis ossificans shows zonation with mature bone at the periphery to less mature areas in the center. (**c**) The center contains a hemorrhagic cyst, akin to a soft tissue aneurysmal bone cyst. (**d**) Immature woven bone can mimic osteosarcoma. (**e,f**) Osteoblastic rimming appears during lesional maturation. (**g**) Malignant transformation of myositis ossificans shows peripheral calcification with the tumor arising with the maturing bony elements. (**h**) Higher magnification reveals atypical osteosarcoma cells producing lace-like malignant osteoid and mitotic activity.

The differential diagnosis includes fibrodysplasia ossificans progressiva (FOP) and osteosarcoma. FOP can be distinguished by its multifocality and specific hypoplasia of the first metacarpal and metatarsal bones. Osteosarcoma can be detected by its "backward" zonation, that is, the most mature bone in the center with woven bone at the periphery, produced by malignant osteoblasts and the presence of definite malignant stroma producing bone, MDM2 positive in low-grade osteosarcoma, can separate parosteal osteosarcoma from myositis ossificans attached to the surface of the bone.

Genetics

No specific genetic abnormalities have been identified. USP6 rearrangements, FISH with probes flanking the USP6 locus on chromosome 17p13 with several partner genes have been identified recently in primary but not in secondary aneurysmal bone cysts (ABC) and in cases that mimic myositis ossificans that are probably soft tissue aneurysmal bone cyst, but not in brown tumor or cherubism, separating it from these entities and placing MO features among those of soft tissue ABC.[115]

Fibrodysplasia ossificans progressiva
Clinical features

This very rare disease of children, FOP, with its progressive ectopic endochondral ossification of the soft tissue, ultimately leads to immobilization of the entire body. The National Institutes of Health website contains useful information: http://ghr.nlm.nih.gov/condition/fibrodysplasia-ossificans-progressiva. Previous terminology includes myositis ossificans progressiva and Munchmeyer disease.

FOP has been estimated to affect approximately 1 per 2 million people.[122,123] The disease is mainly sporadic, since affected patients are often unable to reproduce,[124] but 5% of patients have autosomal dominant transmission.

The signs of disease usually appear by 5 years of age, but can be present at birth.[15,122,123] One-half of the patients have some evidence for the disease by the age of 2 years. Early lesions can be associated with signs of inflammation and systemic symptoms such as low-grade fever and elevated erythrocyte sedimentation rate.[124] These patients typically have skeletal malformations, including short, deviated, and then monophalangic big toes, broad femoral necks, fusion of the cervical vertebrae, hypoplasia of the vertebral bodies, and short first metacarpal bones (short thumbs).[125] The ectopic endochondral ossification has a predilection for the paraspinal and scalp muscles, the jaw muscles, and the extremity musculature in a proximal-to-distal gradient. The patients can develop endochondral ossification of ligaments, tendons, or muscles following the slightest injury, resulting in fusion of the spine, limbs, rib cage, and jawbones, and ultimately leading to immobilization.

The lesions can grow rapidly and can be mistaken for soft tissue tumors and biopsied. Surgical treatment is contraindicated, however, because it is known to exacerbate the ectopic bone formation. The patient usually develops severe scoliosis and ankylosis of all major joints of the axial and appendicular skeleton by early adulthood and is usually confined to a wheelchair by the age of 20 years. Starvation can occur secondary to ankylosis of the jaw, and pneumonia secondary to fixation of the chest wall. Although there is no effective treatment, steroids have been used during the early myositis phase, as well as analgesics, until the acute phase subsides. Other trial medical therapies have been largely unsuccessful.[9,122,126] Future treatment strategies might include inhibition of activin-like kinase signaling.[127]

Imaging studies

FOP demonstrates radiologic features that are similar to myositis ossificans, but much more extensive, frequently beginning in the neck and upper posterior paraspinal regions (Figure 30.27). Associated osseous abnormalities include microdactyly of the big toes (typically the proximal phalanx in >90% of patients and often precedes soft tissue ossification, Figure 30.27) and thumbs, short broad femoral necks, narrowed anteroposterior vertebral body dimension (cervical and lumbar), and enthesophyte formation.

Pathology

By histology, the earliest changes include lymphocyte infiltration and muscle degeneration,[128] followed by a granulation-like tissue and endochondral ossification with woven bone that remodels into mature lamellar bone with marrow elements.[123] The bone formation is central.

The differential diagnosis includes nodular fasciitis in extremity or trunk lesions, cranial fasciitis in skull lesions, and juvenile fibromatosis. The presence of toe abnormalities, multiple lesions, and a tendency to endochondral ossification distinguishes FOP from fasciitis. Other causes for ectopic ossification, such as an entity called progressive osseous heteroplasia or focal cutaneous ossification secondary to acne, burns, hemorrhage, infection, and connective tissue diseases must also be clinically distinguished from FOP. The specific toe anomalies of FOP do not occur in these other clinical entities.

Genetics and biology

The pathogenesis of FOP involves constitutional activation of the bone morphogenetic protein receptor 4, which is an activin-like receptor (ACVR1 [formerly ALK2]) that promotes chondro-osteogenesis in mesenchymal cells.[129] A recurrent mutation leading to ACVR1 protein substitution R206H occurs in most cases of classic FOP.[130] Another ACVR1 mutation leading into altered protein G356D was identified in a Japanese patient with a milder clinical course.[131]

Fibro-osseous pseudotumor of the digits (florid reactive periostitis of the tubular bones of hands and feet, parosteal nodular fasciitis)
Clinical features

This is a heterotopic ossification of the subcutis of the digits, closely related, but less organized than myositis ossificans. It is sometimes considered a subtype of myositis ossificans.

Fibro-osseous pseudotumor of the digits (FOPD) is a variably painful edematous swelling of the proximal phalangeal area of the fingers. It more often involves the index finger, less commonly the middle finger, and rarely the toes. FOPD

Chapter 30: Cartilage- and bone-forming tumors

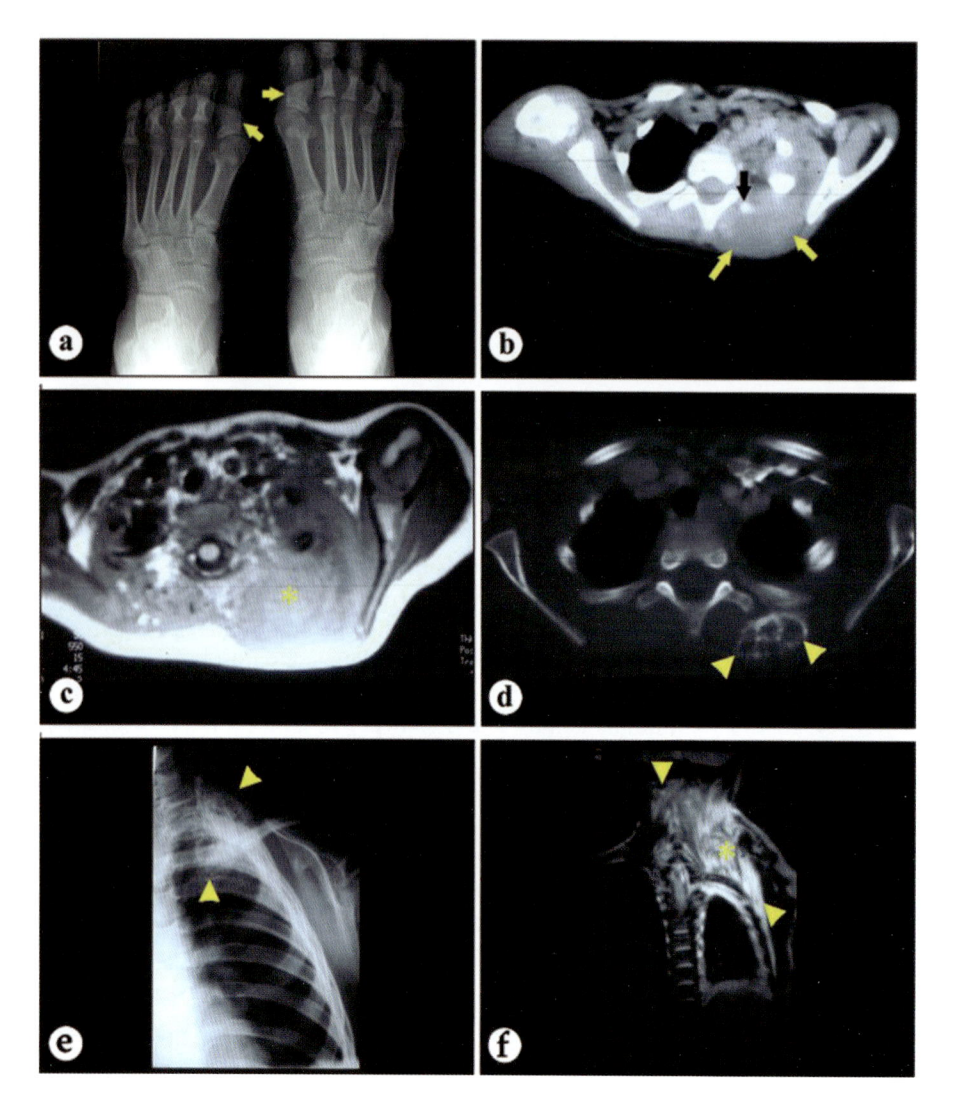

Figure 30.27 Fibrodysplasia ossificans progressiva. (**a**) Radiograph of the feet shows dysplastic changes of the proximal phalanx of the big toes (arrows). (**b**) CT reveals a minimally calcified (small arrow) paraspinal soft tissue mass. (**c,d**) Axial T1-weighted and coronal T2-weighted MR images demonstrate an ill-defined soft tissue mass (•) with extensive surrounding edema (small arrowheads). (**e,f**) Three months later the radiograph and CT show more mature myositis ossificans-like areas (large arrowheads) with a well-defined peripheral rim of ossification.

usually occurs in young adults (age range 10–64 years, median age 40 years), with a female predominance.[132–138]

The duration of these lesions is estimated by radiologic imaging to range from 2 to 6 weeks. A history of trauma can be elicited in approximately 40% of patients.[132–138] Some patients have a history of repetitive manual labor, suggesting minor repetitive trauma as a contributing factor. Signs include a firm mass that does not impair function. Simple excision is diagnostic and curative. Re-excision of residual disease is rarely necessary.

Imaging studies

Imaging can date these lesions, determine their exact location, and help with diagnosis. These lesions are often now considered on a continuum. Radiographs show an ill-defined soft tissue mass overlying the proximal or middle phalanges with evidence of calcification, well separated from the adjacent bone. Periosteal reaction may be apparent and exuberant in florid reactive periostitis.[132,139]

Pathology

Grossly, these are gritty, gray-white, and zonal in approximately one-half of cases. By histology, the zonal organization includes mature woven bone peripherally and immature woven bone centrally; all bone demonstrated osteoblastic rimming.[132–134] FOPD is composed of irregular nodules of fibroblasts, sometimes atypical appearing, and benign bone, within a fibrous to myxoid matrix (Figure 30.28). There is often intramembranous ossification, with both woven and lamellar bone. Bony trabeculae show osteoid seams and are rimmed by the osteoblasts (Figure 30.28). Osteoclasts are uncommon, and no bone marrow elements are apparent.

The differential diagnosis primarily includes osteosarcoma; however, osteoblastic rimming of bony trabeculae (a benign sign) and absence of atypical mitoses in fibro-osseous pseudotumor of the digits are distinguishing features. Other tumors in the differential diagnosis include the spectrum of bone-attached lesions including florid reactive periostitis, bizarre osteochondromatous proliferation (Nora lesion [BPOP]), and

846

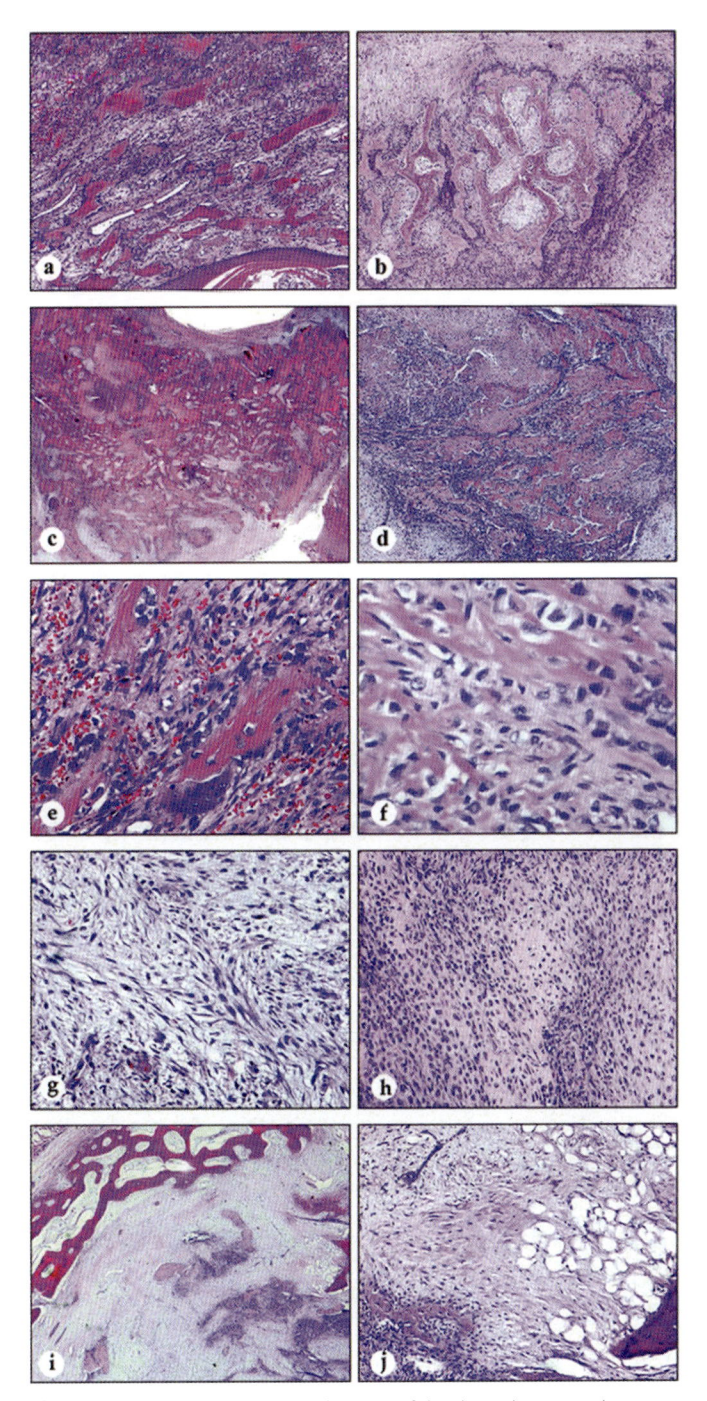

Figure 30.28 Fibro-osseous pseudotumor of the digits shows zonal maturation histologically similar to that in myositis ossificans in one-half of cases. There is mainly intramembranous ossification, so that the lesion is composed of nodular-fasciitis-like stroma, woven bone in various stages of maturation: osteosarcoma-like centrally to the osteoblast and sometimes osteoblast rimming. The periphery matures faster than the center, and it can ossify completely to coupled lamellar bone with fatty marrow; therefore, fat can be seen in these lesions.

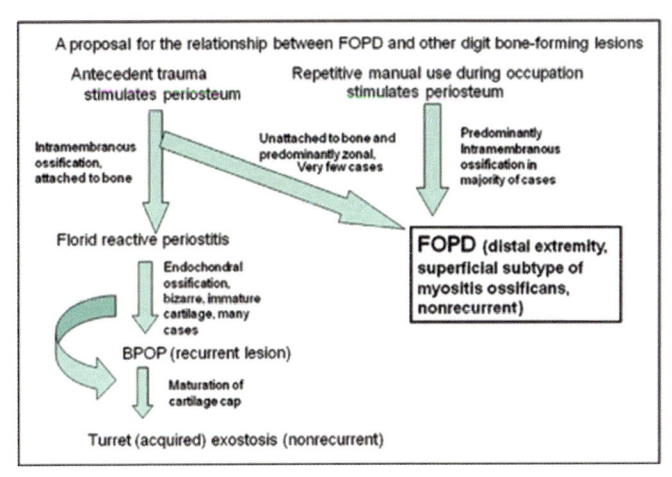

Figure 30.29 The proposed relationship between fibro-osseous pseudotumor of the digit and other bone-forming finger lesions (attached to bone), including the spectrum of florid reactive periostitis, bizarre parosteal osteochondromatous proliferation, and turret exostosis.

turret exostosis. Fibro-osseous pseudotumor is unattached to bone and results in intramembranous ossification with little or no cartilage. The other lesions are parosteal and have a cartilage cap that is bizarre in BPOP or more osteochondroma-like, but no continuity of the marrow space to the host marrow. Subungual exostosis is a distal digit lesion, underneath the nail bed, which has features similar to those of osteochondroma or BPOP, but with fibrocartilage; location distinguishes this entity.[131] A proposed relationship between FOPD, florid reactive periostitis, BPOP, and turret exostosis is delineated in Figure 30.29.

Metaplastic bone

Metaplastic bone can be present in a variety of unrelated soft tissue tumors, including OFT, calcifying aponeurotic fibroma, calcifying fibrous pseudotumor, lipoma (Figure 30.30), liposarcoma, synovial sarcoma, MFH (must be distinguished from osteoid), and MPNST.

Extraskeletal osteosarcoma (extraskeletal osteogenic sarcoma, extraosseous osteosarcoma)

Clinical features

Extraskeletal osteosarcomas constitute 1% to 2% of all soft tissue sarcomas and approximately 4% to 5% of all osteosarcomas.[16,140–145] Mechanical injury (e.g., injection site or fracture), radiation therapy, heterotopic ossification, or stromal metaplasia have been proposed as etiologies. Patients are generally older than 40 years, with variable male or female predominance;[146] the most common locations include the thigh and retroperitoneum. Extraskeletal osteosarcoma can occur in other sites, such as the breast.[147,148]

Prognosis and therapy depend on tumor grade and size.[149] Most extraskeletal osteosarcomas are high grade; therefore adjuvant therapy following surgery is usually clinically indicated, similar to intraosseous osteosarcoma.[143,150] Low-grade

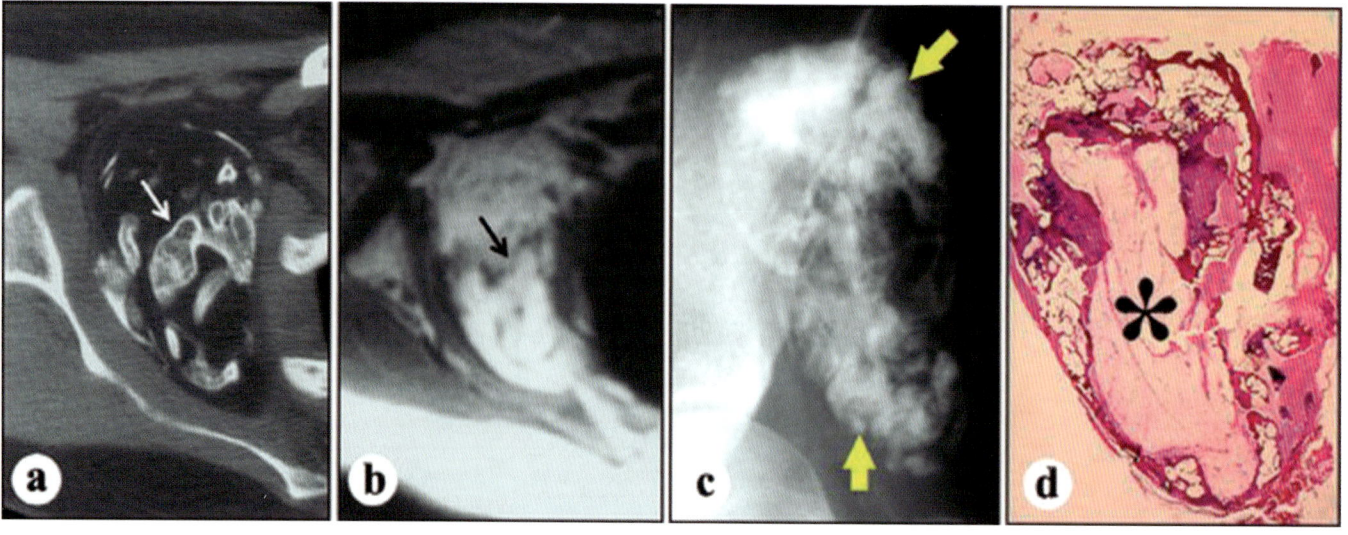

Figure 30.30 (**a**) Metaplastic bone in lipoma from the shoulder CT. (**b**) Axial T1-weighted (right image). MR image shows a lipoma with the same appearance as subcutaneous fat. (**c,d**) Areas of metaplastic mature bone are also seen (arrows); the whole-mount histology section correlates to the imaging, demonstrating a benign fatty tumor (∗) surrounded by bone.

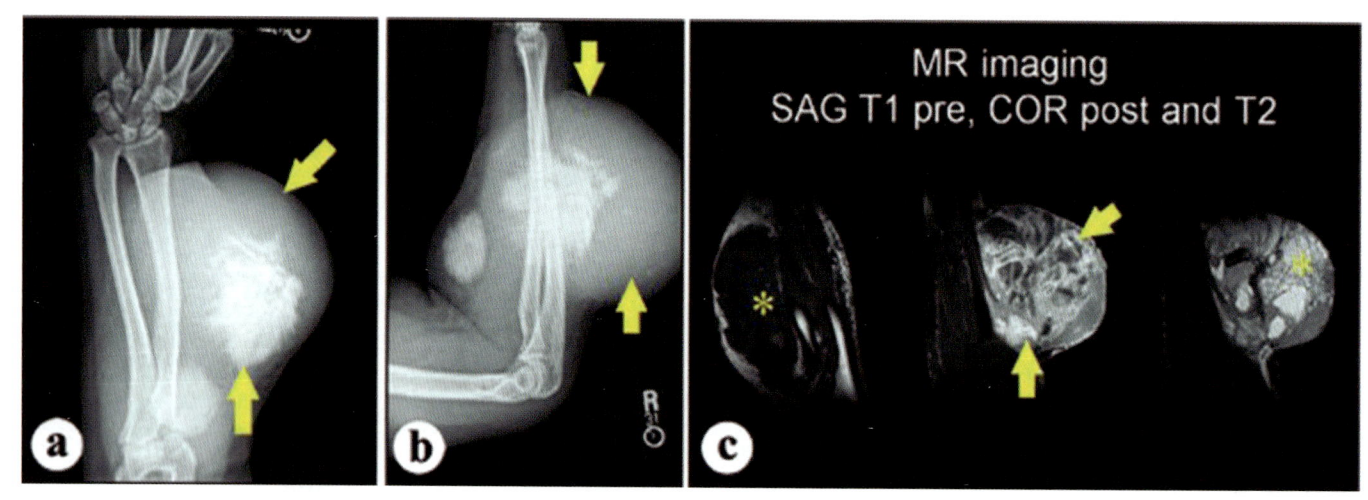

Figure 30.31 Extraskeletal osteosarcoma of the forearm. (**a,b**) The radiograph shows a large, particularly densely calcified soft tissue mass (arrows). (**c**) Multiple MR images reveal low signal intensity on T1-weighting (∗, left image), heterogeneous enhancement on the post-contrast T1-weighted image (arrows, middle image) and heterogeneous high signal intensity on T2-weighting (∗, right image).

forms exist,[145] and extraskeletal osteosarcoma can be secondary to radiation therapy.[146] Metastases are generally to the lung, lymph nodes, bone, and soft tissue. Mortality rates are approximately 80%.[140–154]

Imaging studies

Imaging of extraskeletal osteosarcoma typically reveals a large soft tissue mass with extensive dense amorphous mineralization representing immature osteoid. Unlike myositis ossificans, the calcification is not peripheral. The soft tissue component has a nonspecific attenuation similar to muscle on CT and low to intermediate signal intensity on T1-weighted MR with intermediate to high signal intensity on T2-weighting[155] (Figure 30.31). Radiologic studies are helpful in confirming primary soft tissue origin.

Pathology

Tumor sizes are generally >5 cm, but range from a few to >30 cm. The musculature, subcutis, or dermis can be involved. The lesions are generally well defined. Ossification is sometimes evident radiographically or grossly (Figure 30.32).

By histology, an extraskeletal osteosarcoma is composed of malignant cells of an osteoblastic phenotype with cytologic atypia and mitotic activity that produce bone (osteoid; Figure 30.32). Extraskeletal osteosarcoma is subclassified in a manner similar to intraosseous osteosarcoma.[153,154] The most

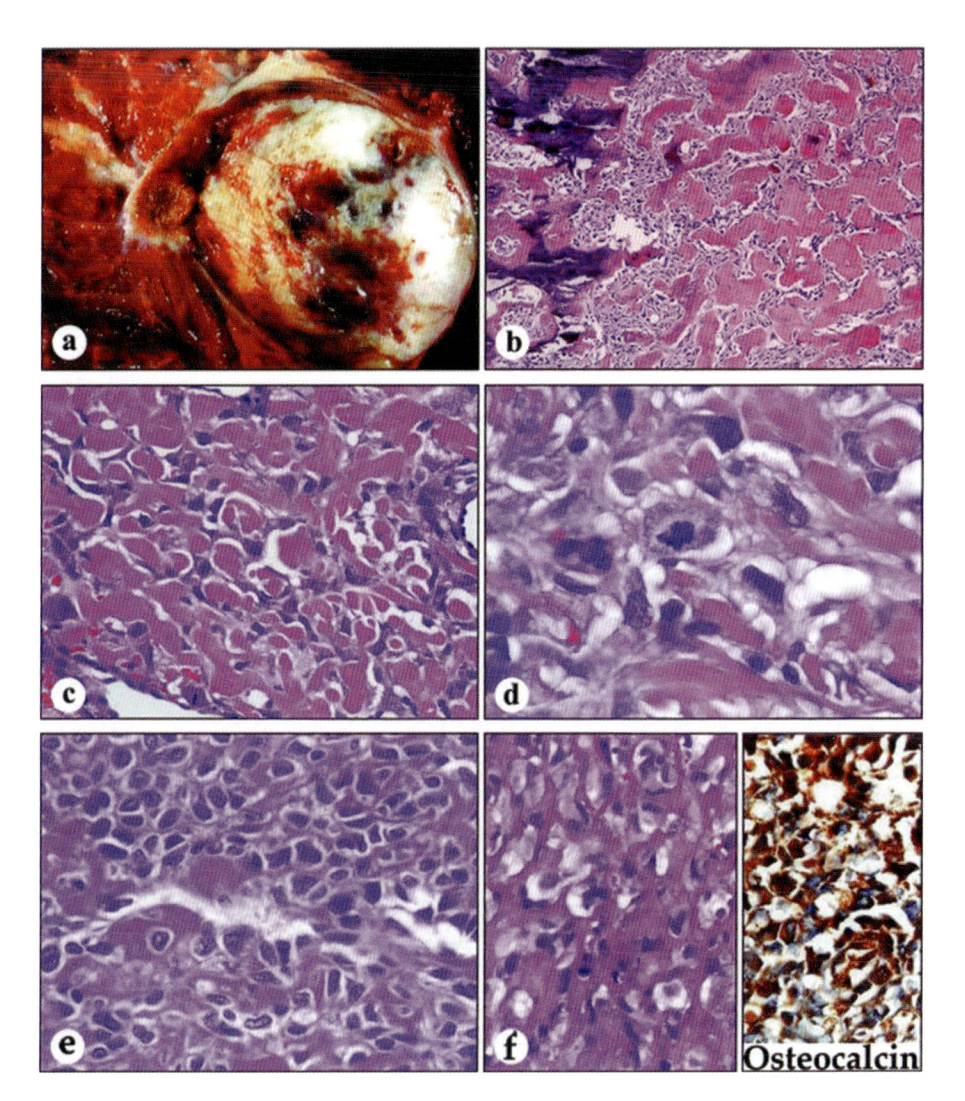

Figure 30.32 (**a**) An extraskeletal osteosarcoma can have a grossly evident bony matrix. (**b–f**) Microscopically there is osteoid produced by malignant osteoblasts. The osteosarcoma tumor bone can be of varying shades of pink and purple. (**d–f**) Lace-like and delicate osteoid is typical, and mitotic activity is present. The background can vary from small round cell to spindled. Osteocalcin immunostaining highlights the osteoblasts and matrix in osteosarcoma.

common subtype is osteoblastic osteosarcoma, which is high grade with marked atypia and mitotic activity. At the other end of the spectrum is the rare well-differentiated, parosteal-like osteosarcoma, characterized by parallel spicules of more mature bone with abundant, relatively bland stroma. High-grade fibroblastic osteosarcoma has lesser amounts of bone and more spindled stroma. The MFH-like type has a storiform, spindle cell pattern and pleomorphism, but it can have any amount of osteoid. The giant-cell-rich type has scattered, benign-appearing osteoclasts in addition to malignant osteoblasts and malignant bone. Small cell osteosarcoma (Figure 30.32) is characterized by small round blue cells with osteoid production, and telangiectatic osteosarcoma has abundant ectatic spaces lined by giant-cell-rich osteoblasts with osteoid.

By immunohistochemistry, osteocalcin, an abundant human bone protein, can be used as a marker for osteoblastic phenotype (Figure 30.32).[156] SATB2, a nuclear transcription factor, may be a more easily interpretable marker for osteoid differentiation in mesenchymal tumors, but experience is limited.[157,158]

The differential diagnosis includes malignant tumors with potential osteoid formation, such as dedifferentiated liposarcoma, metaplastic carcinoma, melanoma, lymphoma, other sarcomas, and benign bone-forming tumors. Proper sampling, histologic search, and radiologic correlation to assess the possibility of an integral lipomatous component are necessary to rule out dedifferentiated liposarcoma. Careful attention to possible epithelial differentiation and immunohistochemistry are helpful for ruling out metaplastic carcinoma or carcinosarcoma.

Genetics

Osteosarcomas have multiple complex chromosomal abnormalities, similar to most pleomorphic and undifferentiated sarcomas.[159] Patients with retinoblastoma, Li–Fraumeni syndrome, Werner syndrome, and Rothmund–Thomson syndrome may develop osteosarcoma.[160] Rarely, osteosarcoma is associated with neurofibromatosis type 1.[161] Lower-grade extraskeletal osteosarcoma has been linked with amplification of 12q, found in adipocytic tumors, as well as M2M2 and CDK4, also found in lipomatous tumors, and MDM2 for low-grade parosteal osteosarcomas.[162,163]

References

General history

1. Lichtenstein L, Goldman RL. Cartilage tumors in soft tissues, particularly in the hand and foot. *Cancer* 1964;17:1203–1208.

2. Jaffe HL. Synovial chondromatosis and other benign articular tumors. In *Tumor and Tumorous Conditions of the Bone and Joints*. Philadelphia: Lea & Febiger; 1958:558–566.

3. Dorfman HD, Czerniak B. Synovial lesions. In *Bone Tumors*. St. Louis, Mo: Mosby; 1998:1041–1086.

4. Reichel PF. Chrondmatose der Kniegelenkskapsel. *Arch Klin Chir* 1900;61:717–724.

5. Stout AP, Verner EW. Chondrosarcoma of the extraskeletal soft tissues. *Cancer* 1953;6:581–590.

6. Enzinger FM, Shiraki M. Extraskeletal myxoid chondrosarcoma: an analysis of 34 cases. *Hum Pathol* 1972;3:421–435.

7. Lichtenstein L, Berstein D. Unusual benign and malignant chondroid tumors of bone: survey of some mesenchymal cartilage tumors and malignant chondroblastic tumors including few multicentric ones as well as many atypical benign chondroblastomas and chondromyxoid fibromas. *Cancer* 1959;12:1142–1157.

8. Park YK, Unni KK, Beabout JW, Hodgson SF. Oncogenic osteomalacia: a clinicopathologic study of 17 bone lesions. *J Korean Med Sci* 1994;9:289–298.

9. Weidner N, Santa Cruz D. Phosphaturic mesenchymal tumors: a polymorphous group causing osteomalacia or rickets. *Cancer* 1987;59:1442–1454.

10. Inclan A. Tumoral calcinosis. *JAMA* 1943;121:490.

11. Becker W. [Calcium gout (calcinosis interstitialis localisate)]. *Medizinische* 1958;40:1589–1590. [Article in German]

12. McCarty D, Kohn N, Faires J. The significance of calcium phosphate crystals in the synovial fluid of arthritic patients: the "pseudogout syndrome." II: Identification of crystals. *Ann Intern Med* 1962;56:738–745.

13. Ling D, Murphy WA, Kyriakos M. Tophaceous pseudogout. *AJR Am J Roentgenol* 1982;138:162–165.

14. Ackerman LV. Extraosseous localized non-neoplastic bone and cartilage formation (so-called myositis ossificans). *J Bone Joint Surg* 1958;40A:279–298.

15. Shore EM, Gannon FH, Kaplan FS. Fibrodysplasia ossificans progressiva: why do some people have two skeletons? *J Clin Rheumatol* 1997;3:84–89.

16. Fine G, Stout AP. Osteogenic sarcoma of the extraskeletal soft tissues. *Cancer* 1956;9:1027–1043.

Extraskeletal (soft tissue) chondroma

17. Dahlin DC, Salvador AH. Cartilaginous tumors of the soft tissues of the hands and feet. *Mayo Clin Proc* 1974;49:721–726.

18. Chung EB, Enzinger FM. Chondroma of soft parts. *Cancer* 1978;41:1414–1424.

19. Humphreys H, Pambakian PH, Fletcher CDM. Soft tissue chondroma: a study of 15 tumors. *Histopathology* 1986;10:147–159.

20. Steiner GC, Meushar N, Norman A, Present D. Intracapsular and paraarticular chondromas. *Clin Orthop Relat Res* 1994;303:231–236.

21. Carney JA. Gastric stromal sarcoma, pulmonary chondroma, and extra-adrenal paraganglioma (Carney's triad): natural history, adrenocortical component and possible familial occurrence. *Mayo Clin Proc* 1999;74:543–552.

22. Hondar Wu HT, Chen W, Lee O, Chang CY. Imaging and pathological correlation of soft tissue chondroma *Clin Imaging* 2005;30:32–36.

23. Shon W, Folpe AL, Fritchie KJ. ERG expression in chondrogenic bone and soft tissue tumours. *J Clin Pathol* 2015;68:125–129.

24. Fetsch JF, Vinh TN, Remotti F, *et al*. Tenosynovial (extraarticular) chondromatosis: an analysis of 37 cases of an underrecognized clinicopathologic entity with a strong predilection for the hands and feet and a high local recurrence rate. *Am J Surg Pathol* 2003;27:1260–1268.

25. Buddingh EP, Naumann S, Nelson M, *et al*. Cytogenetic findings in benign cartilaginous neoplasms. *Cancer Genet Cytogenet* 2003;141:164–168.

26. Sakai Junior N, Abe KT, Formigli LM, *et al*. Cytogenetic findings in 14 benign cartilaginous neoplasms. *Cancer Genet* 2011;204:180–186.

27. Dal Cin P, Qi H, Sciot R, Van Den Berghe H. Involvement of chromosomes 6 and 11 in a soft tissue chondroma. *Cancer Genet Cytogenet* 1997;93:177–178.

28. Fletcher JA, Pinkus GS, Donovan K, *et al*. Clonal rearrangement of chromosome band 6p21 in the mesenchymal component of pulmonary chondroid hamartoma. *Cancer Res* 1992;52:6224–6228.

29. Xiao S, Lux ML, Reeves R, Hudson TJ, Fletcher JA. HMGI(Y) activation by chromosome 6p21 rearrangements in multilineage mesenchymal cells from pulmonary hamartoma. *Am J Pathol* 1997;150:911–918.

Synovial chondromatosis

30. Crotty JM, Monu JUV, Pope TL Jr. Synovial osteochondromatosis. *Radiol Clin North Am* 1996;34:327–342.

31. Sim FH, Dahlin DC, Ivins JC. Extra-articular synovial chondromatosis. *J Bone Joint Surg Am* 1977;59A:492–495.

32. Sviland L, Malcolm AJ. Synovial chondromatosis presenting as painless soft tissue mass: a report of 19 cases. *Histopathology* 1995;27:275–279.

33. Hermann G, Klein M, Abdelwahab IF, Kenan S. Synovial chondrosarcoma arising in synovial chondromatosis of the right hip. *Skeletal Radiol* 1997;26:366–369.

34. Davies RI, Hamilton A, Biggart JD. Primary synovial chondromatosis: a clinicopathologic review and assessment of malignant potential. *Hum Pathol* 1998;29:683–688.

35. Evans S, Boffano M, Chaudhry S, Jeys L, Grimer R. Synovial chondrosarcoma arising in synovial chondromatosis. *Sarcoma* 2014;2014:647939.

36. Murphey MD, Vidal JA, Fanburg-Smith JC, Gajewski DA. From the archives of the AFIP: imaging of synovial chondromatosis with radiologic-pathologic correlation. *Radiographics* 2007;27:1465–1488.

37. McKenzie, G, Raby N, Ritchie O. A pictoral review of primary synovial chondromatosis. *Eur Radiol* 2008;18:2662–2669.

38. Buddingh EP, Krallman P, Neff JR, *et al*. Chromosome 6 abnormalities are recurrent in synovial chondromatosis. *Cancer Genet Cytogenet* 2003;140:18–22.

39. Hocking R, Negrine J. Primary synovial chondromatosis of the subtalar joint affecting two brothers. *Foot Ankle Int* 2003;24:865–867.
40. Hopyan S, Nadesan P, Yu C, Wunder J, Alman BA. Dysregulation of hedgehog signalling predisposes to synovial chondromatosis. *J Pathol* 2005;206:143–150.

Extraskeletal myxoid chondrosarcoma

41. Hachitanda Y, Tsuneyoshi M, Daimaru Y, et al. Extraskeletal myxoid chondrosarcoma in young children. *Cancer* 1988;61:2521–2526.
42. Saleh G, Evans HL, Ro JY, Ayala AG. Extraskeletal myxoid chondrosarcoma: a clinicopathologic study of ten patients with long-term follow-up. *Cancer* 1992;70:2827–2830.
43. Lucas DR, Fletcher CD, Adsay NV, Zalupski MM. High-grade extraskeletal myxoid chondrosarcoma: a high-grade epithelioid malignancy. *Histopathology* 1999;35:201–208.
44. Ramesh K, Gahukamble L, Sarma NH, Al Fituri OM. Extraskeletal myxoid chondrosarcoma with dedifferentiation. *Histopathology* 1995;27:381–382.
45. Meis-Kindblom JM, Bergh P, Gunterberg B, Kindblom LG. Extraskeletal myxoid chondrosarcoma: a reappraisal of its morphologic spectrum and prognostic factors based on 117 cases. *Am J Surg Pathol* 1999;23:636–650.
46. Drilon AD, Popat S, Bhuchar G, et al. Extraskeletal myxoid chondrosarcoma: a retrospective review from 2 referral centers emphasizing long-term outcomes with surgery and chemotherapy. *Cancer* 2008;113:3364–3371.
47. Merimsky O, Bernstein-Molho R, Sagi-Eisenberg R. Targeting the mammalian target of rapamycin in myxoid chondrosarcoma. *Anticancer Drugs* 2008;19:1019–1021.
48. Kapoor N, Shinagare AB, Jagannathan JP, et al. Clinical and radiologic features of extraskeletal myxoid chondrosarcoma including initial presentation, local recurrence, and metastasis. *Radiol Oncol* 2014;48:235–242.
49. Tateishi U, Hasegawa T, Nojima T, Tukegami T, Arai Y. MR features of extraskeletal myxoid chondrosarcoma. *Skeletal Radiol* 2006;35:27–33.
50. Kohashi K, Oda Y, Yamamoto H, et al. SMARCB1/INI1 protein expression in round cell soft tissue sarcomas associated with chromosomal translocations involving EWS: a special reference to SMARCB1/INI1 negative variant extraskeletal myxoid chondrosarcoma. *Am J Surg Pathol* 2008;32:1168–1174.
51. Aigner T, Oliveira AM, Nascimento AB. Extraskeletal myxoid chondrosarcomas do not show a chondrocytic phenotype. *Mod Pathol* 2004;17:214–221.
52. Antonescu CR, Argani P, Erlandson RA, et al. Skeletal and extraskeletal myxoid chondrosarcoma: a comparative clinicopathologic, ultrastructural, and molecular study. *Cancer* 1998;83:1504–1521.
53. O'Hara B, Paetau A, Miettinen M. Keratin subsets and monoclonal antibody HBME-1 in chordoma: immunohistochemical differential diagnosis between tumors simulating chordoma. *Hum Pathol* 1998;29:119–126.
54. Stenman G, Andersson H, Mandahl N, Meis-Kindblom JM, Kindblom LG. Translocation t(9;22)(q22;q12) is a primary cytogenetic abnormality in extraskeletal myxoid chondrosarcoma. *Int J Cancer* 1995;62:398–402.
55. Hirabayashi Y, Ishida T, Yoshida MA, et al. Translocation (9;22)(q22;q12): a recurrent chromosome abnormality in extraskeletal myxoid chondrosarcoma. *Cancer Genet Cytogenet* 1995;81:33–37.
56. Sciot R, Dal Cin P, Fletcher C, et al. t(9;22)(q22–31;q11–12) is a consistent marker of extraskeletal myxoid chondrosarcoma: evaluation of three cases. *Mod Pathol* 1995;8:765–768.
57. Brody RI, Ueda T, Hamelin A, et al. Molecular analysis of the fusion of EWS to an orphan nuclear receptor gene in extraskeletal myxoid chondrosarcoma. *Am J Pathol* 1997;150:1049–1058.
58. Labelle Y, Bussieres J, Courjal F, Goldring MB. The EWS/TEC fusion protein encoded by the t(9;22) chromosomal translocation in human chondrosarcomas is a highly potent transcriptional activator. *Oncogene* 1999;18:3303–3308.
59. Kilpatrick SE, Inwards CY, Fletcher CD, Smith MA, Gitelis S. Myxoid chondrosarcoma (chordoid sarcoma) of bone: a report of two cases and review of the literature. *Cancer* 1997;79:1903–1910.
60. Demicco EG, Wang WL, Madewell JE, et al. Osseous myxochondroid sarcoma: a detailed study of 5 cases of extraskeletal myxoid chondrosarcoma of the bone. *Am J Surg Pathol* 2013;37:752–762.
61. Sjogren H, Meis-Kindblom J, Kindblom L-G, Aman P, Stenman G. Fusion of the EWS-related gene TAF2N to TEC in extraskeletal myxoid chondrosarcoma. *Cancer Res* 1999;59:5064–5067.
62. Ohkura N, Nagamura Y, Tsukada T. Differential transactivation by orphan nuclear receptor NOR1 and its fusion gene product EWS/NOR1: possible involvement of poly(ADP-ribose) polymerase I, PARP-1. *J Cell Biochem* 2008;105:785–800.
63. Agaram NP, Zhang L, Sung YS, Singer S, Antonescu CR. Extraskeletal myxoid chondrosarcoma with non-EWSR1-NR4A3 variant fusions correlate with rhabdoid phenotype and high-grade morphology. *Hum Pathol* 2014;45:1084–1091.
64. Wang WL, Mayordomo E, Czerniak BA, et al. Fluorescence in situ hybridization is a useful ancillary diagnostic tool for extraskeletal myxoid chondrosarcoma. *Mod Pathol* 2008;21:1303–1310.

Extraskeletal mesenchymal chondrosarcoma

65. Guccion JG, Font RL, Enzinger FM, Zimmerman LE. Extraskeletal mesenchymal chondrosarcoma. *Arch Pathol* 1973;95:336–340.
66. Nakashima Y, Unni KK, Shives TC, Swee RG, Dahlin DC. Mesenchymal chondrosarcoma of bone and soft tissue: a study of 111 cases. *Cancer* 1986;57:2444–2453.
67. Rushing EJ, Armonda RA, Ansari Q, Mena H. Mesenchymal chondrosarcoma: a clinicopathologic and flow cytometric study of 13 cases presenting in the central nervous system. *Cancer* 1996;77:1884–1891.
68. Gorelik B, Ziv I, Shohat R, et al. Efficacy of weekly docetaxel and bevacizumab in mesenchymal chondrosarcoma: a new theranostic method combining xenografted biopsies with a mathematical model. *Cancer Res* 2008;68:9033–9040.
69. Dantonello TM, Int-Veen C, Leuschner I, et al. Mesenchymal chondrosarcoma of soft tissues and bone in children, adolescents, and young adults:

experiences of the CWS and COSS study groups. *Cancer* 2008;112:2424–2431.
70. Xu J, Li D, Xie L, Tang S, Guo W. Mesenchymal chondrosarcoma of bone and soft tissue: a systematic review of 107 patients in the past 20 years. *PLoS One* 2015;10:e0122216.
71. Chen Y, Wang X, Guo L, et al. Radiological features and pathology of extraskeletal mesenchymal chondrosarcoma. *Clin Imaging* 2012;36:365–370.
72. Murphey MD, Walter EA, Wilson AJ, et al. Imaging of primary chondrosarcoma: radiologic-pathologic correlation. *Radiographics* 2003;23:1245–1278.
73. Fanburg-Smith JC, Auerbach A, Marwaha JS, et al. Immunoprofile of mesenchymal chondrosarcoma: aberrant desmin and EMA expression, retention of INI1, and negative estrogen receptor in 22 female-predominant central nervous system and musculoskeletal cases. *Ann Diagn Pathol* 2009;14:8–14.
74. Fanburg-Smith JC, Auerbach A, Marwaha JS, Wang Z, Rushing EJ. Reappraisal of mesenchymal chondrosarcoma: novel morphologic observations of the hyaline cartilage and endochondral ossification and beta-catenin, sox 9, and osteocalcin immunostaining. *Hum Pathol* 2010;41:653–662.
75. Wang L, Motoi T, Khanin R, et al. Identification of a novel, recurrent HEY1-NCOA2 fusion in mesenchymal chondrosarcoma based on a genome-wide screen of exon-level expression data. *Genes Chromosomes Cancer* 2012;51:127–139.
76. Naumann S, Krallman PA, Unni KK, et al. Translocation der(13;21)(q10;q10) in skeletal and extraskeletal mesenchymal chondrosarcoma. *Mod Pathol* 2002;15:572–576.
77. Sainati L, Scapinello A, Montaldi A, et al. A mesenchymal chondrosarcoma of a child with the reciprocal translocation (11;22)(q24;q12). *Cancer Genet Cytogenet* 1993;71:144–147.

Phosphaturic mesenchymal tumor
78. Hautmann AH, Hautmann MG, Kölbl O, Herr W, Fleck M. Tumor-induced osteomalacia: an up-to-date review. *Curr Rheumatol Rep* 2015;17:512.
79. Folpe AL, Fanburg-Smith JC, Billings SD, et al. Most osteomalacia-associated mesenchymal tumors are a single histopathologic entity: an analysis of 32 cases and a comprehensive review of the literature. *Am J Surg Pathol* 2004;28:1–30.
80. Zimering MB, Caldarella FA, White KE, Econs MJ. Persistent tumor-induced osteomalacia confirmed by elevated postoperative levels of serum fibroblast growth factor-23 and 5-year follow-up of bone density changes. *Endocr Pract* 2005;11:108–114.
81. Geller JL, Khosravi A, Kelly MH, et al. Cinacalcet in the management of tumor-induced osteomalacia. *J Bone Miner Res* 2007;22:931–937.
82. Nakanishi K, Sakai M, Tanaka H, et al. Whole-body MR imaging in detecting phosphaturic mesenchymal tumor (PMT) in tumor induced hypophosphatemic osteomalacia. *Magn Reson Med Sci* 2013;12:47–52.
83. Clugston E, Gill AC, Graf N, Bonar F, Gill AJ. Use of immunohistochemistry for SSTR2A to support a diagnosis of phosphaturic mesenchymal tumour. *Pathology* 2015;47:173–175.
84. Lee JC, Jeng YM, Su SY, et al. Identification of a novel FN1-FGFR1 genetic fusion as a frequent event in phosphaturic mesenchymal tumour. *J Pathol* 2015;235:539–545.

Tumoral calcinosis
85. McGregor D, Burn J, Lynn K, Robson R. Rapid resolution of tumoral calcinosis after renal transplantation. *Clin Nephrol* 1999;51:54–58.
86. Smack D, Norton SA, Fitzpatrick JE. Proposal for a pathogenesis-based classification of tumoral calcinosis. *Int J Dermatol* 1996;35:265–271.
87. Laskin WB, Miettinen M, Fetsch JF. Calcareous lesions of the distal extremities resembling tumoral calcinosis (tumoral calcinosislike lesions): clinicopathologic study of 43 cases emphasizing a pathogenesis-based approach to classification. *Am J Surg Pathol* 2007;31:15–25.
88. Narchi H. Hyperostosis with hyperphosphatemia: evidence of familial occurrence and association with tumoral calcinosis. *Pediatrics* 1997;99:745–748.
89. Albraham Z, Rozner I, Rozenbaum M. Tumoral calcinosis: report of a case and brief review of the literature. *J Dermatopathol* 1996;23:545–550.
90. Noyez JF, Murphree SM, Chen K. Tumoral calcinosis: a clinical report of eleven cases. *Acta Orthop Belg* 1993;59:249–254.
91. Farzan M, Farhoud AR. Tumoral calcinosis: what is the treatment? Report of two cases of different types and review of the literature. *Am J Orthop (Belle Mead NJ)* 2011;40:E170–E176.
92. Olsen KM, Chew FS. Tumoral calcinosis: pearls, polemics, and alternative possibilities. *Radiographics* 2006;26:871–885.
93. Pakasa NM, Kalengyai RM. Tumoral calcinosis: a clinicopathological study of 111 cases with emphasis of the earliest changes. *Histopathology* 1997;31:18–24.
94. McKee PH, Liomba NG, Hutt MSR. Tumoral calcinosis: a pathological study of fifty-six cases. *Br J Dermatol* 1982;107:669–674.
95. Slavin RE, Wen J, Kumar D, Evans EB. Familial tumoral calcinosis: a clinical, histopathologic, and ultrastructural study with an analysis of its calcifying process and pathogenesis. *Am J Surg Pathol* 1993;17:788–802.
96. Shah A, Miller CJ, Nast CC, et al. Severe vascular calcification and tumoral calcinosis in a family with hyperphosphatemia: a fibroblast growth factor 23 mutation identified by exome sequencing. *Nephrol Dial Transplant* 2014;29:2235–2243.
97. Masi L, Beltrami G, Ottanelli S, et al. Human preosteoblastic cell culture from a patient with severe tumoral calcinosis-hyperphosphatemia due to a new GALNT3 gene mutation: study of in vitro mineralization. *Calcif Tissue Int* 2015;96:438–452.
98. Laleye A, Alao MJ, Gbessi G, et al. Tumoral calcinosis due to GALNT3 C.516-2A >T mutation in a black African family. *Genet Couns* 2008;19:183–192.
99. Folsom LJ, Imel EA. Hyperphosphatemic familial tumoral calcinosis: genetic models of deficient FGF23 action. *Curr Osteoporos Rep* 2015;13:78–87.
100. Topaz O, Indelman M, Chefetz I, et al. A deleterious mutation in SAMD9 causes normophosphatemic familial tumoral calcinosis. *Am J Hum Genet* 2006;79:759–764.

101. Hiramatsu R, Ubara Y, Tajima T, et al. Tumoral calcinosis in a patient with hypoparathyroidism, sensorineural deafness, and renal dysplasia syndrome undergoing hemodialysis. *Clin Case Rep* 2015;3:73–75.

Tophaceous pseudogout

102. Abdelsayed RA, Said-Al-Naief N, Salguerio M, Holmes J, El-Mofty SK. Tophaceous pseudogout of the temporomandibular joint: a series of 3 cases. *Oral Surg Oral Med Oral Pathol Oral Radiol* 2014;117:369–375.

103. Ishida T, Dorfman HD, Bullough PG. Tophaceous pseudogout (tumoral calcium pyrophosphate dihydrate crystal deposition disease. *Hum Pathol* 1995;26:587–593.

104. Tan KB, Scolyer RA, McCarthy SW, et al. Tumoural calcium pyrophosphate dihydrate crystal deposition disease (tophaceous pseudogout) of the hand: a report of two cases including one with a previously unreported associated florid reactive myofibroblastic proliferation. *Pathology* 2008;40:719–722.

105. Athanasou NA, Caughey M, Burge P, Eta L. Deposition of calcium pyrophosphate dihydrate crystals in a soft tissue chondroma. *Ann Rheum Dis* 1991;50:950–952.

106. Schneider I. [Calcium pyrophosphate dihydrate-crystal induced arthropathy] *Z Rheumatol* 2004;63:10–21. [Article in German].

Calciphylaxis

107. Fischer AH, Morris DJ. Pathogenesis of calciphylaxis: study of three cases with literature review. *Hum Pathol* 1995;26:1055–1064.

108. Oh DH, Eulau D, Tokugawa DA, McGuire JS, Kohler S. Five cases of calciphylaxis and a review of the literature. *J Am Acad Dermatol* 1999;40:979–987.

109. Hafner J, Keusch G, Wahl C, et al. Uremic small-artery disease with medial calcification and intimal hyperplasia (so-called calciphylaxis): a complication of chronic renal failure and benefit from parathyroidectomy. *J Am Acad Dermatol* 1995;33:954–962.

110. Oliveira TM, Frazão JM. Calciphylaxis: from the disease to the diseased. *J Nephrol* 2015;28:531–540.

111. Vattikuti R, Towler DA. Osteogenic regulation of vascular calcification: an early perspective. *Am J Physiol Endocrinol Metab* 2004;286:E686–E696.

112. Roverano S, Ortiz A, Henares E, Eletti M, Paira S. Calciphylaxis of the temporal artery masquerading as temporal arteritis: a case presentation and review of the literature. *Clin Rheumatol* 2015;34:1985–1988.

Myositis ossificans

113. Akgun I, Erdogan F, Aydingoz O, Kesmezacar H. Myositis ossificans in early childhood. *J Arthroscop Rel Surg* 1998;15:522–526.

114. Kransdorf MJ, Meis JM. From the archives of the AFIP: extraskeletal osseous and cartilaginous tumors of the extremities. *Radiographics* 1993;13:853–884.

115. Sukov WR, Franco MF, Erickson-Johnson M, et al. Frequency of USP6 rearrangements in myositis ossificans, brown tumor, and cherubism: molecular cytogenetic evidence that a subset of "myositis ossificans-like lesions" are the early phases in the formation of soft-tissue aneurysmal bone cyst. *Skeletal Radiol* 2008;37:321–327.

116. Konishi E, Kusuzaki K, Murata H. Extraskeletal osteosarcoma arising in myositis ossificans. *Skeletal Radiol* 2001;30:39–43.

117. Patel RM, Weiss SW, Folpe AL. Heterotopic mesenteric ossification: a distinctive pseudosarcoma commonly associated with intestinal obstruction. *Am J Surg Pathol* 2006;30:119–122.

118. Kaplan FS, Gannon FH, Hahn GV, Wollner N, Rauner R. Pseudomalignant heterotopic ossification: differential diagnosis and report of two cases. *Clin Orthop Relat Res* 1998;346:134–140.

119. Parikh J, Hyare H, Saifuddin A. The imaging features of post-traumatic myositis ossificans with emphasis on MRI. *Clin Radiol* 2002;57:1058–1066.

120. Wheeler K, Makary R, Berrey H. A case of malignant transformation of myositis ossificans. *Am J Orthop (Belle Mead NJ)* 2014;43(1):E25–E27.

121. Lacout A, Jarraya M, Marcy P-Y, Thariat J, Carlier RY. Myositis ossificans imaging: keys to successful diagnosis. *Indian J Radiol Imaging* 2012;22:35–39.

Fibrodysplasia ossificans progressiva

122. Smith R, Athanasou NA, Vipond SE. Fibrodysplasia (myositis) ossificans progressiva: clinicopathological features and natural history. *Q J Med* 1996;89:445–456.

123. Kaplan FS, Taas JA, Gannon FH, et al. The histopathology of fibrodysplasia ossificans progressiva. *J Bone Joint Surg Am* 1993;75:220–230.

124. Cohen RB, Hahn GV, Tabas JA, et al. The natural history of heterotopic ossification in patients who have fibrodysplasia ossificans progressiva: a study of forty-four patients. *J Bone Joint Surg Am* 1993;75:215–219.

125. Connor JM, Evans DA. Fibrodysplasia ossificans progressiva: the clinical features and natural history of 34 patients. *J Bone Joint Surg Br* 1982;64:76–83.

126. Brantus JF, Meunier PJ. Effects of intravenous etidronate and oral corticosteroids in fibrodysplasia ossificans progressiva. *Clin Orthop* 1998;346:117–120.

127. Faruqi T, Dhawan N, Bahl J, et al. Molecular, phenotypic aspects and therapeutic horizons of rare genetic bone disorders. *Biomed Res Int* 2014;2014:670842.

128. Kaplan FS, Taas JA, Gannon FH, et al. The histopathology of fibrodysplasia ossificans progressiva. *J Bone Joint Surg Am* 1993;75:220–230.

129. Fukuda T, Kohda M, Kanomata K, et al. Constitutively activated ALK2 and increased smad1/5 cooperatively induce BMP signaling in fibrodysplasia ossificans progressiva. *J Biol Chem* 2009;284:7149–7156.

130. Furuya H, Ikezoe K, Wang L, et al. A unique case of fibrodysplasia ossificans progressiva with an ACVR1 mutation, G356D, other than the common mutation (R206H). *Am J Med Genet A* 2008;146A:459–463.

131. Fukuda T, Kanomata K, Nojima J, et al. A unique mutation of ALK2, G356D, found in a patient with fibrodysplasia ossificans progressiva is a moderately activated BMP type I receptor. *Biochem Biophys Res Commun* 2008;377:905–909.

Fibro-osseous pseudotumor

132. Moosavi CA, Al-Nahar LA, Murphey MD, Fanburg-Smith JC. Fibroosseous pseudotumor of the digit: a clinicopathologic study of 43 new cases. *Ann Diagn Pathol* 2008;12:21–28.

133. Dupree WB, Enzinger FM. Fibro-osseous pseudotumor of the digits. *Cancer* 1986;58:2103–2109.

134. Spjut HJ, Dorfman HD. Florid reactive periostitis of the tubular bones of the hands and feet: a benign lesion which may simulate osteosarcoma. *Am J Surg Pathol* 1981;5:423–433.

135. Angervall L, Stener B, Stener I, Ahren C. Pseudomalignant osseous tumor of soft tissue. *J Bone Joint Surg* 1969;51B:654–663.

136. Tang J-B, Gu YQ, Xia RG. Fibro-osseous pseudotumor that may be mistaken for a malignant tumor in the hand: a case report and review of the literature. *J Hand Surg* 1996;21A:714–716.

137. Sleater J, Mullins D, Chun K, Hendricks J. Fibro-osseous pseudotumor of the digit: a comparison to myositis ossificans by light microscopy and immunohistochemical methods. *J Cutan Pathol* 1995;23:373–377.

138. Chaudhry IH, Kazakov DV, Michal M, et al. Fibro-osseous pseudotumor of the digit: a clinicopathological study of 17 cases. *J Cutan Pathol* 2010;37:323–329.

139. Sundanam M, Long L, Rotman M, Howard R, Saboeiro AP. Florid reactive periostitis and bizarre osteochondromatous proliferation: pre-biopsy imaging evolution, treatment and outcome. *Skeletal Radiol* 2001;30(4):192–198.

Extraskeletal osteosarcoma

140. Huvos AG. Osteogenic sarcoma of bones and soft tissue in older persons: a clinicopathologic analysis of 117 patients older than 60 years. *Cancer* 1986;57:1442–1449.

141. Rao U, Cheng A, Didolkar MS. Extraosseous osteogenic sarcoma: clinicopathological study of eight cases and review of the literature. *Cancer* 1978;41:1488–1496.

142. Jensen ML, Schumacher B, Jensen OM, Nielsen OS, Keller J. Extraskeletal osteosarcomas: a clinicopathologic study of 25 cases. *Am J Surg Pathol* 1998;22:588–594.

143. Goldstein-Jackson SY, Gosheger G, Delling G, et al. Extraskeletal osteosarcoma has a favourable prognosis when treated like conventional osteosarcoma. *J Cancer Res Clin Oncol* 2005;131:520–526.

144. Torigoe T, Yazawa Y, Takagi T, Terakado A, Kurosawa H. Extraskeletal osteosarcoma in Japan: multiinstitutional study of 20 patients from the Japanese Musculoskeletal Oncology Group. *J Orthop Sci* 2007;12:424–429.

145. Fang Z, Matsumoto S, Ae K, et al. Postradiation soft tissue sarcoma: a multiinstitutional analysis of 14 cases in Japan. *J Orthop Sci* 2004;9:242–246.

146. Thampi S, Matthay KK, Boscardin WJ, Goldsby R, DuBois SG. Clinical features and outcomes differ between skeletal and extraskeletal osteosarcoma. *Sarcoma* 2014;2014:902620.

147. Silver SA, Tavassoli FA. Primary osteogenic sarcoma of the breast: a clinicopathologic analysis of 50 cases. *Am J Surg Pathol* 1998;22:925–933.

148. Cook PA, Murphy MS, Innis PC, Yu JS. Extraskeletal osteosarcoma of the hand. *J Bone Joint Surg* 1998;80A:725–729.

149. Bane BL, Evans HL, Ro JY, et al. Extraskeletal osteosarcoma: a clinicopathologic review of 26 cases. *Cancer* 1990;66:2762–2770.

150. Berner K, Bjerkehagen B, Bruland ØS, Berner A. Extraskeletal osteosarcoma in Norway, between 1975 and 2009, and a brief review of the literature. *Anticancer Res* 2015;35:2129–2140.

151. Chung EB, Enzinger FM. Extraskeletal osteosarcoma. *Cancer* 1987;60:1132–1142.

152. Sordillo PP, Hajdu SI, Magill GB, Golbey RB. Extraosseous osteogenic sarcoma: a review of 48 patients. *Cancer* 1983;51:727–734.

153. Lee, JSY, Fetsch JF, Wasdhal DA, et al. A review of 32 patients with extraskeletal osteosarcoma. *Cancer* 1995;76:2253–2259.

154. McCarter MD, Lewis JJ, Antonescu CR, Brennan MF. Extraskeletal osteosarcoma: analysis of outcome of a rare neoplasm. *Sarcoma* 2000;4:119–123.

155. McAuley G, Jagannatham J, O'Reagan K, et al. Extraskeletal osteosarcoma: spectrum of imaging findings. *AJR Am J Roentgenol* 2012;198:W31–W37.

156. Fanburg-Smith JC, Bratthauer GL, Miettinen M. Osteocalcin and osteonectin immunoreactivity in extraskeletal osteosarcoma: a study of 28 cases. *Hum Pathol* 1999;30:32–38.

157. Conner JR, Hornick JL. SATB2 is a novel marker of osteoblastic differentiation in bone and soft tissue tumors. *Histopathology* 2013;63:36–49.

158. Ordonez NG. SATB2 is a novel marker of osteoblastic differentiation and colorectal carcinoma. *Adv Anat Pathol* 2014;21:63–67.

159. Mertens F, Fletcher CD, Dal Cin P, et al. Cytogenetic analysis of 46 pleomorphic soft tissue sarcomas and correlation with morphologic and clinical features: a report of the CHAMP Study Group (Chromosomes and MorPhology). *Genes Chromosomes Cancer* 1998;22:16–25.

160. Hauben EI, Arends J, Vandenbroucke JP, et al. Multiple primary malignancies in osteosarcoma patients: incidence and predictive value of ostcosarcoma subtype for cancer syndromes related with osteosarcoma. *Eur J Hum Genet* 2003;11:611–618.

161. Hatori M, Hosaka M, Watanabe M, et al. Osteosarcoma in a patient with neurofibromatosis type 1: a case report and review of the literature. *Tohoku J Exp Med* 2006;208:343–348.

162. Kyriazoglou AI, Vieira J, Dimitriadis E, et al. 12q amplification defines a subtype of extraskeletal osteosarcoma with good prognosis that is the soft tissue homologue of parosteal osteosarcoma. *Cancer Genet* 2012;205:332–336.

163. von Baer A, Ehrhardt A, Baumhoer D, et al. Immunohistochemical and FISH analysis of MDM2 and CDK4 in a dedifferentiated extraskeletal osteosarcoma arising in the vastus lateralis muscle: differential diagnosis and diagnostic algorithm. *Pathol Res Pract* 2014;210:698–703.

Chapter 31

Small round cell tumors

Nick Shillingford
Children's Hospital Los Angeles

Hiroyuki Shimada
Children's Hospital Los Angeles

David M. Parham
University of Southern California Keck School of Medicine

Small round blue cell tumor (SRBCT) is the name given to a group of malignant neoplasms that occur mostly in the pediatric age group. Although this name could also apply to some adult neoplasms, notably small cell carcinoma of the lung and other organs, in soft tissue pathology the term has become widely associated with childhood cancer.[1] The name is derived from the primitive, highly cellular nature of these lesions, which typically present as a sea of dark-blue nuclei on hematoxylin-based stains, an appearance attributable to the scant cytoplasmic volume and the consequently high nuclear to cytoplasmic ratio often seen in the cells of these tumors. Cytoplasmic abundance roughly correlates with cellular differentiation, which is often modest. Although cytoplasmic landmarks allow for cell type identification in some cases, ancillary techniques such as immunohistochemistry, electron microscopy, and molecular/genetic studies have become critical in the diagnosis of these tumors.

Although childhood cancer constitutes a relatively small percentage of all malignancies, it is responsible for a large proportion of pediatric mortality, being second only to accidents as a cause of death in industrialized nations.[2] Pediatric cancer also differs from adult cancer in the greater percentage of cases treated on multi-institutional protocols.[3] Despite success in treatment,[2] more than 100 000 person-years are lost each year to childhood cancers.[3]

Types of round cell tumors and their differential diagnoses

Leukemias and brain tumors constitute the majority of childhood cancers. Although the former can present as solid masses known variably as granulocytic sarcoma, myeloid sarcoma, and chloroma, these lesions do not typically cause extramedullary SRBCT. Table 31.1 lists the most common SRBCT. Of these tumors, neuroblastoma, rhabdomyosarcoma, lymphoma, and Wilms' tumor are most common. This chapter discusses neuroblastoma, Ewing sarcoma/peripheral primitive neuroectodermal tumor (PNET), desmoplastic small round cell tumor (DSRCT), and extrarenal rhabdoid tumor. Rhabdomyosarcoma and lymphoma are discussed elsewhere (Chapters 20 and 35).

Most SRBCT, such as neuroblastoma and Wilms' tumor, are organ-specific blastomas. These embryonal tumors recapitulate the embryogenesis of their organs of origin. Another group, rhabdomyosarcomas and the Ewing sarcoma family of tumors, are primarily soft tissue and bone lesions, the former representing a neoplastic attempt at embryonic muscle development. Often, however, it is impossible to identify the tissue of origin or the differentiated counterpart of the embryonic tissue of small round blue cell tumors at the light microscopic level, necessitating ancillary techniques for diagnosis. These techniques historically included histochemical stains and electron microscopy and

Table 31.1 Pediatric small round blue cell tumors

Neuroblastoma
Rhabdomyosarcoma
Lymphoma
Ewing sarcoma/peripheral primitive neuroectodermal tumor
Desmoplastic small round cell tumor
Melanotic neuroectodermal tumor
Mesenchymal chondrosarcoma
Small cell osteosarcoma
Rhabdoid tumor
Poorly differentiated chordoma
Germ cell tumors
Synovial sarcomas, poorly differentiated monophasic variant
NUT translocation carcinoma
Undifferentiated sarcoma

Organ-specific blastomas
   Wilms' tumor (nephroblastoma)
   Hepatoblastoma
   Sialoblastoma
   Pancreatoblastoma
   Pleuropulmonary blastoma
   Esthesioneuroblastoma

Modern Soft Tissue Pathology, Second Edition, ed. Markku Miettinen. Published by Cambridge University Press. © Cambridge University Press 2016.

now comprise immunohistochemistry and molecular tests, including karyotyping, genetic testing, including reverse transcriptase polymerase chain reaction (RT-PCR), and fluorescence in situ hybridization (CISH, FISH). Gene expression arrays have yielded colorimetric information of prognostic significance and can be used to group tumors into diagnostic categories.[4]

Genetic factors

A variety of genetic factors have been associated with SRBCT. Inherited conditions such as Beckwith–Wiedemann syndrome,[5] Li–Fraumeni syndrome, neurofibromatosis 1 (NF1),[6] familial pleuropulmonary blastoma syndrome, and basal cell nevus syndrome place one at increased risk for small cell tumors such as rhabdomyosarcoma.

Genetic and oncoviral factors implicate several acquired gene rearrangements and epigenetic lesions in the causation of SRBCT. Foremost are translocations, in which gene fusion products alter transcription control and lead to unrestrained cellular proliferation. Perhaps the best known is the t(11;22)(q24;q12), which results in the fusion of the *FLI1* protooncogene at 11q24 with the *EWS* gene at 22q12 and creates a chimeric protein that exhibits abnormal binding to DNA and leads to tumor induction.[7] This cytogenetic aberration led to subsequent discoveries proving that two morphologically disparate lesions, Ewing sarcoma and PNETs, are biologically and clinically identical neoplasms.[8] Translocations serve not only as disease mechanisms but also as diagnostic markers detectable by various molecular techniques.

Mutated genes that predispose to embryonal cancers include *APC*, associated with hepatoblastoma,[9] the *patched* gene, associated with medulloblastoma,[10] and *TP53*, associated with rhabdomyosarcoma.[11] These may represent acquired mutations or occur as a constitutional lesion, as with polyposis coli, basal cell nevus syndrome, or Li–Fraumeni syndrome, respectively.

Epigenetic mechanisms play a causative role in many SRBCT. They produce alterations of the DNA sequence without mutation, and they are responsible for diverse biologic phenomena such as imprinting, X inactivation, and aging.[12,13] Methylation alterations in rhabdomyosarcoma,[14] and Wilms' tumor associated with Beckwith–Wiedemann syndrome indicate that epigenetic aberrations play roles in these tumors. Histone deacetylase and noncoding short RNAs add to this potent epigenetic brew.[15]

The Ewing sarcoma family of tumors
Definition

By definition, Ewing sarcomas are malignant, poorly differentiated, small round cell tumors primarily arising in bone and soft tissue, rarely occurring in diverse organs, and having a propensity for limited neuroepithelial differentiation. They are genetically characterized by fusions of ETS family genes (primarily *FLI1* or *ERG*) and *EWS* (rarely, *FUS* is an alternate). These tumors are the second most common malignant bone tumor and soft tissue tumor in children and adolescents).

Synonyms for this entity include Ewing sarcoma, PNET, peripheral neuroepithelioma, and Askin's tumor (limited to the soft tissue of chest wall).

History

Skeletal tumors of adolescents and young adults, now known as *Ewing sarcoma*, were initially named "endothelioma of bone" by James Ewing in 1918. In that same year, Arthur Purdy Stout reported a sarcoma of the ulnar nerve, which he termed a *peripheral neuroepithelioma*. Although a similar age group was affected, these tumors differed in their histologic appearance and site of origin. Ewing's endothelioma was an undifferentiated small cell skeletal tumor, and Stout's peripheral neuroepithelioma was a soft tissue lesion that displayed definite neural characteristics similar to neuroblastomas. Subsequent bone pathologists debated whether Ewing sarcoma was truly a primary lesion or a metastasis from an undiscovered adrenal neuroblastoma. Stout, however, noted that Ewing sarcomas at times appeared to have vague neural features, similar to peripheral neuroepithelioma.[16,17]

By the 1960s, the question of whether Ewing sarcoma was a primary or metastatic lesion was resolved, but its histogenesis remained obscure. Authors hypothesized that these were of neural, hematopoietic, and mesenchymal origins.[17] In the 1970s, reports appeared of a soft tissue lesion having morphologic features similar to Ewing sarcoma.[18,19] These tumors also affected young adults and were similarly composed of sheets of primitive small cells. By the 1970s, ultrastructural studies showed that both the skeletal[20] and extraskeletal[21] forms of Ewing sarcoma sometimes contained neural organelles.

Reports of soft tissue lesions resembling Stout's peripheral neuroepithelioma, also known as PNET,[22] included the study by Askin and colleagues of a series of tumors that arose in the chest wall of adolescents and young adults.[23] In the early 1980s, Jaffe and others reported PNTs of bone and demonstrated neural proteins by immunohistochemistry.[24] At this juncture, there appeared to be two separate entities, PNET and Ewing sarcoma, that could arise within either bone or soft tissue, although the former appeared more common in soft tissue and the latter in bone.

Cytogenetic studies showed that Ewing sarcoma, Askin's tumor, and PNETs contained the same karyotypic change: a translocation between the long arms of chromosomes 11 and 22, known as t(11;22)(q24;q12).[25,26] Other methods later confirmed that most of these neoplasms had the same chromosomal abnormality and a shared genetic identity, despite their morphologic dissimilarity.[27,28]

Cavazzana and coworkers[29] then demonstrated that neural differentiation could be induced *in vitro* in Ewing sarcoma cells, which began sprouting neurites similar to those described by Stout and Murray in peripheral neuroepithelioma cultures.[30] Cavazzana *et al.* also demonstrated expression of neural proteins indicative of neural differentiation, a finding subsequently verified by others. These experiments showed that Ewing sarcoma and PNET were closely related, if not the

Table 31.2 Ewing sarcoma, racial incidence ratios[32]

USA, Caucasian, SEER data	0.54
USA, African American, and Asian/Native American SEER data	0.12

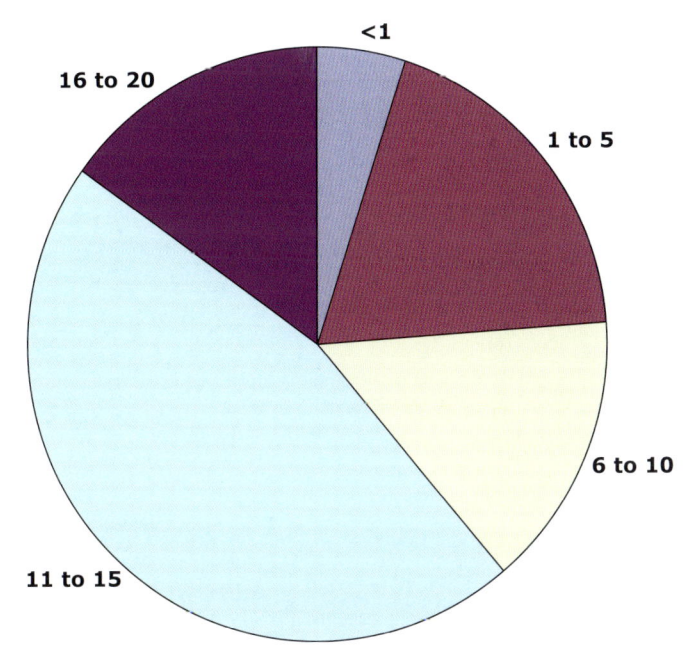

Figure 31.1 Pediatric age distribution, extraosseous Ewing sarcoma

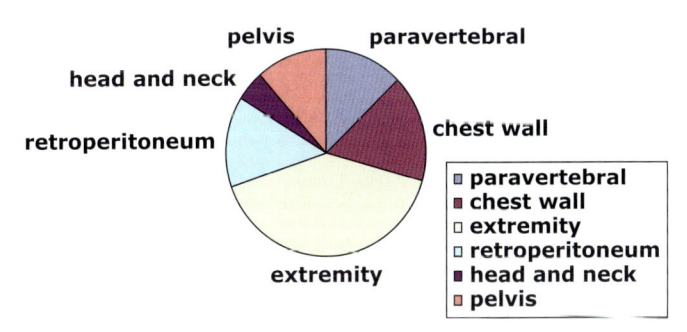

Figure 31.2 Site distribution, extraosseous Ewing sarcoma.

same entity, which became known as *the Ewing sarcoma family of tumors*.[31]

Clinical features

Ewing sarcomas primarily arise in adolescents and young adults and occasionally occur in young children and in older adults (Figure 31.1). There is a striking ethnic predilection, because the disease is distinctly uncommon in non-whites (Table 31.2).[33] Like rhabdomyosarcomas, Ewing sarcomas have a slight predilection for males.[34]

Ewing sarcomas most commonly present as bone lesions and can affect any bone in the skeleton. Increased recognition of the soft tissue lesions has followed standardized reviews,[35] better techniques of diagnosis, and greater awareness of these tumors. Currently Ewing sarcomas comprise the second most common soft tissue malignancies in children and adolescents following rhabdomyosarcoma. For example, in a large series that comprised 1687 childhood sarcomas, 45% of cases were rhabdomyosarcomas, and 23% were PNETs or extraosseous Ewing sarcomas.[36]

A geographic predilection for Ewing sarcoma, independent of racial proclivity, has not been described. Cluster outbreaks,[37] family clusters,[38,39] and an AIDS association raise the possibility of an infectious etiology. Familial clusters have also been reported. No infectious agent has ever been proven in humans, but adenovirus induces PNET-like embryonic neuroepithelial tumors in rodents.[40]

Patients affected with Ewing sarcomas may present with symptoms of inflammation of the affected extremity, resulting in a clinical misdiagnosis of osteomyelitis. It is therefore important for radiologists to recognize the signs of these tumors, particularly those occurring in bones. Otherwise, the most common presentation is with mass lesion, sometimes with a pathologic fracture.

Both axial and extremity bones can be affected by Ewing sarcoma, but diaphyseal tumors predominate.[41] The most common soft tissue locations are the paravertebral region, retroperitoneum, and chest wall, followed by the extremities (Figure 31.2).[42,43] Of note are tumors occurring as renal or skin neoplasms;[44] these have unique features which are different from those of the usual lesions.[45–47] Another unusual but possible location is the epidural space of both the cranium and the spine.[48,49] In soft tissue sites, origin from the peripheral nerves or nerve roots can occur, as in Stout's original description. This phenomenon creates diagnostic and biologic confusion with malignant peripheral nerve sheath tumors (MPNSTs), which may contain Ewing sarcoma-like foci.[50]

Radiologic studies usually determine whether a Ewing sarcoma arises in bone or soft tissue (Figure 31.3). Bony lesions are typically permeative, destructive lesions. There is often a large, associated soft tissue mass that overshadows the bony component. Lesions arising in the kidney are usually large, invasive masses that have no radiographic distinction from other related neoplasms. Paraspinal tumors can present as dumbbell-shaped masses with intraspinal invasion, reminiscent of neuroblastoma.

Gross features

Although lesions limited to the skin and superficial soft tissue can measure <1 cm, Ewing sarcomas are usually large, fleshy masses that can contain extensive necrosis. Untreated tumors can grow to huge lesions that rapidly invade adjacent structures. Bone tumors are typically permeative and destructive. They commonly show periosteal breakthrough and elevation, often with lamination that creates an onion-skin appearance on plain radiographs. Soft tissue lesions can be pseudoencapsulated. The viable portions of the soft tissue masses have a yellow-tan, fleshy appearance. The adjacent muscle is sometimes massively edematous, creating an erroneous impression on magnetic resonance (MR) images of tissue extension.[51]

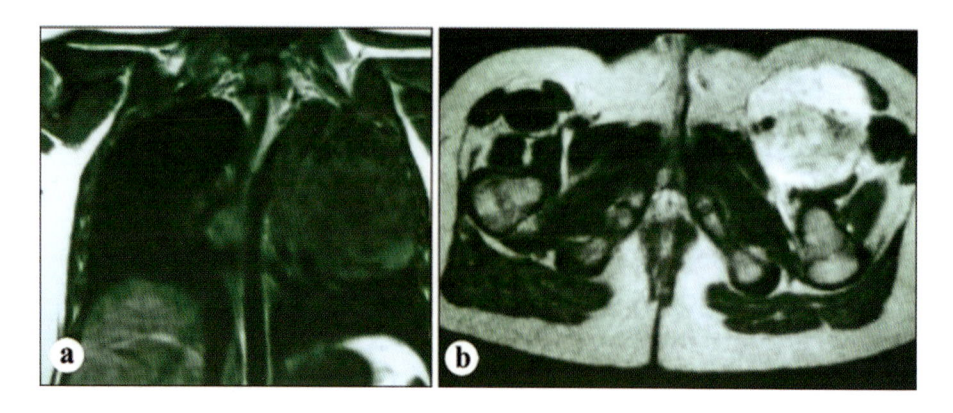

Figure 31.3 Nuclear magnetic images of Ewing sarcoma family of tumors. (**a**) A large chest wall mass protrudes into the left hemithorax, causing a marked mediastinal shift. (**b**) An ovoid soft tissue mass arises from the muscles of the inguinal region. (Images are provided courtesy of Dr. Sue Kaste, St. Jude Children's Research Hospital, Memphis, TN.)

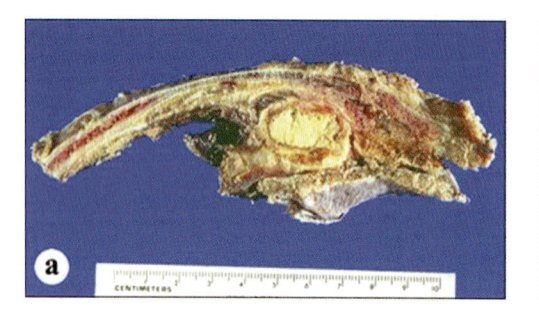

Figure 31.4 Ewing sarcoma after multiagent therapy. (**a**) Gross photograph of a lesion involving rib and thoracic soft tissues, with marked necrosis in soft tissue lesion. (**b**) Photomicrograph of treated Ewing sarcoma of the fibula showing the marrow space with fibrosis and focal hemorrhage (green arrow). No viable residual tumor is identified.

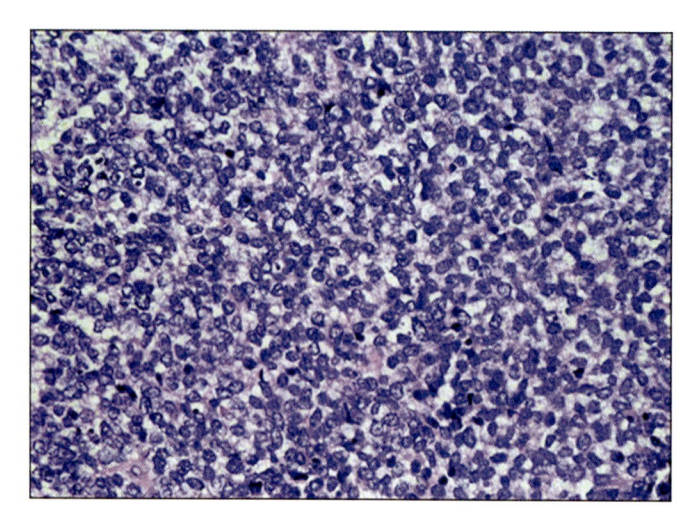

Figure 31.5 Ewing sarcoma is composed of a highly cellular sheet of relatively featureless small cells, with round, dark nuclei and inconspicuous cytoplasm.

After a course of multiagent chemotherapy, there is usually dense fibrosis, hemorrhage, and necrosis, often without macroscopic evidence of tumor (Figure 31.4). Adjacent bone marrow loses its fatty quality and becomes more serous or fibrotic, creating abnormal radiographic images.[52]

Microscopic features

Most Ewing sarcomas present a relatively monotonous pattern of sheets of small blue cells with round to oval, hyperchromatic nuclei, modest amounts of cytoplasm, and inconspicuous cellular boundaries (Figure 31.5). Several subtle cytologic and histologic features have been used to separate these tumors into three major histologic categories: (1) typical Ewing sarcoma, (2) atypical Ewing sarcoma, and (3) PNET.[53] Familiarity with the histologic varieties of the Ewing sarcoma family of tumors aids in their proper diagnosis and distinction from other similar round cell neoplasms.

Typical Ewing sarcoma constitutes the most common category of the Ewing sarcoma family of tumors. There are usually two cell types present: those with round nuclear contours, smooth, even chromatin, and clear, vacuolated cytoplasm, and those with more angulated nuclear contours, hyperchromatic chromatin, and lightly eosinophilic cytoplasm. These have been respectively termed *light cells* and *dark cells* (Figure 31.6). The light and dark cells often intermingle, with no discernible pattern, among diffuse, patternless sheets of tumor cells. Alternatively, the dark cells may present in a streaming pattern among the lighter elements.

Neither light nor dark cells contain conspicuous nucleoli, a property that aids in the distinction of typical Ewing sarcoma from the atypical variant. A somewhat ironic feature of the typical lesions, in view of their rapid, destructive growth, is their paucity of mitotic figures, usually less than 1 per high-power field (HPF). The clear, vacuolated cytoplasm reflects a high content of glycogen demonstrable by periodic-acid–Schiff (PAS) stain (Figure 31.7). Other round cell tumors, such as rhabdomyosarcoma, germinoma, and lymphoblastic lymphoma, can also have glycogen.

Atypical Ewing sarcomas have cytological features that distinguish them from typical tumors, as recounted in Table 31.3. Foremost among these is their pleomorphism, exhibited by irregular nuclear contours and prominent

Table 31.3 Distinguishing features of the Ewing sarcoma family of tumors

Feature	Typical	Atypical	PNET
Nuclear size	Small (roughly the size of histiocytes)	Large (larger than histiocytes)	Small or large
Mitotic index	Low (<1 per 400× field)	High (>1 per 400× field)	Generally high
Nuclear contours	Regular	Irregular	Variable
Rosettes	No	No	Yes
Nucleoli	Inconspicuous	Prominent	Prominent

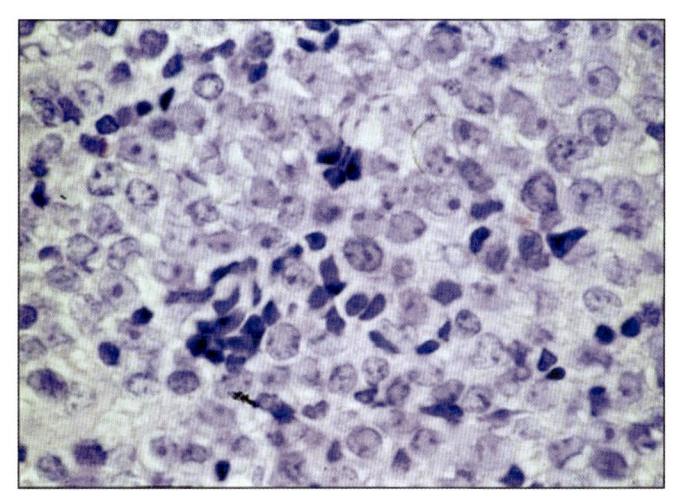

Figure 31.6 At a high magnification, Ewing sarcoma reveals a mixture of cells with smooth, lightly staining nuclear chromatin ("light" cells) and darkly staining nuclear chromatin ("dark" cells).

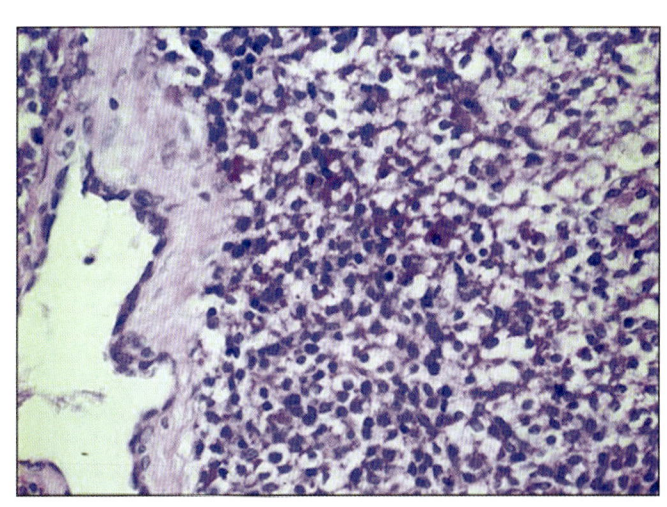

Figure 31.7 Periodic-acid–Schiff stain of Ewing sarcoma, illustrating the cytoplasmic positivity that is indicative of glycogen (PAS/hematoxylin) nuclei and inconspicuous cytoplasm.

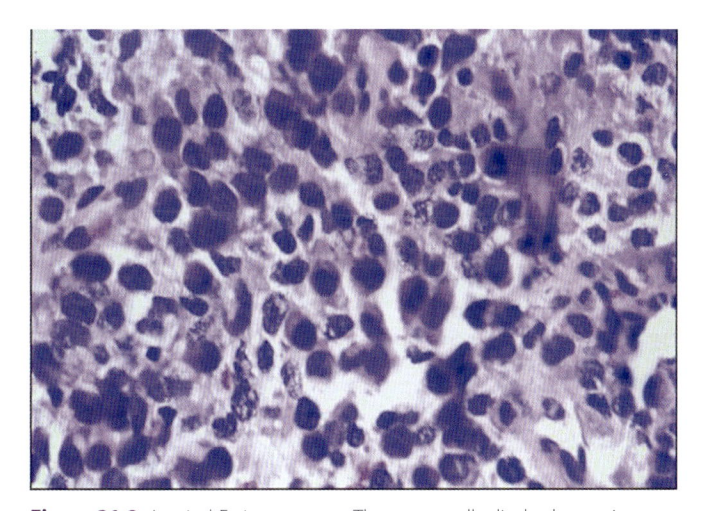

Figure 31.8 Atypical Ewing sarcoma. The tumor cells display larger size, greater nuclear variability, and occasionally prominent nucleoli.

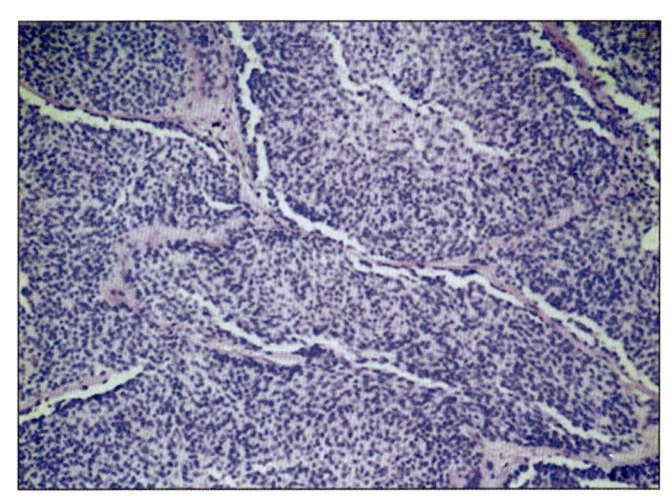

Figure 31.9 Atypical Ewing sarcoma. On low-power examination, this lesion contains lobules of tumor cells invested by a fine fibrovascular stroma, creating a nodular pattern.

nucleoli (Figure 31.8). These characteristics and larger cell size make large cell lymphoma a diagnostic consideration. Some cases display a nodular nested pattern with moderate fibrous stroma on low-power examination, reminiscent of alveolar rhabdomyosarcoma or neuroendocrine tumors (Figure 31.9). Despite their worrisome features, atypical Ewing sarcomas are similar to typical lesions in clinical behavior.[54] They tend to exhibit nascent neural differentiation, however, as shown by neuron-specific enolase (NSE) positivity and ultrastructural characteristics.[53] Newer varieties of Ewing sarcoma have been added to its description. These include adamantinoma-like, sclerosing, spindled, and clear cell anaplastic variants.[55,56]

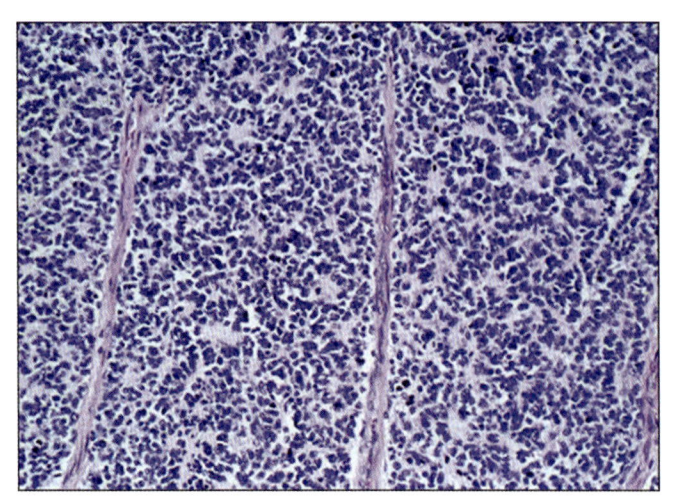

Figure 31.10 Peripheral primitive neuroectodermal tumor (PNET), containing numerous, conspicuous Homer Wright rosettes.

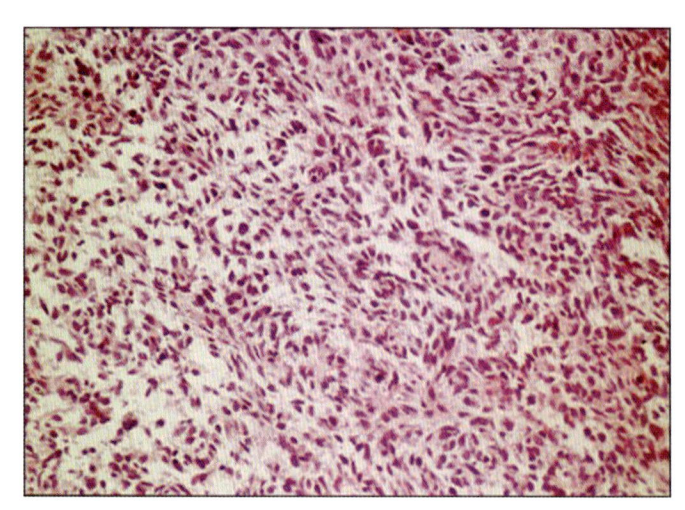

Figure 31.11 PNET arising from the anterior chest wall lesion and containing an admixture of primitive cells and differentiated ganglionic cells.

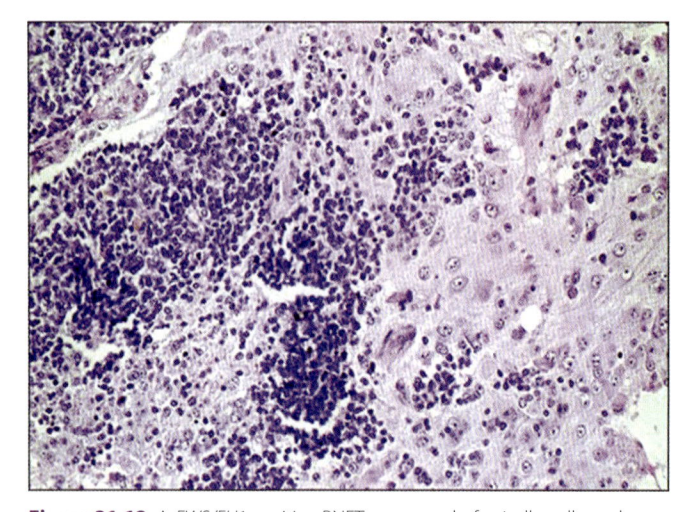

Figure 31.12 A *EWS/FLI1*-positive PNET composed of spindle cells evokes malignant peripheral nerve sheath tumor (H&E, ×50).

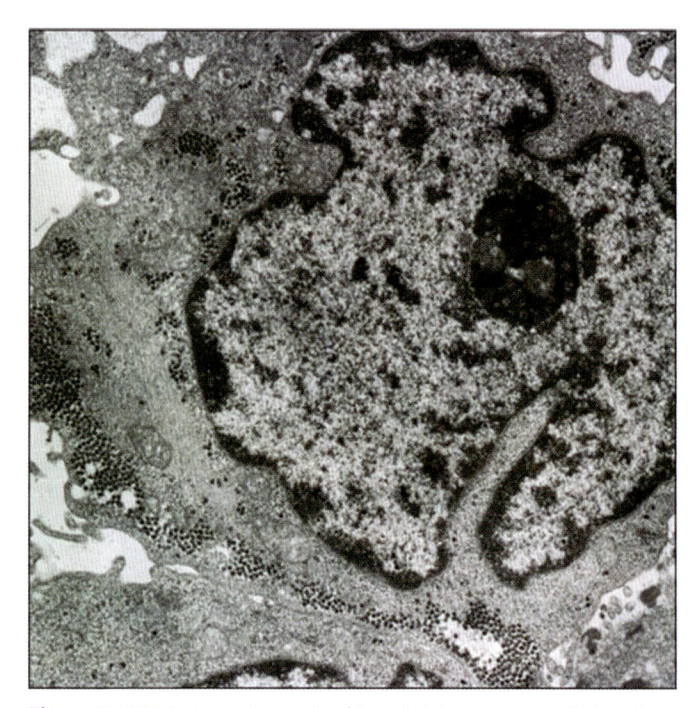

Figure 31.13 Electron micrograph of typical Ewing sarcoma cell. There is a dearth of cytoplasmic organelles, but pools of cytoplasmic glycogen are prominent. Note the simple intercellular junction. magnification: ×13 000)

PNETs are the differentiated form of Ewing sarcoma family tumor. Homer Wright rosettes constitute the major criterion for their diagnosis (Table 31.3). Homer Wright rosettes should be easily recognizable as lightly eosinophilic, fibrillary cores surrounded by circular wreaths of round to oval nuclei (Figure 31.10). Other features include a "lobular" pattern reminiscent of alveolar rhabdomyosarcoma (Figure 31.9),[34] a pericytomatous pattern resembling neuroectodermal or pericytic tumors, and a spindle cell pattern recapitulating nerve sheath lesions (Figure 31.11).[57] In rare tumors, ganglionic differentiation can be strikingly similar to ganglioneuroma (Figure 31.12).[58] The range of potential differentiation shown by PNETs parallels that of neural crest cells and even includes mesenchymal derivatives.

Electron microscopy

Ultrastructurally, a typical Ewing sarcoma displays a dearth of cytoplasmic organelles, principally free ribosomes, occasional mitochondria, and short profiles of endoplasmic reticulum. The presence of primitive intercellular junctions of the zonula occludens type helps to differentiate this tumor from lymphoma.[59] The most striking feature is the prominence of cytoplasmic pools of beta-glycogen, which are sometimes recognizable only as irregular electron-lucent zones in poorly preserved cells (Figure 31.13). Other primitive tumors, such as rhabdomyosarcoma, lymphoblastic lymphoma, and neuroblastoma, can also be glycogen rich.[60]

Atypical Ewing sarcomas have a gradual increase in cytoplasmic diversity. There can be clusters of mitochondria and

Table 31.4 Immunohistochemical markers of the Ewing sarcoma family of tumors

Marker	Comment
CD57 (Leu-7)	Seen in PNET and some atypical Ewing sarcomas
CD99	Characteristic of all Ewing sarcoma family of tumors, but can be negative in unusual examples. Because many tumors are also positive, it cannot be used as a sole marker
Chromogranin	Uncommon in PNET; more typical of neuroblastoma
Cytokeratin	LMW keratins (e.g., as stained with CAM5.2) often seen with Ewing sarcoma; occasional strong positivity in PNET
Desmin	Rare but occasionally seen in PNET, more typical of ectomesenchymoma
FLI1	Characteristic of all Ewing sarcoma family members but can be negative in unusual examples. Has nonspecificity similar to CD99 and therefore cannot be used as a sole marker
Neuron-specific enolase	More common in atypical Ewing sarcoma; usual in PNET
Synaptophysin	More common in PNET; occasional in atypical Ewing sarcoma
NKX2.2	A new marker found to be upregulated on array studies

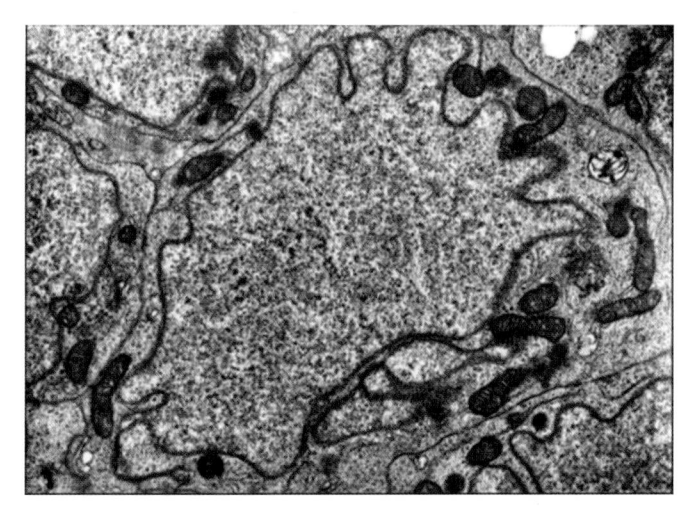

Figure 31.14 Electron micrograph of atypical Ewing sarcoma. Irregularities of the nuclear and cytoplasmic membranes are prominent, with formation of blunt processes, and there are increased numbers of organelles (cf. Figure 31.10; magnification: ×7000). (Courtesy of Ms. Cindy Hastings and Dr. Francine Tryka, University of Arkansas for Medical Sciences, Little Rock, AR.)

globoid concentrations of intermediate filaments akin to those found in rhabdoid tumor.[53,59] Blunted extensions of the cellular periphery can occur and approach the submicroscopic boundary between atypical Ewing sarcoma and PNET (Figure 31.14).[53]

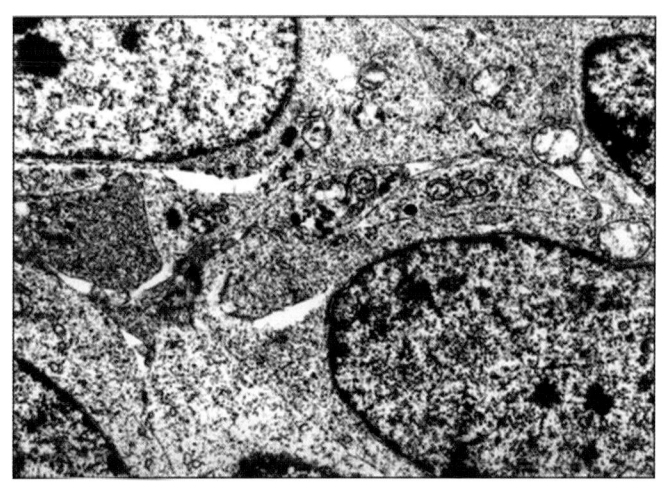

Figure 31.15 Electron micrograph of PNET. Besides the features noted in Figure 31.14, a cytoplasmic process contains somewhat pleomorphic neurosecretory granules (magnification: ×4400). (Courtesy of Ms. Cindy Hastings and Dr. Francine Tryka, University of Arkansas for Medical Sciences, Little Rock, AR.)

The ultrastructural diagnosis of PNET requires intertwining bundles of dendritic processes with discrete neurosecretory granules (Figure 31.15). These granules should contain a central dense core surrounded by an electron-lucent halo wrapped in a trilaminar membrane. They are most commonly found in the cell processes comprising the central portions of Homer Wright rosettes. Primary lysosomes might be mistaken for neurosecretory granules, but they more typically lie within the perikaryon and intermingle with phagolysosomes. Neurosecretory granules of PNETs can be pleomorphic and dumbbell-shaped (Figure 31.15).[53]

Immunohistochemistry

Immunohistochemical features of the Ewing sarcoma family are listed in Table 31.4. PNET exhibits an immunostaining profile that largely parallels other neural tumors (Figure 31.16), such as neuroblastoma.[61] Stains of low-molecular-weight (LMW) keratins and desmosomal proteins are positive in a surprising number of cases[55], paralleling the primitive junctions and tonofilaments seen in these tumors with electron microscopy (Figure 31.17). The epithelial nature of some Ewing sarcomas is underscored by cases that strongly resemble adamantinoma.[62]

A major immunohistochemical feature of the Ewing sarcoma family of tumors is staining for CD99 (Figure 31.18).[63] As recognized by monoclonal antibody clones O13, HBA71, and 12E7, CD99 was originally discovered in immunologic studies of T-cell leukemia,[64] and it was subsequently found to mark most Ewing sarcomas and PNETs.[65,66] It is a sensitive marker, positive in over 95% of these tumors. CD99 is not specific and positivity has been reported in rhabdomyosarcoma, synovial sarcoma, small cell carcinoma, sex cord tumors, and a variety of other cancers.[65] CD99 is also positive in lymphoblastic lymphoma. Lymphoblastic lymphomas, which can arise in bone, can show little or no reactivity to

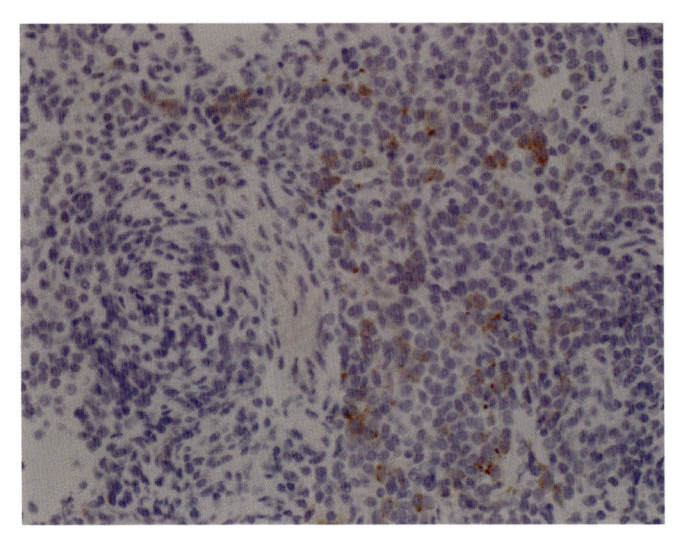

Figure 31.16 Immunohistochemical evidence of neural differentiation in Ewing sarcoma, with cytoplasmic staining for synaptophysin. (Avidin-biotin complex method with hematoxylin counterstain.)

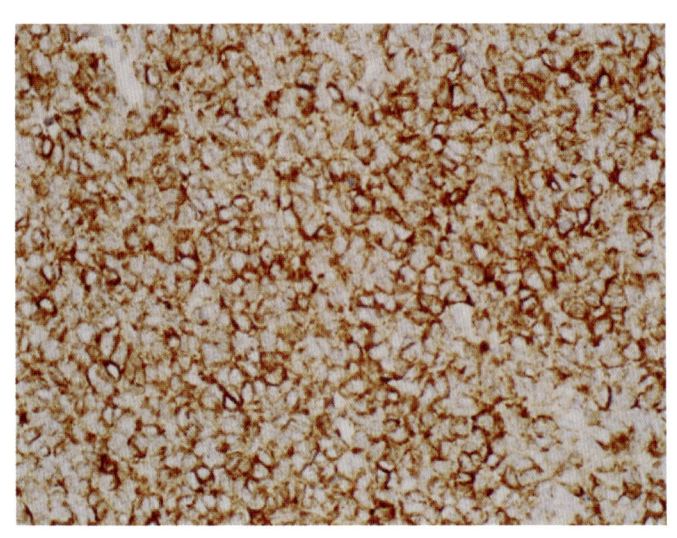

Figure 31.17 Immunohistochemical evidence of epithelial differentiation in Ewing sarcoma, with strong cytokeratin positivity. (AE1/AE3).

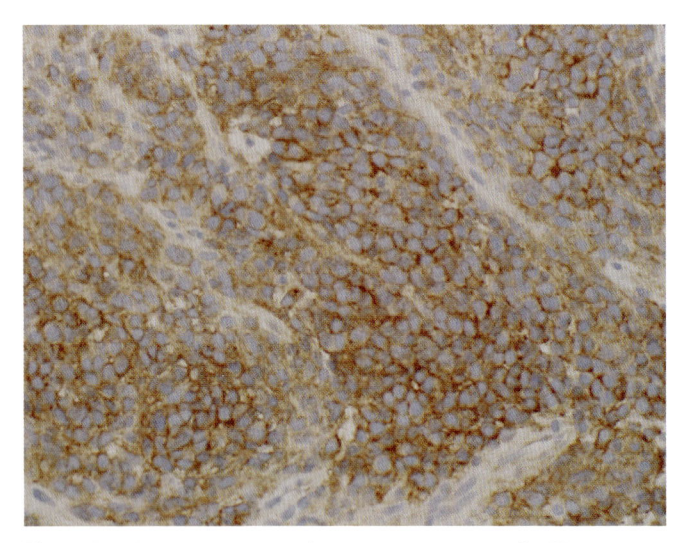

Figure 31.18 Ewing sarcoma with immunostain positivity for CD99. There is diffuse membranous staining.

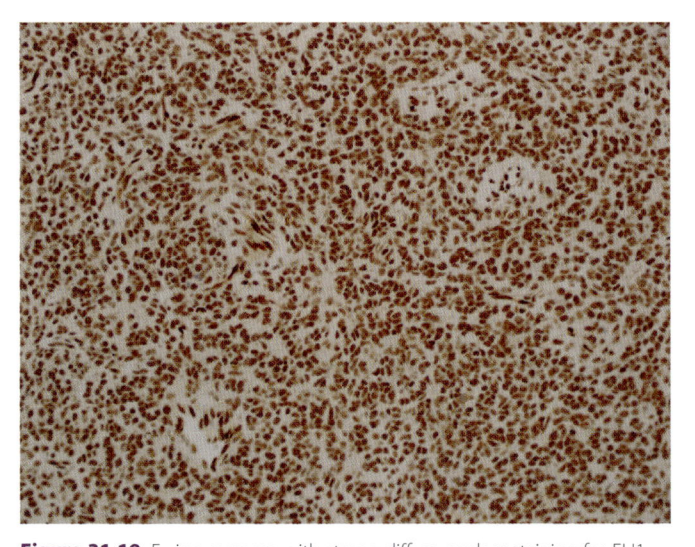

Figure 31.19 Ewing sarcoma with strong diffuse nuclear staining for FLI1.

CD45 (leukocyte common antigen) and strong staining with CD99, creating a diagnostic dilemma.[67,68] It is thus prudent to use CD99 in a panel of other lymphoid markers, such as CD3, terminal deoxynucleotidyl transferase (TDT), or CD43. Conversely, CD99 has not been reported to date in neuroblastomas, so that it remains a useful marker in the distinction between them and PNETs.[65,69] Conversely, a marker that is usually negative in Ewing sarcomas is the neural cell adhesion molecule (CD56, NCAM), typically positive in neuroblastoma.[70–72]

Antibodies against the FLI1 protein have been used to complement CD99 as an ancillary marker for Ewing sarcoma (Figure 31.19).[73] FLI1 gene fusion, as described later, occurs in most Ewing sarcomas, and this leads to overexpression of the protein. Unfortunately, there is similar nonspecificity against endothelial cell elements and T-lymphocytes, but the combination of the two markers might improve their individual reliability.[73]

Gene expression analysis of EWS fusion-positive Ewing sarcomas shows upregulation of NKX2.2, a transcription factor important in neuroendocrine differentiation. It is a sensitive immunohistochemical marker, but it suffers from some lack of specificity. Olfactory neuroblastomas and some small cell carcinomas, synovial sarcomas, mesenchymal chondrosarcomas, and melanomas may be positive.[74]

Rare desmin positivity of PNETs creates potential confusion with rhabdomyosarcoma.[75] Some PNET-like tumors contain myogenous elements and are termed *ectomesenchymomas* (Figure 31.20). Ectomesenchymomas contain a myogenic component recognizable by light microscopy or electron microscopy, in addition to their neural component.[76] Their histogenesis is currently speculative, but the myogenic

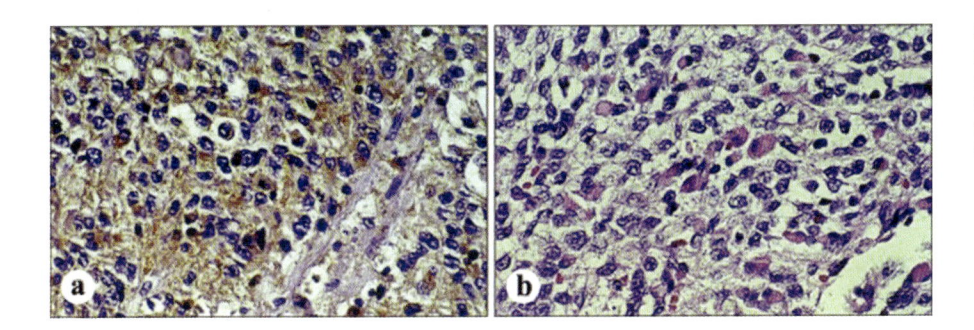

Figure 31.20 Ectomesenchymoma. There is a mixture of neural elements, as evidenced by synaptophysin-positive fibrillary stroma (**a**), and myogenous elements, evidenced by scattered rhabdomyoblasts (**b**). (a: avidin-biotin complex method with hematoxylin counterstain; b: H&E stain.)

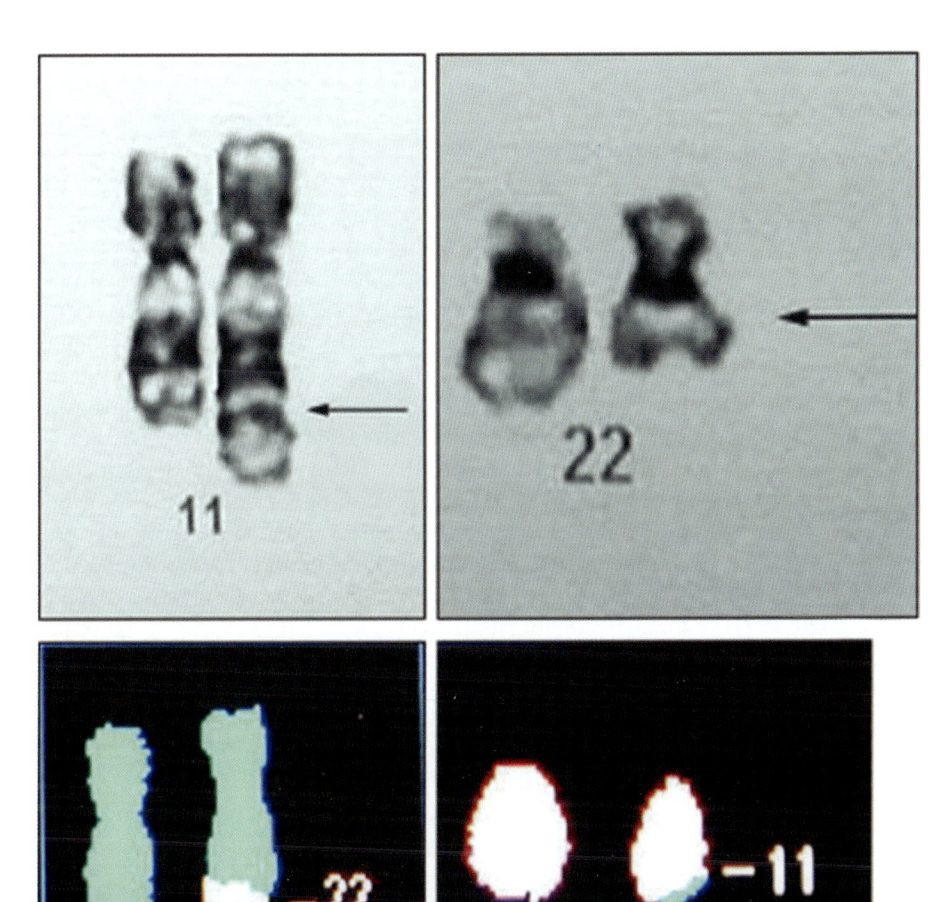

Figure 31.21 t(11;22)(q24;q12) translocation of Ewing sarcoma. The upper panels are taken from a standard G-banded karyotype, and the lower panels are from a spectral karyotype (SKY). In the upper panels, there is a reciprocal translocation involving the long arms of chromosomes 11 and 22 at breakpoints 11q24 and 22q12 (arrows). In the lower panels, chromosomes 11 and 22 are respectively colored blue and white. (Courtesy of Mr. Charles Swanson and Dr. Jeff Sawyer, University of Arkansas Medical Sciences, Little Rock, AR.)

potential of the neural crest has been documented.[77] The Children's Oncology Group currently treats ectomesenchymomas as rhabdomyosarcomas, and their clinical behavior seems to be related to the histologic variety of rhabdomyosarcoma that comprises the myogenic component.[78]

Genetics of the Ewing sarcoma family of tumors

Cytogenetic studies of Ewing sarcomas typically reveal a reciprocal translocation between chromosomes 11 and 22, the t(11;22)(q24;q12) (Figure 31.21). This karyotypic aberration is found in the entire morphologic spectrum of this entity, including typical Ewing sarcomas, atypical Ewing sarcomas, Askin tumors, and PNETs,[8] and it forms the theoretic basis for the biologic relatedness of these lesions.[31,79] Cytogenetic analyses reveal additional cytogenetic abnormalities, such as the der(16)t(1;16)(q21;q13)[80] and trisomy 8 and 12,[81,82] in both Ewing sarcomas and PNETs, but these are not a consistent feature and appear to represent secondary events.[83,84] The frequency of detection of these secondary karyotypic changes ranges from around 40% by karyotyping and FISH[81] to approximately 75% by comparative genomic hybridization.[84]

The t(11;22)(q24;q12) in Ewing sarcoma causes a fusion between the *FLI1* gene, located on chromosome 11q, and the

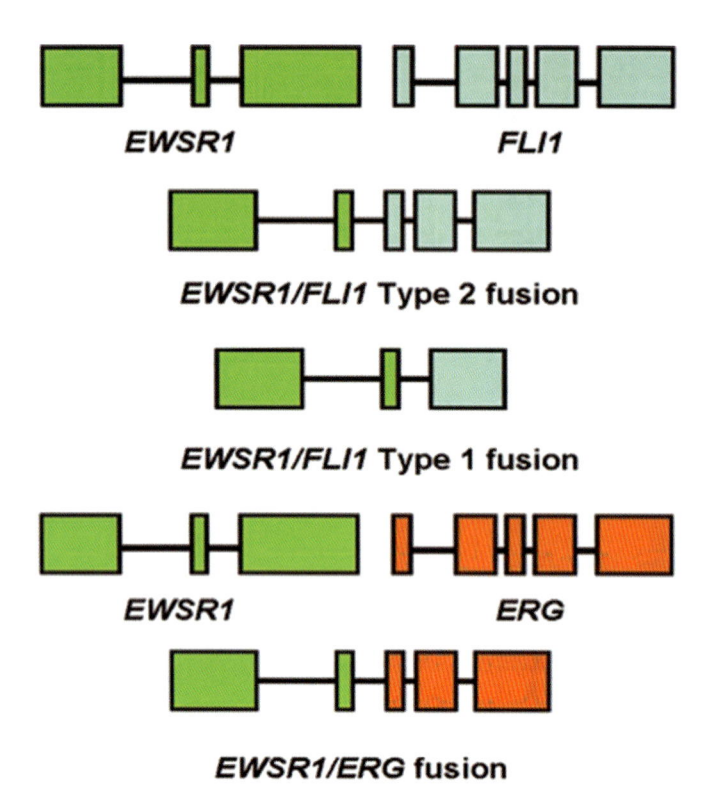

Figure 31.22 Cartoon of genetic fusions resulting from translocation of portions of chromosomes 11 and 22 (EWS/FLI1) and 21 and 22 (EWS/ERG). There is variability in the location of the breakpoints in EWS/FLI1 resulting in different fusion types; the two most common are presented here.

EWS gene on chromosome 22q12 (Figure 31.22).[85] The *FLI1* gene is the human homolog of proto-oncogene that causes murine erythroleukemia, and the *EWS* (from Ewing Sarcoma) gene produces an RNA-binding factor. *FLI1* contains exons identifying it as an ETS family gene, distinguished by a highly conserved DNA-binding region. Following the fusion of the two genes, the RNA-binding region of the EWS protein replaces the FLI1 DNA-binding region and results in abnormal transcription control.[86] The chimeric transcript can produce tumors in mice and can transform cultured cells. Because its antisense RNA represses this tumorigenicity,[87] the abnormal transcript appears to be the key biologic event in the tumorigenesis of Ewing sarcoma.

EWS may fuse with other ETS family genes and produce morphologically identical tumors. The first "alternate partner" discovered for the *EWS* was the *ERG* gene on chromosome 21, creating an *EWS/ERG* fusion resulting from a t(21;22)(q22;q12) (Figure 31.22).[88] Approximately 10% of Ewing sarcomas contain this alternate translocation.[33] Like FLI1, the ERG product contains an ETS motif with a similar DNA transcription region. This translocation produces a fusion product resembling FLI1/EWS and possessing matching tumorigenic properties. Additional alternate translocations for the Ewing sarcoma family of tumors include a t(7;22) that fuses *EWS* with *ETV1*,[89] a t(2;22) that fuses *EWS* with *FEV*,[90] and a t(17;22) that fuses *EWS* with *E1AF*, and the list continues to grow. All of these Ewing sarcoma fusions involve ETS family genes. In rare tumors, *EWS* is substituted by the *FUS* gene, which serves as an alternate partner for *FLI1* or *ERG*.[91] *FUS* and *EWS* are both members of the TET family of genes.

RT-PCR can be used to detect Ewing sarcoma fusions in cellular RNA extracts. FISH with two-colored probes can also be applied to interphase cells,[92] permitting routine use of formalin-fixed, paraffin-embedded tissue. FISH may be a more sensitive method than RT-PCR for the diagnosis of the Ewing sarcoma family of tumors.[93] The EWS/FLI1 molecule exhibits several different fusion points because of so-called combinatorial fusion.[94] The tumorigenic properties of the molecule rely primarily on the polar ends of the fusion as the functional sites. Because of this variability, a relatively long sequence of cDNA is required for the PCR process to span the various breakpoints. The extreme sensitivity of this test in fresh tissue, however, allows for the detection of circulating tumor cells in peripheral blood or cells in bone marrow.[95,96]

Combinatorial fusion results in two major types of EWS/FLI1 fusion proteins, termed *type 1* and *type 2*. De Alava and colleagues suggested that type 1 tumors have a better prognosis.[97] However, this factor did not have prognostic value in more recent studies.[98] There appears to be no clinical significance to date for the alternative fusions, such as the *EWS/ERG*.

Other genetic features of Ewing family sarcomas include losses of p16[99] and amplification of *MDM2* and *CDK4* genes on 12q.[100] One interesting observation, considering the neuroectodermal phenotype of these tumors, is gastrin-releasing peptide expression, also noted in neuroendocrine carcinomas.[101] Recent studies have focused on an array of molecules of potential therapeutic value.[102]

Prognosis and outcome of the Ewing sarcoma family of tumors

Because of improved modern therapy, the overall current prognosis for the Ewing sarcoma family of tumors has improved from their dismal status in previous years. Some relate this primarily to the addition of ifosfamide and etoposide to the therapeutic regimen. At present, the overall 5-year survival rate is approximately 70% for localized tumors.[103] Pulmonary metastases characterize these tumors, in contrast with neuroblastoma, which rarely metastasizes to the lung. Bony and hepatic metastasis also occurs, but lymph node metastases are unusual. Local therapy using postchemotherapy surgery or radiation is important for tumor control.[103]

Although initial studies showed a poor prognosis for PNETs compared with typical Ewing sarcoma,[104] studies based on prospective trials failed to confirm this observation.[105] Another feature of aggressiveness, confirmed in several centers, is histological lack of responsiveness to preoperative (neo-adjuvant) chemotherapy.[106–108] Secondary cytogenetic changes, such as deletions of chromosome 1p, also correlates with adverse tumor behavior.[83] Patients with soft tissue Ewing sarcomas may fare better than those with bony tumors.[109]

Although much has been learned, more research is needed to identify biological factors than reliably predict outcome and response to treatment.[56]

A variety of clinical parameters correlate with aggressive behavior. Foremost among these is the presence of metastatic disease.[110,111] Current trials using agents such as ifosfamide lead to improved survival with localized tumors, but not with metastatic lesions.[112] Another prognostic parameter is tumor size, with lesions >100 mm in diameter having 3-year disease-free survival rates of 32%, compared with 80% for smaller tumors.[108,111] Pelvic tumors show more aggressive behavior than nonpelvic lesions, perhaps because they are also often larger at the time of diagnosis and are nonresectable. Other prognostic factors include completeness of surgical excision and local tumor control,[113,114] types of local treatment and systemic chemotherapy, and serum LDH levels.[108]

Differential diagnosis

The differential diagnosis of Ewing sarcoma depends on patient age, because tumors in older patients can be confused with small cell carcinoma, and lesions in younger patients can be mistaken with rhabdomyosarcoma, neuroblastoma, and Wilms' tumor. Distinction of PNET from Ewing sarcoma is no longer an issue.

Ancillary studies are usually necessary to confirm the diagnosis of the Ewing sarcoma family of tumors. The most important immunohistochemical marker is CD99, because it is almost universally present in Ewing tumors, but is negative in neuroblastomas.[90] FLI1 used in concert with CD99 can enhance the reliability of the immunohistochemistry of these tumors.[73] Molecular genetic methods such as FISH or RT-PCR give turnaround times comparable to immunohistochemistry, and specificity is high, with a few exceptions.[115] Confirmatory genetic testing has now become more critical in most lesions, particularly those arising in unexpected sites.[116]

Undifferentiated Ewing-like sarcomas

There is a growing body of knowledge about "undifferentiated Ewing-like sarcomas," which show a variety of genetic aberrations.[117] These include both cases having EWS partnering with a non-ETS family member and those lacking EWS fusions, but showing a similar genetic profile. The clinical setting and therapeutic responses of these lesions are heterogeneous and at the present time unpredictable. The rarity of these lesions has precluded extensive understanding, so a pragmatic medical approach is often necessary for diagnosis and treatment. These lesions may partner EWS with either transcription factors or nontranscription factors, show typical or atypical Ewing histology, occur either in bone or soft tissue, and show variable CD99 positivity.[118]

EWS–non-ETS sarcomas

In rare tumors, EWS and NFATC2 form a non-ETS fusion in which EWS is fused with the transcription factor. NFATC2 is not part of the ETS family of genes, but shares similar DNA-binding properties. The protein product of the gene activates the same downstream pathway as FLI1 and ERG fusions. Of note, these tumors show recurrent gene amplification, similar to the alternate PAX7-FOXO1 fusion of alveolar rhabdomyosarcoma. Thus, one may discover a large number of rearranged signals on FISH studies.

In EWS-PATZ1 and EWS-SP3 fusion, EWS fuses with a transcription factor gene that is unlike the ETS family genes. Both of these aberrations fuse EWS to a zinc finger fusion partner, similar to that seen in DSRCT. These Ewing-like sarcomas show polyphenotypia similar to DSRCT, and likewise they are more aggressive and drug resistant than Ewing sarcoma. Of note, rarely one may see intra-abdominal tumors with polyphenotypia and desmoplasia, like DSRCTs, but containing an EWS-FL1 fusion.[118]

EWS and SMARCA5 form a variant fusion that links EWS with an epigenetic factor rather than a transcription factor. SMARCA5 is a chromatin remodeling gene, similar to the SMARCB1 that is mutated in rhabdoid tumors. This suggests a relatedness to rhabdoid tumor, but as yet nothing has been proven.

Some have described a Ewing-like sarcoma that contains a fusion of EWS with POU5F1. The latter gene produces a transcription factor that regulates stem cells (OCT4). This lesion thus binds EWS to a transcription factor entirely unlike that of ETS family genes. These lesions show a heterogeneous morphology, with areas containing the nested polygonal and spindle cells, and they show S100 protein positivity. Thus, they more likely represent a form of soft tissue myoepithelial tumor rather than a Ewing-like sarcoma. EWS fusions have also been reported in the former lesion.[118]

CIC-DUX4 sarcomas

Some Ewing-like sarcomas show fusions that are unlike EWS, but transcribe a similar downstream group of genes. These include CIC-DUX4 tumors, which form balanced translocations between chromosomes 4 and 19 or 10 and 19, and the BCOR-CCNB3 fusion tumors, which show a paracentric inversion of the X chromosome.

CIC-DUX4 sarcomas are usually extraskeletal and arise most commonly in the extremities. They show an aggressive course with early metastasis. In particular, lymph node metastasis is relatively frequent, unlike with Ewing sarcoma. They have features more in common with atypical Ewing sarcoma, such as prominent nucleoli, more abundant cytoplasm, and extensive necrosis and mitotic activity (Figure 31.23 and 31.24).[119] CD99 positivity can be variable, and this lesion should be considered in any Ewing-like sarcoma that lacks CD99 staining and EWS fusion. Of note, they are frequently WT1 positive.[119] The CIC gene product contains a binding site for TLE protein, similar to that seen with synovial sarcoma fusions.

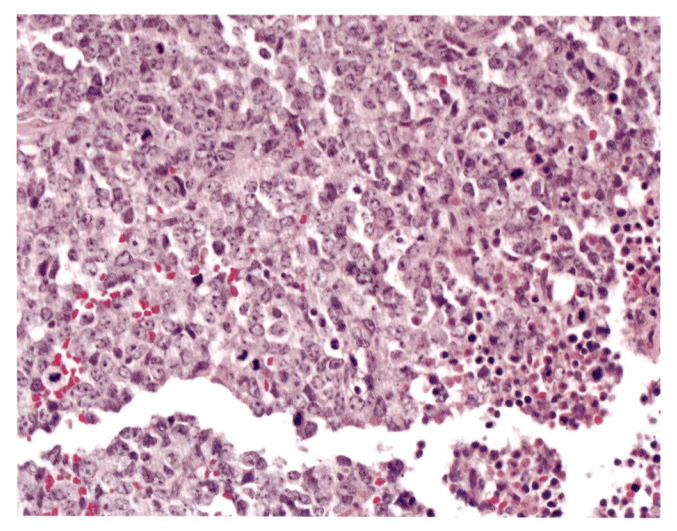

Figure 31.23 *CIC-DUX* sarcoma composed of undifferentiated medium-sized cells with abundant cytoplasm and prominent nucleoli. The features resemble atypical Ewing sarcoma.

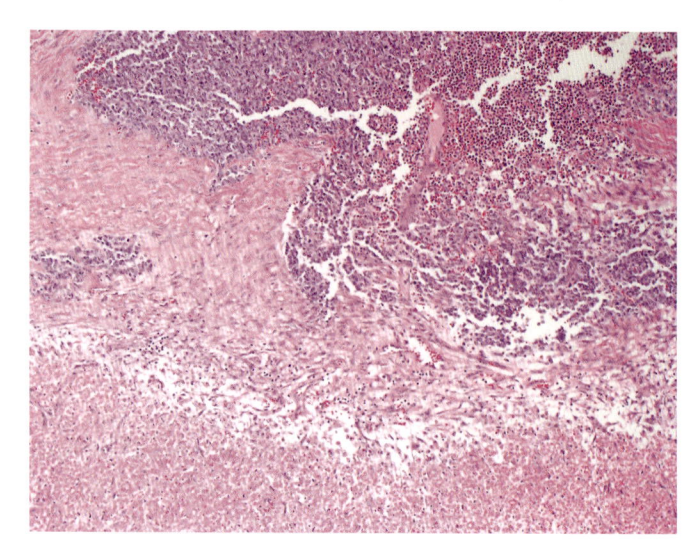

Figure 31.24 *CIC-DUX* sarcoma with extensive tumor necrosis (bottom third of the image).

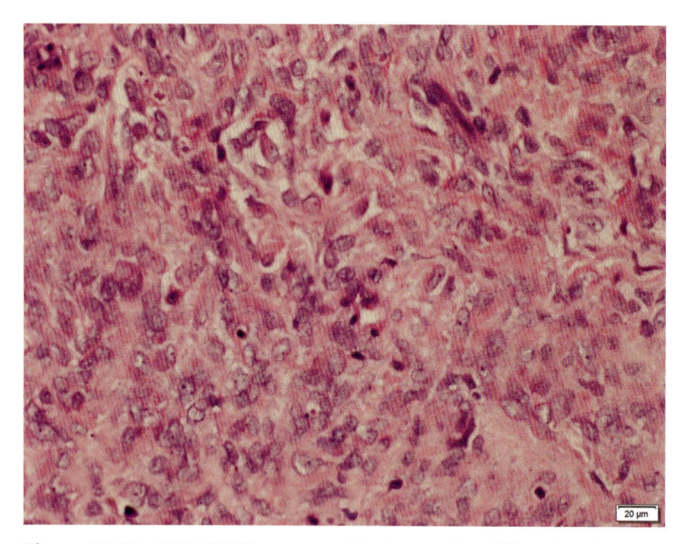

Figure 31.25 *BCOR-CCNB3* sarcoma with clusters of undifferentiated round cells.

DUX4 transcribes a DNA-binding site that upregulates several ETS family genes, similar to those showing *EWS* Ewing-type fusions. Alternate partners for *CIC* have been described in recent papers, such as *CIC-FOXO4(AFX)*. *PAX3-FOXO4* fusion has been reported in alveolar rhabdomyosarcoma.[120] *CIC-DUX4* sarcoma appears to be particularly common in some studies of EWS-negative Ewing-like tumors. In the study by Italiano,[119] two-thirds of *EWS* fusion-negative Ewing-like sarcomas fit into this category. In the series of Graham et al.[121] 5 of 15 undifferentiated SRBCT had a *CIC-DUX4* fusion.

BCOR-CCNB3 sarcomas

BCOR-CCNB3 sarcomas usually arise in bone and resemble classic Ewing sarcomas, both clinically and morphologically.[122] These are rare neoplasms and represent less than 5% of undifferentiated sarcomas. They fuse BCOR, an epigenetic depressor, with cyclin B3. This enhances the ability of cyclin B3 to drive cell cycle events. Also, the resultant loss of function of BCOR causes epigenetic instability.

BCOR-CCNB3 sarcomas are relatively infrequent. Only 10 cases were identified in an archival series of 134 small round cell tumors.[123] Of note, this series included tumors lacking prior molecular testing, so the incidence would likely be higher if one considered only *EWS* fusion-negative small cell sarcomas.

As with other Ewing-like tumors, *BCOR-CCNB3* sarcomas predominantly show the high cellularity typical of SRBCT. *BCOR-CCNB3* sarcomas differ, however, in one aspect – round cells areas frequently merge with areas containing plump spindle cells (Figure 31.25) These areas may also acquire abundant myxoid matrix (Figure 31.26).[123] This "low-grade" appearance may be a diagnostic problem. On limited biopsies, tumors may show more myxoid matrix after therapy. Undifferentiated pleomorphic sarcoma-like features may also appear, and some tumors may acquire osteoid creating more potential confusion.[123]

Immunohistochemistry is helpful in diagnosis, as *BCOR-CCNB3* sarcomas usually show strong nuclear expression of CCNB3 cyclin B3.[124] Other positive markers include BCL2, CD117, BCL6 (weak), and CD99. Desmin and cytokeratin are negative, and beta-catenin shows only cytoplasmic positivity.[123]

Survival of the limited number of reported *BCOR-CCNB3* cases has been good and overlaps with that of Ewing sarcoma. Even metastatic tumors have responded to therapy.[123] BCOR mutation may also occur as a constitutional mutation that leads to a skeletal dysplasia, the oculofaciocardiodental syndrome. This leads to increased osteogenic potential of mesenchymal stem cells and bony overgrowth. Somatic mutations of *BCOR* also occur in AML,

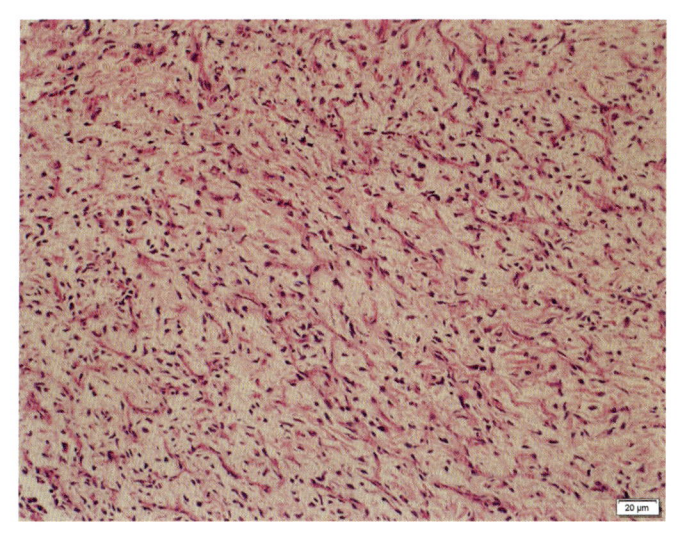

Figure 31.26 Other areas of the *BCOR-CCNB3* sarcoma containing a prominent myxoid matrix.

myelodysplasia, medulloblastoma, endometrial stromal sarcoma, and ossifying fibromyxoid tumor.

Neuroblastoma

Definition

A neuroblastoma is a neuroblastic tumor arising from primitive sympathetic ganglia and containing variably differentiated neural elements and Schwann cells. Neuroblastomas are relatively common lesions in children and constitute the single most common intra-abdominal malignancy and nonhematopoietic, non-CNS pediatric cancer. A significant number of tumors also arise in the posterior mediastinum and the cervical region. After early childhood, neuroblastoma becomes increasingly rare.

History

Dehner recounts the history of neuroblastoma in a historical article of pediatric neoplasia.[125] In it, he credits Virchow with the first discovery of neuroblastoma, in an 1863 report of a small cell abdominal tumor arising in an infant. A useful historical note is provided by Wright, who in 1910[126] assembled a series of case reports describing characteristic features and calling attention to the unique histology of these lesions. Wright's earliest cited case is that of Marchand, who in 1891 reported an adrenal tumor in a 9-month-old child. Wright's paper is of particular interest because of his careful description of ball-like aggregations that became known as *Homer Wright rosettes*. All of his cited cases contained neuroblastic features, although he included one medulloblastoma. All but two of the remaining cases were adrenal primaries.

No historical note on neuroblastoma pathology would be complete without a mention of Willis,[127] who included neuroblastoma among his series of embryonic tumors that recapitulate embryonic, fetal, or early postnatal development. Willis' descriptions carefully notated the phenomenon of differentiation, and he recognized that neuroblastomas and ganglioneuromas represent stages of development of the same basic neoplasm. He also restricted these lesions to the sympathetic nervous system, and pathologists continue to embrace this concept.

In response to a series of articles on adult neuroblastomas, Dehner[128] published a conceptual masterpiece that delineates neuroblastoma from primitive peripheral neuroectodermal tumor and loosely unites them in the overall group of primitive neural tumors. Dehner's ideas have now been confirmed by modern genetic and clinical data.

Several classification systems have linked neuroblastoma histology to clinical outcome. In 1968, Beckwith and Martin[129] published an early system. Other systems include those delineated by Joshi[130] and Shimada.[131] Shimada's classification has stood the test of time and, more recently, it has been modified to constitute the International Classification System.[132]

Beckwith's observations with Perrin are also of note, because he called attention to a precursor lesion, the "neuroblastoma-*in-situ*."[133] These tumorous lesions arise *in utero* and are now of particular interest because of the emphasis by some on prenatal screening.[134] However, there are no data supporting clonality of those so-called "neuroblastoma-*in-situ*" lesions, and the majority of them could be "embryonic rests" rather than tumor precursors.

Finally, the modern era of neuroblastoma discovery has been dominated by molecular genetics, beginning with the observations on tumor ploidy[135] and continuing with seminal observations on *MYCN*[136] amplification. Genetic studies have dominated modern clinical trials and have been refined by the addition of new genetic factors such as telomerase and *TRK* expression.[137] Nevertheless, histology has persisted as an independent factor in clinical outcome.[138,139]

Terminology and synonyms

Peripheral neuroblastic tumors constitute a spectrum of neoplasms which includes undifferentiated neuroblastomas at the most primitive end of the spectrum and mature ganglioneuromas at the most differentiated end, with poorly differentiated neuroblastoma, differentiating neuroblastoma, ganglioneuroblastoma, intermixed, and ganglioneuroma, maturing subtype between the two (Figure 31.27).[140]

In the general sense, neuroblastoma is defined as a neuroblastic tumor in which less than 50% of tumors shows schwannian development and examples are referred to as schwannian stroma-poor tumors. This subgroup includes the undifferentiated, poorly differentiated and differentiating subtypes. The term undifferentiated neuroblastoma is applied to those neuroblastomas in which less than 5% of the cells demonstrate ganglionic differentiation/maturation, which are composed of totally primitive cells and no clearly identifiable neuropil is present within the tumor (Figure 31.28). Immunohistochemistry is often required to clinch the diagnosis in such cases.

Chapter 31: Small round cell tumors

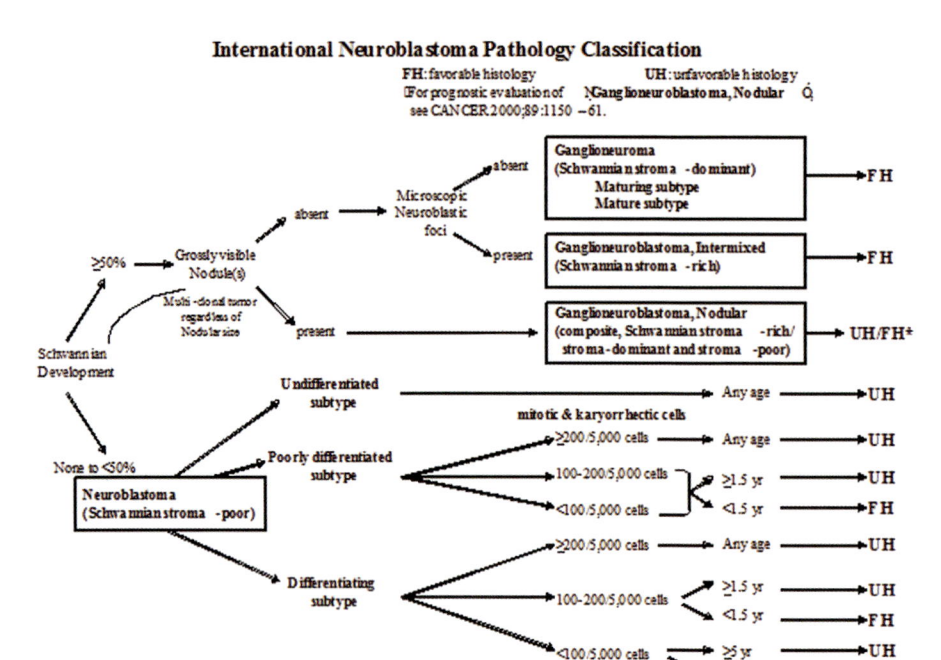

Figure 31.27 International Neuroblastoma Pathology Classification.

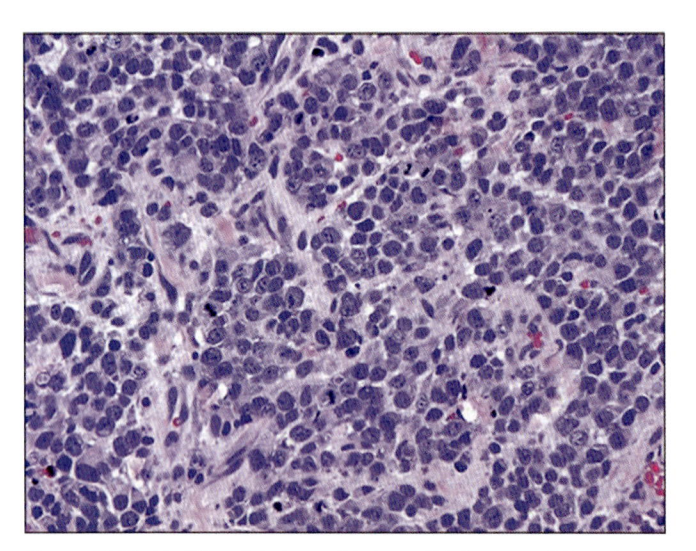

Figure 31.28 Undifferentiated neuroblastoma. This tumor contains only primitive cells with no ganglionic differentiation and no identifiable neuropil. Immunohistochemistry is often required to make the diagnosis in such cases. By definition, all undifferentiated neuroblastomas fall into the unfavorable histology group of the International Neuroblastoma Pathology Classification.

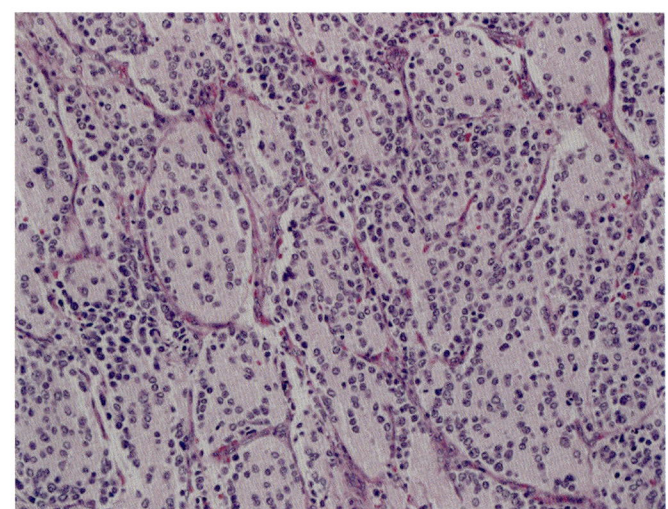

Figure 31.29 Neuroblastoma, poorly differentiated. Note the abundance of neuropil.

Those tumors in which less than 5% of the cells demonstrate neuroblastic differentiation towards ganglion cells (differentiating neuroblasts with both nuclear and cytoplasmic enlargement), but in which neuropil is identified microscopically are designated poorly differentiated (Figures 31.29 and 31.43a and b). In the differentiating subtype, more than 5% of tumor cells demonstrate neuroblastic differentiation/maturation (Figures 31.30 and 31.43c).

Those tumors that show more than 50% schwannian differentiation include the ganglioneuromas and the ganglioneuroblastomas. Ganglioneuromas are composed almost entirely of schwannian stroma with scattered ganglion cells and are termed schwannian stroma-dominant tumors (Figure 31.31). Ganglioneuroma is divided into mature and maturing subtypes. In the mature subtype the ganglion cells are completely mature and are always accompanied by satellite cells. In the maturing subtype there is a minor component of scattered maturing ganglion cells without satellite cells in addition to fully mature ganglion cells.[141] Naked neurites are not detected in the category of ganglioneuroma, since they are incorporated by schwannian stromal cells right after being produced by ganglion cells.

Ganglioneuroblastoma intermixed (schwannian stroma-rich tumors) is the term used to designate those tumors which

Chapter 31: Small round cell tumors

are undergoing tumor maturation towards ganglioneuroma, however the process has not been completed. These tumors demonstrate schwannian development with no macroscopically discernible foci of neuroblastoma; however, there are scattered residual microscopic neuroblastomatous foci or pockets of neuropil with neuroblastic cells in various stages of differentiation including neuroblasts, differentiating neuroblasts (Figure 31.43d). As previously stated, the portion of the ganglioneuromatous (with schwannian stromal development) to residual neuroblastomatous foci should exceed 50% of the total volume in microscopic fields from representative sections of tumor.[141]

In the case of ganglioneuroblastoma nodular, a grossly visible neuroblastomatous nodule is present and is derived from a biologically different clone as the surrounding schwannian stroma-rich (ganglioneuroblastoma, intermixed) or stroma-dominant (ganglioneuroma) tissue (Figures 31.32a and b and 31.33).[141] The nodule is usually hemorrhagic and necrotic (Figure 31.32a); however, in occasional cases the lesion is homogeneously tan-white and fleshy with no significant hemorrhage (Figure 31.32b). The proportion of neuroblastomatous nodule to schwannian stroma-rich/stroma-dominant component varies from case to case (Figure 31.32a and b).[141]

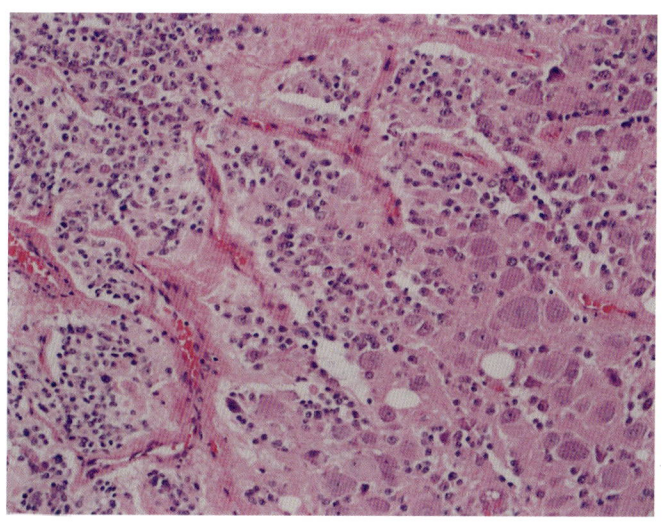

Figure 31.30 Neuroblastoma, differentiating subtype. Note the presence of cells demonstrating ganglionic differentiation/maturation in addition to immature neuroblastic elements and neuropil. By definition, more than 5% of the neoplastic cells should show ganglionic differentiation/maturation in this subtype of neuroblastoma.

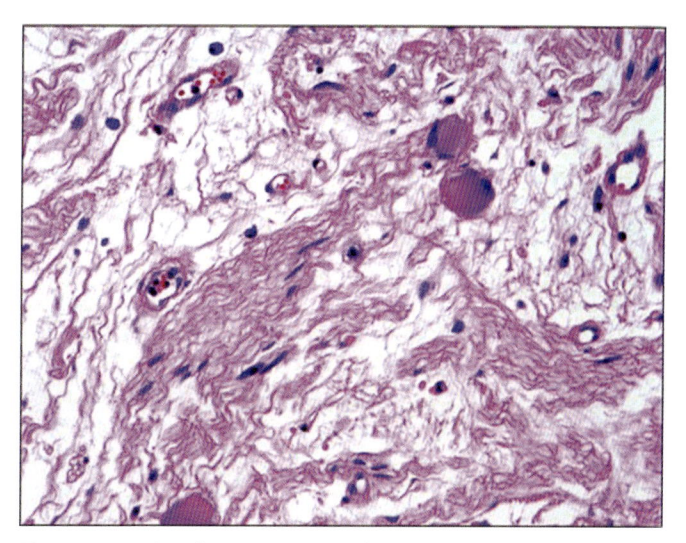

Figure 31.31 Ganglioneuroma, stroma-dominant tumor. This tumor is composed entirely of mature neural elements, including abundant schwannian stroma and scattered mature ganglion cells. Note that the uppermost ganglion cell is accompanied by a satellite cell, a feature of maturity.

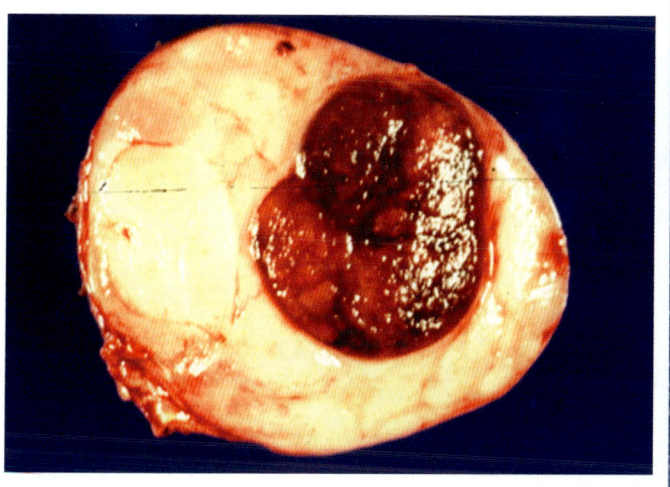

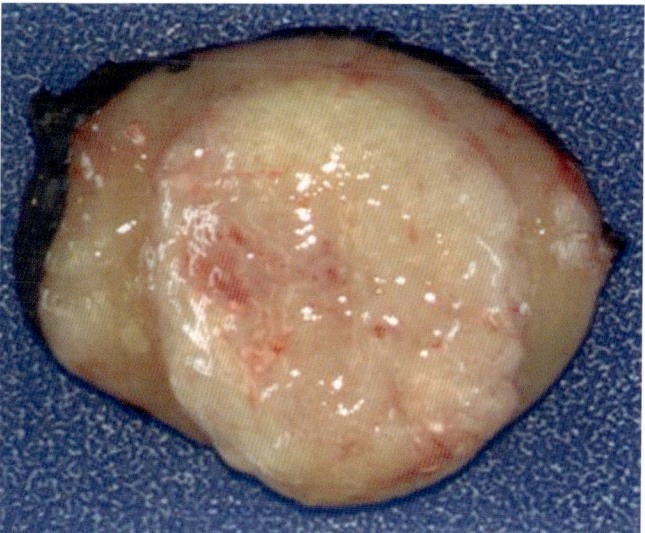

Figure 31.32 Gross photomicrograph of ganglioneuroblastoma, nodular. (**a**) The nodular neuroblastomatous component is hemorrhagic as is often the case. (**b**) The neuroblastomatous component in this case is tan-white and fleshy with significantly less hemorrhage. In both cases the nodule is surrounded by schwannian stroma.

A rare subset of neuroblastoma tentatively designated as "composite neuroblastoma" is composed of two (or more) distinct neuroblastoma clones. One clone has features characteristic of one subtype of stroma-poor neuroblastoma and the other clone has features characteristic of another (Figure 31.34). These tumors are different from ganglioneuroblastoma, nodular (composite, schwannian stroma-dominant/stroma-rich and stroma-poor) and do not have a place in the current International Neuroblastoma Pathology Classification (INPC).[142]

Clinical features of neuroblastoma

Neuroblastoma and the Ewing sarcoma family of tumors share several pathologic features, such as rosettes and primitive histology, and are sometimes referred together as the *primitive neuroectodermal tumor family*.[143] However, these tumors are different clinically, biologically, and molecularly. One of the most distinctive differences between the two is the age of presentation, which is generally below the age of 5 years in the case of neuroblastoma (Figure 31.35), compared with the young adult and adolescent predilection of the Ewing sarcoma family of tumors. Neuroblastoma is the most common solid tumor of children <1 year of age.[144] It typically presents as an abdominal mass, although it can also arise in the sympathetic ganglia of the neck, thorax, and pelvis (Figure 31.36). The adrenal gland is the most common site of origin, and neuroblastoma rightly can be considered a tumor of the sympathetic nervous system, with production of catecholamines, their precursors and their by-products.

Catecholamine secretion is a key diagnostic feature for neuroblastoma. Urine specimens (24-hour) should be tested for homovanillic acid (HVA) and vanillylmandelic acid (VMA) prior to or soon after excision of these tumors, because serum levels drop quickly. The ratio of the precursor molecule, HVA, to the product, VMA, correlates with the degree of tumor differentiation and survival.[145] Other hormonal substances produced by neuroblastoma can cause paraneoplastic syndromes. One such substance, vasoactive intestinal polypeptide, causes watery diarrhea and hypokalemia.[146]

Other paraneoplastic syndromes can result from immune phenomena. In *opsoclonus myoclonus syndrome*, antibodies produced against the tumor are speculated to cause cerebellar ataxia and rapid eye movements. Horner syndrome, Ondine's curse, and a host of other unusual clinical symptoms can also occur.[146] Maternal antibodies in pregnancy serum are capable of tumor cytolysis and possibly represent a mechanism for tumor regression.[147] The histologic presence of lymphocytes

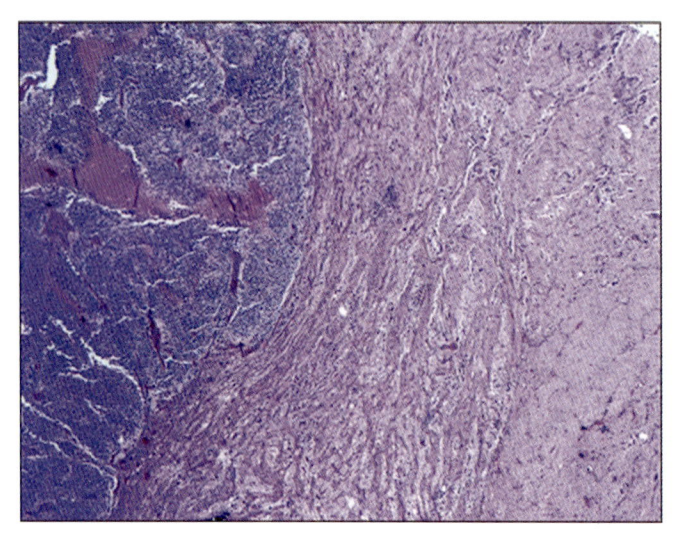

Figure 31.33 Ganglioneuroblastoma, nodular. This otherwise fully differentiated, stroma-rich/stroma-dominant lesion contains a prominent nodular focus of primitive neuroblastoma to the left of the photomicrograph. This was originally a de facto sign of unfavorable histology, but now the nodular focus should be graded independently.

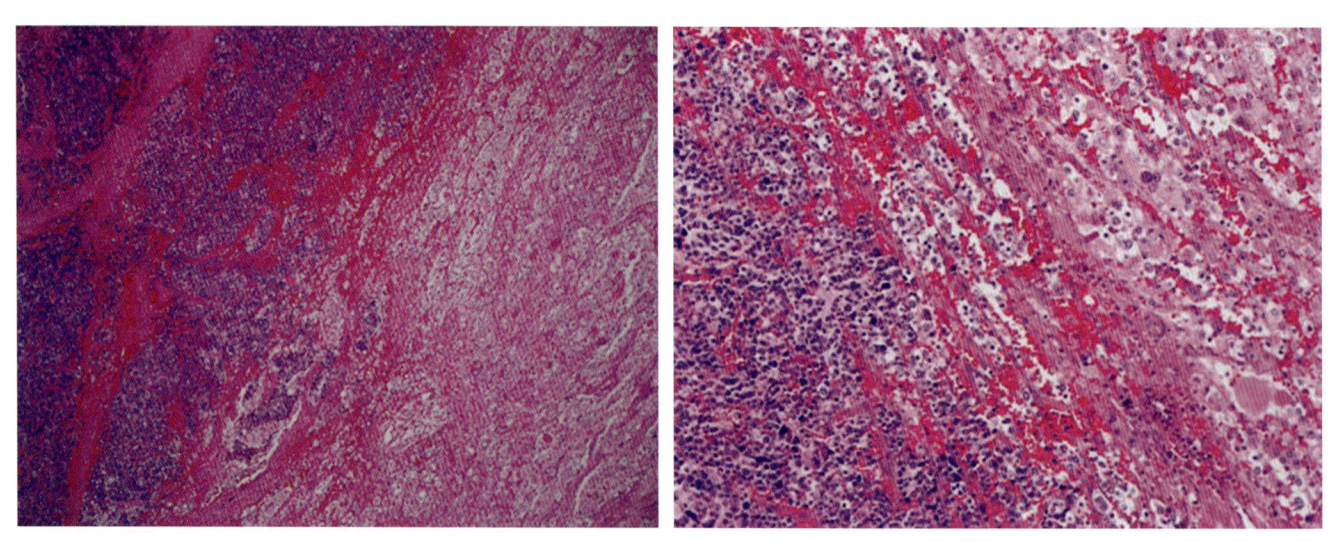

Figure 31.34 Composite neuroblastoma. (**a**) The poorly differentiated neuroblastoma component composed of small round blue cells is seen on the left of the photomicrograph. The right half is composed of differentiating neuroblastoma. (**b**) The same tumor at higher magnification with the poorly differentiated neuroblastoma component on the left and the differentiating neuroblastoma component composed of large maturing neuroblastoma cells with voluminous eosinophilic cytoplasm is seen on the right. (Courtesy of Risa Teshiba, Children's Hospital Los Angeles and Keck Medical School of USC, Los Angeles, California.)

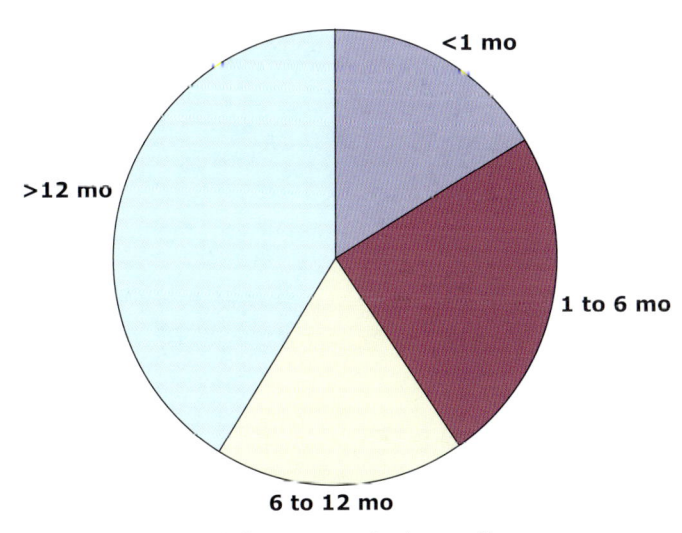

Figure 31.35 Pie graph illustrating age distribution of low-stage neuroblastoma.

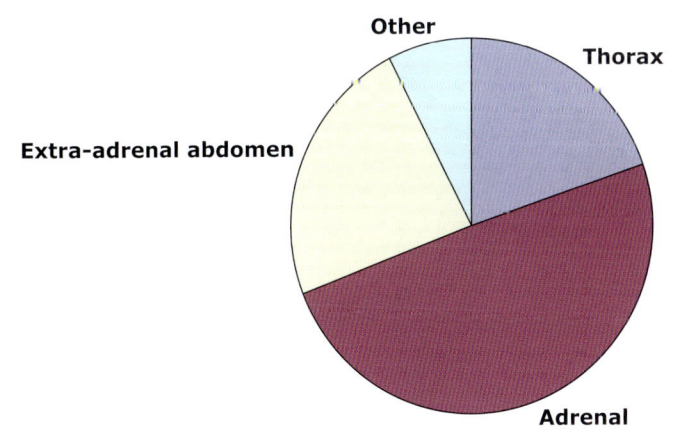

Figure 31.36 Pie graph illustrating neuroblastoma locations.

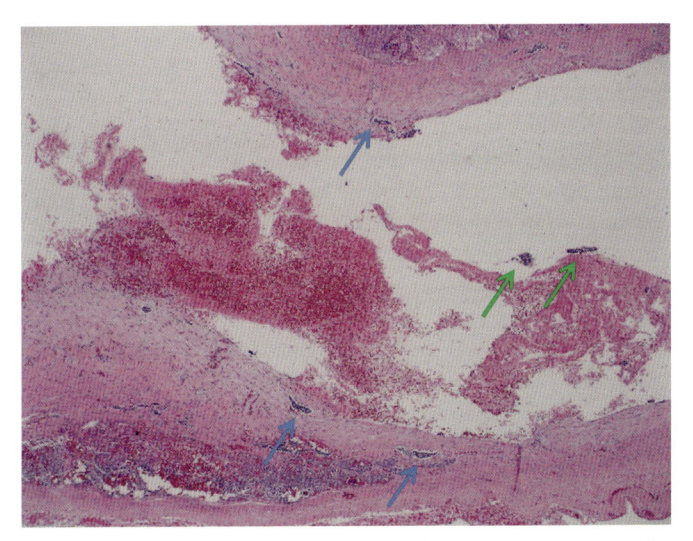

Figure 31.37 Congenital neuroblastoma with large cystic cavity and small nests of residual neuroblastoma *(blue arrows)* embedded in the dense fibroconnective tissue. Small clusters of neuroblastoma cells are seen floating with blood in the cyst cavity *(green arrows)*.

in excised, pretreated tumors is another manifestation of the immunogenicity of these tumors (Figure 31.44).

The clinical behavior of aggressive neuroblastoma differs distinctly from PNET. Neuroblastoma only rarely metastasizes to the lung, and those that do have unusual features.[148] Conversely, aggressive PNETs typically exhibit pulmonary metastases. Both tumors can show bone marrow and bone metastases, but this phenomenon is more typical of neuroblastoma. Neuroblastomas also commonly spread to regional and even distant lymph nodes, an unusual metastatic site in Ewing family sarcomas, except in widespread disease.[149]

Another key characteristic of neuroblastoma is its stippled calcification, easily visible on abdominal radiographs. This feature can be recognizable on prenatal ultrasonography and lead to a diagnosis of a congenital tumor. Most of these congenital tumors are biologically favorable. In some countries, predominantly Japan, in the past widespread population screening using urinary detection of VMA/HVA was performed with the hope of detecting early stage disease before clinical progression. However, this system detected predominantly the aforementioned biologically favorable tumors, which have a tendency to regress spontaneously, hence the effect that this procedure has had on reducing mortality is debatable.

(The combination of metastatic disease in the bone marrow, a calcified suprarenal mass, or catecholamine secretion are confirmatory signs for neuroblastoma.)[150]

The clinical outcome for localized neuroblastoma is generally good, even with intraspinal tumors.[151] As mentioned previously, congenital neuroblastomas often spontaneously involute, leaving cystic lesions with neuroblastic elements in a fibrous shell (Figure 31.37). These have an excellent prognosis. Conversely, metastatic neuroblastoma persists, with a 5-year survival rate of approximately 30% to 35%.[152]

Gross and microscopic features of neuroblastic tumors

Grossly, neuroblastomas typically display a lobulated, encapsulated, ovoid surface that on sectioning is often grossly hemorrhagic and punctuated by flecks of calcification (Figure 31.38). Paravertebral neuroblastomas can invade the intervertebral foramina and encroach on the spinal epidural space in a dumbbell fashion. In adrenal lesions, a small yellow focus of intact glandular tissue can be flattened against the fibrous capsule. After combination chemotherapy, there is often extensive post-treatment effect, with fibrosis, calcification, and necrosis, creating an appearance dissimilar to untreated lesions.

Microscopic diagnosis of neuroblastoma depends on the presence of primitive neural cells that are identical to the migratory neural crest elements that normally invade the fetal adrenal cortex. In fact, adrenal neuroblastic foci can be numerous and large at birth (termed *neuroblastoma in situ*), but they commonly spontaneously involute as the glands mature

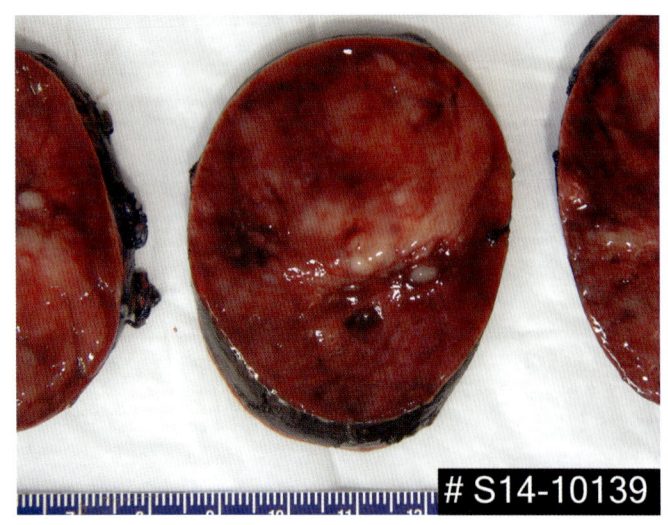

Figure 31.38 Adrenal neuroblastoma from a 5-week-old infant, gross photograph. The adrenal gland is replaced by this 7 cm ovoid, encapsulated mass. The cut surface is fleshy with prominent hemorrhage and focal cavitation. (Courtesy of Harry Kozakewich, Boston Children's Hospital and Harvard Medical School, Boston Massachusetts.)

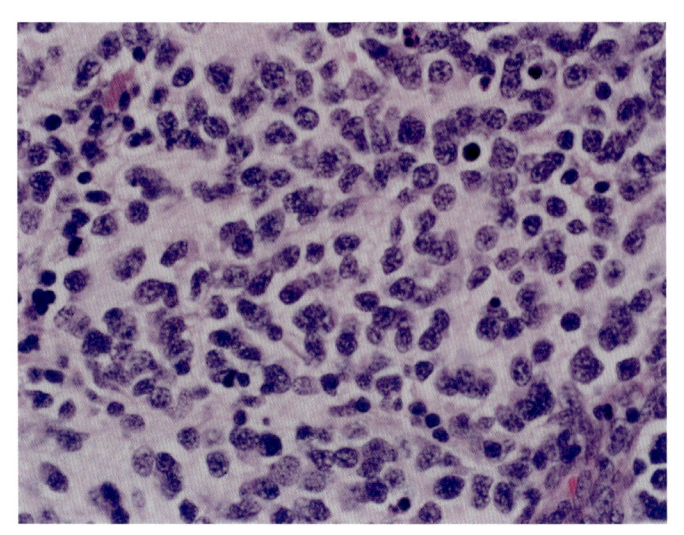

Figure 31.39 Neuroblastoma with neuroblastic cells with the classic salt and pepper nuclei.

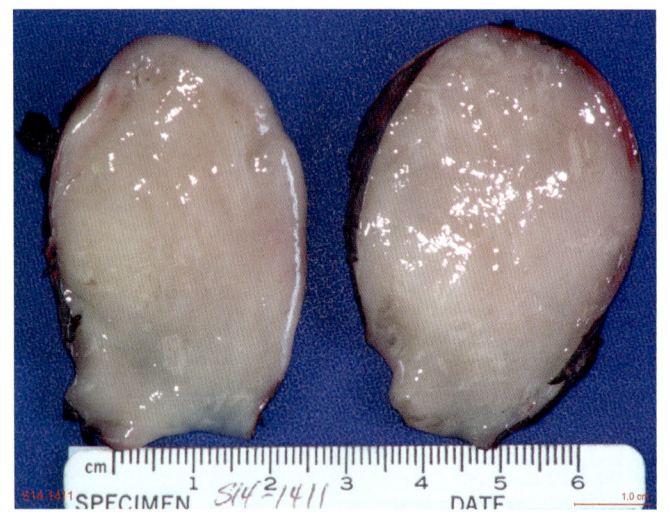

Figure 31.40 Ganglioneuroma, gross appearance. The cut surface of the mass has a homogeneous, pale, fibrous appearance.

(Figure 31.37).[134] Neuroblastic elements in true neuroblastomas are commonly accompanied by aggregates of fibrillary material known as neuropil. Neuroblastic cells may be arranged circumferentially with aggregates of neuropil in the center forming distinct structures known as Homer Wright rosettes (Figure 31.43a). Homer Wright rosettes are composed of lightly eosinophilic, neurofibrillary cores of neuropil surrounded by wreaths of round, hyperchromatic nuclei with coarsely granular chromatin and inconspicuous nucleoli. The granular chromatin is responsible for the classic salt and pepper appearance of the nuclei of neuroblastic cells (Figure 31.39). Neurofibrillary material can be abundant and forms a scaffolding within which oval nuclei are suspended (Figure 31.29). One characteristic feature is the tendency of neuroblasts to undergo varying degrees of cell maturation, characterized by increased cytoplasm with Nissl substance and prominent nucleoli. This process continues to the point of creation of maturing neuroblastic cells which progressively acquire features that more closely resemble non-neoplastic neurons, at which point the combination of more primitive neuroblasts and these differentiating neuroblasts is termed a neuroblastoma, differentiating subtype (Figure 31.30 and Figure 31.43c).

At the fully mature end of the histologic spectrum lies the ganglioneuroma, a benign tumor comprised of mature, neoplastic ganglion cells enmeshed in a prominent spindle cell stroma derived from Schwann cell elements (Figure 31.31). These lesions have a whorled, fibrous cut surface rather than a hemorrhagic one (Figure 31.40). Ganglioneuromas are often diagnosed in the posterior mediastinum rather than the abdomen. They can represent evidence of spontaneous maturation of previous neuroblastomas, a process occasionally noted in exceptionally fortunate patients, or, alternatively, they are seen as metastatic, matured lesions in previously treated children.

Possible ganglioneuromas should be examined carefully during gross inspection for hemorrhagic nodules, which denote nodular foci of neuroblastoma. As stated previously, these composite tumors are referred to as *ganglioneuroblastoma, nodular*, and they are lesions that are capable of metastatic behavior (Figures 31.32 and 31.33). Ganglioneuromas should also be liberally sectioned and examined for microscopic neuroblastic foci and pockets of neuropil with neuroblastic cells in various stages of maturation. These lesions have been termed ganglioneuroblastoma, intermixed, and confer a good prognosis (Figure 31.43d). Lesions containing a few immature cells have been termed "borderline ganglioneuroblastomas" by Joshi et al.,[153] but are now considered "maturing ganglioneuromas."[154]

Table 31.5 The International Neuroblastoma Pathology Classification (the Shimada System)[a]

Good prognosis

Ganglioneuroblastoma, intermixed (stroma-rich) type, any age
Differentiating neuroblastomas with low MKI,[b] 1.5–5 years of age
Undifferentiated/poorly differentiated neuroblastomas with low or intermediate MKI,[b] <1.5 years of age
Ganglioneuroma, any age

Poor prognosis

Ganglioneuroblastomas, nodular type, any age
Neuroblastomas with high MKI,[b] any age
Undifferentiated/poorly differentiated neuroblastomas, age 1.5–5 years
Neuroblastomas with intermediate MKI,[b] age 1.5–5 years
Neuroblastoma, any type, age >5 years

[a] From reference 154
[b] MKI (mitosis/karyorrhexis index): low = <100/5000 cells; intermediate = 100–200/5000 cells; high = >200/5000 cells.

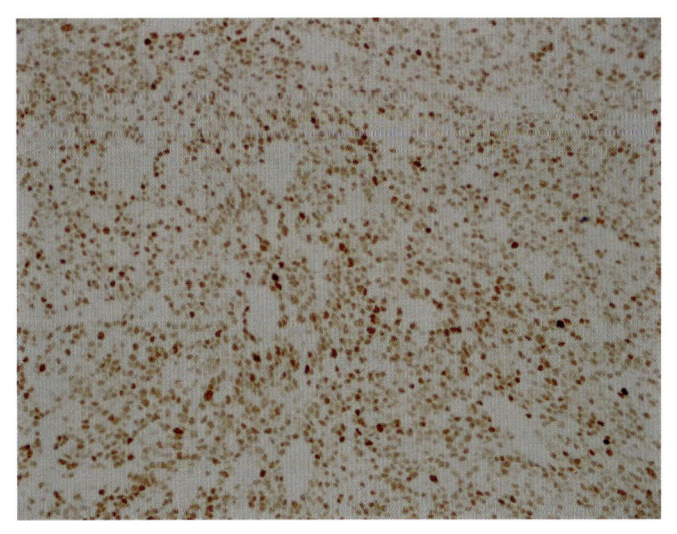

Figure 31.42 The same tumor seen in Figure 31.39 showing n-myc positivity by immunohistochemistry. Note the heterogeneity in staining intensity characteristic of n-myc staining in neuroblastoma and the resulting checkered pattern.

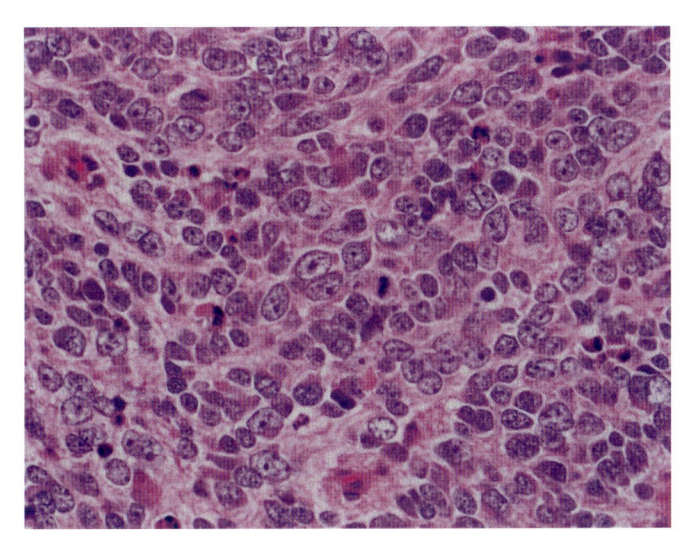

Figure 31.41 Poorly differentiated neuroblastoma with prominent nucleoli and vesicular chromatin. n-myc was amplified in this tumor.

At the other end of the differentiation spectrum lies undifferentiated neuroblastomas (Figure 31.28), which display no evidence of neuroblastic differentiation by light microscopy and require ancillary techniques for diagnosis. These lesions show aggressive features and are thus included in the unfavorable category, independent of other factors.[154] One noteworthy feature of undifferentiated and poorly differentiated neuroblastoma that appears to be associated with poor outcome and correlates with n-myc expression is the presence of prominent nucleoli (Figures 31.41 and 31.42).[155]

Shimada et al. have created a grading scheme for neuroblastomas known as the Shimada classification.[131] This classification is independently predictive of clinical behavior and is now used as an internationally accepted standard.[132] It is presented in Table 31.5 and Figure 31.27, and pertinent histologic features are illustrated in Figure 31.43. Several terms used in this schema require explanation. First, the term *stroma* refers to Schwann cell elements and not neurofibrillary material, so that *stroma-rich* tumors are actually *intermixed ganglioneuroblastomas*, as per the classification created by Shimada et al.[154] Second, the *mitosis/karyorrhexis index* (MKI) is a numeric, semiquantitative value derived from counting the number of mitotic figures and fragmented, karyorrhectic nuclei among 5000 cells. This onerous chore can be greatly simplified by familiarity with estimating the total number of cells in a 40× objective, high-power field (HPF). Used in this manner, the Shimada classification is a reproducible and easily learned technique that is clinically useful and prognostically accurate.

Nodular ganglioneuroblastoma presents another caveat. It was originally defined as a high-risk tumor by the Shimada categorization, but it now appears that behavior depends on the grade and MKI of the neuroblastomatous nodule.[156]

Unusual histologic variants of neuroblastic tumors include "anaplastic neuroblastomas," with prominent large neoplastic cells and bizarre, hyperchromatic nuclei.[157] This finding was thought to have no clinical significance.[153] Neuroblastoma diffusely composed of anaplastic cells is reported to have a poor prognosis; however, clinical significance of focal anaplasia is debatable.[158] Rhabdoid cells with abundant eosinophilic cytoplasm and hyaline inclusions can also be seen.[153] Neural crest differentiation other than neural and schwannian differentiation can occur, particularly melanocytic[159,160] or paragangliomatous.[161] Tumors with the latter features are more commonly classified as composite pheochromocytomas.

Finally, neuroblastic tumors in patients with opsoclonus-myoclonus are often accompanied by a diffuse and extensive lymphocytic infiltration with lymphoid follicles (Figure 31.44).[162]

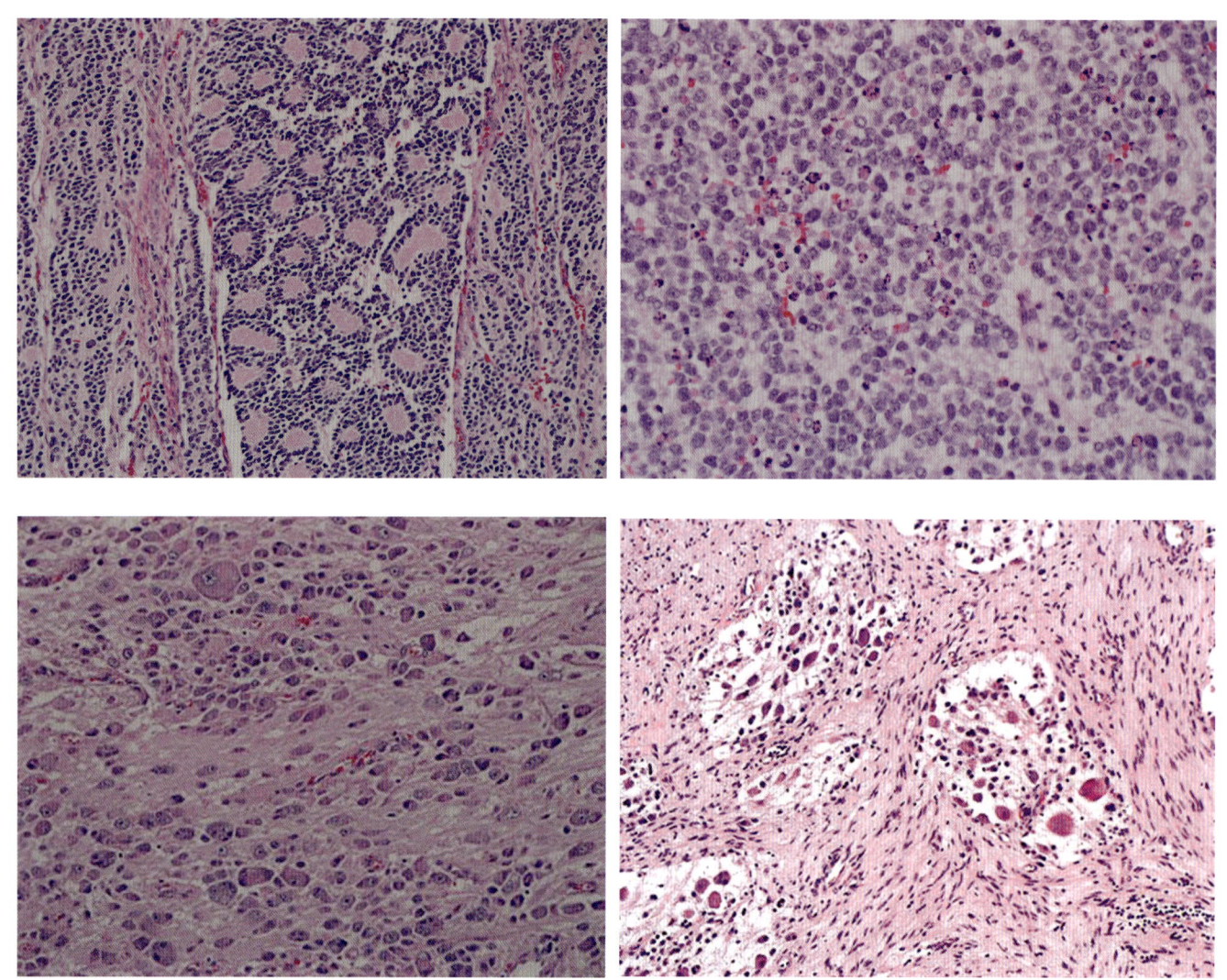

Figure 31.43 Representative images of neuroblastoma classification: (**a**) Poorly differentiated, favorable histology neuroblastoma. The tumor contains Homer Wright rosettes with cores of neurofibrillary material, a sign of incomplete neuroblastic differentiation. There are very few mitoses or karyorrhectic bodies in this lesion. (**b**) Poorly differentiated neuroblastoma, unfavorable histology. There are abundant mitoses and karyorrhectic bodies. (**c**) Differentiating neuroblastoma. This lesion contains neuropil with clusters of predominantly well-differentiated neuroblasts admixed with occasional poorly differentiated ones. (**d**) Intermixed ganglioneuroblastoma. This lesion contains abundant Schwann cell stroma with clusters of neuroblast, differentiated ganglion cells, and naked neurites.

Ancillary diagnostic techniques

Neuroblastomas usually present less of a diagnostic challenge than the Ewing sarcoma family of tumors, because they tend to be more differentiated and tend to have a better defined clinical picture. Undifferentiated forms can occur, older patients can be affected,[163] and paravertebral, nonadrenal primaries can broaden the differential diagnosis, however. In addition, extremely rare examples of PNETs contain differentiated areas similar to ganglioneuroblastoma.[58] In nonadrenal locations and older patients, it is important to rule out PNET, because neuroblastomas are treated differently and display a vastly dissimilar clinical behavior. This can be problematic with undifferentiated tumors, if ancillary techniques are not used.

Ultrastructurally, the neural features of neuroblastomas are prominent and comprise clusters of dense core granules, elongated, intertwining dendritic processes, neurosecretory vesicles, and parallel arrays of microtubules (Figure 31.45). The neurofibrillary material that constitutes the cores of rosettes is composed of intricately entwined neuritic processes (Figure 31.45a), such as appear in cell cultures of these tumors. Although PNETs also contain these structures, experienced electron microscopists note a distinction in their prominence in neuroblastomas that assists in diagnosis (cf. Figure 31.15).[164]

Another key diagnostic feature separating neuroblastoma from other neural neoplasms is the presence of catecholamine secretion, as measured in pre-excisional urine specimens. A biochemical distinction is *MYCN* (or *n-myc*) expression and amplification (discussed later in this chapter), typical of neuroblastoma. The occasional *MYCN* amplification seen in alveolar rhabdomyosarcoma[165,166] excludes use of this phenomenon to differentiate rhabdomyoblastic tumors from neuroblastic ones.

Immunohistochemistry is often of great assistance in the diagnosis of primitive neuroblastic tumors. These lesions are typically positive for neural proteins, such as NSE, synaptophysin (Figure 31.46), neurofilaments, and chromogranin, usually separating them from non-neural embryonal lesions. Unlike PNETs, they are CD99 negative (Figure 31.47), so that this stain can be extremely useful in separating the two lesions.[69] An important observation is that although it is a sensitive marker, NSE is actually relatively nonspecific.[167] Perhaps the most useful marker is tyrosine hydroxylase, which indicates the sympathetic nature of these neoplasms (Figure 31.48). A new marker, PHOX2B[168] is also restricted to sympathetic cells and is constitutionally mutated in patients with Ondine's curse. In undifferentiated neuroblastomas, MYCN immunohistochemical stains may be of benefit (Figure 31.42), although some express MYC (aka c-myc) (Figure 31.49).[155]

Genetics

Neuroblastomas are often typified by three cytogenetic abnormalities: deletion of chromosome 1, double minute (DM) chromosomes, and homogeneous staining regions. The former alteration potentially reflects the loss of a tumor-suppressor gene, and the latter two reflect amplification of *MYCN*.

Alterations of chromosome 1 in neuroblastoma represent the single most common cytogenetic event[169] and usually result in the loss of a large segment of the short arm (1p). The addition of an intact chromosome 1 suppresses the tumorigenicity of neuroblastomas, indicating the tumor-suppressor function of this region. A common locus affected in all tumors is a distal region of 1p, including 1p36.2 and 1p36.3.[137] The exact gene responsible for the tumor-suppressor function of this region remains unproven,[170] but candidates include a p53 homolog, a CDK2 homolog, transcription factors, a transcription elongation factor, and members of the tumor necrosis family.[137] Some evidence suggests CAMTAI as a candidate.[171] Abnormal gene expression affecting multiple loci has been proposed as an alternative hypothesis.[172] Although a strong correlation exists between the presence of losses in 1p and high-risk clinical features such as older age and N-*myc* amplification, it is debatable whether 1p deletions have an independent effect on outcome.[137]

DM chromosomes (Figure 31.50) are short chromosome dyad fragments that are separate from the major chromosomes and are usually multiple. Homogeneous staining regions (hsrs; Figure 31.51) are uniformly staining elongations of chromosomes resulting from insertions of genetic material. Both of these phenomena were noted on cytogenetic studies of neuroblastoma explants that were readily immortalized into cell lines. It became apparent that those tumors that were easily transferred to an *in-vitro* existence were associated with poor clinical outcome *in vivo*, and cytogenetic studies of these cells were noteworthy for both DMs and hsrs. Cloning experiments of these regions revealed that they were composed of amplified segments of *MYCN*, a gene located on chromosome 2p24 and having homology to the *c-myc* proto-oncogene.[137] Thus, the ready immortalization of cells *in vitro* and the aggressive clinical behavior *in vivo* result from multiplied segments of a tumor-causing gene, with high levels of production of the resultant protein. Amplification of *MYCN* typically ranges from 50 to 400 copies of this usually single segment.

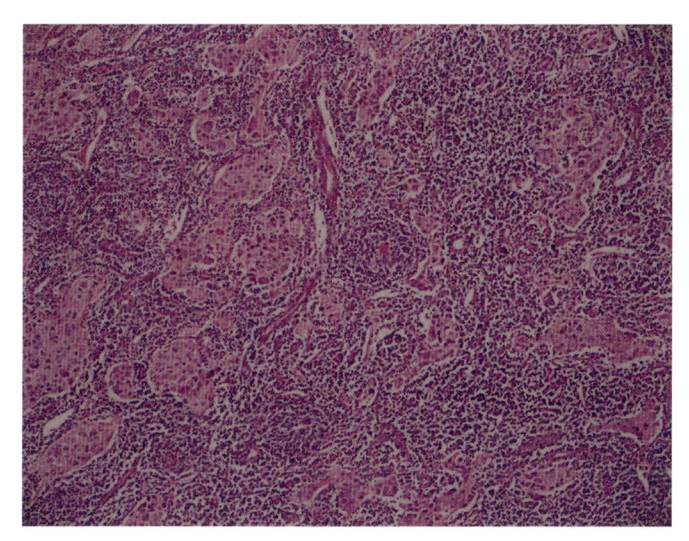

Figure 31.44 Neuroblastoma in a patient with opsoclonus-myoclonus. Note the extensive lymphocytic infiltration and the presence of lymphoid follicles.

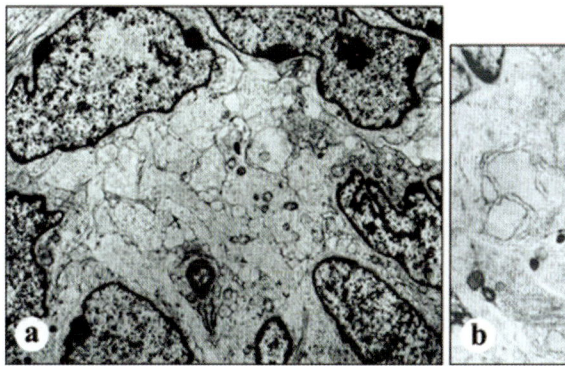

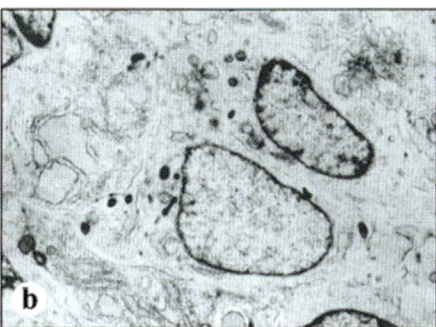

Figure 31.45 Electron micrographs of a neuroblastoma. (**a**) The core of a Homer Wright rosette is composed of tangles of intertwined cellular processes outlined by plasmalemmal membranes. (**b**) Neuroblastoma cells contain cytoplasmic aggregates of dense core granules.

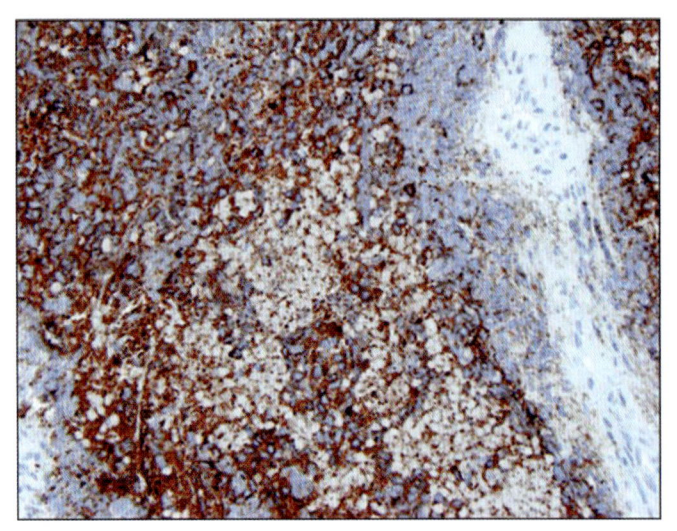

Figure 31.46 Neuroblastoma immunohistochemistry, synaptophysin staining. There is strong brown synaptophysin staining within tumor cell cytoplasm and processes. The fibrous stroma is negative (blue).

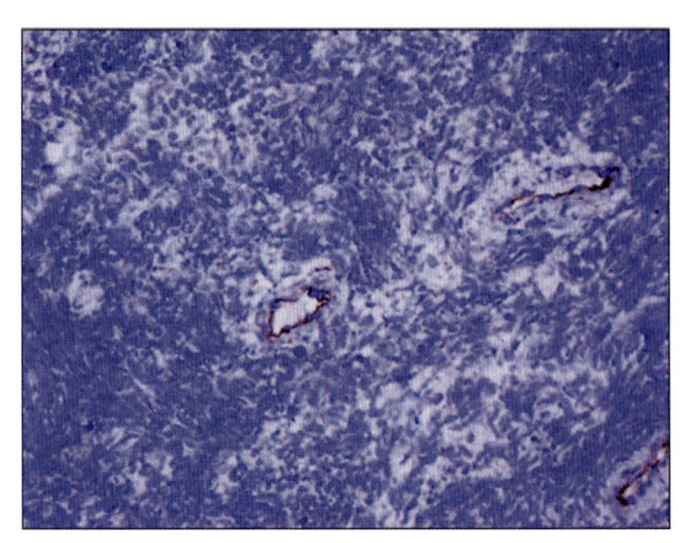

Figure 31.47 Neuroblastoma immunohistochemistry, CD99 staining. There is no evidence of staining in the tumor, with faint endothelial staining as an internal control.

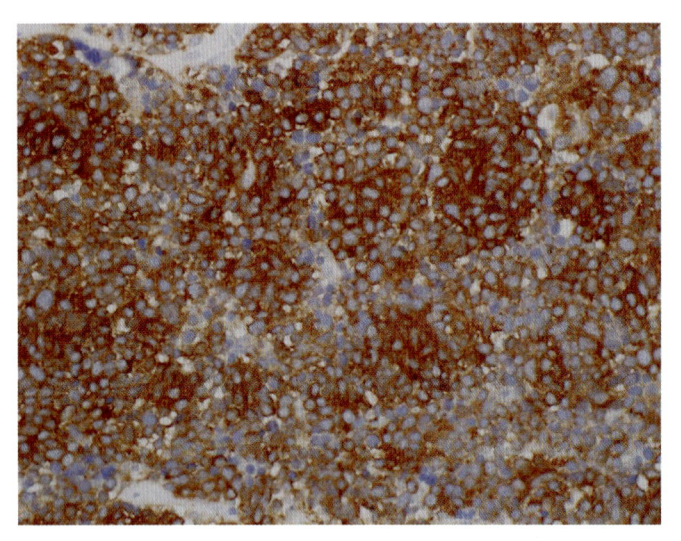

Figure 31.48 Neuroblastoma showing strong tyrosine hydroxylase positivity by immunohistochemistry.

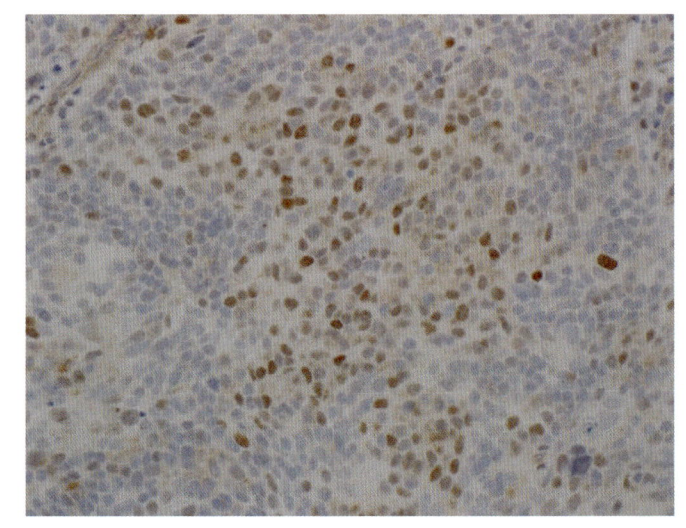

Figure 31.49 In a subset of neuroblastomas with prominent nucleoli, n-myc is not amplified. In a large number of those cases c-myc is overexpressed instead. In this case c-myc protein expression is detected by immunohistochemistry. In c-myc-positive cases the staining is heterogeneous with variable staining giving rise to a checkered pattern.

MYCN is normally expressed in the developing nervous system. The gene product is a DNA-binding protein. It contains a helix-loop-helix/leucine zipper motif that mediates the interactions with DNA and related proteins like Max and Mad. Amplification of *MYCN* causes a perturbation of the balance between its product and the proteins and DNA sequences with which it interacts, resulting in unrestrained cell proliferation. Forced overexpression of *MYCN* in transgenic mice causes neuroblastic tumors, further indicating the relationship between this gene and the tumorigenesis of neuroblastomas.[137]

Detection of *MYCN* amplification has a marked relationship to clinical outcome and is therefore used to stratify patients for treatment purposes. *MYCN* amplification is an unfavorable predictor of outcome, even in young infants.[173] In the past, testing generally required fresh tissue, so portions of tissue would be saved frozen in suspected neuroblastomas, prior to their submersion in fixatives. Initially, Southern blots were used to detect this phenomenon, but now FISH analysis can be done on formalin-fixed paraffin-embedded tissue and obviates the need for frozen tissue.[174] The chromogenic *in situ* hybridization technique has also been adapted for use with paraffin sections.[175] By FISH analysis, *MYCN* amplification is apparent in cells with DMs as multiple, randomly distributed sites of fluorescence (Figure 31.52), whereas hsrs are indicated by the presence of long, contiguous strands of fluorescence.

Another cytogenetic feature that is related to the clinical outcome of neuroblastomas is hyperdiploidy (whole chromosomal gain[s] without structural abnormalities).[135,176]

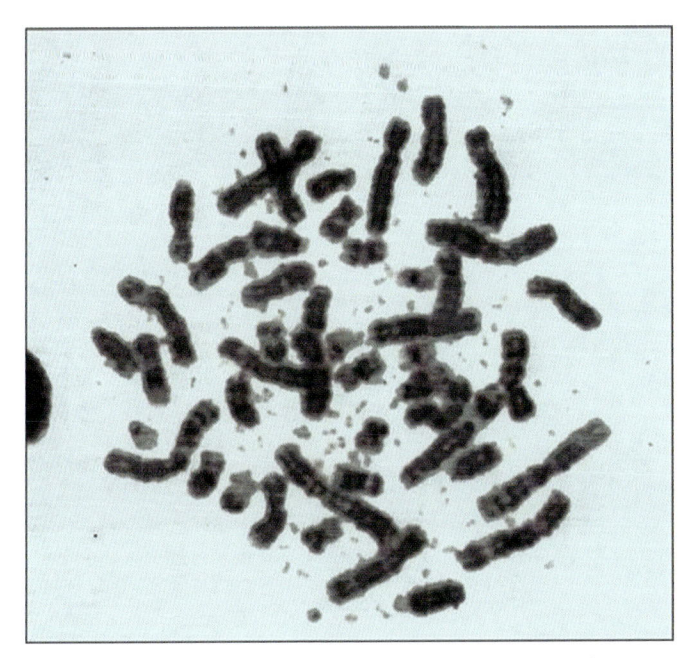

Figure 31.50 Cytogenetics of neuroblastoma. A metaphase spread contains myriad double minutes, seen as small doublets that are interspersed with normal chromosomes. (Courtesy of Mr. Charles Swanson and Dr. Jeffrey Sawyer, Arkansas Children's Hospital, Little Rock, AR.)

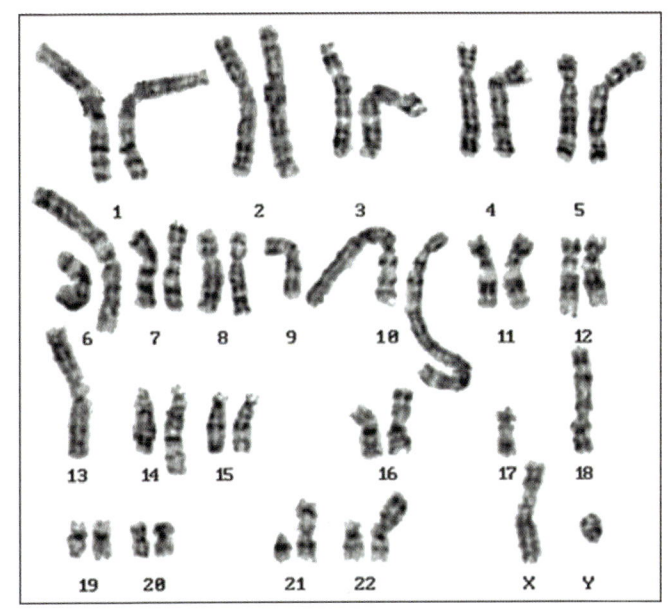

Figure 31.51 Cytogenetics of neuroblastoma. A karyotype contains multiple homogeneous staining regions, particularly prominent in the short and long arms of chromosome 10. These markedly elongated genetic segments represent amplification of the *MYCN* gene and are a marker of a poor prognosis. (Courtesy of Mr. Charles Swanson and Dr. Jeffrey Sawyer, Arkansas Children's Hospital, Little Rock, AR.)

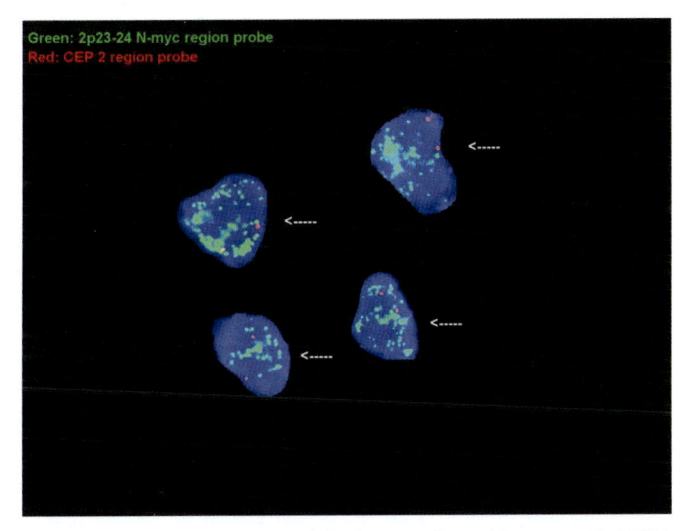

Figure 31.52 Fluorescent *in situ* hybridization of neuroblastoma with *MYCN* amplification. The red dots represent chromosome enumeration probes for chromosome 2. Each cell is seen to contain no more than 2 red signals, each representing one copy of chromosome 2. The green signals represent n-myc region probes. Numerous green signals are present. (Courtesy of Dr. Samuel Wu Children's Hospital Los Angeles and Keck Medical School of USC, Los Angeles, CA.)

Paradoxically, this finding is usually indicative of poor outcome in adult tumors, but it is a favorable feature in neuroblastoma and acute lymphoblastic leukemia, two of the most common pediatric tumors. The presence of hyperdiploidy in these tumors denotes genomic instability that renders them more susceptible to cancer chemotherapy and usually separates them from the poor-prognosis, *MYCN*-amplified group.

Newer genetic features of neuroblastoma to consider include *TRK* and telomerase expression. The tyrosine kinase receptor (Trk) family is a group of receptor proteins that are critical in the transfer of extracellular signals from nerve growth and differentiation factors, or neurotrophins, into the intracellular milieu of neural cells. Three relevant neurotrophin receptors, or Trk proteins, have been identified, TrkA, TrkB, and TrkC. TrkA and TrkC are differentiation-inducing proteins that are inversely related to neuroblastoma stage and N-*myc* amplification.[177] TrkB, conversely, is overexpressed in advanced tumors.[137] *Telomerase* is an enzyme that has been associated with cell immortalization. Increased telomerase expression and telomere length are related to greater neuroblastoma aggressiveness.[178]

Elevated *ALK* gene expression is reported in aggressive neuroblastoma cases. Activating mutations in the *ALK* oncogene are detected in many of the hereditary neuroblastoma cases, as well as some sporadic cases.[179] *ATRX* gene mutation has also been reported in children and young adults with Stage 4 neuroblastoma.[180]

Prognosis and clinical outcome

The prognosis of neuroblastoma is a rather complicated affair related to clinical, histologic, and genetic features. Brodeur and colleagues have identified three distinct genetic subgroups: a low risk group, an intermediate-risk group, and a high-risk group.[181] The low-risk group comprises hyperdiploid tumors with few rearrangements, generally occurring in infants with localized tumors. The intermediate-risk group comprises

Chapter 31: Small round cell tumors

Table 31.6 International Neuroblastoma Staging System[a]

Stage	Definition
1	Localized tumor, complete gross excision, lymph nodes negative[b]
2A	Localized tumor, incomplete gross excision, lymph nodes negative[b]
2B	Localized tumor, with or without complete gross excision, ipsilateral lymph nodes positive[b]
3	Unresectable tumor with midline extension,[c] either by primary tumor or lymph node metastasis
4	Metastatic lesions involving distant lymph nodes, bone, bone marrow, liver, skin, or other (except as defined for Stage 4S)
4S	Localized, Stage 1, 2A, or 2B tumors with dissemination limited to skin, liver, or bone marrow, and in patients <1 year of age

[a] From Reference 150.
[b] Lymph nodes attached to and removed with tumor specimen can be positive in lower stages. To upstage patient, the lymph node must be separately sampled.
[c] Defined by vertebral column.

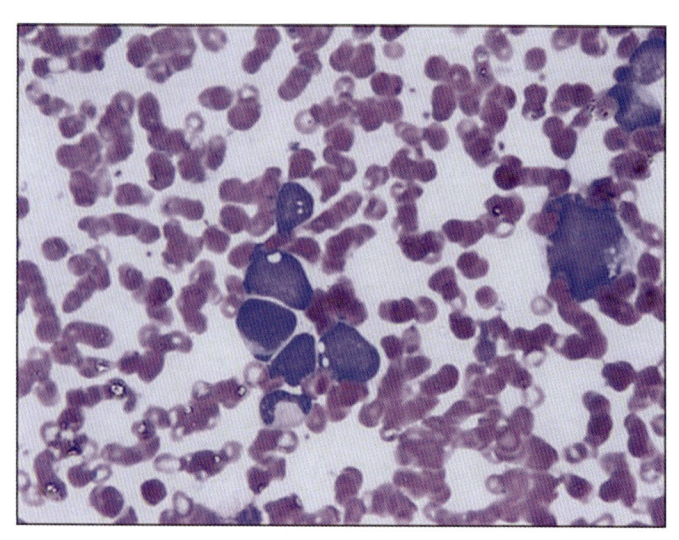

Figure 31.54 Metastatic neuroblastoma, bone marrow. Cohesive primitive cells form well-defined aggregates. Singly dispersed cells contain irregular nuclear and cytoplasmic contours and dwarf adjacent leukocytes (Wright's stain).

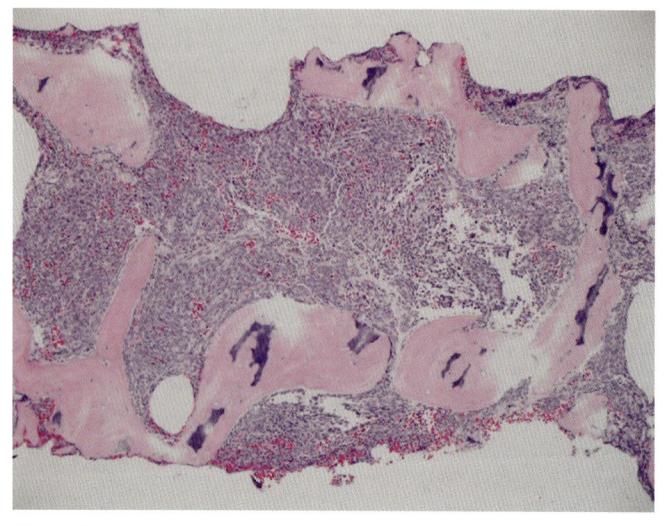

Figure 31.53 Stage 4 neuroblastoma. In this case the bone marrow space is packed with neuroblastoma cells. It is obvious that more than 10% of the nucleated cells seen are neuroblastoma cells, hence Stage 4. Stage 4S cases are defined as those cases in which the tumor cells represent less than 10% of the nucleated cells in children younger than 1 year. Skin and liver metastasis in this age group also constitute Stage 4S and not Stage 4.

near-diploid tumors with no consistent abnormality; these are usually seen in older patients with more advanced disease. The high-risk group comprises lesions with a near-diploid or tetraploid karyotype, with deletions or LOH for 1p36, amplification of *MYCN*, or both, and occurs in older patients with rapidly progressive lesions.

Recent studies report that MYCN protein expression rather than *MYCN* amplification at DNA level is more critical for predicting aggressive behavior of the neuroblastomas.[182]

MYC (aka c-myc) expression is also reported as a new prognostic indicator for a poor prognosis of the patients with neuroblastoma.[155]

Staging is important to management, although low-stage tumors that show high-risk biologic factors must be aggressively treated. Staging is based on the adequacy of excision, extension of disease across the midline, and presence of metastatic tumor, as indicated in Table 31.6.[150] One paradoxical group is Stage 4S, typically composed of infants with apparent metastatic disease, usually to liver, skin or bone marrow, but with relatively good outcome.[183] In the case of bone marrow metastasis, 4S cases are defined as the presence of bone marrow metastasis in a child younger than 1 year of age and with less than 10% of the nucleated cells representing neuroblastoma. Cases that do not meet this criteria are designated Stage 4 (Figure 31.53). Of note is the lesser frequency of *MYCN* amplification in these tumors compared with Stage 4 tumors.[181]

Finally, histologic classification continues to be an independent predictor of outcome in localized, resectable patients.[139] Unfavorable genetic factors and advanced tumor stage occur more frequently with unfavorable histology.[184]

Differential diagnosis of neuroblastoma

Diagnosis of neuroblastoma generally presents less of a problem than that of its less differentiated Ewing sarcoma family cousins. In the presence of a calcified adrenal mass and catecholamine secretion, identification of clusters of primitive cells in fine-needle aspirates or bone marrow is sufficient (Figure 31.54).[150] In patients with primitive, non-adrenal tumors, particularly older children and adolescents, PNET should be strongly considered, requiring the ancillary techniques discussed previously. One particularly

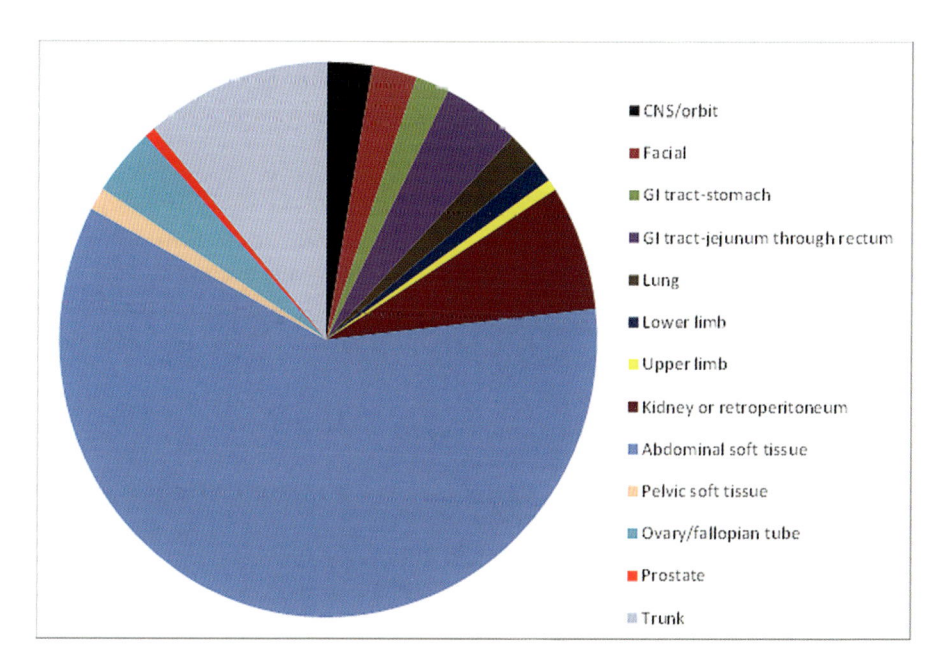

Figure 31.55 Anatomic distribution of desmoplastic small round cell tumor.

Anatomic distribution of desmoplastic small round cell tumor. Christina K Lettieri. *Journal of Cancer Epidemiology.*

important caveat is not to overinterpret *MYCN* amplification as definitive evidence of neuroblastoma in these cases, because this phenomenon also occurs in alveolar rhabdomyosarcoma. Positivity for MyoD or myogenin and the presence of the *PAX/FRKHR* fusion is sufficient to make the latter diagnosis. Another abdominal tumor that should be considered in the differential diagnosis is the DSRCT. This lesion can involve the retroperitoneum and could potentially lead to confusion in the case of large tumors that involve the adrenal. Rosettes have even been described in such tumors.[185]

Desmoplastic small round cell tumor

Definition

A DSRCT is a primitive polyphenotypic neoplasm usually arising in the abdomen and characterized by nests of small cells enmeshed in a dense fibrous stroma. Extra-abdominal examples rarely occur. An *EWS/WT1* fusion genetically characterizes this lesion. Synonyms for this tumor include intra-abdominal desmoplastic small cell tumor, desmoplastic SRBCT, desmoplastic cancer, desmoplastic sarcoma.

History

Gerald *et al.*[186] first described the DSRCT in 1991. Since that seminal description, several clinical, pathologic, and molecular descriptions have appeared. In 1992, Sawyer *et al.*[187] first recognized the characteristic chromosomal translocation, the t(11;22)(p13;q12), and soon thereafter, in 1994, Ladanyi and Gerald first isolated the *EWS/WT1* gene fusion.[188]

Clinical features of DSRCT

DSRCT is predominantly an intra-abdominal lesion that usually occurs in male adolescents.[186] It typically occurs as an intraperitoneal, retroperitoneal, or pelvic mass that can arise in different parts of the abdominal cavity (Figure 31.55). Because of the continuity of the peritoneum with the scrotal sac via the processus vaginalis, DSRCT can also arise as intrascrotal lesions.[189–191] The site diversity of DSRCT has been broadened by the discovery of neoplasms in sites such as the pleura, cranium, kidney, and hand.[192–197] Clinically, these tumors are usually discovered as asymptomatic abdominal masses, similar to ovarian tumors, although complications from intestinal or biliary tract obstruction may signal their presence. Unfortunately, a relatively large percentage exhibit peritoneal spread coincident with or soon after their clinical detection.[198]

Gross and microscopic description

DSRCT are gritty, yellow-gray lesions that have a firm, fibrous cut surface. They are usually nonencapsulated and invade adjacent organs, such as the pancreas. Careful examination for evidence of peritoneal studding should be performed at the time of surgery.

Histologically, fibroplasia is a key diagnostic feature. Nests of small primitive cells punctuate a framework of dense collagen containing scattered fibroblasts and small blood vessels (Figure 31.56). In some areas, the small round cells can fuse into larger aggregates and form ribbons or trabeculae. The small round cell component of these tumors is composed of primitive, cohesive cells with hyperchromatic, round to ovoid nuclei, and scant cytoplasm. Occasionally, scattered tumor cells have an epithelioid appearance, as exemplified by modest amounts of lightly eosinophilic cytoplasm and increased cohesiveness. Other foci can have a vaguely neural appearance, forming ill-defined rosettes.[185] Some tumors have focal pleomorphism, and others contain rhabdoid cells with large amounts of eosinophilic cytoplasm, intracytoplasmic hyaline inclusions, and eccentric nuclei with prominent nucleoli. Because of these rhabdoid cells, DSRCTs must be included among the "pseudo-rhabdoid tumors."

Genetic studies have indicated that some lesions have an atypical histology, with myxoid stroma and anastomosing cords of tumor cells.[200]

Immunohistochemistry and electron microscopy

A defining feature of DSRCTs, which aids in their distinction from rhabdomyosarcomas, is their polyphenotypia. They usually display positivity for keratins, vimentin, desmin,[201] and strong WT1 nuclear positivity (Figure 31.57). One important caveat for WT1 staining with DSRCT is that the monoclonal antibodies used as detection reagents must be directed toward the carboxyl terminal of the molecule.[202] The tumor cells are also positive for epithelial membrane antigen (EMA), NSE, and often for CD57. Cells can be positive for CD99 in some cases, but MyoD and myogenin stains are nonreactive. One should avoid overinterpretation of cytoplasmic MyoD staining as evidence for rhabdomyosarcoma: this pattern is nonspecific. This is important, because DSRCT can resemble alveolar rhabdomyosarcomas in its desmin positivity. Desmin expression often shows a dot-like Golgi pattern in DSRCT. A list of immunohistochemical findings in DSRCT is given in Table 31.7. Another caveat is that occasional cases are keratin negative.[200,203]

Ultrastructural features of DSRCT include intercellular junctions and tonofilaments, representing the epithelial component. Microtubules and neurosecretory granules indicate a neural phenotype.[185] Myogenous organelles such as thick and thin filaments or Z bands are absent.

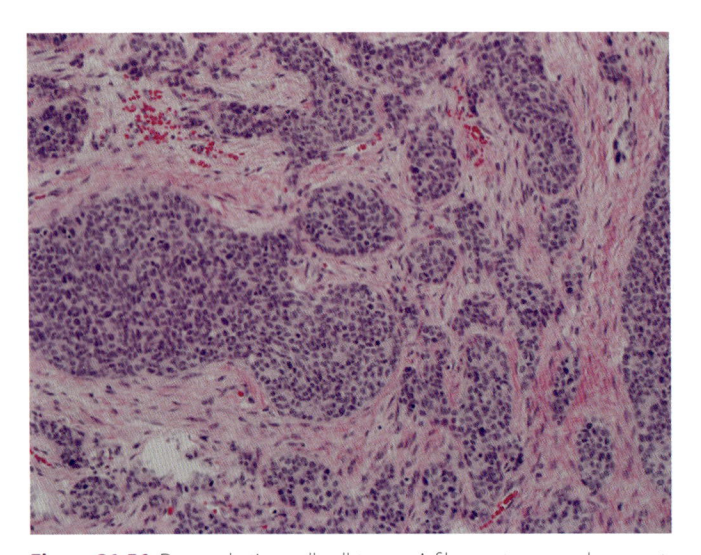

Figure 31.56 Desmoplastic small cell tumor. A fibrous stroma encloses nests and cords of tumor cells with no discernable differentiation.[199]

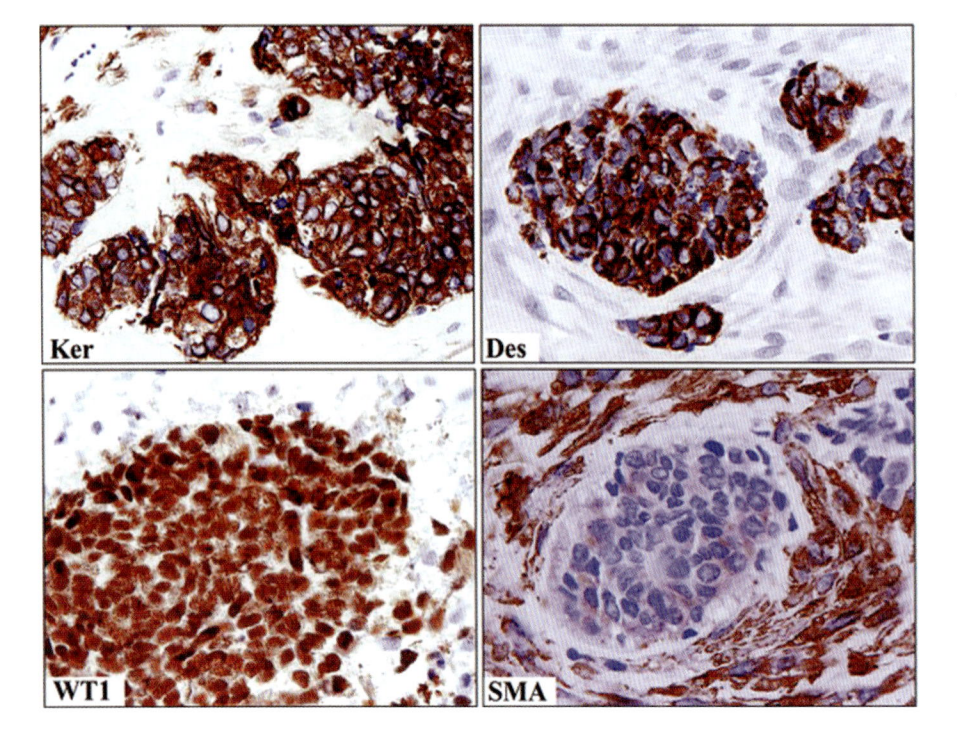

Figure 31.57 Desmoplastic small cell tumor is immunohistochemically positive for keratin cocktail, desmin, and WT1 with a nuclear pattern. There is weak, if any, SMA positivity.

Cytogenetics and molecular biology

DSRCT is cytogenetically defined by a reciprocal translocation, the t(11;22)(p13;q12).[107] This translocation creates a fusion gene that transcribes a chimeric protein containing portions of the WT1 protein and the EWS protein (Figure 31.58).[188] Some lesions previously diagnosed as "extrarenal Wilms' tumor"[204] might represent examples of DSRCT. The fusion protein produced by *WT1/EWS* induces expression of endogenous platelet-derived growth factor-A, a powerful mitogen and chemoattractant.

Table 31.7 Immunohistochemical markers reported to be positive in desmoplastic small cell tumor[a]

Mesenchymal
   Vimentin
   Desmin
   Smooth muscle actin
   Muscle-specific actin

Epithelial
   Keratins
   Epithelial membrane antigen

Neural
   Neuron-specific enolase
   CD57 (Leu7)
   Synaptophysin
   Chromogranin
   NB84

Miscellaneous
   CD15
   WT1
   CD99
   CA125
   Ber-EP4

[a] From reference 201.

WT1/EWS fusion offers a clinically practical means of ancillary diagnosis, accomplished by RT-PCR or FISH[203,205,206] and might be used for confirmation of diagnosis in limited samples.[193] Occasional cases that clinically and phenotypically appear to be DSRCT can show the Ewing sarcoma genotype, *EWS/FLI1* or *EWS/ERG*, by these ancillary studies.[207–209] The effectiveness of chemotherapeutic agents that are typically active against Ewing sarcoma dictates that these latter lesions be treated thusly.

An adolescent leiomyosarcoma like spindle cell tumor with a favorable clinical course has also been shown to possess an *EWS/WT1* fusion.[210] The nosology of this lesion is uncertain.

Prognosis and clinical outcome

DSRCTs are aggressive neoplasms with a penchant for intra- and extraperitoneal dissemination. As with ovarian neoplasms, peritoneal metastases are associated with ascites, serosal adhesions, and intestinal and ureteral obstructions. Modern combination chemotherapy may induce temporary remissions, but the ultimate outcome remains dismal with these aggressive neoplasms,[189,211,212] particularly if chemotherapy and surgery do not produce a remission.[212]

Diagnostic summary and differential diagnosis

The diagnosis of DSRCT should be considered in all intra-abdominal and intrascrotal small cell neoplasms, particularly those with a prominent fibrous stroma. The latter feature, however, does not exclude other small cell neoplasms, such as alveolar rhabdomyosarcoma, Ewing sarcoma, and extrarenal Wilms' tumor. Demonstration of polyphenotypia by expression of epithelial, neural, and mesenchymal markers is key to the diagnosis. If well-formed tubules or glomeruli are present, a diagnosis of extrarenal Wilms' tumor is likely.[204,213] However, one should be aware of considerable morphologic and immunophenotyping overlap.[214] In limited biopsies, demonstration of *EWS/WT1* or t(11;22)(p13;q12) is particularly

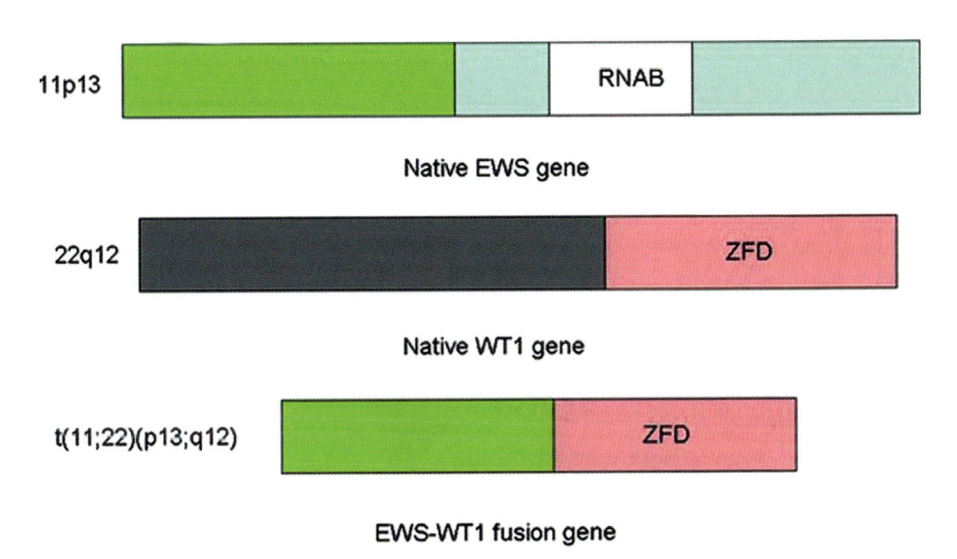

Figure 31.58 Diagram illustrating *EWS/WT1* gene fusion. The sequence transcribing a putative RNA-binding region of the *EWS* gene is replaced by one encoding a zinc finger domain, a DNA-binding region (lowest panel).

helpful. Because of a possible overlap with Ewing sarcoma, RT-PCR, *EWS-FLI1*, and *EWS-ERG*, should also be considered.

Extrarenal rhabdoid tumor of soft tissues

Definition

Rhabdoid tumors are rare lesions that mostly occur in the kidney and CNS. Outside of these locations their rarity increases.

This tumor is a clinically aggressive neoplasm typically arising in infants and characterized by filamentous inclusions within primitive polyphenotypic cells. Genetically, a deletion of the *INI1/SMARCB1* gene on chromosome 22q11 is typically present[215] although exceptions exist.[216]

History

Although eosinophilic rhabdoid cells, filled with whorled eosinophilic filaments, are morphologically highly distinctive, rhabdoid tumor is a problematic entity, because many other tumor types can have cells with rhabdoid cytoplasm.[217,218] Nevertheless, biologic observations indicate that some of these tumors do indeed share genetic alterations that would define a specific disease entity.[219,220]

The rhabdoid tumor was introduced in 1978 by Beckwith and Palmer, who described new categories of aggressive pediatric renal neoplasms based on a cohort originally diagnosed as Wilms' tumors and entered in the first National Wilms' Tumor Study.[221] These tumors included anaplastic Wilms' tumors and the clear cell, hyalinizing, and rhabdomyosarcomatoid variants of Wilms' tumor. The rhabdomyosarcomatoid variant is now known as "rhabdoid tumor," a term coined by Haas and colleagues in 1981.[222]

Subsequent to Beckwith and Palmer's discovery of malignant rhabdoid tumors of the kidney, others described rhabdoid tumors of the soft tissues, liver, brain, genitourinary tract, and skin.[223,224] In retrospect, lesions such as "malignant histiocytoma"[225,226] and "rhabdomyoblastoma,"[227] described prior to Beckwith and Palmer's report, also had rhabdoid features. Reports from Japan and the Intergroup Rhabdomyosarcoma Study, however, found that although some soft tissue lesions could be characterized as discrete entities,[224,228] others were actually different tumors that contained a rhabdoid component.[228,229] Subsequent investigations also indicated that many putative "extrarenal rhabdoid tumors" actually represented examples of other neoplasms, but some were bona fide rhabdoid tumors.[217] Remarkable polyphenotypia was also noted,[217,230,231] similar to that seen in atypical teratoid rhabdoid tumors of the brain.[232]

The recent history of this entity has been dominated by the discovery of a tumor-suppressor gene, *INI1/SMARCB1*, that is often mutated or deleted in rhabdoid tumors of all sites. Lack of expression of this genetic marker has become a standard ancillary feature in the diagnosis of rhabdoid tumor, although caveats are now recognized. Foremost among them is the increasing number of tumors that share this feature.[233]

Terminology and synonyms

The term *composite rhabdoid tumors*, as defined by Wick et al.,[234] refers to neoplasms of variable histology and origin that contain histologic foci that resemble true lesions.

Atypical teratoid/rhabdoid tumors (AT/RT) refers to rhabdoid tumors of the CNS, as defined by Rorke et al.[232]

Clinical features

One prominent feature of rhabdoid tumors of all sites is their predilection for occurring in infants and very young children. Rhabdoid tumors constitute a large percentage of infantile renal tumors,[235] and most of the extrarenal tumors without other apparent histologies also arise in very young patients.[217] Thus, the origin of an "extrarenal rhabdoid tumor" in an older child or adult should raise suspicion that another tumor diagnosis might be in order, and confirmation is necessary.

Extrarenal rhabdoid tumors have been reported in a variety of sites, but variable documentation makes the data difficult to interpret. Nevertheless, hepatic and soft tissue lesions have been well reported. Among favored sites in soft tissue are axial and paravertebral lesions,[217] as well as extremity lesions (Figure 31.59).[226,236] Superficial lesions may arise in the dermis and subcutis,[237] but appropriate imaging studies are essential to rule out a renal primary tumor.[238] Some cutaneous rhabdoid tumors are associated with hamartomas, such as sebaceous nevi.[217,239]

Hypercalcemia has been reported as a paraneoplastic phenomenon with rhabdoid tumors of the kidney.[240,241] This finding has been associated with secretion of parathormone and parathormone-like substances.[242] Another clinical feature described in both renal and extrarenal rhabdoid tumors is the coexistence of primitive neuroectodermal tumors of the CNS.[243,244] Although these brain tumors mistakenly could be considered to be metastases by radiologic reports, they actually represent primitive neuroepithelial tumors, possibly within the spectrum of ATRT. In particular, multiple tumors suggest a constitutional deletion of SMARCB1 and genetic testing and counselling is indicated.[215]

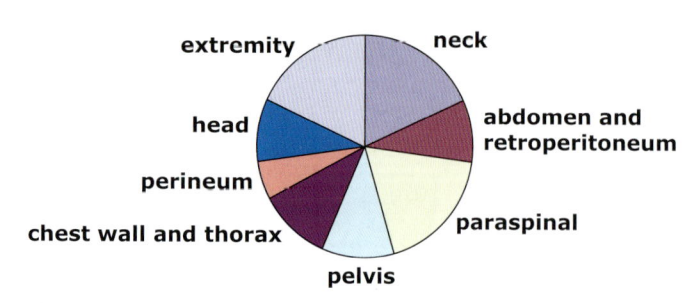

Figure 31.59 Distribution of soft tissue rhabdoid tumors.

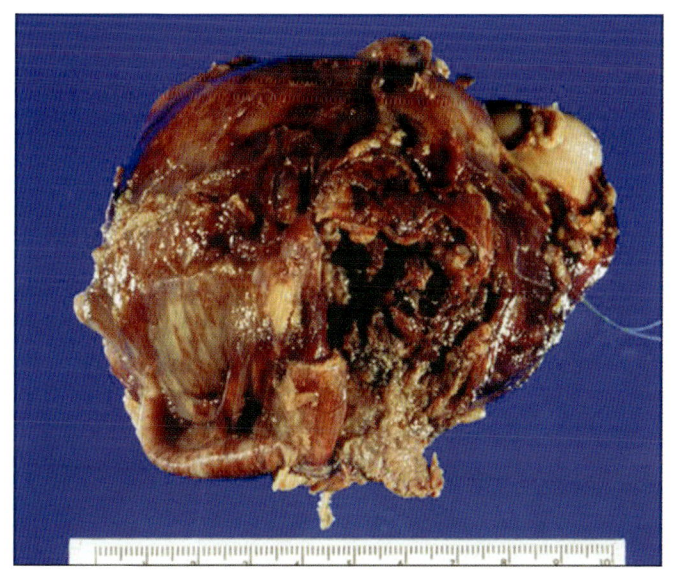

Figure 31.60 Malignant rhabdoid tumor, gross photograph. A bulging soft tissue tumor contains friable, necrotic, hemorrhagic material that exudes from a surgical rupture.

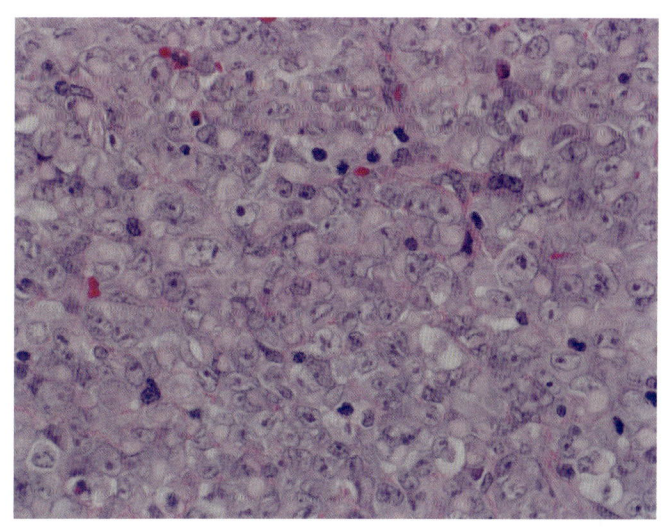

Figure 31.61 Malignant rhabdoid tumor. The neoplasm is composed of cells with large, "owl's-eye" nuclei, eosinophilic cytoplasm, and hyaline intracytoplasmic inclusions.

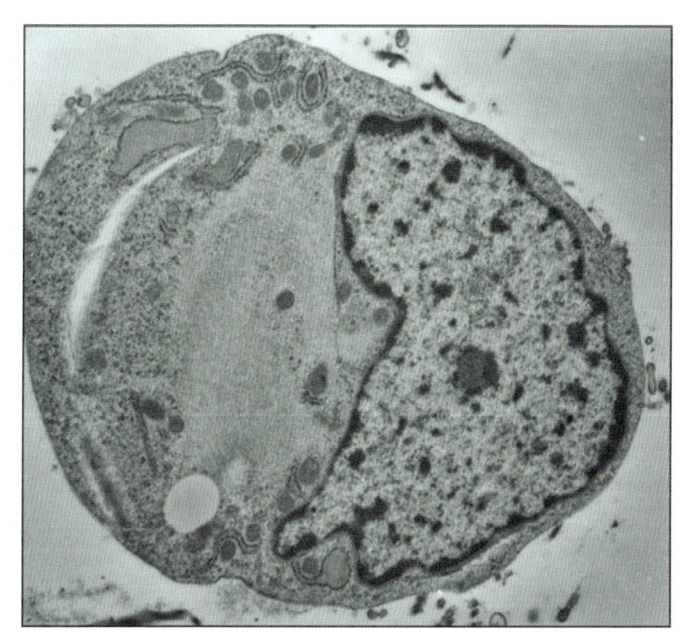

Figure 31.62 Malignant rhabdoid tumor, electron micrograph. A rhabdoid cell contains an eccentric nucleus and a paranuclear cluster of intermediate filaments that entraps cytoplasmic organelles. There is no evidence of differentiation (magnification: ×7000).

Gross and microscopic description

Grossly, the typical invasive, nonencapsulated appearance of rhabdoid tumors, accentuated by a soft, fleshy, focally necrotic cut surface, underscores their malignant nature at any site (Figure 31.60). Because of their invasiveness, rhabdoid tumors are often incompletely excised. Documentation of the surgical margins can be an arduous task, because of the soupy consistency of these tumors.

Microscopically, the rhabdoid cells have eccentric nuclei with prominent nucleoli and abundant cytoplasm with round, hyaline inclusions (Figure 31.61). Some lesions have less abundant cytoplasm, potentially simulating a hematopoietic neoplasm; these tumors should thus be included in the differential diagnosis of small round cell neoplasms. Weeks et al. also described hemangiopericytomatous, spindle cell, and carcinoid-like morphologies in renal rhabdoid tumors,[245] which can be seen in extrarenal examples. The occurrence of paraganglioma-like patterns raises the question of neuroectodermal differentiation, and the gene expression profile seems to support this.[246]

Electron microscopy

The key ultrastructural feature in rhabdoid tumors is the presence of the paranuclear, spherical, intermediate filament aggregates that correspond with the hyaline inclusions seen by light microscopy (Figure 31.62). These inclusions are not specific, and other features could lead to another diagnosis. Tonofilaments indicate possible epithelioid sarcoma or carcinoma, thick and thin filaments characterize rhabdomyosarcoma, and neurosecretory granules and cytoplasmic processes suggest PNET or neuroendocrine carcinoma (Figure 31.14).[217]

Immunohistochemistry

The intermediate filaments in rhabdoid tumors consist of keratins and vimentin (Figure 31.63). The keratin types vary, and immunoreactivity depends on the antibody. Positivity for EMA further confirms the epithelial nature of these lesions (Figure 31.64). A variety of other markers can be positive, including myogenous and neural markers, and the apparent nonspecificity of these stains has led some to postulate that the staining might be artifactual.[245] Nevertheless, this

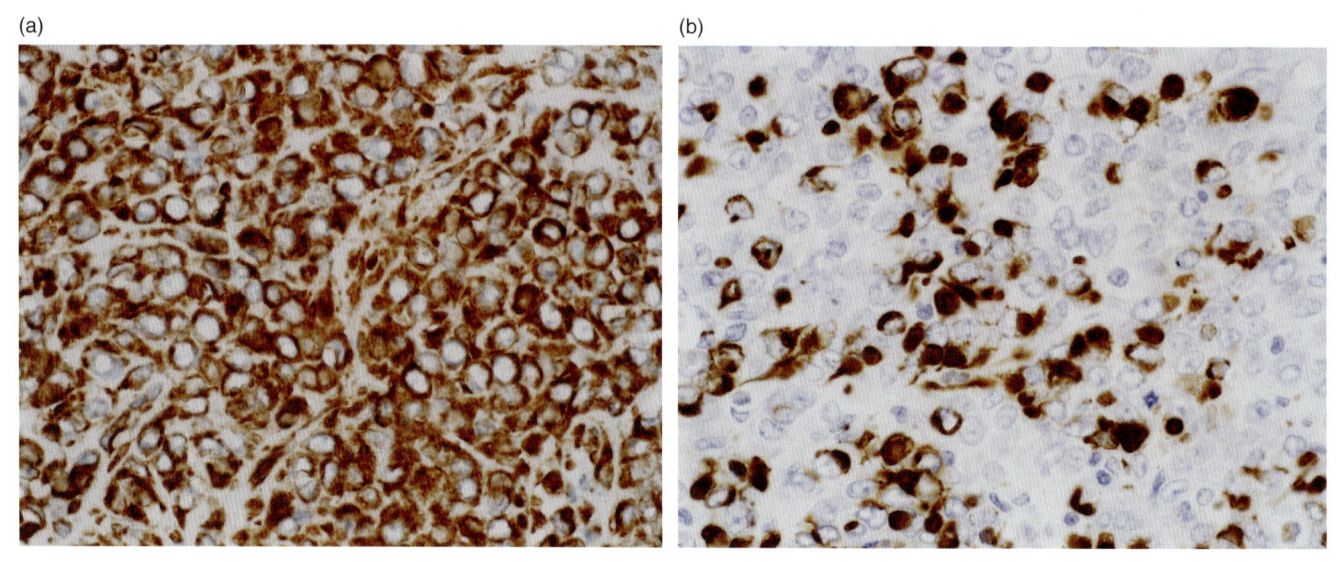

Figure 31.63 Malignant rhabdoid tumor, immunohistochemical stains for vimentin (**a**) and cytokeratin (**b**) coexpression. Note the globular vimentin staining, corresponding to cytoplasmic intermediate filament inclusion. Only a subset of tumor cells are positive for cytokeratin.

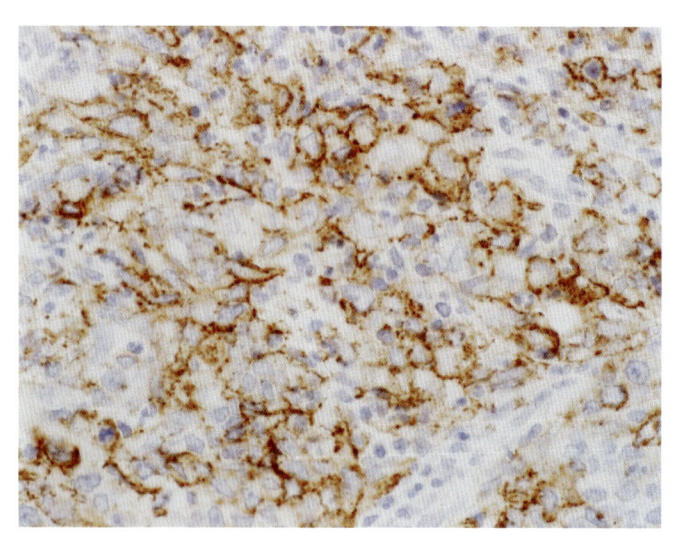

Figure 31.64 Malignant rhabdoid tumor (EMA stain). The epithelial nature of rhabdoid tumor is confirmed by its epithelial membrane positivity, seen here in a typical membranous pattern (avidin-biotin complex method with hematoxylin counterstain).

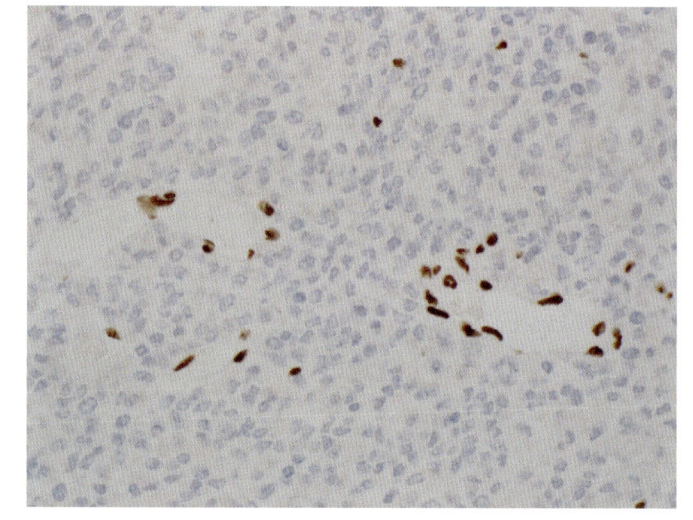

Figure 31.65 Malignant rhabdoid tumor (INI1 stain). There is a diffuse lack of nuclear staining in the tumor cells. In comparison, endothelial cells show strong positivity. (Avidin-biotin complex method with hematoxylin counterstain, monoclonal BAF47).

phenomenon correlates with the diverse organelles that can be observed by careful electron microscopy.[217]

The absence of staining with INI1 immunostain has become recognized as a major diagnostic feature of rhabdoid tumors (Figure 31.65). INI1, as discussed in the next section, is a highly conserved protein that is critical for proper DNA maintenance, so that all normal cells express it. In contrast, rhabdoid tumors usually possess a SMARCB1 mutation or deletion that can be recognized by their lack of staining using anti-INI1 antibodies. This phenomenon has been exploited in recent years as a means of ancillary diagnosis, because other small cell tumors show strong nuclear staining.[247] In initial studies, this appeared to be potentially useful in separating true rhabdoid tumors from composite rhabdoid tumors.[248]

Unfortunately, recent studies suggest that there are exceptions to this rule. The first to be discovered was the negativity of some choroid plexus neoplasms.[249,250] Like CNS atypical teratoid-rhabdoid tumors, these lesions arise in the central-most portions of the brain, and the absence of INI1 expression suggests a possible kinship. Among soft tissue sarcomas, it is now recognized that epithelioid sarcomas can be INI1 negative.[246,251,252] Given the fact that the proximal variant of an epithelioid sarcoma can be a dead ringer for a rhabdoid tumor, both in behavior and in histopathology, one might not be surprised that additional evidence has arisen that suggests that these are possibly related neoplasms. Finally, case reports have indicated that composite rhabdoid tumors of adults on occasion show a lack of INI1 expression.[253,254] This phenomenon suggests that *INI1/SMARCB1* mutation or suppression might develop as a component of tumor progression; a surprising number of other sarcomas have been shown to have INI1 negativity, so careful correlation with other features is still advisable.

Cytogenetics

Early studies of the cytogenetics of rhabdoid tumors were disappointing because of their lack of any apparent karyotypic abnormality,[255] though it was noted that atypical teratoid-rhabdoid tumors of the brain consistently exhibited a monosomy 22.[256,257] Subsequent studies of rhabdoid tumors focused on monosomy 22, with the resultant discovery of a region of common deletion at 22q11 in both renal and extrarenal cases.[258] This region did not seem to involve the *EWS* or *NF2* loci, but rather to overlap with the *BCR* gene, which is involved with chronic myelogenous leukemia and some cases of acute lymphoblastic leukemia.[259] Subsequent studies delimited a putative tumor-suppressor gene locus, termed the *SMARCB1* (also known as *hSNF5/INI1*) gene.[260,261] Genetic studies of chromosome 22 were then used as a means of ancillary diagnosis of extrarenal rhabdoid tumor,[219,220] prior to the development of INI1 immunohistochemistry. Of particular interest has been the discovery of patients with a constitutional mutation of *SMARCB1/INI1*.[219,262] These unfortunate children tend to develop rhabdoid tumors at multiple sites, explaining the phenomenon of separate tumors first described by Bonnin et al.[243] One must consider this possibility when confronted with patients with multiple rhabdoid tumors, because family studies might be indicated.

Another phenomenon of note in rhabdoid tumor cytogenetics is their association with alterations of the short arm of chromosome 11. The mistaken identity of a rhabdoid cell line as Wilms' tumor led to its use in biologic studies of the latter lesion, demonstrating that tumorigenesis could be reversed by implantation of a normal chromosome 11.[263] Genetic alterations in this region can occur in different tumor types and suggest a molecular kinship between rhabdoid tumor and rhabdomyosarcoma.[264,265]

Table 31.8 Tumors reported to contain rhabdoid cells

Rhabdomyosarcoma
Synovial sarcoma
Paraganglioma
Neuroendocrine carcinoma
Melanoma
Transitional cell carcinoma
Epithelioid sarcoma
Desmoplastic small cell tumor
Neuroblastoma
Glioma
Salivary gland tumors
Adenocarcinoma
PNET
Myofibrosarcoma
Myoepithelioma
Merkel cell carcinoma
Leiomyosarcoma

Table 31.9 Nonrhabdoid INI1-deficient tumors of bone and soft tissue

Epithelioid sarcoma
Malignant peripheral nerve sheath tumor
Extraskeletal myxoid chondrosarcoma
Myoepithelial carcinoma
Chordoma
Synovial sarcoma

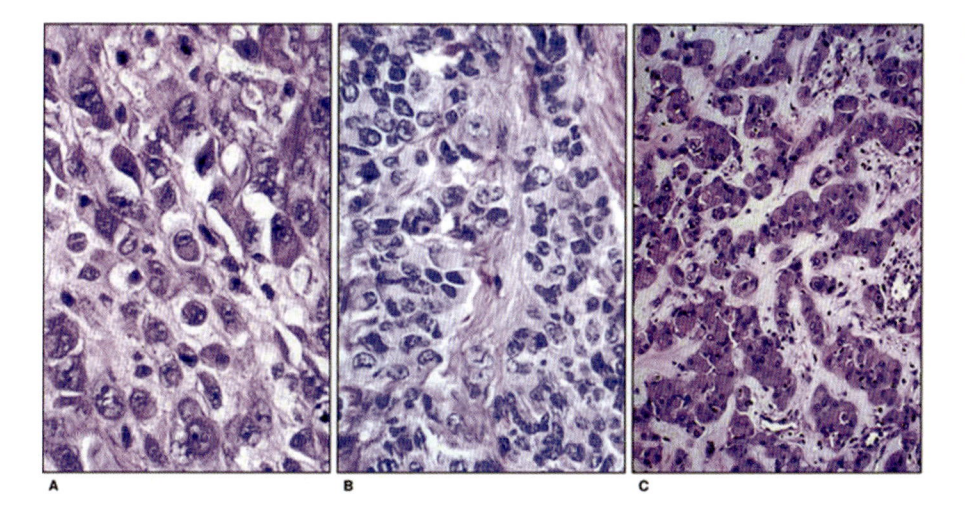

Figure 31.66 "Pseudo-rhabdoid" tumors, including epithelioid sarcoma (**A**), rhabdomyosarcoma (**B**), and neuroendocrine carcinoma (**C**), can contain rhabdoid cells with cytoplasmic inclusions.

Some patients appear to have mutations in genes other than *SMARCB1/INI1*, both as acquired tumors or as constitutional lesions. These patients form a minority of cases to date, but the tumors may express INI1 normally and show no gene deletions, creating potential diagnostic confusion.[266,267] These may involve other members of the SWI/SWF complex.[216]

Differential diagnosis

Rhabdoid tumor has historically been a diagnosis by exclusion, because other neoplasms with rhabdoid cytoplasm must be ruled out (Tables 31.8 and 31.9, Figure 31.66). This is especially true for the extrarenal lesions in adults. Some tumors, such as epithelioid sarcomas, show more aggressive behavior if rhabdoid cells are prominent. MyoD or myogenin expression defines a rhabdomyosarcoma. Melanoma can also have a rhabdoid phenotype,[268] but should be recognized by its positive melanocytic markers. Immunohistochemical or genetic testing for *SMARCB1/INI1* deletions[220] has now become standard for confirmation of diagnosis of rhabdoid tumor, but the existence of occasional nonspecificity or misleading studies must be recognized.[233]

Acknowledgements

The authors acknowledge their keen appreciation and gratitude for the services of Ms. Rossana Desrochers, who ably assisted in the editing of this chapter.

References

1. Triche TJ, Askin FB. Neuroblastoma and the differential diagnosis of small-, round-, blue-cell tumors. *Hum Pathol* 1983;14:569–595.
2. Donaldson SS. Lessons from our children. *Int J Radiat Oncol Biol Phys* 1993;26:739–749.
3. Ross JA, Severson RK, Pollock BH, Robison LL. Childhood cancer in the United States: a geographical analysis of cases from the Pediatric Cooperative Clinical Trials groups. *Cancer* 1996;77:201–207.
4. Kelleher FC, Viterbo A. Histologic and genetic advances in refining the diagnosis of "undifferentiated pleomorphic sarcoma." *Cancers* 2013;5:218–233.
5. Koufos A, Hansen MF, Copeland NG, et al. Loss of heterozygosity in three embryonal tumours suggests a common pathogenetic mechanism. *Nature* 1985;316:330–334.
6. Heyn R, Haeberlen V, Newton WA, et al. Second malignant neoplasms in children treated for rhabdomyosarcoma. Intergroup Rhabdomyosarcoma Study Committee. *J Clin Oncol* 1993;11:262–270.
7. May WA, Lessnick SL, Braun BS, et al. The Ewing's sarcoma EWS/FLI-1 fusion gene encodes a more potent transcriptional activator and is a more powerful transforming gene than FLI-1. *Mol Cell Biol* 1993;13:7393–7398.
8. Whang-Peng J, Triche TJ, Knutsen T, et al. Cytogenetic characterization of selected small round cell tumors of childhood. *Cancer Genet Cytogenet* 1986;21:185–208.
9. Oda H, Imai Y, Nakatsuru Y, Hata J, Ishikawa T. Somatic mutations of the APC gene in sporadic hepatoblastomas. *Cancer Res* 1996;56:3320–3323.
10. Raffel C, Jenkins RB, Frederick L, et al. Sporadic medulloblastomas contain PTCH mutations. *Cancer Res* 1997;57:842–845.
11. Felix CA, Kappel CC, Mitsudomi T, et al. Frequency and diversity of p53 mutations in childhood rhabdomyosarcoma. *Cancer Res* 1992;52:2243–2247.
12. Jones PA, Laird PW. Cancer epigenetics comes of age. *Nat Genet* 1999;21:163–167.
13. Baylin SB. Tying it all together: epigenetics, genetics, cell cycle, and cancer. *Science* 1997;277:1948–1949.
14. Chen B, Dias P, Jenkins JJ, 3rd, Savell VH, Parham DM. Methylation alterations of the MyoD1 upstream region are predictive of subclassification of human rhabdomyosarcomas. *Am J Pathol* 1998;152:1071–1079.
15. Eid JE, Garcia CB. Reprogramming of mesenchymal stem cells by oncogenes. *Semin Cancer Biol* 2014;32:18–31.
16. Dehner LP. Primitive neuroectodermal tumor and Ewing's sarcoma. *Am J Surg Pathol* 1993;17:1–13.
17. Yunis EJ. Ewing's sarcoma and related small round cell neoplasms in children. *Am J Surg Pathol* 1986;10(Suppl 1):54–62.
18. Angervall L, Enzinger FM. Extraskeletal neoplasm resembling Ewing's sarcoma. *Cancer* 1975;36:240–251.
19. Tefft M, Vawter GF, Mitus A. Paravertebral "round cell" tumors in children. *Radiology* 1969;92:1501–1509.
20. Schmidt D, Mackay B, Ayala AG. Ewing's sarcoma with neuroblastoma-like features. *Ultrastruct Pathol* 1982;3:143–151.
21. Mierau GW. Extraskeletal Ewing's sarcoma (peripheral neuroepithelioma). *Ultrastruct Pathol* 1985;9:91–98.
22. Seemayer TA, Thelmo WL, Bolande RP, Wiglesworth FW. Peripheral neuroectodermal tumors. *Perspect Pediatr Pathol* 1975;2:151–172.
23. Askin FB, Rosai J, Sibley RK, Dehner LP, McAlister WH. Malignant small cell tumor of the thoracopulmonary region in childhood: a distinctive clinicopathologic entity of uncertain histogenesis. *Cancer* 1979;43:2438–2451.
24. Jaffe R, Santamaria M, Yunis EJ, et al. The neuroectodermal tumor of bone. *Am J Surg Pathol* 1984;8:885–898.
25. Turc-Carel C, Philip I, Berger MP, Philip T, Lenoir GM. Chromosome study of Ewing's sarcoma (ES) cell lines: consistency of a reciprocal translocation t(11;22)(q24;q12). *Cancer Genet Cytogenet* 1984;12:1–19.
26. Whang-Peng J, Triche TJ, Knutsen T, et al. Chromosome translocation in peripheral neuroepithelioma. *N Engl J Med* 1984;311:584–585.
27. Ladanyi M, Lewis R, Garin-Chesa P, et al. EWS rearrangement in Ewing's sarcoma and peripheral neuroectodermal tumor: molecular detection and correlation with cytogenetic analysis and MIC2 expression. *Diagn Mol Pathol* 1993;2:141–146.
28. Pellin A, Boix J, Blesa JR, et al. EWS/FLI-1 rearrangement in small round cell sarcomas of bone and soft tissue

28. detected by reverse transcriptase polymerase chain reaction amplification. *Eur J Cancer* 1994;30A:827–831.

29. Cavazzana AO, Miser JS, Jefferson J, Triche TJ. Experimental evidence for a neural origin of Ewing's sarcoma of bone. *Am J Pathol* 1987;127:507–518.

30. Stout AP, Murray MA. Neuroepithelioma of radial nerve with study of its behavior in vitro. *Rev Can Biol* 1949;1:651–659.

31. Delattre O, Zucman J, Melot T, et al. The Ewing family of tumors: a subgroup of small-round-cell tumors defined by specific chimeric transcripts. *N Engl J Med* 1994;331:294–299.

32. Worch J, Cyrus J, Goldsby R, et al. Racial differences in the incidence of mesenchymal tumors associated with EWSR1 translocation. *Cancer Epidemiol Biomarkers Prev* 2011;20:449–453.

33. Grier HE. The Ewing family of tumors: Ewing's sarcoma and primitive neuroectodermal tumors. *Pediatr Clin North Am* 1997;44:991–1004.

34. Kissane JM, Askin FB, Foulkes M, Stratton LB, Shirley SF. Ewing's sarcoma of bone: clinicopathologic aspects of 303 cases from the Intergroup Ewing's Sarcoma Study. *Hum Pathol* 1983;14:773–779.

35. Soule EH, Newton W, Jr., Moon TE, Tefft M. Extraskeletal Ewing's sarcoma: a preliminary review of 26 cases encountered in the Intergroup Rhabdomyosarcoma Study. *Cancer* 1978;42:259–264.

36. Harms D. Soft tissue sarcomas in the Kiel Pediatric Tumor Registry. *Curr Top Pathol* 1995;89:31–45.

37. Holman CD, Reynolds PM, Byrne MJ, Trotter JM, Armstrong BK. Possible infectious etiology of six cases of Ewing's sarcoma in Western Australia. *Cancer* 1983;52:1974–1976.

38. Hutter RV, Francis KC, Foote FW, Jr. Ewing's sarcoma in siblings: report of the second known occurrence. *Am J Surg* 1964;107:598–603.

39. Gariepy G, Drouin R, Lemieux N, Richer CL. Ultrastructural, immunohistochemical, and cytogenetic study of a malignant peripheral neuroectodermal tumor in a patient seropositive for human immunodeficiency virus. *Am J Clin Pathol* 1990;93:818–822.

40. Ogawa K. Embryonal neuroepithelial tumors induced by human adenovirus type 12 in rodents. 1. Tumor induction in the peripheral nervous system. *Acta Neuropathol* 1989;77:244–253.

41. Mirra J. *Bone Tumors: Clinical, Radiologic, and Pathologic Correlations*. Philadelphia: Lea & Febiger; 1989.

42. Marina NM, Etcubanas E, Parham DM, Bowman LC, Green A. Peripheral primitive neuroectodermal tumor (peripheral neuroepithelioma) in children: a review of the St. Jude experience and controversies in diagnosis and management. *Cancer* 1989;64:1952–1960.

43. Shimada H, Newton WA, Jr., Soule EH, et al. Pathologic features of extraosseous Ewing's sarcoma: a report from the Intergroup Rhabdomyosarcoma Study. *Hum Pathol* 1988;19:442–453.

44. Rodriguez-Galindo C, Marina NM, Fletcher BD, et al. Is primitive neuroectodermal tumor of the kidney a distinct entity? *Cancer* 1997;79:2243–2250.

45. Sexton CW, White WL. Primary cutaneous Ewing's family sarcoma: report of a case with immunostaining for glycoprotein p30/32 mic2. *Am J Dermatopathol* 1996;18:601–605.

46. Smith LM, Adams RH, Brothman AR, Vanderhooft SL, Coffin CM. Peripheral primitive neuroectodermal tumor presenting with diffuse cutaneous involvement and 7;22 translocation. *Med Pediatr Oncol* 1998;30:357–363.

47. Ehrig T, Billings SD, Fanburg-Smith JC. Superficial primitive neuroectodermal tumor/Ewing sarcoma (PN/ES): same tumor as deep PN/ES or new entity? *Ann Diagn Pathol* 2007;11:153–159.

48. Kaspers GJ, Kamphorst W, van de Graaff M, van Alphen HA, Veerman AJ. Primary spinal epidural extraosseous Ewing's sarcoma. *Cancer* 1991;68:648–654.

49. Mobley BC, Roulston D, Shah GV, Bijwaard KE, McKeever PE. Peripheral primitive neuroectodermal tumor/Ewing's sarcoma of the craniospinal vault: case reports and review. *Hum Pathol* 2006;37:845–853.

50. Meis JM, Enzinger FM, Martz KL, Neal JA. Malignant peripheral nerve sheath tumors (malignant schwannomas) in children. *Am J Surg Pathol* 1992;16:694–707.

51. Hanna SL, Fletcher BD, Kaste SC, Fairclough DL, Parham DM. Increased confidence of diagnosis of Ewing sarcoma using T2-weighted MR images. *Magn Reson Imaging* 1994;12:559–568.

52. Lemmi MA, Fletcher BD, Marina NM, et al. Use of MR imaging to assess results of chemotherapy for Ewing sarcoma. *Am J Roentgenol* 1990;155:343–346.

53. Tsokos M. Peripheral primitive neuroectodermal tumors: diagnosis, classification, and prognosis. *Perspect Pediatr Pathol* 1992;16:27–98.

54. Nascimento AG, Unii KK, Pritchard DJ, Cooper KL, Dahlin DC. A clinicopathologic study of 20 cases of large-cell (atypical) Ewing's sarcoma of bone. *Am J Surg Pathol* 1980;4:29–36.

55. Folpe AL, Goldblum JR, Rubin BP, et al. Morphologic and immunophenotypic diversity in Ewing family tumors: a study of 66 genetically confirmed cases. *Am J Surg Pathol* 2005;29:1025–1033.

56. Pinto A, Dickman P, Parham D. Pathobiologic markers of the Ewing sarcoma family of tumors: state of the art and prediction of behaviour. *Sarcoma* 2011;2011:856190.

57. Arnold MA, Ballester LY, Pack SD, et al. Primary subcutaneous spindle cell Ewing sarcoma with strong S100 expression and EWSR1-FLI1 fusion: a case report. *Pediatr Dev Pathol* 2014;17:302–307.

58. Williams S, Parham DM, Jenkins JJ, 3rd. Peripheral neuroepithelioma with ganglion cells: report of two cases and review of the literature. *Pediatr Dev Pathol* 1999;2:42–49.

59. Llombart-Bosch A, Peydro-Olaya A. Scanning and transmission electron microscopy of Ewing's sarcoma of bone (typical and atypical variants): an analysis of nine cases. *Virchows Arch A Pathol Anat Histopathol* 1983;398:329–346.

60. Triche TJ, Ross WE. Glycogen-containing neuroblastoma with clinical and histopathologic features of Ewing's sarcoma. *Cancer* 1978;41:1425–1432.

61. Llombart-Bosch A, Lacombe MJ, Peydro-Olaya A, Perez-Bacete M, Contesso G. Malignant peripheral

62. Bridge JA, Fidler ME, Neff JR, et al. Adamantinoma-like Ewing's sarcoma: genomic confirmation, phenotypic drift. Am J Surg Pathol 1999;23:159–165.
63. Ambros IM, Ambros PF, Strehl S, et al. MIC2 is a specific marker for Ewing's sarcoma and peripheral primitive neuroectodermal tumors: evidence for a common histogenesis of Ewing's sarcoma and peripheral primitive neuroectodermal tumors from MIC2 expression and specific chromosome aberration. Cancer 1991;67:1886–1893.
64. Ramani P, Rampling D, Link M. Immunocytochemical study of 12E7 in small round-cell tumours of childhood: an assessment of its sensitivity and specificity. Histopathology 1993;23:557–561.
65. Stevenson AJ Chatten J, Bertoni F, et al. CD99 (p30/32MIC2) neuroectodermal/Ewing's sarcoma antigen as an immunohistochemical marker: review of more than 600 tumors and the literature experience. Appl Immunohistochem Mol Morphol 1994;2:231.
66. Fellinger EJ, Garin-Chesa P, Glasser DB, Huvos AG, Rettig WJ. Comparison of cell surface antigen HBA71 (p30/32MIC2), neuron-specific enolase, and vimentin in the immunohistochemical analysis of Ewing's sarcoma of bone. Am J Surg Pathol 1992;16:746–755.
67. Weiss LM, Arber DA, Chang KL. CD45: a review. Appl Immunohistochem Mol Morphol 1993;1:166.
68. Ozdemirli M, Fanburg-Smith JC, Hartmann DP, et al. Precursor B-lymphoblastic lymphoma presenting as a solitary bone tumor and mimicking Ewing's sarcoma: a report of four cases and review of the literature. Am J Surg Pathol 1998;22:795–804.
69. Pappo AS, Douglass EC, Meyer WH, Marina N, Parham DM. Use of HBA 71 and anti-beta 2-microglobulin to distinguish peripheral neuroepithelioma from neuroblastoma. Hum Pathol 1993;24:880–885.
70. Molenaar WM, Muntinghe FL. Expression of neural cell adhesion molecules and neurofilament protein isoforms in skeletal muscle tumors. Hum Pathol 1998;29:1290–1293.
71. Strother DR, Parham DM, Houghton PJ. Expression of the 5.1 H11 antigen, a fetal muscle surface antigen, in normal and neoplastic tissue. Arch Pathol Lab Med 1990;114:593–596.
72. Garin-Chesa P, Fellinger EJ, Huvos AG, et al. Immunohistochemical analysis of neural cell adhesion molecules: differential expression in small round cell tumors of childhood and adolescence. Am J Pathol 1991;139:275–286.
73. Mhawech-Fauceglia P, Herrmann F, Penetrante R, et al. Diagnostic utility of FLI-1 monoclonal antibody and dual-colour, break-apart probe fluorescence in situ (FISH) analysis in Ewing's sarcoma/primitive neuroectodermal tumour (EWS/PNET): a comparative study with CD99 and FLI-1 polyclonal antibodies. Histopathology 2006;49:569–575.
74. Yoshida A, Sekine S, Tsuta K, et al. NKX2.2 is a useful immunohistochemical marker for Ewing sarcoma. Am J Surg Pathol 2012;36:993–999.
75. Parham DM, Dias P, Kelly DR, Rutledge JC, Houghton P. Desmin positivity in primitive neuroectodermal tumors of childhood. Am J Surg Pathol 1992;16:483–492.
76. Kawamoto EH, Weidner N, Agostini RM, Jr., Jaffe R. Malignant ectomesenchymoma of soft tissue: report of two cases and review of the literature. Cancer 1987;59:1791–1802.
77. Le Douarin NM, Ziller C. Plasticity in neural crest cell differentiation. Curr Opin Cell Biol 1993;5:1036–1043.
78. Boue DR, Parham DM, Webber B, Crist WM, Qualman SJ. Clinicopathologic study of ectomesenchymomas from Intergroup Rhabdomyosarcoma Study Groups III and IV. Pediatr Dev Pathol 2000;3:290–300.
79. Navarro S, Cavazzana AO, Llombart-Bosch A, Triche TJ. Comparison of Ewing's sarcoma of bone and peripheral neuroepithelioma: an immunocytochemical and ultrastructural analysis of two primitive neuroectodermal neoplasms. Arch Pathol Lab Med 1994;118:608–615.
80. Douglass EC, Rowe ST, Valentine M, et al. A second nonrandom translocation, der(16)t(1;16)(q21;q13), in Ewing sarcoma and peripheral neuroectodermal tumor. Cytogenet Cell Genet 1990;53:87–90.
81. Maurici D, Perez-Atayde A, Grier HE, et al. Frequency and implications of chromosome 8 and 12 gains in Ewing sarcoma. Cancer Genet Cytogenet 1998;100:106–110.
82. Mugneret F, Lizard S, Aurias A, Turc-Carel C. Chromosomes in Ewing's sarcoma. II. Nonrandom additional changes, trisomy 8 and der(16)t(1;16). Cancer Genet Cytogenet 1988;32:239–245.
83. Hattinger CM, Rumpler S, Strehl S, et al. Prognostic impact of deletions at 1p36 and numerical aberrations in Ewing tumors. Genes Chromosomes Cancer 1999;24:243–254.
84. Armengol G, Tarkkanen M, Virolainen M, et al. Recurrent gains of 1q, 8 and 12 in the Ewing family of tumours by comparative genomic hybridization. Br J Cancer 1997;75:1403–1409.
85. Delattre O, Zucman J, Plougastel B, et al. Gene fusion with an ETS DNA-binding domain caused by chromosome translocation in human tumours. Nature 1992;359:162–165.
86. Ladanyi M. The emerging molecular genetics of sarcoma translocations. Diagn Mol Pathol 1995;4:162–173.
87. Ouchida M, Ohno T, Fujimura Y, Rao VN, Reddy ES. Loss of tumorigenicity of Ewing's sarcoma cells expressing antisense RNA to EWS-fusion transcripts. Oncogene 1995;11:1049–1054.
88. Ida K, Kobayashi S, Taki T, et al. EWS-FLI-1 and EWS-ERG chimeric mRNAs in Ewing's sarcoma and primitive neuroectodermal tumor. Int J Cancer 1995;63:500–504.
89. Jeon IS, Davis JN, Braun BS, et al. A variant Ewing's sarcoma translocation (7;22) fuses the EWS gene to the ETS gene ETV1. Oncogene 1995;10:1229–1234.
90. Peter M, Couturier J, Pacquement H, et al. A new member of the ETS family

fused to EWS in Ewing tumors. *Oncogene* 1997;14:1159–1164.

91. Ishida S, Yoshida K, Kaneko Y, et al. The genomic breakpoint and chimeric transcripts in the EWSR1-ETV4/E1AF gene fusion in Ewing sarcoma. *Cytogenet Cell Genet* 1998;82:278–283.

92. Kumar S, Pack S, Kumar D, et al. Detection of EWS-FLI-1 fusion in Ewing's sarcoma/peripheral primitive neuroectodermal tumor by fluorescence in situ hybridization using formalin-fixed paraffin-embedded tissue. *Hum Pathol* 1999;30:324–330.

93. Bridge RS, Rajaram V, Dehner LP, Pfcifer JD, Perry A. Molecular diagnosis of Ewing sarcoma/primitive neuroectodermal tumor in routinely processed tissue: a comparison of two FISH strategies and RT-PCR in malignant round cell tumors. *Mod Pathol* 2006;19:1–8.

94. Zucman J, Melot T, Desmaze C, et al. Combinatorial generation of variable fusion proteins in the Ewing family of tumours. *EMBO J* 1993;12:4481–4487.

95. West DC, Grier HE, Swallow MM, et al. Detection of circulating tumor cells in patients with Ewing's sarcoma and peripheral primitive neuroectodermal tumor. *J Clin Oncol* 1997;15:583–588.

96. Vermeulen J, Ballet S, Oberlin O, et al. Incidence and prognostic value of tumour cells detected by RT-PCR in peripheral blood stem cell collections from patients with Ewing tumour. *Br J Cancer* 2006;95:1326–1333.

97. de Alava E. Kawai A, Healey JH, et al. EWS-FLI1 fusion transcript structure is an independent determinant of prognosis in Ewing's sarcoma. *J Clin Oncol* 1998;16:1248.

98. Barr FG, Meyer WH. Role of fusion subtype in Ewing sarcoma. *J Clin Oncol* 2010;28:1973–1974.

99. Kovar H, Jug G, Aryee DN, et al. Among genes involved in the RB dependent cell cycle regulatory cascade, the p16 tumor suppressor gene is frequently lost in the Ewing family of tumors. *Oncogene* 1997;15:2225–2232.

100. Ladanyi M, Lewis R, Jhanwar SC, et al. MDM2 and CDK4 gene amplification in Ewing's sarcoma. *J Pathol* 1995;175:211–217.

101. Lawlor ER, Lim JF, Tao W, et al. The Ewing tumor family of peripheral primitive neuroectodermal tumors expresses human gastrin-releasing peptide. *Cancer Res* 1998;58:2469–2476.

102. May WA, Grigoryan R3, Keshelava N, et al. Characterization and drug resistance patterns of Ewing's sarcoma family tumor cell lines. *PloS One* 2013;8:e80060.

103. Rodriguez-Galindo C, Navid F, Khoury J, Krasin M. *Ewing Sarcoma Family of Tumors*. Berlin: Springer; 2006: 181–218.

104. Schmidt D, Herrmann C, Jurgens H, Harms D. Malignant peripheral neuroectodermal tumor and its necessary distinction from Ewing's sarcoma: a report from the Kiel Pediatric Tumor Registry. *Cancer* 1991;68:2251–2259.

105. Parham DM, Hijazi Y, Steinberg SM, et al. Neuroectodermal differentiation in Ewing's sarcoma family of tumors does not predict tumor behavior. *Hum Pathol* 1999;30:911–918.

106. Oberlin O, Patte C, Demeocq F, et al. The response to initial chemotherapy as a prognostic factor in localized Ewing's sarcoma. *Eur J Cancer Clin Oncol* 1985;21:463–467.

107. Picci P, Bohling T, Bacci G, et al. Chemotherapy-induced tumor necrosis as a prognostic factor in localized Ewing's sarcoma of the extremities. *J Clin Oncol* 1997;15:1553–1559.

108. Bacci G, Longhi A, Ferrari S, et al. Prognostic factors in non-metastatic Ewing's sarcoma tumor of bone: an analysis of 579 patients treated at a single institution with adjuvant or neoadjuvant chemotherapy between 1972 and 1998. *Acta Oncol* 2006;45:469–475.

109. Biswas B, Shukla NK, Deo SV, et al. Evaluation of outcome and prognostic factors in extraosseous Ewing sarcoma. *Pediatr Blood Cancer* 2014;61:1925–1931.

110. Aparicio J, Munarriz B, Pastor M, et al. Long-term follow-up and prognostic factors in Ewing's sarcoma: a multivariate analysis of 116 patients from a single institution. *Oncology* 1998;55:20–26.

111. Wexler LH, DeLaney TF, Tsokos M, et al. Ifosfamide and etoposide plus vincristine, doxorubicin, and cyclophosphamide for newly diagnosed Ewing's sarcoma family of tumors. *Cancer* 1996;78:901–911.

112. Jurgens H, Exner U, Gadner H, et al. Multidisciplinary treatment of primary Ewing's sarcoma of bone: a 6-year experience of a European Cooperative Trial. *Cancer* 1988;61:23–32.

113. Ozaki T, Hillmann A, Hoffmann C, et al. Significance of surgical margin on the prognosis of patients with Ewing's sarcoma: a report from the Cooperative Ewing's Sarcoma Study. *Cancer* 1996;78:892–900.

114. Evans RG. The four S's of Ewing's sarcoma. *Int J Radiat Oncol Biol Phys* 1991;21:1671–1673.

115. Thorner P, Squire J, Chilton-MacNeil S, et al. Is the EWS/FLI-1 fusion transcript specific for Ewing sarcoma and peripheral primitive neuroectodermal tumor?: a report of four cases showing this transcript in a wider range of tumor types. *Am J Pathol* 1996;148:1125–1138.

116. Ahmed AA, Nava VE, Pham T, et al. Ewing sarcoma family of tumors in unusual sites: confirmation by rt-PCR. *Pediatr Dev Pathol* 2006;9:488–495.

117. Wei S, Siegal GP. Round cell tumors of bone: an update on recent molecular genetic advances. *Adv Anat Pathol* 2014;21:359–372.

118. Marino-Enriquez A, Fletcher CD. Round cell sarcomas: biologically important refinements in subclassification. *Int J Biochem Cell Biol* 2014;53:493–504.

119. Italiano A, Sung YS, Zhang L, et al. High prevalence of CIC fusion with double-homeobox (DUX4) transcription factors in EWSR1-negative undifferentiated small blue round cell sarcomas. *Genes Chromosomes Cancer* 2012;51:207–218.

120. Sugita S, Arai Y, Tonooka A, et al. A novel CIC-FOXO4 gene fusion in undifferentiated small round cell sarcoma: a genetically distinct variant of Ewing-like sarcoma. *Am J Surg Pathol* 2014;38:1571–1576.

121. Graham C, Chilton-MacNeill S, Zielenska M, Somers GR. The CIC-DUX4 fusion transcript is present in a subgroup of pediatric primitive round cell sarcomas. *Hum Pathol* 2012;43:180–189.

122. Cohen-Gogo S, Cellier C, Coindre JM, et al. Ewing-like sarcomas with BCOR-CCNB3 fusion transcript: a clinical, radiological and pathological retrospective study from the Societe

Francaise des Cancers de L'Enfant. *Pediatr Blood Cancer* 2014;61:2191–2198.

123. Puls F, Niblett A, Marland G, et al. BCOR-CCNB3 (Ewing-like) sarcoma: a clinicopathologic analysis of 10 cases, in comparison with conventional Ewing sarcoma. *Am J Surg Pathol* 2014;38:1307–1318.

124. Pierron G, Tirode F, Lucchesi C, et al. A new subtype of bone sarcoma defined by BCOR-CCNB3 gene fusion. *Nat Genet* 2012;44:461–466.

125. Dehner LP. The evolution of the diagnosis and understanding of primitive and embryonic neoplasms in children: living through an epoch. *Mod Pathol* 1998;11:669–685.

126. Wright J. Neurocytoma or neuroblastoma, a kind of tumor not generally recognized. *J Exp Med* 1910;12:556–561.

127. Willis RA. *The Borderland of Embryology and Pathology*, 2nd edn. London: Butterworths; 1962.

128. Dehner LP. Whence the primitive neuroectodermal tumor? *Arch Pathol Lab Med* 1990;114:16–17.

129. Beckwith JB, Martin RF. Observations on the histopathology of neuroblastomas. *J Pediatr Surg* 1968;3:106–110.

130. Joshi VV, Cantor AB, Altshuler G, et al. Age-linked prognostic categorization based on a new histologic grading system of neuroblastomas: a clinicopathologic study of 211 cases from the Pediatric Oncology Group. *Cancer* 1992;69:2197–2211.

131. Shimada H, Chatten J, Newton WA, Jr., et al. Histopathologic prognostic factors in neuroblastic tumors: definition of subtypes of ganglioneuroblastoma and an age-linked classification of neuroblastomas. *J Natl Cancer Inst* 1984;73:405–416.

132. Shimada H, Ambros IM, Dehner LP, et al. The International Neuroblastoma Pathology Classification (the Shimada system). *Cancer* 1999;86:364–372.

133. Beckwith JB, Perrin EV. In situ neuroblastomas: a contribution to the natural history of neural crest tumors. *Am J Pathol* 1963;43:1089–1104.

134. Acharya S, Jayabose S, Kogan SJ, et al. Prenatally diagnosed neuroblastoma. *Cancer* 1997;80:304–310.

135. Look AT, Hayes FA, Nitschke R, McWilliams NB, Green AA. Cellular DNA content as a predictor of response to chemotherapy in infants with unresectable neuroblastoma. *N Engl J Med* 1984;311:231–235.

136. Brodeur GM, Moley JF. Biology of tumors of the peripheral nervous system. *Cancer Metast Rev* 1991;10:321–333.

137. Maris JM, Matthay KK. Molecular biology of neuroblastoma. *J Clin Oncol* 1999;17:2264–2279.

138. Sano H, Bonadio J, Gerbing RB, et al. International neuroblastoma pathology classification adds independent prognostic information beyond the prognostic contribution of age. *Eur J Cancer* 2006;42:1113–1119.

139. Navarro S, Amann G, Beiske K, et al., European Study Group T. Protocol: Prognostic value of International Neuroblastoma Pathology Classification in localized resectable peripheral neuroblastic tumors: a histopathologic study of localized neuroblastoma European Study Group 94.01 Trial and Protocol. *J Clin Oncol* 2006;24:695–699.

140. Peuchmaur M, d'Amore ES, Joshi VV, et al. Revision of the International Neuroblastoma Pathology Classification: confirmation of favorable and unfavorable prognostic subsets in ganglioneuroblastoma, nodular. *Cancer* 2003;98:2274–2281.

141. Okamatsu C, London WB, Naranjo A, et al. Clinicopathological characteristics of ganglioneuroma and ganglioneuroblastoma: a report from the CCG and COG. *Pediatr Blood Cancer* 2009;53:563–569.

142. Sano H, Gonzalez-Gomez I, Wu SQ, et al. A case of composite neuroblastoma composed of histologically and biologically distinct clones. *Pediatr Dev Pathol* 2007;10:229–232.

143. Dehner LP. Peripheral and central primitive neuroectodermal tumors: a nosologic concept seeking a consensus. *Arch Pathol Lab Med* 1986;110:997–1005.

144. Matthay KK. Neuroblastoma: a clinical challenge and biologic puzzle. *CA Cancer J Clin* 1995;45:179–192.

145. Evans AE, D'Angio GJ, Propert K, et al. Prognostic factors in neuroblastoma. *Cancer* 1987;59:1853–1859.

146. Kelly DR, Joshi VV. Neuroblastoma and related tumors. In *Pediatric Neoplasia: Morphology and Biology*. Parham DM (ed.) Philadelphia: Lippincott-Raven; 1996: 105.

147. Bolande RP. A natural immune system in pregnancy serum lethal to human neuroblastoma cells: a possible mechanism of spontaneous regression. *Perspect Pediatr Pathol* 1992;16:120–133.

148. Graeve JL, de Alarcon PA, Sato Y, Pringle K, Helson L. Miliary pulmonary neuroblastoma: a risk of autologous bone marrow transplantation? *Cancer* 1988;62:2125–2127.

149. Telles NC, Rabson AS, Pomeroy TC. Ewing's sarcoma: an autopsy study. *Cancer* 1978;41:2321–2329.

150. Brodeur GM, Pritchard J, Berthold F, et al. Revisions of the international criteria for neuroblastoma diagnosis, staging, and response to treatment. *J Clin Oncol* 1993;11:1466–1477.

151. Katzenstein HM, Kent PM, London WB, Cohn SL. Treatment and outcome of 83 children with intraspinal neuroblastoma: the Pediatric Oncology Group experience. *J Clin Oncol* 2001;19:1047–1055.

152. Escobar MA, Grosfeld JL, Powell RL, et al. Long-term outcomes in patients with stage IV neuroblastoma. *J Pediatr Surg* 2006;41:377–381.

153. Joshi VV, Silverman JF, Altshuler G, et al. Systematization of primary histopathologic and fine-needle aspiration cytologic features and description of unusual histopathologic features of neuroblastic tumors: a report from the Pediatric Oncology Group. *Hum Pathol* 1993;24:493–504.

154. Shimada H, Ambros IM, Dehner LP, et al. Terminology and morphologic criteria of neuroblastic tumors: recommendations by the International Neuroblastoma Pathology Committee. *Cancer* 1999;86:349–363.

155. Wang LL, Suganuma R, Ikegaki N, et al. Neuroblastoma of undifferentiated subtype, prognostic significance of prominent nucleolar formation, and MYC/MYCN protein expression: a report from the Children's Oncology Group. *Cancer* 2013;119:3718–3726.

156. Umehara S, Nakagawa A, Matthay KK, et al. Histopathology defines prognostic subsets of ganglioneuroblastoma, nodular. *Cancer* 2000;89:1150–1161.

157. Cozzutto C, Carbone A. Pleomorphic (anaplastic) neuroblastoma. *Arch Pathol Lab Med* 1988,112:621–625.

158. Navarro S, Noguera R, Pellin A, *et al.* Pleomorphic anaplastic neuroblastoma. *Med Pediatr Oncol* 2000;35:498–502.

159. Gonzalez-Crussi F, Hsueh W. Bilateral adrenal ganglioneuroblastoma with neuromelanin: clinical and pathologic observations. *Cancer* 1988;61:1159–1166.

160. Tsokos M, Scarpa S, Ross RA, Triche TJ. Differentiation of human neuroblastoma recapitulates neural crest development: study of morphology, neurotransmitter enzymes, and extracellular matrix proteins. *Am J Pathol* 1987;128:484–496.

161. Balazs M. Mixed pheochromocytoma and ganglioneuroma of the adrenal medulla: a case report with electron microscopic examination. *Hum Pathol* 1988;19:1352–1355.

162. Cooper R, Khakoo Y, Matthay KK, *et al.* Opsoclonus-myoclonus-ataxia syndrome in neuroblastoma: histopathologic features-a report from the Children's Cancer Group. *Med Pediatr Oncol* 2001;36:623–629.

163. Blatt J, Gula MJ, Orlando SJ, *et al.* Indolent course of advanced neuroblastoma in children older than 6 years at diagnosis. *Cancer* 1995;76:890–894.

164. Mierau GW, Berry PJ, Malott RL, Weeks DA. Appraisal of the comparative utility of immunohistochemistry and electron microscopy in the diagnosis of childhood round cell tumors. *Ultrastruct Pathol* 1996;20:507–517.

165. Dias P, Kumar P, Marsden HB, *et al.* N-myc gene is amplified in alveolar rhabdomyosarcomas (RMS) but not in embryonal RMS. *Int J Cancer* 1990;45:593–596.

166. Driman D, Thorner PS, Greenberg ML, Chilton-MacNeill S, Squire J. MYCN gene amplification in rhabdomyosarcoma. *Cancer* 1994;73:2231–2237.

167. Leader M, Collins M, Patel J, Henry K. Antineuron specific enolase staining reactions in sarcomas and carcinomas: its lack of neuroendocrine specificity. *J Clin Pathol* 1986;39:1186–1192.

168. Bielle F, Freneaux P, Jeanne-Pasquier C, *et al.* PHOX2B immunolabeling: a novel tool for the diagnosis of undifferentiated neuroblastomas among childhood small round blue-cell tumors. *Am J Surg Pathol* 2012;36:1141–1149.

169. Brodeur GM, Green AA, Hayes FA, *et al.* Cytogenetic features of human neuroblastomas and cell lines. *Cancer Res* 1981;41:4678–4686.

170. Caren H, Ejeskar K, Fransson S, *et al.* A cluster of genes located in 1p36 are down-regulated in neuroblastomas with poor prognosis, but not due to CpG island methylation. *Mol Cancer* 2005;4:10.

171. Henrich KO, Bauer T, Schulte J, *et al.* CAMTA1, a 1p36 tumor suppressor candidate, inhibits growth and activates differentiation programs in neuroblastoma cells. *Cancer Res* 2011;71:3142–3151.

172. Fransson S, Martinsson T, Ejeskar K. Neuroblastoma tumors with favorable and unfavorable outcomes: significant differences in mRNA expression of genes mapped at 1p36.2. *Genes Chromosomes Cancer* 2007;46:45–52.

173. Iehara T, Hosoi H, Akazawa K, *et al.* MYCN gene amplification is a powerful prognostic factor even in infantile neuroblastoma detected by mass screening. *Br J Cancer* 2006;94:1510–1515.

174. Shapiro DN, Valentine MB, Rowe ST, *et al.* Detection of N-myc gene amplification by fluorescence in situ hybridization: diagnostic utility for neuroblastoma. *Am J Pathol* 1993;142:1339–1346.

175. Bhargava R, Oppenheimer O, Gerald W, Jhanwar SC, Chen B. Identification of MYCN gene amplification in neuroblastoma using chromogenic in situ hybridization (CISH): an alternative and practical method. *Diagn Mol Pathol* 2005;14:72–76.

176. Look AT, Hayes FA, Shuster JJ, *et al.* Clinical relevance of tumor cell ploidy and N-myc gene amplification in childhood neuroblastoma: a Pediatric Oncology Group study. *J Clin Oncol* 1991;9:581–591.

177. Tanaka T, Sugimoto T, Sawada T. Prognostic discrimination among neuroblastomas according to Ha-ras/trk A gene expression: a comparison of the profiles of neuroblastomas detected clinically and those detected through mass screening. *Cancer* 1998;83:1626–1633.

178. Ohali A, Avigad S, Ash S, *et al.* Telomere length is a prognostic factor in neuroblastoma. *Cancer* 2006;107:1391–1399.

179. Carpenter EL, Mosse YP. Targeting ALK in neuroblastoma: preclinical and clinical advancements. *Nat Rev Clin Oncol* 2012;9:391–399.

180. Cheung NK, Zhang J, Lu C, *et al.* Association of age at diagnosis and genetic mutations in patients with neuroblastoma. *JAMA* 2012;307:1062–1071.

181. Brodeur GM, Azar C, Brother M, *et al.* Neuroblastoma: effect of genetic factors on prognosis and treatment. *Cancer* 1992;70:1685–1694.

182. Valentijn LJ, Koster J, Haneveld F, *et al.* Functional MYCN signature predicts outcome of neuroblastoma irrespective of MYCN amplification. *Proc Natl Acad Sci USA* 2012;109:19190–19195.

183. Elimam NA, Atra AA, Fayea NY, *et al.* Stage 4S neuroblastoma, a disseminated tumor with excellent outcome. *Saudi Med J* 2006;27:1734–1736.

184. Burgues O, Navarro S, Noguera R, *et al.* Prognostic value of the International Neuroblastoma Pathology Classification in Neuroblastoma (Schwannian stroma-poor) and comparison with other prognostic factors: a study of 182 cases from the Spanish Neuroblastoma Registry. *Virchows Arch* 2006;449:410–420.

185. Ordonez NG. Desmoplastic small round cell tumor. I: A histopathologic study of 39 cases with emphasis on unusual histological patterns. *Am J Surg Pathol* 1998;22:1303–1313.

186. Gerald WL, Miller HK, Battifora H, *et al.* Intra-abdominal desmoplastic small round-cell tumor: report of 19 cases of a distinctive type of high-grade polyphenotypic malignancy affecting young individuals. *Am J Surg Pathol* 1991;15:499–513.

187. Sawyer JR, Tryka AF, Lewis JM. A novel reciprocal chromosome translocation t(11;22)(p13;q12) in an intraabdominal desmoplastic small round-cell tumor. *Am J Surg Pathol* 1992;16:411–416.

188. Ladanyi M, Gerald W. Fusion of the EWS and WT1 genes in the

desmoplastic small round cell tumor. *Cancer Res* 1994;54:2837–2840.

189. Kretschmar CS, Colbach C, Bhan I, Crombleholme TM. Desmoplastic small cell tumor: a report of three cases and a review of the literature. *J Pediatr Hematol Oncol* 1996;18:293–298.

190. Ordonez NG, el-Naggar AK, Ro JY, Silva EG, Mackay B. Intra-abdominal desmoplastic small cell tumor: a light microscopic, immunocytochemical, ultrastructural, and flow cytometric study. *Hum Pathol* 1993;24:850–865.

191. Kawano N, Inayama Y, Nagashima Y, et al. Desmoplastic small round-cell tumor of the paratesticular region: report of an adult case with demonstration of EWS and WT1 gene fusion using paraffin-embedded tissue. *Mod Pathol* 1999;12:729–734.

192. Adsay V, Cheng J, Athanasian E, Gerald W, Rosai J. Primary desmoplastic small cell tumor of soft tissues and bone of the hand. *Am J Surg Pathol* 1999;23:1408–1413.

193. Bian Y, Jordan AG, Rupp M, et al. Effusion cytology of desmoplastic small round cell tumor of the pleura: a case report. *Acta Cytol* 1993;37:77–82.

194. Tison V, Cerasoli S, Morigi F, et al. Intracranial desmoplastic small-cell tumor. Report of a case. *Am J Surg Pathol* 1996;20:112–117.

195. Karavitakis EM, Moschovi M, Stefanaki K, et al. Desmoplastic small round cell tumor of the pleura. *Pediatr Blood Cancer* 2007;49:335–338.

196. Wang LL, Perlman EJ, Vujanic GM, et al. Desmoplastic small round cell tumor of the kidney in childhood. *Am J Surg Pathol* 2007;31:576–584.

197. Finke NM, Lae ME, Lloyd RV, Gehani SK, Nascimento AG. Sinonasal desmoplastic small round cell tumor: a case report. *Am J Surg Pathol* 2002;26:799–803.

198. Schmidt D, Koster E, Harms D. Intraabdominal desmoplastic small-cell tumor with divergent differentiation: clinicopathological findings and DNA ploidy. *Med Pediatr Oncol* 1994;22:97–102.

199. Lettieri CK, Garcia-Filion P, Hingorani P. Incidence and outcomes of desmoplastic small round cell tumor: results from the surveillance, epidemiology, and end results database. *J Cancer Epidemiol* 2014;2014:680126.

200. Zhang J, Dalton J, Fuller C. Epithelial marker-negative desmoplastic small round cell tumor with atypical morphology: definitive classification by fluorescence in situ hybridization. *Arch Pathol Lab Med* 2007;131:646–649.

201. Ordonez NG. Desmoplastic small round cell tumor. II: An ultrastructural and immunohistochemical study with emphasis on new immunohistochemical markers. *Am J Surg Pathol* 1998;22:1314–1327.

202. Murphy AJ, Bishop K, Pereira C, et al. A new molecular variant of desmoplastic small round cell tumor: significance of WT1 immunostaining in this entity. *Hum Pathol* 2008;39:1763–1770.

203. Trupiano JK, Machen SK, Barr FG, Goldblum JR. Cytokeratin-negative desmoplastic small round cell tumor: a report of two cases emphasizing the utility of reverse transcriptase-polymerase chain reaction. *Mod Pathol* 1999;12:849–853.

204. Roberts DJ, Haber D, Sklar J, Crum CP. Extrarenal Wilms' tumors: a study of their relationship with classical renal Wilms' tumor using expression of WT1 as a molecular marker. *Lab Invest* 1993;68:528–536.

205. de Alava E, Ladanyi M, Rosai J, Gerald WL. Detection of chimeric transcripts in desmoplastic small round cell tumor and related developmental tumors by reverse transcriptase polymerase chain reaction: a specific diagnostic assay. *Am J Pathol* 1995;147:1584–1591.

206. Argatoff LH, O'Connell JX, Mathers JA, Gilks CB, Sorensen PH. Detection of the EWS/WT1 gene fusion by reverse transcriptase-polymerase chain reaction in the diagnosis of intra-abdominal desmoplastic small round cell tumor. *Am J Surg Pathol* 1996;20:406–412.

207. Katz RL, Quezado M, Senderowicz AM, et al. An intra-abdominal small round cell neoplasm with features of primitive neuroectodermal and desmoplastic round cell tumor and a EWS/FLI-1 fusion transcript. *Hum Pathol* 1997;28:502–509.

208. Gardner LJ, Ayala AG, Monforte HL, Dunphy CH. Ewing sarcoma/peripheral primitive neuroectodermal tumor: adult abdominal tumors with an Ewing sarcoma gene rearrangement demonstrated by fluorescence in situ hybridization in paraffin sections. *Appl Immunohistochem Mol Morphol* 2004;12:160–165.

209. Ordi J, de Alava E, Torne A, et al. Intraabdominal desmoplastic small round cell tumor with EWS/ERG fusion transcript. *Am J Surg Pathol* 1998;22:1026–1032.

210. Alaggio R, Rosolen A, Sartori F, et al. Spindle cell tumor with EWS-WT1 transcript and a favorable clinical course: a variant of DSCT, a variant of leiomyosarcoma, or a new entity?: report of 2 pediatric cases. *Am J Surg Pathol* 2007;31:454–459.

211. Amato RJ, Ellerhorst JA, Ayala AG. Intraabdominal desmoplastic small round cell tumor. Report and discussion of five cases. *Cancer* 1996;78:845–851.

212. Saab R, Khoury JD, Krasin M, Davidoff AM, Navid F. Desmoplastic small round cell tumor in childhood: the St. Jude Children's Research Hospital experience. *Pediatr Blood Cancer* 2007;49:274–279.

213. Coppes MJ, Wilson PC, Weitzman S. Extrarenal Wilms' tumor: staging, treatment, and prognosis. *J Clin Oncol* 1991;9:167–174.

214. Arnold MA, Schoenfield L, Limketkai BN, Arnold CA. Diagnostic pitfalls of differentiating desmoplastic small round cell tumor (DSRCT) from Wilms tumor (WT): overlapping morphologic and immunohistochemical features. *Am J Surg Pathol* 2014;38:1220–1226.

215. Biegel JA, Busse TM, Weissman BE. SWI/SNF chromatin remodeling complexes and cancer. *Am J Med Genet C Semin Med Genet* 2014;166c:350–366.

216. Hasselblatt M, Gesk S, Oyen F, et al. Nonsense mutation and inactivation of SMARCA4 (BRG1) in an atypical teratoid/rhabdoid tumor showing retained SMARCB1 (INI1) expression. *Am J Surg Pathol* 2011;35:933–935.

217. Parham DM, Weeks DA, Beckwith JB. The clinicopathologic spectrum of putative extrarenal rhabdoid tumors: an analysis of 42 cases studied with immunohistochemistry or electron microscopy. *Am J Surg Pathol* 1994;18:1010–1029.

218. Leong FJ, Leong AS. Malignant rhabdoid tumor in adults–heterogenous tumors with a unique morphological

219. White FV, Dehner LP, Belchis DA, et al. Congenital disseminated malignant rhabdoid tumor: a distinct clinicopathologic entity demonstrating abnormalities of chromosome 22q11. *Am J Surg Pathol* 1999;23:249–256.

220. Simons J, Teshima I, Zielenska M, et al. Analysis of chromosome 22q as an aid to the diagnosis of rhabdoid tumor: a case report. *Am J Surg Pathol* 1999;23:982–988.

221. Beckwith JB, Palmer NF. Histopathology and prognosis of Wilms tumors: results from the First National Wilms' Tumor Study. *Cancer* 1978;41:1937–1948.

222. Haas JE, Palmer NF, Weinberg AG, Beckwith JB. Ultrastructure of malignant rhabdoid tumor of the kidney: a distinctive renal tumor of children. *Hum Pathol* 1981;12:646–657.

223. Balaton AJ, Vaury P, Videgrain M. Paravertebral malignant rhabdoid tumor in an adult: a case report with immunocytochemical study. *Pathol Res Pract* 1987;182:713–718.

224. Tsuneyoshi M, Daimaru Y, Hashimoto H, Enjoji M. Malignant soft tissue neoplasms with the histologic features of renal rhabdoid tumors: an ultrastructural and immunohistochemical study. *Hum Pathol* 1985;16:1235–1242.

225. Gonzalez-Crussi F, Goldschmidt RA, Hsueh W, Trujillo YP. Infantile sarcoma with intracytoplasmic filamentous inclusions: distinctive tumor of possible histiocytic origin. *Cancer* 1982;49:2365–2375.

226. Lemos LB, Hamoudi AB. Malignant thymic tumor in an infant (malignant histiocytoma). *Arch Pathol Lab Med* 1978;102:84–89.

227. Hajdu S. *Pathology of Soft Tissue Tumors*. Philadelphia: Lea & Febiger; 1979.

228. Kodet R, Newton WA, Jr., Hamoudi AB, Asmar L. Rhabdomyosarcomas with intermediate-filament inclusions and features of rhabdoid tumors: light microscopic and immunohistochemical study. *Am J Surg Pathol* 1991;15:257–267.

229. Tsuneyoshi M, Daimaru Y, Hashimoto H, Enjoji M. The existence of rhabdoid cells in specified soft tissue sarcomas: histopathological, ultrastructural and immunohistochemical evidence. *Virchows Arch A Pathol Anat Histopathol* 1987;411:509–514.

230. Tsokos M, Kouraklis G, Chandra RS, Bhagavan BS, Triche TJ. Malignant rhabdoid tumor of the kidney and soft tissues: avidence for a diverse morphological and immunocytochemical phenotype. *Arch Pathol Lab Med* 1989;113:115–120.

231. Fanburg-Smith JC, Hengge M, Hengge UR, Smith JS, Jr., Miettinen M. Extrarenal rhabdoid tumors of soft tissue: a clinicopathologic and immunohistochemical study of 18 cases. *Ann Diagn Pathol* 1998;2:351–362.

232. Rorke LB, Packer RJ, Biegel JA. Central nervous system atypical teratoid/rhabdoid tumors of infancy and childhood: definition of an entity. *J Neurosurg* 1996;85:56–65.

233. Hollmann TJ, Hornick JL. INI1-deficient tumors: diagnostic features and molecular genetics. *Am J Surg Pathol* 2011;35:e47–e63.

234. Wick MR, Ritter JH, Dehner LP. Malignant rhabdoid tumors: a clinicopathologic review and conceptual discussion. *Semin Diagn Pathol* 1995;12:233–248.

235. Chung CJ, Cammoun D, Munden M. Rhabdoid tumor of the kidney presenting as an abdominal mass in a newborn. *Pediatr Radiol* 1990;20:562–563.

236. Kent AL, Mahoney DH, Jr., Gresik MV, Steuber CP, Fernbach DJ. Malignant rhabdoid tumor of the extremity. *Cancer* 1987;60:1056–1059.

237. Dabbs DJ, Park HK. Malignant rhabdoid skin tumor: an uncommon primary skin neoplasm. Ultrastructural and immunohistochemical analysis. *J Cutan Pathol* 1988;15:109–115.

238. Dominey A, Paller AS, Gonzalez-Crussi F. Congenital rhabdoid sarcoma with cutaneous metastases. *J Am Acad Dermatol* 1990;22:969–974.

239. Perez-Atayde AR, Newbury R, Fletcher JA, Barnhill R, Gellis S. Congenital "neurovascular hamartoma" of the skin: a possible marker of malignant rhabdoid tumor. *Am J Surg Pathol* 1994;18:1030–1038.

240. Jayabose S, Iqbal K, Newman L, et al. Hypercalcemia in childhood renal tumors. *Cancer* 1988;61:788–791.

241. Mayes LC, Kasselberg AG, Roloff JS, Lukens JN. Hypercalcemia associated with immunoreactive parathyroid hormone in a malignant rhabdoid tumor of the kidney (rhabdoid Wilms' tumor). *Cancer* 1984;54:882–884.

242. Rousseau-Merck MF, Nogues C, Roth A, et al. Hypercalcemic infantile renal tumors: morphological, clinical, and biological heterogeneity. *Pediatr Pathol* 1985;3:155–164.

243. Bonnin JM, Rubinstein LJ, Palmer NF, Beckwith JB. The association of embryonal tumors originating in the kidney and in the brain: a report of seven cases. *Cancer* 1984;54:2137–2146.

244. Chang CH, Ramirez N, Sakr WA. Primitive neuroectodermal tumor of the brain associated with malignant rhabdoid tumor of the liver: a histologic, immunohistochemical, and electron microscopic study. *Pediatr Pathol* 1989;9:307–319.

245. Weeks DA, Beckwith JB, Mierau GW, Luckey DW. Rhabdoid tumor of kidney: a report of 111 cases from the National Wilms' Tumor Study Pathology Center. *Am J Surg Pathol* 1989;13:439–458.

246. Suzuki A, Ohta S, Shimada M. Gene expression of malignant rhabdoid tumor cell lines by reverse transcriptase-polymerase chain reaction. *Diagn Mol Pathol* 1997;6:326–332.

247. Hoot AC, Russo P, Judkins AR, Perlman EJ, Biegel JA. Immunohistochemical analysis of hSNF5/INI1 distinguishes renal and extra-renal malignant rhabdoid tumors from other pediatric soft tissue tumors. *Am J Surg Pathol* 2004;28:1485–1491.

248. Fuller CE, Pfeifer J, Humphrey P, et al. Chromosome 22q dosage in composite extrarenal rhabdoid tumors: clonal evolution or a phenotypic mimic? *Hum Pathol* 2001;32:1102–1108.

249. Wyatt-Ashmead J, Kleinschmidt-DeMasters B, Mierau GW, et al. Choroid plexus carcinomas and rhabdoid tumors: phenotypic and genotypic overlap. *Pediatr Dev Pathol* 2001;4:545–549.

250. Judkins AR, Burger PC, Hamilton RL, et al. INI1 protein expression distinguishes atypical teratoid/rhabdoid

251. Oda Y, Tsuneyoshi M. Extrarenal rhabdoid tumors of soft tissue: clinicopathological and molecular genetic review and distinction from other soft-tissue sarcomas with rhabdoid features. *Pathol Int* 2006;56:287–295.

252. Modena P, Lualdi E, Facchinetti F, et al. SMARCB1/INI1 tumor suppressor gene is frequently inactivated in epithelioid sarcomas. *Cancer Res* 2005;65:4012–4019.

253. Donner LR, Wainwright LM, Zhang F, Biegel JA. Mutation of the INI1 gene in composite rhabdoid tumor of the endometrium. *Hum Pathol* 2007;38:935–939.

254. Cho YM, Choi J, Lee OJ, et al. SMARCB1/INI1 missense mutation in mucinous carcinoma with rhabdoid features. *Pathol Int* 2006;56:702–706.

255. Douglass EC, Valentine M, Rowe ST, et al. Malignant rhabdoid tumor: a highly malignant childhood tumor with minimal karyotypic changes. *Genes Chromosomes Cancer* 1990;2:210–216.

256. Biegel JA, Rorke LB, Emanuel BS. Monosomy 22 in rhabdoid or atypical teratoid tumors of the brain. *N Engl J Med* 1989;321:906.

257. Bhattacharjee MB, Armstrong DD, Vogel H, Cooley LD. Cytogenetic analysis of 120 primary pediatric brain tumors and literature review. *Cancer Genet Cytogenet* 1997;97:39–53.

258. Schofield DE, Beckwith JB, Sklar J. Loss of heterozygosity at chromosome regions 22q11-12 and 11p15.5 in renal rhabdoid tumors. *Genes Chromosomes Cancer* 1996;15:10–17.

259. Biegel JA, Allen CS, Kawasaki K, et al. Narrowing the critical region for a rhabdoid tumor locus in 22q11. *Genes Chromosomes Cancer* 1996;16:94–105.

260. Versteege I, Sevenet N, Lange J, et al. Truncating mutations of hSNF5/INI1 in aggressive paediatric cancer. *Nature* 1998;394:203–206.

261. Biegel JA. Molecular genetics of atypical teratoid/rhabdoid tumor. *Neurosurg Focus* 2006;20:E11.

262. Janson K, Nedzi LA, David O, et al. Predisposition to atypical teratoid/rhabdoid tumor due to an inherited INI1 mutation. *Pediatr Blood Cancer* 2006;47:279–284.

263. Garvin AJ, Re GG, Tarnowski BI, Hazen-Martin DJ, Sens DA. The G401 cell line, utilized for studies of chromosomal changes in Wilms' tumor, is derived from a rhabdoid tumor of the kidney. *Am J Pathol* 1993;142:375–380.

264. Reid LH, Davies C, Cooper PR, et al. A 1-Mb physical map and PAC contig of the imprinted domain in 11p15.5 that contains TAPA1 and the BWSCR1/WT2 region. *Genomics* 1997;43:366–375.

265. Sabbioni S, Barbanti-Brodano G, Croce CM, Negrini M. GOK: a gene at 11p15 involved in rhabdomyosarcoma and rhabdoid tumor development. *Cancer Res* 1997;57:4493–4497.

266. Fruhwald MC, Hasselblatt M, Wirth S, et al. Non-linkage of familial rhabdoid tumors to SMARCB1 implies a second locus for the rhabdoid tumor predisposition syndrome. *Pediatr Blood Cancer* 2006;47:273–278.

267. Bourdeaut F, Freneaux P, Thuille B, et al. hSNF5/INI1-deficient tumours and rhabdoid tumours are convergent but not fully overlapping entities. *J Pathol* 2007;211:323–330.

268. Bittesini L, Dei Tos AP, Fletcher CD. Metastatic malignant melanoma showing a rhabdoid phenotype: further evidence of a non-specific histological pattern. *Histopathology* 1992;20:167–170.

Chapter 32

Alveolar soft part sarcoma

Markku Miettinen
National Cancer Institute, National Institutes of Health

This rare, histologically distinctive, and histogenetically enigmatic sarcoma with a compartmental pattern, having a predilection for children and young adults, originally was described by Christopherson, Foote, and Stewart in Memorial Sloan-Kettering Cancer Center in 1952.[1] Previously this tumor was often considered a peripheral example of a nonchromaffin paraganglioma or an organoid variant of malignant granular cell tumor. There is no normal cell counterpart known to this tumor.

Clinical features of alveolar soft part sarcoma in soft tissue

Alveolar soft part sarcoma (ASPS) is a very rare tumor, constituting <1% of all soft tissue sarcomas (0.8–0.9%) in two tertiary center series, including one from a large cancer hospital.[2,3] Based on Armed Forces Institute of Pathology (AFIP) statistics from >250 patients, 39% of the cases occur in patients younger than the age of 20 years. Childhood examples have a female predominance, whereas those in the adults >30 years have a predilection for men (Figure 32.1), as was previously observed in one series.[4]

ASPS occurs in a wide variety of soft tissue locations (Figure 32.2). Based on AFIP data, more than one-third of patients present with tumors in the lower extremities, especially the thigh and buttock. Childhood tumors are more common in the head and neck (especially tongue) and upper extremities, whereas examples in adults occur more often in the lower extremities and trunk wall. Ten percent of adult cases occur in visceral locations, most commonly in the gynecologic tract.

Most soft tissue ASPSs are located in deep subfascial tissues. These tumors are not often clinically distinctive. Some examples simulate arteriovenous malformations or other vascular tumors presenting as pulsatile masses.[5] Many patients present with metastatic disease, and metastases have been detected at presentation in >50% of patients in some cancer hospital series.[4] Pulmonary and sometimes brain metastases develop in a high percentage of patients in long-term follow-up. For example, the 5-year survival in Memorial Sloan-Kettering Cancer Center series was 38%, but the expected 25-year survival rate fell to 10%, reflecting continuous development of metastases during long-term follow-up.[2] Among patients with orbital ASPS, only two patients who either died or were followed >10 years died of the tumor.[6] In this series, relatively small tumor size could have been a contributing favorable factor, although tumor sizes were not given. In one case, brain metastasis developed 33 years after surgery.[7] A recent M. D. Anderson Hospital and Tumor Institute series suggested that localized tumors at presentation have a better prognosis, although 10% of these patients developed metastases during a 9-year median follow-up period.[4]

Aggressive surgery is generally thought to be the most important part of the treatment, because these tumors seem to be resistant to sarcoma chemotherapies.[4–8] Surgical treatment should include metastasectomy when possible, because many patients live long after successful surgery. Small tumor size (<5 cm) is a favorable prognostic factor, and two studies have shown a better prognosis in children,[9,10] although long-term follow-up data were scant. Childhood tumors in the tongue also have a good prognosis, probably contributed to by the small tumor size.[11]

Clinical features in visceral sites

The most common primary visceral involvement for ASPS is in the gynecologic tract, where most examples have occurred

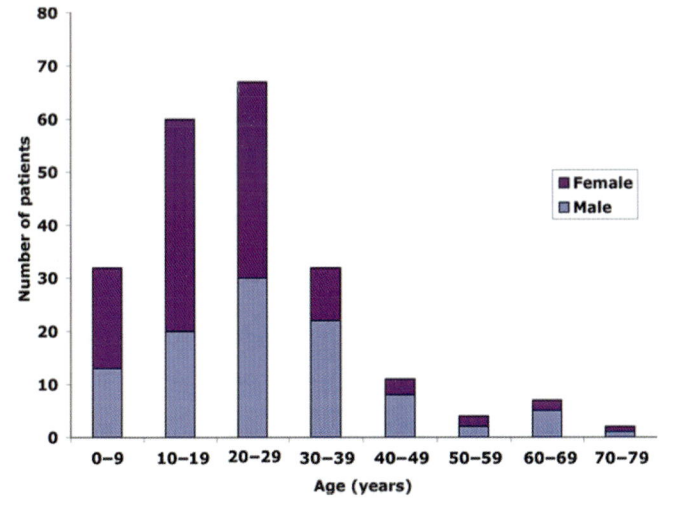

Figure 32.1 Age and sex distribution of 252 alveolar soft part sarcomas (ASPSs) according to AFIP data.

Modern Soft Tissue Pathology, Second Edition, ed. Markku Miettinen. Published by Cambridge University Press. © Cambridge University Press 2016.

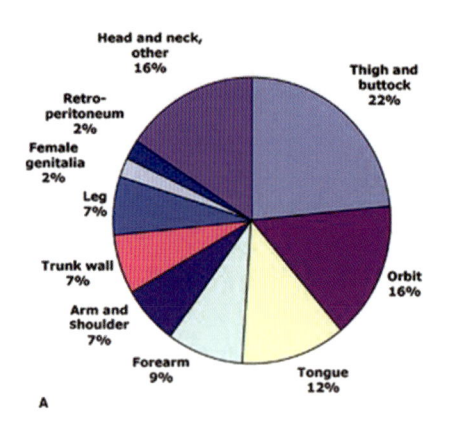

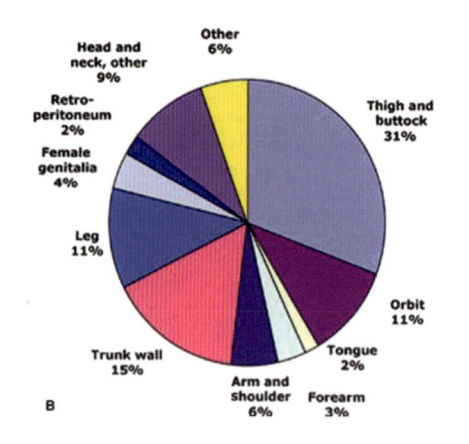

Figure 32.2 Anatomic distribution of ASPS. (**A**) Children (<19 years). (**B**) Adults (≥19 years).

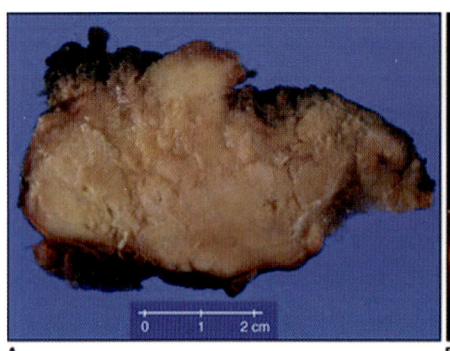

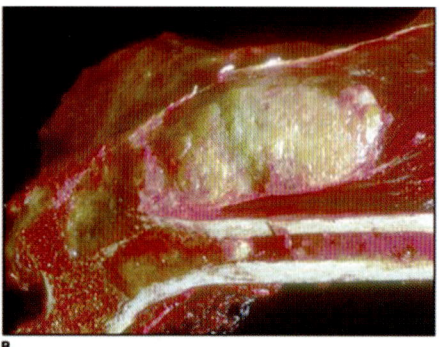

Figure 32.3 Gross appearance of ASPS. (**A**) A fixed specimen shows a yellow-brown surface on sectioning. (**B**) An unfixed sectioned surface of an example from the periarticular upper thigh has a creamy yellowish color.

in the uterus and vagina, usually in premenopausal women, and sometimes in children. Among these sites, occurrence in the uterine cervix seems to be more common than in the vagina and uterine corpus. Three-quarters of the published cases have been <5 cm, and the metastatic rate has been low, although only a few patients have had long-term follow-up.[12,13]

In the gastrointestinal tract, rare cases have been reported in the stomach in patients >50 years of age. One patient survived 10 years.[14,15] There are no reports in the intestines. The author and colleagues have not encountered any ASPSs in the gastrointestinal tract among the large numbers of mesenchymal tumors examined, but have seen one apparent primary example in the liver. Recently, hepatic occurrence has been reported in a 47-year-old man, who died 2 years later with widespread metastases.[16]

Mediastinal primary tumors have been reported in young adults in both the anterior and posterior mediastinum with concomitant pulmonary metastases.[17] Examples thought to be primary pulmonary ASPSs have been reported,[18] and an apparent primary cardiac ASPS was diagnosed in an 11-year-old girl.[19]

Examples thought to be primary osseous ASPS have occurred in the ilium, femur, and fibula.[20–22] One series also pointed out that a significant percentage (23%) of ASPSs have bone involvement, and several were clinically considered bone tumors.[7] The possibility of metastasis from an occult primary tumor should be ruled out before the diagnosis of a primary visceral (especially pulmonary) or osseous ASPS is made.

Pathology

Grossly, the tumor is circumscribed, soft to rubbery, and varies from yellow to brownish and grey; it can be discolored by hemorrhage (Figure 32.3). The tumors in the head and neck are typically smaller (1–3 cm), compared with those located deep in the extremities, which often reach 10 cm or more.

Histologically highly distinctive is a compartmental pattern, often with central empty space leading to an alveolar appearance. The tumor cells have variably eosinophilic cytoplasm (Figure 32.4). The compartments are typically separated by delicate, thin-walled or gaping capillaries, which can appear empty or, less often, hyperemic. In some tumors, several compartmental units form lobules separated by thick fibrous septa, creating a microlobular pattern. Some tumors are composed of very small compartments, resulting in a nearly solid appearance, but clusters of tumor cells are separated by a fine capillary network in this variant as well (Figure 32.5).

Some cases show extensive degenerative changes, such as hyalinization and reactive vascular proliferation, but coagulative necrosis is uncommon. Vascular invasion around the tumor is a common finding.

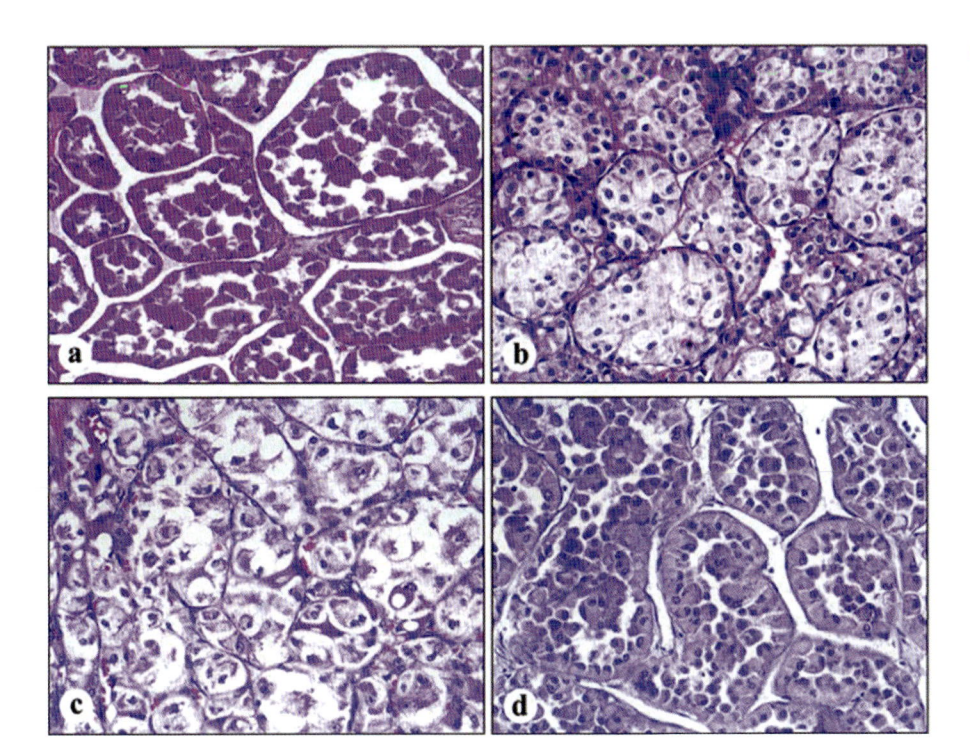

Figure 32.4 Typical compartmental pattern of ASPS with pseudoalveolar spaces.

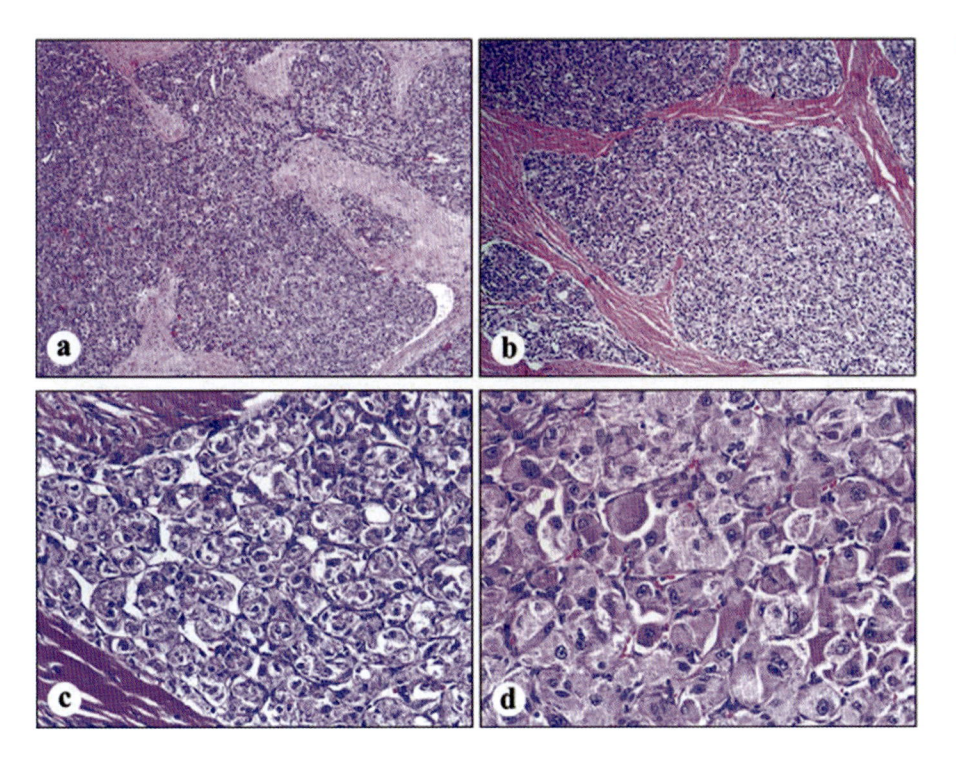

Figure 32.5 Solid variant of ASPS. A vague compartmental pattern resembles that of paraganglioma.

The tumor cell nuclei are usually round and uniform, with a delicate chromatin pattern and a single mildly prominent nucleolus. In rare variants, the tumor cells have focally or extensively clear cytoplasm, and occasional examples have significant nuclear atypia that usually is uncommon in this tumor. Focal calcifications and psammoma bodies occur infrequently (Figure 32.6).

The cytoplasm is rich in glycogen and is therefore periodic-acid–Schiff (PAS) positive. Distinctive PAS-positive, diastase-resistant cytoplasmic crystals are found in 70% to 80% of cases, and the tumor cells, containing large amounts of crystals, are diffusely PAS positive (Figure 32.7). The crystals can be only a focal finding, and multiple sections are sometimes needed for their detection. Rarely, there is focal nuclear pleomorphism

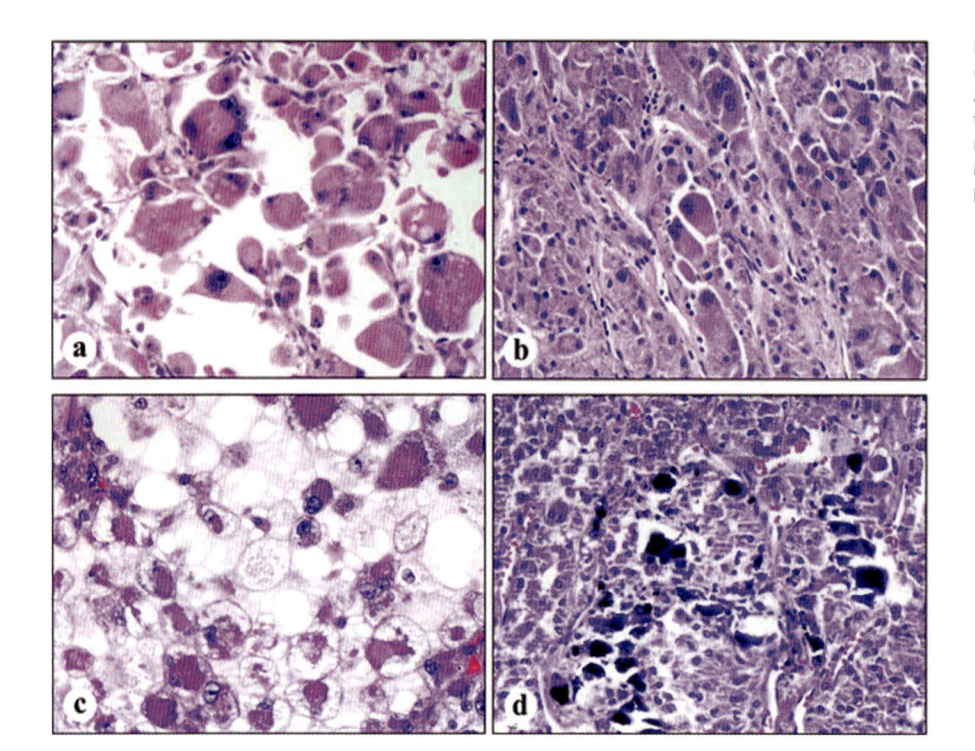

Figure 32.6 Unusual histologic features of ASPS. (**a,b**) Multinucleated tumor giant cells with nuclear atypia and prominent nucleoli are an uncommon finding. (**c**) Cytoplasmic clear cell features resembling those of renal cell carcinoma. (**d**) Microcalcifications, some of which resemble psammoma bodies, are sometimes present.

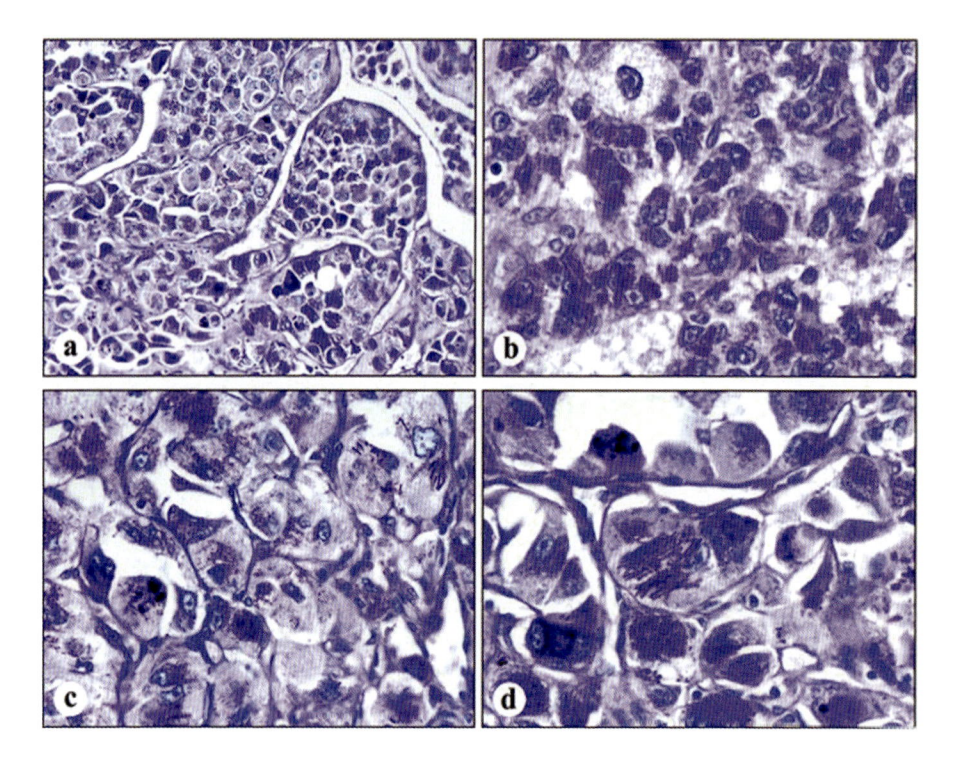

Figure 32.7 PAS-positive cytoplasmic crystals are seen in 70% to 80% of ASPSs. They show a needle-shaped morphology at a high histologic magnification.

with hyperchromasia or nuclear complexity, sometimes with pseudoinclusions. Mitoses and necrosis are distinctly uncommon in this tumor, except in rare examples.[1,2,23]

Immunohistochemistry

Three markers of special interest in this tumor are TFE3, CD147, and desmin. Nuclear positivity for TFE3 transcription factor is typical of this tumor (Figure 32.8). TFE3 is rearranged by an ASPL-TFE3 fusion (see Genetics).[24] This characteristic finding, however, is not totally specific, and most important, it is also seen in renal carcinomas with translocations that involve TFE3, and in some perivascular epithelioid cell tumors (PEComas).

Monocarboxylate transporter 1 (MCT1) and its chaperone protein CD147 antigen are coexpressed and colocalized in the cytoplasmic crystals of ASPS. In addition, MCT1 also shows a membrane-staining pattern. These antigens appear either to be overexpressed and overproduced, or trapped in tumor cells by defective intracellular processing.[25]

Varying desmin positivity has been observed in more than one-half of the cases. The percentage of positive cases in various series has ranged from 0% to 100%, which probably reflects problems previously faced in desmin detection in formalin-fixed tissue, because desmin has been more consistently demonstrated in frozen sections with various antibodies.[26–32]

Smooth muscle actin and muscle actin positivity has been detected in some cases, and sarcomeric actins have been reported. The expression of muscle markers has raised the question of skeletal muscle relationship, and a possible relation to rhabdomyosarcoma has been suggested.[33] Although an initial observation based on one case seemed to indicate MyoD1 positivity and expression by Western blotting,[34] larger series have not confirmed expression of MyoD1 and myogenin, the skeletal muscle cell transcriptional regulators in the tumor cell nuclei. Cytoplasmic MyoD1 positivity is common, however,[35,36] raising the possibility of aberrant cytoplasmic expression of these proteins.

Unlike most other sarcomas, the tumor cells show only limited, if any, vimentin positivity. They can be (weakly) positive for S100 protein, but have been almost uniformly negative for chromogranin, synaptophysin, neurofilaments, various neuropeptides, and keratins.[30,31]

Ultrastructure

Cytoplasmic-membrane-bound rhomboid crystals with internal 70 Å to 100 Å periodicity are a characteristic feature seen in most cases (Figure 32.9).[30,37,38] These crystals contain filamentous structures, possibly actin related.[38] Myofilaments, as expected in true myosarcomas, are not features of this tumor. Electron microscopic studies have ruled out a relationship with paragangliomas and granular cell tumors.[37]

Histogenesis

The histogenesis of this tumor remains obscure, but a relationship with skeletal muscle cells has been suggested, based on expression of desmin and that of myogenic regulators and possibly sarcomeric actin.[33] Carstens[39] noted similar crystals in muscle spindles but was cautious to invoke a histogenetic relationship. It is also possible that the unique genetic makeup of this tumor (see Genetics) has created an unparalleled

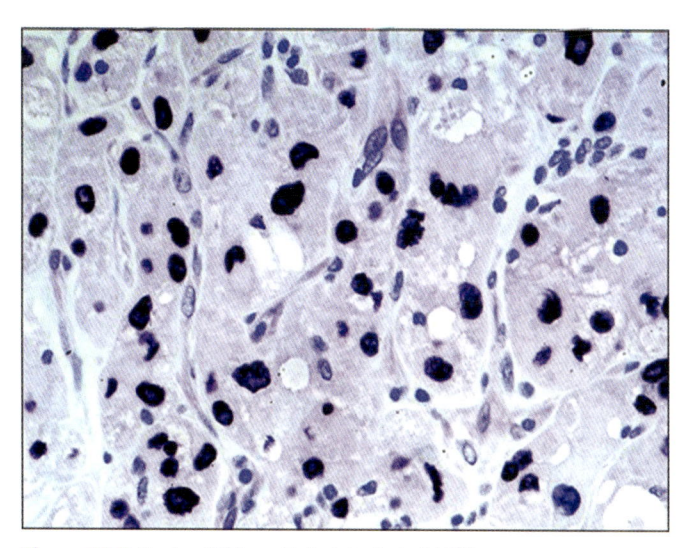

Figure 32.8 Nuclear TFE3 positivity is typical of ASPS.

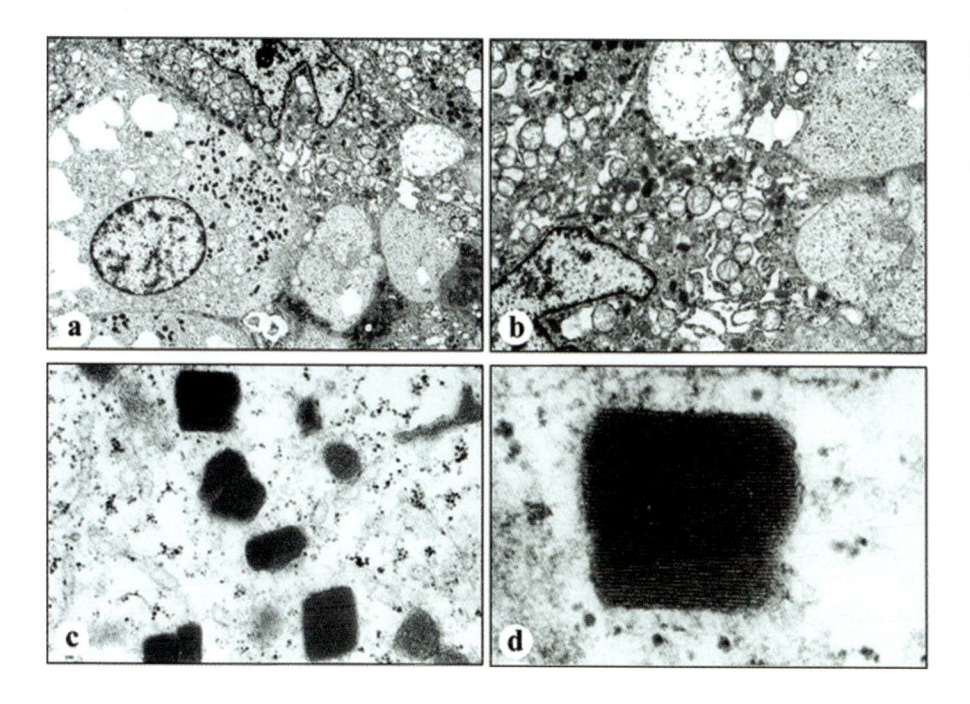

Figure 32.9 Electron microscopy demonstrates that alveolar soft part sarcoma cells contain cytoplasmic crystals that at a high magnification reveal linear lamellar structures with distinct periodicity. The lower magnifications also show abundant mitochondria.

phenotypic pattern not comparable to any normal cell type. No special studies have supported older theories on the relationship of this tumor to granular cell tumor or paraganglioma.

Differential diagnosis

Other tumors with compartmental, granular, or clear cell pattern that enter into the differential diagnosis of ASPS include metastatic melanoma, granular cell tumor, paraganglioma, renal cell carcinoma, adrenocortical carcinoma, and PEComa. Any tumor with a compartmental pattern that displays prominent nuclear atypia and has significant mitotic activity is probably not ASPS.

Melanoma usually shows a more variable histology with a greater nuclear atypia, less regular compartmental pattern, and expression of melanoma markers, at least S100 protein, and often others (GP100/HMB45, and melan A). Granular cell tumors, especially the malignant variety, have a more solid architecture, a tendency for spindling, and are usually strongly S100 protein positive.

Paraganglioma can enter into the differential diagnosis, especially in the head and neck area. It typically has a "biphasic" population composed of chief cells positive for the neuroendocrine markers chromogranin and synaptophysin and sustentacular cells in the periphery of the compartments, positive for S100 protein.

Renal cell carcinoma, primary or metastatic, can be highly vascularized with a compartmental pattern. It is typically keratin and EMA positive. Adrenocortical carcinomas can have a compartmental pattern resembling that of ASPS; however, PAS-positive crystals are absent. In contrast, there is usually immunoreactivity for melan A, inhibin, or both.

PEComa usually presents in the abdomen, especially in the retroperitoneum, pelvis, and gynecologic tract. PEComa is composed of clear cells with a vague compartmental pattern, usually not with alveolar spaces. The tumor cells are by definition variably HMB45 positive, and sometimes also melan A positive, while variably coexpressing smooth muscle markers and estrogen receptors. Notably some PEComas as well as other tumors can have nuclear TFE3 positivity so that this marker cannot be used as the sole identification for ASPS, especially if morphology is unusual. Demonstration of TFE3 gene rearrangement by FISH or a PCR-based translocation test may be required in problem cases.

Genetics

Rearrangement of 17q25 was the recurrent chromosomal aberration initially detected by cytogenetics and FISH.[40–42] Subsequently, a nonbalanced (nonreciprocal) translocation t(X;17)(p11.2q25) was identified with loss of one copy of the terminal 17q and gain of Xp, based on spectral karyotyping or similar techniques, to allow specific identification of chromosomal elements not identifiable by morphology alone.[43,44] Cloning of this translocation revealed fusion of the TFE3 transcription factor gene in Xp11 with a new widely expressed gene named *alveolar soft part sarcoma locus/alveolar soft part sarcoma chromosomal region* (ASPL/ASPSCR1) in 17q25.[45] Recurrent gains in 1q, 8q, and 12q were also seen by CGH; these probably represent secondary chromosomal changes.[46]

The female predominance of ASPS has been explained by the hypothesis that women have a greater likelihood to develop X chromosome-related mutations because of the two copies, and the X chromosomal elements remain active in the situation of an autosomally translocated X chromosome fragment (chromosome 17) free from the normally expected random X chromosome inactivation.[47] The ASPS translocation seems to activate the mesenchymal-epithelial transition factor (MET) receptor tyrosine kinase pathway, and the search for treatment targets for ASPS therefore includes the MET signaling pathway.[48]

Detection of the t(X;17) translocation or the resulting gene fusion is a new gold standard for the definition of this tumor.[49] A small subset of renal carcinomas especially seen in children, but sometimes also in adult patients shows a similar but balanced (reciprocal) translocation, t(X;17)(p11.2q25), among other gene fusions involving TFE3.[50]

References

1. Christopherson WM, Foote FW, Stewart FW. Alveolar soft-part sarcomas; structurally characteristic tumors of uncertain histogenesis. *Cancer* 1952;5:100–111.
2. Lieberman PH, Brennan MF, Kimmel M, et al. Alveolar soft-part sarcoma: a clinico-pathologic study of half a century. *Cancer* 1989;63:1–13.
3. Anderson ME, Hornicek FJ, Gebhardt MC, Raskin KA, Mankin HJ. Alveolar soft part sarcoma: a rare and enigmatic entity. *Clin Orthop Rel Res* 2005;438:144–148.
4. Portera CA Jr, Ho V, Patel SR, et al. Alveolar soft part sarcoma. Clinical course and patterns of metastasis in 70 patients treated at a single institution. *Cancer* 2001;91:585–591.
5. Temple HT, Scully SP, O'Keefe RJ, Rosenthal DI, Mankin HJ. Clinical presentation of alveolar soft-part sarcoma. *Clin Orthop Rel Res* 1994;300:213–218.
6. Font RL, Jurco S 3rd, Zimmerman LE. Alveolar soft-part sarcoma of the orbit: a clinicopathologic analysis of seventeen cases and a review of the literature. *Hum Pathol* 1982;13:569–579.
7. Lillehei KO, Kleinschmidt-DeMasters B, Mitchell DH, Spector E, Kruse CA. Alveolar soft-part sarcoma: an unusually long interval between presentation and brain metastasis. *Hum Pathol* 1993;24:1030–1034.
8. Ogose A, Yazawa Y, Ueda T, et al. Alveolar soft part sarcoma in Japan: multiinstitutional study of 57 patients from the Japanese Musculoskeletal Oncology Group. *Oncology* 2003;65:7–13.
9. Pappo AS, Parham DM, Cain A, et al. Alveolar soft part sarcoma in children and adolescents: clinical features and outcome of 11 patients. *Med Pediatr Oncol* 1996;26:81–84.

10. Kayton ML, Meyers P, Wexler LH, Gerald WL, LaQuaglia MP. Clinical presentation, treatment, and outcome of alveolar soft part sarcoma in children, adolescents, and young adults. *J Pediatr Surg* 2006;41:187–193.

11. Fanburg-Smith JF, Miettinen M, Folpe AL, Weiss SW, Childers EL. Lingual alveolar soft part sarcoma, 14 cases: novel clinical and morphological observations. *Histopathology* 2004;45:526–537.

12. Nielsen GP, Oliva E, Young RH, et al. Alveolar soft-part sarcoma of the female genital tract: a report of nine cases and review of the literature. *Int J Gynecol Pathol* 1995;14:283–292.

13. Radig K, Buhtz P, Roessner A. Alveolar soft part sarcoma of the uterine corpus: report of two cases and review of the literature. *Pathol Res Pract* 1998;194:59–63.

14. Yagihashi S, Yagihashi N, Hase Y, Nagai K, Alquacil-Garcia A. Primary alveolar soft part sarcoma of the stomach. *Am J Surg Pathol* 1991;15:399–406.

15. Yaziji H, Ranaldi R, Verdolini R, et al. Primary alveolar soft part sarcoma of the stomach: a case report and review. *Pathol Res Pract* 2000;196:519–525.

16. Shaddix KK, Fakhre GP, Nields WW, et al. Primary alveolar soft-part sarcoma of the liver: anomalous presentation of a rare disease. *Am Surg* 2008;74:43–46.

17. Flieder DB, Moran C, Suster S. Primary alveolar soft part sarcoma of the mediastinum: a clinicopathological and immunohistochemical study of two cases. *Histopathology* 1997;31:469–473.

18. Sonobe H, Ro JY, Mackay B, et al. Primary pulmonary alveolar soft-part sarcoma. *Int J Surg Pathol* 1994;2:57–62.

19. Luo J, Melnick S, Rossi A, et al. Primary cardiac alveolar soft part sarcoma: a report of the first observed case with molecular diagnostics corroboration. *Pediatr Dev Pathol* 2008;11:142–147.

20. Park YK, Unni KK, Kim YW, et al. Primary alveolar soft part sarcoma of bone. *Histopathology* 1999;35:411–417.

21. Durkin RC, Johnston JO. Alveolar soft part sarcoma involving the ilium: a case report. *Clin Orthop Rel Res* 1999;339:197–202.

22. Aisner SC, Beebe K, Blacksin M, Mirani N, Hameed M. Primary alveolar soft part sarcoma of fibula demonstrating ASPL-TFE3 fusion: a case report and review of the literature. *Skeletal Radiol* 2008;37:1047–1051.

23. Evans HL. Alveolar soft-part sarcoma: a study of 13 typical examples and one with a histologically atypical component. *Cancer* 1985;55:912–917.

Alveolar soft part sarcoma: immunohistochemistry and ultrastructure

24. Argani P, Lai P, Hutchinson B, et al. Aberrant nuclear immunoreactivity for TFE3 in neoplasms with TFE3 gene fusions: a sensitive and specific immunohistochemical assay. *Am J Surg Pathol* 2003;27:750–761.

25. Ladanyi M, Antonescu CR, Drobnjak M, et al. The precrystalline cytoplasmic granules of alveolar soft part sarcoma contain monocarboxylate transporter 1 and CD147. *Am J Pathol* 2002;160:1215–1221.

26. Auerbach HE, Brooks JJ. Alveolar soft part sarcoma: a clinicopathologic and immunohistochemical study. *Cancer* 1987;60:66–73.

27. Persson S, Willems JS, Kindblom LG, Angervall L. Alveolar soft part sarcoma: an immunohistochemical, cytologic and electron-microscopic study and a quantitative DNA analysis. *Virchows Arch A Pathol Anat Histopathol* 1988;412:499–513.

28. Mukai M, Torikata C, Shimoda T, Iri H. Alveolar soft-part sarcoma: assessment of immunohistochemical demonstration of desmin using paraffin sections and frozen sections. *Virchows Arch A Pathol Anat Histopathol* 1989;414:503–509.

29. Matsuno Y, Mukai K, Itabashi M, et al. Alveolar soft-part sarcoma: a clinicopathologic and immunohistochemical study of 12 cases. *Acta Pathol Jpn* 1990;40:199–205.

30. Ordonez NG, Ro JY, Mackay B. Alveolar soft part sarcoma: an ultrastructural and immunocytochemical investigation of its histogenesis. *Cancer* 1989;63:1721–1736.

31. Miettinen M, Ekfors T. Alveolar soft part sarcoma: immunohistochemical evidence for muscle cell differentiation. *Am J Clin Pathol* 1990;93:32–38.

32. Hirose T, Kudo E, Hasegawa T, Abe JI, Hizawa K. Cytoskeletal properties of alveolar soft-part sarcoma. *Hum Pathol* 1990;21:204–211.

33. Foschini MP, Eusebi V. Alveolar soft-part sarcoma: a new type of rhabdomyosarcoma? *Semin Diagn Pathol* 1994;11:58–68.

34. Rosai J, Dias P, Parham DM, Shapiro DN, Houghton P. MyoD1 protein expression in alveolar soft part sarcoma as confirmatory evidence of its skeletal muscle nature. *Am J Surg Pathol* 1991;15:974–981.

35. Wang NP, Bacchi CE, Jiang JJ, McNutt MA, Gown AM. Does alveolar soft part sarcoma exhibit skeletal muscle differentiation?: an immunocytochemical and biochemical study of myogenic regulatory protein expression. *Mod Pathol* 1996;9:496–506.

36. Gomez JA, Amin MB, Ro JY, et al. Immunohistochemical profile of myogenin and Myo D1 does not support skeletal muscle lineage in alveolar soft part sarcoma: a study of 19 tumors. *Arch Pathol Lab Med* 1999;123:503–507.

37. Shipkey FH, Lieberman PH, Foote FW Jr, Stewart FW. Ultrastructure of alveolar soft-part sarcoma. *Cancer* 1964;17:821–830.

38. Mukai M, Torikata C, Iri H, et al. Alveolar soft part sarcoma: an elaboration of a three-dimensional configuration of the crystalloids by digital image processing. *Am J Pathol* 1984;116:398–406.

39. Carstens PHB. Membrane-bound cytoplasmic crystals, similar to those in alveolar soft part sarcoma, in a human muscle spindle. *Ultrastruct Pathol* 1990;14:423–428.

Alveolar soft part sarcoma: genetics

40. Cullinane C, Thorner PS, Greenberg ML, et al. Molecular genetic, cytogenetic, and immunohistochemical characterization of alveolar soft-part sarcoma. Implications for cell of origin. *Cancer* 1992;70:2444–2450.

41. Sciot R, Dal Cin P, De Vos R, et al. Alveolar soft-part sarcoma: evidence for its myogenic origin and for the involvement of 17q25. *Histopathology* 1993;23:439–444.

42. van Echten J, Van Den Berg E, van Baarlen J, et al. An important role for chromosome 17, band q25, in the histogenesis of alveolar soft part

sarcoma. *Cancer Genet Cytogenet* 1995;82:57–61.

43. Heimann P, Devalck C, Debusscher C, Sariban E, Vamos E. Alveolar soft-part sarcoma: further evidence by FISH for the involvement of chromosome band 17q25. *Genes Chromosomes Cancer* 1998;23:194–197.

44. Joyama S, Ueda T, Shimizu K, *et al*. Chromosome rearrangement at 17q25 and Xp11.2 in alveolar soft-part sarcoma: a case report and review of the literature. *Cancer* 1999;86:1246–1250.

45. Ladanyi M, Lui MY, Antonescu CR, *et al*. The der(17)t(X;17)(p11;q25) of human alveolar soft part sarcoma fuses the TFE3 transcription factor gene to ASPL, a novel gene at 17q25. *Oncogene* 2001;20:48–57.

46. Kiuru-Kuhlefelt S, El-Rifai W, Sarlomo-Rikala M, Knuutila S, Miettinen M. DNA copy number changes in alveolar soft part sarcoma: a comparative genomic hybridization study. *Mod Pathol* 1998;11:227–231.

47. Bu X, Bernstein L. A proposed explanation for female predominance in alveolar soft part sarcoma. *Cancer* 2005;103:1245–1253.

48. Tsuda M, Davis IJ, Argani P, *et al*. TFE3 fusions activate MET signaling by transcriptional upregulation, defining another class of tumors as candidates for therapeutic MET inhibition. *Cancer Res* 2007;67:919–929.

49. Aulmann S, Longerich T, Schirmacher P, Mechtersheimer G, Penzel R. Detection of the ASPSCR1-TFE3 gene fusion in paraffin-embedded alveolar soft part sarcomas. *Histopathology* 2007;50:881–886.

50. Argani P, Antonescu CR, Illei PB, *et al*. Primary renal neoplasms with the ASPL-TFE3 gene fusion of alveolar soft part sarcoma: a distinctive tumor entity previously included among renal cell carcinomas of children and adolescents. *Am J Pathol* 2001;159:179–192.

Chapter 33
Pathology of synovia and tendons

Markku Miettinen
National Cancer Institute, National Institutes of Health

This chapter includes infectious and inflammatory diseases of the synovia, tenosynovia, and tendons, with an emphasis on morphology. Also included in the discussion are synovial foreign body reactions related to joint prostheses. The only neoplasm discussed here is the tenosynovial giant cell tumor. Synovial sarcoma is a neoplasm unrelated to synovial tissue, and it is discussed with soft tissue tumors with epithelial differentiation (see Chapter 27). Clear cell sarcoma of tendons and aponeuroses is included among the neuroectodermal neoplasms (see Chapter 25).

Synovial biopsy

Synovial tissue is normally covered by a mildly folded membrane lined by a one- to two-cell layer of specialized fibroblasts and histiocytes; there is no epithelial lining. The stroma contains fat cells and loose connective tissue. There are few if any lymphoid cells or neutrophils in the stroma, but scattered mast cells are sometimes present.

In the past, synovial specimens have mostly included synovectomy samples from patients with rheumatoid arthritis or other surgically treated disease. More recently, diagnostic synovial biopsies have been obtained arthroscopically or blindly, yielding diagnostic information on conditions such as rheumatoid arthritis, granulomatous inflammation, or foreign body arthritis.[1,2] In one study, a blind synovial biopsy of 59 patients with arthritis of unknown cause revealed a specific cause in only three cases (5%). These diagnoses were tuberculotic synovitis, synovial chondromatosis, and villonodular synovitis.[3]

Analysis of synovial fluid

In addition to biopsy (or sometimes instead of it), analysis of synovial fluid gives insight into the nature of joint effusion. Analysis of cell content allows categorization of the process as noninflammatory, inflammatory, purulent, or hemorrhagic. Visual inspection, cell count (i.e., neutrophils, lymphocytes), microscopic analysis for crystals, testing of spontaneous clotting (i.e., fibrin content), and microbial culture should be considered.

Normal synovial fluid is viscous, does not coagulate spontaneously, and contains few cells (<100/μL). Highest cell counts (mainly neutrophils, usually >50 000/μL) are seen in septic arthritis, whereas intermediate counts are seen in inflammatory conditions such as rheumatoid arthritis.

Microscopic analysis of synovial fluid can detect the rhomboid calcium pyrophosphate crystals by native microscopy in pseudogout, and the needle-shaped monosodium urate crystals in gouty arthritis.[4,5] Polarized light with a first-order red (550-nm wavelength) compensator is also helpful in separating calcium pyrophosphate (pseudogout), seen as blue crystals, from monosodium urate (gout), seen as yellow crystals.

Acute purulent synovitis (septic arthritis)
Clinical features

Bacterial infection in the joint cavity is a medical emergency. If untreated, it can lead to rapid destruction of articular cartilage or premature osteoarthrosis and can threaten the functionality of the joint. This is a result of anoxia, acidosis, and proteolytic enzymes released by the inflammatory cells and bacteria. Septic arthritis occurs at all ages, but one-half of all cases are seen in children <3 years of age; older adults are also at risk.[6]

The most commonly involved joints are the knee, hip, and shoulder, and polyarthritis is also possible. The infection can be hematogenous (sepsis related), can spread from a local source from bone (osteomyelitic abscess) or soft tissue, or can be from penetrating trauma by a contaminated object. The most commonly involved bacteria are *Staphylococcus aureus* in adults, and *Haemophilus influenzae* in young children. In young adults, gonococcal synovitis is a consideration. Rapidly initiated systemic antibiotic therapy, based on microbial identification and testing for antibiotic resistance, is the main treatment.

Pathology

The joint region is typically tender, swollen, and warm. The joint cavity is filled with purulent, overwhelmingly neutrophilic exudate containing >50 000 to 100 000 cells/μL. Fibrin clots containing abundant neutrophils are a typical microscopic finding in tissue recovered from the joint cavity. Because bacterial identification by microscopy only is unreliable, the culture material should be obtained from the effusion.

Modern Soft Tissue Pathology, Second Edition, ed. Markku Miettinen. Published by Cambridge University Press. © Cambridge University Press 2016.

The synovial membrane is rarely biopsied in septic arthritis. In fact, when synovial tissue is available, the changes can be surprisingly subtle. The synovial membrane can appear flattened and edematous, and there are numerous neutrophils. The synovial cells sometimes have a smudged, necrobiotic appearance. The most conspicuous inflammatory infiltration is often within the synovial cavity, which contains loose bodies consisting of fibrin and massive infiltration by neutrophils (Figure 33.1).

Lyme disease

Lyme disease is named after the town in southeastern Connecticut where a regional cluster of what was then thought to be juvenile rheumatoid arthritis occurred.[7] Soon this epidemic was found to be caused by the *Borrelia burgdorferi* spirochete. In the United States and Europe, other spirochete species belonging to the *B. burgdorferi* species in the broad sense (i.e., also known by other individual species names) can also be involved. This infection is arthropod borne and usually transmitted by infected mature or nymph-stage deer ticks feeding on humans. Antibiotics (e.g., penicillins, doxycycline) are the main treatment in uncomplicated cases.[8,9]

The initial manifestation is a spherical, enlarging skin lesion, termed *erythema migrans*. Synovitis is one of the most important systemic manifestations, and the central nervous system and heart also can be involved in a systemic infection.

Specific diagnosis of Lyme disease includes serologic testing (ELISA) for *B. burgdorferi*, and confirmatory Western blotting for the corresponding antigens to confirm specificity. Polymerase chain reaction (PCR)-based detection of the organisms from a skin lesion, synovial biopsy, or synovial fluid is also possible, although not routinely performed.[8–10] In some cases, the spirochetes can be demonstrated by the culture of a skin biopsy, but this is generally not successful from synovial tissue.

Histologically, Lyme disease synovitis shows a spectrum of changes related to the duration of infection. Acute inflammation with predominance of neutrophils can be seen (Figure 33.2), or there can be chronic synovitis with villous proliferation, and lymphoplasmacytic infiltration and lymphoid aggregates not unlike those seen in rheumatoid arthritis. Distinctive microscopic changes reported in Lyme disease synovitis include stromal fibrin deposition and small arterial occlusion, with onion-skin-like myoid cell proliferation (microangiopathic lesions).[11,12] Spirochetes have been detected in these lesions using electron microscopy and silver stains, most commonly Dieterle or Warthin–Starry stains.[13,14]

Rheumatoid synovitis

Rheumatoid arthritis is a common polyarthritic inflammatory disease of unknown etiology; however, autoimmune mechanisms are thought to be important factors. The revised diagnostic criteria by the American Rheumatism Association from 1987 are based on clinical signs, serology, or radiology.[15] These criteria were created by analyzing large numbers of parameters in 262 patients with rheumatoid arthritis, and an equal number of control patients with other arthritides. Those criteria are:

(1) Morning stiffness in and around joints lasting at least 1 hour
(2) Soft tissue swelling around at least three joints observed by a physician

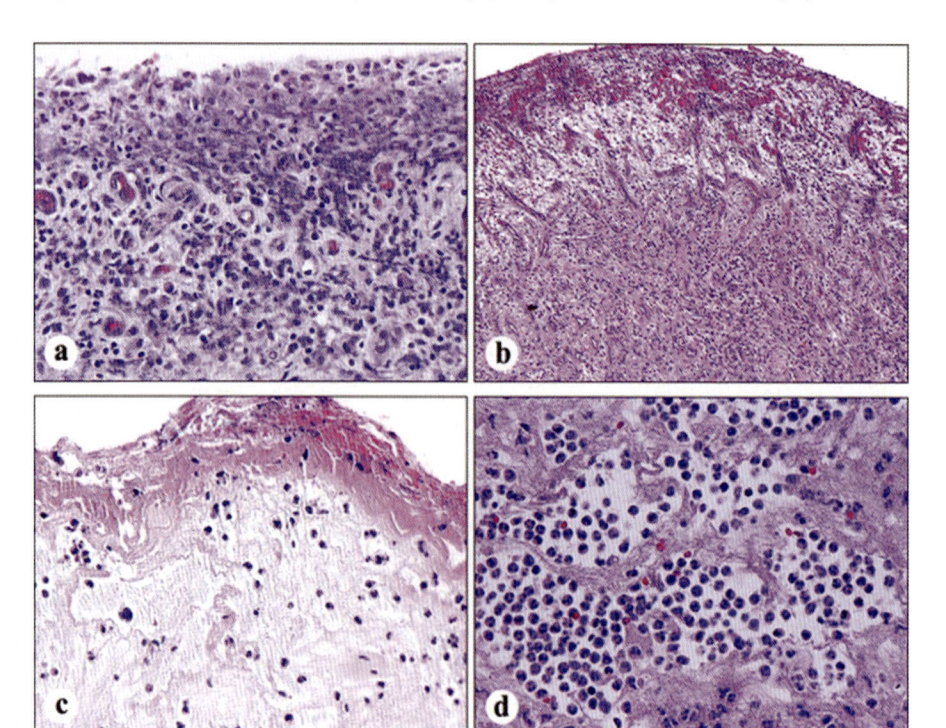

Figure 33.1 (**a**) Acute purulent synovitis shows neutrophilic infiltration in relatively flat synovia. Other inflammatory cells are also present. (**b**) Disrupted synovial surface and vascular granulation tissue lining the synovial cavity. (**c,d**) The most prominent neutrophil tissue infiltration is seen in the fibrinous loose bodies in the synovial cavity.

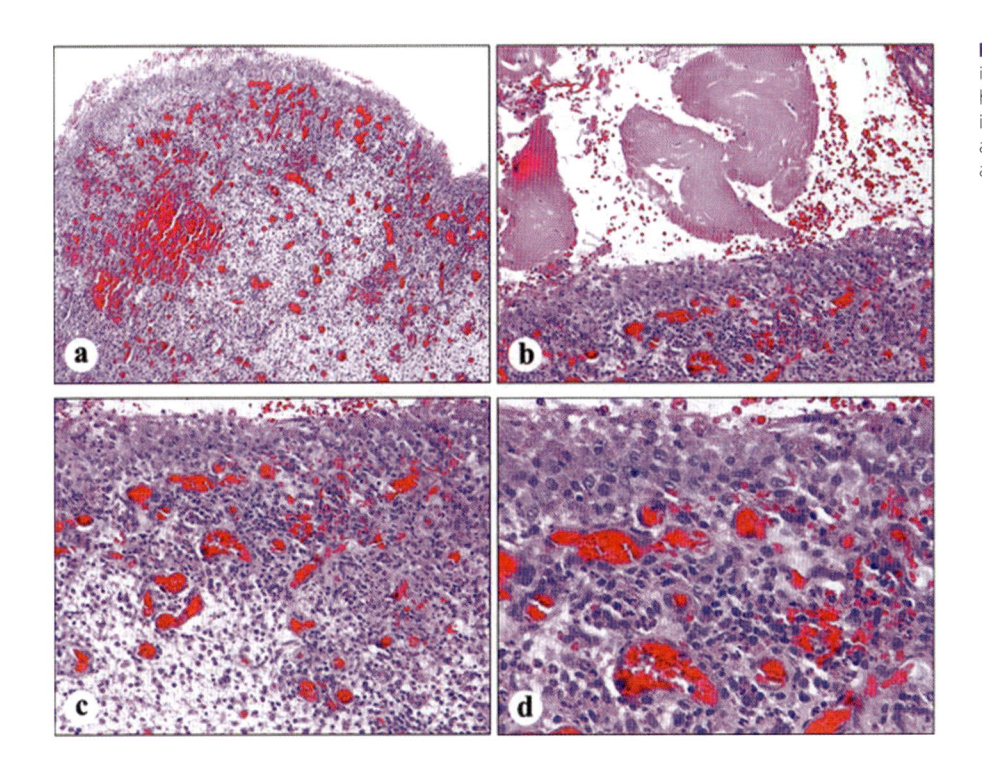

Figure 33.2 Synovial acute and chronic inflammation in Lyme disease. The synovia is hyperemic, containing mixed inflammatory infiltration composed of neutrophils, histiocytes, and plasma cells. (**b**) Fibrinous loose bodies are also present.

(3) Arthritis includes at least one hand joint (wrist, metacarpophalangeal, or proximal interphalangeal joint)
(4) Symmetricity of the arthritis (allowing involvement of different joints of each hand)
(5) Subcutaneous nodules observed by physician (rheumatoid nodules)
(6) Rheumatoid factor present in the serum
(7) Radiologically detected bony erosions or demineralization typical of rheumatoid arthritis.

Pathology

The volume of synovial tissue is markedly increased in active rheumatoid arthritis. The synovium can extend over the articular cartilage, and this abnormal synovial continuation is referred to as *pannus*. It causes ischemic damage to the articular cartilage resulting in cartilage destruction, which is also promoted by the inflammatory process.[16]

The histologic features of rheumatoid synovia vary widely, depending on disease activity. In a quiescent phase, rheumatoid synovia can have minimal inflammation, whereas in the active state, prominent chronic inflammation is present. None of the histologic changes is totally specific. Scoring systems have been proposed, however, showing that the highest levels of chronic inflammatory activity are discriminatory among rheumatoid arthritis and reactive, psoriatic, post-traumatic, and osteoarthritis.[17]

Typical (although not etiologically specific) histologic changes in active chronic rheumatoid arthritis include synovial villous hyperplasia, thickening of the synovial lining, fibrinous exudation at the synovial membrane, and expansion of the synovial stroma by marked chronic inflammation with lymphoid cells.[16–18] The lymphoid infiltration in synovial villi can include germinal centers. Plasma cells are often numerous, sometimes present in sheets and often with Russell bodies (Figure 33.3).

Fibrinous exudation is commonly present and can be extensive, both in the synovial lining and in the synovial fluid. Intra-articular fibrinous loose bodies, grossly whitish granules resembling rice grains, are referred to as *rice bodies*. They vary in size from microscopic (<1 mm) to those >5 mm.

Histologically, they appear nearly acellular, composed of fibrin-like material (Figure 33.4). Rice bodies are commonly seen in active rheumatoid arthritis and are less common in other arthritides. They are generated during the exudative inflammation at the synovial membrane and become subsequently detached, spreading into the joint cavity.[19–24]

Necrobiotic granulomas (rheumatoid nodules) occur in 20% to 25% of patients with rheumatoid arthritis. They usually occur in the subcutis, most commonly on the extensor surface of the proximal forearm and around the joints of the hand and wrist.[25]

The nodules are whitish, with a yellowish center corresponding to the area of central necrosis. Their pathology is similar to that of granuloma annulare or necrobiotic granuloma and is further discussed in Chapter 34.

Synovia in osteoarthritis (degenerative joint disease)

Degenerative joint disease is characterized by degradation of articular cartilage and periarticular bony changes, including localized bony hyperplasia with exophytic excrescences (osteophytes) projecting into the articular cavity, subarticular bone sclerosis, and subchondral ganglion-cyst-like formation.[26]

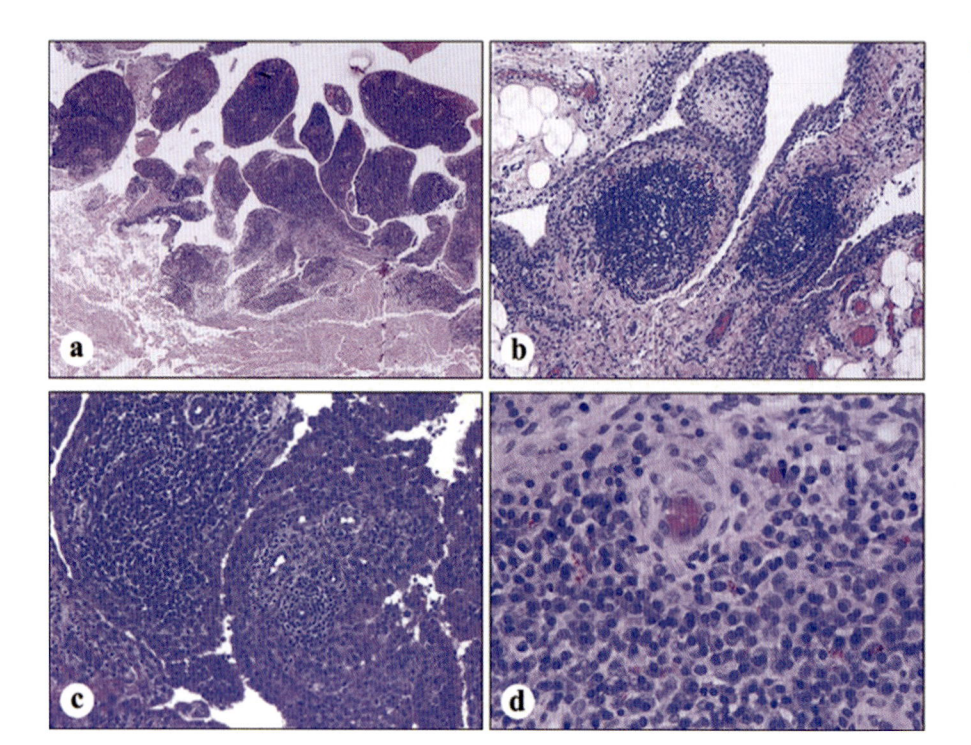

Figure 33.3 Chronic rheumatoid synovitis with villous synovial hyperplasia and markedly expanded stroma containing lymphoplasmacytic infiltration and sheets of plasma cells.

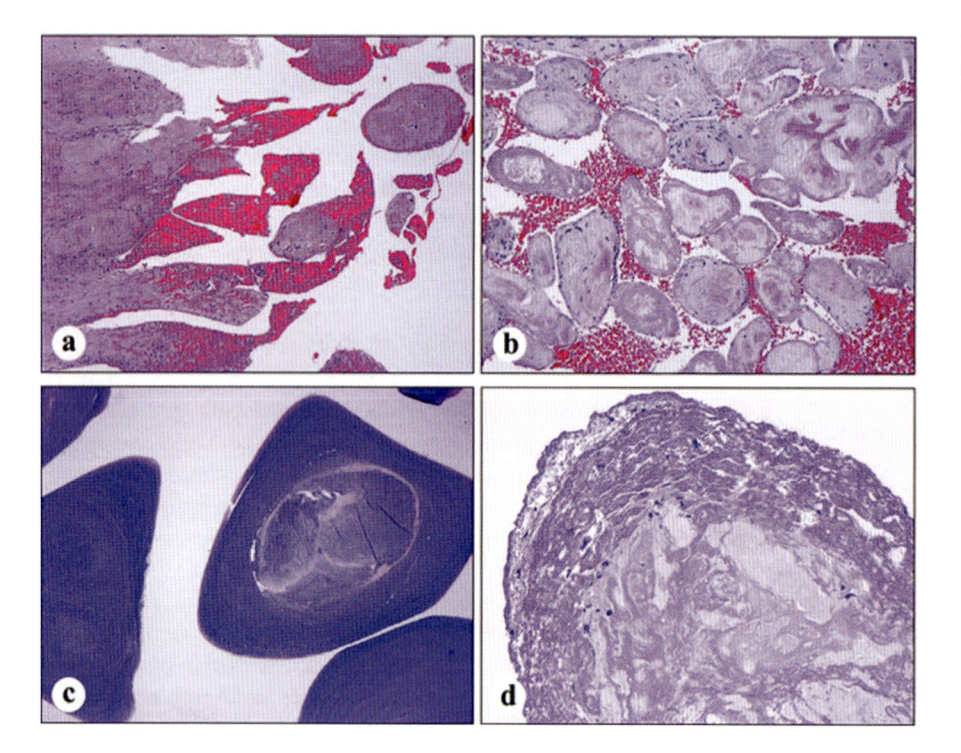

Figure 33.4 Fibrinous bodies (rice bodies) originate from the synovial lining and are shed into the synovial cavity. The bodies vary in size, shape, and tinctorial quality, and they contain very few inflammatory cells.

Histologically, the synovial changes are often mild, and inflammatory infiltration is scant or absent. Typical features include villous hyperplasia of the synovia and thickening of the synovial lining to several layers. Small bone or cartilage fragments (shards) are sometimes present (Figure 33.5). Extensive presence of degenerative fragments is referred to as *detritus synovitis*. A severe degenerative joint disease with prominent bone fragment deposition in the synovia is called *Charcot joint*. This type of disease is often associated with neuropathy or abnormal load as the underlying mechanism.

Synovial villous hyperplasia associated with joint bleeding (hemosiderotic synovitis)

Chronic hemorrhage into the joint cavity for a variety of reasons can lead to hemosiderin deposition in synovial tissue,

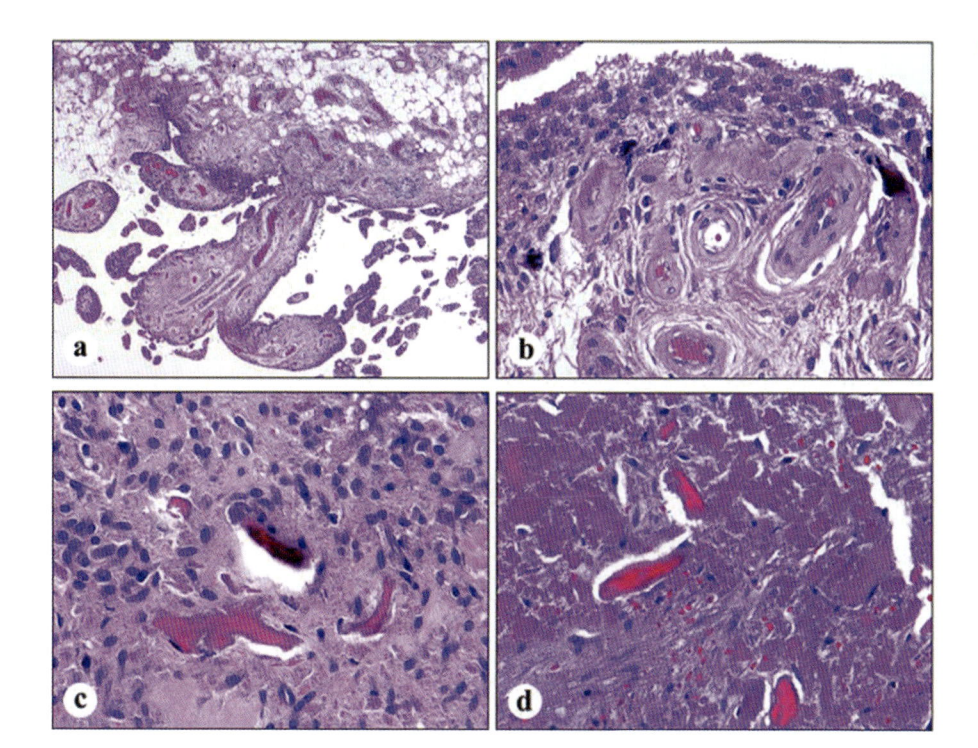

Figure 33.5 (a) Synovia in osteoarthritis shows villous hyperplasia. (b,c) Microscopic bone particles are embedded in the synovia. (d) In more severe cases, amorphous cartilaginous debris can also be present, along with the bone particles.

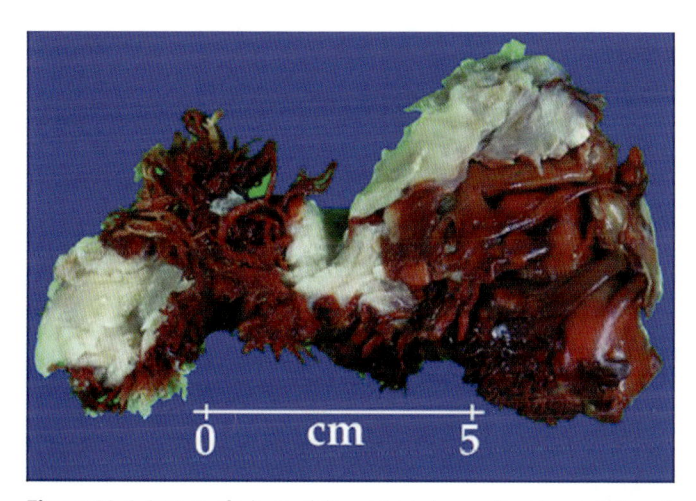

Figure 33.6 Synovia of a hemophiliac patient shows villous hyperplasia and brown discoloration that resembles diffuse synovial giant cell tumor (pigmented villonodular synovitis).

accompanied by synovial hyperplasia. The underlying conditions especially include synovial hemangioma and hemophilia. This condition is sometimes called *hemosiderotic synovitis*, which is a misnomer, because there is no significant inflammatory component.

Grossly, hemosiderotic synovial tissue is brown because of its abundant hemosiderin and villous by its surface proliferation (Figure 33.6). Hemophilia often leads to more severe changes, including the villous brown appearance of synovia that potentially simulates villonodular synovitis (Figure 33.7). A giant-cell-rich stromal cellular proliferation, as would be seen in a true synovial giant cell tumor (villonodular synovitis), is not present, however.

Granulomatous synovitis

The etiology includes various noninfectious and infectious diseases. The evaluation should include histochemical stains for organisms (acid-fast bacilli and fungi), and ideally culture of synovial biopsy or fluid for relevant bacilli and fungi.

The most common noninfectious granulomatous synovitis is one related to sarcoidosis. This is typically associated with abundant presence of non-necrotic epithelioid granulomas with numerous Langerhans giant cells (Figure 33.8). Stains for mycobacteria and fungi are negative; however, supportive laboratory tests and clinical correlation are always needed to establish the diagnosis of sarcoidosis and rule out any infectious cause.[27] It seems to be difficult to find sarcoidosis granulomas in synovial needle biopsies.[28] Those negative findings are probably caused by sampling limitations.

Crohn's disease is another rare cause of non-necrotizing granulomatous synovitis, and in some patients, granulomatous monoarthritis preceded the diagnosis of the inflammatory bowel disease. The reported patients have been children or young adults.[29,30]

Mycobacteria are the most commonly involved organisms in infectious granulomatous synovitis. The most common specific pathogens include *Mycobacterium tuberculosis* and *M. marinum*. The latter infection usually involves the hand, and persons at risk include marine fishermen and pet shop workers. Rarely involved mycobacteria are *M. avium*, *M. kansasii*, and *M. terrae*. Immunocompromised persons are at greater risk.[30] In these infections, the granulomas often contain a necrotizing component (Figure 33.9). Negative stains for acid-fast bacilli never rule out infection, and mycobacterial and fungal culture of synovial fluid or synovial tissue is

Chapter 33: Pathology of synovia and tendons

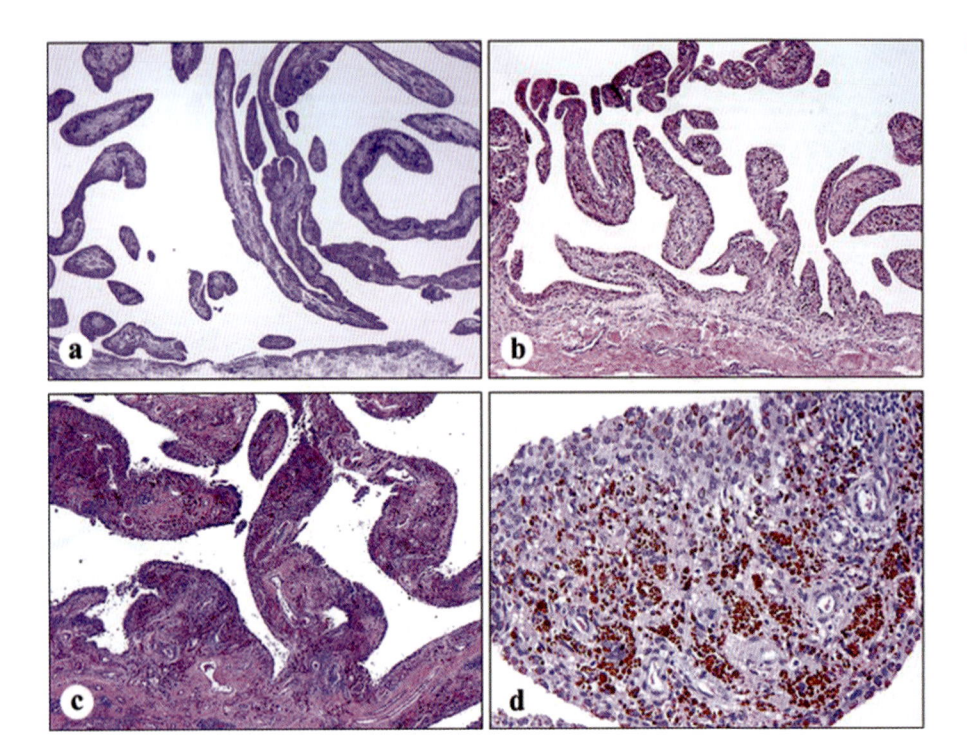

Figure 33.7 Synovial hemosiderosis and villous hyperplasia. The villi are slender and contain aggregates of hemosiderin and fibrosis, but no giant cells or other elements of tenosynovial giant cell tumor.

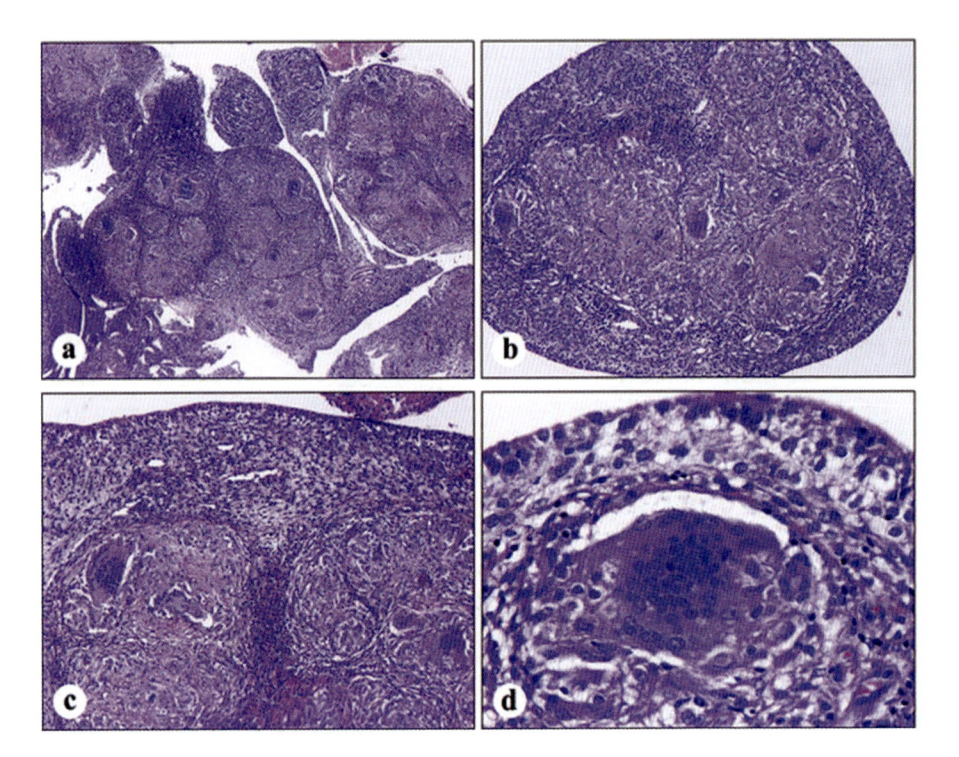

Figure 33.8 Synovia expanded with a florid non-necrotizing granulomatous inflammation in a patient with sarcoidosis.

recommended. Culturally verified *M. marinum* infection seems to be associated with scattered epithelioid granulomas and giant cells (Langerhans and foreign body types), and the inflammatory component is rich in lymphocytes and histiocytes, with a notable paucity of plasma cells.[31,32]

Fungal synovitis caused by *Sporothrix schenkii* has been reported, and in these cases, the probable entry of infection was by a local penetrating trauma with an object contaminated with this common soil fungus. A more superficial trauma can cause similar infection that is limited to the skin and subcutis.[33]

Confirmation of fungal etiology is aided by fungal culture of synovial tissue, and serology for *Sporothrix* antibodies may also be useful. Histologically, there is necrotizing granulomatous inflammation, typically with a strong lymphoid reaction, but a relative paucity of plasma cells, as also seen in some

Chapter 33: Pathology of synovia and tendons

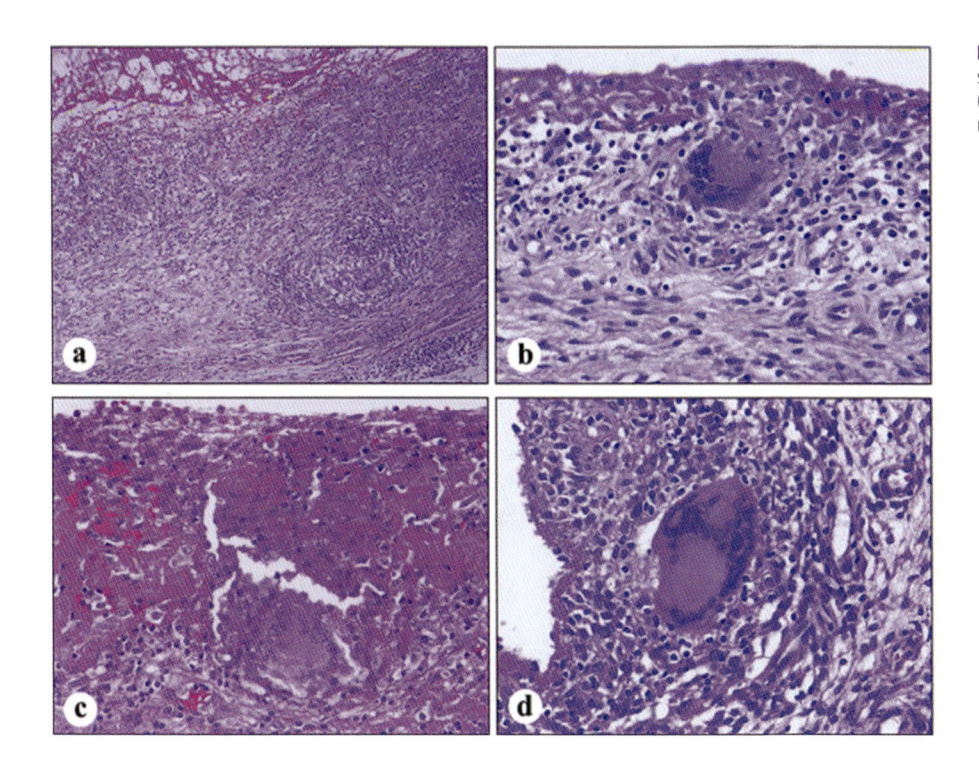

Figure 33.9 Mycobacterial synovitis. (**a**) Fibrinous synovial surface and severe chronic inflammation. (**b–d**) Langerhans-type giant cells and fibrinoid necrosis are present.

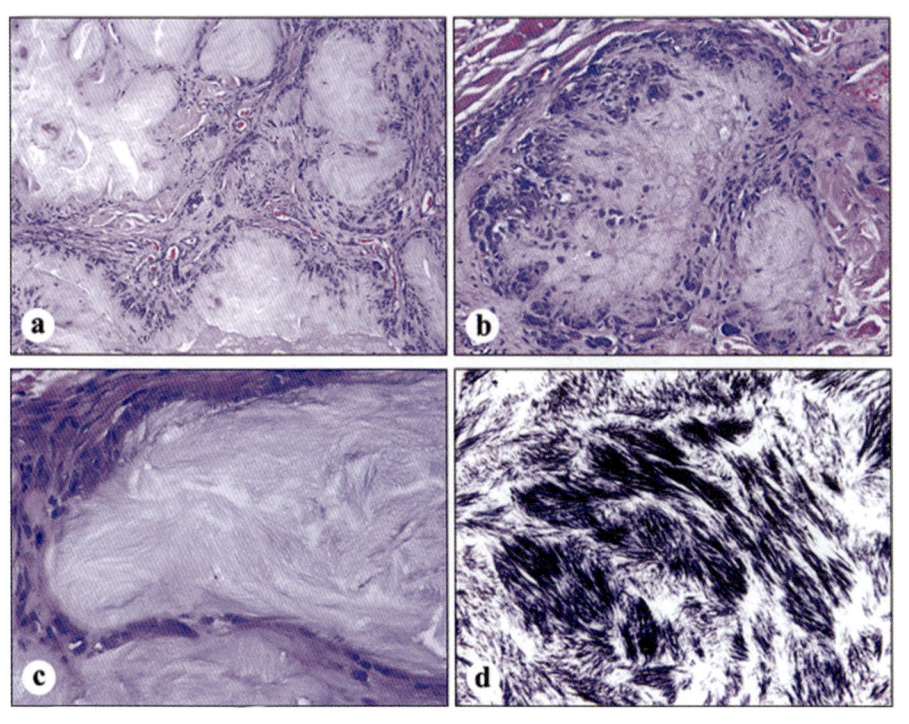

Figure 33.10 (**a–c**) Gouty tophus containing urate crystals surrounded by histiocytic foreign body reaction. (**d**) The crystals are highlighted with urate stain, showing masses of needle-shaped crystals.

mycobacterial synovitides. Long-standing infection can involve bones with a lytic osteitis.[33]

Gouty arthritis

Gout is a metabolic disorder that includes accumulation of uric acid in various tissues due to elevated concentration of circulating uric acid (hyperuricemia). This condition can be due to impaired secretion caused by renal disease or drugs, especially some diuretics. Alternatively, gout can result from increased production of uric acid owing to the breakdown of nucleic acid, most commonly following cell lysis from malignant disease.[34]

Extensive synovial gouty deposits of urate are distinctive as chalky whitish masses. Typical histologic findings include pools of amorphic to crystalline uric acid surrounded by large, sometimes multinucleated histiocytes (Figure 33.10).

Urate stain can be helpful in demonstrating the typical needle-shaped urate crystals, but often the crystals are extracted during histologic processing, resulting in unimpressive histochemical staining.[35,36] Alcohol fixation of the tissue, bypassing water-containing formalin often preserves the urate crystals. Direct microscopic examination of scrape preparations from the grossly observed chalky areas can also be helpful.

Calcium pyrophosphate dihydrate deposition disease

Synovial calcium pyrophosphate dihydrate deposition occurs predominantly in older adults with a female predominance. It most commonly involves the knee joint, symphysis pubis, wrist, shoulder, elbow, and hip joints. Crystalline deposits of calcium pyrophosphate develop in the articular cartilage, meniscal tissue, tendons, and bursal tissue. Patients having these depositions can remain asymptomatic for a long time, but attacks of crystal arthritis not dissimilar to gouty arthritis (pseudogout) can develop. Degenerative joint disease is commonly present, probably caused by crystal arthritis. In some patients, there is an underlying metabolic condition, such as primary hyperparathyroidism or hemochromatosis, and association with diabetes mellitus is common.[37–41]

The crystals can be detected in the synovial fluid by polarization microscopy from patients having arthritis attacks. The presence of pools of dark, basophilic, calcified material containing needle-shaped or rhomboid crystals is the usual histologic finding. There is typically no associated inflammation (Figure 33.11). A foreign body histiocytic reaction is sometimes present in the involved synovia.

Polyethylene wear particles

Polyethylene has been the most commonly used lining material in artificial knee joints and the acetabular cups of prosthetic hip joints. In a total knee prosthesis, both the tibial and femoral surfaces have such a lining. Microscopic wear of this surface occurs during normal knee motion, leading to the release of numerous microparticles into the synovial fluid. The wear can be enhanced by inoptimal positioning of the prosthesis and intra-articular cement fragments. Patients with a long-standing knee prosthesis can have milky synovial fluid with an abundant content of polyethylene microparticles. The recruitment of macrophages by these particles around the prosthesis is thought to be a significant factor in prosthesis loosening and failure through activation of the proteolytic enzymes in the histiocytes.[42–48]

The polyethylene particles are also present in the synovial tissue, where a proliferative reaction with recruitment of macrophages is induced. The polyethylene wear particles can be demonstrated as a polarizing foreign material that can be extracellular or contained in multinucleated histiocytic giant cells (Figure 33.12). Other inflammatory cell content is scant. The polyethylene particles can also cause formation of synovial herniation/cysts, but these are not necessarily associated with prosthesis failure.[49] The polyethylene and other particles can spread to the regional lymph nodes, and the associated nodal enlargement can clinically simulate nodal metastases in patients with pelvic cancer.[50] The wear particles are histochemically positive in oil red O stain.[51]

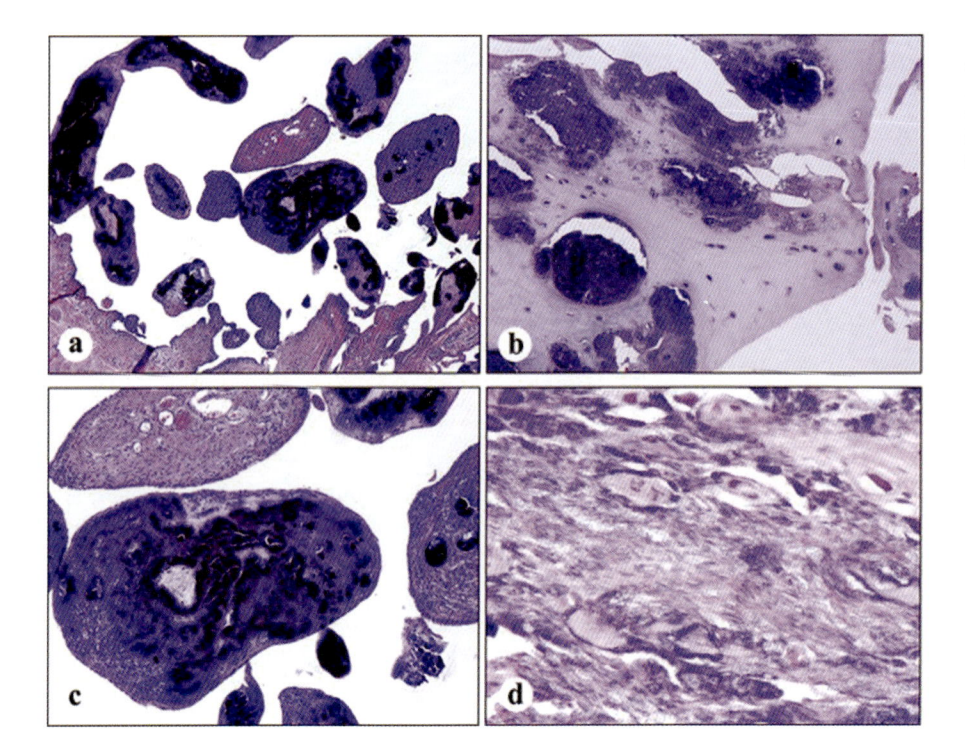

Figure 33.11 (**a,c**) Calcium pyrophosphate deposition is seen as amorphous purple material inside villous synovial elements. (**b**) Calcium pyrophosphate material within metaplastic cartilage and synovial folds. (**d**) Detail of crystals, some of which show rhomboid morphology.

Chapter 33: Pathology of synovia and tendons

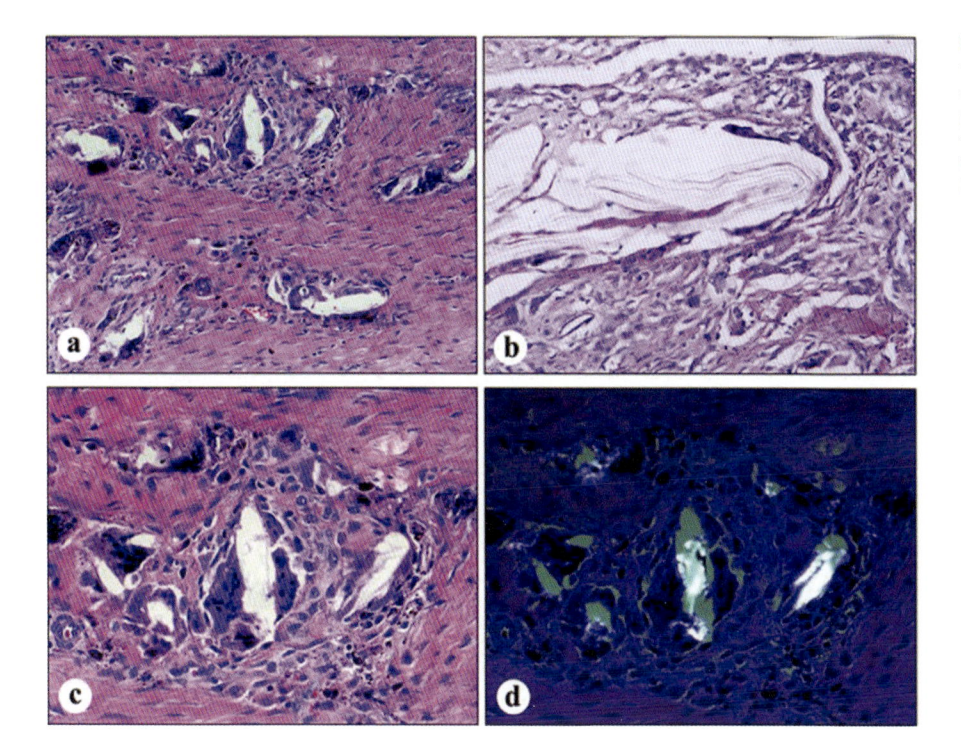

Figure 33.12 (**a**) Polyethylene wear particles in the synovia of a patient with a joint prosthesis covered by polyethylene surfaces. (**b**) Some larger particles also are present. (**c**) Numerous particles of various sizes are present inside or surrounded by foreign body histiocytes. (**d**) The particles are illuminated in polarized light.

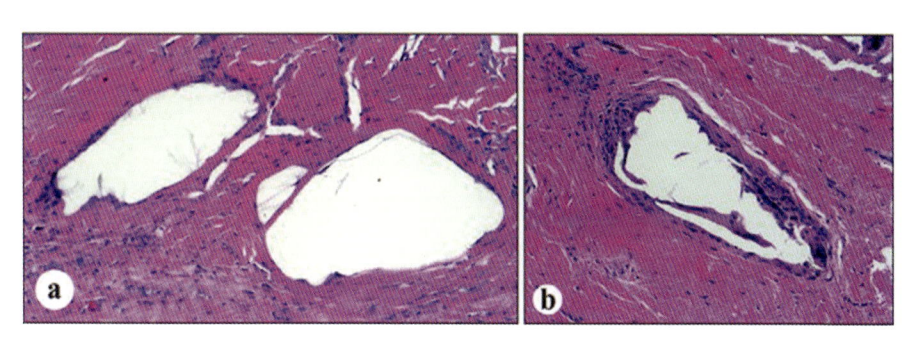

Figure 33.13 Soft tissue around a joint prosthesis contains empty spaces, representing cement surrounded by histiocytic foreign body reaction. Vague granular material in the periphery of the cement spaces is barium sulfate, which is used to make the cement radiopaque and radiologically identifiable.

Bone cement

Bone cement, usually polymerized methyl methacrylate, has been extensively used in the free space between the prosthesis shaft and bone. Small loose fragments of the cement can be deposited around the prosthesis, and such fragments are often seen around the tissue reaction of a failed (loosened) prosthesis.[42,43,52]

The cement itself is not seen because it dissolves during tissue processing, but the sites of these particles are seen as empty, irregularly angled, or rectangular spaces.[42,43] Small amounts of finely granular barium sulfate (used as a radiopaque tracer for implanted cement) can usually be seen in the periphery of such spaces (Figure 33.13).

Silicone (elastomere) deposition

Silicone bone prosthetic parts are used in a limited number of non-weight-bearing joints, most commonly in the wrist, and sometimes in the foot. The articular lining is composed of silicone polymer (elastomere) that sheds silicone microparticles into the joint cavity, leading to their accumulation in the adjacent synovial membranes.[53–55]

Silicone particles can be highlighted as birefringent particles when the condenser is lowered, but they are not polarizing. These particles are typically surrounded by or contained in multinuclear histiocytes (Figure 33.14).

Carbon material

Graphite (originating from pencils) and carbon fibers implanted as tendon repair material are the two types of carbon material seen in synovial tissue.

Carbon fibers have been used as reinforcing elements in artificial ligament transplants. They can be recognized as dark cylindrical elements that tend to be surrounded by fibrosis and separated from each other over time.[56] These fragments are more prominent in unstained sections, because great numbers of fibers are lost during the histologic staining procedure. The carbon fibers are typically surrounded by multinucleated histiocytes (Figure 33.15).

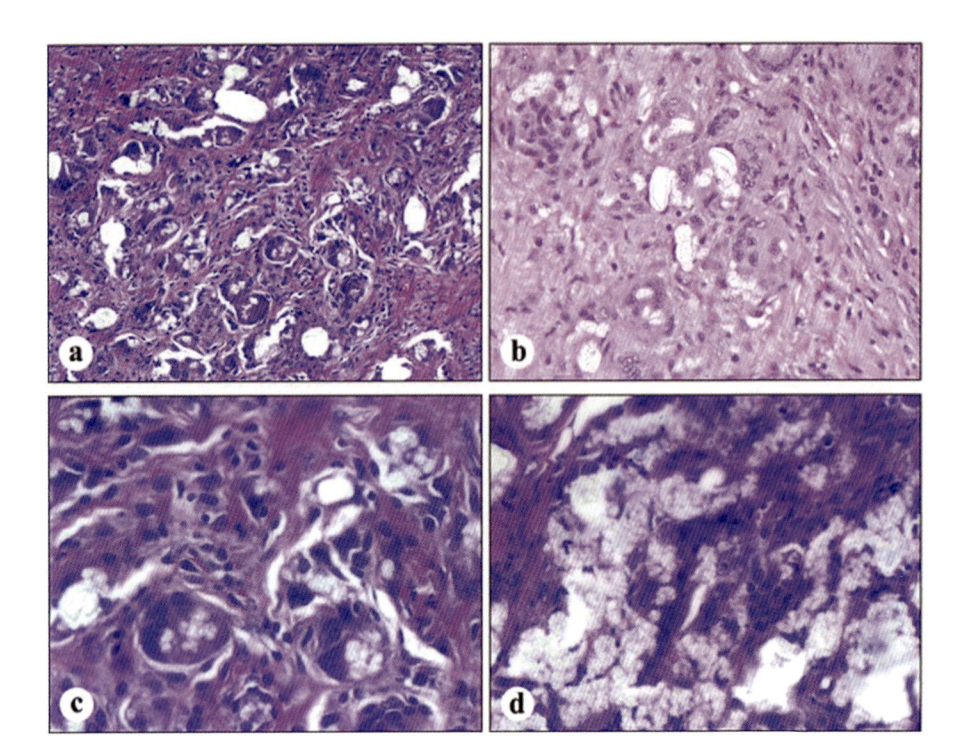

Figure 33.14 (**a–c**) Silicone particles deposited in foreign body histiocytes in the synovia around a silicone bone replacement in the wrist. (**d**) Pools of extracellular silicone material are also present.

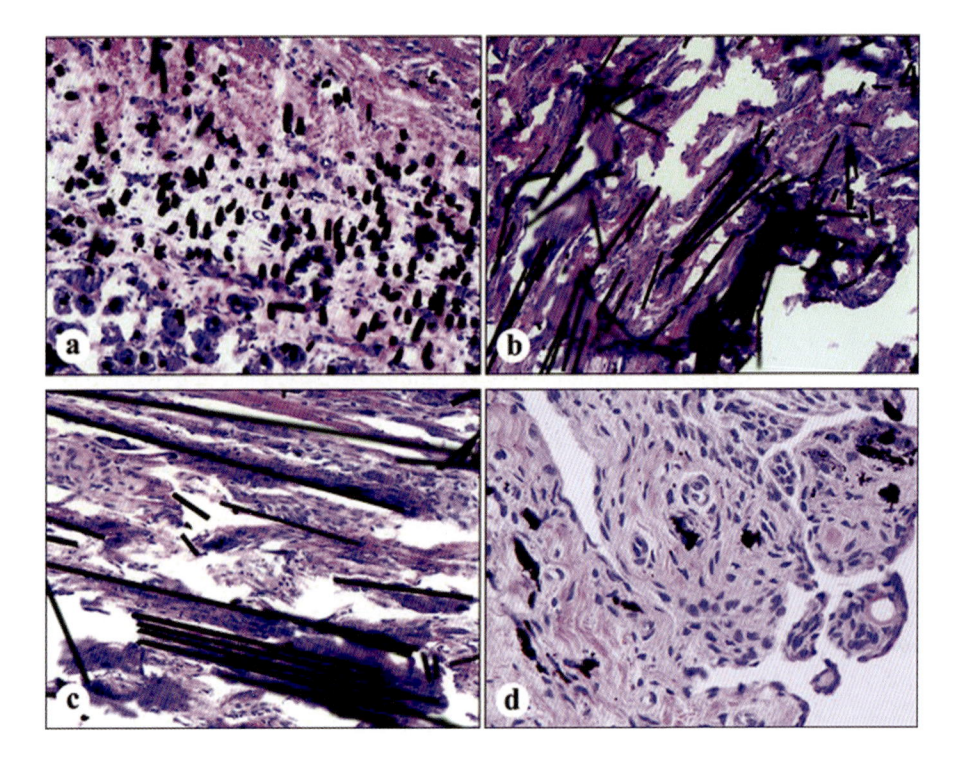

Figure 33.15 Graphite deposition. (**a–c**) Stacks of needle-shaped graphite fibers associated with fibrosis and histiocytic foreign body reaction. (**d**) Small aggregates of graphite in synovia following a pencil-tip injury.

The graphite core of a pencil can be accidentally introduced into the soft tissue of a finger; this core then reaches synovial tissue and becomes deposited in local histiocytes.[57]

Graphite from the pencil core is deposited in the synovia as amorphous globules without much cellular reaction other than histiocytes (Figure 33.15).

Metal fragment deposition

Metal components used in various joint prosthesis types can lead to the release of microscopic titanium-, chromium-, or cobalt-containing metal particles into the synovial fluid and periprosthetic and synovial tissue.[58,59] The metal particles can

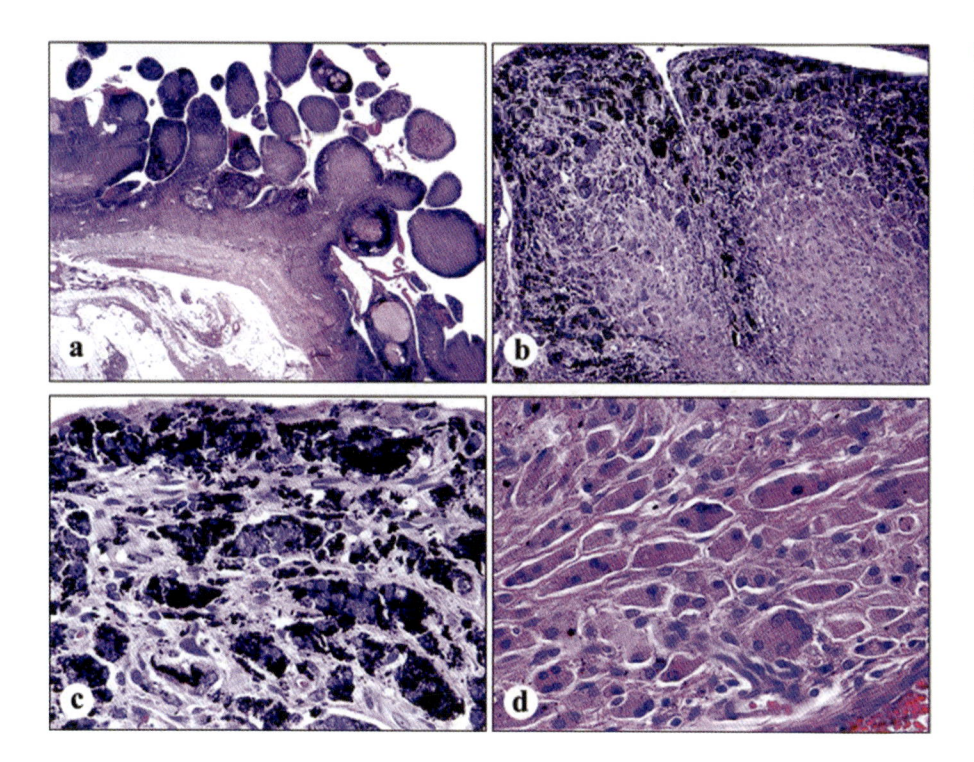

Figure 33.16 Synovia surrounding a failed prosthesis. (**a**) Villous hyperplasia with dark pigment containing metal microparticles. (**b,c**) The particles are concentrated closer to the synovial cavity and are accompanied by histiocytic foreign body reaction. (**d**) Large numbers of granular histiocytes without metal pigment are also present.

also travel into the regional lymph nodes, where they induce a histiocytic reaction.[60]

Metal particles are seen as dark, amorphous flocculent material typically present in densely fibrotic, virtually acellular stroma. An abundant granular histiocytic reaction is typically present (Figure 33.16).

Alkaptonuric ochronosis

Ochronosis refers to brownish (ocher) discoloration of the connective tissues, which generally results from alkaptonuria, an inborn error of metabolism. Secondary weakening of cartilage and ligaments causes orthopedic morbidity.

Pathogenesis and clinical features

This rare hereditary condition is caused by the absence of functional homogentisic acid 1,2 dioxygenase (HGO), an enzyme involved in the catabolism of aromatic amino acids phenylalanine and tyrosine. The pathogenesis is through the loss-of-function germline mutations in the HGO gene, often by compound heterozygous mutations involving both alleles. As a result, homogentisic acid accumulates and subsequently polymerizes in tissues, forming melanin-like pigment. This accumulation results in structural weakening of cartilage, tendons, intervertebral disks, and ligaments by an unknown molecular mechanism. The most common soft tissue manifestation is early-onset degenerative joint disease in large weight-bearing joints, especially the knee. Although the condition is congenital, the tissue damage becomes manifest at an adult age. It has been estimated that one-half of the patients undergo at least one joint replacement by the age of 55 years. Ligament weakening can also cause muscle tears. Some patients develop aortic or mitral valvular disease, and some acquire chronic renal disease.[61–63] Disk herniation and spontaneous tendon rupture are also known, but less common complications.[64–66]

Alkaptonuria (with endogenous ochronosis) is diagnosed based on elevated urinary homogentisic acid levels (normally <10 mM/L). Dark discoloration of urine (especially when alkaline), and joint pain are the most common presenting symptoms. Darkening of the synovial fluid can also be observed. In Caucasians, ochronosis is also observed as a brownish discoloration of skin. Regardless of skin color, there is darkening in the sclerae. Nitisinone, an inhibitor of an enzyme involved in homogentisic acid production, has been shown to have some therapeutic effect.[62]

Pathology

Grossly, ochronosis appears as discoloration by brownish or black pigment, as observed in tendons and ligaments, cartilage, and other pale connective tissues, such as the aorta and heart valves.[67] In microscopic slides, the pigment is most easily detected on gross examination as diffuse brownish discoloration. Microscopically, the pigment is less obvious and is variably masked by transmitted light and filter arrangements.

Microscopically, there is accumulation of brown pigment in the articular cartilage and menisci of knees, and cartilage microfragments are deposited in the synovia (Figure 33.17). Similar pigment can be seen in ruptured tendons, where secondary reactive changes can be minimal (Figure 33.18).

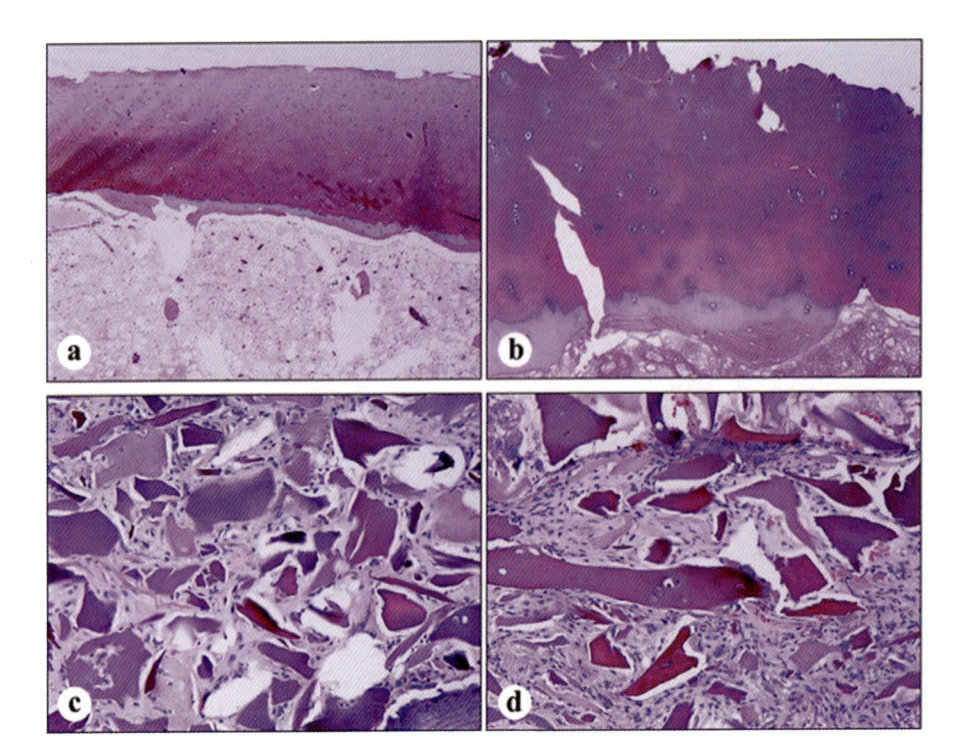

Figure 33.17 (**a,b**) Brown discoloration in the articular cartilage in a patient with alkaptonuric ochronosis. (**c,d**) Numerous brownish cartilaginous shards are embedded in the synovial tissue.

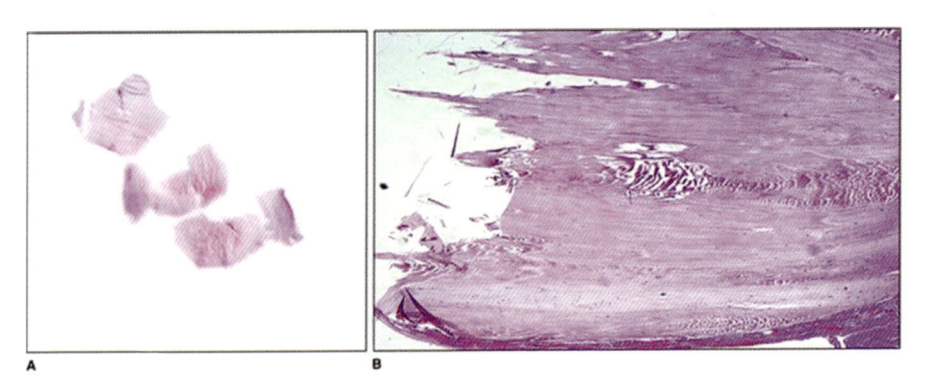

Figure 33.18 (**A**) Ruptured tendon in a patient with ochronosis shows brown discoloration. (**B**) The ruptured tendon shows ragged margins. There is no inflammation.

Tendon rupture

Tendon ruptures show a spectrum of changes that include neovascularization of the tendon adjacent to the rupture, fibrinous necrosis, and granulation tissue formation at the rupture site. An older tendon rupture is more likely to show fibrous scar and metaplastic bone formation (Figure 33.19).

Tenosynovial giant cell tumor (giant cell tumor of tendon sheath, nodular tenosynovitis, pigmented villonodular synovitis, fibroxanthoma, fibrous histiocytoma of tendon sheath)

In 1941, Jaffe et al. outlined the concept of relatedness of tenosynovial giant cell tumor (TSGT), a mainly digital tumor, and pigmented villonodular synovitis, its large joint equivalent.[68] The many synonyms listed in the heading for this section reflect in part this tumor's clinicopathologic heterogeneity and in part the historical differences in understanding of the nature of this tumor. The four clinicopathologic variants of TSGT discussed here are: (1) localized tenosynovial, (2) intra-articular (pigmented villonodular synovitis, diffuse and localized), (3) diffuse tenosynovial, and (4) atypical and malignant tenosynovial giant cell tumor.

The common presence of clonal chromosomal changes supports the neoplastic nature of this tumor, which once was considered a reactive, inflammatory process[69]. Furthermore, similarities in the genetic changes in different subtypes support their mutual relationship.[70,71]

The essential pathogenetic change seems to be translocation-induced overexpression of colony stimulating factor 1 (CSF1) by the lesional cells that typically form a minority of all cells in TSGTs. These growth factor expressing cells recruit the abundant reactive histiocytic population that responds to the growth factor signal by their CSF1 receptor expression.[70–72]

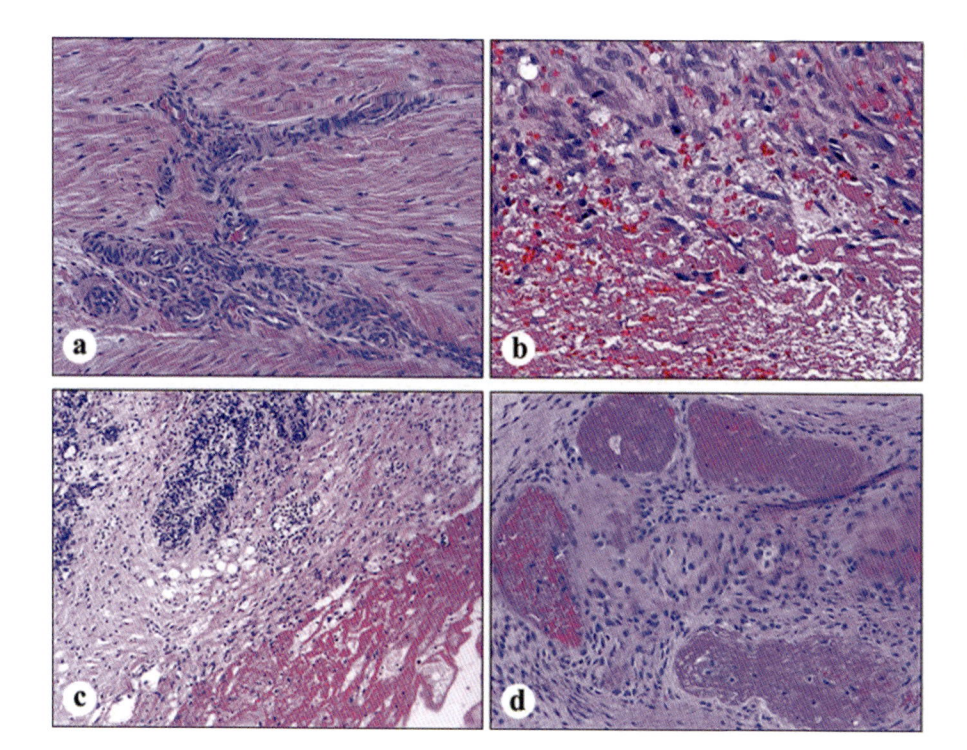

Figure 33.19 (**a**) Tendon near the rupture point shows increased vascularity. (**b,c**) The site of rupture contains fibrin deposition, granulation tissue, and vascular proliferation. (**d**) Vascular thrombosis adjacent to the rupture point.

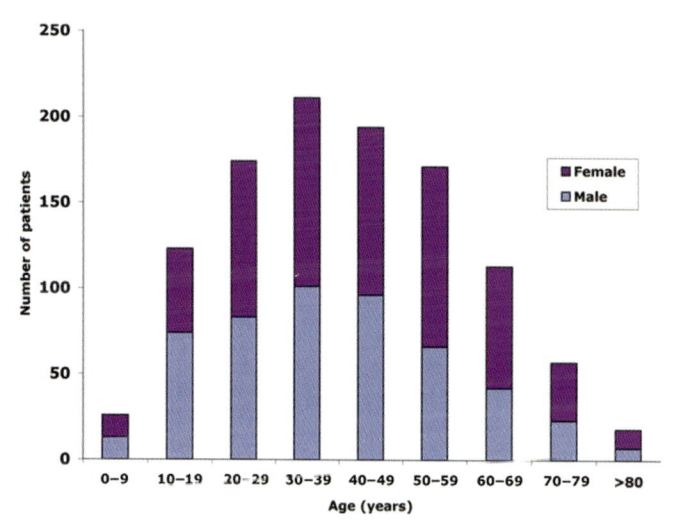

Figure 33.20 Age and sex distribution of 1076 civilian patients with tenosynovial giant cell tumor (TSGT), including all variants.

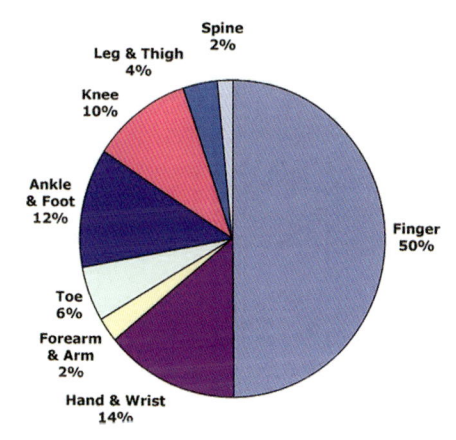

Figure 33.21 Anatomic distribution of 1730 cases of TSGT, including all variants.

The population incidence of all TSGTs has been estimated to be 11 to 20 per 1 million inhabitants in epidemiologic studies.[73,74] In the author's experience, the localized variant makes up approximately 85% of the cases. In one series, 69% were (localized) giant cell tumors, 8% solitary intra-articular, and 23% diffuse intra-articular.[73]

TSGT is usually diagnosed in young to middle-aged adults between 20 and 50 years of age. Some studies have shown a 2:1 female predominance, and a predilection for the right hand has been reported in some studies.[75,76] Based on AFIP statistics of civilian patients, there is only a minor female predominance (Figure 33.20). The age spectrum ranges from preschool age children to rare elderly patients >80 years of age.

When all types are included, TSGT most commonly occurs in the fingers (70–75%). Less frequently it occurs in the hands and wrists, toes, feet, ankles, knees and hip, shoulder, and rarely in the spine and temporomandibular joint (Figure 33.21).

Tenosynovial giant cell tumor, localized type

By definition, this type of TSGT is a circumscribed lesion, which microscopically is not seen to infiltrate fat or skeletal muscle and radiologically is shown to be localized, without extensive longitudinal involvement within the tendon sheath.

Clinical features

The localized type is by far the most common subtype and typically presents as a slowly growing, painless mass varying from 0.5 cm to 3.0 cm (average size <2 cm) in the fingers, but often reaching a larger size (usually ≤5 cm) in the foot and ankle. Less common locations include the hand and wrist, and occurrence in proximal locations and spine is very rare. Erosion of adjacent bones sometimes occurs.[76,77] Some intra-articular TSGTs appear to fall into the category of localized TSGTs by being solitary intra-articular synovial nodules. These tumors usually occur in the knee and rarely in the hip, and they often cause pain, restricted joint motion, and joint effusion.[76]

Localized TSGTs have some tendency to recur. The published frequency of recurrence (most data being from finger tumors) varies widely, but an estimated average is 20% to 25%. Local recurrences can develop several years after the primary excision, but often they occur within 2 years.

Pathology

Clinically, TSGT often forms an irregularly lobulated, subcutaneous nodule (Figure 33.22). On sectioning, the tumor is firm and homogeneous and varies from yellowish to brown.

At low magnification, the tumor is often asymmetric, with an elevated contour on the superficial side and a flat contour on the side facing the tendon and bone (Figure 33.23).

Microscopically, the localized tumors are often partly surrounded by a band of dense fibrosis and lobulated by fibrous septa. The lobules are composed of ovoid and epithelioid mononuclear and multinucleated histiocyte-like cells set in varying amounts of dense collagenous stroma, sometimes in a trabecular pattern (Figure 33.24). The mononuclear cells have a moderate amount of cytoplasm and large, round, sometimes grooved nuclei with a single prominent nucleolus; these are the lesional neoplastic cells (best seen in atypical variants, which are richer in this component). The multinucleated giant cells have an osteoclast-like appearance, vary in numbers, and are often less prominent in highly cellular lesions. The giant cells have a widely varying number of nuclei, from a few to over 50 (Figure 33.25). Their cytoplasm varies from mildly basophilic to eosinophilic, and some giant cells develop into

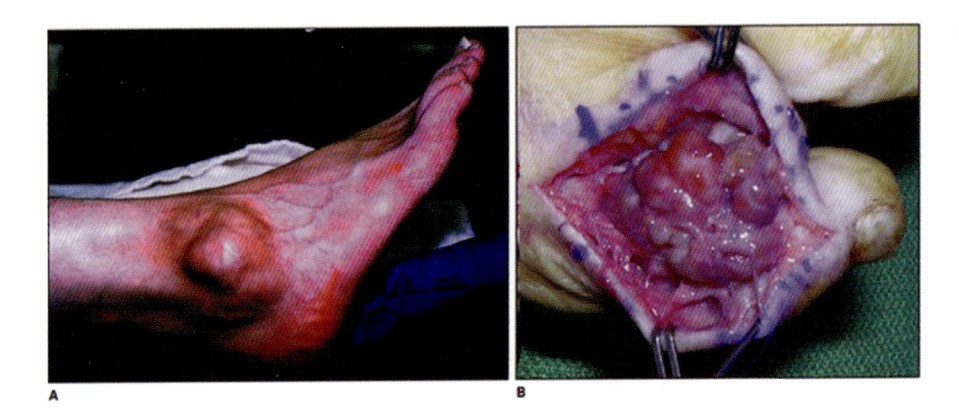

Figure 33.22 (**A**) Clinically a TSGT involving the ankle appears as a multinodular mass underneath intact skin. (**B**) The exposed tumor in the big toe shows a multinodular surface.

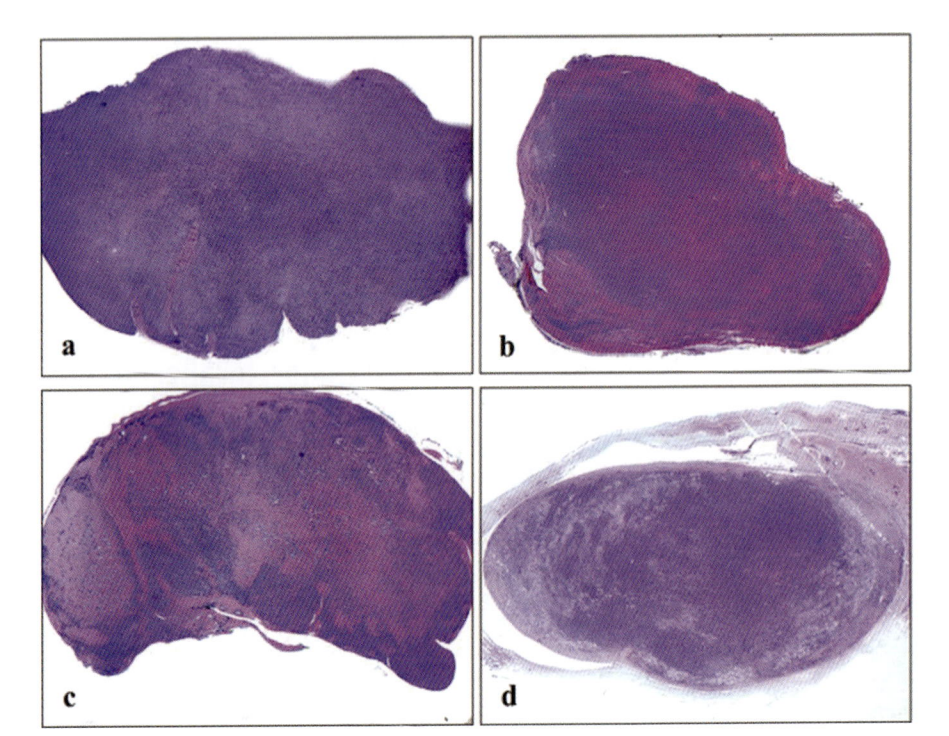

Figure 33.23 (**a–c**) Low-magnification examples of tenosynovial giant cell tumor. Note demarcated nodules. (**d**) This example is seen inside a tenosynovial space.

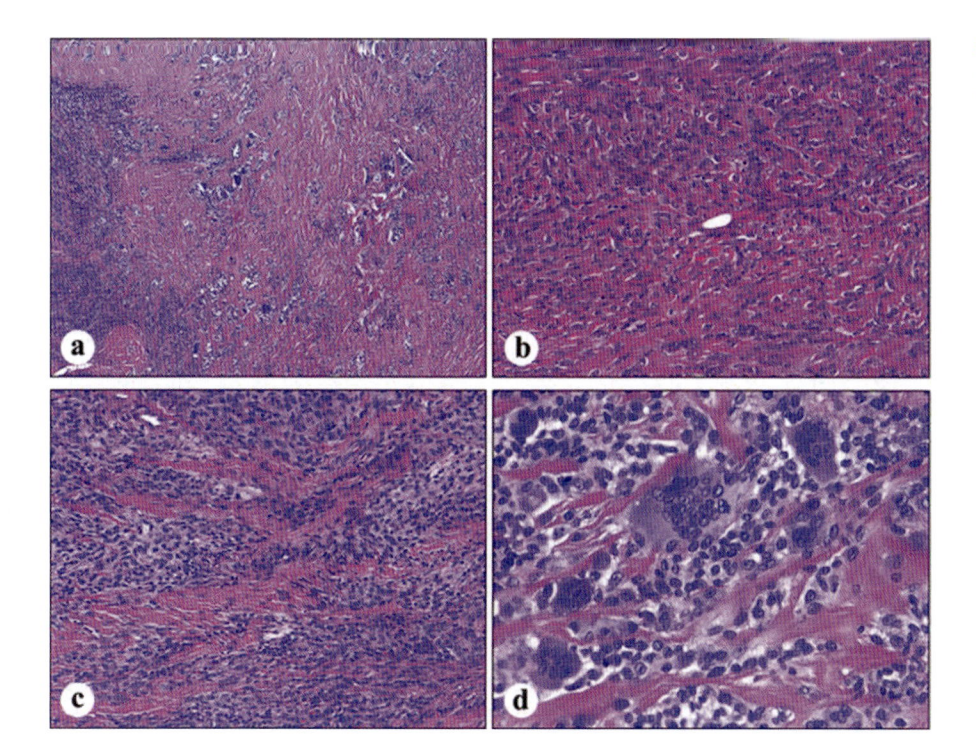

Figure 33.24 TSGT. Note sclerosing collagenous background, and cellular element containing multinucleated histiocytes.

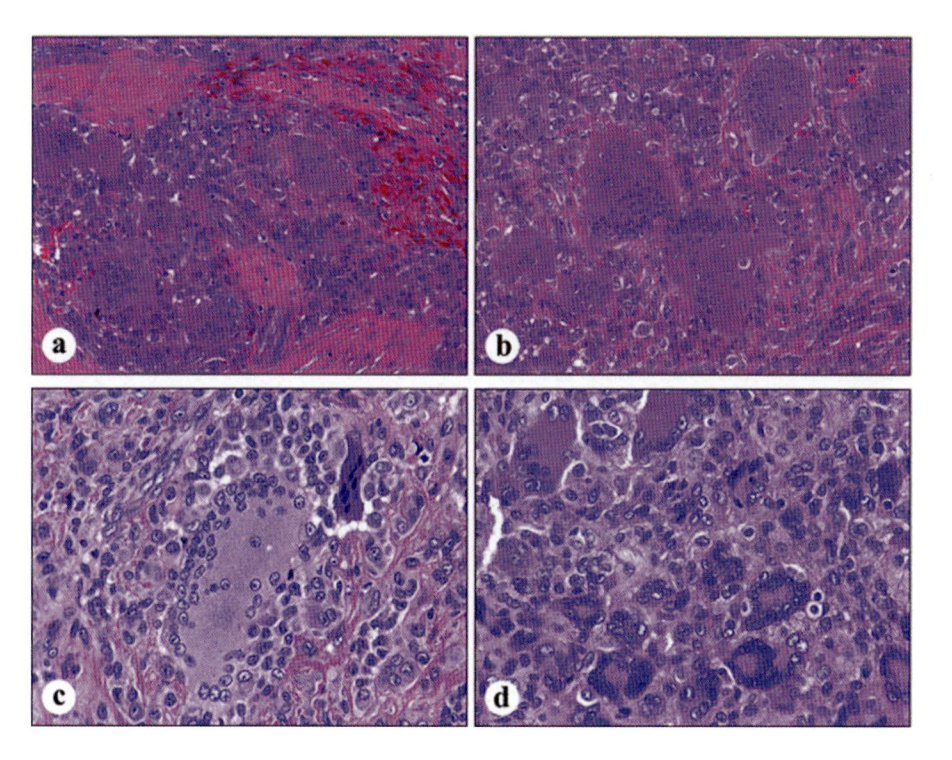

Figure 33.25 Variation in histiocytic morphology in TSGT. (**a**) Small giant cells. (**b**) Giant cells with high numbers of nuclei. (**c**) Giants cells with low and high numbers of nuclei and one extremely large one. (**d**) Giant cells with circular nuclear arrangement.

shrunken, increasingly eosinophilic apoptotic bodies. Foci of xanthoma cells with foamy cytoplasm are usually present, often in the periphery of the lesion; these are accompanied by prominent, branching capillaries. Hyalinized areas also occur, and foci of hemosiderin deposition are seen, especially in the fibrous septa (Figure 33.26). Touton-type giant cells and cholesterol crystals are occasional findings. When involving synovia of the large joints, such as the knee, the intra-articular tumor can be pedunculated and become infarcted and sometimes detached by torsion (Figure 33.27). Cartilaginous metaplasia is a rare finding (Figure 33.28).

Some mitotic activity is almost always present, varying from 2 to >10 mitoses per 10 HPFs (average 5 to 10 HPFs), depending on the degree of cellularity. The higher degree of mitotic activity is not clearly associated with recurrences. Atypical mitoses and nuclear atypia are not present.

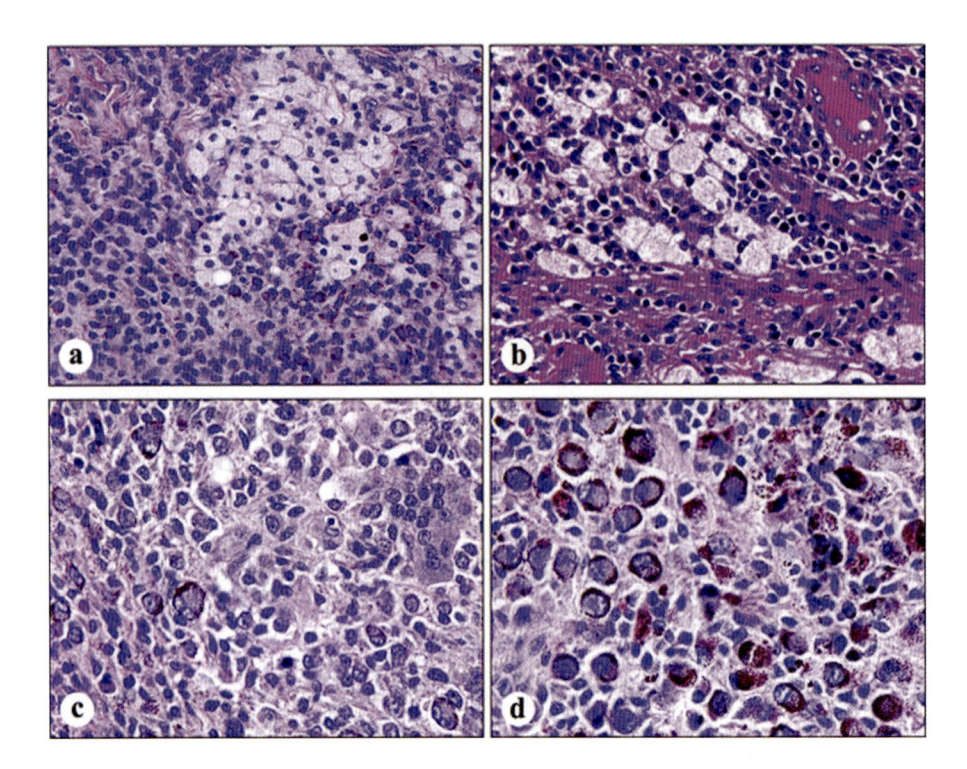

Figure 33.26 Cellular elements of tenosynovial giant cell tumor. (**a,b**) Xanthoma cells are typical focal components, and plasma cells are present in some cases. (**c,d**) Mononuclear epithelioid cells, some of which contain iron pigment, are a minor component.

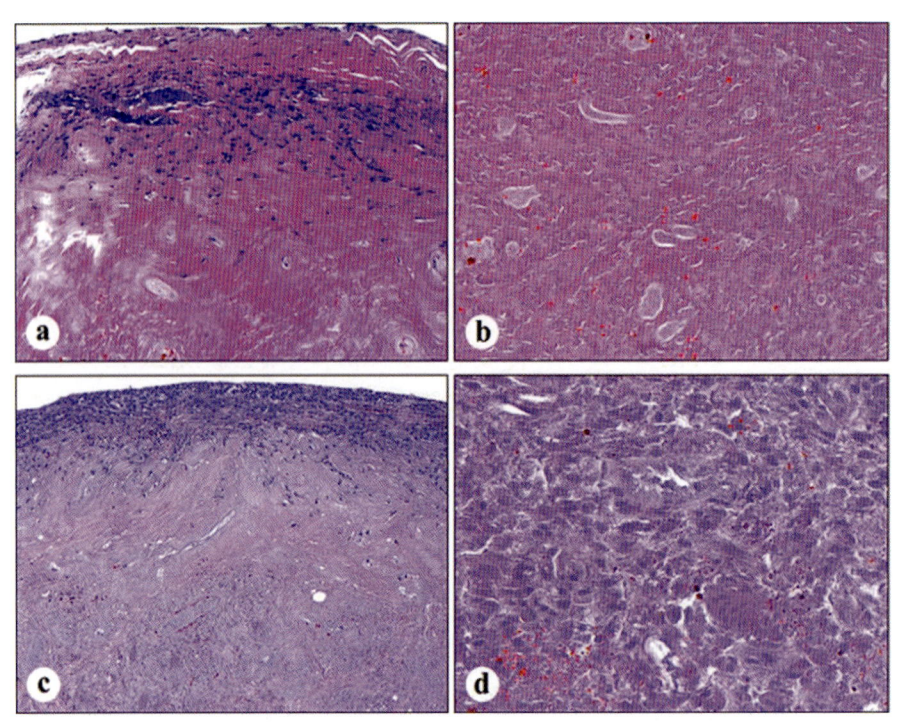

Figure 33.27 Infarcted TSGT. Note the ghosts of mononuclear cells and multinucleated histiocytes.

Tenosynovial giant cell tumor, diffuse type, intra-articular (diffuse pigmented villonodular synovitis)

This locally aggressive disease of large joints most commonly occurs in the knee. Diffuse pigmented villonodular synovitis is a synonym.

Clinical features

The diffuse intra-articular giant cell tumor occurs more in younger patients than does the localized type and has a 2:1 female predominance. It usually presents as a painful monoarticular swelling or effusion of a large joint, most commonly the knee, but sometimes the hip or shoulder and rarely the

Chapter 33: Pathology of synovia and tendons

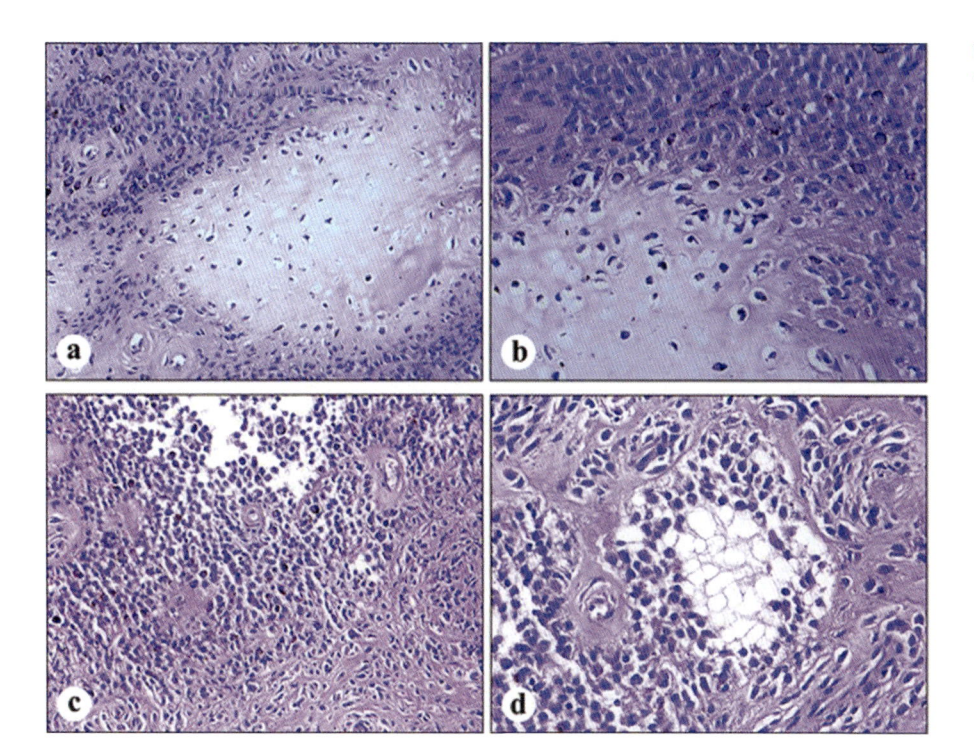

Figure 33.28 An unusual example of TSGT with cartilaginous metaplasia.

The clinical course is often lingering, and recurrences develop cumulatively in 30% to 40% of patients, sometimes 20 years or more after the first surgery. Knee lesions recur more often than those in other large joints.[81,83] Total synovectomy is advocated for treatment, but recurrences nevertheless sometimes develop. In some cases, extensive cartilage destruction requires total arthroplasty and prosthetic joints, especially with hip lesions. Radiation treatment was historically used to control recurrent disease.[79–83]

Pathology

Grossly, the diffuse intra-articular variant transforms the synovia with a villous brown proliferation discolored by hemosiderin deposition. The villous proliferation can coexist with solid, yellow-to-tan tumor nodules and can form complex masses. Extra-articular extension is also possible (Figure 33.29).

Histologically, the villous component is lined with proliferative synovial cells that can involve most of the villi. The villi also can be extensively fibrous, with a rather nonspecific microscopic appearance, except the presence of solid areas of tumor (Figure 33.30). The specific diagnostic elements are often easier to find in the solid tumor areas and the nonvillous component in the base of the synovium. Their composition is generally similar to that of the localized tumors, except that there is more variation, often with greater fibrosis.

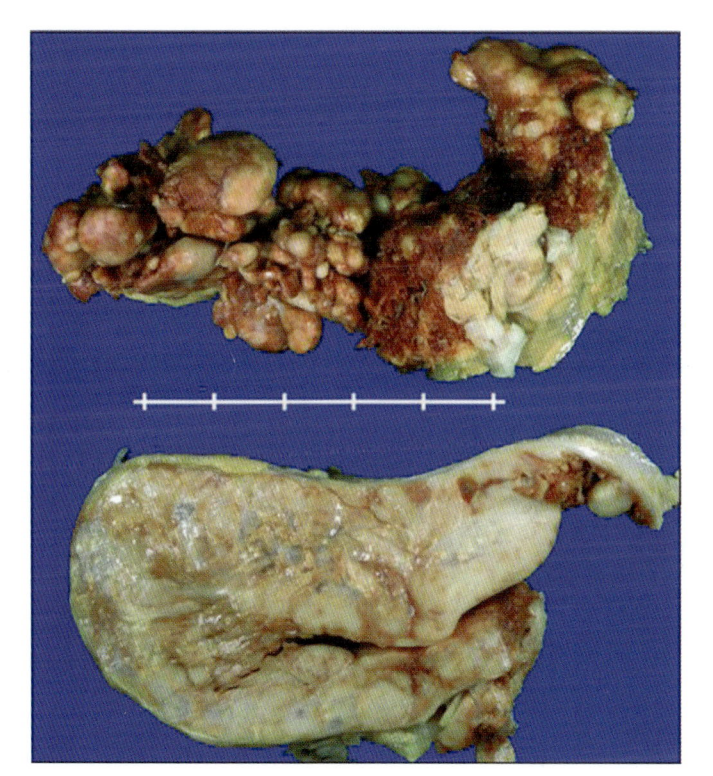

Figure 33.29 Gross appearance of intra-articular diffuse giant cell tumor. The tumor involves the knee synovia in a diffuse, multinodular manner. On sectioning, a large nodule is grayish with yellow and brownish speckles.

elbow or ankle.[78–81] The diffuse nature is best highlighted by MRI studies. The proliferative synovial process can cause extensive cartilage damage and cystic changes in the adjacent bones, especially when the hip and shoulder are involved.[82]

Tenosynovial giant cell tumor, diffuse type, extra-articular

The diffuse extra-articular type is defined by microscopic evidence of invasion into fat or skeletal muscle; this type can also

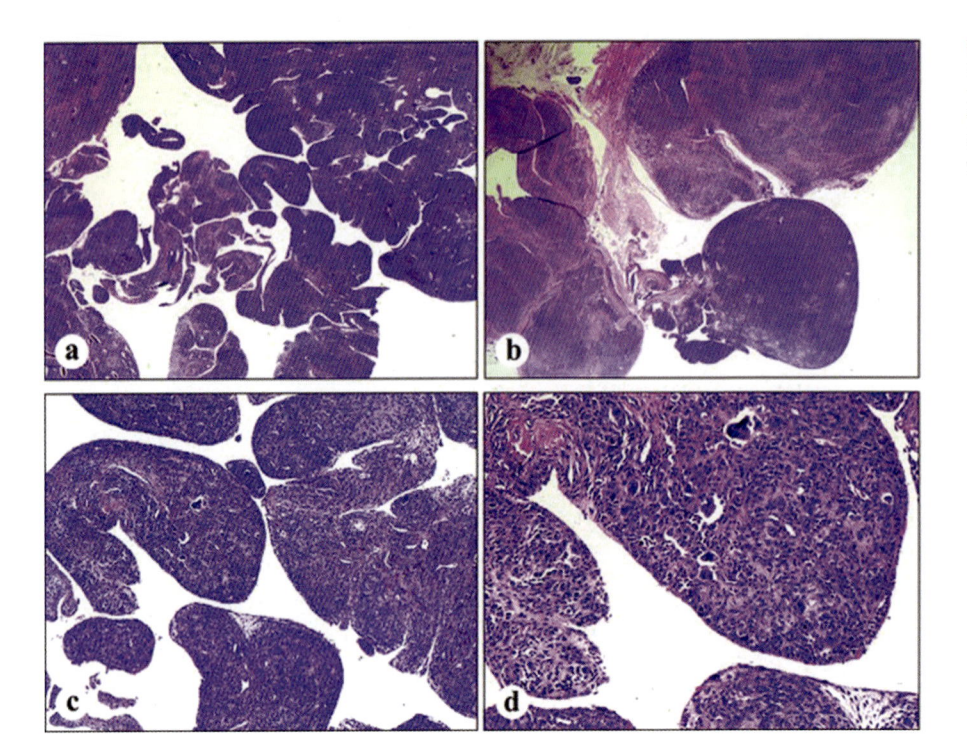

Figure 33.30 Diffuse TSGT forming innumerable villous projections. This process involved the tenosynovial spaces in the length of an entire finger, extending into the hand. Note that numerous multinucleated giant cells are present in the distended villi.

have a concomitant intra-articular component. This designation applies to many TSGTs in proximal locations.

Clinical features

Both the diffuse extra-articular and articular tumors present mainly in young adults with a female predominance, similar to the localized tumors.[84,85] These types overlap because the tumors primarily forming an extra-articular mass can also have an intra-articular component and vice versa. The diffuse tumors typically occur more proximally than do the localized ones and are rare in the digits. In a large series, the wrist, knee, thigh, and foot were the most common locations, and together these joints constituted 60% of cases of extra-articular tumors. These tumors vary from a small nodule to a mass >10 cm. The recurrence rate is much higher than that of the localized tumors and has been estimated as being 33%.[85] Wide excision is therefore optimal whenever possible.

Pathology

The diffuse extra-articular tumors are typically grossly yellowish to brown and can appear circumscribed or infiltrative. On sectioning, they are usually softer than the localized examples, and larger tumors can have cystic changes.

Histologically typical and diagnostic is infiltration into the adjacent adipose tissue and sometimes into the skeletal muscle. The diffuse tumors are in general more cellular and less sclerotic than the localized ones. Other features more frequently seen in the diffuse variant are hemosiderin-pigmented epithelioid tumor cells and microscopic cysts lined by the lesional cells (Figure 33.31). Some tumors contain large numbers of xanthoma cells, and others have lymphocytes and plasma cells, often concentrated in the periphery of the tumor. The mitotic rate varies widely and is generally in a similar range as for the localized type, varying between 1 and 10 mitoses per 10 HPFs.

Immunohistochemical features and histogenesis

The osteoclast-like component shows phenotypic features similar to osteoclasts, is positive for CD68,[86–88] and also expresses calcitonin receptor.[87] The mononuclear component is positive for histiocyte-specific markers, such as CD14, whereas only the giant cells are positive for CD45 (LCA).[86,88] A desmin-positive and actin-negative subpopulation of spindled dendritic tumor cells occurs both in the localized and diffuse types and is probably part of the neoplastic component.[88]

Genetics

Perhaps the most important genetic change is rearrangements of CSF1 at 1p11, and this includes the t(1;2)(p11;q36–37) translocation, in which the CSF1 gene is activated by a translocation that puts this gene under the control of collagen type 6 promoter. The result is massive recruitment of histiocytes in the TSGT, including both localized and diffuse forms.[70,71] The CO6A3-CSF1 fusion is the molecular genetic equivalent of the t(1;2) translocation. It can be detected by PCR or FISH, although probes are not yet commercially available.[89]

Clonal cytogenetic changes have been detected in more than one-half of the localized examples, whereas most diffuse tumors have shown aberrations. The most common recurrent changes in the localized type of tumors have been translocations

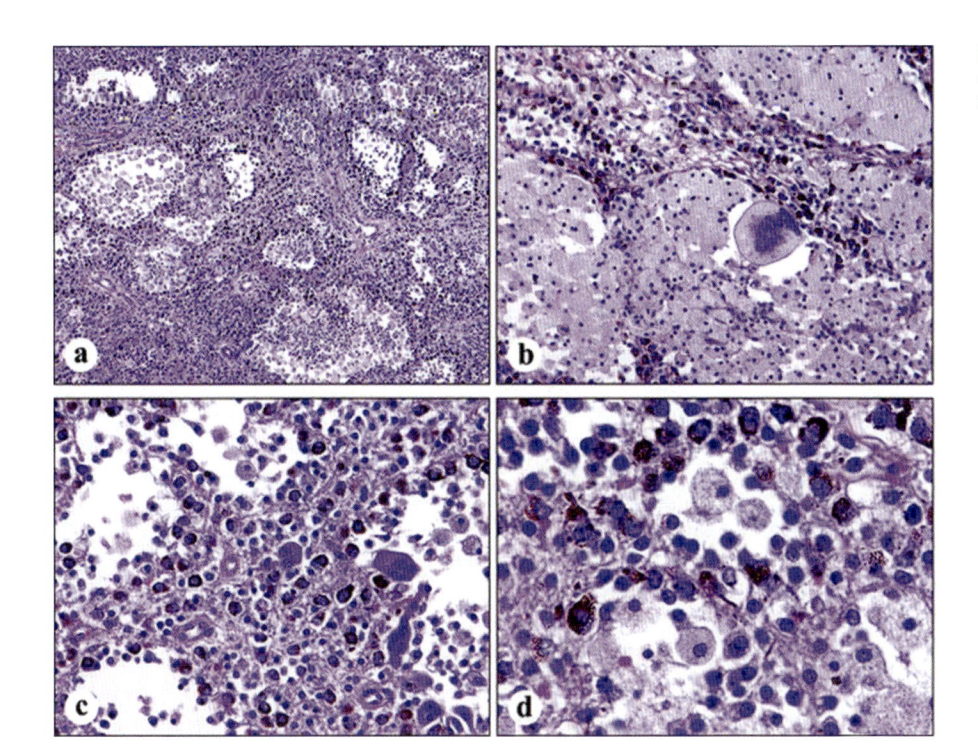

Figure 33.31 Histologically typical of the diffuse extra-articular TSGT are pseudoglandular spaces and iron-pigmented histiocytes.

involving chromosome 1, t(1;2)(p11;q36–37), related to the previously mentioned molecular genetic events.[90,91]

Some diffuse tumors have shown trisomy of chromosomes 5 and 7,[90,92,93] and a subgroup has been identified with involvement of 16q24(22). More complex changes have been seen occasionally in both the localized and diffuse variants.[94] In one study, the localized tumors were reported as diploid in DNA flow cytometric analysis, but diffuse tumors had aneuploidy with an elevated DNA index.[95] This might reflect the more common occurrence of trisomies in the diffuse tumors, as observed in the subsequent cytogenetic studies.[90] Although generally the clonal genetic aberrations support a neoplastic nature, a point has been made that rheumatoid synovia can also have clonal chromosomal changes, such as trisomy 7.[90]

Differential diagnosis

Granulomatous inflammation of the synovia contains epithelioid histiocytes forming clusters and aggregates. The giant cell nuclei are often in a horseshoe configuration, and there is typically an accompanying lymphoid population. The process usually lacks the circumscription typical of giant cell tumors.

True xanthoma has no typical osteoclasts and is composed of histiocytes and prominent cholesterol crystals; the latter are rare in giant cell tumors.

Giant cell tumor of soft tissues is a rare tumor, which by the presence of osteoclast-like cells resembles TGCT (see Chapter 7). These tumors are typically more homogeneously cellular and less collagenous, however. Foci of metaplastic bone can be found in these tumors, which do not occur in tenosynovial locations.

Extraosseous extension of giant cell tumor of bone can be seen near the large joints, especially the knee. These tumors are histologically similar to giant cell tumors of soft tissue, being more homogeneous and richer in giant cells; radiologic studies reveal the parent tumor as a lytic bone lesion, making clinicoradiologic correlation important.

Epithelioid sarcomas can simulate localized TSGTs when they contain numerous osteoclastic giant cells, but they also contain an atypical, eosinophilic epithelioid cell population that can be identified because of keratin and EMA positivity; this differential diagnosis is especially relevant for digital lesions.

Hemosiderotic synovitis resulting from chronic, repetitive intra-articular hemorrhage (seen most often in hemophilia) grossly can simulate diffuse villonodular synovitis as a brown, villous process. Similar to villonodular synovitis, the synovial cells contain extensive hemosiderin deposition; however, the villous projections are narrow and the stroma is paucicellular, fibrous, and hemosiderotic. The stroma also does not have the cellular infiltration.

Foreign body reaction to a prosthetic device can cause diffuse villous synovial hyperplasia. This process contains birefringent foreign body material, however, and often has an inflammatory mononuclear component.

Rheumatoid synovium is sometimes extensively villous, but it typically contains extensive perivascular lymphoplasmacytic infiltration.

Atypical and malignant tenosynovial giant cell tumor

This occurrence is extremely rare, but a small number of well-documented cases have been reported. In some cases, these

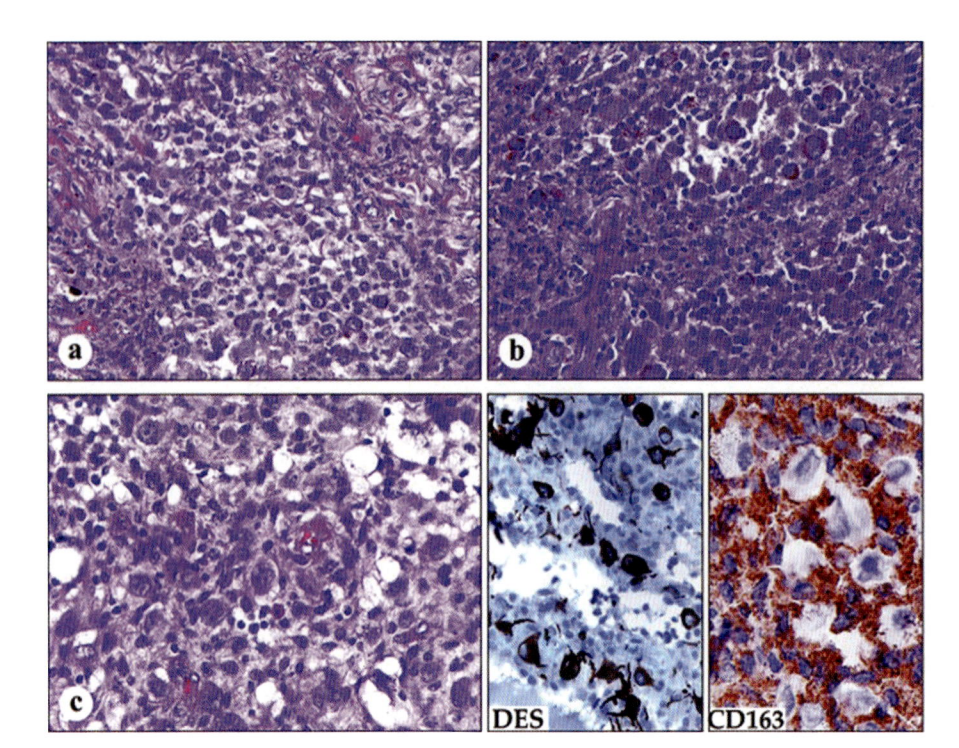

Figure 33.32 (**a, b**) Atypical TSGT. (**c**) Atypical cells with large nuclei and prominent nuclei are present. The atypical cellular component is desmin positive and negative for CD163, which labels the abundant histiocytes.

tumors have been defined based on the observed malignant behavior, and in others on sarcomatous histologic features. In general, the malignant variant can be ascertained by the presence of components diagnostic for TSGTs in connection with an overtly sarcomatous component, with mitotic activity and atypical mitoses. The features of a giant-cell-rich sarcoma are not sufficiently specific.

On average, the reported malignant tumors have been found in older patients than their benign counterparts and have represented the diffuse variants with adipose tissue or skeletal muscle infiltration. The most common location has been the knee joint; local recurrence has developed in most cases and metastases in more than one-half of the patients, one-half of whom died of the tumor. Some malignant TSGTs have arisen *de novo*, and others in the setting of a pre-existing benign TSGT. In some cases, the malignant evolution has been slow, occurring over a span of 20 years. Some patients have developed multiple soft tissue metastases in adjacent regions.[96–102]

Histologically, atypical TSGTs have a multinodular pattern of growth, high cellularity, and relative paucity of giant cells and collagen and the presence of increased number of large cells with prominent nucleoli (Figure 33.32). Although the mitotic rate is often >10 per 10 HPFs, some tumors have counts similar to those found in benign tumors. Some metastasizing tumors have a spindle cell sarcomatous pattern, and others have the features of conventional diffuse TSGTs, but they nevertheless have metastasized. Some patients have received radiation for PVNS, which could have played a role in the sarcomatous transformation.

In a recent series of malignant TSGT, lymph node, lung, and bone metastases developed in three of seven patients, and one patient died of distant metastases. The patients were older than those with benign TSGTs (mean age, 61 years). The tumors were larger (mean, >9 cm), had higher mitotic activity (18 versus 2 mitoses per 10 HPFs), and often had atypical mitoses, tumor necrosis, and a higher Ki67 index (40% versus 13%).[102]

Differential diagnosis

Several specific tumor types potentially occurring in similar locations must be ruled out. They especially include malignant fibrous histiocytoma with giant cells, epithelioid sarcoma, which has deeply eosinophilic epithelioid cells and is positive for epithelial markers (see Chapter 27), clear cell sarcoma of tendons and aponeuroses, which has a melanoma-like immunophenotype (see Chapter 25), and malignant lymphomas.

References

General

1. Breshnihan B. Are synovial biopsies of diagnostic value? *Arthritis Res Ther* 2003;5:271–278.
2. Kroot EJA, Weel AEA, Hazes JMW, et al. Diagnostic value of blind synovial biopsy in clinical practice. *Rheumatology* 2006;45:192–195.
3. Gibson T, Fagg N, Highton J, Wilton M, Dyson M. The diagnostic value of synovial biopsy in patients with arthritis of unknown cause. *Br J Rheumatol* 1985;24:232–241.

4. Teloh HA. Clinical pathology of synovial fluid. *Ann Clin Lab Sci* 1973;5:282–287.
5. Pascual E, Jovani V. Synovial fluid analysis. *Best Pract Res Clin Rheumatol* 2005;19: 371–386.

Infectious arthritis
6. Nade S. Septic arthritis. *Best Pract Res Clin Rheumatol* 2003;17:183–200.
7. Steere AC, Malawista SE, Snydman DR, et al. Lyme arthritis: an epidemic of oligoarticular arthritis in children and adults in three Connecticut communities. *Arthritis Rheum* 1977;20:7–17.
8. Steere AC. Diagnosis and treatment of Lyme arthritis. *Adv Rheumatol* 1997;81:179–194.
9. Wilske B. Epidemiology and diagnosis of Lyme borreliosis. *Ann Med* 2005;37:568–579.
10. Liebling MR, Nishio MJ, Rodriquez A, et al. The polymerase chain reaction for the detection of *Borrelia burgdorferi* in human body fluids. *Arthritis Rheum* 1993;36:665–675.
11. Duray PH. The surgical pathology of human Lyme disease: an enlarging picture. *Am J Surg Pathol* 1987;11 (Suppl 1): 47–60.
12. Duray PH. Histopathology of clinical phases of human Lyme disease. *Rheum Dis Clin North Am* 1989;15:691–710.
13. Johnston YE, Duray PH, Steere AC, et al. Lyme arthritis: spirochetes found in synovial microangiopathic lesions. *Am J Pathol* 1985;118:26–34.
14. de Koning J, Bosma RB, Hoogkamp-Korstanje JAA. Demonstration of spirochetes in patients with Lyme disease with modified silver stain. *J Med Microbiol* 1987;23:261–267.

Rheumatoid arthritis
15. Arnett FC, Edworthy SM, Bloch DA, et al. The American Rheumatism Association 1987 revised criteria for the classification of rheumatoid arthritis. *Arthritis Rheum* 1988;31:315–324.
16. Goldenberg DL, Cohen AS. Synovial membrane histopathology in the differential diagnosis of rheumatoid arthritis, gout, pseudogout, systemic lupus erythematosus, infectious arthritis and degenerative joint disease. *Medicine (Baltimore)* 1978;57: 239–252.
17. Krenn V, Morawietz L, Burmeister GR, et al. Synovitis score: discrimination between chronic low-grade and high-grade synovitis. *Histopathology* 2006;49:358–364.
18. Magalhaes R, Stiehl P, Morawietz L, Berek C, Krenn V. Morphological and molecular pathology of the B cell response in synovitis of rheumatoid arthritis. *Virchows Arch* 2002;441:415–427.
19. Galvez J, Sola J, Ortuno G, et al. Microscopic rice bodies in rheumatoid synovial fluid sediments. *J Rheumatol* 1992;19:1851–1858.
20. Albrecht M, Marinetti GV, Jacox F, Vaughan JH. A biochemical and electron microscopy study of rice bodies from rheumatoid arthritis. *Arthritis Rheum* 1965;8:1053–1063.
21. Berg E, Wainwright R, Barton B, Puchtler H, McDonald T. On the nature of rheumatoid rice bodies: an immunologic, histochemical, and electron microscopic study. *Arthritis Rheum* 1977;20:1343–1349.
22. Popert AJ, Scott DL, Wainwright AC, et al. Frequency of occurrence, mode of development, and significance of rice bodies in rheumatoid joints. *Ann Rheum Dis* 1982;41:109–117.
23. Sugano I, Bagao T, Tajima Y, et al. Variation among giant rice bodies: report of four cases and their clinicopathological features. *Skeletal Radiol* 2000;29:525–529.
24. Steinfeld R, Rock MG, Yonge DA, Cofield RH. Massive subacromial bursitis with rice bodies: report of three cases, one of which was bilateral. *Clin Orthop Relat Res* 1994;201:185–190.
25. Garcia-Patos V. Rheumatoid nodule. *Semin Cutan Med Surg* 2007;26:100–107.
26. Altman R, Asch E, Bloch D, et al. Development of criteria for the classification and reporting of osteoarthritis. *Arthritis Rheum* 1986;29:1039–1049.

Granulomatous synovitis
27. Scott DG, Porto LO, Lovell CR, Thomas GO. Chronic sarcoid synovitis in the Caucasian: an arthroscopic and histological study. *Ann Rheum Dis* 1981;40:121–123.
28. Palmer DG, Schumacher HR. Synovitis with non-specific histological changes in synovium in chronic sarcoidosis. *Ann Rheum Dis* 1984;43:778–782.
29. Frayha R, Stevens MB, Bayless TM. Destructive monoarthritis and granulomatous synovitis as the presenting manifestations of Crohn's disease. *Johns Hopkins Med J* 1975;137:151–155.
30. Hermans PJ, Fieved ML, Descamps CL, Aupaix MA. Granulomatous synovitis and Crohn's disease. *J Rheumatol* 1984;11:710–712.
31. Sutker WL, Lankford LL, Tompsett R. Granulomatous synovitis: the role of atypical mycobacteria. *Rev Infect Dis* 1979;1:729–735.
32. Beckman EN, Pankey GA, McFarland GB. The histopathology of *Mycobacterium marinum* synovitis. *Am J Clin Pathol* 1985;83:457–462.
33. Marrocco GR, Tihen WS, Goodnough CP, Johnson RJ. Granulomatous synovitis and osteitis caused by *Sporothrix schenckii*. *Am J Clin Pathol* 1975;64:345–350.

Crystal-induced arthritis
34. Schumacher HR. Crystal-induced arthritis: an overview. *Am J Med* 1996;100:46S–52S.
35. Darby AJ, Harness NF, Pritchard MS. Demonstration of urate crystals after formalin fixation. *Histopathology* 1998;32:382–383.
36. Shidham V, Chivukula M, Basir Z, Shidman G. Evaluation of crystals in formalin fixed, paraffin embedded tissue sections for the differential diagnosis of pseudogout, gout, and tumoral calcinosis. *Mod Pathol* 2001;14:806–810.
37. Chaplin AJ. Calcium pyrophosphate. Histological characterization of crystals in pseudogout. *Arch Pathol Lab Med* 1976;100:12–15.
38. Skinner M, Cohen AS. Calcium pyrophosphate dihydrate crystal deposition disease. *Arch Intern Med* 1969;123:636–644.
39. Ishida T, Dorfman HD, Bullough PG. Tophaceous pseudogout (tumoral calcium pyrophosphate deposition disease). *Hum Pathol* 1995;26:587–593.
40. Beutler A, Rothfuss S, Clayburne G, Sieck M, Schumacher HR Jr. Calcium pyrophosphate dihydrate crystal deposition in synovium:

relationship to collagen fibers and chondrometaplasia. *Arthritis Rheum* 1993;36:704–715.

41. Fam AG, Topp JR, Stein HB, Little AH. Clinical and roentgenographic aspects of pseudogout: a study of 50 cases and a review. *Can Med Assoc J* 1981;124:545–551.

Synovial foreign body reactions

42. Mirra JM, Marder RA, Amstutz HC. The pathology of failed total joint arthroplasty. *Clin Orthop Relat Res* 1982;170:175–183.

43. Bullough PG, DiCarlo EF, Hansraj KK, Neves MC. Pathologic studies of total joint replacement. *Orthop Clin North Am* 1988;19:611–625.

44. Ewald FC, Sledge CB, Corson JM, Rose RM, Radin EL. Giant cell synovitis associated with failed polyethylene patellar replacements. *Clin Orthop Relat Res* 1976;115:213–219.

45. Willert HG. Reactions of the articular capsule to wear products of artificial joint prostheses. *J Biomed Mater Res* 1977;11:157–164.

46. Goodman S, Lidgren L. Polyethylene wear in knee arthroplasty. *Acta Orthop Scand* 1992;63:358–364.

47. Mohanty M. Cellular basis of failure of joint prosthesis. *Biomed Mater Eng* 1996;6:165–172.

48. Kadoya Y, Kobayashi A, Ohashi H. Wear and osteolysis in total joint replacements. *Acta Orthop Scand Suppl* 1998;278:1–16.

49. Niki Y, Matsumoto H, Otani T, *et al.* Gigantic popliteal synovial cyst caused by wear particles after total knee arthroplasty. *J Arthroplasty* 2003;18:1071–1075.

50. Zaloudek C, Treseler PA, Powell CB. Postarthroplasty histiocytic lymphadenopathy in gynecologic oncology patients: a benign reactive process that clinically may be mistaken for cancer. *Cancer* 1996;78:834–844.

51. Schmalzried TP, Jasty M, Rosenberg A, Harris WH. Histologic identification of polyethylene wear debris using Oil Red O stain. *J Appl Biomater* 1993;4:119–125.

52. Johanson NA, Bullough PG, Wilson PD Jr, Salvati EA, Ranawat CS. The microscopic anatomy of the bone–cement interface in failed total hip arthroplasties. *Clin Orthop Rel Res* 1987;218:123–135.

53. Friedlander GN, Potter GK, Tucker RS, Berlin SJ. Silicone elastomer microshards in fluid from a painful metatarsophalangeal implant site: a case report. *Acta Cytol* 1995;39:586–588.

54. de Heer DH, Owens SR, Swanson AB. The host response to silicone elastomer implants for small joint arthroplasty. *J Hand Surg [Am]* 1995;20:S101–S109.

55. Hirakawa K, Bauer TW, Culver JE, Wilde AH. Isolation and quantitation of debris particles around failed silicone orthopedic implants. *J Hand Surg Am* 1996;21:819–827.

56. Bercovy M, Goutallier D, Voisin MC, *et al.* Carbon-PGLA prostheses for ligament reconstruction: experimental basis and short-term results in man. *Clin Orthop Rel Res* 1985;196:159–168.

57. Johnson FB. Identification of graphite in tissue sections. *Arch Pathol Lab Med* 1980;104:491–492.

58. Buly RL, Huo MH, Salvati E, Brien W, Bansal M. Titanium wear debris in failed cemented total hip arthroplasty: an analysis of 71 cases. *J Arthroplasty* 1992;7:315–323.

59. Matsuda Y, Yamamuro T, Kasai R, Matsusue Y, Okumura H. Severe metallosis due to abnormal abrasion of the femoral head in a dual bearing hip prosthesis: a case report. *Arthroplasty* 1992;7:439–445.

60. Albores-Saavedra J, Vuitch F, Delgado R, Wiley E, Hagler H. Sinus histiocytosis of pelvic lymph nodes after hip replacement: a histiocytic proliferation induced by cobalt-chromium and titanium. *Am J Surg Pathol* 1994;18:83–90.

Ochronosis and pathology of tendons

61. La Du BN Jr. Alcaptonuria and ochronotic arthritis. *Mol Biol Med* 1991;8:31–38.

62. Phornphutkul C, Introne WJ, Perry MB, *et al.* Natural history of alkaptonuria. *N Engl J Med* 2002;347:2111–2121.

63. Keller JM, Macaulay W, Nercessian OA, Jaffe IA. New developments in ochronosis: review of the literature. *Rheumatol Int* 2005;25:81–85.

64. Farzannia A, Shokouhi G, Hadidchi S. Alkaptonuria and lumbar disc herniation. Report of three cases. *J Neurosurg* 2003;98(1 Suppl):87–89.

65. Ando W, Sakai T, Kudawara I, *et al.* Bilateral achilles tendon ruptures in a patient with ochronosis: a case report. *Clin Orthop Relat Res* 2004;424:180–182.

66. Manoj Kumar RV, Rajasekaran S. Spontaneous tendon ruptures in alkaptonuria. *J Bone Joint Surg Br* 2003;85:883–886.

67. Gaines JJ. The pathology of alkaptonuric ochronosis. *Hum Pathol* 1989;20:40–46.

Tenosynovial giant cell tumor: clinicopathologic and immunohistochemical studies

68. Jaffe HL, Lichtenstein J, Sutro C. Pigmented villonodular synovitis, bursitis and tenosynovitis. *Arch Pathol* 1941;31:731–765.

69. Sciot R, Rosai J, Dal Cin P, *et al.* Analysis of 35 cases of localized and diffuse tenosynovial giant cell tumor: a report from the chromosomes and morphology (CHAMP) study group. *Mod Pathol* 1999;12:576–579.

70. West RB, Rubin BP, Miller MA, *et al.* A landscape effect in tenosynovial giant cell tumor from activation of CSF1 expression by a translocation in a minority of tumor cells. *Proc Natl Acad Sci USA* 2006;103:690–695.

71. Cupp JS, Miller MA, Montgomery KD, *et al.* Translocation and expression of CSF1 in pigmented villonodular synovitis, tenosynovial giant cell tumor, rheumatoid arthritis and other reactive synovitides. *Am J Surg Pathol* 2007;31:970–976.

72. Myers BW, Masi AT, Feigenbaum SL. Pigmented villonodular synovitis and tenosynovitis: a clinical and epidemiologic study of 166 cases and literature review. *Medicine* 1980;59:223–238.

73. Monaghan H, Salter DM, Al-Nafussi A. Giant cell tumour of tendon sheath (localised nodular tenosynovitis): clinicopathological features of 71 cases. *J Clin Pathol* 2001;54:404–407.

74. Jones FE, Soule EH, Coventry MB. Fibrous xanthoma of synovium (giant cell tumor of tendon sheath, pigmented nodular synovitis): a study of one hundred and eighteen cases. *J Bone Joint Surg Am* 1969;51:76–86.

75. Ushijima M, Hashimoto H, Tsuneyoshi M, Enjoji M. Giant cell tumor of tendon sheath (nodular tenosynovitis): a study of 207 cases to compare the large joint group with the common digit group. *Cancer* 1986;57:875–884.

76. Rao AS, Vigorita VJ. Pigmented villonodular synovitis (giant cell tumor of the tendon sheath and synovial membrane). *J Bone Joint Surg Am* 1984;66:76–94.

77. Karasick D, Karasick S. Giant cell tumor of tendon sheath: spectrum of radiologic changes. *Skeletal Radiol* 1992;21:219–224.

78. Granowitz SP, D'Antonio J, Mankin HL. The pathogenesis and long-term end results of pigmented villonodular synovitis. *Clin Orthop* 1976;114:335–351.

79. Ushijima M, Hashimoto H, Tsuneyoshi M, Enjoji M. Pigmented villonodular synovitis: a clinicopathologic study of 52 cases. *Acta Pathol Jpn* 1986;36:317–326.

80. Schwartz HS, Unni KK, Pritchard DJ. Pigmented villonodular synovitis: a retrospective review of affected large joints. *Clin Orthop* 1989;247:243–255.

81. Dorwart RH, Genant HK, Johnston WH, Morris JM. Pigmented villonodular synovitis of synovial joints: clinical, pathologic, and radiologic features. *AJR Am J Roentgenol* 1984;143:877–885.

82. Gonzalez Della Valle A, Piccaluga F, Potter HG, Salvati EA, Pusso R. Pigmented villonodular synovitis of the hip: 2- to 23-year follow-up study. *Clin Orthop* 2001;388:187–199.

83. Rowlands CG, Roland B, Hwang WS, Sevick RJ. Diffuse variant tenosynovial giant cell tumor: a rare and aggressive lesion. *Hum Pathol* 1994;25:423–425.

84. de Aubain Somerhausen N, Fletcher CDM. Diffuse-type giant cell tumor: clinicopathologic and immunohistochemical analysis of 50 cases with extraarticular disease. *Am J Surg Pathol* 2000;24:479–492.

85. Wood GS, Beckstead JH, Medeiros LJ, Kempson RL, Warnke RA. The cells of giant cell tumor of tendon sheath resemble osteoclasts. *Am J Surg Pathol* 1988;12:444–452.

86. O'Connell JX, Fanburg JC, Rosenberg AE. Giant cell tumor of tendon sheath and pigmented villonodular synovitis: immunophenotype suggests a synovial cell origin. *Hum Pathol* 1995;26:771–775.

87. Darling JM, Goldring SR, Harada Y, et al. Multinucleated cells in pigmented villonodular synovitis and giant cell tumor of tendon sheath express features of osteoclasts. *Am J Pathol* 1997;150:1383–1393.

88. Folpe AL, Weiss SW, Fletcher CDM, Gown AM. Tenosynovial giant cell tumors: evidence for a desmin-positive dendritic cell subpopulation. *Mod Pathol* 1998;11:939–944.

89. Möller E, Mandahl N, Mertens F, Panagopoulos I. Molecular identification of COL6A3-CSF1 fusion transcripts in tenosynovial giant cell tumors. *Genes Chromosomes Cancer* 2008;47:21–25.

90. Mertens FM, Örndal C, Mandahl N, et al. Chromosome aberrations in tenosynovial giant cell tumors and nontumorous synovial tissue. *Genes Chromosomes Cancer* 1993;6:212–217.

91. Dal Cin P, Sciot R, Samson I, et al. Cytogenetic characterization of tenosynovial giant cell tumors (nodular tenosynovitis). *Cancer Res* 1994;54:3986–3987.

92. Ray RA, Morton CC, Lipinski KK, Corson JM, Fletcher JA. Cytogenetic evidence for clonality in a case of pigmented villonodular synovitis. *Cancer* 1991;67:121–125.

93. Fletcher JA, Henkle C, Atkins L, Rosenberg AE, Morton CC. Trisomy 5 and trisomy 7 are nonrandom aberrations in pigmented villonodular synovitis: confirmation of trisomy 7 in uncultured cells. *Genes Chromosomes Cancer* 1992;4:264–266.

94. Dal Cin P, Sciot R, De Smet L, van Damme B, Van Den Berghe H. A new cytogenetic subgroup in tenosynovial giant cell tumors (nodular tenosynovitis) is characterized by involvement of 16q24. *Cancer Genet Cytogenet* 1996;87:85–87.

95. Abdul-Karim FW, El-Naggar AK, Joyce MJ, Makley JT, Carter JR. Diffuse and localized tenosynovial giant cell tumor and pigmented villonodular synovitis: a clinicopathologic and flow cytometric DNA analysis. *Hum Pathol* 1992;23:729–735.

Malignant tenosynovial giant cell tumor

96. Kahn LB. Malignant giant cell tumor of the tendon sheath. *Arch Pathol* 1973;95:203–208.

97. Castens PHB, Howell RS. Malignant giant cell tumor of tendon sheath. *Virchows Arch A Pathol Anat Histol* 1979;382:237–243.

98. Lynge-Nielsen A, Kiaer T. Malignant giant cell tumor of synovium and locally destructive pigmented villonodular synovitis: ultrastructural and immunohistochemical study and review of the literature. *Hum Pathol* 1989;20:765–771.

99. Choong PFM, Willen H, Nilbert M, et al. Pigmented villonodular synovitis. Monoclonality and metastasis: a case for neoplastic origin? *Acta Orthop Scand* 1995;66:64–68.

100. Bertoni F, Unni KK, Beabout JW, Sim FH. Malignant giant cell tumor of tendon sheaths and joints (malignant pigmented villonodular synovitis). *Am J Surg Pathol* 1997;21:153–163.

101. Layfield LJ, Meloni-Ehrig A, Liu K, Shepard R, Harrelson JM. Malignant giant cell tumor of synovium (malignant pigmented villonodular synovitis): a histopathological and fluorescence in situ hybridization analysis of 2 cases and review of the literature. *Arch Pathol Lab Med* 2000;124:1636–1641.

102. Li CF, Wang JW, Huang WW, et al. Malignant diffuse-type tenosynovial giant cell tumors: a series of 7 cases comparing with 24 benign lesions with review of the literature. *Am J Surg Pathol* 2008;32:587–599.

Chapter 34

Miscellaneous tumor-like lesions, and histiocytic and foreign body reactions

Markku Miettinen
National Cancer Institute, National Institutes of Health

This chapter includes a variety of non-neoplastic lesions that can form tumor-like masses or histologically simulate a neoplasm. Among these are histiocytic reactions to apparently noninfectious stimuli (e.g., granuloma annulare, pulse granuloma) and infections (e.g., mycobacterial pseudotumor, malakoplakia, and xanthogranulomatous reaction to infection). Accumulation or retention of foreign material can cause tumor-like reactive masses (e.g., retained cotton sponge), or microscopically can simulate a tumor, such as mucinous or signet ring cell carcinoma or chordoma (polyvinylpyrrolidone storage in histiocytes). Another example of permanently retained diagnostically used material is *thorotrast*, a historically used radioactive contrast medium associated with fibrous tumors at injection sites and hepatic malignancies. In the case of gadolinium-associated fibrosis, the fibrous reaction is visible, whereas the causative agent is not microscopically detectable. Finally, amyloid tumor is discussed here, including its variants associated with systemic amyloidosis and lymphoplasmacytic lymphoma (immunoglobulin light-chain-derived amyloidosis).

Organizing hematoma

Soft tissue hematoma can form a long-standing and expanding tumor-like mass that clinically, radiologically, and even pathologically sometimes simulates a neoplasm. Such lesions have been variably referred to as *post-traumatic cyst of soft tissue*,[1] *chronic expanding hematoma*,[2] and *ancient hematoma*.[3] *Calcified myonecrosis* is a related term, with this condition also including a significant component of muscular necrosis and often developing after conditions such as compartment syndrome.[4]

Clinical features

Organizing hematoma can develop in a variety of soft tissue sites that most commonly include the thigh, abdominal wall, and chest wall. The tensor fasciae latae muscle in the upper lateral thigh has been the most common location in the reported series.[3] The lesion seems to usually develop in the intermuscular spaces or sometimes in the subcutis. Patients over a wide spectrum of age and of either sex can be afflicted. In many instances, there is a history of a major blunt trauma, such as that associated with a motor vehicle accident. In some cases, such hematomas have developed at the site of previous surgery. The hematoma cavity expands over time, apparently by fluid collection, and it can enlarge and become increasingly symptomatic long after the inciting traumatic event, sometimes more than 20 years later.

Pathology

Grossly, the organizing hematoma often appears as a fusiform or oval saccular mass lined by a thick capsule and containing pasty, dark-brown material consisting of old blood (Figure 34.1), or yellow or brownish fluid. In a lesion of more recent onset, a more blood-like content is sometimes present. The cyst size varies from several centimeters to >20 cm.

The histologic organization depends on the age of the hematoma. In more recent examples, the cavity is filled with recognizable blood and fibrin, and lined with hemorrhagic granulation tissue. In a fully organized older hematoma, three somewhat intermingling layers can be identified. The external

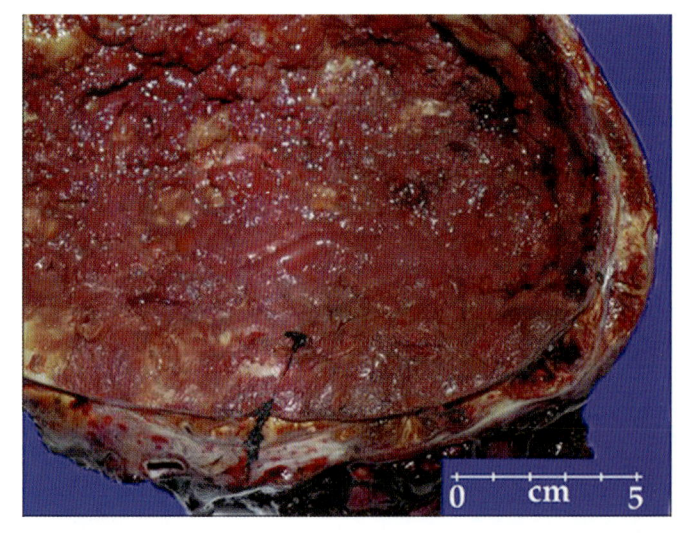

Figure 34.1 Grossly, an organized hematoma forms a sharply circumscribed sac lined by a thick capsule with yellow xanthomatous material. The contents are a paste-like brownish material, an organized old blood clot.

Modern Soft Tissue Pathology, Second Edition, ed. Markku Miettinen. Published by Cambridge University Press. © Cambridge University Press 2016.

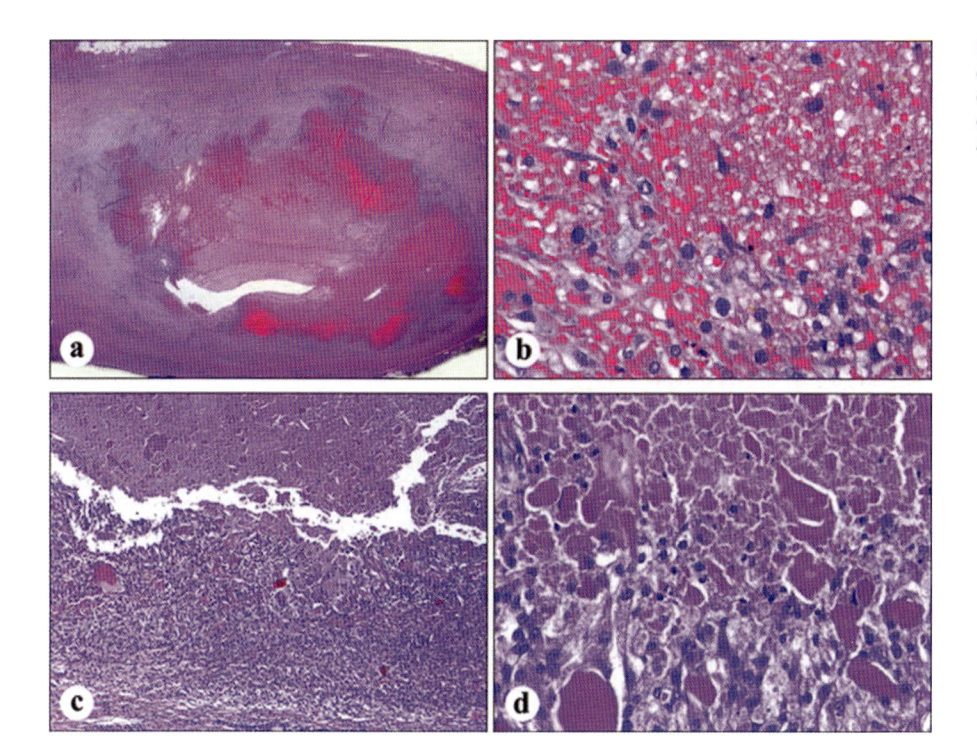

Figure 34.2 (**a,b**) An organized hematoma can contain bloody material admixed with xanthoma cells in the hematoma wall. (**c,d**) An older organizing hematoma contains granular material and sheets of xanthoma cells.

aspect is formed by fascia admixed with a variably thick, collagenous, paucicellular fibrous tissue, sometimes containing focal or extensive dystrophic calcification. Inside this there is a cellular layer, formed by sheets of xanthoma cells and mononuclear histiocytes with granular, eosinophilic cytoplasm, scattered histiocytic giant cells, and often foci of cholesterol crystals. A delicate vascular pattern is present in the cellular layer. Inside the cavity there is blood or acellular amorphous or shard-shaped tan-brown material, representing old disintegrated red blood cells (Figure 34.2).

Immunohistochemically, the histiocytic element shows strong positivity for CD163 and vimentin. Varying numbers of capsular spindle cells are smooth muscle actin positive. The lesional components are negative for keratins and S100 protein.

Differential diagnosis

A complete excision specimen presents little diagnostic challenge, although in some instances, a cellular neoplasm or necrotic tumor might be suspected, based on extensive sheets of histiocytes and old amorphous blood residue. A needle biopsy is often nondiagnostic, and results from such a specimen should be reported with caution, because some sarcomas can contain a large hematoma, masking the neoplastic process in a biopsy. This possibility should be ruled out by clinicoradiologic correlation.

Deep granuloma annulare (pseudorheumatoid nodule, necrobiotic granuloma)

Necrobiotic granuloma of soft tissues is known by several names, all of which refer to the same process. Those associated with rheumatoid arthritis are called *rheumatoid nodules* (see Chapter 33), and those that are not are sometimes called *pseudorheumatoid nodules*. The term *deep granuloma annulare* refers to the similarity of this type to the dermal version of granuloma annulare.

Clinical features

Subcutaneous necrobiotic granulomas unassociated with rheumatoid arthritis most often occur in children, usually younger than age 10. The most common locations are the pretibial lower extremity, hand, and scalp. The lesions vary in size from <1 cm to 4 cm (median, 2 cm). Their clinical behavior is indolent, and spontaneous regression often occurs. Excision is therefore not necessary if the diagnosis can be otherwise ascertained. Even after an excisional biopsy, some of these lesions tend to recur locally.[5–7] A less common occurrence in adults is well documented; according to one series this is usually in younger age groups, with apparent female predominance.[8]

Multiple eruptive cutaneous granuloma annulare lesions often occur in HIV patients. The etiology of this association is unclear, but the role of secondary Epstein–Barr viral infection has been excluded.[9] An association with rheumatoid arthritis should be ruled out, especially in adults. There is a weak association between granuloma annulare and malignancy in older adults.[10]

Pathology

Grossly, there is a pale whitish nodule often punctuated by streaks of yellow geographic necrosis. Histologically distinctive is an eosinophilic to variably mucinous amorphous necrotic center that is nearly acellular or contains residua of inflammatory cells or macrophages (Figure 34.3). The necrotic center is surrounded by perpendicularly oriented histiocytes (Figure 34.4). A narrow rim of lymphocyte-rich chronic inflammation is usually seen between the histiocytes and normal tissue. Special stains for fungi and acid-fast bacilli are negative, and these stains are indicated if there is any doubt about the diagnosis. Some examples contain mitotic activity up to 7 per 10 high-power fields (HPFs; usually <3 per 10 HPFs). This finding does not seem to have any specific significance, however.[11]

Although the histologic spectrum of granuloma annulare and rheumatoid nodule overlap, some differences have been observed. Among them, the mucinous material present in the

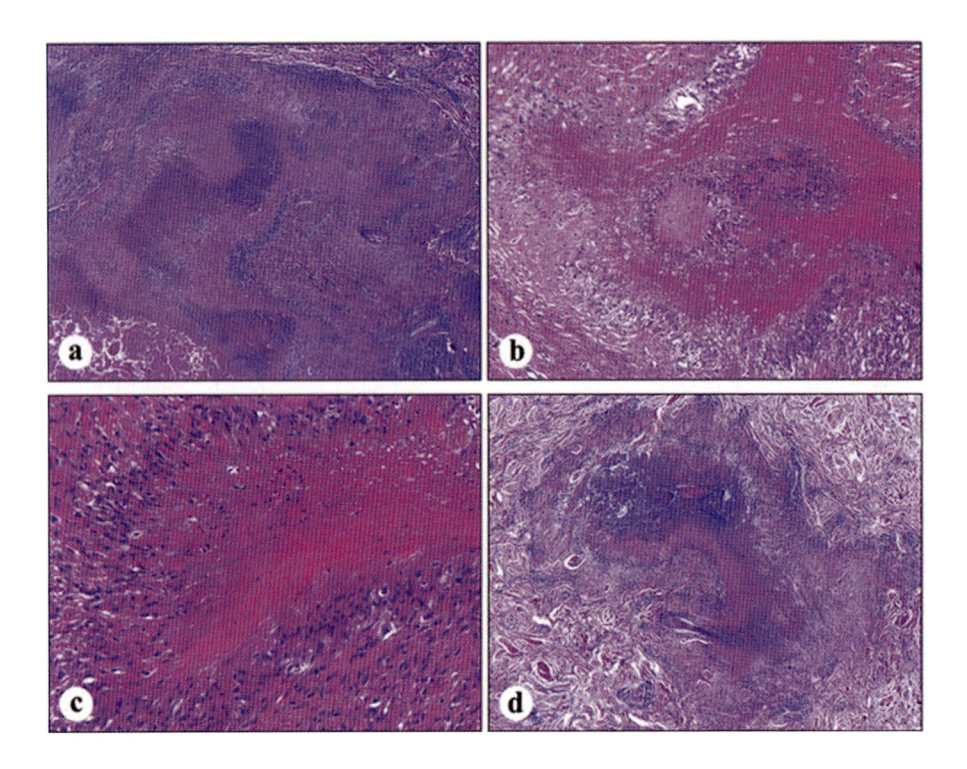

Figure 34.3 A deep granuloma annulare has serpiginous outlines. It contains an inner area of eosinophilic necrosis and is surrounded by palisading histiocytes.

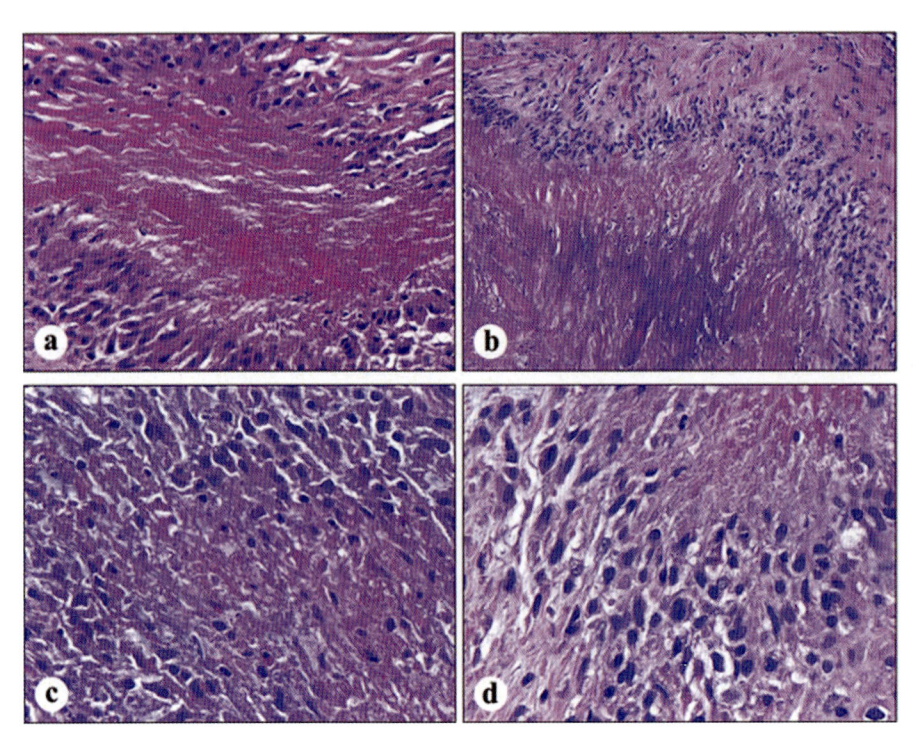

Figure 34.4 The central necrosis is nearly acellular, with small numbers of histiocytes and neutrophils. The histiocytes lining the necrosis are arranged perpendicularly to the interface with necrosis.

central necrosis is more common in granuloma annulare, whereas a rheumatoid nodule more commonly contains a solid eosinophilic, fibrous center.[12] If the distinction cannot be clearly made histologically, clinicoserologic correlation and follow-up studies for rheumatoid disease are indicated.

Differential diagnosis

The necrotizing granuloma pattern somewhat resembles that seen in epithelioid sarcoma. The latter, however, contains mildly atypical epithelioid cells with prominent eosinophilic cytoplasm and shows epithelial markers, both keratin and epithelioid membrane antigen (EMA). In some cases, suture granuloma may simulate necrotic granuloma.[13]

Pulse granuloma (giant cell hyaline angiopathy)

This rare condition, which includes clusters of foreign body giant cells and ring-like arrangements of hyaline material, is generally known by the names *pulse granuloma* or *giant cell hyaline angiopathy*. Neither name is totally accurate, because this condition might not be intravascular, and it is uncertain if legume (pulse) plant parts have any causal role.

Clinical features

Pulse granuloma usually develops in the oral cavity or buccal wall after a dental procedure such as tooth extraction. Food-borne vegetable material, especially from legume plants (e.g., beans, peas, lentils) embedded in tissue, was thought to trigger this granulomatous reaction. This is based on experimental studies in which similar lesions were produced by the introduction of legume material into oral soft tissue. Other causes probably exist, however, because demonstration of vegetable material is successful in only a few cases. Similar granulomas have been occasionally reported in the colon and rectum or their surrounding tissues, associated with diverticula and perforations. Pulse granuloma is a benign inflammatory process.[14–21]

Pathology

Histologic features vary depending on the degree of acute versus chronic inflammation. The process was originally thought to involve vessel walls, but usually there is no vascular involvement. Acute and chronic granulomatous inflammation associated with the vegetable material can occur. In some cases, chronic granulomatous inflammation is present as multiple nodules with a "plexiform" architecture. The nodules contain clusters of foreign body giant cells with irregular nuclear arrangement. Scattered, thin, convoluted, thread-like hyaline rings are present (Figure 34.5). These are typically periodic-acid–Schiff (PAS) stain positive, but are not birefringent on polarization. The hyaline rings therefore cannot be necessarily confirmed as vegetable material.

Mycobacterial pseudotumor

Abundant mycobacterial infestation in histiocytes can create the impression of a spindle cell neoplasm or granular cell tumor. This occurrence is typically associated with an immunosuppressed state, especially HIV/AIDS. The most commonly detected mycobacterium is the *Mycobacterium*

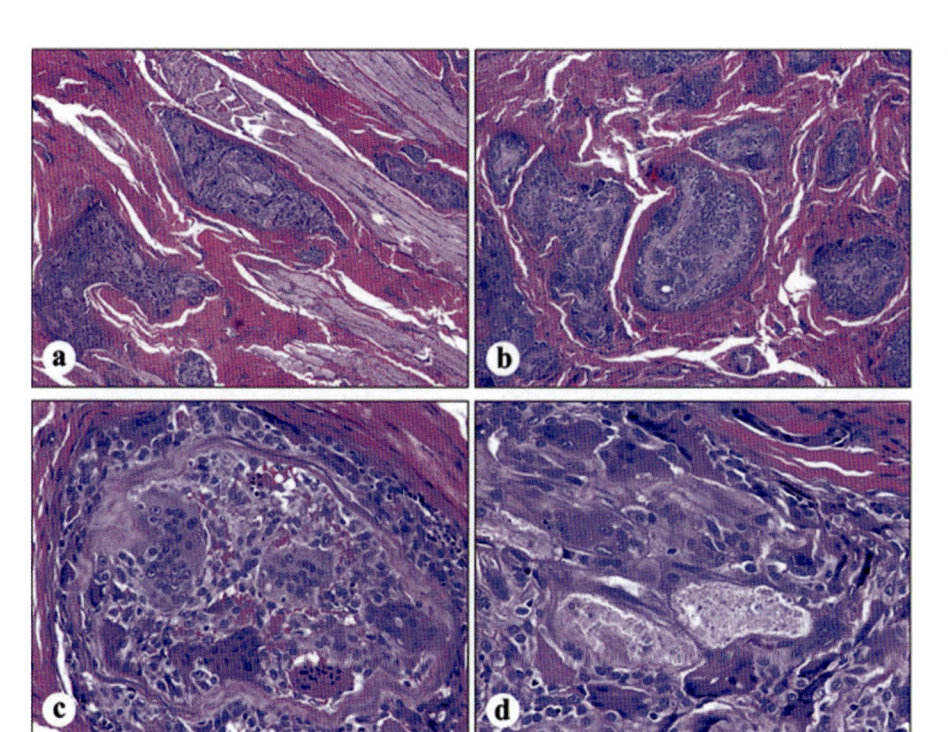

Figure 34.5 Pulse granuloma involving the soft tissue of the cheek is composed of multiple nodules rich in multinucleated histiocytes. Some of the nodules contain a band of eosinophilic material.

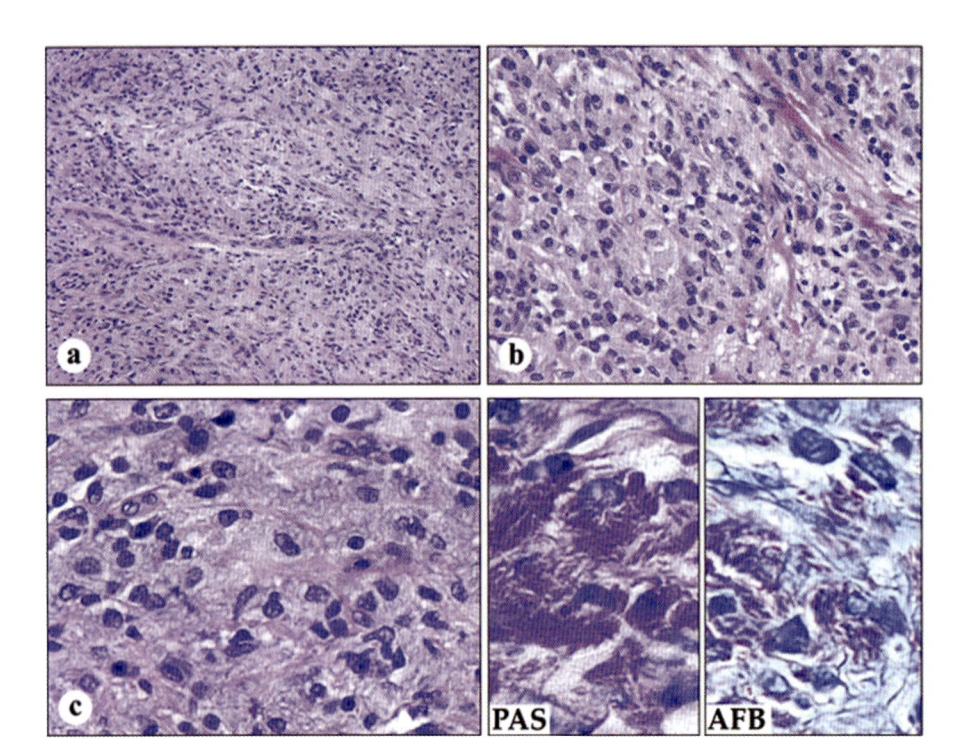

Figure 34.6 Mycobacterial pseudotumor is composed of pale eosinophilic to spindle cells with scattered lymphocytes and plasma cells. The histiocytes contain PAS-positive bacilli that are acid-fast staining (Ziehl–Nielsen) positive.

avium intracellulare. The affected tissues are primarily lymph nodes, but extranodal tissues also can be involved.[22–24]

A similar histologic picture can also occur in so-called histoid leprosy, where lepra bacilli massively infest macrophages, usually in the setting of leprosy resistant to dapsone therapy.[25,26] Histologically, there is a spindle cell proliferation admixed with small numbers of plasma cells and lymphocytes, with a well-developed capillary network. A storiform histologic pattern can resemble the one in fibrohistiocytic tumors. The spindle cells have a variably granular cytoplasm. They are PAS and Ziehl–Neelsen (acid-fast) positive, containing numerous rod-shaped acid-fast bacilli (Figure 34.6). The spindled histiocytes are positive for CD68, and according to one study, the bacteria can be desmin positive, which should not be interpreted as evidence of a myoid tumor.[27] The cytoplasmic bacteria can also be detected in touch imprint preparations in frozen sections.[28]

Malakoplakia

Malakoplakia, literally translated from Greek as "soft plaque," is a rare, histologically distinctive, reactive localized histiocytic proliferation in response to certain bacterial infections. It is strongly associated with infections by Gram-negative bacteria, especially *Escherichia coli*, and less commonly *Klebsiella pneumoniae*, *Proteus* spp., or *Pseudomonas aeruginosa*. Some examples, especially when found in HIV/AIDS or other immunosuppressed patients, are caused by opportunistic bacteria such as *Rhodococcus equi*, and some are negative in bacterial culture. Malakoplakia also occurs in transplant recipients and patients who have received immunosuppressive therapy for other reasons. The precise pathogenesis is not known, but an impaired lysosomal ability to destroy ingested bacteria could be responsible.[29–37]

Malakoplakia has a predilection for older adults, and in the genitourinary system it is more common in women. It is most common in the urogenital tract, especially the kidney and urinary bladder. The respiratory, gynecologic, and gastrointestinal tracts can also be involved. In the gastrointestinal tract it often occurs with colon carcinoma.[35,36] Soft tissue examples are rare, but cutaneous abdominal wall lesions draining to the skin have occurred associated with internal lesions,[32] and solitary soft tissue lesions have been reported in the neck.[30] Treatment of symptomatic examples consists of surgical excision of lesions and antibiotics, but in those cases in which malakoplakia has been an incidental microscopic finding (e.g., with colon cancer), surgery might be sufficient.

Histologically typical are sheets of large, swollen histiocytes with a variably granular cytoplasm (i.e., Hansemann cells). The presence of small intracytoplasmic calcospherites with a central basophilic density, Michaelis–Gutmann bodies (named after the authors who were among the first to describe malakoplakia), is a characteristic feature (Figure 34.7). The histiocytes are often accompanied by large numbers of plasma cells, and variably by neutrophils and lymphocytes. The histiocytes are PAS positive, and they are immunohistochemically positive for CD68, CD163, and CD45.

Xanthogranulomatous inflammation

Extensive deposition of lipid-laden histiocytes can occur in connection with certain infections, especially in the kidney (i.e., xanthogranulomatous pyelonephritis), and in chronic cholecystitis.[38–40]

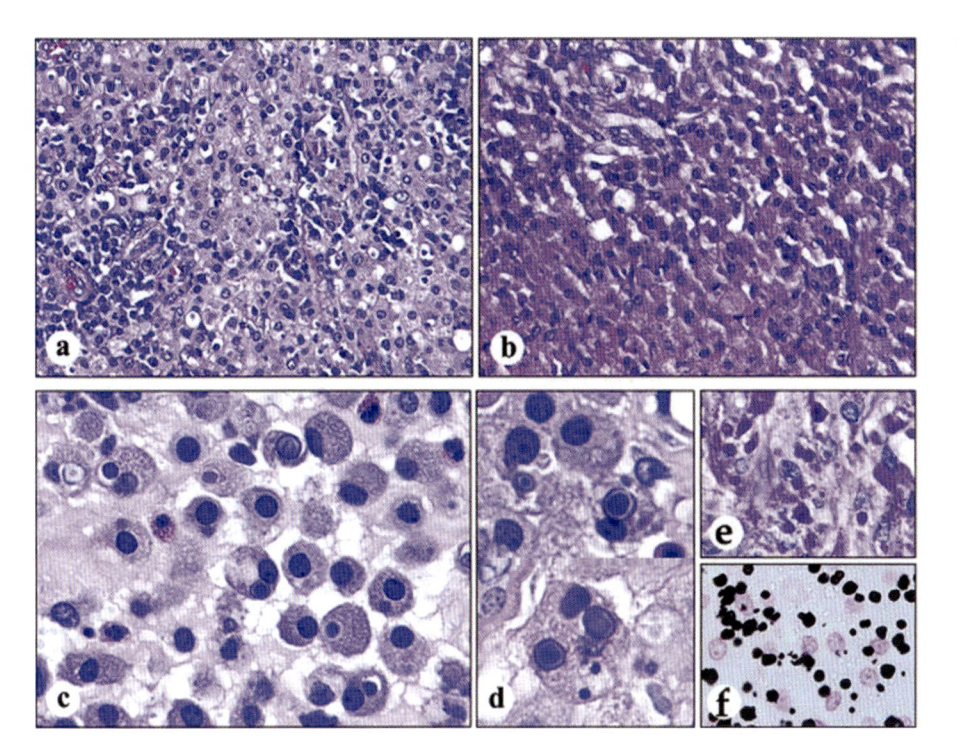

Chapter 34: Miscellaneous tumor-like lesions, and histiocytic and foreign body reactions

Figure 34.7 (**a,b**) A malakoplakia lesion contains masses of swollen histiocytes with variably eosinophilic cytoplasm. (**c,d**) Some of these cells contain Michaelis–Gutmann bodies, target-shaped microcalcifications. (**e,f**) These are positive for PAS and von Kossa stains.

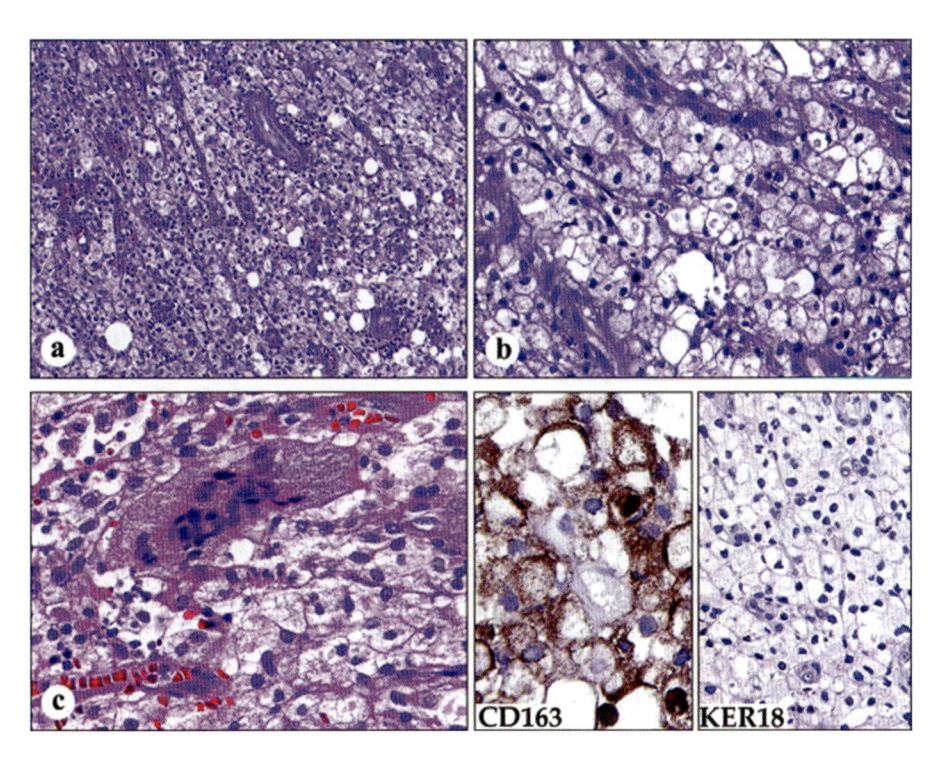

Figure 34.8 (**a,b**) Xanthogranulomatous inflammation contains sheets of lipid-laden histiocytes, and a prominent capillary network is typically present, potentially simulating metastatic renal cell carcinoma. (**c**) The presence of histiocytic giant cells is a clue to the xanthogranulomatous inflammation. The histiocytes are positive for CD163 and negative for keratin 18.

This reaction can also extend in perinephric and retroperitoneal soft tissues. The significance includes the risk of misdiagnosis as a neoplasm, on microscopic analysis, especially as a renal clear cell carcinoma. The histiocytes are positive for histiocyte-specific markers, especially CD163 and CD68, and they are negative for keratins.[41]

Histologically, there are sheets of xanthomatous histiocytes with variably granular cytoplasm, often interspersed with prominent capillaries, which further indicates the similarity of this tumor to renal cell carcinoma. The presence of multinucleated histiocytes is a clue to the reactive nature of this process (Figure 34.8). An early phase of the infection can be

931

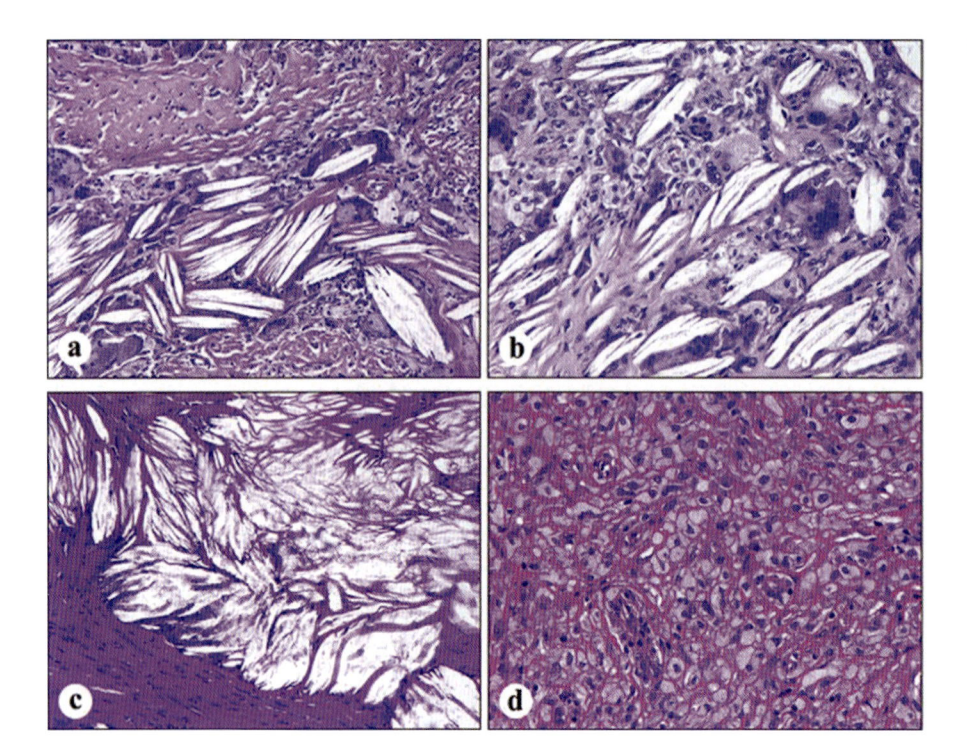

Figure 34.9 (**a–c**) A xanthoma involving the Achilles tendon consists of lipid-laden xanthoma cells and masses of cholesterol crystals infiltrating between the tendinous collagen. (**d**) The xanthomatous histiocytes can form solid sheets.

associated with neutrophilic abscess formation, whereas a later lesion contains plasma cell infiltrates. Because xanthogranulomatous inflammation occasionally can coexist with a carcinoma, this possibility should not be ignored. Extensive judicious sampling and immunohistochemical studies are therefore advisable.

Xanthoma

Xanthoma refers to non-neoplastic mass-forming collections of lipid-rich histiocytes and extracellular lipid material in the skin or deeper soft tissues, or occasionally in internal sites. Most xanthomas are related to inborn errors of lipid metabolism, and many of them are familial. In some cases, a xanthoma can be a clinical warning sign of a lipid disorder with associated cardiovascular or other morbidities. Five xanthoma types are recognized: (1) tendinous xanthoma, (2) xanthoma tuberosum, (3) planar xanthoma (xanthelasma), (4) eruptive xanthoma, and (5) disseminated xanthoma.

Tendinous xanthoma

Occurrence correlates with hypercholesterolemia (familial beta lipoproteinemia). These xanthomas can occur at an early age, even in children younger than 5 years, although more commonly they present in young adults. The Achilles tendon is a common site, and other significant sites include the extensor tendons of the hands. Tendinous xanthomas are located below the subcutis and are recognized as masses movable with the tendons. In the Achilles tendon they can be quite long (>10 cm); their transverse and anteroposterior (AP) dimensions are usually 1 cm to 3 cm.[42–44]

Grossly, these xanthomas are yellow to yellow-gray on sectioning. Microscopically they intermingle with tendinous collagenous bundles. The cellular component is a lipid-laden, polygonal histiocyte (xanthoma cell), and foci of lipids with cholesterol crystals are often present surrounded by collagenous fibrosis (Figure 34.9).

Tuberous xanthoma

These lesions usually occur in the extensor surfaces of extremities, especially the elbows, hands, and knees, and they can occur with tendinous xanthomas, being histologically similar. There is a strong association with familial hypercholesterolemia.

Planar xanthoma (xanthelasma)

These xanthomas form flat lesions in the eyelids. There is a moderate association (50%) with hypercholesterolemia, seen especially in younger patients. *Xanthoma striatum palmare* is a form of planar xanthoma in the creases of the palm, and these lesions are strongly associated with hyperlipidemia.

Eruptive xanthoma

Local, often sudden eruptions on the extensor surfaces of the arms and legs usually signify underlying hyperlipidemia.

Disseminated xanthoma

These rare xanthomas have a predilection for the flexural creases in the arms and axillae, among other places, but they can also occur in mucous membranes, the central nervous system, and bone. There is no association with hyperlipidemia syndromes.[42]

Histiocytic reaction to a ruptured epidermal inclusion cyst

A ruptured epidermal inclusion cyst is often associated with a florid histiocytic reaction, featuring sheets of multinucleated giant cells. If the remnants of the epithelial cyst are not present, and the histiocytic nature of these cells is not recognized, this process can simulate a neoplasm. A diagnostic clue is the presence of lamellar keratin debris, often in microcystic formations (Figure 34.10). Keratin immunostaining can be helpful in highlighting these keratinous elements.

Polyvinylpyrrolidone accumulation

Polyvinylpyrrolidone (PVP), a polymer of vinylpyrrolidone, was used as a plasma expander from the mid-1940s to the 1950s, and in some countries even later for other medical purposes, such as an inert retardant for injected drugs, a detoxifying agent, a component of radiologic contrast medium, or even as a "blood tonic," especially in the Far East. Subsequently, PVP was abandoned, because it became apparent that this polymer (especially when >20 kD) accumulated systemically and permanently, being taken up by histiocytes in the bone marrow, liver, spleen, lymph nodes, skin, and elsewhere.[45,46] Use in subcutaneously administered drug preparations has also led to localized tumor-like accumulations at injection sites.[47] Although PVP accumulation is often asymptomatic and merely detected in surgical specimens removed for a wide variety of unrelated indications, some patients have developed anemia because of extensive bone marrow replacement by PVP-containing histiocytes.[48] Some patients have developed lytic bone lesions, leading to pathologic fractures.[49]

Histologically characteristic are sheets of swollen, rounded histiocytes filled with basophilic, bluish mucoid material, as seen in hematoxylin-eosin (HE) stain. The histiocytes have small eccentric nuclei pushed on the side, somewhat resembling the cells of signet ring cell adenocarcinoma. This likeness especially applies to gastric biopsy specimens containing PVP.[50] The PVP histiocytes have small compacted nuclei and lack atypia, however. Multinucleated foreign body histiocytes

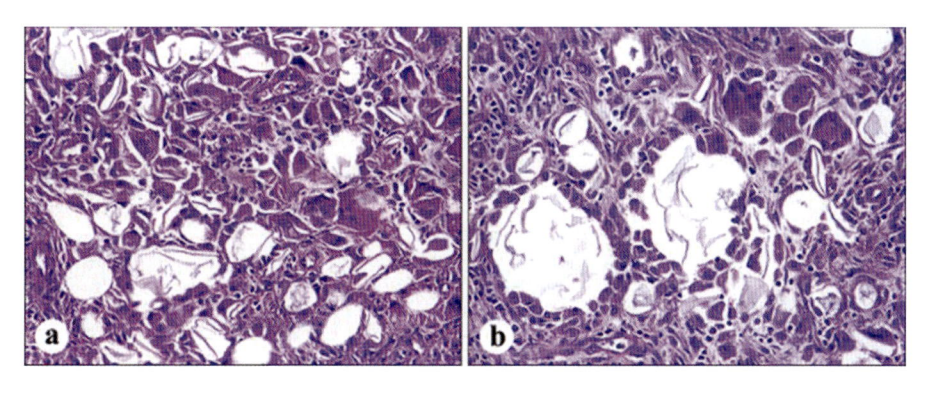

Figure 34.10 Histiocytic reaction to a ruptured epidermal inclusion cyst contains mononuclear and multinucleated histiocytes. Some cystic spaces contain fibrillary remnants of keratin debris.

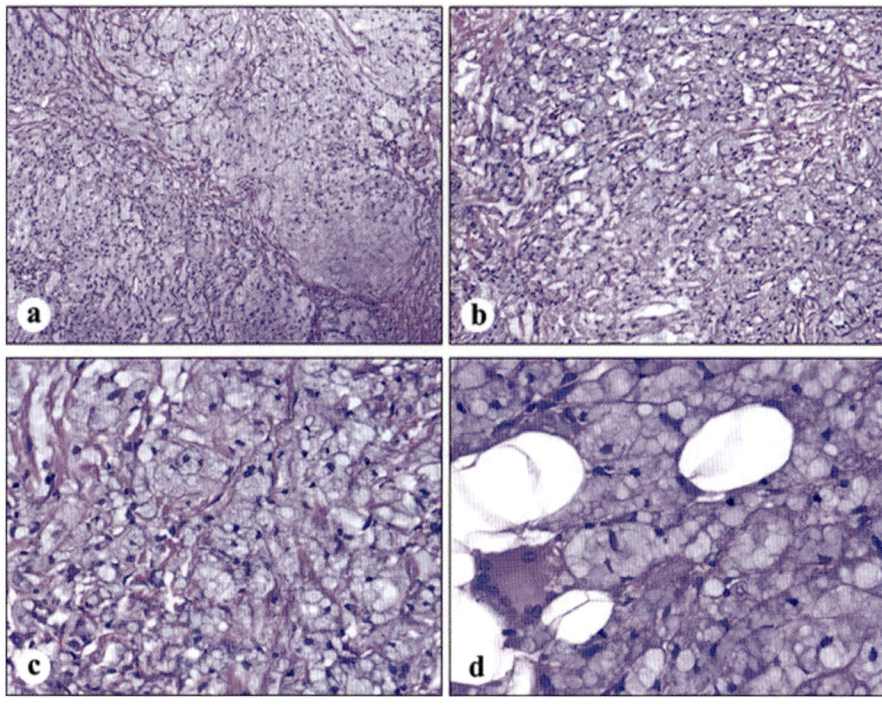

Figure 34.11 Polyvinylpyrrolidone storage in histiocytes can resemble metastatic signet ring cell carcinoma or chordoma. The histiocyte nuclei are typically inconspicuous. Large fields of extracellular mucin-like material can also be present.

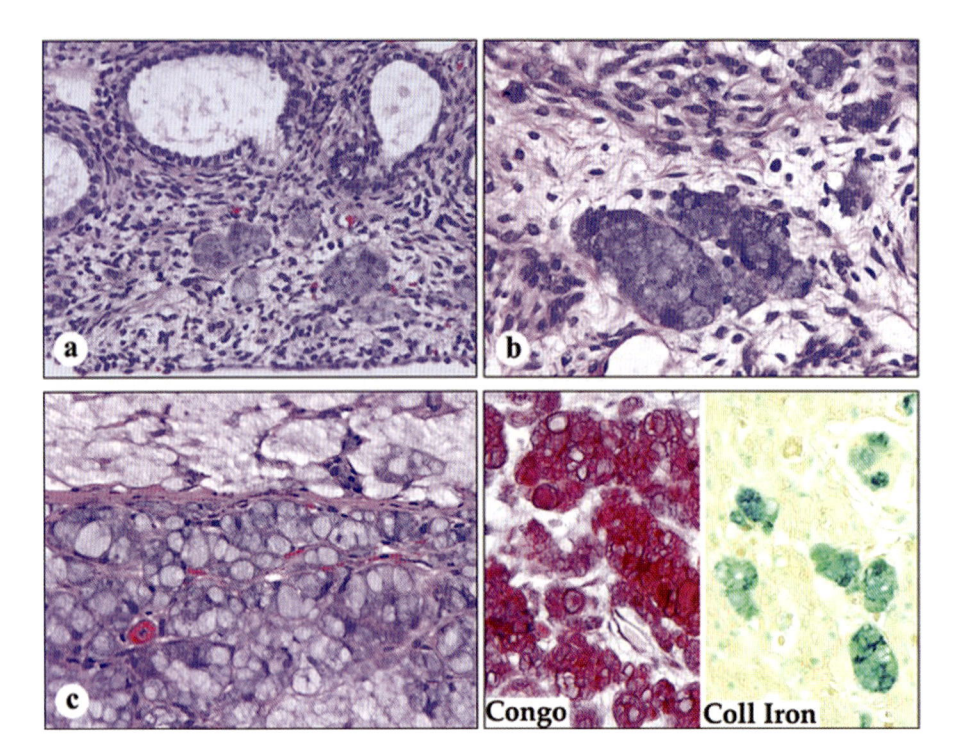

Figure 34.12 Incidental polyvinylpyrrolidone deposition in a hysterectomy specimen. Histiocytes containing the mucin-like material are present in the endometrium, myometrium, and on the peritoneal surface, where they created an appearance not dissimilar to metastatic signet ring cell carcinoma. The histiocytes are positive for Congo and colloidal iron stains.

are often present in small numbers. Diffuse extracellular pools of mucoid PVP material can also be present (Figure 34.11). Any organ can be involved in disseminated PVP accumulation. For example, in a hysterectomy specimen, small clusters of PVP-positive histiocytes can be present, in both the endometrium and the myometrium (Figure 34.12).

Histochemically, the PVP-laden histiocytes are positive for mucicarmine; the name *mucicarminophilic histiocytosis* is sometimes used for this condition. The histiocytes are also positive for colloidal iron, Congo red, Fontana–Masson, Gomori methenamine silver, and Sudan black B, whereas they are negative for PAS and alcian blue.[45,46] The PVP-containing histiocytes are not associated with any cellular inflammatory reaction.

Cotton fiber granuloma (gossypiboma, textiloma)

Cotton fibers cause a foreign body reaction. On a small scale, this phenomenon is observed around small numbers of cotton fibers that are sometimes seen at surgical sites. On HE stain, the fibers are pale-staining, donut-shaped hollow structures, and when cross-sectioned, they appear U-shaped and curved. Histochemically they are positive for Grocott's silver methenamine stain and variably for PAS.[51–56]

Accidentally retained surgical sponges consisting of cotton mesh can form a tumor-like mass, usually in a body cavity (e.g., the abdomen or mediastinum), and this is sometimes referred to as *gossypiboma* (based on the scientific name of the cotton family [*Gossypium*]). Those sponges that have been retained for a shorter period form masses surrounded by a connective tissue capsule containing a recognizable cotton sponge. In some cases, however, the sponge has been retained for long periods, even 40 years.[56] In these cases, the sponge is no longer grossly recognizable, but there is mass that clinically simulates a malignant tumor. Microscopically there are clusters of cotton fibers surrounded by foreign body and xanthomatous histiocytic reaction and hematoma (Figure 34.13).

Thorium oxide deposition (thorotrast)

Thorotrast is the commercial name for a colloidal suspension of oxide of thorium, a heavy metal that emits alpha particles during decay. Thorotrast was used as a radiologic contrast medium especially for carotid angiography until the early 1950s, but was abandoned because of its carcinogenic complications. Thorotrast persists in the body, engulfed by macrophages, especially in the liver, hematopoietic bone marrow, and bone. The continuing alpha-particle radiation is considered the likely cause for a high incidence of hepatocellular carcinoma, other hepatic carcinomas, hepatic angiosarcoma, leukemias, and other malignancies in patients who received thorotrast injections in the past, often 20 to 50 years earlier.[57–64]

Extravasation of thorotrast can form a fibrous mass (*thorotrastoma*) at the injection site, usually in the neck. Histologically, the injection site lesions contain clusters of thorotrast particles amid dense collagenous fibrosis.[65] Cutaneous histiocytic granulomas containing thorotrast particles have also been observed.[63] In the liver, thorotrast particles can be seen especially in the portal areas.

The clustering of these particles suggests that they once were inside histiocytes that subsequently died, probably

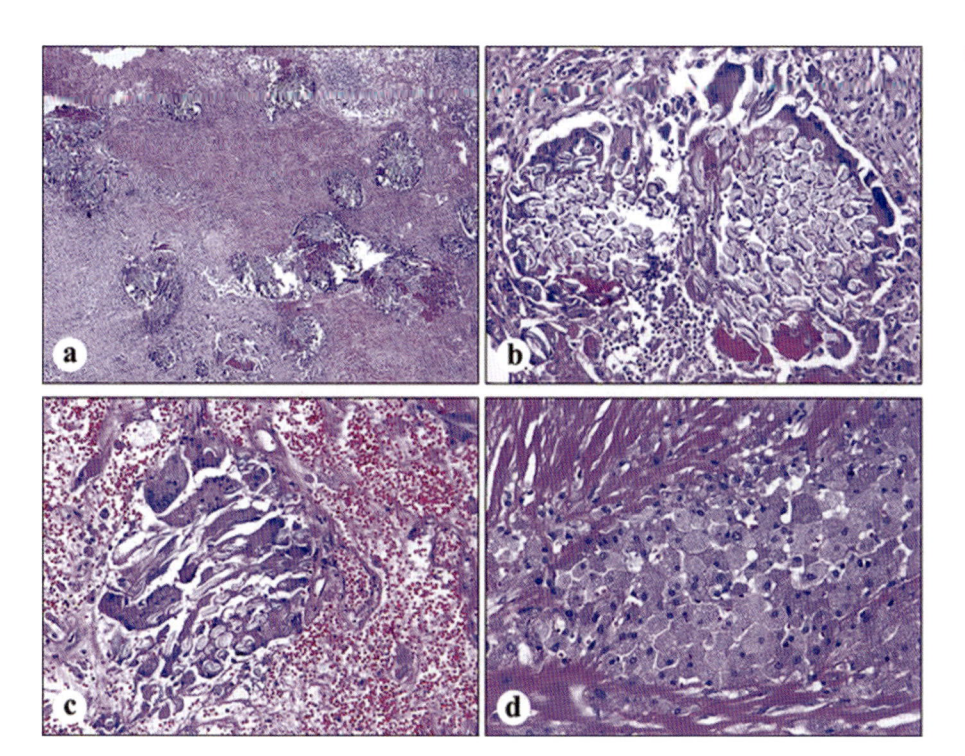

Chapter 34: Miscellaneous tumor-like lesions, and histiocytic and foreign body reactions

Figure 34.13 A long-standing retained surgical sponge creates a tumor-like foreign body reaction, in which rounded foci of preserved cotton fibers are recognizable. Also present are fibrosis, hemorrhage, and sheets of xanthoma cells.

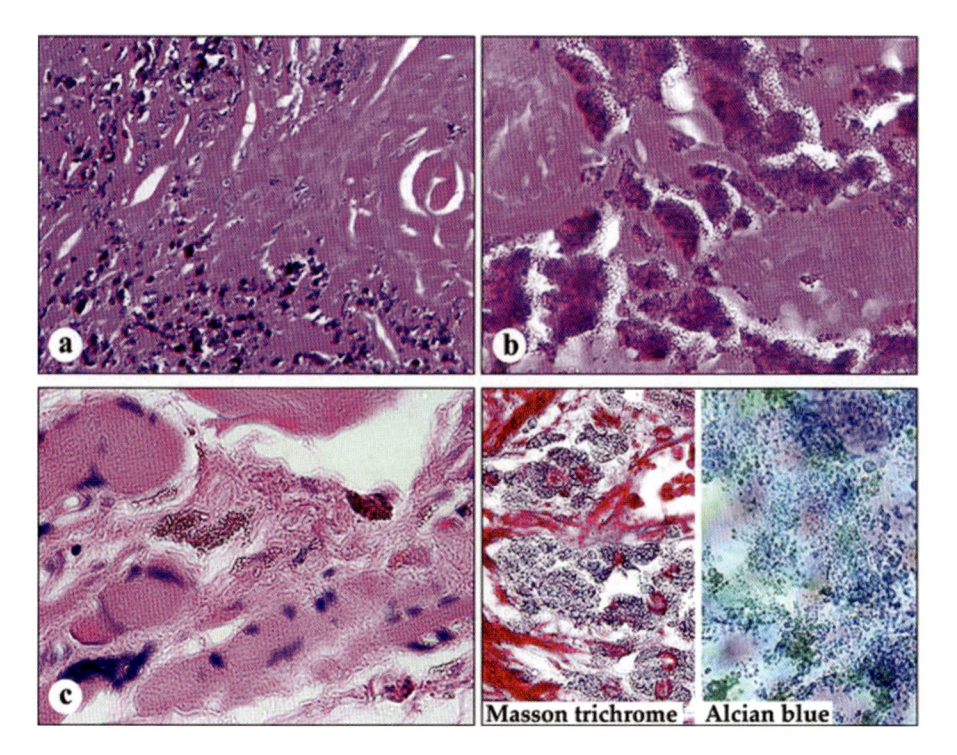

Figure 34.14 (a,b) Thorotrast deposition in soft tissue is associated with acellular fibrous scarring and clusters of thorotrast granules that can appear eosinophilic or golden. The grouping and configuration of the particles suggest that they once were inside histiocytes. (c) Thorotrast granules in the sternocleidomastoid muscle. With Masson trichrome stain, some histiocyte nuclei are still visible. The thorotrast granules do not stain and appear pale. The thorotrast granules are alcian blue positive, because they adsorb acid mucopolysaccharides.

contributed to by continuous alpha-particle radiation. With HE stain, the larger clusters of thorotrast particles often show basophilic staining, whereas the granular particles in the periphery of the clusters appear golden or grayish. With Masson trichrome stain, the granules are pale staining, and remaining histiocyte nuclei are sometimes present. The granules are negative for PAS but appear alcian blue positive (Figure 34.14). The latter effect is a result of the avidity of thorotrast particles for acid mucopolysaccharides. This phenomenon has been used in electron microscopic histochemistry to localize acid mucopolysaccharides with thorotrast particles as a tracer.[66]

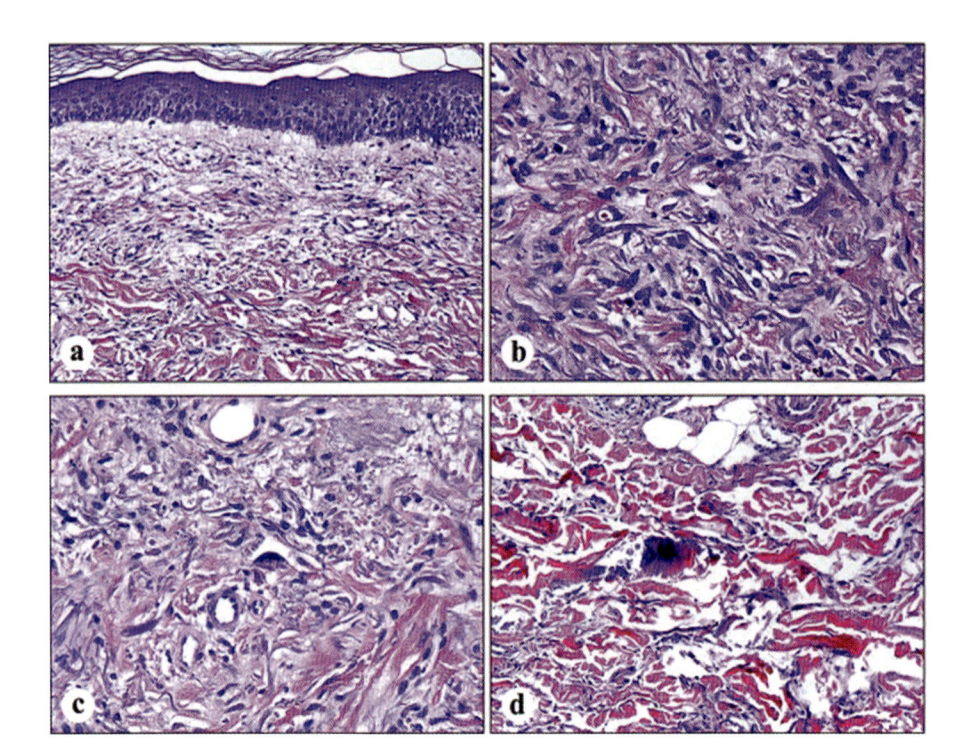

Figure 34.15 (**a–c**) A gadolinium-associated cutaneous fibrosis lesion contains increased numbers of dermal fibroblasts and elastic fibers. (**d**) Focal calcification is also present.

Gadolinium-associated fibrosis (nephrogenic systemic fibrosis)

Originally found to be associated with impaired renal function, this syndrome was subsequently linked to the administration of gadolinium (especially gadodiamide), a contrast agent used for magnetic resonance (MR) imaging. It therefore appears that abnormally long retention of gadodiamide by impaired renal function might be the causative factor.[67–73] Gadodiamide has been found to stimulate fibroblastic collagen and hyaluronic acid synthesis and conversion into myofibroblasts *in vitro*, and this has been proposed as a mechanism of its fibrogenic action.[74]

Clinically, cutaneous indurated plaques and cobblestone-like formations or orange-peel-like skin appears on the extremities and trunk. The process often begins in the distal lower extremities and extends into the trunk wall and upper extremities, but the head and neck are less frequently involved. Internal organ fibrosis involving the mediastinum, pericardium, and diaphragm has been observed in some patients.[68,71]

Histologic changes vary with the age of the lesion, and there is some resemblance to the changes seen in localized scleroderma and eosinophilic fasciitis (although without much inflammation). Early manifestation is a mild increase in dermal fibroblasts and collagen, focal mucin deposition, and increased numbers of thickened elastic fibers that can be seen as glassy structures parallel to the skin surface (Figure 34.15). Foci of multinucleated histiocytes can be present. A lymphoid inflammatory component is not a feature of this disease. Subcutaneous septa are expanded by fibrosis. Gadolinium itself is not seen histologically, but it has been demonstrated by X-ray microanalysis.[75]

Lesions of >20 weeks' duration show more prominent collagen bundles, having possible focal dystrophic calcification with less prominent mucin deposition and histiocytic infiltration.[67]

Amyloid tumor (amyloidoma)

Amyloid is the name given to a heterogeneous group of fibrillary proteins that form insoluble depositions in various tissues. These can be diffuse, often perivascular and extend to tissue stroma, or rarely form mass lesions, *amyloid tumors*. Amyloid tumors occur most frequently in body cavity sites (e.g., the mediastinum or retroperitoneum). They are seen less frequently in the trunk wall and extremities.[76]

Histological analysis shows collections of eosinophilic amorphous amyloid spaced by dense collagen. The amyloid is often surrounded by multinucleated foreign-body-type giant cells (Figure 34.16). Independent of type, amyloid is positive (red) with Congo red stain and shows at least focal apple-green birefringence. Amyloid is also positive for crystal violet (purple metachromasia).

The most common form of amyloid tumor seems to be the one derived from immunoglobulin light-chain material (AL-amyloid) and associated with lymphoplasmacytic lymphoma (usually low-grade), or plasma cell neoplasms. Those cases typically contain clusters of lymphocytes and plasma cells, and the latter can demonstrate immunohistochemical light-chain restriction (Figures 34.17 and 34.18). The small

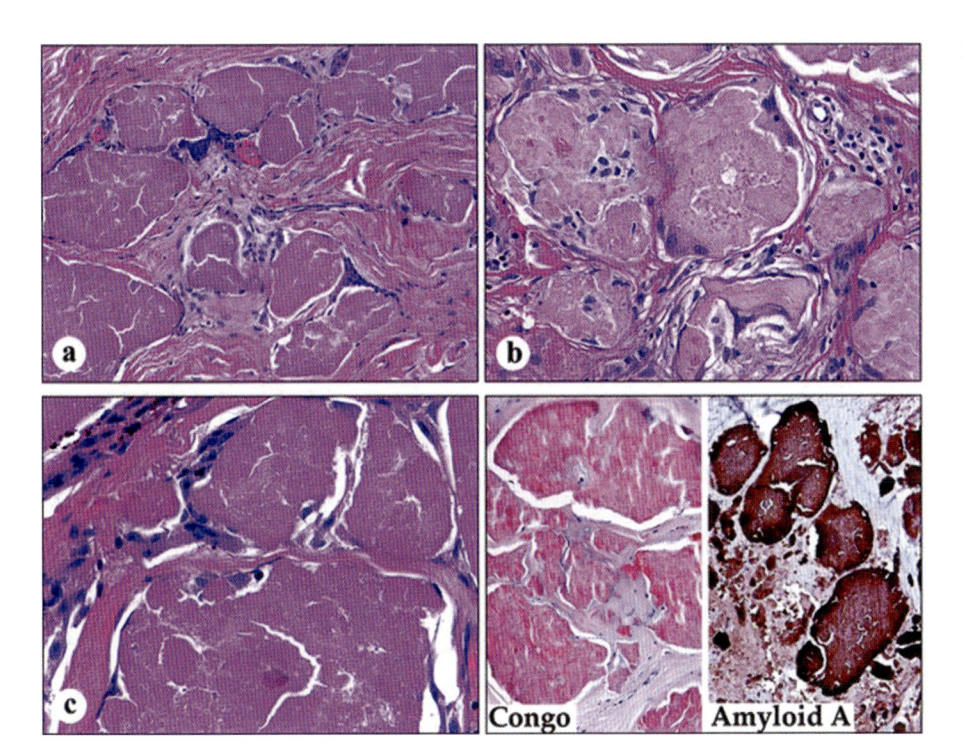

Chapter 34: Miscellaneous tumor-like lesions, and histiocytic and foreign body reactions

Figure 34.16 A tumor-like collection of secondary amyloidosis in soft tissue contains small nodules of amorphous amyloid material focally surrounded by multinucleated histiocytes. The amyloid is positive for Congo stain and amyloid A immunostaining, consistent with secondary amyloidosis.

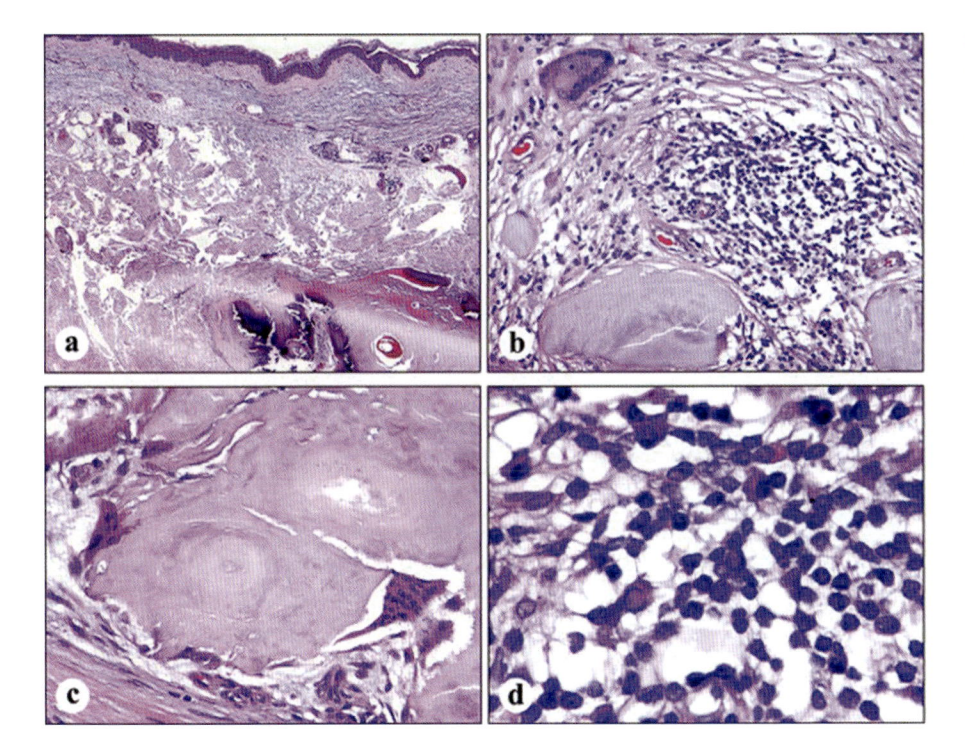

Figure 34.17 (a–c) Amyloid tumor associated with lymphoplasmacytic lymphoma. There are nests of amyloid collection and scattered lymphoid clusters. (d) The lymphoid clusters also contain plasma cells, and some of them show nuclear inclusions (Dutcher bodies).

lymphocytes are mostly B-cells. In these cases, the amyloid is negative for amyloid A precursor protein. Prognosis depends on the course of the underlying plasma cell neoplasia. The most accurate determination of amyloid type is performed by proteomic analysis using laser capture microdissection combined with mass spectrometric analysis.[77]

Amyloid tumor can be a manifestation of systemic secondary amyloidosis related to an underlying condition, such as a chronic inflammatory disease (rheumatoid arthritis). The originating protein is an acute phase protein produced by the liver (serum amyloid precursor). In such cases, the amyloid is immunohistochemically positive for amyloid A precursor

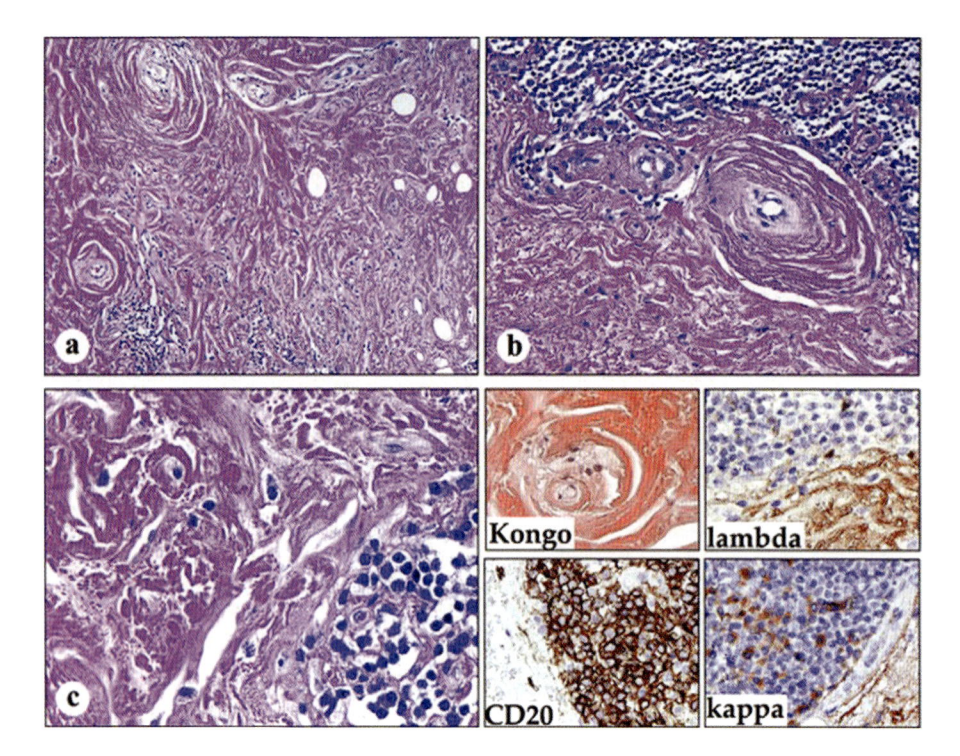

Figure 34.18 Amyloid tumor associated with lymphoplasmacytic lymphoma. There are large fields of amyloid material also seen in blood vessel walls. The amyloid is positive with Congo stain, and the lymphoid population is composed of CD20-positive B-cells. The plasma cells show kappa clonality and are negative for lambda. Note that the amyloid material stains for both light chains.

protein. Rare examples of apparently localized amyloid tumors composed of amyloid A protein have been reported.[78]

Beta-2 microglobulin-derived amyloidosis occurs in chronic hemodialysis patients. In some cases, periarticular tumor-like deposits and lytic bone lesions have developed, and subcutaneous amyloid tumors and tenosynovial (including carpal tunnel) amyloid collection have also been reported.[79,80]

In some patients with carpal tunnel syndrome, amyloid is present in the tenosynovial tissue without signs of systemic amyloidosis. Such amyloid seems to be most commonly derived from transthyretin, and the prognosis is excellent. In some patients, however, tenosynovial amyloid is part of systemic secondary or light-chain-derived amyloidosis. The latter could be related to systemic plasma cell neoplasia.[81,82]

References

Organizing hematoma

1. Sterling A, Butterfield WC, Bonner R Jr, Quigley W, Marjani M. Post-traumatic cysts of soft tissue. *J Trauma* 1977;17:392–396.
2. Reid JD, Kommareddi S, Lankerani M, Park MC. Chronic expanding hematomas: a clinicopathologic entity. *JAMA* 1980;244:2441–2444.
3. Mentzel T, Goodlad JR, Smith MA, Fletcher CD. Ancient hematoma: a unifying concept for a post-traumatic lesion mimicking an aggressive soft tissue neoplasm. *Mod Pathol* 1997;10:334–340.
4. O'Keefe RJ, O'Connell JX, Temple HT, et al. Calcific myonecrosis: a late sequela to compartment syndrome of the leg. *Clin Orthop* 1995;318:205–213.

Deep granuloma annulare (pseudorheumatoid nodule)

5. Felner EI, Steinberg JB, Weinberg AG. Subcutaneous granuloma annulare: a review of 47 cases. *Pediatrics* 1997;100:965–967.
6. Grogg KL, Nascimento AG. Subcutaneous granuloma annulare in childhood: clinicopathological features in 34 cases. *Pediatrics* 2001;107:1–4.
7. McDermott MB, Lind AC, Marley EF, Dehner LP. Deep granuloma annulare (pseudorheumatoid nodule) in children: clinicopathologic study of 35 cases. *Pediatr Dev Pathol* 1998;1:300–308.
8. Barzilai A, Huszar M, Shapiro D, Nass D, Trau H. Pseudorheumatoid nodules in adults: a juxtaarticular form of nodular granuloma annulare. *Am J Dermatopathol* 2005;27:1–5.
9. Toto JR, Chu P, Yen TS, LeBoit PE. Granuloma annulare and human immunodeficiency virus infection. *Arch Dermatol* 1999;135:1341–1346.
10. Li A, Hogan DJ, Sanusi ID, Smoler BR. Granuloma annulare and malignant neoplasms. *Am J Dermatopathol* 2003;25:113–116.
11. Trotter MJ, Crawford RI, O'Connell JX, Tron VA. Mitotic granuloma annulare: a clinicopathologic study of 20 cases. *J Cutan Pathol* 1996;23:537–545.
12. Patterson JW. Rheumatoid nodule and subcutaneous granuloma annulare: a comparative histologic study. *Am J Dermatopathol* 1988;10:1–8.
13. Alguacil-Garcia A. Necrobiotic palisading suture granuloma simulating rheumatoid nodule. *Am J Surg Pathol* 1993;17:920–923.

Pulse granuloma

14. Dunlap CL, Barker BF. Giant-cell hyaline angiopathy. *Oral Surg Oral Med Oral Pathol* 1977;44:587–591.
15. Barker BF, Dunlap CL. Hyaline rings of the oral cavity: the so-called "pulse" granuloma redefined. *Semin Diagn Pathol* 1987;4:237–242.
16. Talacko AA, Radden BG. Oral pulse granuloma: clinical and histopathological features: a review of 62 cases. *Int J Oral Maxillofac Surg* 1988;17:343–346.
17. Talacko AA, Radden BG. The pathogenesis of oral pulse granuloma: an animal model. *J Oral Pathol* 1988;17:99–105.
18. Marcussen LN, Peters E, Carmel D, Mickelborough M, Robinson C. Legume-associated residual cyst. *J Oral Pathol Med* 1993;22:141–144.
19. Martin RW 3rd, Lumadue JA, Corio RL, Kalb RL, Hood AF. Cutaneous giant cell hyaline angiopathy. *J Cutan Pathol* 1993;20:356–358.
20. Pereira TC, Prichard JW, Khalid M, Medich DS, Silverman JF. Rectal pulse granuloma. *Arch Pathol Lab Med* 2001;125:822–823.
21. Zhai J, Maluf HM. Peridiverticular colonic hyaline rings (pulse granuloma): report of two cases associated with perforated diverticula. *Ann Diagn Pathol* 2004;8:375–379.

Mycobacterial pseudotumor

22. Wood C, Nickoloff BJ, Todes-Taylor NR. Pseudotumor resulting from atypical mycobacterial infection: a "histoid" variety of mycobacterium avium-intracellulare complex infection. *Am J Clin Pathol* 1985;83:524–527.
23. Logani S, Lucas DR, Cheng JD, Ioachim HL, Adsay NV. Spindle cell tumors associated with mycobacteria in lymph node of HIV-positive patients: "Kaposi sarcoma with mycobacteria" and "mycobacterial pseudotumor." *Am J Surg Pathol* 1999;23:656–661.
24. Chen KT. Mycobacterial spindle cell pseudotumor of lymph nodes. *Am J Surg Pathol* 1992;16:276–281.
25. Kontochristopoulos DJ, Aroni K, Panteleos DN, Tosca AD. Immunohistochemistry in histoid leprosy. *Int J Dermatol* 1995;34:777–781.
26. Sehgal VN, Srivastava G, Beohar PC. Histoid leprosy: a histopathological reappraisal. *Acta Leprol* 1987;5:125–131.
27. Umlas J, Federman M, Crawford C, et al. Spindle cell pseudotumor due to Mycobacterium avium-intracellulare in patients with acquired immunodeficiency syndrome (AIDS): positive staining of mycobacteria for cytoskeleton filaments. *Am J Surg Pathol* 1991;15:1181–1187.
28. Wolf DA, Wu CD, Medeiros LJ. Mycobacterial pseudotumors of lymph node: a report of two cases diagnosed at the time of intraoperative consultation using touch imprint preparations. *Arch Pathol Lab Med* 1995;119:811–814.

Malakoplakia

29. Damjanov I, Katz SM. Malakoplakia. *Pathol Ann* 1981;16(Part 2):103–126.
30. Douglas-Jones AG, Rodd C, James EMV, Mills RGS. Pre-diagnostic malakoplakia presenting as a chronic inflammatory mass in the soft tissues of the neck. *J Laryngol Otol* 1992;106:173–177.
31. Schmerber S, Lantuejoul S, Lavieille JP, Reyt E. Malakoplakia of the neck. *Arch Otolaryngol Head Neck Surg* 2003;129:1240–1242.
32. Kogulan PK, Smith M, Seidman J, et al. Malakoplakia involving the abdominal wall, urinary bladder, vagina, and vulva: case report and discussion of malakoplakia-associated bacteria. *Int J Gynecol Pathol* 2001;20:403–406.
33. Lowitt MH, Kariniemi AL, Niemi KM, Kao GF. Cutaneous malakoplakia: a report of two cases and review of the literature. *J Am Acad Dermatol* 1996;34:325–332.
34. Bates AW, Dev S, Baithun SI. Malakoplakia and colorectal adenocarcinoma. *Postgrad Med J* 1997;73:171–173.
35. Pillay K, Chetty R. Malakoplakia in association with colorectal carcinoma: a series of four cases. *Pathology* 2002;34:332–335.
36. Yousef GM, Naghibi B. Malakoplakia outside the urinary tract. *Arch Pathol Lab Med* 2007;131:297–300.
37. Kohl SK, Hans CP. Cutaneous malakoplakia. *Arch Pathol Lab Med* 2008;132:113–117.

Xanthogranulomatous inflammation

38. McDonald GS. Xanthogranulomatous pyelonephritis. *J Pathol* 1981;133:203–213.
39. Goodman ZD, Ishak KG. Xanthogranulomatous cholecystitis. *Am J Surg Pathol* 1981;5:653–659.
40. Ladefoged C, Lorentzen M. Xanthogranulomatous cholecystitis: a clinicopathological study of 20 cases and review of the literature. *APMIS* 1993;101:869–875.
41. Cozzutto C, Carbone A. The xanthogranulomatous process: xanthogranulomatous inflammation. *Pathol Res Pract* 1988;183:395–402.

Xanthoma

42. Parker F. Xanthomas and hyperlipidemias. *Am J Acad Dermatol* 1985;13:1–30.
43. Fahey JJ, Stark HH, Donovan WF, Drennan DB. Xanthoma of the Achilles tendon. *J Bone Joint Surg Am* 1973;55:1197–1211.
44. Bulkley BH, Buja M, Ferrans VJ, Bulkley GB, Roberts WC. Tuberous xanthoma in homozygous type II hyperlipoproteinemia. *Arch Pathol* 1975;99;293–300.

Polyvinylpyrrolidone storage disease

45. Wessel W, Schoog M, Winkler E. Polyvinylpyrrolidone (PVP), its diagnostic, therapeutic and technical application and consequences thereof. *Arzneimittelforsch* 1971;21:1468–1482.
46. Kuo TT, Hsueh S. Mucicarminophilic histiocytosis: a polyvinylpyrrolidone (PVP) storage disease simulating signet ring carcinoma. *Am J Surg Pathol* 1984;8:419–428.
47. Soumerai S. Pseudotumors of the arm following injections of procaine polyvinylpyrrolidone: report of two cases. *J Med Soc NJ* 1978;75:407–408.
48. Kuo TT, Hu S, Hung CL, et al. Cutaneous involvement in polyvinylpyrrolidone storage disease: a clinicopathologic study of five patients, including two patients with severe anemia. *Am J Surg Pathol* 1997;21:1361–1367.
49. Kepes JJ, Chen WYK, Jim YF. "Mucoid dissolution" of bones and multiple pathologic fractures in a patient with a past history of intravenous administration of polyvinylpyrrolidone (PVP): a case report. *Bone Miner* 1993;22:33–41.
50. Hewan-Lowe K, Hamners Y, Lyons JM. Polyvinylpyrrolidone storage disease:

a source of error in the diagnosis of signet ring cell gastric adenocarcinoma. *Ultrastruct Pathol* 1994;18:271–278.

Cotton fibers (gossypiboma)

51. Deger RB, Livolsi VA, Noumoff JS. Foreign body reaction (gossypiboma) masking as recurrent ovarian cancer. *Gynecol Oncol* 1995;56:94–96.
52. Moyle H, Hines OJ, McFadden DW. Gossypiboma of the abdomen. *Arch Surg* 1996;131:566–568.
53. Zbar AP, Agrawal A, Saeed IT, Utidjian MR. Gossypiboma revisited: a case report and review of the literature. *J R Coll Surg Edinb* 1998;43:417–418.
54. Rajagopal A, Martin J. Gossypiboma, "A surgeon's legacy": report of a case and review of the literature. *Dis Colon Rectum* 2002;45:119–120.
55. Bani-Hani KE, Gharaibeh KA, Yaghan RJ. Retained surgical sponge (gossypiboma). *Asian J Surg* 2005;28:109–115.
56. Sakayama K, Fujibuchi T, Sugawara Y, et al. A 40-year-old gossypiboma (foreign body granuloma) mimicking a malignant femoral surface tumor. *Skeletal Radiol* 2005; 34:221–224.

Thorium oxide deposition (thorotrast)

57. Van Kaick G, Lieberman D, Lorenz D, et al. Recent results of the German Thorotrast study: epidemiological results and dose effect relationships in thorotrast patients. *Health Phys* 1983;44 (Suppl 1):299–306.
58. Travis LB, Hauptmann M, Gaul LK, et al. Site-specific cancer incidence and mortality after cerebral angiography with radioactive thorotrast. *Radiat Res* 2003;160:691–706.
59. McInroy JF, Gonzales ER, Miglio JJ. Measurement of thorium isotopes and 228Ra in soft tissues and bones in a deceased Thorotrast patient. *Health Phys* 1992;63:54–71.
60. Backer OG, Faber M, Rasmussen H. Local sequelae to carotid angiography with colloid thorium oxide. *Acta Chir Scand* 1958;115:417–421.
61. Lung RJ, Harding RL, Herceg SJ, Schantz JC, Miller SH. Long-term survival with Thorotrast cervical granuloma. *J Surg Oncol* 1978; 10:171–177.
62. Wargoz ES. Cutaneous thorotrast granulomas and chronic lymphocytic leukemia following thorotrast angiography. *Am J Surg Pathol* 1985;9:835–841.
63. DiMarcangelo MT, David ET, Kuroda K. Malignant fibrous histiocytoma induced by thorium. *N Engl J Med* 1990;87:47–49.
64. Polacarz SV, Laing RW, Loomes R. Thorotrast granuloma: an unexpected diagnosis. *J Clin Pathol* 1992;45: 259–261.
65. Wistrow TP, Behbehani AA, Wiebecke B. Thorotrast-induced oro- and hypopharyngeal fibrosis with recurrent bleeding. *J Craniomaxillofac Surg* 1988;16:315–319.
66. Lünsdorf H, Kristen I, Barth E. Cationic hydrous thorium oxide colloids: a useful tool for staining negatively charged surface matrices of bacteria for use in energy-filtered transmission electron microscopy. *BMC Microbiol* 2006;6:59.

Gadolinium-associated fibrosis

67. Cowper SE, Su L, Robin H, Bhawan J, LeBoit PE. Nephrogenic fibrosing dermopathy. *Am J Dermatopathol* 2001;33:383–393.
68. Grobner T. Gadolinium: a specific trigger for the development of nephrogenic fibrosing dermopathy and nephrogenic systemic fibrosis. *Nephrol Dial Transplant* 2006;21:1104–1108.
69. Richmond H, Zwerner J, Kim Y, Fiorentino D. Nephrogenic systemic fibrosis: relationship to gadolinium and response to photopheresis. *Arch Dermatol* 2007;143:1025–1030.
70. Grobner T, Prischl FC. Gadolinium and nephrogenic systemic fibrosis. *Kidney Int* 2007;72:260–264.
71. Moreno-Romero JA, Segura S, Mascaro JM Jr, et al. Nephrogenic systemic fibrosis: a case series suggesting gadolinium as a possible etiologic factor. *Br J Dermatol* 2007;157:783–787.
72. Cowper SE, Rabach M, Girardi M. Clinical and histologic findings in nephrogenic systemic fibrosis. *Eur J Radiol* 2008;66:191–199.
73. Stratta P, Canavese C, Aime S. Gadolinium-enhanced magnetic resonance imaging, renal failure and nephrogenic systemic fibrosis/ nephrogenic fibrosing dermopathy. *Curr Med Chem* 2008;15:1229–1235.
74. Edward M, Quinn JA, Mukherjee S, et al. Gadodiamide contrast agent "activates" fibroblasts: a possible cause of nephrogenic systemic fibrosis. *J Pathol* 2008;214:584–593.
75. High WA, Ayers RA, Chandler J, Zito G, Cowper SE. Gadolinium is detectable within the tissue of patients with nephrogenic systemic fibrosis. *J Am Acad Dermatol* 2007;56:21–26.

Amyloid tumor (amyloidoma)

76. Khrishnan J, Chu WS, Elrod JP, Frizzera G. Tumoral presentation of amyloidosis (amyloidomas) in soft tissues. *Am J Clin Pathol* 1993;100: 135–144.
77. Sun W, Sun J, Zou L, et al. The successful diagnosis and typing of systemic amyloidosis using a microwave-assisted filter-aided fast sample preparation method and LC/MS/MS analysis. *PLoS One* 2015; 10:e0127180.
78. Luo JH, Rotterdam H. Primary amyloid tumor of the breast: a case report and review of the literature. *Mod Pathol* 1997;10:735–738.
79. Casey TT, Stone WJ, DiRaimondo CR, et al. Tumoral amyloidosis of bone of beta 2-microglobulin origin in association with long-term hemodialysis: a new type of amyloid disease. *Hum Pathol* 1986;17:731–738.
80. Maury CP. Beta-2-microglobulin amyloidosis: a systemic amyloid disease affecting primarily synovium and bone in long-term dialysis patients. *Rheumatol Int* 1990;10:1–8.
81. Kyle RA, Eilers SG, Linscheid RL, Gaffey TA. Amyloid localized to tenosynovium at carpal tunnel release. *Am J Clin Pathol* 1989;91:393–397.
82. Kyle RA, Gertz MA, Linke RP. Amyloid localized to tenosynovium at carpal tunnel release: immunohistochemical identification of amyloid type. *Am J Clin Pathol* 1992;97:250–253.

Chapter 35

Lymphoid, myeloid, histiocytic, and dendritic cell proliferations in soft tissues

Markku Miettinen

National Cancer Institute, National Institutes of Health

This chapter includes selected hematopoietic entities that are important in soft tissue pathology because of their presentation in extranodal soft tissues. Practical diagnosis is emphasized, based on histologic and immunohistochemical parameters (Table 35.1). Reactive histiocytic lesions and foreign body reactions are discussed in Chapter 34.

The concepts of soft tissue and cutaneous lymphoma overlap in the literature, because subcutaneous involvement is often included with cutaneous lymphoma. Selected cutaneous lymphoma types are therefore discussed in this chapter. Typically skin-based lymphomas, such as mycosis fungoides, related entities, and some rare T-cell lymphomas are excluded. Discussion of the other entities centers on extranodal, noncutaneous soft tissue manifestations, and genetics emphasizes established clinical tests or newly discovered pathogenesis. For more extensive discussion, the author refers readers to recent review articles and lymphoma classification handbooks.[1–3] Comprehensive reviews on immunophenotyping of lymphomas and other hematopoietic neoplasms are recommended.[4–6]

Soft tissue lymphomas outside the context of cutaneous lymphomas have been reported in older series that predate modern immunophenotypic classification. Morphologically, most examples have been diffuse large cell lymphomas that probably represented large B-cell lymphomas. Localized primary soft tissue lymphomas seem to be distinctly uncommon. In a Mayo Clinic series employing strict criteria for primary extranodal soft tissue lymphoma, there were only 8 cases among 7000 lymphomas (0.1%): 6 in the lower extremities and 2 in the arm. Most of these lymphomas formed large masses (≥5 cm), and systemic disease often developed during follow-up.[7] Based on data extrapolated from an Armed Forces Institute of Pathology (AFIP) series of 75 cases, large B-cell lymphomas dominated, and low-grade lymphomas were primarily follicular. In this series, lymphomas had a predilection for patients >60 years of age. The thigh was the most common site, and other common locations included the chest wall, arm, and leg. Although initial workup showed localized disease, 8 of 33 patients showed systemic disease within 3 months. A majority of patients with large cell lymphomas survived >5 years, however.[8]

Follicular lymphoid hyperplasia

Follicular hyperplasia of unknown etiology can occur in extranodal soft tissue, usually in the subcutis and less commonly in the muscle compartments. Florid follicular hyperplasia is more common in children and young adults.

Histological analysis shows large germinal centers containing both small and large cells; these centers are surrounded by a narrow mantle zone. Heterogeneous cellular composition with small and large cells, follicular polarity (asymmetric distribution of small and large cells), and relative paucity of small cleaved cells supports hyperplasia rather than follicular lymphoma. Nuclear debris-containing macrophages (tingible body macrophages) also supports hyperplasia when present.

Immunohistochemically, the germinal centers are negative for BCL2 and positive for CD20, with scattered T-cells being present (Figure 35.1).[9] Proliferative activity (Ki67 index) is usually higher in reactive hyperplasia than follicular lymphoma.[10] Immunohistochemical parameters of follicular hyperplasia and follicular lymphoma are known to overlap (lymphoma follicles can also be BCL2 negative), so that this distinction can be problematic in some cases.

Molecular genetic demonstration of clonal immunoglobulin heavy-chain rearrangement and the presence of the t(14;18) translocation supports follicular lymphoma when present, but negative results do not rule out lymphoma. Fortunately, isolated superficial follicular lymphoma is usually clinically indolent.[10,11]

Hyaline vascular Castleman disease

This disease is best known to involve the lymph nodes in the body cavities, but a small number of cases have been reported in peripheral soft tissue (the subcutis and deeper). The nodal masses in solitary Castleman disease usually measure 3 cm to 7 cm, but can exceed 10 cm. They are usually located in the mediastinum, and less often in the retroperitoneum, axilla, or neck, and are often incidental radiologic findings. Behavior is benign.[12]

Soft tissue Castleman disease has been reported in young adults involving the subcutis or skeletal muscle of the proximal extremities and chest wall. Tumors have measured 4 cm to

Modern Soft Tissue Pathology, Second Edition, ed. Markku Miettinen. Published by Cambridge University Press. © Cambridge University Press 2016.

Chapter 35: Lymphoid, myeloid, histiocytic, and dendritic cell proliferations

Table 35.1 Summary of the most important immunohistochemical markers applied in the diagnosis of lymphoid and hematopoietic tumors involving soft tissues

Antigen	Synonym or nature of the antigen	Expression in hematopoietic cells	Expression in hematopoietic tumors and tumor-like processes
BC1 (n)	Cyclin D1	Not expressed	Mantle cell lymphoma[a]
BCL2	Mitochondrial membrane protein	T-cells, B-cell subsets	Most small cell lymphomas[a] Variable in large cell lymphomas
BCL6 (n)	B-cell transcription factor	Germinal center B-cells Some CD30-positive cells	Follicular, Burkitt, and large B-cell lymphomas
CD1A	T-cell membrane protein	Langerhans cells Thymocytes	Langerhans cell histiocytosis
CD2		T-lymphocytes	
CD3	T-cell receptor-associated	T-lymphocytes	Most cell lymphomas
CD5	Regulator of B-cell receptor signaling	T-lymphocytes, B-cell subsets	Small lymphocytic lymphoma/CLL, mantle cell lymphoma, most T-cell lymphomas
CD7		T-lymphocytes	T-cell lymphomas (absence can signify abnormal immunophenotype suggestive of T-cell neoplasia)
CD8		Cytotoxic T-lymphocytes, splenic sinusoid lining cells	Panniculitis-like T-cell lymphoma
CD10	Common acute lymphoblastic leukemia antigen (CALLA), peptidase	Follicular B-cells (also fibroblasts)	B-cell lymphomas of follicular origin,[a] Burkitt lymphoma, B-cell lymphoblastic lymphoma
CD15	Carbohydrate determinant	Myeloid cells, monocytes, some T-cells	Differentiated myeloid cells Hodgkin cells[a]
CD19	B-cell antigen receptor-related	Dendritic reticulum cells and B-lymphocytes, except plasma cells	Alternative B-cell marker to the more commonly used CD20
CD20	Cell surface phosphoprotein	B-lymphocytes, except plasma cells	Most B-cell lymphomas except those with plasma cell differentiation
CD21	Complement receptor 3d	Follicular dendritic reticulum cells B-cell subsets	Follicular dendritic cell tumor, some B-cell lymphomas, Hodgkin cells[a]
CD23	IgE receptor	Follicular dendritic cells, B-cell subsets	Follicular dendritic cell tumor, small lymphocytic lymphoma/CLL
CD30	Activation-associated TNF family protein	Activated lymphoid cells	Large cell anaplastic lymphoma,[a] Hodgkin disease
CD34	Hematopoietic progenitor	Hematopoietic stem cells	Extramedullary myeloid tumor (some)[a]
CD35	Complement receptor 3b	Follicular dendritic cells	Follicular dendritic cell tumor
CD38	Surface glycoprotein	Plasma cells	Plasmacytoma
CD43[b]	Sialophorin	Most leukocytes, except B-cells (non–lineage-specific marker)	T-cell lymphomas, some B-cell lymphomas, myeloid infiltrates
CD45[b]	Leukocyte common antigen (LCA)	Nearly all leukocytes (non–lineage-specific marker)	Most hematopoietic tumors
CD56	Neural cell adhesion molecule (NCAM)	Natural killer cells	NK/T-cell lymphoma[a]
CD61	Platelet glycoprotein IIIa (gpIIIa)	Megakaryocytes, platelets	Megakaryocytic component in extramedullary myeloid tumor

Table 35.1 (cont.)

Antigen	Synonym or nature of the antigen	Expression in hematopoietic cells	Expression in hematopoietic tumors and tumor-like processes
CD68	Lysosomal glycoprotein	Monocytes, histiocytes	Histiocytic tumors, mastocytoma[a]
CD79a	B-cell receptor-associated	B-cells, plasma cells	Most B-cell lymphomas, plasmacytoma
CD117	KIT receptor tyrosine kinase	Hematopoietic stem cells Mast cells	Extramedullary myeloid tumor[a] Mastocytoma
CD138	Syndecan-1	Plasma cells	Plasmacytoma[a]
CD163	Erythrocyte scavenge receptor	Histiocytes	Histiocytic infiltrates Juvenile xanthogranuloma, reticulohistiocytoma, histiocytic sarcoma
CD246	Anaplastic lymphoma kinase (ALK)	Not expressed	Subset of anaplastic large cell lymphoma[a]
EMA/MUC1	Membrane glycoprotein	Plasma cells	Plasmacytoma, anaplastic large cell lymphomas[a] (especially ALK positive)
Granzyme B	Cytotoxic T-cell granule protein	Cytotoxic T-cells	Adjunct marker in the diagnosis of anaplastic large cell lymphoma
HGAL	Human-germinal-center-associated lymphoma antigen	Normal germinal center cells	Marker for lymphomas of follicular cell origin: follicular, Burkitt, and majority of DLBCL, LP and classical Hodgkin also positive
MUM1/IRF4 (n)	Transcription factor	Plasma cells	Most diffuse large B-cell lymphomas, plasmacytoma, anaplastic large cell lymphoma
Myeloperoxidase	Lysosomal granule protein	Neutrophils	Extramedullary myeloid tumor, some B-cell acute lymphoblastic leukemias
PAX-5 (n)	B-cell-specific activation factor	B-cells, not plasma cells	B-cell lymphomas, Hodgkin lymphomas
TIA1	T-cell cytotoxic granule protein		Adjunct marker to support anaplastic large cell lymphoma
TDT (n)	Terminal deoxyribonucleotidyl transferase	Thymocytes, immature B- and T-cells	B- and T-lymphoblastic lymphomas,[a] some extramedullary myeloid tumors

[a] Denotes marker expression also present in various nonhematopoietic tumors.
[b] These markers are useful panhematopoietic markers and can be useful in evaluating possible lymphoid/hematopoietic origin of undifferentiated tumors.
(n) = nuclear staining.

6 cm, and in some cases, concomitant mediastinal disease was seen with chest wall involvement. Clinical follow-up has shown an indolent course.[13]

Multicentric Castleman disease is related to the plasma cell type of Castleman disease. It is usually associated with human herpesvirus 8 (HHV8) infection and HIV/AIDS, and some patients have concomitant involvement of Kaposi's sarcoma (also HHV8 associated).[14,15]

Histologically typical of hyaline vascular Castleman disease is a nodular architecture. The center of the nodules contains an atrophic remnant of a germinal center composed of large B-cells and dendritic reticulum cells, surrounded by an expanded mantle zone, often with concentric layers of small B-lymphocytes. The atrophic germinal center typically contains a capillary with sclerosis (Figure 35.2). Dendritic cells may show cytologic dysplasia. Sheets of plasma cells that can show light-chain clonality by immunohistochemistry are characteristic of the plasma cell type of Castleman disease, and the germinal centers may lack the distinctive features seen in the hyaline vascular type.[16]

Immunohistochemically, the germinal center itself shows prominent CD21 positivity, reflecting an abundance of dendritic reticulum cells. The germinal centers are negative for BCL2, whereas the mantle zone and T-cell elements are positive. The germinal center and surrounding mantle cells are CD20 positive, whereas scattered T-cells are present within the germinal center and the mantle zone, and a larger T-cell population is present outside the mantle cell layer (Figure 35.2). T-lymphoblastic populations have been reported and can be prominent in some cases.[17]

Kimura disease

This rare reactive lymphoid infiltration, rich in eosinophils, historically was often confused with epithelioid hemangioma

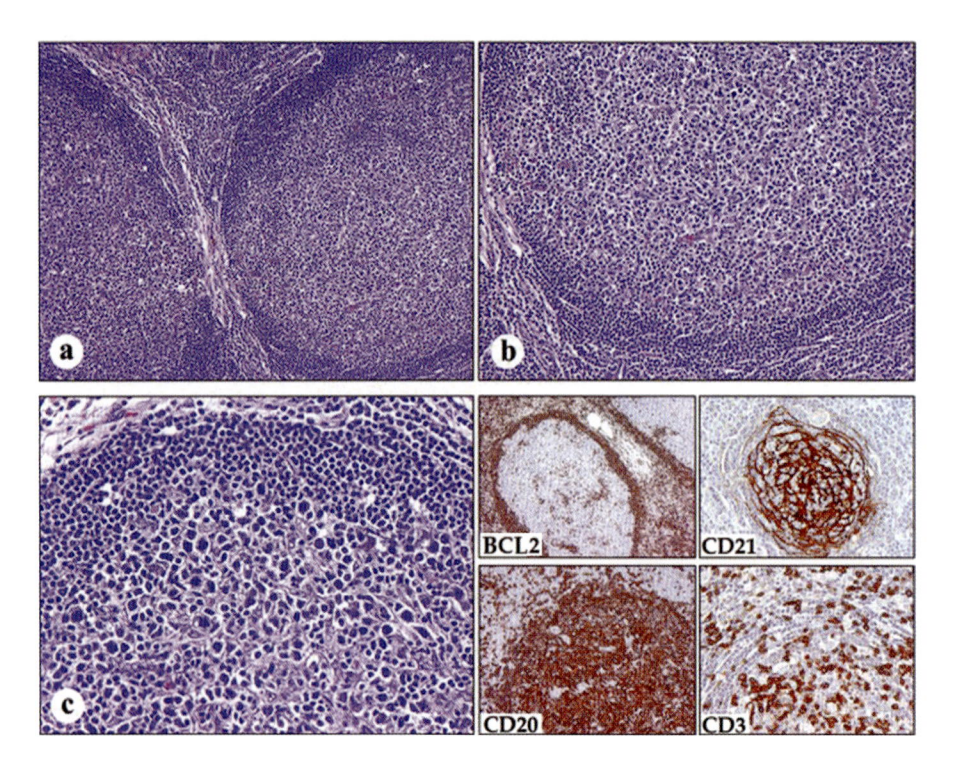

Figure 35.1 Follicular hyperplasia contains large follicles with a mixed cell population and a surrounding mantle zone composed of small lymphocytes. The follicle centers are negative for BCL2, contain a prominent network of CD21-positive dendritic reticulum cells, and are positive for CD20 with scattered CD3-positive T-cells being present.

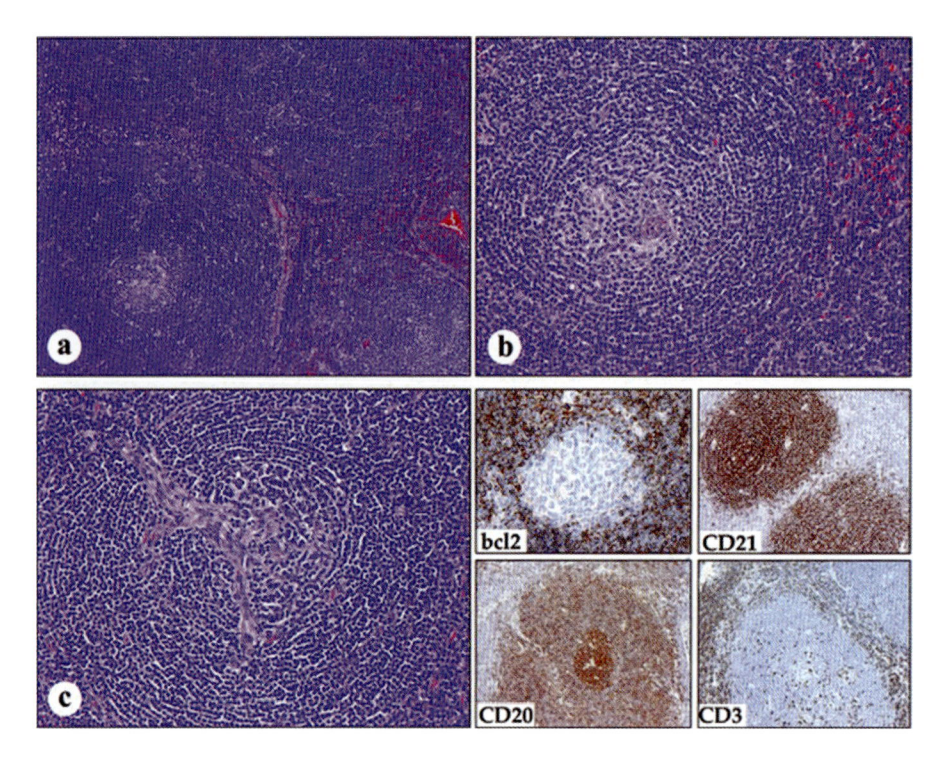

Figure 35.2 Castleman disease has a nodular pattern, and there are small germinal centers surrounded by a prominent mantle zone. Within the germinal center, a sclerosing blood vessel is often present. The germinal centers are BCL2 negative and have a prominent CD21-positive network of dendritic reticulum cells. They are positive for CD20 and negative for CD3.

(also called angiolymphoid hyperplasia with eosinophilia), which is an unrelated entity (see Chapter 21). Epithelioid endothelial cells are not part of Kimura disease. The etiology is unknown.

Occurrence is most common in Oriental male subjects, and there is a predilection for young adults, with cases being reported in children and older adults, however. Cases reported from the United States have predominantly involved non-Oriental persons.

The head and neck is most commonly involved, but nodal involvement has been reported in the groin and axilla. Many patients have subcutaneous nodules and parotid gland and lymph node involvement, and there is often a long history of

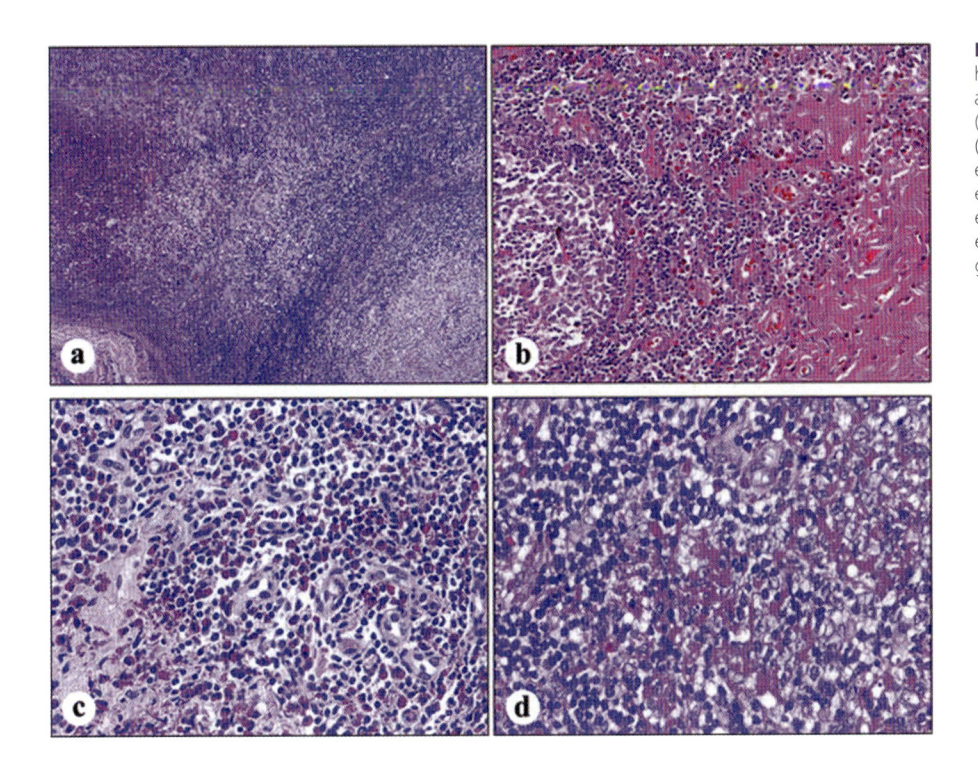

Figure 35.3 (**a**) Kimura disease is a lymphoid hyperplasia that often contains germinal centers and eosinophilic abscesses (left center). (**b**) Prominent fibrous septa are often present. (**c**) There is an admixture of lymphocytes and eosinophilic granulocytes. Prominent high endothelial capillaries are present, but epithelioid endothelial cells are not a feature. (**d**) Amorphous eosinophilic material is often seen within the germinal centers.

the tumor. The mean size of the neck mass is 2 cm to 3 cm, but some are >5 cm. Laboratory studies show elevated levels of blood eosinophils and immunoglobulin E (IgE). The clinical course is benign, although sometimes protracted, and the mass can recur after a local excision.[18–20]

Histologically, there is lymphoid hyperplasia with germinal centers separated by fibrous streaks, and numerous eosinophilic granulocytes are present, sometimes forming microabscesses. The germinal centers often contain amorphous proteinaceous deposits. Numerous thin-walled capillaries are often present, and they can infiltrate the germinal centers (Figure 35.3).

Follicular lymphoma and follicular center cell lymphoma

The most common soft tissue involvement of follicular lymphoma is abdominal tissue: a mesenteric or retroperitoneal mass that typically represents extranodal extension of para-aortic nodal disease, generally part of systemic disease. Clinicopathologic correlation is always required to determine whether a soft tissue follicular lymphoma is localized or part of a systemic process. Furthermore, in peripheral apparently localized disease, follow-up is necessary, because evolution into a systemic follicular lymphoma can occur.

In retroperitoneal disease, the follicles are typically composed of a mixture of large cells (usually a small minority) and small cells with cleaved nuclei (usually a dominant component; Figure 35.4). Significant sclerosis simulating benign fibrosclerosing disease can be present, so that caution is needed in the diagnosis of retroperitoneal fibrosis in small biopsies.

Reports on primary cutaneous follicular lymphoma include cases with both dermal and subcutaneous involvement. These lymphomas occur in older patients than those having follicular hyperplasia, with a median age of 55 to 60 years in major series. There is a predilection for the head and neck, especially the scalp and cheek. Clustered multiple lesions are common, and the clinical course is favorable in most cases. Because these tumors respond to local excision and radiation treatment, systemic therapy is generally unnecessary in asymptomatic patients.[10,11,21–24]

Nodal follicular lymphoma is graded based on the number of centroblasts (large noncleaved cells) per a follicular high-power field (HPF): grade 1, <5; grade 2, 5–15; grade 3, >15. Grade 3 is divided into 3A with small cleaved cells present, and 3B with solid sheets of large cells with small cleaved cells absent.[2,3] Grades 1 and 2 are considered low grade and grade 3, high grade. Grade 3B essentially merges with diffuse large B-cell lymphoma. It is not certain, however, whether clinicopathologic correlation obtained from nodal disease applies to cutaneous or primary extranodal follicular lymphoma.

Immunohistochemical analysis shows that typical follicular lymphoma follicles are positive for CD20, CD10, BCL2 (cytoplasmic), BCL6 (nuclear), and contain large numbers of CD21-positive dendritic cells. Cutaneous and soft tissue examples often differ from the nodal examples and are either negative or only weakly positive for BCL2, however.[10,11,23,25] Human germinal center-associated lymphoma (HGAL) protein is expressed in cutaneous as well as nodal follicular lymphoma and also in lymphocyte predominant and classical Hodgkin lymphomas.[26]

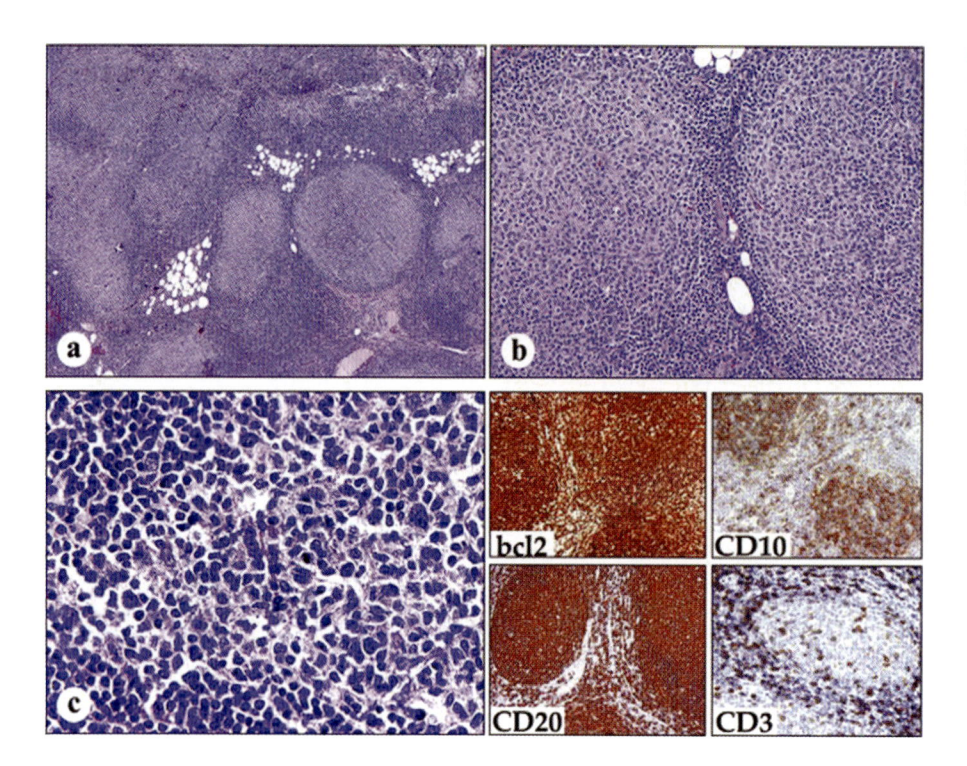

Figure 35.4 Follicular lymphoma follicles involving adipose tissue. Cytologically, they are usually composed of a dominant small cleaved cell population. Immunohistochemically typical is positivity for BCL2, CD10, and CD20, and CD3-positive T-cells are present both between and inside the follicles.

Molecular genetic studies for clonal immunoglobulin heavy-chain (IgH) rearrangement (a dominant band in gel electrophoresis or a dominant peak in capillary electrophoresis) support B-cell lymphoma. The t(14;18) follicular lymphoma translocation involving BCL2 and IgH genes is also typical of this lymphoma. Both tests are consistently positive only in systemic follicular lymphoma and often negative in localized cutaneous disease, however, so that negative molecular results for either test do not rule out lymphoma, if the morphology is diagnostic.[23]

Even if diffuse, however, those lymphomas that contain a cleaved cell (centrocytic) population (small or large) in addition to large noncleaved cells, and express follicular center cell markers (CD10, BCL6), and lack MUM1 expression are included in the category of primary cutaneous follicular center cell lymphomas. These patients have a better prognosis than do those with cutaneous/soft tissue diffuse large B-cell lymphomas.[1] Diffuse follicular center cell lymphomas probably include many cases previously classified as "mixed lymphomas."

Extranodal marginal zone lymphoma

Extranodal marginal zone lymphoma is a low-grade small B-cell lymphoma type that typically occurs in mucosal sites, such as the gastrointestinal tract, salivary gland, lung, skin, and less frequently in other organ systems. In previous terminology, these lymphomas were called *mucosa-associated lymphoid tissue* (MALT) *lymphomas*. Some cases previously designated as lymphoplasmacytic lymphomas and immunocytomas are actually marginal zone lymphomas. The nodal examples in particular were previously called monocytoid B-cell lymphomas. According to the current classification, large cell lymphomas arising from marginal zone lymphomas are classified as diffuse large B-cell lymphomas and excluded from marginal zone lymphomas.

Cutaneous and subcutaneous examples occur mainly in middle-aged adults, with conflicting reports on gender predilection. Isolated cutaneous/subcutaneous disease does not have any significant tendency to systemic progression. Clinical correlation is needed to determine whether a peripheral lesion is the primary focus or of "metastatic" origin from a mucosal or other primary site, although the disease is usually indolent either way. Some marginal zone lymphomas are associated with persistent local infection. In the skin and subcutis, *Borrelia burgdorferi* has been implicated, and in the stomach, *Helicobacter pylori* can be a triggering factor, and their elimination by antibiotics can cause regression of the lymphoma. Local excision and other local therapies are generally sufficient, and the prognosis is excellent.[24,26-32]

Histologically, marginal zone lymphoma is composed of sheets of small lymphoid cells that often have a distinctly clear cytoplasm (monocytoid appearance), but the cells can also resemble small lymphocytes. Plasma cell differentiation can be prominent (Figure 35.5). The nuclei are often notched or mildly cleaved. Plasmacytoid differentiation is common, and Dutcher bodies (intranuclear inclusions) are often seen. Germinal centers can be present and colonized by lesional cells. In the skin, clusters of lymphoma cells can infiltrate sweat gland elements, forming so-called lymphoepithelial lesions, typical of marginal zone lymphomas in mucosal sites. However, some cases may contain only a minor neoplastic population with a prominent reactive component and these may be difficult to recognize.

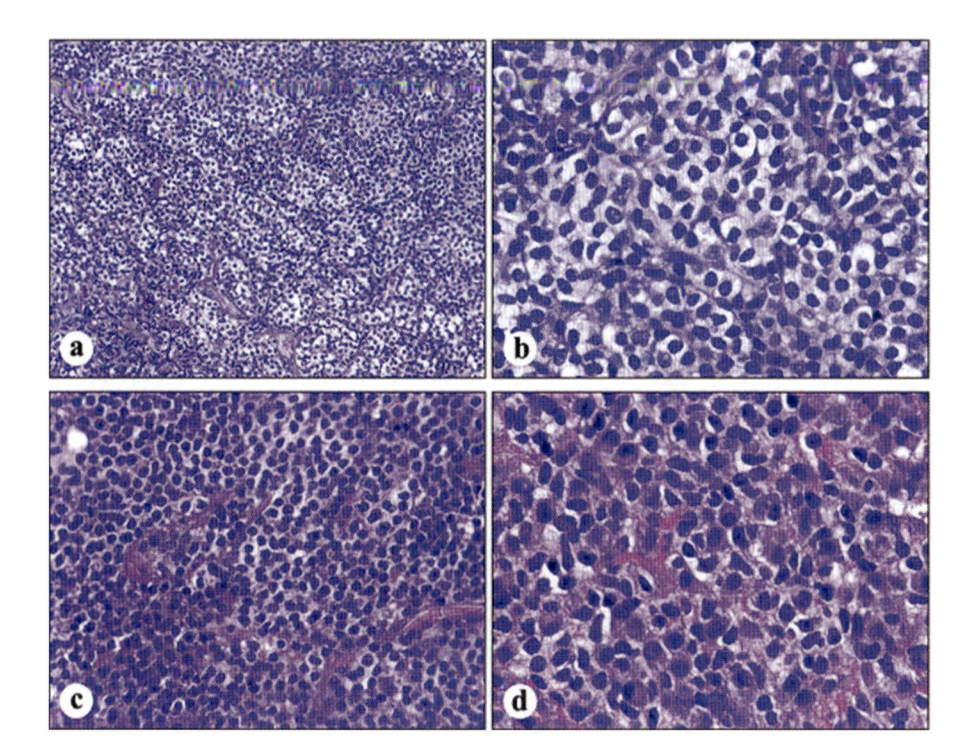

Figure 35.5 (a–c) Extranodal marginal zone lymphoma consists of diffuse sheets of small lymphoid cells that often have delicately cleaved and a moderate amount of clear cytoplasm. (d) Plasma cell differentiation can be prominent.

Immunohistochemically the tumor cells are positive for CD20, and immunoglobulin light-chain restriction can usually be demonstrated, especially in the lesional plasma cells. Otherwise marginal zone lymphomas lack distinctive positive markers. They are usually negative for CD5, CD10, CD23, and BCL1 (cyclin D1), and BCL6, distinct from mantle cell, follicular, and small B-lymphocytic lymphomas (CLL). A small number are positive for CD23.[4,31]

Diffuse large B-cell lymphoma

In soft tissue, diffuse large cell lymphomas are among the most common lymphomas, and most of these are extranodal extensions of nodal lymphomas in sites such as the neck, axilla, and inguinal region. The World Health Organization-European Organization for Research and Treatment of Cancer (WHO-EORTC) 2005 classification of primary cutaneous lymphomas places most diffuse large B-cell lymphomas composed of monotonous large noncleaved cells into the category of "primary cutaneous large B-cell lymphomas of leg type," because they usually present in this location.[1] Other sites can also be involved, but approximately two-thirds of the patients have leg nodules. These nodules are often multiple, and they can involve the skin and subcutis. Clinical staging is necessary to confirm primary soft tissue occurrence. Age-related Epstein–Barr Virus (EBV)-positive diffuse large B-cell lymphoma can also manifest in extranodal soft tissues.[1,24,25,33–35]

There is a predilection for older adults, with a mean age >75 years and a 3:2 female predominance. Many of these lymphomas are rapidly growing and clinically aggressive, and systemic therapy has a significant role. Patients with solitary lesions fare better, however. According to one study, patients with tumors located elsewhere than the legs had a somewhat better prognosis.[35]

Morphology is variable, but monotonous sheets of large cells containing oval to round, noncleaved nuclei with variably prominent nucleoli are typical. Epithelioid cell morphology and a nesting pattern reminiscent of a poorly differentiated carcinoma are not uncommon (Figure 35.6). Rare variants of extranodal large B-cell lymphomas having myxoid or rosette-forming morphology should not be confused with soft tissue sarcomas or neuroblastoma variants.[36,37]

Immunohistochemically typical is positivity for CD20, CD79A, and BCL6 (most cases), whereas CD10 is usually absent. Most leg-type lymphomas express BCL2 and MUM1 transcription factor.[1,34]

Age or immunosuppression-related EBV-positive diffuse large B-cell lymphoma can also form a soft tissue mass. These lymphomas reflect impaired immunosurveillance and can also occur associated with primary immunodeficiency of iatrogenic immunosuppression. Histologic features may include nuclear pleomorphism, including Hodgkin-disease-like features. In general, the latter should always be diagnosed with hesitation in cutaneous and soft tissue sites. Immunohistochemically these lymphomas are positive for pan B-cell markers (CD20, CD79a) and often for MUM1. They are typically negative for follicular center cell markers such as CD20 and BCL6. EBV latent membrane protein (LMP) is generally present, and EBV-virus RNA can be generally demonstrated by EBV-encoded RNA (EBER) *in situ* hybridization.[25]

Precursor B-cell lymphoblastic lymphoma

Extranodal B-cell lymphoblastic lymphoma usually occurs in children or young adults, and it can present as an

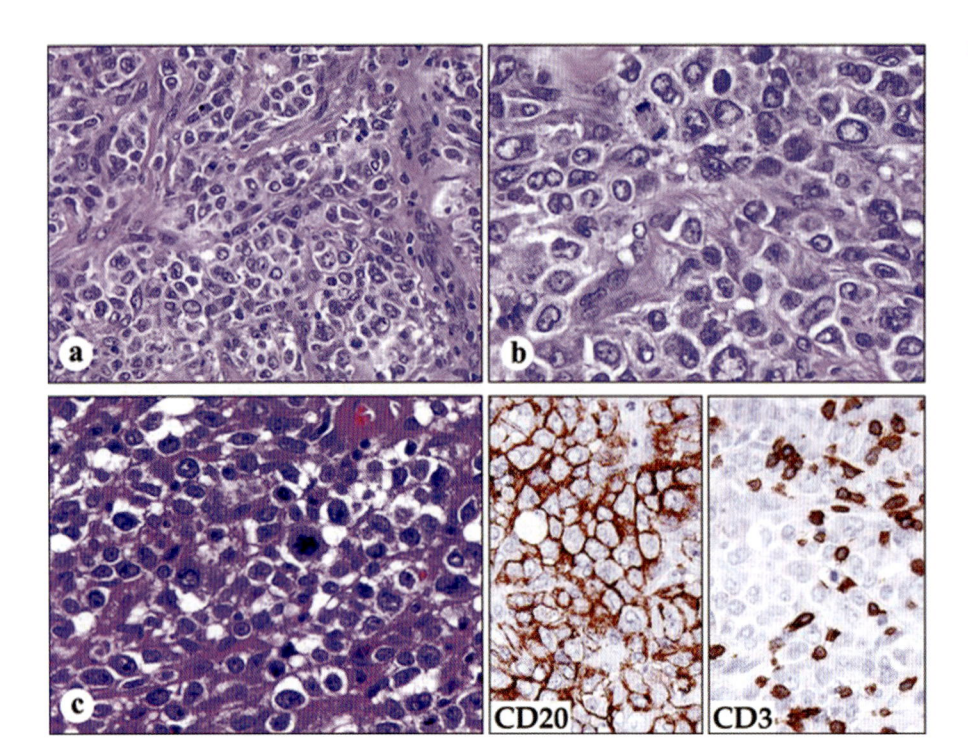

Figure 35.6 Diffuse large cell lymphoma can have a nesting pattern with slightly epithelioid cytology, resembling a carcinoma. The tumor cells are positive for CD20 and negative for CD3, with small lymphocytes being positive.

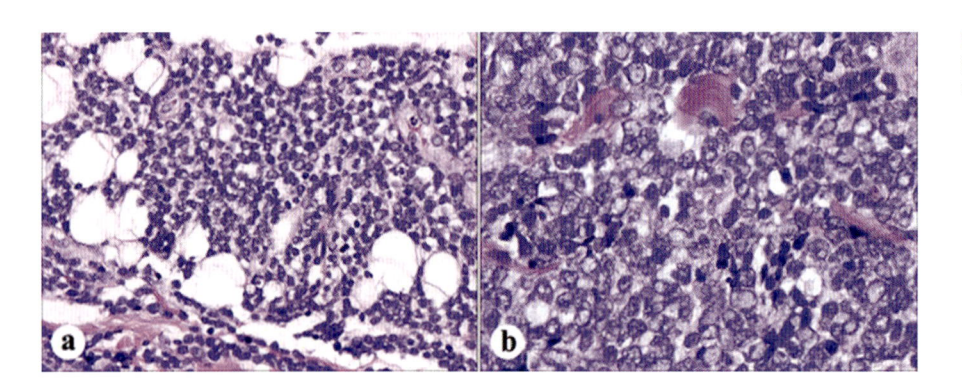

Figure 35.7 Lymphoblastic lymphoma infiltrating in subcutaneous fat. The large cells have vesicular round nuclei and delicate nucleoli.

indolent-appearing solitary cutaneous or soft tissue nodule or mass, especially in the head and trunk, or as a bone tumor, often in the spine. Adequate systemic chemotherapy is necessary to prevent (late) relapses, even in apparently localized peripheral disease.[38,39]

Histologically, precursor B-cell lymphoma cells have large vesicular nuclei. These lymphomas can resemble certain large B-cell lymphomas or blastic extramedullary myeloid tumors in their corded pattern and fibrous stroma (Figure 35.7).

Immunohistochemically, B-cell lymphoblastic lymphoma is distinctive for expression of B-cell markers and terminal deoxyribonucleotidyl transferase (TDT), and usually CD10. CD20 expression varies, however.[12–14] In contrast to blastic extramedullary myeloid tumor (granulocytic sarcoma), precursor B-cell lymphoblastic lymphomas are negative for myeloperoxidase.

Intravascular B-cell lymphoma

This rare manifestation of large cell lymphoma was historically known as *malignant angioendotheliomatosis*, but by the advent of immunohistochemistry, it was recognized as a lymphoma and not a vascular tumor. Although B-cell differentiation is most common, other phenotypes also occur.[39–41] The sole intravascular presentation is thought to stem from the inability of the lymphoma cells to extravasate and home into solid tissues, possibly by the lack of lymphocyte homing receptors, such as the CD11A/CD18 complex, CD29, and CD54.[42,43]

Occurrence is usually in older adults, and systemic intravascular spread causes systemic symptoms, especially those related to the central nervous system. The soft tissue lesions are cutaneous and subcutaneous, and in some cases, vessels of hemangioma or capillary-rich tumors, such as angiolipoma, are involved.[44,45] The clinical course is often aggressive, but

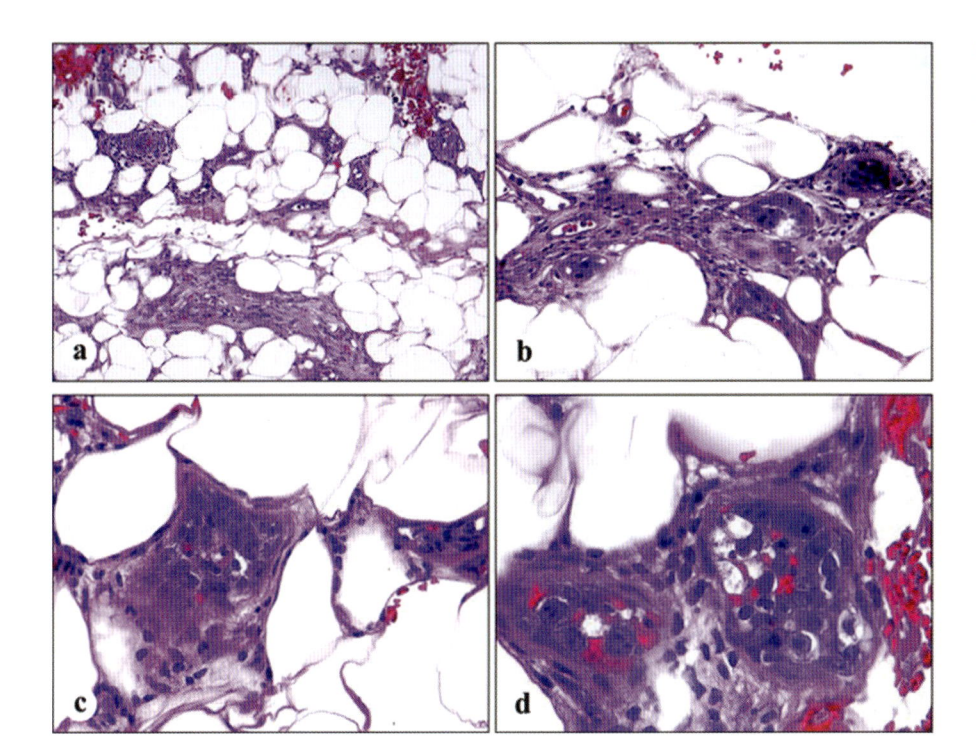

Figure 35.8 Intravascular B-cell lymphoma presenting in the subcutaneous fat contains small clusters of large B-lymphocytes within capillary vessels.

some patients have survived a few years with systemic chemotherapy.[39–41,46]

Pathogenetically this group is heterogeneous and includes B-cell, T-cell, CD30-positive, natural killer (NK)-cell, and HHV8/EBV-associated lymphomas, so that immunophenotyping is necessary for specific diagnosis.[41,46–50] Histologically, there are intravascular clusters of large round cells (Figure 35.8).

Burkitt lymphoma

This highly malignant B-cell lymphoma has a follicular center cell origin. Clinicopathologically, it is divided into three groups: endemic (African Sub-Saharan and Southwestern Pacific), sporadic (Western-worldwide), and in connection with immunosuppression, especially HIV/AIDS and less frequently, transplantation associated.

Endemic Burkitt lymphoma is always associated with EBV infection, and in fact, this virus was originally identified in Burkitt lymphoma cell culture. Only 30% of sporadic cases are EBV-associated. Among non-immunosuppressed patients there is a predilection for children, and it is the most common childhood non-Hodgkin lymphoma. The extranodal pediatric examples typically occur in jawbones in cases in the endemic areas, and in the abdomen, the ileocecal region in the gut, ovaries, and breasts in the nonendemic cases. Some cases, especially in immunosuppressed adults, may present with a soft tissue mass, although this is generally part of systemic disease. Burkitt lymphoma is a clinically aggressive and rapidly proliferating lymphoma that can evolve into acute lymphoblastic leukemia type 3 (ALL-L3), the leukemic equivalent of Burkitt lymphoma.[51–55]

Histologically typical are monotonous sheets of medium-sized to large cells with round nuclei, spaced by variable numbers of nuclear debris containing histiocytes, creating a "starry-sky" pattern (Figure 35.9). Immunohistochemically typical is expression of pan B-cell (CD19, CD20, CD22), follicular center cell markers (CD10, BCL6), and often EBV, detectable by *in situ* hybridization for EBER. High Ki67 index (nearly 100%) is typical of Burkitt lymphoma. Burkitt lymphoma is negative for CD5 and TDT. These lymphomas typically have C-myc activating translocations involving immunoglobulin genes, most commonly the heavy-chain gene with t(8;14)(q24;q32). MYC-immunohistochemistry is typically positive, although this is not pathognomonic.[53–55]

Plasmacytoma/plasmablastic B-cell lymphoma

Plasmacytoma is composed of terminally differentiated B-cells with features of plasma cells, and therefore this tumor can be considered a form of B-cell lymphoma. *Multiple myeloma* is the designation for disseminated plasmacytoma with multifocal bone marrow involvement. It has been estimated that 2–20% of multiple myeloma patients have extramedullary spread, and the frequency is higher with relapsing disease. Autopsy studies show even higher frequency of extramedullary disease, up to 70%. Local growth from adjacent involved bone is more common than pure soft tissue involvement.[56,57] Primary soft tissue plasmacytomas have also been reported and often found to be clinically indolent. It is likely that many of these tumors are actually marginal zone lymphomas with extensive plasma cell differentiation rather than true plasmacytomas.[25] This is

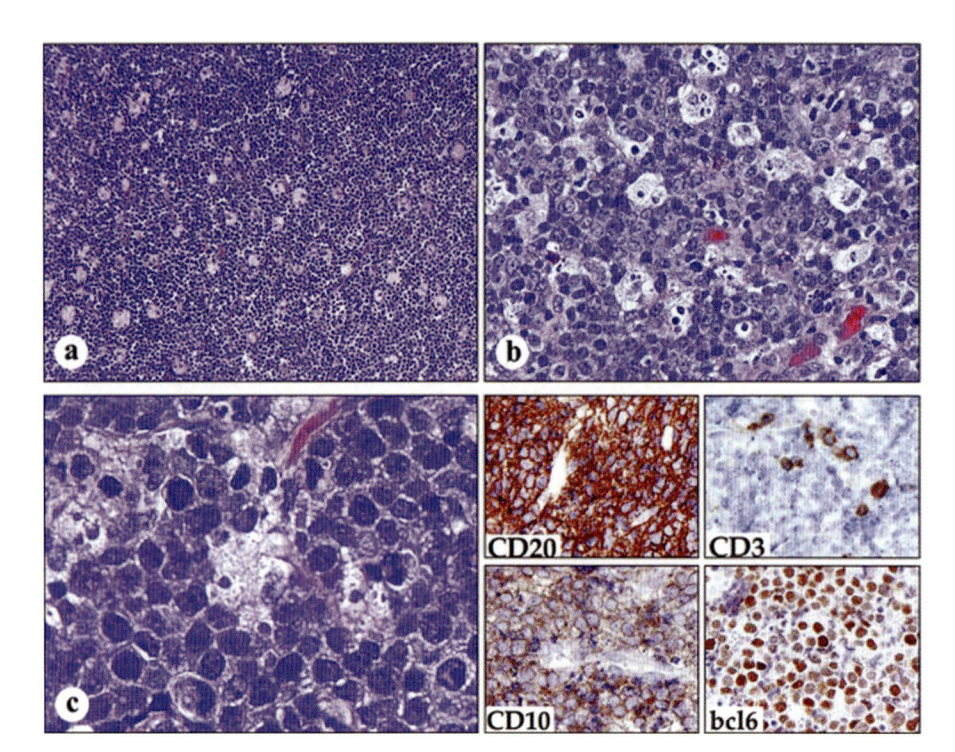

Figure 35.9 Burkitt lymphoma is composed of relatively uniform middle sized lymphocytes and numerous nuclear debris-containing (tingible body) macrophages. Positivity for CD20, CD10, and BCL6 is typical and reflects follicular center B-cell origin.

especially true for plasmacytomas in the respiratory tract and neck of thyroid origin. However, clinical follow-up and possibly bone marrow studies may be indicated to rule out concurrent myeloma or its development in the future.

Plasmablastic B-cell lymphomas are associated with EBV infection, are EBER positive, and often occur in immunosuppressed patients, especially those with HIV/AIDS.[58–60]

Better-differentiated plasmacytomas show plasma cell morphology with eccentric nuclei and abundant mildly basophilic cytoplasm (Figure 35.10). Extranodal marginal zone lymphomas can have extensive plasma cell differentiation and these should not be confused with plasmacytoma, as their prognosis is much better with infrequent dissemination. Poorly differentiated plasmablastic forms of plasmacytoma and plasmablastic B-cell lymphoma seem to be practically inseparable in their morphology and immunohistochemistry.[61]

Immunohistochemically typical is expression of immunoglobulin light chains in a clonal pattern (either kappa or lambda). Other markers expressed include CD38, CD56 (NCAM, variably), CD138 (syndecan), MUM1, and EMA. Plasmacytomas are negative for common B-cell markers such as CD20, variably positive for CD79a, and are negative for CD3, CD30, and BCL6.[4–6,61] One has to consider that CD138 is not specific for plasma cells, but is expressed in a variety of other malignancies, such as many carcinomas.

Panniculitis-like T-cell lymphoma

Clinical features

Panniculitis-like T-cell lymphoma (PTCL) is currently defined as a T-cell lymphoma expressing alpha/beta T-cell receptors (AB-type PTCL). It presents as multiple subcutaneous nodules and plaques in the extremities and trunk. The lesions vary from 1 cm to very large (>10–20 cm) confluent masses. Constitutional lymphoma symptoms (e.g., fever, night sweats, weight loss) are common. Occurrence is from childhood to old age, with the overall median age approximately 40 years and 2:1 female predominance.[62–67]

The patients often have a long history of cutaneous lesions, sometimes thought to be a form of panniculitis, such as erythema nodosum or lupus panniculitis. Prognosis is relatively good, with an 85% disease-specific survival rate in the largest series, but some patients progress to a hemophagocytic syndrome, which is a poor prognostic sign. There is no significant tendency toward nodal involvement.

Gamma-delta (GD)-type PTCL occurs in an older population, with the median age approximately 60 years. It has a more aggressive course, frequently evolving into a hemophagocytic syndrome and with only an 11% survival rate in the largest series.[57] In the current classification, this subtype has been separated from PTCL.[1]

Pathology

Histologically, there is diffuse subcutaneous fat infiltration by a mildly atypical, mitotically active T-cell population involving the fat lobules, often sparing the septa. The neoplastic cells often encircle individual fat cells with reactive histiocytes (Figure 35.11). The cell size varies, with most cases being composed of predominantly small- to medium-sized cells often admixed with large atypical cells. Apoptotic tumoral lymphocytes are often present, as is accompanying plasma cell and eosinophilic infiltration.[53,56]

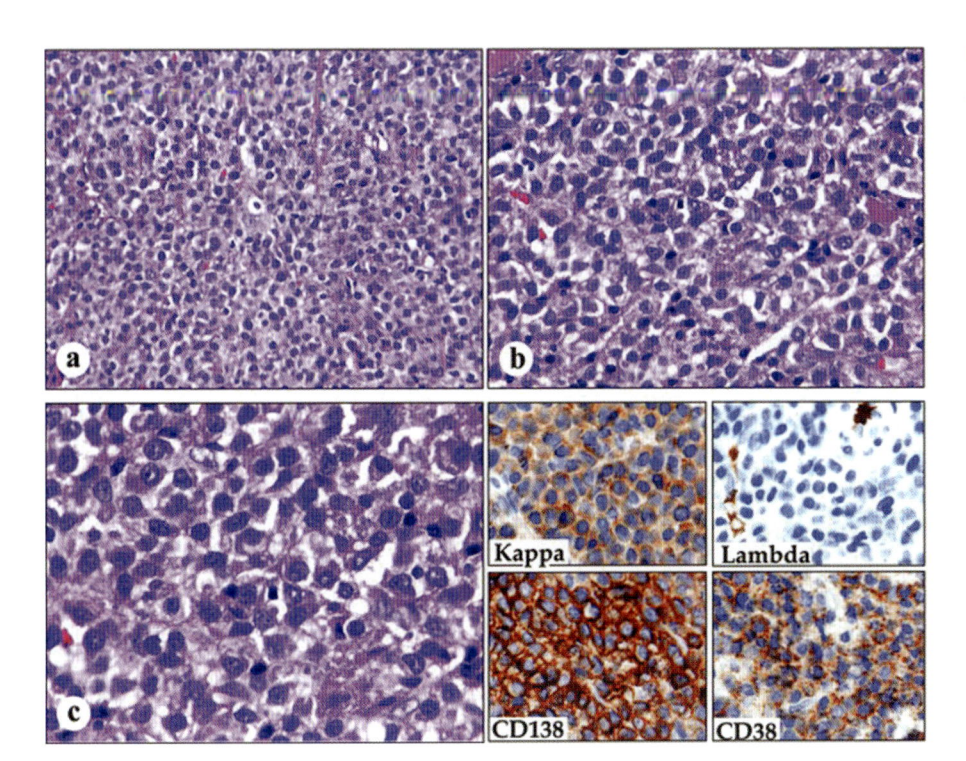

Chapter 35: Lymphoid, myeloid, histiocytic, and dendritic cell proliferations

Figure 35.10 Plasmacytoma is composed of large polygonal cells with resemblance to plasma cells. In this case, the tumor cells show kappa light-chain restriction and are positive for CD138 and CD38.

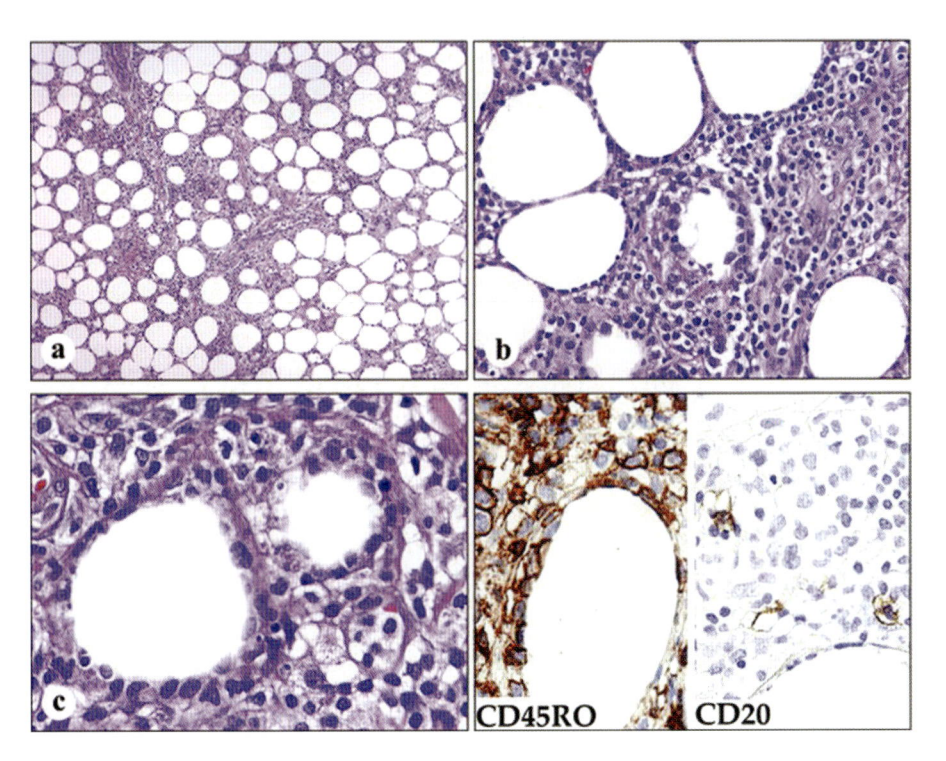

Figure 35.11 Panniculitis-like T-cell lymphoma involves diffusely subcutaneous fat resembling a lobular panniculitis. This example is predominantly composed of small lymphocytes, and these cells often encircle the fat cells. The small cell population is mainly composed of CD45RO-positive T-cells with very few CD20-positive B-cells being present.

Immunohistochemically, the lesional cells are positive for the T-cell antigens CD2, CD3, and CD43, and usually also positive for CD5, CD7, and CD8, but negative for CD4 indicating a cytotoxic (suppressor) T-cell phenotype. Immunoreactivity for the cytotoxic granule antigens, granzyme B, perforin, and TIA1, is typical, although these tumors are generally negative for CD30 and CD56, in contrast to large cell anaplastic and NK/T-cell lymphomas. B-cell markers (e.g., CD20, CD79a), are absent. Clonal T-cell receptor gene (e.g., alpha, beta, gamma, delta) rearrangements are demonstrable in the tumor cells. *In situ* hybridization for EBER is negative.[63–67]

Differential diagnosis

Deep lupus panniculitis has to be considered in the differential diagnosis. Presence of lymphoid follicles and sclerosing fibrous

951

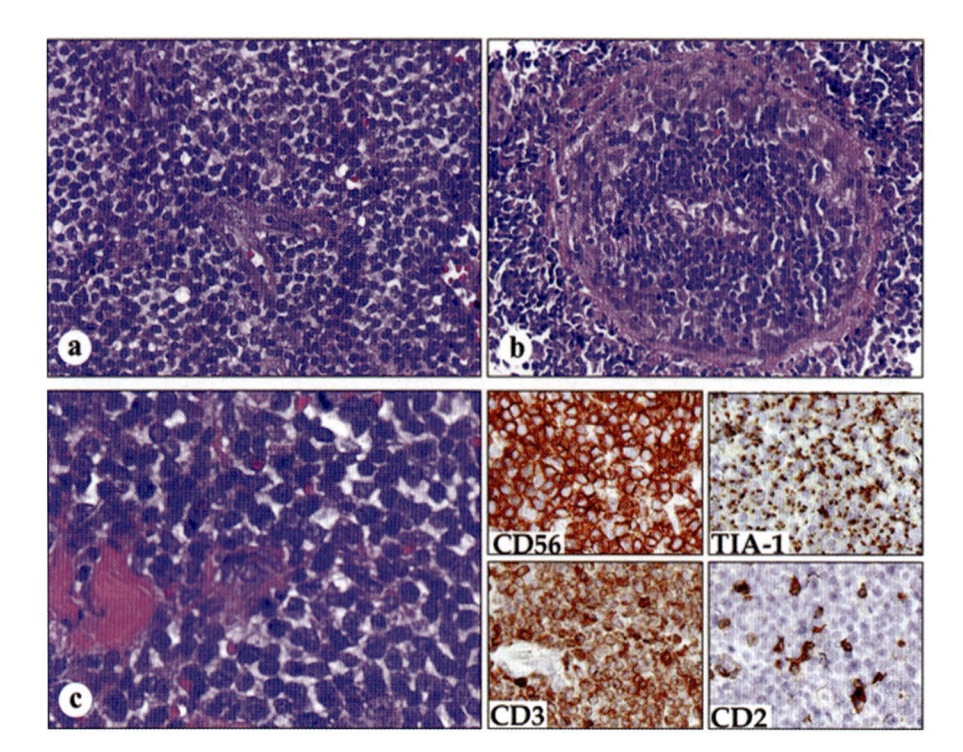

Figure 35.12 (**a,c**) A nasal-type, non-nasal NK/T-cell lymphoma is composed of medium-sized to large lymphoid cells with clear cytoplasm. (**b**) Vascular involvement (angiotropism) is a typical feature. Immunohistochemical findings include positivity for NK cell markers CD56 and TIA-1, whereas expression of T-cell antigens is limited to CD3E, recognized by a polyclonal antibody.

matrix are features of lupus that can be useful in the differential diagnosis, along with clinical setting. Nevertheless, overlap between lupus profundus and panniculitic T-cell lymphoma may exist, requiring clinical follow-up.[68]

Borrelia burgdorferi infection has also been reported with a lobular lymphocytic panniculitis rich in CD8+ and TIA1+ T-cells simulating panniculitic T-cell lymphoma. In this case the lesion differed from lymphoma by being a solitary nodule and positive for *Borrelia burgdorfi* DNA by PCR.[69]

Nasal-type natural killer cell/T-cell lymphoma

The most common presentation of this lymphoma type is in the nasal cavity, but a small number occur in extranodal peripheral soft tissues involving the skin, subcutis, or deep soft tissues, with common multifocality and dissemination. The patients are adults, with a median age of approximately 50 years and significant male predominance in the reported largest series. Prevalence is greater in the Far East, where the nasal NK/T-cell lymphoma also preferentially occurs. NK/T-cell lymphoma is highly aggressive, with most patients dying of their disease in 3 to 6 months.[70,71]

Histologically, NK-cell lymphomas are often angiocentric, involving vessel walls, and are composed of a mixed population of small and large cells, often with irregular nuclear outlines (Figure 35.12).

Immunohistochemically, defining features are CD56 and cytotoxic granule protein (granzyme B, TIA1) positivity, whereas the expression of T-cell markers is essentially limited to the CD3 epsilon chain recognized by polyclonal CD3-antibodies. There is a high association with EBV (positive EBER *in situ* hybridization). T-cell receptor genes are unrearranged (germline configuration), and B-cell markers (CD20) are absent.[64,70,71]

Noncutaneous (systemic) anaplastic large cell lymphoma

This lymphoma, characterized by CD30 positivity (Ki1 antigen, formerly *Ki1 lymphoma*) is a malignant lymphoma of activated T-lymphoid cell phenotype that either expresses or does not express T-cell markers. The T-cell marker-negative "null" variants also have clonal T-cell receptor gene rearrangements supporting their T-cell lineage. Anaplastic large cell lymphoma (ALCL) is presently considered a T-cell lymphoma, and it is not associated with EBV infection. CD30-positive B-cell lymphomas are excluded from ALCL and included among B-cell lymphomas. Also excluded from the ALCL category are peripheral T-cell lymphomas with focal CD30 expression.

ALK expression and ALK gene rearrangements are common, and their presence separates ALCL into ALK-positive and ALK-negative groups that are clinicopathologically different. Most important, ALK-positive tumors occur in younger-aged patients, who have a better prognosis.[72–75] ALCL can present as a soft tissue mass that can clinically and even histologically mimic a soft tissue sarcoma in locations such as the retroperitoneum; most of these tumors have been ALK+.[76–78]

Clinical features

ALK-positive ALCL occurs mainly in children and young adults, with a male predominance. It can present in a wide variety of sites: soft tissue, bone, and organ-based locations (e.g., the gastrointestinal tract). The overall survival rate with systemic chemotherapy is better than that for ALK-negative ALCL. This difference might be related at least partly to a younger patient age.[74–78]

ALK-negative ALCL occurs in older age groups (median age >50 years), with a male predominance. The spectrum of sites is essentially similar to those of ALK-positive ALCLs.[72,73]

Over 100 cases of ALCL have been reported in association with mostly silicone-coated and silicone or saline-filled breast implants. These lymphomas have often occurred long after implantation (>10–20 years) and have varied from minimal tumors to ulcerated nodules. Most have been ALK negative and many were microscopic lesions limited to the seroma fluid around the implant capsule. The reported cases were indolent in most cases, but a small number of patients died of disease. The pathogenesis of implant-capsule-associated ALCL is unclear, but lymphoma-promoting factors may include chronic inflammation or immunosurveillance vacuum in the capsular seroma environment.[79–85]

Pathology

Histologically, ALCL is composed of sheets of cells with large nuclei having complex reniform contours and abundant pale to eosinophilic cytoplasm, thus resembling histiocytes (Figure 35.13). The historical designation *malignant histiocytosis* largely refers to ALCL. Variants with extensive histiocytic infiltration (lymphohistiocytic variant) and those composed of smaller cells (small cell variant) occur with a lower frequency.[72] Examples with a spindle cell pattern or cohesive perivascular arrangements can histologically resemble sarcomas.[86,87]

Immunohistochemically, ALCL cells are, by definition, positive for CD30 and usually also for cytotoxic granule proteins (granzyme B, perforin, TIA-1) in all forms of ALCL. Approximately 50% of cases are positive for ALK. Use of markers other than CD30 is necessary, because CD30 is not specific for ALCL, being present in Hodgkin's disease, some embryonal carcinomas, and some cells in reactive lymphoid infiltrations, among others. Nevertheless, distinction from Hodgkin's lymphoma can be problematic at times, with some overlap between these entities. PAX5 positivity is a feature favoring Hodgkin's disease.[4,5]

ALCL is often, but variably, positive for T-cell markers, but by definition it is negative for B-cell markers, such as CD20, and for PAX5. EMA/MUC1 is preferentially expressed in ALK-positive examples, and the primary cutaneous variant is typically negative.[88] Because ALCL can be negative for CD45 (LCA), CD30 has to be included for evaluation of possible lymphomas.

ALK receptor tyrosine kinase is aberrantly expressed in cases with ALK-involving translocations that activate this signaling pathway. ALK gene rearrangements involving 2p23 are frequently present, and several fusion translocations involving the ALK gene have been detected. Among the most common is t(2;5)(p23;q35), fusing ALK with the nucleophosmin (NPM) gene. Among the other fusion partners are ATIC, clathrin, TPM3, and TPM4.[72–74] Of note, these fusions also occur in

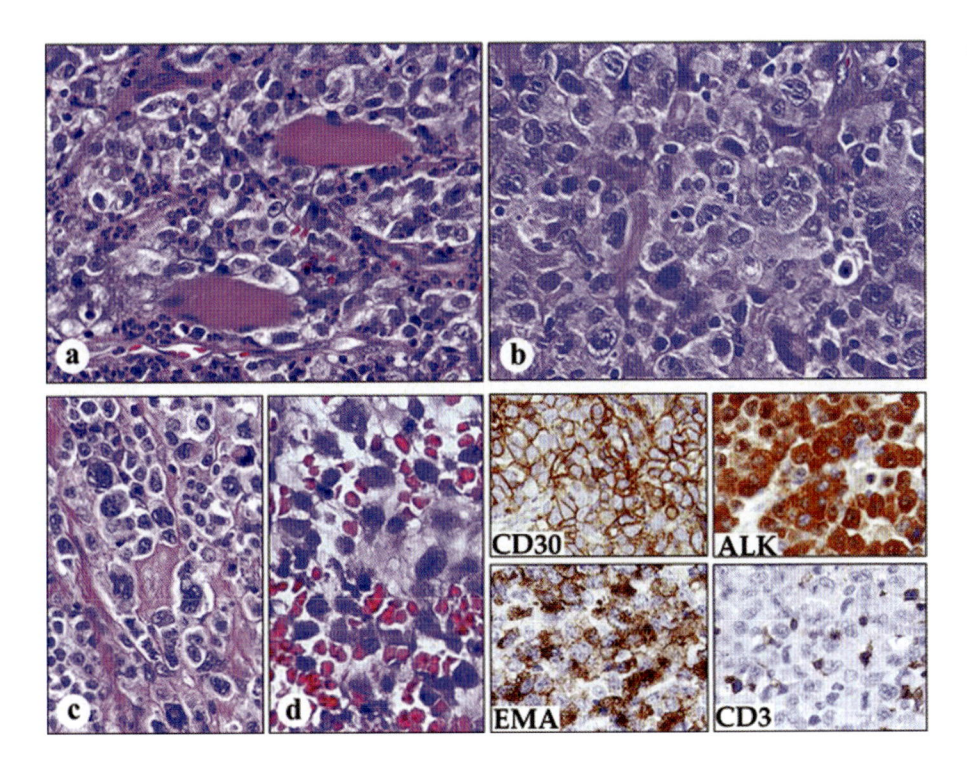

Figure 35.13 Examples of extranodal anaplastic large cell lymphoma. (**a**) Infiltration in skeletal muscle. (**b**) Large tumor cells with abundant mildly eosinophilic cytoplasm. (**c**) Focal pleomorphism. (**d**) Examples with extravasated erythrocytes can simulate an angiosarcoma. Immunohistochemically typical is CD30 positivity (here membranous), cytoplasmic and nuclear ALK positivity, and variable expression of EMA (here positive), and CD3 (here negative).

Primary cutaneous anaplastic large cell lymphoma

Primary cutaneous ALCL is a clinically indolent variant that typically occurs in an older population than does the non-cutaneous or systemic ALCL and has a male predominance. It is generally treated by local excision, often supplemented with radiation therapy, but without systemic chemotherapy. These tumors variably express T-cell antigens, but are typically negative for ALK and EMA. With lymphomatoid papulosis (LP), primary cutaneous ALCL is often included under an umbrella diagnosis: CD30-positive cutaneous lymphoproliferative disorder. LP is a chronic multifocal disease that is generally indolent, but on rare occasions it can progress to cutaneous ALCL. Therefore, LP and primary cutaneous ALCL could be a continuum of a single entity.[72–74]

Noncutaneous systemic ALCL can also involve the skin secondarily. Immunophenotypically, it can often be separated from the cutaneous variant by its positivity for ALK and EMA.

Extramedullary myeloid tumor (myeloid sarcoma, chloroma)

This tumor is an extramedullary analog of acute myeloid leukemia (AML) and occurs in all ages, with some predilection for children and young adults. A wide variety of locations can be involved, including the peripheral soft tissues, body cavities, and internal organs. Most patients have previously diagnosed AML or myelodysplasia, but in some patients, leukemia is diagnosed months or even years later. There have also been some long-term survivors after aggressive therapy who have never had or developed acute leukemia during >15 years of follow-up.[89–94]

The term *chloroma* refers to the often yellowish color of the tumor.

Histologically, the blastic type is composed of sheets of primitive, uniform, medium-sized blastic cells (Figure 35.14) and is the most common variant. In more differentiated forms, there is a component of eosinophilic myelocytes and megakaryocytes (Figure 35.15). The differentiated, megakaryocyte-containing examples can exhibit stromal sclerosis and histologically simulate soft tissue tumors such as sclerosing liposarcoma.[95]

Immunohistochemically, the blastic variant typically expresses the panleukocytic and nonlineage markers CD45 and CD43 and often also CD31 and CD34. Expression of myeloperoxidase is a diagnostic feature, and many examples are also positive for lysozyme (Figure 35.16). Hematolymphoid proliferations negative for CD3, CD20, and CD30 should be tested for myeloperoxidase to rule out an extramedullary myeloid tumor, which otherwise can easily be confused with a non-Hodgkin lymphoma.[92,96] Histochemical demonstration of chloroacetate esterase (Leder stain) was previously also used as a marker for myeloid differentiation in the diagnosis of extramedullary myeloid tumor. Expression of KIT (CD117) is typical.[96]

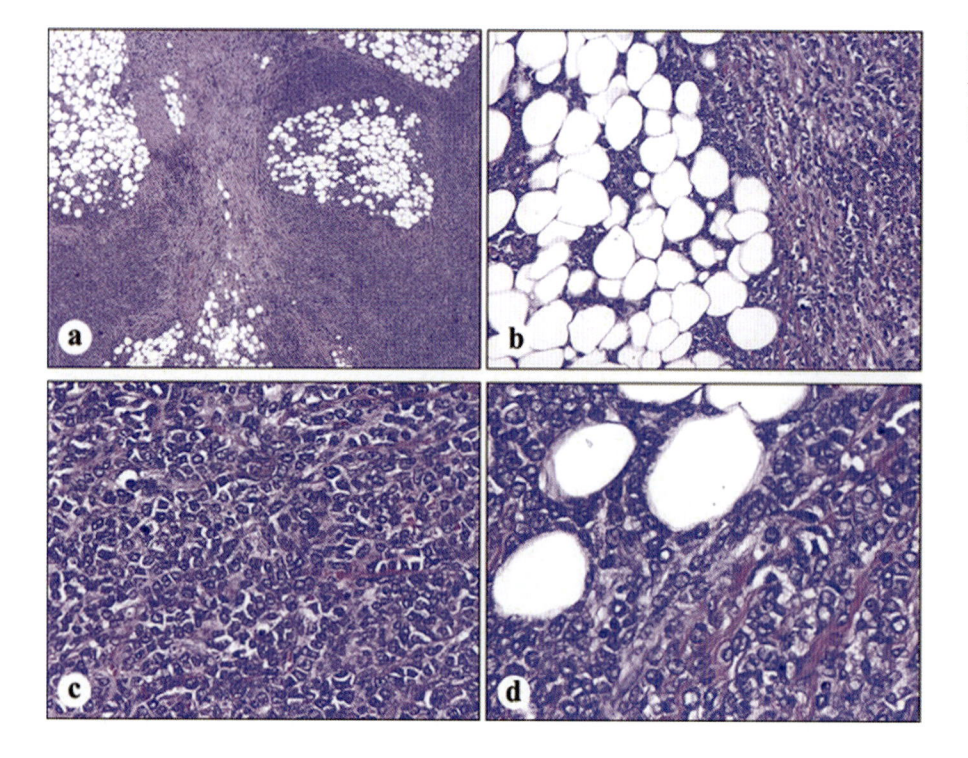

Figure 35.14 Extramedullary myeloid tumor (blastic variant) diffusely involves omental fat. In this case, the tumor cells are uniform and oval with no apparent myeloid differentiation, simulating a large cell lymphoma.

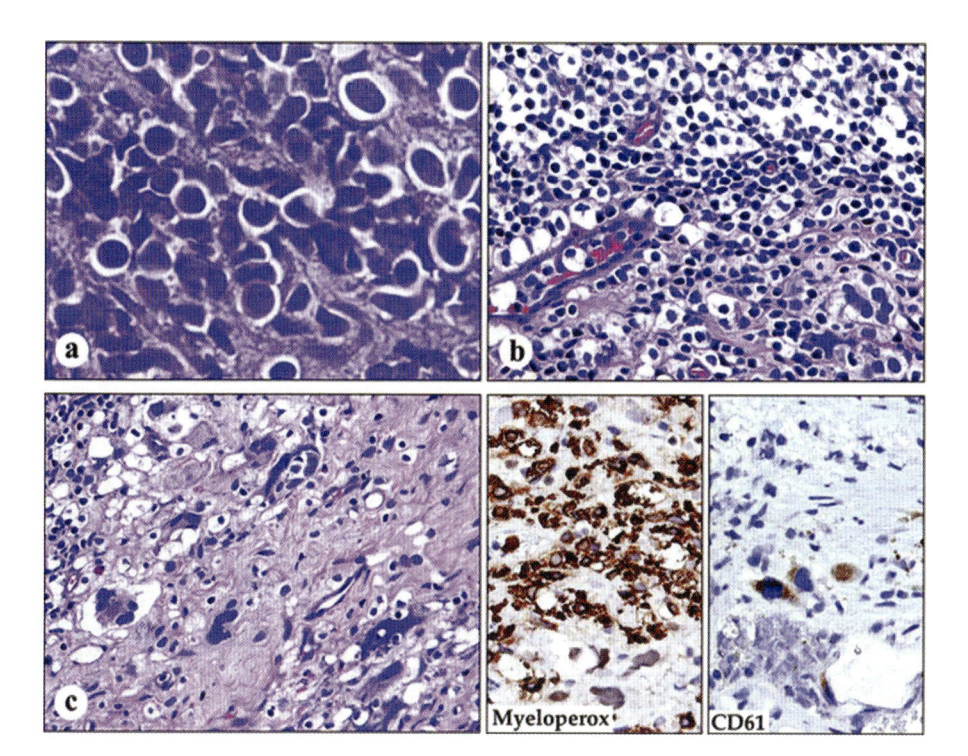

Chapter 35: Lymphoid, myeloid, histiocytic, and dendritic cell proliferations

Figure 35.15 Extramedullary myeloid tumor with differentiation (**a**) Eosinophilic myelocytes are present (below the center). (**b**) This example contains myeloid and megakaryocytic elements. (**c**) Megakaryocytic elements in a sclerosing stroma. The myeloid elements are positive for myeloperoxidase, and megakaryocytes are positive for CD61.

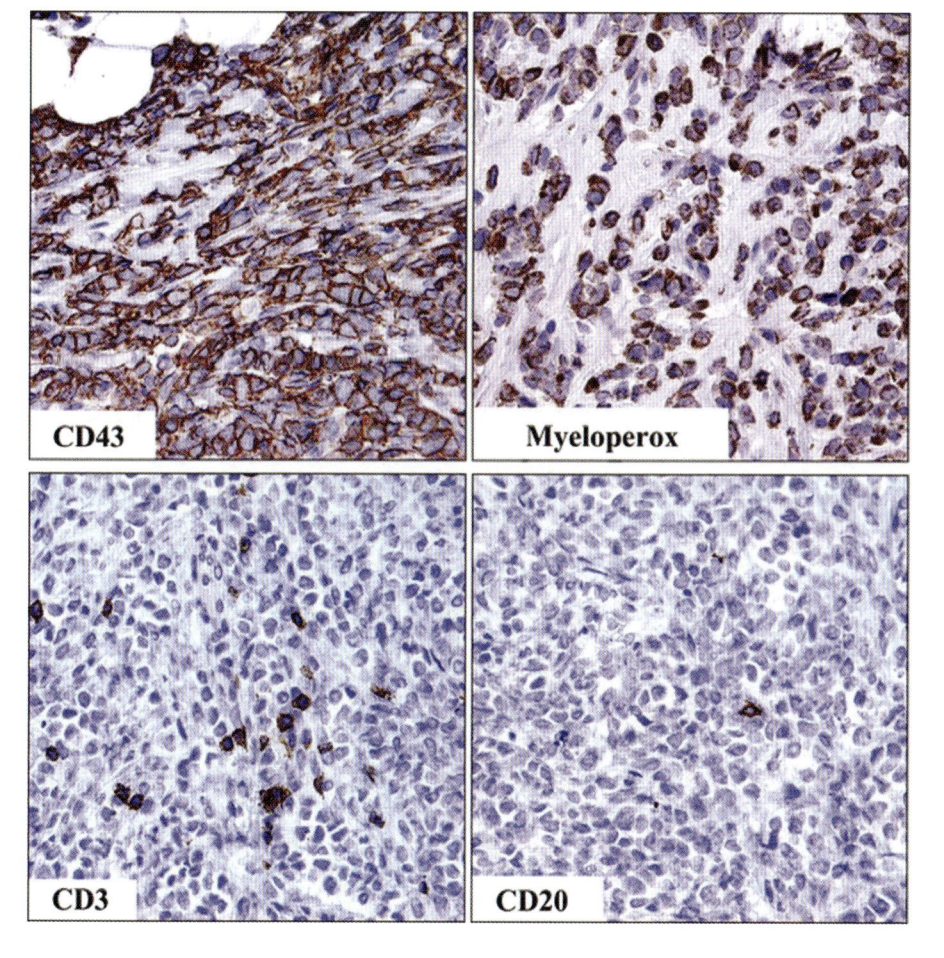

Figure 35.16 Immunohistochemically, a blastic extramedullary myeloid tumor is positive for CD43 and myeloperoxidase, whereas the tumor cells are negative for T- and B-lymphoid markers.

955

Mastocytoma and mastocytosis

Solitary mastocytoma is rare in humans, but is among the common soft tissue tumors in dogs and cats. In humans, it most commonly forms a small <1 cm skin nodule in infants. The prognosis is good, and sometimes the lesion even regresses spontaneously.[97]

Histologically, this tumor forms a sharply demarcated dermal nodule that can be ulcerated. The nodule is composed of sheets of uniform spindled to epithelioid cells.

These cells do not necessarily resemble mature mast cells, but they reveal blue granulation with Giemsa staining. Immunohistochemically, they are positive for KIT and mast cell tryptase, but are negative for CD163, in contrast to juvenile xanthogranuloma (Figure 35.17). Chloroacetate esterase stain is also positive. Infectious lesions such as scabies can elicit a strong mast cell response to simulate mastocytoma, and therefore infection has to be ruled out before diagnosing mastocytoma.[98]

The main diagnostic criteria of systemic mastocytosis is bone marrow or other noncutaneous infiltrates containing >15 mast cells per each cluster. The most common form of systemic mastocytosis is the indolent (urticaria pigmentosa) variant, in which mast cell infiltrations occur in the skin and the bone marrow. In some patients, the lesions are limited to bone marrow. Aggressive forms of systemic mastocytosis have more extensive mast cell infiltrates in the marrow that are associated with anemia and involve internal organs. Serum measurement of mast cell tryptase can assist in determining the tumor load.[97]

Histologically, the bone marrow in indolent systemic mastocytosis contains focal clusters of spindled mast cells with pale cytoplasm. These infiltrates can also contain eosinophils. Immunohistochemically, these cells are similar to those in solitary mastocytoma and are positive for KIT and mast cell tryptase (Figure 35.18). Aberrant expression of CD2 and CD25 (not expressed in normal mast cells) is an adjunct diagnostic criterion.[97]

Systemic mastocytosis is associated with activating either somatic (adult-onset mastocytosis) or germline mutations of KIT exon 17 (familial syndrome with mastocytosis).[99] Some patients also have other neoplasms with KIT-positive cells, especially gastrointestinal stromal tumors (GISTs; see Chapter 17). Solitary mastocytoma has also been reported to harbor KIT mutations.[100]

Rosai–Dorfman disease

Rosai–Dorfman disease (RDD) is a distinctive, chronic, inflammatory histiocyte-rich process of unknown etiology. Originally reported by Rosai *et al.* as a nodal disease predominantly in children, it was named "sinus histiocytosis with massive lymphadenopathy." It has become known as a disease potentially involving nearly every organ system, including soft tissues. Although in general the course is favorable, a few patients have had a fatal course contributed to by extensive multiorgan involvement and immunologic compromise. The most common extranodal sites besides soft tissue include the skin, nasal cavity, eyelids and orbit, and bones.[101] It is not known with certainty whether RDD is a neoplasm or a reactive process, and whether it is a histiocytic or a dendritic cell process has been debated.[102]

According to the RDD registry, soft tissues are involved in at least 9% of patients, and most of these patients have other

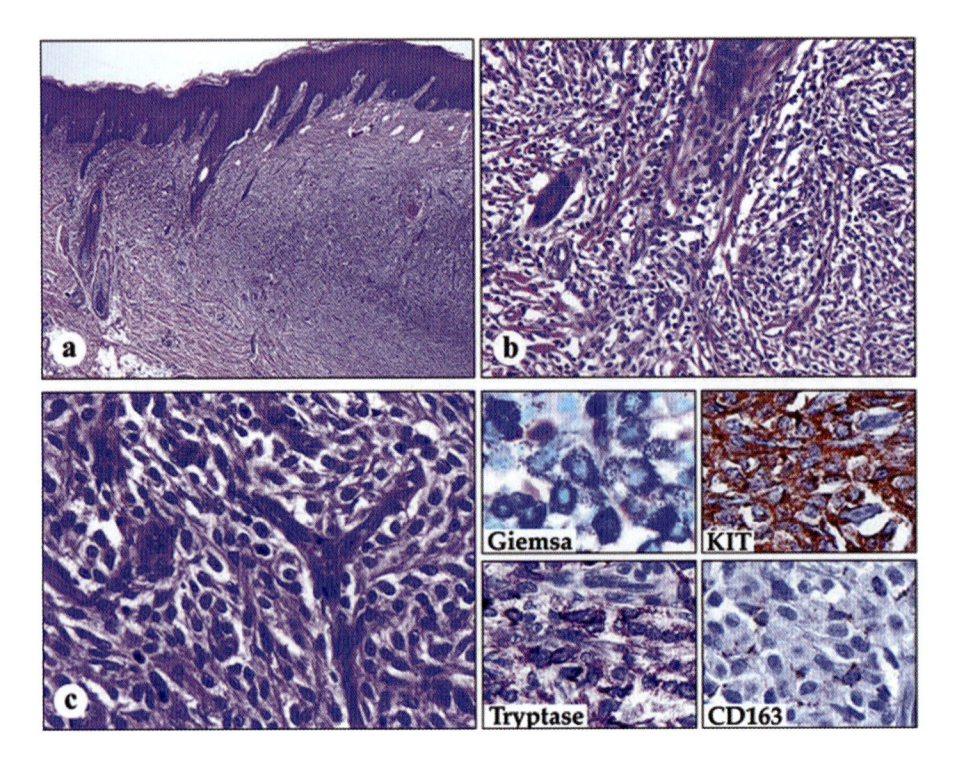

Figure 35.17 Cutaneous mastocytoma forms a mildly elevated cellular dermal infiltrate. This consists of oval to spindled cells. Prominent capillary network is also evident. The tumor cells are positive for Giemsa stain, KIT, and mast cell tryptase, but are negative for CD163.

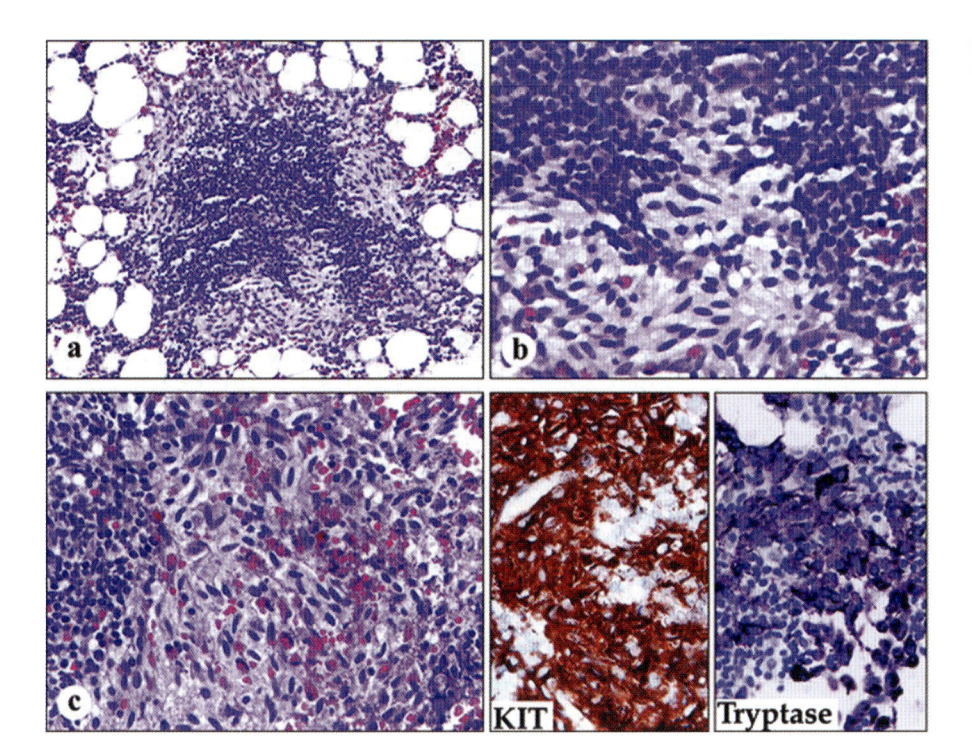

Figure 35.18 Systemic mastocytosis involving bone marrow forms small, pale-staining nests of spindle cells that are often associated with lymphoid infiltration. Immunohistochemically the cells are positive for KIT and mast cell tryptase.

extranodal manifestations. The patients with soft tissue involvement are on average older than those with primarily nodal disease, with the mean age being 46 years in a series of soft tissue RDD.[103] In soft tissues, RDD usually involves subcutaneous tissue in a wide variety of locations in the extremities, trunk wall, and less frequently in the head and neck; multiple masses can be present. The clinical course is marked by common persistent disease and recurrences of soft tissue infiltrations after surgery.[101,103,104]

Cutaneous RDD typically forms papules, plaques, or nodules without any clear anatomic predilection. Occurrence of multiple lesions in different regions is relatively common. Based on RDD registry data, cutaneous involvement does not have a negative impact on prognosis.[101,105]

RDD in soft tissues usually forms an apparently demarcated mass.[103] Microscopically, however, the lesion typically infiltrates the subcutaneous fat in a multinodular manner (Figure 35.19). The lesions measure 1 cm to 10 cm (usually 2–5 cm). The cellular components include dense clusters of B-lymphocytes, scattered T-lymphocytes, plasma cells, and histiocytes, often in clusters or sheets. Distinctive are the large histiocytes with abundant pale to deeply eosinophilic cytoplasm (Figure 35.20). These histiocytes can form syncytial sheets, but in some cases the cell membranes are sharply delineated. Some histiocytes contain intracytoplasmic lymphocytes, a phenomenon called *emperipolesis*.

Immunohistochemically, the large histiocytes are notoriously positive for S100 protein, a significant diagnostic feature.[106–108] They differ from Langerhans histiocytosis cells by being CD1A negative. The histiocytes are also positive for C68 and lysozyme, but are variably positive for CD163. Large clusters of B-cells and scattered T-cells are present (Figure 35.21). Plasma cells are polyclonal and the number of IgG4 plasma cells is similar to that seen in reactive lymph nodes and lower than in IgG4-related disease.[109] A SLC29A3 gene mutation has been reported in familial RDD.[110]

Erdheim–Chester disease

This rare systemic histiocytosis occurs in adults, with a male predilection. It is typically associated with bilateral, often symmetric osteosclerotic lesions involving the diaphyses of the long bones of the lower and sometimes upper extremities. Bone pain is the most common presenting symptom. The histiocytic infiltrations also occur in the orbit, around the aorta and kidneys in the retroperitoneum, and skin, with some patients having xanthoma-like lesions. In the lung, there is pleural and interlobular septal involvement, which can evolve into interstitial fibrosis. Skull base involvement can be complicated by diabetes insipidus, and retroperitoneal lesions can be complicated by obstructive nephropathy. Treatments have included corticosteroids, cytotoxic chemotherapy or radiation.[111–114] More recent treatment options have included alpha-interferon, anakinra, and vemurafenib, based on the presence of BRAF mutation.[111–113] Even if this is a nonmalignant condition, there is moderate, although now decreasing, mortality because of CNS, pulmonary, or other visceral complications.

Reported soft tissue lesions associated with Erdheim–Chester disease include retroperitoneal xanthogranuloma[115] and bilateral breast masses.[116] In addition, orbital and

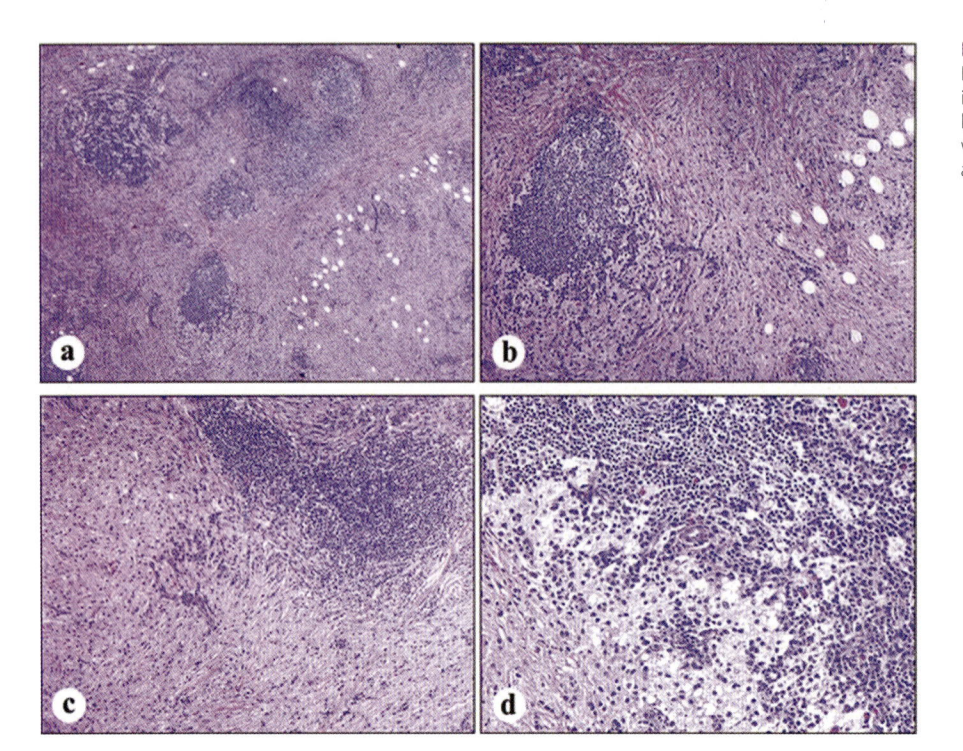

Figure 35.19 (**a,b**) Low magnification of Rosai–Dorfman disease (RDD) shows a process infiltrating in subcutaneous fat. The infiltrate consists of lymphoid foci and mixed histiocyte-rich infiltration with fibrosis. (**c,d**) Pale histiocytes are visible around the lymphoid foci.

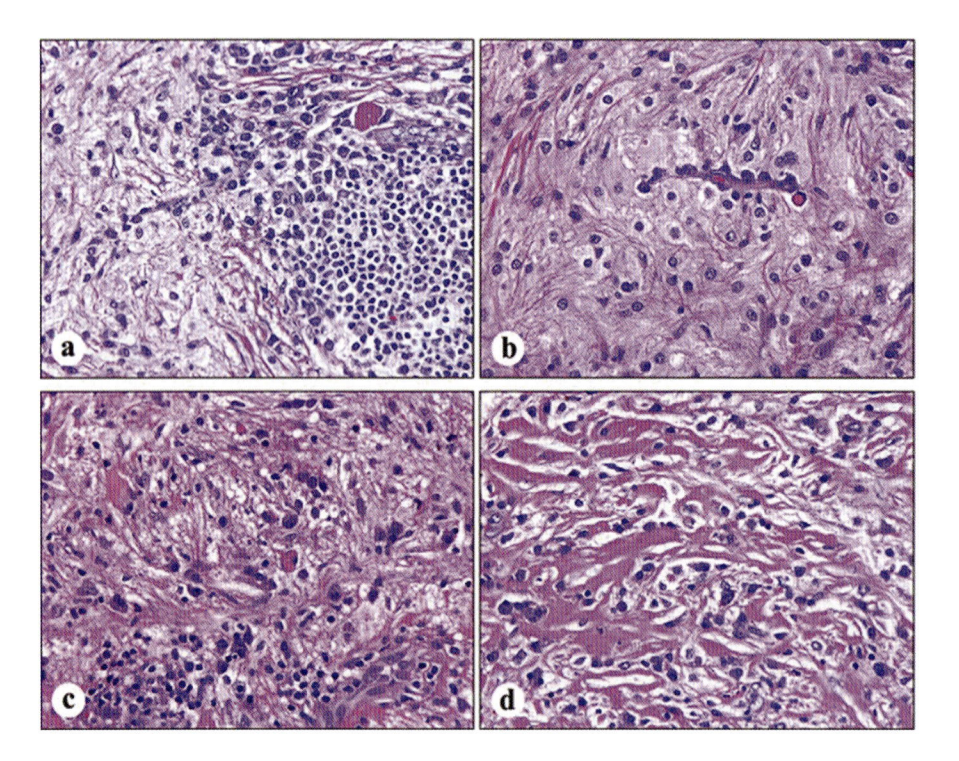

Figure 35.20 (**a,b**) Higher magnification of RDD reveals syncytial masses of pale cytoplasmic histiocytes, lymphoid infiltrate, and plasma cells with Russell bodies. (**c**) The histiocytes can also be sharply delineated. (**d**) Prominent sclerosis can be present.

perirenal histiocytic infiltrates may be biopsied for debulking or diagnostic purposes.

Histologically, there is histiocytic, typically xanthomatous infiltration into the bone marrow. Accompanying changes include spotty fibrosis with lymphoid infiltration of scattered T-cells, clusters of B-cells, and mast cells (Figure 35.22).

Immunohistochemically, large numbers of CD163-positive histiocytes are present (more than expected, based on HE stain). In contrast to Langerhans cell histiocytosis and RDD, the histiocytes are negative for CD1A and S100 protein.[117]

BRAF V600E mutation has been found as a recurrent change and analysis of this mutation can be used for diagnostic purposes, although it is not disease-specific.[118] The reported

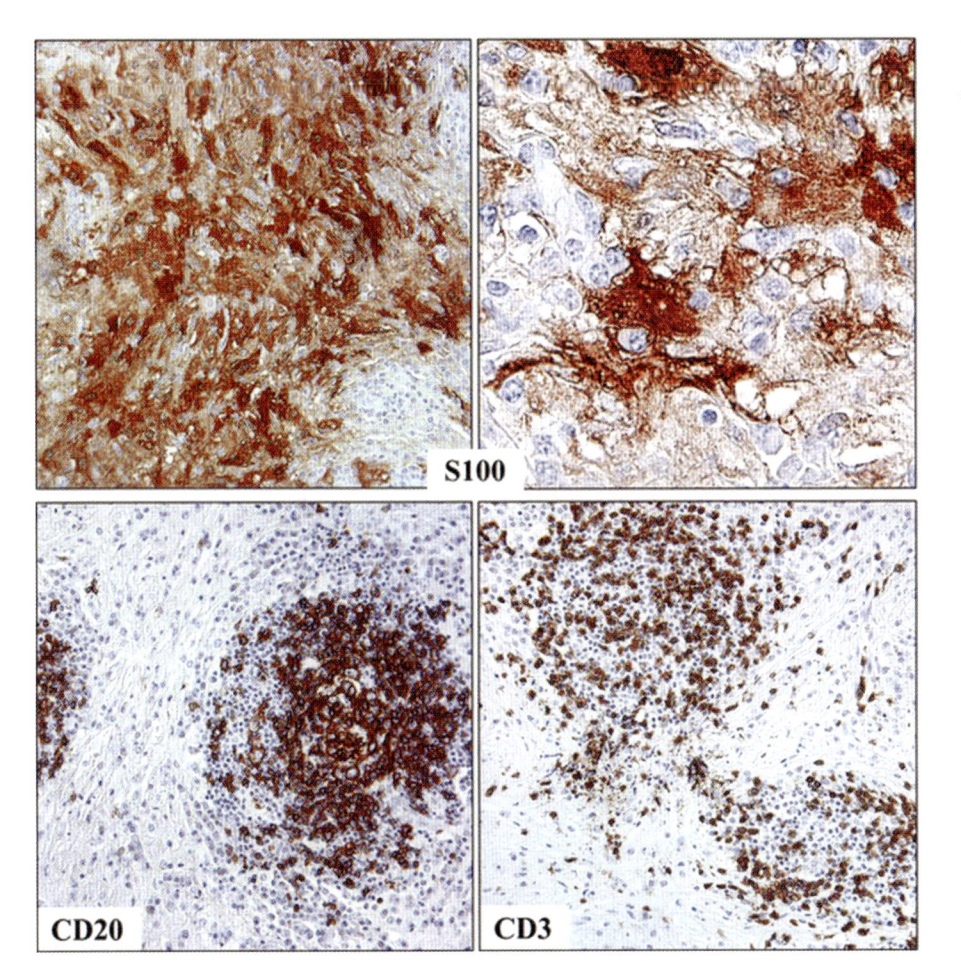

Chapter 35: Lymphoid, myeloid, histiocytic, and dendritic cell proliferations

Figure 35.21 The histiocytic cells in RDD are positive for S100 protein, and some of these cells show dendritic morphology. Clusters of B-lymphocytes and T-lymphocytes are also present.

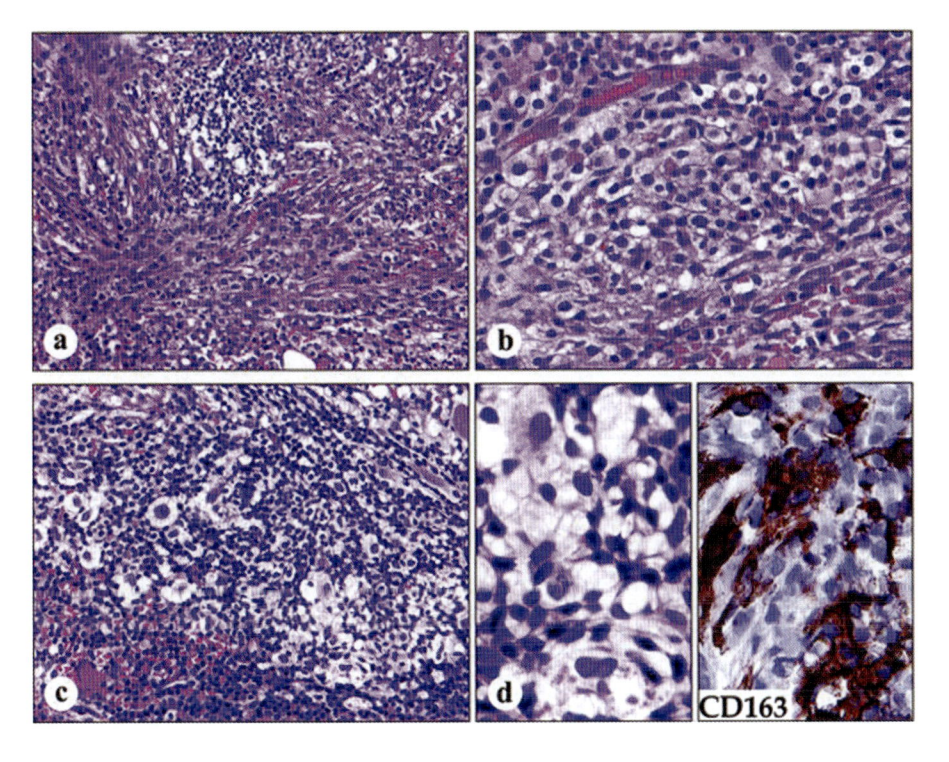

Figure 35.22 (**a**) Erdheim–Chester disease involving bone marrow shows fibrosclerosing cellular infiltrate. (**b**) The infiltrate contains aggregates of pale cytoplasmic histiocytes. (**c**) Lymphoid infiltrates also contain xanthomatous histiocytes. (**d**) The clear cytoplasmic histiocytes have relatively small nuclei with delicate nucleoli. CD163 immunostain reveals numerous positive histiocytes.

chromosomal translocations and positive results on clonality assay also indicate the clonal nature of Erdheim–Chester disease.[119,120]

Langerhans cell histiocytosis

Langerhans cell histiocytosis (LCH) is also known as *Langerhans cell granulomatosis* and *histiocytosis X*.[121] *Hand–Schüller–Christian disease* is the designation for examples of histiocytosis with osseous and skull base involvement associated with diabetes insipidus. The term *Letterer–Siwe disease* has been used for disseminated disease in infants, especially in earlier literature. *Congenital self-healing reticulohistiocytosis (Hashimoto–Pritzker disease)* is a form of LCH in infants with disseminated skin lesions.[122] Co-occurrences of LCH and Erdheim–Chester disease, and LCH and Rosai–Dorfman disease, has been reported.[123]

Clinical features

LCH usually occurs in children and is more common in infancy. Some forms are present in adulthood, including old age. Clinically it is divided into single organ system disease and multisystem disease, and this division has treatment and prognostic implications. The prognosis of patients with solitary disease (mostly bone) is excellent and local excision or bone curettage is generally sufficient, although in some cases the lesions regress spontaneously. Systemic multiorgan involvement (e.g., liver, lungs, bone marrow, spleen) is usually successfully treated with steroids and vinblastine.[124] According to a large cancer hospital series, nearly all mortality in multifocal disease in older children and adults was related to treatment complications and not to the disease itself.[121]

Pathology

The most common soft tissue involvement is extension from a lesion involving a skull bone or rib. Any other bones can be involved, and some examples are limited to lymph nodes. Primary extraskeletal soft tissue masses can also occur independently of bone tumors.[125] Polypoid mass-forming gastrointestinal lesions reported in adults are usually unassociated with systemic disease, whereas involvement in children is often part of systemic disease.[126] Nodal Langerhans histiocytosis can occur in connection with lymphoma, such as Hodgkin's disease.

Histologically, a typical solitary lesion contains a dominant Langerhans cell population. These cells can occur as granuloma-like collections in a connective tissue stroma, as solid sheets, or in a discohesive pattern admixed with varying numbers of eosinophils and sometimes lymphocytes and plasma cells (Figure 35.23). Langerhans cells have complex-shaped, lobulated, or deeply cleaved nuclei. In certain projections, the nuclei appear to have a longitudinal groove, and they often look kidney shaped. They have an open chromatin and a small nucleolus. The cytoplasm is pale eosinophilic. Multinucleated histiocytes are often focally present, and mitotic activity is usually low.

Immunohistochemically, LCH contains a significant S100-protein-positive population, although not all cells are

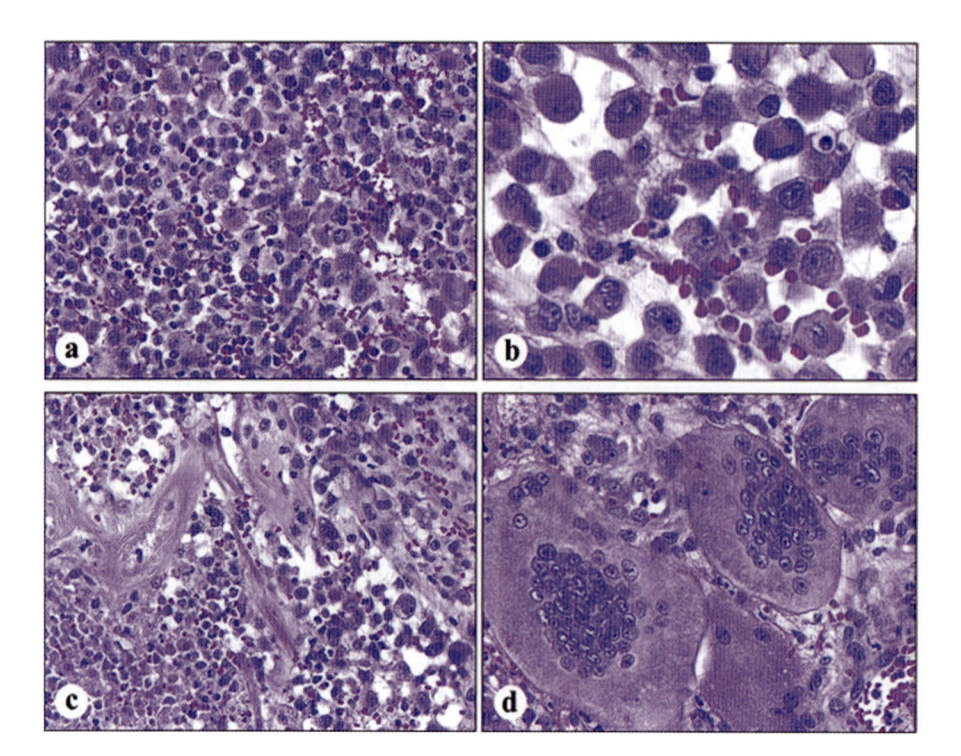

Figure 35.23 (**a**) Langerhans cell histiocytosis contains a mixed cellular infiltrate containing Langerhans cells and variable numbers of eosinophils. (**b**) The Langerhans cells often have complex nuclear outlines, including reniform nuclei. (**c**) Tumor necrosis can be present. (**d**) Multinucleated giant cells are often focally present.

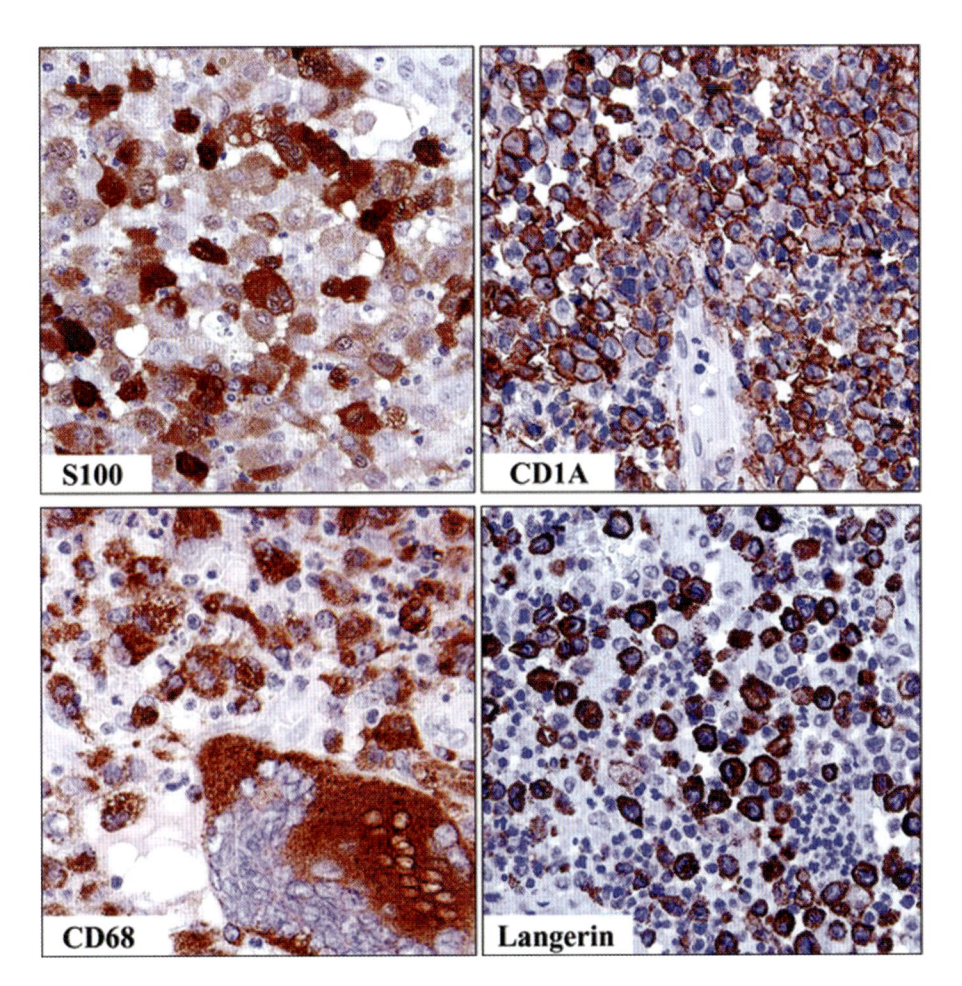

Figure 35.24 Immunohistochemical features of Langerhans cell histiocytosis. There is somewhat variable S100 protein positivity, membrane staining for CD1A, and cytoplasmic positivity for CD68, especially in multinucleated histiocytes. Granular cytoplasmic langerin positivity is a diagnostic feature.

positive. There is widespread cell membrane positivity for CD1A, and granular cytoplasmic positivity for langerin.[127-129] The expression of histiocytic markers, such as CD163 and lysozyme, is seen in some lesional cells, and many of these cells could be nonprincipal cellular components, which are negative for CD1A and langerin (Figure 35.24).

Ultrastructurally characteristic of this disease is the occurrence of Birbeck granules, which are tennis racket-shaped membranous structures that connect to the outer cell membrane. Electron microscopy also reveals complex nuclear outlines.

Somatic BRAF V600E mutations occur in 50% of cases.[130] Tumors negative for this mutation have been found to harbor somatic MAP-kinase (MAP2K1) or ARAF mutations.[131,132] HUMARA clonality assay is also indicative of a clonal proliferation.[133]

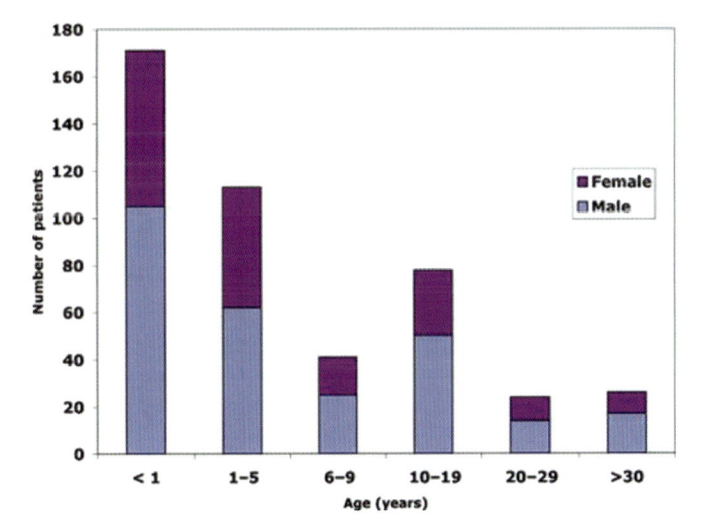

Figure 35.25 Age and sex distribution of 455 patients with juvenile xanthogranuloma.

Juvenile xanthogranuloma

Juvenile xanthogranuloma (JXG) is a true histiomonocytic proliferation typical of young children <2 years of age, with a male predominance. There is a low incidence of JXG in older children and adults (Figure 35.25). JXG most commonly involves the skin, but some lesions occur in deeper soft tissues or the subcutis, or they can be intramuscular (Figure 35.26). Multiple lesions are more common in infants.[134-140]

The cutaneous examples typically form small, solitary elevated papules of a few millimeters in diameter, but some patients have multiple JXGs (Figure 35.27). A few patients have larger nodular lesions measuring 1 cm to 3 cm. Local excision is generally curative. In rare cases, histologic lesions inseparable from JXG are associated with systemic internal organ involvement, especially the spleen and liver, with giant cell hepatitis that has been fatal in some cases.[139] A rare association with myelomonocytic leukemia and neurofibromatosis type 1 (NF1) has been reported.[141] However, in general, cases associated with NF1 do not differ from the sporadic cases.[142] For these reasons, clinical correlation and follow-up studies are recommended, especially in cases of multiple lesions. In some cases, patients with internal organ involvement have been treated with systemic chemotherapy.[138]

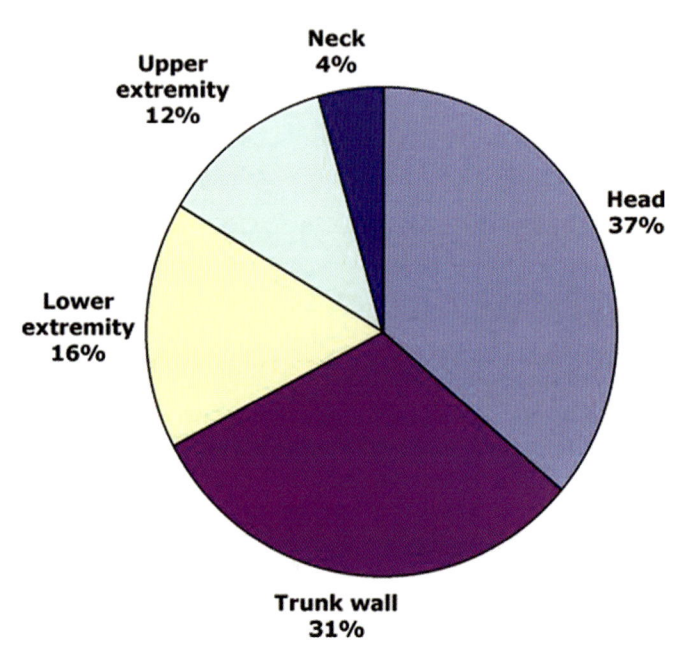

Figure 35.26 Anatomic location of 445 juvenile xanthogranulomas.

Pathology

Grossly, a JXG lesion shows a yellowish surface on sectioning. The cutaneous lesions are often dome-shaped, and the tumor cells often extend close to the epidermis (Figure 35.28). Ulceration is occasionally present. Histologically, there are medium-sized, rounded cells with notched nuclei and a moderate amount of pale cytoplasm, with some of the cells having lipidized cytoplasm with features of xanthoma cells. Multinucleated histiocytes with nuclei in a wreath-like circular pattern (Touton giant cells) or horseshoe pattern (Langerhans giant cells) and scattered eosinophilic granulocytes are typical, although not uniform, lesional components (Figure 35.29).

Immunohistochemically, the main cellular population is positive for the histiocytic markers, lysozyme, CD68, and CD163, Factor XIIIa, and for CD45 (LCA).[138,143] Most examples are negative for S100 protein and CD1A, in contrast to LCH, which is positive for the last two markers.

Reticulohistiocytoma (solitary epithelioid histiocytoma)

This rare cutaneous histiocytic proliferation typically forms a small, asymptomatic cutaneous (or mucosal) papule or nodule of <2 cm to 10 cm. Reticulohistiocytoma occurs in all ages, without a significant gender predilection.

The involved sites include all regions of the trunk wall, the lower part of the head and neck region (i.e., the chin, lip, and neck), and the extremities. In the extremities, there is a predilection for proximal sites such as the upper arm and the thigh, but occurrence in the hands and feet is rare. The duration of the lesion varies, but is often 3 to 4 months. The clinical appearance can resemble a nevus, cyst, dermatofibroma, and basal cell or squamous carcinoma or even melanoma. Simple excision is curative.[144–146]

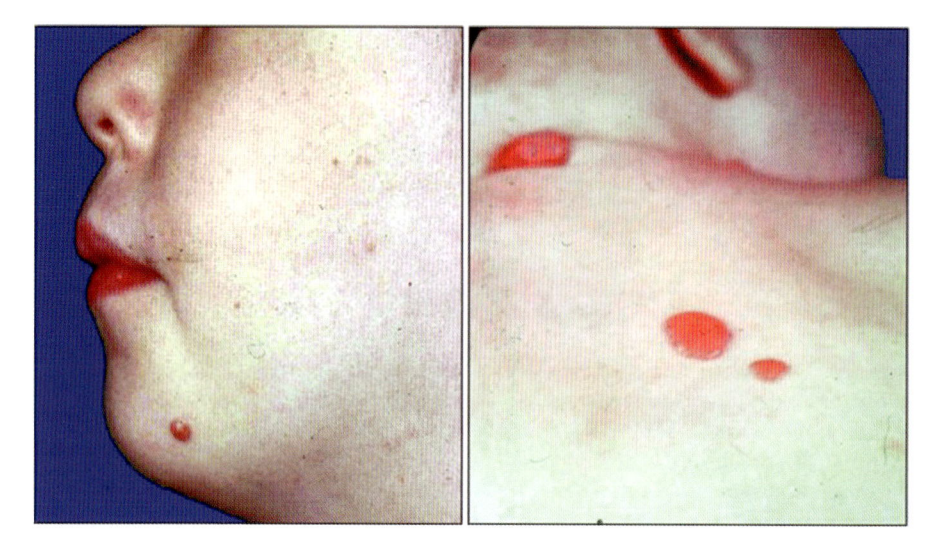

Figure 35.27 Clinically, juvenile xanthogranuloma (JXG) can form a reddish cutaneous papule or multiple papular or plaque-like lesions, usually in the upper body.

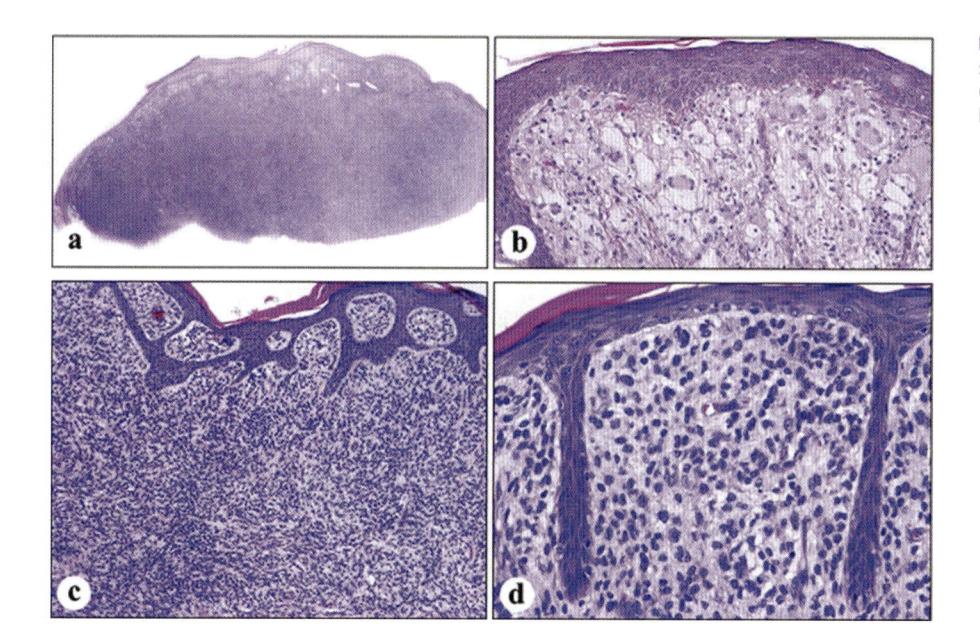

Figure 35.28 (**a**) Low-magnification examples of skin lesions of juvenile xanthogranuloma. (**b**) Numerous Touton-type giant cells. (**c,d**) The lesion can extend into the epidermal junction.

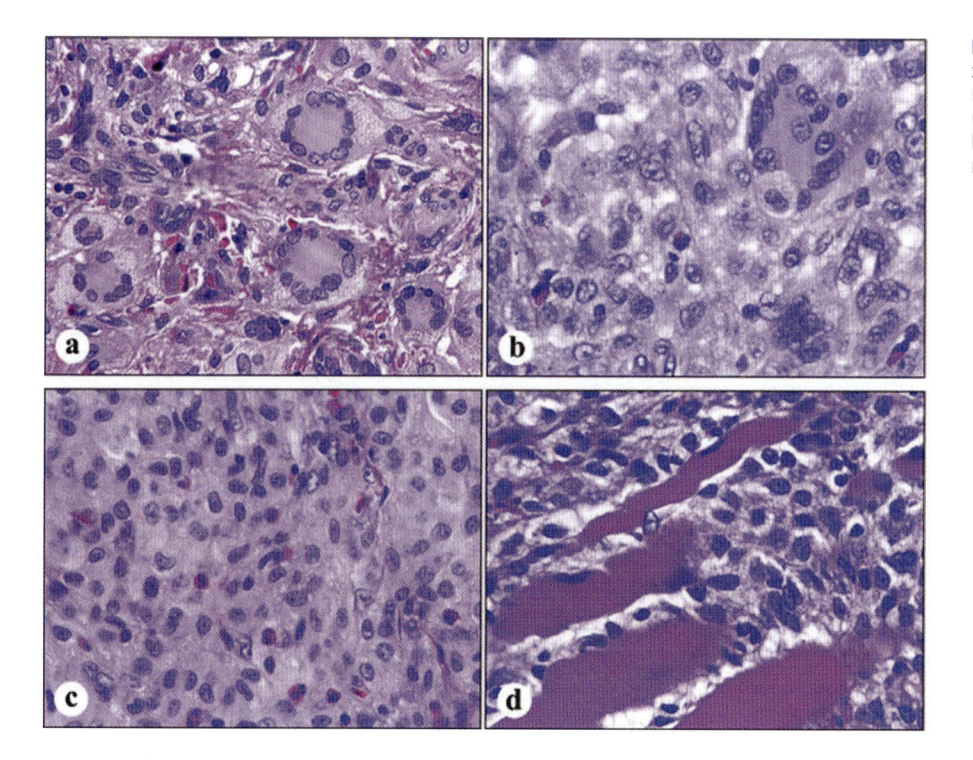

Figure 35.29 Higher magnification of juvenile xanthogranuloma. (**a**) Touton-type giant cells, mononuclear cells, and eosinophils. (**b,c**) The mononuclear cells have a moderate amount of bubbly, mildly eosinophilic cytoplasm. (**d**) Skeletal muscle involvement is possible.

Pathology

At low magnification, a typical example forms a shallow, symmetric dermal elevation, and the extent is often from underneath the epidermis to the mid-dermis (Figure 35.30). Lichenoid epidermal change with a disturbed basal cell layer can be present. The main cell type is a large histiocyte admixed with small numbers of T-lymphocytes, often neutrophils and plasma cells, and occasionally eosinophils. The histiocytes often have a scalloped margin, abundant eosinophilic cytoplasm, and a relatively small, oval, occasionally cleaved nucleus containing a delicate nucleolus. Small numbers of multinucleated histiocytes with randomly oriented nuclei are present in most cases. In some cases, tumor extension to the epidermal junction can histologically resemble the appearance of malignant melanoma (Figure 35.30). Mitotic activity is low (usually no more than 1 per 10 HPFs).

Immunohistochemically, the tumor cells show membranous and cytoplasmic positivity for the histiocytic markers

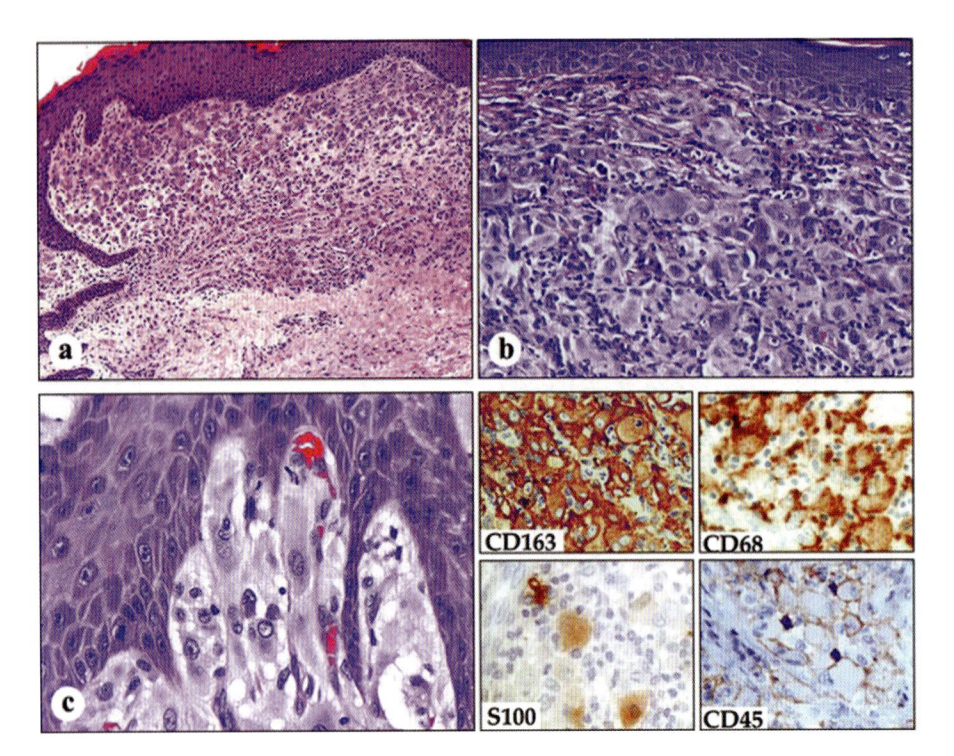

Figure 35.30 (**a,b**) Reticulohistiocytoma forms a cutaneous nodule with dermal infiltration of epithelioid cells. (**c**) The infiltrate can extend into the epidermal junction and simulate melanoma. Note large cells with relatively small nuclei and abundant eosinophilic cytoplasm. Immunohistochemically typical is positivity for CD163 (membranous and cytoplasmic) and CD68. A minor element can be S100 protein positive, and the histiocytes show variable membrane staining for CD45 (LCA).

CD163 and CD68. A subset of tumor cells is positive for lysozyme. Factor XIIIa positivity is typically limited to dendritic cells, but in some cases periepidermal tumor cells show cytoplasmic staining; no lesional cells are positive for CD1A. Nuclear microphthalmia transcription factor positivity is usually present. Consistent with histiomonocytic lineage, there is variable membrane staining for CD31 and CD45. A few tumor cells are S100-protein positive in 25% of cases. There is a low (<1%) Ki67 index. The tumor cells are negative for CD30, CD34, EMA, HMB45, keratins, and SMA.

Differential diagnosis

Lack of Touton-type giant cells and rarity of eosinophils separate reticulohistiocytoma from JXG. The S100-protein-positive cellular component is scant if present, and there is no emperipolesis, in contrast to RDD.

Multicentric reticulohistiocytosis

This very rare syndrome includes destructive polyarthritis and multiple cutaneous nodules, and the clinical presentation can resemble that of rheumatoid arthritis. It occurs in a wide range of ages, with median age from 40 to 45 years and a 3:1 female predominance. Most patients present with polyarthritis followed by the development of cutaneous nodules, often years later; however, both symptoms can present together. Destructive arthritis usually involves the interphalangeal joints, knee, shoulder, elbow, hip, and other joints. The skin nodules usually involve the face and hands. Some patients have associated diseases, such as cancer, tuberculosis, and diabetes. It is uncertain whether there is more than a random association, however.[147,148]

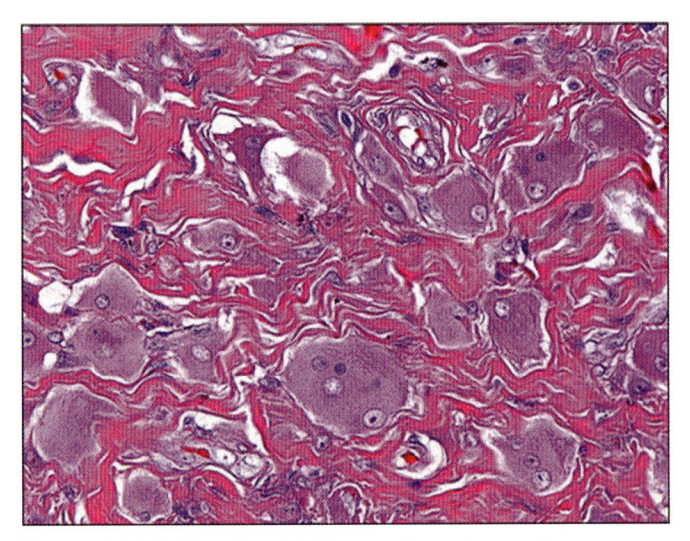

Figure 35.31 Multicentric reticulohistiocytosis typically contains numerous multinucleated histiocytic giant cells, here in a densely collagenous stroma.

Histologic analysis shows a histiocytic infiltration in the dermis that also contains fibrosis, xanthoma cells, and a prominent giant cell component (Figure 35.31). Multinucleated histiocytes also occur in the synovia. The histiocytes are positive for histiomonocytic cell markers and negative for S100 protein.[149]

Crystal-storing histiocytosis

This rare phenomenon is consistently associated with B-cell lymphomas or plasmacytoma. Crystalline deposits of immunoglobulins develop in histiocytes, which can be a prominent

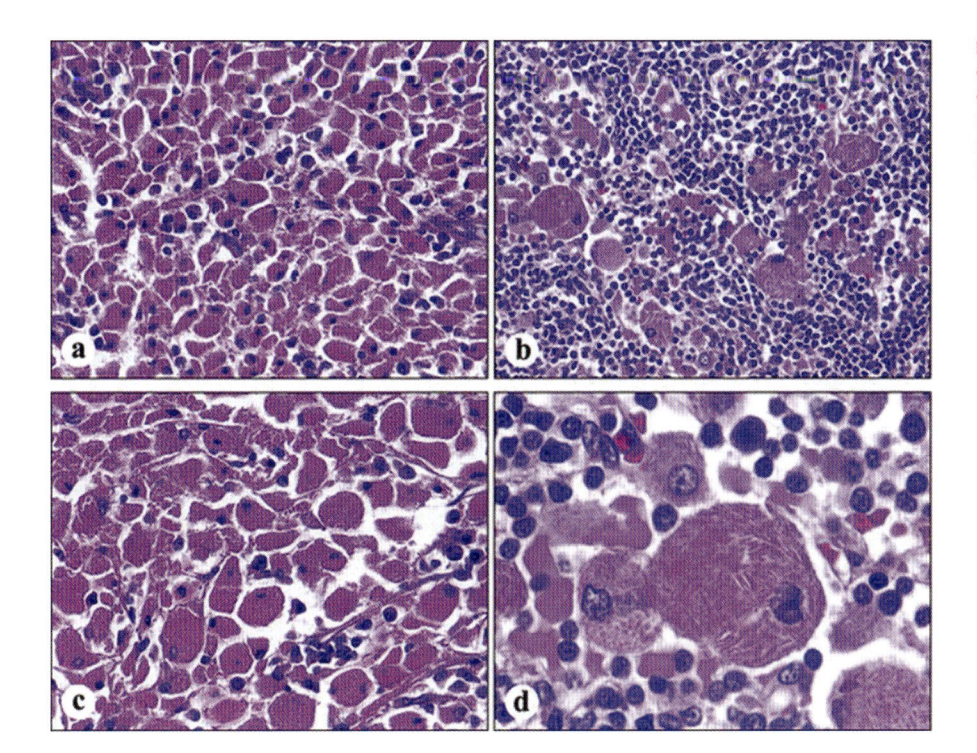

Figure 35.32 (**a–d**) Crystal-storing histiocytosis contains sheets of histiocytes with abundant eosinopholic cytoplasm. (**b,d**) Needle-shaped morphology of the crystals is evident at a higher magnification. A lymphoid component is also present.

component of the tumor, often overshadowing the lymphomatous or plasmacytic component (Figure 35.32). Different lymphomas with plasmacytic differentiation can be involved, most frequently plasmacytoma (myeloma), marginal zone lymphoma, and lymphoplasmacytic lymphoma.[150–152]

Immunohistochemically, the B-cell populations are positive for CD20, whereas the plasma cells are often negative for B-cell markers, but express CD38 and CD138. Most reported lymphomas with the crystal-storing histiocytosis phenomenon have had lambda clonality (Figure 35.33).

Histiocytic sarcoma

True histiocytic malignancies, formerly referred to as *true histiocytic lymphomas*, are now designated as *histiocytic sarcomas*. The previous designation of *malignant histiocytosis* mostly refers to tumors that are now classified as large cell anaplastic lymphoma.[153]

Malignant histiocytic tumors are rare and occur in a broad range of ages, from childhood (second decade) to old age. They occur in both nodal and extranodal locations. The latter can involve the peripheral soft tissues and abdominal sites; the skin/subcutis and gastrointestinal tract seem to be the most common. Tumor behavior varies, and both local recurrences and distant spread are possible. Sites of dissemination include the lymph nodes, lungs, and bone.[154–161] In some cases, histiocytic sarcoma has occurred as a lineage switch from a follicular B-cell lymphoma.[162,163] As development of histiocytic sarcoma is possible from B- or T-cell neoplasia, immunoglobulin or T-cell receptor gene rearrangements can occur and no longer definitively exclude the diagnosis of histiocytic sarcoma.[164] Development in connection with autoimmunolymphoproliferative disease and Rosai–Dorfman disease has also been reported.[165]

Histologically, these tumors are composed of sheets of large cells with typically abundant and variably eosinophilic cytoplasm. The nuclei are often cleaved or complex-shaped (Figure 35.34). Mitotic activity varies, but for most cases it is relatively low, 3 to 5 mitoses per 10 HPFs. Atypical mitoses and nuclear pleomorphism can be present.

Immunohistochemically, the tumor cells are positive for histiocytic markers such as CD163 and CD68.[161] There has been variable S100 protein immunoreactivity, generally no CD1A expression, in contrast to LCH. These tumors are also negative for CD30, melanoma-specific markers, and keratins.

Follicular dendritic cell sarcoma

This rare tumor was originally reported by Monda and Rosai in cervical lymph nodes.[166] It is a neoplastic counterpart of dendritic reticulum cells located in the nodal germinal centers and has also been referred to as *dendritic reticulum cell sarcoma*. The term "dendritic reticulum cell tumor" has also been used due to uncertain biologic potential. WHO classification prefers the above term.[2]

Dendritic cell sarcoma (DCS) occurs in adults of all ages and equally in men and women; the median age is approximately 40 years. Most examples involve the lymph nodes, tonsils, or spleen, but some involve extranodal tissues in the body cavities (mediastinum and retroperitoneum), or the gastrointestinal tract. The examples in the neck are usually smaller

Chapter 35: Lymphoid, myeloid, histiocytic, and dendritic cell proliferations

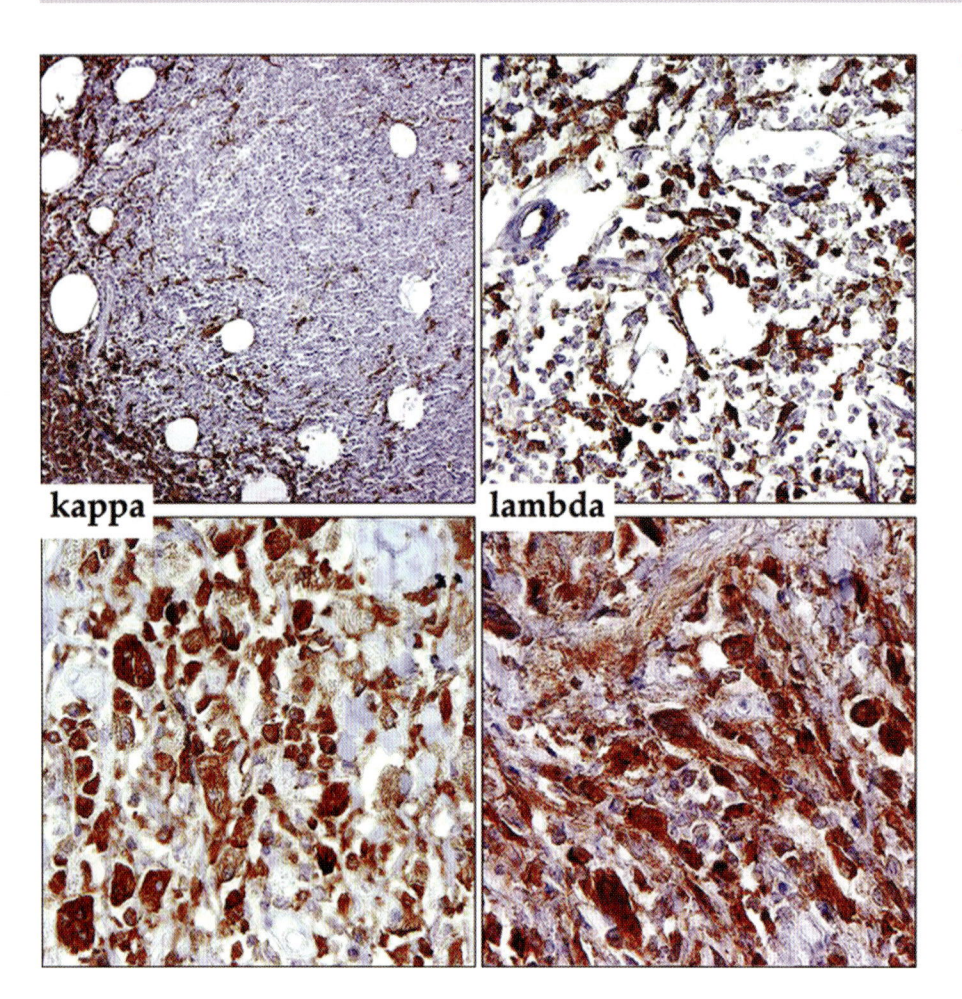

Figure 35.33 Immunohistochemically, crystal-storing histiocytosis shows a lymphoplasmacytic component that is negative for kappa, but positive for lambda light chains (upper panels). Histiocytes are positive for both kappa and lambda light chains (lower panels).

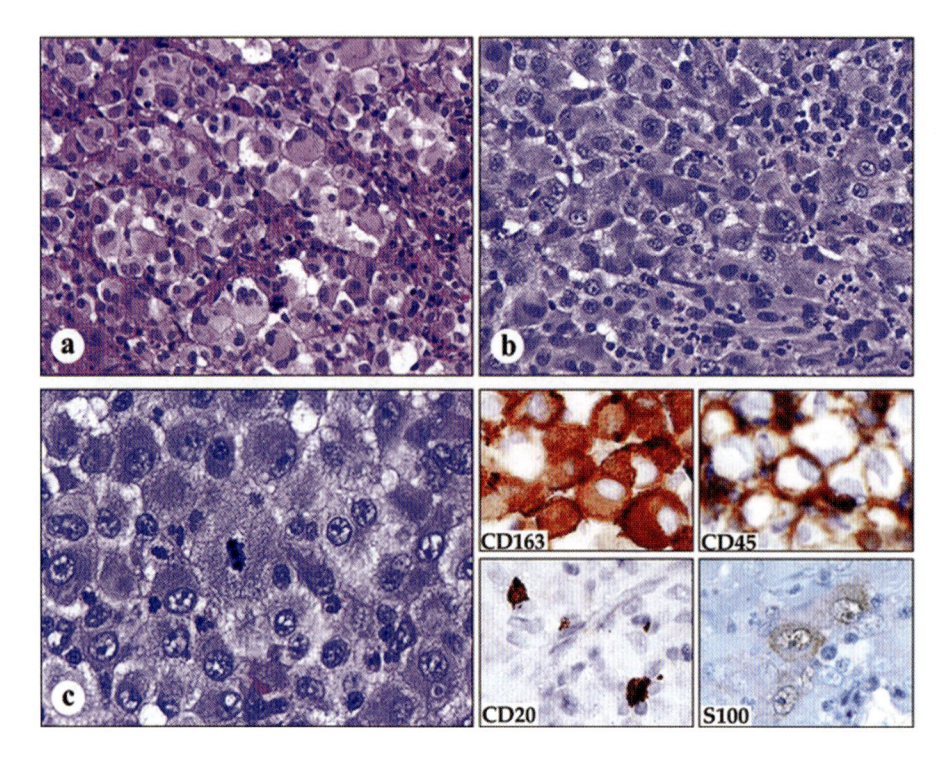

Figure 35.34 Histiocytic sarcoma contains epithelioid histiocytes with large nuclei and prominent nucleoli, and mitotic activity is present. Immunohistochemically, the tumor cells are positive for CD163, show membrane positivity for CD45, and are negative for CD20 with focal S100 protein positivity.

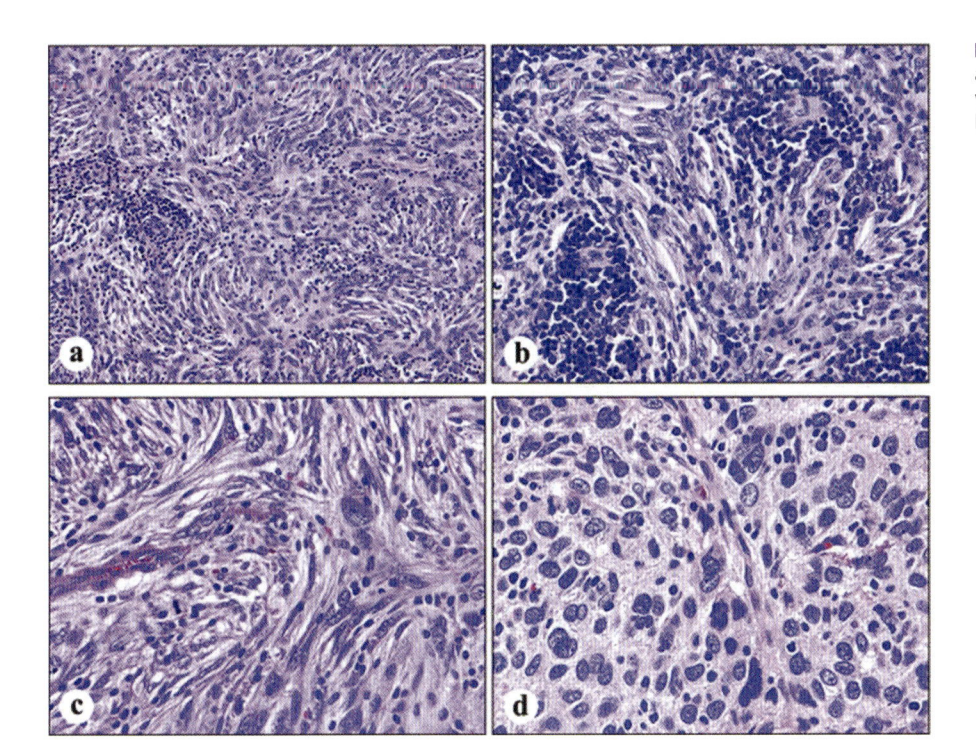

Figure 35.35 Dendritic cell sarcoma often shows a storiform growth pattern. Note intermingling with lymphocytes and perivascular collections of lymphoid cells.

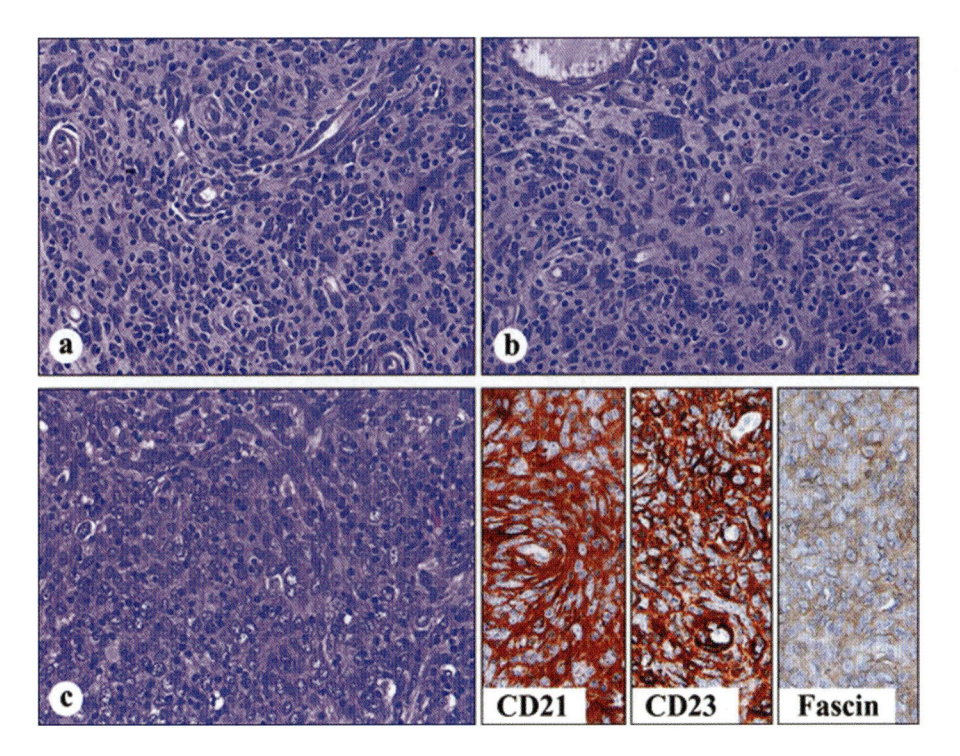

Figure 35.36 Dendritic cell sarcoma consists of oval cells forming syncytial clusters and intermingling with small lymphocytes. The tumor cells are strongly positive for CD21 and CD23 and are also positive for fascin.

(<5 cm), whereas those in the body cavities are often large (>10 cm). Local recurrence and metastasis has been reported in 30% to 40% of patients.[166–172]

Histologically, DC is composed of syncytial sheets of spindled oval cells, often concentrically arranged around blood vessels (Figures 35.35 and 35.36). The tumor cells typically intermingle with lymphocytes. Mitotic activity varies, but is 0 to 5 per 10 HPFs in most cases and can exceed 20 per 10 HPFs. Atypical mitoses and tumor necrosis are present in some cases. Elements of hyaline vascular Castleman disease in connection with the tumor suggest an origin from Castleman disease in some cases.

Immunohistochemically typical is expression of antigens of dendritic reticulum cells: CD21, CD23, and CD35. Because of

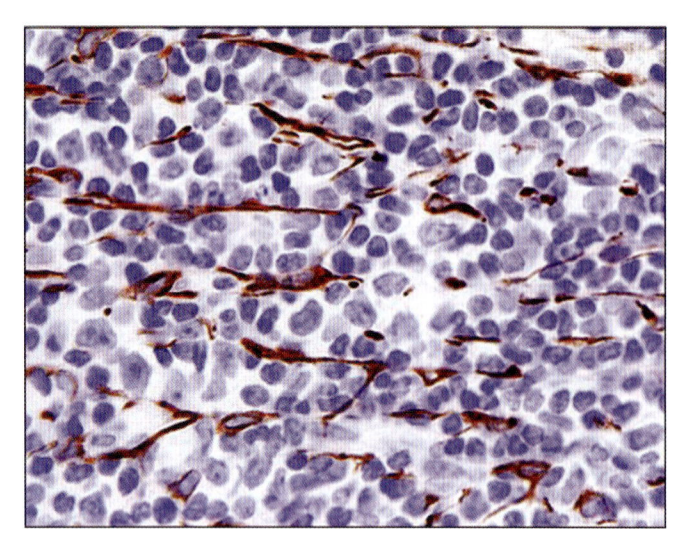

Figure 35.37 Varying numbers of keratin-18-positive bipolar-shaped reticulum cells can be detected in the paracortical areas of normal and reactive lymph nodes.

variable antigen expression, it is advisable to evaluate all these markers. Other markers expressed in DCS, although not specific for it, include fascin, clusterin,[173] and podoplanin.[174,175] In addition, desmoplakin and EMA positivity is common, as is occasional S100 protein positivity (most cases are negative for S100 protein). These tumors are typically negative for keratins, CD1A, and desmin.

Electron microscopy demonstrates prominent cell processes, and desmosomal junctions occur between some processes.

Interdigitating reticulum cell sarcoma

Interdigitating reticulum cells present antigens to T-cells during cell-mediated immunoresponse. They are derived from Langerhans cells of the skin and home into the paracortical area of the lymph nodes. Their immunophenotype includes positivity for S100 protein and leukocyte antigens such as CD11C. They differ from Langerhans cells by being CD1A negative.[102]

Interdigitating reticulum cell sarcoma (IDRCC) is very rare and usually occurs in older adults >50 years in nodal locations such as the neck, axilla, and groin. Morphology varies and can be epithelioid, pleomorphic, or spindled. Immunohistochemically typical is positivity for S100 protein and usually for CD45 and CD11C. Cases that are negative for leukocyte antigens are diagnostically problematic and difficult or impossible to distinguish from metastatic melanoma, which is the main differential diagnosis.[171,176–178] The continuingly small number of reported cases reflect the rarity and ambiguity of this entity.

Fibroblastic reticulum cell sarcoma

A small number of nodal sarcomas have been reported that are thought to recapitulate the features of fibroblastic reticulum cells of the lymph nodes.[178–180] The normal fibroblastic reticulum cells (interstitial reticulum cells) are nodal paracortical, slender, bipolar-shaped spindle cells that are often positive for keratins 8 and 18, and sometimes for desmin (Figure 35.37). The normal keratin-positive reticulum cells should not be confused with metastatic epithelial neoplasms. Fibroblastic reticulum cell sarcomas are identified as primary nodal spindle cell neoplasms that can be keratin and desmin positive.

Before the diagnosis of this keratin-positive sarcoma is made, a thorough clinicopathologic correlation is required to rule out metastatic (sarcomatoid) carcinoma. Potentially desmin-positive tumors, such as angiomatoid (malignant) fibrous histiocytoma, also must be ruled out.

References

General

1. Willemze R, Jaffe ES, Burg G, et al. WHO-EORTC classification of cutaneous lymphomas. *Blood* 2005;105:3768–3785.
2. Swerdlow SH, Campo E, Harris NL, et al. *World Health Organization Classification of Tumours of Haematopoietic and Lymphoid Tissues.* Lyon: IARC Press; 2008.
3. LeBoit PE, Burg G, Weedon D, Sarasin A (eds). *World Health Organization Classification of Tumours: Pathology and Genetics of Skin Tumours.* Lyon: IARC Press; 2006.
4. Chu PG, Chang KL, Arber DA, Weiss LM. Immunophenotyping of hematopoietic neoplasms. *Semin Diagn Pathol* 2000;17:236–256.
5. Hsi ED, Yegappan S. Lymphoma immunophenotyping: a new era in paraffin-section immunohistochemistry. *Adv Anat Pathol* 2001;8:218–239.
6. Falini B, Mason DY. Proteins encoded by genes involved in chromosomal alterations in lymphoma and leukemia: clinical value of their detection by immunocytochemistry. *Blood* 2002;99:409–426.
7. Travis WD, Banks PM, Reiman HM. Primary extranodal soft tissue lymphoma of the extremities. *Am J Surg Pathol* 1987;11:359–366.
8. Lanham GR, Weiss SW, Enzinger FM. Malignant lymphoma: a study of 75 cases presenting in soft tissue. *Am J Surg Pathol* 1989; 13:1–10.

Lymphoid hyperplasia

9. Utz GL, Swerdlow S. Distinction of follicular hyperplasia from follicular lymphoma in B5-fixed tissues: comparison of MT2 and bcl-2 antibodies. *Hum Pathol* 1993;24:1155–1158.
10. Leinweber B, Colli C, Chott A, Kerl H, Cerroni L. Differential diagnosis of cutaneous infiltrates of B lymphocytes with follicular growth pattern. *Am J Dermatopathol* 2004;26:4–13.
11. Goodlad JR, MacPherson S, Jackson R, Batstone P, White J; Scotland, Newcastle Lymphoma Group. Extranodal follicular lymphoma: a clinicopathologic and genetic analysis of 15 cases arising at non-cutaneous sites. *Histopathology* 2004;44:268–276.

12. Keller AR, Hochholzer L, Castleman B. Hyaline-vascular and plasma-cell types of giant lymph node hyperplasia of the mediastinum and other locations. *Cancer* 1972;29:670–683.
13. Kazakov D, Fanburg-Smith JC, Suster S, et al. Castleman disease of the subcutis and underlying skeletal muscle: report of 6 cases. *Am J Surg Pathol* 2004;28:569–577.
14. Waterston A, Bower M. Fifty years of multicentric Castleman's disease. *Acta Oncol* 2004;43:698–704.
15. Carbone A, Gloghini A. KSHV/HHV8-associated lymphomas. *Br J Hematol* 2008;140:13–24.
16. Cronin DM, Warnke RA. Castleman disease: an update on classification and the spectrum of associated lesions. *Adv Anat Pathol* 2009;16:236–246.
17. Ohgami RS, Zhao S, Ohgami JK, et al. TdT+ T-lymphoblastic populations are increased in Castleman disease, in Castleman disease in association with follicular dendritic cell tumors, and in angioimmunoblastic T-cell lymphoma. *Am J Surg Pathol* 2012;36:1619–1628.
18. Kuo TT, Shih LY, Chan HL. Kimura's disease: involvement of regional lymph nodes and distinction from angiolymphoid hyperplasia with eosinophilia. *Am J Surg Pathol* 1988;12:843–854.
19. Li TJ, Chen XM, Wang SZ, et al. Kimura's disease: a clinicopathologic study of 54 Chinese patients. *Oral Surg Oral Med Oral Pathol Oral Radiol Endod* 1996;82:549–555.
20. Chen H, Thompson LD, Aguilera NS, Abbondanzo SL. Kimura disease: a clinicopathologic study of 21 cases. *Am J Surg Pathol* 2004;28:505–513.

Follicular lymphoma/follicular center cell lymphoma

21. Santucci M, Pimpinelli N, Arganini L. Primary cutaneous B-cell lymphoma: a unique type of low-grade lymphoma. clinicopathologic and immunologic study of 83 cases. *Cancer* 1991;67:2311–2326.
22. Aguilera NSI, Tomaszewski MM, Moad JC, et al. Cutaneous follicular center cell lymphoma: clinicopathologic study of 19 cases. *Mod Pathol* 2001;14:828–835.
23. Senff NJ, Hoefnagel JJ, Jansen PM, et al. Reclassification of 300 primary cutaneous B-cell lymphomas according to the new WHO-EORTC classification for cutaneous lymphomas: comparison with previous classifications and identification of prognostic markers. *J Clin Oncol* 2007;25:1581–1587.
24. Vermeer MH, Willemze R. Recent advances in primary cutaneous B-cell lymphomas. *Curr Opin Oncol* 2014;26:230–236.
25. Swerdlow SH, Quintanilla-Martinez L, Willemze R, Kinney MC. Cutaneous B-cell lymphoproliferative disorders: report of the 2011 Society for Hematopathology/European Association for Haematopathology workshop. *Am J Clin Pathol* 2013;139:515–535.
26. Xie X, Sundram U, Natkunam Y, et al. Expression of HGAL in primary cutaneous large B-cell lymphomas: evidence for germinal center derivation of primary cutaneous follicular lymphoma. *Mod Pathol* 2008;21:653–659.

Marginal zone lymphoma

27. Cerroni L, Signoretti S, Höfler H, et al. Primary cutaneous marginal zone B-cell lymphoma: a recently described entity of low-grade malignant cutaneous B-cell lymphoma. *Am J Surg Pathol* 1997;21:1307–1315.
28. Tomaszewski MM, Abbondanzo SL, Lupton GP. Extranodal marginal zone B-cell lymphoma of the skin: a morphologic and immunophenotypic study of 11 cases. *Am J Dermatopathol* 2000;22:205–211.
29. Baldassano MF, Bailey EM, Ferry JA, Harris NL, Duncan LM. Cutaneous lymphoid hyperplasia and cutaneous marginal zone lymphoma: comparison of morphologic and immunophenotypic features. *Am J Surg Pathol* 1999;23:88–96.
30. Servitje O, Gallardo F, Estrach T, et al. Primary cutaneous marginal zone B-cell lymphoma: a clinical, histopathological, immunophenotypic and molecular genetic study of 22 cases. *Br J Dermatol* 2002;147:1147–1158.
31. Cho-Vega JH, Vega F, Rassidakis G, Medeiros LJ. Primary cutaneous marginal zone B-cell lymphoma. *Am J Clin Pathol* 2006;125 Suppl: S38–S49.
32. Dalle S, Thomas L, Balme B, Dumontet C, Thieblemont C. Primary cutaneous marginal zone lymphoma. *Crit Rev Oncol Hematol* 2010;74:156–162.

Large B-cell lymphoblastic and intravascular lymphoma

33. Vermeer MH, Geelen FA, van Haselen CW, et al. Primary cutaneous large B-cell lymphomas of the legs: a distinct type of cutaneous B-cell lymphoma with an intermediate prognosis. *Arch Dermatol* 1996;132:1304–1308.
34. Kodama K, Massone C, Chott A, et al. Primary cutaneous large B-cell lymphomas: clinicopathologic features, classification, and prognostic factors in a large series of patients. *Blood* 2005;106:2491–2497.
35. Grange F, Beylot-Barry M, Courville P, et al. Primary cutaneous diffuse large B-cell lymphoma, leg type: clinicopathologic features and prognostic analysis of 60 cases. *Arch Dermatol* 2007;143:1144–1150.
36. Tse CC, Chan JK, Yuen RW, Ng CS. Malignant lymphoma with myxoid stroma: a new pattern in need of recognition. *Histopathology* 1991;18:31–35.
37. Koo CH, Shin SS, Bracho F, et al. Rosette-forming non-Hodgkin lymphomas. *Histopathology* 1996;29:557–563.
38. Lin P, Jones D, Dorfman DM, Medeiros LJ. Precursor B-cell lymphoblastic lymphoma: a predominantly extranodal tumor with low propensity for leukemic involvement. *Am J Surg Pathol* 2000;24:1480–1490.
39. Maitra A, McKenna RW, Weinberg AG, Schneider NR, Kroft SH. Precursor B-cell lymphoblastic lymphoma: a study of nine cases lacking blood and bone marrow involvement and review of the literature. *Am J Clin Pathol* 2001;115:868–875.
40. Sheibani K, Battifora H, Winberg CD, et al. Further evidence that "malignant angioendotheliomatosis" is an angiotropic large-cell lymphoma. *N Engl J Med* 1986;314:943–948.
41. Wick MR, Mills SE, Scheithauer BW, et al. Reassessment of malignant "angioendotheliomatosis": evidence in favor of its reclassification as "intravascular lymphomatosis." *Am J Surg Pathol* 1986;10:112–123.
42. Stroup RM, Sheibani K, Moncada A, Purdy LJ, Battifora H. Angiotropic (intravascular) large cell lymphoma:

a clinicopathologic study of seven cases with unique clinical presentations. *Cancer* 1990;66:1781–1788.

43. Jalkanen S, Aho R, Kallajoki M, et al. Lymphocyte homing receptors and adhesion molecules in intravascular malignant lymphomatosis. *Int J Cancer* 1989;44:777–782.

44. Ponzoni M, Arrigoni G, Gould VE, et al. Lack of CD29 (beta integrin) and CD54 (ICAM-1) adhesion molecules in intravascular lymphomatosis. *Hum Pathol* 2000;31:220–226.

45. Nixon BK, Kussick SJ, Carlon MJ, Rubin BP. Intravascular large B-cell lymphoma involving hemangiomas: an unusual presentation of a rare neoplasm. *Mod Pathol* 2005;18:1121–1126.

46. Smith ME, Stamatakos MD, Neuhauser TS. Intravascular lymphomatosis presenting within angiolipomas. *Ann Diagn Pathol* 2001;5:103–106.

47. Matsue K, Asada N, Takeuchi M, et al. A clinicopathological study of 13 cases of intravascular lymphoma: experience in a single institution over a 9-yr period. *Eur J Haematol* 2008;80:236–244.

48. Samols MA, Su A, Ra S, et al. Intralymphatic cutaneous anaplastic large cell lymphoma/lymphomatoid papulosis: expanding the spectrum of CD30-positive lymphoproliferative disorders. *Am J Surg Pathol* 2014;38:1203–1211.

49. Liu Y, Zhang W, An J, Li H, Liu S. Cutaneous intravascular natural killer-cell lymphoma: a case report and review of the literature. *Am J Clin Pathol* 2014;142:243–247.

50. Crane GM, Ambinder RF, Shirley CM, et al. HHV-8-positive and EBV-positive intravascular lymphoma: an unusual presentation of extracavitary primary effusion lymphoma. *Am J Surg Pathol* 2014;38:426–432.

Burkitt lymphoma, plasmacytic B-cell lymphoma, and plasmacytoma

51. Wang MB, Strasnick B, Zimmerman MC. Extranodal American Burkitt's lymphoma of the head and neck. *Arch Otolaryngol Head Neck Surg* 1992;118:193–199.

52. Wright D, McKeever P, Carter R. Childhood non-Hodgkin lymphomas in the United Kingdom: findings from the UK Children's Cancer Study Group. *J Clin Pathol* 1997;50:128–134.

53. Bociek RG. Adult Burkitt's lymphoma. *Clin Lymphoma* 2005;6:11–20.

54. Brady G, MacArthur GJ, Farrell PJ. Epstein-Barr virus and Burkitt lymphoma. *J Clin Pathol* 2007;60:1397–1402.

55. Said J, Lones M, Yea S. Burkitt lymphoma and MYC: what else is new? *Adv Anat Pathol* 2014;21:160–165.

56. Bladé J, Fernández de Larrea C, Rosiñol L, et al. Soft-tissue plasmacytomas in multiple myeloma: incidence, mechanisms of extramedullary spread, and treatment approach. *J Clin Oncol* 2011;29:3805–3812.

57. Wirk B, Wingard JR, Moreb JS. Extramedullary disease in plasma cell myeloma: the iceberg phenomenon. *Bone Marrow Transplant* 2013;48:10–18.

58. Delecluse HJ, Anagnostopoulos I, Dallenbach F, et al. Plasmablastic lymphomas of the oral cavity: a new entity associated with the human immunodeficiency virus infection. *Blood* 1997;89:1413–1420.

59. Colomo L, Loong F, Rives S, et al. Diffuse large B-cell lymphomas with plasmablastic differentiation represent a heterogeneous group of disease entities. *Am J Surg Pathol* 2004;28:736–747.

60. Dong HY, Scadden DT, de Leval T, et al. Plasmablastic lymphoma in HIV-positive patients: an aggressive Epstein-Barr virus-associated extramedullary plasmacytic neoplasm. *Am J Surg Pathol* 2005;29:1633–1641.

61. Vega F, Chang CC, Medeiros LJ, et al. Plasmablastic lymphomas and plasmablastic plasma cell myelomas have nearly identical immunophenotypic profiles. *Mod Pathol* 2005;18:806–815.

T-cell and NK-cell lymphomas

62. Salhany KE, Macon WR, Choi JK, et al. Subcutaneous panniculitis-like T-cell lymphoma: clinicopathologic, immunophenotypic, and genotypic analysis of alpha/beta and gamma/delta subtypes. *Am J Surg Pathol* 1998;22:881–893.

63. Kumar S, Knenacs S, Medeiros J, et al. Subcutaneous panniculitic T-cell lymphoma is a tumor of cytotoxic T lymphocytes. *Hum Pathol* 1998;29:397–403.

64. Massone C, Chott A, Metze D, et al. Subcutaneous, blastic natural killer cell (NK), NK/T-cell, and other cytotoxic lymphomas of the skin: a morphologic, immunophenotypic, and molecular study of 50 patients. *Am J Surg Pathol* 2004;28:719–735.

65. Kong YY, Dai B, Kong JC, et al. Subcutaneous panniculitis-like T-cell lymphoma: a clinicopathologic, immunophenotypic, and molecular study of 22 Asian cases according to WHO-EORTC classification. *Am J Surg Pathol* 2008;32:1495–1502.

66. Willemze R, Jansen PM, Cerroni L, et al. Subcutaneous panniculitis-like T-cell lymphoma: definition, classification, and prognostic factors: an EORTC cutaneous lymphoma group study of 83 cases. *Blood* 2008;111:838–845.

67. Huppmann AR, Xi L, Raffeld M, Pittaluga S, Jaffe ES. Subcutaneous panniculitis-like T-cell lymphoma in the pediatric age group: a lymphoma of low malignant potential. *Pediatr Blood Cancer* 2013;60:1165–1170.

68. Arps DP, Patel RM. Lupus profundus (panniculitis): a potential mimic of subcutaneous panniculitis-like T-cell lymphoma. *Arch Pathol Lab Med* 2013;137:1211–1215.

69. Kempf W, Kazakov DV, Kutzner H. Lobular panniculitis due to Borrelia burgdorferi infection mimicking subcutaneous panniculitis-like T-cell lymphoma. *Am J Dermatopathol* 2013;35:e30–e33.

70. Chan JK, Sin VC, Wong KF, et al. Nonnasal lymphoma expressing the natural killer cell marker CD56: a clinicopathologic study of 49 cases of an uncommon aggressive neoplasm. *Blood* 1997;89:4501–4513.

71. Ko YH, Cho EY, Kim JE, et al. NK and NK-like T-cell lymphoma in extranasal sites: a comparative clinicopathological study according to the site and EBV status. *Histopathology* 2004;44:480–489.

Anaplastic large cell lymphoma

72. Stein H, Foss HD, Dürkop H, et al. CD30+ anaplastic large cell lymphoma: a review of its histologic, genetic and clinical features. *Blood* 2000;96:3681–3695.

73. Savage KJ, Harris NL, Vose JM, et al. ALK- anaplastic large-cell lymphoma is clinically and immunophenotypically different from both ALK+ ALCL and peripheral T-cell lymphoma, not

otherwise specified: report from the International Peripheral T-cell Lymphoma Project. *Blood* 2008;111.5496–5504.

74. Amin HN, Lai R. Pathobiology of ALK+ anaplastic large cell lymphoma. *Blood* 2007;110:2259–2267.

75. Medeiros LJ, Elenitoba-Johnson KSJ. Anaplastic large cell lymphoma. *Am J Clin Pathol* 2007;127:707–722.

76. Pant V, Jambhekar NA, Madur P, et al. Anaplastic large cell lymphoma (ALCL) presenting as primary bone and soft tissue sarcoma: a study of 12 cases. *Indian J Pathol Microbiol* 2007;50:303–307.

77. Kounami S, Shibuta K, Yoshiyama M, et al. Primary anaplastic large cell lymphoma of the psoas muscle: a case report and literature review. *Acta Haematol* 2012;127:186–188.

78. Gaiser T, Geissinger E, Schattenberg T, et al. Case report: a unique pediatric case of a primary CD8 expressing ALK-1 positive anaplastic large cell lymphoma of skeletal muscle. *Diagn Pathol* 2012;7:38.

79. Sahoo S, Rosen PP, Feddersen RM, et al. Anaplastic large cell lymphoma arising in a silicone breast implant capsule: a case report and review of the literature. *Arch Pathol Lab Med* 2003;127:115–118.

80. Gauddet G, Friedberg JW, Weng A, Pinkus GS, Freedman AS. Breast lymphoma associated with breast implants: two case reports and review of the literature. *Leuk Lymphoma* 2002;43:115–119.

81. Fritsche FR, Pahl S, Petersen I, et al. Anaplastic large-cell non-Hodgkin's lymphoma of the breast in periprosthetic localization 32 years after treatment for primary breast cancer: a case report. *Virchows Arch* 2006;449:561–564.

82. Roden AC, Macon WR, Keeney GL, et al. Seroma-associated primary anaplastic large-cell lymphoma adjacent to breast implants: an indolent T-cell lymphoproliferative disorder. *Mod Pathol* 2008;21:455–463.

83. Wong AK, Lopategui J, Clancy S, Kulber D, Bose S. Anaplastic large cell lymphoma associated with a breast implant capsule: a case report and review of the literature. *Am J Surg Pathol* 2008;32:1265–1268.

84. Ye X, Shokrollahi K, Rozen WM, et al. Anaplastic large cell lymphoma (ALCL) and breast implants: breaking down the evidence. *Mutat Res Rev Mutat Res* 2014;762C:123–132.

85. Brody GS, Deapen D, Taylor CR, et al. Anaplastic large cell lymphoma (ALCL) occuring in women with breast implants: analysis of 173 cases. *Plast Reconstr Surg* 2014;135:695–705.

86. Chan JK, Buchanan R, Fletcher CD. Sarcomatoid variant of anaplastic large cell Ki-1 lymphoma. *Am J Surg Pathol* 1990;14:983–988.

87. Rekhi B, Sridhar E, Viswanathan S, Shet TM, Jambhekar NA. ALK+ anaplastic large cell lymphoma with cohesive, perivascular arrangements on cytology, mimicking a soft tissue sarcoma: a report of 2 cases. *Acta Cytol* 2010;54:75–78.

88. Ten Berge RL, Snijdewint FGM, von Mensdorff-Pouilly S, et al. MUC1 (EMA) is preferentially expressed by ALK positive anaplastic large cell lymphoma, in the normally glycosylated or only partly hypoglycosylated form. *J Clin Pathol* 2001;54:933–939.

Extramedullary myeloid tumor and mast cell tumors

89. Neiman RS, Barcos M, Berard C, et al. Granulocytic sarcoma: a clinicopathologic study of 61 cases. *Cancer* 1981;48:1426–1437.

90. Meis JM, Butler JJ, Osborne BM, Manning JT. Granulocytic sarcoma in nonleukemic patients. *Cancer* 1986;58:2697–2709.

91. Remstein ED, Kurtin PJ, Nascimento AG. Sclerosing extramedullary hematopoietic tumor in chronic myeloproliferative disorders. *Am J Surg Pathol* 2000;24:51–55.

92. Traweek ST, Arbder DA, Rappaport H, Brynes RK. Extramedullary myeloid tumors: an immunohistochemical and morphological study of 28 cases. *Am J Surg Pathol* 1993;17:1011–1019.

93. Pileri SA, Ascani S, Cox MC, et al. Myeloid sarcoma: clinicopathologic, phenotypic and cytogenetic analysis of 92 adult patients. *Leukemia* 2007;21:340–350.

94. Paydas S, Zorludemir S, Ergin M. Granulocytic sarcoma: 32 cases and review of the literature. *Leuk Lymphoma* 2006;47:2527–2541.

95. Hudock J, Chatten J, Miettinen M. Immunohistochemical evaluation of myeloid leukemia infiltrates (granulocytic sarcomas) in formaldehyde-fixed and paraffin-embedded tissue. *Am J Clin Pathol* 1994;102:55–60.

96. Chen J, Abbondanzo SL. C-kit (CD117) reactivity in extramedullary myeloid tumor/granulocytic sarcoma. *Arch Pathol Lab Med* 2001;125:1448–1452.

97. Carter MC, Metcalfe DD, Komarow HD. Mastocytosis. *Immunol Allergy Clin North Am* 2014;34:181–196.

98. Salces IG, Alfaro J, Sáenz DE Santamaría MC, Sanchez M. Scabies presenting as solitary mastocytoma-like eruption in an infant. *Pediatr Dermatol* 2009;26:486–488.

99. Longley BJ, Metcalfe DD. A proposed classification of mastocytosis incorporating molecular genetics. *Hematol Oncol Clin North Am* 2000;14:697–701.

100. Ma D, Stence AA, Bossler AB, Hackman JR, Bellizzi AM. Identification of KIT activating mutations in paediatric solitary mastocytoma. *Histopathology* 2014;64:218–225.

Rosai–Dorfman disease

101. Foucar E, Rosai J, Dorfman R. Sinus histiocytosis with massive lymphadenopathy (Rosai–Dorfman disease): review of the entity. *Semin Diagn Pathol* 1990;7:19–73.

102. Foucar K, Foucar E. The mononuclear phagocyte and immunoregulatory effector (M-PIRE) system: evolving concepts. *Semin Diagn Pathol* 1990;7:4–18.

103. Montgomery EA, Meis JM, Frizzera G. Rosai–Dorfman disease of soft tissue. *Am J Surg Pathol* 1992;16:122–129.

104. Young PM, Kransdorf MJ, Temple HT, Mousavi F, Robinson PG. Rosai–Dorfman disease presenting as multiple soft tissue masses. *Skeletal Radiol* 2005;34:665–669.

105. Kong Y, Kong J, Shi D, et al. Cutaneous Rosai–Dorfman disease: a clinical and histopathologic study of 25 cases in China. *Am J Surg Pathol* 2007;31:341–350.

106. Bonetti F, Chilosi M, Menestrina F, et al. Immunohistological analysis of Rosai–Dorfman histiocytosis: a disease

of S-100 +CD1-histiocytes. *Virchows Arch* 1987;411:129–135.

107. Miettinen M, Paljakka P, Haveri P, Saxén E. Sinus histiocytosis with massive lymphadenopathy: a nodal and extranodal proliferation of S-100 protein positive histiocytes. *Am J Clin Pathol* 1987;88:270–277.

108. Eisen RN, Buckley PJ, Rosai J. Immunophenotypic characterization of sinus histiocytosis with massive lymphadenopathy (Rosai–Dorfman disease). *Semin Diagn Pathol* 1990;7:74–82.

109. Liu L, Perry AM, Cao W, *et al*. Relationship between Rosai-Dorfman disease and IgG4-related disease: study of 32 cases. *Am J Clin Pathol* 2013;140:395–402.

110. Morgan NV, Morris MR, Cangul H, *et al*. Mutations in SLC29A3, encoding an equilibrative nucleoside transporter ENT3, cause a familial histiocytosis syndrome (Faisalabad histiocytosis) and familial Rosai–Dorfman disease. *PLoS Genet* 2010;6:e1000833.

Erdheim–Chester disease

111. Diamond EL, Dagna L, Hyman DM, *et al*. Consensus guidelines for the diagnosis and clinical management of Erdheim–Chester disease. *Blood* 2014;124:483–492.

112. Haroche H, Arnaud L, Cohen-Aubart F, *et al*. Erdheim–Chester disease. *Rheum Dis Clin North Am* 2013;39:299–311.

113. Munoz J, Janku F, Cohen PR, Kurzrock R. Erdheim–Chester disease: characteristics and management. *Mayo Clin Proc* 2014;89:985–996.

114. Veyssier-Belot C, Cacoub P, Caparros-Lefebvre D, *et al*. Erdheim-Chester disease: clinical and radiologic characteristics of 59 cases. *Medicine* 1996;75:157–169.

115. Eble JN, Rosenberg AE, Young RH. Retroperitoneal xanthogranuloma in a patient with Erdheim–Chester disease. *Am J Surg Pathol* 1994;18:843–848.

116. Provenzano E, Barter SJ, Wright PA, *et al*. Erdheim–Chester disease presenting as bilateral clinically malignant breast masses. *Am J Surg Pathol* 2010;34:584–588.

117. Kenn W, Eck M, Allolio B, *et al*. Erdheim–Chester disease: evidence for a disease entity different from Langerhans cell histiocytosis? Three cases with a detailed radiological and immunohistochemical analysis. *Hum Pathol* 2000;31:734–739.

118. Haroche J, Charlotte F, Arnaud L, *et al*. High prevalence of BRAF V600E mutations in Erdheim–Chester disease but not in other non-Langerhans cell histiocytoses. *Blood* 2012;120:2700–2703.

119. Vencio EF, Jenkins RB, Schiller JL, *et al*. Clonal cytogenetic abnormalities in Erdheim–Chester disease. *Am J Surg Pathol* 2007;31:319–321.

120. Chetritt J, Paradis V, Dargere D, *et al*. Chester-Erdheim disease: a neoplastic disorder. *Hum Pathol* 1999;30:1093–1096.

Langerhans cell histiocytosis

121. Lieberman PH, Jones CR, Steinman RM, *et al*. Langerhans cell (eosinophilic) granulomatosis: a clinicopathologic study encompassing 50 years. *Am J Surg Pathol* 1996;20:519–552.

122. Kapur P, Erickson C, Rakheja D, Carder KR, Hoang MP. Congenital self-healing reticulohistiocytosis (Hashimoto-Pritzker disease): ten-year experience at Dallas Children's Medical Center. *J Am Acad Dermatol* 2007;56:290–294.

123. O'Malley DP, Duong A, Barry TS, *et al*. Co-occurrence of Langerhans cell histiocytosis and Rosai–Dorfman disease: possible relationship of two histiocytic disorders in rare cases. *Mod Pathol* 2010;23:1616–1623.

124. Haupt R, Minkov M, Astigarraga I, *et al*. Langerhans cell histiocytosis (LCH): guidelines for diagnosis, clinical work-up, and treatment for patients till the age of 18 years. *Pediatr Blood Cancer* 2013;60:175–184.

125. Henck ME, Simpson EL, Ochs RH, Eremus JL. Extraskeletal soft tissue masses of Langerhans' cell histiocytosis. *Skeletal Radiol* 1996;25:409–412.

126. Singhi AD, Montgomery EA. Gastrointestinal tract Langerhans cell histiocytosis: a clinicopathologic study of 12 patients. *Am J Surg Pathol* 2011;35:305–310.

127. Emile JF, Wechsler J, Brousse N, *et al*. Langerhans' cell histiocytosis. Definitive diagnosis with the use of monoclonal antibody O10 on routinely paraffin-embedded samples. *Am J Surg Pathol* 1995;19:636–641.

128. Sholl LM, Hornick JL, Pinkus GS, Padera RF. Immunohistochemical analysis of langerin in Langerhans' cell histiocytosis and pulmonary infectious and inflammatory diseases. *Am J Surg Pathol* 2007;31:947–952.

129. Lau SK, Chu PG, Weiss LM. Immunohistochemical expression of langerin in Langerhans cell histiocytosis and non-Langerhans cell histiocytic disorders. *Am J Surg Pathol* 2008;32:615–619.

130. Badalian-Very G, Vergilio JA, Degar BA, *et al*. Recurrent BRAF mutation in Langerhans cell histiocytosis. *Blood* 2010;116:1919–1923.

131. Brown NA, Furtado LV, Betz BL, *et al*. High prevalence of somatic MAP2K1 mutations in BRAF V600E-negative Langerhans cell histiocytosis. *Blood* 2014;124:1655–1658.

132. Nelson DS, Quispel W, Badalian-Very G, *et al*. Somatic activating ARAF mutations in Langerhans cell histiocytosis. *Blood* 2014;123:3152–3155.

133. Willman CL, Busque L, Griffith BB, *et al*. Langerhans'-cell histiocytosis (histiocytosis X): a clonal proliferative disease. *N Engl J Med* 1994;331(3):154–160.

Juvenile xanthogranuloma

134. Cohen BA, Hood A. Xanthogranuloma: a report on clinical and histologic findings in 64 patients. *Pediatr Dermatol* 1989;6:262–266.

135. Janney CG, Hurt MA, Santa Cruz DJ. Deep juvenile xanthogranuloma: subcutaneous and intramuscular forms. *Am J Surg Pathol* 1991;15:150–159.

136. de Graaf JH, Timens W, Tamminga RYJ, Molenaar WM. Deep juvenile xanthogranuloma: a lesion related to dermal indeterminate cells. *Hum Pathol* 1992;23:905–910.

137. Zelger B, Cerio R, Orchard G, Wilson-Jones E. Juvenile and adult xanthogranuloma: a histological and immunohistochemical comparison. *Am J Surg Pathol* 1994;18:126–135.

138. Janssen D, Harms D. Juvenile xanthogranuloma in childhood and adolescence: a clinicopathologic study of 129 patients from the Kiel pediatric tumor registry. *Am J Surg Pathol* 2005;29:21–28.

139. Dehner L. Juvenile xanthogranulomas in the first two decades of life: a clinicopathologic study of 174 cases with cutaneous and extracutaneous manifestations. *Am J Surg Pathol* 2003;27:579–593.

140. Nascimento AG. A clinicopathologic and immunohistochemical comparative study of cutaneous and intramuscular forms of juvenile xanthogranuloma. *Am J Surg Pathol* 1997;21:645–652.

141. Zvulunov A, Barak Y, Metzker A. Juvenile xanthogranuloma, neurofibromatosis, and juvenile chronic myelogenous leukemia: world statistical analysis. *Arch Dermatol* 1995;131(8):904–908.

142. Cambiaghi S, Restano L, Caputo R. Juvenile xanthogranuloma associated with neurofibromatosis. 1: 14 patients without evidence of hematologic malignancies. *Pediatr Dermatol* 2004;21:97–101.

143. Kraus MD, Haley JC, Ruiz, R, et al. "Juvenile" xanthogranuloma: an immunophenotypic study with a reappraisal of its histogenesis. *Am J Dermatopathol* 2001;23:104–111.

Reticulohistiocytoma and multicentric reticulohistiocytosis

144. Purvis WE, Helwig EB. Reticulohistiocytic granuloma ("reticulohistiocytoma") of the skin. *Am J Clin Pathol* 1955;24:1005–1015.

145. Zelger B, Cerio R, Soyer HP, et al. Reticulohistiocytoma and multicentric reticulohistiocytosis. *Am J Surg Pathol* 1994;16:577–584.

146. Miettinen M, Fetsch JF. Reticulohistiocytoma (solitary epithelioid histiocytoma): a clinicopathologic and immunohistochemical study of 44 cases. *Am J Surg Pathol* 2006;30:521–528.

147. Barrow MV, Holubar K. Multicentric reticulohistiocytosis: a review of 33 patients. *Medicine (Baltimore)* 1969;48:287–305.

148. Trotta F, Castellino G, Lo Monaco A. Multicentric reticulohistiocytosis. *Best Pract Res Clin Rheumatol* 2004;18:759–772.

149. Salisbury JR, Hall PA, Williams HC, Mangi MH, Mufti GJ. Multicentric reticulohistiocytosis: detailed immunophenotyping confirms macrophage origin. *Am J Surg Pathol* 1990;14:687–693.

Crystal-storing histiocytosis

150. Kapadia SB, Enzinger FM, Heffner DK, Hyams VJ, Frizzera G. Crystal-storing histiocytosis associated with lymphoplasmacytic neoplasms: report of three cases mimicking adult rhabdomyoma. *Am J Surg Pathol* 1993;17:461–467.

151. Jones D, Bhatia VK, Krausz T, Pinkus GS. Crystal-storing histiocytosis: a disorder occurring in plasmacytic tumors expressing immunoglobulin kappa light chain. *Hum Pathol* 1999;30:1441–1448.

152. Friedman MT, Molho L, Valderrama E, et al. Crystal-storing histiocytosis associated with a lymphoplasmacytic neoplasm mimicking adult rhabdomyoma: a case report and review of the literature. *Arch Pathol Lab Med* 1996;120:1133–1136.

Histiocytic sarcoma

153. Wilson MS, Weiss LM, Gatter KC, et al. Malignant histiocytosis: a reassessment of cases previously reported in 1975 based on paraffin section immunophenotyping studies. *Cancer* 1990;66:530–536.

154. Michgrub S, Kamel OW, Wiley E, et al. Malignant histiocytic neoplasms of the small intestine. *Am J Surg Pathol* 1992;16:11–20.

155. Miettinen M, Fletcher CD, Lasota J. True histiocytic lymphoma of small intestine: analysis of two S-100 protein positive cases with features of interdigitating reticulum cell sarcoma. *Am J Clin Pathol* 1993;100:285–292.

156. Soria C, Orradre JL, Garcia-Almagro D, et al. True histiocytic lymphoma (monocytic sarcoma). *Am J Dermatopathol* 1992;14:511–517.

157. Lauritzen AF, Delsol G, Hansen NE, et al. Histiocytic sarcomas and monoblastic leukemias: a clinical, histologic, and immunophenotypic study. *Am J Clin Pathol* 1994;102:45–54.

158. Kamel OW, Gocke CD, Kell DL, Cleary ML, Warnke RA. True histiocytic lymphoma: a study of 12 cases based on current definition. *Leuk Lymphoma* 1995;18:81–86.

159. Copie-Bergman C, Wotherspoon AC, Norton AJ, Diss TC, Isaacson PG. True histiocytic lymphoma: a morphologic, immunohistochemical, and molecular genetic study of 13 cases. *Am J Surg Pathol* 1998;22:1386–1392.

160. Hornick JL, Jaffe ES, Fletcher CD. Extranodal histiocytic sarcoma: clinicopathologic analysis of 14 cases of a rare epithelioid malignancy. *Am J Surg Pathol* 2004;28:1133–1144.

161. Vos JA, Abbondanzo SL, Barekman CL, et al. Histiocytic sarcoma: a study of five cases including the histiocytic marker CD163. *Mod Pathol* 2005;18:693–704.

162. Wang E, Papalas J, Hutchinson CB, et al. Sequential development of histiocytic sarcoma and diffuse large B-cell lymphoma in a patient with a remote history of follicular lymphoma with genotypic evidence of a clonal relationship: a divergent (bilineal) neoplastic transformation of an indolent B-cell lymphoma in a single individual. *Am J Surg Pathol* 2011;35:457–463.

163. Feldman AL, Arber DA, Pittaluga S, et al. Clonally related follicular lymphomas and histiocytic/dendritic cell sarcomas: evidence for transdifferentiation of the follicular lymphoma clone. *Blood* 2008;111:5433–5439.

164. Takahashi E, Nakamura S. Histiocytic sarcoma: an updated literature review based on the 2008 WHO classification. *J Clin Exp Hematop* 2013;53:1–8.

165. Venkataraman G, McClain KL, Pittaluga S, Rao VK, Jaffe ES. Development of disseminated histiocytic sarcoma in a patient with autoimmune lymphoproliferative syndrome and associated Rosai-Dorfman disease. *Am J Surg Pathol* 2010;34:589–594.

Follicular dendritic reticulum cell tumor

166. Monda L, Warnke R, Rosai J. A primary lymph node malignancy with features suggestive of dendritic reticulum cell differentiation: a report of 4 cases. *Am J Pathol* 1986;122:562–572.

167. Chan JKC, Fletcher CDM, Nayler SJ, Cooper K. Follicular dendritic cell sarcoma: clinicopathologic analysis of 17 cases suggesting a malignant potential higher than currently recognized. *Cancer* 1997;79:294–313.

168. Chan JK. Proliferative lesions of follicular dendritic reticulum cells: an overview including a detailed account of follicular dendritic reticulum cell sarcoma, and neoplasm with

169. Perez-Ordonez B, Erlandson RA, Rosai J. Follicular dendritic reticulum cell tumor: report of 13 additional cases of a distinctive entity. *Am J Surg Pathol* 1996;20:944–955.

170. Kairouz S, Hashash J, Kabbara W, McHayleh W, Tabbara IA. Dendritic cell neoplasms: an overview. *Am J Hematol* 2007;82:924–928.

171. Fonseca R, Yamakawa M, Nakamura S, *et al.* Follicular dendritic reticulum cell sarcoma and interdigitating reticulum cell sarcoma: a review. *Am J Hematol* 1998;59:161–167.

172. Biddle DA, Ro JY, Yonn GS, *et al.* Extranodal follicular dendritic cell sarcoma of the head and neck region: three new cases, with a review of the literature. *Mod Pathol* 2002;15:50–58.

173. Grogg KL, Lae ME, Kurtin PJ, Macon WR. Clusterin expression distinguishes follicular dendritic cell tumors from other dendritic cell neoplasms: report of a novel follicular dendritic cell marker and clinicopathologic data on 12 additional follicular dendritic cell tumors and 6 additional interdigitating dendritic cell tumors. *Am J Surg Pathol* 2004;28:988–998.

174. Yu H, Gibson JA, Pinkus GS, Hornick JL. Podoplanin (D2–40) is a novel marker for follicular dendritic cell tumors. *Am J Clin Pathol* 2007;128:776–782.

175. Yu H, Gibson JA, Pinkus GS, Hornick JL. Podoplanin (D2–40) is a novel marker for follicular dendritic cell tumors. *Am J Clin Pathol* 2007;128:776–782.

Other reticulum cell tumors

176. Weiss L, Berry G, Dorfman RF, *et al.* Spindle cell neoplasms of lymph nodes of probably reticulum cell lineage: true reticulum cell sarcoma? *Am J Surg Pathol* 1990;14:405–414.

177. Gaertner EM, Tsokos M, Derringer GA, *et al.* Interdigitating dendritic cell sarcoma: a report of four cases and review of the literature. *Am J Clin Pathol* 2001;115:589–597.

178. Andriko JW, Kaldjian EP, Tsokos M, Abbondanzo SL, Jaffe ES. Reticulum cell neoplasms of lymph nodes: a clinicopathologic study of 11 cases with recognition of a new subtype derived from fibroblastic reticular cells. *Am J Surg Pathol* 1998;22:1048–1058.

179. Gould VE, Warren WH, Faber LP, Kuhn C, Franke WW. Malignant cells of epithelial phenotype limited to thoracic lymph nodes. *Eur J Cancer* 1990;26:1121–1126.

180. Franke WW, Moll R. Cytoskeletal components of lymphoid organs. I. Synthesis of cytokeratins 8 and 18 and desmin in subpopulations of extrafollicular reticulum cells of human lymph nodes, tonsils, and spleen. *Differentiation* 1987;36:145–163.

Chapter 36

Cytology of soft tissue lesions

Matjaž Šebenik,
John F. Kennedy Hospital, Fort Lauderdale, Florida

Živa Pohar-Marinšek
Institute of Oncology, Ljubljana, Slovenia

Almost 9000 soft tissue sarcomas (and 2500 bone sarcomas) are diagnosed annually in the United States; <1% of the 1 444 920 new cancer cases diagnosed in the United States in 2007.[1] Benign soft tissue tumors outnumber sarcomas by approximately 100:1, increasing the yearly incidence of all soft tissue tumors to about 300 per 100 000.

The use of fine-needle aspiration (FNA) as a primary diagnostic tool in soft tissue lesions is still uncommon, resulting in a lack of experience among many cytopathologists. This, coupled with the ever-present sampling error in larger masses, has led to the avoidance of FNA as the primary diagnostic tool in most medical centers.

The knowledge of general soft tissue surgical pathology cannot be overemphasized. All aspirates should be evaluated with the clinical and imaging data in mind. Cytopathologists, clinicians, and radiologists must work closely together to arrive at the best and most accurate diagnosis. Ancillary techniques, especially cytogenetics, should be employed whenever appropriate.

To give perspective on FNA practice, we list here the most common soft tissue masses aspirated at the Institute of Oncology (IO), in Ljubljana, Slovenia. The FNA clinic at IO opened in 1959 and has a volume of approximately 10 000 annual aspirates in the last 20 years. In 2004, of these, 1818 were aspirates of soft tissues (Table 36.1). To avoid confusion, we will list only the entities that are most common and well defined.

The use of ancillary techniques has greatly improved the accuracy of diagnosis of soft tissue lesions by FNA. Hajdu recommends extreme caution in the interpretation of aspiration smears from soft tissue tumors.[2] Twelve years after this report, Brown reports much more optimistic data about statistical performance characteristics (Table 36.2).[3]

Ancillary techniques
Immunocytochemistry

In the United States, immunocytochemistry (ICC) is performed mostly on cell blocks. At the IO, ICC is performed on alcohol-fixed cytospins (Delaunay's solution). If cytospins are not prepared in advance, ICC can be attempted on destained Pap slides.

Table 36.1 Soft tissue masses aspirated at IO in 2004

Benign lesions	891
Lipoma	406
Benign process, NOS	316
Fat necrosis	50
Ganglion cyst	41
Other cysts	31
Other benign tumors	25
Nodular fasciitis	14
Other	8
Unsatisfactory smears	580
Metastatic carcinoma	155
Other malignant lesions (lymphoma, melanoma)	91
Suspicious for sarcoma	26
Suspicious for other malignant lesions	16
Malignant tumor, NOS	25
Sarcoma, primary	14
Sarcoma, recurrent	20

Table 36.2 Statistical performance characteristics for aspiration smear interpretation

	%
Sensitivity rates	≤95
Specificity for sarcoma	54–98
False-positive rate	0–5
False-negative rate	2–15

Modern Soft Tissue Pathology, Second Edition, ed. Markku Miettinen. Published by Cambridge University Press. © Cambridge University Press 2016.

Table 36.3 Cytogenetic anomalies in soft tissue tumors (only the anomalies with the highest frequencies are listed)

Tumor	Cytogenetic anomaly	Frequency (%)
Alveolar soft part sarcoma	t(X;17)(p11;q25)	>90
Chondrosarcoma, extraskeletal myxoid	t(9;22)(q22;q12)	>75
Dermatofibrosarcoma protuberans	Ring forms of chromosomes 17 and 22	>75
Desmoid fibromatosis	Trisomy 8 and 20	25
Desmoplastic small round cell tumor	t(11;22)(p13;q12)	>75
Endometrial stromal sarcoma	t(7;17)(p15;q21)	70
Fibrosarcoma, congenital	t(12;15)(p13;q25)	>75
Gastrointestinal stromal tumor	Monosomies 14 and 22	>75
Ewing sarcoma/PNET	t(11;22)(q24;q12)	>80
Hibernoma	11q13 rearrangement	>50
Inflammatory myofibroblastic tumor	2p23 rearrangement	50
Lipoblastoma	8q12 rearrangement or polysomy 8	>80
Lipoma, not otherwise specified	12q15 rearrangement	75
Lipoma, spindle cell and pleomorphic	Deletions of 13q	80
Liposarcoma, well-differentiated	Giant marker chromosomes and ring form of chromosome 12	>75
Liposarcoma, myxoid/round cell type	t(12;16)(q13;p11)	>75
Malignant melanoma of soft tissues	t(12;22)(q13;q12)	90
Neuroblastoma, good prognosis	Hyperdiploid, no 1p deletion	90
Neuroblastoma, bad prognosis	Deletion of 1p	>75
Rhabdoid tumor	Deletion of q22	>90
Rhabdomyosarcoma, alveolar	t(2;13)(q35;q14)	>75
Rhabdomyosarcoma, embryonal	Trisomies 2q, 8, and 20	>75
	Loss of heterozygosity at 11p15	
Schwannoma	Deletion of 22q	>80
Synovial sarcoma monophasic	t(X;18)(p11;q11)	>90
Synovial sarcoma biphasic	t(X;18)(p13;q11)	>90

Electron microscopy

The main role for electron microscopy (EM) has been in the diagnosis of small round cell tumors and poorly differentiated neoplasms with specific organelles, such as melanoma and angiosarcoma. This technique has been largely replaced by ICC in recent decades.

Flow cytometry

This technique is used predominantly for immunophenotyping when a lymphoproliferative disorder is suspected, and for DNA analysis when such information has prognostic significance. In addition, flow cytometry (FC) has also been used to detect residual neuroblastoma in the bone marrow.[4]

Cytogenetics

As evident from Table 36.3, quite a few specific and reproducible cytogenetic abnormalities have been identified by routine chromosome analysis, by fluorescence *in situ* hybridization (FISH), and by reverse transcriptase polymerase chain reaction (RT-PCR).

Conventional chromosome analysis is time-consuming and has low sensitivity; however, it does examine the full karyotype.

FISH labels only the portion of the chromosome that needs to be examined. Aspirates provide excellent material with intact cytologic compartments. The best results can be obtained from evenly spread and moderately cellular preparations such as touch preparation, thin-layer preparation, or a cytospin.

An alternative and complementary method for transcript evaluation is RT-PCR. Total RNA isolated from fresh, frozen, or preserved tumor tissue is subjected to reverse transcription. PCR reactions using gene-specific primers allow identification of abnormal fusion transcripts typical of sarcoma translocations.

Adipocytic lesions

Lipoma

Lipoma is the most common soft tissue tumor; it presents mainly in adults and accounts for more than one-half of all soft tissue tumors. Most lipomas are <10 cm and subcutaneous; rare examples are intramuscular. Lipomas are common in trunk and proximal extremities, and rare in the distal extremities.

A smear consists of tissue fragments often associated with capillaries. Single cells are uncommon, and cells are uniform in size and shape. Nuclei are small, with even chromatin (Figure 36.1).[5]

Differential diagnosis

Distinction with normal fat is clinical and not morphologic; lipoma is usually a circumscribed, pseudoencapsulated mass.

In well-differentiated liposarcoma, adipocytes are not uniform, nuclei show at least focal atypia, and multivacuolated lipoblasts are sometimes seen. Deep adipose tissue tumors of the retroperitoneum, mediastinum, and groin should be suspected to be well-differentiated liposarcomas, until proven otherwise.

Myxolipoma of soft tissues

Myxolipomas are lipomas in which a portion of the tumor has been replaced by a mucoid substance. These tumors have a clinical presentation similar to ordinary lipomas.

Aspirates are composed of mature adipose tissue with abundant mucoid material in the background. Thick-walled blood vessels are also present in vascular myxolipomas (Figure 36.2).

Differential diagnosis

Myxoid liposarcoma is deeper seated, and aspirates contain lipoblasts and plexiform capillaries. Myxoma does not contain fat.

Fat necrosis

Fat necrosis is a very common lesion most often found in the breast, where it mimics malignancy, both clinically and on imaging studies. Patients frequently recall a history of trauma to the area.

Smears are often hypocellular, composed of fragments of degenerated fat and individual cells associated with mono- and multinucleated histiocytes with numerous cytoplasmic

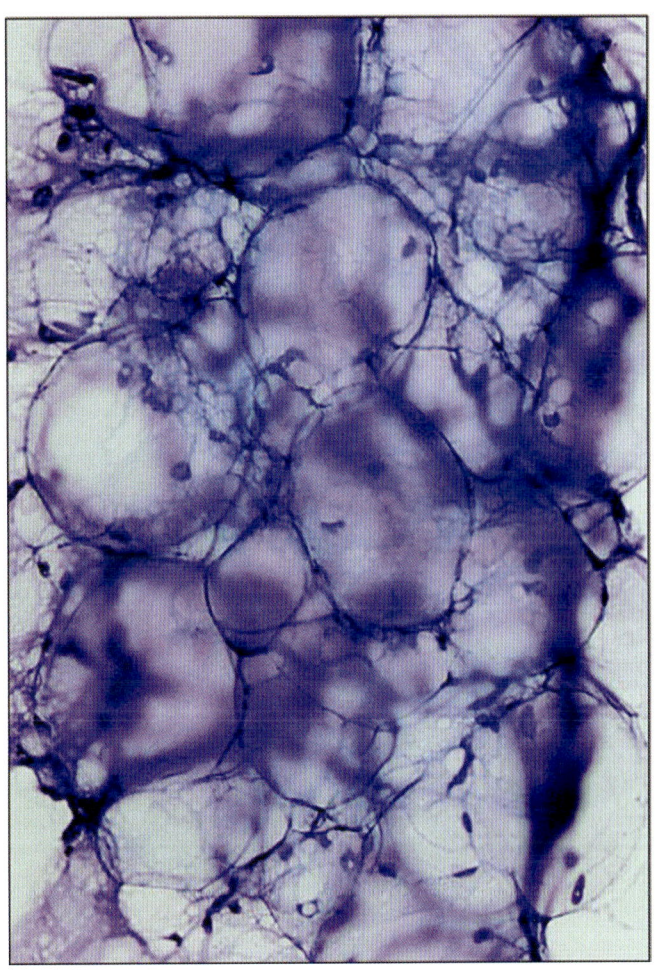

Figure 36.1 Lipoma. The tissue fragment is composed of adipocytes that are uniform in size and shape. Nuclei are very small, and chromatin is fine and even.

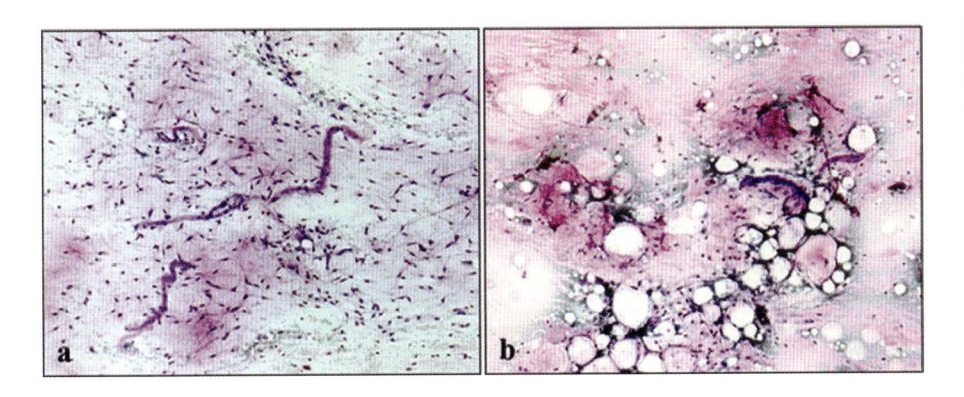

Figure 36.2 Myxolipoma. (**a**) Capillaries, spindle-shaped cells, and abundant mucoid material in the background. (**b**) Mature adipose tissue, capillaries, and mucoid material.

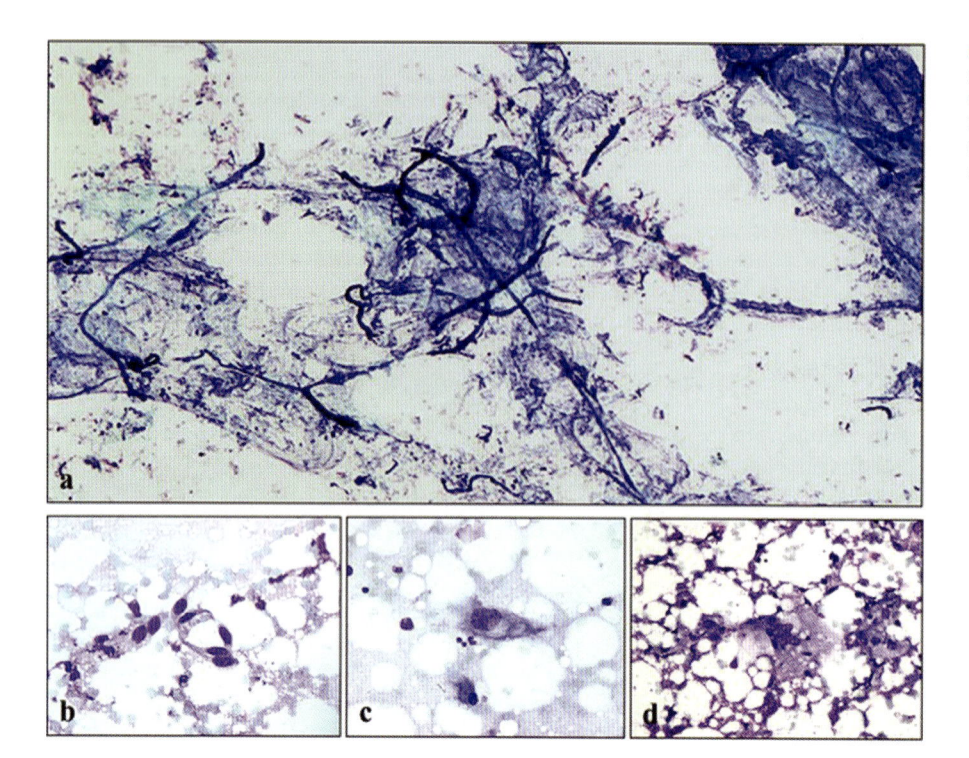

Figure 36.3 Fat necrosis. (**a**) Low power with capillaries, ghosts of lipocytes, and degenerated fat. (**b,c**) Reactive-appearing fibroblasts with enlarged nuclei, but fine chromatin, as can be seen in nodular fasciitis. (**d**) Multinucleated histiocytes with cytoplasmic vacuoles.

vacuoles. Inside the tissue fragments, fat cells are still recognizable; outside fragments they appear as amorphous proteinaceous material. The background is frequently composed of ghosts of lipocytes, old blood, necrotic debris, mixed inflammation, and reactive-appearing fibroblasts. The fibroblasts are often enlarged and pleomorphic, with prominent nucleoli, but fine chromatin, as is seen in nodular fasciitis (Figure 36.3).[6,7]

Differential diagnosis

In ductal carcinoma, especially the lipid-rich variant, the specimen is hypercellular, and nuclei show marked atypia. In silicone granulomas, the vacuoles are larger, often resulting in a signet-ring appearance. In cryptococcosis, there are large cytoplasmic vacuoles that contain refractile budding yeast.

Lipoblastoma

Lipoblastoma is an unusual tumor of childhood. It presents as slowly growing subcutaneous masses in the limbs, especially the lower extremities.

Aspirates contain abundant myxoid material interspersed with a fine network of small capillaries. The background shows numerous bare nuclei and bland-appearing lipoblasts. The latter are small, round, and uniform; eccentric nuclei show minimal atypia with a delicate chromatin and no or inconspicuous nucleoli. Clear cytoplasm shows multiple vacuoles. Intranuclear cytoplasmic invaginations are not uncommon (Figure 36.4).[8]

Differential diagnosis

Myxoid liposarcoma presents in an older population and shows nuclear atypia, more abundant myxoid stroma, and more sheet-like architecture. In lipomas, lipoblasts are not seen, and capillaries are not prominent.

Spindle cell/pleomorphic lipoma

Both entities represent the two ends of a common histologic spectrum and usually present as circumscribed subcutaneous lesions on the neck, back, and shoulder in older men.

Aspirates of spindle cell lipomas usually present as a mixture of mature adipose tissue associated with dispersed or clustered bland spindle cells in a myxoid background (Figure 36.5). The latter sometimes shows eosinophilic collagen/hyaline fibers and mast cells.[9]

Aspirates of pleomorphic lipoma show a mixture of mature adipocytes and a variable number of bizarre pleomorphic giant cells with dark nuclei that on occasion form rings (floret cells). Similar to spindle cell lipomas, the background on occasion shows collagen/eosinophilic fibers and mast cells.[10]

Differential diagnosis

Myxoid liposarcoma is usually deep seated and consists of a branching network of thin capillaries and lipoblasts. Low-grade myxofibrosarcoma often contains coarse vessel fragments. In schwannomas, the spindle cells are positive for S100 protein and negative for CD34. Atypical lipomatous tumor/well-differentiated liposarcoma shows significant adipocytic nuclear atypia and sometimes lipoblasts.

Hibernoma

This relatively uncommon tumor especially occurs in the thigh, back, shoulder, and neck. Most cases present during adulthood.

Chapter 36: Cytology of soft tissue lesions

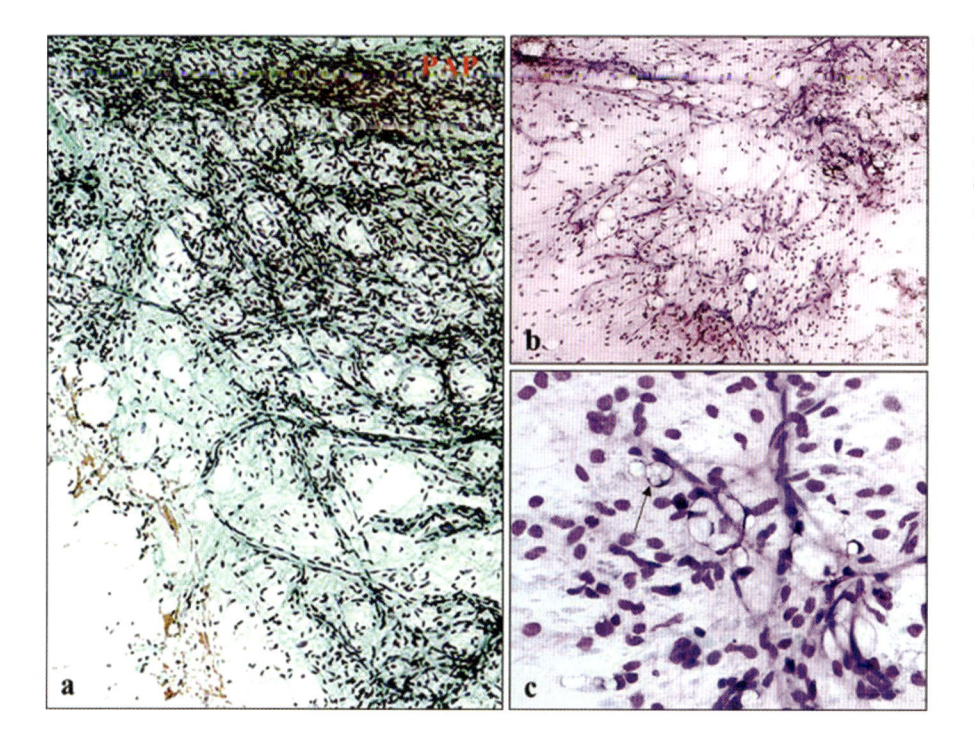

Figure 36.4 Lipoblastoma. (**a**) Low power, highlighting the fine capillary network. (**b**) Capillaries and a bland-appearing lipoblast. (**c**) Lipoblasts are small, round, and uniform. Eccentric nuclei show minimal atypia, delicate chromatin, and inconspicuous or absent nucleoli. Clear cytoplasm shows multiple vacuoles. An intranuclear cytoplasmic vacuole is also seen (arrow).

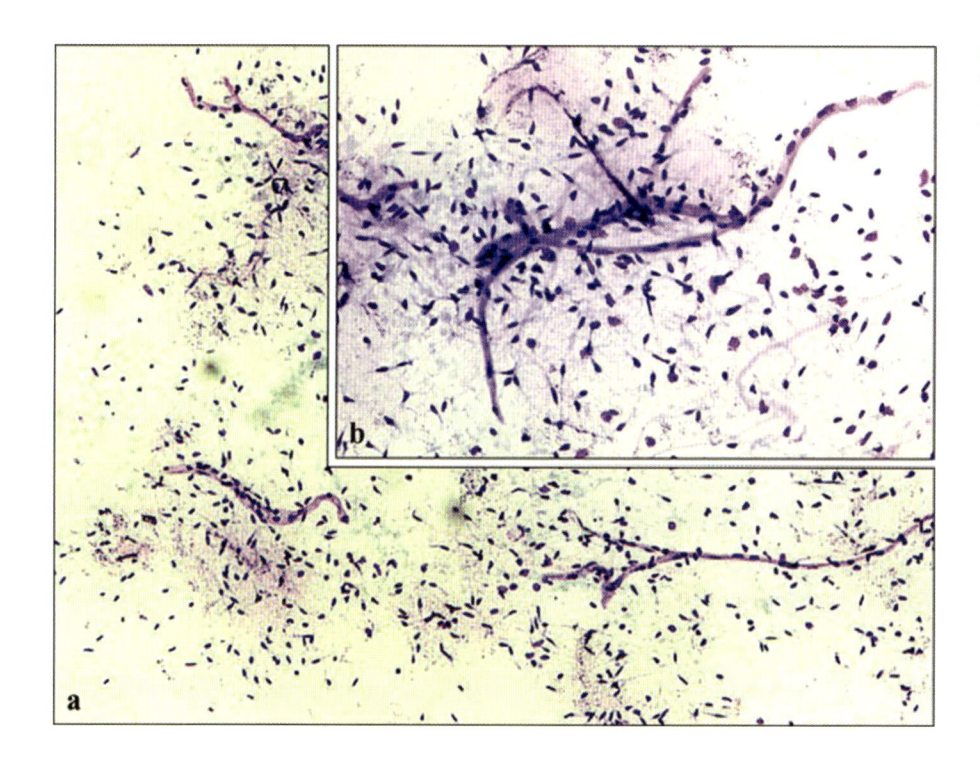

Figure 36.5 Spindle cell lipoma. (**a,b**) Fine capillaries and bland spindle cells in a myxoid background are seen in this case. Mature adipose tissue was seen in other areas (not shown).

Aspirates consist of small tissue fragments admixed with single cells. The cytoplasm shows numerous uniform vacuoles; the nuclei are small and bland and often centrally placed (Figure 36.6). Mature adipocytes are frequently seen in the background.[11]

Differential diagnosis

Granular cell tumors (GCTs) show no adipocytic differentiation and are positive for neuron-specific enolase (NSE) and S100 protein. In adult rhabdomyoma, cells are larger, cytoplasm is denser and fibrillary, and positive for muscle markers.

Atypical lipomatous tumor/well-differentiated liposarcoma

This common soft tissue sarcoma presents in adults, with a peak incidence during the fifth and seventh decades.

979

Chapter 36: Cytology of soft tissue lesions

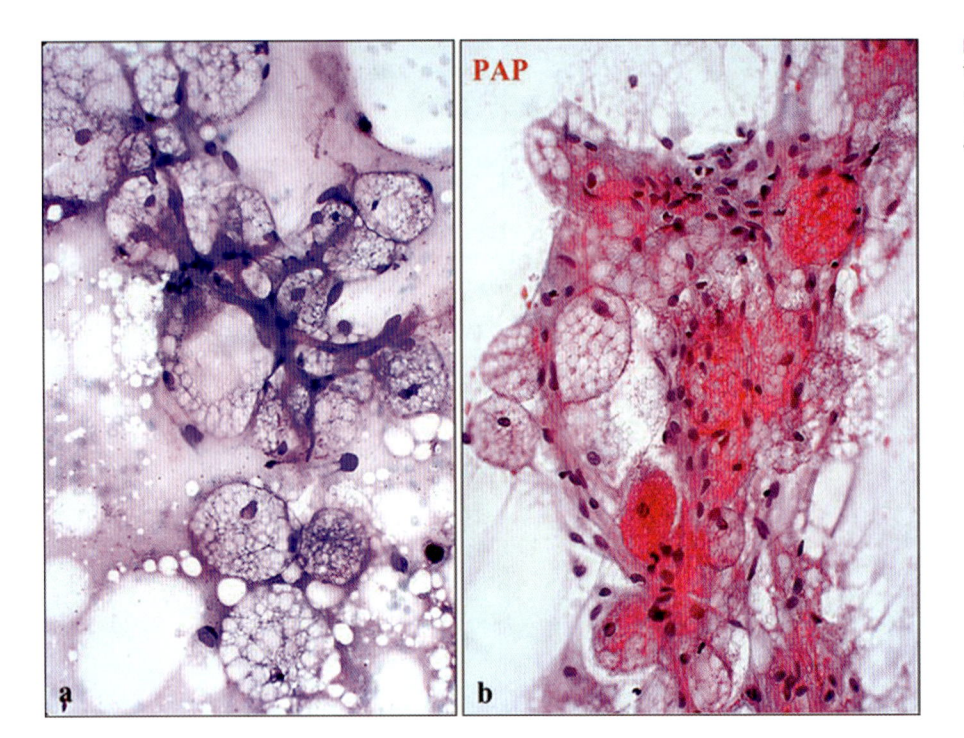

Figure 36.6 Hibernoma. (**a,b**) Small tissue fragments are composed of cells with numerous uniform cytoplasmic vacuoles and small and bland nuclei. Mature adipose tissue was present in other areas (not shown).

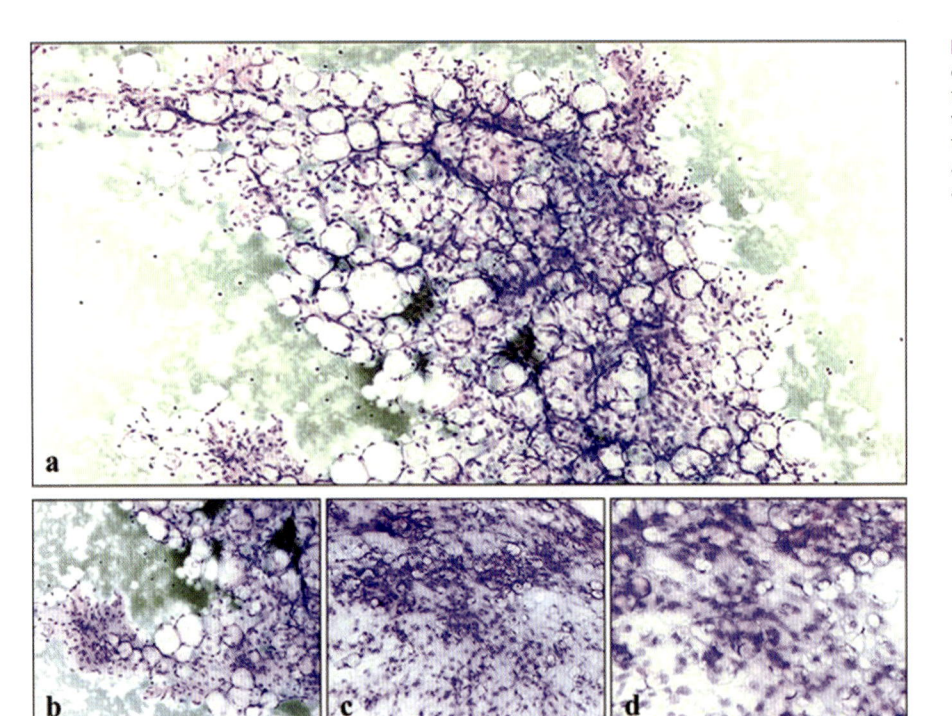

Figure 36.7 Atypical lipomatous tumor/well-differentiated liposarcoma. (**a**) Low power with tissue fragment composed of adipocytes of variable sizes. Note that nuclei are more prominent than in lipoma. (**b**) Minimally atypical adipocytic and stromal nuclei. (**c,d**) Multivacuolated lipoblasts are occasionally seen.

The terms atypical lipomatous tumor and well-differentiated liposarcoma are synonymous. Variation in cell size and adipocytic and septal nuclear atypia are defining features. These tumors do recur, can undergo dedifferentiation, and can then metastasize.

Aspirates yield tissue fragments of variable sizes. Variation in adipocyte size can be very subtle and focal, and is commonly overlooked. Uncommonly, atypical adipocytic and stromal nuclei and rare lipoblasts are seen (Figure 36.7).[12]

Aspirates should always be evaluated with the prior knowledge of the clinical condition of the patient. Because there are very few lipomas (and also very few leiomyomas) in deep locations such as the retroperitoneum, mediastinum, and groin, aspirates of unremarkable fat from these areas should be always viewed as suspicious for well-

Chapter 36: Cytology of soft tissue lesions

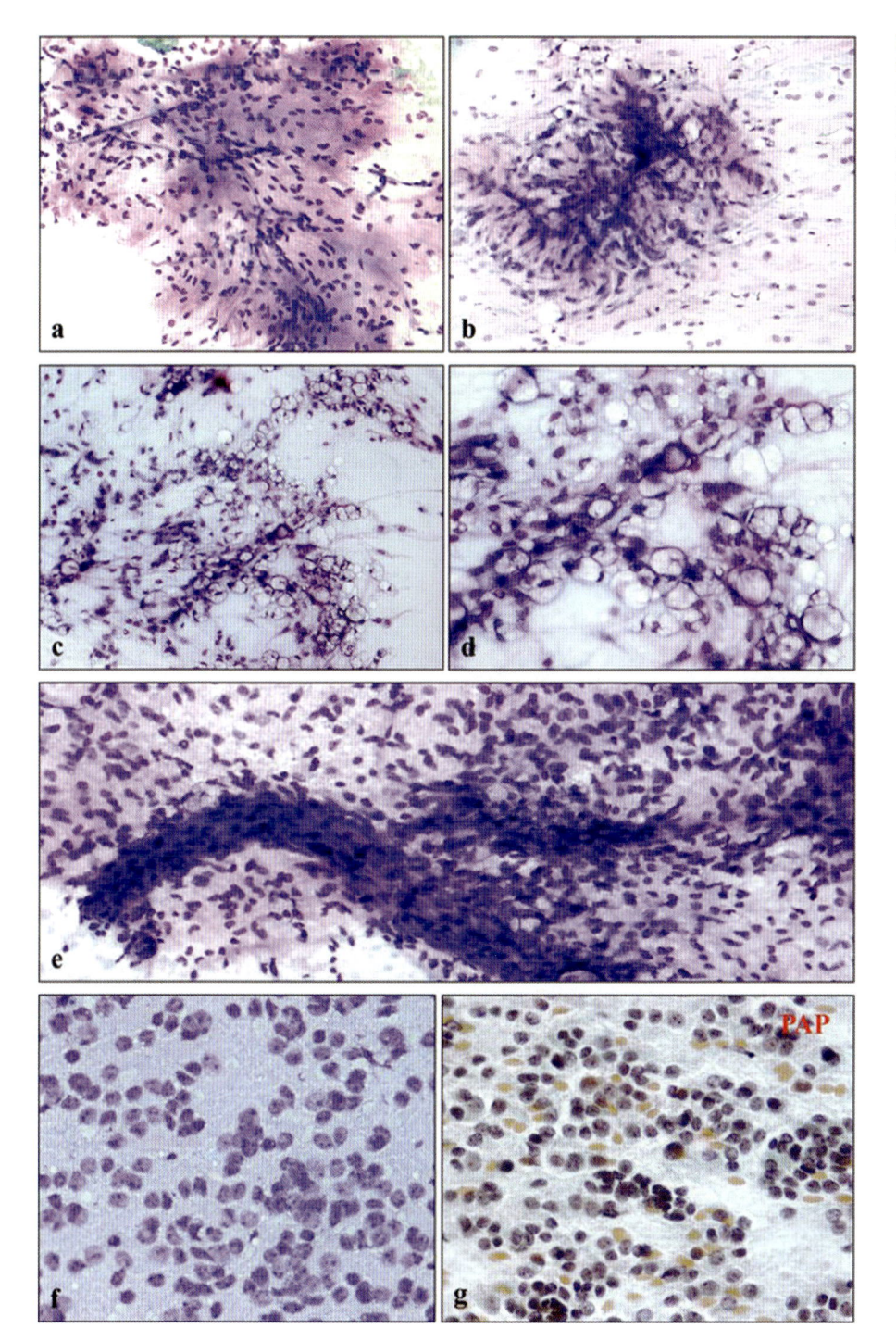

Figure 36.8 Myxoid/round cell liposarcoma. (**a**) Arborizing uniform capillaries and abundant myxoid matrix. (**b**) Capillaries, myxoid matrix, and a few lipoblasts. (**c,d**) Prominent multivacuolated lipoblasts with nuclear atypia. (**e**) Round cells in this case are more prominent in the vicinity of capillaries. (**f,g**) Sheets of round cells with atypical round nuclei, prominent nucleoli, and scanty eccentric cytoplasm. Lipoblasts and myxoid material are present in other areas.

differentiated liposarcomas even if none of the above criteria is seen.

Myxoid/round cell liposarcoma

Both of these tumors share the same chromosomal aberration and represent two ends of the morphologic spectrum of the same entity. A pure myxoid lesion represents the lower-grade end of the spectrum. The grade increases with the higher percentage of the round cell component.

Most tumors occur during the third and fifth decades and are most commonly located in the deep soft tissue of the limbs, especially in the thigh.

Aspirates show abundant myxoid matrix interspersed with arborizing uniform capillaries. Lower-grade lesions present as tissue fragments, whereas higher-grade lesions demonstrate less clustering and a predominance of single cells. Cells consist of lipoblasts and cells varying from round to cuboidal. Lipoblasts show one to a few cytoplasmic vacuoles and eccentric slightly hyperchromatic nuclei (Figure 36.8).[5,12–14]

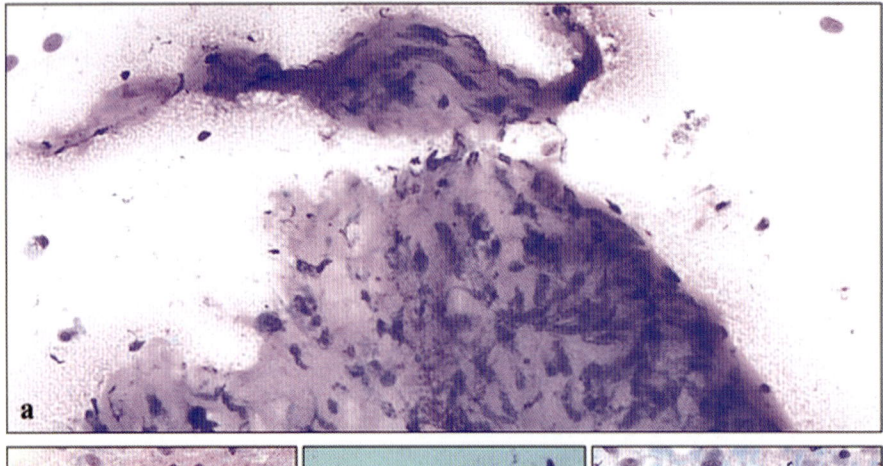

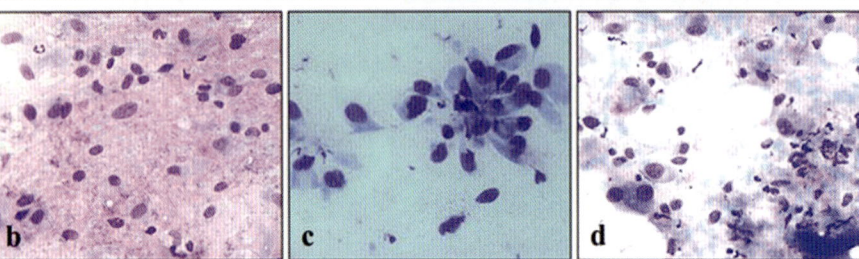

Figure 36.9 Nodular fasciitis. (**a**) Uncommonly seen tissue fragment contains few cells and abundant intercellular matrix. Cells have oval nuclei and moderate amounts of blue cytoplasm. (**b,c**) Commonly seen dispersed cells have round to oval nuclei, bland chromatin with inconspicuous nucleoli, and abundant cytoplasm forming tail-like extensions. (**d**) Proliferative fasciitis is characterized by large basophilic cells with one or two vesicular nuclei with fine chromatin and prominent nucleoli. Abundant basophilic slightly granular cytoplasm lacks the cross-striations typical of rhabdomyoblasts (so-called ganglion-like cells).

Differential diagnosis

A spindle cell lipoma, with abundant myxoid matrix, is typically superficial and grows in the neck or upper torso of middle-aged men.

Intramuscular myxomas are composed of cells with hair-like processes with minimal atypia and only scattered vessel fragments in the myxoid background. Lipoblasts are not seen.

In myxofibrosarcoma, no lipoblasts are seen; however, the nonbranching coarse vessel fragments are present in the background.

The presence and the percentage of round cells are very important and should always be documented. The metastatic potential is increased when 25% or more of the tumor is composed of round cells.[15] Round cells present as sheets, with atypical round nuclei, prominent nucleoli, and scanty eccentric cytoplasm. Lipoblasts and myxoid material can be seen in the background.

(Myo)fibroblastic lesions
Nodular fasciitis

Nodular fasciitis is a common lesion often mistaken for sarcoma. The clinical presentation is crucial: the lesion shows rapid growth, reaching a few centimeters (but not more) in only a few weeks. It is most common during the second to fifth decades. The tumors are superficial and are most commonly found in the upper extremities. They typically involve the fascia, and are not connected to the skin except on the face. Proliferative fasciitis has similar clinical presentation; however, it is characterized by the presence of large basophilic cells resembling ganglion cells.

Aspirates are usually very cellular, with single cells outnumbering cohesive groups, and consist of a polymorphous population of fibroblasts and myofibroblasts. Cells are round to spindle-shaped, polygonal, and stellate. Uniform nuclei have bland chromatin, and nucleoli are usually not prominent. The cytoplasm is abundant, often forming tail-like extensions.

Cells are negative for desmin, but positive for (smooth muscle) actin. The background can contain myxoid material, branching capillaries, red blood cells, and inflammatory cells (Figure 36.9a,b,c).[16]

Differential diagnosis

Proliferative fasciitis can exhibit numerous ganglion-like cells, often with large nucleoli. The myxoid background is less prominent (Figure 36.9d). Proliferative myositis often exhibits multinucleated regenerating muscle fibers. In desmoid fibromatosis, the cell population is more uniform.

Desmoid-type fibromatosis

This benign fibrous tissue proliferation seen mostly in adults is considered "deep fibromatosis." Desmoids usually arise from aponeuroses of muscles of the trunk and extremities. Common sites include abdominal wall and limb girdles. They can attain a large size, and often recur, but do not metastasize.

Superficial fibromatoses arise from fascia or superficial aponeurosis. They grow slowly, remain small, recur less often, and do not metastasize. The superficial group comprises palmar fibromatoses (Dupuytren contracture), plantar fibromatoses (Ledderhose disease), penile fibromatoses (Peyronie's disease), and so-called knuckle pads.

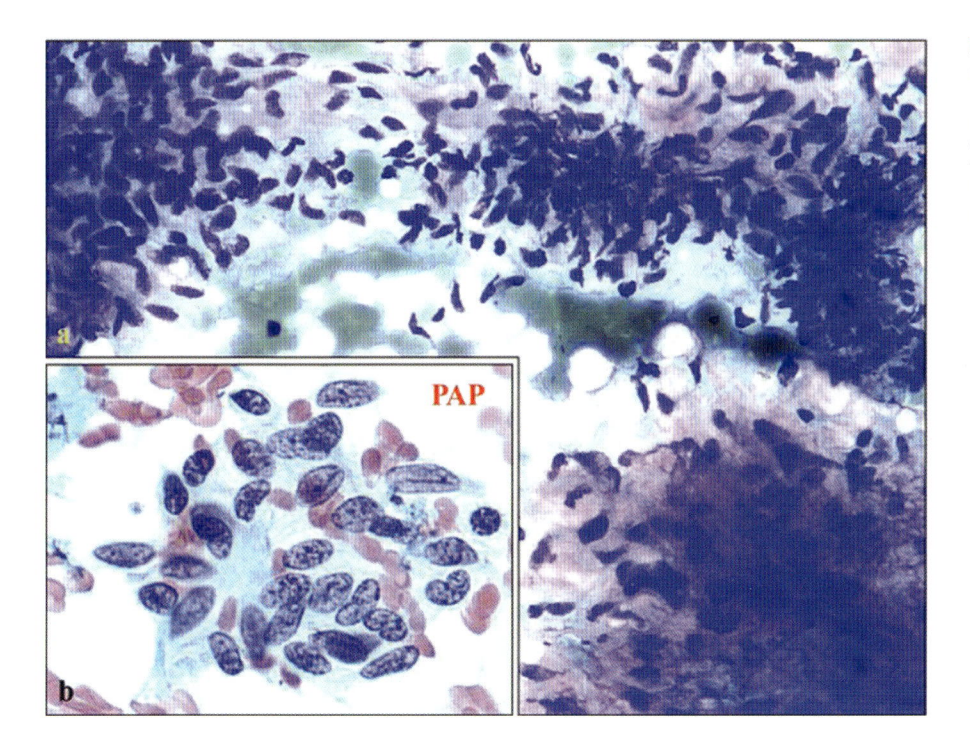

Figure 36.10 Desmoid-type fibromatosis. (**a**) Tissue fragment in this early (proliferative) lesion is moderately cellular. Some of the stellate and polygonal cells exhibit tapered cytoplasmic extensions. (**b**) Round to oval nuclei are "active-appearing" (larger and darker, as seen in late lesions); however, the chromatin is fine, and no mitotic figures are seen.

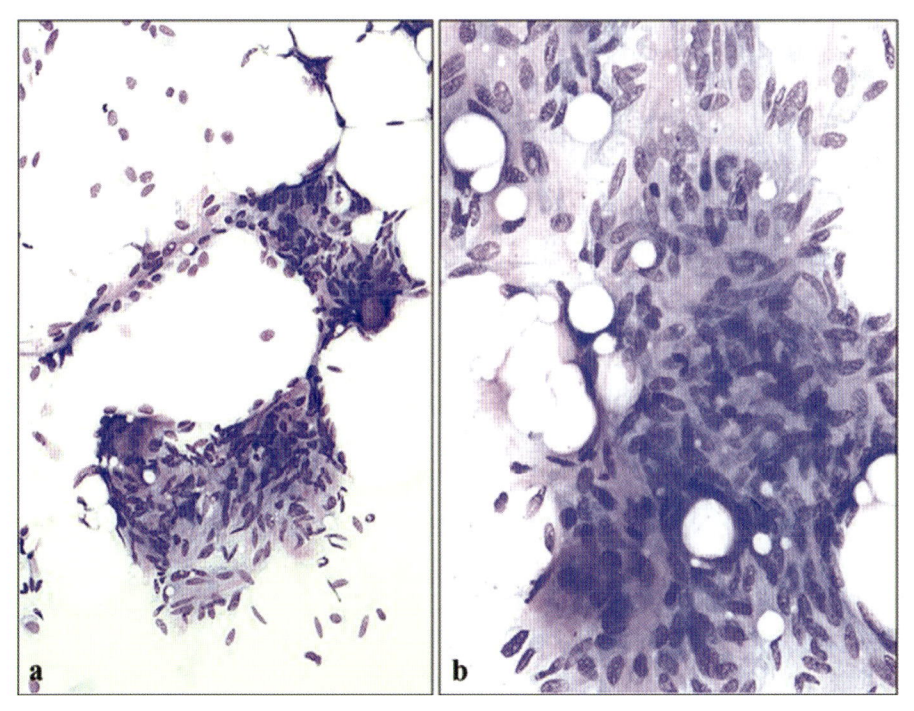

Figure 36.11 Mammary-type myofibroblastoma. (**a**) The cellular aspirate is composed of clustered and dispersed benign spindle-shaped mesenchymal cells. (**b**) Higher power shows moderate amounts of tapered cytoplasm and oval to elongated nuclei with fine chromatin and inconspicuous nucleoli.

Aspirates of superficial or older lesions are moderately cellular. Cells are in small clusters or single, and have tapered cytoplasmic extensions or are stellate and polygonal. Oval nuclei are bland with no mitotic activity. Background consists of rare inflammatory cells.

Aspirates of deep or early (proliferative) lesions are more cellular. The nuclei are slightly bigger and darker (active-appearing), showing minimal atypia but no mitotic figures (Figure 36.10).[17,18]

Mammary-type myofibroblastoma

This uncommon tumor usually presents as a well-circumscribed subcutaneous mass in the breast of adult patients. Some examples occur along putative milk-lines, extending from the axilla to the medial groin.

Aspirates reveal abundant, randomly arranged, single and clustered benign spindle-shaped mesenchymal cells with moderate amounts of tapered cytoplasm and oval to elongated nuclei with fine chromatin and inconspicuous nucleoli (Figure 36.11).

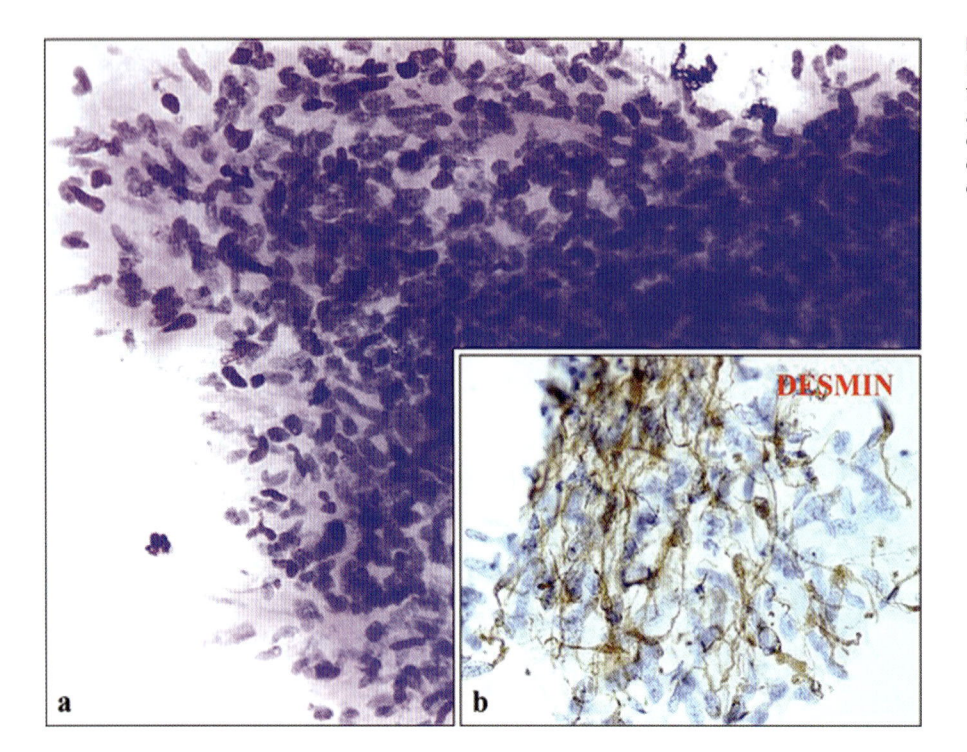

Figure 36.12 (Infantile) myofibroma/myofibromatosis. (**a**) This cellular large tissue fragment is composed of tumor cells and small amounts of intercellular matrix. Cells have uniform oval nuclei without nucleoli and indistinct cytoplasmic borders. (**b**) Cytoplasm is positive for desmin (cytospin preparation).

In one published case, few cells showed nuclear grooves. Cells are positive for CD34 and desmin, and negative for S100 protein and cytokeratin.[19]

Differential diagnosis

Schwannoma is uncommon in the breast. Cases in the authors' experience have been less cellular, with bullet-, comma-, or S-shaped nuclei. The cytoplasm is positive for S100 protein.

(Infantile) myofibroma/myofibromatosis

Solitary myofibroma and multicentric myofibromatosis are more common in male subjects. Many infantile myofibromas are congenital, and most are detected within the first two years of life. Solitary forms appear in the skin, subcutis, and skeletal muscle of the trunk and extremities. Multicentric forms are often found in the deep soft tissues and at visceral locations.

In the authors' single case, the aspirate was very hemorrhagic and sparsely cellular. It contained a few large tissue fragments, with many small oval nuclei in the background. Tissue fragments were highly cellular, containing only tumor cells and small amounts of intercellular matrix. Cells had uniform oval nuclei without nucleoli and indistinct cytoplasmic borders. Cytoplasm was positive for desmin (Figure 36.12). Others have reported similar results.[20,21]

Myxoinflammatory fibroblastic sarcoma

This low-grade sarcoma with myxoid stroma, inflammatory infiltrate, and virocyte-like cells predominantly involves the hands and feet of adults during the fourth and fifth decades. The tumors are often multinodular, measure from 1 cm to 8 cm in diameter, and are located close to the synovium. They practically never metastasize, but do recur.

Aspirates contain all of the characteristic features described in surgical biopsies: myxoid material, spindle-shaped cells with bipolar cytoplasmic extensions, epithelioid cells with globules of extracellular material, and ganglion-like and lipoblast-like giant cells. Only the inflammatory component is scarce (Figure 36.13).[22,23]

Differential diagnosis

Other myxoid tumors often resemble myxoinflammatory fibroblastic sarcoma to some extent. In the authors' experience, however, none of them contains all three components characteristic of this tumor.

Adult fibrosarcoma

Adult fibrosarcoma is a rare malignant tumor composed of atypical fibroblasts with variable collagen production. By definition, no other matrix production is allowed, resulting in a diagnosis of exclusion. Many fibrosarcomas in adults are derived from dermatofibrosarcoma protuberans.

Smears show a uniform population of spindle-shaped cells arranged in small clusters or singly. Nuclei are fusiform, and atypia varies, with dark coarse chromatin and prominent nucleoli in higher-grade lesions. The cytoplasm is elongated (Figure 36.14).[24]

Differential diagnosis

Monophasic fibrous synovial sarcoma (SS) stains with pancytokeratin and epithelial membrane antigen (EMA). MPNSTs stain focally with S100 protein.

Chapter 36: Cytology of soft tissue lesions

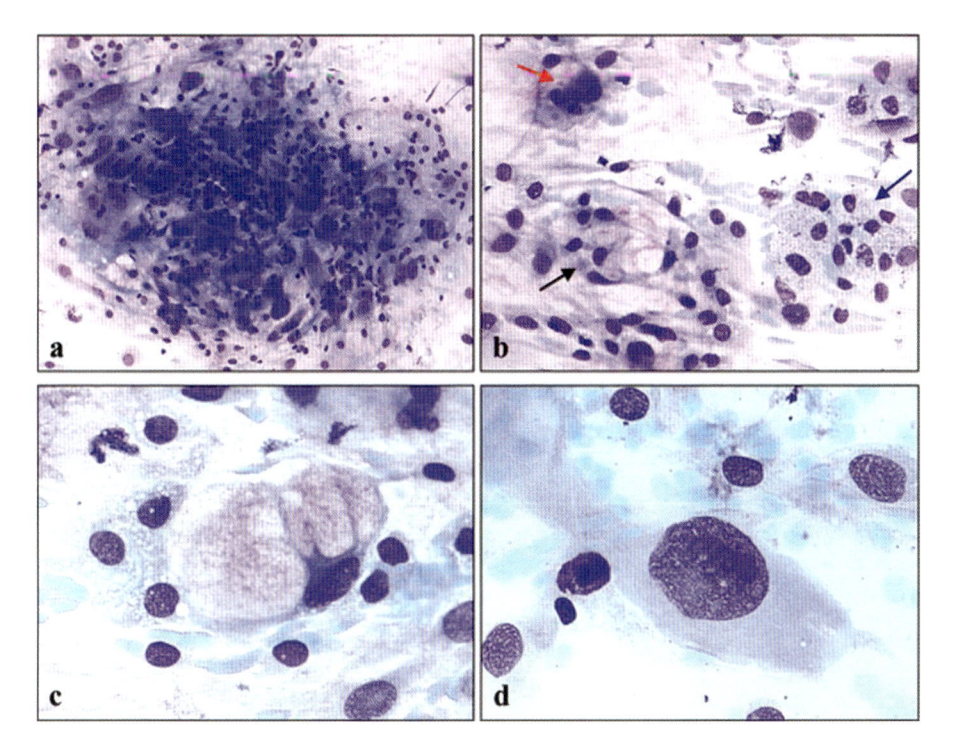

Figure 36.13 Myxoinflammatory fibroblastic sarcoma. (**a**) Mostly dispersed cells and abundant myxoid material. (**b**) Higher power shows numerous spindle-shaped cells with bipolar cytoplasmic extensions *(black arrow)*, few bi- and multinucleated epithelioid cells *(red arrow)* and cells with "bubbly" cytoplasm *(blue arrow)*. (**c**) Lipoblast-like cell. (**d**) Epithelioid cell with large nucleus, fine chromatin, and inconspicuous nucleolus.

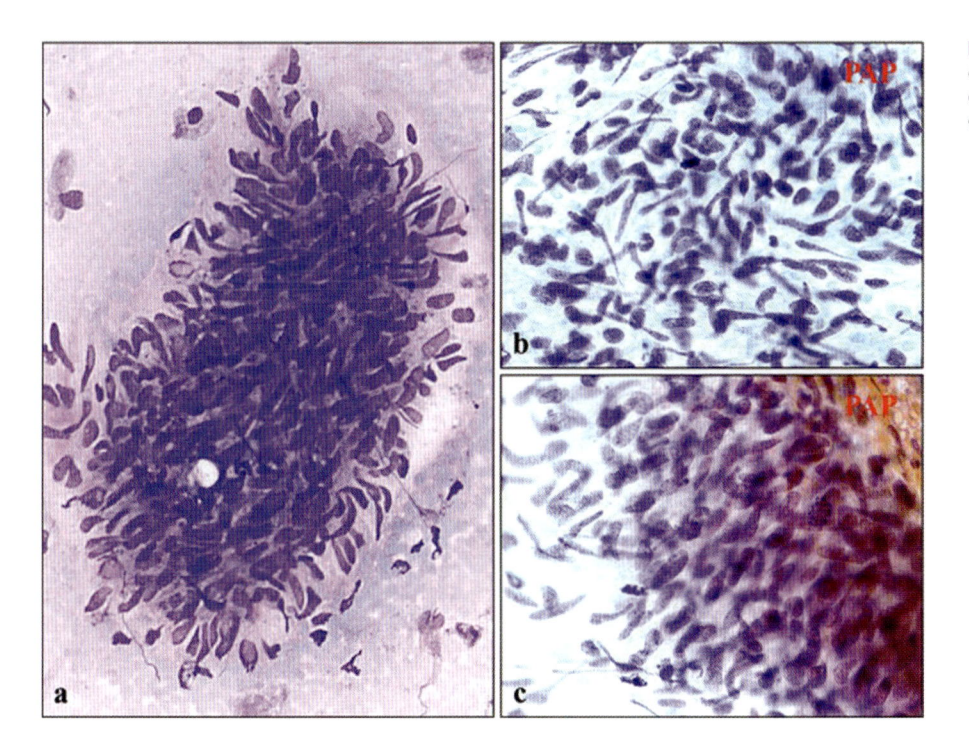

Figure 36.14 Adult fibrosarcoma. (**a**) This small cluster is composed of uniform spindle-shaped cells. (**b,c**) Nuclei are fusiform, chromatin is finely granular, and nucleoli are inconspicuous or absent.

Myxofibrosarcoma (myxoid malignant fibrous histiocytoma)

Myxofibrosarcoma, one of the most common sarcomas, comprises a spectrum of malignant myofibroblastic lesions with variably myxoid stroma, pleomorphism, and a distinct curvilinear vascular pattern.

The tumor is most common during the sixth and eighth decades and presents as a painless, slowly growing mass of the limb girdles, and less often in the trunk, head, and neck. The local recurrence in 50% to 60% of cases is unrelated to the grade and depth of the tumor.

Metastases occur in 20% to 35% of the intermediate- and high-grade tumors, especially if they are deep seated. Lungs and bone are the most common sites.

Aspirates of low-grade tumors show abundant myxoid material and scant mildly atypical cells. With rising grade, there is less myxoid material, but progressively more cells

985

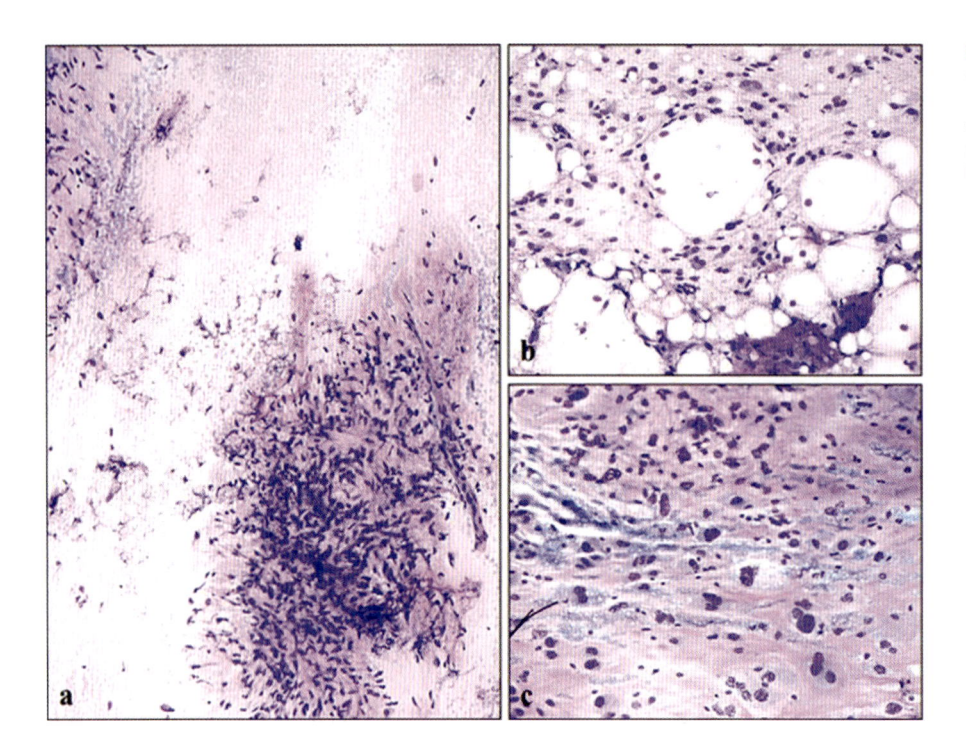

Figure 36.15 Myxofibrosarcoma (myxoid MFH). (**a,b**) Lower-grade lesions show abundant myxoid material and scant mildly atypical cells. (**c**) Higher-grade lesions show less myxoid material, but progressively more cells as the degree of atypia increases.

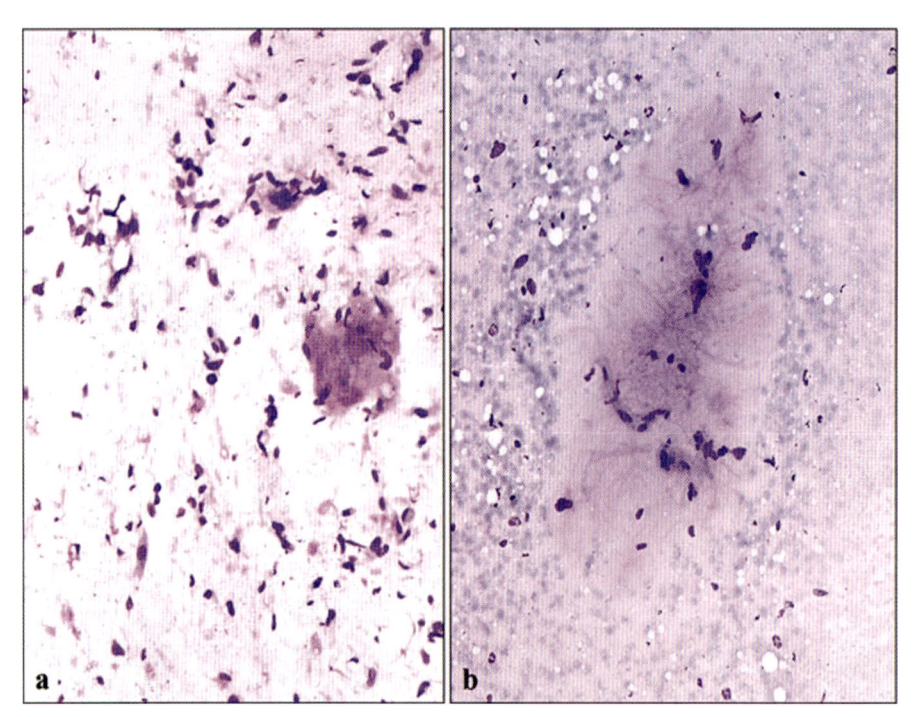

Figure 36.16 Low-grade fibromyxoid sarcoma. (**a,b**) Aspirates are hypocellular, with a monotonous population of fibroblast-like, spindle-shaped, minimally atypical cells, and an abundant myxoid background. Cells are mostly isolated, and the background contains a few bare nuclei, but no vascular fragments.

with a higher degree of atypia (Figure 36.15). High-grade lesions have the appearance of undifferentiated pleomorphic sarcoma.[25,26]

Low-grade fibromyxoid sarcoma

This low-grade variant of fibrosarcoma presents as a deep mass in the subfascial locations of the proximal extremities and trunk in young adults. About 40% of tumors contain collagenous rosettes surrounded by plump fibroblasts (hyalinizing giant cell tumor with giant rosettes) showing the same biologic behavior.

Despite its bland appearance, this tumor recurs in 9%, metastasizes in 6%, and is fatal in 2% of cases. Metastases are encountered even decades after diagnosis, requiring indefinite follow-up.

Aspirates are often hypocellular, with a monotonous population of fibroblast-like, spindle-shaped minimally atypical cells in abundant myxoid background. Cells are found in small clusters or singly. Bare nuclei are common, and vascular fragments are not seen (Figure 36.16).

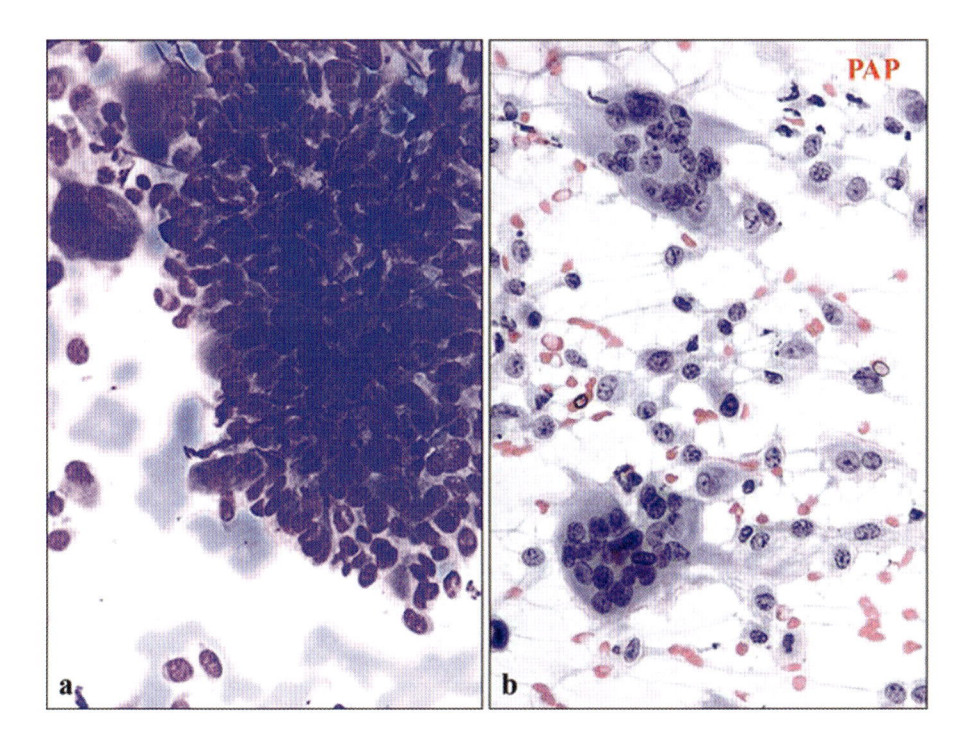

Figure 36.17 Giant cell tumor of tendon sheath. (**a**) The cellular aspirate is composed of loose clusters of spindle-shaped cells and a few multinucleated giant cells. (**b**) The nuclei of the multinucleated and single cells are identical, characterized by a regular round shape, fine chromatin, and single or a few nucleoli.

Differential diagnosis

The differential diagnosis with low-grade myxofibrosarcoma is very difficult. Both tumors are composed of abundant myxoid background and atypical spindle cells. The misdiagnosis has no clinical consequences, however, because both tumors are treated with surgery.

Intramuscular myxoma is usually less cellular, the nuclei show less atypia, and the background has almost no capillary structures.

So-called fibrohistiocytic tumors

Giant cell tumor of the tendon sheath

Giant cell tumor of the tendon sheath presents as a small, circumscribed, lobulated mass in the hand and less often in the arms and legs. The tumor is almost always benign.

Aspirates are cellular, with loose clusters of spindle-shaped cells and varying numbers of multinucleated giant cells, either of which can contain hemosiderin. Spindle-shaped cells have regular, often eccentric nuclei, fine even chromatin, and single nucleoli. Giant cells contain numerous bland nuclei similar to other cells (Figure 36.17).[27,28]

Differential diagnosis

Giant cell tumors of bone rarely occur in the small bones of the hands or feet. The aspirates, similarly composed of mononuclear spindle cell and multinucleated giant cells, show the peripheral adherence of giant cells to spindle cells. Histiocyte-like cells, hemosiderin-laden macrophages, and foamy macrophages are less common.[29]

Deep benign fibrous histiocytoma

Deep benign fibrous histiocytoma usually presents as a slowly growing, painless tumor of deep subcutaneous tissue.

The cellularity of aspirates depends on fibrosis, which is mild in early, but marked in older lesions. Most cells are spindle-shaped, fibroblast-like, admixed with variable amounts of histiocyte-like rounded cells. Atypia is minimal at best. The background consists of degenerated cells, lipophages, sidero-phages, and Touton giant cells in more superficial lesions (Figure 36.18).[30]

Differential diagnosis

In nodular fasciitis, the lesions grow rapidly, clinically simulating sarcoma. Aspirates are more pleomorphic and have few or no histiocytes.

Undifferentiated high-grade pleomorphic sarcoma

Undifferentiated high-grade pleomorphic sarcoma (UPS), also known as malignant fibrous histiocytoma (MFH), is defined as a high-grade sarcoma without a definable line of differentiation. Therefore, UPS is a diagnosis made by exclusion. Contrary to the name used for so many years, MFH shows no true histiocytic differentiation. After generous sampling, careful microscopic examination, and the use of ancillary studies, some tumors diagnosed in the past as MFH would today be categorized as pleomorphic leiomyosarcoma (LMS), rhabdomyosarcoma (RMS), liposarcoma, osteosarcoma, carcinoma, melanoma, or lymphoma. The incidence of MFH/UPS has fallen lately; however, it is still the most common sarcoma in adults >40 years.

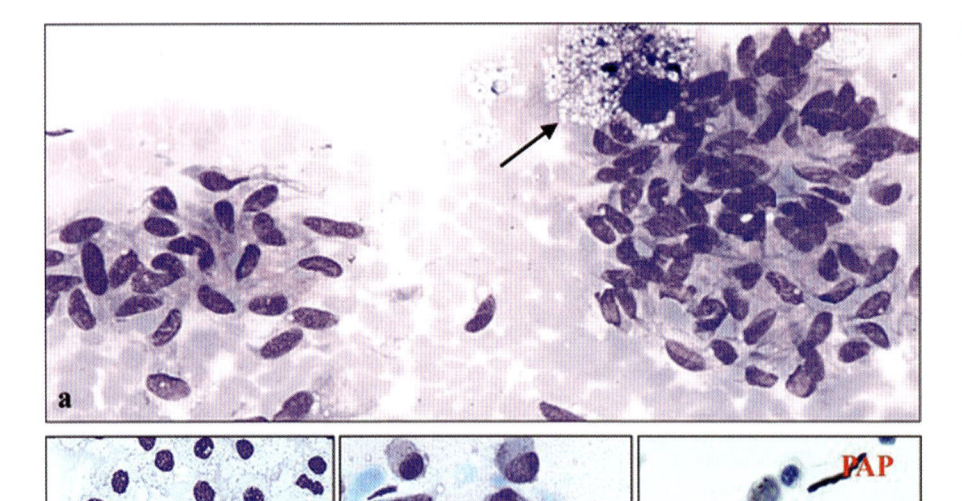

Figure 36.18 Deep benign fibrous histiocytoma. (**a**) Numerous spindle-shaped fibroblast-like cells and a single siderophage *(arrow)*. Atypia is absent or minimal. (**b**) Lipophages. (**c**) Round cells, some with pigment. (**d**) Touton giant cell.

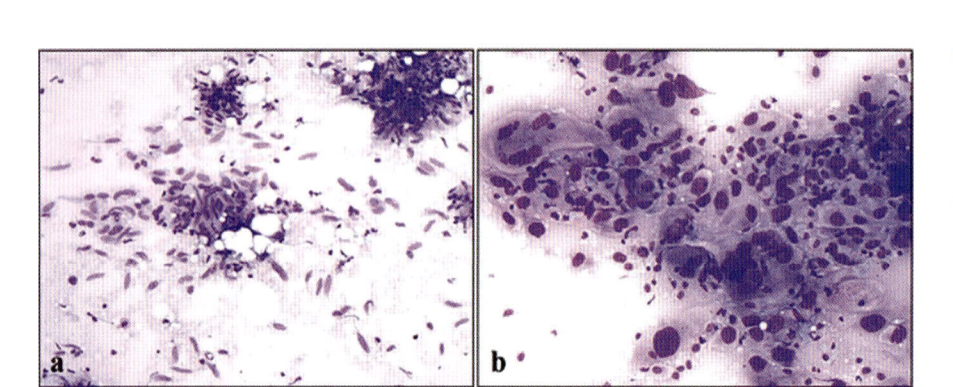

Figure 36.19 Undifferentiated high-grade pleomorphic sarcoma. Cells vary from relatively unremarkable spindle-shaped cells with scanty tapered cytoplasm, as seen in (**a**), to highly pleomorphic bizarre cells with abundant, well-defined, and dense basophilic cytoplasm, as seen in (**b**). Abnormal mitotic figures and necrosis were seen elsewhere (not shown).

UPS is most common during the sixth and seventh decades and is usually deep seated. About 80% of tumors arise in the extremities, especially in thigh, and the rest in the trunk wall and retroperitoneum.

Diagnosis by FNA is difficult owing to sampling errors. Aspirates are cellular, with numerous highly pleomorphic cells, bizarre tumor giant cells, atypical mitotic figures, and necrosis (Figure 36.19). The finding of spindle-shaped, round giant cells, osteoclast-like giant cells, and inflammatory cells is the most consistent feature that allows the identification of storiform/pleomorphic, giant cell, and inflammatory variants of MFH. The myxoid tumors have marked myxoid background, with spindle-shaped cells and less frequently round and giant cells. Unless a specific line of differentiation is provided by adequate sampling or ancillary methods, these aspirates should be designated as pleomorphic high-grade sarcomas. If the cytopathologist is asked for a diagnosis at this point, additional sampling should be requested.[31–33]

Differential diagnosis

If other types of pleomorphic sarcomas are generously sampled they frequently show better-differentiated areas: lipoblasts in pleomorphic liposarcoma, rhabdomyoblasts in pleomorphic RMS, and cells characteristic of classic LMS in dedifferentiated LMS.

Epithelial differentiation in anaplastic carcinoma should be suspected whenever there are small molded groups of tumor cells with an "owls' eyes" appearance. Carcinoma should be confirmed with pancytokeratin.

Most of the cells in anaplastic large cell lymphoma (ALCL) are round with dark-blue cytoplasm focally resembling Reed–Sternberg cells. ALCL should be confirmed with positive staining for CD30, and variably for EMA and ALK.

The diagnosis of melanoma should be confirmed with positive staining for HMB45, S100 protein, MART1, or other markers of melanocytic differentiation.

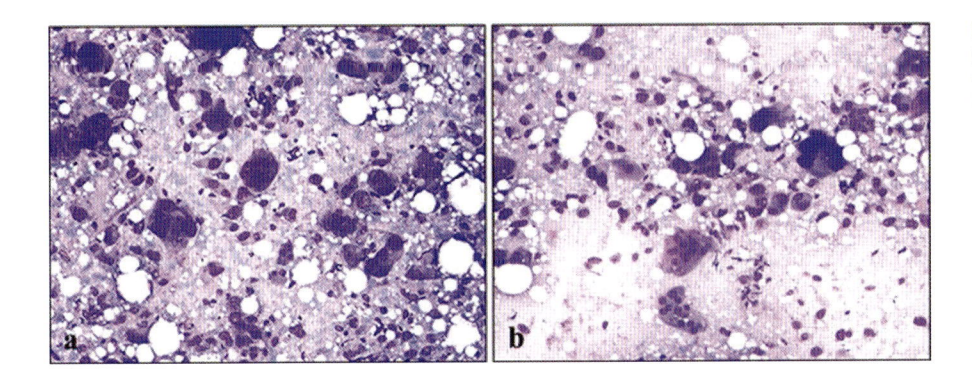

Figure 36.20 Undifferentiated high-grade pleomorphic sarcoma with giant cells. (**a,b**) Numerous osteoclast-like giant cells often lack the cytologic features of malignancy.

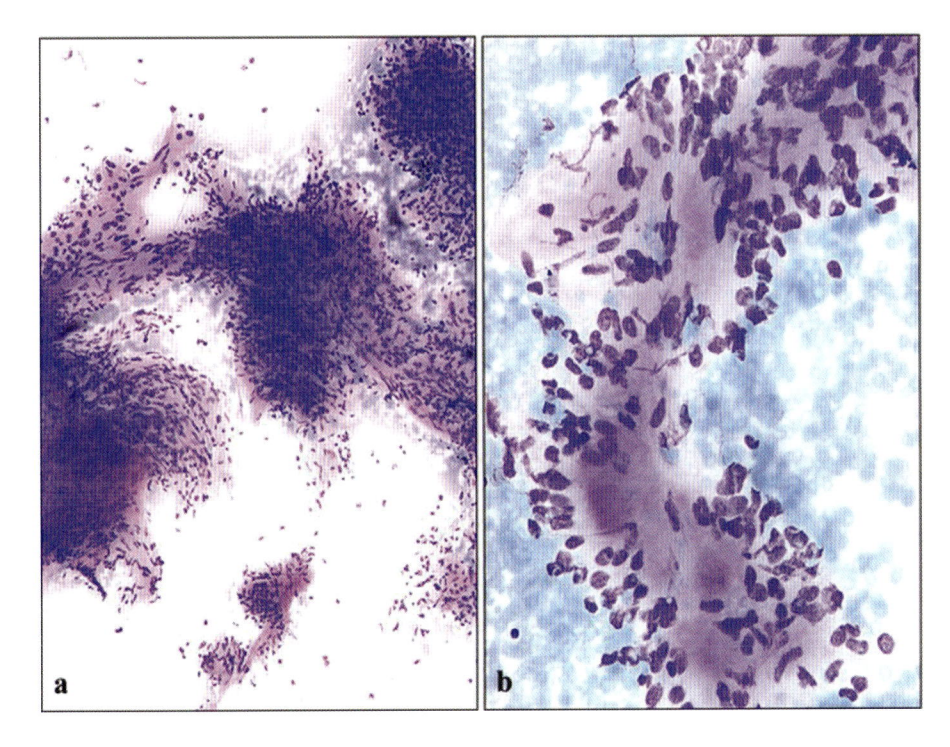

Figure 36.21 Schwannoma. (**a**) Irregular large tissue fragments contain numerous bland, spindle-, comma-, and bullet-shaped nuclei. (**b**) Verocay body with elongated neoplastic nuclei lined up in a palisade. Cytoplasm blends imperceptibly with the adjacent collagen.

UPS with giant cells

UPS with giant cells is defined as MFH with a prominent component of osteoclast-type giant cells. The giant cell component most often lacks the cytologic features of malignancy (Figure 36.20).

Nerve sheath, neuroectodermal and neural tumors

Schwannoma

Schwannoma is a benign tumor of peripheral nerves composed exclusively of Schwann cells, with nerves at the periphery of the tumor. The tumor is most common in adults as a subcutaneous or less often an intramuscular tumor. The anatomic distribution is very wide, most often along the flexor surface of the extremities, neck, posterior spinal roots, and cerebellopontine angle. Aspiration is often painful and can yield nondiagnostic material when the needle hits collagenized areas, or myxoid or cystic degeneration.

Aspirates show numerous bland, spindly, comma- and bullet-shaped cells inside the large irregular tissue fragments. Verocay bodies are occasionally seen (Figure 36.21). Contrary to MPNSTs, schwannomas show diffuse and strong positivity for S100 protein.

Ancient schwannoma

Ancient schwannoma (AS) is a variant featuring prominent regressive changes. AS usually grows deep in the retroperitoneum or mediastinum. Because of its large size, cystic degeneration is common.

Aspirates exhibit marked anisokaryosis and large, dark nuclei, often leading to misinterpretation of these benign tumors as malignant, commonly as MPNSTs or LMSs. Atypical nuclei in AS, however, show degenerated ("smudged") and not crisp chromatin, a complete absence of mitotic figures, and larger nuclear vacuoles (Figure 36.22).[34–37]

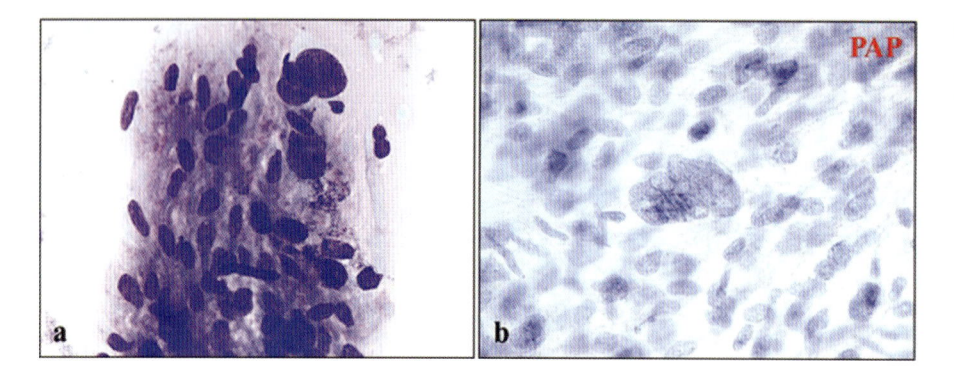

Figure 36.22 Ancient schwannoma. (**a,b**) Nuclei are dark and of varying sizes, often leading to an erroneous diagnosis of malignancy. Chromatin, however, is degenerated ("smudged") and not crisp as it is in MPNSTs (right-hand panel from Figure 36.27, is shown for comparison). Mitotic figures are also not seen.

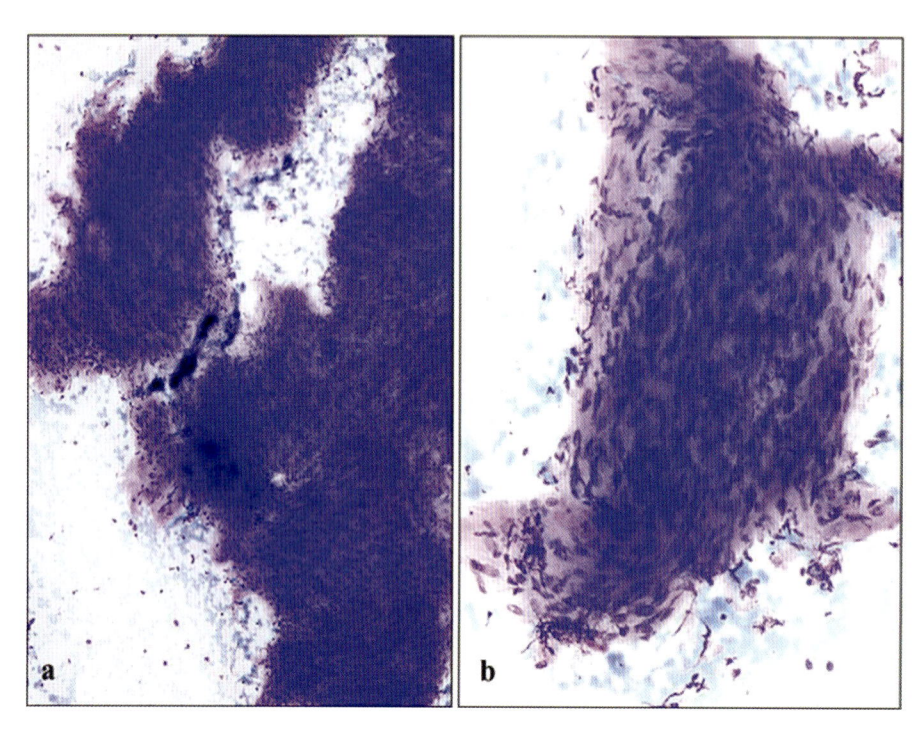

Figure 36.23 Neurofibroma. (**a**) Most of the cells are in cohesive clusters; very few are dispersed. (**b**) Cells are spindle-shaped and have elongated, wavy, irregular nuclei with pointed ends. Chromatin is bland; nucleoli are not seen or are very inconspicuous.

Differential Diagnosis

In MPNSTs, atypia is "real," with crisp, coarser, and darker chromatin. Tumor fragments are homogeneously cellular, and no Antoni A areas can be seen. MPNST is only focally positive for S100 protein.

In LMS, the nuclei are usually plumper, and the cytoplasm stains with desmin and SMA.

Neurofibroma

Neurofibroma (NF) is a tumor of peripheral nerves composed of all nerve components: Schwann cells, nerve axons, fibroblasts, and perineural cells with nerve often penetrating the center of the NF. This tumor presents anywhere in the body. Superficial tumors are small and protrude from the skin. Deeper tumors are larger, often plexiform, and frequently arise in the orbit, neck, back, or in the inguinal region, often in neurofibromatosis type 1 (NF1). Malignant change can occur, especially in patients with NF1.

NF, which is less commonly aspirated than schwannoma, shows variable cellularity owing to the varying content of the collagen fibers. Most cells form cohesive clusters; a few are dispersed. The cells are spindle-shaped, with elongated nuclei and exhibit characteristics of Schwann cells (wavy, irregular nuclei with pointed ends) and fibroblasts. The chromatin is bland, and the nucleoli are not prominent. The background frequently shows a myxoid substance (Figure 36.23).[24,34]

Extracranial meningioma

Extracranial meningiomas (EMs) are rare tumors most often encountered in the skin or soft tissue of the scalp or along the vertebral axis.

Cellular smears are composed of spindle-shaped cells in concentric whorls and scattered psammoma bodies characteristic of meningioma.[38] On occasion, intranuclear cytoplasmic inclusions (Figure 36.24), as well as

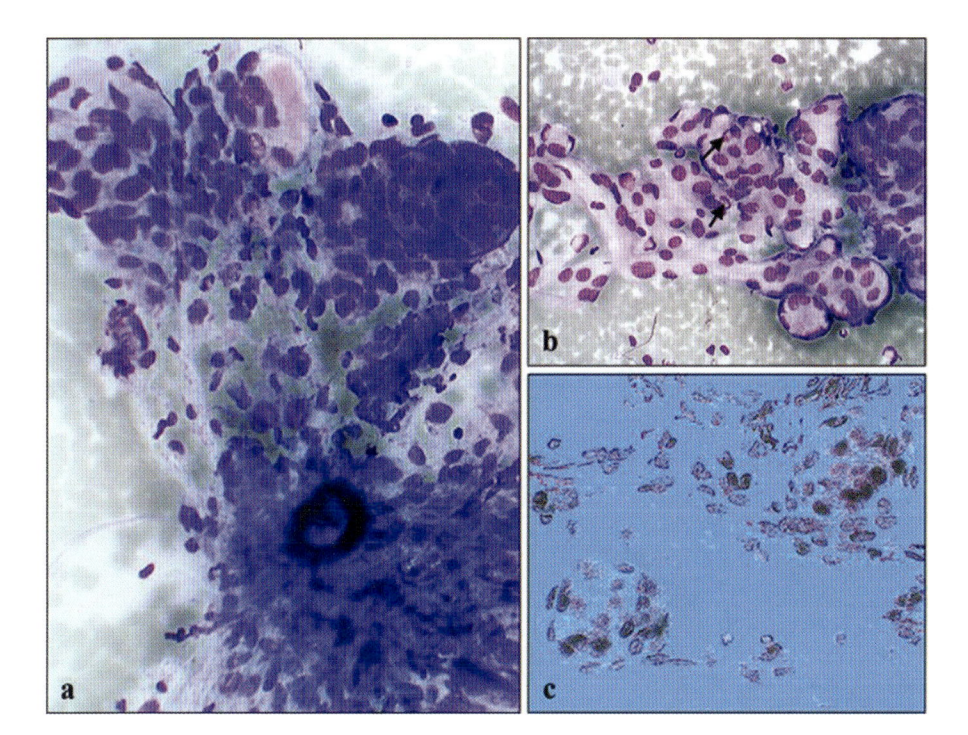

Figure 36.24 Extracranial meningioma. (**a**) Oval-shaped cells in concentric whorls are at the top; a psammoma body is at the bottom. (**b**) Intranuclear cytoplasmic inclusion *(arrows)*. (**c**) Concentric whorls as seen on Pap stain.

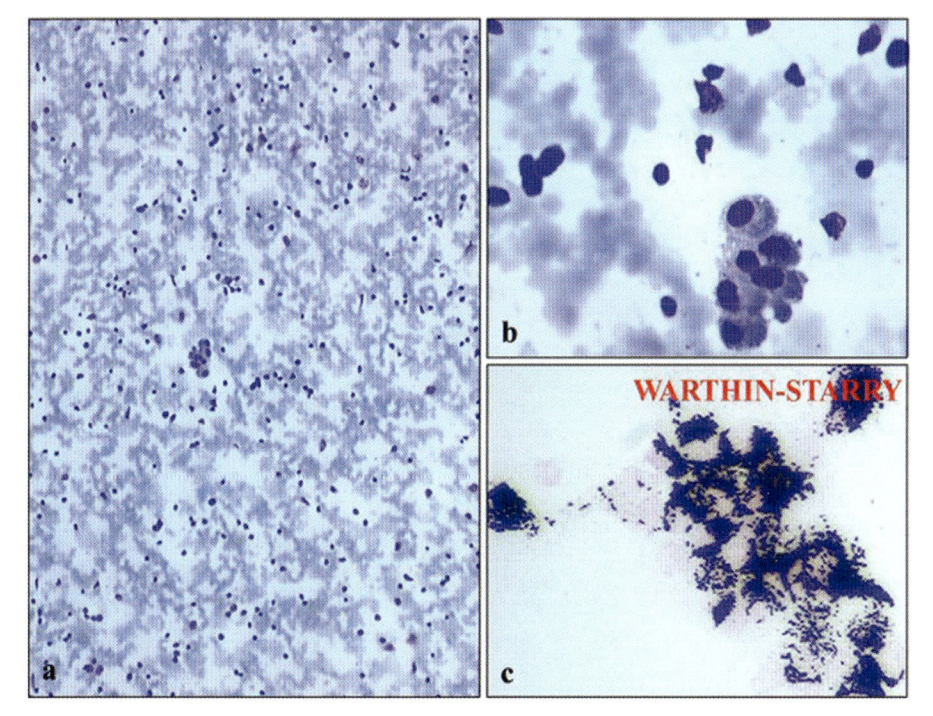

Figure 36.25 Melanotic neuroectodermal tumor of infancy. (**a,b**) Dual population of larger, cuboidal, pigmented epithelial cells, and smaller, rounded, nonpigmented immature neuroblast-like cells. (**c**) A Warthin–Starry stain performed at pH 3.2 highlights melanin granules in the larger cuboidal cells.

binucleated cells with wispy cytoplasmic extensions, are also noted. Immunoperoxidase studies show focal positivity for EMA.[39]

Melanotic neuroectodermal tumor of infancy (retinal anlage tumor, melanotic progonoma)

This uncommon tumor develops during the first year of life and presents as a protruding mass in the upper or lower jaw. Very rarely, the tumor appears in unusual sites, such as the soft tissues of extremities, the long bones, epididymis, mediastinum, and brain.

Aspirates show scant cellularity and are composed of a dual population: larger, cuboidal, pigmented epithelial cells and smaller, rounded, nonpigmented, immature neuroblast-like cells (Figure 36.25).

Myxopapillary ependymoma of soft tissues

This uncommon tumor presents as a long-standing mass of the subcutaneous soft tissue dorsal to the sacrum or coccyx, or

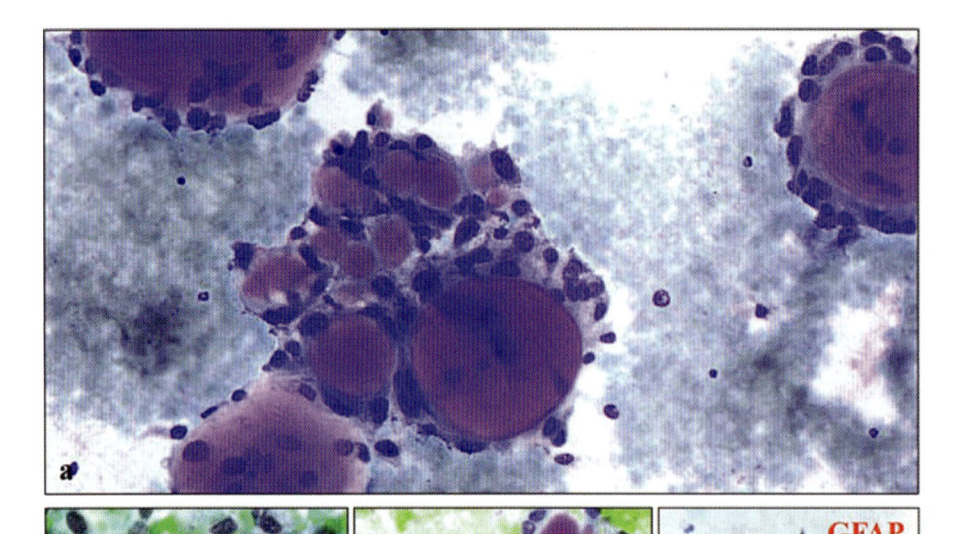

Figure 36.26 Myxopapillary ependymoma. (**a**) Globular structures lie singly or in a grape-like fashion. Their core is acellular, intensely eosinophilic, and mucinous, covered by one to three layers of flattened, cuboidal, or low cylindric mitotically inactive cells. (**b**) The few dispersed cells have nuclei that are oval to spindle-shaped. (**c**) Focal nuclear atypia is frequently seen and has no prognostic value. (**d**) Cytoplasm stains with GFAP (cytospin preparation).

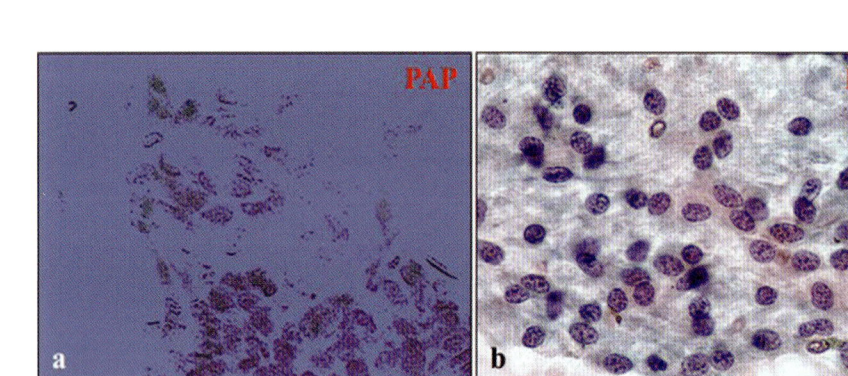

Figure 36.27 MPNST. (**a,b**) Most of the nuclei are oval, with an increased N/C ratio and uneven chromatin. Palisading is not seen, and nuclear membranes are not smooth. Ancient schwannoma (left panel from Figure 36.22) is shown for comparison.

in the deep soft tissue anterior to the sacrum and posterior to the rectum. Except for larger masses, most tumors are encapsulated and can be easily separated from the fascia overlying the sacrum or coccyx.

Cellular aspirates consist of dominant tissue fragments and myxoid material with a moderate number of dissociated cells in the background. Tissue fragments represent the remnants of papillae and are composed of globular structures lying singly or imitating a bunch of grapes. Their core is acellular, intensely eosinophilic, and mucinous, covered by one to three layers of flattened, cuboidal or low cylindrical mitotically inactive cells (Figure 36.26).[40,41]

Differential diagnosis

Adenocarcinoma, especially adenoid cystic carcinoma does not show elongated GFAP-positive ependymal cells. In addition, nuclear atypia is usually, but not always, more prominent in carcinoma.

Malignant peripheral neural sheath tumor (malignant schwannoma)

Most of these tumors are associated with deep-seated nerves and can arise almost anywhere in the body. In neurofibromatosis, the most common sites of origin are the head and neck, retroperitoneum, trunk, and limb girdles. Sporadic cases are more common in the extremities and limb girdles and appear later in life.

The cellularity of aspirates varies and is usually abundant in higher-grade tumors. Similar to the histology, densely cellular fascicles are often mixed with hypocellular myxoid zones, heterologous elements, and necrosis. Cells are single and focally organized in small fascicles.

Most of the cells are spindle-shaped to oval in lower-grade and rounder in higher-grade lesions. The nucleus-to-cytoplasm (N/C) ratio is increased, with eccentric nuclei and tapering cytoplasm sometimes imparting a comma-shaped appearance. Nuclear palisading is rare (Figure 36.27). S100 protein is focally positive in less than one-half of all tumors.[42,43]

Differential diagnosis

Cellular schwannoma is also more diffusely positive for S100 protein. SS stains positively with pancytokeratin, EMA, and CD99.

Skeletal muscle tumors

Rhabdomyoma

A rhabdomyoma is a rare benign tumor with skeletal muscle differentiation. Of all three subtypes (i.e., adult, fetal, and genital), the adult subtype is most common and because of its location in the head and neck region, it is also the most commonly aspirated.

Aspirates are composed of large, rounded or polygonal cells with eosinophilic granulated and often vacuolated cytoplasm. The nuclei are small with prominent nucleoli. Cross-striations, often present on tissue sections, are not seen, but naked nuclei are common in the background (Figure 36.28).[44,45]

Differential diagnosis

In hibernomas, the cells and cytoplasmic vacuoles are smaller.

In GCTs, the cells have no cytoplasmic vacuoles, and the cytoplasm stains diffusely with S100 protein and is negative for muscle markers.

Rhabdomyosarcoma

RMSs occur mostly in children and young adults. Most tumors do not arise in the extremities, where the bulk of skeletal muscle resides. RMSs in children are classified according to prognosis (Table 36.4).[46,47]

Embryonal RMS

These tumors most often occur during the first decade, in the genitourinary tract and head and neck, and less often elsewhere.

Aspirates are moderately cellular; the predominant architectural pattern features large tissue fragments with abundant eosinophilic material and varying numbers of dissociated cells.[48] Most of the cells are primitive and round or spindle-shaped, with round nuclei. Mature rhabdomyoblasts (e.g.,

Table 36.4 Classification of rhabdomyosarcomas according to prognosis

Prognosis/superior prognosis	Frequency (%)
Botryoid RMS	6
Spindle cell RMS	3
Intermediate prognosis:	
Embryonal RMS	49
Poor prognosis:	
Alveolar RMS	31
Undifferentiated sarcoma	
Diffuse anaplasia (alveolar or embryonal)	
Subtypes for which the prognosis cannot presently be evaluated	
RMS with rhabdoid features	

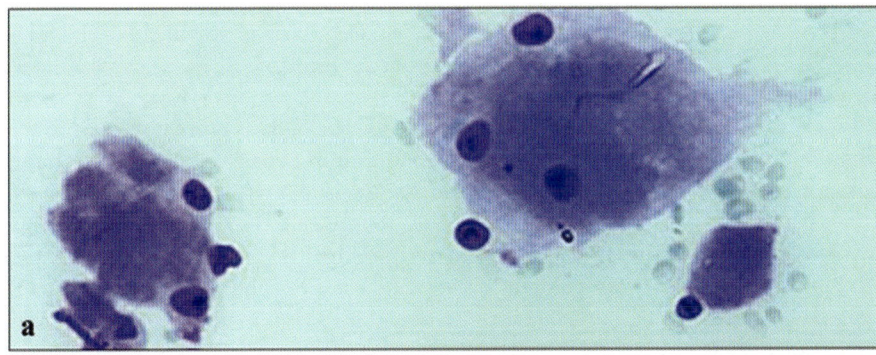

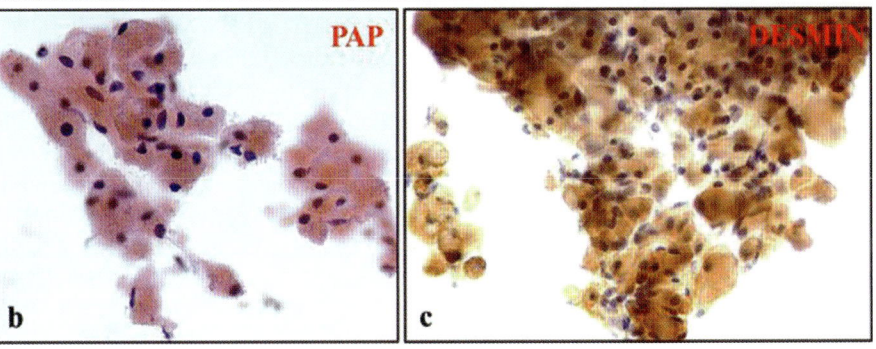

Figure 36.28 Adult rhabdomyoma. (**a,b**) Large, rounded to polygonal cells with granular cytoplasm are characteristic. Nuclei are small; cross-striations often present on tissue sections are not seen. (**c**) Cytoplasm is positive for desmin (cytospin preparation).

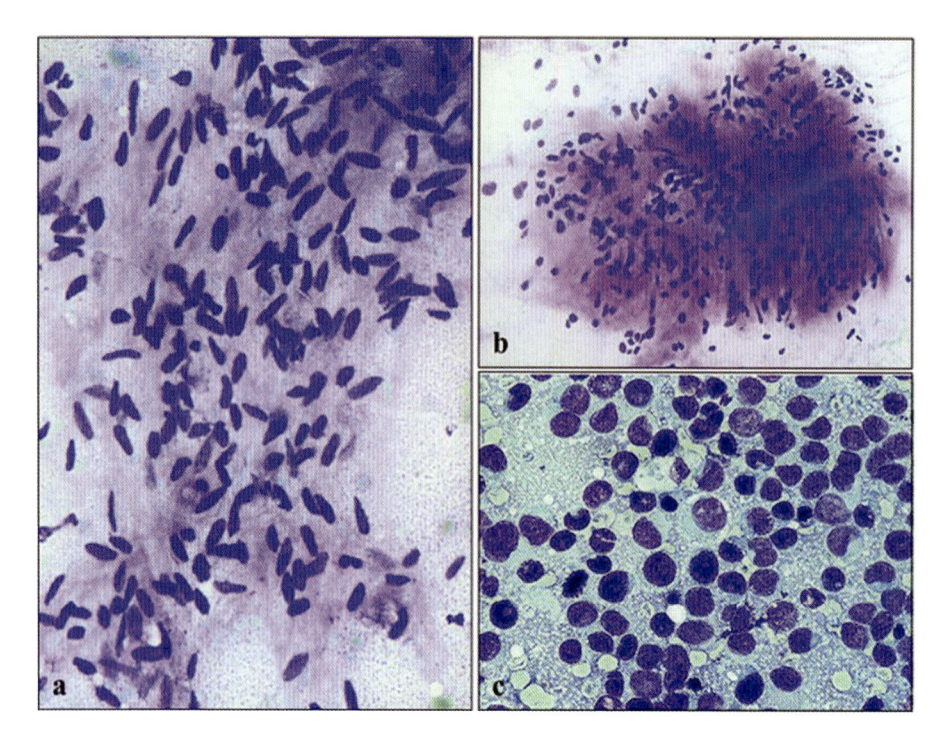

Figure 36.29 Rhabdomyosarcoma. (**a**) The spindle cell type is composed of loose clusters of cells with spindle-shaped to oval, bland-appearing nuclei, scanty cytoplasm, and abundant intercellular substance. (**b**) The embryonal type often presents as tissue fragments with abundant eosinophilic intercellular material and few dissociated cells. Most of the nuclei are round to elongated, and the cytoplasm is indistinct. Mature rhabdomyoblasts are not seen. (**c**) The alveolar type is often composed of numerous small- to medium-sized and oval to pear-shaped dissociated cells. Nuclei exhibit coarse chromatin and prominent nucleoli. The cytoplasm is friable, forming numerous cytoplasmic fragments in the background. Multinucleated cells often seen in tissue sections were seen in other fields (not shown).

tadpole, strap, and ribbon-shaped cells) are rare (Figure 36.29b). An alternative, but less common morphologic picture consists of completely dissociated rhabdomyoblasts in various stages of maturation. The same picture can also be seen in some cases of alveolar RMS. Aspirates of spindle cell variant can present with atypical spindle-shaped cells only and with very rare rhabdomyoblasts. Background consists of variable amounts of myxoid material (Figure 36.29a).

Alveolar RMS

These tumors occur most often in the head and neck and extremities during the second decade. Aspirates are very cellular and are composed of numerous predominantly small to medium, round, oval, or pear-shaped cells with coarse chromatin and prominent nucleoli. The cells are either completely dissociated or make many chance formations. Mitotic figures are common. The cytoplasm is vacuolated and friable. The background consists of numerous naked nuclei and detached cytoplasmic fragments (Figure 36.29c). Multinucleated cells are seen occasionally.

Differential diagnosis

Alveolar RMS versus embryonal RMS: aspirates of RMS exhibit a variety of architectural as well as cytologic pictures even within the same subtype; therefore, a reliable subclassification as either an embryonal or alveolar subtype cannot be achieved by morphology alone. The embryonal subtype can be suggested when large tissue fragments are associated with abundant eosinophilic material and small, tightly packed cells with oval nuclei. When definitive treatment is based solely on FNA smears, all ancillary methods (e.g., ICC and cytogenetic and molecular genetic tests) have to be applied. In the authors' experience, it is possible to differentiate alveolar RMS from embryonal RMS in about 80% of cases when morphology and ancillary methods are combined.[49–51]

Cells in Ewing sarcoma/PNET are smaller and more regular, with less abundant and looser cytoplasm. Multinucleated and spindle-shaped cells are not seen. The cells stain with CD99 and not with skeletal muscle markers.

LMS versus spindle cell RMS: LMS is rare in children and is negative for skeletal muscle markers, such as MyoD1 and myogenin.

The monophasic fibrous type of SS is more common in the lower extremities later in life and is positive for pancytokeratin and EMA.

Desmoplastic small round cell tumor more commonly presents as large intra-abdominal mass in younger males and shows different cytogenetic translocation and immunoprofile.

Pleomorphic rhabdomyosarcoma

Most of these tumors occur during the sixth decade, in deep soft tissues of the lower extremities, especially the thigh. Aspirates are composed of numerous atypical spindle cells and larger pleomorphic and polygonal rhabdomyoblasts with abundant basophilic cytoplasm and multinucleated tumor cells (Figure 36.30).

Differential diagnosis

Other pleomorphic sarcomas are negative for skeletal muscle markers such as MyoD1 or myogenin.

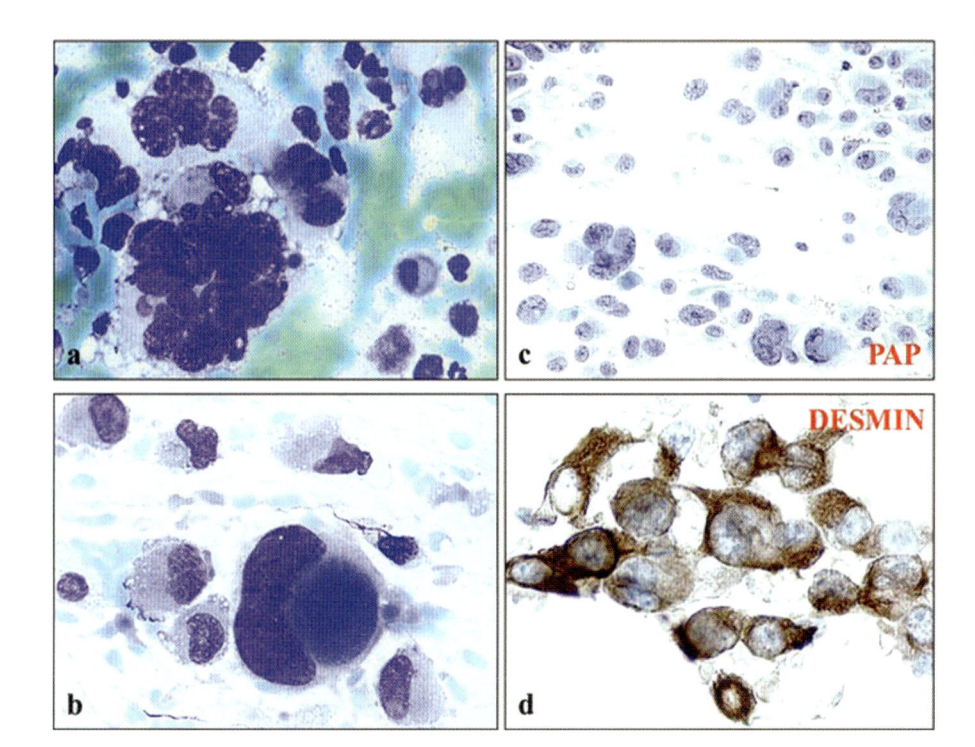

Figure 36.30 Pleomorphic rhabdomyosarcoma. (**a**) In this field, the aspirate consists of multinucleated bizarre tumor cells and numerous, atypical, poorly formed pleomorphic rhabdomyoblasts. Poorly formed rhabdomyoblasts can be easily overlooked, leading to a diagnosis of undifferentiated pleomorphic sarcoma. (**b,c**) In other fields, the rhabdomyoblasts are well formed, with large amounts of dense basophilic cytoplasm. (**d**) The cytoplasm is positive for desmin (cytospin preparation).

Smooth muscle tumors

Leiomyosarcoma

Many LMSs are intra-abdominal, located in the retroperitoneum, where they account for 30% to 50% of all sarcomas. They arise from larger blood vessels, especially from the vena cava, and often grow into large masses >10 cm in diameter. One has to consider that most intraperitoneal sarcomas previously classified as LMS are now considered gastrointestinal stromal tumors (GISTs).

The remaining LMSs are divided into subcutaneous/deep soft tissue, cutaneous, and vascular groups. Subcutaneous or deep soft tissue LMSs arise in the limbs, especially in the thigh in middle-aged patients. Cutaneous LMSs arise in younger adults in limbs, are often painful, and frequently recur. Vascular LMSs arise in older adults adjacent to blood vessels with muscular walls, in particular the inferior vena cava and the large veins of the lower extremity.

Aspirates of low-grade LMSs (most of them are of the spindle cell subtype) are less cellular with deceptively bland appearance. Low-grade malignancy is suggested by anatomic location alone: unless the tumor is attached to the uterus, any deep-seated smooth muscle tumor should be considered malignant, unless proved otherwise.

Morphologic features frequently include nuclear atypia, mitotic activity, and irregular, coarse chromatin. Fascicles of monomorphic spindle cells very often exhibit cigar-shaped nuclei, with perinuclear vacuoles. Chromatin is finely granular, and nucleoli are not prominent. Palisading of nuclei is common, and single cells are rare (Figure 36.31a,b).

Higher-grade LMSs can be epithelioid and pleomorphic. Numerous single cells and few loose clusters of highly atypical cells exhibit coarse chromatin, prominent nucleoli, and variable amounts of cytoplasm. Myxoid or necrotic material is often seen in the background (Figure 36.31c,d,e). In the epithelioid variant, there are many tumors in the differential, including epithelioid GIST, angiosarcoma, MPNST, epithelioid sarcoma, adenocarcinoma, melanoma, and mesothelioma. As in any case of a poorly differentiated tumor, searching for better-differentiated areas can help with the diagnosis.[52–55]

Vascular tumors

Angiosarcoma

An angiosarcoma is a rare malignant vascular neoplasm that can arise in any part of the body. The most common location is in the skin of the head and neck. Other locations include soft tissues and almost any organs.[56]

Published studies describing the cytology of angiosarcomas are mainly case reports. There are only four reports that consider more than ten cases.[57–60] A reliable diagnosis requires an immunocytochemical investigation with CD31, CD34, and Factor VIII, the last of which is the least sensitive. In the epithelioid variant, cytokeratin is usually positive.[61]

Morphologically, angiosarcomas are divided into classic and epithelioid subtypes; the classic subtype can be well, moderately, or poorly differentiated. Both subtypes show variable cellularity on FNA.

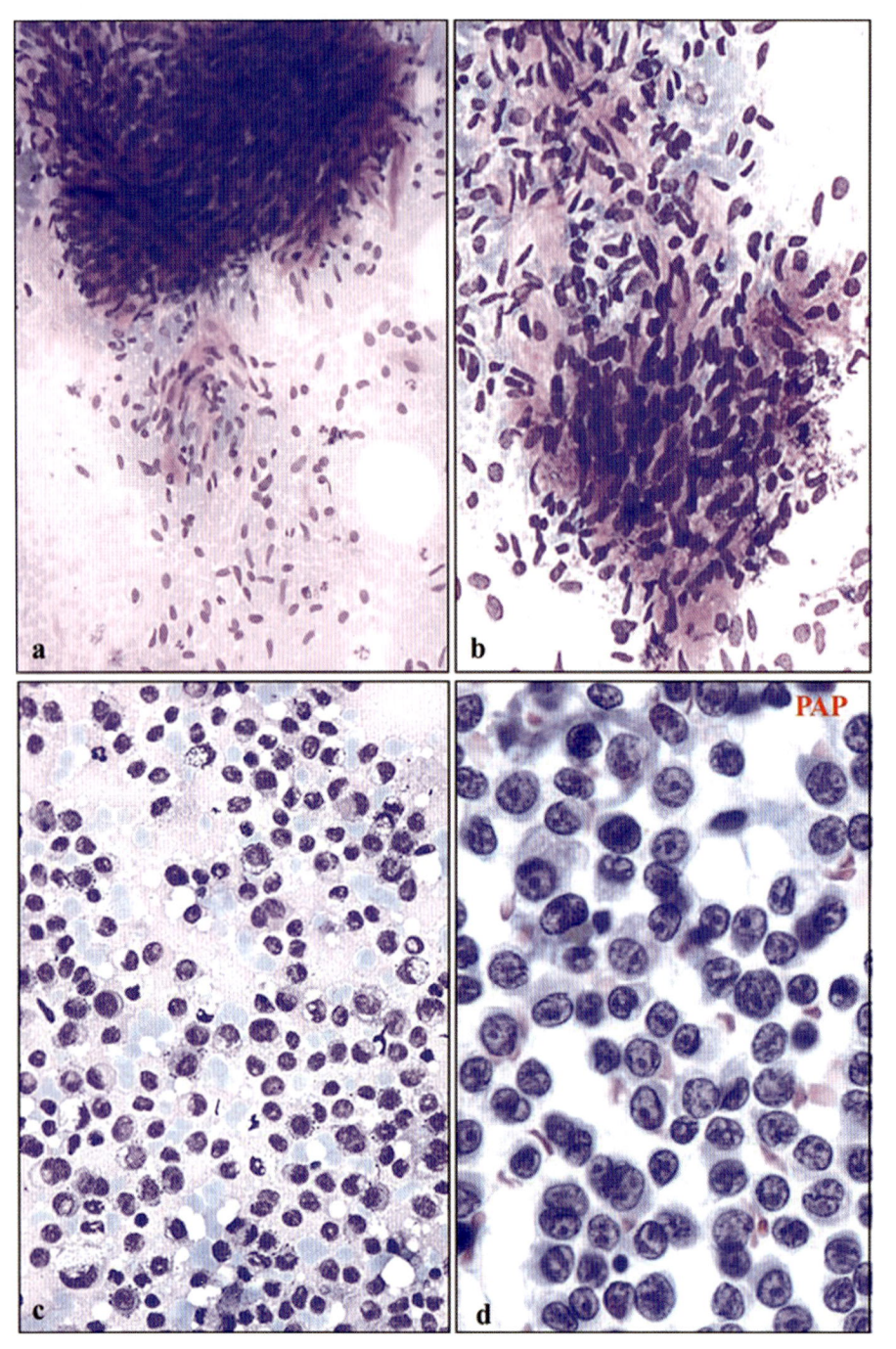

Figure 36.31 Leiomyosarcoma. (**a**) Lower-grade tumors are deceptively bland. Cigar-shaped nuclei with occasional perinuclear vacuoles reveal the smooth muscle origin. Location, not morphologic details as seen on aspirates, reveals the biologic potential, however. Any deep-seated smooth muscle tumor not attached to the uterus is potentially malignant until proved otherwise. (**b**) An intermediate-grade tumor has rounder, more irregular nuclei and coarser chromatin. (**c,d**) A high-grade tumor of epithelioid type shows numerous dispersed round cells with irregular, focally indented nuclei, coarse chromatin, and prominent nucleoli. (**e,f**) A high-grade tumor of pleomorphic type shows numerous mono- and multinucleated pleomorphic cells forming loose clusters. The cytoplasm is positive for SMA (cytospin preparation).

Angiosarcoma, classic type

The variable morphology in the cytology reports reflects the diverse morphology seen in this tumor's histology. In classic angiosarcoma, the authors report a variety of round, oval, spindle, and epithelioid cells. Pleomorphism and hyperchromasia are often pronounced; nucleoli can be conspicuous, large, or multiple (Figure 36.32a,b). Many authors mention nuclear folds, indentations, grooves, and mitotic figures. Cytoplasm, which is scant to moderate, can contain hemosiderin deposits, multiple small vacuoles, or intracytoplasmic lumina. Erythrophagocytosis is seen infrequently. In addition to dissociated cells, pseudoacinar and rosette-like formations, papillary structures and well-formed small vessels can be seen. Most authors call the latter vasoformative structures and consider them to be pathognomonic for angiosarcoma.

Differential diagnosis

Classic low-grade angiosarcomas can be mistaken for benign lesions, especially in smears with low cellularity. The bland spindle cells resemble fibroblasts, and the lesion can be

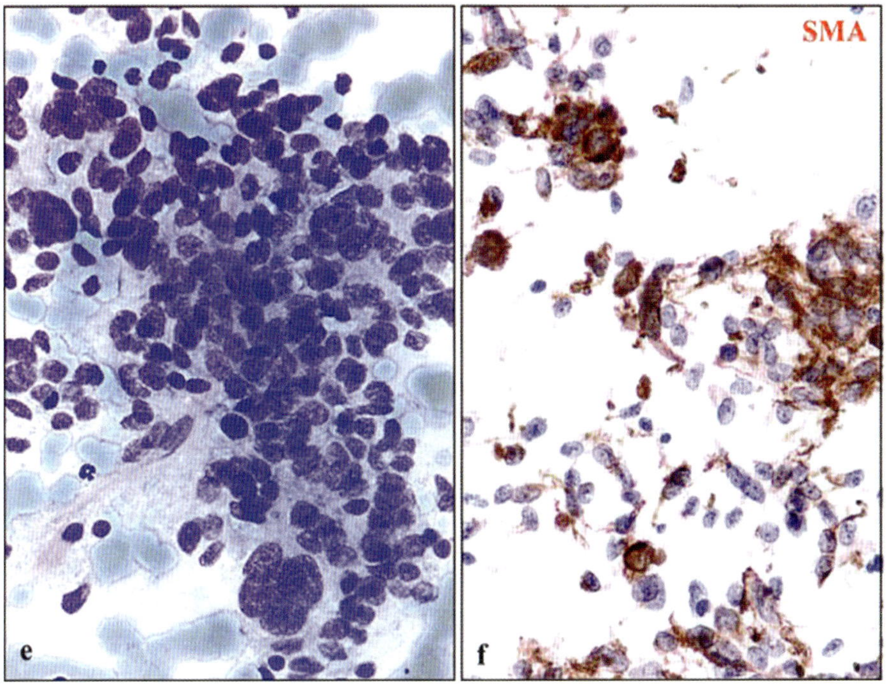

Figure 36.31 (cont.)

mistaken for granulation tissue.[62] When low cellularity is combined with an abundance of blood, hemangioma becomes part of the differential.

In the authors' opinion, it is very difficult to distinguish a low-grade angiosarcoma morphologically from Kaposi's sarcoma or dermatofibrosarcoma protuberans on FNA biopsy. Immunocytochemistry can help; however, because staining for Factor VIII and CD31 is inconsistent in angiosarcomas, this ancillary technique might not be enough.

Classic high-grade angiosarcoma is easily recognized as malignant and most often also as a sarcoma; however, the differentiation from other spindle cell sarcomas can be difficult. Sarcomas in the differential are MPNST, leiomyosarcoma, low-grade myxofibrosarcoma, monophasic SS, and spindle cell liposarcoma. ICC may resolve the problem.

Angiosarcoma, epithelioid

This subset of angiosarcoma is usually deeper and more rapidly growing than the conventional type. These tumors can express both endothelial and epithelial markers (keratins), creating potential confusion with carcinoma.

Smears of epithelioid angiosarcoma are composed of large epithelioid cells averaging three to four times the size of lymphocytes. Single or loosely cohesive cells show moderate to marked nuclear pleomorphism. Nuclei are single, hyperchromatic, and round, with smooth nuclear borders, frequent mitotic figures, and prominent nucleoli (Figure 36.32c). All features described in the classic subtype are also present. Erythrophagocytosis is sometimes seen.[63]

Differential diagnosis
The differential diagnosis includes melanoma, carcinoma, high-grade lymphoma, epithelioid sarcoma, and all epithelioid variants of various sarcomas.[64] ICC may resolve the problem.

Epithelioid hemangioendothelioma

Epithelioid hemangioendothelioma (EHE) arises from deep and superficial soft tissues of the extremities, where it presents as a painful nodule. This rare tumor occurs in all age groups except during early childhood. More than one-half of tumors originate in vessels, usually in small veins.

Aspirates are cellular, with predominantly single cells focally forming small loose aggregates and rosette-like structures. The oval to polygonal cells are monotonous with round to oval, occasionally kidney-shaped nuclei. Binucleation is not uncommon, and mitotic figures are rare. The chromatin is uneven, with one to two nucleoli. Abundant dense eosinophilic cytoplasm can contain a few sharply demarcated vacuoles, which can distort the nucleus (Figure 36.33). Cells stain strongly with endothelial markers CD31 and Factor VIII and are negative for EMA. Up to one-quarter of cases are positive for pancytokeratin.[65,66]

Differential diagnosis
Metastatic malignant melanoma exhibits more prominent nuclear atypia and a higher number of mitotic figures; this entity is positive for melanotic and negative for endothelial markers.

Metastatic carcinoma shows a higher degree of nuclear atypia and a higher number of mitotic figures. Metastatic carcinoma is positive for EMA and negative for endothelial markers.

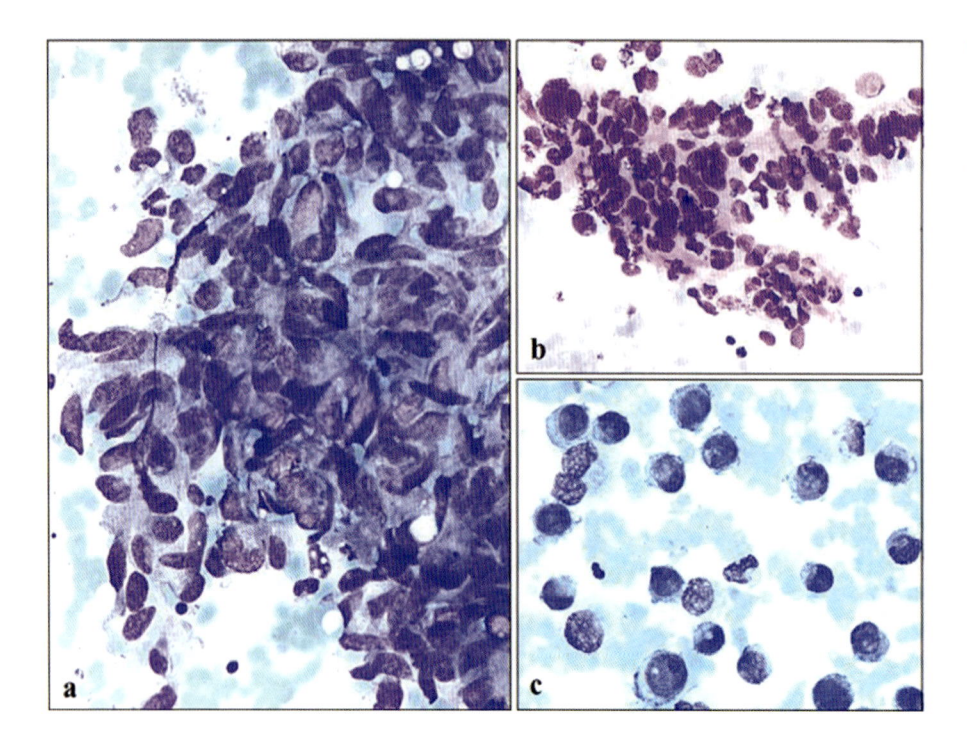

Figure 36.32 Angiosarcoma. (**a,b**) Classic high-grade type. Smear a is composed of moderately atypical spindle cells; smear b of severely atypical pleomorphic cells. (**c**) The epithelioid type is composed of dispersed large epithelioid cells. Nuclei are eccentric, smooth, and hyperchromatic, and nucleoli are prominent. The cytoplasm is scant.

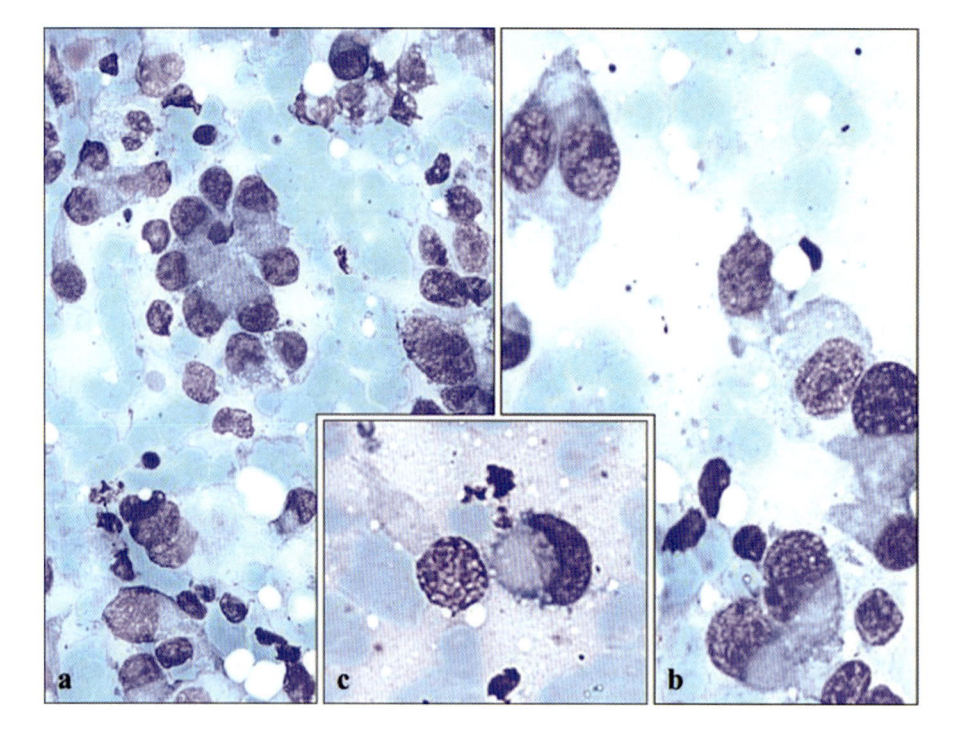

Figure 36.33 Epithelioid hemangioendothelioma. (**a**) Cellular aspirates are composed of predominantly single cells focally forming small, loose aggregates and rosette-like structures. (**b**) Oval to polygonal cells are monotonous, with nuclei that are round to oval. Binucleation is not uncommon, and mitotic figures are rare. The chromatin is uneven, with one to two nucleoli. The cytoplasm is abundant and dense. (**c**) A sharply demarcated intracytoplasmic vacuole distorts the nucleus.

Pericytic (perivascular) tumors

Glomus tumor

Glomus tumors (GTs) are painful and usually <1 cm in diameter. The most common location is the upper extremities, especially the subungual region of the fingers.

Aspirates of variable cellularity show predominantly clustered, medium-sized cells with poorly defined cytoplasmic borders, round or ovoid bland nuclei, and inconspicuous nucleoli. The background is usually hemorrhagic and can contain fibrillary myxoid matrix (Figure 36.34).[67]

Differential diagnosis

Skin adnexal neoplasms are negative for SMA, but positive for cytokeratin, and vascular tumors express endothelial markers.

Chapter 36: Cytology of soft tissue lesions

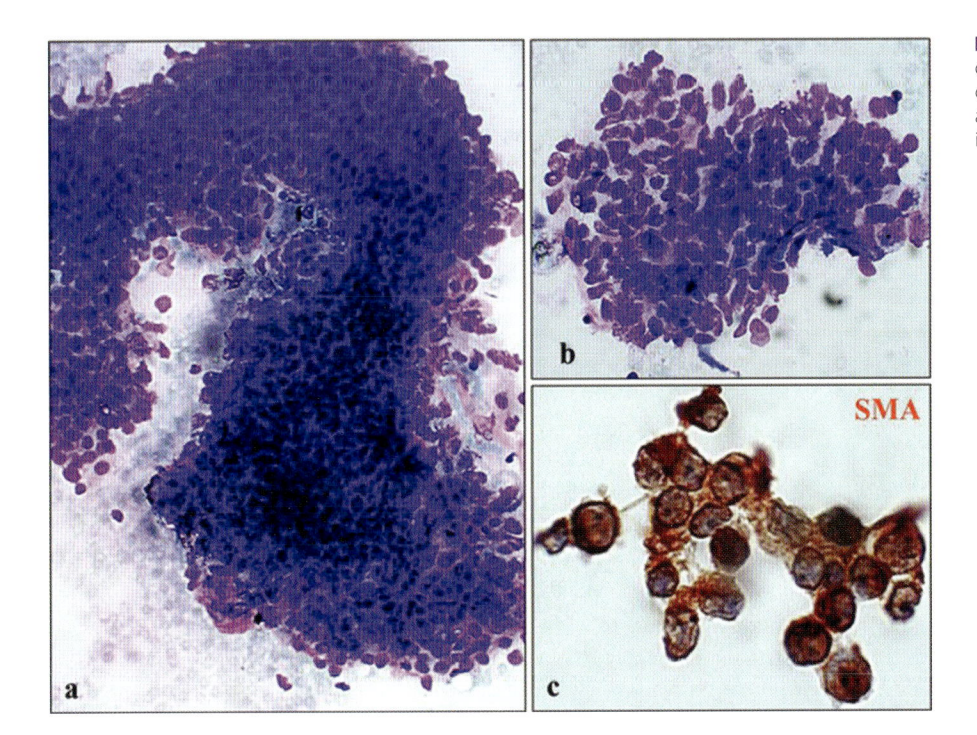

Figure 36.34 Glomus tumor. (**a,b**) Tight clusters of medium-sized cells with poorly defined cytoplasmic borders. Nuclei are round and bland, and nucleoli are inconspicuous. (**c**) The cytoplasm is positive for SMA (cytospin preparation).

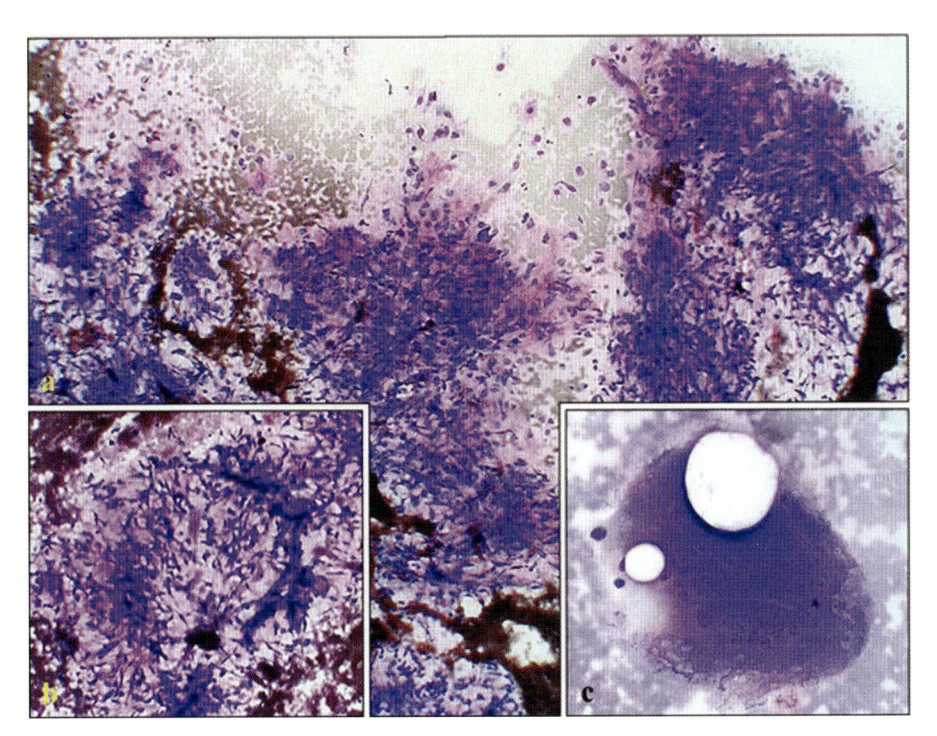

Figure 36.35 Myositis ossificans. (**a,b**) Sheets of spindle-shaped to plump fibroblasts. (**c**) Multinucleated osteoclast.

Chondro-osseous tumors
Myositis ossificans

Myositis ossificans (MO) is a mass-forming non-neoplastic heterotopic ossification most often found in the musculature of the thigh, often associated with a history of trauma.

Aspirates are moderately cellular, composed of sheets of fibroblasts at times associated with scattered skeletal muscle fibers and multinucleated osteoclasts. Fibroblasts vary from spindle-shaped, associated with long cytoplasmic processes, to more plump cells with round to oval nuclei (Figure 36.35).[68,69]

Differential diagnosis

In contrast to many sarcomas, pseudosarcomas such as MOs and NFs show rapid growth and are often associated with pain and tenderness. At low power, pseudosarcomas show tissue

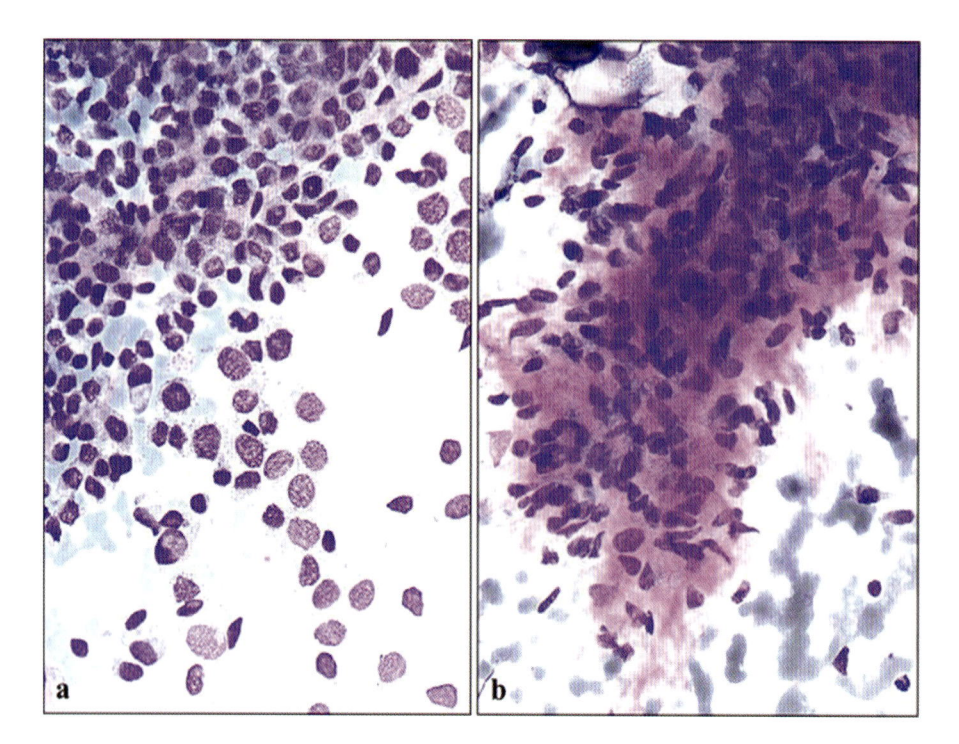

Figure 36.36 Extraskeletal mesenchymal chondrosarcoma. (**a**) Sheets of small blue cells with oval to elongated hyperchromatic nuclei, coarse chromatin, and inconspicuous nucleoli. The cytoplasm is scanty and poorly defined. (**b**) Cartilaginous stroma presents as a fibrillary matrix in this case.

culture morphology. At high power, nuclei are cytologically benign without atypical mitotic figures.

Extraskeletal mesenchymal chondrosarcoma

This uncommon tumor of adolescents and young adults arises in the soft tissue of the orbit, meninges, lower limbs, and in other locations. This tumor often metastasizes to the lung.

Aspirates are cellular, composed of small blue cells admixed with foci of cartilage. Cells are arranged in sheets, with scanty poorly defined cytoplasm and oval or elongated hyperchromatic nuclei with coarse chromatin and one or more small nucleoli. Some malignant cells are spindle-shaped (Figure 36.36); others can be large, polygonal, or even giant cells. The cartilaginous stroma ranges from islands of fibrillar matrix to well-differentiated cartilage. Some smears do not contain cartilaginous stroma.[70,71]

Differential diagnosis

In the absence of a cartilaginous component, the differential with other small blue cell tumors can be difficult. Ewing sarcoma often contains PAS-positive glycogen.

Tumors of uncertain differentiation

Intramuscular myxoma

These tumors are most common in adult women and grow as painless masses of the large muscles of the thigh, shoulder, buttocks, and upper arm, sometimes reaching a size >10 cm. Cellular variant has more prominent vascularity and cellularity.

Aspirates are composed of cells with elongated ovoid to round uniform bland nuclei with attached long hair-like cytoplasmic processes (Figure 36.37a). Background shows prominent myxoid material and sometimes also individual scattered vessel fragments with atrophic muscle fibers.[72–74]

Aspirates of cellular myxomas and intramuscular myxomas are similar, except for the higher cellularity, the presence of larger capillary fragments, and slightly more pleomorphic nuclei in the former (Figure 36.37b,c).

Differential diagnosis

Aspirates of spindle cell lipoma are practically identical to aspirates of cellular myxomas.

Mixed tumor/myoepithelioma

Most mixed tumors/myoepitheliomas arise in the subcutaneous and deep subfascial tissues of the extremities (upper more than lower), and less frequently in the head, neck, and trunk.

Aspirates are moderately cellular, with tumor cells in clusters or occasionally in rows. A glandular arrangement is very rare. Individual cells are spindle-shaped, or rounder and more epithelioid-like. The background consists of an abundant myxoid substance (Figure 36.38).

Angiomyolipoma

Angiomyolipomas (AMLs) are almost always benign and arise in two clinical settings. In tuberous sclerosis they are often bilateral and arise in young adults. When not associated with a clinical syndrome, this tumor is solitary and grows in young and middle-aged women.

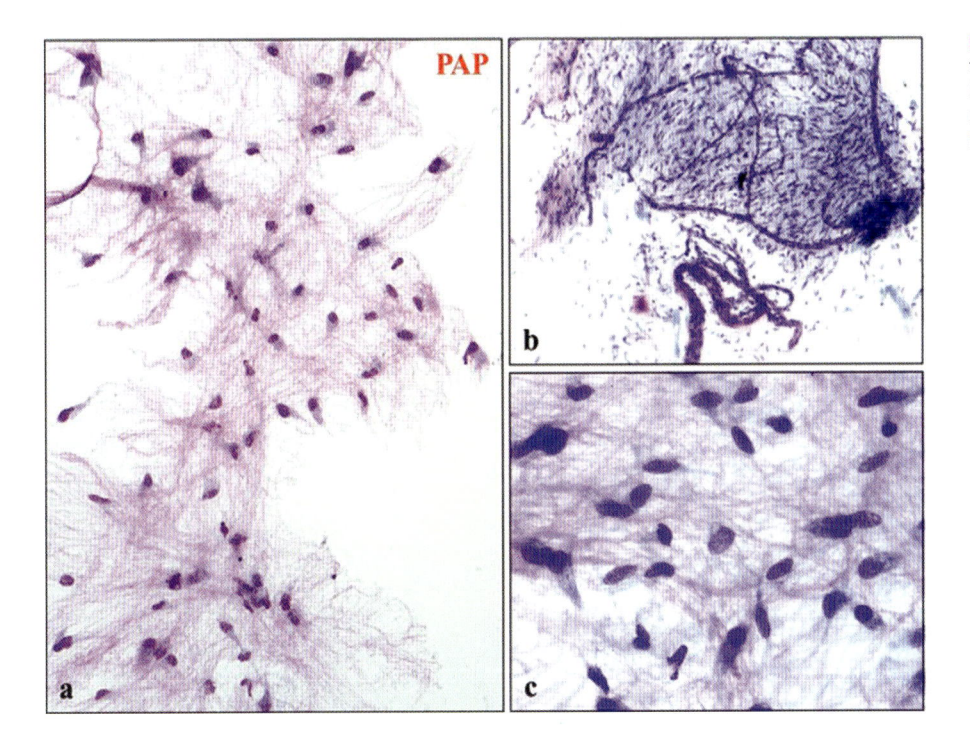

Figure 36.37 Intramuscular myxoma. (**a**) Ovoid to round nuclei are bland and uniform, with long hair-like cytoplasmic processes. (**b,c**) The cellular myxoma is more cellular. Nuclei are slightly pleomorphic, and capillary fragments are present.

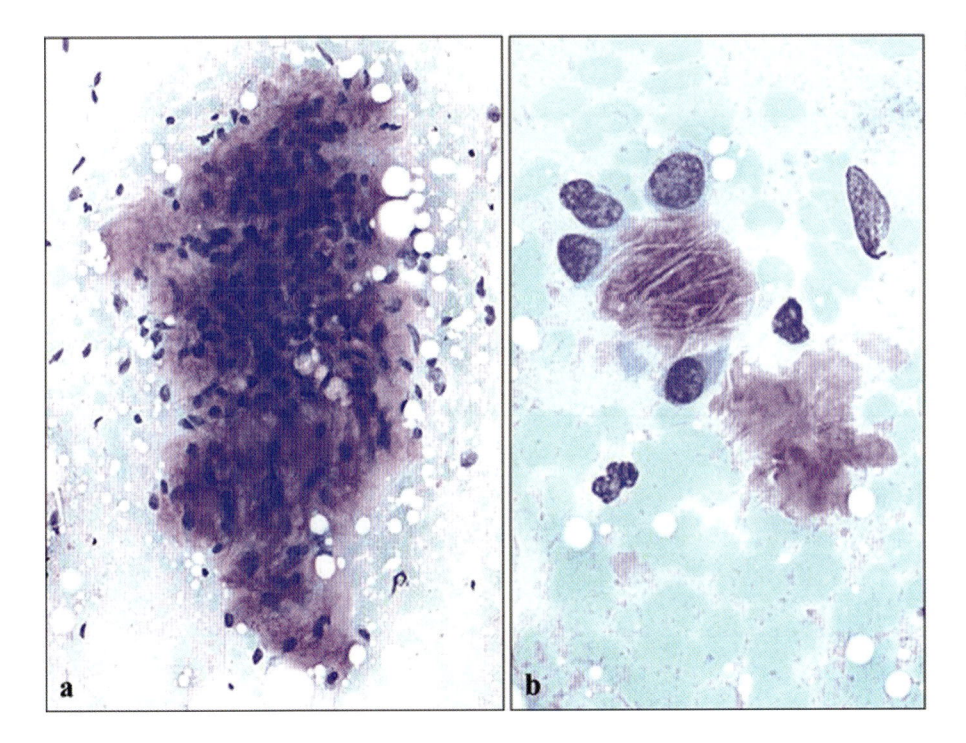

Figure 36.38 Mixed tumor/myoepithelioma. (**a**) This three-dimensional tissue fragment is composed of cells embedded in a myxoid matrix. (**b**) Spindle-shaped to round cells exhibit bland nuclei and inconspicuous nucleoli.

Aspirates are poorly to moderately cellular. The main elements are loose clusters of spindly to epithelioid cells with focal atypia. The chromatin is evenly distributed, and bland nucleoli are inconspicuous or absent. The cytoplasm is delicate and at times finely vacuolated (Figure 36.39). Naked nuclei, adipose tissue, and small vessels can be seen in the background. Cellular atypia and radiologic findings that overlap with renal cell carcinoma are potential diagnostic pitfalls, especially when the epithelioid cells predominate. Positive immunoreactivity of AML with HMB45 is extremely helpful.[75,76] In the authors' experience, however, aspirated material is often too scanty for ICC studies.

Differential diagnosis

Clear cell carcinoma of the kidney usually has abundant clearer cytoplasm, prominent nucleoli, and is negative for HMB45.

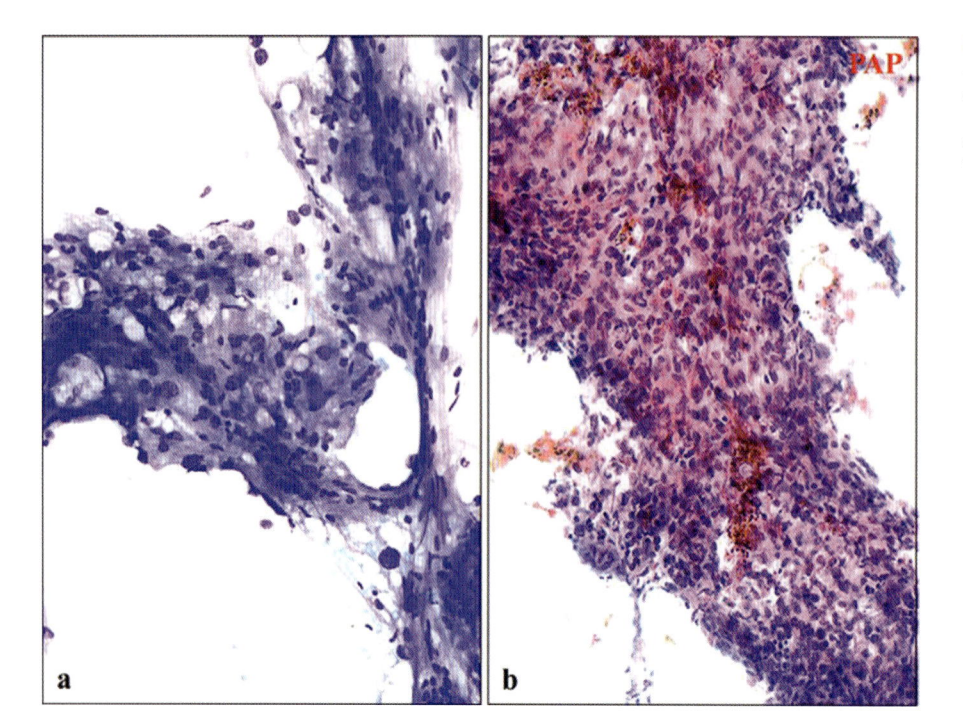

Figure 36.39 Angiomyolipoma. (**a,b**) Clusters of spindle-shaped to epithelioid cells show focal atypia. Chromatin, however, is even and bland, and nucleoli are either inconspicuous or absent. The cytoplasm is delicate and at times finely vacuolated.

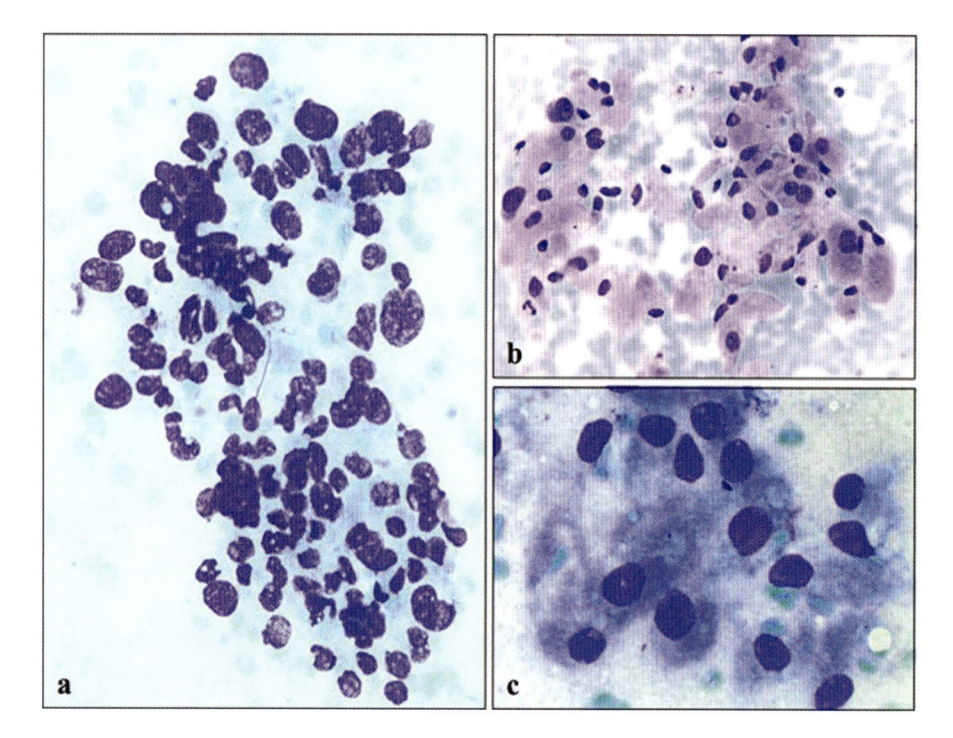

Figure 36.40 Granular cell tumor. (**a**) A cluster of cells with focally enlarged nuclei, coarser chromatin, and prominent nucleoli. Subsequent surgical biopsy revealed a (benign) granular cell tumor. (**b,c**) The cytoplasm is abundant, granular, indistinct, and friable.

Granular cell tumor

GCTs are almost always benign, single, and <5 cm in diameter. The classic location is in the tongue; however, this tumor can arise anywhere.

Aspirates are usually cellular; cells are in cohesive clusters or separate. The cytoplasm is abundant, granular, indistinct, and friable, leading to cytoplasmic granules and stripped nuclei in the background. The nuclei are mostly small and round, with fine chromatin and indistinct nucleoli. A few nuclei are larger, with coarser chromatin and prominent nucleoli (Figure 36.40).[77]

Alveolar soft part sarcomas most often occur in the deep soft tissues of the lower extremities, and are negative for SMA and desmin, and positive for S100 protein. EM shows characteristic crystals.

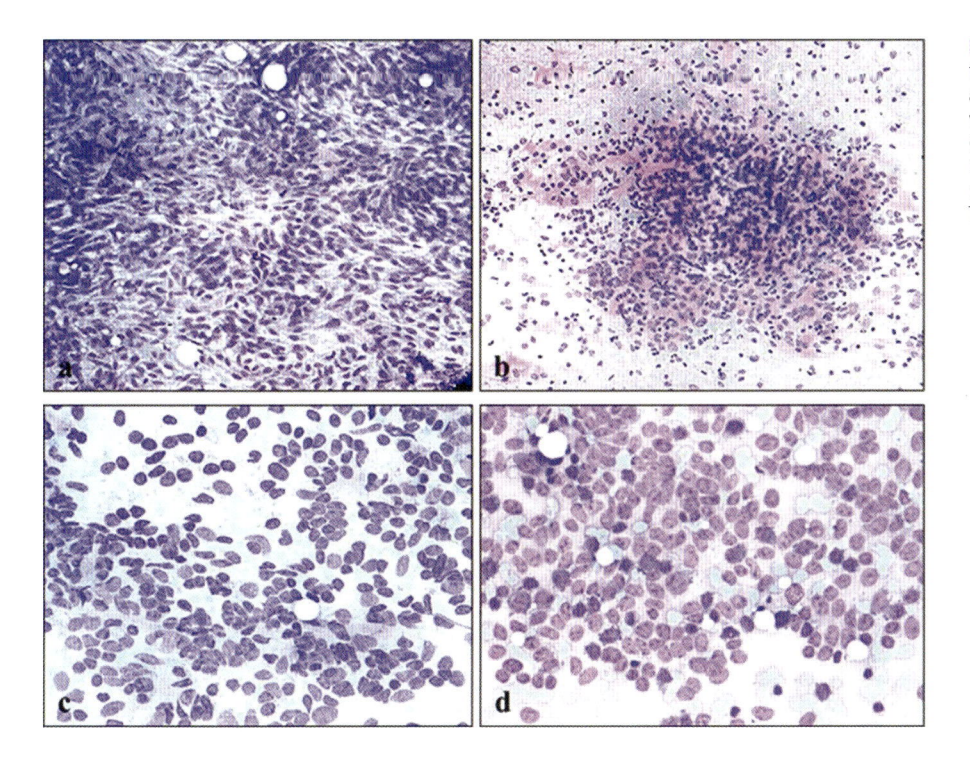

Figure 36.41 Synovial sarcoma. (**a**) Low power of this cellular aspirate exhibits a "school-of-fish" arrangement of cells. (**b**) Cells form loose clusters with mucin and single cells in the background. (**c,d**) Individual cells are round to oval and small to intermediate. Chromatin is finely granular and moderately hyperchromatic; nucleoli are small. The cytoplasm is scant and pale.

Synovial sarcoma

This relatively common sarcoma affects predominantly young adults, shows a wide anatomic distribution, and has no relationship to the synovium. Most SSs arise in the lower extremities, especially near the knee and in the thigh, followed by abdominal wall and other locations. These tumors show slow growth and measure from 3 cm to 10 cm in diameter.

Smears are similar, regardless of the most common histologic subtype.[78,79] They are moderately to highly cellular with a monotonous cell population, occasional naked nuclei, and scanty myxoid or fibrous extracellular matrix. Cells are mostly single or form loose clusters, with a branching network of vessels that may imitate a vascular tumor.

Individual cells are round to oval in shape and small to intermediate in size. The chromatin is finely granular and moderately hyperchromatic. The N/C ratio is high; the cytoplasm is pale and slightly more abundant in oval cells forming tapering extensions. The nucleoli are usually not prominent. Mitotic figures are common, and abnormal ones are rare (Figure 36.41).

In the biphasic variant, small glandular or acinar-like structures are sometimes seen. Comparing them to the histologic subtype, some authors have noted that epithelial cells secreting mucin were restricted to biphasic SS, round cells to poorly differentiated SS, and comma-like nuclei to monophasic fibrous SS.[80–83]

Rare small cell variants present with numerous small round cells and a very high N/C ratio, leading to confusion with pediatric small round cell tumors.[84] An unusual poorly differentiated (pleomorphic) type shows marked cellular pleomorphism with irregular nuclei and rhabdomyoblast-like cells.

In all cases, the ICC and clinical presentation are crucial to the correct diagnosis. SSs are positive for pancytokeratin and EMA in areas of epithelial differentiation and to a smaller extent in the spindle cell areas. Ninety percent express BCL2 (which is nonspecific because it is also positive in other soft tissue tumors), two-thirds express CD99, and one-third express S100 protein. More than 90% of cases are positive for t(X;18)(p11;q11). Patients with the SS18/SSX1 fusion gene product may have a poorer prognosis.[85]

Differential diagnosis

In leiomyosarcoma, the nuclei are blunt ended, and the cytoplasm is denser. In MPNSTs, the nuclei are wavier, S- and bullet-shaped, and there is greater pleomorphism. In about one-fifth of SS cases, poorly differentiated round cells resemble those of Ewing sarcoma/PNET. Ancillary studies are needed for the correct diagnosis. Metastatic spindle cell carcinoma shows greater pleomorphism.

Epithelioid sarcoma

Epithelioid sarcoma (ES) usually presents as firm, slowly growing, painless nodules or plaques along the flexor surfaces of the fingers, hands, wrists, and forearms, followed by the knees, lower legs, and trunk. ES is most common in young men and, as the name implies, shows epithelioid morphology. A proximal variant grows mostly in the perineum, genital tract, and in the pelvis. Tumors are deep seated, occur later in life, show more rhabdoid morphology, and have worse prognosis than the usual (distal or classic) type.

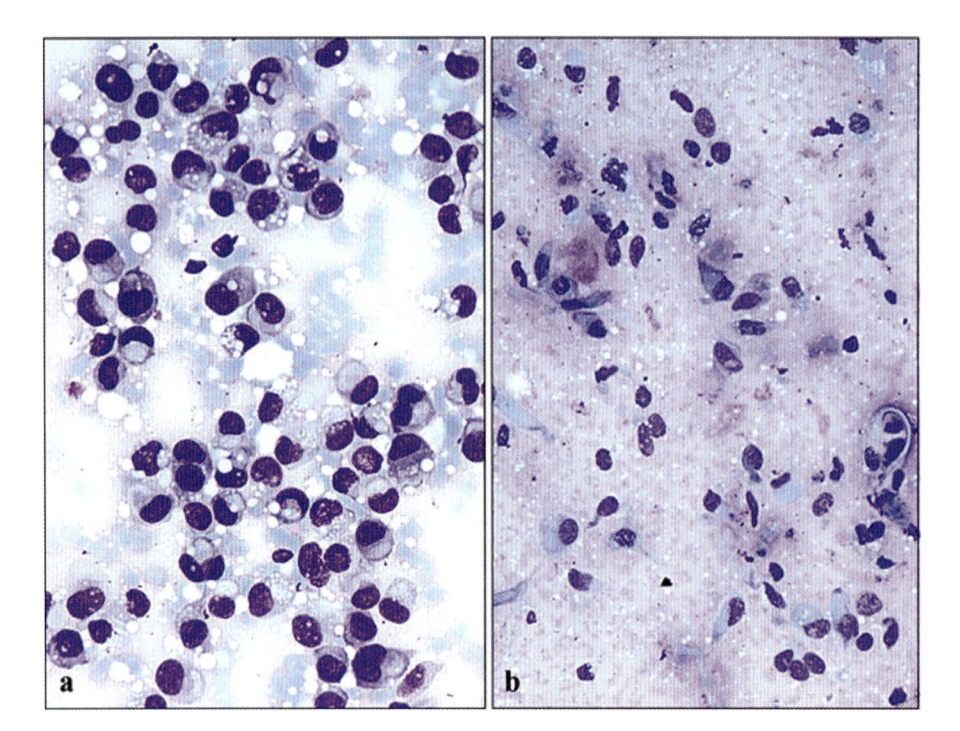

Figure 36.42 Epithelioid sarcoma. (**a**) Classic variant exhibits dispersed, medium to large and predominantly round cells. The eccentric nuclei are round to oval, with a pale perinuclear zone. Nucleoli are focally seen, and cell borders are well defined. (**b**) Fibroma-like variant is composed of spindle-shaped cells and is reminiscent of a fibrohistiocytic benign tumor.

Aspirates of the classic subtype are variably cellular. When tumor cells are scanty, the background is composed of inflammatory cells and necrosis dominates the picture, leading to a false diagnosis of inflammatory (granulomatous) lesion. When the cellularity is adequate, a diagnosis of malignancy is the rule.

Medium to large, predominantly round cells are mostly dispersed, and in a few aspirates they focally form loose clusters. The nuclei are round to oval, eccentric with a pale perinuclear zone in approximately one-third of cases. The nucleoli are prominent and cell borders are well defined. In the authors' experience, aspirates of the proximal type are practically identical to aspirates of the classic type. A fibroma-like variant is composed of spindle cells and is reminiscent of a fibrohistiocytic benign tumor (Figure 36.42).[86,87] The cytoplasm stains with cytokeratin and vimentin.

Differential diagnosis

Metastatic melanoma is more pleomorphic and positive for S100 protein and melanoma markers. Epithelioid leiomyosarcoma is more pleomorphic and is positive for smooth muscle markers. Schwannoma is positive for S100 protein; the nuclei are wavy and both comma- and bullet-shaped. Metastatic adenocarcinoma is more pleomorphic and uncommon along the flexor surfaces of the limbs in young people. Granulomas lack cytologic atypia and immunoreactivity for pancytokeratin. SSs have a similar immunophenotype; however, the cells in SS are smaller, more oval, and less pleomorphic. Nodular hidradenoma cells are more monomorphic and more often positive for S100 protein.

Ewing sarcoma/primitive neuroendocrine tumor

Recent immunohistochemical and cytogenetic studies indicate that Ewing sarcoma and PNET are part of the spectrum of a single entity. These tumors (85%) share the same chromosomal translocation and show characteristic membranous staining with CD99. The term PNET has been used for tumors that show evidence of neuroendocrine differentiation by light microscopy, immunohistochemistry, and electron microscopy.

Most of these tumors develop before the age of 30 years. Common sites for extraskeletal Ewing sarcoma/PNET are the paravertebral region and chest wall, and extremities.

Aspirates of conventional Ewing sarcoma are cellular with dispersed and clustered cells. The cytoplasm has small vacuoles indicative of glycogen. The fragility of the cytoplasm results in ill-defined cytoplasmic borders, abundant tigroid background, and stripped nuclei. Many, but not all authors, describe two distinct cell types. Large cells have abundant fragile cytoplasm, with vacuoles and round bland nuclei with inconspicuous nucleoli. Small cells have scant cytoplasm and more irregular, darker nuclei. The latter are often arranged in small, molded groups within the cell cluster. Atypical Ewing sarcoma has more pronounced cellular and nuclear pleomorphism. Light cells are less common than in conventional Ewing sarcoma. A few cells have thin cytoplasmic processes focally forming rosette-like structures with fibrillary centers (Figure 36.43). Aspirates of PNET are similar to aspirates of atypical Ewing sarcoma; however, the cells with cytoplasmic processes are more numerous, often resulting in rosette-like structures. Large light cells are in the minority.[88–90]

Desmoplastic small round cell tumors (DSRCTs) do not show double cell population, but might present with a background composed of fibroblastic stroma and collagen. ICC and

Chapter 36: Cytology of soft tissue lesions

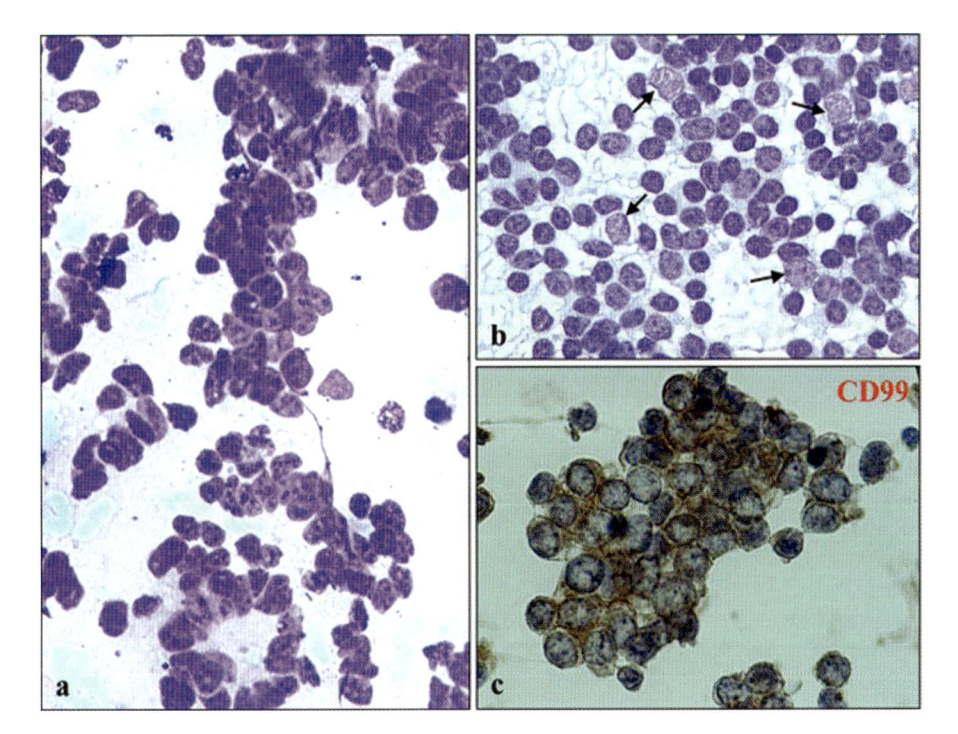

Figure 36.43 Ewing sarcoma/primitive neuroendocrine tumor. (**a**) Atypical Ewing sarcoma is composed predominantly of small cells with irregular, dark nuclei, focally forming small, molded groups. (**b**) Conventional Ewing sarcoma shows a dual population composed of numerous smaller and a few larger cells. Larger cells *(arrows)* possess round bland nuclei, inconspicuous nucleoli, and fragile cytoplasm, resulting in ill-defined cytoplasmic borders and a tigroid background. (**c**) Characteristic membranous staining with CD99 (cytospin preparation).

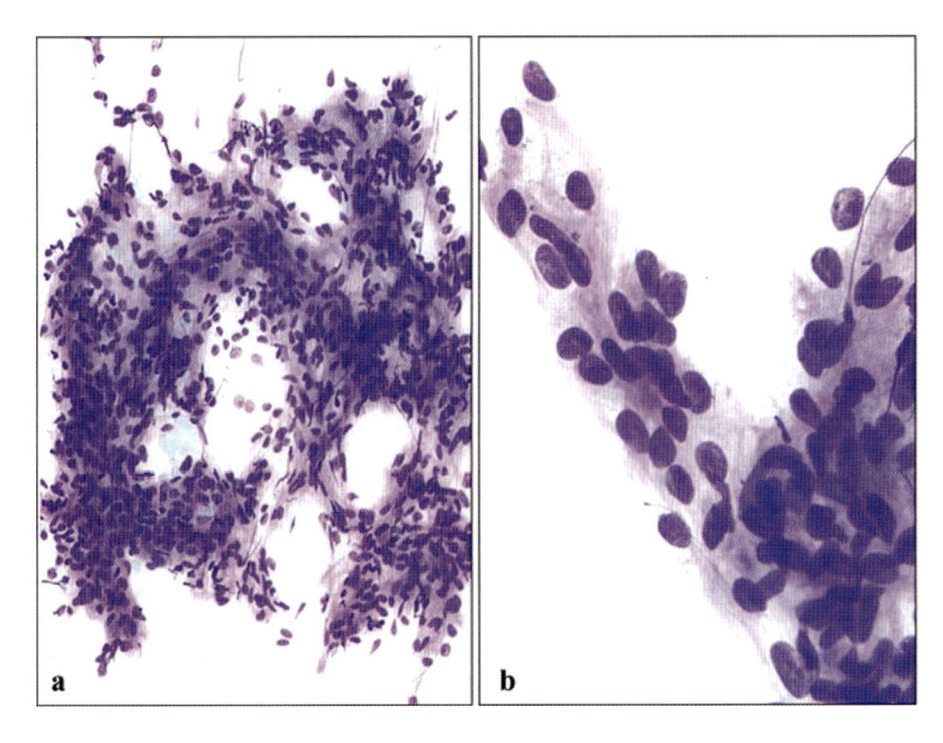

Figure 36.44 Alveolar soft part sarcoma. (**a,b**) A cluster of cells with oval, smooth nuclei, fine chromatin, and few prominent central nucleoli. The cytoplasm is granular and fragile.

gene analysis is helpful when the cellularity is low. FISH, which shows only the gross translocation abnormality, does not discriminate between Ewing sarcoma/PNET and DSRCT. Only a detailed molecular analysis can provide details of the actual rearrangement at the gene sequence level.

Alveolar soft part sarcoma

This uncommon tumor grows as a painless mass of deep soft tissues in the thigh and buttocks in young adults and on occasion also in the head and neck region in children. Metastases to the lung or brain are often the first manifestation.

Aspirates are hemorrhagic, with clustered and single tumor cells. Round cells are large and epithelioid; the nuclei are round with prominent central nucleoli. The cytoplasm is often granular and fragile (Figure 36.44). The background frequently contains stripped nuclei, cytoplasmic fragments, and multinucleated cells. Rod-shaped crystals are almost never seen on

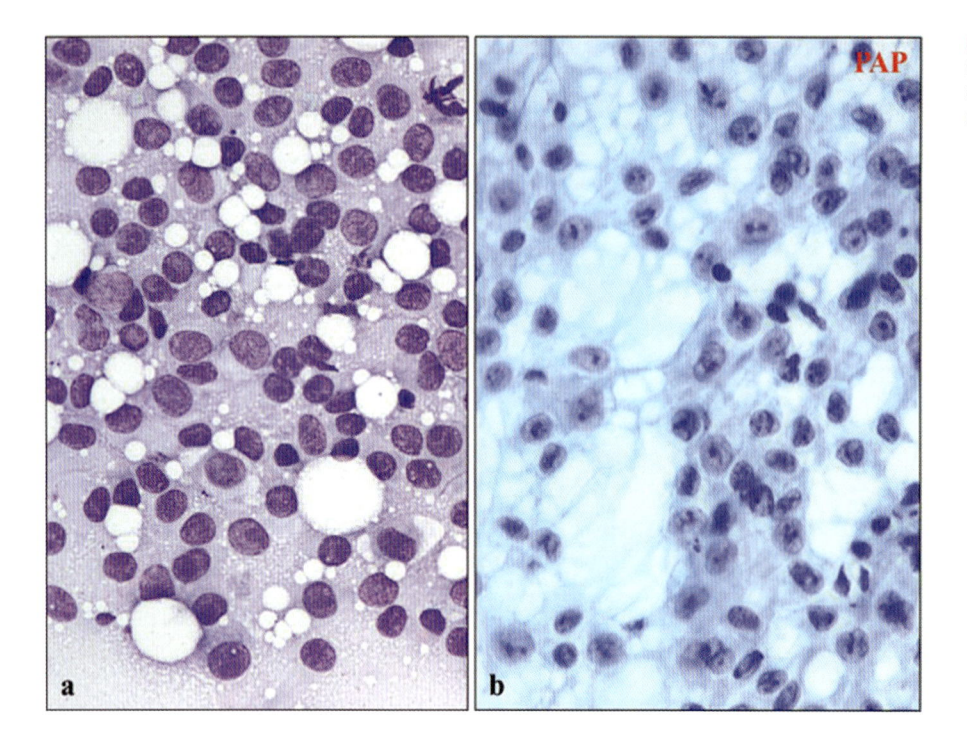

Figure 36.45 Clear cell sarcoma of soft tissue. (**a**) Numerous dispersed, round, obviously malignant cells. (**b**) Eccentric nuclei with prominent nucleoli and abundant cytoplasm.

smears, but can be identified on PAS stain and by electron microscopy.[91,92]

Clear cell sarcoma of soft tissue

This rare tumor, known also as a malignant melanoma of soft parts, is a tumor with melanocytic differentiation typically involving tendons and aponeuroses of the extremities in young adults. The most common anatomic locations are the foot, ankle, and thigh, followed by the wrist and hand.

Aspirates are usually highly cellular, with numerous dispersed and obviously malignant cells. Most of the cells are round to pleomorphic, with round and often eccentric nuclei, prominent and sometimes reddish nucleoli, and abundant eosinophilic cytoplasm (Figure 36.45). Melanin pigment is seen only occasionally. Microacinar structures mimicking adenocarcinoma are sometimes seen.[93–95]

Differential diagnosis

Metastatic malignant melanoma should be ruled out clinically. The characteristic cytogenetic change t(12;22)(q13;q12) has not been found in cutaneous malignant melanoma.

Extraskeletal myxoid chondrosarcoma

This rare sarcoma with no convincing cartilaginous differentiation typically arises in the deep soft tissues of proximal extremities and limb girdles, especially in the thigh and popliteal fossa, followed by the trunk and other sites. Tumors arise in adults.

Of all of the myxoid soft tissue neoplasms, extraskeletal myxoid chondrosarcoma shows the most consistent picture on aspirates, with tissue fragments in the fibrillary myxoid stroma distinct from the granular matrix present in other myxoid entities. Smears are moderately cellular, with plump spindly to oval uniform tumor cells most often arranged in a lace-like fashion, forming loose cords and nests. Cell balls and single cells are also observed. Nuclei are monotonous and round to oval; the chromatin is fine, and the nucleoli are inconspicuous. Nuclear grooves and clefts resulting in chondroblastoma-like nuclei are frequently seen. Scant to moderate cytoplasm is well defined, homogeneous, wispy, and tapered (Figure 36.46).[96]

Differential diagnosis

In mixed tumor/myoepithelioma of soft tissues, the clusters of cells are more compact, and the nuclei do not show grooves.

Desmoplastic small round cell tumor

These tumors primarily affect men during the third decade and usually present as widespread abdominal serosal growth in the retroperitoneum, pelvis, omentum, and mesentery. Infrequently, this tumor grows in the thoracic cavity and paratesticular tissue.

Aspirates are variably cellular, with tumor cells arranged in loose clusters, sheets, and sometimes rosettes. Individual cells are small to medium sized, round to ovoid, with scant cytoplasm and round to ovoid nuclei. The chromatin is finely granular, and the nucleoli are small. Nuclear molding can be prominent at the edges of groups. The background consists of numerous single tumor cells, which are at times admixed with stroma composed of fibroblast-like cells and collagen (Figure 36.47).[97–99]

Chapter 36: Cytology of soft tissue lesions

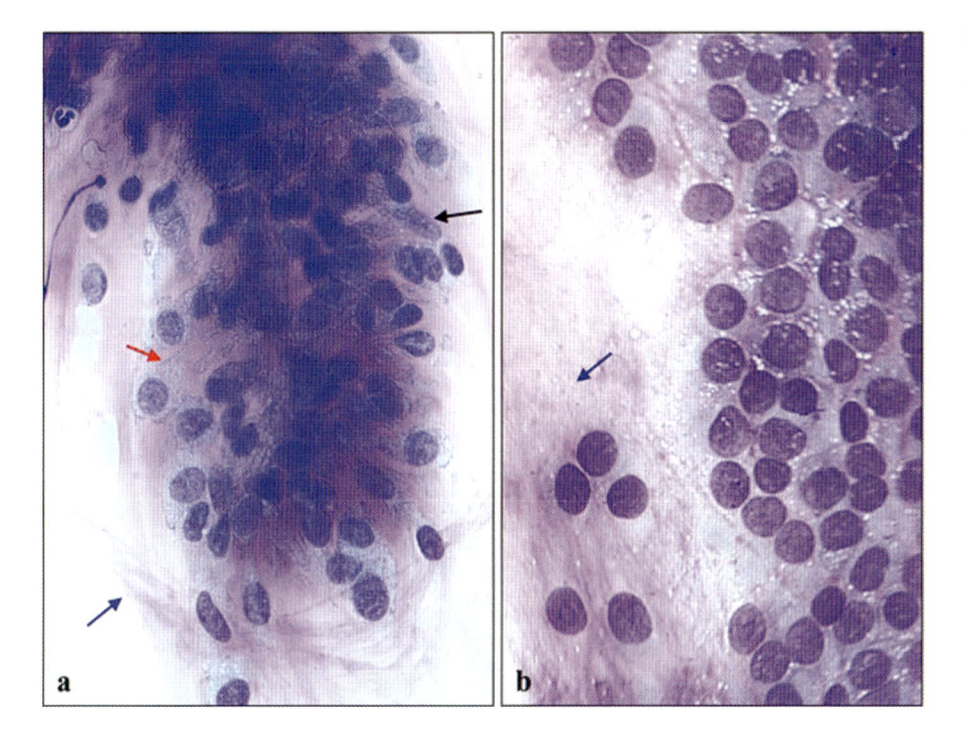

Figure 36.46 Extraskeletal myxoid chondrosarcoma. (**a,b**) Oval tumor cells are arranged in a lace-like fashion. Nuclei are monotonous and round to oval; chromatin is fine, and nucleoli are inconspicuous. A few nuclear grooves are seen (black arrow). The scant to moderate cytoplasm is well defined, wispy, and tapered (red arrow). The background is composed of fibrillary myxoid stroma (blue arrow).

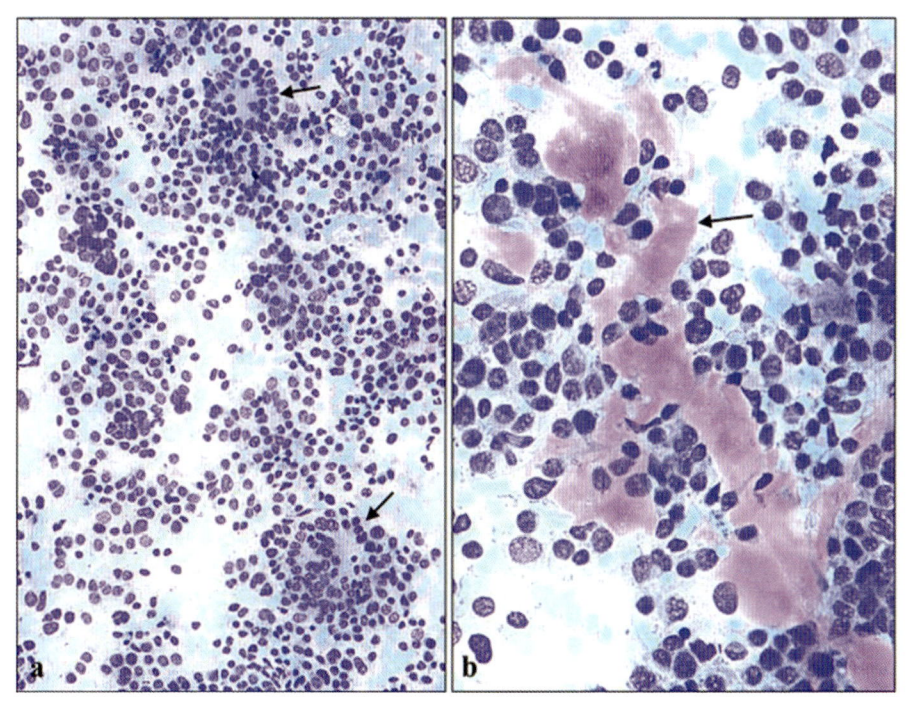

Figure 36.47 Desmoplastic small round cell tumor. (**a**) This cellular smear is composed of loose clusters of cells focally forming rosettes (arrows). (**b**) Individual cells are small to medium in size, with scant cytoplasm and round to ovoid nuclei. Chromatin is finely granular and nucleoli are small. Red fibrillary collagen is seen in the background (arrow).

The tumor cells are positive for low-molecular-weight cytokeratin, neuroendocrine markers, WT1, and often for CD99. Characteristic is also a paranuclear dot-like reactivity for desmin and vimentin corresponding to whorls of filaments seen on electron microscopy.

Differential diagnosis

For discussion of Ewing sarcoma/PNET and alveolar/embryonal RMS differential diagnosis, see the corresponding sections.

Small cell carcinoma is uncommon in younger adults. The cells are more dispersed and more fragile, with a prominent smearing artifact.

Extrarenal malignant rhabdoid tumor

This very aggressive tumor with dismal prognosis affects primarily infants and children. The tumor arises most commonly in the liver, heart, gastrointestinal system, and in deep axial locations such as the neck and paraspinal regions. Owing to significant overlap with other neoplasms, the diagnosis of

1007

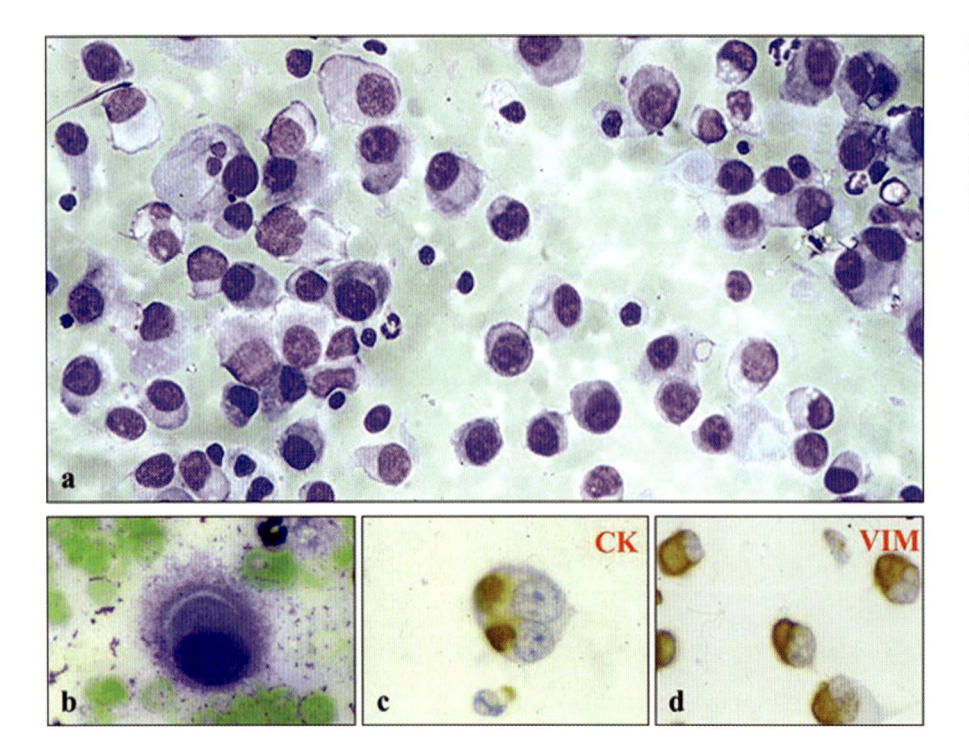

Figure 36.48 Extrarenal malignant rhabdoid tumor. (**a**) Numerous dispersed intermediate to large round cells have abundant cytoplasm. Nuclei are large and usually round, with focally prominent nucleoli. (**b**) A cell with a prominent, round, paranuclear cytoplasmic inclusion. (**c,d**) Cytoplasm stains with pancytokeratin and vimentin (cytospin preparation).

extrarenal malignant rhabdoid tumor requires exclusion of all other lines of differentiation.

Aspirates are composed of numerous predominantly single cells of intermediate to large size. Cells are mostly rounded with an abundant cytoplasm. Nuclei are large, usually round, with prominent nucleoli. Round paranuclear cytoplasmic inclusions are prominent and stain with vimentin and pancytokeratin (Figure 36.48).[100]

Differential diagnosis

RMS shows different anatomic distribution. The cells are more primitive, have no globular hyaline cytoplasmic inclusions, and are desmin positive.

Malignant melanoma is uncommon at an early age. Cells are positive for melanoma markers, the nucleoli are usually more prominent, and inclusions are intranuclear and not cytoplasmic.

Acknowledgments

The authors would like to thank Dr. John S.J. Brooks, Professor at the University of Pennsylvania and Chair of Pathology at Pennsylvania Hospital in Philadelphia, PA; Dr. Edmund S. Cibas, Associate Professor of Pathology at Harvard Medical School and Director of Cytopathology at Brigham and Women's Hospital in Boston, MA; and Dr. B. Hudson Berrey, Professor of Orthopaedics and former Chair of Orthopaedic Surgery at University of Florida College of Medicine in Jacksonville, FL. We are sincerely grateful to all of the aforementioned individuals. Without their inspiration and input, this chapter would have never appeared in the current form.

Notes

All photomicrographs were produced from material aspirated at the Institute of Oncology, Ljubljana, Slovenia, EU. Most photographs were taken by Živa Pohar-Marinšek, MD, PhD.

If not stated otherwise, aspirates were stained with Giemsa's azure eosin methylene blue solution (Merck Cat. No. 1.09204).

References

1. Jemal A, Siegel R, Ward E, et al. Cancer statistics, 2007. *CA Cancer J Clin* 2007;57(1):43–66.
2. Hajdu S. Soft tissue and bone. In *Bibbo M. Comprehensive Cytopathology*, 2nd edn. Philadelphia: WB Saunders; 1997.
3. Brown FM. Soft tissue. In *Cytology: Diagnostic Principles and Clinical Correlates*, 2nd edn. Cibas E, Ducatman B (eds.) Edinburgh: WB Saunders; 2003.
4. Beiske K, Ambros PF, Burchill SA, Cheung IY, Swerts K. Detecting minimal residual disease in neuroblastoma patients: the present state of the art [review]. *Cancer Lett* 2005;228(1–2):229–240.
5. Akerman M, Rydholm A. Aspiration cytology of lipomatous tumors: a 10-year experience at an orthopedic oncology center. *Diagn Cytopathol* 1987;3(4): 295–302.
6. James LP. Cytopathology of mesenchymal repair. *Diagn Cytopathol* 1985;1(2):91–104.
7. De May RM. *The Art and Science of Cytopathology*. Chicago: ASCP Press; 1996.

8. Kloboves-Prevodnik VV, Us-Krasovec M, Gale N, Lamovec J. Cytological features of lipoblastoma: a report of three cases. *Diagn Cytopathol* 2005;33:195–200.

9. Domanski HA, Carlen B, Jonsson K, Mertens F, Akerman M. Distinct cytologic features of spindle cell lipoma: cytologic-histologic study with clinical, radiologic, electron microscopic, and cytogenetic correlations. *Cancer* 2001;93:381–389.

10. Yong M, Raza AS, Greaves TS, Cobb CJ. Fine-needle aspiration of a pleomorphic lipoma of the head and neck: a case report. *Diagn Cytopathol* 2005;32:110–113.

11. Lemos MM, Kindblom LG, Meis-Kindblom JM, et al. Fine-needle aspiration characteristics of hibernoma. *Cancer* 2001;93:206–210.

12. Walaas L, Kindblom LG. Lipomatous tumors: a correlative cytologic and histologic study of 27 tumors examined by fine needle aspiration cytology. *Hum Pathol* 1985;16:6–18.

13. Nemanqani D, Mourad WA. Cytomorphologic features of fine-needle aspiration of liposarcoma. *Diagn Cytopathol* 1999;20:67–69.

14. Szadowska A, Lasota J. Fine needle aspiration cytology of myxoid liposarcoma; a study of 18 tumours. *Cytopathology* 1993;4:99–106.

15. Kempson RL, Fletcher CDM, Evans HL, Hendrickson MR, Sibley RK. *Atlas of Tumor Pathology: Tumors of Soft Tissues*. Washington, DC: Armed Forces Institute of Pathology; 2001.

16. Dahl I, Akerman M. Nodular fasciitis: a correlative cytologic and histologic study of 13 cases. *Acta Cytol* 1981;25:215–223.

17. Raab SS, Silverman JF, McLeod DL, Benning TL, Geisinger KR. Fine needle aspiration biopsy of fibromatoses. *Acta Cytol* 1993;37:323–328.

18. Kilpatrick SE, Geisinger KR. Soft tissue sarcomas: the usefulness and limitations of fine-needle aspiration biopsy [review]. *Am J Clin Pathol* 1998;110:50–68.

19. Odashiro AN, Odashiro Miiji LN, Odashiro DN, Nguyen GK. Mammary myofibroblastoma: report of two cases with fine-needle aspiration cytology and review of the cytology literature. *Diagn Cytopathol* 2004;30:406–410.

20. Jurcić V, Perković T, Pohar-Marinsek Z, Hvala A, Lazar I. Infantile myofibroma in a prematurely born twin: a case report. *Pediatr Dermatol* 2003;20:345–349.

21. Ostrowski ML, Bradshaw J, Garrison D. Infantile myofibromatosis: diagnosis suggested by fine-needle aspiration biopsy. *Diagn Cytopathol* 1990;6:284–288.

22. Pohar-Marinsek Z, Flezar M, Lamovec J. Acral myxoinflammatory fibroblastic sarcoma in FNAB samples: can we distinguish it from other myxoid lesions? *Cytopathology* 2003;14:73–78.

23. García-García E, Rodríguez-Gil Y, Suárez-Gauthier A, et al. Myxoinflammatory fibroblastic sarcoma: report of a case with fine needle aspiration cytology. *Acta Cytol* 2007;51:231–234.

24. Akerman M, Domanski H. The cytological features of soft tissue tumors in fine needle aspiration smears classified according to histotype. In *The Cytology of Soft Tissue Tumors*. Basel, Switzerland: Karger; 2003: 40–41.

25. Merck C, Hagmar B. Myxofibrosarcoma: a correlative cytologic and histologic study of 13 cases examined by fine needle aspiration cytology. *Acta Cytol* 1980;24:137–144.

26. Kilpatrick SE, Ward WG. Myxofibrosarcoma of soft tissues: cytomorphologic analysis of a series. *Diagn Cytopathol* 1999;20:6–9.

27. Dawiskiba S, Eriksson L, Elner A, et al. Diffuse pigmented villonodular synovitis of the temporomandibular joint diagnosed by fine-needle aspiration cytology. *Diagn Cytopathol* 1989;5:301–304.

28. Wakely PE Jr, Frable WJ. Fine-needle aspiration biopsy cytology of giant-cell tumor of tendon sheath. *Am J Clin Pathol* 1994;102:87–90.

29. Gupta K, Dey P, Goldsmith R, Vasishta RK. Comparison of cytologic features of giant-cell tumor and giant-cell tumor of tendon sheath. *Diagn Cytopathol* 2004;30:14–18.

30. Nguyen GK, Neifer R. The cells of benign and malignant hemangiopericytomas in aspiration biopsy. *Diagn Cytopathol* 1985;1:327–331.

31. Walaas L, Angervall L, Hagmar B, Save-Soderbergh J. A correlative cytologic and histologic study of malignant fibrous histiocytoma: an analysis of 40 cases examined by fine-needle aspiration cytology. *Diagn Cytopathol* 1986;2:46–54.

32. Berardo MD, Powers CN, Wakely PE Jr, Almeida MO, Frable WJ. Fine-needle aspiration cytopathology of malignant fibrous histiocytoma. *Cancer* 1997;81:228–237.

33. Klijanienko J, Caillaud JM, Lagace R, Vielh P. Comparative fine-needle aspiration and pathologic study of malignant fibrous histiocytoma: cytodiagnostic features of 95 tumors in 71 patients. *Diagn Cytopathol* 2003;29:320–326.

34. Mooney EE, Layfield LJ, Dodd LG. Fine-needle aspiration of neural lesions. *Diagn Cytopathol* 1999;20:1–5.

35. Yu GH, Sack MJ, Baloch Z, Gupta PK. Difficulties in the fine needle aspiration (FNA) diagnosis of schwannoma. *Cytopathology* 1999;10:186–194.

36. Resnick JM, Fanning CV, Caraway NP, Varma DG, Johnson M. Percutaneous needle biopsy diagnosis of benign neurogenic neoplasms. *Diagn Cytopathol* 1997;16:17–25.

37. Dodd LG, Marom EM, Dash RC, Matthews MR, McLendon RE. Fine-needle aspiration cytology of "ancient" schwannoma. *Diagn Cytopathol* 1999;20:307–311.

38. Rorat E, Yang W, DeLaTorre R. Fine needle aspiration cytology of parapharyngeal meningioma [review]. *Acta Cytol* 1991;35:497–500.

39. Baisden BL, Hamper UM, Ali SZ. Metastatic meningioma in fine-needle aspiration (FNA) of the lung: cytomorphologic finding. *Diagn Cytopathol* 1999;20:291–294.

40. Pohar-Marinsek Z, Frković-Grazio S. Fine needle aspiration (FNA) cytology of primary subcutaneous sacrococcygeal myxopapillary ependymoma. *Cytopathology* 1998;9:415–420.

41. Kulesza P, Tihan T, Ali SZ. Myxopapillary ependymoma: cytomorphologic characteristics and differential diagnosis. *Diagn Cytopathol* 2002;26:247–250.

42. McGee RS Jr, Ward WG, Kilpatrick SE. Malignant peripheral nerve sheath tumor: a fine-needle aspiration biopsy study. *Diagn Cytopathol* 1997;17:298–305.

43. Klijanienko J, Caillaud JM, Lagace R, Vielh P. Cytohistologic correlations of 24 malignant peripheral nerve sheath tumor (MPNST) in 17 patients: the Institut Curie experience. *Diagn Cytopathol* 2002;27:103–108.

44. Bondeson L, Andreasson L. Aspiration cytology of adult rhabdomyoma. *Acta Cytol* 1986;30:679–682.

45. Domanski HA, Dawiskiba S. Adult rhabdomyoma in fine needle aspirates: a report of two cases. *Acta Cytol* 2000;44:223–226.

46. Newton WA Jr, Gehan EA, Webber BL, et al. Classification of rhabdomyosarcomas and related sarcomas: pathologic aspects and proposal for a new classification. An Intergroup Rhabdomyosarcoma Study. *Cancer* 1995;76:1073–1085.

47. Qualman SJ, Coffin CM, Newton WA, et al. Intergroup Rhabdomyosarcoma Study: update for pathologists [review]. *Pediatr Dev Pathol* 1998;1:550–561.

48. Pohar-Marinsek Z, Bracko M. Rhabdomyosarcoma: cytomorphology, subtyping and differential diagnostic dilemmas. *Acta Cytol* 2000;44:524–532.

49. Pohar-Marinsek Z, Us-Krasovec M, Golouh R, Zganec M. Value of image cytometry in the subclassification of rhabdomyosarcoma. *Anal Quant Cytol Histol* 2002;24:212–220.

50. Pohar-Marinsek Z, Anzic J, Jereb B. Topical topic: value of fine needle aspiration biopsy in childhood rhabdomyosarcoma: twenty-six years of experience in Slovenia. *Med Pediatr Oncol* 2002;38:416–420.

51. Pohar-Marinsek Z, Bracko M, Lavrencak J, Us-Krasovec M. DNA ploidy as a prognostic factor in rhabdomyosarcoma: analysis of 35 cases with image cytometry. *Anal Quant Cytol Histol* 2003;25:235–242.

52. Dahl I, Hagmar B, Angervall L. Leiomyosarcoma of the soft tissue: a correlative cytological and histological study of 11 cases. *Acta Pathol Microbiol Scand A* 1981;89:285–291.

53. Tao LC, Davidson DD. Aspiration biopsy cytology of smooth muscle tumors: a cytologic approach to the differentiation between leiomyosarcoma and leiomyoma. *Acta Cytol* 1993;37:300–308.

54. Klijanienko J, Caillaud JM, Lagace R, Vielh P. Fine-needle aspiration of leiomyosarcoma: a correlative cytohistopathological study of 96 tumors in 68 patients. *Diagn Cytopathol* 2003;28:119–125.

55. Domanski HA, Akerman M, Rissler P, Gustafson P. Fine-needle aspiration of soft tissue leiomyosarcoma: an analysis of the most common cytologic findings and the value of ancillary techniques. *Diagn Cytopathol* 2006;34:597–604.

56. Weiss SW, Goldblum JR. Malignant vascular tumors. In *Enzinger's and Weiss's Soft Tissue Tumors*, 4th edn. Strauss M (ed.) St. Louis: Mosby; 2001: 917–954.

57. Liu K, Layfield LJ. Cytomorphologic features of angiosarcoma on fine needle aspiration biopsy. *Acta Cytol* 1999;43:407–415.

58. Boucher LD, Swanson PE, Stanley MW, et al. Cytology of angiosarcoma. Findings in fourteen fine-needle aspiration biopsy specimens and one pleural fluid specimen. *Am J Clin Pathol* 2000;114:210–219.

59. Minimo C, Zakowski M, Lin O. Cytologic findings of malignant vascular neoplasms: a study of twenty-four cases. *Diagn Cytopathol* 2002;26:349–355.

60. Klijanienko J, Caillaud JM, Lagace R, Vielh P. Cytohistologic correlations in angiosarcoma including classic and epithelioid variants: Institut Curie's experience. *Diagn Cytopathol* 2003;29:140–145.

61. Gagner JP, Yim JH, Yang GC. Fine-needle aspiration cytology of epithelioid angiosarcoma: a diagnostic dilemma. *Diagn Cytopathol* 2005;33(33):429–433.

62. Carson KF, Hirschowitz SL, Nieberg RK, Sadeghi S. Pitfalls in the cytologic diagnosis of angiosarcoma of the breast by fine-needle aspiration: a case report. *Diagn Cytopathol* 1994;11:297–299; discussion, 299–300.

63. Jeon YK, Kim HW, Choi HJ, Park IA. Fine needle aspiration cytology of epithelioid angiosarcoma: report of a case with nuclear grooves and indentations. *Acta Cytol* 2004;48:223–228.

64. Vesoulis Z, Cunliffe C. Fine-needle aspiration biopsy of postradiation epithelioid angiosarcoma of breast. *Diagn Cytopathol* 2000;22:172–175.

65. Kilpatrick SE, Koplyay PD, Ward WG, Richards F 2nd. Epithelioid hemangioendothelioma of bone and soft tissue: a fine-needle aspiration biopsy study with histologic and immunohistochemical confirmation. *Diagn Cytopathol* 1998;19:38–43.

66. Pettinato G, Insabato L, De Chiara A, Forestieri P, Manco A. Epithelioid hemangioendothelioma of soft tissue: fine needle aspiration cytology, histology, electron microscopy and immunohistochemistry of a case. *Acta Cytol* 1986;30:194–200.

67. Handa U, Palta A, Mohan H, Punia RP. Aspiration cytology of glomus tumor: a case report. *Acta Cytol* 2001;45:1073–1076.

68. Wong NL. Fine needle aspiration cytology of pseudosarcomatous reactive proliferative lesions of soft tissue. *Acta Cytol* 2002;46:1049–1055.

69. Akerman M. Benign fibrous lesions masquerading as sarcomas: clinical and morphological pitfalls. *Acta Orthop Scand Suppl* 1997;273:37–40.

70. Doria MI Jr, Wang HH, Chinoy MJ. Retroperitoneal mesenchymal chondrosarcoma: report of a case diagnosed by fine needle aspiration cytology. *Acta Cytol* 1990;34:529–532.

71. Trembath DG, Dash R, Major NM, Dodd LG. Cytopathology of mesenchymal chondrosarcomas: a report and comparison of four patients. *Cancer* 2003;99:211–216.

72. Wakely Jr PE, Bos GD, Mayerson J. The cytopathology of soft tissue mxyomas: ganglia, juxtaarticular myxoid lesions, and intramuscular myxoma. *Am J Clin Pathol* 2005;123:858–865.

73. Caraway NP, Staerkel GA, Fanning CV, Varma DG, Pollock RE. Diagnosing intramuscular myxoma by fine-needle aspiration: a multidisciplinary approach. *Diagn Cytopathol* 1994;11:255–261.

74. Akerman M, Rydholm A. Aspiration cytology of intramuscular myxoma: a comparative clinical, cytologic and histologic study of ten cases. *Acta Cytol* 1983;27:505–510.

75. Crapanzano JP. Fine-needle aspiration of renal angiomyolipoma: cytological findings and diagnostic pitfalls in a series of five cases. *Diagn Cytopathol* 2005;32:53–57.

76. Handa U, Nanda A, Mohan H. Fine-needle aspiration of renal angiomyolipoma: a report of four cases. *Cytopathology* 2007;18:250–254.

77. Wieczorek TJ, Krane JF, Domanski HA, et al. Cytologic findings in granular cell

78. Kilpatrick SE, Teot LA, Stanley MW, et al. Fine-needle aspiration biopsy of synovial sarcoma: a cytomorphologic analysis of primary, recurrent, and metastatic tumors. *Am J Clin Pathol* 1996;106:769–775.

79. Akerman M, Ryd W, Skytting B, Scandinavian Sarcoma Group. Fine-needle aspiration of synovial sarcoma. Criteria for diagnosis: retrospective reexamination of 37 cases, including ancillary diagnostics. *Diagn Cytopathol* 2003;28:232–238.

80. Klijanienko J, Caillaud JM, Lagace R, Vielh P. Cytohistologic correlations in 56 synovial sarcomas in 36 patients: the Institut Curie experience. *Diagn Cytopathol* 2002;27:96–102.

81. Viguer JM, Jimenez-Heffernan JA, Vicandi B, Lopez-Ferrer P, Gamallo C. Cytologic features of synovial sarcoma with emphasis on the monophasic fibrous variant: a morphologic and immunocytochemical analysis of bcl-2 protein expression. *Cancer* 1998;84:50–56.

82. Akerman M, Willen H, Carlen B, Mandahl N, Mertens F. Fine needle aspiration (FNA) of synovial sarcoma—a comparative histological-cytological study of 15 cases, including immunohistochemical, electron microscopic and cytogenetic examination and DNA-ploidy analysis. *Cytopathology* 1996;7:187–200.

83. Ewing CA, Zakowski MF, Lin O. Monophasic synovial sarcoma: a cytologic spectrum. *Diagn Cytopathol* 2004;30:19–23.

84. Silverman JF, Landreneau RJ, Sturgis CD, et al. Small-cell variant of synovial sarcoma: fine-needle aspiration with ancillary features and potential diagnostic pitfalls. *Diagn Cytopathol* 2000;23:118–123.

85. Inagaki H, Murase T, Otsuka T, Eimoto T. Detection of SYT-SSX fusion transcript in synovial sarcoma using archival cytologic specimens. *Am J Clin Pathol* 1999;111:528–533.

86. Pohar-Marinsek Z, Zidar A. Epithelioid sarcoma in FNAB smears. *Diagn Cytopathol* 1994;11:367–372.

87. Cardillo M, Zakowski MF, Lin O. Fine-needle aspiration of epithelioid sarcoma: cytology findings in nine cases [review]. *Cancer* 2001;93:246–251.

88. Bakhos R, Andrey J, Bhoopalam N, Jensen J, Reyes CV. Fine-needle aspiration cytology of extraskeletal Ewing's sarcoma. *Diagn Cytopathol* 1998;18:137–140.

89. Renshaw AA, Perez-Atayde AR, Fletcher JA, Granter SR. Cytology of typical and atypical Ewing's sarcoma/PNET. *Am J Clin Pathol* 1996;106:620–624.

90. Das DK. Fine-needle aspiration (FNA) cytology diagnosis of small round cell tumors: value and limitations [review]. *Indian J Pathol Microbiol* 2004;47:309–318.

91. Lopez-Ferrer P, Jimenez-Heffernan JA, Vicandi B, Gonzalez-Peramato P, Viguer JM. Cytologic features of alveolar soft part sarcoma: report of three cases. *Diagn Cytopathol* 2002;27:115–119.

92. Shabb N, Sneige N, Fanning CV, Dekmezian R. Fine-needle aspiration cytology of alveolar soft-part sarcoma. *Diagn Cytopathol* 1991;7:293–298.

93. Creager AJ, Pitman MB, Geisinger KR. Cytologic features of clear cell sarcoma (malignant melanoma) of soft parts: a study of fine-needle aspirates and exfoliative specimens. *Am J Clin Pathol* 2002;117:217–224.

94. Tong TR, Chow TC, Chan OW, et al. Clear-cell sarcoma diagnosis by fine-needle aspiration: cytologic, histologic, and ultrastructural features; potential pitfalls and literature review [review]. *Diagn Cytopathol* 2002;26:174–180.

95. Almeida MM, Nunes AM, Frable WJ. Malignant melanoma of soft tissue: a report of three cases with diagnosis by fine needle aspiration cytology. *Acta Cytol* 1994;38:241–246.

96. Jakowski JD, Wakely PE Jr. Cytopathology of extraskeletal myxoid chondrosarcoma: report of 8 cases. *Cancer* 2007;111:298–305.

97. Caraway NP, Fanning CV, Amato RJ, Ordonez NG, Katz RL. Fine-needle aspiration of intra-abdominal desmoplastic small cell tumor. *Diagn Cytopathol* 1993;9:465–470.

98. Granja NM, Begnami MD, Bortolan J, Filho AL, Schmitt FC. Desmoplastic small round cell tumour: cytological and immunocytochemical features. *Cytojournal* 2005;2:6.

99. Hill DA, Pfeifer JD, Marley EF, et al. WT1 staining reliably differentiates desmoplastic small round cell tumor from Ewing sarcoma/primitive neuroectodermal tumor: an immunohistochemical and molecular diagnostic study. *Am J Clin Pathol* 2000;114:345–353.

100. Pogacnik A, Zidar N. Malignant rhabdoid tumor of the liver diagnosed by fine needle aspiration cytology: a case report. *Acta Cytol* 1997;41:539–543.

Chapter 37

Surgical management of soft tissue sarcoma: histologic type and grade guide surgical planning and integration of multimodality therapy

Charlotte Ariyan
Memorial Sloan-Kettering Cancer Center, New York

Samuel Singer
Memorial Sloan-Kettering Cancer Center, New York

The primary treatment of sarcoma is surgical resection; however, as the understanding of this rare disease has improved, it is now clear that different sarcoma subtypes have varying growth patterns and metastatic potential. Although surgery continues to play a central role in the treatment of sarcomas, pathologic subtyping has allowed characterization of sarcomas that have an improved sensitivity to chemotherapy or radiation therapy. This has translated into a refinement of the surgical approach, and the addition of multidisciplinary treatment in an adjuvant or neo-adjuvant fashion. This chapter focuses on the surgical management of sarcoma, which is driven by the pathologic subtype.

Clinical evaluation and biopsy

Soft tissue sarcomas (STSs) are often asymptomatic and are usually found incidentally. A survey on the clinical presentation demonstrated that most patients present with a mass, only one-third of which cause pain symptoms.[1] There can be an antecedent history of trauma, which is usually not related, but often draws attention to a mass. There is often a hesitation to diagnose sarcoma; up to 20% of patients have been found to have a delay of 6 or more months between presentation to a physician and tissue diagnosis.[1]

The differential diagnosis of a mass includes a hematoma, benign lipoma, lymphoma, germ cell tumor, and sarcoma. A history of an enlarging or changing lesion should raise the suspicion of a malignancy. On physical examination it is important to differentiate between a soft, mobile lesion and one that is fixed or invading local structures.

Regardless of the features, any persistent mass should be imaged. A CT scan is ideal for examining sarcomas of the chest and abdomen, whereas an MRI demonstrates the muscle and fat planes in the extremities with greater clarity. Sarcomas often do not differentially take up fluoro-2-deoxy-D-glucose (FDG) based on malignant potential, and there is little role for positron emission tomography (PET) scans. Features such as heterogeneity on imaging, invasion into fascial planes of muscles, or enlarging size warrant a biopsy.

A biopsy of the soft tissue mass is usually necessary to determine the specific histologic type of sarcoma and grade of malignancy. A treatment plan can then be designed that is tailored to a particular lesion's predicted pattern of local growth, risk of metastasis, and likely sites of distant spread. Ideally, the initial diagnostic procedure should be performed at the center where the patient will be treated. Not only does this facilitate proper placement of the biopsy site, but it also avoids the complications and diagnostic difficulties that can arise if such biopsy samples are mishandled. In the authors' opinion, it is not appropriate for patients with suspected STSs to undergo biopsy in the primary care setting.

Biopsy is performed by fine-needle aspiration (FNA), core-needle biopsy, or incisional or excisional biopsy. An FNA rarely provides enough tissue for histology and cytogenetics, and the authors favor either a carefully planned incisional biopsy or percutaneous core-needle biopsy.

A core biopsy can be performed easily in the clinic with 95% accuracy.[2] This technique is especially useful for large, deep-seated tumors in the pelvis or paraspinal area, because it results in negligible contamination of tissue planes and can provide sufficient information to avoid an open surgical biopsy. One limitation of a core biopsy is that it can miss a small high-grade region in a heterogeneous tumor. This limitation can be overcome by taking three to four core biopsy samples through the same percutaneous approach, and sampling an area of the tumor that, based on its imaging characteristics, most likely represents a high-grade region.

An incisional biopsy can provide 1 cm³ to 2 cm³ of tissue; however, this requires surgery for the patient. In addition, if not properly performed, this procedure runs the risk of making the definitive surgery more morbid for the patient. Incisions for these open biopsies should be oriented along the longitudinal axis of the limb, and meticulous dissection and hemostasis should be employed to avoid flaps and hematomas, which can compromise the definitive surgical resection. A poorly planned and executed biopsy can spread sarcoma cells into the surrounding tissue, requiring more extensive soft tissue resections and

Modern Soft Tissue Pathology, Second Edition, ed. Markku Miettinen. Published by Cambridge University Press. © Cambridge University Press 2016.

functional loss for the patient. Complex reconstructive procedures to achieve surgical wound closure could even be required.

Excisional biopsy is recommended only for small cutaneous or subcutaneous tumors, <3 cm in size. The surgeon again should avoid contamination of surrounding tissue planes and must obtain complete hemostasis. In addition, use of flaps or dissection that would complicate the definitive resection should be avoided.

The most important information obtained from the biopsy is the grade and histologic subtype. Together, this information can be integrated with the size, site, and depth obtained from the imaging and physical examination, as well as patient age and other comorbidities, to derive an optimal treatment plan for an individual patient.

Surgical principles
General principles in sarcoma resection

A properly executed surgical resection that obtains negative margins remains the most important part of the overall treatment of sarcoma. In general, the scope of the excision is dictated by the size, site, and pathology of the tumor, its anatomic relation to normal structures, and the degree of function that would be lost after resection. Although most sarcomas often have a pseudocapsule, the surgeon should avoid using this as a plane of dissection because invariably, viable sarcoma cells extend into the surrounding normal tissue planes well beyond the pseudocapsule. This is particularly true of all high-grade sarcomas and for certain histologic types of low-grade sarcoma. These sarcoma types are best resected with a 1-cm to 2-cm normal tissue margin or intact fascial barrier beyond the lesion, as long as this can be achieved with minimal functional loss.

A marginal excision is when the border of tissue is close (<0.1–0.2 mm), and should be performed only when limited by neurovascular or bony structures. Patients with anticipated resections that would either be marginal or severely limiting functionally might benefit from chemotherapy or radiation therapy, as discussed later in this chapter.

Many patients have an initial biopsy or even resection with positive margins performed at an outside institution, and then present to a tertiary care center for care. This delay and need for a second surgery, although morbid for the patient, does not necessarily translate into a worse outcome for the patient. Patients therefore should be aggressively referred for a definitive operation.[3]

Reconstructive options

Researchers have continued to develop reconstructive techniques over the past three decades, permitting the repair of nerves and vessels and the transfer of muscle and soft tissue to surgically resected defects. This has contributed to the decreased need for amputation and has allowed more extensive operations to be performed with the knowledge that function can be preserved, and defects and dead space can be filled, to enhance wound healing and outcome.

A sarcoma resection that required a nerve resection was previously thought to be best treated with amputation; however, it is now evident that the nerve can either be resected or autografted. Studies on nerve autografts have mixed results, in that a majority of patients with a long segment of reconstruction do not recover enough function to ambulate without the assistance of a brace.[4] Removal of the sciatic, peroneal, or tibial nerves without reconstruction results in a good quality of life and functional outcome, as reported by patients. Although all patients who had a sciatic nerve resection require an ankle/foot orthosis for ambulation, all patients stated a preference for this over amputation.[5] In the upper extremities, removal of the median, ulnar, or radial nerve often can be reconstructed with tendon transfers to restore function.[6]

Venous or prosthetic grafts can be used for vascular reconstruction with acceptable long-term patency rates.[7,8] There is a slightly increased incidence of morbidity when vascular resection is required (e.g., infection, deep venous thrombosis [DVT], edema), which more likely reflects the magnitude of the operation rather than the vascular reconstruction itself.[9]

Soft tissue defects can be replaced with skin or combination skin-muscle flaps. Skin grafts are the most readily applicable for wound coverage. Full-thickness skin grafts provide the most cosmetic repair because they include the full thickness of the dermis for texture and pliability, as well as the epidermal elements of the dermis to incorporate oil glands for a soft and moist skin surface. A small full-thickness skin graft can easily be taken from the flexion creases in the skin such as the neck, antecubital area, or groin to provide cosmetically appealing coverage. The major limitation to the use of full-thickness skin grafts is the smaller amount of skin area that can be harvested from these areas. Occasionally, larger areas of full-thickness skin graft can be harvested from the lower abdomen. The donor site soft tissue is then resected, with the donor site closed as in an abdominoplasty.

Conversely, a partial-thickness skin graft can easily be taken from an extremity or trunk, and if a large surface area is required, this can be meshed to increase the coverage area. The amount of surface area of a surgical wound that can be covered is limited only by the amount of surface area available from the trunk or extremities. The resulting wound at the donor site heals with re-epithelialization, usually requiring about 2 weeks to complete, but leaves scars that are the size and pattern of the grafts.

For the skin graft to survive at the recipient site, there must be adequate nutrient blood supply for passive diffusion in the initial few days, until new capillary ingrowth begins after 72 hours. Therefore, a skin graft cannot be placed directly on bare bone (without viable periosteum), ischemic tissue, or tissue treated in the past with high doses of radiation.

Myocutaneous flaps such as the rectus abdominus, pectoralis major, latissimus dorsi, and sternocleidomastoid rely on the preservation of at least one of the various muscular blood supplies to provide nutrient flow to the accompanying skin and soft tissue. The flap can then be rotated to cover the defect of interest, while the donor site can be closed directly or with a skin graft if a large area of skin is required. Because of the vascularity and bulk, these flaps are ideal for reconstructions of complicated, previously irradiated wounds. In fact, the pectoralis major flap was originally devised for coverage in patients after radical resections of oropharyngeal carcinomas that had been previously treated with full-course radiation therapy.[10] The flaps can be harvested and surgically transferred expeditiously and have a high rate of success, even in patients that are treated with intraoperative brachytherapy.[11,12]

Occasionally, however, resection of the sarcoma requires such extensive resections of anatomic blocks of muscle and tissue that it compromises the vascular supply of the remaining local tissue, precluding the use of local flaps. The only reliable option that is available then is the use of a microvascular free flap. Free flaps entail the transfer of an entire muscular/skin pedicle with its donor vascular pedicle to a new site of surgical defect, where it is attached to the recipient vascular supply. This requires a microvascular anastomosis at the recipient site and is therefore a more skill- and time-intensive procedure. The advantage to free flap reconstructions is that less bulky, more cosmetically appropriate flaps, such as the radial arm free flap, can be fashioned to cover wounds of the face and hands. More complex reconstructive needs can be met by expanding the flap to incorporate bone, such as the fibular free flap, to repair bone and soft tissue defects of the upper extremities and face. The additional time and complexity of reconstruction with a free flap might be too risky for a patient with multiple medical comorbidities. Centers with extensive expertise and experience with various levels of reconstruction have demonstrated a free flap success rate of 98%, in appropriately selected patients, however.[13]

The advances and successes of these reconstructive techniques over these past three decades have expanded the indications of wound coverage from skin grafts to free flaps. This in itself has expanded the possibility of extensive and more complicated resections to increase the chances of cancer control, with the knowledge that the wounds can be reliably repaired.

Site-specific surgical therapy

Surgical strategies based on specific extremity/truncal subtypes

Of >3000 extremity sarcomas removed at the Memorial Sloan-Kettering Cancer Center (MSKCC), the most common subtypes in the extremities are liposarcoma and malignant fibrous histiocytoma (MFH), as shown in Figure 37.1. The most common truncal sarcomas are desmoids, MFH, and liposarcomas, as shown in Figure 37.2.

Primary surgical therapy

Extremity and truncal sarcomas with a low risk of distant spread are the atypical lipomatous tumors/well-differentiated liposarcoma, dermatofibrosarcoma protuberans (DFSP), and desmoids. If the anatomic location allows, these lesions should be treated with wide surgical excision and rarely require additional treatment.

Atypical lipomatous tumors and well-differentiated liposarcomas in the extremities share a similar biologic behavior and should undergo complete excision with negative margins. Often 1-mm to 10-mm margins are acceptable and dictated by proximity to important major neurovascular structures.

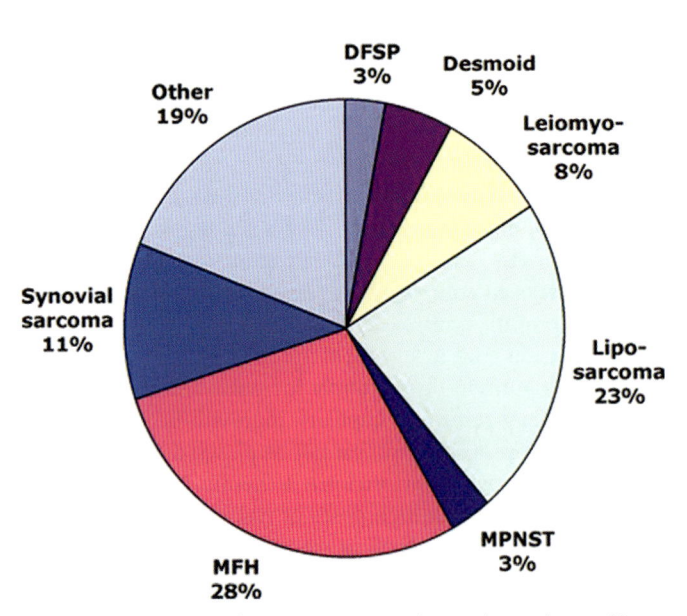

Figure 37.1 Incidence of extremity sarcoma subtypes observed over a 25-year period at MSKCC. Of a total of 3039 cases, liposarcoma and MFH are the most common.

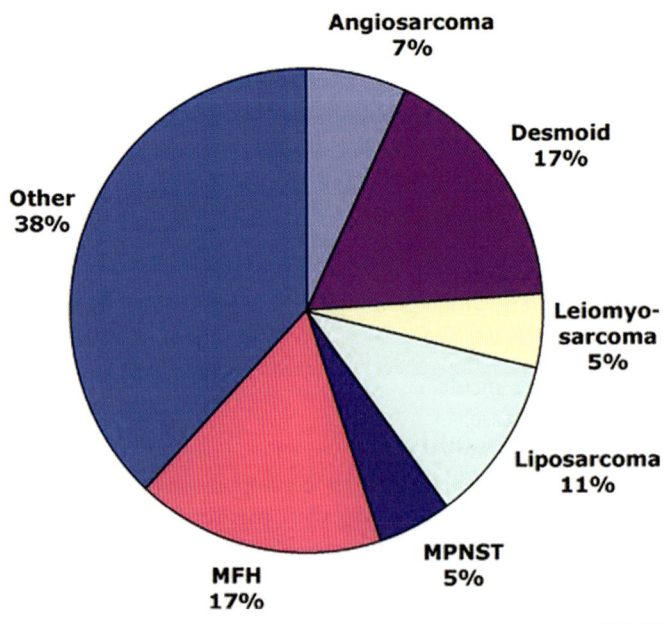

Figure 37.2 Subtypes in 776 cases of truncal sarcoma over 25 years at MSKCC. The most common subtypes are desmoid, MFH, and liposarcoma. The "other" group includes subtypes such as cystosarcoma, sarcoma NOS, fibrosarcoma, and rhabdomyosarcoma.

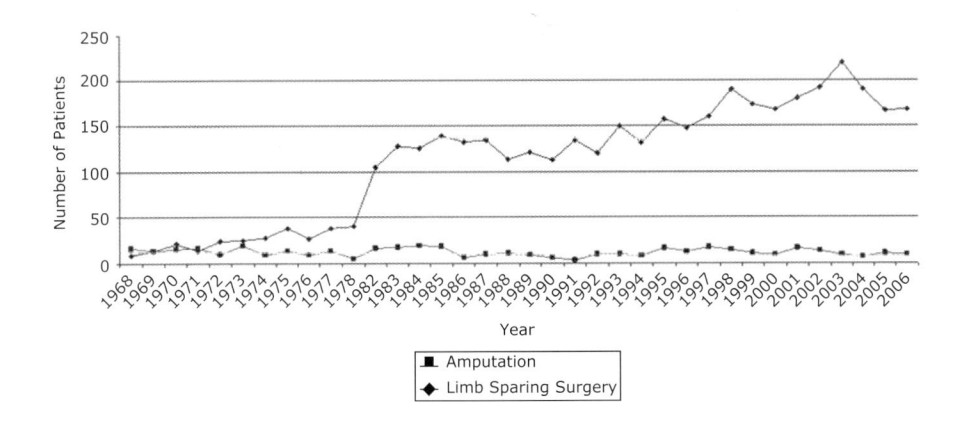

Figure 37.3 Incidence of amputations in patients at MSKCC from 1968 to 2006. Note the increase in the number of limb-sparing surgeries performed since the 1980s. This coincides with clinical trials, which supported the use of multimodality therapy to avoid amputation. Although the number of amputations has remained relatively stable, the number of amputations per procedure performed has markedly decreased.

As long as complete excision is achieved, even patients with microscopically positive margins will not have a recurrence or will have late recurrences (after 5–10 years). The efficacy of the surgery in these lesions is seen in a study of 91 patients who had atypical lipomatous tumors or well-differentiated liposarcoma. There was not a single recurrence in 10 years of follow-up for those patients that did not have a sclerosing component and had a negative-margin resection. Analysis of the entire cohort revealed a 78% local recurrence-free survival at 10 years, with distant metastatic disease developing in only one patient who had dedifferentiated sarcoma in the recurrence.[14] Most of these patients therefore can be managed with surgery alone without the use of adjuvant radiation.

DFSP tends to have papillary-like projections that grow along tissue planes into the adjacent fat tissue that can be difficult to detect in normal subcutaneous tissue. This likely contributes to the high recurrence rate observed after inadequate surgery. Whenever possible, the surgical resection must include 2 cm to 3 cm of skin and normal subcutaneous tissue beyond the gross visible or palpable mass, and careful histologic analysis should confirm clean lateral margins. In addition, the underlying fascia must be removed en bloc with the specimen because this provides a barrier through which most DFSP will not penetrate. With proper surgery, the rate of local recurrence is <5% at 5 years.[15] Patients with microscopic positive margins should be re-excised, and adjuvant radiation is rarely indicated for these lesions. Frozen-section assessment of margins is exceedingly difficult, even for an experienced sarcoma pathologist and should be avoided. The more aggressive form of DFSP, the fibrosarcomatous variant, is associated with a higher incidence of local recurrence and a 1% risk of metastatic disease.[15]

Desmoid tumors (desmoid fibromatoses) of the extremities are deep seated, whereas superficial fibromatoses can often be followed. The management of desmoids is complicated, because they do not have metastatic potential yet they can be locally invasive. Extremity desmoids should be treated with a wide excision (WE) to obtain negative margins, yet preserve functional outcomes. In a series of 189 extremity desmoids, the local recurrence rate at 5 years for patients with a negative margin was 22%. In patients with a positive margin, the recurrence rate was 24%; however, some of those patients had adjuvant radiation.[16] This provides evidence that debilitating operations should not be performed in an effort to obtain negative margins. The role of radiation therapy for extremity desmoids remains controversial, but radiation should be used judiciously. Desmoids are low-grade tumors that never metastasize, and even when they do recur locally, they can remain stable for many years even without treatment.

Role of amputation

Over the years, the incidence of amputation has diminished for extremity sarcoma, as shown in Figure 37.3. This is partly secondary to a prospective, randomized trial that was published in 1982 that did not demonstrate a survival advantage for amputation. Forty-three patients, all undergoing chemotherapy, consented to either amputation or WE with postoperative radiation therapy. Analysis at 5 years demonstrated an increased local recurrence rate in patients with WE; however, there was no difference in disease-specific survival (DSS) rates at 5 years (71% versus 78%, p = 0.75).[17] The efficacy of limb-sparing surgery (LSS) in managing sarcomas has been confirmed in follow-up studies from later periods.[18] Currently, amputation for extremity sarcoma is performed only for extensive lesions in which limb-sparing resection would still leave substantial residual disease or result in a functionless limb.

Role of radiation

Radiation is an effective modality to reduce the risk of local recurrence when used as an adjuvant to surgery. Alternatively, radiation can be employed prior to surgery to shrink a tumor away from vital structures and optimize the surgical margin in patients who present with advanced local disease. The use of radiation is associated with an increased risk of limb edema, fibrosis, bone fracture, wound complications, and even formation of a radiation-associated sarcoma in the future, however. The use of radiation therefore has been appropriately scrutinized and defined by prospective randomized trials.

Extremity/truncal radiation can be delivered by brachytherapy (BRT) catheters or external beam radiation therapy

Chapter 37: Surgical management of soft tissue sarcoma

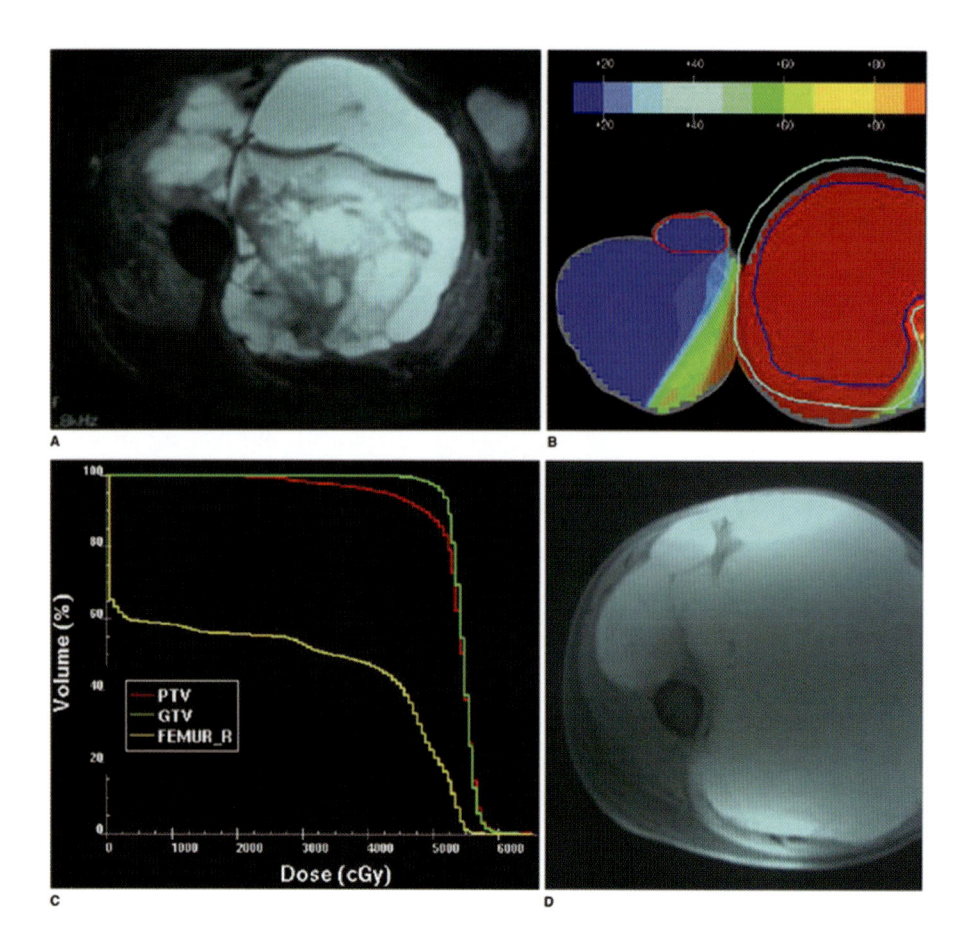

Figure 37.4 Treatment planning in IMRT is the (**A**) pretreatment sarcoma of the extremity. The IMRT treatment mapping is demonstrated (**B** and **C**) with sparing of the circumference of the bone. The postradiation film is shown in (**d**).

(EBRT). A prospective randomized trial compared the use of BRT catheters, placed at the time of surgical resection, to surgical resection alone. BRT was delivered by loading the catheters with iridium-192 at a dose of 42 to 45 gray (Gy) over 4 to 6 days in the postoperative period. Although BRT increased the rate of local control at 5 years in the patients with high-grade tumors (89% versus 66% respectively, p = 0.0025), it was ineffective in patients with low-grade tumors.[19] EBRT, in contrast, improved local control in high-grade and low-grade lesions.[20] DSS was not altered by radiation with either BRT or EBRT.[21,22] Therefore, adjuvant radiation therapy in the form of BRT or EBRT should be considered for patients with a high-grade sarcoma that has been excised with close margins <1 cm as a means to limit local recurrence. Patients with low-grade sarcomas, or high-grade sarcomas <5 cm, can be treated with primary surgery alone if adequate margins (1–2 cm) are obtained, because these tumors have a low local recurrence rate.[22]

There is an association between the timing of the administration of radiation and the development of complications. Preoperative radiation delivers a small targeted amount of radiation to the tumor mass, yet it compromises healing and the rate of wound complications after definitive surgery is higher.[20] In addition, pathologic analysis of the tumor specimen can be more difficult because of treatment effect. In this situation, the treating physician must often rely on the histologic features present in a core or incisional biopsy. Postoperative radiation encompasses the field beyond the incision, and therefore is a larger radiation dose. There is increased morbidity with radiation (defined as limb edema and fibrosis), but the local recurrence and progression-free survival rate is not different between pre- and postoperative radiation at 3 years of follow-up.[20] The risk of increased infection with preoperative radiation therapy, primarily in the lower extremities, must be balanced with the risk of an increased radiation dose and long-term fibrosis of the wound when radiation is delivered in the postoperative time period.

A new therapy evolving in the treatment of sarcoma is intensity-modulated radiation therapy (IMRT). This type of radiation allows a conformal delivery of radiation to intensify in areas of tumor and avoids vital structures such as neurovascular structures and bone. A demonstration of the treatment planning for a patient receiving IMRT is shown in Figure 37.4. Sarcoma resections that involve periosteal stripping or circumferential radiation of the bone are associated with a high risk of bone fractures, reaching 29% at 5 years.[23] A recent study from MSKCC has demonstrated the efficacy of IMRT. Although this report has a relatively short follow-up period (2 years), in this high-risk group of patients where 30% of patients had periosteal stripping and IMRT, the fracture rate was only 6.4%.[24]

Chemotherapy in extremity/truncal sarcomas

Historically, the chemotherapy used in the treatment of sarcoma was the anthracycline, doxorubicin. The enthusiasm for this therapy was in part demonstrated by a randomized trial of adjuvant treatment in patients with extremity sarcoma. There was an improvement in overall survival and DSS rates at 3 years.[25] Later follow-up studies of these patients did not demonstrate a significant survival benefit, however.[26] This was followed by a series of trials that used doxorubicin in combination with a variety of agents. Most of these trials were relatively small, contained heterogeneous groups of sarcoma patients, and had no standardization based on sarcoma subtype. A meta-analysis of the most carefully performed adjuvant trials accumulated 1568 patients in 14 trials that involved doxorubicin alone or in combination with other agents. This group was composed of primarily high-grade (67%), resected (76% negative margin) patients. Slightly more patients did not have radiation (51%) versus those that did receive radiation (47%). There was a statistically significant improvement in the local and distant recurrence-free intervals; however, the overall survival rate at 10 years did not reach statistical significance (p = 0.12). A similar result was demonstrated when only the subgroup of patients with large (>8 cm) and high-grade tumors was analyzed. In the subgroup of patients with an extremity sarcoma, a 7% absolute benefit in overall survival was demonstrated at 10 years (p = 0.029).[27] Despite the limitations of a retrospective meta-analysis, this work suggests that chemotherapy with anthracyclines alone is of limited benefit for patients with STS. Since 1990, anthracyclines have been combined with alkylating agents in an attempt to improve efficacy, and recently, several studies have tried to define the subgroups of patients that are most likely to benefit from combination chemotherapy.

Ifosfamide is an alkylating agent originally demonstrated to have efficacy in the treatment of advanced or metastatic sarcoma.[28-30] In a randomized trial in which patients with metastatic or unresectable sarcoma were treated with doxorubicin and dacarbazine alone or in combination with ifosfamide, there was a small but significant increase in time to disease progression (6 months with ifosfamide versus 4 months with standard chemotherapy), and an increase in overall survival (13 versus 12 months, respectively).[30] There have been multiple additional trials that include ifosfamide, which are difficult to interpret again because of small sample size and absence of stratification based on sarcoma subtype. Because therapies for treatment of sarcoma are limited, however, studies have continued to investigate further whether there is a role for this drug in the treatment of sarcoma.

A recent study used ifosfamide in an adjuvant fashion in 104 patients with high-grade extremity sarcoma after optimal surgery. One arm of the study underwent observation, whereas the other arm underwent treatment with adjuvant epidoxorubicin and ifosfamide. After 2 years the trial was stopped, because at interim analysis there was a significant increase in disease-free survival (48 months with chemotherapy versus 16 months with observation, p = 0.04) and overall survival (75 versus 46 months, respectively, p = 0.03).[31] On longer follow-up, however, the difference in survival rates no longer persisted.[32]

Although the role of ifosfamide in adjuvant use remains unclear, there have been data to support its role in a neo-adjuvant approach. Although the studies are relatively small and are retrospective in nature, they have identified those subtypes of sarcoma that might be more chemosensitive. In addition, neo-adjuvant chemotherapy allows the physician to monitor the response *in vivo* to therapy and to avoid subsequent chemotherapy for those patients who progress after the first two cycles of chemotherapy.

Neo-adjuvant chemotherapy is currently offered to all patients with large (≥10 cm), deep, high-grade extremity tumors, round cell and pleomorphic liposarcomas >5 cm, and synovial sarcomas >5 cm. The use of neo-adjuvant ifosfamide in combination with doxorubicin and mesna resulted in a 21% improvement in DSS rate at 3 years in all patients with extremity subtypes of high-grade, deep tumors, >10 cm. In this study, the most commonly represented subtypes were MFH, liposarcoma, synovial sarcoma, and leiomyosarcoma.[33]

The role of neo-adjuvant chemotherapy has been further clarified in liposarcomas. There are three biologic groups of liposarcoma, encompassing five subtypes: (1) well-differentiated/dedifferentiated, (2) myxoid/round cell, and (3) pleomorphic. The high-grade subtypes with a high risk of systemic recurrence are myxoid/round cell (>5% round cell component) and pleomorphic.[34-38] A recent retrospective analysis of patients with extremity liposarcomas from MSKCC and UCLA compared patients who received chemotherapy (in an adjuvant or neo-adjuvant manner) to a similar cohort of patients who did not receive chemotherapy. The doxorubicin regimens were alone or in combination with methotrexate, cyclophosphamide, or cisplatin. All the ifosfamide regimens included doxorubicin with mesna, cisplatin, or dacarbazine. There was no improvement in survival demonstrated with doxorubicin therapy; however, ifosfamide combined with doxorubicin therapy demonstrated an improved DSS at 5 years compared with no chemotherapy (92% versus 65% respectively, p = 0.0003).[39] This translates into a 14% survival advantage at 5 years for lesions ≤10 cm, and a 31% survival benefit for lesions >10 cm. Based on the results of this retrospective study, and in the absence of any prospective randomized data, neo-adjuvant doxorubicin and ifosfamide chemotherapy is offered to young patients with large (>5 cm), extremity round cell or pleomorphic liposarcomas. Dedifferentiated liposarcoma of the extremities is not particularly responsive to chemotherapy and has a substantially smaller risk of distant metastasis than other high-grade extremity sarcomas, and so neo-adjuvant chemotherapy is not indicated for this subtype.

Synovial sarcoma predominates in the extremities of young adults, and unfortunately is associated with a high risk of distant disease and death, with 34% of patients dying of disease in 5 years. The risk of mortality increases with tumors >5 cm or with bone or neurovascular involvement.[40] Data to support systemic chemotherapy in patients with synovial cell sarcoma come from a joint study of patients from MSKCC and UCLA. Patients were analyzed in a retrospective cohort that included the period of treatment with ifosfamide. Most of the patients were treated with LSS and radiation. The patients were divided into two groups: those that received additional systemic chemotherapy with ifosfamide and those that underwent surgery alone. At a median follow-up of 58 months, treatment with ifosfamide was associated with an improvement in DSS over observation (88% versus 67%, p = 0.01).[41] Based on the improvement demonstrated, given the lack of randomized trial data, the authors now offer chemotherapy to patients with extremity synovial sarcomas >5 cm and acceptable comorbidities.

Sarcoma subtypes requiring multimodality treatment

Certain sarcoma subtypes tend to have ill-defined patterns of growth that expand along fascial planes or even present with skip areas. Pathologic knowledge of these subtypes warrants an aggressive surgical treatment with 2-cm margins and vigilance for sites of multifocal disease.

Myxofibrosarcomas are one such tumor that predominates in the extremities of older patients. The mass can have satellite nodules or fascial extension well beyond the primary location of the tumor.[42,43] This partly explains the high recurrence rate for these lesions, as high as 57% at 21 months of follow-up in some series.[44] For this reason, careful assessment of the area around the palpable tumor is paramount in planning the resection, to encompass all areas at risk with a wide margin. CT or MRI imaging of these lesions can often assist in demonstrating the extension of the tumor along fascial planes, as well as sites of unusual distant spread such as bone, soft tissue, and mesentery.[45] These tumors are often relatively resistant to radiation and chemotherapy, making surgery a critical component of local control. High-grade, >10-cm myxofibrosarcomas of the extremities or trunk have a high incidence of distant failure, and patients with these tumors should be considered good candidates for neo-adjuvant chemotherapy, provided that they have acceptable comorbidities. Most high-grade, >5-cm, and deep extremity myxofibrosarcomas require treatment with adjuvant radiation therapy.

Angiosarcoma is an aggressive tumor that mandates a wide excision, because it can be multifocal at presentation and difficult to obtain negative margins. Angiosarcoma of the liver is a rare phenomenon noted to occur in patients with vinyl chloride exposure.[46,47] Patients exposed to radiation have an increased risk of developing angiosarcoma, and there is an increased incidence in angiosarcoma of the breast and pelvis after radiation for malignancy.[48,49] Patients with chronic lymphedema are also at increased risk of developing extremity angiosarcoma (Stewart–Treves syndrome).[50] The overall survival is poor, with ranges reported from 21% at 2 years[51] to 40% at 5 years,[49] so that the utility of radiation, a potential causative factor, in reducing a local recurrence must be balanced with the development of an aggressive secondary malignancy. In patients with extensive and multifocal angiosarcoma at presentation, paclitaxel-based neo-adjuvant chemotherapy can be beneficial for reducing systemic risk and enhancing local control prior to definitive surgery.[52,53]

Malignant peripheral nerve sheath tumors (MPNSTs) can arise sporadically or when associated with neurofibromatosis 1 (NF1) from neurofibromas. This presents a difficult problem in patients with NF1. Any mass that expands rapidly or shows evidence of central necrosis on imaging should be removed. MPNSTs also tend to grow in an infiltrative pattern and, when arising from a major motor nerve, they should not be enucleated from the nerve, but the motor nerve must be sacrificed. Thus, the potential for functional nerve loss has to be discussed with the patient and accepted or, in some cases, nerve grafts or functional tendon grafts can be performed to minimize the functional deficit. Often the route of extension is beyond the palpable mass, up the fascicle of a nerve bundle; surgery therefore should include at least a 2-cm soft tissue resection margin with resection of 3 cm to 4 cm of the involved nerve beyond any palpable or visible disease whenever possible. Survival rates for patients with MPNSTs treated aggressively parallel those of other high-grade sarcomas.[54] High-grade, >10-cm MPNSTs of the extremities or trunk have a high incidence of distant failure, and patients with acceptable comorbidities should be considered candidates for neo-adjuvant chemotherapy. Most patients with deep, high-grade, >5-cm extremity MPNSTs will also require the addition of adjuvant radiation therapy.

Histologic subtypes requiring primary chemotherapy

Chemotherapy is usually indicated as primary treatment for Ewing sarcoma/primitive neuroectodermal tumor and rhabdomyosarcoma. Ewing sarcoma affects children in the bone and occurs most frequently in soft tissue at any age. Historically, the survival was poor, on the order of 10% to 20% with surgery or radiation therapy alone. New regimens of multiagent chemotherapy with ifosfamide, etoposide, and vincristine, doxorubicin, cyclophosphamide, and actinomycin D (VACA) are associated with a 5-year survival rate of 72%.[55] Typically, the chemotherapy is given in a neo-adjuvant fashion for four to six cycles, followed by surgery and further adjuvant therapy.

Rhabdomyosarcoma is the most common solid tumor of children.[56] There are five types: embryonic, alveolar, botryoid, pleomorphic, and anaplastic. The pleomorphic subtype has occurred in adults. The primary treatment for this tumor is also chemotherapy with vincristine, doxorubicin, and cyclophosphamide (VAC) first, followed by surgical resection and adjuvant chemotherapy. This has improved the 5-year survival rate from 20% to >70%.[57] An algorithm of treatment for

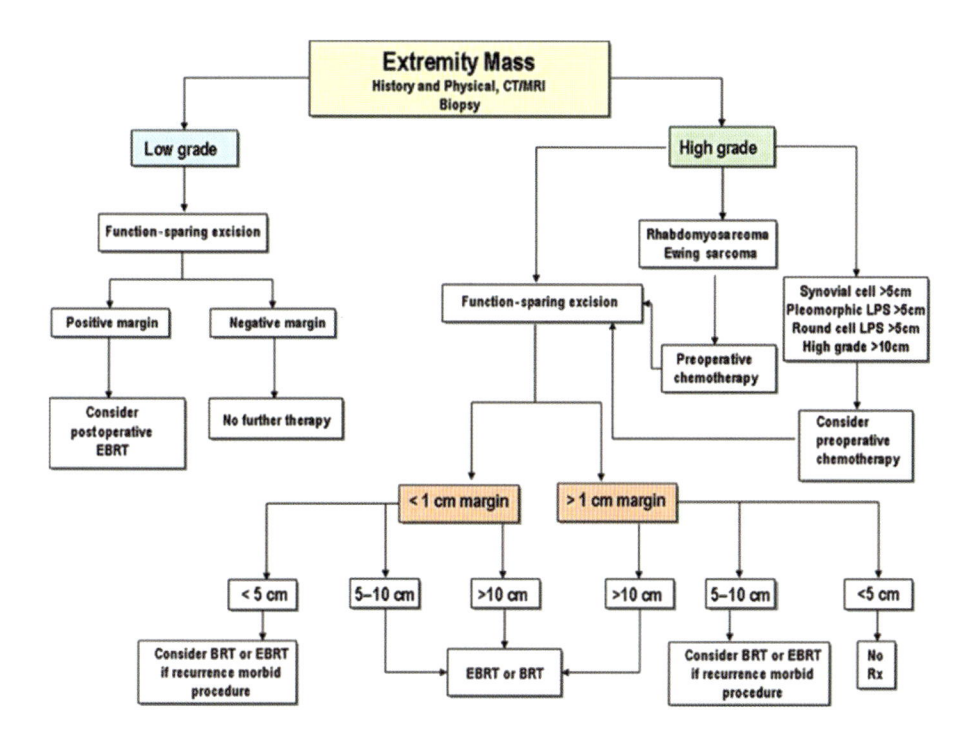

Figure 37.5 Algorithm for treatment in extremity sarcoma.

extremity lesions is demonstrated in Figure 37.5. Patients with rhabdomyosarcoma and Ewing sarcoma often harbor micrometastatic disease that is not detectable with conventional imaging modalities and should be treated with chemotherapy immediately after diagnosis, because once clinically detectable metastatic disease develops, patients are largely incurable. For this reason, extensive procedures to resect the primary tumor are not indicated and only delay potentially curative chemotherapy.

Retroperitoneal sarcoma

Subtypes and goals of surgery

The most common subtypes presenting as retroperitoneal sarcomas are liposarcomas and leiomyosarcomas, as shown in Figure 37.6. Retroperitoneal sarcomas are often detected late, because the abdominal cavity accommodates significant growth without specific symptoms. Once found, the lesions are usually large (>10 cm) and encroach on major abdominal structures.[58]

On presentation, there is no need for biopsy unless one thinks that other diseases besides sarcoma are in the differential diagnosis. Additional tumors to consider are lymphomas, angiomyolipomas, pancreatic tail tumors, and metastatic germ cell tumors. Primary treatment necessitates an aggressive surgical resection. The surgeon who performs the first definitive surgery must remove all disease and should not hesitate to excise and remove adjacent organs or viscera. If the mass does not encase the renal vessels or invade the renal hilum, a parenchyma-sparing capsular stripping can be performed without compromising DSS.[59] Patients with the best survival rate have had a surgical resection with negative margins. For example, in one study of over 500 patients, an R0 resection

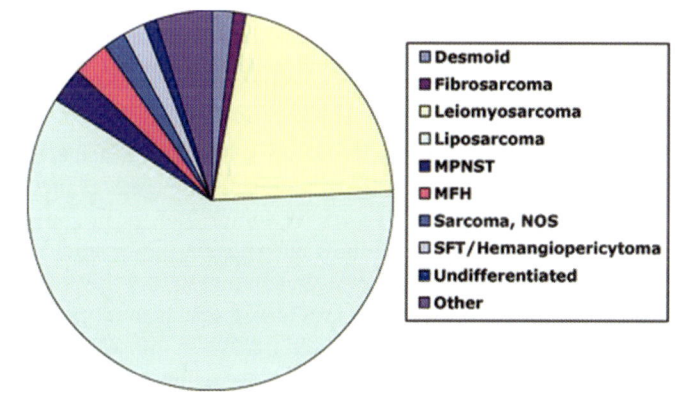

Figure 37.6 Subtypes of sarcoma in the retroperitoneum seen over a 25-year period at MSKCC. Note that out of the 684 cases, most of the retroperitoneal tumors are either liposarcomas or leiomyosarcomas.

correlated with a median survival of 103 months, whereas those that had had an R2 resection had a median survival of 18 months.[58]

Patients with retroperitoneal liposarcomas can be further risk stratified based on subtypes. In a study of 177 patients with retroperitoneal liposarcomas, histologic subtype was the most important predictor of DSS. Patients with the well-differentiated and dedifferentiated subtype had an 83% and 20% DSS at 5 years, respectively. The local recurrence rate for patients with well-differentiated liposarcoma was 31% at 3 years compared with 83% at 3 years for dedifferentiated liposarcoma. The dedifferentiated subtype had a 30% risk of distant disease at 3 years. In contrast, patients with well-differentiated liposarcomas rarely developed distant metastases (1% risk of distant disease at 3 years).[59]

Radiation in retroperitoneal sarcomas

Retroperitoneal sarcomas tend to recur locally, with an incidence of 20% to 50%, depending on length of follow-up.[58,60,61] Median survival after local recurrence is only 28 months;[58] therefore, consideration has been given to both chemotherapy and radiation therapy in retroperitoneal sarcomas. Although there are few data to support the use of chemotherapy, radiation, which addresses the local treatment failure, has been addressed.

Radiation of the retroperitoneum is a difficult task because the abdominal cavity is difficult to penetrate, and radiation cannot be delivered without risk to the surrounding vessels and intestine. Small studies suggested high-dose radiation improves local control.[62,63] A prospective trial of preoperative radiation in patients undergoing a R0 or R1 resection demonstrated an improvement to 60% local recurrence-free survival in intermediate or high-grade lesions, at a median follow-up of 40 months.[64] Further multi-institutional randomized trials are needed to define the benefit of preoperative radiation therapy for the treatment of retroperitoneal sarcoma; however, these trials have closed secondary to poor accrual. Newer techniques of EBRT such as IMRT with dose painting might enable administration of radiation therapy preoperatively to retroperitoneal sarcomas, with less toxicity to surrounding organs and vital structures.

Locally recurrent retroperitoneal sarcoma

Recurrence in retroperitoneal sarcomas is more difficult to treat, in part secondary to prior contamination of tissue planes, as well as the intrinsically aggressive tumor biology in some cases. Resection must encompass all palpable tumor and all potential microscopic foci present in adjacent tissues traversed during previous surgical procedures (e.g., fascial planes opened up in dissection, skin flaps, and drain tracks). In difficult cases, it might be possible for the surgeon to remove the entire recurrent tumor only by dissecting it off nerve, artery, and vein. This usually fails to provide local control, unless adjuvant radiation therapy is administered (either by EBRT/IMRT or BRT techniques), or if neo-adjuvant radiation therapy or chemotherapy is given before the tumor is excised. The other alternative is to resect the nerve, artery, or vein to achieve a clean margin.

Gastrointestinal stromal tumors
Treatment of primary disease

Gastrointestinal stromal tumors (GISTs) occur as submucosal or pedunculated lesions arising in the gastrointestinal tract, most commonly the stomach. On CT or MRI they appear as well-circumscribed, low-density lesions, with possible areas of enhancement or necrosis. If a mass is suspected to be a GIST, it should be removed for both diagnosis and treatment. Biopsy is not necessary because it risks rupturing the mass. Complete surgical resection offers the best chance of cure.

Factors that contribute to a risk of recurrence are site of primary GIST, size, and mitotic index. For example, a distal small bowel GIST >5 cm with a mitotic index >5 per 50 high-power fields (HPFs) has an 85% chance of disease progression, whereas a tumor <2 cm with ≤5 mitoses per HPF has a negligible chance of recurrence, regardless of site.[65,66]

It is now known that the majority (80%) of GIST tumors have an activating mutation of the KIT receptor tyrosine kinase.[67] Exon 11 mutations, the most common, are associated with a good response to imatinib, while exon 9 mutations tend to occur in lesions of the small bowel and colon, which are less sensitive to imatinib.[68,69] A small percentage (5–10%) of GISTs harbor a mutation in platelet-derived growth factor receptor alpha (PDGFRA),[70] some of which could be resistant to imatinib, but the new multikinase inhibitors such as sunitinib might have efficacy.

An American College of Surgeons Oncology Group (ACO-SOG) intergroup trial randomized patients with primary GIST tumors >3 cm to 1 year of imatinib at 400 mg daily or placebo after complete surgical resection. At the interim analysis, there was a significant improvement in recurrence-free survival with imatinib treatment, although no difference was seen in DSS.[71] Longer follow-up will help to clarify if imatinib prevents or merely delays the onset of disease recurrence. It might be that therapy initiated at the time of disease recurrence is equally effective.

Surgical treatment of advanced GIST

Therapy with imatinib can produce dramatic responses. For example, 70% of patients in a randomized trial of treatment with high-dose imatinib in metastatic disease were alive at 2 years, and 50% were alive without disease progression.[72] This raises the question of whether surgery should be used in conjunction with imatinib therapy to reduce tumor burden to delay or prevent the emergence of resistant disease. One important factor in determining surgical intervention is that while on imatinib, resistant clones can develop in as little as 6 months, with a median time to drug resistance of 20 months.[72] Most patients with advanced disease will achieve maximal response to imatinib by 6 months and at that time should be evaluated by an experienced oncologic surgeon for consideration of complete resection of all clinically detectable disease.

A recent study of 40 patients with metastatic GIST supports surgical intervention for those patients with disease that responds, or has no more than one focus of resistance, to imatinib. Patients with multifocal resistance, however, have a poor 1-year survival rate (36% alive at 1 year) and should not proceed to surgical debulking, but should be considered for a clinical trial with a second-line tyrosine kinase inhibitor.[73]

Summary

STSs represent a heterogeneous group of tumors with varying malignant potential. Surgery with an adequate margin of resection remains the mainstay of treatment and often is sufficient for most low-grade lesions and many small high-grade sarcomas. Radiation reduces the incidence of recurrence in the extremities but must be balanced with the long-term side effects and risk of secondary malignancies. Chemotherapy has efficacy in rhabdomyosarcoma and Ewing sarcoma, and in some histologic types of high-grade extremity sarcomas (e.g., round cell and pleomorphic liposarcoma, synovial sarcoma). For other histologic subtypes, chemotherapy should be reserved for patients with high-risk (>10 cm) high-grade sarcomas.

In patients with retroperitoneal and visceral sarcomas, complete surgical resection is a critical determinant of outcome. Most patients' disease tends to recur locally and this, in contrast to extremity disease, is often difficult to control with additional surgery, eventually leading to death from progressive locoregional disease. Most retroperitoneal and visceral sarcomas are poorly responsive to conventional chemotherapy, so chemotherapy is generally used only for palliation of disease that can no longer be completely removed surgically. The role of radiation remains controversial, but seems to provide minimal benefit with substantial toxicity when used in the adjuvant setting. Carefully targeted radiation therapy might best be used in select patients with locally advanced retroperitoneal sarcoma in a neo-adjuvant fashion to improve resectability. GIST tumors represent the success story in which a rational target was identified, the constitutively active KIT receptor tyrosine kinase, and effective therapy with a selective tyrosine kinase inhibitor was instituted. As physicians continue to learn more about the biology and behavior of each sarcoma subtype through improved molecular techniques, opportunities for effective treatment will continue to evolve.

References

1. Lawrence W Jr, Donegan WL, Natarajan N, et al. Adult soft tissue sarcomas: a pattern of care survey of the American College of Surgeons. *Ann Surg* 1987;205(4):349–359.
2. Heslin MJ, Lewis JJ, Woodruff JM, Brennan MF. Core needle biopsy for diagnosis of extremity soft tissue sarcoma. *Ann Surg Oncol* 1997;4(5):425–431.
3. Lewis JJ, Leung D, Espat J, Woodruff JM, Brennan MF. Effect of reresection in extremity soft tissue sarcoma. *Ann Surg* 2000;231(5):655–663.
4. Melendez M, Brandt K, Evans GR. Sciatic nerve reconstruction: limb preservation after sarcoma resection. *Ann Plast Surg* 2001;46(4):375–381.
5. Brooks AD, Gold JS, Graham D, et al. Resection of the sciatic, peroneal, or tibial nerves: assessment of functional status. *Ann Surg Oncol* 2002;9(1):41–47.
6. Saint-Cyr M, Langstein HN. Reconstruction of the hand and upper extremity after tumor resection. *J Surg Oncol* 2006;94(6):490–503.
7. Baxter BT, Mahoney C, Johnson PJ, et al. Concomitant arterial and venous reconstruction with resection of lower extremity sarcomas. *Ann Vasc Surg* 2007;21(3):272–279.
8. Adelani MA, Holt GE, Dittus RS, Passman MA, Schwartz HS. Revascularization after segmental resection of lower extremity soft tissue sarcomas. *J Surg Oncol* 2007;95(6):455–460.
9. Ghert MA, Davis AM, Griffin AM, et al. The surgical and functional outcome of limb-salvage surgery with vascular reconstruction for soft tissue sarcoma of the extremity. *Ann Surg Oncol* 2005;12(12):1102–1110.
10. Ariyan S. The functional pectoralis major musculocutaneous island flap for head and neck reconstruction. *Plast Reconstr Surg* 1990;86(4):807–808.
11. Ross DA, Hundal JS, Son YH, et al. Microsurgical free flap reconstruction outcomes in head and neck cancer patients after surgical extirpation and intraoperative brachytherapy. *Laryngoscope* 2004;114(7):1170–1176.
12. Duman H, Evans GR, Reece G, et al. Brachytherapy: reconstructive options and the role of plastic surgery. *Ann Plast Surg* 2000;45(5):477–480.
13. Bui DT, Cordeiro PG, Hu QY, et al. Free flap reexploration: indications, treatment, and outcomes in 1193 free flaps. *Plast Reconstr Surg* 2007;119(7):2092–2100.
14. Kooby DA, Antonescu CR, Brennan MF, Singer S. Atypical lipomatous tumor/well-differentiated liposarcoma of the extremity and trunk wall: importance of histological subtype with treatment recommendations. *Ann Surg Oncol* 2004;11(1):78–84.
15. Bowne WB, Antonescu CR, Leung DH, et al. Dermatofibrosarcoma protuberans: a clinicopathologic analysis of patients treated and followed at a single institution. *Cancer* 2000;88(12):2711–2720.
16. Merchant NB, Lewis JJ, Woodruff JM, Leung DH, Brennan MF. Extremity and trunk desmoid tumors: a multifactorial analysis of outcome. *Cancer* 1999;86(10):2045–2052.
17. Rosenberg SA, Tepper J, Glatstein E, et al. The treatment of soft-tissue sarcomas of the extremities: prospective randomized evaluations of (1) limb-sparing surgery plus radiation therapy compared with amputation and (2) the role of adjuvant chemotherapy. *Ann Surg* 1982;196(3):305–315.
18. Williard WC, Hajdu SI, Casper ES, Brennan MF. Comparison of amputation with limb-sparing operations for adult soft tissue sarcoma of the extremity. *Ann Surg* 1992;215(3):269–275.
19. Pisters PW, Harrison LB, Leung DH, et al. Long-term results of a prospective randomized trial of adjuvant brachytherapy in soft tissue sarcoma. *J Clin Oncol* 1996;14(3):859–868.
20. O'Sullivan B, Davis AM, Turcotte R, et al. Preoperative versus postoperative radiotherapy in soft-tissue sarcoma of the limbs: a randomised trial. *Lancet* 2002;359(9325):2235–2241.
21. Yang JC, Chang AE, Baker AR, et al. Randomized prospective study of the benefit of adjuvant radiation therapy in the treatment of soft tissue sarcomas of the extremity. *J Clin Oncol* 1998;16(1):197–203.
22. Baldini EH, Goldberg J, Jenner C, et al. Long-term outcomes after function-sparing surgery without radiotherapy for soft tissue sarcoma of the extremities and trunk. *J Clin Oncol* 1999;17(10):3252–3259.

23. Lin PP, Schupak KD, Boland PJ, Brennan MF, Healey JH. Pathologic femoral fracture after periosteal excision and radiation for the treatment of soft tissue sarcoma. *Cancer* 1998;82(12):2356–2365.

24. Alektiar KM, Hong L, Brennan MF, Della-Biancia C, Singer S. Intensity modulated radiation therapy for primary soft tissue sarcoma of the extremity: preliminary results. *Int J Radiat Oncol Biol Phys* 2007;68(2):458–464.

25. Rosenberg SA, Tepper J, Glatstein E, et al. Prospective randomized evaluation of adjuvant chemotherapy in adults with soft tissue sarcomas of the extremities. *Cancer* 1983;52(3):424–434.

26. Chang AE, Kinsella T, Glatstein E, et al. Adjuvant chemotherapy for patients with high-grade soft-tissue sarcomas of the extremity. *J Clin Oncol* 1988;6(9):1491–1500.

27. Sarcoma Meta-Analysis Collaboration. Adjuvant chemotherapy for localised resectable soft-tissue sarcoma of adults: meta-analysis of individual data. *Lancet* 1997;350(9092):1647–1654.

28. Elias A, Ryan L, Sulkes A, et al. Response to mesna, doxorubicin, ifosfamide, and dacarbazine in 108 patients with metastatic or unresectable sarcoma and no prior chemotherapy. *J Clin Oncol* 1989;7(9):1208–1216.

29. Elias AD, Eder JP, Shea T, et al. High-dose ifosfamide with mesna uroprotection: a phase I study. *J Clin Oncol* 1990;8(1):170–178.

30. Antman KH, Elias A, Ryan L. Ifosfamide and mesna: response and toxicity at standard- and high-dose schedules. *Semin Oncol* 1990;17(2 Suppl 4):68–73.

31. Frustaci S, Gherlinzoni F, De Paoli A, et al. Adjuvant chemotherapy for adult soft tissue sarcomas of the extremities and girdles: results of the Italian randomized cooperative trial. *J Clin Oncol* 2001;19(5):1238–1247.

32. Frustaci S, De Paoli A, Bidoli E, et al. Ifosfamide in the adjuvant therapy of soft tissue sarcomas. *Oncology* 2003;65(Suppl 2):80–84.

33. Grobmyer SR, Maki RG, Demetri GD, et al. Neo-adjuvant chemotherapy for primary high-grade extremity soft tissue sarcoma. *Ann Oncol* 2004;15(11):1667–1672.

34. Chang HR, Hajdu SI, Collin C, Brennan MF. The prognostic value of histologic subtypes in primary extremity liposarcoma. *Cancer* 1989;64(7):1514–1520.

35. Gebhard S, Coindre JM, Michels JJ, et al. Pleomorphic liposarcoma: clinicopathologic, immunohistochemical, and follow-up analysis of 63 cases: a study from the French Federation of Cancer Centers Sarcoma Group. *Am J Surg Pathol* 2002;26(5):601–616.

36. Antonescu CR, Tschernyavsky SJ, Decuseara R, et al. Prognostic impact of P53 status, TLS-CHOP fusion transcript structure, and histological grade in myxoid liposarcoma: a molecular and clinicopathologic study of 82 cases. *Clin Cancer Res* 2001;7(12):3977–3987.

37. McCormick D, Mentzel T, Beham A, Fletcher CD. Dedifferentiated liposarcoma: clinicopathologic analysis of 32 cases suggesting a better prognostic subgroup among pleomorphic sarcomas. *Am J Surg Pathol* 1994;18(12):1213–1223.

38. Henricks WH, Chu YC, Goldblum JR, Weiss SW. Dedifferentiated liposarcoma: a clinicopathological analysis of 155 cases with a proposal for an expanded definition of dedifferentiation. *Am J Surg Pathol* 1997;21(3):271–281.

39. Eilber FC, Eilber FR, Eckardt J, et al. The impact of chemotherapy on the survival of patients with high-grade primary extremity liposarcoma. *Ann Surg* 2004;240(4):686–695; discussion 695–697.

40. Lewis JJ, Antonescu CR, Leung DH, et al. Synovial sarcoma: a multivariate analysis of prognostic factors in 112 patients with primary localized tumors of the extremity. *J Clin Oncol* 2000;18(10):2087–2094.

41. Eilber FC, Brennan MF, Eilber FR, et al. Chemotherapy is associated with improved survival in adult patients with primary extremity synovial sarcoma. *Ann Surg* 2007;246(1):105–113.

42. Mentzel T, Calonje E, Wadden C, et al. Myxofibrosarcoma: clinicopathologic analysis of 75 cases with emphasis on the low-grade variant. *Am J Surg Pathol* 1996;20(4):391–405.

43. Angervall L, Kindblom LG, Merck C. Myxofibrosarcoma: a study of 30 cases. *Acta Pathol Microbiol Scand A* 1977;85A(2):127–140.

44. Huang HY, Lal P, Qin J, Brennan MF, Antonescu CR. Low-grade myxofibrosarcoma: a clinicopathologic analysis of 49 cases treated at a single institution with simultaneous assessment of the efficacy of 3-tier and 4-tier grading systems. *Hum Pathol* 2004;35(5):612–621.

45. Waters B, Panicek DM, Lefkowitz RA, et al. Low-grade myxofibrosarcoma: CT and MRI patterns in recurrent disease. *AJR Am J Roentgenol* 2007;188(2):W193–W198.

46. Mundt KA, Dell LD, Austin RP, et al. Historical cohort study of 10 109 men in the North American vinyl chloride industry, 1942–72: update of cancer mortality to 31 December 1995. *Occup Environ Med* 2000;57(11):774–781.

47. Lee FI, Smith PM, Bennett B, Williams DM. Occupationally related angiosarcoma of the liver in the United Kingdom 1972–1994. *Gut* 1996;39(2):312–318.

48. Strobbe LJ, Peterse HL, van Tinteren H, Wijnmaalen A, Rutgers EJ. Angiosarcoma of the breast after conservation therapy for invasive cancer, the incidence and outcome: an unforseen sequela. *Breast Cancer Res Treat* 1998;47(2):101–109.

49. Cha C, Antonescu CR, Quan ML, Maru S, Brennan MF. Long-term results with resection of radiation-induced soft tissue sarcomas. *Ann Surg* 2004;239(6):903–909; discussion 909–910.

50. Stewart FW, Treves N. Classics in oncology: lymphangiosarcoma in postmastectomy lymphedema: a report of six cases in elephantiasis chirurgica. *CA Cancer J Clin* 1981;31(5):284–299.

51. Naka N, Ohsawa M, Tomita Y, et al. Prognostic factors in angiosarcoma: a multivariate analysis of 55 cases. *J Surg Oncol* 1996;61(3):170–176.

52. Skubitz KM, Haddad PA. Paclitaxel and pegylated-liposomal doxorubicin are both active in angiosarcoma. *Cancer* 2005;104(2):361–366.

53. Fata F, O'Reilly E, Ilson D, et al. Paclitaxel in the treatment of patients with angiosarcoma of the scalp or face. *Cancer* 1999;86(10):2034–2037.

54. Vauthey JN, Woodruff JM, Brennan MF. Extremity malignant peripheral nerve sheath tumors (neurogenic

55. Grier HE, Krailo MD, Tarbell NJ, et al. Addition of ifosfamide and etoposide to standard chemotherapy for Ewing's sarcoma and primitive neuroectodermal tumor of bone. *N Engl J Med* 2003;348(8):694–701.

56. Linet MS, Ries LA, Smith MA, Tarone RE, Devesa SS. Cancer surveillance series: recent trends in childhood cancer incidence and mortality in the United States. *J Natl Cancer Inst* 1999;91(12):1051–1058.

57. Paulino AC, Okcu MF. Rhabdomyosarcoma. *Curr Probl Cancer* 2008;32(1):7–34.

58. Lewis JJ, Leung D, Woodruff JM, Brennan MF. Retroperitoneal soft-tissue sarcoma: analysis of 500 patients treated and followed at a single institution. *Ann Surg* 1998;228(3):355–365.

59. Singer S, Antonescu CR, Riedel E, Brennan MF. Histologic subtype and margin of resection predict pattern of recurrence and survival for retroperitoneal liposarcoma. *Ann Surg* 2003;238(3):358–370; discussion 370–371.

60. Fabre-Guillevin E, Coindre JM, Somerhausen Nde S, et al. Retroperitoneal liposarcomas: follow-up analysis of dedifferentiation after clinicopathologic reexamination of 86 liposarcomas and malignant fibrous histiocytomas. *Cancer* 2006;106(12):2725–2733.

61. Karakousis CP, Velez AF, Gerstenbluth R, Driscoll DL. Resectability and survival in retroperitoneal sarcomas. *Ann Surg Oncol* 1996;3(2):150–158.

62. Tepper JE, Suit HD, Wood WC, et al. Radiation therapy of retroperitoneal soft tissue sarcomas. *Int J Radiat Oncol Biol Phys* 1984;10(6):825–830.

63. Sindelar WF, Kinsella TJ, Chen PW, et al. Intraoperative radiotherapy in retroperitoneal sarcomas: final results of a prospective, randomized, clinical trial. *Arch Surg* 1993;128(4):402–410.

64. Pawlik TM, Pisters PW, Mikula L, et al. Long-term results of two prospective trials of preoperative external beam radiotherapy for localized intermediate- or high-grade retroperitoneal soft tissue sarcoma. *Ann Surg Oncol* 2006;13(4):508–517.

65. Dematteo RP, Gold JS, Saran L, et al. Tumor mitotic rate, size, and location independently predict recurrence after resection of primary gastrointestinal stromal tumor (GIST). *Cancer* 2008;112(3):608–615.

66. Raut CP, DeMatteo RP. Prognostic factors for primary GIST: prime time for personalized therapy? *Ann Surg Oncol* 2008;15(1):4–6.

67. Hirota S, Isozaki K, Moriyama Y, et al. Gain-of-function mutations of c-kit in human gastrointestinal stromal tumors. *Science* 1998;279(5350):577–580.

68. Heinrich MC, Corless CL, Demetri GD, et al. Kinase mutations and imatinib response in patients with metastatic gastrointestinal stromal tumor. *J Clin Oncol* 2003;21(23):4342–4349.

69. Antonescu CR, Sommer G, Sarran L, et al. Association of KIT exon 9 mutations with nongastric primary site and aggressive behavior: KIT mutation analysis and clinical correlates of 120 gastrointestinal stromal tumors. *Clin Cancer Res* 2003;9(9):3329–3337.

70. Heinrich MC, Corless CL, Duensing A, et al. PDGFRA activating mutations in gastrointestinal stromal tumors. *Science* 2003;299(5607):708–710.

71. Dematteo RP, Owzar K, Maki RG, et al. Adjuvant imatinib mesylate increases recurrence-free survival (RFS) in patients with completely localized primary gastrointestinal stromal tumor (GIST): North American Intergroup Phase III Trial ACOSOG Z9001. *Proc Am Soc Clin Oncol* 2007;Abstract No(10079).

72. Verweij J, Casali PG, Zalcberg J, et al. Progression-free survival in gastrointestinal stromal tumours with high-dose imatinib: randomised trial. *Lancet* 2004;364(9440):1127–1134.

73. DeMatteo RP, Maki RG, Singer S, et al. Results of tyrosine kinase inhibitor therapy followed by surgical resection for metastatic gastrointestinal stromal tumor. *Ann Surg* 2007;245(3):347–352.

Chapter 38

Medical oncology of soft tissue sarcomas

Robert G. Maki
Memorial Sloan-Kettering Cancer Center, New York

Systemic therapies, including classic cytotoxic chemotherapy drugs and newer small molecule oral inhibitor of kinases and other proteins, are used commonly in the metastatic setting for patients with soft tissue sarcomas. Conversely, the use of chemotherapy in the adjuvant setting remains controversial, because the benefit for most sarcomas is small. Exceptions to this rule include rhabdomyosarcoma and Ewing sarcoma of soft tissue, in which adjuvant or neo-adjuvant chemotherapy is a critical component of treatment in most patients, as well as in gastrointestinal stromal tumor (GIST), where the small molecule oral kinase inhibitor (SMOKI) imatinib is the standard of care for most high-risk GISTs.

It is increasingly appreciated that systemic therapy must be tailored to the specific type of sarcoma being treated. In this respect, a working relationship with an expert sarcoma pathologist is paramount to the optimal treatment of patients with primary or metastatic sarcoma. If chemotherapy is to have the same impact as radiation therapy and surgery in the management of sarcomas, effective drugs must be identified that help to improve the cure rate for patients with primary tumors and unseen microscopic metastatic disease for each specific sarcoma subtype. This section reviews the treatment of GIST, which represents what we hope to achieve for all sarcoma subtypes, and in fact in cancer in general. Also discussed are the use of systemic agents in the adjuvant and metastatic settings for other sarcomas. A brief discussion of simultaneous chemotherapy and radiotherapy concludes this chapter.

Gastrointestinal stromal tumors (GISTs): a paradigm of modern systemic therapy for solid tumors

GIST is one of the most common sarcomas, and it can vary in size and clinical outcome from an incidental finding during an operation to life-threatening metastatic disease.[1] Surgery is the standard of care for primary disease, and the oral drug imatinib is the first-line standard of care for metastatic disease,[2] a remarkable change of events for a tumor that is essentially impervious to standard cytotoxic chemotherapy. Sunitinib was approved in the United States in early 2006 for GISTs refractory to imatinib, and regorafenib in 2013 for GISTs refractory to imatinib and sunitinib.[3,4] Remarkably, the mutation status of the kitten receptor tyrosine kinase *(KIT)* or platelet-derived growth factor receptor alpha *(PDGFRA)* within the GIST leads to specific sensitivity patterns to imatinib, sunitinib, regorafenib and other tyrosine kinase inhibitors (Table 38.1).[5,6] The latest developments in the world of GIST involve recognition and development of agents to combat the Darwinian evolution of these tumors, which develop multiple unique secondary mutations (among other resistance mechanisms), rendering tumors resistant to the present generation of tyrosine kinase inhibitors. (See Figure 38.1 for a clinical image of the development of resistance in a previously responding tumor). The data regarding these remarkable findings are highlighted in the sections that follow. Because tyrosine kinase inhibitors first showed utility in metastatic GIST, treatment of metastatic disease is discussed first. This section is followed by a discussion of the adjuvant therapy of primary GIST, seeking to improve the cure rate for this frequently aggressive tumor.

Imatinib as the first-line therapy for metastatic GIST

GIST is essentially impervious to cytotoxic chemotherapy, even when applied directly to the peritoneum after removal

Table 38.1 *KIT* or *PDGFRA* mutation status predicts imatinib sensitivity

Mutation type	Frequency in study population (n = 127)	Median event-free survival (months)
Exon 11 *KIT*	67%	22.6
Exon 9 *KIT*	18%	6.6
Any *PDGFRA* receptor mutation	7%	Not examined
Unmutated *KIT* and *PDGFRA*	5%	2.7

Key: PDGFRA, platelet-derived growth factor receptor alpha.
From reference 5.

Modern Soft Tissue Pathology, Second Edition, ed. Markku Miettinen. Published by Cambridge University Press. © Cambridge University Press 2016.

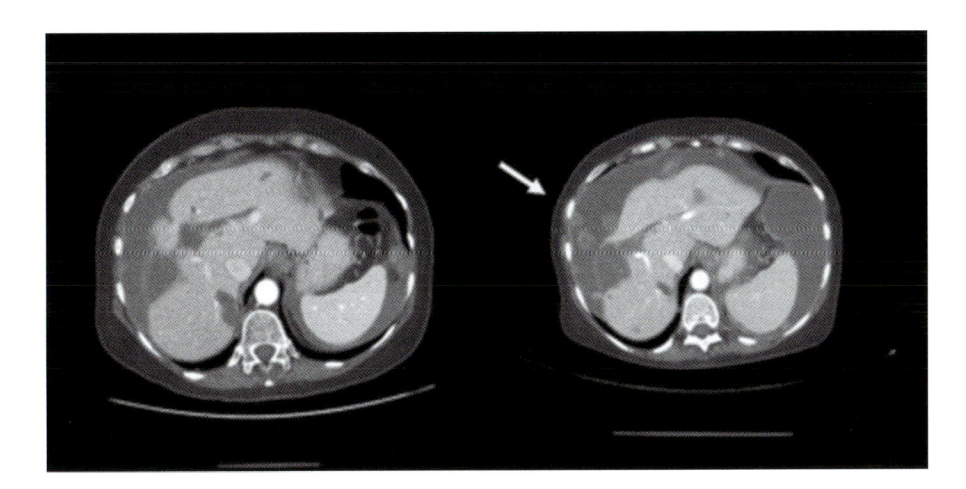

Figure 38.1 "Tumor within a tumor" developing on therapy with sunitinib. Examination of the tumor material from these areas of progressing GIST metastases often demonstrates secondary mutations in *KIT*, indicating selection of a resistant tumor clone. (Left panel) CT scan with IV contrast before therapy. (Right panel) CT scan with IV contrast after 6 months of treatment. Multiple lesions are developing in the area of previously necrotic and responsive tumor just outside the liver parenchyma.

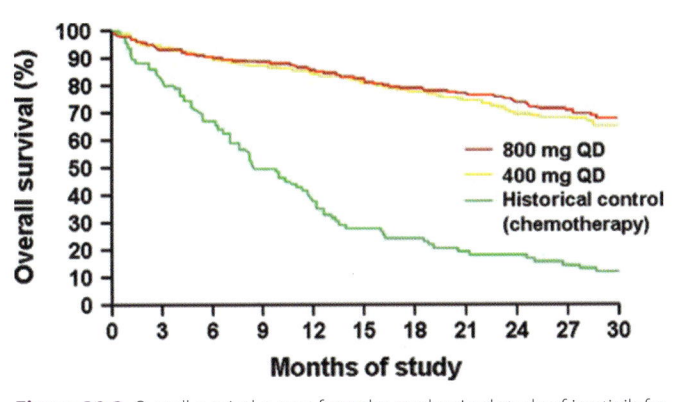

Figure 38.2 Overall survival curves from the randomized study of imatinib for metastatic GIST. Red line: imatinib 400 mg oral twice daily. Yellow line: imatinib 400 mg oral once daily. Green line: historical controls who received cytotoxic chemotherapy. The survival of patients on imatinib is substantially improved. (From Verweij, J., P. G. Casali, et al. [2004]. Progression-free survival in gastrointestinal stromal tumours with high-dose imatinib: randomised trial. *Lancet* 364(9440): 1127–1134; with permission.)

of metastatic disease.[7] On the basis of phase I, II, and III studies, imatinib became the standard of care for metastatic disease in patients with recurrent GIST. From both American and European/Australasian phase III studies, a starting dose of 400 mg orally daily was determined to be as effective as a starting dose of 800 mg orally daily. In the European/Australasian study, 800 mg orally daily of imatinib led to a longer time to progression; however, patients who crossed over from 400 mg to 800 mg at the time of progression fared as well as those who received 800 mg orally daily from the study outset.[2] Because the higher starting dose is associated with greater toxicity, the lower dose is the preferred starting dose for patients with metastatic disease. The survival for patients with metastatic GIST is clearly improved compared with the pre-imatinib era (Figure 38.2).

KIT mutation status is associated with tumor response in metastatic disease. The exon 11 mutation in *KIT* predicts a much better outcome for patients than does an exon 9 mutation, for unclear reasons, and patients with GIST that has no *KIT* mutation fare worst of all (Table 38.1).[4] *PDGFRA* is also mutated in a small proportion of GISTs, and mutations in *PDGFRA* are associated with at least some imatinib sensitivity.[8]

Finally, for GISTs without *KIT* mutations, these usually younger patients have been found to have loss of expression of the succinate dehydrogenase complex of proteins (SDH). These tumors are usually slower growing, and occasionally respond to second- and third-line agents against KIT.

KIT inhibitors after imatinib failure

Approximately one patient in six with metastatic GIST shows overt resistance to imatinib from the start of therapy (no response within the first 3–6 months of treatment).[2] The median time to progression on imatinib is approximately 2 years. The patterns of progression after resistance to imatinib are quite different from those seen with carcinomas developing resistance to cytotoxic chemotherapy. Rather than observing multiple lesions growing after a good response to a cytotoxic drug, only one or two lesions growing in size are often observed when resistance to imatinib develops. Another curious finding is the apparent reactivation of tumor(s) when imatinib resistance develops. Although the size of a lesion does not change, intravenous (IV) contrast highlights the lesion and shows nodules within a mass that were not present previously (Figure 38.1).[9] The "tumor-in-a-tumor" phenomenon is the clearest evidence of clonal selection of a subset of GIST cells, in that the areas found to show renewed IV contrast uptake are viable and show secondary mutations in *KIT* that are associated with imatinib resistance in patient tumor samples[10] and *in vitro* models.

In relatively rapid succession, studies of sunitinib and regorafenib led to their approval by many regulatory agencies for treatment of imatinib-resistant GIST (or patients intolerant of imatinib).[3,4] In the phase III sunitinib study, an 8% response rate for sunitinib and no responses in the placebo arm were observed.[3] Time to progression and overall survival were superior for the sunitinib arm, indicating that even with a cross-over there was an advantage to using sunitinib as early as possible after progression on imatinib.

In a similar fashion, regorafenib was tested versus placebo in a phase III study after earlier development, and demonstrated a median progression-free survival (PFS) of 4.6 months versus 0.9 months for placebo. The study was also considered a success as people on placebo were rapidly crossed over to the active drug and did not have inferior survival. Response rate was 4.5% for regorafenib versus 1.5% for placebo, indicating most of the disease control came from tumor stabilization rather than shrinking, as best as could be determined.[4]

Responses of GIST to other tyrosine kinase inhibitors are also dictated by *KIT* or *PDGFRA* mutation status

KIT mutation status also determines the sensitivity to sunitinib, as it does imatinib. Secondary mutations are a common means of GISTs developing resistance to imatinib.[10]

People who have GISTs with exon 9 *KIT* mutations or wild-type *KIT* had a greater chance to achieve a partial response on sunitinib and had improved progression-free and overall survival.[11] The median PFS for patients with exon 11 mutations (most with secondary mutations as well) was inferior to those people who had GIST with exon 9 mutations (and no secondary mutation in *KIT*), p <0.001. Overall survival was 12.3 months for people with exon 11 *KIT* mutations, and 26.9 months for those with exon 9 mutations. Data for the people with unmutated *KIT/PDGFRA* were similar to the exon 9 patients.

These clinical data correlate well with *in vitro* testing of GISTs bearing mutated *KIT* kinases or transfection experiments in which mutated *KIT* genes are introduced into KIT null cells and assayed for the median inhibitory concentrations for each drug related to each primary or secondary mutations. These data are summarized in Figure 38.3 below, which shows the sensitivity pattern of GISTs bearing secondary *KIT* exon 13, 14, 17, and 18 mutations. Data are shown for imatinib, sunitinib, and sorafenib, which is a close relative of regorafenib approved in other cancers. Given the heterogeneity of mutations within a single person's GIST, it is not surprising the response rate was not higher for newer-generation KIT inhibitors, the topic of future basic science and clinical studies.

Palliative surgery for metastatic GIST

Surgery is also useful for selected patients with metastatic disease.[12] As noted previously, patterns of resistance to imatinib are different from those seen with other diseases, likely reflecting the selective pressure of imatinib on one or at most a few tumor genes. The tumor-within-a-tumor phenomenon, or progression of a limited number of tumor masses, is commonly a sign of progression on imatinib.[9,13] Each site within a progressing lesion might have a unique mutation, which complicates therapy after the development of imatinib resistance.[9,12–14] Not surprisingly, patients with a single progressive lesion or limited progressive disease fare better than those with multifocal disease progression. Thus, selected patients might benefit from surgery for progressing lesions, employing imatinib in the postsurgical setting to maintain pressure on the pathways that allow GIST tumor cells to survive.[15] For patients with multifocal disease progression, practitioners typically change systemic therapy, because such patients usually show further evidence of tumor progression on subsequent imaging studies. Although patients with imatinib-responsive disease might fare even better in surgical series than those who have developed resistance, there is no clear proven benefit to early surgery versus late surgery, because there can be a significant lead time bias for those patients who have achieved their best overall result before developing resistance.

Adjuvant therapy for GIST

Given the findings on the sensitivity of GIST to imatinib, the question of its utility is often raised. Although disease-free survival is improved while patients are actually on therapy, there are presently no data to indicate that imatinib improves the cure rate of patients with resection of primary disease.[15]

The first study examining adjuvant imatinib for high-risk GIST (i.e., >10-cm disease, ruptured tumor, or satellite implants near a primary tumor) was completed by the American College of Surgeons Oncology Group (ACOSOG). More than 100 evaluable patients were accrued on study Z9000 from September 2001 to September 2003. Patients received imatinib at a starting dose of 400 mg orally daily for 48 weeks and were then followed for recurrence, showing data that were at least on the surface far better than historical controls of GIST patients from prior retrospective analyses.[1,16] The Z9000 study set the stage for ACOSOG study Z9001, examining imatinib 400 mg daily versus placebo for 48 weeks for any GIST >3 cm in the greatest dimension.[17] This study completed accrual in 2007 and showed a significant difference in recurrence-free survival (RFS) at the end of 1 year of therapy (3% recurrence versus 17% for those on imatinib or placebo, respectively). The RFS advantage did not translate to a survival advantage, since people failing primary therapy all received imatinib in the metastatic setting. These data were sufficient for approval of imatinib in the adjuvant setting.

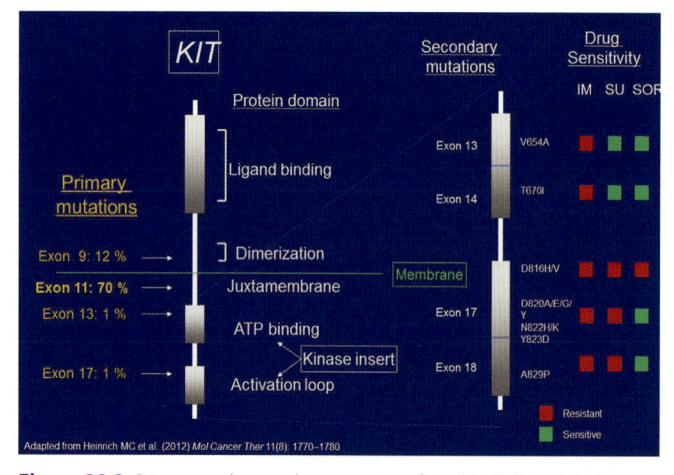

Figure 38.3 Primary and secondary mutations found in GIST and their sensitivity to newer KIT inhibitors. (Adapted from Heinrich MC et al. (2012) *Mol Cancer Ther* 11(8): 1770–1780.)

Joensuu et al. pursued the adjuvant question therapy further with what serves as the contemporary standard for care of higher-risk GIST in the adjuvant setting.[18] In the SSG XVIII study, imatinib for 1 year versus 3 years were compared in people with GIST who had at least one high-risk feature of their GIST (per NIH consensus criteria of the time): (1) longest dimension over 10 cm, (2) mitotic count over 10 mitoses per 50 high-power microscopic fields, (3) tumor diameter greater than 5 cm and mitotic count over 5, or (4) tumor rupture before surgery or at surgery. The RFS, landmark 5-year RFS, and overall survival for people assigned to 3 years of imatinib were all superior to that of people receiving 1 year of imatinib. At 5 years, the overall survival was 92% versus 82% for longer versus shorter exposure to imatinib, respectively, thus defining 3 years of adjuvant therapy for higher-risk GIST. This study has led to general acceptance by regulatory authorities of the use of 3 years of imatinib in the adjuvant setting as well. A newer study will examine 3 versus 5 years of imatinib for the highest-risk patients.

Therapy for metastatic GIST resistant to tyrosine kinase inhibitors

Given the relatively short time to disease progression in patients receiving sunitinib or regorafenib after progression on imatinib, new agents are needed to control metastatic disease and to move into the setting of newly metastatic disease or even into the adjuvant setting. A variety of agents have been tested with variable results, not compelling enough to move forward, however, including alternate kinase inhibitors such as dasatinib, ponatinib, everolimus[19] as well as hsp90 inhibitors such as retaspimycin.[20] The latter agent failed in a phase III study of kinase-inhibitor-refractory GIST, calling into question the pharmacokinetics of these agents with respect to the ability to stabilize or shrink metastatic disease.

The rapid pace and increasing sophistication of basic and translational research in sarcoma research gives us ample opportunity to manage metastatic GIST in the near future. Work in immunotherapy that may be relevant to GIST and other cancers is rapidly altering the spectrum of options for cancer in general, as are evolving trends in epigenetics and more efficient and effective design of drugs against essentially arbitrary targets. With this improved arsenal it is fairly clear that the management of GIST will continue to improve in years to come despite the heterogeneity evident in every primary tumor.

Other soft tissue sarcomas

Although GIST is the most common of the malignancies treated as a soft tissue sarcoma, there are at least 50 other soft tissue malignancies that can be fatal.[21] Each is distinct in its biologic behavior, such as metastatic pattern and sensitivity to specific chemotherapy agents (Figure 38.4). Understanding the most common subtypes of sarcoma (i.e., liposarcoma, leiomyosarcoma, high-grade undifferentiated pleomorphic sarcoma [UPS, formerly called malignant fibrous histiocytoma or MFH], and synovial sarcoma) makes it possible to manage many of the situations faced by clinical oncologists. The issues regarding adjuvant chemotherapy are described first, followed by a discussion of systemic therapy for metastatic disease, with an emphasis on a few special management issues at the end of the chapter.

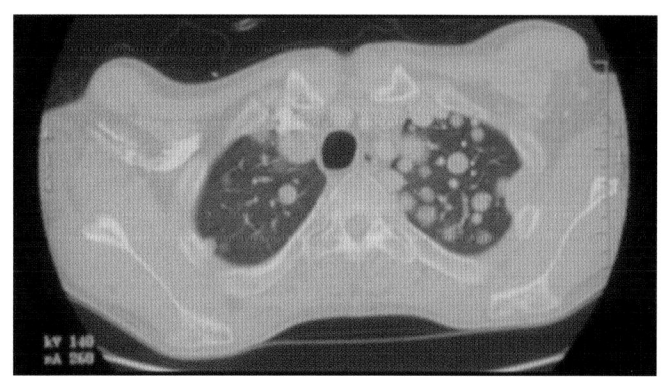

Figure 38.4 Metastatic alveolar soft part sarcoma. The tumor characteristically metastasizes early as multiple small round nodules that can occupy a large volume of the thorax without causing significant symptoms.

Adjuvant and neo-adjuvant chemotherapy for soft tissue sarcomas

Surgery and radiation therapy form the backbone of therapy for local control of soft tissue sarcoma.[22] Nearly one-half of people with primary non-GIST sarcomas will develop distant metastases despite good local disease control. Can chemotherapy after primary surgery reduce the risk of distant recurrence and thus increase the cure rate for people with primary disease?

Approximately 20 studies of adjuvant therapy for soft tissue sarcoma have been performed to date. Because anthracyclines are the most active agents for metastatic sarcomas, they have been universally employed in adjuvant trials, alone or in combination. From the first adjuvant studies performed at the National Cancer Institute (NCI) onward, only a few have accrued large (several hundred) patients and have lacked statistical power to detect small changes in overall survival.

Two more contemporary studies highlight the difficulties with the use of adjuvant chemotherapy in unselected people with primary soft tissue sarcomas. On one hand, the largest study of adjuvant chemotherapy versus placebo, conducted by the EORTC, demonstrated no improvement in overall survival with the use of doxorubicin and ifosfamide chemotherapy.[23] In fact, the numerical result for 5-year survival was worse for people receiving chemotherapy than those who did not. Conversely, the largest meta-analysis, which did not include the EORTC study, showed an overall survival benefit for adjuvant chemotherapy, in particular that employing ifosfamide-based therapy.[24] Primary criticisms of meta-analyses are those of lack of review of the primary histological diagnosis, which often will change after review at an expert institution, as well as ongoing concern of publication bias. Beyond general problems with selection bias that can even occur in randomized studies, staging and dose intensity also affect the ability to draw conclusions from individual studies. By enrolling patients with lower-risk tumors (as was the case in some of the protocols in the past) it is possible to artificially "improve" the overall outcome, regardless of chemotherapy. The author's conclusion is that if there is a benefit to adjuvant or neo-adjuvant chemotherapy, it is small, and that such benefit has to be discussed on a patient-by-patient basis, to balance the potential toxicity of therapy with any potential benefit.

In contrast to most typical soft tissue sarcoma diagnoses in adults, such as UPS or leiomyosarcoma, chemotherapy has been very successful in the management of predominantly pediatric sarcomas such as rhabdomyosarcoma, Ewing sarcoma, and osteosarcoma. Adults with these diagnoses should be treated according to the outlines for these diagnoses in children, although it is often impossible to maintain the dose intensity of therapies in adults that are achieved in children, e.g. with the use of weekly vincristine in patients with rhabdomyosarcoma. Although a detailed description of all the adjuvant and neo-adjuvant studies in soft tissue sarcoma is beyond the scope of this text, the largest individual studies of neo-adjuvant and adjuvant chemotherapy are discussed in the next section.

It is increasingly appreciated that metastatic soft tissue sarcomas are heterogeneous in their chemotherapy sensitivity patterns. In the future it may be most effective to examine adjuvant therapy for specific subtypes of sarcoma, using the successful application of chemotherapy in Ewing sarcoma, osteosarcoma, and rhabdomyosarcoma as an example. There are some situations, such as uterine leiomyosarcoma, where cooperative group studies have begun to focus on adjuvant therapy in a manner similar to that of the pediatric cooperative groups, although no clear benefit of adjuvant chemotherapy has been seen to date in the treatment of uterine leiomyosarcoma, the most common diagnosis in that anatomic site.[25]

When data are not present or are conflicting, it may be useful to discuss the risk of recurrence in the light of potential benefit or toxicity of the proposed therapy. The use of nomograms that are more or less particular to specific clinical situations may be helpful in this regard.[26–28]

Limb perfusion and hyperthermia for locoregional disease

Although tumor necrosis factor (TNF) has not been made available in the United States for testing, it is used in combination with chemotherapy in limb perfusion protocols. *Limb perfusion* involves isolation of an extremity with a tourniquet and application of a bypass pump to circulate blood from the venous back to the arterial side of the circulation of the extremity, at the same time circulating high doses of chemotherapy and TNF. Several chemotherapeutic agents have been investigated for limb perfusion, such as melphalan, dactinomycin, and doxorubicin. Melphalan and TNF seem to be a particularly useful combination. Eggermont and colleagues described in detail their experience of this technique.[29] After isolation of the extremity, a total of 246 patients with otherwise unresectable sarcomas received a melphalan perfusion with a dose of TNF ten times the lethal dose for humans, under mild hyperthermic conditions. Both components of the regimen appear important; omitting TNF led to a decrease in tissue dose of melphalan, probably from vascular effects. Surgery to remove residual tumor was performed 2 to 4 months after limb perfusion, when feasible. Overall, 71% of patients had successful limb salvage at a median of 3 years follow-up. More recent studies have shown that lower doses of TNF might be as effective as higher doses.[30] It is important to recognize, however, that limb perfusion requires expertise and specialized equipment. Complications include TNF leak and subsequent shock, chronic soft tissue damage, persistent edema, thrombosis, and infection.

Again, particularly in Europe, hyperthermia has also been employed in other ways to enhance the effects of chemotherapy. Neo-adjuvant chemotherapy with etoposide, ifosfamide, and doxorubicin was examined with or without hyperthermia supplied by an external electromagnetic field (a so-called phased array).[31] The investigators of this 341 patient study noted a 29% response rate with hyperthermia compared to 13% for the chemotherapy-only cohort, and improved 2-year local control rate. Overall survival was also better in the hyperthermia group, data which have led to the approval of hyperthermia for primary sarcoma patients in Germany. This approach has not been adopted more generally, perhaps because of the specialized equipment and training involved, and hyperthermia remains investigational in the United States.

Combined chemoradiotherapy for control of localized disease

Larger primary tumors present risks of both local and distant recurrence. Chemoradiation therapy, in the form of interdigitated chemotherapy and radiation, was first examined in 48 patients with localized, high-grade, large-extremity soft tissue sarcomas.[32] Therapy included three courses of doxorubicin, ifosfamide, mesna, and dacarbazine (MAID) and two 22 Gy courses of radiation (11 fractions each) for a total preoperative radiation dose of 44 Gy. A 16 Gy boost (in eight fractions) was give to patients with microscopically positive surgical margins. Five-year actuarial local control, distant-metastasis-free survival, and overall survival rates for the chemoradiation group were all impressive, at 92%, 75%, and 87%, respectively. The therapy was quite toxic, however, and included wound-healing complications in the lower limbs (29% of subjects).[32] One patient died of chemotherapy-associated marrow failure.

A follow-up multicenter study examined 64 patients and showed good local control, but also significant toxicity, with three patients (5%) having experienced fatal grade 5 toxicities consisting of myelodysplasia in two and sepsis in one. Moreover, another 53 patients (83%) experienced a variety of grade 4 toxicities, and 5 patients required amputation.[33] Recurrence-free and overall survival were also somewhat lower than the single institution study. The combination of "sandwich" chemotherapy and radiotherapy is used at some institutions, while either preoperative or postoperative radiation therapy (without concurrent or interdigitated chemotherapy) remains the standard of care at most institutions, based on a trial of preoperative versus postoperative radiation therapy for extremity/truncal soft tissue sarcomas.[34]

Management of recurrent or metastatic soft tissue sarcomas

Approximately one-half of patients with soft tissue sarcomas will die of metastatic or locally advanced disease. The most active chemotherapeutic options are unfortunately of limited value and are associated with serious and occasionally life-threatening side effects. Approximately 20% to 25% of patients with metastatic sarcoma are alive 2 years after diagnosis using such therapy. Although patients can remain free of symptoms for a relatively long period, sometimes even years, progression is inevitable. Surgical resection can cure a small percentage of patients with metastatic disease, in particular those tumors that recur as a small number of lung metastases more than a year or two after initial presentation. Radiation therapy can be palliative in patients with local complications of metastatic disease, such as spinal cord compression. Maximizing the quantity and quality of life with unresectable or metastatic soft tissue sarcoma requires an appreciation for the natural history of the specific sarcoma subtype, as well as an understanding of the limitations and possible utility of the therapeutic options at hand.

Systemic therapy for advanced disease

There are few agents with significant activity in metastatic soft tissue sarcomas (Table 38.2). Doxorubicin remains the mainstay of chemotherapy for advanced sarcoma. Although older studies with less stringent response criteria indicated an overall response rate of 30%,[35] more recent studies indicate a response rate in the 10% to 20% range, depending on the variety of histologies included in the study.[36] Studies with polyethylene-glycol-coated (PEGylated) liposomal doxorubicin (Doxil/Caelyx) showed response rates similar to doxorubicin with fewer side effects, making it a good option for patients with poor performance status who need systemic therapy.[37]

Ifosfamide is approximately as effective as doxorubicin in metastatic sarcomas (Table 38.2).[38] Ifosfamide doses as large as 14 to 18 g/m² or more over 1 to 2 weeks are sometimes given, thanks to the use of mesna, which protects the bladder from what would otherwise be a high risk of hemorrhagic cystitis. Ifosfamide appears more effective than cyclophosphamide in at least one clinical trial and remains the standard of care for

Table 38.2 Traditional cytotoxic chemotherapy agents and combinations for soft tissue sarcoma

Agent(s)	Dose	Schedule
Doxorubicin	60–90 mg/m²	Bolus or IVCI over 3–4 d q3wk
Ifosfamide		
24-h continuous dosing	5 g/m²	24-h IVCI with mesna q3–4wk
High dose	2–4 g/m²/d	Bolus or IVCI with mesna for 3–4 d q3–4wk
Dacarbazine		
Routine dosing	1000 mg/m²	IV over 60 min (sometimes longer) every 3–4 weeks
AI or AIM		
Doxorubicin	50–90 mg/m²	Bolus or divided over 2–3 d by bolus or IVCI q3–4wk
Ifosfamide (with mesna)	5–10 g/m²	Daily × 3 d or IVCI q3–4wk with mesna
Gemcitabine combinations		
Gemcitabine & Dacarbazine	1800 mg/m² + 500 mg/m² respectively	IV every 2 weeks
Gemcitabine & Docetaxel	900 mg/m² (d1 d8) + 75 mg/m² (d8 only), respectively	IV d1, d8 every 21 days

Key: IVCI, intravenously by continuous infusion.

an agent other than doxorubicin for many metastatic sarcomas. While cyclophosphamide causes more myelotoxicity overall, there is more renal and central nervous system toxicity with ifosfamide, in particular in people over the age of 60. It is also worth noting that patients who fail on a lower dose of ifosfamide occasionally respond to a higher dose of ifosfamide. Patients with synovial sarcoma and myxoid-round cell liposarcoma appear to be more responsive to ifosfamide than patients with other sarcomas.

If doxorubicin and ifosfamide are both active, then the combination is expected to be better. The best test of this concept to date was that from the EORTC, comparing doxorubicin 75 mg/m² per cycle to the same dose of doxorubicin with ifosfamide 10 g/m² per cycle, the so-called MD Anderson schedule of administration.[39] The response rate for the combination was approximately twice that of doxorubicin alone (26% versus 14%), and PFS was a median of nearly 3 months better (7.4 versus 4.6 months). Combination therapy was more toxic, with 46% of patients experiencing febrile neutropenia with the combination versus 13% for doxorubicin alone. A total of 18% stopped treatment for toxicity with the combination versus 3% for the single agent. Despite stopping therapy in a number of patients, overall survival showed borderline statistical improvement for patients receiving the combination (14.3 versus 12.8 months, p = 0.076). Thus, for patients needing a response, the combination is justified, while for less symptomatic patients in need of therapy, single agents are reasonable; there does not appear to be significant synergy between doxorubicin and ifosfamide.

Dacarbazine (also called DTIC) is an FDA-approved drug for metastatic sarcomas. Dacarbazine has frequently been used in combination chemotherapy with doxorubicin, and the author uses this combination for leiomyosarcomas in need of a response, given the low response rate of leiomyosarcomas to ifosfamide. In addition, a randomized phase II trial showed improvement in PFS and overall survival compared to dacarbazine alone, and in particular for leiomyosarcoma this represents a good standard of care after failure of doxorubicin-based therapy.[40] Temozolomide, an oral agent, has activity similar to dacarbazine, but has not been examined in randomized studies in soft tissue sarcoma.

In a manner similar to gemcitabine and dacarbazine, gemcitabine and docetaxel proved to be superior to gemcitabine alone, with respect to both progression-free and overall survival in a group of patients with unselected soft tissue sarcomas.[41] Responses were noted in pleomorphic liposarcoma and pleomorphic rhabdomyosarcoma, in addition to UPS and leiomyosarcoma, suggesting the major benefit for this combination rests with the use in highly aneuploid sarcomas. It is unclear for a given patient if the toxicity of the addition of docetaxel merits its addition, and patients must be monitored for signs of pneumonitis, edema, and neuropathy with this combination.

Pazopanib is the first oral kinase inhibitor to be approved for use in refractory sarcoma patients, by virtue of responses seen in phase II, as well as a large phase III study.[42] In the phase III trial pazopanib 800 mg oral daily was compared to placebo, without the possibility of cross-over. The response rate was 6%, very similar to the low response rate of agents already approved for hepatocellular carcinoma and renal cell carcinoma, while PFS was 4.6 months for the drug versus 1.6 months for placebo. There was a numerically greater median overall survival of 12.5 months versus 10.7 months for placebo, but this was not statistically significant. These data were sufficient for the approval of pazopanib for use in refractory soft tissue sarcoma.

The novel minor groove binding agent trabectedin (ecteinascidin, ET-743) has been approved for use in Europe, but not in the United States, by virtue of a randomized trial showing statistically superior PFS (3.3 months for treatment every 3 weeks versus 2.3 months for weekly administration), p = 0.04, with a numerically greater median overall survival (13.9 versus 11.9 months, p = 0.19 in this randomized phase II study).[43] Myxoid liposarcoma appears particularly sensitive to trabectedin.[44] Trabectedin covalently attaches to DNA in its minor groove, and also has effects on transcription that remain a topic of research. Both trabectedin and an antimicrotubule agent approved in breast cancer, eribulin, showed activity against leiomyosarcoma and liposarcomas in preliminary data from randomized trials presented in 2015, but are not approved for use in the United States as of 2015.

Dose intensification to achieve better responses

A dose–response effect for doxorubicin and for ifosfamide has been observed in older studies of the single agents. The use of hematopoietic growth factors has allowed the study of higher dosages of chemotherapy in sarcoma. Other toxicities ultimately intervene and limit the ability to give higher doses. Doxorubicin is limited by its cardiac toxicity and mucositis at very high doses, and the nephrotoxicity and central nervous system toxicity of ifosfamide limit even higher doses than are given today. One logical step to examine regarding dose intensity is to employ stem cell support with very high doses of standard chemotherapy. These studies indicate occasional long-term responders with Ewing sarcoma, osteosarcoma, or rhabdomyosarcoma, but these are few and far between. As a result, high-dose therapy and stem cell rescue remain highly investigational.[45] The pursuit of new agents with better activity against specific sarcoma subtypes thus remains a high priority for medical and pediatric oncologists.

Response by site and histology: examples beyond GIST

Among the important factors to consider when examining activity of a specific systemic therapy for sarcomas are the criteria used to determine the response (which have changed over time), the reality of improved imaging over time, and critically, the distribution of soft tissue sarcoma subtypes enrolled in the study. Pediatric soft-tissue sarcomas are known for their relative sensitivity to chemotherapy (most commonly Ewing sarcoma and rhabdomyosarcoma). For adult sarcomas, synovial sarcoma and round cell liposarcomas are more responsive to chemotherapy than other subtypes. GIST, alveolar soft part sarcoma, and extraskeletal myxoid chondrosarcoma are among the diagnoses notorious for their resistance to standard cytotoxic chemotherapy. An unfavorable distribution of the subtypes of sarcoma in a given study affects the observed outcome.

Perhaps somewhat surprisingly, the disease site is also a factor that determines outcome. For example, subjects with large low-grade liposarcomas of the extremities show lower relapse rates than patients with low-grade liposarcomas of the retroperitoneum; it is very uncommon to achieve local control of a retroperitoneal sarcoma, if patients are followed long enough. It is also important to recall that most tumors formerly called gastrointestinal leiomyosarcomas are now considered GISTs, which are essentially unresponsive to standard chemotherapy.

It is evident that specific subtypes of soft tissue sarcoma demonstrate unique biologic behavior. As diagnosis and classification of sarcomas improve, these unique features might become more evident. Discussed here are a few examples of site or histology that can affect clinical decision-making.

Angiosarcomas

Angiosarcomas are one end of a spectrum of vascular tumors, ranging from benign hemangiomas to hemangioendotheliomas, which can have highly variable behavior, to fully fledged angiosarcomas, with the ability to metastasize early and widely.[46] Perhaps their derivation from endoderm accounts for their biologic properties that are somewhat different from mesodermally derived sarcomas. They have an insidious nature with a high propensity to recur locoregionally, and are one of the few sarcomas that can metastasize to lymph nodes.[47] They can metastasize to nearly any site or organ, including bone and lung, as well as more unusual locations such as heart and brain. They are unique among sarcomas in their responsiveness to taxanes, which have activity similar to that of anthracyclines.[48–50]

Uterine sarcomas

Uterine sarcomas are unusual, representing approximately 5% of uterine malignancies. The uterus is rather unique in that three different sarcomatous or sarcoma-like entities can arise from this organ, including leiomyosarcoma, endometrial stromal sarcoma, and malignant mixed müllerian tumor (MMMT), the most common of which is carcinosarcoma. MMMT is the most common of the three of these entities (but given the understanding that MMMT behaves more like carcinoma than sarcoma, is not discussed further here). Akin to pediatric sarcomas, a series of studies examined chemotherapy responsiveness of specific uterine sarcoma subtypes[51,52]. This is increasingly important, given the recognized genetic differences between low- and high-grade/undifferentiated endometrial stromal sarcomas (with recognized characteristic chromosomal translocations) and leiomyosarcomas (with aneuploidy and mutations in *TP53* and *CDKN2A* commonly, among other genes).

The combination of gemcitabine and docetaxel is particularly active against uterine leiomyosarcoma,[53] from the apparent synergy of the combination, these data have been put into question from an (underpowered) randomized study of gemcitabine versus gemcitabine-docetaxel in leiomyosarcomas specifically.[54] Furthermore, endometrial stromal sarcomas express estrogen and progesterone receptors, and responses to hormonal therapy are seen more in this subtype than with any other sarcoma subtype.[55] Conversely, the frequency of positive estrogen or progesterone receptor staining in uterine leiomyosarcomas is far greater than the observed response rate to hormonal therapy for leiomyosarcoma. For example, 0 of 19 patients with leiomyosarcoma responded to tamoxifen in a prospective phase II study of different uterine sarcoma subtypes.[56]

"Pediatric" sarcomas in adults

Osteogenic sarcoma, Ewing sarcoma (in soft tissue or bone), and rhabdomyosarcoma all occur in adults, although at lower frequencies compared with children. Pediatric sarcomas are distinctive in their sensitivity to chemotherapy and improvement in overall survival with the use of adjuvant chemotherapy for each of these diseases.[57,58] There remains controversy as to whether to treat soft tissue extraskeletal osteogenic sarcoma with adjuvant chemotherapy like its bone counterpart, owing to the poor response rates to chemotherapy in patients with metastatic disease.[59]

Rhabdomyosarcoma and Ewing sarcoma therapy focuses on the combination of vincristine, doxorubicin, and cyclophosphamide (or dactinomycin instead of doxorubicin in rhabdomyosarcoma), as well as the combination of ifosfamide and etoposide, active in all pediatric sarcomas.[58,60] There is still a debate over whether adults do worse than pediatric patients with the same stage of disease. Adults do not tolerate the aggressive chemotherapy typically used in children. Adults also present with advanced-stage disease relative to children or adolescents. Adult medical oncologists are indebted to pediatric oncologists, who accrue a large percentage of their patients to protocols examining new therapeutics. Adults with a diagnosis of a sarcoma typically seen in children should be included in pediatric protocols whenever possible, to help to determine appropriate care for patients with these rare diagnoses.

Aggressive fibromatosis (desmoid tumor)

Desmoid tumors are part of a family of myofibroblastic fibromatoses that are remarkable for their slow progression and rather unremarkable histology. Surgery remains the primary treatment of choice for these lesions, which cannot truly be called sarcomas owing to their lack of metastatic potential.

For patients with negative margins after surgery, most authors think that postoperative radiation is not recommended. For patients with positive microscopic margins, however, the role of adjuvant radiation remains an open question. For example, one study reported a local control rate of 61% for patients with primary tumors with positive microscopic margins treated with surgery alone.[61] Other studies reported lower local control rates.[62] It is important to note that a positive microscopic margin does not necessarily foreordain failure, and increasingly a watch-and-wait approach is being taken, especially for tumors that are relatively static over time.[63,64] When adjuvant radiation is indicated (e.g., for some primary lesions or most recurrent tumors), the usual dose is on the order of 50 Gy. In advanced cases, desmoids can still cause significant morbidity in proximal extremity lesions and can be fatal if they arise in the retroperitoneum or head and neck areas, because in such sites local control might not be possible.

Desmoid tumors, although superficially similar, have differences in their outcomes and sensitivity to systemic therapy based on their site of origin. For example, desmoids arising in the abdominal wall during pregnancy often dissipate postpartum. It is therefore not surprising that some desmoids express estrogen receptors (ER), more typically ER beta than ER alpha.[65] There are anecdotal accounts of responses to hormonal manipulation such as tamoxifen or aromatase inhibitors. There are also well-documented responses of desmoids to sulindac and other nonsteroidal anti-inflammatory compounds. It is also worth noting that these responses typically occur very slowly, sometimes only over months or even years. Results with any therapy have to be considered in light of the phenomenon of spontaneous improvement, seen in a significant percentage of desmoids,[66] in particular those associated with pregnancy.

The basis for systemic chemotherapy for desmoid stems from a study from the MD Anderson Cancer Center, in which the combination of cyclophosphamide, vincristine, doxorubicin, and dacarbazine (CyVADIC) was investigated.[67,68] Responses have been reported to single-agent doxorubicin chemotherapy as well as combination chemotherapy at either standard or relatively low doses.[69] As with other interventions for desmoid tumors, responses can be slow. In the author's experience, those desmoids that are growing are more likely to respond than those that have been static for a prolonged period. It is difficult to decide when to stop therapy, because complete responses are unusual.

It is increasingly clear that multitargeted kinase inhibitors such as sorafenib have significant activity in desmoid tumors.[70] The finding that hopefully safer, nongenotoxic agents have such activity, along with the understanding that desmoids have alterations in either *APC* (usually patients with familial adenomatous polyposis and desmoids of the root of the mesentery) or *CTNNB1* (beta-catenin), raises hope that more effective therapy based on the molecular biology of desmoid tumors is in the offing soon.[71,72]

Hence, although for primary resectable disease, surgery alone appears to be the optimal approach, in advanced cases, or in cases in which surgery would be morbid, a trial of kinase-targeted therapy, hormonal therapy, or nonsteroidal anti-inflammatories can be considered for slowly changing disease. Observation is also reasonable, because some patients demonstrate regression without any therapy. For the symptomatic patient, the author favors chemotherapy (e.g., PEGylated liposomal doxorubicin), or in relatively dire circumstances, radiation, to try to gain control of the tumor. The author and colleagues do not favor acetic acid tumor injection or other local therapies, which have generated morbidity without clinical benefit in their experience, but it is entirely possible that novel local therapies will emerge in the coming decade as well.

Future directions

A variety of frontiers will have direct impact on the treatment of sarcomas before the year 2020. Chief among these are immunotherapy, epigenetic therapy, and the increasing ability to target chemically nearly any protein in the cell with increased specificity. Of note, despite research on immune checkpoint inhibitors in other cancers leading to their approval in several diagnoses, such agents have not been made available for study until 2015, but data are accumulating quickly. Failures have been seen in drug development for reasons beyond data in sarcomas, for example with IGF1 receptor inhibitors in lung cancer, which masked to the pharmaceutical industry the activity of such agents in Ewing sarcoma. It will take cycles of interest and support to ultimately see new agents in any of these realms to an accepted standard in patient care. Certainly the demonstration of cancer germline antigens in synovial sarcoma and myxoid/round cell liposarcoma, and the ability to target such proteins with T cells, gives hope that increasingly sophisticated immunological approaches may be brought to bear on the issue of metastatic disease, as well as cancer heterogeneity.[73,74] Epigenetic therapy may become interesting in part due to activity seen in hematological malignancies coupled to the very deep relationship of hematopoietic stem cells with the aortic endothelium or its precursors during development.[75]

The explosion of data regarding the many previously unrecognized functions of RNA within the cell, e.g. small inhibitory RNA molecules,[76] along with the rapid developments in nanotechnology and genomics also gives hope that active alterations of miRNA profiles of sarcoma cells will become possible. The discovery of drugs that read through premature stop codons in patients with Duchenne muscular dystrophy causing increased expression of the muscular protein dystrophin[77] encourages researchers to think that similar tools could be employed in that percentage of tumors with nonsense mutations (mutations yielding a stop codon) in important cancer-related genes such as *TP53*. With improved understanding of the cancer cell, thanks to basic scientists and translational biologists, it is eminently clear that more will be learned about the survival and apoptotic cascades that could be in the patient's favor for sarcomas as well as other cancers.

References

1. Gold, JS, Van Der Zwan SM, Gönen M, et al. Outcome of metastatic GIST in the era before tyrosine kinase inhibitors. *Ann Surg Oncol* 2007;14(1):134–142.
2. Verweij J, Casali PG, Zalcberg J, et al. Progression-free survival in gastrointestinal stromal tumours with high-dose imatinib: randomised trial. *Lancet* 2004;364(9440):1127–1134.
3. Demetri GD, van Oosterom, AT, Garrett CR, et al. 2006 Efficacy and safety of sunitinib in patients with advanced gastrointestinal stromal tumour after failure of imatinib: a randomised controlled trial. *Lancet* 2006;368(9544):1329–1338.
4. Demetri GD, Reichardt P, Kang YK, et al. Efficacy and safety of regorafenib

for advanced gastrointestinal stromal tumours after failure of imatinib and sunitinib (GRID): an international, multicentre, randomised, placebo-controlled, phase 3 trial. *Lancet* 2013;381(9863):295–302.

5. Heinrich MC, Corless CL, Demetri GD, et al. Kinase mutations and imatinib response in patients with metastatic gastrointestinal stromal tumor. *J Clin Oncol* 2003;21(23):4342–4349.

6. Heinrich MC, Marino-Enriquez, A, Presnell A, et al. Sorafenib inhibits many kinase mutations associated with drug-resistant gastrointestinal stromal tumors. *Mol Cancer Ther* 2012;11(8):1770–1780.

7. Eilber FC, Rosen G, Forscher C, et al. Surgical resection and intraperitoneal chemotherapy for recurrent abdominal sarcomas. *Ann Surg Oncol* 1999;6(7):645–650.

8. Corless CL, Schroeder A, Griffith D, et al. *PDGFRA* mutations in gastrointestinal stromal tumors: frequency, spectrum and in vitro sensitivity to imatinib. *J Clin Oncol* 2005;23(23):5357–5364.

9. Shankar S, van Sonnenberg E, Desai J, et al. Gastrointestinal stromal tumor: new nodule-within-a-mass pattern of recurrence after partial response to imatinib mesylate. *Radiology* 2005; 235(3):892–898.

10. Antonescu CR, Besmer P, Guo T, et al. Acquired resistance to imatinib in gastrointestinal stromal tumor occurs through secondary gene mutation. *Clin Cancer Res* 2005;11(11):4182–4190.

11. Heinrich MC, Maki RG, Corless CL, et al. Primary and secondary kinase genotypes correlate with the biological and clinical activity of sunitinib in imatinib-resistant gastrointestinal stromal tumor. *J Clin Oncol* 2008;26(33):5352–5359.

12. von Mehren M, Randall RL, Benjamin RS, et al. Gastrointestinal stromal tumors, version 2.2014. *J Natl Comp Canc Netw* 2014;12(6):853–862.

13. Wardelmann E, Merkelbach-Bruse S, Pauls K, et al. Polyclonal evolution of multiple secondary *KIT* mutations in gastrointestinal stromal tumors under treatment with imatinib mesylate. *Clin Cancer Res* 2006;12(6):1743–1749.

14. Wardelmann E, Thomas N, Merkelbach-Bruse S, et al. Acquired resistance to imatinib in gastrointestinal stromal tumours caused by multiple *KIT* mutations. *Lancet Oncol* 2005;6(4):249–251.

15. DeMatteo RP, Maki RG, Singer S, et al. Results of tyrosine kinase inhibitor therapy followed by surgical resection for metastatic gastrointestinal stromal tumor. *Ann Surg* 2007;245(3):347–352.

16. DeMatteo RP, Ballman KV, Antonescu CR, et al. Long-term results of adjuvant imatinib mesylate in localized, high-risk, primary gastrointestinal stromal tumor: ACOSOG Z9000 intergroup phase 2 trial. *Ann Surg* 2013;258(3):422–429.

17. DeMatteo RP, Ballman KV, Antonescu CR, et al. Adjuvant imatinib mesylate after resection of localised, primary gastrointestinal stromal tumour: a randomised, double-blind, placebo-controlled trial. *Lancet* 2009;373(9669):1097–1104.

18. Joensuu H, Erikson M, Sundby Hall K, et al. One vs three years of adjuvant imatinib for operable gastrointestinal stromal tumor: a randomized trial. *JAMA* 2012;307(12):1265–1272.

19. Schöffski P, Reichardt P, Blay J-Y, et al. A phase I-II study of everolimus (RAD001) in combination with imatinib in patients with imatinib-resistant gastrointestinal stromal tumors. *Ann Oncol* 2010;21(10):1990–1998.

20. Wagner AJ, Chugh R, Rosen LS, et al. A phase I study of the HSP90 inhibitor retaspimycin hydrochloride (IPI-504) in patients with gastrointestinal stromal tumors or soft-tissue sarcomas. *Clin Cancer Res* 2013;19(21):6020–6029.

21. Fletcher CDM, Bridge JA, Hogendoorn PCW, Mertens F (eds). *WHO Classification of Tumours of Soft Tissue and Bone*, 4th edn. Lyon: IARC; 2013.

22. Brennan MF, Antonescu CR, Maki RG. *Management of Soft Tissue Sarcomas*. New York: Springer; 2013.

23. Woll PJ, Reichardt P, Le Cesne A, et al. Adjuvant chemotherapy with doxorubicin, ifosfamide, and lenograstim for resected soft-tissue sarcoma (EORTC 62931): a multicentre randomised controlled trial. *Lancet Oncol* 2012;13(10):1045–1054.

24. Pervaiz N, Colterjohn N, Farrokhyar F, et al. A systematic meta-analysis of randomized controlled trials of adjuvant chemotherapy for localized resectable soft-tissue sarcoma. *Cancer* 2008;113(3):573–581.

25. Hensley ML, Wathen JK, Maki RG, et al. Adjuvant therapy for high-grade, uterus-limited leiomyosarcoma: results of a phase 2 trial (SARC 005). *Cancer* 2013;119(8):1555–1561.

26. Kattan MW, Leung DH, Brennan M. Postoperative nomogram for 12-year sarcoma-specific death. *J Clin Oncol* 2002;20(3):791–796.

27. Zivanovic O, Jacks LM, Iasonos A, et al. A nomogram to predict postresection 5-year overall survival for patients with uterine leiomyosarcoma. *Cancer* 2012;118(3):660–669.

28. Gronchi A, Miceli R, Shurell E, et al. Outcome prediction in primary resected retroperitoneal soft tissue sarcoma: histology-specific overall survival and disease-free survival nomograms built on major sarcoma center data sets. *J Clin Oncol* 2013;31(13):1649–1655.

29. Eggermont AM, ten Hagen TL. Tumor necrosis factor-based isolated limb perfusion for soft tissue sarcoma and melanoma: ten years of successful antivascular therapy. *Curr Oncol Rep* 2003;5(2):79–80.

30. Bonvalot S, Laplanche A, Lejeune F, et al. Limb salvage with isolated perfusion for soft tissue sarcoma: could less TNF-alpha be better? *Ann Oncol* 2005;16(7):1061–1068.

31. Issels RD, Lindner LH, Verweij J, et al. Neo-adjuvant chemotherapy alone or with regional hyperthermia for localised high-risk soft-tissue sarcoma: a randomised phase 3 multicentre study. *Lancet Oncol* 2010;11(6):561–570.

32. DeLaney TF, Spiro IJ, Suit HD, et al. Neoadjuvant chemotherapy and radiotherapy for large extremity soft-tissue sarcomas. *Int J Radiat Oncol Biol Phys* 2003;56(4):1117–1127.

33. Kraybill WG, Harris J, Spiro IJ, et al. Phase II study of neoadjuvant chemotherapy and radiation therapy in the management of high-risk, high-grade, soft tissue sarcomas of the extremities and body wall: Radiation Therapy Oncology Group Trial 9514. *J Clin Oncol* 2006;24(4):619–625.

34. O'Sullivan B, Davis AM, Turcotte R, et al. Preoperative versus postoperative radiotherapy in soft-tissue sarcoma of the limbs: a randomised trial. *Lancet* 2002;359(9325):2235–2241.

35. O'Bryan RM, Luce JK, Talley RW, et al. Phase II evaluation of adriamycin in human neoplasia. *Cancer* 1973;32(1):1–8.
36. Lorigan P, Verweij J, Papai Z, et al. Phase III trial of two investigational schedules of ifosfamide compared with standard-dose doxorubicin in advanced or metastatic soft tissue sarcoma: a European Organisation for Research and Treatment of Cancer Soft Tissue and Bone Sarcoma Group Study. *J Clin Oncol* 2007;25(21):3144–3150.
37. Judson I, Radford JA, Harris M, et al. Randomised phase II trial of pegylated liposomal doxorubicin (DOXIL/CAELYX) versus doxorubicin in the treatment of advanced or metastatic soft tissue sarcoma: a study by the EORTC Soft Tissue and Bone Sarcoma Group. *Eur J Cancer* 2001;37(7):870–877.
38. Nielsen OS, Judson I, van Hoesel Q, et al. Effect of high-dose ifosfamide in advanced soft tissue sarcomas: a multicentre phase II study of the EORTC Soft Tissue and Bone Sarcoma Group. *Eur J Cancer* 2000;36(1):61–67.
39. Judson I, Verweij J, Gelderblom H, et al. Doxorubicin alone versus intensified doxorubicin plus ifosfamide for first-line treatment of advanced or metastatic soft-tissue sarcoma: a randomised controlled phase 3 trial. *Lancet Oncol* 2014;15(4):415–423.
40. García-Del-Muro X, López-Pousa A, Maurel J, et al. Randomized phase II study comparing gemcitabine plus dacarbazine versus dacarbazine alone in patients with previously treated soft tissue sarcoma. *J Clin Oncol* 2011;29:2528–2533.
41. Maki, RG, Wathen JK, Patel SR, et al. Randomized phase II study of gemcitabine and docetaxel compared with gemcitabine alone in patients with metastatic soft tissue sarcomas. *J Clin Oncol* 2007;25(19):2755–2763.
42. van der Graaf WT, Blay J-Y, Chawla SP, et al. Pazopanib for metastatic soft-tissue sarcoma (PALETTE): a randomised, double-blind, placebo-controlled phase 3 trial. *Lancet* 2012;379(9829):1879–1886.
43. Demetri GD, Chawla SP, von Mehren M, et al. Efficacy and safety of trabectedin in patients with advanced or metastatic liposarcoma or leiomyosarcoma after failure of prior anthracyclines and ifosfamide: results of a randomized phase II study of two different schedules. *J Clin Oncol* 2009;27(25):4188–4196.
44. Grosso F, Jones RL, Demetri GD, et al. Efficacy of trabectedin (ecteinascidin-743) in advanced pretreated myxoid liposarcomas: a retrospective study. *Lancet Oncol* 2007;8(7):595–602.
45. Seynaeve C, Verweij J. High-dose chemotherapy in adult sarcomas: no standard yet. *Semin Oncol* 1999;26(1):119–133.
46. Deyrup AT, Weiss SW. Grading of soft tissue sarcomas: the challenge of providing precise information in an imprecise world. *Histopathology* 2006;48(1):42–50.
47. Fong Y, Coit DG, Woodruff JM, Brennan MF. Lymph node metastasis from soft tissue sarcoma in adults: analysis of data from a prospective database of 1772 sarcoma patients. *Ann Surg* 1993;217(1):72–77.
48. Casper ES, Waltzman RJ, Schwartz GK, et al. Phase II trial of paclitaxel in patients with soft-tissue sarcoma. *Cancer Invest* 1998;16(7):442–446.
49. Fury MG, Antonescu CR, Van Zee KJ, Brennan MF, Maki RG. A 14-year retrospective review of angiosarcoma: clinical characteristics, prognostic factors, and treatment outcomes with surgery and chemotherapy. *Cancer J* 2005;11(3):241–247.
50. Italiano A, Cioffi A, Penel N, et al. Comparison of doxorubicin and weekly paclitaxel efficacy in metastatic angiosarcomas. *Cancer* 2012;118(13):3330–3336.
51. Maki RG, D'Adamo DR, Keohan ML, et al. Phase II study of sorafenib in patients with metastatic or recurrent sarcomas. *J Clin Oncol* 2009;28(19):3133–3140.
52. Agulnik M, Yarber JL, Okuno SH, et al. An open-label, multicenter, phase II study of bevacizumab for the treatment of angiosarcoma and epithelioid hemangioendotheliomas. *Ann Oncol* 2013;24(1):257–263.
53. Hensley ML, Maki RG, Venkatraman E, et al. Gemcitabine and docetaxel in patients with unresectable leiomyosarcoma: results of a phase II trial. *J Clin Oncol* 2002;20(12):2824–2831.
54. Pautier P, Floquet A, Penel N, et al. Randomized multicenter and stratified phase II study of gemcitabine alone versus gemcitabine and docetaxel in patients with metastatic or relapsed leiomyosarcomas. *Oncologist* 2012;17(9):1213–1220.
55. Keen CE, Philip G. Progestogen-induced regression in low-grade endometrial stromal sarcoma. Case report and literature review. *Br J Obstet Gynaecol* 1989;96(12):1435–1439.
56. Wade K, Quinn MA, Hammond I, Williams K, Cauchi M. Uterine sarcoma: steroid receptors and response to hormonal therapy. *Gynecol Oncol* 1990;39(3):364–367.
57. Goorin AM, Schwartzentruber DJ, Devidas M, et al. Presurgical chemotherapy compared with immediate surgery and adjuvant chemotherapy for nonmetastatic osteosarcoma: Pediatric Oncology Group Study POG-8651. *J Clin Oncol* 2003;21(8):1574–1580.
58. Womer RB, West DC, Krailo MD, et al. Randomized controlled trial of interval-compressed chemotherapy for the treatment of localized Ewing sarcoma. *J Clin Oncol* 2012;30(33):4148–4154.
59. Bane BL, Evans HL, Ro JY, et al. Extraskeletal osteosarcoma: a clinicopathologic review of 26 cases. *Cancer* 1990;65(12):2762–2770.
60. Crist WM, Anderson JR, Meza JL, et al. Intergroup rhabdomyosarcoma study-IV: results for patients with nonmetastatic disease. *J Clin Oncol* 2001;19(12):3091–3102.
61. Spear MA, Jennings LC, Mankin HJ, et al. Individualizing management of aggressive fibromatoses. *Int J Radiat Oncol Biol Phys* 1998;40(3):637–645.
62. Ballo MT, Zagars GK, Pollack A. Radiation therapy in the management of desmoid tumors. *Int J Radiat Oncol Biol Phys* 1998;42(5):1007–1014.
63. Lev D, Kotilingam D, Wei C, et al. Optimizing treatment of desmoid tumors. *J Clin Oncol* 2007;25(13):1785–1791.
64. Gronchi A, Colombo C, Le Péchoux C, et al. Sporadic desmoid-type fibromatosis: a stepwise approach to a non-metastasising neoplasm. *Ann Oncol* 2014;25(3):578–583.
65. Deyrup AT, Tretiakova M, Montag AG. Estrogen receptor-beta expression in extraabdominal fibromatoses: an analysis of 40 cases. *Cancer* 2006;106(1):208–213.

66. Bonvalot S, Ternès N, Fiore M, *et al.* Spontaneous regression of primary abdominal wall desmoid tumors: more common than previously thought. *Ann Surg Oncol* 2013;20(13):4096–4102.

67. Patel SR, Evans HL, Benjamin RS. Combination chemotherapy in adult desmoid tumors. *Cancer* 1993;72(11):3244–3247.

68. Patel SR, Benjamin RS. Desmoid tumors respond to chemotherapy: defying the dogma in oncology. *J Clin Oncol* 2006;24(1):11–12.

69. Weiss AJ, Horowitz S, Lackman RD. Therapy of desmoid tumors and fibromatosis using vinorelbine. *Am J Clin Oncol* 1999;22(2):193–195.

70. Gounder MM, Lefkowitz RA, Keohan ML, *et al.* Activity of sorafenib against desmoid tumor/deep fibromatosis. *Clin Cancer Res* 2011;17(12):4082–4090.

71. Alman BA, Li C, Pajerski ME, Diaz-Cano S, Wolfe HJ. Increased beta-catenin protein and somatic APC mutations in sporadic aggressive fibromatoses (desmoid tumors). *Am J Pathol* 1997;151(2):329–334.

72. Tejpar S, Nollet F, Li C, *et al.* Predominance of beta-catenin mutations and beta-catenin dysregulation in sporadic aggressive fibromatosis (desmoid tumor). *Oncogene* 1999;18(47):6615–6620.

73. Jungbluth AA, Antonescu CR, Busam KJ, et al. Monophasic and biphasic synovial sarcomas abundantly express cancer/testis antigen NY-ESO-1 but not MAGE-A1 or CT7. *Int J Cancer* 2001;94:252–256.

74. Robbins PF, Morgan RA, Feldman SA, *et al.* Tumor regression in patients with metastatic synovial cell sarcoma and melanoma using genetically engineered lymphocytes reactive with NY-ESO-1. *J Clin Oncol* 2011;29(7):917–924.

75. Bertrand JY, Chi NC, Santoso B, *et al.* Haematopoietic stem cells derive directly from aortic endothelium during development. *Nature* 2010;464(7285):108–111.

76. Wu SY, Lopez-Bernstein G, Calin GA, Sood AK. RNAi therapies: drugging the undruggable. *Sci Transl Med* 2014;6(240):240ps7.

77. Welch EM, Barton ER, Zhuo J, *et al.* PTC124 targets genetic disorders caused by nonsense mutations. *Nature* 2007;447(7140):87–91.

Index

Locators in *italics* refer to Figures.

AAM *see* aggressive angiomyxoma
abdominal endometriosis 506–507
  histopathology 507, *508*
abdominal ependymoma 711–713, *713*
abdominal inflammatory myofibroblastic tumor (IMT) 289
  epithelioid variant *290*, 291
abdominal wall desmoid tumor 239–241
abscesses 33–34
aCGH (array-CGH) 100–101, 123, 143
acquired elastotic hemangioma 575, *577*
acquired progressive lymphangioma (benign lymphangioendothelioma) 585–587
  histopathology *586*, 586–587
acquired tufted angioma 566–567, *568*
acral arteriovenous tumor (arteriovenous hemangioma) 26, 561–562, *562*
acral myxoinflammatory fibroblastic sarcoma 352–356
  appearance *354*
  cytology 984, *985*
  differential diagnosis 354–355, 984
  histopathology 354, *355*
  pseudolipoblasts *356*
  Reed–Sternberg-like cells *356*
acroangiodermatitis (pseudo-Kaposi's sarcoma) 580, *580*
actins 49–50, 546
acute myeloid leukemia 954
ACVR1 (activin-like receptor) 845
adamantinoma *770*, 770
adamantinoma-like Ewing sarcoma 99

adenocarcinoma *see also* carcinoma
  aggressive digital papillary 764–765, *765–766*
  keratins 65
  pancreatic 809
  pulmonary 792–793, 805, *806*
adenofibroma 505
adenomatoid tumor 777, *778*
adenomatous polyposis coli gene (*APC*) 137–138, 245–246
adenosarcoma 505
adipose tissue pathology *see also* lipoblastoma; lipodystrophy; lipogranuloma; lipoma; lipomatoses; liposarcoma
  cytology 977–982
  imaging 19–25
  immunohistochemistry 46, 57, 72
adrenal gland
  cortical tumors
    carcinoma 60, 815, *816*
    myelolipoma 390, *391*
  medullary tumors *see* pheochromocytoma
adult fibrosarcoma 336, 351–352
  cytology 984, *985*
  spindle cells 352–353
adult granulosa cell tumor 505–506
  appearance *506*
  histopathology 505–506, *506*
  immunohistochemistry 506, *507*
adult myofibroma 267, *270–271*
adult rhabdomyoma 529, 529–530
adults with "pediatric" sarcomas 1031
AFH *see* angiomatoid fibrous histiocytoma
AFX *see* atypical fibroxanthoma
age, and tumor type 19
aggressive angiomyxoma (AAM)
  female 498–501
    appearance 498, *499*
    histopathology 498–499, *500*
    male 498

aggressive digital papillary adenocarcinoma 764–765, *765, 766*
aggressive fibromatosis *see* desmoid fibromatosis
AIDS *see* HIV/AIDS
AJCC staging system 5
AKT/mTOR signaling pathway 128, 136
  inhibitors 133, 518
ALCL *see* anaplastic large cell lymphoma
alcohol fixation 8
ALK/ALK (anaplastic lymphoma kinase)
  in ALCL 952–954
  as an IHC marker 70, 291
  gene fusion 124, 127, 292, 953–954
  in inflammatory myofibroblastic tumor 127, 291–292
  in neuroblastoma 130–131, 877
alkaptonuric ochronosis 913, *914*
allele-specific PCR (AS-PCR) 141
alpha smooth muscle actin 50, 546
Alport syndrome 450
ALT (atypical lipomatous tumor) *see* well-differentiated liposarcoma
altitude (and carotid body paraganglioma) 735
alveolar rhabdomyosarcoma (alveolar sarcoma) 539–541
  cytology 994, 994
  genetics 129, 134–135, 539–541, 542, 878–879
  histopathology 539, *540–541*
  leukemoid variant 540
  imaging 539
  immunohistochemistry 546, *547*
  mixed with embryonal form 544
alveolar soft part sarcoma (ASPS) 895–900

  appearance 896, 896
  clinical features
    soft tissue 895
    visceral sites 895–896
  cytology *1005*, 1005–1006
  differential diagnosis 737, 900
  electron microscopy 899, *899*
  genetics 129, 900
  histogenesis 899–900
  histopathology 896–898, *897–898*
  immunohistochemistry 898–899, *899*
  metastatic disease *1027*
American Joint Committee of Cancer (AJCC) staging system 5
AML *see* angiomyolipoma
amplification of genes 98–99, 101, 130–131, 611, 875–876
amputation 1015
amputation (traumatic) neuroma 30, 639, *640*
amyloid tumor (amyloidoma) 936–938, *937–938*
anaplastic large cell lymphoma (ALCL)
  noncutaneous (systemic) 952–954, *953*
  primary cutaneous 954
anaplastic lymphoma kinase *see* *ALK*/ALK
anaplastic rhabdomyosarcoma 543–544, *544*
anaplastic spindle cell rhabdomyosarcoma 542
ancient hematoma *see* organizing hematoma
ancient schwannoma 655, 989–990, *990*
aneuploidy 98–99
  hyperdiploidy 548, 876–877
  trisomy 237, 246, 294, 506, 921
aneurysmal bone cyst 104, 844–845

Index

aneurysmal fibrous histiocytoma 224–225, *224*
angioblastic meningioma (meningeal SFT) 324–325, 331
angioblastoma of Nakagawa (acquired tufted angioma) 566–567, *568*
angioendothelioma *see* papillary intralymphatic angioendothelioma
angioendotheliomatosis, reactive 579–580
angiofibroma, cellular (vulvar) 496–497, *497–498*
angiofibroma, giant cell 330, *331*
angiofibroma, juvenile nasopharyngeal 263–264, *263–265*
angiogenesis 553
angiography 13, 29
angiokeratoma 559–560, *561*
angioleiomyoma 446–447
 histopathology 446–447, *448–449*
angiolipoma 388–390
 histopathology 389, *389*
 cellular variant 389–390, *390*
 infiltrative *see* intramuscular hemangioma
angiolymphoid hyperplasia with eosinophilia *see* epithelioid hemangioma
angiomatoid fibrous histiocytoma (AFH) 283–288
 appearance *284–285*
 histopathology 284–285, *286*
 cysts *286*
 hemorrhagic appearance *286–287*
 resembling lymph node metastasis *285*
 variations *287–288*
angiomatosis 562–564
 histopathology 563–564, *565*
 imaging 27, *29*
angiomatosis, bacillary 578–579
 histopathology 578, *579*
 Warthin–Starry stain of *Bartonella* 579
angiomatous lesions *see* angiomatosis; angiosarcoma; hemangioendothelioma; hemangioma; hemangiopericytoma; lymphangioma
angiomyofibroblastoma, vulvar 494–496, *495–496*
angiomyofibroblastoma-like tumors in men 496–497, *497*
angiomyolipoma (AML) 513–518
 appearance *514*
 cytology 1000–1001, *1002*
 differential diagnosis 517
 genetics 517–518
 histopathology 515–517, *515*
  epithelioid *517*, 517
  lipomatous component *515*, 515–516
  smooth muscle component 516–517, *516*
  vascular component *516*, 516
 immunohistochemistry 59, *59*, *60*, *61*, 517
angiomyoma (angioleiomyoma) 446–447
 histopathology 446–447, *448–449*
angiomyomatous hamartoma of lymph node 444, *445*
angiomyxoma, aggressive female 498–501
 appearance 498, *499*
 histopathology 498–499, *500*
 male 498
angiomyxoma, superficial (cutaneous myxoma) 306–310
 clinical features 306–307, 309
 differential diagnosis 307–308
 genetics 309–310
 histopathology 307, *308–309*
angiosarcoma (AS) 593, 601–611, 1031
 appearance
  breast *605*
  deep soft tissue *606*
  lymphedematous arm *604*
  scalp *603*
 breast parenchyma 604–605
 cardiovascular 606–607
 in children 606, 608
 cutaneous (face and scalp) 602–603
 cytology 995–997, *998*
 of deep soft tissues 605, 608
 differential diagnosis 27, 605, 608–610, 996–997
 etiology 3–4, 593
 foreign-body-associated 607–608
 genetics 611
 in germ cell tumors 607
 hepatic 3–4, 606
 histopathology 608
  breast parenchyma *605*
  epithelioid 608, *609–610*
  hepatic *606*
  lymphoid infiltration *609*
  poorly differentiated *610*
  radiation-induced 603, *604*
  splenic *607*
  well-differentiated *608*
 imaging 27
 immunohistochemistry 45, 66, 610–611, *611*
 in lymphedematous extremities 4, 603–604
 in nerve sheath tumors 607, 659, 683
 postradiation 603
 rare sites 607
 splenic 606
 treatment 1018
angulated bodies 673
anoctamin-1 484–485
anthracyclines *see* doxorubicin
Antoni A/Antoni B areas in schwannoma 655
aorticopulmonary paraganglioma 734
APC (adenomatous polyposis coli) 137–138, 245–246
appearance *see* gross anatomy/morphology
appendiceal GISTs 481
array-CGH (aCGH) 100–101, 123, 143
arsenic 4, 798
arteriovenous hemangioma 26, 561–562, *562*
arthritis *see also* pseudogout
 gout 841, 903, *907*, 909–910
 osteoarthritis 905–906, *907*
 rheumatoid 904–905, *906*
 septic 903–904, *904*
AS *see* angiosarcoma
asbestos 782–783
 asbestos fibers *782*
 ferruginous bodies 782–783, *783*
Askin's tumor (chest wall Ewing sarcoma) 856 *see also* Ewing sarcoma family of tumors
ASPS *see* alveolar soft part sarcoma
ASPSCR1 (ASPL)-TFE3 gene fusion 129, 900
ATF1/ATF gene fusion 126, 287–288, 707–708
atypical decubital fibroplasia (ischemic fasciitis) 195
 histopathology 195, *197*
atypical fibrous histiocytoma 226–228, *227*
atypical fibroxanthoma (AFX) 364–367
 appearance 365
 histopathology 365–366, *366*
 immunohistochemistry 366, *367*
atypical glomus tumor 626, *630*
atypical lipomatous tumor (ALT) *see* well-differentiated liposarcoma
atypical mesothelial hyperplasia 790
atypical tenosynovial giant cell tumor 921–922, *922*
atypical teratoid/rhabdoid tumor 882, 885
B-cell lymphoma
 Burkitt lymphoma 949, *950*
 and crystal-storing histiocytosis 964–965, *965–966*
 diffuse large 947, *948*
 extranodal marginal zone 946–947, *947*
 follicular 945–946, *946*
 intravascular 948–949, *949*
 lymphoblastic 947–948, *948*
 plasmablastic 949–950, *951*
bacillary angiomatosis 578–579
 histopathology 578, *579*
 Warthin–Starry stain of *Bartonella* 579
BAF47 protein 74, 133–134 *see also* INI1/INI1 (*SMARCB1*)
Baker (popliteal) cyst 17, 31–32, *33*
BAP1/BAP1 (BRCA1-associated protein) 794
Bartonella infections
 bacillary angiomatosis 578–579, *579*
 verruga peruana 579
 Warthin–Starry stain 579
basal cell keratins 67–68
basal cell nevus syndrome (Gorlin syndrome) 529
basement membrane proteins 72
BCL2 protein *70*
BCOR gene 124, 866–867
BCOR-CCNB3 fusion sarcoma 866–867, *866–867*
Beckwith–Wiedemann syndrome 535, 856
Bednar tumor (pigmented variant of DFSP) 342, *342*
benign fibrous histiocytoma (BFH, dermatofibroma) 220–228
 appearance 220–221, *221*, *224*
 clinical features 220
 cytology 987, *988*
 differential diagnosis 224–226, 344–345, 987
 genetics 228
 histopathology 221–222, *222–223*
  aneurysmal variant *224*, 224–225
  atypical variant *227*
  cellular variant *225–226*
  epithelioid variant *226*
  lipidized variant *227*
 immunohistochemistry 222–224, *225*, *227*, *223*
 metastasis 221
 recurrence 220
 variants 224
  aneurysmal 224–225, *224*
  atypical 226–228, *227*
  cellular 225–226, *225*
  epithelioid 225, *226*
  granular cell 228
  lipidized 225–226, *227*

Index

benign lymphangioendothelioma (acquired progressive lymphangioma) 585–587
histopathology *586*, 586–587
benign metastasizing leiomyoma 464–465, *465*
benign Triton tumor (neuromuscular hamartoma/choristoma) 530, *530*, 638, *638*
benign tumors
classification 2
distinguishing from malignant tumors 11–12, 19
epidemiology 3, 11
beta-2 microglobulin-derived amyloidosis 938
beta-catenin 128, 138, 631
immunohistochemistry of desmoids *245*
BFH *see* benign fibrous histiocytoma
biological therapies *see* targeted therapies
biopsy 6–7
fine-needle aspiration 975, 1012
of synovia 903
tissue handling 8
types of 1012–1013
biphasic mesothelioma 786–787, *787*
biphasic synovial sarcoma 68, 747–748, *747–748*
Birbeck granules 961
bladder carcinoma 67–68, 811–813
bladder inflammatory myofibroblastic tumor (IMT) 288–289, *292*
bladder leiomyoma 452
bladder paraganglioma 729–734, *733–734*
blue nevus *see* cellular blue nevus
bone cement *911*, 911
bone imaging 11, 13
bone-forming tumors 825 *see also* calciphylaxis; fibrodysplasia ossificans progressiva; fibro-osseous pseudotumor of the digits; metaplastic bone; myositis ossificans; odontogenic myxoma; osteosarcoma, extraskeletal; phosphaturic mesenchymal tumor; tophaceous pseudogout; tumoral calcinosis
differential diagnosis 836–839, 842–847, 849
Borrelia burgdorferi infection
Lyme disease 904, *905*
lymphoma 952
botryoid rhabdomyosarcoma 533–534

histopathology 535, *537*
brachytherapy 1015–1016
brachyury (transcription factor) 766
BRAF mutations 487, 702, 958–961
brain tumors
angiosarcoma 289
atypical teratoid/rhabdoid tumor 882, 885
inflammatory myofibroblastic tumor 289
metastatic small cell carcinoma of the lung 806
branchial anlage mixed tumor (ectopic hamartomatous thymoma) 766–768
histopathology 768–769, *768*
immunohistochemistry 768, *769*
breast angiosarcoma 604–605, *605*
breast carcinoma
metastatic 73, 803–805, *805*
sarcomatoid 805, *805*
breast fibromatosis 239
breast implants
ALCL 953
silicone granuloma 409, *410*
breast myofibroblastoma 218–220
cytology *983*, 983–984
histopathology 218–219, *220*
breast smooth muscle hamartoma 445
Brooke–Spiegler syndrome 763
brown fat 398
brown fat tumor *see* hibernoma
Burkitt lymphoma 949, *950*
bursal fluid collections 32–33

cadherins 70
café au lait spots *653*
Cajal cells 475, *484*
calcific uremic arteriolopathy *see* calciphylaxis
calcification *see also* calciphylaxis; tumoral calcinosis
hemangioma 25
imaging 11, 13
neuroblastoma 871
phosphaturic mesenchymal tumor *836*
synovial chondromatosis 34, *35*
synovial sarcoma 749
calcified myonecrosis 926
calcifying aponeurotic fibroma 256–257
histopathology 257, *259–260*
calcifying fibrous pseudotumor 253–254
histopathology 254, *255*
calciphylaxis 840–841

histopathology 841, *843*
imaging 841, *842*
calcium pyrophosphate deposition disease *see* pseudogout
Call–Exner bodies 505
CALLA (common acute lymphoblastic leukemia antigen, CD10) 70–71
calponin 51
calretinin 72, 752
capillary hemangioma 26 *see also* infantile hemangioma; lobular capillary hemangioma
acquired elastotic hemangioma 575, *577*
acquired tufted angioma 566–567, *568*
cherry hemangioma 575, *576*
imaging 26, *28*
intramuscular *28*, 564
microvenular hemangioma 575, *576*
carbon fibers/graphite 911–912, *912*
carcinogens (chemical) 3–4, 606, 782–783, 798
carcinoid tumors 53, 806
carcinoma
adrenocortical 60, 815, *816*
aggressive digital papillary adenocarcinoma 764–765, *765*, *766*
breast
metastatic 73, 803–805, *805*
sarcomatoid 805, *805*
compared with mesothelioma 792–793, *792–793*
gastrointestinal 808–809
hepatocellular 809, *809*
immunohistochemistry 50, 58, 65, 812, 818–819
lung
adenocarcinoma/large cell 792–793, 805, *806*
sarcomatoid 807, *808*
small cell 53–54, 802–803, 806, *806–807*
Merkel cell *see* Merkel cell carcinoma
myoepithelial 50, 702
ovarian 815–816
pancreatic 809
prostatic 125, 813–815, *814–815*
renal cell *see* renal cell carcinoma
squamous cell 758, 803, *803–804*
thyroid 184, 816–817, *817*
unknown primary 818–819
urothelial 67–68, 811–813
carcinosarcoma *see* malignant mixed müllerian tumor; sarcomatoid carcinoma
cardiac angiosarcoma 606–607

cardiac inflammatory myofibroblastic tumor 289, *291*, 291
cardiac lipoma 385
cardiac myxoma 313–315
histopathology 314, *314–315*
degenerative features *316*
glandular epithelial differentiation 314, *316*
cardiac paraganglioma 734
cardiac rhabdomyoma *523*, 523
cardiac synovial sarcoma 746
Carney complex 306, 308–310, 661
Carney triad 482, 825
Carney–Stratakis syndrome 482
carotid body paraganglioma 734–735, *737*
appearance *735*
histopathology 735–737, *736*
immunohistochemistry *54*
carpal tunnel syndrome 938
Carrion's disease (verruga peruana) 579
cartilage-associated lesions 825 *see also* chondroma of soft parts; extraskeletal mesenchymal chondrosarcoma; extraskeletal myxoid chondrosarcoma; synovial chondromatosis
alkaptonuric ochronosis 913, *914*
differential diagnosis 827, 830–832, 834
immunohistochemistry 48, 57
lipoma with metaplastic cartilage and bone 382–383, *383*
Castleman disease 941–943, *944*
catecholamines
neuroblastoma 870
pheochromocytoma 723, 726
cauda equina paraganglioma 738–740, *739*
cavernous hemangioma 558–559
appearance *559*
histopathology 559, *560*
imaging 25–26, *26–27*
CBN *see* cellular blue nevus
CCNB3-BCOR fusion sarcoma 866–867, *866–867*
CD10 (CALLA) 70–71
CD31 (PECAM-1) 41–44, 610
CD34 (hematopoietic progenitor cell antigen) 44–47
CD34-positive fibroblasts 182
dermatofibrosarcoma protuberans 46, *345*
GISTs 47, 485, *485–486*
superficial acral fibromyxoma 312
CD56 (NCAM) 56, 806

1038

Index

CD57 (Leu7) 56
CD68 62
CD99 71–72, *71*, 861–862, *862*, 865, *876*
CD117 (KIT)
  immunohistochemistry 74, 484, *485*, 611
  mutations
    in GISTs 131–132, 475, 486–487
    in mastocytosis 956
    and TKI sensitivity 486, 1024–1026
  NGS profile *145*
  as a therapeutic target 132, 487–488
CD163 62
CDK4/CDK4 (cyclin-dependent kinase 4) 101, 134
CDKN1a (cyclin-dependent kinase inhibitor 1a) (p21) 135
CDKN2a/*CDKN2b* (cyclin-dependent kinase inhibitors 2a/2b) (p14ARF/p15a/p16) 77, 134–135, 682
cDNA microarrays 143–144
cell culture media 93
  for transport 93
cell cycle markers 75–77
cell cycle regulatory pathways 134–136
cellular angiofibroma (vulvar) 496–497, *497–498*
cellular angiolipoma 389–390, *390*
cellular blue nevus (CBN) 694–695
  atypical/malignant 694–695
  differential diagnosis 694–695, 707
  histopathology 694, *695*
  malignant 696
  immunohistochemistry *59*, *61*, 694
cellular fibrous histiocytoma 225–226, *225*
cellular neurofibroma 645–646, *646*, 652
cellular plexiform schwannoma 657
cellular schwannoma 655–656, *657*
central nervous system tumors *see* brain tumors; spinal tumors
centromere-specific FISH probes 97
cerebriform fibrous proliferation in Proteus syndrome *273*
cervicovaginal myofibroblastoma 493
  histopathology 493, *493*
  immunohistochemistry 493, *494*
CGH (comparative genomic hybridization) 100–101, 123, 143

Charcot joint 906
chemoradiotherapy 1028–1029
chemotherapy *see also* targeted therapies
  adjuvant/neoadjuvant 1017–1018, 1027–1028, 1028–1029
  for advanced/metastatic disease 1029–1030
  high-dose therapy 1030
  and hyperthermia 1028
  limb perfusion therapy 1028
  response by histologic type/anatomic location 1030–1032
  rhabdomyosarcoma and Ewing sarcoma 547–548, 1018–1019
  in adults 1031
cherry hemangioma 575, *576*
chief cells 723
childhood tumors 19, 249, 277
  angiomatoid fibrous histiocytoma 283–288
  angiosarcoma 606, 608
  calcifying aponeurotic fibroma 256–257
  calcifying fibrous pseudotumor 253–254, *255*
  cellular plexiform schwannoma 657
  chemotherapy 547–548, 1018–1019
  cranial fasciitis 182, 249–251, *250*
  fibrodysplasia ossificans progressiva 845, *846*
  fibromatosis colli 251, *251–252*
  fibrous hamartoma of infancy 265–266, *265–266*
  fibrous umbilical polyp 251–253, *252*
  Gardner fibroma 207, *262*, 262–263
  giant cell fibroblastoma 342–343, *343*, *344*, *343*
  gingival fibromatosis 253, *253–254*
  gingival granular cell tumor 676, *677*
  gliomatosis 710–711, *711*
  grading system 4–5
  infantile diffuse fibromatosis 260–261, *261*
  infantile digital fibroma 254–256, *256*, *256*, *257*
  infantile fibrosarcoma 292–294, *293–294*
  infantile hemangioma (and related forms) 554–557, *556*
  infantile myofibroma/myofibromatosis 266–268, 269–270, 984, *984*

inflammatory myofibroblastic tumor *see* inflammatory myofibroblastic tumor
juvenile hyaline fibromatosis 253, 269–272, *271–272*
juvenile nasopharyngeal angiofibroma 263–265
juvenile xanthogranuloma *63*, 224, 961–962, *962–963*
Kaposi's sarcoma 612–613
Langerhans cell histiocytosis 960–961, *960–961*
leiomyosarcoma 459
lipoblastoma 19, 21, 405–407, *406–407*, 978, *979*
lipofibromatosis 257–260, *260–261*, 283
lipoma (spinal cord) 384–385
liposarcoma 430
lymphangiomatosis 587, *587–588*
melanotic neuroectodermal tumor of infancy 708–709, *710*, 991, *991*
mortality 855
nasal glioma 710
neuromuscular choristoma 530, *530*, 638, *638*
neurothekeoma 224, 277–282, *279–281*
papillary intralymphatic angioendothelioma 593–594, *593–595*
plexiform fibrohistiocytic tumor 224, 281–283, 368
Proteus syndrome 272–273, *272*
rhabdoid tumor *see* rhabdoid tumor, extrarenal
rhabdomyoma
  cardiac 523, *523*
  fetal 528, *528–529*
rhabdomyomatous mesenchymal hamartoma 528–530, *530*
rhabdomyosarcoma *see* rhabdomyosarcoma
smooth muscle hamartoma of skin 443–444, *444*
CHK2 (checkpoint kinase 2) 135
chloroma (extramedullary myeloid tumor) 954
  histopathology 954–955, *954*
  immunohistochemistry 954, *955*
  sclerosing 424
choanal polyps 265
chondro-osseous metaplasia in lipoma 382–383, *383*
chondroid lipoma 402–404
  appearance *402*
  histopathology *402–403*, *402–403*
chondroid syringoma *see* mixed tumor

chondroma of soft parts 825–827
  appearance 826, *826*
  histopathology 826–827, *826*
  imaging 825, *826*
chondromatosis, synovial 825, 827–830
  appearance 827–828, *828*
  histopathology 828–830, *829*
  imaging 34, *35*, 827, *828*
  secondary 827, *829*, 829–830
chondrosarcoma *see* extraskeletal mesenchymal chondrosarcoma; extraskeletal myxoid chondrosarcoma
CHOP (*DDIT3*) gene fusion 126–127, *127*, 434–435
chordoma 765–766
  appearance *766*
  histopathology 765–766, *767*
  immunohistochemistry 66, 766, *767*
choristoma, neuromuscular 530, *530*, 638, *638*
chromaffin paraganglioma *see* pheochromocytoma
chromogenic *in situ* hybridization (CISH) 99–100
chromogranin A 54
chromosomes *see* cytogenetics
chromothripsis 101–102
CIC-DUX4 fusion sarcoma 865–866, *866*, *866*
circumscribed storiform collagenoma (sclerotic fibroma) 211, *214*
cirsoid aneurysm (arteriovenous hemangioma) 26, 561–562, *562*
CISH (chromogenic *in situ* hybridization) 99–100
citric acid cycle dysfunction 138, 724–725
classification
  hemangioma 554
  myxofibrosarcoma 356–357
  neuroblastoma 867–870, 873, *874*
  paraganglioma 723
  rhabdomyosarcoma 531, 993
  skeletal muscle tumors 527
  soft tissue tumors 1–2
clear cell renal carcinoma *810*
clear cell rhabdomyosarcoma 544, *544*
clear cell sarcoma, gastrointestinal 708
  histopathology 708, *709*
clear cell sarcoma of tendons and aponeuroses 704–708
  cytology 1006, *1006*
  differential diagnosis 702, 707, 1006
  genetics 707–708

1039

Index

clear cell sarcoma of tendons and aponeuroses (cont.)
   histopathology 705–707, *706–707*
   immunohistochemistry 56, 59, 707, *708*
clear cell "sugar" tumor 519–520
   histopathology 519, *520*
clonal evolution 94, 123, 145
clonality assay (HUMARA) 138–139
CNS tumors *see* brain tumors; spinal tumors
COBRA-FISH (combined binary ratio FISH) 100
coccygeal glomus bodies 624
COL1A1-PDFGB gene fusion 128–129, 345
collagenase 93
collagenous fibroma 208–209
   appearance 208, *210*
   histopathology 208, *210–211*
   immunohistochemistry 208, *211*
collagens 182
   collagen IV 72
   in palisaded myofibroblastoma 216, 218, *219*
colon
   carcinoma 808
   FAP 137–138
   GIST 480–481
   leiomyoma 451–452
   leiomyosarcoma 457
   lipoma 385
   nerve sheath tumor 661, *662*
   PEComa 521, *522*
colony stimulating factor 1 gene (*CSF1*) 914, 920–921
combined binary ratio FISH (COBRA-FISH) 100
comparative genomic hybridization (CGH) 100–101, 123, 143
composite rhabdoid tumor 882
computed tomography (CT) 13, 25
   cavernous hemangioma 26
   chondroma of soft parts *826*
   extraskeletal myxoid chondrosarcoma *830*
   fibrodysplasia ossificans *846*
   lipomatous lesions *19, 23*
   lymphangioma *28*
   myositis ossificans *12, 843*
   synovial chondromatosis *34, 35*
   tumor within a tumor *1025*
congenital fibromatosis *see* infantile myofibroma/myofibromatosis
congenital fibrosarcoma *see* infantile fibrosarcoma
congenital hemangioma 556–557
congenital self-healing reticulohistiocytosis 960
contrast media 13–14

gadolinium-associated fibrosis *936*, 936
   thorotrast reaction 3, 934–935, *935*
copy number variation 98–99, 143 *see also* amplification of genes
core biopsy 6, 1012
cotton fiber granuloma 934, *935*
Cowden's disease 211
cranial fasciitis 182, 249–251
   histopathology *250*, 250–251
   imaging *250*
cranial nerve schwannoma 133, 654, 659–660
CREB1-EWSR1 gene fusion 126, 287–288, 708
CREB3L1/CREB3L1 127, 348–349
CREB3L2/CREB3L2 126–127, 348–349, 351
crizotinib 127
Crohn's disease 907
crystal deposition disease *see* pseudogout
crystal-storing histiocytosis 964–965, *965–966*
CSF1 gene translocation (colony stimulating factor 1) 914, 920–921
CT *see* computed tomography
CTNNB1 (beta-catenin gene) 128, 138, 246, 631
culture media 93
   for transport 93
cutaneous anaplastic large cell lymphoma 954
cutaneous angiosarcoma 602–603
   appearance *603*
   histopathology
     epithelioid *609*
     lymphoid infiltration *609*
     poorly differentiated *610*
     well-differentiated *608*
cutaneous follicular lymphoma 945–946
cutaneous leiomyosarcoma 455–456, *456*
cutaneous mucinosis, focal 310, *311*
cutaneous myxoid cyst 305–306, *306*, 307
cutaneous myxoma *see* superficial angiomyxoma
cutaneous neurofibroma 644–645
   appearance (NF1 patient) *653*
   histopathology 645, *645–646*
cutaneous PEComa 522
cutaneous pleomorphic liposarcoma 435
cutaneous sarcomatoid carcinoma 803–804, *803*
cyclin-dependent kinase inhibitors (CDKNs) 76–77
   CDKN1a (p21) 135

CDKN2a/CDKN2b (p14ARF/p15a/p16) 77, 134–135, 682
CYLD gene 763
cylindroma 763–764, *764*
cystic hygroma 582, *584*
cysts/cystic masses *see also* multicystic peritoneal mesothelioma
   Baker (popliteal) cyst 17, 31–32, *33*
   cutaneous myxoid cyst 305–306, 307, *306*
   ganglion cyst 32, *33*, 303–305, *304–305*
   imaging 14, 31–34
   schwannoma *656*
cytogenetics 92, 976–977
   CGH/aCGH 100–101, 123, 143
   chromosomal abnormalities
     amplification 98–99, 101, 130–131, 611, 875–876
     fusion *see* fusion genes
     hyperdiploidy 548, 876–877
     multiple/complex 103, 108–110, 123, 134
     translocations 103–108, 117, 120
     trisomy 237, 246, 294, 506, 921
   conventional 8
     cell culture and harvest 8, 93–94
     G/GTW-banding stains 94
     image analysis 94
     pros and cons 95
     specimen preparation 93
     specimen selection 92
     specimen transport 92–93
   FISH 95
     in Ewing sarcomas 97, 864
     MYCN amplification 98–99, 877
     probes 97–99, *99–100*, 119
     pros and cons 102
     specimen preparation 98, 140–141, 976
     specimen selection 95–97
     technical variations 99–100
     nomenclature 94–96
cytokeratins *see* keratins
cytokines 362
cytology
   adult fibrosarcoma 984, *985*
   alveolar soft part sarcoma *1005*, 1005–1006
   ancillary techniques 975–977
   angiomyolipoma 1000–1001, *1002*
   angiosarcoma 995–997, *998*
   benign fibrous histiocytoma 987, *988*
   clear cell sarcoma 1006, *1006*
   desmoid fibromatosis 982–983, *983*
   desmoplastic small round cell tumor 1006–1007

   epithelioid hemangioendothelioma 997, *998*
   epithelioid sarcoma 1003–1004, *1004*
   Ewing sarcoma/PNET 1004–1005, *1005*
   fat necrosis 977–978, *978*
   fibromyxoid sarcoma (low-grade) 986–987, *986*
   fine-needle aspiration 975
   glomus tumor 998, *999*
   granular cell tumor *1002*, 1002
   hibernoma 978–979, *980*
   leiomyosarcoma 995, *996*
   lipoblastoma 978, *979*
   lipoma 977, *977*
     myxolipoma 977, *977*
     pleomorphic/spindle cell 978, *979*
   liposarcoma
     myxoid/round cell 981–982, *981*
     well-differentiated 979–981, *980*
   malignant fibrous histiocytoma
     giant cell 989
     pleomorphic 988, *988*
   malignant peripheral nerve sheath tumor 992, 992–993
   melanotic neuroectodermal tumor of infancy 991
   meningioma 990–991, *991*
   mesenchymal chondrosarcoma, extraskeletal 1000, *1000*
   myoepithelioma/mixed tumor cytology 1000, *1001*
   myofibroblastoma, breast 983, 983–984
   myositis ossificans 999, 999–1000
   myxofibrosarcoma 985–986, *986*
   myxoid chondrosarcoma, extraskeletal 1006, *1007*
   myxoinflammatory fibroblastic sarcoma 984, *985*
   myxoma, intramuscular 1000, *1001*
   myxopapillary ependymoma 991–992, *992*
   neurofibroma 990, *990*
   nodular fasciitis 982, *982*
   rhabdoid tumor, extrarenal 1007–1008, *1008*
   rhabdomyoma 993, 993
   rhabdomyosarcoma 993–994, *994–995*
   schwannoma 989, 989
     ancient 989–990, *990*
   synovial sarcoma 1003, *1003*
   tenosynovial giant cell tumor 987, 987

Index

D2-40 (podoplanin antibody) 47
Dabska tumor *see* papillary intralymphatic angioendothelioma
dacarbazine 1030
DDIT3 (*CHOP*) gene fusion 126–127, *127*, 434–435
deciduoid mesothelioma 785–786, *786*
dedifferentiated leiomyosarcoma 453–454, *454*
dedifferentiated liposarcoma (DDLS) 7, 416, 424–430
  appearance 425, *426*
  clinical features 424–425
  differential diagnosis 424, 429–430, 487
  genetics 108, 430
  histopathology 425–428
    leiomyosarcomatous differentiation *429*
    meningothelial-like whorls 426–427, *427–428*
    pleomorphic forms *428*
    resembling inflammatory MFH *426*
    resembling various sarcoma types *427*
    rhabdomyosarcomatous differentiation *429*
  imaging 6, 24, *25*
  immunohistochemistry 429
deep (benign) fibrous histiocytoma 220, 224, 987
deep granuloma annulare 927–929, *928*
deep plexiform schwannoma 657
deep soft tissue angiosarcoma 605, *606*, 608
definition of soft tissue tumors 1
degenerative joint disease (osteoarthritis) 905–906, *907*
denaturing gradient gel electrophoresis (DGGE) 142
denaturing high-performance liquid chromatography (DHPLC) 142
dendritic reticulum cell sarcoma 965–968, *967*
dermatofibroma *see* benign fibrous histiocytoma
dermatofibroma with monster cells (atypical fibrous histiocytoma) 226–228, *227*
dermatofibrosarcoma protuberans (DFSP) 336–346
  appearance 338, *338*
  Bednar tumor (pigmented variant) 342
  clinical features 336–338, 342

differential diagnosis 224, 344–345
genetics 128–129, 345
giant cell fibroblastoma 342–343
histopathology 338, 338–340, *340*
  Bednar tumor *342*
  borders *339*
  fat infiltration *339*
  fibrosarcomatous transformation 341–342, *341*
  giant cell fibroblastoma 342–343, *343–344*
  myoid nodules *341*
  myxoid variant *340*
immunohistochemistry 46, 343–345, *345*
treatment
  imatinib 129, 345–346
  surgery 337, 1015
dermatomyofibroma 212, *215*
desmin 50, 486, 546, 899
desmoid fibromatosis (desmoid tumor) 238–246
  appearance 242, *242*
  clinical features 238–240
  cytology 982–983, *983*
  differential diagnosis 184–185, 244–245, 487
  genetics
    APC 137–138, 245–246
    CTNNB1 138, 246
    other 240, 246
  histopathology 242, *243–244*
    dilated blood vessels *243*
    keloid-like fibers *243*
    mesenteric desmoids *245*
    skeletal muscle infiltration *244*
  imaging 34–35, *36*
  immunohistochemistry 242–244, *245*
  multiple desmoids 240
  site of occurrence 238
    abdominal wall 239–241
    extra-abdominal *36*, 239–241
    intra-abdominal 240–241, *245*, 487
  terminology 233
  treatment 241
    hormonal 1032
    radiotherapy (adjuvant) 1031
    surgery 239–241, 1015
    TKIs 241–242, 1032
desmoplastic fibroblastoma 208–209
desmoplastic melanoma 702–704
  histopathology 702–703, *703*
  immunohistochemistry 704, *704*
desmoplastic mesothelioma 788–789, *790*, *788*

desmoplastic small round cell tumor (DSRCT) 879–882
  clinical features 882
  cytology 1006–1007
  differential diagnosis 806, *808*, 881–882
  genetics 126–156, 881
  histopathology 879–880, *880*
  immunohistochemistry 75, 880, *880*
DFSP *see* dermatofibrosarcoma protuberans
diabetic scleroderma 207
diagnosis 6–9, 17–19 *see also* imaging; immunohistochemistry
diffuse angiokeratoma in Fabry disease 560
diffuse extra-articular tenosynovial giant cell tumor 919–921, *921*
diffuse intra-articular tenosynovial giant cell tumor 918–921
  appearance *919*, 919
  histopathology 919, *920*
diffuse large cell B-cell lymphoma 947, *948*
diffuse neurofibroma 648, *651*, 652 *see also* pigmented neurofibroma
diffuse symmetric lipomatosis 386–387
digital mucoid cyst 305–306, *306*, *307*
digital papillary adenocarcinoma 764–765, *765–766*
dioxin 3
disseminated xanthoma 932
DNA extraction from formalin-fixed specimens 140–141
DNA microarrays (DNA chips) 143–144
DNA sequencing
  next-generation (NGS) 123, 144–145
  pyrosequencing 144
  Sanger (dideoxy) method 144
docetaxel 1030–1031
Doppler ultrasound 14
double minutes (dmins) 130
  neuroblastoma 875, *877*
doxorubicin 1017, 1029
  combination therapies 1029–1030
  high-dose therapy 1030
DSRCT *see* desmoplastic small round cell tumor
duodenal gangliocytic paraganglioma 737–738, *739*
duodenal GIST *see* small intestinal GISTs
Dupuytren's contracture *see* palmar fibromatosis

DUX4-CIC fusion sarcoma 865–866, *866*, *866*
dysgerminoma 818

E-cadherin 70
ear *see also* vestibular schwannoma
  jugulotympanic paraganglioma 737, *738*
  meningioma 716
EBV *see* Epstein–Barr virus
eccrine acrospiroma (hidradenoma) 762, *763*
eccrine spiradenoma 762–763, *764*
ectomesenchymoma 545–546, 862–863, *863*
ectopic hamartomatous thymoma (branchial anlage mixed tumor) 766–768
  histopathology *768–769*, 768
  immunohistochemistry 768, *769*
ectopic meningioma *see* meningioma, ectopic
EGLN1/PHD2 (prolyl hydroxylase) mutations 726
EGLN2/PHD1 (prolyl hydroxylase) mutations 726
EHE *see* epithelioid hemangioendothelioma
elastofibroma 204–206
  appearance 204–205, *205*
  histopathology 205–206, *205–206*
  imaging 22, 34
electron microscopy 8, 976
  alveolar soft part sarcoma 899, *899*
  desmoplastic small round cell tumor 880
  Ewing sarcomas/PNETs
    atypical form 860–861, *861*
    PNET 861, *861*
    typical form 860, *860*
  granular cell tumor 673
  hemangioma 555
  Langerhans cell histiocytosis 961
  leiomyosarcoma 455
  malignant fibrous histiocytoma
    myxoid 357–358
    pleomorphic 362
  Merkel cell carcinoma 801
  mesothelioma 789, *790*
  MPNST 680
  neuroblastoma 874, *875*
  perineurioma 668, 671
  pheochromocytoma 728
  rhabdoid tumor 883, *883*
  rhabdomyoma 529

Index

electron microscopy (cont.)
  schwannoma 655
    melanotic 663
EMA *see* epithelial membrane antigen
embolization of tumors 13
embryonal rhabdomyosarcoma (ERMS) 532, 533–535
  appearance 535
  botryoid 533–535
  cytology 993–994, *994*
  genetics 108, 535
  histopathology 534–535, *536*
    botryoid subtype 535, *537*
  immunohistochemistry 51, *53*, *547*
  mixed with alveolar form 544
EMC *see* extraskeletal myxoid chondrosarcoma
emperipolesis 957
endometrial carcinoma *see* malignant mixed müllerian tumor
endometrial stromal sarcoma (ESS) 501–503
  differential diagnosis 464
  genetics 123, 503
  histopathology 501, *502*
  hormonal therapy 1031
  immunohistochemistry 70, *502, 503*
endometriosis 506–507
  histopathology 507, *508*
endothelial biology 553–554
endothelial markers 41–48
Enneking surgical staging system 6, 18
eosinophilic fasciitis 195–197, *198*
ependymoma
  abdominal 711–713
  cytology 991–992, *992*
  histopathology
    abdominal *713*
    myxopapillary 711, *712*
  immunohistochemistry 712–713
  myxopapillary 711, 740, 991–992
epidemiology 2–3, 11, 975
  aggressive angiomyxoma 498
  alveolar soft part sarcoma 895
  angioleiomyoma 446
  angiolipoma *388*
  angiomatoid fibrous histiocytoma 283
  angiomyolipoma 513
  angiosarcoma 601, *602*
  atypical fibroxanthoma 364
  benign fibrous histiocytoma 220
  calcifying aponeurotic fibroma 257
  cellular blue nevus 694
  clear cell sarcoma of tendons *705*

collagenous fibroma 208
dermatofibrosarcoma protuberans *337*, 342
desmoids 238–240
elastofibroma 204
epithelioid sarcoma 754
Ewing sarcomas/PNETS 857
extremity sarcoma *1014*
fasciitis
  cranial *249*
  intravascular 185
  ischemic 195
  necrotizing 198
  nodular 182
  proliferative 191
fibromyxoid sarcoma *346*
Gardner fibroma 262
giant cell angiofibroma 330
giant cell fibroblastoma 342
GISTs 474, 482
glomus tumor *625*
granular cell tumor 672
hemangioendothelioma 596
hemangioma
  epithelioid 569
  hobnail 565
  infantile 554
  intramuscular *563*
  lobular capillary 557
  spindle cell *572*
hibernoma 398–399
juvenile xanthogranuloma 961
Kaposi's sarcoma 612–613
leiomyoma 446, *450*
leiomyosarcoma 456, *457*, 457, 466
lipoblastoma *405*
lipoma 379, *388*, 392–393
liposarcoma
  dedifferentiated 425
  myxoid 430
  pleomorphic 436
  well-differentiated 417
lymphangioma 582
lymphangiomyoma(tosis) 518
lymphedema in the obese patient 410
malignant fibrous histiocytoma
  inflammatory 367
  myxoid (myxofibrosarcoma) 357
  pleomorphic 362
melanoma 696, 702
meningioma, ectopic *714*
Merkel cell carcinoma 798
mesodermal stromal polyps 491
mesothelioma 782–783
MPNST 678, 683–684
myofibroblastoma
  breast 218
  palisaded 214
myofibroma 267

myxoinflammatory fibroblastic sarcoma *354*
myxoma
  cardiac 313
  cutaneous 306
  intramuscular *299*
  juxtaarticular *302*
  nerve sheath 665
neuroblastoma 870
neurofibroma 645, *650*
neuroma, palisaded encapsulated 642
neurothekeoma 277
nuchal-type fibroma 207
oral giant cell fibroma 213
oral irritation fibroma 212–213
ossifying fibromyxoid tumor 317
palmar fibromatosis 233
paraganglioma 726, *728*, 735, 737
penile fibromatosis 237
pheochromocytoma 726
plantar fibromatosis 233
retroperitoneal fibrosis 199
rhabdomyoma 528
rhabdomyosarcoma *532*
schwannoma 654, 656
solitary fibrous tumor 324, 330
superficial angiomyxoma 306
synovial sarcoma 744
tendon sheath fibroma 209
tenosynovial giant cell tumor 915, 918, 920
truncal sarcoma *1014*
epidermal inclusion cyst, ruptured *933*, 933
epididymal tumors
  adenomatoid tumor 777
  melanotic neuroectodermal tumor of infancy 709
epigenetics (epimutations) 116, 136
  SDH-deficient GIST 483
  SDH-deficient paraganglioma 725
  small round blue cell tumors 856
  as a therapeutic target 1032
epithelial cell markers *see* epithelial membrane antigen; keratins
epithelial membrane antigen (EMA) 69
  perineurioma 668, *671*, 671
  rhabdoid tumor *884*
epithelial tumors of unknown histogenesis *see* epithelioid sarcoma; synovial sarcoma
epithelioid angiomyolipoma *517*, 517
epithelioid angiosarcoma 608, *609–610*
  cytology 997, *998*

epithelioid fibrous histiocytoma 225, *226*
epithelioid gastric GISTs 479, *480*
epithelioid hemangioendothelioma (EHE) 595–599
  cytology 997, *998*
  differential diagnosis 570, 997
  hepatic 596–597
  histopathology 596–597, *597–598*
    hepatic 597, *598*
    pulmonary 597, *599*
    sarcoma-like (pseudomyogenic) *601*
  immunohistochemistry 47, 597–598, *600*
  pulmonary 596–597
  sarcoma-like (pseudomyogenic) 599–601
epithelioid hemangioma 104, 554, 569–570
  appearance *570*
  histopathology 570, *571–572*
epithelioid histiocytoma, solitary (reticulohistiocytoma) 962–964, *964*
epithelioid inflammatory myofibroblastic tumor (epithelioid myofibroblastic sarcoma) *290*, 291
epithelioid leiomyoma/leiomyosarcoma *463*, 463
epithelioid mesothelioma 785, *786*
epithelioid nerve sheath tumor (benign) 664, *665*
epithelioid rhabdomyosarcoma 544–545, *545*
epithelioid sarcoma 753–758
  appearance *755*
  clinical features 754–755
  cytology 1003–1004, *1004*
  differential diagnosis 756–758, 1004
  genetics 758
  histopathology 755, *756–757*
    proximal-type 758, *759*
  immunohistochemistry 47, 74, 755, *757*, 758
  proximal-type 758
epithelioid sarcoma-like hemangioendothelioma 599–601, *601*
epithelioid schwannoma 657, *658*, 661
Epstein–Barr virus (EBV)
  Burkitt lymphoma 949
  diffuse large B-cell lymphoma 947
  leiomyosarcoma 3–4, 467–468, *468*

ER-positive tumors *see* estrogen receptor (ER)-positive tumors
Erdheim–Chester disease 957–960, *959*
ERG/ERG (ETS-related gene)
 as an IHC marker 47–48, 610
 fusion with *EWSR1* 864
 fusion with *FUS* 127, 864
eribulin 1030
ERMS *see* embryonal rhabdomyosarcoma
eruptive xanthoma 932
erythema migrans 904
esophageal tumors
 GISTs 475–476, *477*
 intramural leiomyoma 449–451, *451*, 451
 leiomyomatosis 450
ESS *see* endometrial stromal sarcoma
estrogen receptor (ER)-positive tumors 72–73
 desmoids 1032
 leiomyoma 72–73
  abdominal *73*, 459–460, *459*
  benign metastasizing (pulmonary nodules) 464–465, *465*
  uterine 459–463, *461–462*
  vulvar 459
 and response to hormonal therapy 1031
etiology 3–4
 angiosarcoma 3–4, 593
 atypical fibroxanthoma 364
 EBV-associated tumors 3–4, 467–468, 947, 949
 Kaposi's sarcoma 3, 612
 Merkel cell carcinoma 798, 801–802
 mesothelioma 782–783
ETS gene/transcription factor family 47–48, 125–127, 863–864 *see also* ERG/ ERG; *ETV6*; *FLI1*
ETV6 (ETS variant gene 6), fusion with *NTRK3* 124, 127–128, 294
everolimus 133
Ewing sarcoma family of tumors 856–867
 adamantinoma-like 99
 appearance 857–858, *858*
 BCOR-CCNB3 sarcoma 866–867
 chemotherapy 1018, 1031
 CIC-DUX4 sarcoma 865–866
 clinical features 857, 1004
 cytology 1004–1005, *1005*
 differential diagnosis 753, 865, 1004–1005
 electron microscopy
  atypical form 860–861, *861*
  PNET 861, *861*

 typical form 860, *860*
 genetics 104, 856
  BCOR-CCNB3 fusion 866
  CIC-DUX4 fusion 865–866
  EWSR1/ETS family fusions 107–108, 123, 125–126, 856, 863–864
  EWSR1/non-ETS family fusions 864–865
  FISH probes 97, 864
  t(11;22) (q24;q12) translocation *106*, 863
 histopathology 858, *858*
  atypical form 858–859, *859*
  BCOR-CCNB3 fusion sarcoma 866, *866–867*
  CIC-DUX4 fusion sarcoma *866*
  PNET *860*, 860
  typical form 858, *859*
 history 856–857
 imaging 857, *858*
 immunohistochemistry 71–72, 861–863, 866
  CD99 *71*, 861–862, *862*, 865
  chromogranin 54
  FLI1 48, *862*, 862
  keratins 66, *862*
  synaptophysin 53, *862*
 prognosis 864–865
Ewing-like sarcoma, undifferentiated 865
EWSR1 gene fusions/ translocations (EWS RNA-binding protein 1) 104, 124, 126
 ATF1-CREB1 126, 287–288, 707–708
 DDIT3 126–127
 ETS genes (*FLI1*, *ERG*) 107–108, 123, 125–126, 863–864
 activity of the fusion products 136, 856
 and prognosis 864
 FISH probes 97, 864
 NFATC2 865
 NR4A3 126, 832
 PATZ1 865
 PBX1 126, 760–761
 POU5F1 126, 760–761, 865
 SMARCA5 865
 SP3 865
 WT1 126–156, 881
excisional biopsy 1013
external beam radiation therapy (EBRT) 1015–1016
extra-abdominal desmoids 36, 239–241
extragastrointestinal gastrointestinal stromal tumors (EGISTs) 481–482
extramedullary myeloid tumor *see* myeloid tumor, extramedullary (myeloid sarcoma)

extranodal marginal zone lymphoma 946–947, *947*
extrarenal rhabdoid tumor *see* rhabdoid tumor, extrarenal
extraskeletal chondroma *see* chondroma of soft parts
extraskeletal mesenchymal chondrosarcoma (MCHSA) 832–834
 appearance *833*
 cytology 1000, *1000*
 histopathology 832–833, *833*
 imaging 832, *833*
 immunohistochemistry 57, 833–834, *834*
extraskeletal myxoid chondrosarcoma (EMC) 830–832
 appearance *830*
 cytology 1006, *1007*
 genetics 126, 832
 histopathology 830, *831–832*
 hybrid with synovial sarcoma 750
 imaging *830*, 830
 immunohistochemistry 54, 57, 831
extraskeletal osteosarcoma *see* osteosarcoma, extraskeletal
extremity angiosarcoma (Stewart– Treves syndrome) 4, 603–604, *604*
eye *see* orbital tumors
eyelids, xanthelasma 932

Fabry disease 560
Factor XIIIa 62
falciform ligament PEComa 521
familial adenomatous polyposis (FAP) 137–138, 238, 240, 245–246, 263
familial cancer syndromes 117–119 *see also individual syndromes*
familial cylindromatosis 763
familial GIST syndrome 486–487
familial glomuvenous malformation 630–631
familial infiltrative fibromatosis 137
fascicular sign 30
fasciitis 181
 cranial 182, 249–251, *250*
 eosinophilic 195–197, *198*
 intravascular 182, 185–189, *190*
 ischemic 195, *197*
 necrotizing 197–199, *198–199*
 nodular *see* nodular fasciitis
 proliferative 190–194, *193–194*, 982
fasciitis ossificans (panniculitis ossificans) 182–183
fat atrophy (lipodystrophy) 407, *408*

fat necrosis (lipogranuloma) 407–409, *408*, 977–978, *978*
FDG-PET (positron emission tomography) 12
female genital tract tumors 491
 see also ovarian tumors; uterine tumors; vaginal tumors; vulvar tumors
 differential diagnosis 464, 492–493, 497, 500–503
ferruginous bodies 782–783, *783*
FET proteins 125 *see also* EWSR1 fusions/translocations; FUS gene fusion
fetal rhabdomyoma 528–529
 histopathology *528*, 528
FFCC (French Federation of Cancer Centers) grading system 4–5
FH/fumarate hydratase 138
 fumarase-deficient leiomyoma 461, *463*
fibroblast biology 181–182
fibroblast growth factors (FGFs) 182, 834, 837
fibroblastic proliferations 181, 249, 277, 336 *see also individual lesions*
fibroblastic reticulum cells
 normal 968
 sarcoma 968
fibroblastic sarcoma *see* acral myxoinflammatory fibroblastic sarcoma
fibrocartilaginous pseudotumor, nuchal 207–208, *208*
fibrodysplasia ossificans progressiva 845
 imaging 845, *846*
fibrohistiocytic tumor *see* plexiform fibrohistiocytic tumor
fibroids (uterine leiomyoma) 459–463
 histopathology *461–462*, 461
fibrolipoma 382
 neural 385–386, *386–387*
fibrolipomatous hamartoma of nerve 385–386, *386–387*
fibroma *see also* myofibroblastoma; neurofibroma
 calcifying aponeurotic 256–257
 collagenous 208–209, *210–211*
 dermatomyofibroma 212, *215*
 elastofibroma 22, 34, 204–206, *205–206*
 Gardner fibroma 207, 262, *262–263*
 infantile digital fibroma/ fibromatosis 254–256, *256–257*
 nuchal-type 206–207, *207*
 oral
  giant cell 213, *218*

Index

fibroma (cont.)
  irritation 212–213, 217
  skin
    pleomorphic 211–212, 215
    sclerotic 211, 214
  tendon sheath 209–211, 213–214
fibromatosis see also desmoid fibromatosis (desmoid tumor); infantile myofibroma/ myofibromatosis; neurofibromatosis type 1; neurofibromatosis type 2
  familial infiltrative 137
  gingival 253, 253–254
  imaging 34–35, 36
  infantile diffuse fibromatosis 260–261, 261
  juvenile hyaline 253, 269–272, 271–272
  lipofibromatosis 257–260, 260–261, 283
  palmar/plantar 233–237
    palmar histopathology 234–235, 235
    plantar histopathology 235–236, 236–237
  penile 237–238, 238
  terminology 233, 249, 260–261, 266
fibromatosis colli 251
  appearance 251
  histopathology 251, 252
fibromyxoid sarcoma (low-grade) (LGFMS) 346–349 see also sclerosing epithelioid fibrosarcoma
  appearance 347
  clinical features 346
  cytology 986–987, 986
  differential diagnosis 347–348, 987
  genetics 101, 348–349
  histopathology 346–347, 347–348
    giant hyaline rosette variant 350
    higher-grade variant 349
  hyalinizing spindle cell tumor with giant rosettes 347
  immunohistochemistry 74, 347
fibromyxoma see superficial acral fibromyxoma
fibro-osseous pseudotumor of the digits 845–847, 847
fibrosarcoma see also dermatofibrosarcoma protuberans; myofibrosarcoma; myxofibrosarcoma
  adult 336, 351–352
    cytology 984, 985
    spindle cells 352–353
  infantile 292–294
    appearance 293

histopathology 293–294, 293–294
  inflammatory 288
  sclerosing epithelioid 74, 336, 350–351, 351
fibrosarcoma-like lipomatous neoplasm 423
fibrosarcomatous transformation of DFSP 341–342, 341
fibrous dysplasia of bone 300, 301–302
fibrous hamartoma of infancy 265–266, 265–266
fibrous histiocytoma of tendon sheath see tenosynovial giant cell tumor
fibrous mesothelioma see solitary fibrous tumor
fibrous umbilical polyp 251–253, 252
fibroxanthoma see atypical fibroxanthoma; tenosynovial giant cell tumor
FICTION technique 99
filariasis 603
fine-needle aspiration (FNA) 975, 1012
FISH see fluorescence in situ hybridization
fixatives 8
  in FISH 98
  recovery of DNA/RNA from fixed specimens 140–141
FKHR (FOXO1A)-PAX gene fusion 129
FLI1/FLI1 (Freund's leukemia integration site)
  as an IHC marker 48, 611, 862, 862
  fusion with EWSR1 107–108, 125, 863–864
    activity of fusion products 136, 856
    and prognosis 864
florid reactive periostitis of the tubular bones of hands and feet (fibro-osseous pseudotumor of the digits) 845–847, 847
flow cytometry 976
fluid collections 32–33
fluorescence in situ hybridization (FISH) 95
  in Ewing sarcomas 97, 864
  MYCN amplification 98–99, 877
  probes 97–99, 99–100, 119
  pros and cons 102
  specimen preparation 98, 140–141, 976
  specimen selection 95–97
  technical variations 99–100
FNA (fine-needle aspiration) 975, 1012

focal cutaneous mucinosis 310, 311
focal myositis 194, 531, 531
focal tumoral calcium pyrophosphate dihydrate deposition disease (tophaceous pseudogout) 838–839
  histopathology 839, 841
  imaging 839, 840
follicular dendritic (reticulum) cell sarcoma 965–968, 967
follicular lymphoid hyperplasia 941, 944
follicular lymphoma 945–946, 946
foot see also acral myxoinflammatory fibroblastic sarcoma; plantar fibromatosis
  hemangioma 562
  infantile digital fibroma 254–256, 256–257
  Morton neuroma 30, 31, 638, 639
  pleomorphic hyalinizing angiectatic tumor 369–372, 371, 371
  Proteus syndrome 272
Fordyce angiokeratoma 560
foreign body reactions 3
  angiosarcoma 607–608
  cotton fiber granuloma 934, 935
  PVP accumulation 933–934, 933–934
  to silicone
    ALCL 953
    granuloma 409, 410
    synovial 911, 912
  synovial
    to bone cement 911, 911
    to carbon fibers/graphite 911–912, 912
    to metal fragments 912–913, 913
    to polyethylene wear particles 910, 911
    to silicone 911, 912
formalin fixatives 8
Fournier's gangrene 198
FOXL2 mutations 506
FOXO1A (FKHR)-PAX gene fusion 129, 535, 539–540, 542, 547
French Federation of Cancer Centers (FFCC) grading system 4–5
frozen tissue specimens 8
fumarate hydratase (FH) 138
  fumarase-deficient leiomyoma 461, 463
fungal synovitis 908–909
FUS (TLS) gene fusions 124, 126–127
  with CREB3L1 or CREB3L2 127, 348–349, 351

with DDIT3 126–127, 434–435
  with FLI1 or ERG 127, 864
fusion genes 103, 106–107, 117–129
  ALK 124, 127, 292, 953–954
  ASPSCR1-TFE3 129, 900
  BCOR 124, 866
  CIC-DUX4 865
  COL1A1-PDGFB 128–129, 345
  ETV6-NTRK3 124, 127–128, 294
  EWSR1 see EWSR1 gene fusions/translocations
  FUS 124, 864
    FUS-CREB3L1/FUS-CREB3L2 127, 348–349, 351
    FUS-DDIT3 126–127, 434–435
  HMGA2 124, 382
  JAZF1-SUZ12 123, 503
  NAB2-STAT6 330
  NCOA2 125, 834
  PAX-FOXO1A (FKHR) 129, 535, 539–540, 542, 547
  PHF1 125, 318
  PLAG1 125, 407
  ROS1 125, 127
  SS18-SSX 123, 128, 136, 753
  TFE3 125, 129, 599, 900
  TFK 126
  TMPRSS2 125
  USP6 (TRE17) 125
  WWTR-CAMTA1 598–599
  YAP1-TFE3 599

gadolinium contrast agent 13–14
gadolinium-associated fibrosis 936, 936
gallium scans 11–12
gangliocytic (duodenal) paraganglioma 737–738, 739
ganglion cyst 32, 303–305
  histopathology 304–305, 304–305
  imaging 33
ganglion of distal interphalangeal joint 305–306, 306
ganglioneuroblastoma 868–869, 869–870, 872, 874
ganglioneuroma 872
  appearance 872
  gastrointestinal/mucosal 639
  histopathology 868, 869, 872, 872
  mucosal 641
  immunohistochemistry 55
  with MPNST/pheochromocytoma 686, 732
ganglioneuromatosis 639
Gardner fibroma 262–263
  differential diagnosis 207, 263
  histopathology 262, 262

1044

Index

immunohistochemistry 262, 262
Gardner syndrome
    (adenomatous polyposis
    with desmoids) 137–138,
    238, 240, 245–246
gastric GIST 476–479
    appearance 477, 477–478
    genetics
        KIT and PDGFRA
            mutations 486–487
        succinate dehydrogenase-
            deficient 482–483
    histopathology 478–479
        epithelioid type 479, 480
        hypercellular type 479, 479
        palisaded-vacuolated type
            478, 479
        sarcomatous type 479, 479
        sclerosing spindle cell type
            478, 478
    origin of omental GISTS 482
gastric glomus tumor 624–626
gastrointestinal autonomic nerve
    tumor (GANT) see
    gastrointestinal stromal
    tumor (GIST)
gastrointestinal carcinoma
    808–809
gastrointestinal clear cell
    sarcoma 708, 709
gastrointestinal leiomyoma
    448–449
    appearance (esophageal) 450,
        451
    intramural 449–451, 451
    muscularis mucosae 451–452,
        452
gastrointestinal leiomyosarcoma
    457
gastrointestinal lipoma 385
gastrointestinal mucosal
    (ganglio)neuroma 639
gastrointestinal nerve sheath
    tumor 661, 662
gastrointestinal
    neurofibromatosis 653
gastrointestinal paraganglioma
    (duodenal gangliocytic)
    737–738, 739
gastrointestinal PEComa 521,
    522
gastrointestinal schwannoma
    487, 660–661
    appearance 660, 660
    histopathology 660–661, 661
gastrointestinal stromal tumor
    (GIST) 474
    appearance 477, 477–478,
        480
    clinical features 474–476,
        479, 481–483
    differential diagnosis 292, 457,
        481–482, 487
    familial syndromes 482–483,
        486–487

genetics
    KIT and PDGFRA
        mutations 131–132, 475,
        486–487, 1024–1026
    other 108, 487
    SDH-deficient form 483
histopathology 475
    esophageal 477
    gastric 478–479, 478–480
    in NF1 patients 484
    post-treatment 488
    rectal 481, 482
    SDH-deficient tumors 483,
        483
    small intestinal 479–480, 481
imaging of "tumor within a
    tumor" 1025
immunohistochemistry 70,
    484–486, 485
    CD34 47, 485–486
    desmin 50, 486
    heavy caldesmon 52, 486
    keratins 66, 486
    KIT 74, 484, 485
    S100 57
    SDH-deficient tumors 483
location 474–475
    appendix 481
    colon 480–481
    esophagus 475–476
    extragastrointestinal 481–482
    rectum 481
    small intestines/duodenum
        476, 479, 483–484
    stomach 476–479
in NF1 patients 483–484, 653
succinate dehydrogenase-
    deficient 482–483, 488,
    1025
surgery 487, 1020
    palliative 1026
targeted therapies
    future directions 1027
    SDH-deficient tumors 488,
        1025
    TKIs, adjuvant 1020, 1026–
        1027
    TKIs, first line 132, 487–488,
        1024–1026
    TKIs, resistance 132, 486,
        1020, 1024–1025, 1025,
        1027
terminology 474
gastrointestinal synovial sarcoma
    746
gastrointestinal vascular
    proliferation related to
    intestinal intussusception
    580, 581
GCT see granular cell tumor
gel electrophoresis 142
gemcitabine 1030–1031
gene expression profiling 123,
    143–144, 977
genetics 115 see also epigenetics

aggressive angiomyxoma 501
alveolar soft part sarcoma 129,
    900
analytical techniques 92
    cDNA microarrays 143–144
    CGH/aCGH 100–101, 123,
        143
    clonality assay (HUMARA)
        138–139
    conventional cytogenetics 8,
        92–95
    for detection of gene
        amplification 98–99, 130
    DNA sequencing 123, 144–
        145
    FISH (and variants) 95–100,
        102, 119
    PCR (and variants) 119–
        123, 141–143, 977
    specimen preparation and
        handling 8, 92–93, 95–97,
        140–141
anaplastic large cell lymphoma
    952–954
aneurysmal bone cyst 104,
    844–845
angioleiomyoma 447
angiomatoid fibrous
    histiocytoma 287–288
angiomyolipoma 517–518
angiosarcoma 611
atypical fibroxanthoma 366
benign fibrous histiocytoma
    228
Burkitt lymphoma 949
Carney complex 309–310
cell cycle regulatory pathways
    134–136
cellular angiofibroma 497
chondroma 827
chromosomal abnormalities
    amplification 98–99, 101,
        130–131, 611, 875–876
    fusion see fusion genes
    hyperdiploidy 548, 876–877
    multiple/complex 103, 108–
        110, 123, 134
    translocations 103–108, 117,
        120
    trisomy 237, 246, 294, 506,
        921
clear cell sarcoma 707–708
collagenous fibroma 209
cylindroma 763
dermatofibrosarcoma
    protuberans 128–129, 345
desmoid fibromatosis
    APC 137–138, 245–246
    CTNNB1 138, 246
    other 240, 246
desmoplastic small round cell
    tumor 126–156, 881
elastofibroma 206
endometrial stromal sarcoma
    123, 503

epithelioid sarcoma 758
Erdheim–Chester disease
    958–960
Ewing sarcomas/PNETs see
    Ewing sarcoma family of
    tumors, genetics
fasciitis, nodular 185
fibrodysplasia ossificans
    progressiva 845
fibromyxoid sarcoma 101,
    348–349
follicular lymphoma 946
gingival fibromatosis 253
GISTs
    KIT and PDGFRA
        mutations 131–132, 475,
        486–487, 1024–1026
    other 108, 487
    SDH-deficient form 483
glomangiopericytoma 631
glomus tumor 630–631
granulosa cell tumor 506
growth factor signaling
    pathways 136–138
hemangioendothelioma 598–
    599, 601
hemangioma 104, 575, 587
hemangiopericytoma 330, 333
hereditary cancer syndromes
    117–119 see also
    individual syndromes
infantile fibrosarcoma 294
inflammatory myofibroblastic
    tumor 107, 127, 292
juvenile hyaline fibromatosis
    271–272
Langerhans cell histiocytosis
    961
leiomyoma 447, 450–451, 462–
    463
leiomyosarcoma 455
lipoblastoma 407
lipoma 131, 382
    angiolipoma 390
    chondroid 404
    hibernoma 402
    lipomatosis 387
    spindle cell/pleomorphic
        394
liposarcoma 101
    dedifferentiated 108, 430
    myxoid 106, 126–127, 434–
        435
    pleomorphic 437–438
    well-differentiated 102, 131,
        416–423
malignant fibrous
    histiocytoma 134, 359,
    362–364
malignant mixed müllerian
    tumor 505
mastocytosis 956
melanoma 702, 704
melanotic neuroectodermal
    tumor of infancy 709

1045

genetics (cont.)
  mesenchymal chondrosarcoma (extraskeletal) 834
  mesothelioma 794
  miRNA 116–117, 129, 136
  mixed tumor 760–761
  MPNST 108–110, 117, 682
  myxofibrosarcoma 359
  myxoid chondrosarcoma (extraskeletal) 126, 832
  myxoinflammatory fibroblastic sarcoma 355–356
  myxoma 301–303, 314–315
  neuroblastoma 130–131, 875–877
    MYCN amplification 99, 130, 875–876, *877*
    and prognosis 876–878
  neurofibromatosis type 1 (NF1) 108–110, 132–133, 637, 653
  neurofibromatosis type 2 (NF2) 133, 637
  nomenclature
    cytogenetics 94–96
    of individual genes 146–153
  ossifying fibromyxoid tumor 318
  osteosarcoma 849
  palmar/plantar fibromatosis 237
  paraganglioma 724–726
  PEComa 513, 517–518, 523
  perineurioma 669, 671
  phosphaturic mesenchymal tumor 837
  plexiform fibrohistiocytic tumor 283
  proliferative myositis 194–195
  Proteus syndrome 272–273
  rhabdoid tumor 133, 882, 885–886
  rhabdomyoma 529–530
  rhabdomyosarcoma 548
    alveolar 129, 134–135, 539–541, *542*, 878–879
    embryonal 108, 535
    epithelioid 545
    mixed alveolar and embryonal 544
    pleomorphic 542
    and prognosis 547
    spindle cell 538
  schwannoma 660–661
  sclerosing epithelioid fibrosarcoma 351
  solitary fibrous tumor 330
  superficial angiomyxoma 309–310
  synovial chondromatosis 830
  synovial sarcoma 123, 128, 753
  TCA cycle dysfunction 138, 724–725
  tendon sheath fibroma 211

tenosynovial giant cell tumor 105, 914, 920–921
tumoral calcinosis 838
vascular malformations 587
germ cell tumors
  with angiosarcoma 607
  dysgerminoma 818
  seminoma 48, 817–818, *818*
GFAP (glial fibrillary acidic protein) 73–74, 712–713, *712*
giant cell angiofibroma 330, *331*
giant cell fibroblastoma 342–343, *343*
giant cells 344
giant cell fibroma of the oral cavity 213, *218*
giant cell hyaline angiopathy (pulse granuloma) 929, *929*
giant cell malignant fibrous histiocytoma (MFH) 368–369, 989
  cytology 989
  histopathology 369, *370*
giant cell tumor of soft parts 282–283, 367–368, *369*
giant cell tumor of tendon sheath *see* tenosynovial giant cell tumor
Giemsa stain 94
gingival fibromatosis 253
  appearance *253*
  histopathology 253, *254*
gingival granular cell tumor (gingival epulis of newborn) 676, *677*
  histopathology 676, *677*
  immunohistochemistry 676, *677*
GIST *see* gastrointestinal stromal tumor
Gleevec *see* imatinib mesylate
glial fibrillary acidic protein (GFAP) 73–74, 712–713, *712*
glioma, nasal 710
gliomatosis 710–711, *711*
glomangioma 628
glomangiomyoma 629
glomangiopericytoma 631–632, *632*
glomangiosarcoma 626, *630*
glomeruloid hemangioma 564–565, *566*
glomulin 630–631
glomus bodies 624, *624*
glomus jugulare tumor (jugulotympanic paraganglioma) 737, *738*
glomus tumor 624–631
  appearance 625, *626*
  atypical/malignant 626
  clinical features 624–625
  cytology 998, *999*

differential diagnosis 628–629
genetics 630–631
histopathology 625–626, *627–628*
  atypical/malignant 630
  glomangioma 628
  glomangiomyoma 629
  glomus bodies 624
  intramuscular 629
immunohistochemistry 626–627, *631*
glycogen
  in alveolar soft part sarcoma 897, *898*
  in Ewing sarcoma cells 859–860
Gorlin syndrome (nevoid basal cell carcinoma syndrome) 529
gossypiboma (cotton fiber granuloma) 934, *935*
gout 841, 903, *907*, 909–910 *see also* pseudogout
grading of tumors 4–5, 75–76
granular cell fibrous histiocytoma 228
granular cell tumor (GCT) 672–674
  appearance 673, *675*
  cytology *1002*, 1002
  differential diagnosis 673–674
  histopathology 672–673, *673*, *675*
    malignant 676
    pseudoepitheliomatous hyperplasia 674
  immunohistochemistry 56–57, 673, *675*
  malignant 675–676
granular cell tumor (GCT), gingival 676, *677*
  histopathology 676, *677*
  immunohistochemistry 676, *677*
granuloma annulare, deep 927–929, *928*
granulomatous synovitis 907–909
  mycobacterial 907–908, *909*
  in sarcoidosis 907
granulosa cell tumor 505–506
  appearance *506*
  histopathology 505–506, *506*
  immunohistochemistry 506, *507*
gross anatomy/morphology 7
  alveolar soft part sarcoma 896, *896*
  angiomatoid fibrous histiocytoma 284–285
  angiomyofibroblastoma-like tumor (male) 497
  angiomyolipoma 514
  angiosarcoma
    breast 605
    deep soft tissue 606
    lymphedematous arm 604

scalp 603
atypical fibroxanthoma 365
benign fibrous histiocytoma 220–221, *221*
  aneurysmal variant 224
  cellular variant 225
chondroma 826, *826*
chordoma 766
collagenous fibroma 208, *210*
dermatofibrosarcoma protuberans 338, *338*
desmoids 242, *242*
elastofibroma 204–205, *205*
epithelioid sarcoma 755
Ewing sarcomas/PNETs 857–858, *858*
fasciitis
  intravascular 186–187, *190*
  necrotizing 198, *198*
  nodular 183, *185*
  proliferative 192–200, *193*
fibromatosis colli 251
fibromyxoid sarcoma 347
gingival fibromatosis 253
GISTs 477, 477–478, *480*
glomus tumor 625, *626*
granular cell tumor 673, *675*
granulosa cell tumor 506
hemangioendothelioma
  epithelioid 596, *597*
  PILA 593
hemangioma 559, 564, 570, 573
hemangiopericytoma 331
hibernoma 399, *400*
infantile digital fibroma 254–255, *256*
infantile fibrosarcoma 293
inflammatory myofibroblastic tumor 289, *289*
juvenile hyaline fibromatosis 271
juvenile nasopharyngeal angiofibroma 263–264, *263*
juvenile xanthogranuloma 962
Kaposi's sarcoma 613
leiomyoma
  esophageal 450, *451*
  retroperitoneal 459
leiomyomatosis, peritoneal 465, *466*
leiomyosarcoma 452–453, *453*
  uterine 467
  vena cava 458
lipoblastoma 406
lipoma 379, *380*
  chondroid 402
  lipoma arborescens 383
  myelolipoma 391
  neural fibrolipoma 386
  spindle cell lipoma 394
  thymolipoma 385
liposarcoma

dedifferentiated 425, *426*
well-differentiated *419*, 419
lymphangioma 584
lymphangiomyoma(tosis) 518–519, *518*
malignant mixed müllerian tumor 504, *504*
Merkel cell carcinoma 799, *799*
mesenchymal chondrosarcoma (extraskeletal) 833
mesothelioma, malignant 783
  peritoneal 784
  pleural 784
MPNST 678, *679*, *684*
multicystic peritoneal mesothelioma 779
myxofibrosarcoma 358
myxoid chondrosarcoma (extraskeletal) 830
myxoinflammatory fibroblastic sarcoma 354
myxoma
  cardiac 314
  intramuscular 300, *300*
neuroblastoma 871, *872*
  ganglioneuroma *872*
neurofibroma
  cutaneous 653
  intraneural 648
  plexiform 649
neurofibromatosis type 1 (café au lait spots) 653
organizing hematoma 926, *926*
ossifying fibromyxoid tumor 319
palisaded myofibroblastoma 218
palmar fibromatosis 233–234
papillary endothelial hyperplasia 577
papillary intralymphatic angioendothelioma 593
paraganglioma
  carotid body 735
  retroperitoneal 729
PEComa (renal capsule) 520
penile fibromatosis 237
pheochromocytoma 726
phosphaturic mesenchymal tumor 835
plantar fibromatosis 233–234
Proteus syndrome 272
renal cell carcinoma (cutaneous metastases) 809
rhabdoid tumor *883*, 883
rhabdomyoma, oral 529
rhabdomyosarcoma 531–532, *535*
schwannoma 654–655, *654*
  gastrointestinal 660, *660*
  melanotic 662, *662*
solitary fibrous tumor 326, *326*
superficial acral fibromyxoma 311

superficial angiomyxoma 306–307
synovial chondromatosis 827–828, *828*
synovial sarcoma 746, *746*
tendon sheath fibroma 213
tenosynovial giant cell tumor
  diffuse intra-articular *919*, 919
  localized *916*, 916
tumoral calcinosis *838*, 838
growth factors/growth factor receptors 136
  FGF 182, 834, 837
  HGF/MET 130, 136–137
  IGF/IGF1R 136–137
  PDGFA receptor 132, 475, 1024
  PDGFB/PDGFRB 128–129, 242
  as therapeutic targets 137
  VEGF/VEGFR 47, 553, 555, 593–594
Gs alpha mutations 301 302
gynecological tumors 491 *see also* ovarian tumors; uterine tumors; vaginal tumors; vulvar tumors
  differential diagnosis 464, 492–493, 497, 500–503

HAART (highly active antiretroviral therapy) 407
hamartin (*TSC2*) 517–518
hamartoma
  angiomyomatous hamartoma of lymph node 444, *445*
  fibrolipomatous hamartoma of nerve 385–386, *386–387*
  fibrous hamartoma of infancy 265–266, *265–266*
  meningiothelial 714–715, *715*
  neuromuscular 530, *530*, 638, *638*
  rhabdomyomatous mesenchymal hamartoma 528–530, *530*
  smooth muscle hamartoma of skin 443–444, *444–445*
hand *see also* acral myxoinflammatory fibroblastic sarcoma; chondroma of soft parts; palmar fibromatosis
  calcifying aponeurotic fibroma 256–257
  digital mucoid cyst 305–306, *306*, *307*
  digital papillary adenocarcinoma 764–765, *765–766*
  fibro-osseous pseudotumor of the digits 845–847, *847*

infantile digital fibroma 254–256, *256–257*
lipomatosis of the nerve 385–386, *386–387*
pacinian neuroma 640–641, *642*
sclerotic lipoma 395–396, *398*
Hand–Schüller–Christian disease 960
Hashimoto–Pritzker disease 960
heart *see entries at* cardiac
heavy caldesmon (HCD) 50–51, 486
hedgehog signaling pathway 529
hemangioendothelioma
  appearance
    epithelioid 596, *597*
    PILA 593
  cytology (epithelioid) *997*, 998
  differential diagnosis 570, 997
  epithelioid 570, 595–599, 997
  epithelioid sarcoma-like 599–601
  genetics 598–599, 601
  hepatic 596–597
  histopathology
    epithelioid 596–597, *597–598*
    epithelioid sarcoma-like *601*, 601
    hepatic *597*, 598
    kaposiform 567–568, *568–569*
    PILA 593, *594*
    pulmonary 597, *599*
    retiform 594–595, *595*
  imaging 27
  immunohistochemistry
    epithelioid 47, 597–598, *600*
    epithelioid sarcoma-like *601*, 601
    kaposiform 568
    PILA 593–594, *594*
  kaposiform 567–568
  papillary intralymphatic angioendothelioma (PILA) 593–594
  pulmonary 596–597
  retiform 594–595
hemangioma 554–575 *see also* angiokeratoma; kaposiform hemangioendothelioma
  acquired elastotic 575
  acquired tufted angioma 566–567
  angiomatosis 27, 562–564
  appearance *559*, *564*, *570*, *573*
  arteriovenous 26, 561–562
  cavernous 25–26, 558–559
  cherry 575
  classification 554
  compared with vascular malformations/telangiectasia 554

congenital forms 556–557
differential diagnosis 381–382, 557–558, 570
epithelioid 104, 554, 569–570
genetics 104, 575, 587
glomeruloid 564–565
histopathology
  acquired elastotic 577
  acquired tufted angioma 567, *568*
  angiomatosis 563–564, *565*
  arteriovenous 562
  cavernous 559, *560*
  cherry 576
  epithelioid 570, *571–572*
  glomeruloid 565, *566*
  hobnail 565–566, *567*
  infantile (juvenile capillary) 555, *556*
  intramuscular 562, *564*
  intravascular pyogenic granuloma 557, *559*
  lobular capillary 557–558, *558*
  microvenular 576
  spindle cell 573–575, *573–574*
  venous 563
  verrucous 560–561, *561*
hobnail 565–566
imaging 25–27
  angiomatosis 27, *29*
  capillary 26, *28*
  cavernous 25–26, *26–27*
  immunohistochemistry 555, 566, 570, 572, 574
infantile 554–555
intramuscular 381–382, 562
lobular capillary 557–558
microvenular 575
spindle cell 571–575
venous 562
verrucous 560–561
hemangiopericytoma (HPC) 330–333
  clinical features 331, *333*
  differential diagnosis 330, 333
  genetics 330, 333
  histopathology 331–332, *332*
  glomangiopericytoma 631, *632*
  lipomatous *332*
  imaging 27, *30*
  immunohistochemistry 333
  lipomatous 332–333
  not pericytic 330–331
  sinonasal (glomangiopericytoma) 631–632
hematoma 926–927
  appearance 926, *926*
  histopathology 926–927, *927*
  imaging 33
hematopoiesis, extramedullary 390

Index

hematopoietic cell markers 942
hematopoietic progenitor cell antigen *see* CD34
hemophilia, synovial pathology 906–907, 907–908
hemosiderin
    in angiomatoid fibrous histiocytoma 284–285, 287
    in benign fibrous histiocytoma 222
    in pigmented villonodular synovitis 21, 34
hemosiderotic fibrolipomatous tumor 404, 404–405
hemosiderotic synovitis 906–907, 907–908
Henderson–Jones disease *see* synovial chondromatosis
hepatic angiomyolipoma 513–514, 514
hepatic angiosarcoma 3–4, 606, 606
hepatic hemangioendothelioma 596–597, 598
hepatic PEComa (falciform ligament) 521
hepatocellular carcinoma 809, 809
hepatocyte growth factor (HGF)/MET receptor 130, 136–137
herbicides 3
hereditary cancer syndromes 117–119 *see also individual syndromes*
hereditary gingival fibromatosis *see* gingival fibromatosis
hereditary leiomyomatosis and renal cell cancer 445–446, 461, 462–463
hernia sac lipoma 380–381, 385
herpes simplex virus 313
heterotopic ossification *see* myositis ossificans
HHF-35 (actin antibody) 49, 546
HHV8 virus (human herpesvirus 8) 3, 136, 612
hibernoma 396–402
    appearance 399, 400
    cytology 978–979, 980
    histopathology 399–400, 400–401
    brown fat 398
    imaging 14, 21–22
hidradenoma (eccrine acrospiroma) 762, 763
high-performance liquid chromatography (HPLC) 142
highly active antiretroviral therapy (HAART) 407
histiocytic markers 58, 61–62
histiocytic proliferations
    crystal-storing histiocytosis 964–965, 965–966

deep granuloma annulare 927–929, 928
dendritic reticulum cell sarcoma 965–968, 967
Erdheim-Chester disease 957–960, 959
fibroblastic reticulum cell sarcoma 968
interdigitating reticulum cell sarcoma 58, 702, 968
juvenile xanthogranuloma 63, 224, 961–962, 962–963
Langerhans cell histiocytosis 960–961, 960–961
malakoplakia 930, 931
malignant (histiocytic sarcoma) 965, 966
multicentric reticulohistiocytosis 964, 964
mycobacterial pseudotumor 929–930, 930
pulse granuloma 929, 929
PVP accumulation 933–934, 933–934
reticulohistiocytoma 962–964, 964
Rosai-Dorfman disease 58, 956–957, 958–959
ruptured epidermal inclusion cyst 933, 933
xanthogranulomatous inflammation 930–932, 931
histiocytosis X *see* Langerhans cell histiocytosis
histogenesis 2, 527, 899–900
histoid leprosy 930
histopathology *see under individual lesions*
HIV/AIDS
    bacillary angiomatosis 578
    granuloma annulare 927
    Kaposi's sarcoma 612
    lipodystrophy 407, 408
    mycobacterial pseudotumor 929–930, 930
HMB45 monoclonal antibody 58–59
HMGA2 (*HMGIC*) gene
    in aggressive angiomyxoma 501
    fusions 124, 382
    in lipoma/lipomatosis 382, 387
hobnail hemangioma 565–566, 567
Hodgkin's disease 367, 953
Hoffa disease 384
Homer Wright rosettes
    neuroblastoma 867, 872, 874
    PNETs 860, 860
homogeneous staining regions (hsrs) 130
    neuroblastoma 875, 877
homovanillic acid 870

hormonal therapy 1031–1032
HPC *see* hemangiopericytoma
HPV16 (human papillomavirus 16) 803
HRAS oncogene 115–116
human herpesvirus 8 (HHV8) 3, 136, 612
human milk fat globule protein *see* epithelial membrane antigen
human papillomavirus 16 (HPV16) 803
HUMARA clonality assay 138–139
hyaline vascular Castleman disease 941–943, 944
hyalinizing spindle cell tumor with giant rosettes 347, 350
hypercalcemia 837, 882
hypercellular spindle cell GISTs 479, 479
hypercholesterolemia, lesions associated with *see* xanthoma
hyperdiploidy 548, 876–877
hyperphosphatemia 837–838
hyperthermia therapy 1028
hypoglycemia 324

iatrogenic gingival hyperplasia 253
iatrogenic Kaposi's sarcoma 613
IDH1/IDH2 genes (isocitrate dehydrogenase) 138, 575
idiopathic retroperitoneal fibrosis *see* retroperitoneal fibrosis
ifosfamide 1017, 1029–1030
    high-dose therapy 1030
IGF/IGF1R (insulin-like growth factor/receptor) 136–137
IgG4 sclerosing disease 201, 292
IHC *see* immunohistochemistry
imaging 11–36, 1012
    angiomatosis 27, 29
    angiosarcoma 27
    calciphylaxis 841, 842
    chondroma 825, 826
    collagenous fibroma 210
    cranial fasciitis 250
    cystic masses 31–34
    elastofibroma 22, 34
    Ewing sarcomas/PNETs 857, 858
    fibrodysplasia ossificans progressiva 845, 846
    fibromatosis 34–35, 36
    ganglion cyst 32, 33
    GIST "tumor within a tumor" 1025
    hemangioma 25–27, 26–28
    hemangiopericytoma 27, 30
    lipomatous lesions 19–25
    hibernoma 14, 21–22

lipoblastoma 19, 21
lipoma 15, 20–21, 23
liposarcoma 6, 17, 22–25, 23–25
lymphangioma 27, 28
malignant fibrous histiocytoma 16–20, 21
mesenchymal chondrosarcoma (extraskeletal) 832, 833
myositis ossificans 12, 842, 843
myxoid chondrosarcoma (extraskeletal) 830, 830
myxoma 300
neuroblastoma 871
neurogenic tumors 27–31
    Morton neuroma 30, 31
    MPNST/BPNST 11–12, 31, 32
    in NF1 22
    schwannoma 20, 31
nonspecific appearance 35–36
osteosarcoma (extraskeletal) 848, 848
phosphaturic mesenchymal tumor 834, 835
pros and cons of different modalities 11–15
PVNS 21, 34
rhabdomyosarcoma 535, 539
synovial chondromatosis 34, 35, 827, 828
synovial sarcoma 18
tophaceous pseudogout 839, 840
tumoral calcinosis 838, 838
imatinib mesylate
    for desmoids 241–242
    for DFSP 129, 345–346
    for GISTs 132
        as adjuvant therapy 1020, 1026–1027
        as primary therapy 487–488, 1024–1025
        resistance 486–488, 1020, 1024–1025
immunocytochemistry *see* immunohistochemistry
immunoFISH 99
immunohistochemistry (IHC) 41–77, 975
    adenomatoid tumor 777
    adrenocortical carcinoma 60, 815
    alveolar soft part sarcoma 898–899, 899
    anaplastic large cell lymphoma 953–954
    angioleiomyoma 447
    angiomatoid fibrous histiocytoma 285
    angiomyolipoma 59, 59, 60, 61, 517
    angiosarcoma 45, 66, 608–610, 611

Index

atypical fibroxanthoma 366, 367
benign fibrous histiocytoma 222–225, 223, 227
branchial anlage mixed tumor 768, 769
breast carcinoma 73, 803–804, 805
  sarcomatoid 805
Burkitt lymphoma 949
calcifying fibrous pseudotumor 254
Castleman disease 943
CD antigens *see entries at* CD
cell cycle markers 75–77
cellular blue nevus 59, 61, 694
chordoma 66, 766, 767
clear cell sarcoma, gastrointestinal 708
clear cell sarcoma of tendons and aponeuroses 56, 59, 707, 708
clear cell "sugar" tumor 520
collagenous fibroma 208, 211
crystal-storing histiocytosis 965, 966
dendritic reticulum cell sarcoma 965–968
dermatofibrosarcoma protuberans 46, 343–345, 345
desmoid fibromatosis, 242–244, 245
desmoplastic small round cell tumor 75, 880, 880
diffuse large cell B-cell lymphoma 947
ectomesenchymoma 545
endometrial stromal sarcoma 70, 502, 503
endometriosis 507
endothelial/multispecific markers 41–48
ependymoma 712–713
epithelial markers 62–69
epithelioid sarcoma 47, 74, 755, 757, 758
Erdheim–Chester disease 958
Ewing sarcomas/PNETs *see* Ewing sarcoma family of tumors, immunohistochemistry
fasciitis
  intravascular 187–189
  nodular 183–184, 188
  proliferative 193
female genital tract stromal tumors 492–493, 496–497, 499
fibromyxoid sarcoma (low-grade) 74, 347
fibrous umbilical polyp 252–253
follicular hyperplasia 941
follicular lymphoma 945
Gardner fibroma 262, 262

giant cell angiofibroma 330
GIST *see* gastrointestinal stromal tumor (GIST), immunohistochemistry
glomangiopericytoma 631
glomus tumor 626–627, 631
granular cell tumor 56–57, 673, 675
granular cell tumor, gingival 676, 677
granulosa cell tumor 506, 507
hemangioendothelioma
  epithelioid 47, 597–598, 600
  epithelioid sarcoma-like 601, 601
  kaposiform 568
  PILA 593–594, 595
hemangioma 555, 566, 570, 572, 574
hemangiopericytoma 333
hematoma 927
hematopoietic cell markers 942
hepatocellular carcinoma 809
histiocytic markers 58, 61–62
histiocytic sarcoma 965
infantile digital fibroma 256
infantile fibrosarcoma 294
inflammatory myofibroblastic tumor 70, 291, 291
juvenile nasopharyngeal angiofibroma 265
juvenile xanthogranuloma 63, 962
juxtaglomerular cell tumor 633–634
Kaposi's sarcoma 46, 48, 616, 616
Langerhans cell histiocytosis 960–961, 961
leiomyoma 49–51, 73, 447, 450, 459, 460
leiomyosarcoma 47, 49–51, 66, 454–455, 455
lipofibromatosis 258–259
lipoma 394, 397, 401, 403–404
liposarcoma 46, 57, 423, 429, 431, 436
lung carcinoma
  adenocarcinoma/large cell 792–793, 805, 806
  small cell 53–54, 54, 806, 807
lymphangioma 585
lymphangiomyoma(tosis) 59, 59, 519
lymphoblastic lymphoma 948
lymphoid markers 942
malignant fibrous histiocytoma 62, 63, 71
  pleomorphic MFH (undifferentiated sarcoma) 71, 360, 362
malignant mixed müllerian tumor 75, 505, 505
marginal zone lymphoma 947

mastocytoma/mastocytosis 956
melanoma 56, 58–61, 60, 74, 699–701, 700–701
  desmoplastic 704, 704
melanotic neuroectodermal tumor of infancy 709
meningioma, ectopic 66, 69, 715–718
Merkel cell carcinoma 55, 802, 802
mesenchymal chondrosarcoma (extraskeletal) 57, 833–834, 834
mesothelioma
  malignant 67, 70, 72, 789, 789, 792–794
  multicystic peritoneal (benign) 779
  sarcomatoid 789, 790
metastatic carcinoma (in general/unknown primary) 65, 812, 818–819
mixed tumor/myoepithelioma 74, 759, 761
MPNST 54, 57, 678–679, 681–682, 685–686
multispecific markers 56
muscle cell markers 48–52, 546–547
myeloid tumor, extramedullary 954, 955
myofibroblastoma
  cervicovaginal 493, 494
  palisaded 218
myofibroma/myofibromatosis 267
myopericytoma 633
myxoid chondrosarcoma (extraskeletal) 54, 57, 831
myxoinflammatory fibroblastic sarcoma 354
myxoma 300, 307
  cardiac 314
  nerve sheath 666, 667
nasal type NK/T-cell lymphoma 952
nerve sheath tumors
  epithelioid 664
  MPNST 54, 57, 678–679, 681–682, 685–686
  myxoma 666, 667
neural/neuroendocrine markers 52–57, 73–74
neuroblastoma *see* neuroblastoma, immunohistochemistry
neurofibroma 56, 647, 647, 650–652
neuroma (palisaded encapsulated) 644, 644
neurothekeoma 277–280, 281
ossifying fibromyxoid tumor 57, 318

osteosarcoma 74–75, 849
ovarian carcinoma 815–816
palmar/plantar fibromatosis 236
panniculitis-like T-cell lymphoma 951
papillary intralymphatic angioendothelioma 593–594, 595
paraganglioma *see* paraganglioma, immunohistochemistry
PEComa 521–522, 522
penile myointimoma 190, 191
perineurioma 668, 671, 671
pheochromocytoma 728
phosphaturic mesenchymal tumor 836
plasmacytoma 950
plexiform fibrohistiocytic tumor 282
prostatic carcinoma 815, 815
renal cell carcinoma 811, 812–813
reticulohistiocytoma 963–964
retroperitoneal fibrosis 200
rhabdoid tumor 883–885, 884
rhabdomyoma 50, 528–529
rhabdomyosarcoma *see* rhabdomyosarcoma (RMS), immunohistochemistry
Rosai–Dorfman disease 58, 957, 959
schwannoma 56, 73–74, 655
  cellular 656
  epithelioid 657, 658
  gastrointestinal 661
  melanotic 664
sclerosing epithelioid fibrosarcoma 74, 351
seminoma 818
solitary fibrous tumor 70, 328–329, 329, 330
squamous cell carcinoma 803
superficial acral fibromyxoma 312, 312
superficial angiomyxoma 307
synovial sarcoma *see* synovial sarcoma, immunohistochemistry
tenosynovial giant cell tumor 920
thyroid carcinoma 817
urothelial carcinoma 67–68, 811–813
immunosuppressed patients
  bacillary angiomatosis 578
  benign fibrous histiocytoma 220
  Burkitt lymphoma 949
  EBV-associated leiomyosarcoma 3–4, 467–468, 468

1049

immunosuppressed patients (cont.)
   EBV-associated lymphoma 947
   Kaposi's sarcoma 612–613
   malakoplakia 930
   Merkel cell carcinoma 798
immunotherapy 1032
implants 3
   carbon fibers/graphite 911–912, *912*
   metal fragments 912–913, *913*
   polyethylene wear particles 910, *911*
   silicone
      breast implants 409, *410*, 953
      joint prostheses 911, *912*
imprinting 725
IMT *see* inflammatory myofibroblastic tumor
incidence 2–3
incisional biopsy 1012–1013
inclusion body fibromatosis *see* infantile digital fibroma
infantile diffuse fibromatosis 260–261, *261*
infantile digital fibroma 254–256
   appearance 254–255, *256*
   histopathology 255–256, *256–257*
infantile fibrosarcoma 292–294
   appearance 293
   histopathology *293–294*, 293–294
infantile hemangioma 554–555
   histopathology 555, *556*
infantile myofibroma/myofibromatosis 266–268
   cytology 984, *984*
   histopathology 267, *269–270*
infiltrative angiolipoma (intramuscular hemangioma) 381–382
inflammatory fibrosarcoma 288
inflammatory liposarcoma (lymphocyte-rich) 421, 421–422, *422*
inflammatory malignant fibrous histiocytoma 336, 367
   differential diagnosis 367
   histopathology 367, *368*
inflammatory myofibroblastic tumor (IMT) 288–292
   appearance 289, *289*
   clinical features 288–289
   differential diagnosis 291–292, 487
   genetics *107*, 127, 292
   histopathology 289–291, *290*
      cardiac variant *291*
      epithelioid variant *290*, 291
   immunohistochemistry 70, *291*, 291

no oncogenic viruses 291–292
terminology 288
treatment 127
inflammatory myxohyaline tumor of distal extremities (myxoinflammatory fibroblastic sarcoma) 352–356
   appearance 354
   cytology 984, *985*
   differential diagnosis 354–355, 984
   histopathology 354, *355*
      pseudolipoblasts 356
      Reed–Sternberg-like cells 356
inflammatory pseudotumor 288
inhibin 506
INI1/INI1 (*SMARCB1*/SMARCB1) (SWI/SNF complex) 128, 133–134, 545, 758, 882, 885
   IHC marker 74, 545, 755, *884*, 884–885
insulin-like growth factor/receptor (IGF/IGF1R) 136–137
intensity-modulated radiation therapy (IMRT) 1016, *1016*
interdigitating reticulum cell sarcoma 58, 702, 968
intermetatarsal compression neuritis *see* Morton neuroma
intermuscular lesions 18, 20
intra-abdominal desmoid fibromatosis 240–241, *245*, 487
intra-articular lesions 18–20
intralymphangial stromal myosis *see* endometrial stromal sarcoma
intramuscular glomus tumor 629
intramuscular hemangioma 381–382, 562
   appearance 564
   histopathology 562, *564*
intramuscular leiomyosarcoma 456
intramuscular lesions 18, 20
intramuscular lipoma 381
intramuscular myxoma 299–302
   appearance 300, *300*
   cytology 1000, *1001*
   histopathology 300, *301–302*
intramuscular nodular fasciitis 183, *188*
intraneural neurofibroma 646–647 *see also* plexiform neurofibroma
   appearance 648
   histopathology 647, *648–649*
   atypia 649

intraneural perineurioma 668, *668*
intravascular B-cell lymphoma 948–949, *949*
intravascular bronchioloalveolar tumor (pulmonary epithelioid hemangioendothelioma) 596–597, *599*
intravascular fasciitis 182, 185–189, *190 see also* penile myointimoma
   histopathology 187, *190*
intravascular pyogenic granuloma 557, *559*
intravenous leiomyomatosis 463–464, *464*
intussusception, florid vascular proliferation 580, *581*
irritation fibroma of the oral cavity 212–213, *217*
ischemic fasciitis 195
   histopathology 195, *197*
isocitrate dehydrogenase (IDH) 138, 575

jaw, myxoma 315–317, *317*
JAZF1-SUZ12 gene fusion 123, 503
jugulotympanic paraganglioma 737, *738*
juvenile aponeurotic fibroma (calcifying fibroma) 256–257
   histopathology 257, *259–260*
juvenile capillary hemangioma *see* infantile hemangioma
juvenile diffuse fibromatosis 259–260
juvenile hyaline fibromatosis 253, 269–272
   appearance *271*
   histopathology 271, *272*
juvenile nasopharyngeal angiofibroma 263–265
   appearance 263–264, *263*
   histopathology *264*, 264–265
juvenile rhabdomyoma 528, *528*
juvenile variant DFSP (giant cell fibroblastoma) 342–343, *343–344*
juvenile xanthogranuloma (JXG) 224, 961–962
   appearance *962*
   histopathology 962, *963*
   immunohistochemistry *63*, 962
juxtaarticular myxoma 302–303
   histopathology 303, *303–304*
juxtaglomerular cell tumor 633–634, *634*
JXG *see* juvenile xanthogranuloma

kaposiform hemangioendothelioma 567–568, *568–569*

Kaposi's sarcoma 611–616
   appearance *613*
   classic form 612
   differential diagnosis 218, 616
   endemic form 612–613
   epidemic form (AIDS-associated) 612
   etiology (HHV8 virus) 3, 612
   histopathology 613–616, *614–615*
      nodular lesions 614–615, *614*
      patch stage *613*
      pulmonary infiltrates *612*
      sarcomatoid *615*
   iatrogenic form 613
   immunohistochemistry *46*, *48*, 616, *616*
karyotype 94–95
Kasabach–Merritt syndrome 556–557, 567
KBA62 (IHC marker) 61
keloids 201, *202*
keratins 62–65, 744
   in angiosarcoma 66, 611
   antibodies used in IHC 68–69
   in branchial anlage mixed tumor 768, *769*
   of complex epithelia
      K5/K14 67–69
      K17 68
   in epithelioid sarcoma 755
   in GIST 66, 486
   in hemangioendothelioma 598
   of internal squamous epithelia/urothelium (K4/K13) 67–68
   K3/K12 68
   K6/K16 68
   K15 68
   of keratinizing squamous epithelia (K2/K9/K11) 68
   of keratinizing stratified epithelia
      K1 68
      K10 68
   in leiomyosarcoma 455
   in melanoma 700–701
   in Merkel cell carcinoma 802
   in metastatic carcinomas 65, 819
   in mixed tumor 759
   of simple nonstratified epithelia
      K7 67, 611
      K8/K18 65–69, 486, 598, 611, 700–701, 802
      K19 67, 598
      K20 67
   in synovial sarcoma 66–67, 750, *751*
Ki67 (MK167) 75–76, 329
kidney tumors *see also* angiomyolipoma; renal cell carcinoma, metastatic

glomus tumor 625
hereditary leiomyomatosis and renal cell cancer 445–446, 461, *462–463*
juxtaglomerular cell tumor 633–634, *634*
renal capsule PEComa *520*, 520
synovial sarcoma 746
Wilms' tumor 856, 882
Kimura disease 943–945, *945*
KIT/KIT (CD117)
　immunohistochemistry 74, 484, *485*, 611
　mutations
　　in GISTs 131–132, 475, 486–487
　　in mastocytosis 956
　　and TKI sensitivity 486, 1024–1026
　NGS profile *145*
　as a therapeutic target 132, 487–488
Knudson, Alfred, 'two-hit hypothesis' 108 *109*, 116, 137–138
Krebs cycle dysfunction 138, 724–725

labia majora
　angiomyofibroblastoma 494–496
　cellular angiofibroma 496
LAM see lymphangiomyoma/lymphangiomyomatosis
laminin 72
Langerhans cell histiocytosis (LHH) 960–961
　histopathology 960, *960*
　immunohistochemistry 960–961, *961*
Langerhans cells 58
lapatinib 133
large B-cell lymphoma, diffuse 947, *948*
large cell carcinoma of the lung (metastatic) 792–793, 805, *806*
large cell lymphoma, anaplastic see anaplastic large cell lymphoma (ALCL)
laryngeal inflammatory myofibroblastic tumor 289
laryngeal liposarcoma 418–419
laryngeal sarcomatoid carcinoma 804
Ledderhose disease see plantar fibromatosis
leiomyoma 443
　angioleiomyoma 446–447
　appearance
　　esophageal 450, *451*
　　retroperitoneal *459*
　differential diagnosis 443, 465, 467

epithelioid, gynecological 463
ER-positive 72–73
　abdominal 459
　benign metastasizing 464–465
　uterine 459–463
　vulvar 459
gastrointestinal 448–449
　intramural 449–451
　muscularis mucosae 451–452
genetics
　angioleiomyoma 447
　gastrointestinal 450–451
　uterine 462–463
histopathology
　abdominal uterine-type 459, *460*
　angioleiomyoma 446–447, *448–449*
　benign metastasizing 465
　epithelioid 463
　gastrointestinal 450, *451–452*
　lipoleiomyoma 404, *460*
　pilar 446, *446*
　scrotal *449*
　uterine 461–462, *461*
immunohistochemistry 49–51, 73, 459, *460*
　angioleiomyoma 447
　gastrointestinal 450
　pilar 444–446
　scrotal 447–448
　visceral (ER-negative) 452
leiomyomatosis
　esophageal 450
　hereditary leiomyomatosis and renal cell cancer 445–446, 461, *462–463*
　intravenous 463–464, *464*
　peritoneal 465–466, *466*
leiomyosarcoma (LMS) 443, 452–455, 995
　appearance 452–453, *453*
　uterine *467*
　vena cava *458*
chemotherapy 1031
in children 459
cutaneous 455–456
cytology 995, *996*
differential diagnosis 443, 457, 467, 482
EBV-associated 3–4, 467–468, *468*
electron microscopy 455
epithelioid, gynecological 463
gastrointestinal 457
genetics 455
histopathology *453–454*, 453–454
cutaneous *456*
EBV-associated *468*
inflammatory 455
pleomorphic *454*
vascular form *458*

immunohistochemistry 47, 49–51, 66, 454–455, *455*
other sites 458
retroperitoneal 456–457
subcutaneous/intramuscular 456
uterine 466–467, 1031
venous 457
leprosy 930
Letterer–Siwe disease 960
LGFMS see fibromyxoid sarcoma (low-grade)
LHFP (lipoma HMGA2/HMGIC fusion partner gene) 382
LHH see Langerhans cell histiocytosis
Li–Fraumeni syndrome 135
limb perfusion therapy 1028
lip, mucosal neuroma 639, *641*
lipid-rich histiocytes see xanthogranulomatous inflammation; xanthoma
lipid-rich rhabdomyosarcoma (RMS) 544
lipidized fibrous histiocytoma 225–226, *227*
lipoblastoma 405–407
　appearance *406*
　cytology 978, *979*
　histopathology 406–407, *406–407*
　imaging *19*, 21
lipodystrophy (fat atrophy) 407, *408*
lipofibromatosis 257–260
　differential diagnosis 259–260, 283
　histopathology 258, *260–261*
lipogranuloma
　fat necrosis 407–409, *408*, 977–978, *978*
　sclerosing 409, *409*
lipoleiomyoma 404, *460*
lipoma 379
　angiolipoma 388–390
　appearance 379, *380*
　　chondroid *402*
　　hibernoma 399, *400*
　　lipoma arborescens *383*
　　myelolipoma *391*
　　neural fibrolipoma *386*
　　spindle cell lipoma *394*
　　thymolipoma *385*
　atypical see well-differentiated liposarcoma
　with chondro-osseous metaplasia 382–383
　chondroid 402–404
　clinical features 379, 384–386, 388, 390, 392, 398–399, 402
　cytology 977, *977*
　　hibernoma 978–979, *980*
　　myxolipoma 977, *977*

spindle cell/pleomorphic 978, *979*
differential diagnosis 380–382, 394, 401–402, 404, 977–979
genetics 131, 382
　angiolipoma 390
　chondroid 404
　hibernoma 402
　lipomatosis 387
　spindle cell/pleomorphic 394
hemosiderotic fibrolipomatous tumor 404–405
hibernoma 21–22, 396–402, 978–979
histopathology 379–380, *380–381*
　angiolipoma 389–390, *389*
　angiolipoma, cellular 389–390, *390*
　chondro-osseous metaplasia 383
　chondroid lipoma 402–403, *402 403*
　fibrolipoma 382
　hemosiderotic fibrolipomatous tumor *404*
　hibernoma 399–400, *400–401*
　intramuscular *381*
　lipoma arborescens *384*
　myelolipoma 390, *391*
　myxolipoma *382*
　neural fibrolipoma *387*
　nevus lipomatosus superficialis *383*
　pleomorphic lipoma 393–394, *397*
　sclerotic lipoma *398*
　spindle cell lipoma 393, *394–396*
　thymolipoma 385, *386*
imaging *15*, *20–21*, *23*
　hibernoma *14*, 21–22
immunohistochemistry 394, *397*, 401, 403–404
lipomatoses 386–387
lipomatosis of the nerve (neural fibrolipoma) 385–386
lumbosacral/spinal cord-associated 384–385
myelolipoma 390
myolipoma 404
myxolipoma 977
nevus lipomatosus superficialis 383
sclerotic 395–396
spindle cell/pleomorphic 391–394, 978
thymolipoma 385
visceral 384–385
lipoma arborescens (synovial lipomatosis) 383–384, *383–384*

Index

lipoma HMGA2/HMGIC fusion partner gene (*LHFP*) 382
lipoma-like well-differentiated liposarcoma *419–420*, 419
lipomatoses 386–387
lipomatosis of the nerve 385–387, *386–387*
lipomatous hemangiopericytoma 332–333, *332*
lipomatous hypertrophy of the interatrial septum 385
liposarcoma 416
  appearance
    dedifferentiated 425, *426*
    myxoid *431*, 431
    pleomorphic 435, *436*
    well-differentiated *419*, 419
  chemotherapy 1017
  clinical features 417–419, 424–425, 430–431, 435
  cytology
    myxoid/round cell 981–982, *981*
    well-differentiated 979–981, *980*
  dedifferentiated (DDLS) 7, 24, 108, 416, 424–430
  differential diagnosis 381, 401–402, 424, 429–430, 432–433
  genetics 101
    dedifferentiated 108, 430
    myxoid *106*, 126–127, 434–435
    pleomorphic 437–438
    well-differentiated *102*, 131, 416–423
  histopathology 419, 425–428, 431, 435–436
    DDLS meningothelial-like whorls 426–427, *427–428*
    DDLS pleomorphic forms *428*
    DDLS resembling inflammatory MFH *426*
    DDLS resembling various sarcoma types *427*
    DDLS rhabdomyosarcomatous type *429*
    myxoid hibernoma-like *433*
    myxoid lymphangioma-like *433*
    myxoid postradiation therapy *435*
    myxoid round cell *434*
    myxoid typical forms *432*
    pleomorphic *437*
    pleomorphic epithelioid form *438*
    spindle cell *423*
    WDLS inflammatory type *421–422*, 421–422
    WDLS lipoma-like *419–420*, 419
    WDLS myxoid type *422*, 422–423
    WDLS sclerosing type 420–421, *421*
  imaging 22–25
    dedifferentiated 6, *25*
    myxoid *17*, *24*
    well-differentiated *23*
  immunohistochemistry 46, 57, 423, 429, 431, 436
  myxoid 24, 126–127, 416, 430–435, 981–982
  pleomorphic 24, 416, 435–438
  round cell 24, 431, 981–982
  spindle cell 423
  surgery 1014–1015, 1019–1020
  well-differentiated (WDLS) 22–24, 131, 381, 416–424, 979–981, 1014–1015
liver tumors
  angiomyolipoma 513–514, *514*
  angiosarcoma 3–4, 606, *606*
  hemangioendothelioma 596–597, *598*
  hepatocellular carcinoma 809, *809*
  PEComa in the falciform ligament 521
LMS *see* leiomyosarcoma
lobular capillary hemangioma 557–558
  histopathology 557–558, *558*
  intravascular pyogenic granuloma 557, *559*
localized fibrous tumor *see* solitary fibrous tumor
localized interdigital neuritis *see* Morton neuroma
localized tenosynovial giant cell tumor 915–917, 920–921
  histopathology 916–918, *916–917*
  cartilaginous metaplasia *919*
  giant cell morphology *917*
  infarction *918*
loss of heterozygosity (LOH) 143
low-grade fibromyxoid sarcoma *see* fibromyxoid sarcoma (low-grade)
lumbosacral lipoma 384–385
lung tumors
  clear cell "sugar" tumor 519–520, *520*
  differential diagnosis of mesothelioma 792–793
  glomus tumor 625
  hemangioendothelioma 596–597, *599*
  inflammatory myofibroblastic tumor 288–289
  large cell carcinoma, metastatic 792–793, 805, *806*
  leiomyoma 452
    benign metastasizing 464–465, *465*
  lipoma 385
  liposarcoma 424–425
  lymphangiomyomatosis 518–519, *519*
  meningiothelial nodules 717, *717*
  sarcomatoid carcinoma 807, *808*
  small cell carcinoma, metastatic 53–54, 802–803, 806, *806–807*
  synovial sarcoma 746
lupus panniculitis 951–952
Lyme disease 904, *905*
lymph node tumors
  angiomyomatous hamartoma 444, *445*
  endemic Kaposi's sarcoma in children 613
  palisaded myofibroblastoma 213–218
lymphangioendothelioma, benign 585–587
  histopathology 586, *586–587*
lymphangioendotheliomatosis with thrombocytopenia 585–586
lymphangioma 581–585
  acquired progressive (benign lymphangio endothelioma) 585–587
  appearance 584, *587*
  differential diagnosis 585, 587
  diffuse (lymphangiomatosis) 587–588
  histopathology 584–585, *584–585*
  benign lymphangioendothelioma 586, *586–587*
  imaging *27, 28*
lymphangioma circumscriptum 582
lymphangiomatosis 587, 587–588
lymphangiomyoma/lymphangiomyomatosis (LAM) 518–519
  appearance 518–519, *518*
  histopathology 519, *519*
  immunohistochemistry 59, *59*, *519*
lymphedema
  angiosarcoma in lymphedematous extremities 4, 603–604, *604*
  in the obese patient 409–411, *411*
lymphoblastic lymphoma 947–948, *948*
lymphoma 941
  anaplastic large cell
    cutaneous 954
    noncutaneous 952–954, *953*
  Burkitt 949, *950*
  and crystal-storing histiocytosis 964–965, *965–966*
  diffuse large cell B-cell 947, *948*
  extranodal marginal zone 946–947, *947*
  follicular 945–946, *946*
  genetics 946, 949, 952–954
  Hodgkin's disease 367, 953
  intravascular B-cell 948–949, *949*
  lymphoplasmacytic, and amyloid tumor 936–937
  nasal-type NK/T-cell *952*, 952
  panniculitis-like T-cell 950–952, *951*
  plasmablastic B-cell 949–950, *951*
  precursor B-cell lymphoblastic 947–948, *948*
lymphomatoid papulosis 954
lymphoplasmacytic lymphoma, and amyloid tumor 936–937
Lyon's hypothesis 139
lysosomal proteins 62

M-FISH (multicolor FISH) 100
macroscopic appearance *see* gross anatomy/morphology
Maffucci syndrome 573, *573*, 575
magnetic resonance imaging (MRI) 13–14
  angiomatous lesions 25–26, *27–28*, *30*
  cartilaginous tumors 34, *35*, *825*, *826*, *828*, *833*
  collagenous fibroma *210*
  elastofibroma 22
  fibromatosis 34–35, *36*
  lipomatous lesions 6, *15*, 23–25
  lymphangioma *28*
  myxoma *300*
  neurogenic lesions *20*, 22, 30–31, *31–32*
  popliteal cyst *33*
  PVNS *21*, 34
  rhabdomyosarcoma *535*, *539*
malakoplakia 930, *931*
male genital tumors *see* penile fibromatosis; penile myointimoma; scrotal tumors; testicular tumors
malignant eccrine spiradenoma 763
malignant fibrous histiocytoma (MFH)
  clinical features 357, 362, 367, 369
  cytology
    giant cell MFH *989*
    pleomorphic MFH 988, *988*
  differential diagnosis 348, 358–359, 364, 366–367, 988

electron microscopy 357–358, 362
genetics 134, 359, 362–364
giant cell 368–369, 989
histopathology
　giant cell MFH 369, *370*
　inflammatory MFH 367, *368*
　myxoid MFH 357, *359–361*
　pleomorphic MFH 362, *363–364*
imaging 16–20, *21*
immunohistochemistry 62, *63*
　pleomorphic MFH *71*, 360, 362
inflammatory 336, 367
myxoid 336, 356–359, 985–986, 1018
pleomorphic 336, 359–364, 987–988
malignant glomus tumor 626, *630*
malignant granular tumor 675–676, *676*
malignant histiocytic tumors (histiocytic sarcoma) 965, *966*
malignant histiocytosis *see* anaplastic large cell lymphoma
malignant melanoma *see* melanoma, metastatic
malignant mesothelioma *see* mesothelioma, malignant
malignant mixed müllerian tumor 503–505
　appearance *504*, 504
　histopathology 504, 504–505
　immunohistochemistry 75, *505*, 505
malignant mixed tumor
　myoepithelial carcinoma 50, 702
　myoepithelioma 761–762
malignant perineurioma 671–672
malignant peripheral nerve sheath tumor (MPNST) (malignant schwannoma) 676–682
　with angiosarcoma 607, 683
　appearance 678, *679*, *684*
　cytology 992, 992–993
　differential diagnosis 680–684, 702, 993
　epithelioid differentiation 684–686
　with ganglioneuroma or pheochromocytoma 686
　genetics 108–110, 117, 682
　glandular epithelial differentiation 54, 683–684
　granular cell tumor 675–676, *675–676*
　histopathology 678, *680*
　　epithelioid differentiation *685*, 685

geographic necrosis *681*
glandular differentiation *684*
in a neurofibroma *679*
with pheochromocytoma *686*, 732
rhabdomyosarcomatous differentiation (Triton tumor) *683*
vascular proliferation *681*
imaging 11–12, 29, 31, *32*
immunohistochemistry 54, 57, 678–679, *681–682*, 685–686
surgery 1018
transformed neurofibroma 652
Triton tumor 683
malignant rhabdoid tumor *see* rhabdoid tumor, extrarenal
malignant tenosynovial giant cell tumor (TSGT) 921–922, *922*
MALT lymphoma (extranodal marginal zone lymphoma) 946–947, *947*
mammary-type myofibroblastoma *see* breast myofibroblastoma
marginal zone lymphoma 946–947, *947*
MART-1 (melan A) 58, *60*
Masson tumor (papillary endothelial hyperplasia) 577–578, *577–578*
mastocytoma 956, *956*
mastocytosis 956, *957*
Mazabraud syndrome 300
McCune-Albright syndrome 300
MCHSA *see* extraskeletal mesenchymal chondrosarcoma
MDM2/MDM2 (murine double minute)
　amplification 101, 131
　liposarcoma 416–424, 430
　and p53 135–136
medallion-like dermal dendrocytic hamartoma 345
media for cell culture 93
　for transport 93
mediastinal fibrosis 200
mediastinal liposarcoma, well-differentiated 418
mediastinal paraganglioma 734
mediastinal thymolipoma 385–386, *385*
Meissner-like bodies 56, *651*
melan A (MART-1) 58, *60*
melanocytes 56
melanoma, desmoplastic 702–704
　histopathology 702–703, *703*
　immunohistochemistry *704*, 704
melanoma, metastatic 695–702

differential diagnosis 9, *697*, 701–702
genetics 702
histopathology 696–698
　osteoclastic giant cells *699*
　pleomorphism *699*
　resembling hematopoietic tumors *700*
　various patterns 697–698
immunohistochemistry 58–61, *60*, *74*, 699–701, *700–701*
S100 56, 699, *701*
melanoma of soft parts *see* clear cell sarcoma of tendons and aponeuroses
melanotic neuroectodermal tumor of infancy (melanotic progonoma) 708–709, 991
cytology *991*
histopathology 709, *710*
melanotic schwannoma 661–664
appearance 662, *662*
differential diagnosis 664, 702
histopathology 662–663, *663–664*
melphalan 1027–1028
MEN2a/MEN2b syndromes (multiple endocrine neoplasia)
mucosal neuroma/ganglioneuroma 639, *641*
paraganglioma 725–726
meningeal hemangiopericytoma 331
meningeal solitary fibrous tumor 324–325, 331
meningioma, ectopic 713
cytology 990–991, *991*
differential diagnosis 718
histopathology
　paraspinal 716, *716*
　pulmonary *717*
　scalp 714–715, *715*
　sinonasal *716*, 716
immunohistochemistry 66, 69, 715–718
metastatic 717
paraspinal 716
pulmonary 717
scalp 714–715
sinonasal, ear and orbital 716–717
meningiothelial hamartoma 714–715, *715*
meniscal cyst 32–33
Merkel cell carcinoma 798–803
appearance *799*, 799
clinical features 798–799
differential diagnosis 802–803
etiology 798, 801–802
histopathology 799–801, *800–801*
immunohistochemistry 55, *802*, 802

Merkel cell polyomavirus (MCV) 801–802
merlin (NF2 protein) 133, 659–660
mesenchymal chondrosarcoma, extraskeletal (MCHSA) 832–834
appearance *833*
cytology 1000, *1000*
histopathology 832–833, *833*
imaging 832, *833*
immunohistochemistry 57, 833–834, *834*
mesenchymoma 1
mesna 1029
mesodermal stromal polyp 491–493, *492*
mesothelial proliferations (benign or precancerous)
adenomatoid tumor 777, *778*
differential diagnosis 777, 789–792, *791*
multicystic peritoneal mesothelioma 778–780
　appearance *779*
　cellular nodules 789, *791*
　histopathology *779–780*, 779
well-differentiated papillary mesothelioma 780–781, *781*, 789–790, *791*
mesothelioma, malignant 782–794
appearance *783*
peritoneal *784*
pleural *784*
differential diagnosis 789
vs benign/reactive mesothelial proliferation 777, 789–792, *791*
vs carcinoma 789, *792–793*, 792–793
vs mesenchymal tumors 793
vs vascular tumors 793–794
electron microscopy 789, *790*
epidemiology 782–783
etiology 782–783
genetics 794
histopathology
　biphasic 786–787, *787*
　deciduoid 785–786, *786*
　desmoplastic 788–790, *788*
　epithelioid 785, *786*
　sarcomatoid 787–788, *787–788*
　tubulopapillary 784–785, *784*
　tubulopapillary adenomatoid-like *785*
immunohistochemistry 67, 70, *72*, 789, *789*, 792–794
sarcomatoid 789, *790*
peritoneal 783, *784*, 792–793
pleural 783, *784*
MET (hepatocyte growth factor receptor) 130, 136–137

metabolism of cancer cells 138
metal foreign bodies 3, 607–608
    synovial reaction to prosthetic fragments 912–913, *913*
metal-containing fixatives 8
metaplasia
    bone 847, *848*
    lipoma with metaplastic cartilage and bone 382–383, *383*
metastatic disease
    adrenocortical carcinoma 60, 815, *816*
    angiomatoid fibrous histiocytoma 283
    benign fibrous histiocytoma 221
    benign metastasizing leiomyoma 464–465, *465*
    breast carcinoma 73, 803–805, *805*
    endometrial stromal sarcoma 501
    gastrointestinal carcinoma 808–809
    GISTs 476, 481, 483, 1024–1027
    hepatocellular carcinoma 809, *809*
    IHC profiles of carcinomas 65, 812, 818–819
    liposarcoma 424–425, 431
    lung carcinoma
        large cell 792–793, 805, *806*
        small cell 53–54, 802–803, 806, *806–807*
    melanoma *see* melanoma, metastatic
    meningioma 717
    ovarian carcinoma 815–816
    prostatic carcinoma 813–815, *814*
    renal cell carcinoma *see* renal cell carcinoma, metastatic
    seminoma 817–818, *818*
    squamous cell carcinoma 758, 803, *803–804*
    thyroid carcinoma 816–817, *817*
    treatment 1029–1030
    unknown primary 818–819
    urothelial carcinoma 67–68, 811–813
MFH *see* malignant fibrous histiocytoma
Mibelli angiokeratoma 560
Michaelis–Gutman bodies *931*
microphthalmia transcription factor (MITF) 58, 60
microRNA (miRNA) 116–117, 129, 136
microsatellite markers 143
microvenular hemangioma 575, *576*
midostaurin 128
Milroy disease 603
mitochondrial conditions 386, *387*

mitosis/karyorrhexis index (MKI) 873
mixed tumor *see also* branchial anlage mixed tumor; malignant mixed müllerian tumor
    myoepithelial carcinoma 50, 702
    myoepithelioma 758–761
        cytology 1000, *1001*
        histopathology 759, *760*
        immunohistochemistry 58, 74, 759, *761*
        malignant 761–762
MLS *see* myxoid liposarcoma
molecular genetics *see* genetics
monomorphous round cell RMS *see* alveolar rhabdomyosarcoma
monophasic (spindle cell) synovial sarcoma 52, 748–749, *749*, *751*
Morton neuroma 638
    histopathology *639*
    imaging 30, *31*
mouth tumors
    giant cell fibroma 213, *218*
    gingival fibromatosis 253, *253–254*
    irritation fibroma 212–213, *217*
    mucosal neuroma 639, *641*
    pulse granuloma 929, *929*
    rhabdomyoma 529, *529–530*
    solitary fibrous tumor 324
MPNST *see* malignant peripheral nerve sheath tumor
MRI *see* magnetic resonance imaging
mTOR signaling pathway 128, 136
    inhibitors 133, 518
MUC4 protein 74
mucin, production by myxomas 299
mucinosis, focal cutaneous mucinosis 310, *311*
mucocele 308, *310*
mucosa-associated lymphoid tissue (MALT) lymphoma (extranodal marginal zone lymphoma) 946–947, *947*
mucosal ganglioneuroma 639, *641*
mucosal neuroma 639, *641*
müllerian adenosarcoma 505
müllerian mixed tumor *see* malignant mixed müllerian tumor
multicentric Castleman disease 943
multicentric reticulohistiocytosis 964, *964*
multicolor FISH (M-FISH) 100
multicystic peritoneal mesothelioma

(multilocular peritoneal inclusion cyst) 778–780
    appearance *779*
    cellular nodules 789, *791*
    histopathology 779–780, *779*
multinodular (myxoid) myofibroblastoma *see* neurothekeoma
multiple endocrine neoplasia types 2a and 2b (MEN2a/MEN2b)
    mucosal neuroma/ganglioneuroma 639, *641*
    paraganglioma 725–726
multiple hamartoma syndrome 211
multiple myeloma (disseminated plasmacytoma) 949
multiplex PCR 141, 143
muscle
    IHC markers 48–52
    myogenesis 531, *534*
muscle-specific actin (HHF-35) 49, 546
muscularis mucosae leiomyoma 451–452, *452*
musculoaponeurotic fibromatosis *see* desmoid fibromatosis
MYC/MYCN/MYC/MYCN
    gene amplification 130
        angiosarcoma 611
        FISH 98–99, *99*, 877
        neuroblastoma 99, 130, 875–876, *877*
        rhabdomyosarcoma 131
    gene translocation in Burkitt lymphoma 949
    MYC/MYCN expression in neuroblastoma
        as an IHC marker 873, 875, *876*
        and prognosis 878
mycobacterial pseudotumor 929–930, *930*
mycobacterial synovitis 907–908, *909*
myeloid tumor, extramedullary (myeloid sarcoma) 954
    histopathology 954–955, *954*
    immunohistochemistry 954, *955*
    sclerosing 424
myelolipoma 390, *391*
    histopathology 390, *391*
myeloperoxidase 954
myocutaneous flaps 1013–1014
MyoD1 (myogenic regulatory factor) 51–52, 546, 899
myoepithelial carcinoma 50, 702
myoepithelioma 758–761
    cytology 1000, *1001*
    histopathology 759, *760*
    immunohistochemistry 74, 759, *761*
    malignant 761–762

myofibroblastic sarcoma (myofibrosarcoma) 352, *353*
myofibroblastoma
    breast 218–220
        cytology 983, *983–984*
        histopathology 218–219, *220*
    cervicovaginal 493, *493*
    palisaded 213–218, *218*
        histopathology 216–218, *219*
myofibroblasts 49–50, 181–182
myofibroma/myofibromatosis
    adult 267, *270–271*
    infantile 266–268
        cytology 984, *984*
        histopathology 267, *269–270*
myofibrosarcoma (myofibroblastic sarcoma) 352, *353*
myogenesis 531, *534*
myogenic regulatory factor (MyoD1) 51–52, 546, 899
myogenic sarcoma 364
myogenin 51–52, 546, *547*
myoglobin 52, 546
myolipoma 404
myopericytoma 632–633, *633–634*
myosin 50
myositis
    focal 194, 531, *531*
    proliferative 194–195, *195*
myositis ossificans 841–845
    cytology 999, *999–1000*
    histopathology 842–843, *844*
    imaging 12, 842, *843*
myxofibrosarcoma (myxoid MFH) 336, 356–359
    appearance *358*
    classification 356–357
    clinical features 357
    cytology 985–986, *986*
    differential diagnosis 348, 358–359
    genetics 359
    histopathology 357
        high-grade *361*
        intermediate-grade *360–361*
        low-grade *359–360*
    imaging *16–20*
    treatment 1018
myxoid chondrosarcoma (extraskeletal, EMC) 830–832
    appearance *830*
    cytology 1006, *1007*
    genetics 126, 832
    histopathology 830, *831–832*
    hybrid with synovial sarcoma 750
    imaging 830, *830*
    immunohistochemistry 54, 57, 831
myxoid cyst, cutaneous 305–306, *306*, *307*
myxoid lipoma (myxolipoma) 380, *382*, 977, *977*

Index

myxoid liposarcoma (MLS) 416, 430–435
  appearance *431*, 431
  clinical features 430–431
  cytology 981–982, *981*
  differential diagnosis 301, 432–433, *982*
  genetics *106*, 126–127, 434–435
  histopathology 431, *432*
    hibernoma-like *433*
    lymphangioma-like *433*
    postradiation therapy *435*
    round cell form *434*
  imaging *17*, 24, *24*
  immunohistochemistry 431
myxoid malignant fibrous histiocytoma *see* myxofibrosarcoma
myxoid smooth muscle uterine tumors 463
myxoid well-differentiated liposarcoma 422, *422–423*
myxoinflammatory fibroblastic sarcoma *see* acral myxoinflammatory fibroblastic sarcoma
myxolipoma (myxoid lipoma) 380, *382*, 977, *977*
myxoma 299
  cardiac 313–315
    glandular epithelial differentiation 314, *316*
    histopathology 314, *314–316*
  Carney complex 306, 308–310
  cutaneous 306–310
    histopathology *307*, *308–309*
  differential diagnosis 300–301, 307–308, 312–313, 317, 666
  intramuscular 299–302
    appearance 300, *300*
    cytology 1000, *1001*
    histopathology 300, *301–302*
  juxtaarticular 302–303, *303–304*
  Mazabraud/McCune–Albright syndromes 300
  nerve sheath 665–666
    histopathology 666, *666–667*
    immunohistochemistry 666, *667*
    not neurothekeoma 277, 280–281, *281*, 665–666
  odontogenic 315–317, *317*
myxomatous tumors, cystic appearance 34
myxopapillary ependymoma 711, 740
  cytology 991–992, *992*
  histopathology 711, *712*

N-cadherin 70
NAB2-STAT6 gene fusion 330

nail bed tumors (glomus tumor) 624
nasal glioma 710
nasal glomangiopericytoma 631–632, *632*
nasal-type NK/T-cell lymphoma 952, *952*
nasopharyngeal angiofibroma, juvenile 263–265
  appearance 263–264, *263*
  histopathology *264*, 264–265
NB84 (neuroblastoma marker) 55, 55–56
NCAM (CD56) 56, 806
NCOA2 gene fusion 125, 834
neck paraganglioma 734–737
  appearance *735*
  histopathology 735–737, *736*
  immunohistochemistry *54*, 737
necrobiotic granuloma pseudorheumatoid nodule (deep granuloma annulare) 927–929, *928*
  rheumatoid nodule 905
necrotizing fasciitis 197–199
  appearance 198, *198*
  histopathology 199, *199*
necrotizing granuloma 758
  in renal cell carcinoma 811
needle biopsy *see* core biopsy; fine-needle aspiration
nephrogenic systemic fibrosis (gadolinium-associated) 936, *936*
nerve growth factor receptor p75 56
nerve resection 1013
nerve sheath cellular anatomy 637–638
nerve sheath myxoma 665–666
  histopathology 666, *666–667*
  immunohistochemistry 666, *667*
  not neurothekeoma 277, 280–281, *281*, 665–666
nerve sheath tumor-associated angiosarcoma 607, 659, 683
nerve sheath tumors *see also* granular cell tumor; malignant peripheral nerve sheath tumor (MPNST); nerve sheath myxoma; neurofibroma; neuroma; perineurioma; schwannoma
  differential diagnosis 345, 652, 656, 661, 664
  imaging 11–12, 29, 31
  epithelioid 664, *665*
  hybrid 660
nested PCR 123, 141
neural fibrolipoma 385–386, *386–387*

neural and neuroendocrine IHC markers 52–57
neurilemmoma *see* schwannoma
neuroblastoma 867–879
  appearance 871, *872*
    ganglioneuroma *872*
  catecholamines 870
  classification 867–870, *873*, *874*
  clinical features 870–871
  differential diagnosis 874, 878–879
  electron microscopy 874, *875*
  genetics 130–131, 875–877
    MYCN amplification 99, 130, 875–876, *877*
    and prognosis 876–878
  histopathology 871–873
    in bone marrow 878
    composite 870, *870*
    congenital *871*
    differentiating 868, *869*, *874*
    ganglioneuroblastoma 868–869, *869–870*, *872*, *874*
    ganglioneuroma 868, *869*, *872*, 872
    ganglioneuroma, mucosal 639, *641*
    lymphocytic infiltration in opsoclonus-myoclonus 875
    neuropil/Homer Wright rosettes 868, *872*, *874*
    nucleoli *873*
    poorly differentiated 868, *868*, 873–874
    salt and pepper nuclei *872*
    staging 878
    undifferentiated 867, *868*
  history 867
  imaging 871
  immunohistochemistry 53, 875, *876*
    MYC/MYCN *873*, 875, *876*
    NB84 55, 55–56
    neurofilaments 55
    synaptophysin 53, *54*, *876*
    tyrosine hydroxylase 875, *876*
  prognosis 871, 876–878
  staging 878
neuroblastoma-*in-situ* 867
neuroblastoma-like schwannoma 657–658, *659*
neuroectodermal tumors *see also* cellular blue nevus; ependymoma; gastrointestinal clear cell sarcoma; melanoma; meningioma
  gastrointestinal clear cell sarcoma 708, *709*
  gliomatosis 710–711, *711*
  melanotic neuroectodermal tumor of infancy 708–709, *710*, 991, *991*

nasal glioma 710
neuroendocrine tumors 806 *see also* Merkel cell carcinoma; pheochromocytoma; small cell carcinoma of the lung (metastatic)
neurofibroma 644–652
  appearance
    cutaneous *653*
    intraneural *648*
    plexiform *649*
  cellular 645–646, 652
  cutaneous 644–645
  cytology 990, *990*
  differential diagnosis 345, 652
  diffuse 648, 652
  genetics 108–110
  histopathology
    cellular *646*
    cutaneous 645, *645–646*
    diffuse 648, *651*
    intraneural 647, *648–649*
    pigmented *651*
    plexiform *650*
  hybrid tumors 660
  imaging 30–31
  immunohistochemistry 56, 647, *647*, 650–652
  intraneural 646–647
  malignant transformation 652
  pigmented 648–650
  plexiform 31, 647
neurofibromatosis type 1 (NF1) 19, 653
  café au lait spots *653*
  genetics 108–110, 132–133, 637, 653
  GIST 483–484, *484*, 653
  glomus tumor 631
  imaging 22
  MPNST 652, 676
  pheochromocytoma 725–726
neurofibromatosis type 2 (NF2) 133, 637, 659–660, 669
neurofibrosarcoma *see* malignant peripheral nerve sheath tumor (MPNST)
neurofilament (NF) proteins 54–55
neuroma
  Morton 30, *31*, 638, *639*
  mucosal 639, *641*
  pacinian 640–641, *642*
  palisaded encapsulated 642–644, *643–644*
  traumatic 30, 639, *640*
neuromuscular hamartoma (choristoma) 530, *530*, 638, *638*
neuropil 868, *872*, *874*
neurosecretory granules in PNET cells 861, *861*
neurothekeoma 277–282
  clinical features 277–278

Index

neurothekeoma (cont.)
  differential diagnosis 224, 280–282
    not nerve sheath myxoma 277, 280–281, *281*, 665
    histopathology 278, *279–281*
      fascicular component *280*
      mixed multinodular *279*
    immunohistochemistry 277–280, *281*
  myxoid *see* nerve sheath myxoma
nevoid basal cell carcinoma syndrome (Gorlin syndrome) 529
nevus flammeus (port wine stain) 575–576
nevus lipomatosus superficialis 383, *383*
next-generation sequencing (NGS) 123, 144–145
NF1 *see* neurofibromatosis type 1
NF2 *see* neurofibromatosis type 2
NFATC2-EWSR1 gene fusion 865
NGS (next-generation sequencing) 123, 144–145
NK/T-cell lymphoma 952, *952*
NKX2.2 (IHC marker) 862
nodular fasciitis 182–185
  appearance 183, *185*
  clinical features 182–183
  cranial fasciitis 182, 249–251
  cytology *982*, 982
  differential diagnosis 184–185, 224, 251, 982
  genetics 185
  histopathology 183, *185–186*
    cranial fasciitis *250*, 250–251
    early/late forms *186*
    intramuscular 183, *188*
    pseudosarcomatous *187*
    repair variant 183, *185*, *187*
  imaging (cranial fasciitis) *250*
  immunohistochemistry 183–184, *188*
  terminology 182
nodular mesothelial hyperplasia 789, *791*
nodular tenosynovitis *see* tenosynovial giant cell tumor
noninvoluting congenital hemangioma 556–557
NPM-ALK gene fusion 127
NR4A3-EWSR1 gene fusion 126, 832
NTRK3-ETV6 gene fusion 124, 127–128, 294
nuchal fibrocartilaginous pseudotumor 207–208, *208*
nuchal-type fibroma 206–207, *207*
nuclear medicine 11–12, 29

nutlins 131
obese patients, massive localized lymphedema 409–411, *411*
ochronosis 913, *914*
odontogenic myxoma 315–317, *317*
OFT *see* ossifying fibromyxoid tumor
omental GISTs 482
oncogenes 115, 130–131 *see also* ALK/ALK; MYC/MYCN/MYC/MYCN
  IDH1/IDH2 138, 575
  KIT 131–132, 145, 486–487
  MDM2 101, 131, 135–136, 416–424, 430
  RAS family 115–116
oncogenic osteomalacia 834–837
oncogenic viruses *see* viruses, oncogenic
oncomir *see* microRNA
Ondine's curse 875
oral tumors
  giant cell fibroma 213, *218*
  gingival fibromatosis 253, *253–254*
  irritation fibroma 212–213, *217*
  mucosal neuroma 639, *641*
  pulse granuloma *929*, 929
  rhabdomyoma 529, *529–530*
  solitary fibrous tumor 324
orbital tumors
  giant cell angiofibroma 330
  hemangioma 559
  meningioma 716
  solitary fibrous tumor 224, 324
organizing hematoma 926–927
  appearance 926, *926*
  histopathology 926–927, *927*
  imaging 33
Ormond's disease (retroperitoneal fibrosis) 199–201
  histopathology 200, *200–201*
Oroya fever (verruga peruana) 579
ossifying fasciitis (panniculitis ossificans) 182–183
ossifying fibromyxoid tumor (OFT) 317–320
  appearance/X-ray image *319*
  genetics 318
  histopathology 317–318, *319–320*
  immunohistochemistry 57, *318*
osteoarthritis 905–906, *907*
osteoclastic giant cells
  in melanoma 699
  in MFH *370*, 989
  in PMT *836*
osteolipoma 382–383, *383*

osteomalacia, oncogenic 834–837
osteosarcoma, extraskeletal 847–849
  histopathology 848–849, *849*
  imaging 848, *848*
  immunohistochemistry 74–75, 849
ovarian tumors *see also* adult granulosa cell tumor
  carcinoma 815–816
  dysgerminoma 818
  germ cell tumors with angiosarcoma 607

p15(*CDKN2b*) 135
p16 (*CDKN2a*) 77, 134–135, 682
p21 (*CDKN1a*) 135
p27 76
p53 (*TP53*)
  atypical fibroxanthoma 366
  cell cycle regulation 135–136
  immunohistochemistry 76
  rhabdomyosarcoma 548
  as a therapeutic target 131
p75 (nerve growth factor receptor) 56
pacinian neuroma 640–641, *642*
paclitaxel 1018
palisaded encapsulated neuroma 642–644
  histopathology 642–643, *643*
  immunohistochemistry 644, *644*
palisaded myofibroblastoma 213–218
  appearance *218*
  histopathology 216–218, *219*
palisaded-vacuolated type GIST *478*, 479
palmar fibromatosis
  clinical features 233
  differential diagnosis 237
  genetics 237
  histopathology 234–235, *235*
  immunohistochemistry 236
  treatment 234
pancreatic adenocarcinoma 809
panniculitis-like T-cell lymphoma 950–952, *951*
panniculitis ossificans (fasciitis ossificans) 182–183, 841
papillary endothelial hyperplasia (Masson tumor) 577–578, *577–578*
papillary hemangioma 565
papillary intralymphatic angioendothelioma (PILA, Dabska tumor) 593–594
  appearance *593*
  histopathology 593, *594*
  immunohistochemistry 595
papillary mesothelioma, well-differentiated 780–781, *781*, 789–790, *791*

parachordoma 759–761
paraffinoma 409
paraganglia *723*, 723
paraganglioma 723–740 *see also* ganglioneuroma; neuroblastoma
  appearance *729*, *735*
  Carney triad 482
  cauda equina 738–740
  classification 723
  differential diagnosis 737, 740, 900
  gangliocytic (duodenal) 737–738
  genetics 724–726
  histopathology
    carotid body 735–737, *736*
    cauda equina 739, *739*
    combined pheochromocytoma and ganglioneuroma *732*
    gangliocytic duodenal 738, *739*
    jugulotympanic *738*
    pheochromocytoma 726, *727*
    pheochromocytoma-like retroperitoneal *730*
    retroperitoneal 728–729, *729–731*
    urinary bladder *733*, 733
  immunohistochemistry
    cauda equina 739–740
    gangliocytic duodenal 738
    neck 54, 737
    pheochromocytoma 728
    retroperitoneal *55*, 732
    urinary bladder 734, *734*
  jugulotympanic 737
  malignant 726–728
  neck/carotid body 734–737
  pheochromocytoma 723, 726–728
    with MPNST *686*, 686, *732*
  retroperitoneal (extra-adrenal) 723–724, 728–729
  syndromic forms 724–726
  thoracic 734
  urinary bladder 729–734
paraneoplastic syndromes
  in DSCRT 882
  in neuroblastoma 870–871
  oncogenic osteomalacia 834–837
paraspinal tumors
  meningioma 716, *716*
  paraganglioma 734
paratesticular serous papillary carcinoma 793, *793*
paratesticular spindle cell rhabdomyosarcoma 535–539
parosteal lesions 182, 845–847
PAS (periodic-acid–Schiff) positive cells

Index

in alveolar soft part sarcoma 897, *898*
in Ewing sarcoma *859–860*
PATZ1-EWSR1 gene fusion 865
PAX-FOXO1A (FKHR) gene fusion 129, 535, 539–540, 542, 547
pazopanib 1030
PBX1-EWSR1 gene fusion 126, 760–761
PCR *see* polymerase chain reaction
PDGFB-COL1A1 gene fusion 128–129, 345
PDGFRA (platelet-derived growth factor alpha receptor) 132, 475, 486, 1024
 targeted therapy *see* imatinib mesylate
PDGFRB (platelet-derived growth factor beta receptor) 242
PECAM-1 (CD31) 41–44, 610
PEComa (perivascular epithelioid cell tumor) 513 *see also* angiomyolipoma; lymphangiomyoma/ lymphangiomyomatosis
 abdominal sites 521–522
 appearance *520*
 clear cell "sugar" tumor 519–520
 cutaneous 522
 differential diagnosis 465, 517, 523, 900
 genetics 513, 517–518, 523
 histopathology 521, *521–522*
  clear cell "sugar" tumor 519, *520*
 immunohistochemistry 521–522, *522*
 renal capsule 520
 uterine 520–521
pediatric tumors *see* childhood tumors
pelvic desmoids 240
pelvic lipomatosis 386–387
penile fibromatosis 237–238, *238*
penile myointimoma 189–190, *191*
periarticular lesions 18, 20 *see also* synovial sarcoma
perineurioma 666–672
 differential diagnosis 668–672
 genetics 669, 671
 histopathology
  intraneural *668*, 668
  retiform *670*
  sclerosing 668, *669*
  soft tissue (other) 670, *670*
 hybrid tumors 660
 immunohistochemistry 668, *671*, 671
 intraneural 668

malignant 671–672
perineural cells 637
retiform (reticular) 670
sclerosing 668–670
soft tissue (other) 670–671
periodic-acid–Schiff (PAS) positive cells
 in alveolar soft part sarcoma 897, *898*
 in Ewing sarcoma *859–860*
periosteal fasciitis 182, 845–847
peripheral nerve sheath tumors (PNST) *see also* granular cell tumor; malignant peripheral nerve sheath tumor (MPNST); nerve sheath myxoma; neurofibroma; neuroma; perineurioma; schwannoma
 differential diagnosis 345, 652, 656, 661, 664
 imaging 11–12, 29, 31
peritoneal leiomyomatosis 465–466, *466*
peritoneal mesothelioma 783
 appearance *784*
 differential diagnosis 792–793
perivascular epithelioid cell tumor *see* PEComa
pesticides 3
Peyronie's disease (penile fibromatosis) 237–238, *238*
phagolysosomes 673
pheochromocytoma 723, 726–728
 histopathology 726, *727*
 immunohistochemistry 728
 malignant 726–728
 with MPNST 686, *686*, 732
 syndromic 725–726
pheochromocytoma of adrenal scaled score (PASS) 726–727
PHF1 gene fusion 125, 318
phleboliths 25, 574
phosphaturic mesenchymal tumor (PMT) of mixed connective tissue variant 834–837
 appearance *835*
 histopathology *835–837*, 835–836
 imaging 834, *835*
 immunohistochemistry 836
photography of tumors 7
PHOX2B (IHC marker) 875
PI3K/AKT/mTOR signaling pathway 128, 133, 136, 518
pigmented DFSP (Bednar tumor) 342, *342*
pigmented neurofibroma 648–650, *651*

pigmented villonodular synovitis (PVNS) *see* tenosynovial giant cell tumor
pilar leiomyoma (piloleiomyoma) 444–446, *446*
PLAG1 gene fusion 125, 407
planar xanthoma (xanthelasma) 932
plantar fibromatosis
 clinical features 233–234
 differential diagnosis 237
 genetics 237
 histopathology 235–236, *236–237*
 immunohistochemistry 236
 treatment 234
plaque-like CD34-positive dermal fibroma 345
plasmacytoma 949–950, *951*
plastic induration of the penis (penile fibromatosis) 237–238, *238*
platelet-derived growth factor alpha receptor (PDGFRA) 132, 475, 486, 1024
 targeted therapy *see* imatinib mesylate
platelet-derived growth factor beta (PDGFB)
 gene fusion with *COL1A1* 128–129, 345
 PDGFB receptor in desmoids 242
pleomorphic fibroma of skin 211–212, *215*
pleomorphic hyalinizing angiectatic tumor 369–372, *371*
pleomorphic leiomyosarcoma 453–454, *454*
pleomorphic lipoma 391–394
 appearance *394*
 cytology 978, *979*
 histopathology 393–394, *397*
 immunohistochemistry 394, *397*
pleomorphic liposarcoma 24, 416, 435–438
 appearance 435, *436*
 histopathology 435–436, *437*
 epithelioid form *438*
pleomorphic malignant fibrous histiocytoma (MFH) (pleomorphic undifferentiated sarcoma) 336, 359–364, 987–988
 clinical features 362
 cytology 988, *988*
 differential diagnosis 364, 988
 genetics 134, 362–364
 histopathology 362, *363–364*
 immunohistochemistry 71, 360, 362

pleomorphic rhabdomyosarcoma (RMS) 541–542
 cytology 994, *995*
 histopathology 542, *543*
 immunohistochemistry *543*
pleomorphic undifferentiated sarcoma *see* pleomorphic malignant fibrous histiocytoma
pleural mesothelioma 783, *784*
 see also mesothelioma, malignant
pleural plaque *791*
pleural solitary fibrous tumor 324
plexiform fibrohistiocytic tumor 282–283
 differential diagnosis 224, 281–283, 368
 histopathology 282–283, *282*
plexiform neurofibroma 31, 647
 appearance *649*
 histopathology *650*
plexiform schwannoma 656–657, *658*, 661
plywood fibroma (sclerotic fibroma) 211, *214*
PMT *see* phosphaturic mesenchymal tumor (PMT) of mixed connective tissue variant
PNET (primitive neuroectodermal tumor) *see* Ewing sarcoma family of tumors
PNL2 (IHC marker) 61
PNST (peripheral nerve sheath tumors) *see also* granular cell tumor; malignant peripheral nerve sheath tumor (MPNST); nerve sheath myxoma; neurofibroma; neuroma; perineurioma; schwannoma
 differential diagnosis 345, 652, 656, 661, 664
 imaging 11–12, 29, 31
podoplanin (D2-40) 47
POEMS syndrome 564–565
polyethylene wear particles 910, *911*
polymerase chain reaction (PCR) 141–142
 allele-specific PCR 141
 amplification of microsatellite markers 143
 multiplex PCR 141, 143
 nested PCR 123, 141
 RACE 119–123, 141
 real-time PCR 141–143
 RT-PCR 119–123, 141, 977
polyvinylpyrrolidone (PVP) accumulation *933–934*, 933–934

Index

poorly differentiated neuroblastoma 868, 868, 873–874
poorly differentiated synovial sarcoma 750, 750
popliteal (Baker) cyst 17, 31–32, 33
port wine stain 575–576
positron emission tomography (PET) 12
postradiation angiosarcoma/atypical vascular lesions 603, 604
post-traumatic cyst of soft tissue see organizing hematoma
POU5F1-EWSR1 gene fusion 126, 760–761, 865
PR (progesterone receptor) 73, 1031
precursor B-cell lymphoblastic lymphoma 947–948, 948
primary neuroendocrine carcinoma of the skin see Merkel cell carcinoma
primitive neuroectodermal tumor (PNET) see Ewing sarcoma family of tumors
PRKAR1A (protein kinase A) gene 309–310, 315
progesterone receptor (PR) 73, 1031
progressive angioma (acquired tufted angioma) 566–567, 568
proliferative fasciitis 190–194
  appearance 192–200, 193
  cytology 982
  histopathology 192–193, 193–194
proliferative myositis 194–195, 195
prostatic carcinoma 125, 813–815
  histopathology 814, 814
  immunohistochemistry 815, 815
prosthetics (foreign body reactions) 3
  bone cement 911, 911
  carbon fibers/graphite 911–912, 912
  metal fragments 912–913, 913
  polyethylene wear particles 910, 911
  silicone 911, 912
protein kinase A (PRKAR1A gene) 309–310, 315
Proteus syndrome 272–273
  gyriform lesions on the soles of the feet 272
  histopathology 272, 273
proto-oncogenes 115 see also oncogenes
proximal-type epithelioid sarcoma 758, 759
psammoma bodies 663, 716, 792

psammomatous melanotic schwannoma see melanotic schwannoma
pseudogout 910
  histopathology 903, 910, 910
  tophaceous (focal) form 839, 841
  tophaceous (focal) 838–839, 840
pseudo-Kaposi's sarcoma (acroangiodermatitis) 580, 580
pseudolipoblasts 356, 359
pseudomyogenic hemangioendothelioma 599–601, 601
pseudorheumatoid nodule (deep granuloma annulare) 927–929, 928
pseudosarcomatous fasciitis see nodular fasciitis
pseudosarcomatous fibrous histiocytoma 226–228, 227
pseudosarcomatous stromal polyp 492, 492
PTCH1 gene 529
pulmonary tumors see lung tumors
pulse granuloma 929, 929
PVP (polyvinylpyrrolidone) accumulation 933–934, 933–934
pyogenic granuloma (lobular capillary hemangioma) 557–558
  histopathology 557–558, 558
  intravascular 557, 559
pyrosequencing 144

quantitative-PCR (Q-PCR, real-time PCR) 141–143

RACE (rapid amplification of cDNA ends) 119–123, 141
radiation-induced tumors 3
  angiosarcoma 603, 604
  mesothelioma 783
radiography 11, 29
  calciphylaxis 842
  chondroma of soft parts 826
  cranial fasciitis 250
  fibrodysplasia ossificans 846
  mesenchymal chondrosarcoma 833
  myositis ossificans 12, 842, 843
  neuroblastoma 871
  osteosarcoma 848
  phosphaturic mesenchymal tumor 835
  pseudogout 840
  synovial chondromatosis 34, 35, 828–829
  TSGT 34

tumoral calcinosis 838
radionuclide imaging 11–12, 29
radiotherapy
  adjuvant/neoadjuvant 1015–1016
  chemoradiotherapy 1028–1029
  desmoid tumors 1031
  retroperitoneal sarcoma 1020
  adverse effects 3, 1015–1016
rapidly involuting congenital hemangioma 556–557
RAS family oncogenes 115–116
RB1/RB1 (retinoblastoma) 76, 116, 134
reactive mesothelial proliferation 789–792, 791
reactive vascular proliferations 578
  acroangiodermatitis 580, 580
  angioendotheliomatosis 579–580
  bacillary angiomatosis 578–579, 579
  florid proliferation related to intestinal intussusception 580, 581
  superficial vascular pseudoaneurysm 580, 581
  verruga peruana 579
real-time PCR 141–143
receptor tyrosine kinases see tyrosine kinase receptors
reconstructive surgery 1013–1014
rectal GISTs 481, 482
recurrent digital fibroma see infantile digital fibroma
Reed–Sternberg-like cells 356
regorafenib 1025–1026
Reichel disease see synovial chondromatosis
renal cell carcinoma, metastatic 809–811
  appearance 809
  genetics 130
  histopathology 810–811, 810
  sarcomatoid 813
  immunohistochemistry 811, 812–813
  sarcomatoid 811
renal tumors see also angiomyolipoma
  glomus tumor 625
  hereditary leiomyomatosis and renal cell cancer 445–446, 461, 462–463
  juxtaglomerular cell tumor 633–634, 634
  renal capsule PEComa 520, 520
  synovial sarcoma 746
  Wilms' tumor 856, 882
reticulohistiocytoma 962–964, 964

reticulohistiocytosis, multicentric 964, 964
reticulum cells 968
  dendritic reticulum cell sarcoma 965–968, 967
  fibroblastic reticulum cell sarcoma 968
  interdigitating reticulum cell sarcoma 58, 702, 968
retiform hemangioendothelioma 594–595, 595
retiform (reticular) perineurioma 670, 670
retinal anlage tumor (melanotic neuroectodermal tumor of infancy) 708–709, 991
  cytology 991
  histopathology 709, 710
retinoblastoma gene/product (RB1/RB1) 76, 116, 134
retroperitoneal fibrosis 199–201
  histopathology 200, 200–201
retroperitoneal leiomyoma, ER-positive 73, 459, 459
retroperitoneal leiomyosarcoma 456–457
retroperitoneal liposarcoma, well-differentiated 417–418, 424
retroperitoneal paraganglioma (extra-adrenal) 723–724, 728–729
  histopathology 728–729
  combined pheochromocytoma and ganglioneuroma 732
  nuclear atypia 731
  organoid patterns 729–731
  pheochromocytoma-like 730
  pigmented 731
  immunohistochemistry 55, 729, 732
retroperitoneal sarcoma 1019–1021
retroperitoneal schwannoma 654, 654
reverse transcription PCR (RT-PCR) 119–123, 141, 977
rhabdoid rhabdomyosarcoma 544–545, 545
rhabdoid tumor, extrarenal 882–886
  appearance 883, 883
  cytology 1007–1008, 1008
  differential diagnosis 758, 886, 1008
    pseudo-rhabdoid tumors 885
  electron microscopy 883, 883
  genetics 133, 882, 885–886
  histopathology 883, 883
  history 882
  immunohistochemistry 883–885, 884

rhabdomyoma 527–528
  adult 529, 529–530
  cardiac 523
  cytology 993, 993
  differential diagnosis 527–530, 993
  fetal 528–529
  genetics 529–530
  histopathology
    adult 529, 529
    cardiac 523
    fetal 528, 528
    genital (vaginal) 495
    juvenile 528, 528
  immunohistochemistry 50, 529
  fetal 528
  vaginal 493–494, 530
rhabdomyoma-like rhabdomyosarcoma (RMS) 544, 545
rhabdomyomatous mesenchymal hamartoma 528–530, 530
rhabdomyosarcoma (RMS) 531–548
  alveolar 129, 134–135, 539–541, 878–879, 994
  anaplastic 543–544
  anaplastic spindle cell 542
  appearance 531–532, 535
  chemotherapy 547–548, 1018–1019, 1031
  classification 531, 993
  clear cell 544
  clinical features 531, 533–534, 536, 539, 541–542
  cytology 994, 995
    alveolar 994, 994
    embryonal 993–994, 994
  differential diagnosis 292, 527, 535, 539, 541–542, 545, 878–879, 994
  ectomesenchymoma 545–546, 862–863
  embryonal 51, 108, 532, 533–535, 993–994
  epithelioid/rhabdoid 544–545
  genetics 548
    alveolar 129, 134–135, 539–541, 542, 878–879
    embryonal 108, 535
    epithelioid 545
    mixed alveolar and embryonal 544
    pleomorphic 542
    and prognosis 547
    spindle cell 538
  histopathology 532
    alveolar 539, 540–541
    alveolar leukemoid variant 540
    anaplastic 544
    botryoid 535, 537
    clear cell 544
    ectomesenchymoma 863

  embryonal 534–535, 536
  and embryonic myogenesis 531, 534
  pleomorphic 542, 543
  rhabdoid 545
  rhabdomyoma-like 545
  sclerosing 536–537, 538
  spindle cell 536, 537
imaging 535, 539
immunohistochemistry 49–52, 51, 57, 546–547
  alveolar 546, 547
  ectomesenchymoma 545
  embryonal 51, 53, 547
  fusion status markers 546–547
  pleomorphic 543
  spindle cell 538
lipid-rich 544
miRNA 117
mixed alveolar and embryonal 544
oncogenesis 527
pleomorphic 541–542, 994
prognosis 547–548, 993
rhabdomyoma-like 544
spindle cell/sclerosing 535–539
staging 547–548
treatment 531–532, 547–548, 1018–1019
  in adults 1031
rheumatoid arthritis 904–905, 906
rheumatoid nodule (necrobiotic granuloma) 905
rice bodies 905, 906
RMS see rhabdomyosarcoma
RNA
  extraction from formalin-fixed specimens 140–141
  miRNA 116–117, 129, 136
  in therapies 1032
RNA-seq technique 123
ROS1 gene fusion 125, 127
Rosai–Dorfman disease (RDD) 956–957
  histopathology 957, 958
  immunohistochemistry 58, 957, 959
rosettes
  giant hyaline rosettes in low-grade fibromyxoid sarcoma 350
  Homer Wright rosettes
    neuroblastoma 867, 872, 874
    PNETs 860, 860
    schwannoma 657–658, 659
round cell liposarcoma 24, 431
  cytology 981–982, 981
  histopathology 434
RT-PCR (reverse transcription PCR) 119–123, 141, 977
RTKs see tyrosine kinase receptors

S100 protein 56–58, 277
  melanoma, metastatic 56, 699, 701

  synovial sarcoma 57, 752, 752
SAF see superficial acral fibromyxoma
salivary gland tumors 74
SAM see superficial angiomyxoma
Sanger DNA sequencing 144
sarcoidosis, synovitis 907
sarcoma
  epidemiology 2–3, 11, 975
  etiology 3–4
  grading 4–5, 75–76
  histogenesis 2, 899–900
  staging 5–6, 15–17
sarcoma botryoides (botryoid rhabdomyosarcoma) 533–534
  histopathology 535, 537
sarcomatoid carcinoma
  breast 805, 805
  lung 807, 808
  renal cell 811, 813
    chromophobe 813
  squamous cell 803, 803–804
    laryngeal 804
sarcomatoid mesothelioma 787–788, 787–788
sarcomatous spindle cell GIST 479, 479
SATB2 protein 74–75
scalp angiosarcoma 602–603, 603
scalp meningioma 714–715, 715
scapular tip tumors see elastofibroma
scars (keloids) 201, 202
Schwann cells 637
schwannoma 654 see also neurofibromatosis type 2
  ancient 655, 989–990
  appearance 654–655, 654
    gastrointestinal 660, 660
    melanotic 662, 662
  cellular 655–656
  cellular plexiform 657
  clinical feature 654, 656–657
  conventional 654–655
  cytology 989, 989
    ancient schwannoma 990
  differential diagnosis 652, 656, 661, 664, 702, 990
  electron microscopy 655, 663
  epithelioid 657, 661
  gastrointestinal 487, 660–661
  genetics 637, 661
  histopathology 655, 655
    cellular 656, 657
    cystic 656
    gastrointestinal 660–661, 661
    melanotic 662–663, 663–664
    plexiform 657, 658
    rosettes 659
  hybrid tumors 660

  imaging 20, 31
  immunohistochemistry 56, 73–74, 655
    cellular 656
    epithelioid 657, 658
    gastrointestinal 661
    melanotic 664
  malignant see malignant peripheral nerve sheath tumor (MPNST)
  malignant transformation 658–659
    angiosarcoma 607, 659
  melanotic 661–664, 702
  plexiform 656–657, 661
  with rosettes 657–658
  vestibular 133, 654, 659–660
schwannomatosis 654, 660
schwannomin (merlin) 133, 659–660
sciatic nerve resection 1013
scintigraphy 11–12, 29
sclerosing epithelioid fibrosarcoma 350–351
  as form of LGFMS 336, 350
  histopathology 351, 351
  immunohistochemistry 74, 351
sclerosing extramedullary myeloid tumor 424
sclerosing hemangioma see benign fibrous histiocytoma
sclerosing lipogranuloma 409, 409
sclerosing liposarcoma 420–421, 421
sclerosing perineurioma 668–670, 669
sclerosing rhabdomyosarcoma 536–537, 538, 539
sclerotic fibroma of skin 211, 214
sclerotic lipoma 395–396, 398
scrotal tumors
  adenomatoid 777
  aggressive angiomyxoma 498
  angiomyofibroblastoma-like tumor 496–497, 497
  leiomyoma 447–448, 449
  liposarcoma 418
  melanotic neuroectodermal tumor of infancy 709
  necrotizing fasciitis (Fournier's gangrene) 198
  paratesticular spindle cell rhabdomyosarcoma 535–539
  serous papillary carcinoma 793, 793
  smooth muscle hyperplasia 444
SDH (succinate dehydrogenase) 138, 724
SDH-deficient tumors
  GISTs 482–483, 483
  treatment 488, 1025

Index

SDH-deficient tumors (cont.)
  paraganglioma 724–725
segmental neurofibromatosis
  type 1 (NF1) 653
seminoma 48, 817–818, *818*
senile hemangioma (cherry
  hemangioma) 575, *576*
septic arthritis 903–904, *904*
serous carcinoma 792–793, *792–793*
sex cord stromal tumors *see*
  granulosa cell tumor
SFT *see* solitary fibrous tumor
Shimada classification
  (neuroblastoma) 867–870, 873, *874*
Shulman syndrome (eosinophilic
  fasciitis) 195–197, *198*
signet ring cell-like schwannoma
  660
signet ring cells in breast
  carcinoma 805
silicone
  breast implants
    ALCL 953
    granuloma 409, *410*
  joint prostheses 911, *912*
simian virus 40 (SV40) 783
single-strand conformation
  polymorphism (SSCP)
  assay 142
sinonasal hemangiopericytoma
  (glomangiopericytoma)
  631–632, *632*
sinonasal leiomyoma 452
sinonasal meningioma 716–717, *716*
sinus histiocytosis with massive
  lymphadenopathy *see*
  Rosai–Dorfman disease
sinusoidal hemangioma 559
sirolimus 518
skeletal muscle tumors 527 *see
  also* myositis;
  rhabdomyoma;
  rhabdomyosarcoma
  classification 527
  neuromuscular choristoma
    530, *530*, 638, *638*
  rhabdomyomatous
    mesenchymal hamartoma
    528–530, *530*
skin grafts 1013
small cell carcinoma of the lung
  (metastatic) 802–803, 806
  immunohistochemistry 53–54, *54*, 806, *807*
  intracranial *806*
small intestinal GISTs 476, 479
  appearance *480*
  histopathology 479–480, *481*
  in NF1 patients 483–484, *484*
small round blue cell tumors
  (SBRCT) 855–856 *see
  also* desmoplastic small

round cell tumor; Ewing
  sarcoma family of
  tumors; neuroblastoma;
  rhabdoid tumor,
  extrarenal
SMARCA5-EWSR1 gene fusion
  865
SMARCB1/SMARCB1 *see INI1*/
  INI1
smooth muscle actin (SMA) 50,
  546
smooth muscle tumors 443 *see
  also* leiomyoma;
  leiomyomatosis;
  leiomyosarcoma
  angiomyomatous hamartoma
    of lymph node 444, *445*
  hamartoma of skin 443–444,
    444–445
  myxoid (uterine) 463
  scrotal smooth muscle
    hyperplasia 444
  of uncertain malignant
    potential 468–469
soft part chondroma *see*
  chondroma of soft parts
soft tissue tumors, definition and
  classification 1–2
solitary circumscribed neuroma
  see palisaded
  encapsulated neuroma
solitary epithelioid histiocytoma
  (reticulohistiocytoma)
  962–964, *964*
solitary fibrous tumor (SFT)
  324–330
  appearance 326, *326*
  clinical features 324–326, 330
  differential diagnosis 329–330,
    793
  genetics 330
  giant cell angiofibroma variant
    330
  histopathology 326–327, *327–328*
    giant cell angiofibroma 330,
      331
    malignant *328*
    uncertain potential *329*
  immunohistochemistry 70,
    328–329, *329*, 330
  orbital 224, 324
  terminology 324, 777
Sox10 (transcription factor) 58
SP3-EWSR1 gene fusion 865
spectral karyotyping (SKY) 100
spider cells (cardiac
  rhabdomyoma) 523
spina bifida 384
spinal tumors *see also* paraspinal
  tumors
  paraganglioma 738–740, *739*
  spinal cord-associated lipoma
    384–385
spindle cell angiosarcoma 608

spindle cell carcinoma
  breast 805, 805
  squamous cell carcinoma 803–804
spindle cell hemangioma 571–575
  histopathology *573–575*, 573–574
spindle cell lipoma 391–394
  cytology 978, *979*
  histopathology 393, *394–396*
    pleomorphic variant 393–394, *397*
    pseudoangiomatoid variant
      396
  immunohistochemistry 394, *397*
spindle cell liposarcoma *423*, 423
spindle cell rhabdomyosarcoma
  535–539
  histopathology 536, *537*
  immunohistochemistry *538*
spindle cell synovial sarcoma
  748–749, *749*
splenic angiosarcoma 606, *607*
split-fat sign 18, *20*, 30–31
squamous cell carcinoma 758,
  803, *803*–804
  laryngeal 804
SS18-SSX gene fusion 123, 128,
  136, 753
staging of tumors 5–6, 15–17
  neuroblastoma 878
  rhabdomyosarcoma 547–548
STAT6-NAB2 gene fusion 330
stem cells (precursor cells) 2
sternocleidomastoid tumors *see*
  branchial anlage mixed
  tumor
sternomastoid tumor *see*
  fibromatosis colli
Stewart–Treves syndrome
  (angiosarcoma in
  lymphedematous
  extremities) 4, 603–604,
  604
stomach GIST *see* gastric GIST
stomach glomus tumor 624–626
storage of tissue samples 8
strawberry nevus *see* infantile
  hemangioma
stromatosis *see* endometrial
  stromal sarcoma
STUMP (smooth muscle tumors
  of uncertain malignant
  potential) 468–469
subcutaneous lesions 20
subungual tumors (glomus
  tumor) 624
succinate dehydrogenase (SDH)
  138, 724
succinate dehydrogenase (SDH)-
  deficient tumors
  GISTs 482–483, *483*
  treatment 488, 1025

paraganglioma 724–725
sunitinib 1025–1026
superficial acral fibromyxoma
  (SAF) 310–313
  appearance *311*
  differential diagnosis 307,
    312–313
  histopathology 311–312, *312*
  immunohistochemistry *312*,
    312
superficial angiomyxoma (SAM)
  306–310
  clinical features 306–307,
    309
  differential diagnosis 307–308
  genetics 309–310
  histopathology 307, *308–309*
superficial cervicovaginal
  myofibroblastoma 493
  histopathology 493, *493*
  immunohistochemistry 493,
    494
superficial plexiform
  schwannoma 657
superficial vascular
  pseudoaneurysm 580, *581*
surgery 1012–1021
  adjuvant/neoadjuvant
    chemotherapy 1017–1018, 1027–1029
  adjuvant/neoadjuvant
    radiotherapy 1015–1016,
    1020, 1031
  amputation 1015
  desmoids 239–241, 1015
  DFSP 337, 1015
  excisional margins 1013
  GISTs 487, 1020
    palliative 1026
  liposarcoma 1014–1015,
    1019–1020
  for metastatic disease 1029
  MPNST 1018
  myxofibrosarcoma 1018
  palmar/plantar fibromatosis
    234
  preoperative work up 1012–1013
  reconstruction 1013–1014
  retroperitoneal sarcoma 1019–1021
surgical sponges, retained 934,
  935
surgical staging system
  (Enneking) 6, 18
sustentacular cells 57, 723, 728
SUZ12-JAZF1 gene fusion 123,
  503
sweat gland tumors
  eccrine acrospiroma 762, *763*
  eccrine spiradenoma 762–763,
    764
SWI/SNF complex *see INI1*/INI1
  (*SMARCB1*/SMARCB1)
synaptophysin 53–54, 876

Index

synovial chondromatosis 825, 827–830
  appearance 827–828, *828*
  histopathology 828–830, *829*
  imaging *34, 35*, 827, *828*
  secondary 827, 829–830, *829*
synovial hemangioma 26–27
synovial lipomatosis (lipoma arborescens) 383–384, *383–384*
synovial pathology (other than tumors)
  acute purulent synovitis (septic arthritis) 903–904, *904*
  Baker (popliteal) cyst *17*, 31–32, *33*
  biopsy 903
  foreign body reactions
    bone cement *911*, 911
    carbon fibers/graphite 911–912, *912*
    metal fragments 912–913, *913*
    polyethylene wear particles 910, *911*
    silicone 911, *912*
  gout *841*, 903, *907*, 909–910
  granulomatous synovitis 907–909
    mycobacterial 907–908, *909*
    in sarcoidosis 907
  hemosiderotic synovitis 906–907, *907–908*
  Lyme disease 904, *905*
  osteoarthritis 905–906, *907*
  pseudogout 903, 910, *910*
    tophaceous 838–839, *840–841*
  rheumatoid 904–905, *906*
  synovial fluid analysis 903
synovial sarcoma 744–753
  appearance 746, *746*
  chemotherapy 1017–1018
  clinical features 744–746
  cytology 1003, *1003*
  differential diagnosis *18*, 753, *793*, 1003
    MPNST 680–684
  genetics 123, 128, 753
  histopathology 746–747
    biphasic *747–748*, 747–748
    calcification *749*
    monophasic spindle cell 748–749, *749*
    poorly differentiated *750*, 750
  hybrid with extraskeletal myxoid chondrosarcoma 750
  imaging *18*
  immunohistochemistry 750–753, *752*, 1003
    biphasic *68*, *751*
    calretinin *72*, *752*
    INI1 *74*

    keratins 66–67, 750, *751*
    monophasic *52*, *751*
    S100 *57*, *752*, 752
  prognosis 745–746, 750
  visceral 746
synovial villous hyperplasia 906–907, *907*, *908*
systemic calcinosis *see* calciphylaxis

T-cell lymphomas
  cutaneous ALCL 954
  nasal-type NK/T-cell *952*, 952
  noncutaneous ALCL 952–954, *953*
  panniculitis-like 950–952, *951*
TAF15 protein 126
tamoxifen 1031–1032
target sign 30
targeted therapies 137
  for advanced disease 1030
  for *ALK*-rearranged tumors 127
  for desmoids 241–242, 1032
  for DFSP *129*, 345–346
  for *ETV6-NTRK3* gene fusion tumors 128
  future directions 1027, 1032
  for GISTs 1027
    adjuvant therapy 1020, 1026–1027
    first-line therapy 132, 487–488, 1024–1026
    resistance 132, 486, 1020, 1024–1025, *1025*, 1027
    SDH-deficient tumors 488, 1025
  MDM2 inhibitors 131
  mTOR pathway inhibitors 133, 518
  for vestibular schwannoma 133
targetoid hemosiderotic hemangioma (hobnail hemangioma) 565–566, *567*
taxanes 1018, 1030–1031
telangiectasia 554, 575 *see also* angiokeratoma
  port wine stain 575–576
  venous lake 577
telomerase 877
tendinous xanthoma 932, *932*
tendon pathology (other than tumors)
  amyloid 938
  ochronosis 913, *914*
  rupture 914–915, *914*
tendon sheath fibroma 209–211
  appearance *213*
  histopathology 209–210, *213–214*
tenosynovial giant cell tumor (TSGT) 914–922
  appearance
    diffuse intra-articular *919*, *919*

    localized *916*, 916
  atypical and malignant 921–922
  clinical features 915–916, 918–920
  cytology *987*, 987
  differential diagnosis 368, 921–922, 987
  diffuse extra-articular 919–920
  diffuse intra-articular 918–919
  genetics 105, 914, 920–921
  histopathology
    atypical/malignant *922*, 922
    diffuse extra-articular 920, *921*
    diffuse intra-articular 919, *920*
    localized type 916–917, *916–918*
    localized type with cartilaginous metaplasia *919*
    localized type infarction *918*
  imaging *21*, *34*
  immunohistochemistry 920
  localized 915–917
testicular tumors
  adenomatoid tumor 777
  seminoma *48*, 817–818, *818*
  of the tunica vaginalis 783, 790–793, *793*
TFE3/TFE3
  gene fusion 125
    TFE3-ASPSCR1 (ASPL) *129*, 900
    TFE3-YAP1 599
  as IHC marker 898
TFG gene fusion 126
thalidomide 553
thoracic paraganglioma 734
thorotrast contrast material 3, 934–935, *935*
thymolipoma 385–386, *385*
thymoma *see* ectopic hamartomatous thymoma (branchial anlage mixed tumor)
thyroid carcinoma 184, 816–817, *817*
tibial tumors, adamantinoma 770, *770*
tissue handling procedures 8
TKIs *see* tyrosine kinase inhibitors
TLE1 (transducin-like enhancer of split 1) 752, *752*
TLS *see FUS* gene fusions
TMPRSS2 gene fusion 125
TNM staging system 5
  rhabdomyosarcoma 548
tophaceous pseudogout 838–839
  histopathology 839, *841*
  imaging 839, *840*
torticollis 251

Touton-type giant cells
  in BFH *223*, 988
  in juvenile xanthogranuloma 963
TP53 see p53 (*TP53*)
trabectedin 1030
trabecular carcinoma of the skin *see* Merkel cell carcinoma
transcriptomics 123, 143–144, 977
transgelin 50
transitional cell carcinoma 811–813
translocations 103–108, 117, 120
  see also fusion genes
transplantation *see* immunosuppressed patients
transport of specimens 92–93
traumatic neuroma 30, 639, *640*
TRE17 (USP6) gene fusion 125
tricarboxylic acid (TCA) cycle dysfunction 138, 724–725
trichodiscoma 307
trisomy 237, 246, 294, 506, 921
Triton tumor
  benign 530, *530*, 638, 638
  malignant 683, *683*
TRKs *see* tyrosine kinase receptors
TSGT *see* tenosynovial giant cell tumor
TTF1 (thyroid transcription factor 1) 789, 805–806, *817*
tuberous sclerosis complex (TSC) 513, 517–518, 523
tuberous xanthoma 932
tubulopapillary mesothelioma 784–785, *784*
  adenomatoid-like *785*
tumor migration 240
tumor necrosis factor (TNF) 1028
tumor suppressor genes 116
  APC 137–138, 245–246
  FH 138
  INI1 (SMARCB1) 133–134, 758, 882, 885
  NF1 108–110, 132–133, 637, 653
  NF2 133, 637, 659–660, 669
  RB1 76, 116, 134
  SDH 138
  TP53 see p53 (*TP53*)
tumoral calcinosis 837–838
  appearance *838*, 838
  histopathology 838, *839*
  imaging *838*, 838
tunica vaginalis tumors 783, 790–793, *793*
turban tumor 763
Turner syndrome 562, 582
"two-hit hypothesis" (Knudson) 108–109, 116, 137–138

Index

tyrosinase 58–60
tyrosine hydroxylase 875, *876*
tyrosine kinase inhibitors (TKIs) 137
  crizotinib 127
  in desmoids 241–242, 1032
  in DFSP 129, 345–346
  in GISTs 132
    adjuvant 1020, 1026–1027
    first-line therapy 132, 487–488, 1024–1026
    resistance 132, 486, 1020, 1024–1025, *1025*, 1027
  imatinib *see* imatinib mesylate
  lapatinib 133
  midostaurin 128
  pazopanib 1030
  regorafenib 1025–1026
  sunitinib 1025–1026
tyrosine kinase receptors (TKRs/RTKs) 127, 877 *see also* *ALK*/ALK (anaplastic lymphoma kinase); CD117 (KIT)
  IGFR 136–137
  MET 130, 136–137
  NTRK3 127–128, 294
  PDGFRA 132, 475, 1024
  PDGFRB 242
  ROS1 125, 127
  VEGFR 47, 553, 555, 593–594

UICC-TNM staging system 5
ultra-deep targeted sequencing 145
ultrasound (US) 14–15, 25, 32
ultrastructure *see* electron microscopy
ultraviolet radiation 364, 798
umbilical polyp 251–253, *252*
undifferentiated pleomorphic sarcoma (UPS) *see* malignant fibrous histiocytoma (MFH)
undifferentiated pleomorphic sarcoma with giant cells *see* giant cell malignant fibrous histiocytoma
unknown primary 818–819
ureteral obstruction 200
urinary bladder inflammatory myofibroblastic tumor 288–289, 292

urinary bladder leiomyoma 452
urinary bladder paraganglioma 729–734, *733–734*
urogenital solitary fibrous tumor 325
uroplakin 2 (UP II) 811–813
urothelial carcinoma 67–68, 811–813
USP6 (*TRE17*) gene fusion 125
uterine tumors *see also* endometrial stromal sarcoma
  adenomatoid tumor 777, *778*
  chemotherapy 1031
  epithelioid smooth muscle tumors 463, *463*
  inflammatory myofibroblastic tumor 289
  leiomyoma 72–73, 459–463
    histopathology *461–462*, 461
  leiomyomatosis, intravenous 463–464, *464*
  leiomyosarcoma 466–467, *467*, 1031
  malignant mixed müllerian tumor (carcinosarcoma) 503–505, *504*
    histopathology 504–505, *504*
  myxoid 463
  PEComa 520–521, *522*

vagal body paraganglioma 735
vaginal tumors
  cervicovaginal myofibroblastoma 493, *493*
  mesodermal stromal polyp 491–493, *492*
  rhabdomyoma 493–494, *495*, 530
vanillylmandelic acid 870
vascular endothelial growth factor/receptor (VEGF/VEGFR) 553
  VEGFR2 555
  VEGFR3 47, 593–594
vascular malformations 554
vascular resection/reconstruction 1013
vascular tumors
  benign *see* hemangioma; lymphangioma; papillary endothelial hyperplasia; reactive vascular proliferations; telangiectasia
  borderline *see* hemangioendothelioma
  endothelial biology 553–554
  endothelial cell markers 41–48
  genetic syndromes 587
  malignant *see* angiosarcoma; Kaposi's sarcoma; venous leiomyosarcoma
  vasculogenesis 553–554
  VEGF/VEGFR *see* vascular endothelial growth factor/receptor
vena cava leiomyosarcoma 457, *458*
venous hemangioma 562, *563*
venous lake 577
venous leiomyosarcoma 457, *458*, *458*
venous stasis, acroangiodermatitis 580, *580*
Verocay bodies *643*, 655, *989*
verrucous hemangioma 560–561, *561*
verruga peruana 579
vestibular schwannoma 133, 654, 659–660
vimentin 75
viruses, oncogenic 3
  absent from IMT 291–292
  EBV 3–4, 467–468, 947, 949
  herpes simplex 313
  HHV8 3, 136, 612
  HPV16 803
  Merkel cell polyomavirus 801–802
  SV40 783
Von Hippel–Lindau disease type 2 (VHL-2) 725–726
von Recklinghausen's disease *see* neurofibromatosis type 1 (NF1)
von Willebrand factor (vWF) 47, 610
vulvar tumors
  angiomyofibroblastoma 494–496, *495*, *496*
  cellular angiofibroma 496–497, *497–498*
  leiomyoma 459
  mesodermal stromal polyp 491–493, *492*
  PEComa 522

Warburg effect 138
well-differentiated liposarcoma (WDLS) 416–424
  appearance *419*, 419
  clinical features 417–419
  cytology 979–981, *980*
  differential diagnosis 381, 424
  genetics 102, 131, 416–423
  histopathology 419
    inflammatory *421–422*, 421–422
    lipoma-like *419–420*, 419
    myxoid 422, 422–423
    sclerosing 420–421, *421*
    spindle cell 423
  imaging 22–24, *23*
  immunohistochemistry 423
  spindle cell 423
  surgery 1014–1015
well-differentiated papillary mesothelioma 780–781, *781*, 789–790, *791*
whole-chromosome probes (WCPs) 97
Wilms' tumor 856, 881–882
Wnt signaling pathway 128, 138
WT1/WT1 (Wilms' tumor protein)
  as an IHC marker 75, *76*, 880
  fusion with *EWSR1* 126–156, 881
WWTR-CAMTA1 gene fusion 598–599

X chromosome inactivation, basis of clonality assay 139
X-rays *see* radiography
xanthogranuloma, juvenile *see* juvenile xanthogranuloma
xanthogranulomatous inflammation 930–932, *931*
xanthoma 226, 932
  disseminated 932
  eruptive 932
  planar (xanthelasma) 932
  tendinous 932, *932*
  tuberous 932

YAP1-TFE3 gene fusion 599

Zellballen 723